ATEC Ass...

5150 E. 65th St.

Chemistry Division

849-4990

MICROBIOLOGY

An Introduction

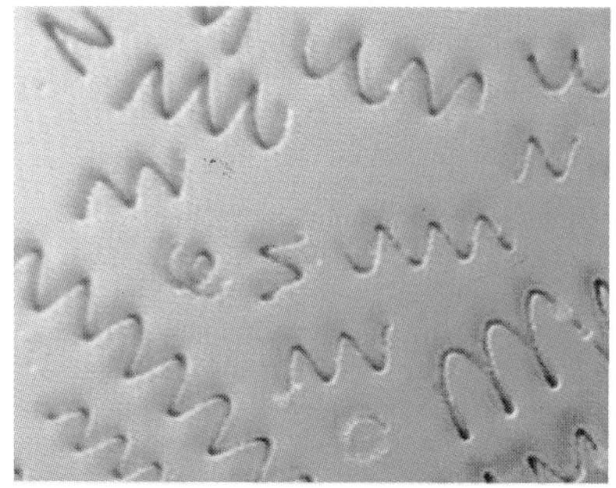

OF RELATED INTEREST

FROM THE BENJAMIN/CUMMINGS SERIES IN THE LIFE SCIENCES

GENERAL BIOLOGY

N.A. Campbell
Biology, Second Edition (1990)

PLANT BIOLOGY

M.G. Barbour, J.H. Burk, and W.D. Pitts
Terrestrial Plant Ecology, Second Edition (1987)

J. Mauseth
Plant Anatomy (1988)

L. Taiz and E. Zeiger
Plant Physiology (1991)

BIOCHEMISTRY AND CELL BIOLOGY

W.M. Becker and D.W. Deamer
The World of the Cell, Second Edition (1991)

C. Mathews and K.H. van Holde
Biochemistry (1990)

G.L. Sackheim
Chemistry for Biology Students, Fourth Edition (1990)

MOLECULAR BIOLOGY AND GENETICS

E.J. Ayala and J.A. Kieger, Jr.
Modern Genetics, Second Edition (1984)

L.E. Hood, I.L. Weissman, W.B. Wood, and J.H. Wilson
Immunology, Second Edition (1984)

R. Schief
Genetics and Molecular Biology (1986)

J.D. Watson, N.H. Hopkins, J.W. Roberts, J.A. Steitz, and A.M. Weiner
Molecular Biology of the Gene, Fourth Edition (1987)

G. Zubay
Genetics (1987)

MICROBIOLOGY

E. Alcamo
Fundamentals of Microbiology, Third Edition (1991)

R.M. Atlas and R. Bartha
Microbial Ecology: Fundamentals and Applications
Second Edition (1987)

J. Cappuccino and N. Sherman
Microbiology: A Laboratory Manual, Third Edition (1992)

T.R. Johnson and C. Case
Laboratory Experiments in Microbiology, Brief Edition
Third Edition (1992)

EVOLUTION, ECOLOGY, AND BEHAVIOR

D.D. Chiras
Environmental Science, Third Edition (1991)

R.J. Lederer
Ecology and Field Biology (1984)

M. Lerman
Marine Biology: Environment, Diversity, and Ecology (1986)

D. McFarland
Animal Behavior (1985)

E. Minkoff
Evolutionary Biology (1983)

R. Trivers
Social Evolution (1985)

ANIMAL BIOLOGY

E.N. Marieb
Essentials of Human Anatomy and Physiology
Third Edition (1991)

E.N. Marieb
Human Anatomy and Physiology, Second Edition (1992)

E.N. Marieb and J. Mallatt
Human Anatomy (1992)

L.G. Mitchell, J.A. Mutchmor, W.D. Dolphin
Zoology (1988)

A.P. Spence
Basic Human Anatomy, Third Edition (1991)

MICROBIOLOGY

An Introduction

FOURTH EDITION

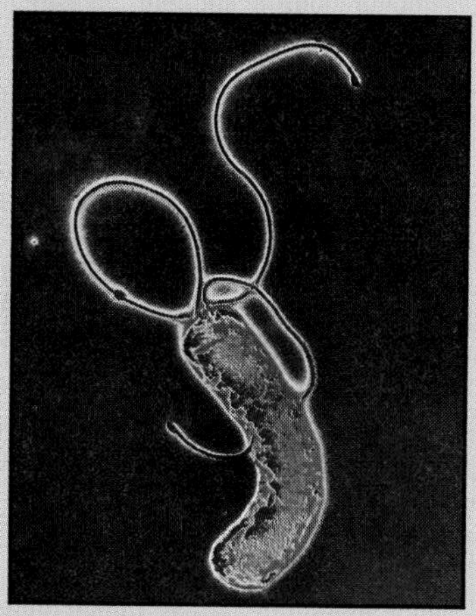

Gerard J. Tortora
BERGEN COMMUNITY COLLEGE

Berdell R. Funke
NORTH DAKOTA STATE UNIVERSITY

Christine L. Case
SKYLINE COLLEGE

THE BENJAMIN/CUMMINGS PUBLISHING COMPANY, INC.

Redwood City, California • Menlo Park, California • Reading, Massachusetts
New York • Don Mills, Ontario • Wokingham, U.K. • Amsterdam • Bonn
Sydney • Singapore • Tokyo • Madrid • San Juan

Sponsoring Editor: Edith Beard Brady
Editorial Assistant: Sissy Hodge
Developmental Manager: Robin J. Heyden
Developmental Editor: Shelley Parlante
Production Manager: Glenda Epting
Art and Design Manager: Michele Carter
Text and Cover Designer: Paula Schlosser
Production Supervisors: Larry Olsen and Brian Jones
Copy Editor: Yvonne Strong
Photo Editor: Cecilia Mills
Photo Researcher: Sarah Bendersky
Art Coordinator: Kevin Richardson
Artists: Ben Turner Graphics, Georg Klatt
Composition and Film: York Graphic Services, Inc.
Printing and Binding: Rand McNally & Company
Cover photograph: © Will and Deni McIntyre/Allstock

Photograph acknowledgments appear following the Glossary.

Library of Congress Cataloging-in-Publication Data

Tortora, Gerard J.
Microbiology: an introduction/Gerard J. Tortora, Berdell R. Funke, Christine L. Case.—4th ed.
p. cm.
Includes bibliographical references and index.
ISBN 0-8053-8480-4
1. Microbiology. I. Funke, Berdell R. II. Case, Christine L., 1948– . III. Title.
QR41.2.T67 1991
576—dc20 91-26219
CIP

ISBN 0-8053-8480-4

3 4 5 6 7 8 9 10—RN—95 94 93 92

The Benjamin/Cummings Publishing Company, Inc.
390 Bridge Parkway
Redwood City, California 94065

ABOUT THE AUTHORS

Gerard J. Tortora

Jerry Tortora is a professor of biology and teaches microbiology and human anatomy and physiology at Bergen Community College in Paramus, New Jersey, where he has been a faculty member for the past 24 years. He received his M.A. in biology from Montclair State College in 1965. He belongs to numerous biology/microbiology organizations, such as the American Society of Microbiology (ASM), American Association for the Advancement of Science (AAAS), National Education Association (NEA), New Jersey Education Association (NJEA), and the Metropolitan Association of College and University Biologists (MACUB). Jerry is the author of a number of best-selling biology textbooks.

Berdell R. Funke

Bert Funke received his Ph.D., M.S., and B.S. in microbiology from Kansas State University. He has been a professor of microbiology for 27 years at North Dakota State University. Currently he is teaching microbiology, pathogenic microbiology, food microbiology, and parasitology. Bert is an active member of the ASM and the AAAS. As a research scientist in the Experiment Station at North Dakota State, he has published numerous papers in soil microbiology and food microbiology.

Christine L. Case

Chris Case is a registered microbiologist and a professor of microbiology at Skyline College in San Bruno, California, where she has taught for the past 21 years. She received her Ed.D. in curriculum and instruction from Nova University and her M.A. in microbiology from San Francisco State University. Presently she is an active member of the ASM, the Northern California Association of Microbiologists (NCASM), Northern California Chapter of the Society for Industrial Microbiology (NCA–SIM), and the Scientific Research Society of North America (Sigma Xi). She served as president of NCA–SIM in 1990–91. In addition to teaching, Chris contributes regularly to the professional literature, develops innovative instructional methodologies, and maintains a personal and professional commitment to conservation and the importance of science in society. Chris is also an avid photographer, and many of her photographs appear in this volume.

PREFACE

Microbiology: An Introduction, fourth edition, is a comprehensive text for students in a wide variety of programs, including allied-health sciences, biological sciences, environmental studies, animal science, forestry, agriculture, home economics, and liberal arts. It is a beginning text, assuming no previous study of biology or chemistry.

During the ten years since the publication of the first edition, this book has been used at more than 800 colleges and universities by over 350,000 students. We have been gratified to hear from instructors and students alike that the book has become a favorite among their textbooks—a learning tool that is both effective and enjoyable.

Features of the Revision

Our primary goal for the fourth edition of *Microbiology: An Introduction* was to update the book throughout, making certain it reflects the important new discoveries of the past few years. We also wanted to give our book a fresh look with new photomicrographs, new color illustrations and photographs, and a more inviting design. At the same time, we wanted to maintain a manageable length and the emphasis on the clear coverage of fundamental principles of microbiology. Every page of the book was scrutinized with these goals in mind. The following list highlights the resulting major changes in the fourth edition:

- One of the noticeable changes for this edition is the use of full color for many of the illustrations and photographs. Moreover, many of the illustrations have been redrawn. Color is used in the illustrations to highlight key concepts and to focus attention on the key elements of the discussion. For example, molecules such as DNA and ATP are the same color throughout the book to help students follow pathways and see mechanisms. Colored electron micrographs were selected to highlight structures without distorting natural perspective. Color should help students relate new information to the professional environment by providing accurate examples of organisms and stains.

- A new chapter (Chapter 9) entitled Recombinant DNA and Biotechnology has been added to reflect the increasing importance of this topic, which has been called the technology of the 1990s. This chapter describes the principles employed and their numerous applications and includes many new figures at a level commensurate with the background of the average undergraduate student.

- The immunology chapters (Chapters 17 to 19) have again been thoroughly revised, expanded, and reorganized in response to the many changes that have taken place recently in this dynamic, rapidly evolving field. The discussion of AIDS, for example, has been rewritten in response to current research findings and applications.

- In the chapters dealing with diseases (Chapters 21 to 26), many new photographs have been added, and the majority of these are in color. All discussions of diseases have been carefully scrutinized and updated. The treatment of oral diseases has been expanded greatly, and several new illustrations have been added in response to user comments.

- Many of the MMWR and Microbiology in the News boxes have been replaced with more timely articles. Microbiology Highlights boxes contain interesting information relevant to the study of microbiology that we hope will further motivate students.

- Chapter 5, Microbial Metabolism, has been completely revised to further clarify the basic principles.

- Chapter 8, Microbial Genetics, has been carefully revised to include updated information on molecular biology.

Features Retained from the Earlier Editions

We have retained in this new edition the features that made the previous editions so popular. These include:

- **An appropriate balance between microbiological fundamentals and applications and between medical applications and other applied areas.** As in previous editions, basic principles are given greater emphasis than applications, and health-related applications are emphasized. Applications are integrated throughout the text, and considerable attention is devoted to microorganisms in habitats outside the human body. We hope that all students will gain an appreciation of the fascinating diversity of microbial life, the central roles of microorganisms in nature, and the importance of microorganisms in our daily lives.

- **An illustration program carefully developed to support the text.** Included are both state-of-the-art electron micrographs that dramatically show microbial structures and light micrographs that more closely resemble what is usually seen in a microbiology laboratory. As with previous editions, quality line drawings support text discussions. In the fourth edition, we have improved many illustrations and added new diagrams to further enhance the visual quality of the text.

- **Straightforward presentation of complex topics.** Each section of the text has been revised with the student in mind to maintain the clarity of explanation for which our book has been known.

- **Phonetic pronunciations.** Throughout the text, phonetic pronunciations are provided in parentheses for the majority of genus and species names for microorganisms. The pronunciations are typically provided at the point where the terms are first introduced. A comprehensive, consolidated list of all phonetic pronunciations cited in the book appears following the Appendices.

- **Several appendices at the end of the book heighten its usefulness.** Appendix A is the Classification of Bacteria According to *Bergey's Manual of Systematic Bacteriology*. Appendix B, Word Roots Used in Microbiology, provides basic rules of pronunciation and some phonetic pronunciations for genus and species names used in the text. Appendix C is a Most Probable Numbers (MPN) Table, and Appendix D describes Methods for Taking Clinical Samples. Appendix E, Biochemical Pathways, provides more elaborate illustrations of these key biochemical concepts. Appendix F provides background on Exponents, Exponential Notation, Logarithms, and Generation Time. A Glossary following the appendices provides definitions of all important terms used in the text.

Course Sequences

We have organized the book in what we feel is a useful fashion, but we recognize that the material might be effectively presented in a number of alternative sequences. For those who wish to follow another sequence, we have made each chapter as independent as possible and have included numerous cross-references. Thus, the Survey of the Microbial World, Part Two, could be studied at the beginning of a course, immediately after Chapter 1. Applied Microbiology and the Environment, Part Five, could follow Parts One and Two. Or, Chapter 8, Microbial Genetics, and Chapter 9, Recombinant DNA and Biotechnology, can be covered with Part Five. Since Chapters 7 and 20 both deal with the control of microbial growth, they could be covered together. The material on Microorganisms and Human Disease, Part Four, readily lends itself to rearrangement or selective coverage. The various diseases are organized into chapters according to the host organ-system affected. The *Instructor's Guide* provides detailed guidelines for organizing the disease material in several alternative ways.

Supplementary Materials

All the ancillaries to the text have been revised by their original authors:

- ***Study Guide***
  The *Study Guide*, by Berdell Funke, will help students master and review major concepts and facts from the text. Each chapter of the *Study Guide* begins with a chapter summary organized by the text headings. Important terms are printed in boldface type and are defined, and important figures and tables from the text are included. Following the summary is an extensive self-testing section containing matching questions, fill-in questions, and an answer key.

- ***Instructor's Guide***
  The *Instructor's Guide*, by Christine Case, includes many practical suggestions for using the text in a course. Suggested course outlines and corresponding text pages provide maximum flexibility in organizing your course. Important fourth edition changes are highlighted for each chapter to facilitate the transition from the third to the fourth edition. Also included are answers to study questions from the text. Special case studies of clinical issues ask students to be "medical detectives" by solving the problem presented.

- ***Software***
  Benjamin/Cummings Testing Software is available for IBM PC and Macintosh personal computers.

- ***Transparencies***
  Acetate overhead transparencies of 200 two-color and full-color line drawings from the text are available from the publisher to qualified adopters.

- ***Transparency Masters***
  Transparency masters for all the line drawings from the text are available in a reproducible format from the publisher to qualified adopters.

- ***Slides***
  Two slide sets to complement instructors' lectures are available to qualified adopters. The topics of the slide sets are (1) Microbial Agents of Disease and (2) Human Immunodeficiency Virus.

- ***Lab Manual***

  Laboratory Experiments in Microbiology: Brief Edition (third edition, 1992), by Ted Johnson and Christine Case, is a very successful lab manual that has been updated throughout with new exercises on genetic engineering, transformation, bioremediation, and new safety guidelines for the handling of blood and body fluids according to CDC guidelines. The manual is organized so that lab report forms immediately follow each exercise. A detailed *Instructor's Guide* for the lab manual facilitates preparation for laboratory sessions.

Acknowledgments

In the preparation of this textbook, we have benefited from the guidance and advice of a large number of microbiology instructors across the country. R.L. Bernstein, San Francisco State University, provided an early draft of Chapter 9, Recombinant DNA and Biotechnology, and many hours of content expertise. Many dedicated teachers participated in focus groups to help plan the revision. Reviewers offered constructive criticism and valuable suggestions at various stages of manuscript preparation. Contributors, focus group attendees, and reviewers are listed on the following pages. We gratefully acknowledge our debt to these individuals.

We offer special thanks to the staff at Benjamin/Cummings for their dedication to excellence. Shelley Parlante's careful attention to detail served to keep information clear. The scientific and editorial contributions of Jane Reece were invaluable. Larry Olsen and Brian Jones expertly guided our book through the production process. Edith Beard Brady, our sponsoring editor, and Cecilia Mills, our photo editor, were instrumental throughout the revision process. And, we have enduring appreciation for our students, whose comments and suggestions both provide insight and remind us of their needs. This text is for them.

Gerard J. Tortora
Berdell R. Funke
Christine L. Case

CONSULTANTS FOR *MICROBIOLOGY: AN INTRODUCTION,* FOURTH EDITION

FOCUS GROUP PARTICIPANTS

Les Albin, Austin Community College
Clementine de Angelis, Tarrant County Junior College, South
Louis Debetaz, Angelina College
Joe Harber, San Antonio College
Sally Jackson, Baylor University
Elaine Johnson, San Francisco City College
Diana Kaftan, Diablo Valley College
Carolyn Mohr, St. Mary's College of California
Remo Morelli, San Francisco State University
Luis A. Rodriguez, San Antonio Community College
John Searle, College of San Mateo
Helen Sowers, California State University, Hayward
Hideo Yonenaka, San Francisco State University
Shanna Yonenaka, San Francisco State University

REVIEWERS OF THE FOURTH EDITION

Kenneth L. Anderson, California State University, Los Angeles
Walter Appelgren, Northern Arizona University
Kostia Bergman, Northeastern University
R.L. Bernstein, San Francisco State University
David Berryhill, North Dakota State University
Frank Binder, Marshall University
Carol J. Burger, Virginia Polytechnic Institute and State University
Russell J. Centanni, Boise State University
Bruce Cochrane, University of South Florida
William Coleman, University of Hartford
Irene Cotton, Lorain County Community College
Jacqueline Dushensky, Hudson Valley Community College
Cindy Erwin, City College of San Francisco
Joseph J. Gauthier, University of Alabama, Birmingham
Diana Kaftan, Diablo Valley College
Juhee Kim, California State University, Long Beach
John Lammert, Gustavus Adolphus College
Thomas Maier, University of Colorado, Denver
Joel Ostroff, Brevard Community College
Helen Oujesky, University of Texas, San Antonio
Kay Pauling, Foothill College
Marian Price, Onondaga Community College
Ralph Rascati, Kennesaw State College
Lisa Steiner, Massachusetts Institute of Technology
Sheldon Steiner, University of Kentucky, Lexington
Mary Lou Tortorello, Cornell University
James E. Urban, Kansas State University
Brian J. Wilkinson, Illinois State University, Normal
Shanna Yonenaka, San Francisco State University

REVIEWERS OF PREVIOUS EDITIONS

Wendall E. Allen, East Carolina University
Lucia Anderson, Queensborough Community College
Kenneth L. Anderson, California State University, Los Angeles
Barry Batzing, SUNY College at Cortland
Jeff Becker, University of Tennessee
Lois Beishir, Antelope Valley College
Harold Bendigkeit, De Anza College
Kostia Bergman, Northeastern University
Richard Bernstein, San Francisco State University
L.J. Berry, University of Texas
Russell F. Bey, University of Minnesota
Frank L. Binder, Marshall University
Robert B. Boley, University of Texas
Russell J. Centanni, Boise State University
J. John Cohen, University of Colorado, Denver
William Coleman, University of Hartford
Mary Lynne Perille Collins, University of Wisconsin, Milwaukee
Richard Davis, West Valley College
Norman Epps, University of Guelph
Cindy Erwin, City College of San Francisco
David Filmer, Purdue University
Randy Firstman, College of the Sequoias
Roger Furbee, Middlesex County College
David Gabrielson, North Dakota State University
James G. Garner, C.W. Post College
Joseph J. Gauthier, University of Alabama, Birmingham
Blanche Griggs, Georgia State University
Rebecca Halyard, Clayton State College
Joan Handley, University of Kansas
Bettina Harrison, University of Massachusetts, Boston
Diana S. Herson, University of Delaware
Ronald Hochede, City College of San Francisco
John G. Holt, Iowa State University
Robert Janssen, University of Arizona
Ted R. Johnson, St. Olaf College
Wallis L. Jones, DeKalb Community College
Diana Kaftan, Diablo Valley College
Ken Keudell, Western Illinois University
Juhee Kim, California State University, Long Beach
Alan Konopka, Purdue University
Walter Koostra, University of Montana
Robert I. Krasner, Providence College
John Lammert, Gustavus Aldolphus College
John Lewis, San Bernardino Valley College
Peter Ludovici, University of Arizona
Thomas Maier, University of Colorado, Denver
John C. Makemson, Florida International University
Eleanor K. Marr, Dutchess Community College
William Matthai, Tarrant County Junior College
J.R. Milam, University of Florida, Gainesville
Robert Mitchell, Community College of Philadelphia
Frank Mittermeyer, Elmhurst College
Henry Mulcahy, Suffolk University
Roger Nichols, Weber State College
Elinor O'Brien, Boston College
J. Dennis O'Malley, University of Arkansas
Dennis Opheim, Texas A & M University
Helen Oujesky, University of Texas, San Antonio
Robert Pengra, South Dakota State University
Jane Phillips, University of Wisconsin, Madison
Christine Pootjes, Pennsylvania State University, University Park
Joseph R. Powell, Florida Junior College
Ralph Rascati, Kennesaw College
Robert Satterfield, College of Du Page
Violet Schirone, Suffolk County College
Robert Schoenhoff, Jefferson Community College
Michael H. Scholla, E.I. DuPont de Nemours & Company
Louis Shainberg, Mount San Antonio College
Gregory L. Shipley, Texas A & M University
Jeffrey J. Sich, Youngstown State University
Josephine Smith, Montgomery County Community College
Cynthia V. Sommer, University of Wisconsin, Milwaukee
Lisa Steiner, Massachusetts Institute of Technology
Sheldon Steiner, University of Kentucky, Lexington
Bernice Stewart, Prince Georges Community College
Teresa Ann Thomas, Southwestern College
Dan Trubovitz, Pasadena City College
Robert Twarog, University of North Carolina, Chapel Hill
James E. Urban, Kansas State University
Pat Vary, Northern Illinois University
Roberta Wald, Portland Community College
James White, Prairie State College
Mary K. Wicksten, Texas A & M University
Fred Williams, Iowa State University
Brian J. Wilkinson, Illinois State University, Normal
Shanna Yonenaka, San Francisco State University

A Student's Guide to Microbiology

CHAPTER 1

The Microbial World and You

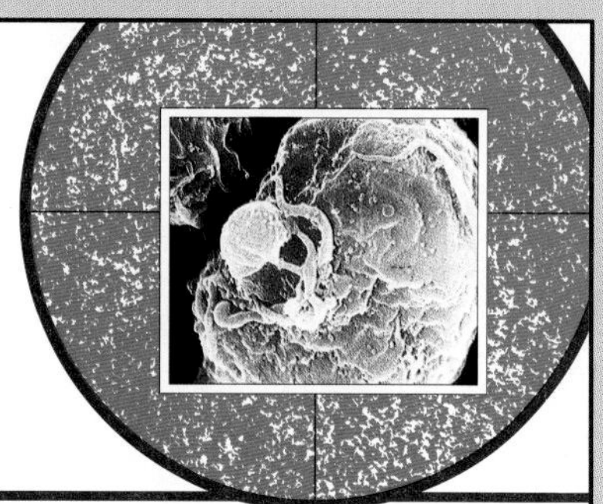

LEARNING OBJECTIVES

- Identify the contributions to microbiology made by Anton van Leeuwenhoek, Robert Hooke, Louis Pasteur, Robert Koch, Joseph Lister, Paul Ehrlich, Alexander Fleming, and Edward Jenner.
- Compare the theories of spontaneous generation and biogenesis.
- Recognize scientific genus and specific epithet names.
- List the major groups of organisms studied in microbiology.
- List at least four beneficial activities of microorganisms.
- Define normal flora.
- Define immunology, microbial ecology, microbial genetics, microbial physiology, molecular biology, and virology.
- List applications of recombinant DNA, biotechnology, and bioremediation.

Part One photo, page 1: Although billions of microorganisms are present in, on, and around us, they are too small to see without special instruments such as this compound light microscope.

In the summer of 1985, newspapers announced that a well-known actor was suffering from AIDS (acquired immunodeficiency syndrome). The American public had heard about the disease, but until then AIDS had not been perceived as much of a health threat by the majority of Americans, perhaps because the victims were primarily from a group outside the mainstream of American society—male homosexuals. While the actor sought treatment in hospitals in France, Americans became increasingly aware of the illness that would end the life of one of their screen idols as well as the lives of over 8000 other Americans that same year.

The general public was just beginning to worry about what would become the most frightening epidemic of the century. Yet medical researchers had been gathering information about the d
for several years. The same concep
that had helped them identify ot
means of transmission, and treat
applied to the study of AIDS.

AIDS came to public attention i
from Los Angeles that a few youn
had died of a previously rare t
known as *Pneumocystis* (nü-mō-s
These men had experienced a sever
immune system, which normally fi
ease. Soon these cases were correl
sual number of occurrences of a ra
Kaposi's sarcoma, among young
Similar increases in such rare dis
among hemophiliacs and intraven

By the end of 1990, nearly 160
United States had been diagnose
and more than 50% of them had di
disease. A great many more people
for the presence of the AIDS virus i

2

Learning objectives open each chapter
Each chapter begins with learning objectives, directing the student to the chapter's important points for mastery.

The authors effectively use examples throughout the text to illustrate the study of microorganisms. Chapter 1, for example, begins with a discussion of the impact of AIDS during the past decade as an example of how microorganisms affect our lives and what microbiologists do.

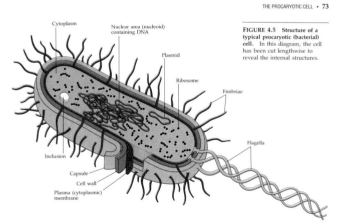

THE PROCARYOTIC CELL • 73

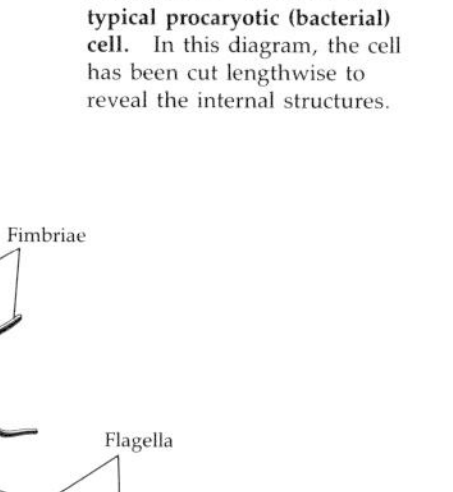

FIGURE 4.5 Structure of a typical procaryotic (bacterial) cell. In this diagram, the cell has been cut lengthwise to reveal the internal structures.

from Chapter 2 that the D forms of amino acids are unusual. Since only encapsulated *B. anthracis* cause anthrax, it is possible that the capsule prevents phagocytosis.

Another function of the sticky glycocalyx is to allow a bacterium to attach to various surfaces in order to survive in its natural environment. Through attachment, bacteria can grow on diverse surfaces such as rocks in fast-moving streams, plant roots, human teeth and tissues, and even other bacteria. *Streptococcus mutans* (mū'tans), an important cause of dental caries, attaches itself to the surface of teeth by a glycocalyx. The capsule of *Klebsiella pneumoniae* (kleb-sē-el'lä nü-mō'nē-ī) prevents phagocytosis and allows this bacterium to adhere to and colonize the respiratory tract.

S. mutans may use its capsule as a source of nutrition by breaking it down and utilizing the sugars when energy stores are low. A glycocalyx can protect a cell against dehydration. Also, its viscosity may inhibit the movement of nutrients from the cell.

FLAGELLA

Some procaryotic cells carry **flagella** (singular, *flagellum*, meaning whip), which are long filamentous appendages that propel bacteria (see Figure 4.5).

Bacterial cells have four arrangements of flagella (Figure 4.6)—**monotrichous** (single polar flagellum), **amphitrichous** (single flagellum at each end of the cell), **lophotrichous** (two or more flagella at one or both poles of the cell), and **peritrichous** (flagella distributed over the entire cell).

A flagellum has three basic parts (Figure 4.6e). The long outermost region, the *filament*, is constant in diameter and contains the globular (roughly spherical) protein *flagellin* arranged in several chains that intertwine and form a helix around a hollow core. Flagellar proteins serve to identify certain pathogenic bacteria. In most bacteria, filaments are not covered by a membrane or sheath, as in eucaryotic cells. The filament is attached to a slightly wider *hook*, which consists of a different protein. The third portion of a flagellum is the *basal body*, which anchors the flagellum to the cell wall and plasma membrane.

The basal body is composed of a small central rod inserted into a series of rings. Gram-negative bacteria contain two pairs of rings. The outer pair of rings is anchored to various portions of the cell wall, and the inner pair of rings is anchored to the plasma membrane. In gram-positive bacteria, only the inner pair is present. As you will see later in the chapter, the flagella (and cilia) of eucaryotic cells are more complex than those of procaryotic cells.

Bacteria with flagella are motile. That is, they have the ability to move on their own. Each procaryotic flagellum is a semirigid, helical rotor that moves the cell

Text and illustrations are fully integrated
The integration of text and illustrations is a key to the success of *Microbiology: An Introduction*, fourth edition. These pages from Chapter 4 highlight the illustration program. Notice the clearly rendered diagram in Figure 4.5 with key structures labeled and in color. The procaryotic cell reappears in Figure 4.6 as an orienting figure.

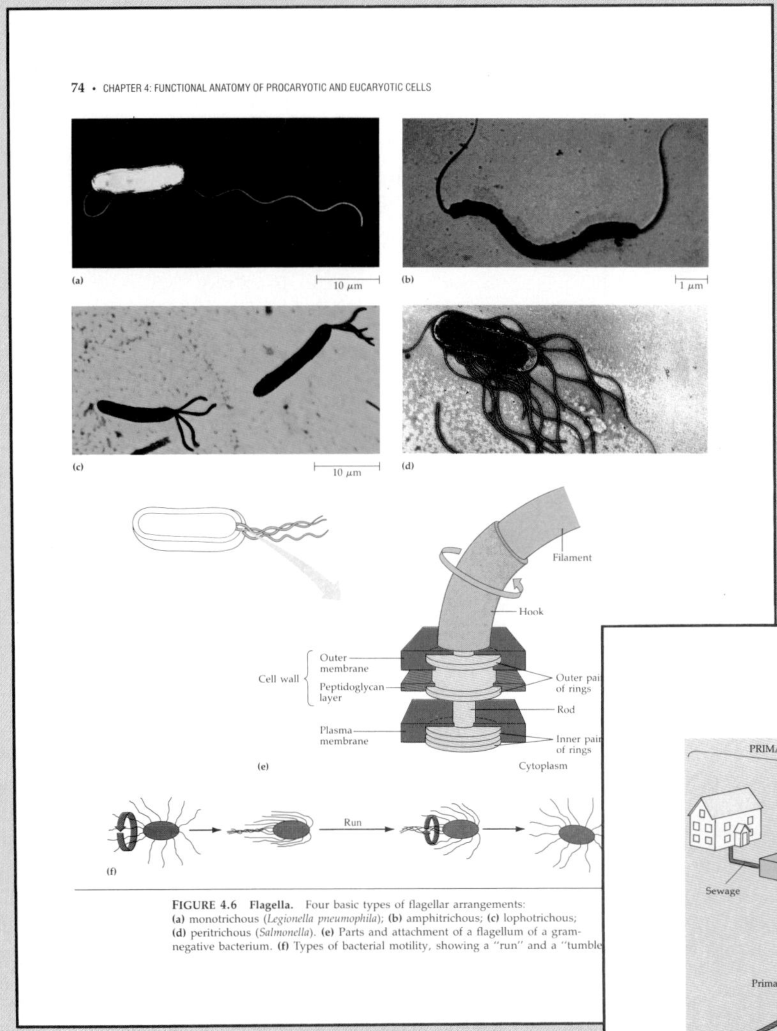

74 • CHAPTER 4: FUNCTIONAL ANATOMY OF PROCARYOTIC AND EUCARYOTIC CELLS

(a) 10 μm (b) 1 μm

(c) 10 μm (d)

Filament

Hook

Cell wall

Outer membrane

Peptidoglycan layer

Outer pai of rings

Rod

Plasma membrane

Inner pai of rings

(e) Cytoplasm

Run

(f)

FIGURE 4.6 Flagella. Four basic types of flagellar arrangements: **(a)** monotrichous (*Legionella pneumophila*); **(b)** amphitrichous; **(c)** lophotrichous; **(d)** peritrichous (*Salmonella*). **(e)** Parts and attachment of a flagellum of a gram-negative bacterium. **(f)** Types of bacterial motility, showing a "run" and a "tumble

Text and illustrations are fully integrated

Figure 4.6: The color photographs (a) through (d) illustrate examples of the flagellum structure diagrammed in part (e). Note the orientation diagram. Part (f) shows the use of the flagellum.

AQUATIC MICROBIOLOGY AND SEWAGE TREATMENT • 689

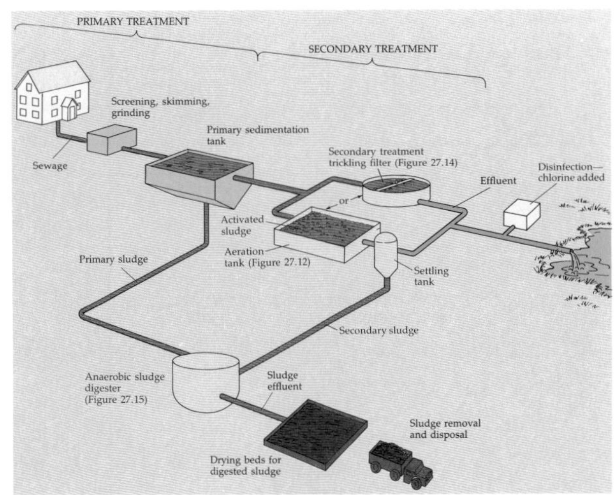

FIGURE 27.10 Steps of typical sewage waste treatment. A particular system would use either activated sludge aeration tanks or trickling filters, not both, as shown in this figure. The sludge is disposed of in landfills or agricultural land. Microbial activity occurs aerobically in trickling filters or in activated sludge aeration tanks and anaerobically in the anaerobic sludge digester.

water is initially aerated to provide a relatively high level of dissolved oxygen and is seeded with bacteria if necessary. The filled bottles are then incubated in the dark for five days at 20°C, and the decrease in dissolved oxygen is determined by a chemical or electronic testing method. The more oxygen that is used up as the bacteria degrade the organic matter in the sample, the greater the BOD—which is usually expressed in milligrams of oxygen per liter of water. The amount of oxygen that normally can be dissolved in water is only about 10 mg/liter. Typical BOD values of waste water may be twenty times this amount. If this waste water enters a lake, for example, bacteria in the lake begin to consume the organic matter responsible for the high BOD, rapidly depleting the oxygen in the lake water.

Secondary Treatment

After primary treatment, the greater part of the BOD remaining in the sewage is in the form of dissolved organic matter. **Secondary treatment,** which is primarily biological, is designed to remove most of this organic matter and reduce the BOD (Figure 27.10). In this process, the sewage undergoes strong aeration to encourage the growth of aerobic bacteria and other microorganisms that oxidize the dissolved organic

Color supports the learning process

The illustration program includes many step-by-step diagrams that help the reader understand important processes. This example from Chapter 27 illustrates the steps in sewage waste treatment.

A Student's Guide to Microbiology

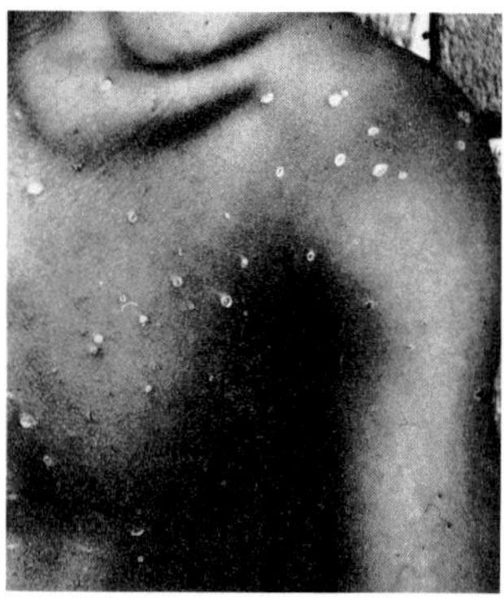

(a)

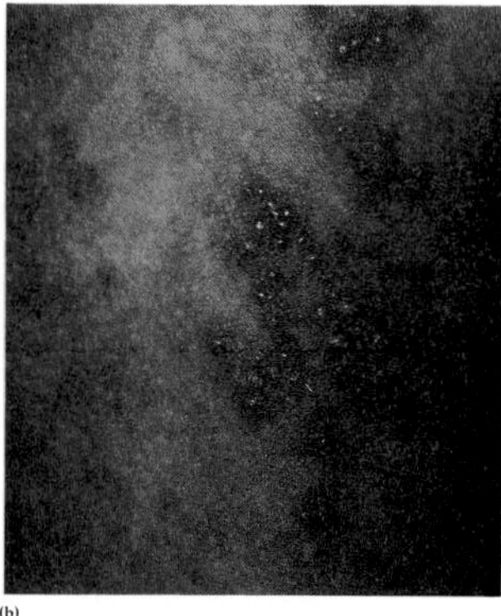

(b)

FIGURE 21.8 Typical lesions associated with (a) chickenpox and (b) shingles (herpes zoster), shown affecting the back of this patient.

The eradication of smallpox was possible because there are no animal host reservoirs for the disease. Once an effective vaccine became available, eradication was accomplished by a concerted vaccination effort coordinated by the World Health Organization.

Currently, the smallpox virus collections in laboratories have been the most likely sources of new infections. The risk of such infection is not merely a hypothetical concern as there have already been several laboratory-associated infections, one of which caused death. Today, only two sites maintain the smallpox virus, one in the United States and one in the USSR.

CHICKENPOX (VARICELLA) AND SHINGLES (HERPES ZOSTER)

Chickenpox (varicella) is a relatively mild childhood disease. After gonorrhea, it is the most common reportable infectious disease in the United States. It is probably greatly underreported, and more than 2 million cases probably occur each year in the United States. Disease summaries of the Centers for Disease Control show that about 100 deaths per year, usually from encephalitis (infection of the brain), are attributed to chickenpox.

Chickenpox is acquired by infection of the respiratory system, and the infection localizes in skin cells after about two weeks. The infected skin is vesicular for three to four days. During that
with pus, rupture, and form a scab
ure 21.8a). Lesions are mostly co
throat, and lower back. The vesi
appear in the mouth and throat.
occurs in adults—which is not fr
high incidence in childhood grant
persons—it is a more severe disea
mortality rate.

Reye's syndrome is an occasio
tion of chickenpox, influenza, an
diseases. A few days after the init
ceded, the patient persistently v
signs of brain dysfunction. Coma,
of the liver, and death can follow.
age in survivors is from brain swell
blood circulation. At one time, th
ported cases approached 90%, bu
declining with improved care and i
when the disease is recognized a
Reye's syndrome affects children a
exclusively. The use of aspirin to lo

New color photographs are effective

The illustration program for the fourth edition includes many new full-color photographs to enable the reader to better visualize the effects of diseases caused by microbes.

MMWR

MORBIDITY AND MORTALITY WEEKLY REPORT

AIDS: The Risk to Health Care Workers

With the advent of the AIDS epidemic, health care workers are understandably concerned about the risk of contracting AIDS after exposure to body fluids of infected patients. However, when precautions are observed, the risk to workers is very small, even for those treating AIDS patients.

Understanding the Risk

The first protection for health care workers is a clear understanding of how AIDS can (and cannot) be transmitted in the course of their work. To date, direct inoculation of infected material is the only proven method of transmission in the health care environment. Infected materials that can transmit AIDS are blood, semen, vaginal secretions, and breast milk. The most common route of transmission is through accidental needle sticks. However, inoculation is also possible if infected material contacts mucous membranes or a break in the health care worker's skin. There is no evidence of HIV transmission by aerosols, the fecal-oral route, mouth-to-mouth or casual contact, or via environmental surfaces such as floors, walls, chairs, and toilets. Saliva, tears, cerebrospinal fluid, amniotic fluid, and urine do not pose the same risk as other body fluids.

The number of health care workers who have become infected and the sources of their infections are being carefully monitored by the CDC. In the United States, by the end of 1990, 37 health care workers who denied other risk factors and who were exposed to blood from patients with HIV had contracted HIV infections, 29 of these following needlestick injuries. Six cases had extensive contact with blood of infected patients and did not observe routine barrier precautions. Additionally, two workers in industrial laboratories producing large quantities of highly concentrated HIV have been reported to have laboratory-acquired HIV infections.

In addition to monitoring all health care workers who have actually contracted HIV, the CDC and others also study workers who have been exposed to infected materials, to determine the magnitude of risk. The largest study followed 4000 health care workers who had been exposed to HIV, 1200 by needlesticks. The study found that the probability of infection following a needlestick injury with blood containing HIV is 0.4%. The probability of infection from mucous membrane exposure to blood is 10 times lower. The probability of acquiring hepatitis B from a needlestick injury with blood containing HBV is from 6 to 30%.

Precautions

In 1985, the CDC developed the strategy of "universal precautions," which should be followed in **all** health care settings. These are:

Gloves. Disposable gloves should be used for direct exposure to infected blood, other body fluids, and tissues. Double-gloving is recommended during invasive surgical procedures. Personnel should not work when they have open skin lesions, weeping dermatitis, or cutaneous wounds.

Gowns, Masks, and Goggles. Masks and protective eyewear are recommended when splashes are expected, such as during airway manipulation, endoscopy, and dental procedures, and in the laboratory.

Needles. To minimize the risk of needle sticks, needles should not be resheathed and should be put in a puncture-proof container for sterilization and disposal.

Disinfection. Routine housekeeping in health care settings should include washing floors, walls, and other areas not normally associated with disease transmissions with 1 : 100 dilution of household bleach. A 1 : 10 dilution is recommended for disinfecting a spill.

Prophylactic Treatment After Exposure

Avoidance of exposure is the health care worker's first line of defense. Admittedly, however, accidental exposure cannot always be prevented. Promising new evidence points to the possibility that exposed persons may be able to reduce their risk by prophylactic use of AZT. The National Institutes of Health and the University of California at San Francisco are developing recommendations for use of AZT.

The Risk to Patients

Transmission of HBV and HIV from health care workers to patients during invasive dental procedures (tooth extractions) has been documented. Restrictions on the procedures that can be performed by health care workers with HIV infection have also been considered by the American Medical Association, the American Dental Association, the CDC, and other organizations. Their recommendations state that the risk of HIV transmission from health care workers to patients is greatest during invasive procedures and that decisions regarding restriction of patient care by infected workers who perform such procedures should be made on an individual basis.

Adapted from *MMWR* 38:(S-2) (5/12/89); *MMWR* 39(S-6) (6/23/89); *MMWR* 39:489 (7/27/90).

Box program

An innovative box program supports the text with interesting stories and clinical examples of microbiology in action. The three types of boxes are Microbiology in the News, Microbiology Highlights, and the newsworthy Morbidity and Mortality Weekly Report boxes. This example, AIDS: The Risk to Health Care Workers, illustrates the updated material found throughout the fourth edition.

End-of-chapter material reinforces key ideas

Extensive end-of-chapter material supports the text, including a conceptually organized Study Outline, two types of Study Questions, and updated suggestions for Further Reading.

STUDY OUTLINE

Bacterial Groups (pp. 273–293)

1. *Bergey's Manual* divides bacteria into sections based on Gram-stain reaction, cellular morphology, oxygen requirements, and nutritional properties.
2. In some cases, the sections include families and orders, and some bacteria are included as genera of uncertain affiliation.
3. Spirochetes are long, thin, helical cells that move by means of an axial filament.
4. Aerobic/microaerophilic, motile, helical/vibrioid gram-negative bacteria move by means of one or more polar flagella.
5. Gram-negative aerobic rods and cocci have polar flagella, if flagellated, and can utilize a wide variety of organic compounds.
6. Facultatively anaerobic gram-negative rods have flagella, if motile, and include the enterics, Vibrionaceae, Pasteurellaceae, and the genus *Gardnerella*.
7. Members of the anaerobic gram-negative straight, curved, and helical rods can be found in humans.
8. Dissimilatory sulfate-reducing or sulfur-reducing bacteria are anaerobes that are important in the sulfur cycle.
9. Anaerobic gram-negative cocci are normal flora of the human mouth.
10. Rickettsias and chlamydias are obligate intracellular parasites.
11. Mycoplasmas are bacteria that lack cell walls.
12. Gram-positive cocci include the catalase-positive *Staphylococcus* and catalase-negative *Streptococcus*.
13. Endospore-forming gram-positive rods and cocci may be aerobic, facultatively anaerobic, or anaerobic.
14. The diverse group of regular, nonsporing, gram-positive rods includes *Lactobacillus* and *Listeria*.
15. Irregular, nonsporing, gram-positive rods include the irregular-staining corynebacteria.
16. Pathogenic species of mycobacteria are acid-fast.
17. Nocardioform bacteria may be acid-fast; they form short filaments.
18. Bacteria with unusual morphologies are discussed in this book in terms of the following groups: budding and/or appendaged; gliding, nonfruiting; gliding, fruiting; budding; sheathed.
19. The chemoautotrophic bacteria play important roles in the cycles of elements in the environment.
20. Extreme halophiles, acidophiles, thermophiles, and methane-producing bacteria are included in the archaeobacteria.
21. Photosynthetic purple and green bacteria are included in the group of anoxygenic phototrophic bacteria; they do not produce molecular oxygen.
22. Cyanobacteria produce molecular oxygen during photosynthesis; they are oxygenic phototrophs.
23. Actinomycetes produce mycelia and reproduce by external spores.

STUDY QUESTIONS

Review

1. The following outline is a key that can be used to identify the medically important groups of bacteria. Fill in the name of the group indicated by the key.

Name of Group and Representative Genus

I. Gram-positive
  A. Endospore-forming ______
  B. Nonsporing
    1. Cocci ______
    2. Rods
      a. Regular ______
      b. Irregular ______
      c. Acid-fast ______
  C. Mycelium produced
    1. Acid-fast ______
    2. Produce chains of conidia ______
II. Gram-negative
  A. Cells are helical or curved
    1. Axial filament ______
    2. No axial filament
      a. Aerobic
      b. Anaerobic
  B. Cells are rods or cocci
    1. Aerobic, nonfermenting
    2. Facultatively anaerobic
    3. Anaerobic
  C. Intracellular parasites
III. Lacking cell wall

2. Compare and contrast each of the
  (a) Cyanobacteria and algae
  (b) Actinomycetes and fungi
  (c) *Bacillus* and *Lactobacillus*
  (d) *Pseudomonas* and *Escherichia*
  (e) *Leptospira* and *Spirillum*
  (f) *Veillonella* and *Bacteroides*
  (g) *Rickettsia* and *Chlamydia*
  (h) *Thermoplasma* and *Mycoplasma*

3. Matching:

I. Gram-positive
  A. Nitrogen-fixing
II. Gram-negative
  A. Phototrophic
    1. Anoxygenic
    2. Oxygenic
  B. Chemoautotrophic
    1. Oxidize inorganics, such as NO_2^-
    2. Reduce CO_2 to CH_4
  C. Chemoheterotrophic
    1. Move via a slime layer
    2. Form myxospores
    3. Reduce sulfate to H_2S
      a. Anaerobic
      b. Thermophilic
    4. Long filaments, found in sewage
    5. Form projections from the cell
III. Unusual cell wall (lacking peptidoglycan)

(a) Archaeobacteria
(b) Cyanobacteria
(c) *Cytophaga*
(d) *Desulfovibrio*
(e) *Frankia*
(f) *Hyphomicrobium*
(g) Methanogenic bacteria
(h) Myxobacteria
(i) *Nitrobacter*
(j) Purple bacteria
(k) *Sphaerotilus*
(l) *Sulfolobus*

Challenge

1. Place each section listed in Table 11.1 in the appropriate division:
  (a) Typical gram-positive
  (b) Typical gram-negative
  (c) Wall-less
  (d) Unusual walls
2. Where are each of the following classified in *Bergey's Manual?* Why are they described more than once?
  (a) *Gardnerella*
  (b) Nocardioforms
  (c) *Halobacterium*
3. Identify the genus that best fits each description given below:
  (a) This gram-negative genus is well-suited to degrade hydrocarbons in an oil spill.
  (b) This gram-positive genus presents the greatest source of bacterial damage to the beekeeping industry.
  (c) This gram-negative genus is most probably able to oxidize arsenic and asbestos compounds and to clean up polluted water.
  (d) This group can produce a fuel used for home heating and for generating electricity.

FURTHER READING

Balows, A., H.G. Trüper, M. Dworkin, W. Harder, and K.H. Schliefer, eds. *The Prokaryotes: A Handbook on the Biology of Bacteria*, 2nd ed., 3 volumes. New York: Springer-Verlag, 1991. Discusses characteristics and provides information on cultivation, isolation, and identification of bacteria.

Baron, E.J., and S.M. Finegold. *Diagnostic Microbiology*, 8th ed. St. Louis: Mosby, 1990. A good reference for procedures used in the clinical microbiology laboratory.

Carr, N.G., and B.A. Whitton. *The Biology of Cyanobacteria. Botanical Monographs*, Vol. 19. Berkeley: University of California Press, 1982. A comprehensive reference on morphology, metabolism, and reproduction of cyanobacteria.

Goodfellow, M., M. Mordarski, and S. Williams. *Biology of the Actinomycetes*. New York: Academic Press, 1984. A survey of the biology, ecology, and pathogenicity of this diverse group of bacteria.

Harwood, C.S., and E. Canale-Parola. "Ecology of spirochetes." *Annual Review of Microbiology* 38:161–192, 1984. Describes the role of spirochetes in decomposition and in their associations with humans and other organisms.

Holt, J.G., ed. *Bergey's Manual of Determinative Bacteriology*, 9th ed. Baltimore: Williams & Wilkins, 1992. The standard reference for identification of bacteria.

Holt, J.G., ed. *Bergey's Manual of Systematic Bacteriology*, 1st ed., 4 volumes. Baltimore: Williams & Wilkins, 1984–1989. These four volumes are the standard reference for identification and classification of bacteria.

Kloos, W.E. "Natural populations of the genus *Staphylococcus*." *Annual Review of Microbiology* 34:559–592, 1980. Illustrates the use of modern technology to determine a species; includes a discussion of staphylococci that live on birds and mammals.

Maniloff, J. "Evolution of wall-less prokaryotes." *Annual Review of Microbiology* 37:477–499, 1983. Mycoplasmas and archaeobacteria are examples in this discussion of evolution; information comes from advances in molecular biology.

Shapiro, J.A. "Bacteria as multicellular organisms." *Scientific American* 258(6):82–89, June 1988. Describes bacterial colonies that function as multicellular organisms.

Starr, M.P., H. Stolp, H.G. Truper, A. Balows, and H.G. Schlegel. *The Prokaryotes: A Handbook on Habitats, Isolation and Identification of Bacteria*, 2nd ed., 2 volumes. New York: Springer-Verlag, 1989. Discusses characteristics and provides information on cultivation, isolation, and identification of bacteria.

BRIEF CONTENTS

PART ONE

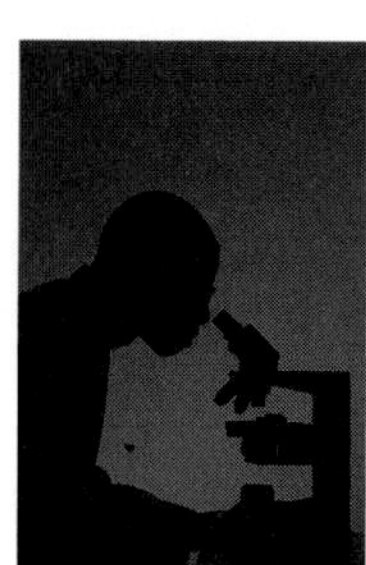

FUNDAMENTALS OF MICROBIOLOGY 1

1 The Microbial World and You 2
2 Chemical Principles 22
3 Observing Microorganisms Through a Microscope 52
4 Functional Anatomy of Procaryotic and Eucaryotic Cells 69
5 Microbial Metabolism 101
6 Microbial Growth 141
7 Control of Microbial Growth 167
8 Microbial Genetics 190
9 Recombinant DNA and Biotechnology 227

PART TWO

SURVEY OF THE MICROBIAL WORLD 249

10 Classification of Microorganisms 250
11 Bacteria 273
12 Fungi, Algae, Protozoans, and Multicellular Parasites 296
13 Viruses 332

PART THREE

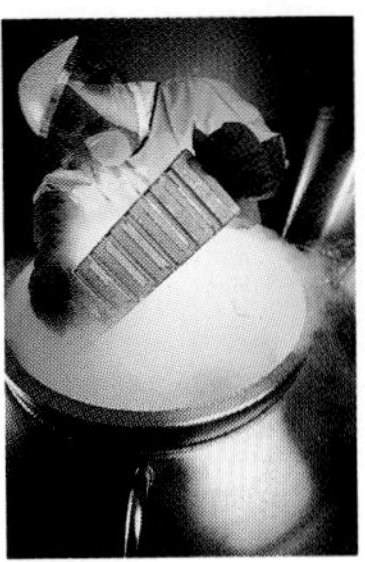

INTERACTION BETWEEN MICROBE AND HOST 367

14 Principles of Disease and Epidemiology 368
15 Mechanisms of Pathogenicity 392
16 Nonspecific Defenses of the Host 407
17 Specific Defenses of the Host: The Immune Response 426
18 Practical Applications of Immunology 449
19 Disorders Associated with the Immune System 467
20 Antimicrobial Drugs 492

PART FOUR

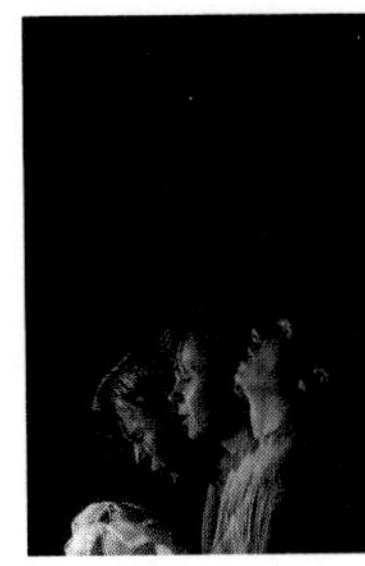

MICROORGANISMS AND HUMAN DISEASE 517

21 Microbial Diseases of the Skin and Eyes 518
22 Microbial Diseases of the Nervous System 539
23 Microbial Diseases of the Cardiovascular System 560
24 Microbial Diseases of the Respiratory System 589
25 Microbial Diseases of the Digestive System 617
26 Microbial Diseases of the Urinary and Reproductive Systems 650

PART FIVE

APPLIED MICROBIOLOGY AND THE ENVIRONMENT 671

27 Soil and Water Microbiology 672
28 Applied and Industrial Microbiology 700

APPENDICES

A Classification of Bacteria According to *Bergey's Manual of Systematic Bacteriology* 723
B Word Roots Used in Microbiology 729
C Most Probable Numbers (MPN) Table 733
D Methods for Taking Clinical Samples 734
E Biochemical Pathways 736
F Exponents, Exponential Notation, Logarithms, and Generation Time 742

PRONUNCIATION OF SCIENTIFIC NAMES 744
GLOSSARY 749
PHOTOGRAPH ACKNOWLEDGMENTS 770
INDEX 773

CONTENTS

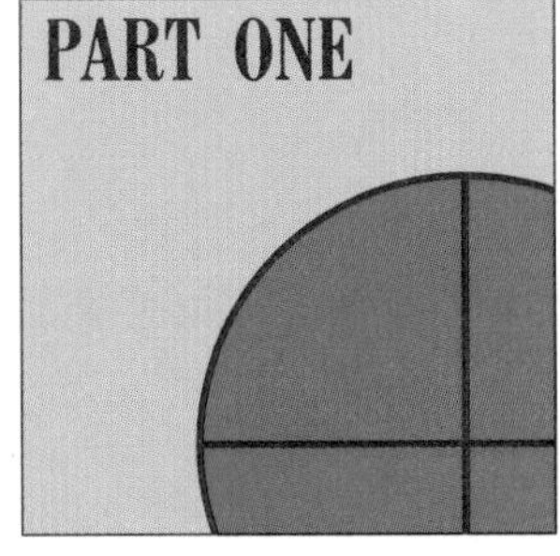

PART ONE

FUNDAMENTALS OF MICROBIOLOGY 1

1 The Microbial World and You 2

Microbes in Our Lives 4
A Brief History of Microbiology 6
THE FIRST OBSERVATIONS 6
THE DEBATE OVER SPONTANEOUS GENERATION 6
THE GOLDEN AGE OF MICROBIOLOGY 8
THE BIRTH OF MODERN CHEMOTHERAPY: DREAMS OF A "MAGIC BULLET" 11
MODERN DEVELOPMENTS IN MICROBIOLOGY 13
Naming and Classifying Microorganisms 14
The Diversity of Microorganisms 14
BACTERIA 14
FUNGI 15
PROTOZOANS 15
ALGAE 15

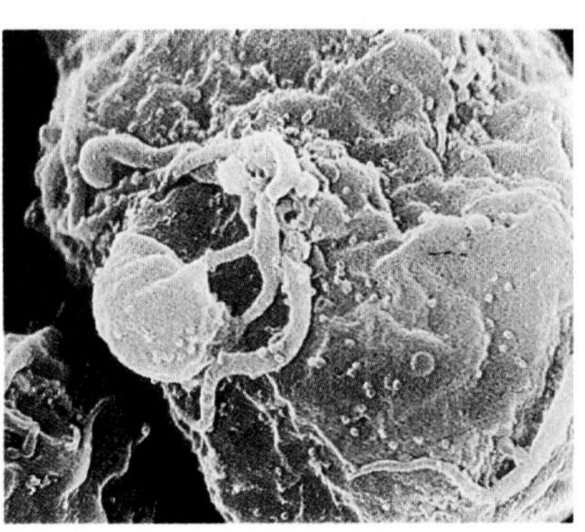

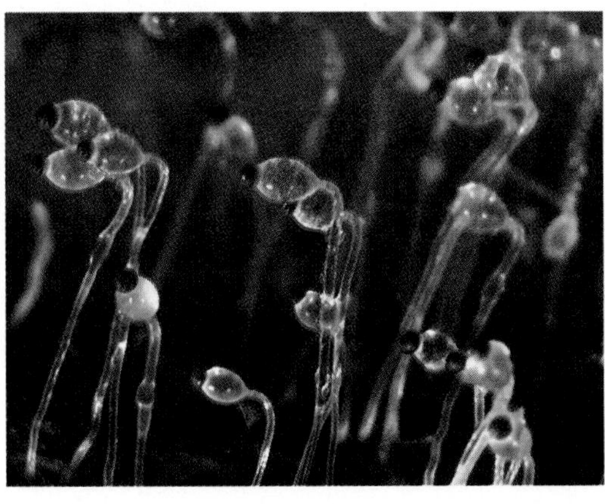

VIRUSES 15
MULTICELLULAR ANIMAL PARASITES 15
Microbes and Human Welfare 17
RECYCLING VITAL ELEMENTS 17
SEWAGE TREATMENT: USING MICROBES TO RECYCLE WATER 17
USING MICROBES TO CLEAN UP TOXIC DUMPS 17
INSECT PEST CONTROL BY MICROORGANISMS 17
MODERN INDUSTRIAL MICROBIOLOGY AND GENETIC ENGINEERING 18
Microbes and Human Disease 18
Study Outline 19
Study Questions 20
Further Reading 21
Microbiology Highlights: What Makes Sourdough Bread Different? 5

2 Chemical Principles 22

Structure of Atoms 22
- CHEMICAL ELEMENTS 23
- ELECTRONIC CONFIGURATIONS 24

How Atoms Form Molecules: Chemical Bonds 24
- IONIC BONDS 26
- COVALENT BONDS 27
- HYDROGEN BONDS 27
- MOLECULAR WEIGHT AND MOLES 27

Chemical Reactions 29
- ENERGY OF CHEMICAL REACTIONS 29
- SYNTHESIS REACTIONS 29
- DECOMPOSITION REACTIONS 29
- EXCHANGE REACTIONS 30
- THE REVERSIBILITY OF CHEMICAL REACTIONS 30
- HOW CHEMICAL REACTIONS OCCUR 30

IMPORTANT BIOLOGICAL MOLECULES 31

Inorganic Compounds 32
- WATER 32
- ACIDS, BASES, AND SALTS 33
- ACID–BASE BALANCE 33

Organic Compounds 34
- FUNCTIONAL GROUPS 35
- MACROMOLECULES 36
- CARBOHYDRATES 36
- LIPIDS 37
- PROTEINS 41
- NUCLEIC ACIDS 46
- ADENOSINE TRIPHOSPHATE (ATP) 47

Study Outline 48

Study Questions 50

Further Reading 51

Microbiology in the News: Bacterial Banqueters Attend Oil Spill 38

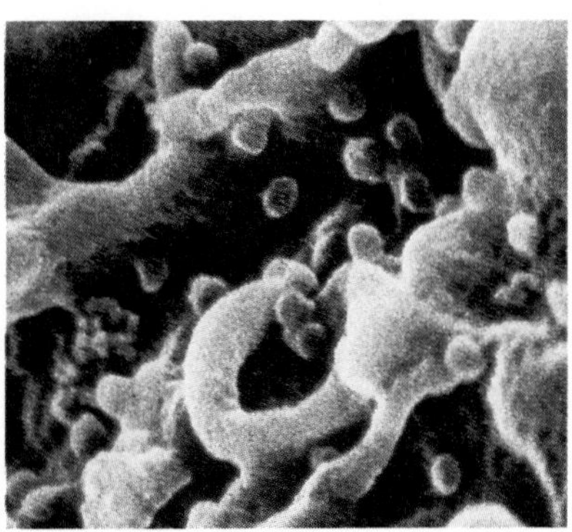

3 Observing Microorganisms Through a Microscope 52

Units of Measurement 52

Microscopy: The Instruments 53
- COMPOUND LIGHT MICROSCOPY 53
- DARKFIELD MICROSCOPY 55
- PHASE-CONTRAST MICROSCOPY 57
- FLUORESCENCE MICROSCOPY 57
- ELECTRON MICROSCOPY 58

Preparation of Specimens for Light Microscopy 60
- PREPARING SMEARS AND STAINING 60
- SIMPLE STAINS 60
- DIFFERENTIAL STAINS 63
- SPECIAL STAINS 65

Study Outline 66

Study Questions 67

Further Reading 68

Microbiology Highlights: *Bdellovibrio*, Predator Extraordinaire 62

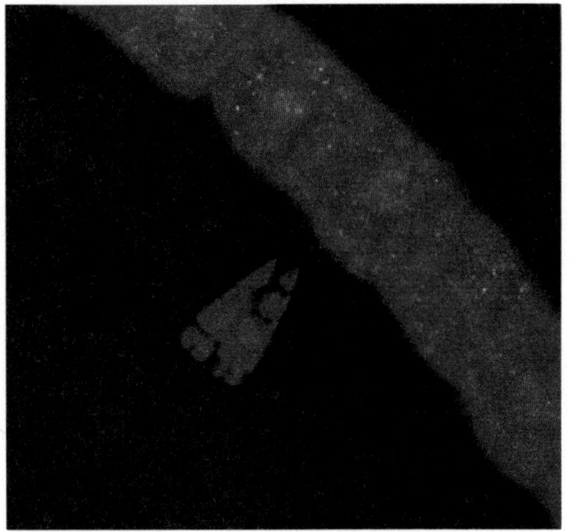

4 Functional Anatomy of Procaryotic and Eucaryotic Cells 69

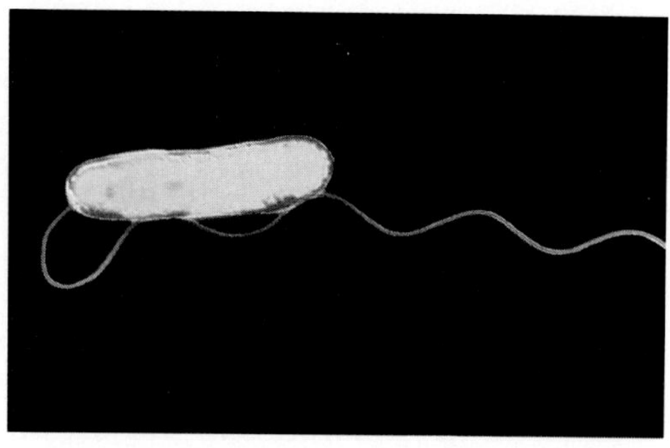

The Procaryotic Cell 71
Size, Shape, and Arrangement of Bacterial Cells 71
Structures External to the Cell Wall 72
GLYCOCALYX 72
FLAGELLA 73
AXIAL FILAMENTS 75
FIMBRIAE AND PILI 76
The Cell Wall 77
COMPOSITION AND CHARACTERISTICS 77
ATYPICAL CELL WALLS 79
DAMAGE TO THE CELL WALL 80
Structures Internal to the Cell Wall 81
PLASMA (CYTOPLASMIC) MEMBRANE 81
MOVEMENT OF MATERIALS ACROSS MEMBRANES 83
CYTOPLASM 86
NUCLEAR AREA 86
RIBOSOMES 86
INCLUSIONS 87
ENDOSPORES 88
The Eucaryotic Cell 89
Flagella and Cilia 90
The Cell Wall and Glycocalyx 90
The Plasma (Cytoplasmic) Membrane 90
Cytoplasm 91
Organelles 92
NUCLEUS 92
ENDOPLASMIC RETICULUM 93
RIBOSOMES 93
GOLGI COMPLEX 93
MITOCHONDRIA 93
CHLOROPLASTS 95
LYSOSOMES 95
CENTRIOLES 95
Evolution of Eucaryotes 95
Study Outline 96
Study Questions 99
Further Reading 100
Microbiology Highlights: Microbial Magnets 76

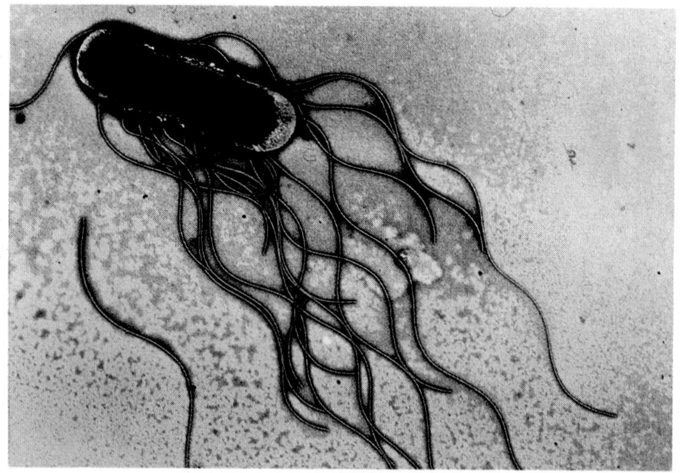

5 Microbial Metabolism 101

Catabolic and Anabolic Reactions 101
Enzymes 102
NAMING ENZYMES 103
ENZYME COMPONENTS 103
MECHANISM OF ENZYMATIC ACTION 104
FACTORS INFLUENCING ENZYMATIC ACTIVITY 105
FEEDBACK INHIBITION 108
Energy Production 109
OXIDATION-REDUCTION 109
GENERATION OF ATP 110
Biochemical Pathways of Energy Production 110
Carbohydrate Catabolism 111
GLYCOLYSIS 111
ALTERNATIVES TO GLYCOLYSIS 114
DEFINITION OF RESPIRATION 114
THE KREBS CYCLE 114
ELECTRON TRANSPORT CHAIN 116
CHEMIOSMOTIC MECHANISM OF ATP GENERATION 117
SUMMARY OF AEROBIC RESPIRATION 118
ANAEROBIC RESPIRATION 120
FERMENTATION 120
LIPID CATABOLISM 122
PROTEIN CATABOLISM 124

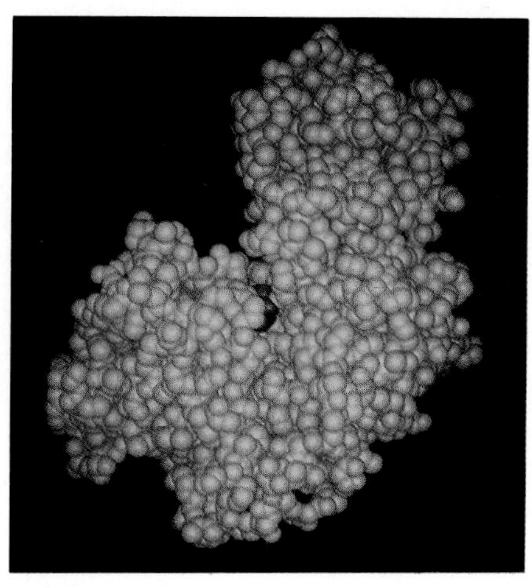

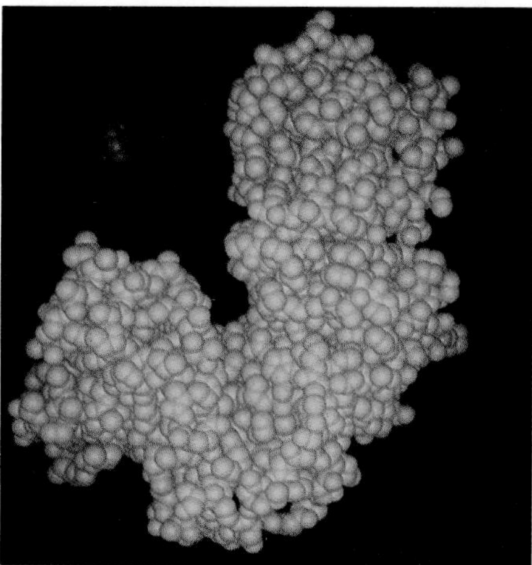

Photosynthesis 125
THE LIGHT REACTIONS: PHOTOPHOSPHORYLATION 125
THE DARK REACTIONS: CALVIN–BENSON CYCLE 125
NUTRITIONAL PATTERNS AMONG ORGANISMS 125
SUMMARY OF ENERGY PRODUCTION MECHANISMS 130
Biochemical Pathways of Energy Utilization (Anabolism) 131
BIOSYNTHESIS OF POLYSACCHARIDES 131
BIOSYNTHESIS OF LIPIDS 131
BIOSYNTHESIS OF AMINO ACIDS AND PROTEINS 132
BIOSYNTHESIS OF PURINES AND PYRIMIDINES 132
Integration of Metabolism 132
Study Outline 135
Study Questions 138
Further Reading 140
Microbiology Highlights: What Is Fermentation? 120
Microbiology Highlights: Photosynthesis Without Chlorophyll 128

6 Microbial Growth 141

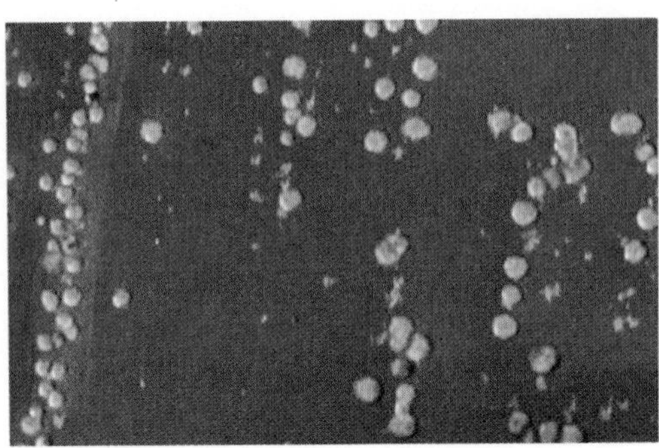

Requirements for Growth 141
PHYSICAL REQUIREMENTS 142
CHEMICAL REQUIREMENTS 145
Culture Media 148
CHEMICALLY DEFINED MEDIA 149
COMPLEX MEDIA 150
ANAEROBIC GROWTH MEDIA AND METHODS 150
SPECIAL CULTURE TECHNIQUES 151
SELECTIVE AND DIFFERENTIAL MEDIA 152
ENRICHMENT CULTURE 153
OBTAINING PURE CULTURES 153
Preserving Bacterial Cultures 155
Growth of Bacterial Cultures 155
BACTERIAL DIVISION 155
GENERATION TIME 156
LOGARITHMIC REPRESENTATION OF BACTERIAL POPULATIONS 156
PHASES OF GROWTH 157
MEASUREMENT OF MICROBIAL GROWTH 158
ESTIMATIONS OF BACTERIAL NUMBERS BY INDIRECT METHODS 162
Study Outline 163
Study Questions 164
Further Reading 166
Microbiology Highlights: Hydrothermal Bacteria 143

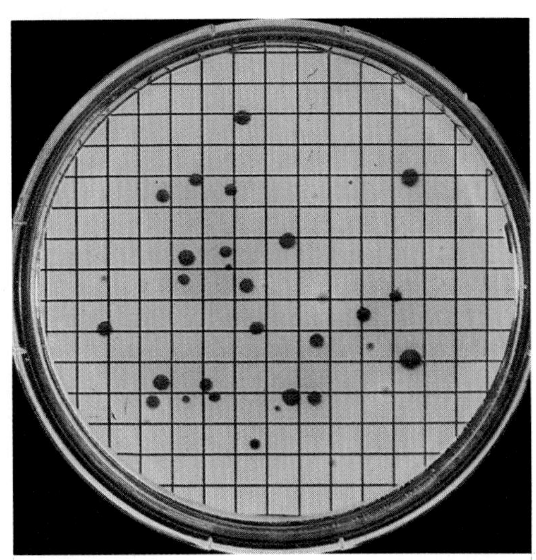

7 Control of Microbial Growth 167

Conditions Influencing Microbial Control 168
TEMPERATURE 169
TYPES OF MICROBES 169
PHYSIOLOGICAL STATE OF THE MICROBE 169
ENVIRONMENT 169
Actions of Microbial Control Agents 169
ALTERATION OF MEMBRANE PERMEABILITY 169
DAMAGE TO PROTEINS AND NUCLEIC ACIDS 169
Rate of Microbial Death 170
Physical Methods of Microbial Control 170
HEAT 170
FILTRATION 173
LOW TEMPERATURE 173
DESICCATION 174
OSMOTIC PRESSURE 175
RADIATION 175
Chemical Methods of Microbial Control 176
PRINCIPLES OF EFFECTIVE DISINFECTION 176
EVALUATING A DISINFECTANT 177
TYPES OF DISINFECTANTS 177
Study Outline 185
Study Questions 187
Further Reading 189
MMWR: Contaminated Povidone—Iodine Solution—Texas 180

8 Microbial Genetics 190

STRUCTURE AND FUNCTION OF THE GENETIC MATERIAL 190
Genotype and Phenotype 191
DNA and Chromosomes 191
DNA Replication 192
THE RATE OF DNA REPLICATION 197
THE FLOW OF GENETIC INFORMATION 197
RNA and Protein Synthesis 197
TRANSCRIPTION 197
TRANSLATION 199
The Genetic Code 202
REGULATION OF GENE EXPRESSION IN BACTERIA 203
Repression, Induction, and Attenuation 203
REPRESSION AND INDUCTION 203
OPERON MODEL 203
ATTENUATION 206
MUTATION: CHANGE IN THE GENETIC MATERIAL 207
Types of Mutations 207
CHEMICAL MUTAGENS 208
RADIATION 210
Frequency of Mutation 210
Identifying Mutants 211
Identifying Chemical Carcinogens 212
GENETIC TRANSFER AND RECOMBINATION 214
Transformation in Bacteria 214
Conjugation in Bacteria 216
Transduction in Bacteria 218
Recombination in Eucaryotes 218
PLASMIDS 220
TRANSPOSONS 221
Genes and Evolution 222
Study Outline 222
Study Questions 225
Further Reading 226
Microbiology Highlights: Stress Brings Out Hidden Talents—Even in Bacteria 205

9 Recombinant DNA and Biotechnology 227

The Advent of Recombinant-DNA Technology 227
Overview of Recombinant-DNA Procedures 228
Restriction Enzymes 231
Cloning a Gene with a Plasmid Vector 231
Alternative Ways To Get Foreign DNA into Cells 232
Sources of DNA 233
GENE LIBRARIES 233
SYNTHETIC DNA 234
Identifying the Desired Clone in a Library 234
Making a Gene Product 235
Genetically Engineered Products for Medical Therapy 236
Obtaining Information from DNA for Basic Research and Medical Applications 236
Applications of Recombinant-DNA Technology to Agriculture 241
Safety Issues 243
The Future of Genetic Engineering 243
Study Outline 243
Study Questions 245
Further Reading 246
Microbiology in the News: Bioremediation of PCBs 240

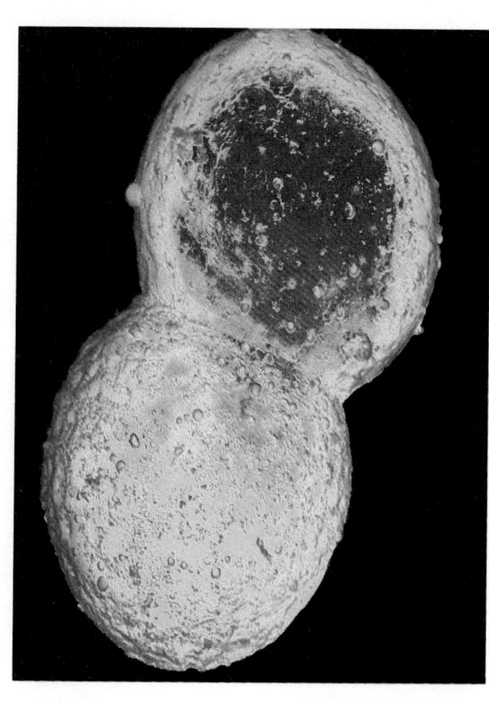

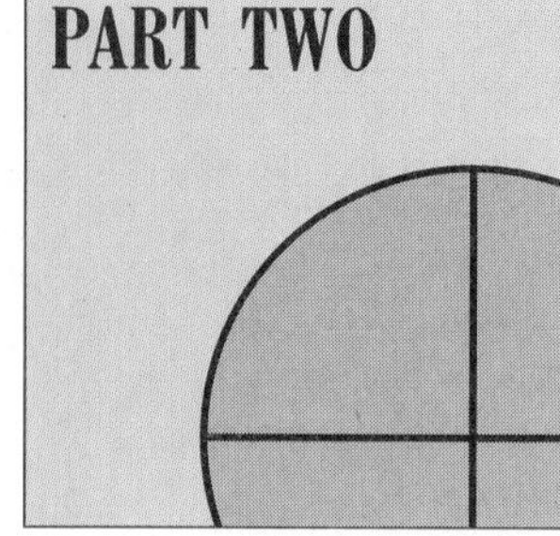

PART TWO SURVEY OF THE MICROBIAL WORLD 249

10 Classification of Microorganisms 250

Phylogenetic Relationships 250
- A FIVE-KINGDOM SYSTEM 251
- A THREE-KINGDOM SYSTEM 251

Classification of Organisms 254
- SCIENTIFIC NOMENCLATURE 254
- THE TAXONOMIC HIERARCHY 255
- A PHYLOGENETIC HIERARCHY 256
- CLASSIFICATION OF BACTERIA 257
- CLASSIFICATION OF VIRUSES 257

Criteria for Classification and Identification of Microorganisms 257
- MORPHOLOGICAL CHARACTERISTICS 258
- DIFFERENTIAL STAINING 259
- BIOCHEMICAL TESTS 259
- SEROLOGY 260
- PHAGE TYPING 261
- AMINO ACID SEQUENCING 262
- PROTEIN ANALYSIS 262
- BASE COMPOSITION OF NUCLEIC ACIDS 265
- NUCLEIC ACID HYBRIDIZATION 265
- FLOW CYTOMETRY 267
- GENETIC RECOMBINATION 267
- NUMERICAL TAXONOMY 268

Study Outline 270

Study Questions 271

Further Reading 271

Microbiology in the News: Mass Deaths of Dolphins Worry Experts 264

11 Bacteria 273

Bacterial Groups 273
- SPIROCHETES 275
- AEROBIC/MICROAEROPHILIC, MOTILE, HELICAL/VIBRIOID GRAM-NEGATIVE BACTERIA 276
- GRAM-NEGATIVE AEROBIC RODS AND COCCI 277
- FACULTATIVELY ANAEROBIC GRAM-NEGATIVE RODS 280
- ANAEROBIC, GRAM-NEGATIVE, STRAIGHT, CURVED, AND HELICAL RODS 282
- DISSIMILATORY SULFATE-REDUCING OR SULFUR-REDUCING BACTERIA 282
- ANAEROBIC GRAM-NEGATIVE COCCI 283
- RICKETTSIAS AND CHLAMYDIAS 283
- MYCOPLASMAS 285
- GRAM-POSITIVE COCCI 285
- ENDOSPORE-FORMING GRAM-POSITIVE RODS AND COCCI 286
- REGULAR NONSPORING GRAM-POSITIVE RODS 287
- IRREGULAR NONSPORING GRAM-POSITIVE RODS 288
- MYCOBACTERIA 288
- NOCARDIOFORMS 288
- GLIDING, SHEATHED, AND BUDDING AND/OR APPENDAGED BACTERIA 289
- CHEMOAUTOTROPHIC BACTERIA 290
- ARCHAEOBACTERIA 291
- PHOTOTROPHIC BACTERIA 291
- ACTINOMYCETES 293

Study Outline 294

Study Questions 294

Further Reading 295

Microbiology Highlights: Why Microbiologists Study Termites 278

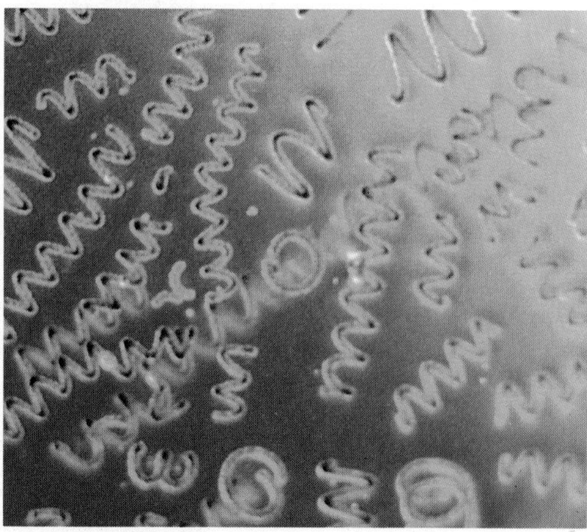

12 Fungi, Algae, Protozoans, and Multicellular Parasites 296

FUNGI 296
Characteristics of Fungi 297
VEGETATIVE STRUCTURES 297
REPRODUCTIVE STRUCTURES 299
NUTRITIONAL ADAPTATIONS 301
Medically Important Phyla of Fungi 302
DEUTEROMYCOTA 302
ZYGOMYCOTA 304
ASCOMYCOTA 304
BASIDIOMYCOTA 304
Fungal Diseases 304
Economic Effects of Fungi 306

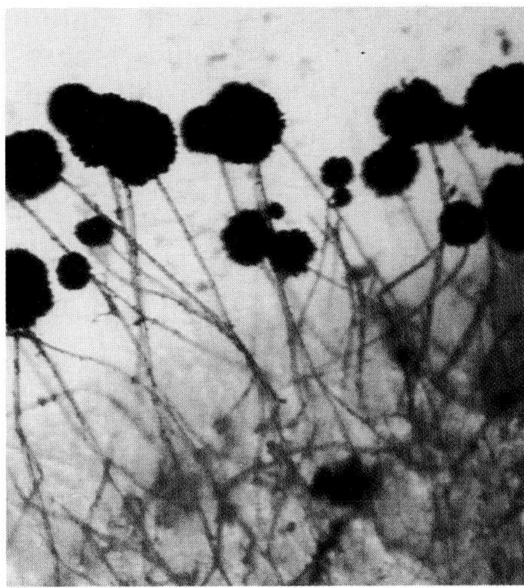

ALGAE 306
Phyla of Algae 307
Structure and Reproduction 310
Roles of Algae in Nature 310
LICHENS 310
SLIME MOLDS 312
PROTOZOANS 313
Protozoan Biology 313
NUTRITION 314
REPRODUCTION 314
ENCYSTMENT 315
Medically Important Phyla of Protozoans 315
SARCODINA 315
MASTIGOPHORA 315
CILIATA 316
SPOROZOA 316
HELMINTHS 319
Helminth Biology 319
REPRODUCTION 319
LIFE CYCLE 319
Platyhelminthes 319
TREMATODES 319
CESTODES 320
Aschelminthes 321
NEMATODES 322
Arthropods as Vectors 325
Study Outline 328
Study Questions 330
Further Reading 331
MMWR: *Aedes albopictus*—A New Vector 324

13 Viruses 332

General Characteristics of Viruses 333
HOST RANGE 334
SIZE 335
Viral Structure 335
NUCLEIC ACID 335
CAPSID AND ENVELOPE 335
GENERAL MORPHOLOGY 336
Taxonomy of Viruses 339
Isolation, Cultivation, and Identification of Viruses 340
GROWTH OF BACTERIOPHAGES IN THE LABORATORY 341
GROWTH OF ANIMAL VIRUSES IN THE LABORATORY 341
VIRAL IDENTIFICATION 343
Viral Multiplication 345
MULTIPLICATION OF BACTERIOPHAGES 345
MULTIPLICATION OF ANIMAL VIRUSES 348
Effects of Animal Viral Infection on Host Cells 355
Viruses and Cancer 356
TRANSFORMATION OF NORMAL CELLS INTO TUMOR CELLS 357
ACTIVATION OF ONCOGENES 358
DNA-CONTAINING ONCOGENIC VIRUSES 358
RNA-CONTAINING ONCOGENIC VIRUSES 359
Latent Viral Infections 359
Slow Viral Infections 359
Unconventional Agents of Disease 360
Plant Viruses and Viroids 360
Study Outline 362
Study Questions 364
Further Reading 365
MMWR: AIDS: The Risk to Health Care Workers 344

PART THREE INTERACTION BETWEEN MICROBE AND HOST 367

14 Principles of Disease and Epidemiology 368

Pathology, Infection, and Disease 369
Normal Flora 369
RELATIONSHIPS BETWEEN THE NORMAL FLORA AND HOST 370
OPPORTUNISTIC ORGANISMS 370
Etiology of Infectious Diseases 371
KOCH'S POSTULATES 371
EXCEPTIONS TO KOCH'S POSTULATES 372
Classifying Infectious Diseases 373
OCCURRENCE OF DISEASE 373
SEVERITY OR DURATION OF DISEASE 374
EXTENT OF HOST INVOLVEMENT 374
Spread of Infection 374
RESERVOIRS 374
TRANSMISSION OF DISEASE 375
PORTALS OF EXIT 379
Nosocomial (Hospital-Acquired) Infections 380
MICROORGANISMS IN THE HOSPITAL 380
COMPROMISED HOST 382
CHAIN OF TRANSMISSION 382
CONTROL OF NOSOCOMIAL INFECTIONS 383

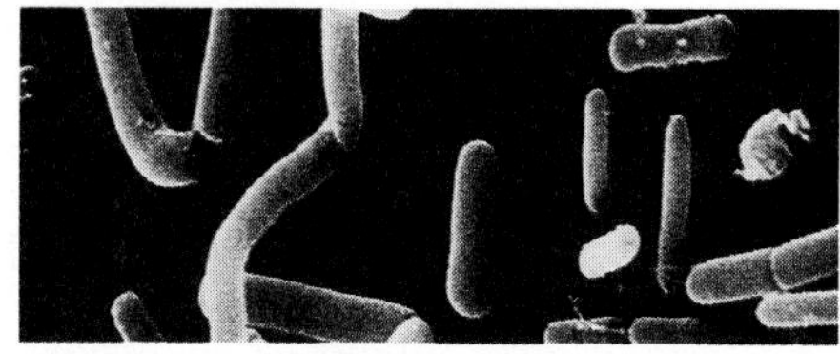

Patterns of Disease 383
PREDISPOSING FACTORS 383
DEVELOPMENT OF DISEASE 383
Epidemiology 384
DESCRIPTIVE EPIDEMIOLOGY 386
ANALYTICAL EPIDEMIOLOGY 386
EXPERIMENTAL EPIDEMIOLOGY 386
CASE REPORTING 386
CENTERS FOR DISEASE CONTROL (CDC) 386
Study Outline 387
Study Questions 389
Further Reading 391
MMWR: Postsurgical Infections Associated with Breaks in Aseptic Techniques—California, Illinois, Maine, and Michigan, 1990 381

15 Mechanisms of Pathogenicity 392

Entry of a Microorganism into the Host 392
- MUCOUS MEMBRANES 392
- SKIN 393
- PARENTERAL ROUTE 393
- PREFERRED PORTAL OF ENTRY 393
- NUMBERS OF INVADING MICROBES 393
- ADHERENCE 393

How Pathogens Penetrate Host Defenses 395
- CAPSULES 395
- COMPONENTS OF THE CELL WALL 396
- ENZYMES 396

Damage to Host Cells 397
- DIRECT DAMAGE 397
- TOXINS 397

Plasmids, Lysogeny, and Pathogenicity 401

Pathogenic Properties of Nonbacterial Microorganisms 402
- VIRUSES 402
- FUNGI, PROTOZOANS, HELMINTHS, AND ALGAE 403

Study Outline 404

Study Questions 406

Further Reading 406

Microbiology Highlights: Are Ulcers an Infectious Disease? 398

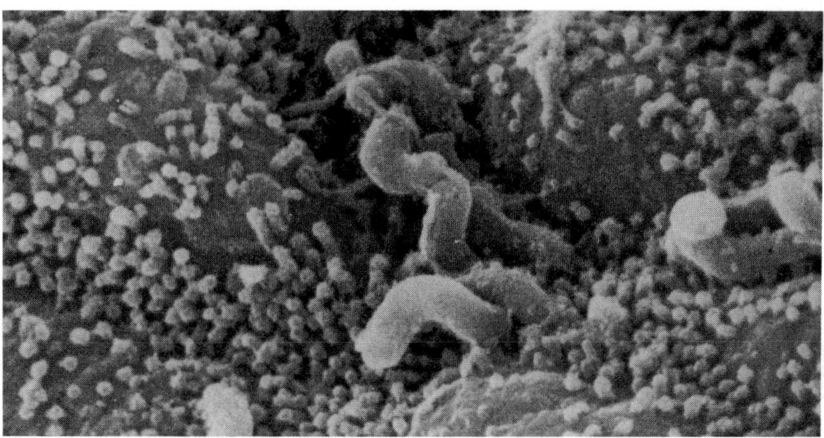

16 Nonspecific Defenses of the Host 407

Skin and Mucous Membranes 407
- MECHANICAL FACTORS 408
- CHEMICAL FACTORS 409

Phagocytosis 410
- FORMED ELEMENTS IN BLOOD 411
- ACTIONS OF PHAGOCYTIC CELLS 413
- MECHANISM OF PHAGOCYTOSIS 414

Inflammation 415
- VASODILATION AND INCREASED PERMEABILITY OF BLOOD VESSELS 415
- PHAGOCYTE MIGRATION 416
- REPAIR 417

Fever 417

Antimicrobial Substances 418
- THE COMPLEMENT SYSTEM 418
- INTERFERONS (IFNs) 421

Study Outline 423

Study Questions 424

Further Reading 425

Microbiology Highlights: Neutrophil Defect Leads to Periodontal Disease 410

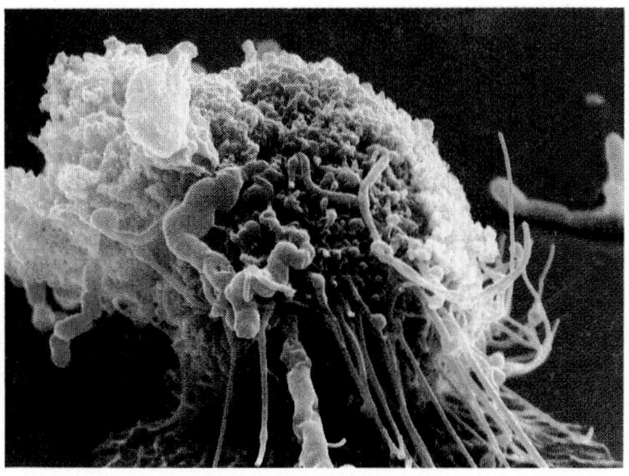

17 Specific Defenses of the Host: The Immune Response 426

Types of Acquired Immunity 427
NATURALLY ACQUIRED IMMUNITY 427
ARTIFICIALLY ACQUIRED IMMUNITY 427
The Duality of the Immune System 428
THE HUMORAL IMMUNE SYSTEM 428
THE CELL-MEDIATED IMMUNE SYSTEM 429
Antigens and Antibodies 429
THE NATURE OF ANTIGENS 429
THE NATURE OF ANTIBODIES 430
B Cells and Humoral Immunity 433
B-CELL AND ANTIGEN INTERACTIONS 433
ANTIBODY–ANTIGEN BINDING 435
MOLECULAR BASIS OF THE DIVERSITY OF ANTIGEN RECEPTORS 436
MONOCLONAL ANTIBODIES 436
T Cells and Cell-Mediated Immunity 439
THE COMPONENTS OF CELL-MEDIATED IMMUNITY 439
THE CELL-MEDIATED IMMUNE RESPONSE 441
Clonal Selection 443
Immunological Memory 443
Study Outline 445
Study Questions 447
Further Reading 448
Microbiology in the News: Parasite Outwits the Immune System 440

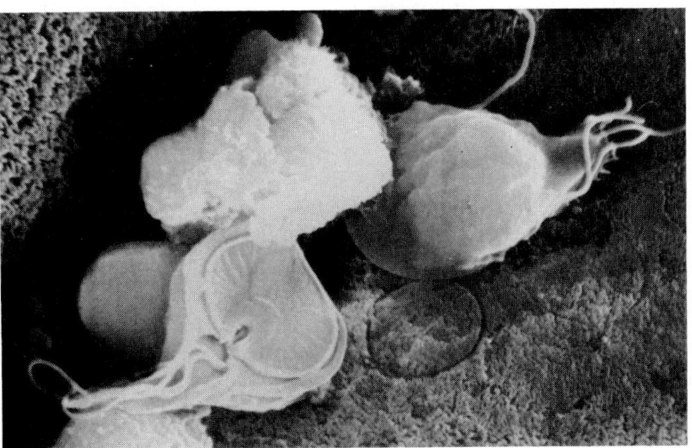

18 Practical Applications of Immunology 449

Vaccines 449
CHARACTERISTICS OF VACCINES 450
NEW VACCINE DEVELOPMENT 451
Diagnostic Immunology 453
PRECIPITATION REACTIONS 453
AGGLUTINATION REACTIONS 456
COMPLEMENT FIXATION REACTIONS 457
NEUTRALIZATION REACTIONS 458
IMMUNOFLUORESCENCE AND FLUORESCENT-ANTIBODY TECHNIQUES 461
ENZYME-LINKED IMMUNOSORBENT ASSAY (ELISA) 463
The Future of Vaccines and Diagnostic Immunology 463
Study Outline 464
Study Questions 465
Further Reading 466
Microbiology in the News: Molecular Biology: The Latest Weapon Against Malaria 454

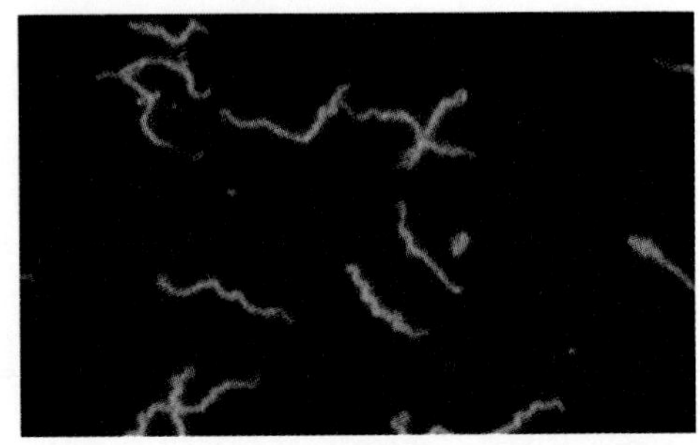

19 Disorders Associated with the Immune System 467

Hypersensitivity 467
Type I (Anaphylaxis) Reactions 468
- SYSTEMIC ANAPHYLAXIS 469
- LOCALIZED ANAPHYLAXIS 470
- PREVENTION OF ANAPHYLACTIC REACTIONS 470

Type II (Cytotoxic) Reactions 471
- THE ABO BLOOD GROUP SYSTEM 471
- THE RH BLOOD GROUP SYSTEM 471
- DRUG-INDUCED CYTOTOXIC REACTIONS 473
- AUTOIMMUNE DISORDERS (TYPE II REACTIONS) 474

Type III (Immune Complex) Reactions 474
- AUTOIMMUNE DISORDERS (TYPE III REACTIONS) 475

Type IV (Cell-Mediated) Reactions 475
- CAUSES OF TYPE IV REACTIONS 475
- CELL-MEDIATED HYPERSENSITIVITY REACTIONS OF THE SKIN 475
- AUTOIMMUNE DISORDERS (TYPE IV REACTIONS) 476
- LOSS OF IMMUNOLOGICAL TOLERANCE 476
- MAJOR HISTOCOMPATIBILITY COMPLEX 476

Transplantation 478
- TYPES OF TRANSPLANTS 478
- IMMUNOSUPPRESSION 480

Natural Immune Deficiencies 480
Acquired Immunodeficiency Syndrome (AIDS) 480
- THE ORIGIN OF AIDS 480
- THE HIV INFECTION 480
- MODES OF TRANSMISSION 482
- VACCINES AND TREATMENTS 486
- THE FUTURE OF AIDS 486

Immune Response to Cancer 486
- IMMUNOLOGICAL SURVEILLANCE 486
- IMMUNOLOGICAL ESCAPE 486
- IMMONOTHERAPY 487

Study Outline 488
Study Questions 490
Further Reading 491
Microbiology in the News: New Weapons Against AIDS 484

20 Antimicrobial Drugs 492

Historical Development 492
Criteria for Evaluating Antimicrobial Drugs 494
Spectrum of Activity 494
Action of Antimicrobial Drugs 496
- INHIBITION OF CELL WALL SYNTHESIS 496
- INHIBITION OF PROTEIN SYNTHESIS 497
- INJURY TO THE PLASMA MEMBRANE 498
- INHIBITION OF NUCLEIC ACID SYNTHESIS 498
- INHIBITION OF ENZYMATIC ACTIVITY 498

Survey of Commonly Used Antimicrobial Drugs 499
- ANTIBACTERIAL SYNTHETICS 499
- ANTIFUNGAL SYNTHETICS 502
- ANTIBIOTICS 502
- ANTIFUNGAL DRUGS 506
- ANTIVIRAL DRUGS 507
- ANTIPROTOZOAN AND ANTIHELMINTHIC DRUGS 509

Tests for Microbial Susceptibility to Chemotherapeutic Agents 509
- DISK-DIFFUSION METHOD 510
- BROTH DILUTION TESTS 510

Effectiveness of Chemotherapeutic Agents 510
- DRUG RESISTANCE 510
- EFFECTS OF COMBINATIONS OF DRUGS 511
- THE FUTURE OF CHEMOTHERAPEUTIC AGENTS 511

Study Outline 512
Study Questions 514
Further Reading 515
Microbiology in the News: Antibiotics in Animal Feed Linked to Human Disease 495

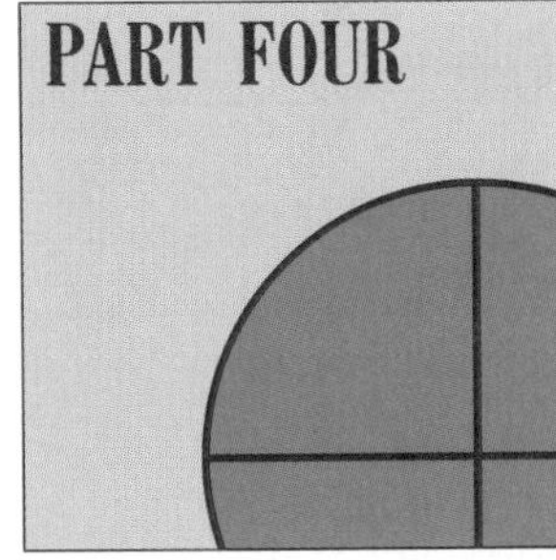

PART FOUR MICROORGANISMS AND HUMAN DISEASE 517

21 Microbial Diseases of the Skin and Eyes 518

Structure and Function of the Skin 518
Normal Flora of the Skin 519
DISEASES OF THE SKIN 520
Bacterial Diseases of the Skin 520
STAPHYLOCOCCAL SKIN INFECTIONS 520
STREPTOCOCCAL SKIN INFECTIONS 522
INFECTIONS BY PSEUDOMONADS 523
ACNE 523
Viral Diseases of the Skin 524
WARTS 524
SMALLPOX (VARIOLA) 524
CHICKENPOX (VARICELLA) AND SHINGLES (HERPES ZOSTER) 525
HERPES SIMPLEX 526
MEASLES (RUBEOLA) 527
RUBELLA 528
Fungal Diseases of the Skin 530
CUTANEOUS MYCOSES 530
SUBCUTANEOUS MYCOSES 532
CANDIDIASIS 533
DISEASES OF THE EYE 533
Conjunctivitis 533
Neonatal Gonorrheal Ophthalmia 533
Inclusion Conjunctivitis 533
Trachoma 533
Herpetic Keratitis 534
Acanthamoeba Keratitis 534
Study Outline 535
Study Questions 537
Further Reading 538
MMWR: Scabies in Health Care Facilities, Iowa 529

22 Microbial Diseases of the Nervous System 539

Structure and Function of the Nervous System 539
Bacterial Diseases of the Nervous System 541
BACTERIAL MENINGITIS 541
LISTERIOSIS 543
TETANUS 543
BOTULISM 544
LEPROSY 546
Viral Diseases of the Nervous System 547
POLIOMYELITIS 547
RABIES 549
ARTHROPOD-BORNE ENCEPHALITIS 551
Fungal Disease of the Nervous System 553
CRYPTOCOCCUS NEOFORMANS MENINGITIS (CRYPTOCOCCOSIS) 553
Protozoan Diseases of the Nervous System 553
AFRICAN TRYPANOSOMIASIS 554
NAEGLERIA MICROENCEPHALITIS 554
Nervous System Diseases Caused by Unconventional Agents 554

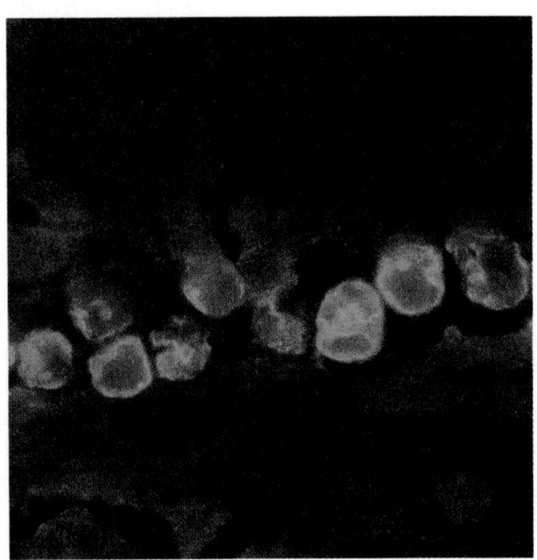

Study Outline 556
Study Questions 558
Further Reading 559
MMWR: Human Rabies in San Francisco 550

23 Microbial Diseases of the Cardiovascular System 560

Structure and Function of the Cardiovascular System 560
Structure and Function of the Lymphatic System 561
Bacterial Diseases of the Cardiovascular System 562
SEPTICEMIA 562
PUERPERAL SEPSIS 562
BACTERIAL ENDOCARDITIS 563
RHEUMATIC FEVER 563
TULAREMIA 564
BRUCELLOSIS (UNDULANT FEVER) 565
ANTHRAX 566
GANGRENE 566
SYSTEMIC DISEASES CAUSED BY BITES AND SCRATCHES 567
PLAGUE 568
RELAPSING FEVER 569
LYME DISEASE (LYME BORRELIOSIS) 569
TYPHUS 570
Viral Diseases of the Cardiovascular System 573
BURKITT'S LYMPHOMA 573
INFECTIOUS MONONUCLEOSIS 574
YELLOW FEVER 575
DENGUE 575
Protozoan Diseases of the Cardiovascular System 575
TOXOPLASMOSIS 575
AMERICAN TRYPANOSOMIASIS (CHAGAS' DISEASE) 577
MALARIA 578
Helminthic Diseases of the Cardiovascular System 581
SCHISTOSOMIASIS 582
SWIMMER'S ITCH 583
Study Outline 583
Study Questions 586
Further Reading 588
MMWR: Babesiosis–Connecticut 578

24 Microbial Diseases of the Respiratory System 589

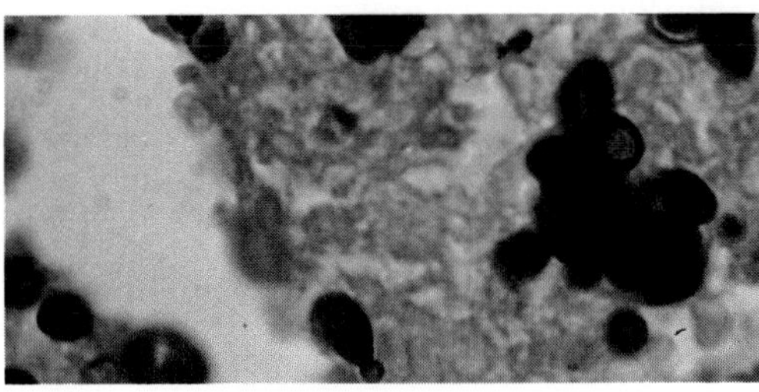

Structure and Function of the Respiratory System 589
Normal Flora of the Respiratory System 590
DISEASES OF THE UPPER RESPIRATORY SYSTEM 591
Bacterial Diseases of the Upper Respiratory System 592
STREPTOCOCCAL PHARYNGITIS (STREP THROAT) 592
SCARLET FEVER 592
DIPHTHERIA 593
CUTANEOUS DIPHTHERIA 594
OTITIS MEDIA 594
Viral Disease of the Upper Respiratory System 594
COMMON COLD (ACUTE CORYZA) 594
DISEASES OF THE LOWER RESPIRATORY SYSTEM 595
Bacterial Diseases of the Lower Respiratory System 595
WHOOPING COUGH (PERTUSSIS) 595
TUBERCULOSIS 596
BACTERIAL PNEUMONIAS 599
PSITTACOSIS (ORNITHOSIS) 601
CHLAMYDIAL PNEUMONIA 603
Q FEVER 603
Viral Diseases of the Lower Respiratory System 604
VIRAL PNEUMONIA 604
INFLUENZA (FLU) 604
Fungal Diseases of the Lower Respiratory System 606
HISTOPLASMOSIS 606
COCCIDIOIDOMYCOSIS 608
BLASTOMYCOSIS (NORTH AMERICAN BLASTOMYCOSIS) 609
OTHER FUNGI INVOLVED IN RESPIRATORY DISEASE 609
Protozoan Disease of the Lower Respiratory System 609
PNEUMOCYSTIS PNEUMONIA 609
Study Outline 612
Study Questions 614
Further Reading 616
MMWR: Legionellosis from a Decorative Fountain 602

25 Microbial Diseases of the Digestive System 617

Structure and Function of the Digestive System 617
Normal Flora of the Digestive System 618
Bacterial Diseases of the Mouth 618
DENTAL CARIES (TOOTH DECAY) 618
PERIODONTAL DISEASE 621
Bacterial Diseases of the Lower Digestive System 622
STAPHYLOCOCCAL FOOD POISONING (STAPHYLOCOCCAL ENTEROTOXICOSIS) 622
SALMONELLOSIS (*SALMONELLA* GASTROENTERITIS) 624
TYPHOID FEVER 625
BACILLARY DYSENTERY (SHIGELLOSIS) 626
CHOLERA 626
VIBRIO PARAHAEMOLYTICUS GASTROENTERITIS 627
ESCHERICHIA COLI GASTROENTERITIS (TRAVELER'S DIARRHEA) 628
CAMPYLOBACTER GASTROENTERITIS 628
HELICOBACTER GASTROENTERITIS 629
YERSINIA GASTROENTERITIS 629
CLOSTRIDIUM PERFRINGENS GASTROENTERITIS 629
BACILLUS CEREUS GASTROENTERITIS 630
Viral Diseases of the Digestive System 630
MUMPS 630
CYTOMEGALOVIRUS (CMV) INCLUSION DISEASE 630
HEPATITIS 631
VIRAL GASTROENTERITIS 636
Fungal Diseases of the Digestive System 637

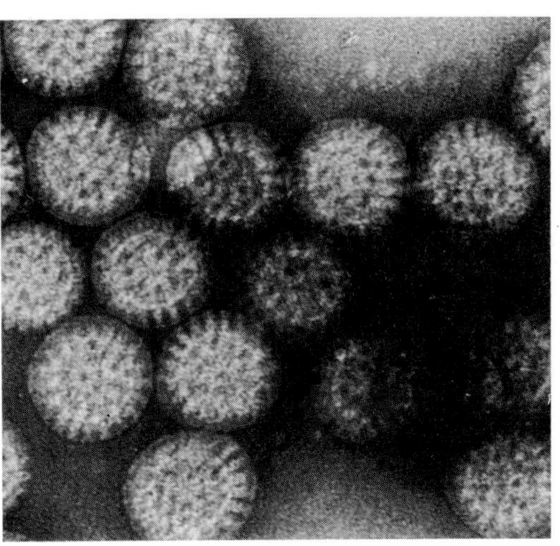

ERGOT POISONING 637
AFLATOXIN POISONING 637
Protozoan Diseases of the Digestive System 637
GIARDIASIS 637
BALANTIDIASIS (BALANTIDIAL DYSENTERY) 638
AMOEBIC DYSENTERY (AMOEBIASIS) 638
CRYPTOSPORIDIOSIS 638
Helminthic Diseases of the Digestive System 639
TAPEWORM INFESTATIONS 639
HYDATID DISEASE 642
NEMATODE INFESTATIONS 642
Study Outline 644
Study Questions 648
Further Reading 649
MMWR: Foodborne Hepatitis A in Seven States 634

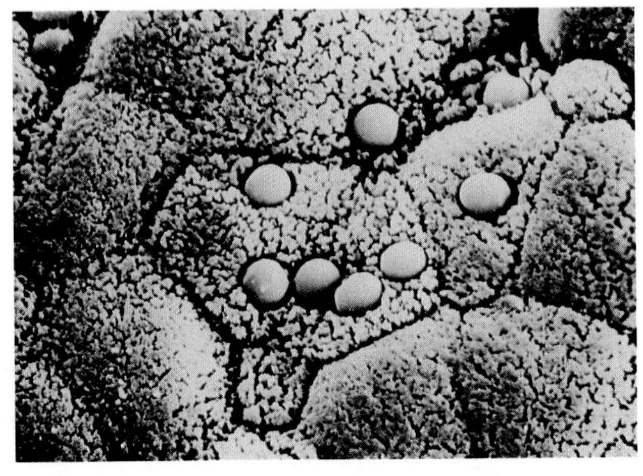

26 Microbial Diseases of the Urinary and Reproductive Systems 650

Structure and Function of the Urinary System 650
Structure and Function of the Reproductive System 651
Normal Flora of the Urinary and Reproductive Systems 652
DISEASES OF THE URINARY SYSTEM 653
Bacterial Diseases of the Urinary System 653
CYSTITIS 653
PYELONEPHRITIS 653
LEPTOSPIROSIS 653
GLOMERULONEPHRITIS 654
DISEASES OF THE REPRODUCTIVE SYSTEM 654
Bacterial Diseases of the Reproductive System 654
GONORRHEA 655
NONGONOCOCCAL URETHRITIS (NGU) 658
SYPHILIS 659
GARDNERELLA VAGINITIS 662
LYMPHOGRANULOMA VENEREUM 662
CHANCROID (SOFT CHANCRE) 663
Viral Diseases of the Reproductive System 663
GENITAL HERPES 663
GENITAL WARTS 664
AIDS 664
Fungal Disease of the Reproductive System 664
CANDIDIASIS 664
Protozoan Disease of the Reproductive System 666
TRICHOMONIASIS 666
Study Outline 667
Study Questions 669
Further Reading 670
MMWR: Antibiotic-Resistant Gonococci and Natural Selection 657

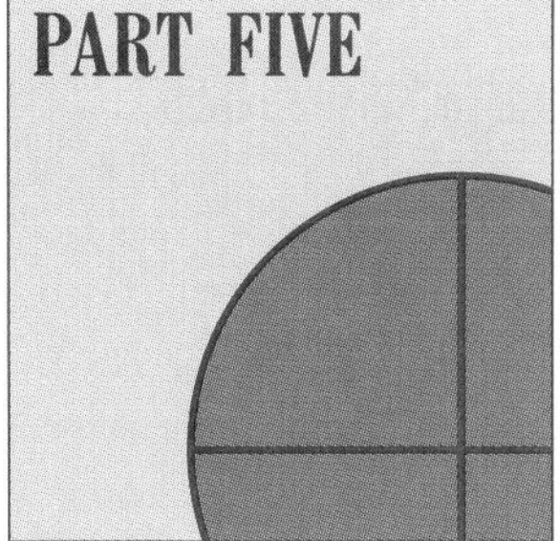

PART FIVE

APPLIED MICROBIOLOGY AND THE ENVIRONMENT 671

27 Soil and Water Microbiology 672

SOIL MICROBIOLOGY AND CYCLES OF THE ELEMENTS 673
The Components of Soil 673
MINERALS 673
ORGANIC MATTER 673
WATER AND GASES 673
ORGANISMS 673
Microorganisms and Biogeochemical Cycles 676
THE CARBON CYCLE 676
THE NITROGEN CYCLE 677
OTHER BIOGEOCHEMICAL CYCLES 681
Degradation of Synthetic Chemicals in the Soil 681
AQUATIC MICROBIOLOGY AND SEWAGE TREATMENT 682
Aquatic Microorganisms 682
FRESHWATER MICROBIAL FLORA 682
SEAWATER MICROBIAL FLORA 683
The Role of Microorganisms in Water Quality 683
WATER POLLUTION 683
TESTS FOR WATER PURITY 685
WATER TREATMENT 686
SEWAGE TREATMENT 688
SOLID MUNICIPAL WASTE 694
Study Outline 695
Study Questions 697
Further Reading 699
Microbiology in the News: Bacteria Contribute to the Greenhouse Effect 675

28 Applied and Industrial Microbiology 700

FOOD MICROBIOLOGY 700
Food Preservation and Spoilage 700
CANNING 701
PASTEURIZATION 703
ASEPTIC PACKAGING 703
LOW-TEMPERATURE PRESERVATION 704
RADIATION AND FOOD PRESERVATION 704
CHEMICAL PRESERVATIVES 705
Foodborne Infections and Microbial Intoxications 706
The Role of Microorganisms in Food Production 706
CHEESE 707
OTHER DAIRY PRODUCTS 708
NONDAIRY FERMENTATIONS 710
ALCOHOLIC BEVERAGES AND VINEGAR 710
MICROORGANISMS AS A FOOD SOURCE 712
INDUSTRIAL MICROBIOLOGY 712
Fermentation Technology 713
IMMOBILIZED ENZYMES AND ORGANISMS 714
Industrial Products 714
AMINO ACIDS 714
CITRIC ACID 716
ENZYMES 716
VITAMINS 717
PHARMACEUTICALS 717
URANIUM AND COPPER 717
MICROORGANISMS 717
Alternative Energy Sources Using Microorganisms 718
Industrial Microbiology and the Future 719
Study Outline 719
Study Questions 721
Further Reading 722
Microbiology Highlights: Microbial Miner 715

APPENDICES

A Classification of Bacteria According to *Bergey's Manual of Systematic Bacteriology* 723

B Word Roots Used in Microbiology 729

C Most Probable Numbers (MPN) Table 733

D Methods for Taking Clinical Samples 734

E Biochemical Pathways 736

F Exponents, Exponential Notation, Logarithms, and Generation Time 742

PRONUNCIATION OF SCIENTIFIC NAMES 744
GLOSSARY 749
PHOTOGRAPH ACKNOWLEDGMENTS 770
INDEX 773

PART ONE

Fundamentals of Microbiology

CHAPTER 1

The Microbial World and You

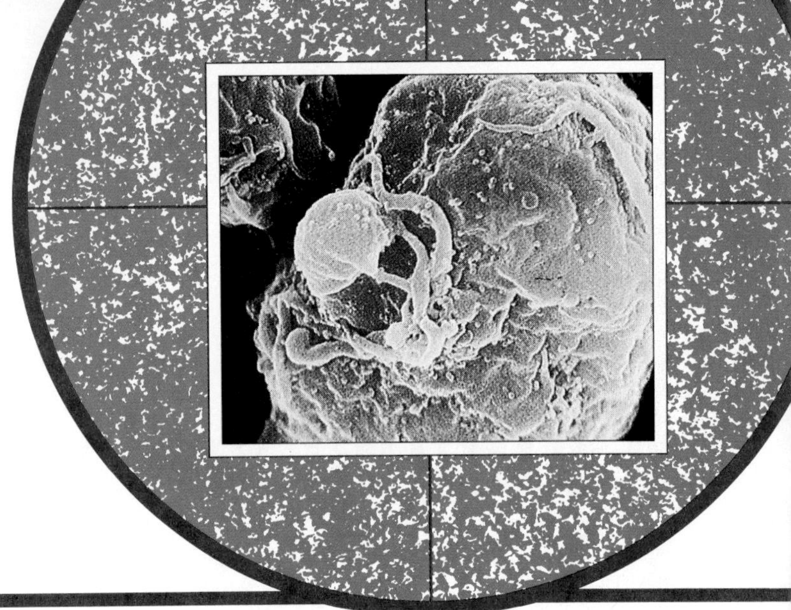

LEARNING OBJECTIVES

- Identify the contributions to microbiology made by Anton van Leeuwenhoek, Robert Hooke, Louis Pasteur, Robert Koch, Joseph Lister, Paul Ehrlich, Alexander Fleming, and Edward Jenner.
- Compare the theories of spontaneous generation and biogenesis.
- Recognize scientific genus and specific epithet names.
- List the major groups of organisms studied in microbiology.
- List at least four beneficial activities of microorganisms.
- Define normal flora.
- Define immunology, microbial ecology, microbial genetics, microbial physiology, molecular biology, and virology.
- List applications of recombinant DNA, biotechnology, and bioremediation.

Part One photo, page 1: Although billions of microorganisms are present in, on, and around us, they are too small to see without special instruments such as this compound light microscope.

In the summer of 1985, newspapers announced that a well-known actor was suffering from AIDS (acquired immunodeficiency syndrome). The American public had heard about the disease, but until then AIDS had not been perceived as much of a health threat by the majority of Americans, perhaps because the victims were primarily from a group outside the mainstream of American society—male homosexuals. While the actor sought treatment in hospitals in France, Americans became increasingly aware of the illness that would end the life of one of their screen idols as well as the lives of over 8000 other Americans that same year.

The general public was just beginning to worry about what would become the most frightening epidemic of the century. Yet medical researchers had been gathering information about the devastating disease for several years. The same concepts of microbiology that had helped them identify other diseases, their means of transmission, and treatments were being applied to the study of AIDS.

AIDS came to public attention in 1981 with reports from Los Angeles that a few young homosexual men had died of a previously rare type of pneumonia known as *Pneumocystis* (nü-mō-sis'tis) pneumonia. These men had experienced a severe weakening of the immune system, which normally fights infectious disease. Soon these cases were correlated with an unusual number of occurrences of a rare form of cancer, Kaposi's sarcoma, among young homosexual men. Similar increases in such rare diseases were found among hemophiliacs and intravenous drug users.

By the end of 1990, nearly 160,000 people in the United States had been diagnosed as having AIDS, and more than 50% of them had died as a result of the disease. A great many more people had tested positive for the presence of the AIDS virus in their blood. As of

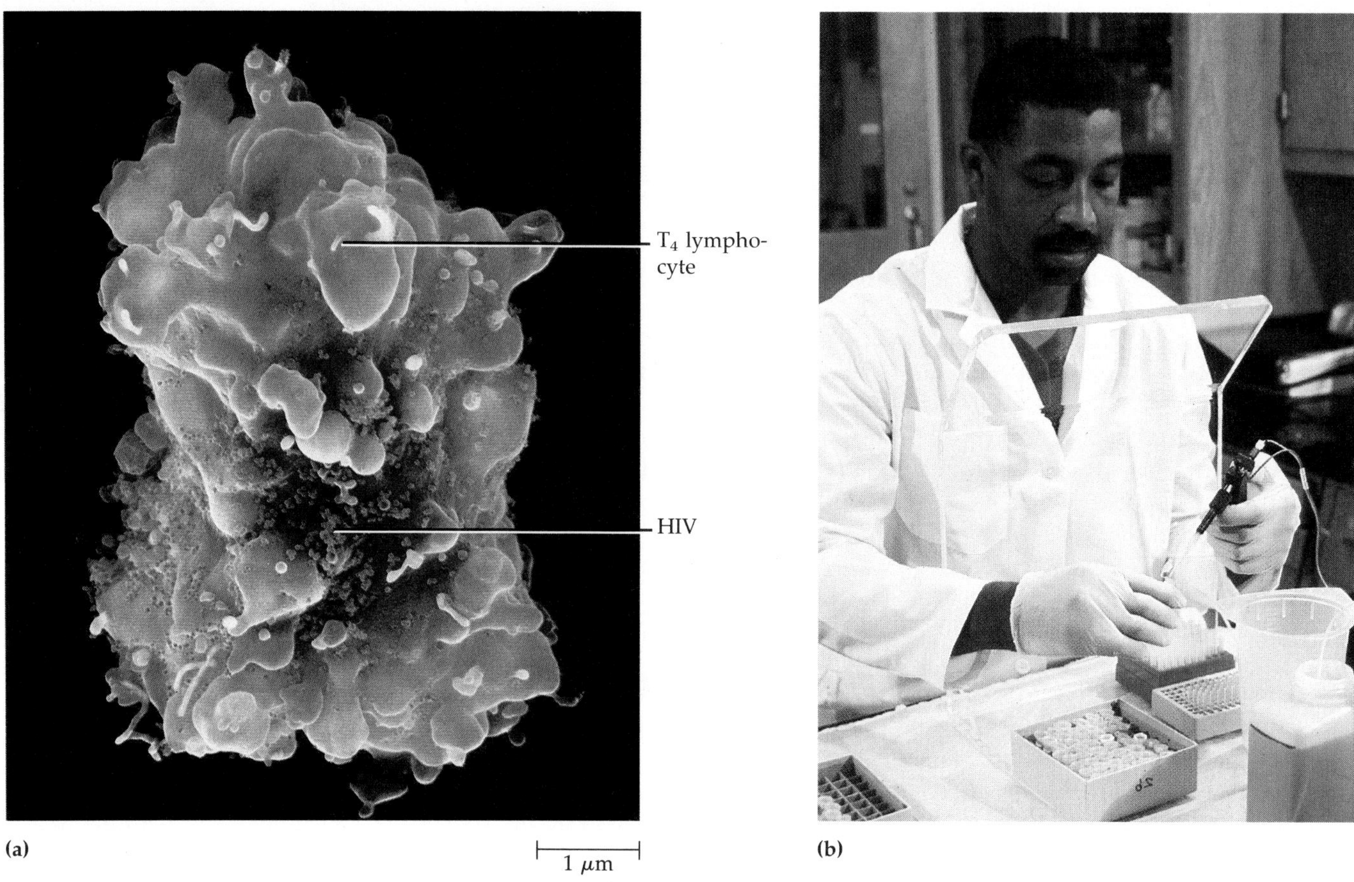

FIGURE 1.1 Human immunodeficiency virus (HIV). **(a)** This scanning electron micrograph shows HIV (black area), the virus that causes AIDS, surrounding one of its target cells, a white blood cell called a T4 lymphocyte. **(b)** Laboratory technician performing the standard enzyme-linked immunosorbent assay (ELISA).

1992, health officials estimated that AIDS cases in the United States total about 365,000, and more than 1.5 million Americans carried the virus. In 1991, the World Health Organization estimated that 1,300,000 people worldwide had AIDS. Scientists studying AIDS are still uncertain whether all individuals carrying the virus will eventually develop the disease or whether some may be able to live normally without any symptoms.

Researchers quickly discovered that the cause of AIDS was a previously unknown virus (Figure 1.1). The virus, now called **human immunodeficiency virus (HIV),** causes disease in an unusual way. Instead of directly causing disease symptoms—as do such viruses as those causing mumps, chickenpox, hepatitis, and smallpox—HIV destroys certain cells of the immune system, the body's defense system. Sickness and death result from microorganisms or cancerous cells that might otherwise have been defeated by the body's natural defenses. So far, the disease has been inevitably fatal once symptoms develop.

By studying disease patterns, medical researchers found that HIV could spread through sexual intercourse, from contaminated needles, from infected mothers to their fetuses before birth, and in blood transfusions—in short, by transmission of bodily fluids from one person to another. Since 1985, blood used for transfusions has been carefully checked for the presence of HIV, and it is now quite unlikely for the virus to be spread through this route.

The majority of AIDS sufferers are in the sexually active age group, and since heterosexual partners of AIDS sufferers are at high risk of infection, public health officials are concerned that teenagers are at high risk of contracting AIDS. Forty-six percent of U.S. teenage girls with AIDS became infected through heterosexual intercourse, as compared with 30% of all U.S. women with AIDS. In several U.S. cities, AIDS is now the major cause of death of young men and women.

The first drug that has been used to treat AIDS (although it does not provide a cure) is zidovudine (AZT). Other drugs are in clinical trials. In addition, considerable effort is being given to developing a vaccine against AIDS. Public health officials have also focused on prevention through education.

In the months and years to come, microbiological techniques will be applied intensively to help scientists learn more about the structure of this deadly virus, how it is transmitted, how it grows in cells and causes disease, how drugs can be directed against it, and whether an effective vaccine can be developed.

AIDS poses one of this century's most formidable health threats, but it is not the first serious epidemic of a sexually transmitted disease. Syphilis was also once a fatal epidemic disease. As recently as 1941, syphilis caused an estimated 14,000 deaths in the United States. With few drugs available for treatment and no vaccines to prevent it, efforts to control the disease focused mainly on restraining sexual behavior and on the use of condoms. The eventual development of drugs to treat syphilis contributed significantly to preventing the spread of the disease. Reported cases of syphilis dropped from a record high of 575,000 in 1943 to 42,128 in 1990.

Just as microbiological techniques helped researchers in the fight against syphilis, in 1976 they helped scientists discover the cause of Legionnaires' disease—a tiny, airborne bacterium not detectable by standard laboratory tests. Though the microorganism was unfamiliar to microbiologists, it had, in fact, caused several unexplained epidemics of pneumonia. Once it had been carefully documented, the bacterium was assigned the scientific name *Legionella pneumophila* (lē-jä-nel'lä nü-mō'fi-lä); the disease itself was later renamed *legionellosis.*

In 1978, toxic shock syndrome (TSS) caught the attention of the public for the first time when a number of otherwise healthy young women suddenly died. Microbiologists were able to show that the cause of death was **toxins** (poisonous substances) produced by the bacterium *Staphylococcus aureus* (staf-i-lō-kok'kus ô'rē-us), a microorganism carried on human skin (see Figure 11.10). In their research on how such a common bacterium could become so toxic, one thing scientists noticed was that all the victims used tampons during menstruation. In 1985, a Harvard research team found that certain tampon fibers absorb magnesium that is normally present in the vagina. When the amount of magnesium in the vagina is reduced, *S. aureus* produces large amounts of toxin. The association of TSS with a particular brand of tampon resulted in the recall of that brand and the publication of cautions about the use of tampons in general.

In 1975, a cluster of disease cases first diagnosed as rheumatoid arthritis was reported in young people living near the city of Lyme, Connecticut. Today this disease is known as **Lyme disease.** By 1982, microbiologists determined that the disease is caused by a bacterium (*Borrelia burgdorferi* [bôr-rel'ē-ä burg-dôr'fėr-ē]) transmitted to humans by ticks that live on field mice and deer. These ticks are so small that their bites go unnoticed. Within a few weeks of the tick bite, a rash may appear at the site, although some people never develop one. The typical rash resembles a bull's-eye, but there are many variations. The rash often is accompanied by joint stiffness, fever and chills, headache, stiff neck, nausea, and low back pain. In the next stage, some patients may develop cardiac abnormalities or neurological problems. In the third stage, arthritis, usually affecting the larger joints, is the principal complication. Lyme disease responds well to antibiotics, especially if they are given early in the course of the disease.

All the diseases we have mentioned are caused by viruses or bacteria, two types of microorganisms. This book introduces you to the enormous variety of these microscopic organisms. It shows you how microbiologists use specific techniques and procedures to study the microbes that cause such diseases as AIDS, legionellosis, TSS, and Lyme disease. You will also learn about the body's responses to microbial infection and the ways certain drugs combat microbial diseases. Finally, you will learn about the many beneficial roles that microbes play in the world around us.

Microbes in Our Lives

For many people, the words *germ, microbe,* and *microorganism* bring to mind a group of tiny creatures that do not quite fit any of the categories in that old question, "Is it animal, vegetable, or mineral?" Microorganisms are minute living things that individually are too small to be seen with the naked eye. The group includes bacteria (see Chapter 11), fungi (yeasts and molds), protozoans, and microscopic algae (Chapter 12). It also includes viruses, those noncellular entities sometimes regarded as being at the border of life and nonlife (see Chapter 13). Figure 1.6 on page 16 gives you an idea of the diversity of the microorganisms that you will learn more about in later chapters.

We tend to associate these small organisms only with uncomfortable infections, major diseases such as AIDS, or such common inconveniences as spoiled food. However, the majority of microorganisms make crucial contributions to the welfare of the world's inhabitants by helping to maintain the balance of living organisms and chemicals in our environment. Marine and freshwater microorganisms form the basis of the food chain in oceans, lakes, and rivers. Soil microbes

MICROBIOLOGY *HIGHLIGHTS*

What Makes Sourdough Bread Different?

Imagine being a miner during the California gold rush. You've just made bread dough from your last supplies of flour and salt when someone yells, "Gold!" Temporarily forgetting your hunger, you run off to the gold fields. Many hours later, you return. The dough has been rising longer than usual, but you are too cold, tired, and hungry to start a new batch. Later you find that your bread tastes different from previous batches; it is slightly sour. During the gold rush, miners baked so many sour loaves that they were nicknamed "sourdoughs."

Conventional bread is made from flour, water, sugar, salt, shortening, and a living microbe, yeast. The yeast belongs to the Kingdom Fungi and is named *Saccharomyces cerevisiae.* When flour is mixed with water, an enzyme in the flour breaks its starch into two sugars, maltose and glucose. After the ingredients for the bread are mixed, the yeast metabolizes the sugars and produces alcohol (ethanol) and carbon dioxide as waste products. This metabolic process is called fermentation. The dough rises as carbon dioxide bubbles get trapped in the sticky matrix. The alcohol evaporates during baking. It and the carbon dioxide gas form spaces that remain in the bread.

Originally, breads were leavened by wild yeast from the air, which had been trapped in the dough. Later, bakers kept a starter culture of yeast—dough from the last batch of bread—to leaven each new batch of dough. Sourdough bread is made with a special sourdough starter culture that is added to flour, water, and salt. Perhaps the most famous sourdough bread made today comes from San Francisco, where a handful of bakeries have continuously cultivated their starters for more than 100 years and have meticulously maintained the starters to keep out unwanted microbes that can produce different and undesirable flavors. After bakeries in other areas made several unsuccessful attempts to match the unique flavor of San Francisco sourdough, rumors attributed the taste to a unique local climate or contamination from bakery walls. T.F. Sugihara and L. Kline from the United States Department of Agriculture set out to debunk these rumors and to determine the microbiological reason for the bread's different taste so that it could be made in other areas.

The U.S.D.A. workers found that sourdough is eight to ten times more acidic than conventional bread because of the presence of lactic and acetic acids. These acids account for the sour flavor of the bread. The workers isolated and identified the yeast in the starter as *Saccharomyces exiguus,* a unique yeast that does not ferment maltose and thrives in the acidic environment of this dough. The sourdough question was not answered, however, because the yeast did not produce the acids and did not use maltose. Sugihara and Kline searched the starter for a second agent capable of fermenting maltose and producing the acids. The bacterium they isolated, so carefully guarded all those years, was classified into the genus *Lactobacillus.* Many members of this genus are used in dairy fermentations and are part of the normal flora of humans and other mammals. Analyses of cell structure and genetic composition showed that the sourdough bacterium is genetically different from other previously characterized lactobacilli. It has been given the name *Lactobacillus sanfrancisco.*

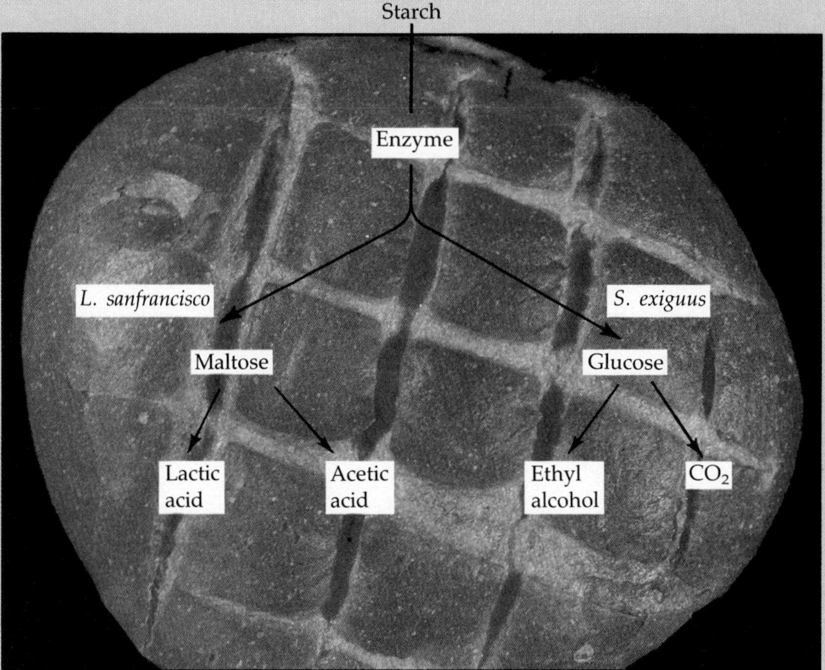

help to break down wastes and incorporate nitrogen gas from the air into organic compounds, thereby recycling chemical elements in the soil, water, and air. Certain bacteria and algae play important roles in photosynthesis, a food- and oxygen-generating process that is critical to life on Earth. Humans (and many other animals) depend on the bacteria in our intestines for digestion and synthesis of some vitamins that our bodies require, including some B vitamins for metabolism and vitamin K for blood clotting.

Microorganisms also have commercial applications. They are used in the synthesis of such chemical products as acetone, organic acids, enzymes, alcohols, and many drugs. Our food industry frequently uses microbes in producing vinegar, sauerkraut, pickles, alcoholic beverages, green olives, soy sauce, buttermilk, cheese, yogurt, and bread.

Recently, it has also become possible to manipulate bacteria and other microbes to produce substances that they normally do not synthesize. Through this new technology, called *genetic engineering*, bacteria can produce important therapeutic substances such as insulin, human growth hormone, and interferon.

Though only a minority of microorganisms are **pathogenic** (disease-producing), practical knowledge of microbes is necessary for medicine and the related health sciences. For example, hospital workers must be able to protect patients from common bacteria that are normally harmless but that pose a threat to the sick and injured.

Today, we understand that microorganisms are found almost everywhere. Yet, not long ago, before the invention of the microscope, microbes were unknown to scientists. Thousands of people died in devastating epidemics whose causes were not understood. Food spoilage often could not be controlled, and entire families died because vaccinations and antibiotics were not available to fight infection.

We can get an idea of how our current concepts of microbiology developed by looking at a few of the historic breakthroughs in microbiology that have changed our lives.

A Brief History of Microbiology

Now that we have generally examined how microbes relate to our everyday lives, we will take a look at how the field of microbiology developed from its beginning hundreds of years ago to its current high-technology status.

THE FIRST OBSERVATIONS

One of the most important discoveries in the history of biology occurred in 1665 with the help of a relatively crude microscope. An Englishman, Robert Hooke, reported to the world that life's smallest structural units were "little boxes," or "cells," as he called them. Hooke was able to see individual cells with a microscope that he had developed, an improved version of a compound microscope (one that uses two sets of lenses). Hooke's discovery marked the beginning of the cell theory—the theory that *all living things are composed of cells.* Subsequent investigations into the structure and functions of cells were based on this theory.

Though Hooke's microscope was capable of showing protozoans and probably bacteria, he lacked the staining techniques (see Chapter 3) that would have allowed him to see such small microbes clearly. The Dutch merchant and amateur scientist Anton van Leeuwenhoek was probably the first to actually observe microorganisms through magnifying lenses. Between 1674 and 1723, he wrote a series of letters to the Royal Society of London describing the "animalcules" he saw through his simple, single-lens microscope. Van Leeuwenhoek's detailed drawings of "animalcules" in rain water, in liquid in which peppercorns had soaked, and in material taken from teeth scrapings have since been identified as representations of bacteria and protozoans (Figure 1.2).

THE DEBATE OVER SPONTANEOUS GENERATION

After van Leeuwenhoek discovered the "invisible" world of microorganisms, the scientific community became interested in the origins of these tiny living things. Until the second half of the nineteenth century, many scientists and philosophers believed that some forms of life could arise spontaneously from nonliving matter; this process was called **spontaneous generation.** People thought that toads, snakes, and mice could be born of moist soil; that flies could emerge from manure; and that maggots, the larvae of flies, could arise from decaying corpses.

Evidence Pro and Con

A strong opponent of spontaneous generation, the Italian physician Francesco Redi, set out in 1668 (even before van Leeuwenhoek's discovery of microscopic life) to demonstrate that maggots do not arise spontaneously from decaying meat, as was commonly believed. Redi filled three jars with decaying meat and sealed them tightly. Then he arranged three other jars similarly but left them open. Maggots appeared in the open vessels after flies entered the jars and laid their eggs. But the sealed containers showed no signs of maggot larvae. Still, Redi's antagonists were not convinced; they claimed that fresh air was needed for spontaneous generation. So Redi set up a second experiment, in which three jars were covered with a fine net instead of being sealed. No larvae appeared in the gauze-covered jars, even though air was present. Maggots appeared only if flies were allowed to leave their eggs on the meat.

Redi's results were a serious blow to the long-held belief that large forms of life could arise from nonlife. However, many scientists still believed that small organisms, such as van Leeuwenhoek's "animalcules," were simple enough to be generated from nonliving materials.

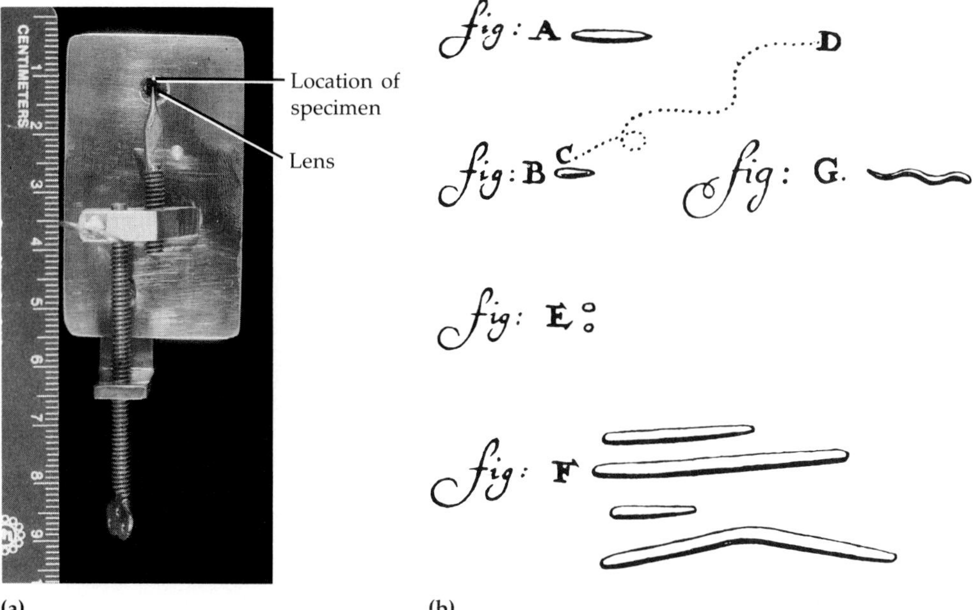

FIGURE 1.2 Anton Van Leeuwenhoek's research with the microscope. **(a)** A replica of a simple microscope made by Anton van Leeuwenhoek to observe living organisms too small to be seen with the naked eye. The specimen was placed on the tip of the adjustable point and viewed from the other side through the tiny round lens. The highest magnification possible with his microscopes was about 300× (times). **(b)** Some of van Leeuwenhoek's drawings of bacteria, made in 1683. He was the first to see the microorganisms we now call bacteria and protozoans.

The case for the spontaneous generation of microorganisms was strengthened in 1745 when John Needham, an Englishman, found that even after he heated nutrient fluids (chicken broth and corn broth) before pouring them into covered flasks, the cooled solutions were soon teeming with microorganisms. Needham claimed that the microbes developed spontaneously from the fluids. Twenty years later, Lazzaro Spallanzani, an Italian scientist, suggested that microorganisms from the air probably had entered Needham's solutions after they were boiled. Spallanzani showed that nutrient fluids heated *after* being sealed in a flask did not develop microbial growth. Needham responded by claiming that the "vital force" necessary for spontaneous generation had been destroyed by the heat and was kept out of the flasks by the seals.

This intangible "vital force" was given all the more credence shortly after Spallanzani's experiment when Laurent Lavoisier showed the importance of oxygen to life. Spallanzani's observations were criticized on the grounds that there was not enough oxygen in the sealed flasks to support microbial life.

The Theory of Biogenesis

The issue was still unresolved in 1858, when the German scientist Rudolf Virchow challenged spontaneous generation with the concept of **biogenesis,** the claim that living cells can arise only from preexisting living cells. Arguments about spontaneous generation continued until 1861, when the issue was resolved experimentally by Louis Pasteur, the French scientist. Pasteur demonstrated that microorganisms are indeed present in the air and can contaminate seemingly sterile solutions, but air itself does not create microbes.

Pasteur designed a series of ingenious and persuasive experiments to prove his points. He began by filling several short-necked flasks with beef broth and boiling them. Some were then left open and allowed to cool. In a few days, these flasks were found to be contaminated with microbes. The other flasks, sealed after boiling, remained free of microorganisms. From these results, Pasteur reasoned that microbes in the air were the agents responsible for contaminating nonliving matter such as the broths in Needham's flasks.

Pasteur's next step was to place broth in open-ended long-necked flasks; he then bent the necks into S-shaped curves (Figure 1.3). The contents of these flasks were then boiled and cooled. The broth in the flasks did not decay and showed no signs of life after days, weeks, and even months. Pasteur's unique design allowed air to pass into the flask, but the curved neck trapped any airborne microorganisms that might contaminate the broth. (Some of these original vessels, which were later sealed, are on display at the Pasteur Institute in Paris. They still show no sign of contamination more than 100 years later.)

Pasteur showed that microorganisms can be present in nonliving matter—on solids, in liquids, and in the air. Furthermore, he demonstrated conclusively that microbial life can be destroyed by heat and that methods can be devised to block the access of airborne microorganisms to nutrient environments. These discoveries form the basis of the **aseptic techniques** (techniques to prevent contamination by unwanted microorganisms) that are now the standard practice in laboratory and many medical procedures. Modern aseptic techniques are among the first and most important things that a beginning microbiologist learns.

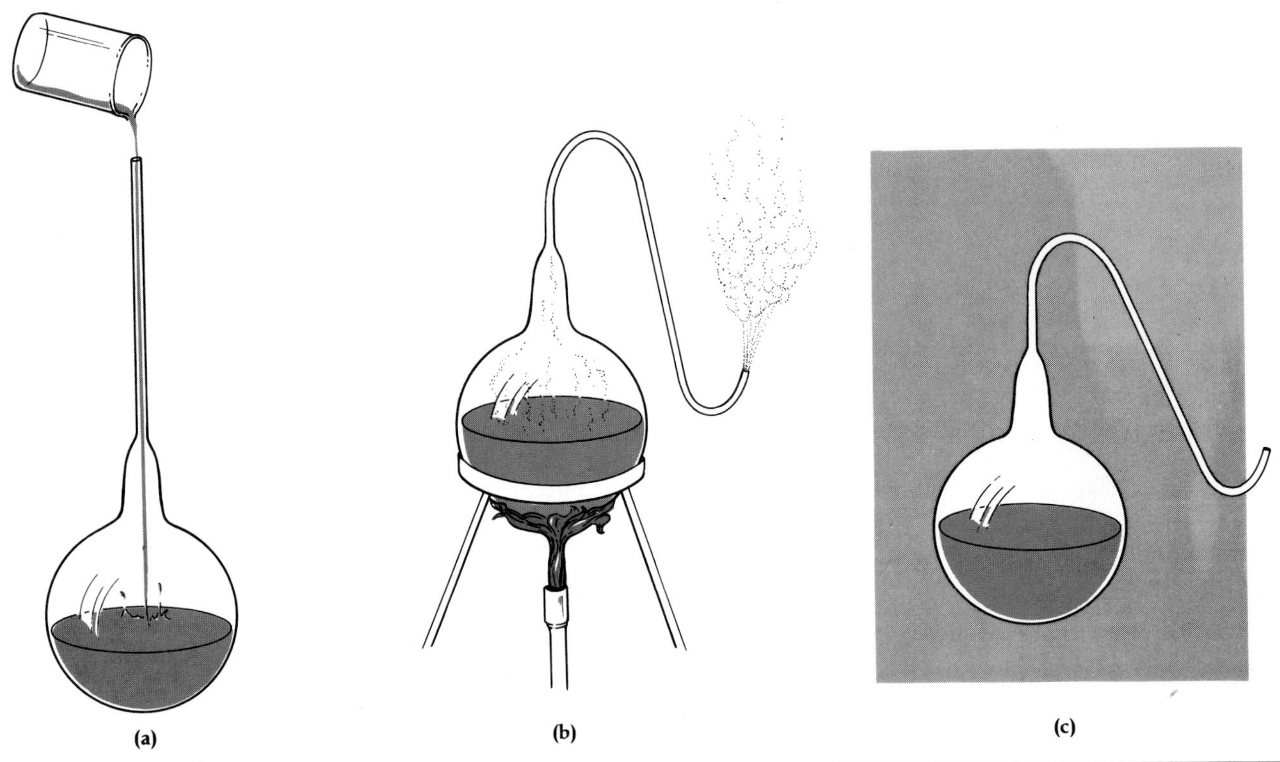

FIGURE 1.3 Pasteur's experiment disproving the theory of spontaneous generation. **(a)** Pasteur first poured nutrient broth into a long-necked flask. **(b)** Next he heated the neck of the flask and bent it into an S-shaped curve; then he boiled the solution for several minutes. **(c)** Microorganisms did not appear in the cooled broth, even after long periods.

Pasteur's work provided evidence that microorganisms cannot originate from mystical forces present in nonliving materials. Rather, any appearance of "spontaneous" life in nonliving solutions can be attributed to microorganisms that were already present in the air or in the fluids themselves. Scientists now believe that a form of spontaneous generation probably did occur on the primitive Earth when life first began, but they agree that this does not happen under our present environmental conditions.

THE GOLDEN AGE OF MICROBIOLOGY

For about 60 years, beginning with the work of Pasteur, there was an explosion of discoveries in microbiology. The period from 1857 to 1914 has been appropriately named the **Golden Age of Microbiology.** During this period, rapid advances, spearheaded mainly by Pasteur and Robert Koch, led to the establishment of microbiology as a science. These years saw discoveries of the agents of many diseases and the role of immunity in the prevention and cure of disease. During this productive period, microbiologists studied the chemical activities of microorganisms, improved the techniques for performing microscopy and culturing microorganisms, and developed vaccines and surgical techniques. Some of the major events that occurred during the Golden Age of Microbiology are listed in Figure 1.4.

Fermentation and Pasteurization

One of the key steps that established the relationship between microorganisms and disease occurred when a group of French merchants asked Pasteur to find out why wine and beer soured. They hoped to develop a method that would prevent spoilage when those beverages were shipped long distances. At the time, many scientists believed that air converted the sugars in these fluids into alcohol. Pasteur found instead that microorganisms called yeasts convert the sugars to alcohol in the absence of air. This process, called **fermentation** (Chapter 5), is used to make wine and beer. Souring and spoilage are caused by different microorganisms called bacteria. In the presence of air, bacteria change the alcohol in the beverage into vinegar (acetic acid).

Pasteur's solution to the spoilage problem was to heat the beer and wine just enough to kill most of the bacteria; the process does not greatly affect the flavor

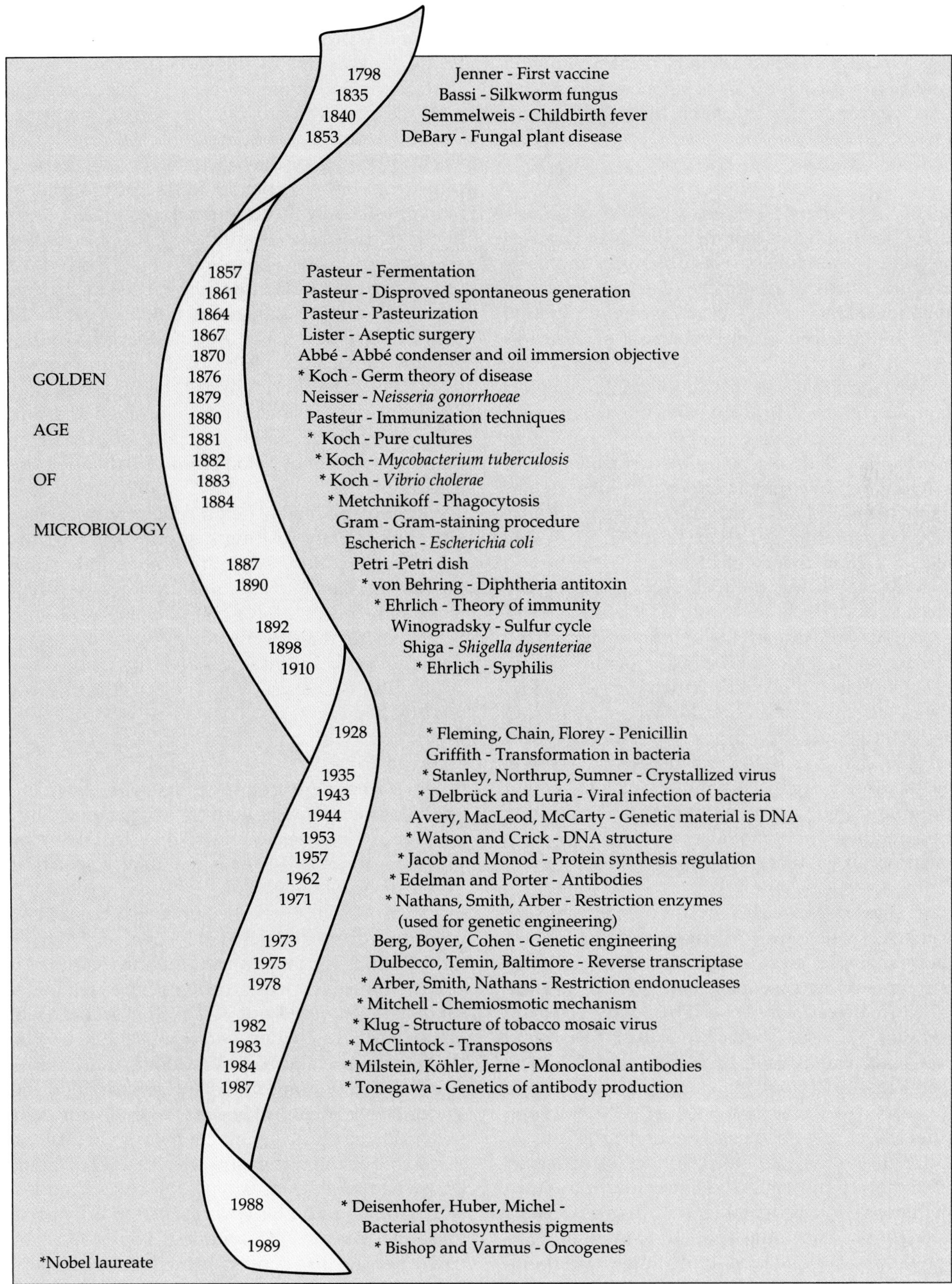

1798 Jenner - First vaccine
1835 Bassi - Silkworm fungus
1840 Semmelweis - Childbirth fever
1853 DeBary - Fungal plant disease
GOLDEN AGE OF MICROBIOLOGY
1857 Pasteur - Fermentation
1861 Pasteur - Disproved spontaneous generation
1864 Pasteur - Pasteurization
1867 Lister - Aseptic surgery
1870 Abbé - Abbé condenser and oil immersion objective
1876 * Koch - Germ theory of disease
1879 Neisser - *Neisseria gonorrhoeae*
1880 Pasteur - Immunization techniques
1881 * Koch - Pure cultures
1882 * Koch - *Mycobacterium tuberculosis*
1883 * Koch - *Vibrio cholerae*
1884 * Metchnikoff - Phagocytosis
Gram - Gram-staining procedure
Escherich - *Escherichia coli*
1887 Petri -Petri dish
1890 * von Behring - Diphtheria antitoxin
* Ehrlich - Theory of immunity
1892 Winogradsky - Sulfur cycle
1898 Shiga - *Shigella dysenteriae*
1910 * Ehrlich - Syphilis
1928 * Fleming, Chain, Florey - Penicillin
Griffith - Transformation in bacteria
1935 * Stanley, Northrup, Sumner - Crystallized virus
1943 * Delbrück and Luria - Viral infection of bacteria
1944 Avery, MacLeod, McCarty - Genetic material is DNA
1953 * Watson and Crick - DNA structure
1957 * Jacob and Monod - Protein synthesis regulation
1962 * Edelman and Porter - Antibodies
1971 * Nathans, Smith, Arber - Restriction enzymes (used for genetic engineering)
1973 Berg, Boyer, Cohen - Genetic engineering
1975 Dulbecco, Temin, Baltimore - Reverse transcriptase
1978 * Arber, Smith, Nathans - Restriction endonucleases
* Mitchell - Chemiosmotic mechanism
1982 * Klug - Structure of tobacco mosaic virus
1983 * McClintock - Transposons
1984 * Milstein, Köhler, Jerne - Monoclonal antibodies
1987 * Tonegawa - Genetics of antibody production
1988 * Deisenhofer, Huber, Michel - Bacterial photosynthesis pigments
1989 * Bishop and Varmus - Oncogenes
*Nobel laureate

FIGURE 1.4 Milestones in microbiology.

of the beverage. This process, called **pasteurization,** is now commonly used to kill potentially harmful bacteria in milk as well as in some alcoholic drinks. Showing the connection between spoilage of food and microorganisms was a key step in establishing the relationship between disease and microbes.

The Germ Theory of Disease

As we have seen, the fact that many kinds of diseases are related to microorganisms was unknown until relatively recently. Before the time of Pasteur, effective treatments for many diseases were discovered by trial and error, but the causes of the diseases were unknown.

The realization that yeasts play a crucial role in fermentation was the first link between the activity of a microorganism and physical and chemical changes in organic materials. This discovery alerted scientists to the possibility that microorganisms might have similar relationships with plants and animals—specifically, that microorganisms might cause disease. This idea was called the **germ theory of disease.**

The germ theory was a difficult concept for many people to accept because for centuries it was believed that disease was punishment for an individual's crimes or misdeeds. When the inhabitants of an entire village became ill, people often blamed the disease on foul odors from sewage or on poisonous vapors from swamps. Most people born in Pasteur's time found it inconceivable that "invisible" microbes could travel through the air to infect plants and animals or remain on clothing and bedding to be transmitted from one person to another. But gradually, scientists accumulated the information needed to support the new germ theory.

In 1865, Pasteur was called upon to help fight silkworm disease, which was ruining the silk industry throughout Europe. Years earlier, in 1835, Agostino Bassi, an amateur microscopist, had proved that another silkworm disease was caused by a fungus. Using data provided by Bassi, Pasteur found that the more recent infection was caused by a protozoan, and he developed a method for recognizing afflicted silkworm moths.

In the 1860s, Joseph Lister, an English surgeon, applied the germ theory to medical procedures. Lister was aware that in the 1840s, the Hungarian physician Ignaz Semmelweis had demonstrated that physicians who went from one obstetrical patient to another without disinfecting their hands transmitted puerperal (childbirth) fever. Lister had also heard of Pasteur's work connecting microbes to animal diseases. Disinfectants were unknown at the time, but Lister knew that carbolic acid (now called phenol) kills bacteria, so he began soaking surgical dressings in a solution of it. The practice so reduced the incidence of infections and deaths that other surgeons quickly adopted it. Lister's technique was one of the earliest medical attempts to control the infections caused by microorganisms. In fact, his findings proved that microorganisms cause surgical wound infections.

The first proof that bacteria actually cause disease came from Robert Koch in 1876. Koch, a brilliant German physician, was Pasteur's young rival in the race to discover the cause of anthrax, a disease that was destroying cattle and sheep in Europe. Koch discovered rod-shaped bacteria now known as *Bacillus anthracis* (bä-sil'lus an-thrā'sis) in the blood of cattle that had died of anthrax. He cultured the bacteria on nutrients and then injected samples of the culture into healthy animals. When these animals became sick and died, Koch isolated the bacteria in their blood and compared them with the bacteria originally isolated. He found that the two sets of blood cultures contained the same bacteria.

Koch thus established a sequence of experimental steps for directly relating a specific microbe to a specific disease. These steps are known today as **Koch's postulates** (see Figure 14.3). During the past 100 years, these same criteria have been invaluable in investigations proving that specific microorganisms cause many diseases. Koch's postulates, their limitations, and their application to disease will be discussed in greater detail in Chapter 14.

Vaccination

Often a treatment or preventive procedure is developed before scientists know why it works. The smallpox vaccine is an example of this. In 1798, almost 70 years before Koch established that a specific microorganism causes anthrax, Edward Jenner, a young British physician, embarked on an experiment to find a protection from smallpox.

Smallpox epidemics were greatly feared. The disease periodically swept through Europe, killing thousands, and it wiped out 90% of the American Indians on the East Coast when European settlers first brought the infection to the New World.

When a young milkmaid told Jenner that she couldn't get smallpox because she already had been sick from cowpox—a much milder disease—he decided to put the girl's story to a test. First Jenner collected scrapings from cowpox blisters. Then he made inoculations with the cowpox material by scratching the arm of a healthy volunteer with a pox-contaminated needle. The scratch turned into a raised bump. In a few days, the person became mildly sick but recovered and never again contracted either cowpox or smallpox. The process was called *vaccination,* from the Latin word *vacca,* for cow. The protection from disease provided by vaccination (or by recovery from the disease itself) is called **immunity.** We will discuss the mechanisms of immunity in Chapter 17.

Years after Jenner's experiment, in about 1880, Pasteur discovered why vaccinations work. He found that the bacterium that causes chicken cholera lost its ability to cause disease (lost its *virulence,* or became *avirulent*) after it was grown in the laboratory for long periods. However, it—and other microorganisms with decreased virulence—was able to induce immunity against subsequent infections by its virulent counterparts. The discovery of this phenomenon provided a clue to Jenner's successful experiment with cowpox. Both cowpox and smallpox are caused by viruses. Even though cowpox virus is not a laboratory-produced derivative of smallpox virus, it is so closely related to the smallpox virus that it can induce immunity to both viruses. Pasteur used the term *vaccine* for cultures of avirulent microorganisms used for preventive inoculation.

Jenner's experiment was the first time in a Western culture that a living viral agent—the cowpox virus—was used to produce immunity. Physicians in ancient China had immunized patients by removing scales from drying pustules of a person suffering from a mild case of smallpox, grinding the scales to a fine powder, and inserting the powder into the nose of the person to be protected.

Some vaccines are still produced from avirulent microbial strains that stimulate immunity to the related virulent strain. Other vaccines are made from killed virulent microbes, from isolated components of virulent microorganisms, or by genetic engineering techniques.

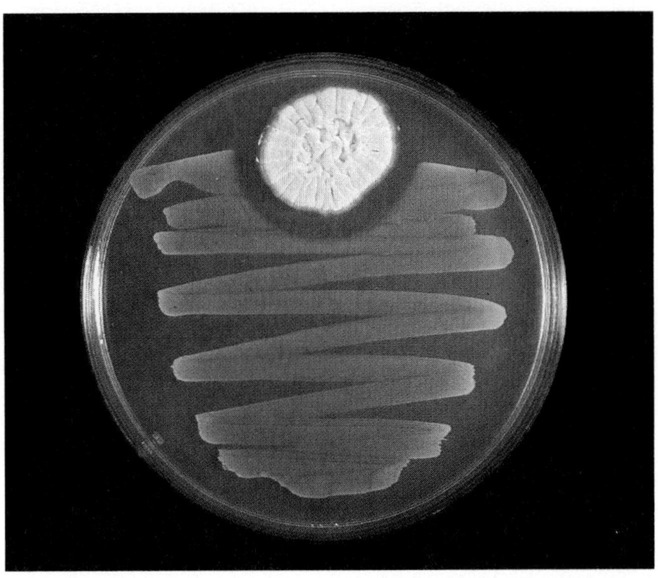

FIGURE 1.5 Inhibition of bacterial growth around the mold *Penicillium chrysogenum.* Colonies of the bacteria *Staphylococcus aureus* do not grow in the vicinity of the contaminating *Penicillium* colony near the top of the photo. The reason is that the mold secretes penicillin.

THE BIRTH OF MODERN CHEMOTHERAPY: DREAMS OF A "MAGIC BULLET"

After the relationship between microorganisms and disease was established, medical microbiologists next focused on the search for substances that could destroy pathogenic microorganisms without damaging the infected animal or human. Treatment of disease by using chemical substances is called **chemotherapy.** (The term also commonly refers to chemical treatment of noninfectious diseases, such as cancer.) The chemotherapeutic agents prepared from chemicals in the laboratory are called **synthetic drugs.** Chemicals produced naturally by bacteria and fungi that act against other microorganisms are called **antibiotics.** The success of chemotherapy is based on the fact that some chemicals are more poisonous to microorganisms than to the hosts infected by the microbes.

The First Synthetic Drugs

Paul Ehrlich, a German physician, was the imaginative thinker who fired the first shot in the chemotherapy revolution. As a medical student, Ehrlich speculated about a "magic bullet" that could hunt down and destroy a pathogen without harming the infected host. Ehrlich launched a search for such a bullet. In 1910, after testing hundreds of substances, he found a chemotherapeutic agent called *salvarsan,* an arsenic derivative effective against syphilis. Prior to this discovery, the only known chemical in Europe's medical arsenal was an extract from the bark of a South American tree, *quinine,* which had been used by Spanish conquistadors to treat malaria.

By the late 1930s, researchers had developed several other synthetic drugs that could destroy microorganisms. Most of these drugs were derivatives of dyes. In addition, *sulfa drugs* were discovered at about the same time.

A Fortunate Accident—Antibiotics

In contrast to the sulfa drugs, which were deliberately developed from a series of industrial chemicals, the first antibiotic was discovered by accident. Alexander Fleming, a Scottish physician and bacteriologist, almost tossed out some culture plates that had been contaminated by mold. Fortunately, he took a second look at the curious pattern of growth on the contaminated plates. There was a clear area around the mold where the bacterial culture had stopped growing (Figure 1.5). Fleming was looking at a mold that could inhibit the growth of a bacterium. The mold was later identified

TABLE 1.1 Selected Nobel Prizes Awarded for Research in Microbiology

NOBEL LAUREATE	YEAR OF PRESENTATION	COUNTRY OF BIRTH	CONTRIBUTION
Emil A. von Behring	1901	Germany	Developed a diphtheria antitoxin
Robert Koch	1905	Germany	Cultured tuberculosis microbe
Paul Ehrlich	1908	Germany	Developed theories on immunity
Elie Metchnikoff	1908	Russia	Described phagocytosis, the intake of solid materials by cells
Alexander Fleming, Ernst Chain, and Howard Florey	1945	Scotland England	Discovered penicillin
Selman A. Waksman	1952	United States	Discovered streptomycin
Hans A. Krebs	1953	Germany	Discovered chemical steps of the Krebs cycle in carbohydrate metabolism
John F. Enders, Thomas H. Weller, and Frederick C. Robbins	1954	United States	Cultured polio virus in cell cultures
James D. Watson, Frances H.C. Crick, and Maurice A.F. Wilkins	1962	United States England New Zealand	Determined the structure of DNA
François Jacob, Jacques Monod, and André Lwoff	1965	France	Described how protein synthesis is regulated in bacteria
Robert Holley, Har Gobind Khorana, and Marshall W. Nirenberg	1968	United States India United States	Discovered the genetic code for amino acids
Max Delbrück, Alfred D. Hershey, and Salvador E. Luria	1969	Germany United States Italy	Described the mechanism of viral infection of bacterial cells
Gerald M. Edelman and Rodney R. Porter	1972	United States England	Described the nature and structure of antibodies
Renato Dulbecco, Howard Temin, and David Baltimore	1975	United States	Discovered reverse transcriptase and described how RNA viruses could cause cancer
Daniel Nathans, Hamilton Smith, and Werner Arber	1978	United States United States Switzerland	Described the action of restriction enzymes (now used in genetic engineering)
Peter Mitchell	1978	Great Britain	Described the chemiosmotic mechanism for ATP synthesis
Paul Berg	1980	United States	Performed experiments in gene splicing (genetic engineering)
Walter Gilbert	1980	United States	Discovered method of DNA sequencing
Aaron Klug	1982	South Africa	Described the structure of TMV
Barbara McClintock	1983	United States	Discovered transposons
César Milstein, Georges J.F. Köhler, and Niels Kai Jerne	1984	Argentina Germany Denmark	Developed technique for producing monoclonal antibodies
Susumu Tonegawa	1987	Japan	Described genetics of antibody production

(Continued)

TABLE 1.1 *(Continued)*

NOBEL LAUREATE	YEAR OF PRESENTATION	COUNTRY OF BIRTH	CONTRIBUTION
Johann Deisenhofer, Robert Huber, and Hartmut Michel	1988	Germany	Described the structure of bacterial photosynthesis pigments
J. Michael Bishop and Harold E. Varmus	1989	United States	Discovered cancer-causing genes called oncogenes
Joseph E. Murray and E. Donnall Thomas	1990	United States	Performed first successful transplants by using immunosuppressive agents

as *Penicillium notatum* (pe-ni-sil'lē-um nō-tä'tum), and in 1928 Fleming named the mold's active inhibitor *penicillin*.

Penicillin is an antibiotic produced by a fungus. The enormous usefulness of penicillin was not apparent until the 1940s, when it was finally mass produced and tested clinically. Since then, many other antibiotics have been discovered.

Unfortunately, antibiotics and other chemotherapeutic drugs are not without problems. Many antimicrobial chemicals are too toxic to humans for practical use; they kill the pathogenic microbes, but they also have damaging side effects on the infected host. For reasons we shall discuss later, toxicity to humans is a particular problem in the development of drugs for treating viral diseases. Because viral growth is so closely linked with the life processes of the normal host cell, there are very few successful antiviral drugs.

Another major problem associated with antimicrobial drugs is the emergence and spread of new varieties of microorganisms that are resistant to antibiotics. The quest to solve this and other problems requires sophisticated research techniques and correlated studies that were never dreamed of in the days of Koch and Pasteur.

MODERN DEVELOPMENTS IN MICROBIOLOGY

The groundwork laid during the Golden Age of Microbiology provided the basis for several monumental achievements during the twentieth century (Table 1.1). New branches of microbiology were developed, including immunology and virology. Most recently, the development of a set of new methods called recombinant DNA technology has revolutionized research and practical applications in all areas of microbiology.

Immunology

Immunology, the study of immunity, actually dates back to Jenner's development of the first vaccine in 1798. Since that time, knowledge about the immune system has accumulated steadily, and it expanded rapidly in the twentieth century. Vaccines are now available for numerous diseases, including measles, mumps, polio, rubella, and hepatitis B. The smallpox vaccine is so effective that not a single case of the disease has been reported worldwide since 1977. In 1960, interferon, a substance generated by the body's own immune system, was discovered. Interferon inhibits replication of viruses and has triggered considerable research related to treatment of viral diseases and cancer. One of this century's biggest challenges for immunologists is learning how the immune system might be stimulated to ward off HIV, the virus responsible for AIDS.

Virology

The study of viruses, **virology,** actually had its origin during the Golden Age of Microbiology. In 1892, Dmitri Iwanowski reported that the organism that caused mosaic disease of tobacco was so small that it passed through filters fine enough to stop all known bacteria. At the time, Iwanowski was not aware that the organism in question was a virus in the sense that we now understand the term. In 1935, Wendell Stanley showed that the organism, called tobacco mosaic virus (TMV), was fundamentally different from other microbes and so simple and homogeneous that it could be crystallized like a chemical compound. Stanley's work facilitated the study of viral structure and chemistry. Since the development of the electron microscope in the 1940s, microbiologists have been able to observe the structure of viruses in detail, and today much is known about their structure and activity.

Recombinant DNA Technology

The set of methods called **recombinant DNA technology** had its origins in two related fields. The first, **microbial genetics,** studies the mechanisms by which microorganisms inherit traits. The second, **molecular biology,** specifically studies how genetic information is carried in molecules of DNA and how DNA directs the synthesis of proteins.

Although molecular biology encompasses all organisms, much of our knowledge of how genes cause specific traits has been revealed through experiments with bacteria. Until the 1930s, all genetic research was based on the study of plant and animal cells. But in the 1940s, scientists turned to unicellular organisms, primarily bacteria, which have several advantages for genetic and biochemical research. For one thing, bacteria are less complex than plants and animals. For another, the life cycles of many bacteria require less than an hour, so scientists can cultivate very large numbers of individuals for study in a relatively short time.

Once science turned to the study of unicellular life, breakthroughs in genetics began to appear rapidly. DNA was established as the hereditary material in 1944 by Oswald Avery, Colin MacLeod, and Maclyn McCarty. Then, in 1953, James Watson and Francis Crick proposed a model for the structure and replication of DNA. The early 1960s witnessed a further explosion of discoveries relating to the way DNA controls protein synthesis. François Jacob and Jacques Monod discovered messenger RNA in 1961 and, later, made the first major discoveries about regulation of gene function in bacteria. During the same period, scientists were able to break the genetic code and thus understand the biochemical signals that DNA transmits.

Because of the knowledge gained in such research, microorganisms can now be genetically engineered to manufacture large amounts of human hormones and other urgently needed medical substances. In the late 1960s, Paul Berg showed that fragments of human or animal DNA that code for important proteins can be attached to bacterial DNA. The resulting hybrid was the first example of **recombinant DNA.** When recombinant DNA is inserted into bacteria, the bacteria can be used to make large quantities of the desired protein.

Naming and Classifying Microorganisms

The system of nomenclature (naming) for organisms in use today was established in 1735 by Carolus Linnaeus. Scientific names are latinized because Latin was the language traditionally used by scholars. Scientific nomenclature assigns each organism two names—the **genus** is the first name and is always capitalized; the **specific epithet (species)** follows and is not capitalized. The organism is referred to by both the genus and the specific epithet, and both words are underlined or italicized. By custom, after a scientific name has been mentioned once, it can be abbreviated with the initial of the genus followed by the specific epithet.

Scientific names can, among other things, describe the organism, honor a researcher, or identify the habitat of the species. For example, consider *Staphylococcus aureus,* a bacterium that is commonly found on the skin of humans and, as we saw earlier, is involved in TSS. *Staphylo-* describes the clustered arrangement of the cells; *coccus* indicates that they are shaped like spheres. The specific epithet, *aureus,* is Latin for golden, the color of many colonies of this bacterium. The genus of the bacterium *Escherichia coli* (esh-ėr-i'kē-ä kō'lī or kō'lē) is named for a scientist, Theodor Escherich, whereas its specific epithet, *coli,* reminds us that *E. coli* live in the colon, or large intestine.

Before the existence of microbes was known, all organisms were grouped into either the animal kingdom or the plant kingdom. When microscopic organisms with characteristics of animals or plants were discovered late in the seventeenth century, a new system of classification was needed. In 1866, Ernst Heinrich Haeckel, a German zoologist, proposed a classification system with three kingdoms: Animalia, Plantae, and Protista. The Protista consisted of single-celled microorganisms, such as bacteria, some fungi, algae, and protozoans.

In 1969, Robert Whittaker of Cornell University devised a five-kingdom classification system that is based on cellular organization and nutritional patterns of organisms. The five-kingdom system is widely accepted. It groups all organisms into the following kingdoms:

- Procaryotae or Monera (eubacteria and archaeobacteria)
- Protista (slime molds, protozoans, and some algae)
- Fungi (unicellular yeasts, multicellular molds, and mushrooms)
- Plantae (some algae and all mosses, ferns, conifers, and flowering plants)
- Animalia (including, among others, sponges, worms, insects, and vertebrates)

Classification will be discussed in more detail in Part Two. For now, however, let us simply introduce the major groups of microorganisms studied in this text.

The Diversity of Microorganisms

BACTERIA

Bacteria are very small, relatively simple, single-celled organisms whose genetic material is not enclosed in a special nuclear membrane (Figure 1.6a). For this reason, bacteria are called **procaryotes,** from Greek words meaning prenucleus. Bacteria make up the kingdom that Whittaker calls Monera; others, including microbiologists, call it Procaryotae. (Criteria for classifying bacteria will be discussed in Part Two.)

Bacterial cells generally appear in one of several shapes. **Bacillus** (rodlike), **coccus** (spherical or ovoid), and **spiral** (corkscrew or curved) are the most common shapes, but some bacteria are star-shaped or square (see Figures 4.1 through 4.4). Individual bacteria may form pairs, chains, clusters, or other groupings; such formations are usually characteristic of a particular species.

Bacteria are enclosed in cell walls that are largely composed of a substance called *peptidoglycan.* (By contrast, cellulose is the main substance of plant cell walls.) Bacteria generally reproduce by dividing into two equal daughter cells; this process is called *binary fission.* For nutrition, most bacteria use organic chemicals, which in nature can be derived from other, dead organisms or from a living host. Some bacteria can manufacture their own food by photosynthesis, and some can derive nutrition from inorganic substances. Many bacteria can "swim" by using moving appendages called *flagella.*

FUNGI

Fungi are **eucaryotes**—organisms whose cells have a distinct nucleus that contains the cell's genetic material and is surrounded by a special envelope called the nuclear membrane. They make up the Kingdom Fungi (see Figure 10.1, p. 252). Fungi may be unicellular or multicellular. Large multicellular fungi, such as mushrooms, may look somewhat like plants, but they cannot carry out photosynthesis, as most plants can. True fungi have cell walls composed primarily of a substance called *chitin.* The unicellular forms of fungi, *yeasts,* are oval microorganisms that are larger than bacteria. The most typical fungi are *molds.* Molds form *mycelia,* which are long filaments that branch and intertwine (Figure 1.6b). The cottony growths sometimes found on bread and fruit are mold mycelia. Fungi can reproduce sexually or asexually. They obtain nourishment by absorbing solutions of organic material from their environment—whether soil, water, or an animal or plant host.

PROTOZOANS

Protozoans are unicellular, eucaryotic microbes. They belong to the Kingdom Protista (Figure 1.6c). Protozoans are classified according to their means of locomotion. Amoebas move by using extensions of their cytoplasm called *pseudopods* (false feet). Other protozoans have flagella or numerous shorter appendages called *cilia.* They have a variety of shapes and live as free entities or parasites (organisms that derive nutrients from living hosts) that absorb or ingest organic compounds from their environment. Protozoans can reproduce sexually or asexually.

ALGAE

Algae are photosynthetic eucaryotes with a wide variety of shapes and both sexual and asexual reproductive forms (Figure 1.6d). The algae of interest to microbiologists are members of the Kingdom Protista and are usually unicellular. The cell walls of many algae, like those of plants, are composed of cellulose. They are abundant in fresh and salt water, in soil, and in association with plants. As photosynthesizers, algae need light and air for food production and growth, but they do not generally require organic compounds from the environment. As a result of photosynthesis, algae produce oxygen and carbohydrates, which are utilized by other organisms, including animals. Thus they play an important role in the balance of nature.

VIRUSES

Viruses (Figure 1.6e) are very different from the other microbial groups mentioned here. They are so small that most can be seen only with an electron microscope, and they are not cellular. Structurally very simple, a virus particle contains a core made of only one type of nucleic acid, either DNA (deoxyribonucleic acid) or RNA (ribonucleic acid). This core is surrounded by a protein coat. Sometimes the coat is encased by an additional layer, a lipid membrane called an envelope. All living cells have RNA *and* DNA, can carry out chemical reactions, and can reproduce as self-sufficient units. Viruses can reproduce only inside the cells of other organisms. Thus, all viruses are parasites of other forms of life.

MULTICELLULAR ANIMAL PARASITES

Although multicellular parasites are not strictly microorganisms, they are of medical importance and so will be discussed in this text. The two major groups of parasites are the flatworms and the roundworms, collectively called **helminths** (see Chapter 12). During some stages of their life cycle, the helminths are microscopic in size. Laboratory identification of these organisms includes many of the same techniques used for the identification of microbes.

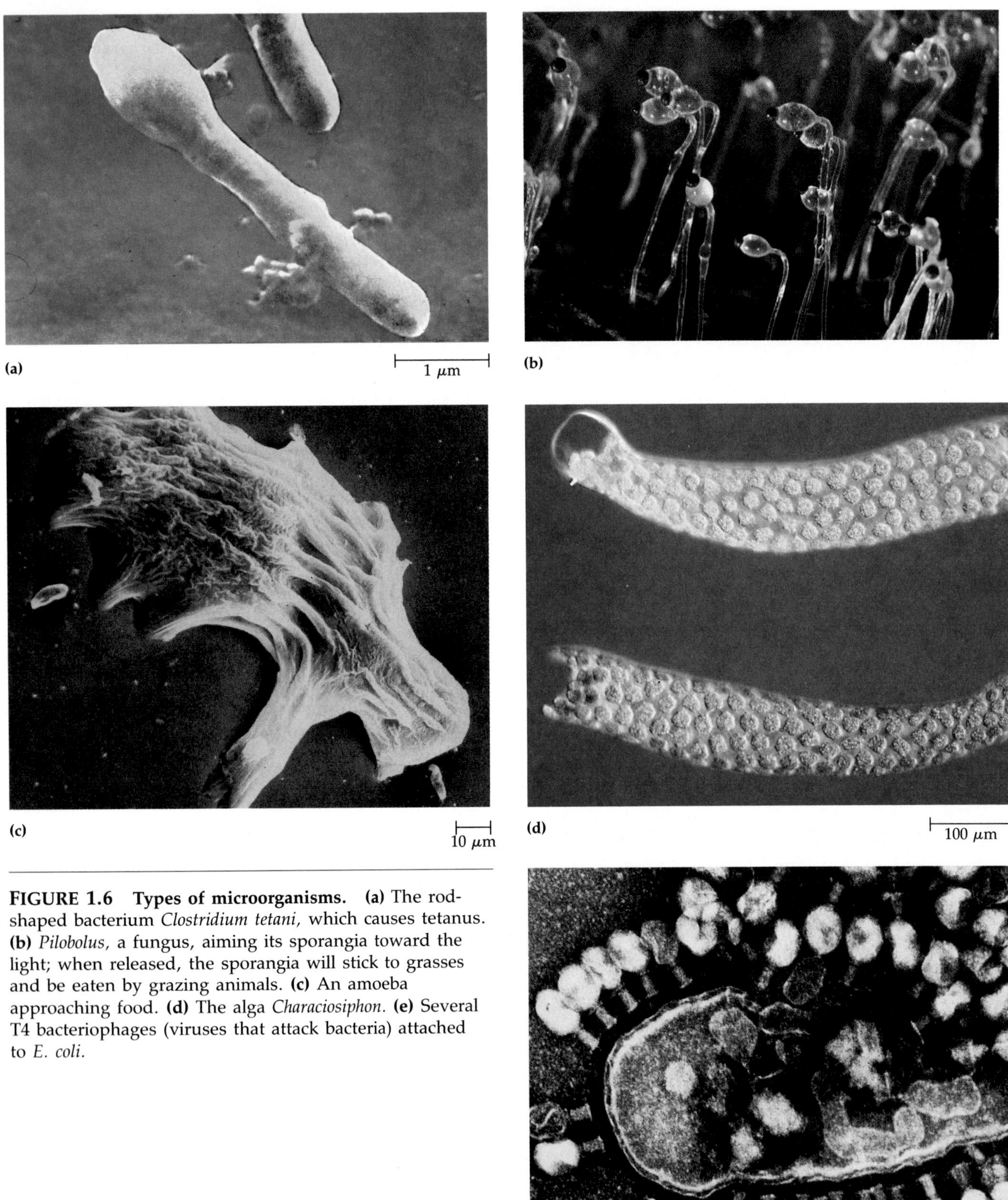

FIGURE 1.6 Types of microorganisms. **(a)** The rod-shaped bacterium *Clostridium tetani,* which causes tetanus. **(b)** *Pilobolus,* a fungus, aiming its sporangia toward the light; when released, the sporangia will stick to grasses and be eaten by grazing animals. **(c)** An amoeba approaching food. **(d)** The alga *Characiosiphon.* **(e)** Several T4 bacteriophages (viruses that attack bacteria) attached to *E. coli.*

Microbes and Human Welfare

As we mentioned earlier, only a minority of all microorganisms are pathogenic. Microbes that cause food spoilage, such as soft spots on fruits and vegetables, decomposition of meats, rancidity of fats and oils, and ropiness of bread, are also a minority. The vast majority of microbes benefit humans as well as other plants and animals in numerous ways. The following sections will outline some of these beneficial activities. In later chapters, we will discuss these activities in greater detail.

RECYCLING VITAL ELEMENTS

Discoveries made by two microbiologists in the 1880s have formed the basis for today's understanding of the biochemical cycles that support life on Earth. Martinus Beijerinck and Sergei Winogradsky were the first to show how bacteria help recycle vital elements between the soil and the atmosphere. **Microbial ecology,** the study of the relationship between microorganisms and their environment, originated with the work of Beijerinck and Winogradsky. Today, microbial ecology has branched out and includes the study of how microbial populations interact in various environments with plants and animals. Among the concerns of microbial ecologists are water pollution and toxic chemicals in the environment.

The chemical elements nitrogen, carbon, oxygen, sulfur, and phosphorus are essential for life and are available only in limited, though large, amounts. Nitrogen, oxygen, and carbon (as carbon dioxide) exist as gases in the atmosphere. Sulfur and phosphorus are stored in the Earth's crust. For the most part, microorganisms convert these elements into forms that can be used by plants and animals.

The **nitrogen cycle** is one example of how microorganisms recycle an element (see Chapter 27). Nitrogen is a major constituent of all living cells and is present in proteins and in nucleic acids (DNA and RNA). Although 79% of the atmosphere is gaseous nitrogen, animals and plants cannot use it in that form. Certain soil bacteria and cyanobacteria (Monera) in water combine gaseous nitrogen with other elements (*nitrogen fixation*) to form compounds that other organisms can use. When other bacteria help to decompose dead plants and animals, nitrogen-containing chemicals are released into the soil; still other microbes then convert the soil nitrogen back into nitrogen gas.

In the **carbon cycle,** green plants and algae remove carbon dioxide from the air and use sunlight to convert the carbon dioxide into food by the process of photosynthesis (see Chapter 27). When this food is consumed by other organisms, including humans, carbon dioxide is produced in the process called respiration. Microorganisms, mostly bacteria, play another key role in the carbon cycle by returning carbon dioxide to the atmosphere when they decompose organic wastes and dead plants and animals. Algae, higher plants, and certain bacteria (cyanobacteria) also participate in the **oxygen cycle** by recycling oxygen to the air during photosynthesis. The nitrogen, carbon, and oxygen cycles, as well as those involving sulfur and phosphorus, will be discussed in greater detail in Chapters 5 and 27.

SEWAGE TREATMENT: USING MICROBES TO RECYCLE WATER

With our growing awareness of the need to preserve the environment, we are conscious of our responsibility to recycle precious water and prevent the pollution of rivers and oceans. One major pollutant is sewage, which consists of human excrement, waste water, industrial wastes, and surface run-off. Sewage is about 99.9% water with a few hundredths of one percent suspended solids. The remainder is a variety of dissolved materials.

Sewage plant treatments remove the undesirable materials and harmful microorganisms. Treatments combine various physical and chemical processes with action by beneficial microbes. Large solids such as paper, wood, glass, gravel, and plastic are removed from sewage; left behind are liquid and organic materials that bacteria convert into such by-products as carbon dioxide, nitrates, phosphates, sulfates, ammonia, hydrogen sulfide, and methane. (We shall discuss sewage treatment in detail in Chapter 28.)

USING MICROBES TO CLEAN UP TOXIC DUMPS

In 1988, scientists began using microbes to clean up toxic wastes produced by various industrial processes. For example, some bacteria can actually turn toxins into energy sources that they consume; others produce enzymes that break down toxins into less harmful substances. By using bacteria in these ways—a process known as **bioremediation**—it is possible to remove toxins from underground wells, chemical spills, toxic waste sites, and oil spills.

INSECT PEST CONTROL BY MICROORGANISMS

Insects cause devastating crop damage besides spreading diseases. Insect pest control is therefore important for agriculture as well as for the prevention of human disease.

The bacterium *Bacillus thuringiensis* (bä-sil'lus thür-in-jē-en'sis) has been used extensively in the United States to control such pests as alfalfa caterpillars, bollworms, corn borers, cabbageworms, tobacco budworms, and fruit tree leaf rollers. It is incorporated into a dusting powder that is applied to the crops these insects eat. The bacterium produces protein crystals that are toxic to the digestive systems of the insects.

By using microbial rather than chemical insect control, farmers can avoid harming the environment. Many chemical insecticides, such as DDT, remain in the soil as toxic pollutants and are eventually incorporated into the food chain.

MODERN INDUSTRIAL MICROBIOLOGY AND GENETIC ENGINEERING

Earlier in this chapter, we touched on the commercial use of microorganisms to produce some common foods and chemicals. This is called **biotechnology.** One recent extension of biotechnology that has been attracting considerable attention is **genetic engineering,** the use of recombinant DNA technology to expand the potential of bacteria as miniature biochemical factories. **Single-cell protein (SCP),** a microbe-made food, is another extension of industrial microbiology.

Genetic Engineering

Genetic engineering promises important potential uses in agriculture, but to date most of its contributions to human welfare have been in the medical area. Recombinant DNA techniques have been used thus far to produce a number of natural proteins that are otherwise expensive and difficult to isolate and purify. Such proteins have great potential for medical use. Among them are insulin, a hormone used to lower blood sugar level in diabetics; human growth hormone, required for growth during childhood; interferon, an antiviral (and possibly anticancer) substance; factor VIII, a clotting substance missing in the blood of most hemophiliacs; tissue plasminogen activator, a substance used to dissolve blood clots in coronary arteries; beta-endorphin, a substance that suppresses pain; hepatitis B protein, which is used in a vaccine against hepatitis B; and monoclonal antibodies, which are used to diagnose and treat cancer and assist in AIDS research.

Scientists are using recombinant DNA techniques in attempts to develop vaccines against several microorganisms, including those that cause herpes, AIDS, and malaria. Just recently, they have begun to use recombinant DNA techniques in the diagnosis and treatment of cancer in its early stages. It is possible that certain genetic diseases, such as diabetes and hemophilia, will eventually be eliminated by the insertion of correct DNA sequences directly into unhealthy human cells.

Single-Cell Protein

The demand for food steadily increases along with the world's growing population. In response to this demand, scientists are investigating several new food sources. One such source is SCP—the protein (and other nutrients) produced by microorganisms cultivated on industrial and domestic wastes.

The advantage of SCP over plant crops is that microorganisms grow rapidly and can produce a high protein yield, estimated to be 15 times greater than the amount of soybeans (a high-protein food) and 50 times greater than the amount of corn grown during the same time. However, the high nucleic acid (DNA) content of SCP may elevate blood levels of uric acid and result in kidney stones, gout, allergic reactions, and gastrointestinal problems. In addition, SCP is not very tasty and in some cases might have to be supplemented with essential amino acids to provide a balanced protein source. At present, SCP is used primarily in animal feed.

Microbes and Human Disease

We all live in a microbial world, from birth until death, and we all have a variety of microorganisms on and inside our bodies. These microorganisms make up the *normal flora* (Figure 1.7). Although the normal flora usually benefit us, under some circumstances they can cause us illness or infect people we contact.

When is a microbe a welcome part of a healthy human, and when is it a harbinger of disease? The distinction between health and disease is in large part a balance between the natural defenses of the body and the disease-producing properties of microorganisms.

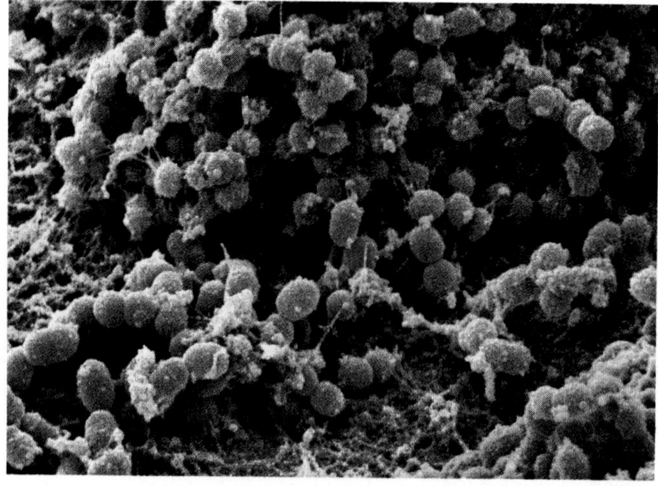

FIGURE 1.7 Several bacteria found as members of the normal flora inside the human mouth.

Whether our bodies overcome the offensive tactics of a particular microbe depends on our **resistance.** Important natural resistance is provided by the barrier of the skin, mucous membranes, cilia, stomach acid, and the antimicrobial chemicals such as interferon. Microbes can be destroyed by white blood cells, inflammation, fever, and our powerful immune systems. Sometimes, when our natural defenses are not strong enough to overcome an invader, they have to be supplemented by antibiotics or other drugs.

In the chapters that follow, you will become acquainted with the principles and tools that microbiologists use to understand the many interactions between microbes and their hosts.

STUDY OUTLINE

Microbes in Our Lives (pp. 4–6)

1. Living things too small to be seen with the naked eye are called microorganisms.
2. Microorganisms are important in the maintenance of an ecological balance on Earth.
3. Some microorganisms live in humans and other animals and are needed to maintain the animal's health.
4. Some microorganisms are used to produce foods and chemicals.
5. Some microorganisms cause disease.

A Brief History of Microbiology (pp. 6–14)

THE FIRST OBSERVATIONS (p. 6)

1. Robert Hooke observed that plant material was composed of "little boxes"; he introduced the term *cell* (1665).
2. Hooke's observations were the groundwork for development of the cell theory, the concept that all living things are composed of cells.
3. Anton van Leeuwenhoek, using a simple microscope, was the first to observe microorganisms (1673).

THE DEBATE OVER SPONTANEOUS GENERATION (pp. 6–8)

1. Until the mid-1880s, many people believed in spontaneous generation, the idea that living organisms could arise from nonliving matter.
2. Francesco Redi demonstrated that maggots appear on decaying meat only when flies are able to lay eggs on the meat (1668).
3. John Needham claimed that microorganisms could arise spontaneously from heated nutrient broth (1745).
4. Lazzaro Spallanzani repeated Needham's experiments and suggested that Needham's results were due to microorganisms in the air entering his broth (1765).
5. Rudolf Virchow introduced the concept of biogenesis—living cells can arise only from preexisting cells (1858).
6. Louis Pasteur demonstrated that microorganisms are in the air everywhere and offered proof of biogenesis (1861).
7. Pasteur's discoveries led to the development of aseptic techniques used in laboratory and medical procedures to prevent contamination by microorganisms that are in the air.

THE GOLDEN AGE OF MICROBIOLOGY (pp. 8–11)

1. Rapid advances in the science of microbiology were made between 1857 and 1914.

Fermentation and Pasteurization (pp. 8–10)

1. Pasteur found that yeast ferments sugars to alcohol and that bacteria can oxidize the alcohol to acetic acid.
2. A heating process called pasteurization is used to kill bacteria in some alcoholic beverages and milk.

The Germ Theory of Disease (p. 10)

1. Agostino Bassi (1834) and Pasteur (1865) showed a causal relationship between microorganisms and disease.
2. Joseph Lister introduced the use of a disinfectant to clean surgical dressings in order to control infections in humans (1860s).
3. Robert Koch proved that microorganisms transmit disease. He used a sequence of procedures called Koch's postulates (1876), which are used today to prove that a particular microorganism causes a particular disease.

Vaccination (pp. 10–11)

1. In a vaccination, immunity (resistance to a particular disease) is conferred by inoculation with a vaccine.
2. In 1798, Edward Jenner demonstrated that inoculation with cowpox material provides humans with immunity from smallpox.
3. About 1880, Pasteur discovered that avirulent bacteria could be used as a vaccine for chicken cholera; he coined the word *vaccine.*
4. Modern vaccines are prepared from living avirulent microorganisms or killed pathogens, from isolated components of pathogens, and by recombinant DNA techniques.

THE BIRTH OF MODERN CHEMOTHERAPY: DREAMS OF A "MAGIC BULLET" (pp. 11–13)

1. Chemotherapy is the chemical treatment of a disease.
2. Two types of chemotherapeutic agents are synthetic drugs (chemically prepared in the laboratory) and antibiotics (substances produced naturally by bacteria and fungi that inhibit the growth of other microorganisms).
3. Paul Ehrlich introduced an arsenic-containing chemical called salvarsan to treat syphilis (1910).
4. Alexander Fleming observed that the mold (fungus) *Penicillium* inhibited the growth of a bacterial culture. He named the active ingredient penicillin (1928).
5. Penicillin has been used clinically as an antibiotic since the 1940s.
6. Researchers are tackling the problem of drug-resistant microbes.

MODERN DEVELOPMENTS IN MICROBIOLOGY (pp. 13–14)

1. The study of AIDS, analysis of interferon action, and the development of new vaccines are among the current research interests in immunology.
2. New techniques in molecular biology and electron microscopy have provided tools for advancement of our knowledge of virology.
3. The development of recombinant DNA technology has helped advance all areas of microbiology.

Naming and Classifying Microorganisms (p. 14)

1. In a nomenclature system designed by Carolus Linnaeus (1735), each living organism is assigned two names.
2. The two names consist of a genus and specific epithet, both of which must be underlined or italicized.
3. In the five-kingdom system, all organisms are classified into Procaryotae (or Monera), Protista, Fungi, Plantae, or Animalia.

The Diversity of Microorganisms (pp. 14–16)

BACTERIA (pp. 14–15)

1. Bacteria are one-celled organisms. Because they have no nucleus, the cells are described as procaryotic.
2. The three major basic shapes of bacteria are bacillus, coccus, and spiral.
3. Most bacteria have a peptidoglycan cell wall; they divide by binary fission; and they may possess flagella.
4. Bacteria can use a wide range of chemical substances for their nutrition.

FUNGI (p. 15)

1. Fungi (mushrooms, molds, and yeasts) have eucaryotic cells (with a true nucleus). Most fungi are multicellular.
2. Fungi obtain nutrients by absorbing organic material from their environment.

PROTOZOANS (p. 15)

1. Protozoans are unicellular eucaryotes and are classified according to their means of locomotion.
2. Protozoans obtain nourishment by absorption or ingestion through specialized structures.

ALGAE (p. 15)

1. Algae are unicellular or multicellular eucaryotes that obtain nourishment by photosynthesis.
2. Algae produce oxygen and carbohydrates that are used by other organisms.

VIRUSES (p. 15)

1. Viruses are noncellular entities that are parasites of cells.
2. Viruses consist of a nucleic acid core (DNA or RNA) surrounded by a protein coat. An envelope may surround the coat.

MULTICELLULAR ANIMAL PARASITES (p. 15)

1. The principal groups of multicellular animal parasites are flatworms and roundworms, collectively called helminths.
2. The microscopic stages in the life cycle of helminths are identified by traditional microbiologic procedures.

Microbes and Human Welfare (pp. 17–18)

1. Microorganisms degrade dead plants and animals and recycle chemical elements to be used by living plants and animals.
2. Bacteria are used to decompose organic matter in sewage.
3. Bioremediation processes use bacteria to clean up toxic wastes.
4. Bacteria that cause diseases in insects are being used as biological controls of insect pests. Biological controls are specific for the pest and do not harm the environment.
5. Using recombinant DNA, bacteria can produce important human proteins, such as insulin, beta-endorphin, and hepatitis B vaccine.
6. Microorganisms can be used to help produce foods. They are also food sources (single-cell protein) themselves.

Microbes and Human Disease (pp. 18–19)

1. Everyone has microorganisms in and on the body; these make up the *normal flora.*
2. The disease-producing properties of a species of microbe and the host's resistance are important factors in determining whether a person will contract a disease.

STUDY QUESTIONS

Review

1. How did the idea of spontaneous generation come about?
2. Some proponents of spontaneous generation believed that air is necessary for life. They thought that Spallanzani did not really disprove spontaneous generation because he hermetically sealed his flasks to keep air out. How did Pasteur's experiments address the air question without allowing the microbes in the air to ruin his experiment?
3. Briefly state the role played by microorganisms in each of the following.
   (a) Biological control of pests
   (b) Recycling of elements
   (c) Normal flora
   (d) Sewage treatment
   (e) Human insulin production
   (f) Vaccine production

4. Match the following people to their contribution toward the advancement of microbiology.

____ Ehrlich	**(a)** First to observe bacteria
____ Fleming	**(b)** First to observe cells in plant material and name them
____ Hooke	**(c)** Disproved spontaneous generation
____ Koch	**(d)** Proved that microorganisms can cause disease
____ Lister	**(e)** Discovered penicillin
____ Pasteur	**(f)** Used the first synthetic chemotherapeutic agent
____ Van Leeuwenhoek	**(g)** First to employ disinfectants in surgical procedures

5. Match the following microorganisms to their descriptions.

____ Algae	**(a)** Not composed of cells
____ Bacteria	**(b)** Cell wall made of chitin
____ Fungi	**(c)** Cell wall made of peptidoglycan
____ Protozoans	**(d)** Cell wall made of cellulose; photosynthetic
____ Viruses	**(e)** Complex cell structure lacking a cell wall

6. The genus name of a bacterium is "erwinia" and the specific epithet is "carotovora." Write the scientific name of this organism correctly. Using this name as an example, explain how scientific names are chosen.

7. Into which field of microbiology would the following scientists best fit?

Researcher Who	Field
____ Studies biodegradation of toxic wastes.	**(a)** Immunology
____ Studies the causative agent of AIDS.	**(b)** Microbial ecology
____ Studies the production of human proteins by bacteria.	**(c)** Microbial genetics
____ Studies the symptoms of AIDS.	**(d)** Microbial physiology
____ Studies the production of TSS toxin by *S. aureus.*	**(e)** Molecular biology
____ Studies the life cycle of *B. burgdorferi.*	**(f)** Virology

Challenge

1. How did the theory of biogenesis lead the way for the germ theory of disease?

2. Since the germ theory of disease was not demonstrated until 1876, why did Semmelweis (1840) and Lister (1867) argue for the use of aseptic techniques?

3. Find at least three supermarket products made by microorganisms. (*Hint:* The label will state the scientific name of the organism or include the word *culture, fermented,* or *brewed.*)

FURTHER READING

Brock, T.D., ed. *Milestones in Microbiology.* Washington, D.C.: American Society for Microbiology, 1971. The writings of Jenner, Lister, Fleming, and others, with instructive editorial comments.

De Kruif, P. *Microbe Hunters.* New York: Harcourt Brace Jovanovich, 1966. The stories of van Leeuwenhoek, Koch, Pasteur, and others, presented in an interesting narrative.

Habicht, G.S., G. Beck, and J.L. Benach. "Lyme disease." *Scientific American* 257(1):78–83, July 1987. A summary of the disease, its causes, and its effects on the human body.

Jaret, P. "Our immune system: the war within." *National Geographic* 169(6):702–734, June 1986. An exciting look at AIDS, malaria, and conception through color-enhanced electron micrographs.

Margulis, L., and K.V. Schwartz. *Five Kingdoms: An Illustrated Guide to the Phyla of Life on Earth,* 2nd ed. New York: W.H. Freeman, 1988. A discussion of the classification of living organisms and descriptions of phyla.

"Microbes for hire." *Science 85* 6(6):30–46, July/August 1985. Three articles that discuss and illustrate technological uses of microbes in large-scale production of foods, vaccines, and drugs.

Rosebury, T. *Life on Man.* New York: Viking Press, 1969. A humorous yet scientific account of the role of microbes on the human body.

"What science knows about AIDS." *Scientific American* 259(4), October 1988. Ten articles cover the spread of the virus, the search for prevention of and cures for AIDS, and social issues concerning the disease.

CHAPTER 2

Chemical Principles

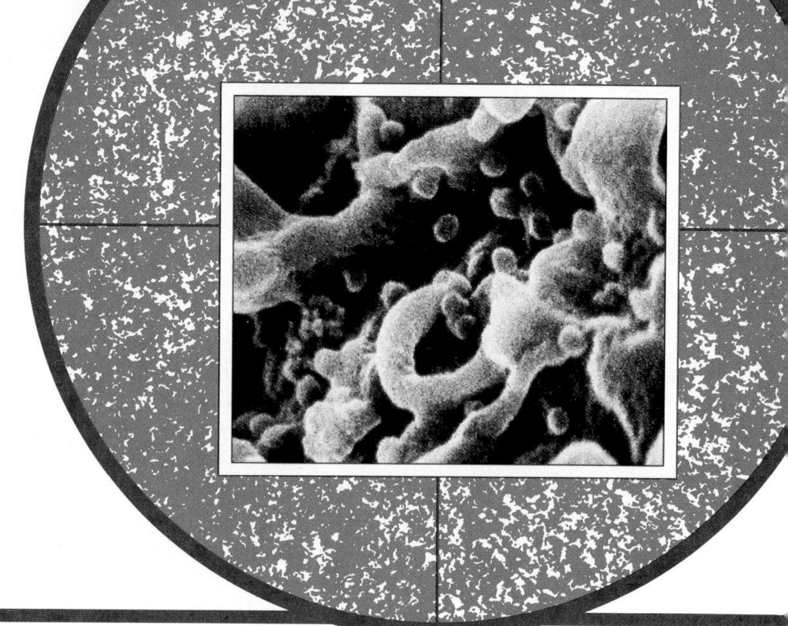

LEARNING OBJECTIVES

- Discuss the structure of an atom and its relation to the chemical properties of elements.
- Define ionic bond, covalent bond, hydrogen bond, molecular weight, mole.
- Diagram three basic types of chemical reactions.
- Identify the role of enzymes in chemical reactions.
- List several properties of water that are important to living systems.
- Distinguish between inorganic and organic molecules.
- Define acid, base, salt, pH.
- Identify the building blocks of carbohydrates, simple lipids, phospholipids, proteins, and nucleic acids.
- Identify the role of ATP in cellular activities.

We can see a tree rot and smell milk when it sours, but we might not realize what is happening on a microscopic level. In both cases, microbes are conducting chemical operations. The tree rots when microorganisms induce the hydrolysis of wood. (Hydrolysis is a chemical reaction that involves the splitting of water.) Milk turns sour from the production of lactic acid by bacteria. Most of the activities of microorganisms are the result of a series of chemical reactions.

Like all organisms, microorganisms must use nutrients to make chemical building blocks that are used for growth and for all the other functions essential to life. For most microorganisms, synthesizing these building blocks requires them to break down nutrient substances and use the energy released to assemble the molecular fragments into new substances. These chemical reactions take place minute by minute in countless microenvironments that in many cases are smaller than the head of a pin.

All matter—whether air, rock, or a living organism—is made up of small units called **atoms.** Atoms interact with each other and in certain combinations to form **molecules.** Living cells are made up of molecules, some of which are very complex. The science of the interaction of atoms and molecules is called **chemistry.**

The chemistry of microbes is the microbiologist's concern. To understand the changes that occur in microorganisms and the changes microbes make in the world around us, you will need to know how molecules are formed and how they interact.

Structure of Atoms

Atoms are the smallest units of matter that enter into chemical reactions. All atoms have a centrally located

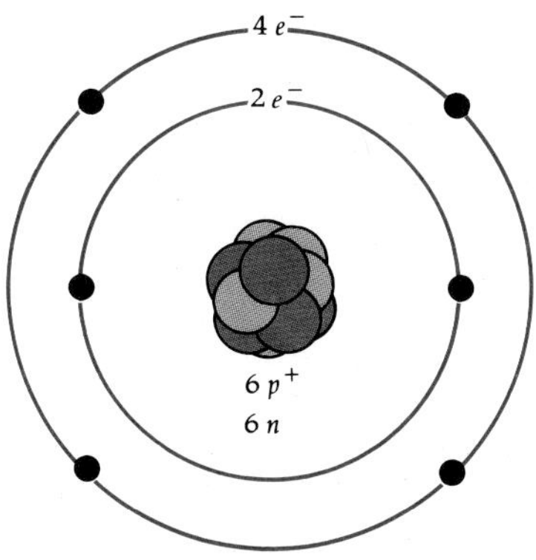

FIGURE 2.1 Atomic structure. Illustrated here is a very simplified version of a carbon atom (C). It contains six protons (p^+, dark red) and six neutrons (n, light red) in its centrally located nucleus. Six electrons (e^-, black) orbit the nucleus.

nucleus and particles called **electrons** that move around the nucleus in patterns called electronic configurations (Figure 2.1). The nuclei of most atoms are stable—that is, they do not change spontaneously—and nuclei do not participate in chemical reactions. The nucleus is made up of positively (+) charged particles called **protons** and uncharged (neutral) particles called **neutrons.** The nucleus, therefore, bears a net positive charge. Neutrons and protons have approximately the same weight, which is about 1840 times that of an electron. The charge on electrons is negative (−), and in all atoms the number of protons is equal to the number of electrons. Because the total positive charge of the nucleus equals the total negative charge of the electrons, each atom is electrically neutral.

The number of protons in an atomic nucleus ranges from one (in a hydrogen atom) to more than 100 (in the largest atoms known). Atoms are often listed by their **atomic number,** the number of protons in the nucleus (Table 2.1). The total number of protons and neutrons in an atom is its approximate **atomic weight.**

TABLE 2.1 The Elements of Life*

ELEMENT	SYMBOL	ATOMIC NUMBER	APPROXIMATE ATOMIC WEIGHT
Hydrogen	H	1	1
Carbon	C	6	12
Nitrogen	N	7	14
Oxygen	O	8	16
Fluorine	F	9	19
Sodium	Na	11	23
Magnesium	Mg	12	24
Silicon	Si	14	28
Phosphorus	P	15	31
Sulfur	S	16	32
Chlorine	Cl	17	35
Potassium	K	19	39
Calcium	Ca	20	40
Vanadium	V	23	51
Chromium	Cr	24	52
Manganese	Mn	25	55
Iron	Fe	26	56
Cobalt	Co	27	59
Nickel	Ni	28	59
Copper	Cu	29	64
Zinc	Zn	30	65
Arsenic	As	33	75
Selenium	Se	34	79
Molybdenum	Mo	42	96
Tin	Sn	50	119
Iodine	I	53	127

*Hydrogen, carbon, nitrogen, and oxygen are the most abundant chemical elements in living organisms.

CHEMICAL ELEMENTS

All atoms with the same atomic number behave the same way chemically and are classified as the same **chemical element.** Each element has its own name and a one- or two-letter symbol, usually derived from the English or Latin name for the element. For example, the symbol for the element hydrogen is H, and the symbol for carbon is C. The symbol for sodium is Na—the first two letters of its Latin name, *natrium*—to distinguish it from nitrogen, N, and from sulfur, S. There are 92 naturally occurring elements. However, only about 26 elements are commonly found in living things; these are listed in Table 2.1. The elements most abundant in living matter are hydrogen, carbon, nitrogen, and oxygen.

Most elements have several **isotopes**—atoms with differing numbers of neutrons in their nuclei. All isotopes of an element have the same number of protons in their nuclei, but their atomic weights differ because of the difference in the number of neutrons. For example, in a natural sample of oxygen, all the atoms will contain eight protons. However, 99.76% of the atoms will have eight neutrons, 0.04% will contain nine neutrons, and the remaining 0.2% will contain ten neutrons. Therefore, the three isotopes composing a natural sample of oxygen will have atomic weights of 16, 17, and 18, although all will have the atomic number 8. Atomic numbers are written as a subscript to the left of an element's chemical symbol. Atomic weights are written as a superscript above the atomic number. Thus, natural oxygen isotopes are designated as $^{16}_{8}O$, $^{17}_{8}O$, and $^{18}_{8}O$. Rare isotopes of certain elements are extremely useful in biological research, medical diagnosis, and treatment of some disorders.

ELECTRONIC CONFIGURATIONS

In an atom, electrons are arranged in **electron shells,** which are regions corresponding to different **energy levels.** The arrangement is called an **electronic configuration.** Shells are layered outward from the nucleus, and each shell can hold a characteristic maximum number of electrons—two electrons in the innermost shell (lowest energy level), eight electrons in the second shell, and eight electrons in the third shell, if it is the atom's outermost (valence) shell. The fourth, fifth, and sixth electron shells can each accommodate 18 electrons, although there are some exceptions to this generalization. Table 2.2 shows the electronic configurations for atoms of some elements found in living organisms.

There is a tendency for the outermost shell to be filled with the maximum number of electrons. An atom can give up, accept, or share electrons with other atoms to fill this shell. The chemical properties of atoms are largely a function of the number of electrons in the outermost electron shell. When its outer shell is filled, the atom is chemically stable, or inert: it does not tend to react with other atoms. Helium (atomic number 2) and neon (atomic number 10) are examples of atoms that have filled outer shells. Helium has two electrons in the first and only shell, and neon has two electrons in the first shell and eight electrons in the second shell, its outermost shell.

When an atom's outer electron shell is only partially filled, the atom is chemically unstable. Such an atom reacts with other atoms, and this reaction depends, in part, on the degree to which the outer energy levels are filled. Note the number of electrons in the outer energy levels of the atoms in Table 2.2. We will see later how the number correlates with the chemical reactivity of the elements.

How Atoms Form Molecules: Chemical Bonds

When the outermost energy level of an atom is not completely filled by electrons, it may be thought of as having either unfilled spaces or extra electrons in that energy level, depending on whether it is easier for the atom to gain or lose electrons. For example, an atom of oxygen, with two electrons in the first energy level and six in the second, has two unfilled spaces in the second electron shell; an atom of magnesium has two extra electrons in its outermost shell. The most chemically stable configuration for any atom is to have its outermost shell filled, as do the inert gases. Therefore, for these two atoms to attain that state, oxygen has to gain two electrons, and magnesium has to lose two electrons. All atoms tend to combine so that the extra electrons in the outermost shell of one atom fill the spaces of the outermost shell of the other atom; in the cases of oxygen and magnesium, the outermost shell of each atom would have eight electrons when each element combines with another element whose electrons fill the outer shell.

The **valence,** or combining capacity, of an atom is the number of extra or missing electrons in its outermost electron shell. For example, hydrogen has a valence of 1 (one unfilled space, or one extra electron), oxygen has a valence of 2 (two unfilled spaces), carbon has a valence of 4 (four unfilled spaces, or four extra electrons), and magnesium has a valence of 2 (two extra electrons). Valence may also be viewed as the bonding capacity of an element. Hydrogen can form one chemical bond with another atom, oxygen can form two chemical bonds with other atoms, carbon can form four chemical bonds, and magnesium can form two chemical bonds.

Basically, atoms gain chemical stability by achieving the full complement of electrons in their outermost energy shells. They do this by combining to form molecules, which are made up of atoms of one or more elements. A molecule that contains at least two different kinds of atoms, such as H_2O, the water molecule, is called a **compound.** In H_2O, the subscript 2 indicates that there are two atoms of hydrogen and one atom of oxygen. Molecules hold together because the valence electrons of the combining atoms form attractive forces, called **chemical bonds,** between the atomic nuclei. Because energy is required for chemical bond formation, each chemical bond possesses a certain amount of potential chemical energy. Atoms of a given element can form only a specific number of chemical bonds, based on their electron configurations.

In general, atoms form bonds in one of two ways—by gaining or losing electrons from their outer electron shell or by sharing outer electrons. When atoms have gained or lost outer electrons, the chemical bond is called an ionic bond. When outer electrons are shared, the bond is called a covalent bond. Although we will discuss ionic and covalent bonds separately, the kinds of bonds actually found in molecules do not belong

TABLE 2.2 Electronic Configurations for the Atoms of Some Elements Found in Living Organisms

ELEMENT	FIRST ELECTRON SHELL	SECOND ELECTRON SHELL	THIRD ELECTRON SHELL	DIAGRAM	NUMBER OF VALENCE (OUTERMOST) SHELL ELECTRONS	NUMBER OF UNFILLED SPACES	MAXIMUM NUMBER OF BONDS FORMED
Hydrogen	1			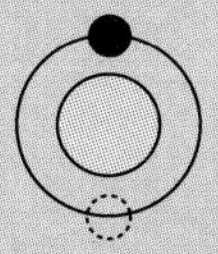	1	1	1 (by either losing or gaining 1 electron)
Carbon	2	4		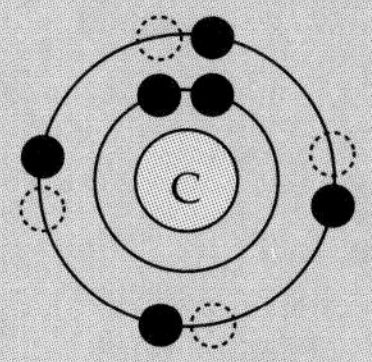 C	4	4	4 (by either losing or gaining 4 electrons)
Nitrogen	2	5		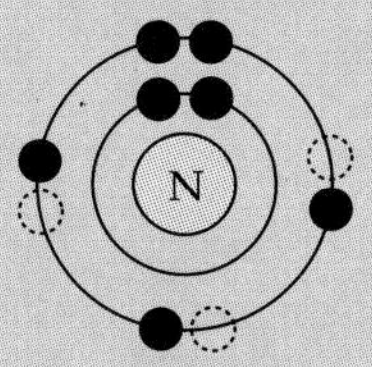 N	5	3	3 (by gaining 3 electrons)
Oxygen	2	6		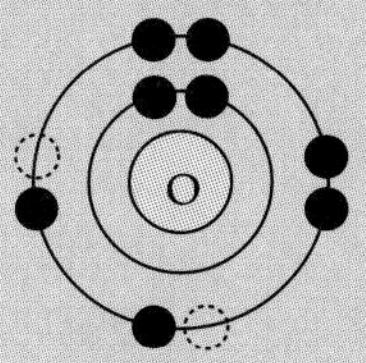 O	6	2	2 (by gaining 2 electrons)
Magnesium	2	8	2	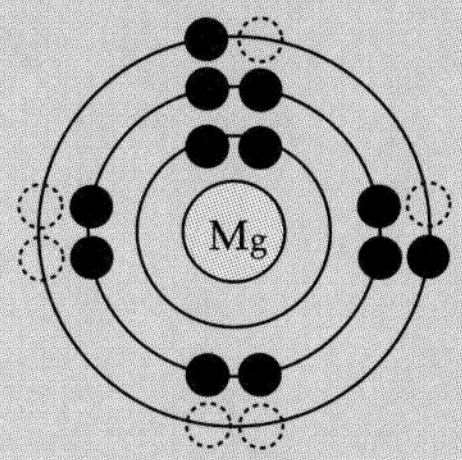 Mg	2	6	2 (by losing 2 electrons)
Phosphorus	2	8	5	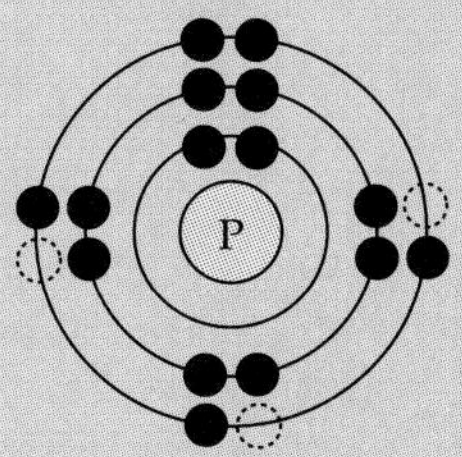 P	5	3	3 (by gaining 3 electrons)
Sulfur	2	8	6	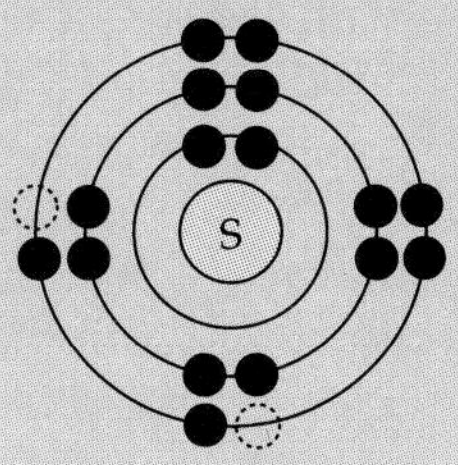 S	6	2	2 (by gaining 2 electrons)

Key
● electron
◌ unfilled space
nucleus

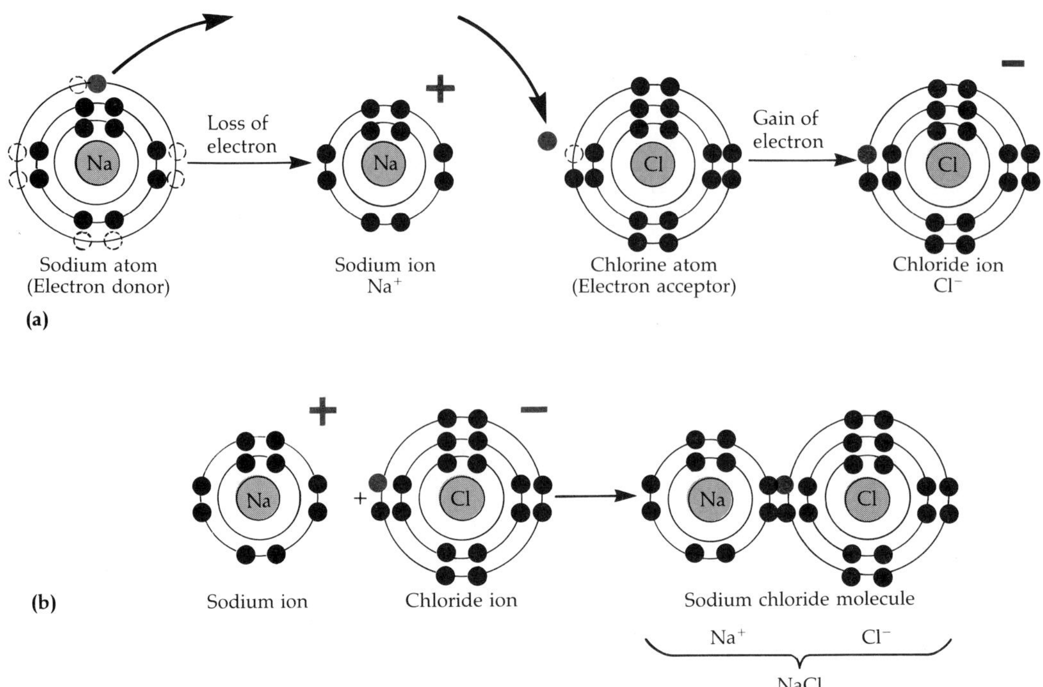

FIGURE 2.2 Ionic bond formation. **(a)** A sodium atom (Na), left, loses one electron to an electron acceptor and forms a sodium ion (Na$^+$). A chlorine atom (Cl), right, accepts one electron from an electron donor to become a chloride ion (Cl$^-$). **(b)** The sodium and chloride ions are attracted because of their opposite charges and are held together by an ionic bond to form a molecule of sodium chloride.

entirely to either category. Instead, bonds range from the highly ionic to the highly covalent.

IONIC BONDS

Atoms are electrically neutral when the number of positive charges (protons) equals the number of negative charges (electrons). But when an isolated atom gains or loses electrons, this balance is upset. If the atom gains electrons, it acquires an overall negative charge; if the atom loses electrons, it acquires an overall positive charge. Such a negatively or positively charged atom (or group of atoms) is called an **ion.**

Consider the following examples (Figure 2.2). Sodium (Na) has 11 protons and 11 electrons, with one electron in its outer electron shell. Sodium tends to lose the single outer electron; it is an *electron donor* (Figure 2.2a). When sodium donates an electron to another atom, it is left with 11 protons and only 10 electrons and so has an overall charge of +1. This positively charged sodium atom is called a sodium ion and is written as Na$^+$. Chlorine (Cl) has a total of 17 electrons, 7 of them in the outer electron shell. Because this outer shell can hold 8 electrons, chlorine tends to pick up an electron that has been lost by another atom; it is an *electron acceptor* (Figure 2.2a). By accepting an electron, chlorine totals 18 electrons. However, it still has only 17 protons in its nucleus. The chloride ion therefore has a charge of −1 and is written as Cl$^-$ (Figure 2.2a).

The opposite charges of the sodium ion (Na$^+$) and chloride ion (Cl$^-$) attract each other. The attraction, an ionic bond, holds the two atoms together, and a molecule is formed (Figure 2.2b). The formation of this molecule, called sodium chloride (NaCl) or table salt, is a common example of ionic bonding. Thus, an **ionic bond** is an attraction between ions of opposite charge that holds them together to form a stable molecule. Put another way, an ionic bond is an attraction between atoms in which one atom loses electrons and another atom gains electrons. Strong ionic bonds, such as those that hold Na$^+$ and Cl$^-$ ions together in salt crystals, have only limited importance in living cells. But the weaker ionic bonds formed in aqueous (water) solutions are important in biochemical reactions of microbes and other organisms. For example, weaker ionic bonds assume a role in certain antigen—antibody reactions—that is, reactions in which molecules produced by the immune system (antibodies) combine with foreign substances (antigens) in order to combat infection.

In general, an atom whose outer electron shell is less than half filled will lose electrons and form positively charged ions, called **cations.** Examples of cations are the potassium ion (K$^+$), calcium ion (Ca^{2+}), and sodium ion (Na$^+$). The symbol for an ion is the chemical abbreviation followed by the ion's number of positive (+) or negative (−) charges. When an atom's outer electron shell is more than half filled, the atom will gain electrons and form negatively charged ions, called **anions.** Examples are the iodide ion (I$^-$), chloride ion (Cl$^-$), and sulfide ion (S^{2-}).

Hydrogen is an example of an atom whose outer level is exactly half filled. The first energy level of a hydrogen atom can hold two electrons, but it contains only one electron. Hydrogen can lose its electron and become a positive ion (H^+), which is precisely what has happened when a hydrogen ion combines with a chloride ion to form hydrochloric acid (HCl). Hydrogen can gain an electron and become a negative ion (H^-), which can form ionic bonds with positive ions, but H^- ions are not very significant in living systems. Hydrogen atoms can also form a more important and altogether different kind of bond, a covalent bond.

COVALENT BONDS

A **covalent bond** is a chemical bond formed by two atoms sharing one or more pairs of electrons. Covalent bonds are stronger and far more common in organisms than are true ionic bonds. In the hydrogen molecule, H_2, two hydrogen atoms share a pair of electrons. Each hydrogen atom has its own electron plus one electron from the other atom (Figure 2.3a). The shared pair of electrons actually orbits the nuclei of both atoms. Therefore, the outer electron shells of both atoms are filled. When only one pair of electrons is shared between atoms, a *single covalent bond* is formed. For simplicity, a single covalent bond is expressed as a single line between the atoms (H—H). When two pairs of electrons are shared between atoms, a *double covalent bond* is formed, expressed as two single lines (=) (Figure 2.3b). A *triple covalent bond*, expressed as three single lines (≡), occurs when three pairs of electrons are shared (Figure 2.3c).

The principles of covalent bonding that apply to atoms of the same element also apply to atoms of different elements. Methane (CH_4) is an example of covalent bonding between atoms of different elements (Figure 2.3d). The outer electron shell of the carbon atom can hold eight electrons but has only four. Each hydrogen atom can hold two electrons but has only one. Consequently, in the methane molecule, the carbon atom gains four hydrogen electrons to complete its outer shell, and each hydrogen atom completes its pair by sharing one electron from the carbon atom. Each outer electron of the carbon atom orbits both the carbon nucleus and a hydrogen nucleus. Each hydrogen electron orbits both its own nucleus and the carbon nucleus.

Elements such as hydrogen and carbon, whose outer electron shells are half-filled, form covalent bonds quite easily. In fact, in living organisms, carbon almost always forms covalent bonds; it almost never becomes an ion. However, many atoms whose outer electron shells are almost full will also form covalent bonds. An example is oxygen. It is not necessary for now to know why some atoms tend to form covalent bonds rather than ionic bonds. But it is important to understand the basic principles of bond formation because chemical reactions are nothing more than the making or breaking of bonds between atoms. *Remember:* Covalent bonds are formed by the *sharing* of electrons between atoms. Ionic bonds are formed by *attractions* between atoms that have lost or gained electrons and are therefore positively or negatively charged.

HYDROGEN BONDS

Another chemical bond of special importance to all organisms is the **hydrogen bond,** in which a hydrogen atom that is covalently bonded to one oxygen or nitrogen atom is attracted to another oxygen or nitrogen atom. Such bonds are weak and do not bind atoms into molecules. However, they do serve as bridges between different molecules or between various portions of the same molecule.

When hydrogen combines with atoms of oxygen or nitrogen, the relatively large positive nucleus of these larger atoms attracts the hydrogen electron more strongly than does the small hydrogen nucleus. Thus, in a molecule of water (H_2O), all the electrons tend to be closer to the oxygen nucleus than to the hydrogen nuclei. The oxygen portion of the molecule thus has a slightly negative charge, and the hydrogen portion of the molecule has a slightly positive charge (Figure 2.4a). When the positively charged end of one molecule is attracted to the negative end of another molecule, a hydrogen bond is formed (Figure 2.4b). This attraction can also occur between hydrogen and other atoms of the same molecule, especially in large molecules. Because nitrogen and oxygen have unshared pairs of electrons, they are the elements most frequently involved in hydrogen bonding.

Hydrogen bonds are considerably weaker than either ionic or covalent bonds; they have only about 5% of the strength of covalent bonds. Consequently, hydrogen bonds are formed and broken relatively easily. This property accounts for the temporary bonding that occurs between certain atoms of large and complex molecules, such as proteins and nucleic acids. Even though hydrogen bonds are relatively weak, large molecules containing several hundred of these bonds have considerable strength and stability.

MOLECULAR WEIGHT AND MOLES

You have seen that bond formation results in the creation of molecules. Molecules are often discussed in terms of units of measure called molecular weight and moles. The **molecular weight** of a molecule is the sum of the atomic weights of all its atoms. To relate the molecular level to the laboratory level, we use a unit called the mole. One **mole** of a substance is its molecu-

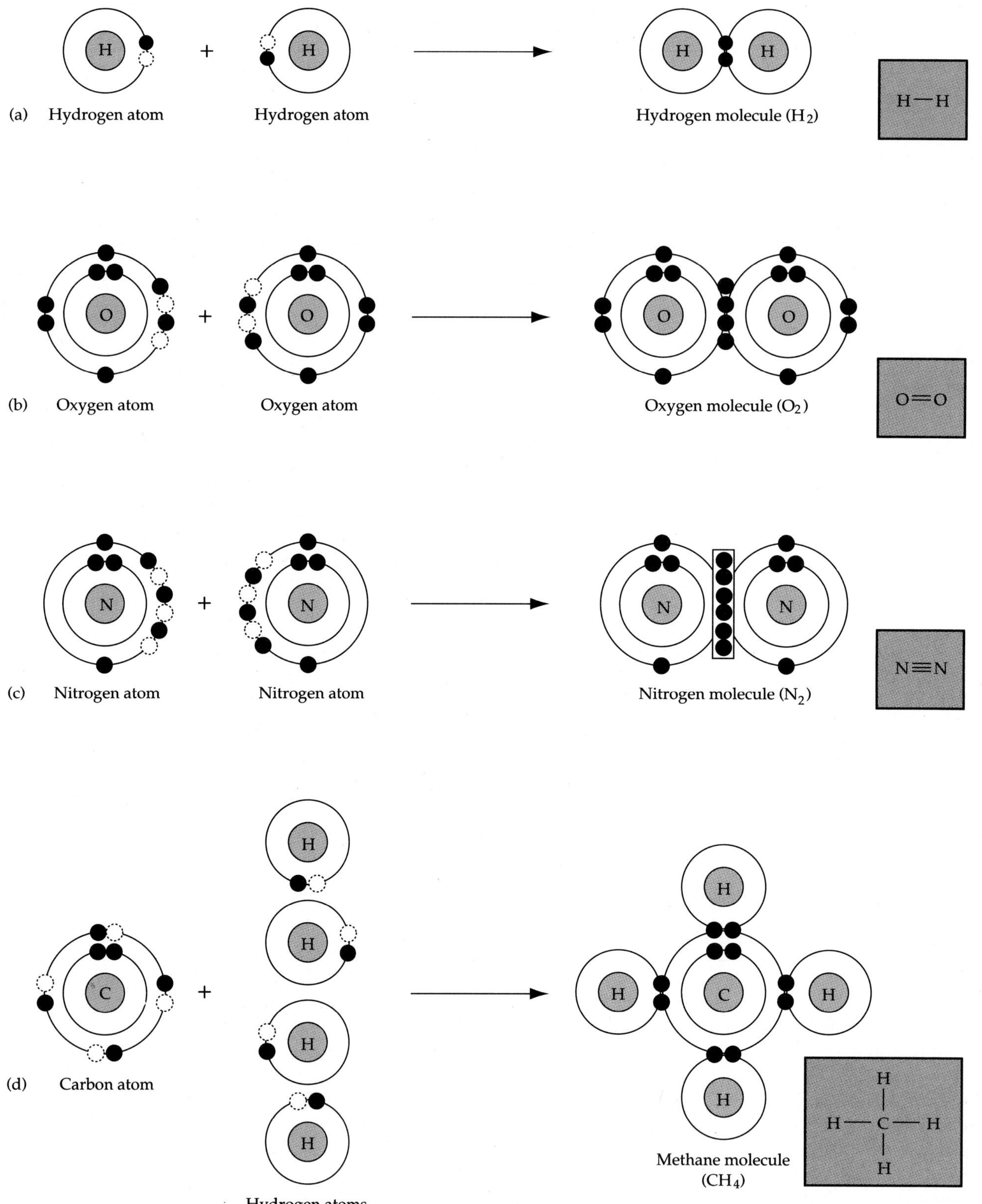

FIGURE 2.3 Covalent bond formation. **(a)** Single covalent bond between two hydrogen atoms. **(b)** Double covalent bond between two oxygen atoms. **(c)** Triple covalent bond between two nitrogen atoms. **(d)** Single covalent bonds between four hydrogen atoms and a carbon atom, forming the methane molecule.

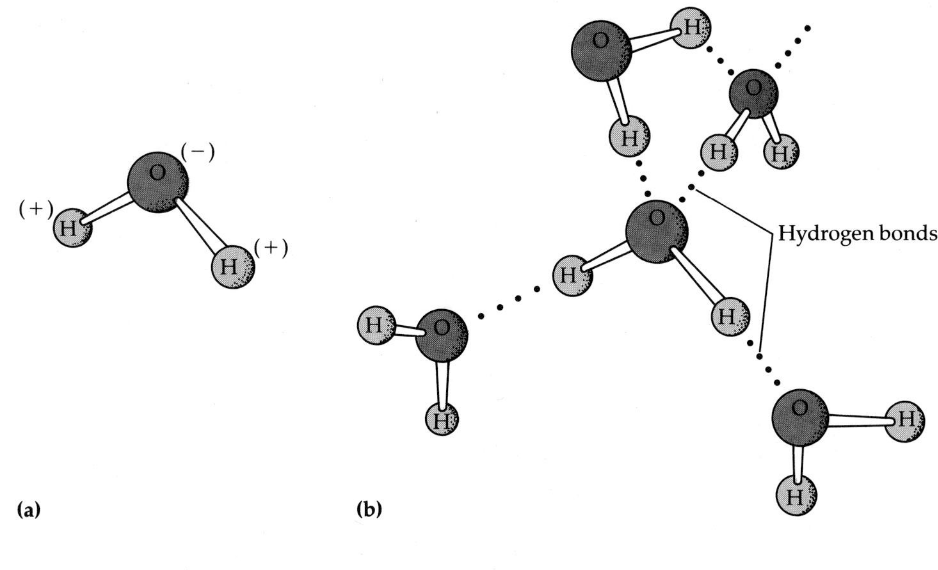

FIGURE 2.4 Hydrogen bond formation in water. (a) In a water molecule, the electrons of the hydrogen atoms are strongly attracted to the oxygen atom. Therefore, the part of the water molecule containing the oxygen atom has a slightly negative charge, and the part containing hydrogen atoms has a slightly positive charge. **(b)** In a hydrogen bond between water molecules, the hydrogen of one water molecule is attracted to the oxygen of another water molecule. Many water molecules may be attracted to each other by hydrogen bonds (black dots).

lar weight expressed in grams. For example, one mole of water weighs 18 grams because the molecular weight of H_2O is 18 ($2 \times 1 + 16$). The word *mole* can also be applied to atoms or even to ions. Thus, one mole of hydrogen ions is equal to 1 gram.

To determine the number of moles in a certain quantity of a substance, the mass of the substance (in grams) is divided by the molecular weight of the substance. For example, to calculate the number of moles in 500 grams of glucose ($C_6H_{12}O_6$), first calculate the number of grams in one mole ($6 \times 12 + 12 \times 1 + 6 \times 16 = 180$). Then divide 500 grams by 180 grams per mole to get 2.78 moles.

Chemical Reactions

As we said earlier, **chemical reactions** involve the making or breaking of bonds between atoms. After a chemical reaction, the total number of atoms remains the same, but there are new molecules with new properties because the atoms are rearranged.

ENERGY OF CHEMICAL REACTIONS

Some change of energy occurs whenever bonds between atoms are formed or broken during chemical reactions. This energy is called **chemical energy.** When a chemical bond is formed, energy is required. Such a chemical reaction that requires energy is called an **endergonic reaction,** meaning that energy is directed inward. When a bond is broken, energy is released. A chemical reaction that releases energy is an **exergonic reaction,** meaning that energy is directed outward.

In this section, we will look at three basic types of chemical reactions common to all living cells. By becoming familiar with these reactions, you will be able to understand the specific chemical reactions we will discuss later, particularly in Chapter 5.

SYNTHESIS REACTIONS

When two or more atoms, ions, or molecules combine to form new and larger molecules, the reaction is called a **synthesis reaction.** To synthesize means to put together, and a synthesis reaction *forms new bonds.* Synthesis reactions can be expressed in the following way:

A	+	B	$\longrightarrow$	AB
Atom, ion, or molecule A		Atom, ion, or molecule B	Combine to form	New molecule AB

The combining substances, A and B, are called the *reactants;* the substance formed by the combination, AB, is the *product.* The arrow indicates the direction in which the reaction proceeds.

Pathways of synthesis reactions in living organisms are collectively called anabolic reactions, or simply **anabolism.** The combining of sugar molecules to form starch, and of amino acids to form proteins, are two examples of anabolism.

DECOMPOSITION REACTIONS

The reverse of a synthesis reaction is a **decomposition reaction.** To decompose means to break down into smaller parts, and in a decomposition reaction, *bonds are broken.* Typically, decomposition reactions split large molecules into smaller molecules, ions, or atoms. A decomposition reaction occurs in this way:

$$\underset{\text{Molecule AB}}{AB} \xrightarrow[\text{down into}]{\text{Breaks}} \underset{\text{Atom, ion, or molecule A}}{A} + \underset{\text{Atom, ion, or molecule B}}{B}$$

Decomposition reactions that occur in living organisms are collectively called catabolic reactions, or simply **catabolism.** An example of catabolism is the breakdown of sucrose (table sugar) into simpler sugars, glucose and fructose, during digestion. Bacterial decomposition of petroleum is discussed on p. 38.

EXCHANGE REACTIONS

All chemical reactions are based on synthesis and decomposition. Many reactions, such as **exchange reactions,** are actually part synthesis and part decomposition. An exchange reaction works in this way:

$$AB + CD \xrightarrow[\text{to form}]{\text{Recombine}} AD + BC$$

First, the bonds between A and B and between C and D are broken in a decomposition process. New bonds are then formed between A and D and between B and C in a synthesis process. For example, an exchange reaction occurs when sodium hydroxide (NaOH) and hydrochloric acid (HCl) react to form table salt (NaCl) and water (H_2O) as follows:

$$NaOH + HCl \longrightarrow NaCl + H_2O$$

THE REVERSIBILITY OF CHEMICAL REACTIONS

All chemical reactions are, in theory, reversible; that is, they can occur in either direction. In practice, however, some reactions do this more easily than others. A chemical reaction that is readily reversible (meaning, the end product can revert to the original molecules) is termed a **reversible reaction** and is indicated by two arrows, as shown here:

$$A + B \underset{\text{Breaks down into}}{\overset{\text{Combines to form}}{\rightleftharpoons}} AB$$

Some reversible reactions occur because neither the reactants nor the end products are very stable. Other reactions will reverse only under special conditions:

$$A + B \underset{\text{Water}}{\overset{\text{Heat}}{\rightleftharpoons}} AB$$

Whatever is written above or below the arrows indicates the special condition under which the reaction in that direction occurs. In this case, A and B react to produce AB only when heat is applied, and AB breaks down into A and B only in the presence of water.

HOW CHEMICAL REACTIONS OCCUR

The **collision theory** explains how chemical reactions occur and how certain factors affect the rates of those reactions. The basis of the collision theory is that all atoms, ions, and molecules are continuously moving and are thus continuously colliding with one another. The energy transferred by the particles in the collision can disrupt their electron structures enough so that chemical bonds are broken or new ones are formed.

Factors Affecting Chemical Reactions

Several factors determine whether a collision will cause a chemical reaction—the velocities of the colliding particles, their energy, and their specific chemical configurations. Up to a point, the higher the particles' velocities, the greater the probability that their collision will cause a reaction. Also, each chemical reaction requires a specific level of energy. But even if colliding particles possess the minimum energy needed for reaction, no reaction will take place unless the particles are properly oriented toward each other.

Let us assume that molecules of substance AB (the reactant) are to be converted to molecules of substances A and B (the products). In a given population of molecules of substance AB, at a specific temperature, some molecules will possess relatively little energy; the majority of the population will possess an average amount of energy; and a small portion of the population will have high energy. If only the high-energy AB molecules are able to react and be converted to A and B molecules, then only relatively few molecules at any one time possess enough energy to react in a collision. The collision energy required for a chemical reaction is its **activation energy,** which is the amount of energy needed to disrupt the stable electronic configuration of any specific molecule so that the electrons can be rearranged.

The **reaction rate,** the frequency of collisions containing sufficient energy to bring about a reaction, depends on the number of reactant molecules at or above the activation energy level. One way to increase the reaction rate of a substance is to raise its temperature. By causing the molecules to move faster, heat increases both the frequency of collisions and the number of molecules that attain activation energy. The number of collisions also increases when pressure is increased or the reactants are more concentrated and the distance between molecules is thereby decreased.

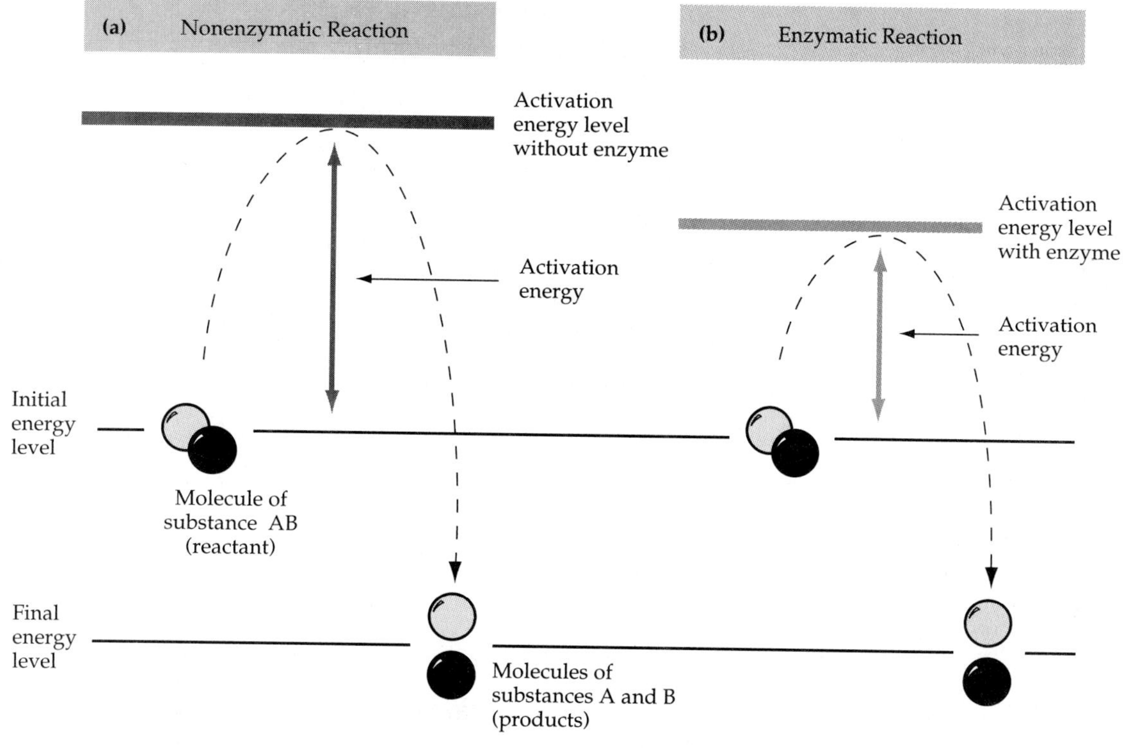

FIGURE 2.5 Energy requirements of a chemical reaction. **(a)** Without an enzyme; **(b)** with an enzyme. In (b), the enzyme lowers the activation energy level. Thus, more molecules of substance AB are converted to substances A and B because more molecules of substance AB possess the activation energy needed for the reaction.

Enzymes and Chemical Reactions

Living systems contain large protein molecules called **enzymes** that act as catalysts. A **catalyst** is a substance that changes the rate of a chemical reaction, usually by increasing it. As catalysts, enzymes typically accelerate chemical reactions. The three-dimensional enzyme molecule has an *active site,* a region that will interact with a specific chemical substance (see Figure 5.3). The chemical substance (reactant) an enzyme acts on is referred to as the enzyme's **substrate.**

The enzyme orients the substrate into a position that increases the probability of a reaction. The **enzyme-substrate complex** formed by the temporary binding of enzyme and reactants enables the collisions to be more effective and lowers the activation energy of the reaction (Figure 2.5). The enzyme therefore speeds up the reaction by increasing the number of A and B molecules that attain sufficient activation energy to react.

An enzyme is able to accelerate a reaction without the need for an increase in temperature. This ability is crucial to living systems because a significant temperature increase would destroy cellular proteins. The crucial function of enzymes, therefore, is to speed up biochemical reactions at a temperature that is compatible with the normal functioning of the cell.

Important Biological Molecules

Biologists and chemists divide compounds into two principal classes, inorganic and organic. **Inorganic compounds** are defined as molecules, usually small, that typically lack carbon and in which ionic bonds may play an important role. Inorganic compounds include water, oxygen, carbon dioxide, and many salts, acids, and bases.

Organic compounds always contain carbon and hydrogen. Carbon is a unique element in the chemistry of life because it has four electrons in its outer shell and four unfilled spaces. It can combine with a variety of atoms, including other carbon atoms, to form straight or branched chains and rings (see Figure 2.9). Carbon chains form the basis of many organic compounds in living cells, including sugars, amino acids,

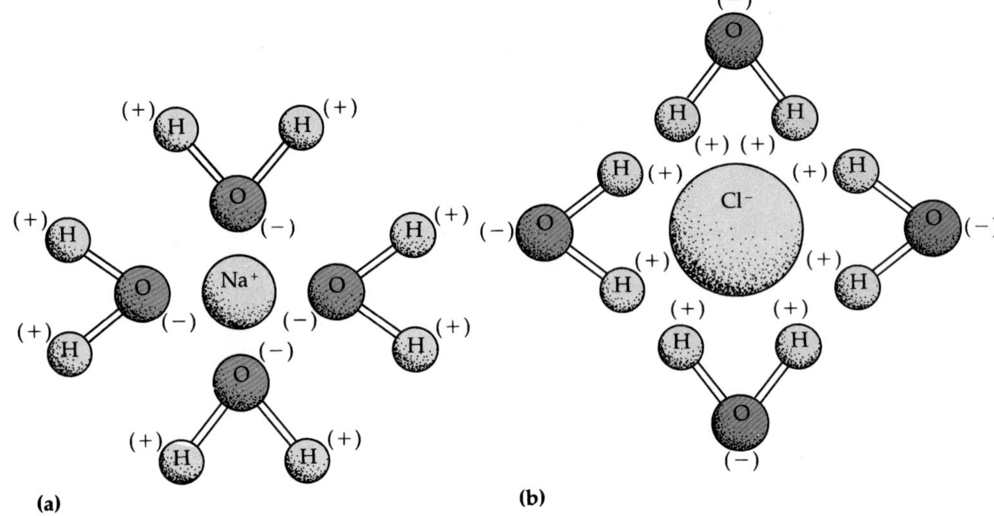

FIGURE 2.6 How water acts as a solvent for sodium chloride (NaCl). (a) The positively charged sodium ion (Na^+) is attracted to the negative part of the water molecule. **(b)** The negatively charged chloride ion (Cl^-) is attracted to the positive part of the water molecule. In the presence of water molecules, the bonds between the Na^+ and Cl^- ions are disrupted, and the NaCl dissolves in the water.

and vitamins. Organic compounds are held together mostly or entirely by covalent bonds. Some organic molecules, such as polysaccharides, proteins, and nucleic acids, are very large and usually contain thousands of atoms. Such giant molecules are called *macromolecules.* In the following section, we will discuss inorganic and organic compounds that are essential for cells.

Inorganic Compounds

WATER

All living organisms require a wide variety of inorganic compounds for growth, repair, maintenance, and reproduction. Water is one of the most important, as well as one of the most abundant, of these compounds, and it is particularly vital to microorganisms. Outside the cell, nutrients are dissolved in water, which facilitates their passage through cell membranes. And inside the cell, water is the medium for most chemical reactions. In fact, water is by far the most abundant component of almost all living cells. Water makes up 5% to 95% or more of each cell, the average being between 65% and 75%. Simply stated, no organism can survive without water.

Water has structural and chemical properties that make it particularly suitable for its role in living cells. As we discussed, the total charge on the water molecule is neutral, but the oxygen region of the molecule has a slightly negative charge and the hydrogen region has a slightly positive charge (see Figure 2.4a). Any molecule having such an unequal distribution of charges is called a **polar molecule.** The polar nature of water gives it four characteristics that make it a useful medium for living cells.

First, every water molecule is capable of forming four hydrogen bonds with nearby water molecules (see Figure 2.4b). This property results in a strong attraction between water molecules. Because of this strong attraction, a great deal of heat is required to separate water molecules from each other to form water vapor; thus, water has a relatively high boiling point (100°C). Since water has such a high boiling point, it exists in the liquid state on most of the Earth's surface. Furthermore, the hydrogen bonding between water molecules affects the density of water, depending on whether it occurs as ice or a liquid. For example, the hydrogen bonds in the crystalline structure of water (ice) make it more spacious. As a result, ice has fewer molecules than an equal volume of liquid water. This makes its crystalline structure less dense than liquid water. It is for this reason that ice floats and can serve as an insulating layer on the surfaces of lakes and streams that harbor living organisms.

Second, the polarity of water makes it an excellent dissolving medium or **solvent.** Many polar substances **dissociate** (separate) into individual molecules in water—that is, dissolve—because the negative part of the water molecule is attracted to the positive part of the molecules in the **solute** (dissolving substance), and the positive part of the water molecules is attracted to the negative part of solute molecules. Substances (such as salts) that are composed of atoms (or groups of atoms) held together by ionic bonds tend to dissociate into separate cations and anions in water. Thus, the polarity of water allows molecules of many different substances to separate and become surrounded by water molecules (Figure 2.6).

Third, polarity accounts for water's characteristic role as a reactant or product in many chemical reactions. Its polarity facilitates the splitting and rejoining of hydrogen (H^+) and hydroxide (OH^-) ions. Water is a key reactant in the digestive processes of organisms, whereby larger molecules are broken down into smaller ones. Water molecules are also involved in synthetic reactions—water is an important source of the hydrogen and oxygen that are incorporated into numerous organic compounds in living cells.

Finally, the relatively strong hydrogen bonding between water molecules (see Figure 2.4b) makes water an excellent temperature buffer. A given quantity of water requires a great gain of heat to increase its temperature and a great loss of heat to decrease its temperature, as compared with many other substances. Normally, heat absorption by molecules increases their kinetic energy and thus increases their rate of motion and their reactivity. In water, however, heat absorption first breaks hydrogen bonds rather than increasing the rate of motion. Therefore, much more heat must be applied to raise the temperature of water than to raise the temperature of a non-hydrogen-bonded liquid. The reverse is true as water cools. Thus, water more easily maintains a constant temperature than other solvents and tends to protect a cell from fluctuations in environmental temperatures.

ACIDS, BASES, AND SALTS

As we saw in Figure 2.6, when inorganic salts such as sodium chloride (NaCl) are dissolved in water, they undergo **dissociation,** or **ionization.** That is, they break apart into ions. Substances called acids and bases show similar behavior.

An **acid** can be defined as a substance that dissociates into one or more hydrogen ions (H^+) and one or more negative ions (anions). Thus, an acid can also be defined as a proton (H^+) donor. A **base** dissociates into one or more positive ions (cations) plus one or more negatively charged hydroxide ions (OH^-) that can accept, or combine with, protons. Thus, sodium hydroxide (NaOH) is a base because it dissociates to release hydroxide ions, which have a strong attraction for protons and are among the most important proton acceptors. A **salt** is a substance that dissociates in water into cations and anions, neither of which is H^+ or OH^-. Figure 2.7 shows common examples of each type of compound and how they dissociate in water.

ACID–BASE BALANCE

An organism must maintain a fairly constant balance of acids and bases to remain healthy. In the aqueous environment within organisms, acids dissociate into hydrogen ions (H^+) and anions. Bases, on the other hand, dissociate into hydroxide ions (OH^-) and cations. The more hydrogen ions that are free in a solution, the more acid the solution is. Conversely, the more hydroxide ions that are free in a solution, the more basic, or alkaline, it is.

Biochemical reactions—that is, reactions in living systems—are extremely sensitive to even small changes in the acidity or alkalinity of the environments in which they occur. In fact, H^+ and OH^- ions are involved in almost all biochemical processes, and the functions of a cell are modified greatly by any deviation from its narrow band of normal H^+ and OH^- concentrations. For this reason, the acids and bases that are continually formed in an organism must be kept in balance.

It is convenient to express the amount of H^+ in a solution by a logarithmic pH scale, which ranges from 0 to 14 (Figure 2.8). The **pH** of a solution is calculated by $-\log_{10}[H^+]$, the negative logarithm to the base 10 of the hydrogen ion concentration, determined in moles per liter $[H^+]$. For example, if the H^+ concentration of a solution is 1.0×10^{-4} moles/liter, or 10^{-4}, its pH equals $-\log_{10}10^{-4} = -(-4) = 4$; this is about the pH of wine. In the laboratory, you will usually measure the pH of a solution with a pH meter or with chemical test papers.

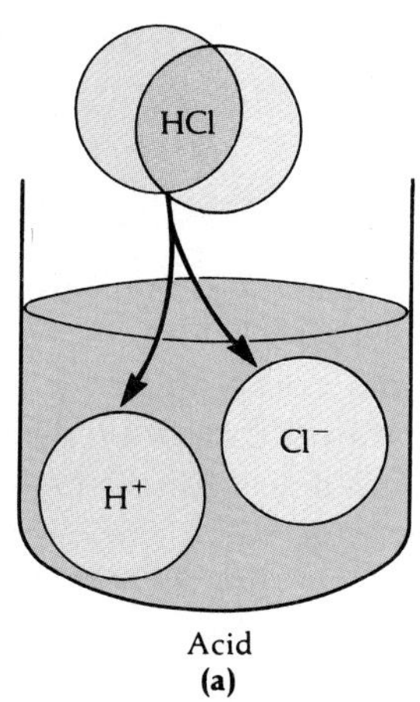

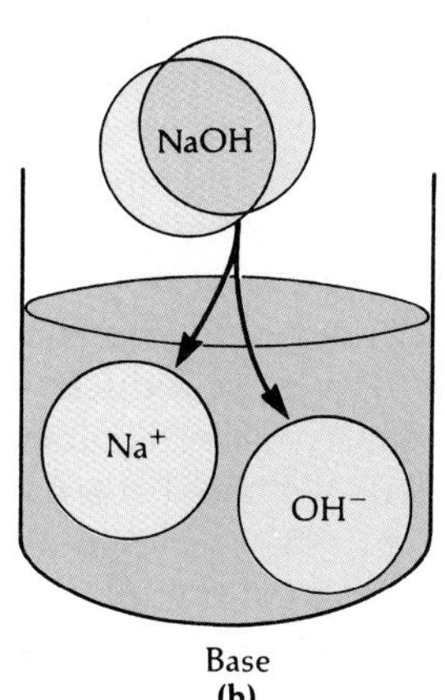

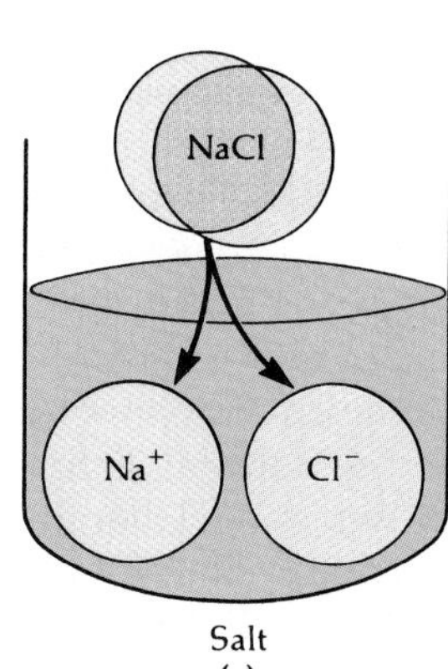

FIGURE 2.7 Acids, bases, and salts. **(a)** In water, hydrochloric acid (HCl) dissociates into H^+ ions and Cl^- ions. **(b)** Sodium hydroxide (NaOH), a base, dissociates into OH^- ions and Na^+ ions in water. **(c)** In water, table salt (NaCl) dissociates into positive ions (Na^+) and negative ions (Cl^-), neither of which are H^+ or OH^- ions.

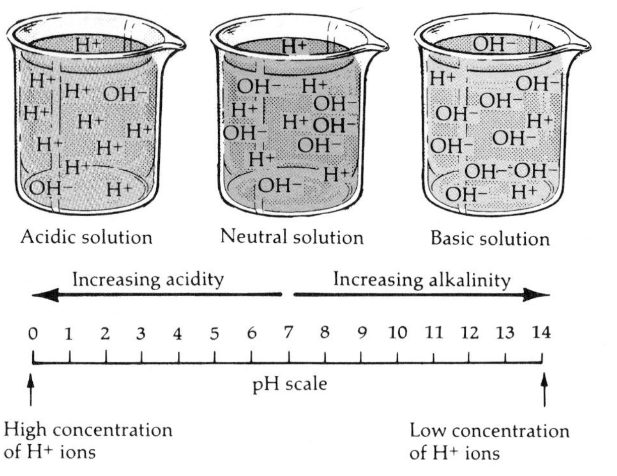

FIGURE 2.8 The pH scale. The concentrations of H^+ and OH^- ions are equal at pH 7, which is the neutral point. As the pH values decrease from 14 to 0, the H^+ concentration increases. Thus, the lower the pH, the more acidic the solution; the higher the pH, the more basic the solution. If the pH value of a solution is below 7, the solution is acidic; if the pH is above 7, the solution is basic (alkaline).

TABLE 2.3 pH Values of Some Human Body Fluids and Common Substances

FLUID OR SUBSTANCE	pH
Stomach acid	1.5 to 2.5
Lemon juice	2.2 to 2.4
Carbonated soft drinks	3.0 to 3.5
Tomato juice	4.2
Urine	5.0 to 7.8
Saliva	6.35 to 6.85
Milk	6.6 to 6.9
Distilled (pure) water	7.0
Blood	7.35 to 7.45
Eggs	7.6 to 8.0
Milk of magnesia	10.0 to 11.0
Limewater	12.3

Acidic solutions contain more H^+ ions than OH^- ions and have a pH lower than 7. If a solution has more OH^- ions than H^+ ions, it is a **basic,** or **alkaline, solution.** In pure water, a small percentage of the molecules are dissociated into H^+ and OH^- ions, so it has a pH of 7. Because the concentrations of H^+ and OH^- are equal, this pH is said to be the pH of a **neutral solution.**

The pH scale is logarithmic, so a change of one whole number represents a tenfold change from the previous concentration. Thus, a solution of pH 1 has ten times more H^+ ions than a solution of pH 2 and has 100 times more H^+ ions than a solution of pH 3.

The pH values of some human body fluids and other common substances are shown in Table 2.3.

Keep in mind that the pH of a solution can be changed. We can increase its acidity by adding substances that will increase the concentration of hydrogen ions. As a living organism takes up nutrients, carries out chemical reactions, and excretes wastes, its balance of acids and bases tends to change, and the pH fluctuates. Fortunately, organisms possess natural **pH buffers,** compounds that help keep the pH from changing drastically. But the pH in our environment's water and soil can be altered by waste products from organisms, pollutants from industry, or fertilizers used in agricultural fields or gardens. When bacteria are grown in a laboratory medium, they excrete waste products such as acids that can alter the pH of the medium. If this effect were to continue, the medium would become acidic enough to inhibit bacterial enzymes and cause the death of the bacteria. To prevent this problem, pH buffers are added to the culture medium. One very effective pH buffer for some culture media uses a mixture of K_2HPO_4 and KH_2PO_4 (see Table 6.4).

Different microbes function best within different pH ranges, but most organisms grow best in environments with a pH between 6.5 and 8.5. Fungi are the microbes that are best able to tolerate acidic conditions, whereas the procaryotes called cyanobacteria tend to do well in alkaline habitats. *Propionibacterium acnes* (prō-pē-on-ē-bak-ti'rē-um ak'nēz), a bacterium that contributes to acne, has its natural environment on human skin, which tends to be slightly acidic, with a pH of about 4. *Thiobacillus ferrooxidans* (thī-ō-bä-sil'lus fer-rō-oks'i-danz) is a bacterium that grows on elemental sulfur and produces sulfuric acid (H_2SO_4). Its optimum growth pH is from 1 to 3.5. The sulfuric acid produced by this bacterium in mine water is important in dissolving uranium and copper from low-grade ore. (See the box in Chapter 28.)

Organic Compounds

Inorganic compounds, excluding water, constitute about 1% to 1.5% of living cells. These relatively simple components, whose molecules have only a few atoms, cannot be used by cells to perform complicated biological functions. Organic molecules, whose carbon atoms can combine in an enormous variety of ways with other carbon atoms and atoms of other elements, are relatively complex and thus are capable of more complicated biological functions. In the formation of organic molecules, carbon's four outer electrons can participate in up to four covalent bonds, and carbon atoms can bond to each other to form straight-chain, branched-chain, or ring structures (Figure 2.9).

When single bonds hold a carbon atom to four other atoms, those four atoms form a skeleton of a three-dimensional shape called a **tetrahedron** (Figure 2.10). Carbon atoms can also bond via double or triple bonds; the molecular shape then is not tetrahedral.

In addition to carbon, the most common elements in organic compounds are hydrogen (which can form one bond), oxygen (two bonds), and nitrogen (three bonds). Sulfur (two bonds) and phosphorus (five bonds) appear less often. Other elements are found, but only in a relatively few organic compounds. If you look back at Table 2.1, you will see that the elements most abundant in organic compounds are the same as those most abundant in living cells.

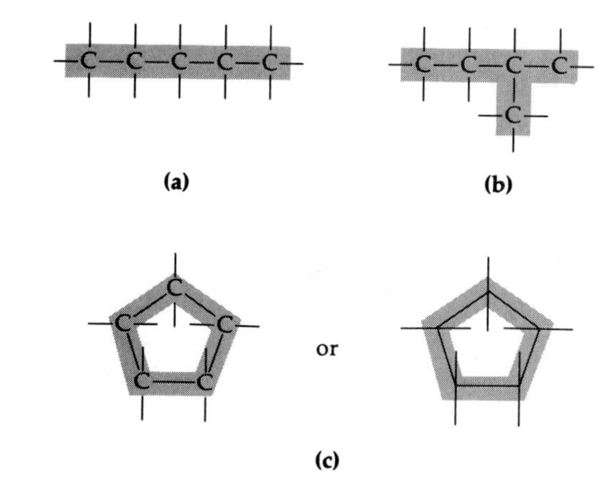

FIGURE 2.9 Various bonding patterns of carbon. **(a)** Straight-chain structures. **(b)** Branched-chain structures. **(c)** Ring. In ring structures, the carbon atoms at the corners are often not shown.

FUNCTIONAL GROUPS

The chain of carbon atoms in a molecule is called the **carbon skeleton**; there is a huge number of combinations possible for carbon skeletons. Most of these carbons are bonded to hydrogen atoms. The bonding of other elements with carbon and hydrogen forms characteristic **functional groups** (Table 2.4). Functional groups are specific groups of atoms that are responsible for most of the characteristic chemical properties and many of the physical properties of a particular organic compound. Thus, functional groups help us to classify organic compounds. For example, the —OH group is present in each of the following molecules:

Methanol

Ethanol

Isopropanol

Toluol

Because the characteristic reactivity of the molecules is based on the —OH group, they are grouped together in a class called alcohols. The —OH group is called the *hydroxyl group* and is not to be confused with the *hydroxide ion* (OH^-) of bases. The hydroxyl group of alcohols does not ionize at neutral pH; it is covalently bonded to a carbon atom.

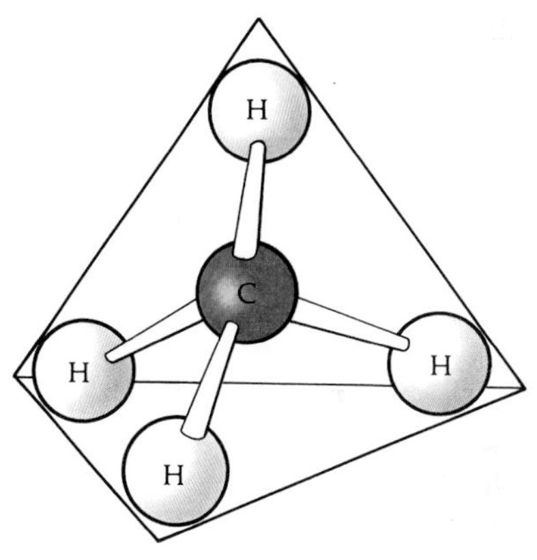

FIGURE 2.10 A tetrahedral carbon compound. Organic molecules containing carbon atoms with four single bonds have this distinctive shape because of the directions of the four bonds. Here, the carbon atom is bonded to four hydrogen atoms to form the tetrahedral methane molecule (CH_4).

When a class of compounds is characterized by a certain functional group, the letter *R* can be used to stand for the remainder of the molecule. For example, alcohols in general may be written R—OH.

Frequently, more than one functional group is found in a single molecule. For example, an amino acid molecule contains both amino and carboxyl groups. The amino acid glycine is shown below.

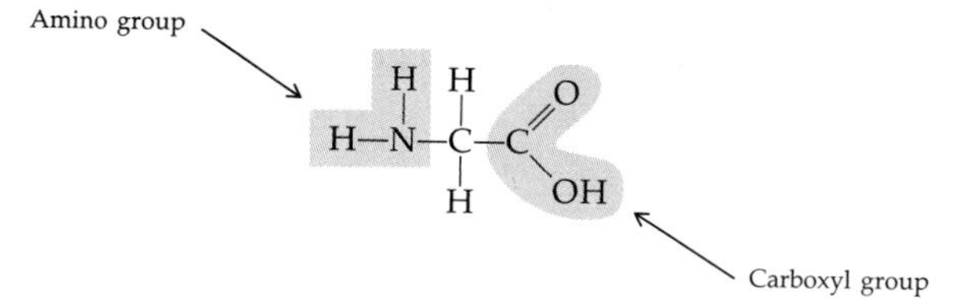

TABLE 2.4 Representative Functional Groups and the Compounds in Which They Are Found

FUNCTIONAL GROUP	NAME OF GROUP	CLASS OF COMPOUNDS
R—O—H	Hydroxyl	Alcohol
R—C(=O)—H	Carbonyl (terminal)*	Aldehyde
R—C(=O)—R	Carbonyl (internal)*	Ketone
R—C(H)(H)—NH_2	Amino	Amine
R—C(=O)—O—R′	Ester	Ester
R—C(H)(H)—O—C(H)(H)—R′	Ether	Ether
R—C(H)(H)—SH	Sulfhydryl	Sulfhydryl
R—C(=O)—OH	Carboxyl	Organic acid

*A "terminal" carbonyl group is at an end of a molecule. In contrast, an "internal" carbonyl is at least one C atom removed from both ends.

Most of the organic compounds found in living organisms are quite complex; a large number of carbon atoms form the skeleton, and many functional groups are attached. In organic molecules, it is important that each of the four bonds of carbon is satisfied (attached to another atom) and each of the attaching atoms has its characteristic number of bonds satisfied. Because of this, such molecules are chemically stable.

MACROMOLECULES

Small organic molecules can be combined into the very large molecules called **macromolecules.** Macromolecules are usually **polymers,** large molecules formed by covalent bonding of many repeating small molecules called **subunits** or **monomers.** When two monomers join together, the reaction usually involves the elimination of a hydrogen atom from one monomer and a hydroxyl group from the other; these combine to produce water:

$$X{-}OH + H{-}Y \longrightarrow X{-}Y + H_2O$$

This type of exchange reaction is called **dehydration synthesis** or **condensation** because a molecule of water is released. Such macromolecules as carbohydrates, lipids, proteins, and nucleic acids are assembled in the cell essentially by dehydration synthesis. However, other molecules must also participate to provide energy for bond formation. ATP, the cell's chief energy provider, is discussed at the end of this chapter.

CARBOHYDRATES

The **carbohydrates** are a large and diverse group of organic compounds that includes sugars and starches. The carbohydrates perform a number of major functions in living systems. For instance, one type of sugar (deoxyribose) is a building block of deoxyribonucleic acid (DNA), the molecule that carries hereditary information. Other sugars help to form the cell walls of bacterial cells. Simple carbohydrates are used in the synthesis of amino acids and fats or fatlike substances, which are used to build structures and provide an emergency source of energy. Macromolecular carbohydrates function as food reserves. The principal function of carbohydrates, however, is to fuel cell activities with a ready source of energy.

Carbohydrates are made up of carbon, hydrogen, and oxygen atoms. The ratio of hydrogen to oxygen atoms is always 2:1 in simple carbohydrates. This ratio can be seen in the formulas for the carbohydrates ribose ($C_5H_{10}O_5$), glucose ($C_6H_{12}O_6$), and sucrose ($C_{12}H_{22}O_{11}$). Although there are exceptions, the general formula for carbohydrates is $(CH_2O)_n$, where n indicates that there are three or more CH_2O units. Carbohydrates can be divided into three major groups on the basis of size—monosaccharides, disaccharides, and polysaccharides.

Monosaccharides

Simple sugars are called **monosaccharides;** each molecule contains from three to seven carbon atoms. The number of carbon atoms in the molecule of a simple sugar is indicated by the prefix in its name. For example, simple sugars with three carbons are called trioses. There are also tetroses (four-carbon sugars), pentoses (five-carbon sugars), hexoses (six-carbon sugars), and heptoses (seven-carbon sugars). Pentoses and hexoses are extremely important to living organisms. Deoxyribose is a pentose found in DNA, the genetic material of the cell. Glucose, a common hexose, is the main energy-supplying molecule of living cells.

Disaccharides

Disaccharides are formed when two monosaccharides bond in a dehydration synthesis reaction (Figure 2.11). For example, molecules of two monosaccharides, glucose and fructose, combine to form a molecule of the

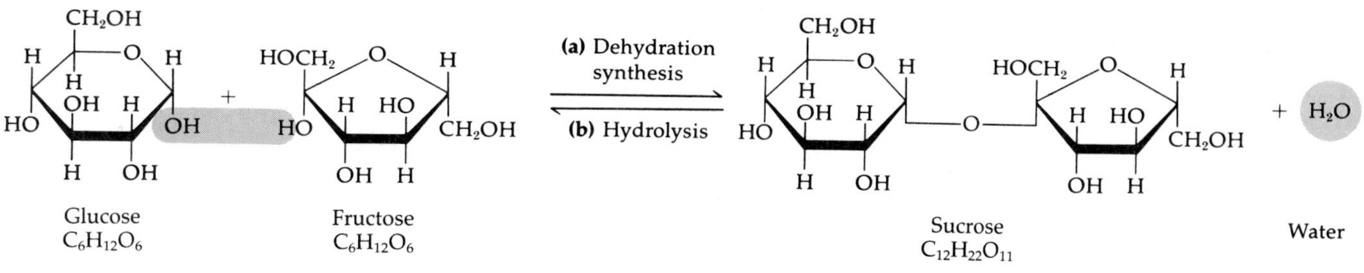

FIGURE 2.11 Dehydration synthesis and hydrolysis. **(a)** In dehydration synthesis (left to right), the monosaccharides glucose and fructose combine to form a molecule of the disaccharide sucrose. A molecule of water is lost in the reaction. **(b)** In hydrolysis (right to left), the sucrose molecule breaks down into the smaller molecules glucose and fructose. For the hydrolysis reaction to proceed, water must be added to the sucrose.

disaccharide sucrose (table sugar) and a molecule of water:

$$\underset{\text{Glucose (monosaccharide)}}{C_6H_{12}O_6} + \underset{\text{Fructose (monosaccharide)}}{C_6H_{12}O_6} \longrightarrow \underset{\text{Sucrose (disaccharide)}}{C_{12}H_{22}O_{11}} + \underset{\text{Water}}{H_2O}$$

Similarly, the dehydration synthesis of the monosaccharides glucose and galactose forms the disaccharide lactose (milk sugar).

It may seem odd that glucose and fructose should have the same chemical formula since they are different monosaccharides. The positions of the oxygens and carbons vary in the two different molecules (see Figure 2.11), and consequently the molecules have different physical and chemical properties. Two molecules with the same chemical formula but different structures and properties are called **isomers.**

Disaccharides can be broken down into smaller, simpler molecules when water is added. This chemical reaction, the reverse of dehydration synthesis, is called **hydrolysis,** which means to split water. A molecule of sucrose, for example, may be hydrolyzed (digested) into its components of glucose and fructose by reacting with the H^+ and OH^- of water. This reaction is also represented in Figure 2.11.

Polysaccharides

Carbohydrates in the third major group, the **polysaccharides,** consist of eight or more monosaccharides joined through dehydration synthesis. Polysaccharides often have side chains branching off the main structure and are classified as macromolecules. Like disaccharides, polysaccharides can be split apart into their constituent sugars through hydrolysis. Unlike monosaccharides and disaccharides, however, they usually lack the characteristic sweetness of sugars such as fructose and sucrose and usually are not soluble in water.

One important polysaccharide is *glycogen,* which is composed of glucose subunits and is synthesized as a storage material by animals and some bacteria. *Cellulose,* another important glucose polymer, is the main component of the cell walls of plants and most algae. The polysaccharide *dextran,* which is produced as a sugary slime by certain bacteria, is used in a blood plasma substitute. *Starch* is a polymer of glucose produced by plants and used as food by humans.

LIPIDS

If lipids were suddenly to disappear from the Earth, all living cells would collapse in a pool of fluid, because lipids are essential to the structure and function of membranes that separate living cells from their environment. **Lipids** are a second major group of organic compounds found in living matter. Like carbohydrates, they are composed of atoms of carbon, hydrogen, and oxygen, but lipids lack the 2:1 ratio between hydrogen and oxygen atoms. Lipids are a very diverse group of compounds, their common characteristic being that they are very *nonpolar* molecules; unlike water, they do not have a positive and a negative end (pole). Hence, most are insoluble in water but dissolve readily in nonpolar solvents, such as ether and chloroform. Lipids function in energy storage and provide part of the structure of membranes and cell walls.

Simple Lipids

Simple lipids, called *fats* or *triglycerides,* contain an alcohol called *glycerol* and a group of compounds known as *fatty acids.* Glycerol molecules have three carbon atoms to which are attached three hydroxyl (—OH) groups (Figure 2.12a). Fatty acids consist of long hydrocarbon chains (composed only of carbon and hydrogen atoms) ending in a carboxyl (—COOH, organic acid) group (Figure 2.12b). Most common fatty acids contain an even number of carbon atoms.

MICROBIOLOGY *IN THE NEWS*

Bacterial Banqueters Attend Oil Spill

Readers of science fiction have long realized that beings from another planet might have quite a different chemical makeup from earthlings and might be able to eat, drink, and breathe the substances we find deadly. Such aliens could be invaluable in helping clean up pollutants such as oil and mercury, which harm plants, animals, and humans. Fortunately, however, we need not wait for a visit from outer space to find creatures whose unusual chemistry can be harnessed for environmental cleanup. Although many bacteria have dietary requirements similar to ours—that's why they cause food spoilage—others metabolize (or process chemically) the substances we might expect at a banquet of extraterrestrials—heavy metals, sulfur, nitrogen gas, petroleum, and even polychlorinated biphenyls (PCBs) and mercury.

Bacteria have several advantages as pollution fighters. They can extract pollutants that have combined with soil and water and hence cannot be simply shoveled away. In addition, they may chemically alter a harmful substance so that it becomes harmless or even beneficial. Bacteria that are able to degrade many pollutants are naturally present in soil and water. However, their small numbers make them inefficient to deal with large-scale contamination. Scientists are now working to improve the efficiency of natural pollution fighters and, in some cases, are altering organisms by genetic engineering to give them exactly the right chemical appetites.

One of the most promising successes occurred on an Alaskan beach following the *Exxon Valdez* oil spill. Several naturally occurring bacteria in the genus *Pseudomonas* are able to degrade oil for their carbon and energy requirements. In the presence of air, they remove two carbons at a time from a large petroleum molecule.

The bacteria degrade the oil too slowly to be helpful cleaning up an oil spill. However, scientists hit upon a very simple way to speed them up—with no need for genetic engineering. They simply dumped ordinary nitrogen and phosphorus plant fertilizers onto the beach (a process called bioremediation). The number of oil-degrading bacteria increased compared with that on unfertilized control beaches, and the test beach is now free of oil.

Another group of bacteria is being investigated for its ability to clean up mercury contamination. Mercury is contained in such common substances as discarded leftover paint and can leak into soil and water from garbage dumps. One species of bacteria that is common in the environment, *Desulfovibrio desulfuricans*, actually makes the mercury more dangerous by adding a methyl group, converting it into the highly toxic substance methyl mercury. Methyl mercury in ponds or marshes sticks to small organisms such as plankton, which are eaten by larger organisms, which in turn are eaten by fish. Fish and human poisonings have been attributed to the ingestion of methyl mercury.

However, other bacteria, such as species of *Pseudomonas,* may offer the solution. To avoid mercury poisoning, these bacteria first convert methyl mercury to mercuric ion:

$$\underset{\text{Methyl mercury}}{CH_3Hg} \longrightarrow \underset{\text{Methane}}{CH_4} + \underset{\text{Mercuric ion}}{Hg^{2+}}$$

They then convert the positively charged mercuric ion to the relatively harmless elemental form by adding electrons, which they take from hydrogen atoms.

$$\underset{\text{Mercuric ion}}{Hg^{2+}} + \underset{\text{Hydrogen atoms}}{2H} \xrightarrow{2e^-} \underset{\text{Elemental mercury}}{Hg} + \underset{\text{Hydrogen ions}}{2H^+}$$

These bacteria work too slowly in nature for cleaning up manmade toxic spills, but scientists are experimenting with bioaugmentation and other techniques to increase their effectiveness. Unlike some forms of environmental cleanup, in which dangerous substances are removed from one place only to be dumped in another, bacterial cleanup eliminates the toxic substance and often returns a harmless or useful substance to the environment.

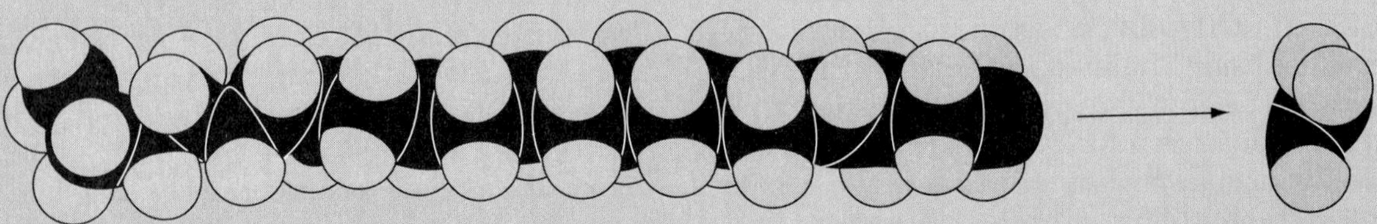

Typical saturated hydrocarbon found in petroleum

Two-carbon unit can be metabolized in cell

FIGURE 2.12 Structural formulas. **(a)** Glycerol; **(b)** lauric acid, a fatty acid. In the condensed formula for lauric acid, $COOH(CH_2)_{10}CH_3$, the subscript 10 means that the molecule has ten CH_2 units. **(c)** The chemical combination of a molecule of glycerol with three fatty acid molecules, in this case lauric acid, forms **(d)** one molecule of fat (triglyceride) and three molecules of water in a dehydration synthesis reaction. The addition of three water molecules to a fat forms glycerol and three fatty acid molecules in a hydrolysis reaction.

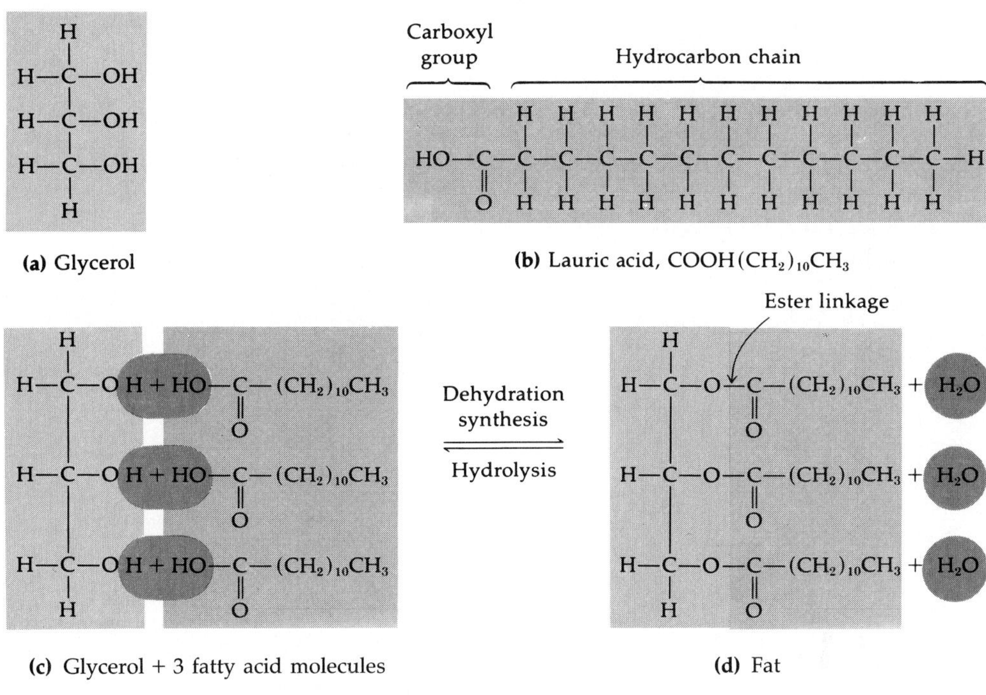

A fat molecule (Figure 2.12d) is formed when a molecule of glycerol combines with one to three fatty acid molecules to form a monoglyceride, diglyceride, or triglyceride (Figure 2.12c). In the reaction, one to three molecules of water are formed (dehydration), depending on the number of fatty acid molecules reacting. The chemical bond formed where the water molecule is removed is an *ester linkage.* In the reverse reaction, hydrolysis, a fat molecule is broken down into its component fatty acid and glycerol molecules.

Because the fatty acids that form lipids have different structures, there is a wide variety of lipids. For example, three molecules of fatty acid A might combine with a glycerol molecule, or one molecule each of fatty acids A, B, and C might unite with a glycerol molecule. Simple lipids serve as energy storage materials.

The terms *saturated fat* and *unsaturated fat* refer to the structure of the hydrocarbon chains of the fatty acids in the fat molecule. If there are no double bonds between carbon atoms in the hydrocarbon chain, then the carbon skeleton contains the maximum number of hydrogen atoms. Such a fat is said to be *saturated,* meaning that it is saturated with hydrogen atoms (Figure 2.13a). Most animal fats are saturated fats. If there is one or more double bonds, which reduces the number of hydrogen atoms that can be attached to the carbon skeleton, then the fat molecule is referred to as *unsaturated.* At each point where a double bond does occur in the hydrocarbon chain of an unsaturated fat, the fatty acid has a kink (bend) in its structure. Plant fats are usually unsaturated.

Complex Lipids

Complex lipids contain such elements as phosphorus, nitrogen, and sulfur in addition to the carbon, hydrogen, and oxygen found in simple lipids. The complex lipids called *phospholipids* are made up of glycerol, two fatty acids, and, in place of a third fatty acid, a phosphate group bonded to one of several organic groups (Figure 2.13). Phospholipids are the lipids that build membranes; they are thus essential to a cell's survival. Phospholipids have polar as well as nonpolar regions (see Figures 2.13 and 4.11). When placed in water, phospholipid molecules twist and orient themselves in such a way that all polar (hydrophilic) portions will orient themselves toward the polar water molecules, with which they then form hydrogen bonds (Figure 2.13b). (*Hydrophilic* means water-loving.) Polar portions consist of a phosphate group and glycerol. In contrast to the polar regions, all nonpolar (hydrophobic) parts of the phospholipid make contact only with the nonpolar portions of neighboring molecules. (*Hydrophobic* means water-fearing.) Nonpolar portions consist of fatty acids. This characteristic behavior makes phospholipids particularly suitable to be a major component of the membranes that enclose cells. Phospholipids enable the membrane to act as a barrier that separates the contents of the cell from the water-based environment in which it lives.

Some complex lipids are useful in identifying certain bacteria. For example, the cell wall of *Mycobacterium tuberculosis* (mī-kō-bak-ti'rē-um tü-bėr-kū-lō'sis), the bacterium that causes tuberculosis, is distin-

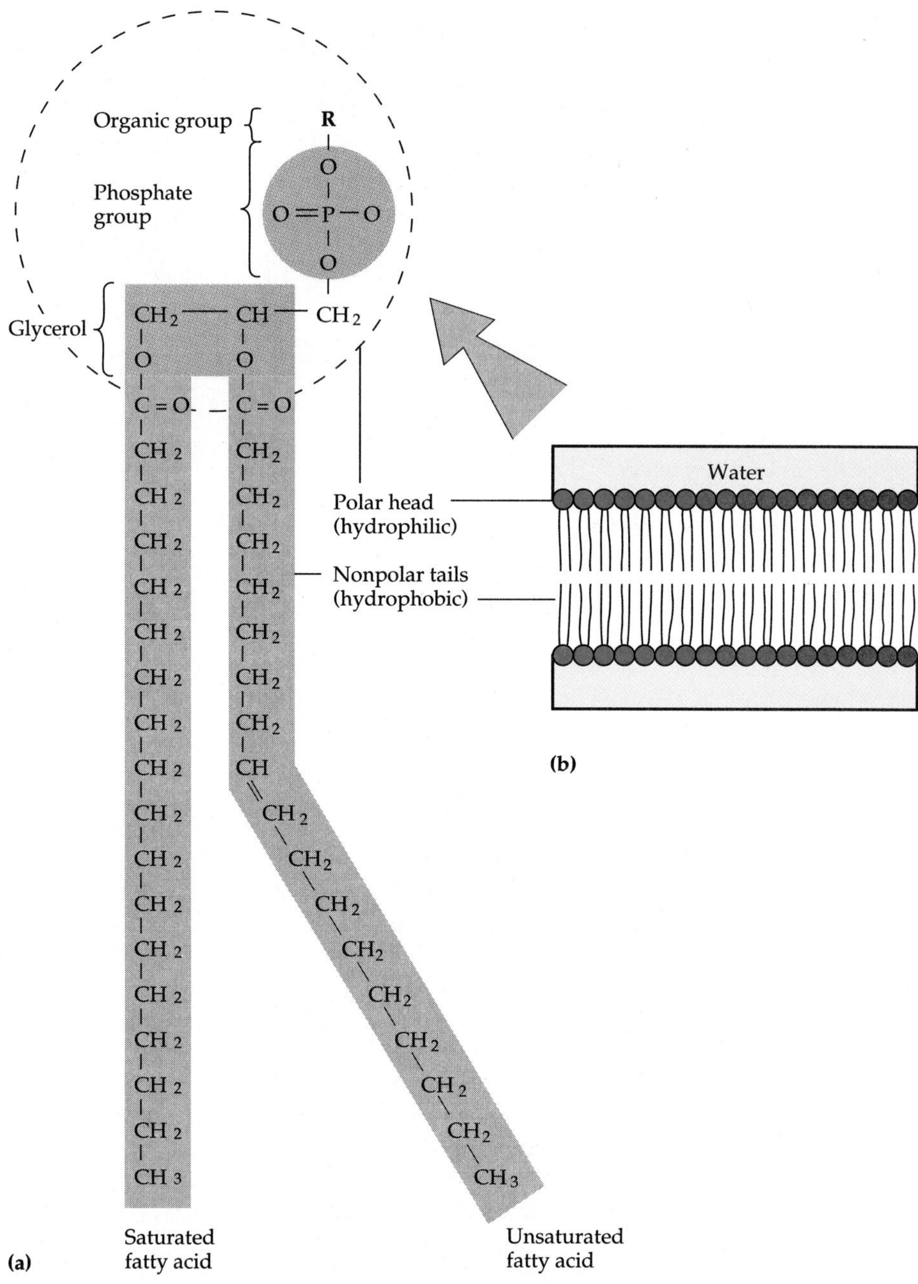

FIGURE 2.13 Phospholipids. **(a)** Structure of a phospholipid. The fatty acids and the R group (at top) show the structure of the polar heads in diagram (b); this structure may vary with the particular phospholipid. **(b)** Orientation of phospholipids in water.

guished by its lipid-rich content. The cell wall contains complex lipids such as waxes and glycolipids (lipids with carbohydrates attached) that give the bacterium distinctive staining characteristics. Cell walls rich in such complex lipids are characteristic of all members of the genus *Mycobacterium*.

Steroids

Steroids are structurally very different from the lipids described previously. Figure 2.14 shows the structure of the steroid cholesterol with the characteristic ring structure of steroids. The —OH group in cholesterol makes it a *sterol*. Sterols are important constituents of the plasma membranes of animal cells and of one

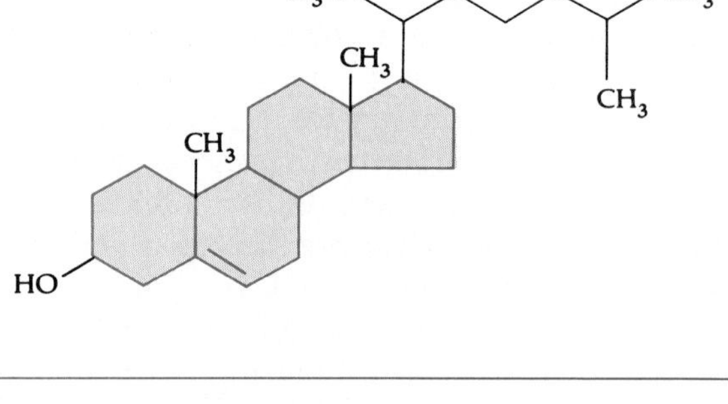

FIGURE 2.14 Cholesterol, a steroid. Note the four "fused" carbon rings, which are characteristic of steroid molecules. The hydrogen atoms attached to the carbons at the corners of the rings have been omitted.

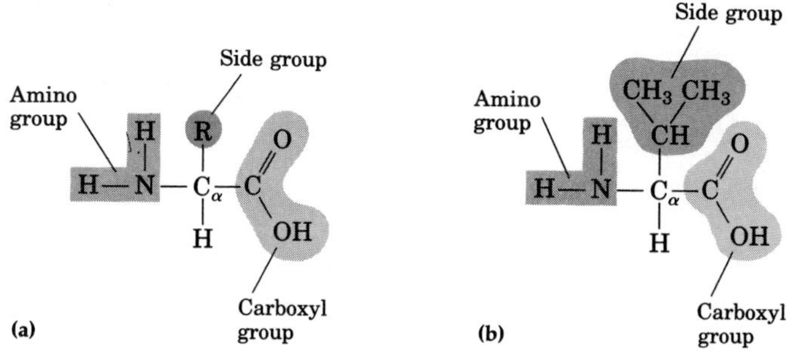

FIGURE 2.15 Amino acids. (a) General structural formula for an amino acid. The alpha-carbon (C_α) is shown in the center. The letter R can stand for any of a number of groups of atoms. Different amino acids have different R groups, also called side groups. **(b)** Structural formula for the amino acid valine. Note the side group.

group of bacteria (mycoplasmas), and they are also found in fungi and plants. Animals synthesize steroid hormones and vitamin D from sterols.

PROTEINS

Proteins are organic molecules that contain carbon, hydrogen, oxygen, and nitrogen. Some also contain sulfur. If you were to separate and weigh all the groups of organic compounds in a living cell, the proteins would tip the scale. Hundreds of different proteins can be found in any single cell, and together they make up 50% or more of the cell's dry weight.

Proteins are essential ingredients in all aspects of cell structure and function. We have already mentioned enzymes, the proteins that catalyze biochemical reactions. But proteins have other vital functions as well. *Carrier proteins* help to transport certain chemicals into and out of cells. Other proteins, such as the *bacteriocins* produced by many bacteria, kill other bacteria. Toxins produced by certain disease-causing microorganisms are also proteins. Some proteins play a key role in contraction of animal muscle cells and movement of microbial and other types of cells. Other proteins are integral parts of cell structures such as walls, membranes, and cytoplasmic components. Others, such as the hormones of certain organisms, have regulatory functions. Then there are proteins that play a key role in vertebrate immune systems—antibodies, for instance, are proteins.

Amino Acids

Just as the monosaccharides are the building blocks of larger carbohydrate molecules and fatty acids and glycerol are the building blocks of fats, **amino acids** are the building blocks of proteins. Amino acids contain at

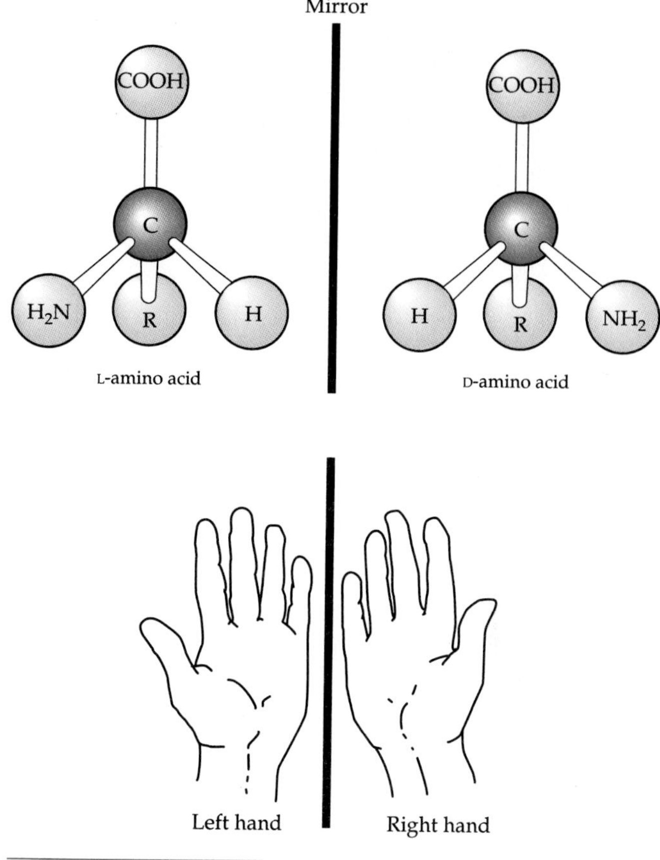

FIGURE 2.16 The L- and D-isomers of an amino acid, shown with ball-and-stick models. The two isomers, like left and right hands, are mirror images of each other and cannot be superimposed on one another. (Try it!)

least one carboxyl (—COOH) group and one amino (—NH_2) group attached to the same carbon atom, called an alpha-carbon (written C_α) (Figure 2.15). Such amino acids are called *alpha-amino acids.* Also attached to the alpha-carbon is a side group (R group), which is the amino acid's distinguishing factor. The side group can be a hydrogen atom, an unbranched or branched chain of atoms, or a ring structure that is cyclic (all carbon) or heterocyclic (when an atom other than carbon is included in the ring). Figure 2.15b illustrates the structural formula of a branched-chain side group—in this case, the amino acid valine. The side group can contain functional groups, such as the sulfhydryl group (—SH), the hydroxyl group (—OH), or additional carboxyl or amino groups. These side groups and the carboxyl and alpha-amino groups affect the total structure of a protein, as will be described later. The structures and standard abbreviations of the 20 amino acids found in proteins are shown in Table 2.5.

Amino acids exist in either of two configurations called *stereoisomers,* designated by D and L. These two configurations are mirror images, corresponding to

TABLE 2.5 Twenty Amino Acids Found in Proteins

AMINO ACID (ABBREVIATION)	STRUCTURAL FORMULA	CHARACTERISTIC OF R GROUP
Glycine (Gly)	$H-\underset{NH_2}{\overset{H}{C}}-COOH$	Hydrogen atom
Alanine (Ala)	$CH_3-\underset{NH_2}{\overset{H}{C}}-COOH$	Unbranched chain
Valine (Val)	$H_3C-\underset{CH_3}{CH}-\underset{NH_2}{\overset{H}{C}}-COOH$	Branched chain
Leucine (Leu)	$H_3C-\underset{CH_3}{CH}-CH_2-\underset{NH_2}{\overset{H}{C}}-COOH$	Branched chain
Isoleucine (Ile)	$H_3C-CH_2-\underset{CH_3}{CH}-\underset{NH_2}{\overset{H}{C}}-COOH$	Branched chain
Serine (Ser)	$HO-CH_2-\underset{NH_2}{\overset{H}{C}}-COOH$	Hydroxyl (—OH) group
Threonine (Thr)	$H_3C-\underset{OH}{CH}-\underset{NH_2}{\overset{H}{C}}-COOH$	Hydroxyl (—OH) group
Cysteine (Cys)	$HS-CH_2-\underset{NH_2}{\overset{H}{C}}-COOH$	Sulfhydryl (—SH) group
Methionine (Met)	$H_3C-S-CH_2-CH_2-\underset{NH_2}{\overset{H}{C}}-COOH$	Sulfhydryl (—SH) group
Glutamic acid (Glu)	$HOOC-CH_2-CH_2-\underset{NH_2}{\overset{H}{C}}-COOH$	Additional carboxyl (—COOH) group, acidic

(Continued)

"right-handed" (D) and "left-handed" (L) three-dimensional shapes (Figure 2.16). The amino acids found in proteins are always the L-isomers (except for glycine, the simplest amino acid, which does not have stereoisomers). However, D-amino acids occasionally occur in nature—for example, in certain bacterial cell walls and antibiotics. (Many other kinds of organic molecules also can exist in D and L forms. One example is the sugar glucose, which occurs in nature as D-glucose.)

Although only 20 different amino acids occur naturally in proteins, a single protein molecule can contain from 50 to hundreds of amino acid molecules, which can be arranged in an infinite number of ways to make proteins of different lengths, compositions, and structures. The number of proteins is practically endless, and every living cell produces many different proteins.

Peptide Bonds

Amino acids bond between the carboxyl (—COOH) group of one amino acid and the amino (—NH_2) group of another. For the formation of every bond between

TABLE 2.5 *(Continued)*

AMINO ACID (ABBREVIATION)	STRUCTURAL FORMULA	CHARACTERISTIC OF R GROUP
Aspartic acid (Asp)	H | HOOC—CH_2—C—COOH | NH_2	Additional carboxyl (—COOH) group, acidic
Lysine (Lys)	H | H_2N—CH_2—CH_2—CH_2—CH_2—C—COOH | NH_2	Additional amino (—NH_2) group, basic
Arginine (Arg)	H | H_2N—C—NH—CH_2—CH_2—CH_2—C—COOH ‖ | NH NH_2	Additional amino (—NH_2) group, basic
Asparagine (Asn)	H | H_2N—C—CH_2—C—COOH ‖ | O NH_2	Additional amino (—NH_2) group, basic
Glutamine (Gln)	H | H_2N—C—CH_2—CH_2—C—COOH ‖ | O NH_2	Additional amino (—NH_2) group, basic
Phenylalanine (Phe)	H | (benzene ring)—CH_2—C—COOH | NH_2	Cyclic
Tyrosine (Tyr)	H | HO—(benzene ring)—CH_2—C—COOH | NH_2	Cyclic
Histidine (His)	H | HC=C—CH_2—C—COOH | | | N NH NH_2 \\ / C | H	Heterocyclic
Tryptophan (Trp)	H | (benzene ring)—C—CH_2—C—COOH ‖ | N NH_2 | H	Heterocyclic
Proline (Pro)	CH_2 / \\ H_2C CH_2 | | N——C—COOH | | H H	Heterocyclic

two amino acids, one water molecule is released (dehydration). The bonds between amino acids are called **peptide bonds.**

In the example in Figure 2.17, a peptide bond between two amino acids is formed by dehydration synthesis. The carboxyl group of one amino acid supplies an OH^- and the amino group of the other releases an H^+ for the formation of water. The peptide bond is between the carbon atom of the carboxyl group of one amino acid and the nitrogen atom of the amino group of another amino acid. The resulting compound is called a *dipeptide* because it consists of two amino acids

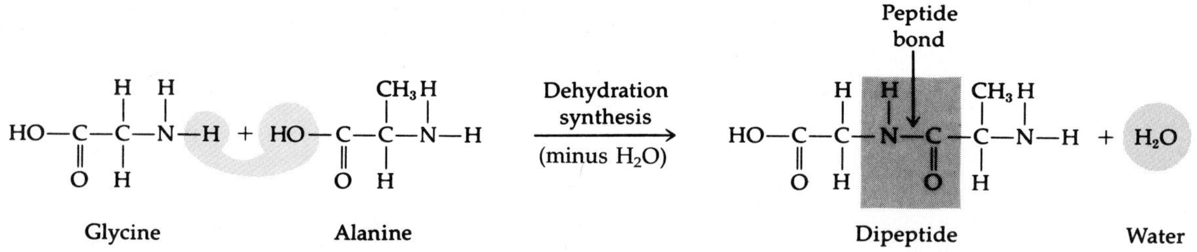

FIGURE 2.17 Peptide bond formation by dehydration synthesis. The amino acids glycine and alanine combine to form a dipeptide. The newly formed bond between the nitrogen atom of glycine and the carbon atom of alanine is called a peptide bond.

joined by a peptide bond. Adding another amino acid to a dipeptide would form a *tripeptide.* Further additions of amino acids would produce a long, chain-like molecule called a *polypeptide,* a protein molecule.

Levels of Protein Structure

Proteins vary tremendously in structure. Different proteins have different architectures and different three-dimensional shapes. This variation in structure is directly related to their diverse functions.

When a cell makes a protein, the polypeptide chain folds spontaneously to assume a certain shape. One reason for folding of the polypeptide is that some parts of a protein are attracted to water and other parts are repelled by it. In practically every case, the function of a protein depends on its ability to recognize and bind to some other molecule. As an example, an enzyme binds specifically with its substrate. A hormonal protein binds to a receptor on a cell whose function it will alter. An antibody binds to a foreign substance (antigen) that has invaded the body. The unique shape of a protein permits it to interact with specific other molecules in order to carry out specific functions.

Proteins are described in terms of four levels of organization—primary, secondary, tertiary, and quaternary. The *primary structure* is the unique order (the sequence) in which the amino acids are linked together (Figure 2.18a). This sequence is genetically determined. Alterations in sequence can have profound metabolic effects. For example, a single incorrect amino acid in a blood protein can produce the deformed hemoglobin molecule characteristic of sickle-cell anemia. But proteins do not exist as long, straight chains. Each polypeptide chain folds and coils in specific ways into a relatively compact structure with a characteristic three-dimensional shape.

A protein's *secondary structure* is the localized, repetitious twisting or folding of the polypeptide chain. This aspect of a protein's shape comes from hydrogen bonds joining the atoms of peptide bonds at different locations along the polypeptide chain. The two types of secondary protein structures are clockwise spirals called helices and pleated sheets, which form from roughly parallel portions of the chain (Figure 2.18b). Both structures are held together by hydrogen bonds between oxygen and nitrogen atoms that are part of the polypeptide's backbone.

Tertiary structure refers to the overall three-dimensional structure of a polypeptide chain (Figure 2.18c). The folding is not repetitive or predictable, as in secondary structure. Whereas secondary structure involves hydrogen bonding between atoms of the amino and carbonyl groups involved in the peptide bonds, tertiary structure involves interactions between various amino acid side groups in the polypeptide chain. Thus, the tertiary structure of a protein is elaborately irregular. Hydrogen bonds and other relatively weak interactions play an important role in tertiary structure. In addition, sulfhydryl groups (—SH) on two amino acid subunits can form a covalent, disulfide link (—S—S—) by removal of the hydrogen atoms. Such a reaction, in which electrons (in this case belonging to hydrogen atoms) are removed from a molecule is called an oxidation reaction.

Some proteins have a *quaternary structure,* which consists of an aggregation of two or more individual polypeptide chains (subunits) that operate as a single functional unit. Figure 2.18d shows a hypothetical protein consisting of two identical polypeptide chains. More commonly, proteins have two or more kinds of polypeptide subunits. The bonds that hold a quaternary structure together are basically the same as those that maintain tertiary structure. The overall shape of a protein may be globular (compact and roughly spherical, as in the figure) or fibrous (threadlike).

If a protein encounters a hostile environment in terms of temperature, pH, or salt concentrations, it may unravel and lose its characteristic shape. This process is called **denaturation** (see Figure 5.5). As a result of denaturation, the protein is no longer functional. This is discussed in more detail in Chapter 5 with regard to denaturation of enzymes.

The proteins we have been discussing are *simple proteins,* which contain only amino acids. *Conjugated*

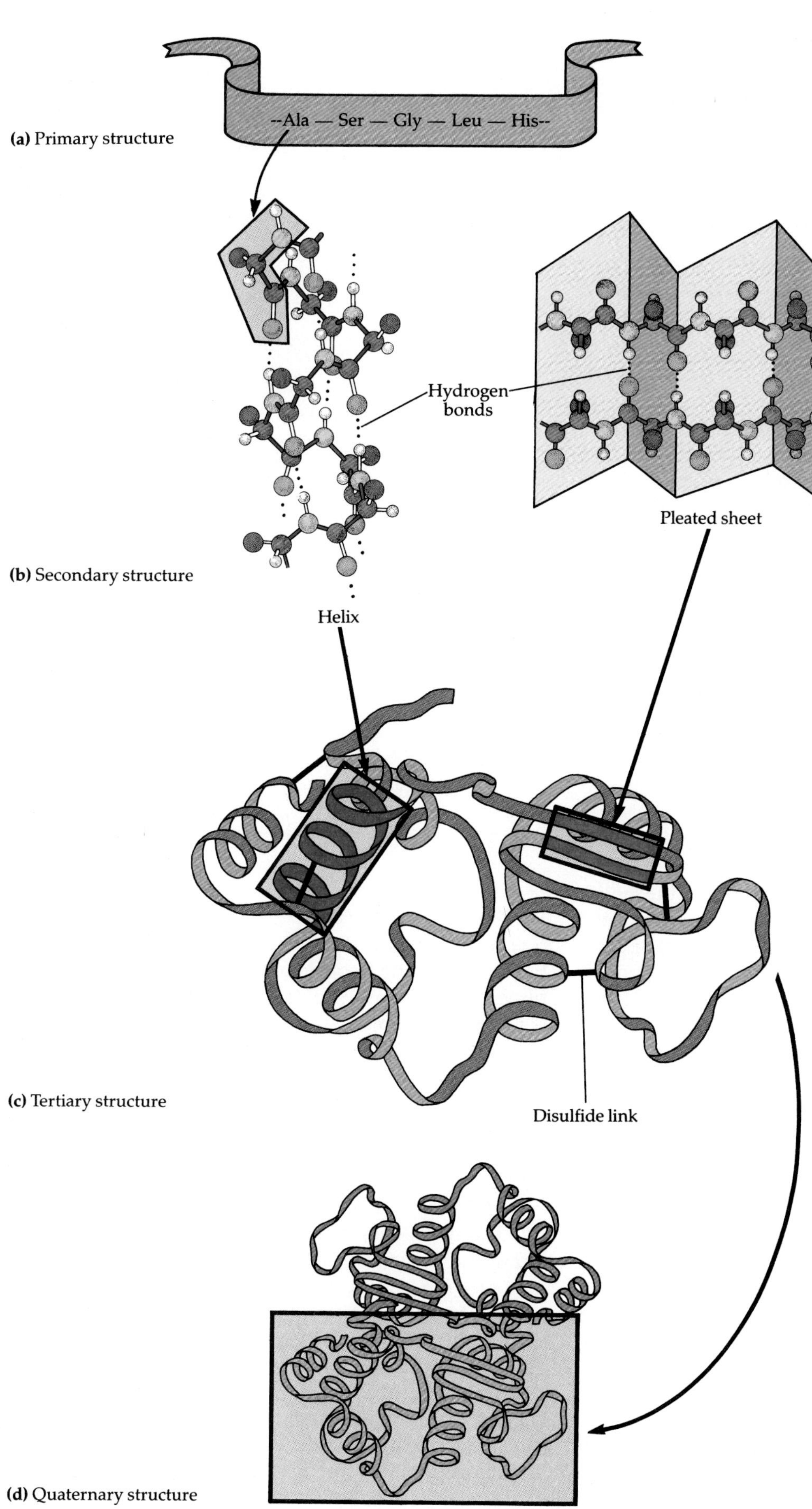

FIGURE 2.18 Protein structure. In this figure, polypeptide chains are shown as ribbons. **(a)** Primary structure, the amino acid sequence. Part of the polypeptide chain has been expanded to illustrate this. Each amino acid is designated by its three-letter abbreviation, as given in Table 2.5. **(b)** Secondary structures, helix and pleated sheet. Each ball represents an atom or R group of an amino acid. One amino acid is outlined in black. **(c)** Tertiary structure, the overall three-dimensional folding of a polypeptide chain. **(d)** Quaternary structure, the relationship between several polypeptide chains that make up a protein. Shown here is the quaternary structure of a hypothetical protein composed of two identical polypeptide chains.

proteins are combinations of amino acids with other organic or inorganic components. Conjugated proteins are named by their non-amino-acid component. Thus, glycoproteins contain sugars, nucleoproteins contain nucleic acids, metalloproteins contain metal atoms, lipoproteins contain lipids, and phosphoproteins contain phosphate groups. An example of a microbial phosphoprotein is phospholipase C. Synthesized by the bacterium *Clostridium perfringens* (klôs-tri'dē-um pėr-frin'jens), this enzyme breaks down red blood cells and induces some of the symptoms of gas gangrene.

NUCLEIC ACIDS

In 1944, three American microbiologists, Oswald Avery, Colin MacLeod, and Maclyn McCarty, discovered that a substance called **deoxyribonucleic acid (DNA)** is the substance of which genes are made. Nine years later, James Watson and Francis Crick, working with molecular models and X-ray information supplied by Maurice Wilkins and Rosalind Franklin, determined the physical structure of DNA. In addition, Crick suggested a mechanism for DNA replication and how it works as the hereditary material. DNA, and another substance called **ribonucleic acid (RNA),** are together referred to as **nucleic acids** because they were first discovered in the nuclei of cells.

DNA

According to the model proposed by Watson and Crick, a DNA molecule consists of two long strands wrapped around each other to form a **double helix** (Figure 2.19). The double helix looks like a twisted ladder. Just as amino acids are the structural units of proteins, so nucleotides are the structural units of nucleic acids. Each strand of the DNA double helix is composed of many nucleotides.

Each **nucleotide** of DNA is composed of three parts—a nitrogen-containing base, a pentose (five-carbon) sugar called **deoxyribose,** and a phosphate group (phosphoric acid), (Figure 2.19). The nitrogen-containing bases are cyclic compounds made up of carbon, hydrogen, oxygen, and nitrogen atoms. The bases are named adenine (A), thymine (T), cytosine (C), and guanine (G). A and G are double-ring structures called **purines,** whereas T and C are single-ring structures referred to as **pyrimidines.**

Nucleotides are named according to their nitrogenous base. Thus, a nucleotide containing thymine is called a **thymine nucleotide,** one containing adenine is called an **adenine nucleotide,** and so on. The term **nucleoside** refers to the combination of a purine or pyrimidine plus a pentose sugar; it does not contain a phosphate group.

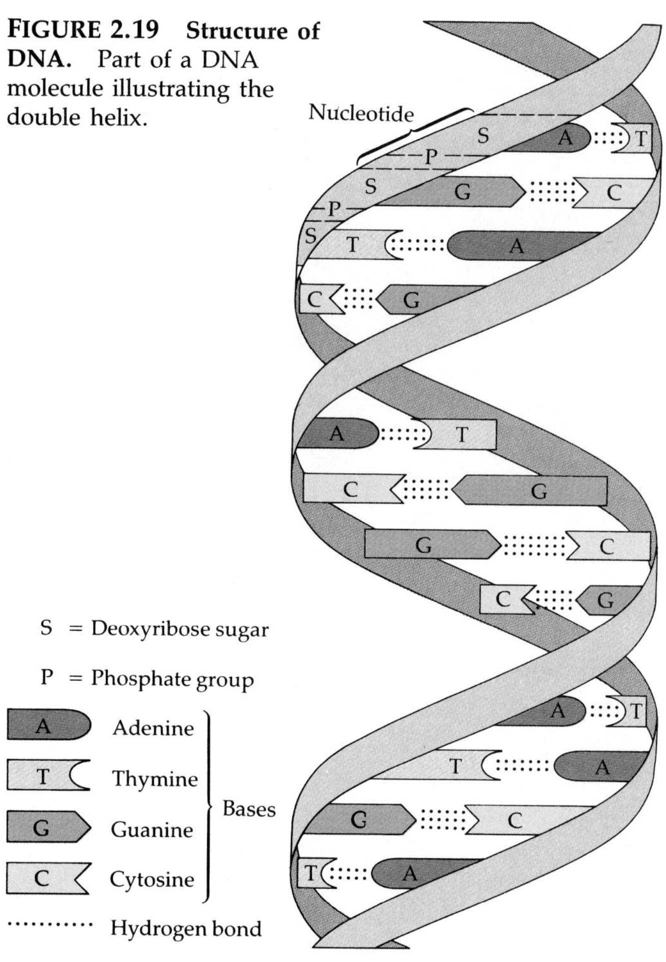

FIGURE 2.19 Structure of DNA. Part of a DNA molecule illustrating the double helix.

Each strand of DNA composing the double helix has a "backbone" consisting of alternating sugar and phosphate groups. The deoxyribose of one nucleotide is joined to the phosphate group of the next. (Refer to Figure 8.3 to see how nucleotides are bonded.) The nitrogen-containing bases make up the rungs of the ladder. Note that the purine A is always paired with the pyrimidine T and that purine G is always paired with the pyrimidine C. The bases are held together by hydrogen bonds; A—T is held by two hydrogen bonds, and G—C is held by three.

The order in which the nitrogen-base pairs occur along the backbone is extremely specific and in fact contains the genetic instructions for the organism. A certain segment of a nucleotide chain constitutes a gene, and a single DNA molecule may contain thousands of genes. Genes determine all hereditary traits, and they control all the activities that take place within cells.

A very important consequence of nitrogen-containing base pairing is that, if the sequence of bases

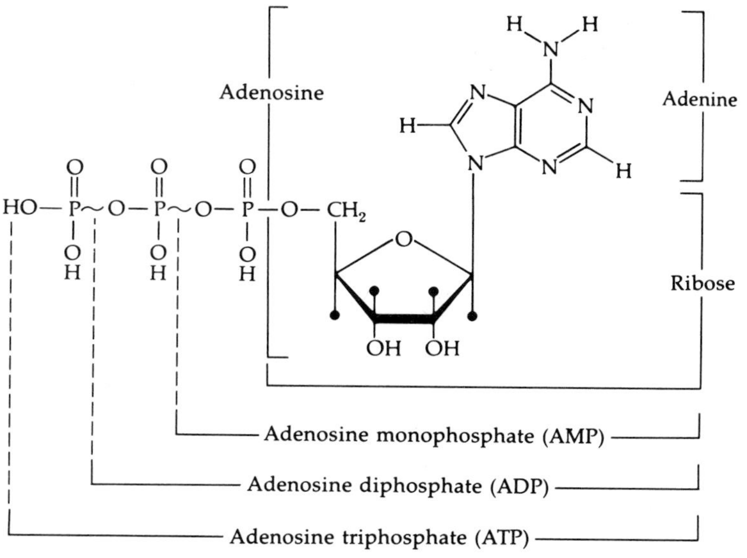

FIGURE 2.20 Structure of ATP. High-energy phosphate bonds are indicated by wavy lines. When ATP breaks down to ADP and inorganic phosphate, a large amount of chemical energy is released for use in other chemical reactions.

of one strand is known, then the sequence of the other strand is also known. For example, if one strand has the sequence . . . ATGC . . . , then the other strand has the sequence . . . TACG . . . . Since the sequence of bases of one strand is determined by the sequence of bases of the other, the bases are said to be **complementary.** The actual transfer of information becomes possible because of DNA's unique structure and will be discussed further in Chapter 8.

RNA

RNA, the second principal kind of nucleic acid, differs from DNA in several respects. Whereas DNA is double-stranded, RNA is usually single-stranded. The five-carbon sugar in the RNA nucleotide is **ribose,** which has one more oxygen atom than deoxyribose. Also, one of RNA's bases is uracil (U) instead of thymine. The other three bases (A, G, C) are the same as in DNA. Three major kinds of RNA have been identified in cells. These are referred to as **messenger RNA, ribosomal RNA,** and **transfer RNA.** As we shall see in Chapter 8, each type of RNA has a specific role in protein synthesis.

ADENOSINE TRIPHOSPHATE (ATP)

Adenosine triphosphate (ATP) is the principal energy-carrying molecule of all cells and is indispensable to the life of the cell. It stores the chemical energy released by some chemical reactions, and it provides the energy for reactions that require energy. ATP consists of an adenosine unit, composed of adenine and ribose, with three phosphate groups (abbreviated Ⓟ) attached (Figure 2.20). In other words, it is an adenine nucleotide (also called adenosine monophosphate, or AMP) with two extra phosphate groups. ATP is called a high-energy molecule because it releases a large amount of usable energy when it loses its terminal phosphate group to become **adenosine diphosphate (ADP).** This reaction can be represented as follows:

ATP	$\rightleftharpoons$	ADP	+	Ⓟ	+ Energy
Adenosine triphosphate		Adenosine diphosphate		Inorganic phosphate	

Because the supply of ATP at any particular time is limited, a mechanism exists to replenish it. The addition of a phosphate group to ADP manufactures more ATP. Because energy is required to manufacture ATP, the reaction can be represented as follows:

ADP	+	Ⓟ	+ Energy	$\rightleftharpoons$	ATP
Adenosine diphosphate		Phosphate			Adenosine triphosphate

The energy required to attach the terminal phosphate group to ADP is supplied by the cell's various decomposition reactions, particularly the decomposition of glucose. ATP can be stored in every cell, where its potential energy is not released until needed.

STUDY OUTLINE

Introduction (p. 22)

1. The interaction of atoms and molecules is called chemistry.
2. The metabolic activities of microorganisms involve complex chemical reactions.
3. Nutrients are broken down by microbes to obtain energy and to make new cells.

Structure of Atoms (pp. 22–24)

1. Atoms are the smallest units of chemical elements that enter into chemical reactions.
2. Atoms consist of a nucleus, which contains protons and neutrons, and electrons that move around the nucleus.
3. The atomic number is the number of protons in the nucleus; the total number of protons and neutrons is the atomic weight.

CHEMICAL ELEMENTS (pp. 23–24)

1. Atoms with the same atomic number and same chemical behavior are classified as the same chemical element.
2. Chemical elements are designated by letter abbreviations called chemical symbols.
3. There are about 26 elements commonly found in living cells.
4. Atoms that have the same atomic number (are of the same element) but different atomic weights are called isotopes.

ELECTRONIC CONFIGURATIONS (p. 24)

1. In an atom, electrons are arranged around the nucleus in electron shells.
2. Each shell can hold a characteristic maximum number of electrons.
3. Chemical properties of an atom are largely due to the number of electrons in its outermost shell.

How Atoms Form Molecules: Chemical Bonds (pp. 24–29)

1. Molecules are made up of two or more atoms; molecules consisting of at least two different kinds of atoms are called compounds.
2. Atoms form molecules in order to fill their outermost electron shells.
3. Attractive forces that bind the atomic nuclei of two atoms together are called chemical bonds.
4. The combining capacity of an atom—the number of chemical bonds the atom can form with other atoms—is its valence.

IONIC BONDS (pp. 26–27)

1. A positively or negatively charged atom or group of atoms is called an ion.
2. A chemical attraction between ions of opposite charge is called an ionic bond.
3. To form an ionic bond, one ion is an electron donor; the other ion is an electron acceptor.

COVALENT BONDS (p. 27)

1. In a covalent bond, atoms share pairs of electrons.
2. Covalent bonds are stronger than ionic bonds and are far more common in organisms.

HYDROGEN BONDS (p. 27)

1. A hydrogen bond exists when a hydrogen atom covalently bonded to one oxygen or nitrogen atom is attracted to another oxygen or nitrogen atom.
2. Hydrogen bonds form weak links between different molecules or between parts of the same large molecule.

MOLECULAR WEIGHT AND MOLES (pp. 27–29)

1. The molecular weight is the sum of the atomic weights of all the atoms in a molecule.
2. A mole of an atom, ion, or molecule is equal to its atomic or molecular weight expressed in grams.
3. The number of moles of a substance equals its mass in grams divided by its molecular weight.

Chemical Reactions (pp. 29–31)

1. Chemical reactions are the making or breaking of chemical bonds between atoms.

ENERGY OF CHEMICAL REACTIONS (p. 29)

1. A change of energy occurs during chemical reactions.
2. Endergonic reactions require energy; exergonic reactions release energy.
3. In a synthesis reaction, atoms, ions, or molecules are combined to form a larger molecule.
4. In a decomposition reaction, a larger molecule is broken down into its component molecules, ions, or atoms.
5. In an exchange reaction, two molecules are decomposed, and their subunits are used to synthesize two new molecules.
6. The products of reversible reactions can readily revert back to form the original reactants.

HOW CHEMICAL REACTIONS OCCUR (pp. 30–31)

1. For a chemical reaction to take place, the reactants must collide with each other.
2. The minimum energy of collision that can produce a chemical reaction is called its activation energy.
3. Specialized proteins called enzymes accelerate chemical reactions in living systems by lowering the activation energy.

IMPORTANT BIOLOGICAL MOLECULES (pp. 31–47)

Inorganic Compounds (pp. 32–34)

1. Inorganic compounds are usually small, ionically bonded molecules.
2. Water, and many common acids, bases, and salts are examples of inorganic compounds.

WATER (pp. 32–33)

1. Water is the most abundant substance in cells.
2. Because water is a polar molecule, it is an excellent solvent.
3. Water is a reactant in many of the decomposition reactions of digestion.
4. Water is an excellent temperature buffer.

ACIDS, BASES, AND SALTS (p. 33)

1. An acid dissociates into H^+ ions and anions.
2. A base dissociates into OH^- ions and cations.
3. A salt dissociates into negative and positive ions, neither of which is H^+ or OH^-.

ACID–BASE BALANCE (pp. 33–34)

1. The term pH refers to the concentration of H^+ in a solution.
2. A solution with a pH of 7 is neutral; a pH below 7 indicates acidity; a pH above 7 indicates alkalinity.
3. A pH buffer, which stabilizes the pH inside a cell, can be used in culture media.

Organic Compounds (pp. 34–47)

1. Organic compounds always contain carbon and hydrogen.
2. Carbon atoms form up to four bonds with other atoms.
3. Organic compounds are mostly or entirely covalently bonded, and many of them are large molecules.

FUNCTIONAL GROUPS (pp. 35–36)

1. A chain of carbon atoms forms a carbon skeleton.
2. The letter *R* may be used to denote a particular functional group of atoms within an organic molecule.
3. Functional groups of atoms are responsible for most of the properties of organic molecules.
4. Frequently encountered classes of molecules are R—OH (alcohols), R—COOH (organic acids), H_2N—R—COOH (amino acids).

MACROMOLECULES (p. 36)

1. Small organic molecules may combine into very large molecules called macromolecules.
2. Monomers usually bond together by dehydration synthesis or condensation reactions that form water and a polymer.

CARBOHYDRATES (pp. 36–37)

1. Carbohydrates are compounds consisting of atoms of carbon, hydrogen, and oxygen, with hydrogen and oxygen in a 2:1 ratio.
2. Carbohydrates include sugars and starches.
3. Carbohydrates can be divided into three types, monosaccharides, disaccharides, and polysaccharides.
4. Monosaccharides contain from three to seven carbon atoms.
5. Isomers are two molecules with the same chemical formula but different structures and properties—for example, glucose ($C_6H_{12}O_6$) and fructose ($C_6H_{12}O_6$).
6. Monosaccharides may form disaccharides and polysaccharides by dehydration synthesis.
7. Polysaccharides and disaccharides may be broken down by hydrolysis, a reaction involving the splitting of water molecules.

LIPIDS (pp. 37–41)

1. Lipids are a diverse group of compounds distinguished by their insolubility in water.
2. Simple lipids (fats) consist of a molecule of glycerol and three molecules of fatty acids.
3. A saturated fat has no double bonds between carbon atoms in the fatty acids; an unsaturated fat has one or more double bonds.
4. Phospholipids are complex lipids consisting of glycerol, two fatty acids, and phosphate.
5. Steroids have carbon ring systems with functional hydroxyl and carbonyl groups.

PROTEINS (pp. 41–46)

1. Amino acids are the building blocks of proteins.
2. Amino acids consist of carbon, hydrogen, oxygen, nitrogen, and sometimes sulfur.
3. Twenty amino acids occur naturally.
4. By linking amino acids, peptide bonds (formed by dehydration synthesis) allow the formation of polypeptide chains.
5. Proteins have four levels of structure—primary (sequence of amino acids), secondary (regular coils or pleats), tertiary (overall three-dimensional structure of a polypeptide), and quaternary (two or more polypeptide chains).
6. Conjugated proteins consist of amino acids combined with other organic or inorganic compounds.

NUCLEIC ACIDS (p. 46)

1. Nucleic acids—DNA and RNA—are macromolecules consisting of repeating nucleotides.
2. A nucleotide is composed of a pentose, a phosphate group, and a nitrogenous base. A nucleoside is composed of a pentose and a nitrogenous base.
3. A DNA nucleotide consists of deoxyribose (a pentose) and one of these nitrogenous bases: thymine or cytosine (pyrimidines) or adenine or guanine (purines).
4. DNA consists of two strands of nucleotides wound in a double helix. The strands are held together by hydrogen bonds between purine and pyrimidine nucleotides: A—T and G—C.
5. An RNA nucleotide consists of ribose (a pentose) and one of these nitrogenous bases: cytosine, guanine, adenine, or uracil.

ADENOSINE TRIPHOSPHATE (ATP) (p. 47)

1. ATP stores chemical energy for various cellular activities.
2. When the bond to ATP's terminal phosphate group is broken, energy is released.
3. The energy from decomposition reactions is used to regenerate ATP from ADP and phosphate.

STUDY QUESTIONS

Review

1. What is a chemical element?
2. Diagram the electronic configuration of a carbon atom.
3. How does ${}^{14}_{6}C$ differ from ${}^{12}_{6}C$?
4. What type of bonding exists between water molecules?
5. What type of bonds will hold the following atoms together?
   (a) Li^+ and Cl^- ions in LiCl
   (b) Carbon and oxygen atoms in CO_2
   (c) Oxygen atoms in O_2
   (d) A carbon atom from glutamic acid and a nitrogen atom from lysine in:

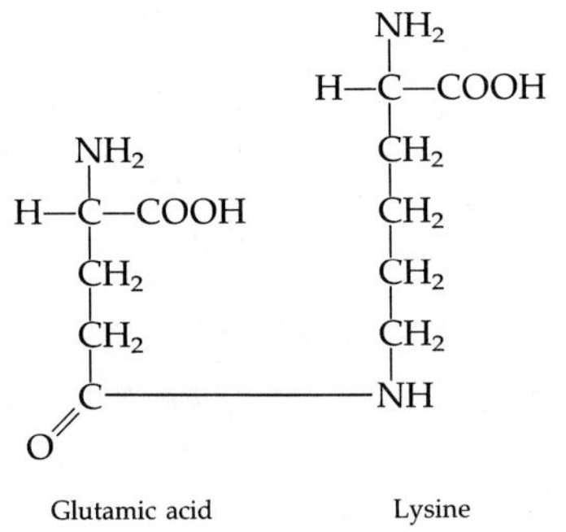

Glutamic acid: NH_2—H—C—COOH, CH_2, CH_2, C(=O) — Lysine: NH_2—H—C—COOH, CH_2, CH_2, CH_2, CH_2, NH

Glutamic acid Lysine

   (e) A hydrogen atom of one nucleotide to nitrogen or oxygen atoms of another nucleotide in:

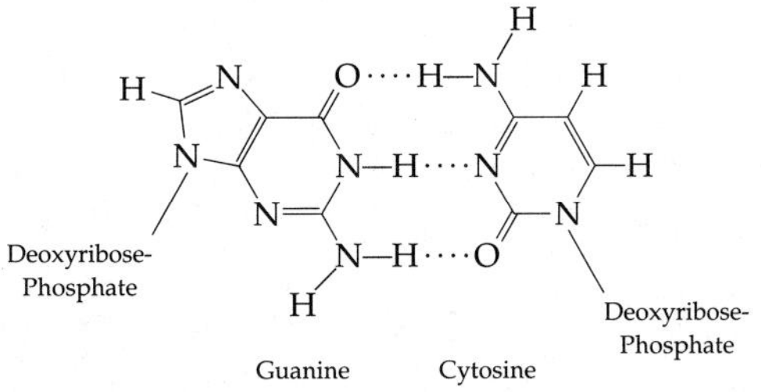

Deoxyribose-Phosphate Guanine Cytosine Deoxyribose-Phosphate

6. Classify the following inorganic molecules as an acid, base, or salt. Their dissociation products are shown to help you.
   (a) $HNO_3 \rightarrow H^+ + NO_3^-$
   (b) $H_2SO_4 \rightarrow 2H^+ + SO_4^{2-}$
   (c) $NaOH \rightarrow Na^+ + OH^-$
   (d) $MgSO_4 \rightarrow Mg^{2+} + SO_4^{2-}$
7. Vinegar, pH 3, is how many times more acidic than pure water, pH 7?
8. Calculate the molecular weight of $C_6H_{12}O_6$.
9. How many moles are in 360 grams of glucose?
10. Classify the following types of chemical reactions.
    (a) Glucose + Fructose → Sucrose + H_2O
    (b) Lactose → Glucose + Galactose
    (c) $NH_4Cl + H_2O \rightarrow NH_4OH + HCl$
    (d) ATP → ADP + Ⓟ
11. Bacteria use the enzyme urease to obtain nitrogen in a form they can use from urea in this reaction:

$$CO(NH_2)_2 + H_2O \longrightarrow 2NH_4^+ + CO_2$$

Urea Ammonium ion Carbon dioxide

    What purpose does the enzyme serve in this reaction? What type of reaction is this?
12. Classify the following as subunits of either a carbohydrate, lipid, protein, or nucleic acid.
    (a) $CH_3—(CH_2)_7—CH{=}CH—(CH_2)_7—COOH$
    Oleic acid
    (b) NH_2—H—C—COOH, CH_2, OH
    Serine
    (c) $C_6H_{12}O_6$
    (d) Thymine nucleotide
13. Add the appropriate functional group(s) to this ethyl group to produce each of the following compounds: ethanol, acetic acid, acetaldehyde, ethanolamine, diethyl ether.

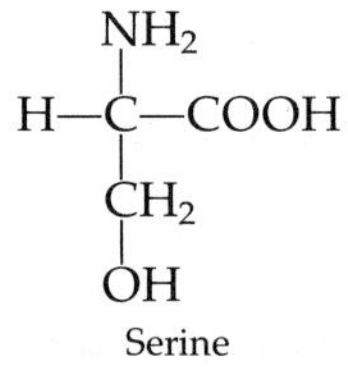

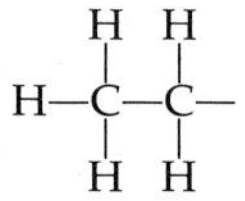

$$H—\underset{H}{\overset{H}{C}}—\underset{H}{\overset{H}{C}}—$$

14. Water plays an important role in these reactions:

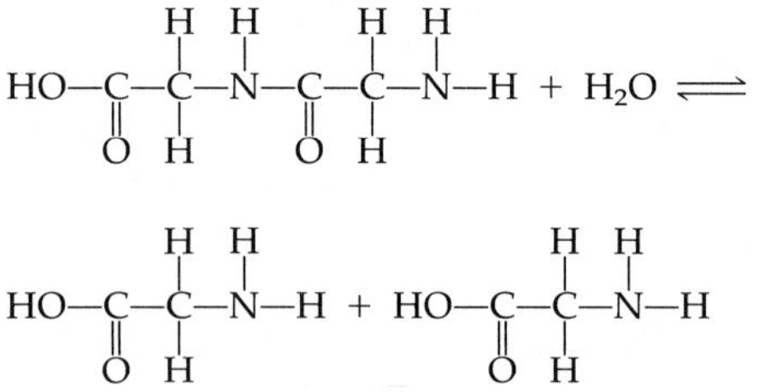

$$HO—\underset{O}{C}—\underset{H}{\overset{H}{C}}—\overset{H}{N}—\underset{O}{C}—\underset{H}{\overset{H}{C}}—\overset{H}{N}—H + H_2O \rightleftharpoons$$

$$HO—\underset{O}{C}—\underset{H}{\overset{H}{C}}—\overset{H}{N}—H + HO—\underset{O}{C}—\underset{H}{\overset{H}{C}}—\overset{H}{N}—H$$

    (a) What direction is the hydrolysis reaction (left to right or right to left)?
    (b) What direction is the dehydration synthesis reaction?
    (c) Circle the atoms involved in the formation of water.
    (d) Identify the peptide bond.
15. The energy-carrying property of the ATP molecule is due to energy dynamics that favor the breaking of bonds between ______________. What type of bonds are these? ______________.

16. The diagram at right shows a protein molecule. Indicate the regions of primary, secondary, and tertiary structure. Does this protein have quaternary structure?
17. Draw a simple lipid, and show how it could be modified to a phospholipid.
18. ATP is an energy-storage compound. Where does it get this energy from?

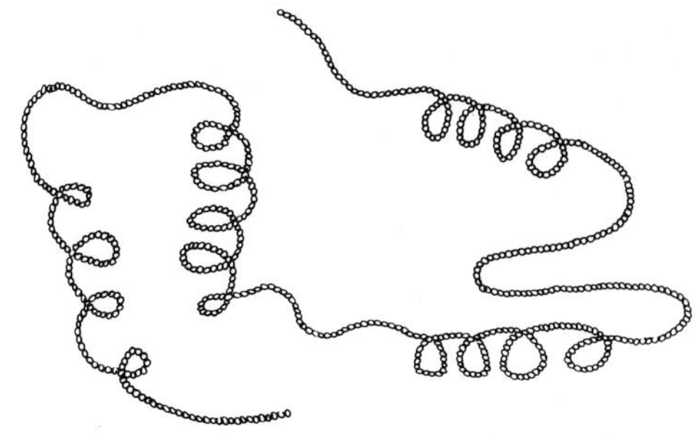

Challenge

1. When you blow bubbles into a glass of water, the following reactions take place.

$$H_2O + CO_2 \xrightarrow{A} H_2CO_3 \xrightarrow{B} H^+ + HCO_3^-$$

   (a) What type of reaction is *A*?
   (b) What does reaction *B* tell you about the type of molecule H_2CO_3 is?

2. What are the common structural characteristics of ATP and DNA molecules?
3. *Thiobacillus ferrooxidans* was responsible for destroying buildings in the Midwest by causing changes in the earth. The original rock, which contained lime ($CaCO_3$) and pyrite (FeS_2), expanded as the bacterial metabolism caused gypsum ($CaSO_4$) crystals to form. How did *T. ferrooxidans* bring about the change from lime to gypsum?

FURTHER READING

Chang, R. *Chemistry*, 4th ed. New York: McGraw-Hill, 1990. An introductory chemistry textbook with examples of biological and environmental chemistry.

Hakomori, S. "Glycosphingolipids." *Scientific American* 254(5):44–53, May 1986. Discusses lipid structure and the role of lipids in cancer.

Harold, F.M. *The Vital Force: A Study of Bioenergetics*. New York: W.H. Freeman, 1986. Chapter 4 covers energy production in bacteria, including photosynthesis, respiration, and translocation. Chapter 5 describes energy utilization such as movement and transport. Chapter 6 uses metabolic examples to illustrate evolution of eucaryotic cells.

Karplus, M., and J.A. McCammon. "The dynamics of proteins." *Scientific American* 254(4):42–51, April 1986. Relates the constant motion of proteins to their biological functions by using X-ray crystallography.

Mathews, C.K., and K.E. van Holde. *Biochemistry*. Redwood City, Calif.: Benjamin/Cummings, 1990. A biochemistry text with detailed descriptions of organic molecules and their functions in cells.

"Molecules of Life." *Scientific American* 253(4), October 1985. Entire issue devoted to biochemistry; includes articles on RNA, DNA, and proteins.

Sharon, N. "Carbohydrates." *Scientific American* 243(5):90–116, November 1980. Illustrates new information on the functions of carbohydrates other than as carbon and energy sources.

Watson, J.D. *The Double Helix*, ed. Gunther S. Stent. New York: W.W. Norton, 1980. Original papers and Watson's account of the struggle to decipher the structure of DNA give insight into creative scientific processes. Includes commentaries and reviews by other scientists.

CHAPTER 3

Observing Microorganisms Through a Microscope

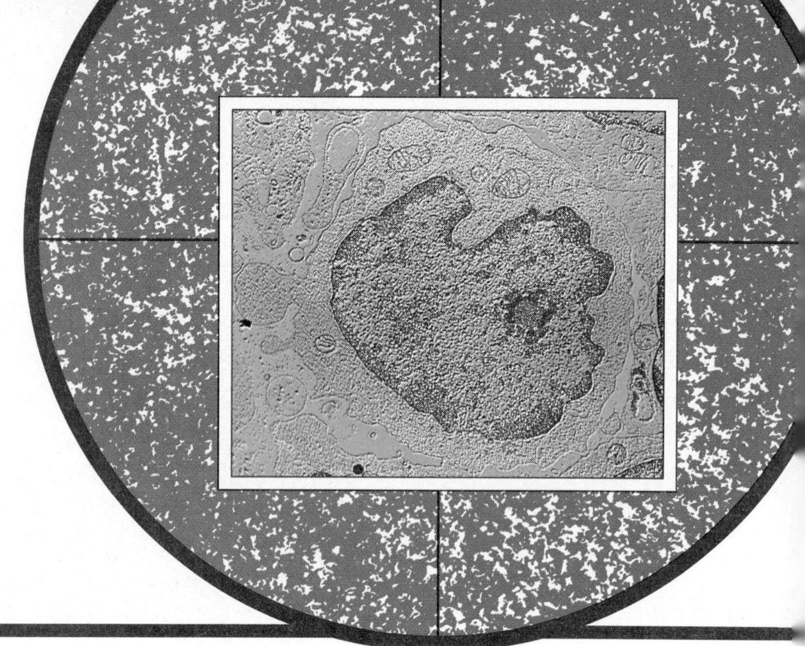

LEARNING OBJECTIVES

- List the units of measurement used for microorganisms, and know their equivalents.
- Diagram the path of light through a compound microscope.
- Define resolution and total magnification.
- Cite an advantage of darkfield, phase-contrast, and fluorescence microscopy, and compare each with brightfield illumination.
- Explain how electron microscopy differs from light microscopy.
- Differentiate between an acidic dye and a basic dye.
- Compare simple, differential, and special stains.
- List the steps in preparing a Gram stain, and describe the appearance of a gram-positive and gram-negative cell after each step.

Microorganisms are too small to be seen with the naked eye, so they must be observed with a microscope (*micro* means small, *skopein* means to see). In Anton van Leeuwenhoek's day, looking through a microscope meant looking through rainbow rings and shadows and multiple images. Modern microbiologists, however, have access to microscopes that produce, with great clarity, magnifications from ten to thousands of times greater than those of van Leeuwenhoek's single lens (see Figure 1.2a). This chapter will describe how different types of microscopes function and the advantages of each type.

Some microbes are more readily visible than others. Many have to undergo several staining procedures before cell walls, membranes, and structures lose their opaque or colorless natural state. The last part of this chapter will explain the methods of preparing specimens for light microscopy.

You may be wondering how we are going to sort out, measure, and count the specimens we will be studying. This chapter opens with a discussion of how to use the metric system for measuring microbes.

Units of Measurement

Because microorganisms and their component parts are so very small, they are measured in units that are unfamiliar to many of us in everyday life. When measuring microorganisms, we use the metric system. The standard unit of length in the metric system is the meter (m). A major advantage of the metric system is that the units are related to each other by factors of 10. Thus, 1 m equals 10 decimeters (dm) or 100 centimeters (cm) or 1000 millimeters (mm). Units in the U.S. system of measure do not have the advantage of conversion by a factor of 10. For example, we have to use 3 ft or 36 in. to equal 1 yd.

Microorganisms and their structural components are measured in even smaller units, such as micrometers, nanometers, and angstroms. A **micrometer (μm),** formerly known as a micron (μ), is equal to 0.000001 m (10^{-6} m). The prefix *micro* indicates that the unit following it should be divided by one million, or 10^6 (see Exponential Notation in Appendix G). A **nanometer (nm),** formerly known as a millimicron (mμ), is equal to 0.000000001 m (10^{-9} m). The prefix *nano* tells us that the unit after it should be divided by one billion (10^9). An **angstrom (Å)** is equal to 0.0000000001 m (10^{-10} m). The angstrom is no longer an official unit of measure; however, because of its widespread presence in scientific literature, you should be familiar with it. Its accepted equivalent is 0.1 nm. Table 3.1 presents the basic metric units of length and some of their U.S. equivalents. In Table 3.1, you can compare the microscopic units of measurement with the commonly known macroscopic units of measurement, such as centimeters, meters, and kilometers. If you look ahead to Figure 3.2 on page 55, you will see the relative sizes of various organisms on the metric scale.

Microscopy: The Instruments

The simple microscope used by Anton van Leeuwenhoek in the seventeenth century had only one lens and was similar to a magnifying glass. Contemporaries of van Leeuwenhoek, such as Robert Hooke, built microscopes with multiple lenses, called compound microscopes. In fact, a Dutch spectacle-maker, Zaccharias Janssen, is credited with making the first compound microscope around 1600. However, these early compound microscopes were of poor quality. It was not until about 1830 that a significantly better microscope was developed, by Joseph Jackson Lister (the father of Joseph Lister). Various improvements on Lister's microscope were made that resulted in the development of the modern compound microscope, the kind used in microbiology laboratories today.

COMPOUND LIGHT MICROSCOPY

A modern **compound light microscope** has two sets of lenses, objective and ocular, and uses visible light as its source of illumination (Figure 3.1a). Using the compound light microscope, we can examine very small specimens as well as some of their fine detail, or **ultrastructure.** A series of finely ground lenses forms a clearly focused image that is many times larger than the specimen itself (Figure 3.1b). This magnification is achieved when light rays from an **illuminator,** the light source, are passed through a **condenser,** which directs the light rays through the specimen. From here, light rays pass into the **objective lens,** the lens closest to the specimen. The image of the specimen is magnified again by the **ocular lens,** or **eyepiece.**

We can calculate the total magnification of a specimen by multiplying the objective lens magnification (power) by the ocular lens magnification (power). Most microscopes used in microbiology have several objective lenses, including 10× (low power), 40× (high power), and 100× (oil immersion). Most oculars magnify specimens by 10×. Multiplying the magnification of a specific objective lens with that of the ocular, we see that the total magnifications would be 100× for low power, 400× for high power, and 1000× for oil immersion. Some compound light microscopes can achieve a magnification of 2000× with the oil immersion lens.

Resolution, or **resolving power,** is the ability of the lenses to distinguish fine detail and structure. Specifically, it refers to the ability of the lenses to distinguish between two points at a specified distance apart. For example, if a microscope has a resolving power of 0.4 nm, it is capable of distinguishing two points as separate objects if they are at least 0.4 nm apart. A

TABLE 3.1 Metric Units of Length and U.S. Equivalents

METRIC UNIT	MEANING OF PREFIX	METRIC EQUIVALENT	U.S. EQUIVALENT
1 kilometer (km)	*kilo* = 1000	1000 m = 10^3	3280.84 ft or 0.62 mi; 1 mi = 1.61 km
1 meter (m)		Standard unit of length	39.37 in. or 3.28 ft or 1.09 yd
1 decimeter (dm)	*deci* = 1/10	0.1 m = 10^{-1} m	3.94 in.
1 centimeter (cm)	*centi* = 1/100	0.01 m = 10^{-2} m	0.394 in.; 1 in. = 2.54 cm
1 millimeter (mm)	*milli* = 1/1000	0.001 m = 10^{-3} m	
1 micrometer (μm)	*micro* = 1/1,000,000	0.000001 m = 10^{-6} m	
1 nanometer (nm)	*nano* = 1/1,000,000,000	0.000000001 m = 10^{-9} m	

FIGURE 3.1 The compound light microscope. **(a)** Principal parts and their functions. **(b)** The path of light (bottom to top). **(c)** Refraction. Because the refractive indexes of the glass microscope slide and immersion oil are the same, the light rays do not refract when passing from one to the other when the oil immersion objective lens is used.

Arm

Condenser Focuses light through specimen

Fine adjustment knob

Coarse adjustment knob

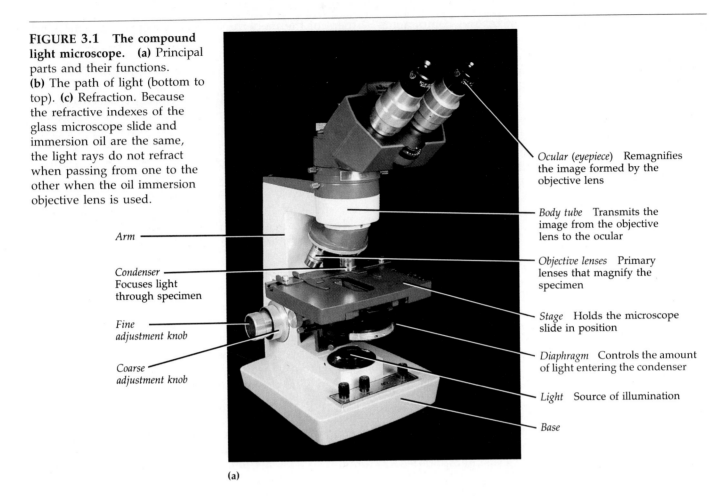

Ocular (eyepiece) Remagnifies the image formed by the objective lens

Body tube Transmits the image from the objective lens to the ocular

Objective lenses Primary lenses that magnify the specimen

Stage Holds the microscope slide in position

Diaphragm Controls the amount of light entering the condenser

Light Source of illumination

Base

(a)

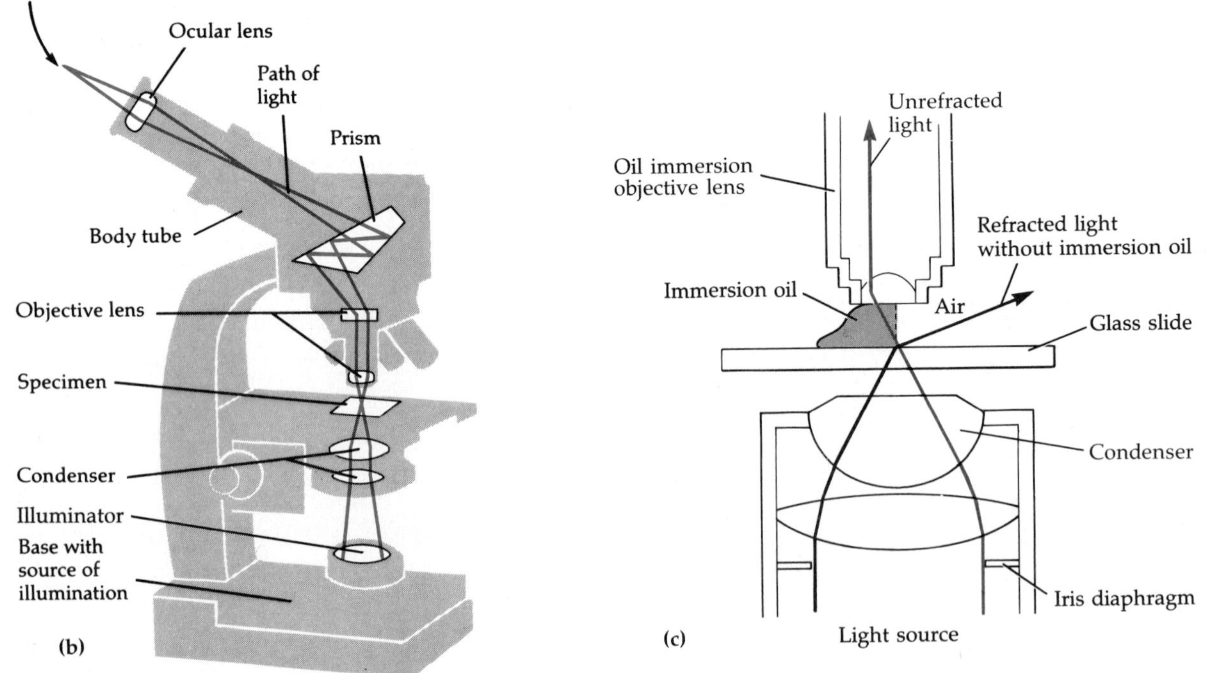

general principle of microscopy is that, the shorter the wavelength of light used in the instrument, the greater the resolution. The white light used in a compound light microscope has a relatively long wavelength and cannot resolve structures smaller than 0.2 μm. This fact and other practical considerations limit the magnification achieved by even the best compound light microscopes to about 2000×.

Figure 3.2 shows various specimens that can be resolved by the human eye, light microscope, and electron microscope.

To obtain a clear, finely detailed image under a compound light microscope, specimens must be made to contrast sharply with their medium. To attain such contrast, we must change the refractive index of specimens from that of their medium. The **refractive index** is a measure of the relative velocity at which light passes through a material. We change the refractive index of specimens by staining them, a procedure we will discuss shortly. Light rays move in a straight line through a single medium. After staining, when light rays pass through the two materials (the specimen and its medium) with different refractive indexes, the rays change direction (refract) at the boundary between the materials and increase the image's contrast between the specimen and the medium. As the light rays travel away from the specimen, they spread out to resolve points that are close together in the image. These light rays enter the objective lens, and the image is magnified.

To achieve high magnification (100×) with good resolution, the lens must be small. Although we want light traveling through the specimen and medium to refract differently, we do not want to lose light rays after they have passed through the stained specimen. To preserve the direction of light rays at the highest magnification, immersion oil is placed between the glass slide and the oil immersion objective lens. The immersion oil has the same refractive index as glass, so the oil becomes part of the optics of the glass of the microscope. Unless immersion oil is used, light rays are refracted as they enter the air from the slide, and the objective lens would have to be increased in diameter to capture them. The oil has the same effect as increasing the objective diameter; therefore, it improves the resolving power of the lenses. If oil is not used with an oil immersion objective lens, the image becomes fuzzy, with poor resolution.

Under usual operating conditions, the field of vision in a compound light microscope is brightly illuminated (Figure 3.3a). By focusing the light, the condenser produces a **brightfield** illumination (Figure 3.4a).

It is not always desirable to stain a specimen. However, an unstained cell has little contrast with its surroundings and is therefore difficult to see. Unstained cells are more easily observed with the modified compound microscopes described in the next section.

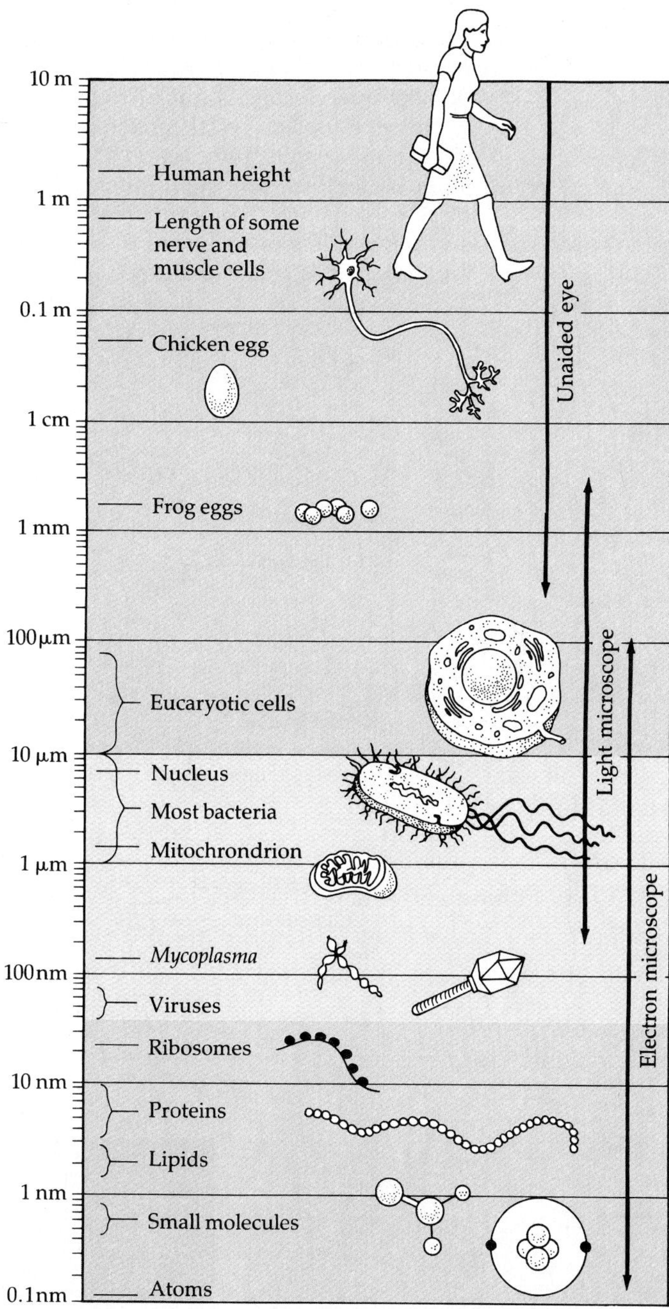

FIGURE 3.2 Relationships between the sizes of various specimens and the resolution of the human eye, light microscope, and electron microscope. The area in yellow shows the size range of the organisms we will be studying in this book.

DARKFIELD MICROSCOPY

A **darkfield microscope** is used for examining live microorganisms that either are invisible in the ordinary light microscope, cannot be stained by standard meth-

FIGURE 3.3 Brightfield, darkfield, and phase-contrast microscopy. **(a)** The path of light in brightfield microscopy. **(b)** The path of light in darkfield microscopy. The darkfield microscope is similar to the brightfield microscope, except that a darkfield microscope uses a special condenser that contains an opaque disc that eliminates all light in the center of the beam. The only light that reaches the specimen comes in at an angle; thus, only light diffracted by the specimen (colored arrows) reaches the objective lens. **(c)** Phase-contrast microscopy. Light rays are diffracted differently and travel different pathways to reach the eye of the viewer. Diffracted light rays are indicated in color; undiffracted light rays are indicated in black.

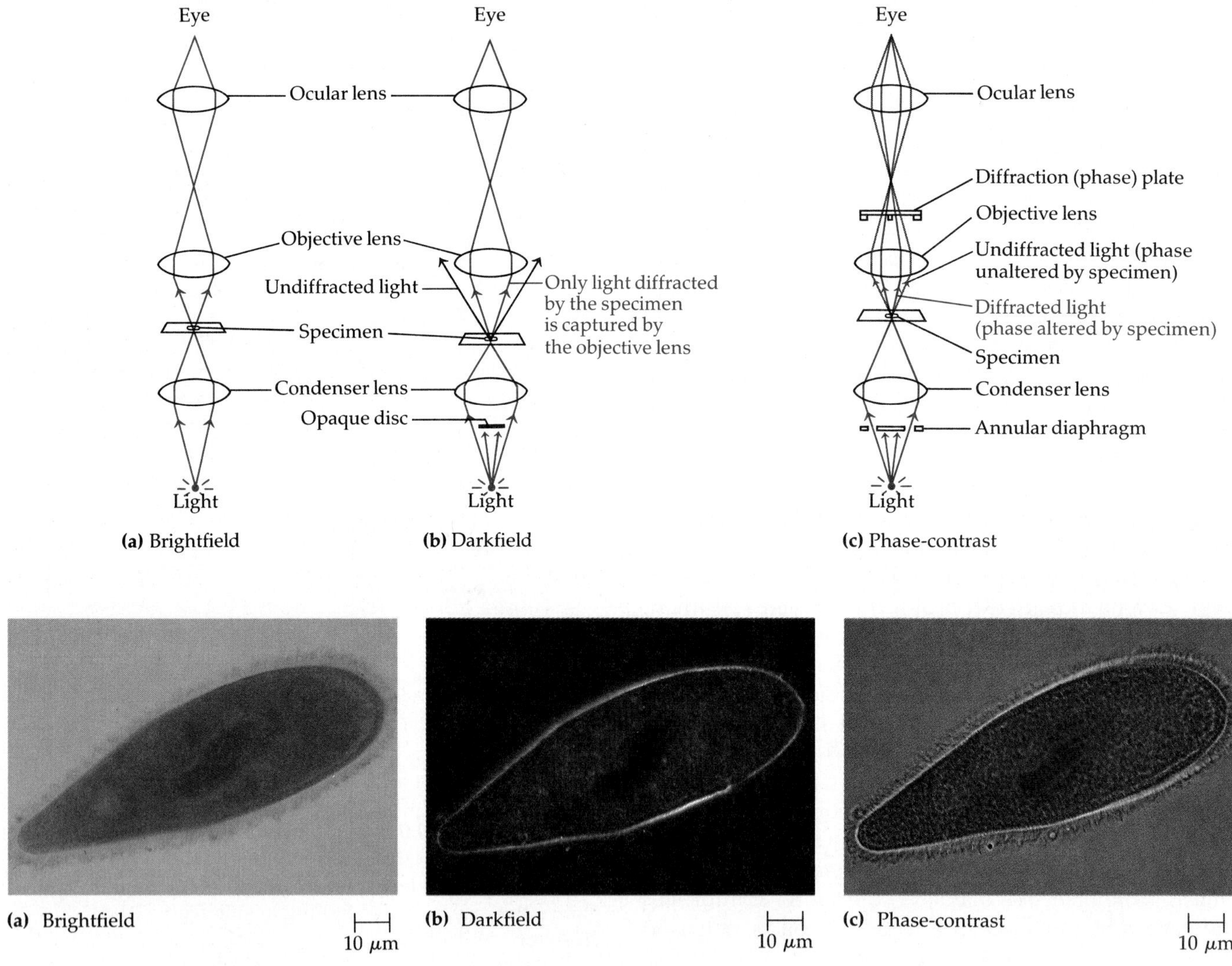

FIGURE 3.4 Types of images produced by a compound light microscope. Shown here is the same *Paramecium* specimen using different microscopic techniques. **(a)** Brightfield illumination shows internal structures and the outline of the transparent sheath. **(b)** Against the black background seen with darkfield microscopy, edges of the cell are bright, some internal structures seem to sparkle, and the sheath (external covering) is almost invisible. **(c)** Phase-contrast microscopy shows greater differentiation among the internal structures and also shows the sheath wall. The wide light band around each cell is an artifact resulting from this type of microscopy—a "phase halo."

ods, or are so distorted by staining that their characteristics then cannot be identified. Instead of the normal condenser, a darkfield microscope uses a darkfield condenser that contains an opaque disc. The disc blocks light that would enter the objective directly. Only light that is diffracted (bent) by the specimen enters the objective lens (Figure 3.3b). Because there is no direct background light, the specimen appears light against a black background—the dark field (Figure 3.4b). This technique is frequently used to examine unstained microorganisms suspended in liquid. One use for darkfield microscopy is the examination of very thin spirochetes, such as *Treponema pallidum* (tre-pō-nē'mä pal'li-dum), the causative agent of syphilis.

PHASE-CONTRAST MICROSCOPY

Another way to observe microorganisms is with a **phase-contrast microscope.** This technique is especially useful because it permits the detailed examination of internal structures in *living* microorganisms. In addition, it is not necessary to fix (attach the microbes to the microscope slide) or stain the specimen—procedures that could harm or kill the microorganisms. The principle of phase-contrast microscopy is based on slight variations in refractive index. As rays pass from the light source through the specimen, their velocity may be altered by differences in the thickness and physical properties of various portions of the specimen. Light rays passing through the specimen are diffracted (bent) differently and travel different pathways (out of phase with one another) to reach the eye of the viewer (Figure 3.3c). These phase differences are seen through the microscope as different degrees of brightness. Details of the internal structures of the specimen also become more sharply defined in phase-contrast microscopy. The internal details of a cell appear as degrees of brightness against a dark background (see Figure 3.4c).

A phase-contrast microscope uses a special condenser that contains an annular (ring-shaped) diaphragm. The diaphragm allows a ring of light to pass through the condenser, focusing light on the specimen and a ring-shaped diffraction (phase) plate in the objective lens. The diffracted and undiffracted rays are then brought in phase with each other to produce the image that meets the eye.

FLUORESCENCE MICROSCOPY

Fluorescence microscopy takes advantage of the fluorescence of substances. Fluorescent substances absorb short wavelengths of light (ultraviolet) and give off light at a longer wavelength that can be seen by the use of special light filters. Some organisms fluoresce naturally under special lighting (Figure 3.5); if the specimen to be viewed does not naturally fluoresce, it is stained with one of a group of fluorescent dyes called *fluorochromes.* When microorganisms stained with a fluorochrome are examined under a fluorescent microscope with an ultraviolet or near-ultraviolet light source, they appear as luminescent, bright objects against a dark background.

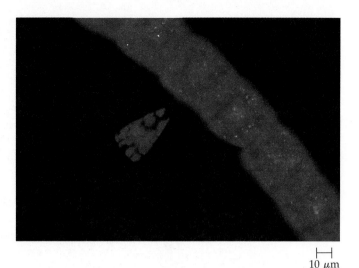

FIGURE 3.5 Fluorescence microscopy. Fluorescence microscopy has been used to visualize this triangular diatom (*Licmorphora*) and strand of an alga (*Spongomorphora*). Under ultraviolet light, the chlorophyll in these organisms fluoresces.

Fluorochromes have special attractions for different microorganisms. For example, the fluorochrome auramine O, which glows yellow when exposed to ultraviolet light, is strongly absorbed by *Mycobacterium tuberculosis* (mī-kō-bak-ti'rē-um tü-bėr-kū-lō'sis), the bacterium that causes tuberculosis. When the dye is applied to a sample of material suspected of containing the bacterium, the bacterium is detected by the appearance of bright yellow organisms against a dark background. *Bacillus anthracis,* the causative agent of anthrax, appears apple green when stained with another fluorochrome, fluorescein isothiocyanate (FITC).

The principal use of fluorescence microscopy is a diagnostic technique called the **fluorescent-antibody technique,** or **immunofluorescence. Antibodies** are natural defense molecules that are produced by humans and many animals in reaction to a foreign substance, or **antigen.** Fluorescent antibodies for a particular antigen are obtained as follows: An animal is injected with a specific antigen, such as a bacterium; the animal then begins to produce specific antibodies against that antigen. After a sufficient time, the antibodies are removed from the serum of the animal. Next, as you can see in Figure 3.6, a fluorochrome is chemically combined with the antibodies. These fluo-

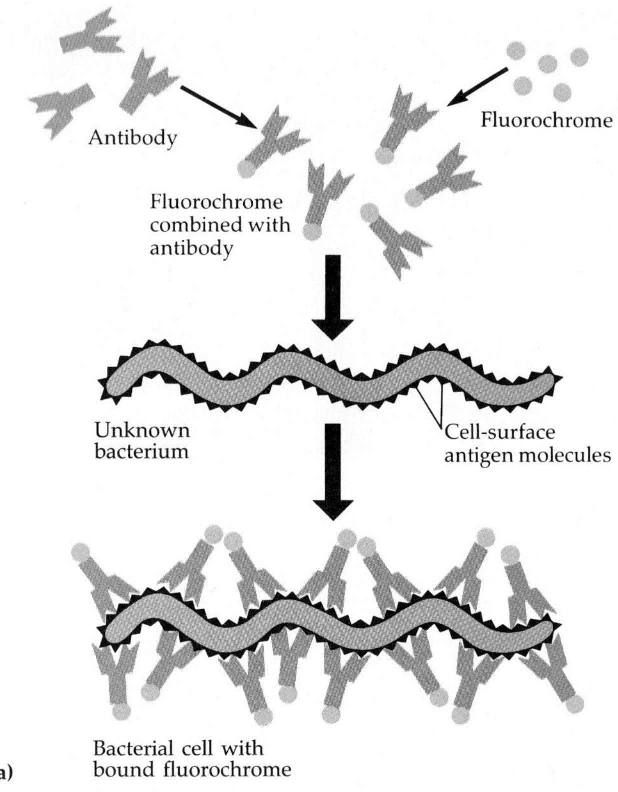

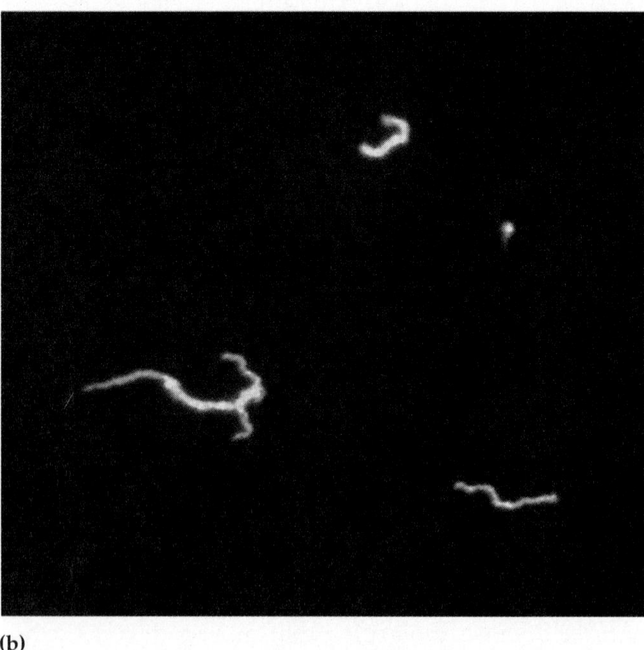

FIGURE 3.6 The principle of immunofluorescence. **(a)** A fluorochrome is combined with an antibody against a specific bacterium. When the preparation is added to bacterial cells on a microscope slide, the antibody attaches to the bacterial cells and the cells fluoresce when illuminated with ultraviolet light. **(b)** In the FTA-ABS test for syphilis, *Treponema pallidum* shows up as light regions against a darker background.

rescent antibodies are then added to a microscope slide containing an unknown bacterium. If this unknown bacterium is the same bacterium that was injected into the animal, the fluorescent antibodies bind to antigens on the surface of the bacterium, causing it to fluoresce.

This technique can detect bacteria or other disease-producing microorganisms even within cells, tissues, or other clinical specimens (Figure 3.6b). It is especially useful in diagnosing syphilis and rabies. We will say more about antigen-antibody reactions and immunofluorescence in Chapter 18.

ELECTRON MICROSCOPY

Objects smaller than 0.2 μm, such as viruses or the internal structures of cells, must be examined with an **electron microscope.** In electron microscopy, a beam of electrons is used instead of light. Free electrons travel in waves. The resolving power of the electron microscope is far greater than that of the other microscopes we have mentioned. The better resolution of electron microscopes is due to the shorter wavelengths of electrons; the wavelengths of electrons are about 100,000 times smaller than the wavelengths of visible light. Thus, electron microscopes are used to examine structures too small to be resolved with light microscopes.

Instead of using glass lenses, an electron microscope uses electromagnetic lenses to focus a beam of electrons through an evacuated tube onto a specimen. There are two types of electron microscopes, the transmission electron microscope and the scanning electron microscope.

Transmission Electron Microscopy

In the **transmission electron microscope,** a finely focused beam of electrons from an electron gun passes through a specially prepared, ultrathin section of the specimen (Figure 3.7a). The beam is focused on a small area of the specimen by an electromagnetic condenser lens that performs roughly the same function as the condenser of a light microscope—to direct the beam of electrons in a straight line to illuminate the specimen.

Electron microscopes use electromagnetic lenses rather than glass lenses, as in a light microscope. Electromagnetic lenses control illumination, focus, and magnification. Instead of being placed on a glass slide, as in light microscopes, the specimen is usually placed on a copper mesh grid. The beam of electrons passes through the specimen and then through an electromagnetic objective lens, which magnifies the image. Finally, the electrons are focused by an electromagnetic projector lens, rather than by an ocular lens as in a light microscope, onto a fluorescent screen or photographic plate. The final image, called a **transmission electron micrograph,** appears as many light and dark

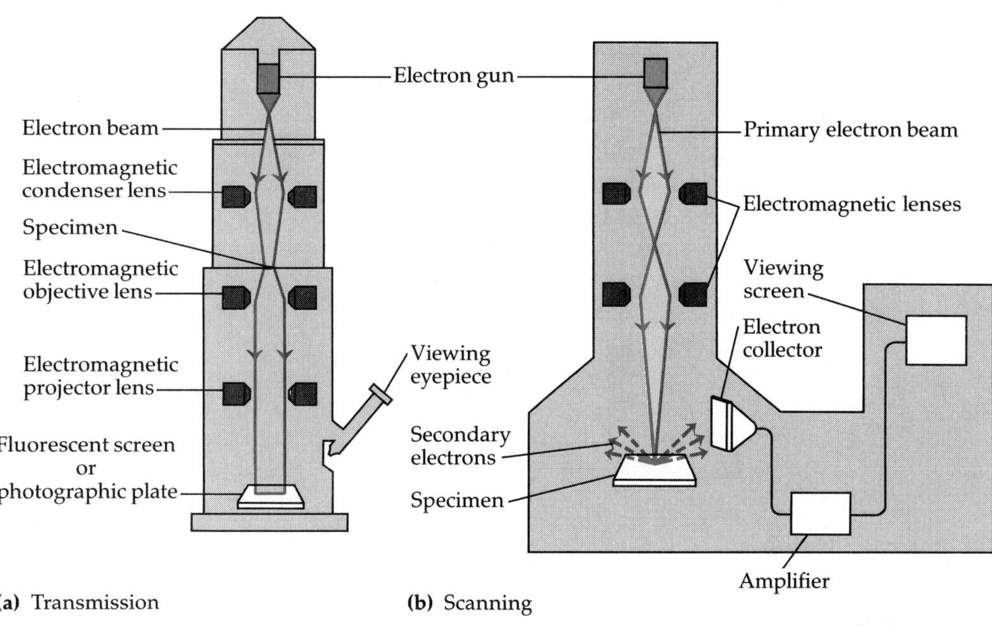

FIGURE 3.7 Electron microscopes. **(a)** In a transmission electron microscope, electrons pass through the specimen and are scattered. Magnetic lenses focus the image onto a fluorescent screen or photographic plate. **(b)** In a scanning electron microscope, the electrons sweep across the specimen and knock electrons from the specimen surface. The electrons are picked up by a collector, amplified, and transmitted onto a viewing screen or photographic plate.

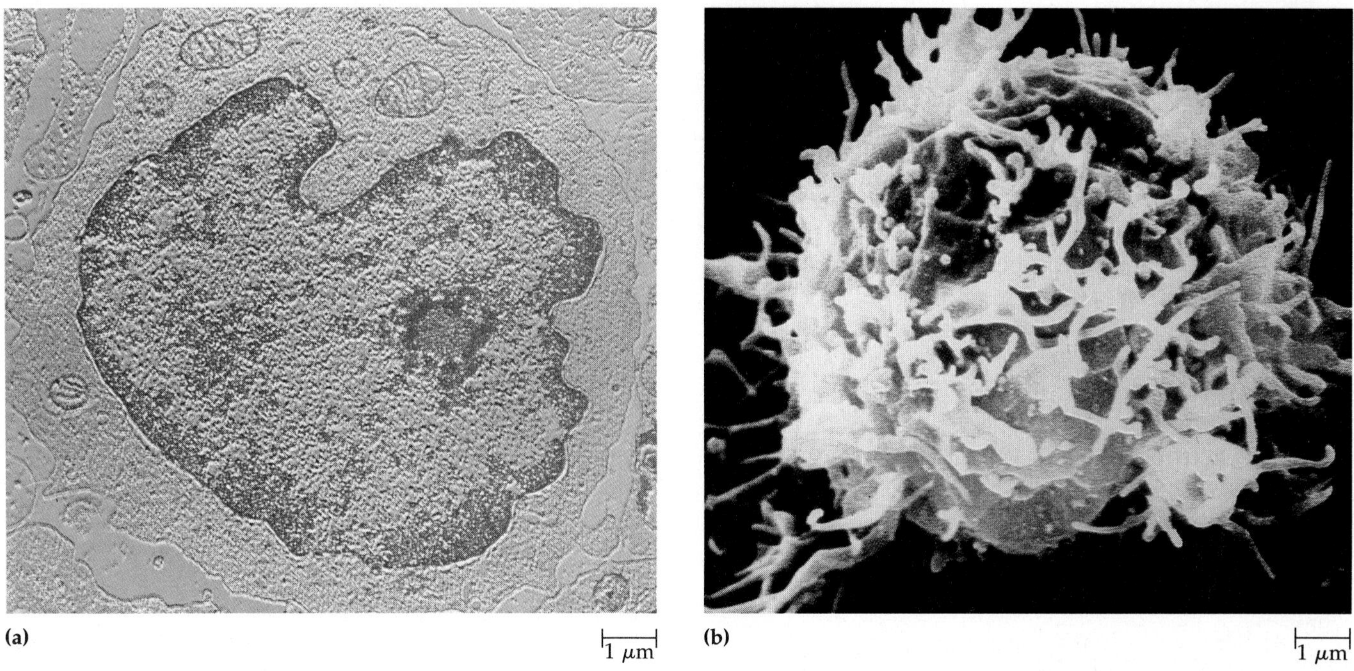

FIGURE 3.8 Lymphocyte viewed through a transmission electron microscope and a scanning electron microscope. **(a)** This transmission electron micrograph shows a thin slice of a lymphocyte, a type of white blood cell. This type of microscopy allows one to see the internal structures present in the slice. **(b)** In a scanning electron micrograph, surface structures can be seen, as demonstrated in this view of a lymphocyte. Note the three-dimensional appearance of this cell, in contrast to the two-dimensional appearance of the cell in (a).

areas, depending on how many electrons are absorbed by different areas of the specimen (Figure 3.8a).

In practice, the transmission electron microscope can resolve objects as close as 2.5 nm, and objects are generally magnified 10,000× to 100,000×. Because most microscopic specimens are so thin, the contrast between their ultrastructures and the background is weak. Contrast can be greatly enhanced by use of a "stain" that absorbs electrons and produces a darker image in the stained region. Salts of various heavy metals, such as lead, osmium, tungsten, and uranium, are commonly used as stains. These metals can be fixed onto the specimen (positive staining) or used to increase the electron opacity of the surrounding field (negative staining). Negative staining is useful for the study of the very smallest specimens, such as virus particles, bacterial flagella, and protein molecules.

Transmission electron microscopy has the advantage of high resolution and is extremely valuable for examining different layers of specimens. However, it does have certain disadvantages. Because electrons have limited penetrating power, only a very thin section of a specimen (about 100 nm) can be studied effectively. Thus, the specimen has no three-dimensional aspect. In addition, specimens must be fixed, dehydrated, and viewed under a high vacuum. These treatments not only kill the specimen but also cause some shrinkage and distortion.

Scanning Electron Microscopy

The **scanning electron microscope** overcomes the problem of sectioning associated with a transmission electron microscope. A scanning electron microscope provides striking three-dimensional views of specimens (Figure 3.8b). In scanning electron microscopy, an electron gun produces a finely focused beam of electrons called the primary electron beam (Figure 3.7b). These electrons pass through electromagnetic lenses and are directed over the surface of the specimen. The primary electron beam knocks electrons out of the surface of the specimen, and the secondary electrons thus produced are transmitted to an electron collector, amplified, and used to produce an image on a viewing screen or photographic plate. The scanning electron microscope is especially useful in studying the surface structures of intact cells and viruses. In practice, it can resolve objects as close as 20 nm, and objects are generally magnified 1000× to 10,000×.

The various types of microscopy just described are summarized in Table 3.2.

Preparation of Specimens for Light Microscopy

Because most microorganisms appear almost colorless when viewed through a standard light microscope, one often must prepare them for observation. One of the ways this can be done is by staining (coloring). Here we will discuss several different staining procedures.

PREPARING SMEARS AND STAINING

Most studies of the shapes and cellular arrangements of microorganisms are made with stained preparations. **Staining** simply means coloring the microorganisms with a dye that emphasizes certain structures. Before the microorganisms can be stained, however, they must be attached, or **fixed,** to the microscope slide; otherwise, the stain might wash them from the slide.

When a specimen is fixed, a thin film of material containing the microorganisms is spread over the surface of the slide. This film, called a **smear,** is allowed to air dry. In most staining procedures the slide is then slowly passed through the flame of a Bunsen burner several times, smear side up. Air drying and flaming fix the microorganisms to the slide. This fixing procedure usually kills them. Stain is applied and then washed off with water; then the slide is blotted with absorbent paper. The stained microorganisms are now ready for microscopic examination.

Stains are salts composed of a positive and a negative ion, one of which is colored and is known as the chromophore. The color of so-called **basic dyes** is in the positive ion. In **acidic dyes,** it is in the negative ion. Bacteria are slightly negatively charged at pH 7.0. Thus, the colored positive ion in a basic dye is attracted to the negatively charged bacterial cell. Basic dyes, which include crystal violet, methylene blue, and safranin, are more commonly used than acidic dyes. Acidic dyes are not attracted to most types of bacteria because the dye's negative ions are repelled by the negatively charged bacterial surface, so the stain colors the background instead. This preparation of colorless bacteria against a colored background is called **negative staining.** It is valuable in the observation of overall cell shapes, sizes, and capsules because the cells are made highly visible against a contrasting dark background (see Figure 3.10a). Distortions of cell size and shape are minimized because heat fixing is not necessary and the cells do not pick up the stain. Examples of acidic dyes are eosin, nigrosin, and India ink.

To apply acidic or basic dyes, microbiologists use three kinds of staining techniques—simple, differential, and special.

SIMPLE STAINS

A **simple stain** is an aqueous or alcohol solution of a single basic dye. Although different dyes bind specifically to different parts of cells, the primary purpose of a simple stain is to highlight the entire microorganism so that cellular shapes and structures are visible. The stain is applied to the fixed smear for a certain length of time and then washed off, and the slide is dried and examined. Occasionally, a chemical is added to the solution to intensify the stain. Such an additive is called a **mordant.** One function of a mordant is to increase the affinity of a stain for a biological specimen. Another is to coat a structure (such as a flagellum) to make it

TABLE 3.2 Summary of Various Types of Microscopes

MICROSCOPE TYPE	*DISTINGUISHING FEATURES*	*TYPICAL IMAGE*	*PRINCIPAL USES*
Brightfield	Uses visible light as a source of illumination; cannot resolve structures smaller than 0.2 μm; specimen appears against a bright background.	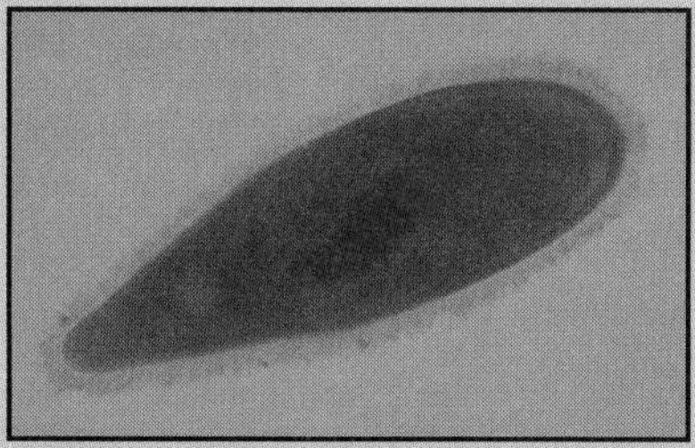	Commonly used to observe various stained (killed) specimens and to count microbes; does not resolve very small specimens, such as viruses. Instrument is inexpensive and simple to use.
Darkfield	Uses a special condenser with an opaque disc that blocks light from entering the objective directly; light diffracted by specimen enters the objective, and the specimen appears light against a black background.	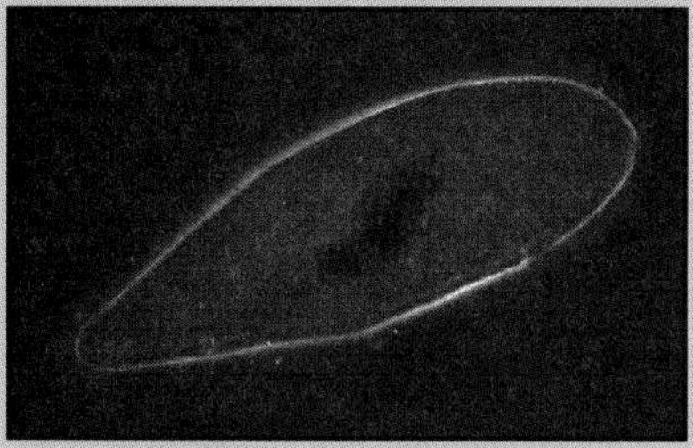	Commonly used to examine living microorganisms that are invisible in brightfield microscopy, do not stain easily, or are distorted by staining; frequently used to detect *Treponema pallidum.*
Phase-contrast	Uses a special condenser and diffraction plate to diffract light rays so that they are out of phase with one another; the specimen appears as different degrees of brightness and contrast.	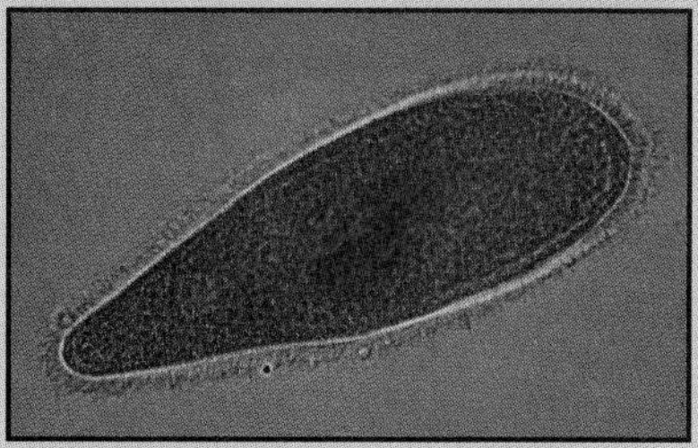	Commonly used to provide detailed examination of the internal structures of living specimens; no staining is required.
Fluorescence	Uses an ultraviolet or near-ultraviolet source of illumination that causes fluorescent compounds in a specimen to emit light.		Principal use is for fluorescent-antibody techniques (immunofluorescence) to identify and detect microorganisms in tissues or clinical specimens rapidly.
Electron	Uses a beam of electrons instead of light; because of the shorter wavelength of electrons, structures smaller than 0.2 μm can be resolved.	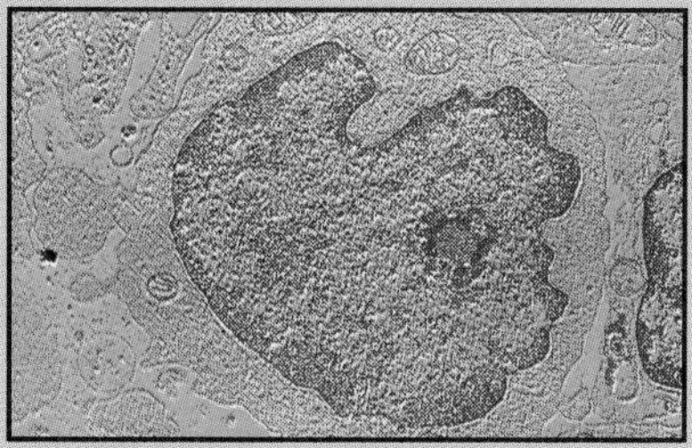 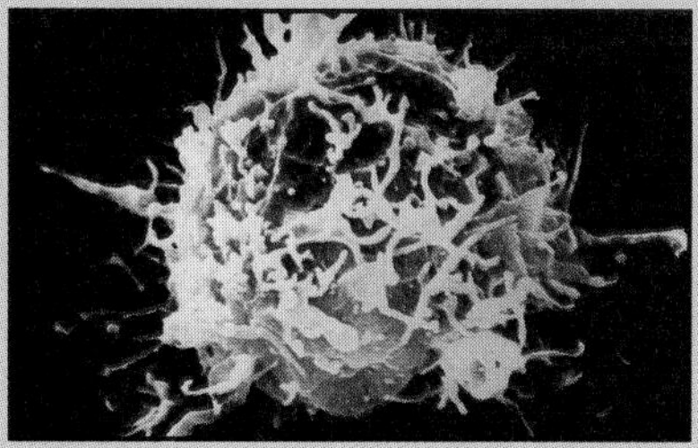	A transmission electron microscope is used to examine viruses or the internal ultrastructure in thin sections of cells (usually magnified 10,000× to 100,000×); the image produced is not three dimensional. A scanning electron microscope is used to study the surface features of cells and viruses (usually magnified 1000× to 10,000×); the image produced is three dimensional.

MICROBIOLOGY *HIGHLIGHTS*

Bdellovibrio, Predator Extraordinaire

Even through a microscope, appearances can be deceptive and first impressions can be quite misleading. For example, at first glance, the bacterium *Bdellovibrio* looks like a perfectly ordinary 1-μm-long curved, gram-negative rod. *Bdellovibrio* cells exhibit the normal runs and tumbles of flagellated bacteria, although they move about ten times faster than other bacteria of the same size. However, more patient observation or use of a microscope equipped with a video camera reveals that *Bdellovibrio*'s behavior is unlike that of any other known bacteria. A unique microbial drama occurs when *Bdellovibrio* cells are placed in a suspension with other bacteria, such as *E. coli.* Apparently drawn by chemical attractants, *Bdellovibrio* speeds toward and rams the larger, slower *E. coli.* The two cells become attached and rotate rapidly. As the two bacteria whirl about, the *Bdellovibrio* paralyzes the *E. coli* by an unknown mechanism that most probably involves one of *Bdellovibrio*'s protein products. *Bdellovibrio* kills *E. coli* without breaking down its contents.

Next, in preparation for moving into its newly acquired home, *Bdellovibrio* adds structural components to the outer membrane of *E. coli.* This activity, which stabilizes the outer membrane, is another ability unique to *Bdellovibrio.* After it penetrates *E. coli*'s cell wall and detaches its flagellum, *Bdellovibrio* snuggles into the periplasmic space (that is, the space between the plasma membrane and the cell wall) and prepares for a feast (see figure).

Macromolecules leak from *E. coli*'s cytoplasm into the periplasmic space. *Bdellovibrio* digests the macromolecules and grows into a long, coiled filament. In about two or three hours, the cytoplasmic contents of the host (including its DNA) are digested, and the *Bdellovibrio* filament fragments into as many as 20 individual curved rods. Each rod develops a flagellum and is released when the host cell is broken open by a *Bdellovibrio* enzyme. The new *Bdellovibrio* cells scurry away to feed on other bacteria.

Bdellovibrio was once regarded as simply a parasite. However, thorough studies have shown that *Bdellovibrio* does not merely siphon off its host's resources and energy, it actually digests the cell's contents as food, so *Bdellovibrio* is now considered to be a bacterial predator. The ecological importance of *Bdellovibrio* is that it, along with protozoans and bacteriophages, helps to control the population growth of soil bacteria.

In addition, now that its habits are known, *Bdellovibrio* may have a new career as a bacterial exterminator. Researchers led by Richard Whiting at the U.S. Department of Agriculture are investigating the use of *Bdellovibrio* to control *Salmonella* (a gram-negative bacterium) in chickens and eggs, where it is one of the most common causes of food poisoning in the United States. *Salmonella* can infect the bird's ovaries and enter the egg before it is laid, or it can penetrate the eggshell from bird droppings. Although *Salmonella* is destroyed by cooking, foods that contain raw eggs (e.g., mayonnaise) are a problem. Of the 91,678 cases of foodborne infection reported in the United States from 1983 through 1987, roughly half were caused by *Salmonella* associated with eggs. For example, one outbreak was traced to a blender used to make scrambled eggs. *Salmonella* left in the blender contaminated other foods throughout the day and caused numerous cases of food poisoning.

Since *Bdellovibrio* cells prey on gram-negative bacteria, the U.S.D.A. researchers hope that if *Bdellovibrio* can be introduced into chickens with food or water, it will be able to survive and kill salmonellae. As harmful as they are to other bacteria, *Bdellovibrio* cells are safe for humans. If the U.S.D.A. project succeeds, picnickers need not even be aware of the violent microbiological drama acted out in defense of their chicken salad.

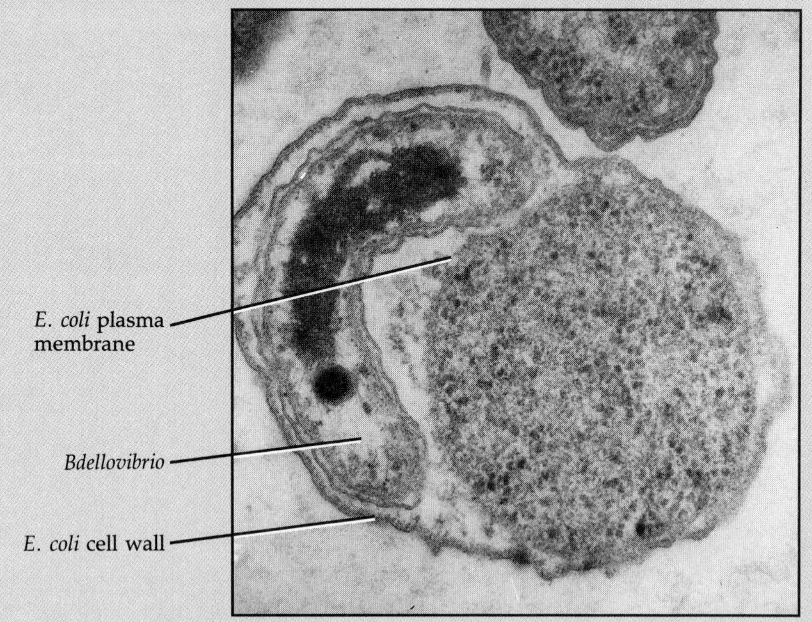

Bdellovibrio. A *Bdellovibrio* bacterium has penetrated the cell wall of an *E. coli* cell. The *Bdellovibrio,* having killed the prey cell, utilizes the *E. coli*'s contents for its elongation and multiplication.

thicker and easier to see after it is stained with a dye. Some of the simple stains commonly used in the laboratory are methylene blue, carbolfuchsin, crystal violet, and safranin.

DIFFERENTIAL STAINS

Unlike simple stains, **differential stains** react differently with different kinds of bacteria and thus can be used to distinguish among them. The differential stains most frequently used for bacteria are the Gram stain and the acid-fast stain.

Gram Stain

The **Gram stain** was developed in 1884 by the Danish bacteriologist Hans Christian Gram. It is one of the most useful staining procedures because it divides bacteria into two large groups, gram-positive and gram-negative.

In this procedure, the heat-fixed smear is covered with a basic purple dye, usually crystal violet (Figure 3.9a). Since the purple stain imparts its color to all cells, it is referred to as a **primary stain.** After a short time, the purple dye is washed off, and the smear is covered with iodine, a mordant (Figure 3.9b). When the iodine is washed off, both gram-positive and gram-negative bacteria appear dark violet or purple. Next, the slide is washed with ethanol or an ethanol–acetone solution (Figure 3.9c). This solution is a **decolorizing agent,** which removes the purple from the cells of some species but not from others. The alcohol is rinsed off, and the slide is then stained with safranin, a basic

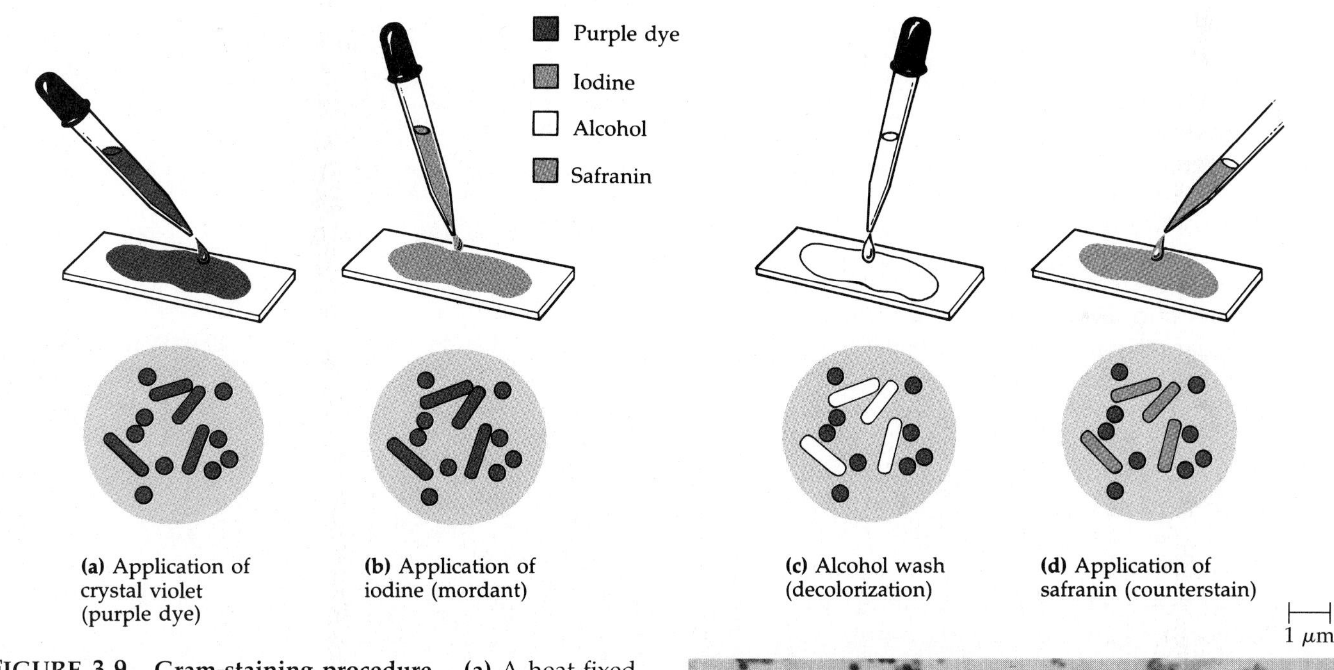

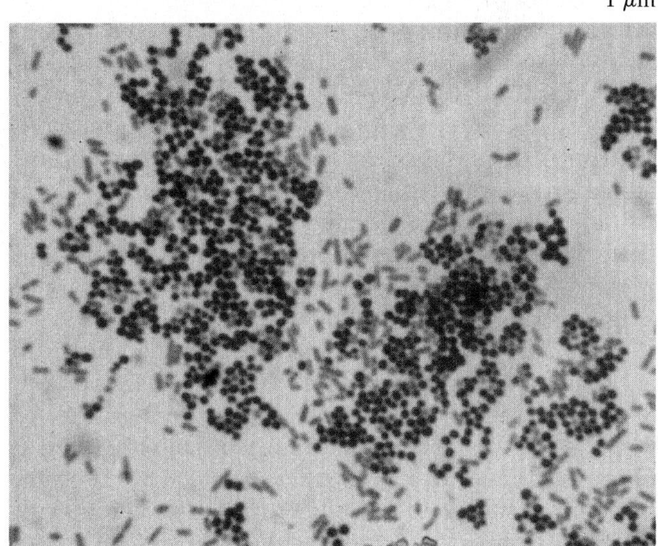

FIGURE 3.9 Gram-staining procedure. (a) A heat-fixed smear of gram-positive cocci and gram-negative rods is first covered with a basic purple dye (primary stain) such as crystal violet, and then the dye is washed off. **(b)** Then the smear is covered with iodine (a mordant) and washed off. At this time, both gram-positive and gram-negative bacteria are purple. **(c)** The slide is washed with ethanol or an alcohol–acetone solution (a decolorizer) and then washed with water. Now the gram-positive cells are purple and the gram-negative cells are colorless. **(d)** In the final step, safranin is added as a counterstain, and the slide is washed, dried, and examined microscopically. Gram-positive bacteria retain the purple dye, even through the alcohol wash. Gram-negative bacteria appear pink because they pick up the safranin counterstain. **(e)** Photomicrograph of gram-stained bacteria. *Staphylococcus aureus* (purple) are gram-positive, and *Escherichia coli* (red) are gram-negative.

red dye (Figure 3.9d). The smear is washed again, blotted dry, and examined microscopically.

The purple dye and the iodine combine with each bacterium and color it dark violet or purple. Bacteria that retain this color after the alcohol has attempted to decolorize them are classified as **gram-positive** (see Figure 3.9e). Bacteria that lose the dark violet or purple color after decolorization are classified as **gram-negative** (see Figure 3.9e). Because gram-negative bacteria are colorless after the alcohol wash, they are no longer visible. It is for this reason that the basic dye safranin is applied; it turns the gram-negative bacteria pink. Stains such as safranin that have a contrasting color to the primary stain are called **counterstains.** Because gram-positive bacteria retain the original purple stain, they are not affected by the safranin counterstain.

As you will see in the next chapter, different kinds of bacteria react differently to the Gram stain, probably because structural differences in their cell walls affect the retention or escape of a combination of crystal violet and iodine, called the crystal violet–iodine (CV–I) complex. Among other differences, gram-positive bacteria have a thicker peptidoglycan cell wall than gram-negative bacteria (see Chapter 4). In addition, gram-negative bacteria contain a layer of lipopolysaccharide as part of their cell wall. When applied to both gram-positive and gram-negative cells, crystal violet and then iodine readily enter the cells. Inside the cells, the crystal violet and iodine combine to form CV–I. This complex is larger than the crystal violet molecule that entered the cells, and, because of its size, it cannot be washed out of the intact peptidoglycan layer of gram-positive cells by alcohol. Consequently, gram-positive cells retain the color of the crystal violet dye. In gram-negative cells, however, the alcohol wash disrupts the outer lipopolysaccharide layer, and the CV–I complex is washed out through the thin layer of peptidoglycan. As a result, gram-negative cells are rendered colorless until counterstained with safranin, after which they are pink. In summary, the gram-positive cells retain the dye and remain purple. The gram-negative cells do not retain the dye. They are colorless until counterstained with a red dye, after which they appear pink.

The Gram method is one of the most important staining techniques in medical microbiology. But Gram staining results are not universally applicable because some bacterial cells stain poorly or not at all. The Gram reaction is most consistent when it is used on young, growing bacteria.

In many cases, the Gram reaction of a bacterium provides valuable information for the treatment of disease. For example, gram-positive bacteria tend to be killed easily by penicillin and sulfonamide drugs. Gram-negative bacteria resist these drugs, but they are much more susceptible to such drugs as streptomycin, chloramphenicol, and tetracycline. Thus, Gram-stain

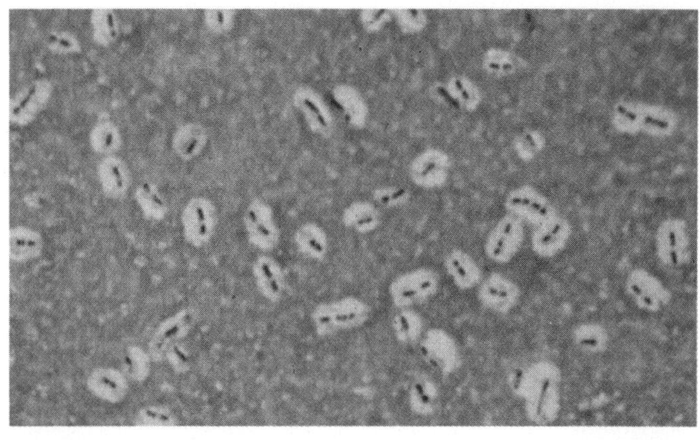

(a)

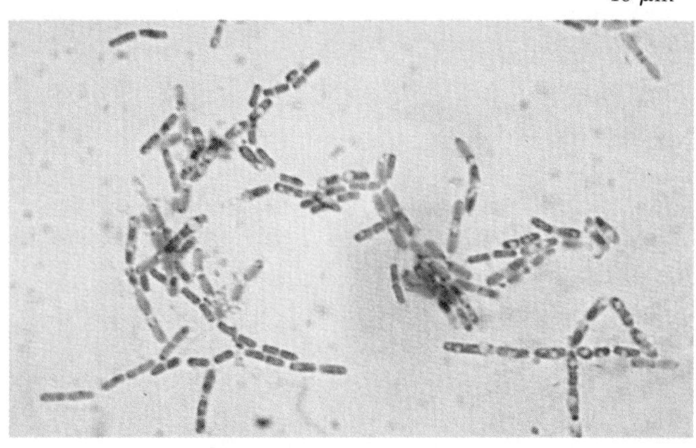

(b)

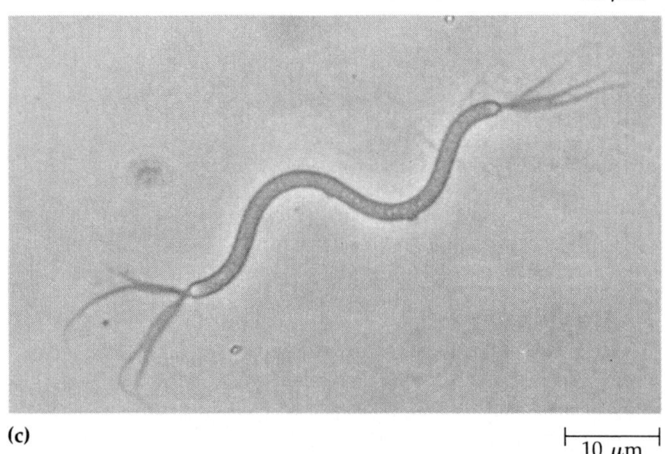

(c)

FIGURE 3.10 Negative staining of capsules and staining of endospores and flagella. **(a)** India ink provides a dark background, so the capsules of these bacteria, *Klebsiella pneumoniae,* show up as light areas surrounding the stained cells. **(b)** Endospores are seen as green areas in the centers of these rod-shaped cells of the bacterium *Bacillus cereus,* using the Schaeffer-Fulton spore stain. The light areas contain storage deposits. **(c)** Flagella are shown as wavy extensions from one end of these cells of the bacterium *Spirillum volutans.* In relation to the bodies of the cells, the flagella are much thicker than they normally are because layers of the stain accumulated from treatment with a mordant.

identification of a bacterium can help determine which drug will be most effective against a disease.

Acid-Fast Stain

Another important differential stain (one that divides bacteria into distinctive groups) is the **acid-fast stain,** which binds strongly only to bacteria that have a waxy material in their cell walls. Microbiologists use this stain to identify all bacteria in the genus *Mycobacterium,* including the two important disease producers *Mycobacterium tuberculosis* and *Mycobacterium leprae* (lep'rī). This stain is also used to identify the disease-producing strains of the genus *Nocardia* (nō-kär'dē-ä).

In the acid-fast staining procedure, the red dye carbolfuchsin is applied to a fixed smear, and the slide is gently heated for several minutes. Heating enhances penetration and retention of the dye. Then the slide is cooled and washed with water. The smear is next treated with acid-alcohol, a decolorizer, which removes the red stain from bacteria that are not acid-fast. The acid-fast microorganisms retain the red color since the carbolfuchsin is more soluble in the cell wall waxes than in the acid-alcohol. In non-acid-fast bacteria, whose cell walls lack the waxy components, the carbolfuchsin is rapidly removed during decolorization, leaving the cells colorless. The smear is then stained with a methylene blue counterstain. Non-acid-fast cells appear blue after application of the counterstain.

TABLE 3.3 Summary of Various Stains and Their Uses

STAIN	PRINCIPAL USES
Simple (methylene blue, carbolfuchsin, crystal violet, safranin)	Aqueous or alcohol solution of a single basic dye. (Sometimes a mordant is added to intensify the stain.) Used to highlight microorganisms to determine cellular shapes and arrangements.
Differential	React differently with different kinds of bacteria in order to distinguish them.
Gram	Divides bacteria into two large groups, gram-positive and gram-negative. Gram-positive bacteria retain the crystal violet stain and appear purple. Gram-negative bacteria do not retain the crystal violet stain and remain colorless until counterstained with safranin and then appear pink.
Acid-fast	Used to distinguish *Mycobacterium* species and some species of *Nocardia.* Acid-fast bacteria, once stained with carbolfuchsin and treated with acid-alcohol, remain red because they retain the carbolfuchsin stain. Non-acid-fast bacteria, when stained and treated the same way and then stained with methylene blue, appear blue because they lose the carbolfuchsin stain but retain the methylene blue stain.
Special	Used to color and isolate various structures, such as capsules, endospores, and flagella; sometimes used as a diagnostic aid.
Negative	Used to demonstrate the presence of capsules. Since capsules do not accept most stains, the capsules appear as halos around bacterial cells and stand out against a dark background.
Endospore	Used to detect the presence of endospores in seven genera of bacteria. When malachite green is applied to a heat-fixed smear of bacterial cells, the stain penetrates the endospores and stains them green. When safranin (red) is then applied, it stains the remainder of the cells red or pink.
Flagella	Used to demonstrate the presence of flagella. A mordant is used to build up the diameters of flagella until they become visible microscopically when stained with carbolfuchsin.

SPECIAL STAINS

Special stains are used to color and isolate specific parts of microorganisms, such as endospores and flagella, and to reveal the presence of capsules.

Negative Staining for Capsules

Many microorganisms contain a gelatinous covering called a **capsule,** which we will discuss in our examination of the procaryotic cell in Chapter 4. In medical microbiology, demonstrating the presence of a capsule is a means of determining the organism's **virulence,** the degree to which a pathogen can cause disease. Capsule staining is more difficult than other types of staining procedures because capsular materials are soluble in water and may be dislodged or removed during rigorous washing. To demonstrate the presence of capsules, a microbiologist can mix the bacteria in a solution containing a fine colloidal suspension of colored particles (usually India ink or nigrosin) to provide a dark background and then can stain the bacteria with a simple stain, such as safranin (Figure 3.10a). Because of their chemical composition, capsules do not accept most biological dyes, such as safranin, and thus appear as halos surrounding each stained bacterial cell. The use of India ink illustrates a negative-staining technique; negative stains do not penetrate the cell capsule, and thus they provide a contrast between the capsule and the surrounding dark medium.

Endospore (Spore) Staining

An **endospore** is a special resistant, dormant structure formed within a cell that protects the microorganism from adverse environmental conditions. Although endospores are relatively uncommon in bacterial cells, they can be formed by seven genera of bacteria. Endospores cannot be stained by ordinary methods, such as simple staining and Gram staining, because the dyes do not penetrate the wall of the endospore.

The most commonly used endospore stain is the *Schaeffer-Fulton endospore stain.* Malachite green, the primary stain, is applied to a heat-fixed smear and heated to steaming for about five minutes. The heat helps the stain to penetrate the endospore wall. Then the preparation is washed for about 30 seconds with water to remove the malachite green stain from all of the cells' parts except the endospores. Next, safranin, a counterstain, is applied to the smear to stain portions of the cell other than endospores. In a properly prepared smear, the endospores appear green within red or pink cells. Because endospores are highly refractive, they can be detected under the light microscope but cannot be differentiated from inclusions of stored material (Figure 3.10b).

Flagella Staining

Bacterial **flagella** are structures of locomotion too small to be seen with light microscopes. A tedious and delicate staining procedure uses a mordant and the stain carbolfuchsin to build up the diameters of the flagella until they become visible under the light microscope. Microbiologists use the number and arrangement of flagella as diagnostic aids. Figure 3.10c depicts stained flagella.

A summary of stains is presented in Table 3.3. In the next chapter we will take a closer look at the structure of microbes and how they protect, nourish, and reproduce themselves.

STUDY OUTLINE

Units of Measurement (pp. 52–53)

1. The standard unit of length is the meter (m).
2. Microorganisms are measured in micrometers, μm (10^{-6} m), nanometers, nm (10^{-9} m), and angstroms, Å (10^{-10} m).

Microscopy: The Instruments (pp. 53–60)

COMPOUND LIGHT MICROSCOPY (pp. 53–55)

1. The most common microscope used in microbiology is the compound light microscope, which uses two sets of lenses, ocular and objective.
2. We calculate the total magnification of an object by multiplying the magnification of the objective lens by the magnification of the ocular lens.
3. The compound light microscope uses visible light.
4. The maximum resolving power (ability to distinguish two points) of a compound light microscope is 0.2 μm; maximum magnification is 2000×.
5. Specimens are stained to increase the difference between the refractive indexes of the specimen and the medium.
6. Immersion oil is used with the oil immersion lens to reduce light loss between the slide and the lens.
7. Brightfield illumination is used for stained smears.
8. Unstained cells are more productively observed using darkfield, phase-contrast, or fluorescence microscopy. These types of microscopy use modified compound microscopes.

DARKFIELD MICROSCOPY (pp. 55–57)

1. The darkfield microscope shows a light silhouette of an organism against a dark background.
2. It is most useful to detect the presence of extremely small organisms.

PHASE-CONTRAST MICROSCOPY (p. 57)

1. The phase-contrast microscope uses a special condenser to enhance differences in the refractive indexes of the cell's parts and its surroundings.
2. It allows the detailed observation of living organisms.

FLUORESCENCE MICROSCOPY (pp. 57–58)

1. In fluorescence microscopy, specimens are first stained with fluorochromes and then viewed through a compound microscope by using an ultraviolet (or near-ultraviolet) light source.
2. The microorganisms appear as bright objects against a dark background.
3. Fluorescence microscopy is used primarily in a diagnostic procedure called fluorescent antibody technique.

ELECTRON MICROSCOPY (pp. 58–60)

1. A beam of electrons, instead of light, is used with an electron microscope.
2. Electromagnets, instead of glass lenses, control focus, illumination, and magnification.
3. Thin sections of organisms can be seen in an electron micrograph produced using a transmission electron microscope. *Magnification:* 10,000× to 100,000×. *Resolving power:* 2.5 nm.
4. Three-dimensional views of the surfaces of whole microorganisms can be obtained with a scanning electron microscope. *Magnification:* 1000× to 10,000×. *Resolving power:* 20 nm.

Preparation of Specimens for Light Microscopy (pp. 60–66)

PREPARING SMEARS AND STAINING (p. 60)

1. Staining means to color with a dye to make some structures more visible.

2. Fixing uses air and heat to attach microorganisms to a slide.
3. A smear is a thin film of material used for microscopic examination.
4. Bacteria are negatively charged, and the colored positive ion of a basic dye will stain bacterial cells.
5. The colored negative ion of an acidic dye will stain the background of a bacterial smear; a negative stain is produced.

SIMPLE STAINS **(pp. 60–63)**

1. A simple stain is an aqueous or alcohol solution of a single basic dye.
2. It is used to make cellular shapes and arrangements visible.
3. A mordant may be used to improve bonding between the stain and the specimen.

DIFFERENTIAL STAINS **(pp. 63–65)**

1. Differential stains, such as the Gram stain and acid-fast stain, divide bacteria into groups according to their reactions to the stains.
2. The Gram-stain procedure uses a purple stain, iodine as a mordant, an alcohol decolorizer, and a red counterstain.
3. Gram-positive bacteria retain the purple stain after the decolorization step; gram-negative bacteria do not and thus appear pink from the counterstain.
4. Acid-fast bacteria, such as members of the genera *Mycobacterium* and *Nocardia,* retain carbolfuchsin after acid-alcohol decolorization and appear red; non-acid-fast bacteria take up the methylene blue counterstain and appear blue.

SPECIAL STAINS **(pp. 65–66)**

1. Stains such as the endospore stain and flagella stain color only certain parts of microbes.
2. Negative staining is used to make microbial capsules visible.

STUDY QUESTIONS

Review

1. Fill in the following blanks.
   1 μm = __________ m
   1 __________ = 10^{-9} m
   1 Å= __________ m
   1 μm = __________ nm = __________ Å
2. Label the parts of the compound light microscope below.

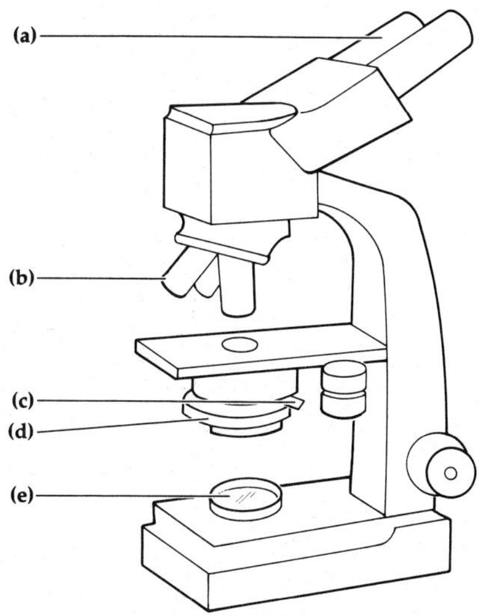

3. Calculate the total magnification of the nucleus of a cell being observed through a compound light microscope with a 10× ocular lens and the oil immersion lens.
4. Which type of microscope would be best to use to observe the following?
   (a) A stained bacterial smear
   (b) Unstained bacterial cells where the cells are small and no detail is needed
   (c) Unstained live tissue where it is desirable to see some intracellular detail
   (d) A sample that emits light when illuminated with ultraviolet light
   (e) Intracellular detail of a cell that is 1 μm long
5. An electron microscope differs from a light microscope in that __________ focused by __________ is used instead of light, and the image is viewed not through the ocular lenses but on __________.
6. The maximum magnification of a compound microscope is __________; of an electron microscope, __________. The maximum resolution of a compound microscope is __________; of an electron microscope, __________.
7. One advantage of a scanning electron microscope over a transmission electron microscope is __________.
8. Acidic dyes stain the (cells/background) in a smear and are used for (negative/simple) stains.
9. Basic dyes stain the (cells/background) in a smear and are used for (negative/simple) stains.
10. Why do basic dyes stain bacterial cells? Why don't acidic dyes stain bacterial cells?
11. When is it most appropriate to use:
    (a) A simple stain? **(c)** A negative stain?
    (b) A differential stain? **(d)** A flagella stain?
12. Why is a mordant used in the Gram stain? In the flagella stain?

13. What is the purpose of a counterstain in the acid-fast stain?

14. What is the purpose of a decolorizer in the Gram stain? In the acid-fast stain?

15. Choose from the following terms to fill in the blanks: counterstain, decolorizer, mordant, primary stain. In the endospore stain, safranin is the ______________. In the Gram stain, safranin is the ________.

16. Fill in the following table regarding the Gram stain.

Steps	Appearance after this step of	
	Gram-positive cells	Gram-negative cells
Crystal violet		
Iodine		
Alcohol-acetone		
Safranin		

Challenge

1. In a Gram stain, one step could be omitted and you could still differentiate between gram-positive and gram-negative cells. What is that one step?

2. Using a good compound light microscope with a resolving power of 0.3 μm, a 10× ocular lens, and a 100× oil immersion lens, would you be able to discern two objects separated by 3 μm? 0.3 μm? 300 nm? 3000 Å?

3. Why isn't the Gram stain used on acid-fast bacteria? If you did Gram stain acid-fast bacteria, what would their Gram reaction be? What is the Gram reaction of non-acid-fast bacteria?

FURTHER READING

Balows, A., et al., eds. *Manual of Clinical Microbiology,* 5th ed. Washington, D.C.: American Society for Microbiology, 1991. Contains chapters on microscopy, specimen preparation, and staining.

Boatman, E.S., et al. "Today's microscopy." *BioScience* 37(6):384–394, 1987. Provides a background in fundamentals and then describes recent advances in computer enhancement and laser and acoustic microscopy.

Branson, D. *Methods in Clinical Bacteriology.* Springfield, Ill.: Thomas, 1972. Good stepwise directions for staining procedures and preparation of stains. Also see *Procedure Manual for Clinical Bacteriology* (1982), same author and publisher.

Ford, B.J. *Single Lens: The Story of the Simple Microscope.* New York: Harper & Row, 1985. An interesting history of the simple microscope and its impact on biology.

Gray, P., ed. *Encyclopedia of Microscopy and Microtechnique.* Melbourne, Fla.: Krieger, 1982. An illustrated encyclopedia of terminology and techniques.

Howells, M.R., J. Kirz, and W. Sayre. "X-ray microscopes." *Scientific American* 264(2): 88–94, February 1991. This new microscopy technique provides three-dimensional images of cells and tissues.

Wickramasinghe, H.K. "Scanned-probe microscopes." *Scientific American* 261(4):98–105, October 1989. Reports on new microscopes capable of resolutions of 1 nm that allow observations of atoms and molecular interactions, such as the attachment of a virus to a cell.

CHAPTER 4

Functional Anatomy of Procaryotic and Eucaryotic Cells

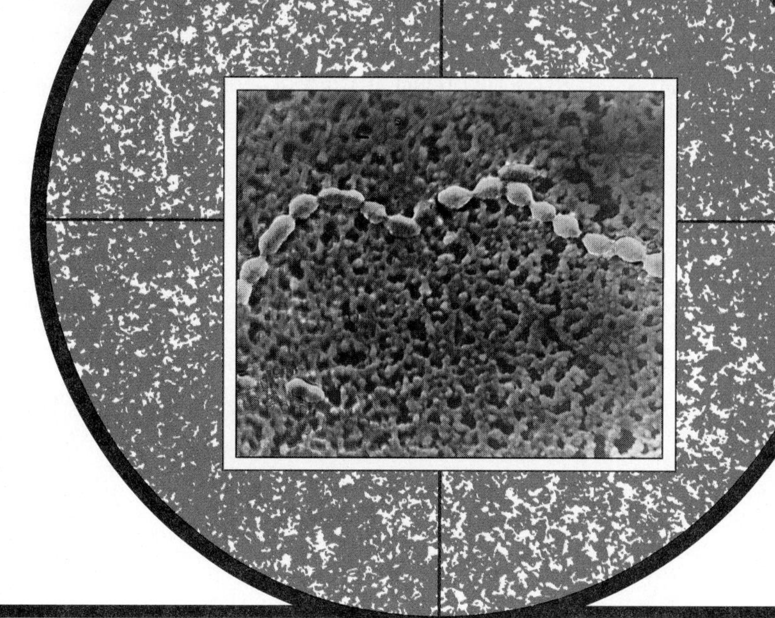

LEARNING OBJECTIVES

- Identify the three basic shapes of bacteria.
- Compare and contrast the cell walls of gram-positive bacteria, gram-negative bacteria, archaeobacteria, and mycoplasmas.
- Describe the structure, chemistry, and functions of the procaryotic plasma membrane.
- Define simple diffusion, osmosis, facilitated diffusion, active transport, and group translocation.
- Identify the functions of procaryotic cell structures.
- Compare and contrast the overall cell structure, glycocalyx, cell wall, flagella, nucleus, and ribosomes of procaryotes and eucaryotes.
- Explain what an organelle is.
- Describe the functions of the endoplasmic reticulum, Golgi complex, mitochondria, chloroplasts, and lysosomes.

Despite their complexity and variety, all living cells can be divided into two groups, procaryotes and eucaryotes, based on their ultrastructure as seen with the electron microscope. Plants and animals are entirely composed of eucaryotic cells. In the microbial world, bacteria are procaryotes. Other cellular microbes—fungi (yeasts and molds), protozoans, and algae—are eucaryotes.

Viruses, as noncellular elements with some cell-like properties, do not fit into any organizational scheme of living cells. They are genetic particles that replicate but are unable to perform the usual chemical activities of living cells. Viral structure and activity will be discussed in Chapter 13. In this chapter, we will concentrate on describing procaryotic and eucaryotic cells.

Procaryotes and eucaryotes are chemically similar, in the sense that they both contain nucleic acids, proteins, lipids, and carbohydrates. They use the same kinds of chemical reactions to metabolize food, build proteins, and store energy. It is primarily the *structure* of cell walls, membranes, and *organelles* (specialized cellular structures that have specific functions) that distinguishes procaryotes from eucaryotes.

The chief distinguishing characteristics of **procaryotic cells** (from the Greek for prenucleus) are:

1. Their genetic material (DNA) is not enclosed within a membrane.
2. They lack other membrane-bounded organelles.
3. Their DNA is not associated with histone proteins (special chromosomal proteins found in eucaryotes).
4. Their cell walls almost always contain the complex polysaccharide peptidoglycan.

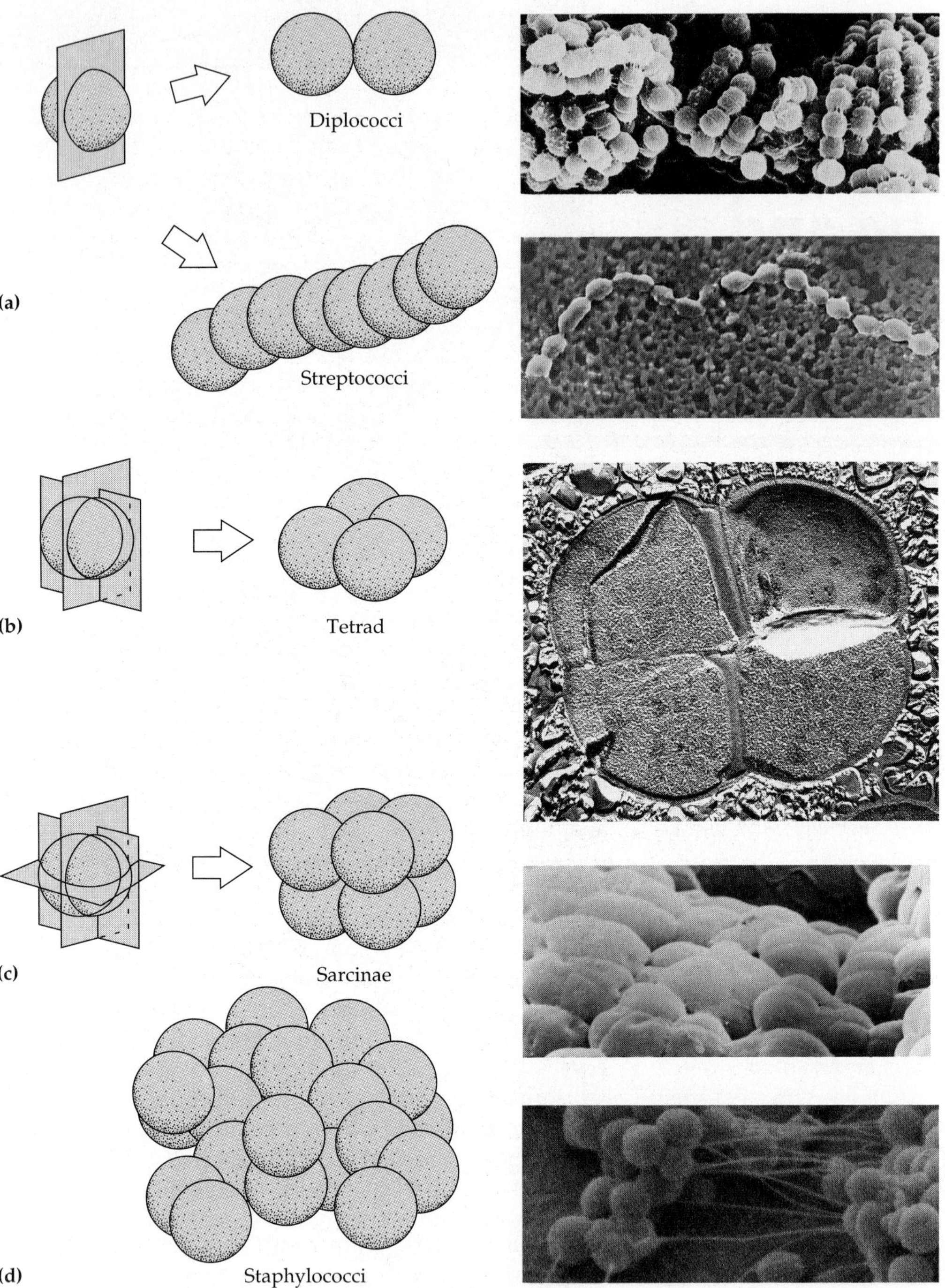

FIGURE 4.1 Arrangements of cocci. The number of planes in which the cell divides determines the arrangement of cells. Shown are diagrams (left) and corresponding photos (right). **(a)** Division in one plane produces diplococci and streptococci. **(b)** Division in two planes produces tetrads. **(c)** Division in three planes produces sarcinae, and **(d)** division in multiple planes produces staphylococci.

5. They usually divide by **binary fission.** During this process, the DNA is copied and the cell splits into two cells. Binary fission involves fewer structures and processes than eucaryotic mitosis and cell division.

Eucaryotic cells (from the Greek for true nucleus) have linear structures of DNA called chromosomes; these are found in the cell's nucleus, which is separated from the cytoplasm by a nuclear membrane. The DNA of eucaryotic chromosomes is consistently associated with chromosomal proteins called histones and nonhistones. Eucaryotes also have a mitotic apparatus (various cellular structures that participate in a type of nuclear division called mitosis) and a number of organelles, including mitochondria, endoplasmic reticulum, and sometimes chloroplasts. We will elaborate on the particular characteristics and functions of these organelles later in this chapter.

THE PROCARYOTIC CELL

The members of the procaryotic world make up a vast heterogeneous group of very small unicellular organisms. This group includes eubacteria, or true bacteria, and archaeobacteria. Although eubacteria and archeobacteria look similar, they are different in chemical composition, as will be described later. The thousands of species of bacteria are differentiated by many factors, including morphology (shape), chemical composition (often detected by staining reactions), nutritional requirements, biochemical activities, and source of energy (sunlight or chemicals).

Size, Shape, and Arrangement of Bacterial Cells

There are a great many sizes and shapes among bacteria. Most bacteria fall within a range from 0.20 to 2.0 μm in diameter and from 2 to 8 μm in length. They have a few basic shapes—spherical **coccus** (plural, *cocci,* meaning berries), rod-shaped **bacillus** (plural, *bacilli,* meaning little staffs), and **spiral.**

Cocci are usually round but can be oval, elongated, or flattened on one side. When cocci divide to reproduce, the cells can remain attached to one another. Cocci that remain in pairs after dividing are called **diplococci** (Figure 4.1a). Those that divide and remain attached in chainlike patterns are called **streptococci** (Figure 4.1a). Those that divide into two planes and remain in groups of four are known as **tetrads** (Figure 4.1b). Those that divide in three planes and remain attached in cubelike groups of eight are called **sarcinae** (Figure 4.1c). Those that divide in multiple planes and form grapelike clusters or broad sheets are called **staphylococci** (Figure 4.1d). These group characteristics are frequently helpful in the identification of certain cocci.

Bacilli divide only across their short axis, so there are fewer groupings of bacilli than of cocci. **Diplobacilli** appear in pairs after division (Figure 4.2b), and **streptobacilli** occur in chains (Figure 4.2c). Some bacilli look like straws. Others have tapered ends, like cigars. Still others are oval and look so much like cocci that they are called **coccobacilli** (Figure 4.2d). However, most bacilli appear as single rods (Figure 4.2a).

The term *bacillus* has two meanings in microbiology. As we have just used it, it refers to a bacterial shape. When capitalized and italicized, it refers to a specific genus. For example, the bacterium *Bacillus anthracis* is the causative agent of anthrax.

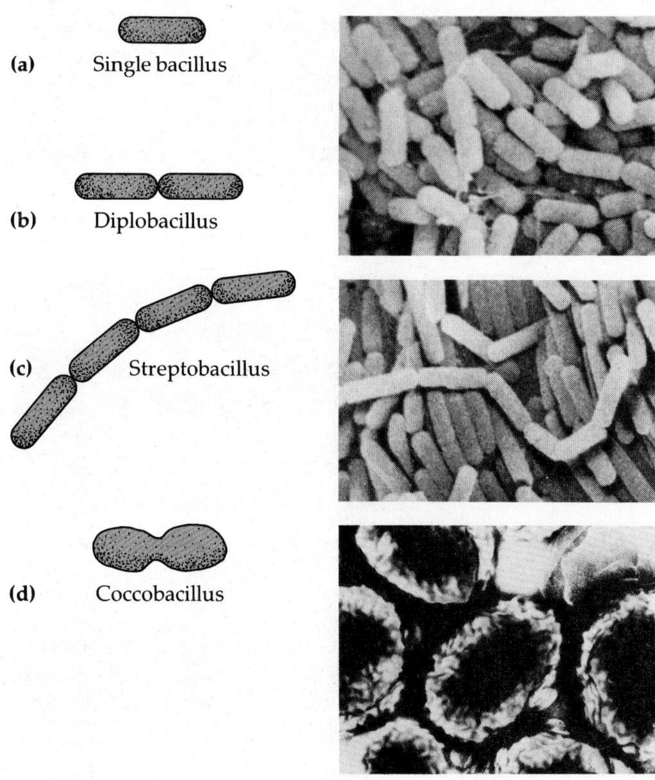

FIGURE 4.2 Bacilli. Shown are diagrams (left) and corresponding photos (right). **(a)** Single bacilli. **(b)** Diplobacilli. In the photo, a few joined pairs of bacilli serve as examples of diplobacilli. **(c)** Streptobacilli. **(d)** Coccobacilli.

Spiral bacteria have one or more twists; they are never straight. Curved bacteria that look like commas are called **vibrios** (Figure 4.3a). Others, called **spirilla,** have a helical shape, like a corkscrew, and fairly rigid bodies (Figure 4.3b). Yet another group of spirals are helical and flexible; they are called **spirochetes** (Figure 4.3c). Unlike the spirilla, which have outside appendages called flagella, spirochetes move by means of an axial filament, which resembles a flagellum but is contained under an external flexible sheath.

In addition to the three basic shapes, there are star-shaped cells (genus *Stella*) and recently discovered square, flat cells (halophilic archaeobacteria) and triangular cells *(Haloarcula)* (Figure 4.4).

The shape of a bacterium is determined by heredity. Genetically most bacteria are **monomorphic;** that is, they maintain a single shape. However, a number of environmental conditions can alter that shape. If the shape is altered, identification becomes difficult. Moreover, some bacteria, such as *Rhizobium* (rī-zō'bē-um) and *Corynebacterium* (kô-rī-nē-bak-ti'rē-um), are genetically **pleomorphic,** which means they can have many shapes, not just one.

The structure of a typical procaryotic (bacterial) cell is shown in Figure 4.5. We will discuss its components according to the following organization: (1) structures external to the cell wall, (2) the cell wall, and (3) structures internal to the cell wall.

(a) Vibrio

(b) Spirillum

(c) Spirochete

FIGURE 4.3 Spiral bacteria. Diagrams (left) and corresponding photos (right) of **(a)** vibrio, **(b)** spirillum, and **(c)** spirochete.

Structures External to the Cell Wall

Among the structures external to the procaryotic cell wall are the glycocalyx, which can take the form of a capsule or a slime layer, flagella, axial filaments, and pili.

GLYCOCALYX

Glycocalyx is the general term used for substances that surround cells. The bacterial glycocalyx is a viscous (sticky), gelatinous polymer that is external to the cell wall and is composed of polysaccharide, polypeptide, or both. Its chemical composition varies widely with the species. For the most part, it is made inside the cell and excreted to the cell surface. If the substance is organized and is firmly attached to the cell wall, the glycocalyx is described as a **capsule** (see Figure 15.2).

The presence of a capsule can be determined by using negative staining, such as the India ink method mentioned in Chapter 3 (see Figure 3.10a). If the substance is unorganized and only loosely attached to the cell wall, the glycocalyx is described as a **slime layer.**

Capsules are important in contributing to bacterial virulence (the degree to which a pathogen causes disease) in certain species. Capsules often protect pathogenic bacteria from phagocytosis by the cells of the host. Phagocytosis, which is discussed in Chapter 16, is a process by which certain white blood cells engulf and destroy microbes (see Figure 16.8). *Streptococcus pneumoniae* (strep-tō-kok'kus nü-mō'nē-ī) causes pneumonia when the cells are protected by a polysaccharide capsule. Unencapsulated *S. pneumoniae* cells are readily phagocytized and cannot cause pneumonia. *Bacillus anthracis* produces a capsule of D-glutamic acid. Recall

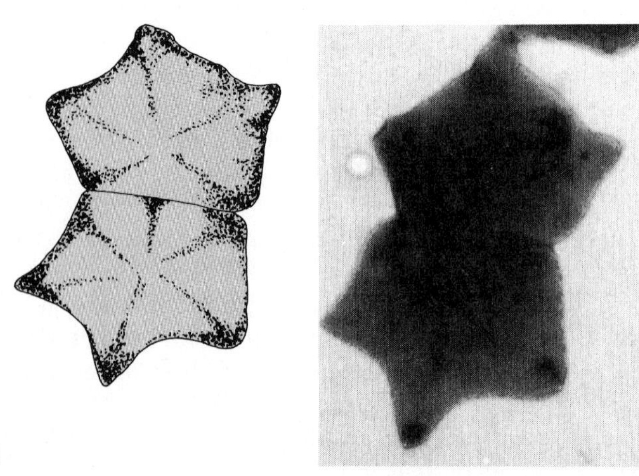

(a)

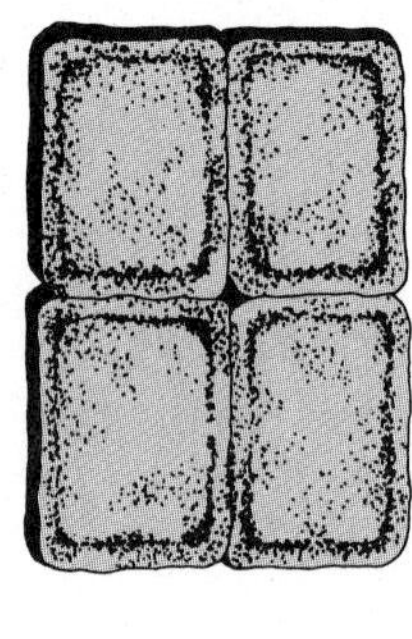

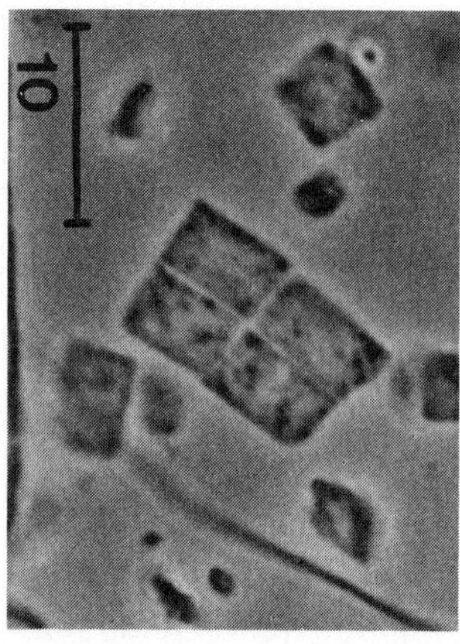

(b)

FIGURE 4.4 Star-shaped and square cells. **(a)** *Stella* (star-shaped). **(b)** Halophilic archaeobacteria (square cells).

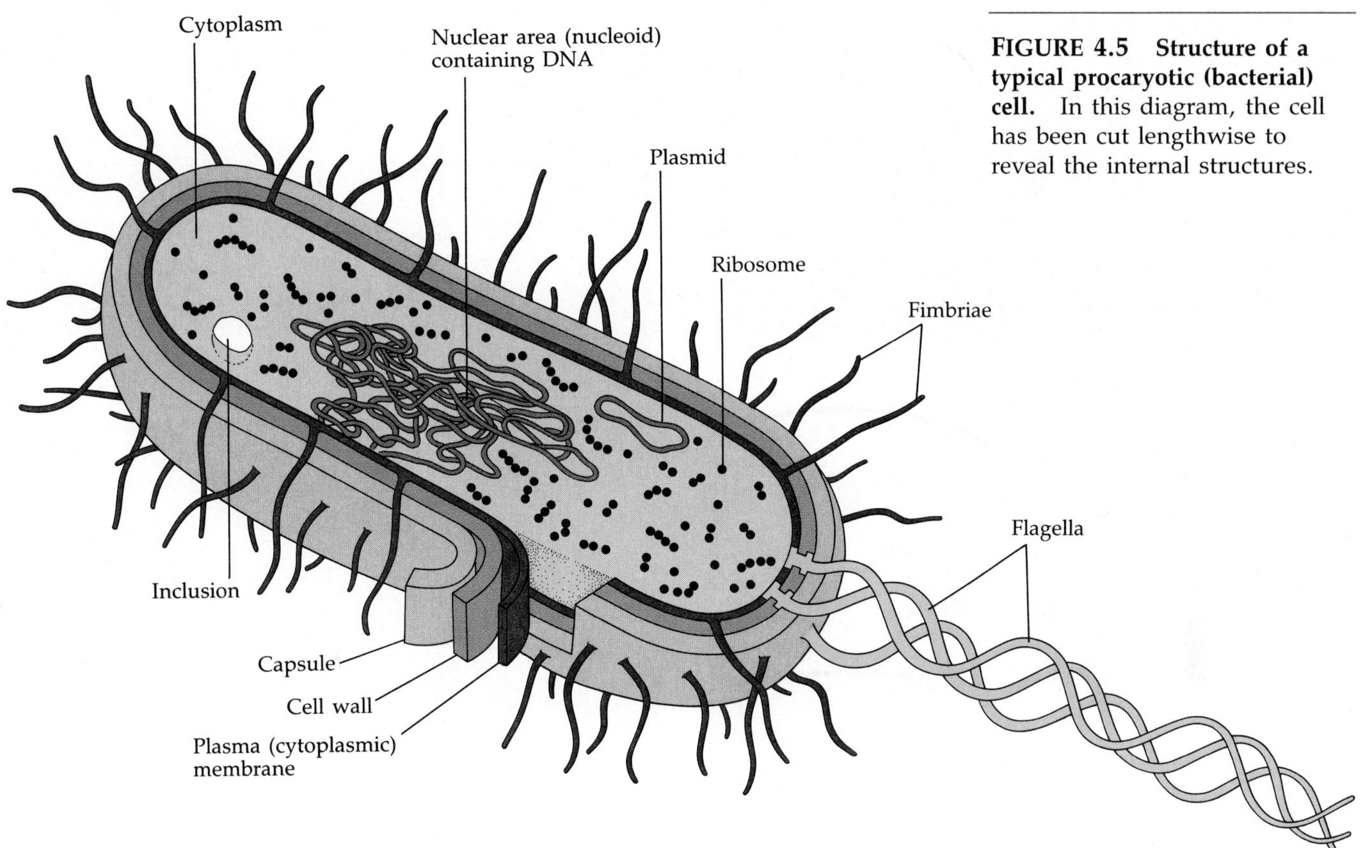

FIGURE 4.5 Structure of a typical procaryotic (bacterial) cell. In this diagram, the cell has been cut lengthwise to reveal the internal structures.

from Chapter 2 that the D forms of amino acids are unusual. Since only encapsulated *B. anthracis* cause anthrax, it is possible that the capsule prevents phagocytosis.

Another function of the sticky glycocalyx is to allow a bacterium to attach to various surfaces in order to survive in its natural environment. Through attachment, bacteria can grow on diverse surfaces such as rocks in fast-moving streams, plant roots, human teeth and tissues, and even other bacteria. *Streptococcus mutans* (mū'tans), an important cause of dental caries, attaches itself to the surface of teeth by a glycocalyx. The capsule of *Klebsiella pneumoniae* (kleb-sē-el'lä nü-mō'nē-ī) prevents phagocytosis and allows this bacterium to adhere to and colonize the respiratory tract.

S. mutans may use its capsule as a source of nutrition by breaking it down and utilizing the sugars when energy stores are low. A glycocalyx can protect a cell against dehydration. Also, its viscosity may inhibit the movement of nutrients from the cell.

FLAGELLA

Some procaryotic cells carry **flagella** (singular, *flagellum,* meaning whip), which are long filamentous appendages that propel bacteria (see Figure 4.5).

Bacterial cells have four arrangements of flagella (Figure 4.6)—**monotrichous** (single polar flagellum), **amphitrichous** (single flagellum at each end of the cell), **lophotrichous** (two or more flagella at one or both poles of the cell), and **peritrichous** (flagella distributed over the entire cell).

A flagellum has three basic parts (Figure 4.6e). The long outermost region, the *filament,* is constant in diameter and contains the globular (roughly spherical) protein *flagellin* arranged in several chains that intertwine and form a helix around a hollow core. Flagellar proteins serve to identify certain pathogenic bacteria. In most bacteria, filaments are not covered by a membrane or sheath, as in eucaryotic cells. The filament is attached to a slightly wider *hook,* which consists of a different protein. The third portion of a flagellum is the *basal body,* which anchors the flagellum to the cell wall and plasma membrane.

The basal body is composed of a small central rod inserted into a series of rings. Gram-negative bacteria contain two pairs of rings. The outer pair of rings is anchored to various portions of the cell wall, and the inner pair of rings is anchored to the plasma membrane. In gram-positive bacteria, only the inner pair is present. As you will see later in the chapter, the flagella (and cilia) of eucaryotic cells are more complex than those of procaryotic cells.

Bacteria with flagella are motile. That is, they have the ability to move on their own. Each procaryotic flagellum is a semirigid, helical rotor that moves the cell

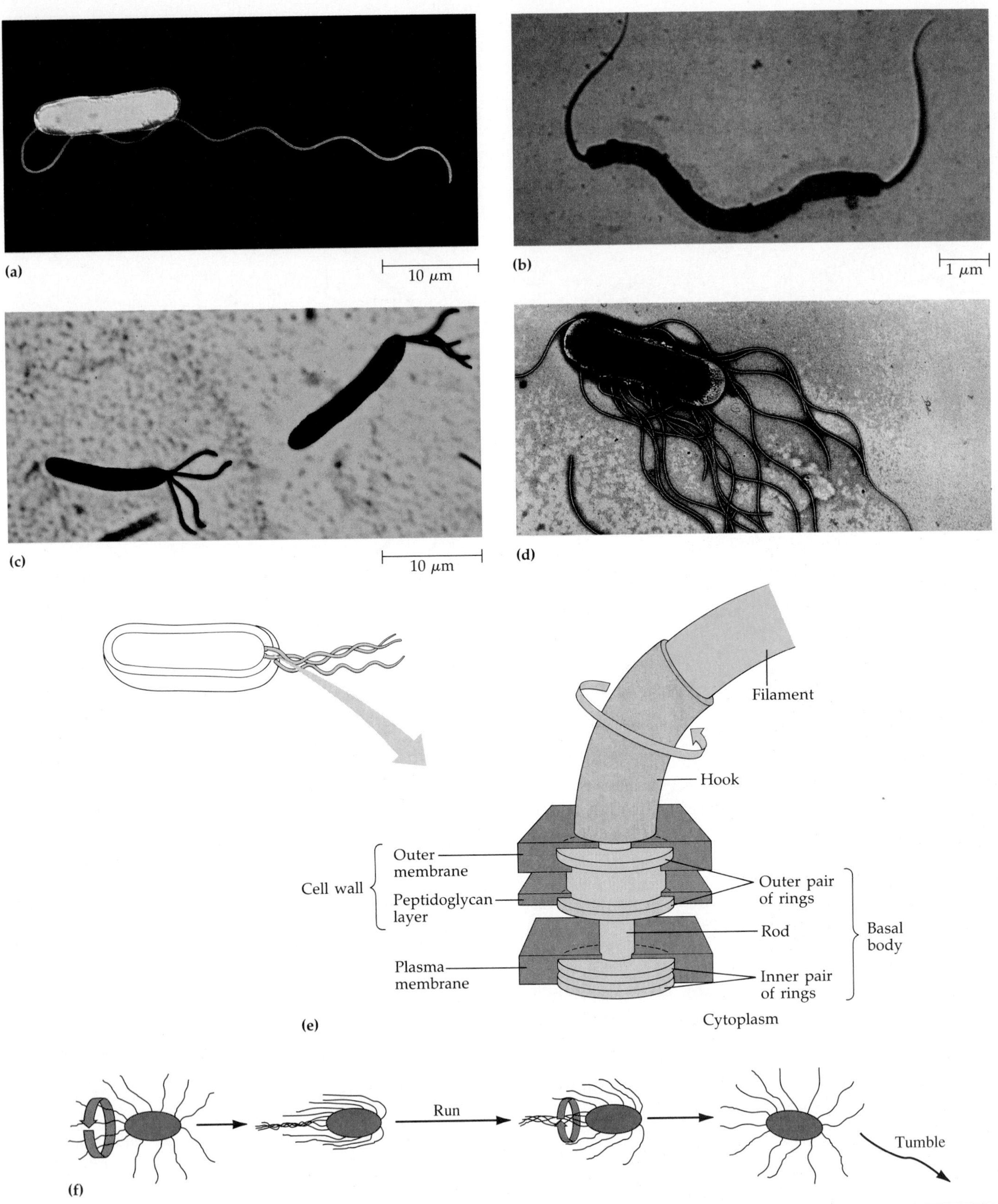

FIGURE 4.6 Flagella. Four basic types of flagellar arrangements: **(a)** monotrichous (*Legionella pneumophila*); **(b)** amphitrichous; **(c)** lophotrichous; **(d)** peritrichous (*Salmonella*). **(e)** Parts and attachment of a flagellum of a gram-negative bacterium. **(f)** Types of bacterial motility, showing a "run" and a "tumble."

by rotating from the basal body. The rotation of a flagellum is either clockwise or counterclockwise around its long axis. Eucaryotic flagella, by contrast, undulate in a wavelike motion. The movement of a procaryotic flagellum results from rotation of its basal body and is similar to the movement of the shaft of an electric motor. As the flagella rotate, they form a bundle that pushes against the surrounding liquid and propels the bacterium (Figure 4.6f). Although the exact mechanochemical basis for this biological "motor" is not completely understood, we know that it depends on the cell's continuous generation of energy.

Bacterial cells can alter the speed and direction of rotation of flagella and thus are capable of various patterns of motility (Figure 4.6f). When a bacterium moves in one direction for a length of time, the movement is called a "run" or "swim." "Runs" are interrupted by periodic, abrupt, random changes in direction called "tumbles." Then, a "run" resumes again. "Tumbles" are caused by a reversal of flagellar rotation. Some species of bacteria endowed with large numbers of flagella (*Proteus* [prō'tē-us], for example) can "swarm," or show rapid wavelike growth across a solid culture medium.

One advantage of motility is that it enables a bacterium to move toward a favorable environment or away from an adverse one. The movement of a bacterium toward or away from a particular stimulus is called **taxis.** Such stimuli include chemicals **(chemotaxis)** and light **(phototaxis).** Motile bacteria contain receptors in various locations, such as in or just under the cell wall. These receptors pick up chemical stimuli, such as oxygen, ribose, and galactose. In response to the stimuli, information is passed to the flagella. If the chemotactic signal is positive, called an *attractant,* the bacteria move toward the stimulus with many runs and few tumbles. If the chemotactic signal is negative, called a *repellent,* the frequency of tumbles increases as the bacterium moves away from the stimulus (see the box on p. 76).

AXIAL FILAMENTS

Spirochetes are a group of bacteria that have unique structure and motility. One of the best-known spirochetes is *Treponema pallidum,* the causative agent of syphilis. Another spirochete, one that has been in the news recently, is *Borrelia burgdorferi,* the causative agent of Lyme disease. Spirochetes move by means of **axial filaments,** bundles of fibrils that arise at the ends of the cell beneath the outer sheath and spiral around the cell (Figure 4.7). Axial filaments, which are anchored at one end of the spirochete, have a structure

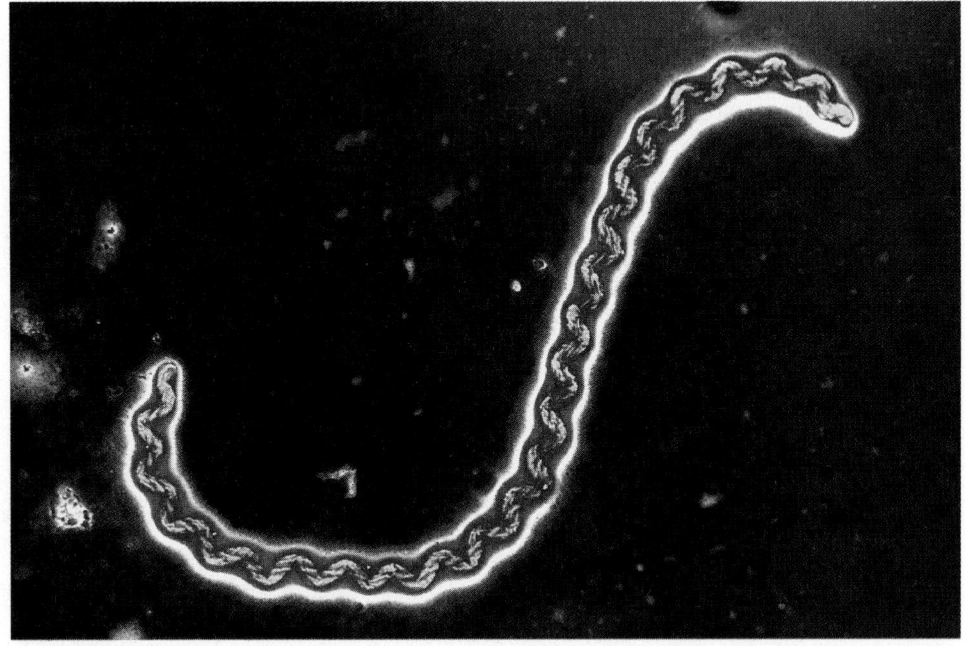

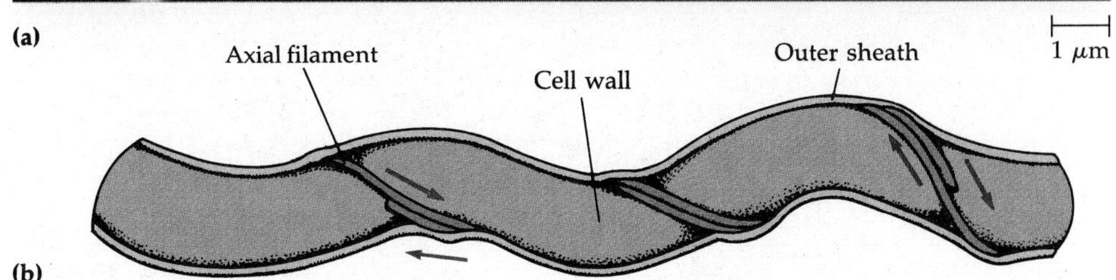

FIGURE 4.7 Axial filaments. **(a)** Photomicrograph of the spirochete *Leptospira,* showing an axial filament. **(b)** Diagram of axial filaments wrapping around part of a spirochete.

MICROBIOLOGY *HIGHLIGHTS*

Microbial Magnets

For nearly 100 years, biologists have suspected that birds can use the Earth's magnetic field for orientation during flight. During the 1970s, magnetotaxis (response to a magnetic field) was demonstrated in some birds as well as in honeybees and some bacteria. Around 1980, magnetic material was found in these birds and honeybees, as well as in some dolphins.

In 1975, R.P. Blakemore reported that sulfide-rich mud contained highly motile bacteria that "... swam in the same geographic direction even when the microscope was turned around, moved to another location, or covered with a pasteboard box. To my astonishment, when a magnet was brought near the microscope, the hundreds of swimming cells instantly turned and rushed away from the end of the magnet! They were always attracted by the end that also attracted the north-seeking end of a compass needle and they were repelled by its opposite end. Their swimming speed was very fast, on the order of 100 μm per second, and the entire population swerved in unison as the magnet was moved about nearby."*

Magnetotactic bacteria synthesize magnetite, iron oxide (Fe_3O_4), and store it in inclusions called magnetosomes, as shown in the photo. The magnetosomes act like magnets and cause the bacterial cells to align on the Earth's geomagnetic field. All bacteria with magnetosomes are gram-negative and have fimbriae or a polysaccharide glycocalyx to provide attachment. These bacteria are found in sediments throughout the world. In the Northern Hemisphere, north-seeking bacteria predominate, and in the Southern Hemisphere, south-seeking bacteria predominate.

Aquaspirillum magnetotacticum (see photo) metabolizes organic molecules for carbon and energy and requires a microaerophilic environment. It was found that when *A. magnetotacticum* are separated from an attachment site they swim along what appear to be geomagnetic lines. They move downward, toward either the north or south pole, until reaching a suitable attachment site.

How these bacteria synthesize magnetite is presently being studied, and the function of magnetite in the cell's metabolism is not yet clear. The formation of ferric iron (Fe^{3+}) from ferrous iron (Fe^{2+}) could provide adenosine triphosphate (ATP) for the cell, but the estimated amount of ATP is so small that the magnetite must serve another function. In vitro, the magnetosomes can decompose hydrogen peroxide, which forms in cells in the presence of oxygen and is usually degraded by the enzyme catalase before it reaches toxic levels. Researchers speculate that the magnetosomes protect the cell against hydrogen peroxide accumulation in vivo.

The synthesis and function of magnetosomes are of interest to microbiologists to further our understanding of cells and may provide a model to explain the formation of similar iron oxides in birds, insects, and other animals. The ecology of magnetic microbes may also provide a tool for scientists studying the Earth's magnetic field. Evidence suggests that the Earth's magnetic field may have reversed or shifted several times. Examination of the orientation of fossils of magnetotactic bacteria preserved in rock or sediment may shed light on the movement of the continents and magnetic poles in relation to one another.

In 1987, researchers discovered other bacteria that live in the bottom sediments of lakes and bays and that produce megnetite. The Tokyo Institute of Agriculture and TDK Corporation are developing culture methods to obtain large quantities of magnetite from bacteria. They hope to use the magnetic particles in the production of magnetic tapes for recording sound and data.

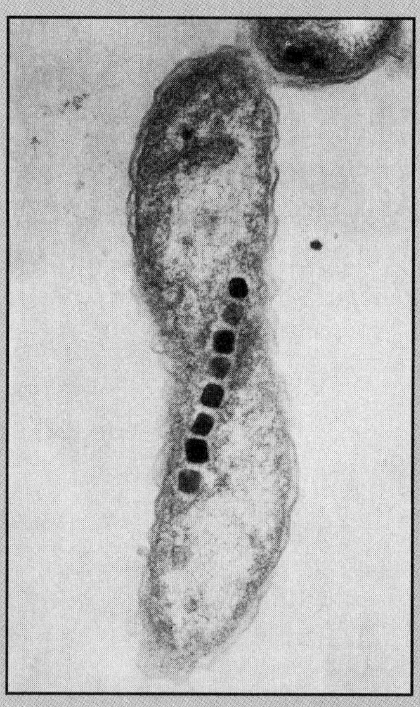

Aquaspirillum magnetotacticum, showing chain of magnetosomes. Outer membrane of gram-negative wall is also visible.

*Reproduced, with permission, from the *Annual Review of Microbiology,* Volume 36, © 1982 by Annual Reviews Inc. and Richard P. Blakemore. (Richard P. Blakemore, "Magnetotactic bacteria." *Annual Review of Microbiology* 36:217–238, 1982.)

similar to that of flagella. The rotation of the filaments produces an opposing movement of the outer sheath that propels the spirochetes by causing them to move like corkscrews.

FIMBRIAE AND PILI

Many gram-negative bacteria contain hairlike appendages that are shorter, straighter, and thinner than flagella and are used for attachment rather than for motility. These structures consist of a protein called *pilin* arranged helically around a central core and are divided into two types, *fimbriae* and *pili*, with very different functions. (Some microbiologists use the two terms interchangeably to refer to all such structures, but we distinguish between them.)

Fimbriae can occur at the poles of the bacterial cell, or they can be evenly distributed over the entire surface of the cell. They can number anywhere from a few to several hundred per cell (Figure 4.8). Like the glycocalyx, fimbriae allow a cell to adhere to surfaces, including the surfaces of other cells. For example, fimbriae associated with the bacterium *Neisseria gonorrhoeae* (nī-se′rē-ä go-nôr-rē′ī), the causative agent of gonorrhea, help the microbe to colonize mucous membranes. Once colonization occurs, the bacteria are capable of causing the disease. When fimbriae are absent (because of genetic mutation), colonization cannot occur, and no disease ensues.

Pili are usually longer than fimbriae and number only one or two per cell. Pili function to join bacterial cells prior to the transfer of DNA from one cell to another. For this reason, they are sometimes also called **sex pili** (see Chapter 8).

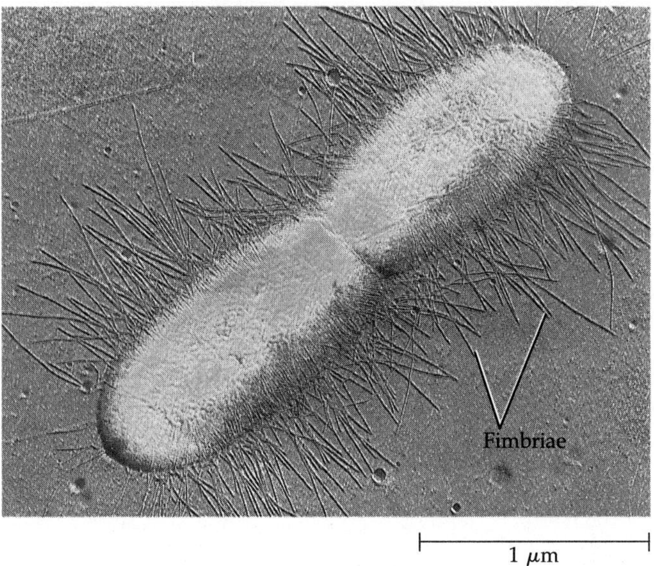

FIGURE 4.8 Fimbriae. The fimbriae seem to bristle from this cell of *E. coli,* which is beginning to divide.

The Cell Wall

The **cell wall** of the bacterial cell is a complex, semirigid structure that is responsible for the characteristic shape of the cell. The cell wall surrounds the underlying, fragile plasma (cytoplasmic) membrane and protects it and the internal parts of the cell from adverse changes in the surrounding environment (see Figure 4.5). Almost all procaryotes have cell walls.

The major function of the cell wall is to prevent bacterial cells from rupturing when the osmotic pressure inside the cell is greater than that outside the cell. It also helps maintain the shape of a bacterium and serves as a point of anchorage for flagella. As the volume of a bacterial cell increases, there is a corresponding extension of the plasma membrane and cell wall. Clinically, the cell wall is important because it contributes to the ability of some species to cause disease and is the site of action of some antibiotics. In addition, the chemical composition of the cell wall is used to differentiate major types of bacteria.

Although some eucaryotes, including plants, algae, and fungi, contain cell walls, their cell walls differ chemically from those of procaryotes, are simpler in structure, and are less rigid.

COMPOSITION AND CHARACTERISTICS

The bacterial cell wall is composed of a macromolecular network called *peptidoglycan (murein),* which is present either alone or in combination with other substances. Peptidoglycan is a mucopolysaccharide consisting of a repeating disaccharide attached to chains of four or five amino acids. The component monosaccharides, called N-acetylglucosamine (NAG) and N-acetylmuramic acid (NAM) (from *murus,* meaning wall), are related to glucose. The structural formulas for NAG and NAM are shown in Figure 4.9.

The various components of peptidoglycan are assembled in the cell wall (see Figure 4.10a, right). Alter-

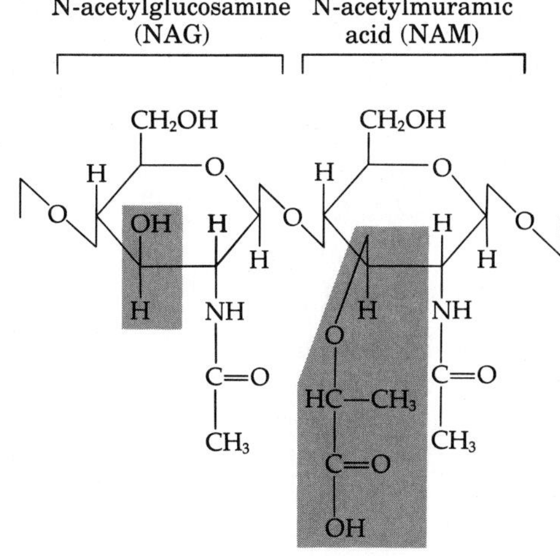

FIGURE 4.9 N-acetylglucosamine (NAG) and N-acetylmuramic acid (NAM) joined as in peptidoglycan. The colored areas show the differences between the two molecules. The linkage between them is called a β-1,4 linkage.

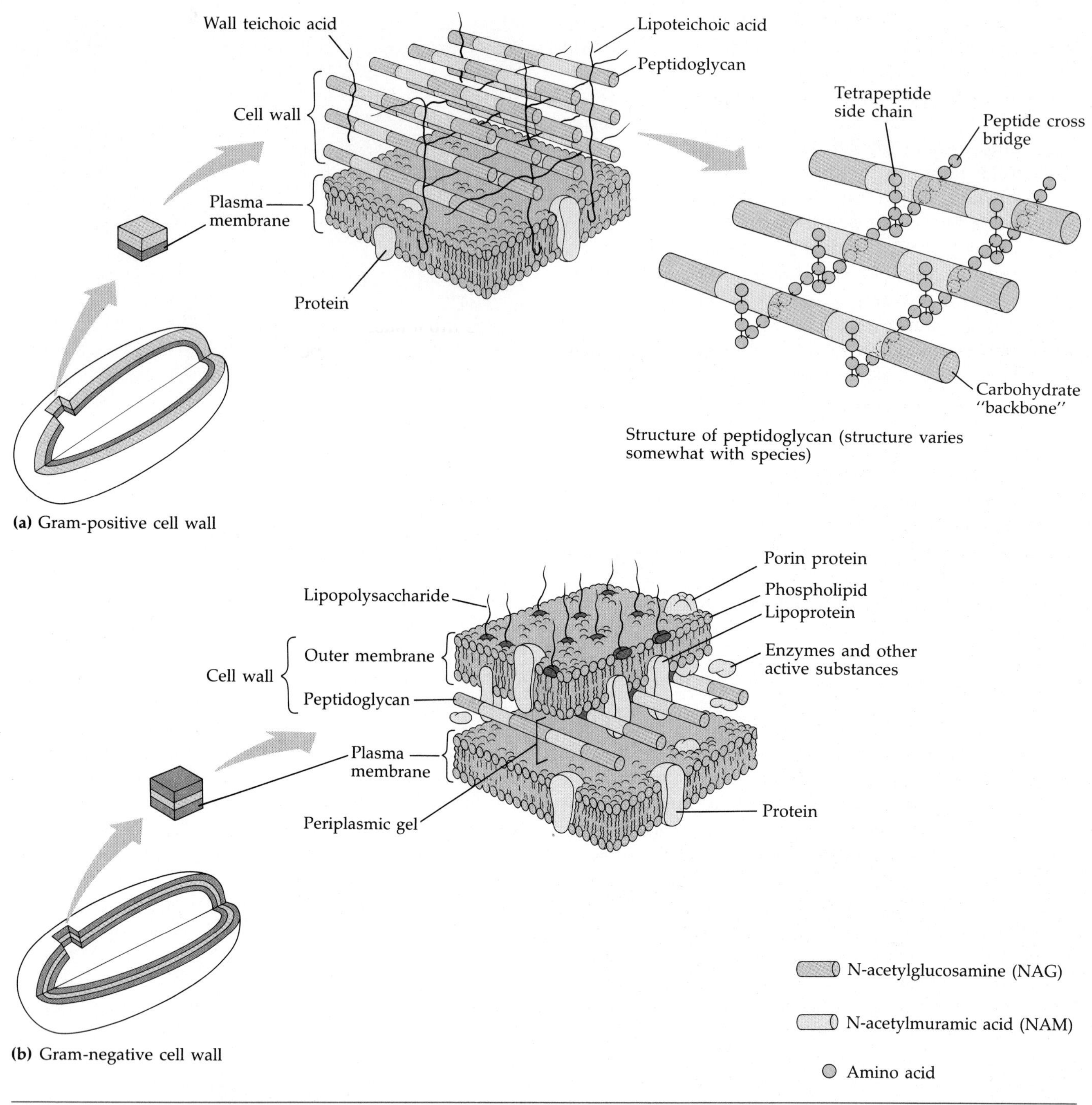

FIGURE 4.10 Bacterial cell wall. **(a)** Gram-positive cell wall showing the structure of peptidoglycan. Together the carbohydrate backbone (glycan portion) and amino acids (peptide portion) make up peptidoglycan. The frequency of peptide cross bridges and the number of amino acids in these bridges vary with the species of bacterium. **(b)** Gram-negative cell wall.

nating NAM and NAG molecules are linked in rows of 10 to 65 sugars to form a carbohydrate "backbone" (the glycan portion of peptidoglycan). Adjacent rows are linked by polypeptides (the peptide portion of peptidoglycan) to form a lattice that surrounds and protects the entire cell. Although the structure of the polypeptide link varies, it always includes *tetrapeptide side chains*. A tetrapeptide side chain consists of four amino acids. The amino acids occur in an alternating pattern of D and L forms (see Figure 2.16). This is unique since the amino acids found in other proteins are L forms. Parallel tetrapeptide side chains may be directly bonded to each other or linked by a *peptide cross bridge,* consisting of one to five amino acids.

In most gram-positive bacteria, the cell wall consists of many layers of peptidoglycan, forming a thick, rigid structure (Figure 4.10a). (By contrast, gram-negative cell walls contain only a thin layer of peptidoglycan.) In addition, the cell walls of gram-positive bacteria contain **teichoic acids**, which consist primarily of an alcohol (such as glycerol or ribitol) and phosphate. There are two classes of teichoic acids: *lipoteichoic acid,* which spans the peptidoglycan layer and is linked to the plasma membrane, and *wall teichoic acid,* which is linked to the peptidoglycan layer. Because of their negative charge (from the phosphate groups), teichoic acids may bind and regulate the movement of cations (positive ions) into and out of the cell. They may also assume a role in cell growth. When the cell wall grows, enzymes called *autolysins* are needed. These enzymes separate components of the wall to allow insertion of new subunits. Teichoic acids regulate the activity of autolysins, thus preventing extensive wall breakdown and possible *cell lysis* (rupturing). Teichoic acids may also be involved in storing phosphorus. Finally, teichoic acids provide much of the wall's antigenic specificity and thus make it possible to identify bacteria by serological means (see Chapter 10).

The cell walls of gram-positive streptococci are covered with various polysaccharides that allow them to be grouped into medically significant types. The cell walls of acid-fast bacteria, such as *Mycobacterium* consist of as much as 60% mycolic acid, a waxy lipid, while the rest is peptidoglycan. These bacteria can be stained with the Gram stain and are considered gram-positive.

The cell walls of gram-negative bacteria consist of one or a very few layers of peptidoglycan and an outer membrane (see Figure 4.10b). The peptidoglycan is bonded to lipoproteins in the outer membrane and is embedded in a soft material, the *periplasmic gel,* which contains a high concentration of degradative enzymes and transport proteins. Gram-negative cell walls do not contain teichoic acids. Because the cell walls of gram-negative bacteria contain only a small amount of peptidoglycan, they are more susceptible to mechanical breakage.

The *outer membrane* of the gram-negative cell consists of lipoproteins, lipopolysaccharides (LPS), and phospholipids (Figure 4.10b). Lipoproteins are composed of protein covalently linked to lipid. The lipoproteins firmly bind the outer membrane to the underlying peptidoglycan layer via covalent bonds. The outer membrane has several specialized functions. Its strong negative charge is an important factor in evading phagocytosis and the action of complement, two components of the defenses of the host (discussed in detail in Chapter 16).

The outer membrane also provides a barrier to certain antibiotics (for example, penicillin), lysozyme (an enzyme that can break down the cell walls of all gram-positive bacteria but of only a few gram-negative bacteria), detergents, heavy metals, bile salts, digestive enzymes, and certain dyes.

However, the outer membrane does not provide a barrier to all substances in the environment since nutrients must pass through to sustain the metabolism of the cell. Part of the permeability of the outer membrane is due to proteins in the membrane, called **porins**, that form channels. Porins are nonspecific and permit the passage of small molecules up to a molecular weight of about 800. The permeability of the outer membrane is also related to membrane proteins called **specific channel proteins**. These proteins permit the passage of specific substances only (for example, vitamin B_{12}, iron, nucleotides, and maltose). They can also make bacteria vulnerable to attack by providing attachment sites for viruses and bacteriocins (proteins produced by some bacteria that inhibit or kill closely related species).

The lipopolysaccharide component of the outer membrane provides two important characteristics of gram-negative bacteria. The polysaccharide portion of the lipopolysaccharide is composed of sugars, called *O polysaccharides,* that function as antigens and are useful for distinguishing species of gram-negative bacteria (for example, the over 1400 *Salmonella* [sal-mōn-el'lä] species) by serological means. This role is comparable to that of teichoic acids in gram-positive cells. The lipid portion of the lipopolysaccharide, called *lipid A,* is referred to as *endotoxin* and is toxic when in the host's bloodstream or gastrointestinal tract. It causes fever and shock. The nature and importance of endotoxins and other bacterial toxins will be discussed in Chapter 15.

A comparison of some characteristics of gram-positive and gram-negative bacteria is presented in Table 4.1.

ATYPICAL CELL WALLS

Among procaryotes, there are cells that naturally have no walls or have very little wall material. These include members of the genus *Mycoplasma* (mī-kō-plaz'mä) and related organisms. Mycoplasmas are the smallest known bacteria that can grow and reproduce outside living host cells. Because they have no cell walls, they pass through most bacterial filters and were first mistaken for viruses. Their plasma membranes are unique

TABLE 4.1 Some Comparative Characteristics of Gram-Positive and Gram-Negative Bacteria

CHARACTERISTIC	GRAM-POSITIVE	GRAM-NEGATIVE
Gram reaction	Retain crystal violet dye and stain dark violet or purple	Can be decolorized to accept counterstain (safranin) and stain red
Peptidoglycan layer	Thick (multilayered)	Thin (single-layered)
Teichoic acids	Present in many	Absent
Periplasmic gel	Absent	Present
Outer membrane	Absent	Present
Lipopolysaccharide (LPS) content	Virtually none	High
Lipid and lipoprotein content	Low (acid-fast bacteria have lipids linked to peptidoglycan)	High (due to presence of outer membrane)
Flagellar structure	2 rings in basal body	4 rings in basal body
Toxins produced	Primarily exotoxins	Primarily endotoxins
Resistance to physical disruption	High	Low
Cell wall disruption by lysozyme	High	Low (requires pretreatment to destabilize the outer membrane)
Susceptibility to penicillin and sulfonamide	High	Low
Susceptibility to streptomycin, chloramphenicol, and tetracycline	Low	High
Inhibition by basic dyes	High	Low
Susceptibility to anionic detergents	High	Low
Resistance to sodium azide	High	Low
Resistance to drying	High	Low

among bacteria in having lipids called *sterols*, which are thought to help protect them from osmotic lysis.

Archaeobacteria may lack walls or may have unusual walls composed of polysaccharides and proteins but not peptidoglycan. These walls do, however, contain a substance similar to peptidoglycan that contains N-acetyltalosaminuronic acid instead of NAM but lacks the D amino acids found in bacterial cell walls. The substance is called *pseudomurein*.

Other atypical bacterial cells are the **L forms** (so named after the Lister Institute, where they were discovered). These are tiny mutant bacteria with defective cell walls. Certain chemicals and antibiotics like penicillin induce many bacteria to produce L forms. Although some L forms can revert to the original bacterial form, others are stable. L forms tend to contain just enough cell wall material to prevent lysis from occurring in dilute solutions.

Recall from Chapter 3 that the mechanism for the Gram stain is related to the structure and chemical composition of the cell wall. It might serve you well to review that mechanism before continuing to the next section.

DAMAGE TO THE CELL WALL

Chemicals that damage bacterial cell walls, or interfere with their synthesis, often do not harm the cells of an animal host because the bacterial cell wall is made of chemicals unlike those in eucaryotic cells. Thus, cell wall synthesis is the target for some antimicrobial drugs. One way that the cell wall can be damaged is by exposure to the enzyme *lysozyme*. This enzyme occurs naturally in some eucaryotic cells and is a constituent of tears, mucus, and saliva. Lysozyme is particularly active on the major cell wall components of most gram-positive bacteria, making them vulnerable to rupture or lysis. The lysozyme catalyzes the hydrolysis of the bonds between the sugars in the polysaccharide chain of peptidoglycan. This act is analogous to cutting the steel supports of a bridge with a cutting torch. The gram-positive cell wall is almost completely destroyed by lysozyme. The cellular contents that remain surrounded by the plasma membrane may remain intact if osmotic lysis does not occur; this wall-less cell is termed a **protoplast.** Typically, the protoplast is spherical and is still capable of carrying on metabolism.

When lysozyme is applied to gram-negative cells, usually the wall is not destroyed to the same extent as in gram-positive cells; some of the outer membrane also remains. In this case, the cellular contents, plasma membrane, and remaining outer wall layer are called a **spheroplast**, also a spherical structure. For lysozyme to exert its effect on gram-negative cells, the cells are first treated with ethylenediaminetetraacetic acid (EDTA), a substance that weakens ionic bonds in the outer membrane and damages it, giving the lysozyme access to the peptidoglycan layer.

Protoplasts and spheroplasts burst in pure water or very dilute salt or sugar solutions because the water molecules from the surrounding fluid rapidly move into and enlarge the cell, which has a much lower internal concentration of water. This rupturing is called **osmotic lysis** and will be discussed in detail shortly.

Certain antibiotics, such as penicillin, destroy bacteria by interfering with the formation of the peptide cross bridges of peptidoglycan, thus preventing the formation of a functional cell wall. Most gram-negative bacteria are not as susceptible to penicillin as gram-positive bacteria are because the outer membrane of gram-negative bacteria forms a barrier that inhibits the entry of this and other substances, and gram-negative bacteria have fewer peptide cross bridges. However, gram-negative bacteria are quite susceptible to some β-lactam antibiotics that penetrate the outer membrane better than penicillin. Antibiotics will be discussed in more detail in Chapter 20.

Structures Internal to the Cell Wall

Thus far, we have discussed the procaryotic cell wall and structures external to it. We will now look inside the procaryotic cell and discuss the structures and functions of the plasma membrane and other components within the cytoplasm of the cell.

PLASMA (CYTOPLASMIC) MEMBRANE

The **plasma (cytoplasmic) membrane** or **inner membrane** is a thin structure lying inside the cell wall and enclosing the cytoplasm of the cell (see Figure 4.5). The plasma membrane of procaryotes consists primarily of phospholipids (see Figure 2.13), which are the most abundant chemicals in the membrane, and proteins. Eucaryotic plasma membranes also contain carbohydrates and sterols, such as cholesterol. Because they lack sterols, procaryotic plasma membranes are less rigid than eucaryotic membranes. One exception to this is the wall-less procaryote called *Mycoplasma*, which contains membrane sterols.

Structure

In electron micrographs, procaryotic and eucaryotic plasma membranes (and the outer membranes of gram-negative bacteria) look like two-layered structures; there are two dark lines with a light space between the lines (Figure 4.11a). The phospholipid molecules are arranged in two parallel rows, called a *phospholipid bilayer* (Figure 4.11b). Each phospholipid molecule contains a polar head, composed of a phosphate group and glycerol that is hydrophilic (water-loving) and soluble in water, and nonpolar tails, composed of fatty acids that are hydrophobic (water-fearing) and are insoluble in water (Figure 4.11c). The polar heads are on the two surfaces of the phospholipid bilayer, and the nonpolar tails are in the interior of the bilayer.

The protein molecules in the membrane can be arranged in a variety of ways. Some, called *peripheral proteins,* are easily removed from the membrane by mild treatments and are thought to lie at the inner or outer surface of the membrane. They may function as enzymes that catalyze chemical reactions, as a "scaffold" for support, and as mediators of changes in membrane shape during movement. Other proteins, called *integral proteins,* can be removed from the membrane only after disrupting the bilayer (by using detergents, for example). Some integral proteins are believed to penetrate the membrane completely. Some of these proteins contain channels through which substances enter and exit the cell. Studies have demonstrated that the phospholipid and protein molecules in membranes are not static but move quite freely within the membrane surface. This movement is most probably associated with the many functions performed by the plasma membrane. Since the fatty acid tails cling together, phospholipids in the presence of water form a self-sealing bilayer so that breaks and tears in the membrane will heal themselves. The membrane must be about as viscous as olive oil to allow membrane proteins to move freely enough to perform their functions without destroying the structure of the membrane. This dynamic arrangement of phospholipid and protein is referred to as the *fluid mosaic model.*

Functions

The most important function of the plasma membrane is to serve as a selective barrier through which materials enter and exit the cell. In this function, plasma membranes are **selectively permeable** (sometimes called **semipermeable**). This term indicates that certain molecules and ions pass through the membrane, but others are restricted. The permeability of the membrane depends on several factors. Large molecules (such as proteins) cannot pass through the plasma membrane. This may be because these molecules are larger than the channels in integral proteins. But

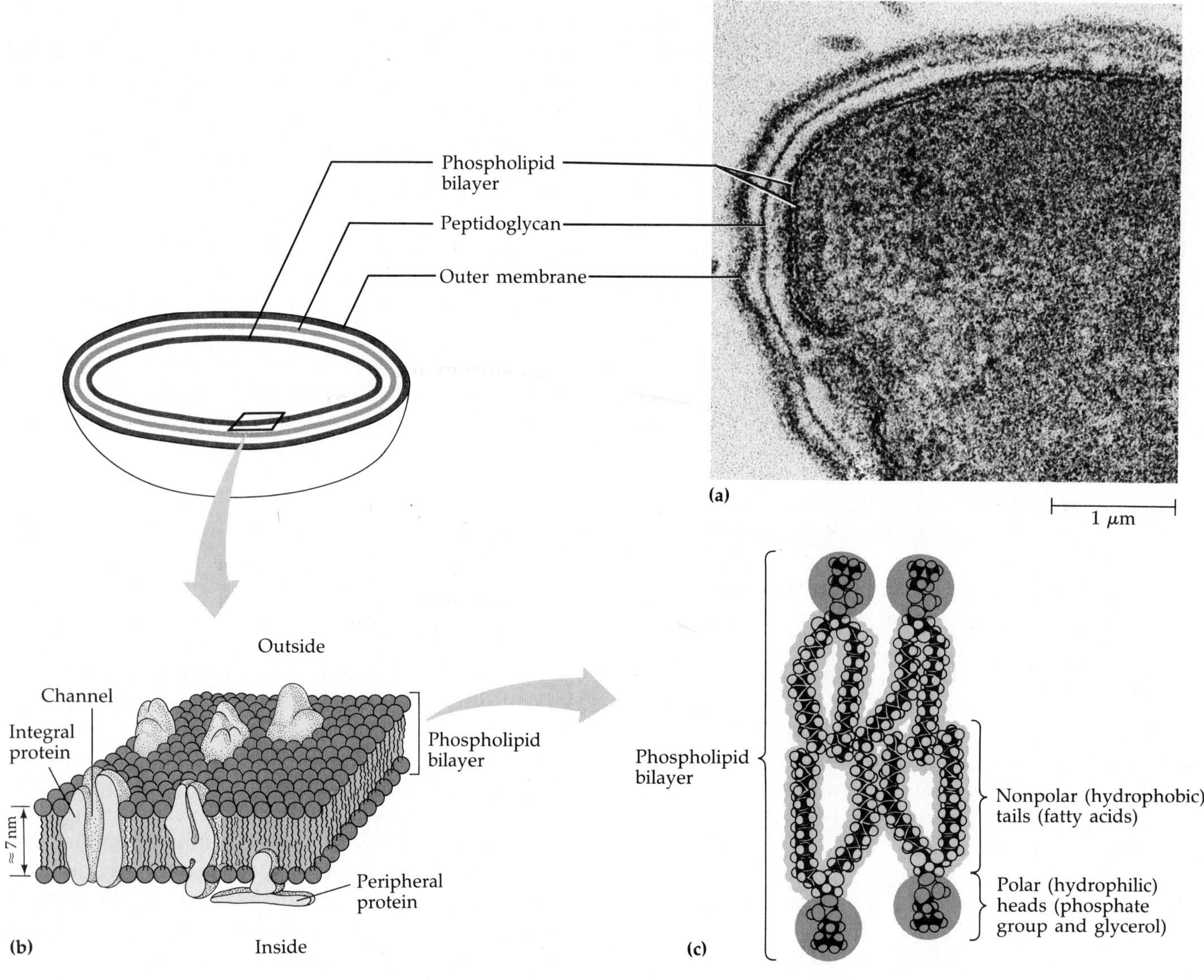

FIGURE 4.11 Plasma membrane. **(a)** Electron micrograph showing the phospholipid bilayer forming the plasma membrane of the bacterium *Bacillus brevis*. Layers of the cell wall can be seen outside the plasma membrane. **(b)** Drawing of a membrane showing phospholipid bilayer and proteins. The outer membrane of gram-negative bacteria is also a phospholipid bilayer. **(c)** Space-filling models of several molecules as they are arranged in the phospholipid bilayer.

smaller molecules (such as water, oxygen, carbon dioxide, and some simple sugars) usually pass through easily. Ions penetrate the membrane very slowly. Substances that dissolve easily in lipids (such as oxygen, carbon dioxide, and nonpolar organic molecules) enter and exit more easily than other substances because the membrane consists mostly of phospholipids. The movement of materials across plasma membranes also depends on carrier molecules, which will be described shortly.

Plasma membranes are also important to the breakdown of nutrients and the production of energy. The plasma membranes of bacteria contain enzymes capable of catalyzing the chemical reactions that break down nutrients and produce ATP. In some bacteria, pigments and enzymes involved in photosynthesis (the conversion of light energy into chemical energy) are found in infoldings of the plasma membrane that extend into the cytoplasm. These membrane layers are called **chromatophores** or **thylakoids** (Figure 4.12).

When bacteria are viewed with an electron microscope, it appears that the plasma membranes often contain one or more large, irregular folds called **mesosomes.** Many functions have been proposed for meso-

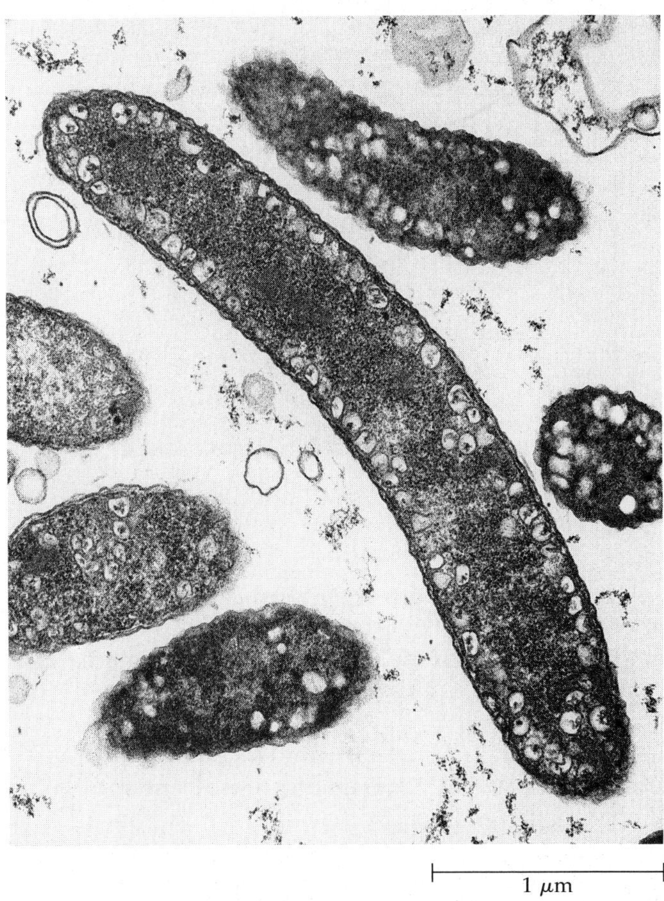

FIGURE 4.12 Chromatophores. Membranous structures that are sites of photosynthesis can be seen in this transmission electron micrograph of *Rhodospirillum rubrum,* a purple bacterium.

somes. However, recent evidence indicates that they are artifacts (structures that appear in microscopic preparations as a result of the method of preparation), not true cell structures. It is thought that they are folds in the plasma membrane that develop when cells are dehydrated prior to electron microscopy. When a new electron microscopy technique made it possible to view cells without dehydration, no mesosomes were seen.

Destruction of the Plasma Membrane by Antimicrobial Agents

Because the plasma membrane is vital to the bacterial cell, it is not surprising that it is the site at which several antimicrobial agents exert their effect. In addition to the chemicals that damage the cell wall and hence indirectly expose the membrane to injury, many compounds specifically damage plasma membranes. These compounds include certain alcohols and quaternary ammonium compounds, which are used as disinfectants. By disrupting the membrane's phospholipids, a group of antibiotics known as the *polymyxins* cause leakage of intracellular contents and subsequent cell death. This mechanism is discussed in Chapter 20.

MOVEMENT OF MATERIALS ACROSS MEMBRANES

When the concentration of a substance is stronger on one side of a membrane than on another, a *concentration gradient* (difference) exists. If the substance can cross the membrane, it will move to the more dilute side until the concentrations are equal or until other forces stop its movement.

Materials move across plasma membranes of both procaryotic and eucaryotic cells by two kinds of processes, passive and active. In *passive processes,* substances cross the membrane from an area of high concentration to an area of low concentration (with the concentration gradient), without any expenditure of energy (ATP) by the cell. Examples of passive processes are simple diffusion, osmosis, and facilitated diffusion, to be discussed shortly. In *active processes,* the cell must use energy (ATP) to move substances from areas of low concentration to areas of high concentration (against the concentration gradient). Examples of active processes are active transport and group translocation, which we will also discuss shortly.

Simple Diffusion

Simple diffusion is the net (overall) movement of molecules or ions from an area of high concentration to an area of low concentration (Figure 4.13). The movement from areas of high to low concentration continues until the molecules or ions are evenly distributed. The point of even distribution is called *equilibrium.* Cells rely on diffusion to transport certain small molecules, such as oxygen and carbon dioxide, across their cell membranes.

Facilitated Diffusion

When **facilitated diffusion** occurs, the substance (glucose, for example) to be transported combines with a *carrier protein* in the plasma membrane. Such carriers are sometimes called *permeases.* The carrier can transport a substance across the membrane from an area of high concentration to one of low concentration. Carrier proteins bind to the substance they transport on the outer surface of the plasma membrane and, by a poorly understood mechanism, transport the substance through the membrane to the inner surface and release it into the cytoplasm. One proposed mechanism for facilitated diffusion is that the carrier protein remains in place in the membrane but undergoes a change in shape that transports substances from one side of the membrane to the other (Figure 4.14). Facilitated diffusion is similar to simple diffusion, in that the

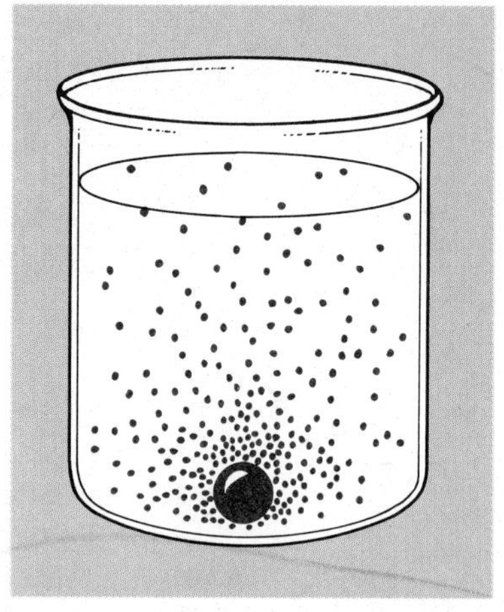

(a)

(b)

FIGURE 4.13 The principle of simple diffusion. (a) The molecules of dye in the pellet are diffusing into the water from an area of high dye concentration to areas of low dye concentration. **(b)** Potassium permanganate in the process of diffusing.

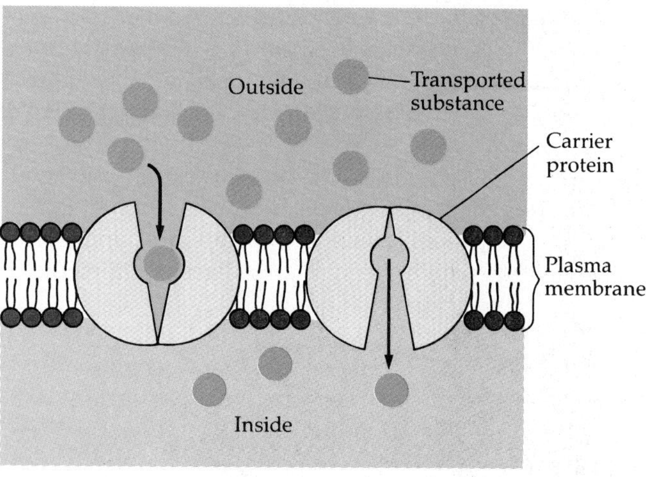

FIGURE 4.14 Facilitated diffusion. Carrier proteins in the plasma membrane transport molecules across the membrane from an area of high concentration to one of low concentration (with the concentration gradient). The carrier molecule probably undergoes a change in shape to transport the substance. The process does not require ATP.

cell does not need to expend energy because the substance moves from a high to a low concentration. The process differs from simple diffusion in its use of carriers.

In some cases, molecules that bacteria need are too large to be transported into the cells by the methods just described. Most bacteria, however, produce enzymes that can break down large molecules into simpler ones (such as proteins into amino acids, and polysaccharides into simple sugars). Such enzymes, which are released by the bacteria into the surrounding medium, are appropriately called *extracellular enzymes.* Once the enzymes degrade the large molecules, the subunits are transported by permeases into the cell. For example, specific carriers retrieve DNA bases, such as the purine guanine, from extracellular media and bring them into the cell's cytoplasm.

Osmosis

Osmosis is the net movement of solvent molecules across a selectively permeable membrane from an area in which the solvent molecules are highly concentrated to an area of low concentration. In living systems, the chief solvent is water.

Osmosis may be demonstrated with the apparatus shown in Figure 4.15a and b. A sack constructed from cellophane, which is a selectively permeable membrane, is filled with a solution of 20% sucrose (table sugar). The opening of the cellophane sack is plugged with a rubber stopper, through which a glass tubing is fitted. The cellophane sack is placed into a beaker containing distilled water. Initially, the concentrations of water on either side of the membrane are different. Because of the sucrose molecules, the concentration of water is lower inside the cellophane sack than outside. Therefore, water moves from the beaker (where its concentration is higher) into the cellophane sack (where its concentration is lower).

There is no movement of sugar out of the cellophane sack into the beaker, however, because the cellophane is impermeable to molecules of sugar—the sugar molecules are too large to go through the pores of the membrane. As water moves into the cellophane sack, the sugar solution becomes increasingly dilute, and, because the cellophane sack is at maximum expansion as a result of its increased volume of water, water begins to move up the glass tubing. In time, the water that has accumulated in the cellophane sack and the glass tubing exerts a downward pressure that forces water molecules out of the cellophane sack and back into the beaker. This movement of water through

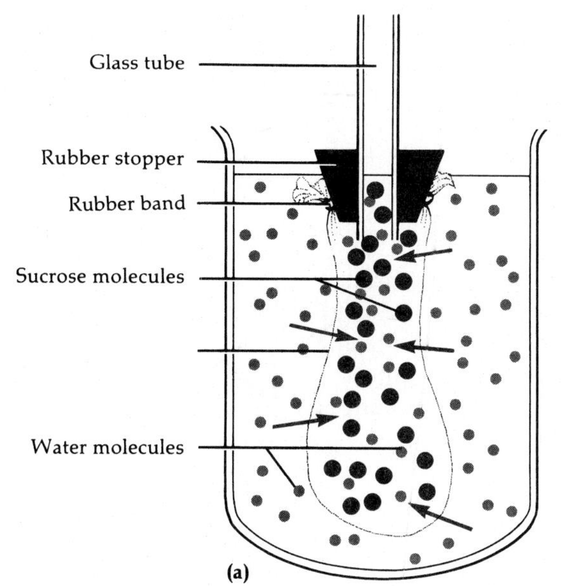

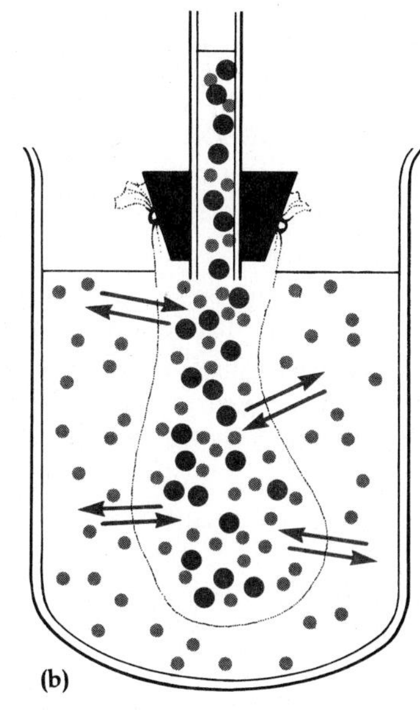

FIGURE 4.15 The principle of osmosis. (a) Setup at the beginning of the experiment. Water molecules start to move from the beaker into the sack along the concentration gradient. **(b)** Setup at equilibrium. The osmotic pressure exerted by the solution in the sack pushes water molecules from the sack back into the beaker to balance the rate of water entry into the sack. The final height of solution in the glass tube in (b) is a measure of the osmotic pressure. **(c)–(e)** Effects of various solutions on bacterial cells.

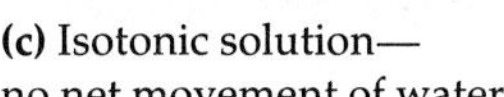

(c) Isotonic solution—
no net movement of water

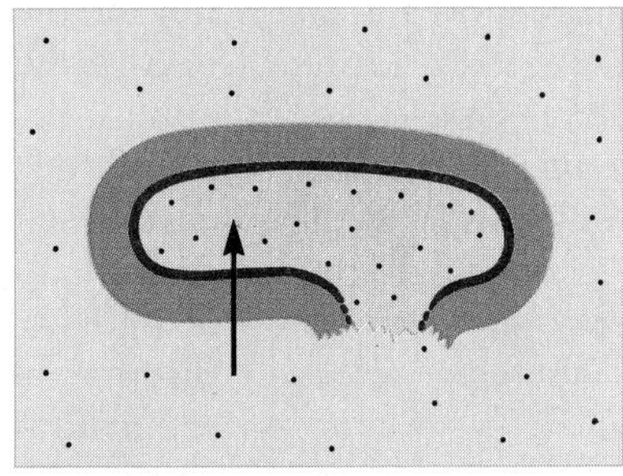

(d) Hypotonic solution—
water moves into the cell and
may cause the cell to burst if the wall
is weak or damaged (osmotic lysis)

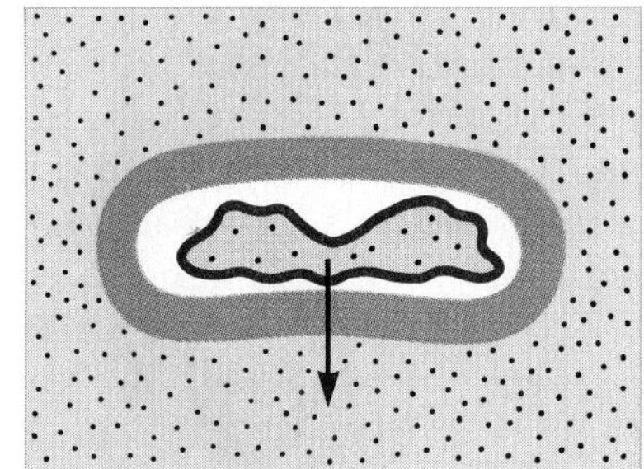

(e) Hypertonic solution—
water moves out of the cell,
causing it to shrink (plasmolysis)

a selectively permeable membrane produces a pressure called osmotic pressure. **Osmotic pressure** is the pressure required to prevent the movement of pure water (water with no solutes) into a solution containing some solutes. In other words, osmotic pressure is the pressure needed to stop the flow of water across the selectively permeable membrane (cellophane). When water molecules leave and enter the cellophane sack at the same rate, equilibrium is reached.

A bacterial cell may be subjected to three kinds of osmotic solutions—isotonic, hypotonic, or hypertonic. An **isotonic solution** (*iso* means equal) is one in which the overall concentrations of solutes are the same on both sides of the membrane. Water leaves and enters the cell at the same rate (no net change); the cell's contents are in equilibrium with the solution outside the cell wall (Figure 4.15c).

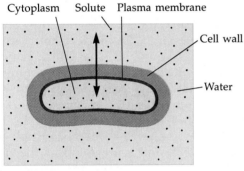

Earlier we mentioned that lysozyme and certain antibiotics damage bacterial cell walls, causing the cells to rupture, or lyse. Such rupturing occurs because bacterial cytoplasm usually contains such a strong concentration of solutes that, when the wall is weakened or removed, additional water enters the cell by osmosis. The damaged (or removed) cell wall cannot constrain the swelling of the cytoplasmic membrane, and the membrane bursts. This is an example of osmotic lysis caused by immersion in a hypotonic solution. A **hypotonic solution** (*hypo* means under or less) outside the cell is a medium whose concentration of solutes is lower than that inside the cell. Most bacteria live in hypotonic solutions, and swelling is contained by the cell wall. Cells with weak cell walls, such as gram-negative bacteria, may burst or undergo osmotic lysis as a result of excessive water intake (Figure 4.15d).

A **hypertonic solution** (*hyper* means above) is a medium having a higher concentration of solutes than the cell has. Most bacterial cells placed in a hypertonic solution shrink and collapse or plasmolyze because water leaves the cells by osmosis (Figure 4.15e). Keep in mind that the terms *isotonic, hypotonic,* and *hypertonic* describe the concentration of solutions outside the cell *relative* to the concentration inside the cell.

Active Transport

In performing **active transport,** the cell *uses energy* in the form of adenosine triphosphate (ATP) to move substances across the plasma membrane. The movement of a substance in active transport is usually from outside to inside, even though the concentration might be much higher inside the cell. Like facilitated diffusion, active transport depends on carrier proteins in the plasma membrane (see Figure 4.14). There appears to be a different carrier for each transported substance or group of closely related transported substances.

Group Translocation

In active transport, the substance that crosses the membrane is not altered by transport across the membrane. In **group translocation,** a special form of active transport that occurs exclusively in procaryotes, the substance is chemically altered during transport across the membrane. Once the substance is altered and inside the cell, the plasma membrane becomes impermeable to it, so it remains inside the cell. This is important because a cell can accumulate various substances even though they may be in low concentrations outside the cell. Group translocation requires energy supplied by high-energy phosphate compounds, such as phosphoenolpyruvic acid (PEP) found in *E. coli.*

One example of group translocation is the transport of the sugar glucose, which is often used in growth media for bacteria. During transport of the glucose molecule across the membrane by a specific carrier protein, a phosphate group is added to the sugar. The phosphorylated form of glucose, which cannot be transported out, can then be used in the cell's metabolic pathways.

Simple diffusion and facilitated diffusion are useful mechanisms for transporting substances into cells when the concentration of the substances is greater outside the cell. However, when a bacterial cell is in an environment in which nutrients are in low concentration, the cell must utilize active transport and group translocation to accumulate the needed substances.

Some eucaryotic cells (those without cell walls) can use two additional active transport processes, not found in bacteria, called phagocytosis and pinocytosis. Both processes are described in later sections.

CYTOPLASM

For a procaryotic cell, the term **cytoplasm** refers to the internal matrix of the cell contained inside the plasma membrane (see Figure 4.5). Cytoplasm is about 80% water and contains primarily proteins (enzymes), carbohydrates, lipids, inorganic ions, and many low-molecular-weight compounds. Inorganic ions are present in much higher concentrations in cytoplasm than in most media. Cytoplasm is thick, aqueous, semitransparent, and elastic. The major structures in the cytoplasm are DNA, particles called ribosomes, and reserve deposits called inclusions.

Procaryotic cytoplasm lacks certain features of eucaryotic cytoplasm, such as a cytoskeleton and cytoplasmic streaming. These features will be described later.

NUCLEAR AREA

The **nuclear area,** or **nucleoid,** of a bacterial cell (see Figure 4.5) contains a single long circular molecule of double-stranded DNA, the **bacterial chromosome** (see Figure 8.1a). This is the cell's genetic information, its DNA, carrying all the information required for the cell's structures and functions. Unlike the chromosomes of eucaryotic cells, bacterial chromosomes do not include histones and are not surrounded by a nuclear envelope (membrane). The nuclear area can be spherical, elongated, or dumbbell-shaped. In actively growing bacteria, as much as 20% of the cell volume is occupied by DNA, since such cells presynthesize nuclear material for future cells. The chromosome is attached to the plasma membrane. Proteins in the plasma membrane are believed to be responsible for replication of the DNA and segregation of the new chromosomes to daughter cells in cell division (see Figure 6.10).

Bacteria often contain, in addition to the bacterial chromosome, small circular, double-stranded DNA molecules called **plasmids** (see the F factor in Figure 8.25a). These are extrachromosomal genetic elements; that is, they are not connected to the main bacterial chromosome, and they replicate independently of chromosomal DNA. Research indicates that plasmids are associated with plasma membrane proteins. Plasmids usually contain from five to 100 genes, which are generally not crucial for the survival of the bacterium under normal environmental conditions, and plasmids may be gained or lost without harming the cell. Under certain conditions, however, plasmids are an advantage to cells. Plasmids may carry genes for such activities as antibiotic resistance, tolerance to toxic metals, production of toxins, and synthesis of enzymes. Plasmids can be transferred from one bacterium to another. In fact, plasmid DNA is used for gene manipulation in biotechnology.

RIBOSOMES

All eucaryotic and procaryotic cells contain **ribosomes,** which function as the sites of protein synthesis. Cells that have high rates of protein synthesis, such as those that are actively growing, have a great number of ribosomes. The cytoplasm of a procaryotic cell contains tens of thousands of these very small structures, which give the cytoplasm a granular appearance (see Figure 4.5).

Ribosomes are composed of two subunits, each subunit being composed of protein and a type of RNA called ribosomal RNA (rRNA). Procaryotic ribosomes differ from eucaryotic ribosomes in the number of proteins and rRNA molecules they contain and are somewhat smaller and less dense than ribosomes of eucaryotic cells. Accordingly, procaryotic ribosomes are called 70S ribosomes, and those of eucaryotic cells are known as 80S ribosomes (Figure 4.16). The letter *S* refers to Svedberg units, which indicate the relative rate of sedimentation during ultra-high-speed centrifugation. Sedimentation rate is a function of the size, weight, and shape of a particle. The subunits of a 70S ribosome are a small 30S subunit containing one molecule of rRNA and a larger 50S subunit containing two molecules of rRNA.

Several antibiotics, such as streptomycin, neomycin, and tetracyclines, exert their antimicrobial effects by inhibiting protein synthesis on the ribosomes. Because of differences in procaryotic and eucaryotic ribosomes, the microbial cell can be killed by the antibiotic while the eucaryotic host cell is unaffected.

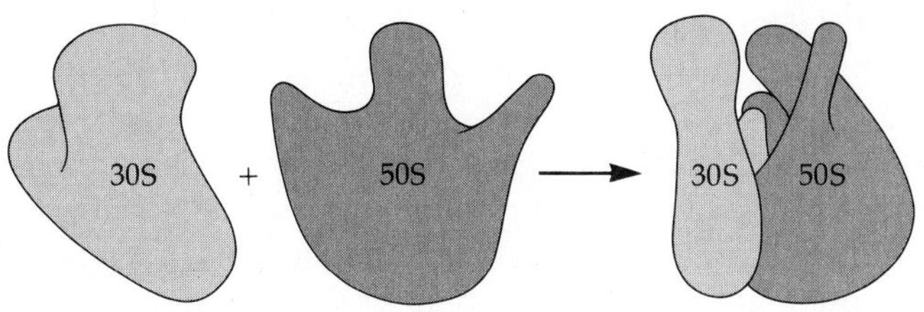

FIGURE 4.16 Procaryotic ribosome. A small 30S subunit **(a)** and a large 50S subunit **(b)** make up the complete 70S procaryotic ribosome **(c)**.

INCLUSIONS

Within the cytoplasm of procaryotic (and eucaryotic) cells are several kinds of reserve deposits, known as **inclusions.** Some inclusions are common to a wide variety of bacteria, whereas others are limited to a small number of species and therefore serve as a basis for identification. Among the more prominent bacterial inclusions are the following.

Metachromatic Granules

These inclusions sometimes stain red with certain blue dyes, such as methylene blue, and are collectively known as **volutin**. Volutin represents a reserve of inorganic phosphate (polyphosphate) that can be used in the synthesis of ATP. It is generally formed by cells that grow in phosphate-rich environments. Metachromatic granules are found in algae, fungi, and protozoans, as well as in bacteria. These granules are quite large and are characteristic of *Corynebacterium diphtheriae* (kô-rī-nē-bak-ti'rē-um dif-thi'rē-ī), the causative agent of diphtheria; thus, they have diagnostic significance.

Polysaccharide Granules

These inclusions typically consist of glycogen and starch, and their presence can be demonstrated when iodine is applied to the cells. In the presence of iodine, glycogen granules appear reddish brown and starch granules appear blue.

Lipid Inclusions

Lipid inclusions appear in various species of *Mycobacterium, Bacillus, Azotobacter* (ä-zō-tō-bak'tėr), *Spirillum* (spī-ril'lum), and other genera. A lipid that is commonly found as a storage material and unique to bacteria is the polymer *poly-β-hydroxybutyric acid.* Lipid inclusions are revealed by use of fat-soluble dyes, such as Sudan dyes.

Sulfur Granules

Certain bacteria, for example, the "sulfur bacteria," which belong to the genus *Thiobacillus,* derive energy by oxidizing sulfur and sulfur-containing compounds. These bacteria may deposit sulfur granules in the cell, where they serve as an energy reserve.

Carboxysomes

These are polyhedral and hexagonal inclusions that contain the enzyme ribulose 1,5-diphosphate carboxylase. Bacteria that use carbon dioxide as their sole source of carbon require this enzyme for carbon dioxide fixation during photosynthesis. Among the bacteria containing carboxysomes are nitrifying bacteria, cyanobacteria, and thiobacilli.

Gas Vacuoles

These are hollow cavities found in many aquatic procaryotes, including cyanobacteria, anoxygenic photosynthetic bacteria, and halobacteria. Each vacuole consists of rows of several individual *gas vesicles,* which are hollow cylinders covered by protein. The function of gas vacuoles is to maintain buoyancy so that the cells can remain at the depth in the water appropriate for them to receive sufficient amounts of oxygen, light, and nutrients.

For a discussion of inclusions in magnetotactic bacteria, see the box on p. 76.

ENDOSPORES

When essential nutrients are depleted, or when water is unavailable, certain gram-positive bacteria, such as those of the genera *Clostridium* and *Bacillus,* form specialized "resting" cells called **endospores.** Unique to bacteria, endospores are highly durable dehydrated cells with thick walls and additional layers. They are formed internal to the bacterial cell membrane. When released into the environment, they can survive extreme heat, lack of water, and exposure to many toxic chemicals and radiation. For example, 7500-year-old endospores of *Thermoactinomyces vulgaris* (thėr-mō-ak-tin-ō-mī'sēs vul-ga'ris) from the freezing muds of Elk Lake in Minnesota have germinated when rewarmed and placed in a nutrient medium. Although true endospores are found in gram-positive bacteria, one gram-negative species, *Coxiella burnetii* (käks-ē-el'lä bėr-ne'tē-ē), the cause of Q fever, forms endosporelike structures that resist heat and chemicals and can be stained with endospore stains.

The process of endospore formation within a vegetative (parent) cell is known as **sporulation** or **sporogenesis** (Figure 4.17a and b). It is not clear what biochemical events trigger this process. In the first observable stage of sporogenesis, a newly replicated bacterial chromosome and a small portion of cytoplasm are isolated by an ingrowth of the plasma membrane called a *spore septum.* The spore septum becomes a double-layered membrane that surrounds the chromosome and cytoplasm. This structure, entirely enclosed within the original cell, is called a *forespore.* Thick layers of peptidoglycan are laid down between the two membrane layers. Then a thick *spore coat* of protein forms around the outside membrane. It is this coat that is responsible for the resistance of endospores to many harsh chemicals.

The diameter of the endospore may be the same as, smaller than, or larger than the diameter of the vegetative cell. Depending on the species, the endospore might be located *terminally* (at one end), *subterminally* (near one end), or *centrally* inside the vegetative cell. When the endospore matures, the vegetative cell wall dissolves (lyses), killing the cell, and the endospore is freed.

Most of the water present in the forespore cytoplasm is eliminated by the time sporogenesis is complete, and endospores do not carry out metabolic reactions. The highly dehydrated endospore core contains only DNA, small amounts of RNA, ribosomes, enzymes, and a few important small molecules. The latter include a strikingly large amount of an organic acid called *dipicolinic acid* (found in the cytoplasm), which is accompanied by a large amount of calcium ions. These

FIGURE 4.17 Endospores. **(a)** Sporogenesis, the process of endospore formation. **(b)** Electron micrograph of an endospore in *Bacillus sphaericus.*

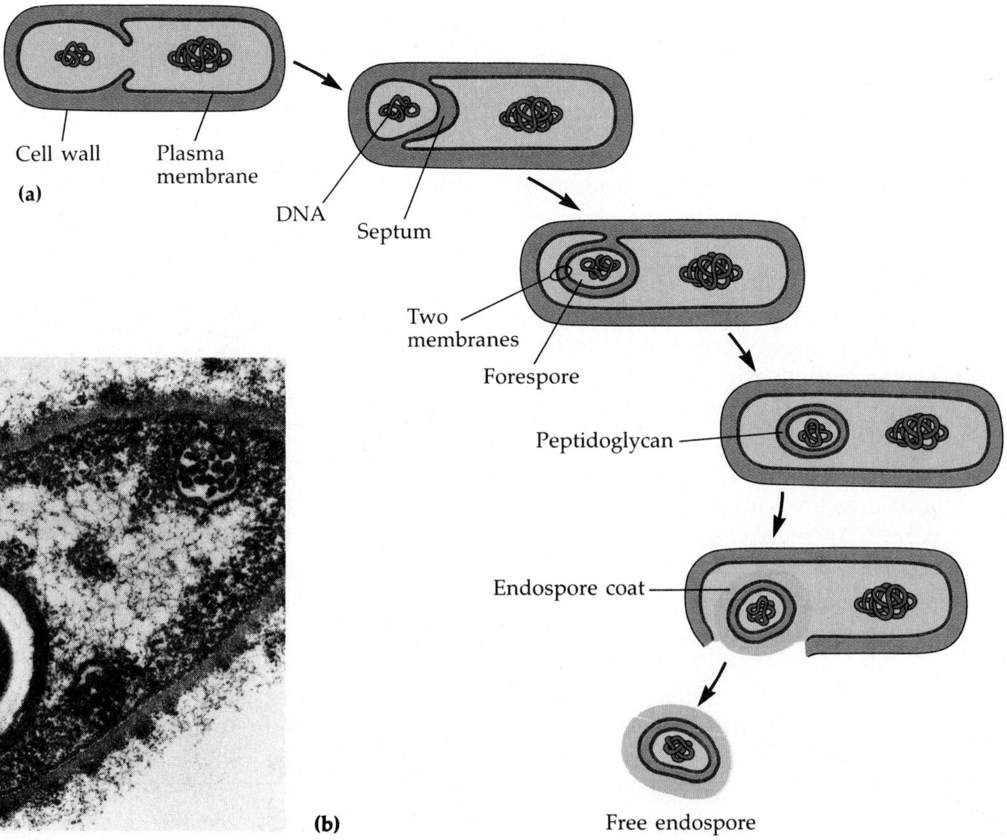

(a) Cell wall; Plasma membrane; DNA; Septum; Two membranes; Forespore; Peptidoglycan; Endospore coat; Free endospore

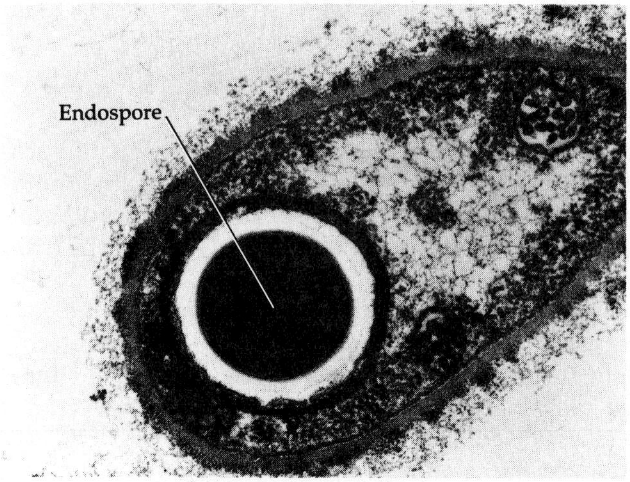

(b) Endospore

cellular components will be essential for resuming metabolism later.

Endospores can remain dormant for thousands of years. An endospore returns to its vegetative state by a process called **germination.** Germination is triggered by physical or chemical damage to the endospore's coat. The endospore's enzymes then break down the extra layers surrounding the endospore, water enters, and metabolism resumes. Because one vegetative cell forms a single endospore, which, after germination, remains one cell, sporogenesis in bacteria is *not* a means of reproduction. There is no increase in the number of cells.

Endospores are important from a clinical viewpoint and in the food industry because they are resistant to processes that normally kill vegetative cells. Such processes include heating, freezing, desiccation, use of chemicals, and radiation. Whereas most vegetative cells are killed by temperatures above 70°C, endospores can survive in boiling water for several hours or more. Endospores of thermophilic (heat-loving) bacteria can survive in boiling water for 19 hours. Endospore-forming bacteria are a problem in the food industry because they are likely to survive underprocessing, and if conditions for growth occur, some species produce toxins and disease. Special methods used to control organisms that produce endospores are discussed in Chapter 7.

As noted in Chapter 3, endospores are difficult to stain for detection. Thus, a specially prepared stain must be used along with heat. (The Schaeffer–Fulton endospore stain is commonly used.)

Having examined the functional anatomy of the procaryotic cell, we will now look at the functional anatomy of the eucaryotic cell.

THE EUCARYOTIC CELL

As mentioned earlier, eucaryotic organisms include algae, protozoans, fungi, higher plants, and animals. The eucaryotic cell (Figure 4.18) is typically larger and structurally more complex than the procaryotic cell. By comparing the structure of the procaryotic cell in Fig-

FIGURE 4.18 Highly schematic diagram of a composite eucaryotic cell.

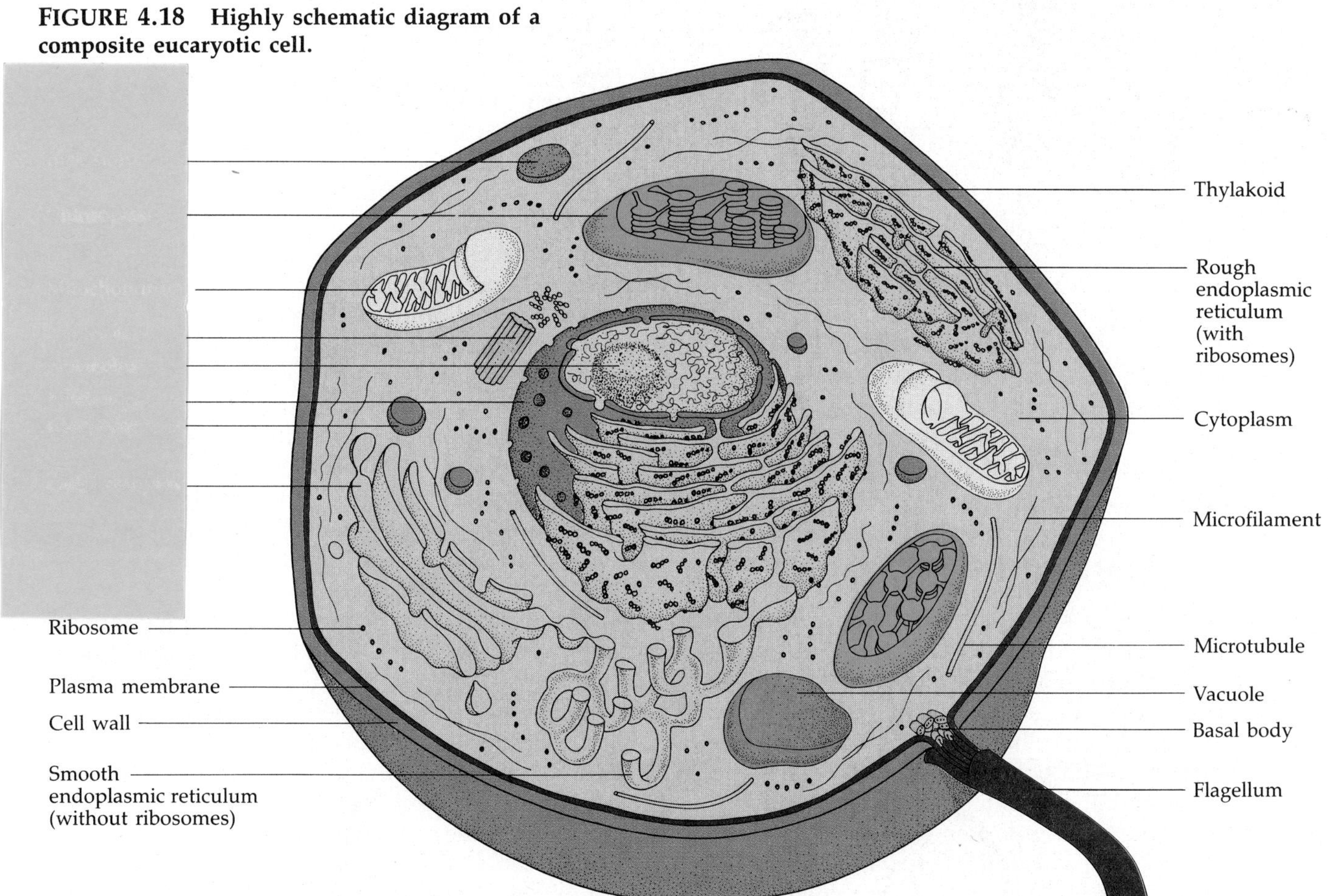

ure 4.5 with that of the eucaryotic cell, we can see the differences between the two types of cells.

To review briefly, eucaryotic cells contain membrane-bounded organelles; procaryotic cells do not. **Organelles** are specialized structures that perform specific functions. Although both types of cells carry on the same basic functions, procaryotic cells generally do not localize these functions in specific organelles. The genetic material (DNA) of procaryotic cells is not membrane-bounded or associated in a regular way with protein. The genetic material of eucaryotic cells *is* membrane-bounded, organized into chromosomes, and closely associated with histones and other proteins.

The following discussion of eucaryotic cells will parallel our discussion of procaryotic cells by starting with structures that extend external to the cell. The principal differences between procaryotic and eucaryotic cells are summarized in Table 4.2 at the end of the discussion.

Flagella and Cilia

Many types of eucaryotic cells have projections that are used for cellular locomotion or for moving substances along the surface of the cell. These projections contain cytoplasm and are enclosed by the plasma membrane. If the projections are few and are long in relation to the size of the cell, they are called **flagella.** If the projections are numerous and short, resembling hairs, they are called **cilia**. Both flagella and cilia consist of nine pairs of microtubules arranged in a ring, plus two single microtubules in the center of the ring. Euglenoid algae use a flagellum for locomotion, whereas protozoans, such as *Paramecium* (păr-ä-mē'sē-um), use cilia for locomotion (Figure 4.19a and b). A procaryotic flagellum rotates, but a eucaryotic flagellum moves in a wavelike manner (Figure 4.19c). To help keep foreign material out of the lungs, ciliated cells of the human respiratory system move the material along the surface of the cells in the bronchial tubes and trachea toward the throat and mouth (see Figure 16.3).

The Cell Wall and Glycocalyx

Some eucaryotic cells have cell walls, although these walls are generally much simpler than those of procaryotic cells. Most algae have cell walls consisting of the polysaccharide *cellulose* (as do all plants). Cell walls of some fungi also contain cellulose, but in most fungi the principal structural component of the cell wall is the polysaccharide *chitin,* a polymer of N-acetylglucosamine (NAG) units. (Chitin is also the main structural component of the exoskeleton of crustaceans and insects.) The cell walls of yeasts contain the polysaccharides *glucan* and *mannan.* In eucaryotes that lack a cell wall, the plasma membrane may be the outer covering; however, cells that have direct contact with the environment may have coatings outside the plasma membrane. Protozoans do not have a typical cell wall; instead, they have a flexible outer covering called a *pellicle.*

In other eucaryotic cells, including animal cells, the plasma membrane is covered by a **glycocalyx** ("sugar coat"), a layer of material containing substantial amounts of sticky carbohydrates. Some of these carbohydrates are covalently bonded to proteins and lipids in the plasma membrane, forming glycoproteins and glycolipids that anchor the glycocalyx to the cell. The glycocalyx strengthens the cell surface, helps attach cells together, and may contribute to cell-to-cell recognition.

An important clinical consideration is that eucaryotic cells do not contain peptidoglycan, the framework of the procaryotic cell wall. This is significant medically because antibiotics, such as penicillins and cephalosporins, act against peptidoglycan and therefore do not affect human eucaryotic cells.

The Plasma (Cytoplasmic) Membrane

In eucaryotic cells that lack a cell wall, the **plasma membrane** is the external covering of the cell. In function and basic structure, the eucaryotic and procaryotic plasma membranes are very similar. There are, however, differences in the proteins. In eucaryotes, carbohydrates serve as receptor sites that assume a role in such functions as cell-to-cell recognition. These carbohydrates also provide adherence for bacteria. Eucaryotic plasma membranes also contain *sterols,* complex lipids not found in procaryotic plasma membranes (with the exception of the mycoplasmas). Sterols seem to be associated with the ability of the membranes to resist lysis due to increased osmotic pressure.

Substances can cross eucaryotic and procaryotic plasma membranes by simple diffusion, facilitated diffusion, osmosis, or active transport. Group translocation does not occur in eucaryotic cells. However, eucaryotic cells can utilize an additional mechanism called **endocytosis.** This occurs when a segment of the plasma membrane surrounds a particle or large molecule, encloses it, and brings it into the cell. Endocytosis is one of the ways that viruses can enter animal cells.

Two very important types of endocytosis are *phagocytosis* and *pinocytosis.* During phagocytosis, cellular projections called pseudopods engulf particles and bring them into the cell. Phagocytosis is used by white blood cells to destroy bacteria and foreign substances (see Figure 16.8 and further discussion in Chapter 16).

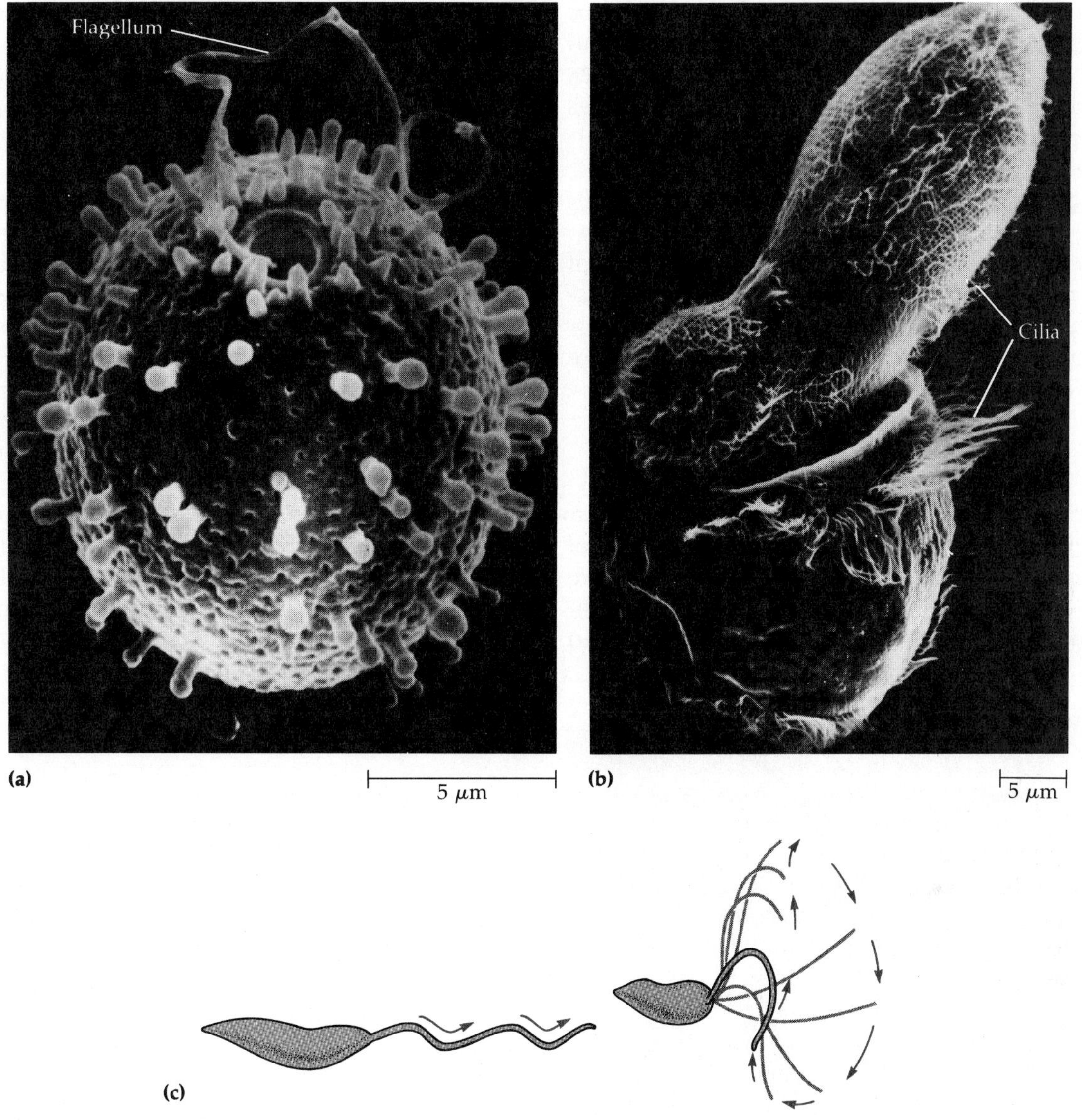

FIGURE 4.19 Eucaryotic flagella and cilia. **(a)** The freshwater euglenoid *Trachelomonas* sp. Its long, whiplike flagellum is emerging from the cell body at the top of this scanning electron micrograph. **(b)** Ciliate eats ciliate. *Didinium nasutum,* the lower cell with two rows of cilia around its body, has begun to engulf a cilia-covered cell of *Paramecium multimicronucleatum.* **(c)** Movement of a eucaryotic flagellum.

In pinocytosis, the plasma membrane folds inward, bringing extracellular fluid into the cell along with whatever substances are dissolved in the fluid.

Cytoplasm

The **cytoplasm** of the eucaryotic cell encompasses the matrix inside the cell membrane and outside the nucleus (see Figure 4.18). The cytoplasm is the substance in which various cellular components are found. A major difference between eucaryotic and procaryotic cytoplasm is that eucaryotic cytoplasm has a complex internal structure, consisting of exceedingly small rods called *microfilaments* and *intermediate filaments* and cylinders called *microtubules.* Together, they form the **cytoskeleton.** The cytoskeleton and other cytoplasmic components are held together by a three-dimensional membrane of fine filaments called the *microtrabecular lattice.* It provides support and shape, organizes chemical reactions that occur in the cytoplasm, and assists in transporting substances through the cell and even moving the entire cell, as in phagocytosis. The movement of eucaryotic cytoplasm from one part of the cell to another, which helps to distribute nutrients and move the cell over a surface, is called *cytoplasmic streaming*. Another difference between procaryotic and eucaryotic cytoplasm is that many of the important enzymes found in the cytoplasmic fluid of procaryotes are sequestered in the organelles of eucaryotes.

Organelles

Organelles are specialized structures characteristic of eucaryotic cells. They include the nucleus, endoplasmic reticulum, Golgi complex, mitochondria, chloroplasts, lysosomes, centrioles, and ribosomes that are larger and denser than procaryotic ribosomes.

NUCLEUS

The most characteristic eucaryotic organelle is the nucleus (see Figure 4.18). The **nucleus** is usually spherical or oval (Figure 4.20), is frequently the largest structure in the cell, and contains almost all of the cell's hereditary information (DNA). Some DNA is also found in mitochondria and chloroplasts of photosynthetic organisms.

The nucleus is separated from the cytoplasm by a double membrane called the **nuclear envelope.** Each of the two membranes resembles the plasma membrane in structure. Minute *pores* in the nuclear membrane allow the nucleus to communicate with the membranous network in the cytoplasm, called the endoplasmic reticulum (see next section). Substances entering and exiting the nucleus are believed to pass through the tiny pores. Within the nuclear envelope is a gelatinous fluid called **nucleoplasm.** One or more spherical bodies called **nucleoli** are also present (singular is *nucleolus*). These structures are a center for the synthesis of ribosomal RNA, an essential constituent of ribosomes (see next section).

Finally, there is the DNA, which is combined with a number of proteins, including several basic proteins called **histones** and nonhistones. The combination of about 165 base pairs of DNA and 9 molecules of histones is referred to as a *nucleosome*. When the cell is not

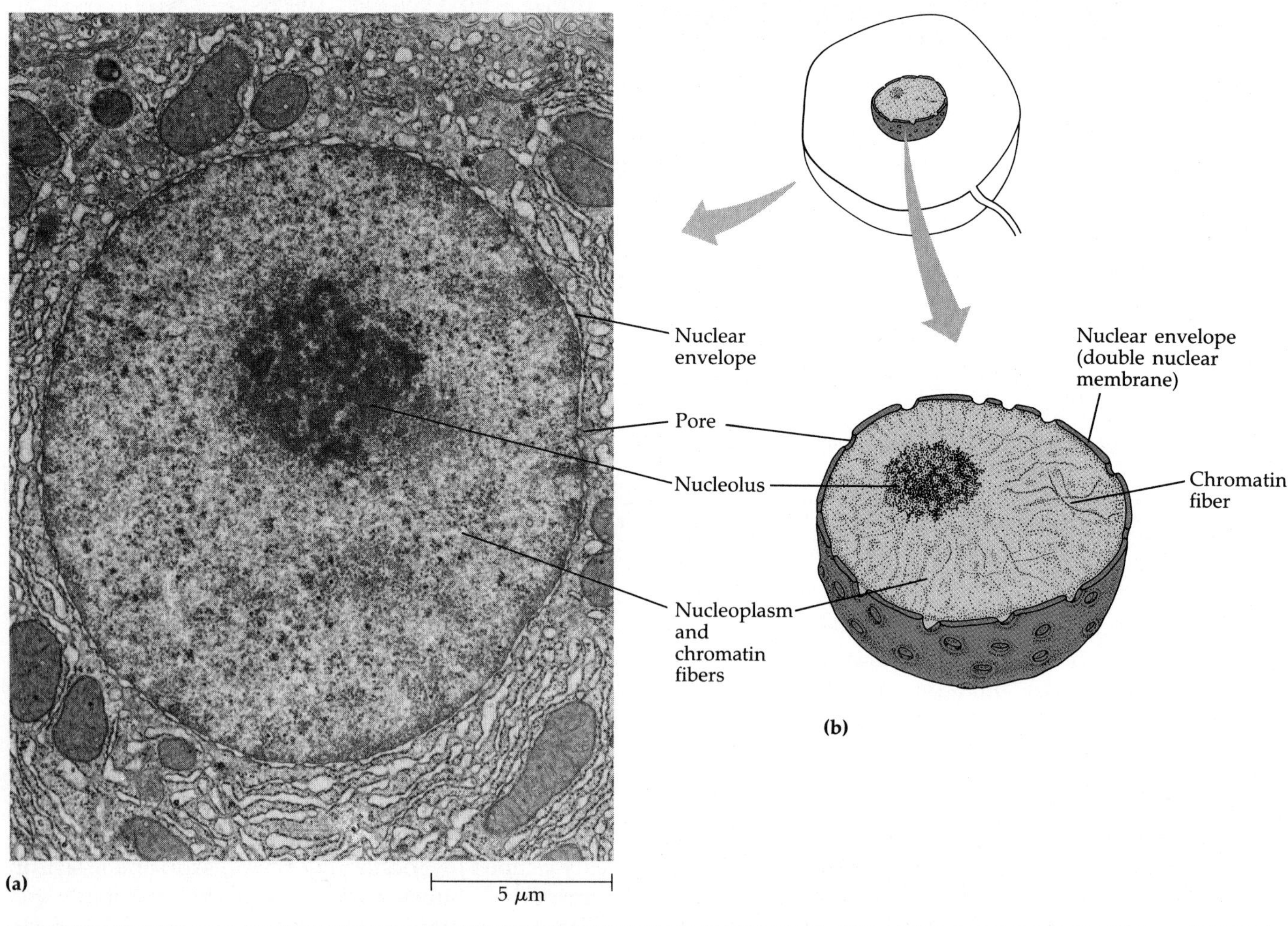

FIGURE 4.20 Eucaryotic nucleus. **(a)** Transmission electron micrograph of a nucleus. **(b)** Diagram of a nucleus.

reproducing, the DNA and its associated proteins appear as a threadlike mass called **chromatin.** Prior to nuclear division, the chromatin coils into shorter and thicker rodlike bodies called **chromosomes.** Nucleosomes are the basic structural units of chromosomes. Procaryotic "chromosomes" do not undergo this process, do not have histones, are not formed of nucleosomes, and are not enclosed in a nuclear envelope.

Eucaryotic cells divide by two elaborate mechanisms called mitosis and meiosis (see Chapter 8). Neither occurs in procaryotic cells.

ENDOPLASMIC RETICULUM

Within the cytoplasm, there is a system consisting of pairs of parallel membranes that enclose narrow vesicles or sacs of various shapes. This system, not present in procaryotes, is known as the **endoplasmic reticulum,** or **ER** (Figure 4.21).

The ER is a network of tubes running throughout the cytoplasm that is continuous with both the plasma membrane and the nuclear membrane (see Figure 4.18). It is thought that the ER provides a surface area for chemical reactions, a pathway for the transportation of molecules within the cell, and a storage area for synthesized molecules. The ER plays a role in both lipid synthesis and protein synthesis.

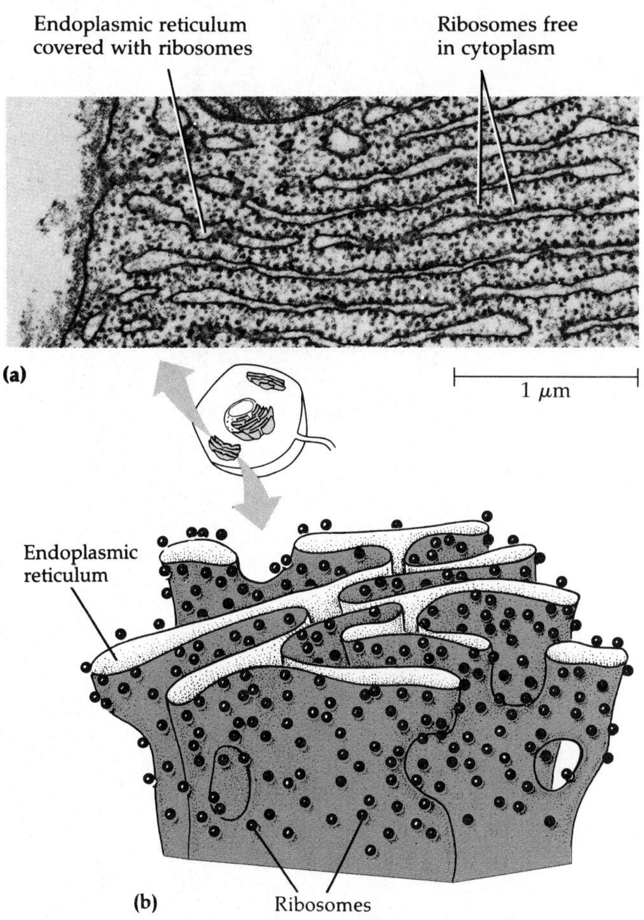

FIGURE 4.21 Endoplasmic reticulum and ribosomes. **(a)** Transmission electron micrograph of endoplasmic reticulum and ribosomes in cross section. **(b)** Three-dimensional drawing of endoplasmic reticulum and ribosomes.

RIBOSOMES

Attached to the outer surface of some of the ER are ribosomes, which are also found free in the cytoplasm (Figure 4.21). As in procaryotes, ribosomes are the sites of protein synthesis in the cell.

The ribosomes of the eucaryotic ER and cytoplasm are somewhat larger and denser than those of procaryotic cells. These eucaryotic ribosomes are 80S ribosomes. An 80S ribosome consists of a large 60S subunit containing three molecules of rRNA and a smaller 40S subunit with one molecule of rRNA. Chloroplasts and mitochondria contain 70S ribosomes, which may indicate their evolution from procaryotes. This hypothesis is discussed later in this chapter. The role of ribosomes in protein synthesis will be discussed in more detail in Chapter 8.

GOLGI COMPLEX

Another organelle found in the cytoplasm of eucaryotic cells is the **Golgi complex.** This structure usually consists of four to eight flattened sacs (**cisternae**) whose expanded ends separate from the complex and are known as secretory vesicles. The cisternae are stacked one on the other (Figure 4.22). The Golgi complex is sometimes connected to the ER. One function of the Golgi complex is to package and secrete (release from the cell) certain proteins and lipids. Another function is to synthesize carbohydrates and to combine carbohydrates and proteins to form complexes called *glycoproteins.* Carbohydrates and glycoproteins are also secreted from the cell in vesicles.

MITOCHONDRIA

Spherical or rod-shaped organelles called **mitochondria** appear throughout the cytoplasm of eucaryotic cells (see Figure 4.18). A mitochondrion consists of a double membrane similar in structure to the plasma membrane (Figure 4.23). The outer mitochondrial membrane is smooth, but the inner mitochondrial membrane is arranged in a series of folds called **cristae.** The center of the mitochondrion is a semifluid substance called the *matrix.* Because of the nature and arrangement of the cristae, the inner membrane provides an enormous surface area on which chemical reactions can occur. Some proteins that function in cellular respiration, including the enzyme that makes

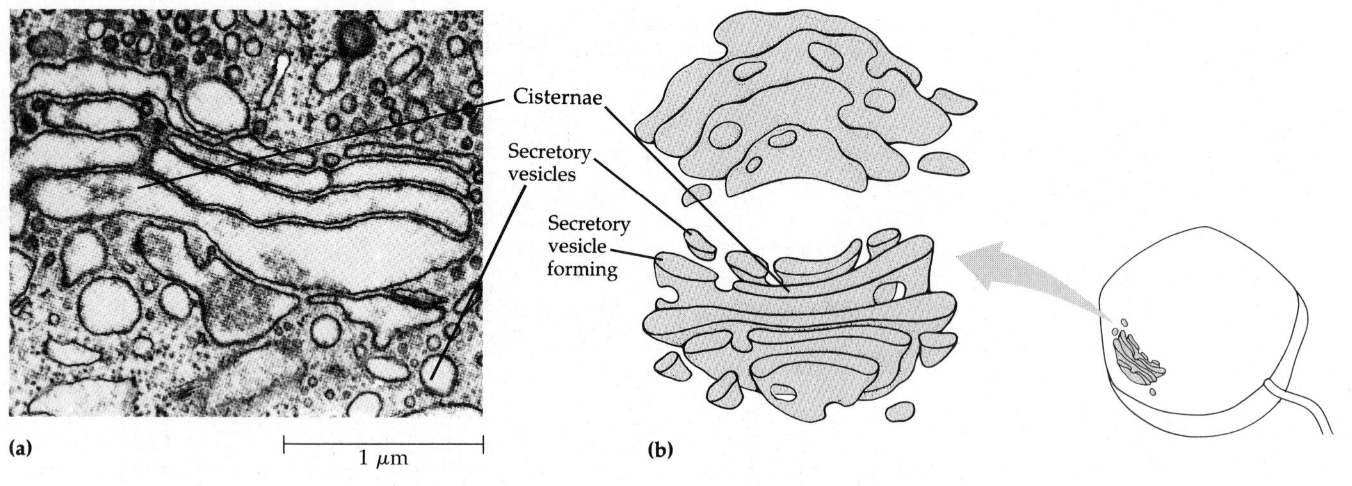

FIGURE 4.22 Golgi complex. **(a)** Transmission electron micrograph of a Golgi complex in cross section. **(b)** Three-dimensional drawing of a Golgi complex.

ATP, are located on the cristae of the inner mitochondrial membrane, and many of the metabolic steps involved in cellular respiration are concentrated in the matrix (see Chapter 5). Mitochondria are frequently called the "powerhouses of the cell" because of their central role in the production of ATP.

Mitochondria contain 70S ribosomes and some DNA of their own, as well as the machinery necessary to replicate, transcribe, and translate the information encoded by their DNA. In addition, mitochondria can reproduce more or less on their own by growing and dividing in two.

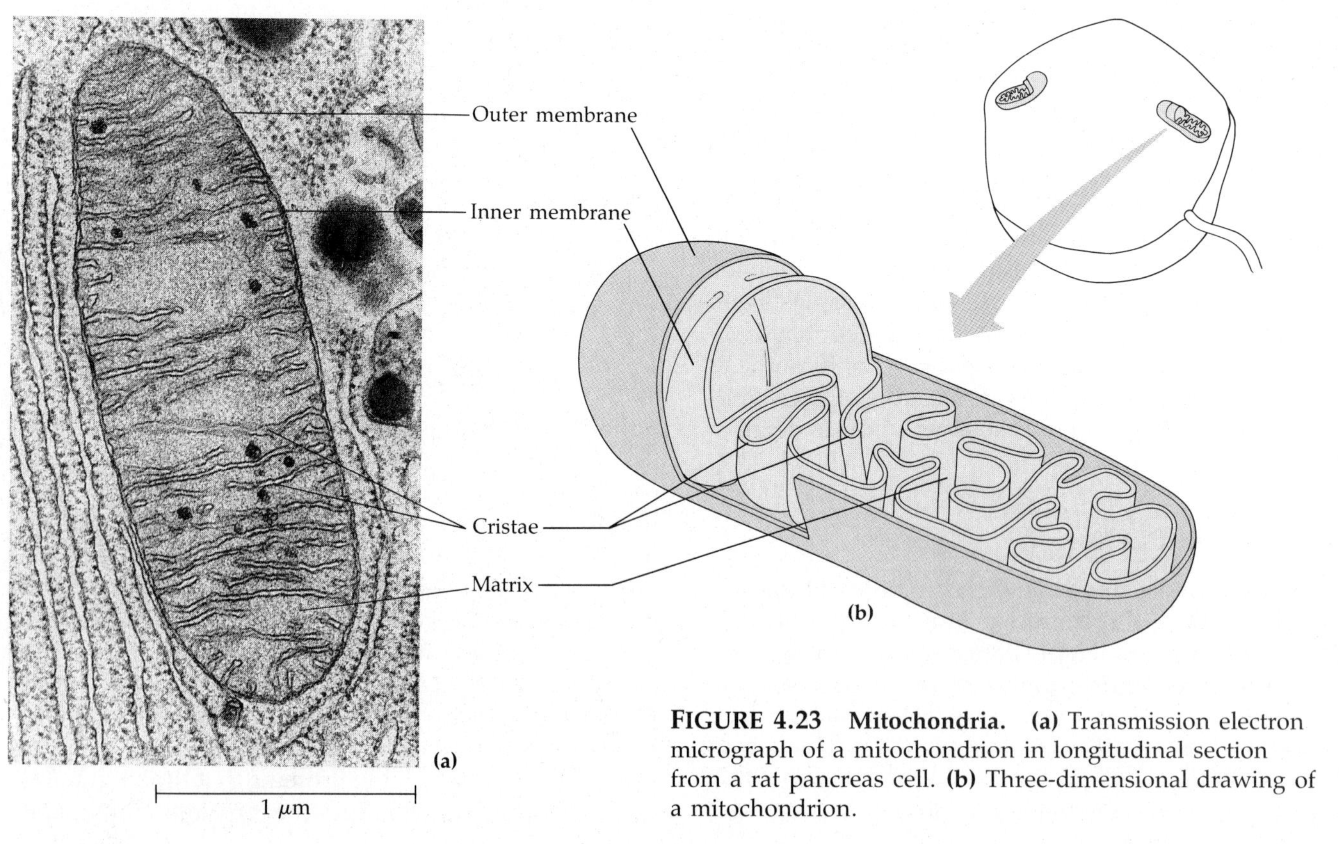

FIGURE 4.23 Mitochondria. **(a)** Transmission electron micrograph of a mitochondrion in longitudinal section from a rat pancreas cell. **(b)** Three-dimensional drawing of a mitochondrion.

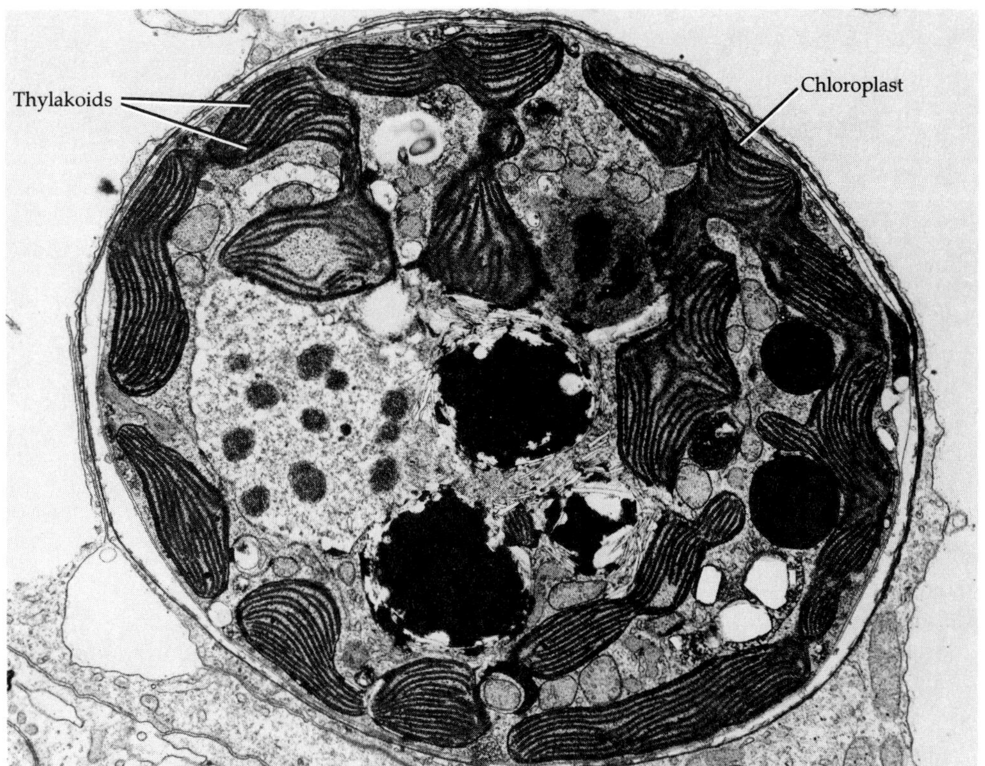

FIGURE 4.24 Chloroplasts. Transmission electron micrograph of chloroplasts in a cell of the alga *Gymnodynium microadriaticum,* which lives within cells of a coral. Photosynthesis occurs in the chloroplasts, providing nutrients for both the alga and the coral. The light-trapping pigments are located on the thylakoids.

CHLOROPLASTS

Algae and green plants contain a unique organelle called a **chloroplast** (Figure 4.24). A chloroplast is a membrane-bounded structure that contains both the pigment chlorophyll and the enzymes required for the light-gathering phases of photosynthesis (see Chapter 5). The chlorophyll is contained in flattened membrane sacs called *thylakoids* (see Figure 4.18). Stacks of thylakoids are called *grana* (the singular is *granum*). Like mitochondria, chloroplasts contain 70S ribosomes, DNA, and enzymes involved in protein synthesis. They are capable of multiplying on their own within the cell. Interestingly, the way both chloroplasts and mitochondria multiply—by increasing in size and then dividing in two—is strikingly reminiscent of bacterial multiplication.

LYSOSOMES

As viewed under the electron microscope, **lysosomes,** which are formed from Golgi complexes, appear to be membrane-enclosed spheres in eucaryotic cytoplasm. Unlike mitochondria, lysosomes have only a single membrane and lack detailed structure (see Figure 4.18). But they contain powerful digestive enzymes capable of breaking down many kinds of molecules. Moreover, these enzymes can also digest bacteria that enter the cell. Human white blood cells, which use phagocytosis to ingest bacteria, contain large numbers of lysosomes.

CENTRIOLES

Located near the nucleus are a pair of cylindrical structures, the **centrioles** (see Figure 4.18). Each centriole is a ring of nine evenly spaced bundles, each of which consists of three microtubules. The two centrioles are situated so that the long axis of one is at right angles to the long axis of the other. Centrioles play a role in eucaryotic cell division and as basal bodies in the formation of cilia and flagella.

The principal differences between procaryotic and eucaryotic cells are presented in Table 4.2. Most of the differences have been discussed in this chapter; a few will be treated in greater detail in subsequent chapters.

Evolution of Eucaryotes

At present, there are two major views on the origin of eucaryotes. According to the **autogenous hypothesis,** specialized internal membranes that derived from a procaryotic plasma membrane evolved into organelles characteristic of eucaryotic cells. According to this view, the nuclear envelope, endoplasmic reticulum, Golgi complex, and organelles bounded by single membranes (such as lysosomes) evolved this way.

TABLE 4.2 Principal Differences Between Procaryotic and Eucaryotic Cells

CHARACTERISTIC	*PROCARYOTIC*	*EUCARYOTIC*
Size of cell	Typically 0.20–2 μm in diameter	Typically 10–100 μm in diameter
Nucleus	No nuclear membrane or nucleoli	True nucleus, consisting of nuclear membrane and nucleoli
Membrane-bounded organelles	Absent	Present; examples include lysosomes, Golgi complex, endoplasmic reticulum, mitochondria, and chloroplasts
Flagella	Consist of two protein building blocks	Complex; consist of multiple microtubules
Glycocalyx	Present as a capsule or slime layer	Present in some cells that lack a cell wall
Cell wall	Usually present; chemically complex (typical bacterial cell wall includes peptidoglycan)	When present, chemically simple
Plasma membrane	No carbohydrates and generally lacks sterols	Sterols and carbohydrates that serve as receptors present
Cytoplasm	No cytoskeleton or cytoplasmic streaming	Cytoskeleton; cytoplasmic streaming
Ribosomes	Smaller size (70S)	Larger size (80S); small size (70S) in organelles
Chromosome (DNA) arrangement	Single circular chromosome; lacks histones	Multiple linear chromosomes with histones
Cell division	Binary fission	Mitosis
Sexual reproduction	No meiosis; transfer of DNA fragments only	Involves meiosis

These organelles are sometimes referred to as the *endomembrane system.* The other view, the **endosymbiotic hypothesis,** proposes that the forerunners of eucaryotic cells were associations of small, symbiotic (living-together) procaryotic cells living within larger procaryotic cells. This hypothesis focuses primarily on the origin of mitochondria and chloroplasts (which multiply similarly to bacteria). The proposed ancestors of mitochondria are believed to have been oxygen-requiring heterotrophic bacteria. Heterotrophs obtain food molecules by eating other organisms or their by-products. Chloroplasts are believed to be descendants of photosynthetic procaryotes. Photosynthetic organisms use light energy to produce food molecules. It has been suggested that oxygen-requiring heterotrophs and photosynthetic procaryotes both gained entry into the large procaryotic cell as undigested prey or internal parasites. It is possible that the origin of eucaryotes from procaryotes took place by a combination of the two mechanisms. The nuclear envelope and other cellular membranes may have evolved as outlined in the autogenous hypothesis, whereas mitochondria and chloroplasts may have evolved by the endosymbiotic mechanism.

Our next concern is to examine microbial metabolism. In Chapter 5, you will learn about the importance of enzymes to microbes and the ways microbes produce and utilize energy.

STUDY OUTLINE

Introduction (pp. 69–70)

1. Procaryotic and eucaryotic cells are similar in their chemical composition and chemical reactions.
2. Procaryotic cells lack membrane-bounded organelles (including a nucleus).
3. Peptidoglycan is found in procaryotic cell walls and not in eucaryotic cell walls.

THE PROCARYOTIC CELL (pp. 71–89)

1. Bacteria are unicellular, and most of them multiply by binary fission.
2. Bacterial species are differentiated by morphology, chemical composition, nutritional requirements, biochemical activities, and source of energy.

Size, Shape, and Arrangement of Bacterial Cells (pp. 71–72)

1. Most bacteria are from 0.20 to 2.0 μm in diameter and from 2 to 8 μm in length.
2. The three basic bacterial shapes are coccus (spheres), bacillus (rods), and spiral (twisted).
3. Pleomorphic bacteria can assume several shapes.

Structures External to the Cell Wall (pp. 72–77)

GLYCOCALYX (pp. 72–73)

1. The glycocalyx (capsule or slime layer) is a gelatinous polysaccharide and/or polypeptide covering.
2. Capsules may protect pathogens from phagocytosis.
3. Capsules provide adherence to surfaces, prevent desiccation, and may provide nutrients.

FLAGELLA (pp. 73–75)

1. Flagella are relatively long filamentous appendages consisting of a filament, hook, and basal body.
2. Procaryotic flagella rotate to push the cell.
3. Motile bacteria exhibit taxis—positive taxis is movement toward an attractant and negative taxis is movement away from a repellent.

AXIAL FILAMENTS (pp. 75–76)

1. Spiral cells that move by means of an axial filament are called spirochetes.
2. Axial filaments are similar to flagella, except that they wrap around the cell.

FIMBRIAE AND PILI (pp. 76–77)

1. Fimbriae and pili are short, thin appendages.
2. Fimbriae help cells adhere to surfaces.
3. Pili join cells for the transfer of DNA from one cell to another.

The Cell Wall (pp. 77–81)

COMPOSITION AND CHARACTERISTICS (pp. 77–79)

1. The cell wall surrounds the plasma membrane and protects the cell from changes in osmotic pressure.
2. The bacterial cell wall consists of peptidoglycan, a polymer consisting of NAG and NAM and short chains of amino acids.
3. Gram-positive cell walls consist of many layers of peptidoglycan and also contain teichoic acids.
4. Gram-negative bacteria have a lipoprotein–lipopolysaccharide–phospholipid outer membrane surrounding a thin peptidoglycan layer.
5. The outer membrane protects the cell from phagocytosis and penicillin, lysozyme, and other chemicals.
6. Porins are proteins that permit small molecules to pass through the outer membrane; specific channel proteins allow other molecules to move through the outer membrane.
7. The lipopolysaccharide component of the outer membrane consists of sugars that function as antigens and lipid A, which is an endotoxin.

ATYPICAL CELL WALLS (pp. 79–80)

1. *Mycoplasma* is a bacterial genus that naturally lacks cell walls.
2. Archaeobacteria have pseudomurein; they lack peptidoglycan.
3. L forms are mutant bacteria with defective cell walls.

DAMAGE TO THE CELL WALL (pp. 80–81)

1. In the presence of lysozyme, gram-positive cell walls are destroyed and the remaining cellular contents are referred to as a protoplast.
2. In the presence of lysozyme, gram-negative cell walls are not completely destroyed and the remaining cellular contents are referred to as a spheroplast.
3. Protoplasts and spheroplasts are subject to osmotic lysis.
4. Antibiotics such as penicillin interfere with cell wall synthesis.

Structures Internal to the Cell Wall (pp. 81–89)

PLASMA (CYTOPLASMIC) MEMBRANE (pp. 81–83)

1. The plasma membrane encloses the cytoplasm and is a phospholipid bilayer with protein (fluid mosaic).
2. The plasma membrane is selectively permeable.
3. Plasma membranes carry enzymes for metabolic reactions, such as nutrient breakdown, energy production, and photosynthesis.
4. Mesosomes—irregular infoldings of the plasma membrane—are now considered artifacts.
5. Plasma membranes can be destroyed by alcohols and polymyxins.

MOVEMENT OF MATERIALS ACROSS MEMBRANES (pp. 83–86)

1. Movement across the membrane may be by passive processes, in which materials move from higher to lower concentration and no energy is expended by the cell.
2. In simple diffusion, molecules and ions move until equilibrium is reached.
3. In facilitated diffusion, substances are transported by permeases across membranes from high to low concentration.
4. Osmosis is the movement of water from high to low concentration across a selectively semipermeable membrane until equilibrium is reached.
5. In active transport, materials move from low to high concentrations by permeases, and the cell must expend energy.
6. In group translocation, energy is expended to modify chemicals and transport them across the membrane.

CYTOPLASM (p. 86)

1. Cytoplasm is the fluid component inside the plasma membrane.
2. The cytoplasm is mostly water, with inorganic and organic molecules, DNA, ribosomes, and inclusions.

NUCLEAR AREA (p. 86)

1. The nuclear area contains the DNA of the bacterial chromosome. Bacteria can also contain plasmids, which are extrachromosomal DNA circles.

RIBOSOMES (pp. 86–87)

1. The cytoplasm of a procaryote contains numerous 70S ribosomes; ribosomes consist of rRNA and protein.
2. Protein synthesis occurs at ribosomes; this can be inhibited by certain antibiotics.

INCLUSIONS (p. 87)

1. Inclusions are reserve deposits found in procaryotic and eucaryotic cells.
2. Among the inclusions found in bacteria are metachromatic granules (inorganic phosphate), polysaccharide granules (usually glycogen or starch), lipid inclusions, sulfur granules, carboxysomes (ribulose 1,5-diphosphate carboxylase), and gas vacuoles.

ENDOSPORES (pp. 88–89)

1. Endospores are resting structures formed by some bacteria for survival during adverse environmental conditions.
2. The process of endospore formation is called sporulation; the return of an endospore to its vegetative state is called germination.

THE EUCARYOTIC CELL (pp. 89–96)

Flagella and Cilia (p. 90)

1. Flagella are few and long in relation to cell size; cilia are numerous and short.
2. Flagella and cilia are used for motility, and cilia also move substances along the surface of the cells.
3. Both flagella and cilia contain a "nine pairs + two pairs" arrangement of microtubules.

The Cell Wall and Glycocalyx (p. 90)

1. The cell walls of most algae and some fungi consist of cellulose.
2. The main material of fungal cell walls is chitin.
3. Yeast cell walls consist of glucan and mannan.
4. Animal cells are surrounded by a glycocalyx; this makes the cell stronger and provides a means of attachment to other cells.

The Plasma (Cytoplasmic) Membrane (pp. 90–91)

1. Like the procaryotic plasma membrane, the eucaryotic plasma membrane is a phospholipid bilayer containing proteins.
2. Eucaryotic plasma membranes contain carbohydrates attached to the proteins and sterols not found in procaryotic cells (except *Mycoplasma* cells).
3. Eucaryotic cells can move materials across the plasma membrane by the passive processes used by procaryotes, in addition to active transport and endocytoses (phagocytosis and pinocytosis).

Cytoplasm (pp. 91–92)

1. The cytoplasm of eucaryotic cells includes everything inside the plasma membrane and external to the nucleus.
2. The chemical characteristics of the cytoplasm of eucaryotic cells resemble that of the cytoplasm of procaryotic cells.
3. Eucaryotic cytoplasm has a cytoskeleton and exhibits cytoplasmic streaming.

Organelles (pp. 92–95)

1. Organelles are specialized membrane-bounded structures in the cytoplasm.
2. Organelles are characteristic of eucaryotic cells.
3. The nucleus, which contains DNA in the form of chromosomes, is the most characteristic eucaryotic organelle.
4. The nuclear membrane is connected to a system of parallel membranes in the cytoplasm, called the endoplasmic reticulum.
5. The endoplasmic reticulum provides a surface for chemical reactions, serves as a transporting network, and stores synthesized molecules.
6. 80S ribosomes are found in the cytoplasm or attached to the endoplasmic reticulum.
7. The Golgi complex consists of cisternae. It functions in secretion, carbohydrate synthesis, and glycoprotein formation.
8. Mitochondria are the primary sites of ATP production. They contain small 70S ribosomes and DNA, and they multiply by fission.
9. Chloroplasts contain chlorophyll and enzymes for photosynthesis. Like mitochondria, they contain 70S ribosomes and DNA and multiply by fission.
10. Lysosomes are formed from Golgi complexes. They store powerful digestive enzymes.
11. A pair of cylindrical structures called centrioles that are involved in cell division are near the nucleus.

Evolution of Eucaryotes (pp. 95–96)

1. Two hypotheses offer explanations of the origin of eucaryotes from procaryotic ancestors.
2. The autogenous hypothesis states that organelles (for example, the nucleus and Golgi complex) evolved from internal membranes derived from the plasma membrane.
3. According to the endosymbiotic hypothesis, organelles (for example, mitochondria and chloroplasts) evolved from symbiotic procaryotes living inside other procaryotic cells.

STUDY QUESTIONS

Review

1. Diagram each of the following flagellar arrangements:
   (a) Lophotrichous
   (b) Monotrichous
   (c) Peritrichous

2. Endospore formation is called ________. It is initiated by ________. Formation of a new cell from an endospore is called ________. This process is triggered by ________.

3. Draw the bacterial shapes listed in a, b, and c. Show how d, e, and f are special conditions of a, b, and c, respectively.
   (a) Spiral
   (b) Bacillus
   (c) Coccus
   (d) Spirochete
   (e) Streptobacilli
   (f) Staphylococci

4. List three differences between procaryotic and eucaryotic cells.

5. Match the structures to their functions.

____ Cell wall	(a) Attachment to surfaces
____ Endospore	(b) Cell wall formation
____ Fimbriae	(c) Motility
____ Flagella	(d) Protection from osmotic lysis
____ Glycocalyx	(e) Protection from phagocytes
____ Pili	(f) Resting
____ Plasma membrane	(g) Protein synthesis
____ Ribosomes	(h) Selectively permeable
	(i) Transfer of genetic material

6. Of what value is each of the following to the cell?
   (a) Metachromatic granules
   (b) Polysaccharide granules
   (c) Lipid inclusions
   (d) Sulfur granules
   (e) Carboxysomes
   (f) Gas vacuoles

7. Why is an endospore called a resting structure? Of what advantage is an endospore to a bacterial cell?

8. Explain what would happen in the following experiments.
   (a) A suspension of bacteria is placed in distilled water.
   (b) A suspension of bacteria is placed in distilled water with lysozyme.
   (c) A suspension of bacteria is placed in an aqueous solution of lysozyme and 10% sucrose.
   (d) A suspension of gram-negative bacteria is placed in distilled water with penicillin.

9. Compare and contrast the following:
   (a) Simple diffusion and facilitated diffusion
   (b) Active transport and facilitated diffusion
   (c) Active transport and group translocation

10. Why are mycoplasmas resistant to antibiotics that interfere with cell wall synthesis?

11. Answer the following questions using the diagrams below, which represent cross sections of bacterial cell walls.
    (a) Which diagram represents a gram-positive bacterium? How can you tell?
    (b) Explain how the Gram stain works to distinguish between these two types of cell walls.
    (c) Why does penicillin have no effect on most gram-negative cells?
    (d) How do essential molecules enter cells through each wall?

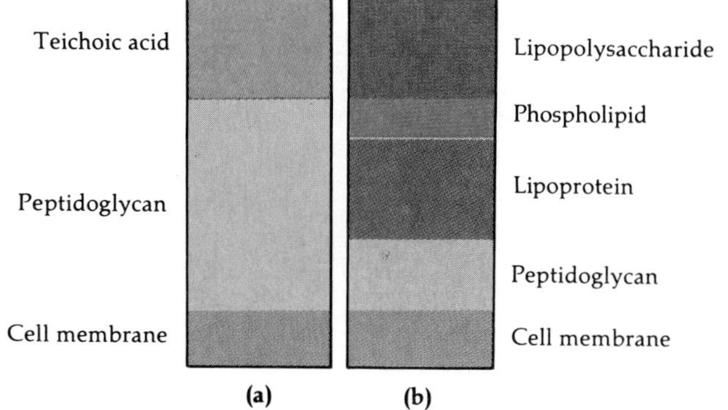

12. Starch is readily metabolized by many cells, but a starch molecule is too large to cross the plasma membrane. How does a cell get the glucose molecules from a starch polymer? How does the cell get these glucose molecules across the plasma membrane?

13. Match the following characteristics of eucaryotic cells with their functions.

____ Chloroplasts	(a) Intracellular transport
____ Endoplasmic reticulum	(b) Photosynthesis
____ Golgi complex	(c) ATP production
____ Lysosomes	(d) Digestive enzyme storage
____ Mitochondria	(e) Secretion

14. Eucaryotic cells might have evolved from early procaryotic cells living in close association. What do you know about eucaryotic organelles that would support this theory?

15. What process would a eucaryotic cell use to ingest a procaryotic cell? To ingest a virus?

Challenge

1. Why can procaryotic cells be smaller than eucaryotic cells and still carry on all the functions for life?

2. Within a three-day period at a large hospital, five patients undergoing hemodialysis developed fever and chills. *Pseudomonas aeruginosa* and *Klebsiella pneumoniae* were isolated from three of the patients. *P. aeruginosa, K. pneumoniae,* and *Enterobacter agglomerans* were isolated from the dialysis system. Why do all three bacteria cause similar symptoms?

3. Two types of procaryotic cells have been distinguished: eubacterial and archaeobacterial. How do these cells differ from each other? How are they similar?

FURTHER READING

Adler, J. "The sensing of chemicals by bacteria." *Scientific American* 235(4):40–47, April 1976. Describes the chemotactic response of bacteria using molecules that detect the presence of chemicals in the environment.

Allen, M.M. "Cyanobacterial cell inclusions." *Annual Review of Microbiology* 38:1–25, 1984. Discusses the effects of starvation and feeding on carboxysomes, gas vacuoles, lipid inclusions, and other cell inclusions.

Benz, R. "Structure and function of porins from gram-negative bacteria." *Annual Review of Microbiology* 42:359–393, 1988. Describes the chemical structure of porins and their role in allowing materials into a bacterial cell.

Costerton, J.W., R.T. Irvin, and K.-J. Cheng. "The bacterial glycocalyx in nature and disease." *Annual Review of Microbiology* 35:299–324, 1981. A thorough discussion of the functions of the glycocalyx.

Ferris, F.G., and T.J. Beveridge. "Functions of bacterial cell surface structure." *BioScience* 35:172–177, 1985. Includes ultrastructures and functions of the cell wall, capsules, flagella, pili, and a recently discovered structure, spinae.

Henrichsen, J. "Twitching motility." *Annual Review of Microbiology* 37:81–93, 1983. Describes a newly observed locomotion that is not related to flagella.

Moir, A., and D.A. Smith. "The genetics of bacterial spore germination." *Annual Review of Microbiology* 44:531–553, 1990. Provides a detailed description of the events of sporulation and germination.

Thomas, L. *The Lives of a Cell: Notes of a Biology Watcher*. New York: Viking Press, 1974. Thoughtful and entertaining essays by a physician and researcher, including essays on evolution of eucaryotic cells and the importance of microbes.

CHAPTER 5

Microbial Metabolism

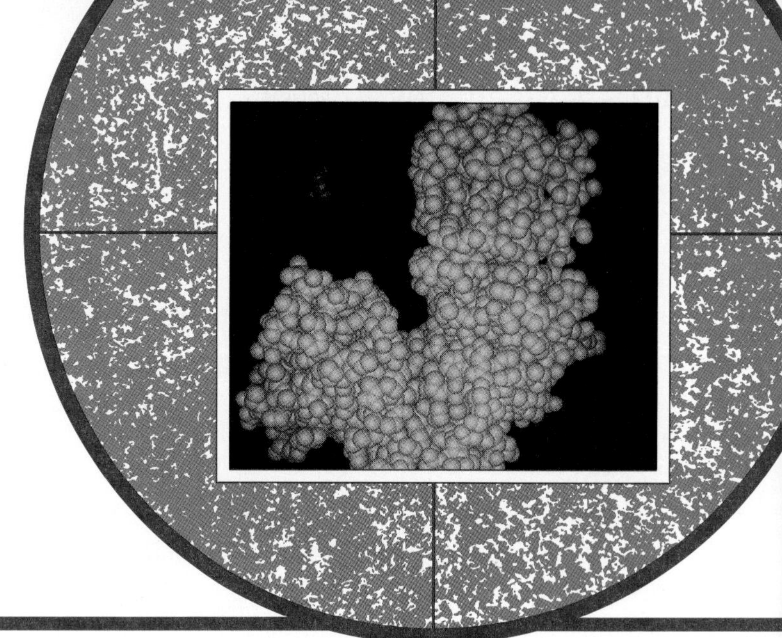

LEARNING OBJECTIVES

- Define metabolism, and describe the fundamental differences between anabolism and catabolism.
- Describe the mechanism of enzymatic action.
- List the factors that influence enzymatic activity.
- Explain what is meant by oxidation-reduction.
- List and provide examples of three types of phosphorylation reactions that generate ATP.
- Describe the chemical reactions of glycolysis.
- Compare and contrast aerobic and anaerobic respiration, cyclic and noncyclic photophosphorylation, and the light reaction and dark reaction of photosynthesis.
- Explain the products of the Krebs cycle.
- Describe the chemiosmotic model for ATP generation.
- Describe the chemical reactions of and list some products of fermentation.
- Categorize the various nutritional patterns among organisms.
- Describe the major types of anabolism and their relationship to catabolism.

Now that you are familiar with the structure of procaryotic cells, we can discuss the activities that allow these microbes to thrive. The life support activity of even the most structurally simple organism involves a large number of complex biochemical reactions. Most, although not all, of the biochemical processes of bacteria also occur in eucaryotic microbes and in the cells of multicellular organisms, including humans. However, the reactions that are unique to bacteria are fascinating because they allow microorganisms to do things we cannot do. For example, some bacteria (the chemoautotrophs) can grow on diets of such inorganic substances as carbon dioxide, iron, sulfur, hydrogen gas, and ammonia.

This chapter examines some representative chemical reactions that either produce energy (the catabolic reactions) or use energy (the anabolic reactions) in microorganisms. We will also look at how these various reactions are integrated within the cell.

Catabolic and Anabolic Reactions

We use the term **metabolism** to refer to the sum of all chemical reactions within a living organism. Because chemical reactions either release or require energy, metabolism can be viewed as an energy-balancing act. Accordingly, metabolism can be divided into two classes of chemical reactions—those that release energy and those that require energy. In living cells, the chemical reactions that release energy are generally the ones involved in **catabolism,** the breakdown of complex organic compounds into simpler ones. These reactions are called catabolic, or *degradative*, reactions. On the other hand, the energy-requiring reactions are mostly involved in **anabolism,** the building of complex

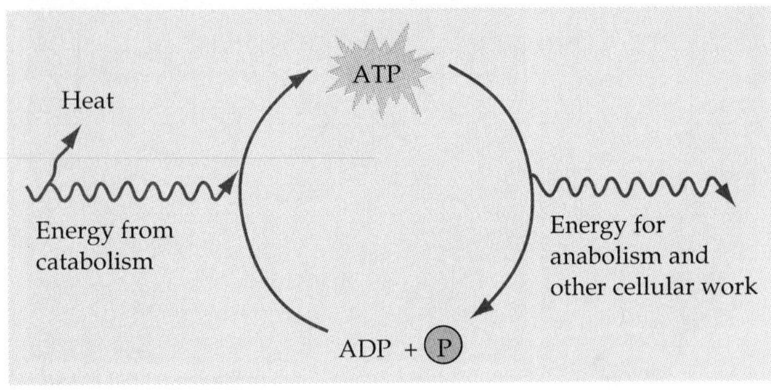

FIGURE 5.1 The relationship between anabolism, catabolism, and ATP.

organic molecules from simpler ones. These reactions are called **anabolic** or *biosynthetic* reactions. Anabolic processes often involve dehydration synthesis reactions (reactions that release water) and require energy to form new chemical bonds. Examples of anabolic processes are the formation of proteins from amino acids, nucleic acids from nucleotides, and polysaccharides from simple sugars. These biosynthetic reactions generate the materials for cell growth. Catabolic reactions are generally hydrolytic reactions (reactions that use water to break chemical bonds). Chemical bonds store energy; when they are broken, chemical energy is released. An example of catabolism occurs when cells break down sugars into carbon dioxide and water.

Catabolic reactions furnish the energy needed to drive anabolic reactions. This coupling of energy-requiring and energy-releasing reactions is made possible through the molecule adenosine triphosphate (ATP). (You can review its structure in Figure 2.20.) ATP stores energy derived from catabolic reactions and releases it later to drive anabolic reactions and perform other cellular work. Recall from Chapter 2 that a molecule of ATP consists of an adenine, a ribose, and three phosphate groups. When the terminal phosphate group is split from ATP, adenosine diphosphate (ADP) is formed, and energy is released to drive anabolic reactions. Using Ⓟ to represent a phosphate group, we can write this reaction as (ATP → ADP + Ⓟ + energy). Then, the energy from catabolic reactions is used to combine ADP and a Ⓟ to resynthesize ATP (ADP + Ⓟ + energy → ATP). Thus, anabolic reactions are coupled to ATP breakdown, and catabolic reactions are coupled to ATP synthesis. This concept of coupled reactions is very important. We will see why by the end of this chapter. For now, you should know that the chemical composition of a living cell is constantly changing; some molecules are being broken down while others are being synthesized. This balanced flow of chemicals and energy maintains the life of a cell.

The role of ATP in coupling anabolic and catabolic reactions is shown in Figure 5.1. Only part of the energy released in catabolism is actually available for cellular functions. Part of the energy is lost to the environment as heat. Because the cell must use energy to maintain life, it has a continuous need for new external sources of energy.

Before we discuss how cells produce energy, let us first consider the principal properties of a group of proteins involved in almost all biologically important chemical reactions. These proteins, the enzymes, were described briefly in Chapter 2. It is important to understand that a cell's metabolic pathways are determined by its enzymes, which are in turn determined by the cell's genetic makeup.

Enzymes

We indicated in Chapter 2 that chemical reactions occur when chemical bonds are formed or broken. In order for reactions to take place, atoms, ions, or molecules must collide. Whether a collision produces a reaction depends on the speed of the particles, the amount of energy required to trigger the reaction (called **activation energy**), and the specific configuration of the particles. The physiological temperature and pressure of organisms are too low for chemical reactions to occur quickly enough to maintain the life of the organism. Raising the temperature and pressure and the number of reacting molecules can increase the frequency of collisions and the rate of chemical reactions. However, such changes could damage or kill the organism. The living cell's solution to this problem is a class of proteins called **enzymes.** Enzymes can speed up chemical reactions in several ways. For example, an enzyme may bring two reactant molecules close together and may properly orient them to react. Whatever the method, the result is that the enzyme lowers the activation energy for the reaction without increasing the temperature or pressure inside the cell.

Substances that can speed up a chemical reaction without themselves being altered are called **catalysts.** In living cells, **enzymes** serve as biological catalysts. As catalysts, enzymes are specific. Each acts on a specific substance, called the enzyme's **substrate** (or substrates when there are two or more reactants), and each catalyzes only one reaction. For example, sucrose (table sugar) is the substrate of the enzyme sucrase, which catalyzes the hydrolysis of sucrose to glucose and fructose. The specificity of enzymes is made possible by their structures. Enzymes are generally large globular proteins that range in molecular weight from about 10,000 to several million. Each of the thousands of known enzymes has a characteristic three-dimensional shape with a specific surface configuration as a result of its primary, secondary, and tertiary structures (see

Figure 2.18). The unique configuration of each enzyme enables it to "find" the correct substrate from among the large number of diverse molecules in the cell.

Enzymes are extremely efficient. Under optimum conditions, they can catalyze reactions at rates 10^8 to 10^{10} times (up to 10 billion times) higher than those of comparable reactions without enzymes. The **turnover number** (maximum number of substrate molecules an enzyme molecule converts to product each second) is generally between 1 and 10,000 and can be as high as 500,000. For example, the enzyme lactate dehydrogenase, which removes hydrogen atoms from lactic acid, has a turnover number of 1000, whereas the enzyme DNA polymerase I, which participates in the synthesis of DNA, has a turnover number of 15.

Enzymes are subject to various cellular controls. Two main types are control of enzyme *synthesis* (see Chapter 8) and control of enzyme *activity* (how *much* enzyme is present vs. how *active* it is). Many enzymes exist in the cell in both active and inactive forms. The rate at which the inactive form becomes active or the active form becomes inactive is determined by the cellular environment.

NAMING ENZYMES

The names of enzymes usually end in *-ase*. All enzymes can be grouped into six classes, according to the type of chemical reaction they catalyze (Table 5.1). Enzymes within each of the major classes are named according to the more specific types of reactions they carry out. For example, the class called oxidoreductases is involved with oxidation–reduction reactions (described shortly). During some oxidation reactions, a substrate loses hydrogen; in others, a substrate gains oxygen. Enzymes in the oxidoreductase class that remove hydrogen are called *dehydrogenases;* those that add oxygen are called *oxidases*. As you will see later, dehydrogenase and oxidase enzymes have even more specific names, such as lactate dehydrogenase and cytochrome oxidase, depending on the specific substrates on which they act.

ENZYME COMPONENTS

Although some enzymes consist entirely of proteins, most consist of both a protein portion called an **apoenzyme** and a nonprotein component called a **cofactor.** Together, the apoenzyme and cofactor form a **holoenzyme,** or whole enzyme (Figure 5.2). If the cofactor is removed, the apoenzyme will not function. The cofactor can be a metal ion or a complex organic molecule called a **coenzyme.**

Coenzymes may assist the enzyme by accepting atoms removed from the substrate or donating atoms required by the substrate. Some coenzymes act as electron carriers, removing electrons from the substrate and donating them to other molecules in subsequent reactions. Many coenzymes are derived from vitamins (Table 5.2). Two of the most important coenzymes in cellular metabolism are **nicotinamide adenine dinucleotide (NAD^+)** and **nicotinamide adenine dinucleotide phosphate ($NADP^+$).** Both compounds contain derivatives of the B vitamin nicotinic acid (niacin), and both function as electron carriers. The flavin coenzymes, such as **flavin mononucleotide (FMN)** and **flavin adenine dinucleotide (FAD),** contain derivatives of the B vitamin riboflavin and are also electron carriers. Another important coenzyme, *coenzyme A (CoA)*, contains a derivative of pantothenic acid, another B vitamin. This coenzyme plays an important role in the synthesis

TABLE 5.1 Enzyme Classification Based on Type of Chemical Reaction

CLASS	*TYPE OF CHEMICAL REACTION*	*EXAMPLES*
Oxidoreductase	Oxidation-reduction in which oxygen and hydrogen are gained or lost	Cytochrome oxidase, lactate dehydrogenase
Transferase	Transfer of functional groups, such as an amino group, acetyl group, or phosphate group	Acetate kinase, alanine deaminase
Hydrolase	Hydrolysis (addition of water)	Lipase, sucrase
Lyase	Removal of groups of atoms without hydrolysis	Oxalate decarboxylase, isocitrate lyase
Isomerase	Rearrangement of atoms within a molecule	Glucose-phosphate isomerase, alanine racemase
Ligase	Joining of two molecules (using energy usually derived from breakdown of ATP)	Acetyl-CoA synthetase, DNA ligase

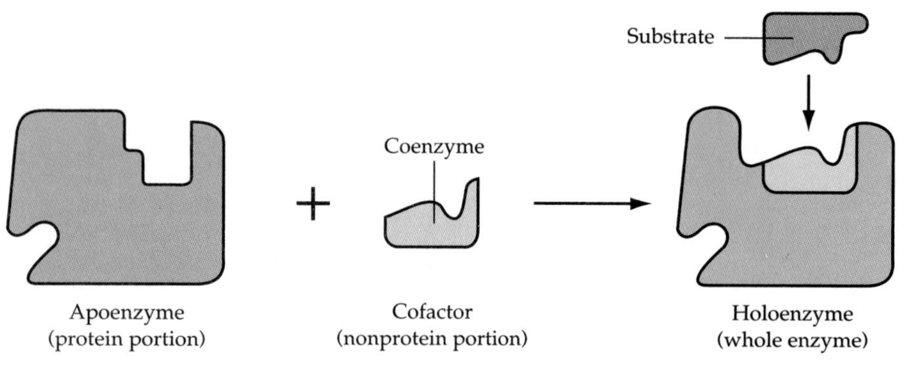

FIGURE 5.2 Components of a holoenzyme. Many enzymes require both an apoenzyme (protein portion) and a cofactor (nonprotein portion) to become active. The cofactor can be a metal ion or an organic molecule, called a coenzyme (as shown here). The apoenzyme and cofactor together make up the holoenzyme, or complete enzyme. The substrate is the reactant acted upon by the enzyme.

and breakdown of fats and in a series of oxidizing reactions called the Krebs cycle. We will come across all of these coenzymes in our discussion of metabolism.

As noted earlier, some cofactors are metal ions, including iron, copper, magnesium, manganese, zinc, calcium, and cobalt. Such cofactors may help catalyze a reaction by forming a bridge between the enzyme and a substrate. For example, magnesium (Mg^{2+}) is required by many phosphorylating enzymes (enzymes that transfer a phosphate group from ATP to another substrate). The Mg^{2+} can form a link between the enzyme and the ATP molecule. Most trace elements required by living cells are probably used in some such way to activate cellular enzymes.

MECHANISM OF ENZYMATIC ACTION

Although scientists do not completely understand how enzymes lower the activation energy, the general sequence of events in enzyme action is as follows (Figure 5.3a):

1. The surface of the substrate contacts a specific region of the surface of the enzyme molecule, called the **active site.**
2. A temporary intermediate compound forms, called an **enzyme–substrate complex.**
3. The substrate molecule is transformed by the rearrangement of existing atoms, the breakdown of the substrate molecule, or combination with another substrate molecule.
4. The transformed substrate molecules, the products of the reaction, are released from the enzyme molecule because they no longer fit in the active site of the enzyme.
5. The unchanged enzyme, now freed, reacts with other substrate molecules.

As a result of these events, an enzyme speeds up a chemical reaction.

As mentioned earlier, enzymes have *specificity* for particular substrates. For example, a specific enzyme

TABLE 5.2 Selected Vitamins and Their Functions

VITAMIN	FUNCTION
Vitamin B_1 (thiamine)	Part of coenzyme cocarboxylase; has many functions, including the metabolism of pyruvic acid
Vitamin B_2 (riboflavin)	Coenzyme in flavoproteins; active in electron transfers
Niacin (nicotinic acid)	Part of NAD molecule; active in electron transfers
Vitamin B_6 (pyridoxine)	Coenzyme in amino acid metabolism
Vitamin B_{12} (cyanocobalamin)	Coenzyme (methyl cyanocobalamide) involved in transfer of methyl groups; active in amino acid metabolism
Pantothenic acid	Part of coenzyme A molecule; involved in metabolism of pyruvic acid and lipids
Biotin	Involved in carbon dioxide fixation reactions, fatty acid synthesis
Folic acid	Coenzyme used in synthesis of purines and pyrimidines
Vitamin E	Needed for cellular and macromolecular syntheses
Vitamin K	Coenzyme used in electron transport (napthoquinones and quinones)

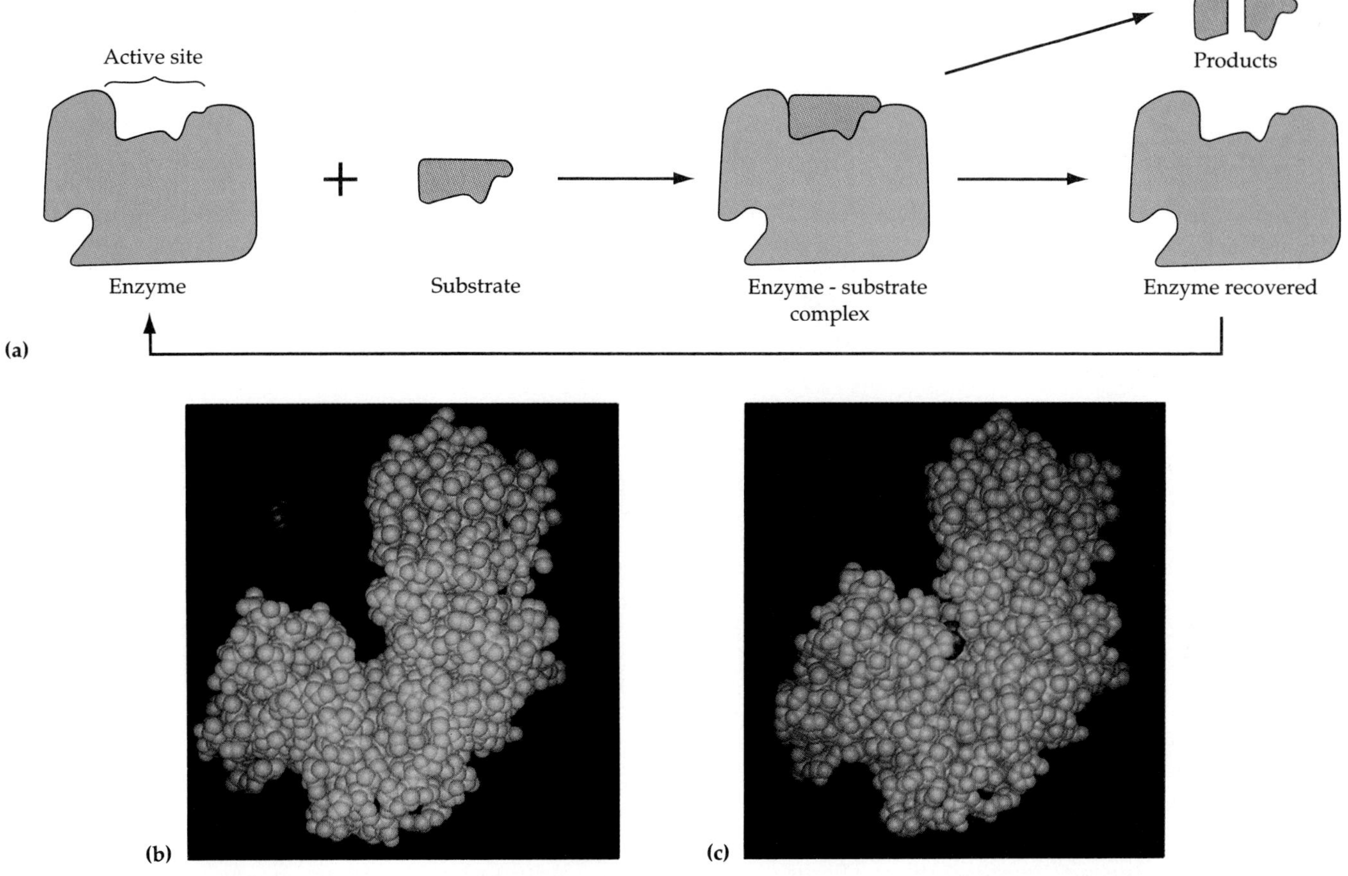

FIGURE 5.3 Mechanism of enzymatic action. (a) The substrate contacts the active site on the enzyme to form an enzyme–substrate complex. The substrate is then transformed into products, and the enzyme is recovered unchanged. In the example shown, the transformation into products involves a breakdown of the substrate into two products. Other transformations, however, may occur. **(b)** Molecular model of the enzyme (blue) and substrate (red). The active site of the enzyme can be seen here as a groove on the surface of the protein. **(c)** As they meet, the enzyme and substrate change shape slightly to fit together more tightly.

may be capable of hydrolyzing a peptide bond only between two specific amino acids. Other enzymes are capable of hydrolyzing starch but not cellulose; even though both starch and cellulose are polysaccharides composed of glucose subunits, the orientations of the subunits in the two polysaccharides differ. Enzymes have this specificity because the three-dimensional shape of the active site fits the substrate somewhat as a lock fits with its key. However, the active site and substrate are flexible, and they change shape somewhat as they meet to fit together more tightly (Figure 5.3b). The substrate is usually much smaller than the enzyme, and relatively few of the enzyme's amino acids make up the active site.

A certain compound can be a substrate for a number of different enzymes that catalyze different reactions, so the fate of a compound depends on the enzyme that acts upon it. Glucose 6-phosphate, a molecule that is important in cell metabolism, can be acted upon by at least four different enzymes, and each reaction will give a different product.

FACTORS INFLUENCING ENZYMATIC ACTIVITY

Several factors influence the activity of an enzyme. Among the more important are temperature, pH, substrate concentration, and inhibitors.

Temperature

The rate of most chemical reactions increases as the temperature increases. Molecules move more slowly at low than at higher temperatures and may not have enough energy to cause a chemical reaction. For enzymatic reactions, however, elevation beyond a certain temperature drastically reduces the rate of reaction

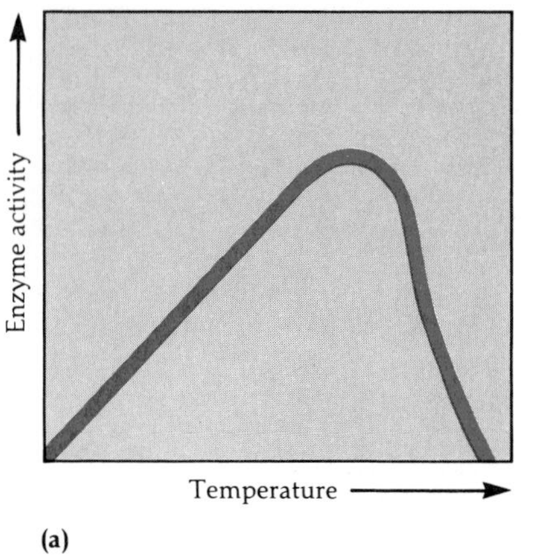

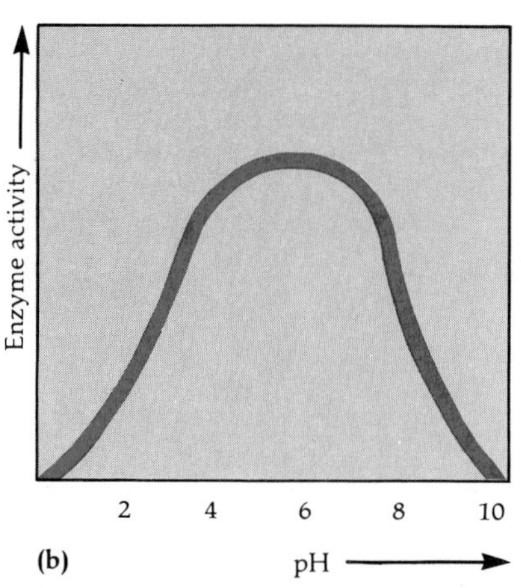

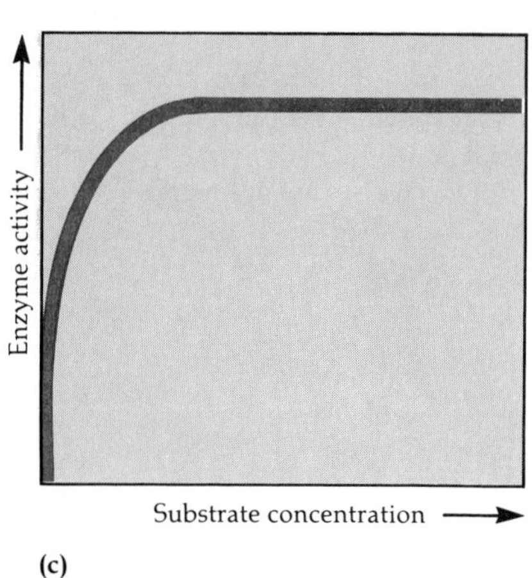

FIGURE 5.4 Factors that influence enzymatic activity, plotted for a hypothetical enzyme. **(a)** Temperature. The enzymatic activity (rate of reaction catalyzed by the enzyme) increases with increasing temperature until the enzyme, a protein, is denatured by heat and inactivated. At this point, the reaction rate falls steeply. **(b)** pH. The level at which the enzyme illustrated is most active is about pH 6. **(c)** Substrate concentration. With increasing concentration of substrate molecules, the rate of reaction increases until the active sites on all the enzyme molecules are filled, at which point the maximum rate of reaction is reached.

(Figure 5.4a). This decrease is due to the enzyme's **denaturation,** the loss of its characteristic three-dimensional structure (tertiary configuration) (Figure 5.5). Denaturation of a protein involves breakage of hydrogen bonds and other noncovalent bonds.

A common example of denaturation is the transformation of uncooked egg white (a protein called albumin) to a hardened state by heat. As might be expected, denaturation of an enzyme changes the arrangement of the amino acids in the active site, altering its shape and causing the enzyme to lose its catalytic ability and vital biological activity. In some cases, denaturation is partially or fully reversible. However, if denaturation continues until the enzyme has lost its solubility and coagulates (as with cooked albumin), the process is irreversible and the enzyme cannot regain its original properties. Enzymes can also be denatured by concentrated acids, bases, heavy-metal ions (such as lead, arsenic, or mercury), alcohol, and ultraviolet radiation.

pH

Most enzymes have a pH optimum at which their activity is characteristically maximal. Above or below this pH value, enzyme activity, and therefore the reaction rate, declines (Figure 5.4b). When the H^+ concentration (pH) in the medium is changed, many of the enzyme's amino acids are affected and the protein's three-dimensional structure is altered. Extreme changes in pH can cause denaturation.

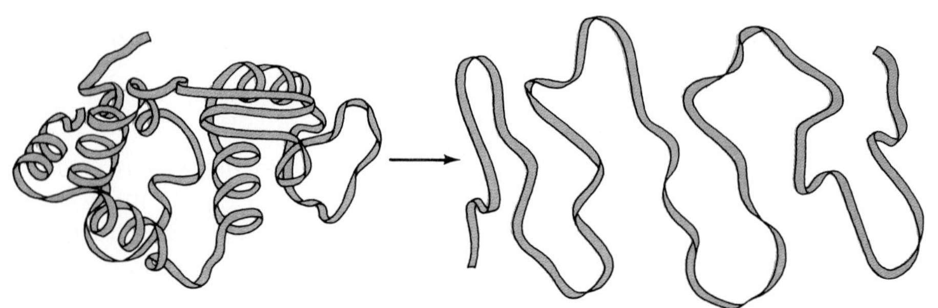

FIGURE 5.5 Denaturation of a protein. Breaking the noncovalent bonds (such as hydrogen bonds) that hold the active protein in its three-dimensional shape makes the denatured protein no longer functional.

Substrate Concentration

There is a maximum rate at which a certain amount of enzyme can catalyze a specific reaction. Only when the concentration of substrate(s) is extremely high can this maximum rate be attained. Under conditions of high substrate concentration, the enzyme is said to be **saturated;** that is, its active site is always occupied by substrate or product molecules. In this condition, a further increase in substrate concentration will not affect the reaction rate because all active sites are already in use (Figure 5.4c). If a substrate's concentration exceeds a cell's saturation level for a particular enzyme, the rate of reaction can be increased only if the cell produces additional enzyme molecules. However, under normal cellular conditions, enzymes are not saturated with substrate(s). At any given time, many of the enzyme molecules are inactive for lack of substrate; thus, the rate of reaction is likely to be influenced by the substrate concentration.

Inhibitors

As you will see in later chapters, an effective way to control the growth of bacteria is to control their enzymes. Certain poisons, such as cyanide, arsenic, and mercury, combine with enzymes and prevent them from functioning. As a result, the cells stop functioning and die.

Enzyme inhibitors are classified according to their mechanism of action as competitive inhibitors and noncompetitive inhibitors. **Competitive inhibitors** fill the active site of an enzyme and compete with the normal substrate for the active site. The competitive inhibitor is able to do this because its shape and chemical structure are similar to those of the normal substrate (Figure 5.6b). However, unlike the substrate, it does not undergo any reaction to form products. Some competitive inhibitors bind irreversibly to amino acids in the active site, preventing any further interactions with the substrate. Others bind reversibly, alternately occupying and leaving the active site. These slow the enzyme's interaction with the substrate. Reversible competitive inhibition can be overcome by increasing the substrate concentration. As active sites become available, more substrate molecules than competitive inhibitor molecules are available to attach to the active sites of enzymes. One good example of a competitive inhibitor is sulfanilamide (a sulfa drug), which inhibits the enzyme whose normal substrate is para-aminobenzoic acid (PABA).

NH_2 — O=S=O — (benzene ring) — NH_2 HO—C(=O) — (benzene ring) — NH_2

Sulfanilamide   PABA

PABA is an essential nutrient used by many bacteria in the synthesis of folic acid, a vitamin that functions as a coenzyme. When sulfanilamide is administered to bacteria, the enzyme that normally converts PABA to folic acid combines instead with the sulfanilamide. Folic acid is not synthesized, and the bacteria cannot grow. Since human cells do not use PABA to make their folic acid, sulfanilamide selectively kills bacteria but does not harm human cells.

Noncompetitive inhibitors do not compete with the substrate for the enzyme's active site; instead, they interact with another part of the enzyme (Figure 5.6c). In this process, called **allosteric** ("other space") **inhibition,** an enzyme's activity is reduced because of a change in shape caused by binding of an inhibitor at a site other than the substrate's binding site. The change in shape can be either reversible or irreversible. In some cases, allosteric interactions can activate an enzyme rather than inhibit it. Another type of noncompetitive inhibition can operate on enzymes that require metal ions for their activity. Certain chemicals can bind or tie up the metal ion activators and thus prevent an enzymatic reaction. Cyanide can bind the iron in iron-containing enzymes, and fluoride can bind calcium or magnesium. Substances such as cyanide and fluoride are sometimes called *enzyme poisons* because they permanently inactivate enzymes.

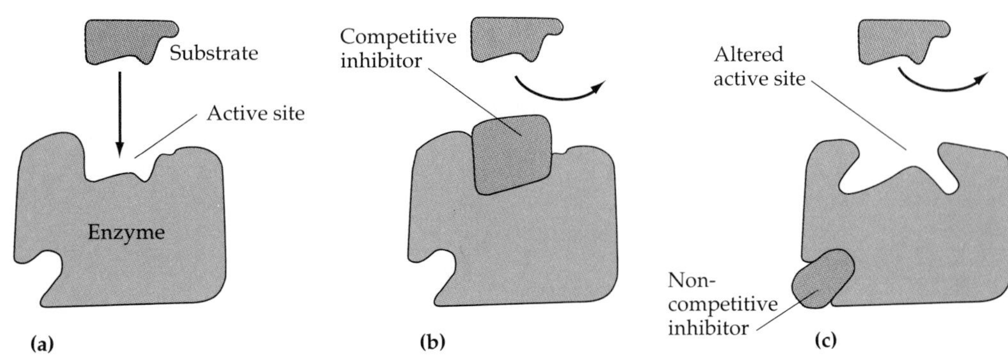

FIGURE 5.6 Enzyme inhibitors. (a) Uninhibited enzyme and its normal substrate. **(b)** Competitive inhibitor. **(c)** One type of noncompetitive inhibitor, causing allosteric inhibition.

FEEDBACK INHIBITION

Allosteric inhibitors play a role in a kind of biochemical control called **feedback inhibition** (or **end-product inhibition**). This control mechanism stops the cell from wasting chemical resources by making more of a substance than it needs. In some metabolic reactions, several steps are required for the synthesis of a particular chemical compound, called the *end-product*. The process is similar to an assembly line, with each step catalyzed by a separate enzyme (Figure 5.7). In many biosynthetic pathways, the final product can allosterically inhibit the activity of one of the enzymes earlier in the pathway. This phenomenon is feedback inhibition.

Feedback inhibition generally acts on the first enzyme in a metabolic pathway (similar to shutting down an assembly line by stopping the first worker). Because the enzyme is inhibited, the product of the first enzymatic reaction in the pathway is not synthesized. Because that unsynthesized product would normally be the substrate for the second enzyme in the pathway, the second reaction stops immediately as well. Thus, even though only the first enzyme in the pathway is inhibited, the entire pathway shuts down and no new end-product is formed. By inhibiting the first enzyme in the pathway, the cell also keeps metabolic intermediates from accumulating. As the existing end-product is used up by the cell, the first enzyme's allosteric site will more often remain unbound, and the pathway will resume activity.

The bacterium *Escherichia coli* can be used to demonstrate feedback inhibition in the synthesis of the amino acid isoleucine, which is required for its growth. In this metabolic pathway, five steps are taken to enzymatically convert the amino acid threonine to isoleucine (Figure 5.8). If isoleucine is added to the *E. coli* medium, it inhibits the first enzyme in the pathway and the bacteria stop synthesizing isoleucine. This condition is maintained until the supply of isoleucine is depleted. This type of feedback inhibition is also involved in regulating the cells' production of other amino acids, as well as vitamins, purines, and pyrimi-

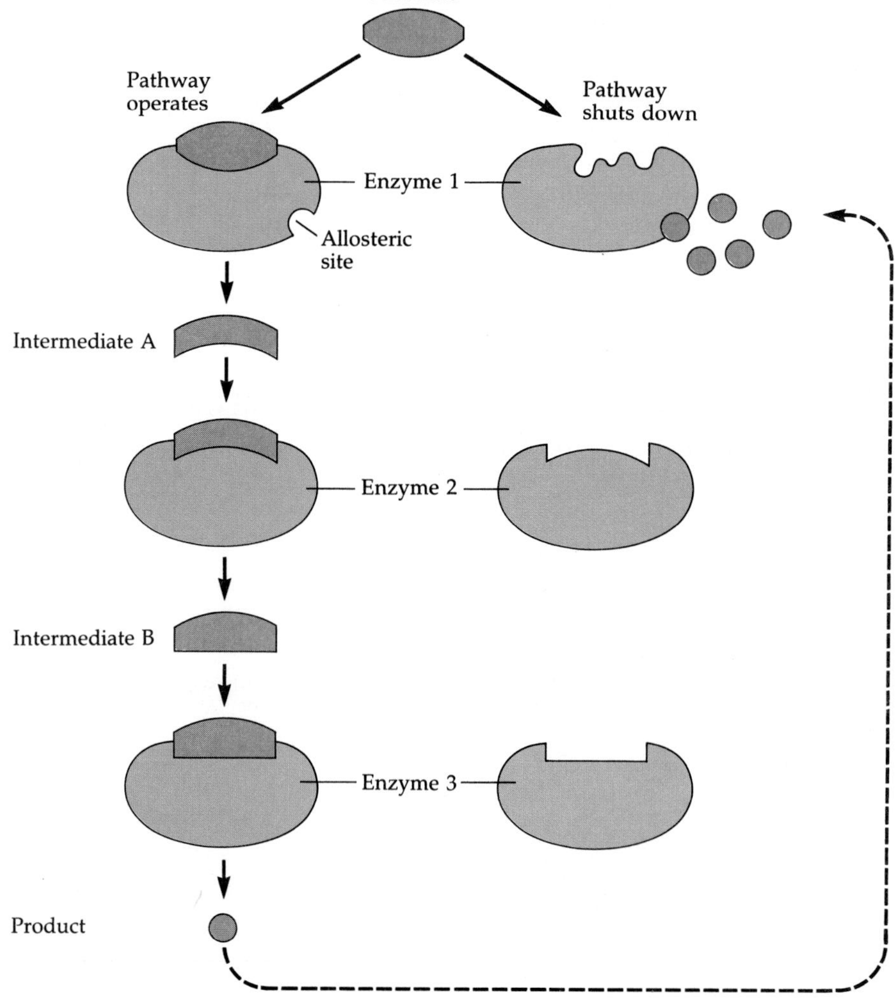

FIGURE 5.7 Feedback inhibition. A series of enzymes makes a product that inhibits the first enzyme in the series, thus shutting down the entire pathway when sufficient product has been made.

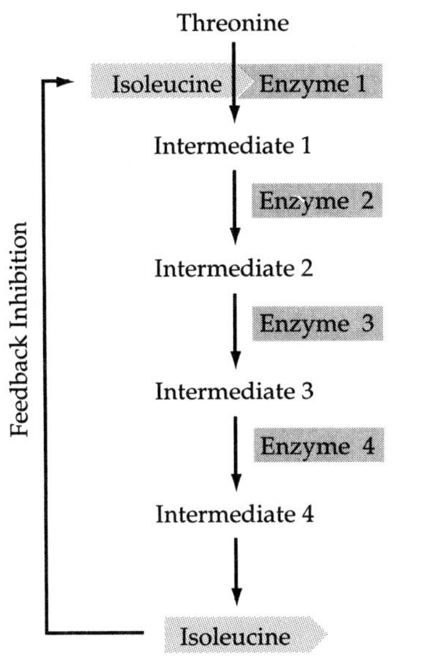

FIGURE 5.8 Feedback inhibition in *E. coli.* Shown here is a sequence of reactions in which threonine is converted to isoleucine. If isoleucine is added to the culture medium, *E. coli* no longer synthesizes isoleucine because the isoleucine combines with the enzyme that catalyzes the first step of the synthesis. Thus, isoleucine inhibits the synthesis of more isoleucine.

dines. In the following section on energy production methods, you will see the key roles enzymes play in both catabolic and anabolic reactions.

Energy Production

Nutrient molecules, like all molecules, have energy associated with the electrons that form bonds between their atoms. When this energy is spread throughout the molecule, it is difficult for the cell to use it. Various reactions in catabolic pathways, however, concentrate the energy into the bonds of ATP, which serves as a convenient energy carrier. ATP is generally referred to as having "high-energy" bonds. Actually, a better term is probably "unstable" bonds. Although there is not an exceptionally large amount of energy in these bonds, it can be released quickly and easily. In a sense, ATP is similar to a highly flammable liquid such as kerosene. Although a large log might eventually burn to produce more heat than a cup of kerosene, the kerosene is easier to ignite and provides heat more quickly and conveniently. In a similar way, the "high-energy" unstable bonds of ATP provide the cell with readily available energy for anabolic reactions.

Before discussing the catabolic pathways, we will consider two general aspects of energy production—the concept of oxidation–reduction and the mechanisms of ATP generation.

OXIDATION–REDUCTION

Oxidation is the removal of electrons (e^-) from an atom or molecule and is often an energy-producing reaction. Figure 5.9 shows an example of an oxidation in which molecule A loses an electron to molecule B. Molecule A is oxidized (meaning it has lost one or more electrons), while molecule B has undergone **reduction** (meaning that it has gained one or more electrons). Oxidation and reduction reactions are always coupled. In other words, each time one substance is oxidized, another is simultaneously reduced. The pairing of these reactions is called **oxidation–reduction.**

In many cellular oxidations, electrons and protons (hydrogen ions, H^+) are removed at the same time; this is equivalent to the removal of hydrogen atoms, because a hydrogen atom is made up of one proton and one electron (see Table 2.2). Because most biological oxidations involve the loss of hydrogen atoms, they are also called **dehydrogenation** reactions. Figure 5.10 shows an example of a biological oxidation. An organic molecule is oxidized by the loss of two hydrogen atoms, and a molecule of NAD^+ is reduced. As you may recall, NAD^+ assists enzymes by accepting hydrogen atoms removed from the substrate, in this case the organic molecule. As shown in Figure 5.10, NAD^+ accepts two electrons and one proton. One proton (H^+) is left over and is released into the surrounding medium. The reduced coenzyme, NADH, contains more energy than NAD^+. This energy can be used to generate ATP in later reactions.

An important point to remember about biological oxidation–reduction reactions (also called *redox* reactions) is that cells use them in catabolism to extract energy from nutrient molecules. Cells take nutrients, some of which serve as energy sources, and degrade them from highly reduced compounds (with many hydrogen atoms) to highly oxidized compounds. For example, when a cell oxidizes a molecule of glucose ($C_6H_{12}O_6$) to CO_2 and H_2O, the energy in the glucose molecule is removed in a stepwise manner and ulti-

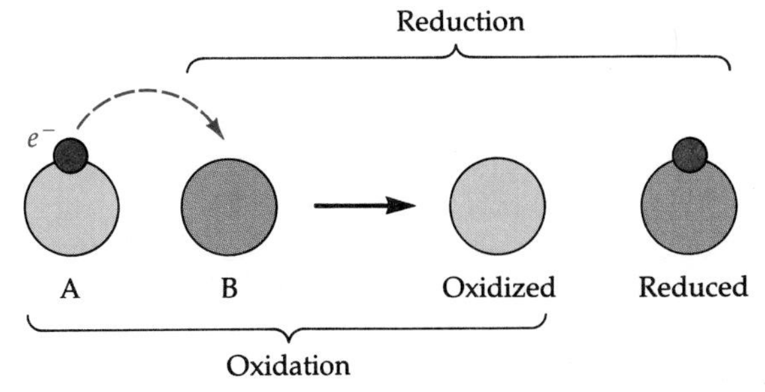

FIGURE 5.9 Oxidation–reduction. An electron is transferred from molecule A to molecule B. In the process, molecule A is oxidized and molecule B is reduced.

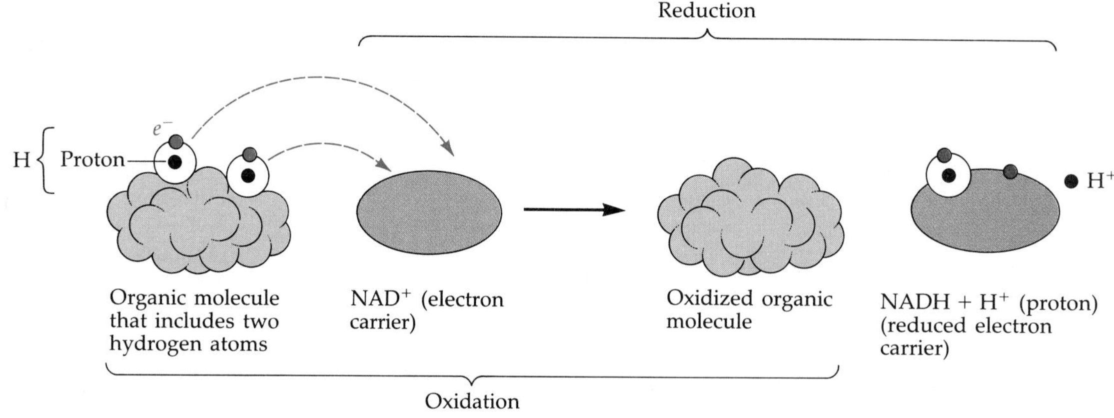

FIGURE 5.10 Representative biological oxidation. Two electrons and two protons (altogether equivalent to two hydrogen atoms) are transferred from an organic substrate molecule to a coenzyme, NAD^+. NAD^+ actually receives one hydrogen atom and one electron, while one proton is released into the medium. NAD^+ is reduced to NADH, which is a more energy-rich molecule.

mately trapped by ATP, which can then serve as an energy source for energy-requiring reactions. Compounds such as glucose that have many hydrogen atoms are highly reduced compounds, containing a large amount of potential energy. Thus, glucose is a valuable nutrient for organisms.

GENERATION OF ATP

Much of the energy released during oxidation–reduction reactions is trapped within the cell by the formation of ATP. Specifically, a phosphate group, Ⓟ, is added to ADP with the input of energy to form ATP:

$$\underbrace{\text{Adenosine—Ⓟ} \sim \text{Ⓟ}}_{\text{ADP}} + \text{Energy} + \text{Ⓟ} \longrightarrow \underbrace{\text{Adenosine—Ⓟ} \sim \text{Ⓟ} \sim \text{Ⓟ}}_{\text{ATP}}$$

The symbol ~ designates a "high-energy" bond, i.e., one that can readily be broken to release usable energy. The high-energy bond that attaches the third Ⓟ in a sense contains the energy stored in the above reaction. When this Ⓟ is removed, usable energy is released. The addition of Ⓟ to a chemical compound is called **phosphorylation**. Organisms use three mechanisms of phosphorylation to generate ATP from ADP.

In **substrate-level phosphorylation** ATP is generated when a high-energy Ⓟ is directly transferred from a phosphorylated compound (a substrate) to ADP. Generally, the Ⓟ has acquired its energy during an earlier reaction in which the substrate itself was oxidized. The following example shows only the carbon skeleton and the Ⓟ of a typical substrate:

$$\text{C—C—C} \sim \text{Ⓟ} + \text{ADP} \longrightarrow \text{C—C—C} + \text{ATP}$$

When **oxidative phosphorylation** occurs, electrons transferred from organic compounds to electron carriers (usually to NAD^+) are passed through a series of different electron carriers to molecules of oxygen (O_2) or other inorganic molecules. This process occurs in the plasma membrane of procaryotes and in the inner mitochondrial membrane of eucaryotes. The series of electron carriers used in oxidative phosphorylation is called an **electron transport chain** (see Figure 5.14). The transfer of electrons from one electron carrier to the next releases energy, some of which is used to generate ATP from ADP through a process called *chemiosmosis*, to be described shortly.

The third mechanism of phosphorylation, **photophosphorylation,** occurs only in photosynthetic cells, which contain light-trapping pigments such as chlorophylls. In photosynthesis, organic molecules, especially sugars, are synthesized with the energy of light from the energy-poor building blocks carbon dioxide and water. Photophosphorylation starts this process by converting light energy to the chemical energy of ATP and NADPH, which, in turn, are used to synthesize organic molecules. As in oxidative phosphorylation, an electron transport chain is involved.

Biochemical Pathways of Energy Production

Organisms release and store energy from organic molecules by a series of controlled reactions rather than in a single burst. If the energy were released all at once, as a large amount of heat, it could not be readily used to drive chemical reactions and would, in fact, damage the cell. To extract energy from organic compounds and store it in chemical form, organisms pass electrons from one compound to another through a series of oxidation–reduction reactions.

A sequence of enzymatically catalyzed chemical reactions occurring in a cell is called a **biochemical pathway**. Let us write out a hypothetical pathway that converts starting-material A to end-product F in a series of five steps:

$$\underset{\text{Starting material}}{A} \xrightarrow{NAD^+ \;\curvearrowright\; NADH + H^+} B \longrightarrow$$

$$C \xrightarrow{ADP + \text{(P)} \;\curvearrowright\; ATP} D \rightleftharpoons E \xrightarrow[CO_2 \quad H_2O]{O_2} \underset{\text{End-product}}{F}$$

The first step is the conversion of molecule A to molecule B. The curved arrow indicates that coupled to that reaction is the reduction of coenzyme NAD^+ to NADH; the electrons and protons come from molecule A. Similarly, the two arrows in the third step show a coupling of two reactions. As C is converted to D, ADP is converted to ATP; the energy needed comes from C as it transforms into D. The reaction converting D to E is readily reversible, as indicated by the double arrow. In the fifth step, the curved arrow leading from O_2 indicates that O_2 is a reactant in the reaction. The curved arrows leading to CO_2 and H_2O indicate that these substances are secondary products produced in the reaction, in addition to F, the end-product that (presumably) interests us the most. Secondary products such as CO_2 and H_2O shown here are sometimes called "by-products" or "waste products." Keep in mind that almost every reaction in a biochemical pathway is catalyzed by a specific enzyme; sometimes the name of the enzyme is printed near the arrow.

Carbohydrate Catabolism

Most microorganisms oxidize carbohydrates as their primary source of cellular energy. Carbohydrate catabolism, the breakdown of carbohydrate molecules to produce energy, is therefore of great importance in cell metabolism. Glucose is the most common carbohydrate energy source used by cells. Microorganisms can also catabolize various lipids and proteins for energy production, as you will see later.

To produce energy from glucose, microorganisms use two general processes, respiration and fermentation. Both processes usually start with the same first step, glycolysis, but follow different subsequent pathways (Figure 5.11). Before examining the details of respiration and fermentation, we will first look at a general overview of the processes.

As shown in Figure 5.11, the respiration of glucose typically occurs in three principal stages: glycolysis, the Krebs cycle, and the electron transport chain.

1. Glycolysis is the oxidation of glucose to pyruvic acid with the production of some ATP and energy-containing NADH.
2. The Krebs cycle is the oxidation of a derivative of pyruvic acid to carbon dioxide with the production of some ATP, energy-containing NADH, and another reduced electron carrier, $FADH_2$.
3. In the electron transport chain, NADH and $FADH_2$ are oxidized, contributing the electrons they have carried from the substrates to a "cascade" of redox reactions involving a series of additional electron carriers. Energy from these reactions is used to generate a considerable amount of ATP. In respiration, most of the ATP is generated in the third step.

Since respiration involves a long series of oxidation–reduction reactions, the entire process can be thought of as involving a flow of electrons from the energy-rich glucose molecule to the relatively energy-poor CO_2 and H_2O molecules. The coupling of ATP production to this flow is somewhat analogous to the production of electric power by using energy provided by a flowing stream. Carrying the analogy further, you could imagine a stream flowing down a gentle slope during gycolysis and the Krebs cycle, supplying energy to turn two old-fashioned waterwheels. Then the stream rushes down a steep slope in the electron transport chain, supplying energy for a large modern power plant. In a similar way, glycolysis and the Krebs cycle generate a small amount of ATP and also supply the electrons that generate a great deal of ATP at the electron transport chain stage.

The initial stage of fermentation usually is also glycolysis (Figure 5.11). However, once glycolysis has taken place, the pyruvic acid is converted into one or more different products, depending on the type of cell. These products might include alcohol (ethanol) and lactic acid. Unlike respiration, there is no Krebs cycle or electron transport chain in fermentation. Accordingly, the ATP yield is much lower.

GLYCOLYSIS

Glycolysis, the oxidation of glucose to pyruvic acid, is usually the first stage in carbohydrate catabolism.

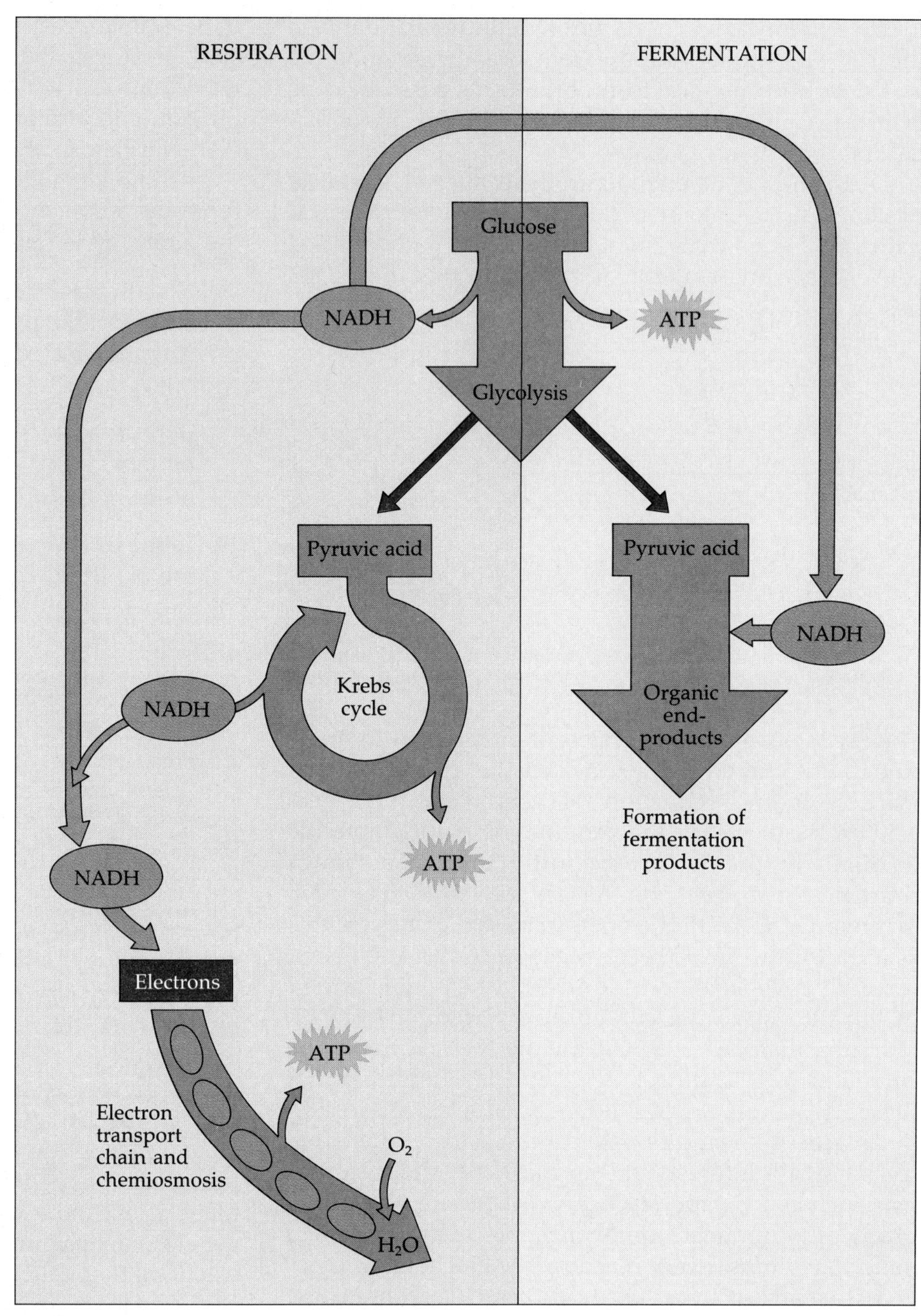

FIGURE 5.11 Overview of respiration and fermentation. NADH carries electrons to the electron transport chain, where a great deal of ATP is produced. In the Krebs cycle, electrons are actually carried by NADH and another carrier, $FADH_2$. In fermentation, electrons carried by NADH are incorporated into organic wastes.

Most microorganisms use this pathway; in fact, it occurs in most living cells.

Glycolysis is also called the **Embden–Meyerhof pathway.** The word *glycolysis* means splitting of sugar, and this is exactly what happens. The enzymes of glycolysis catalyze the splitting of glucose, a six-carbon sugar, into two three-carbon sugars. These sugars are then oxidized, releasing energy, and their atoms are rearranged to form two molecules of pyruvic acid. During glycolysis NAD^+ is reduced to NADH, and there is a net production of two ATP molecules by substrate-level phosphorylation. Glycolysis does not require oxygen; it can occur under either aerobic or anaerobic conditions. This pathway is a series of ten chemical reactions, each catalyzed by a different enzyme. The steps are outlined in Figure 5.12; see also Appendix E for a more detailed representation of glycolysis. To summarize the process, glycolysis consists of two basic stages, a preparatory stage and an energy-conserving stage.

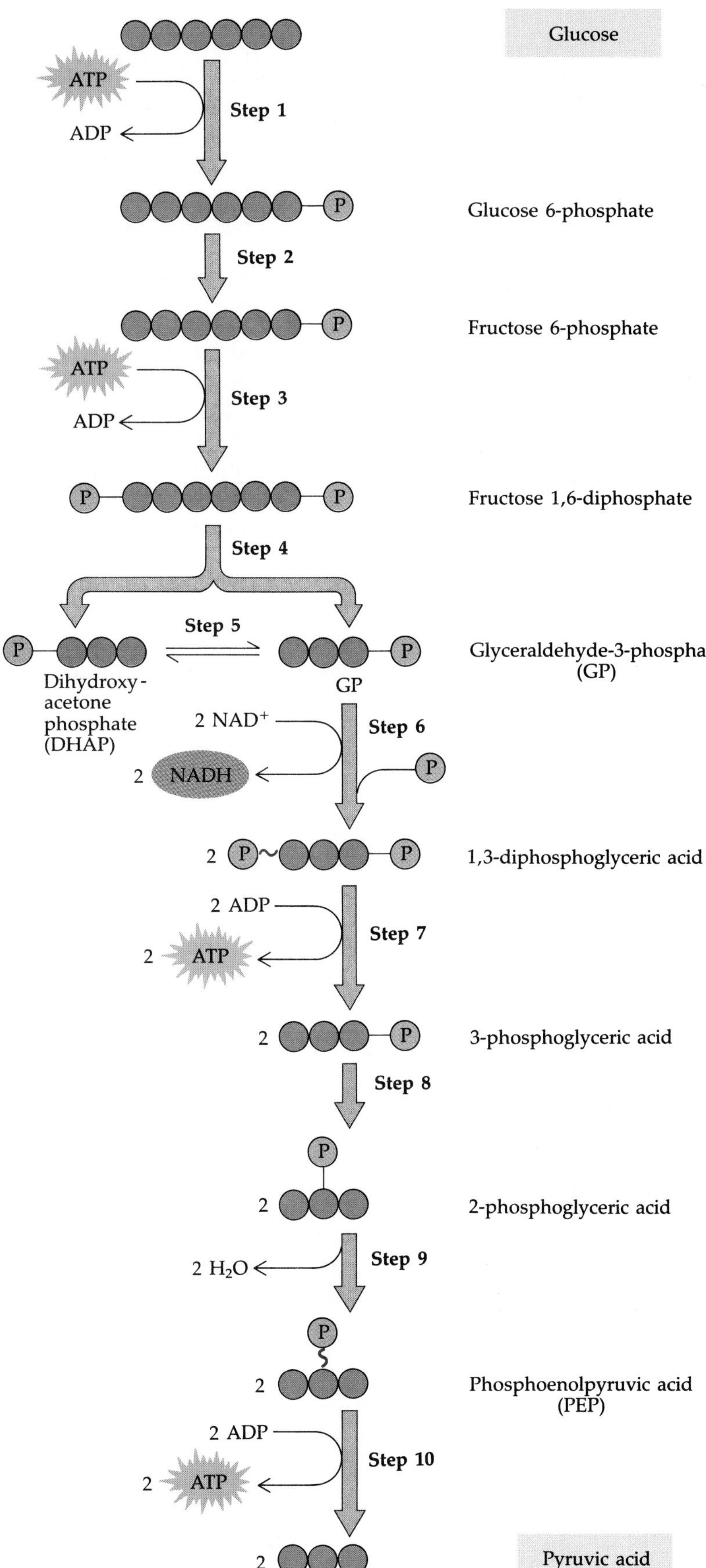

Step 1: Glucose enters the cell and is phosphorylated. A molecule of ATP is invested. The product of this reaction is glucose 6-phosphate.

Step 2: Glucose 6-phosphate is rearranged to form fructose 6-phosphate.

Step 3: The Ⓟ from another ATP is required to produce fructose 1,6-diphosphate, still a 6-carbon compound. (Note the total investment of 2 ATP molecules up to this point.)

Step 4: An enzyme cleaves (splits) the sugar into two 3-carbon molecules, dihydroxyacetone phosphate (DHAP) and glyceraldehyde 3-phosphate (GP).

Step 5: Dihydroxyacetate phosphate and glyceraldehyde 3-phosphate can each be converted to the other.

Step 6: The next enzyme converts glyceraldehyde phosphate to another three-carbon compound, 1,3-diphosphoglyceric acid. Since every dihydroxyacetone phosphate molecule can be converted to glyceraldehyde 3-phosphate, and each glyceraldehyde 3-phosphate to 1,3-diphosphoglyceric acid, the result is two molecules of 1,3-diphosphoglyceric acid for each initial molecule of glucose. Glyceraldehyde 3-phosphate is oxidized by the transfer of two hydrogen atoms to NAD^+ to form NADH. The enzyme couples this reaction with the creation of a high-energy bond between the sugar and a Ⓟ. The three-carbon sugar now has two Ⓟ groups.

Step 7: The high-energy Ⓟ is moved to ADP, forming ATP, the first ATP production of glycolysis. (Since the sugar splitting in step 4, all products are doubled. Therefore, this step actually repays the earlier investment of two ATP molecules.)

Step 8: An enzyme relocates the remaining Ⓟ of 3-phosphoglyceric acid to form 2-phosphoglyceric acid in preparation for the next step.

Step 9: By the loss of a water molecule, 2-phosphoglyceric acid is converted to phosphoenolpyruvic acid (PEP). In the process, the phosphate bond is upgraded to a high-energy bond.

Step 10: This high-energy Ⓟ is transferred from PEP to ADP, forming ATP. For each initial glucose molecule, the result of this step is *two* molecules of ATP and two molecules of a 3-carbon compound called pyruvic acid.

FIGURE 5.12 Outline of the reactions of glycolysis. A more detailed version is presented in Appendix E.

1. First, in the preparatory stage, two molecules of ATP are used as a six-carbon glucose molecule is phosphorylated, restructured, and split into two three-carbon compounds, glyceraldehyde 3-phosphate and dihydroxyacetone phosphate.
2. Then, in the energy-conserving stage, the two three-carbon molecules are oxidized in several steps to two molecules of pyruvic acid. In these reactions, two molecules of NAD^+ are reduced to NADH and four molecules of ATP are formed by substrate-level phosphorylation.

Because two molecules of ATP were needed to get glycolysis started and four molecules of ATP are generated by the process, *there is a net gain of two molecules of ATP for each molecule of glucose that is oxidized.*

ALTERNATIVES TO GLYCOLYSIS

Pentose Phosphate Pathway

Many bacteria have another pathway in addition to glycolysis for the oxidation of glucose. The most common alternative is the **pentose phosphate pathway** (or **hexose monophosphate shunt**), which operates simultaneously with glycolysis. The pentose phosphate pathway is a cyclic pathway that provides a means for the breakdown of five-carbon sugars (pentoses) as well as glucose (see the detailed figure in Appendix E). A key feature of this pathway is that it produces important intermediate pentoses that act as precursors in the synthesis of (1) nucleic acids, (2) glucose from carbon dioxide in photosynthesis, and (3) certain amino acids. The pathway is an important producer of the reduced coenzyme NADPH from $NADP^+$. Like NAD^+, NADP carries an electron and a hydrogen atom. Twelve molecules of $NADPH^+$ can be produced from each molecule of glucose if the glucose is completely oxidized. NADPH is then used in various biosynthetic reactions in the cell. Unlike glycolysis, the pentose phosphate pathway yields a net gain of only one molecule of ATP for each molecule of glucose oxidized. Bacteria that use the pentose phosphate pathway include *Bacillus subtilis* (su'til-us), *E. coli, Leuconostoc mesenteroides* (lü-kō-nos'tok mes-en-ter-oi'dēz), and *Enterococcus faecalis* (fē-kāl'is).

Entner–Doudoroff Pathway

The **Entner–Doudoroff pathway** is still another pathway for the oxidation of glucose to pyruvic acid. From each molecule of glucose, two molecules of NADPH and one molecule of ATP are produced for use in cellular biosynthetic reactions (see Appendix F for a more detailed representation). Bacteria that have the enzymes for the Entner–Doudoroff pathway can metabolize glucose without either glycolysis or the pentose phosphate pathway. The Entner–Doudoroff pathway is found in some gram-negative bacteria, including *Rhizobium, Pseudomonas* (sū-dō-mō'nas), and *Agrobacterium* (ag-rō-bak-ti'rē-um). It is generally not found among gram-positive bacteria. Tests for the ability to oxidize glucose by this pathway are sometimes used to identify *Pseudomonas* in the clinical laboratory.

DEFINITION OF RESPIRATION

After glucose has been broken down to pyruvic acid, the pyruvic acid can be channeled into the next step of either fermentation (to be described later) or respiration (see Figure 15.11). **Respiration** is defined as an ATP-generating process in which molecules are oxidized and the final electron acceptor is (almost always) an inorganic molecule. An essential feature of respiration is the operation of an electron transport chain. In **aerobic respiration** the final electron acceptor is O_2, and in **anaerobic respiration** the final electron acceptor is an inorganic molecule other than molecular oxygen or, rarely, an organic molecule. For now we will describe respiration as it typically occurs in an aerobic cell.

THE KREBS CYCLE

Pyruvic acid cannot enter the Krebs cycle directly. In a preparatory step, it must lose one molecule of CO_2 and become a two-carbon compound (Figure 5.13, top). This process is called **decarboxylation.** The two-carbon compound, called an *acetyl group,* attaches to coenzyme A through a high-energy bond; the resulting complex is known as *acetyl coenzyme A* (*acetyl CoA*). During this reaction, pyruvic acid is also oxidized and NAD^+ is reduced to NADH.

Remember that the oxidation of one glucose molecule produces two molecules of pyruvic acid, so for each molecule of glucose, two molecules of CO_2 are lost in this preparatory step, two molecules of NADH are produced, and two molecules of acetyl CoA are formed. Once the pyruvic acid has undergone decarboxylation and its derivative (the acetyl group) has attached to CoA, the resulting acetyl CoA is ready to enter the Krebs cycle.

The **Krebs cycle,** also called the *tricarboxylic acid (TCA) cycle* or *citric acid cycle,* is a series of biochemical reactions in which the large amount of potential chemical energy stored in acetyl CoA is released step by step. In this cycle, a series of oxidations and reductions transfer that potential energy, in the form of electrons, to electron carrier coenzymes, chiefly NAD^+. The pyruvic acid derivatives are oxidized; the coenzymes are reduced.

As acetyl CoA enters the Krebs cycle, CoA detaches from the acetyl group. The two-carbon acetyl group combines with a four-carbon compound called oxaloacetic acid to form the six-carbon citric acid. This

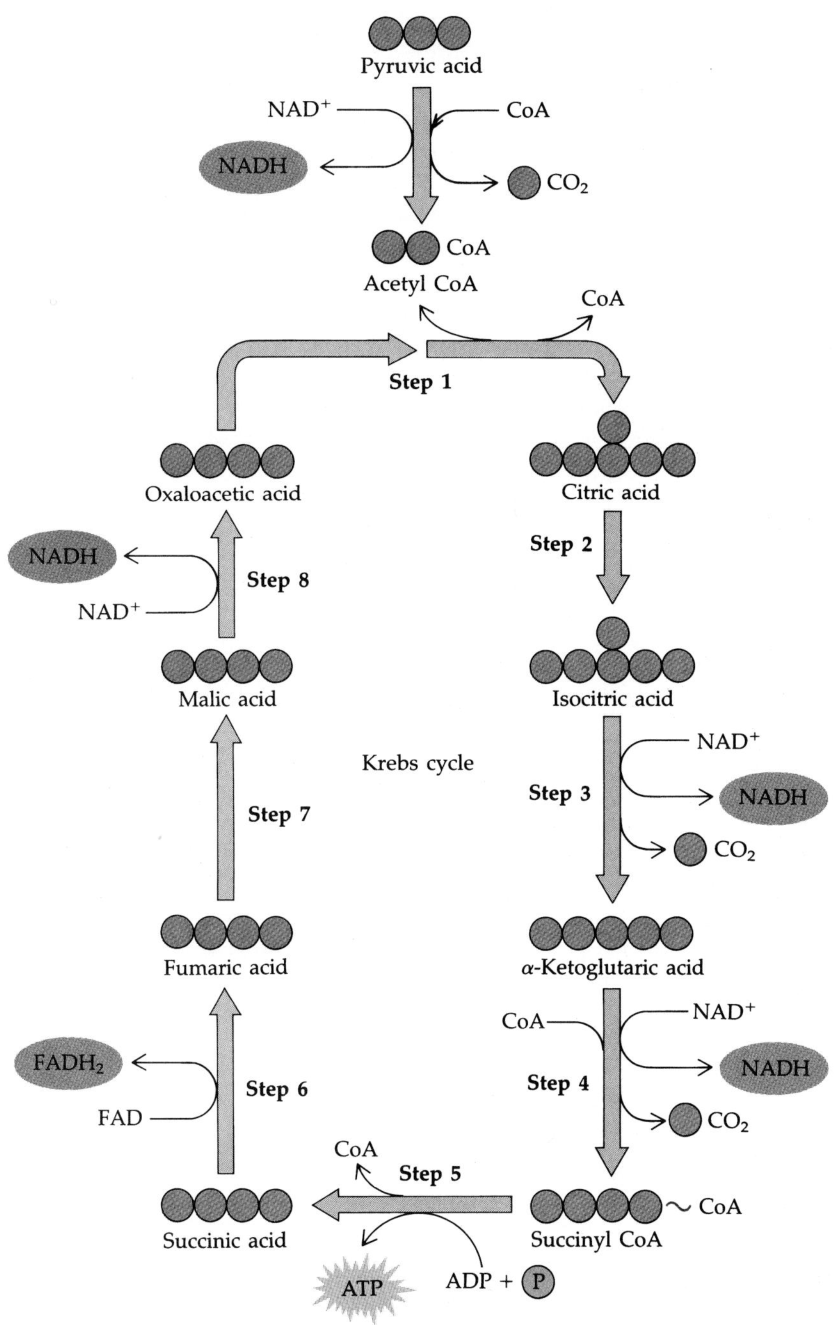

FIGURE 5.13 The Krebs cycle. In the preparatory step, pyruvic acid is oxidized, losing one carbon atom as CO_2 and reducing NAD^+ to NADH. The resulting two-carbon acetyl molecule is joined to CoA.

Step 1: A turn of the cycle begins as enzymes strip off the CoA portion from acetyl CoA and combine the remaining two-carbon acetyl group with oxaloacetic acid. Adding the acetyl group produces the six-carbon molecule citric acid.

Steps 2 to 4: Oxidations generate NADH. Step 2 is a rearrangement. Steps 3 and 4 combine oxidations and decarboxylations. They dispose of two carbon atoms that came from oxaloacetic acid. The carbons are released as CO_2, and the oxidations generate NADH from NAD^+. During the second oxidation (step 4), CoA is added into the cycle, forming the compound succinyl CoA.

Step 5: ATP is produced by substrate-level phosphorylation. CoA is removed from succinyl CoA, leaving succinic acid.

Steps 6 to 8: Enzymes rearrange chemical bonds, producing three different molecules before regenerating oxaloacetic acid. In step 6, an oxidation produces $FADH_2$. In step 8, a final oxidation generates NADH and converts malic acid to oxaloacetic acid, which is ready to enter another round of the Krebs cycle.

synthesis reaction requires energy, which is provided by the cleavage of the high-energy bond between the acetyl group and CoA. The formation of citric acid is thus the first step in the Krebs cycle. The major chemical reactions of this cycle are outlined in Figure 5.13; a more detailed representation of the Krebs cycle is provided in Appendix E. Keep in mind that each reaction is catalyzed by a specific enzyme.

The chemical reactions of the Krebs cycle fall into several general categories. One of these is decarboxylation. For example, in step 3, isocitric acid, a six-carbon compound, is decarboxylated to the five-carbon compound called α-ketoglutaric acid. Another decarboxylation takes place in step 4. Since one decarboxylation has taken place in the preparatory step and two in the Krebs cycle, all three carbon atoms in pyruvic acid are eventually released as CO_2 by the Krebs cycle. This represents the conversion to CO_2 of all six carbon atoms contained in the original glucose molecule.

Another general category of Krebs cycle chemical reactions is oxidation–reduction. For example, in step 3, two hydrogen atoms are lost during the conversion of the six-carbon isocitric acid to a five-carbon compound. In other words, the six-carbon compound is oxidized. Hydrogen atoms are also released in the Krebs cycle in steps 4, 6, and 8 and are picked up by the coenzymes NAD^+ and FAD. Since NAD^+ picks up two electrons but only one additional proton, its reduced form is represented as NADH; however, FAD picks up two complete hydrogen atoms and is reduced to $FADH_2$.

If we look at the Krebs cycle as a whole, we see that for every two molecules of acetyl CoA that enter the cycle four molecules of CO_2 are liberated by decarboxylation, six molecules of NADH and two molecules of $FADH_2$ are produced by oxidation–reduction reactions, and two molecules of ATP are generated by substrate-level phosphorylation. Many of the intermediates in the Krebs cycle also play a role in other pathways, especially in amino acid biosynthesis (discussed later in this chapter).

The CO_2 produced in the Krebs cycle is liberated into the atmosphere as a gaseous by-product of aerobic respiration. (Humans produce CO_2 from the Krebs cycle in most cells of the body and discharge it through the lungs during exhalation.) The reduced coenzymes NADH and $FADH_2$ are the most important products of the Krebs cycle because they contain most of the energy originally stored in glucose. During the next phase of respiration, a series of reductions indirectly transfers the energy stored in those coenzymes to ATP. These reactions are collectively called the electron transport chain.

ELECTRON TRANSPORT CHAIN

An **electron transport chain** consists of a sequence of carrier molecules that are capable of oxidation and reduction. As electrons are passed through the chain, there is a stepwise release of energy, which is used to drive the chemiosmotic generation of ATP, to be described shortly. The final oxidation is irreversible. In eucaryotic cells, the electron transport chain is contained in the inner membrane of mitochondria; in procaryotic cells, it is found in the plasma membrane.

There are three classes of carrier molecules in electron transport chains. The first are **flavoproteins.** These proteins contain flavin, a coenzyme derived from riboflavin (vitamin B_2), and are capable of performing alternating oxidations and reductions. One important flavin coenzyme is flavin mononucleotide (FMN). The second class of carrier molecules are **cytochromes,** proteins with an iron-containing group (heme) capable of existing alternately as a reduced form (Fe^{2+}) and an oxidized form (Fe^{3+}). The cytochromes involved in electron transport chains include cytochrome *b* (cyt*b*), cytochrome c_1 (cytc_1), cytochrome *c* (cyt*c*), cytochrome *a* (cyt*a*), and cytochrome a_3 (cyta_3). The third class of carriers is known as **ubiquinones** (or *coenzyme Q*), symbolized Q; these are small nonprotein carriers.

The electron transport chains of bacteria are somewhat diverse, in that the particular carriers used by a bacterium and the order in which they act may differ from those of other bacteria and from those of eucaryotic mitochondrial systems. Even a single bacterium may have several types of electron transport chains. However, keep in mind that all electron transport chains achieve the same basic goal, that of releasing energy as electrons are transferred from higher-energy compounds to lower-energy compounds. Much is known about the electron transport chain in the mitochondria of eucaryotic cells, so this is the one we will describe.

The first step in the mitochondrial electron transport chain involves the transfer of high-energy electrons from NADH to FMN, the first carrier in the chain (Figure 5.14). This transfer actually involves the passage of a hydrogen atom with two electrons to FMN, which then picks up an additional H^+ from the surrounding aqueous medium. As a result of the first transfer, NADH is oxidized to NAD^+, and FMN is reduced to $FMNH_2$. In the second step in the electron transport chain, $FMNH_2$ passes $2H^+$ to the other side of the mitochondrial membrane and passes two electrons to Q. As a result, $FMNH_2$ is oxidized to FMN. Q also picks up an additional $2H^+$ from the surrounding aqueous medium and releases it on the other side of the membrane.

The next part of the electron transport chain involves the cytochromes. Electrons are passed successively from Q to cyt*b*, cytc_1, cyt*c*, cyt*a*, and cyta_3. Each cytochrome in the chain is reduced as it picks up electrons and is oxidized as it gives up electrons. The last cytochrome, cyta_3, passes its electrons to molecular oxygen (O_2), which becomes negatively charged and then picks up protons from the surrounding medium to form H_2O.

Note that in Figure 5.14, $FADH_2$, derived from the Krebs cycle, is another source of electrons. However, $FADH_2$ adds its electrons to the electron transport chain at a lower level than NADH. Because of this, the electron transport chain produces about one-third less energy for ATP generation when $FADH_2$ donates electrons than when NADH is involved.

An important feature of the electron transport chain is the presence of some carriers, such as FMN and Q, that accept and release protons as well as electrons, and other carriers, such as cytochromes, that transfer electrons only. Electron flow down the chain is accompanied at several points by the active transport (pumping) of protons from the matrix side of the inner mitochondrial membrane to the opposite side of the membrane. The result is a buildup of protons on

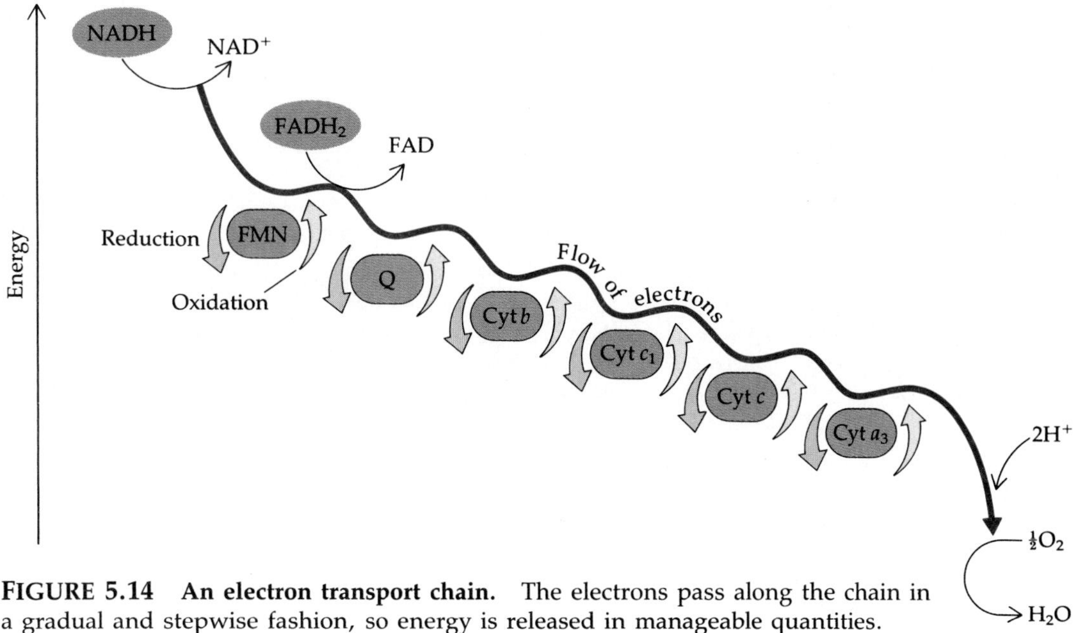

FIGURE 5.14 An electron transport chain. The electrons pass along the chain in a gradual and stepwise fashion, so energy is released in manageable quantities.

one side of the membrane. Just as water behind a dam stores energy that can be used to generate electricity, this buildup of protons provides energy for the generation of ATP by the chemiosmotic mechanism.

CHEMIOSMOTIC MECHANISM OF ATP GENERATION

The chemiosmotic mechanism of ATP synthesis—**chemiosmosis**—was first proposed by British biochemist Peter Mitchell in 1961. Although it took a number of years to gain full acceptance, it is now recognized as a major milestone in the history of biochemistry.

What is chemiosmosis, and how does it drive the energy-requiring synthesis of ATP from ADP? To understand the answers to these questions, we need to recall several concepts that were introduced in Chapter 4 as part of the section Movement of Materials Across Membranes (pp. 83–86). Recall that substances diffuse passively across membranes from regions of high concentration to regions of low concentration; this diffusion yields energy. Recall also that the movement of substances *against* such a concentration gradient *requires* energy and that, in such an active transport of molecules or ions across biological membranes, the required energy is usually provided by ATP. In chemiosmosis, the energy released when a substance moves along a gradient is used to *synthesize* ATP. The "substance" in this case refers to protons. In cellular respiration, chemiosmosis is responsible for most of the ATP that is generated. The steps of chemiosmosis are as follows (Figure 5.15):

1. As energetic electrons from NADH pass down the electron transport chain, some of the carriers in the chain pump (actively transport) protons across the membrane. Such carrier molecules are called *proton pumps*.
2. The phospholipid membrane is normally impermeable to protons, so this one-directional pumping establishes a proton gradient (a difference in the concentrations of protons on the two sides of the membrane). In addition to a concentration gradient, there is an electrical charge gradient. The excess H^+ on one side of the membrane makes that

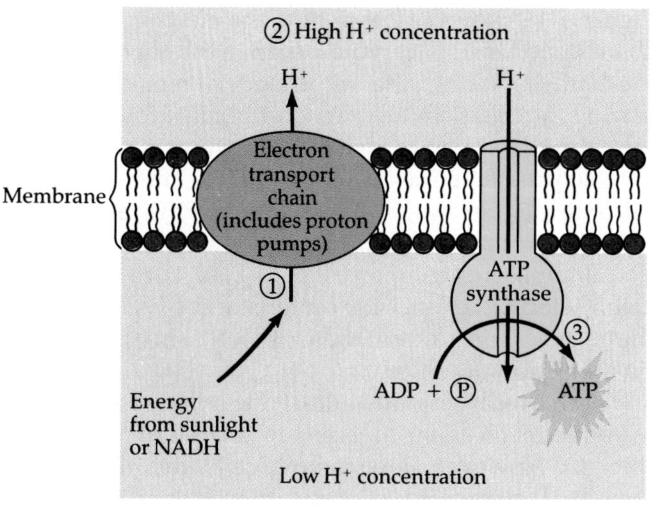

FIGURE 5.15 Chemiosmosis.

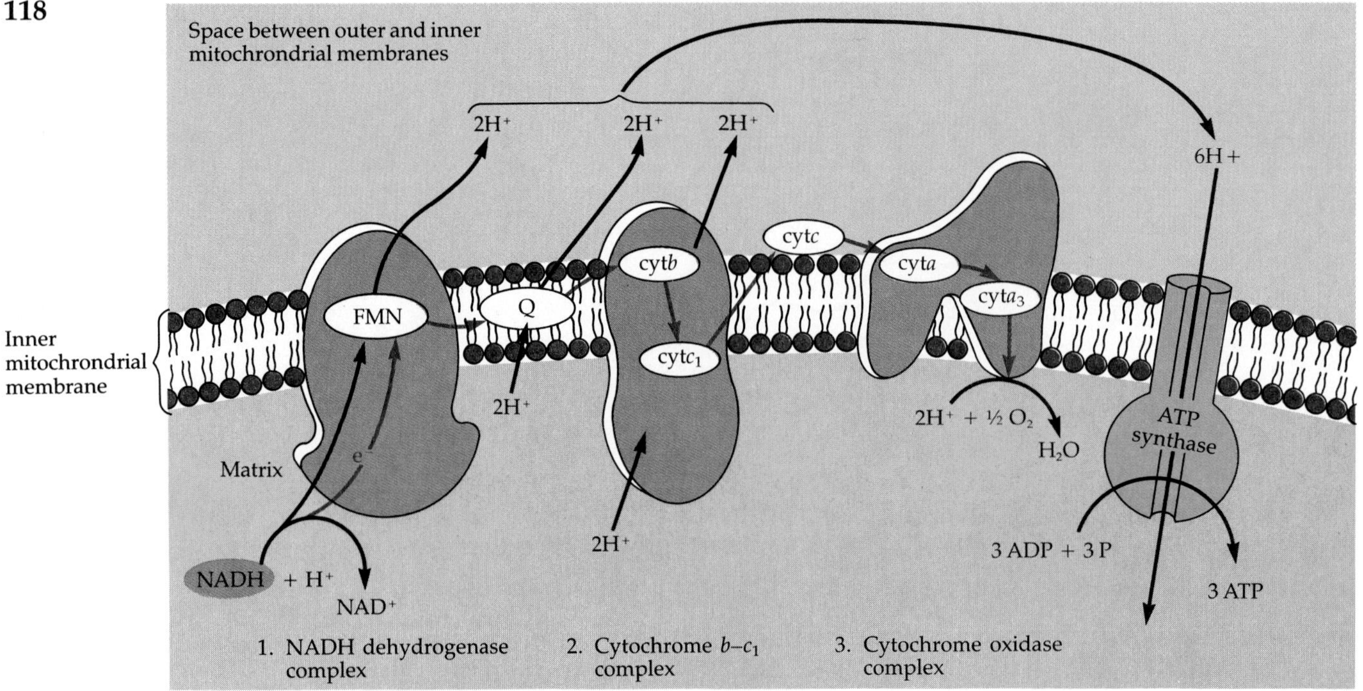

FIGURE 5.16 Electron transport and the chemiosmotic generation of ATP. Electron carriers are organized into three complexes, and protons (H^+) are pumped across the membrane at three points. In a eucaryotic cell, they are pumped from the matrix side of the mitochondrial membrane to the opposite side. The flow of electrons is shown in red.

side positively charged compared with the other side. The resulting electrochemical gradient has potential energy, called the *proton motive force.*

3. The protons on the side of the membrane with the higher proton concentration can diffuse across the membrane only through special protein channels that contain an enzyme called *adenosine triphosphatase (ATP synthase).* When this flow occurs, energy is released and is used by the enzyme to synthesize ATP from ADP and Ⓟ.

Figure 5.16 shows in detail how the electron transport chain operates in eucaryotes to drive the chemiosmotic mechanism. Within the mitochondrial membrane, the carriers of the electron transport chain are organized into three complexes, with Q transporting electrons between the first and second complex, and cyt*c* transporting them between the second and third complex. Three components of the system pump protons: the first and second complexes and Q. At the end of the chain, electrons join with protons and oxygen (O_2) in the matrix fluid to form water (H_2O). Thus, O_2 is the final electron acceptor.

Both procaryotic and eucaryotic cells use the chemiosmotic mechanism to generate energy for ATP production. However, in eucaryotic cells, the inner mitochondrial membrane contains the electron transport carriers at the ATP synthase, whereas in most procaryotes, the plasma membrane does so. An electron transport chain also operates in photophosphorylation and is located in the thylakoid membrane of cyanobacteria and of eucaryotic chloroplasts.

SUMMARY OF AEROBIC RESPIRATION

The various electron transfers in the electron transport chain generate about 34 molecules of ATP from each molecule of glucose oxidized: approximately three from each of the ten molecules of NADH (a total of 30) and approximately two from each of the two molecules of $FADH_2$ (a total of 4). To arrive at the total number of ATP molecules generated for each molecule of glucose, the 34 from chemiosmosis are added to those generated in glycolysis and the Krebs cycle. In aerobic respiration among procaryotes, a total of 38 molecules of ATP can be generated from one molecule of glucose. Note that four of those ATPs come from substrate-level phosphorylation in glycolysis and the Krebs cycle. Table 5.3 on p. 120 provides a detailed accounting of the ATP yield during procaryotic aerobic respiration.

Aerobic respiration among eucaryotes produces a total of only 36 molecules of ATP. There are fewer ATPs than in procaryotes because some energy is lost when electrons are shuttled across the mitochondrial membranes that separate glycolysis (in the cytoplasm) from the electron transport chain. There is no such separation in procaryotes. We can now summarize the

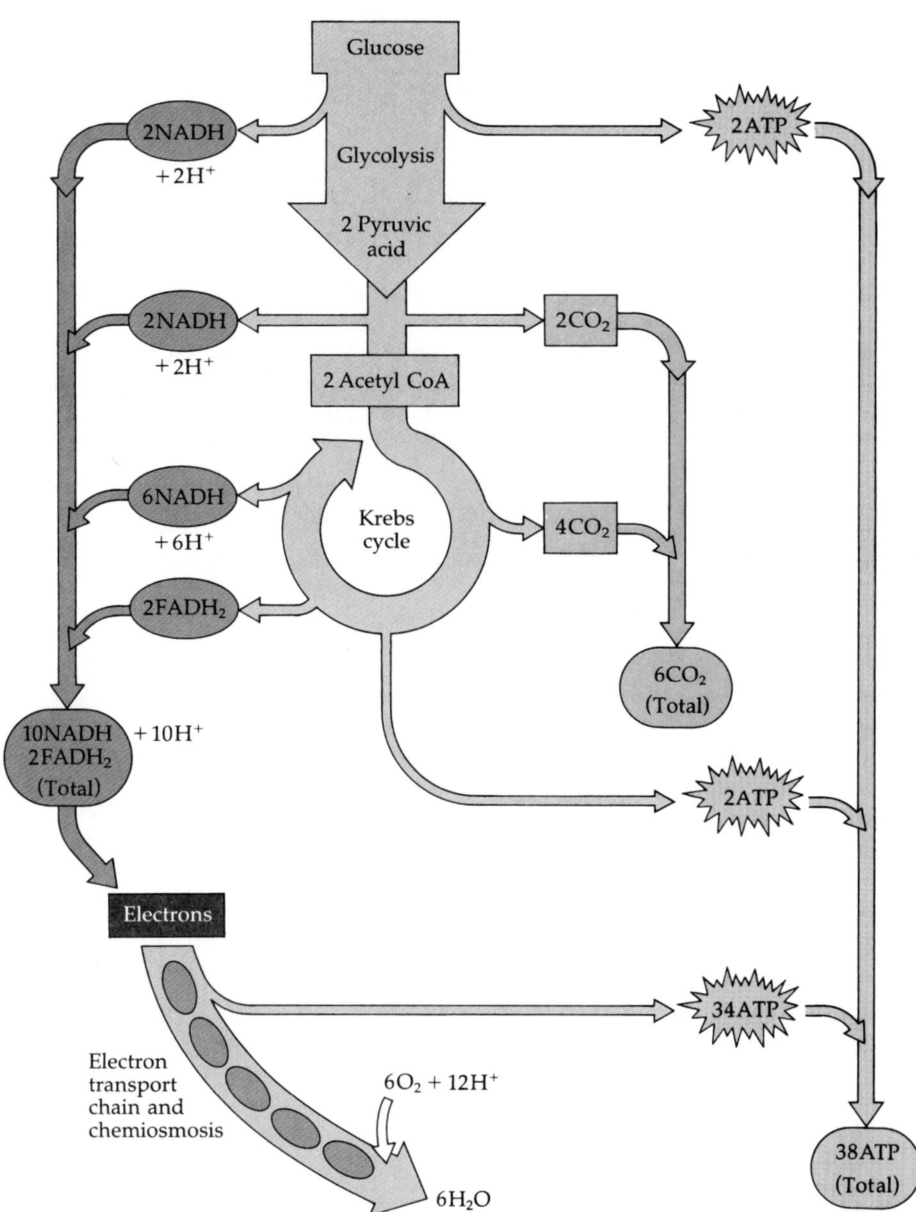

FIGURE 5.17 Summary of aerobic respiration. Glucose is broken down completely to carbon dioxide and water, and ATP is generated. This process has three major phases—glycolysis, the Krebs cycle, and the electron transport chain. The key event in the process is that electrons are picked up from intermediates of glycolysis and the Krebs cycle by NAD^+ or FAD and are carried by NADH or $FADH_2$ to the electron transport chain. NADH is also produced in the conversion of pyruvic acid to acetyl CoA. Most of the ATP generated by aerobic respiration is made by the chemiosmotic mechanism during the electron transport chain phase; this is called oxidative phosphorylation.

TABLE 5.3 Summary of ATP Produced During Procaryotic Aerobic Respiration of One Glucose Molecule

SOURCE	*ATP YIELD (METHOD)*
Glycolysis	
1. Oxidation of glucose to pyruvic acid	2ATP (substrate-level phosphorylation)
2. Production of 2NADH	6ATP (oxidation phosphorylation in electron transport chain)
Preparatory Step	
1. Formation of acetyl CoA produces 2NADH	6ATP (oxidative phosphorylation in electron transport chain)
Krebs Cycle	
1. Oxidation of succinyl CoA to succinic acid	2GTP (equivalent of ATP; substrate-level phosphorylation)
2. Production of 6NADH	18ATP (oxidative phosphorylation in electron transport chain)
3. Production of 2FADH	4ATP (oxidative phosphorylation in electron transport chain)
Total:	38ATP

overall reaction for aerobic respiration in procaryotes as follows:

$$\underset{\text{Glucose}}{C_6H_{12}O_6} + \underset{\text{Oxygen}}{6O_2} + 38ADP + 38Ⓟ \longrightarrow \underset{\text{Carbon dioxide}}{6CO_2} + \underset{\text{Water}}{6H_2O} + 38ATP$$

A summary of the various stages of aerobic respiration is presented in Figure 5.17.

ANAEROBIC RESPIRATION

In anaerobic respiration, the final electron acceptor is an inorganic substance other than oxygen (O_2). Some bacteria, such as *Pseudomonas* and *Bacillus*, can use a nitrate ion (NO_3^-) as a final electron acceptor; it is reduced to nitrite ion (NO_2^-), nitrous oxide (N_2O), or nitrogen gas (N_2). Other bacteria, such as *Desulfovibrio* (dē-sul-fō-vib′rē-ō), use sulfate (SO_4^{2-}) as the final electron acceptor to form hydrogen sulfide (H_2S). Still other bacteria use carbonate (CO_3^{2-}) to form methane (CH_4). (See the box on p. 128.) Anaerobic respiration by bacteria using nitrate and sulfate as final acceptors is essential for the nitrogen and sulfur cycles that occur in nature. The amount of ATP generated in anaerobic respiration varies with the organism and the pathway. Since only part of the Krebs cycle operates under anaerobic conditions, the ATP yield is never as high as in aerobic respiration, but it can be nearly as high.

FERMENTATION

After glucose has been broken down into pyruvic acid, the pyruvic acid can be completely broken down in respiration, as previously described, or can be converted to an organic product in fermentation (see Figure 5.11). **Fermentation** can be defined in several ways (see the box on this page), but we will define it here as a process that:

1. Releases energy from sugars or other organic molecules, such as amino acids, organic acids, purines, and pyrimidines.
2. Does not require oxygen (but sometimes can occur in its presence).
3. Does not require use of the Krebs cycle or an electron transport chain.
4. Uses an organic molecule as the final electron acceptor.
5. Produces only small amounts of ATP (only one or two ATP molecules for each molecule of starting material) because much of the original glucose energy remains in the chemical bonds of the organic end-products, such as lactic acid or ethanol.

MICROBIOLOGY HIGHLIGHTS

What Is Fermentation?

To many people, fermentation simply means the production of alcohol. Grains and fruits are fermented to produce beer and wine. If a food soured, you might say it was "off" or fermented. Here are some definitions of fermentation. They range from informal, general usage to more scientific definitions. Fermentation is:

1. Any process that produces alcoholic beverages or acidic dairy products (general use).
2. Any spoilage of food by microorganisms (general use).
3. Any large-scale microbial process occurring with or without air (common definition used in industry).
4. Any energy-releasing metabolic process that takes place only under anaerobic conditions (becoming more scientific).
5. All metabolic processes that release energy from a sugar or other organic molecule, do not require oxygen or an electron transport system, and use an organic molecule as the final electron acceptor. This is the definition we will use in this book.

During fermentation, electrons are transferred (along with protons) from reduced coenzymes (NADH, NADPH) to pyruvic acid or its derivatives (Figure 5.18a). Those final electron acceptors are reduced to the end-products shown in Figure 5.18b. In the process, NAD^+ and $NADP^+$ are regenerated and can enter another round of glycolysis. An essential function of the second stage of fermentation is to ensure a steady supply of NAD^+ and $NADP^+$ so that glycolysis can continue. In fermentation, ATP is generated only during glycolysis.

Various microorganisms can ferment various substrates; the end-products depend on the particular microorganism, the substrate, and the enzymes that are present and active. Chemical analyses of these end-products are useful in identifying microorganisms. We next consider two of the more important processes, lactic acid fermentation and alcohol fermentation.

Lactic Acid Fermentation

During glycolysis, the first phase of **lactic acid fermentation,** a molecule of glucose is oxidized to two molecules of pyruvic acid (Figure 5.19a; see also Figure 5.12). This oxidation generates the energy with which the two molecules of ATP are formed. In the next step, the two molecules of pyruvic acid are reduced by two molecules of NADH to form two molecules of lactic acid. Because lactic acid is the end-product of the reaction, it undergoes no further oxidation, and most of the energy produced by the reaction remains stored in the lactic acid. Thus, this fermentation yields only a small amount of energy. The overall reaction is:

$$\text{Glucose} + 2\text{ADP} + 2\text{Ⓟ} \longrightarrow 2 \text{ Lactic acid} + 2\text{ATP}$$

Two important genera of lactic acid bacteria are *Streptococcus* and *Lactobacillus* (lak-tō-bä-sil'lus). Because these microbes produce only lactic acid, they are referred to as **homolactic** (or *homofermentative*). Lactic acid fermentation can result in food spoilage. However, the process can also produce yogurt from milk, sauerkraut from fresh cabbage, and pickles from cucumbers.

Alcohol Fermentation

Alcohol fermentation also begins with the glycolysis of a molecule of glucose to yield two molecules of pyruvic acid and two molecules of ATP. In the next reaction, the two molecules of pyruvic acid are converted to two molecules of acetaldehyde and two molecules of CO_2 (Figure 5.19b). The two molecules of acetaldehyde are next reduced by two molecules of NADH to form two molecules of ethanol. Again, alcohol fermentation is a low-energy-yield process because most of the energy contained in the initial glucose molecule remains in the ethanol, the end-product. Alcohol fermentation is carried out by a number of bacteria and yeasts. The ethanol and carbon dioxide produced by the yeast *Saccharomyces* (sak-ä-rō-mī'sēs) are waste products for yeast cells but are useful to humans. Ethanol made by yeasts is the alcohol in alcoholic beverages, and carbon dioxide made by yeasts causes bread dough to rise (see the box in Chapter 1, p. 5).

FIGURE 5.18 Fermentation. (a) Overview of fermentation. The first step is glycolysis, the conversion of glucose to pyruvic acid. In the second step, the reduced coenzymes from glycolysis or its alternatives (NADH, NADPH) donate their electrons and hydrogen ions to pyruvic acid or a derivative to form an organic by-product. **(b)** End-products of various microbial fermentations.

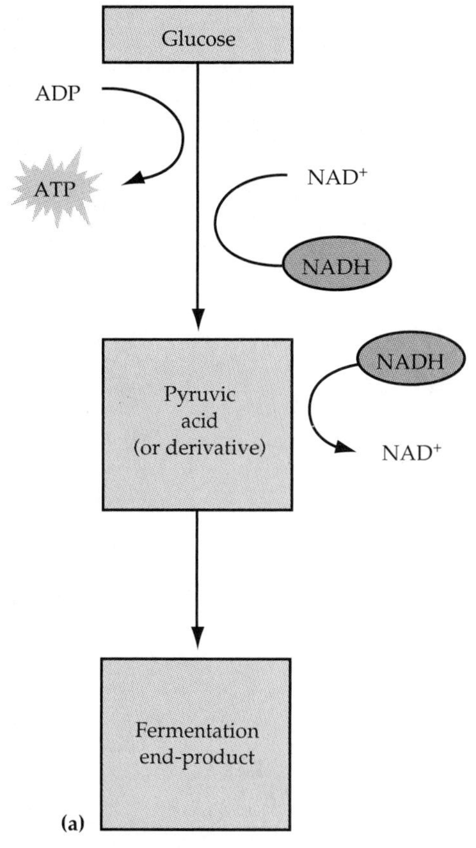

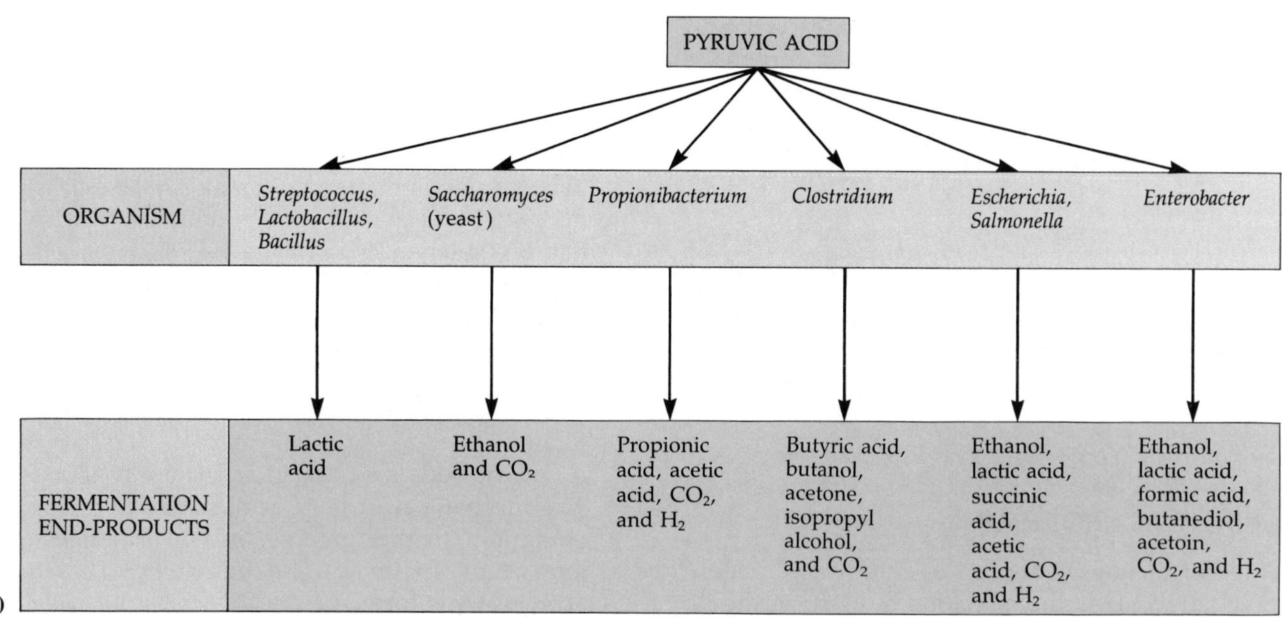

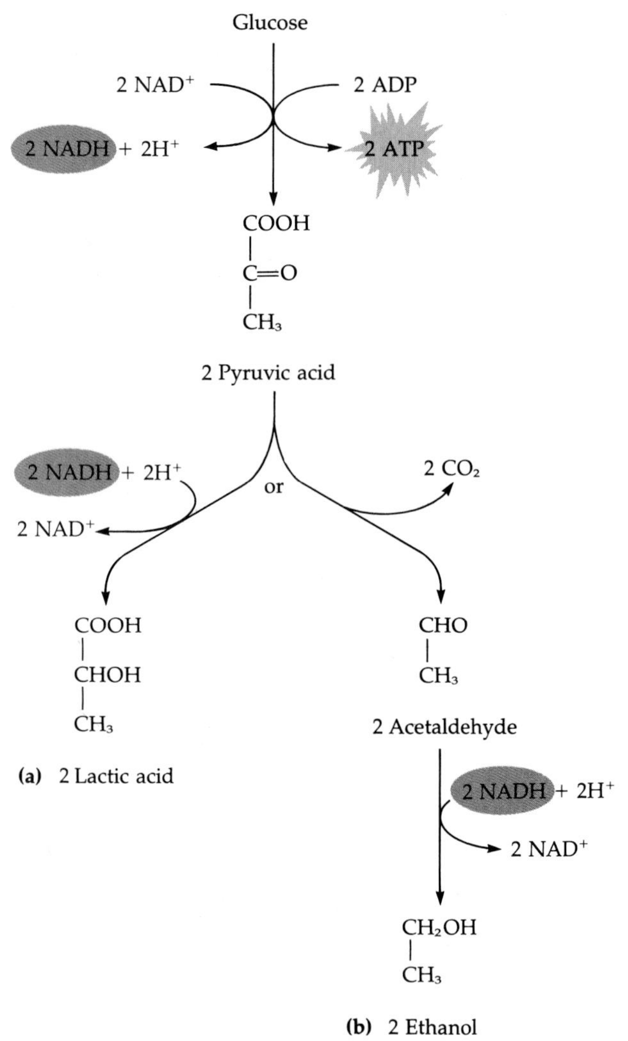

FIGURE 5.19 Types of fermentation. (a) Lactic acid fermentation. **(b)** Alcohol fermentation.

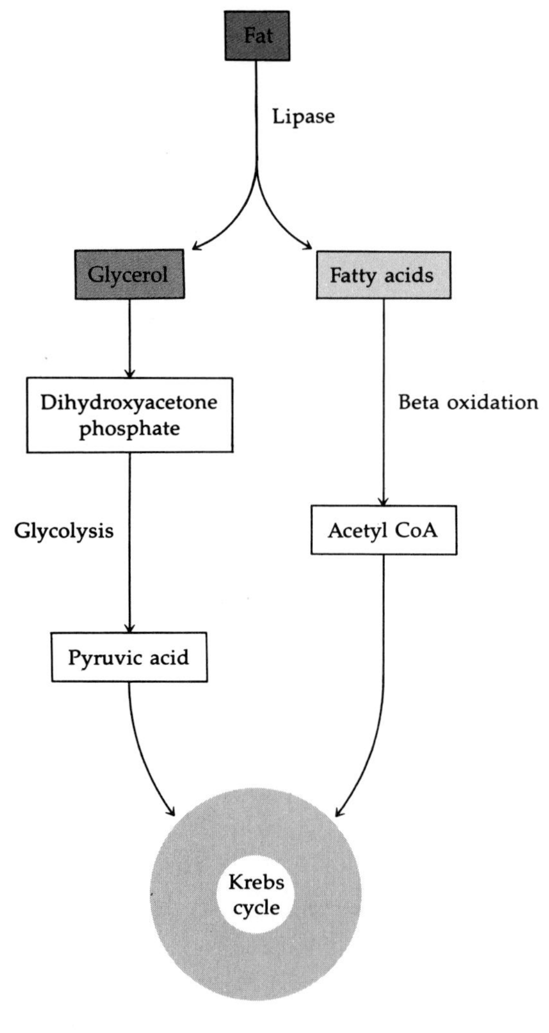

FIGURE 5.20 Lipid catabolism.

Organisms that produce lactic acid as well as other acids or alcohols are known as **heterolactic** (or *heterofermentative*) and often use the pentose phosphate pathway. The overall reaction for a typical heterolactic fermentation is:

$$\text{Glucose} + \text{ADP} + \textcircled{P} \longrightarrow \text{Lactic acid} + \text{Ethanol} + CO_2 + \text{ATP}$$

Table 5.4 lists some of the various microbial fermentations used by industry to convert inexpensive raw materials into useful end-products. A summary of aerobic respiration, anaerobic respiration, and fermentation is given in Table 5.5.

LIPID CATABOLISM

Thus far in our discussion of energy production, we have emphasized the oxidation of glucose, the principal energy-supplying carbohydrate. However, microbes also oxidize lipids and proteins, and the oxidations of all these nutrients are related.

Fats, as you recall, are lipids consisting of fatty acids and glycerol. Microbes produce extracellular enzymes called **lipases** that break fats down into their fatty acid and glycerol components. Each component is then metabolized separately.

Many types of microbes can convert glycerol into dihydroxyacetone phosphate, one of the three-carbon intermediates formed during glycolysis (see Figure 5.12). The dihydroxyacetone phosphate is then metabolized by glycolysis and the Krebs cycle, as shown in Figure 5.20.

Fatty acids are catabolized somewhat differently. The mechanism of fatty acid oxidation is called **beta oxidation.** In this process, carbon fragments from a long chain of fatty acid are removed two at a time and acetyl CoA is formed from them. As the molecules of

TABLE 5.4 Some Industrial Fermentations

FERMENTATION PRODUCT	COMMERCIAL USE	STARTING MATERIAL	MICROORGANISM
Ethanol	Beer	Malt extract	*Saccharomyces cerevisiae* (yeast, a fungus)
	Wine	Grape or other fruit juices	*Saccharomyces ellipsoideus* (yeast, a fungus)
	Fuel	Agricultural wastes	*Saccharomyces cerevisiae*
Acetic acid	Vinegar	Ethanol	*Acetobacter* (bacterium)
Lactic acid	Cheese, yogurt	Milk	*Lactobacillus, Streptococcus* (bacteria)
	Rye bread	Grain, sugar	*Lactobacillus bulgaricus* (bacterium)
	Sauerkraut	Cabbage	*Lactobacillus plantarum* (bacterium)
	Summer sausage	Meat	*Pediococcus* (bacterium)
Propionic acid and carbon dioxide	Swiss cheese	Milk	*Propionibacterium freudenreichii* (bacterium)
Acetone and butanol	Pharmaceutical, industrial uses	Molasses	*Clostridium acetobutylicum* (bacterium)
Glycerol	Pharmaceutical, industrial uses	Molasses	*Saccharomyces cerevisiae*
Citric acid	Flavoring	Molasses	*Aspergillus* (fungus)
Methane	Fuel	Acetic acid	*Methanosarcina* (bacterium)
Sorbose	Vitamin C (ascorbic acid)	Sorbitol	*Acetobacter* (bacterium)

TABLE 5.5 Comparison of Aerobic Respiration, Anaerobic Respiration, and Fermentation

ENERGY-PRODUCING PROCESS	GROWTH CONDITIONS	FINAL HYDROGEN (ELECTRON) ACCEPTOR	TYPE OF PHOSPHORYLATION USED TO GENERATE ATP	ATP MOLECULES PRODUCED PER GLUCOSE MOLECULE
Aerobic respiration	Aerobic	Molecular oxygen (O_2)	Substrate-level and oxidative	36 or 38*
Anaerobic respiration	Anaerobic	Usually an inorganic substance (such as NO_3^-, SO_4^{2-}, or CO_3^{2-}), but not molecular oxygen (O_2)	Substrate-level and oxidative	Variable (fewer than 38, but more than 2)
Fermentation	Aerobic or anaerobic	An organic molecule	Substrate-level	2

*In procaryotic aerobic respiration, 38 ATP molecules are produced; in eucaryotic aerobic respiration, 36 ATP molecules are produced.

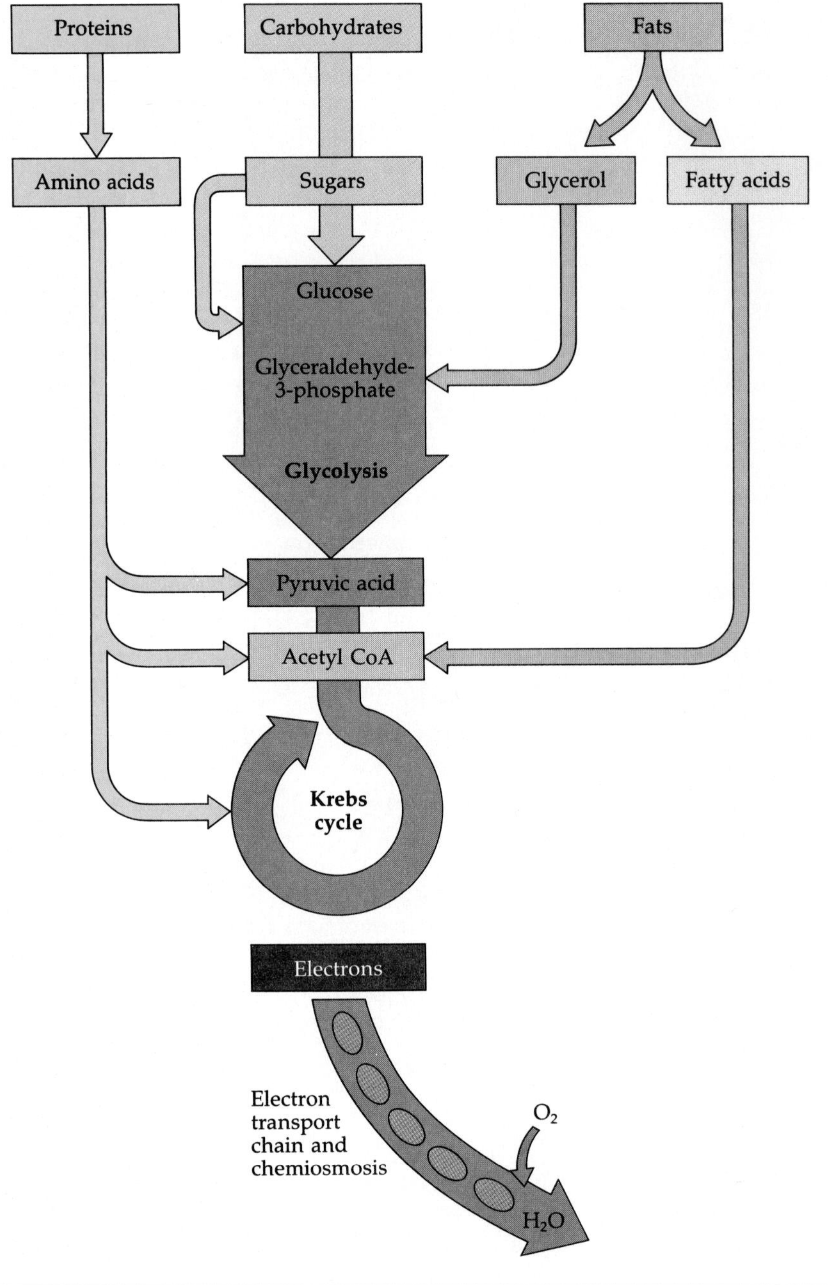

FIGURE 5.21 Catabolism of various organic food molecules. Carbohydrates, fats, and proteins can all be sources of electrons and protons for respiration. These food molecules enter glycolysis or the Krebs cycle at various points. Glycolysis and the Krebs cycle are catabolic funnels through which high-energy electrons from all kinds of organic molecules flow on their energy-releasing pathways.

acetyl CoA form, they enter the Krebs cycle, as do the acetyl CoA molecules formed by the oxidation of pyruvic acid (see Figure 5.13). Thus, the Krebs cycle functions in the oxidation of glycerol, fatty acids, and glucose and is the stage at which all three pathways converge. Beta oxidation of petroleum is discussed in the box in Chapter 2, page 38.

PROTEIN CATABOLISM

Proteins are too large to pass unaided through plasma membranes. Microbes produce extracellular *proteases* and *peptidases*, which break down proteins into their component amino acids, which can cross the membrane. However, before amino acids can be catabolized, they must be converted to substances that can enter the Krebs cycle. In one such conversion, called **deamination,** the amino group of an amino acid is removed and converted to an ammonium ion (NH_4^+), which can be excreted from the cell. The remaining organic acid can enter the Krebs cycle. Other conversions involve **decarboxylation** (removal of —COOH) and **dehydrogenation.** These conversions are complex; the important thing to remember is that various methods can prepare amino acids to enter the Krebs cycle.

A summary of the interrelation of carbohydrate, lipid, and protein catabolism is shown in Figure 5.21.

Photosynthesis

In all of the metabolic pathways just discussed, organisms obtain energy for cellular work by oxidizing organic compounds. But where do organisms obtain these organic compounds? Some, including animals and many microbes, feed on matter produced by other organisms. For example, bacteria may catabolize compounds from dead plants and animals or may obtain nourishment from a living host. Other organisms synthesize complex organic compounds from simple inorganic substances.

The major mechanism for doing so is a process called **photosynthesis,** which is used by plants and many microbes. Essentially, photosynthesis is the conversion of light energy from the sun into chemical energy. The chemical energy is then used to convert CO_2 from the atmosphere to more reduced carbon compounds, primarily sugars. The word *photosynthesis* summarizes the process: *photo* means light, and *synthesis* refers to the assembly of organic compounds. This synthesis of sugars by using carbons from CO_2 gas is also called **carbon fixation.** Continuation of life as we know it on Earth depends on the recycling of carbon in this way. Cyanobacteria, algae, and green plants all contribute to this vital recycling with photosynthesis.

Photosynthesis can be summarized as follows:

$$6CO_2 + 12H_2O + \text{Light energy} \longrightarrow C_6H_{12}O_6 + 6O_2 + 6H_2O$$

In the course of photosynthesis, electrons are taken from the hydrogen atoms of water, an energy-poor molecule, and incorporated into sugar, an energy-rich molecule. The energy boost is supplied by light energy, although indirectly.

Photosynthesis takes place in two stages. In the first stage, called the **light reactions,** light energy is used in a process that converts ADP + Ⓟ to ATP. In addition, in the predominant form of the light reactions, the electron carrier NADP is reduced to NADPH. The coenzyme NADPH, like NADH, is an energy-rich carrier of electrons. In the second stage, called the **dark (light-independent) reactions,** these electrons are used along with energy from ATP to reduce CO_2 to sugar.

The Light Reactions: Photophosphorylation

Photophosphorylation was introduced earlier in this chapter as one of the three ways that ATP is formed. Recall that photophosphorylation occurs only in photosynthetic cells. In this mechanism, light energy is absorbed by chlorophyll molecules in the photosynthetic cell, exciting some of their electrons. The chlorophyll principally used by green plants, algae, and cyanobacteria is *chlorophyll a*. It is located in the membranous thylakoids of chloroplasts in algae and green plants (see Figure 4.24) and in the thylakoids found in the photosynthetic structures of cyanobacteria. Other bacteria use *bacteriochlorophylls*.

The excited electrons jump from the chlorophyll to the first of a series of carrier molecules, an electron transport chain similar to that used in respiration. As electrons are passed along the series of carriers, protons are pumped across the membrane and ADP is converted to ATP by chemiosmosis. In **cyclic photophosphorylation** the electron eventually returns to chlorophyll (Figure 5.22a). In **noncyclic photophosphorylation,** which is the more common process, the electrons released from chlorophyll do not return to chlorophyll but become incorporated into NADPH (Figure 5.22b). The electrons lost from chlorophyll are replaced by electrons from water (H_2O) or another oxidizable compound, such as hydrogen sulfide (H_2S). To summarize: the products of noncyclic photophosphorylation are ATP (formed by chemiosmosis using energy released in an electron transport chain), O_2 (from water molecules), and NADPH (in which the hydrogen electrons and protons were derived ultimately from water) (Figure 5.22b).

The Dark Reactions: Calvin–Benson Cycle

The dark reactions are so called because no light is directly required for them to occur. They include a complex cyclic pathway called the **Calvin–Benson cycle,** in which CO_2 is "fixed," that is, used to synthesize sugars (Figure 5.23; see also Appendix E). For the net synthesis of one molecule of a three-carbon sugar called glyceraldehyde 3-phosphate, the cycle of reactions must occur three times, fixing three molecules of CO_2 (note that this sugar is also a key intermediate in glycolysis). These three turns of the cycle require the input of nine molecules of ATP and six molecules of NADPH. To make glucose, a six-carbon sugar, three more turns of the cycle are required. Thus, to make one molecule of glucose, the cycle must turn six times and requires an investment of $6CO_2$, 18ATP, and 12NADPH.

Nutritional Patterns Among Organisms

We have looked in detail at some of the energy-generating metabolic pathways that are used by animals and plants as well as many microbes. Microbes are distinguished by their great metabolic diversity, however, and some can sustain themselves on inorganic sub-

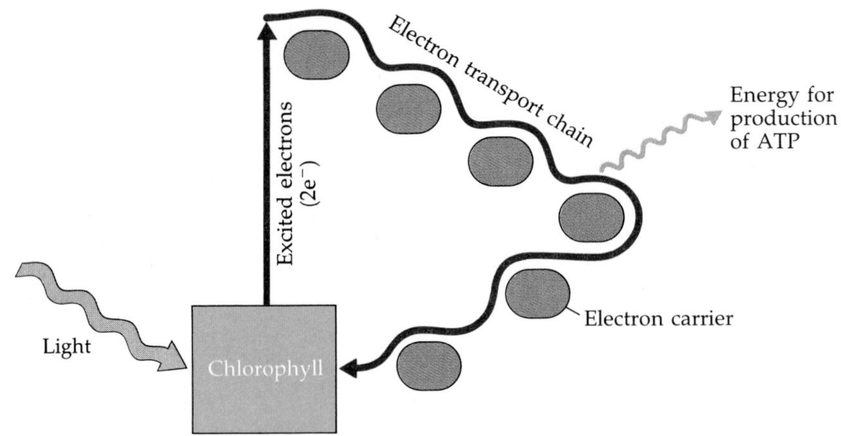

(a) Cyclic photophosphorylation

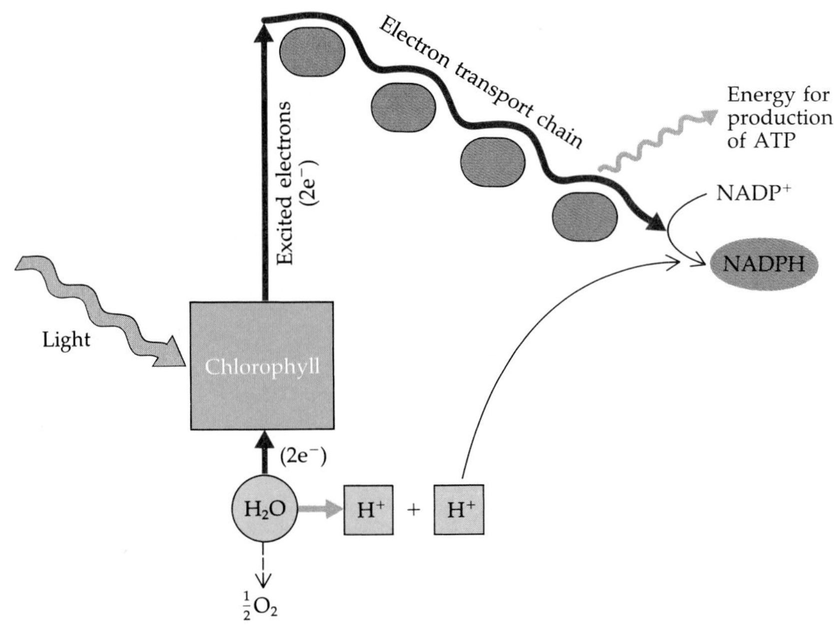

(b) Noncyclic photosphosphorylation

FIGURE 5.22
Photophosphorylation. **(a)** In cyclic photophosphorylation, electrons released from chlorophyll by light return to chlorophyll after passage along the electron transport chain. The energy from electron transfer is converted to ATP. **(b)** In noncyclic photophosphorylation, electrons released from chlorophyll are replaced by electrons from water. The chlorophyll electrons are passed along the electron transport chain to the electron acceptor $NADP^+$. $NADP^+$ combines with electrons and with hydrogen ions from water, forming NADPH.

stances by using pathways that are unavailable to either plants or animals. All organisms, including microbes, can be classified metabolically according to their *nutritional pattern,* i.e., their source of energy and their source of carbon.

First considering the energy source, we can generally classify organisms as phototrophs or chemotrophs. **Phototrophs** use light as their primary energy source, whereas **chemotrophs** depend on oxidation–reduction reactions of inorganic or organic compounds for energy. For their principal carbon source, **autotrophs** (self-feeders) use carbon dioxide and **heterotrophs** (feeders on others) require an organic carbon source. Autotrophs are also referred to as *lithotrophs* (rock eating), and heterotrophs are referred to as *organotrophs.*

If we combine the energy and carbon sources, we derive the following nutritional classifications for organisms: *photoautotrophs, photoheterotrophs, chemoautotrophs,* and *chemoheterotrophs.* This nutritional classification of microbes is summarized in Table 5.6. Almost all of the medically important microorganisms discussed in this book are chemoheterotrophs. Typically, infectious organisms catabolize substances obtained from the host.

Photoautotrophs

Photoautotrophs use light as a source of energy and carbon dioxide as their chief source of carbon. They include photosynthetic bacteria (green sulfur and purple sulfur bacteria and cyanobacteria), algae, and green plants. In the photosynthetic reactions of cyano-

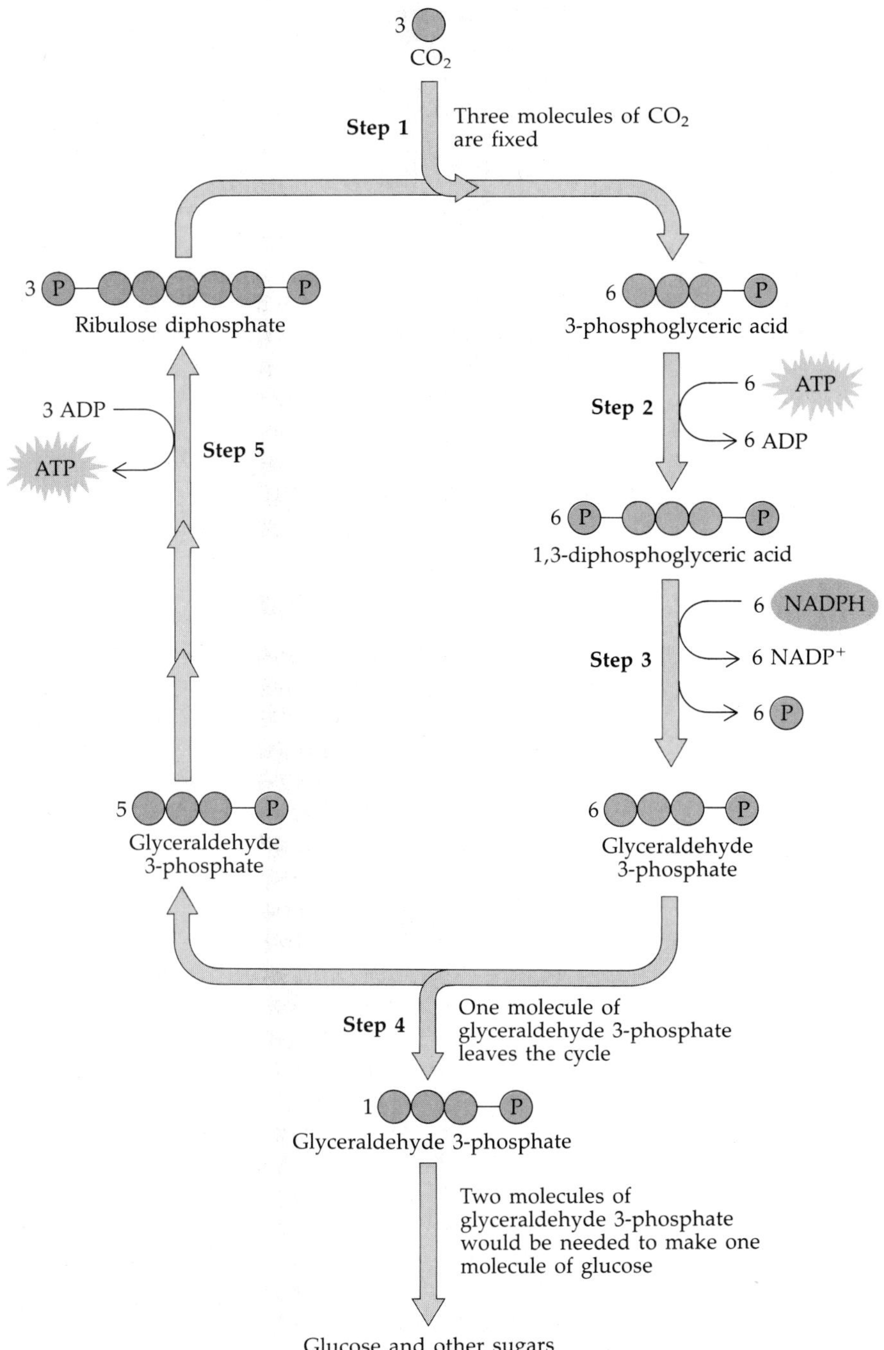

FIGURE 5.23 Simplified version of the Calvin–Benson cycle. To make one molecule of glucose, the cycle must turn six times so that there is a total investment of 6 molecules of CO_2, 18 molecules of ATP, and 12 molecules of NADPH. A more detailed version is presented in Appendix E.

bacteria, algae, and green plants the hydrogen atoms of water are used to reduce carbon dioxide, and oxygen gas is given off. Because this photosynthetic process produces O_2, it is sometimes called **oxygenic.**

In addition to the cyanobacteria (see Figure 11.19), there are several other families of photosynthetic procaryotes. Each is classified according to the way it reduces CO_2. These bacteria cannot use H_2O to reduce CO_2 and cannot carry on photosynthesis when oxygen is present (they must have an *anaerobic* environment). Consequently, their photosynthetic process does not produce O_2 and is called **anoxygenic.** Two of the families are photoautotrophs—the green sulfur and purple sulfur bacteria. The **green sulfur bacteria,** such as *Chlorobium* (klô-rō'bē-um) use sulfur (S), sulfur compounds (such as hydrogen sulfide, H_2S), or hydrogen gas (H_2) to reduce carbon dioxide and form organic compounds. Applying the energy from light and the appropriate enzymes, these bacteria oxidize sulfide (S^{2-}) or sulfur (S) to sulfate (SO_4^{2-}), or hydrogen gas to

MICROBIOLOGY *HIGHLIGHTS*

Photosynthesis Without Chlorophyll

The interesting archaeobacterium *Halobacterium* lives where very little else can grow. This bacterium is found in salt lakes, in the salt licks on ranches, in salt flats, or in any other environment with a concentration of salt that is from five to seven times greater than that of the ocean. Halobacteria are easy to detect because they turn their environment red. These bacteria cannot ferment carbohydrates and do not contain chlorophyll, so it was assumed that all their energy comes from oxidative phosphorylation. The exciting discovery of a new system of photophosphorylation arose through the study of the plasma membrane of *Halobacterium halobium*.

Researchers found that the plasma membrane of *H. halobium* fragments into two fractions (red and purple) when the cell is broken down and its components are sorted. The red fraction, which comprises most of the membrane, contains cytochromes, flavoproteins, and other parts of the electron transport chain that carries out oxidative phosphorylation. The purple fraction is more interesting. This purple membrane occurs in distinct patches of hexagonal lattices within the plasma membrane. The purple color comes from a protein that makes up 75% of the purple membrane. This protein is similar to the retinal pigment in the rod cells of the human eye, rhodopsin, so the protein was named bacteriorhodopsin. At the time it was discovered, its function was not known.

During the 1970s, further study of *H. halobium* suggested some startling explanations. When starved cells in anaerobic, dark environments were exposed to either light or oxygen, ATP synthesis increased. *H. halobium* can grow in the presence of either light or oxygen but cannot grow when neither is present.

This unexpected result suggested that *Halobacterium* can obtain energy by using either of two systems, one that operates in the presence of oxygen (oxidative phosphorylation) and one that operates in the presence of light (some kind of photophosphorylation). It was found that the rate of ATP synthesis by *H. halobium* is highest when the cells receive light that is between 550 and 600 nm in wavelength; this range exactly corresponds to the absorption spectrum of bacteriorhodopsin.

Researchers hypothesized that bacteriorhodopsin, like chlorophyll-containing systems, acts as a proton pump to create a proton gradient across a cell membrane; in this case, the gradient is created across the purple membrane. The proton gradient can do cellular work—drive the synthesis of ATP (see figure) or transport solutes.

Halobacterium synthesizes the purple membrane only when it is in low oxygen concentrations and in the presence of light. When *H. halobium* is exposed to optimum light for ATP synthesis, it keeps swimming in one direction. When the light wavelength is changed, the bacteria swim randomly, reversing direction up to 20 times per second. It is clear that the membrane provides the bacterium with energy when oxygen concentrations in its highly salinic environment become too low to support oxidative phosphorylation; that condition occurs frequently. Cells can accumulate solutes by group translocation to prevent water loss.

Until the discovery of the halobacterial system, chlorophyll-containing systems were the only ones known to generate ATP by using sunlight. Use of the pigment bacteriorhodopsin by *Halobacterium* is the simplest form of photophosphorylation known. Bacteriorhodopsin is itself an intriguing discovery. Its similarity to rhodopsin is the basis of much conjecture about the evolutionary relationship between halobacteria and eucaryotic rhodopsin.

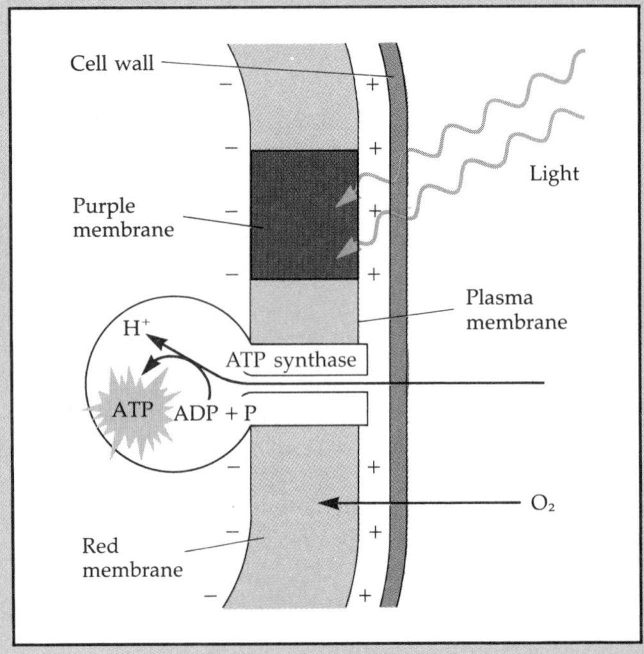

TABLE 5.6 Nutritional Classification of Organisms

NUTRITIONAL TYPE	ENERGY SOURCE	CARBON SOURCE	EXAMPLES
*Photoautotroph	Light	Carbon dioxide (CO_2)	Photosynthetic bacteria (green sulfur and purple sulfur bacteria), cyanobacteria, algae, plants (some green and purple sulfur bacteria may grow chemotrophically)
*Photoheterotroph	Light	Organic compounds	Purple nonsulfur and green nonsulfur bacteria (some purple nonsulfur bacteria may grow photoautotrophically and chemotrophically)
Chemoautotroph	Usually electrons from inorganic compounds	Carbon dioxide (CO_2)	Hydrogen, sulfur, iron, and nitrifying bacteria
Chemoheterotroph	Usually electrons from organic compounds	Organic compounds	Most bacteria and all fungi, protozoans, and animals

*The electron source for noncyclic photophosphorylation for photosynthetic bacteria is H_2, H_2S, or S; for algae and plants the source is H_2O; for purple nonsulfur and green nonsulfur bacteria, it is organic compounds; for cyanobacteria the source is usually H_2O.

water (H_2O). The **purple sulfur bacteria,** such as *Chromatium* (krō-mā'tē-um), also use sulfur, sulfur compounds, or hydrogen gas to reduce carbon dioxide. They are distinguished from the green sulfur bacteria by their type of chlorophyll and location of stored sulfur.

The chlorophylls used by these photosynthetic bacteria are called *bacteriochlorophylls*, and they absorb light at longer wavelengths than that absorbed by chlorophyll *a*. Bacteriochlorophylls of green sulfur bacteria are found in vesicles called *chlorosomes* (or *chlorobium vesicles*) underlying and attached to the plasma membrane. In the purple sulfur bacteria, the bacteriochlorophylls are located in invaginations of the plasma membrane (*intracytoplasmic membranes*).

Several characteristics that distinguish eucaryotic photosynthesis from procaryotic photosynthesis are presented in Table 5.7. See the box on p. 128 for a discussion of an exceptional photosynthetic system that exists in *Halobacterium*. The system does not use chlorophyll.

Photoheterotrophs

Photoheterotrophs use light as a source of energy but cannot convert carbon dioxide to sugar; rather, they use organic compounds, such as alcohols, fatty acids, other organic acids, and carbohydrates, as sources of carbon. They are anoxygenic. Among the photoheterotrophs are the **green nonsulfur bacteria,** such as

TABLE 5.7 Comparison of Eucaryotic and Selected Procaryotic Photosynthesis

	EUCARYOTES	PROCARYOTES		
CHARACTERISTIC	ALGAE, PLANTS	CYANOBACTERIA	GREEN SULFUR BACTERIA	PURPLE SULFUR BACTERIA
Substance that reduces CO_2	H atoms of H_2O	H atoms of H_2O	Sulfur, sulfur compounds, H_2 gas	Sulfur, sulfur compounds, H_2 gas
Oxygen production	Oxygenic	Oxygenic (and anoxygenic)	Anoxygenic	Anoxygenic
Light-trapping pigment	Chlorophyll *a*	Chlorophyll *a*	Bacteriochlorophyll *a*	Bacteriochlorophyll *a* or *b*
Site of photosynthesis	Chloroplasts with thylakoids	Thylakoids	Chlorosomes	Intracytoplasmic membrane
Environment	Aerobic	Aerobic (and anaerobic)	Anaerobic	Anaerobic

Chloroflexus (klô-rō-flex'us), and **purple nonsulfur bacteria,** such as *Rhodopseudomonas* (rō-dō-sū-dō-mō'nas).

Chemoautotrophs

Chemoautotrophs use the electrons from reduced inorganic compounds as a source of energy, and they use CO_2 as their principal source of carbon (see Figure 27.1, the carbon cycle). Inorganic sources of energy for these organisms include hydrogen sulfide (H_2S) for *Beggiatoa* (bej-jē-ä-tō'ä); elemental sulfur (S) for *Thiobacillus thiooxidans;* ammonia (NH_3) for *Nitrosomonas* (nī-trō-sō-mō'näs); nitrite ions (NO_2^-) for *Nitrobacter* (nī-trō-bak'tėr); hydrogen gas (H_2) for *Hydrogenomonas* (hī-drō-je-nō-mō'näs); and ferrous iron (Fe^{2+}) for *Thiobacillus ferrooxidans* (see the box in Chapter 28, p. 715). The energy derived from the oxidation of these inorganic compounds is eventually stored in ATP, which is produced by oxidative phosphorylation.

Chemoheterotrophs

When we discuss photoautotrophs, photoheterotrophs, and chemoautotrophs, it is easy to categorize the energy source and carbon source because they occur as separate entities. However, in chemoheterotrophs, the distinction is not so clear because the energy source and carbon source are usually the same organic compound—glucose, for example. Chemoheterotrophs specifically use the electrons from hydrogen atoms in organic compounds as their energy source.

Heterotrophs are further classified according to their source of organic molecules—**saprophytes** live on dead organic matter, and **parasites** derive nutrients from a living host. Most bacteria, and all fungi, protozoans, and animals, are chemoheterotrophs.

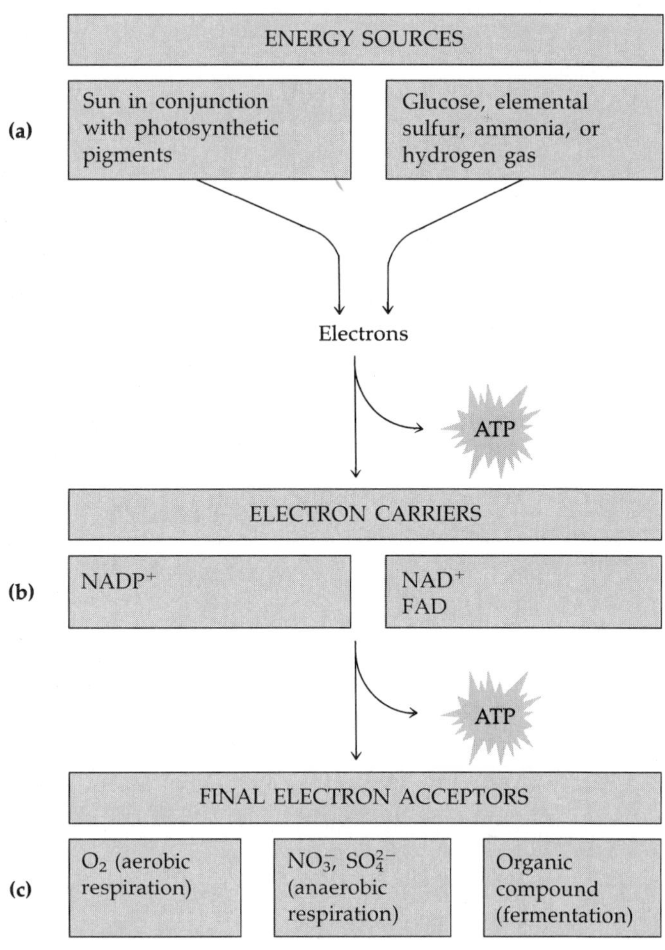

FIGURE 5.24 ATP production. The production of ATP requires **(a)** an energy source, **(b)** the transfer of electrons to an electron carrier during an oxidation–reduction reaction, and **(c)** the transfer of electrons to a final electron acceptor.

SUMMARY OF ENERGY PRODUCTION MECHANISMS

In the living world, energy passes from the sun to photosynthetic organisms and then to other organisms in the form of the potential energy contained in the bonds of chemical compounds. To obtain energy in a usable form, a cell must have an electron (or hydrogen) donor, which serves as an initial energy source within the cell. Electron donors can be as diverse as photosynthetic pigments, glucose or other organic compounds, elemental sulfur, ammonia, or hydrogen gas (Figure 5.24a). Next, electrons removed from the chemical energy sources are transferred to electron carriers, such as the coenzymes NAD^+, $NADP^+$, and FAD (Figure 5.24b). This transfer is an oxidation–reduction reaction; the initial energy source is oxidized as this first electron carrier is reduced. During this phase, some ATP is produced. In the third stage, electrons are transferred from electron carriers to their final electron acceptors in further oxidation–reduction reactions (Figure 5.24c).

In aerobic respiration, oxygen (O_2) serves as the final electron acceptor. In anaerobic respiration, inorganic substances other than oxygen, such as nitrate ions (NO_3^-) or sulfate ions (SO_4^{2-}), serve as the final electron acceptors. In fermentation, organic compounds serve as the final electron acceptors. In aerobic and anaerobic respiration, a series of electron carriers called an electron transport chain releases energy that is used by the mechanism of chemiosmosis to synthesize ATP. Regardless of their energy sources, all organisms use similar oxidation–reduction reactions to transfer electrons and similar mechanisms to use the energy released to produce ATP.

We will next consider how cells use ATP pathways for the synthesis of organic compounds such as carbohydrates, lipids, proteins, and nucleic acids.

Biochemical Pathways of Energy Utilization (Anabolism)

Up to now we have been considering energy production. Through the oxidation of organic molecules, organisms produce energy by aerobic respiration, anaerobic respiration, and fermentation. Much of this energy is given off as heat. The complete metabolic oxidation of glucose to carbon dioxide and water is considered a very efficient process, but about 45% of the energy of glucose is lost as heat. Cells use the remaining energy, which is trapped in the bonds of ATP, in a variety of ways. Microbes use ATP to provide energy for the transport of substances across plasma membranes—the process called active transport that we discussed in Chapter 4. Microbes also use some of their energy for flagellar motion (also discussed in Chapter 4). Most of the ATP, however, is used in the production of new cellular components. This production is a continuous process in cells, and, in general, is faster in procaryotic cells than in eucaryotic cells.

Autotrophs build their organic compounds by fixing carbon dioxide in the Calvin–Benson cycle (see Figure 5.23). This requires both energy (ATP) and electrons (from the oxidation of NADPH). Heterotrophs, by contrast, must have a ready source of organic compounds for biosynthesis, the production of needed cellular components usually from simpler molecules. The cells use these compounds as both the carbon source and the energy source. We will next consider the biosynthesis of a few representative classes of biological molecules—carbohydrates, lipids, amino acids, purines, and pyrimidines. As we do so, keep in mind that synthesis reactions require a net input of energy.

BIOSYNTHESIS OF POLYSACCHARIDES

Microorganisms synthesize sugars and polysaccharides. The carbon atoms required to synthesize glucose are derived from the intermediates produced during processes such as glycolysis and the Krebs cycle and from lipids or amino acids. After synthesizing glucose (or other simple sugars), bacteria may assemble it into the more complex polysaccharides. One such example is glycogen. For bacteria to build glucose into glycogen, glucose units must be phosphorylated and linked. The product of glucose phosphorylation is glucose 6-phosphate. Such a process involves the expenditure of energy, usually in the form of ATP, and dehydration (loss of water). The key intermediate in the synthesis of glycogen in bacteria is called *adenosine diphosphoglucose (ADPG)*. It is formed when a molecule of ATP is added to glucose 6-phosphate (Figure 5.25). Once the building block is synthesized, it is linked with similar units through dehydration synthesis to form glycogen.

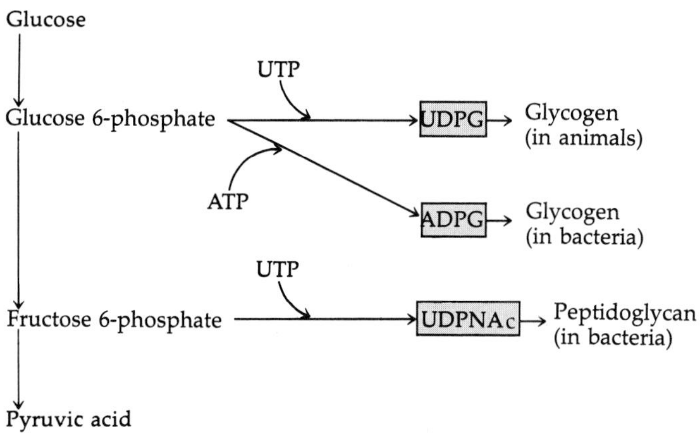

FIGURE 5.25 Biosynthesis of polysaccharides.

Using a nucleotide called uridine triphosphate (UTP) as a source of energy, animals synthesize glycogen and many other carbohydrates from *uridine diphosphoglucose, UDPG* (Figure 5.25). A compound related to UDPG, called *UDP-N-acetylglucosamine (UDPNAc)*, is a key starting material in the biosynthesis of peptidoglycan, the substance that forms bacterial cell walls. UDPNAc is formed from fructose 6-phosphate, and the reaction also uses UTP.

BIOSYNTHESIS OF LIPIDS

Because lipids vary in chemical composition, they are synthesized by a variety of routes. Cells synthesize fats by joining glycerol and fatty acids. The glycerol portion of the fat is derived from dihydroxyacetone phosphate, an intermediate formed during glycolysis. Fatty acids, which are long-chain hydrocarbons (hydrogen linked to carbon), are built up when two-carbon fragments of acetyl CoA are successively added to each other (Figure 5.26). As with polysaccharide synthesis, the building units of fats and other lipids are linked via dehydration reactions that require energy, not always in the form of ATP. In the synthesis of phospholipids, for example, the energy is derived from the hydrolysis of a nucleotide called cytidine triphosphate (CTP).

As mentioned previously, the most important role of lipids is as structural components of biological membranes, and most membrane lipids are phospholipids. A lipid of a very different structure, cholesterol, is also found in plasma membranes of eucaryotic cells. Waxes are lipids that are important components of the cell wall of acid-fast bacteria. Other lipids, such as carotenoids, provide the red, orange, and yellow pigments of some microorganisms. Some lipids form portions of chlorophyll molecules. Lipids also function in energy storage. Recall that the breakdown products of lipids after biological oxidation feed into the Krebs cycle.

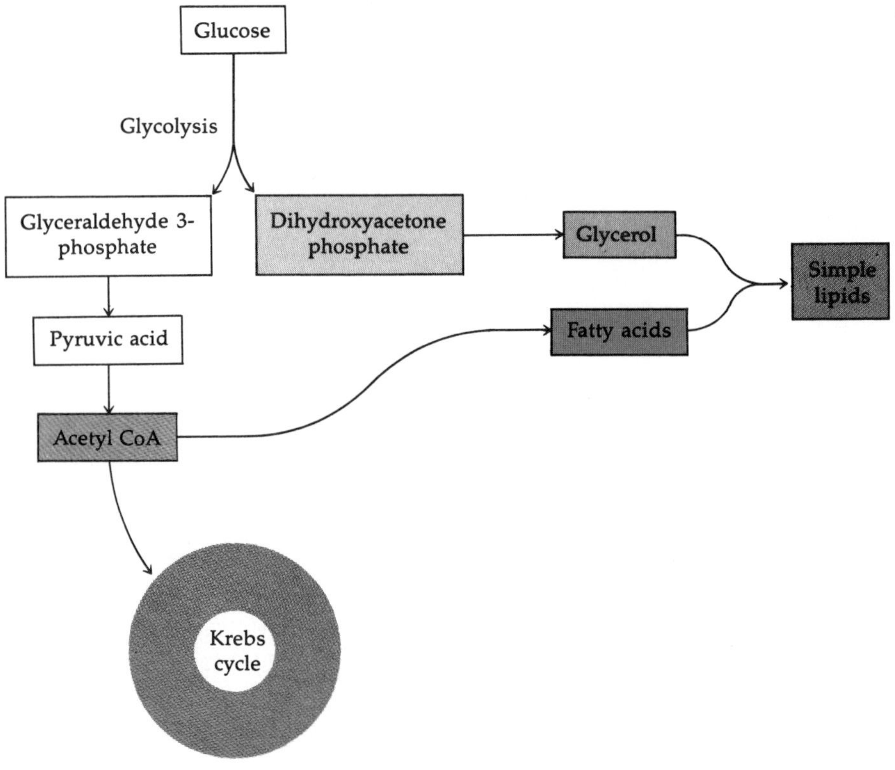

FIGURE 5.26 Biosynthesis of simple lipids.

BIOSYNTHESIS OF AMINO ACIDS AND PROTEINS

Amino acids are required for protein biosynthesis. Some microbes, such as *E. coli,* contain the enzymes necessary to use starting materials, such as glucose and inorganic salts, for the synthesis of all the amino acids they need. Organisms with the necessary enzymes can synthesize all amino acids directly or indirectly from intermediates of carbohydrate metabolism (Figure 5.27a). Other microbes require that the environment provide some preformed amino acids.

One important source of the *precursors* (intermediates) used in amino acid synthesis is the Krebs cycle. Adding an amine group to pyruvic acid or to an appropriate organic acid of the Krebs cycle converts the acid into an amino acid. This process is called *amination.* If the amine group comes from a preexisting amino acid, the process is called *transamination* (Figure 5.27b). Other precursors for amino acid biosynthesis are derived from the pentose phosphate pathway and from the Entner–Doudoroff pathway.

Most amino acids within cells are destined to be building blocks for protein synthesis. Proteins play major roles in the cell as enzymes, structural components, and toxins, to name just a few uses. The joining of amino acids to form proteins involves dehydration synthesis and requires energy in the form of ATP. The mechanism of protein synthesis involves genes and will be discussed in Chapter 8.

BIOSYNTHESIS OF PURINES AND PYRIMIDINES

As you may recall from Chapter 2, the informational molecules DNA and RNA consist of repeating units called *nucleotides*, each of which consists of a purine or pyrimidine, a pentose (five-carbon sugar), and a phosphate group. The five-carbon sugars of nucleotides are derived from either the pentose phosphate pathway or the Entner–Doudoroff pathway. Certain amino acids—aspartic acid, glycine, and glutamine—made from intermediates produced during glycolysis and in the Krebs cycle participate in the biosyntheses of purines and pyrimidines (Figure 5.28). The carbon and nitrogen atoms derived from these amino acids form the purine and pyrimidine rings, and the energy for synthesis is provided by ATP. DNA contains all the information necessary to determine the specific structures and functions of cells. Both RNA and DNA are required for protein synthesis. In addition, such nucleotides as ATP, NAD^+, and $NADP^+$ assume roles in stimulating and inhibiting the rate of cellular metabolism. The synthesis of DNA and RNA from nucleotides will be discussed in Chapter 8.

Integration of Metabolism

We have seen thus far that the metabolic processes of microbes produce energy from light, inorganic compounds, and organic compounds. Reactions also occur

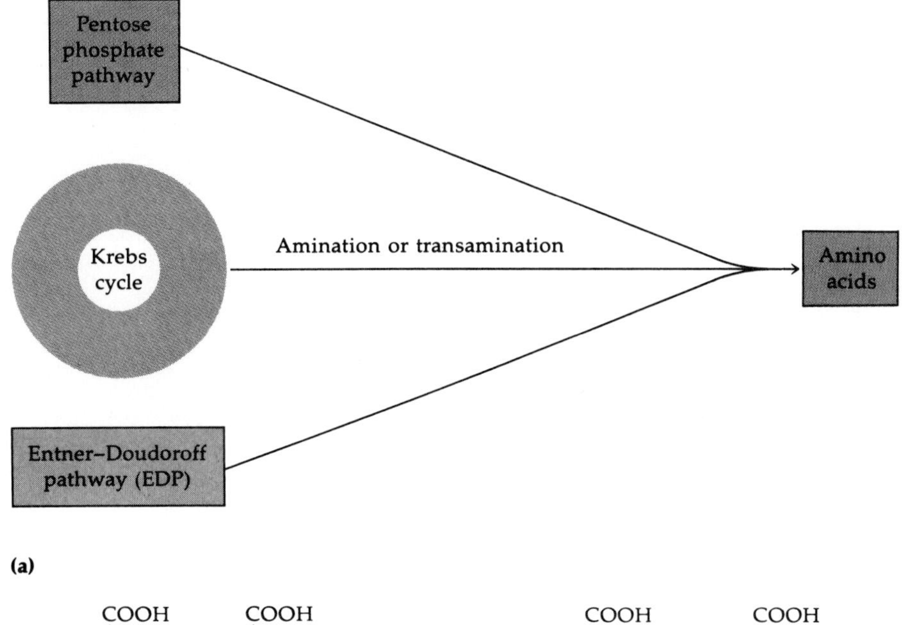

(a)

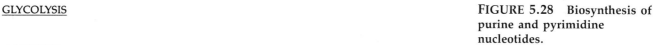

COOH		COOH		COOH		COOH
CH_2		CH_2		CH_2		CH_2
CH_2	+	C=O	$\rightleftharpoons$ Transamination	CH_2	+	H—C—NH_2
H—C—NH_2		COOH		C=O		COOH
COOH				COOH		
Glutamic acid		Oxaloacetic acid		α-Ketoglutaric acid		Aspartic acid

(b)

FIGURE 5.27 Biosynthesis of amino acids. **(a)** Pathways of amino acid biosynthesis through amination or transamination of intermediates of carbohydrate metabolism. **(b)** Transamination, a process by which new amino acids are made with the amine groups from old amino acids. Glutamic acid and aspartic acid are both amino acids. The other two compounds are intermediates in the Krebs cycle.

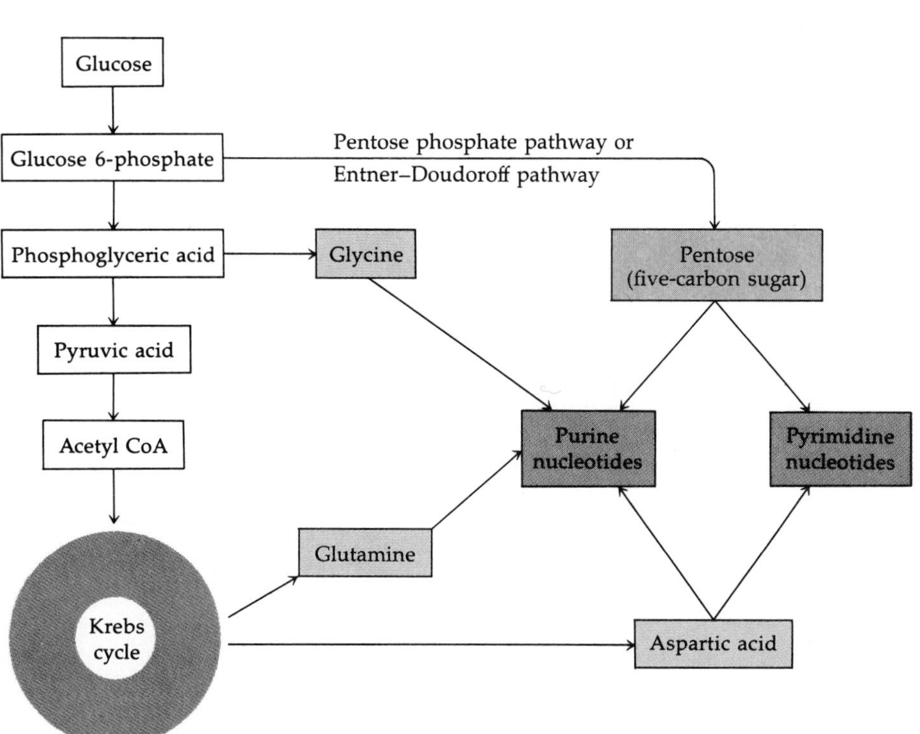

FIGURE 5.28 Biosynthesis of purine and pyrimidine nucleotides.

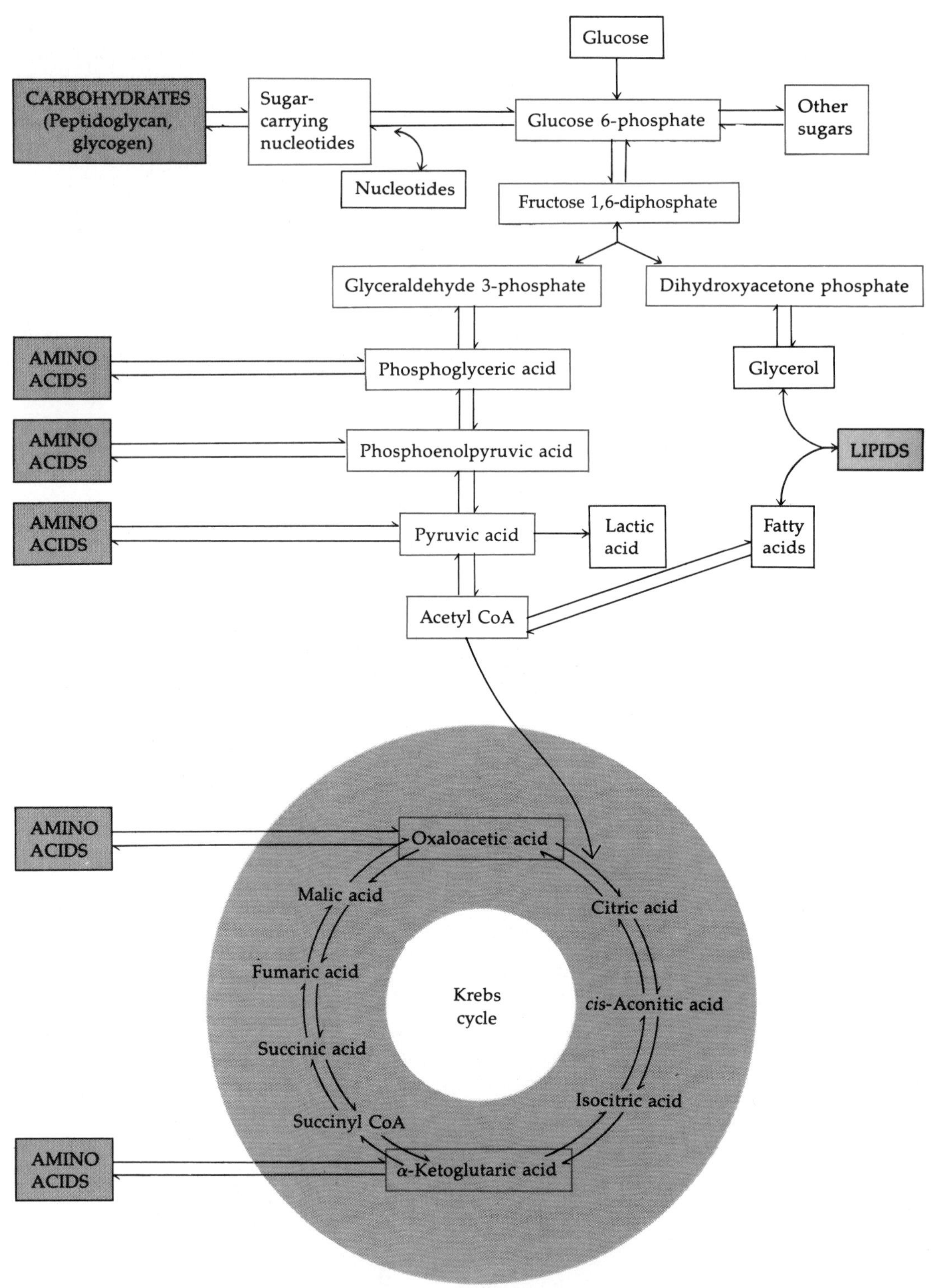

FIGURE 5.29 Integration of metabolism. Key intermediates are outlined in color. Although not indicated in the figure, amino acids and ribose are used in the synthesis of purine and pyrimidine nucleotides (see Figure 5.28). The double arrows indicate the amphibolic pathways.

in which energy is utilized for biosynthesis. With such a variety of activity, it might be imagined that anabolic and catabolic reactions occur independently of each other in space and time. Actually, anabolic and catabolic reactions are joined through a group of common intermediates (identified as key intermediates in Figure 5.29). Both anabolic and catabolic reactions also share some biochemical pathways, such as the Krebs cycle. For example, reactions in the Krebs cycle not only participate in the oxidation of glucose but also produce intermediates that can be converted to amino acids. Biochemical pathways that function in both anabolism and catabolism are called **amphibolic pathways,** meaning that they are dual-purpose.

Amphibolic pathways bridge the reactions that lead to the breakdown and synthesis of carbohydrates, lipids, proteins, and nucleotides. Such pathways enable simultaneous reactions to occur in which the breakdown product formed in one reaction is used in another reaction to synthesize a different compound, and vice versa. Since various intermediates are common to both anabolic and catabolic reactions, mechanisms exist that regulate synthesis and breakdown pathways and allow these reactions to occur simultaneously. One such mechanism involves the use of different coenzymes for opposite pathways. For example, NAD^+ is involved in catabolic reactions, whereas $NADP^+$ is involved in anabolic reactions. Enzymes can also coordinate anabolic and catabolic reactions by accelerating or inhibiting the rates of biochemical reactions. The energy stores of a cell can also affect the rates of biochemical reactions. For example, if ATP begins to accumulate, an enzyme shuts down glycolysis. This control helps to synchronize the rates of glycolysis and the Krebs cycle. Thus, if citric acid consumption increases, either because of a demand for more ATP or because anabolic pathways are draining off intermediates of the citric acid cycle, glycolysis accelerates and meets the demand.

STUDY OUTLINE

Catabolic and Anabolic Reactions (pp. 101–102)

1. The sum of all chemical reactions within a living organism is known as metabolism.
2. Catabolism refers to chemical reactions that result in the breakdown of more complex organic molecules into simpler substances. Catabolic reactions usually release energy.
3. Anabolism refers to chemical reactions in which simpler substances are combined to form more complex molecules. Anabolic reactions usually require energy.
4. The energy of catabolic reactions is used to drive anabolic reactions.
5. The energy for chemical reactions is stored in ATP.

Enzymes (pp. 102–109)

1. Enzymes are proteins produced by living cells that catalyze chemical reactions.
2. Enzymes are generally globular proteins with characteristic three-dimensional shapes.
3. Enzymes catalyze chemical reactions by lowering the activation energy.
4. Enzymes are efficient, are able to operate at relatively low temperatures, and are subject to various cellular controls.

NAMING ENZYMES (p. 103)

1. Enzymes' names usually end in *-ase*.
2. The six classes of enzymes are defined on the basis of the types of reactions they catalyze.

ENZYME COMPONENTS (pp. 103–104)

1. Most enzymes are holoenzymes, consisting of a protein portion (apoenzyme) and a nonprotein portion (cofactor).
2. The cofactor can be a metal ion (iron, copper, magnesium, manganese, zinc, calcium, or cobalt) or a complex organic molecule known as a coenzyme (NAD, NADP, FMN, FAD, and coenzyme A).

MECHANISM OF ENZYMATIC ACTION (pp. 104–105)

1. When an enzyme and substrate combine, the substrate is transformed, and the enzyme is recovered.
2. Enzymes are characterized by specificity, which is a function of their active sites.

FACTORS INFLUENCING ENZYMATIC ACTIVITY (pp. 105–107)

1. At high temperatures, enzymes undergo denaturation and lose their catalytic properties; at low temperatures, the reaction rate decreases.
2. The pH at which enzymatic activity is maximal is known as the pH optimum.
3. Within limits, enzymatic activity increases as substrate concentration increases.
4. Competitive inhibitors compete with the normal substrate for the active site of the enzyme. Noncompetitive inhibitors act on other parts of the apoenzyme or on the cofactor and decrease the enzyme's ability to combine with the normal substrate.

FEEDBACK INHIBITION (pp. 108–109)

1. Feedback inhibition occurs when the end-product of a pathway inhibits an enzyme's activity in the pathway.

Energy Production (pp. 109–110)

OXIDATION–REDUCTION (pp. 109–110)

1. Oxidation is the removal of one or more electrons from a substrate. Protons (H^+) are often removed with the electrons.
2. Reduction of a substrate refers to its gain of one or more electrons.
3. Each time a substance is oxidized, another is simultaneously reduced.
4. NAD^+ is the oxidized form; NADH is the reduced form.
5. Glucose is a reduced molecule; energy is released during a cell's oxidation of glucose.

GENERATION OF ATP (p. 110)

1. Energy released during certain metabolic reactions can be trapped to form ATP from ADP and Ⓟ (phosphate). Addition of a Ⓟ to a molecule is called phosphorylation.
2. During substrate-level phosphorylation, a high-energy Ⓟ from an intermediate in catabolism is added to ADP.
3. During oxidative phosphorylation, energy is released as electrons are passed to a series of electron acceptors (an electron transport chain) and finally to O_2 or another inorganic compound.
4. During photophosphorylation, energy from light is trapped by chlorophyll, and electrons are passed through a series of electron acceptors. The electron transfer releases energy used for the synthesis of ATP.

Biochemical Pathways of Energy Production (pp. 110–111)

1. Series of enzymatically catalyzed chemical reactions called biochemical pathways store energy in and release energy from organic molecules.

Carbohydrate Catabolism (pp. 111–124)

1. Most of a cell's energy is produced from the oxidation of carbohydrates.
2. Glucose is the most commonly used carbohydrate.
3. The two major types of glucose catabolism are respiration, in which glucose is completely broken down, and fermentation, in which it is partially broken down.

GLYCOLYSIS (pp. 111–114)

1. The most common pathway for the oxidation of glucose is glycolysis. Pyruvic acid is the end-product.
2. Two ATP and two NADH molecules are produced from one glucose molecule.

ALTERNATIVES TO GLYCOLYSIS (p. 114)

1. The pentose phosphate pathway is used to metabolize five-carbon sugars; one ATP and 12 NADPH molecules are produced from one glucose molecule.
2. The Entner–Doudoroff pathway yields one ATP and two NADPH molecules from one glucose molecule.

DEFINITION OF RESPIRATION (p. 114)

1. During respiration, organic molecules are oxidized. Energy is generated from the electron transport chain.
2. In aerobic respiration, O_2 functions as the final electron acceptor.
3. In anaerobic respiration, the final electron acceptor is an inorganic molecule other than O_2.

THE KREBS CYCLE (pp. 114–116)

1. Decarboxylation of pyruvic acid produces one CO_2 molecule and one acetyl group.
2. Two-carbon acetyl groups are oxidized in the Krebs cycle. Electrons are picked up by NAD^+ and FAD for the electron transport chain.
3. From one molecule of glucose, oxidation produces six molecules of NADH, two molecules of $FADH_2$, and two molecules of ATP.
4. Decarboxylation produces six molecules of CO_2.

ELECTRON TRANSPORT CHAIN (pp. 116–117)

1. Electrons are brought to the electron transport chain by NADH.
2. The electron transport chain consists of carriers, including flavoproteins, cytochromes, and ubiquinones.

CHEMIOSMOTIC MECHANISM OF ATP GENERATION (pp. 117–118)

1. Protons being pumped across the membrane generate a proton motive force, as electrons move through a series of acceptors or carriers.
2. Energy produced from movement of the protons back across the membrane is used by ATPase to make ATP and ADP and Ⓟ.
3. In eucaryotes, electron carriers are located on the inner mitochondrial membrane; in procaryotes, electron carriers are in the plasma membrane.

SUMMARY OF AEROBIC RESPIRATION (pp. 118–120)

1. In aerobic procaryotes, 38 ATP molecules can be produced from complete oxidation of a glucose molecule in glycolysis, the Krebs cycle, and the electron transport chain.
2. In eucaryotes, 36 ATP molecules are produced from complete oxidation of a glucose molecule.

ANAEROBIC RESPIRATION (p. 120)

1. The final electron acceptors in anaerobic respiration include NO_3^-, SO_4^{2-}, and CO_3^{2-}.
2. The total ATP yield is less than aerobic respiration because only part of the Krebs cycle operates under anaerobic conditions.

FERMENTATION (pp. 120–122)

1. Fermentation releases energy from sugars or other organic molecules by oxidation.
2. O_2 is not required in fermentation.
3. Two ATP molecules are produced by substrate-level phosphorylation.

4. Electrons removed from the substrate reduce NAD^+.
5. The final electron acceptor is an organic molecule.
6. In lactic acid fermentation, pyruvic acid is reduced by NADH to lactic acid.
7. In alcohol fermentation, acetaldehyde is reduced by NADH to produce ethanol.
8. Heterolactic fermenters can use the pentose phosphate pathway to produce lactic acid and ethanol.

LIPID CATABOLISM (pp. 122–124)

1. Lipases hydrolyze lipids into glycerol and fatty acids.
2. Glycerol is catabolized by conversion to dihydroxyacetone phosphate, and fatty acids are catabolized by beta oxidation.
3. Catabolic products can be further broken down in glycolysis and the Krebs cycle.

PROTEIN CATABOLISM (pp. 124)

1. Before amino acids can be catabolized, they must be converted to various substances that enter the Krebs cycle.
2. Transamination, decarboxylation, and dehydrogenation reactions convert the amino acids to be catabolized.

Photosynthesis (pp. 125–130)

1. Photosynthesis is the conversion of light energy from the sun into chemical energy; the chemical energy is used for carbon fixation.

THE LIGHT REACTIONS: PHOTOPHOSPHORYLATION (p. 125)

1. Chlorophyll *a* is used by green plants, algae, and cyanobacteria; it is found in thylakoid membranes.
2. Electrons from chlorophyll pass through an electron transport chain, where ATP is produced by chemiosmosis.
3. In cyclic photophosphorylation, the electrons return to the chlorophyll.
4. In noncyclic photophosphorylation, the electrons are used to reduce NADP and electrons are returned to chlorophyll from H_2O or H_2S.
5. When H_2O is oxidized by green plants, algae, and cyanobacteria, O_2 is produced.

THE DARK REACTIONS: CALVIN–BENSON CYCLE (p. 125)

1. CO_2 is used to synthesize sugars.
2. Synthesis of glyceraldehyde 3-phosphate requires three molecules of CO_2, nine molecules of ATP, and six molecules of NADPH.

NUTRITIONAL PATTERNS AMONG ORGANISMS (pp. 125–130)

1. Photoautotrophs obtain energy by photophosphorylation and fix carbon from CO_2 via the Calvin–Benson cycle to synthesize organic compounds.
2. Cyanobacteria are oxygenic phototrophs. Green sulfur bacteria and purple sulfur bacteria are anoxygenic phototrophs.
3. Photoheterotrophs use light as an energy source and an organic compound for their carbon source or electron donor.
4. Chemoautotrophs use inorganic compounds as their energy source and carbon dioxide as their carbon source.
5. Chemoheterotrophs use complex organic molecules as their carbon and energy sources.

SUMMARY OF ENERGY PRODUCTION MECHANISMS (p. 130)

1. Sunlight is converted to chemical energy in oxidation–reduction reactions carried on by phototrophs. Chemotrophs can use this chemical energy.
2. In oxidation–reduction reactions, energy is derived from the transfer of electrons.
3. To produce energy, a cell needs an electron donor (organic or inorganic), a system of electron carriers, and a final electron acceptor (organic or inorganic).

Biochemical Pathways of Energy Utilization (Anabolism) (pp. 131–132)

BIOSYNTHESIS OF POLYSACCHARIDES (p. 131)

1. Glycogen is formed from ADPG.
2. UDPNAc is the starting material for the biosynthesis of peptidoglycan.

BIOSYNTHESIS OF LIPIDS (p. 131)

1. Lipids are synthesized from fatty acids and glycerol.
2. Glycerol is derived from dihydroxyacetone phosphate, and fatty acids are built from acetyl CoA.

BIOSYNTHESIS OF AMINO ACIDS AND PROTEINS (p. 132)

1. Amino acids are required for protein biosynthesis.
2. All amino acids can be synthesized either directly or indirectly from intermediates of carbohydrate metabolism, particularly from the Krebs cycle.

BIOSYNTHESIS OF PURINES AND PYRIMIDINES (p. 132)

1. The sugars composing nucleotides are derived from either the pentose phosphate pathway or the Entner–Doudoroff pathway.
2. Carbon and nitrogen atoms from certain amino acids form the backbones of the purines and pyrimidines.

Integration of Metabolism (pp. 132–135)

1. Anabolic and catabolic reactions are integrated through a group of common intermediates.
2. Such integrated pathways are referred to as amphibolic pathways.

STUDY QUESTIONS

Review

1. Define metabolism.

2. Distinguish between catabolism and anabolism. How are these processes related?

3. Using the diagrams below, show
   (a) Where the substrate will bind.
   (b) Where the competitive inhibitor will bind.
   (c) Where the noncompetitive inhibitor will bind.
   (d) Which of those four elements could be the inhibitor in feedback inhibition?

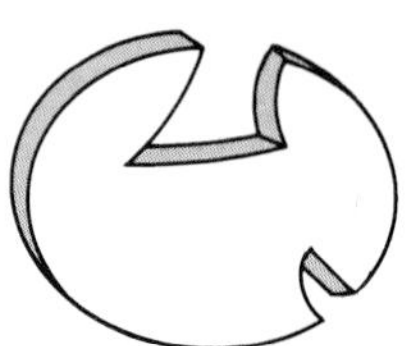
Enzyme

Substrate

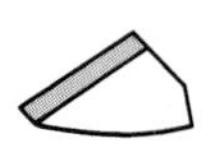
Competitive inhibitor

Noncompetitive inhibitor

4. What will the effect of the reactions in Question 3 be?

5. Why are most enzymes active at one particular temperature? Why are enzymes less active below this temperature? What happens above this temperature?

6. Which substances in each of the following reactions are being oxidized? Which are being reduced?

$$\underset{\text{(a) Acetaldehyde}}{\mathrm{H{-}\overset{}{C}(=O){-}CH_3}} + NADH + H^+ \longrightarrow \underset{\text{Ethanol}}{\mathrm{H{-}C(H)(OH){-}CH_3}} + NAD^+$$

$$\underset{\text{(b) Succinic acid}}{\mathrm{COOH{-}CH_2{-}CH_2{-}COOH}} + FAD \longrightarrow \underset{\text{Fumaric acid}}{\mathrm{COOH{-}CH{=}CH{-}COOH}} + FADH_2$$

7. What is the fate of pyruvic acid in an organism that uses respiration? Fermentation?

8. List four compounds that can be made from pyruvic acid by an organism that uses fermentation only.

9. Fill in the table below with the carbon source and energy source of each type of organism.

Organism	Carbon Source	Energy Source
Photoautotroph		
Photoheterotroph		
Chemoautotroph		
Chemoheterotroph		

10. There are three mechanisms for the phosphorylation of ADP to produce ATP. Write the name of the mechanism that describes each of the following reactions in the table below.

ATP Generated By	Reaction
	An electron, liberated from chlorophyll by light, is passed down an electron transport chain.
	Cytochrome *c* passes two electrons to cytochrome *a*.
	$\mathrm{CH_2{=}C(O \sim Ⓟ){-}COOH}$ (Phosphoenolpyruvic acid) $\longrightarrow$ $\mathrm{CH_3{-}C({=}O){-}COOH}$ (Pyruvic acid)

11. Define *oxidation–reduction*, and differentiate between the following terms:
    (a) Aerobic and anaerobic respiration
    (b) Respiration and fermentation
    (c) Cyclic and noncyclic photophosphorylation

12. The pentose phosphate pathway produces only one ATP. List four advantages of this pathway for the cell.

Use the following diagrams for Questions 13 through 21.

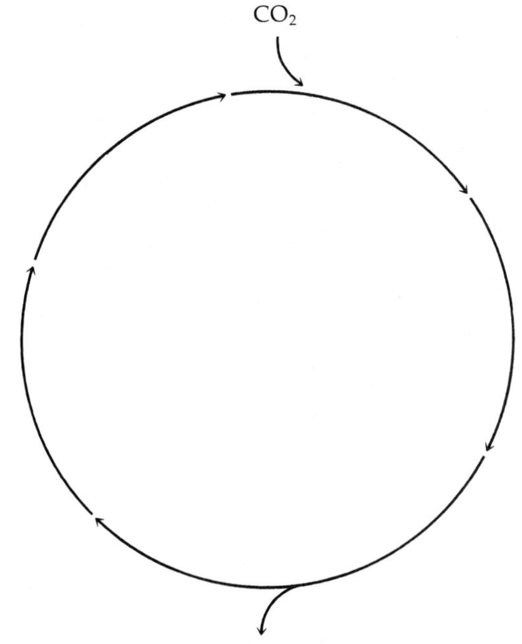

(a)

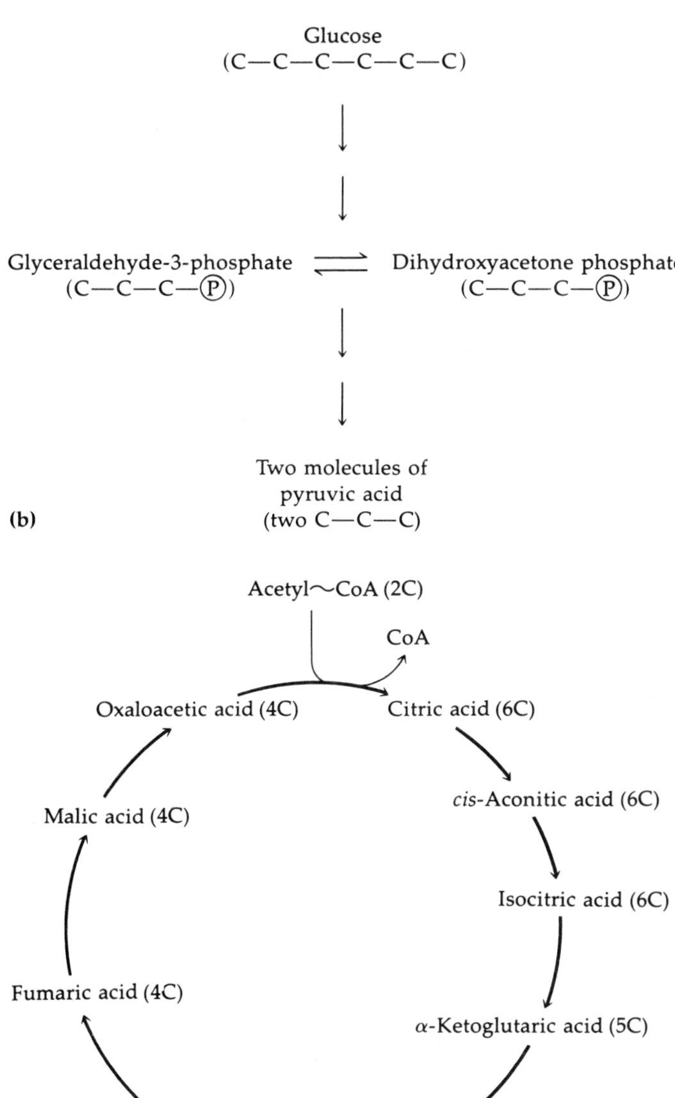

13. Name the pathways diagrammed in a, b, and c.

14. Show where glycerol is catabolized and where fatty acids are catabolized.

15. Show where the amino acid glutamic acid is catabolized:

$$HOOC—CH_2—CH_2—\underset{NH_2}{\overset{H}{C}}—COOH$$

16. Show how these pathways are related.

17. Where is ATP required in pathways a and b?

18. Where is CO_2 released in pathways a and b?

19. Show where a long-chain hydrocarbon such as petroleum is catabolized.

20. Where is NADH (or $FADH_2$ or NADPH) used and produced in these pathways?

21. Identify four places where anabolic and catabolic pathways are integrated.

22. Explain how ATP is a key intermediate in metabolism.

23. An enzyme and substrate are combined. The rate of reaction begins as shown in this graph. To complete the graph, show the effect of increased substrate concentration on a constant enzyme concentration. Show the effect of increased temperature.

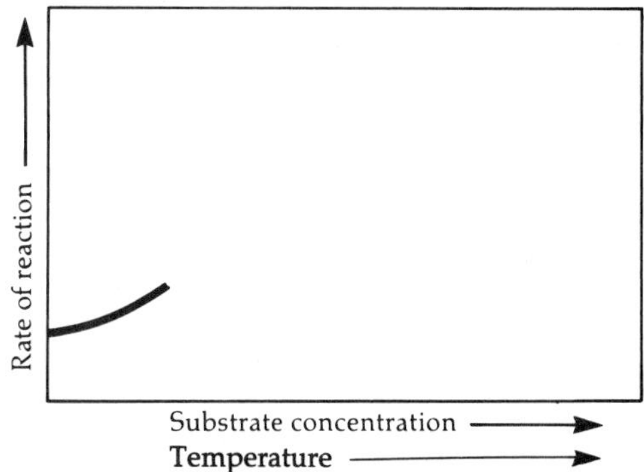

Challenge

1. Write your own definition of the chemiosmotic mechanism of ATP generation. On Figure 5.16, mark the following:
   (a) The acidic side of the membrane
   (b) The side with a positive electrical charge
   (c) Potential energy
   (d) Kinetic energy
2. Explain why, even under ideal conditions, *Streptococcus* grows slowly.
3. The graph at right shows the normal rate of reaction of an enzyme and its substrate (black) and the rate when an excess of competitive inhibitor is present (red). Explain why the graph appears as it does.

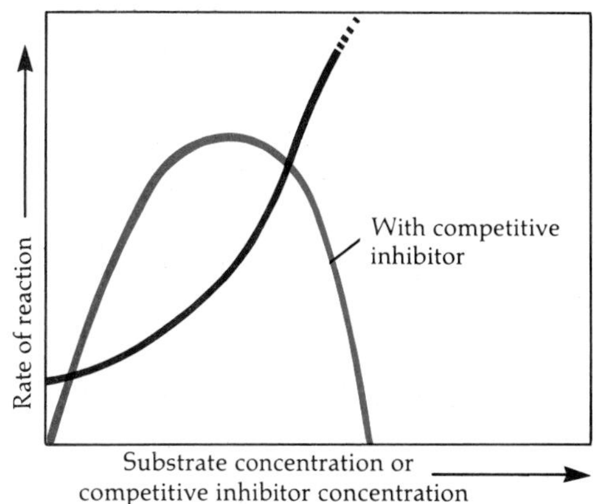

4. Why must NADH be reoxidized? How does this happen in an organism that uses respiration? Fermentation?
5. Compare and contrast carbohydrate catabolism and energy production in the following bacteria:
   (a) *Pseudomonas*, an aerobic chemoheterotroph
   (b) *Spirulina*, an oxygenic photoautotroph
   (c) *Ectothiorhodospira*, an anoxygenic photoautotroph
6. How much ATP could be obtained from the complete oxidation of one molecule of glucose? From one molecule of butterfat containing one glycerol and three 12-carbon chains?
7. The chemoautotroph *Thiobacillus* can obtain energy from oxidation of arsenic ($As^{3+} \rightarrow As^{5+}$). How does this reaction provide energy? How can this bacterium be put to use by humans?

FURTHER READING

Blakemore, R.P., and R.B. Frankel. "Magnetic navigation in bacteria." *Scientific American* 245(6):58–65, December 1981. Describes synthesis and function of magnets in bacterial cells.

Lehninger, A.L. *Bioenergetics: The Molecular Basis of Biological Energy Transformation*, 2nd ed. Redwood City, Calif.: Benjamin/Cummings, 1975. A classic reference that provides clear explanations of bioenergetics; predates the chemiosmotic theory, however.

Mathews, C.K., and K.E. van Holde. *Biochemistry*. Redwood City, Calif: Benjamin/Cummings, 1990. A biochemistry test with chapters on metabolic pathways of energy production and biosynthesis.

Neidhardt, F.C., J.L. Ingraham, and M. Schaechter. *Physiology of the Bacterial Cell*. Sunderland, Mass.: Sinauer, 1990. Includes chapters on all aspects of microbial metabolism, with emphasis on the success of bacteria in a variety of environments.

Nicholls, D.G. *Bioenergetics: An Introduction to the Chemiosmotic Theory*. New York: Academic Press, 1982. An introductory text on this theory.

Oliver, D. "Protein secretion in *Escherichia coli*." *Annual Review of Microbiology* 39:615–648, 1985. A summary of the processes by which membrane proteins and extracellular proteins are synthesized and excreted from cells.

Proton Pumping. *BioScience* 35:14–48, January 1985. Includes seven articles, with excellent illustrations, describing proton chemistry and photosynthetic, halobacterial, and eucaryotic membrane systems.

Stryer, L. *Biochemistry*, 3rd ed. New York: W.H. Freeman, 1988. The latest edition of a clearly written and widely used textbook.

Youvan, D.C., and B.L. Marrs. "Molecular mechanisms of photosynthesis." *Scientific American* 256(6):42–48, June 1987. Describes modern techniques from physics and molecular genetics used to identify the electron transport chain in bacterial (cyclic) photophosphorylation.

CHAPTER 6

Microbial Growth

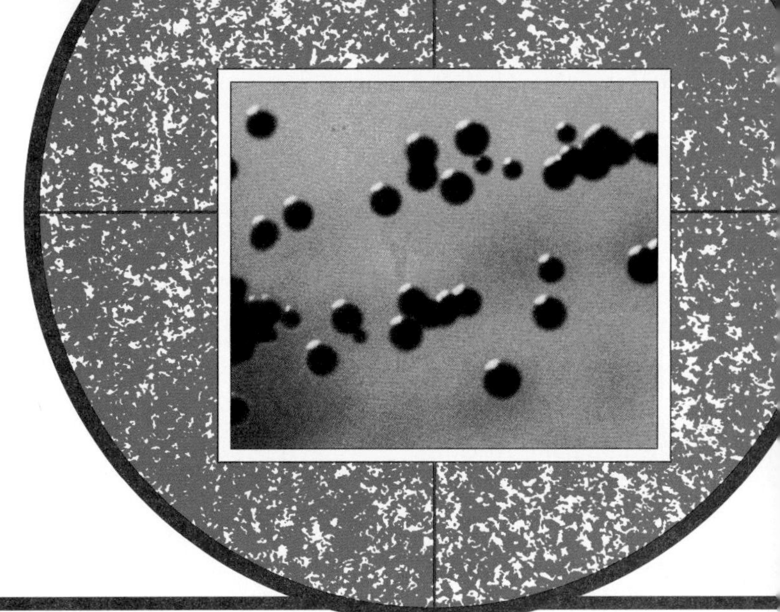

LEARNING OBJECTIVES

- Define bacterial growth, including binary fission.
- Classify microbes into three principal groups on the basis of preferred temperature range.
- Explain the importance of pH and osmotic pressure to microbial growth.
- In general terms, list the chemical requirements for growth.
- Explain how microbes are classified on the basis of oxygen requirements.
- Distinguish between chemically defined and complex media.
- Justify the use of the following culture methods: selective and differential media, anaerobic techniques, living host cells, candle jars.
- Describe how pure cultures can be isolated by using streak plates.
- Compare the phases of microbial growth and their relation to generation time.
- Describe several direct and indirect measurements of microbial growth.

When we talk about microbial growth, we are really referring to the *number* of cells, not the *size* of the cells. Microbes that are "growing" are increasing in number, accumulating into *clumps* of hundreds, *colonies* of hundreds of thousands, or *populations* of billions. For the most part, we are not concerned with the growth of individual cells. Although individual cells approximately double in size during their lifetime, this is not a very significant change compared with the size increases observed during the lifetime of plants and animals.

Requirements for Growth

The requirements for microbial growth can be divided into two main categories, physical and chemical. Physical aspects include temperature, pH, and osmotic pressure. Chemical requirements include water, sources of carbon and nitrogen, minerals, oxygen, and organic growth factors.

Microbial populations can become incredibly large in a very short time. By understanding the conditions necessary for microbial growth, we can predict how quickly microorganisms will grow in various situations, and we can determine how to control the growth of microbes that cause serious disease and food spoilage. At the same time, we can learn how to encourage the growth of helpful microbes and those we wish to study. The manager of a sewage treatment plant and the manager of a brewery do not seem to have much in common, but both would be interested in promoting rapid microbial activity.

We will examine in this chapter the physical and chemical requirements for microbial growth, the various kinds of culture media, bacterial division, the phases of microbial growth, and the methods of mea-

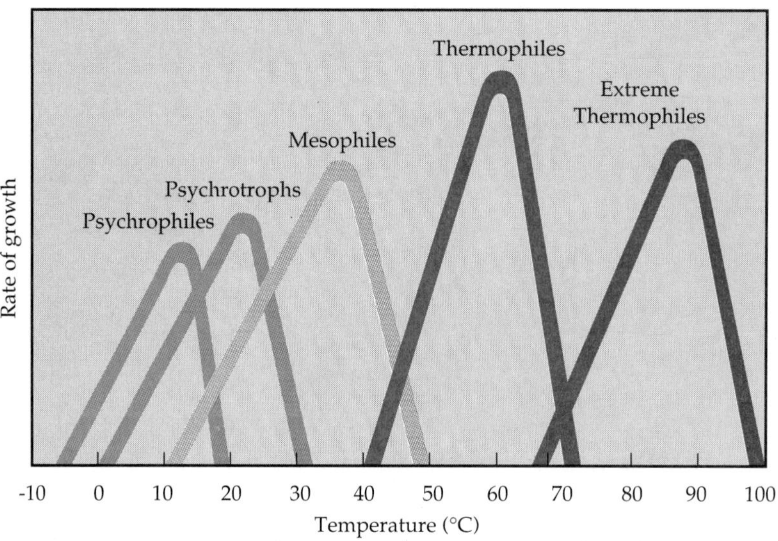

FIGURE 6.1 Typical growth responses to temperature by psychrophiles, mesophiles, and thermophiles. The optimum growth (fastest reproduction) is represented by the peak of the curve. Note that the reproductive rate drops off very quickly at temperatures only a little above the optimum. At either extreme of the temperature range, the reproductive rate is much lower than the rate at the optimum temperature.

suring microbial growth. In later chapters we will see how unwanted microbial growth can be controlled.

PHYSICAL REQUIREMENTS

Temperature

Most microorganisms grow well at the same temperatures favored by humans. However, certain bacteria are capable of growing in extreme cold or extreme heat, temperatures that would certainly hinder the survival of most higher organisms.

Microorganisms are divided into three groups on the basis of their preferred range of temperature. These are the **psychrophiles** (cold-loving microbes), **mesophiles** (moderate-temperature-loving microbes), and **thermophiles** (heat-loving microbes). Most bacteria grow only within a limited range of temperatures, and their maximum and minimum growth temperatures are about 30°C apart. They grow poorly at some temperatures within their range and well at others. Each bacterial species grows at particular minimum, optimum, and maximum temperatures. The **minimum growth temperature** is the lowest temperature at which the species will grow. The **optimum growth temperature** is the temperature at which the species grows best. The **maximum growth temperature** is the highest temperature at which growth is possible. By graphing the growth response over a temperature range, we can see that the optimum growth temperature is usually near the top of the range; above that temperature, the rate of growth drops off rapidly (Figure 6.1). This presumably happens because the high temperature has inactivated necessary enzymatic systems of the cell.

The ranges and maximum growth temperatures that define bacteria as psychrophiles, mesophiles, or thermophiles are not rigidly defined and vary from reference to reference. Psychrophiles, for example, were originally considered simply as organisms capable of growing at 0°C. However, there seem to be two fairly distinct groups of organisms capable of growth at that temperature. One group, composed of psychrophiles in the strictest sense, can grow at 0°C but has an optimal growth temperature of about 15°C. Most of these are so sensitive to higher temperatures that they will not even grow in a reasonably warm room (20°C). Such organisms are found mostly in the oceans' depths or in certain Arctic regions and seldom cause problems in food preservation. The other group that can grow at 0°C has higher optimal temperatures, usually between 20° and 30°C. It has been proposed that these organisms that grow fairly well at refrigerator temperatures be called **psychrotrophs.** Organisms of this type are much more common and are the most likely to be encountered in low-temperature food spoilage. It is the term we will use for microorganisms responsible for such spoilage.

Refrigeration is the most common method of preserving household food supplies. It is based on the principle that microbial reproductive rates decrease at low temperatures. Although microbes usually survive even subfreezing temperatures (they might become entirely dormant), they gradually decline in number. Some species decline faster than others. Psychrotrophs actually do not grow well at low temperatures, except in comparison with other organisms, but given time they are able to slowly degrade the food. This spoilage might take the form of mold mycelium, slime on food surfaces, or off tastes or colors in foods. The temperature inside a properly set refrigerator will greatly slow the growth of most spoilage organisms and will entirely prevent the growth of all but a few pathogenic bacteria.

MICROBIOLOGY HIGHLIGHTS

Hydrothermal Bacteria

Until humans explored the ocean floor, scientists believed that only a few forms of life could survive in that extremely cold, completely dark, oxygen-poor environment. Then, in 1977, the first manned vehicle able to penetrate to the bottom of the deepest oceans carried two scientists 2600 meters below the surface at the Galápagos Rift, about 350 km northeast of the Galápagos Islands. There, amid the vast expanse of barren basalt pillows, the scientists found unexpectedly rich oases of life, including mollusks, crustaceans, and worms. How do such creatures survive in these harsh conditions?

Ecosystem of the Hydrothermal Vents

Life at the surface of the world's oceans depends on photosynthetic organisms, such as bacteria, plankton, and green plants, to harness the Sun's energy and make carbohydrates. At the ocean floor, where no light penetrates, photosynthesis is not possible. The scientists found that the so-called primary producers at the ocean floor are chemoautotrophic bacteria. Using such inorganic compounds as hydrogen sulfide (H_2S) as a source of energy and carbon dioxide (CO_2) as a source of carbon, the chemoautotrophs create an environment that supports higher life forms. Hydrothermal vents in the sea floor supply the H_2S and CO_2. As superheated water from within the Earth rises through fractures in the Earth's crust called vents, it reacts with surrounding rock and dissolves metal ions, sulfides, and CO_2. The ecosystems of these vents depend on an abundance of sulfur compounds in the hot water. The concentration of sulfide (S^{2-}) is three times greater than the concentration of molecular oxygen in and around the vents. Such high sulfide concentrations are toxic to many organisms, but not to the creatures that inhabit this exotic environment.

Mats of bacteria grow along the sides of the vents, where temperatures can be as high as 350°C. This is the highest temperature that any organism is known to tolerate. Above the vent, where temperatures are about 30°C, the concentration of bacteria is about four times greater than that in water farther from the vents, and the growth rate of bacteria is equal to that found in productive, sunlit coastal waters. The bacteria in and around the vents create an environment in which bacteria consumers, such as clams, mussels, and worms, and decomposers can exist.

A Rare Relationship

One of the most astonishing discoveries in the hydrothermal ecosystem is of worms that live with some species of chemoautotrophic bacteria in mutually beneficial relationships called *endosymbioses,* in which the bacteria live inside the animal. Each worm species appears to harbor a single bacterial species. Between 5 and 30 bacteria live inside a single cell, and each is enclosed by a host cell membrane.

One such relationship involves *Riftia,* a tubeworm that can grow up to one meter in length and has no mouth or gut. The photograph shows hundreds of such tubeworms at a hydrothermal vent on the Galápagos Rift. *Riftia* exists in an unusual endosymbiotic relationship, as it actually receives food from the parasitic bacteria it hosts. *Riftia's* tentacles absorb metal ions, sulfides, CO_2, and O_2 from the plume of hot water. These elements are transported through the worm's circulatory system, so that the bacteria are supplied with CO_2 and H_2S. (Although hydrogen sulfide is usually toxic to animals, *Riftia* appears to have in its blood a protein that binds the sulfide so that it cannot inhibit metabolism.) The bacteria fix the carbon from CO_2 into glycogen by using energy obtained from the oxidation of sulfide to sulfate: $CO_2 + H_2S \rightarrow \text{glycogen} + SO_4^{2-}$. *Riftia's* lysosomes digest the bacteria and release glycogen for the animal. Thus, it is ultimately a chemical reaction—the oxidation of H_2S by its symbiotic bacteria—and not the sun that gives *Riftia* its energy.

Tubeworms and mussels at the Galápagos vent.

The mesophile, with an optimum growth temperature of between 25° and 40°C, is the most common type of microbe. Organisms that have adapted to live in the bodies of higher animals usually have an optimum temperature close to that of their host. The optimum temperature for many pathogenic bacteria is about 37°C, and incubators for clinical cultures are usually set at about this temperature. The mesophiles include most of our common spoilage and disease organisms.

A very interesting group of microorganisms are the thermophiles, those capable of growth at high temperatures. Many of these organisms have an optimum growth temperature between 50° and 60°C. This is about the temperature of the water from a hot-water tap. Such temperatures can also be reached in soil on which sunlight falls and in thermal waters, such as the hot springs in Yellowstone National Park.

Some of these organisms, **extreme thermophiles,** grow at temperatures well above 90°C, near the boiling point of water. Remarkably, many thermophiles are incapable of growth at temperatures below about 45°C. Endospores formed by thermophilic bacteria are unusually heat-resistant. They survive the usual heat treatment given canned goods but will not grow at normal storage temperatures. Although elevated temperatures may cause surviving endospores to germinate and grow, these thermophilic bacteria are not a public health problem. Thermophiles are important in organic compost piles, in which the temperature can rise rapidly to 50° or 60°C.

pH

As you will recall from Chapter 2, pH refers to the acidity or alkalinity of a solution. Most bacteria grow best in a narrow range of pH near neutrality, between pH 6.5 and 7.5. Very few bacteria grow at an acidic pH below about 4.0. That is the reason why a number of foods, such as sauerkraut, pickles, and many cheeses, are preserved by the acids of bacterial fermentation. Nonetheless, some bacteria, **acidophiles,** are remarkably tolerant of acidity. One chemoautotrophic bacterium, which is found in the drainage water from coal mines and which oxidizes sulfur to form sulfuric acid, can survive at a pH of 1 (see the box in Chapter 28). Molds and yeasts will grow over a greater pH range than bacteria will, but the optimum pH of molds and yeasts is generally below that of bacteria, usually about pH 5 to 6. Alkalinity also inhibits microbial growth but is rarely used to preserve foods.

When bacteria are cultured in the laboratory, they often produce acids that eventually interfere with their own growth. To neutralize the acids and maintain the proper pH, chemicals called **buffers** are included in the growth medium. The peptones and amino acids in some media act as buffers, and many media also contain phosphate salts. Phosphate salts have the advantage of exhibiting their buffering effect in the pH growth range of most bacteria. They are also nontoxic; in fact, they provide phosphorus, an essential nutrient element.

Osmotic Pressure

Microbes obtain almost all of their nutrients in solution from the surrounding water. They therefore require water for growth and are actually about 80% to 90% water. High osmotic pressures have the effect of removing necessary water from a cell. When a microbial cell is in a solution that has a higher concentration of solutes than in the cell (hypertonic), the cellular water passes out through the cytoplasmic membrane to the high salt concentration. (See the discussion of osmosis in Chapter 4, and review Figure 4.15 for the three types of solution a cell may encounter.) This osmotic loss of water causes **plasmolysis,** or shrinkage of the cell (Figure 6.2). What is so important about this phenomenon is that the growth of the cell is inhibited as the cytoplasmic membrane pulls away from the cell wall. Thus, the addition of salts (or other solutes) to a solution, and the resulting increase in osmotic pressure, can be used to preserve foods. Salted fish, honey, and sweetened condensed milk are preserved largely by this mechanism; the high salt or sugar concentrations draw water out of any microbial cells that are present and thus prevent their growth. These effects of osmotic pressure are roughly related to the number of molecules in a volume of solution. A low-molecular-weight compound such as sodium chloride has a greater antimicrobial effect per gram than does, for example, sucrose, which has a somewhat higher molecular weight.

Some bacteria, **extreme halophiles,** have adapted so well to high salt concentrations that they actually require them for growth. Bacteria from such saline waters as the Dead Sea often require nearly 30% salt, and the inoculating loop (a device for handling bacteria in the laboratory) used to transfer them must first be dipped into a saturated salt solution. More common are **facultative halophiles,** which do not require high salt concentrations but are able to grow at salt concentrations up to 2%, a concentration that inhibits the growth of many other bacteria. A few species of facultative halophiles can even tolerate 15% salt. A halophile is described in the box in Chapter 5, p. 128.

Most microbes, however, must be grown in a medium that is nearly all water. For example, the concentration of agar (a complex polysaccharide isolated from algae) used to solidify microbial growth media is usually about 1.5%. If markedly higher concentrations are used, the growth of some bacteria can be inhibited by the increased osmotic pressure.

If the osmotic pressure is unusually low (hypotonic), such as in distilled water, for example, water

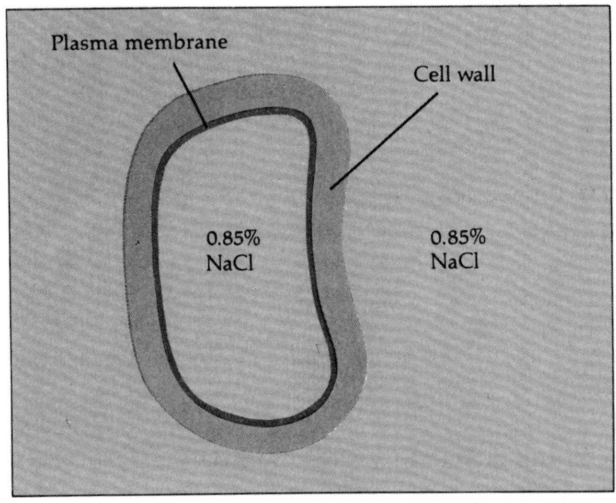

Normal cell

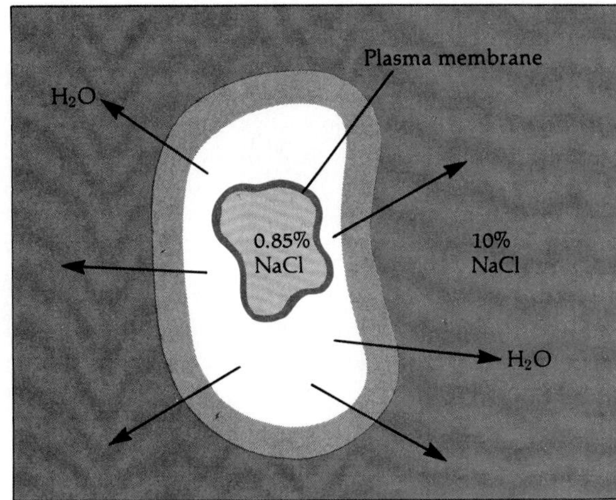

Plasmolyzed cell

FIGURE 6.2 Plasmolysis. If the concentration of solutes, such as sodium chloride, is higher in the surrounding medium (hypertonic), then water tends to leave the cell. The cell membrane shrinks away from the cell wall (an action called plasmolysis), and cell growth is inhibited.

tends to enter the cell rather than leave it. Some microbes that have a relatively weak cell wall may actually be lysed by such treatment. Red blood cells, which do not have the tough cell wall of bacteria, can be handled only in weak saline solutions that approximate the osmotic pressures of their normal habitats in the body.

CHEMICAL REQUIREMENTS

Carbon

Besides water, one of the most important requirements for microbial growth is a source of carbon. Carbon is the structural backbone of living matter; it is needed for all the organic compounds that make up a living cell. As we saw in Chapter 2, its four valences allow it to be used in constructing extended and complicated organic molecules (see Figures 2.9 and 2.10). One half of the weight of a typical bacterial cell, exclusive of water, is carbon.

Chemoheterotrophs get most of their carbon from the source of their energy, namely, organic materials such as proteins, carbohydrates, and lipids. Chemoautotrophs and photoautotrophs derive their carbon from carbon dioxide.

Because the utilization of carbon is such a basic trait of life on Earth, one way to test for life on another planet is to look for a sign that carbon is being metabolized. If carbon compounds are being formed or broken down, living organisms might be present.

Nitrogen, Sulfur, and Phosphorus

In addition to carbon, other elements are needed by microbes for the synthesis of cellular material. For example, protein synthesis requires considerable amounts of nitrogen as well as some sulfur. The syntheses of DNA and RNA also require nitrogen and some phosphorus, as does the synthesis of ATP, the molecule so important for storage and transfer of chemical energy within the cell. Together, nitrogen, sulfur, and phosphorus constitute about 18% of the dry weight of the cell; of that amount, 15% is nitrogen.

Organisms use nitrogen primarily to form the amino group of the amino acids of proteins. Many bacteria meet this requirement by decomposing protein-containing material and reincorporating the amino acids into newly synthesized proteins and other nitrogen-containing compounds. Other bacteria use nitrogen from ammonium ions (NH_4^+), which are already in the reduced form and are usually found in organic cellular material. Still other bacteria are able to derive nitrogen from nitrates (compounds that dissociate to give the nitrate ion, NO_3^-, in solution).

Some important bacteria, including the cyanobacteria (blue-green algae), use gaseous nitrogen (N_2) directly from the atmosphere. The process by which they absorb N_2 is called **nitrogen fixation.** Some organisms that can use this method are free-living, mostly in the soil, but others live in symbiosis with the roots of certain plants. *Symbiosis* is a close association of two or more different species of organisms. The most important symbiotic nitrogen-fixing bacteria belong to the genera *Rhizobium* and *Bradyrhizobium* (brad-ē-rī-zō'bē-um) and are associated with such legumes as clover, soybeans, alfalfa, beans, and peas. The nitrogen fixed in the symbiosis is used by both the plant and the bacterium. This reaction greatly increases soil fertility and is therefore very important in agriculture.

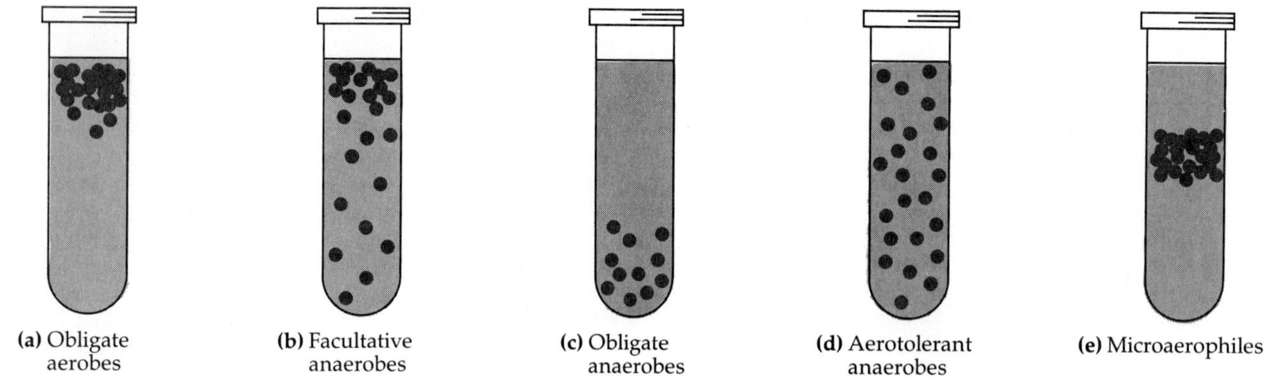

FIGURE 6.3 Effect of oxygen concentration on the growth of various bacteria in tubes of solid medium. **(a)** Obligate aerobes—growth occurs only in the short distance to which the oxygen diffuses into the medium. **(b)** Facultative anaerobes—growth is best near the surface, where oxygen is available, but occurs throughout the tube. **(c)** Obligate anaerobes—oxygen is toxic, and there is no growth near the surface. **(d)** Aerotolerant anaerobes—growth occurs evenly throughout the tube but is not better at the surface because the organisms do not use oxygen. **(e)** Microaerophiles, aerobic organisms that do not tolerate atmospheric concentrations of oxygen—growth occurs only in a narrow band of optimal oxygen concentration.

Sulfur is used to synthesize sulfur-containing amino acids and vitamins such as thiamine and biotin. The box on p. 143 describes an unusual ecosystem that is based on a supply of hydrogen sulfide (H_2S) that would be toxic to most organisms. Important natural sources of sulfur include the sulfate ion (SO_4^{2-}), hydrogen sulfide, and the sulfur-containing amino acids. Phosphorus is essential to synthesize nucleic acids and the phospholipids of cell membranes. Among other places, it is also found in the energy bonds of ATP. An important source of phosphorus is the phosphate ion (PO_4^{3-}). Potassium, magnesium, and calcium are also examples of elements that microorganisms require, often as cofactors for enzymes (Chapter 5).

Trace Elements

Microbes require very small amounts of other mineral elements, such as iron, copper, molybdenum, and zinc; these are referred to as **trace elements.** Most are essential for activity of certain enzymes, usually as cofactors. Although these components are sometimes added to a laboratory medium, they are usually assumed to be naturally present in tap water and other components of media. Even most distilled waters contain adequate amounts, but tap water is sometimes specified to ensure that these trace minerals will be present in culture media.

Oxygen

We are accustomed to thinking of molecular oxygen (O_2) as a necessity of life, but it is actually a poisonous gas. Molecular oxygen did not exist in the atmosphere during most of our planet's history. It is speculated that life could not have arisen if oxygen had been present. However, many current forms of life have metabolic systems that require oxygen for aerobic respiration. As we have seen, oxygen combines with hydrogen atoms that have been stripped from organic compounds and forms water. This process yields a great deal of energy while neutralizing a potentially toxic gas—a very neat solution, all in all.

Microbes that use molecular oxygen, called **aerobes,** produce more energy from nutrients than do microbes that do not use oxygen. Organisms that require oxygen to live are called **obligate aerobes** (Figure 6.3a).

Obligate aerobes are at a disadvantage, in that oxygen is poorly soluble in water and much of the environment is oxygen poor. Therefore, many of the aerobic bacteria have developed, or retained, the ability to continue growing in the absence of oxygen. Such organisms, including some that exist in your intestinal tract, are called **facultative anaerobes** (Figure 6.3b). In other words, facultative anaerobes can use oxygen when it is present but are able to continue growth by using fermentation or anaerobic respiration when oxygen is not available. However, their efficiency in producing energy decreases in the absence of oxygen. Examples of facultative anaerobes include the familiar *E. coli* and many yeasts.

The ability, or inability, of the organisms to ferment different sugars is often used in laboratory identification of microbes. Also important in identification of microbes may be differing abilities to substitute other electron acceptors, such as nitrates, for oxygen (anaerobic respiration).

Obligate anaerobes (Figure 6.3c) are bacteria that are unable to use molecular oxygen for energy-yielding reactions. In fact, most are harmed by it. The genus *Clostridium,* which contains the species that cause tetanus and botulism, is the most familiar example. These bacteria do use oxygen atoms in cellular materials; the atoms are usually obtained from water.

Understanding how organisms can be harmed by oxygen requires a brief discussion of the toxic forms of oxygen.

1. **Singlet oxygen** is normal molecular oxygen (O_2) that has been boosted into a higher-energy state and is extremely reactive. It is formed most commonly by the action of visible light. It is present in phagocytic cells, specialized cells that ingest and eliminate small foreign bodies such as bacteria. It plays a role in the destruction of foreign cells after they are ingested by the phagocytes.

2. **Superoxide free radicals** ($O_2^- \cdot$) are formed in small amounts during the normal respiration of organisms that use oxygen as a final electron acceptor, forming water (see Figure 5.17). In the presence of oxygen, obligate anaerobes also appear to form some superoxide free radicals. These free radicals are so toxic to cellular components that all organisms attempting to grow in atmospheric oxygen must produce an enzyme, **superoxide dismutase,** to neutralize them. Their toxicity is caused by their great instability, which leads them to avidly steal an electron from a neighboring molecule, which in turn becomes a radical and steals an electron, and so on. Unless this reaction series is quickly neutralized, extreme damage to the cell results. Aerobic bacteria, facultative anaerobes growing aerobically, and aerotolerant anaerobes (discussed later) produce this enzyme (Table 6.1). Using superoxide dismutase, they convert the superoxide free radical into molecular oxygen (O_2) and hydrogen peroxide (H_2O_2):

$$O_2^- \cdot + O_2^- \cdot + 2H^+ \longrightarrow H_2O_2 + O_2$$

3. The hydrogen peroxide produced in this reaction contains the peroxide anion O_2^{2-} and is also toxic. In the following chapter we will encounter it as the active principle in the antimicrobial agents hydrogen peroxide and benzoyl peroxide. Because the hydrogen peroxide produced during normal aerobic respiration is toxic, microbes have developed enzymes to neutralize it. The most familiar of these is **catalase,** which converts it into water and oxygen:

$$2H_2O_2 \longrightarrow 2H_2O + O_2$$

Catalase is easily detected by its action on hydrogen peroxide. When a drop of hydrogen peroxide is added to a colony of bacterial cells producing catalase, oxygen bubbles are released. This colony is designated as a catalase-positive organism. Anyone who has put hydrogen peroxide on a wound will recognize that cells in human tissue also contain catalase. The other enzyme that breaks down hydrogen peroxide is **peroxidase;** it differs from catalase in that its reaction does not produce oxygen:

$$H_2O_2 + 2H^+ \longrightarrow 2H_2O$$

4. The **hydroxyl free radical** ($OH \cdot$) is another intermediate form of oxygen and probably the most reactive. It is formed in the cellular cytoplasm by ionizing radiation. It is also produced by a reaction between superoxide free radicals and peroxide, so most aerobic respiration produces some hydroxyl radicals.

Obligate anaerobes usually produce neither superoxide dismutase nor catalase. Since aerobic conditions probably lead to an accumulation of superoxide free radicals in their cytoplasm, they are extremely sensitive to oxygen. The ability to form endospores has probably aided the survival of these oxygen-sensitive bacteria.

Aerotolerant anaerobes (Figure 6.3d) cannot use oxygen for growth, but they tolerate it fairly well. On the surface of a solid medium, they will grow without

TABLE 6.1 Oxygen-Related Enzymes of Bacteria

TYPE OF MICROBE	CATALASE	SUPEROXIDE DISMUTASE	RELATION TO OXYGEN
Aerobes and facultative anaerobes	Present	Present	Oxygen tolerated, all harmful forms neutralized
Aerotolerant anaerobes	Absent	Present	Oxygen tolerated because harmful forms partially neutralized
Obligate anaerobes	Absent	Absent	Oxygen is toxic, harmful forms not neutralized

the special techniques (discussed later) required by obligate anaerobes. Many of the aerotolerant bacteria characteristically ferment carbohydrates to lactic acid. As lactic acid accumulates, it inhibits the growth of aerobic competitors and establishes an ecological niche for the lactic acid producers. A common example of lactic-acid-producing aerotolerant anaerobes are the lactobacilli used in the production of many acidic fermented foods, such as pickles and cheese. In the laboratory, they are handled and grown much like any other bacteria, but they make no use of the oxygen in the air. These bacteria can tolerate oxygen because they possess superoxide dismutase (Table 6.1) or an equivalent system that neutralizes the toxic forms of oxygen discussed above.

A few bacteria are **microaerophilic.** They are aerobic; they do require oxygen. However, they grow only in oxygen concentrations lower than those in air. In a test tube of solid nutrient medium, they grow only at a depth where small amounts of oxygen have diffused into the medium; they do not grow at the oxygen-rich surface (Figure 6.3e). This limited tolerance is probably due to their sensitivity to superoxide free radicals and peroxides, which they produce in lethal concentrations under oxygen-rich conditions.

Organic Growth Factors

Organic growth factors are essential organic compounds that the organism is unable to synthesize; they must be directly obtained from the environment. A group of organic growth factors for humans is vitamins. Most vitamins function as coenzymes, the accessories required by certain enzymes in order to function. Many bacteria can synthesize all their own vitamins and are not dependent on outside sources. However, some bacteria lack the enzymes needed for the synthesis of certain vitamins, and for them those vitamins are organic growth factors. Other organic growth factors required by some bacteria are amino acids, purines, and pyrimidines.

Culture Media

A nutrient material prepared for the growth of microorganisms in a laboratory is called a **culture medium.** Some bacteria can grow well on just about any culture medium; others require special media, and still others cannot grow on any nonliving medium yet developed. The microbes that grow and multiply in or on a culture medium are referred to as a **culture.**

Suppose we want to grow a culture of a particular microorganism, perhaps the microbes from a particular clinical specimen. What criteria must the culture medium meet? First, it must contain the right nutrients for the particular microorganism we want to grow. It should also contain sufficient moisture, a properly adjusted pH, and a suitable level of oxygen, perhaps none at all. The medium must initially be **sterile**—that is, it must initially contain no living microorganisms—so that the culture will contain only the microorganisms we add to the medium (and their offspring). Finally, the growing culture should be incubated at the proper temperature.

TABLE 6.2 Chemically Defined Medium for the Growth of a Chemoautotrophic Bacterium That Is Able To Use Ammonium Ions for Energy

CONSTITUENT	CONCENTRATION (GRAMS/LITER)
Ammonium sulfate $[(NH_4)_2SO_4]$	0.5
Sodium bicarbonate $[NaHCO_3]$*	0.5
Sodium phosphate, dibasic $[Na_2HPO_4]$	13.5
Potassium phosphate, monobasic $[KH_2PO_4]$	0.7
Magnesium sulfate $[MgSO_4 \cdot 7H_2O]$	0.1
Ferric chloride $[FeCl_3 \cdot 6H_2O]$	0.014
Calcium chloride $[CaCl_2 \cdot 2H_2O]$	0.18
Water	1 L

*The $NaHCO_3$ is a source of carbon dioxide (the carbon source) in solution.
Source: D. Pramer, and E.L. Schmidt, *Experimental Soil Microbiology.* Minneapolis: Burgess, 1964.

A wide variety of media are available for the growth of microorganisms in the laboratory. Most of these, which are available from commercial sources, have premixed components and require only the addition of water and sterilization. Media are constantly being developed or revised for use in the isolation and identification of bacteria that are of interest to researchers in such areas as food, water, and clinical microbiology.

When it is desirable to grow bacteria on a solid medium, a solidifying agent such as agar is added to the medium. **Agar** is a complex polysaccharide derived from a marine alga, and it has long been used as a thickener in foods such as jellies, soups, and ice cream.

Agar has some very important properties that make it valuable to microbiology, and no really satisfactory substitute has ever been found. Few microbes can degrade agar, so it remains solid. Also important is the fact that it melts at about the boiling point of water but remains liquid until the temperature drops to about 40°C. For laboratory use, agar is held in water baths at about 50°C. At this temperature, it does not injure most bacteria when it is poured over a bacterial inoculum. Once the agar has solidified, it can be incubated at temperatures approaching 100°C before it liq-

TABLE 6.3 Chemically Defined Medium for the Growth of a Less Fastidious Chemoheterotroph, Such As *E. coli*

CONSTITUENT	*CONCENTRATION (GRAMS/LITER)*
Glucose	5.0
Ammonium phosphate, monobasic [$NH_4H_2PO_4$]	1.0
Sodium chloride [NaCl]	5.0
Magnesium sulfate [$MgSO_4 \cdot 7H_2O$]	0.2
Potassium phosphate, dibasic [K_2HPO_4]	1.0
Water	1 L

uefies; this property is particularly useful when thermophilic bacteria are being grown.

Agar media are usually contained in test tubes or Petri plates. The test tubes are called *slants* when they are allowed to solidify with the tube held at an angle so that a large surface for growth is available. When the agar solidifies in a vertical tube, it is called a *deep*. Petri plates, named for their inventor, are shallow dishes with a lid that nests over the bottom to prevent contamination.

CHEMICALLY DEFINED MEDIA

When a medium is being prepared for microbial growth, consideration must be given to providing an energy source as well as sources of carbon, nitrogen, sulfur, phosphorus, and any necessary growth factors that the organism is unable to synthesize. A **chemically defined medium** is one whose exact chemical composition is known. Table 6.2 shows a chemically defined medium used for the growth of a chemoautotrophic organism capable of extracting energy from the oxidation of ammonium ions to nitrite ions. For a chemoheterotroph, the chemically defined medium must contain organic growth factors that serve as a source of carbon and energy. For example, as shown in Table 6.3, glucose is included in the medium for growing the chemoheterotroph *E. coli.*

As Table 6.4 demonstrates, many organic growth factors must be provided in the chemically defined medium used to cultivate species of *Lactobacillus*. Organisms that require many growth factors are described as *fastidious.* Organisms of this type, such as *Lactobacillus,* are sometimes used in tests that determine the concentration of a particular vitamin in a substance. For such *microbiological assays* (tests), the growth medium to which the test substance and the bacterium will be added contains all the growth requirements of the bacterium except the vitamin being assayed. When the medium, test substance, and bacterium are combined, the growth of bacteria will be proportional to the amount of vitamin in the test substance. This growth is reflected by the amount of lactic acid produced. The more lactic acid, the more the *Lactobacillus* cells have been able to grow, so the more vitamin is present.

TABLE 6.4 Chemically Defined Medium for the Growth of a Fastidious Chemoheterotrophic Bacterium, Such As *Lactobacillus* Species

CONSTITUENT	*CONCENTRATION (GRAMS/LITER)*
Carbon and Energy Sources	
Glucose	10.0
Sodium acetate	6.0
Salts (added as two solutions of mixed salts)	
Ammonium chloride [NH_4Cl]	2.5
Ammonium sulfate [$(NH_4)_2SO_4$]	2.5
Potassium phosphate, monobasic [KH_2PO_4]	0.50
Potassium phosphate, dibasic [K_2HPO_4]	0.50
Sodium chloride [NaCl], Ferrous sulfate [$FeSO_4$], and Manganese sulfate [$MnSO_4$], each	0.005
Amino Acids	
Arginine, cystine, glycine, histidine, hydroxyproline, proline, tryptophan, and tyrosine, each	0.20
Alanine, aspartic and glutamic acids, isoleucine, leucine, lysine, methionine, norleucine, phenylalanine, serine, threonine, and valine, each	0.10
Glutamine	0.025
Purines	
Adenine, guanine, and uracil, each	0.010
Vitamins	
Pyridoxamine-HCl	0.0004
Riboflavin, thiamine, niacin, and pantothenic acid, each	0.0002
Para-aminobenzoic acid	0.00004
Folic acid	0.00002
Biotin	0.0000002
Water	1 L

Source: K. Thimann, *The Life of Bacteria*, 2nd ed. New York: Macmillan, 1963.

COMPLEX MEDIA

Chemically defined media are usually reserved for laboratory experimental work or the growth of autotrophic bacteria. Most heterotrophic bacteria and fungi are routinely grown on **complex media,** whose exact chemical composition varies slightly from batch to batch. These complex media are made up of nutrients such as extracts from yeasts, meat, or plants or digests of proteins from these and other sources. Table 6.5 gives one widely used recipe.

In complex media such as this one, the energy, carbon, nitrogen, and sulfur requirements of the growing microorganisms are met largely by protein. Protein is a large, relatively insoluble molecule that few microorganisms can utilize directly, but a partial digestion by acids or enzymes reduces the protein to shorter chains of amino acids called *peptones.* These small, soluble fragments can be digested by the bacteria.

The vitamins and other organic growth factors are provided by meat extracts or yeast extracts. The soluble vitamins and minerals from the meats or yeasts are dissolved in the extracting water, which is then evaporated so that these factors are concentrated. (These extracts also supplement the organic nitrogen and carbon compounds.) Yeast extracts are particularly rich in the B vitamins. If this type of medium is in liquid form, it is called **nutrient broth.** When agar is added, it is called **nutrient agar.** (This terminology sometimes confuses students, who may think that agar itself is a nutrient—it is not.)

ANAEROBIC GROWTH MEDIA AND METHODS

The cultivation of anaerobic bacteria poses a special problem. Because anaerobes might be killed by exposure to oxygen, special media called **reducing media** must be used. These media contain ingredients, such as sodium thioglycolate, that chemically combine with dissolved oxygen and deplete the oxygen in the culture medium. To routinely grow and maintain pure cultures of obligate anaerobes, microbiologists use reducing media stored in ordinary, tightly capped test tubes. These media are heated shortly before use, so that any absorbed oxygen is driven off.

When the culture must be grown in Petri plates so that individual colonies can be observed, special jars are used that hold several Petri plates in an oxygen-free atmosphere (Figure 6.4). The oxygen is removed by the following process: A packet of chemicals (sodium bicarbonate and sodium borohydride) in the jar is moistened with a few milliliters of water, and the jar is sealed. Hydrogen and carbon dioxide are produced by the reaction of the chemicals with the water. A palladium catalyst in the jar combines the oxygen in the jar with the hydrogen produced by the chemical reaction, and water is formed. As a result, the oxygen disappears in a short time. Moreover, the carbon dioxide that is produced aids the growth of many anaerobic bacteria.

TABLE 6.5 Composition of Nutrient Agar, a Complex Medium for the Growth of Heterotrophic Bacteria

CONSTITUENT	CONCENTRATION (GRAMS/LITER)
Peptone	5.0
Beef extract	3.0
Sodium chloride	8.0
Agar	15.0
Water	1 L

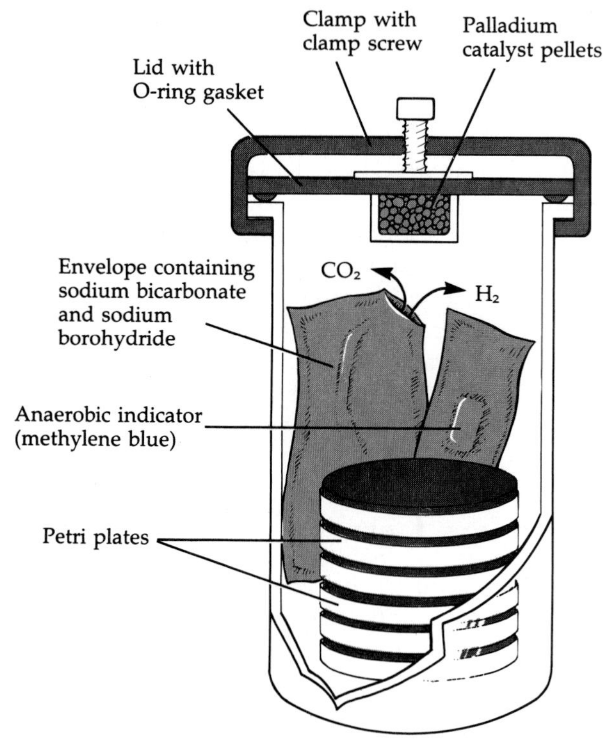

FIGURE 6.4 Anaerobic container used in cultivation of anaerobic bacteria on Petri plates. When water is mixed with the chemical packet containing sodium bicarbonate and sodium borohydride, hydrogen and carbon dioxide are generated. Reacting on the surface of a palladium catalyst in a screened reaction chamber, the hydrogen and the atmospheric oxygen in the jar combine to form water. The oxygen is thus removed. An anaerobic indicator is also in the container. It contains methylene blue, which is blue when oxidized and turns colorless when the oxygen is removed from the container.

Researchers working with anaerobes on a regular basis use anaerobic glove boxes, which are transparent chambers equipped with air locks and filled with inert gases. The technicians are able to manipulate the equipment by inserting their hands into airtight rubber gloves fitted to the wall of the chamber (Figure 6.5).

Colonies of anaerobes can also be grown in deep test tubes instead of Petri plates. The natural atmosphere in the test tube is replaced with an inert gas, such as nitrogen. A bacterial inoculum can be mixed with melted nutrient medium in the tube, and the tube, tightly capped, is rolled on horizontal rollers. The nutrient medium solidifies against the interior walls of the test tube (usually referred to as **roll tubes**), and colonies that appear there can be counted and picked, as on a shallow layer of agar in a Petri plate.

SPECIAL CULTURE TECHNIQUES

Some bacteria have never been successfully grown on artificial laboratory media. *Mycobacterium leprae,* the leprosy bacillus, can be grown on the foot pads of mice but is now usually grown in armadillos, which have a relatively low body temperature that matches the requirements of the microbe. Another example is the syphilis spirochete, although certain specialized strains of the latter have been grown on laboratory media. With few exceptions, the obligate intracellular parasites, such as the rickettsias and the chlamydia bacteria, do not grow on artificial media. They, like viruses, can reproduce only in a living host cell.

Many clinical laboratories have special *carbon dioxide incubators* in which to grow aerobic bacteria that require concentrations of CO_2 higher or lower than that found in the atmosphere. Desired carbon dioxide levels are maintained by electronic controls. High levels of carbon dioxide are also obtained with simple *candle jars* or CO_2-generating packets (Figure 6.6). Cultures are placed in a large sealed jar containing a lighted candle, which consumes oxygen. The candle stops burning when the atmosphere in the jar has a low concentration of oxygen (but one that is still adequate for the growth of aerobic bacteria) and has reached the high CO_2 concentration required by certain clinically important bacteria, such as those causing brucellosis, gonorrhea, and meningococcal meningitis. Candle jars are still used occasionally, but commercially available chemical packets are now used more often to generate

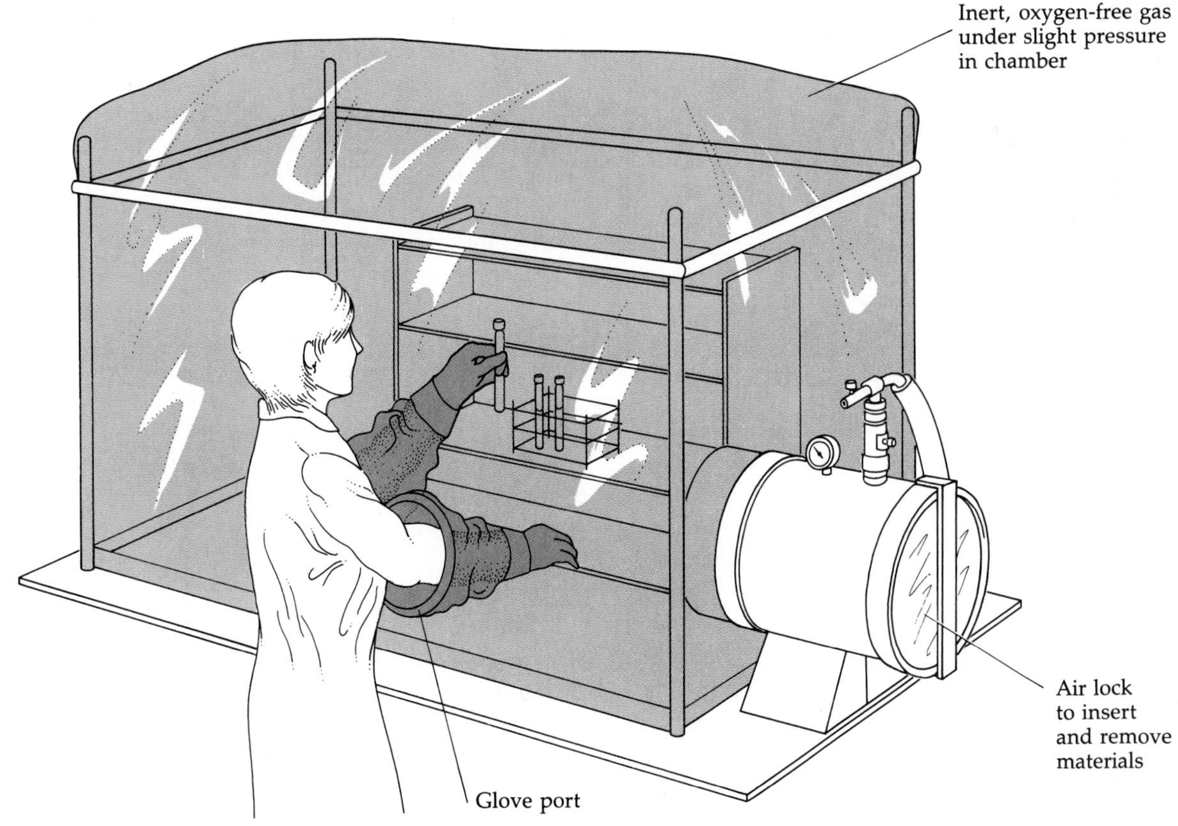

FIGURE 6.5 Anaerobic chamber. Microbiologists who work regularly with oxygen-sensitive anaerobes often use an anaerobic chamber. The transparent enclosure is filled with an inert, oxygen-free gas. Organisms and materials enter and leave through an air lock, and manipulations are made by means of glove ports.

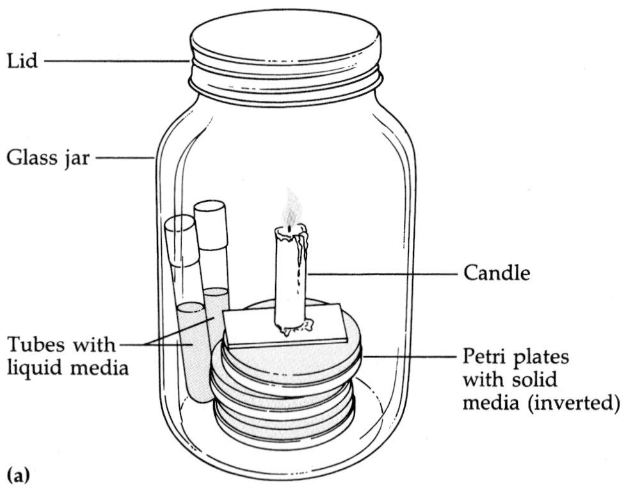

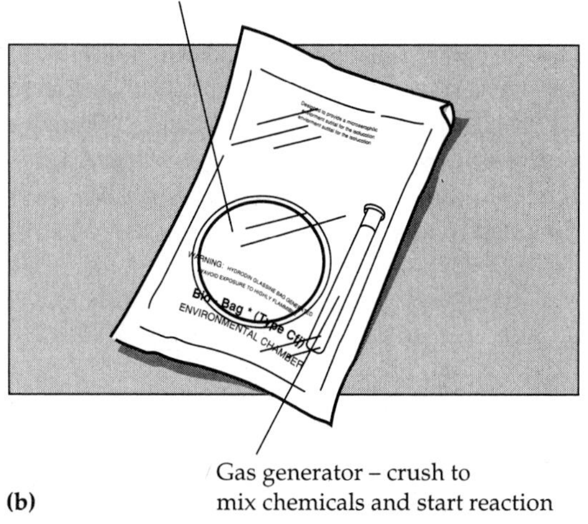

FIGURE 6.6 Candle jar. **(a)** Plates and tubes inoculated with, for example, *Neisseria meningitidis* are placed in a jar with a lighted candle, and the jar is sealed. The candle will go out when the atmosphere contains approximately 10% carbon dioxide. **(b)** Bag with Petri plate and CO_2 gas generator, which produces CO_2 when crushed.

carbon dioxide atmospheres in containers. When only one or two Petri plates of cultures are to be incubated, clinical laboratory investigators often use small plastic bags with self-contained chemical gas generators that are activated by crushing the packet or moistening it with a few milliliters of water. These are sometimes specially designed to provide precise carbon dioxide and oxygen concentrations in the bag for culture of organisms such as the microaerophilic *Campylobacter* bacteria.

SELECTIVE AND DIFFERENTIAL MEDIA

In clinical and public health microbiology, it is frequently necessary to detect the presence of specific microorganisms associated with disease or poor sanitation. For this task, selective and differential media are used. **Selective media** are designed to suppress the growth of unwanted bacteria and encourage the growth of the desired microbes. For example, bismuth sulfite agar is one medium used to isolate the typhoid bacterium, the gram-negative *Salmonella typhi* (tī'fē), from feces. Bismuth sulfite inhibits gram-positive bacteria and most gram-negative intestinal bacteria, other than *S. typhi*, as well. Sabouraud's dextrose agar, which has a pH of 5.6, is used to isolate fungi that outgrow most bacteria at this pH. Dyes such as brilliant green selectively inhibit gram-positive bacteria, and this dye is the basis of a medium called Brilliant Green agar that is used to isolate the gram-negative *Salmonella*.

Differential media make it easier to distinguish colonies of the desired organism from other colonies growing on the same plate. Similarly, pure cultures of microorganisms have identifiable reactions with differential media in tubes or plates. Blood agar (which contains red blood cells) is a dark, reddish-brown medium that microbiologists often use to identify bacterial species that destroy red blood cells. These species, such as *Streptococcus pyogenes* (pī-äj'en-ēz), the bacterium that causes strep throat, show a clear ring around their colonies where they have lysed the surrounding blood cells. (See Figure 6.7.)

Sometimes, selective and differential media are used together. Suppose we want to isolate the common bacterium *Staphylococcus aureus*, found in the nasal passages. One of the characteristics of this organism is its tolerance for high concentrations of sodium chloride; another characteristic is its ability to ferment the carbohydrate mannitol to form acid. There is an isolation medium called mannitol salt agar medium that combines features of both selective and differential media. Mannitol salt agar contains 7.5% sodium chloride, which will discourage the growth of competing organisms and thus *select for* (favor the growth of) *S. aureus*. This salty medium also contains a pH indicator that changes color if the mannitol in the medium is fermented to acid; the mannitol-fermenting colonies of *S. aureus* are thus differentiated from colonies of bacteria that do not ferment mannitol. Bacteria that grow at the high salt concentration *and* ferment mannitol to acid can be readily identified by the color change. These are probably colonies of *S. aureus*, and their identification can be confirmed with additional tests.

Another medium that is both selective and differential is MacConkey agar. This medium contains bile salts and crystal violet, which inhibit the growth of gram-positive bacteria. Because this medium also contains lactose, gram-negative bacteria that can grow on lactose can be differentiated from similar bacteria that cannot. The ability to distinguish between lactose fer-

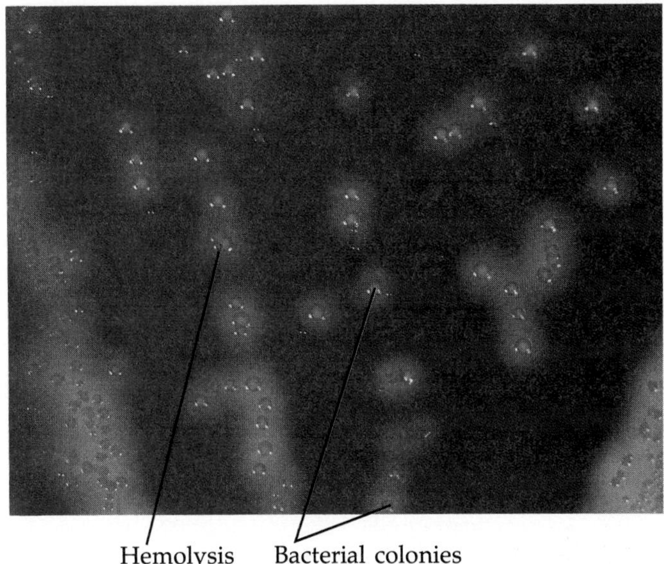

FIGURE 6.7 Blood agar, a differential medium containing red blood cells. The bacteria have lysed the red blood cells (hemolysis), so clear areas have formed around the colonies.

TABLE 6.6 Culture Media

TYPE OF MEDIA	PURPOSE
Chemically defined	Grow chemoautotrophs, photoautotrophs, and for microbiological assays
Complex	Most chemoheterotrophic organisms
Reducing	Obligate anaerobes
Selective	Suppress unwanted microbes, encourage desired microbes
Differential	Distinguish colonies of desired microbes from others
Enrichment	Similar to selective media but designed to increase numbers of desired microbes to detectable levels

menters (red or pink colonies) and nonfermenters (colorless colonies) is useful in distinguishing between the pathogenic *Salmonella* bacteria and other related bacteria. Figure 6.8 shows the appearance of bacterial colonies on several differential media.

ENRICHMENT CULTURE

Because bacteria present in small numbers can be missed, especially if other bacteria are present in much larger numbers, it is sometimes necessary to resort to an **enrichment culture.** This is often the case for soil or fecal samples. The medium for an enrichment culture is usually liquid and provides nutrients and environmental conditions that favor the growth of a particular microbe but are not suitable for the growth of other types of microbes. In this sense, it is also a selective medium, but it is designed to increase very small numbers of desired microbes to detectable levels.

Let us assume that we want to isolate from a soil sample a microbe that is capable of growth on phenol and is present in much smaller numbers than other species. If the soil sample is placed in a liquid enrichment medium, in which phenol is the only source of carbon and energy, microbes unable to metabolize phenol will not grow. The culture medium is allowed to incubate for a few days, and then a small amount of it is transferred into another flask of the same medium. After a series of such transfers, the surviving population will consist of bacteria capable of metabolizing phenol. The bacteria are given time to grow in the medium between transfers; this is the enrichment stage. Any nutrients in the original inoculum are rapidly diluted out with the successive transfers. When the last dilution is streaked onto a solid medium of the same composition, only those colonies of organisms that are capable of using phenol should grow.

Table 6.6 summarizes the purposes of the main types of culture media.

OBTAINING PURE CULTURES

Most infectious materials, such as pus, sputum, and urine, contain several kinds of bacteria; so do samples of soil, water, or foods. If these materials are plated out onto the surface of a solid medium, colonies will form that are exact copies of the same organism. A **colony,** or **clone,** theoretically arises from a single spore or vegetative cell or from a group of the same microorganisms attached to one another in clumps or chains. Microbial colonies often have a distinctive appearance that distinguishes one microbe from another. The bacteria must be distributed widely enough so that the colonies that grow are visibly separated from each other.

Most bacteriological work requires pure cultures, or clones, of bacteria. The isolation method most commonly used to get pure cultures is the **streak plate method** (Figure 6.9). A sterile inoculating loop is dipped into a mixed culture that contains more than one type of microbe and is streaked in a pattern (that shown in the figure is only one example of such a pattern), over the surface of the nutrient medium. As the pattern is traced, bacteria are rubbed off the loop onto the medium in paths of fewer and fewer cells. The last cells to be rubbed off the loop are far enough apart to

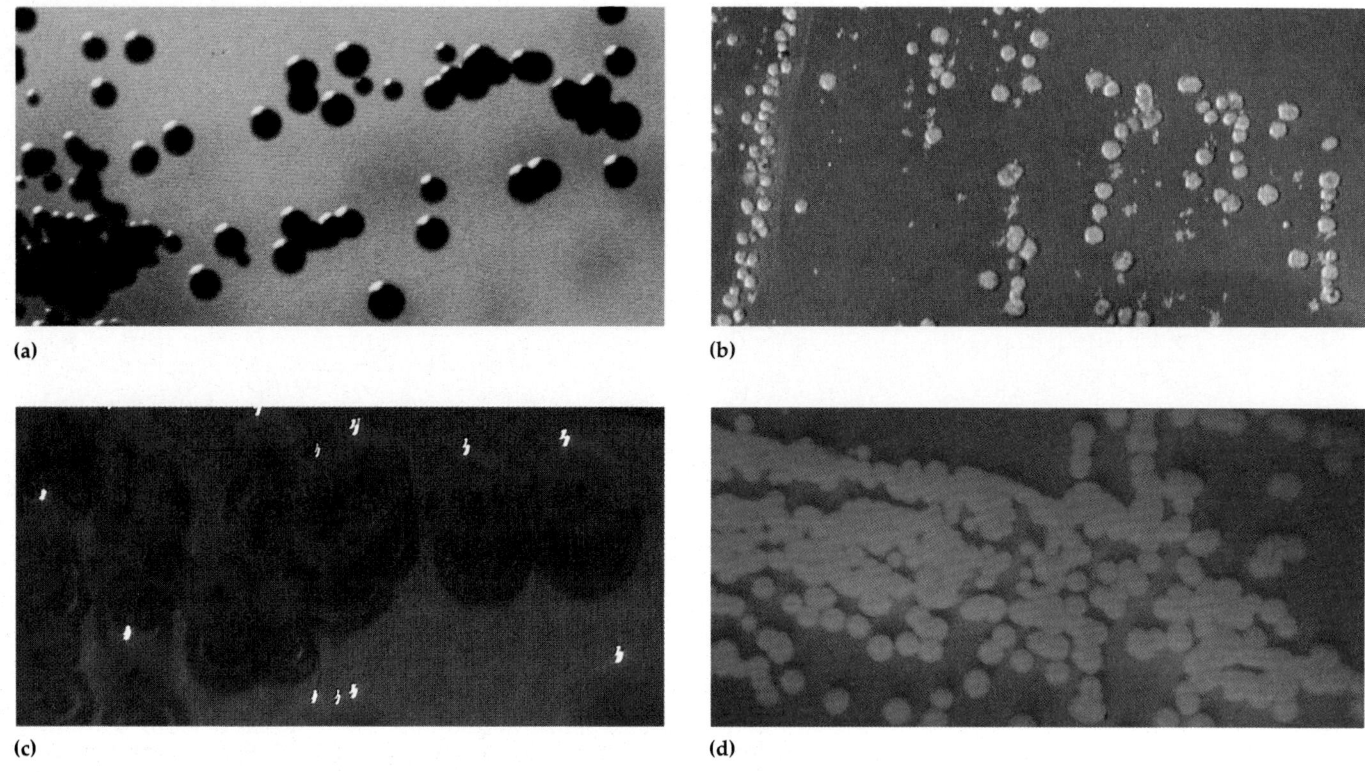

FIGURE 6.8 Bacterial colonies on several differential media. (a) *Staphylococcus aureus* on tellurite-glycine media. **(b)** *Escherichia coli* on EMB. **(c)** *Enterobacter aerogenes* on EMB. **(d)** The pigment produced by *Pseudomonas aeruginosa* on PSP agar fluoresces under ultraviolet light.

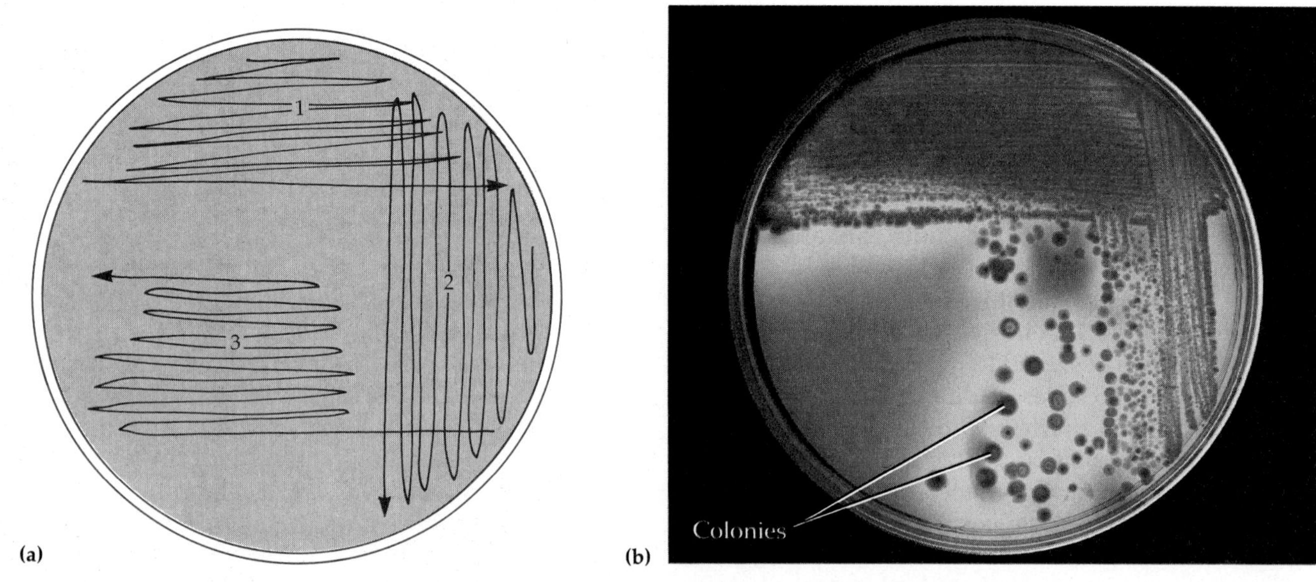

FIGURE 6.9 Streak plate method for isolation of pure cultures of bacteria. (a) The direction of streaking is indicated by arrows. Streak series 1 is made from the original bacterial mixture. The inoculating loop is sterilized following each streak series. In series 2 and 3 the loop picks up bacteria from the previous series, diluting out the number of cells each time. There are numerous variants of such patterns. **(b)** In series 3 of this example, note that well-isolated colonies of two different types of bacteria have been obtained.

grow into isolated colonies. These colonies can be picked up with an inoculating loop and transferred to a test tube of nutrient medium to form a pure culture containing only one type of bacterium.

The streak plate method works well when the organism to be isolated is present in large numbers relative to the population. However, sometimes the microbe to be isolated is present only in very small numbers. The soil microbe that is capable of using the disinfectant phenol as a nutrient is one such example. Its numbers must be greatly increased by enrichment before streak plate isolation.

Preserving Bacterial Cultures

Refrigeration can be used for short-term storage of bacterial cultures. To preserve microbial cultures for long periods, two common methods are deep-freezing and lyophilization. **Deep-freezing** is a process in which a pure culture of microbes is placed in a suspending liquid and quick-frozen at temperatures ranging from −50° to −95°C. The culture can usually be thawed and used up to several years later. During **lyophilization (freeze-drying),** a suspension of microbes is quickly frozen at temperatures ranging from −54° to −72°C, and the water is removed by a high vacuum (sublimation). While under vacuum, the container is sealed by a high-temperature torch. The remaining powderlike residue that contains the surviving microbes can be stored for years. The microbes can be revived at any time by hydration with suitable liquid nutrient medium.

Growth of Bacterial Cultures

An essential part of microbiology is to be able to graphically represent the enormous populations resulting from their growth. It is also necessary to be able to determine microbial numbers by counting or indirectly, by their metabolic activity.

BACTERIAL DIVISION

As we mentioned at the beginning of this chapter, bacterial growth refers to an increase in bacterial numbers, not an increase in the size of the individual cells. Bacteria normally reproduce by binary fission (Figure 6.10).

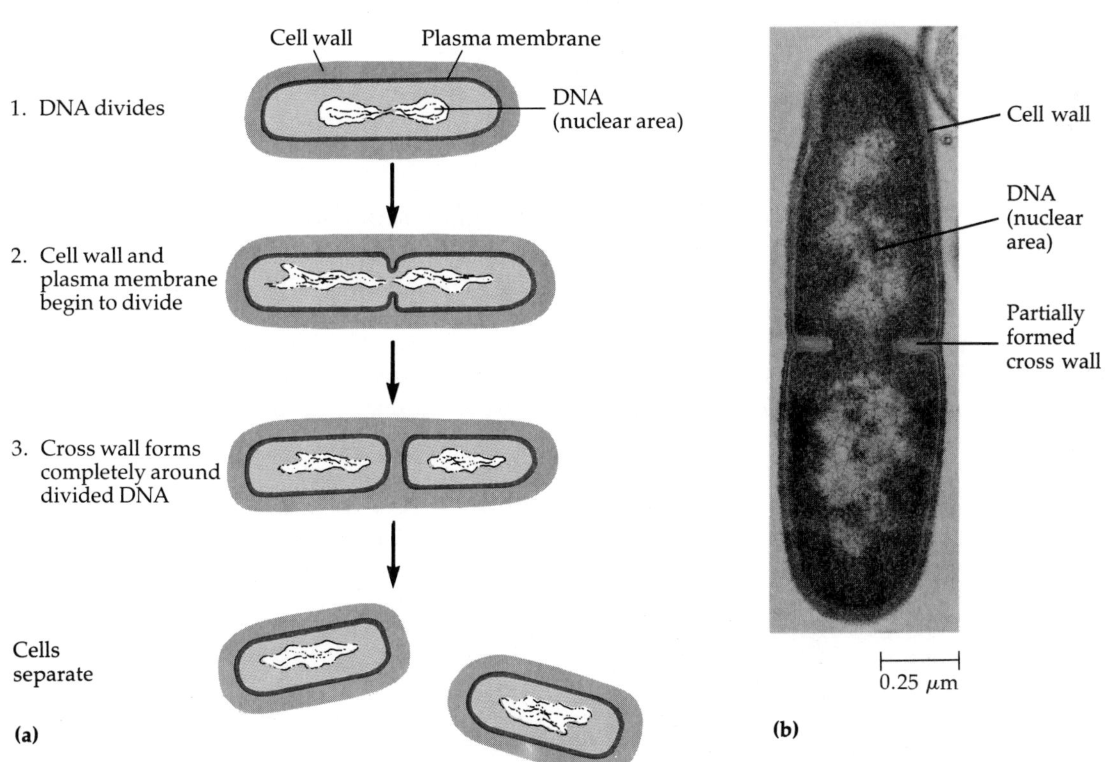

FIGURE 6.10 Binary fission in bacteria. (a) Diagram of sequence of cell division. **(b)** Electron micrograph of a thin section of a cell of *Bacillus licheniformis* starting to divide.

The first step in division is cell elongation and duplication of the chromosomal DNA. The cell wall and cell membrane then begin to grow inward from all sides at a point between the two regions of the chromosomal DNA. Eventually, the ingrowing cell walls meet, and two individual cells are formed, each of which is essentially identical to the parent cell.

A few bacterial species **bud;** that is, they form a small initial outgrowth that enlarges until its size approaches that of the parent cell, and then it separates. Some filamentous bacteria (certain actinomycetes) reproduce by producing chains of spores carried externally at the tips of the filaments. A few filamentous species simply fragment, and the fragments initiate the growth of new cells.

GENERATION TIME

For purposes of calculating the generation time of bacteria, we will consider only reproduction by binary fission, which is by far the most common method. As can be seen in Figure 6.11, one cell's division produces two cells, two cells' division produces four cells, and so on. When the arithmetic number of cells in each generation is expressed as a power of 2, the exponent tells the number of doublings (generations) that have occurred. (You might at this point wish to refer to the discussion of exponential notation in Appendix F.)

The time required for a cell to divide (and its population to double) is called the **generation time.** It varies considerably among organisms. Most bacteria have a generation time from one to three hours; others require more than 24 hours per generation. If binary fission continues unchecked, an enormous number of cells will be produced. If a doubling occurred every 20 minutes—which is the case for *E. coli* under favorable conditions—in 20 generations a single initial cell would increase to over 1 million cells. This would require a little less than 7 hours. In 30 generations, or 10 hours, the population would be 1 billion, and in 24 hours it would be a number trailed by 21 zeroes. It is difficult to graph population changes of such enormous magnitude by using arithmetic numbers. This is why logarithmic scales rather than arithmetic scales are generally used to graph bacterial growth. Understanding logarithmic representations of bacterial populations requires some use of mathematics and is necessary for anyone studying microbiology.

LOGARITHMIC REPRESENTATION OF BACTERIAL POPULATIONS

To illustrate the difference between logarithmic and arithmetic graphing of bacterial populations, let us graph 20 bacterial generations both logarithmically and arithmetically. In 5 generations (2^5), there would be 64 cells; in 10 generations (2^{10}), there would be 1024 cells, and so on. (If your calculator has a y^x key and a *log* key, you can duplicate the calculations in the third column.)

GENERATION NUMBER	ARITHMETIC NUMBER OF CELLS	LOG_{10} OF ARITHMETIC NUMBER OF CELLS
0	1	0
5 (2^5) =	32	1.51
10 (2^{10}) =	1024	3.01
15 (2^{15}) =	32768	4.52
16 (2^{16}) =	65536	4.82
17 (2^{17}) =	131072	5.12
18 (2^{18}) =	262144	5.42
19 (2^{19}) =	524288	5.72
20 (2^{20}) =	1048576	6.02

In Figure 6.12, note that the arithmetically plotted line (solid) does not clearly show the population changes in the early stages of the growth curve at this scale. In fact, the first 10 generations essentially do not leave the baseline. Furthermore, another one or two arithmetic generations graphed to the same scale would greatly increase the height of the graph and take the line off the page.

The dashed line in Figure 6.12 shows how these plotting problems can be avoided by graphing the $\log_{10}$ of the population numbers. The $\log_{10}$ of the population

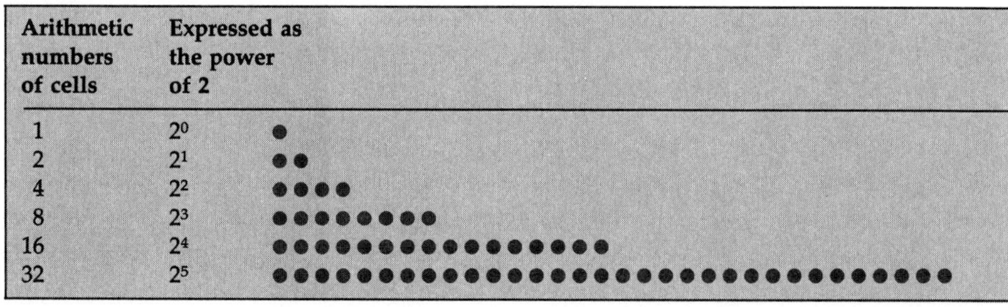

FIGURE 6.11 Cell division. When the arithmetic number of cells in each generation is expressed as a power of 2, the exponent tells the number of doublings that have occurred.

Arithmetic numbers of cells	Expressed as the power of 2	
1	2^0	●
2	2^1	●●
4	2^2	●●●●
8	2^3	●●●●●●●●
16	2^4	●●●●●●●●●●●●●●●●
32	2^5	●●●●●●●●●●●●●●●●●●●●●●●●●●●●●●●●

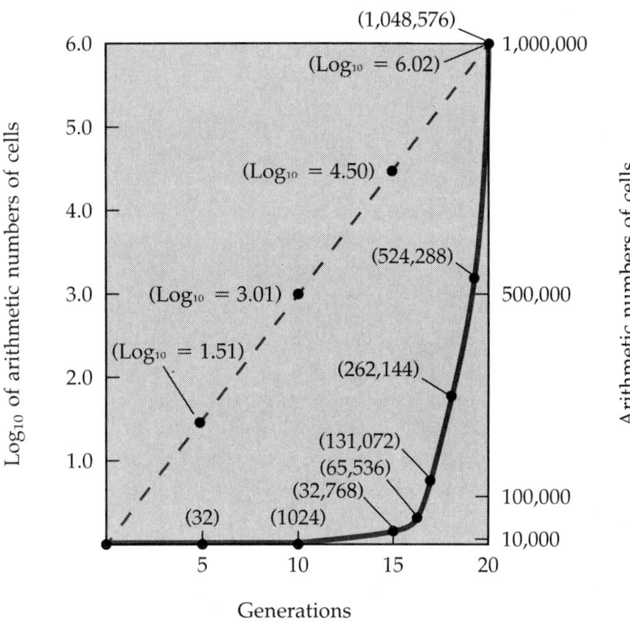

FIGURE 6.12 Growth curve for an exponentially increasing population, plotted logarithmically and arithmetically.

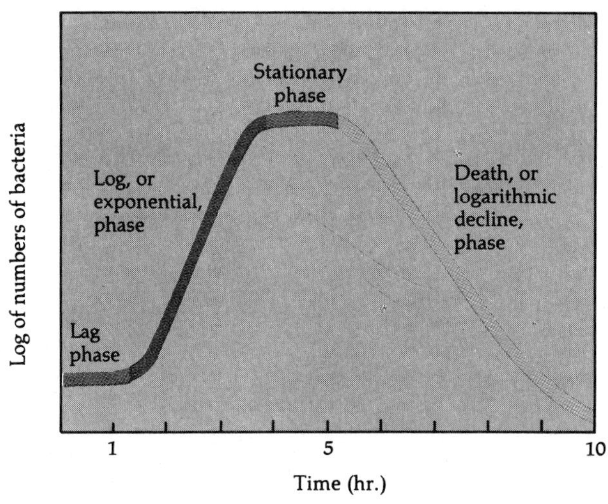

FIGURE 6.13 Bacterial growth curve, showing the four typical phases of growth.

is plotted at 5, 10, 15, and 20 generations. Note that a straight line is formed and that a thousand times this population (1,000,000,000, or $\log_{10}$ 9.0) could be accommodated in relatively little extra space. However, this advantage is obtained at the cost of distorting our "common sense" perception of the actual situation. We are not accustomed to thinking in logarithmic relationships, but it is necessary for a proper understanding of graphs of microbial populations.

PHASES OF GROWTH

When a few bacteria are inoculated into a liquid growth medium and the population is counted at intervals, it is possible to plot a **bacterial growth curve** that shows the growth of cells over time (Figure 6.13). There are four basic phases of growth.

Lag Phase

For a while, there is little or no change in the number of cells because they do not immediately reproduce in a new medium. This period of little or no cell division is called the **lag phase,** and it can last for an hour or several days. During this time, however, the cells are not dormant. The microbial population has intense metabolic activity, in particular, DNA and enzyme synthesis. (The situation is analogous to a factory newly equipped to produce automobiles—there is considerable initial activity but no immediate increase in the automobile population.)

Log Phase

Eventually, the cells begin to divide and enter a period of growth, or logarithmic increase, called the **log phase** or **exponential growth phase.** Cellular reproduction is most active during this period, and generation time reaches a constant minimum. There is apparently a characteristic minimum generation time—a maximum rate of doubling—genetically determined for each species. Because the generation time is constant, a logarithmic plot of growth during the log phase is a straight line. The log phase is the time when cells are most active metabolically.

However, during their log phase of growth, microorganisms are particularly sensitive to adverse conditions. Radiation and many antimicrobial drugs—for example, the antibiotic penicillin—exert their effect by interfering with some important step in the growth process and are therefore most harmful to cells during this phase.

Stationary Phase

If exponential growth continues unchecked, startlingly large numbers of cells could arise. For example, a single bacterium at a weight of 9.5×10^{-13} g per cell dividing every 20 minutes for only 25.5 hours can theoretically produce an enormously large population, equivalent in weight to that of an 80,000-ton aircraft carrier. In reality, this does not happen. Eventually, the growth rate slows, the number of microbial deaths balances the number of new cells, and the population stabilizes. The metabolic activities of individual surviving cells also slow at this stage. This period of equilibrium is called the **stationary phase.**

The reasons for cessation of exponential growth are not always clear. In addition to the exhaustion of nutrients and accumulation of waste products, harmful changes in pH and temperature may also play a role. In a specialized apparatus called a *chemostat,* it is possible to keep a population in the exponential growth phase indefinitely by continually draining off spent medium and adding fresh medium.

Death Phase

Usually, the number of deaths soon exceeds the number of new cells formed, and the population enters the **death phase,** or **logarithmic decline phase.** This continues until the population is diminished to a tiny fraction of the cells, or the population might die out entirely. Some species pass through the entire series of phases in only a few days; others retain some surviving cells almost indefinitely. Microbial death will be discussed further in the next chapter.

MEASUREMENT OF MICROBIAL GROWTH

There are a number of ways to measure bacterial growth. Some methods measure cell numbers; other methods measure the population's total mass, which is often directly proportional to cell numbers. Population numbers are usually recorded as the number of cells in a milliliter of liquid or in a gram of solid material. Because bacterial populations are usually very large, most methods of counting them are based on direct or indirect counts of very small samples; calculations then determine the size of the total population. Assume, for example, that a millionth of a milliliter (10^{-6} ml) of sour milk is found to contain 70 bacterial cells. Then there must be 70 times 1 million cells, or 70 million cells, per milliliter.

The problem in this method is that it is not practical to measure out a millionth of a milliliter of liquid or a millionth of a gram of food. Therefore, the procedure is done indirectly, in a series of dilutions. For example, if we add one milliliter of milk to 99 milliliters of water, each milliliter of this dilution now has one-hundredth as many bacteria as each milliliter of the original sample had. By making a series of such dilutions, we can readily estimate the number of bacteria in our original sample. To count microbial populations in foods (such as hamburger), a homogenate of one part food to nine parts water is finely ground in a food blender. Samples can then be transferred with a pipette for further dilutions or cell counts.

Plate Counts

The most frequently used method for the measurement of bacterial populations is the **plate count.** An important advantage of this method is that it measures the number of viable cells. One disadvantage may be that it takes some time, usually 24 hours or more, for visible colonies to form. This can be a serious problem in some applications, such as quality control of milk, when it is not possible to hold a particular lot for this amount of time.

The plate count is based on the assumption that each bacterium grows and divides to produce a single colony. It is also assumed that the original inoculum is homogeneous and that no aggregates of cells are present.

When a plate count is performed, it is important that only a limited number of colonies develop in the plate. When too many colonies are present, some cells are overcrowded and do not develop; these conditions cause inaccuracies in the count. Generally, only plates with from 25 to 250 colonies are counted. To ensure that some colony counts will be within this range, the original inoculum is diluted several times in a process called **serial dilution** (Figure 6.14a).

Serial dilutions. Let us say, for example, that a milk sample has 10,000 bacteria per milliliter. If 1 ml of this sample were plated out, there would theoretically be 10,000 colonies formed in the Petri plate of medium. Obviously, this would not be countable. If 1 ml of this sample were transferred to a tube containing 9 ml of sterile water, each milliliter of fluid in this tube would now contain 1000 bacteria. If 1 ml of this sample were inoculated into a Petri plate, there would still be too many potential colonies to count on a plate. Therefore, another serial dilution could be made. One milliliter containing 1000 bacteria would be transferred to a second tube of 9 ml of water. Each milliliter of this tube would now contain only 100 bacteria, and if 1 ml of the contents of this tube were plated out, potentially 100 colonies would be formed.

Pour plates and spread plates. A plate count is done by either the pour plate or the spread plate method. The **pour plate method** follows the procedure shown in Figure 6.14b. Dilutions of the bacterial suspension in the amounts of 1.0 ml or 0.1 ml are introduced into a Petri dish. The nutrient medium, in which the agar is kept liquid by holding it in a water bath at about 50°C, is poured over the sample, which is then mixed into the medium by gentle agitation of the plate. When the agar solidifies, the plate is incubated. With the pour plate technique, colonies will grow within the nutrient agar (from cells suspended in the nutrient medium as the agar solidifies) as well as on the surface of the agar plate. This technique has some drawbacks since some relatively heat-sensitive microorganisms may be damaged by the melted agar and will therefore be unable to form colonies. Also, when certain differential media are used, the distinctive appearance of the colony on the surface is essential for diagnostic purposes. Colonies that form beneath the surface of a pour plate are not satisfactory for such

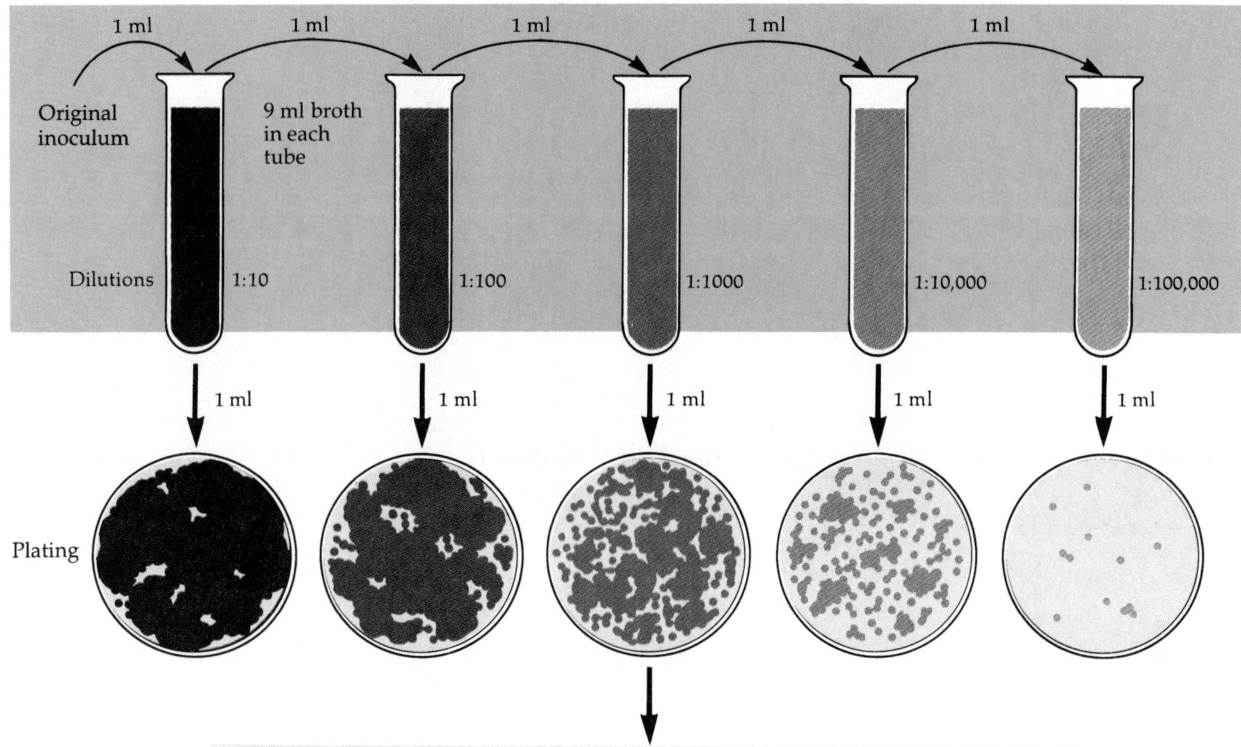

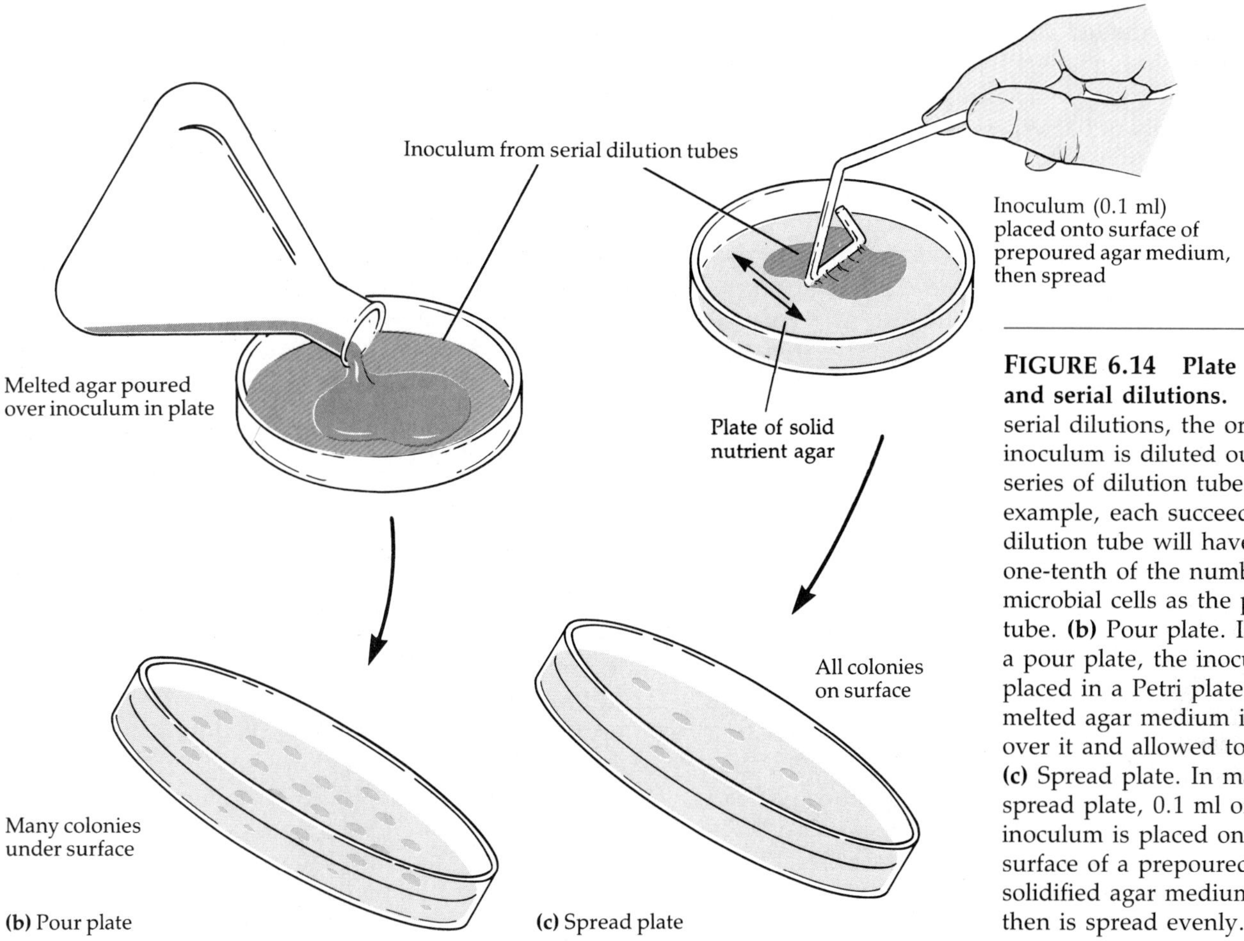

(a) Serial dilutions

Inoculum from serial dilution tubes

Melted agar poured over inoculum in plate

Inoculum (0.1 ml) placed onto surface of prepoured agar medium, then spread

Plate of solid nutrient agar

All colonies on surface

Many colonies under surface

(b) Pour plate

(c) Spread plate

FIGURE 6.14 Plate counts and serial dilutions. (a) In serial dilutions, the original inoculum is diluted out in a series of dilution tubes. In our example, each succeeding dilution tube will have only one-tenth of the number of microbial cells as the preceding tube. **(b)** Pour plate. In making a pour plate, the inoculum is placed in a Petri plate. The melted agar medium is poured over it and allowed to solidify. **(c)** Spread plate. In making a spread plate, 0.1 ml of inoculum is placed on the surface of a prepoured plate of solidified agar medium and then is spread evenly.

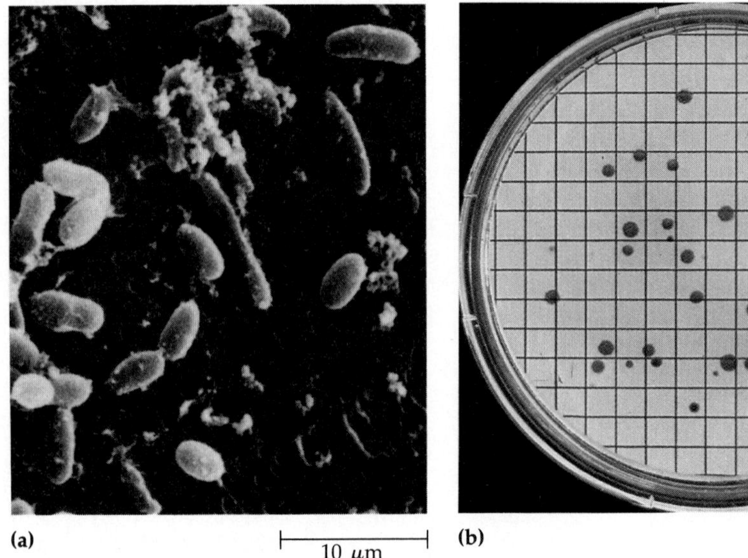

FIGURE 6.15 Counting bacteria by filtration. (a) The bacteria in 100 ml of water were sieved out onto the surface of a membrane filter. **(b)** Such a filter, with the bacteria much more widely spaced, was placed on a pad saturated with liquid nutrient medium, and the individual bacteria grew into visible colonies. There are 27 colonies visible, so we would record 27 bacteria per 100 ml of water sample.

tests. To avoid these problems, the **spread plate method** is frequently used instead (Figure 6.14c). A 0.1 ml inoculum is added to the surface of a prepoured, solidified agar medium. The inoculum is then spread uniformly over the surface of the medium with a specially shaped sterilized glass rod. This method positions all the colonies on the surface and avoids contact of the cells with melted agar.

Filtration

When the quantity of bacteria is very small, as in lakes or relatively pure streams, bacteria can be counted by **filtration methods** (Figure 6.15). One hundred milliliters or more of water are passed through a thin membrane filter whose pores are too small to pass bacteria. Thus, the bacteria are sieved out and retained on the surface of the filter. This filter is then transferred to a plate of nutrient agar, where colonies arise from the bacteria on the filter's surface. This method is applied frequently to coliform bacteria, which are indicators of fecal pollution of food or water (Chapter 27). The colonies formed by these bacteria are distinctive when a differential nutrient medium is used.

Most Probable Number Method

Another method for the determination of the number of bacteria in a sample is the **most probable number (MPN) method.** Figure 6.16 illustrates the MPN technique. This statistical estimating method is based on the fact that the greater the number of bacteria in a sample, the more dilution is needed to reduce the density to the point where no bacteria are left to grow in the replicated tubes in a dilution series. The MPN method is most useful when the microbes being counted will not grow on solid media (such as the chemoautotrophic nitrifying bacteria). It is also useful when growth of bacteria in a liquid differential medium is used to identify them (such as coliform bacteria in water testing). The MPN is only a statement that there is a 95% chance that the bacterial population falls within a certain range and that the MPN is statistically the most probable number.

Direct Microscopic Count

In the method known as the **direct microscopic count,** a measured volume of a bacterial suspension must be placed inside a defined area on a microscope slide. For example, in the *Breed count method* for milk bacteria, a 0.01 ml sample is spread over a marked square centimeter of slide and stained. By using an oil immersion objective, you can inspect many microscope fields and average the number of bacteria seen in each field.

Since the area of the viewing field of the objective can be determined, the average number of bacteria per viewing field can be obtained. By solving the following equation,

$$\frac{\text{Average number of bacteria per viewing field}}{\text{Area of the viewing field}} = \frac{\text{Number of bacteria applied to the slide}}{1\ \text{mm}^2}$$

you can determine the number of bacteria in the marked square centimeter on the slide, and you can then determine the number of bacteria in the 0.01-ml sample applied to the slide.

A specially designed slide called a *Petroff–Hausser counter* is also used in direct microscopic counts (Figure 6.17). A shallow well of known volume is indented into the surface of a microscope slide and covered with a thin glass inscribed with squares of known area. The well is filled with the microbial suspension. The average number of bacteria in each of a series of these

Volume of inoculum for each set of 5 tubes	Tubes of nutrient medium (sets of 5 tubes)	Number of positive tubes in set
10 ml		5
1 ml		3
0.1 ml		1

(a)

FIGURE 6.16 Most probable number (MPN) determination. **(a)** In this example, there are three sets of tubes and five tubes in each set. The first set of five tubes each receives 10 ml of the inoculum, such as a sample of water. The second set of five tubes each receives 1 ml of the sample, and the third set, 0.1 ml each. In our example, there were enough bacteria in the sample so that all five tubes in the set receiving 10 ml showed bacterial growth and were recorded as positive. In the next set of tubes, which received only one-tenth as much inoculum, only three tubes were positive. In the third set of tubes, which received one-hundredth as much inoculum, only one tube was positive. **(b)** Tables of MPN allow us to calculate the microbial numbers in a sample that are statistically likely to lead to such a result. The number of positive tubes is recorded for each set: 5, 3, and 1. If we look up this combination in an MPN table (see Appendix C for complete MPN table), we find that the MPN index per 100 ml is 110. Statistically, this means that 95% of the water samples that give this result contain between 40 and 300 bacteria, with 110 bacteria being the most frequent number.

Combination of Positives	MPN Index/ 100 mL	95% Confidence Limits	
		Lower	Upper
4-2-0	22	9.0	56
4-2-1	26	12	65
4-3-0	27	12	67
4-3-1	33	15	77
4-4-0	34	16	80
5-0-0	23	9.0	86
5-0-1	30	10	110
5-0-2	40	20	140
5-1-0	30	10	120
5-1-1	50	20	150
5-1-2	60	30	180
5-2-0	50	20	170
5-2-1	70	30	210
5-2-2	90	40	250
5-3-0	80	30	250
5-3-1	110	40	300
5-3-2	140	60	360

(b)

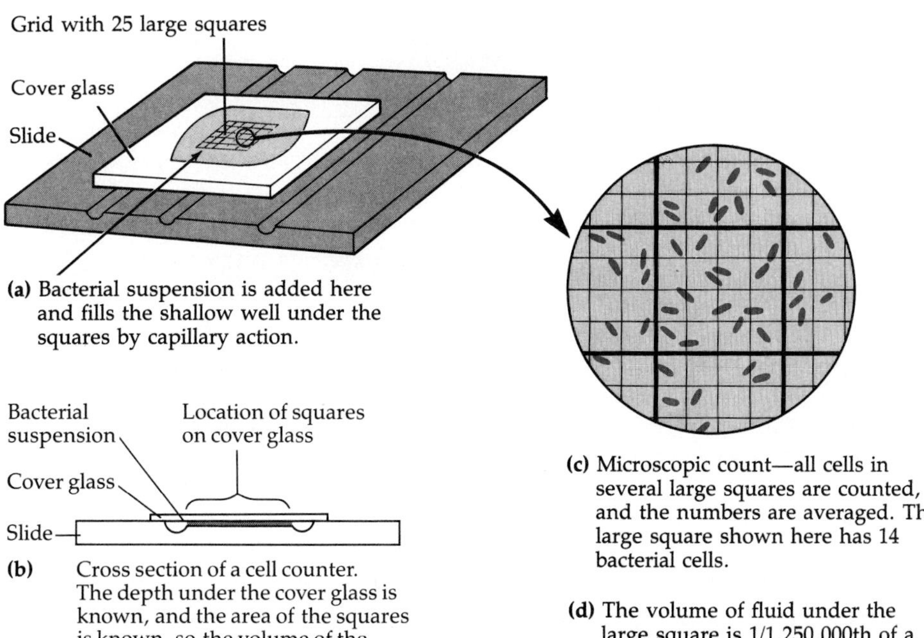

Grid with 25 large squares
Cover glass
Slide
(a) Bacterial suspension is added here and fills the shallow well under the squares by capillary action.
Bacterial suspension
Location of squares on cover glass
Cover glass
Slide
(b) Cross section of a cell counter. The depth under the cover glass is known, and the area of the squares is known, so the volume of the bacterial suspension under the squares can be calculated (depth × area).
(c) Microscopic count—all cells in several large squares are counted, and the numbers are averaged. The large square shown here has 14 bacterial cells.
(d) The volume of fluid under the large square is 1/1,250,000th of a milliliter. If it contains 14 cells, as shown here, then there are 14 times 1,250,000 cells in a milliliter.

FIGURE 6.17 Direct microscopic count of bacteria with a Petroff-Hausser cell counter. The average number of cells within a large square multiplied by a factor of 1,250,000 gives the number of bacteria per milliliter.

squares is calculated and then multiplied by the factor that produces the count per milliliter. Motile bacteria are difficult to count by this method, and, as happens with other microscopic methods, dead cells are about as likely to be counted as live ones. In addition to these disadvantages, a rather high concentration of cells is required, about 10 million bacteria per milliliter. The chief advantage of microscopic counts is that no incubation time is required, and they are usually reserved for applications in which time is the primary consideration. This advantage also holds for *electronic cell counters,* which automatically count the number of cells in a measured volume of liquid. These instruments are available in some research laboratories and hospitals.

ESTIMATIONS OF BACTERIAL NUMBERS BY INDIRECT METHODS

It is not always necessary to count microbial cells to estimate their numbers. In science and industry, microbial numbers and activity are determined by some of the following indirect means as well.

Turbidity

For some types of experimental work, estimating turbidity is a practical way of monitoring bacterial growth. As bacteria multiply in a liquid medium, the medium becomes **turbid,** or cloudy with cells.

The instrument used to measure turbidity is a spectrophotometer, or colorimeter. In the spectrophotometer, a beam of light is transmitted through a bacterial suspension to a photoelectric cell (Figure 6.18). As bacterial numbers increase, less light will reach the photoelectric cell. This change of light will register on the instrument's scale as the *percentage of transmission.* Also printed on the instrument's scale is a logarithmic expression called the *absorbance* (sometimes called *optical density* or *OD),* a value derived from the percentage of transmission. The absorbance is used to plot bacterial growth. When the bacteria are in logarithmic growth or decline, the graph of the absorbance versus time will form an approximately straight line. If absorbance readings are matched with plate counts of the same culture, this correlation can be used in future estimations of bacterial numbers made directly by turbidity.

More than a million cells per milliliter must be present for the first traces of turbidity to be visible. About 10 to 100 million cells per milliliter are needed to make a suspension turbid enough to be read on a spectrophotometer. Therefore, turbidity is not a useful measure of contamination of liquids by relatively small numbers of bacteria.

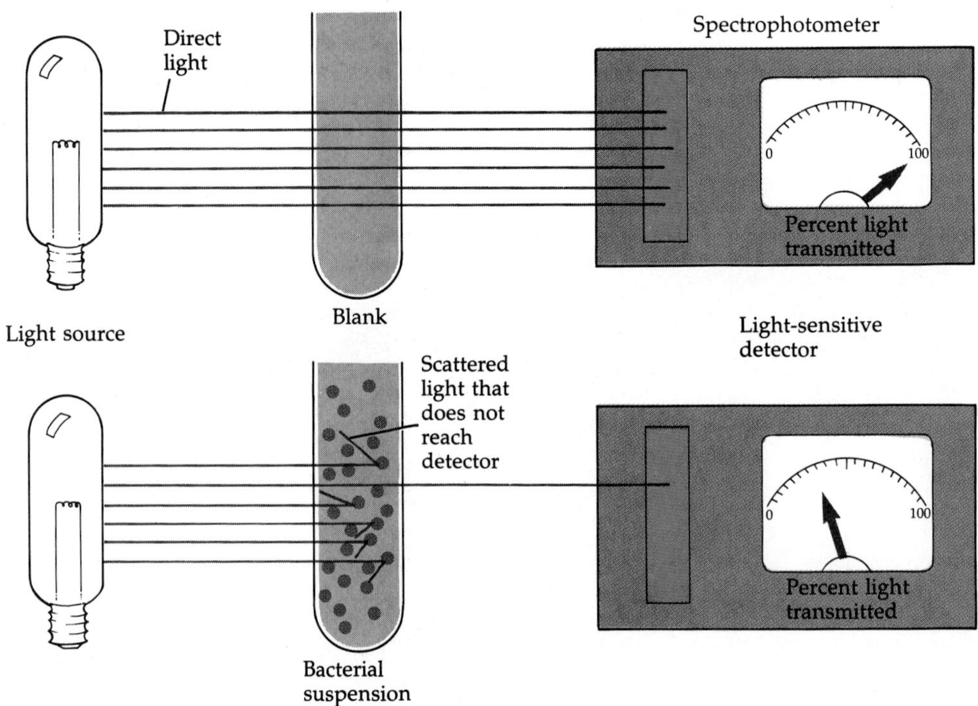

FIGURE 6.18 Turbidity estimation of bacterial numbers. The amount of light striking the light-sensitive detector is inversely proportional to the number of bacteria under standardized conditions. The less light transmitted, the more bacteria in the sample.

Metabolic Activity

Another indirect way to estimate bacterial numbers is by the measurement of the **metabolic activity** of the population. Rather than attempting to count bacterial colonies or estimate turbidity, we measure the amount of a certain metabolic product and assume that it is in direct proportion to the number of bacteria present. The metabolic product might be acid or CO_2. Bacterial numbers can also be estimated by a reduction test, which measures oxygen uptake directly or indirectly. To a medium such as milk, a dye that changes color in the presence or absence of oxygen is added and the filled tube of milk is tightly capped. Methylene blue, for example, is blue in the presence of oxygen and colorless in its absence. The bacteria use the oxygen as they metabolize the milk. The faster the dye loses color, the faster the oxygen is being depleted and the more bacteria are presumed to be present in the milk. Reduction tests are frequently used in microbiology teaching laboratories, but they lack accuracy and are seldom used in commercial applications.

Dry Weight

For filamentous organisms, such as molds, the usual measuring methods are less satisfactory. A plate count would not measure this filamentous increase in biomass. In plate counts of molds, the number of asexual spores is counted instead, and this is often not a good measure of growth. One of the better measurements of growth of filamentous organisms is by *dry weight.* In this procedure, the fungus is removed from the growth medium, filtered to remove extraneous material, placed in a weighing bottle, and dried in a desiccator. For bacteria, the same basic procedure is followed. It is customary to remove the bacteria from the culture medium by centrifugation.

You now have a basic understanding of the requirements for, and measurements of, microbial growth. In Chapter 7, we will look at ways in which this growth is controlled in laboratories, hospitals, industry, and our homes.

STUDY OUTLINE

Requirements for Growth (pp. 141–148)

1. The growth of a population is an increase in the number of cells or in mass.
2. Requirements for microbial growth are divided into physical and chemical needs.

PHYSICAL REQUIREMENTS (pp. 142–145)

1. On the basis of growth range of temperature, microbes are classified as psychrophiles (cold-loving), mesophiles (moderate-temperature-loving), and thermophiles (heat-loving).
2. The minimum growth temperature is the lowest temperature at which a species will grow; the optimum growth temperature is the temperature at which it grows best; and the maximum growth temperature is the highest temperature at which growth is possible.
3. Most bacteria grow best at a pH between 6.5 and 7.5.
4. In a hypertonic solution, microbes undergo plasmolysis; halophiles can tolerate high salt concentrations.

CHEMICAL REQUIREMENTS (pp. 145–148)

1. All organisms require a carbon source; chemoheterotrophs use an organic molecule, and autotrophs typically use carbon dioxide.
2. Nitrogen is needed for protein and nucleic acid synthesis. Nitrogen can be obtained from decomposition of proteins or from NH_4^+ or NO_3^-; a few bacteria are able to fix N_2.
3. On the basis of oxygen requirements, organisms are classified as obligate aerobes, facultative anaerobes, obligate anaerobes, aerotolerant anaerobes, and microaerophiles.
4. Aerobes, facultative anaerobes, and aerotolerant anaerobes must have the enzymes superoxide dismutase ($2O_2^{2-} + 2H^+ \rightarrow O_2 + H_2O_2$) and either catalase ($2H_2O_2 \rightarrow 2H_2O + O_2$) or peroxidase ($H_2O_2 + 2H^+ \rightarrow 2H_2O$).
5. Other chemicals required for microbial growth include sulfur, phosphorus, trace elements, and, for some microorganisms, organic growth factors.

Culture Media (pp. 148–155)

1. A culture medium is any material prepared for the growth of bacteria in a laboratory.
2. Microbes that grow and multiply in or on a culture medium are known as a culture.
3. Agar is a common solidifying agent for a culture medium.

CHEMICALLY DEFINED MEDIA (p. 149)

1. A chemically defined medium is one in which the exact chemical composition is known.

COMPLEX MEDIA (p. 150)

1. A complex medium is one in which the exact chemical composition is not known.

ANAEROBIC GROWTH MEDIA AND METHODS (pp. 150–151)

1. Reducing media chemically remove molecular oxygen (O_2) that might interfere with the growth of anaerobes.

2. Petri plates can be incubated in an anaerobic jar or anaerobic glove box.

SPECIAL CULTURE TECHNIQUES (pp. 151–152)

1. Some parasitic and fastidious bacteria must be cultured in living animals or in cell cultures.
2. CO_2 incubators or candle jars are used to grow bacteria requiring an increased CO_2 concentration.

SELECTIVE AND DIFFERENTIAL MEDIA (pp. 152–153)

1. By inhibiting unwanted organisms with salts, dyes, or other chemicals, selective media allow growth of only the desired microbe.
2. Differential media are used to distinguish between different organisms.

ENRICHMENT CULTURE (p. 153)

1. An enrichment culture is used to encourage the growth of a particular microorganism in a mixed culture.

OBTAINING PURE CULTURES (pp. 153–155)

1. Colonies are the progeny of a single cell that have grown in or on a culture medium.
2. Pure cultures are usually obtained by the streak plate method.

Preserving Bacterial Cultures (p. 155)

1. Microbes can be preserved for long periods of time by deep-freezing or lyophilization (freeze-drying).

Growth of Bacterial Cultures (pp. 155–163)

BACTERIAL DIVISION (pp. 155–156)

1. The normal reproductive method of bacteria is binary fission, in which a single cell divides into two identical cells.
2. Some bacteria reproduce by budding, aerial spore formation, or fragmentation.

GENERATION TIME (p. 156)

1. The time required for a cell to divide or a population to double is known as the generation time.

LOGARITHMIC REPRESENTATION OF BACTERIAL POPULATIONS (pp. 156–157)

1. Bacterial division occurs according to a logarithmic progression (2 cells, 4 cells, 8 cells, etc.).

PHASES OF GROWTH (pp. 157–158)

1. During the lag phase, there is little or no change in the number of cells, but metabolic activity is high.
2. During the log phase, the bacteria multiply at the fastest rate possible under the conditions provided.
3. During the stationary phase, there is an equilibrium between cell division and death.
4. During the death phase, the number of deaths exceeds the number of new cells formed.

MEASUREMENT OF MICROBIAL GROWTH (pp. 158–162)

1. A standard plate count reflects the number of viable microbes and assumes that each bacterium grows into a single colony.
2. A plate count may be done by either the pour plate or spread plate methods.
3. In filtration, bacteria are retained on the surface of a membrane filter and then are transferred to a culture medium to grow and subsequently be counted.
4. The most probable number (MPN) method can be used for microbes that will grow in a liquid medium; it is a statistical estimation.
5. In a direct microscopic count, the microbes in a measured volume of a bacterial suspension are counted with the use of a specially designed slide.

ESTIMATIONS OF BACTERIAL NUMBERS BY INDIRECT METHODS (pp. 162–163)

1. A spectrophotometer is used to determine turbidity by measuring the amount of light that passes through a suspension of cells.
2. An indirect way of estimating bacterial numbers is by measurement of the metabolic activity of the population, for example, acid production or oxygen consumption.
3. For filamentous organisms such as fungi, measuring dry weight is a convenient method of growth measurement.

STUDY QUESTIONS

Review

1. Describe binary fission.
2. What physical and chemical properties of agar make it useful in culture media?
3. Draw a typical bacterial growth curve. Label and define each of the four phases.
4. Macronutrients (needed in relatively large amounts) are often listed as CHONPS. What does each of these letters indicate, and why are they needed by the cell?
5. Most bacteria grow best at pH ________ .
6. Why can high concentrations of salt or sugar be used to preserve food?
7. Explain five bacterial categories, based on their requirements for oxygen.
8. Define the following terms and explain the importance of each enzyme.
   (a) Catalase
   (b) Hydrogen peroxide
   (c) Peroxidase
   (d) Superoxide free radical
   (e) Superoxide dismutase

9. Which of these curves best depicts the log phase of:
   (a) A thermophile incubated at room temperature?
   (b) *Listeria monocytogenes* growing in a human?
   (c) A psychrophile when incubated at 9°C?

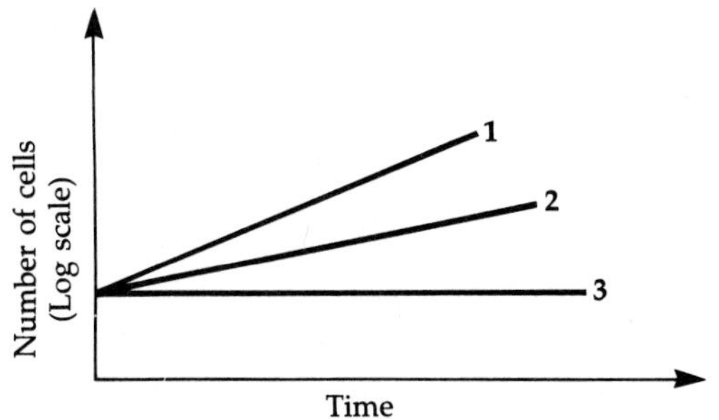

10. *Clostridium* can be cultured in an anaerobic incubator or in the presence of atmospheric oxygen if thioglycolate is added to the nutrient broth. Compare these two techniques. Using terms from Question 8, explain why elaborate culture techniques are used for *Clostridium.*

11. Seven methods of measuring microbial growth were explained in this chapter. Categorize each as a direct or indirect method.

12. By deep-freezing, bacteria can be stored without harm for extended periods. Why do refrigeration and freezing preserve foods?

13. A pastry chef inoculated a cream pie with six *S. aureus* cells. If *S. aureus* has a generation time of 60 minutes, how many cells would be in the cream pie after seven hours?

14. Nitrogen and phosphorus added to beaches following an oil spill encourage growth of natural oil-degrading bacteria. Explain why the bacteria do not grow if nitrogen and phosphorus are not added.

15. A normal fecal sample has a high concentration of coliform bacteria. A patient with typhoid fever will excrete *Salmonella* along with normal bacteria. Feces from a suspected case of typhoid fever are inoculated on an S-S agar plate. After proper incubation, coliform bacteria form red colonies, and *Salmonella* form colorless colonies with a black center. Is S-S agar selective? Differential? How can you tell?

16. Differentiate between complex and chemically defined media.

17. Draw the following growth curves for *E. coli,* starting with 100 cells with a generation time of 30 min at 35°C.
   (a) The cells are incubated for 5 hours at 35°C.
   (b) After 5 hours, the temperature is changed to 20°C for 2 hours.
   (c) After 5 hours at 35°C, the temperature is changed to 5°C for 2 hours followed by 35°C for 5 hours.

Challenge

1. *E. coli* was incubated with aeration in a nutrient medium containing two carbon sources, and the growth curve below was made from this culture.
   (a) Explain what happened at the time marked *x*.
   (b) Which substrate provided "better" growth conditions for the bacteria? How can you tell?

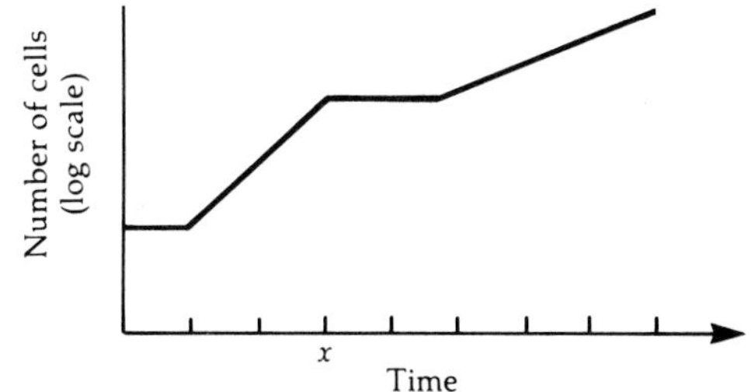

2. *Clostridium* and *Streptococcus* are both catalase-negative. *Streptococcus* grows by fermentation. Why is *Clostridium* killed by oxygen, while *Streptococcus* is not?

3. Design an enrichment medium and procedure for growing an endospore-forming bacterium that fixes nitrogen and uses cellulose for its carbon source.

4. Most laboratory media contain a fermentable carbohydrate and peptone because the majority of bacteria require carbon, nitrogen, and energy sources in these forms. How are these three needs met by glucose–minimal salts medium? (See Table 6.4.)

5. Flask A contains yeast cells in glucose–minimal salts broth, incubated at 30°C with aeration. Flask B contains yeast cells in glucose–minimal salts broth, incubated at 30°C in an anaerobic jar. The yeasts are facultative anaerobes.
   (a) Which culture produced the most ATP?
   (b) Which culture produced the most alcohol?
   (c) Which culture had the shortest generation time?
   (d) Which culture had the greatest cell mass?
   (e) Which culture had the highest absorbance?

FURTHER READING

Atlas, R.M., and R. Bartha. *Microbial Ecology: Fundamentals and Applications,* 2nd ed. Redwood City, Calif.: Benjamin/Cummings, 1987. Discusses the interrelationships of microbes with their environments; includes enumeration and isolation techniques.

Difco Manual: Dehydrated Culture Media and Reagents for Microbiology, 10th ed. Detroit: Difco Laboratories, 1984. Formulas and applications of media from the manufacturer.

Dworkin, M. *Developmental Biology of the Bacteria.* Redwood City, Calif.: Benjamin/Cummings, 1985. Describes life cycles, cell structure, and ecology of selected bacteria, including endospore-formers, cyanobacteria, and actinomycetes.

Geesey, G.G., and D.C. White. "Determination of bacterial growth and activity at solid–liquid interfaces." *Annual Review of Microbiology* 44:579–602, 1990. Describes methods for examining the growth of bacteria as they grow attached to rocks or pipes.

Ingraham, J.L., O. Maaloe, and F.C. Niedhardt. *Growth of the Bacterial Cell.* Sunderland, Mass.: Sinauer Associates, 1983. Provides comprehensive coverage of cell growth.

Krieg, N.R., and P.S. Hoffman. "Microaerophily and oxygen toxicity." *Annual Review of Microbiology* 40:107–130, 1986. Offers seven explanations for microaerophily.

Sutter, V., et al. *Wadsworth Anaerobic Bacteriology Manual,* 4th ed. Belmont, Calif.: Starr Publishing Co., 1986. Describes methods for isolation and cultivation of anaerobes.

CHAPTER 7

Control of Microbial Growth

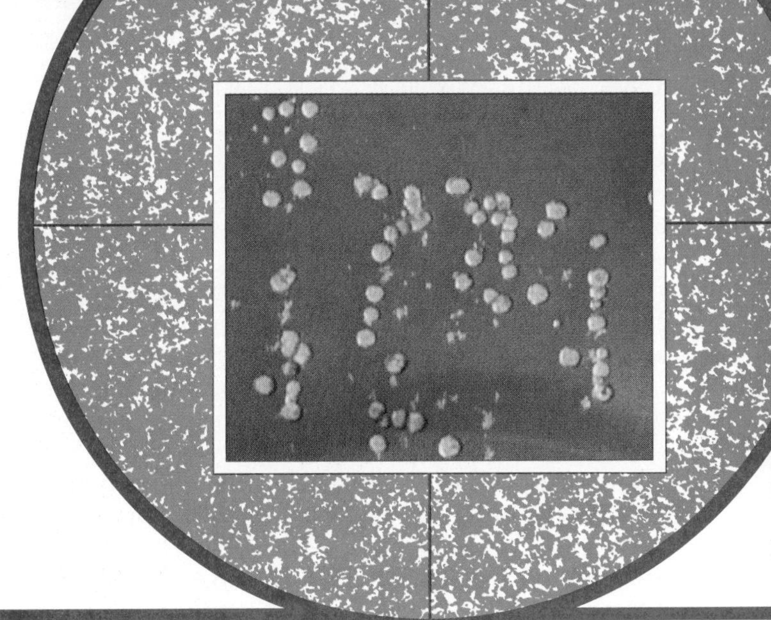

LEARNING OBJECTIVES

- Define the key terms related to the destruction or suppression of microbial growth: sterilization, disinfection, antisepsis, germicide, bacteriostasis, asepsis, degerming, sanitization.
- Explain how microbial growth is affected by the type of microbe, its physiological state, and the environmental conditions.
- Describe the effects of microbial control agents on cellular structures.
- Describe the patterns of microbial death caused by treatments with microbial control agents.
- Describe the physical methods of microbial control.
- Describe the factors related to effective disinfection and their effects on the evaluation of a disinfectant.
- Describe the methods of use and preferred uses for disinfectants.

The scientific control of microbial growth began only about 100 years ago. Prior to that time, epidemics periodically killed thousands of people. In some hospitals, 25% of delivering mothers died of infections carried by the hands and instruments of attending nurses and physicians. Ignorance of microbes was such that, during the American Civil War, a surgeon might have cleaned his scalpel on his bootsole between incisions.

Ignatz Semmelweis (1816–1865), a Hungarian physician working in Vienna, and Joseph Lister (1827–1912), an English physician, first introduced the concept of microbial control. At the obstetrics ward in the Vienna General Hospital, Semmelweis required all personnel to wash their hands in chlorinated lime; that procedure significantly lowered the infection rate. Lister, meanwhile, read about Pasteur's work with microbes and concluded that the number of infected surgical wounds (sepsis) could be decreased through procedures that prevented the access of microbes to the wound. This system, known as **aseptic surgery,** included the heat sterilization of surgical instruments and, following surgery, the application of phenol (carbolic acid) to wounds.

We have come a long way in controlling microbial growth since the time of Semmelweis and Lister. Today's procedures are far more sophisticated and effective. They are used not only to control disease organisms but also to curb microbial growth that results in food spoilage. This chapter will discuss how microbial growth can be controlled by physical methods and chemical agents. Physical methods include the use of heat, low temperatures, desiccation, osmotic pressure, filtration, and radiation. Chemical agents include several groups of substances that destroy microbes or limit microbial growth on body surfaces and on inanimate objects. In this chapter, discussion will be limited to agents that prevent microbial infection. In Chapter 20, we will discuss methods for the control of microbes once infection has occurred. Chemotherapy is one such method.

Before we begin our survey of physical and chemical methods used to control microbial growth, it would be useful to define the important terms related to control. Microbial control is needed to prevent the transmission of infection, contamination, and spoilage. Microbial control does not always mean killing the microbes; some situations require that they simply be inhibited or removed. A variety of agents and procedures have been developed; for different situations, each has its own application and range of controlling effects. Table 7.1 contains a list of terms related to the destruction or suppression of microbial growth.

Conditions Influencing Microbial Control

Usually, if we want to sterilize an object, such as a laboratory flask, we do not bother to identify the species of microbes on it. Instead, we launch a general attack calculated to be strong enough to kill the most resistant microbial forms that could possibly be present—endospores. At any time, the flask could carry a multitude of microorganisms in different phases of their growth cycles, with different metabolic states and even with different microenvironments. Thus, it is possible that an antimicrobial agent will not kill all the microbial population immediately. The time it takes to reduce or eliminate microbial growth on any object depends, in part, on the number and species of

TABLE 7.1 Terminology Related to the Control of Microbial Growth

TERM	*DEFINITION*	*COMMENTS*
Terms Related to Destruction of Organisms		
Sterilization	The process of destroying all forms of microbial life on an object or in a material. This includes the destruction of endospores—the most resistant form of microbial life. *Sterilization* is absolute; there are no degrees of sterilization.	Typical temperatures required: moist heat, 121°C for 15 minutes; dry heat, 170°C for 120 minutes. Other methods are ionizing radiation and gases such as ethylene oxide.
Disinfection	The process of destroying vegetative pathogens but not necessarily endospores or viruses. Usually, a *disinfectant* is a chemical applied to an object or a material. Disinfectants tend to reduce or inhibit growth; they usually do not sterilize.	Term usually applied to use of liquid chemical solutions on surfaces or to elimination of pathogens in water—for example, by chlorination.
Antisepsis	Chemical disinfection of the skin, mucous membranes, or other living tissues.	Term especially applied to treatment of wounds. *Antisepsis* is a specific kind of *disinfection.*
Germicide (*cide*, kill)	A chemical agent that rapidly kills microbes but not necessarily their endospores.	A *bactericide* kills bacteria, a *sporicide* kills endospores, a *fungicide* kills fungi, a *virucide* kills viruses, and an *amoebicide* kills amoebas.
Terms Related to Suppression of Organisms		
Bacteriostasis (*stasis*, halt)	A condition in which bacterial growth and multiplication are inhibited, but the bacteria are not killed. If the bacteriostatic agent is removed, bacterial growth and multiplication may resume. *Fungistasis* refers to the inhibition of fungal growth.	Refrigeration is bacteriostatic; many chemicals such as dyes are bacteriostatic rather than bactericidal.
Asepsis (*asepsis*, without infection)	The absence of pathogens from an object or area. Aseptic techniques are designed to prevent the entry of pathogens into the body. Whereas *surgical asepsis* is designed to exclude all microbes, *medical asepsis* is designed to exclude microbes associated with communicable diseases.	Air filtration, ultraviolet lights, personnel masks, gloves, and gowns, and instrument sterilization are all factors in achieving *asepsis.*
Degerming	The removal of transient microbes from the skin by mechanical cleansing or by the use of an antiseptic.	For routine injections, alcohol swabs are often used; before surgery, iodine-containing products are often used.
Sanitization	The reduction of pathogens on eating utensils to safe public health levels by mechanical cleansing or chemicals.	Any chemicals must be compatible with safety and palatability of foods.

microbes on the object. It also depends on the following factors.

TEMPERATURE

Everyone is familiar with the use of ice or mechanical refrigeration to control microbial growth. Like all chemical reactions, the biochemical reactions required for growth are slowed considerably by low temperatures. Chemical disinfectants are also often inhibited by low temperatures. Because their activity is due to temperature-dependent chemical reactions, disinfectants tend to work somewhat better in warm solutions. You might have noticed that directions on disinfectant containers frequently specify that a warm solution be used.

TYPES OF MICROBES

Many disinfectants and antiseptics tend to have a greater effect on gram-positive bacteria, as a group, than on gram-negative bacteria, although this differentiation is by no means as clear-cut as it is with antibiotic activity. A certain group of gram-negative bacteria, the pseudomonads (genus *Pseudomonas*), is of special interest (see Chapter 11). Pseudomonads are unusually resistant to chemical activity and will even grow actively *in* some disinfectants and antiseptics (see box on p. 180). These bacteria are also resistant to many antibiotics. This resistance to antimicrobials is probably related to the characteristics of the *porins* (structural openings in the wall) of pseudomonads (see Chapter 4). Because they are difficult to eliminate and are common in the environment, pseudomonads are very troublesome in a hospital setting.

The mycobacteria are another group of non–endospore-forming organisms that exhibit greater than normal resistance to disinfectants. This group includes *Mycobacterium tuberculosis*, the microbe that causes tuberculosis, and a number of mycobacteria that are pathogenic in persons with depressed immune systems (opportunistically pathogenic). Special tuberculocidal tests have been developed to evaluate the effectiveness of chemical antimicrobials acting on the mycobacteria.

The endospores of bacteria and cysts of protozoans are also highly resistant to chemical agents, as are some viruses. Their resistance to chlorine disinfection in water treatment is a particularly important factor in public health efforts.

PHYSIOLOGICAL STATE OF THE MICROBE

Microorganisms that are actively growing tend to be more susceptible to chemical agents than older cells. This might be because an actively metabolizing and reproducing cell has more points of vulnerability than does an older, possibly dormant cell. When microorganisms have formed endospores, the endospores are generally more resistant than the vegetative cells. The endospores of the resistant strains of *Clostridium botulinum* (bo-tū-lī'num), for example, can withstand boiling for several hours.

ENVIRONMENT

Organic matter frequently interferes with the action of chemical agents. In hospitals, the presence of organic matter in vomit and feces influences the selection of disinfectants because some disinfectants are more effective than others under these conditions. Even bacteria found in food tend to be protected by the organic matter of the food. For example, bacteria in cream are protected by proteins and fats and can survive more heat than bacteria in skim milk. Heat is much more effective in killing bacteria under acidic conditions than at a neutral pH. Some disinfectants also work best at an acidic pH.

Let us now examine the mechanisms of action of control agents, that is, the way various agents actually kill or inhibit microbes.

Actions of Microbial Control Agents

ALTERATION OF MEMBRANE PERMEABILITY

The plasma membrane, located just inside the cell wall, is the target of many control agents. This membrane actively controls the passage of nutrients into the cell and the elimination of wastes from the cell. Damage to the lipids or proteins of the plasma membrane by antimicrobial agents, such as quaternary ammonium compounds, typically causes cellular contents to leak into the surrounding medium and interferes with the growth of the cell. Several types of chemical agents and antibiotics work at least partially in this manner.

DAMAGE TO PROTEINS AND NUCLEIC ACIDS

Bacteria are sometimes thought of as "little bags of enzymes." Enzymes, which are primarily protein, are vital to all cellular activities. You might recall that the functional properties of proteins are due to their three-dimensional shape. This shape is maintained by chemical bonds that link adjoining portions of the amino acid chain as it folds back and forth upon itself. Some of those bonds are hydrogen bonds, which are susceptible to breakage by heat or certain chemicals; breakage results in denaturation of the protein. Covalent bonds, which are stronger, are also subject to attack. For example, disulfide bridges, which play an important role in protein structure by joining amino acids with exposed sulfhydryl (—SH) groups, can be broken by certain chemicals or sufficient heat.

Because DNA and RNA carry the genetic message, damage to them by radiation or chemicals is frequently lethal to the cell; this damage prevents both replication and normal functioning.

Rate of Microbial Death

When bacterial populations are heated or treated with antimicrobial chemicals, they usually die at a constant rate. For example, say that a population of 1 million microbes has been treated for one minute and 90% of the population has died. We are now left with 100,000 microbes. If the population is treated for another minute, 90% of *those* microbes die, and we are left with 10,000 survivors. In other words, for each minute that the treatment is applied, 90% of the remaining population is killed (Table 7.2). If the death curve is plotted logarithmically, the death rate is seen to be constant, as shown by the straight line in Figure 7.1. Obviously, the more microbes there are at the beginning, the longer it will take to kill them all.

Now that you have an understanding of the basic principles of microbial control, we will examine some representative physical and chemical methods.

TABLE 7.2 Example of Microbial Death Rate

TIME (MINUTES)	DEATHS PER MINUTE	NUMBER OF SURVIVORS
0	0	1,000,000
1	900,000	100,000
2	90,000	10,000
3	9,000	1,000
4	900	100
5	90	10
6	9	1

Source: O. Rahn, *Physiology of Bacteria*. New York: McGraw-Hill, 1932.

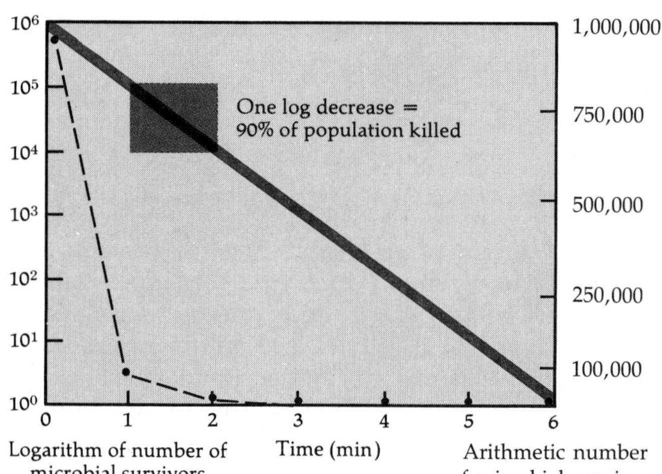

FIGURE 7.1 Microbial death curve. The curve is plotted logarithmically (solid line) and arithmetically (broken line). In this case, the cells are dying at a rate of 90% each minute.

Physical Methods of Microbial Control

HEAT

Probably the most common method by which microbes are killed is heat. A visit to any supermarket will demonstrate that canned goods—in which microorganisms have been killed by moist heat—easily outnumber the foods preserved by other methods. Laboratory media and glassware and hospital instruments are also usually sterilized by heat.

Heat is not only the most widely applicable and effective agent for sterilization but also the most economical and easily controlled. Heat appears to kill microbes by denaturing their enzymes. In sterilization by heat, the degree of heat resistance of the bacteria must be considered. Heat resistance varies among different microbes; these differences can be expressed through the concept of thermal death point. **Thermal death point (TDP)** is the lowest *temperature* required to kill all of the microorganisms in a liquid suspension in 10 minutes.

Another factor to be considered in sterilization is the length of time required for the material to be rendered sterile. This is expressed as **thermal death time (TDT),** the minimal length of *time* in which all bacteria in a liquid culture will be killed at a given temperature. Both TDP and TDT are useful guidelines that indicate the severity of treatment required to kill a given population of bacteria.

Decimal reduction time (DRT or **D value)** is a third concept related to the bacteria's degree of heat resistance. It is the time, in minutes, in which 90% of the population of bacteria at a given temperature will be killed (Figure 7.1). DRT is especially useful in the canning industry.

The heat used in sterilization can be applied in the form of moist heat or dry heat. Dry heat kills by oxidation effects. A simple analogy is the slow charring of

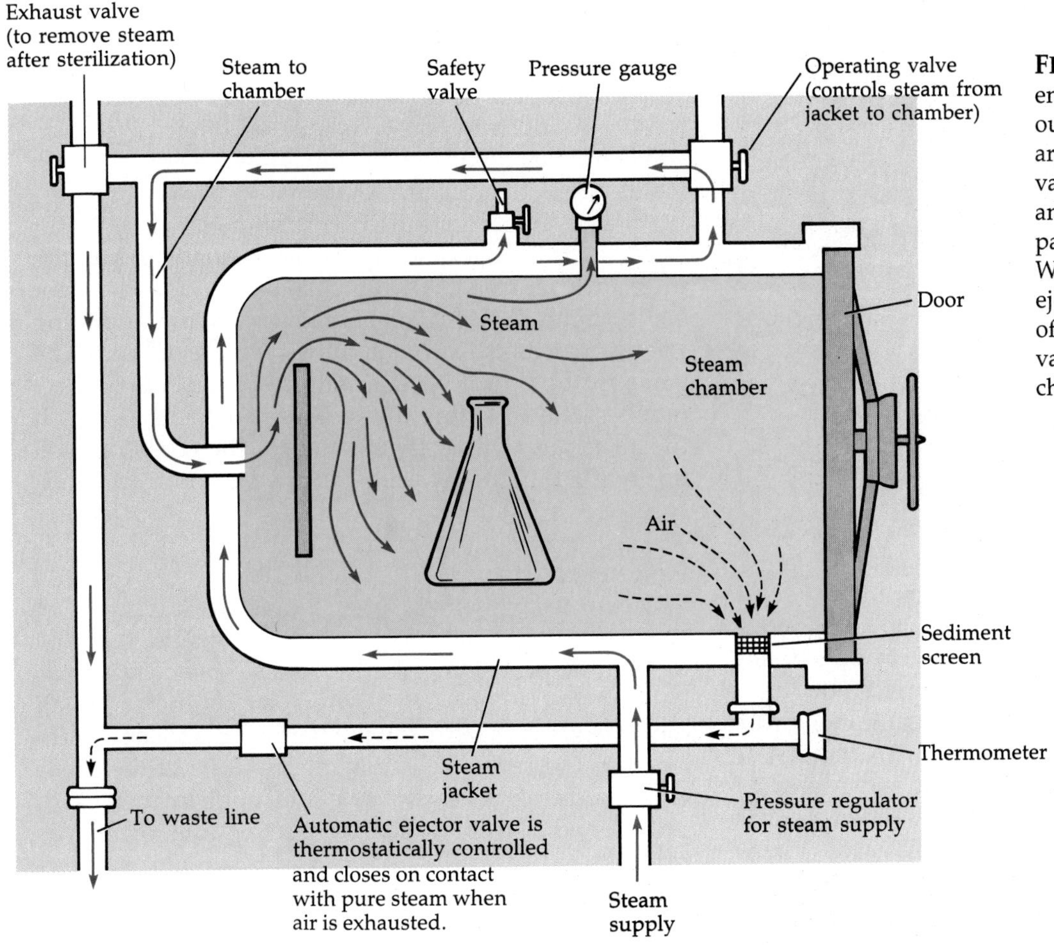

FIGURE 7.2 Autoclave. The entering steam forces the air out of the bottom (black arrows). The automatic ejector valve remains open as long as an air–steam mixture is passing out of the waste line. When all the air has been ejected, the higher temperature of the pure steam closes the valve, and the pressure in the chamber increases.

paper in a heated oven, even when the temperature remains below the ignition point of paper. Moist heat kills microorganisms more quickly because the water hastens the breaking of hydrogen bonds that hold proteins in their three-dimensional structure.

Moist Heat

One type of moist heat sterilization is boiling (100°C at sea level), which kills vegetative forms of bacterial pathogens, many viruses, and fungi and their spores within about 10 minutes. Free-flowing (unpressurized) steam is equivalent in temperature to boiling water. Endospores and some viruses, however, are not destroyed so quickly. One type of hepatitis virus, for example, can survive up to 30 minutes of boiling, and some bacterial endospores have resisted boiling temperatures for more than 20 hours. Boiling is therefore not always a reliable sterilization procedure. However, boiling for a few minutes will kill most pathogens and generally make food or water safe to eat or drink.

Reliable sterilization with moist heat requires temperatures above that of boiling water. These high temperatures are most commonly achieved by steam under pressure in an **autoclave** (Figure 7.2). As the preferred method of sterilization, autoclaving is used unless the material to be sterilized can be damaged by heat or moisture.

The higher the pressure in the autoclave, the higher the temperature. For example, when free-flowing steam at a temperature of 100°C is placed under a pressure of 1 atmosphere above sea level pressure, that is, about 15 pounds of pressure per square inch (psi), the temperature rises to 121°C. Increasing the pressure to 20 pounds per square inch raises the temperature to 126°C. The relationship between temperature and pressure is shown in Table 7.3. Sterilization in an autoclave is most effective when the organisms either contact the steam directly or are contained in a small volume of aqueous (primarily water) liquid. Under these conditions, steam at a pressure of about 15 pounds per square inch (121°C) will kill *all* organisms and their endospores in about 15 minutes.

Autoclaving is used to sterilize culture media, instruments, dressings, intravenous equipment, applicators, solutions, syringes, transfusion equipment, and numerous other items that can withstand high temperature and pressure. Large industrial autoclaves are called *retorts,* but the same principle is used for the common household pressure cooker in the home canning of foods.

TABLE 7.3 Relationship Between Pressure and Temperature of Steam at Sea Level*

PRESSURE (PSI IN EXCESS OF ATMOSPHERIC PRESSURE)	TEMPERATURE (°C)
0 psi	100
5 psi	110
10 psi	116
15 psi	121
20 psi	126
30 psi	135

*At higher altitudes, the gauge pressure would be higher for a particular temperature.

Heat requires extra time to reach the center of solid materials, such as canned meats, in which heat is conducted without the efficient convection currents in liquids. Heating large containers also requires extra time. Table 7.4 shows the different time requirements for various container sizes. It is important to remember that, to sterilize the surface of a solid, steam has to actually contact it. This is not necessary when sterilizing aqueous liquids, but with dry glassware, bandages, and the like, care must be taken to ensure that contact is made. For example, aluminum foil is impervious to steam and should not be used to wrap materials that are to be sterilized; paper should be used instead. Care should also be taken to avoid trapping air in the bottom of a dry container because trapped air will not be replaced by steam, which is lighter than air. The trapped air is the equivalent of a small hot-air oven, which, as we shall see shortly, requires a higher temperature and longer time. Air-trapping containers should be placed in a tipped position so that the steam will force out the air. Products that do not permit penetration by moisture, such as mineral oil or petroleum jelly, are not sterilized by the same treatments that would sterilize aqueous solutions.

TABLE 7.4 Effect of Container Size on Sterilization Times for Liquid Solutions*

CONTAINER SIZE	LIQUID VOLUME	STERILIZATION TIME (MINUTES)
Test tube: 18 × 150 mm	10 ml	15
Erlenmeyer flask: 125 ml	95 ml	15
Erlenmeyer flask: 2000 ml	1500 ml	30
Fermentation bottle: 9000 ml	6750 ml	70

*Sterilization times in the autoclave include the time needed for the contents of the containers to reach sterilization temperatures. For smaller containers this is only 5 minutes or less, but for a 9000-ml bottle it might be as much as 70 minutes. A container is usually not filled past 75% of its capacity.

Steam under pressure fails to sterilize when the air is not completely exhausted. This is usually due to the premature closing of the autoclave's automatic ejector valve (Figure 7.2). Anyone familiar with home canning knows that the steam must flow vigorously out of the valve in the lid for several minutes to entrain and remove all the air before the pressure cooker is sealed. If the air is not completely exhausted, the container will not reach the temperature expected for a given pressure.

Pasteurization

You may recall from Chapter 1 that Louis Pasteur, in the early days of microbiology, found a practical method of preventing the spoilage of beer and wine. Pasteur used mild heating, which was sufficient to kill the organisms that caused the particular spoilage problem without seriously damaging the taste of the product. The same principle was later applied to milk to produce what we now call pasteurized milk. Milk was first pasteurized to eliminate the tuberculosis bacterium. Many relatively heat-resistant *(thermoduric)* organisms survive pasteurization, but these are unlikely to cause disease or cause refrigerated milk to spoil in a short time.

In the classic pasteurization treatment of milk, the milk was exposed to a temperature of about 63°C for 30 minutes. Most milk pasteurization today uses higher temperatures, at least 72°C, but for only about 15 seconds. This treatment, known as **high-temperature short-time (HTST) pasteurization,** is applied as the milk flows continuously past a heat exchanger. In addition to killing pathogens, HTST pasteurization lowers total bacterial counts, so the milk keeps well under refrigeration. Milk can also be sterilized—something quite different from pasteurization—by **ultra-high-temperature (UHT)** treatments so that it can be stored without refrigeration. This is more useful in parts of the world with poor refrigeration facilities. In the United States, sterilization is sometimes used on the small containers of coffee creamers found in restaurants. To avoid imparting a cooked taste to the milk, it was necessary to devise a UHT system in which the liquid milk never touches a surface hotter than the milk itself while being heated by steam. The milk falls in a thin film through a chamber of superheated steam and reaches 140°C in less than a second. It is held for 3 seconds in a holding tube and then cooled in a vacuum chamber, where the steam flashes off. With this process, in less than 5 seconds the milk temperature rises from 74°C to 140°C and drops back to 74°C.

The heat treatments we have just discussed illustrate the concept of **equivalent treatments**—as the temperature is increased, much less time is needed to kill the same number of microbes. For example, the destruction of highly resistant endospores might take 70 minutes at 115°C, whereas only 7 minutes would be needed at 125°C. Both treatments yield the same result. The concept of equivalent treatments also explains why classic pasteurization temperatures of 63°C for 30 minutes, HTST temperatures of 72°C for 15 seconds, and UHT temperatures of 140°C for less than a second can have similar effects.

Dry Heat Sterilization

One of the simplest methods of dry heat sterilization is *direct flaming*. You will use this procedure many times in the microbiology laboratory when you sterilize inoculating loops. To sterilize the inoculating loop, you heat the wire to a red glow, which is 100% effective. A similar principle is used in *incineration*, an effective way to sterilize and dispose of contaminated paper cups, bags, and dressings.

Another form of dry heat sterilization is **hot-air sterilization.** Items to be sterilized by this procedure are placed in an oven. Generally, a temperature of about 170°C maintained for nearly two hours ensures sterilization. The longer period and higher temperature (relative to moist heat) are required because the heat in water is more readily transferred to a cool body than is the heat in air. For example, imagine the different effects of immersing your hand in boiling water at 100°C (212°F) and of holding it in a hot-air oven at the same temperature for the same time.

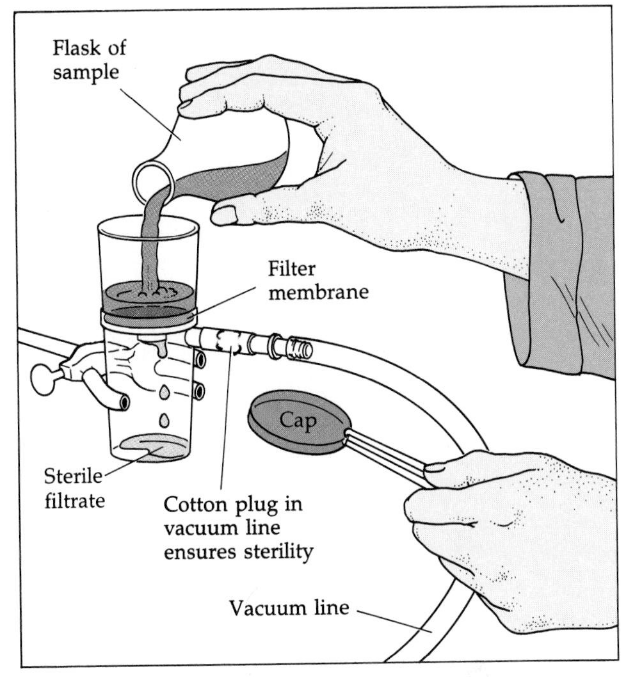

FIGURE 7.3 Filter sterilization with a disposable, presterilized plastic unit. The sample is placed into the upper chamber and forced through the membrane filter by a vacuum in the lower chamber. Pores in the membrane filter are smaller than the bacteria, so bacteria are retained on the filter. The sterilized sample can then be decanted from the lower chamber. Similar equipment with removable filter disks is used to count bacteria in samples (see Figure 6.15).

FILTRATION

Filtration is the passage of a liquid or gas through a screenlike material with pores small enough to retain microorganisms (this is often the same apparatus used for counting; see Figure 6.15). A vacuum that is created in the receiving flask aids gravity in pulling the liquid through the filter. Filtration is used to sterilize heat-sensitive materials, such as some culture media, enzymes, vaccines, and antibiotic solutions.

Some operating theaters and rooms occupied by burn patients receive filtered air to lower the numbers of airborne microbes. **High-efficiency particulate air (HEPA) filters** remove almost all microorganisms larger than about 0.3 μm in diameter.

In the early days of microbiology, hollow candle-shaped filters of unglazed porcelain were used to filter liquids. The long and indirect passageways through the walls of the filter adsorbed the bacteria. Unseen pathogens that passed through such filters (causing such diseases as rabies) were called *filterable viruses.*

In recent years, **membrane filters,** composed of such substances as cellulose esters or plastic polymers, have become popular for industrial and laboratory use (see Figures 6.15 and 7.3). These filters are only 0.1 mm thick. The pores of a membrane filter are closely matched in size. In some brands, plastic film is irradiated so that very uniform holes, where the radiation particles have passed, are etched in the plastic. The pores of membrane filters include 0.22- and 0.45-μm sizes, intended for bacteria. Some very flexible bacteria, such as spirochetes, or the wall-less mycoplasma, will sometimes pass through such filters, however. Filters are available that range down to only 0.01 μm, a size that will retain viruses and even some large protein molecules.

LOW TEMPERATURE

The effect of low temperatures on microbes depends on the particular microbe and the intensity of the application. For example, at temperatures of ordinary refrigerators (0° to 7°C), the metabolic rate of most microbes is so reduced that they cannot reproduce or synthesize toxins. In other words, ordinary refrigeration has a bacteriostatic effect. Yet, psychrotrophs do grow slowly at refrigerator temperatures and will alter

TABLE 7.5 Summary of Physical Methods Used to Control Microbial Growth

METHOD	MECHANISM OF ACTION	COMMENT	PREFERRED USE
Heat			
1. Moist heat sterilization			
a. Boiling or flowing steam	Denaturation	Kills vegetative bacterial and fungal pathogens and many viruses within 10 minutes; less effective on endospores.	Dishes, basins, pitchers, various equipment.
b. Autoclaving	Denaturation	Very effective method of sterilization; at about 15 pounds of pressure (121°C), all vegetative cells and their endospores are killed in about 15 minutes at sterilizing temperature.	Microbiological media, solutions, linens, utensils, dressings, equipment, and other items that can withstand temperature and pressure.
2. Dry heat sterilization			
a. Direct flaming	Burning contaminants to ashes	Very effective method of sterilization.	Inoculating loops.
b. Incineration	Burning to ashes	Very effective method of sterilization.	Paper cups, dressings, animal carcasses, bags, and wipes.
c. Hot-air sterilization	Oxidation	Very effective method of sterilization, but requires temperature of 170°C for about 2 hours.	Empty glassware, instruments, needles, and glass syringes.
3. Pasteurization	Denaturation	Heat treatment for milk (72°C for about 15 seconds) that kills all pathogens and some nonpathogens.	Milk, cream, and certain alcoholic beverages (beer and wine).
Filtration	Separation of bacteria from suspending liquid	Passage of a liquid or gas through a screenlike material that traps microbes; most filters in use consist of cellulose acetate or nitrocellulose.	Useful for sterilizing liquids (toxins, enzymes, vaccines) that are destroyed by heat.

(Continued)

the appearance and taste of foods after a considerable time. Surprisingly, some bacteria can grow at temperatures several degrees below freezing. One-third of the population of some vegetative bacteria might survive a year of freezing, while other species might have very few survivors. Rapidly attained subfreezing temperatures tend to render microbes dormant but do not necessarily kill them. Slow freezing is most harmful to bacteria; the ice crystals that form and grow disrupt the cellular and molecular structure of the bacteria.

DESICCATION

To grow and multiply, microbes require water. In the absence of water—a condition known as **desiccation**—microbes are not capable of growth or reproduction but can remain viable for years. Then, when water is made available to them, they can resume their growth and division. This ability is used in the laboratory when microbes are preserved by lyophilization, a process described in Chapter 6.

The resistance of vegetative cells to desiccation varies with the species and the organism's environment. For example, the gonorrhea bacterium can withstand dryness for only about an hour, but the tuberculosis bacterium can remain viable for months. A normally susceptible bacterium is much more resistant if it is embedded in mucus, pus, or feces. Viruses are generally resistant to desiccation, but not as resistant as endospores. This ability of certain dried microbes and endospores to remain viable is important in a hospital setting. Dust, clothing, bedding, and dressings

TABLE 7.5 *(Continued)*

METHOD	MECHANISM OF ACTION	COMMENT	PREFERRED USE
Low Temperature			
1. Refrigeration	Decreased chemical reactions and possible changes in proteins	Has a bacteriostatic effect.	Food, drug, and culture preservation.
2. Deep-freezing (see Chapter 6)	Decreased chemical reactions and possible changes in proteins	An effective method for preserving microbial cultures, in which cultures are quick-frozen between −50° and −95°C.	Food, drug, and culture preservation.
3. Lyophilization (see Chapter 6)	Decreased chemical reactions and possible changes in proteins	Most effective method for long-term preservation of microbial cultures; water removed by high vacuum at low temperature.	Food, drug, and culture preservation.
Desiccation	Disruption of metabolism	Involves removing water from microbes; primarily bacteriostatic.	Food preservation.
Osmotic pressure	Plasmolysis	Results in loss of water from microbial cells.	Food preservation.
Radiation			
1. Ionizing	Destruction of DNA by gamma rays and high-energy electron beams	Not widespread in routine sterilization.	Used for sterilizing pharmaceuticals and medical and dental supplies.
2. Nonionizing	Damage to DNA by UV light	Radiation not very penetrating.	Practical application is the UV (germicidal) lamp.

might contain infectious microbes in dried mucus, urine, pus, and feces.

Nomadic peoples such as some Native Americans used desiccation to preserve foods. They found that if meat is sliced thin and sun dried (made into jerky), it can be preserved for long periods. Other examples include raisins and similarly dried fruits.

OSMOTIC PRESSURE

The use of high concentrations of salts and sugars to preserve food is based on the effects of *osmotic pressure.* High concentrations of these substances create a hypertonic environment that causes water to leave the microbial cell (see Chapters 4 and 6). This process resembles preservation by desiccation, in that both methods deny the cell the moisture it needs for growth. As water leaves the microbial cell, the plasma membrane shrinks away from the cell wall (plasmolysis), and the cell stops growing, although it might not immediately die. The principle of osmotic pressure is used in the preservation of foods. For example, concentrated salt solutions are used to cure meats, and thick sugar solutions are used to preserve fruits.

As a general rule, molds and yeasts are much more capable than bacteria of growing in materials with low moisture or high osmotic pressures. This property of molds, sometimes combined with their ability to grow under acidic conditions, is the reason fruits and grains are spoiled by molds rather than by bacteria. It is also part of the reason why molds are able to form "mildew" growth on a damp wall or a shower curtain.

RADIATION

Radiation has various effects on cells, depending on its wavelength, intensity, and duration. There are two types of sterilizing radiation, ionizing and nonionizing.

Ionizing radiation, such as *gamma rays* or *high-energy electron beams,* has a wavelength shorter than that of nonionizing radiation, less than about 1 nm. Therefore, it carries much more energy (Figure 7.4). Gamma rays are emitted by radioactive cobalt, and electron beams are produced by accelerating electrons to high energies in special machines. Gamma rays penetrate deeply but may require hours for sterilization of large masses; high-energy electron beams have much lower penetrating power but usually require only a few seconds of exposure. The principal effect of ionizing radiation is the ionization of water, which forms highly reactive hydroxyl radicals (see the discussion of

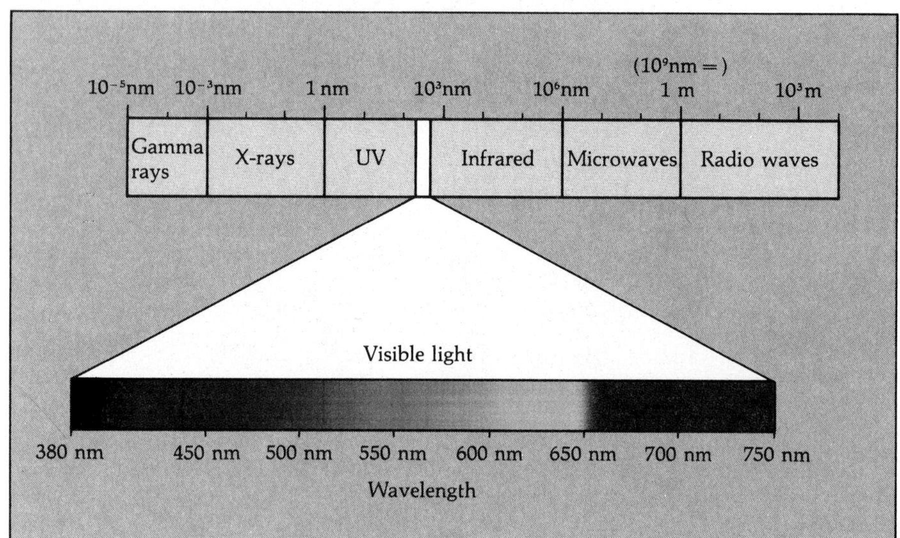

FIGURE 7.4 Radiant energy spectrum.

toxic forms of oxygen in Chapter 6). These radicals react with cellular organic components, especially DNA, and kill the cell.

The food industry has recently renewed its interest in the use of radiation for food preservation. (This subject will be discussed more fully in Chapter 28.) Low-level ionizing radiation, used for years in many countries, has been approved recently in the United States for certain purposes. However, public apprehension about the safety of radiation has limited its use.

Ionizing radiation, especially high-energy electron beams, is increasingly used for the sterilization of pharmaceuticals and disposable dental and medical supplies, such as plastic syringes, surgical gloves, suturing materials, and catheters. Radiation has been replacing gases for sterilizing these items. Sterilizing gases will be discussed later in this chapter.

Nonionizing radiation has a wavelength longer than that of ionizing radiation, usually greater than about 1 nm. A good example of nonionizing radiation is ultraviolet (UV) light. UV light damages the DNA of exposed cells. It causes bonds to form between adjacent thymines in DNA chains (see Figure 8.18). These *thymine dimers* inhibit correct replication of the DNA during reproduction of the cell. The UV wavelengths most effective for killing microorganisms are about 260 nm; these wavelengths are specifically absorbed by cellular DNA. UV radiation is also used to control microbes in the air. A UV or "germicidal" lamp is commonly found in hospital rooms, nurseries, operating rooms, and cafeterias. UV light is also used to sterilize vaccines, serum, and toxins, and occasionally municipal waste waters and drinking waters. A major drawback of UV light as a sterilizer is that the radiation is not very penetrating, so the organisms to be killed must be directly exposed to the rays. Organisms protected by solids and such coverings as paper, glass, and textiles are not affected. Another potential problem is that UV light can damage the eyes and prolonged exposure to UV light can cause burns and skin cancer.

Sunlight contains some UV light, but the most effective (shorter) wavelengths are screened out by the atmosphere. The antimicrobial effect of sunlight is due almost entirely to the formation of singlet oxygen in the cytoplasm (see Chapter 6). Many pigments produced by bacteria provide protection from sunlight.

Microwaves do not have much direct effect on microorganisms but kill them indirectly by heating the medium, such as food.

Table 7.5 summarizes the physical methods of microbial control.

Chemical Methods of Microbial Control

Chemical agents are used to control the growth of microbes on living tissue and inanimate objects. Unfortunately, few chemical agents achieve sterility; most of them merely reduce microbial populations to safe levels or remove vegetative forms of pathogens from objects. A common problem in disinfection is the selection of an agent that will kill all organisms in the shortest time without damaging the contaminated material. Just as there is no single physical method of microbial control that can be used in every situation, there is no one disinfectant that will be appropriate for all circumstances.

PRINCIPLES OF EFFECTIVE DISINFECTION

To select a disinfectant for a particular job, we must first understand its action. For example, what are the properties of the disinfectant? What is it designed to do? Simply by reading the label, we can learn a great deal about a disinfectant's properties. We should also remember that the concentration of a disinfectant will affect its action. A disinfectant should always be diluted exactly as specified by the manufacturer. Solutions that are too weak may be ineffective or be bacteriostatic instead of bactericidal. Solutions that are too strong can be dangerous to humans who come in contact with them.

Also to be considered is the nature of the material being disinfected. Are organic materials present that might interfere with the action of the disinfectant? The presence of organic materials and the pH of the medium in which the microbes are present might determine whether a chemical control agent is lethal or only inhibitory to the microbes.

Another very important consideration is whether the disinfectant will easily make contact with the microbes. An area might have to be scrubbed and rinsed before the disinfectant is applied. In general, disinfection is a gradual process. Thus, to be effective, a disinfectant might have to be left on a surface for several hours.

A final factor to keep in mind is that the higher the temperature at which the disinfectant is applied, the more effective it usually is.

EVALUATING A DISINFECTANT

Phenol Coefficient

There is an obvious need to compare the effectiveness of disinfectants and antiseptics. Phenol, which was once a widely used disinfectant, is still used as a standard for these comparisons. However, some antimicrobials cannot be compared with phenol, particularly if their action is bacteriostatic or if they have a prolonged residual activity on the skin. Nonetheless, the **phenol coefficient test** is still used and serves to illustrate the general principles involved in all such comparisons. For this test, three test organisms are used: *Staphylococcus aureus* (a gram-positive bacterium), *Salmonella typhi* (a gram-negative bacterium), and *Pseudomonas aeruginosa* (ā-rü-ji-nō'sä) (a gram-negative organism often resistant to antimicrobials). In broth cultures under standard conditions, all three test organisms are exposed to the test chemical for a specified time. If the test chemical must be used at a greater concentration, or for a longer period, than that required for phenol to achieve comparable effects, the test chemical's phenol coefficient will be less than 1. This rating indicates that the test chemical is weaker than phenol. A coefficient greater than 1 indicates that the chemical is more active than phenol. For example, if Brand X at a dilution concentration of 1:200 is effective under the specified conditions, and if phenol is equally effective at a concentration of 1:100, then Brand X is more efficient than phenol by a factor of 200/100 and has a phenol coefficient of 2. The result of a phenol coefficient test is usually confirmed by the use-dilution test.

The **use-dilution test** is sometimes considered the official evaluation of a disinfectant. In this test, standardized preparations of several test bacteria are added to a series of tubes containing increasingly strong concentrations of the test disinfectant. The tubes are then incubated, and growth, or lack of it, is recorded. The more highly diluted the chemical can be and still be effective, the higher its rating.

Many specialized tests can be used to determine specific effects of disinfectants, such as their sporicidal, tuberculocidal, or fungicidal effectiveness.

Filter Paper Method

The *filter paper method* is often used in teaching laboratories to evaluate the efficacy of a chemical agent. A disk of filter paper is soaked with a chemical and placed on the surface of an agar plate that has been previously inoculated and incubated with the test organism. After incubation, if the chemical agent is effective, a clear zone representing inhibition of growth can be observed around the disk (Figure 7.5).

TYPES OF DISINFECTANTS

Phenol and Phenolics

Phenol (carbolic acid), the substance first used by Joseph Lister in his operating room, is rarely used as an antiseptic or disinfectant because it irritates the skin and has a disagreeable odor. It is often used in throat lozenges for its local anesthetic effect but has no antimicrobial effect at the low concentrations used. At concentrations above 1%, however, phenol has an antibacterial effect. The structure of a phenol molecule is shown in Figure 7.6a.

Derivatives of phenol, called **phenolics,** contain a molecule of phenol that has been chemically altered to reduce its irritating qualities or increase its antibacterial activity in combination with a soap or detergent. Phenolics exert antimicrobial activity by injuring plasma membranes, inactivating enzymes, and denaturing proteins. Phenolics are frequently used as disinfectants because they remain active in the presence of organic compounds, they are stable, and they persist for long periods after application. For these reasons, phenolics are suitable agents for disinfecting pus, saliva, and feces. The addition of halogens such as chlorines to phenolics usually increases their antimicrobial activity.

One of the most frequently used phenolics is derived from coal tar, a group of chemicals called *cresols.*

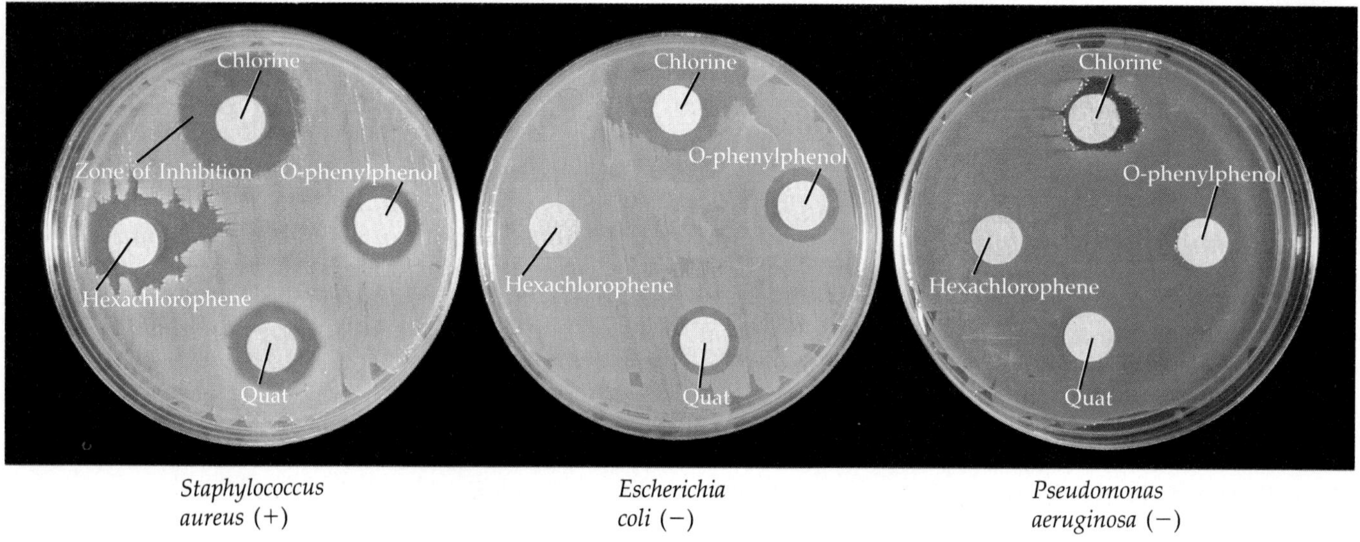

Staphylococcus aureus (+) *Escherichia coli* (−) *Pseudomonas aeruginosa* (−)

FIGURE 7.5 Evaluation of disinfectants by the filter paper method. The paper disks are soaked in a solution of the disinfectant and are placed on the surface of a nutrient medium, on which a culture of test bacteria has been spread to produce uniform growth. Note that the gram-positive *Staphylococcus aureus* was the only one susceptible to the inhibitory effects of all the disinfectants. The two gram-negative bacteria were relatively less affected. Only sodium hypochlorite (chlorine bleach) was effective against them; however, note that the zones of inhibition were smaller than for *Staphylococcus*. You can see that hexachlorophene is effective only against gram-positive bacteria. O-phenylphenol and the quaternary ammonium compound ("quat") were also most effective against gram-positive bacteria.

A very important cresol is *o-phenylphenol* (Figures 7.5 and 7.6b), the main ingredient in most formulations of Lysol®. Cresols are very good surface disinfectants.

Another phenolic, used in the past much more widely than now, is *hexachlorophene* (Figure 7.6c). This phenolic was originally an ingredient in soaps and lotions (such as pHisoHex®) used for surgical scrubs and hospital microbial control procedures, cosmetic soaps, deodorants, feminine hygiene sprays, and even toothpaste. As a bacteriostatic agent, hexachlorophene is especially effective against gram-positive staphylococci and streptococci (Figure 7.5), which can cause infections of the skin. Moreover, the hexachlorophene persists on the skin for long periods. For these reasons, it was once heavily used to control staphylococcal and streptococcal infections in hospital nurseries. However, in 1972, it was found that *excessive* use of hexachlorophene, such as bathing infants with it several times a day, could lead to neurological damage. Hexachlorophene is still used in hospitals to control nosocomial (hospital-acquired) infections and as a scrub for hospital personnel. However, it is no longer casually used by the general public. At present, a prescription is required for the purchase of a 3% or stronger solution of hexachlorophene.

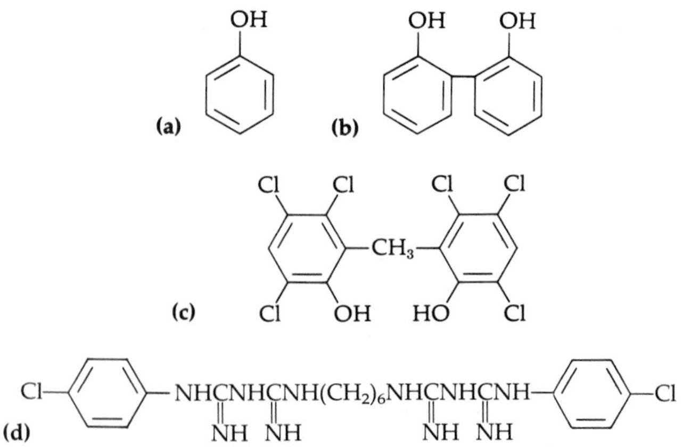

FIGURE 7.6 Structure of phenolics and chlorheximide. **(a)** Phenol. **(b)** O-Phenylphenol. **(c)** Hexachlorophene. **(d)** Chlorhexidine.

Chlorhexidine

Chlorhexidine, which is not a phenol, although its structure and application resemble hexachlorophene (Figure 7.6d), is frequently used for disinfection of skin

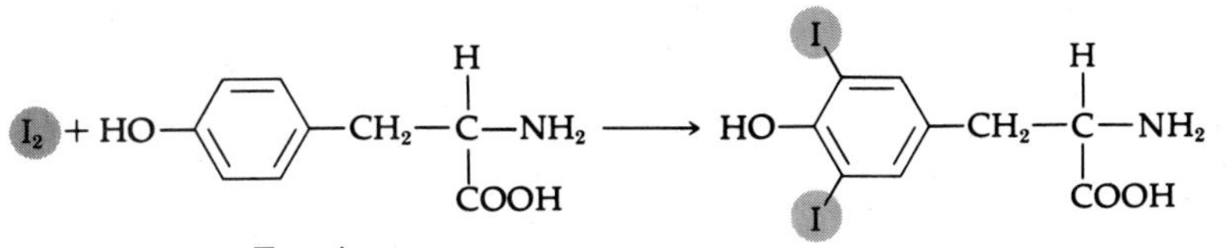

$$I_2 + HO-C_6H_4-CH_2-CH(NH_2)-COOH \longrightarrow HO-C_6H_2I_2-CH_2-CH(NH_2)-COOH$$

Tyrosine Diiodotyrosine

FIGURE 7.7 A proposed explanation of the mechanism for the activity of iodine. Iodine reacts with tyrosine, an amino acid found in protein. The diiodotyrosine formed inhibits the function of the protein in the cell. Iodine probably also reacts with other cellular constituents.

and mucous membranes as an alternative to hexachlorophene. It is combined with a detergent or alcohol for surgical hand scrubs and preoperative skin preparation in patients. In such applications, it works more rapidly than hexachlorophene and is equally persistent on the skin. The skin does not absorb it, and no toxicity has been reported. Its killing effect is related to damage to the plasma membrane. It is effective against most vegetative bacteria but is not sporicidal. The only viruses affected are certain enveloped (lipophilic) types (see Chapter 13).

Halogens

The *halogen elements,* particularly iodine and chlorine, are effective antimicrobial agents, both alone (as I_2 or Cl_2 in solution) and as constituents of inorganic or organic compounds. *Iodine* (I_2) is one of the oldest and most effective antiseptics. It is effective against all kinds of bacteria, many endospores, various fungi, and some viruses. One proposed mechanism for the activity of iodine is that it combines with the amino acid tyrosine, a common component of many enzymes and other cellular proteins (Figure 7.7). As a result, microbial protein function is inhibited. Iodine also reacts with other amino acids and apparently with the fatty acids of the plasma membranes.

Iodine is available as a **tincture,** that is, in solution in aqueous alcohol, and as an iodophor. An **iodophor** is a combination of iodine and an organic molecule, usually a detergent in which the iodine is released slowly. Iodophors have the antimicrobial activity of iodine, but they do not stain and are less irritating. The preparations most commonly seen commercially are Betadine® and Isodine®. These are *povidone-iodines;* povidone is a surface-active iodophor that improves the wetting action. The main use of iodines is for skin disinfection and wound treatment. *Pseudomonas* can survive for long periods in iodophores (see the box, p. 180).

Chlorine (Cl_2), as a gas or in combination with other chemicals, is another widely used disinfectant. Its germicidal action is caused by the hypochlorous acid (HOCl) that forms when chlorine is added to water.

$$(1)\quad \underset{\text{Chlorine}}{Cl_2} + \underset{\text{Water}}{H_2O} \rightleftharpoons \underset{\text{Hydrogen ion}}{H^+} + \underset{\text{Chloride ion}}{Cl^-} + \underset{\text{Hypochlorous acid}}{HOCl}$$

$$(2)\quad \underset{\text{Hypochlorous acid}}{HOCl} \rightleftharpoons \underset{\text{Hydrogen ion}}{H^+} + \underset{\text{Hypochlorite ion}}{OCl^-}$$

Exactly how hypochlorous acid exerts its killing power is not known. It is a strong oxidizing agent that prevents much of the cellular enzyme system from functioning. Hypochlorous acid is the most effective form of chlorine because it is neutral in electrical charge and diffuses as rapidly as water through the cell wall. Because of its negative charge, the hypochlorite ion (OCl^-) cannot enter the cell freely.

A liquid form of compressed chlorine gas is used extensively for disinfecting municipal drinking water, water in swimming pools, and sewage. Several compounds of chlorine are also effective disinfectants. For example, solutions of *calcium hypochlorite* [$Ca(OCl)_2$] are used to disinfect dairies, barns, slaughterhouses, and restaurant eating utensils. Another chlorine compound, *sodium hypochlorite* (NaOCl), is used as a household disinfectant and a bleach (Clorox®) and as a disinfectant in dairies, food-processing establishments, and hemodialysis systems (see Figure 7.5). When the quality of drinking water is in question, household bleach can provide a rough equivalent of municipal chlorination. After two drops of bleach are added to a liter of water (four drops if the water is cloudy) and the mixture has sat for 30 minutes, the water is considered safe for drinking under emergency conditions.

Another group of chlorine compounds, the *chloramines,* consist of chlorine and ammonia. They are used as disinfectants, antiseptics, or sanitizing agents. Chloramines are very stable compounds that release chlorine over long periods. They are relatively effective in organic matter, but they have the disadvantages of acting more slowly and being less effective purifiers than many other chlorine compounds. Chloramines are used to sanitize glassware and eating utensils and to treat dairy and food-manufacturing equipment. Ammonia is usually mixed with chlorine in municipal

MMWR

MORBIDITY AND MORTALITY WEEKLY REPORT

Contaminated Povidone–Iodine Solution—Texas

From December 29, 1988, to January 21, 1989, *Pseudomonas cepacia* was isolated from peritoneal fluid of four patients and blood of two patients at a children's hospital in Texas. Three patients who were receiving peritoneal dialysis for renal failure developed peritonitis. Two intensive-care unit (ICU) patients who were not on dialysis and whose blood cultures grew the organism had no symptoms attributable to *P. cepacia* bacteremia.

The hospital personnel recovered *P. cepacia* in pure culture from three previously opened 1-gallon containers of Clinidine®, a povidone–iodine (iodophor) solution. This solution was being used by the peritoneal dialysis staff to disinfect tops of multidose vials of dialysis fluid additives, peritoneal fluid administration set connectors, and ports of peritoneal dialysis systems. Clinidine was also being used by ICU staff for antisepsis of skin before venipuncture and to disinfect the tops of blood culture bottles. In further investigations by CDC and Food and Drug Administration (FDA), *P. cepacia* was isolated from two unopened bottles and one opened bottle of the same lot number being used in another health-care facility.

On February 9, 1989, the company initiated a voluntary recall of the implicated lot.

Comment: In this particular outbreak, three patients developed peritonitis and three had pseudoinfections associated with probable contamination of a povidone–iodine solution. This is the third instance of contamination of an iodophor solution reported to CDC. In 1980, a cluster of *P. cepacia* pseudobacteremias in seven northeastern U.S. hospitals was associated with a contaminated iodophor solution (from another manufacturer) used to disinfect the tops of blood-culture bottles before inoculation. In 1982, a cluster of *P. aeruginosa* peritonitis cases in peritoneal dialysis patients was associated with a contaminated iodophor solution (from a third manufacturer) being used as a peritoneal catheter disinfectant in a hospital. *P. cepacia* and *P. aeruginosa,* organisms commonly found in water, may colonize water distribution pipes or filters in plants that manufacture iodine solutions. Once affixed to the inner surface of pipes, *P. cepacia* and *P. aeruginosa* could be protected from the bactericidal effect of the iodophor solution, probably by a glycocalyx film. Bacteria that come free from the pipes end up in bottles of iodophor destined for clinical use. These bacteria can survive up to 15 months in iodophors.

New types of plastic and metals are being sought that prevent this bacterial biofilm from developing. Bacteria that adhere to plastic and metal are also a source of infection for people with implants such as heart valves and pacemakers. New materials that prevent bacterial adhesion would be beneficial to these people as well.

Source: MMWR 38:133–134(3/3/89).

water treatment systems to form chloramines. The chloramines control taste and odor problems caused by the reaction of chlorine with other nitrogenous compounds in the water. Because chloramines are less effective as germicides, sufficient chlorine must be added to ensure a residual of chlorine in the form of HOCl. (Chloramines are toxic to aquarium fish, but pet shops sell chemicals to neutralize them.)

Alcohols

Alcohols effectively kill bacteria and fungi but not endospores and nonenveloped viruses. The mechanism of action of alcohol is usually protein denaturation, but alcohol can also disrupt membranes and dissolve many lipids, including the lipid component of enveloped viruses. Alcohols have the advantage of acting and then evaporating rapidly and leaving no residue. In a quick swabbing of the skin (degerming) before an injection, most of the microbial control activity comes from a simple *wiping away* of dirt and microorganisms. However, alcohols are unsatisfactory when applied to wounds. They cause coagulation of a layer of protein under which bacteria continue to grow.

Two of the most commonly used alcohols are ethanol and isopropanol. The recommended optimum concentration of *ethanol* is 70%, but concentrations between 60% and 95% seem to kill as fast (Table 7.6). Pure ethanol is less effective than aqueous solutions (ethanol mixed with water) because denaturation requires water. *Isopropanol,* often sold as rubbing alcohol, is slightly superior to ethanol as an antiseptic and disinfectant. Moreover, it is less volatile, less expensive, and more easily obtained than ethanol.

Ethanol and isopropanol are often used to enhance the effectiveness of other chemical agents. For example, an aqueous solution of Zephiran™ (described shortly) kills about 40% of the population of a test organism in two minutes, whereas a tincture of Zephiran™ kills about 85% in the same period. You can compare the effectiveness of tinctures and aqueous solutions in Figure 7.10.

Heavy Metals and Their Compounds

Several heavy metals, such as silver, mercury, and copper, can be germicidal or antiseptic. The ability of very small amounts of heavy metals, especially silver and copper, to exert antimicrobial activity is referred to as **oligodynamic action** (*oligo* means few). This action can be seen when we place a coin or other clean piece of metal containing silver or copper on an inoculated Petri plate. Extremely small amounts of metal diffuse from the coin and inhibit the growth of bacteria for some distance around the coin (Figure 7.8). This effect is produced by the action of heavy metal ions on microbes. When the metal ions combine with the —SH groups on cellular proteins, denaturation results.

Silver is used as an antiseptic in a 1% *silver nitrate* solution. The solution is bactericidal for most organisms. Many states require that the eyes of newborns be treated with a few drops of silver nitrate to guard against a gonococcal infection of the eyes called gonorrheal ophthalmia neonatorum, which the infants might have contracted as they passed through the birth canal. In recent years, antibiotics have been replacing silver nitrate for this purpose.

Inorganic mercury compounds, such as *mercuric chloride,* probably have the longest history of use as disinfectants. They have a very broad spectrum of activity; their effect is primarily bacteriostatic. However, their use is now limited because of their toxicity, corrosiveness, and ineffectiveness in organic matter. Organic mercury compounds, such as Mercurochrome and Merthiolate, are less irritating and less toxic than inorganic mercury compounds. Mercurochrome and Merthiolate are antiseptics used on skin and mucous membranes. Unfortunately, bacteria resume growth if the compounds are removed (washed away).

Copper in the form of *copper sulfate* is used chiefly to destroy green algae (algicide) that grow in reservoirs, swimming pools, and fish tanks. If the water does not contain excessive organic matter, copper sulfate is effective in concentrations of one part per million of water. To prevent mildew, copper compounds are sometimes included in paint.

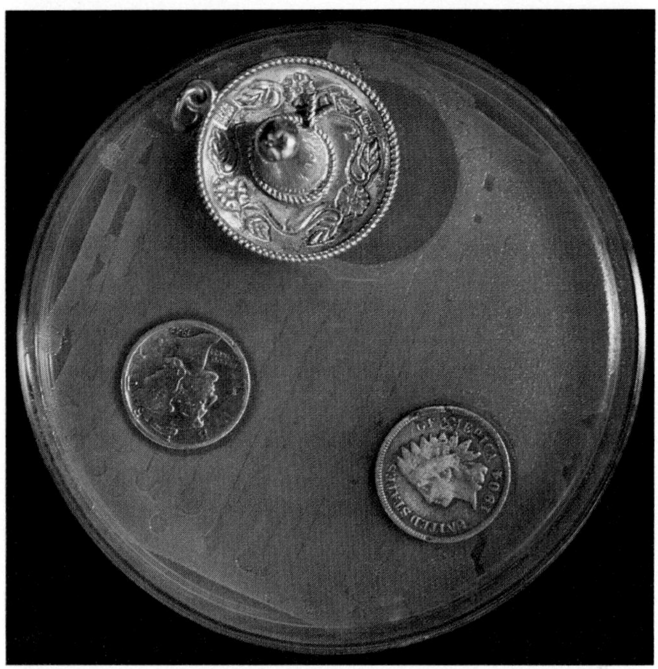

FIGURE 7.8 Oligodynamic action of heavy metals. Clear zones where bacterial growth has been inhibited are seen around the sombrero charm (pushed aside), the dime, and the penny. The charm and the dime contain silver; the penny contains copper.

TABLE 7.6 Germicidal Action of Various Concentrations of Ethanol in Aqueous Solution Against *Streptococcus pyogenes**

PERCENTAGE OF ETHANOL	SECONDS 10	20	30	40	50
100	+	+	+	+	+
95	–	–	–	–	–
90	–	–	–	–	–
80	–	–	–	–	–
70	–	–	–	–	–
60	–	–	–	–	–
50	+	+	–	–	–
40	+	+	+	+	+

*+ = growth, microbes not killed; – = no growth.
Source: H.E. Morton, "Relationship of concentration and germicidal efficiency of ethyl alcohol." *Annals of the New York Academy of Sciences* 53:191–196, 1950.

Another metal used as an antimicrobial is zinc. *Zinc chloride* is a common ingredient in mouthwashes, and *zinc oxide* is probably the most widely used antifungal in paints.

Surface-Active Agents

Surface-active agents, or **surfactants,** can decrease surface tension among molecules of a liquid. Such agents include soaps and detergents. Soap has little value as an antiseptic, but it does have an important function in the mechanical removal of microbes through scrubbing. The skin normally contains dead cells, dust, dried sweat, microbes, and oily secretions from oil glands. Soap breaks the oily film into tiny droplets, a process called **emulsification,** and the water and soap together lift up the emulsified oil and debris and float them away as the lather is washed off. In this sense, soaps are good degerming and emulsifying agents. Many so-called deodorant soaps contain compounds such as *triclocarban* that strongly inhibit gram-positive bacteria.

Acid-anionic surface-active sanitizers are very important in the dairy industry's cleaning of utensils and equipment. Their sanitizing ability is related to the negatively charged portion (anion) of the molecule, which reacts with the plasma membrane. These sanitizers act on a wide spectrum of microbes, including troublesome thermoduric bacteria. They are nontoxic, noncorrosive, and fast-acting.

Quaternary Ammonium Compounds

The most widely used surface-active agents are the cationic detergents, especially the **quaternary ammonium compounds (quats).** Their cleansing ability is related to the positively charged portion—the cation—of the molecule. Their name comes from the fact that they are modifications of the four-valence ammonium ion, NH_4^+ (Figure 7.9). Quaternary ammonium compounds are strongly bactericidal against gram-positive bacteria and somewhat less active against gram-negative bacteria.

Quats are also fungicidal, amoebicidal, and virucidal against enveloped viruses. They do not kill endospores or tuberculosis bacteria, but they are bacteriostatic. Their chemical mode of action is unknown, but they most probably affect the plasma membrane. They change the cell's permeability and cause the loss of essential cytoplasmic constituents, such as potassium.

Two popular quats are Zephiran™, a brand name of benzalkonium chloride (see Figure 7.5), and Cepacol®, a brand name of cetylpyridinium chloride. They are strongly antimicrobial, colorless, odorless, tasteless, stable, easily diluted, and nontoxic, except at high concentrations. If your mouthwash bottle fills with foam when shaken, the mouthwash probably contains a quat. However, organic matter interferes with their activity, and they are rapidly neutralized by soaps and anionic detergents.

Anyone associated with medical applications of quats should remember that certain bacteria, such as some species of *Pseudomonas,* not only survive in quaternary ammonium compounds but actively grow in them. This resistance occurs not only to the disinfectant solution but also to the moistened gauze and bandages, whose fibers tend to neutralize the quats. This resistance is another example of the peculiar and troublesome characteristics of the pseudomonads. Their ability to metabolize unusual substrates for growth permits pseudomonads to grow in unlikely places, such as in simple saline solutions and even distilled water. They are using as nutrients such things as traces of soap residues and cap-liner adhesives.

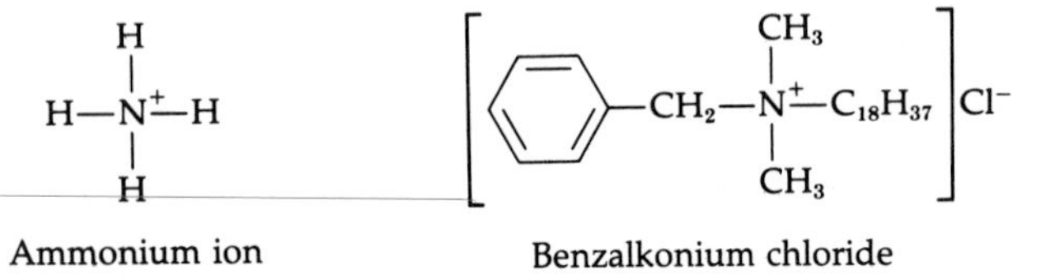

Before we move on to the next group of chemical agents, refer to Figure 7.10, which compares the effectiveness of some of the antimicrobials we have discussed so far.

H
H—N$^+$—H
H
Ammonium ion

[C$_6$H$_5$—CH_2—N$^+$(CH_3)$_2$—$C_{18}H_{37}$] Cl^-
Benzalkonium chloride

FIGURE 7.9 The ammonium ion and a quaternary ammonium compound, benzalkonium chloride (Zephiran™). Note how other groups replace the hydrogens of the ammonium ion.

Organic Acids and Derivatives

A number of organic acids are used as preservatives to control mold growth. *Sorbic acid* (often as the salt *potassium sorbate*) is used to inhibit mold growth in acidic foods, such as cheese. *Benzoic acid* (or its salt *sodium benzoate*) is an antifungal that is effective at low pH values and has wide use in soft drinks and other acidic foods. The *parabens,* such as *methylparaben* and *propylparaben,* are often used to inhibit mold growth in liquid cosmetics and shampoos. The parabens are derivatives of benzoic acid but work at a neutral pH. *Calcium propionate* prevents mold growth in bread.

The activity of these organic acids is not related to their acidity but to their ability to inhibit enzymatic and metabolic activity. Organic acids are listed on the labels of many foods and a few cosmetic preparations. By and large, the body metabolizes these organic acids readily, so their use is considered quite safe.

Aldehydes

Aldehydes are among the most effective antimicrobials. Two examples are formaldehyde and glutaraldehyde. They inactivate proteins by forming covalent crosslinks with a number of organic functional groups on proteins (—NH_2, —OH, —COOH, and —SH). *Formaldehyde gas* is an excellent disinfectant. However, it is more commonly available as *formalin,* a 37% aqueous solution of formaldehyde gas. Formalin was once used extensively to preserve biological specimens and inactivate bacteria and viruses in vaccines. However, it is considered carcinogenic and is little used at present, except for embalming corpses.

Glutaraldehyde is a chemical relative of formaldehyde that is less irritating and more effective than formaldehyde. Glutaraldehyde is used to sterilize hospital instruments, including respiratory-therapy equipment. When used in a 2% solution (Cidex™), it is bactericidal, tuberculocidal, and virucidal in 10 minutes and sporicidal in 3 to 10 hours. It is probably the only liquid chemical disinfectant that can be considered a possible sterilizing agent. However, such a long

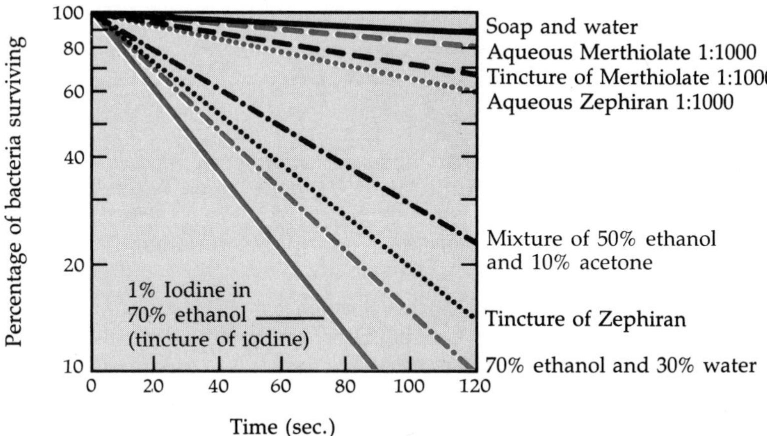

FIGURE 7.10 Comparison of the effectiveness of various antiseptics. The steeper the downward slope of the killing curve of the antiseptic, the more effective it is. A 1% iodine in 70% ethanol solution is the most effective; soap and water are the least effective. Note that tinctures are more effective than aqueous solutions of the same antiseptic.

exposure time is required for sporicidal activity that chemical agents cannot be relied on as sterilants.

Gaseous Chemosterilizers

Gaseous chemosterilizers are chemicals that sterilize in a closed chamber (similar to an autoclave). A gas suitable for this method is *ethylene oxide:*

$$H_2C{-}CH_2 \text{ (bridged by O)}$$

Its activity depends on the denaturation of proteins: The proteins' labile hydrogens, such as —SH, —COOH, or —OH, are replaced by alkyl groups *(alkylation),* such as —CH_2CH_2OH. Ethylene oxide kills all microbes and endospores. It is toxic and explosive in its pure form, so it is usually mixed with a nonflammable gas, such as carbon dioxide or nitrogen. One of its principal advantages is that it is highly penetrative. Although materials to be sterilized with ethylene oxide must be exposed to it for 4 to 18 hours, its remarkable penetrating power is one reason why ethylene oxide was chosen to sterilize spacecraft sent to land on the Moon and other planets. Using heat to sterilize the electronic gear on these vehicles was not practical.

Because of their ability to sterilize without heat, gases like ethylene oxide are also widely used on medical supplies and equipment. Examples include disposable sterile plasticware, such as syringes and Petri plates, textiles, sutures, lensed instruments, artificial heart valves, heart–lung machines, and mattresses. Many large hospitals have ethylene oxide chambers, some large enough to sterilize mattresses, as part of their sterilizing equipment. Propylene oxide and beta-propiolactone are also used for gaseous sterilization:

$H_3C{-}CH{-}CH_2$ (bridged by O) — Propylene oxide

$H_2C{-}CH_2$ / $O{-}C{=}O$ — Beta-propiolactone

All these gases may be hazardous by causing mutations and are suspected carcinogens, especially beta-propiolactone. For this reason, there has been concern about exposure of hospital workers to ethylene oxide from such sterilizers; certainly, it is wise to minimize exposures.

Oxidizing Agents

Oxidizing agents exert antimicrobial activity by oxidizing cellular components of the treated microbes. Examples of oxidizing agents are ozone and hydrogen peroxide. *Ozone* (O_3) is a highly reactive form of oxygen that is generated by high-voltage electrical discharges. It is responsible for the air's rather fresh odor after a lightning storm, in the vicinity of electric sparking, or around an ultraviolet light. Ozone is often used to supplement chlorine in the disinfection of water. It helps neutralize tastes and odors. Although ozone is a more effective killing agent, its residual activity is difficult to maintain in water, and it is more expensive than chlorine.

Hydrogen peroxide is an antiseptic found in many household medicine cabinets and in hospital supply rooms. It is not a good antiseptic for open wounds because it is quickly broken down to water and gaseous oxygen by the action of the enzyme catalase, which is present in human cells (Chapter 6). Although its usefulness as an antiseptic is limited, hydrogen peroxide is effectively used to disinfect inanimate objects, an application where it is even sporicidal. On a nonliving surface, the normally protective enzymes of aerobic bacteria and facultative anaerobes are overwhelmed by the high concentrations of peroxide used. The food industry is increasing its use of hydrogen peroxide for aseptic packaging (Chapter 28). The packaging materials pass through a hot solution of the chemical before being assembled into a container. Many wearers of contact lenses are familiar with disinfection by hydro-

TABLE 7.7 Summary of Chemical Agents Used to Control Microbial Growth

CHEMICAL AGENT	*MECHANISM OF ACTION*	*PREFERRED USE*	*COMMENT*
Phenol and Phenolics			
1. Phenol	Disruption of plasma membrane, denaturation, inactivation of enzymes.	Still used as a standard for the effectiveness of other disinfectants *(phenol coefficient)*.	Seldom used as a disinfectant or antiseptic because of its irritating qualities and disagreeable odor.
2. Phenolics	Disruption of plasma membrane, denaturation, inactivation of enzymes.	Environmental surfaces, instruments, skin surfaces, and mucous membranes.	Derivatives of phenol that are reactive even in the presence of organic material; examples include o-phenylphenol and hexachlorophene.
Chlorhexidine	Disruption of plasma membrane.	Skin degerming, especially for surgical scrubs.	Bactericidal to gram-positives and gram-negatives. Nontoxic, persistent.
Halogens	Iodine inhibits protein function and is a strong oxidizing agent; chlorine forms the strong oxidizing agent hypochlorous acid, which alters cellular components.	Effective antiseptic available as a tincture and an iodophor; chlorine gas is used to disinfect water; chlorine compounds are used to disinfect dairy equipment, eating utensils, household items, and glassware.	Iodine and chlorine may act alone or as components of inorganic and organic compounds.
Alcohols	Denaturation and lipid dissolution.	Thermometers and other instruments; in a quick swabbing of the skin with alcohol before an injection, most of the disinfecting action probably comes from a simple wiping away (cleansing) of dirt and some microbes; usually another antiseptic is used in addition.	Bactericidal and fungicidal, but not effective against endospores, nonenveloped viruses; commonly used alcohols are ethanol and isopropanol.
Heavy Metals and Their Compounds	Denature enzymes and other essential proteins.	Silver nitrate is used to prevent gonococcal eye infections; Mercurochrome and Merthiolate disinfect skin and mucous membranes; copper sulfate is an algicide.	Heavy metals such as silver and mercury are germicidal or antiseptic.
Surface-Active Agents			
1. Soaps and acid-anionic detergents	Mechanical removal of microbes through scrubbing.	Skin degerming and emulsification of debris.	Many deodorant soaps contain antimicrobials such as triclocarban.
2. Acid-anionic detergents	Not certain; may involve enzyme inactivation or disruption.	Sanitizers in dairy and food-processing industry.	Wide spectrum of activity, nontoxic, noncorrosive, fast-acting.
3. Cationic detergents (quaternary ammonium compounds)	Enzyme inhibition, protein denaturation, and disruption of plasma membranes.	Antiseptic for skin, instruments, utensils, rubber goods.	Bactericidal, bacteriostatic, fungicidal, and virucidal against enveloped viruses; examples of quats are Zephiran™ and Cepacol®.

(Continued)

TABLE 7.7 ***(Continued)***

CHEMICAL AGENT	*MECHANISM OF ACTION*	*PREFERRED USE*	*COMMENT*
Organic Acids	Metabolic inhibitors, mostly affecting molds; action not related to their acidity.	Sorbic acid and benzoic acid effective at low pH; parabens much used in cosmetics, shampoos; calcium propionate used in bread; all are mainly antifungals.	Widely used to control molds and some bacteria in foods and cosmetics.
Aldehydes	Protein inactivation	Formalin (37% aqueous solution of formaldehyde) for embalming corpses; glutaraldehyde (Cidex™) is less irritating than formaldehyde and is used for sterilization of medical equipment.	Very effective antimicrobials; glutaraldehyde is considered a liquid sterilant.
Gaseous Sterilants	Denaturation	Excellent sterilizing agent, especially for objects that would be damaged by heat.	Ethylene oxide is most commonly used.
Oxidizing Agents	Oxidation	Contaminated surfaces; some deep wounds, in which they are very effective against oxygen-sensitive anaerobes.	Ozone is gaining attention as a potential replacement for chlorination; hydrogen peroxide is a poor antiseptic but a good disinfectant.

gen peroxide. A platinum catalyst in the lens-disinfecting kit destroys residual hydrogen peroxide so it does not persist on the lens, where it might be an irritant.

Zinc peroxide and hydrogen peroxide are useful in irrigation of deep wounds, where the oxygen released makes an environment that inhibits the growth of anaerobic bacteria.

Benzoyl peroxide is another useful treatment for wounds infected by anaerobic pathogens, but it is probably much more familiar as the main ingredient in over-the-counter medications for treating acne, which is caused by an anaerobic bacterium.

A summary of chemical agents that control microbial growth is presented in Table 7.7.

The compounds discussed in this chapter are not generally useful in the treatment of diseases. Because antibiotics are used in chemotherapy, antibiotics and the pathogens against which they are active will be discussed together, in Chapter 20.

STUDY OUTLINE

Introduction (pp. 167–168)

1. Ignatz Semmelweis introduced handwashing with chlorinated lime by hospital personnel to reduce infections.
2. Joseph Lister introduced aseptic surgery, including heat sterilization of instruments and disinfection of wounds.
3. Microbial growth can be controlled by physical and chemical methods.
4. Control of microbial growth can prevent infections and food spoilage.
5. Sterilization is the process of destroying all microbial life on an object.
6. Disinfection is the process of reducing or inhibiting microbial growth.

Conditions Influencing Microbial Control (pp. 168–169)

1. A general rule of disinfection is to try to kill the most resistant microbes that can be found on the object to be disinfected.
2. The number and species of microorganisms present will affect the rate of disinfection.

TEMPERATURE (p. 169)

1. Biochemical reactions occur more rapidly at warm temperatures.
2. Disinfectant activity is enhanced by warm temperatures.

TYPES OF MICROBES (p. 169)

1. Gram-positive bacteria are generally more susceptible to disinfectants than gram-negative bacteria.
2. Pseudomonads can even grow in some disinfectants and antiseptics.

PHYSIOLOGICAL STATE OF THE MICROBE (p. 169)

1. An actively growing microorganism tends to be less resistant to chemical agents than an older microorganism.
2. Endospores are more resistant to chemical agents and physical methods.

ENVIRONMENT (p. 169)

1. Organic matter and pH level frequently interfere with the actions of chemical control agents.
2. Examples of such organic matter include vomit, feces, pus, and food.

Actions of Microbial Control Agents (pp. 169–170)

ALTERATION OF MEMBRANE PERMEABILITY (p. 169)

1. The susceptibility of the plasma membrane is due to its lipid and protein components.
2. Certain chemical control agents damage the plasma membrane by altering its permeability.

DAMAGE TO PROTEINS AND NUCLEIC ACIDS (pp. 169–170)

1. Some microbial control agents damage cellular proteins by breaking hydrogen and covalent bonds.
2. Other agents interfere with DNA and RNA replication and protein synthesis.

Rate of Microbial Death (p. 170)

1. Bacterial populations subjected to heat or antimicrobial chemicals die at a constant rate.
2. Such a death curve can be plotted logarithmically.

Physical Methods of Microbial Control (pp. 170–176)

HEAT (pp. 170–173)

1. Heat is frequently used to eliminate microorganisms; it is economical and easily controlled.
2. The mechanism or action by which heat kills microbes is denaturation of enzymes.
3. Thermal death point (TDP) is the lowest temperature at which all the bacteria in a liquid culture will be killed in 10 minutes.
4. Thermal death time (TDT) is the length of time required to kill all bacteria in a liquid culture at a given temperature.
5. Decimal reduction time (DRT) is the length of time in which 90% of a bacterial population will be killed at a given temperature.
6. Boiling (100°C) kills many vegetative cells and viruses within 10 minutes.
7. Autoclaving (steam under pressure) is the most effective method of moist heat sterilization. The steam must directly contact the material to be sterilized.
8. In pasteurization, a high temperature is used for a short time (72°C for 15 seconds) to destroy pathogens without altering the flavor of the food.
9. Methods of dry heat sterilization include direct flaming, incineration, and hot-air sterilization.
10. Different methods that produce the same effect (reduction in microbial growth) are called equivalent treatments.

FILTRATION (p. 173)

1. Filtration is the passage of a liquid or gas through a filter with pores small enough to retain microbes.
2. Microbes can be removed from air by high-efficiency particulate air filters.
3. Membrane filters composed of nitrocellulose or cellulose acetate are commonly used to filter out bacteria, viruses, and even large proteins.

LOW TEMPERATURE (pp. 173–174)

1. The effectiveness of low temperatures depends on the particular microorganism and the intensity of the application.
2. Most microorganisms do not reproduce at ordinary refrigerator temperatures (0° to 7°C).
3. Many microbes survive (but do not grow) at the subzero temperatures used to store foods.

DESICCATION (pp. 174–175)

1. In the absence of water, microorganisms cannot grow but can remain viable.
2. Viruses and endospores can resist desiccation.

OSMOTIC PRESSURE (p. 175)

1. Microorganisms in high concentrations of salts and sugars undergo plasmolysis.
2. Molds and yeasts are more capable of growing in materials with low moisture or high osmotic pressure than bacteria are.

RADIATION (pp. 175–176)

1. The effects of radiation depend on its wavelength, intensity, and duration.
2. Ionizing radiation (gamma rays and high-energy electron beams) has a high degree of penetration and exerts its effect primarily by ionizing water and forming highly reactive hydroxyl radicals.
3. Ultraviolet radiation, a form of nonionizing radiation, has a low degree of penetration and causes cell damage by making DNA thymine dimers that interfere with the DNA replication; the most effective germicidal wavelength is 260 nm.
4. Microwaves can kill microbes indirectly as materials get hot.

Chemical Methods of Microbial Control (pp. 176–185)

1. Chemical agents are used on living tissue (as antiseptics) and on inanimate objects (as disinfectants).
2. Few chemical agents achieve sterility.

PRINCIPLES OF EFFECTIVE DISINFECTION (pp. 176–177)

1. Careful attention should be paid to the properties and concentration of the disinfectant to be used.

2. The presence of organic matter, degree of contact with microorganisms, and temperature should also be considered.

EVALUATING A DISINFECTANT (p. 177)

1. The phenol coefficient is the comparison of one chemical's disinfecting action with that of phenol, applied for the same length of time on the same organism under identical conditions.
2. In the use-dilution test, a series of tubes contain increasing concentrations of disinfectant; the more the chemical can be diluted and still be effective, the higher its rating.
3. In the filter paper method, a disk of filter paper is soaked with a chemical and placed on an inoculated agar plate; a clear zone of inhibition indicates effectiveness.

TYPES OF DISINFECTANTS (pp. 177–185)

Phenol and Phenolics (pp. 177–178)

1. Phenolics exert their action by injuring plasma membranes, inactivating enzymes, and denaturing proteins.
2. Common phenolics are cresols and hexachlorophene.

Chlorhexidine (pp. 178–179)

1. Chlorhexidine damages plasma membranes of vegetative cells.

Halogens (pp. 179–180)

1. Some halogens (iodine and chlorine) are used alone or as components of inorganic or organic solutions.
2. To inactivate enzymes and other cellular proteins, iodine combines with the amino acid tyrosine.
3. Iodine is available as a tincture (in solution with alcohol) or as an iodophor (combined with an organic molecule).
4. The germicidal action of chlorine is based on the formation of hypochlorous acid when chlorine is added to water.
5. Chlorine is used as a disinfectant in gaseous form (Cl_2) or in the form of a compound, such as calcium hypochlorite, sodium hypochlorite, and chloramines.

Alcohols (p. 180)

1. Alcohols exert their action by denaturing proteins and dissolving lipids.
2. In tinctures, they enhance the effectiveness of other antimicrobial chemicals.
3. Aqueous ethanol (60% to 95%) and isopropanol are used as disinfectants.

Heavy Metals and Their Compounds (pp. 180–181)

1. Silver, mercury, copper, and zinc are used as germicidals or antiseptics.
2. They exert their antimicrobial action through oligodynamic action. When heavy metal ions combine with —SH groups, proteins are denatured.

Surface-Active Agents (pp. 181–182)

1. Surface-active agents decrease the tension between molecules that lie on the surface of a liquid; soaps and detergents are examples.
2. Soaps have limited germicidal action but assist in the removal of microorganisms through scrubbing.
3. Acid-anionic detergents are used to clean dairy equipment.

Quaternary Ammonium Compounds (p. 182)

1. Quats are cationic detergents attached to NH_4^+.
2. By disrupting plasma membranes, they allow cytoplasmic constituents to leak out of the cell.
3. They are most effective against gram-positive bacteria.

Organic Acids and Derivatives (p. 182)

1. Sorbic acid, benzoic acid, and propionic acid inhibit fungal metabolism. They are used as food preservatives.

Aldehydes (pp. 182–183)

1. Aldehydes such as formaldehyde and glutaraldehyde exert their antimicrobial effect by inactivating proteins.
2. They are among the most effective chemical disinfectants.

Gaseous Chemosterilizers (p. 183)

1. Ethylene oxide is the gas most frequently used for sterilization.
2. It penetrates most materials and kills all microorganisms by protein denaturation.

Oxidizing Agents (pp. 183–185)

1. Ozone and peroxide are used as antimicrobial agents.
2. They exert their effect by oxidizing molecules inside cells.

STUDY QUESTIONS

Review

1. Name the cause of cell death resulting from damage to the following:
   (a) Cell wall (c) Proteins
   (b) Plasma membrane (d) Nucleic acids

2. The thermal death time for a suspension of *B. subtilis* endospores is 30 minutes in dry heat and less than 10 minutes in an autoclave. Which type of heat is more effective? Why?

3. If pasteurization does not achieve sterilization, why is food treated by pasteurization?

4. Thermal death point is not considered an accurate measure of the effectiveness of heat sterilization. List three factors that can alter the thermal death point.

5. The antimicrobial effect of gamma radiation is due to ________. The antimicrobial effect of ultraviolet radiation is due to ________.

6. A bacterial culture was in log phase. At time x, an antibacterial compound was added to the culture. Which line indicates addition of a bactericidal compound? A bacteriostatic compound? How can you tell? Explain why the viable count does not immediately drop to zero in line a.

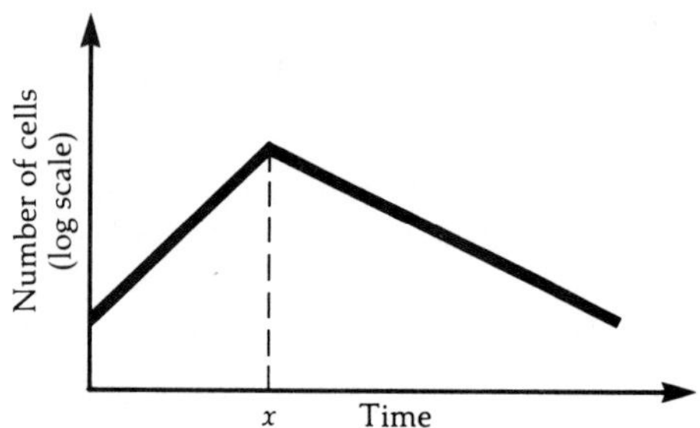

7. Fill in the following table.

Method of Sterilization	Temp.	Time	Type of Heat	Pref'd Use	Type of Action
Autoclaving					
Hot air					
Pasteurization					

8. How do the examples in Question 7 illustrate the concept of equivalent treatments?

9. Heat-labile solutions such as glucose–minimal salts broth can be sterilized by ________.

10. How do salts and sugars preserve foods? Why are these considered physical rather than chemical methods of control? Name one food that is preserved with sugar, and one preserved with salt. How do you account for the occasional growth of *Penicillium* mold in jelly, which is 50% sucrose?

11. List five factors to consider before selecting a disinfectant.

12. Give the method of action and at least one standard use of each of the following types of disinfectants:
    - **(a)** Phenolics
    - **(b)** Iodine
    - **(c)** Chlorine
    - **(d)** Alcohol
    - **(e)** Heavy metals
    - **(f)** Aldehydes
    - **(g)** Ethylene oxide
    - **(h)** Oxidizing agents

13. The phenol coefficients against *S. aureus* for ethanol and isopropanol are 0.039 and 0.054, respectively. Which is the more effective antimicrobial agent? What is a phenol coefficient?

14. The use-dilution values for two disinfectants tested under the same conditions are Disinfectant A—1:2; Disinfectant B—1:10,000. If both disinfectants are designed for the same purpose, which would you select?

15. A large hospital washes burn patients in a stainless steel tub. After each patient, the tub is cleaned with a quat. It was noticed that 14 of 20 burn patients acquired *Pseudomonas* infections after being bathed. Provide an explanation for this high rate of infection.

Challenge

1. The filter paper method was used to evaluate three disinfectants. The results were as follows:

Disinfectant	Zone of Inhibition
X	0 mm
Y	5 mm
Z	10 mm

   (a) Which disinfectant was the most effective against the organism?
   (b) Can you determine whether compound Y was bactericidal or bacteriostatic?

2. Why is each of the following bacteria often resistant to disinfectants?
   (a) *Mycobacterium*
   (b) *Pseudomonas*
   (c) *Bacillus*

3. *Entamoeba histolytica* and *Giardia lamblia* were isolated from the stool sample of a 45-year-old man, and *Shigella sonnei* was isolated from the stool sample of an 18-year-old woman. Both patients experienced diarrhea and severe abdominal cramps, and, prior to onset of digestive symptoms, both had been treated by the same chiropractor.

   The chiropractor had administered colonic irrigations (enemas) to these patients. The device used for this treatment was a gravity-dependent apparatus using 12 liters of tap water. There were no check valves to prevent backflow, so all parts of the apparatus could have become contaminated with feces during each colonic treatment. The chiropractor provided colonic treatment to four or five patients per day. Between patients, the adaptor piece that is inserted into the rectum was placed in a "hot-water sterilizer."

   What two errors were made by the chiropractor?

4. Between March 9 and April 12, five chronic peritoneal dialysis patients at one hospital became infected with *Pseudomonas aeruginosa*. Four patients developed peritonitis (inflammation of the abdominal cavity), and one developed a skin infection at the catheter insertion site. All patients with peritonitis had low-grade fever, cloudy peritoneal fluid, and abdominal pain. All patients had permanent indwelling peritoneal catheters, which the nurse wiped with gauze soaked with an iodophor

(Prepodyne®) solution each time the catheter was connected to or disconnected from the machine tubing. Aliquots of the iodophor were transferred from stock bottles to small in-use bottles.

Cultures from the dialysate concentrate and the internal areas of the dialysis machines were negative; iodophor from a small in-use plastic container yielded a pure culture of *P. aeruginosa*.

What improper technique led to this infection?

5. Eleven patients received injections of methylprednisolone and lidocaine to relieve the pain and inflammation of arthritis at the same orthopedic surgery office. All the patients developed septic arthritis caused by *Serratia marcescens*. Unopened bottles of methyprednisolone from the same lot numbers tested sterile; the methylprednisolone was preserved with a quat. Cotton balls were used to wipe multiple-use injection vials before the medication was drawn into a disposable syringe. The site of injection on each patient was also wiped with a cotton ball. The cotton balls were soaked in benzalkonium chloride, and fresh cotton balls were added as the jar was emptied. Opened methylprednisolone and the jar of cotton balls contained *S. marcesens*.

   How was the infection transmitted? What part of the routine procedure caused the contamination?

FURTHER READING

Balows, A., ed. *Manual of Clinical Microbiology*, 5th ed. Washington, D.C.: American Society for Microbiology, 1991. Includes chapters on preventing contamination in clinical laboratories.

Block, S.S., ed. *Disinfection, Sterilization and Preservation*, 4th ed. Philadelphia: Lea and Febiger, 1991. A comprehensive reference on principles and procedures for disinfection and sterilization.

Castle, M., and E. Ajemian. *Hospital Infection Control: Principles and Practice*, 2nd ed. New York: Delmar, 1987. An excellent reference for practical surveillance and maintenance of asepsis.

Collins, C.H., P.M. Lyne, and J.M. Grange, eds. *Microbiological Methods*, 6th ed. Stoneham, Mass.: Butterworths, 1989. This manual includes chapters on safety in microbiology and sterilization and disinfection.

Miller, B.M., ed. *Laboratory Safety: Principles and Practices*. Washington, D.C.: American Society for Microbiology, 1986. Contains a chapter on sterilization and disinfection.

Phillips, G.B., and W.S. Miller, eds. *Industrial Sterilization*. Durham, N.C.: Duke University Press, 1973. Methods and applications for the use of gas, formaldehyde, and radiation sterilization.

Russell, A. *The Destruction of Bacterial Spores*. New York: Academic Press, 1982. A discussion of physical and chemical control of endospores in industry and hospitals.

CHAPTER 8

Microbial Genetics

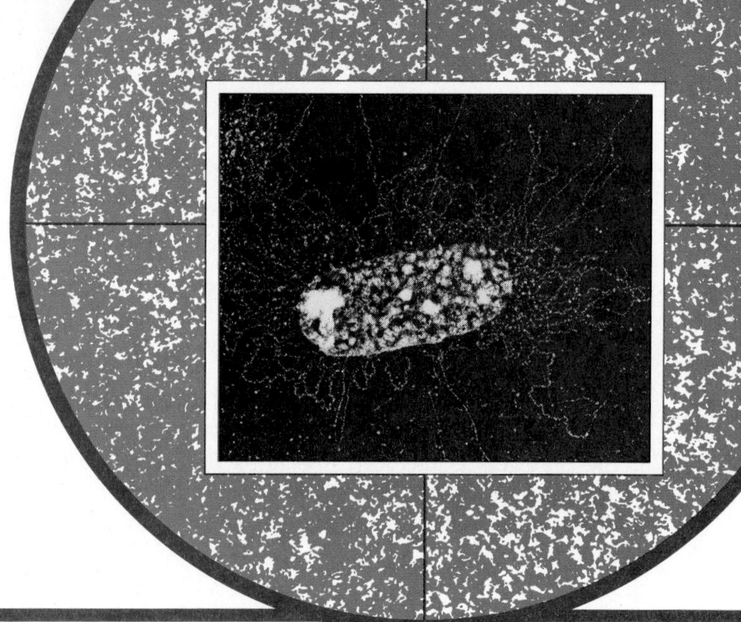

LEARNING OBJECTIVES

- Define genetics, chromosome, gene, genetic code, genotype, phenotype, mutagen, genetic recombination.
- Describe how DNA serves as genetic information.
- Describe DNA replication.
- Describe protein synthesis, including transcription and translation.
- Explain the regulation of gene expression in bacteria by induction, repression, attenuation, and catabolite repression.
- Classify mutations by type.
- Outline methods for direct and indirect selection of mutants.
- Compare mechanisms of genetic transfer in bacteria.
- Define plasmid and transposon and discuss their functions.
- Relate mechanisms for genetic change to microbial evolution.

Virtually all the microbial traits you have read about in earlier chapters are controlled or influenced by the microbe's heredity. The inherited traits of microbes include their shape and structural features (morphology), their biochemical reactions (metabolism), their ability to move or behave in other ways, and their ability to interact with other organisms—perhaps causing disease. Individual organisms transmit these characteristics to their offspring through genes, which are the units of the hereditary material (DNA) that contain the information that determines these characteristics.

Chromosomes are the cellular structures that physically carry hereditary information; the chromosomes contain the genes. **Genetics** is the science of heredity. It includes the study of what genes are, how they carry information, how they are replicated and passed to future generations of cells or passed between organisms, and how the expression of their information within an organism determines the particular characteristics of that organism.

STRUCTURE AND FUNCTION OF THE GENETIC MATERIAL

Genes consist of DNA. We saw in Chapter 2 that DNA is a macromolecule composed of repeating units called *nucleotides*. You may recall that each nucleotide consists of a nitrogenous base—adenine (A), thymine (T), cytosine (C), or guanine (G)—deoxyribose (a pentose sugar), and a phosphate group. Review the general arrangement of the nucleotides in a DNA molecule in Figure 2.19. The DNA within a cell exists as long strands of nucleotides twisted together in pairs to form a double helix. Each strand has a string of alternating sugar and phosphate groups, its *sugar–phosphate backbone*, and to each sugar is attached a nitrogenous base. The two strands are held together by hydrogen bonds between their nitrogenous bases. The bases are paired in a specific way. Adenine always pairs with thymine, and cytosine always pairs with guanine. Because of the specific base pairing, if the base sequence of one DNA strand is known, then the sequence of the other strand is also known. The two strands of DNA are thus *complementary*. A good analogy to complementary sequences in DNA strands is the relationship between a positive photograph and its negative. The complementary structure of DNA helps explain how DNA stores and transmits genetic information, as we shall see later.

The genetic language uses an alphabet with only four letters—the four kinds of nitrogenous bases in DNA (or RNA). But 1000 of these four bases, the number contained in an average gene, can be arranged in 4^{1000} different ways. This astronomically large number explains how genes can be varied enough to provide all the information a cell needs to grow and perform its functions. The genetic code, which determines how a nucleotide sequence is converted into the amino acid sequence of a protein, is discussed in more detail later in this chapter.

Although an individual hydrogen bond is weak, the numerous hydrogen bonds along even a short stretch of DNA provide sufficient bonding energy to make the double helix a stable structure. In spite of this stability, the two strands of DNA do separate under the influence of special proteins. This allows the two separated strands of DNA to serve as templates for replication, which leads to the production of two daughter DNA molecules.

The specific sequence of nucleotides contained in the DNA is usually duplicated accurately by enzymes each time the cell divides. This accurate duplication allows DNA to carry genetic information from cell to cell and from generation to generation. The information in the sequence of nucleotides determines the characteristics of the cell and transfers these characteristics to subsequent generations of cells. Later in this chapter, we will describe several important experiments that clearly identified DNA as the genetic material of cells.

A **gene** can be defined as a segment of DNA (a sequence of nucleotides in DNA) that codes for a functional product. The final product can be a molecule of ribosomal RNA (rRNA) or transfer RNA (tRNA). Usually, however, the final product is a protein. Each gene has a unique base sequence that codes for a unique protein. The DNA of a cell therefore specifies the complete collection of proteins to be found in the cell.

Proteins are so important that much of a cell's machinery is concerned with translating the genetic message of genes into specific proteins. The DNA sequence is *transcribed* (copied) to produce a specific molecule of RNA called messenger RNA (mRNA). The information encoded in the mRNA is *translated* into a specific sequence of amino acids that forms a protein.

When the ultimate molecule that a gene codes for (a protein, rRNA, or tRNA) has been produced, we say that the gene has been *expressed*. Thus, gene expression involves the transcription of DNA to produce specific molecules of RNA. If the RNA is mRNA, the information encoded in it is then translated into protein. This process can be symbolized as follows:

$$\text{DNA} \xrightarrow{\text{Transcription}} \text{mRNA} \xrightarrow{\text{Translation}} \text{Protein}$$

DNA can change, or *mutate*. Although many genetic mutations cause harm or kill the cell, others create new characteristics that give the offspring inheriting the mutation a better chance of survival in new environments. Thus, mutation is in the long run an advantage that contributes to the successful evolution of species. We will discuss various types of mutation later in this chapter.

Genotype and Phenotype

The **genotype** of an organism is its genetic makeup, the information that codes for all the particular characteristics of the organism. The genotype represents the *potential* properties but not the properties themselves. **Phenotype** refers to the *actual, expressed* properties, such as the organism's ability to perform a particular chemical reaction. The phenotype, then, is the manifestation of the genotype.

In molecular terms, an organism's genotype is its collection of genes, its entire DNA. What constitutes the organism's phenotype in molecular terms? In a sense, an organism's phenotype is its collection of proteins. Most of a cell's properties derive from the structures and functions of its proteins. In microbes, most proteins are either enzymatic (catalyzing particular reactions) or structural (participating in large functional complexes such as membranes or ribosomes). Even phenotypes that depend on structural macromolecules other than protein (such as lipids or polysaccharides) rely indirectly on proteins. For instance, the structure of a complex lipid or a polysaccharide molecule results from the catalytic activities of enzymes that synthesize, process, and degrade them. Thus, although it is not completely accurate to say that phenotypes are due only to proteins, it is a useful simplification.

DNA and Chromosomes

What is the relationship of DNA to chromosomes? Evidence available so far suggests that the DNA in each

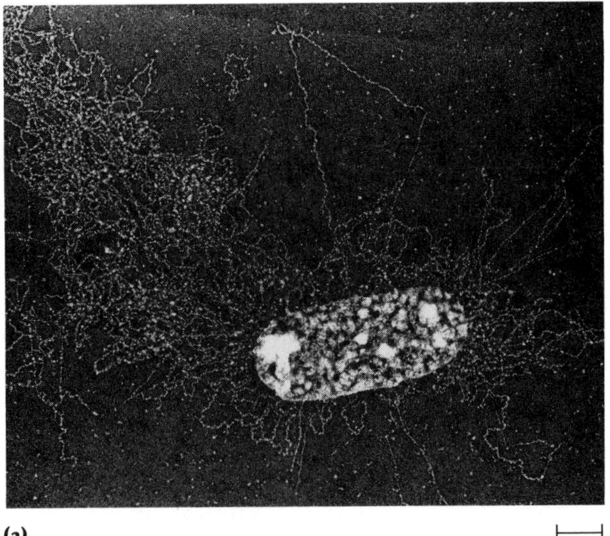

FIGURE 8.1 Chromosomes. **(a)** Electron micrograph of a procaryotic chromosome. The tangled mass and looping strands of DNA emerging from this disrupted cell of *E. coli* are part of its single chromosome. **(b)** Electron micrograph of a eucaryotic chromosome. This human chromosome is one of 46 chromosomes found in a normal human cell. The individual fibers are strands of DNA with attached proteins.

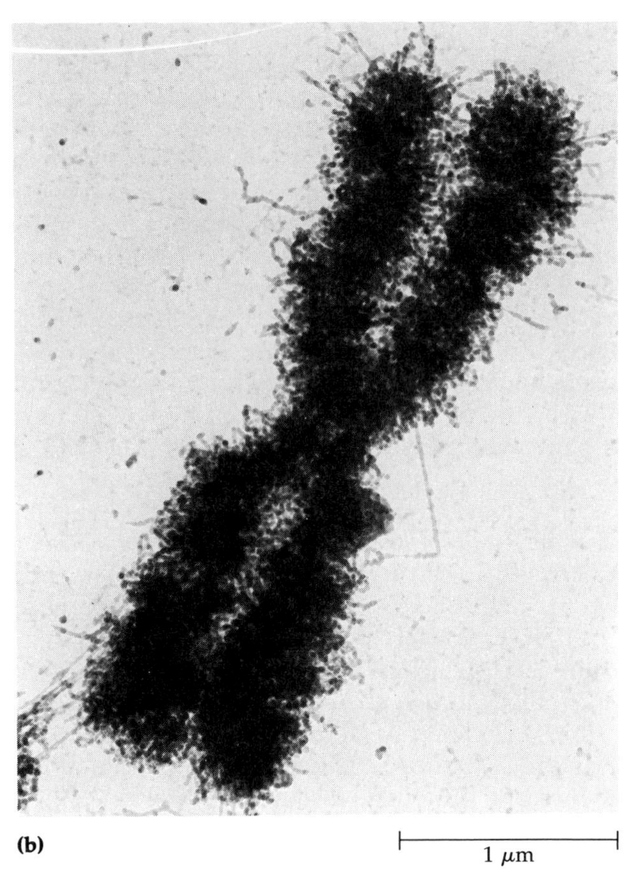

chromosome—even in the large, complex chromosomes of eucaryotes—is one long double helix. Various proteins that function in DNA replication and expression are bound to the DNA.

Bacteria typically have a single circular chromosome consisting of a single circular molecule of DNA (Figure 8.1a). The chromosome is looped and folded and attached at one or several points to the plasma membrane. The DNA of *E. coli*, the best studied bacterium, has about 4 million base pairs and is about 1 mm long—1000 times longer than the entire cell. However, DNA is very thin and is tightly packed inside the cell, so that this twisted, coiled macromolecule takes up only about 10% of the cell's volume.

Eucaryotic chromosomes (Figure 8.1b) contain DNA that is even more highly coiled (condensed) than procaryotic DNA. Eucaryotic chromosomes contain much more protein than procaryotic chromosomes do. In eucaryotic cells, a group of proteins known as histones form complexes around which DNA is wound. The structure and function of a eucaryotic chromosome are also influenced by diverse nonhistone chromosomal proteins, which help determine many tissue-specific and species-specific phenotypes. The detailed structure of the eucaryotic chromosome, and the precise arrangement of DNA with proteins, is still under investigation. Researchers believe that an understanding of the detailed structure is likely to reveal how the cell turns genes on and off to produce crucial proteins when needed. This regulation of gene expression governs the differentiation of eucaryotic cells into the different types of cells found in multicellular organisms as well as an individual cell's ongoing activities.

We will next examine more closely how DNA functions as the genetic material in replication and protein synthesis.

DNA Replication

In DNA replication, one "parental" double-stranded DNA molecule is converted to two identical "daughter" molecules. The complementary structure of the nitrogenous base sequences in the DNA molecule provides the key to understanding DNA replication. Because the bases that comprise the two strands of double-helical DNA are complementary, one strand can act as a template for the production of the other strand.

When DNA replicates, the two strands of parental DNA unwind and separate from each other in one small DNA segment after another. Free nucleotides present in the cytoplasm of the cell match up to the exposed bases of the single-stranded parental DNA. Where thymine is present on the original strand, only

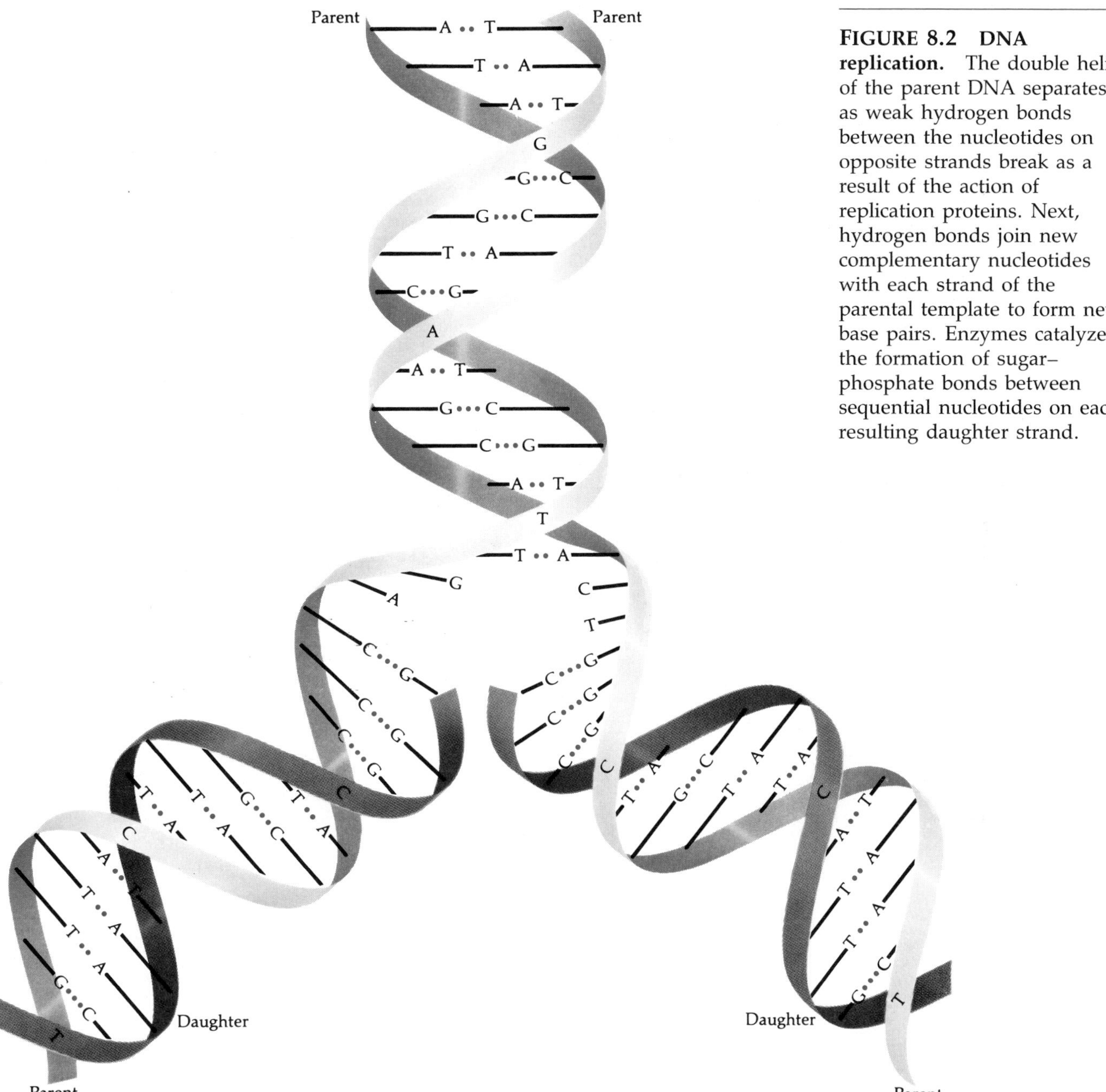

FIGURE 8.2 DNA replication. The double helix of the parent DNA separates as weak hydrogen bonds between the nucleotides on opposite strands break as a result of the action of replication proteins. Next, hydrogen bonds join new complementary nucleotides with each strand of the parental template to form new base pairs. Enzymes catalyze the formation of sugar–phosphate bonds between sequential nucleotides on each resulting daughter strand.

adenine can fit into place on the new strand; where guanine is present on the original strand, only cytosine can fit into place, and so on. Once aligned, the newly added nucleotide is joined to the growing DNA strand by an enzyme called DNA polymerase. Then the parental DNA unwinds a bit further to allow the addition of the next nucleotides. The point at which replication is occurring is called the **replication fork** (Figure 8.2).

As the replication fork moves along the parental DNA, each of the unwound single strands combines with new nucleotides. The original strand and this newly synthesized daughter strand then rewind. Because each new double-stranded DNA molecule contains one original strand (conserved) and one new strand, the process of replication is referred to as **semiconservative replication.**

Before discussing DNA replication in more detail, let us take a closer look at the structure of DNA (Figure 8.3). Note in the inset of Figure 8.3 that the carbon atoms of the sugar component of each nucleotide are numbered 1′ (pronounced "one-prime") to 5′. Although the two strands of DNA are complementary, their backbones have different chemical senses of direction in terms of the orientation of the sugar groups. Each phosphate group in the sugar–phosphate backbone of DNA attaches the 5′ carbon of one sugar to the

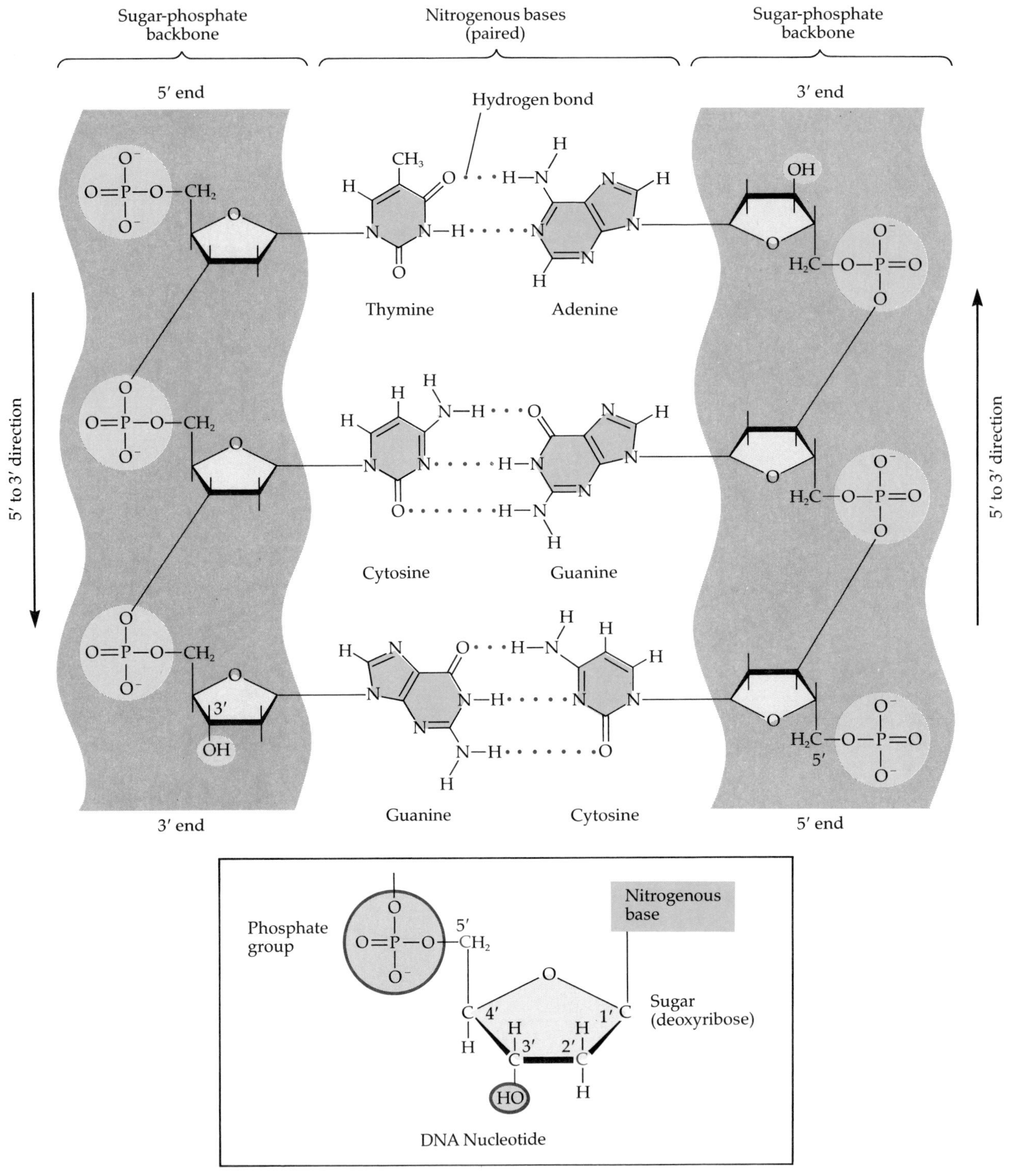

FIGURE 8.3 Chemical structure of DNA.

3′ carbon of the next sugar. (The nitrogenous base attaches to the 1′ carbon.) If the sugar–phosphate backbone is broken by the action of an enzyme or the DNA is a linear molecule, a hydroxyl group is found at one end of each DNA strand. The end of the DNA strand that has the hydroxyl attached to the 3′ carbon of the sugar is called the 3′ end of the DNA strand; the end having a phosphate group attached to the 5′ carbon of the sugar is called the 5′ end. Because of the way in which the two strands of DNA fit together, if one strand has the chemical sense of direction 5′ → 3′, the other must be reversed, 3′ → 5′. The two strands of the

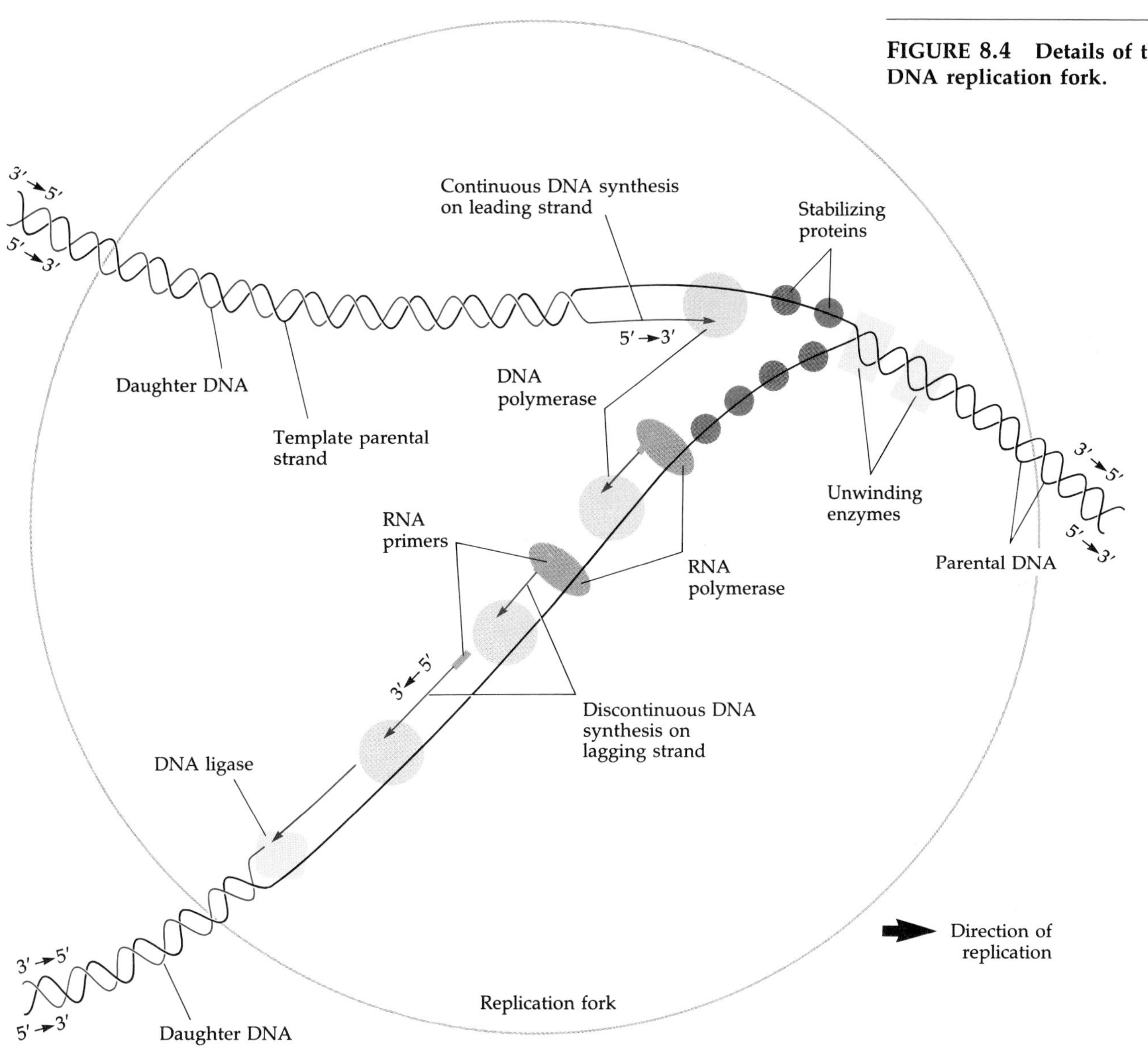

FIGURE 8.4 Details of the DNA replication fork.

DNA double helix are therefore said to be *antiparallel.*

The antiparallel structure of DNA affects the replication process. DNA polymerases can join new nucleotides to the growing daughter strand only by adding them to its 3′ end. That is, a DNA strand grows only in the 5′ → 3′ direction. Therefore, as the site of the replication fork moves along the parental DNA, the two new DNA strands must grow in slightly different fashions. One new DNA strand, called the *leading strand,* is synthesized continuously in the 5′ → 3′ direction (from a template parental strand running 3′ → 5′). In contrast, the *lagging strand* of new DNA is synthesized *discontinuously* in fragments of about 1000 nucleotides, which must later be joined to make one continuous strand (Figure 8.4). Thus the *overall* growth of the lagging strand in the 3′ → 5′ direction is achieved by assembling short stretches that have each been synthesized in the usual 5′ → 3′ direction.

The process of DNA synthesis requires many proteins and enzymes, including several types of DNA polymerase molecules. At the site of the replication fork, the parental double helix is unwound by enzymes. Stabilizing proteins prevent the rewinding of the single-stranded parental DNA. DNA polymerase cannot initiate synthesis of a DNA strand entirely on its own, but it can attach additional nucleotides to the free 3′ end of an existing DNA or RNA strand to elongate the strand. While the leading strand of DNA is being synthesized by DNA polymerase, an RNA polymerase starts the synthesis of each DNA fragment of the lagging strand with a short stretch of RNA nucleotides, called an *RNA primer.* DNA synthesis catalyzed by DNA polymerase can then proceed in the 5′ → 3′ direction from the RNA primers. The RNA is later digested away by the *5′ → 3′ exonuclease* activity that is part of the function of DNA polymerase. Another en-

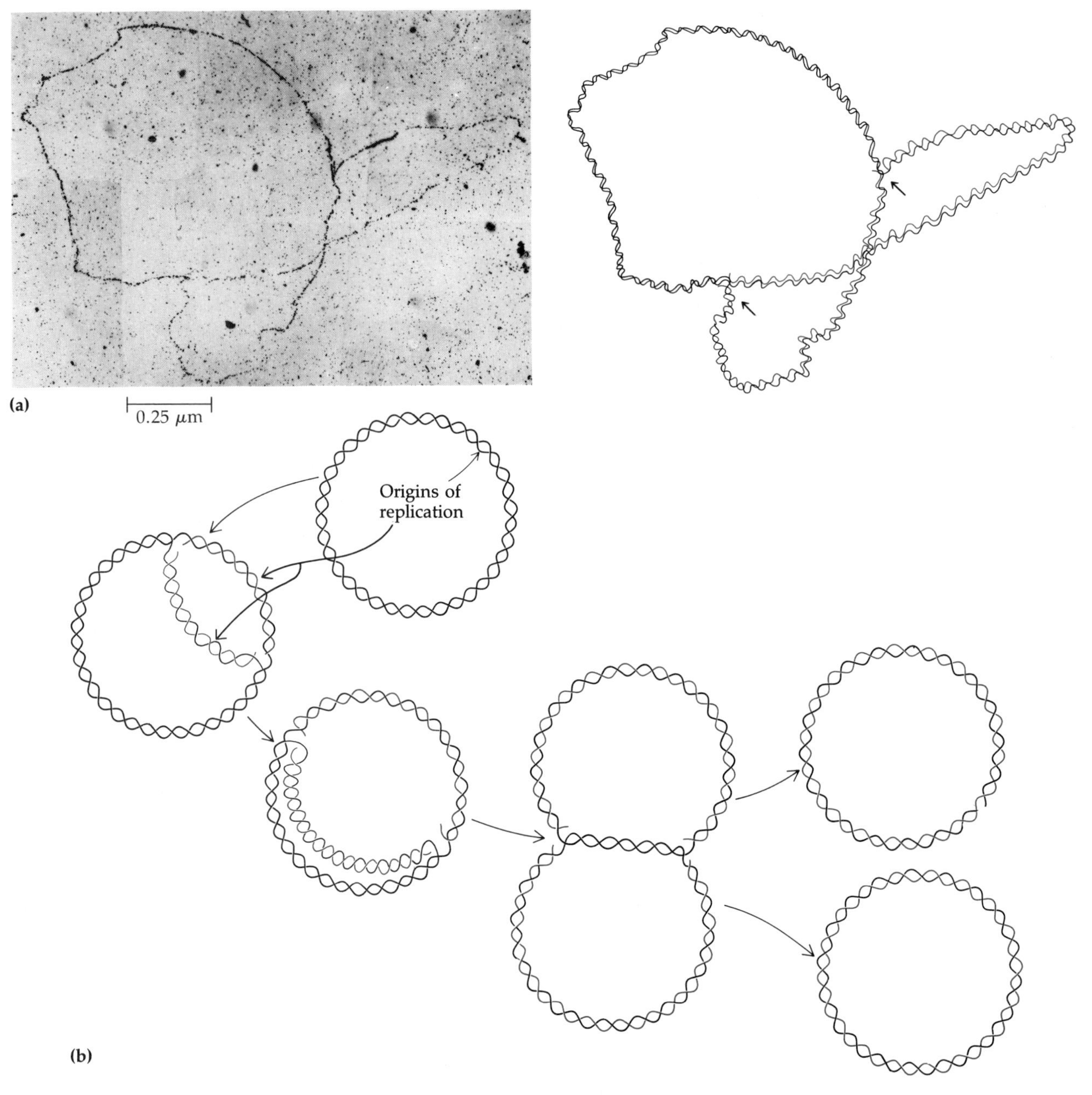

FIGURE 8.5 Replication of bacterial DNA. (a) Autoradiograph of the replication of an *E. coli* chromosome (corresponding diagram at right). The arrows point to the two replication forks. The chromosome is about one-third replicated. Note that one of the new helices is crossed over the other one. **(b)** Diagrammatic representation of the bidirectional replication of a circular DNA molecule. The new strand is shown in red.

zyme, *DNA ligase*, joins the discontinuous fragments with covalent bonds to make a continuous lagging strand. Each newly synthesized DNA strand forms a double helix with its parental strand. The two daughter DNA double helices that are the final result of the process are indistinguishable from each other and from the original parental DNA.

In bacteria, the replication process begins at a unique site on the bacterial chromosome. It is known that DNA replication of some bacteria, such as *E. coli*, goes *bidirectionally* around the chromosome (Figure 8.5). Two replication forks move in opposite directions away from the origin of replication. Because the bacterial chromosome is a closed loop, the replication forks eventually meet when the replication of the chromosome is completed. Much evidence exists to show an association between the bacterial cell membrane and the site on the DNA where replication begins. The DNA–membrane attachment in procaryotes might serve to ensure that in cell division each daughter cell receives one copy of the DNA molecule, that is, one complete chromosome.

THE RATE OF DNA REPLICATION

DNA synthesis is a surprisingly fast process, about 1000 nucleotides per second in *E. coli* growing at 37°C. At first glance, this speed seems improbable, considering that nucleotide substrates must be synthesized and then must diffuse to the replication fork. Furthermore, several attempts are probably made by wrong nucleotides to pair at each position before the correct bases pair up. Nevertheless, the speed and specificity of DNA replication are governed by the same principles that guide all chemical reactions.

Under some conditions, namely log phase growth in a rich nutrient medium (see Figure 6.12), *E. coli* can grow faster than the two replication forks can complete the circular chromosome. Under these conditions, the cell initiates *multiple* replication forks at the origin on the chromosome; a new pair of forks begins before the last pair has finished. In this way, the overall rate of DNA synthesis matches the rate at which the cell divides. Similarly, when the cell's growth greatly slows, the initiation of DNA synthesis at the origin of replication may be delayed. The rate at which each replication fork moves is generally constant (at a stable temperature). However, by regulating how often replication is initiated, the cell controls its overall rate of DNA synthesis to match its rate of growth and cell division.

THE FLOW OF GENETIC INFORMATION

DNA replication makes possible the flow of genetic information from one generation to the next. As shown in Figure 8.6, the DNA of a cell replicates before cell division, so that each daughter cell receives a chromosome identical to the parent's. Within each metabolizing cell, the genetic information contained in DNA also flows in another way: It is transcribed into RNA and then translated into protein. We describe the processes of transcription and translation in the next section.

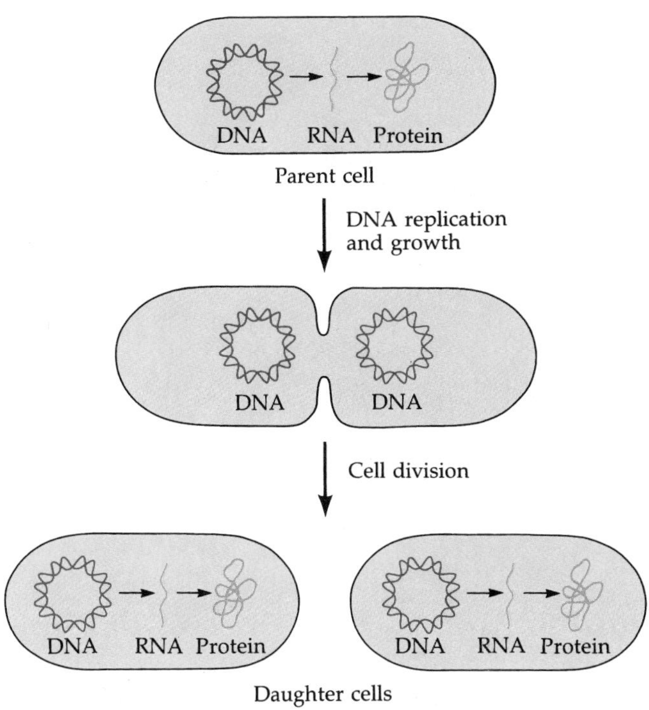

FIGURE 8.6 Flow of genetic information. Genetic information can be transferred between generations of cells and within a cell. The cell represented in this diagram is a bacterium, which has a single circular chromosome.

RNA and Protein Synthesis

How is the information in DNA used in making the proteins that control cell activities? In a process called transcription, genetic information in DNA is copied, or transcribed, into a complementary base sequence of RNA. The cell then uses the information encoded in this RNA message to synthesize specific proteins through the process of translation. We will now take a closer look at these two processes as they occur in a bacterial cell.

TRANSCRIPTION

Transcription can be defined as the synthesis of a complementary strand of RNA from a DNA template. As we discussed earlier, there are three kinds of RNA in bacterial cells—messenger RNA, ribosomal RNA, and transfer RNA. Ribosomal RNA forms an integral part of ribosomes, the cellular machinery for protein synthesis. Transfer RNA is also involved in protein synthesis, as we will see later. **Messenger RNA (mRNA)** carries the coded information for making specific proteins.

A strand of mRNA is synthesized during transcription, using a specific gene, a portion of the cell's DNA, as a template (Figure 8.7). In other words, the genetic information stored in the sequence of nitrogenous bases of DNA is rewritten so that the same information appears in the base sequence of mRNA. As in DNA replication, a G in the DNA template dictates a C in the mRNA being made; a C in the DNA template dictates a

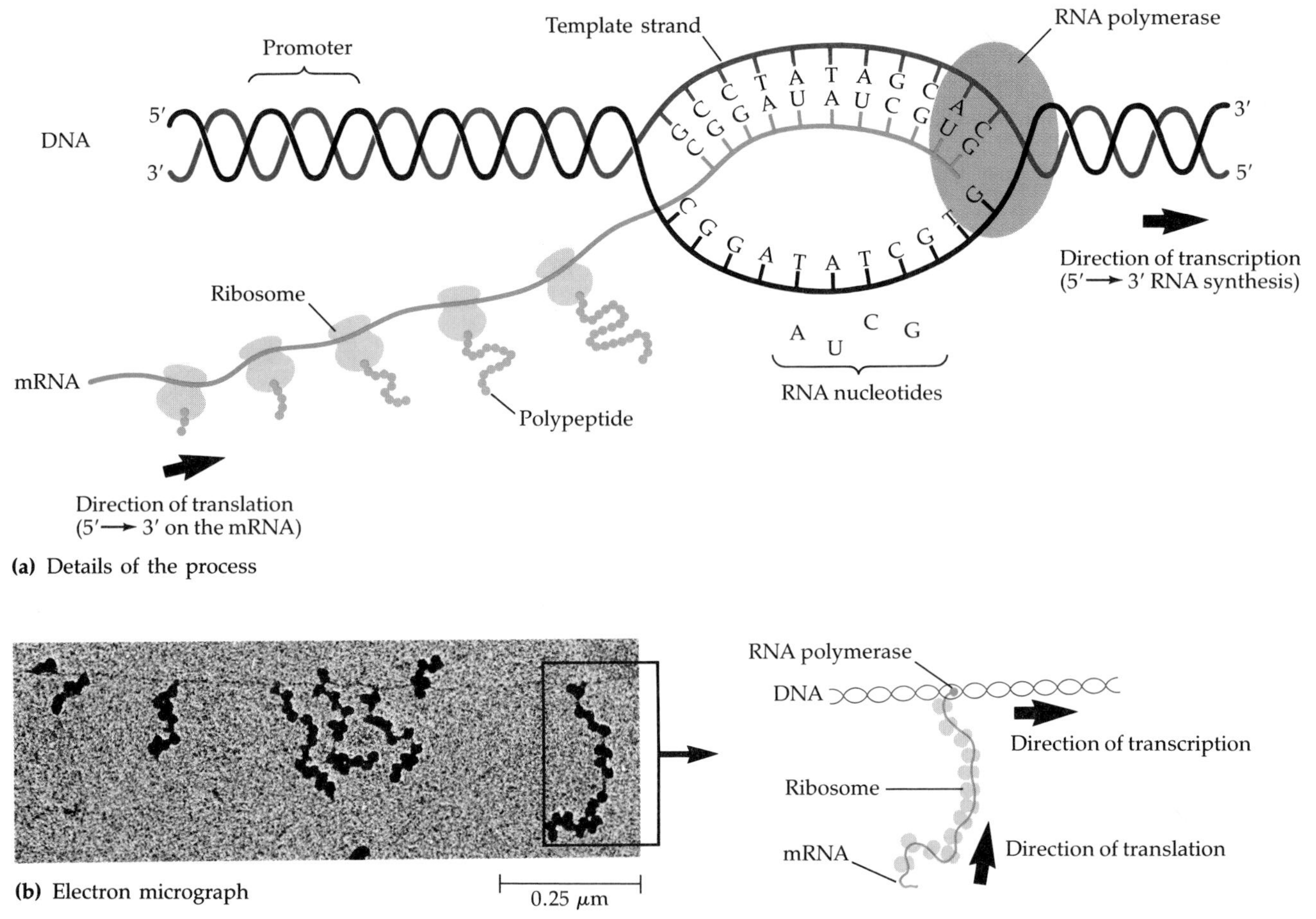

FIGURE 8.7 Simultaneous transcription and translation in bacteria. **(a)** Diagram of a partly uncoiled DNA helix acting as a template for RNA synthesis. A molecule of RNA polymerase binds to a promoter, a nucleotide sequence that signals the start of a gene. Transcription takes place on the temporarily unwound region of DNA, where specific basepairing between the template strand of DNA and RNA nucleotides determines the correct sequence of mRNA. The free end of completed mRNA has begun to be translated by ribosomes. In procaryotes, transcription and translation can be coupled in this way. **(b)** Electron micrograph of transcription from a single gene. Many molecules of mRNA are being synthesized simultaneously, starting at the promoter site. The longest mRNA molecules were the first to begin at the promoter. Note the ribosomes attached to the newly forming mRNA. The newly synthesized polypeptides are not visible here.

G in the mRNA; and a T in the DNA template dictates an A in the mRNA. However, an A in the DNA template dictates a uracil (U) in the mRNA because RNA contains U instead of T. (U has a chemical structure slightly different from T, but it base-pairs in the same way.) If, for example, the template portion of DNA has the base sequence ATGCAT, then the newly synthesized mRNA strand will have the complementary base sequence UACGUA.

The process of transcription requires an enzyme called *RNA polymerase* and a supply of RNA nucleotides. Transcription begins after RNA polymerase binds to the DNA at a site called a **promoter.** Only one of the two DNA strands serves as the template for RNA synthesis for a given gene. Like DNA, RNA is synthesized in the $5' \rightarrow 3'$ direction. The endpoint for transcription of the gene is signaled by a **terminator** region in the DNA. At this region, RNA polymerase and newly formed single-stranded mRNA are released from the DNA. In procaryotic cells, the translation of mRNA into protein can begin even before transcription is finished. Both transcription and translation take place together in the cytoplasm of the cell.

In eucaryotic organisms, on the other hand, transcription takes place in the nucleus. The mRNA must be completed and exported to the cytoplasm before translation can begin. In addition, before it leaves the nucleus, the RNA undergoes a further processing step. In eucaryotic cells the regions of genes that code for proteins are often interrupted by noncoding DNA. Thus, eucaryotic genes are composed of **exons**, the regions of DNA *expressed*, and **introns**, the *intervening* regions of DNA that do not code for protein (Figure 8.8). In the nucleus of a eucaryotic cell, RNA polymerase synthesizes a long, continuous RNA product from the entire gene, including all exons and introns. This long RNA is then processed by other enzymes, which remove the intron-derived RNA and splice together the exon-derived RNA. In most organisms RNA splicing is carried out in the nucleus by a complex of proteins and RNA called the **spliceosome.** The resulting RNA leaves the nucleus and becomes the messenger RNA of the cytoplasm.

In summary, in both procaryotes and eucaryotes, the genetic information for protein synthesis is stored in DNA and passed to mRNA during transcription. The mRNA then acts as the source of information for protein synthesis.

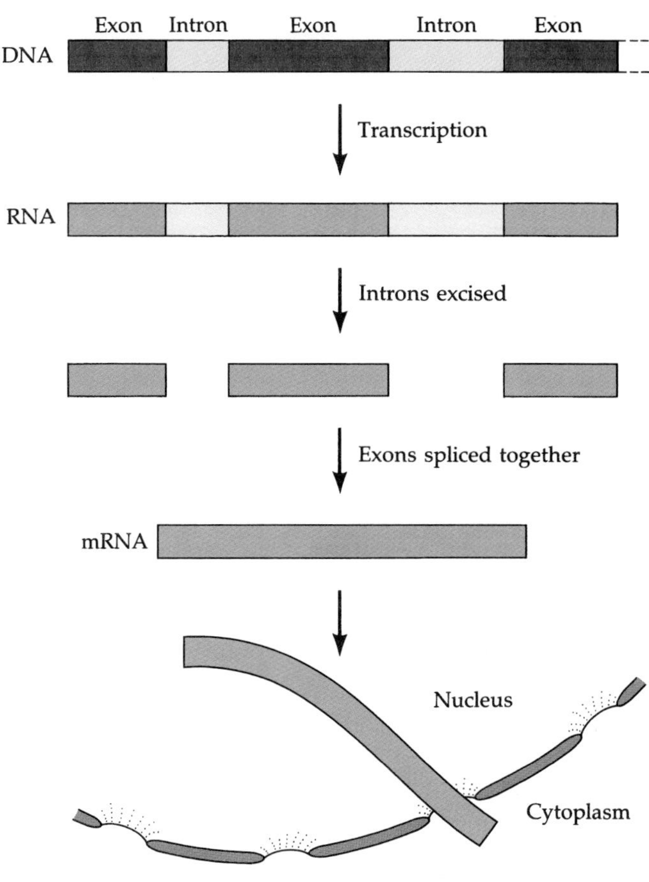

FIGURE 8.8 RNA processing in eucaryotic cells. A gene composed of exons and introns is transcribed to RNA by RNA polymerase. Processing enzymes in the nucleus cut and splice the RNA. After further modification, the mature mRNA travels to the cytoplasm, where it directs protein synthesis.

TRANSLATION

The process in which the nitrogenous-base sequence of mRNA dictates the amino acid sequence of a protein is called **translation**. In bacteria, which lack a membrane-enclosed nucleus, translation and transcription both take place in the cytoplasm. In eucaryotic organisms, translation begins when mRNA enters the cytoplasm. The events involved in translation are described in this section and shown in Figure 8.9 and Figure 8.10.

First, the 5′ end of the mRNA molecule becomes associated with a ribosome, the site of protein synthesis (Figure 8.9a). Ribosomes consist of the special type of RNA called **ribosomal RNA (rRNA)** and proteins. Each ribosome consists of two subunits. At the start of translation, the two subunits come together with the mRNA and several other components of the process.

In solution in the cytoplasm are 20 different amino acids that participate in protein synthesis (see Table 2.5). (These amino acids can be synthesized by the cell or taken up from the external medium.) However, before the appropriate amino acids can be joined together to form a protein, they must be *activated* by attaching to **transfer RNA (tRNA)** (Figure 8.10a). For each different amino acid, there is a different type of tRNA. During amino acid activation, a specific amino acid attaches to its specific tRNA. This attachment is accomplished with an *amino acid activating enzyme* and energy from ATP (Figure 8.10b).

Now we are ready to see how the mRNA actually determines the order in which amino acids are linked together to form a protein. Each set of three nucleotides of mRNA, called a **codon,** specifies (codes for) a single amino acid. For example, the sequence

AUGCCAGGCAAA

contains four codons, specifying the amino acids methionine (AUG) – proline (CCA) – glycine (GGC) – lysine (AAA).

The tRNA molecules "read" the coded message on the mRNA. On one special part of each tRNA molecule is a set of three nucleotides, called an **anticodon,** that is complementary to the codon for the amino acid carried by the tRNA (Figure 8.10c). During translation, the anticodon of a molecule of tRNA hydrogen-bonds to its complementary codon on mRNA. For example, a tRNA with anticodon AGC pairs with the mRNA codon UCG. The pairing of codon and anticodon occurs only where mRNA is attached to a ribosome.

After the first tRNA, with its amino acid, attaches to mRNA at the *start codon* (Figure 8.9b), a second tRNA molecule, with its amino acid, moves into posi-

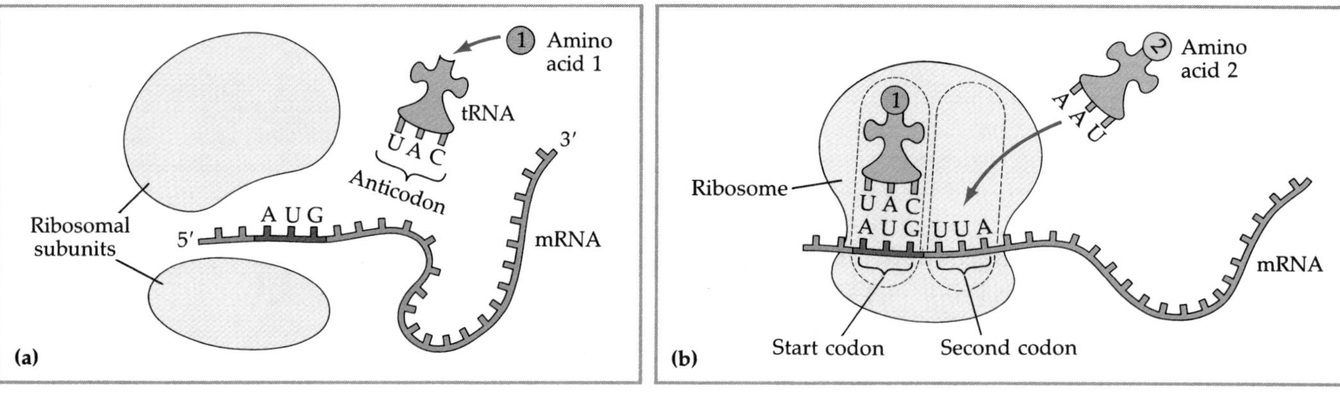

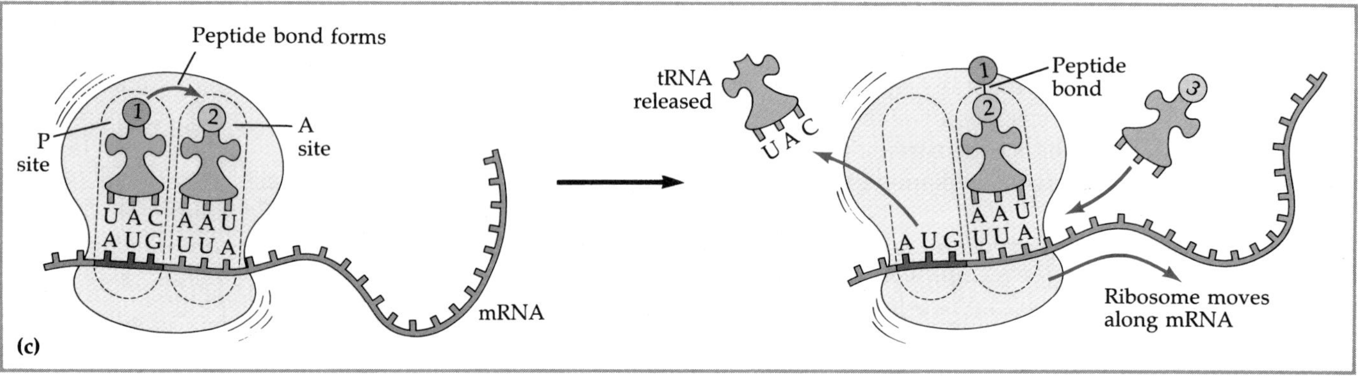

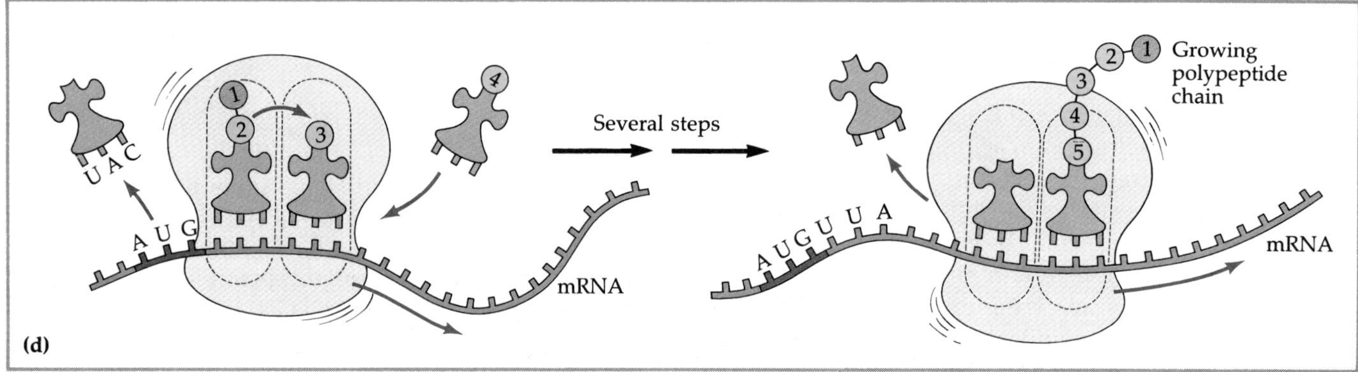

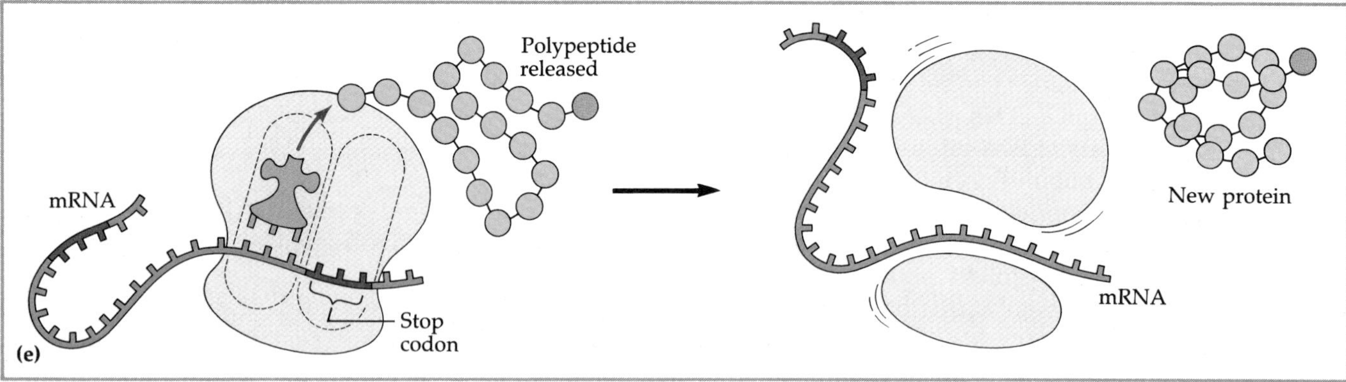

FIGURE 8.9 Translation. **(a)** Components needed to begin translation. **(b)** On the assembled ribosome, a tRNA carrying the first amino acid is paired with the start codon on the mRNA. A tRNA carrying the second amino acid approaches. **(c)** The place on the ribosome where the first tRNA sits is called the P site. In the A site next to it, the second codon of the mRNA pairs with a tRNA carrying the second amino acid. The first amino acid joins to the second by a peptide bond, and the first tRNA is released. (Nucleotide bases are labeled only for the first two codons.) **(d)** The ribosome moves along the mRNA until the second tRNA is in the P site, and the process continues. **(e)** When the ribosome reaches the stop codon, the polypeptide is released, forming a new protein. Meanwhile, the last tRNA is released, and the ribosome comes apart.

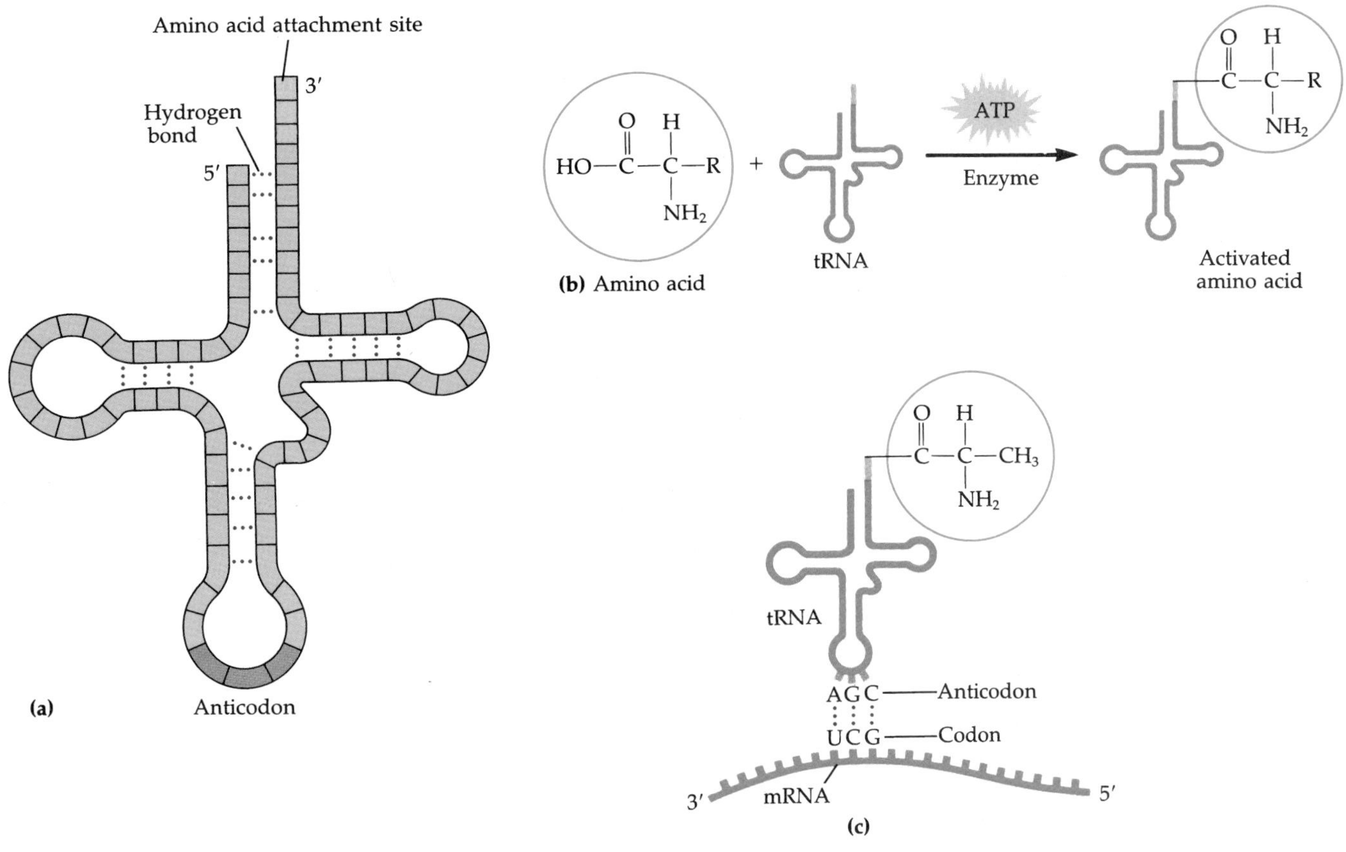

FIGURE 8.10 Transfer RNA. **(a)** Structure of tRNA shown in two-dimensional form. Each "box" represents a nucleotide. Note the regions of hydrogen bonding between base pairs and the loops of unpaired bases. **(b)** Amino acid activated by attachment to tRNA. **(c)** The anticodon of tRNA pairs with its complementary codon on an mRNA strand. The tRNA shown here carries the amino acid alanine. Codons are always read in the 5′ → 3′ direction; hence, the codon for alanine is read G–C–U.

tion on the second codon. The two amino acids are joined by a peptide bond, and the first tRNA molecule leaves the ribosome (Figure 8.9c). (The sites on the ribosome where the tRNA molecules base-pair with the mRNA are called the *A site* and the *P site*.) The detached tRNA can now pick up another molecule of its specific amino acid. Each tRNA can be used over and over again. As the proper amino acids are brought into line one by one, peptide bonds form between the amino acids, and a polypeptide chain is formed (Figure 8.9d).

A special *stop codon* in the mRNA (called a *nonsense codon* because it does not specify any amino acid) signals the end of a polypeptide chain, and the chain is released from the ribosome (Figure 8.9e). The ribosome then comes apart into its two subunits.

Before the ribosome moving along the mRNA completes translation of that message, another ribosome, and then another, can attach and begin translation of the same mRNA molecule; there can be a number of ribosomes attached at different positions to the same mRNA molecule (see Figure 8.7b). In this way, a single mRNA strand can be translated simultaneously into several identical protein molecules. The combination of an mRNA strand with several attached ribosomes is called a **polyribosome.**

In summary, the synthesis of a protein requires that the genetic information in DNA first be transferred to a molecule of mRNA by a process called transcription. Then, in a process called translation, the mRNA directs the assembly of amino acids into a polypeptide. The mRNA attaches to a ribosome, and tRNA delivers the amino acids to mRNA according to the complementary base-pairings of codons and anticodons. The amino acids are then linked to form polypeptides. Recall that proteins can be composed of one or more polypeptide chains, each of which may range in length from 50 to several hundred amino acids.

Because a typical polypeptide has about 300 amino acids, the typical mRNA must have about 300 codons and thus be at least 900 nucleotides long. The DNA of *E. coli* is long enough to contain about 4000 genes and can therefore specify about 4000 different kinds of proteins. At any one time, there are only a few copies of some proteins in a cell, but there are several thousand copies of other proteins. Table 8.1 lists the main components of transcription and translation.

The Genetic Code

The **genetic code** is the set of rules that relates the nitrogenous base sequence of DNA, the corresponding codons of mRNA, and the amino acids that the codons

TABLE 8.1 Major Components of Protein Synthesis

	TRANSCRIPTION	*TRANSLATION*
Messenger RNA (mRNA)	Synthesized as copy of information in DNA	Acts as coded message specifying amino acid sequence of proteins
Monomer subunits	RNA nucleotides (4)	Amino acids (20)
Proteins		
Enzymes	RNA polymerase	Amino-acid activating enzymes; ribosomal catalysis of peptide bond formation
Structural proteins	None	Ribosomal proteins
Other RNAs		
Transfer RNA (tRNA)	tRNA and rRNA are themselves made by transcription from DNA	tRNAs with amino acids attached
Ribosomal RNA (rRNA)		rRNAs, structural components of ribosomes
Chemical energy	From RNA nucleotide triphosphates (ATP, GTP, CTP, UTP)	From ATP and GTP

FIGURE 8.11 The genetic code. The three nucleotides in an mRNA codon are designated, respectively, as the first position, second position, and third position of the codon, going in the $5' \rightarrow 3'$ direction on the mRNA. Each set of three nucleotides specifies a particular amino acid, represented by a three-letter abbreviation (see Table 2.5). The codon AUG (which specifies the amino acid methionine) is the start of protein synthesis. The word *End* stands for the nonsense codons that serve as signals to terminate protein synthesis.

Second Position

First Position	U		C		A		G		Third Position
U	UUU	Phe	UCU		UAU	Tyr	UGU	Cys	U
	UUC		UCC	Ser	UAC		UGC		C
	UUA	Leu	UCA		UAA	End	UGA	End	A
	UUG		UCG		UAG	End	UGG	Trp	G
C	CUU		CCU		CAU	His	CGU		U
	CUC	Leu	CCC	Pro	CAC		CGC	Arg	C
	CUA		CCA		CAA	Gln	CGA		A
	CUG		CCG		CAG		CGG		G
A	AUU	Ile	ACU		AAU	Asn	AGU	Ser	U
	AUC		ACC	Thr	AAC		AGC		C
	AUA		ACA		AAA	Lys	AGA	Arg	A
	AUG	Met	ACG		AAG		AGG		G
G	GUU		GCU		GAU	Asp	GGU		U
	GUC	Val	GCC	Ala	GAC		GGC	Gly	C
	GUA		GCA		GAA	Glu	GGA		A
	GUG		GCG		GAG		GGG		G

stand for (Figure 8.11). Codons are written in terms of their base sequence in mRNA. Note that there are 64 possible codons but only 20 amino acids. This means that most amino acids are signaled by several alternative codons; this situation is referred to as the *degeneracy* of the code. For example, leucine has 6 codons and alanine has 4 codons.

Of the 64 codons, 61 are sense codons and 3 are nonsense codons. **Sense codons** code for amino acids, and **nonsense codons** do not. Rather, they signal the end of the protein molecule's synthesis. The nonsense codons are UAA, UAG, and UGA. The start codon that initiates the synthesis of the protein molecule is AUG, which is also the codon for methionine. The initiating methionine is often removed later, so not all proteins begin with methionine.

REGULATION OF GENE EXPRESSION IN BACTERIA

In Chapter 5 we learned that the bacterial cell carries out an enormous number of metabolic reactions. Some of these reactions are concerned with biosynthesis and are known as anabolic reactions. Others are concerned with degradation and are known as catabolic reactions. The common feature of all metabolic reactions is that they are catalyzed by enzymes.

We have seen that genes, through transcription and translation, direct the synthesis of proteins, many of which serve as enzymes. These are the very enzymes used for cellular metabolism. Therefore, the genetic machinery and metabolic machinery of a cell are integrated and interdependent. We will next look at how gene expression, and therefore metabolism, is regulated. Because protein (enzyme) synthesis requires a tremendous expenditure of energy (ATP), the regulation of protein synthesis is important to the cell's energy economy. The cell conserves energy by making only the proteins needed at a particular time.

Enzymes that are constantly produced at a fixed rate in a cell are called *constitutive.* They are produced from genes that are effectively turned on all the time. The production of other enzymes is regulated so that they are present only when needed.

Repression, Induction, and Attenuation

Mechanisms called repression, induction, and attenuation regulate the transcription of messenger RNAs and consequently the synthesis of enzymes from the mRNAs. These genetic control mechanisms regulate the formation and amounts of enzymes in the cell, not the activities of the enzymes.

REPRESSION AND INDUCTION

The regulatory mechanism that inhibits gene expression and decreases the synthesis of enzymes is called **repression.** Repression usually is a response to the presence of an end-product of a metabolic pathway; there is a decrease in the rate of synthesis of the enzymes leading to the formation of that product. Repression is mediated by regulatory proteins called *repressors,* which block the ability of RNA polymerase to initiate transcription from the repressed genes.

Induction is the process that turns on the transcription of a gene or genes. A substance that acts to induce transcription of a gene is called an *inducer,* and enzymes that are synthesized in the presence of inducers are *inducible enzymes.* The genes required for lactose metabolism in *E. coli* are a well-known example of an inducible system. One of these genes codes for the enzyme β-galactosidase, which splits the substrate lactose into two simple sugars, glucose and galactose. (β is the Greek letter "beta" and refers to the type of linkage that joins the glucose and galactose.) If *E. coli* is placed in a medium in which no lactose is present, the organism is found to contain almost no β-galactosidase enzyme. However, when lactose is added to the medium, the bacterial cells produce a large quantity of the enzyme. Lactose is converted in the cell to the related compound allolactose, which is the inducer for these genes; the presence of lactose thus indirectly induces the cells to synthesize more enzyme. This response, which is under genetic control, is termed **enzyme induction.** An example of induction is given in the box on page 205.

OPERON MODEL

Details of the control of gene expression by induction and repression are described by the operon model. In 1961, François Jacob and Jacques Monod formulated this general model to account for the regulation of protein synthesis. They based their model on studies of the induction of the enzymes of lactose catabolism in *E. coli.* These enzymes include β-galactosidase, which breaks the disaccharide lactose into glucose and galactose; β-galactoside permease, which is involved in the transport of lactose into the cell; and thiogalactoside transacetylase, which metabolizes certain disaccharides other than lactose.

The genes for the three enzymes involved in lactose uptake and utilization are next to each other on the bacterial chromosome and are regulated together (Figure 8.12a). These genes are called **structural genes** because they determine the structures of proteins and

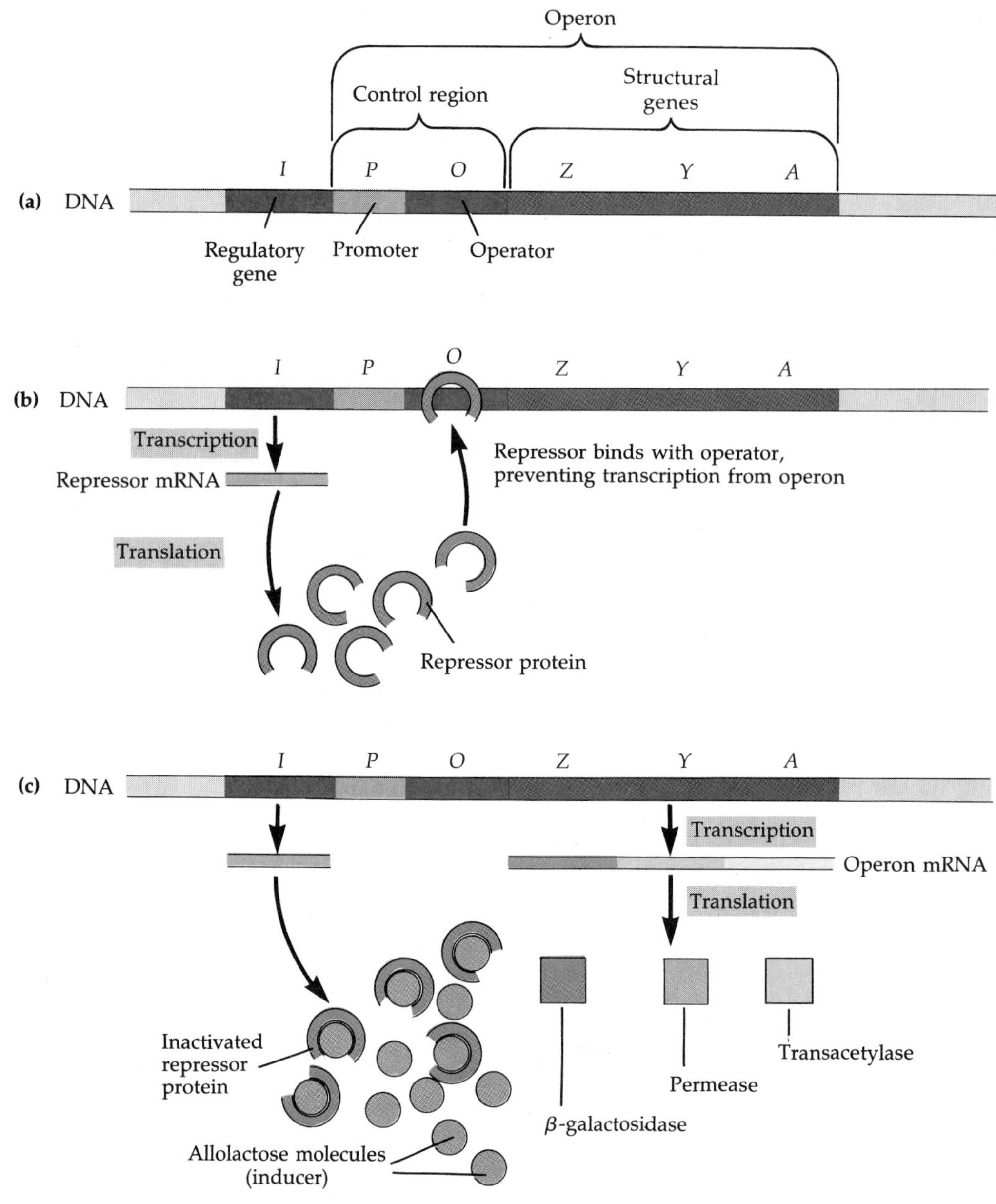

FIGURE 8.12 The *lac* operon. **(a)** This segment of the DNA molecule of *E. coli* shows the genes involved in lactose catabolism. In addition to the promoter and operator sites, the lactose operon contains three structural genes, which code for the proteins β-galactosidase, β-galactoside permease, and thiogalactoside transacetylase. For this operon the regulatory gene (*I*) is located next to the promoter (more often it is some distance away from the operon). **(b)** Control of the operon in the absence of lactose (repression). **(c)** Control of the operon in the presence of lactose (induction). When the inducer binds to the repressor protein, the repressor can no longer block transcription.

to distinguish them from an adjoining control region on the DNA. The combination of the three structural genes and the adjoining control region is called the *lac* operon. When lactose is introduced into the culture medium, the *lac* structural genes are all transcribed and translated rapidly and simultaneously. We will now see how this regulation occurs.

In the control region of the *lac* operon are two relatively short segments of DNA called regulatory sites. One, the *operator,* is like a traffic light that gives a go or stop signal for transcription of the structural genes. The other is the *promoter,* which is the region of DNA where RNA polymerase initiates transcription. A set of operator and promoter sites and the structural genes they control are what define an **operon.**

Near the *lac* operon on the bacterial DNA is a regulatory gene called the *I* gene, which codes for a *repressor protein.*

MICROBIOLOGY HIGHLIGHTS

Stress Brings Out Hidden Talents—Even in Bacteria

Stressed out? Imagine being threatened by sharp enamel teeth and pushed by a tidal wave of saliva down a tunnel into a hydrochloric acid bath. This is what happens to the food-borne pathogen *Salmonella* when it moves from the environment into the human mouth and from there to the stomach. If any salmonellae are still alive, they get pushed and shoved by the muscular stomach into the intestine to be doused with digestive enzymes and bile.

Bacteria encounter many environmental changes as they move from the free-living state into a host. In addition, they may be challenged by the host's many defenses. Change and challenge create stress for the organism. To survive, many bacteria have stress-induced genes whose products offer needed protection under adverse conditions.

Recent work has shown that when they are stressed salmonellae make special proteins that allow them to survive the inhospitable digestive tract by seeking safety inside the cells of the host. These proteins are not virulence factors (factors that cause disease); they simply defend the bacterium. (*Salmonella*'s ability to cause disease is encoded by genes on its virulence plasmid.)

Salmonellae first stick to epithelial cells lining the digestive tract by means of the LPS portion of the cell wall. The bacteria are loosely attached and can move away. However, contact with the epithelial cell glycocalyx induces *Salmonella* to produce fimbriae that attach it permanently. The cell then takes in the bacterium by endocytosis, and the cell's cytoskeleton forms an actin basket that appears to cuddle the *Salmonella* (see the photo) and move it through the epithelial cell and into underlying cells.

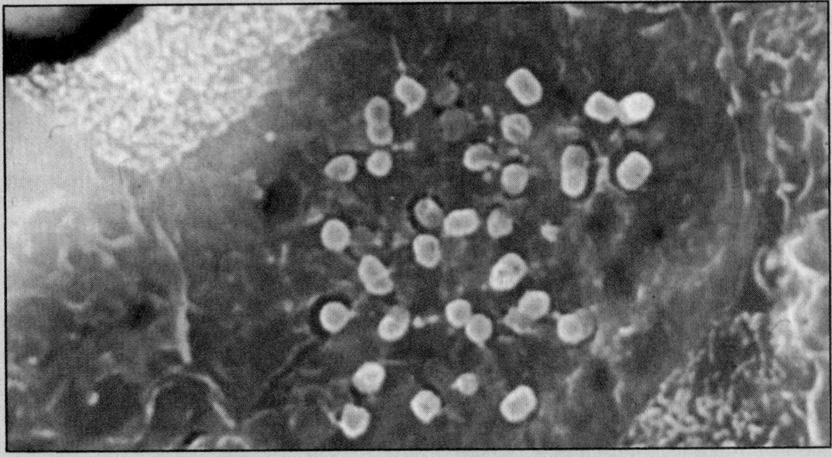

Salmonella entering epithelial cells.

Among the host's cells are macrophages that specialize in engulfing bacteria, digesting them, and signaling the body to produce antibodies against the remaining bacteria. Salmonellae "intentionally" enter macrophages in their search for a safe place to grow. Inside a macrophage, *Salmonella*'s *proP* gene is induced to make its product, which, in turn, activates special stress-induced proteins that combat the onslaught of lysosomal enzymes and oxidizing agents inside the macrophage.

Salmonella infections are a major public health problem. Different species cause illnesses ranging from salmonellosis, characterized by fever, cramping, diarrhea, and sometimes death, to typhoid fever. There are 40,000 cases of salmonellosis with 500 deaths reported annually in the United States alone. Researchers hope that an understanding of how *Salmonella*'s host-induced genes are activated will lead to the ability to control these genes with drugs against their regulatory systems.

An understanding of how stress-induced genes keep *Salmonella* alive may also help researchers construct live vaccines from salmonellae that lack virulence factors. Live vaccines are more effective than dead ones. They can also be given orally, thus eliminating the need for sterile syringes, which are frequently not available in underdeveloped countries. In 1990, a new live attenuated vaccine against *Salmonella typhi* (typhoid fever) was licensed for use. This new vaccine, called Vivotif, was shown to reduce infection by 67%; its mechanisms of action are not known. Researchers are now trying to develop a strain of *Salmonella* that lacks virulence genes but contains survival genes, to provide a live vaccine that is 100% effective.

When lactose is absent (Figure 8.12b), the repressor protein binds tightly to the operator site. This binding prevents the RNA polymerase enzyme from transcribing the adjacent structural genes; consequently, no mRNA is made and no enzymes are synthesized. But when lactose is present (Figure 8.12c), some of it diffuses into the cells and is converted into the inducer allolactose. The inducer binds the repressor protein and alters it so that it cannot bind to the operator site. In the absence of an operator-bound repressor, the RNA polymerase enzyme can transcribe the structural genes into mRNA, which is then translated into en-

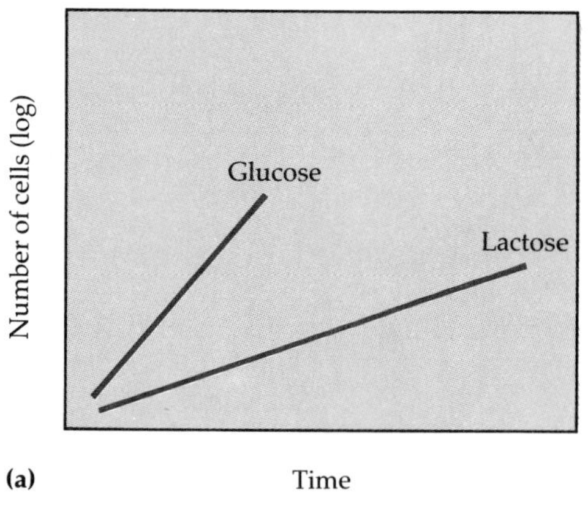

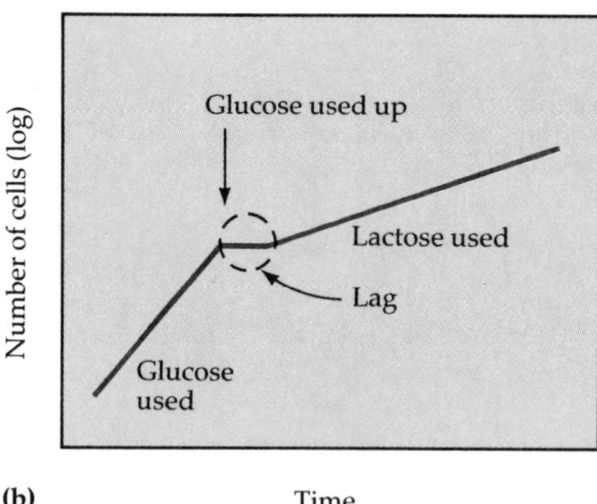

FIGURE 8.13 Growth rate of *E. coli* on glucose and lactose. The steeper the straight line, the faster the growth. **(a)** Bacteria growing on glucose as the sole carbon source grow faster than on lactose. **(b)** Bacteria growing in a medium containing glucose and lactose first consume the glucose and then, after a short lag time, the lactose. During the lag time, the intracellular cyclic AMP increases, the *lac* operon is transcribed, more lactose is transported into the cell, and β-galactosidase is synthesized to break down lactose.

zymes. This is why, in the presence of lactose, enzymes are produced. Lactose is said to induce enzyme synthesis, and the *lac* operon is called an inducible operon.

In repressible operons, the repressor requires the presence of a small molecule called a **corepressor,** which helps it bind to the operator gene. Without the corepressor, the repressor is unable to bind. The enzymes involved in several biosynthetic pathways are regulated in this manner.

Remember that many genes, perhaps from 60% to 80%, are not regulated but are constitutive. Usually these genes code for enzymes that the cell needs in fairly constant amounts for its major life processes. The enzymes of glycolysis are examples.

Regulation of the lactose operon also depends on the level of glucose in the medium, which in turn controls the intracellular level of the small molecule **cyclic AMP,** a substance that is derived from ATP and serves as a cellular alarm signal. The cell prefers glucose to lactose as a carbon source because glucose is more directly metabolized as a carbon and energy source. Enzymes that metabolize glucose are constitutive, and cells grow at their maximal rate with glucose as their carbon source (Figure 8.13a). When glucose is no longer available, the cell responds by producing cyclic AMP. The increased level of cyclic AMP in the cell triggers a series of events that, in the presence of lactose, result in transcription of the structural genes in the lactose operon. As a result, the cell can grow on lactose (Figure 8.13b). Thus, transcription of the *lac* operon requires both the presence of lactose and the absence of glucose.

Cyclic AMP is an example of an *alarmone,* an alarm signal that the cell uses to respond to environmental or nutritional stress. In this case, the stress is the lack of glucose. Similar mechanisms involving cyclic AMP allow the cell to grow on other sugars. Inhibition of the metabolism of alternative carbon sources by glucose is termed **catabolic repression** or the **glucose effect.**

Table 8.2 summarizes the components of regulation of the lactose operon and provides additional information about how cyclic AMP works.

ATTENUATION

A control mechanism that allows mRNA synthesis to begin but then prematurely terminates it is called **attenuation.** For example, cells of *E. coli* growing in a medium lacking any amino acids contain the enzymes necessary for the synthesis of all 20 amino acids needed to make proteins. But introducing an amino acid such as tryptophan to the culture greatly decreases the synthesis of enzymes needed to make that amino acid. This response is adaptive and conserves resources because the enzymes are not needed when the end-product of the biosynthetic pathway (tryptophan, in the case just described) is available to the cell.

The mechanism of attenuation involves the coupled processes of transcription and translation that were shown in Figure 8.7a. RNA polymerase begins transcribing the genes that code for the biosynthetic enzymes needed to make tryptophan, for example, but in the presence of tryptophan it soon reaches a site on the newly made RNA called the *attenuator,* where it stops transcribing.

TABLE 8.2 Components of Gene Expression of the Lactose Operon

	COMPONENT	FUNCTION
Proteins	RNA polymerase	Binds to the promoter site on DNA and catalyzes the synthesis of mRNA
	lac repressor	Binds to the operator site on DNA and the inducer allolactose
	CAP (catabolite activator protein)*	Binds to the promoter site and cyclic AMP
Small molecules	Lactose	Forms the inducer allolactose inside the cell
	Cyclic AMP	Binds to CAP
Sites in DNA	Promoter site	Attachment site for RNA polymerase and CAP
	Operator site	Attachment site for repressor protein (when not bound by inducer)

*CAP is a protein that is always present in small amounts in the cytoplasm; it is active only when bound to cyclic AMP.

The attenuator serves as a termination site only when tryptophan is present. Why is that? It turns out that there are repeated codons that specify tryptophan in the leading portion of mRNA synthesized before RNA polymerase reaches the attenuator. If enough tryptophan is present in the cell, translation of these codons takes place, and transcription and translation both proceed smoothly until RNA polymerase reaches the attenuator region, where it terminates transcription. In the absence of tryptophan, translation stalls, and the leading portion of the mRNA folds up in such a way that the attenuator cannot function. In that case, synthesis of the complete mRNA for the tryptophan operon can be completed, and all the tryptophan biosynthetic enzymes are made. With these enzymes available, tryptophan is restored to the cell in large amounts. Attenuation is an elegant form of repression that regulates the genes that specify amino acid biosynthesis in bacteria.

Mutation: Change in the Genetic Material

A **mutation** is a change in the base sequence of DNA. We can reasonably expect that a change in the base sequence of a gene will sometimes cause a change in the product encoded by that gene. When the gene for an enzyme mutates, for example, the enzyme encoded by the gene may become inactive or less active because its amino acid sequence has changed. Such a change in genotype may be disadvantageous or even lethal if the cell loses a phenotypic trait it needs. However, a mutation can be beneficial if, for instance, the altered enzyme encoded by the mutant gene has a new activity that benefits the cell.

Many simple mutations are neutral: the change in DNA base sequence causes no change in the activity of the product encoded by the gene. Neutral mutations commonly occur when one nucleotide is substituted for another in the DNA, especially at a location corresponding to the third position of the mRNA codon. Because of the degeneracy of the genetic code, the resulting new codon might still code for the same amino acid. Even if the amino acid is changed, there may be no change in protein function if the amino acid is in a nonvital portion of the protein or is chemically very similar to the original amino acid.

Types of Mutations

The most common type of mutation involving single base pairs is **base substitution** or **point mutation,** in which a single base at one point in the DNA is replaced with a different one. Then, when the DNA replicates, the result is a substituted base pair (Figure 8.14). For example, A-T might be substituted for G-C, or C-G for G-C. If a base substitution occurs in a portion of the DNA molecule that codes for a protein, then the mRNA transcribed from the gene will carry an incorrect base at some position. When the mRNA is translated into protein, the incorrect base can cause the insertion of an incorrect amino acid in the protein. Thus, the base substitution in DNA can result in an amino acid substitution in the synthesized protein. This is known as a **missense mutation** (Figure 8.15a and b).

By creating a stop (nonsense) codon in the middle of an mRNA molecule, some base substitutions effec-

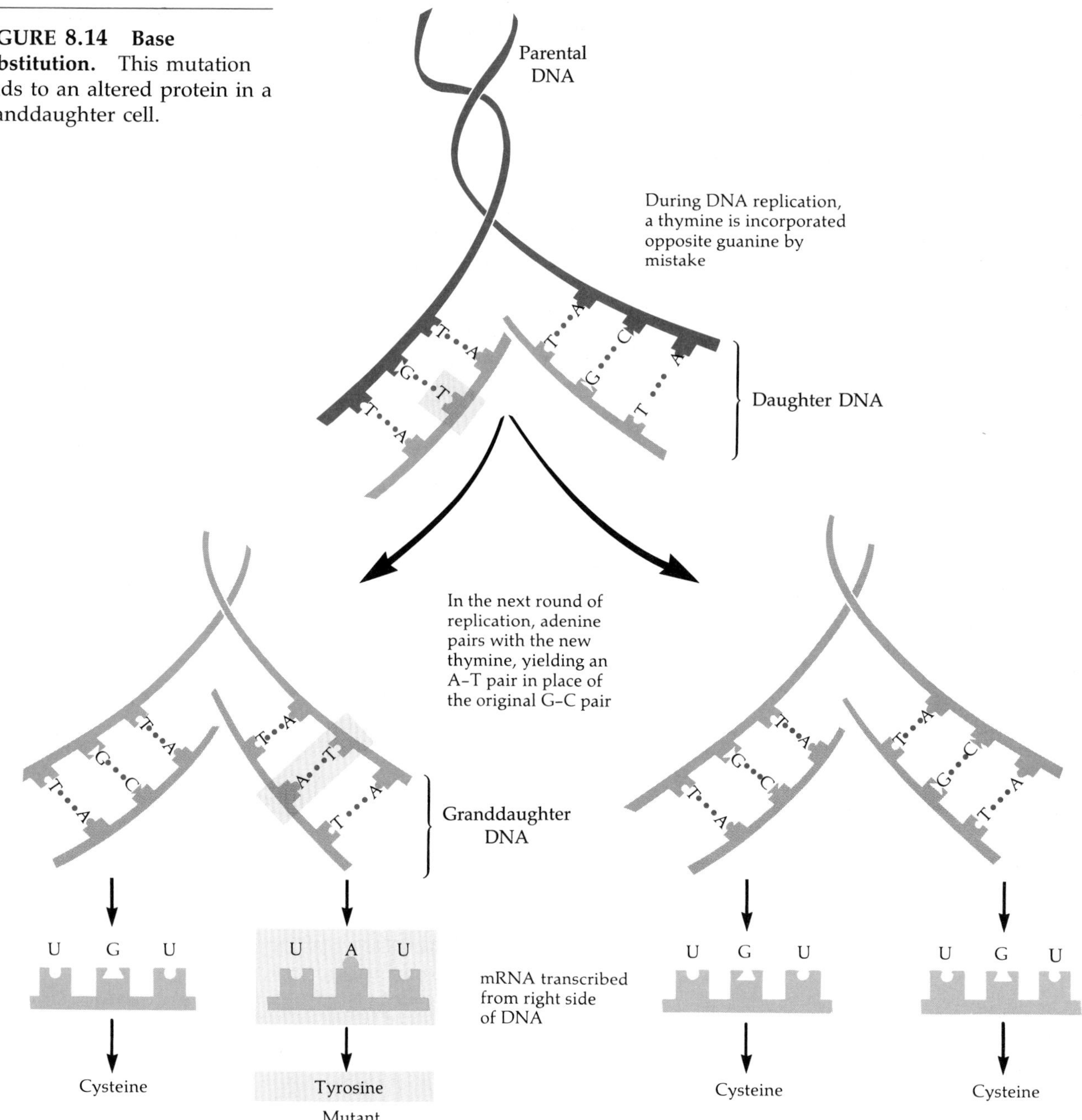
Parental DNA
During DNA replication, a thymine is incorporated opposite guanine by mistake
Daughter DNA
In the next round of replication, adenine pairs with the new thymine, yielding an A–T pair in place of the original G–C pair
Granddaughter DNA
mRNA transcribed from right side of DNA
U G U
U A U
Cysteine
Tyrosine
Mutant

FIGURE 8.14 Base substitution. This mutation leads to an altered protein in a granddaughter cell.

tively prevent the synthesis of a functional protein. Only a fragment of the protein is synthesized. A base substitution thus resulting in a nonsense codon is called a **nonsense mutation** (Figure 8.15c).

Besides base-pair mutations, there are also changes in DNA called **frameshift mutations.** Here, one or a few nucleotide pairs are deleted or inserted in the DNA (Figure 8.15d). This can shift the "translational reading frame," that is, the three-by-three grouping of nucleotides recognized as codons by the tRNAs during translation. For example, inserting one nucleotide pair in the middle of a gene causes many amino acids downstream from the site of the original mutation to change. Frameshift mutations almost always result in a long stretch of missense and an inactive protein produced for the mutated gene. In most cases, a nonsense codon will eventually be generated and thereby terminate translation.

Base substitutions and frameshift mutations may occur spontaneously because of occasional mistakes made during DNA replication. These **spontaneous mutations** are mutations that occur without known intervention of mutation-causing agents. Agents in the environment, such as certain chemicals and radiation, that directly or indirectly bring about mutations are called **mutagens.**

CHEMICAL MUTAGENS

Among the many chemicals known to be mutagenic is nitrous acid. Figure 8.16 shows how exposure of DNA to nitrous acid can convert the base adenine (A) to a

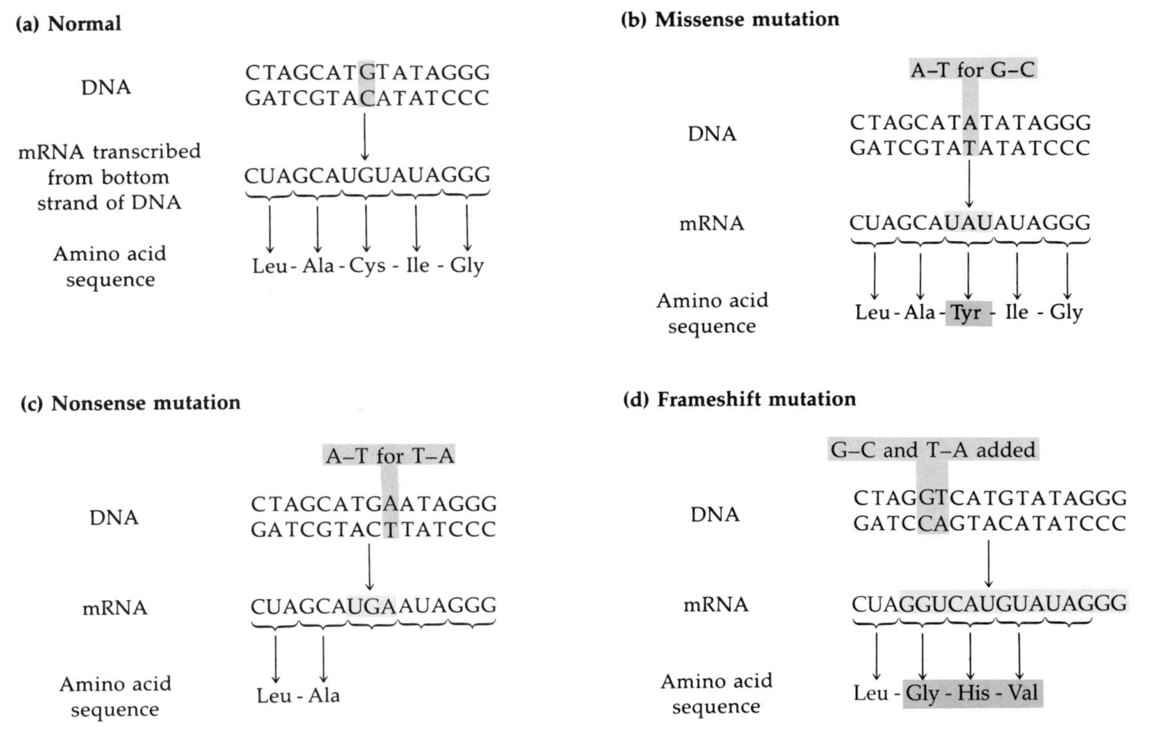

FIGURE 8.15 Types of mutations. **(a)** Normal DNA molecule. **(b)** Missense mutation. **(c)** Nonsense mutation. **(d)** Frameshift mutation.

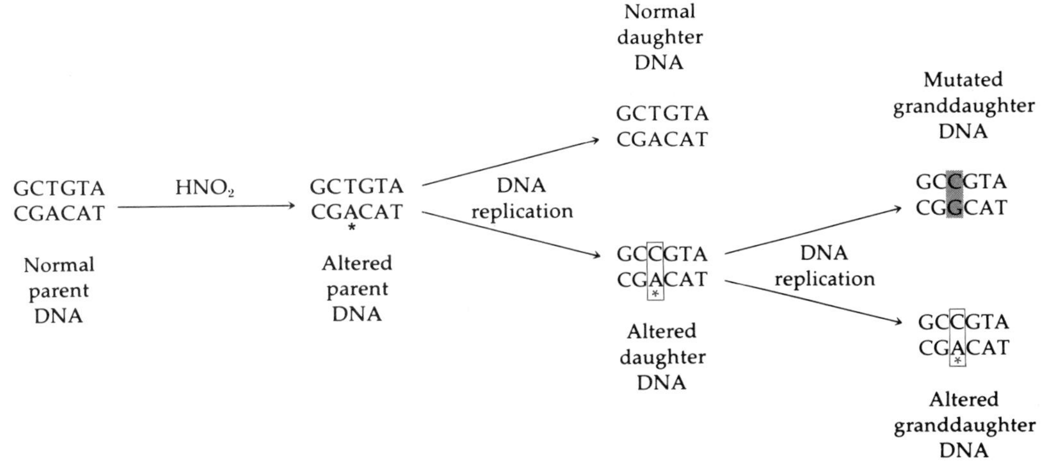

FIGURE 8.16 Mutagenesis by nitrous acid (HNO_2). The nitrous acid alters an adenine (asterisk), with the result that it pairs with cytosine instead of thymine.

form that no longer pairs with thymine (T) but instead pairs with cytosine (C). When DNA containing such modified adenines replicates, one daughter DNA molecule will have a base-pair sequence different from that of the parent DNA. Eventually, some A-T base pairs of the parent will have been changed to G-C base pairs in a granddaughter cell. Nitrous acid is thus an effective mutagen that makes a specific base-pair change in DNA. Like all mutagens, it alters DNA at random locations.

Other chemical mutagens are **base analogs;** examples are 2-aminopurine and 5-bromouracil. These molecules are structurally similar to normal nitrogenous bases, but they have slightly altered basepairing properties. The 2-aminopurine molecule (Figure 8.17a) is incorporated into DNA in place of adenine (A) but can

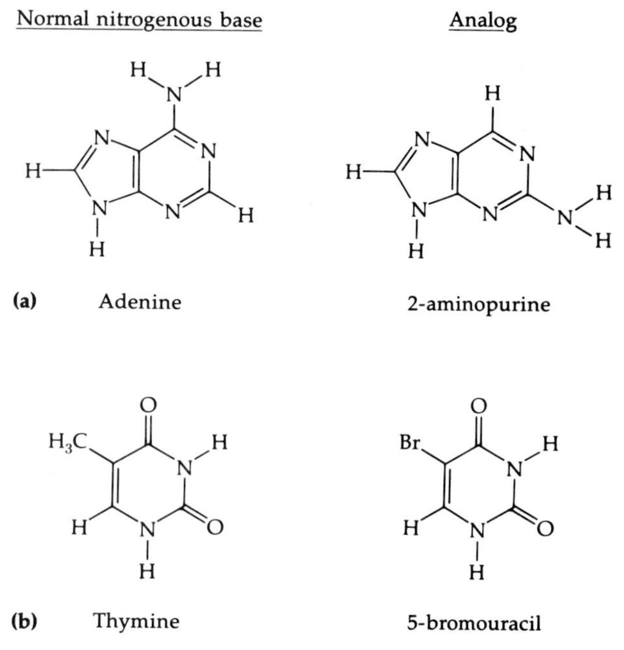

FIGURE 8.17 Base analogs and the nitrogenous bases they replace. **(a)** 2-Aminopurine and adenine. **(b)** 5-Bromouracil and thymine.

sometimes pair with cytosine (C). The 5-bromouracil molecule (Figure 8.17b) is incorporated into DNA in place of thymine (T) but often pairs with guanine (G). When base analogs are given to growing cells, the analogs are randomly incorporated into cellular DNA in place of the normal bases. Then, during DNA replication, the analogs cause mistakes in basepairing. The wrongly paired bases will be copied during subsequent replication of the DNA, resulting in base-pair substitutions in the progeny cells. Some antiviral and antitumor drugs are base analogs.

Still other mutagens cause small deletions or insertions, which can result in frameshifts. For instance, under certain conditions, benzpyrene, which is present in smoke and soot, is an effective **frameshift mutagen.** Likewise, aflatoxin, produced by *Aspergillus flavus* (a-spėr-jil'lus flā'vus), a mold that grows on peanuts and grain, is a frameshift mutagen, as are the acridine dyes used experimentally against herpesvirus infections. Frameshift mutagens usually have the right size and chemical properties to slip between the stacked base pairs of the DNA double helix. It is believed that frameshift mutagens work by slightly offsetting the two strands of DNA, leaving a gap in one strand or a bulge in the other. When the staggered DNA strands are copied during DNA synthesis, one or more base pairs can be inserted or deleted in the new double-stranded DNA. Interestingly, frameshift mutagens are often potent carcinogens.

RADIATION

X rays and gamma rays are forms of radiation that are potent mutagens because of their ability to *ionize* atoms and molecules. The penetrating rays of ionizing radiation cause electrons to pop out of their usual shells. The affected electrons bombard other molecules and cause more damage, and many of the resulting ions and free radicals (molecular fragments with unpaired electrons) are very reactive. Some of these ions can combine with bases in DNA, resulting in errors in DNA replication and repair that produce mutations. An even more serious outcome is the breaking of covalent bonds in the sugar–phosphate backbone of DNA. This causes physical breaks in chromosomes.

Another form of mutagenic radiation is ultraviolet light (UV), a nonionizing component of ordinary sunlight. The most important effect of UV light on DNA is the formation of harmful covalent bonds between certain bases. Adjacent thymines in a DNA strand can crosslink to form thymine dimers (Figure 8.18a). Such dimers, unless repaired, may cause serious damage or death to the cell because the cell cannot properly transcribe or replicate such DNA.

Bacteria and other organisms have enzymes that can repair radiation damage. Some of these enzymes also participate in DNA replication and recombination. One type of repair is illustrated in Figure 8.18b. Enzymes cut out the distorted crosslinked thymines by opening a wide gap. They then fill in the gap with newly synthesized DNA that is complementary to the undamaged strand. By this means, the original base-pair sequence is restored. The last step is the covalent sealing of the DNA backbone by the enzyme DNA ligase. Occasionally, such a repair process makes an error, and the original base-pair sequence is not properly restored. The result of this error is a mutation.

Frequency of Mutation

The **mutation rate** is the probability that a gene will mutate when a cell divides. The rate is usually stated as a power of 10, and because mutations are very rare, the exponent is always a negative number. For example, if there is 1 chance in 10,000 that a gene will mutate when the cell divides, the mutation rate is 1/10,000, which is expressed as 10^{-4}. Spontaneous mistakes in DNA replication occur at a very low rate because the replication machinery is remarkably faithful. Perhaps only once in 10^9 replicated base pairs does an error occur (a mutation rate of 10^{-9}). Because the average gene has about 10^3 base pairs, the spontaneous rate of mutation is about once in 10^6 (a million) replicated genes. Mutations usually occur more or less randomly along a chromosome. The occurrence of random mutations at low frequency is an essential aspect

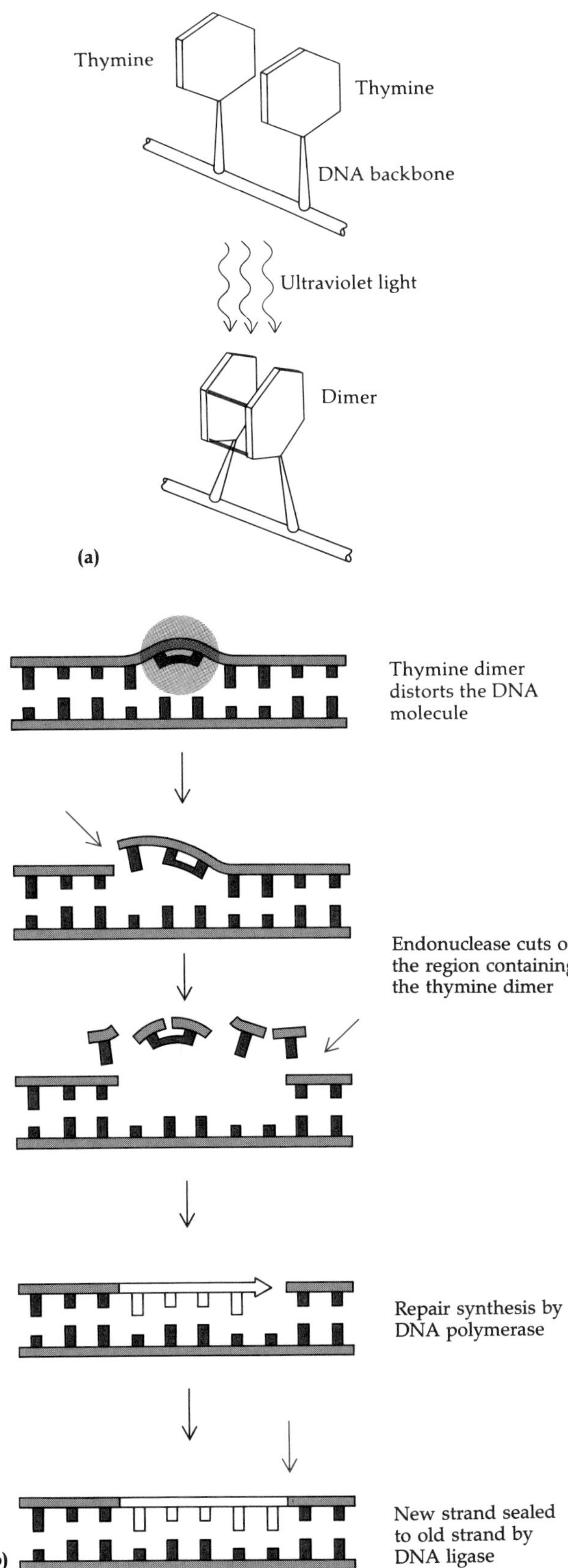
Thymine
Thymine
DNA backbone
Ultraviolet light
Dimer
(a)
Thymine dimer distorts the DNA molecule
Endonuclease cuts out the region containing the thymine dimer
Repair synthesis by DNA polymerase
New strand sealed to old strand by DNA ligase
(b)

FIGURE 8.18 Mutagenesis by ultraviolet light. **(a)** Ultraviolet light crosslinks adjacent thymines. **(b)** Mechanism of repair of crosslinked thymines.

in the adaptation of species to their environments, for evolution requires that genetic diversity be generated randomly and at a low rate. For example, in a bacterial population of significant size, say, greater than 10^7 cells, a few new mutant cells will always be produced in every generation. Most mutations either are harmful, and likely to die out with the individual cell, or are neutral. However, a few may be beneficial. For example, a mutation that confers antibiotic resistance is beneficial if a population of bacteria is regularly exposed to antibiotics. Once such a trait has made its appearance through mutation, cells carrying the mutated gene are more likely than other cells to survive and reproduce. Soon most of the cells in the population will have the gene; an evolutionary change will have occurred, although on a small scale.

A mutagen will usually increase the spontaneous rate 10 to 1000 times. In other words, a mutagen will cause a mutation rate of 10^{-5} to 10^{-3} per gene per cell per generation. Mutagens are used experimentally to enhance the production of mutant cells for research on the genetic properties of microorganisms and for commercial use.

Identifying Mutants

Mutants can be detected by selecting or testing for an altered phenotype. Whether or not a mutagen is used, mutant cells with specific mutations are always rare compared with other cells in the population. The problem is one of detecting a rare event.

Experiments are done more easily with bacteria than with other organisms because bacteria reproduce rapidly and large numbers of organisms (more than 10^9 per milliliter of broth) can easily be used. Furthermore, because bacteria generally have only one copy of each gene per cell, the effects of a mutated gene are not masked by the presence of a normal version of the gene, as in many eucaryotic organisms.

Positive (direct) selection involves the detection of mutant cells by rejection of the unmutated parent cells. For example, suppose we were trying to find mutant bacteria that are resistant to penicillin. When the bacterial cells are plated on a medium containing penicillin, the mutant can be identified directly. The few cells in the population that are resistant (mutants) will grow and form colonies. The normal, penicillin-sensitive, parental cells cannot grow.

For identifying mutations in other kinds of genes, it may be necessary to use **negative (indirect) selection.** This is made possible by the **replica-plating technique.** In replica plating, about 100 bacterial cells are first inoculated onto an agar plate. This plate, called the master plate, contains a rich, nonselective medium on which all cells will grow. After several hours of incubation, each cell reproduces to form a colony. Then a pad of sterile velvet is pressed over the master plate (Figure 8.19). Some of the cells from each colony adhere to the velvet. Next, the velvet is pressed down onto two (or more) sterile plates. One plate contains a rich medium, and one contains a minimal medium on which the original, nonmutant bacteria can grow (for example, glucose and inorganic salts). A mutant colony that grew on the rich medium of the master plate but that has a new requirement for a growth factor will not grow on the minimal medium, which lacks that growth factor. Cells taken from the mutant colony on the master plate can then be tested on minimal media supplemented with various growth factors. These tests determine precisely which factor is required.

The replica-plating technique is a very effective means of isolating mutants that require one or more new growth factors. Any mutant microorganism possessing a nutritional requirement not possessed by the parent is known as an **auxotroph.** For example, an auxotroph may be lacking an enzyme needed for synthesizing a particular amino acid and will therefore require that amino acid as a growth factor in its nutrient medium.

Identifying Chemical Carcinogens

Many known mutagens have been found to be carcinogens. A **carcinogen** is any substance that causes cancer in animals, including humans. In recent years, chemicals in the environment, the work place, and the diet

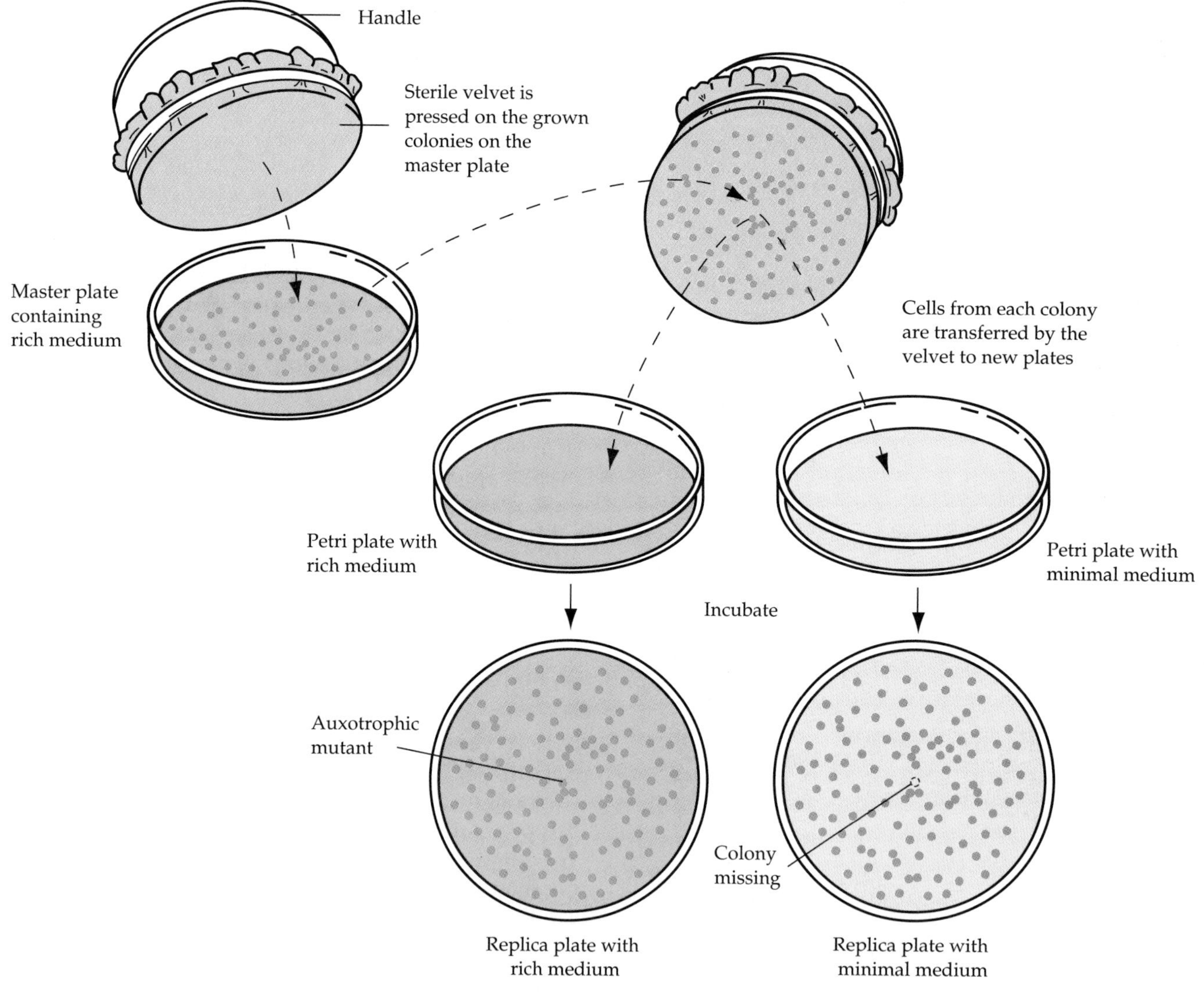

FIGURE 8.19 Replica plating. This technique transfers bacteria quickly and easily from colonies on a master plate to a different medium in another plate. The procedure permits the identification of auxotrophic mutants, which form colonies on the complete medium of the master plate but are unable to grow on the minimal medium of the replica plate.

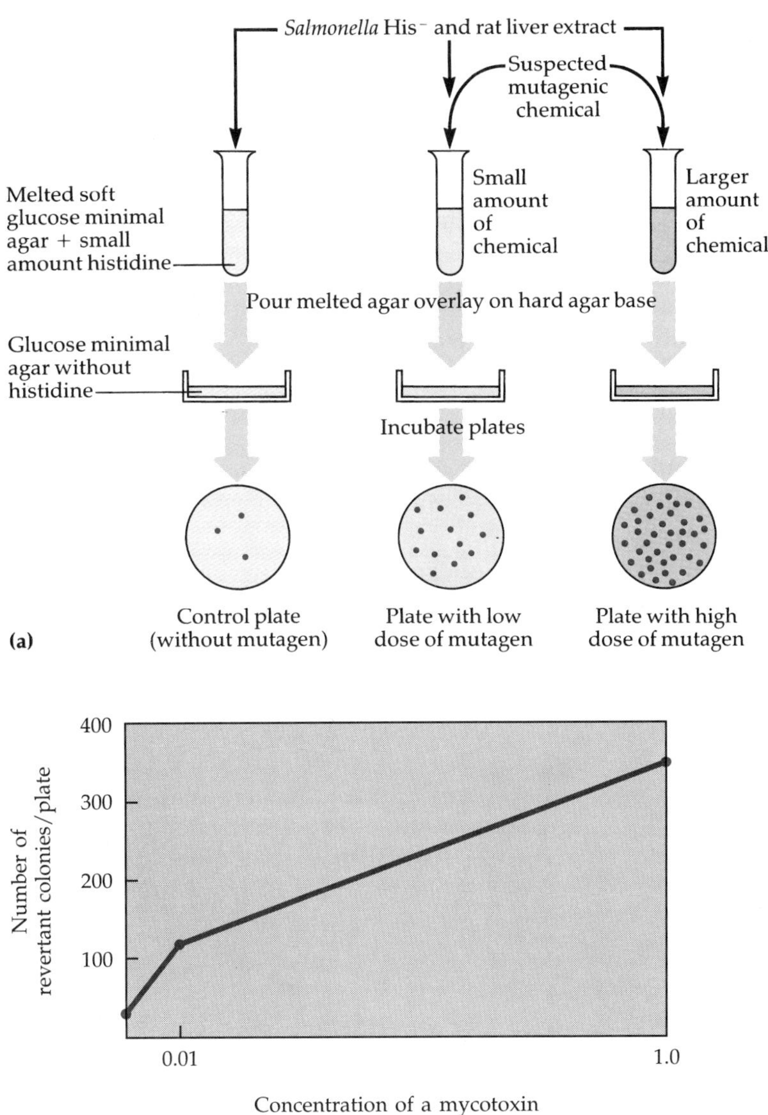

FIGURE 8.20 Ames test for suspected carcinogen. **(a)** Mutant *Salmonella* bacteria unable to synthesize histidine (His^-) are mixed with a potentially mutagenic chemical and rat liver extract, which provides a source of mammalian activating enzymes. The test bacteria are then plated on growth medium with a very small amount of histidine, to start the bacteria growing. (Some growth is necessary for mutagenesis.) Control His^- bacteria, which have not been treated with the mutagen, are inoculated onto another plate containing the same medium. Only bacteria whose His^- phenotype has mutated back (reverted) to His^+ (able to synthesize histidine) can grow into colonies. The control plate may show a few spontaneous His^+ revertants. The test plates will show an increase in the number of His^+ revertants if the test chemical is indeed a mutagen and potential carcinogen. **(b)** Quantitative results can be expressed as a mutational dose–response curve. This graph shows the effect of increasing concentrations of a mutagenic mycotoxin.

have been implicated as causes of cancer in humans. The usual subjects of tests to determine potential carcinogens are animals, and the testing procedures are both time consuming and expensive. Now there are faster and less expensive procedures for the preliminary screening of potential carcinogens. One of these, called the **Ames test,** uses bacteria as carcinogen indicators.

The Ames test is based on the observation that exposure of mutant bacteria to mutagenic substances may cause new mutations that reverse the effect (the phenotype) of the original mutation. Specifically, the test measures the reversion of histidine auxotrophs of *Salmonella* (His^- cells—mutants that have lost the ability to synthesize the amino acid histidine) to histidine-synthesizing cells (His^+) after treatment with a mutagen. Bacteria are incubated in both the presence and absence of the substance being tested (Figure 8.20). Since many chemicals must be activated (transformed chemically) by animal enzymes for mutagenic or carcinogenic activity to be exhibited, the chemical to be tested and the mutant bacteria are incubated together with rat liver extract, a rich source of activation enzymes. If the substance being tested is mutagenic, it will cause reversion of His^- bacteria to His^+ at a rate higher than the spontaneous reversion rate. The number of observed revertants provides an indication of the degree to which a substance is mutagenic and therefore possibly carcinogenic.

The test can be used in many ways. Several potential mutagens can be qualitatively tested by spotting the individual chemicals on small paper disks on a single plate inoculated with bacteria. Not just pure chemicals but also mixtures such as wine, blood, smoke condensates, and extracts of foods can be tested for mutagenic substances.

About 90% of the substances found by the Ames test to be mutagenic have also been shown to be carcinogenic in animals. Furthermore, the more mutagenic substances have generally been found to be more carcinogenic.

Genetic Transfer and Recombination

Genetic recombination refers to the exchange of genes between two DNA molecules to form new combinations of genes on a chromosome. Figure 8.21 shows one type of recombinational event occurring between two pieces of DNA, which we shall regard as chromosomes for the sake of simplicity. We have called one chromosome A and the other B. If these two chromosomes break and rejoin as shown (a process called **crossing over,**) some of the genes carried by these chromosomes are shuffled. The original chromosomes have *recombined,* so that each now carries a portion of the other chromosome's genes.

If A and B represent DNA from different individuals, how are they brought close enough to recombine? In eucaryotes, recombination generally takes place during meiosis, when chromosomes originating from the organism's two parents pair (a process called synapsis). Thus, the gametes resulting from meiosis contain recombined DNA. Genetic recombination in eucaryotes is an ordered process that usually occurs as part of the sexual cycle of the organism. In bacteria, genetic recombination can happen in a number of ways, which we will discuss in the following sections.

Like mutation, genetic recombination contributes to a population's genetic diversity, which is the source of variation in evolution. In highly evolved organisms such as present-day microorganisms, recombination is more likely than mutation to be beneficial because recombination is less likely to destroy the function of a gene and may bring together combinations of genes that enable the organism to carry out a valuable new function.

Genetic material can be transferred between bacteria in several ways. In all of the mechanisms, the transfer involves a **donor cell** that gives a portion of its total DNA to a **recipient cell.** Once transferred, part of the donor's DNA is usually incorporated into the recipient's DNA; the remainder is degraded by cellular enzymes. The recipient cell that incorporates donor DNA into its own DNA is called a **recombinant.** The transfer of genetic material between bacteria is by no means a frequent event. It may occur in only 1% or less of an entire population. Let us now examine the specific types of genetic transfer in detail.

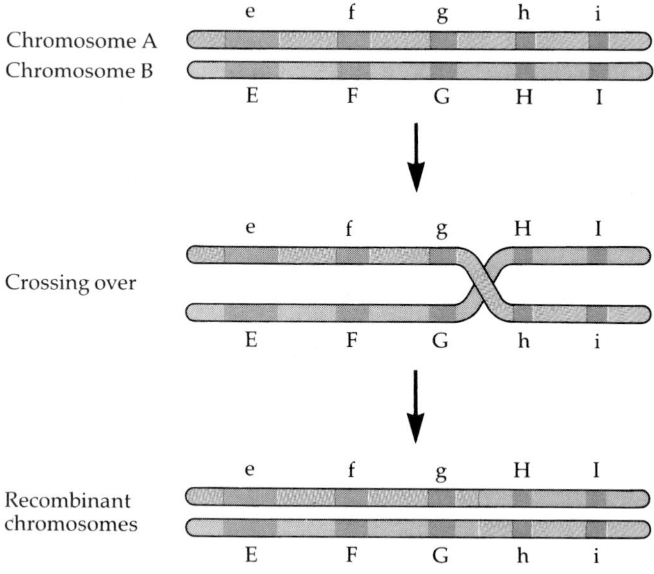

FIGURE 8.21 Genetic recombination between two related chromosomes. Chromosomes A and B each carry one copy of genes E through I. The chromosomes cross over by breaking and rejoining, sometimes in more than one location. The result is two recombinant chromosomes, each of which carries genes originating from both chromosomes.

Transformation in Bacteria

During the process of **transformation,** genes are transferred from one bacterium to another as "naked" DNA in solution. This process was first demonstrated more than 60 years ago, although it was not understood at that time. Not only did transformation show that genetic material could be transferred from one bacterial cell to another, but study of this phenomenon eventually led to the conclusion that DNA is the genetic material. The initial experiment on transformation was performed by Frederick Griffith in England in 1928 while he was working with two strains of *Streptococcus pneumoniae.* One, a virulent strain, has a polysaccharide capsule and causes pneumonia; the second, an avirulent strain, lacks the capsule and does not cause disease.

Griffith was interested in determining whether injections of heat-killed bacteria of the virulent strain could be used to vaccinate mice against pneumonia. As he expected, injections of the killed bacteria did not make mice ill, nor did injections of live avirulent bacteria. However, when the dead virulent bacteria were mixed with live avirulent bacteria and injected into the mice, many of the mice died. In the blood of the dead

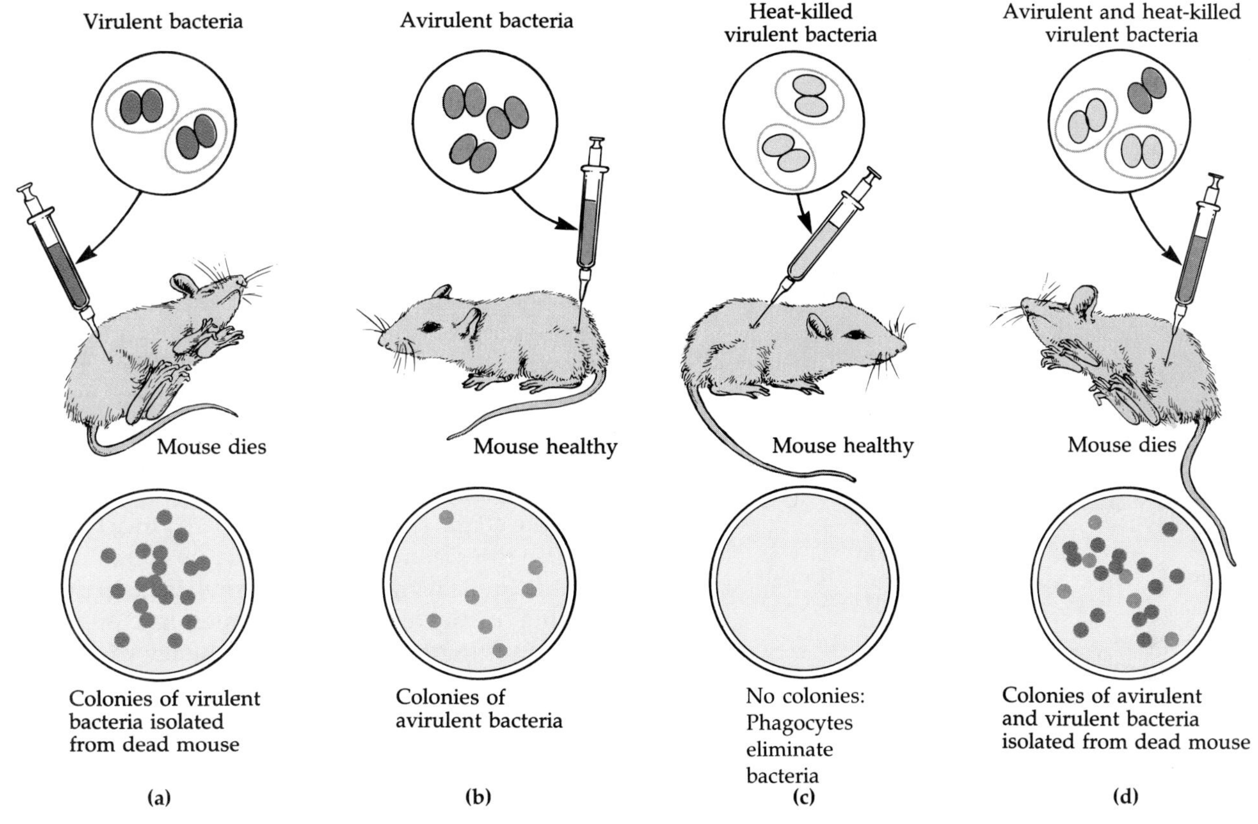

FIGURE 8.22 Transformation. Griffith's experiment demonstrating genetic transformation. Some material from the heat-killed virulent bacteria transformed the living avirulent bacteria into virulent bacteria, which killed the mouse. Avirulent bacteria lack capsules and are readily destroyed by the host; therefore, few show up as colonies on the medium in (b) and (d).

mice, Griffith found living, encapsulated, virulent bacteria. Hereditary material (genes) from the dead bacteria had entered the live cells and changed them genetically so that their offspring were encapsulated, virulent forms (Figure 8.22).

Subsequent investigations based on Griffith's research revealed that bacterial transformation could be carried out without mice. A broth (liquid culture medium) was inoculated with live avirulent bacteria. Dead virulent bacteria were then added to the broth. After incubation, the culture was found to contain living bacteria that were encapsulated and virulent. The avirulent bacteria had been transformed; they had acquired a new hereditary trait by incorporating genes from the killed virulent bacteria.

The next step was to extract various chemical components from the killed cells to determine which component caused the transformation. These crucial experiments were performed by Oswald T. Avery and his associates Colin M. MacLeod and Maclyn McCarty in the United States. After years of research, they announced in 1944 that the component responsible for transforming harmless *S. pneumoniae* into virulent strains was DNA.

Since the time of Griffith's experiment, considerable information has been gathered about transformation. In nature, some bacteria, perhaps after death and cell lysis, release their DNA into the environment. Other bacteria can then encounter the DNA and, depending on the particular species and growth conditions, take up fragments of DNA into their cytoplasm and recombine the DNA into their own chromosomes. A recipient cell having this new combination of genes is a kind of hybrid, or recombinant cell (Figure 8.23). All the descendants of such a recombinant cell will be identical to it. Transformation occurs naturally among very few genera of bacteria. These include *Bacillus, Hemophilus* (hē-mä'fi-lus), *Neisseria, Acinetobacter* (a-si-ne'tō-bak-tėr), and certain strains of *Streptococcus* and *Staphylococcus.*

Transformation works best when the donor and recipient cells are very closely related and when the recipient cells are in the late log phase of growth. Even though only a small portion of a cell's DNA is trans-

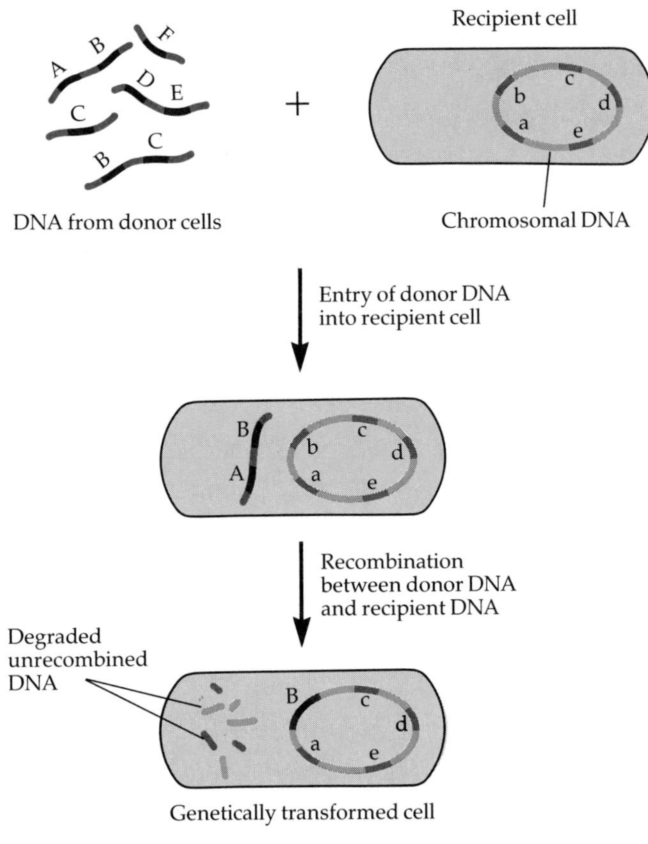

FIGURE 8.23 Mechanism of genetic transformation.

Conjugation differs from transformation in two major ways. First, conjugation requires that there be direct cell-to-cell contact. Second, the conjugating cells must generally be of opposite "mating type"—donor cells must carry the plasmid, and recipient cells usually do not. The plasmid carries genes that code for the synthesis of *sex pili,* projections from the donor's cell surface that contact the recipient and help to bring the two cells into direct contact. In the process of conjugation, the plasmid is replicated during transfer of a single-stranded copy of the plasmid DNA to the recipient, where the complementary strand is synthesized.

Because most experimental work on conjugation has been done with *E. coli,* we will describe the process in this organism. In *E. coli,* the *F factor* was the first plasmid observed to transfer between cells during conjugation (Figure 8.24). Donors carrying F (F^+ cells) transfer the plasmid to recipients (F^- cells), which become F^+ cells as a result (Figure 8.25a). In some cells carrying F, the F factor integrates into the chromosome, forming an **Hfr (high frequency of recombination) cell** (Figure 8.25b). During conjugation between an Hfr and an F^- cell, the Hfr cell's chromosome, with its integrated F factor, replicates, and a parental strand of the chromosome is transferred to the recipient cell (Figure 8.25c). Replication of the Hfr chromosome begins within the F factor, and a small piece of the F factor

ferred to the recipient, it is still a very large molecule that must pass through the recipient cell wall and membrane. When a recipient cell is in a physiological state in which it can take up the donor DNA, it is said to be **competent**. Competence may be related to alterations in the cell wall that make it permeable to the large DNA molecule.

The well-understood and widely used bacterium *E. coli* is not naturally competent for transformation. However, a simple laboratory treatment enables *E. coli* to readily take up DNA. The discovery of this treatment has enabled *E. coli* to be used for genetic engineering, as will be discussed in Chapter 9.

Conjugation in Bacteria

Conjugation is another mechanism by which genetic material is transferred from one bacterium to another.

Conjugation is mediated by *plasmids,* which are circular pieces of DNA that replicate independently from the cell's chromosome. They are like minichromosomes, but they differ from bacterial chromosomes in that the genes they carry are usually not essential for the growth of the cell. The plasmids responsible for conjugation are transmissible between cells during conjugation.

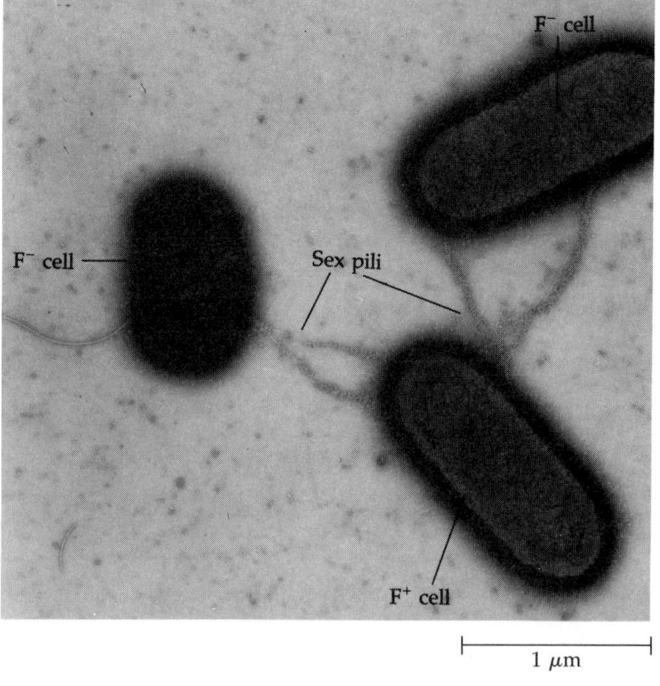

FIGURE 8.24 Bacterial conjugation. The DNA-donating F^+ *E. coli* cell extends sex pili to two F^- recipients. Later the F^+ and F^- cells will come into direct contact, and DNA will be transferred from one to the other.

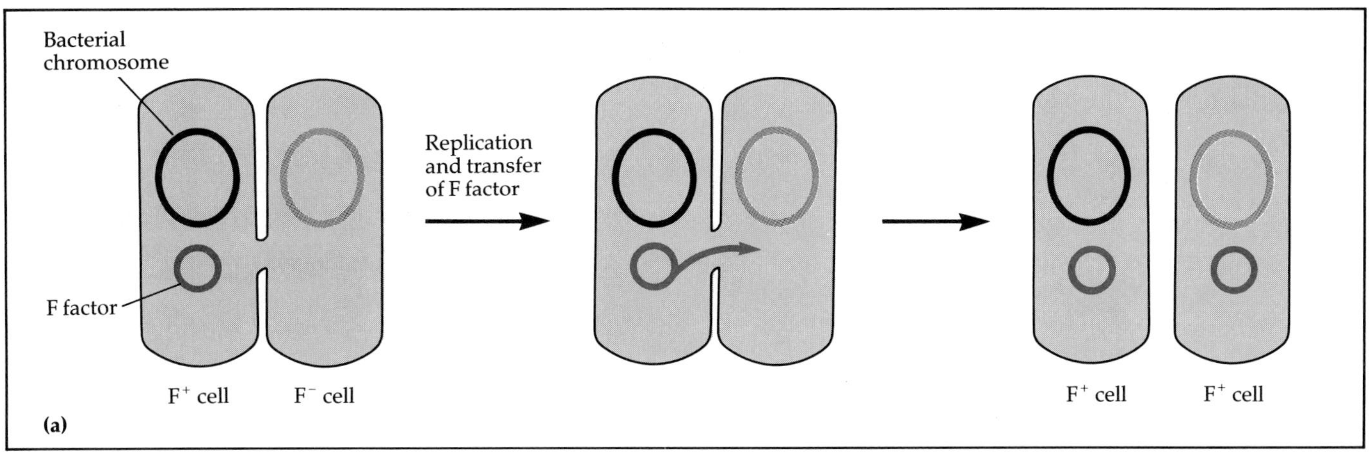

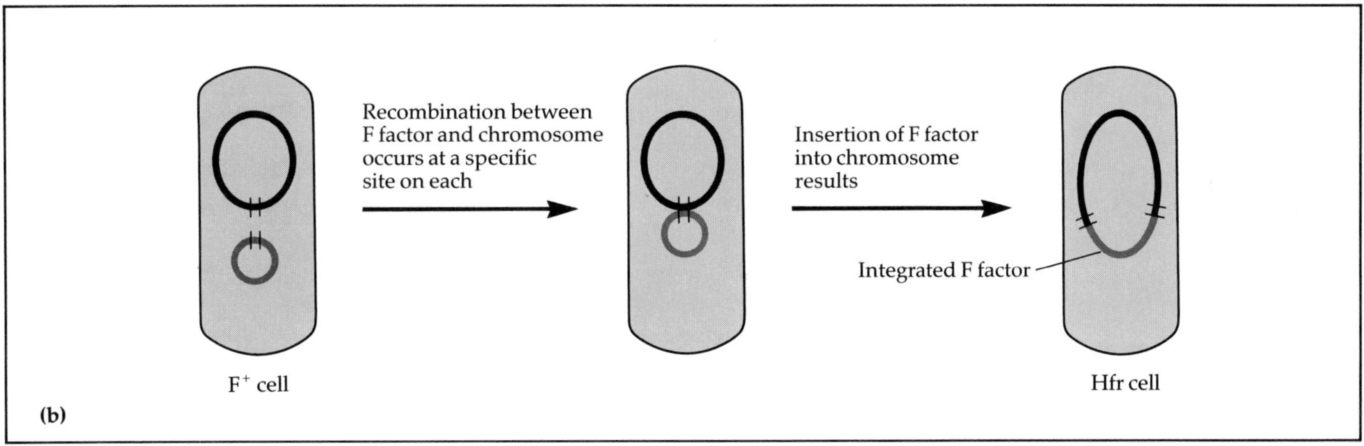

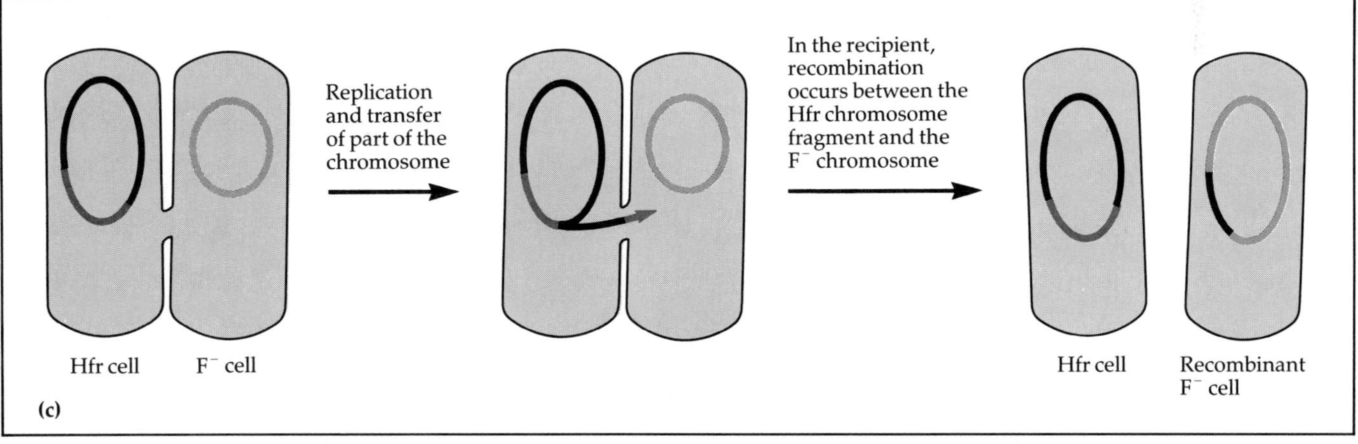

FIGURE 8.25 Conjugation in *E. coli*. **(a)** When an F factor (a plasmid) is transferred from donor (F^+) to recipient (F^-), the F^- cell is converted into an F^+ cell. **(b)** Integration of an F factor into the bacterial chromosome makes the cell an Hfr cell. **(c)** An Hfr donor passes a portion of its chromosome into an F^- cell. A recombinant F^- cell results.

tor leads the chromosomal genes into the F^- cell. Most of the integrated F factor enters the recipient cell last, if at all. Usually, the chromosome breaks before it is completely transferred. Once within the recipient cell, donor DNA can recombine with the recipient's DNA, as occurs in transformation. Donor DNA that is not integrated is degraded.

By the process of conjugation between an Hfr cell and an F^- cell, the F^- cell can acquire new versions of chromosomal genes (just as in transformation).

Transduction in Bacteria

A third mechanism of genetic transfer between bacteria is **transduction.** In this process, bacterial DNA is transferred from the donor cell to the recipient cell inside a virus that infects bacteria, called a **bacteriophage** (or simply, **phage**). (Phages will be discussed further in Chapter 13.)

To understand how transduction works, we shall consider the life cycle of one type of transducing phage of *E. coli;* this phage carries out a process called **generalized transduction.** In the process of infection, the phage attaches to the bacterial cell wall and injects its DNA into the bacterium (Figure 8.26a). The phage DNA acts as a template for the synthesis of new phage DNA and also directs the synthesis of phage protein coats (Figure 8.26b). During phage development inside the infected cell, the bacterial chromosome breaks apart and at least some fragments of the bacterial chromosome happen to be packaged inside phage protein coats (Figure 8.26c). (Even plasmid DNA or DNA of another virus that is inside the cell can be given phage protein coats.) The resulting phage particles then carry bacterial DNA instead of phage DNA. When the released phage particles later infect a new population of bacteria, bacterial genes will be transferred to the newly infected cells at low frequency (Figure 8.26d).

Transduction of cellular DNA by a virus can lead to recombination between the DNA of the first host cell and the DNA of the second host cell (Figure 8.26e). Transduction is thus another way that bacteria can acquire new genotypes, in addition to mutation, transformation, and conjugation. The process of generalized transduction just described is typical of bacteriophages such as phage P1 of *E. coli* and phage P22 of *Salmonella*.

All genes contained within a bacterium infected by a generalized transducting phage are equally likely to be packaged in a phage coat and transferred. There is another type of transduction, called **specialized transduction,** in which only certain bacterial genes are transferred. Specialized transduction will be discussed in Chapter 13.

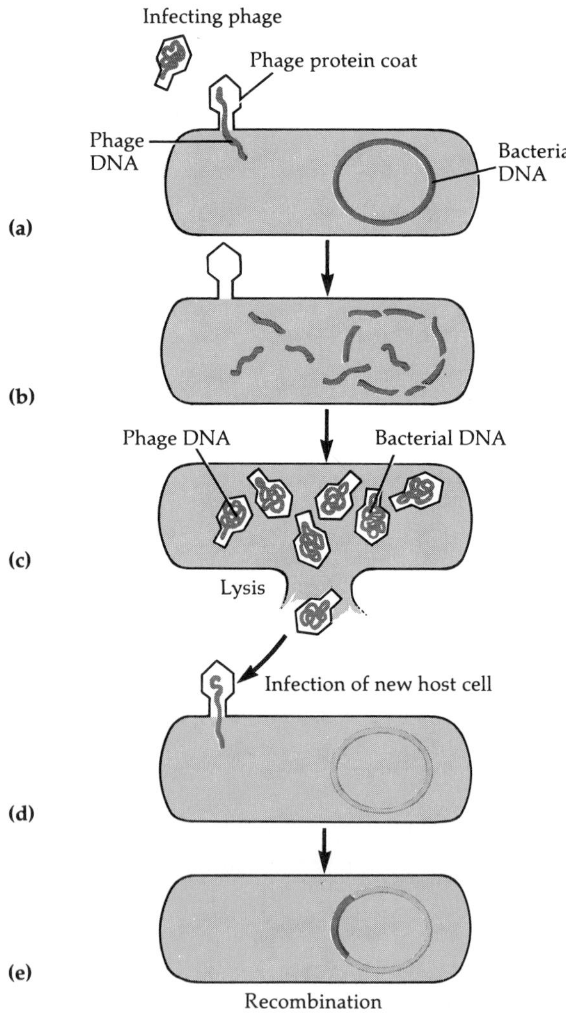

FIGURE 8.26 Transduction by bacteriophage. Shown here is generalized transduction, in which any bacterial DNA can be transferred from one cell to another.

Recombination in Eucaryotes

In transformation, conjugation, and transduction in bacteria, only a small segment of the DNA of the donor cell enters and recombines with the DNA of the recipient cell. In eucaryotes, genetic recombination is the result of a sexual reproductive process. In contrast to the processes in bacteria, sexual reproduction is a regular event for most eucaryotes and is necessary for the survival of many species.

Each species of eucaryotic organism has a characteristic **chromosome number,** that is, the number of chromosomes in each cell. For example, the human chromosome number is 46, meaning that each ordinary body cell (*somatic cell*) has 46 chromosomes. These chromosomes fall into two sets of 23 chromosomes. Each set is derived from one parent and carries genes for virtually all the activities of human cells.

Sexual reproduction in eucaryotes involves the fusion of the nuclei of two parent cells that are haploid (Figure 8.27). A **haploid cell** is one that contains only one of each chromosome; the number of chromosomes in such a cell is called the haploid number, symbolized *N*. In humans, the haploid number of chromosomes is 23. In higher organisms the haploid cells that fuse are called **gametes.** (In animals and plants, the male gametes are sperm and the female gametes are ova.) Fusion of the gametes produces a fertilized ovum or **zygote**, which has received half its chromosomes from one parent and half from the other (Figure 8.27a). The

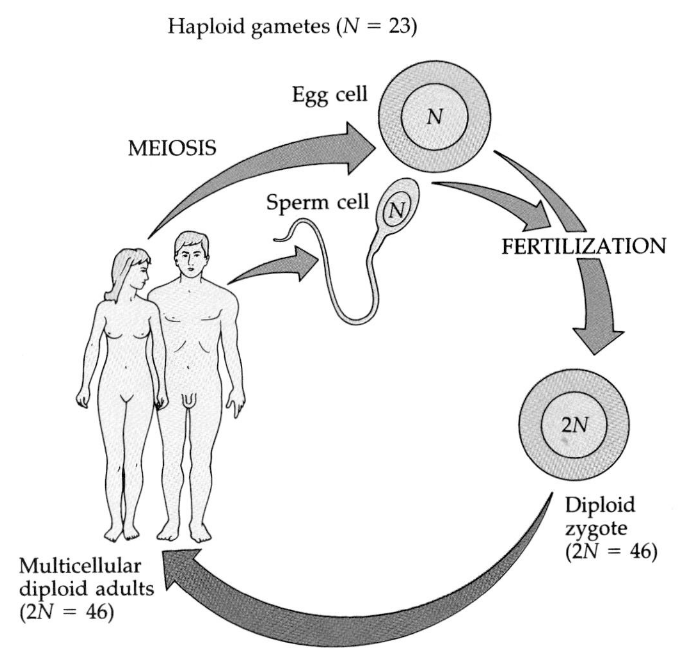

(a) Humans

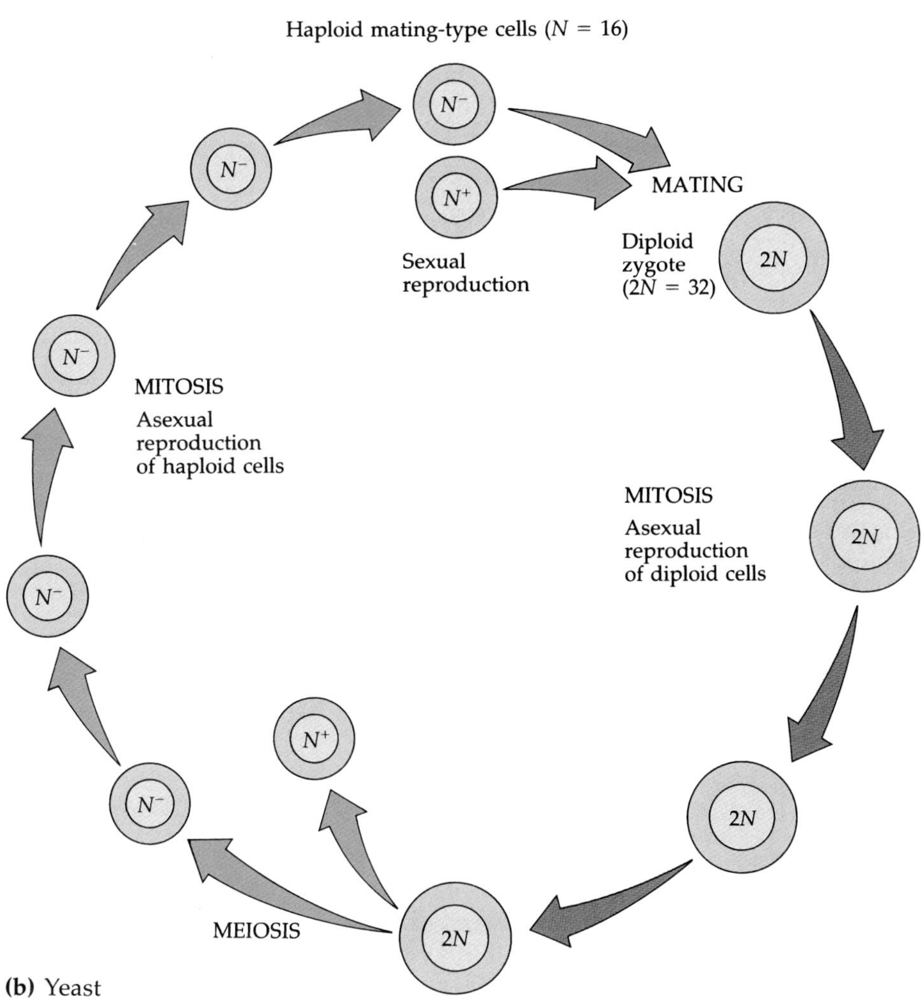

(b) Yeast

FIGURE 8.27 Sexual reproduction in eucaryotes. **(a)** Human reproductive cycle. **(b)** Reproductive cycle of the yeast *S. cerevisiae.* Plus and minus signs are used to distinguish the opposite mating types. (Meiosis actually produces four haploid cells because the chromosomes are duplicated during the process.)

zygote is called a **diploid cell** because it contains two sets of chromosomes. The diploid number (2*N*) is twice the haploid number (*N*). The two sets of chromosomes are said to be *homologous;* i.e., for each chromosome in one set, there is, in the other set, a corresponding chromosome that carries genes for the same traits in the same order (exceptions exist for chromosomes that determine sex, in humans the X and Y chromosomes). Subsequent growth and development of the zygote produces a mature multicellular organism capable of sexual reproduction. Most cells in this mature organism are produced by **mitosis,** a eucaryotic cell division process in which the chromosomes are duplicated and then one copy of each is distributed to each of two daughter cells.

Many eucaryotic microorganisms also form haploid cells of opposite mating types that can fuse to form a zygote. However, eucaryotic microorganisms are often capable of asexual reproduction as well. In the yeast *Saccharomyces cerevisiae* (se-ri-vis'ē-ī), for example, asexual reproduction, in which diploid or haploid cells simply divide by mitosis, can go on for many generations. In yeasts, periods of asexual reproduction alternate with sexual reproduction, in which haploid cells join to form a zygote (Figure 8.27b, p. 219).

Eventually, for sexual reproduction to recur in either microbes or higher organisms, some of the diploid cells must give rise to haploid cells. Diploid cells produce haploid cells by a special cell division process called **meiosis,** in which the homologous chromosomes are separated and a single chromosome from each pair goes to each daughter cell. In the early stages of meiosis the homologous chromosomes come together in pairs, and portions of homologous chromosomes can be exchanged by crossing over (Figure 8.21). Crossing over is in fact genetic recombination; and in eucaryotic cells it can often be observed microscopically.

In eucaryotic microorganisms, meiosis may be followed by a period of asexual reproduction in which generations of haploid organisms are produced by mitosis. In both higher organisms and microbes, another round of sexual reproduction is initiated when two haploid mating-type cells fuse and their nuclei unite to form a new zygote. Since these two haploid cells are usually from different individuals, new genetic combinations result. The regular events of crossing over in meiosis, segregation of chromosomes into haploid cells, and formation of diploid zygotes generate enormous diversity in every generation.

PLASMIDS

Plasmids are gene-containing circular pieces of DNA about 1% to 5% the size of the bacterial chromosome. They are found mainly in bacteria but also in some eucaryotic microorganisms, such as *S. cerevisiae.* Recall that the F factor is a plasmid that carries genes for sex pili and for transfer of the plasmid to another cell. Although plasmids are usually dispensable, under certain conditions genes carried by plasmids can be crucial to the survival and growth of the cell. **Dissimilation plasmids** code for enzymes that catalyze the catabolism of certain unusual sugars and hydrocarbons. Some species of *Pseudomonas* can actually use such substances as toluene, camphor, and hydrocarbons of petroleum as primary carbon and energy sources because they have catabolic enzymes encoded by genes carried on plasmids. Other plasmids code for proteins that enhance the pathogenicity of a bacterium. The strain of *E. coli* that causes infant diarrhea and traveler's diarrhea carries plasmids that code for toxin production and for bacterial attachment to intestinal cells. Without these plasmids, *E. coli* is a harmless resident of the large intestine. With them, it is pathogenic. Still other plasmids contain genes for the synthesis of **bacter-**

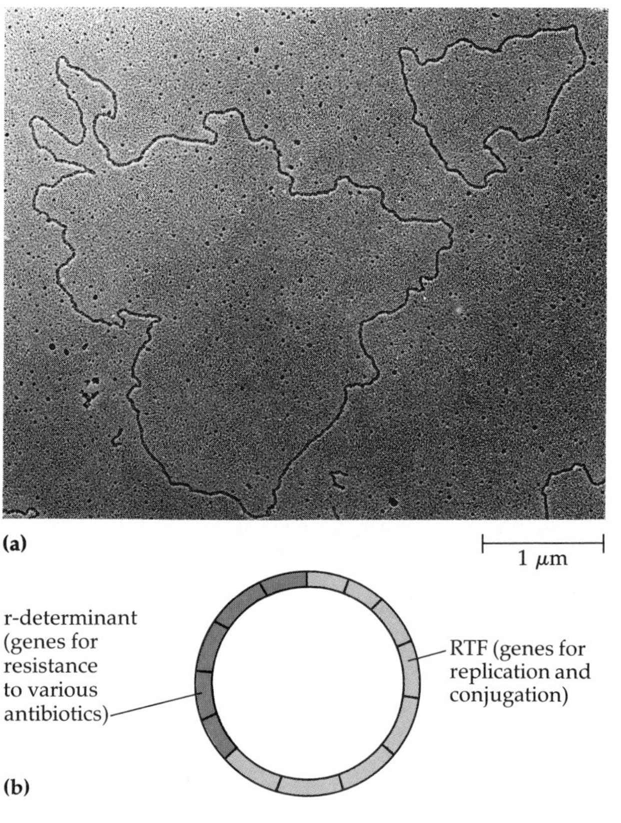

FIGURE 8.28 R factors, a type of plasmid. (a) Electron micrograph showing two different resistance (R) factors. The larger loop of DNA is a plasmid isolated from the bacterium *Bacteroides fragilis* and encodes resistance to the antibiotic clindamycin. The smaller loop of DNA is a plasmid from *E. coli* and encodes resistance to tetracycline. **(b)** Diagram of an R factor with multiple resistance determinants.

iocins, toxic proteins that kill other bacteria. These plasmids have been found in many genera, and they are useful markers for the identification of certain bacteria in clinical laboratories.

Resistance factors (R factors) are plasmids of great medical importance. R factors carry genes that confer upon their host cell resistance to antibiotics, heavy metals, or cellular toxins. Many R factors contain two groups of genes. One group is called the **resistance transfer factor (RTF)** and includes genes for plasmid replication and conjugation. The other group, the **r-determinant,** has the resistance genes; it codes for the production of enzymes that inactivate certain drugs or toxic substances (Figure 8.28). Different R factors, when present in the same cell, can recombine to produce R factors with new combinations of genes on their r-determinants.

R factors present very serious problems for the treatment of infectious diseases by using antibiotics. The widespread use of antibiotics in medicine and agriculture (many animal feeds contain antibiotics) has led to the preferential survival (selection) of bacteria that have R factors. So populations of resistant bacteria grow larger and larger. The transfer of resistance between bacterial cells of a population, and even between bacteria of different genera, also contributes to the problem. (See the boxes in Chapter 20, p. 495, and Chapter 26, p. 657.)

Plasmids are an important tool for genetic engineering, which will be discussed in Chapter 9.

TRANSPOSONS

Transposons (also called *transposable genetic elements*) are small segments of DNA that can move (be "transposed") from one region of the DNA molecule to another. In the popular press, they are often called "jumping genes." These pieces of DNA measure from 700 to 40,000 base pairs long.

Barbara McClintock of the United States discovered transposons in corn, but transposons occur in all organisms and have been studied most thoroughly in microorganisms. Their wide occurrence is not surprising, because they can insert themselves into almost any stretch of DNA. They may move from one site on a chromosome to another site on the same chromosome or to another chromosome or plasmid. As you might imagine, frequent movement of transposons could wreak havoc within a cell. For example, as transposons move about on chromosomes, they may insert themselves *inside* genes, inactivating them. Fortunately, transposition occurs relatively rarely. The frequency of transposition is comparable with the spontaneous mutation rate that occurs in bacteria, that is, from 10^{-5} to 10^{-7} per generation.

(a) Insertion sequence "IS1"

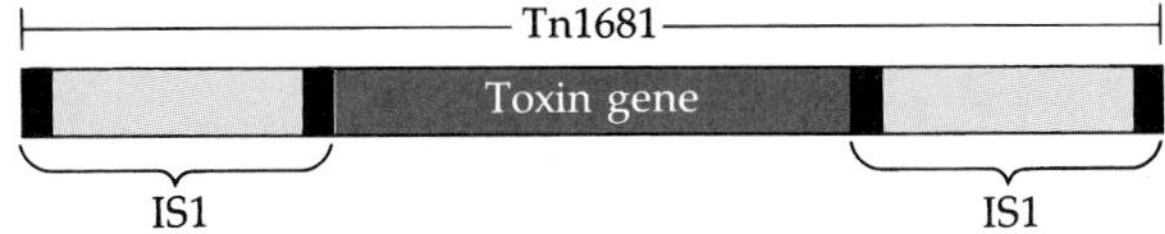

(b) Complex transposon "Tn1681"

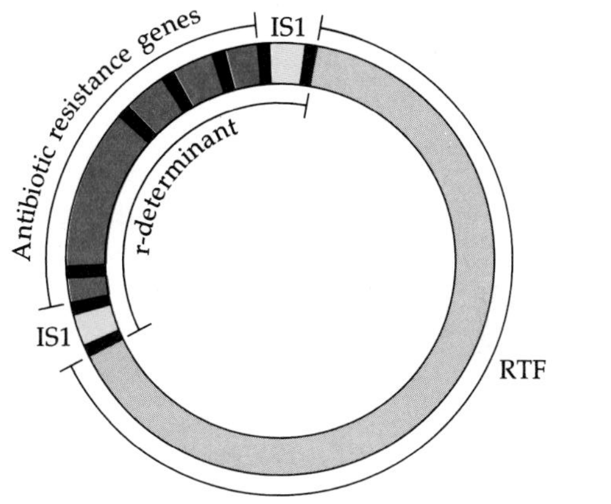

(c) R plasmid "R1" with r-determinant

FIGURE 8.29 Transposons and insertion sequences. **(a)** The simplest type of transposon is the insertion sequence (IS)—a segment of DNA that contains a gene for transposase (the enzyme that catalyzes transposition) and is bounded at each end by inverted repeat (IR) sequences that the transposase recognizes as sites of recombination. ISI is one example of an insertion sequence. **(b)** Complex transposons carry other genetic material in addition to transposase genes. The example shown here, Tn1681, carries the gene for a bacterial toxin and has complete copies of the insertion sequence ISI at each end. **(c)** A large transposon makes up a major portion of the plasmid RI, which has two parts. One part, the RTF, contains the genes needed for plasmid replication and transfer of the plasmid by conjugation; the other, the r-determinant, carries genes for resistance to five different antibiotics. The presence of an insertion sequence at each end of the group of antibiotic resistance genes makes this part of the plasmid a transposon.

All transposons contain the information for their own transposition. As shown in Figure 8.29a, the simplest transposons, also called **insertion sequences (IS),** contain only a gene that codes for an enzyme (transposase, which catalyzes the cutting and ligating of DNA that occurs in transposition) and recognition sites. Recognition sites are short regions of DNA the enzyme recognizes as recombination sites between the transposon and the chromosome.

Complex transposons also carry other genes not connected with the transposition process. For example, bacterial transposons may contain genes for toxins or for antibiotic resistance (Figure 8.29b). R factors are frequently made up of a collection of transposons (Figure 8.29c).

Transposons with antibiotic resistance genes are of particular practical interest, but there is no limitation on the kinds of genes that transposons can have. Thus, transposons provide a powerful natural mechanism for the movement of genes from one chromosome to another. Furthermore, since they may be carried between cells on plasmids or viruses, they can also spread from one organism—or even species—to another. Transposons are thus a potentially powerful mediator of evolution in organisms.

Genes and Evolution

We have now seen how gene activity can be controlled by the cell's internal regulatory mechanisms and how genes themselves can be altered or rearranged by mutation, transposition, and recombination. All these processes provide diversity in the descendants of cells, and diversity is the driving force of evolution. The different kinds of microorganisms that exist today are the result of a long history of evolution. Microorganisms have continuously changed by altering their genetic properties and adapting to many different habitats.

STUDY OUTLINE

STRUCTURE AND FUNCTION OF THE GENETIC MATERIAL (pp. 190–203)

1. Genetics is the study of what genes are, how they carry information, how their information is expressed, and how they are replicated and passed to subsequent generations or other organisms.
2. DNA in cells exists as a double-stranded helix; the two strands are held together by hydrogen bonds between specific nitrogenous base pairs: A-T and C-G.
3. A gene is a segment of DNA, a sequence of nucleotides, that codes for a functional product, usually a protein.
4. The genetic information in a region of DNA is transcribed to produce RNA; mRNA is translated into proteins.

Genotype and Phenotype (p. 191)

1. Genotype is the genetic composition of an organism—its entire DNA.
2. Phenotype is the expression of the genes—the proteins of the cell and the properties they confer on the organism.

DNA and Chromosomes (pp. 191–192)

1. The DNA in a chromosome exists as one long double helix associated with various proteins that regulate genetic activity.
2. Bacterial DNA is circular; the chromosome of *E. coli*, for example, contains about 4 million base pairs and is approximately 1000 times longer than the cell.
3. The DNA of eucaryotic chromosomes is more condensed and is associated with much protein, including histones.

DNA Replication (pp. 192–197)

1. During DNA replication, the two strands of the double helix separate at the replication fork, and each strand is used as a template by DNA polymerases to synthesize two new strands of DNA according to the rules of nitrogenous basepairing.
2. The result of DNA replication is two new strands of DNA, each having a base sequence complementary to one of the original strands.
3. Because each double-stranded DNA molecule contains one original and one new strand, the replication process is called semiconservative.
4. DNA is synthesized in one chemical direction called $5' \rightarrow 3'$. At the replication fork, one daughter strand is synthesized continuously, the other, discontinuously.
5. DNA synthesis of new fragments on the lagging, or discontinuous, strand moves in the opposite direction from the movement of the replication fork. Completion of the lagging strand requires RNA primers, RNA polymerase, and DNA ligase, as well as DNA polymerase.

THE RATE OF DNA REPLICATION (p. 197)

1. DNA can be synthesized at 1000 nucleotides per second in *E. coli* grown at 37°C.
2. Because multiple replication forks occur in log phase, the generation time for bacteria can be shorter than the time it takes for the bacterial chromosome to replicate.

THE FLOW OF GENETIC INFORMATION (p. 197)

1. Each daughter bacterium receives a chromosome identical to the parent's.
2. Information contained in the DNA is transcribed into RNA and translated into proteins.

RNA and Protein Synthesis (pp. 197–202)

TRANSCRIPTION (pp. 197–199)

1. During transcription, the enzyme RNA polymerase synthesizes a strand of RNA from one strand of double-stranded DNA, which serves as a template.
2. RNA is synthesized from nucleotides containing the bases A, C, G, and U, which pair with the bases of the DNA sense strand.

3. The starting point for transcription, where RNA polymerase binds to DNA, is the promoter site; the region of DNA that is the end point of transcription is the terminator site.
4. RNA is synthesized in the $5' \rightarrow 3'$ direction.
5. In procaryotes, transcription occurs in the cytoplasm; in eucaryotes, transcription occurs in the nucleus.
6. After transcription in eucaryotes, enzymes remove intron RNA, and the spliceosome joins the exon RNA together to form mRNA.

TRANSLATION (pp. 199–202)

1. In procaryotes, translation can begin before transcription is completed.
2. In eucaryotes, translation begins after the mRNA enters the cytoplasm.
3. Translation is the process in which the information in the nucleotide-base sequence of mRNA is used to dictate the amino acid sequence of a protein.
4. The mRNA associates with ribosomes, which consist of rRNA and protein.
5. Three-base segments of mRNA that specify amino acids are called codons.
6. Specific amino acids are attached to molecules of tRNA. Another portion of the tRNA has a base triplet called an anticodon.
7. The base-pairing of codon and anticodon at the ribosome results in specific amino acids being brought to the site of protein synthesis.
8. The ribosome moves along the mRNA strand as amino acids are joined to form a growing polypeptide.
9. A polyribosome is an mRNA with several ribosomes attached to it.

The Genetic Code (pp. 202–203)

1. The genetic code refers to the relationship between the nucleotide-base sequence of DNA, the corresponding codons of mRNA, and the amino acids for which the codons code.
2. The genetic code is degenerate; that is, most amino acids are coded for by more than one codon.
3. Of the 64 codons, 61 are sense codons (which code for amino acids), and 3 are nonsense codons (which do not code for amino acids and are stop signals for translation).
4. The start codon, AUG, codes for methionine.

Regulation of Gene Expression in Bacteria (pp. 203–207)

1. Regulating protein synthesis at the gene level is energy efficient because proteins are only synthesized as they are needed.
2. Constitutive enzymes are always present in a cell.

Repression, Induction, and Attenuation (pp. 203–207)

1. For these gene regulatory mechanisms, the control is aimed at mRNA synthesis.

REPRESSION AND INDUCTION (p. 203)

1. Repression controls the synthesis of one or several (repressible) enzymes.
2. When cells are exposed to a particular end-product, the synthesis of enzymes related to that product decreases.
3. In the presence of certain chemicals (inducers), cells synthesize more enzymes. This process is called induction.
4. An example of induction is the production of β-galactosidase by *E. coli* in the presence of lactose, so lactose can be metabolized.

OPERON MODEL (pp. 203–206)

1. The formation of enzymes is determined by structural genes. In bacteria, a group of coordinately regulated structural genes with related metabolic functions and the promoter and operator sites that control their transcription are called an operon.
2. In the operon model for an inducible system, a regulatory gene codes for repressor protein.
3. When the inducer is absent, the repressor binds to the operator and no mRNA is synthesized.
4. When the inducer is present, it binds to the repressor so that it cannot bind to the operator; thus, mRNA is made and enzyme synthesis is induced.
5. In repressible systems, the repressor requires a corepressor in order to bind to the operator site; thus, the corepressor controls enzyme synthesis.
6. Constitutive genes produce proteins regardless of how much substrate is present. Examples are genes for most of the enzymes of glycolysis.
7. In catabolite repression, transcription of structural genes for catabolic enzymes (such as β-galactosidase) is induced by the absence of glucose and the presence of an alternative substrate (such as lactose).

ATTENUATION (pp. 206–207)

1. Attenuation is a regulatory mechanism that causes premature termination of mRNA synthesis. It was discovered in studies of the operon for tryptophan synthesis.
2. Near the start of the operon for tryptophan synthesis are a number of tryptophan codons.
3. When tryptophan is present at a high level, the coupling of transcription and translation that occurs in bacteria facilitates premature termination of transcription at the attenuator site.
4. When tryptophan is at a low level, the attenuator does not function and the mRNA is completed and translated into enzymes for making more tryptophan.

Mutation: Change in the Genetic Material (pp. 207–214)

1. A mutation is a change in the nitrogenous-base sequence of DNA; that change causes a change in the product coded for by the mutated gene.
2. Many mutations are neutral, many are disadvantageous, and some are beneficial.

Types of Mutations (pp. 207–210)

1. A base substitution occurs when one base pair in DNA is replaced with a different base pair.

2. In a frameshift mutation, one or a few base pairs are deleted or added to DNA.
3. Alterations in DNA can result in missense mutations (which cause amino acid substitutions) or nonsense mutations (which create stop codons).
4. Mutagens are agents in the environment that cause permanent changes in DNA.
5. Some mutations occur without the presence of a mutagen; these are called spontaneous mutations.

CHEMICAL MUTAGENS (pp. 208–210)

1. Chemical mutagens include base-pair mutagens (for example, nitrous acid), base analogs (for example, 2-aminopurine and 5-bromouracil), and frameshift mutagens (for example, benzpyrene).

RADIATION (p. 210)

1. Ionizing radiation causes the formation of ions and free radicals that react with DNA; base substitutions or breakage of the sugar–phosphate backbone results.
2. Ultraviolet radiation is nonionizing; it causes bonding between adjacent thymines.
3. Damage to DNA caused by ultraviolet radiation can be repaired by enzymes that cut out and replace the damaged portion of DNA.

Frequency of Mutation (pp. 210–211)

1. Mutation rate is the probability that a gene will mutate when a cell divides; the rate is expressed as 10 to a negative power.
2. Mutations usually occur randomly along a chromosome.
3. A low rate of spontaneous mutations is beneficial in providing the genetic diversity needed for evolution.

Identifying Mutants (pp. 211–212)

1. Mutants can be detected by selecting or testing for an altered phenotype.
2. Positive selection involves the selection of mutant cells and rejection of nonmutated cells.
3. Replica plating is used for negative selection—to detect, for example, auxotrophs that have nutritional requirements not possessed by the parent (nonmutated) cell.

Identifying Chemical Carcinogens (pp. 212–214)

1. The Ames test is a relatively inexpensive and rapid test for identifying possible chemical carcinogens.
2. The test assumes that a mutant cell can revert to a normal cell in the presence of a mutagen and that many mutagens are carcinogens.
3. Histidine auxotrophs of *Salmonella* are exposed to an enzymatically treated potential carcinogen, and reversions to the nonmutant state are selected.

GENETIC TRANSFER AND RECOMBINATION (pp. 214–222)

1. Genetic recombination, which refers to the rearrangement of genes from separate groups of genes, usually involves DNA from different organisms; it contributes to genetic diversity.
2. In crossing over, genes from two chromosomes are recombined into one chromosome containing some genes from each original chromosome.
3. Mechanisms for genetic transfer in bacteria involve a portion of the cell's DNA being transferred from donor to recipient.
4. When some of the donor's DNA has been integrated into the recipient's DNA, the resultant cell is called a recombinant.

Transformation in Bacteria (pp. 214–216)

1. During this process, genes are transferred from one bacterium to another as "naked" DNA in solution.
2. This process was first demonstrated in *Streptococcus pneumoniae*.
3. It occurs naturally among a few genera of bacteria.

Conjugation in Bacteria (pp. 216–217)

1. This process requires contact between living cells.
2. One type of genetic donor cell is an F^+; recipient cells are F^-. F^+ cells contain plasmids called F factors; these are transferred to the F^- cell during conjugation.
3. When the plasmid becomes incorporated into the chromosome, the cell is called an Hfr (high-frequency recombinant).
4. During conjugation, an Hfr can transfer chromosomal DNA to an F^-. Usually, the Hfr chromosome breaks before it is fully transferred.

Transduction in Bacteria (p. 218)

1. In this process, DNA is passed from one bacterium to another in a bacteriophage and is then incorporated into the recipient's DNA.
2. In generalized transduction, any bacterial genes can be transferred.

Recombination in Eucaryotes (pp. 218–222)

1. In eucaryotes, genetic recombination is associated with sexual reproduction.
2. Sexual reproduction involves the fusion of haploid gametes to form a diploid zygote.
3. Haploid gametes are produced through meiosis, in which portions of chromosomes can be exchanged in the recombination process of crossing over.
4. In eucaryotic microorganisms, sexual reproduction often alternates with asexual reproduction.

PLASMIDS (pp. 220–221)

1. Plasmids are self-replicating circular molecules of DNA carrying genes that are not usually essential for survival of the cell.
2. There are several types of plasmids, including the F factor, dissimilation plasmids, plasmids carrying genes for toxins or bacteriocins, and resistance factors.

TRANSPOSONS (pp. 221–222)

1. Transposons are small segments of DNA that can move from one region of a chromosome to another region of

the same chromosome or to a different chromosome or a plasmid.
2. Transposons are found in the main chromosomes of organisms, in plasmids, and in the genetic material of viruses. They vary from simple (insertion sequences) to complex.
3. Complex transposons can carry any type of gene, including antibiotic-resistance genes, and are thus a natural mechanism for moving genes from one chromosome to another.

Genes and Evolution (p. 222)

1. Diversity is the precondition of evolution.
2. Genetic mutation and recombination provide a diversity of organisms, and the process of natural selection allows the growth of those best adapted for a given environment.

STUDY QUESTIONS

Review

1. Briefly describe the components of DNA, and explain its functional relationship to RNA and protein.
2. Draw a diagram showing a portion of a chromosome undergoing replication.
   (a) Identify the replication fork.
   (b) What is the role of DNA polymerase? Of RNA polymerase?
   (c) What is the role of DNA ligase?
   (d) Why is the lagging strand synthesized discontinuously?
   (e) How does this process represent semiconservative replication?
3. Below is a code for a strand of DNA.

   (3′) A T A T T A C T T T G C T T G G A C T (5′)
   1 2 3 4 5 6 7 8 9 10 11 12 13 14 15 16 17 18 19
   ATAT = Promoter sequence

   (a) Write the code for the complementary strand of DNA.
   (b) Using the genetic code provided in Figure 8.11, identify the sequence of amino acids coded for by mRNA transcribed from this strand of DNA (the complementary strand is the antisense strand).
   (c) What would be the effect if C was substituted for T at base 10?
   (d) What would be the effect if A was substituted for G at base 11?
   (e) What would be the effect if G was substituted for T at base 14?
   (f) What would be the effect if C was inserted between bases 9 and 10?
   (g) How would ultraviolet radiation affect this strand of DNA?
   (h) Identify a nonsense sequence in this strand of DNA.
4. Describe translation, and be sure to include the following terms: ribosome, rRNA, amino acid activation, tRNA, anticodon, and codon.
5. Contrast the structures of a bacterial chromosome and a eucaryotic chromosome.
6. Explain how you would find an antibiotic-resistant mutant by direct selection and how you would find an antibiotic-sensitive mutant by indirect selection.
7. Match the following examples of mutagens.

____ A mutagen that is incorporated into DNA in place of a normal base	(a) Frameshift mutagen
____ A mutagen that causes the formation of highly reactive ions	(b) Base analog
____ A mutagen that alters adenine so that it base-pairs with cytosine	(c) Base-pair mutagen
____ A mutagen that causes insertions	(d) Ionizing radiation
____ A mutagen that causes the formation of pyrimidine dimers	(e) Nonionizing radiation

8. Describe the principle of the Ames test for identifying chemical carcinogens.
9. Differentiate between transformation and transduction.
10. Define *plasmids*, and explain the relationship between F plasmids and conjugation.
11. Use this metabolic pathway to answer the questions below:

    $$\text{Substrate } A \xrightarrow{\text{enzyme } a} \text{Intermediate } B \xrightarrow{\text{enzyme } b} \text{End-product } C$$

    (a) If enzyme *a* is inducible and is not being synthesized at present, a ________ protein must be bound tightly to the ________ site. When the inducer is present, it will bind to the ________ so that ________ can occur.
    (b) If enzyme *a* is repressible, end-product C, called a ________, causes the ________ to bind to the ________. What causes derepression?
    (c) If enzyme *a* is constitutive, what effect, if any, will the presence of *A* or *C* have on it?

12. What is the advantage to a bacterium of having an attenuator on the mRNA for enzymes used to synthesize an amino acid?
13. How does a decrease in glucose result in transcription of new catabolic genes?
14. Define the following terms:
    (a) Genotype
    (b) Phenotype
    (c) Recombination
15. Which sequence is the best target for damage by UV radiation: AGGCAA, CTTTGA, GUAAAU? Why aren't all bacteria killed when they are exposed to sunlight?
16. You are provided with cultures with the following characteristics.

    Culture 1: F^+, genotype A^+ B^+ C^+
    Culture 2: F^-, genotype A^- B^- C^-

    (a) Indicate the possible genotypes of a recombinant cell resulting from the conjugation of cultures 1 and 2.
    (b) Indicate the possible genotypes of a recombinant cell resulting from conjugation of the two cultures after the F^+ has become an Hfr.
17. Why are semiconservative replication and degeneracy of the genetic code advantageous to the survival of species?
18. Why are mutation and recombination important in the process of natural selection and the evolution of organisms?

Challenge

1. Base analogs and ionizing radiation are used in the treatment of cancer. These mutagens can cause cancer, so how do you suppose they are used to treat the disease?
2. Replication of the *E. coli* chromosome takes from 40 to 45 minutes, but the organism has a generation time of 26 minutes. How does the cell have time to make complete chromosomes for each daughter cell? For each granddaughter cell?
3. Chloroquine and erythromycin are used to treat microbial infections. Chloroquine acts by fitting between base pairs in the DNA molecule. Erythromycin binds in front of the A site on the 50S subunit of a ribosome.
    (a) What steps in protein synthesis are inhibited by each drug?
    (b) Which drug is more effective against bacteria? Why?
    (c) Which drug will have effects on the host's cells? Why?
    (d) Use the index to find out the disease for which chloroquine is primarily used. Why is it more effective than erythromycin for treating this disease?
4. *Pseudomonas* has a plasmid containing the *mer* operon, which includes the gene for mercuric reductase. This enzyme catalyzes the reduction of the mercuric ion Hg^{2+} to the uncharged form of mercury, Hg^0. Hg^{2+} is quite toxic to cells; Hg^0 is not.
    (a) What do you suppose is the inducer for this operon?
    (b) The protein encoded by one of the *mer* genes binds Hg^{2+} in the periplasmic gel and brings it into the cell. Why would a cell bring in a toxin?
    (c) What is the value of the *mer* operon to *Pseudomonas?*

FURTHER READING

Cech, T.R. "RNA as an enzyme." *Scientific American* 255(5):64–75, November 1986. Describes how eucaryotic RNA works to remove exons to produce an RNA molecule. Poses the question of whether proteins are the only enzymes.

Devoret, R. "Bacterial tests for potential carcinogens." *Scientific American* 241(2):40–49, August 1979. Describes short tests using bacteria instead of animals to detect carcinogenic chemicals and the effects of carcinogens on DNA.

Howard-Flanders, P. "Inducible repair of DNA." *Scientific American* 245(5):72–80, November 1981. Describes the roles of two enzymes that respond to an "SOS" for DNA damage.

Radman, M., and R. Wagner. "The high fidelity of DNA duplication." *Scientific American* 259(2):40–46, August 1988. A discussion of the roles of DNA polymerase, exonucleases, and repair enzymes in making perfect copies of DNA.

Stahl, F.W. "Genetic recombination." *Scientific American* 256(2):91–101, February 1987. Describes current studies in genetic recombination using a lambda-phage–*E. coli* model.

Steitz, J.A. "'Snurps.'" *Scientific American* 258(6):56–63, June 1988. Describes how introns are removed from eucaryotic RNA by small nuclear ribonucleoproteins.

Stewart, G.J., and C.A. Carlson. "The biology of natural transformation." *Annual Review of Microbiology* 40:211–235, 1986. An overview of information on transformation acquired from laboratory studies and a discussion of the importance of transformation in nature.

Watson, J.D., N.H. Hopkins, J.W. Roberts, J.A. Steitz, and A. M. Weiner. *Molecular Biology of the Gene*, 4th ed. Redwood City, Calif.: Benjamin/Cummings, 1988. A recent, authoritative textbook that discusses topics of microbial genetics in great detail; very clear explanations.

Weinberg, R.A. "Finding the anti-oncogene." *Scientific American* 259(3):44–51, September 1988. Describes the first known growth-suppressing or anticancer gene.

CHAPTER 9

Recombinant DNA and Biotechnology

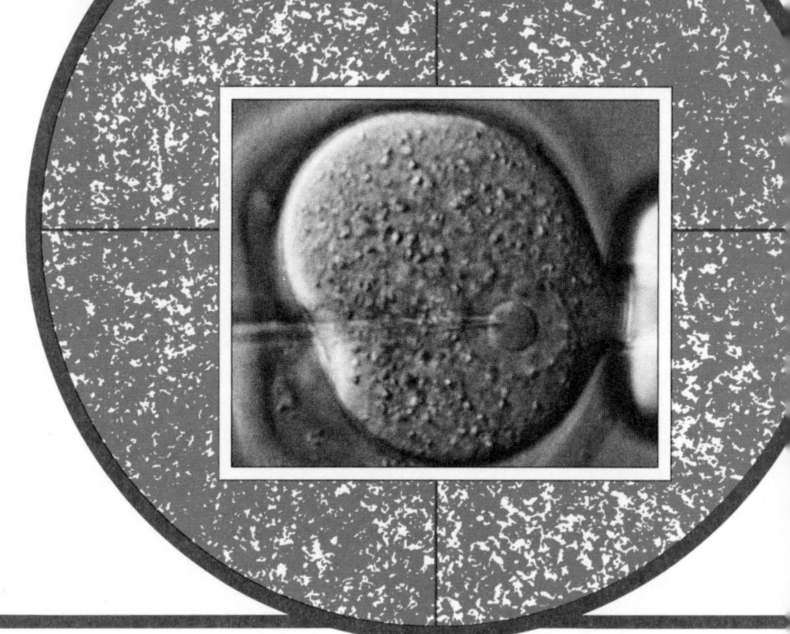

LEARNING OBJECTIVES

- Define recombinant DNA, cDNA, genetic engineering, vector, and protoplast fusion.
- Describe restriction enzymes, and outline how they are used to make recombinant DNA.
- Describe how a gene library is made.
- Outline genetic engineering using transformation.
- Outline genetic engineering using transduction.
- Outline genetic engineering with *Agrobacterium*.
- Define DNA probes, and provide three examples of their use.
- Diagram the Southern blot procedure.
- Outline the polymerase chain reaction, and provide an example of its use.
- List at least five applications of genetic engineering.

During the nineteenth century, scientists showed that microorganisms were responsible for many useful functions, including the long-practiced production of bread, beer, wine, and cheese. As the science of microbiology developed, microbiologists learned how to isolate and grow microorganisms in pure culture. Once they had obtained pure strains, they could select the ones that were more efficient than others in brewing beer, for example—this was a logical extension of the reasoning long used in selecting desirable breeds of animals or strains of plants. Much more recently, scientists working with antibiotic-producing microbes discovered that they could create new strains by exposing the microbes to mutagenic radiation or chemicals. Some of these variant strains produced more, or better, antibiotics. This mutational approach was responsible for many advances in the industrial production of antibiotics following World War II.

During this period, scientists discovered recombination of DNA in microbes. Recombination occurs naturally in microorganisms, generally between the DNAs of closely related strains (see Figure 8.21, p. 214). In bacteria, the DNA of a donor strain may be brought into contact with the DNA of a recipient strain by transformation, conjugation, or transduction. Recombination is more likely to be a beneficial form of genetic change than are mutations, most of which are harmful to the bacterium and are not useful commercially.

The Advent of Recombinant-DNA Technology

During the 1970s and 1980s, the practical application of microorganisms expanded almost beyond imagination

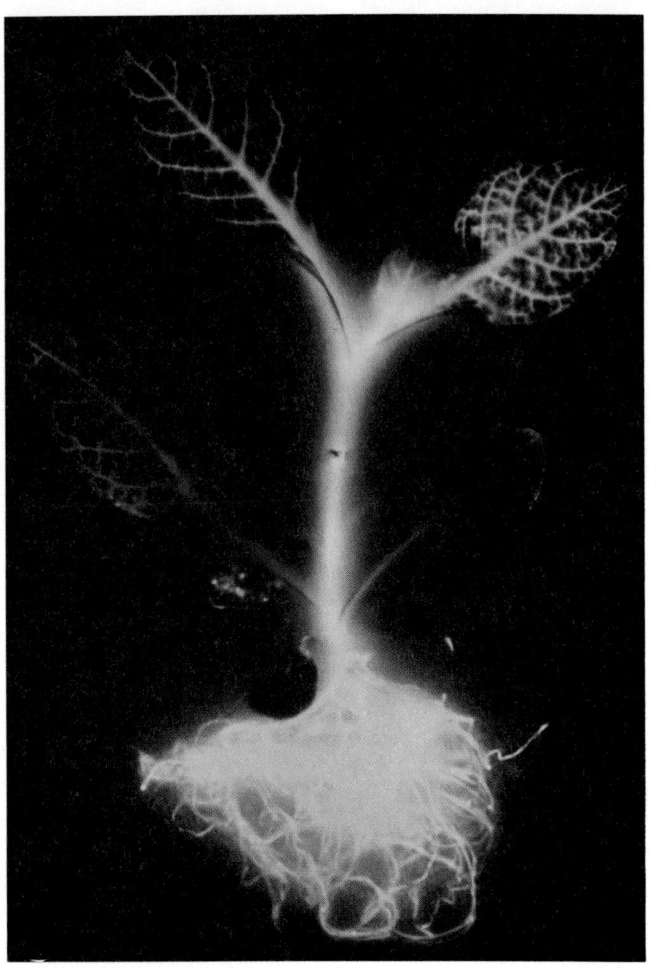

FIGURE 9.1 Tobacco plant expressing a firefly gene.

with the development of new, artificial techniques for making recombinant DNA. Although natural recombination makes it possible for closely related organisms to exchange genes, the new techniques make it possible to transfer genes between completely unrelated species. So very powerful are these techniques that the term **recombinant DNA** is now widely understood to refer to DNA that has been *artificially manipulated* to combine genes from two different sources. A gene from a vertebrate animal, such as a human, can be inserted into the DNA of a bacterium, or a gene from a virus into a yeast. In many cases, the recipient can then be made to express the gene, which may code for a commercially useful product. For example, bacteria with genes for human insulin are now being used to produce insulin for treating diabetes. Also, a vaccine for hepatitis B is being made by yeast carrying a gene for part of the hepatitis virus; the yeast produce a viral protein. Scientists hope that such a protein may prove to be an effective vaccine, eliminating the need to use whole viruses as in conventional vaccines. In a lighter vein, scientists working to refine recombinant-DNA procedures in plants have succeeded in transferring a firefly gene to a tobacco plant and getting it to function (Figure 9.1).

The new recombinant-DNA techniques can also be used to make thousands of copies of the same DNA molecule—to *amplify* DNA, thus generating sufficient DNA for various kinds of experimentation and analysis. This has found practical application in criminology. DNA from a few human cells of blood or semen found at a crime scene can be amplified and then compared with the DNA of suspects. Since each person's DNA is unique (except for identical twins), the procedure has great potential usefulness.

Artificial gene manipulation is popularly known as **genetic engineering.** In fact, the term **biotechnology,** which had been generally defined to include all industrial applications of microorganisms, has increasingly become identified in the public mind as the industrial use of genetically engineered microorganisms.

Genetic engineering was made possible by the discovery of bacterial enzymes that can be used to cut DNA from different sources into pieces that are easy to recombine in vitro (*in vitro* means "in glass," that is, in a test tube rather than inside a living organism). Also essential was the development of methods for inserting this recombinant DNA into a cell. If a mosquito carrying the virus for yellow fever bites and infects a human, the mosquito is known as a "disease vector" because it can transmit the virus from one host to another. The term **vector** has been adopted to describe a self-replicating DNA molecule that is used as a carrier to transmit a gene from one organism into another. It is also known as a **cloning vector.**

Overview of Recombinant-DNA Procedures

Figure 9.2 gives an overview of some of the procedures typically used in genetic engineering and some promising applications. The gene of interest is first inserted into the vector DNA in vitro. In our example, the vector is a plasmid. Next, this recombinant vector DNA is put into a cell where it can multiply. The DNA molecule chosen as a vector must be a self-replicating type, such as a plasmid or a viral genome. The cell containing the recombinant vector is then grown in culture to form a **clone** of many genetically identical cells (see Chapter 6, p. 153), each carrying a copy of the vector. This cell clone therefore contains many copies of the foreign gene. This is why DNA vectors are often called gene-cloning vectors, or simply cloning vectors. The final step can go in two alternative directions, as the figure shows. From the cell clone, the genetic engineer

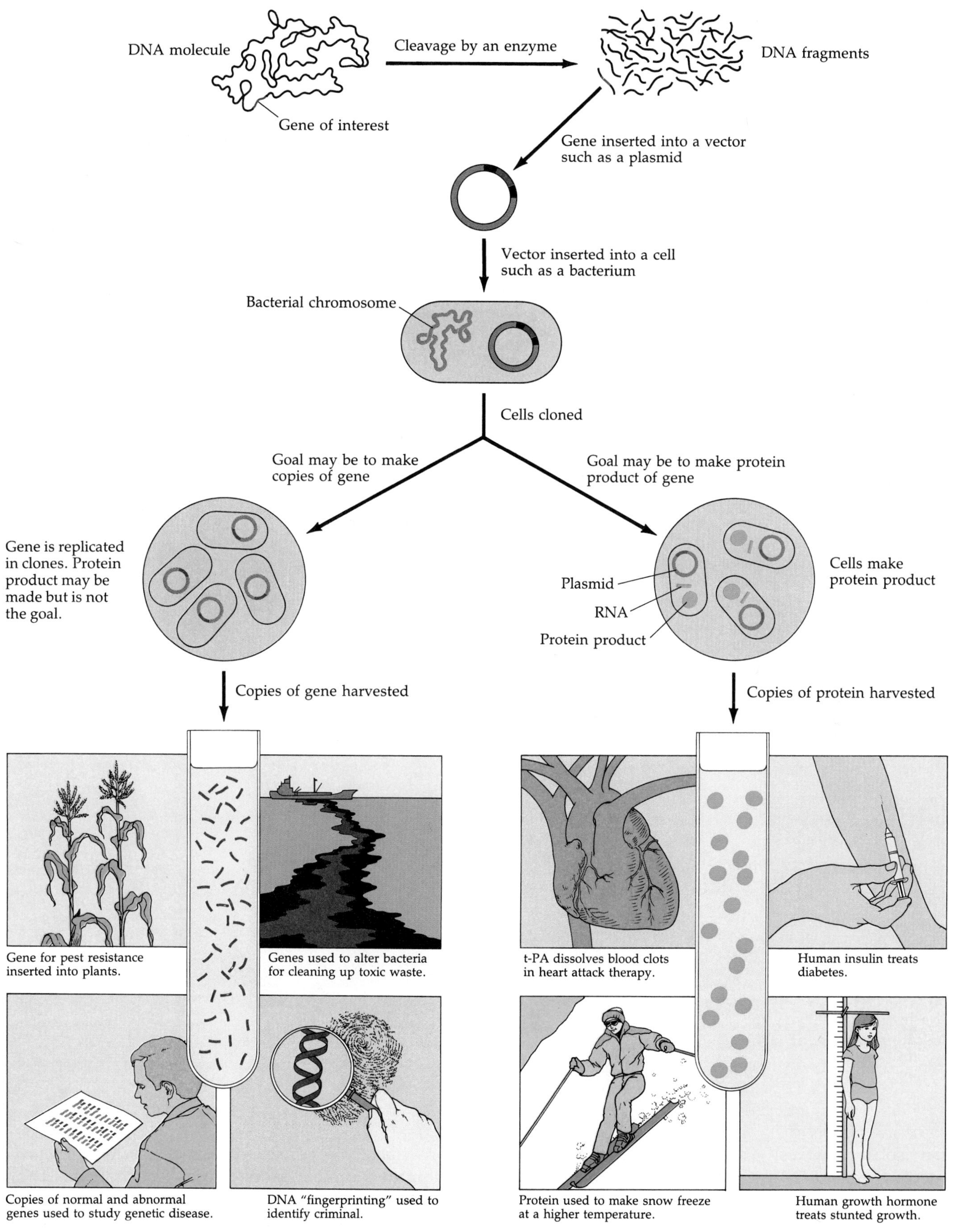
DNA molecule
Gene of interest
Cleavage by an enzyme
DNA fragments
Gene inserted into a vector
such as a plasmid
Vector inserted into a cell
such as a bacterium
Bacterial chromosome
Cells cloned
Goal may be to make
copies of gene
Goal may be to make protein
product of gene
Gene is replicated
in clones. Protein
product may be
made but is not
the goal.
Plasmid
RNA
Protein product
Cells make
protein product
Copies of gene harvested
Copies of protein harvested
Gene for pest resistance
inserted into plants.
Genes used to alter bacteria
for cleaning up toxic waste.
Copies of normal and abnormal
genes used to study genetic disease.
DNA "fingerprinting" used to
identify criminal.
t-PA dissolves blood clots
in heart attack therapy.
Human insulin treats
diabetes.
Protein used to make snow freeze
at a higher temperature.
Human growth hormone
treats stunted growth.

FIGURE 9.2 Typical genetic engineering procedures with examples of possible applications.

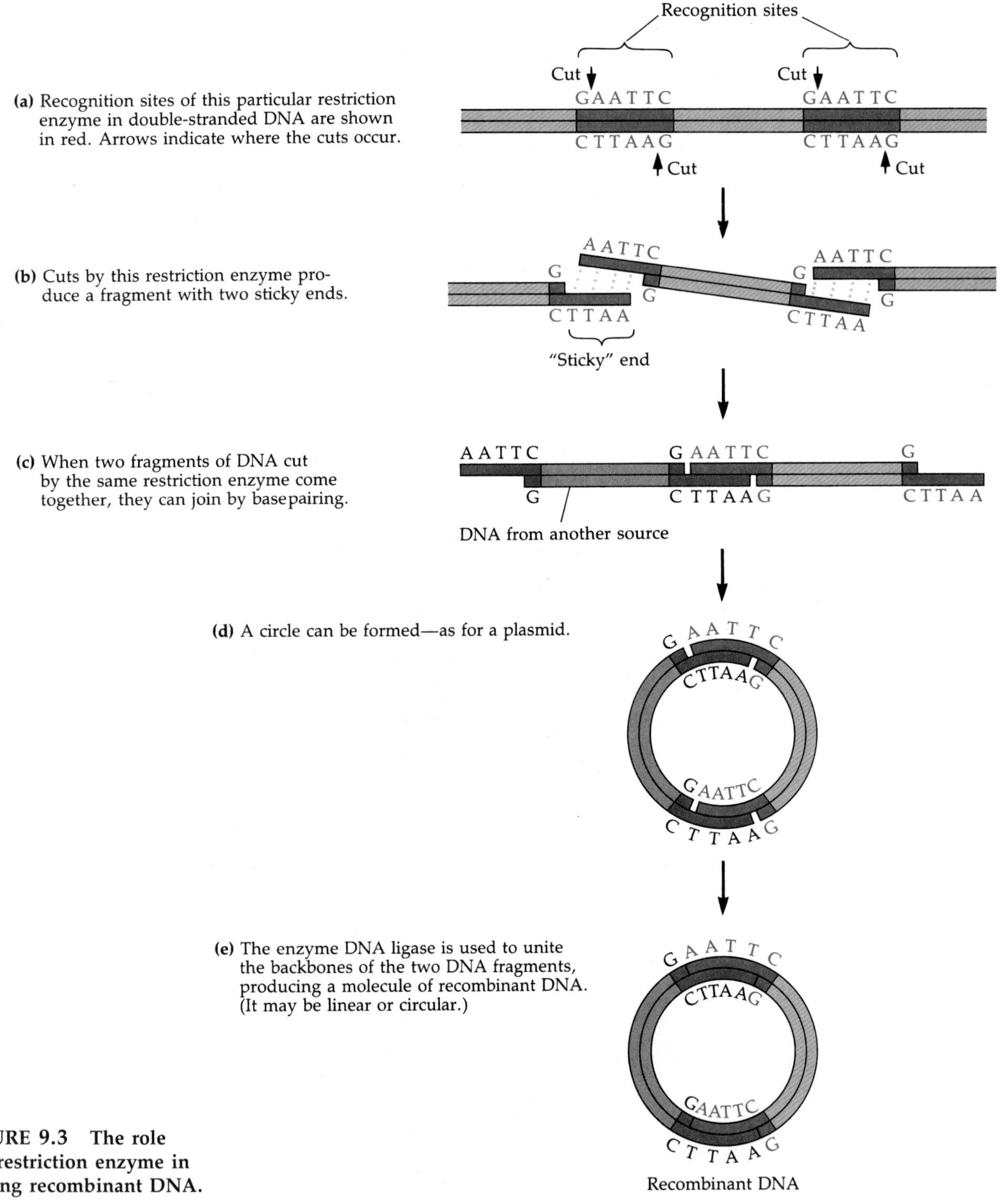

Recognition sites
Cut
GAATTC
CTTAAG
Cut
(a) Recognition sites of this particular restriction enzyme in double-stranded DNA are shown in red. Arrows indicate where the cuts occur.
AATTC
G
CTTAA
"Sticky" end
(b) Cuts by this restriction enzyme produce a fragment with two sticky ends.
(c) When two fragments of DNA cut by the same restriction enzyme come together, they can join by basepairing.
DNA from another source
(d) A circle can be formed—as for a plasmid.
(e) The enzyme DNA ligase is used to unite the backbones of the two DNA fragments, producing a molecule of recombinant DNA. (It may be linear or circular.)
Recombinant DNA

FIGURE 9.3 The role of a restriction enzyme in making recombinant DNA.

may isolate ("harvest") large quantities of the gene of interest, which may then be used for a variety of purposes. The gene may even be inserted into another vector, for introduction into another kind of cell (such as a plant or animal cell).

Alternatively, if the gene of interest is expressed (transcribed and translated) in the cell clone, its protein product can be harvested and used for a variety of possible purposes. The advantages of genetic engineering for obtaining such proteins is illustrated by one of its early successes, somatostatin. This human hormone, which is involved in limiting growth, is now

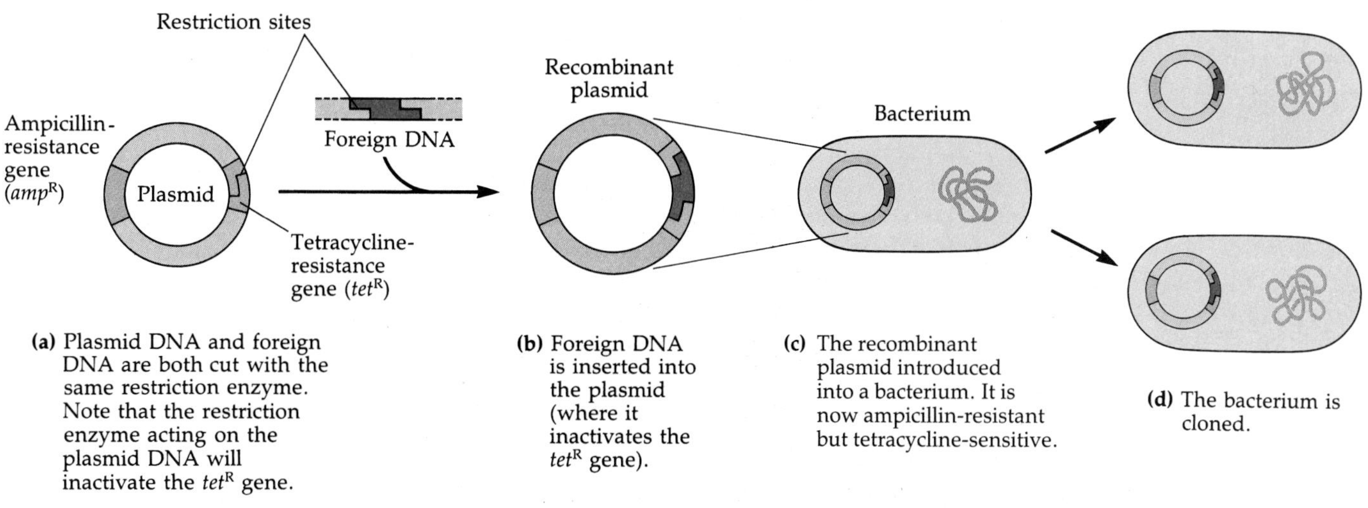

Restriction sites
Ampicillin-resistance gene (amp^R)
Plasmid
Foreign DNA
Tetracycline-resistance gene (tet^R)
Recombinant plasmid
Bacterium
(a) Plasmid DNA and foreign DNA are both cut with the same restriction enzyme. Note that the restriction enzyme acting on the plasmid DNA will inactivate the tet^R gene.
(b) Foreign DNA is inserted into the plasmid (where it inactivates the tet^R gene).
(c) The recombinant plasmid introduced into a bacterium. It is now ampicillin-resistant but tetracycline-sensitive.
(d) The bacterium is cloned.

FIGURE 9.4 Insertion of foreign gene into a bacterium by using a plasmid as the vector.

being produced in *Escherichia coli* for research purposes at the rate of 10,000 somatostatin molecules per cell. At one time 500,000 sheep brains were needed to produce 5 mg of animal somatostatin for experimental purposes. By contrast, only 8 liters of a genetically engineered bacterial culture are now required to obtain the equivalent amount of the human hormone.

Restriction Enzymes

Genetic engineering had its technical roots in the discovery of a special class of DNA-cutting enzymes that exist in many bacteria. These enzymes, called restriction enzymes, were first isolated in 1970. However, their role in nature had actually been observed earlier, when it was found that certain bacteriophages had a restricted host range. If these phages were used to infect bacteria other than their usual hosts, "restriction" enzymes in the new host destroyed almost all the phage DNA. The purified forms of these bacterial enzymes are used by today's genetic engineers.

What is important for genetic engineering is that a **restriction enzyme** is an enzyme that recognizes and cuts only one particular sequence of nucleotide bases in DNA and that it cuts this sequence in the same way each time. Many restriction enzymes make *staggered* cuts in the two strands of a DNA molecule, i.e., cuts that are not directly opposite each other (Figure 9.3). Note in Figure 9.3a that the red base sequences on the two strands are the same, but running in opposite directions. Staggered cuts leave stretches of single-stranded DNA at the ends of the DNA fragments (Figure 9.3b). These are called *sticky ends*, because they can "stick" to complementary stretches of single-stranded DNA by base-pairing. There are hundreds of restriction enzymes known, each producing DNA fragments with characteristic ends.

As shown in Figure 9.3c and d, if two fragments of DNA from different sources have been produced by the same restriction enzyme, the two pieces will have identical sets of sticky ends and can be spliced (recombined) in vitro. The result is a linear or circular recombinant molecule. The sticky ends first join spontaneously by hydrogen bonding (base-pairing). Then the enzyme DNA ligase is used to covalently link the backbones of the DNA pieces (Figure 9.3e).

Cloning a Gene with a Plasmid Vector

Figure 9.4 diagrams the cloning of a DNA fragment by using a plasmid vector. Once the desired DNA fragment has been spliced into the plasmid vector, the entire plasmid can be introduced into a host microbe. In nature, plasmids are usually transferred between closely related microbes by cell-to-cell contact, such as conjugation. In genetic engineering, the plasmid must be inserted by transformation. If the host cell is one like *E. coli* or a yeast that does not naturally take up DNA by transformation, it can be made to do so by chemical treatment. Plasmids can be versatile; some can survive in both bacteria and yeasts and can be used for introducing recombinant DNA into either host. If the vector is bacteriophage DNA rather than a plasmid, the recombinant vector can be inserted into host cells either by transformation or, once enclosed in a phage coat, by transduction.

In addition to showing the general procedure for gene cloning using a plasmid, Figure 9.4 illustrates a common trick for identifying bacterial clones that have acquired recombinant plasmids. The plasmid used as vector carries two antibiotic-resistance genes, one conferring resistance to ampicillin (amp^R) and the other to tetracycline (*tet*^{R}). Bacteria carrying the nonrecombinant version of the plasmid will be resistant to both antibiotics. However, bacteria carrying recombinant plasmids will be resistant only to ampicillin, because the plasmid's restriction site is in the middle of the tetracycline-resistance gene. When the foreign DNA inserts there, it inactivates the *tet*^{R} gene. So cells carrying recombinant plasmids will grow on media containing ampicillin but not on media containing tetracycline.

Alternative Ways To Get Foreign DNA into Cells

Genetic engineers have alternatives to vectors for transferring genes between two cells that would not ordinarily recombine their genes. One is **protoplast fusion,** the joining of two cells whose cell walls have been removed. As outlined in Figure 9.5a for two bacteria, the cell walls are enzymatically removed and the resulting protoplast is kept in a solution with an osmotic pressure high enough to keep the cell intact. At a low but significant frequency, two such protoplasts will fuse; the addition of polyethylene glycol raises the frequency of fusion. In the new hybrid cell, the DNA derived from the two "parent" cells may undergo natural recombination. This method is especially valuable in genetic manipulation of plant cells (Figure 9.5b).

Removal of the cell wall to form protoplasts can also be used in another technique in which DNA is introduced into bacteria, fungi, or plant cells without using vectors. In this technique, an electric current is used to form microscopic pores in the cell membranes of the protoplasts. The DNA, which may be the product of gene cloning, enters the cells through the pores. The method is also directly applicable to animal cells, which lack cell walls.

A remarkable way of introducing foreign DNA into plant cells directly through the thick cellulose wall is literally to shoot it in with a "gun." Microscopic particles of tungsten are coated with DNA and fired through the plant cell wall. Some of the cells express the violently introduced DNA as if it were their own genes. As shown in Figure 9.6, DNA can also be injected directly into an animal cell by using a glass micropipette that has been drawn out until its diame-

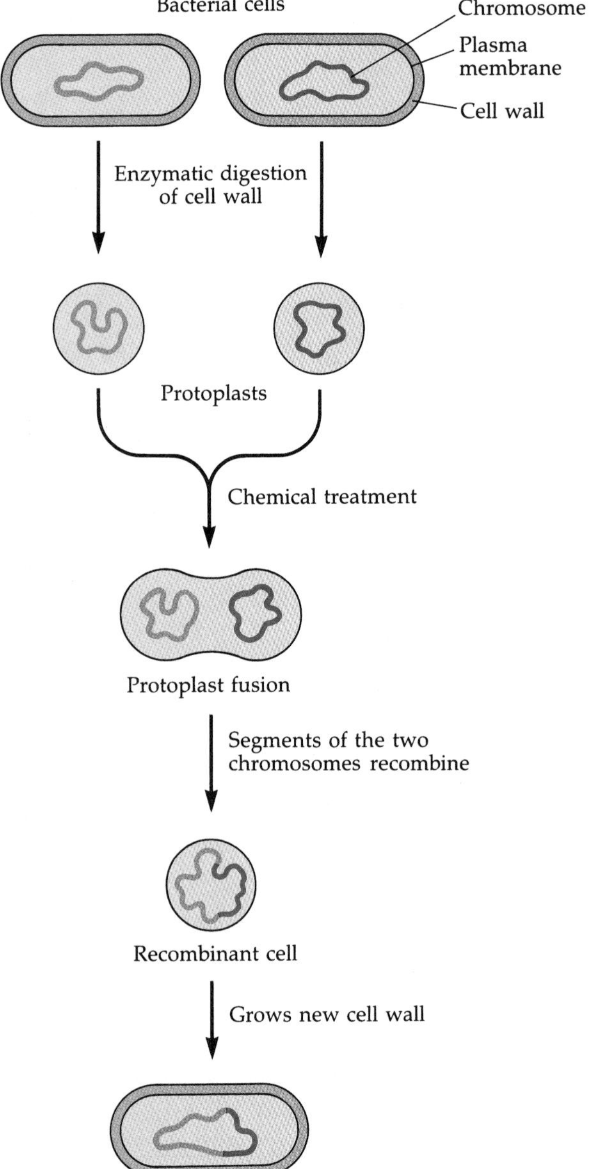

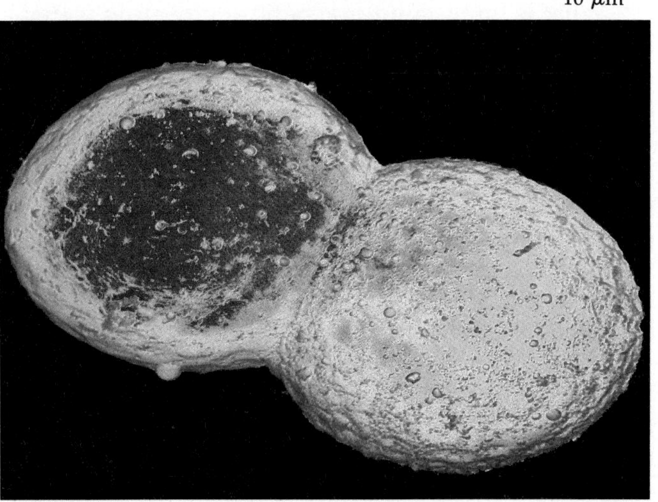

FIGURE 9.5 Protoplast fusion. **(a)** Diagram of protoplast fusion with bacterial cells. **(b)** Two protoplast leaf cells from a tobacco plant are shown undergoing fusion. The cellulose wall was removed, leaving only the delicate plasma membrane to bind the cell contents together, allowing exchange of DNA.

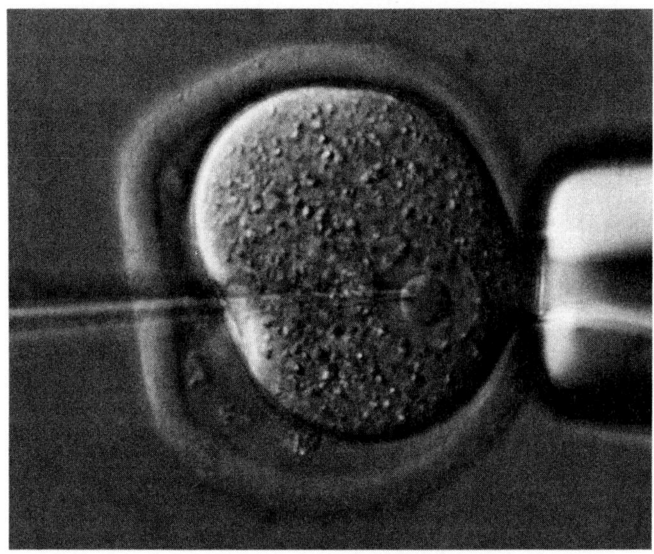

FIGURE 9.6 Injection of foreign DNA into a fertilized mouse egg. The egg is first immobilized by applying mild suction to the large, blunt holding pipette (right). Several hundred copies of the gene of interest are then injected through the sharp end of the micropipette (left).

ter is bacterial in scale. Whatever the means by which foreign DNA is introduced into a cell, in order to survive it must either be present on a self-replicating vector or be inserted into one of the cell's chromosomes by natural recombination.

Sources of DNA

Let's backtrack now to ask where the genetic engineer gets the genes he or she is interested in. There are three main kinds of sources: gene libraries containing natural copies of genes, gene libraries containing "cDNA" copies of genes made from mRNA, and artificial synthesis of DNA.

GENE LIBRARIES

Researchers interested in genes from a particular organism usually start by cutting up the entire genome with restriction enzymes, splicing as many as possible of the genome fragments into copies of a plasmid or phage vector and then introducing the recombinant vectors into bacterial cells. The goal is to make a collection of clones large enough to ensure that there is at least one clone for every gene in the organism. (This is a necessity because isolating genes directly from the initial enzymatic digestion of the DNA is usually not practical.) This collection of clones containing different DNA fragments is called a **gene library;** each "book" is a bacterial or phage strain that contains a fragment of the genome (Figure 9.7). Such libraries are essential for maintaining and retrieving DNA clones; they can even be purchased commercially.

The other form of gene library is one in which the clones contain **complementary DNA (cDNA),** which is DNA synthesized by using mRNA as a template. Recall that the genes of eucaryotic cells generally contain both exons (stretches of DNA that code for protein) and introns (intervening stretches of DNA that do not code for protein). When the RNA transcript of such a gene is converted to mRNA, the introns are removed.

In cloning genes of eucaryotic cells, it is desirable to use a version of the gene that lacks introns, the extraneous genetic material. A gene that includes introns may be too large to work with easily. In addition, if such a gene is put into a bacterial cell, the bacterium will not have the machinery to remove the introns from the RNA transcript and therefore will not be able to make the correct protein product. However, an artificial gene that contains only exons can be produced by using an enzyme called *reverse transcriptase* to synthesize cDNA from an mRNA template (Figure 9.8). After the cDNA is produced, the mRNA can be removed. This synthesis is the reverse of the normal DNA-to-RNA transcription process. The cDNA method is most commonly applied in cases when a specific type of eu-

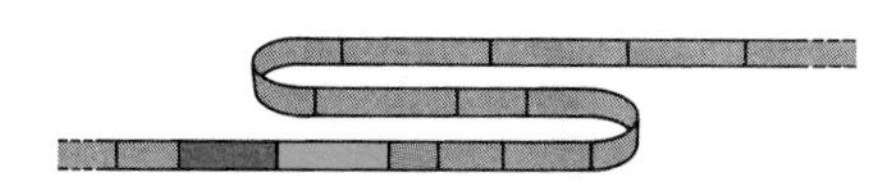

(a) Genome is to be stored in library (short section shown)

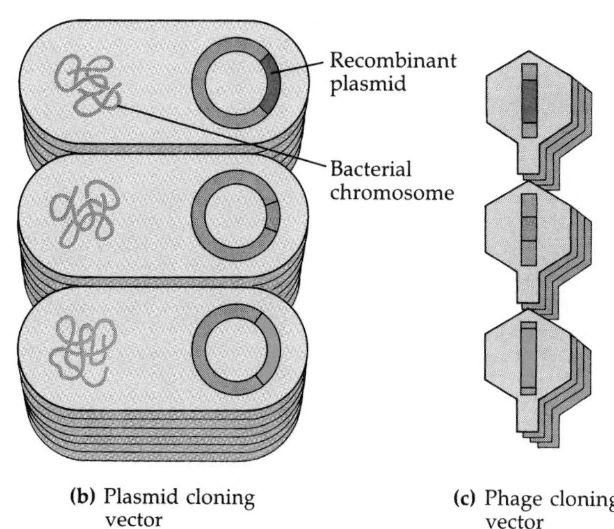

(b) Plasmid cloning vector

(c) Phage cloning vector

FIGURE 9.7 Gene libraries. A gene library is a collection of a large number of DNA fragments from a genome **(a).** Each fragment, containing about one gene, is carried by a vector, either a plasmid carried within a bacterial cell **(b)** or a phage **(c).**

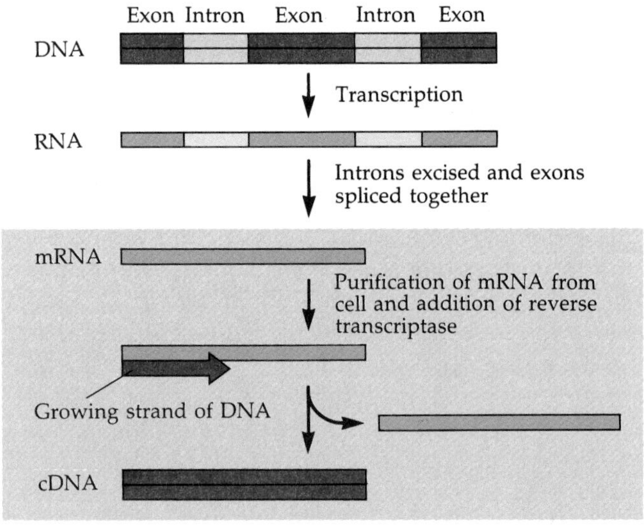

FIGURE 9.8 Making complementary DNA (cDNA) for a eucaryotic gene.

caryotic cell contains large amounts of a particular mRNA, such as mRNA for hemoglobin in immature red blood cells. A difficulty with the method is that long molecules of mRNA may not get completely reverse-transcribed into DNA: the reverse transcription often aborts, forming only parts of the desired gene.

SYNTHETIC DNA

Under certain circumstances, genes can be made in vitro with the help of DNA-synthesizing machines (Figure 9.9). A keyboard on the machine is used to enter the sequence of nucleotides desired, much as letters are entered into a word processor to compose a sentence. A microprocessor controls the synthesis of the DNA from stored supplies of nucleotides and the other necessary reagents. A chain of about 40 nucleotides can be synthesized by this method. To make an entire gene, ten or more such chains must be linked together.

Identifying the Desired Clone in a Library

A gene library is of little use without some convenient method for locating a clone containing the gene of interest. One method for finding the desired clone is shown in Figure 9.10. Short segments of single-stranded DNA, consisting of a sequence of nucleotides unique to the gene we are seeking, are synthesized. These small DNA molecules, called **probes**, are radioactively labeled so that they can be located later. Mean-

FIGURE 9.9 DNA synthesis machine. Short sequences of DNA can be synthesized by instruments such as this one. The necessary chemical components are placed in containers in the machine, and the desired DNA base sequences are typed on the keyboard.

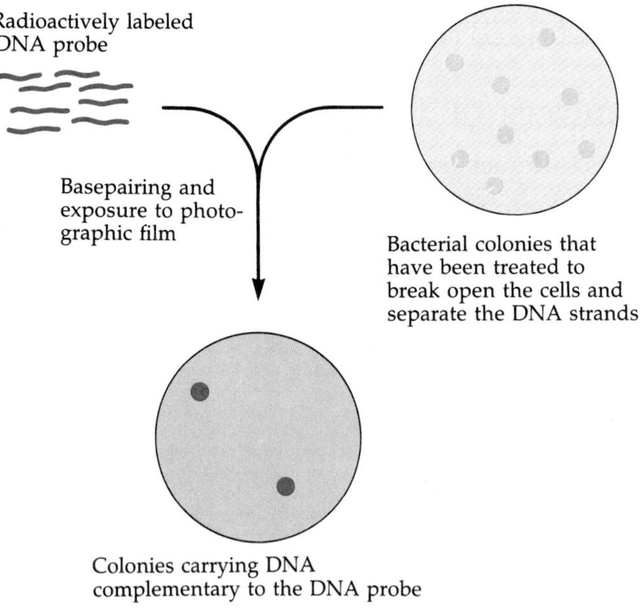

FIGURE 9.10 Use of a DNA probe to identify a cloned gene of interest. This technique makes use of the fact that nucleic acids of complementary sequence will basepair (hybridize). Here the probe is DNA tagged with a radioactive isotope. It is used to find the particular bacterial colonies that carry recombinant plasmids containing the gene of interest.

while, bacteria from the library that carry different fragments of foreign DNA are grown into colonies on a plate of nutrient medium. After incubation, a replica plate is made (see Figure 8.19). The plate is treated to gently break open the bacterial cells and separate their DNA into single strands. Then the radioactively labeled probe is added to the plate. Any bacterial colony containing DNA that base-pairs with the probe can be identified by exposure to photographic film. Living cells from the same clone can then be obtained from the untreated replica plate.

An analogous approach based on protein, rather than DNA, recognition can be used to identify colonies that have made the protein product of the foreign gene. Here the "probe" consists of radioactively labeled antibodies that specifically attach to molecules of the protein product (antibodies are discussed in Chapter 17). Colonies producing the desired protein can be identified by the radioactive tag.

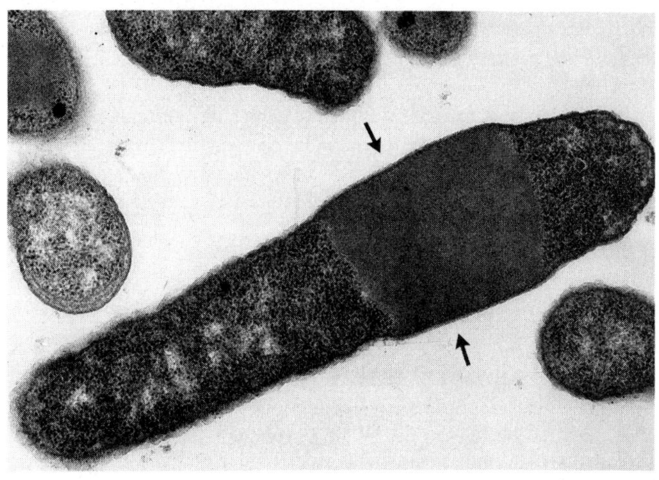

FIGURE 9.11 ***E. coli*** **genetically engineered to produce human insulin.** The inclusion body of human insulin is indicated by the arrows in this transmission electron micrograph.

Making a Gene Product

We have just seen that one way of identifying cells carrying a particular gene is by assaying for the gene product. Such products are themselves, of course, a frequent objective of genetic engineering. Most of the earliest work in genetic engineering made use of *E. coli* to synthesize the gene products (Figure 9.11). *E. coli* has the advantage that researchers are very familiar with this easily grown organism and with its genetics. It also has several disadvantages. Like other gram-negative bacteria, it produces endotoxins as part of its outer layer. Since endotoxins cause fever and shock in animals, their accidental presence in products intended for use in humans would be a serious problem. Another disadvantage of *E. coli* is that it does not usually secrete protein products. To obtain a product, cells must usually be broken open and the product purified from the resulting "soup" of cell components.

Recovering the product from such a mixture is expensive when done on an industrial scale. It is more economical to have an organism secrete the product so that it can be recovered continuously from the growth medium. One approach has been to link the product to a natural *E. coli* protein that the bacterium does secrete. As we will discuss later, this approach has been used to produce insulin. Certain gram-positive bacteria, such as *Bacillus subtilis,* are more likely to secrete their products and are often preferred industrially for that reason.

Another microbe that shows promise as a vehicle for the expression of genetically engineered genes is baker's yeast, *Saccharomyces cerevisiae.* Its genome is only about four times larger than that of *E. coli* and is probably the best understood eucaryotic genome. Yeasts may carry plasmids, and their cell walls can readily be removed to introduce plasmids carrying engineered genes. As eucaryotic cells, yeasts may be more successful in expressing foreign eucaryotic genes than bacteria. Furthermore, yeasts are likely to continuously secrete the product. Because of all these factors, yeasts have become the workhorse of eucaryotic cells. Yeasts also have a psychological advantage in the marketplace. Bacteria and viruses are, unfairly, associated in the public's mind with diseases, whereas yeasts have a much more benign image, thanks to their association with baking, brewing, and wine-making.

Animal viruses have also been used in making engineered gene products, primarily in the field of vaccine production. For example, scientists have been able to insert genes for the surface proteins of pathogenic microbes into the generally harmless vaccinia virus. The result is a sort of "sheep in wolf's clothing," a virus that has the external proteins of a pathogen but does not cause disease. When an animal host is infected with the engineered virus, the host's immune system recognizes these proteins as foreign and, in response, develops an immunity that can protect it against the actual pathogen. Because the vaccinia virus is unusually large and has room for several extra genes, a genetically engineered vaccinia virus might theoretically be used as a vaccine for several diseases simultaneously.

Mammalian cells in culture, even human cells, can be used much like bacteria to produce genetically engineered products. Scientists have developed effective methods for growing mammalian cells in culture as hosts for growing viruses (see Chapter 13). In genetic engineering, mammalian cells are often the best suited to make protein products for medical use; these products include hormones, lymphokines (which regulate cells of the immune system), and interferon (a natural antiviral substance that is also used to treat some cancers). Using mammalian cells to make foreign gene products on an industrial scale often requires a preliminary step of cloning the gene in bacteria. Consider the example of colony-stimulating factor (CSF). CSF, a protein produced naturally in tiny amounts by white blood cells, is valuable because it stimulates the growth of certain cells that protect against infection. To produce huge amounts of CSF industrially, the gene is first inserted into a plasmid, and bacteria are used to make multiple copies of the plasmid (as in Figure 9.4). The recombinant plasmids are then inserted into mammalian cells that are grown in huge tanks.

Plant cells can also be grown in culture, altered by recombinant-DNA techniques, and then used to generate genetically engineered plants. Such plants may prove useful as sources of valuable plant products, such as alkaloids (the painkiller codeine, for example) and the isoprenoids that are the basis of synthetic rubber. We will return to the topic of genetic engineering of plants later in this chapter.

Genetically Engineered Products for Medical Therapy

An extremely valuable pharmaceutical product is the hormone insulin, which is a small protein. In the mammalian body, this protein is produced by the pancreas and controls the body's uptake of glucose from blood. For many years insulin-deficient diabetics have controlled their disease by the injection of insulin obtained from the pancreases of slaughtered animals. Obtaining this insulin is an expensive process, and the insulin from animals is not as effective as human insulin.

Because of the value of human insulin and the small size of the protein, the production of human insulin by recombinant-DNA techniques was an early goal for the pharmaceutical industry. To produce the hormone, synthetic genes were first constructed for each of the two short polypeptide chains that make up the insulin molecule. The small size of these chains, only 21 and 30 amino acids long, made it possible to use synthetic genes. Following the strategy discussed earlier, each of the two synthetic genes was inserted into a plasmid vector and linked to the end of a gene coding for the bacterial enzyme β-galactosidase, so that the insulin polypeptide was coproduced with the enzyme and was secreted with it. Two different *E. coli* bacterial cultures were used, one to produce each of the insulin polypeptide chains. The polypeptides were then recovered from the bacteria, separated from the β-galactosidase, and chemically joined to make human insulin. This accomplishment was one of the early commercial successes of genetic engineering and illustrates a number of the principles and procedures we have discussed in this chapter.

Another human protein hormone that is now being produced commercially by genetic engineering is somatotropin, which is also known as human growth hormone. Some individuals do not produce adequate amounts of somatotropin and are stunted in growth. In the past, growth hormone to correct this deficiency had to be obtained from human pituitary glands at autopsy (growth hormone from other animals is not effective in humans). This practice was not only expensive but also dangerous as on several occasions neurological diseases were transmitted along with the hormone. Human growth hormone produced by genetically engineered *E. coli* is therefore an important product of modern biotechnology.

Earlier we mentioned a recombinant-DNA approach to vaccines by using vaccinia virus engineered to carry genes for surface proteins of a pathogen. A related approach already in use is to harvest such proteins from engineered microbial cells and then to use the purified protein as a vaccine. The term for a vaccine consisting only of a limited protein portion of a pathogen is *subunit vaccine.* Unlike conventional vaccines, which are killed or weakened whole microorganisms, subunit vaccines do not contain extraneous material that could cause undesirable side effects. Table 9.1 lists other important genetically engineered products for medical therapy.

Obtaining Information from DNA for Basic Research and Medical Applications

Recombinant-DNA technology can be used to make products, but this is not its only important application. Because of its ability to make multiple copies of DNA, it can serve as a sort of DNA printing press. Once a large amount of a particular piece of DNA is available, various analytic techniques, which we will discuss in this section, can be used to "read" the information contained in the DNA. To date, the most important achievements made possible by recombinant-DNA technology have been in basic molecular biology. Only 25 years ago, research into the detailed structures and functions of genes was extremely difficult. Now, even the complex genes of humans and other mammals are open to study.

As with all biological research, we can expect new discoveries in molecular genetics to lead to practical applications of value. Already, recently developed techniques for DNA analysis are being used for *genetic*

TABLE 9.1 Some Products of Genetic Engineering

PRODUCT	COMMENTS
Pharmaceutical Products	
Tissue plasminogen activator (t-PA)	Dissolves the fibrin of blood clots; therapy for heart attacks; produced by mammalian cell culture
Erythropoietin (EPO)	Treatment of anemia; produced by mammalian cell culture
Human insulin	Therapy for diabetics; better tolerated than insulin extracted from animals; produced by *E. coli*
Interleukin-2 (IL-2)	Possible treatment for cancer; stimulates the immune system; produced by *E. coli*
Alpha-interferon and gamma-interferon	Possible cancer and virus-disease therapy; produced by *E. coli*, *S. cerevisiae*
Tumor necrosis factor (TNF)	Causes disintegration of tumor cells; produced by *E. coli*
Human growth hormone (somatotropin)	Corrects growth deficiencies in children; produced by *E. coli*
Epidermal growth factor (EGF)	Heals wounds, burns, ulcers; produced by *E. coli*
Prourokinase	Anticoagulant; therapy for heart attacks; produced by *E. coli*
Factor VIII	Treatment of hemophilia; improves clotting; produced by mammalian cell culture
Colony-stimulating factor (CSF)	Counteracts effects of chemotherapy; improves resistance to infectious disease such as AIDS; treatment of leukemia; produced by *E. coli*, *S. cerevisiae*
Superoxide dismutase	Minimizes damage caused by oxygen free radicals during reperfusion of oxygen-deprived tissues; produced by *S. cerevisiae*
Monoclonal antibodies	Possible therapy for cancer and transplant rejection; used in diagnostic tests; produced by mammalian cell culture (from fusion of cancer cell and antibody-producing cell)
Hepatitis B vaccine	Produced by *S. cerevisiae* that carries hepatitis-virus gene on a plasmid (the hepatitis virus is difficult to grow)
Agricultural Products	
Pseudomonas syringae, ice-minus bacterium	Lacks normal protein product that initiates undesirable ice formation on plants
Pseudomonas fluorescens bacterium	Has toxin-producing gene from insect pathogen *Bacillus thuringiensis;* toxin kills root-eating insects that ingest bacteria
Rhizobium meliloti bacterium	Modified for enhanced nitrogen fixation
Animal Husbandry Products	
Porcine growth hormone (PGH)	Improves weight gain in swine; produced by *E. coli*
Bovine growth hormone (BGH)	Improves weight gain and milk production in cattle; produced by *E. coli*
Food Production Products	
Rennin	Causes formation of milk curds for dairy products; produced by *Aspergillus niger*
Cellulase	Enzymes that degrade cellulose to make animal feedstocks; produced by *E. coli*
Recreation Products	
Snomax®	A protein that causes water to freeze at higher than normal temperatures; used for making snow at ski resorts; produced by *Pseudomonas syringae*

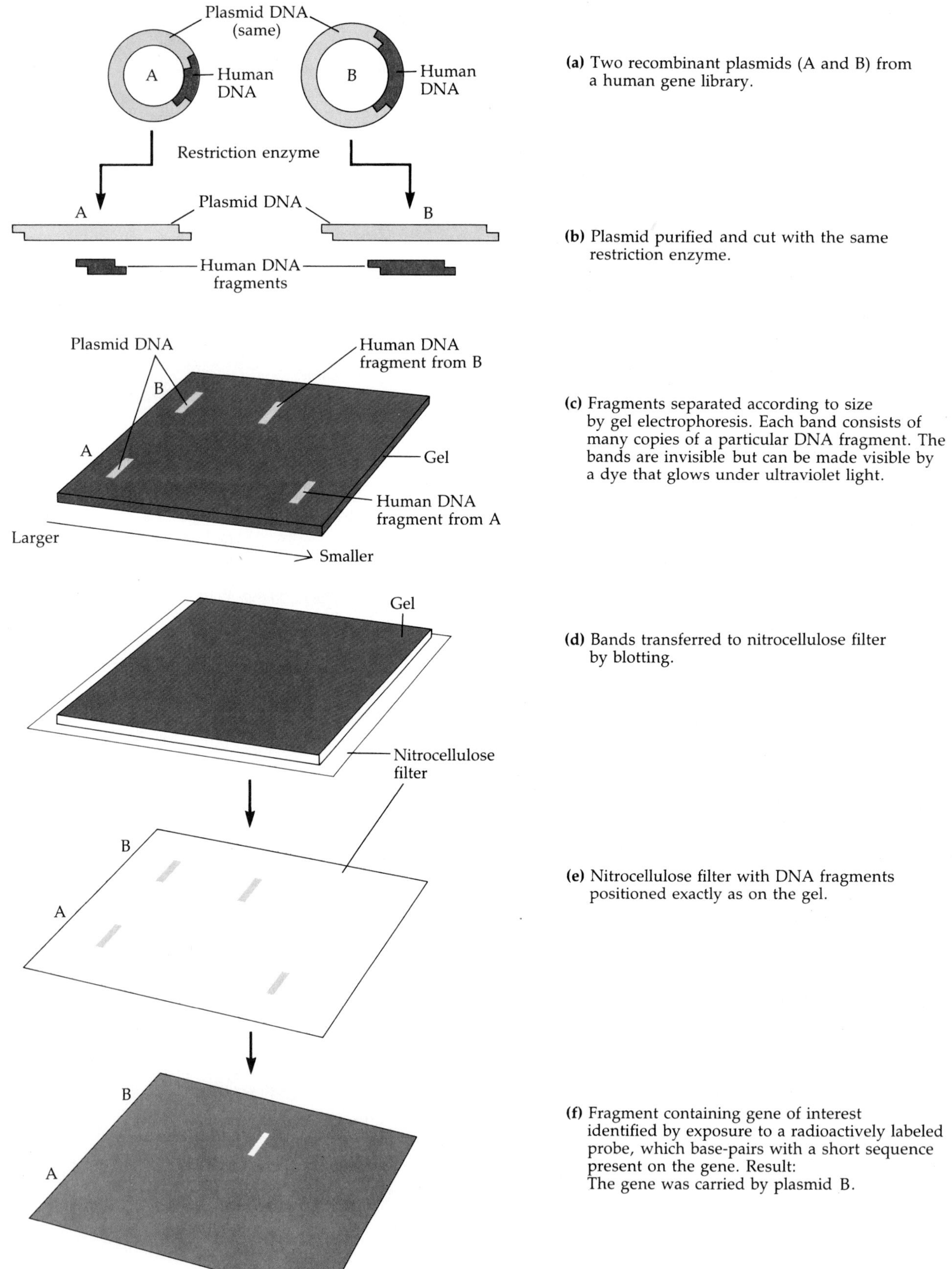

(a) Two recombinant plasmids (A and B) from a human gene library.

(b) Plasmid purified and cut with the same restriction enzyme.

(c) Fragments separated according to size by gel electrophoresis. Each band consists of many copies of a particular DNA fragment. The bands are invisible but can be made visible by a dye that glows under ultraviolet light.

(d) Bands transferred to nitrocellulose filter by blotting.

(e) Nitrocellulose filter with DNA fragments positioned exactly as on the gel.

(f) Fragment containing gene of interest identified by exposure to a radioactively labeled probe, which base-pairs with a short sequence present on the gene. Result: The gene was carried by plasmid B.

FIGURE 9.12 Southern blotting.

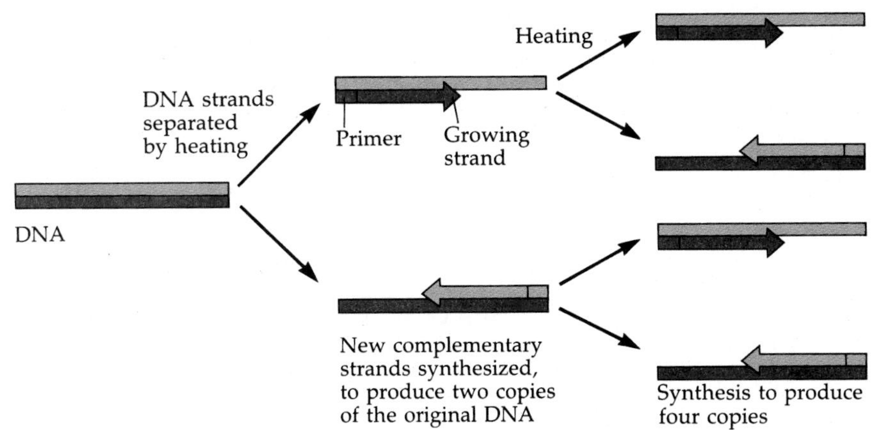

FIGURE 9.13 Polymerase chain reaction.

fingerprinting, which is based on the fact that the DNA of every individual is unique. In contrast to conventional fingerprinting, genetic fingerprinting requires only a few cells from the person to be identified. Furthermore, since the DNA of each species of microbe is also unique, microbiologists can use similar fingerprinting techniques to identify microorganisms, including pathogens. DNA analysis techniques are also being used to pinpoint genetic abnormalities that cause or contribute to various diseases. These techniques are also contributing to advances in *gene therapy,* the replacing of abnormal genes with normal genes in a living individual.

Let us suppose, by using the techniques described in earlier sections, we have created a human gene library, using a bacterial plasmid as the cloning vector. Further assume that we have used a DNA probe to identify two bacterial clones that seem to carry a gene we are interested in. To verify that we have the right gene and to isolate a fragment of DNA containing it, we can now use a method known as **Southern blotting** (Figure 9.12); (the technique's name comes from the scientist who developed it, E. M. Southern, not from geography). We treat the DNA plasmids from each candidate clone with the same restriction enzyme and separate the resulting fragments by size employing a technique called *gel electrophoresis* (see p. 262). In the simplest case, the DNA of each recombinant plasmid will produce only two bands on the gel, one corresponding to the original vector DNA and one to the fragment of human DNA that was inserted in it. We then transfer the DNA fragments onto a special filter by blotting. Next we bathe the filter with a solution of radioactively tagged probes of DNA complementary to the gene we want. After we have seen which band has the gene, we can recover copies of that DNA fragment in pure form from gels simply by cutting out the band and soaking it in solvent.

For further study of the gene—for example, to compare it with other known genes—we may want to do **DNA sequencing,** that is, to determine the sequence of nucleotides in the DNA. This used to require a laborious procedure suitable only for sequencing very short stretches of DNA. In recent years, though, DNA sequencing has become much easier. Automated machines now make it possible to quickly determine the sequence of nucleotides in lengthy molecules of DNA, such as the entire genome of a virus.

Analytical techniques such as Southern blotting and DNA sequencing require multiple copies of the DNA to be analyzed. These can be made by relatively slow and laborious cloning. However, the recent development of the **polymerase chain reaction (PCR)** technique has revolutionalized the amplification of DNA. Starting with just one gene-sized piece of DNA, PCR can be used to make literally billions of copies in only a few hours (Figure 9.13). In the PCR technique, a solution containing the target piece of DNA is first heated in a test tube to separate the two strands of DNA; these will serve as the initial templates for DNA synthesis. With this DNA is a supply of DNA's four nucleotides, for assembly into new DNA, and the enzyme for catalyzing the synthesis, DNA polymerase (see Chapter 8). Short pieces of nucleic acid called primers are also present, to help start the reaction. Each newly synthesized DNA strand serves in turn as a template for more new DNA. As a result, the process proceeds exponentially. After each cycle of synthesis, the DNA is heated to convert all the new DNA into single strands. PCR is made possible by the use of DNA polymerase taken from a thermophilic bacterium; this enzyme can survive the heating phase without being destroyed. Thirty cycles, just one day's work, will increase the amount of target DNA by more than a billion times.

The applications of PCR are potentially of great

MICROBIOLOGY *IN THE NEWS*

Bioremediation of PCBs

Some of the most important experiments in using microbes to protect human health are taking place not in medical facilities but in drag strips, landfills, and sludge heaps. These experiments are aimed at using both natural and genetically engineered bacteria to degrade toxic wastes that would otherwise pose a permanent—and growing—threat to the health of humans and to the environment.

Each year in the United States, 14,000 industrial plants generate 265 metric tons of hazardous wastes, 80% of which make their way into landfills. One of the best known pollutants is the class of compounds known as the polychlorinated biphenyls (PCBs), which do pose a severe problem because of their toxicity and persistence in the environment. Until their production was banned in 1979, PCBs were used in capacitors, hydraulic fluids, carbonless paper, and transformers (and are still present in transformers manufactured before 1979). If PCBs are inhaled, they cause nausea, eye and nose irritation, and headaches. PCBs may cause immune dysfunction and teratogenesis (malformation of the fetus). They have been shown to be carcinogenic in rats.

PCBs are degraded very slowly in nature. In water, PCBs decrease photosynthesis, inhibit heterotrophic bacteria, and are stored in the fats of fish and other aquatic animals. If these animals are eaten or used as fertilizers, the PCBs are passed on and further concentrated. Milk cows that feed on pasture that has been contaminated by PCB-containing water or fertilizer concentrate PCBs in their milk.

The danger is great enough to justify large-scale cleanup efforts. For example, to remove PCBs from the Hudson River, the Department of Environmental Conservation plans to dredge contaminated soil from a 40-mile stretch of river and haul the soil to a landfill site. However, even this costly project won't remove the PCBs from the ecosystem but will just move them to land, where they again may find their way into bodies of water.

Researchers hope that such bioremediation will remove PCBs permanently. Bioremediation is the controlled application of a biodegradation process. In its simplest form, this could involve spraying appropriate bacteria onto a contaminated area. Microbial ecologists at General Electric have sprayed PCB-degrading *Pseudomonas* strains on a drag strip where PCBs were used to hold dust down. The bacteria reduced the PCBs by 25%, but took nearly six months to do so. Scientists are now looking for ways to improve the speed and efficiency of the process.

PCBs consist of two linked benzene rings with chlorine atoms attached to some or all of the carbon atoms. For example,

```
 Cl        H H        Cl
   \      /   \      /
    C==C       C--C
   /    \     //   \\
H--C     C--C        C--H
   \\   //    \     /
    C--C       C==C
   /    \     /    \
 Cl      H H        Cl
```

A consortium of species may be needed to degrade an entire PCB molecule. Some bacteria can remove the chlorine atoms. Another bacterium may open the rings. A third species may be needed to produce acetaldehyde, which can be completely degraded to CO_2 and H_2O by many bacteria. Using recombinant-DNA techniques, researchers are trying to produce a single bacterium that makes all the necessary enzymes for degradation.

Even with genetically engineered bacteria, degradation that follows surface spraying of bacteria is slow, and PCBs deep in the soil are not available to the bacteria. Removal of the soil for treatment will be necessary. Soils will be excavated, piled in lined trenches, and inoculated with the desired bacteria. Pipes laid through the piles can be used for aeration and addition of more bacteria and fertilizer (see the figure). To prevent the release of PCBs into the air and to increase the temperature, the pile must be covered with plastic until degradation is complete. After a few days or weeks, the soil can be returned or used as any other clean soil.

Vacuum heap treatment. A vacuum applied to the pipes draws air into the soil. The pipes can also be used to add bacteria and nutrients.

medical importance. For example, the genes causing genetic diseases such as hemophilia, cystic fibrosis, and muscular dystrophy can be increased to detectable levels for research and diagnosis. Another medical use still in the experimental stage is for the identification of infectious viruses, such as the AIDS virus, in situations where they would otherwise be undetectable. In one dramatic case, PCR has been used to amplify DNA of the AIDS virus in tissue samples from a British sailor who died in 1959 of a then mysterious cause. Thus, this sailor has been established as having the earliest documented case of AIDS, over 30 years after his death. Because of the power of the PCR technique, machines that perform it automatically are becoming common fixtures in research laboratories.

DNA probes like the ones used to screen gene libraries are a promising tool for rapid identification of microorganisms. For use in medical diagnosis, these probes are derived from the DNA of a pathogenic microbe and are cloned with an attached tag, such as a radioactive label. The probe then serves as a diagnostic test by combining with the DNA of the pathogen to locate it in body tissue or, perhaps, in food. Probes are also being used in nonmedical areas of microbiology, for example, for locating and identifying specific microbes in soil. We will discuss probes further in Chapter 10.

Applications of Recombinant-DNA Technology to Agriculture

It has always been a time-consuming process to select for plants that are genetically superior. Performing plant crosses in the conventional way is laborious and involves waiting for the planted seed to mature for harvesting. Plant breeding has been revolutionized by the use of plant cells grown in culture. Clones of plant cells, including ones that have been genetically altered by recombinant-DNA techniques, can be grown in large numbers. These cells can then be induced to regenerate whole plants, from which seeds can be harvested.

Recombinant DNA can be introduced into plant cells in several ways. Previously we mentioned protoplast fusion and the use of DNA-coated "bullets." The most elegant method, however, makes use of a plasmid called the **Ti plasmid** (*Ti* stands for *tumor inducing*), which occurs naturally in the bacterium *Agrobacterium tumefaciens*. This bacterium infects plants, where the Ti plasmid causes the formation of a tumorlike growth called a crown gall (Figure 9.14). A part of the Ti plasmid, called T-DNA, integrates into the genome of the infected plant. The T-DNA stimulates local cellular growth (the crown gall), and at the same time it causes production of certain products that are used by the bacteria as a source of nutritional carbon and nitrogen.

For plant scientists, the attraction of the Ti plasmid is that it provides a vehicle for the introduction of genetically engineered DNA into a plant (Figure 9.15). The scientist can insert foreign genes into the T-DNA, put the recombinant plasmid back into the *Agrobacterium* cell, and use the bacterium to insert the recombinant Ti plasmid into a plant cell. The plant cell with the foreign gene can then be used to generate a new plant. With luck, the new plant will express the foreign gene.

A noteworthy accomplishment of this approach has been the introduction of resistance to the herbicide glyphosate into plants. Normally, this herbicide kills both weeds and useful plants by inhibiting an enzyme necessary for making certain essential amino acids. It happens that *Salmonella* bacteria have this enzyme, and some *Salmonella* bacteria have a mutant enzyme that is resistant to the herbicide. When the DNA for this enzyme is introduced into a crop plant, the crop becomes resistant to the herbicide, which then kills only the weeds.

Perhaps the most exciting potential use of genetic engineering in plants has to do with nitrogen fixation, the ability to convert the nitrogen gas in the air to compounds that living cells can use. The availability of

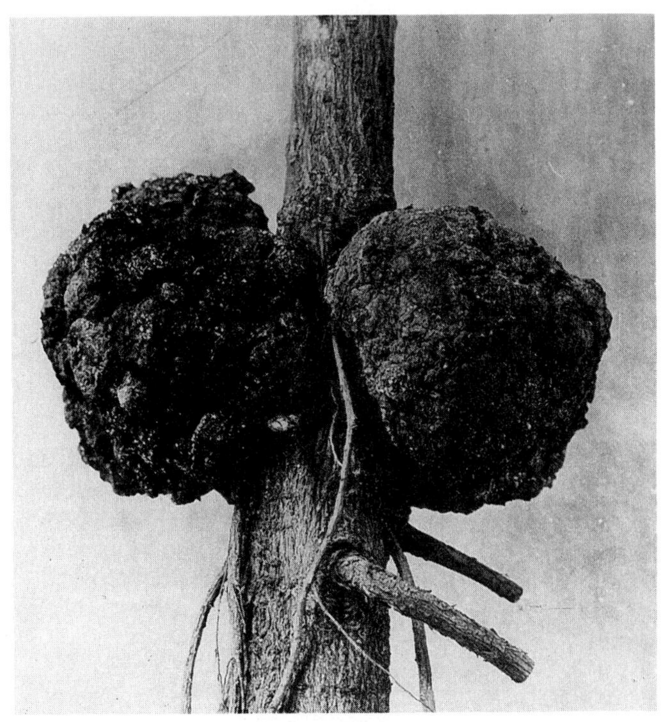

FIGURE 9.14 Crown gall disease in a pecan tree. The tumorlike growth is stimulated by a gene on the Ti plasmid carried by the bacterium *A. tumefaciens*, which has infected the tree.

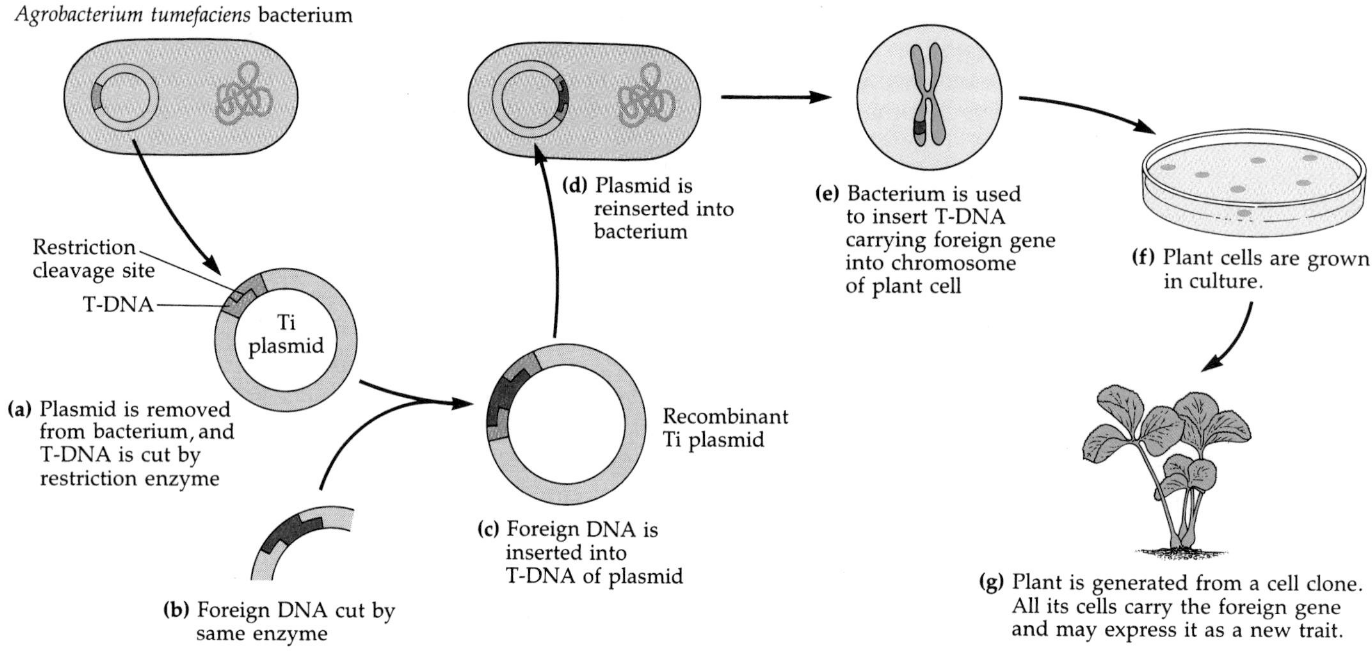

Agrobacterium tumefaciens bacterium
Restriction cleavage site
T-DNA
Ti plasmid
(a) Plasmid is removed from bacterium, and T-DNA is cut by restriction enzyme
(b) Foreign DNA cut by same enzyme
(c) Foreign DNA is inserted into T-DNA of plasmid
Recombinant Ti plasmid
(d) Plasmid is reinserted into bacterium
(e) Bacterium is used to insert T-DNA carrying foreign gene into chromosome of plant cell
(f) Plant cells are grown in culture.
(g) Plant is generated from a cell clone. All its cells carry the foreign gene and may express it as a new trait.

FIGURE 9.15 Using the Ti plasmid as a vector for genetic engineering in plants.

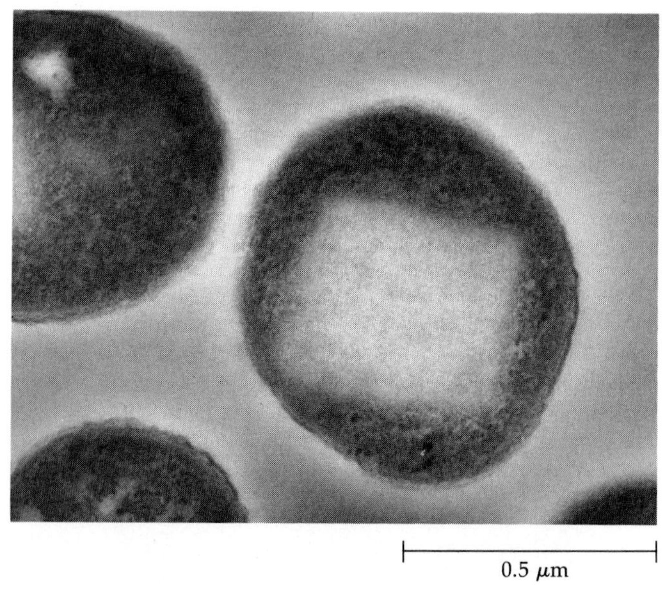

0.5 μm

FIGURE 9.16 Production of an insect toxin by genetic engineering techniques. Root-dwelling *Pseudomonas fluorescens* bacteria have been genetically engineered to produce the insecticidal toxin normally produced by *Bacillus thuringiensis*. The crystal of toxin can be seen in this cell of *P. fluorescens*, which is capable of producing the toxin in much larger amounts than *B. thuringiensis* can. When root-eating insect larvae ingest the engineered bacteria, they are killed by the toxin. The use of the engineered bacteria is several times more effective than the conventional method of applying *B. thuringiensis* spores and toxin to the plant leaves.

such nitrogen-containing nutrients is usually the main factor limiting crop growth. But in nature, only certain bacteria have genes for carrying out this process, although some plants, such as alfalfa, benefit from a symbiotic relationship with these microbes. Already, some species of the symbiotic bacterium *Rhizobium* have been genetically engineered for enhanced nitrogen fixation. In the future it may be possible to design *Rhizobium* strains that can colonize such crop plants as corn and wheat, perhaps eliminating their requirement for nitrogen fertilizer. The ultimate goal would be to introduce functioning nitrogen-fixation genes directly into the plants. Although this goal is not achievable at our present state of knowledge, work toward it will continue because of the potential for dramatically increasing the world's food supply.

Another example of a genetically engineered bacterium now in agricultural use is *Pseudomonas fluorescens* that has been engineered to produce a toxin normally produced by *Bacillus thuringiensis* (Figure 9.16). This toxin kills certain plant pathogens, such as the European corn borer. The genetically altered *Pseudomonas*, which produces much more toxin than *B. thuringiensis*, can be added to plant seeds and in time will enter the vascular system of the growing plant. Its toxin is ingested by the feeding insect larvae and kills them.

Animal husbandry has also benefited from genetic engineering. We have seen how one of the early commercial products of genetic engineering was human growth hormone. By similar methods, it is possible to manufacture bovine growth hormone (BGH). When BGH is injected into cattle, it increases their weight gain. In dairy cows, it also causes a 10% increase in milk production. Whether the procedure will become a standard practice in the dairy industry will depend on factors such as consumer acceptance.

Table 9.1, p. 237, lists these and several other genetically engineered products for use in agriculture and animal husbandry.

Safety Issues

There will always be concern about the safety of any new technology, and genetic engineering of microbes is certainly no exception. One reason for possible concern is that it is nearly impossible to prove that something is entirely safe under all conceivable conditions. People worry that the same techniques that can alter a microbe to make it useful could also inadvertently make it pathogenic to humans or otherwise dangerous to living organisms. Therefore, laboratories engaged in recombinant-DNA research have to meet rigorous standards of control to avoid accidentally releasing genetically engineered microbes into the environment or exposing humans to any risk of infection. To reduce risk further, microbiologists engaged in genetic engineering often delete from the microbes' genomes certain genes that are essential for growth in environments outside the laboratory. Finally, recombinant-DNA-carrying microbes intended for use in the environment (in agriculture, for example) may be engineered to contain "suicide genes." These are genes that eventually turn on to produce a toxin that kills the microbes, thus ensuring that the microbes do not survive in the environment for very long after they have accomplished their task.

The Future of Genetic Engineering

Like the invention of the microscope, the development of recombinant-DNA techniques is bringing about profound changes in science, agriculture, and human health care. With this technology only a little more than twenty years old, it is difficult to predict exactly what all the changes will be. However, it is likely that, within another twenty years, many of the treatments and diagnostic methods discussed in this book will have been replaced by far more powerful techniques based on the unprecedented ability to manipulate DNA precisely. One monumental project involving much of the new technology is the *human genome project*, which is currently under way. The goal of this project is to map the 100,000 or so genes in human DNA and to sequence the entire genome, approximately 3 billion nucleotide pairs. Technically, the project has been compared to the reconstruction of the contents of several sets of shredded encyclopedias. To reach the basic goal and to interpret the data will probably require cooperation among thousands of workers all over the world over several decades. Even before it is completed, however, the project should be of immense value to our understanding of biology. It will also be of great medical benefit, especially for the diagnosis and possibly for the repair of genetic diseases. James Watson, codiscoverer of the structure of DNA, has said that, "A more important set of instruction books will never be found by human beings. When finally interpreted, the genetic messages encoded within our DNA molecules will provide the ultimate answers to the chemical underpinnings of human existence."

STUDY OUTLINE

The Advent of Recombinant-DNA Technology (pp. 227–228)

1. Closely related organisms can exchange genes in natural recombination.
2. Genes can be transferred between unrelated species via laboratory manipulations called genetic engineering.
3. Recombinant DNA refers to DNA that has been artificially manipulated to combine genes from two different sources.
4. Biotechnology includes all industrial applications of microorganisms as well as industrial uses of genetically engineered cells.
5. A DNA molecule used to carry a desired gene from one organism to another is called a vector.

Overview of Recombinant-DNA Procedures (pp. 228–231)

1. A desired gene is inserted into a DNA vector such as a plasmid or viral genome.
2. The vector inserts the DNA into a new cell, which is grown to form a clone.
3. Large quantities of the gene or the gene product can be harvested from the clone.

Restriction Enzymes (p. 231)

1. A restriction enzyme recognizes and cuts only one particular nucleotide sequence in DNA.
2. Some restriction enzymes produce sticky ends, short

stretches of single-stranded DNA at the ends of the DNA fragments.
3. Fragments of DNA produced by the same restriction enzyme will spontaneously join by hydrogen-bonding. DNA ligase can covalently link the DNA backbones.

Cloning a Gene with a Plasmid Vector (pp. 231–232)

1. A plasmid containing a new gene can be inserted into a cell by transformation.
2. A bacteriophage DNA vector can be inserted into a cell by transduction.
3. Antibiotic-resistance markers on a plasmid vector are used to identify cells containing the engineered vector.

Alternative Ways to Get Foreign DNA into Cells (pp. 232–233)

1. Plant, fungal, and bacterial protoplasts can be formed by enzymatically removing their cell walls.
2. Protoplast fusion is the joining of cells whose cell walls have been removed.
3. Pores made in protoplasts and animal cells by electric current can provide entrance for new pieces of DNA.
4. Foreign DNA can be introduced into plant cells by shooting DNA-coated tungsten particles into the cells.
5. Foreign DNA can be injected into animal cells by using a fine glass micropipette.

Sources of DNA (pp. 233–234)

1. Gene libraries can be made by cutting up an entire genome with restriction enzymes and inserting the fragments into bacterial plasmids or phages.
2. cDNA made from mRNA by reverse transcription can be cloned in gene libraries.
3. Synthetic DNA can be made in vitro by a DNA-synthesizing machine.

Identifying the Desired Clone in a Library (pp. 234–235)

1. Bacteria from the gene library are grown on a plate of nutrient medium. After replica plating, one plate is treated to lyse the cells and separate the DNA into single strands.
2. A short piece of radioactively labeled DNA called a probe is added to the plate.
3. Bacterial colonies containing DNA that base-pairs with the probe are identified by the radioactive label.
4. A radioactively labeled antibody against a protein product can be used to identify colonies carrying the gene encoding that protein.
5. The desired colony is selected from the replica plate.

Making a Gene Product (pp. 235–236)

1. *E. coli* is used to produce proteins by genetic engineering because it is easily grown and its genetics are well understood.
2. Efforts must be made to ensure that *E. coli*'s endotoxin does not contaminate a product intended for human use.
3. To recover the product, *E. coli* must be lysed or the gene must be linked to a gene that produces a naturally secreted protein.
4. Yeasts can be genetically engineered and are likely to continuously secrete the gene product.
5. Animal viruses can be engineered to carry a gene for a pathogen's surface protein. When the virus is used as a vaccine, the host develops an immunity to the pathogen.
6. Mammalian cells can be engineered to produce proteins such as insulin for medical use.
7. Plant cells can be engineered and used to produce plants with new properties.

Genetically Engineered Products for Medical Therapy (p. 236)

1. Synthetic genes linked to the β-galactosidase gene in a plasmid vector were inserted into *E. coli*, allowing *E. coli* to produce and secrete the two polypeptides used to make human insulin.
2. Cells can be engineered to produce a pathogen's surface protein, which can be used as a subunit vaccine.

Obtaining Information from DNA for Basic Research and Medical Applications (pp. 236–241)

1. Recombinant-DNA techniques can be used to increase understanding of DNA, for genetic fingerprinting, and for gene therapy.
2. Southern blotting can be used to locate the desired DNA fragment in a bacterial clone.
3. DNA sequencing machines are used to determine the nucleotide base sequence in a gene.
4. The polymerase chain reaction (PCR) is used to make multiple copies of a desired piece of DNA enzymatically.
5. PCR can be used to increase the amounts of DNA in samples to detectable levels. This may allow sequencing of genes, diagnosis of genetic diseases, or detection of viruses.
6. DNA probes can be used to quickly identify a pathogen in body tissue or food.

Applications of Recombinant-DNA Technology to Agriculture (pp. 241–243)

1. Cells from plants with desirable characteristics can be cloned to produce many identical cells. These cells can then be used to produce whole plants from which seeds can be harvested.
2. Plant cells can be engineered by using the Ti plasmid vector.
3. The tumor-producing T genes are replaced with desired genes, and the recombinant DNA is inserted into *Agrobacterium*. The bacterium naturally transforms its plant hosts.
4. *Rhizobium* has been engineered for enhanced nitrogen fixation.

5. *Pseudomonas* has been engineered to produce *Bacillus thuringiensis* toxin against insects.
6. Bovine growth hormone is being produced by *E. coli.*

Safety Issues (p. 243)

1. Strict safety standards are employed to avoid accidental release of genetically engineered microorganisms.
2. Some microbes used in genetic engineering have been altered so that they cannot survive outside of the laboratory.
3. Microorganisms that are intended for use in the environment may be engineered to contain suicide genes so the organisms do not persist in the environment.

The Future of Genetic Engineering (p. 243)

1. Genetic engineering techniques may provide new treatments for disease and new diagnostic tools.
2. Genetic engineering techniques are being used to map the human genome through the human genome project.
3. This will provide tools for diagnosis and possibly repair of genetic diseases.

STUDY QUESTIONS

Review

1. Compare and contrast the following terms:
   (a) cDNA and gene
   (b) Restriction fragment and gene
   (c) DNA probe and gene
   (d) DNA polymerase and DNA ligase
   (e) Recombinant DNA and cDNA
2. How are each of the following used in genetic engineering?
   (a) Plasmid
   (b) Viral genome
   (c) Antibiotic-resistance genes
   (d) Radioactively labeled antibody
3. Suppose you want multiple copies of a gene you have synthesized. How would you obtain the necessary copies by cloning? By PCR?
4. Describe a genetic engineering experiment in two or three sentences. Use the following terms: intron, exon, DNA, mRNA, cDNA, RNA polymerase, reverse transcriptase.
5. The polymerase chain reaction involves heating (to 90°C for 10 seconds) to break the hydrogen bonds in DNA, cooling to allow a primer to hydrogen-bond to the single strands, and incubation to allow DNA polymerase to build complementary strands of DNA. This procedure is repeated 30 times an hour to get large quantities of the desired DNA. Why did the use of DNA polymerase from the bacterium *Thermus aquaticus* allow researchers to add the necessary reagents to tubes in a preprogrammed heating block?
6. You are attempting to insert a gene for salt-water tolerance into a plant by using the Ti plasmid. In addition to the desired gene, you add a gene for tetracycline-resistance (*tet*) to the plasmid. What is the purpose of the *tet* gene?
7. List at least two examples of the use of genetic engineering in medicine. In agriculture.
8. Some commonly used restriction enzymes are listed below. The cutting site is indicated by ↓. Indicate which enzymes produce sticky ends. Of what value are sticky ends in making recombinant DNA?

Enzyme	Bacterial source	Recognition sequence
Bam HI	*Bacillus amyloliquefaciens*	G↓G A T C C C C T A G↑G
Eco RI	*Escherichia coli*	G↓A A T T C C T T A A↑G
Hae III	*Haemophilus aegyptius*	G G↓C C C C↑G G
Hin dIII	*Haemophilus influenzae*	A↓A G C T T T T C G A↑A

9. Suppose your biotechnology company is trying to make a subunit vaccine consisting of a surface protein from hepatitis C virus (HCV). Place the following steps for this experiment in the correct order. (Which one of the items listed is not appropriate?)
   (a) Culture *E. coli* on nutrient agar containing tetracycline.
   (b) Determine the chemical structure of the HCV surface protein.
   (c) Determine the nucleotide sequence coding for the HCV surface protein.
   (d) Digest HCV genome with restriction enzyme *Eco* RI.
   (e) Digest plasmid DNA with *Eco* RI.
   (f) Digest plasmid DNA with *Pst* I.
   (g) Separate HCV genome fragments by electrophoresis.

(h) Isolate a plasmid containing a tetracycline-resistance gene.
(i) Locate surface protein gene by using a radioactively labeled DNA probe.
(j) Mix plasmid and HCV surface protein gene with DNA ligase.
(k) Remove surface protein from gel.
(l) Synthesize short chain of DNA that is complementary to surface protein gene.
(m) Test colonies with radioactively labeled antibodies against HCV surface protein.
(n) Transform *E. coli* with recombinant plasmid.

Challenge

1. The following gel electrophoresis patterns were obtained from *Eco* RI digests of various DNA molecules from a transformation experiment. Can you conclude from these data that transformation occurred? Explain why or why not.

Vector
Vector + new gene
New gene
Original cell's DNA
Transformed cell's DNA
Origin

2. Using this map of plasmid pMICRO, give the number of restriction fragments that would result from digesting pMICRO with *Eco* RI, *Hin* dIII, and both enzymes together. What is the smallest fragment that would contain the tetracycline-resistance gene?

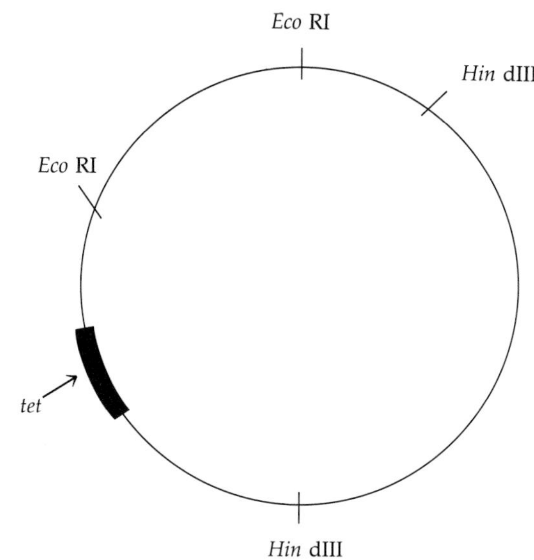

3. Design an experiment using vaccinia virus to make a vaccine against the AIDS virus (HIV).

FURTHER READING

Anderson, W.F., and E.G. Diacumakos. "Genetic engineering in mammalian cells." *Scientific American* 245(1):106–121, July 1981. Describes methods of inserting genes into mammalian cells with the goal of curing inherited diseases.

Antébi, E., and D. Fishlock. *Biotechnology: Strategies for Life.* Cambridge, Mass.: MIT Press, 1986. This oversized book includes hundreds of color photographs and illustrations to show genetic engineering of cells.

Barton, J.H. "Patenting life." *Scientific American* 264(3):40–46, March 1991. This article discusses the legal issues of ownership of genetically engineered cells.

Binns, A.N., and M.F. Thomashow. "Cell biology of *Agrobacterium* infection and transformation of plants." *Annual Review of Microbiology* 42:575–606, 1988. This article describes T DNA integration and the host range of *Agrobacterium.*

Biotechnology. *Science* 219:611ff, 1983. A special issue on biotechnology, including discussions of genetic engineering and industrial microbiology.

Biotechnology. This journal is published monthly. Each issue contains articles on new products, methods, and current research.

Chilton, M.D. "A vector for introducing new genes into plants." *Scientific American* 248(6):50–59, June 1983. Describes the use of the *Agrobacterium* Ti plasmid for genetic engineering.

Industrial Microbiology. *Scientific American* 245(3), September 1981. An issue devoted entirely to the topic of industrial microbiology and genetic engineering.

Microbes for Hire. *Science85* 6(6):30–46, July/August 1985. A special issue on the use of microbes in biotechnology.

Mullis, K.B. "The unusual origin of the polymerase chain reaction." *Scientific American* 262(4):56–65, April 1990. An interesting article by the discoverer of PCR.

Neufeld, P.J., and N. Colman. "When science takes the witness stand." *Scientific American* 262(5):46–53, May 1990. Illustrates separation of DNA fragments by electrophoresis and analysis by Southern blotting in describing use of DNA analysis in courts of law.

Old, R.W., and S.B. Primrose. *Principles of Gene Manipulation: An Introduction to Genetic Engineering,* 4th ed. Boston: Blackwell Scientific, 1989. Volume 2 in the series *Studies in Microbiology* covers transformation and other genetic engineering tools.

Präve, P., ed. *Basic Biotechnology.* New York: VCH Publishers, 1987. This text includes control of metabolic pathways, product recovery, and bioreactors.

Weaver, R.F. "Changing life's genetic blueprint." *National Geographic Magazine* 166(6):818–847, December 1984. Well illustrated with diagrams and photographs explaining the basics of genetic engineering.

PART TWO

Survey of the Microbial World

CHAPTER 10

Classification of Microorganisms

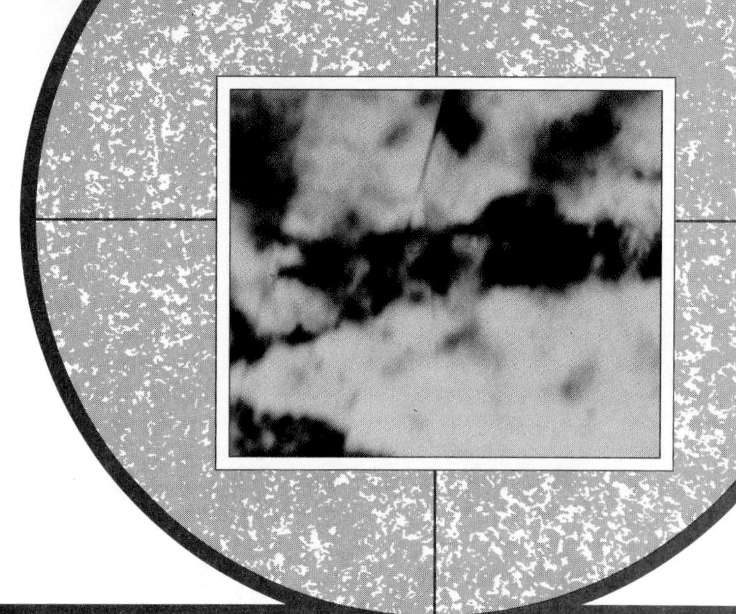

LEARNING OBJECTIVES

- Define taxonomy and binomial nomenclature.
- List the major characteristics used to differentiate among kingdoms in the five-kingdom system and the three-kingdom system.
- Identify the characteristics of the Kingdom Procaryotae that differentiate it from other kingdoms.
- Describe the divisions used in *Bergey's Manual.*
- Compare and contrast classification and identification.
- Describe ten methods of classification and identification of microorganisms.

Part Two photo, p. 249: In a health care setting, correct identification of a pathogen can make the difference between successful and unsuccessful treatment.

The science of classification, especially the classification of living forms, is called **taxonomy,** from the Greek words for law and order. The objective of taxonomy is to classify living organisms, that is, to establish the relationships between one group of organisms and another and to differentiate between them. A taxonomic system enables us to classify organisms that have not yet been studied in detail. That is, a heretofore unknown organism can be identified and then grouped or classified with other organisms that have similar characteristics. For example, very little is known about bacteria that live at great depths in the ocean, but data suggest that these bacteria are related to the known archaeobacteria.

The practical value of taxonomy is to provide a method of identifying organisms already classified. For example, when a bacterium suspected of causing a specific disease is isolated from a patient, characteristics of that isolate are matched to lists of characteristics of previously classified bacteria to identify the isolate. After the bacterium has been identified, drugs can be selected that affect that bacterium.

Taxonomy is a basic and necessary tool for scientists. It provides a universal language of communication and a common reference for identification. Modern taxonomy is an exciting and dynamic field; new techniques in molecular biology and genetics are providing new insights into classification and evolution.

In this chapter, you will learn the various classification systems, the different criteria used for classification, and tests that are used to identify bacteria that are already classified.

Phylogenetic Relationships

Charles Darwin's (1809–1882) theory of evolution explains that the many similarities among organisms are a result of their descent from common ancestors. The arrangement of organisms into taxonomic categories, collectively called **taxa,** reflects degrees of relatedness among organisms. That is, the hierarchy of taxa shows evolutionary or **phylogenetic** (from a common ancestor) relationships.

The most fundamental taxonomic or phylogenetic division is that between procaryotes and eucaryotes. One widely held theory (discussed in Chapter 4) is that eucaryotic cells evolved from procaryotic cells living inside one another (as endosymbionts). In fact, the similarities between procaryotic cells and eucaryotic organelles (Table 10.1) provide striking evidence for this endosymbiotic relationship. (Recall that procaryotes do not have a nucleus separated from the cytoplasm by a membrane, as eucaryotes do. Instead, they have a nuclear region in which the genetic material is more or less localized.) The oldest known fossils are the remains of organisms that lived more than 3.5 billion years ago. They are procaryotes. Eucaryotic cells evolved more recently, about 1.4 billion years ago.

A FIVE-KINGDOM SYSTEM

In 1969, Robert H. Whittaker founded the **five-kingdom system** of biological classification, which shows procaryotes to be the ancestors to all eucaryotes (Figure 10.1). In this system, all procaryotes are included in the Kingdom **Procaryotae** (or Monera). The four eucaryotic kingdoms are distinguished according to nutritional requirements, patterns of development, tissue differentiation, and possession of 9 + 2 flagella. (Recall that eucaryotic flagella and cilia are composed of nine outer pairs plus an inner pair of microtubules, whereas procaryotic flagella are composed of repeating flagellin subunits.)

Simple eucaryotic organisms, mostly unicellular, are grouped as the **Protista.** Protists generally have flagella at some time during their life cycle. The multicellular protists lack tissue organization. The Kingdom Protista includes funguslike water molds, slime molds, protozoans, and the primitive eucaryotic algae.

Fungi, plants, and animals comprise the three kingdoms of more complex eucaryotic organisms, most of which are multicellular.

Fungi include the unicellular yeasts, multicellular molds, and macroscopic species such as mushrooms. To obtain raw materials for vital functions, a fungus absorbs dissolved organic matter through its plasma membranes. The cells of a multicellular fungus are commonly joined to form thin tubes called hyphae. The hyphae are divided into multinucleated units by cross walls that have holes so cytoplasm can flow between the cell-like units. Most fungi lack flagella. Fungi develop from spores or from fragments of hyphae.

The Kingdom **Plantae** (plants) includes some algae and all mosses, ferns, conifers, and flowering plants. All members of this kingdom are multicellular. To obtain energy, a plant uses photosynthesis; this process converts carbon dioxide and water into organic molecules used by the cell.

The kingdom of multicellular organisms called **Animalia** (animals) includes sponges, various worms, insects, and animals with backbones (vertebrates). Animals obtain nutrients and energy by ingesting organic matter through a mouth of some kind.

A THREE-KINGDOM SYSTEM

In 1978, Carl R. Woese proposed a three-kingdom system of classification (Figure 10.2). The grouping of all bacteria into Procaryotae in the five-kingdom system had been based on microscopic observations. Woese's proposal was based on modern techniques in molecular biology and biochemistry that revealed that there are actually two types of procaryotic cells. This discovery was based on the observation that ribosomes are not the same in all cells (as you will recall from Chapter 4). In particular, comparing the sequences of nucleotides in ribosomal RNA (rRNA) from different kinds of cells shows that there are three distinctly different cell groups, the eucaryotes and two different types of procaryotes—the **eubacteria,** or "true bacteria," and the **archaeobacteria** (from *archaios,* meaning ancient).*

*Debate over the proper spelling of the term *archaeobacteria* currently exists. *Archaeobacteria* is used in *Bergey's Manual of Systematic Bacteriology;* other references use *archaebacteria.*

TABLE 10.1 A Comparison of Procaryotic and Eucaryotic Cells and Organelles

	PROCARYOTIC CELLS	EUCARYOTIC CELLS	MITOCHONDRIA AND CHLOROPLASTS (EUCARYOTIC ORGANELLES)
DNA	Circular	Linear	Circular
Histones	No	Yes	No
Ribosomes	70S	80S	70S
Growth	Binary fission	Mitosis	Binary fission

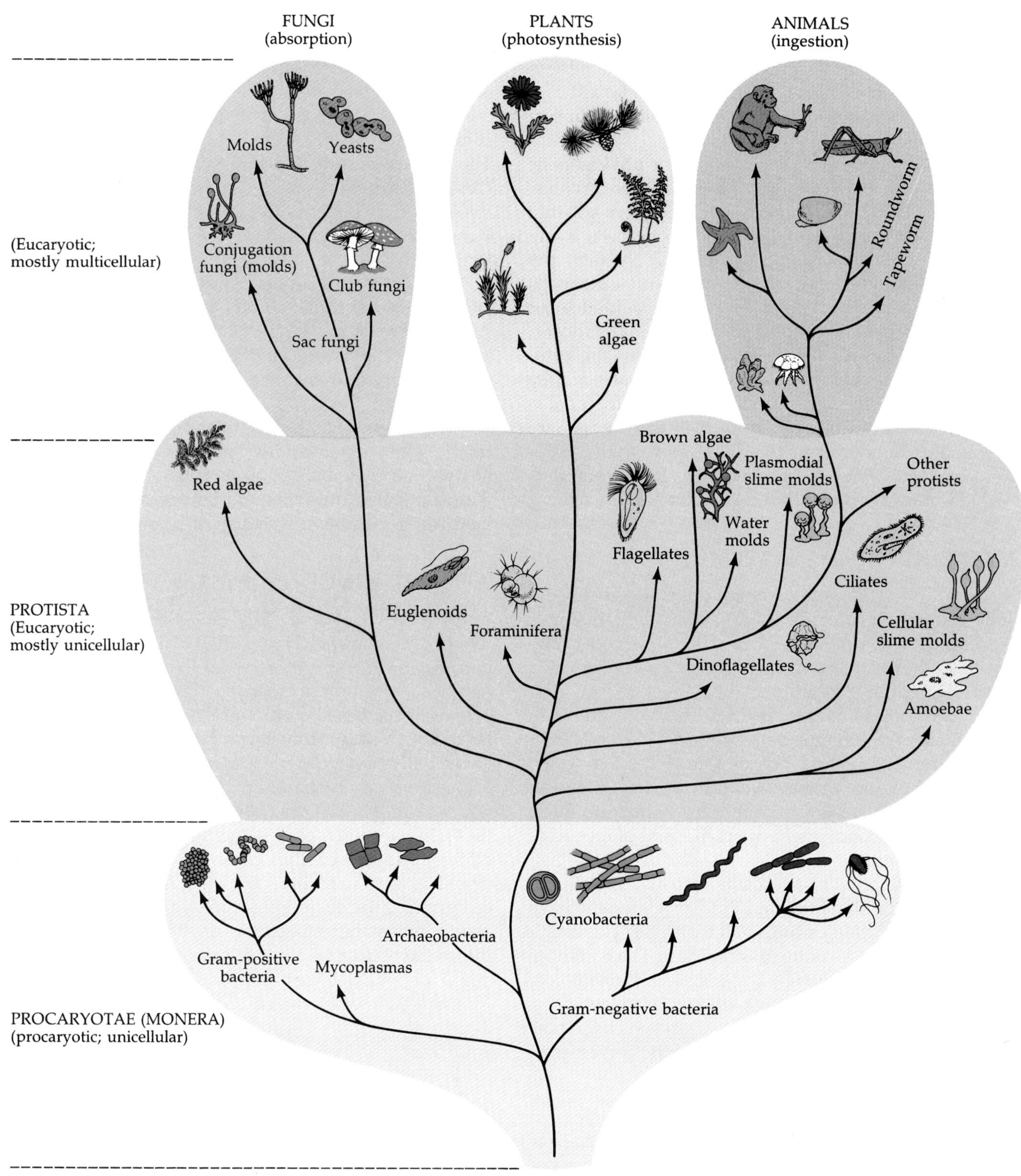

FIGURE 10.1 Five-kingdom system. This is a commonly accepted system of classification.

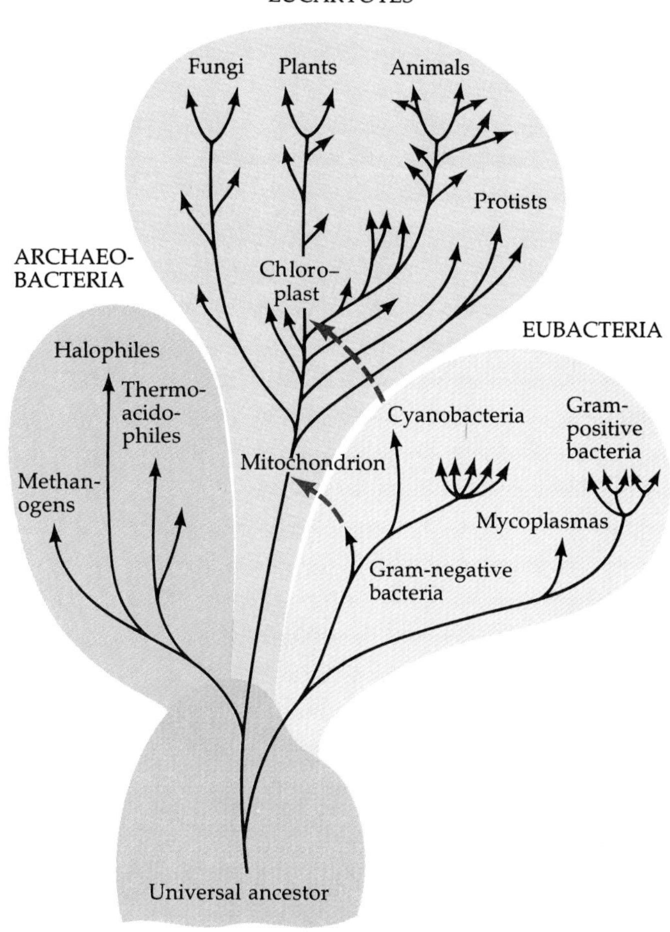

FIGURE 10.2 Three-kingdom system. This system of classification can be used to show three kingdoms or three superkingdoms.

Woese believed, therefore, that the archaeobacteria and the eubacteria (the Procaryotae in the five-kingdom system), while similar in appearance, should form their own separate branches on the evolutionary tree. All eucaryotes (divided into Protista, Fungi, Plantae, and Animalia in Whittaker's system), he believed, should be grouped into a third kingdom. Other workers have suggested raising the three kingdoms to the status of *superkingdoms*. In this scheme, animals, plants, fungi, and protists would then be kingdoms in the eucaryote superkingdom. Differences in membrane lipid structure, transfer RNA molecules, and sensitivity to antibiotics also support classification of the three cell types into separate kingdoms (Table 10.2).

Archaeobacteria differ from eubacteria in a number of other ways. For example, their cell walls never contain peptidoglycan, they often live in extreme environments, and they carry out unusual metabolic processes. They include three groups:

1. The methanogens, strict anaerobes that produce methane (CH_4) from carbon dioxide and hydrogen.
2. Extreme halophiles, which require high concentrations of salt for survival.
3. Thermoacidophiles, which normally grow in hot, acidic environments.

The evolutionary relationship of the three kingdoms is the subject of current research by Woese and others. Originally, archaeobacteria were thought to be the most primitive group, while eubacteria were assumed to be more closely related to eucaryotes. How-

TABLE 10.2 Some Characteristics of Archaeobacteria, Eubacteria, and Eucaryotes

	ARCHAEOBACTERIA	*EUBACTERIA*	*EUCARYOTES*
Cell type	Procaryotic	Procaryotic	Eucaryotic
Cell wall	Varies in composition; contains no peptidoglycan	Contains peptidoglycan	Varies in composition; contains carbohydrates
Membrane lipids	Composed of branched carbon chains attached to glycerol by ether linkage	Composed of straight carbon chains attached to glycerol by ester linkage	Composed of straight carbon chains attached to glycerol by ester linkage
Start signal for protein synthesis	Methionine	Formylmethionine	Methionine
Antibiotic sensitivity	No	Yes	No
rRNA loop*	Lacking	Present	Lacking
Common arm of tRNA**	Lacking	Present	Present

*Binds to ribosomal protein; found in all eubacteria.
**A sequence of bases on tRNA found in all eucaryotes and eubacteria: guanine-thymine-pseudouridine-cytosine-guanine.

ever, recent studies of the DNA sequences of the genes for RNA polymerase show more similarity between the archaeobacteria and eucaryotes than between the archaeobacteria and eubacteria. In light of new information, the researchers believe that the eubacterial kingdom branched off from the other two quite early (Figure 10.2); that is, Eubacteria is the *older* kingdom.

Analysis of ribosomal structure suggests that eucaryotes actually evolved from archaeobacteria. An archaeobacterium may have provided the original host in which endosymbiotic eubacteria developed into organelles. Biochemical similarities between archaeobacteria and eucaryotes (Table 10.2) suggest that the original eucaryotic cell inhabited by symbiotic eubacteria was an archaeobacterium such as *Thermoplasma* (thėr-mō-plaz′mä), whose chromosome contains proteins similar to eucaryotic histones. An ancestor of the modern eubacteria *Bdellovibrio* (del-lō-vib′rē-ō) or *Daptobacter* (dap-tō-bak′tėr), which can invade and reproduce within other bacteria, may have been the original symbiont. Acceptance of the three-kingdom system depends, at least in part, on answering the question about the origin of eucaryotes. At present, the five-kingdom system is accepted by most biologists as the most useful taxonomic tool.

Taxonomists are searching for a natural classification system that depicts phylogenetic relationships. As you can see, taxonomy has developed into a tool for clarifying the evolution of organisms as well as their relationships to each other.

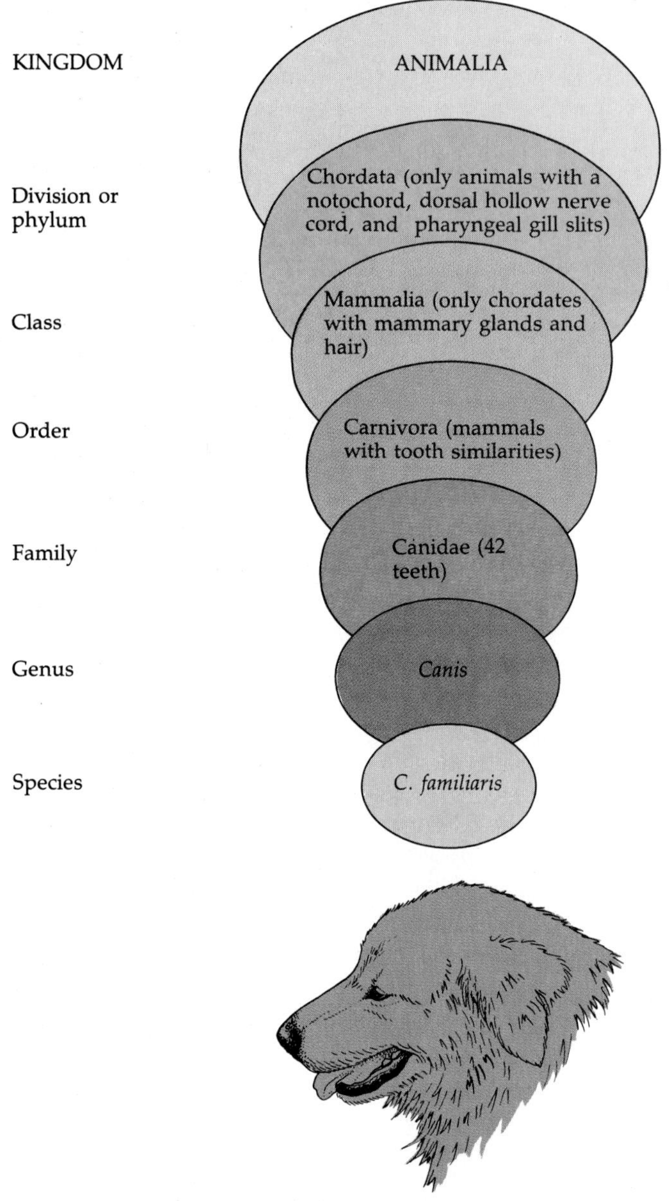

Classification of Organisms

Living organisms are grouped according to similar characteristics (classification), and each organism is assigned a unique scientific name. The rules for classifying and naming, used by biologists worldwide, are discussed below.

SCIENTIFIC NOMENCLATURE

In a world inhabited by millions of living organisms, biologists must be sure they know exactly which organism is being discussed. We cannot use common names because the same name is often used for many different organisms in different regions. For example, there are two different plants with the common name Spanish moss, and neither one is a moss. Three different animals are referred to as a gopher. Because common names are rarely specific and can often be misleading, a system of scientific names, referred to as scientific nomenclature, was developed in the eighteenth century by Carolus Linnaeus.

This system gives every organism two names, usually derived from either Latin or Greek. These are the **genus** name and **specific epithet (species),** and both names are printed underlined or italicized. The genus name is always capitalized and is always a noun. The species name is written in lowercase and is usually an adjective. Because this system gives two names to each organism, the system is called **binomial nomenclature.**

Let us consider a few examples. Your own genus and specific epithet are *Homo sapiens* (hō′mō sā′pē-ens); the noun, or genus, means *man,* and the adjective, or specific epithet, means *wise.* One bacterium that causes pneumonia is called *Klebsiella pneumoniae;* the genus honors Edwin Klebs, a colleague of Robert Koch, and the specific epithet describes a diseased condition. A mold that contaminates bread is called *Rhizopus nigricans* (rī′zō-pus nī′gri-kans). *Rhizo-* (root)

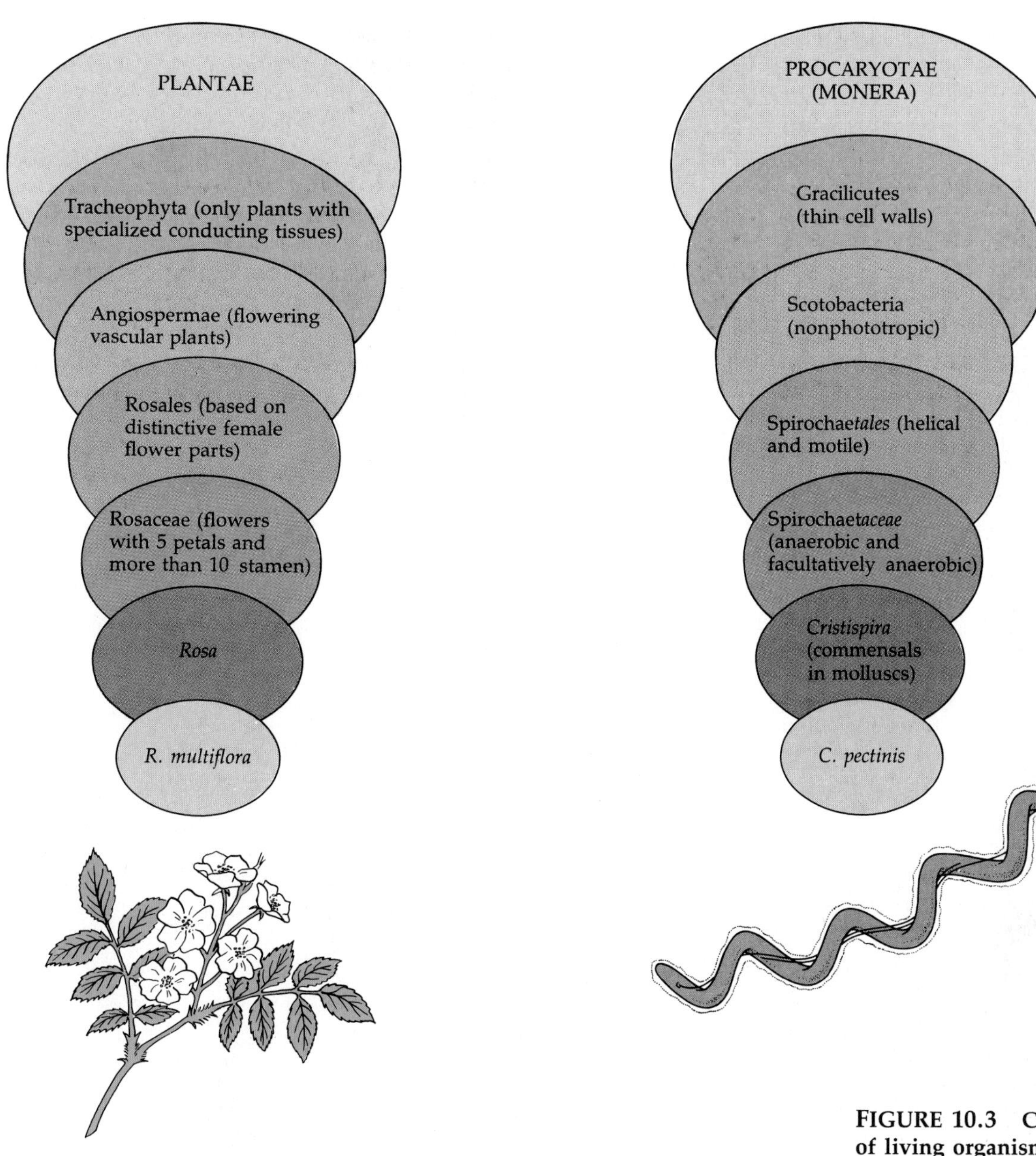

FIGURE 10.3 Classification of living organisms.

describes rootlike structures on the fungus; *nigr-* (black) identifies the color of its spore sacs. The genus and specific epithet are spelled out the first time they are used *(Klebsiella pneumoniae),* but thereafter the genus name may be abbreviated *(K. pneumoniae).*

Rules for assigning names to newly classified bacteria and for assigning bacteria to taxa are established by the International Committee on Systematic Bacteriology and are published in the *Bacteriological Code.* Descriptions of bacteria and evidence for their classifications are published in the *International Journal of Systematic Bacteriology* before being incorporated into a reference called *Bergey's Manual of Systematic Bacteriology.* According to the *Bacteriological Code,* scientific names are to be taken from Latin (a genus name can be taken from Greek) or latinized by the addition of the appropriate suffix. Suffixes for order and family are specified as *-ales* and *-aceae,* respectively (Figure 10.3).

THE TAXONOMIC HIERARCHY

All organisms can be grouped into a series of subdivisions that make up the taxonomic hierarchy. For eucaryotic organisms, a species is a group of closely related organisms that breed among themselves. (Bacterial species will be discussed shortly.) A genus (plural, *genera*) consists of species that differ from each other in certain ways but are related by descent. For example, *Quercus* (kwer'kus), the genus name for oak, consists of all types of oak trees (white oak, red oak, bur oak, velvet oak, and so on). Even though each species of oak differs from every other species, they are all related genetically. Just as a number of species make up a genus, related genera make up a **family.** A group of similar families constitutes an **order,** and a group of similar orders makes up a **class.** Related classes, in turn, make up a **division.** (In zoology, the comparable term **phylum** is used.) Thus, a particular organism (or species) has a genus name and specific epithet and belongs to a family, order, class, and phylum or division. All phyla or divisions that are related to each other make up a **kingdom** (Figure 10.3).

A PHYLOGENETIC HIERARCHY

Grouping organisms according to common properties implies that a group of organisms evolved from a common ancestor. Each species retains some of the characteristics of the ancestor. Some of the information used to classify and to determine evolutionary relationships in higher organisms comes from fossils. Bones, shells, or stems that contain mineral matter or have left imprints in rock that was once mud are examples of fossils.

The structures of most microorganisms are not readily fossilized. Some notable exceptions are:

1. A marine protist whose fossilized colonies form the White Cliffs of Dover, England.
2. Stromatolites, the fossilized remains of microbial communities that flourished between 0.5 and 2 billion years ago.
3. The fossilized cyanobacteria found in 3- to 3.5-billion-year-old rocks in Western Australia. These are the oldest known fossils. Some fossils of procaryotes are shown in Figure 10.4.

FIGURE 10.4 Fossilized procaryotes. (a) Filamentous procaryotes from the Early Precambrian (ca. 3500 million years ago) of Western Australia. **(b)** and **(c)** Coccoid cyanobacteria from the Late Precambrian (ca. 850 million years ago) of central Australia.

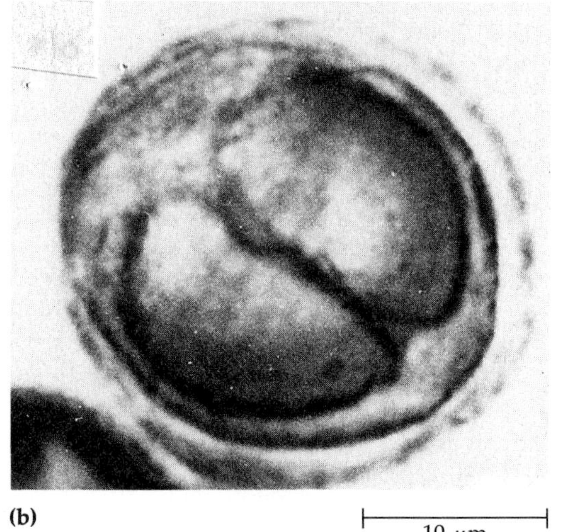

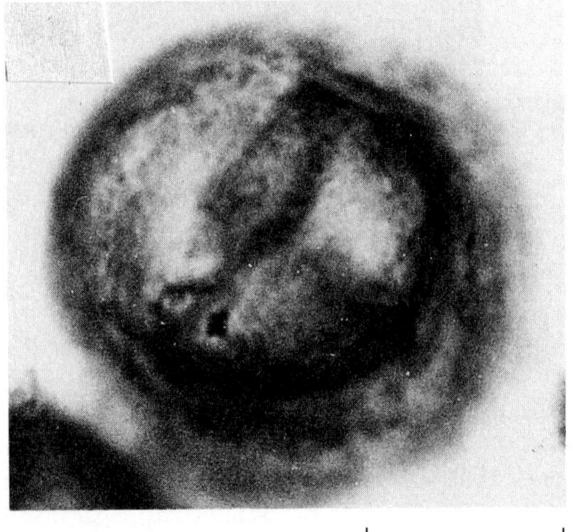

Since fossil evidence is not available for the majority of procaryotes, their phylogeny must be based on other kinds of evidence. Conclusions from DNA hybridization studies (discussed shortly) of selected orders and families of eucaryotes are in agreement with the fossil records. This has encouraged workers to use DNA and RNA hybridization and RNA sequencing to gain an understanding of the evolutionary relationships among procaryotic groups.

CLASSIFICATION OF BACTERIA

The taxonomic classification scheme for bacteria may be found in *Bergey's Manual of Systematic Bacteriology.* In *Bergey's Manual,* bacteria are divided into four divisions. Three divisions consist of eubacterial cells, and the fourth division consists of the archaeobacteria. Each division is divided into classes; in all, there are seven classes of bacteria. The divisions and classes are listed in Table 10.3. Classes are divided into orders; orders, into families; families, into genera; and genera, into species.

A bacterial species is defined somewhat differently than a eucaryotic species, which is a group of closely related organisms that can interbreed. Unlike reproduction in eucaryotic organisms, cell division in bacteria is not directly tied to sexual conjugation, which is infrequent and does not always need to be species-specific. A **bacterial species,** therefore, is defined simply as a population of cells with similar characteristics. (The types of characteristics will be discussed later in this chapter.) The members of a bacterial species are essentially indistinguishable from each other but are distinguishable from members of other species, usually on the same basis of several features. In some cases, pure cultures of the same species are not identical in all ways. Each such group is called a **strain,** which is a group of cells all derived from a single cell. Strains are identified by numbers, letters, or names that follow the specific epithet.

Bergey's Manual provides a reference for identifying bacteria in the laboratory as well as a classification scheme for bacteria. Suggested evolutionary relationships are given for most groups. Information used to build (and modify) the phylogenic models comes from analyses of nucleotide sequences in DNA and RNA, from DNA hybridizations, and from chemical analyses of cellular components. (All these techniques will be described later in this chapter.) A scheme of the suggested evolutionary relationships of bacteria is shown in Figure 10.5.

Not all bacteria have yet been surveyed, and those needing further investigation are temporarily placed in descriptive categories or sections, such as "Endosymbionts" and "Dissimilatory Sulfate- or Sulfur-Reducing Bacteria."

TABLE 10.3 Divisions and Classes in the Kingdom Procaryotae (Monera) Identified by Common Names

DIVISION	CLASS
Typical gram-negative cell wall	Nonphotosynthetic bacteria Anaerobic photosynthetic bacteria Cyanobacteria
Typical gram-positive cell wall	Rods and cocci Actinomycetes and related organisms
Wall-less procaryotes	Mycoplasmas
Unusual walls	Archaeobacteria

CLASSIFICATION OF VIRUSES

Viruses are not classified as part of any of the five kingdoms discussed earlier in this chapter. Viruses are not composed of cells, and they use the anabolic machinery of host cells to multiply inside living cells. Because a viral genome (genetic material) can direct biosynthesis inside a host cell, and because some viral genomes can become incorporated into the host genome, it is possible that viruses are more closely related to their hosts than to other viruses.

Viruses are obligatory intracellular parasites, so they must have evolved after a suitable host cell had evolved. There are two hypotheses on the origin of viruses: (1) they arose from independently replicating strands of nucleic acids (such as plasmids), and (2) they developed from degenerative cells that, through many generations, gradually lost the ability to survive independently but could survive when associated with another cell. Viruses will be discussed in Chapter 13.

Criteria for Classification and Identification of Microorganisms

A classification scheme provides a list of characteristics and a means for comparison to aid in the identification of an organism. Once an organism is identified, it can be placed into a previously devised classification scheme. Microorganisms are *identified* for practical purposes—for example, to determine an appropriate treatment for an infection. They are not necessarily identified by the same techniques by which they are classified. Most identification procedures are easily performed in a laboratory and require as few procedures or tests as possible.

Bergey's Manual of Determinative Bacteriology has been a widely used reference since publication of the

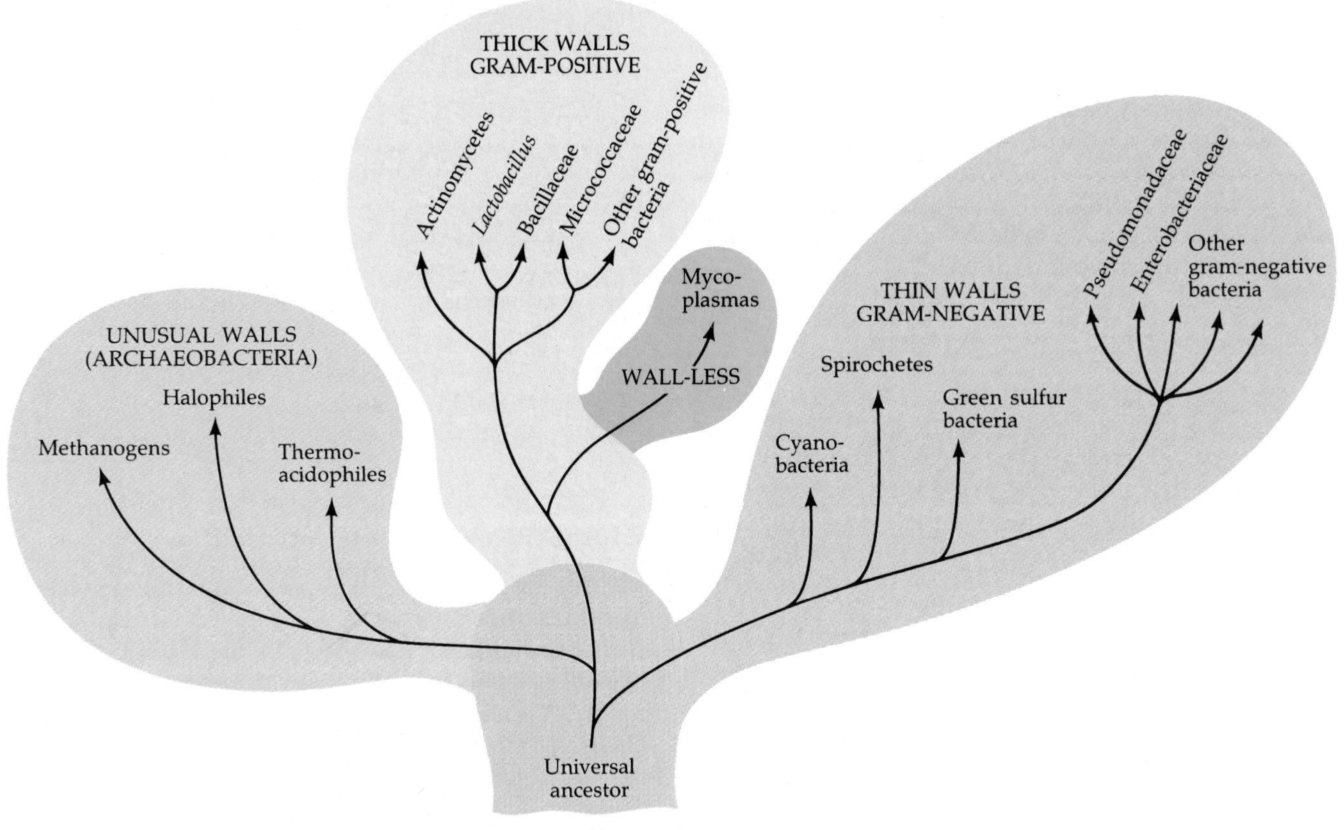

FIGURE 10.5 Phylogenetic relationships of procaryotes. Pathways indicate possible major lines of descent of bacterial groups. The DNA base composition of the mycoplasmas is more similar to that of the gram-positive bacteria than to that of the gram-negative division.

first edition in 1923. *Bergey's Manual of Determinative Bacteriology* (9th ed., 1992) does not classify bacteria according to evolutionary relatedness but provides identification *(determining)* schemes based on such criteria as cell wall composition, morphology, differential staining, oxygen requirements, and biochemical testing.

In 1984, the first edition of *Bergey's Manual of Systematic Bacteriology* was published. In recent years, microbiologists have been acquiring information on relationships among bacteria and are constructing identification schemes that reflect evolutionary or *systematic* relationships used in *Bergey's Manual of Systematic Bacteriology*.

Bergey's Manual of Systematic Bacteriology is divided into four volumes: Volume 1 includes wall-less eubacteria and some gram-negative eubacteria. Volume 2 describes gram-positive eubacteria. Volume 3 includes some gram-negative eubacteria (photosynthetic bacteria; chemolithotrophic bacteria; budding, appendaged, and sheathed bacteria; gliding and fruiting bacteria); and the archaeobacteria. Volume 4 describes the actinomycetes.

Medical microbiology (the branch of microbiology dealing with human pathogens) has dominated interest in microbes, and this interest is reflected in many identification schemes. However, to put the pathogenic properties of bacteria in perspective, of the more than 1800 species listed in the *Approved Lists of Bacterial Names*, fewer than 200 are human pathogens.

We will next discuss several criteria and methods for the classification of microorganisms and the routine identification of some. In addition to properties of the organism itself, the source and habitat of a bacterial isolate will be considered as part of the classification and identification processes.

MORPHOLOGICAL CHARACTERISTICS

Higher organisms are frequently classified according to observed anatomical detail. However, many microorganisms appear too similar to be classified by their structures. Through a microscope, there is a great deal of similarity between organisms that might differ in metabolic or physiological properties. There are literally hundreds of bacterial species that are small rods or small cocci.

Cell morphology tells us little about phylogenetic relationships. However, morphological characteristics are still useful in identifying bacteria. For example, differences in such structures as endospores or flagella can be helpful.

DIFFERENTIAL STAINING

One of the first steps in the identification of bacteria is differential staining. As we have seen from earlier chapters, most bacteria are either gram-positive or gram-negative. Other differential stains, such as the acid-fast stain, can be useful for a more limited group of microorganisms. Recall that these stains are based on the chemical composition of the cell walls and are therefore not useful in identifying either the wall-less bacteria or the archaeobacteria with unusual walls.

BIOCHEMICAL TESTS

Enzymatic activities are widely used to differentiate among bacteria. Even closely related bacteria can usually be separated into distinct species by subjecting them to biochemical tests, such as their ability to ferment an assortment of selected carbohydrates.

Enteric, gram-negative bacteria are a large heterogeneous group of microbes whose natural habitat is the intestinal tract of humans and animals. All members of the family Enterobacteriaciae are oxidase-negative. Among the enteric bacteria are members of the genera *Escherichia, Enterobacter* (en-te-rō-bak'tėr), *Shigella* (shi-gel'lä), *Citrobacter* (sit'rō-bak-tėr), and *Salmonella. Escherichia, Enterobacter,* and *Citrobacter* can be distinguished from *Salmonella* and *Shigella* according to their ability to ferment lactose—the former ferment lactose to produce acid and gas, whereas the latter do not. Further biochemical testing, as shown in Figure 10.6, can differentiate between the genera.

The time needed to identify bacteria can be reduced considerably by the use of selective and differential media or by rapid identification methods. Recall from Chapter 6 that selective media contain ingredients that suppress the growth of competing organisms and encourage the growth of desired ones, and differential media allow the desired organism to form a colony that is somehow distinctive.

Rapid identification tools are manufactured for groups of medically important bacteria, such as the enterics. Such tools are designed to perform several

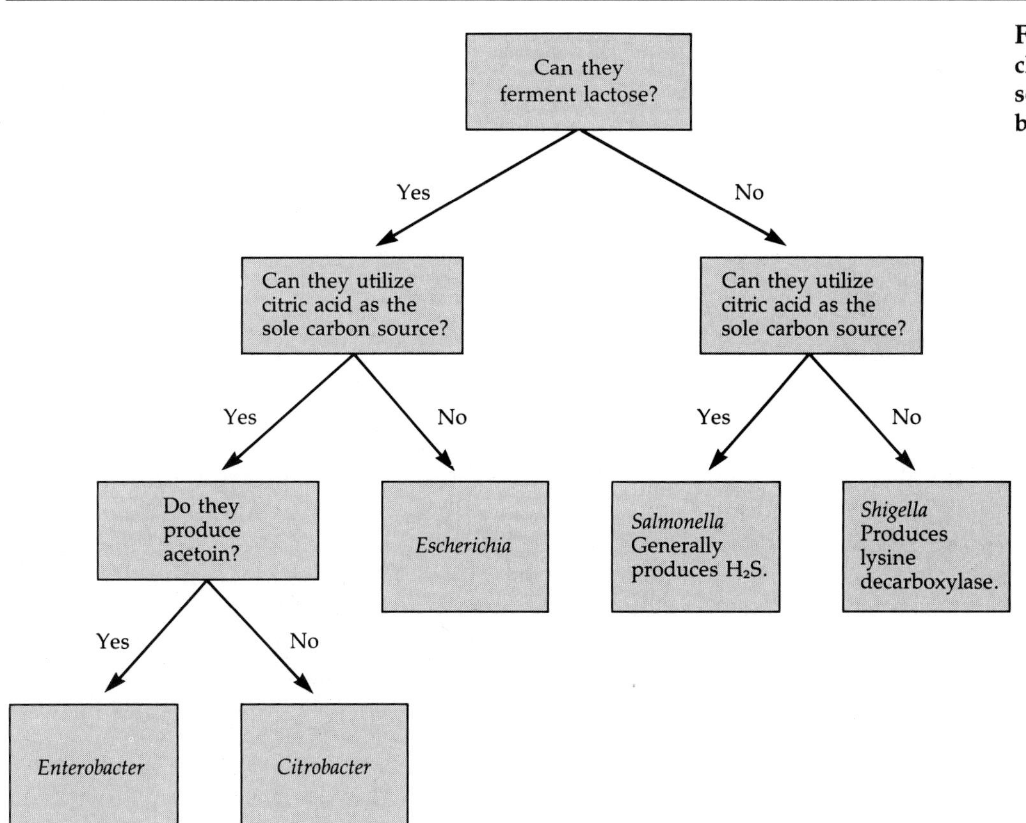

FIGURE 10.6 Metabolic characteristics used to identify selected genera of enteric bacteria.

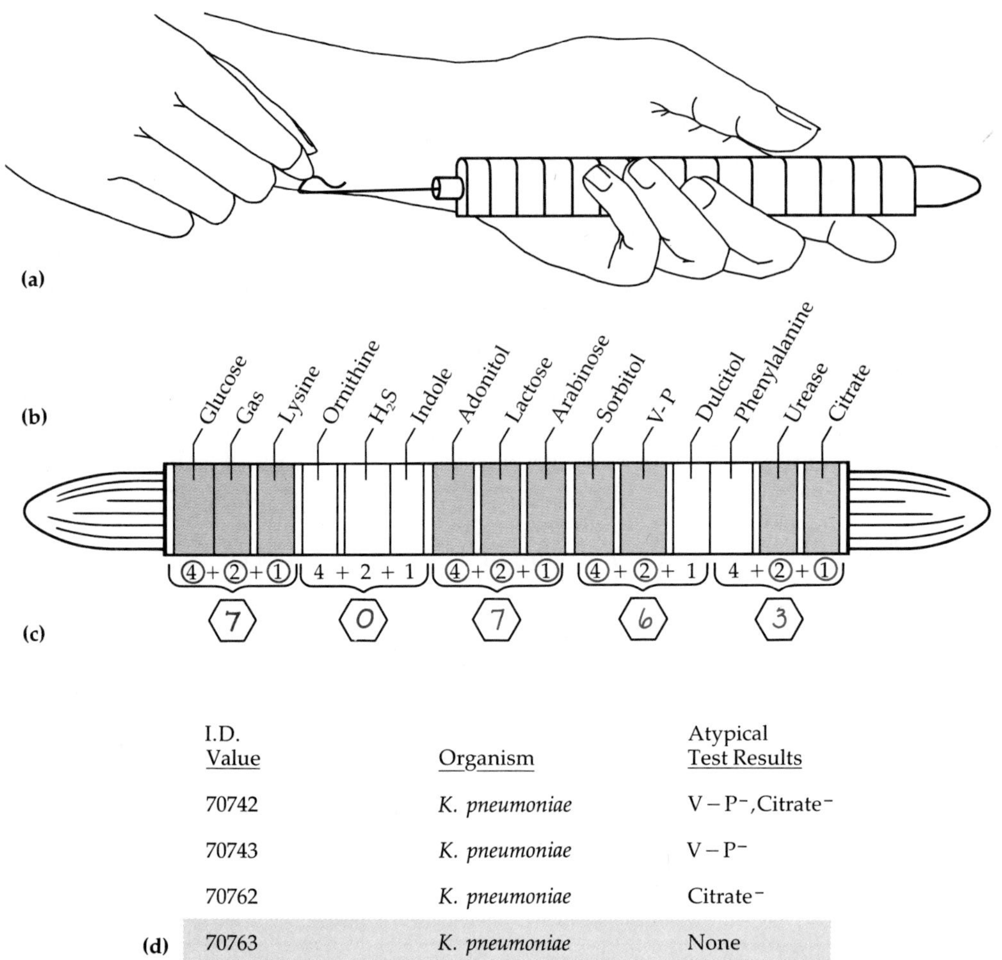

I.D. Value	Organism	Atypical Test Results
70742	*K. pneumoniae*	V – P⁻,Citrate⁻
70743	*K. pneumoniae*	V – P⁻
70762	*K. pneumoniae*	Citrate⁻
70763	*K. pneumoniae*	None

FIGURE 10.7 One type of rapid identification method for bacteria. Enterotube® II from Roche Diagnostics. **(a)** One tube containing media for 15 tests is inoculated. **(b)** After incubation, the tube is observed for positive results (colored boxes). **(c)** The value for each positive test is circled, and the numbers from each group of tests are added to give the I.D. value. **(d)** Comparing the I.D. value with a computerized listing shows that the organism in the tube is *Klebsiella pneumoniae.* Different strains of *K. pneumoniae* may produce different test results, which are listed in the Atypical Test Results column.

biochemical tests simultaneously and can identify bacteria within 4 to 24 hours. The results of each test are assigned a number (which varies with the tool being used) based on the relative reliability and importance of each test.

In the example shown in Figure 10.7, an unknown enteric bacterium is inoculated into a tube that is designed to perform 15 biochemical tests. After incubation, results in each compartment are recorded. Positive results are marked on the scoring form. Note that each test is assigned a value; the number derived from all the tests is called the I.D. value. Fermentation of glucose is important, and a positive reaction is valued at 4, as compared with the production of acetoin (V–P test), where a positive reaction is assigned the value of 2. A computerized interpretation of the simultaneous test results is essential and is provided by the manufacturer. In this example, the test results indicate that the bacterium is *K. pneumoniae,* whether the strain is V–P-positive or V–P-negative and citrate-positive or citrate-negative. *Bergey's Manual* does not evaluate the relative importance of each biochemical test and does not always describe strains. In clinical diagnosis of a disease, it is important that a particular species and even a particular strain be identified in order to proceed with proper treatment.

SEROLOGY

Serology is the science that deals with the blood serum and particularly with immune responses evident in serum (see Chapter 18). Microorganisms are antigenic; that is, injected microorganisms are capable of stimulating antibody formation in animals. For example, the immune system of a rabbit injected with killed typhoid bacteria (antigens) responds by producing antibodies against typhoid bacteria. The antibodies are proteins that circulate in the blood and combine in a highly specific way with the bacteria that caused their production. **Antisera** (singular, **antiserum**) are commercially available solutions of such antibodies used in the identification of many medically important bacteria. If an unknown bacterium is isolated from a patient, it can be tested against known antisera and often identified quickly.

In a procedure called a **slide agglutination test**, the unknown bacterium is placed in a drop of saline on a slide and mixed with a drop of known antiserum. The same bacterium is mixed with other different antisera on different slides. The bacteria agglutinate (clump) when mixed with antibodies that are produced to combine with that strain or species. A positive test is therefore determined by agglutination. A slide agglutination test is illustrated in Figure 10.8.

These serological techniques can differentiate not only among microbial species but also among strains within species. Because closely related bacteria produce some of the same antigens, serological testing is also useful in screening bacterial isolates for possible similarities. If an antiserum reacts with proteins from different bacterial strains or species, these bacteria can be tested further for relatedness.

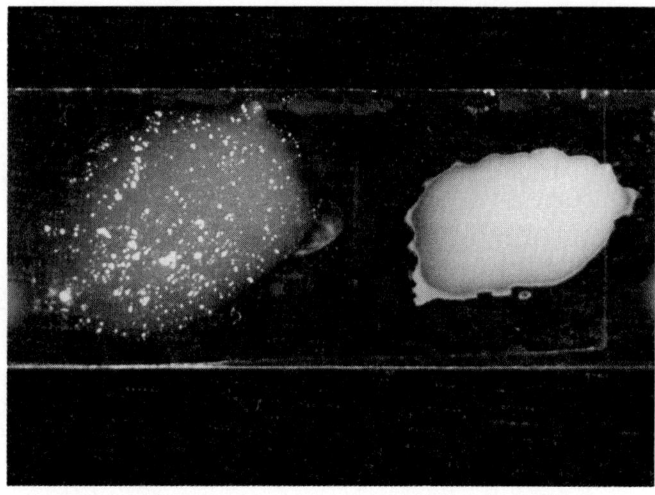

(a) Positive test **(b)** Negative test

FIGURE 10.8 Slide agglutination test. (a) Positive test. The grainy appearance is due to the clumping (agglutination) of the bacteria. Agglutination results when the bacteria are mixed with antibodies that are produced to react with the same strain. **(b)** Negative test. The bacteria are still evenly distributed in the saline and antiserum.

PHAGE TYPING

Phage typing, like serological testing, looks for similarities among bacteria. Both techniques are useful in tracing the origin and course of a disease outbreak. **Phage typing** is a test for determining which phages a bacterium is susceptible to. Recall from Chapter 8 that bacteriophages (phages) are bacterial viruses and that they usually cause lysis of the bacterial cells they infect. They are highly specialized, in that they usually infect only members of a particular species or even particular strains within a species. One bacterial strain might be susceptible to two different phages, whereas another strain of the same species might be susceptible to those two phages plus a third phage. Bacteriophages will be discussed further in Chapter 13.

To see how phage typing can help researchers trace the origin and spread of a specific strain of a bacterium, let us consider the following example. Years ago, there was a very high incidence of hospital-associated (nosocomial) staphylococcal infections. Moreover, the chain of transmission was unknown in many cases. Although such outbreaks still occur, they are less frequent because their sources can be traced by phage typing. In one version of this procedure, a plate of bacteria growing on an agar medium is marked off in small squares, and a drop of each different phage type to be used in the test is placed on the bacteria (Figure 10.9). In squares in which the phages are able to infect and lyse the bacterial cells, clearings in the bacterial growth (called plaques) appear. Such a test might show that bacteria isolated from a surgical wound have the same pattern of phage sensitivity as those isolated from the operating surgeon or surgical nurses. This establishes that the surgeon or a nurse is the source of infection.

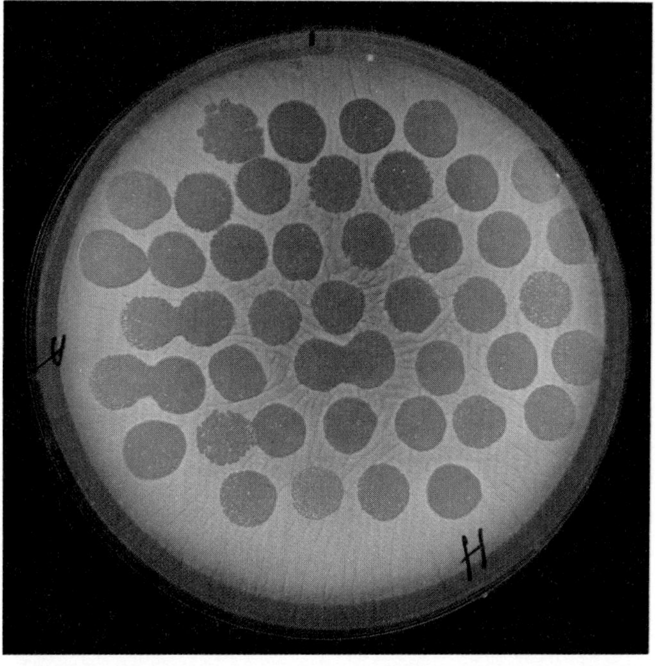

FIGURE 10.9 Phage typing of a strain of *Salmonella typhi*. The tested strain was grown over the entire plate. Plaques, or areas of lysis, were produced by bacteriophages, indicating that the strain was sensitive to infection by these phages. Accordingly, the strain is identified as *Salmonella typhi* phage type A.

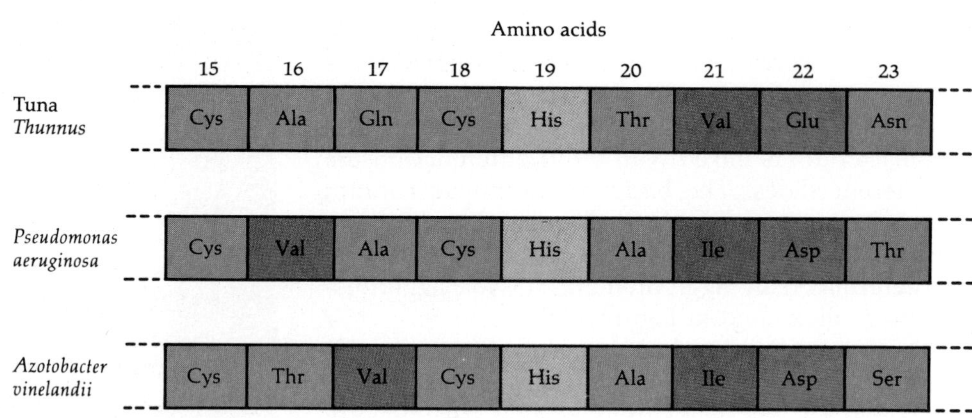

FIGURE 10.10 Amino acid sequences for a fragment of cytochrome *c* from tuna and from two species of bacteria. Three-letter abbreviations indicate the amino acids. Amino acids of the same chemical type are shown in the same color (for example, Glu and Asp are both "acidic amino acids," having an extra —COOH group). The similarity among the three sequences is striking, although the two bacteria are clearly more closely related to each other than to tuna.

AMINO ACID SEQUENCING

A great deal of information can be derived from the study of the amino acid sequences of proteins. The sequence of amino acids in a protein directly reflects the base sequence of the DNA that encodes the protein. The longer the time between the evolution of two organisms, the more changes in the DNA sequence (mutations) encoding the same protein in the two organisms. Thus, the comparison of amino acid sequences from proteins of two different organisms can help determine the similarity in the DNA sequence and therefore the evolutionary relatedness of the species. The more similar the proteins, the more closely related are the organisms. The phylogenetic tree derived for animals from traditional methods, such as fossils and comparative anatomy, is being reinforced today by such a comparison of amino acid sequences.

The sequencing of amino acids in several important proteins, such as cytochrome *c*, has been useful in the study of phylogenetic relationships among many different organisms (Figure 10.10). Amino acid sequencing is limited to use with groups of organisms that share a common protein. Also, it cannot be assumed that all organisms that share a protein are related. Although it is unlikely, it is not impossible for two unrelated organisms to evolve the same solution to a problem—for example, the production of a certain enzyme to transport a needed nutrient.

PROTEIN ANALYSIS

Amino acid sequencing focuses on the sequence of one or a few known proteins in different organisms. Protein analysis compares the pattern of all proteins found in the cells of different organisms. Since a cell's proteins are products of its genes, each species is expected to produce a unique set of proteins. This concept is the basis for making profiles of the proteins in a cell.

Protein profiles are obtained by **polyacrylamide gel electrophoresis (PAGE).** In this process, proteins from lysed cells are dissolved in a detergent and a sample of the resulting solution is placed on a thin slab of polyacrylamide gel. In a process called **electrophoresis**, an electric current is passed through the gel. While the charge is applied, different proteins migrate through the gel at different rates, which depend on the size and charge of each protein. After the gel is stained, it shows a characteristic pattern of the cells' proteins (Figure 10.11).

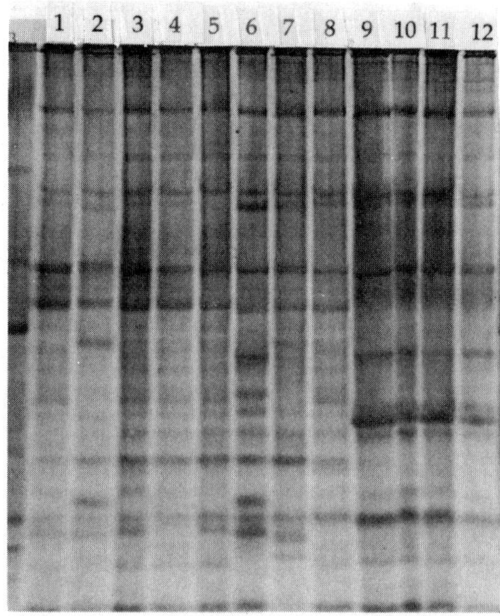

FIGURE 10.11 PAGE patterns. Twelve strains of *Bifidobacterium* were isolated from humans and other mammals. All 12 strains produced identical results from biochemical tests. The PAGE patterns, however, reveal that there are two species. Lanes 1–8 are a new species, *B. globosum;* lanes 9–12 are *B. pseudolongum.*

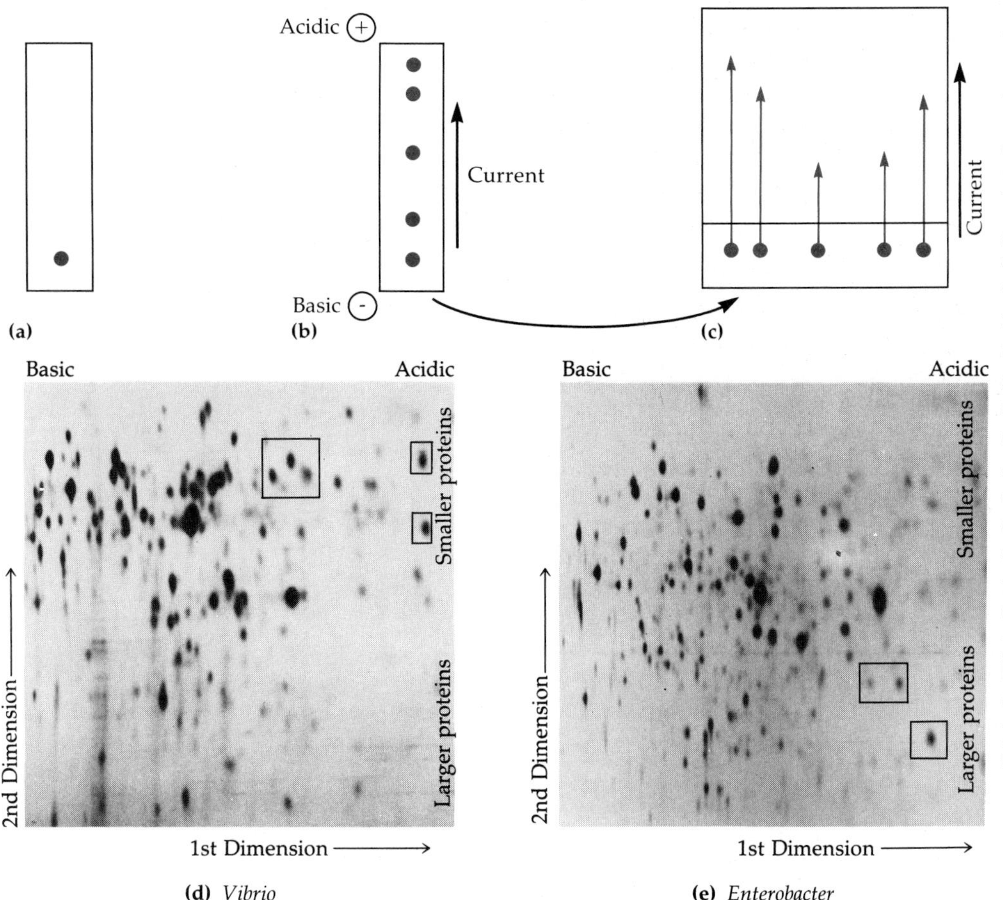

FIGURE 10.12 Two-dimensional polyacrylamide gel electrophoresis. **(a)** A sample of cell extract is placed at one end of a gel strip. **(b)** During electrophoresis, the buffer establishes a pH gradient in the gel, and charged proteins migrate through the gel to their isoelectric points. **(c)** The gel strip is then placed on a larger gel slab and submitted to a second electrophoresis at right angles to the first. In this case, a detergent is present that eliminates the differences in charge, so the migration rates of the proteins depend only on their sizes. **(d)** *Vibrio* and **(e)** *Enterobacter* protein patterns. Proteins shown in boxes are not found in both species. (Note that proteins are present in different amounts as shown by the sizes of spots.)

We assume that identical organisms have identical protein profiles, that the profiles of closely related organisms show only minor differences, and that the profiles of distantly related organisms show major differences. As procedures and materials are developed further, this method may provide a useful tool for identification as well as classification. By comparing the protein profile of the unknown cell with a set of standard profiles, a laboratory technician might be able to identify an organism within 24 hours of its isolation.

PAGE analysis is now being used to compare one group of proteins, the cytochromes, in different organisms. Although the presence of certain cytochromes is affected by the environment, analysis of bacterial cytochromes has disclosed some patterns that may be useful for classifying organisms. The following table shows the cytochromes that are common to three metabolic categories of organisms. The letters correspond to the cytochromes that are present in these cells.

ORGANISMS	***CYTOCHROMES***
Gram-positive chemoheterotrophs	b, c, a, a_3, o
Gram-negative chemoheterotrophs	b, c, d, o, a
Gram-negative chemoautotrophs	b, c, a, a_3, o, a

A limitation of the PAGE process is that different proteins may not be resolved from one another because they migrate to the same position. However, Patrick O'Farrell has developed a more sensitive version called **two-dimensional PAGE,** shown in Figure 10.12. This process differs from the PAGE process because the protein solution is exposed to electrophoresis twice, once to separate proteins by charge and again to separate them by size. Proteins from lysed cells are mixed with a special pH buffer, and a sample of the mixture is placed on a polyacrylamide gel (Figure 10.12a). When exposed to the electric field, the buffer molecules arrange themselves in a pH gradient (basic to acidic), and the proteins migrate through the gel until each reaches its **isoelectric point**—the pH at which the protein is no longer charged and therefore no longer migrates (Figure 10.12b). This process is called **isoelectric focusing.** The gel is then placed on a larger polyacrylamide slab gel and exposed to an electric field in a detergent buffer at a 90° angle to the electric field in the isoelectric focusing gel. This process separates the proteins at each pH by size (Figure 10.12c). The gel is stained to make the proteins visible. It is unlikely that two proteins will have both the same isoelectric point and the same size. Comparison of protein patterns from two bacteria shows differences be-

MICROBIOLOGY *IN THE NEWS*

Mass Deaths of Dolphins Worry Experts

During a normal year, ten or twelve dead dolphins are found washed ashore on beaches between New Jersey and Virginia. However, in one recent year the death toll climbed to more than 200, and scientists believe that hundreds more may have died offshore. Since this represents a large part of the Atlantic dolphin population, some experts are worried that this population may ultimately be destroyed. Moreover, alarmed swimmers have asked whether they themselves are also in danger from whatever is killing the dolphins. The U.S. Office of Naval Research is continuing to fund marine mammal research at colleges and universities to find the cause of the dolphins' death.

Large numbers of opportunistic pathogens, including *Edwardsiella tarda* and 55 species of *Vibrio,* were found in the dead dolphins. These bacteria are a part of the normal flora of dolphins and of coastal waters and can cause disease only if the animals' immune system, its normal defense against infection, has been weakened. To find the ultimate cause of the deaths, scientists must find out what has weakened the animals' immune systems.

One possibility is a chemical spill. High levels of polychlorinated biphenyls (PCBs), which are chemical pollutants, have been found in the 53 dolphins tested. Daniel Martineau of Cornell University says that PCBs are strong immunosuppressants. Another possibility is a virus infection that affects the immune system. The cause of thousands of seal deaths in northern Europe in April 1988 is now known to be a previously undiscovered virus. Could a new virus be infecting dolphins as well?

Marine mammal researchers examine a Pacific Bottlenosed dolphin at Marine World/Africa USA in Vallejo, California.

Information Is Scarce

Such questions are the concern of veterinary microbiology, which until recently has been a neglected branch of medical microbiology. Although the diseases of such animals as cattle, chickens, and mink have been studied—partly because of their availability to researchers—the microbiology of wild animals, especially marine mammals, is a newly emerging field. Gathering samples of animals that live in the open ocean and performing bacteriological analyses on them are very difficult. Currently, the animals being studied are those that live in captivity and those that come onto the shore to breed, such as the northern fur sea lion.

The scientists are identifying bacteria in marine mammals by using conventional test batteries and rapid-identification test kits developed for human microbiology. The bacteria are compared with species described in *Bergey's Manual* to assign names or to identify them, although they are not always a perfect match. Perhaps new species of bacteria will be found in the marine mammals.

Veterinary microbiologists hope that increased study of the microbiology of wild animals, including marine mammals, not only will promote improved wildlife management but also will provide models for the study of human diseases.

Key to selected species of human pathogens isolated from marine animals.

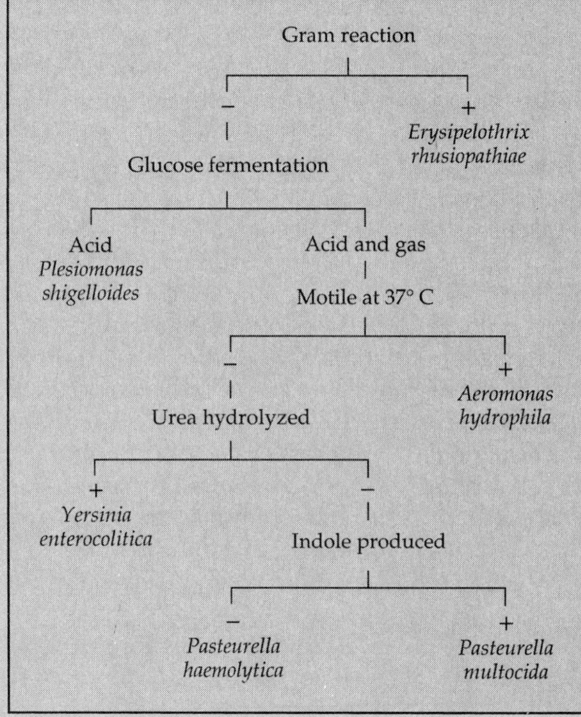

tween the two species (Figures 10.12d and 10.12e). If enough research had been done to constitute a library of these protein patterns, researchers would be able to identify unknown bacteria by comparing their two-dimensional PAGE protein patterns with those of known bacteria.

BASE COMPOSITION OF NUCLEIC ACIDS

A classification technique that has come into wide use among taxonomists because it has a good possibility of at least suggesting evolutionary relationships is the determination of the nitrogenous-base composition of DNA. This base composition is usually expressed as percentage of guanine plus cytosine (G + C). The base composition of a single species is theoretically a fixed property; thus, a comparison of the G + C content in different species can reveal the degree of species relatedness. As we saw in Chapter 8, each guanine (G) in DNA has a complementary cytosine (C). Similarly, each adenine (A) in the DNA has a complementary thymine (T). Therefore, the percentage of DNA bases that are G-C pairs also tells us the percentage that are A-T pairs (G-C + A-T = 100%). Two organisms that are closely related and hence have many identical or similar genes will have similar amounts of the various bases in their DNA. However, if there is a difference of more than 10% in their percentage of G-C pairs (for example, if one bacterium's DNA contains 40% G-C and another bacterium has 60% G-C), then these two organisms are probably not related. Of course, two organisms that have the same percentage of G-C are not necessarily closely related; other supporting data are needed to draw conclusions about evolutionary relationships.

Actually determining the entire *sequence* of bases in an organism's DNA is now possible with modern biochemical methods, but this is currently impractical for all but the smallest microorganisms (viruses). However, the use of restriction enzymes allows the comparison of base sequences of different organisms in two ways. Restriction enzymes cut a molecule of DNA everywhere a specific base sequence occurs, producing restriction fragments (as discussed in Chapter 9). For example, the enzyme *Eco* R1 cuts DNA at the arrows in the sequence

```
    ↓
...GAATTC...
...CTTAAG...
        ↑
```

In the first technique, using restriction enzymes, the DNA from two microorganisms is treated with the same restriction enzyme, and the restriction fragments that are produced are separated by electrophoresis on a thin layer of agar. A comparison of the number and sizes of restriction fragments that are produced from different organisms provides information about their genetic similarities and differences—the more similar the patterns, the more closely related the organisms are expected to be (Figure 10.13).

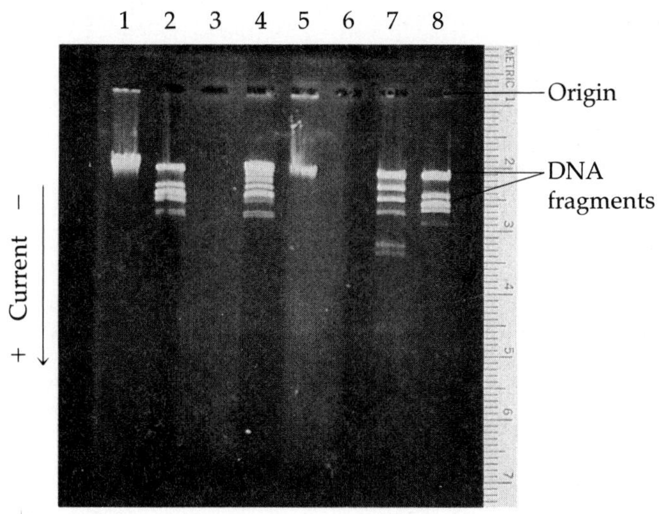

FIGURE 10.13 Plasmids from six different bacteria were digested with the same restriction enzyme. Each digest was put in a different well (origin) in the agarose gel. An electric current was then applied to the gel to separate the fragments by size and electric charge. The DNA was made visible by staining with ethidium bromide, which fluoresces under ultraviolet light. Comparison of the lanes shows that the DNA (and therefore the bacteria) in lanes 2 and 8 are identical.

Restriction enzymes can also be used to facilitate the determination of the base sequence of specific genes. Although the techniques are still somewhat tedious, base sequences can be determined for entire genes or gene fragments cut from the organism's DNA by restriction enzymes. As mentioned earlier, the sequencing of bases in the genes that code for ribosomal RNA has provided useful information about evolutionary relationships among bacteria. We assume that closely related genera have a similar base sequence that differs from that of distantly related genera. An indirect determination of the degree of similarity between the DNA nucleotide sequences of two organisms can be made with the method of nucleic acid hybridization.

NUCLEIC ACID HYBRIDIZATION

If a double-stranded molecule of DNA is subjected to heat, the complementary strands will separate as the hydrogen bonds between the bases break. If the single strands are then cooled slowly, they will reunite to form a double-stranded molecule identical to the original double strand. (This reunion occurs because the

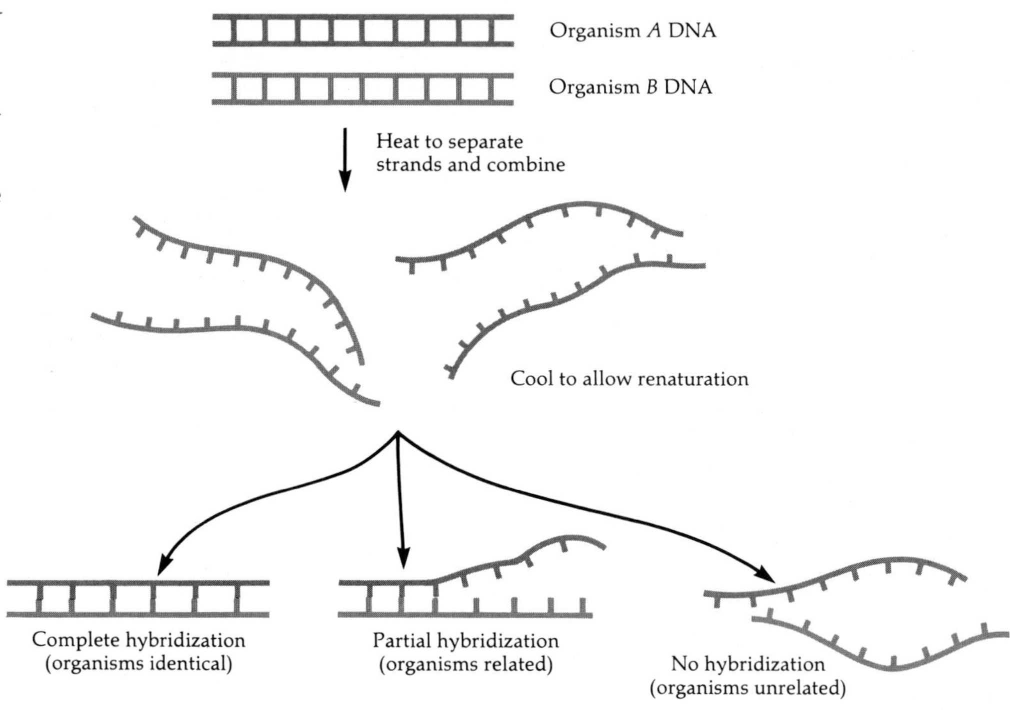

FIGURE 10.14 DNA hybridization. The greater the amount of pairing between DNA strands from different organisms (hybridization), the more closely the organisms are related.

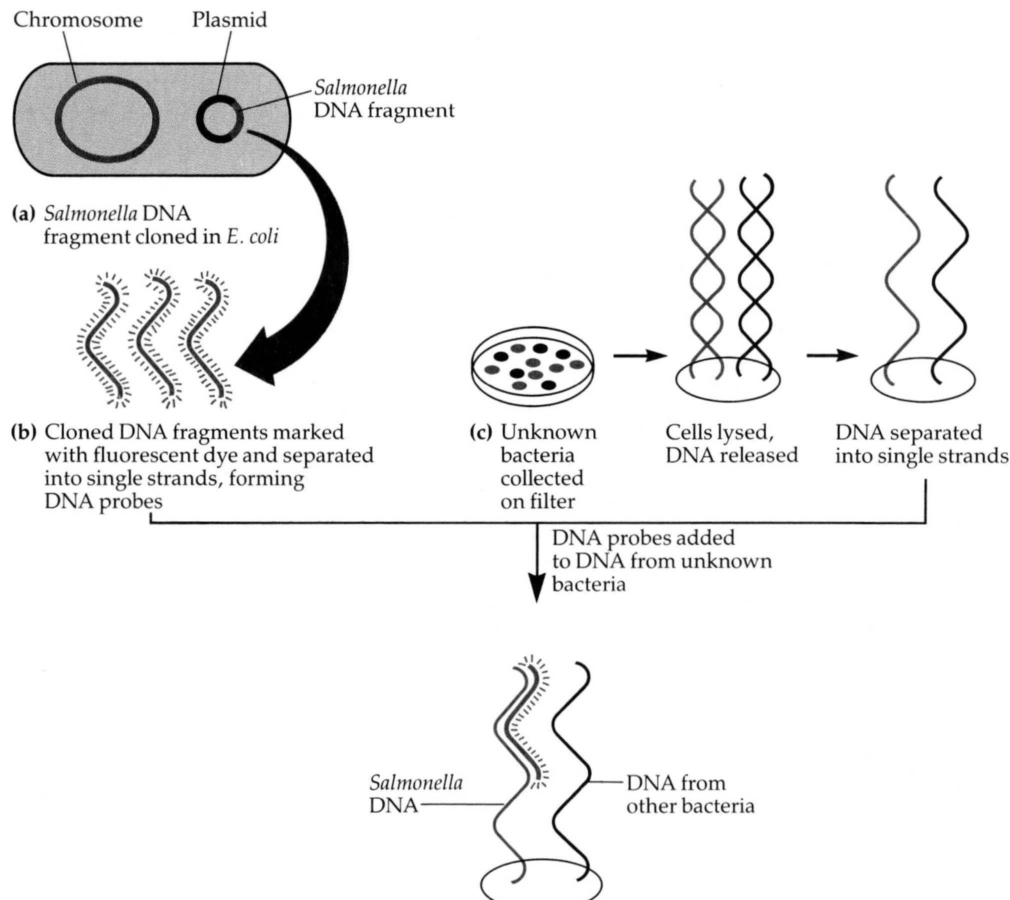

FIGURE 10.15 DNA probes used to identify bacteria. **(a)** and **(b)** Method of producing *Salmonella* DNA probes. **(c)** and **(d)** DNA probes are used to identify *Salmonella* in a sample of food.

single strands have complementary sequences.) When this technique is applied to separated DNA strands from two different organisms, it is possible to determine the extent of similarity between the base sequences of the two organisms. This method is known as **nucleic acid hybridization.** The procedure assumes that, if two species are similar or related, a major portion of their nucleic acid sequences will also be similar. The procedure measures the ability of DNA strands from one organism to hybridize (bind through complementary basepairing) with the DNA strands of another organism (Figure 10.14). The greater the degree of hybridization, the greater the degree of relatedness.

Recall from Chapter 8 that RNA is single-stranded and is transcribed from one strand of DNA; a particular strand of RNA, therefore, is complementary to the strand of DNA from which it was transcribed and will hybridize with that separated strand of DNA. DNA–RNA hybridization can thus be used to determine relatedness between DNA from one organism and RNA from another organism in the same way that DNA–DNA hybridization is used.

Nucleic acid hybridization tools called **DNA probes** are being developed for rapid identification of bacteria. For example, DNA extracted from *Salmonella* is broken into fragments by using a restriction enzyme, and a specific fragment is selected as the probe for *Salmonella.* This fragment must be able to hybridize with the DNA of all *Salmonella* strains but not with the DNA of closely related enteric bacteria. The chosen DNA fragment is cloned in a plasmid in *E. coli* (Figure 10.15a), producing hundreds of specific *Salmonella* DNA fragments. These fragments are tagged with radioactive isotopes or a fluorescent dye (Figure 10.15b) and are separated into single strands of DNA (Figure 10.15c). The resulting DNA probes can then be mixed with single-stranded DNA from a food sample suspected of containing *Salmonella* (Figure 10.15d). If *Salmonella* is present, the DNA probes will hybridize with the *Salmonella's* DNA (Figure 10.15d), and hybridization can be detected by the radioactivity or fluorescence of the probes.

TABLE 10.4 A Summary of Classification and Identification Criteria

	USED FOR	
CRITERION	**CLASSIFICATION**	**IDENTIFICATION**
Morphological characteristics	No (yes for cyanobacteria)	Yes
Differential staining	Yes (for cell wall type)	Yes
Biochemical testing	No	Yes
Serology	No	Yes
Phage typing	No	Yes
Amino acid sequencing	Yes	No
Protein analysis	Yes	No
Base composition of nucleic acids	Yes	No
Nucleic acid hybridization	Yes	Yes (DNA probes)
Flow cytometry	No	Yes
Genetic recombination	Yes	Yes (being developed)
Numerical taxonomy	Yes	No

FLOW CYTOMETRY

Flow cytometry can be used to identify bacteria in a sample without culturing the bacteria. In a *flow cytometer,* a moving fluid containing bacteria is forced through a small opening. The simplest method detects the presence of bacteria by detecting the difference in electrical conductivity between cells and the surrounding medium. If the fluid passing through the opening is illuminated by a laser, the scattering of light provides information about the cell size, shape, density, and surface, which is analyzed by a computer (See Figure 18.13). Fluorescence can be used to detect naturally fluorescent cells, such as *Pseudomonas,* or cells tagged with fluorescent dyes.

Milk is a vehicle for transmission of disease. A test that uses flow cytometry to detect *Listeria* in milk saves time because the bacteria do not have to be cultured for identification. Antibodies against *Listeria* can be labeled with a fluorescent dye and added to the milk to be tested. The milk is passed through the flow cytometer, which records the light scattering of the antibody-labeled cells.

GENETIC RECOMBINATION

Genetic recombination also might provide information useful to the classification and identification of bacteria. Recall from Chapter 8 that genetic recombination (that is, the rearrangement of genes to form new combinations) can occur by transformation, transduction, or conjugation. The use of transformation in identification of a microorganism is based on the fact that recombination between bacterial chromosomes is likely to occur only between closely related organisms. This technique is limited to those organisms that can take up a strand of DNA.

A test to identify *Neisseria gonorrhoeae* uses transformation. DNA is extracted from bacteria suspected

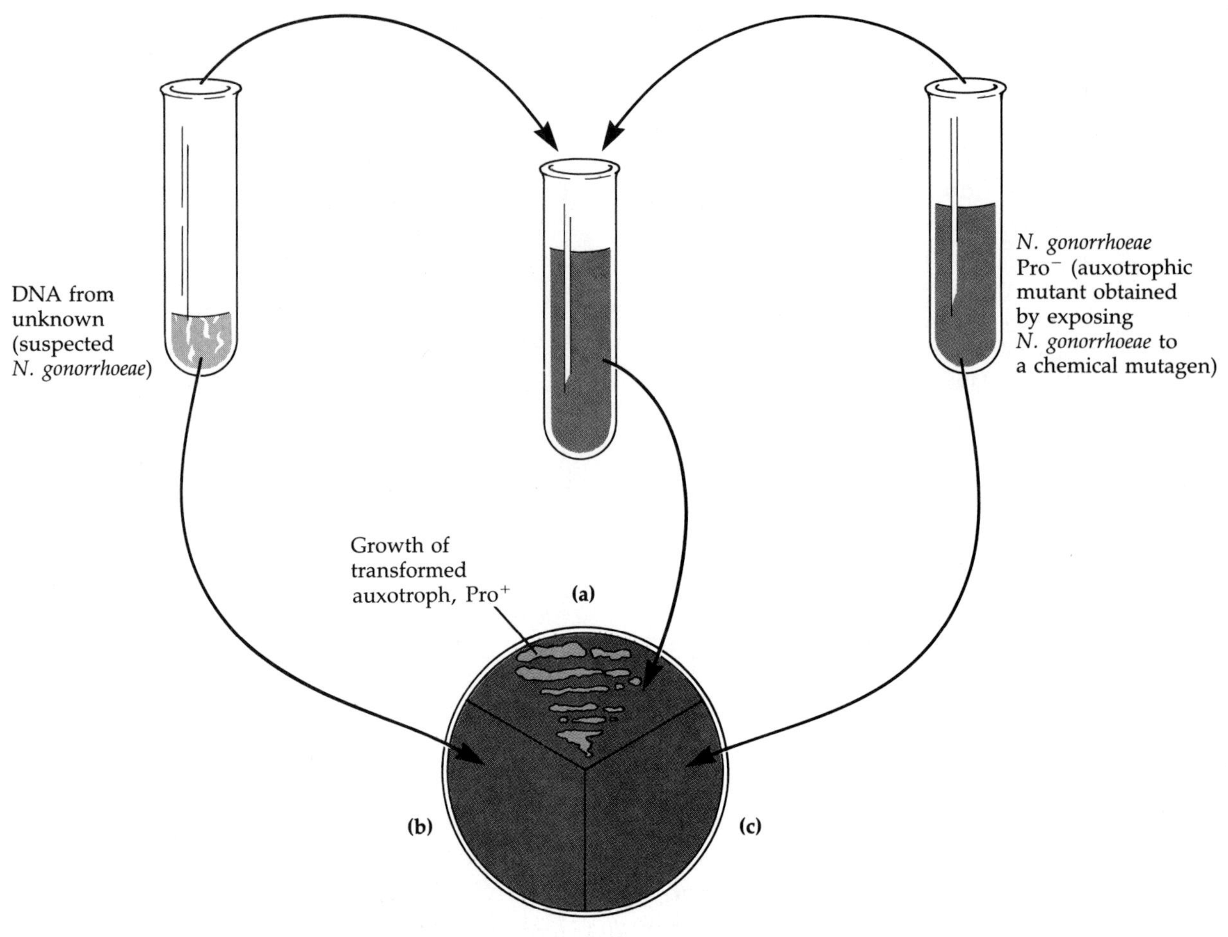

FIGURE 10.16 Using transformation to identify bacteria. If the unknown species and the auxotroph are the same species, transformation can occur, and **(a)** the resultant recombinant cell will grow. Lack of growth of the DNA alone **(b)** and the auxotroph **(c)** serves as a control.

to be *N. gonorrhoeae*. The DNA is then incubated with an auxotrophic *N. gonorrhoeae* mutant. After incubation, samples are inoculated onto a minimal medium that will not support growth of the mutant (Figure 10.16). Only transformed cells—those that have exchanged genetic material with the DNA of the normal *N. gonorrhoeae* and have acquired the necessary wild-type gene—will grow. A relatively high level of growth suggests that the unknown bacterium is *N. gonorrhoeae*.

The use of genetic recombination for identification is desirable because few media are used and a pure culture of the unknown organism is not necessary.

NUMERICAL TAXONOMY

Numerical taxonomy (Figure 10.17) has been used to infer evolutionary relationships between organisms. Many microbial characteristics are listed, such as the ability to use a certain carbohydrate, to form a pigment, or to produce a certain acid, and the presence or absence of each characteristic is scored for each organism. Amino acid sequences, the percentage of G-C pairs, and the percentage of hybridization observed between two nucleic acid strands can be listed. Tests must be selected carefully because each test is weighted equally. That is, a positive reaction is assigned a value of 1, and a negative result, 0, unlike the rapid identification methods, in which a relative weighting can be assigned to each test.

A computer then matches the characteristics of each organism against those of other organisms. The greater the number of characteristics shared by two organisms, the closer the taxonomic relationship is assumed to be. A match of 90% similarity or higher usually indicates a single taxonomic unit, or species. The computer can also generate a dendrogram (Figure 10.17c) that illustrates phylogenetic relationships de-

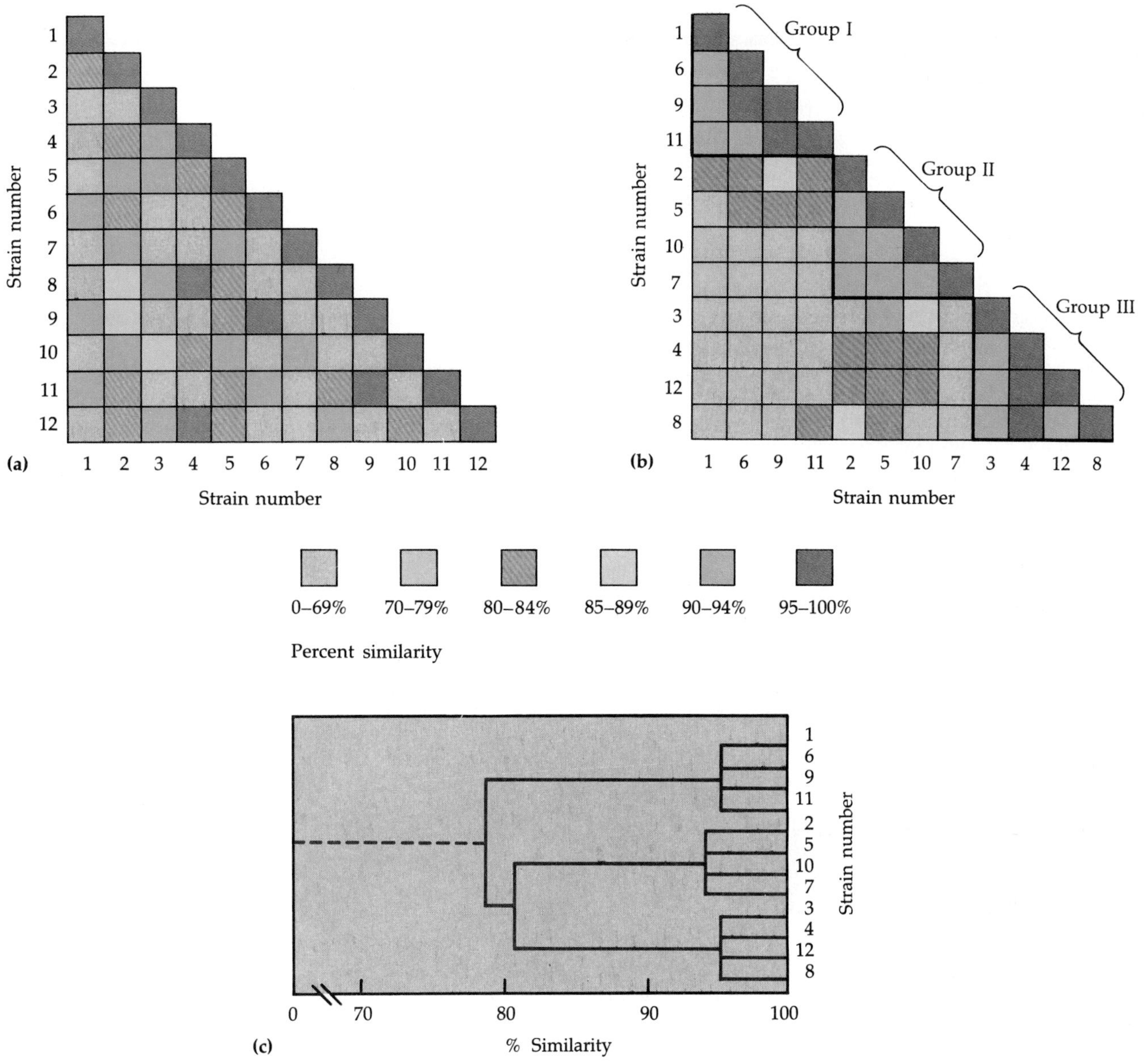

FIGURE 10.17 Numerical taxonomy. With the aid of a computer, a similarity matrix (array) can be constructed to show relationships among different organisms. As many as 100 different tests may be run on each organism. Each organism is then given a numerical score. The score is shown on a matrix **(a)**. The closer the scores of two organisms, the more similar they are. **(b)** In another matrix, organisms with similar test results are grouped together. Groups I, II, and III represent three species, each containing four strains. **(c)** A dendrogram constructed from the similarity values shows that species 1, 6, 9, and 11 are closely related to each other but not to the other organisms tested. The remaining strains are related to each other; they form a family that is divided into two genera, each consisting of four species.

rived from similarities from hybridization and other genetic experiments.

Morphological characteristics, differential staining, and biochemical testing were the only identification tools available just a few years ago. Technological advancements are making it possible to use nucleic acid and protein analysis techniques, once reserved for classification, for routine identification. (A practical example is described in the box in Chapter 20, p. 495.) A summary of taxonomic criteria and their uses is provided in Table 10.4.

STUDY OUTLINE

Phylogenetic Relationships (pp. 250–254)

1. Taxonomy is the science of the classification of organisms, with the goal of showing relationships among organisms.
2. Taxonomy provides a means to identify organisms.

A FIVE-KINGDOM SYSTEM (p. 251)

1. Living organisms can be classified into five kingdoms.
2. Procaryotic organisms are placed in the Kingdom Monera (Procaryotae).
3. Eucaryotic organisms may be classified into the Kingdoms Protista, Fungi, Plantae, or Animalia.
4. Protista are mostly unicellular organisms.
5. Fungi are absorptive heterotrophs that develop from spores.
6. Multicellular photoautotrophs are placed in the Kingdom Plantae.
7. Multicellular ingestive heterotrophs are classified as Animalia.

A THREE-KINGDOM SYSTEM (pp. 251–254)

1. Another proposal divides living organisms into three kingdoms.
2. In this system, plants, animals, fungi, and protists belong to a single kingdom.
3. Eubacteria (with peptidoglycan) form a second kingdom.
4. Archaeobacteria (with unusual cell walls) form a third kingdom.

Classification of Organisms (pp. 254–257)

SCIENTIFIC NOMENCLATURE (pp. 254–255)

1. According to scientific nomenclature, each organism is assigned two names (binomial nomenclature), a genus and a specific epithet, or species.
2. Rules for the assignment of names to bacteria are published in the *Bacteriological Code.*

THE TAXONOMIC HIERARCHY (pp. 255–256)

1. A eucaryotic species is a group of organisms that interbreeds but does not breed with individuals of another species.
2. Similar species are grouped into a genus; similar genera are grouped into a family; families, into an order; orders, into a class; classes, into a division or phylum; and phyla, into a kingdom.

A PHYLOGENETIC HIERARCHY (pp. 256–257)

1. Organisms are grouped into taxa according to phylogenetic (from a common ancestor) relationships.
2. Some of the information for eucaryotic relationships is obtained from the fossil record.
3. Procaryotic relationships are determined by molecular biology techniques.

CLASSIFICATION OF BACTERIA (p. 257)

1. *Bergey's Manual of Systematic Bacteriology* is the standard reference on bacterial classification.
2. In *Bergey's Manual,* bacteria are divided into four divisions.
3. A division is divided into classes; classes are composed of orders; orders consist of families; a family is divided into genera; and a genus is composed of species.
4. A group of bacteria derived from a single cell is called a strain.
5. Closely related strains constitute a species.

CLASSIFICATION OF VIRUSES (p. 257)

1. Viruses are not placed in a kingdom. They are not composed of cells and are incapable of growth without a host cell.

Criteria for Classification and Identification of Microorganisms (pp. 257–269)

1. *Bergey's Manual of Systematic Bacteriology* and *Bergey's Manual of Determinative Bacteriology* are references for laboratory identification of bacteria.
2. Morphological characteristics are useful in the identification of microorganisms, especially when aided by differential staining techniques.
3. The possession of various enzymes as determined by biochemical tests is used in the identification of microorganisms.
4. Serological tests, involving the reactions of microorganisms with specific antibodies, are useful in determining the identity of strains and species as well as relationships between organisms.
5. Phage typing is the identification of bacterial species and strains by the determination of their susceptibility to various phages.
6. The sequences of amino acids in proteins of related organisms are similar.
7. Related organisms have identical proteins; this characteristic can be ascertained by PAGE "fingerprints."
8. The percentage of G-C pairs in the nucleic acid of cells can be used in the classification of organisms.
9. The number and sizes of DNA fragments produced by restriction enzymes are used to determine genetic similarities.
10. Single strands of DNA, or of DNA and RNA, from related organisms will hydrogen-bond to form a double-stranded molecule; this bonding is called nucleic acid hybridization.
11. Flow cytometry measures physical and chemical characteristics of cells.
12. Transformation of an auxotrophic mutant by the DNA of an unknown bacterium can be used in identification of the unknown bacterium.
13. Grouping many different characteristics to show relatedness is called numerical taxonomy.

STUDY QUESTIONS

Review

1. What is taxonomy?
2. List and define the five kingdoms used in the five-kingdom system of classification.
3. What characterizes the Kingdom Procaryotae, and what organisms are placed in this kingdom?
4. Discuss the evidence that supports a three-kingdom system of classification.
5. What is binomial nomenclature?
6. Why is binomial nomenclature preferable to the use of common names?
7. Put the following terms in the correct sequence from the most general to the most specific: order, class, genus, kingdom, species, division, family.
8. Define species.
9. List the eleven bases discussed in this chapter for the classification of microorganisms. Separate your list into those tests used primarily for taxonomic classification and those used primarily for identification of microorganisms already classified.
10. Higher organisms are arranged into taxonomic groups on the basis of evolutionary relationships. Why is this type of classification only being developed for bacteria?
11. Describe the use of a DNA probe and transformation for:
    (a) Rapid identification of an unknown bacterium.
    (b) Determining which of a group of bacteria are most closely related.
12. Using *Bergey's Manual,* identify the organism with the following characteristics: gram-negative rod; facultative anaerobe; motile; oxidase-negative; fermentative; acid and gas from lactose; indole produced; methyl red positive; V–P$^-$; citrate negative; H_2S not produced; does not grow in KCN broth.
13. Can you tell which of these organisms are most closely related? Are any two the same species? On what did you base your answer?

Characteristic	A	B	C	D
Morphology	Rod	Coccus	Coccus	Rod
Gram reaction	+	–	–	+
Glucose utilization	Fermentative	Oxidative	Fermentative	Fermentative
Cytochrome oxidase	Present	Present	Absent	Absent
G-C Moles %	48–52	23–40	30–40	49–53

Challenge

1. Here is some additional information on the organisms in Question 13:

Organisms	% DNA Hybridization
A and B	5–15
A and C	5–15
A and D	70–90
B and C	10–20
B and D	2–5

Which of these organisms are most closely related? Compare this answer with your response to Question 13.

2. In *Bergey's Manual,* the gram-positive cocci include the genera *Micrococcus* and *Staphylococcus.* The reported G + C content of *Micrococcus* is 66 to 75 moles %, and of *Staphylococcus,* 30 to 40 moles %. According to this information, would you conclude that these two genera are closely related?
3. Compare and contrast *Bergey's Manual of Systematic Bacteriology* and *Bergey's Manual of Determinative Bacteriology.*

FURTHER READING

DeLong, E.F., G.S. Wickham, and N.R. Pace. "Phylogenetic stains." *Science* 243:1360–1363, March 10, 1989. Describes the potential for using RNA probes to classify and identify single cells.

Goodfellow, M., and R.G. Board, eds. *Microbiological Classification and Identification.* New York: Academic Press, 1980. Comprehensive coverage of modern identification techniques, including genetics, immunology, and protein electrophoresis.

Holt, J.G., ed. *Bergey's Manual of Systematic Bacteriology*, 1st ed., 4 volumes. Baltimore: Williams and Wilkins, 1984. A discussion of bacterial classification that provides a perspective on the relative importance of classification and identification is in the front of each volume.

Isenberg, H.D., ed. *Clinical Microbiology Procedures Handbook.* Washington, D.C.: American Society for Microbiology, 1991. Provides useful methods for standard bacteriological tests for identification and includes rapid methods and molecular biology techniques.

Knoll, A.H. "The distribution and evolution of microbial life in the late Proterozoic Era." *Annual Review of Microbiology* 39:391–417, 1985. Describes fossil and chemical evidence for early life and provides living models for early life.

Margulis, L. *Early Life.* Boston: Science Books International, 1982. A good book that explains hypotheses and experiments pertaining to the evolution of procaryotes, eucaryotes, fermentation, and respiration.

Schleifer, K.H., and E. Stackebrandt. "Molecular systematics of procaryotes." *Annual Review of Microbiology* 37:143–187, 1983. Describes modern methods of taxonomy for rapid identification and for classifying bacteria.

Woese, C.R. "Archaebacteria." *Scientific American* 244(6):98–122, June 1981. Gives detailed descriptions of the archaeobacteria and comparisons with eubacteria and eucaryotes.

CHAPTER 11

Bacteria

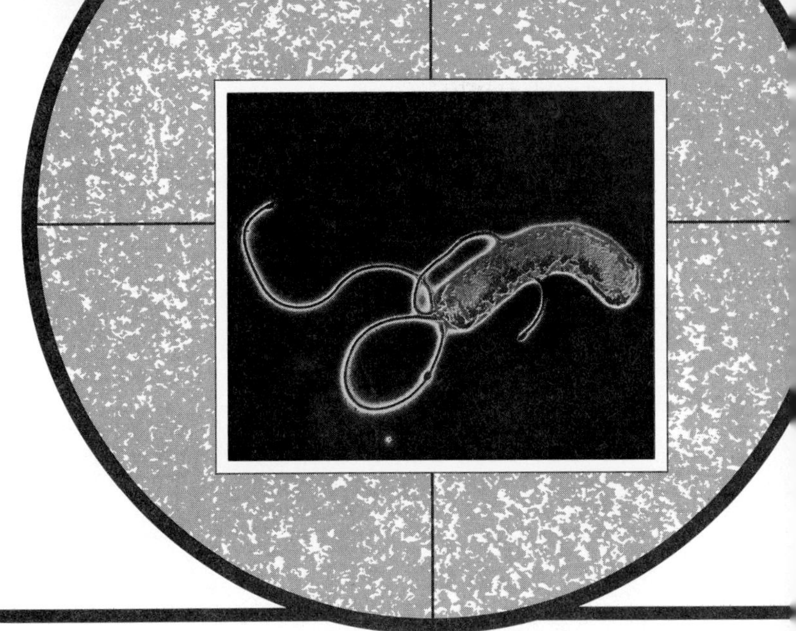

LEARNING OBJECTIVES

- List at least six characteristics used to classify and identify bacteria according to *Bergey's Manual of Systematic Bacteriology.*
- List two major characteristics of each section of *Bergey's Manual of Systematic Bacteriology* described in this chapter.
- Identify the sections in *Bergey's Manual* that contain species of medical importance.

Bacteria are the most metabolically diverse of all microorganisms. Relatively few species of bacteria cause disease in humans or in other organisms, and without the activities of the many other kinds of bacteria, all life on Earth would cease. Indeed, one widely accepted theory suggests that eucaryotic organisms probably evolved from bacterialike organisms. Because bacteria are relatively simple in structure and many are readily cultured and controlled in the laboratory, microbiologists have devoted intensive study to their life processes. From such studies, we have learned that the basic processes of life are the same for all organisms.

Bacterial Groups

This chapter will introduce you to many of the more common bacteria. All bacteria share the property of procaryotic cellular organization, but it is difficult to correlate structural and functional characteristics to show evolutionary relationships. As we mentioned in the previous chapter, the most widely accepted taxonomic classification for bacteria is *Bergey's Manual of Systematic Bacteriology.* The most recent revision of *Bergey's Manual* divides bacteria into four divisions (or phyla) according to the characteristics of cell walls (see Table 10.3). Each division is subdivided into *sections* according to such characteristics as Gram stain reaction, cell shape, cell arrangements, oxygen requirements, motility, and nutritional and metabolic properties. Sections are named according to these characteristics—for example, *spirochetes, gram-negative aerobic rods and cocci, anoxygenic photosynthetic bacteria,* and *endospore-forming rods and cocci.* Each section consists of a number of genera. In some sections, genera are grouped into families and orders; in other sections, they are not (see Table 11.1 and Appendix A).

TABLE 11.1 Summary of Selected Characteristics of Bacterial Groups, from *Bergey's Manual of Systematic Bacteriology*, 1st Edition

NAME OF SECTION	IMPORTANT GENERA	HABITAT	GRAM REACTION	SPECIAL FEATURES
Spirochetes	*Treponema, Borrelia, Leptospira*	Aquatic; animal parasites	–	Helical morphology; motility by axial filaments; several important pathogens
Aerobic/microaerophilic, motile, helical/vibrioid, gram-negative bacteria	*Spirillum, Azospirillum, Campylobacter, Bdellovibrio*	Soil and aquatic environments; human intestinal tract and oral cavity	–	Helical morphology; motility by flagella, not axial filaments; vibrioids do not have a complete turn. Includes nitrogen-fixing bacteria and some pathogens
Nonmotile (or rarely motile), gram-negative, curved bacteria*	*Spirosoma, Meniscus*	Aquatic and sedimentary environments	–	Uncommon; mostly aquatic, not pathogenic; form S-shapes, C-shapes, rings
Gram-negative; aerobic rods and cocci	*Pseudomonas, Legionella, Neisseria, Brucella, Bordetella, Francisella, Rhizobium, Agrobacterium*	Soil; water; animal parasites	–	Organisms of medical, industrial, and environmental importance
Facultatively anaerobic, gram-negative rods	*Escherichia Salmonella, Shigella, Klebsiella, Yersinia, Vibrio, Enterobacter, Hemophilus, Gardnerella, Pasteurella*	Soil; plants; animal intestinal tracts	–	Many important pathogens
Anaerobic, gram-negative straight, curved, or helical rods	*Bacteroides, Fusobacterium*	Animals and insects	–	Obligate anaerobes, mostly of intestinal tract; some common in mouth and genital tract
Dissimilatory sulfate-reducing or sulfur-reducing bacteria	*Desulfovibrio*	Anaerobic sediments	–	Reduce oxidized forms of sulfur to H_2S
Anaerobic gram-negative cocci	*Veillonella*	Mostly animal intestinal tracts	–	Nonmotile anaerobes
Rickettsias and chlamydias	*Rickettsia, Coxiella, Chlamydia*	Parasites of arthropods and animals	–	Obligate intracellular bacteria; many important pathogens
Mycoplasmas	*Mycoplasma*	Parasites of animals, plants, insects	–	Pleomorphic; lack cell walls; some pathogens
Endosymbionts*	*Holospora, Blattabacterium*	Intracellular symbionts of protozoa, insects, helminths, plants, etc.	–	Assorted bacteria that live symbiotically in protozoa, insects, and fungi
Gram-positive cocci	*Staphylococcus, Streptococcus*	Soils; skin and mucous membranes of animals	+	Some important pathogens

(Continued)

*Not discussed in this chapter. See Appendix A.

TABLE 11.1 ***(Continued)***

NAME OF SECTION	IMPORTANT GENERA	HABITAT	GRAM REACTION	SPECIAL FEATURES
Endospore-forming rods and cocci	*Bacillus, Clostridium*	Soil; animal intestinal tract	+	*Bacillus,* aerobic or facultative anaerobes; *Clostridium,* anaerobic
Regular, nonsporing, gram-positive rods	*Lactobacillus, Listeria*	Dairy products; genital and oral cavities, animal feces	+	*Lactobacillus* forms lactic acid from carbohydrates, important industrially; *Listeria,* an animal pathogen
Irregular, nonsporing, gram-positive rods	*Gardnerella, Corynebacterium, Propionibacterium, Actinomyces*	Human pathogens, soil organisms	+	Pleomorphic morphology; several important pathogens
Mycobacteria	*Mycobacterium*	Soil; plants; animals	+	*Mycobacterium* is an important pathogen; acid-fast
Nocardioforms	*Nocardia*	Soil and animals	+	Form branched filaments; reproduce by fragmentation; often acid-fast; some pathogens
Budding and/or appendaged bacteria	*Hyphomicrobium, Caulobacter*	Mostly aquatic, some soil	–	Possess prosthecae; *Hyphomicrobium* reproduces by budding; *Caulobacter* is stalked
Sheathed bacteria	*Sphaerotilus*	Aquatic cause of sewage treatment problems	–	Cells encased in hollow sheath
Nonphotosynthetic, nonfruiting, gliding bacteria	*Cytophaga, Beggiatoa*	Aquatic	–	*Cytophaga* degrades cellulose; *Beggiatoa* oxidizes H_2S
Gliding, fruiting bacteria	*Myxococcus*	Dung, soil	–	Cells aggregate to form a fruiting body
Aerobic chemoautotrophic bacteria	*Nitrosomonas, Nitrobacter, Thiobacillus*	Soil	–	Nitrifying and sulfur-oxidizing bacteria; agriculturally and environmentally important
Archaeobacteria	*Methanobacterium, Halobacterium, Sulfolobus*	Anaerobic sediments; environments of extreme temperature and osmotic pressure	Varies	Not related to other bacterial groups; no peptidoglycan in cell walls; methane producers useful in sewage treatment
Anoxygenic photosynthetic bacteria	*Chromatium, Rhodospirillum, Chlorobium*	Anaerobic sediments	–	Includes the green and purple sulfur and nonsulfur bacteria; green and purple sulfur bacteria use H_2S as an electron donor and release sulfur
Oxygenic photosynthetic bacteria (cyanobacteria)	*Chroococcus, Anabaena*	Aquatic	–	Produce oxygen during photosynthesis; many species fix atmospheric nitrogen
Actinomycetes	*Streptomyces, Frankia, Micromonospora*	Soil; some aquatic	+	Common in soil; branching filaments with reproductive conidiospores; *Frankia* involved in nitrogen-fixing symbiosis with plants

This chapter briefly describes the principal bacterial groups according to the basic organization of *Bergey's Manual.* Our discussion emphasizes bacteria considered to be of practical importance, those important in medicine, or those that illustrate biologically unusual or interesting principles.

SPIROCHETES

The spirochetes are found in contaminated water, in sewage, soil, and decaying organic matter, and within the bodies of humans and animals. One of the first microorganisms described by Anton van Leeuwen-

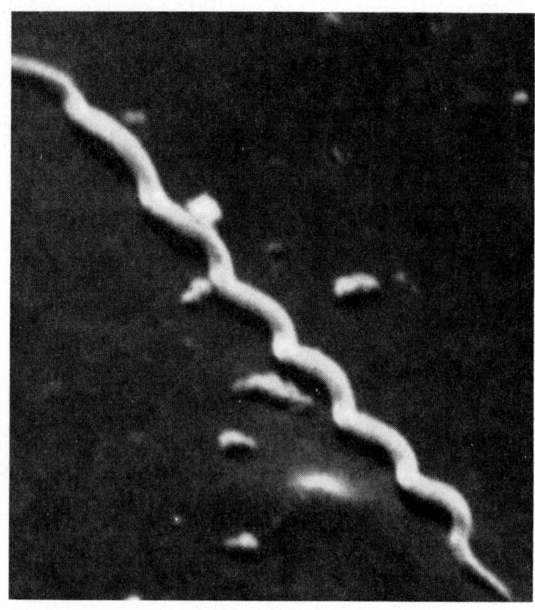

FIGURE 11.1 ***Treponema pallidum.*** Spirochetes such as *T. pallidum,* the causative agent of syphilis, are helical and have axial filaments under their outer sheath that allow them to move by a corkscrewlike rotation.

hoek in the 1600s was a large spirochete taken from saliva and tooth scrapings. These bacteria are typically coiled, like a metal spring. Some are tightly coiled; others resemble a spring that is stretched.

All spirochetes are actively motile and achieve their motility by means of two or more *axial filaments.* The axial filaments are enclosed in the space between an outer sheath and the body of the cell.

One end of each axial filament is attached near a pole of the cell (see Figure 4.7). By rotating its axial filament, the cell rotates in the opposite direction to move through liquids like a corkscrew. The rotating flagella of flagellated bacteria also conform to this motion (see Chapter 4), which is very efficient in moving bacteria through liquids. To a bacterium, water is as viscous as molasses or tar is to an animal. However, a bacterium can typically move about 100 times its body length in a second (or about 50 μm/sec), while a large fish such as a tuna can move only about 10 times its body length in this time.

The spirochetes can be aerobic, facultatively anaerobic, or anaerobic; they do not have flagella or endospores.

The spirochetes include a number of important pathogenic bacteria. The best known is the genus *Treponema,* which includes *Treponema pallidum,* the cause of syphilis (Figure 11.1). Members of the genus *Borrelia* cause relapsing fever and Lyme disease, serious diseases that are usually transmitted by ticks or lice.

Leptospirosis is a disease usually spread to humans by water contaminated by *Leptospira* (lep-tō-spī'rä) species. The bacteria are excreted in the urine of such animals as dogs, rats, and swine, so domestic dogs and cats are routinely immunized against it. The cells of *Leptospira* typically have hooked ends (see Figure 26.4).

AEROBIC/MICROAEROPHILIC, MOTILE, HELICAL/VIBRIOID GRAM-NEGATIVE BACTERIA

A distinctly helical shape does not serve as the universal criterion for classification of a bacterium as a spirochete. Some helical bacteria are not included with the spirochetes because they lack an axial filament. They use flagella instead. These bacteria possess a single flagellum at one or both poles or sometimes tufts of flagella at these locations (Figure 11.2a). Spiral bacteria, unlike spirochetes, are rigid helices or curved rods. These bacteria tend to be microaerophilic.

Most spiral bacteria are harmless aquatic organisms, such as *Spirillum volutans* (vol'ū-tans). This is an unusually large bacterium, up to 60 μm long. The box in Chapter 4, p. 76, describes a member of the genus *Aquaspirillum* that is able to respond to magnetic fields.

Agricultural microbiologists have been interested in members of the genus *Azospirillum* (ā-zō-spī-ril'lum), a soil bacterium that grows in close association with the roots of many plants, especially tropical grasses. It utilizes nutrients excreted by the plants and in return fixes nitrogen from the atmosphere. This nitrogen fixation is most significant in some tropical grasses and in sugar cane, although the organism can be isolated from the root system of many temperate climate plants, such as corn.

The spiral bacteria also include pathogenic organisms. One species, *Campylobacter fetus* (kam-pī-lō-bak'tėr fē'tus), causes abortion in domestic animals. Another species, *C. jejuni* (jē-jū'nē), causes outbreaks of foodborne intestinal disease (see Chapter 25). *Helicobacter* (hē'lik-ō-bak-tėr) causes ulcers in humans (see Figure 11.2b and the box in Chapter 15, p. 398.

Sometimes, helical bacteria do not make a complete turn or twist; these are **vibrioids** (see the box in Chapter 3, p. 62). A particularly interesting example is the comma-shaped bacterium *Bdellovibrio,* which attacks other bacteria. It reproduces within the periplasmic space of the gram-negative bacteria that it preys upon. However, if there are no prey bacteria available, it follows a different reproductive cycle. The cell elongates into a tight spiral, which then fragments almost

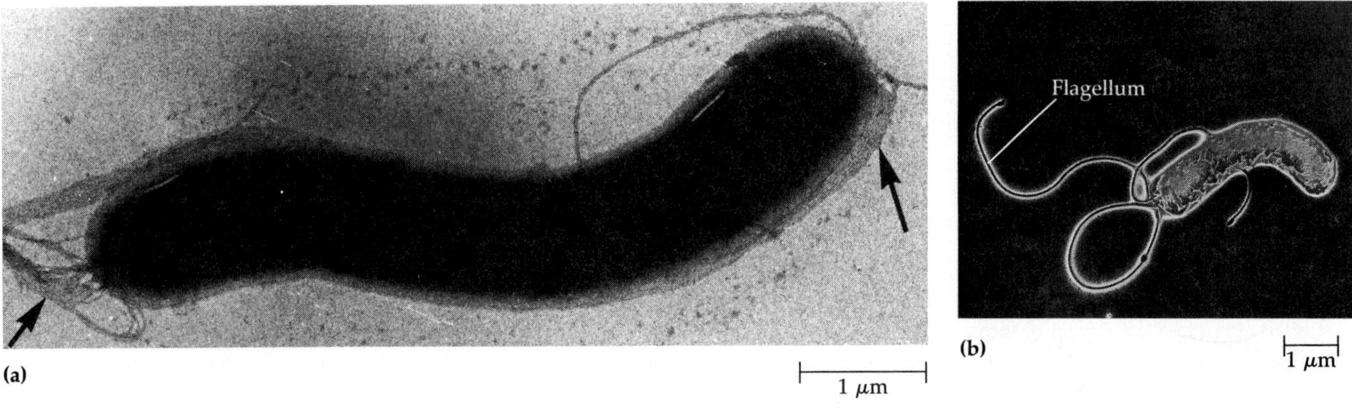

FIGURE 11.2 Helical/vibrioid bacteria. (a) Spiral cell of *Aquaspirillum bengal,* which is motile by means of flagella. The arrows point to the tufts of flagella at each end of the cell. **(b)** *Helicobacter pylori,* an example of a helical bacterium that does not make a complete twist, resulting in a comma shape.

simultaneously into a number of individual flagellated cells.

For a discussion of an especially interesting group of spirochetes, see the box on p. 278.

GRAM-NEGATIVE AEROBIC RODS AND COCCI

This group contains many microbial species of medical, industrial, and environmental importance. Some of the more interesting are those of the genus *Pseudomonas* (Figure 11.3). The common name for these bacteria is *pseudomonads.* These organisms are rod-shaped and have polar flagella. Many species excrete extracellular, water-soluble pigments that diffuse into their media. One species, *Pseudomonas aeruginosa,* produces a characteristically soluble, blue-green pigmentation. Under the right conditions, particularly in weakened hosts, this organism can infect the urinary tract, burns, and wounds and can cause septicemia, abscesses, and meningitis. Other pseudomonads produce soluble fluorescent pigments that glow when illuminated by ultraviolet light.

Pseudomonads are very common in soil and other natural environments and are generally no threat to a healthy individual. These bacteria are less efficient than some other heterotrophic bacteria in utilizing many of the common nutrients, but they have compensating characteristics. For example, many are psychrophilic (or psychrotrophic) and grow at refrigerator temperatures; these impart off tastes and colors to foods. Pseudomonads are also capable of synthesizing an unusually large number of enzymes and probably contribute significantly to the decomposition of chemicals, such as pesticides, that are added to soils.

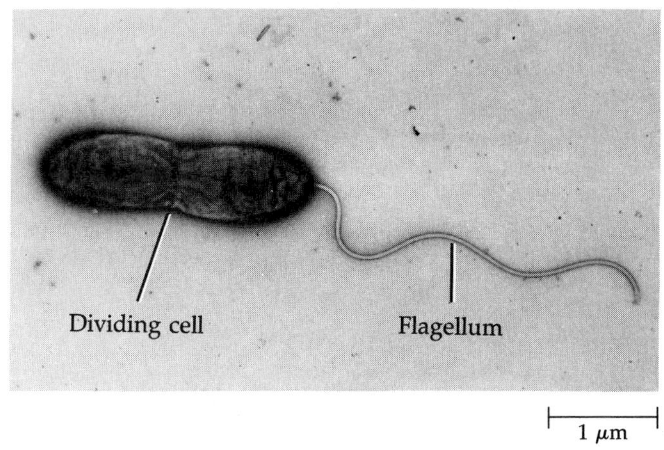

FIGURE 11.3 ***Pseudomonas andropogonis.*** Note the single polar flagellum.

In hospitals and other places where pharmaceuticals are prepared, the ability of pseudomonads to grow on minute traces of unusual carbon sources, such as soap residues or cap liner adhesives found in a solution, has been unexpectedly troublesome (see the box in Chapter 7, p. 180). They are even capable of growth in some antiseptics, such as quarternary ammonium compounds. Their resistance to most antibiotics has also been a source of medical concern. This resistance is related to the characteristics of the cell wall porins (see Chapter 4). Some of these considerations will be discussed in more detail later. Many pseudomonad genes that code for unusual metabolic characteristics, including resistance to antibiotics, are carried on plasmids (see Chapter 8).

MICROBIOLOGY HIGHLIGHTS

Why Microbiologists Study Termites

Although termites are famous for their ability to eat wood, causing damage to wooden structures and recycling cellulose in the soil, they are unable to digest the wood that they eat. To break down the cellulose, termites enlist the help of a variety of microorganisms. Some termites, for example, dig tunnels in the wood, then inoculate the tunnels with fungi that grow on the wood. These termites then eat the fungi, not the wood itself.

What microbiologists find more interesting are the termites that contain, within their digestive tracts, symbiotic microorganisms that digest the cellulose that the termites chew and swallow. Even more fascinating to microbiologists is the fact that these microorganisms themselves can survive only because of even smaller symbionts that live on and within them, without which they would not even be able to move. By studying how a single termite survives, microbiologists have begun to gain an entirely new understanding of symbiosis.

The termite's dependence on nitrogen-fixing bacteria to supply its nitrogen and on protozoans such as *Trichonympha sphaerica* to digest cellulose is an example of endosymbiosis, a symbiotic relationship with an organism that lives inside the body of the host organism (in this case, within the hindgut of the termite).

Theoretically, the protozoan produces the cellulolytic enzymes that digest the cellulose. The picture is more complicated than this, however, for *T. sphaerica* itself is unable to digest cellulose without the aid of bacteria that live within its body; in other words, the protozoan has its own endosymbionts.

Certain hindgut ciliates such as *T. sphaerica* also demonstrate another form of symbiosis—ectosymbiosis, a symbiotic relationship with organisms that live outside its body. Recent advances in microscopy have shown that these ciliates are covered by precise rows consisting of thousands of bacteria, either rods or spirochetes. If these bacteria are killed, the protozoan is unable to move. Evidently, the protozoan does not use its cilia to move; instead, the rows of bacteria move the protozoan like rows of oarsmen in a boat.

The protozoan *Mixotricha,* for example, has rows of spirochetes on its surface. As shown in part (a) of the figure, the end of each spirochete abuts against a swelling known as a bracket. The spirochetes undulate in unison, thereby creating waves of motion along *Mixotricha*'s surface.

In 1978, Sidney L. Tamm observed rod-shaped bacteria aligned in grooves that covered the surface of devescovinids, another group of termite-hindgut protozoans. Each rod has twelve flagella that overlap the flagella of the adjacent bacteria to form a continuous filament along the groove (see part b of the figure). The bacteria beat their flagella to create coordinated waves along all these rows of filaments, thus propelling the protozoan.

An important question remains: Do the bacteria detect a food source and move the protozoan toward it, or does the protozoan force the bacteria to respond to its commands?

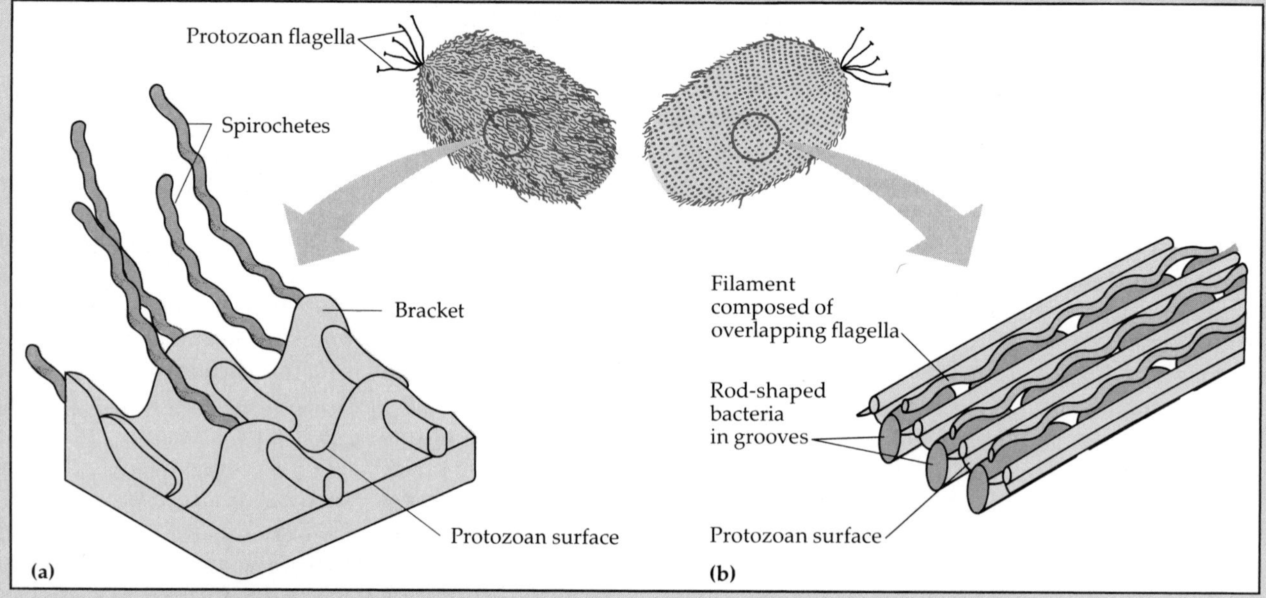

Arrangements of bacteria on the surfaces of two protozoans. **(a)** Spirochetes (red) attached to brackets over the surface of a *Mixotricha* protozoan align themselves and move in unison. **(b)** On this devescovinid protozoan, the flagella from one rod-shaped bacterium (yellow) overlap the next to form a continuous filament.

Although pseudomonads are classified as aerobic, like many other bacteria they are capable of substituting nitrate for oxygen as a terminal electron acceptor. This process, anaerobic respiration, yields almost as much energy as aerobic respiration yields (see Chapter 5). Pseudomonads cause important losses of valuable nitrogen in fertilizer and soil. Nitrate is the form of fertilizer nitrogen most easily used by plants. Under anaerobic conditions, as in waterlogged soil, pseudomonads eventually convert nitrate into nitrogen gas (N_2), which is lost to the atmosphere (see Chapter 27).

Several genera of gram-negative aerobic rods and cocci are of medical importance.

Legionella is a recently discovered genus, which now contains six species. Originally isolated during a search for the cause of a pneumonia now known as legionellosis, these bacteria do not grow on usual laboratory isolation media or stain with the usual histological staining techniques. After intensive effort, they were isolated through tests of guinea pigs inoculated with infected tissue taken from patients. A specially buffered charcoal–yeast extract agar has since been developed for use in their isolation and growth. It has been demonstrated that this is a relatively common organism in streams, and it colonizes such habitats as warm-water supply lines in hospitals and water in cooling towers of air conditioning systems (see the box in Chapter 24, p. 602).

Neisseria is a genus of non–endospore-forming diplococci. These bacteria are aerobic or facultatively anaerobic, parasitic on human mucous membranes, and able to grow well only at temperatures near body temperature. Pathogenic *Neisseria* species include the gonococcus bacterium *Neisseria gonorrhoeae*, the causative agent of gonorrhea (Figure 11.4), and *N. meningitidis* (me-nin-ji'ti-dis), the agent of meningococcal meningitis (see Chapter 22).

Members of the genus *Moraxella* (mô-raks-el'lä) are strictly aerobic coccobacilli. The coccobacilli are egg-shaped, a structure intermediate between cocci and rods. *Moraxella lacunata* (la-kü-nä'tä) is implicated in conjunctivitis (pink eye), an inflammation of the conjunctiva, the membrane that covers the eye and lines the eyelids.

Brucella (brü-sel'lä) is a small nonmotile coccobacillus. All species of *Brucella* are obligate parasites of mammals and cause brucellosis. Of medical interest is the unusual ability of *Brucella* to survive phagocytosis, an important element of our defense against bacteria (see Chapter 16).

Bordetella (bôr-de-tel'lä) is a nonmotile rod. Virulent forms have capsules. Found only in humans, *Bordetella pertussis* (pėr-tus'sis) is the primary cause of whooping cough.

Francisella (fran-sis-el'lä) is a genus of small, pleomorphic bacteria that grow only on complex media enriched with blood or tissue extracts. *Francisella tularensis* (tü-lä-ren'sis) causes tularemia.

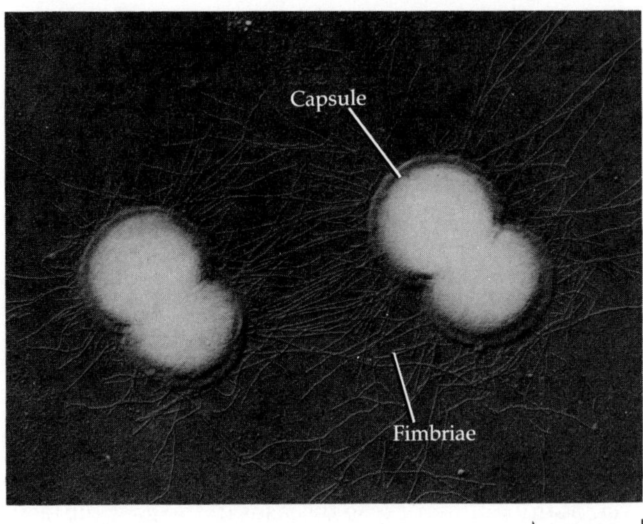

FIGURE 11.4 The gram-negative coccus *Neisseria gonorrhoeae*. Note the paired arrangement (diplococcus). The fimbriae serve as organs of attachment to the mucous membrane and contribute to the pathogenicity of the organism, which causes gonorrhea.

Thus far in our discussion of this group of bacteria, most genera have been largely of medical importance. There are also several genera of industrial and environmental importance.

Rhizobium and *Bradyrhizobium* are especially important to agriculture. These organisms cause formation of nodules on the roots of leguminous plants, such as beans or clover. They live within these nodules in a symbiotic relationship with the plants and fix nitrogen from the air (see Chapter 27).

Some nitrogen-fixing bacteria, such as *Azotobacter* and *Azomonas* (ā-zō-mō'nas), are free-living in the soil. They are frequently used in laboratory demonstrations of nitrogen fixation. However, to fix agriculturally significant amounts of nitrogen, they would require energy sources, such as carbohydrates, that are in limited supply in soil.

Agrobacterium tumefaciens (tü-mē-fā'shens) is a plant pathogen that causes a disease of plants called crown gall. The tumorlike gall grows when *A. tumefaciens* inserts a plasmid containing bacterial genetic information into the plant's chromosomal DNA (see Figure 9.14). For this reason, microbial geneticists are deeply interested in this organism. They hope that eventually it may be a vehicle for inserting useful genetic information into plants—perhaps even the ability to fix nitrogen into plants such as wheat and corn.

Acetobacter (ä-sē-tō-bak'tėr) and *Gluconobacter* (glü-kon-ō-bak'tėr) are industrially important aerobic organisms that convert ethanol to vinegar. *Zoogloea* (zō-ō-glē'ä) will be discussed in Chapter 27 in the context of aerobic sewage treatment processes, such as the activated sludge system. As they grow, *Zoogloea* organisms form unusual fluffy, slimy masses that are essential to the proper operation of such systems.

FACULTATIVELY ANAEROBIC GRAM-NEGATIVE RODS

From a medical viewpoint, facultatively anaerobic gram-negative rods are a very important group of bacteria. Many of them cause diseases of the gastrointestinal tract as well as diseases of other organs. Discussed here are three families, Enterobacteriaceae, Vibrionaceae, and Pasteurellaceae.

Enterobacteriaceae (Enterics)

The family **Enterobacteriaceae,** or **enterics,** as they are commonly called, includes a group of bacteria that inhabit the intestinal tracts of humans and other animals. Some species are permanent residents, others are found in only a fraction of the population, and still others are present only as agents of disease conditions. Most enterics are active fermenters of glucose and other carbohydrates.

Because of the clinical importance of enterics, there are many techniques for their isolation and identification. An identification key to some enterics is shown in Figure 10.6, and a modern tool using 15 biochemical tests is shown in Figure 10.7. Biochemical tests are especially important in the detection of foodborne contamination by *Salmonella.* Enterics can also be distinguished from each other according to antigens present on their surfaces (see Chapter 18).

Enterics include motile as well as nonmotile species; those that are motile have peritrichous flagella. Many enterics have fimbriae (see Figure 4.8) that help them adhere to surfaces of mucous membranes. Specialized sex pili aid in the exchange of genetic information between cells, which often includes antibiotic resistance (see Figure 8.24).

Enterics, like most bacteria, produce proteins called *bacteriocins* that cause the lysis of closely related species of bacteria. Bacteriocins produced by some enterics have been most thoroughly studied. Bacteriocins might help maintain the ecological balance of various enterics in the intestine.

Among the important genera included as enterics are *Escherichia, Salmonella, Shigella, Klebsiella, Serratia* (ser-rä'tē-ä), *Proteus* (prō'tē-us), *Yersinia* (yėr-sin'ē-ä), *Erwinia* (ėr-wi'nē-ä), and *Enterobacter.*

Escherichia. The facultative anaerobe *E. coli* is one of the most common inhabitants of the intestinal tract and is probably the most familiar organism in microbiology. As you may remember from earlier chapters, a great deal is known about the biochemistry and genetics of *E. coli,* and it continues to be an important tool for basic biological research. Its presence in water or food is also important as an indication of fecal contamination (see Chapter 27). *E. coli* is not usually considered to be pathogenic. However, it can be a common cause of urinary tract infections, and certain strains produce enterotoxins that commonly cause traveler's diarrhea (see Chapter 25).

Salmonella. Almost all members of this genus are potentially pathogenic. Because of this, there are extensive biochemical and serological tests to clinically isolate and identify salmonellae. They are common inhabitants of the intestinal tracts of many animals, especially poultry and cattle. Under unsanitary conditions, they can contaminate food.

Typhoid fever, caused by *Salmonella typhi,* is the most severe illness caused by any member of the genus *Salmonella.* A less severe gastrointestinal disease caused by other salmonellae is called *salmonellosis.* Salmonellosis is one of the most common forms of foodborne illness.

Although many members of the genus *Salmonella* have specieslike names, such as *S. typhimurium* (tī-fi-mür'ē-um) and *S. dublin* (dub'lin), no individual species are recognized. Instead, this genus is taxonomically divided into about 2000 **serovars** (serotypes). That is, they are differentiated by serological means. Serovars can be further differentiated by special biochemical or physiological properties into **biovars,** or biotypes.

When the salmonellae are injected into appropriate animals, their flagella, capsules, and cell walls cause the animal to form antibodies that are specific for each of these structures. (This topic is more fully discussed in Chapter 17.) These specific antibodies, which are available commercially, can be used to differentiate *Salmonella* serovars by a system known as the Kauffmann–White scheme. The Kauffmann–White scheme designates an organism by numbers and letters that correspond to specific antigens on the organism's capsule, body, and flagella. For example, the antigenic formula of the bacterium *S. typhimurium* is 1,4,[5],12,i,1,2. Some salmonellae are named only by their antigenic formulas.

Shigella. Species of *Shigella* are responsible for a disease called bacillary dysentery or shigellosis. These organisms are second only to *E. coli* as a cause of traveler's diarrhea. Some strains of *Shigella* can cause a life-threatening dysentery.

Klebsiella. *Klebsiella pneumoniae* is a major cause of septicemia in pediatric wards and is also a cause of one form of pneumonia, which is contracted especially by persons suffering from chronic alcoholism.

Serratia. *Serratia marcescens* (mär-ses'sens), which is distinguished by its production of red pigment, has been used as an experimental organism for tests of air dispersal of bacteria for biological warfare. *S. marcescens* has become increasingly important in recent years because of its connection with hospital-acquired (nosocomial) infections. The organism causes urinary and respiratory tract infections and has been found on catheters, in saline irrigation solutions, and in other supposedly sterile solutions.

Proteus. *Proteus* organisms are very actively motile. They are implicated in infections of the urinary tract and infections of wounds and in infant diarrhea (Figure 11.5).

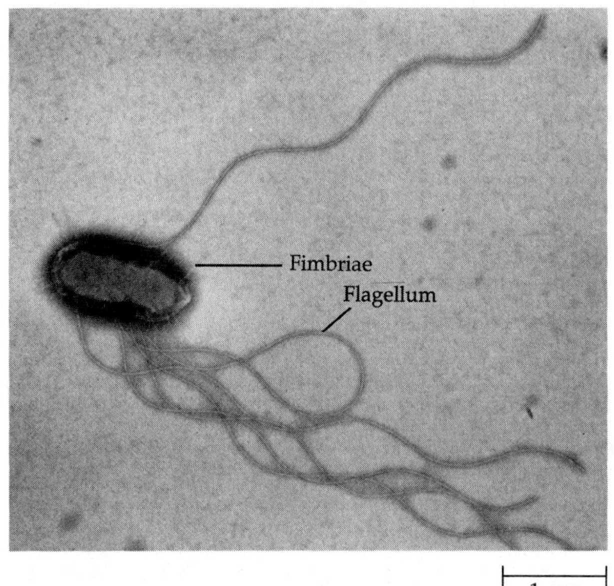

FIGURE 11.5 ***Proteus mirabilis,*** **a gram-negative rod.** Note the flagella distributed about the body, and the short fimbriae.

Yersinia. *Yersinia pestis* (pes'tis) is the causative agent of bubonic plague ("black death"). Urban rats in some parts of the world and ground squirrels in the American Southwest are carriers of these organisms. Fleas usually transmit the organisms among animals and to humans, although contact with animals and respiratory droplets from infected persons can be involved in transmission.

Erwinia. *Erwinia* species are primarily plant pathogens; some cause plant soft-rot diseases. These species produce enzymes that hydrolyze the pectin between individual plant cells and cause the plants to rot.

Enterobacter. Two *Enterobacter* species, *E. cloacae* (klō-ā'kī) and *E. aerogenes* (ā-rä'jen-ēz), can cause urinary tract infections and nosocomial infections. They are widely distributed in humans and animals, as well as in water, sewage, and soil (see Chapter 27).

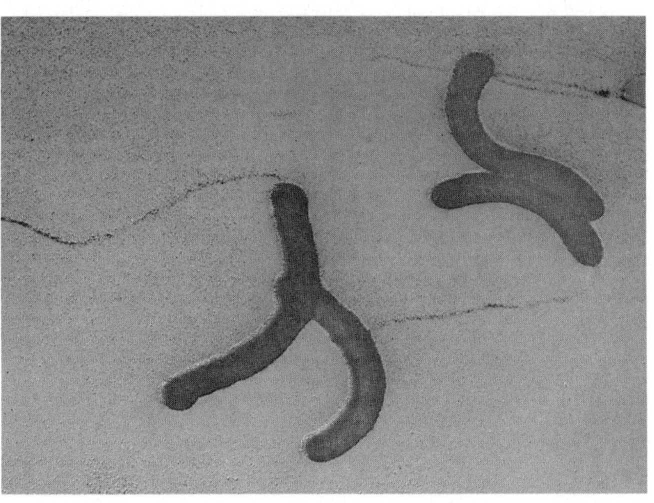

FIGURE 11.6 ***Vibrio cholerae.*** Note the characteristic slight curvature of the rods.

Vibrionaceae

This is a family of gram-negative, facultatively anaerobic rods, many of which are slightly curved. They are found mostly in aquatic habitats. *Vibrio* (vib'rē-ō) is the most important genus in the family.

Vibrio. Members of the genus *Vibrio* are rods that are frequently slightly curved (Figure 11.6). Although most are nonpathogenic, one important pathogen is *Vibrio cholerae* (kol'ėr-ī), the causative agent of cholera. The disease is characterized by a profuse and watery diarrhea. *V. parahaemolyticus* (pa-rä-hē-mō-li'ti-kus) causes a less serious form of gastroenteritis. *V. parahaemolyticus* usually inhabits coastal salt waters and is transmitted to humans, mostly by raw or undercooked shellfish.

Pasteurellaceae

The third medically important family among the gram-negative facultatively anaerobic rods includes *Pasteurella* (pas-tyėr-el'lä), *Hemophilus,* and *Actinobacillus* (ak-tin-o-bä-sil'lus).

Pasteurella. This genus is primarily known as a pathogen of domestic animals. It causes septicemia in cattle, fowl cholera in chickens and other fowl, and pneumonias in several types of animals. The best-known species is *Pasteurella multocida* (mul-tō'si-dä), which can be transmitted to humans by dog and cat bites.

Hemophilus. *Hemophilus,* also spelled *Haemophilus,* is a very important genus of pathogenic bacteria that receives less attention than it should. These organisms commonly inhabit the mucous membranes of the upper respiratory tract, mouth, vagina, and intestinal tract. The best-known species that affects humans is *Hemophilus influenzae* (in-flü-en'zī), named long ago after the erroneous conclusion that it was responsible for influenza.

The name *Hemophilus* is derived from the bacteria's requirement for blood in their culture medium. They are unable to synthesize important parts of the cytochrome system needed for respiration. They obtain these substances from the heme fraction, known as the X factor, of blood hemoglobin. In some formulations, such as chocolate agar, the blood is heated, which causes it to turn a dark brown. The culture medium must also supply the cofactor nicotinamide adenine dinucleotide (NAD^+ or $NADP^+$), which is known as V factor. Clinical laboratories use tests for the requirement of X and V factors to identify isolates as *Hemophilus* species.

Hemophilus is responsible for several important diseases. It is the most common cause of meningitis in young children and is a common cause of earaches. Other clinical conditions caused by *H. influenzae* include epiglottitis (a sore throat complication in which the epiglottis becomes infected), septic arthritis in children, bronchitis, and pneumonia.

Gardnerella

Gardnerella vaginalis (gärd-nė-rel'lä va-jin-al'is), a bacterium that causes one of the most common forms of vaginitis, is not presently assigned to any family. In 1955, researchers named it *Hemophilus vaginalis,* and it was placed with *Hemophilus* in *Bergey's Manual of Determinative Bacteriology,* 8th edition, because it did not seem to fit anywhere else. Its pleomorphic (highly irregular) form and occasional gram-positive reaction have caused it to be classified at times with the genus *Corynebacterium,* which will be discussed later in this chapter. The organism has a cell wall that is structurally similar to that of a gram-positive cell. However, the wall is quite thin, which leads to a gram-negative staining reaction. The present editions of *Bergey's Manual* places this genus in the group of gram-negative rods that we are now discussing. However, *Bergey's Manual* also includes *Gardnerella* in a group of irregular, nonsporing gram-positive rods, which we will discuss later in this chapter.

There are at least five genera in *Bergey's Manual* that have a Gram-staining reaction that does not match their cell wall anatomy, and they are similarly grouped in two different taxonomic sections. Such dual classification is unheard of in botany or zoology today. It illustrates the difficulty of applying taxonomy to bacteria. The types of evolutionary criteria used to classify plants and animals are not so evident among bacteria. For example, although a bat and a bird superficially seem closely related, it can be determined that the bat is a mammal and the bird occupies another evolutionary branch more closely related to reptiles. Such relationships between bacteria may eventually become clearer as we learn to detect similarities and differences in ribosomal RNA and chromosomal DNA of bacteria. A footnote in Appendix A lists the other bacterial genera that, for one reason or another, have been placed in two different sections.

Anaerobic, Gram-Negative, Straight, Curved, and Helical Rods

Among the anaerobic gram-negative bacteria, the genus *Bacteroides* (bak-tė-roi'dēz) is a large group of microbes that live in the human intestinal tract (Figure 11.7a). Some *Bacteroides* species also reside in the oral cavity, genital tract, and upper respiratory tract. *Bacteroides* organisms are non–endospore-forming and nonmotile. Infections due to *Bacteroides* often result from puncture wounds or surgery and are a frequent cause of peritonitis.

Another genus of gram-negative anaerobic bacteria is *Fusobacterium* (fü-sō-bak-ti'rē-um). These microbes are long and slender, with pointed rather than blunt ends (Figure 11.7b). In humans, they are found most often in the gums and are responsible for dental abscesses.

Dissimilatory Sulfate-Reducing or Sulfur-Reducing Bacteria

The sulfur-reducing bacteria are obligately anaerobic organisms that use oxidized forms of sulfur, such as sulfates (SO_4^{2-}) or elemental sulfur (S^0), rather than oxygen as electron acceptors. The product of this reduction is hydrogen sulfide (H_2S). The activity of these bacteria releases millions of tons of H_2S into the atmosphere every year. H_2S is an essential part of the sulfur cycle and is discussed in Chapter 27.

These bacteria are found in anaerobic muds and sediments and also in the intestinal tracts of humans and animals. *Desulfovibrio* is the best-known genus. Still other sulfur bacteria, which we will discuss later in

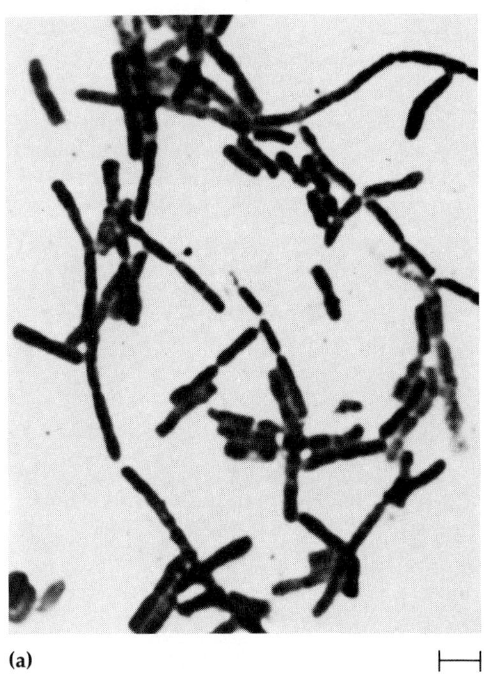

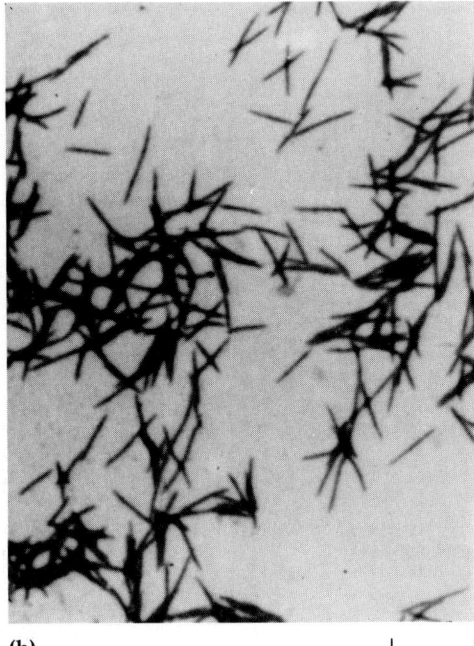

FIGURE 11.7 Gram-negative anaerobes. (a) Chains of *Bacteroides hypermegas.* **(b)** *Fusobacterium.* Note the slender shape with pointed ends.

this chapter, are able to use the H_2S either as part of photosynthesis or as an autotrophic energy source. Altogether, the sulfur bacteria of different types are some of the most interesting microorganisms known; their means of energy production are physiologically unique and ecologically crucial.

ANAEROBIC GRAM-NEGATIVE COCCI

The cells of anaerobic gram-negative cocci typically occur in pairs, but they can occur singly, in clusters, or in chains. They are nonmotile and non–endospore-forming. Bacteria of the genus *Veillonella* (vi-lo-nel'lä) are found as part of the normal flora of the mouth and are components of dental plaque.

RICKETTSIAS AND CHLAMYDIAS

Both rickettsias and chlamydias are obligate intracellular parasites, which means that they can reproduce only within a host cell. In this respect, they are similar to viruses. In fact, they are even smaller than some of the largest viruses. However, in morphological and in limited biochemical aspects, they resemble bacteria and are therefore classified as such. A comparison of rickettsias, chlamydias, and viruses appears in Table 13.1, p. 333.

Rickettsias (Figure 11.8a) are rod-shaped bacteria or coccobacilli that have a high degree of pleomorphism. They are gram-negative and nonmotile and divide by binary fission. Rickettsias range in length from 1 to 2 μm. One distinguishing feature of most rickettsias is that they are transmitted to humans by insects and ticks. The one exception is *Coxiella burnetii,* which causes Q fever. Although cattle ticks harbor the organism, it is most commonly transmitted by aerosols or contaminated milk. A sporogenic cycle has been reported for *C. burnetii.* This cycle might explain the bacterium's relatively high resistance to pasteurization temperatures and antimicrobial chemicals.

Diseases caused by rickettsias include epidemic typhus, caused by *Rickettsia prowazekii* (ri-ket'sē-ä prou-wä-ze'kē-ē) and transmitted by lice; endemic murine typhus (not to be confused with typhoid fever), caused by *R. typhi* (tī'fē) and transmitted by rat fleas; and Rocky Mountain spotted fever, caused by *R. rickettsii* (ri-ket'sē-ē) and transmitted by ticks. In humans, rickettsial infections damage the permeability of capillaries; severe infections can cause the cardiovascular system to collapse. Rickettsias are usually cultivated in the yolk sac of chicken embryos.

Chlamydias (Figure 11.8b) are coccoid bacteria that range in size from 0.2 to 1.5 μm. They are gram-negative and nonmotile and, unlike most rickettsias, do not require insects or ticks for transmission. They are transmitted by interpersonal contact or by airborne respiratory routes. The developmental cycle of chlamydias is perhaps their most distinguishing characteristic. First, the microbe's infectious form, called the **elementary body,** attaches to a host cell. The host cell phagocytizes the elementary body and houses it in a vacuole. (This cycle differs from that of the rickettsias,

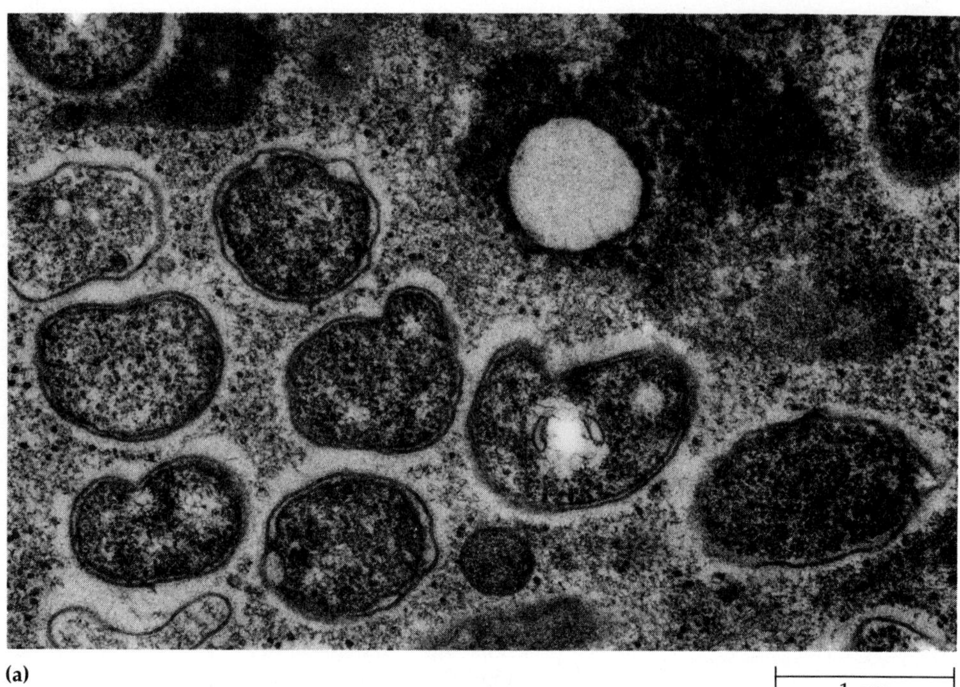

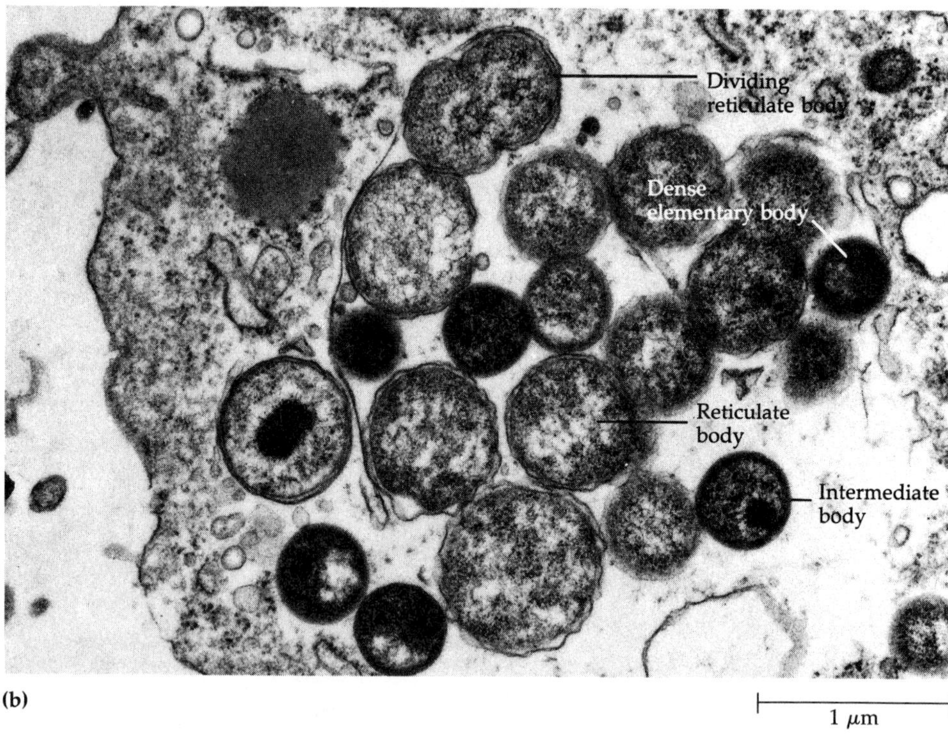

FIGURE 11.8 Rickettsias and chlamydias. **(a)** Electron micrograph of a slice of tick tissue infected by *Rickettsia prowazekii*, showing cross sections of several rickettsias in the tissue. **(b)** Electron micrograph of *Chlamydia psittaci* in a slice of the cytoplasm of a host cell. The dense, dark, relatively small elementary bodies have thin walls similar to those of other gram-negative bacteria. These are the infective forms. The chlamydias reproduce within the host cell as reticulate bodies, one of which is shown dividing. The intermediate bodies shown are transitional stages during the development of elementary bodies into reticulate bodies.

which usually multiply in the cytoplasm of the host cell.) Within the host cell, the chlamydia's elementary body reorganizes and develops into a larger, less infective **reticulate body.** This body then divides successively and eventually condenses to produce the smaller infectious elementary bodies. Finally, these elementary bodies are released from the host cell and spread to infect surrounding host cells.

There are only three species of chlamydias. *Chlamydia trachomatis* (kla-mi'dē-ä trä-kō'mä-tis) is the causative agent of trachoma, a common cause of blindness in humans. *C. trachomatis* seems to be the primary causative agent of nongonococcal urethritis, which might now be the most common sexually transmitted disease in the United States, and lymphogranuloma venereum, another sexually transmitted disease.

C. psittaci (sit'tä-sē) is the causative agent of psittacosis (ornithosis). A third species, *C. pneumoniae,* has recently been identified as the cause of a mild pneumonia. Chlamydias can be cultivated in laboratory animals, cell cultures, or the yolk sac of chicken embryos.

MYCOPLASMAS

Mycoplasmas are bacteria that do not form cell walls. They should not be confused with L forms (see Chapter 4), which are mutants of wall-forming bacteria that fail to form normal cell walls. The differences between mycoplasmas and L forms are summarized in Table 11.2.

The majority of *Mycoplasma* species are aerobes or facultative anaerobes. Because they lack cell walls, they are highly pleomorphic (Figure 11.9). They can produce filaments that resemble fungi; hence their name (*myco* means fungus). *Mycoplasma* cells are very small, ranging in size from 0.1 to 0.25 μm. The most significant human pathogen among mycoplasmas is *M. pneumoniae,* the causative agent of primary atypical pneumonia (commonly called "walking" pneumonia).

Also included in this group of bacteria are *Spiroplasma* (spī-ro-plaz'ma) species, cells with a tight corkscrew morphology that are serious plant pathogens and are common parasites of plant-feeding insects. *Ureaplasma* (ū-rē-ä-plaz'mä) species, so called because they can enzymatically split the urea in urine, are occasionally associated with urinary tract infections. *Thermoplasma* species, which were originally isolated from spontaneously heating coal refuse piles, are harmless organisms occasionally found in household hot water systems. This genus is taxonomically grouped with the archaeobacteria as well as with the mycoplasmas.

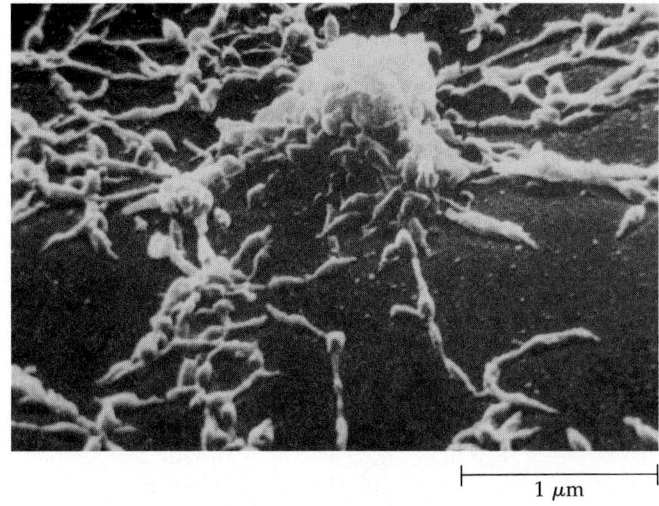

FIGURE 11.9 ***Mycoplasma pneumoniae.*** Bacteria such as *Mycoplasma pneumoniae* have no cell wall, and their morphology is irregular (pleomorphic). The filament in the center shows a series of bulges; the organism reproduces by fragmentation of the filaments at these sites.

TABLE 11.2 Comparison Between Mycoplasmas and L Forms

PROPERTY	MYCOPLASMAS	L FORMS
Stability	Cannot revert to cell-wall-containing type	May revert to cell-wall-containing type
Growth medium	Do not require high osmotic pressure to maintain cellular integrity	Require high osmotic pressure to maintain cellular integrity
Plasma membrane	High sterol content (a few exceptions)	No sterols
Reaction to penicillin	No adverse effect	Reproduction is inhibited

Mycoplasmas can be grown on artificial media that provide them with sterols, if necessary, and other special nutritional or physical requirements. Colonies are less than 1 mm in diameter and have a characteristic "fried-egg" appearance when viewed through a microscope. Because of the small colony size on artificial media, cell culture methods are often more satisfactory.

GRAM-POSITIVE COCCI

Most gram-positive cocci of medical importance are members of the genera *Staphylococcus* and *Streptococcus.*

Staphylococcus

Staphylococci typically occur in grapelike clusters (Figure 11.10). The most important staphylococcal species is *Staphylococcus aureus,* named for its yellow-pigmented colonies (*aureus* means golden). Members of this species are aerobes or facultative anaerobes.

Understanding the characteristics of the staphylococci may help in understanding the reasons for their pathogenicity, which takes many forms. They grow comparatively well under conditions of high osmotic pressure and low moisture, which partially explains why they can grow and survive in nasal secretions (many of us carry the bacteria in our noses) and on the skin. This ability also explains how *S. aureus* can grow in certain foods with high osmotic pressures (such as

FIGURE 11.10 ***Staphylococcus aureus.*** Note the grapelike clusters of these gram-positive cocci.

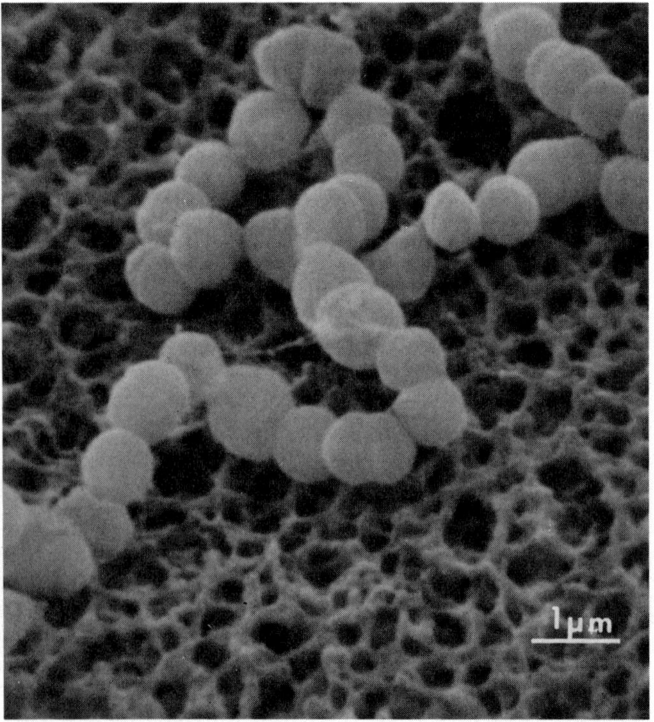

FIGURE 11.11 ***Streptococcus mutans.*** Most streptococci form chains, as shown in this scanning electron micrograph.

ham and other cured meats) or in low-moisture foods that tend to inhibit the growth of other organisms. Furthermore, the yellow pigment probably confers some protection from the antimicrobial effects of sunlight.

S. aureus produces many toxins that contribute to the bacterium's pathogenicity by increasing its ability to invade the body or damage tissue. The infection of surgical wounds by *S. aureus* is a common problem in hospitals, as we will see in Chapter 14. The bacterium's ability to develop resistance quickly to such antibiotics as penicillin contributes to its danger in hospitals. *S. aureus* is the agent of toxic shock syndrome, a severe infection causing high fever and vomiting and sometimes death. *S. aureus* also produces an *enterotoxin* that causes vomiting and nausea when ingested, one of the most common causes of food poisoning.

Streptococcus

Members of the genus *Streptococcus* are spherical gram-positive bacteria. They are probably responsible for more illnesses and cause a greater variety of diseases than any other group of bacteria. Among the diseases caused by streptococci are scarlet fever, pharyngitis (sore throat), and pneumococcal pneumonia. They are also important as lactic acid producers in many dairy products.

Streptococci typically appear in chains that can contain as few as four to six cocci or as many as 50 or more (Figure 11.11). One species, *Streptococcus pneumoniae,* is usually found only in pairs (see Figure 23.8). Streptococci do not use oxygen, although most are aerotolerant. A few are obligately anaerobic.

One basis for the classification of streptococci is their action on blood agar. *Alpha-hemolytic* species produce a substance called alpha-hemolysin that reduces hemoglobin (red) to methemoglobin (green). This reduction causes a greenish zone to surround the colony. *Beta-hemolytic* species produce a hemolysin that forms a clear zone of hemolysis on blood agar (see Figure 6.7). Some species have no apparent effect on red blood cells. These are referred to as *nonhemolytic* or, sometimes less logically, as gamma-hemolytic.

Like staphylococci, pathogenic streptococci produce a number of extracellular substances that contribute to disease. Among them are products that destroy phagocytic cells (which are essential to the body's defenses), enzymes that digest connective tissue of the host and spread the infection, and enzymes that lyse fibrin. Fibrin is a fibrous protein deposited when blood clots that limits movement of pathogens in infected areas.

ENDOSPORE-FORMING GRAM-POSITIVE RODS AND COCCI

The formation of endospores by bacteria is important both to medicine and to the food industry because of

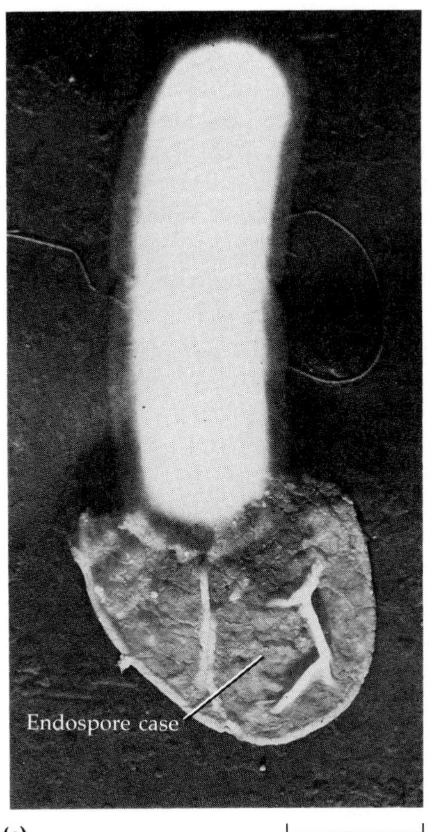

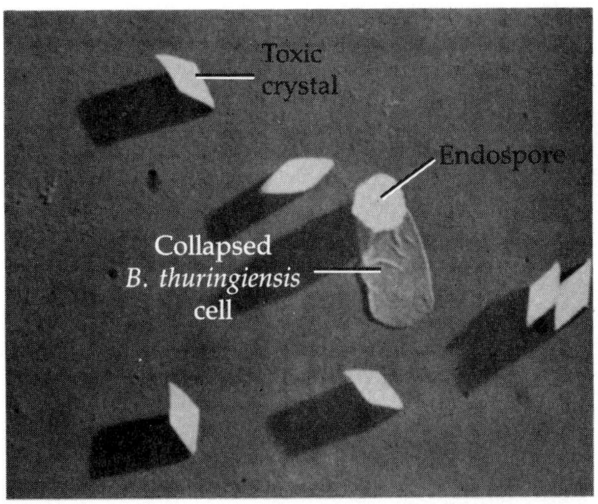

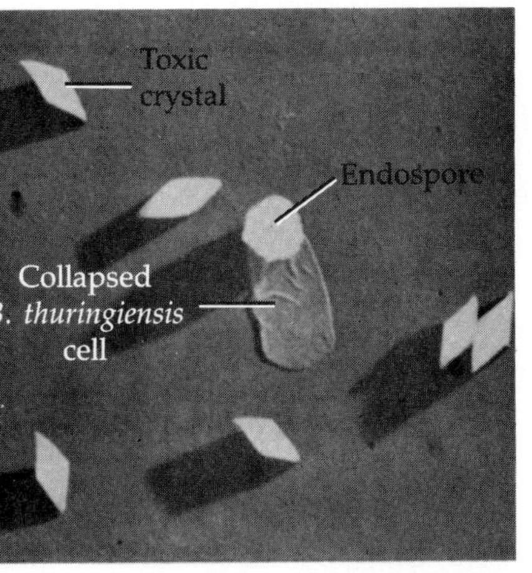

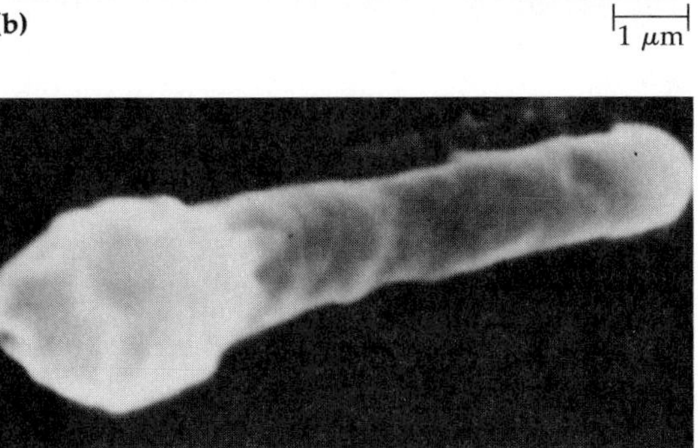

FIGURE 11.12
Representative gram-positive, endospore-forming rods. **(a)** A germinating cell of *Bacillus cereus.* **(b)** *Bacillus thuringiensis;* the diamond-shaped crystal shown next to the endospore is toxic to insects that ingest it. **(c)** *Clostridium botulinum,* showing the terminal location of the endospore.

the endospores' resistance to heat and many chemicals. The majority of endospore-forming rods and cocci are gram-positive. With respect to oxygen requirements, they can be strict aerobes, facultative anaerobes, obligate anaerobes, or microaerophiles. The two important genera are *Bacillus* and *Clostridium* (Figure 11.12).

Bacillus anthracis causes anthrax, a disease of cattle, sheep, and horses that can be transmitted to humans. The anthrax bacillus is a nonmotile facultative anaerobe ranging in length from 4 to 8 μm. It is one of the largest bacterial pathogens. Smears of *B. anthracis* from growth in tissue show a capsule. The endospores are centrally located.

Bacillus thuringiensis is probably the best-known microbial insect pathogen (Figure 11.12b). It produces intracellular crystals of toxic glycoproteins when it sporulates. Commercial preparations containing endospores and crystalline toxin of this bacterium are sold in gardening supply shops to be sprayed on plants. When ingested by an insect, the toxin quickly causes paralysis of the insect's gut, and the insect ceases to feed. In some insects the ingested endospores germinate, and bacterial growth is a factor in the death of the insect. Attempts to use this bacterium in mosquito control are also promising.

Members of the genus *Clostridium* are obligate anaerobes. They vary in length from 3 to 8 μm, and in most species the cells containing endospores appear swollen (Figure 11.12c). Some clostridial endospores can withstand temperatures of 120°C for 15 minutes. Diseases associated with clostridia include tetanus, or lockjaw, caused by *C. tetani* (te'tan-ē); botulism, caused by *C. botulinum;* and gas gangrene, caused by *C. perfringens* (pėr-frin'jens) and other clostridia. *C. perfringens* is also the cause of a common form of food-borne diarrhea.

REGULAR NONSPORING GRAM-POSITIVE RODS

A chief representative of nonsporing gram-positive rods is the genus *Lactobacillus.* Lactobacilli lack a cytochrome system and are unable to use oxygen as an electron acceptor. Instead, these aerotolerant rods produce lactic acid from simple carbohydrates. The acidity inhibits competing bacteria, but lactobacilli grow well in acidic environments. In humans, lactobacilli are located in the vagina, intestinal tract, and oral cavity. Common industrial uses of lactobacilli are in the production of sauerkraut, pickles, buttermilk, and yogurt. Typically, a succession of lactobacilli, each more acid-

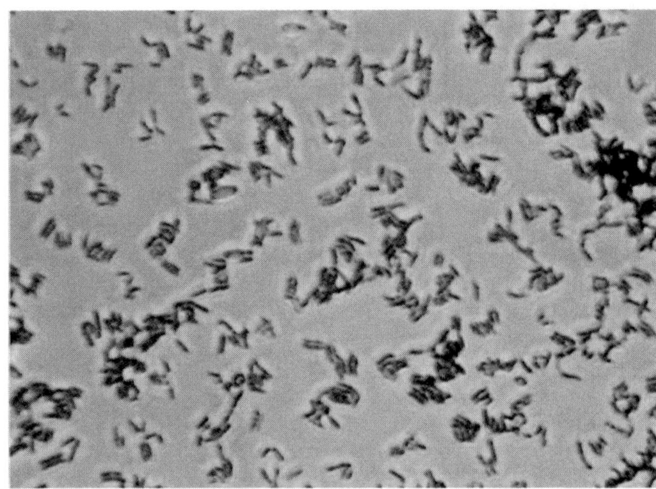

FIGURE 11.13 ***Corynebacterium xerosis.*** Note the irregular, club-shaped morphology.

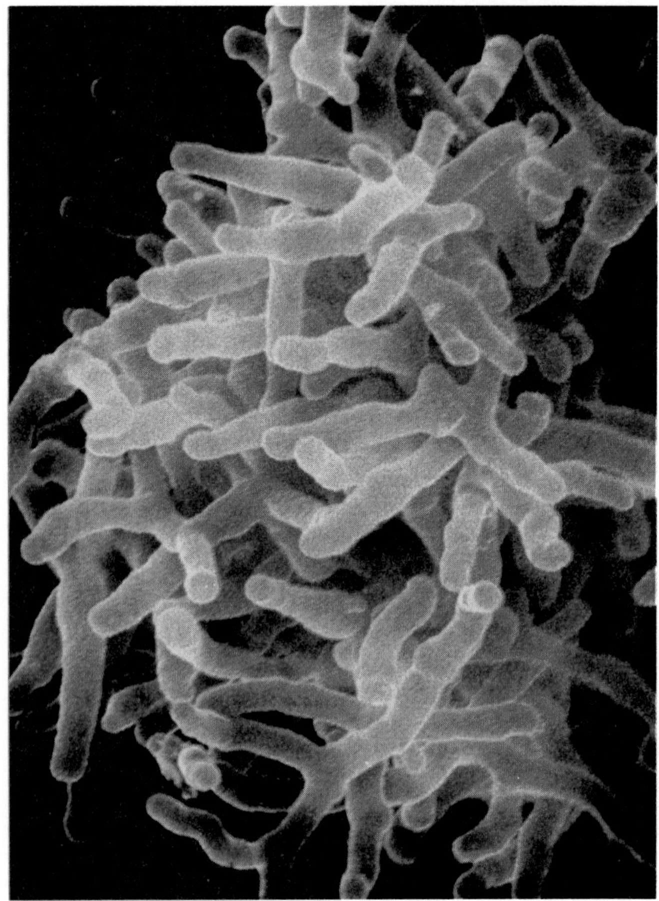

FIGURE 11.14 ***Actinomyces sp.*** Note the branched filamentous morphology.

tolerant than the other, participates in these lactic acid fermentations.

A pathogen in this group, *Listeria monocytogenes* (lis-te'rē-ä mo-nō-sī-tô'je-nēz), can contaminate foods; this problem has led to recalls of dairy products. *L. monocytogenes* survives within phagocytic cells and is capable of growth at refrigeration temperatures. The organism poses the threat of stillbirth or serious damage to the fetus if it infects a pregnant woman.

IRREGULAR NONSPORING GRAM-POSITIVE RODS

The organisms in this group are often grouped under the general term **corynebacteria** (*coryne* means club-shaped). They tend to be pleomorphic (highly irregular in morphology); their morphology often varies with the age of the cells (Figure 11.13). They may be aerobic, anaerobic, or microaerophilic. The best-known and most widely studied species is *Corynebacterium diphtheriae,* the causative agent of diphtheria. One related species, *Propionibacterium acnes,* is commonly found on human skin and is implicated in acne.

The genus *Actinomyces* (ak-tin-ō-mī'sēs) consists of anaerobes that are found in the mouth and throat of humans and animals. They occasionally form filaments (Figure 11.14) that can fragment into coryneform cells. One species, *Actinomyces israelii* (is-rā'lē-ē), causes actinomycosis, a tissue-destroying disease usually affecting the head, neck, or lungs.

MYCOBACTERIA

Mycobacteria are aerobic, non–endospore-forming, nonmotile, rod-shaped organisms. Their name (*myco* means fungus) was suggested by their occasional exhibition of filamentous growth. Most of the pathogenic species are acid-fast. A number of species of *Mycobacterium* are found in the soil. Others are important pathogens: *Mycobacterium tuberculosis* causes tuberculosis, and *Mycobacterium leprae* causes leprosy.

NOCARDIOFORMS

The best-known genus in this group is *Nocardia.* These organisms morphologically resemble *Actinomyces;* however, they are aerobic. To reproduce, they form rudimentary filaments, which fragment into short rods. The structure of their cell wall resembles that of the mycobacteria; therefore, they are often acid-fast. *Nocardia* species are common in soil. Some species, such as *Nocardia asteroides* (as-tėr-oi'dēz), occasionally cause chronic, difficult-to-treat pulmonary nocardiosis (see Chapter 24). *N. asteroides* is also one of the causative agents of mycetoma, a localized destructive infection of the feet or hands.

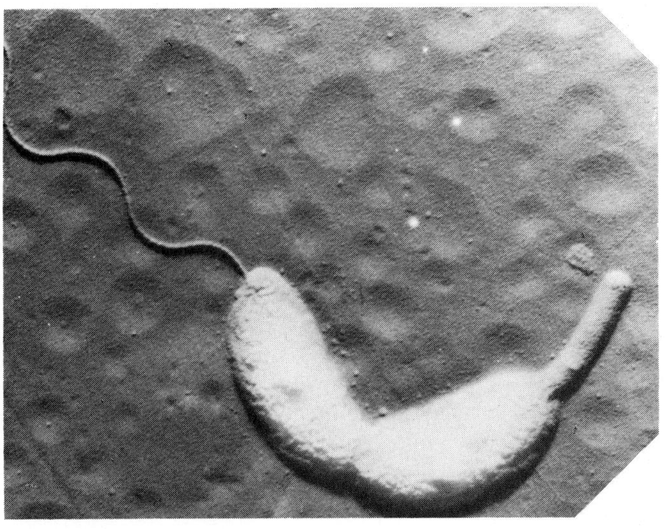

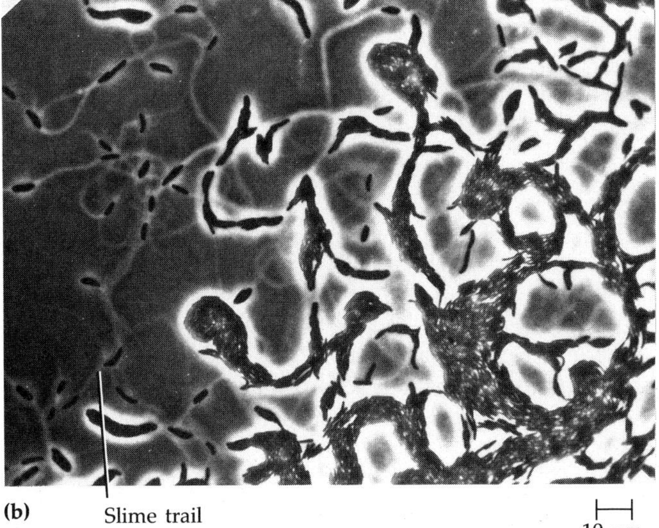

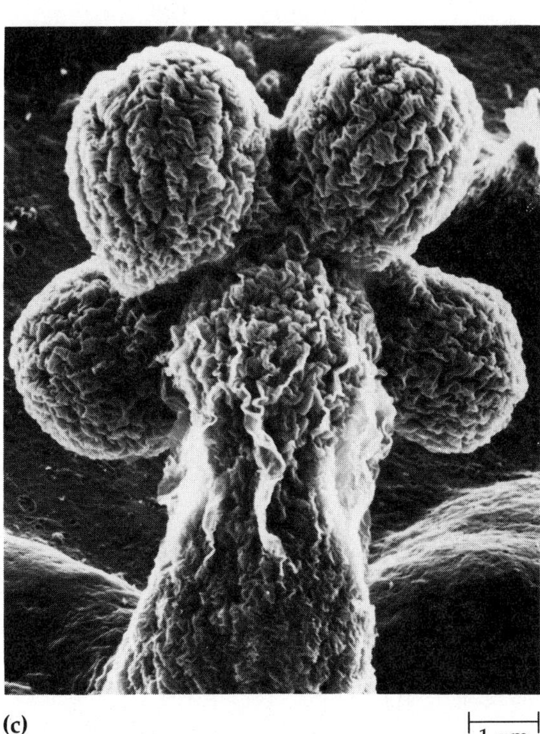

FIGURE 11.15 Gliding or appendaged bacteria. **(a)** *Caulobacter*. The stalked cell is dividing, and a flagellated swarmer cell is being formed. **(b)** Slime trails of the gliding, fruiting myxobacterium *Myxococcus fulvus*. Under appropriate conditions, these bacteria aggregate and form a stalk bearing a fruiting body similar to that shown in c. **(c)** *Stigmatella aurantiaca*, an example of a gliding, fruiting bacterium. Numerous gliding vegetative cells have aggregated to form the stalk shown here, on which four spores (fruiting bodies) have formed. When growth conditions improve, these spores can germinate into new vegetative cells with gliding motility.

GLIDING, SHEATHED, AND BUDDING AND/OR APPENDAGED BACTERIA

Appendaged Bacteria

This is an unusual group of bacteria that is linked taxonomically by the presence of **prosthecae.** Prosthecae are protrusions, such as stalks and buds. Most of these organisms are found in low-nutrient aquatic environments such as lakes. One of the best studied is the genus *Caulobacter* (kô-lō-bak'tėr), which has stalks that anchor the organisms to surfaces (Figure 11.15a). This tends to increase their nutrient uptake because they are exposed to a continuously changing flow of water and because the stalk increases the surface-to-volume ratio of the cell. Also, if the surface to which they anchor is a living host, these bacteria can use the host's excretions as nutrients. When the nutrient concentration is exceptionally low, the size of the stalk increases, evidently to provide an even greater surface area for nutrient absorption.

Another unusual characteristic of *Caulobacter* is that reproduction by fission does not result in two essentially identical cells. When a stalked *Caulobacter* cell divides, the result is one stalked cell and one flagellated swarmer cell (see Figure 11.15a). The swarmer cell is highly motile and confers the advantage of movement to new locations. After a period of motility, the swarmer cell loses its flagellum and becomes another stalked cell. Because of this unusual reproductive cycle, biologists studying the mechanics and genetics of cellular differentiation have found *Caulobacter* to be a useful research tool.

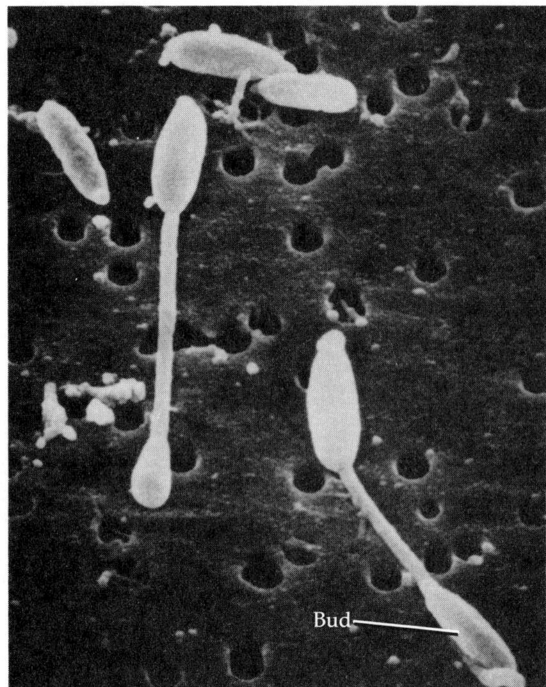

(a)

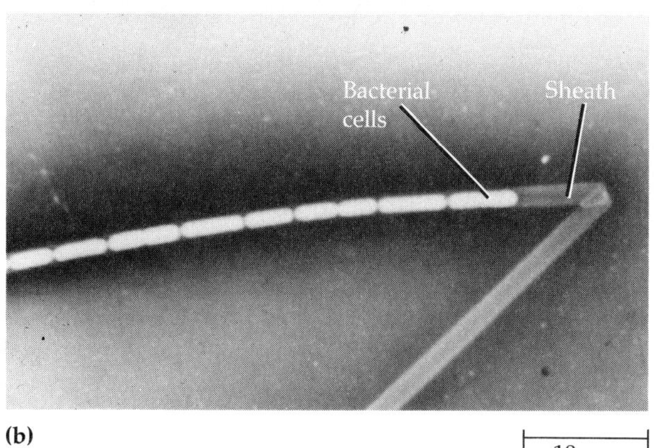

(b)

FIGURE 11.16 Budding or sheathed bacteria. (a) The budding bacterium *Hyphomicrobium* sp. **(b)** The sheathed bacterium *Sphaerotilus natans*. This organism, which is found in sewage and aquatic environments, forms elongated sheaths in which the bacteria live. The bacteria have flagella (not visible here) and are capable of swimming free of the sheath eventually.

Gliding, Nonfruiting Bacteria

Gliding bacteria are classified as such on the basis of their method of motility. Embedded in a layer of slime, they glide over surfaces, often leaving a visible trail. Members of the genus *Cytophaga* (sī-täf'äg-ä) are important cellulose degraders in the soil. Another gliding bacterium is *Beggiatoa*. This organism morphologically resembles certain cyanobacteria, but it is not photosynthetic. Nutritionally, it uses hydrogen sulfide (H_2S) as an autotrophic energy source and accumulates granules of sulfur within the cytoplasm. Its affinity for H_2S allows it to have a symbiotic relationship with rice. Rice grows in muds in which bacteria such as *Desulfovibrio* produce H_2S, which the plant is sensitive to. The *Beggiatoa* cells lower the amounts of toxic H_2S and, in return, benefit from low levels of oxygen that the plant transports to the root area.

Gliding, Fruiting Bacteria

This group of bacteria, the *myxobacteria* (*myxa* means slime), have a remarkable life cycle. Large numbers of vegetative, gram-negative cells converge upon a single point. They move by gliding, leaving a slime trail (Figure 11.15b). An important source of nutrition is the lysis of bacteria they encounter. Where the moving cells aggregate, they differentiate and form a stalked fruiting body that contains resting cells called *myxospores* (Figure 11.15c). Under proper conditions, the myxospores germinate and form new vegetative gliding cells. This is probably the most complex life cycle among procaryotic cells.

Budding Bacteria

Budding bacteria do not divide by fission into nearly identical cells. Members of the genus *Hyphomicrobium* (hī-fō-mī-krō'bē-um) and a few other genera instead form buds (Figure 11.16a). The process in principle resembles the asexual reproductive processes of many yeasts. The parent cell retains its identity, while the bud increases in size until it separates as a complete new cell.

Sheathed Bacteria

Sheathed bacteria, such as *Sphaerotilus natans* (sfe-rä'ti-lus nā'tans), are found in fresh water and in sewage. These gram-negative bacteria with polar flagella form a hollow, filamentous sheath to live in (Figure 11.16b). They are the cause of bulking, an important problem in sewage treatment (see Chapter 27).

CHEMOAUTOTROPHIC BACTERIA

These organisms are of great importance to the environment and to agriculture. They are autotrophs capable of using inorganic chemicals as energy sources and carbon dioxide as the only source of carbon. They possess great synthesizing ability. Especially important for agriculture are the nitrifying bacteria. The energy sources of the genera *Nitrobacter* and *Nitrosomonas* are reduced nitrogenous compounds, such as ammonium (NH_4^+) and nitrite (NO_2^-); the bacteria convert these compounds into nitrates (NO_3^-). Nitrate is a nitrogen form that is mobile in soil and is therefore likely to be encountered and used by plants. *Thiobacillus* and other sulfur-oxidizing bacteria are parts of the important sulfur cycle. These bacteria are capable of obtaining energy by oxidizing the reduced forms of sulfur, such as hydrogen sulfide (H_2S) or elemental sulfur (S^0), into sulfates (SO_4^{2-}).

ARCHAEOBACTERIA

This exceptionally interesting group of bacteria includes microbes that are highly unusual in morphology and in their environmental niches. Prominent among them are bacteria that are extreme halophiles, surviving in very high concentrations of salt (Figure 11.17a). Others tolerate extremes of heat (Figure 11.17b) and acidity. The obligately anaerobic, methane-producing bacteria (methanogens) are of considerable economic importance. The archaeobacteria are grouped together because of their related ribosomal RNA sequences. Also, their cell walls lack the peptidoglycan common to most other bacterial groups.

Examples of extreme halophiles include *Halobacterium* (hal-ō-bak-ti′rē-um) (see the box in Chapter 5, p. 128) and *Halococcus* (hal-ō-kok′kus), which live in high concentrations of sodium chloride (NaCl) and actually require such environmental conditions for growth. When an inoculating loop is used to transfer these bacteria from their normal habitat, the loop must first be dipped into a concentrated NaCl solution so that the cells do not lyse from lowered osmotic pressure. Other archaeobacteria thrive in acidic, sulfur-rich hot springs. Such an organism is *Sulfolobus* (sul-fō-lō′bus), which has a pH optimum of about 2 and a temperature optimum of more than 70°C.

The methane-producing bacteria are used in sewage treatment processes (see Chapter 27). These archaeobacteria derive energy from combining hydrogen (H_2) with CO_2 to form methane (CH_4). An essential part of the treatment of sewage sludge is encouraging the growth of these organisms, in anaerobic digestion tanks, to convert sewage sludge into CH_4.

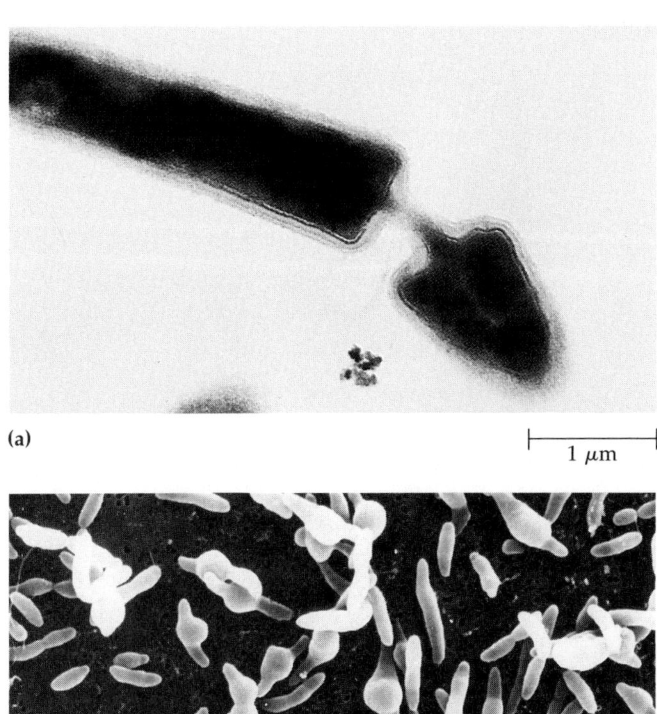

FIGURE 11.17 Archaeobacteria. Examples of bacteria capable of growth under extreme environmental conditions. **(a)** This halophilic archaeobacterium, *Thermotoga neopolitana,* is capable of growth in saturated salt brines, where as much salt as possible has been dissolved. **(b)** This methane-producing organism, *Methanopyrus,* was discovered growing near a hot vent in deep oceanic sediments at a temperature of 110°C.

PHOTOTROPHIC BACTERIA

There are three groups of procaryotes that use light as an energy source. These **phototrophic** bacteria are the purple bacteria, the green bacteria, and the blue-green bacteria (cyanobacteria). The energetics of phototrophic bacteria were discussed in Chapter 5.

Purple and Green Phototrophic Bacteria

The purple and green bacteria (which are not necessarily these colors) are generally anaerobic (Figure 11.18). Their habitat is usually the deep sediments of lakes and ponds. Like plants, algae, and the cyanobacteria, purple and green bacteria carry out photosynthesis. However, unlike plantlike photosynthesis, the photo-

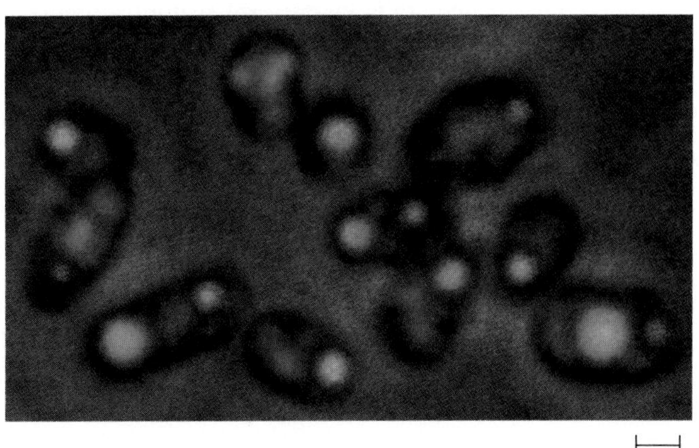

FIGURE 11.18 Purple sulfur bacteria. This phase contrast micrograph of cells of the genus *Chromatium* shows the intracellular sulfur granules as multicolored objects.

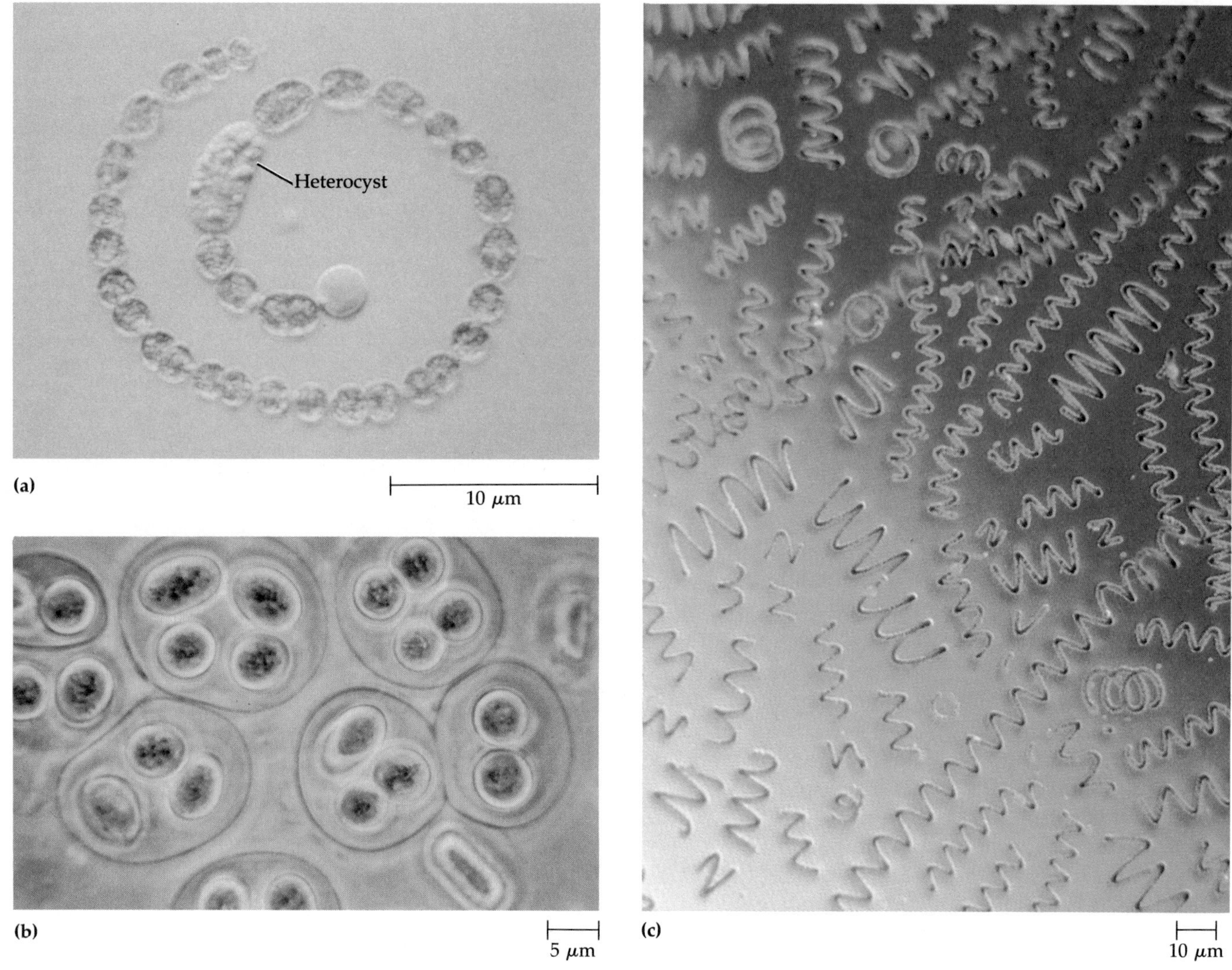

FIGURE 11.19 Cyanobacteria. **(a)** Light micrograph of a filamentous cyanobacterium, *Anabaena spiroides.* Note the heterocysts in which nitrogen-fixing activity is located. **(b)** A nonfilamentous cyanobacterium, *Gleocapsa,* as seen with a phase-contrast microscope. Groups of 2, 3, or 4 cells are held together by the glycocalyx. **(c)** Scanning electron micrograph of *Anabaena spiroides,* showing helical chains of rod-shaped cells.

synthesis of purple or green phototrophic bacteria is **anoxygenic**—it does not produce oxygen. Plants, algae, and the cyanobacteria produce oxygen (O_2) from water (H_2O) as they carry out photosynthesis, as in equation (1):

$$(1)\ 2H_2O + CO_2 \xrightarrow{\text{light}} (CH_2O) + H_2O + O_2$$

The **purple sulfur** and **green sulfur bacteria** use reduced sulfur compounds, such as hydrogen sulfide (H_2S), instead of water, and they produce granules of sulfur (S) rather than oxygen, as shown in equation (2):

$$(2)\ 2H_2S + CO_2 \xrightarrow{\text{light}} (CH_2O) + H_2O + 2S^0$$

At one time, an important question in biology concerned the source of the oxygen produced by plant photosynthesis; was it from the CO_2 or from the H_2O? Until the advent of radioisotope tracers, which traced the oxygen in water and carbon dioxide and finally settled the question, the comparison of equations (1) and (2) above was the only evidence that the oxygen source was indeed H_2O.

Other phototrophs, the **purple nonsulfur** and **green nonsulfur bacteria,** use organic compounds, such as acids and carbohydrates, for the photosynthetic reduction of carbon dioxide. Morphologically, the photosynthetic bacteria are very diverse, with spirals, rods, cocci, and even budding forms.

Cyanobacteria

The **cyanobacteria** are essentially aerobic. They carry out oxygen-producing photosynthesis, much as plants and the eucaryotic algae do (see Chapter 12). Some, however, are capable of using reduced sulfur compounds for anoxygenic photosynthesis. A considerable number of cyanobacteria are capable of fixing nitrogen. Specialized cells called *heterocysts* contain enzymes that fix nitrogen gas (N_2) into ammonium (NH_4^+) for use by the growing cell (Figure 11.19a). *Gas vacuoles* are found in many species that grow in water. The gas vacuole is a series of chambers or gas vesicles surrounded by a protein wall that is permeable to air but not to water. Gas vacuoles provide buoyancy that helps the cell to move to favorable environments. Cyanobacteria that are motile move about by gliding.

The cyanobacteria are morphologically varied. They have unicellular forms that divide by simple fission (Figure 11.19b), colonial forms that divide by multiple fission, and filamentous forms (Figure 11.19c) that reproduce by fragmentation of the filaments. The filamentous forms usually exhibit some differentiation of cells, which are often bound together within an envelope or sheath.

The cyanobacteria, especially those that fix nitrogen, are extremely important to the environment. They occupy environmental niches similar to those occupied by the eucaryotic algae described in Chapter 12 (see Roles of Algae in Nature, p. 310), but the ability of cyanobacteria to fix nitrogen makes them even more adaptable. The environmental role of the cyanobacteria is presented more fully in Chapter 27, in the discussion of eutrophication.

ACTINOMYCETES

Actinomycetes are filamentous bacteria. Superficially, their morphology resembles that of the filamentous fungi; however, the filaments of actinomycetes consist of procaryotic cells with diameters much smaller than those of molds. Some actinomycetes further resemble molds by using externally carried asexual spores for reproduction.

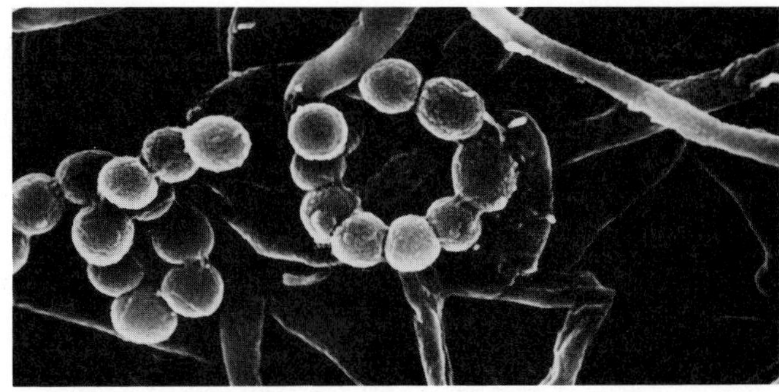

FIGURE 11.20 ***Streptomyces.*** Scanning electron micrograph of a typical *Streptomyces* species. Reproduction is by means of the conidiospores carried in long spiral chains at the end of the branched filaments of the bacterium.

Actinomycetes are very common inhabitants of soil, where mycelia (a filamentous habit of growth) have advantages. The organism can bridge water-free gaps between soil particles to move to a new nutritional site. This morphology also gives the organism a much higher surface-area-to-volume ratio and improves its nutritional efficiency in the highly competitive soil environment.

One genus, *Frankia* (frank'ē-ä), causes nitrogen-fixing nodules to form in alder tree roots, much as *Rhizobium* cause nodules on the roots of legumes (see Chapter 27). The best-known genus of actinomycetes is *Streptomyces* (strep-tō-mī'sēs), which is one of the bacteria most commonly isolated from soil (Figure 11.20). The reproductive asexual spores of *Streptomyces,* conidiospores, are formed at the ends of aerial filaments. If each conidiospore lands on a suitable substrate, it is capable of germinating into a new colony. These organisms are strict aerobes. They often produce extracellular enzymes that enable them to utilize proteins, polysaccharides such as starch or cellulose, and many other organic materials found in soil. *Streptomyces* produces a gaseous compound called *geosmin,* which gives fresh soil its typical musty odor. Species of *Streptomyces* are valuable because they produce most of our commercial antibiotics (see Chapter 20).

In the next chapter, we will turn our attention to algae, fungi, protozoans, and helminths.

STUDY OUTLINE

Bacterial Groups (pp. 273–293)

1. *Bergey's Manual* divides bacteria into sections based on Gram-stain reaction, cellular morphology, oxygen requirements, and nutritional properties.
2. In some cases, the sections include families and orders, and some bacteria are included as genera of uncertain affiliation.
3. Spirochetes are long, thin, helical cells that move by means of an axial filament.
4. Aerobic/microaerophilic, motile, helical/vibrioid gram-negative bacteria move by means of one or more polar flagella.
5. Gram-negative aerobic rods and cocci have polar flagella, if flagellated, and can utilize a wide variety of organic compounds.
6. Facultatively anaerobic gram-negative rods have flagella, if motile, and include the enterics, Vibrionaceae, Pasteurellaceae, and the genus *Gardnerella.*
7. Members of the anaerobic gram-negative straight, curved, and helical rods can be found in humans.
8. Dissimilatory sulfate-reducing or sulfur-reducing bacteria are anaerobes that are important in the sulfur cycle.
9. Anaerobic gram-negative cocci are normal flora of the human mouth.
10. Rickettsias and chlamydias are obligate intracellular parasites.
11. Mycoplasmas are bacteria that lack cell walls.
12. Gram-positive cocci include the catalase-positive *Staphylococcus* and catalase-negative *Streptococcus.*
13. Endospore-forming gram-positive rods and cocci may be aerobic, facultatively anaerobic, or anaerobic.
14. The diverse group of regular, nonsporing, gram-positive rods includes *Lactobacillus* and *Listeria.*
15. Irregular, nonsporing, gram-positive rods include the irregular-staining corynebacteria.
16. Pathogenic species of mycobacteria are acid-fast.
17. Nocardioform bacteria may be acid-fast; they form short filaments.
18. Bacteria with unusual morphologies are discussed in this book in terms of the following groups: budding and/or appendaged; gliding, nonfruiting; gliding, fruiting; budding; sheathed.
19. The chemoautotrophic bacteria play important roles in the cycles of elements in the environment.
20. Extreme halophiles, acidophiles, thermophiles, and methane-producing bacteria are included in the archaeobacteria.
21. Photosynthetic purple and green bacteria are included in the group of anoxygenic phototrophic bacteria; they do not produce molecular oxygen.
22. Cyanobacteria produce molecular oxygen during photosynthesis; they are oxygenic phototrophs.
23. Actinomycetes produce mycelia and reproduce by external spores.

STUDY QUESTIONS

Review

1. The following outline is a key that can be used to identify the medically important groups of bacteria. Fill in the name of the group indicated by the key.

	Name of Group and Representative Genus
I. Gram-positive	
A. Endospore-forming	________
B. Nonsporing	
1. Cocci	________
2. Rods	
a. Regular	________
b. Irregular	________
c. Acid-fast	________
C. Mycelium produced	
1. Acid-fast	________
2. Produce chains of conidia	________
II. Gram-negative	
A. Cells are helical or curved	
1. Axial filament	________
2. No axial filament	
a. Aerobic	________
b. Anaerobic	________
B. Cells are rods or cocci	
1. Aerobic, nonfermenting	________
2. Facultatively anaerobic	________
3. Anaerobic	________
C. Intracellular parasites	________
III. Lacking cell wall	________

2. Compare and contrast each of the following:
   (a) Cyanobacteria and algae
   (b) Actinomycetes and fungi
   (c) *Bacillus* and *Lactobacillus*
   (d) *Pseudomonas* and *Escherichia*
   (e) *Leptospira* and *Spirillum*
   (f) *Veillonella* and *Bacteroides*
   (g) *Rickettsia* and *Chlamydia*
   (h) *Thermoplasma* and *Mycoplasma*

3. Matching:
   I. Gram-positive
      A. Nitrogen-fixing
   II. Gram-negative
      A. Phototrophic
         1. Anoxygenic
         2. Oxygenic
      B. Chemoautotrophic
         1. Oxidize inorganics, such as NO_2^-
         2. Reduce CO_2 to CH_4
      C. Chemoheterotrophic
         1. Move via a slime layer
         2. Form myxospores
         3. Reduce sulfate to H_2S
            a. Anaerobic
            b. Thermophilic
         4. Long filaments, found in sewage
         5. Form projections from the cell
   III. Unusual cell wall (lacking peptidoglycan)

   (a) Archaeobacteria
   (b) Cyanobacteria
   (c) *Cytophaga*
   (d) *Desulfovibrio*
   (e) *Frankia*
   (f) *Hyphomicrobium*
   (g) Methanogenic bacteria
   (h) Myxobacteria
   (i) *Nitrobacter*
   (j) Purple bacteria
   (k) *Sphaerotilus*
   (l) *Sulfolobus*

Challenge

1. Place each section listed in Table 11.1 in the appropriate division:
   (a) Typical gram-positive
   (b) Typical gram-negative
   (c) Wall-less
   (d) Unusual walls
2. Where are each of the following classified in *Bergey's Manual?* Why are they described more than once?
   (a) *Gardnerella*
   (b) Nocardioforms
   (c) *Halobacterium*
3. Identify the genus that best fits each description given below:
   (a) This gram-negative genus is well-suited to degrade hydrocarbons in an oil spill.
   (b) This gram-positive genus presents the greatest source of bacterial damage to the beekeeping industry.
   (c) This gram-negative genus is most probably able to oxidize arsenic and asbestos compounds and to clean up polluted water.
   (d) This group can produce a fuel used for home heating and for generating electricity.

FURTHER READING

Balows, A., H.G. Trüper, M. Dworkin, W. Harder, and K.H. Schliefer, eds. *The Prokaryotes: A Handbook on the Biology of Bacteria,* 2nd ed., 3 volumes. New York: Springer-Verlag, 1991. Discusses characteristics and provides information on cultivation, isolation, and identification of bacteria.

Baron, E.J., and S.M. Finegold. *Diagnostic Microbiology,* 8th ed. St. Louis: Mosby, 1990. A good reference for procedures used in the clinical microbiology laboratory.

Carr, N.G., and B.A. Whitton. *The Biology of Cyanobacteria. Botanical Monographs,* Vol. 19. Berkeley: University of California Press, 1982. A comprehensive reference on morphology, metabolism, and reproduction of cyanobacteria.

Goodfellow, M., M. Mordarski, and S. Williams. *Biology of the Actinomycetes.* New York: Academic Press, 1984. A survey of the biology, ecology, and pathogenicity of this diverse group of bacteria.

Harwood, C.S., and E. Canale-Parola. "Ecology of spirochetes." *Annual Review of Microbiology* 38:161–192, 1984. Describes the role of spirochetes in decomposition and in their associations with humans and other organisms.

Holt, J.G., ed. *Bergey's Manual of Determinative Bacteriology,* 9th ed. Baltimore: Williams & Wilkins, 1992. The standard reference for identification of bacteria.

Holt, J.G., ed. *Bergey's Manual of Systematic Bacteriology,* 1st ed., 4 volumes. Baltimore: Williams & Wilkins, 1984–1989. These four volumes are the standard reference for identification and classification of bacteria.

Kloos, W.E. "Natural populations of the genus *Staphylococcus.*" *Annual Review of Microbiology* 34:559–592, 1980. Illustrates the use of modern technology to determine a species; includes a discussion of staphylococci that live on birds and mammals.

Maniloff, J. "Evolution of wall-less prokaryotes." *Annual Review of Microbiology* 37:477–499, 1983. Mycoplasmas and archaeobacteria are examples in this discussion of evolution; information comes from advances in molecular biology.

Shapiro, J.A. "Bacteria as multicellular organisms." *Scientific American* 258(6):82–89, June 1988. Describes bacterial colonies that function as multicellular organisms.

Starr, M.P., H. Stolp, H.G. Truper, A. Balows, and H.G. Schlegel. *The Prokaryotes: A Handbook on Habitats, Isolation and Identification of Bacteria,* 2nd ed., 2 volumes. New York: Springer-Verlag, 1989. Discusses characteristics and provides information on cultivation, isolation, and identification of bacteria.

CHAPTER 12

Fungi, Algae, Protozoans, and Multicellular Parasites

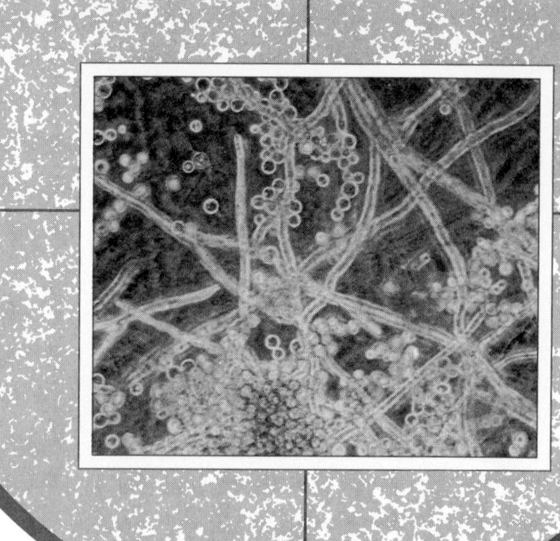

LEARNING OBJECTIVES

- Differentiate between asexual and sexual reproduction, and use as examples the processes of a fungus and a protozoan.
- List the defining characteristics of the four phyla of fungi described in this chapter.
- Describe the outstanding characteristics of the six phyla of algae discussed in this chapter.
- List the distinguishing characteristics of lichens, and describe their nutritional needs.
- Compare and contrast cellular slime molds and plasmodial slime molds.
- Describe the outstanding characteristics of the four phyla of protozoans discussed in this chapter.
- List the characteristics of the three classes of parasitic helminths, and give an example of each class.
- Define intermediate host, definitive host, and arthropod vector.

In this chapter, we will examine the eucaryotic microorganisms—fungi, algae, protozoans, and parasitic helminths, or worms. (See Table 12.1 for a comparison of their characteristics.) Although most adult helminths are not microscopic, they do have microscopic stages in their development, and some of these stages cause disease; therefore, microbiologists study helminths. Microbiologists also study arthropods (jointed-legged animals) because some of them transmit microbial diseases.

FUNGI

The study of fungi is called **mycology.** The fungi include yeasts, molds, and fleshy fungi. Yeasts are unicellular organisms. Molds are multicellular filamentous organisms, such as mildews, rusts, and smuts. Fleshy fungi are multicellular, filamentous organisms that produce a thick (fleshy) reproductive body. The fleshy fungi include mushrooms, puffballs, and coral fungi.

All fungi are chemoheterotrophs, requiring organic compounds for energy and carbon. Fungi are aerobic or facultatively anaerobic; no strictly anaerobic fungi are known. The majority of fungi are saprophytes in soil and water; there they primarily decompose plant material. Like bacteria, fungi contribute significantly to the decomposition of matter and recycling of nutrients. By using extracellular enzymes, such as cellulases and pectinases, fungi are the primary decomposers of the hard parts of plants, which cannot be digested by most animals. Table 12.2 lists the basic differences between fungi and bacteria. Of the more than 100,000 species of fungi, only about 100 are pathogenic for humans and other animals. However, thousands of fungi are pathogenic to plants. Virtually every eco-

TABLE 12.1 Major Differences Among Fungi, Algae, Protozoans, and Helminths (All Are Eucaryotes)

	FUNGI	*ALGAE*	*PROTOZOANS*	*HELMINTHS*
Kingdom	Fungi	Protista and Plantae	Protista	Animalia
Nutritional type	Chemoheterotroph	Photoautotroph	Chemoheterotroph	Chemoheterotroph
Multicellular	All, except yeasts	Some	None	All
Cellular arrangement	Unicellular, filamentous, fleshy (such as mushrooms)	Unicellular, colonial, filamentous; tissues	Unicellular	Tissues and organs
Food acquisition	Absorptive	Absorptive	Absorptive; cytostome	Mouth (ingestive); absorptive
Characteristic features	Sexual and asexual spores	Pigments	Motility; some form cysts	Many have elaborate life cycles, including egg, larva, and adult
Embryo	None	Some	None	All

nomically important plant is attacked by one or more of the fungi.

Characteristics of Fungi

Yeast identification, like bacterial identification, uses biochemical tests. However, multicellular fungi are identified on the basis of physical appearance, including colony characteristics and reproductive spores.

VEGETATIVE STRUCTURES

Fungal colonies are described as **vegetative** structures because they are composed of the cells involved in catabolism and growth.

Molds and Fleshy Fungi

The **thallus** (body) of a mold or fleshy fungus consists of long filaments of cells joined together; these filaments are called **hyphae** (singular, **hypha**). In most molds, the hyphae contain crosswalls called **septa** (singular, **septum**), which divide the hyphae into distinct, uninucleate (one-nucleus) cell-like units. These hyphae are called **septate hyphae** (Figure 12.1a). In a few classes of fungi, however, the hyphae contain no septa and appear as long, continuous cells with many nuclei. These hyphae are called **coenocytic hyphae** (Figure 12.1b). Even in fungi with septate hyphae, there are usually openings in the septa that make the cytoplasm of adjacent "cells" continuous; therefore, these fungi are actually coenocytic organisms, too. The hyphae of a thallus grow by elongating at the tips (see

TABLE 12.2 Comparison of Selected Features of Fungi and Eubacteria

	FUNGI	*EUBACTERIA*
Cell type	Eucaryotic with well-defined nuclear membrane	Procaryotic
Cell membrane	Sterols present	Sterols absent, except in *Mycoplasma*
Cell wall	Glucans; mannans; chitin (no peptidoglycan)	Peptidoglycan
Spores	Produce a wide variety of sexual and asexual reproductive spores	Endospores (not for reproduction); some asexual reproductive spores
Metabolism	Limited to heterotrophic; aerobic, facultatively anaerobic	Heterotrophic, chemoautotrophic, photoautotrophic; aerobic, facultatively anaerobic, anaerobic
Sensitivity to antibiotics	Often sensitive to polyenes, imidazoles, and griseofulvin	Often sensitive to penicillins, tetracyclines, and aminoglycosides

Source: After B.D. Davis et al. *Microbiology*, 4th ed. Philadelphia: J.B. Lippincott, 1990, p. 746.

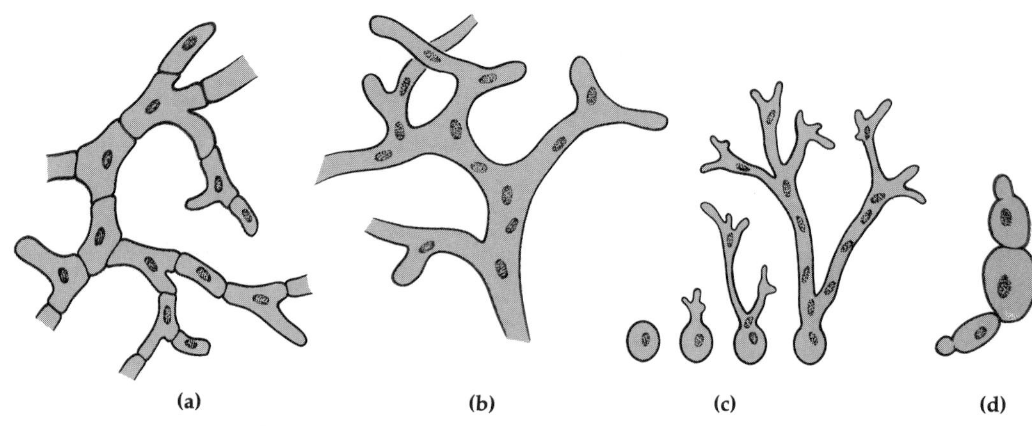

FIGURE 12.1 Vegetative structures of fungi. (a) Septate hyphae have crosswalls dividing the hyphae into cell-like units. **(b)** Coenocytic hyphae lack crosswalls. **(c)** Hyphae grow by elongating at the tips. **(d)** Pseudohyphae are short chains of cells formed by some yeasts.

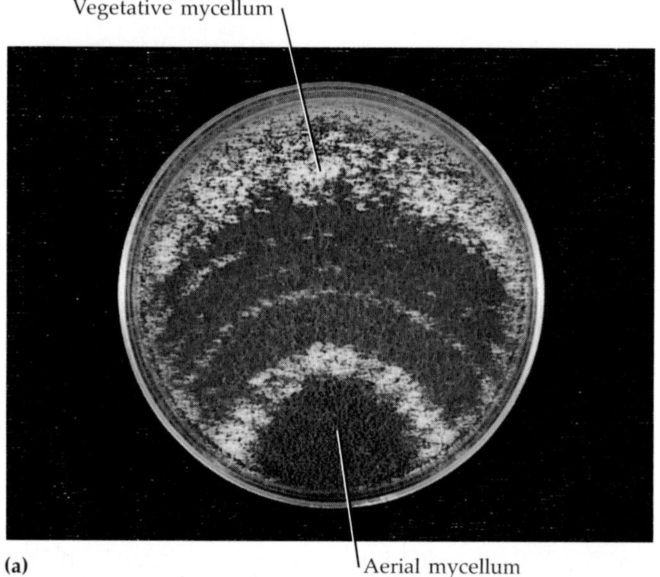

FIGURE 12.2 ***Aspergillus niger*** **grown on agar. (a)** Colony on glucose agar plate. **(b)** Aerial mycelia bearing reproductive spores.

Figure 12.1c). Each part of a hypha is capable of growth, and, when a fragment breaks off, it can elongate to form a new hypha. In the laboratory, fungi are usually grown from fragments obtained from a fungal thallus (Figure 12.2a).

When environmental conditions are suitable, the hyphae grow, intertwine, and form a mass called a **mycelium,** which is visible to the naked eye. The portion of the mycelium concerned with obtaining nutrients is called the **vegetative mycelium;** the portion concerned with reproduction is the **reproductive** or **aerial mycelium,** so called because it projects above the surface of the medium on which the fungus is growing (Figure 12.2b). The aerial mycelium often bears reproductive spores, which we will discuss later.

Yeasts

Yeasts are nonfilamentous, unicellular fungi that are typically spherical or oval. Like molds, yeasts are widely distributed in nature; they are frequently found as a white powdery coating on fruits and leaves. Because most yeasts are colonies of unicellular organisms, they do not reproduce as a unit. Instead, the colony grows as the number of yeast cells increases. This increase usually happens by **budding.** In budding, the parent cell forms a protuberance (bud) on its outer surface. As the bud elongates, the parent cell's nucleus divides, and one nucleus migrates into the bud. Cell wall material is then laid down between the bud and parent cell, and the bud eventually breaks away (Figure 12.3).

One yeast cell can in time produce up to 24 daughter cells by budding. Some yeasts produce buds that fail to detach themselves; these buds form a short chain of cells called a **pseudohypha** (see Figure 12.1d). A few types of yeast grow by fission. During fission, the parent cell itself elongates, its nucleus divides, and two daughter cells are produced. Increases in the number of yeast cells on a solid medium produce a colony similar to a bacterial colony.

Yeasts are capable of facultative anaerobic growth. As you may recall from Chapter 5, yeasts can use oxygen or an organic compound as the final electron acceptor; this is a valuable attribute. If given access to oxygen, yeasts perform aerobic respiration to metabolize carbohydrates to carbon dioxide and water. Denied oxygen, they ferment carbohydrates and produce ethanol and carbon dioxide. This fermentation is the basis of the brewing, wine-making, and baking industries. Species of *Saccharomyces* produce ethanol, used for brewing beverages, and carbon dioxide, used for raising dough.

Dimorphic Fungi

Some fungi, most notably the pathogenic species, exhibit **dimorphism,** that is, two forms of growth. Such fungi can grow either as a mold or as a yeast. The moldlike forms produce vegetative and aerial mycelia. The yeastlike forms reproduce by budding. Frequently, dimorphism is temperature-dependent: At 37°C the fungus is yeastlike, and at 25°C it is moldlike. The appearance of the dimorphic fungus shown in Figure 12.4 changes with CO_2 concentration.

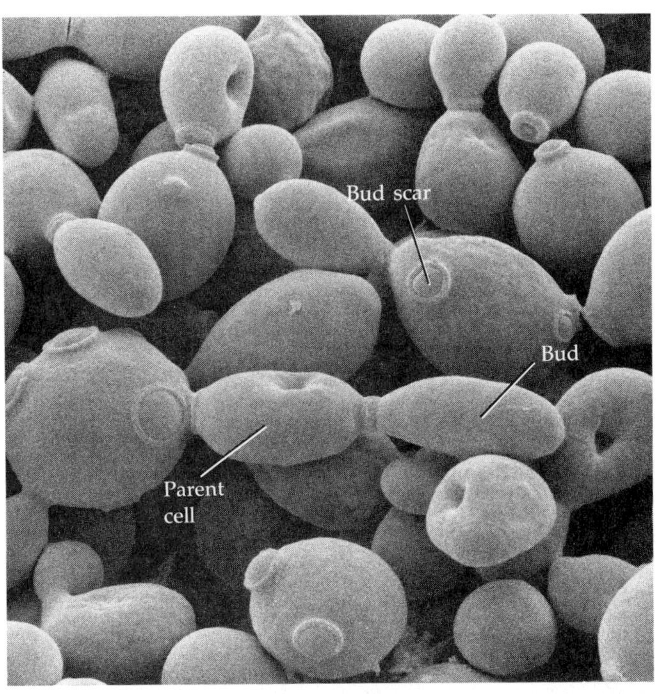

FIGURE 12.3 Bakers' yeast, *Saccharomyces cerevisiae.* Various stages of budding.

REPRODUCTIVE STRUCTURES

Viewed with a microscope, the hyphae of fungi often look alike. When fungi are identified, the reproductive structures, or **spores,** must be examined.

Reproduction in fungi occurs by spore formation. These spores, however, are quite different from bacterial endospores. Bacterial endospores (see Chapter 4) are formed so the bacterial cell will survive adverse environmental conditions. A single vegetative bacterial cell forms one endospore, which eventually germinates to produce a single vegetative bacterial cell. Formation of an endospore by bacteria is not reproduction because it does not increase the total number of bacterial cells. But after a mold forms a spore, the spore detaches from the parent and germinates into a new mold (Figure 12.1c). Unlike the bacterial endospore, this is a true reproductive spore—one organism produces many spores. Although fungal spores can survive for extended periods in dry or hot environments, they do not exhibit the extreme tolerance and longevity of bacterial endospores.

Spores are formed from the aerial mycelium in a great variety of ways, depending on the species. Fungal spores can be asexual or sexual. **Asexual spores** are formed by the aerial mycelium of one organism. When these spores germinate, they become organisms that are genetically identical to the parent. **Sexual spores** result from the fusion of nuclei from two opposite mating strains of the same species of fungus. Organisms that grow from sexual spores will have genetic characteristics of both parental strains. Because spores are of

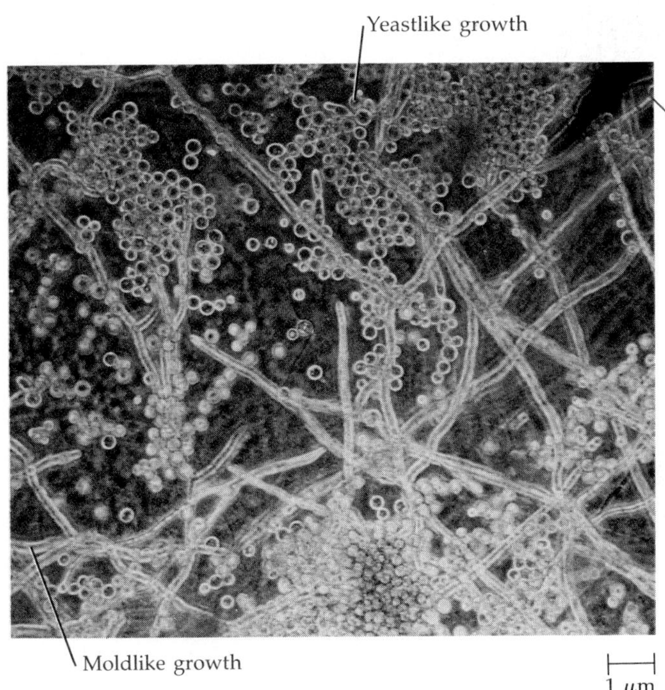

FIGURE 12.4 Dimorphism. Dimorphism in the fungus *Mucor rouxii* depends on CO_2 concentration. On the agar surface, *Mucor* exhibits yeastlike growth, but in the agar it is moldlike.

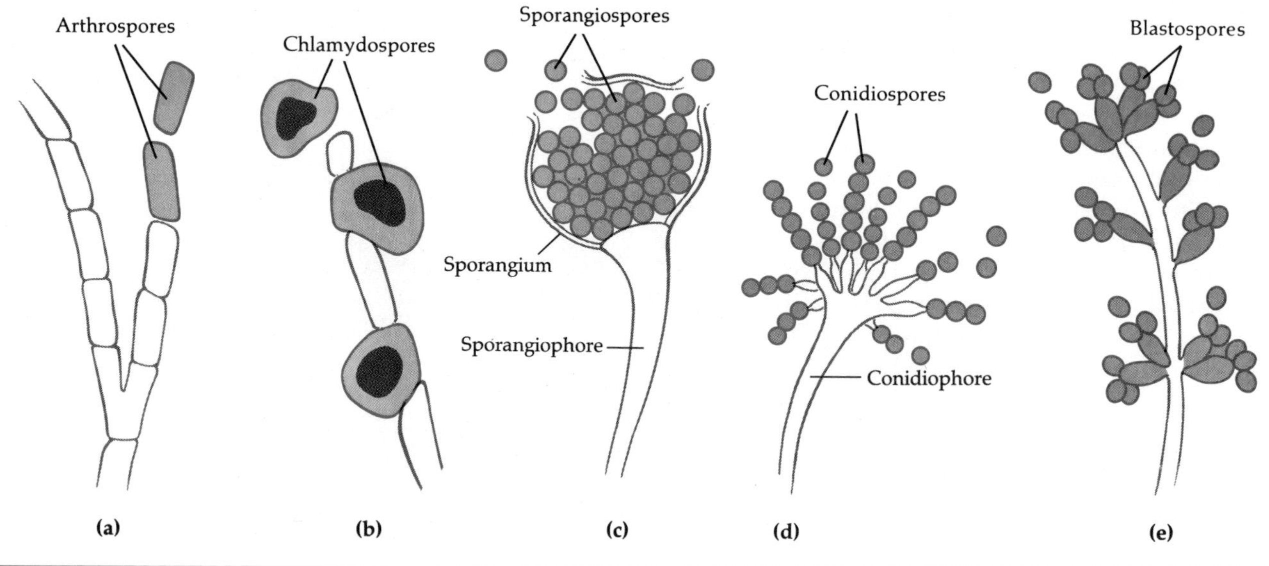

FIGURE 12.5 Representative asexual spores. **(a)** Fragmentation of hyphae results in formation of arthrospores. **(b)** Chlamydospores are thick-walled cells within the hyphae. **(c)** Sporangiospores are formed within a sporangium (spore sac). **(d)** Conidiospores are arranged in chains at the end of a conidiophore. **(e)** Blastospores are formed from buds of the parent cell.

key importance in the identification of fungi, we will next look at some of the various types of asexual and sexual spores.

Asexual Spores

Asexual spores are produced by an individual fungus through mitosis and subsequent cell division; there is no fusion of the nuclei of cells. Several types of asexual spores are produced by fungi. One type, an **arthrospore,** is formed by the fragmentation of a septate hypha into single, slightly thickened cells (see Figures 12.5a and 24.15b. One species that produces such spores is *Coccidioides immitis* (kok-sid-ē-oi′dēz im′mi-tis). Another type of asexual spore is a **chlamydospore,** a thick-walled spore formed by rounding and enlargement within a hyphal segment (Figure 12.5b). A fungus that produces chlamydospores is the yeast *Candida albicans* (kan′did-ä al′bi-kanz). A **sporangiospore** is an asexual spore formed within a sac (**sporangium**) at the end of an aerial hypha called a **sporangiophore.** The sporangium can contain hundreds of sporangiospores (Figure 12.5c). Such spores are produced by *Rhizopus.* A fourth principal type of asexual spore is a **conidiospore,** which is a unicellular or multicellular spore that is not enclosed in a sac (Figure 12.5d). Conidiospores are produced in a chain at the end of a **conidiophore.** Such spores are produced by *Penicillium.* A fifth type of asexual spore, a **blastospore,** consists of a bud coming off the parent cell (Figure 12.5e). Such spores are found in some yeasts.

Sexual Spores

A fungal sexual spore results from sexual reproduction, consisting of three phases:

1. A haploid nucleus of a donor cell (+) penetrates the cytoplasm of a recipient cell (−).
2. The (+) and (−) nuclei fuse to form a diploid zygote nucleus.
3. By meiosis, the diploid nucleus gives rise to haploid nuclei (sexual spores), some of which may be genetic recombinants.

Fungi produce sexual spores less frequently than asexual spores. Often, the sexual spores are produced only under special circumstances. The plant-pathogenic ascomycetes produce ascospores only when host plants are at the end of a growing season. This suggests that the ascospores are initiated by changes in moisture, temperature, or nutrient availability. The ascospores survive until the plant hosts are growing again. The sexual spores produced by fungi are the criterion used to group the fungi into several phyla.

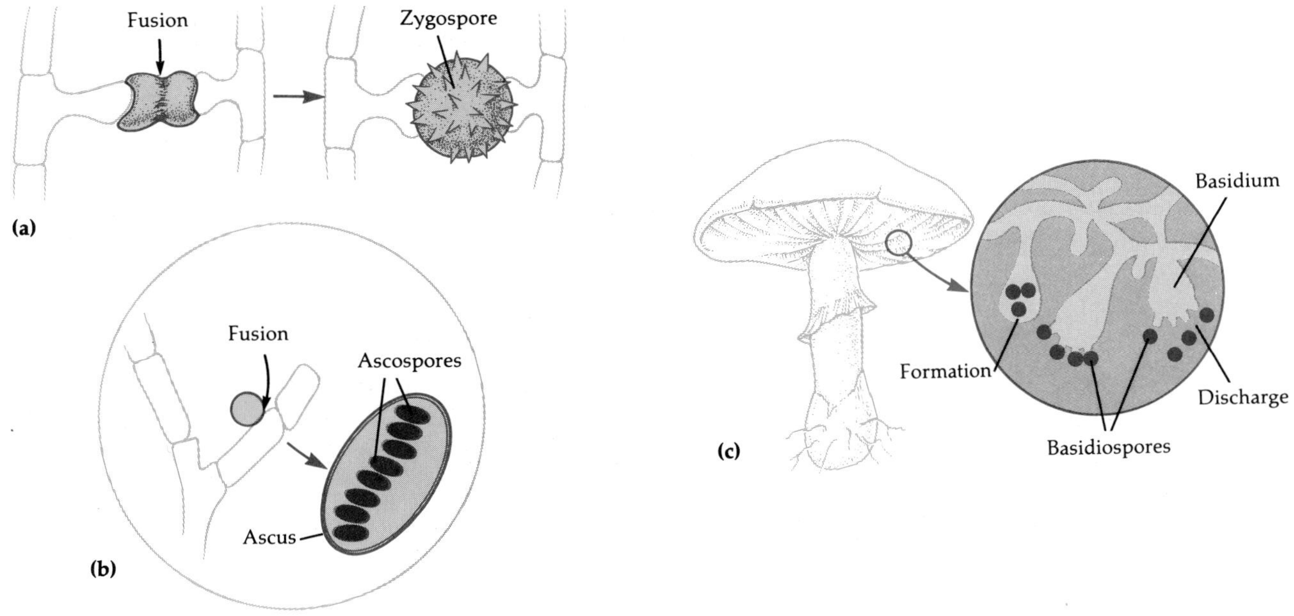

FIGURE 12.6 Sexual spores. **(a)** Zygospores, characteristic of the Phylum Zygomycota, are produced from the fusion of two cells that are morphologically alike. **(b)** Ascospores, produced by the Ascomycota, are formed within an ascus. **(c)** Basidiospores, produced only by Basidiomycota, are formed on the tip of a pedestal called a basidium.

One kind of sexual spore is a **zygospore,** a large spore enclosed in a thick wall (Figure 12.6a). This type of spore results when the nuclei of two cells that are morphologically similar to each other fuse. Such spores are produced by fungi of the Phylum Zygomycota. A second type of sexual spore is an **ascospore,** a spore resulting from the fusion of the nuclei of two cells that can be morphologically similar or dissimilar. These spores are produced in a saclike structure called an **ascus** (Figure 12.6b). There are usually two to eight ascospores in an ascus. Such spores are produced by the Phylum Ascomycota. Another type of sexual spore is the **basidiospore,** a spore formed externally on a base pedestal called a **basidium** (Figure 12.6c). There are usually four basidiospores per basidium. Basidiospores are produced only by the Phylum Basidiomycota, which includes the mushrooms. Mature basidiospores are often discharged explosively when a raindrop or an animal touches the mushroom.

NUTRITIONAL ADAPTATIONS

Fungi are generally adapted to environments that would be hostile to bacteria; thus, they can survive in places where microbial growth is not expected. Fungi are chemoheterotrophs, and, like bacteria, they absorb nutrients rather than ingesting them, as animals do. However, fungi differ from bacteria in certain environmental requirements and in the following nutritional characteristics.

1. Fungi usually grow better in an acidic pH (5.0), which is too acidic for the growth of most common bacteria.
2. Most molds are aerobic, so they grow on surfaces rather than throughout a substrate. Yeasts are facultative anaerobes.
3. Most fungi are more resistant to osmotic pressures than bacteria are; most fungi are therefore able to grow in high sugar or salt concentrations.
4. Fungi are capable of growing on substances with a very low moisture content, generally too low to support the growth of bacteria.
5. Fungi require somewhat less nitrogen for growth than bacteria.
6. Fungi are capable of using complex carbohydrates, such as lignin (wood), that most bacteria cannot metabolize.

These characteristics allow fungi to grow on such unlikely substrates as painted walls or shoe leather. The undesirable effects of these adaptations will be discussed shortly. Some of the beneficial uses of fungi are discussed in Chapter 28.

TABLE 12.3 Characteristics of Some Parasitic Fungi

PHYLUM	GROWTH CHARACTERISTICS	ASEXUAL SPORE TYPES	HUMAN PATHOGENS
Zygomycota	Nonseptate hyphae	Sporangiospores	*Rhizopus*
			Mucor
Ascomycota	Septate hyphae	Conidiospores	*Allescheria boydii*
	Dimorphic	Conidiospores	*Aspergillus**
			*Blastomyces dermatitidis**
			*Histoplasma capsulatum**
	Septate hyphae, strong affinity for keratin	Conidiospores	*Microsporum**
		Arthrospores	*Trichophyton**
Basidiomycota	Septate hyphae include the fleshy fungi (mushrooms), rust and smuts, and plant pathogens; yeastlike encapsulated cells	Conidiospores	*Cryptococcus neoformans**
Deuteromycota	Septate hyphae	Conidiospores	*Epidermophyton*
		Chlamydospores	*Cladosporium werneckii*
	Dimorphic	Conidiospores	*Sporothrix schenckii*
		Arthrospores	*Coccidioides immitis*
	Yeastlike, pseudohyphae	Chlamydospores	*Candida albicans**
	Dimorphic	Chlamydospores Arthrospores Conidiospores	*Paracoccidioides brasiliensis*
	Septate hyphae, produce melaninlike pigments	Conidiospores	*Phialophora* *Fonsecaea*
	Yeastlike pseudohyphae	Arthrospores	*Trichosporon*
	Unknown	Unknown	*Pneumocystis carinii***

*Imperfect name. Refer to Table 12.4 for current classification and names.
**Classification is unclear; also listed with protozoa.

Medically Important Phyla of Fungi

This section provides an overview of medically important phyla of fungi. The actual diseases they cause will be studied in Part Four, Chapters 21–26. Note that not all fungi cause disease. There are other phyla of fungi not discussed here, because they contain no pathogens.

The genera named in the following phyla include many that are readily found as contaminants in foods and in laboratory bacterial cultures. Although these genera are not all of primary medical importance, they are typical examples of their respective groups. Table 12.3 gives a specific list of human pathogens.

DEUTEROMYCOTA

The Deuteromycota are also known as the Fungi Imperfecti. These fungi are "imperfect" because they have not yet been found to produce sexual spores. Members of this phylum produce the asexual chlamydospores, arthrospores, and conidiospores; budding also occurs. Deuteromycota have septate hyphae.

Most of the pathogenic fungi are, or once were, classified as Deuteromycota. This phylum might be described as a "holding category" in which fungi are placed until sexual spores are observed and the fungus can be properly classified. Table 12.4 lists some well-known Fungi Imperfecti that have recently been classi-

TABLE 12.3 Characteristics of Some Parasitic Fungi

CLINICAL NOTES	TYPE OF MYCOSIS	HABITAT	REFER TO CHAPTER
Opportunistic pathogen	Systemic	Ubiquitous	24
Opportunistic pathogen	Systemic	Ubiquitous	24
Primary cause of maduromycosis	Subcutaneous	Soil	21
Opportunistic pathogen	Systemic	Ubiquitous	24
Inhalation	Systemic	Unknown	24
Inhalation	Systemic	Soil	24
Tinea capitis (ringworm)	Cutaneous	Soil, animals	21
Tinea pedis (athlete's foot)	Cutaneous	Soil, animals	21
Inhalation	Systemic	Soil, bird feces	22
Tinea cruris (jock itch), tinea unguium (of fingernails or toenails)	Cutaneous	Soil, humans	21
Tinea nigra	Superficial	Ubiquitous	—
Puncture wound	Subcutaneous	Soil	21
Inhalation	Systemic	Soil	24
Opportunistic pathogen	Cutaneous, systemic, mucocutaneous	Human normal flora	21, 26
Inhalation	Systemic	Presumably soil	—
Chromomycosis	Subcutaneous	Soil, plant debris	—
Chromomycosis	Subcutaneous	Soil, plant debris	—
White piedra	Superficial	Soil (?), humans	—
Pneumonia in immunosuppressed individuals	Systemic	Humans (?)	24

fied. Note that the generic names are changed with reclassification. When different species with a deuteromycete genus are observed to have morphologically different sexual spores, the species might be reclassified into two or more different genera, as was done with *Penicillium*.

A Case of Mistaken Identity

Pneumocystis pneumonia is the most common opportunistic infection in immunosuppressed individuals and is a significant cause of death in AIDS patients. The cause of this infection, *Pneumocystis carinii*, was not considered a human pathogen until the 1970s.

Pneumocystis lacks structures that can be easily used for identification (Figure 12.7), and its taxonomic position has been uncertain since its discovery in 1908.

TABLE 12.4 Several Genera of Reclassified Imperfect Fungi

IMPERFECT NAME	RECLASSIFIED TO PHYLUM	PERFECT NAME
Aspergillus	Ascomycota	*Sartorya, Eurotium, Emericella*
Blastomyces	Ascomycota	*Ajellomyces*
Candida	Ascomycota	*Pichia*
Cryptococcus	Basidiomycota	*Filobasidiella*
Histoplasma	Ascomycota	*Emmonsiella, Gymnoascus*
Microsporum	Ascomycota	*Nannizia*
Penicillium	Ascomycota	*Talaromyces, Carpenteles*
Trichophyton	Ascomycota	*Arthroderma*

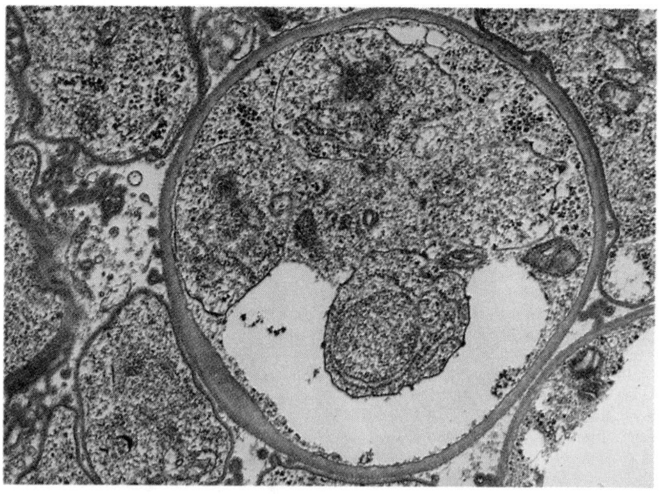

FIGURE 12.7 Pneumocystis carinii. The disease caused by this organism occurs only in patients with severely depressed immune systems.

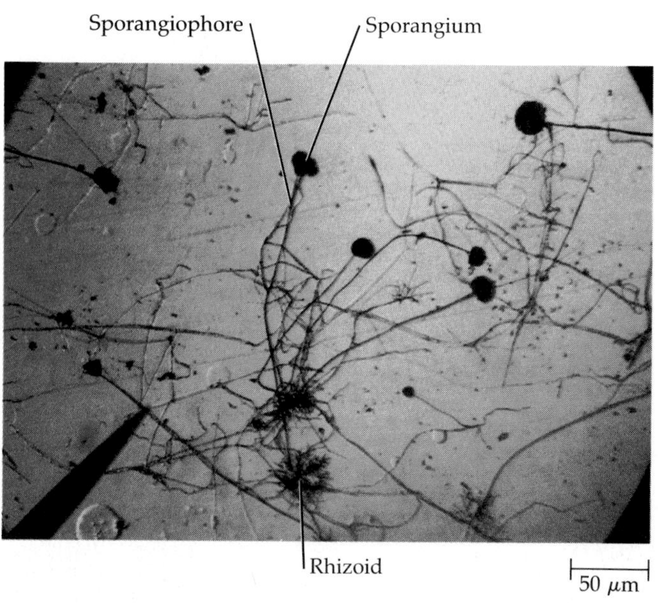

FIGURE 12.8 ***Rhizopus nigricans.*** Note the sporangia at the tops of the sporangiophores.

Although it was provisionally classified as a protozoan, recent studies comparing its ribosomal RNA sequence with those of other protozoa, *Euglena*, cellular slime molds, plants, mammals, and fungi have shown that it might actually be a member of the Kingdom Fungi. Researchers have not been able to culture *Pneumocystis* or develop treatments for *Pneumocystis* pneumonia. Perhaps as researchers start thinking of this organism as a fungus, appropriate culture methods and treatments will result. If *Pneumocystis* is reclassified, it will be placed in the Deuteromycota until sexual spores are observed.

ZYGOMYCOTA

The **Zygomycota,** or conjugation fungi, are saprophytic molds that have coenocytic hyphae. An example is *Rhizopus nigricans*, the common black bread mold (Figure 12.8). The asexual spores of *Rhizopus* are sporangiospores. The dark sporangiospores inside the sporangium give *Rhizopus* its descriptive common name. When the sporangium breaks open, the sporangiospores are dispersed. If they fall on a suitable medium, they will germinate into a new mold thallus. The sexual spores are zygospores.

ASCOMYCOTA

The **Ascomycota,** or sac fungi, include molds with septate hyphae and yeasts. They are called sac fungi because their sexual spores are ascospores produced in an ascus. Their asexual spores are usually conidiospores produced in long chains from the conidiophore. The arrangements of conidiospores in *Penicillium* and *Aspergillus* are shown in Figure 12.9. The term *conidia* means dust, and these spores freely detach from the chain at the slightest disturbance and float in the air like dust.

BASIDIOMYCOTA

The **Basidiomycota,** or club fungi, also possess septate hyphae. The common name is derived from the shape of the basidium that bears the sexual basidiospores. Some of the basidiomycota produce asexual conidiospores. Representative basidiomycetes are shown in Figure 12.10.

Fungal Diseases

Any fungal infection is called a **mycosis.** Mycoses are generally chronic (long-lasting) infections because fungi grow slowly. Mycoses are divided into five groups, according to the level of infected tissue and mode of entry into the host. They are classified as systemic, subcutaneous, cutaneous, superficial, or opportunistic. Characteristics of the medically important fungi are summarized in Table 12.3.

Systemic mycoses are fungal infections deep within the body. They are not restricted to any particular region of the body but can affect a number of tissues and organs. Systemic, or deep, mycoses are usually caused by saprophytic fungi that live in the soil. Inhalation of spores is the route of transmission; these infections typically begin in the lungs and then spread

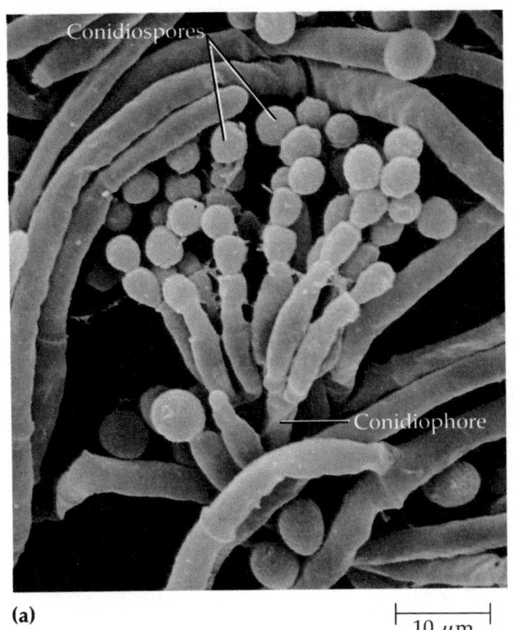

(a)

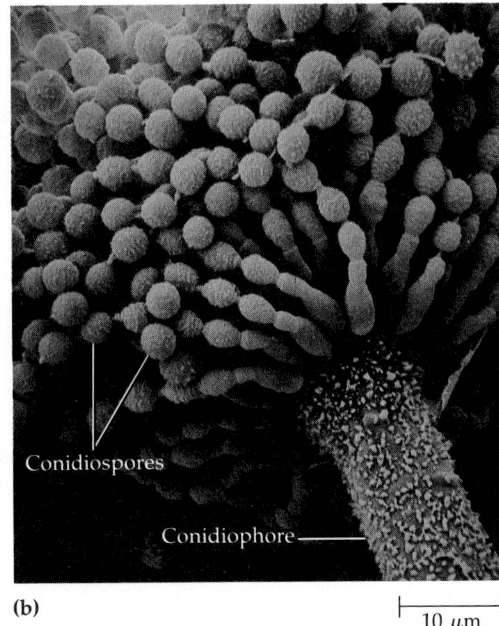

(b)

FIGURE 12.9 Conidiospores. The arrangement of conidiospores is useful in identification of fungi. **(a)** *Penicillium chrysogenum* produces conidiospores from a branched conidiophore. **(b)** Conidiospores of *Aspergillus flavus* are produced from the enlarged terminal end (vesicle) of the conidiophore.

(a)

(b)

(c)

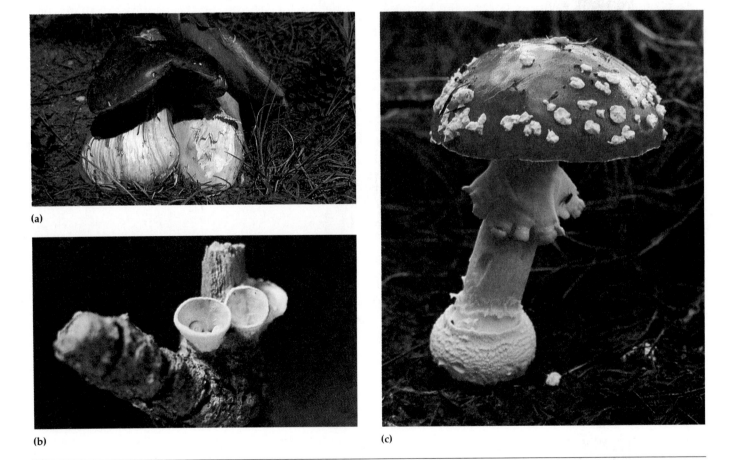

FIGURE 12.10 Representative basidiomycota. **(a)** Two fruiting bodies of the edible basidiomycete *Boletus edulis*. **(b)** Bird's nest fungus growing on a twig. The basidiospores seen in one of the cups will pop out when the cup is hit by a raindrop. **(c)** *Amanita muscaria* produces deadly amanitin toxin.

to other body tissues. They are not contagious from animal to human or from human to human. Two systemic mycoses, histoplasmosis and coccidioidomycosis, are discussed in Chapter 24.

Subcutaneous mycoses are fungal infections beneath the skin and are caused by saprophytic fungi that live in soil and on vegetation. Infection occurs by direct implantation of spores or mycelial fragments into a puncture wound in the skin.

Fungi that infect only the epidermis, hair, and nails are called **dermatophytes,** and their infections are called **dermatomycoses** or **cutaneous mycoses** (see Figure 21.14). Dermatophytes secrete *keratinase,* an enzyme that degrades keratin. Keratin is a protein found in hair, skin, and nails. Infection is transmitted from human to human or from animal to human by direct contact or by contact with infected hairs and epidermal cells (as from barber shop clippers or shower room floors).

The fungi that cause **superficial mycoses** are localized along hair shafts and in superficial (surface) epidermal cells. These infections are prevalent in tropical climates.

An **opportunistic pathogen** is generally harmless in its normal habitat but can become pathogenic in a host who is seriously debilitated or traumatized, who is under treatment with broad-spectrum antibiotics, or whose immune system is suppressed by drugs or by an immune disorder. AIDS patients are quite susceptible to opportunistic pathogens. A person's normal flora can become opportunistic pathogens under these conditions.

Mucormycosis is an opportunistic mycosis caused by *Rhizopus* and *Mucor* (mū-kôr); the infection occurs mostly in patients with ketoacidosis resulting from diabetes mellitus, leukemia, or treatment with immunosuppressive drugs. **Aspergillosis** is also an opportunistic mycosis; it is caused by *Aspergillus* (see Figure 12.2). This disease occurs in persons who have debilitating lung diseases or cancer and have inhaled *Aspergillus* spores. The mycosis **candidiasis** (see Figure 21.15) is most frequently caused by *Candida albicans* and may occur as vulvovaginal candidiasis during pregnancy. **Thrush,** a mucocutaneous candidiasis, is an inflammation of the mouth and throat; it frequently occurs in newborns.

Some fungi cause disease by producing toxins. These toxins will be discussed in Chapter 15.

Economic Effects of Fungi

Fungi can have undesirable effects for industry and agriculture because of their nutritional adaptations. As most of us have observed, mold spoilage of fruits, grains, and vegetables is relatively common, but bacterial spoilage of such foods is not. There is little moisture on the unbroken surfaces of such foods, and the interiors of fruits are too acidic for many bacteria to be able to grow there. Jams and jellies also tend to be acidic, and they have a high osmotic pressure from the sugars they contain. These factors all discourage bacterial growth but readily support the growth of mold. A paraffin layer on top of a jar of homemade jelly helps deter mold growth because molds are aerobic and the paraffin layer keeps out the oxygen. However, fresh meats and certain other foods are such good substrates for bacterial growth that bacteria will not only outgrow molds but will actively suppress mold growth in these foods.

The ability of fungi to grow at low moisture levels is of particular importance in their role as plant pathogens. Bacterial plant pathogens are far less common than fungal pathogens. In fact, the fungus that caused the great potato blight in Ireland during the early 1800s, *Phytophthora infestans* (fī-tof'thô-rä in-fes'tans), was one of the first microorganisms to be associated with a disease.

The spreading chestnut tree, of which Longfellow wrote, no longer grows in this country except in a few widely isolated sites. An important fungal blight killed virtually all of them. First seen in the United States in 1904, the ascomycete *Cryphonectria parasitica* (krip-tō-nek'trē-ä par-ä-si'ti-kä) was introduced from China. The tree roots live and put forth shoots regularly, but the shoots are just as regularly killed by the fungus. Another devastating fungal plant disease that was also imported to this country is Dutch elm disease, caused by *Ceratocystis ulmi* (sē-rä-tō-sis'tis ul'mē). Carried from tree to tree by a bark beetle, the fungus blocks the afflicted tree's circulation.

ALGAE

Algae are familiar as the large brown kelp in coastal waters, the green scum in a puddle, and the green stains on soil or on rocks. Some algae are unicellular, others form chains of cells (*filamentous*), and a few have plantlike bodies called *thalli* (singular, thallus). Algae are mostly aquatic, although some are found in soil or on trees when sufficient moisture is available there. Unusual algal habitats include the hair of both the sedentary South American sloth and the polar bear. Water is necessary for physical support, reproduction, and diffusion of nutrients. Generally, algae are found in cool temperate waters, although the large floating mats of the brown alga *Sargassum* (sär-gas'sum) are found in the subtropical Sargasso Sea. Some species of brown algae grow in antarctic waters.

Algae are photoautotrophs; hence, they are found

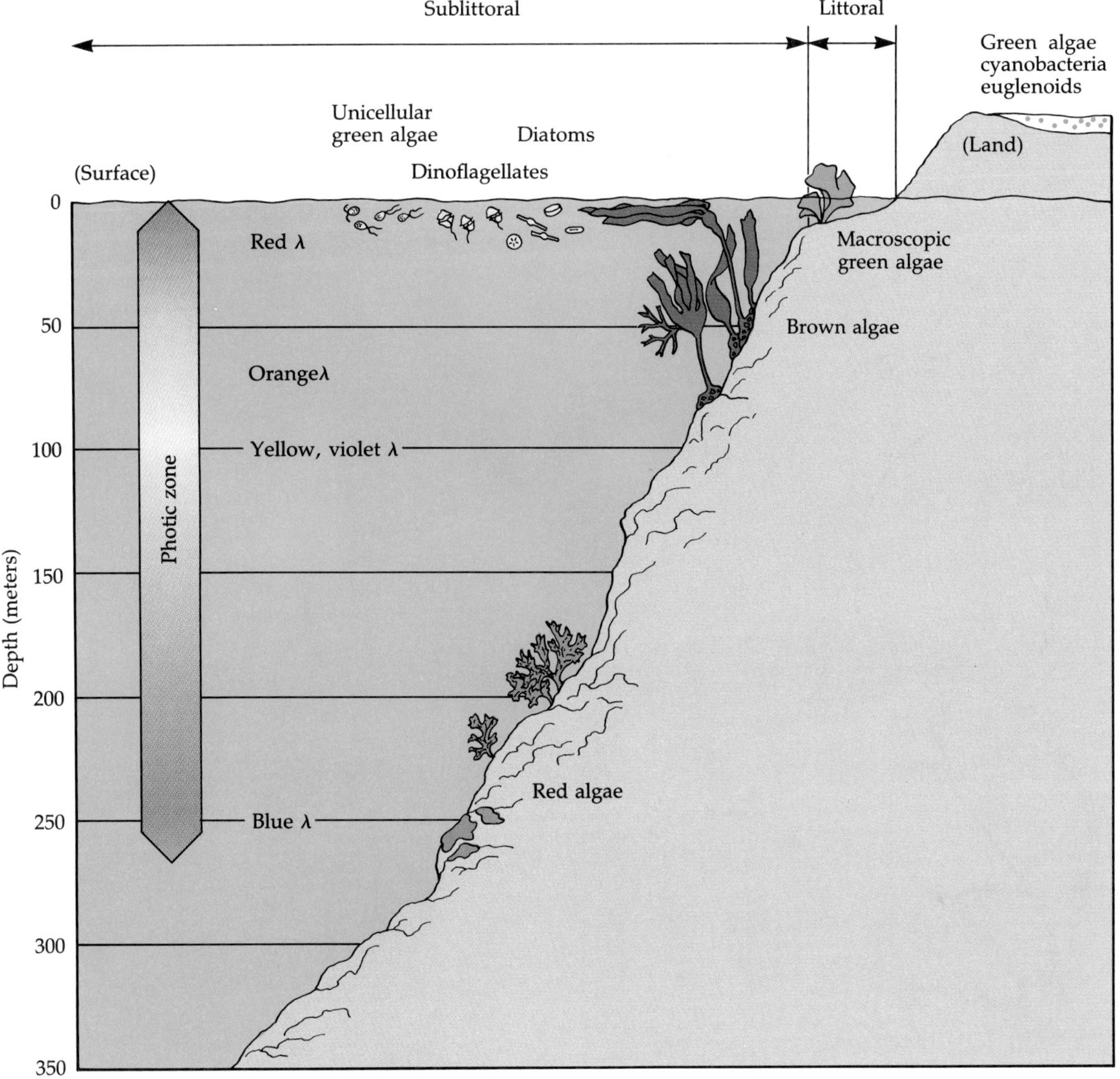

FIGURE 12.11 Algal habitats. Unicellular and filamentous algae can be found on land; they exist as plankton in aquatic environments. Multicellular green, brown, and red algae require a suitable attachment site, adequate water for support, and light of the appropriate wavelengths.

throughout the photic (light) zone of water. Their locations depend on the availability of nutrients, wavelengths of light, and type of substratum. Red and brown algae require rocks, snails, or even other algae to which their holdfasts can attach. Probable locations for representative marine algae are shown in Figure 12.11.

Chlorophyll *a* (light-trapping pigment) and accessory pigments involved in photosynthesis are responsible for the distinctive colors of many algae. Some algae exist as single cells; others exist as colonies of hundreds or thousands of cells. Algae are classified as protists or plants according to their structures, pigments, and other qualities (Table 12.5). Following are descriptions of some phyla of algae.

Phyla of Algae

Dinoflagellates (Figure 12.12a) are unicellular **planktonic,** or free-floating, algae. Their cell walls consist of many individual plates made of cellulose and silica. Marine dinoflagellates in the genus *Gonyaulax* (gō-nē-ō'laks) produce neurotoxins that cause **paralytic shellfish poisoning.** This disease spreads to humans when large numbers of dinoflagellates are eaten by mollusks—such as mussels or clams—that in turn are eaten by humans. Large concentrations of *Gonyaulax* give the ocean a deep red color, from which the name *red tide* originates. Mollusks should not be harvested for human consumption during a red tide. A similar disease called **ciguatera** occurs when the dino-

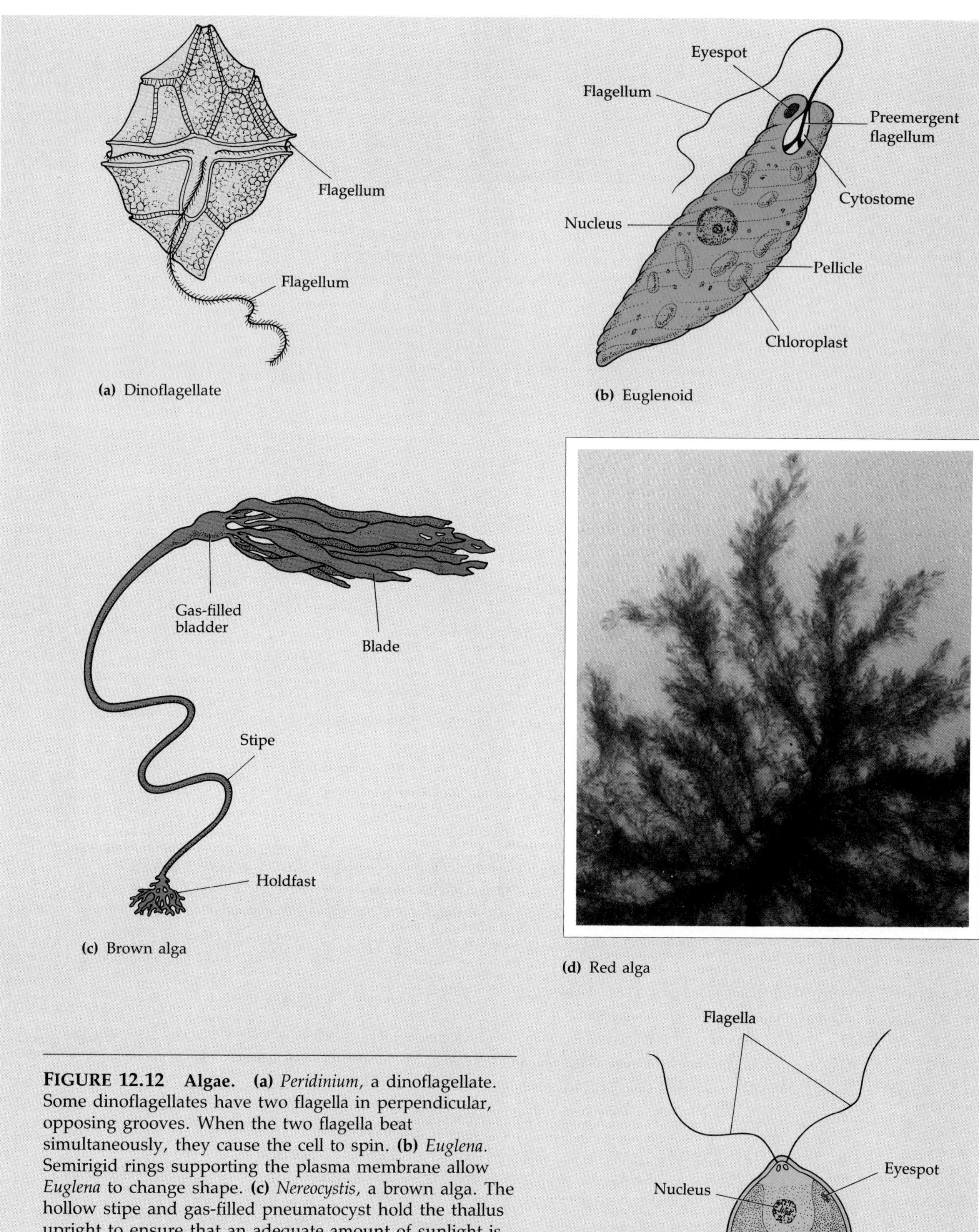

FIGURE 12.12 Algae. **(a)** *Peridinium,* a dinoflagellate. Some dinoflagellates have two flagella in perpendicular, opposing grooves. When the two flagella beat simultaneously, they cause the cell to spin. **(b)** *Euglena.* Semirigid rings supporting the plasma membrane allow *Euglena* to change shape. **(c)** *Nereocystis,* a brown alga. The hollow stipe and gas-filled pneumatocyst hold the thallus upright to ensure that an adequate amount of sunlight is received for growth. **(d)** *Microcladia,* a red alga. The delicately branched red algae get their color from phycobiliprotein accessory pigments. **(e)** *Chlamydomonas,* a green alga. Two whiplike flagella propel this cell.

TABLE 12.5 Characteristics of Selected Phyla of Algae

	DINOFLAGELLATES	EUGLENOIDS	DIATOMS	BROWN ALGAE	RED ALGAE	GREEN ALGAE
Classification	Protist	Protist	Protist	Protist	Protist	Often in Plant Kingdom
Color	Brownish	Green	Brownish	Brownish	Reddish	Green
Cell wall	Cellulose and silica	None; protein pellicle	Pectin and silica	Cellulose and alginic acid	Cellulose	Cellulose
Flagella	2	1	In gametes	In gametes	No	In unicellular forms and in gametes
Cell arrangement	Unicellular	Unicellular	Unicellular	Multicellular	Most are multicellular	Unicellular and multicellular
Nutrition	Autotrophic; some heterotrophic	Autotrophic; some heterotrophic	Autotrophic	Autotrophic	Autotrophic	Autotrophic
Pigments	Chlorophyll *a* and *c*, carotene, xanthins	Chlorophyll *a* and *b*, carotene	Chlorophyll *a* and *c*, carotene, xanthophylls	Chlorophyll *a* and *c*, xanthophylls	Chlorophyll *a* and *d*, phycobiliproteins	Chlorophyll *a* and *b*
Sexual reproduction	In a few (?)	No	Yes	Yes	Yes	Yes
Storage material	Starch	Glucose polymer	Oil	Carbohydrate (laminarin)	Starchlike	Starch

flagellate *Gambierdiscus toxicus* (gam′bē-ėr-dis-kus toks′i-kus) passes up the food chain and is concentrated in large fish. Ciguatera is endemic (constantly present) in the South Pacific Ocean and Caribbean Sea.

Euglenoids (Figure 12.12b) are unicellular, flagellated algae. Euglenoids have a semirigid plasma membrane called a *pellicle,* and they move by means of an anterior flagellum. Some euglenoids are facultative chemoheterotrophs. In the dark, they ingest organic matter through the *cytostome*. Euglenoids are frequently studied with protozoans because they lack a cell wall and can move and feed using the flagellum.

All euglenoids except one have a red *eyespot* at the anterior end. This carotenoid-containing organelle senses light and directs the cell in the appropriate direction by using the *preemergent flagellum*.

Diatoms (Figure 12.13) are unicellular or filamentous algae with complex cell walls that consist of pectin and a layer of silica. The two parts of the wall fit together like the halves of a Petri plate. The distinctive patterns of the walls are a useful tool in diatom identification. Diatoms store energy captured through photosynthesis in the form of oil. Much of the world's petroleum was formed from diatoms that lived over 300 million years ago.

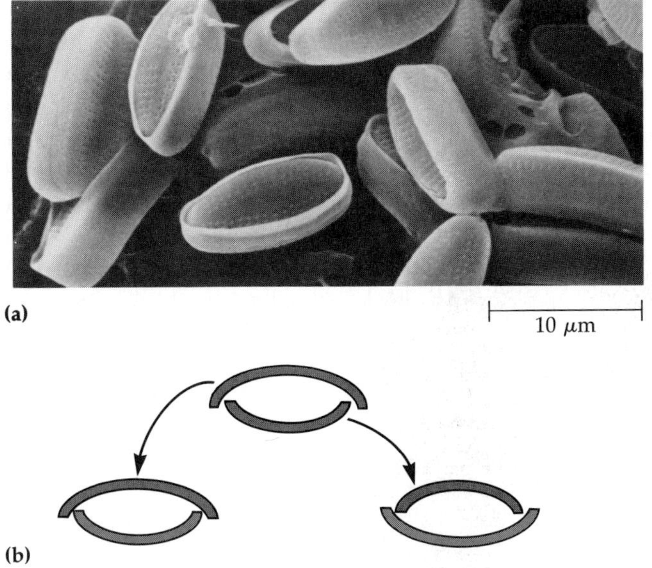

FIGURE 12.13 Diatoms. **(a)** In this scanning electron micrograph of *Achinanthes minutissima,* note how the two parts of the cell wall fit together. **(b)** Asexual reproduction. During mitosis, each daughter cell retains one-half of the cell wall from the parent (red) and must synthesize the remaining half (gray).

The **brown algae** or **kelp** (see Figure 12.12c) are macroscopic; some reach lengths of 50 meters. Most brown algae are found in coastal waters. Brown algae have a phenomenal growth rate. Some grow at rates exceeding 20 cm per day and therefore can be harvested regularly. *Algin,* a thickener used in many foods, such as ice cream and cake decorations, is extracted from their cell walls. Algin is also used in the production of a wide variety of nonfood goods, including rubber tires and hand lotion. Most **red algae** (Figure 12.12d) have delicately branched thalli and live at greater ocean depths than other algae. The thalli of a few red algae form crustlike coatings on rocks and shells. The red pigments allow red algae to absorb the blue light that penetrates deepest into the ocean. Agar is extracted from the red alga *Gelidium* (jel-id'ē-um). Another gelatinous material, called carrageen, comes from a species of red algae called Irish moss. Carrageen can be a thickening ingredient in evaporated milk, ice cream, and pharmaceuticals. **Green algae** (see Figure 12.12e) can be classified as plants because they have cellulose cell walls, contain chlorophyll *a* and *b*, and store starch, as plants do. Green algae are believed to have given rise to terrestrial plants. Most green algae are microscopic. Some filamentous kinds form the grass-green scum in ponds.

Structure and Reproduction

The body of a multicellular alga is called a **thallus.** Thalli of the larger multicellular algae consist of branched **holdfasts** (which anchor the alga to a rock), stemlike and often hollow **stipes,** and leaflike **blades** (Figure 12.12c). The cells covering the thallus can carry on photosynthesis. The thallus lacks conductive tissue (xylem and phloem) characteristic of vascular plants; algae absorb nutrients from the water over their entire surface. The stipe is not lignified or woody, so it does not offer the support of a plant's stem. Instead, the pressure of surrounding water supports the algal thallus; some algae are also buoyed by a floating, gas-filled bladder.

All algae can reproduce asexually. Algae with thalli and filamentous forms can fragment; each piece is capable of forming a new thallus or filament. When a unicellular alga divides, its nucleus divides (mitosis) and then the two nuclei move to opposite parts of the cell. The cell then divides into two complete cells (cytokinesis).

Sexual reproduction occurs in some algae. In some species, asexual reproduction may occur for several generations and then, under different conditions, the same species reproduce sexually. Other species alternate generations so that the offspring resulting from sexual reproduction reproduce asexually, and the next generation then reproduces sexually.

Roles of Algae in Nature

Algae are an important part of any aquatic food chain because they fix carbon dioxide into organic molecules that can be consumed by chemoheterotrophs. Using the energy produced in photophosphorylation, algae convert the carbon dioxide in the atmosphere into carbohydrates. Molecular oxygen (O_2) is a by-product of their photosynthesis. The top few meters of any body of water contain planktonic algae. As three-fourths of the Earth is covered with water, it is estimated that 80% of the Earth's photosynthesis is carried on by planktonic algae.

Seasonal changes in nutrients, light, and temperature cause fluctuations in algal populations; periodic increases in numbers of planktonic algae are called **blooms.** Blooms of dinoflagellates are responsible for seasonal red tides. Blooms of a few species indicate that the water in which they grow is polluted because these algae thrive in high concentrations of organic materials that exist in sewage or industrial wastes. When algae die, the decomposition of the large numbers of cells associated with an algal bloom depletes the level of dissolved oxygen in the water. (This phenomenon is discussed in Chapter 27.)

When diatoms and other planktonic organisms that grew millions of years ago died and were buried by sediments, the organic molecules they contained did not decompose to be returned to the carbon cycle as CO_2. Heat and pressure resulting from the Earth's geologic movements altered the oil stored in the cells as well as the cell membranes. Oxygen and other elements were eliminated, leaving a residue of hydrocarbons in the form of petroleum and natural gas deposits.

Many unicellular algae are symbionts in animals. The giant clam, *Tridacna* (trī-dak'nä), has evolved special organs that host dinoflagellates. As the clam sits in shallow water, the algae proliferate in these organs when they are exposed to the sun. The algae release glycerol into the clam's bloodstream, thus supplying the clam's carbohydrate requirement. Evidence suggests that the clam gets essential proteins by phagocytizing old algae.

LICHENS

A **lichen** is a combination of a green alga (or cyanobacterium) and a fungus. The two organisms exist in a *mutualistic* relationship, in which each partner benefits. The lichen is very different from either the alga or fungus growing alone, and if the partners are separated, the lichen no longer exists. There are approximately 20,000 species of lichens occupying quite di-

(a) Crustose lichen

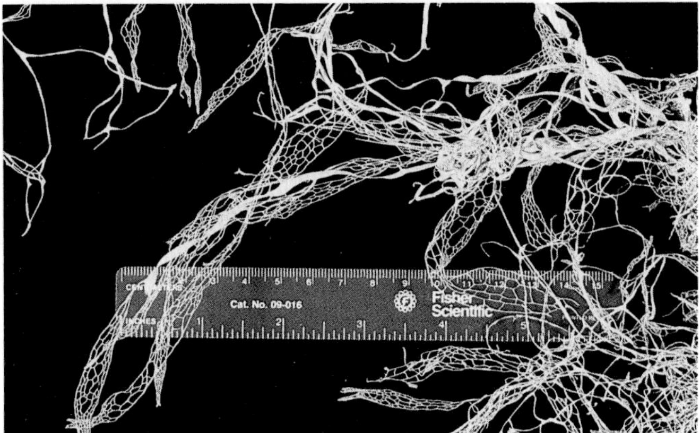

(c) Fruticose lichen

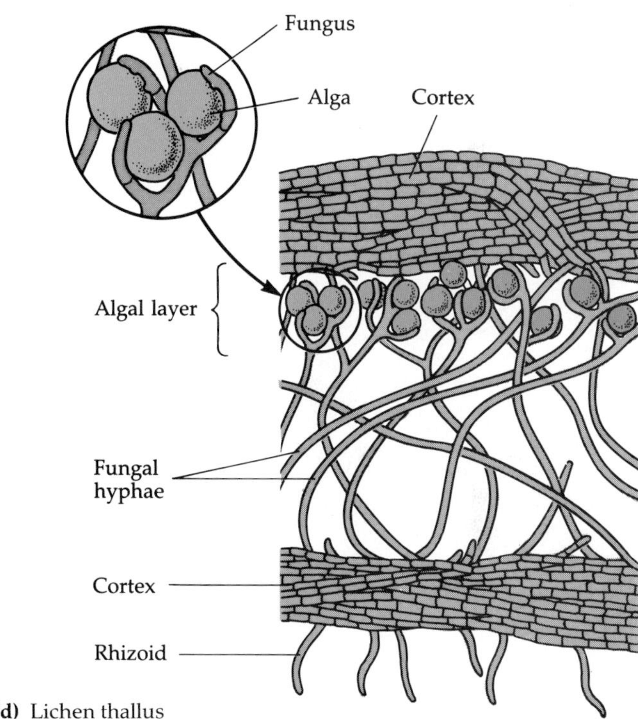

(d) Lichen thallus

FIGURE 12.14 Lichens. **(a)** Crustose. **(b)** Foliose. **(c)** Fruticose. **(d)** Cross section of a lichen thallus. The medulla is composed of fungal hyphae and surrounds the algal layer. The protective cortex is a layer of irregularly organized fungal hyphae that covers the surface and sometimes the bottom of the lichen.

verse habitats. Because lichens can inhabit areas in which neither fungi nor algae could survive alone, lichens are often the first life forms to colonize on newly exposed soil or rock. Lichens secrete organic acids that chemically weather rock, and they accumulate nutrients needed for plant growth. They are some of the slowest-growing organisms on Earth.

Lichens can be grouped into three morphologic categories. **Crustose lichens** (see Figure 12.14a) grow flush or encrusting onto the substratum; **foliose lichens** (Figure 12.14b) are more leaflike; and **fruticose lichens** (Figure 12.14c) have fingerlike projections. The lichen's thallus, or body, forms when fungal hyphae (**medulla**) grow around algal cells (Figure 12.14d). Fungal hyphae project below the lichen body to form **rhizoids,** or holdfasts. Fungal hyphae form a **cortex,** or protective covering, over the algal layer and sometimes under it as well. After incorporation into a lichen thallus, the alga continues to grow by mitosis in the algal layer of the lichen, and the growing hyphae can incorporate new algal cells. In a lichen, the fungus reproduces sexually.

When the algal partner is cultured separately in vitro, about 1% of the carbohydrates produced during photosynthesis are released into the culture medium; however, when the alga is associated with a fungus, the algal plasma membrane is more permeable, and up to 60% of the products of photosynthesis are released to the fungus or are found as end-products of fungus metabolism. It is clear that the fungus benefits from its

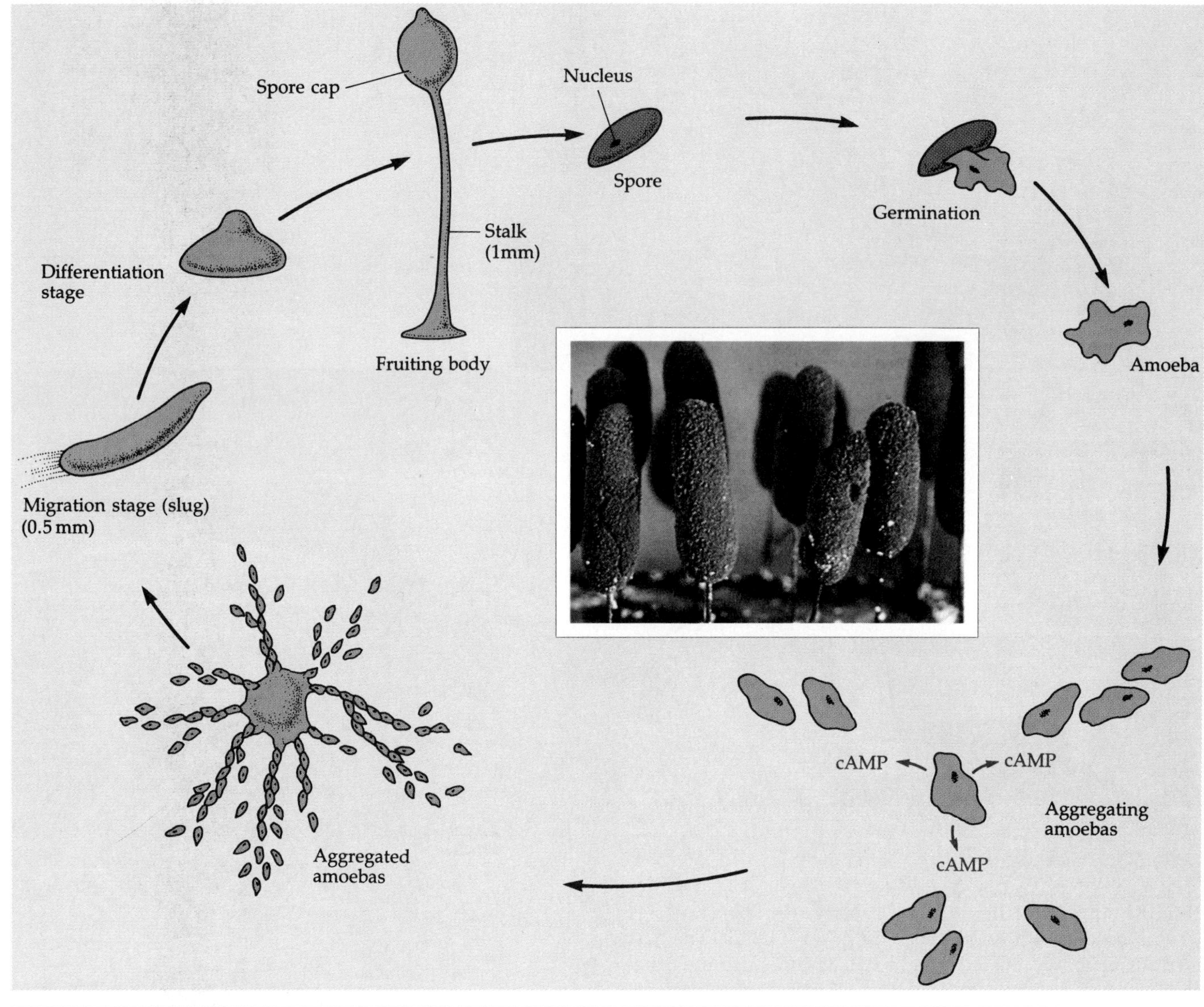

FIGURE 12.15 Generalized life cycle of a cellular slime mold. Inset shows fruiting bodies of an *Arcyria* slime mold.

association. The alga, which appears to be giving up valuable nutrients, is compensated by the protection from desiccation and the holdfast that the fungus provides.

Populations of lichens readily incorporate cations (positively charged ions) into their thalli. The concentrations and types of cations in the atmosphere can be determined by chemical analyses of lichen thalli. The presence or absence of species that are quite sensitive to pollutants can be used to ascertain air quality. A 1985 study in the Cuyahoga Valley (Ohio) revealed that 81% of the 172 lichen species that were present in 1917 are gone. Since this area is severely affected by air pollution, the inference is that air pollutants, primarily sulfur dioxide (the major contributor to acid rain), caused the death of sensitive species.

SLIME MOLDS

Slime molds have both fungal and animal characteristics and have membrane-bounded nuclei. Thus, they are classified as protists. Cellular slime molds (Figure 12.15) are typical eucaryotic cells that resemble amoebas. The amoeboid cells ingest other microorganisms and bacteria by phagocytosis. Cellular slime molds are of interest to biologists who study cellular migration and aggregation. When conditions are unfavorable, large numbers of amoeboid cells aggregate to form a single structure. This aggregation happens because some individual amoebas produce the chemical cyclic AMP (cAMP), toward which the other amoebas migrate. The aggregated amoebas are enclosed in a slimy sheath called a *slug*. The slug migrates as a unit toward light. After a period of hours, the slug ceases to migrate and becomes vertically oriented. Some of the amoeboid cells form a stalk; others swarm up the stalk to form a spore cap, and most of these differentiate into spores. When spores are released under favorable conditions, they germinate to form single amoebas.

In 1973, a Dallas resident discovered a pulsating red blob in his backyard. The news media claimed that a "new life form" had been found. For some people, the "creature" evoked a spine-chilling recollection of an old science fiction movie. Before imaginations got carried away too far, biologists calmed everyone's worst fears (or highest hopes). The amorphous mass was merely a plasmodial slime mold, they explained. Its unusually large size (46 cm in diameter), however, startled even scientists.

Plasmodial (acellular) slime molds were first scientifically reported in 1729. They belong to a separate phylum. A plasmodial slime mold exists as a mass of protoplasm with many nuclei (it is multinucleated). This mass of protoplasm is called a **plasmodium** (Figure 12.16a). The entire plasmodium moves as a giant amoeba. It engulfs organic debris and bacteria. Biologists have found that musclelike proteins forming structures called microfilaments account for the movement of the plasmodium. When plasmodial slime molds are grown in laboratories, a phenomenon called **cytoplasmic streaming** is observed. During cytoplasmic streaming, the protoplasm within the plasmodium moves and changes both its speed and direction so that the oxygen and nutrients are evenly distributed.

The plasmodium grows as long as there is enough food and moisture. When either is in short supply, the plasmodium separates into many groups of protoplasm; each of these groups forms a stalked sporangium, in which spores (a resistant, resting form of the slime mold) develop (Figure 12.16b). Nuclei within these spores undergo meiosis and form uninucleate haploid amoebalike cells. When conditions improve, these spores germinate, fuse to form diploid cells, and develop into a multinucleate plasmodium.

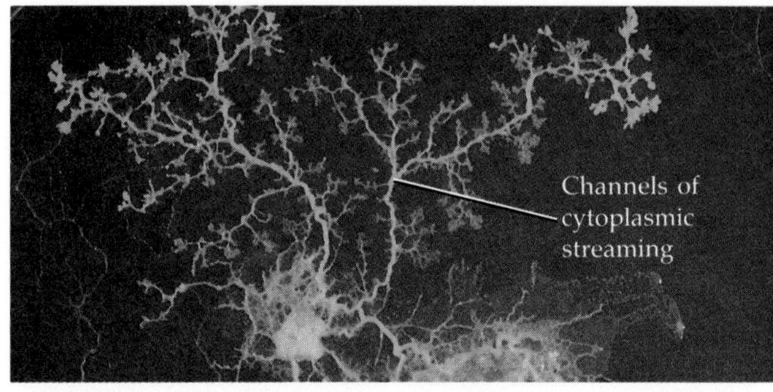

(a)

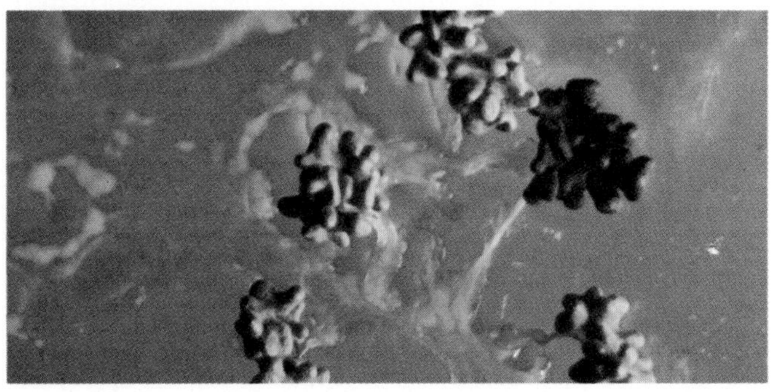

(b)

FIGURE 12.16 Acellular slime mold. **(a)** Plasmodium. **(b)** Sporangia.

PROTOZOANS

Protozoans are one-celled, eucaryotic organisms that belong to the Kingdom Protista. Among the protozoans are many variations on this cell structure, as we shall see. Protozoans inhabit water and soil and feed upon bacteria and small particulate nutrients. Some protozoans are part of the normal flora of animals. For example, the box on p. 278 in Chapter 11 describes protozoans that live in the intestinal tracts of termites to help them digest cellulose. Of the nearly 20,000 species of protozoans, relatively few cause disease.

Protozoans are classified into phyla on the basis of their means of motility. We will restrict our study to the four phyla of protozoans that include disease-causing species. One such phylum, the Sarcodina, consists of amoebas (Figure 12.17a). The amoebas move by extending usually blunt, lobelike projections of the cytoplasm called **pseudopods.** Any number of pseudopods can flow from one side of the amoeba cell, and the rest of the cell will flow toward the pseudopods.

Another phylum, the Mastigophora, or flagellates, possesses flagella. Flagella are capable of whiplike movements that pull the cell through the medium (Figure 12.17b). Most flagellates have one or two flagella, but some of the parasitic species have as many as eight. Species in the phylum Ciliata have projections called cilia that are similar to but shorter than flagella. The cilia are in a precise arrangement over the cell (Figure 12.17c) and move to propel the cell through its medium. The species in the fourth phylum we will study, the Sporozoa, are incapable of independent movement.

Protozoan Biology

The term *protozoa* refers to "first animal" to generally describe their nutrition. In addition to getting food,

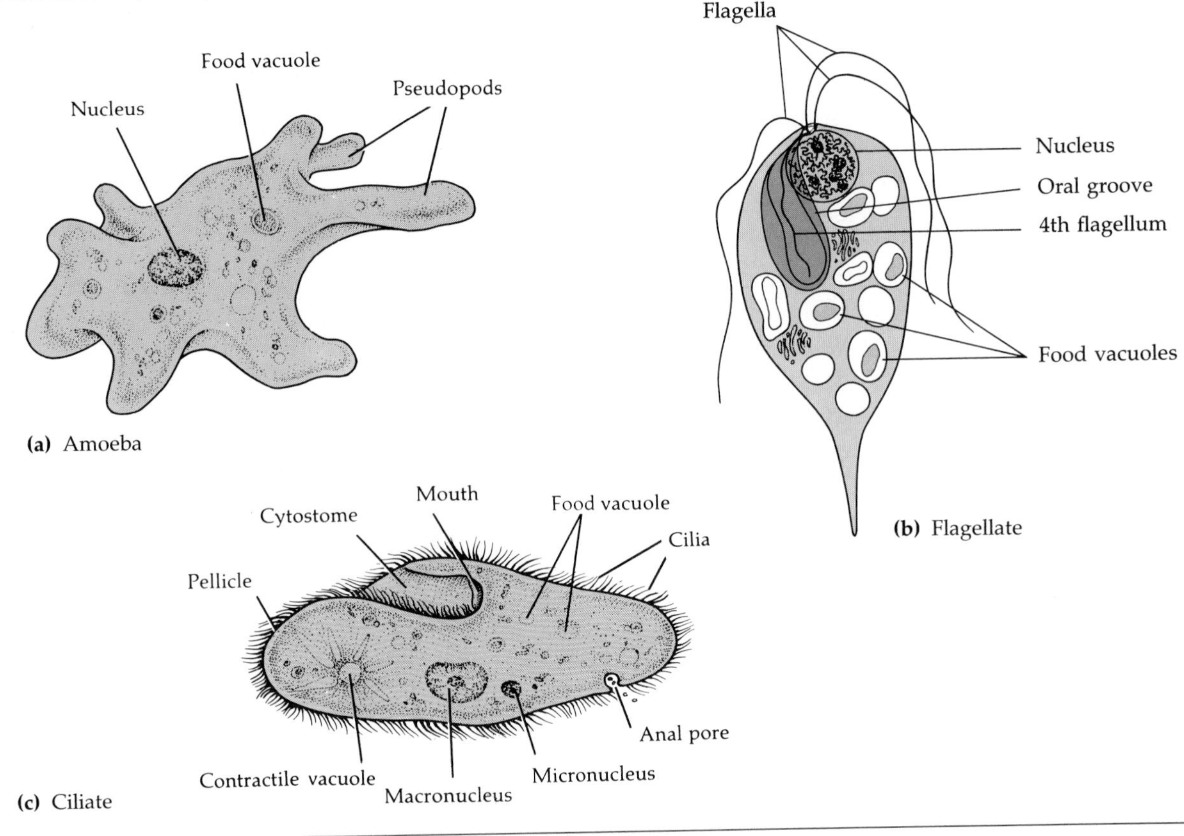

FIGURE 12.17 Examples of protozoans. (a) Amoeba. To move and to engulf food, amoebas extend cytoplasmic portions called pseudopods. Once surrounded by the pseudopods, the food is in a food vacuole. **(b)** ***Chilomastix,*** a flagellate. The fourth flagellum is used to move food into the oral groove where food vacuoles are formed. **(c)** ***Paramecium,*** a ciliate. Rows of cilia cover the cells of ciliates. The cilia are moved in unison for locomotion and to bring food particles to the protozoan. *Paramecium* has specialized structures for ingestion (mouth), elimination of wastes (anal pore), and regulation of osmotic pressure (contractile vacuoles). The macronucleus is involved with protein synthesis and other ongoing cellular activities. The micronucleus functions in sexual reproduction.

protozoans must reproduce, and parasitic species must be able to get from one host to another.

NUTRITION

Protozoans are mostly aerobic heterotrophs, although many intestinal protozoans are capable of anaerobic growth. One chlorophyll-containing flagellate, *Euglena* (ū-glē'nä) (Figure 12.12b), can also grow in the dark as a heterotroph and is often included along with the protozoans.

All protozoans live in areas with a large supply of water. Some protozoans transport food across the plasma membrane. However, some have a protective covering called the **pellicle** and require specialized structures to take in food. The ciliates take in food by waving their cilia toward a mouthlike opening called the **cytostome.** The amoebas engulf food by surrounding it with pseudopods and phagocytizing it. In the protozoans, digestion takes place in membrane-bounded **vacuoles,** and waste may be eliminated through the plasma membrane or through a specialized **anal pore.**

REPRODUCTION

Protozoans reproduce asexually by fission, budding, or schizogony. **Schizogony** is *multiple fission.* The nucleus undergoes multiple divisions before the cell divides. After many nuclei are formed, a small portion of cytoplasm concentrates around each nucleus, and separation of the single cell into daughter cells follows.

Sexual reproduction has been observed in some protozoans. The ciliates reproduce sexually by **conjugation** (Figure 12.18), which is very different from the bacterial process of the same name. During protozoan conjugation, two cells fuse and a haploid nucleus (the micronucleus) from each cell migrates to the other cell.

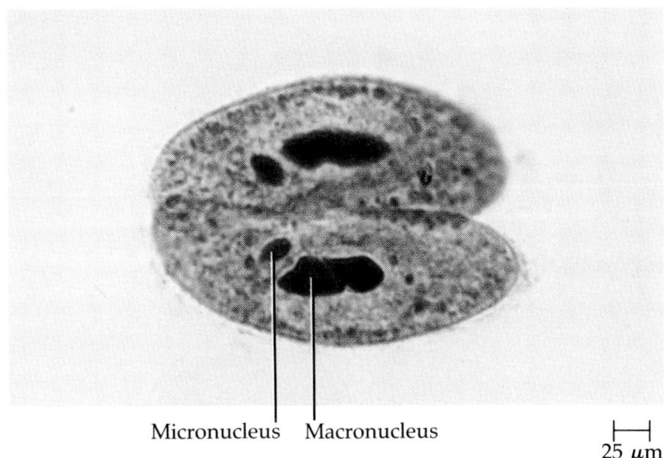

FIGURE 12.18 ***Paramecium* conjugation.** Sexual reproduction in ciliates is by conjugation. Each cell has two nuclei, a micronucleus and a macronucleus. The micronucleus is haploid and is specialized for conjugation. One micronucleus from each cell will migrate to the other cell during conjugation.

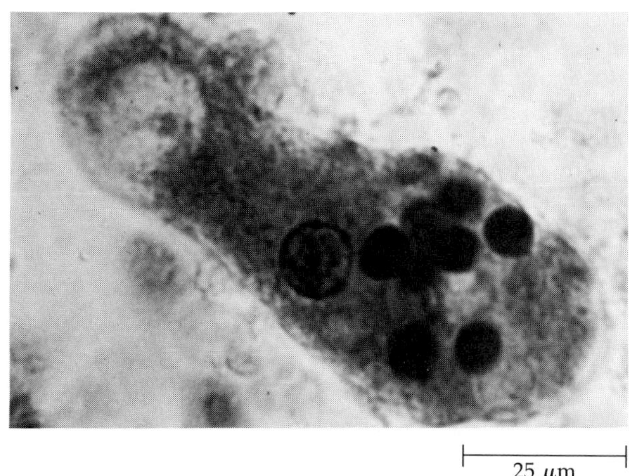

FIGURE 12.19 ***Entamoeba histolytica* in the vegetative form.** The dark circles inside the amoeba are engulfed red blood cells from its human host.

That haploid nucleus fuses with the haploid nucleus within the cell. The parent cells separate, each now a fertilized cell. When the cells later divide, they produce daughter cells with recombined DNA. Some protozoans produce **gametes** or **gametocytes,** haploid sex cells. In reproduction, two gametes fuse to form a diploid zygote.

ENCYSTMENT

Under certain adverse conditions, some protozoans are capable of producing a protective capsule called a **cyst.** A cyst permits the organism to survive when food, moisture, or oxygen are lacking, when temperatures are not suitable, or when toxic chemicals are present. A cyst also enables a parasitic species to survive outside a host. This is important because parasitic protozoans can have life cycles that involve more than one host.

Medically Important Phyla of Protozoans

The phyla Sarcodina, Mastigophora, Ciliata, and Sporozoa include disease-causing species.

SARCODINA

A well-known parasitic amoeba is *Entamoeba histolytica* (en-tä-mē'bä his-tō-li'ti-kä), the causative agent of amoebic dysentery (Figure 12.19). The primary food of *E. histolytica* is red blood cells. *E. histolytica* is transmitted from human to human through ingestion of cysts that are excreted in the feces.

MASTIGOPHORA

As we mentioned previously, members of the Phylum Mastigophora move by means of flagella. Their outer membrane is a tough, flexible pellicle. Flagellates are typically spindle-shaped, with flagella projecting from the front end. Food is ingested through a cytostome. Some flagellates have an **undulating membrane,** which appears to consist of highly modified flagella (Figure 12.20).

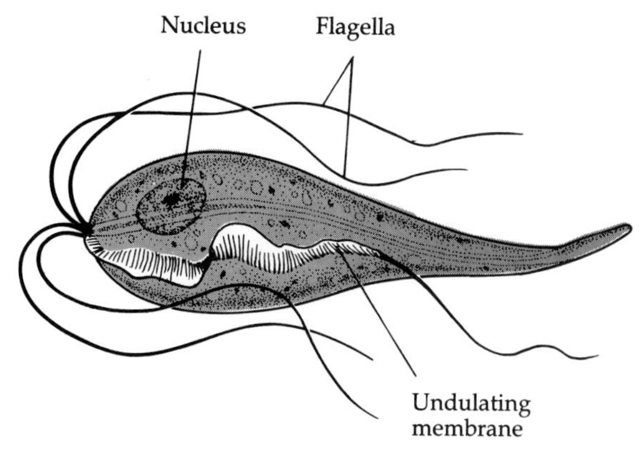

FIGURE 12.20 ***Trichomonas vaginalis.*** This flagellate is the cause of urinary and genital tract infections. It has a small undulating membrane. This flagellate does not have a cyst stage.

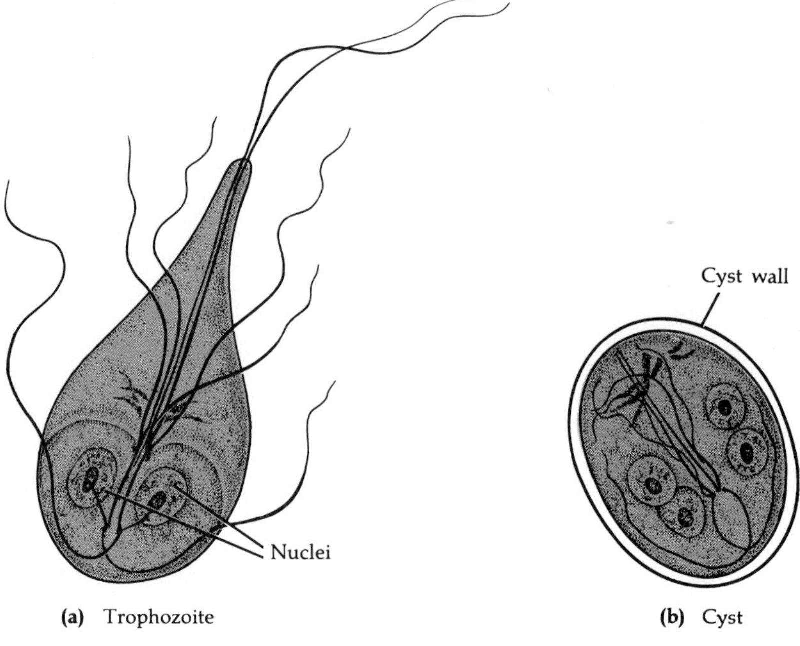

FIGURE 12.21 ***Giardia lamblia.*** **(a)** The trophozoite of this intestinal parasite has eight flagella and two prominent nuclei, giving it a distinctive appearance. **(b)** The cyst provides protection in the environment before being ingested by a new host.

An example of a flagellate that is a human parasite is *Giardia lamblia* (jē-är′dē-ä lam′lē-ä) (see Figures 12.21 and 25.16, which show the vegetative form, or **trophozoite,** as well as the cyst stage of *G. lamblia*). The parasite is found in the small intestine of humans and other mammals. It is excreted in the feces and survives as a cyst before being ingested by the next host. Diagnosis of giardiasis, the disease caused by *G. lamblia,* is often based on identification of cysts in feces.

Another parasitic flagellate is *Trichomonas vaginalis* (trik-ō-mōn′as va-jin-al′is), shown in Figure 12.20. *T. vaginalis* does not have a cyst stage and must be transferred from host to host quickly before desiccation occurs. *T. vaginalis* is found in the vagina and in the male urinary tract. It is usually transmitted by sexual intercourse but can be transmitted by toilet facilities or towels.

The hemoflagellates are transmitted by the bites of blood-sucking insects and are found in the circulatory system of the bitten host. To survive in this viscous fluid, hemoflagellates have long, slender bodies and an undulating membrane. The genus *Trypanosoma* (tri-pa-nō-sō′mä) includes the species that cause African sleeping sickness (*T. brucei gambiense* [brüs′ē gam-bē-ens′], which is transmitted by the tsetse fly. *T. cruzi* [kruz′ē]; see Figure 23.17), the causative agent of Chagas' disease, is transmitted by the "kissing bug," so called because it bites on the face. After entering the insect, the trypanosome rapidly multiplies by fission. If the insect then defecates while biting a human, it can release trypanosomes that can contaminate the bite wound.

CILIATA

The only ciliate that is a human parasite is *Balantidium coli* (bal-an-tid′ē-um kō′lī), the causative agent of a severe, though rare, type of dysentery. When cysts are ingested by the host, they enter the colon, into which the trophozoites are released. The trophozoites feed on bacteria and fecal debris as they multiply, and cysts are passed out with feces.

SPOROZOA

Sporozoans are not motile in their mature forms and are obligate intracellular parasites. They have complex life cycles that ensure their survival and transmission from host to host. An example of a sporozoan is *Plasmodium* (plaz-mō′dē-um), the causative agent of malaria. A vaccine against malaria is now being tested on humans (see the box in Chapter 18, p. 454).

Plasmodium grows by sexual reproduction in the *Anopheles* (an-of′el-ēz) mosquito, in human liver cells, and in red blood cells (Figure 12.22). When an *Anopheles* carrying the infective stage of *Plasmodium* (called a **sporozoite**) bites a human, sporozoites can be injected into the human (Figure 12.22a). The sporozoites are carried by the blood to the liver. They undergo schizogony in liver cells and produce thousands of progeny called **merozoites**. Merozoites enter the bloodstream and infect red blood cells (Figure 12.22b). The young trophozoite looks like a ring in which the nucleus and cytoplasm are visible. This is called a **ring stage**. The ring stage enlarges and divides repeatedly, and the red blood cells eventually rupture and release more mero-

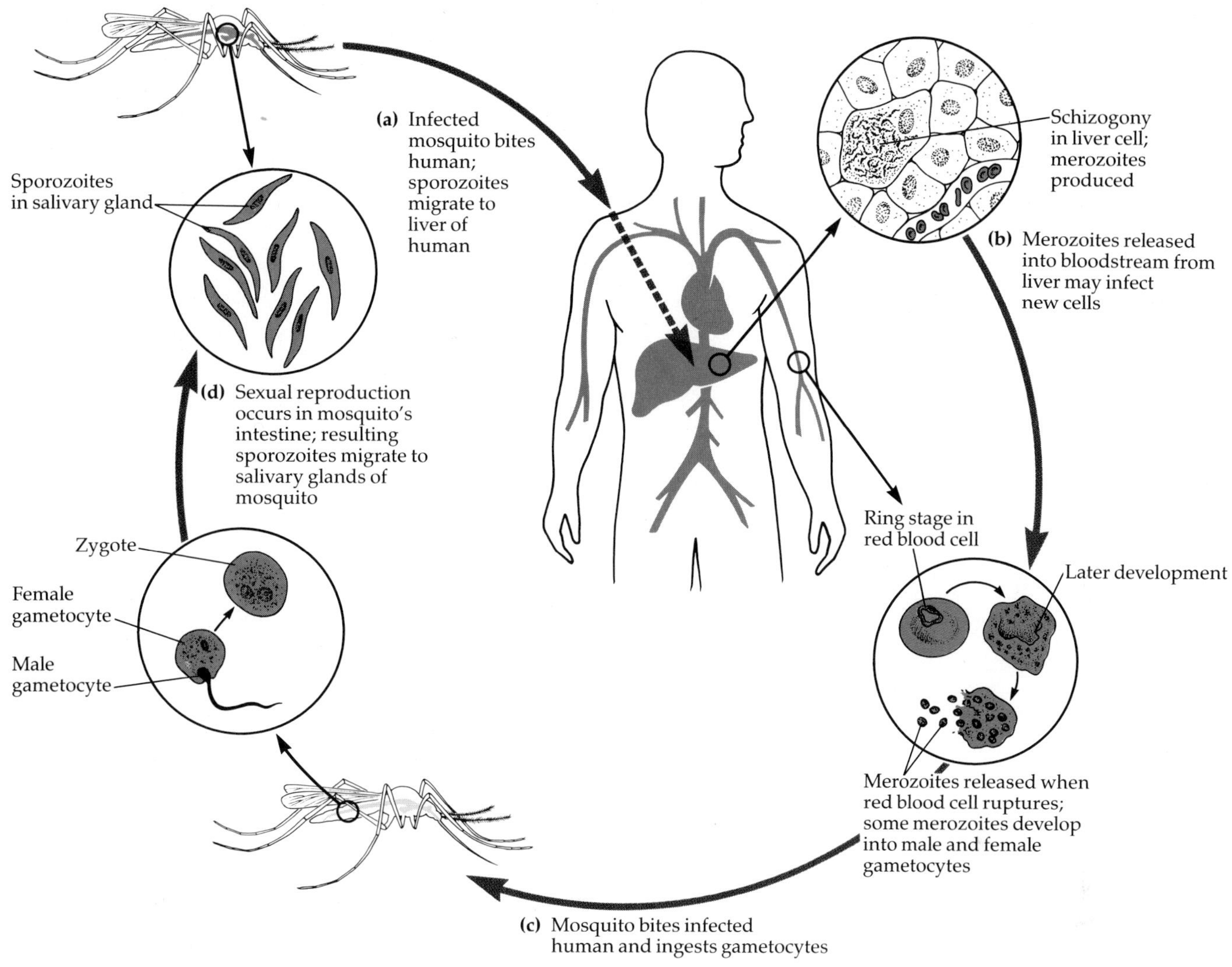

FIGURE 12.22 Life cycle of *Plasmodium vivax*. Asexual reproduction (schizogony) of the parasite takes place in the liver and in the red blood cells of a human host. Sexual reproduction occurs in the intestine of an *Anopheles* mosquito after the mosquito has ingested gametocytes.

zoites. Upon release of the merozoites, their waste products, which cause fever and chills, are also released. Most of the merozoites infect new red blood cells and perpetuate their cycle of asexual reproduction. However, some develop into male and female sexual forms (gametocytes). Even though the gametocytes themselves cause no further damage, they can be picked up by the bite of another *Anopheles* mosquito (Figure 12.22c); they then enter the mosquito's intestine and begin their sexual cycle. Here the male and female gametocytes unite to form a zygote. The zygote forms an oocyst, in which cell division occurs, and asexual sporozoites are formed. When the cyst ruptures, the sporozoites migrate to the salivary glands of the mosquito (Figure 12.22d). They can then be injected into a new human host by the biting mosquito.

Malaria is diagnosed in the laboratory by microscopic observation of thick blood smears for the presence of *Plasmodium* (see Figure 23.24). A peculiar characteristic of malaria is that the interval between periods of fever caused by release of merozoites is always the same for a given species of *Plasmodium* and is always a multiple of 24 hours. The reason and mechanism for such precision has intrigued scientists. After all, why should a parasite need a biological clock? Frank Hawking and his coworkers have demonstrated that *Plasmodium's* development is regulated by the host's body temperature, which normally fluctuates over a 24-hour period. The parasite's careful timing ensures that gametocytes are mature at night, when mosquitoes are feeding, to enable transmission of the parasite to a new host. Another sporozoan parasite of red blood cells is

TABLE 12.6 Some Representative Parasitic Protozoans

PHYLUM	HUMAN PATHOGENS	DISTINGUISHING FEATURES	DISEASE	SOURCE OF HUMAN INFECTIONS	SEE FIGURE	REFER TO CHAPTER
Sarcodina (amoebas)	*Acanthamoeba*	Pseudopods	Keratitis	Water	—	21
	Entamoeba histolytica	Pseudopods	Amoebic dysentery	Fecal contamination of drinking water	12.19	25
	Naegleria fowleri	Some flagellated forms	Microencephalitis	Water in which people swim	22.14	22
Mastigophora (flagellates)	*Giardia lamblia*	Two nuclei, eight flagella	Giardial enteritis	Fecal contamination of drinking water	12.21 25.16	25
	Trichomonas vaginalis	No encysting stage	Urethritis; vaginitis	Contact with vaginal-urethral discharge	12.20 26.15	26
	T.b. gambiense, T.b. rhodesiense	Undulating membrane	African trypanosomiasis	Bite of tsetse fly	—	22
	Trypanosoma cruzi		Chagas' disease	Bite of *Triatoma* (kissing bug)	23.19	23
Ciliata	*Balantidium coli*	Only parasitic ciliate of humans	Balantidial dysentery	Fecal contamination of drinking water	—	25
Sporozoa	*Babesia microti*		Babesiosis	Domestic animals, ticks	—	Box, Ch. 23
	Cryptosporidium	Complex life cycles may require more than one host	Diarrhea	Humans, other animals, water	25.18	25
	Isospora		Coccidiosis	Domestic animals	—	Table 19.7
	Plasmodium		Malaria	Bite of *Anopheles* mosquito	12.22 23.22	23
	*Pneumocystis carinii**		Pneumonia	Humans	12.7 24.17	24
	Toxoplasma gondii		Toxoplasmosis	Cats, other animals; congenital	23.18	23
	Microsporidia	Unknown	Diarrhea, kerato-conjunctivitis	Other animals	—	—

*Classification is uncertain, also listed with fungi.

Babesia microti, is described in the box in Chapter 23, p. 578.

Toxoplasma gondii (toks-ō-plaz'mä gon'dē-ē) is another intracellular parasite of humans. The life cycle of this sporozoan involves domestic cats. The trophozoites, called *tachyzoites,* reproduce sexually and asexually in an infected cat, and oocysts, each containing eight sporozoites, are passed out with feces. If the oocysts are ingested by humans or other animals, the sporozoites emerge as trophozoites, which can reproduce in the tissues of the new host (see Figure 23.18). *T. gondii* is dangerous to pregnant women, as it can cause congenital infections in utero. Tissue examination and observation of *T. gondii* is used for diagnosis. Antibodies may be detected by ELISA and indirect fluorescent-antibody tests (see Chapter 18).

Cryptosporidium (krip-tō-spô-ri'dē-um), an intracellular parasite of mammals, was first reported as a cause of human disease in 1976. It is now thought that *Cryptosporidium* may be responsible for up to 30% of the diarrheal illness in underdeveloped countries. In AIDS patients and other immunosuppressed persons, *Cryptosporidium* can cause respiratory and gallbladder infections and may be a major cause of death. The organism, which lives inside the cells lining the small intestine, can be transmitted to humans through the feces of cows, rodents, dogs, and cats. Waterborne and hospital-acquired infections have also been reported. Inside the host cell, each *Cryptosporidium* organism forms four oocysts, each containing four sporozoites. When the oocyst ruptures, sporozoites may infect new cells in the host or be released with the feces. The disease is diagnosed by acid-fact staining or fluorescent-antibody tests.

The microsporidial protozoans, another group of intracellular parasites, have been reported since 1984 as the cause of a number of human diseases, most notably in AIDS patients, including chronic diarrhea and keratoconjunctivitis (inflammation of the conjunctiva near the cornea). Microsporidial protozoans are obligate intracellular parasites that lack mitochondria. They have been assigned to the Phylum Sporozoa by some workers, but their taxonomic position is unclear.

Table 12.6 lists some typical parasitic protozoans and the diseases they cause.

HELMINTHS

There are a number of parasitic animals that spend part or all of their lives in humans. Most of these animals belong to two phyla. Platyhelminthes (flatworms) and Aschelminthes (roundworms). These worms are commonly called **helminths.** There are also free-living species in these phyla, but we will limit our discussion to the parasitic species.

Helminth Biology

Helminths are multicellular eucaryotic animals, generally possessing digestive, circulatory, nervous, excretory, and reproductive systems. Parasitic helminths must be highly specialized to live inside their hosts. The following generalizations distinguish parasitic helminths from their free-living relatives:

1. They either *lack* a digestive system or have a greatly simplified one. They can absorb nutrients from the host's food, body fluids, and tissues.
2. Their nervous system is *reduced.* They do not need an extensive nervous system because they do not have to search for food or respond to their environment. The environment within a host is fairly constant.
3. Their means of locomotion is either *reduced* or *completely lacking.* Because they are transferred from host to host, they do not have to search for a suitable habitat.
4. Their reproductive system is often *more* complex; it produces more fertilized eggs, by which a suitable host is infected.

REPRODUCTION

Adult helminths may be **dioecious**—male reproductive organs are in one individual, and female reproductive organs are in another. In those species, reproduction occurs only when two adults of the opposite sex are in the same host.

Adult helminths may also be **hermaphroditic**—one animal has both male and female reproductive organs. Two hermaphrodites may copulate and simultaneously fertilize each other. A few types of hermaphrodites fertilize themselves.

LIFE CYCLE

The life cycle of parasitic helminths can be extremely complex, involving a succession of hosts. The term **definitive host** is given to the organism that harbors the adult, sexually mature parasite. One or more **intermediate hosts** might be necessary for completion of each *larval* (developmental) stage of the parasite.

Platyhelminthes

Members of the Phylum Platyhelminthes (flatworms) are flattened from top to bottom. They have what is called an *incomplete digestive system.* This type of digestive system has only one opening (mouth), through which food enters and wastes leave. The classes of parasitic flatworms include the trematodes and cestodes.

TREMATODES

Trematodes, or **flukes,** have flat, leaf-shaped bodies with a ventral sucker and an oral sucker (Figure 12.23). The suckers hold the organism in place and suck fluids

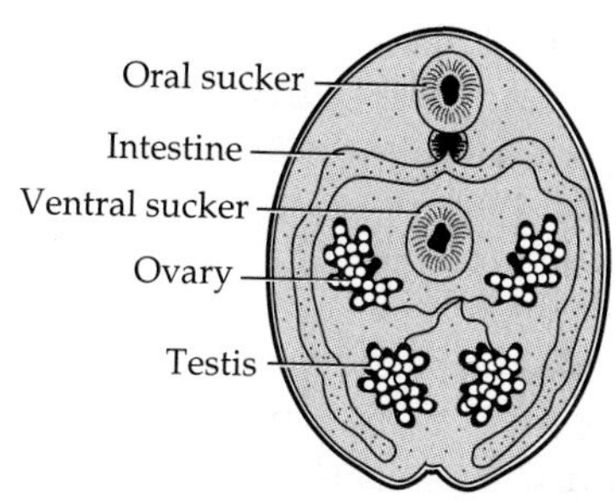

FIGURE 12.23 Adult fluke. This generalized diagram of the anatomy of an adult fluke shows the oral and ventral suckers. The suckers attach the fluke to the host. The mouth is located in the center of the oral sucker. Flukes are hermaphroditic; each animal contains both testes and ovaries.

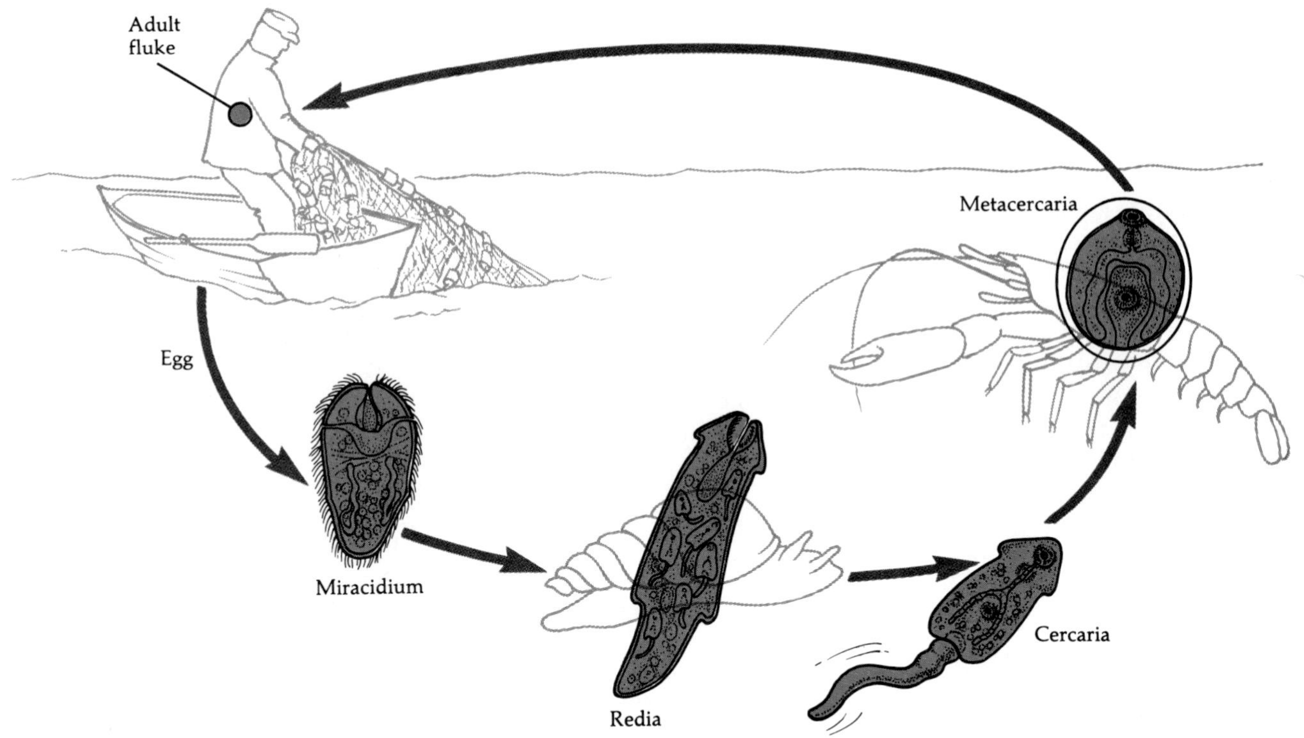

FIGURE 12.24 Life cycle of *Paragonimus westermanni.* The free-swimming miracidium invades the first intermediate host, a snail. Two generations of rediae develop within the snail, and the rediae give rise to cercariae. The cercariae leave the snail and penetrate the second intermediate host, a crayfish. Here they encyst as metacercariae. When a human eats raw or undercooked crayfish, the metacercariae are released to develop into adults.

from the host. Flukes can also obtain food by absorbing it through their nonliving outer covering, called the cuticle. Flukes are named according to the tissue of the definitive host in which the adults live (for example, lung fluke, liver fluke, blood fluke).

To exemplify a fluke's life cycle, let us look at the lung fluke, *Paragonimus westermanni* (păr-ä-gōn'e-mus we-ster-man'nē), shown in Figure 12.24. The adult lung fluke lives in the bronchioles of humans and other mammals and is approximately 6 mm wide by 12 mm long. The hermaphroditic adults liberate eggs into the bronchi. Because sputum that contains eggs is frequently swallowed, the eggs are usually excreted in feces from the host.

To continue the life cycle, the eggs must be excreted into a body of water. Inside the egg, a miracidial larva, or **miracidium,** then develops. When the egg hatches, the larva enters a suitable snail. Only certain species of aquatic snails can be the intermediate host. Inside the snail, the lung fluke undergoes asexual reproduction and produces **rediae.** Each redia develops into a **cercaria** that bores out of the snail and penetrates the cuticle of a crayfish. The parasite encysts as a **metacercaria** in the muscles and other tissues of the crayfish. When the crayfish is eaten by a human, the metacercaria is freed in the human's small intestine. It bores out and wanders around until it penetrates the lungs, enters the bronchioles, and develops into an adult lung fluke.

In a laboratory diagnosis, sputum and feces are examined microscopically for eggs. Infection results from eating undercooked crayfish, and the disease could be prevented by thoroughly cooking crayfish.

The cercariae of the blood fluke *Schistosoma* (shis-tō-sō'mä) are not ingested. Instead, they burrow through the skin of the human host and enter the circulatory system. The adults are found in certain abdominal and pelvic veins. The disease schistosomiasis is a major world health problem; it will be discussed further in Chapter 23.

CESTODES

Cestodes, or **tapeworms,** are intestinal parasites. Their structure is shown in Figure 12.25. The head, or **scolex,** has suckers for attaching to the intestinal mucosa of the host. Some species also have small hooks for attaching to the host. Tapeworms do not ingest the tissues of their hosts; in fact, they completely lack a digestive system. To obtain nutrients from the small intestine, they absorb food through their cuticle. The body consists of segments called **proglottids.** Proglottids are continually produced by the neck region of the scolex, as long as the scolex is attached and alive. Each proglottid contains both male and female reproductive organs. The proglottids farthest away from the scolex are the mature ones containing fertilized eggs.

Humans as Definitive Hosts

The adults of *Taenia saginata* (te'nē-ä sa-ji-nä'tä), the beef tapeworm, live in humans and can reach a length of 6 meters. The scolex is about 2 mm in length and is followed by a thousand or more proglottids. The feces of an infected human contain mature proglottids, and each proglottid contains thousands of eggs. As the proglottids wriggle away from the fecal material, they increase their chances of being ingested by an animal that is grazing. Upon ingestion by cattle, the larvae hatch from the eggs and bore through the intestinal wall. The larvae migrate to muscle (meat), in which they encyst as **cysticerci.** When the cysticerci are ingested by humans, all but the scolex is digested. The scolex anchors itself in the small intestine and begins producing proglottids.

Diagnosis is based on the presence of mature proglottids and eggs in feces. Cysticerci can be seen macroscopically in meat; their presence is referred to as "measly beef." Inspecting beef that is intended for human consumption for "measly" appearance is one way to prevent infections by beef tapeworm. Another method of prevention is to avoid the use of untreated human sewage as fertilizer in grazing pastures.

Humans as Intermediate Hosts

Humans are the intermediate hosts for *Echinococcus granulosus* (ē-kīn-ō-kok'kus gra-nū-lō'sus), shown in Figure 12.26. Dogs and cats, both wild and domestic, are the definitive hosts for this minute (2 to 8 mm) tapeworm. Eggs are excreted with feces and are ingested by deer or humans. Humans can become infected by contaminating their hands with dog feces or saliva from a dog's tongue. The eggs hatch in the human's small intestine, and the larvae migrate to the liver or lungs. The host forms a cyst around the larvae in these organs. This cyst, called a **hydatid cyst,** contains "brood capsules," in which thousands of scoleces (singular, *scolex*) might be produced. In the wild, the cysts might be in a deer that is eaten by a wolf. The scoleces would be able to attach themselves in the wolf's intestine and produce proglottids.

Diagnosis of hydatid cysts is frequently made only on autopsy, although X rays can detect the cysts (Figure 25.21).

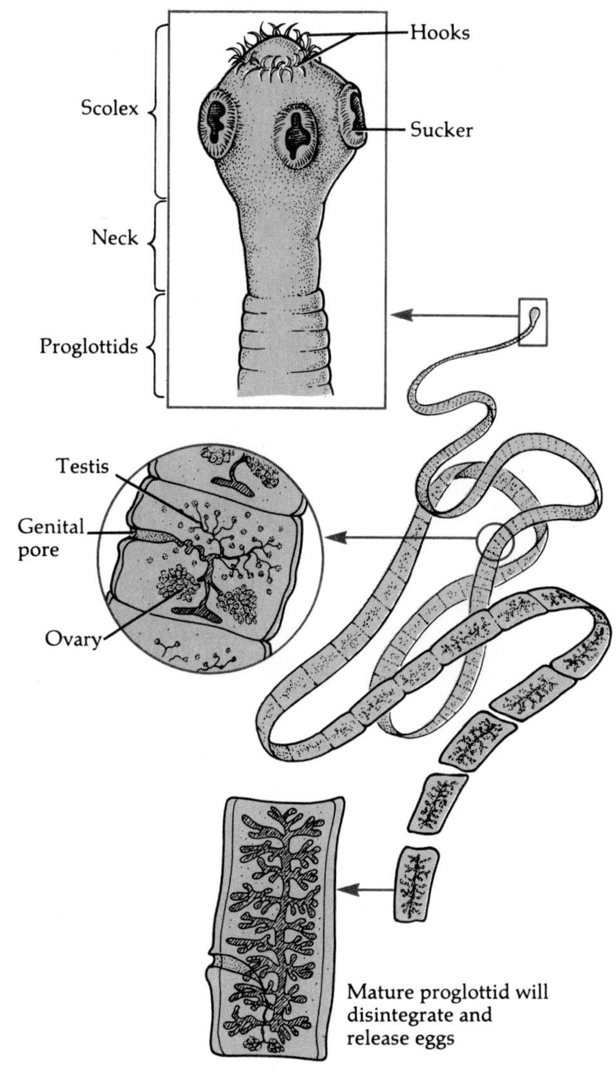

FIGURE 12.25 General anatomy of an adult tapeworm. The scolex consists of suckers and hooks that attach to the host's tissues. The body lengthens as new proglottids form at the neck. Each proglottid contains both testes and ovaries.

Aschelminthes

Members of the Phylum Aschelminthes, the roundworms, are cylindrical and tapered at each end. Roundworms have a *complete* digestive system, consisting of a mouth, an intestine, and an anus. Most species are dioecious. Males are smaller than females and have one or two hardened *spicules* on their posterior ends. Spicules are used to guide sperm to the female's genital pore. Parasites of humans belong only to the class consisting of nematodes.

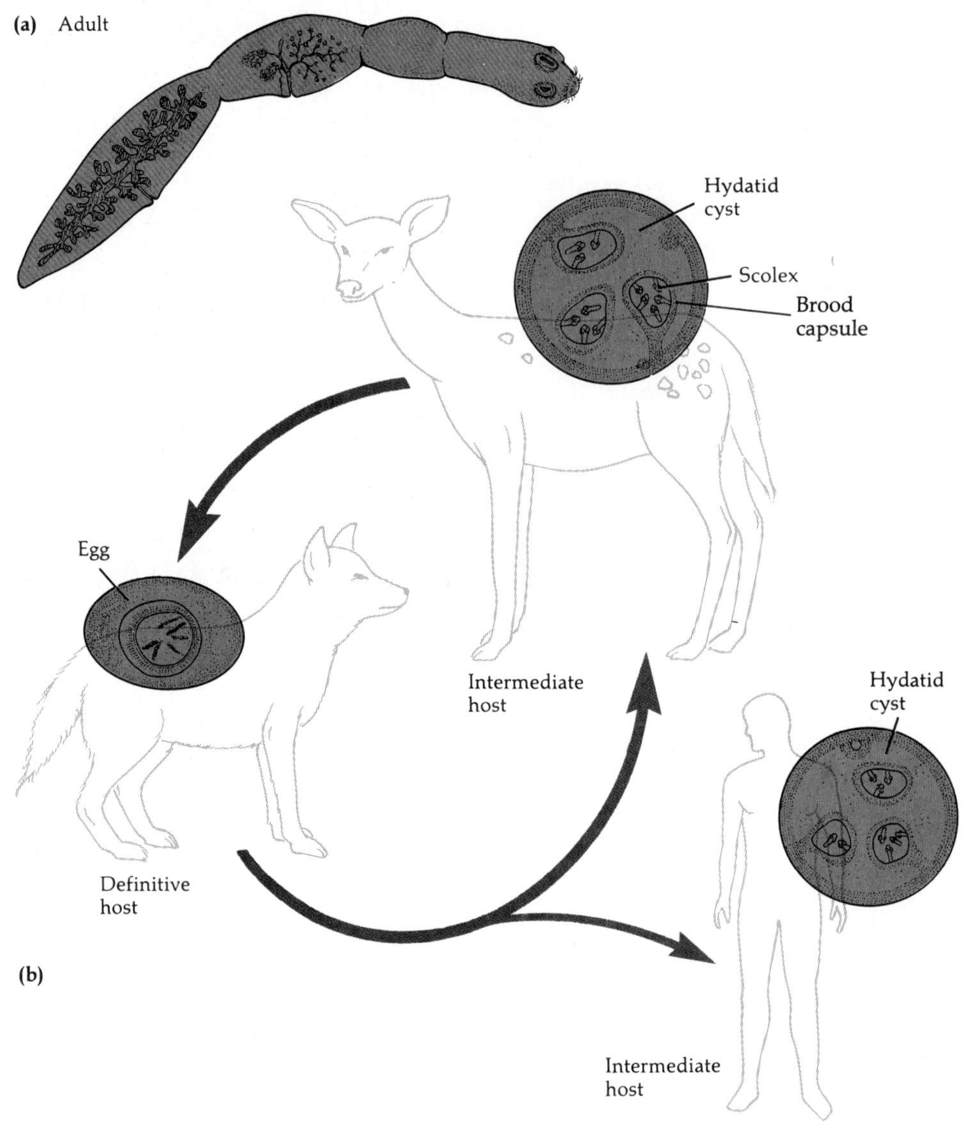

FIGURE 12.26 ***Echinococcus granulosus.*** This tiny tapeworm is found in the intestines of dogs, cats, wolves, and foxes. **(a)** Adult. **(b)** Life cycle. Eggs are excreted from the definitive host and ingested by an intermediate host, such as a deer. In the intermediate host, the eggs hatch, and the larvae form hydatid cysts in the host's tissues. A hydatid cyst is a fluid-filled sac containing many scoleces. For the cycle to be complete, the cysts must be ingested by a definitive host that eats the intermediate host. A human serving as the intermediate host is a dead end for the parasite unless the human is eaten by an animal.

NEMATODES

Some species of nematodes are free-living in soil and water, and others are parasites on plants and animals. Parasitic nematodes do not have the succession of larval stages exhibited by flatworms. Some nematodes pass their entire life cycle, from egg to mature adult, in a single host.

Nematode infections of humans can be divided into two categories—those in which the egg is infective and those in which the larva is infective.

Eggs Infective for Humans

The pinworm *Enterobius vermicularis* (en-te-rō'bē-us ver-mi-kū-lar'is) spends its entire life in a human host (Figure 12.27). Adult pinworms are found in the large intestine. From there, the female pinworm migrates to the anus to deposit her eggs in the perianal skin. The eggs can be ingested by the host or by another person exposed through contaminated clothing. Pinworm infections are diagnosed by the Graham sticky-tape method. A piece of tape is placed on the perianal skin in such a way that the sticky side is exposed to pick up eggs that were deposited earlier. The tape is microscopically examined for the presence of eggs adhering to it.

Ascaris lumbricoides (as'kar-is lum-bri-koi'dēz) is a large nematode (30 cm in length). It is dioecious with **sexual dimorphism;** that is, the male and female worms look distinctly different, the male being smaller with a curled tail. The adult *Ascaris* lives in the small intestines of humans and domestic animals (such as pigs and horses); it feeds primarily on semidigested food. Eggs, excreted with feces, can survive in the soil

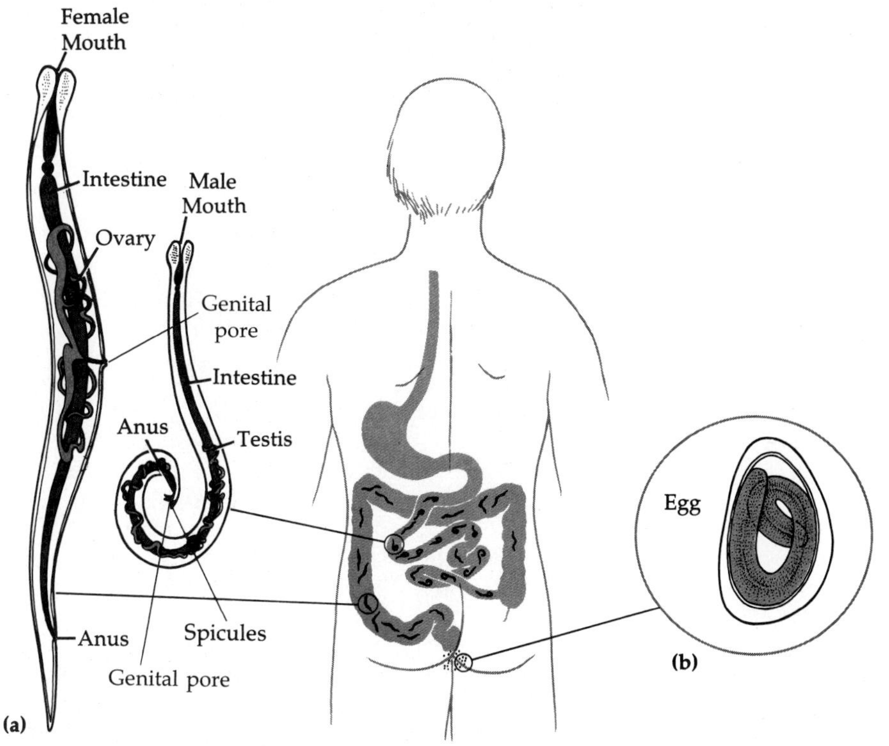

FIGURE 12.27 ***Enterobius vermicularis.*** **(a)** Adult pinworms, which live in the intestines of humans. Roundworms have a complete digestive system with a mouth, intestine, and anus. Most roundworms are dioecious, and the female (left) is often distinctly larger than the male (right). **(b)** Eggs, which the female deposits on the perianal skin at night.

for long periods until accidentally ingested by another host. The eggs hatch in the intestine of the host, mature in the lungs, and from there migrate to the intestines.

Diagnosis is frequently made when the adult worms are excreted with feces. Prevention of the disease in humans is managed by proper sanitary habits. The life cycle of *Ascaris* in pigs can be interrupted by keeping pigs in areas free of fecal material.

Larvae Infective for Humans

Adult hookworms, *Necator americanus* (ne-kā′tôr ä-me-ri-ka′nus), live in the small intestines of humans (Figure 12.28); the eggs are excreted with feces. The larvae hatch in the soil and feed on bacteria there. A larva enters its host by penetrating the host's skin. It then enters a blood or lymph vessel, which carries it to the lungs. It is coughed up in sputum, swallowed, and finally carried to the small intestine. Diagnosis is based on the presence of eggs in feces. People can avoid hookworm infections by wearing shoes.

Trichinella spiralis (trik-in-el′lä spī-ra′lis) infections, called *trichinosis,* are usually acquired by eating encysted larvae in pork or bear meat. In the human digestive tract, the larvae are freed from the cysts. They mature into adults in the intestine and sexually reproduce there. Eggs develop in the female, and she gives birth to live nematodes. The larvae enter lymph and blood vessels in the intestines and migrate from there

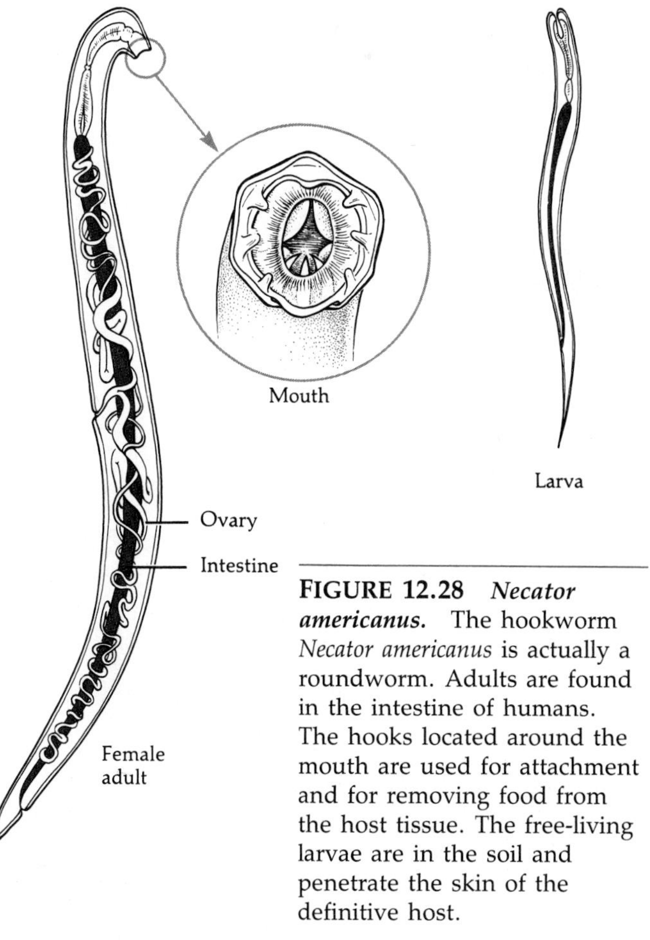

FIGURE 12.28 ***Necator americanus.*** The hookworm *Necator americanus* is actually a roundworm. Adults are found in the intestine of humans. The hooks located around the mouth are used for attachment and for removing food from the host tissue. The free-living larvae are in the soil and penetrate the skin of the definitive host.

MMWR

MORBIDITY AND MORTALITY WEEKLY REPORT

Aedes albopictus—A New Vector

In August 1985, an infestation of *Aedes albopictus* (Asian tiger mosquito), an exotic mosquito known to transmit epidemic dengue fever in its native Asia, was discovered in Harris County, Texas. In the spring of 1986, efforts were begun to determine the distribution of this mosquito vector in the United States. The figure shows the present distribution of *A. albopictus*-positive counties.

A. albopictus commonly breeds in standing water in tires stored outdoors. The mosquito was probably introduced into the United States in used-tire casings imported from Asia, the source of several million imported used tires each year. As of January 1, 1988, used-tire casings must be clean and dry and be treated by an approved fumigation procedure. However, as shown in the figure, *A. albopictus* is now entrenched in the United States, and, in many areas, it appears to be replacing the native mosquito, *A. aegypti.* As yet, *A. albopictus* has not been proved to cause the spread of any viral disease in the Americas, but it represents a public health concern because of its potential to increase the efficiency of dengue virus and of certain other pathogenic viruses. Laboratory studies have shown it to be a more efficient vector of dengue virus than *A. aegypti* and a competent vector of California encephalitis virus and yellow fever virus. Once introduced into an area where these viruses are present, *A. albopictus* could spread the viruses into areas previously free of them.

Dengue in the Americas

Worldwide there are 100 million cases of dengue per year. Dengue is now one of the leading causes of morbidity and mortality among Southeast Asian children. Dengue has been endemic in the Americas for over 200 years. Presently, the disease appears to be following a pattern similar to that in Southeast Asia 30 years ago. When dengue fever first breaks out in a geographical area where there have been no recent occurrences, it causes only a mild flulike illness. However, subsequent outbreaks occur as gastrointestinal hemorrhage and shock called dengue hemorrhagic fever (DHF). For example, in 1978, Mexico reported its first cases of dengue in many years, and, in 1984, the first DHF cases associated with epidemic dengue were reported.

Since 1981, all four dengue virus serotypes have been imported in the United States. The map of the Americas shows the progression of dengue over the past 10 years.

Preventive Measures

Scientists have investigated the mechanism by which dengue fever progresses from a mild illness in the first outbreak to DHF in subsequent outbreaks. A person's first infection sensitizes the immune system, so that it reacts more rapidly and vigorously to the second infection. However, it is thought that, during the second infection, antibodies and live viruses form immune complexes that attach to macrophages and other cells, actually increasing the number of viruses that infect a cell. This large number of viruses per cell results in massive tissue destruction. Because immunity actually increases the severity of the disease, immunization appears to be impractical as a preventive measure. Instead, the most effective methods of preventing the spread of the disease appear to be general mosquito control measures.

Sources: MMWR 38:440, 445–446 (6/30/89); *MMWR* 39:127–128, 133 (3/2/90).

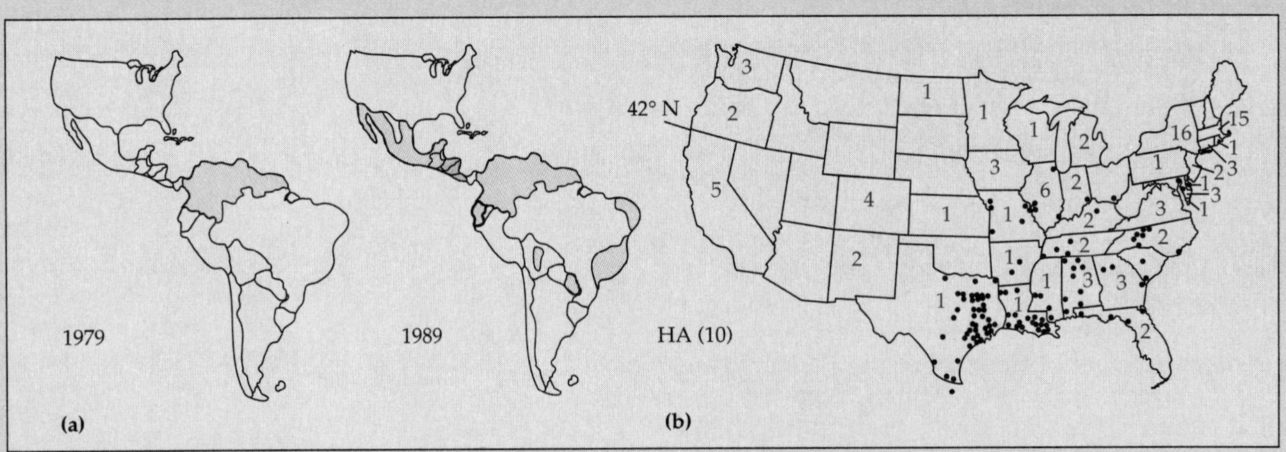

(a) Worldwide progression of dengue fever, 1979–1989. **(b)** Distribution of *A. albopictus* in the United States and confirmed cases of dengue fever (1989). • = Counties with confirmed *A. albopictus* infestation. Numbers indicate the number of confirmed dengue fever cases by state (total = 94).

TABLE 12.7 Representative Parasitic Helminths

PHYLUM	CLASS	HUMAN PARASITES	INTERMEDIATE HOST	DEFINITIVE HOST	STAGE PASSED TO HUMANS	DISEASE	LOCATION IN HUMANS	SEE FIGURE
Platyhelminthes	Trematodes	*Paragonimus westermanni*	Freshwater snails and crayfish	Humans; lungs	Metacercaria in crayfish	Paragonimiasis (lung fluke)	Lungs	12.24
		Schistosoma	Freshwater snails	Humans	Cercariae through skin	Schistosomiasis	Veins	23.25 23.26 23.27
	Cestodes	*Taenia saginata*	Cattle	Humans; small intestine	Cysticerci in beef	Tapeworm	Small intestine	25.21 25.22
		Echinococcus granulosus	Humans	Dogs and other animals; intestines	Eggs from other animals	Hydatid cyst	Lungs	12.26 25.23
Aschelminthes	Nematodes	*Ascaris lumbricoides*	—	Humans; small intestine	Ingestion of eggs	Ascariasis	Small intestine	25.25
		Enterobius vermicularis	—	Humans; large intestine	Ingestion of eggs	Pinworm	Large intestine	12.27a
		Necator americanus	—	Humans; small intestine	Penetration of larvae through skin	Hookworm	Small intestine	12.28 25.24
		Trichinella spiralis	—	Humans, swine, and other mammals; small intestine	Ingestion of larvae in meat, especially pork	Trichinosis	Muscles	25.26
		Anisakidae	Marine fish and squid	Marine mammals	Ingestion of larvae in fish	Anisakiasis ("wriggly worms")	Gastrointestinal tract	

throughout the body. They encyst in muscles and other tissues and remain there until ingested by another host (see Figure 25.24).

Diagnosis of trichinosis is made by microscopic examination for larvae in a muscle biopsy. Trichinosis can be prevented by thoroughly cooking meat prior to consumption.

Four genera of roundworms called anisakine or wriggly worms can be transmitted to humans from infected fish and squid. The anisakine larvae are in the fish's intestine and migrate to the muscle during refrigerated storage. These roundworms are not transmitted by fresh fish. Freezing or thorough cooking will kill the larvae. Table 12.7 lists representative parasitic helminths of each phylum and class and the diseases they cause.

ARTHROPODS AS VECTORS

Arthropods are jointed-legged animals. With nearly 1 million species, this is the largest phylum in the animal kingdom. We will briefly describe arthropods here because a few suck the blood of humans and other animals and can transmit microbial diseases while doing so. Arthropods that carry disease-causing microorganisms are called **vectors.** Scabies, a disease caused by an arthropod, is described in the box in Chapter 21 on page 529.

Representative classes of arthropods include the following:

1. Arachnida (eight legs): spiders, mites, ticks
2. Crustacea (four antennae): crabs, crayfish
3. Chilopoda (two legs per segment): centipedes
4. Diplopoda (four legs per segment): millipedes
5. Insecta (six legs): bees, flies

Table 12.8 lists those arthropods that are important vectors, and Figures 12.29, 12.30, and 12.31 illustrate some of them. These insects and ticks reside on an animal only when they are feeding. An exception to this is the louse, which spends its entire life on its host and cannot survive long away from a host. Among in-

TABLE 12.8 Important Arthropod Vectors of Human Diseases

CLASS	ORDER	VECTOR	SEE FIGURE	DISEASE	REFER TO CHAPTER
Arachnida	Mites and ticks	*Dermacentor* (tick)	—	Rocky Mountain spotted fever	23
		Ixodes (tick)	12.30	Lyme disease, babesiosis	23
		Ornithodorus	—	Relapsing fever	23
Insecta	Sucking lice	*Pediculus* (human louse)	12.31a	Epidemic typhus	23
	Fleas	*Xenopsylla* (rat flea)	12.31b	Endemic murine typhus, plague	23
	True flies	*Chrysops* (deer fly)	12.31c	Tularemia	23
		Aedes (mosquito)	—	Dengue fever, yellow fever	23
		Anopheles (mosquito)	12.29b	Malaria	23
		Culex (mosquito)	—	Arboviral encephalitis	22
		Glossina (tsetse fly)	—	African trypanosomiasis	22
	True bugs	*Triatoma* (kissing bug)	12.30d	Chagas' disease	23

(a)

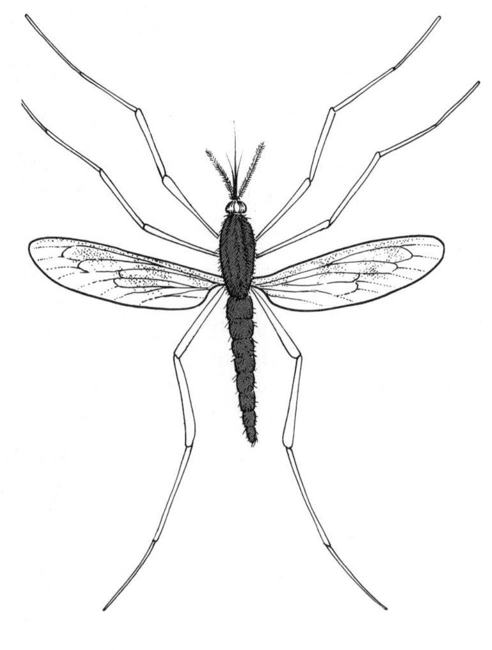

(b)

FIGURE 12.29 Mosquitoes. **(a)** Female sucking blood from human skin. Mosquitoes transmit several diseases from person to person, including the viral diseases yellow fever and dengue fever. **(b)** *Anopheles* transmits malaria.

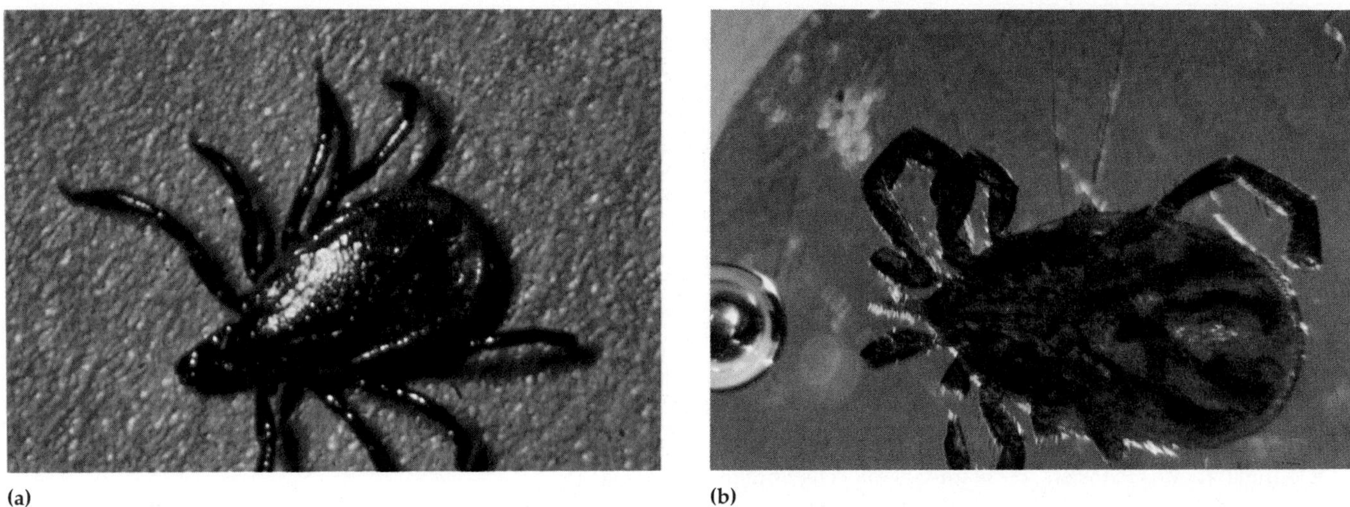

FIGURE 12.30 Ticks. **(a)** *Ixodes dammini* transmits Lyme disease on the East Coast. **(b)** *I. pacificus* is the Lyme disease vector on the West Coast.

sect vectors, only the adult female flies (including mosquitoes) bite.

Some vectors are just a mechanical means of transport for a pathogen. For example, houseflies are attracted to decaying organic matter in which to lay their eggs. A housefly can then pick up a pathogen on its feet or body from feces and transport the pathogen to food.

Some parasites multiply in their vectors. When this happens, the parasites can accumulate in the vector's feces or saliva. Large numbers of parasites can then be deposited in the host while the vector is feeding there. The spirochete that causes Lyme disease is transmitted by ticks in this manner, and the dengue fever virus is transmitted in the same way by mosquitos (see the box on page 324).

Plasmodium is an example of a parasite that requires that its vector also be the host. *Plasmodium* can sexually reproduce only in the gut of an *Anopheles* mosquito (Figure 12.29b). *Plasmodium* is introduced into a human host with the mosquito's saliva, which acts as an anticoagulant to keep blood flowing.

To eliminate vector-borne diseases (such as African sleeping sickness), health workers focus on eradicating the vectors.

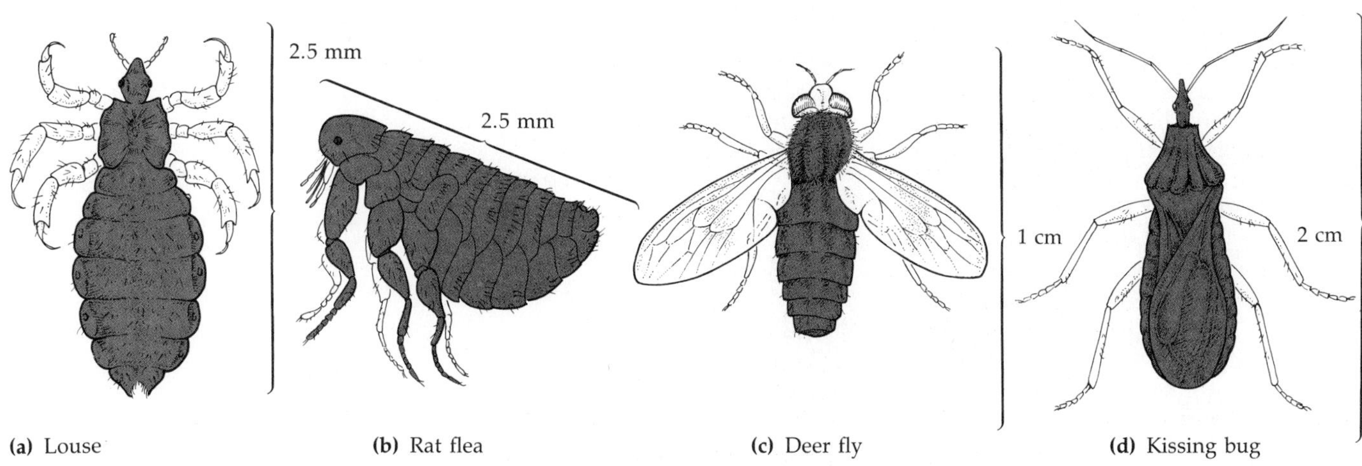

FIGURE 12.31 Arthropod vectors. **(a)** The human louse, *Pediculus.* **(b)** The rat flea, *Xenopsylla.* **(c)** The deer fly, *Chrysops.* **(d)** The kissing bug, *Triatoma.* See Table 12.8 for the diseases these vectors carry.

STUDY OUTLINE

FUNGI (pp. 296–306)

1. The Kingdom Fungi includes yeasts, molds, and fleshy fungi (mushrooms).
2. Fungi are aerobic or facultatively anaerobic chemoheterotrophs.
3. Most fungi are decomposers, and a few are parasites of plants and animals.

Characteristics of Fungi (pp. 297–303)

VEGETATIVE STRUCTURES (pp. 297–299)

1. A fungal thallus consists of filaments of cells called hyphae; a mass of hyphae is called a mycelium.
2. Yeasts are unicellular fungi that reproduce by budding and occasionally by fission.
3. Buds that do not separate from the mother cell form pseudohyphae.
4. Dimorphic fungi are yeastlike at 37°C and moldlike at 25°C.

REPRODUCTIVE STRUCTURES (pp. 299–301)

1. The following spores can be produced asexually: arthrospores, chlamydospores, sporangiospores, conidospores, and blastospores.
2. Fungi are classified according to the type of sexual spore that they form.
3. Sexual spores include zygospores, ascospores, and basidiospores.

NUTRITIONAL ADAPTATIONS (p. 301)

1. Fungi can grow in acidic, low-moisture, aerobic environments.
2. They are able to metabolize complex carbohydrates.

Medically Important Phyla of Fungi (pp. 302–304)

1. Deuteromycota reproduce asexually by one or more of the following: chlamydospores, arthrospores, conidiospores, or budding.
2. Deuteromycota are called the Fungi Imperfecti because sexual spores have not yet been seen.
3. The classification of *Pneumocystis* as a fungus or protozoan is currently being studied.
4. The Zygomycota have coenocytic hyphae and produce sporangiospores and zygospores.
5. The Ascomycota have septate hyphae and produce ascospores and frequently conidiospores.
6. Basidiomycota have septate hyphae and produce basidiospores; some produce conidiospores.

Fungal Diseases (pp. 304–306)

1. Systemic mycoses are fungal infections deep within the body and affect many tissues and organs.
2. Subcutaneous mycoses are fungal infections beneath the skin.
3. Cutaneous mycoses affect keratin-containing tissues such as hair, nails, and skin.
4. Superficial mycoses are localized on hair shafts and superficial skin cells.
5. Opportunistic mycoses are caused by normal flora or fungi that are not usually pathogenic.
6. Opportunistic mycoses include mucormycosis, caused by some zygomycetes; aspergillosis, caused by *Aspergillus;* and candidiasis, caused by *Candida.*
7. Opportunistic mycoses can infect any tissues. However, they are usually systemic.

Economic Effects of Fungi (p. 306)

1. Mold spoilage of fruits, grains, and vegetables is more common than bacterial spoilage of these products.
2. Many fungi cause diseases in plants (for example, in potatoes, chestnuts, and elms).

ALGAE (pp. 306–310)

1. Algae are photoautotrophs that produce oxygen.
2. Algae are classified as plants or protists according to their structures and pigments.

Phyla of Algae (pp. 307–310)

1. Unicellular dinoflagellates have cellulose and silica cell walls; some produce a neurotoxin that causes paralytic shellfish poisoning and ciguatera.
2. Euglenoids have a semirigid cell membrane and one flagellum.
3. Diatoms have pectin and silica cell walls.
4. Multicellular algae include the brown algae, red algae, and green algae.
5. Brown algae (kelp) are harvested for algin.
6. Red algae grow deeper in the ocean than other algae because their red pigments can absorb the blue light that penetrates to deeper levels.
7. Green algae have cellulose and chlorophyll *a* and *b* and store starch.

Structure and Reproduction (p. 310)

1. The thallus (or body) of multicellular algae usually consists of a stipe, holdfast, and blades.
2. Algae reproduce asexually by cell division and fragmentation.
3. Many algae reproduce sexually.

Roles of Algae in Nature (p. 310)

1. Algae are the primary producers in aquatic food chains.
2. Planktonic algae produce most of the molecular oxygen in the Earth's atmosphere.
3. Petroleum is the fossil remains of planktonic algae.
4. Unicellular algae are symbionts in such animals as *Tridacna.*

LICHENS (pp. 310–312)

1. A lichen is a symbiotic combination of an alga and a fungus.
2. The alga photosynthesizes, providing carbohydrates for the lichen; the fungus provides a holdfast.
3. Lichens colonize habitats that are unsuitable for either the alga or fungus alone.
4. Lichens may be classified on the basis of morphology as crustose, foliose, or fruticose.
5. In the lichen thallus, the alga reproduces by cell division, and the fungus forms sexual spores.

SLIME MOLDS (pp. 312–313)

1. Cellular slime molds resemble amoebas and ingest bacteria by phagocytosis.
2. A plasmodial (acellular) slime mold is a multinucleated mass of protoplasm that engulfs organic debris and bacteria as it moves.

PROTOZOANS (pp. 313–319)

1. Protozoans are unicellular eucaryotes in the Kingdom Protista.
2. Protozoans are found in soil and water and as normal flora in animals.
3. Protozoans are classified by their means of locomotion: members of the Phylum Sarcodina move by amoeboid motion; the Mastigophora (flagellates) use flagella for motility; the Ciliata possess cilia; and the Sporozoa lack a means of locomotion and are obligate parasites.

Protozoan Biology (pp. 313–315)

1. Most protozoans are heterotrophs and feed on bacteria and particulate organic material.
2. Protozoans are complex cells that have a pellicle, a cytostome, and an anal pore.
3. Asexual reproduction is by fission, budding, or schizogony.
4. Sexual reproduction is by conjugation.
5. During protozoan conjugation, two haploid nuclei fuse to produce a zygote.
6. Some protozoans can produce a cyst for protection during adverse environmental conditions.

Medically Important Phyla of Protozoans (pp. 315–319)

1. Parasitic sarcodinae that cause amoebic dysentery are found in the genus *Entamoeba.*
2. Parasitic flagellates (Mastigophora) include the following: *Giardia lamblia,* causing an intestinal infection called giardiasis; *Trichomonas vaginalis,* causing genitourinary infections that may be transmitted by coitus; and *Trypanosoma,* which are found in the blood of human hosts and are transmitted by blood-sucking insects.
3. The only ciliate that is a parasite of humans is *Balantidium coli,* the cause of one form of dysentery.
4. *Plasmodium* is the sporozoan that causes malaria.
5. Asexual reproduction of *Plasmodium* occurs in red blood cells and the liver of humans.
6. Sexual reproduction of *Plasmodium* takes place in the intestine of the female *Anopheles* mosquito.
7. *Toxoplasma gondii* is a sporozoan that infects humans; it can be transmitted to a fetus in utero.
8. *Cryptosporidium* causes respiratory and diarrheal diseases in immunosuppressed patients.
9. Microsporidia cause diarrhea and keratoconjunctivitis in AIDS patients.

HELMINTHS (pp. 319–325)

1. Some parasitic flatworms belong to the Phylum Platyhelminthes.
2. Parasitic roundworms belong to the Phylum Aschelminthes.

Helminth Biology (p. 319)

1. Helminths are multicellular animals; a few are parasites of humans.
2. The anatomy and life cycles of parasitic helminths are modified for parasitism.
3. Helminths can be hermaphroditic or dioecious.
4. The adult stage of a parasitic helminth is found in the definitive host.
5. Each larval stage of a parasitic helminth requires an intermediate host.

Platyhelminthes (pp. 319–321)

1. Flatworms are dorsoventrally flattened animals that have an incomplete digestive system.
2. Adult trematodes or flukes have an oral and ventral sucker with which they attach to and feed on host tissue.
3. Eggs of trematodes hatch into free-swimming miracidia that enter the first intermediate host; two generations of rediae develop in the first intermediate host; the rediae become cercariae that bore out of the first intermediate host and penetrate the second intermediate host; cercariae encyst as metacercariae in the second intermediate host; after they are ingested by the definitive host, the metacercariae develop into adults.
4. A cestode, or tapeworm, consists of a scolex (head) and proglottids.
5. Humans serve as the definitive host for the beef tapeworm, and cattle are the intermediate host.
6. Humans serve as the intermediate host for *Echinococcus granulosus;* the definitive hosts are dogs, wolves, and foxes.

Aschelminthes (pp. 321–325)

1. Members of the Phylum Aschelminthes are roundworms with a complete digestive system.
2. Parasitic members of Aschelminthes are in the class consisting of nematodes.

3. The Aschelminthes that infect humans with their eggs are *Enterobius vermicularis* (pinworm) and *Ascaris lumbricoides*.
4. The Aschelminthes that infect humans with their larvae are *Necator americanus* (hookworm), *Trichinella spiralis*, and anisakine worms.

ARTHROPODS AS VECTORS (pp. 325–327)

1. Jointed-legged animals, including ticks and insects, belong to the Phylum Arthropoda.
2. Arthropods that carry diseases are called vectors.
3. Elimination of vector-borne diseases is best done by the control or eradication of the vectors.

STUDY QUESTIONS

Review

1. Contrast the mechanism of conidiospore and ascospore formation by *Penicillium*.
2. Fill in the following table.

	Spore Type(s)	
Phylum	Sexual	Asexual
Zygomycota		
Ascomycota		
Basidiomycota		
Deuteromycota		

3. Fungi are classified into phyla on the basis of ______.
4. The following is a list of fungi, their methods of entry into the body, and sites of infections they cause. Categorize each type of mycosis as cutaneous, opportunistic, subcutaneous, superficial, or systemic.

Genus	Method of Entry	Site of Infection	Mycosis
Blastomyces	Inhalation	Lungs	
Sporothrix	Puncture	Ulcerative lesions	
Microsporum	Contact	Fingernails	
Trichosporon	Contact	Hair shafts	
Aspergillus	Inhalation	Lungs	

5. A mixed culture of *Escherichia coli* and *Penicillium chrysogenum* is inoculated onto the following culture media. On which media would you expect each to grow? Why?
   (a) 0.5% peptone in tap water
   (b) 10% glucose in tap water
6. What is the role of the alga in a lichen? What is the role of the fungus?
7. Briefly discuss the importance of lichens in nature. Briefly discuss the importance of algae.
8. Transmission of helminthic parasites to humans usually occurs by ______.
9. Complete the following table.

Phylum	Cell Wall Composition	Special Features/ Importance
Dinoflagellates		
Euglenoids		
Diatoms		
Red algae		
Brown algae		
Green algae		

   Indicate which phyla consist primarily of unicellular forms. Which phylum would you include in the plant kingdom? Why? Into which kingdom would you place the others?
10. Differentiate between cellular and plasmodial slime molds. How does each survive adverse environmental conditions?
11. Why is it significant that *Trichomonas* does not have a cyst stage? Name a protozoan parasite that does have a cyst stage.
12. Protozoans are classified on the basis of ______.
13. Recall the life cycle of *Plasmodium*. Where does asexual reproduction occur? Where does sexual reproduction occur? Identify the definitive host. Identify the vector.
14. To what phylum and class does this animal belong?

   List two characteristics that put it in this phylum. Name the body parts. What is the name of the encysted larva of this animal?
15. Most nematodes are dioecious. What does this term mean? To what phylum do nematodes belong?
16. Vectors can be divided into three major types, according to the roles they play for the parasite. List the three types of vectors and a disease transmitted by each.

Challenge

1. A generalized life cycle of the liver fluke *(Clonorchis sinensis)* is shown below. Identify the intermediate host(s). Identify the definitive host(s). To what phylum and class does this animal belong?

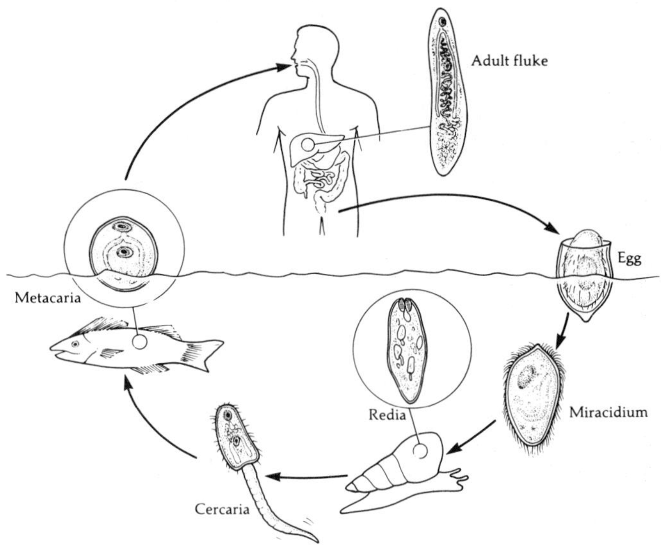

2. The size of a cell is limited by its surface-to-volume ratio; that is, if the volume becomes too great, internal heat cannot be dissipated and nutrients and wastes cannot be efficiently transported. How do plasmodial slime molds manage to circumvent the surface-to-volume rule?

3. The life cycle of the fish tapeworm *Diphyllobothrium* is similar to that of *Taenia saginata,* except that the intermediate host is fish. Describe the life cycle and method of transmission to humans. Why are freshwater fish more likely to be a source of tapeworm infestation than marine fish?

4. Microorganisms don't care what they are called. Why, then, might it be important to classify *Pneumocystis?*

5. *T. b. gambiense* (part a, below) is the causative agent of African sleeping sickness. To what phylum and class does it belong? Part b shows a generalized life cycle for *T. b. gambiense.* Identify the host and vector of this parasite.

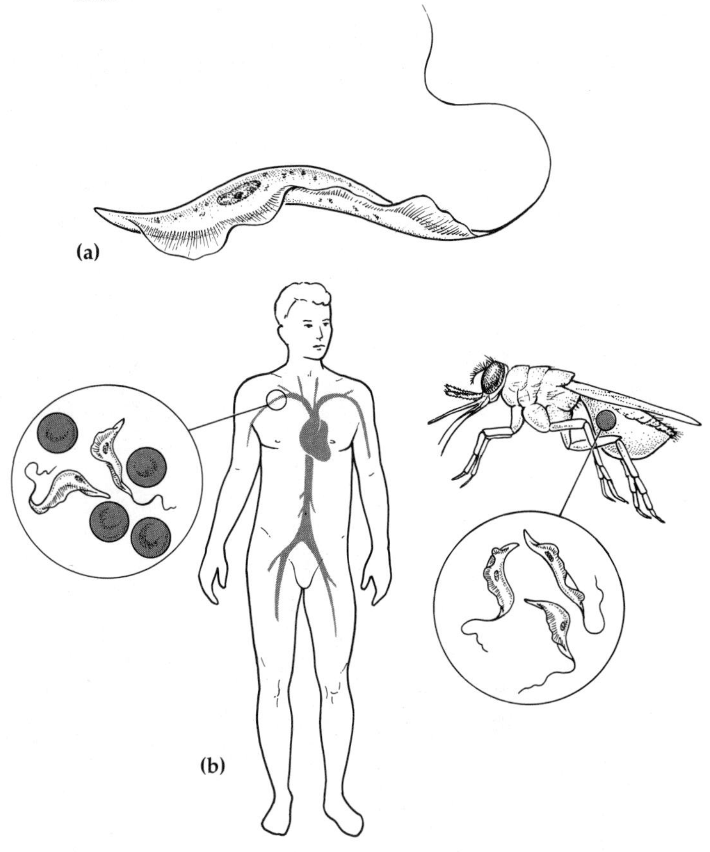

FURTHER READING

Ahmadjian, V., and M. Hale, eds. *The Lichens.* New York: Academic Press, 1974. An excellent text covering the biology, ecology, and antibiotic production by lichens.

Alexopoulos, C.J., and C.W. Mims. *Introductory Mycology,* 3rd ed. New York: Macmillan, 1979. A classic textbook on the biology of fungi.

Barnett, H.L., and B.B. Hunter. *Illustrated Genera of Imperfect Fungi,* 4th ed. New York: Macmillan, 1987. Good line drawings facilitate identification of saprophytic molds.

Barnett, J.A., R.W. Payne, and D. Yarrow. *Yeasts: Characteristics and Identification,* 2nd ed. New York: Cambridge University Press, 1990. A complete key to yeasts, based on morphologic and biochemical characteristics.

Bold, H.C., and M.J. Wynne. *Introduction to the Algae,* 2nd ed. Englewood Cliffs, N.J.: Prentice-Hall, 1985. Covers biology of freshwater and marine unicellular and multicellular algae.

Donelson, J.E., and M.J. Turner. "How the trypanosome changes its coat." *Scientific American* 252(2):44–51, February 1985. In studying the trypanosome's variable surface proteins, scientists may have a model for how gene expression is controlled.

Walsh, T.J., and P.A. Pizzo. Nosocomial fungal infections: a classification for hospital-acquired fungal infections and mycoses arising from endogenous flora or reactivation. *Annual Review of Microbiology* 42:517–545, 1988. Discusses the fungal infections that are becoming more common in chronically ill and immunosuppressed patients.

CHAPTER 13

Viruses

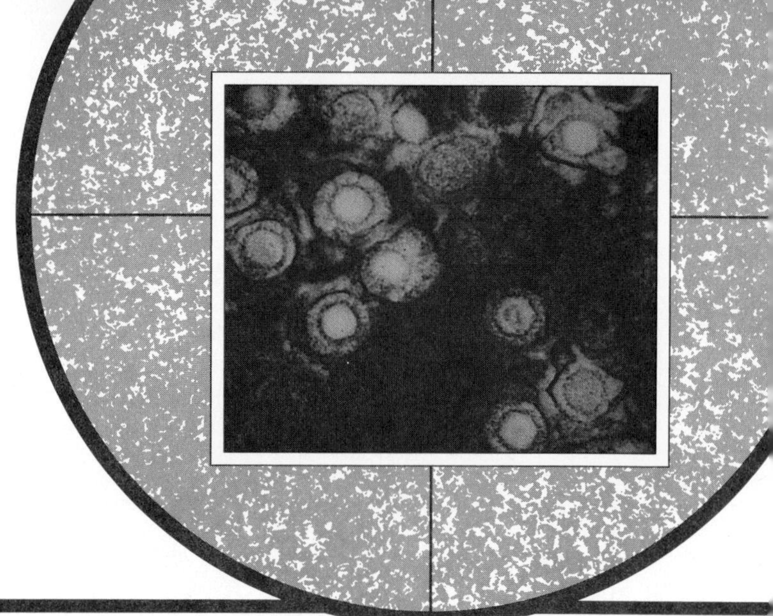

LEARNING OBJECTIVES

- Define virus, viroid, prion, transformed cell, and oncogene.
- Describe the chemical composition of a typical virus.
- Classify viruses according to their morphology, and list other criteria used to classify viruses.
- Explain how viruses are cultured.
- Describe the lytic cycle of T-even bacteriophages.
- Compare and contrast the multiplication cycle of DNA- and RNA-containing animal viruses.
- Differentiate between slow viral infections and latent viral infections.
- List the effects of animal viral infections on host cells.
- Describe four mechanisms of oncogene activation.
- Explain what a tumor is, and distinguish between malignant and benign tumors.
- Discuss the relationship of DNA- and RNA-containing viruses to cancer.

In 1892, in an experiment to isolate the cause of tobacco mosaic disease, Dmitri Iwanowski filtered the sap of diseased tobacco plants through a porcelain filter that was designed to retain bacteria. He expected to find the microbe trapped in the filter. Instead, he found that the infectious agent had passed through the minute pores of the filter. When he injected healthy plants with the filtered fluid, they contracted tobacco mosaic disease.

Iwanowski still believed the infectious agent to be a bacterium that was small enough to pass through the filter. But later, other scientists, led by Martinus Beijerinck, observed that the behavior of this agent was different from that of bacteria. In the early 1900s, a distinction was finally made between bacteria and viruses, the filterable agents that cause tobacco mosaic disease and many other diseases. (*Virus* is the Latin word for poison.) In 1935, an American chemist, Wendell M. Stanley, isolated the tobacco mosaic virus (TMV), making it possible for the first time to carry out chemical and structural studies on a purified virus. At about the same time, the invention of the electron microscope made it possible to see viruses for the first time.

We now know that viruses are found as parasites in all types of organisms. They can reproduce only within cells. Many diseases of humans are known to be caused by viruses; some diseases of agriculturally important animals and plants are also known to be caused by viruses. Viruses infect fungi, bacteria, and protists as well.

Advances in molecular biology techniques in the 1980s have led to the discovery of new human viruses. Hepatitis C virus and a pestivirus causing severe pediatric diarrhea were discovered in 1989. Public health officials are concerned about health risks posed by these new viruses, and they are also concerned that the ease of world travel and changing environments will spread viruses into new areas. For example, the potential of epidemics caused by the dengue fever virus in the United States because of the introduction of a suitable vector is a public health concern (see the box in Chapter 12, p. 324).

General Characteristics of Viruses

The question of whether viruses are living organisms has an ambiguous answer. Life can be defined as a complex set of processes resulting from the actions of proteins specified by nucleic acids. The nucleic acids of living cells are in action all the time. Because viruses are inert outside of living host cells, in this sense they are not considered to be living organisms. However, once viruses enter a host cell, the viral nucleic acids become active, and viral multiplication results. In this sense, viruses are alive when they multiply in the host cells they infect. From a clinical point of view, viruses can be considered alive because they cause infection and disease, just as pathogenic bacteria, fungi, and protozoans do. Depending on one's viewpoint, a virus may be regarded as an exceptionally complex aggregation of nonliving chemicals or as an exceptionally simple living microorganism.

How, then, do we define a virus? Viruses were originally distinguished from other infectious agents because they are especially small (filterable) and because they are **obligatory intracellular parasites**—that is, they absolutely require living host cells in order to multiply. However, both of these properties are shared by certain small bacteria, such as some rickettsias. Viruses and bacteria are compared in Table 13.1.

The truly distinctive features of viruses are now known to relate to their simple structural organization and composition and their mechanism of multiplication. Accordingly, **viruses** are entities that:

1. Contain a single type of nucleic acid, either DNA or RNA.
2. Contain a protein coat (sometimes itself enclosed by an envelope of lipids, proteins, and carbohydrates) that surrounds the nucleic acid.
3. Multiply inside living cells by using the synthesizing machinery of the cell.
4. Cause the synthesis of specialized structures that can transfer the viral nucleic acid to other cells.

Viruses have few or no enzymes of their own for metabolism—for example, they lack enzymes for protein synthesis and ATP generation. To multiply, viruses must take over the metabolic machinery of the host cell. This fact has considerable medical significance for the development of antiviral drugs because most drugs that would interfere with viral multiplication would also interfere with the functioning of the host cell and hence are too toxic for clinical use. (Antiviral drugs are discussed in Chapter 20.) However, the

TABLE 13.1 A Comparison of Viruses and Bacteria

	BACTERIA			*VIRUSES*
	Typical Bacteria	*Rickettsias*	*Chlamydias*	
Intracellular parasite	−	+	+	+
Plasma membrane	+	+	+	−
Binary fission	+	+	+	−
Filterable through bacteriological filters	−	−	+	+
Possess both DNA and RNA	+	+	+	−
ATP-generating metabolism	+	+	−	−
Ribosomes	+	+	+	−
Sensitive to antibiotics	+	+	+	−
Sensitive to interferon	−	−	−	+

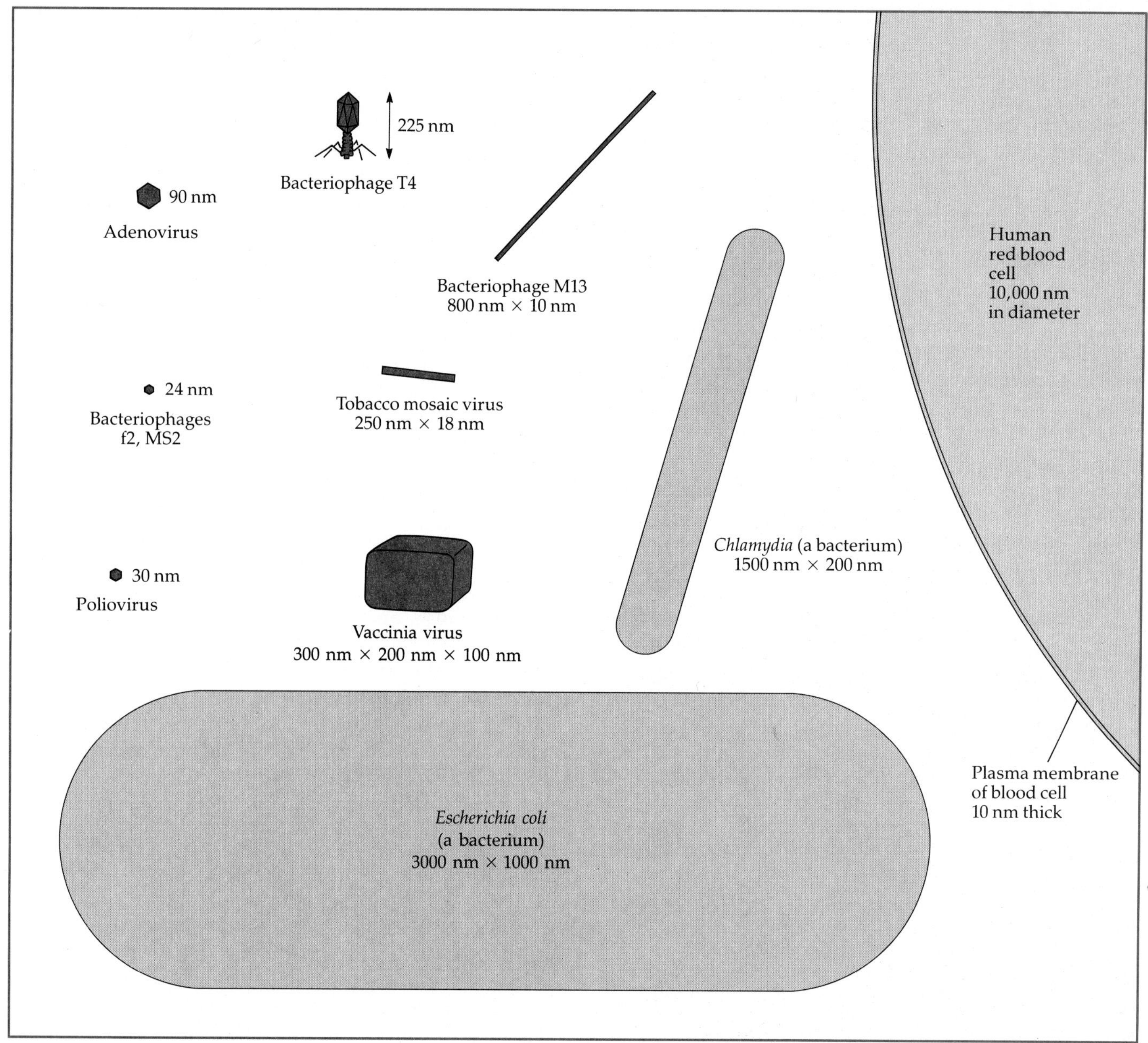

FIGURE 13.1 Virus sizes. Sizes of several viruses (red) and bacteria (gray) are compared with a human red blood cell, shown to the right of the microbes. Dimensions are given in nanometers (nm) and are either diameters or length by width.

presence of lipids in the coverings of some viruses makes these viruses sensitive to disinfection (outside of a host cell). For example, a 1:10 dilution of household chlorine bleach can inactivate HIV (AIDS virus) outside the body. Viral lipids may also be damaged by lipid solvents, such as ether, and by emulsifying agents, such as bile salts and detergents.

HOST RANGE

The **host range** of a virus is the spectrum of host cells that the virus can infect. Viruses multiply only in cells of particular species and thus are divided into three main classes—*animal viruses, bacterial viruses* (bacteriophages), and *plant viruses.* (Protists and fungi can also be hosts for viruses, but in this chapter we are concerned mainly with animal viruses and bacterial viruses.) Within each class, each virus is usually able to infect cells of certain species only.

The particular host range of a virus is determined by the virus's requirements for its specific attachment to the host cell and the availability within the potential host of cellular factors required for viral multiplication. For the virus to infect the host cell, the outer surface of the virus must chemically interact with specific receptor sites on the surface of the cell. The two complementary components are held together by weak bonds, such as hydrogen bonds. For some bacteriophages, the receptor site is part of the cell wall of the host; in other cases, it is part of the fimbriae or flagella. For animal viruses, the receptor sites are on the plasma membranes of the host cells.

SIZE

The sizes of viruses were first estimated by filtration through membranes of known pore diameters. Viral sizes are determined today by ultracentrifugation and by electron microscopy, which seems to produce the most accurate results. Viruses vary considerably in size. Although most are quite a bit smaller than bacteria, some of the larger viruses (such as the smallpox virus) are about the same size as some very small bacteria (such as the mycoplasmas, rickettsias, and chlamydias). Viruses range from 20 to 300 nm in diameter. The comparative sizes of several viruses and bacteria are shown in Figure 13.1.

Viral Structure

A **virion** is a complete, fully developed viral particle composed of nucleic acid surrounded by a coat that protects it from the environment and serves as a vehicle of transmission from one host cell to another.

NUCLEIC ACID

As we noted earlier, the core of a virus contains a single kind of nucleic acid, either DNA or RNA, which is the genetic material. The percentage of nucleic acid in relation to protein is about 1% for the influenza virus and about 50% for certain bacteriophages. The total amount of nucleic acid varies from a few thousand nucleotides (or pairs) to as many as 250,000 nucleotides. (*E. coli*'s chromosome consists of approximately four million nucleotide pairs.)

In contrast to procaryotic and eucaryotic cells, in which DNA is always the primary genetic material (and RNA plays an auxiliary role), a virus can have either DNA or RNA but never both. The nucleic acid of a virus can be single-stranded or double-stranded. Thus, there are viruses with the familiar double-stranded DNA, with single-stranded DNA, with double-stranded RNA, and with single-stranded RNA. Depending on the virus, the nucleic acid can be linear or circular. In some viruses (such as the influenza virus), the nucleic acid is in several separate segments.

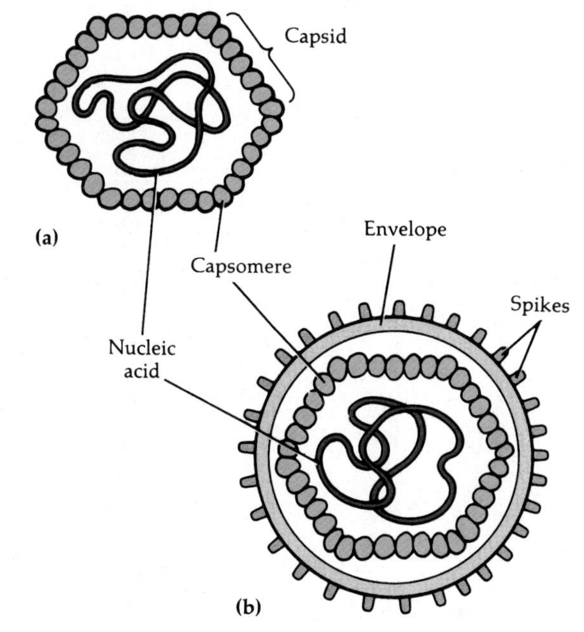

FIGURE 13.2 General structure of two types of viruses. **(a)** Naked virus. **(b)** Enveloped virus with spikes.

CAPSID AND ENVELOPE

The nucleic acid of a virus is surrounded by a protein coat called the **capsid** (Figure 13.2a). The structure of the capsid is ultimately determined by the viral nucleic acid and accounts for most of the mass of a virus, especially of small ones. Each capsid is composed of protein subunits called **capsomeres.** In some viruses, the proteins composing the capsomeres are of a single type; in other viruses, several types of protein may be present. Individual capsomeres are often visible in electron micrographs (see Figure 13.4b, for an example). The arrangement of capsomeres is characteristic of a particular type of virus.

In some viruses, the capsid is covered by an **envelope** (Figure 13.2b), which usually consists of some combination of lipids, proteins, and carbohydrates. Little is known about the molecular organization of these envelopes. Some animal viruses are released from the host cell by an extrusion process that coats the virus with a layer of the host cell's plasma membrane; that layer becomes the viral envelope. In many cases, the envelope contains proteins determined by viral nucleic acid and materials derived from the normal host cell components.

Depending on the virus, envelopes may or may not be covered by **spikes,** which are carbohydrate–protein complexes that project from the surface of the envelope. Some viruses attach to host cells by means of spikes. Spikes are such a reliable characteristic of some viruses that they can be used as a means of iden-

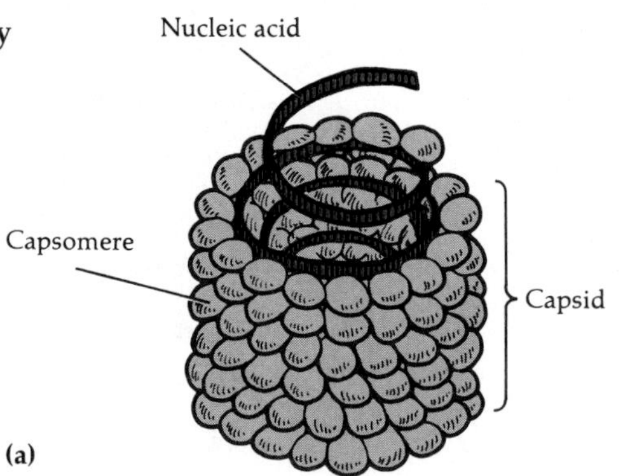

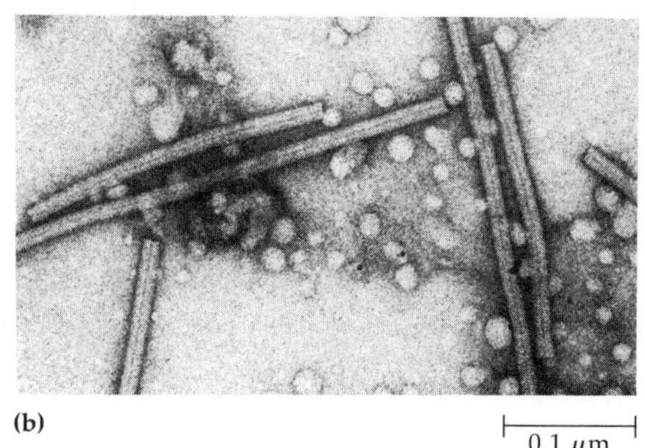

FIGURE 13.3 Morphology of a helical virus. **(a)** Diagram of a portion of a tobacco mosaic virus. Several rows of capsomeres have been removed to reveal the nucleic acid. **(b)** Electron micrograph of a tobacco mosaic virus showing helical rods.

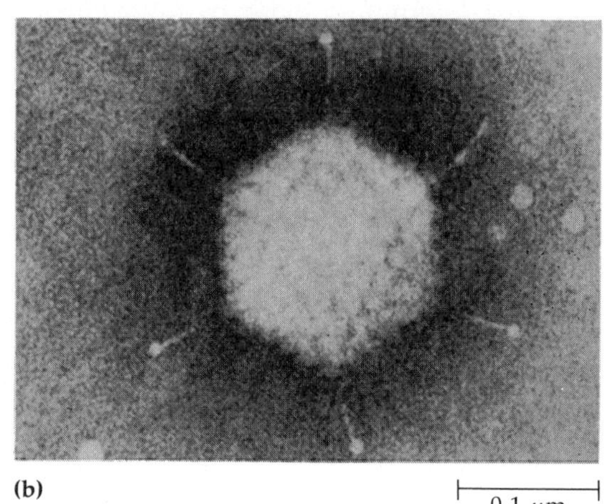

FIGURE 13.4 Morphology of a naked polyhedral virus in the shape of an icosahedron. **(a)** Diagram of an icosahedron. **(b)** Electron micrograph of an adenovirus. Individual capsomeres in the protein coat are visible.

tification. The ability of certain viruses, such as the influenza virus, to clump red blood cells is associated with spikes. Such viruses bind to red blood cells and form bridges between them. The resulting clumping is called **hemagglutination** and is the basis for several useful laboratory tests.

Viruses whose capsids are not covered by an envelope are known as **naked viruses** or **nonenveloped viruses** (Figure 13.2a). The capsid of a naked virus protects the nucleic acid from nuclease enzymes in biological fluids and promotes the virus's attachment to susceptible host cells.

When the host has been infected by a virus, the host immune system produces antibodies (proteins that react with the virus) to inactivate that virus and stop the infection. Some viruses can escape antibodies because regions of the genes that code for these viruses' surface proteins are susceptible to mutations. Mutant viruses alter their surface proteins so that the antibodies are not able to react with them. Influenza virus frequently undergoes such changes in its spikes. This is the reason why you can get influenza more than once. Although you may have produced antibodies to one influenza virus, the virus will mutate and be able to infect you again.

GENERAL MORPHOLOGY

Viruses may be classified into several morphological types on the basis of their capsid architecture as revealed by electron microscopy and a technique called x-ray crystallography.

Helical viruses. Helical viruses resemble long rods that may be rigid or flexible. Surrounding the nucleic acid, their capsid is a hollow cylinder with a helical structure. An example of a helical virus that is a rigid rod is the tobacco mosaic virus (Figure 13.3). Another is bacteriophage M13.

Polyhedral viruses. Many animal, plant, and bacterial viruses are polyhedral viruses; that is, they are many-sided. The capsid of most polyhedral viruses is in the shape of an *icosahedron,* a regular polyhedron with 20 triangular faces and 12 corners (Figure 13.4a). The capsomeres of each face form an equilateral triangle. An example of a polyhedral virus in the shape of

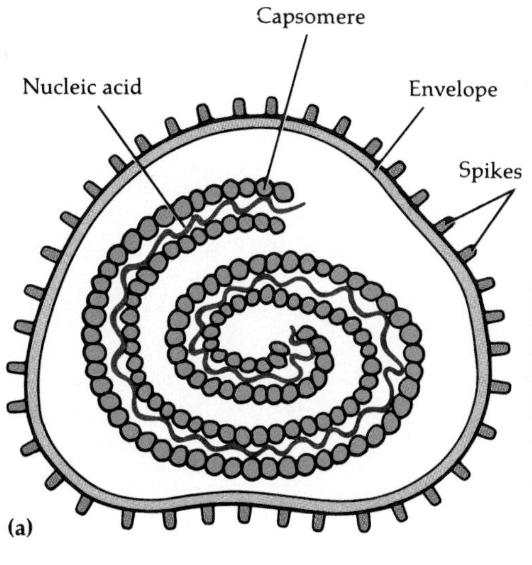

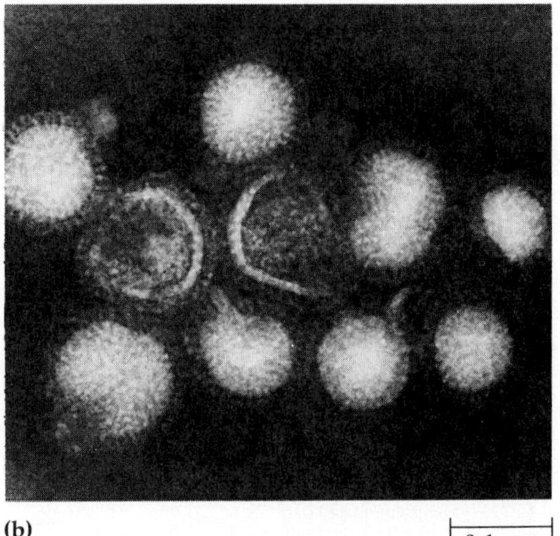

FIGURE 13.5 Morphology of an enveloped helical virus. **(a)** Diagram of an enveloped helical virus. **(b)** Electron micrograph of influenza viruses. Note the halo of spikes projecting from the outer surface of each envelope (see Chapter 24).

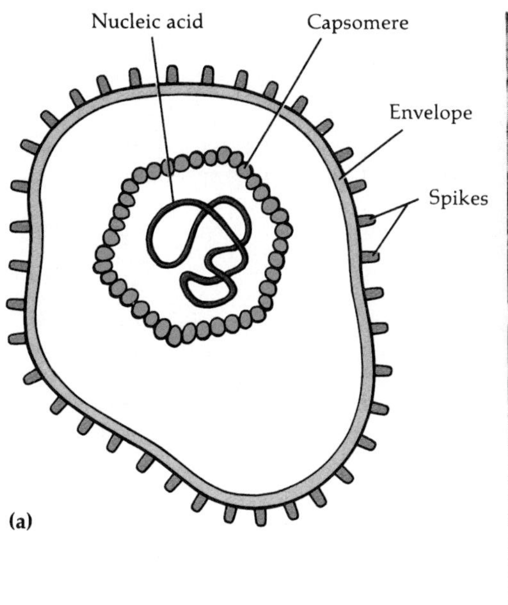

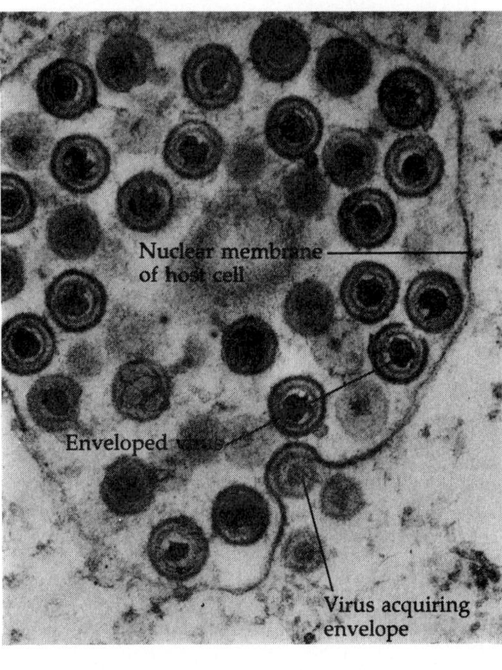

FIGURE 13.6 Morphology of an enveloped polyhedral (icosahedral) virus. **(a)** Diagram of an enveloped polyhedral virus. **(b)** Electron micrograph of a group of herpes simplex viruses. To the lower right, a virus particle is acquiring its envelope as it buds out through the nuclear membrane of a host cell.

an icosahedron is the adenovirus (Figure 13.4b). Another icosahedral virus is the poliovirus.

Enveloped viruses. As we noted earlier, the capsid of some viruses is covered by an envelope. Enveloped viruses are roughly spherical but highly pleomorphic (variable in shape) because the envelope is not rigid. When helical or polyhedral viruses are enclosed by envelopes, they are called *enveloped helical* and *enveloped polyhedral viruses*. An example of an enveloped helical virus is the influenza virus (Figure 13.5). An example of an enveloped polyhedral (icosahedral) virus is the herpes simplex virus (Figure 13.6).

Complex viruses. Some viruses, particularly bacterial viruses, have very complicated structures and are called **complex viruses.** Examples of complex viruses are poxviruses, which do not contain clearly identifiable capsids but have several coats around the nucleic acid (Figure 13.7a). Certain bacteriophages have capsids to which additional structures are attached (Figure 13.7b and c). If you take a close look at the bacteriophages shown in Figure 13.7b and c, you will note that the capsid (head) is polyhedral and the tail is helical. The head contains the nucleic acid. Later in this chapter, we will discuss the functions of the other structures, such as the tail sheath, tail fibers, plate, and pin.

TABLE 13.2 Classification of Animal Viruses by Nucleic Acid Replication

NUCLEIC ACID CATEGORY	VIRAL GROUP AND SPECIFIC EXAMPLES	MORPHOLOGICAL CLASS	DIMENSIONS OF CAPSID (DIAMETER IN NM)	CLINICAL OR SPECIAL FEATURES
Single-stranded DNA viruses	Parvoviruses	Naked polyhedral	18–25	Very small viruses, most of which depend on coinfection with adenoviruses for growth; probably infect only rats, mice, and hamsters.
Double-stranded DNA viruses	Papovaviruses (papilloma, polyoma, simian)	Naked polyhedral	40–57	Small viruses that induce tumors; the human wart virus (papilloma) and certain viruses that produce cancer in animals (polyoma and simian) belong to this family. Refer to Chapters 21 and 26.
	Adenoviruses	Naked polyhedral (Figure 13.4b)	70–80	Medium-sized viruses that cause various respiratory infections in humans; some cause tumors in animals.
	Herpesviruses (herpes simplex [HHV 1 and HHV 2], herpes zoster [HHV 3])	Enveloped polyhedral (Figure 13.8a)	150–250	Medium-sized viruses that cause various human diseases, such as fever blisters, chickenpox, shingles, and infectious mononucleosis; implicated in a type of human cancer called Burkitt's lymphoma. Refer to Chapters 21, 23, and 26.
	Poxviruses (variola, cowpox, vaccinia)	Enveloped complex	200–350	Very large, complex, brick-shaped viruses that cause diseases such as smallpox (variola), molluscum contagiosum (wartlike skin lesion), cowpox, and vaccinia; vaccinia virus gives immunity to smallpox. Refer to Chapter 21.
Sense strand RNA viruses	Picornaviruses (poliovirus, rhinovirus, hepatitis A virus)	Naked polyhedral	18–38	Smallest RNA-containing viruses; at least 70 human enteroviruses are known, including the polio-, coxsackie-, hepatitis A, and echoviruses; more than 100 rhinoviruses exist and are the most common cause of colds. Refer to Chapters 22, 23, 24, and 25.
	Togaviruses (arboviruses, alpha virus, flavivirus, rubella, pestivirus, hepatitis C virus)	Enveloped polyhedral	40–60	Included are many viruses transmitted by arthropods (arboviruses); diseases include eastern equine encephalitis (EEE), St. Louis encephalitis (SLE), yellow fever, and dengue. Rubella virus, which is not an arbovirus, is transmitted by the respiratory route. Refer to Chapters 21, 22, 23, and 26.
Antisense strand RNA viruses	Orthomyxoviruses (influenza A, B, C)	Enveloped helical (Figure 13.9a)	80–200	Medium-sized viruses with a spiked envelope; have the ability to agglutinate red blood cells; cause influenza. Refer to Chapter 24.
	Paramyxoviruses (measles, mumps)	Enveloped helical	150–300	Morphologically similar to orthomyxoviruses, but generally larger; cause parainfluenza, measles, mumps. Refer to Chapters 21 and 25.
	Coronaviruses	Enveloped helical	80–130	Associated with upper respiratory tract infections and the common cold. Refer to Chapter 24.
	Rhabdoviruses (rabies)	Enveloped helical (Figure 22.18)	70 × 180	Bullet-shaped viruses with a spiked envelope; cause rabies and Newcastle disease of chickens. Refer to Chapter 22.

(Continued)

TABLE 13.2 *(Continued)*

NUCLEIC ACID CATEGORY	VIRAL GROUP AND SPECIFIC EXAMPLES	MORPHOLOGICAL CLASS	DIMENSIONS OF CAPSID (DIAMETER IN NM)	CLINICAL OR SPECIAL FEATURES
	Arenaviruses	Enveloped helical	50–300	Viruses contain RNA-containing granules; cause slow viral infections and hemorrhagic fever.
	Filoviruses	Enveloped helical	80 × 14,000	Ebola and Marburg viruses cause hemorrhagic fever. Others cause asymptomatic infection in primate handlers.
Double-stranded RNA viruses	Reoviruses	Naked polyhedral	60–80	Involved in mild respiratory infections and infantile gastroenteritis; causes Colorado tick fever.
Reverse transcription viruses	Retroviruses (lentiviruses)	Enveloped helical (Figure 13.9c)	100–120	Includes all RNA tumor viruses. Double-stranded RNA viruses; cause leukemia and tumors in animals; cause AIDS. Refer to Chapter 19.
	Hepadnaviruses (hepatitis B)	Enveloped polyhedral	42	Double-stranded DNA virus; cause hepatitis; cause liver tumors. Refer to Chapter 25.

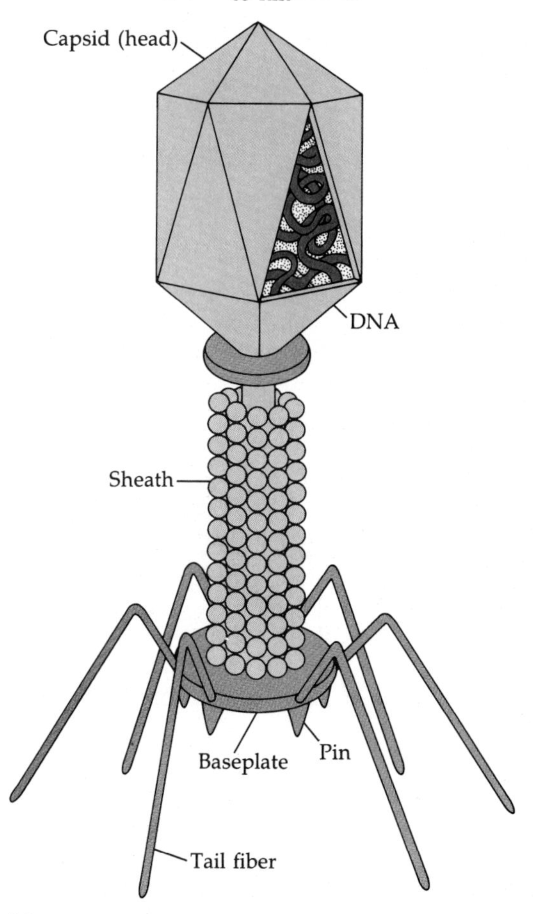

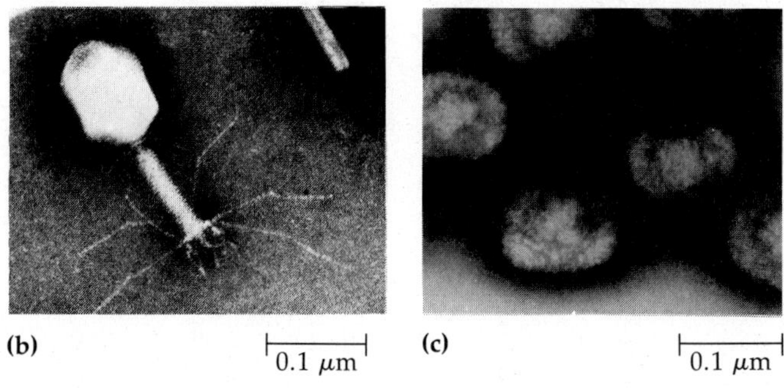

FIGURE 13.7 Morphology of complex viruses. **(a)** Diagram of a T-even bacteriophage. **(b)** T4 bacteriophage. **(c)** Electron micrograph of vaccinia virus, a strain of poxvirus used to vaccinate humans against smallpox.

Taxonomy of Viruses

For our purposes in this chapter, we are grouping viruses as animal viruses, bacterial viruses, or plant viruses, according to host range. This classification is convenient, but it is not scientifically acceptable.

In the oldest system, an animal virus was classified according to the organs that were affected by the diseases it caused. This is a classification by symptomatology. In this system, poliovirus might be considered a nervous system virus; herpesvirus, a skin virus; and

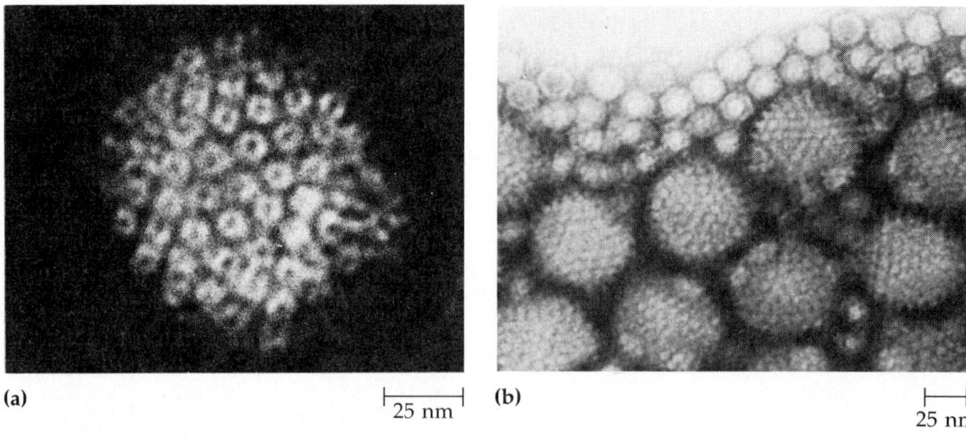

FIGURE 13.8 DNA-containing viruses. **(a)** Negatively stained capsid of a herpesvirus. The individual capsomeres are clearly visible. **(b)** Negatively stained viruses that have been concentrated in a centrifuge gradient. The larger capsids are an adenovirus, and the smaller "adeno-associated particles" are a parvovirus.

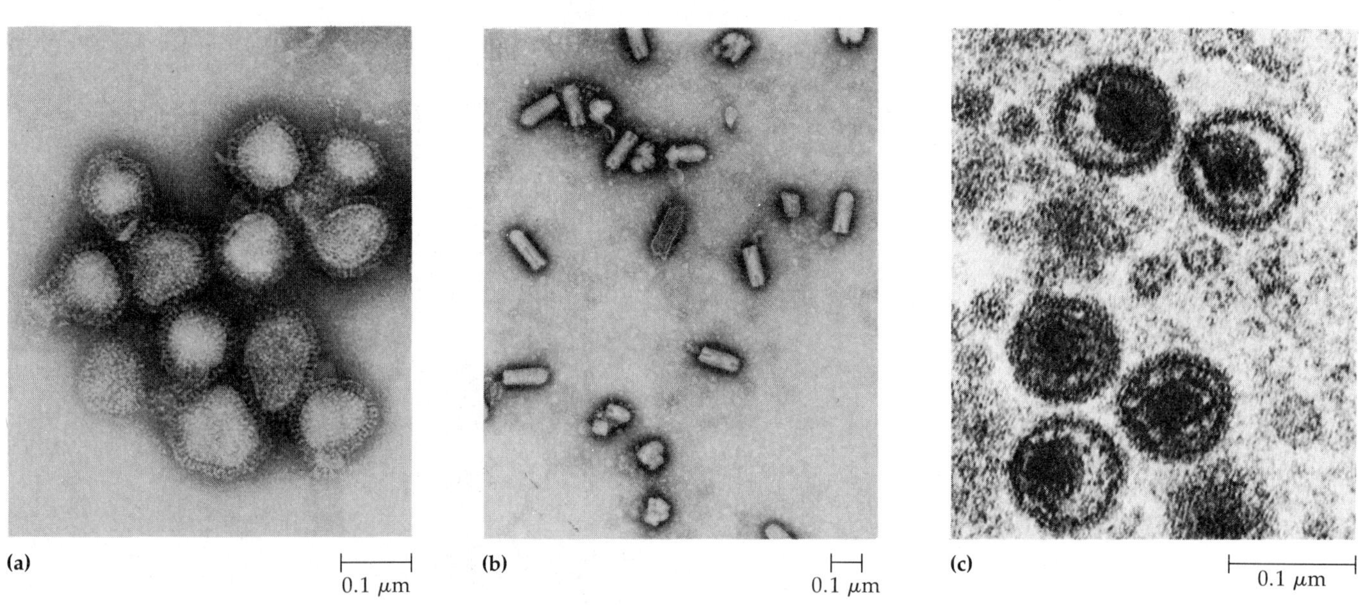

FIGURE 13.9 RNA-containing viruses. **(a)** Envelopes of A2 influenza virus. The spikes projecting from each envelope are visible. **(b)** Particles of vesicular stomatitis virus (VSV). This is a rhabdovirus, a rod-shaped virus similar to rabies. **(c)** Mouse mammary tumor virus, a B-type retrovirus.

hepatitis virus, a liver virus. However, as we learned that the same virus can cause more than one disease, this scheme became unsatisfactory.

In the past few decades, hundreds of viruses have been isolated from plants, animals, and humans. As the list grows, the problem of viral classification becomes more and more complex. Present classification systems are based on such factors as the type of nucleic acid, the mechanism of nucleic acid replication, morphological class, presence of an envelope, size of capsid, and number of capsomeres. A summary of the classification of animal viruses based on these morphological, biochemical, and physical properties is presented in Table 13.2. Other classification schemes also take into account susceptibility to physical and chemical agents, immunologic properties, site of multiplication (nucleus or cytoplasm), and natural methods of transmission.

Figure 13.8 shows electron micrographs of DNA-containing animal viruses, and Figure 13.9 shows electron micrographs of RNA-containing animal viruses.

Isolation, Cultivation, and Identification of Viruses

The fact that viruses cannot multiply outside a living host cell complicates their detection, enumeration, and

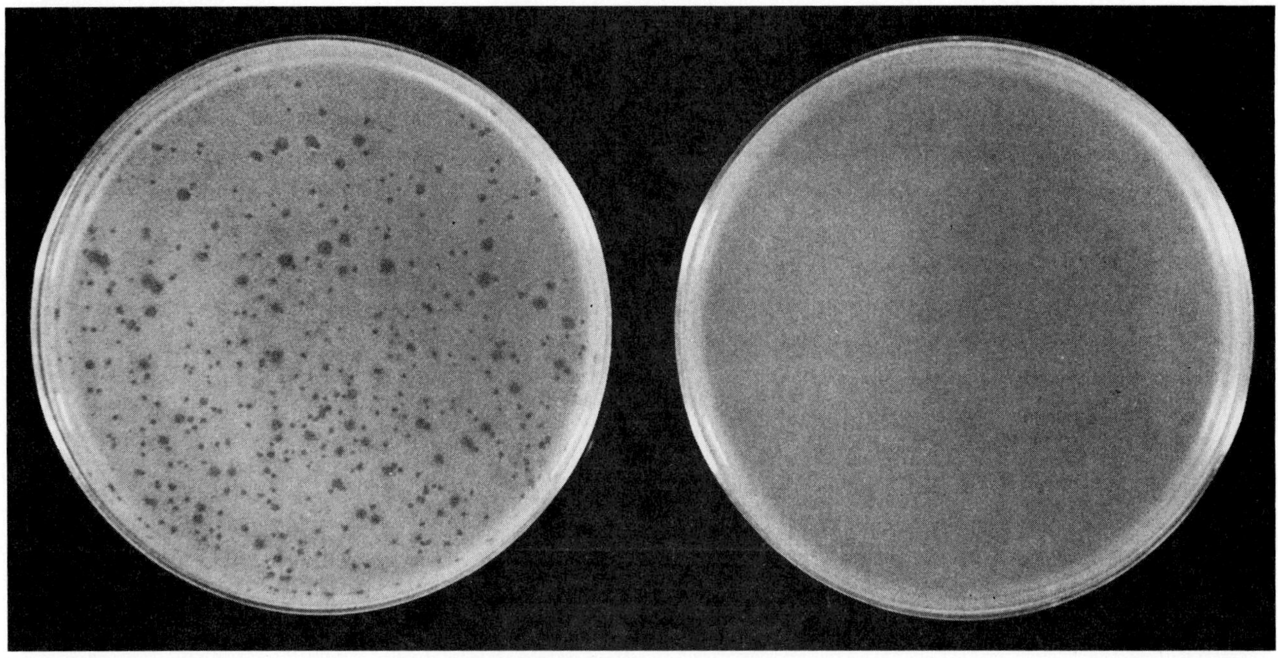

FIGURE 13.10 Viral plaque formed by bacteriophage. In the Petri plate on the left, clear viral plaques of varying sizes have been formed by the bacteriophage λ on a lawn of *E. coli.* For comparison, the Petri plate on the right contains a bacterial culture without phage.

identification. It is necessary to provide viruses with living cells instead of a fairly simple chemical medium. Living plants and animals are difficult and expensive to maintain, and disease-causing viruses that grow only in higher primates and human hosts cause additional complications. However, viruses that use bacterial cells as a host (bacteriophages) are rather easily grown on bacterial cultures. This is one reason why so much of our understanding of viral multiplication has come from bacteriophages.

GROWTH OF BACTERIOPHAGES IN THE LABORATORY

Bacteriophages can be grown either in suspensions of bacteria in liquid media or in bacterial cultures on solid media. The use of solid media makes possible the *plaque method* for detecting and counting viruses. Only simple materials and equipment are needed for this procedure, which was first developed for use with bacteriophages. A sample of bacteriophage is mixed with host bacteria and melted agar. The agar containing bacteriophage and host bacteria is then poured into a Petri plate containing a hardened layer of agar growth medium. The virus–bacteria mixture solidifies into a thin top layer that contains a layer of bacteria approximately one cell thick. Each virus infects a bacterium, multiplies, and releases several hundred new viruses. These newly produced viruses infect other bacteria in the immediate vicinity, and more new viruses are produced. Following several virus multiplication cycles, all the bacteria in the area surrounding the original virus are destroyed. This produces a number of clearings, or **plaques,** visible against a "lawn" of bacterial growth on the surface of the agar (Figure 13.10). While the plaques form, uninfected bacteria elsewhere in the Petri plate multiply rapidly and produce a turbid background.

Each plaque theoretically corresponds to a single virus in the initial suspension. Because a single plaque can arise from more than one virion, and because some virions may not be infectious, the concentrations of viral suspensions measured by the number of plaques are usually given in terms of **plaque-forming units (pfu).**

GROWTH OF ANIMAL VIRUSES IN THE LABORATORY

In Living Animals

Some animal viruses can be cultured only in *living animals,* such as mice, rabbits, and guinea pigs. Most experiments to study the immune system's response to viral infections must also be performed in virally in-

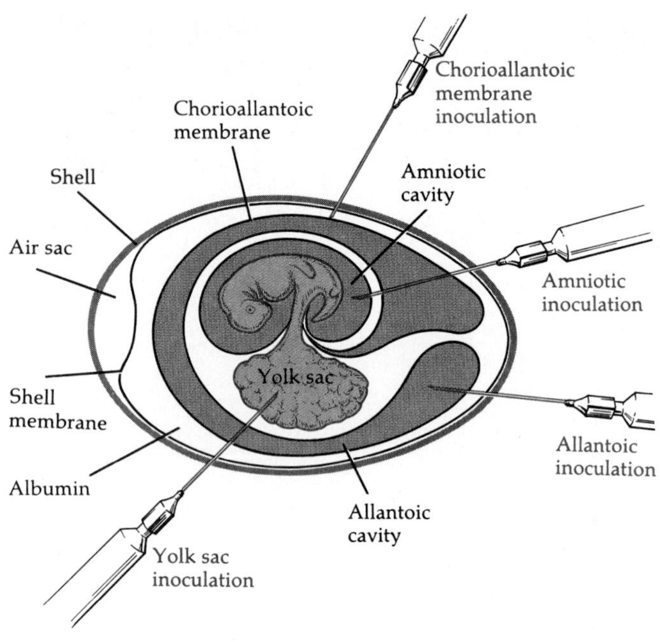

FIGURE 13.11 Inoculation of embryonated eggs. The injection site determines the membrane on which the viruses will grow.

fected live animals. Animal inoculation may be used as a diagnostic procedure for identifying and isolating a virus from a clinical specimen. After the animal is inoculated with the specimen, the animal is observed for signs of disease or is killed so that infected tissues can be examined for the virus.

Some human viruses cannot be grown in animals, or can be grown but do not cause disease. The lack of animal models for AIDS has slowed our understanding of the disease process and prevented experimentation with drugs that inhibit growth of the virus in vivo. Chimpanzees can be infected with one strain of human immunodeficiency virus (HIV-1), but since they do not show symptoms of the disease, they cannot be used to study effects of viral growth and disease treatments. AIDS vaccines are presently being tested in humans, but the disease progresses so slowly in humans that it can take years to determine the effectiveness of these vaccines. In 1986 simian AIDS (SAIDS), an immune deficiency disease of green monkeys, was reported, followed in 1987 by feline AIDS (FAIDS), an immune deficiency disease of domestic cats. These diseases are not caused by the HIV group, since viruses are generally host specific. They are, however, caused by closely related retroviruses, and the diseases develop within a few months, thus providing a model for study of viral growth in different tissues. In 1990 a way to infect mice with human AIDS was found when mice were genetically engineered to produce human T cells and human gamma globulin.

In Embryonated Eggs

If the virus will grow in an *embryonated egg*, this can be a fairly convenient and inexpensive form of animal host for many animal viruses (Figure 13.11). A hole is drilled in the shell of the embryonated egg, and a viral suspension or suspected virus-containing tissue is injected into the fluid of the egg. There are several membranes in the egg on which the virus can be made to grow, and the virus is injected into the proper location in the egg. Viral growth is signaled by the death of the embryo, by embryo cell damage, or by the formation on the membranes of the egg of typical pocks or lesions that result from viral growth. This method was once the most widely used method of viral isolation and growth and is still used to grow viruses for some vaccines. You may be asked if you are allergic to eggs before receiving a vaccination since egg proteins may be present in the viral vaccine preparations. (Allergy is discussed in Chapter 19.)

In Cell Culture

Cell cultures (sometimes called *tissue cultures,* although that is not the best term) have replaced embryonated eggs as growth media for many viruses. Cell cultures consist of cells grown in culture media in the laboratory. Because these cultures are generally rather homogeneous collections of cells and can be propagated and handled much like bacterial cultures, they are more convenient to work with than whole animals or embryonated eggs.

Cell culture lines are readily started by treatment of a slice of animal tissue with enzymes that separate the individual cells (Figure 13.12). These cells are suspended in a solution that provides the osmotic pres-

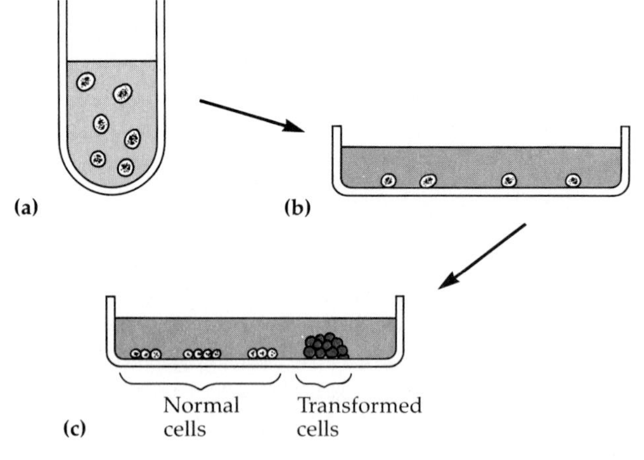

FIGURE 13.12 Cell cultures. (a) A tissue is treated with enzymes to separate the cells. **(b)** Cells are suspended in culture medium. **(c)** Normal cells or primary cell lines grow in a monolayer across the glass or plastic container. Transformed cells or continuous cell cultures do not grow in a monolayer.

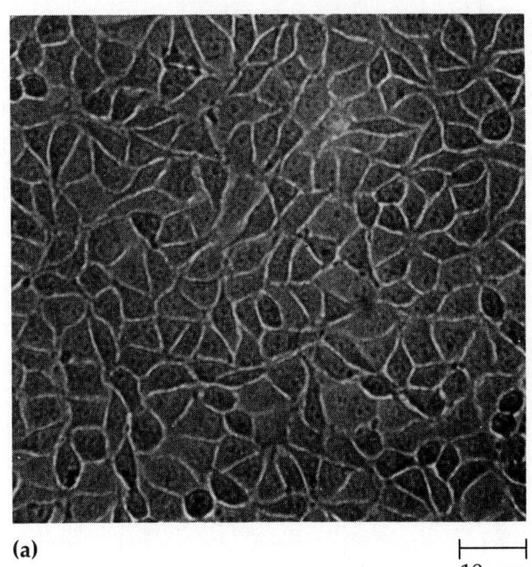

(a)

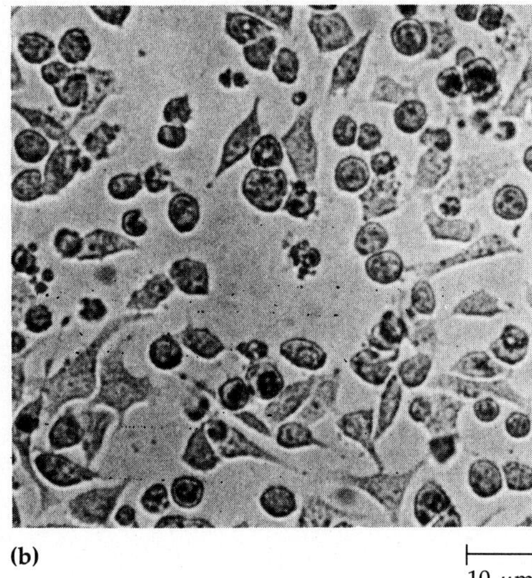

(b)

FIGURE 13.13 Cytopathic effect of viruses. **(a)** A monolayer of uninfected mouse L cells. **(b)** The same cells 24 hours after infection with VSV (see Figure 13.9). Note the separation and "rounding up" of the cells.

sure, nutrients, and growth factors needed for the cell to grow. The normal cells tend to adhere to the glass or plastic container and reproduce to form a monolayer. Transformed (or cancerous) cells do not form a single layer. Viruses infecting such a monolayer sometimes cause the cells of the monolayer to deteriorate as they multiply. This tissue deterioration is called **cytopathic effect (CPE)** and is illustrated in Figure 13.13. CPE can be detected and counted in much the same way as plaques caused by bacteriophages on a lawn of bacteria.

Primary cell lines, derived from tissue slices, tend to die out after only a few generations. Certain cell lines, called **diploid cell lines,** developed from human embryos can be maintained for about 100 generations and are widely used for culturing viruses that require a human host. Cell lines developed from embryonic human cells are used to culture rabies virus for a rabies vaccine called human diploid culture vaccine (see Chapter 22).

When viruses are routinely grown in a laboratory, **continuous cell lines** are used. These are "transformed" cells that can be maintained through an indefinite number of generations, and they are sometimes called "immortal" cell lines (see the discussion of transformation at the end of this chapter). One of these, the HeLa cell line, was isolated from the cancer of a woman who died in 1951. After years of laboratory cultivation, many such cell lines have lost almost all the original characteristics of the cell, but these changes have not interfered with the use of the cells for viral propagation. In spite of the success of cell culture in viral isolation and growth, there are still some viruses that have never been successfully cultivated in cell culture.

The idea of cell culture dates back to the end of the last century, but it was not a practical laboratory technique until the development of antibiotics in the years following World War II. A major problem with cell culture is that the cell lines must be kept free of microbial contamination. The maintenance of cell culture lines requires trained technicians with considerable experience working on a full-time basis. Because of these difficulties, most hospital laboratories and many state health laboratories do not isolate and identify viruses in clinical work. Instead, the tissue or serum samples are sent to central laboratories that specialize in this work.

VIRAL IDENTIFICATION

The identification of viral isolates is not an easy task. For one thing, viruses cannot be seen at all without the use of an electron microscope. Serological methods are the most commonly used means of identification. In these tests, the virus is detected and identified by its reaction with antibodies. Antibodies, which are proteins produced by animals in response to exposure to a virus, are highly specific for the virus that causes their formation. We will discuss antibodies in detail in Chapter 17 and a number of immunological tests for identifying viruses in Chapter 18. Later in this chapter, we will describe how viruses can be identified by observation of their effects on host cells.

Using modern biochemical methods, molecular biologists can identify and characterize purified viruses by isolating the particular nitrogenous base sequences of the virus's nucleic acid. For DNA viruses, restriction enzymes are used to cut DNA into specific nitrogenous base sequences, or fragments (see Chap-

AIDS: The Risk to Health Care Workers

With the advent of the AIDS epidemic, health care workers are understandably concerned about the risk of contracting AIDS after exposure to body fluids of infected patients. However, when precautions are observed, the risk to workers is very small, even for those treating AIDS patients.

Understanding the Risk

The first protection for health care workers is a clear understanding of how AIDS can (and cannot) be transmitted in the course of their work. To date, direct inoculation of infected material is the only proven method of transmission in the health care environment. Infected materials that can transmit AIDS are blood, semen, vaginal secretions, and breast milk. The most common route of transmission is through accidental needle sticks. However, inoculation is also possible if infected material contacts mucous membranes or a break in the health care worker's skin. There is no evidence of HIV transmission by aerosols, the fecal-oral route, mouth-to-mouth or casual contact, or via environmental surfaces such as floors, walls, chairs, and toilets. Saliva, tears, cerebrospinal fluid, amniotic fluid, and urine do not pose the same risk as other body fluids.

The number of health care workers who have become infected and the sources of their infections are being carefully monitored by the CDC. In the United States, by the end of 1990, 37 health care workers who denied other risk factors and who were exposed to blood from patients with HIV had contracted HIV infections, 29 of these following needlestick injuries. Six cases had extensive contact with blood of infected patients and did not observe routine barrier precautions. Additionally, two workers in industrial laboratories producing large quantities of highly concentrated HIV have been reported to have laboratory-acquired HIV infections.

In addition to monitoring all health care workers who have actually contracted HIV, the CDC and others also study workers who have been exposed to infected materials, to determine the magnitude of risk. The largest study followed 4000 health care workers who had been exposed to HIV, 1200 by needlesticks. The study found that the probability of infection following a needlestick injury with blood containing HIV is 0.4%. The probability of infection from mucous membrane exposure to blood is 10 times lower. The probability of acquiring hepatitis B from a needlestick injury with blood containing HBV is from 6 to 30%.

Precautions

In 1985, the CDC developed the strategy of "universal precautions," which should be followed in **all** health care settings. These are:

Gloves. Disposable gloves should be used for direct exposure to infected blood, other body fluids, and tissues. Double-gloving is recommended during invasive surgical procedures. Personnel should not work when they have open skin lesions, weeping dermatitis, or cutaneous wounds.

Gowns, Masks, and Goggles. Masks and protective eyewear are recommended when splashes are expected, such as during airway manipulation, endoscopy, and dental procedures, and in the laboratory.

Needles. To minimize the risk of needle sticks, needles should not be resheathed and should be put in a puncture-proof container for sterilization and disposal.

Disinfection. Routine housekeeping in health care settings should include washing floors, walls, and other areas not normally associated with disease transmissions with 1:100 dilution of household bleach. A 1:10 dilution is recommended for disinfecting a spill.

Prophylactic Treatment After Exposure

Avoidance of exposure is the health care worker's first line of defense. Admittedly, however, accidental exposure cannot always be prevented. Promising new evidence points to the possibility that exposed persons may be able to reduce their risk by prophylactic use of AZT. The National Institutes of Health and the University of California at San Francisco are developing recommendations for use of AZT.

The Risk to Patients

Transmission of HBV and HIV from health care workers to patients during invasive dental procedures (tooth extractions) has been documented. Restrictions on the procedures that can be performed by health care workers with HIV infection have also been considered by the American Medical Association, the American Dental Association, the CDC, and other organizations. Their recommendations state that the risk of HIV transmission from health care workers to patients is greatest during invasive procedures and that decisions regarding restriction of patient care by infected workers who perform such procedures should be made on an individual basis.

Adapted from *MMWR* 38/(S-2) (5/12/89); *MMWR* 39(S-6) (6/23/89); *MMWR* 39:489 (7/27/90).

ter 10). Fragments of closely related viruses have similar electrophoretic patterns. Using more sophisticated methods, researchers can even determine detailed nitrogenous base sequences of entire viral genomes. However, these laboratory research methods are not practical for the routine identification of clinical samples.

Viral Multiplication

The nucleic acid in a virion contains only a few of the genes needed for the synthesis of new viruses. These include genes for the virion's structural components, such as the capsid proteins, and genes for a few of the enzymes used in the viral life cycle. These enzymes are synthesized and functional only when the virus is within the host cell. Viral enzymes are almost entirely concerned with replicating or processing viral nucleic acid. Enzymes needed for protein synthesis, ribosomes, tRNA, and energy production are supplied by the host cell and used for synthesis of virus proteins, including viral enzymes. Although the smallest naked virions do not contain any preformed enzymes, the larger virions may contain one or a few enzymes, which usually function in helping the virus to penetrate the host cell or to replicate its own nucleic acid.

Thus, for a virus to multiply, it must invade a host cell and take over the host's metabolic machinery. A single virus can give rise to several or even thousands of similar viruses in a single host cell. This process drastically changes the host cell and often causes its death.

MULTIPLICATION OF BACTERIOPHAGES

Although the means by which a virus enters and exits a host cell may vary, the basic mechanism of viral multiplication is similar for animal viruses, plant viruses, and bacteriophages. The best-understood viral life cycles are those of the bacteriophages (phages). Because the so-called *T-even bacteriophages* (T2, T4, and T6) have been studied most extensively, we will first describe the multiplication of T-even bacteriophages in their host, *E. coli.*

T-Even Bacteriophages

The T-even bacteriophages have large, complex, naked virions, whose characteristic head-and-tail structure is shown in Figures 13.7a and b and 13.14. The length of DNA contained in these bacteriophages is only about 6% of that contained in *E. coli;* the bacteriophage has enough DNA for over 100 genes. The multiplication cycle of these phages, like that of all viruses, can be divided into several distinct stages—attachment, penetration, biosynthesis of viral components, maturation, and release.

Attachment of phage to host cell. After a chance collision between phage particles and bacteria, **attachment,** or **adsorption,** occurs. During this process, an attachment site on the virus attaches to a complementary receptor site on the bacterial cell. This attachment is a chemical interaction in which weak bonds are formed between the attachment and receptor sites. T-even bacteriophages use fibers at the end of the tail as attachment sites. The complementary receptor sites are on the bacterial cell wall (Figure 13.14a). Other phages attach to flagella or fimbriae.

Penetration. After attachment, the T-even bacteriophage injects its DNA (nucleic acid) into the bacterium. To do this, the bacteriophage's tail releases an enzyme, *phage lysozyme,* which breaks down a portion of the bacterial cell wall. During the process of **penetration,** the tail sheath of the phage contracts, and the tail core is driven through the cell wall. When the tip of the core reaches the plasma membrane, the DNA from the bacteriophage's head passes through the tail core and through the plasma membrane and enters the bacterial cell. The capsid of most bacteriophages remains outside the bacterial cell (Figure 13.14b).

Biosynthesis of viral components. Once the bacteriophage DNA has reached the cytoplasm of the host cell, biosynthesis of viral nucleic acid and protein occurs. Host protein synthesis is stopped by (a) virus-induced degradation of host DNA, (b) virus proteins that interfere with transcription, or (c) repression of translation.

Initially, the phage uses the host cell's nucleotides and several of its enzymes to synthesize many copies of phage DNA. Soon after, biosynthesis of viral proteins begins. Any RNA transcribed in the cell is mRNA transcribed from phage DNA for biosynthesis of phage enzymes and capsid protein. The host cell's ribosomes, enzymes, and amino acids are used for translation (Figure 13.14c). There are genetic controls that regulate when different regions of phage DNA are transcribed into mRNA during the multiplication cycle. For example, there are early messages that are translated into early phage proteins, the enzymes used in the synthesis of phage DNA. Also, there are late messages that are translated into late phage proteins for the synthesis of capsid proteins. This control mechanism is mediated by RNA polymerase.

For several minutes following infection, complete phages cannot be found in the host cell. Only separate components can be detected—DNA and proteins. The period during viral multiplication when complete, infective virions are not yet present is called the **eclipse period.**

Maturation. In the next sequence of events, **maturation** occurs. In this process, bacteriophage DNA and

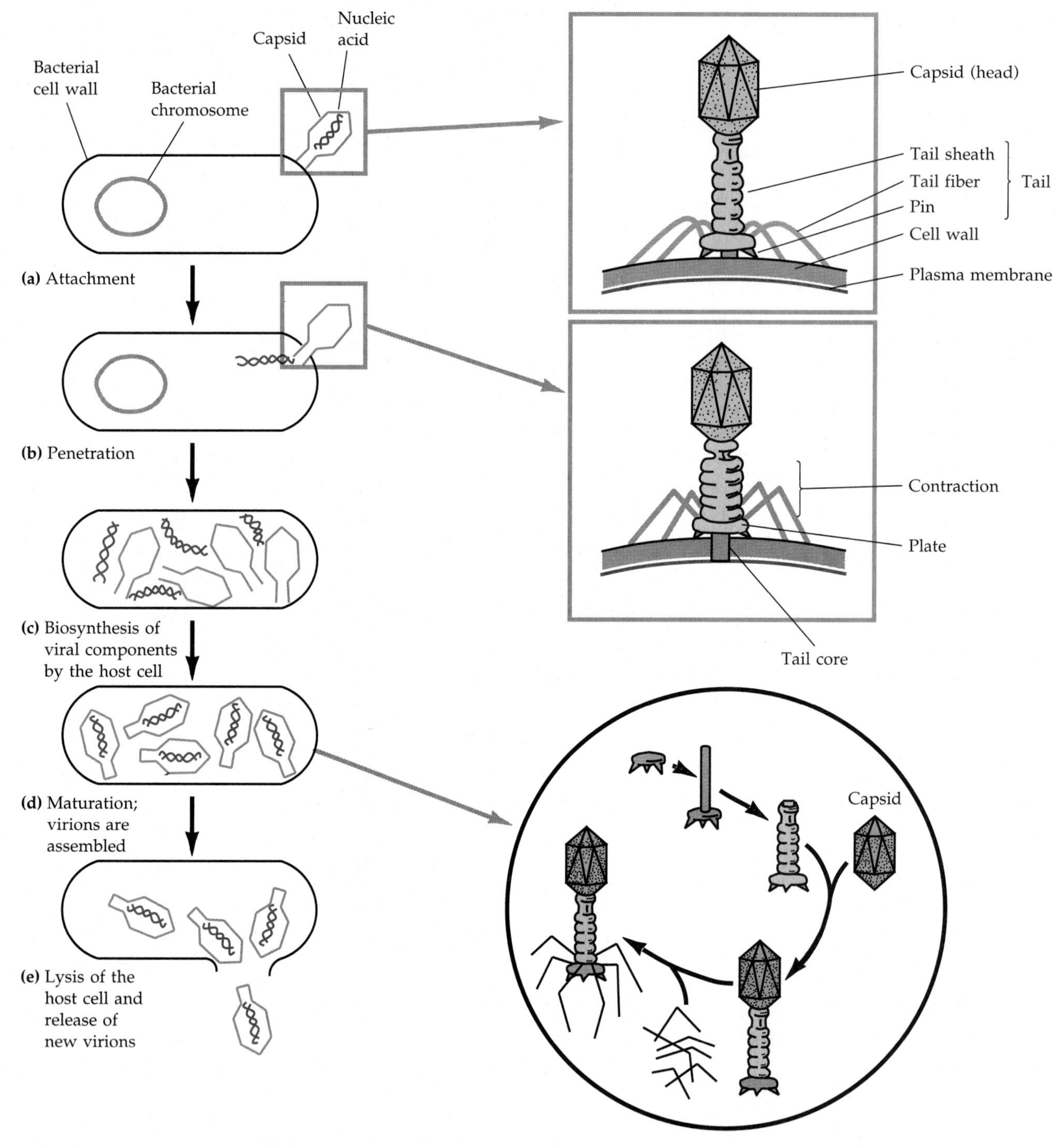

FIGURE 13.14 Multiplication cycle of T-even bacteriophage.

capsids are assembled into complete virions. The assembly process is guided by the products of certain viral genes in a step-by-step sequence. The phage heads and tails are separately assembled from protein subunits, the head is packaged with phage DNA, and the tail is attached (Figure 13.14d). (For many simpler viruses, the nucleic acid and capsid proteins assemble spontaneously to form virions, without the intervention of other phage gene products.)

Release. The final stage of viral multiplication is the **release** of virions from the host cell. The term **lysis** is generally used for this stage in the multiplication of T-even phages because in this case the plasma membrane actually breaks open (lyses). Lysozyme, whose code is provided by a phage gene, is synthesized within the cell. This enzyme causes a breakdown of the bacterial cell wall, and the newly produced bacteriophages are released from the host cell (Figure 13.14e).

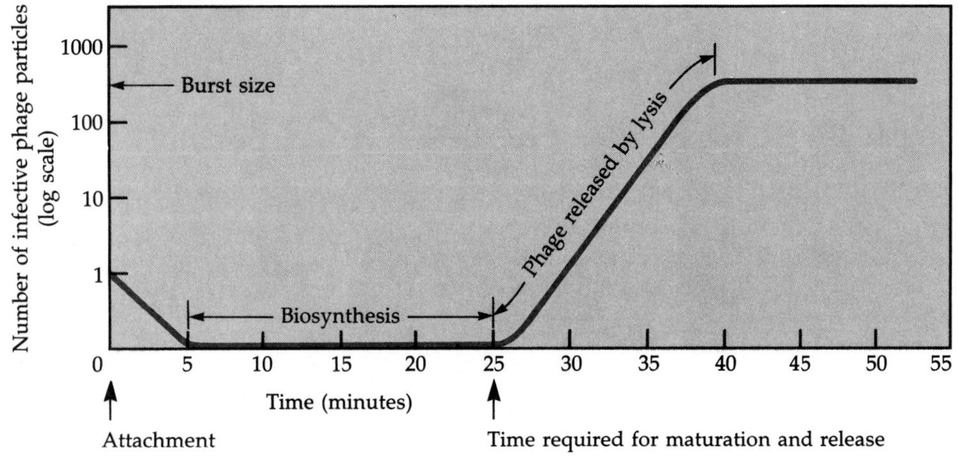

FIGURE 13.15 A bacteriophage one-step growth curve. No new infective phage particles are found in a culture until after biosynthesis and maturation have taken place.

The released bacteriophages infect other susceptible cells in the area, and the viral multiplication cycle is repeated within those cells.

The time elapsed from phage attachment to release is known as **burst time** and averages from 20 to 40 minutes. The number of newly synthesized phage particles released from a single cell is referred to as **burst size** and usually ranges from about 50 to 200.

The various stages involved in the multiplication of phages can be demonstrated experimentally in what is known as a *one-step growth experiment* (Figure 13.15). In this procedure, a phage suspension is diluted until a sample containing only a few phage particles can be obtained. These particles are then introduced into a culture of host cells. Periodically, samples of phage particles are removed from the culture and inoculated onto a plate culture of susceptible host cells; the plaque method is used to determine the number of infective phage particles on this culture (see Figure 13.10). A few minutes after attachment, there are no infective particles present. However, phage nucleic acid is found inside the infected cells, and capsid proteins will be synthesized. The time required for maturation is the interval between the appearance of phage nucleic acid and the synthesis of mature phages. After a few minutes, the number of infective phage particles found in subcultures begins to rise. The burst size is determined once the number of infective phage particles remains constant, indicating that no further phage multiplication will occur.

Lysogeny

In the sequence of events just described for the multiplication of T-even bacteriophages, the release of the phages causes lysis and death of the host cell. Thus, such a sequence is called a **lytic cycle.** Some phages can either proceed through a lytic cycle or incorporate their DNA into the host cell's DNA. In the latter state, the phage remains latent and does not cause lysis of the host cell. Such a state is called **lysogeny,** and such phages are called *lysogenic phages* or *temperate phages.* The participating bacterial host cells are known as *lysogenic cells.*

Let us go through the steps in lysogeny for bacteriophage λ (lambda), a well-studied temperate phage (Figure 13.16). Upon penetration into an *E. coli* cell, the linear phage DNA forms a circle. This circle can multiply and be transcribed, leading to the production of new phage and to cell lysis (lytic cycle). Alternatively, the circle can recombine with and become part of the circular bacterial DNA (lysogenic cycle). The inserted phage DNA is now called a **prophage.** Most of the prophage genes are repressed by two repressor proteins that are the products of phage genes. These repressors stop transcription of all the other phage genes by binding to operators. Thus, the phage genes that would otherwise direct the synthesis and release of new virions are turned off, in much the same way that the genes of the *E. coli lac* operon are turned off by the *lac* repressor (see Chapter 8).

Every time the host cell's machinery replicates the bacterial chromosome, it also replicates the prophage DNA. The prophage remains latent within the progeny cells. However, a rare spontaneous event or the action of ultraviolet light or certain chemicals can lead to the excision (popping-out) of the phage DNA—and initiation of the lytic cycle.

There are three important outcomes of lysogeny. First, the lysogenic cells are immune to reinfection by the same phage. (However, the host cell is not immune to infection by other phage types.) The second outcome of lysogeny is that the host cell may exhibit new properties. For example, the bacterium *Corynebacterium diphtheriae,* which causes diphtheria, is a pathogen whose disease-producing properties are re-

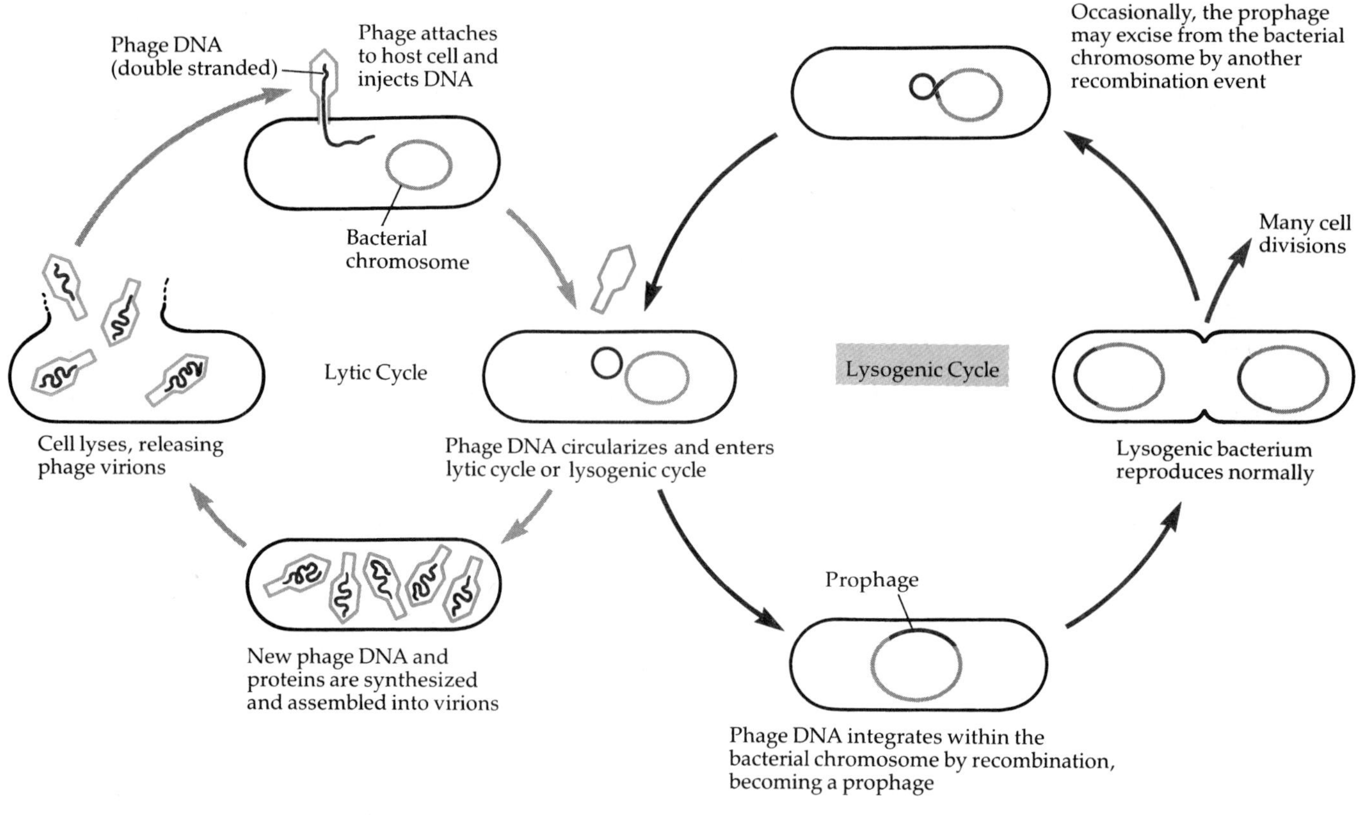
Phage DNA
(double stranded)
Phage attaches
to host cell and
injects DNA
Bacterial
chromosome
Occasionally, the prophage
may excise from the bacterial
chromosome by another
recombination event
Many cell
divisions
Lytic Cycle
Lysogenic Cycle
Cell lyses, releasing
phage virions
Phage DNA circularizes and enters
lytic cycle or lysogenic cycle
Lysogenic bacterium
reproduces normally
Prophage
New phage DNA and
proteins are synthesized
and assembled into virions
Phage DNA integrates within the
bacterial chromosome by recombination,
becoming a prophage

FIGURE 13.16 Lysogenic cycle of bacteriophage λ in *E. coli.*

lated to the synthesis of a toxin. The organism is capable of producing toxin only when it carries a temperate phage because the prophage carries the gene coding for the toxin. As another example, only streptococci carrying a temperate phage are capable of producing the toxin associated with scarlet fever. The toxin produced by *Clostridium botulinum,* which causes botulism, is encoded by a prophage gene.

The third possible outcome of lysogeny is **specialized transduction.** Recall from Chapter 8 that bacterial genes can be picked up in a phage coat and transferred to another bacterium in a process called generalized transduction (see Figure 8.26). Any bacterial genes can be transferred by generalized transduction because the host chromosome is broken down into fragments, any of which can be packaged into a phage coat. In specialized transduction, however, only certain bacterial genes can be transferred.

Specialized transduction is mediated by a lysogenic phage, which packages bacterial DNA *along with* its own DNA in the same capsid. When a prophage is excised from the host chromosome, adjacent genes from either side may remain attached to the phage DNA. The genes that remain attached to bacteriophage λ are genes for galactose utilization and biotin synthesis. A phage particle newly synthesized from this DNA may thus contain one or more of the bacterial genes along with phage DNA. Upon lysogenizing a new host cell, the phage can confer new characteristics to the cell; a cell infected by bacteriophage λ, for example, could gain the ability to metabolize galactose or synthesize biotin. In Figure 13.17, bacteriophage λ has picked up the *gal* gene for galactose fermentation from its galactose-positive host. The phage carries this gene to a galactose-negative cell, which can become galactose-positive.

Certain animal viruses are able to undergo processes very similar to lysogeny. Animal viruses that can remain latent in cells for long periods without multiplying or causing disease may become inserted in a host chromosome or remain separate from host DNA in a repressed state (as some lysogenic phages). Cancer-causing viruses may also be lysogenic, as will be discussed later in this chapter.

Multiplication of Animal Viruses

The multiplication of animal viruses follows the basic pattern of bacteriophage multiplication but has several notable differences. Animal viruses differ from phages

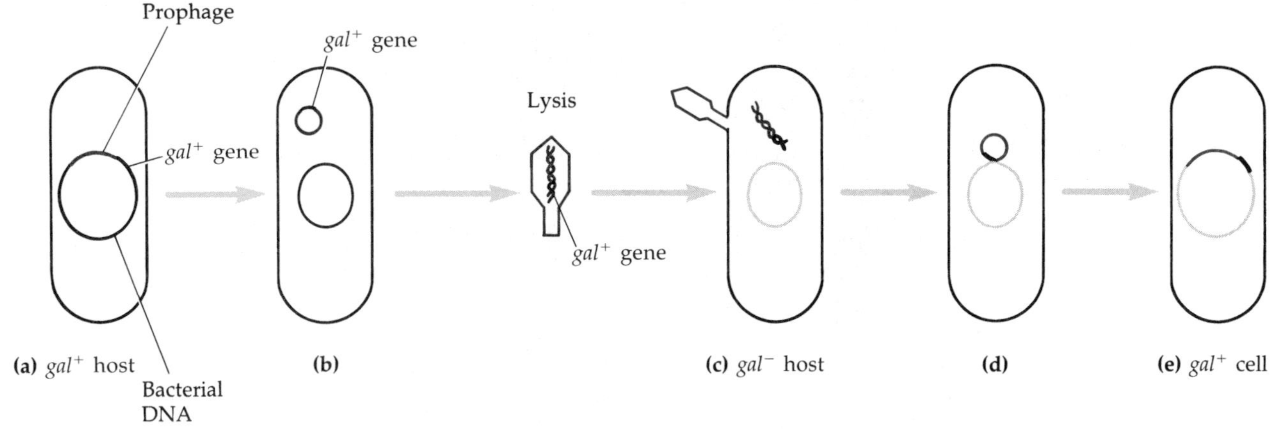

FIGURE 13.17 Specialized transduction. When a prophage is excised from its host chromosome, it can take with it a bit of the adjacent DNA from the bacterial chromosome. **(a)** Prophage in galactose-utilizing *E. coli.* **(b)** Excised phage genome carries the adjacent host gene. **(c)** Phage carries the *gal*$^+$ gene from the original host's chromosome to its new (*gal*$^-$) host. **(d)** Along with the prophage, the bacterial *gal*$^+$ gene becomes integrated into the new host's DNA. **(e)** Lysogenic cell can now metabolize galactose.

in their mechanism of entering the host cell. Also, once the virus is inside, the synthesis and assembly of the new viral components are somewhat different, partly because of the differences between procaryotic cells and eucaryotic cells. Animal viruses may have certain types of enzymes not found in phages. Finally, the mechanisms of maturation and release, and the effects on the host cell, differ in animal viruses and phages.

In the following discussion of the multiplication of animal viruses, we will first consider the processes that are shared by both DNA- and RNA-containing animal viruses. These processes are attachment, penetration, and uncoating. Then we will examine how DNA- and RNA-containing viruses differ with respect to their processes of biosynthesis and release from host cells.

Attachment

Like bacteriophages, animal viruses have attachment sites that attach to complementary receptor sites on the host cell's surface. However, the receptor sites of animal cells are proteins and glycoproteins of the plasma membrane (Figure 13.18a). Moreover, animal viruses do not possess appendages like the tail fibers of some bacteriophages. The attachment sites of animal viruses are distributed over the surface of the virus. The sites themselves vary from one group of viruses to another. In adenoviruses, which are naked icosahedral viruses, the attachment sites are small fibers at the corners of the icosahedron (see Figure 13.4b). In many of the enveloped viruses, such as the orthomyxoviruses, the attachment sites are spikes located on the surface of the envelope (see Figure 13.5b). As soon as one spike attaches to a host receptor, additional receptor sites on the same cell migrate to the virus. Attachment is completed when many sites are bound. Moreover, attachment is enhanced by the presence of cations.

Receptor sites are inherited characteristics of the host. Consequently, the receptor for a particular virus can vary from person to person. This could account for the individual differences in susceptibility to a particular virus. Understanding the nature of attachment can lead to the development of drugs that prevent viral infections. Monoclonal antibodies (discussed in Chapter 17) that combine with a virus's attachment site or the cell's receptor site may soon be used to treat some viral infections.

Penetration

Following attachment, penetration occurs. Penetration of enveloped animal viruses occurs by **endocytosis.** Endocytosis is an active cellular process by which nutrients and other molecules are brought into a cell. A cell's plasma membrane continuously folds inward to form vesicles. These vesicles contain elements that originate outside the cell and are brought into the interior of the cell to be digested. If a virion attaches to a small outfolding (called a microvillus) on the plasma membrane of a potential host cell, the host cell will enfold the virion into a fold of plasma membrane, forming a vesicle (Figure 13.18b). Once the virion is enclosed within the vesicle, its viral envelope is destroyed. The capsid is digested when the cell attempts

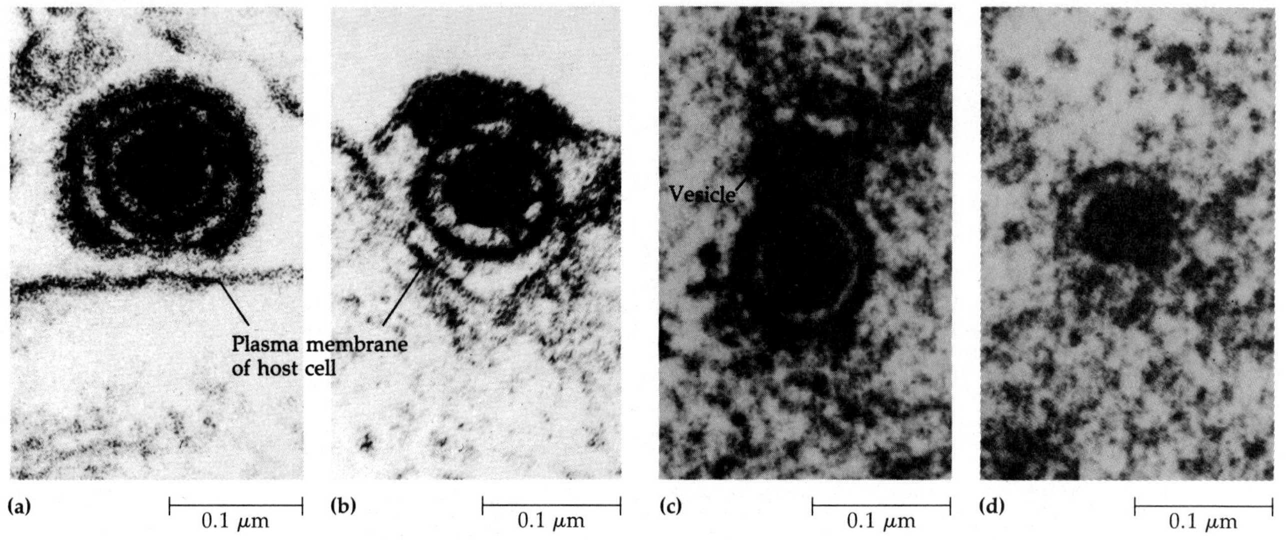

FIGURE 13.18 Entry of herpes simplex virus into an animal cell. **(a)** Attachment of the viral envelope to the plasma membrane. **(b)** Formation of a vesicle around the virus, which results in loss of the envelope. **(c)** Unenveloped capsid entering the cytoplasm of the cell from the vesicle. **(d)** Digestion of the capsid leaves only the nucleic acid core.

to digest the vesicle's contents. The viral nucleic acid is then released into the cytoplasm of the host cell (Figure 13.18c).

An alternative method of penetration called **fusion** has been hypothesized—the viral envelope fuses with the plasma membrane and releases the capsid into the host cell's cytoplasm.

Uncoating

Uncoating is the separation of the viral nucleic acid from its protein coat. It is a poorly understood process, which apparently varies with the type of virus. Uncoating occurs only in animal viruses; recall that phage DNA is inserted directly into the host cell. Some animal viruses accomplish uncoating by the action of lysosomal enzymes contained inside phagocytic vacuoles and coated vesicles of the host cell. These enzymes degrade the proteins of the viral capsid. The uncoating of poxviruses is completed by a specific enzyme encoded by the viral DNA and synthesized soon after infection. For other viruses, uncoating appears to be exclusively caused by enzymes in the host cell cytoplasm (Figure 13.18d). For at least one virus, the poliovirus, uncoating seems to begin while the virus is still attached to the host cell's plasma membrane.

Biosynthesis for DNA-Containing Viruses

Generally, DNA-containing viruses replicate their DNA in the nucleus of the host cell by using viral enzymes, and they synthesize their capsid and other proteins in the cytoplasm by using host cell enzymes. Then the proteins migrate into the nucleus and are joined with the newly synthesized DNA to form virions. These virions are transported along the endoplasmic reticulum into the host cell's membrane for release. Herpesviruses, papovaviruses, and adenoviruses all follow this pattern of biosynthesis (see Table 13.2). Poxviruses are an exception, since all components are synthesized in the cytoplasm.

Adenoviruses. Adenoviruses cause acute respiratory diseases ("common colds"). They are named after adenoids, from which they were first isolated. To show an example of the multiplication of a DNA virus, the sequence of events in adenovirus is shown in Figure 13.19.

Poxviruses. All diseases caused by poxviruses, including smallpox and cowpox, include skin lesions. *Pox* refers to pus-filled sacs. Viral multiplication is started by viral transcriptase; the viral components are synthesized and assembled in the cytoplasm of the host cell.

Herpesvirus. Nearly 100 herpesviruses are known. They are named after the spreading *(herpetic)* appearance of cold sores. Species of human herpesviruses (HHV) include those that cause cold sores (HHV 1 and

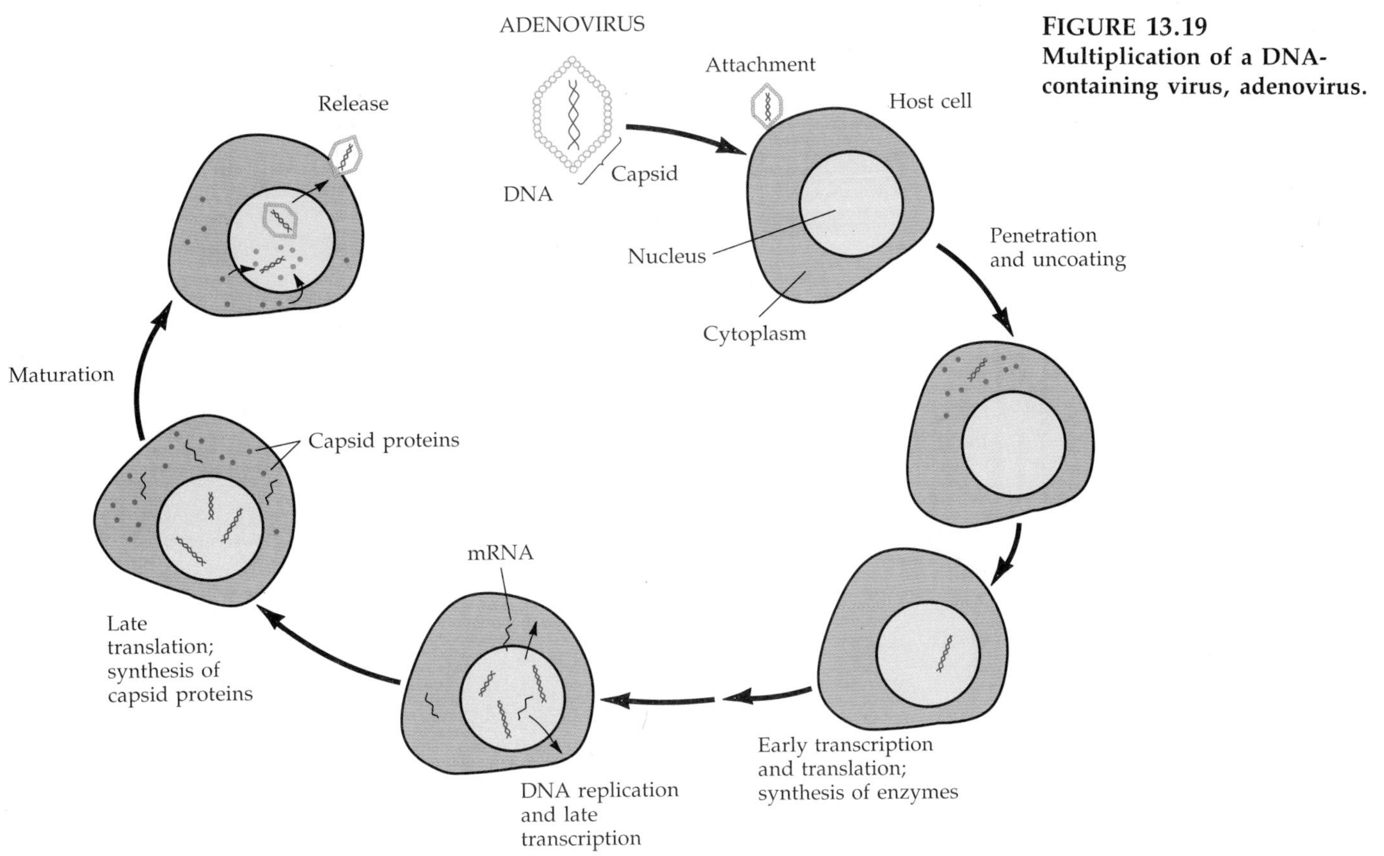

FIGURE 13.19
Multiplication of a DNA-containing virus, adenovirus.

HHV 2), chickenpox (HHV 3 or varicella-zoster virus), and infectious mononucleosis (HHV 4 or Epstein–Barr virus).

Papovaviruses. Papovaviruses are named for *pa*pillomas (warts), *po*lyomas (tumors), and *va*cuolation (cytoplasmic vacuoles produced by some of these viruses). Warts are caused by a papovavirus. Some of the papovaviruses are capable of transforming cells and causing cancer. Viral DNA is replicated in the host cell's nucleus along with host cell chromosomes. Host cells may proliferate, resulting in a tumor.

After attachment, penetration, and uncoating, the viral DNA is released into the nucleus of the host cell. Next occurs the transcription of a portion of the viral DNA—the "early" genes. Translation follows. The products of these genes are enzymes required for the multiplication of viral DNA. In most DNA viruses, early transcription is carried out with the host's transcriptase (RNA polymerase); poxviruses, however, contain their own transcriptase. Sometime after the initiation of DNA multiplication, transcription and translation of the remaining "late" viral genes occur. This leads to the synthesis of capsid proteins, which occurs in the cytoplasm of the host cell. After the capsid proteins migrate into the nucleus of the host cell, maturation occurs—the viral DNA and capsid proteins assemble to form complete viruses.

Biosynthesis for RNA-Containing Viruses

The multiplication of RNA viruses is essentially the same as that of DNA viruses, except that several different mechanisms of mRNA formation occur among different groups of RNA viruses. Although the details of these mechanisms are beyond the scope of this text, for comparative purposes we will trace the multiplication cycles of some representative RNA viruses. (Figure 13.20 summarizes the multiplication cycles of the four major types of RNA viruses.) Multiplication of RNA viruses takes place in the host cell's cytoplasm. The major differences among the multiplication processes of these viruses lie in how mRNA and viral RNA are produced. Once viral RNA and viral proteins are synthesized, maturation occurs by similar means among all animal viruses. This will be discussed shortly.

Picornavirus. Picornaviruses, such as poliovirus, are single-stranded RNA viruses. They are the smallest viruses; hence, the prefix *pico* (small) plus *RNA* gives these viruses their name. The RNA within the virion is identified as a + or **sense strand** because it can act as mRNA. After attachment, penetration, and uncoating

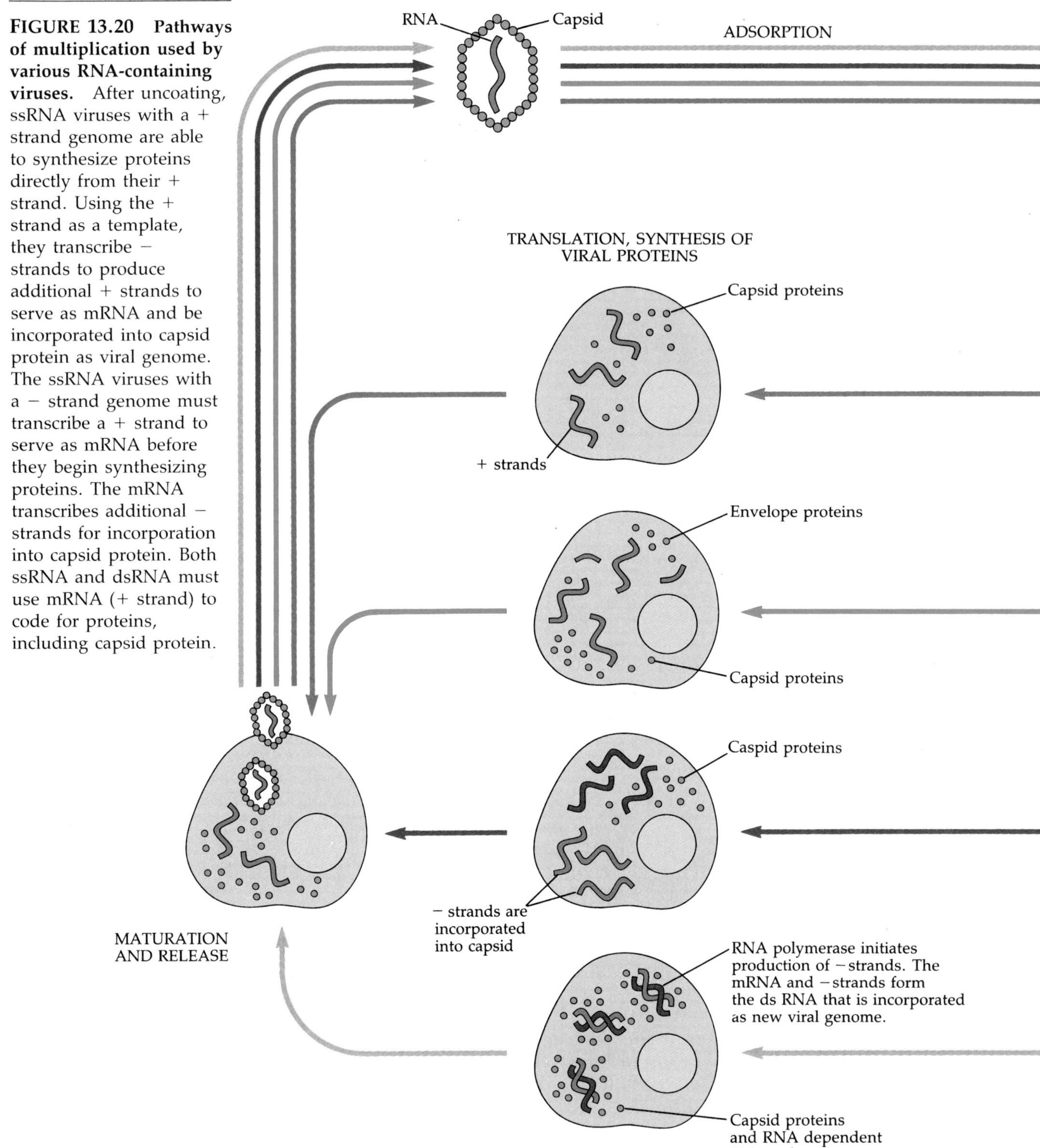

FIGURE 13.20 Pathways of multiplication used by various RNA-containing viruses. After uncoating, ssRNA viruses with a + strand genome are able to synthesize proteins directly from their + strand. Using the + strand as a template, they transcribe – strands to produce additional + strands to serve as mRNA and be incorporated into capsid protein as viral genome. The ssRNA viruses with a – strand genome must transcribe a + strand to serve as mRNA before they begin synthesizing proteins. The mRNA transcribes additional – strands for incorporation into capsid protein. Both ssRNA and dsRNA must use mRNA (+ strand) to code for proteins, including capsid protein.

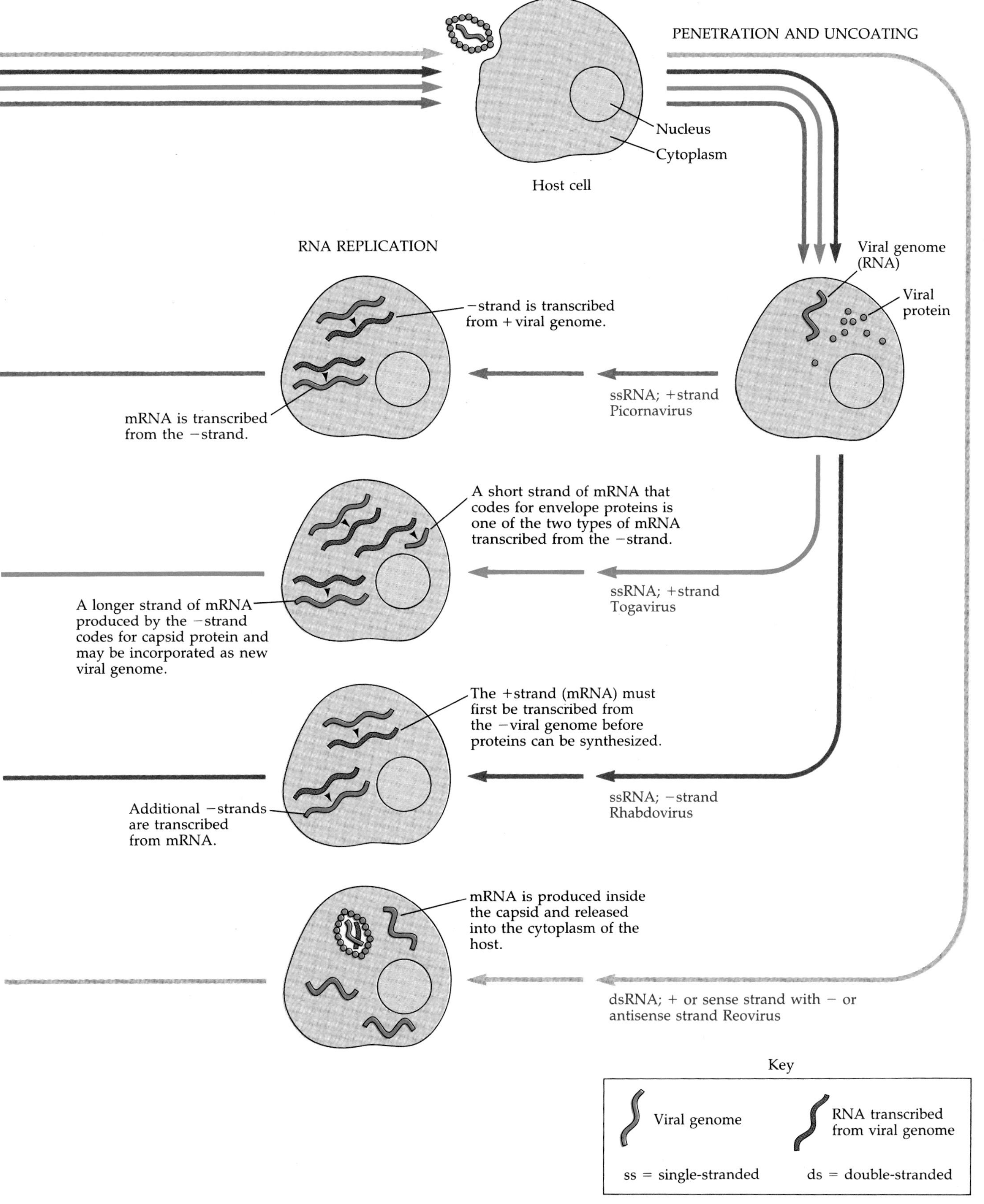
PENETRATION AND UNCOATING
Nucleus
Cytoplasm
Host cell
RNA REPLICATION
Viral genome (RNA)
Viral protein
−strand is transcribed from + viral genome.
mRNA is transcribed from the −strand.
ssRNA; +strand Picornavirus
A short strand of mRNA that codes for envelope proteins is one of the two types of mRNA transcribed from the −strand.
A longer strand of mRNA produced by the −strand codes for capsid protein and may be incorporated as new viral genome.
ssRNA; +strand Togavirus
The +strand (mRNA) must first be transcribed from the −viral genome before proteins can be synthesized.
Additional −strands are transcribed from mRNA.
ssRNA; −strand Rhabdovirus
mRNA is produced inside the capsid and released into the cytoplasm of the host.
dsRNA; + or sense strand with − or antisense strand Reovirus
Key
Viral genome
RNA transcribed from viral genome
ss = single-stranded
ds = double-stranded

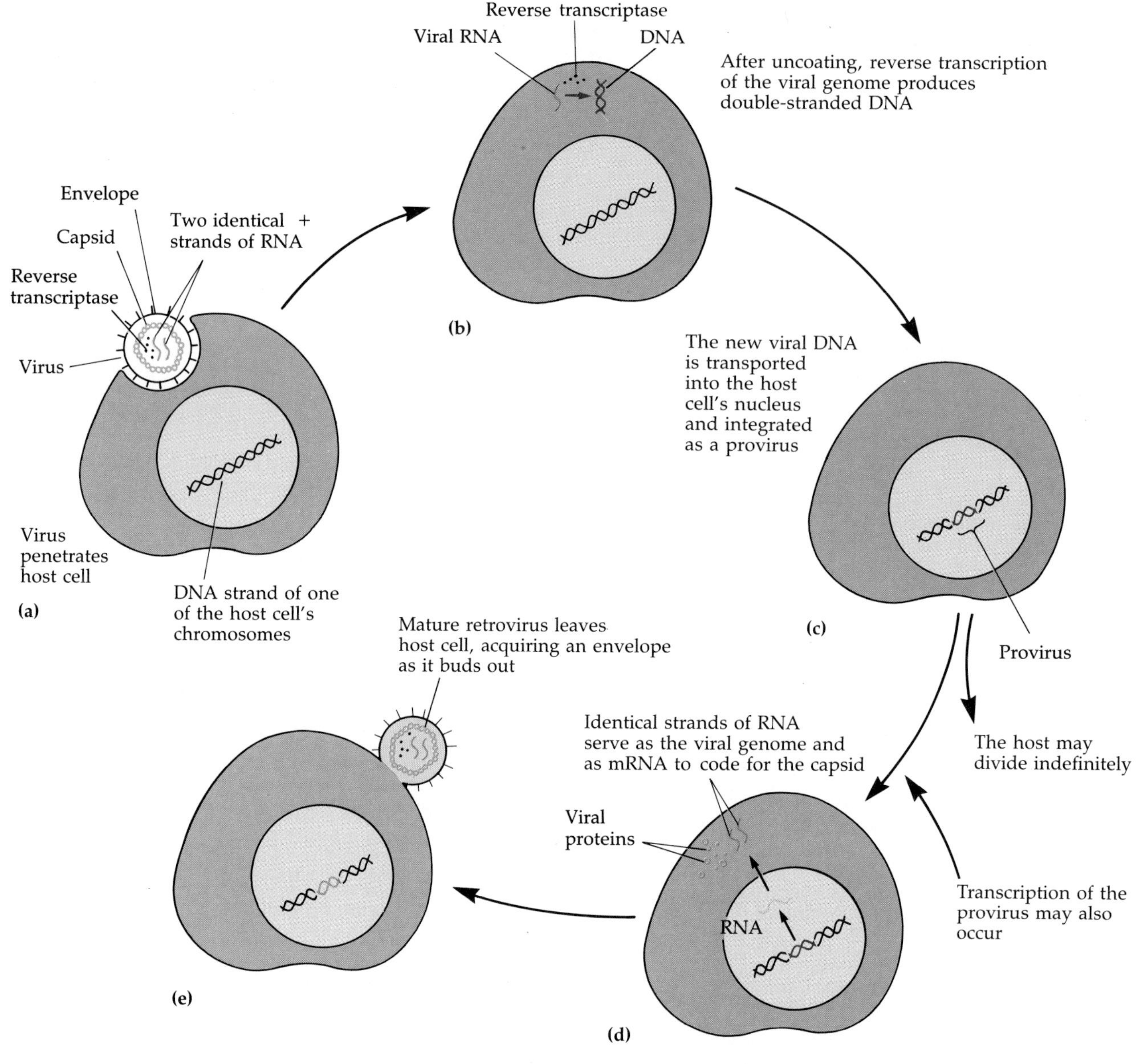

FIGURE 13.21 Multiplication and inheritance processes of a retrovirus. Retrovirus may become a provirus that replicates in a latent state, and it may produce new retroviruses.

are completed, the single-stranded viral RNA is translated into two principal proteins, which inhibit the host cell's synthesis of RNA and protein and which form an enzyme called *RNA-dependent RNA polymerase.* This enzyme catalyzes the synthesis of another strand of RNA, which is complementary in base sequence to the original infecting strand. This new strand is called a − or **antisense strand.** It serves as a template to produce additional + strands. The + strands may serve as mRNA for the translation of capsid protein, may become incorporated into capsid protein to form a new virus, or may serve as a template for continued RNA multiplication. Once viral RNA and viral protein are synthesized, maturation occurs.

Togavirus. Togaviruses, which include arthropod-borne viruses (see Chapter 22), also contain a single + strand of RNA. Togaviruses are enveloped viruses; their name is from the Latin word for covering, *toga.* Keep in mind that these are not the only enveloped viruses. After a − strand is made from the + strand, two types of mRNA are transcribed from the − strand.

One type of mRNA is a short strand that codes for envelope proteins. The other, longer strand serves as mRNA for capsid proteins and can become incorporated into a capsid.

Rhabdovirus. Rhabdoviruses, such as rabiesvirus, are usually bullet-shaped. *Rhabdo* is from the Greek word for rod, which is not really an accurate description of their morphology. They contain a single − strand of RNA. They also contain an RNA-dependent RNA polymerase that uses the − strand as a template from which to produce a + strand. The + strand serves as mRNA and as a template for synthesis of new viral RNA.

Reovirus. Reoviruses are found in the respiratory and enteric (digestive) systems of humans. They either are not pathogenic or are of such low virulence that they were not associated with any diseases when first discovered; they were considered orphan viruses. Their name comes from the words *respiratory, enteric,* and *orphan.*

After the capsid containing the double-stranded RNA enters a host cell, mRNA is produced inside the capsid and is released into the cytoplasm. There it is used to synthesize more viral proteins. One of the newly synthesized viral proteins acts as RNA-dependent RNA polymerase to produce more − strands of RNA. The mRNA (+) and − strands form the double-stranded RNA that is then surrounded by capsid proteins.

Retrovirus. Many retroviruses infect vertebrates. One group of retroviruses, the human immunodeficiency viruses (HIV-1 and HIV-2), causes AIDS (see the box on p. 344 and in Chapter 19, p. 484). The retroviruses that cause cancer will be discussed later in this chapter.

The formation of mRNA and RNA for new retrovirus virions is quite interesting. These viruses carry their own polymerase, which uses the RNA of the virus to synthesize a complementary strand of DNA, which in turn is replicated to form double-stranded DNA (Figure 13.21a and b). This enzyme also degrades the original viral RNA. The enzyme is an RNA-dependent DNA polymerase called **reverse transcriptase,** so called because it carries out a reaction (RNA → DNA) that is exactly the reverse of the familiar transcription of DNA → RNA. The name *retrovirus* is derived from the enzyme name *reverse transcriptase.* The formation of complete viruses requires that DNA be transcribed back into the RNA that will serve as mRNA for viral-protein synthesis and be incorporated into new virions. However, before transcription can take place, the viral DNA must be integrated into the DNA of a host cell chromosome (Figure 13.21c). In this integrated state, the viral DNA is called a **provirus.** Unlike a prophage, the provirus never comes out of the chromosome.

Once the provirus is integrated into the host cell's DNA, several things can happen. Sometimes the provirus simply remains in a latent state and replicates when the DNA of the host cell replicates. In other cases, the provirus becomes transcribed and produces new viruses, which may infect adjacent cells (Figure 13.21d and e). The provirus can also convert the host cell into a tumor cell; possible mechanisms will be discussed later.

Maturation and Release

The first step in viral maturation is the assembly of the protein capsid; this assembly is usually a spontaneous process. The capsids of many of the RNA-containing animal viruses are enclosed by an envelope consisting of protein, lipid, and carbohydrate, as noted earlier. Examples of such viruses include orthomyxoviruses and paramyxoviruses. The envelope protein is synthesized by the virus and is incorporated into the plasma membrane of the host cell. The envelope lipid and carbohydrate are synthesized by host cell enzymes and are present in the plasma membrane. The envelope actually develops around the capsid by a process called **budding** (Figure 13.22).

After the sequence of attachment, penetration, uncoating, and biosynthesis of viral nucleic acid and protein, the assembled capsid-containing nucleic acid pushes through the plasma membrane. As a result, a portion of the plasma membrane, now the envelope, adheres to the virus. This extrusion of a virus from a host cell is one method of release. Budding does not immediately kill the host cell, and in some cases the host cell survives.

Naked viruses are released through ruptures in the host cell plasma membrane. In contrast to budding, this type of release usually results in the death of the host cell.

Effects of Animal Viral Infection on Host Cells

Infection of a host cell by an animal virus usually kills the host cell. Death can be caused by the accumulation of large numbers of multiplying viruses, by the effects of viral proteins on the permeability of the host cell's plasma membrane, or by inhibition of host DNA, RNA, or protein synthesis. The different abnormalities that can lead to damage or death of a host cell are known as *cytopathic effects (CPE)* (see Table 15.4). These effects are frequently used as tools for the diagnosis of many viral infections.

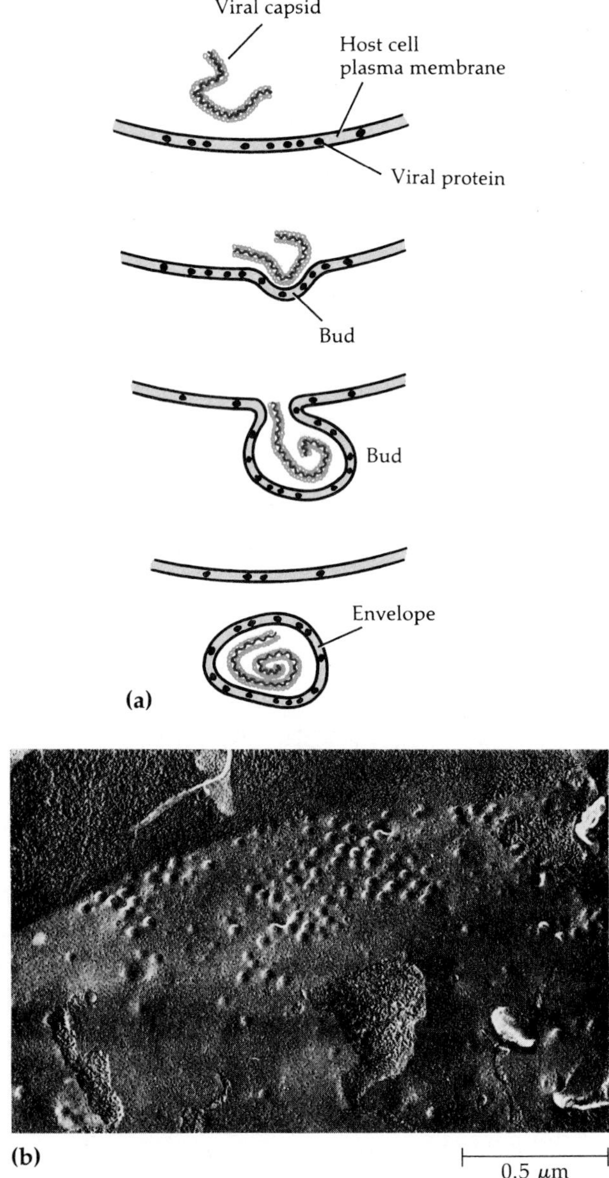

FIGURE 13.22 Budding of an enveloped virus. **(a)** Diagram of the budding process. **(b)** The small "bumps" seen on this freeze-fractured plasma membrane are Sindbis virus particles caught in the act of budding out from an infected cell.

One type of CPE is an **inclusion body,** some abnormal body of material in a cell. Some inclusion bodies arise as a result of the accumulation of assembled or unassembled viruses in the nucleus or cytoplasm of the host cell. Other inclusion bodies arise at sites of earlier viral synthesis but do not contain assembled viruses or their components. Inclusion bodies are important because their presence can help to identify the virus causing an infection. For example, rabiesvirus produces inclusion bodies (Negri bodies) in the cytoplasm of nerve cells, and their presence in the brain tissue of animals suspected of being rabid has been used as a diagnostic tool for rabies. Diagnostic inclusion bodies are also associated with measles virus, vaccinia virus, smallpox virus, herpesvirus, and adenoviruses (Figure 13.23).

As a result of viral infection, some host cells exhibit another type of CPE. At times, several adjacent infected cells fuse to form giant cells called **polykaryocytes.** These are produced from infections by paramyxoviruses. Other viral infections result in chromosomal damage to the host cell, most often chromosomal breakage. Some viral infections result in changes in the host cell's functions with no visible changes in the infected cells. For example, if a virus changes the production level of a hormone, the cells normally affected by that hormone will not function properly.

One very significant response of some virus-infected cells is their production of a substance called **interferon.** Viral infection induces cells to produce interferon, but it is the host cell's DNA that codes for the interferon. This substance protects neighboring uninfected cells from viral infection. More will be said about interferon in Chapter 16.

Viruses and Cancer

A number of types of cancer are now known to be caused by viruses. Research on cancer-causing viruses has led to breakthroughs in our general understanding of cancer. Molecular biological research shows that the mechanisms of the diseases are similar, even when a virus is not the cause of the cancer.

Cancer is not a single disease but is a name given to many diseases. The human body contains more than 100 different kinds of cells, each of which can malfunction in its own distinctive way to cause cancer. When cells multiply in an uncontrolled way, the excess tissue that develops is called a tumor. A cancerous tumor is known as a malignant tumor, whereas a noncancerous tumor is called a benign tumor. (Tumors are generally named by attachment of the suffix *-oma* to the name of the tissue from which the tumor arises.)

Not all cancerous cells form solid tumors. When the malignant cells are those that give rise to white blood cells, the resulting cancer, characterized by an excess of white cells in the cardiovascular system, is called leukemia.

The relationship between cancers and viruses was first demonstrated in 1908, when Wilhelm Ellerman and Olaf Bang were trying to isolate the causative agent of chicken leukemia. They found that leukemia could be transferred to healthy chickens by cell-free

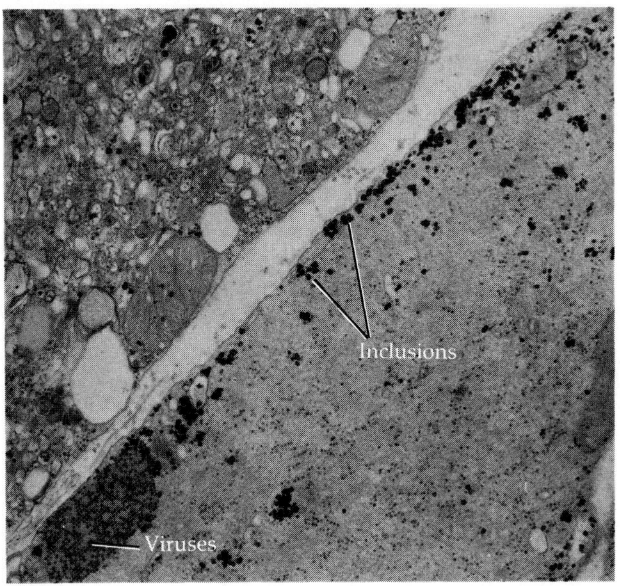

FIGURE 13.23 Viral inclusion bodies. The dark spots in the sections of two mouse muscle cells shown here are intranuclear bodies of coxsackie virus A4. Notice the large crystalline array of virus particles in the lower left.

filtrates that contained viruses. Three years later, Francis Rous found that a chicken **sarcoma** (cancer of connective tissue) can be similarly transmitted. Virus-induced **adenocarcinomas** (cancers of glandular epithelial tissue) in mice were discovered in 1936. At that time, it was clearly shown that mouse mammary gland tumors are transmitted from mother to offspring through the mother's milk.

The viral cause of cancer can often go unrecognized, for several reasons. First, most of the particles of some viruses infect cells but do not induce cancer. Second, cancer might not develop until long after viral infection. Third, cancers do not seem to be contagious, as viral diseases usually are.

TRANSFORMATION OF NORMAL CELLS INTO TUMOR CELLS

Almost anything that can alter the genetic material of a cell has the potential to make a normal cell cancerous. It is now known that these cancer-causing alterations to cellular DNA affect parts of the genome called **oncogenes.** Oncogenes were first identified in cancer-causing viruses and were thought to be a part of the normal viral genome. However, J. Michael Bishop and Harold E. Varmus of the United States received the 1989 Nobel Prize in Medicine for proving that the cancer-inducing genes carried by viruses are actually derived from animal cells. In 1976, Bishop and Varmus showed that the cancer-causing *src* gene in avian sarcoma viruses is derived from a normal part of chicken genes. At that time, they suggested that the *src* gene was involved in cell growth.

All nucleated cells have oncogenes, which are capable of transforming normal cells into cancerous cells. Under normal conditions, these genes probably code for proteins that are necessary to the cell's growth, but mutations can interfere with the normal regulation of oncogenes or cause oncogenes to code for abnormal proteins. Oncogenes that function normally are called **proto-oncogenes** to differentiate them from their abnormally activated counterparts. The wide occurrence of oncogenes suggests that they are not only normal but vital for cell functions.

Essentially the same oncogenes have been isolated from a wide variety of eucaryotes, from yeasts to humans, and from oncogenic viruses as well. More than 20 oncogenes have been isolated from animal cells; some of these are associated with human cancers. Many cancers are leukemias, diseases characterized by the production of large numbers of white blood cells. Different types of white blood cells proliferate in the different forms of leukemia. Note that most of the oncogenes listed are also found in viruses. Because the same oncogenes are isolated from different types of cancer, it is thought that the total number of different oncogenes is less than 100. Furthermore, oncogenes can be grouped by nucleotide sequence and protein product to form a much smaller number of classes.

The most common types of oncogene products isolated so far are protein kinases. Normal **protein kinases** remove a phosphate group from ATP and add it to the amino acids *serine* or *threonine* within a protein. Phosphorylation modifies a protein's function. The abnormal protein kinases isolated from cancerous cells attach a phosphate group to the amino acid *tyrosine.* The exact role of the phosphorylated proteins in transforming normal cells into cancerous cells is unknown. However, hypotheses being investigated relate their activity to loss of normal controls on cell shape and cell growth.

Oncogenes can be activated by a variety of agents, including mutagenic chemicals, high-energy radiation, and viruses. Viruses capable of inducing tumors in animals are called **oncogenic viruses.** Approximately 20% of cancers are known to be virus-induced.

Both DNA- and RNA-containing viruses are capable of inducing tumors in animals. When this occurs, the tumor cells are **transformed** in such a way that they acquire properties that are distinct from properties of uninfected cells or from infected cells that do not form tumors. An outstanding feature of all oncogenic viruses is that their genetic material integrates into the host cell's DNA and replicates along with the host cell's chromosome. This mechanism is similar to the phenomenon of lysogeny in bacteria, and it can alter the host cell's characteristics in the same way.

Transformed cells also lose a property called **contact inhibition.** Normal animal cells in tissue culture move about randomly by amoeboid movement and divide repeatedly until they come into contact with each other. Then, both movement and cell division stop. This phenomenon is known as contact inhibition. Transformed cells in cell culture do not exhibit contact inhibition but instead form tumorlike cell masses. Transformed cells sometimes produce tumors when injected into susceptible animals.

After being transformed by viruses, many tumor cells contain a virus-specific antigen on their cell surface, called **tumor-specific transplantation antigen (TSTA)**, or an antigen in their nucleus, called the **T antigen.** Transformed cells tend to be less round than normal cells and tend to exhibit certain chromosomal abnormalities, such as unusual numbers of chromosomes and fragmented chromosomes.

ACTIVATION OF ONCOGENES

Cancer appears to be a *multistep* process. A number of abnormalities must accumulate in the cell, and at least two oncogenes must be activated to abnormal functioning for malignancy to result. At present, four mechanisms for the activation of oncogenes are proposed.

1. A **single mutation** such as a nitrogenous base-pair substitution resulting from exposure to a chemical mutagen or radiation could alter the amino acid sequence of a gene product. For example, the *ras* oncogene in malignant cells differs from the normal *ras* proto-oncogene in only one nucleotide. It is thought that the mutated *ras* gene increases a cell's response to growth-regulating hormones.

   A new protein resulting from a mutation may contribute directly to transformation. Alternatively, the mutation might affect a gene whose product normally regulates an oncogene. If that oncogene codes for a cell growth factor and is no longer properly regulated, the growth factor might be produced in an uncontrolled way and lead to cancer.

2. **Transduction** of oncogenes by a virus could remove the oncogenes from normal cellular controls and place them under control of viral regulatory proteins. The viral proteins could cause an oncogene product to be produced at an abnormal time or in abnormal amounts. Some human oncogenes that have been isolated from viruses are listed in Table 13.3.

3. **Translocation** of oncogenes from one chromosomal locus to another can occur during normal cellular activities and could place the oncogenes into a site in which normal controls are not active. In Burkitt's lymphoma, the *myc* oncogene is moved from its normal position on chromosome 8 to chromosome 14, and its product is increased because it is now regulated by an antibody-regulating gene. It is thought that the continual presence of the *myc* gene product might allow cells to replicate indefinitely.

4. **Gene amplification** could cause the cell to produce unusually large amounts of oncogenic products. A gene is amplified when an unknown mechanism causes the gene to be duplicated several times. In one central nervous system cancer (glioma), more than 50 copies of a particular gene were found in a malignant cell.

Of these four major mechanisms of oncogene activation, viruses are or may be involved in the last three.

DNA-CONTAINING ONCOGENIC VIRUSES

Oncogenic viruses are found within several groups of DNA-containing viruses. These groups include adenoviruses, herpesviruses, poxviruses, and papovaviruses. Among the papovaviruses are the papillomaviruses that cause benign warts in humans and other animals, polyomaviruses that cause several kinds of tumors when injected into newborn mice, and simian virus 40 (SV40), which was originally isolated from cell cultures being used to cultivate polioviruses for vaccine production. Papillomaviruses are frequently found in cancerous cervical cells, providing evidence that they may cause uterine (cervical) cancer.

During polyomavirus and SV40 infection, there is an increase in the host cell's synthesis of DNA. This increase is followed by the appearance of TSTA and T antigen. When the host cell is transformed, viral DNA is integrated into the host cell's DNA as a provirus. This mechanism, similar to lysogeny in bacteria, results in the transformed cells' distinctive properties noted earlier.

Among the herpesviruses that affect humans are the herpes simplex virus, the herpes zoster virus, and the Epstein–Barr (EB) virus. The EB virus has an attraction for lymphocytes (a type of white blood cell) and has the potential for transforming lymphocytes into highly proliferating cells. This virus, in addition to being the cause of infectious mononucleosis, has been implicated as the causative agent of two human cancers, Burkitt's lymphoma and nasopharyngeal carcinoma. Burkitt's lymphoma is a rare cancer of the lymphatic system; it affects mostly children in certain areas of Africa. Nasopharyngeal carcinoma, a cancer of the nose and throat, is found worldwide. Some researchers have also suggested that EB virus is involved in Hodgkin's disease, a cancer of the lymphatic system.

It is estimated that about 90% of the population of the United States carries the latent stage of the EB virus in their lymphocytes but have no disease. This latent stage is indicated by the presence of antibodies to EB virus in blood serum. Although the EB virus leads to no apparent symptoms in healthy persons, it can cause infectious mononucleosis, mostly in teenagers. TSTA and T antigen specific for the EB virus have been found in specimens from persons with either Burkitt's lymphoma or nasopharyngeal cancer. Moreover, DNA from EB virus is always found in malignant cells from these cancers.

The role of EB virus in human cancer was accidentally demonstrated in 1985 when a 12-year-old boy received a bone marrow transplant. David was born without a functioning immune system and was kept from all microbes inside a sterile "bubble" his entire life—hence he was known as "the bubble boy." (The lymphocytes that provide immunity are produced in bone marrow.) Several months after the transplant, he developed signs of infectious mononucleosis (caused by EB virus), and a few months after that he died of cancer. An autopsy revealed that the EB virus had been unwittingly introduced into the boy with the bone marrow transplant. The case confirmed suspicions that EB virus can induce cancer in immunosuppressed individuals.

One type of herpes simplex virus (HHV 1) produces cold sores. Another type, HHV 2, is associated with more than 90% of genital herpes infections. Women with cervical cancer have more HHV 2 antibodies than asymptomatic patients have, so it has been suggested that HHV 2 might be associated with cervical cancer.

Another virus implicated in cancer is hepatitis B virus (HBV). HBV is an enveloped DNA virus with a small, circular genome. The DNA from HBV can become incorporated into the host genome. Its presence has been associated with liver cancer.

RNA-Containing Oncogenic Viruses

Among the RNA viruses, only the retroviruses seem to be oncogenic. Human T cell leukemia viruses (HTLV 1 and HTLV 2) are retroviruses that are associated with leukemia and lymphoma in humans. (T cells are a type of white blood cell involved in immune responses to disease.) HTLV 1 has been isolated from adult T cell leukemia (ATL) patients. Antibodies against the virus are found in people living in regions where the disease is endemic (constantly present), but antibodies are not found in people outside these areas. The presence of antibodies shows that HTLV infections occur in the same geographic areas as ATL, which suggests that the viral infection induces the cancer.

Sarcoma viruses of cats, chickens, and rodents and the mammary tumor viruses of mice are also retroviruses. Another retrovirus, feline leukemia virus (FeLV), causes leukemia in cats and is transmissible between cats. There is a test to detect the virus in cats' serum.

The ability of retroviruses to induce tumors is related to their production of a reverse transcriptase by the mechanism described earlier (see Figure 13.21). The provirus, which is the double-stranded DNA molecule synthesized from the viral RNA, becomes integrated into the host cell's DNA; new genetic material is thereby introduced into the host's genome, and this is the key reason why retroviruses can contribute to cancer. Some retroviruses contain oncogenes; others contain promoters that turn on oncogenes or other cancer-causing factors.

Latent Viral Infections

A virus can remain in equilibrium with the host and not actually produce disease for a long period, often many years. The classic example of such a **latent viral infection** is the infection of the skin by herpes simplex virus, which produces cold sores. This virus can inhabit the host's nerve cells but cause no damage until it is activated by such stimuli as fever and sunburn—hence the term *fever blister*.

It is interesting that in some individuals, viral production occurs but the symptoms never appear. Even though a large percentage of the human population carries the herpes simplex virus, only 10% to 15% of people carrying the virus exhibit the disease. The virus of some latent infections can exist in a lysogenic state within host cells.

The varicella-zoster virus (also called human herpes virus 3) can also exist in a latent state. Chickenpox (varicella) is a childhood disease in which the viral agent is indistinguishable from the viral agent (herpes zoster) that causes shingles, an adult neurological disorder. Shingles occurs only in a small fraction of the population who have had chickenpox. Many researchers believe that shingles may be caused by reactivation of the virus that has remained latent in certain nerve cells for years.

Slow Viral Infections

The term **slow viral infection** was coined in the 1950s to refer to a disease process that occurs gradually over a long period and was thought to be caused by a virus. (The term does not imply that viral multiplication is unusually slow.) Typically, slow viral infections are fatal.

TABLE 13.3 Examples of Slow Viral Infections Caused by Conventional Viruses

DISEASE	HOST	PRIMARY EFFECT	VIRUS
Subacute sclerosing panencephalitis	Human	Mental deterioration	Measles virus (paramyxovirus)
Progressive encephalitis	Human	Rapid mental deterioration	Rubella virus (togavirus)
Progressive multifocal leukoencephalopathy	Human	Brain degeneration	Papovavirus
Progressive pneumonia	Sheep	Degeneration of lungs	Retrovirus
Lymphocytic choriomeningitis	Mouse	Degeneration of kidneys, brain, liver	Arenavirus
Aleutian mink disease	Mink	Kidney failure	Parvovirus
Visna	Sheep	Complete paralysis	Retrovirus
Encephalopathy	Human	Brain degeneration	HIV (retrovirus)

A number of diseases called slow viral infections have in fact been shown to be caused by conventional viruses. For example, several years after causing measles, the measles virus can be responsible for a rare form of encephalitis called *subacute sclerosing panencephalitis (SSPE).* Slow viral infection is apparently different from latent viral infection in that, in most slow viral infections, detectable infectious virus gradually builds up over a long period, rather than appearing suddenly.

Several examples of slow viral infections caused by conventional viruses are listed in Table 13.3.

Unconventional Agents of Disease

Other diseases called slow viral infections have not been found to have a viral cause and are now thought to be caused by unconventional agents. Ironically, many scientists have abandoned the term "slow viral infection" for truly viral diseases such as SSPE (instead using terms such as "persistent viral infection" for SSPE); some medical scientists now reserve "slow viral infection" for diseases that are apparently *not* caused by conventional viruses. The prototype of these latter diseases is scrapie, a neurologic disease of sheep, for which the infectious agent appears to be pure protein. The mystery of how an infectious agent can lack nucleic acid has prompted several hypotheses, including the following:

1. The infectious agent might actually be a conventional virus, but one whose nucleic acid is extraordinarily difficult to detect.
2. The infectious agent might be an undetectably small nucleic acid enclosed in a *host* protein. The nucleic acid might not encode protein at all but might induce disease by somehow directly disrupting the host cell, perhaps by interacting in the host DNA. Although the agent might physically resemble a virus, it would be protected from the host immune system because its coat of host protein would not be recognized as foreign. This hypothetical infectious agent has been called a **virino.**
3. The infectious agent might really be pure protein. If so, how would it replicate? In the past, the unlikely possibility of protein self-replication was raised. Now, however, there is evidence that, at least for scrapie, a major portion of the infectious agent is a protein whose gene (PrP, for prion protein) is found in normal host DNA. The PrP gene is located on chromosome 2 in humans. Questions remain, including the questions of how the gene is turned on, how its product acts as an infectious disease agent, and whether other genes and proteins are involved. The postulated infectious protein agent has been named a **prion,** for **prot**einaceous **in**fectious particle. (The name would more logically be "proin," but the coiner thought that "prion" sounded better.)

All diseases thought to be caused by unconventional agents are neurologic diseases. In addition to scrapie, these include Creutzfeldt–Jakob disease, kuru, and Gerstmann–Sträussler syndrome. (Neurologic diseases are discussed in Chapter 22.)

Plant Viruses and Viroids

Plant viruses resemble animal viruses in many respects; plant viruses are morphologically similar to ani-

TABLE 13.4 Classification of Some Major Plant Viruses According to Morphological, Chemical, and Physical Properties

VIRUS	MORPHOLOGICAL CLASS	NUCLEIC ACID*	DIMENSIONS OF CAPSID (NM)	METHOD OF TRANSMISSION	SHOWS SIMILARITIES TO
Tobacco mosaic virus	Helical	ssRNA	300 (length)	Wounds	Picornavirus
Brome grass mosaic virus	Polyhedral	ssRNA	23 (diameter)	Pollen	Picornavirus
Potato yellow dwarf virus	Bullet-shaped	ssRNA	380 (length)	Leafhoppers and aphids	Rhabdovirus
Wound tumor virus	Polyhedral	segmented dsDNA	70 (diameter)	Leafhoppers	Reovirus
Cauliflower mosaic virus	Polyhedral	dsDNA	50 (diameter)	Aphids	Papillomavirus

*ss = single-stranded; ds = double-stranded.

mal viruses, and they have similar types of nucleic acid (Table 13.4). In fact, some plant viruses can multiply inside insect cells. Plant viruses cause many diseases of economically important crops, including tomatoes (tomato spotted wilt virus), corn and sugarcane (wound tumor virus), and potatoes (potato yellow dwarf virus). Viruses can cause color change, deformed growth, wilting, and stunted growth in their plant hosts. Some hosts, however, remain symptomless and only serve as reservoirs of infection.

Plant cells are generally protected from disease by an impermeable cell wall. Viruses must enter through wounds or be assisted by other plant parasites, including nematodes, fungi, and, most often, insects that suck the plant's sap. Once one plant is infected, it can spread infection to other plants in its pollen and seeds.

In laboratories, plant viruses are cultured in protoplasts (plant cells with the cell walls removed) and in insect cell cultures.

Some plant diseases are caused by **viroids,** short pieces of naked RNA, only 300 to 400 nucleotides long, with no protein coat. The nucleotides are often internally paired, so the molecule has a closed, folded, three-dimensional structure that presumably helps protect it from attack by cellular enzymes. The RNA does not code for any proteins. Thus far, viroids have been conclusively identified as pathogens only of plants. Infections by viroids, such as potato spindle tuber viroid (Figure 13.24), result in losses of millions of dollars in crop damage.

Current research on viroids shows a similarity in base sequences with introns. Recall from Chapter 8 that introns are sequences of genetic material that do not code for polypeptides. This observation has led to the hypothesis that viroids evolved from introns—which leads to speculation that future work may discover animal viroids.

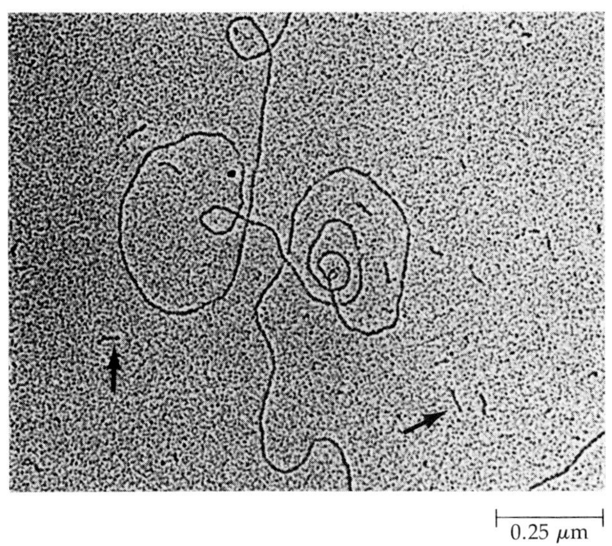

FIGURE 13.24 Potato spindle tuber viroid (PSTV). The arrows point to the viroids, which are adjacent to much longer viral nucleic acid from bacteriophage T7, for comparison in this electron micrograph.

STUDY OUTLINE

General Characteristics of Viruses (pp. 333–335)

1. Depending on one's viewpoint, viruses may be regarded as exceptionally complex aggregations of nonliving chemicals or as exceptionally simple living microbes.
2. Viruses contain a single type of nucleic acid (DNA or RNA) and a protein coat, sometimes enclosed by an envelope composed of lipids, proteins, and carbohydrates.
3. Viruses are obligatory intracellular parasites. They multiply by using the host cell's synthesizing machinery to cause the synthesis of specialized elements that can transfer the viral nucleic acid to other cells.
4. A virion is a complete, fully developed viral particle composed of nucleic acid surrounded by a coat.

HOST RANGE (pp. 334–335)

1. Host range refers to the spectrum of host cells in which a virus can multiply.
2. Depending on its host range, a virus is generally classified as an animal virus, bacterial virus (bacteriophage), or plant virus. A virus can infect only certain species within each class.
3. Host range is determined by the specific attachment site on the host cell's surface and the availability of host cellular factors.

SIZE (p. 335)

1. Viral size is determined by filtration through membrane filters, ultracentrifugation, and electron microscopy.
2. Viruses range from 20 to 300 nm in diameter.

Viral Structure (pp. 335–339)

NUCLEIC ACID (p. 335)

1. The proportion of nucleic acid in relation to protein in viruses ranges from about 1% to about 50%.
2. Viruses contain either DNA or RNA, never both, and the nucleic acid may be single- or double-stranded, linear or circular, or divided into several separate molecules.

CAPSID AND ENVELOPE (pp. 335–336)

1. The protein coat surrounding the nucleic acid of a virus is called the capsid.
2. The capsid is composed of subunits, the capsomeres, which can be a single type of protein or several types.
3. The capsid of some viruses is enclosed by an envelope consisting of lipids, proteins, and carbohydrates.
4. Some envelopes are covered with carbohydrate–protein complexes called spikes.
5. Viruses without envelopes are called naked viruses.

GENERAL MORPHOLOGY (pp. 336–339)

1. Helical viruses (for example, tobacco mosaic virus) resemble long rods, and their capsids are hollow cylinders surrounding the nucleic acid.
2. Polyhedral viruses (for example, adenovirus) are many-sided. Usually the capsid is an icosahedron.
3. Enveloped viruses are covered by an envelope and are roughly spherical but highly pleomorphic. There are also enveloped helical viruses (for example, influenza virus) and enveloped polyhedral viruses (for example, herpes simplex virus).
4. Complex viruses have complex structures. For example, many bacteriophages have a polyhedral capsid with a helical tail attached.

Taxonomy of Viruses (pp. 339–340)

1. Classification of viruses is based on type of nucleic acid, morphological class, size of capsid, and number of capsomeres.
2. Other classification schemes take into account the virus's susceptibility to microbial control agents, immunological properties, site of multiplication, and method of transmission.

Isolation, Cultivation, and Identification of Viruses (pp. 340–345)

1. Viruses must be grown in living cells.
2. The easiest viruses to grow are bacteriophages.

GROWTH OF BACTERIOPHAGES IN THE LABORATORY (p. 341)

1. The plaque method mixes bacteriophages with host bacteria and nutrient agar.
2. After several viral multiplication cycles, the bacteria in the area surrounding the original virus are destroyed; the area of lysis is called a plaque.
3. Each plaque can originate with a single viral particle or with more than one; the concentration of viruses is given as plaque-forming units.

GROWTH OF ANIMAL VIRUSES IN THE LABORATORY (pp. 341–343)

1. Cultivation of some animal viruses requires whole animals.
2. Simian AIDS and feline AIDS provide models for study of human AIDS.
3. Some animal viruses can be cultivated in embryonated eggs.
4. Cell cultures are cells growing in culture media in the laboratory.
5. Primary cell lines and embryonic diploid cell lines grow for a short time in vitro.
6. Continuous cell lines can be maintained in vitro indefinitely.
7. Viral growth can cause cytopathic effects in the cell culture.

VIRAL IDENTIFICATION (pp. 343–345)

1. Serological tests are used most often to identify viruses.
2. Viruses may be identified by restriction enzyme fragments and nucleic acid base sequencing.

Viral Multiplication (pp. 345–355)

1. Viruses do not contain enzymes for energy production or protein synthesis.
2. For a virus to multiply, it must invade a host cell and direct the host's metabolic machinery to produce viral enzymes and components.

MULTIPLICATION OF BACTERIOPHAGES (pp. 345–348)

T-Even Bacteriophages (pp. 345–347)

1. The T-even bacteriophages that infect *E. coli* have been studied extensively.
2. In attachment, sites on the phage's tail fibers attach to complementary receptor sites on the bacterial cell.
3. In penetration, phage lysozyme opens a portion of the bacterial cell wall, the tail sheath contracts to force the tail core through the cell wall, and phage DNA enters the bacterial cell. The capsid remains outside.
4. In biosynthesis, transcription of phage DNA produces mRNA coding for proteins necessary for phage multiplication. Phage DNA is replicated, and capsid proteins are produced. During the eclipse period, separate phage DNA and protein can be found.
5. During maturation, phage DNA and capsids are assembled into complete viruses.
6. During release, phage lysozyme breaks down the bacterial cell wall, and the multiplied phages are released.
7. The time from phage adsorption to release is called burst time (20 to 40 minutes). Burst size, the number of newly synthesized phages produced from a single infected cell, ranges from 50 to 200.

Lysogeny (pp. 347–348)

1. During a lytic cycle, a phage causes the lysis and death of a host cell.
2. Some viruses can either cause lysis or have their DNA incorporated as a prophage into the DNA of the host cell. The latter situation is called lysogeny.
3. Prophage genes are regulated by a repressor coded for by the prophage. The prophage is replicated each time the cell divides.
4. Exposure to certain mutagens can lead to excision of the prophage and initiation of the lytic cycle.
5. Because of lysogeny, lysogenic cells become immune to reinfection with the same phage, and the host cell can exhibit new properties.
6. A lysogenic phage can transfer bacterial genes from one cell to another through transduction. Any genes can be transferred in generalized transduction, and specific genes can be transferred in specialized transduction.

MULTIPLICATION OF ANIMAL VIRUSES (pp. 348–355)

1. Animal viruses attach to the plasma membrane of the host cell.
2. Penetration of enveloped viruses occurs by endocytosis.
3. Animal viruses are uncoated by viral or host cell enzymes.
4. The DNA of most DNA viruses is released into the nucleus of the host cell. Transcription of viral DNA and translation produce viral DNA and, later, capsid protein. Capsid protein is synthesized in the cytoplasm of the host cell.
5. DNA viruses include herpesvirus, poxvirus, papovavirus, and adenovirus.
6. Multiplication of RNA viruses occurs in the cytoplasm of the host cell. RNA-dependent RNA polymerase synthesizes a double-stranded RNA.
7. Picornavirus + strand RNA acts as mRNA and directs the synthesis of RNA-dependent RNA polymerase.
8. Togavirus + strand RNA acts as a template for RNA-dependent RNA polymerase, and mRNA is transcribed from a new − RNA strand.
9. Rhabdovirus − strand RNA is a template for viral RNA-dependent RNA polymerase, which transcribes mRNA.
10. mRNA is produced inside the capsid of reoviruses.
11. Retroviruses carry reverse transcriptase (RNA-dependent DNA polymerase), which transcribes DNA from RNA.
12. After maturation, viruses are released. One method of release (and envelope formation) is budding. Naked viruses are released through ruptures in the host cell membrane.

Effects of Animal Viral Infection on Host Cells (pp. 355–356)

1. Cytopathic effects (CPE) are abnormalities that lead to damage or death of a host cell.
2. Cytopathic effects include inclusion bodies, polykaryocytes, and altered function.
3. Interferon is produced by virus-infected cells and protects neighboring cells from viral infection.

Viruses and Cancer (pp. 356–359)

1. An excess of tissue due to unusually rapid cell multiplication is called a tumor. Tumors are malignant (cancerous) or benign (noncancerous). Metastasis refers to the spread of cancer to other parts of the body.
2. Tumors are usually named by attachment of the suffix *-oma* to the name of the tissue from which the tumor arises.
3. The earliest relationship between cancer and viruses was demonstrated in the early 1900s, when chicken leukemia and chicken sarcoma were transferred to healthy animals by cell-free filtrates.

TRANSFORMATION OF NORMAL CELLS INTO TUMOR CELLS (pp. 357–358)

1. Eucaryotic cells have proto-oncogenes that code for proteins necessary for the cells' normal growth. When activated to oncogenes, these genes transform normal cells into cancerous cells.
2. Viruses capable of producing tumors are called oncogenic viruses.
3. Several DNA viruses and retroviruses are oncogenic.
4. The genetic material of oncogenic viruses becomes integrated into the host cell's DNA.
5. Transformed cells lose contact inhibition, contain virus-specific antigens (TSTA and T antigen), exhibit chromosomal abnormalities, and can produce tumors when injected into susceptible animals.

ACTIVATION OF ONCOGENES (p. 358)

1. A single mutation can result in the production of a protein required for transformation.
2. Transduction of oncogenes could result in oncogene products being made in abnormal amounts or at the wrong time.
3. Translocation of oncogenes could remove normal controls.
4. Gene amplification causes unusually large amounts of oncogene products.

DNA-CONTAINING ONCOGENIC VIRUSES (pp. 358–359)

1. Oncogenic viruses are found among adenoviruses, herpesviruses, poxviruses, and papovaviruses.
2. The EB virus, a herpesvirus, causes infectious mononucleosis and has been implicated in Burkitt's lymphoma and nasopharyngeal carcinoma. One type of herpes simplex virus (HHV 2), associated with over 90% of genital herpes infections, might be implicated in cervical cancer.

RNA-CONTAINING ONCOGENIC VIRUSES (p. 359)

1. Among the RNA viruses, only retroviruses seem to be oncogenic.
2. HTLV 1 and HTLV 2 have been associated with human leukemia and lymphoma.
3. The virus's ability to produce tumors is related to the production of reverse transcriptase. The DNA synthesized from the viral RNA becomes incorporated as a provirus into the host cell's DNA.
4. A provirus can remain latent, can produce viruses, or can transform the host cell.

Latent Viral Infections (p. 359)

1. A latent viral infection is one in which the virus remains in the host cell for long periods without producing an infection.
2. Examples are cold sores and shingles.

Slow Viral Infections (pp. 359–360)

1. Slow viral infections are disease processes that occur over a long period and are generally fatal.
2. Some of these diseases are caused by conventional viruses; others apparently are not.

Unconventional Agents of Disease (p. 360)

1. The "slow viral infections" not caused by conventional viruses—such as scrapie, Creutzfeldt–Jakob disease, and certain other neurologic diseases—may be caused by unconventional infectious agents that are pure protein, without nucleic acid (prions).
2. Another hypothesized agent of such diseases is the virino, a tiny piece of nucleic acid associated with host protein.

Plant Viruses and Viroids (pp. 360–361)

1. Plant viruses must enter plant hosts through wounds or with invasive parasites, such as insects.
2. Some plant viruses also multiply in insect (vector) cells.
3. Viroids are infectious pieces of RNA that cause some plant diseases, such as potato spindle tuber viroid disease.

STUDY QUESTIONS

Review

1. Viruses were first detected because they are filterable. What do we mean by the term *filterable,* and how could this property have helped their detection before invention of the electron microscope?

2. Why do we classify viruses as obligatory intracellular parasites?

3. List the four properties that define a virus. What is a virion?

4. Describe the four morphological classes of viruses, then diagram and give an example of each.

5. Describe how bacteriophages are detected and enumerated by the plaque method.

6. Explain how animal viruses are cultured in each of the following:
   (a) Whole animals
   (b) Embryonated eggs
   (c) Cell cultures

7. Why are continuous cell lines of more practical use than primary cell lines for culturing viruses? What is unique about continuous cell lines?

8. Describe the multiplication of a T-even bacteriophage. Be sure to include the essential features of attachment, penetration, biosynthesis, maturation, and release.

9. *Streptococcus pyogenes* produces erythrogenic toxin and is capable of causing scarlet fever only when it is lysogenic. What does this mean?

10. Describe the principal events of attachment, penetration, uncoating, biosynthesis, maturation, and release of an enveloped DNA-containing virus.

11. Recall from Chapter 1 that Koch's postulates are used to determine the etiology of a disease. Why is it difficult to determine the etiology of
    (a) a viral infection such as influenza?
    (b) cancer?

12. Assume that this strand of RNA is the nucleic acid for an RNA-containing animal virus: UAGUCAAGGU.
    (a) Describe the steps of RNA replication for a virus that contains a + strand of RNA.
    (b) Describe the steps of RNA replication for a virus that contains a − strand of RNA.
    (c) Describe the steps of RNA replication for a virus that contains double-stranded RNA.
    (d) Describe the steps of RNA replication for a virus that contains reverse transcriptase.
13. Provide an explanation for the chronic recurrence of cold sores in some people. (*Note:* The cold sores almost always return at the same site.)
14. Slow viral infections such as ________ might be caused by ________ that are ________.
15. What are cytopathic effects?
16. Which viruses have been proven to cause cancer? How was this proven?
17. The DNA of DNA-containing oncogenic viruses can become integrated into the host DNA. When integrated, the DNA is called a ________. How does this process result in transformation of the cell? Describe the changes of transformation. How can an RNA-containing virus be oncogenic?
18. Why was the discovery of SAIDS and FAIDS important?
19. Contrast viroids and unconventional agents. Name a disease thought to be caused by each.
20. Plant viruses cannot penetrate intact plant cells because ________; therefore, they enter cells by ________. Plant viruses can be cultured in ________.

Challenge

1. Discuss the arguments for and against the classification of viruses as living organisms.
2. In some viruses, capsomeres function as enzymes as well as structural supports. Of what advantage is this to the virus?
3. Discuss four mechanisms of oncogene activation, and discuss how viruses may be implicated.
4. Prophages and proviruses have been described as being similar to bacterial plasmids. What similar properties do they exhibit? How are they different?

FURTHER READING

Desrosiers, R.C. "Simian immunodeficiency viruses." *Annual Review of Microbiology* 42:607–625, 1988. A discussion of the prevalence and diversity of these viruses and their use as a model for human AIDS.

Diener, T.O. "Viroids and their interactions with host cells." *Annual Review of Microbiology* 36:239–258, 1982. Comprehensive coverage of the structure, pathogenesis, and possible origin of viroids by their discoverer.

Girard, M. "The Pasteur Institute's contributions to the field of virology." *Annual Review of Microbiology* 42:745–763, 1988. A history of virology featuring research on rabies, yellow fever, and polio.

Hogle, J.M., M. Chow, and D.J. Filman. "The structure of poliovirus." *Scientific American* 256(3):42–49, March 1987. Describes virus assembly and provides a mechanism for protein coat changes that allow a virus to escape the host's immune response.

Hunter, T. "The proteins of oncogenes." *Scientific American* 251(2):70–79, August 1984. Explains what the products of oncogenes do.

Oldstone, M.B.A. "Viral attenuation of cell function." *Scientific American* 261(2):42–48, August 1989. Suggestions that some human glandular disorders may be due to viral interference with a cell's production of hormones.

Prusiner, S.B. "Prions." *Scientific American* 251(4):50–59, October 1984. The scientist who discovered and named prions poses intriguing and as yet unanswered questions regarding the origin and nature of the prion and its genome.

Prusiner, S.B. "Scrapie prions." *Annual Review of Microbiology* 43:345–374, 1989. This updated work by the discoverer of prions includes descriptions of human prion diseases such as Creutzfeldt–Jakob disease.

Roizman, B., and A.E. Sears. "An inquiry into the mechanisms of herpes simplex virus latency." *Annual Review of Microbiology* 41:543–571, 1987. Describes host cell changes and provides models for establishment and maintenance of latency.

Varmus, H. "Reverse transcription." *Scientific American* 257(3):56–64, September 1987. Presents the action of this enzyme and its role in retroviruses, hepatitis B virus, and evolution.

Watson, J.D., N.H. Hopkins, J.W. Roberts, J.A. Steitz, and A.M. Weiner. *Molecular Biology of the Gene,* 4th ed. Redwood City, Calif.: Benjamin/Cummings, 1988. Chapter 7 discusses phages and Chapter 24 gives an excellent overview of the molecular biology of eucaryotic viruses.

Weinberg, R.A. "A molecular basis of cancer." *Scientific American* 249(5):126–141, November 1983. Describes two methods of oncogene activation, by point mutation and by viruses.

White, D.O., and F.J. Fenner. *Medical Virology,* 3rd ed. New York: Academic Press, 1986. A comprehensive textbook on the biology of medically important viruses.

PART THREE

Interaction Between Microbe and Host

CHAPTER 14

Principles of Disease and Epidemiology

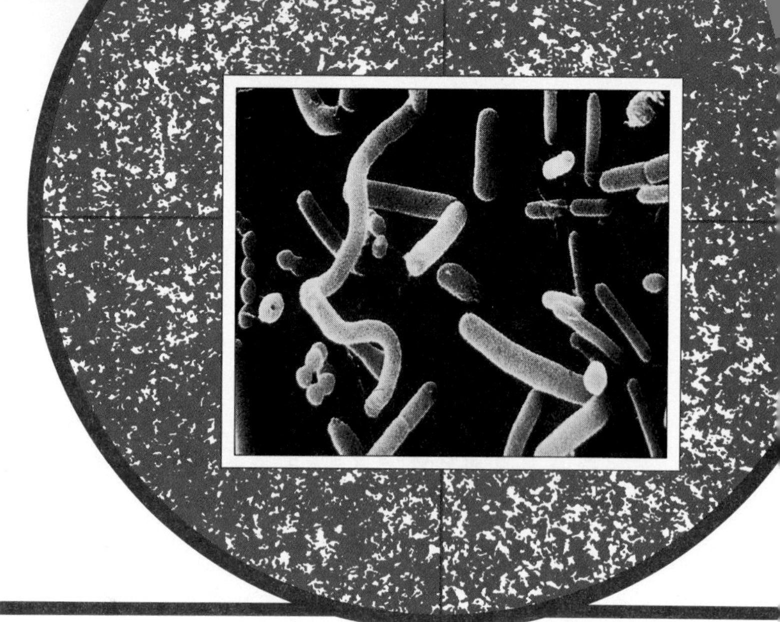

LEARNING OBJECTIVES

- Define normal flora; compare commensalism, mutualism, and parasitism, and give one example of each.
- List Koch's postulates.
- Define a reservoir of infection; contrast human, animal, and nonliving reservoirs, and give one example of each.
- Define nosocomial infections and explain their importance.
- Explain four methods of transmission of disease.
- Categorize diseases according to incidence.
- Define epidemiology, and describe three methods of epidemiologic investigation.
- Differentiate between a communicable and a noncommunicable disease.
- Define pathogen, etiology, infection, host, disease, and acute, chronic, subacute, and latent disease.
- Identify four predisposing factors for disease.
- Put the following terms in proper sequence in terms of the pattern of disease: period of decline, period of convalescence, period of illness, crisis, prodromal period, period of incubation.

Part Three photo, p. 367: Samples of the AIDS virus are frozen for later use by researchers working to develop new treatments and vaccines.

Now that you have a basic understanding of the structures and functions of microorganisms and some idea of the variety of microorganisms that exist, we can consider how the human body and various microorganisms interact in terms of health and disease.

We all have our defense mechanisms to keep us healthy. For instance, unbroken skin and mucous membranes are effective barriers against microbial invasion. Within the body, certain cells and certain specialized proteins called antibodies can work together to destroy microbes. In spite of our defenses, however, we are still susceptible to **pathogens** (disease-causing microorganisms). Some bacteria can invade our tissues and resist our defenses by producing capsules or enzymes. Other bacteria release toxins that can seriously affect our health. A rather delicate balance exists between our defenses and the disease-producing mechanisms of microorganisms. When our defenses resist these disease-producing capabilities, we maintain our health. But when the disease-producing capability overcomes our defenses, disease results. After the disease has become established, an infected person may recover completely, suffer temporary or permanent damage, or die. The outcome depends on many factors.

In Part Three, we examine some of the principles of infection and disease, the mechanisms of pathogenicity, the body's defenses against disease, and the ways that microbial diseases can be prevented by immunization and controlled by drugs. This first chapter discusses the general principles of disease, starting with a discussion of the meaning and scope of pathology. In the last section of this chapter, Epidemiology, you will learn how these principles are useful in studying and controlling disease.

Pathology, Infection, and Disease

Pathology is the scientific study of disease (*pathos* means suffering; *logos* means science). Pathology is first concerned with the cause, or **etiology,** of disease. Second, it deals with **pathogenesis,** the manner in which a disease develops. Third, pathology is concerned with the structural and functional changes brought about by disease and with its final effects on the body.

Although the terms *infection* and *disease* are sometimes used interchangeably, they differ somewhat in meaning. **Infection** is the invasion or colonization of the body by pathogenic microorganisms. **Disease** occurs when an infection results in any change from a state of health. Disease is an abnormal state in which part or all of the body is not properly adjusted or is not capable of carrying on its normal functions. An infection may exist in the absence of detectable disease. For example, the body may be infected with the virus that causes AIDS, but there may be no symptoms of the disease.

The presence of a particular type of microorganism in a part of the body where it is not normally found is also called an infection—and may lead to disease. For example, although large numbers of *E. coli* are normally present in the healthy intestine, their infection of the urinary tract usually results in disease.

Few microorganisms are pathogenic. In fact, the presence of some microorganisms can even benefit the host. Therefore, before we discuss the role of microorganisms in causing disease, let us examine the relationship of the normal flora to the human body.

Normal Flora

Animals, including humans, are generally germ-free in utero. (Viruses and some bacteria may infect a fetus in the uterus.) At birth, however, normal and characteristic microbial populations begin to establish themselves. Just before a woman gives birth, lactobacilli in her vagina multiply rapidly. The newborn's first contact with microorganisms is usually with these lactobacilli, and they become the predominant organisms in the newborn's intestine. More microorganisms are introduced to the newborn's body from the environment when breathing begins and feeding starts. After birth, *E. coli* and other bacteria acquired from foods begin to inhabit the large intestine. These microorganisms remain there throughout life and, in response to altered environmental conditions, may increase or decrease in number and contribute to disease.

Many other normally harmless microorganisms establish themselves inside other parts of the normal adult body and on its surface. It is estimated that a typical human body contains 1×10^{13} body cells yet harbors 1×10^{14} bacterial cells. This gives you an idea of the abundance of normal organisms residing in the human body. The microorganisms that establish more or less permanent residence (colonize) but that do not produce disease under normal conditions are known as **normal flora** or **normal microbiota** (Figure 14.1).

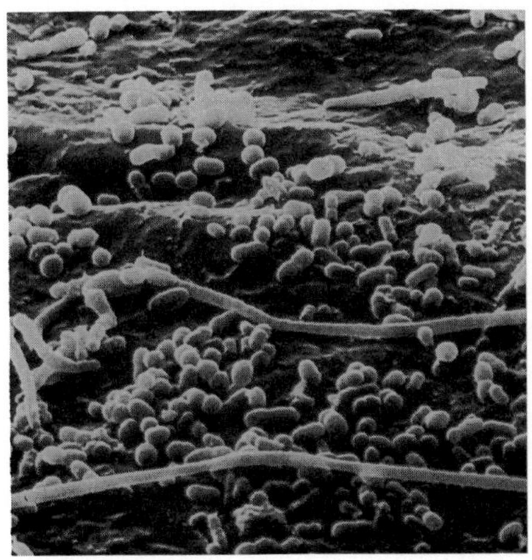
(a)

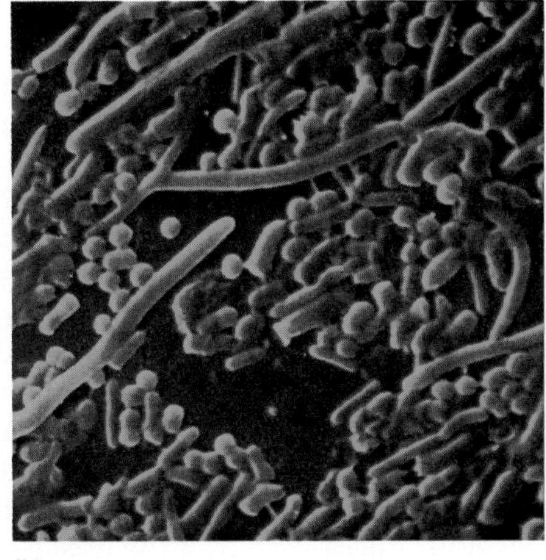
(b)

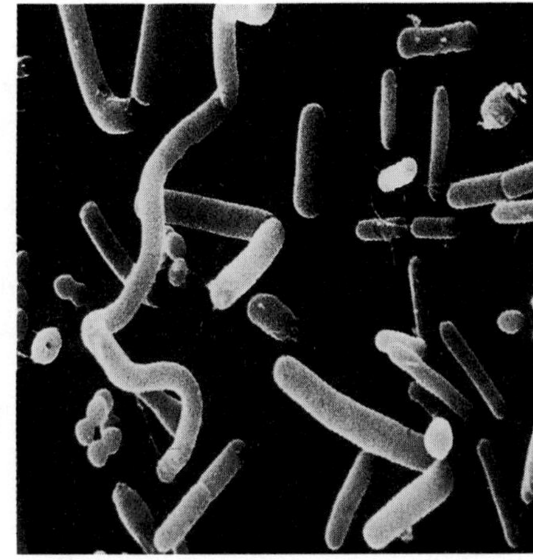
(c)

FIGURE 14.1 Representative normal flora for different regions of the body. **(a)** Bacteria on the surface of the skin. **(b)** Plaque on enamel near the gums. The bacteria that cause dental plaque, while part of the normal flora, will cause gum disease and caries unless removed frequently. **(c)** Bacteria of the large intestine.

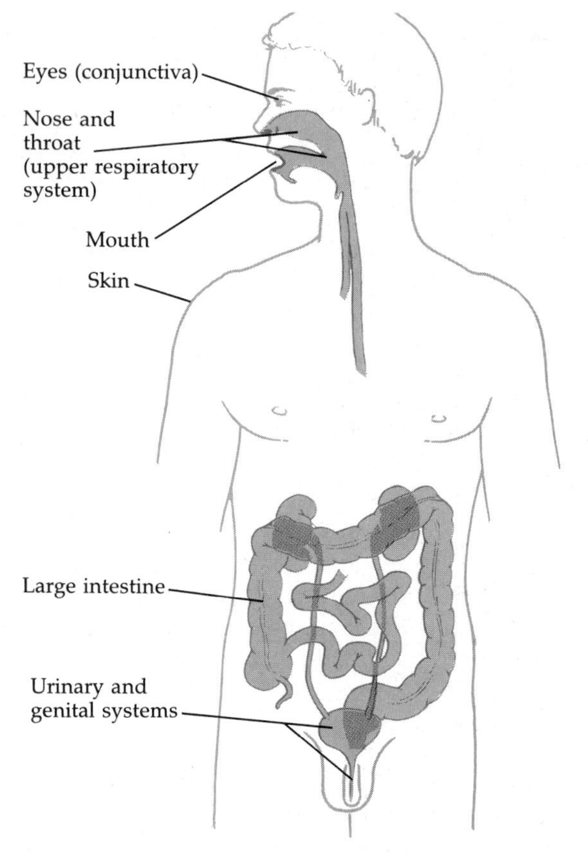

FIGURE 14.2 Locations of the normal flora on and in the human body. Various species of microbes are found in all the named regions.

Others, called the **transient flora,** may be present for several days, weeks, or months and then disappear. Microorganisms are not found throughout the entire human body but are localized in certain regions, as shown in Figure 14.2.

The principal normal floras in different regions of the body and some distinctive features of each region are listed in Table 14.1. The normal flora is also discussed more specifically in Part Four.

Relationships Between the Normal Flora and Host

Once established, the normal flora can benefit the host by preventing the overgrowth of harmful microorganisms, a phenomenon called **microbial antagonism**. When this balance between the normal flora and pathogenic microbes is upset, disease can result. For example, the normal bacterial flora of the adult human vagina maintains a local pH of from 3.5 to 4.5. The presence of the normal flora inhibits overgrowth of the yeast *Candida albicans*, which cannot grow under these conditions and is normally present in small numbers in the vagina. If the bacterial population is eliminated by antibiotics, excessive douching, or deodorants, the pH of the vagina reverts to nearly neutral, and *C. albicans* can flourish and become the dominant microorganism there. This condition can lead to a form of vaginitis (vaginal infection).

The relationship between the normal flora and the host is called **symbiosis** (living together). In the symbiotic relationship called **commensalism,** one of the organisms is benefited and the other is unaffected. Many of the microorganisms that make up our normal flora are commensals; these include the corynebacteria that inhabit the surface of the eye and certain saprophytic mycobacteria that inhabit the ear and external genitals. These bacteria live on secretions and sloughed-off cells, and they bring no apparent benefit or harm to the host.

Mutualism is a type of symbiosis that benefits both organisms. For example, the large intestine contains bacteria, such as *E. coli*, that synthesize vitamin K and some B vitamins. These vitamins are absorbed by the blood and distributed for use by body cells. In exchange, the large intestine provides nutrients for the bacteria so that they can survive. In still another kind of symbiosis, one organism is benefited at the expense of the other; this relationship is called **parasitism.** Many disease-causing bacteria are parasites.

Opportunistic Organisms

Although it is convenient to categorize symbiotic relationships by type, we must keep in mind that under certain conditions the relationship can change. For example, given the proper circumstances, a mutualistic organism, such as *E. coli*, can become harmful. *E. coli* is generally harmless as long as it remains in the large intestine. But if it gains access to other body sites, such as the urinary bladder, lungs, spinal cord, or wounds, it may cause urinary tract infections, pulmonary infections, meningitis, or abscesses, respectively.

Opportunists are potentially pathogenic organisms that ordinarily do not cause disease in their normal habitat in a healthy person. For example, organisms that gain access through broken skin or mucous membranes can cause opportunistic infections. Tooth decay and gum disease are caused by bacteria that are considered members of the normal flora of the mouth. Or, if the host is already weakened or compromised by infection, microbes that are usually harmless can cause disease. The disease AIDS is often accompanied by a common opportunistic infection, *Pneumocystis* pneumonia, caused by the opportunistic organism *Pneumocystis carinii* (see Figures 12.7 and 24.17). This secondary infection can develop in AIDS sufferers because

TABLE 14.1 Representative Members of the Normal Flora by Body Region*

REGION	PRINCIPAL COMPONENTS OF THE FLORA	COMMENT
Skin	*Propionibacterium acnes, Staphylococcus epidermidis, Staphylococcus aureus, Corynebacterium xerosis, Pityrosporum* spp. (fungus), *Candida* spp. (fungus)	Most of the microbes in direct contact with skin do not become residents because secretions from sweat and oil glands have antimicrobial properties.
Eyes (conjunctiva)	*S. epidermidis, S. aureus,* diphtheroids	The conjunctiva, a continuation of the skin or mucous membrane, contains basically the flora found on the skin.
Nose and throat (upper respiratory system)	*S. aureus, S. epidermidis,* and aerobic diphtheroids in the nose; *S. epidermidis, S. aureus,* diphtheroids, *Streptococcus pneumoniae, Hemophilus,* and *Neisseria* in the throat	Although some of the normal flora are potential pathogens, their ability to cause disease is reduced by microbial antagonism.
Mouth	Various species of *Streptococcus, Lactobacillus, Actinomyces, Bacteroides, Fusobacterium, Treponema, Corynebacterium,* and *Candida* (fungus)	Abundant moisture, warmth, and constant presence of food make the mouth an ideal environment that supports very large and diverse microbial populations on the tongue, cheeks, teeth, and gums.
Large intestine	*Bacteroides, Fusobacterium, Lactobacillus, Enterococcus, Bifidobacterium, Escherichia coli, Enterobacter, Citrobacter, Proteus, Klebsiella, Shigella, Candida* (fungus)	The large intestine contains the largest numbers of resident flora in the body because of its available moisture and nutrients.
Urinary and genital systems	*Staphylococcus epidermidis,* aerobic micrococci, *Enterococcus, Lactobacillus,* aerobic diphtheroids, *Pseudomonas, Klebsiella,* and *Proteus* in urethra; lactobacilli in vagina; aerobic diphtheroids, *Streptococcus, Staphylococcus, Bacteroides, Clostridium, Candida albicans* (fungus), and *Trichomonas vaginalis* (protozoan) in vagina	The lower urethra in both sexes has a resident population; the vagina has its acid-tolerant population of microbes because of the nature of its secretions.

*This table lists some flora that are not discussed in this chapter; they will be discussed in Part Four. Unless indicated, the organisms are bacteria.

their immune systems are suppressed. Prior to the AIDS epidemic, this type of pneumonia was rare.

In addition to the usual symbionts, many people carry other microorganisms that are generally regarded as pathogenic but that may not cause disease in those people. Among the pathogens that are frequently carried in normal individuals are echoviruses ("echo" comes from *e*nteric *c*ytopathogenic *h*uman *o*rphan), which can cause intestinal diseases, and adenoviruses, which can cause respiratory diseases. *Neisseria meningitidis,* which often resides benignly in the respiratory tract, can cause meningitis, a disease that inflames the coverings of the brain and spinal cord. *Streptococcus pneumoniae,* a normal resident of the nose and throat, can cause a type of pneumonia.

Etiology of Infectious Diseases

Some diseases, such as polio, have a well-known etiology. Some have an etiology that is not completely understood—for example, the relationship between chronic fatigue syndrome and EB virus. For still others, such as Creutzfeldt–Jakob disease, the etiology is unknown. Of course, not all diseases are caused by microorganisms. For example, the disease hemophilia is an *inherited (genetic) disease*; osteoarthritis and cirrhosis are considered *degenerative diseases*. There are several other categories of disease, but here we will discuss only *infectious diseases,* those caused by microorganisms. To see how microbiologists determine the etiology of an infectious disease, we will discuss in greater detail the work of Robert Koch that was introduced in Chapter 1.

KOCH'S POSTULATES

In the historical overview of microbiology presented in Chapter 1, we briefly discussed Koch's famous postulates. Koch, you might recall, was a German physician who played a major role in establishing that microorganisms cause specific diseases. In 1877, Koch published some early papers on anthrax, a disease of cattle that can also occur in humans. Koch demonstrated that certain bacteria, today known as *Bacillus anthracis,* were always present in the blood of animals that had the disease and were not present in healthy animals.

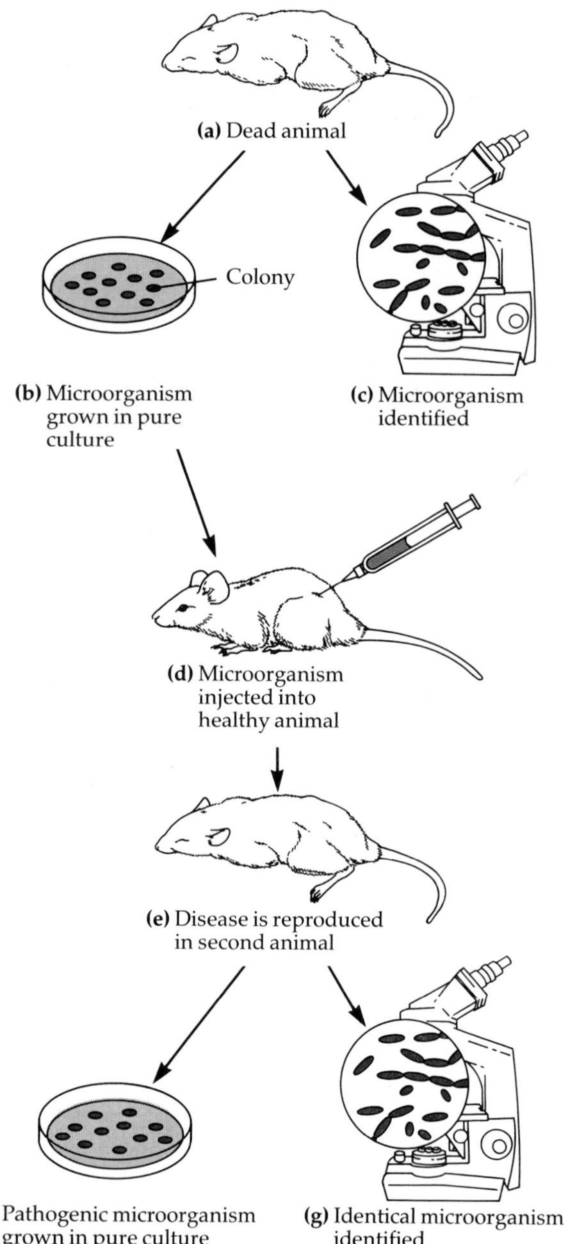

FIGURE 14.3 Koch's postulates. **(a)** Microorganisms from a dead animal are **(b)** grown in pure culture and **(c)** identified. **(d)** A sample of the pure culture is inoculated into a healthy animal, and **(e)** the disease is reproduced in the animal. **(f)** Pathogenic microorganisms are again grown in pure culture and **(g)** identified, confirming that they are identical with the original pathogen.

He knew that the mere presence of the bacteria did not prove that they had caused the disease; the bacteria could have been there as a result of the disease. Thus he experimented further. He took a sample of blood from a diseased animal and injected it into a healthy animal. The second animal developed the same disease and died. He repeated this procedure many times, always with the same results. (A key criterion in the validity of any scientific proof is that experimental results be repeatable.) Koch also cultivated the microorganism in fluids outside the animal's body, and he demonstrated that the bacterium would cause anthrax even after many culture transfers. In short, he showed that a specific infectious disease (anthrax) is caused by a specific microorganism (*B. anthracis*) that can be isolated and cultured on artificial media.

Koch later used the same methods to show that the bacterium *Mycobacterium tuberculosis* is the causative agent of tuberculosis.

Koch's research provides a framework for the study of the etiology of any infectious disease. Today, we refer to Koch's experimental requirements as **Koch's postulates** (Figure 14.3). They are summarized as follows:

1. The same pathogen must be present in every case of the disease.
2. The pathogen must be isolated from the diseased host and grown in pure culture.
3. The pathogen from the pure culture must cause the disease when it is inoculated into a healthy, susceptible laboratory animal. In exclusively human diseases, this step is not morally acceptable, but it sometimes happens accidentally in a laboratory or through other circumstances. The case of the "bubble boy" (see Chapter 13, p. 359) is an example where a pathogen accidentally introduced into a human confirmed the cause of a disease.
4. The pathogen must be isolated from the inoculated animal and must be shown to be the original organism.

EXCEPTIONS TO KOCH'S POSTULATES

Although Koch's postulates are useful in determining the causative agent of most bacterial diseases, there are some exceptions. For instance, it is known that the bacterium *Treponema pallidum* is the causative agent for syphilis, but virulent strains have never been cultured on artificial media. The causative agent of leprosy, *Mycobacterium leprae,* has never been grown on artificial media. Moreover, many rickettsial and viral pathogens cannot be cultured on artificial media because they multiply only within cells.

The discovery of microorganisms that cannot grow on artificial media has made necessary some modifications of Koch's postulates and the use of alternative methods of culturing and detecting particular microbes. For example, when investigators searching for the microbial cause of legionellosis (Legionnaires' disease) were unable to isolate the microbe directly from a victim, they took the alternative step of inoculating a victim's lung tissue into guinea pigs. These guinea pigs developed the disease's pneumonialike symptoms, whereas guinea pigs inoculated with tissue from an unafflicted person did not. Then tissue samples from the diseased guinea pigs were cultured in yolk sacs of chick embryos—a method, discussed in Chapter 13, that reveals the growth of extremely small microbes. After the embryos were incubated, electron microscopy revealed rod-shaped bacteria in the chick embryos. Finally, modern immunologic techniques (which will be discussed in Chapter 18) were used to show that the bacteria in the chick embryos were the same bacteria as those in the guinea pigs and in afflicted humans.

In a number of situations, a human host exhibits certain signs and symptoms that are associated only with a certain pathogen and its disease. For example, the pathogens responsible for diphtheria and tetanus give rise to distinguishing signs and symptoms that can be produced by no other microbe. They are unequivocally the only organisms that produce their respective diseases. But some infectious diseases are not quite so clear-cut. For example, nephritis (inflammation of the kidneys) can involve any of several different pathogens, all of which give rise to the same signs and symptoms. Thus, it is often difficult to know which particular microorganism is causing a disease. Other infectious diseases that sometimes have poorly defined etiologies are pneumonia, meningitis, and peritonitis (inflammation of the peritoneum, the membrane that lines the abdomen and pelvis and covers the organs within them).

Some pathogens can cause several disease conditions. *Mycobacterium tuberculosis*, for example, is implicated in diseases of the lungs, skin, bones, and internal organs. *Streptococcus pyogenes* can cause sore throat, scarlet fever, skin infections (erysipelas), puerperal fever, and osteomyelitis (inflammation of bone), among other diseases. When clinical signs and symptoms are used together with laboratory methods, these infections can usually be distinguished from infections of the same organs by other pathogens.

Classifying Infectious Diseases

Each disease affecting the body alters body structures and functions in particular ways, and these alterations are usually indicated by several kinds of evidence. For example, the patient may experience certain **symptoms**—changes in body function—such as pain and malaise (a vague feeling of body discomfort). These *subjective* changes are not apparent to the observer. The patient can also exhibit **signs,** which are *objective* changes that the physician can observe and measure. Frequently evaluated signs include lesions (changes produced in tissues by disease), swelling, fever, and paralysis. Sometimes a specific group of symptoms or signs always accompanies a particular disease. Such a group is called a **syndrome.** The diagnosis of a disease is made by evaluation of the signs and symptoms together with the results of laboratory tests.

Diseases are often classified in terms of how they behave within a host and within a given population. Any disease that spreads from one host to another, either directly or indirectly, is said to be a **communicable disease.** Chickenpox, measles, genital herpes, typhoid fever, and tuberculosis are examples. Chickenpox and measles are also examples of **contagious diseases** that are *easily* spread from one person to another. A **noncommunicable disease** is not spread from one host to another. These diseases are caused by microorganisms that normally inhabit the body and only occasionally produce disease or by microorganisms that reside outside the body and produce disease only when introduced into the body. An example is tetanus. *Clostridium tetani* produces disease only when it is introduced into the body via abrasions or wounds.

OCCURRENCE OF DISEASE

To understand the full scope of a disease, we should know something about its occurrence. The **incidence** of a disease is the fraction of a population that contracts it during a particular period. The **prevalence** of a disease is the fraction of the population having the disease at a specified time. Knowing the incidence and prevalence of a disease in different populations (for example, in populations representing different geographical regions or different racial groups) permits epidemiologists to estimate the range of the disease's occurrence and its tendency to affect some groups of people more than others.

Frequency of occurrence is another criterion that is used in the classification of diseases. If a particular disease occurs only occasionally, it is called a **sporadic disease.** Typhoid fever in the United States is such a disease. A disease constantly present in a population is called an **endemic disease;** an example of such a disease is the common cold. If many people in a given area acquire a certain disease in a relatively short period, it is called an **epidemic disease.** Influenza is an example of a disease that often achieves epidemic status. Public health authorities consider AIDS near or at epidemic status at this time (see Figure 19.12). Some authorities consider gonorrhea and certain other sexually transmitted diseases to be epidemic at this time as well. An epidemic disease that occurs worldwide is called a **pandemic disease.** We experience pandemics of influenza from time to time. Some authorities also consider AIDS to be pandemic.

SEVERITY OR DURATION OF DISEASE

Another useful way of defining the scope of a disease is in terms of its severity or duration. An **acute disease** is one that develops rapidly but lasts only a short time. A good example is influenza. A **chronic disease** develops more slowly, and the body's reactions may be less severe, but the disease is likely to be continual or recurrent for long periods. Infectious mononucleosis, tuberculosis, syphilis, and leprosy fall into this category. A disease that is intermediate between acute and chronic is described as a **subacute disease.** An example is subacute sclerosing panencephalitis, a rare brain disease characterized by diminished intellectual function and loss of nervous function. A **latent disease** is one in which the causative agent remains inactive for a time but then becomes active to produce symptoms of the disease. An example is shingles, one of the diseases caused by the varicella-zoster virus.

EXTENT OF HOST INVOLVEMENT

Infections can also be classified according to the extent to which the host's body is affected. A **local infection** is one in which the invading microorganisms are limited to a relatively small area of the body. Examples of local infections are boils and abscesses. In a **systemic (generalized) infection,** microorganisms or their products are spread throughout the body by the blood or lymphatic system. Measles is an example of a systemic infection. Very frequently, agents of a local infection enter a blood or lymph vessel and spread to other parts of the body. We refer to this condition as a **focal infection.** Focal infections can arise from infections in the teeth, tonsils, or sinuses. The presence of bacteria in the blood is known as **bacteremia,** and if the bacteria actually multiply in the blood, the condition is called **septicemia. Toxemia** refers to the presence of toxins in blood (as occurs in tetanus), and **viremia** refers to the presence of viruses in blood.

The state of host resistance also determines the extent of infections. A **primary infection** is an acute infection that causes the initial illness. A **secondary infection** is one caused by an opportunist microbe after the primary infection has weakened the body's defenses. Secondary infections of the skin and respiratory tract are common and are sometimes more dangerous than the primary infections. *Pneumocystis* pneumonia as a consequence of AIDS is an example of a secondary infection. Streptococcal bronchopneumonia following influenza is an example of a secondary infection more serious than the primary infection. An **inapparent (subclinical) infection** is one that does not cause any noticeable illness. Polio and hepatitis A, for example, can be carried by persons who never develop the illness.

Spread of Infection

Now that you have an understanding of the concept of the normal flora, the etiology of infectious diseases, and the types of infectious diseases, we will examine the source of pathogens and how diseases are transmitted.

RESERVOIRS

For a disease to perpetuate itself, there must be a continual source of the disease organisms. This source can be either a living organism or an inanimate object that provides a pathogen with adequate conditions for survival and multiplication and an opportunity for transmission. Such a source is called a **reservoir of infection.** These reservoirs may be human, animal, or nonliving (Figure 14.4).

Human Reservoirs

The principal living reservoir of human disease is the human body itself. Many people harbor pathogens and transmit them directly or indirectly to others. People with signs and symptoms of a disease may transmit the disease. In addition, some people can harbor pathogens and transmit them to others without exhibiting any signs of illness. These people, called **carriers,** are important living reservoirs of infection. Some carriers have inapparent infections for which no signs or symptoms are ever exhibited. Other people, for example, those with latent diseases, carry a disease during its symptom-free stages—during the incubation period (before symptoms appear) or during the convalescent period (recovery). Human carriers play an important role in the spread of such diseases as AIDS, diphtheria, typhoid fever, hepatitis, gonorrhea, amoebic dysentery, and streptococcal infections.

Animal Reservoirs

Both wild (sylvatic) and domestic animals are living reservoirs of microorganisms that can cause human diseases. Diseases that occur primarily in wild and domestic animals and can be transmitted to humans are called **zoonoses.** Some types of influenza (transmitted by wild birds), rabies (transmitted by bats, skunks, foxes, dogs, and cats), and Rocky Mountain spotted fever (transmitted by ticks) are examples of zoonoses. Other representative zoonoses are presented in Table 14.2.

About 150 zoonoses are known. The transmission of zoonoses to humans can occur via one of many

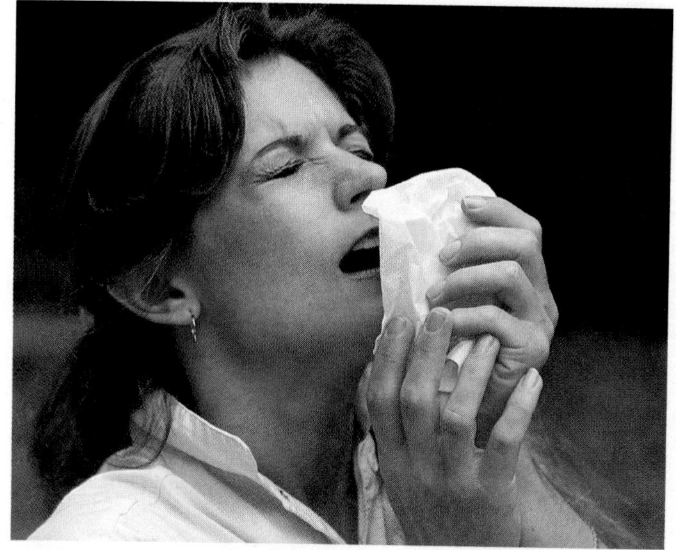
(a)

(b)

(c)

FIGURE 14.4 Reservoirs of disease. **(a)** Human. **(b)** Animal. **(c)** Nonliving.

routes—by direct contact with infected animals, by contamination of food and water, by contact with contaminated hides, fur, or feathers, by consumption of infected animal products, or by insect vectors (insects that transmit pathogens).

Nonliving Reservoirs

The two major nonliving reservoirs of infectious disease are soil and water. Soil harbors such pathogens as fungi, which cause mycoses, and *Clostridium botulinum*, the bacterial agent that causes botulism. Water that has been contaminated by the feces of humans and other animals is a reservoir for several pathogens, most notably those responsible for gastrointestinal diseases.

TRANSMISSION OF DISEASE

The causative agents of disease can be transmitted from the reservoir of infection to a susceptible host by three principal routes: (1) contact, (2) vehicle, and (3) vectors.

(a)

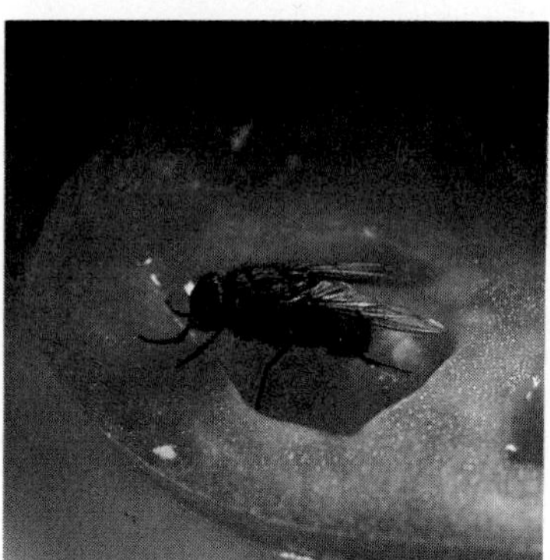
(c)

FIGURE 14.5 Means of transmission of disease. **(a)** Contact transmission (direct). **(b)** Vehicle transmission (foodborne). **(c)** Vector transmission (mechanical).

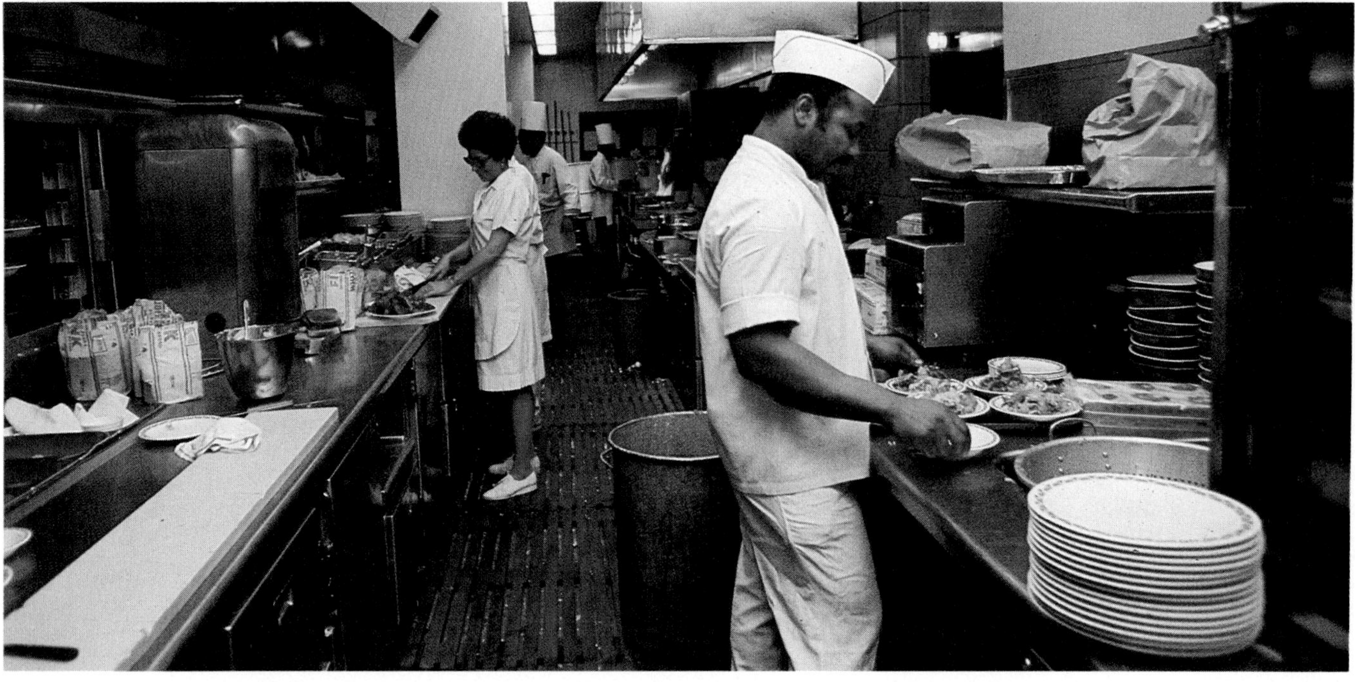
(b)

Contact Transmission

Contact transmission is the spread of the agent of disease by direct contact, indirect contact, or droplet transmission (Figure 14.5a). *Direct contact transmission* refers to the direct transmission of an agent by physical contact between its source and a susceptible host; no intermediate object is involved. This is also known as *person-to-person transmission*. The most common forms of direct contact transmission are touching, kissing, and sexual intercourse. Among the diseases that can be transmitted by direct contact are viral respiratory tract diseases (common cold and influenza), staphylococcal infections, hepatitis A, measles, scarlet fever, smallpox, and sexually transmitted diseases (syphilis, gonorrhea, and genital herpes). Direct contact is also one way of spreading AIDS and infectious mononucleosis. To guard against person-to-person transmission, health personnel use gloves and other protective measures (Figure 14.6). Potential pathogens can also be transmitted by direct contact from animals (or animal products) to humans. Examples are the pathogens causing rabies and anthrax.

Indirect contact transmission occurs when the agent of disease is transmitted from its reservoir to a susceptible host by means of a nonliving object. The general term for any nonliving object involved in the spread of an infection is a **fomite.** Examples of fomites are tissues, handkerchiefs, towels, bedding, diapers, drinking cups, eating utensils, toys, money, and thermometers. Contaminated syringes serve as fomites in the transmission of AIDS and hepatitis B. Other fomites may transmit diseases such as tetanus.

Droplet transmission is a third type of contact transmission in which microbes are spread in *droplet nuclei* (mucus droplets) that travel only short distances (Figure 14.7). These droplets are discharged into the air by coughing, sneezing, laughing, or talking and travel less than one meter from the reservoir to the host. Disease agents that travel such short distances are not regarded as airborne (discussed shortly). Examples of diseases spread by droplet transmission are influenza, pneumonia, and whooping cough.

Vehicle Transmission

Vehicle transmission refers to the transmission of disease agents by a medium, such as water, food, and air (Figure 14.5b). Other media include blood and other body fluids, drugs, and intravenous fluids. A hepatitis outbreak caused by vehicle transmission is described in the box in Chapter 25 (p. 634). Here, we will discuss water, food, and air as vehicles of transmission.

In *waterborne transmission*, pathogens usually are spread by water contaminated with untreated or poorly treated sewage. Diseases transmitted via this route include cholera, waterborne shigellosis, and leptospirosis. In *foodborne transmission*, pathogens are generally transmitted in foods that are incompletely cooked, poorly refrigerated, or prepared under unsanitary conditions. Foodborne pathogens cause diseases such as food poisoning and tapeworm infestation. *Airborne transmission* refers to the spread of agents of infection by droplet nuclei in dust that travel more than one meter from the reservoir to the host. In one type of airborne transmission, microbes are spread by droplets, which may be discharged in a fine spray from the mouth and nose during coughing and sneezing (see Figure 14.7) and which are small enough to remain airborne for prolonged periods. The virus that causes measles and the bacterium that causes tuberculosis can be transmitted via airborne droplets. Dust particles can harbor various pathogens. Staphylococci and streptococci can survive on dust and be transmitted by the airborne route. Spores produced by certain fungi are also transmitted by the airborne route and can cause such diseases as histoplasmosis, coccidioidomycosis, and blastomycosis (see Chapter 24).

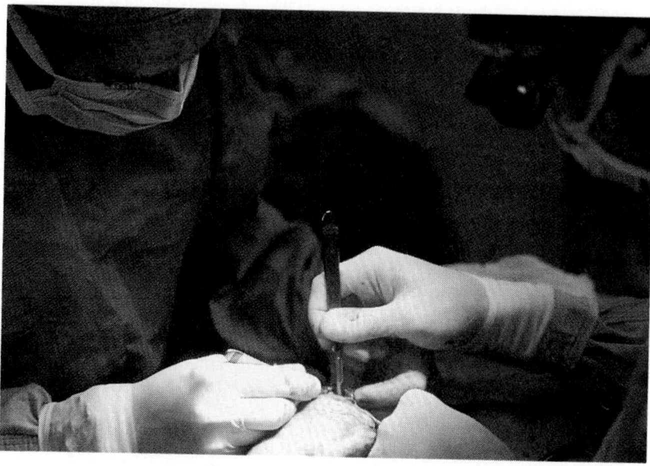

FIGURE 14.6 Health workers use masks, gloves, and other protective measures to avoid transmission of pathogens.

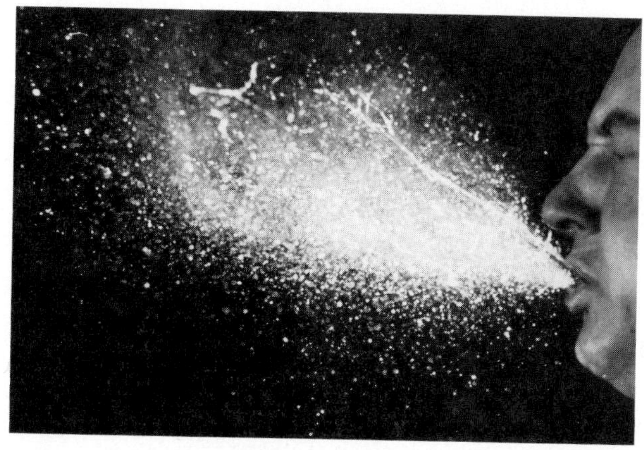

FIGURE 14.7 Droplet transmission. This high-speed photograph shows the spray of small droplets that come from the mouth during a sneeze.

Vectors

Arthropods are the most important group of disease **vectors**—animals that carry pathogens from one host to another. (Insects and other arthropod vectors were discussed in Chapter 12.) Arthropod vectors transmit disease by two general methods. *Mechanical transmission* is the passive transport of the pathogens on the insect's feet or other body parts (Figure 14.5c). If the insect makes contact with a host's food, pathogens can be transferred to the food and later swallowed by the host. Houseflies, for instance, are capable of transferring the pathogens of typhoid fever and bacillary dysentery (shigellosis) from the feces of infected persons to food.

TABLE 14.2 Selected Zoonoses That Can Be Transmitted to Humans

DISEASE	CAUSE	RESERVOIR	METHOD OF TRANSMISSION	REFER TO CHAPTER
Viral				
Influenza (some types)	Orthomyxovirus	Swine	Direct contact	24
Rabies	Rhabdovirus	Bats, skunks, foxes, dogs, cats	Direct contact (bite)	22
Western equine encephalitis	Arbovirus	Horses, birds	*Culex* mosquito bite	22
Yellow fever	Arbovirus	Monkeys	*Aedes* mosquito bite	23
Bacterial				
Anthrax	*Bacillus anthracis*	Domestic livestock	Direct contact with contaminated hides or animals; air; food	23
Brucellosis	*Brucella* spp.	Domestic livestock	Direct contact with contaminated milk, meat, or animals	23
Bubonic plague	*Yersinia pestis*	Rodents	Flea bites	23
Cat-scratch fever	Gram-negative bacillus	Domestic cats	Direct contact	23
Leptospirosis	*Leptospira*	Wild mammals, domestic dogs and cats	Direct contact with urine, soil, water	26
Lyme disease	*Borrelia burgdorferi*	Field mice, deer	Tick bites	23
Pneumonic plague	*Yersinia pestis*	Rodents	Direct contact	23
Psittacosis (Ornithosis)	*Chlamydia psittaci*	Birds, especially parrots	Direct contact	24
Q fever	*Coxiella burnetii*	Domestic livestock	Inhalation, tick bites (between animals), direct contact with contaminated milk	24
Rocky Mountain spotted fever	*Rickettsia rickettsii*	Rodents	Tick bites	23
Salmonellosis	*Salmonella* spp.	Poultry, rats, turtles	Ingestion of contaminated food, water	25
Tularemia	*Francisella tularensis*	Wild and domestic mammals, especially wild rabbits	Direct contact with infected animals; deerfly bites	23
Typhus fever	*Rickettsia typhi*	Rodents	Flea bites	23
Fungal				
Ringworms	*Trichophyton* *Microsporum* *Epidermophyton*	Domestic mammals	Direct contact; fomites	21
Protozoan				
Chagas' disease	*Trypanosoma cruzi*	Wild mammals	Kissing bug bite	23
Malaria	*Plasmodium* spp.	Monkeys	*Anopheles* mosquito bite	23

(Continued)

TABLE 14.2 (Continued)

DISEASE	ETIOLOGY	RESERVOIR	METHOD OF TRANSMISSION	REFER TO CHAPTER
Toxoplasmosis	*Toxoplasma gondii*	Cats and other mammals	Direct contact with infected tissues or fecal material	23
Helminthic				
Hydatid cyst	*Echinococcus granulosus*	Dogs	Direct contact with fecal material	12
Tapeworm (beef)	*Taenia saginata*	Cattle	Ingestion of contaminated beef	25
Trichinosis	*Trichinella spiralis*	Pigs, bears	Ingestion of contaminated meat	25

Biological transmission is an active process and is more complex. The arthropod bites an infected person or animal and ingests some of the infected blood. The pathogens then reproduce in the vector, and the increase in the number of pathogens increases the possibility that they will be transmitted to another host. Some parasites reproduce in the gut of the arthropod; these can be passed with feces. If the arthropod defecates or vomits while biting a potential host, the parasite can enter the wound. Other parasites reproduce in the vector's gut and migrate to the salivary gland. These are directly injected into a bite. Some protozoan and helminthic parasites use the vector as a host for a developmental stage in their life cycle.

Table 14.3 lists a few important arthropod vectors and the diseases they transmit.

PORTALS OF EXIT

In the next chapter, you will learn how microorganisms enter the body through a preferred route or portal of entry, such as mucous membrane or skin. For now, since we are concerned mainly with the spread of disease throughout a population, it is important to know that pathogens also have definite routes of exit, called **portals of exit**. In general, portals of exit are related to the part of the body that has been infected.

The most common portals of exit are the respiratory and gastrointestinal tracts. For example, many pathogens living in the respiratory tract exit in discharges from the mouth and nose; these discharges are expelled during coughing or sneezing. These microorganisms are found in droplets formed from mucus.

TABLE 14.3 Representative Arthropod Vectors and the Diseases They Transmit

DISEASE	CAUSATIVE AGENT	ARTHROPOD VECTOR	REFER TO CHAPTER
Malaria	*Plasmodium* spp.	*Anopheles* (mosquito)	23
African trypanosomiasis	*Trypanosoma brucei gambiense* and *Trypanosoma brucei rhodesiense*	*Glossina* sp. (tsetse fly)	22
American trypanosomiasis	*Trypanosoma cruzi*	*Triatoma* sp. (kissing bug)	23
Yellow fever	Arbovirus (yellow fever virus)	*Aedes* (mosquito)	23
Dengue	Arbovirus (dengue fever virus)	*Aedes aegypti* (mosquito)	23
Arthropod-borne encephalitis	Arbovirus (encephalitis virus)	*Culex* (mosquito)	22
Epidemic typhus	*Rickettsia prowazekii*	*Pediculus humanus* (louse)	23
Endemic murine typhus	*Rickettsia typhi*	*Xenopsylla cheopis* (rat flea)	23
Rocky Mountain spotted fever	*Rickettsia rickettsii*	*Dermacentor andersoni* and other species (tick)	23
Plague	*Yersinia pestis*	*Xenopsylla cheopis* (rat flea)	23
Relapsing fever	*Borrelia* sp.	*Ornithodorus* spp. (soft ticks)	23
Lyme disease	*Borrelia burgdorferi*	*Ixodes spp.* (ticks)	23
Babesiosis	*Babesia microti*	*Ixodes dammini*	Box Ch. 23

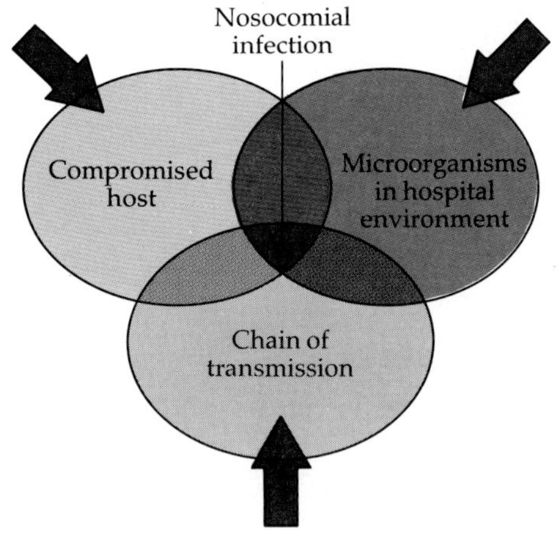

FIGURE 14.8 Nosocomial infections. Interaction of three principal factors that contribute to nosocomial infections.

Pathogens that cause tuberculosis, whooping cough, pneumonia, scarlet fever, meningococcal meningitis, measles, mumps, smallpox, and influenza are discharged through the respiratory route. Other pathogens exit from the gastrointestinal tract in feces or saliva. Feces may be contaminated with pathogens associated with salmonellosis, cholera, typhoid fever, shigellosis, amoebic dysentery, and poliomyelitis. Saliva can also contain pathogens, such as the rabies virus.

Another important route of exit is the urogenital tract. Microbes responsible for sexually transmitted diseases are found in secretions from the penis and vagina. Urine can also contain the pathogens responsible for typhoid fever and brucellosis, which can exit via the urinary tract. Skin or wound infections are other portals of exit. Drainage from wounds can spread infections to another person directly or by contact with a contaminated fomite. Infected blood can be removed and reinjected by biting insects and contaminated needles and syringes to spread infection within the population. Examples of diseases transmitted by biting insects are yellow fever, Rocky Mountain spotted fever, tularemia, and malaria. AIDS and hepatitis B may be transmitted by contaminated needles and syringes.

Nosocomial (Hospital-Acquired) Infections

Infections acquired as a result of a hospital stay are called **nosocomial infections.** (The word *nosocomial* is derived from the Greek word for hospital; the term also includes infections acquired in nursing homes and other health-related facilities.) The Centers for Disease Control (a public health service) estimates that from 5% to 15% of all hospital patients acquire some type of nosocomial infection. It is estimated that about 20,000 people die of nosocomial infections each year.

Nosocomial infections result from the interaction of several factors: (1) microorganisms in the hospital environment, (2) the compromised (or weakened) status of the host, and (3) the chain of transmission in the hospital. Figure 14.8 illustrates how it is not simply the presence of any single one of these factors but rather the interaction of all three factors that poses a significant risk of nosocomial infection.

MICROORGANISMS IN THE HOSPITAL

Although every effort is made to kill or check the growth of microorganisms in the hospital, the hospital environment is a major reservoir for a variety of pathogens. One reason is that the bacteria of the normal

TABLE 14.4 Bacteria Involved in More Than 70% of All Nosocomial Infections

BACTERIA	*PERCENTAGE OF TOTAL INFECTIONS*	*INFECTIONS CAUSED*
Enterobacteria (*E. coli*, *Klebsiella* species, *Proteus* species, *Enterobacter* species, and *Serratia marcescens*)	Over 40	Urinary tract infections, peritonitis, bacteremia, pneumonia, septicemia, gastrointestinal inflammations, and neonatal (in newborns) meningitis
Staphylococcus aureus	11	Urinary and respiratory tract infections and endocarditis (inflammation of the lining of the heart)
Enterococcus	10	Urinary tract surgical wound infections and endocarditis
*Pseudomonas aeruginosa**	9	Burn and surgical wound infections, septicemia, and pneumonia

*See the box in Chapter 7, p. 180.

MMWR

Morbidity and Mortality Weekly Report

Postsurgical Infections Associated with Breaks in Aseptic Techniques—California, Illinois, Maine, and Michigan, 1990

In May and June 1990, the Hospital Infections Program in CDC's Center for Infectious Diseases received reports of four clusters of postsurgical infections occurring in patients after a variety of surgical procedures.

California. During an 8-day period, five patients at one hospital developed *Staphylococcus aureus* surgical wound infections (SWI) following clean surgical procedures. All patients developed fever and SWI within 12–72 hours of surgery. All *S. aureus* isolates had the same phage type. An epidemiological investigation identified use of an intravenous anesthetic, propofol, and attendance by one anesthesiologist as risk factors. A throat culture of the implicated anesthesiologist grew *S. aureus*; the isolate had the same phage type as that recovered from the patients' wounds.

Illinois. During a 5-day period, four patients who underwent different surgical procedures at one hospital developed *Candida albicans* bloodstream infections and/or endophthalmitis. An epidemiologic investigation identified receipt of propofol and preparation by one anesthesiologist as risk factors for infection.

Maine. During a 2-day period, two patients who each underwent different surgical procedures at one hospital developed fever and hypertension within 2 hours following surgery. Both patients recovered after aggressive supportive therapy. An epidemiological investigation identified receipt of propofol and preparation by one nurse anesthetist as risk factors for the reactions. The same syringe and propofol preparation were used for the two patients; cultures and endotoxin assays of the propofol solution at the time of the second patient's reactions grew *Moraxella osloensis* and detected endotoxin.

Michigan. During a 2-week period, 13 of 56 patients at one hospital developed postoperative *S. aureus* bacteremia and/or SWI; all patient isolates had the same phage type. Epidemiologic studies identified receipt of propofol and preparation by one nurse anesthetist as risk factors for infection. Cultures of the hands of the implicated nurse anesthetist grew *S. aureus*; phage typing is pending. A review of anesthesia procedures revealed that propofol remaining at the completion of one surgery was used during the next surgical procedure.

Comment: For at least four reasons, the preliminary results of these investigations suggest that propofol was contaminated during manipulation after receipt from the manufacturer and not contaminated at the time of manufacture. First, at each of the hospitals investigated, different lots of propofol were used, and cultures and endotoxin assays of previously unopened ampules from each hospital were negative. Second, at each hospital, cases were associated only with propofol that was administered via a 60-ml syringe in a pump and prepared by a specific anesthetist/anesthesiologist. Third, aseptic technique was not observed during preparation of the propofol for use; syringes were usually used on multiple patients. Fourth, since propofol is delivered over a longer period than most anesthetics, contaminating microorganisms could proliferate during use and between use in different patients. Growth studies performed at CDC show that when propofol is inoculated with low numbers (10^1–10^2 cfu/ml) of *S. aureus*, the organisms rapidly proliferate to large numbers (10^5–10^6 cfu/ml) within 24 hours at 33°C.

The investigation of the current clusters suggests that severe, life-threatening complications may occur in patients as a consequence of breaks in health care workers' aseptic technique in combination with the use of a drug that is capable of supporting the rapid growth of microorganisms. These outbreaks underscore the importance of aseptic technique and infection control in anesthesia practice.

Source: *MMWR* 39:426–427; 433 (6/29/90).

flora from the human body are opportunists and present a particularly strong danger to hospital patients. In fact, most of the bacteria that cause nosocomial infections do not cause disease in healthy people but are pathogenic only for individuals whose defenses have been weakened by illness or therapy.

In the past, most nosocomial infections were caused by gram-positive microbes. At one time, the gram-positive *Staphylococcus aureus* was the primary cause of nosocomial infections. Although antibiotic-resistant strains of that organism are still significant factors, the major causes today are gram-negative bacteria, such as *E. coli* and *Pseudomonas aeruginosa*, especially the antibiotic-resistant strains. *P. aeruginosa* has the ability to cause opportunistic skin infections, especially in surgical and burn patients. The major bacteria involved in nosocomial infections are summarized in Table 14.4.

In addition to the threat of opportunists in the hospital, some microorganisms become resistant to antimicrobial drugs, which are commonly used in hospitals. For example, *P. aeruginosa* and other such gram-negative bacteria tend to be difficult to control with antibiotics because of their R factors, which carry genes that determine resistance to antibiotics (see Chapter 8). As the R factors recombine, new and multiple resistance factors are produced. These strains become part of the flora of patients and hospital personnel and become progressively more resistant to antibiotic therapy. In this way, people become part of the reservoir (and chain of transmission) for antibiotic-resistant strains of bacteria. Usually, if the host's resistance is high, the new strains are not much of a problem. But if disease, surgery, or trauma has weakened the host's defenses, secondary infections may be difficult to treat.

COMPROMISED HOST

A **compromised host** is one whose resistance to infection is impaired by disease, therapy, or burns. Two principal conditions can compromise the host: (1) broken skin or mucous membranes and (2) a suppressed immune system.

As long as the skin and mucous membranes remain intact, they provide formidable physical barriers against most pathogens. Burns, surgical wounds, trauma (such as accidental wounds), injections, invasive diagnostic procedures, intravenous therapy, and urinary catheters (used to drain urine) can all break the first line of defense and make an individual more susceptible to disease in hospitals. Burn patients are especially susceptible to nosocomial infections because their skin is no longer an effective barrier to microorganisms.

The risk of infection is also related to invasive procedures, such as administering anesthesia, which may alter respiration and contribute to pneumonia, and tracheotomy, in which an incision is made into the trachea to assist breathing (see the box on p. 381).

In healthy individuals, white blood cells called T lymphocytes provide resistance to disease by killing pathogens directly, mobilizing phagocytes and other lymphocytes, and secreting chemicals that kill pathogens. White blood cells called B lymphocytes, which develop into antibody-producing cells, also protect against infection. Antibodies provide immunity by such actions as neutralizing toxins, inhibiting attachment of a pathogen to host cells, and helping to lyse pathogens. Drugs, radiation therapy, steroid therapy, burns, diabetes, leukemia, kidney disease, stress, and malnutrition can all adversely affect the actions of T and B lymphocytes and compromise the host. In addition, the AIDS virus destroys certain T cells.

TABLE 14.5 The Principal Kinds of Nosocomial Infections

TYPE OF INFECTION	*COMMENT*
Urinary tract infection	Most common, usually accounting for about 50% of all nosocomial infections. Typically related to urinary catheterization.
Surgical wound infection	Ranks second in incidence (about 25%). It is estimated that from 5% to 12% of all surgical patients develop postoperative infections; the percentage can reach 30% for certain surgeries, such as colon surgery and amputations.
Lower respiratory infection	Nosocomial pneumonias rank third in incidence (about 12%) and have high mortality rates. Most of these pneumonias are related to respiratory devices that aid breathing or administer medications.
Bacteremia	Bacteremias account for about 6% of hospital infections. Intravenous catheterization is implicated in nosocomial infections of the bloodstream, particularly infections caused by bacteria and fungi.
Cutaneous infection	Among the least common of all nosocomial infections. However, newborns have a high rate of susceptibility to skin and eye infections.

A summary of the principal kinds of nosocomial infections is presented in Table 14.5.

CHAIN OF TRANSMISSION

Given the variety of pathogens (and potential pathogens) in the hospital and the compromised state of the host, routes of transmission are a constant concern. The principal routes of transmission of nosocomial infections are direct contact transmission from hospital staff to patient and from patient to patient, and indirect contact transmission through fomites and the hospital's ventilation system (airborne transmission).

Because hospital personnel are often in direct contact with patients, they can often transmit disease (see the box in Chapter 21, p. 529). For example, a physician or nurse may transmit her or his flora to a patient when changing a dressing, or a kitchen worker who carries *Salmonella* can infect a food supply.

Certain areas of a hospital are reserved for specialized care. These include the burn, hemodialysis, recovery, intensive care, and oncology units. Unfortunately, these units also group patients together and provide environments for the epidemic spread of nosocomial infections from patient to patient.

Many diagnostic and therapeutic hospital procedures provide a fomite route of transmission. The urinary catheter used to drain urine from the bladder is a fomite in many nosocomial infections. Intravenous catheters, which pass into a vein through the skin to provide fluids, nutrients, or medication, can also transmit nosocomial infections. Respiratory aids can introduce contaminated fluids into the lungs. Needles may introduce pathogens into muscle or blood, and surgical dressings can become infected and promote disease.

CONTROL OF NOSOCOMIAL INFECTIONS

Control measures aimed at preventing nosocomial infections vary from one institution to another, but certain procedures are generally implemented. It is important to reduce the number of pathogens to which patients are exposed by using aseptic techniques, handling contaminated materials carefully, insisting on frequent and conscientious handwashing, educating personnel about basic infection control measures (see the box on page 381), and using isolation rooms and wards. Tubs used to bathe patients should be disinfected between uses so that bacteria from the previous patient will not contaminate the next one. Respirators and humidifiers provide both a suitable growth environment for some bacteria and a method of airborne transmission. These sources of nosocomial infections must be kept scrupulously clean and disinfected, and materials used for bandages and intubation (insertion of tubes into organs, such as the trachea) should be single-use disposable or sterilized prior to use. Packaging used to maintain sterility should be removed aseptically. Physicians can help improve patients' resistance to infection by prescribing antibiotics only when necessary, avoiding invasive procedures if possible, and minimizing the use of immunosuppressive drugs.

Accredited hospitals should have an infection control committee. Most hospitals at least have an infection control nurse or epidemiologist (an individual who studies disease in populations). The role of these personnel is to identify problem sources, such as antibiotic-resistant strains of bacteria and improper sterilization techniques. The infection control officer should make periodic examinations of equipment to determine the amount of microbial contamination. Samples should be taken from tubing, catheters, respirator reservoirs, and other equipment.

Patterns of Disease

A definite sequence of events usually occurs during infection and disease. As you already know, there must be a reservoir of infection as a source of pathogens for an infectious disease to occur. Next, the pathogen must be transmitted by either direct contact, indirect contact, or vectors to a susceptible host. This event is followed by invasion, in which the microorganism enters the host and multiplies. Following invasion, the microorganism injures the host through a process called pathogenesis (which you will learn more about in the next chapter). The extent of injury depends on the extent to which host cells are damaged, either directly or by toxins. Despite the effects of all these factors, the occurrence of disease ultimately depends on the resistance of the host to the offensive weapons of the pathogen.

PREDISPOSING FACTORS

Certain predisposing factors also affect the occurrence of disease. A **predisposing factor** is one that makes the body more susceptible to a disease and may alter the course of the disease. Gender is sometimes a predisposing factor. For example, females have a higher incidence of urinary tract infections than males. Males have higher rates of pneumonia and meningitis. Other aspects of genetic background may play a role as well. Individuals with sickle-cell anemia, for instance, are actually more resistant to malaria than are other individuals.

Climate and weather seem to have some effect on the incidence of infectious diseases. In temperate regions, the incidence of respiratory diseases increases during the winter. This increase may be related to the fact that when people stay indoors and have closer contact with each other, the spread of respiratory pathogens is facilitated. Other predisposing factors include inadequate nutrition, fatigue, age, environment, habits, life-style or occupation, preexisting illness, chemotherapy, and emotional disturbances. It is often difficult to know the exact relative importance of different predisposing factors.

DEVELOPMENT OF DISEASE

Once a microorganism does overcome the defenses of the host, development of the disease follows a certain sequence of steps that tends to be similar whether the disease be acute or chronic (Figure 14.9).

Period of Incubation

The period of incubation is the time interval between the actual infection and the first appearance of any signs or symptoms. In some diseases, the incubation

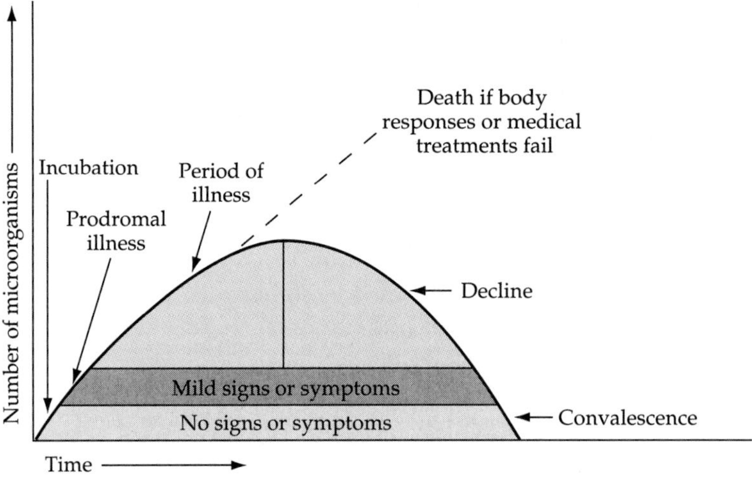

FIGURE 14.9 Stages of disease.

period is always the same; in others, it is quite variable. The time of incubation depends on the specific microorganism involved, its virulence (degree of pathogenicity), the number of infecting microorganisms, and the resistance of the host. Table 15.1 in the next chapter lists the incubation periods for a number of microbial diseases.

Prodromal Period

The prodromal period is a relatively short period that follows the period of incubation in some diseases. The prodromal period is characterized by early, mild symptoms of disease, such as general aches and malaise, and certain very specific symptoms, such as Koplik's spots in measles.

Period of Illness

During the period of illness, the disease is most acute. The person exhibits overt signs and symptoms of disease, such as fever, chills, muscle pain (myalgia), sensitivity to light (photophobia), sore throat (pharyngitis), lymph node enlargement (lymphadenopathy), and gastrointestinal disturbances. During the period of illness, the number of white blood cells may increase or decrease. Generally, the patient's immune response and other defense mechanisms overcome the pathogen and the period of illness ends. When the disease is not successfully overcome (or treated medically), the patient dies during this period.

Period of Decline

During the period of decline, the signs and symptoms subside. The fever decreases, and the feeling of malaise diminishes. During this phase, which may take from less than 24 hours to several days, the patient is vulnerable to secondary infections.

Period of Convalescence

During the period of convalescence, the person regains strength and the body returns to its prediseased state. Recovery has occurred.

We all know that, during the period of illness, people can serve as reservoirs of disease and can easily spread infections to other persons. However, you should also know that persons can spread infection during incubation and convalescence. This is especially true of diseases such as typhoid fever and cholera, where the convalescing person carries the pathogenic microorganism for months or even years.

Epidemiology

In today's crowded, overpopulated world, where frequent travel and mass production and distribution of foods and other goods are a way of life, diseases can spread rapidly. A contaminated food or water supply, for example, can affect many thousands of people very quickly. It is desirable to identify the causative agent so that a disease can be effectively controlled and treated. It is also desirable to understand the mode of transmission and geographical distribution of the disease. The science that deals with when and where diseases occur and how they are transmitted in the human population is called **epidemiology.**

Modern epidemiology began in 1854 when John Snow, a British physician, reasoned that cholera was transmitted by contaminated water. He questioned some victims of a cholera epidemic in London and found that most of them obtained drinking water from the Broad Street well. He also determined that the incidence of cholera was higher among people who drank from the London section of the Thames than among the population using cleaner water upstream. Even

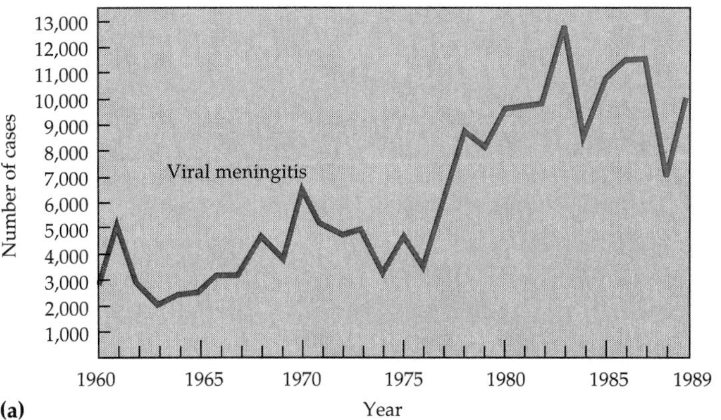

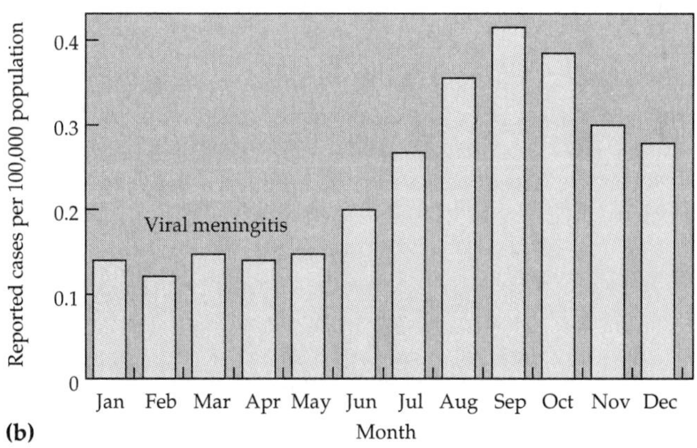

FIGURE 14.10 Epidemiological graphs. Epidemiologists use various types of graphs to describe patterns of disease. **(a)** This graph of the number of viral meningitis cases reported from 1960 to 1989 simply shows the annual occurrence of the disease during that period. **(b)** This bar graph looks at viral meningitis from a different perspective, allowing epidemiologists to begin drawing some conclusions about the disease. It records the incidence by month for one year (1989). Over a period of years, researchers saw that the incidence was consistently highest during the hot months. These data helped them realize that the disease was often contracted by waterborne transmission during swimming. Note that this graph records the number of cases per 100,000 people, rather than the total number of cases. **(c)** Annual incidence of toxic shock syndrome among menstruating women from 1979 to 1990. This graph shows how public education can help in disease control. Once women learned how tampons were involved in the disease and decreased their use, the incidence of the disease fell sharply.

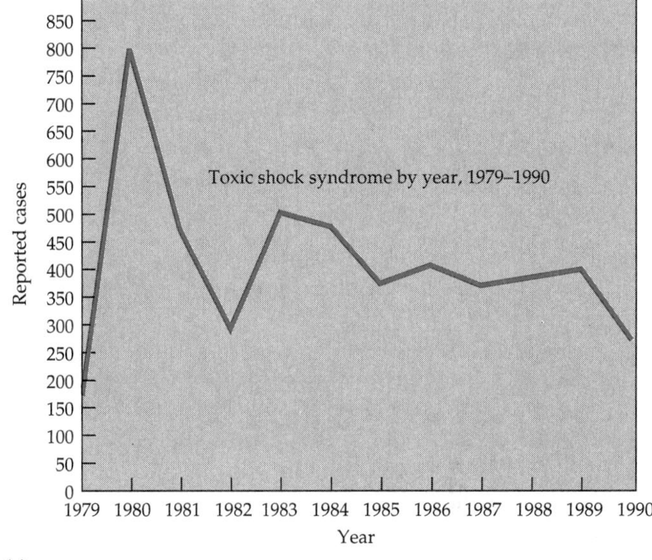

before Pasteur's work, Snow used epidemiologic investigation to improve public health. Today, epidemiologists study specified groups of individuals to discover the causes and prevent the spread of diseases.

An epidemiologist not only determines the etiology of a disease but also identifies other possibly important factors and patterns concerning the persons affected. An important part of the epidemiologist's work is assembling and analyzing such data as age, sex, occupation, personal habits, socioeconomic status, history of immunization, the presence of any other diseases, and the common history of affected individuals (such as eating the same food or visiting the same doctor's office). Also important for prevention of future outbreaks is knowledge of the site at which a susceptible host came into contact with the agent of infection. In addition, the epidemiologist considers the period during which the disease occurs, either on a seasonal basis (to indicate whether the disease is prevalent during the summer or winter) or on a yearly basis (to indicate the effects of hygiene or immunization).

Figure 14.10 shows graphs indicating the incidence of selected diseases. Such graphs provide information about whether disease outbreaks are sporadic or epidemic and, if epidemic, how the disease might have spread. By establishing the frequency of a disease in a population and identifying the factors responsible for its transmission, an epidemiologist provides physicians with information that is important in determining the prognosis and treatment of a disease. Epidemiologists also evaluate how effectively a disease is being controlled in a community—by a vaccination program, for example. Finally, epidemiologists can provide data to help in evaluating and planning overall health care for a community.

Epidemiologists use three basic types of investigations when analyzing the occurrence of a disease—descriptive, analytical, and experimental.

DESCRIPTIVE EPIDEMIOLOGY

Descriptive epidemiology entails collecting all data that describe the occurrence of the disease under study. Relevant information usually includes information about the affected persons and the place and period in which the disease occurred.

Such a study is generally *retrospective* (looking backward after the episode has ended). In other words, the epidemiologist backtracks to the cause and source of the disease (see the box in Chapter 22, p. 550). The search for the cause of toxic shock syndrome is an example of a fairly recent retrospective study. In the initial phase of an epidemiological study, retrospective studies are more common than *prospective* (looking forward) studies, in which an epidemiologist chooses a group of persons who are free of a particular disease to study. The group's subsequent disease experiences are then recorded for a given period. Prospective studies were used to test the Salk polio vaccine in 1954 and 1955.

ANALYTICAL EPIDEMIOLOGY

Analytical epidemiology analyzes a particular disease to determine its probable cause. This study can be done in two ways. With the *case control method*, the epidemiologist looks for factors that might have preceded the disease. A group of persons who have the disease is compared with another group of persons who are free of the disease. For example, one group with meningitis and one without the disease might be matched by age, sex, socioeconomic status, and location. These statistics are compared to determine which of all the possible factors—genetic, environmental, nutritional, and so forth—might be responsible for the meningitis. With the *cohort method*, the epidemiologist studies two populations, one that has had contact with the agent causing a disease and another that has not (both groups are called *cohort groups*). For example, a comparison of one group composed of people who have received blood transfusions and one composed of people who have not could reveal an association between blood transfusions and the incidence of hepatitis B virus. (See the box in Chapter 24, p. 602.)

EXPERIMENTAL EPIDEMIOLOGY

Experimental epidemiology begins with a hypothesis about a particular disease; experiments to test the hypothesis are then conducted with a group of people. One such hypothesis could be the assumed effectiveness of a drug. A group of infected individuals is selected and divided randomly so that some receive the drug and others receive a placebo, a substance that has no effect. If all other factors are kept constant between the two groups, and if it is determined that those people who received the drug recovered more rapidly than those who received the placebo, it can be concluded that the drug was the experimental factor (variable) that made the difference.

CASE REPORTING

We noted earlier in this chapter that establishing the chain of transmission for a disease is extremely important. Once known, the chain can be interrupted in order to slow down or stop the spread of the disease.

An effective way to establish the chain of transmission is case reporting, a procedure that requires health personnel to report specified diseases to local, state, and national health officials (see the box in Chapter 26, p. 657). Examples of such diseases include AIDS, measles, gonorrhea, tetanus, and typhoid fever. Case reporting provides epidemiologists with an approximation of the incidence and prevalence of a disease. This information helps officials decide whether or not to investigate a given disease.

Case reporting provided epidemiologists with valuable leads regarding the origin and spread of AIDS. In fact, one of the first clues about AIDS came from reports of young males with Kaposi's sarcoma, formerly a disease of older males. Using these reports, epidemiologists began various studies of the patients. If an epidemiological study shows that a large enough segment of the population is affected by a disease, an attempt is then made to isolate and identify its causative agent. Identification is accomplished by a number of different microbiological methods. Very frequently, identification of the causative agent provides valuable information regarding the reservoir for the disease.

Once the chain of transmission is discovered, it is possible to apply control measures to stop the disease from spreading. These might include elimination of the source of infection, isolation and segregation of contaminated persons, development of vaccines, and, as in the case of AIDS, education.

CENTERS FOR DISEASE CONTROL (CDC)

Epidemiology is a major concern of state and federal public health departments. The Centers for Disease Control (CDC), a U.S. Public Health Service branch located in Atlanta, Georgia, is a central source of epidemiological information in the United States.

The CDC issues a publication called *Morbidity and Mortality Weekly Report (MMWR)*, which is read by microbiologists, physicians, and other hospital and public health personnel. The *MMWR* contains data on the incidence of specific notifiable diseases (morbidity) and on the deaths from these diseases (mortality); these data are usually organized by state. (Notifiable diseases are those for which physicians are required by law to report cases to the Public Health Service.) Publication articles include reports of disease outbreaks, case histories of special interest, and summaries of the status of particular diseases during a recent period. These articles often include recommendations for procedures for diagnosis, immunization, and treatment. A number of graphs and other data in this book are from *MMWR*, and some of the boxes are direct excerpts from this publication.

In the next chapter, we consider the mechanisms of pathogenicity. We will discuss in more detail the methods by which microorganisms enter the body and cause disease, the effects of disease on the body, and the means by which pathogens leave the body.

STUDY OUTLINE

Introduction (p. 368)

1. Pathogenic microorganisms have special properties that allow them to invade the human body or produce toxins.
2. When the microorganism overcomes the body's defenses, a state of disease results.

Pathology, Infection, and Disease (p. 369)

1. Pathology is the scientific study of disease.
2. Pathology is concerned with the etiology (cause) of disease, pathogenesis (development), and effects of disease.
3. Infection is the invasion and growth of pathogens in the body.
4. A host is an organism that shelters and supports the growth of pathogens.
5. Disease is an abnormal state in which part or all of the body is not properly adjusted or is not capable of carrying on normal functions.

Normal Flora (pp. 369–371)

1. Animals, including humans, are usually germ-free in utero.
2. Microorganisms begin colonization in and on the surface of the body soon after birth.
3. Microorganisms that establish permanent colonies inside or on the body without producing disease make up the normal flora.
4. The transient flora is composed of microbes that are present for various periods and then disappear.

RELATIONSHIPS BETWEEN THE NORMAL FLORA AND HOST (p. 370)

1. The normal flora can prevent pathogens from causing an infection; this phenomenon is known as microbial antagonism.
2. The normal flora and the host exist in a symbiosis (living together).
3. The three types of symbiosis are commensalism (one organism benefits and the other is unaffected), mutualism (both organisms benefit), and parasitism (one organism benefits and one is harmed).

OPPORTUNISTIC ORGANISMS (pp. 370–371)

1. Opportunists (opportunistic pathogens) do not cause disease under normal conditions but cause disease under special conditions.

Etiology of Infectious Diseases (pp. 371–373)

KOCH'S POSTULATES (pp. 371–372)

1. Koch's postulates are a method for establishing that specific microbes cause specific diseases.
2. Koch's postulates state the following requirements: (1) the same pathogen must be present in every case of the disease; (2) the pathogen must be isolated in pure culture; (3) the pathogen isolated from pure culture must cause the same disease in a healthy, susceptible laboratory animal; and (4) the pathogen must be reisolated from the inoculated laboratory animal.

EXCEPTIONS TO KOCH'S POSTULATES (pp. 372–373)

1. Koch's postulates are modified to establish etiologies of diseases caused by viruses and some bacteria, which cannot be grown on artificial media.
2. Some diseases, such as tetanus, have unequivocal signs and symptoms.
3. Some diseases, such as pneumonia and nephritis, may be caused by a variety of microbes.
4. Some pathogens, such as *S. pyogenes*, cause several different diseases.

Classifying Infectious Diseases (pp. 373–374)

1. A patient may exhibit symptoms (subjective changes in body functions) and signs (measurable changes), which are used by a physician to make a diagnosis (identification of the disease).
2. A specific group of symptoms or signs that always accompanies a specific disease is called a syndrome.
3. Communicable diseases are transmitted directly or indirectly from one host to another.
4. Noncommunicable diseases are caused by microorganisms that normally grow outside the human body and are not transmitted from one host to another.
5. A contagious disease is one that is easily spread from one person to another.

OCCURRENCE OF DISEASE (pp. 373–374)

1. Disease occurrence is reported by incidence (number of people with the disease) and prevalence (incidence at a particular time).
2. Diseases are classified by frequency of occurrence—sporadic, endemic, epidemic, and pandemic.

SEVERITY OR DURATION OF DISEASE (p. 374)

1. The scope of a disease may be defined as acute, chronic, subacute, or latent.

EXTENT OF HOST INVOLVEMENT (p. 374)

1. A local infection affects a small area of the body; a systemic infection is spread throughout the body through the circulatory system.
2. A secondary infection can occur after the host is weakened from a primary infection.
3. An inapparent, or subclinical, infection does not cause any signs of disease in the host.

Spread of Infection (pp. 374–380)

RESERVOIRS (pp. 374–375)

1. A continual source of infection is called a reservoir of infection.
2. People who have a disease or are carriers of pathogenic microorganisms are human reservoirs of infection.
3. Zoonoses (diseases that occur in wild and domestic animals) can be transmitted to humans from animal reservoirs of infection.
4. Some pathogenic microorganisms grow in nonliving reservoirs, such as soil and water.

TRANSMISSION OF DISEASE (pp. 375–379)

1. Transmission by direct contact involves close physical contact between the source of the disease and a susceptible host.
2. Transmission by fomites (inanimate objects) constitutes indirect contact.
3. Transmission by saliva or mucus is called droplet infection.
4. Transmission by a medium such as water, food, and air is called vehicle transmission.
5. Airborne transmission refers to pathogens carried on water droplets or dust for a distance greater than one meter.
6. Arthropod vectors carry pathogens from one host to another by both mechanical and biological transmission.

PORTALS OF EXIT (pp. 379–380)

1. Just as pathogens have preferred portals of entry, they also have definite portals of exit.
2. Three common portals of exit are the respiratory tract via coughing or sneezing, the gastrointestinal tract via saliva or feces, and the genital tract via secretions from the vagina and penis.
3. Arthropods and syringes provide a portal of exit for microbes in blood.

Nosocomial (Hospital-Acquired) Infections (pp. 380–383)

1. Nosocomial infections are defined as any infections that are acquired during the course of stay in a hospital, nursing home, or other health care facility.
2. From 5% to 15% of all hospitalized patients acquire nosocomial infections.

MICROORGANISMS IN THE HOSPITAL (pp. 380–382)

1. Members of normal flora are often responsible for nosocomial infections when they are introduced into the body through such medical procedures as surgery and catheterization.
2. Opportunistic, drug-resistant gram-negative bacteria are the most frequent causes of nosocomial infections.

COMPROMISED HOST (p. 382)

1. Patients with burns, surgical wounds, and suppressed immune systems are the most susceptible to nosocomial infections.

CHAIN OF TRANSMISSION (pp. 382–383)

1. Nosocomial infections are transmitted by direct contact between staff and patient and between patients.
2. Fomites such as catheters, syringes, and respiratory aids can transmit nosocomial infections.

CONTROL OF NOSOCOMIAL INFECTIONS (p. 383)

1. Aseptic techniques can prevent nosocomial infections.
2. Hospital infection control personnel are responsible for overseeing proper storage and handling of equipment and supplies.

Patterns of Disease (pp. 383–384)

PREDISPOSING FACTORS (p. 383)

1. A predisposing factor is one that makes the body more susceptible to disease or alters the course of a disease.
2. Examples include climate, age, fatigue, and inadequate nutrition.

DEVELOPMENT OF DISEASE (pp. 383–384)

1. The period of incubation is the time interval between the actual infection and the first appearance of signs and symptoms.
2. The prodromal period is characterized by the appearance of the first mild signs and symptoms.
3. During the period of illness, the disease is at its height and all disease signs and symptoms are apparent.
4. During the period of decline, the signs and symptoms subside.
5. During the period of convalescence, the body returns to its prediseased state, and health is restored.

Epidemiology (pp. 384–387)

1. The science of epidemiology is the study of the transmission, incidence, and frequency of disease.
2. Data about infected persons are collected and analyzed in descriptive epidemiology.
3. In analytical epidemiology, a group of infected persons is compared with an uninfected group.
4. Controlled experiments designed to test hypotheses are performed in experimental epidemiology.
5. Case reporting provides data on incidence and prevalence to local, state, and national health officials.

6. The Centers for Disease Control (CDC) is the main source of epidemiologic information in the United States.

7. The CDC publishes *Morbidity and Mortality Weekly Report* to provide information on morbidity (incidence) and deaths (mortality).

STUDY QUESTIONS

Review

1. Differentiate between the following pairs of terms.
   (a) Etiology and pathogenesis
   (b) Infection and disease
   (c) Communicable disease and noncommunicable disease

2. What is meant by the normal flora? How does it differ from the transient flora?

3. Define symbiosis. Differentiate between commensalism, mutualism, and parasitism, and give an example of each.

4. What is a reservoir of infection? Match the following diseases with their reservoirs.

____ Influenza	**(a)** Nonliving
____ Rabies	**(b)** Human
____ Botulism	**(c)** Animal

5. Describe how Koch's postulates establish the etiology of many infectious diseases. Why don't Koch's postulates apply to all infectious diseases?

6. Describe the various ways diseases can be transmitted in each of the following categories. Name one disease transmitted by each method.
   (a) Transmission by direct contact
   (b) Transmission by indirect contact
   (c) Transmission by arthropod vectors
   (d) Droplet transmission
   (e) Vehicle transmission
   (f) Airborne transmission

7. List four predisposing factors to disease.

8. Indicate whether each of the conditions described is typical of subacute, chronic, or acute infections.
   (a) Patient experiences rapid onset of malaise; symptoms last five days.
   (b) Patient experiences cough and breathing difficulty for months.
   (c) Patient has no apparent symptoms and is a known carrier.

9. Of all the hospital patients with infections, one-third do not enter the hospital with an infection. How do they acquire these infections? What is the method of transmission of these infections? What is the reservoir of infection?

10. Differentiate between an endemic and epidemic state of infectious disease.

11. What is epidemiology? What is the role of the Centers for Disease Control (CDC)?

12. Distinguish between symptoms and signs as signals of disease.

13. How can a local infection become a systemic infection?

14. Why are some organisms that constitute the normal flora described as commensals, whereas others are described as mutualistic?

15. Put the following in the correct order to describe the pattern of disease: period of convalescence, crisis, prodromal period, period of decline, period of incubation, period of illness.

Challenge

1. Ten years before Robert Koch published his work on anthrax, Anton De Bary showed that potato blight was caused by the fungus *Phytophthora infestans*. Why do you suppose we use Koch's postulates instead of something called "De Bary's postulates"?

2. Three days before a nurse developed meningococcemia, she assisted with intubation of a patient with an *N. meningitidis* infection. Of the 24 medical personnel involved, only this nurse became ill. The nurse recalled that she had exposure to nasopharyngeal secretions and did not receive antibiotic prophylaxis. What two mistakes did the nurse make? How is meningitis transmitted?

3. The following is a case history of a 49-year-old man. Identify each period in the pattern of disease that he experienced. On February 7, he handled a parakeet with a respiratory illness. On March 9, he experienced intense pain in the legs, followed by severe chills and headaches. On March 16, he had chest pains, cough, and diarrhea, and his temperature was 40°C. Appropriate antibiotics were administered on March 17, and his fever subsided within 12 hours. He continued taking antibiotics for 14 days. (*Note*: The disease is psittacosis. Can you find the etiology?)

4. Florence Nightingale gathered the following data in 1855.

Population Sampled	Deaths from Contagious Diseases
Englishmen	0.2%
English soldiers (in England)	18.7%
English soldiers (in Crimean War)	42.7%
English soldiers (in Crimean War) after Nightingale's sanitary reforms	2.2%

Discuss how Nightingale employed the three basic types of epidemiologic investigation. The contagious diseases were primarily cholera and typhus; how are these diseases transmitted and prevented?

5. Name the method of transmission of the following diseases.
   - **(a)** Malaria
   - **(b)** Tuberculosis
   - **(c)** Nosocomial infections
   - **(d)** Salmonellosis
   - **(e)** Streptococcal pharyngitis
   - **(f)** Mononucleosis
   - **(g)** Measles
   - **(h)** Hepatitis A
   - **(i)** Tetanus
   - **(j)** Hepatitis B
   - **(k)** Chlamydial urethritis

6. This graph shows the incidence of typhoid fever in the United States from 1955 to 1989. Mark the graph to show when this disease occurred sporadically and epidemically. What appears to be the endemic level? What would have to be shown to indicate a pandemic of this disease? How is typhoid fever transmitted?

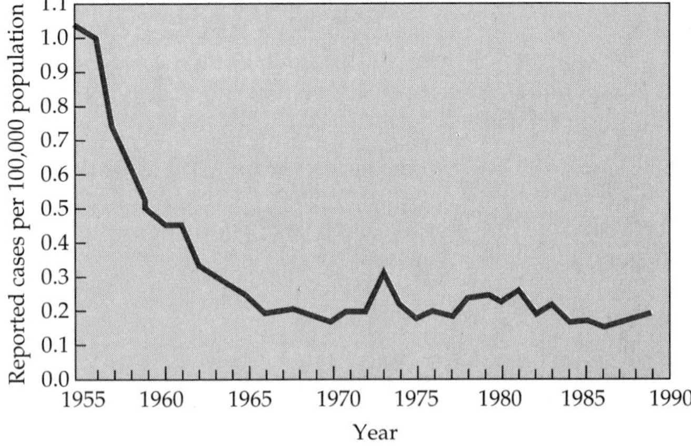

7. Three patients in a large hospital acquired infections of *Pseudomonas cepacia* during their stay. All three patients received cryoprecipitate, which is prepared from blood that has been frozen in a standard plastic blood transfer pack. The transfer pack is then placed in a water bath to thaw. What is the probable origin of the nosocomial infections? What characteristics of *Pseudomonas* would allow it to be involved in this type of infection?

8. During 1989, 21% of the patients in a large hospital acquired *Clostridium difficile* diarrhea and colitis during their hospital stay. These patients required longer hospital stays than uninfected patients. Epidemiological studies provided the following information. What is the most likely mode of transmission of this bacterium in hospitals? How can transmission be prevented?

   Rate of infection for patients

Single room	7%
Double room	17%
Triple room	26%

   Rate of environmental isolation of *C. difficile*

Bedrail	10%
Commode	1%
Floor	18%
Call Button	6%
Toilet	3%

   Hands of hospital personnel after patient contact that were culture-positive for *C. difficile*

Used gloves	0%
Did not use gloves	59%
Had *C. difficile* before patient contact	3%
Washed with nondisinfectant soap	40%
Washed with disinfectant soap	3%
Did not wash hands	20%

9. *Mycobacterium avium* is prevalent in AIDS patients. In an effort to determine the source of this infection, hospital water systems were sampled. The water contained chlorine.

   Percentage of samples with *M. avium*

Hot water	
February	88%
June	50%
Cold water	
February	22%
June	11%

   What is the usual method of transmission for *Mycobacterium*? What is a probable source of infection in hospitals? How can nosocomial infections be prevented?

FURTHER READING

Chesney, P.J., M.S. Bergdoll, and J.P. Davis. "The disease spectrum, epidemiology, and etiology of toxic-shock syndrome." *Annual Review of Microbiology* 38:315–338, 1984. Includes the history, pathogenesis, and method of action of toxic-shock toxin.

Epidemiology Issue. *Science* 234:921, 951–979, November 21, 1986. Three articles (on AIDS, salmonellosis, leishmaniasis, and malaria) about the tools and techniques used to identify and control an epidemic.

Jaret, P. "The disease detectives." *National Geographic* 179(1): 114–140, January 1991. Introduces scientists who are currently working on treatments and vaccines for cholera, Lassa fever, Lyme disease, and AIDS.

Laurence, J. "The AIDS virus." *Scientific American* 256(1): 47–56, January 1987. A summary of the search to identify the cause of AIDS.

Mandell, G.L., R.G. Douglas, and J.E. Bennett, eds. "Nosocomial infections." In *Principles and Practice of Infectious Disease,* 3rd ed. New York: Wiley, 1989. Discusses specific nosocomial infections, such as pneumonia.

Mausner, J.S., and S. Kramer. *Mausner and Bahn Epidemiology,* 2nd ed. Philadelphia: Saunders, 1985. A comprehensive reference on the collection and interpretation of epidemiological data.

McEvedy, C. "The bubonic plague." *Scientific American* 258(2):118–123, February 1988. A historical account of plague with hypotheses about the occurrence of epidemics.

Mills, J., and H. Masur. "AIDS-related infections." *Scientific American* 263(2):48–57, August 1990. A discussion of the causes and treatments of opportunistic infections affecting AIDS patients.

Roueche, B. *The Medical Detectives,* 2 vols. New York: Truman Talley Books. Vol. 1 by Times Books, Volume 2 by Washington Square Press, 1984. True accounts describing the scientific investigations of epidemic diseases. Written so the reader identifies the etiology with the epidemiologic investigator.

Symposium Papers. *Respiratory Care* 32(2):81–124, February 1989. Five articles on the spread of infections in hospitals.

CHAPTER 15

Mechanisms of Pathogenicity

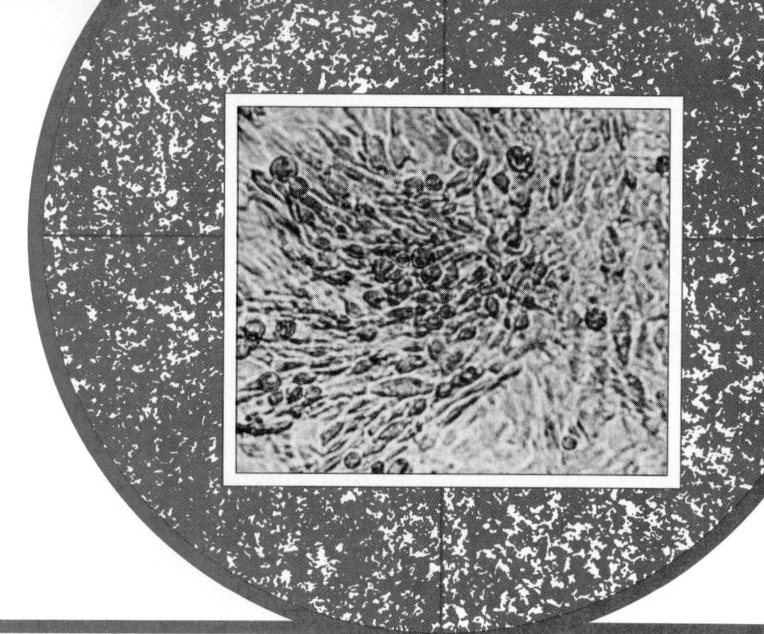

LEARNING OBJECTIVES

- Define portal of entry, pathogenicity, virulence, and LD_{50}.
- Explain how adherence, capsules, cell wall components, and enzymes contribute to pathogenicity.
- Compare the effects of hemolysins, leukocidins, coagulase, kinases, hyaluronidase, and collagenase.
- Contrast the nature and effects of exotoxins and endotoxins.
- Outline the mechanisms of action of diphtherotoxin, botulinum toxin, tetanus toxin, choleragen, and lipid A.
- List seven cytopathic effects of viral infections.
- Discuss the causes of symptoms in fungal, protozoan, and helminthic diseases.

Now that you have a basic understanding of how microorganisms cause disease, we will take a look at some of the specific properties of microorganisms that contribute to **pathogenicity,** the ability to cause disease in a host, and **virulence,** the degree of pathogenicity. Keep in mind that many of the properties contributing to microbial pathogenicity and virulence are unclear or unknown. We do know, however, that if the microbes' attack overpowers the host defenses, disease results.

Entry of a Microorganism into the Host

Most pathogens must gain access to the host, adhere to host tissues, penetrate host defenses, and damage the host tissue to cause disease. Some microbes, such as those that cause dental caries and acne, can cause disease without penetrating the body. Pathogens can gain entrance to the human body and other hosts through several avenues, which we call **portals of entry.**

MUCOUS MEMBRANES

Many bacteria and viruses gain access to the body by penetrating mucous membranes lining the respiratory tract, gastrointestinal tract, genitourinary tract, and conjunctiva, a delicate membrane that covers the eyeball and lines the eyelids. Most pathogens enter through the mucous membranes of the gastrointestinal and respiratory tracts. The respiratory tract is the easiest and most frequently traveled portal of entry for infectious microorganisms. Microbes are inhaled into the nose or mouth in drops of moisture and dust particles. Some diseases that are commonly contracted via the respiratory tract include the common cold, pneumonia, tuberculosis, influenza, measles, and smallpox.

Microorganisms can gain access to the gastrointestinal tract in food and water. Bringing contaminated fingers to the mouth is another common mode of entry. Most microbes that enter the body in these ways are destroyed by hydrochloric acid (HCl) and enzymes in the stomach or by bile and enzymes in the small intestine. Those that survive can cause disease. Microbes in the gastrointestinal tract can cause poliomyelitis, hepatitis A, typhoid fever, amoebic dysentery, giardiasis, bacillary dysentery (shigellosis), and cholera. These pathogens are then eliminated with feces and can be transmitted to other hosts via water, food, or contaminated fingers.

An important pathogen capable of penetrating mucous membranes of the genitourinary tract is *Treponema pallidum*, the causative agent of syphilis.

SKIN

The skin is one of the largest organs of the body and is a significant factor in defense against disease. When unbroken, the skin is impenetrable by most microorganisms, although even then it does not provide complete protection. Some microbes gain access to the body through openings in the skin, such as hair follicles and sweat gland ducts. It has been demonstrated that the hookworm *Necator americanus* actually bores through intact skin, and some fungi grow on the keratin in skin or infect the skin itself.

PARENTERAL ROUTE

Other microorganisms gain access to the body when they are deposited directly into the tissues beneath the skin or into mucous membranes when these barriers are penetrated or injured. This route is called the **parenteral route.** Punctures, injections, bites, cuts, wounds, surgery, and splitting due to swelling or drying can all establish parenteral routes.

PREFERRED PORTAL OF ENTRY

Even after microorganisms have entered the body, they do not necessarily cause disease. Occurrence of disease depends on several factors, only one of which is the portal of entry. Many pathogens have a preferred portal of entry, and using the portal is a prerequisite to their being able to cause disease. If they gain access to the body by another portal, disease might not occur. For example, the bacteria of typhoid fever, *Salmonella typhi,* produce all the signs and symptoms of the disease when swallowed (preferred route), but if the same bacteria are rubbed on the skin no reaction (or only a slight inflammation) occurs. Streptococci that are inhaled (preferred route) can cause pneumonia; those that are swallowed generally do not produce signs or symptoms. Some pathogens, such as the microorganism that causes plague, can initiate disease from more than one portal of entry. The preferred portals of entry for some common pathogens are given in Table 15.1.

NUMBERS OF INVADING MICROBES

If only a few microbes enter the body, it is likely that they will be overcome by the host's defenses. However, if large numbers of microbes gain entry, the stage is probably set for disease. Thus, the likelihood of disease increases as the number of pathogens increases.

The virulence of a microbe or the potency of its toxin is often expressed as the **LD_{50} (lethal dose for 50% of hosts),** the number of microbes in a dose that will kill 50% of inoculated test animals under normal conditions. The dose required to produce a demonstrable infection in 50% of the test animals is called the **ID_{50} (infectious dose for 50% of hosts).**

ADHERENCE

Once pathogens gain entry to a host, almost all of them have some means of attaching themselves to host tissues. For most pathogens, this attachment, called **adherence,** is a necessary step in pathogenicity. (Of course, nonpathogens also have structures for attachment.) The attachment between pathogen and host takes place by means of surface molecules on the pathogen called **adhesins** or **ligands** that bind specifically to complementary surface receptors on the cells of certain host tissues (Figure 15.1). (Nonpathogenic microbes that colonize the body also have adhesins. The presence or absence of complementary cell receptors in various host tissues determines which hosts or tissues are colonized.) Adhesins may be located on a microbe's glycocalyx or on other microbial surface structures, such as fimbriae (see Chapter 4).

The majority of adhesins on the microorganisms we have studied so far are glycoproteins or lipoproteins. The receptors on host cells are typically sugars, such as mannose. Adhesins on different strains of the same species of pathogen can vary in structure. Different cells of the same host can also have different receptors that vary in structure. If adhesins, receptors, or both can be altered to interfere with adherence, it is possible to prevent or control infection.

The following examples illustrate the diversity of adhesins. *Streptococcus mutans,* a bacterium that plays a key role in tooth decay, attaches to the surface of teeth by its glycocalyx. An enzyme, *glucosyltransferase,* produced by *S. mutans,* converts glucose (derived from sucrose or table sugar) into a sticky polysaccharide

TABLE 15.1 Causative Agents for Some Common Diseases, by Portal of Entry

PORTAL OF ENTRY	CAUSATIVE AGENT*	DISEASE	INCUBATION PERIOD
Respiratory tract	*Corynebacterium diphtheriae*	Diphtheria	2–5 days
	Neisseria meningitidis	Meningococcal meningitis	1–7 days
	Streptococcus pneumoniae	Pneumococcal pneumonia	Variable
	*Mycobacterium tuberculosis***	Tuberculosis	Variable
	Bordetella pertussis	Whooping cough (pertussis)	12–20 days
	Myxovirus	Influenza	18–36 hours
	Paramyxovirus	Measles (rubeola)	11–14 days
	Togavirus	German measles (rubella)	2–3 weeks
	Epstein–Barr virus (herpesvirus)	Infectious mononucleosis	2–6 weeks
	Varicella-zoster virus (herpesvirus)	Chickenpox (varicella)	14–16 days
	Poxvirus	Smallpox (variola)	12 days
	Coccidioides immitis (fungus)	Coccidioidomycosis (primary infection)	1–3 weeks
	Histoplasma capsulatum (fungus)	Histoplasmosis	5–18 days
Gastrointestinal tract	*Shigella* species	Bacillary dysentery (shigellosis)	1–2 days
	Brucella species	Brucellosis (undulant fever)	6–14 days
	Vibrio cholerae	Cholera	1–3 days
	Salmonella enteritidis, Salmonella typhimurium, Salmonella choleraesuis	Salmonellosis	7–22 hours
	Salmonella typhi	Typhoid fever	14 days
	Hepatitis A virus (picornavirus)	Hepatitis A	15–50 days
	Paramyxovirus	Mumps	2–3 weeks
	Poliovirus	Poliomyelitis	4–7 days
	Trichinella spiralis (helminth)	Trichinosis	2–28 days
Genitourinary tract	*Neisseria gonorrhoeae*	Gonorrhea	3–8 days
	Treponema pallidum	Syphilis	9–90 days
Skin or parenteral route	*Clostridium perfringens*	Gas gangrene	1–5 days
	Clostridium tetani	Tetanus	3–21 days
	Leptospira interrogans	Leptospirosis	2–20 days
	Yersinia pestis	Plague	2–6 days
	Rickettsia rickettsii	Rocky Mountain spotted fever	3–12 days
	Hepatitis B virus**	Hepatitis B	6 weeks–6 months
	Rhabdovirus	Rabies	10 days–1 year
	Togavirus	Yellow fever	3–6 days
	Plasmodium species (protozoan)	Malaria	2 weeks

*All causative agents are bacteria, unless indicated otherwise. For viruses, only the viral group is given, except where the virus has a name different from the disease it causes.
**These pathogens can also cause disease after entering the body via the gastrointestinal tract.

called dextran, which forms the glycocalyx. *Actinomyces* cells have fimbriae that adhere to the glycocalyx of *S. mutans,* further contributing to plaque. Enteropathogenic strains of *E. coli* (those responsible for gastrointestinal disease) have adhesins on fimbriae that adhere only to specific kinds of cells in certain regions of the small intestine. After adhering, *Shigella* and *E. coli* induce endocytosis as a vehicle to enter host cells and then multiply within them. *Neisseria gonorrhoeae,* the causative agent of gonorrhea, also has fimbriae containing adhesins, which in this case permit attachment to cells with appropriate receptors in the genitourinary tract, eyes, and pharynx. *Staphylococcus aureus,* which can cause skin infections, binds to skin by a mechanism of adherence that resembles viral attachment (discussed in Chapter 13). Attachment of *Helicobacter* to the stomach is described in the box, p. 398.

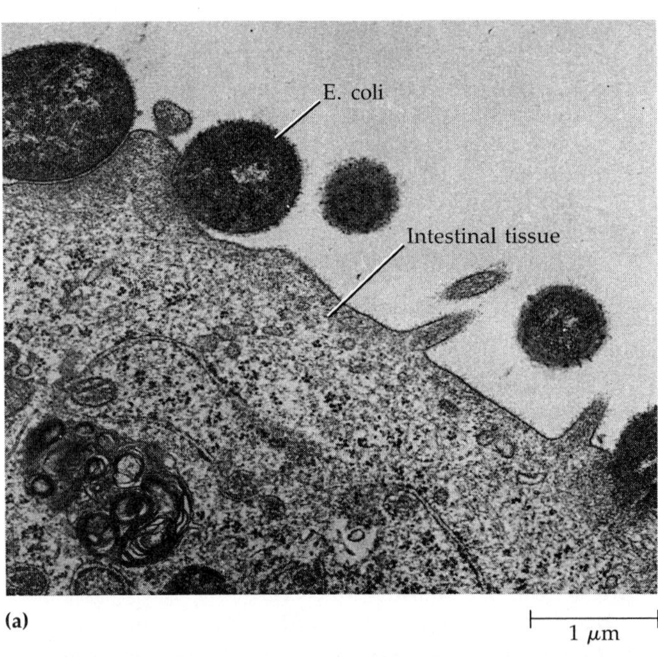

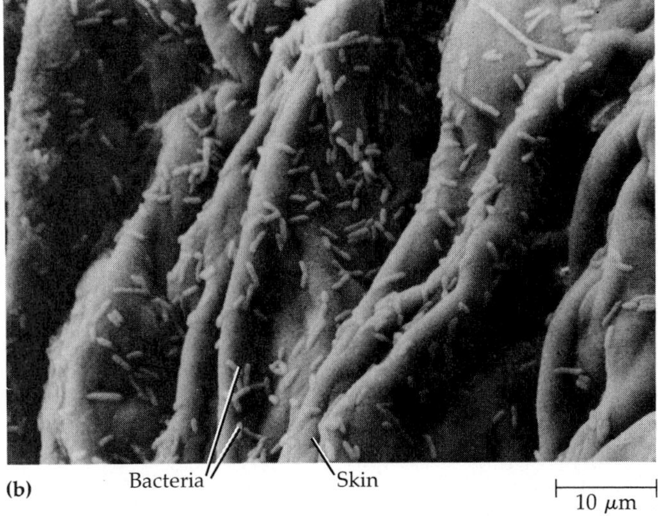

FIGURE 15.1 Adherence. (a) Selective attachment of a pathogenic strain of *E. coli* to the intestinal tissue of a rabbit. **(b)** Scanning electron micrograph of bacteria adhering to the skin of a salamander.

How Pathogens Penetrate Host Defenses

Although some pathogens can cause damage on the surface of tissues, most must penetrate tissues to cause disease. Here we will consider several factors that contribute to a microorganism's ability to invade a host.

CAPSULES

We noted in Chapter 4 that some bacteria make glycocalyx material that forms capsules around their cell walls (Figure 15.2); this property increases the virulence of the species. The capsule resists the host's defenses by impairing phagocytosis, a process by which certain cells of the body engulf and destroy microbes (to be discussed in Chapter 16). The chemical nature of the capsule appears to prevent the phagocytic cell from

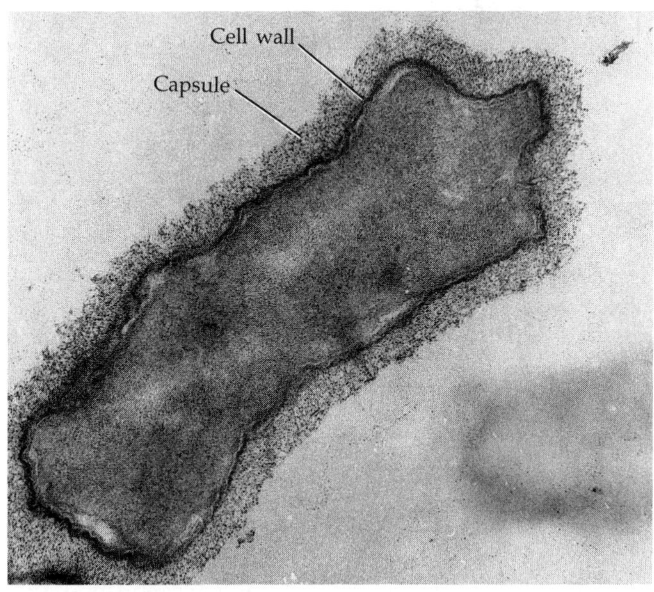

FIGURE 15.2 A bacterial capsule. Electron micrograph of a thin section through cells of an encapsulated strain of *Enterobacter aerogenes*. The capsule appears as a thick gray layer of material outside the dark cell wall. Capsules play an important role in the resistance of some pathogenic bacteria to phagocytosis.

adhering to the bacterium. However, the human body is capable of producing antibodies against the capsule, and when these antibodies are present on the capsule surface, the encapsulated bacteria are easily destroyed by phagocytosis.

One bacterium that owes its virulence to the presence of a polysaccharide capsule is *Streptococcus pneumoniae,* the causative agent of pneumococcal pneumonia (see Figure 24.9). Some strains of this organism have capsules, and others do not. Strains with capsules are virulent, but strains without capsules are avirulent because they are susceptible to phagocytosis. Other bacteria that produce capsules related to virulence are *Klebsiella pneumoniae,* a causative agent of bacterial pneumonia; *Hemophilus influenzae,* a cause of pneumonia and meningitis in children; *Bacillus anthracis,* the cause of anthrax; and *Yersinia pestis,* the causative agent of bubonic plague. Keep in mind that capsules are not the only cause of virulence. Many nonpathogenic bacteria produce capsules, and the virulence of some pathogens is not related to the presence of a capsule.

COMPONENTS OF THE CELL WALL

The cell walls of certain bacteria contain chemical substances that contribute to virulence. For example, *Streptococcus pyogenes* produces a heat-resistant and acid-resistant protein called **M protein.** This protein is found on both the cell surface and fimbriae. The M protein mediates attachment of the bacterium to epithelial cells of the host and helps the bacterium resist phagocytosis by white blood cells. The protein thereby increases the virulence of the microorganism. Immunity to *S. pyogenes* depends on the body's production of an antibody specific to M protein.

The waxes that make up the cell wall of *Mycobacterium tuberculosis* also increase virulence by resisting digestion by phagocytes. In fact, *M. tuberculosis* can even multiply inside phagocytes.

ENZYMES

The virulence of some bacteria is thought to be aided by the production of extracellular enzymes (exoenzymes) and related substances. These chemicals can break cells open, dissolve materials between cells, and form or dissolve blood clots, among other functions. However, the importance of some of these enzymes to bacterial virulence has not been proved conclusively.

Substances called **leukocidins,** produced by some bacteria, can destroy neutrophils, a type of white blood cell (leukocyte) that is very active in phagocytosis. Leukocidins are also active against macrophages (phagocytic cells) present in tissue. Among the bacteria that secrete leukocidins are staphylococci and streptococci. Leukocidins produced by streptococci degrade lysosomes within leukocytes and thereby cause the death of the white blood cell (also see the box in Chapter 16, p. 410). Hydrolytic enzymes released from the leukocytic lysosomes can damage other cellular structures and thus intensify streptococcal lesions. This type of damage to white blood cells decreases host resistance.

Hemolysins are another group of enzymes that are produced by bacteria and might contribute to their virulence. Hemolysins cause the lysis of red blood cells. Bacteria produce a number of hemolysins that differ in their ability to lyse different kinds of red blood cells (humans, sheep, and rabbits, for example) and the type of lysis they cause. Important producers of hemolysins are staphylococci, streptococci, and *Clostridium perfringens* the most common causative agent of gas gangrene.

Coagulases are bacterial enzymes that coagulate (clot) the fibrinogen in blood. Fibrinogen, a plasma protein produced by the liver, is converted by coagulases into fibrin, the threads that form a blood clot. The fibrin clot caused by coagulases may protect the bacterium from phagocytosis and isolate it from other defenses of the host. Coagulases are produced by some members of the genus *Staphylococcus;* they may be involved in the walling-off process in boils produced by staphylococci. However, some staphylococci that do not produce coagulases are still virulent. A capsule, and not coagulases, may be the more important factor in the virulence of these bacteria.

Bacterial kinases are another group of enzymes that may contribute to bacterial virulence. By breaking down fibrin, the kinases dissolve clots formed by the body to isolate the infection. One of the better-known kinases is **streptokinase (fibrinolysin),** which is produced by such streptococci as *Streptococcus pyogenes.* Another kinase, **staphylokinase,** is produced by such staphylococci as *Staphylococcus aureus.* Injected directly into the blood, streptokinase has been used successfully to dissolve some type of blood clots in cases of heart attacks due to obstructed coronary arteries.

Hyaluronidase is yet another enzyme secreted by certain bacteria, such as streptococci, and possibly related to microbial virulence. It hydrolyzes hyaluronic acid, a type of polysaccharide that holds together certain cells of the body, particularly cells in connective tissue. It is believed that this digesting action is involved in the tissue blackening of infected wounds and helps the microorganism spread from its initial site of infection. Hyaluronidase is also produced by some clostridia that cause gas gangrene. For therapeutic use, hyaluronidase may be mixed with a drug to promote the spread of the drug through a body tissue. Another enzyme, **collagenase,** produced by several species of *Clostridium,* facilitates the spread of gas gangrene. Collagenase breaks down the protein collagen, which forms the connective tissue of muscles and other body organs and tissues.

Other bacterial substances believed to contribute to virulence are *necrotizing factors,* which cause the death of body cells; *hypothermic factors,* which decrease body temperature; *lecithinase,* which destroys the plasma membrane, especially around red blood cells; *protease,* which breaks down proteins, especially in muscle tissue; and *siderophores,* which scavenge iron from the host's body fluids.

Damage to Host Cells

When a microorganism invades a body tissue, it initially encounters phagocytes of the host. If the phagocytes are successful in destroying the invader, no further damage is done to the host. But if the pathogen overcomes the host defense, then the microorganism can damage host cells in three basic ways: (1) by causing direct damage in the *immediate vicinity* of the invasion; (2) by producing toxins, transported by blood and lymph, that damage sites *far removed* from the original site of invasion; and (3) by inducing hypersensitivity reactions. This latter mechanism is considered in detail in Chapter 19. For now, we will discuss only the first two mechanisms.

DIRECT DAMAGE

Once a pathogen is attached to host cells, it can pass through them to invade other tissues. During this invasion, the pathogens metabolize and multiply to kill host cells. Some bacteria, such as *E. coli, Shigella, Salmonella,* and *Neisseria gonorrhoeae,* can induce host epithelial cells to engulf them by a process that resembles phagocytosis. These pathogens can also be extruded from the host cells by a reverse phagocytosis process in order to enter other host cells (see the box in Chapter 8, p. 205). Some bacteria can also penetrate host cells by excreting enzymes and by their own motility; such penetration can itself damage the host cell. Most damage by bacteria, however, is done by toxins.

TOXINS

Toxins are poisonous substances that are produced by certain microorganisms. They are frequently the primary factor that contributes to the pathogenic properties of those microbes. The capacity of microorganisms to produce toxins is called **toxigenicity.** Toxins transported by the blood or lymph can cause serious, and sometimes fatal, effects. Some toxins produce fever, cardiovascular disturbances, diarrhea, and shock. Toxins can also inhibit protein synthesis, destroy blood cells and blood vessels, and disrupt the nervous system by causing spasms. Of the 220 known bacterial toxins, nearly 40% cause disease by damaging eucaryotic cell membranes. The term **toxemia** refers to symptoms caused by toxins in the blood. Toxins are of two types, exotoxins and endotoxins.

Exotoxins

Exotoxins are produced inside some bacteria as part of their growth and metabolism and are released into the surrounding medium. Exotoxins are proteins, and many are enzymes that catalyze only certain biochemical reactions. Most bacteria that produce exotoxins are gram-positive. The genes for most (perhaps all) exotoxins are carried on bacterial plasmids or phages. Because exotoxins are soluble in body fluids, they can easily diffuse into the blood and are rapidly transported throughout the body.

Exotoxins work by destroying particular parts of the host's cells or by inhibiting certain metabolic functions. Exotoxins are among the most lethal substances known. Only 1 mg of the botulinum exotoxin is enough to kill 1 million guinea pigs. Fortunately, only a few bacterial species produce such potent exotoxins.

Diseases caused by bacteria that produce exotoxins are often caused by minute amounts of exotoxins, not by the bacteria themselves. It is the exotoxins that produce the specific signs and symptoms of the disease. Thus, exotoxins are disease-specific. For example, the infection of a wound by *Clostridium tetani* need be no larger or more painful than a pin prick, yet the organisms in a wound that size can produce enough tetanus toxin to kill an unvaccinated human.

The body produces antibodies called **antitoxins** that provide immunity to exotoxins. When exotoxins are inactivated by heat, formaldehyde, iodine, or other chemicals, they no longer cause the disease but are still able to stimulate the body to produce antitoxins. Such altered exotoxins are called **toxoids.** When toxoids are injected into the body as a vaccine, they stimulate antitoxin production so that immunity is produced to disease. Diphtheria and tetanus are common diseases that can be prevented by toxoid vaccination.

Here we will briefly describe a few of the more notable exotoxins (antitoxins will be discussed further in Chapter 18). Exotoxins may be grouped into three principal types, based on their mode of action: (1) **cytotoxins,** which kill host cells or affect their functions; (2) **neurotoxins,** which interfere with normal nerve impulse transmission; and (3) **enterotoxins,** which affect cells lining the gastrointestinal tract.

Diphtheria toxin. *Corynebacterium diphtheriae* produces the diphtheria toxin (diphtherotoxin) only when it is infected by a lysogenic phage carrying the *tox* gene. This cytotoxin inhibits protein synthesis in eucaryotic cells. The mechanism by which it does this is

MICROBIOLOGY HIGHLIGHTS

Are Ulcers an Infectious Disease?

Most people think of ulcers as an unfortunate consequence of a high-stress, fast-paced lifestyle. However, recent investigations into the cause of ulcers have begun to focus on the surprising role played by a spiral bacterium. The role of bacteria in stomach disease was first suspected in 1906, when spirilla were found in the stomachs of patients with stomach cancer. In 1983, 77 years later, a curved rod was found in the stomachs of nearly all patients with chronic gastritis and rarely in healthy stomachs. Gastritis, inflammation of the mucosa lining the stomach, is the condition that precedes the development of most ulcers. The bacterium has been named *Helicobacter pylori*.

Indirect evidence supports the hypothesis that bacteria cause ulcers. Ulcers are holes in the mucosa that develop when portions are actually digested away. They have long been associated with stress, genetic factors, and smoking and have generally been treated by acid-suppressive therapy and lifestyle changes. However, even with such therapy, ulcers often recur. The possible role of bacteria was underlined by the recent discovery that, if acid-suppressive therapy is combined with antibiotics used to eliminate bacteria, ulcers usually do not recur. Furthermore, the antibiotics and bismuth compounds most effective in treating ulcers are the same drugs to which *Helicobacter* is most sensitive. Finally, in various studies, *H. pylori* has been isolated from nearly 100% of ulcer patients.

The links between *Helicobacter* and chronic gastritis, which precedes the development of intestinal and stomach ulcers, are also very strong. In one study, analysis of sera taken from a group of patients with gastritis showed antibodies against *Helicobacter*. Such antibodies are evidence of current or prior infection by the organism. In addition, *Helicobacter* is always associated with gastritis lesions (see the photo) when tissue samples are examined. Finally, in an attempt to fulfill Koch's postulates, two healthy individuals were inoculated with *H. pylori*. The bacteria colonized the stomach, acute gastritis developed, and the infections were then successfully treated with antibiotics.

The pathogenesis is not known. The bacteria apparently use projections from their cell wall to adhere to the stomach's mucus-secreting cells. It is thought that they survive the highly acidic environment by hydrolyzing urea to make ammonia, which neutralizes stomach acid in the immediate vicinity of the bacteria. They also suppress acid production by damaging the stomach cells. Then, when the patient's immune response succeeds in suppressing the bacteria, acid secretion returns to normal and the acid digests the damaged cells, leading to chronic gastritis and ulcers.

The method of transmission of *Helicobacter* has not been established. In one study, household members of gastritis patients did not develop antibodies against *Helicobacter*, suggesting that the bacterium is not transmitted by contact. However, a study of clinical staff involved in obtaining biopsy samples shows that the staff did develop antibodies, indicating transmission of *Helicobacter* from patients. These studies provide evidence that exchange of bacteria of the intestinal and skin floras, which is normal among household members, does not transmit *Helicobacter* but that direct contact with stomach contents does.

It is not yet clear how bacterial infection is related to the other conditions, such as smoking, stress, and genetics, that are associated with the development of ulcers. Bacteria may cause the initial lesion, or they may be able to grow only on mucosa made susceptible by these other factors. It would be valuable to have data on the prevalence of *Helicobacter* in normal individuals, in contacts of gastritis patients, and in patient follow-up. However, such studies are hampered by the difficulty of testing for *Helicobacter*. At present, culturing *Helicobacter* requires obtaining a sample of the gastric mucosa by inserting biopsy forceps down the esophagus. In taking the sample, it is possible to miss areas colonized by bacteria, or the local anaesthetic required for taking the sample or disinfectant on the forceps can kill the bacteria. Noninvasive tests based on the presence of bacterial urease are being developed.

Understanding the role of *Helicobacter* in gastritis could lead to treatment of gastritis and reduced incidence of ulcers. Furthermore, chronic gastritis is thought to be a predisposing condition to cancer of the stomach. Early treatment of gastritis might prevent stomach cancer.

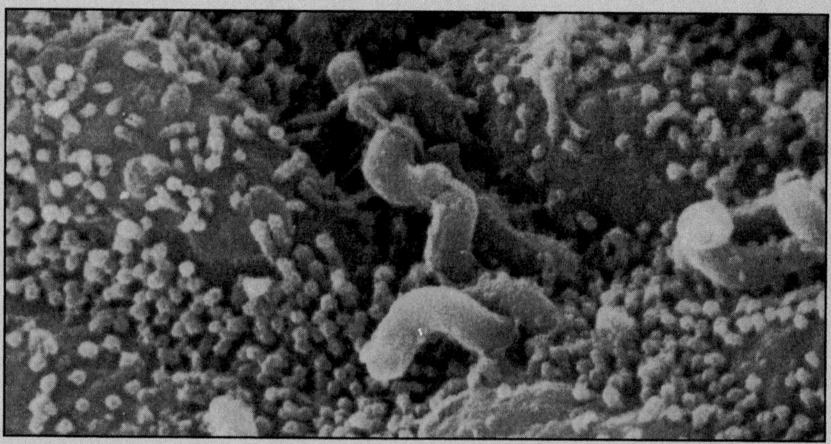

Helicobacter pylori attached to stomach mucosa.

an excellent example of how an exotoxin interacts with host cells (Figure 15.3). Diphtherotoxin is a protein consisting of two different polypeptides, designated A (active) and B (binding). (Recall from Chapter 2 that polypeptides are long, chainlike molecules of amino acids.) Although only polypeptide A is the active component that causes symptoms in the host, polypeptide B is required for polypeptide A to be active. Polypeptide B binds to surface receptors on the host cell and causes the transport of the entire protein across the plasma membrane into the cell. Once this is accomplished, polypeptide A inhibits protein synthesis within the target cell. (This model also applies to the cholera exotoxin, which is described shortly.)

Erythrogenic toxins. *Streptococcus pyogenes* has the genetic material to synthesize three types of cytotoxins, designated as A, B, and C. These erythrogenic toxins damage blood capillaries under the skin and produce a red skin rash. Scarlet fever, produced by *S. pyogenes,* is named for this characteristic rash.

Botulinum toxin. Botulinum toxin is produced by *Clostridium botulinum*. Although the toxin production is associated with germination of endospores and with growth of vegetative cells, little of it appears in the medium until it is released by lysis late in growth. Botulinum toxin is a neurotoxin. It acts at the neuromuscular junction (the junction between nerve cell and muscle cell) and prevents the transmission of impulses from the nerve cell to muscle. The toxin accomplishes this by binding to the nerve cell and inhibiting the release of a neurotransmitter called acetylcholine. As a result, botulinum toxin causes paralysis in which muscle tone is lacking (flaccid paralysis). *C. botulinum* produces eight different types of botulinum toxin, and each possesses a different potency.

Tetanus toxin. *Clostridium tetani* produces tetanus neurotoxin, also known as tetanospasmin. This toxin reaches the central nervous system and binds to nerve cells that control contraction of various skeletal muscles. These nerve cells normally send inhibiting impulses that prevent random contractions and terminate completed contractions. The binding of tetanospasmin blocks the inhibitory nerve impulses to one skeletal muscle while the opposing muscle is contracting. The result is uncontrollable muscle contraction, producing the convulsive symptoms (spasmodic contractions) of tetanus, or "lockjaw."

Vibrio enterotoxin. *Vibrio cholerae* produces an enterotoxin called **choleragen.** Like diphtherotoxin, choleragen consists of two polypeptides, A (active) and B (binding). The B component binds to plasma membranes of epithelial cells lining the small intestine, and the A component induces the formation of cyclic AMP from ATP in the cytoplasm (see Chapter 8). As a result, epithelial cells discharge large amounts of fluids and electrolytes (ions). Normal muscular contractions are disturbed, leading to severe diarrhea that may be accompanied by vomiting. *Heat-labile enterotoxin* (so called because it is more sensitive to heat than are most toxins) produced by some strains of *E. coli* has an action identical to that of choleragen.

Staphylococcal enterotoxin. *Staphylococcus aureus* produces an enterotoxin that affects the intestines in the same way as choleragen. A strain of *S. aureus* also produces enterotoxins that result in the symptoms associated with toxic shock syndrome, discussed in Chapter 1. A summary of diseases produced by exotoxins is presented in Table 15.2.

Endotoxins

Endotoxins differ from exotoxins in several ways. Endotoxins are part of the outer portion of the cell wall of

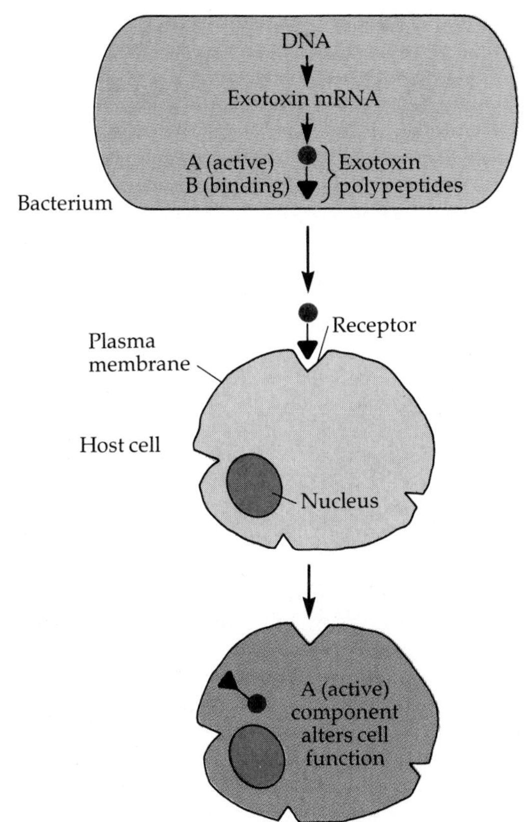

FIGURE 15.3 Exotoxin. Proposed model for mechanism of action of diphtherotoxin.

TABLE 15.2 Diseases Caused by Exotoxins

DISEASE	*BACTERIUM*	*MECHANISM*
Botulism	*Clostridium botulinum*	Neurotoxin prevents transmission of nerve impulses; flaccid paralysis results
Tetanus	*Clostridium tetani*	Neurotoxin blocks nerve impulse that permits relaxation of one skeletal muscle while the opposing muscle is contracting
Gas gangrene and food poisoning	*Clostridium perfringens* and other species of *Clostridium*	One exotoxin (cytotoxin) causes massive red blood cell destruction (hemolysis); another exotoxin (enterotoxin) is related to food poisoning and causes diarrhea
Diphtheria	*Corynebacterium diphtheriae*	Cytotoxin inhibits protein synthesis, especially in nerve, heart, and kidney cells
Scalded skin syndrome, food poisoning, and toxic shock syndrome	*Staphylococcus aureus*	One exotoxin causes skin layers to separate and slough off (scalded skin); another exotoxin (enterotoxin) produces diarrhea and vomiting; still another exotoxin produces symptoms associated with TSS
Cholera	*Vibrio cholerae*	Enterotoxin induces diarrhea
Scarlet fever	*Streptococcus pyogenes*	Cytotoxins cause vasodilation that results in the characteristic rash
Traveler's diarrhea	Enterotoxigenic *Escherichia coli* and *Shigella* spp.	Enterotoxin causes excessive secretion of ions and water; diarrhea results

most gram-negative bacteria. Recall from Chapter 4 that gram-negative bacteria have an outer membrane surrounding the peptidoglycan layer of the cell wall. This outer membrane consists of lipoproteins, phospholipids, and lipopolysaccharides (LPS) (see Figure 4.10). The lipid portion of LPS, called **lipid A,** is the endotoxin. Thus, endotoxins are lipopolysaccharides, whereas exotoxins are proteins.

Endotoxins exert their effects when gram-negative bacteria die and their cell walls undergo lysis, thus liberating the endotoxin. Antibiotics used to treat diseases caused by gram-negative bacteria can lyse the bacterial cells; this reaction releases endotoxin and may lead to an immediate worsening of the symptoms, but the condition usually improves as the endotoxin breaks down. All endotoxins produce the same signs and symptoms, regardless of the species of microorganism, although not to the same degree. Responses by the host include fever, weakness, generalized aches, and, in some cases, shock. Endotoxins can also induce miscarriage and prevent blood from clotting.

The fever (pyrogenic response) caused by endotoxins is believed to occur as depicted in Figure 15.4. When gram-negative bacteria are ingested by phagocytes and degraded in vacuoles, the lipopolysaccharides (LPS) of the bacterial cell wall are released. The lipopolysaccharides cause macrophages to produce small protein molecules called **interleukin-1 (IL-1),** for-

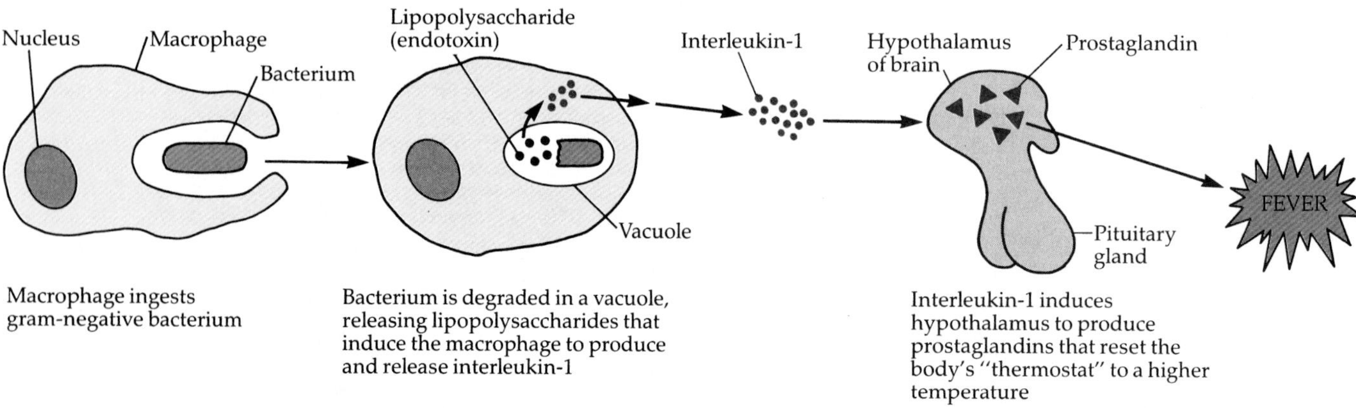
Nucleus
Macrophage
Bacterium
Lipopolysaccharide (endotoxin)
Vacuole
Interleukin-1
Hypothalamus of brain
Prostaglandin
Pituitary gland
FEVER
Macrophage ingests gram-negative bacterium
Bacterium is degraded in a vacuole, releasing lipopolysaccharides that induce the macrophage to produce and release interleukin-1
Interleukin-1 induces hypothalamus to produce prostaglandins that reset the body's "thermostat" to a higher temperature

FIGURE 15.4 Pyrogenic response. The postulated mechanism by which endotoxins cause fever.

TABLE 15.3 Comparison of Exotoxins and Endotoxins

PROPERTY	EXOTOXIN	ENDOTOXIN
Bacterial source	Mostly from gram-positive bacteria	Almost exclusively from gram-negative bacteria
Relation to microorganism	Metabolic product of growing cell	Present in LPS of outer membrane of cell wall and released only with destruction of cell
Chemistry	Protein or short peptide	Lipid portion (lipid A) of LPS of outer membrane
Heat stability	Unstable; can usually be destroyed at 60° to 80°C (except staphylococcal enterotoxin)	Stable; can withstand autoclaving (121°C for one hour)
Toxicity (power to cause disease)	High	Low
Immunology (relation to antibodies)	Can be converted to toxoids; neutralized by antitoxin	Not easily neutralized by antitoxin; therefore, effective toxoids cannot be made
Pharmacology (effect on body)	Specific for a particular cell structure or function in the host (mainly affects cell functions, nerves, and gastrointestinal tract)	General, such as fever, weaknesses, aches, and shock; all produce the same effects
Lethal dose	Small	Considerably larger
Representative diseases	Gas gangrene, tetanus, botulism, diphtheria, scarlet fever	Typhoid fever, urinary tract infections, and meningococcal meningitis

merly called *endogenous pyrogen,* which are carried via the blood to the hypothalamus, a temperature control center in the brain. IL-1 induces the hypothalamus to release lipids called prostaglandins that reset the thermostat in the hypothalamus at a higher temperature. The result is a fever. Bacterial cell death caused by lysis or antibiotics can also produce fever by this mechanism. Both aspirin and acetaminophen reduce fever by inhibiting the synthesis of prostaglandins. The function of fever in the body is discussed in Chapter 16.

Shock caused by gram-negative bacteria is called **septic** or **endotoxic shock.** The shock produced by endotoxins, like fever, is related to secretion of a substance by macrophages. Phagocytosis of gram-negative bacteria causes the phagocytes to secrete a polypeptide called **tumor necrosis factor (TNF),** or **cachectin.** This substance binds to many tissues in the body and alters their metabolism in a number of ways. One effect of TNF is damage to blood capillaries; their permeability is increased, and they lose large amounts of fluid. The result is a drop in blood pressure that results in shock. Low blood pressure has serious effects on the kidneys, lungs, and gastrointestinal tract. TNF also alters fat metabolism so that blood glucose level increases (hyperglycemia).

Endotoxins do not promote the formation of effective antitoxins. Antibodies are produced, but they tend not to counter the effect of the toxin; sometimes, in fact, they actually enhance its effect.

Representative microorganisms that produce endotoxins are *Salmonella typhi* (the causative agent of typhoid fever), *Proteus* species (frequently the causative agents of urinary tract infections), and *Neisseria meningitidis* (the causative agents of meningococcal meningitis). A comparison of exotoxins and endotoxins appears in Table 15.3.

Plasmids, Lysogeny, and Pathogenicity

Recall from Chapters 4 and 8 that plasmids are small circular DNA molecules that are not connected to the main bacterial chromosome and are capable of autonomous replication. One group of plasmids, called resistance (R) factors, is responsible for the resistance of some microorganisms to antibiotics. In addition, a plasmid may carry the information that determines a microbe's pathogenicity. Some example of virulence factors that are encoded by plasmid genes include tetanospasmin, heat-labile enterotoxin, and staphylococcal enterotoxin. Other examples include dextransucrase, an enzyme produced by *Streptococcus mutans* that is involved in tooth decay; adhesins and coagulase produced by *Staphylococcus aureus;* and a type of fimbria specific to enteropathogenic strains of *E. coli.*

In Chapter 13, we noted that some bacteriophages (viruses that infect bacteria) can incorporate their DNA into the bacterial chromosome and thus remain latent and not cause lysis of the bacterium. Such a state is called *lysogeny,* and the bacteriophage DNA that has been incorporated into the chromosome of its host bacterial cell is called a prophage. One outcome of lysog-

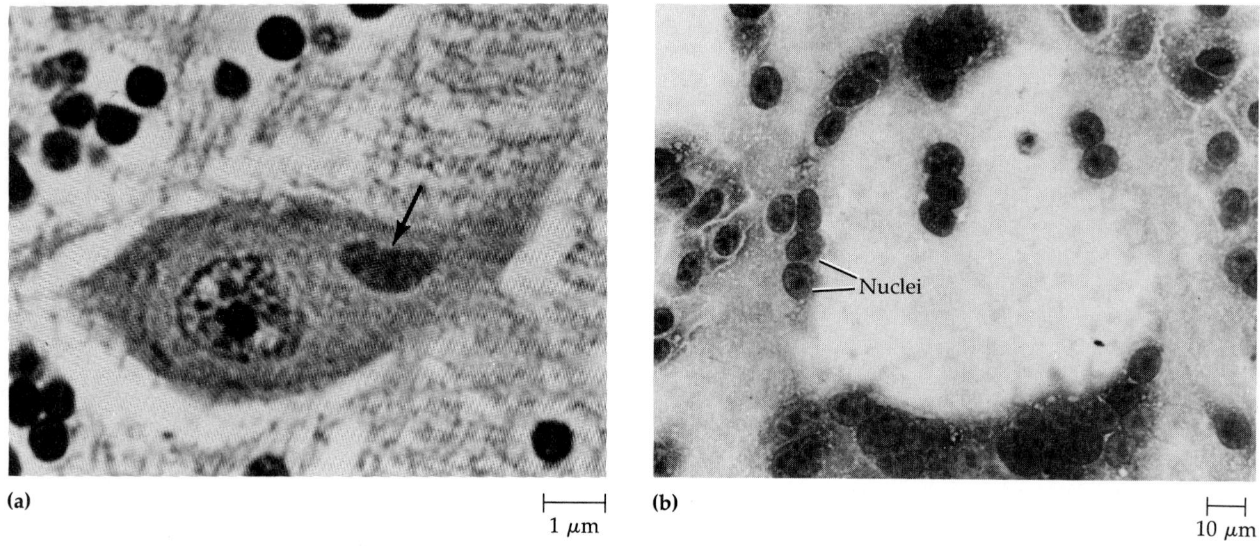

FIGURE 15.5 Effects of viruses on cells. (a) Cytoplasmic inclusion body (arrow) in a human brain cell from a person who died of rabies. **(b)** The relatively clear area in the center is a giant cell formed in a culture of cells infected with measles virus. The numerous dark, oval bodies around the cell's edges are its multiple nuclei.

eny is that the host bacterial cell and its progeny may exhibit new properties coded for by the bacteriophage DNA. Among the bacteriophage genes that contribute to pathogenicity are the genes for diphtherotoxin, erythrogenic toxins, staphylococcal enterotoxin and pyrogenic toxin, botulinum neurotoxin, and the capsule produced by *Streptococcus pneumoniae*.

Pathogenic Properties of Nonbacterial Microorganisms

VIRUSES

Viruses use the host cell's metabolism to interfere with normal cell reactions or to produce substances that can cause observable changes in infected cells. These observable changes, called *cytopathic effects (CPE)*, vary with the virus. One difference is the point in the viral infection cycle at which the effects occur. Some viral infections result in early changes in the host cell; in other infections, changes are not seen until a much later stage. Moreover, some viruses cause **cytocidal effects** (changes resulting in cell death), whereas others are noncytocidal. Some viruses are cell-specific; that is, they attack only particular cells of the body. For example, the AIDS virus attacks cells with a specific receptor, such as T lymphocytes. A virus can produce one or more of the cytopathic effects listed below. (Some of these effects have already been discussed in Chapter 13; others will be discussed in more detail in later chapters.)

1. At some stage in their multiplication, cytocidal viruses cause the macromolecular synthesis within the host cell to stop. Some viruses, such as herpes simplex virus, irreversibly stop mitosis.
2. Infection by a cytocidal virus begins cell destruction by causing the release of enzymes from lysosomes. These enzymes cause autolysis of the cell.
3. Inclusion bodies (Figure 15.5a) are granules found in the cytoplasm or nucleus of some infected cells. These granules are viral parts—nucleic acids or proteins in the process of being assembled into virions. The granules vary in size, shape, and staining properties, according to the virus. Inclusion bodies are characterized by their ability to stain with an acidic stain (acidophilic) or with a basic stain (basophilic).
4. Some viral infections cause host cells to fuse, producing multinucleated syncytia or "giant" cells (Figure 15.5b).
5. Many viral infections induce antigenic changes on the surface of the infected cells. These antigenic changes elicit a host antibody response against the infected cell, and thus they target the cell for destruction by the host's immune system.
6. Some viruses induce chromosomal changes in the host cell. Frequently oncogenes (cancer-causing genes) may be contributed or activated by a virus.
7. Most normal cells cease growing in vitro when they come close to another cell, a phenomenon

TABLE 15.4 Cytopathic Effects of Selected Viruses

VIRUS	CYTOPATHIC EFFECT
Poliovirus	Cell death
Papovavirus	Acidophilic inclusion bodies in nucleus
Adenovirus	Basophilic inclusion bodies in nucleus
Rhabdovirus	Acidophilic inclusion bodies in cytoplasm
Measles, cytomegalovirus	Acidophilic inclusion bodies in nucleus and cytoplasm
Measles	Cell fusion
Polyoma virus	Transformation
AIDS virus (HIV)	Destruction of T lymphocytes

known as *contact inhibition*. Viruses capable of causing cancer *transform* host cells, as discussed in Chapter 13. Transformation results in an abnormal, spindle-shaped cell that does not recognize contact inhibition. (Figure 15.6). Loss of contact inhibition results in unregulated cell growth.

Some representative viruses that cause cytopathic effects are presented in Table 15.4. In subsequent chapters, we will discuss the pathological properties of viruses in more detail.

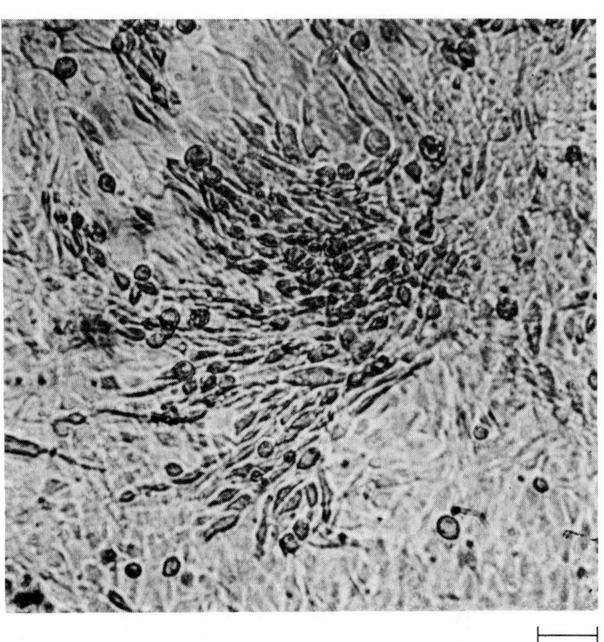

FIGURE 15.6 Transformed cells in culture. In the center of this micrograph is a cluster of chick embryo cells transformed by Rous sarcoma virus. Such a concentration of transformed cells in a cell culture is called a *focus;* it results from multiplication of a single cell infected with a transforming virus. Note how the transformed cells of the focus appear dark, in contrast to the monolayer of light, flat, normal cells around them. This appearance is caused by their spindle shapes and uninhibited growth on top of one another; there is no contact inhibition.

FUNGI, PROTOZOANS, HELMINTHS, AND ALGAE

Although fungi cause disease, they do not have a well-defined set of virulence factors. For example, *Cryptococcus neoformans* (krip-tō-kok'kus nē-ō-fôr'manz) is a fungus that causes a type of meningitis; it produces a capsule that helps it to resist phagocytosis. Other fungi have metabolic products that are toxic to human hosts. In such cases, however, the toxin is only an indirect cause of disease, as the fungus is already growing in or on the host. Fungal infection can also provoke an allergic response in the host.

The human disease called *ergotism,* which was common in Europe during the Middle Ages, is caused by a toxin produced by an ascomycete plant pathogen, *Claviceps purpurea* (kla'vi-seps pur-pū-rē'ä), that grows on grains. The toxin is contained in **sclerotia** (Figure 15.7), highly resistant portions of the fungus's mycelia that can detach. The toxin itself, **ergot,** is an alkaloid that can cause hallucinations resembling those produced by LSD; in fact, ergot is a natural source of LSD. Ergot also constricts capillaries and can cause gangrene of the extremities through prevention of proper blood circulation. Although *Claviceps* still occasionally occurs on grains, modern milling usually removes the sclerotia.

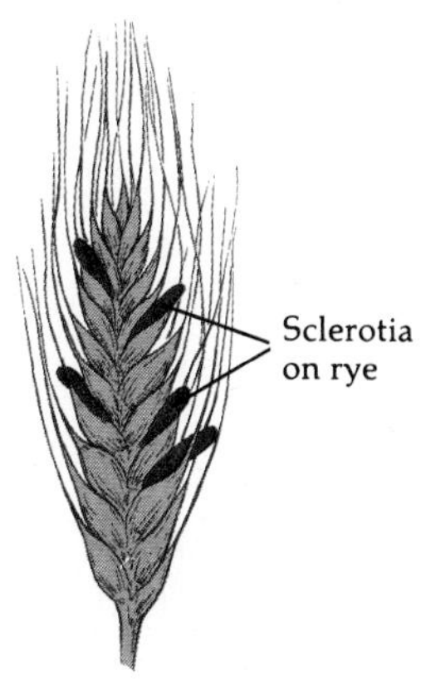

FIGURE 15.7 ***Claviceps purpurea.*** The sclerotia of *Claviceps purpurea* are visible on this rye flower. *Claviceps* is a natural source of the drug LSD.

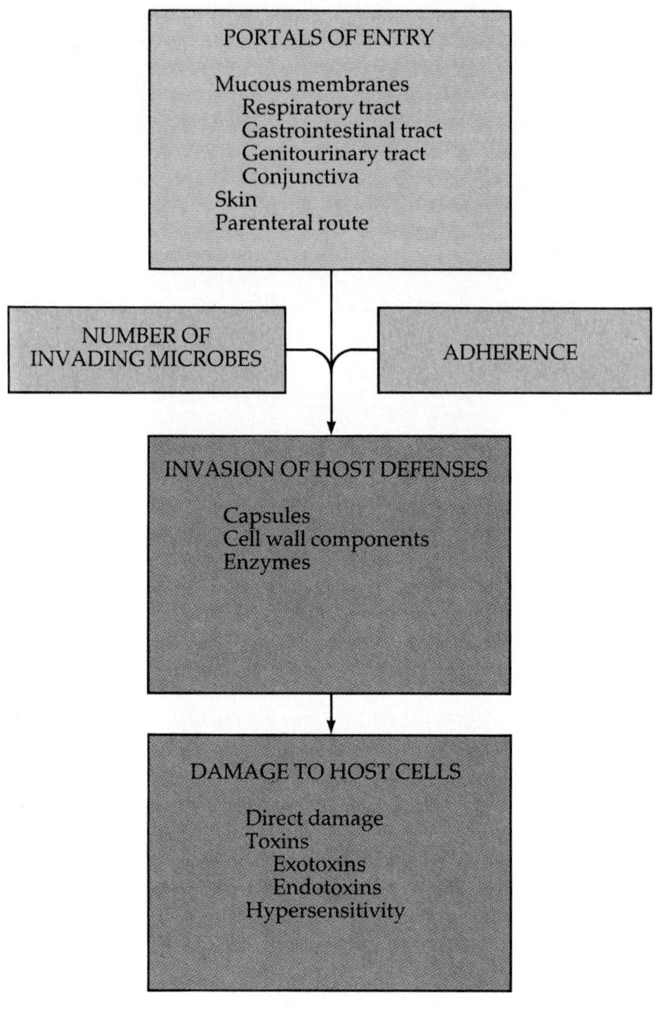

FIGURE 15.8 Mechanisms of pathogenicity. A summary of how microorganisms cause disease.

A number of other toxins are produced by fungi that grow on grains or other plants. For example, peanut butter is occasionally recalled because of excessive amounts of **aflatoxin,** a toxin that has carcinogenic properties. Aflatoxin is produced by the growth of the mold *Aspergillus flavus*. It is believed that the toxin might be altered in the animal's body to a mutagenic compound.

A few mushrooms produce toxins called **mycotoxins** (toxins produced by fungi). Examples are **phalloidin** and **amanitin,** produced by *Amanita phalloides* (am-an-ī'ta fal-loi'dēz), commonly known as the death angel. These neurotoxins are so potent that ingestion of the *Amanita* mushroom may result in death.

The presence of protozoans or helminths often produces disease symptoms in the host. Some of these organisms actually use host tissues for their own growth or produce large parasitic masses that block the flow of lymph; the resulting cellular damage evokes the symptoms. An example is the roundworm *Wuchereria bancrofti* (vū-kėr-ār'ē-ä ban-krof'tē), the causative agent of elephantiasis. Waste products of the metabolism of these parasites can also contribute to the symptoms of a disease.

A few species of algae produce neurotoxins. For example, some genera of dinoflagellates such as *Gonyaulax,* are important because they produce a neurotoxin called **saxitoxin.** Although mollusks that feed on the dinoflagellates that produce saxitoxin show no symptoms of disease, people who eat the mollusks develop symptoms similar to botulism. Recall from Chapter 12 that extensive dinoflagellate growth is responsible for a situation called red tide. Public health agencies frequently prohibit human consumption of mollusks during red tides.

Specific diseases caused by fungi, protozoans, and helminths, along with the pathological properties of these organisms, are discussed in detail in later chapters.

In the next chapter, we will examine a group of nonspecific defenses of the host against disease. But before proceeding, examine Figure 15.8 very carefully. It summarizes some key concepts of the mechanisms of pathogenicity we have discussed in this chapter.

STUDY OUTLINE

Introduction (p. 392)

1. Pathogenicity is the ability of a pathogen to produce a disease by overcoming the defenses of the host.
2. Virulence is the degree of pathogenicity.

Entry of a Microorganism into the Host (pp. 392–395)

1. The specific route by which a particular pathogen gains access to the body is called its portal of entry.

MUCOUS MEMBRANES (pp. 392–393)

1. Many microorganisms can penetrate mucous membranes of the conjunctiva and the respiratory, gastrointestinal, and genitourinary tracts.
2. Microorganisms that are inhaled with droplets of moisture and dust particles gain access to the respiratory tract.
3. The respiratory tract is the most frequently used portal of entry.
4. Microorganisms enter the gastrointestinal tract via food, water, and contaminated fingers.

SKIN (p. 393)

1. Most microorganisms cannot penetrate intact skin; they enter hair follicles and sweat ducts.
2. Some fungi infect the skin itself.

PARENTERAL ROUTE (p. 393)

1. Some microorganisms can gain access to tissues by inoculation through the skin and mucous membranes in bites, injections, and other wounds.
2. This route of penetration is called the parenteral route.

PREFERRED PORTAL OF ENTRY (p. 393)

1. Many microorganisms can cause infections only when they gain access through their specific portal of entry.

NUMBERS OF INVADING MICROBES

1. Virulence can be expressed as LD_{50} (lethal dose for 50% of the inoculated hosts) or ID_{50} (infectious dose for 50% of the inoculated hosts).

ADHERENCE (pp. 393–395)

1. Surface projections on a pathogen called adhesins (ligands) adhere to complementary receptors on the host cells.
2. Ligands can be glycoproteins or lipoproteins and are frequently associated with fimbriae.
3. Mannose is the most common receptor.

How Pathogens Penetrate Host Defenses (pp. 395–397)

CAPSULES (pp. 395–396)

1. Some pathogens have capsules that prevent them from being phagocytized.

COMPONENTS OF THE CELL WALL (p. 396)

1. Proteins in the cell wall can facilitate adherence or prevent a pathogen from being phagocytized.
2. Some microbes can reproduce inside phagocytes.

ENZYMES (pp. 396–397)

1. Leukocidins destroy neutrophils and macrophages.
2. Hemolysins lyse red blood cells.
3. Local infections can be protected in a fibrin clot caused by the bacterial enzyme coagulase.
4. Bacteria can spread from a focal infection by means of kinases (which destroy blood clots), hyaluronidase (which destroys a mucopolysaccharide that holds cells together), and collagenase (which hydrolyzes connective tissue collagen).

Damage to Host Cells (pp. 397–401)

DIRECT DAMAGE (p. 397)

1. Host cells can be destroyed when pathogens metabolize and multiply inside the host cells.

TOXINS (pp. 397–401)

1. Poisonous substances produced by microorganisms are called toxins; toxemia refers to symptoms caused by toxins in the blood.
2. The ability to produce toxins is called toxigenicity.

Exotoxins (pp. 397–399)

1. Exotoxins are produced by bacteria and released into the surrounding medium.
2. Exotoxins, not the bacteria, produce the disease symptoms.
3. Antibodies produced against exotoxins are called antitoxins.
4. Cytotoxins include diphtherotoxin (which inhibits protein synthesis) and erythrogenic toxins (which damage capillaries).
5. Neurotoxins include botulinum toxin (which prevents nerve transmission) and tetanus toxin (which prevents inhibitory nerve transmission).
6. *Vibrio* choleragen and staphylococcal enterotoxin are enterotoxins, which induce fluid and electrolyte loss from host cells.

Endotoxins (pp. 399–401)

1. Endotoxins are lipopolysaccharides—the lipid A component of the cell wall of gram-negative bacteria.
2. Bacterial cell death, antibiotics, and antibodies may cause release of endotoxins.
3. Endotoxins cause fever (by inducing release of interleukin-1) and shock (because of TNF-induced decrease in blood pressure).

Plasmids, Lysogeny, and Pathogenicity (pp. 401–402)

1. Plasmids may carry genes for antibiotic resistance, toxins, capsules, and fimbriae.
2. Lysogeny can result in bacteria with virulence factors, such as toxins or capsules.

Pathogenic Properties of Nonbacterial Microorganisms (pp. 402–404)

VIRUSES (pp. 402–403)

1. Signs of viral infections are called cytopathic effects (CPE).
2. Some viruses cause cytocidal effects (cell death), and others cause noncytocidal effects.
3. Cytopathic effects include the stopping of mitosis, lysis, formation of inclusion bodies, cell fusion, antigenic changes, chromosomal changes, and transformation.

FUNGI, PROTOZOANS, HELMINTHS, AND ALGAE (pp. 403–404)

1. Symptoms of fungal infections can be caused by capsules, toxins, and allergic responses.
2. Symptoms of protozoan and helminthic diseases can be caused by damage to host tissue or by the metabolic waste products of the parasite.

STUDY QUESTIONS

Review

1. List three portals of entry, and describe how microorganisms gain access through each.
2. Compare pathogenicity with virulence.
3. How would drugs that bind each of the following affect pathogenicity:
   (a) Mannose on human cell membranes
   (b) *Neisseria gonorrhoeae* fimbriae
   (c) *Streptococcus pyogenes* M protein
4. Define cytopathic effects, and give five examples.
5. Compare and contrast the following aspects of endotoxins and exotoxins: bacterial source, chemistry, toxicity, and pharmacology. Give an example of each toxin.
6. How are capsules and cell wall components related to pathogenicity? Give specific examples.
7. Describe how hemolysins, leukocidins, coagulase, kinases, and hyaluronidase might contribute to pathogenicity.
8. Describe the factors contributing to the pathogenicity of fungi, protozoans, and helminths.
9. Which of the following genera is the most infectious?

Genus	ID_{50}
Escherichia	10^8 cells
Salmonella	10^5 cells
Shigella	200 cells
Treponema	52 cells

10. The LD_{50} of botulinum toxin is 0.000025 μg. The LD_{50} of *Salmonella* toxin is 200 μg. Which of these is the more potent toxin? How can you tell from the LD_{50} values?
11. Food poisoning can be divided into two categories, food infection and food intoxication. On the basis of toxin production by bacteria, explain the difference between these two categories.

Challenge

1. The cyanobacterium *Microcystis aeruginosa* produces a peptide that is toxic to humans. According to the graph below, when is this bacterium most toxic?

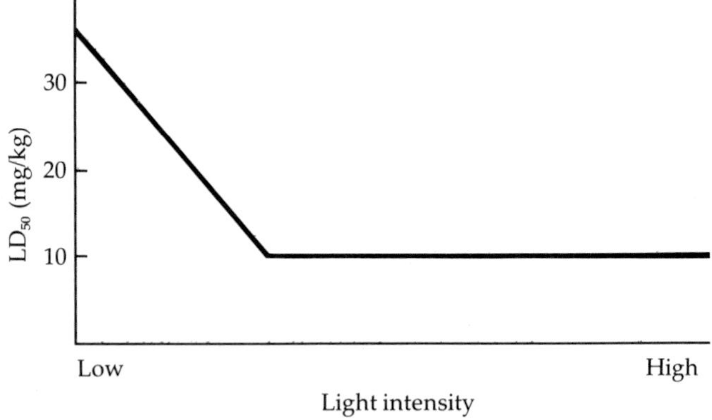

2. Explain whether each of the following examples is a food infection or intoxication. What is the probable etiological agent in each case?
   (a) Eighty-two people who ate shrimp at a dinner in Port Allen, Louisiana, developed diarrhea, cramps, weakness, nausea, chills, headache, and fever from 4 hours to 2 days after eating.
   (b) Two people in Vermont who ate barracuda caught in Florida developed malaise, nausea, blurred vision, breathing difficulty, and numbness from 3 to 6 hours after eating.
3. Washwater containing *Pseudomonas* was sterilized and used to wash cardiac catheters. Three patients undergoing cardiac catheterization developed fever, chills, and hypotension. The water and catheters were sterile. Why did the patients show these reactions?

FURTHER READING

Fischetti, V.A. "Streptococcal M protein." *Scientific American* 264(6):58–65, June 1991. Describes the mechanism by which M protein repels phagocytes and allows antibodies to attach to a highly variable region.

Goodenough, U.W. "Deception by pathogens." *American Scientist* 79:344–355, July–August 1991. Discussion of bacteria and viruses that use camouflage, mimicry, and distraction.

Habicht, G.S., G. Beck, and J.L. Benach. "Lyme disease." *Scientific American* 251(1):78–83, July 1987. Describes how investigators discovered the cause of this new disease since the first reports in 1975. Also describes the actions of bacterial LPS and interleukin-1 in causing symptoms.

Moberg, C.L., and Z.A. Cohn. "René Jules Dubos." *Scientific American* 264(5):66–74, May 1991. This biography recaps Dubos's work on the relationship between bacterial virulence and disease.

CHAPTER 16

Nonspecific Defenses of the Host

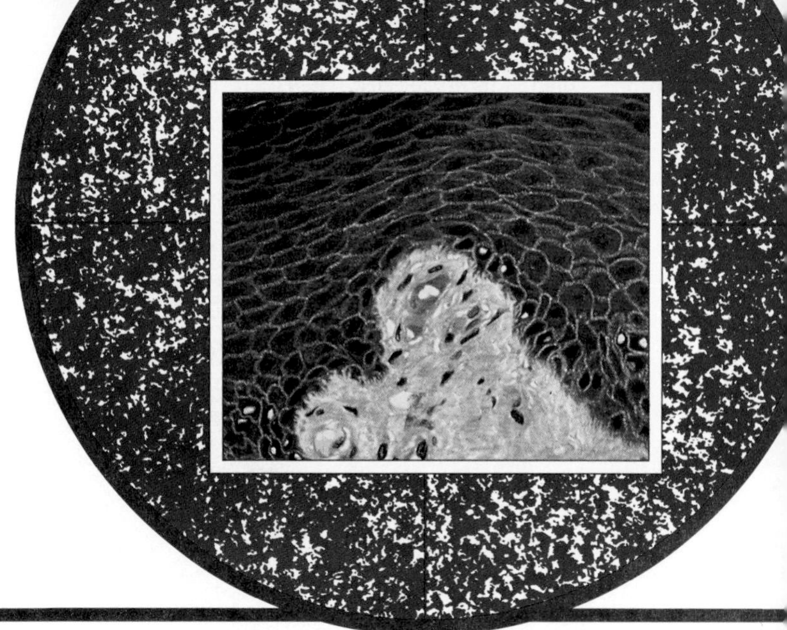

LEARNING OBJECTIVES

- Define resistance and susceptibility.
- Define nonspecific resistance.
- Describe the role of the skin and mucous membranes in nonspecific resistance, and differentiate between mechanical and chemical factors.
- Define phagocytosis, and include the stages of adherence and ingestion.
- Classify phagocytic cells, and describe the roles of granulocytes and monocytes.
- Describe the stages of inflammation and their relation to nonspecific resistance.
- Discuss the role of fever in nonspecific resistance.
- Discuss the role of interferon.
- Discuss the function of complement.

From our discussion to this point, you can see that pathogenic microorganisms are endowed with special properties that enable them to cause disease if given the right opportunity. If microorganisms never encountered resistance from the host, we would constantly be ill and would eventually die of various diseases, but in most cases our body defenses prevent this from happening. Some of these defenses are designed to keep out microorganisms altogether. Other defenses remove the microorganisms if they do get in, and still others combat them if they remain inside. Our ability to ward off disease through our defenses is called **resistance.** Vulnerability or lack of resistance is known as **susceptibility.**

In discussing resistance, we will divide our body defenses into two general kinds, nonspecific and specific. *Nonspecific resistance* refers to defenses that protect us from *any* pathogen, regardless of species. *Specific resistance,* or *immunity,* is the defense that the body offers against a particular pathogen. Specific defenses are based on the production of specific proteins, called antibodies, and special cells of the immune system. Chapter 17 deals with specific resistance. In this chapter, we consider nonspecific resistance.

Contributing to nonspecific resistance are the skin and mucous membranes, phagocytosis, inflammation, fever, and the production of antimicrobial substances other than antibodies. These are the main topics of this chapter.

Skin and Mucous Membranes

The skin and mucous membranes are the body's first line of defense against pathogens. This function results from both mechanical and chemical factors.

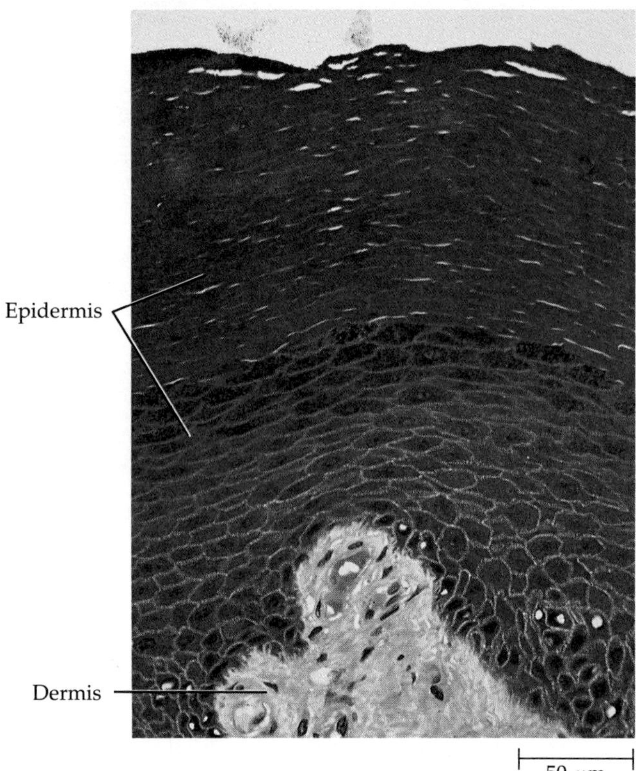

FIGURE 16.1 Section through human skin. The thin layer at the top of the photo contains keratin. This layer and the darker purple cells beneath it make up the epidermis. The lighter purple material near the bottom part of the photo is the dermis.

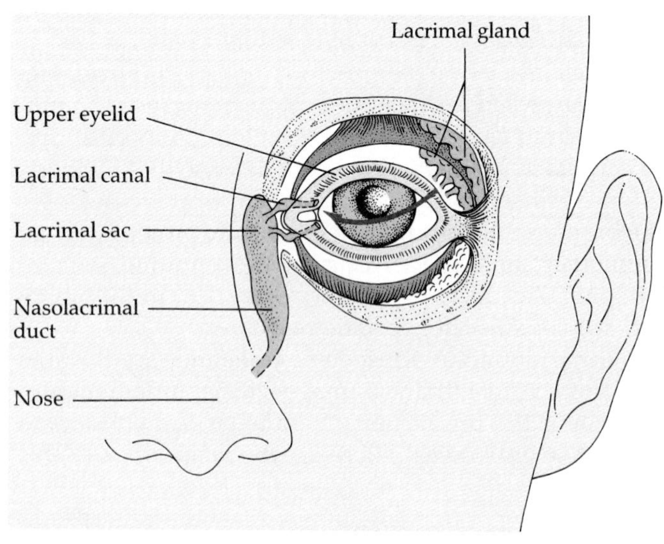

FIGURE 16.2 The lacrimal apparatus. The washing action of the tears, shown by the arrow, prevents microorganisms from settling on the surface of the eyeball. Tears produced by the lacrimal glands pass across the surface of the eyeball into two small holes that convey the tears into the lacrimal canals, lacrimal sac, and nasolacrimal duct.

MECHANICAL FACTORS

The intact skin is the human body's largest organ in terms of surface area. It consists of two distinct portions, the dermis and the epidermis (Figure 16.1). The *dermis,* the skin's inner, thicker portion, is composed of connective tissue. The *epidermis,* the outer, thinner portion, is in direct contact with the external environment. The epidermis consists of layers of continuous sheets of tightly packed epithelial cells with little or no material between the sheets. Two types of cells in the epidermis, Langerhans and Granstein cells, participate in immunity by assisting helper T cells and suppressor T cells, respectively. (These cells are discussed in Chapter 17.) The top layer of epidermal cells is dead and contains a waterproofing protein called **keratin.**

If we consider the closely packed cells, continuous layering, and presence of keratin, we can see why the intact skin provides such a formidable physical barrier to the entrance of microorganisms. The intact surface of healthy epidermis is rarely, if ever, penetrated by microorganisms. But when the epithelial surface is broken, a subcutaneous (below-the-skin) infection often develops. The bacteria most likely to cause such an infection are the staphylococci that normally inhabit the epidermis, hair follicles, and sweat and oil glands of the skin. Infections of the skin and underlying tissues frequently occur as a result of burns, cuts, stab wounds, or other conditions that break the skin. Moreover, when the skin is moist, as in hot, humid climates, skin infections are quite common, especially fungus infections such as athlete's foot.

Mucous membranes also consist of an epithelial layer and an underlying connective tissue layer. Mucous membranes line the entire gastrointestinal, respiratory, urinary, and reproductive tracts. The epithelial layer of a mucous membrane secretes a fluid called *mucus,* which prevents the tracts from drying out. Some pathogens that can thrive on the moist secretions of a mucous membrane are able to penetrate the membrane if the microorganism is present in sufficient numbers. *Treponema pallidum, Mycobacterium tuberculosis,* and *Streptococcus pneumoniae* are such pathogens. This penetration may be facilitated by toxic substances produced by the microorganism, prior injury by viral infection, or mucosal irritation. Although mucous membranes do inhibit the entrance of many microorganisms, they offer less protection than the skin.

Besides the physical barrier presented by the skin and mucous membranes, several other mechanical factors help protect certain epithelial surfaces. One such mechanism that protects the eyes is the *lacrimal apparatus* (Figure 16.2), a group of structures that manufactures and drains away tears. The lacrimal glands, located toward the upper, outermost portion of each eye socket, produce the tears and pass them under the

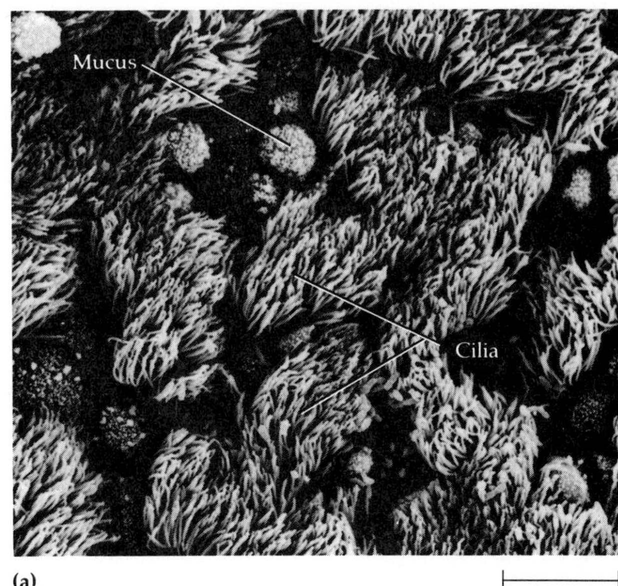

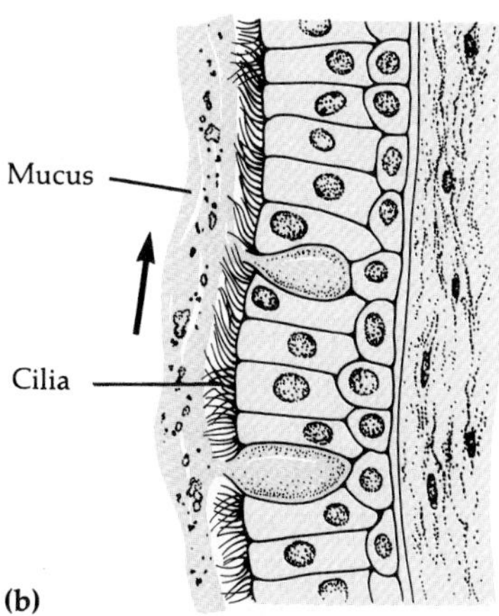

FIGURE 16.3 Mucous membrane of the trachea (windpipe). **(a)** Micrograph of cilia and mucus-secreting goblet cells, from which blobs of mucus are emerging. **(b)** Action of the ciliary escalator.

upper eyelid. From here, tears pass toward the corner of the eye near the nose and into two small holes that lead through tubes (lacrimal canals) to the nose. The tears are spread over the surface of the eyeball by blinking. Normally, the tears evaporate or pass into the nose as fast as they are produced. This continual washing action helps to keep microorganisms from settling on the surface of the eye. If an irritating substance or large numbers of microorganisms come in contact with the eye, the lacrimal glands start to secrete heavily, and the tears accumulate more rapidly than they can be carried away. This excessive production is a protective mechanism because the excess tears dilute and wash away the irritating substance or microorganisms.

In a cleansing action very similar to that of tears, saliva is produced by the salivary glands. Saliva helps to dilute the numbers of microorganisms and wash them from both the surface of the teeth and the mucous membrane of the mouth. This helps prevent colonization by microbes.

The respiratory and gastrointestinal tracts have many mechanical forms of defense. Mucus is slightly viscous (thick); thus, it traps many of the microorganisms that enter the respiratory and gastrointestinal tracts. The mucous membrane of the nose also has mucus-coated hairs that filter inhaled air and trap microorganisms, dust, and pollutants. The cells of the mucous membrane of the lower respiratory tract are covered with cilia (Figure 16.3). By moving synchronously, these cilia propel inhaled dust and microorganisms that have become trapped in mucus upward toward the throat. This so-called **ciliary escalator** keeps the mucous blanket moving toward the throat at a rate of 1 to 3 cm per hour; coughing and sneezing speed up the escalator. (Some substances in cigarette smoke are toxic to cilia and can seriously impair the functioning of the ciliary escalator.) Microorganisms are also prevented from entering the lower respiratory tract by a small lid of cartilage called the epiglottis, which covers the larynx (voicebox) during swallowing.

The cleansing of the urethra by the flow of urine is another mechanical factor that prevents microbial colonization in the urinary system. Vaginal secretions likewise move microorganisms out of the female body.

CHEMICAL FACTORS

Mechanical factors alone do not account for the high degree of resistance of skin and mucous membranes to microbial invasion. Certain chemical factors also play important roles.

Sebaceous (oil) glands of the skin produce an oily substance called **sebum** that prevents hair from drying and becoming brittle. Sebum also forms a protective film over the surface of the skin. One of the components of sebum is unsaturated fatty acids, which inhibit the growth of certain pathogenic bacteria and fungi. The low pH of the skin, between pH 3 and 5, is caused in part by the secretion of fatty acids and lactic acid. The skin's acidity probably discourages the growth of many other microorganisms.

Bacteria that live as commensals on the skin decompose sloughed-off skin cells, and these organic molecules and the end-products of their metabolism produce body odor. As you will see in Chapter 21, certain bacteria commonly found on the skin metabolize sebum, and this metabolism forms free fatty acids that cause the inflammatory response associated with acne. Isotretinoin (Accutane™), a derivative of vitamin A that prevents sebum formation, is a treatment for a very severe type of acne called cystic acne.

The sweat glands of the skin produce perspiration, which helps to maintain body temperature, eliminate certain wastes, and flush microorganisms from the surface of the skin. Perspiration also contains **lysozyme,** an enzyme capable of breaking down cell walls of gram-positive bacteria and, to a lesser extent, gram-negative bacteria (see Figure 4.10). Lysozyme is also found in tears, saliva, nasal secretions, and tissue fluids, where it exhibits its antimicrobial activity. Alexander Fleming was actually studying lysozyme in 1929 when he accidentally discovered the antimicrobial effects of penicillin (Figure 1.5).

Gastric juice is produced by the glands of the stomach. It is a mixture of hydrochloric acid, enzymes, and mucus. The very high acidity of gastric juice (pH 1.2 to 3.0) is sufficient to preserve the usual sterility of the stomach. This acidity destroys bacteria and most bacterial toxins except those of *Clostridium botulinum* and *Staphylococcus aureus.* However, many enteric pathogens are protected by food particles and are able to enter the intestines via the gastrointestinal tract. (The box in Chapter 15, p. 398, describes a bacterium that neutralizes stomach acid so it can grow in the stomach.) Blood also contains antimicrobial chemicals. For example, iron-binding proteins, called **transferrins,** inhibit bacterial growth by reducing the amount of available iron.

We will next discuss phagocytosis, commonly regarded as the body's second line of defense against infection.

Phagocytosis

Phagocytosis (from the Greek words for *eat* and *cell*) is the ingestion of a microorganism or any particulate matter by a cell. We have previously mentioned phagocytosis as the method of nutrition of certain protozoans. In this chapter, phagocytosis is discussed as a means by which cells in the human body counter infection. The human cells that perform this function are collectively called **phagocytes.** All are types of white blood cells or derivatives of white blood cells. Before we look at phagocytes specifically, let us consider the various components of the blood more generally.

MICROBIOLOGY *HIGHLIGHTS*

Neutrophil Defect Leads to Periodontal Disease

What happens when the body creates white blood cells that fail to function properly? Since these cells are a key line of defense against infection and disease, it is clear that malfunctioning white blood cells would leave us more vulnerable. Indeed, recent dental research indicates that an unusual periodontal disease called *localized juvenile periodontitis* (*LJP*) attacks children whose bodies produce faulty neutrophils. Neutrophils, which leave the blood to go to damaged tissue, normally prevent the spread of bacteria into gums and roots.

Periodontal disease is normally an adult problem in which infections of the soft tissues and bones that support the teeth cause loosening and eventual loss of the teeth. Unlike normal periodontal problems, however, LJP attacks children and adolescents. It affects the tissues that support the incisors and first molars. If the disease is left untreated, there is serious bone loss and the affected teeth fall out by the time the victim is 20 years old.

According to the latest findings, the disease-causing agent in LJP is the gram-negative bacterium *Actinobacillus actinomycetemcomitans,* which is the only plaque bacterium that kills neutrophils. It produces a leukotoxin that binds to the neutrophil's membrane and kills it very quickly. Although *A. actinomycetemcomitans* is present in the mouths of about one-fifth of the population, it can cause the disease only in people who have reduced neutrophil function.

This reduced function is caused by depressed chemotaxis—that is, the neutrophils do not respond to the chemicals that should alert them to the presence of the bacteria, and therefore they do not approach and attack the bacteria. It appears that the neutrophils are deficient in receptors, the membrane proteins that bind them to the attracting chemicals. This deficiency seems to be hereditary.

Traditionally, LJP has been treated by killing the bacteria with antibiotics and surgically removing infected tissue. Although the affected tissues regenerate, the disease often recurs. Twenty-five percent of the patients suffer a relapse within one year.

Now that the relationship of LJP to the neutrophils' chemotaxis deficiency has been recognized, new methods of treatment deserve recognition. A chemical known as Bestatin, which is purified from the bacterium *Streptomyces olivoreticuli,* is able to double the chemotactic activity of neutrophils from LJP patients and therefore is potentially useful in treating LJP. One theory is that Bestatin increases the density of chemotactic receptors. So far, Bestatin has been tested only in vitro.

FORMED ELEMENTS IN BLOOD

Blood consists of a fluid called **plasma,** which contains **formed elements,** that is, cells and cell fragments (Table 16.1). Of the cells listed in Table 16.1, those that concern us at present are the **leukocytes,** or white blood cells (Figure 16.4).

During many kinds of infections, especially bacterial infections, there is an increase in the total number of white blood cells; this increase is called *leukocytosis.* During the active stage of infection, the leukocyte count might double, triple, or quadruple, depending on the severity of the infection. Diseases that might cause such an elevation in the leukocyte count are meningitis, infectious mononucleosis, appendicitis, pneumococcal pneumonia, and gonorrhea. Other diseases, such as salmonellosis, brucellosis, pertussis, and some viral and rickettsial infections, may cause a *decrease* in the leukocyte count (called *leukopenia*). Leukopenia may be related to impaired white blood cell production or the effect of increased sensitivity of white blood cell membranes to damage by complement, antimicrobial plasma proteins discussed later in the chapter. Leukocyte increase or decrease can be detected by a **differential white blood cell count,** which is a calculation of the percentage of each kind of white cell in a sample of white blood cells. The percentages in a normal differential white blood cell count are

TABLE 16.1 Formed Elements in Blood

TYPE OF CELL	*NUMBERS PER CUBIC MILLIMETER (mm^3)*	*FUNCTION*
Erythrocytes (*Red blood cells*)	4.8 to 5.4 million	Transport of O_2 and CO_2
Leukocytes (White blood cells)	5000 to 9000	
A. Granulocytes		
1. Neutrophils (PMNs) (60% to 70% of leukocytes)		Phagocytosis
2. Basophils (0.5% to 1%)		Production of heparin and histamine
3. Eosinophils (2% to 4%)		Production of toxic proteins against certain parasites; some phagocytosis
B. Monocytes (3% to 8%)		Phagocytosis (when they mature into macrophages)
C. Lymphocytes (20% to 25%)		Antibody production (B lymphocytes); cell-mediated immunity (T lymphocytes)*
Thrombocytes (Platelets)	250,000 to 400,000	Blood clotting

*Discussed in Chapter 17.

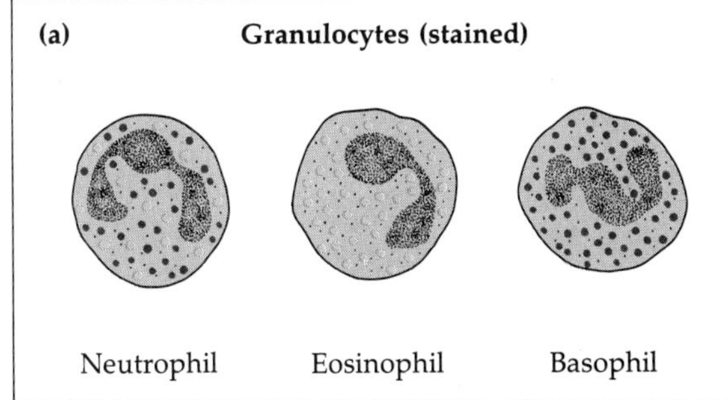

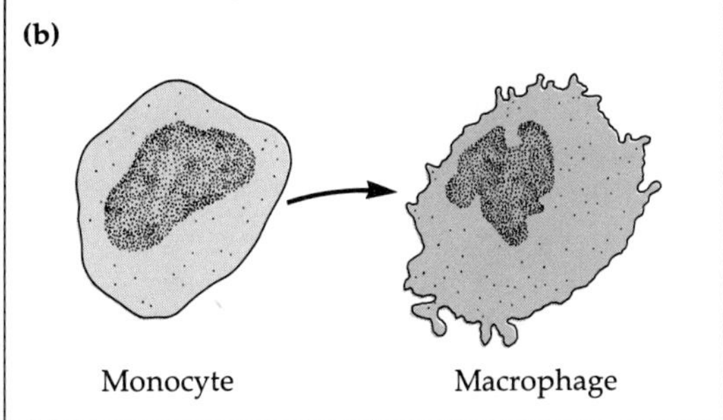

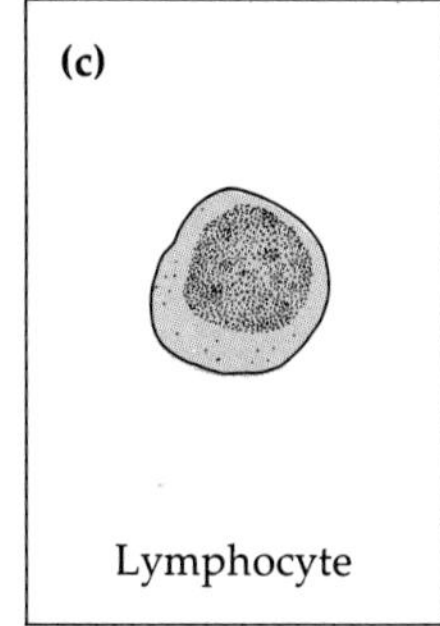

FIGURE 16.4 Principal kinds of leukocytes. (a) Granulocytes can be distinguished by staining. **(b)** Monocytes mature into actively phagocytic macrophages. Neutrophils, eosinophils, and macrophages are the phagocytic cells of the nonspecific defense system. **(c)** Lymphocytes are important in specific immunity.

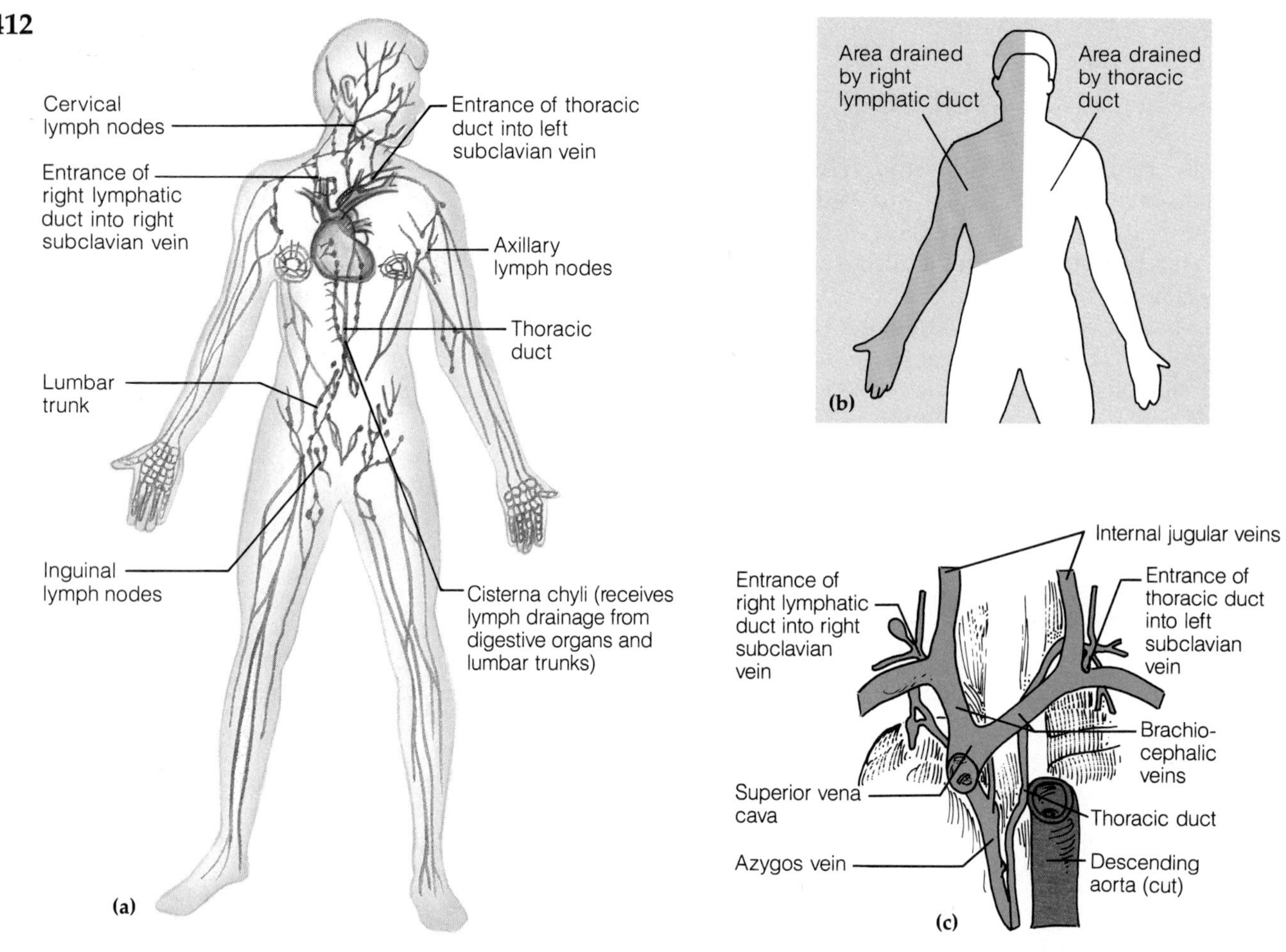

FIGURE 16.5 Components of the lymphatic system. **(a)** Fluid circulating between tissue cells is picked up by lymph vessels. The fluid is then called lymph. Microorganisms can enter the lymph vessels. Macrophages in the lymph nodes remove microorganisms from the lymph. **(b)** Lymph is transported into larger lymph vessels, to be collected in the lymphatic duct and thoracic duct. **(c)** Eventually, the lymph is returned to the blood at the subclavian veins, just before the blood enters the heart.

shown in parentheses in the first column of Table 16.1.

Leukocytes are divided into three categories: granulocytes, monocytes, and lymphocytes. **Granulocytes** include some phagocytic types. They owe their name to the presence of large granules in their cytoplasm that can be seen under a light microscope. They are differentiated into three types on the basis of how the granules stain. The granules of **neutrophils** stain red and blue with a mixture of acidic and basic dyes, respectively; those of **basophils** stain blue with the basic dye methylene blue; and those of **eosinophils** stain red with the acid dye eosin.

Neutrophils are also commonly called *polymorphonuclear leukocytes* (*PMNs*), or *polymorphs*. The term *polymorphonuclear* refers to the fact that the nuclei of neutrophils contain from two to five lobes. Neutrophils, which are highly phagocytic and motile, are active in the initial stages of an infection. They have the ability to leave the blood, enter an infected tissue, and destroy microbes and foreign particles (see the box on p. 410).

The role of basophils is not clear. However, they release substances, such as histamine, that are important in inflammation and allergic responses. They also release heparin, an anticoagulant.

Eosinophils are somewhat phagocytic and also have the ability to leave the blood. Their major function is to produce toxic proteins against certain parasites. Their number increases significantly during certain parasitic worm infections and hypersensitivity (allergy) reactions.

Monocytes lack granules in their cytoplasm and are not actively phagocytic until they leave circulating blood, enter body tissues, and mature into **macrophages.** In fact, the maturation and proliferation of macrophages (along with lymphocytes) is one factor responsible for the swelling of lymph nodes during an infection.

Lymphocytes are not phagocytic but play a key

role in specific immunity (see Chapter 17). They occur in lymphoid tissues of the lymphatic system in the tonsils, lymph nodes, spleen, thymus gland, thoracic duct, bone marrow, appendix, Peyer's patches of the small intestine, and lymph nodes in the respiratory, gastrointestinal, and reproductive tracts (Figure 16.5). They also circulate in the blood.

ACTIONS OF PHAGOCYTIC CELLS

When an infection occurs, both granulocytes (especially neutrophils) and monocytes migrate to the infected area. During this migration, monocytes enlarge and develop into actively phagocytic macrophages (Figure 16.6). Because these cells leave the blood and migrate through tissue to infected areas, they are called *wandering macrophages.* Some macrophages, called *fixed macrophages* or *histiocytes,* are located in certain tissues and organs of the body. Fixed macrophages are found in the liver (Kupffer's cells), lungs (alveolar macrophages), nervous system (microglial cells), bronchial tissue, spleen, lymph nodes, bone marrow, and the peritoneal cavity surrounding abdominal organs. The various macrophages of the body constitute the **mononuclear phagocytic (reticuloendothelial) system.**

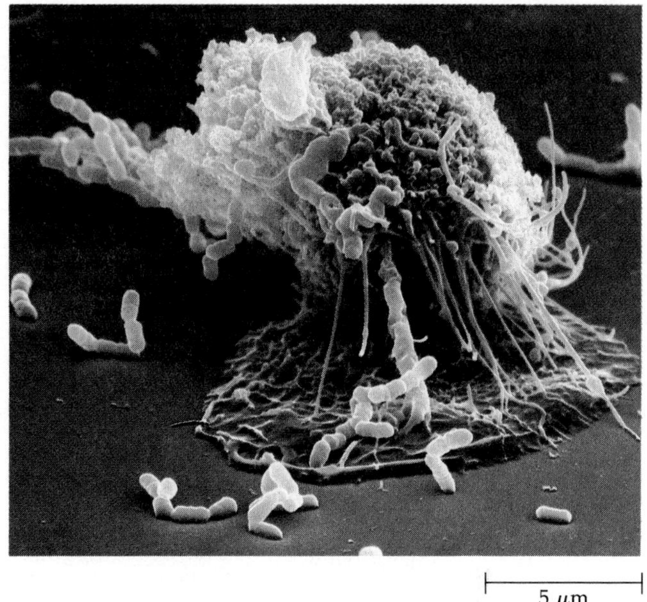

FIGURE 16.6 Electron micrograph of a macrophage engulfing a rod-shaped bacterium.

During the course of an infection, there is a shift in the type of white blood cell that predominates in the bloodstream. Granulocytes, especially neutrophils, predominate during the initial phase of infection, at which time they are actively phagocytic; this dominance is indicated by their number in a differential white blood cell count. However, as the infection progresses, the macrophages predominate; they scavenge and phagocytize remaining living bacteria and dead or dying bacteria. The increased number of monocytes (which develop into macrophages) will also be reflected in a differential count. As blood and lymph that contain microorganisms pass through organs with fixed macrophages, cells of the mononuclear phagocytic system remove the microorganisms by phagocytosis. The mononuclear phagocytic system also disposes of worn-out blood cells.

Figure 16.7 summarizes the different types of phagocytic cells and their functions.

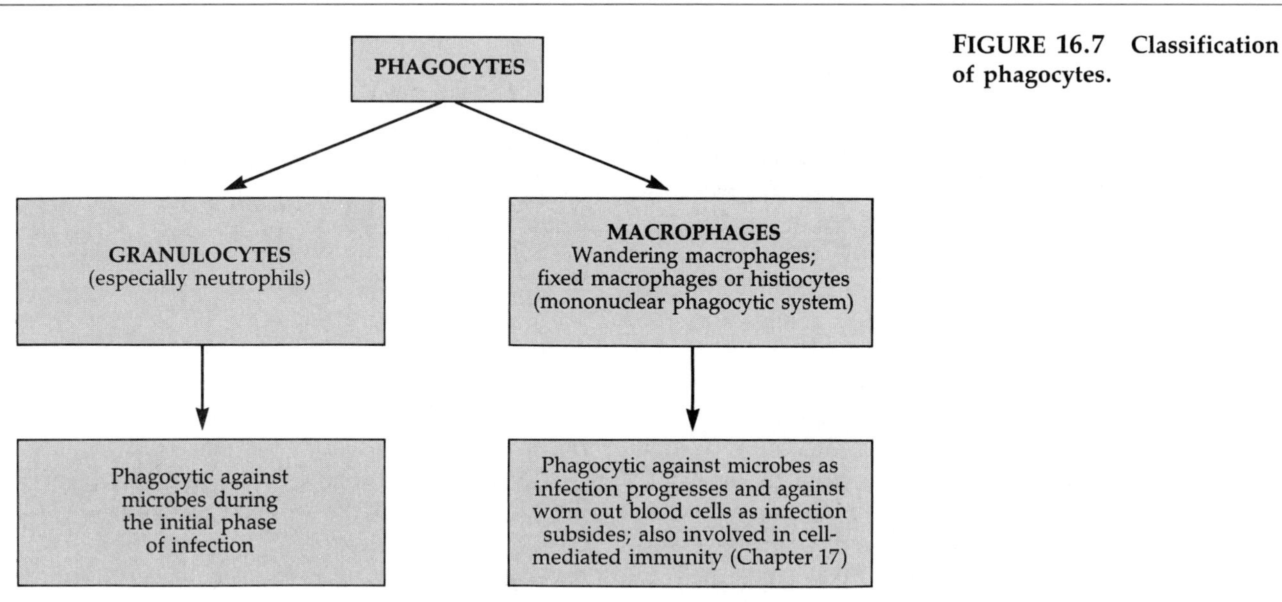

FIGURE 16.7 Classification of phagocytes.

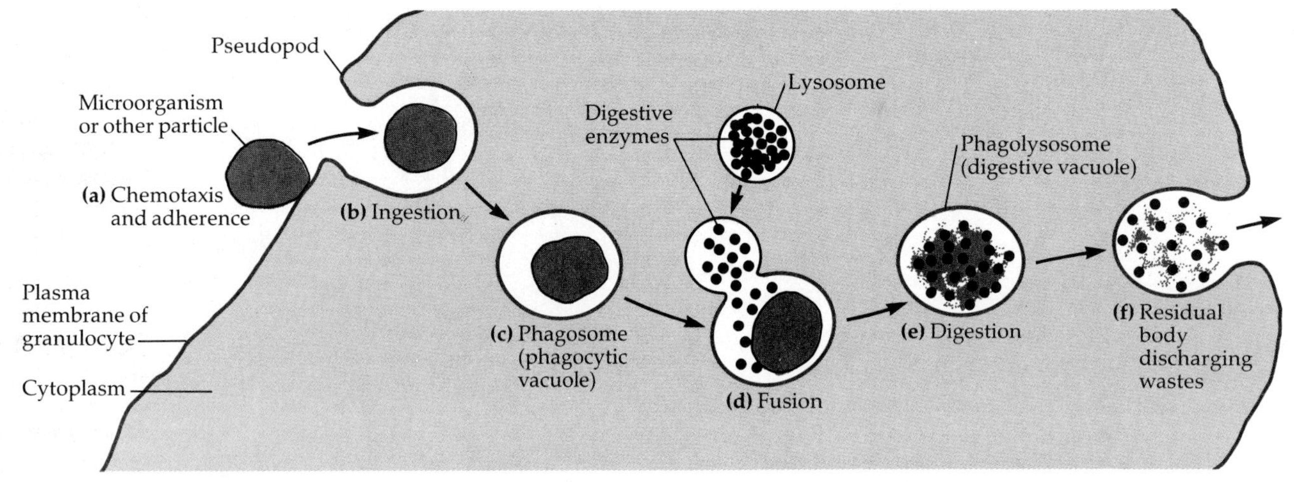

FIGURE 16.8 Phagocytosis. The drawing shows the mechanism within a macrophage. **(a)** Chemotaxis and adherence. **(b)** Ingestion. **(c)** Formation of the phagosome. **(d)** Fusion of the phagosome with a lysosome forms a phagolysosome. **(e)** Digestion—destruction of the ingested microorganism. **(f)** Residual body containing indigestible material.

MECHANISM OF PHAGOCYTOSIS

How does phagocytosis occur? For convenience of study, we will divide phagocytosis into four phases—chemotaxis, adherence, ingestion, and digestion.

Chemotaxis

Chemotaxis is the chemical attraction of phagocytes to microorganisms. (The mechanism of chemotaxis is discussed in Chapter 4.) Among the chemotactic chemicals that attract phagocytes are microbial products, components of white blood cells and damaged tissue cells, and peptides derived from complement, a system of host defense discussed later in this chapter.

Adherence

As it pertains to phagocytosis, **adherence** is the attachment of the phagocyte's plasma membrane to the surface of the microorganism or other foreign material (Figure 16.8a). In some instances, adherence occurs easily, and the microorganism is readily phagocytized. However, adherence can be hampered by the presence of large capsules or M protein. As mentioned in Chapter 15, the M protein of *Streptococcus pyogenes* inhibits the attachment of phagocytes to their surfaces and makes adherence more difficult. Organisms with large capsules include *Streptococcus pneumoniae* and *Klebsiella pneumoniae.* Heavily encapsulated microorganisms like these can be phagocytized only if the phagocyte traps the microorganism against a rough surface, such as a blood vessel, blood clot, or connective tissue fiber, from which the microbe cannot slide away.

Microorganisms can be more readily phagocytized if they are first coated with certain plasma proteins that promote the attachment of the microorganisms to the phagocyte. This coating process is called **opsonization.** The proteins that act as *opsonins* include some components of the complement system and antibody molecules (described later in this chapter and in Chapter 17).

Ingestion

Following adherence, **ingestion** occurs. During the process of ingestion, the plasma membrane of the phagocyte extends projections called pseudopods that engulf the microorganism (Figure 16.8b). Once the microorganism is surrounded, the pseudopods meet and fuse, surrounding the microorganism with a sac called a *phagosome* or *phagocytic vacuole.*

Digestion

In this phase of phagocytosis, the phagosome pinches off from the plasma membrane and enters the cytoplasm (Figure 16.8c). Within the cytoplasm, it contacts lysosomes that contain digestive enzymes and bactericidal substances (see Chapter 4). Upon contact, the phagosome and lysosome membranes fuse (Figure 16.8d) to form a single, larger structure called a *phagolysosome,* or *digestive vacuole* (Figure 16.8e). The contents of the phagolysosome take only 10 to 30 minutes to kill most types of bacteria. It is assumed that microbial destruction occurs because of the contents of the lysosomes.

Lysosomal enzymes that attack microbial cells directly include lysozyme, which hydrolyzes peptidoglycan in bacterial cell walls. A variety of other enzymes, such as lipases, proteases, ribonuclease, and deoxyribonuclease, hydrolyze other macromolecular components of microorganisms. The hydrolytic enzymes are most active at about pH 4, which is the phagolysosome's usual pH because of lactic acid produced by the phagocyte. Lysosomes also contain enzymes that can produce toxic oxygen products such as superoxide radical ($O_2^-\cdot$), hydrogen peroxide (H_2O_2), singlet oxygen (O_2), and hydroxyl radical ($OH\cdot$). Other enzymes can make use of these toxic oxygen products in killing ingested microorganisms. For example, the enzyme myeloperoxidase converts chloride (Cl^-) ions and hydrogen peroxide into highly toxic hypochlorous acid (HOCl). The acid contains hypochlorous ions, which are found in household bleach and account for its antimicrobial activity.

Once enzymes have digested the contents of the phagolysosome brought into the cell by ingestion, the phagolysosome moves toward the cell boundary and discharges its wastes. At this point, the phagolysosome containing indigestible material is called a *residual body* (Figure 16.8f).

Not all phagocytized microorganisms are killed by lysosomal enzymes. The toxins of some microbes, such as toxin-producing staphylococci, can actually kill the phagocytes (see the box on p. 410 for another example involving a plaque bacterium). Other microorganisms, such as *Mycobacterium tuberculosis,* can multiply within the phagolysosome and eventually destroy the phagocyte. Still others, such as the causative agents of tularemia and brucellosis, can remain dormant in phagocytes for months or years at a time.

Phagocytosis, in addition to providing a nonspecific defense of the host, plays a role in immunity. Macrophages help T and B lymphocytes perform vital functions in immunity. We will discuss how phagocytosis supports immunity in more detail in Chapter 17.

In the next section, we will see how phagocytosis often occurs as part of another nonspecific mechanism of resistance, inflammation.

Inflammation

Damage to the body's tissues triggers a defensive response called **inflammation.** The damage can be caused by microbial infection, physical agents (such as heat, radiant energy, electricity, or sharp objects), or chemical agents (acids, bases, and gases). Inflammation is usually characterized by four symptoms—redness, pain, heat, and swelling. Sometimes a fifth symptom, loss of function, is present; its occurrence depends on the site and extent of damage. In apparent contradiction to the symptoms observed, however, the inflammatory response is beneficial. Inflammation has the following functions: (1) to destroy the injurious agent, if possible, and to remove it and its by-products from the body; (2) if destruction is not possible, to limit the effects on the body by confining or walling off the injurious agent and its by-products; and (3) to repair or replace tissue damaged by the injurious agent or its by-products.

For purposes of our discussion, we will divide the process of inflammation into three stages—vasodilation and increased permeability of blood vessels, phagocyte migration, and repair.

Vasodilation and Increased Permeability of Blood Vessels

Immediately following tissue damage, blood vessels dilate in the area of damage, and their permeability increases (Figure 16.9a and b). **Vasodilation** is an increase in the diameter of blood vessels. Vasodilation increases blood flow to the damaged area and is responsible for the redness (erythema) and heat associated with inflammation.

Increased permeability permits defensive substances normally retained in the blood to pass through the walls of the blood vessels and enter the injured area. The increase in permeability, which permits fluid to move from the blood into tissue spaces, is responsible for the **edema** (swelling) of inflammation. The pain of inflammation can be caused by nerve damage, irritation by toxins, or the pressure of edema.

Vasodilation and the increase in permeability of blood vessels are caused by chemicals released by damaged cells in response to injury. One such substance is **histamine,** a chemical present in many cells of the body, especially in mast cells in connective tissue, circulating basophils, and blood platelets. Histamine is released in direct response to the injury of cells that contain it; it is also released in response to stimulation by certain components of the complement system (to be discussed later). Phagocytic granulocytes attracted to the site of injury can also produce chemicals that cause the release of histamine.

Kinins are another group of substances that cause vasodilation and increase the permeability of blood vessels. These chemicals are present in blood plasma and, once activated, also attract phagocytic granulocytes—chiefly neutrophils—to the injured area.

Prostaglandins, substances released by damaged cells, and **leukotrienes** (substances related to prostaglandins), produced by mast cells and basophils, are other chemicals that cause vasodilation.

Vasodilation and the increase in permeability of blood vessels also help to deliver clotting elements of blood into the injured area. The blood clots that form around the site of activity prevent the microorganism (or its toxins) from spreading to other parts of the body. As a result, there may be a localized collection of pus in a cavity formed by the breakdown of body tis-

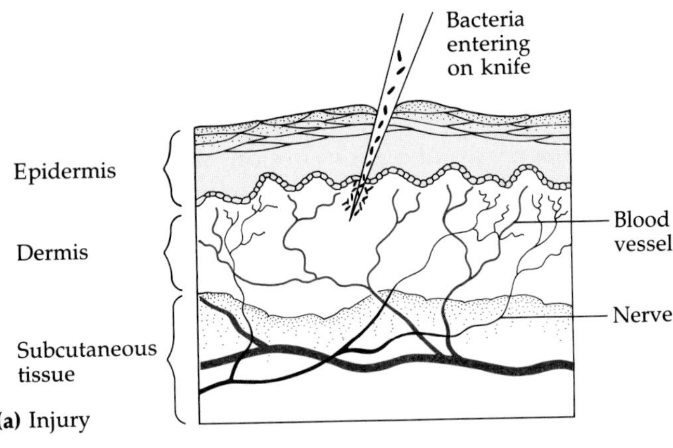

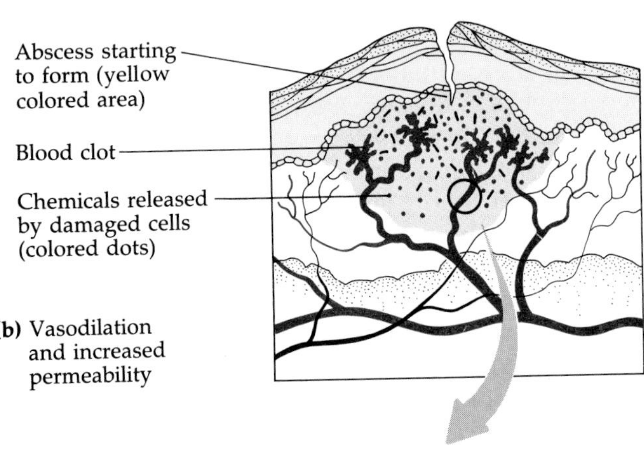

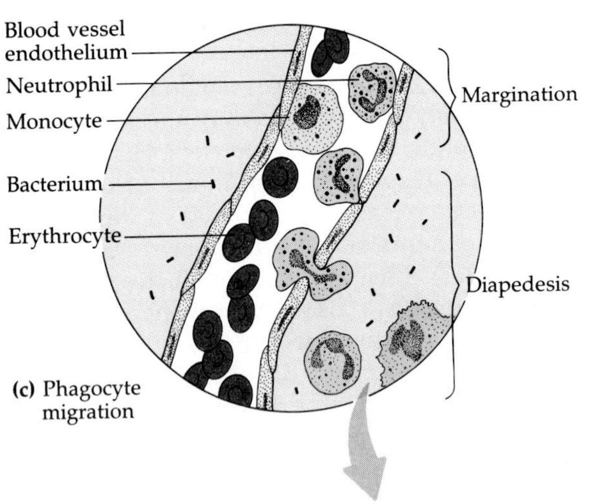

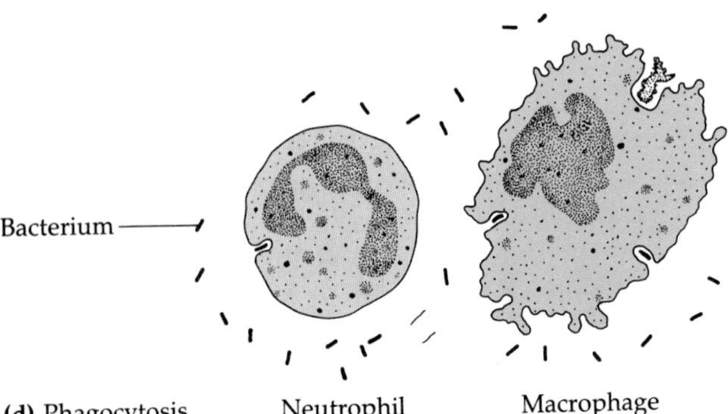

FIGURE 16.9 Aspects of the inflammatory response. **(a)** Injury to otherwise healthy skin. **(b)** Vasodilation and increased permeability. **(c)** Phagocyte migration. **(d)** Phagocytosis of bacteria and cellular debris by macrophages and neutrophils. Macrophages develop from monocytes.

sues. This focus of infection is called an **abscess** (Figure 16.9b). Common abscesses include pustules and boils.

The next stage in inflammation involves the migration of phagocytes to the injured area.

PHAGOCYTE MIGRATION

Generally, within an hour after the process of inflammation is initiated, phagocytes appear on the scene (Figure 16.9c). As the flow of blood gradually decreases, phagocytes (both neutrophils and monocytes) begin to stick to the inner surface of the endothelium (lining) of blood vessels. This sticking process is called **margination.** Then the collected phagocytes begin to squeeze between the endothelial cells of the blood vessel to reach the damaged area. This migration, which resembles amoeboid movement, is called **diapedesis,** or cell walking; the migratory process can take as little as two minutes.

As mentioned earlier, certain chemicals attract neutrophils to the site of injury (chemotaxis). These include chemicals produced by microorganisms and even other neutrophils; other chemicals are kinins and components of the complement system. The availability of a steady stream of neutrophils is ensured by the production and release of additional granulocytes from bone marrow.

As the inflammatory response continues, monocytes follow the granulocytes into the infected area. Once the monocytes are contained in the tissue, they undergo changes in biological properties and become wandering macrophages. The granulocytes predominate in the early stages of infection but tend to die off rapidly. Macrophages enter the picture during a later stage of the infection, once granulocytes have accomplished their function. They are several times more phagocytic than granulocytes and are large enough to phagocytize tissue that has been destroyed, granulocytes that have been destroyed, and invading microorganisms (Figure 16.9d).

After granulocytes or macrophages engulf large numbers of microorganisms and damaged tissue, they themselves eventually die. After a few days, an area containing dead phagocytes and damaged tissue forms in the inflamed tissue. This collection of dead cells and various body fluids is called **pus.** Pus formation usually continues until the infection subsides. At times, the pus pushes to the surface of the body or into an internal cavity for dispersal. On other occasions, the pus remains even after the infection is terminated. In this case, the pus is gradually destroyed over a period of days and is absorbed by the body.

REPAIR

The final stage of inflammation is tissue repair, the process by which tissues replace dead or damaged cells. Repair begins during the active phase of inflammation, but it cannot be completed until all harmful substances have been removed or neutralized at the site of injury. The ability of a tissue to regenerate, or repair itself, depends on the type of tissue. For example, skin has a high capacity for regeneration, whereas cardiac muscle tissue does not regenerate at all.

A tissue is repaired when its stroma or parenchyma produces new cells. The *stroma* is the supporting connective tissue, and the *parenchyma* is the functioning part of the tissue. For example, the capsule around the liver that encloses and protects it is part of the stroma since it is not involved in the functions of the liver; liver cells (hepatocytes) that perform the functions of the liver are part of the parenchyma. If only parenchymal cells are active in repair, a perfect or near-perfect reconstruction of the tissue occurs. A familiar example of perfect reconstruction is a minor skin cut, in which parenchymal cells are more active in repair. However, if repair cells of the stroma of the skin are more active, scar tissue is formed.

The process of inflammation is summarized in Figure 16.10.

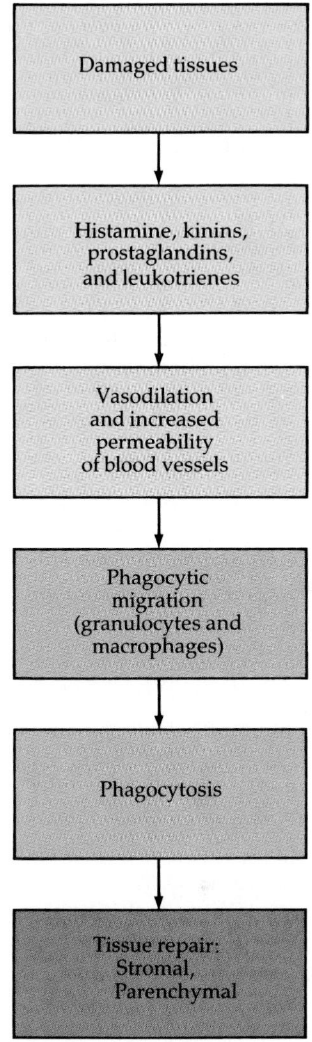

FIGURE 16.10 Summary of the process of inflammation.

Fever

Inflammation is a local response of the body to microbial invasion. There are also systemic, or overall, responses, one of the most important being fever, an abnormally high body temperature. The most frequent cause of fever is infection from bacteria (and their toxins) or viruses.

Body temperature is controlled by a part of the brain called the hypothalamus. The hypothalamus is sometimes called the body's thermostat, and it is normally set at 37°C (98.6°F). It is believed that certain substances affect the hypothalamus by setting it at a higher temperature. Recall from Chapter 15 that, when phagocytes ingest gram-negative bacteria, the lipopolysaccharides (LPS) of the cell wall (endotoxins) are released, causing the phagocytes to release interleukin-1 (endogenous pyrogens). Interleukin-1 causes the hypothalamus to release prostaglandins that reset the hypothalamic thermostat at a higher temperature, thereby causing fever (see Figure 15.4).

Assume that the body is invaded by pathogens and the thermostat setting is increased to 39°C (102.2°F). To adjust to the new thermostat setting, the body responds with blood vessel constriction, increased rate of metabolism, and shivering, all of which raise body temperature. Even though body temperature is climbing higher than normal, the skin remains cold, and shivering occurs. This condition, called a *chill,* is a definite sign that temperature is rising. When body temperature reaches the setting of the thermostat, the chill disappears. But the body will continue to maintain its temperature at 39°C until the interleukin-1 is eliminated. The thermostat is then reset at 37°C. As the infection subsides, heat-losing mechanisms such as vasodilation and sweating go into operation. The skin becomes warm and the person begins to sweat. This phase of the fever, called the *crisis*, indicates that body temperature is falling.

Up to a certain point, fever is considered a defense against disease. Interleukin-1 helps step up the production of T lymphocytes. High body temperature intensifies the effect of interferon (an antiviral protein that we will discuss shortly). It is believed to inhibit the growth of some microorganisms by decreasing the amount of iron available to them. Also, because the high temperature speeds up the body's reactions, it may help body tissues to repair themselves more quickly.

Antimicrobial Substances

The body produces certain antimicrobial substances in addition to the chemical factors mentioned earlier. Among the most important of these are the proteins of the complement system and the interferons.

THE COMPLEMENT SYSTEM

Complement is a defensive system consisting of serum proteins that participate in lysis of foreign cells, inflammation, and phagocytosis. The system can be activated by an immune reaction in the classical pathway or by direct interaction with a bacterium in the alternative pathway. Complement is nonspecific, since the same proteins can be activated in response to any foreign cell. However, in the classical pathway, complement assists, or complements, specific immunity.

Components

The complement system consists of a group of at least 20 different interacting proteins found in normal serum. Proteins of the complement system make up about 5% of the serum proteins in vertebrates. The major components of the classical pathway are designated by a complex numbering system ranging from C1 through C9 (C stands for complement). The C1 protein also has three subcomponents, C1q, C1r, and C1s. The alternative pathway consists of proteins called factor B and factor D along with the C3 and C5 through C9 proteins active in the classical pathway.

Pathways of Activation

The proteins of the classical and alternative pathways act in an ordered sequence, or *cascade*, which, except for C4, follows the numerical designations. In a series of steps, each protein activates the next one in the series, usually by cleaving (splitting) it. The fragments of the cleaved proteins have new physiological or enzymatic functions. For example, one fragment might cause blood vessel dilation, whereas another fragment serves as part of the enzyme that cleaves the next protein in the series.

C3 plays a central role in the complement system. Its activation triggers several mechanisms that contribute to microbial destruction. As you can see in Figure 16.11, either of two pathways can activate C3. The classical pathway is initiated by the binding of antibodies to antigen (described in Chapter 17). The antigen could consist of bacteria or other foreign cells. Once a pair of antibodies recognizes and attaches to the antigen, the complement protein C1 (which actually consists of three subcomponents) binds to two or more adjacent antibodies, thus activating C1 (Figure 16.12a).

Next, activated C1 in turn activates C2 and C4. It does this by functioning as a proteolytic (protein-splitting) enzyme and splitting the C2 and C4 proteins. C2 is split into fragments called C2a and C2b, and C4 into fragments called C4a and C4b (Figure 16.12b). Then C2b and C4b combine to form another proteolytic enzyme, which in turn activates C3 by splitting it into two fragments, C3a and C3b.

The alternative pathway indicated in Figure 16.11 does not involve antibodies; it is initiated by the interaction between certain polysaccharides and the proteins of the pathway. Factors B and D react with C3 to produce low levels of C3b in the serum. The alternative pathway indicated in Figure 16.11 does not involve antibodies; it is initiated when C3b, factor B, and factor D combine with certain polysaccharides. Most of these polysaccharides are contained in the cell walls of certain bacteria and fungi, although they also include molecules on the surface of some foreign mammalian red blood cells. The alternative pathway is of particular importance in combating enteric gram-negative bacteria. The outer membrane of the bacteria's cell wall contains a lipopolysaccharide that is an endotoxin (lipid A), triggering the alternative pathway. Note that this pathway does not involve C1, C2, or C4.

Consequences of Complement Activation

How does the complement system contribute to microbial destruction? Both the classical and alternative pathways lead to the cleavage of C3 into two fragments, C3a and C3b. These fragments induce three processes that are destructive to microorganisms—cytolysis, inflammation, and opsonization.

Cytolysis. The main function of the complement system is to destroy foreign cells by damaging their plasma membranes, causing the cellular contents to leak out. This is called **cytolysis** and is accomplished as follows (Figure 16.12c). C3b initiates a sequence of reactions involving C5, C6, C7, C8, and C9, which are known collectively as the **membrane attack complex.** The activated components of these proteins, with C9 proteins possibly playing a key role, attack the invad-

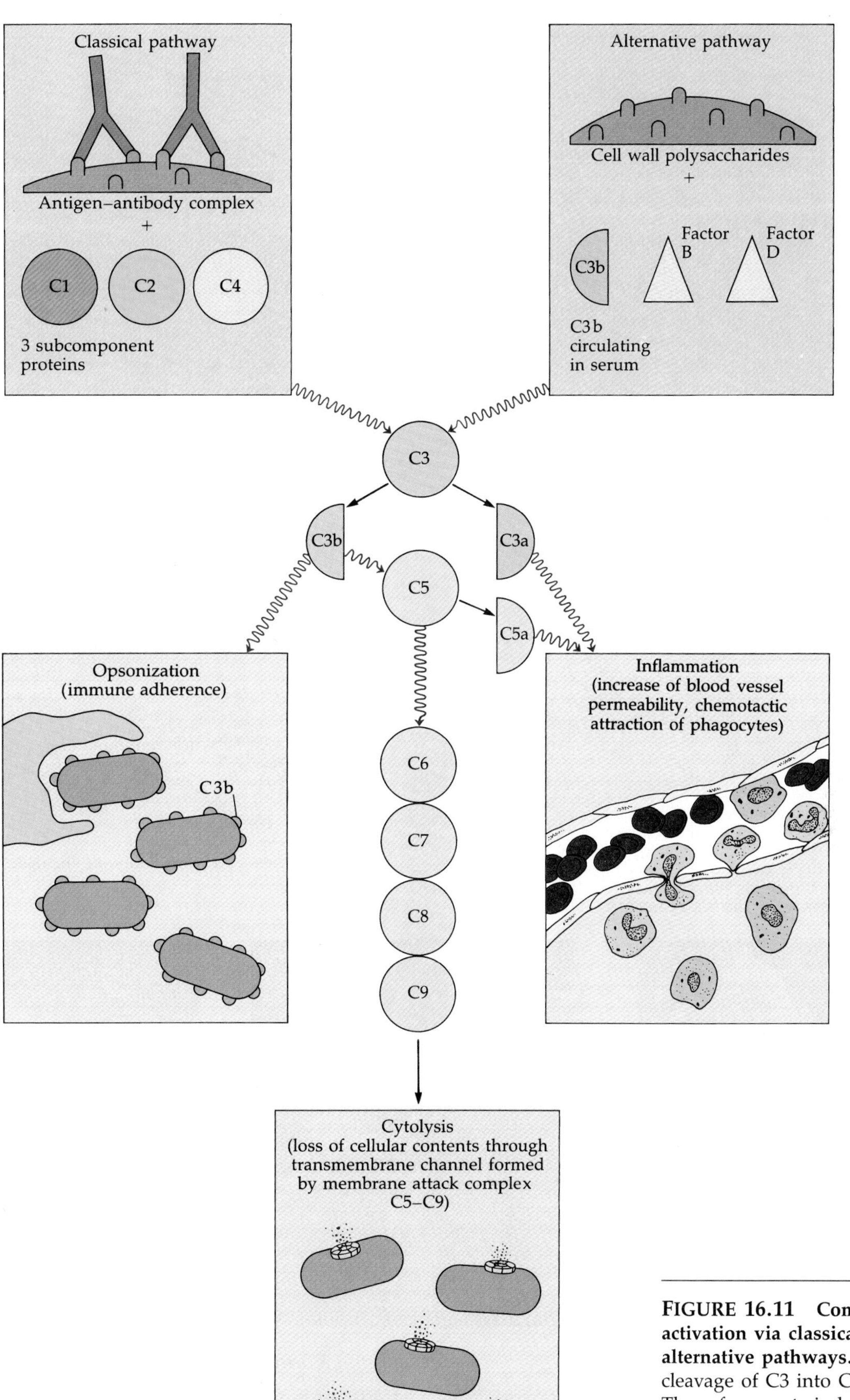

FIGURE 16.11 Complement activation via classical and alternative pathways. Note the cleavage of C3 into C3a and C3b. These fragments induce three kinds of consequences destructive to microorganisms—opsonization, inflammation, and cytolysis.

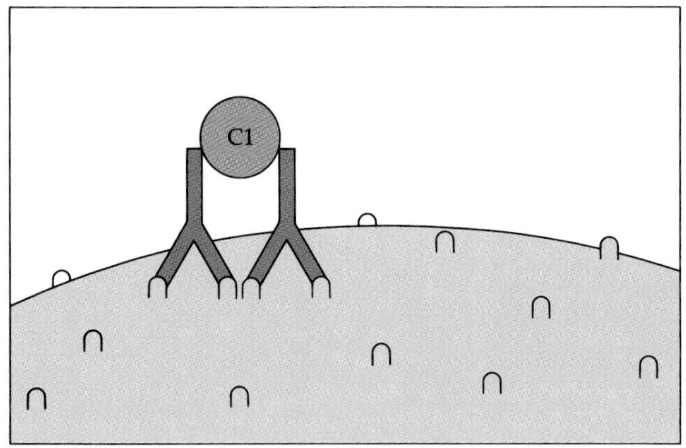

(a) Once antibodies recognize and attach to the antigen, complement protein C1 binds to two adjacent antibodies.

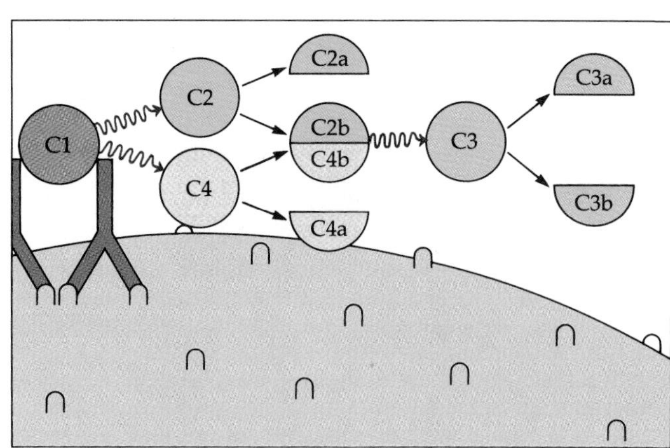

(b) C1 acts as an enzyme that splits the C2 and C4 proteins into fragments. Fragments C2b and C4b combine to form another enzyme, which splits C3 into two fragments. The active fragment is called C3b.

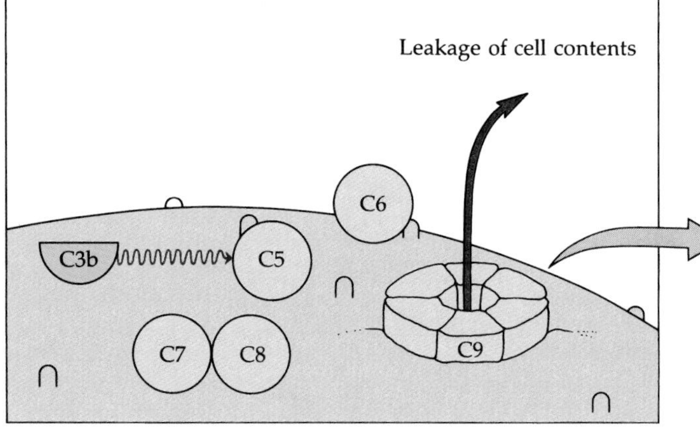

(c) C3b initiates a series of reactions involving C5–C9, collectively called the membrane attack complex. This complex forms circular transmembrane channels (lesions) in the antigenic cell's membrane, with C9 proteins possibly playing a key role. The result is leakage of the cell's contents—cytolysis.

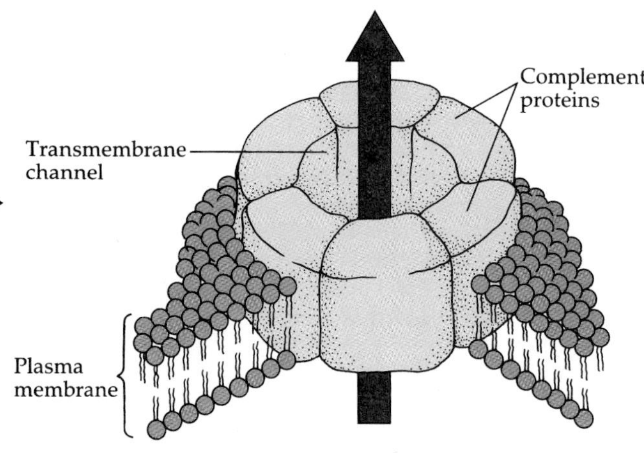

(d) Enlarged view of a transmembrane channel.

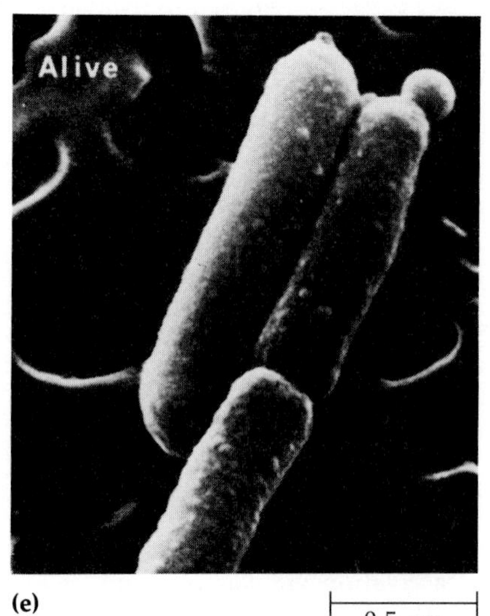

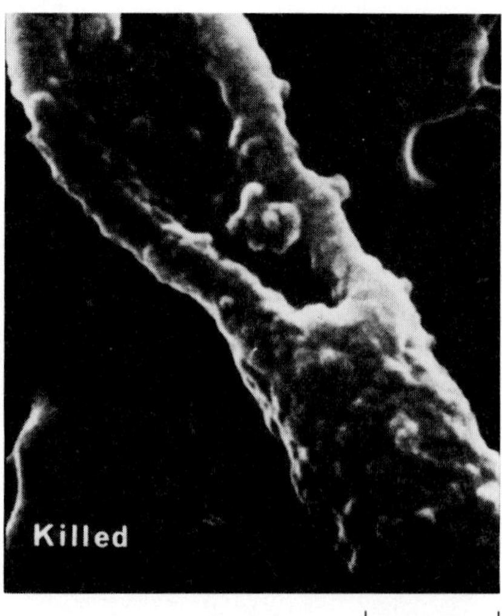

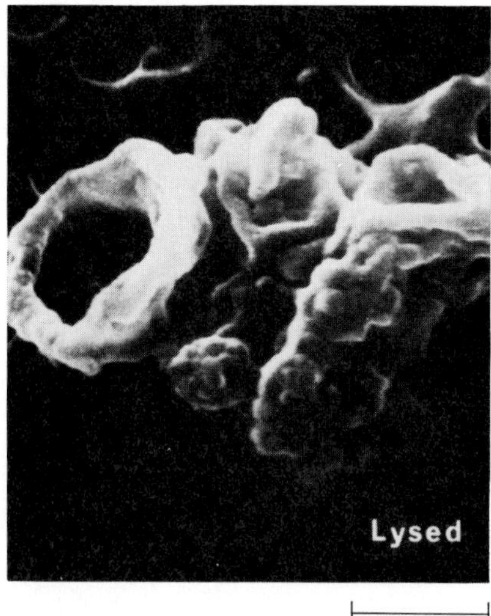

(e)

FIGURE 16.12 Cytolysis caused by complement activation. This figure illustrates just a few key steps in the sequence that leads to holes in a plasma membrane. As you can see, the process is complex.

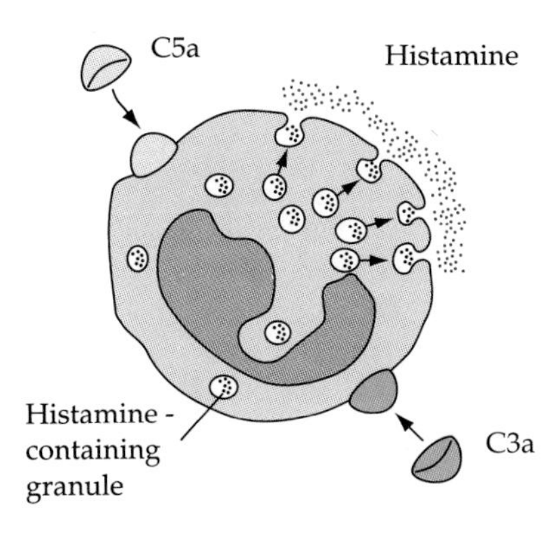

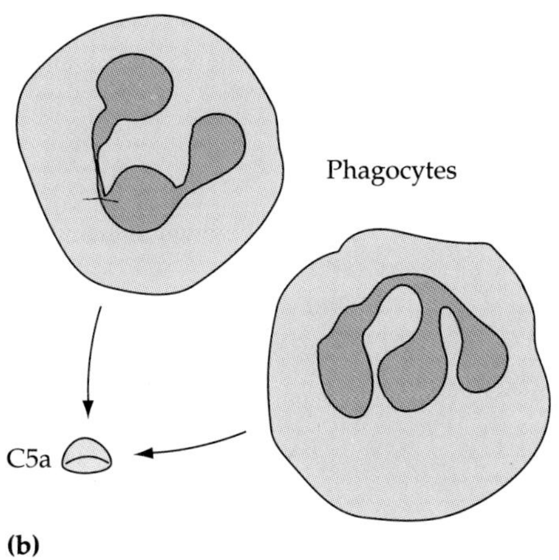

FIGURE 16.13 Stimulation of inflammatory response by complement. **(a)** C3a and C5a bind to mast cells, basophils, and platelets to trigger the release of histamine, which increases blood vessel permeability. **(b)** C5a functions as a chemotactic factor that attracts phagocytes to the site of complement activation.

ing cell's membrane and produce circular lesions, called **transmembrane channels** (Figure 16.12d), that lead to loss of ions and eventual cytolysis. The utilization of the complement components in this process is called **complement fixation;** it forms the basis of an important clinical laboratory test that will be explained in Chapter 18. Figure 16.12e illustrates the effects of complement on a microbial cell.

Inflammation. C3a, a cleavage product from C3, and C5a, a cleavage product from C5, can contribute to the development of acute *inflammation* (Figure 16.13). C3a and C5a bind to mast cells, basophils, and blood platelets to trigger the release of histamine, which increases blood vessel permeability. C5a also functions as a powerful chemotactic factor that attracts phagocytes to the site of complement fixation.

Opsonization. When bound to the surface of a microorganism, C3b can interact with special receptors on phagocytes to promote phagocytosis (Figure 16.14). This phenomenon is called *opsonization* or immune adherence. In the process, C3b functions as an opsonin by coating the microorganism and promoting attachment of the phagocyte to the microbe.

Once complement is activated, its destructive capabilities usually cease very quickly in order to minimize destruction of the host cells. This is accomplished by various regulatory proteins found in the host's blood and on certain cells, such as blood cells. They bring about the breakdown of activated complement and function as inhibitors and destructive enzymes.

In addition to its importance in defense, the complement system assumes a role in causing disease as a result of inherited deficiencies. For example, deficiencies of C1, C2, or C4 cause collagen vascular disorders that result in hypersensitivity (anaphylaxis); deficiency of C3, though rare, results in increased susceptibility to bacterial infections; and C5 through C9 defects result in increased susceptibility to *Neisseria meningitidis* and *N. gonorrhoeae* infections.

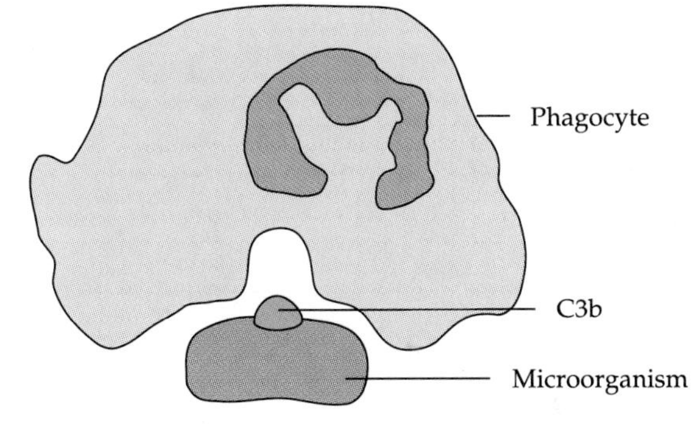

FIGURE 16.14 Opsonization by complement.

INTERFERONS (IFNs)

Because viruses depend on their host cells to provide many functions of viral multiplication, it is difficult to inhibit viral multiplication without at the same time affecting the host cell itself. One way the infected host counters viral infections is with interferons. **Interferons** are a class of similar antiviral proteins produced by certain animal cells after viral stimulation. One of the principal functions of interferons is to interfere with viral multiplication.

One of the most interesting features of interferons is that they are host-cell-specific but not virus-specific. This means that interferon produced by human cells protects human cells but will produce little antiviral activity for cells of other species, such as mice or chick-

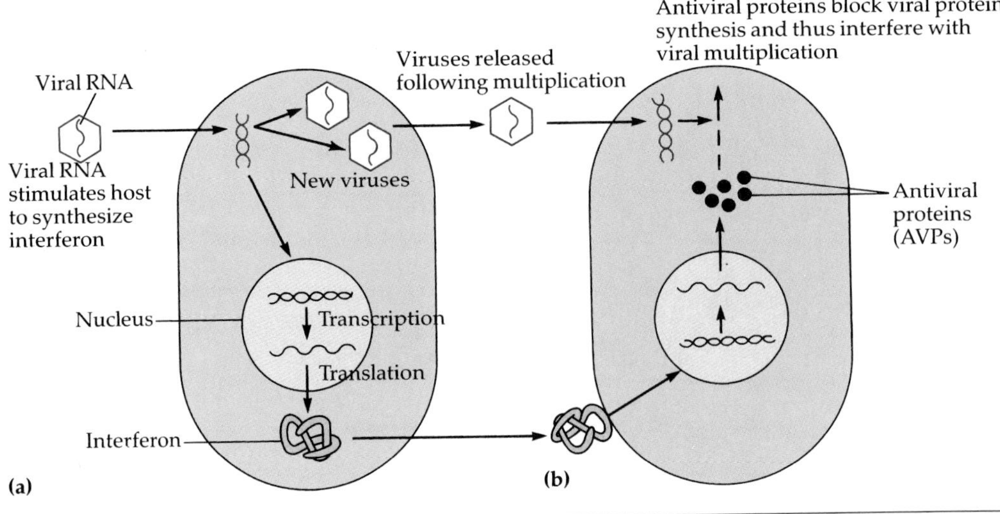

FIGURE 16.15 Antiviral action of interferon. (a) A virus-infected cell produces interferon. **(b)** In an uninfected neighboring cell, interferon reacts with plasma or the nuclear membrane receptors and induces synthesis of antiviral proteins (AVPs).

ens. However, the interferon of a species is active against a number of different viruses.

Not only do different animal species produce different interferons, but also different types of cells in an animal produce different interferons. Human interferons are of three principal types: (1) *alpha interferon* (α-IFN), (2) *beta interferon* (β-IFN), and (3) *gamma interferon* (γ-IFN). There are also various subtypes of interferon within each of the principal groups. In the human body, interferon is produced by fibroblasts in connective tissue, by lymphocytes, and by other leukocytes. The three types of interferon produced by these cells can each have a slightly different effect on the body.

All interferons are small proteins, with molecular weights between 15,000 and 30,000. They are quite stable at low pH and fairly resistant to heat.

Produced by virus-infected host cells only in very small quantities, interferon diffuses to uninfected neighboring cells. It reacts with plasma or nuclear membrane receptors, inducing the uninfected cells to manufacture mRNA for the synthesis of **antiviral proteins (AVPs)** (Figure 16.15). These proteins are enzymes that disrupt various stages of viral multiplication. For example, one AVP inhibits translation of viral mRNA by blocking initiation of protein synthesis. Another inhibits polypeptide elongation. Still another is involved in destroying viral mRNA before translation.

The low concentrations at which interferon inhibits viral multiplication are nontoxic to uninfected cells. Because of its beneficial properties, interferon would seem to be an ideal antiviral substance. But certain problems do exist. For one thing, interferon is effective for only short periods. It typically plays a major role in infections that are acute and short-term, such as colds and influenza. Another problem is that it has no effect on viral multiplication in cells already infected.

The importance of interferon in protecting the body against viruses, as well as its potential as an anticancer agent, has made its production in large quantities a top public health priority. Several groups of scientists have successfully applied recombinant DNA technology in inducing certain species of bacteria to produce interferon. (This technique is described in Chapter 9.) The interferons produced with recombinant DNA techniques, called recombinant interferons (rINFs), are important for two reasons: They are pure, and they are plentiful.

Clinical trials to determine the anticancer effects of interferon were begun in 1981. At present, it appears that large doses of rINFs have only limited effects against some tumors and no effect on others. The clinical trials have also revealed that high doses of interferon can have side-effects that range from minor to quite serious, including fatigue, malaise, loss of appetite, fever, chills, pains in the joints, mental confusion, seizures, and cardiac complications. Several studies are also focusing on the use of rIFNs to treat AIDS, hepatitis, genital herpes, influenza, and the common cold.

Although glowing predictions about interferon's anticancer properties have not been justified by the results of clinical trials to date, research is continuing, for several reasons. First, there are many subtypes of interferon, and it might be possible to find the right one with definitive anticancer properties. Second, some laboratory studies indicate that interferon might work together with other chemotherapeutic agents so that the effects of interferon are enhanced. Encouraging results have been obtained by use of interferon in combination with doxorubicin (used to treat a variety of blood cancers and solid tumors) or cimetidine (used to treat ulcers), for example. Third, it has been observed that patients who did not respond well to either specific chemotherapy or subsequent treatment with interferon showed improvement when returned to the original chemotherapy.

In Chapter 17, we will discuss specific immune responses. We will also look at the principal factors that contribute to immunity.

STUDY OUTLINE

Introduction (p. 407)

1. The ability to ward off disease through body defenses is called resistance.
2. Lack of resistance is called susceptibility.
3. Nonspecific resistance refers to all body defenses that protect the body from any kind of pathogen.
4. Specific resistance refers to defenses (antibodies) against specific microorganisms.

Skin and Mucous Membranes (pp. 407–410)

MECHANICAL FACTORS (pp. 408–409)

1. The structure of intact skin and the waterproof protein keratin provide resistance to microbial invasion.
2. Some pathogens, if present in large numbers, can penetrate mucous membranes.
3. The lacrimal apparatus protects the eyes from irritating substances and microorganisms.
4. Saliva washes microorganisms from teeth and gums.
5. Mucus traps many microorganisms that enter the respiratory and gastrointestinal tracts; in the lower respiratory tract, the ciliary escalator moves mucus up and out.
6. The flow of urine moves microorganisms out of the urinary tract, and vaginal secretions move microorganisms out of the vagina.

CHEMICAL FACTORS (pp. 409–410)

1. Sebum contains unsaturated fatty acids, which inhibit the growth of pathogenic bacteria. Some bacteria commonly found on the skin can metabolize sebum and cause the inflammatory response associated with acne.
2. Perspiration washes microorganisms off the skin.
3. Lysozyme is found in tears, saliva, nasal secretions, and perspiration.
4. The high acidity (pH 1.2 to 3.0) of gastric juice prevents microbial growth in the stomach.
5. Normal flora prevent the growth of many pathogens.

Phagocytosis (pp. 410–415)

1. Phagocytosis is the ingestion of microorganisms or particulate matter by a cell.
2. Phagocytosis is performed by phagocytes, certain types of white blood cells or derivatives of them.

FORMED ELEMENTS IN BLOOD (pp. 411–413)

1. Blood consists of plasma (fluid) and formed elements (cells and cell fragments).
2. Leukocytes (white blood cells) are divided into three categories—granulocytes (neutrophils, basophils, and eosinophils), lymphocytes, and monocytes.
3. During many infections, the number of leukocytes increases (leukocytosis); some infections are characterized by leukopenia (decrease in leukocytes).

ACTIONS OF PHAGOCYTIC CELLS (p. 413)

1. Among the granulocytes, neutrophils are the most important phagocytes.
2. Enlarged monocytes become wandering macrophages and fixed macrophages.
3. Fixed macrophages are located in selected tissues and are part of the mononuclear phagocytic system.
4. Granulocytes predominate during the early stages of infection, whereas monocytes predominate as the infection subsides.

MECHANISM OF PHAGOCYTOSIS (pp. 414–415)

1. Chemotaxis is the process by which phagocytes are attracted to microorganisms.
2. The phagocyte then adheres to the microbial cells; adherence may be facilitated by opsonization—coating the microbe with plasma proteins.
3. Pseudopods of phagocytes engulf the microorganism and enclose it in a phagosome to complete ingestion.
4. Many phagocytized microorganisms are killed by lysosomal enzymes and oxidizing agents.

Inflammation (pp. 415–417)

1. Inflammation is a bodily response to cell damage; it is characterized by redness, pain, heat, swelling, and sometimes loss of function.

VASODILATION AND INCREASED PERMEABILITY OF BLOOD VESSELS (pp. 415–416)

1. The release of histamine, kinins, and prostaglandins causes vasodilation and increased permeability of blood vessels.
2. Blood clots can form around an abscess to prevent dissemination of the infection.

PHAGOCYTE MIGRATION (pp. 416–417)

1. Phagocytes have the ability to stick to the lining of the blood vessels (margination).
2. They also have the ability to squeeze through blood vessels (diapedesis).
3. Pus is the accumulation of damaged tissue and dead microbes, granulocytes, and macrophages.

REPAIR (p. 417)

1. A tissue is repaired when the stroma (supporting tissue) or parenchyma (functioning tissue) produces new cells.
2. Stromal repair by fibroblasts produces scar tissue.

Fever (pp. 417–418)

1. Fever is an abnormally high body temperature produced in response to a bacterial or viral infection.
2. Bacterial endotoxins and interleukin-1 can induce fever.
3. A chill indicates a rising body temperature; crisis (sweating) indicates that the body's temperature is falling.

Antimicrobial Substances (pp. 418–423)

THE COMPLEMENT SYSTEM (pp. 418–421)

1. The complement system consists of a group of serum proteins that activate one another to destroy invading microorganisms.
2. C1 binds to antigen–antibody complexes to eventually activate C3 protein. Factor B, factor D, and C3b bind to certain cell wall polysaccharides to activate C3b.
3. C3 activation can result in cell lysis, inflammation, and opsonization.

INTERFERONS (IFNs) (pp. 421–423)

1. Interferons are antiviral proteins produced in response to viral infection.
2. There are three types of human interferon, α-IFN, β-IFN, and γ-IFN. Recombinant interferon has also been produced.
3. Interferon's mode of action is to induce uninfected cells to produce antiviral proteins (AVPs) that prevent viral replication.
4. Interferons are host-cell-specific but not virus-specific.

STUDY QUESTIONS

Review

1. Define the following terms.
   (a) Resistance
   (b) Susceptibility
   (c) Nonspecific resistance
2. Describe the mechanical factors of the skin and mucous membranes that assume a role in nonspecific resistance.
3. Describe the chemical factors of the skin and mucous membranes that assume a role in nonspecific resistance.
4. Define phagocytosis.
5. Compare the structures and functions of granulocytes and monocytes in phagocytosis.
6. How do fixed and wandering macrophages differ?
7. Diagram the following processes that result in phagocytosis: margination, diapedesis, adherence, and ingestion.
8. Define inflammation, and list its characteristics.
9. Explain how kinins could be responsible for symptoms of the common cold.
10. Why is inflammation beneficial to the body?
11. How is fever related to nonspecific defense?
12. What is the importance of chill and crisis during fever?
13. What is interferon? Discuss its role in nonspecific resistance.
14. What is complement?
15. Summarize the major outcomes of complement activation.
16. How does complement function in the alternative pathway?

Challenge

1. Recent evidence reveals that people with rhinovirus infections of the nose and throat have an 80-fold increase in kinins and no increase in histamine. What do you expect for rhinoviral symptoms? What disease is caused by rhinoviruses?
2. Why do serum levels of transferrin increase during an infection? What can a bacterium do to respond to high levels of transferrin?
3. How is the complement system activated by bacterial endotoxin in the bloodstream? How does endotoxic shock result in massive host cell destruction?
4. A variety of drugs with the ability to reduce inflammation are available. Comment on the danger of misuse of these anti-inflammatory drugs.
5. A hematologist often performs a differential white blood cell count on a blood sample. A differential white blood cell count determines the relative numbers of white blood cells. Why are these numbers important? What do you think the hematologist would find in a differential white blood cell count of a patient with mononucleosis? With neutropenia? With eosinophilia?

FURTHER READING

Beaman, L., and B.L. Beaman. "The role of oxygen and its derivatives in microbial pathogenesis and host defense." *Annual Review of Microbiology* 38:27–48, 1984. Describes the neutrophil's "respiratory burst" using myeloperoxidase and singlet oxygen to kill microbes.

Bretscher, M.S. "How animal cells move." *Scientific American* 257(6):72–90, December 1987. Describes the movement of phagocytes, fibroblasts, and cancer cells.

Edelson, R.L., and J.M. Fink. "The immunologic function of skin." *Scientific American* 256(6):46–53, June 1985. Discusses the role of T cells in skin.

Joiner, K.A. "Complement evasion by bacteria and parasites." *Annual Review of Microbiology* 42:201–230, 1988. Identifies mechanisms of bacterial blocking of complement and bacterial use of complement to access cells.

McNabb, P.C., and T.B. Tomasi. "Host defense mechanisms at mucosal surfaces." *Annual Review of Microbiology* 35:477–496, 1981. Describes various host defenses, including the normal microbial flora and IgA antibodies, and how they prevent the growth of pathogens.

Old, L.J. "Tumor necrosis factor." *Scientific American* 258(5):59–75, May 1988. Discusses how bacterial (LPS) endotoxin stimulates production of tumor-destroying proteins in animals.

Ross, G.E., ed. *Immunobiology of the Complement System: An Introduction for Research and Clinical Medicine.* New York: Academic Press, 1986. A compilation of information on complement.

Shafer, W.M., and R.F. Rest. "Interactions of gonococci with phagocytic cells." *Annual Review of Microbiology* 43:121–145, 1989. A discussion of why gonococci can be bound to or engulfed by PMNs and not be killed.

Taylor-Papadimitrious, J. *Interferons: Their Impact on Biology and Medicine.* New York: Oxford University Press, 1985. Coverage includes natural production and mechanisms of action of interferons and clinical treatment with interferons.

Trager, W. "The biochemistry of resistance to malaria." *Scientific American* 244(3):154–164, March 1981. Describes the evolution of sickle-cell anemia and how this disease offers protection from malaria.

CHAPTER 17

Specific Defenses of the Host: The Immune Response

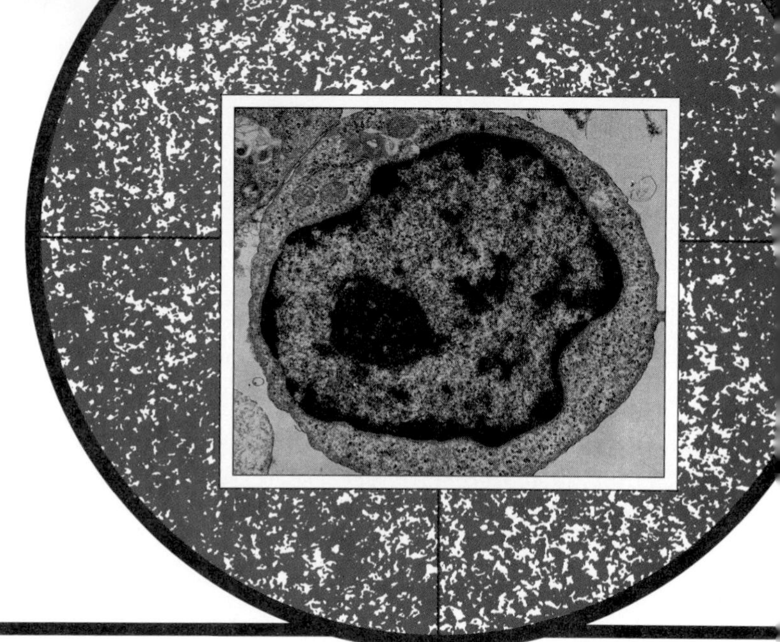

LEARNING OBJECTIVES

- Contrast the four kinds of acquired immunity.
- Define immunity, antigen, anamnestic response, lymphokines, and monoclonal antibody.
- Describe the clonal selection theory.
- Explain what an antibody is, and describe the structural and chemical characteristics of antibodies.
- Describe the genetic mechanism for antibody diversity.
- Provide at least one function for each of the five classes of antibodies.
- Compare and contrast humoral and cell-mediated immunity.
- Differentiate among the types of T cells.

In the last chapter, we discussed nonspecific host defenses, including the intact skin and mucous membranes, phagocytosis, inflammation, and fever. These are integral components of the body's resistance to disease. In addition to these nonspecific defenses, we have an **innate resistance** to certain illnesses. For example, all humans are resistant to many infectious animal diseases, such as canine distemper and chicken and hog cholera. However, resistance to human diseases, such as measles, can vary from person to person. Although the illness is often relatively mild for persons of European ancestry (possibly because many generations have been exposed to the measles virus), the disease caused many deaths among the Polynesians who were first exposed to measles by European explorers. A person's resistance to disease also depends on genetic factors, age, nutritional status, and general health.

In this chapter, we will study the immune system, another aspect of the body's defenses. In contrast to resistance, immunity involves a *specific* defensive response when the host is invaded by foreign organisms or other foreign substances. Invading organisms might be pathogenic bacteria, viruses, fungi, protozoans, or helminths; foreign materials include pollen, insect venom, and transplanted tissue. The body recognizes these substances as not belonging to itself, and it develops an immune response to inactivate or destroy them. Body cells that become cancerous are also recognized as foreign and eliminated—but if they evade this defense and become established as solid tumors, they evade the immune system. We will see in this chapter and the next two chapters that the immune system is essential to our survival, but the immune system can also cause harm when its efforts are misdirected.

Types of Acquired Immunity

Acquired immunity refers to the protection an organism develops against certain types of microbes or foreign substances. Acquired immunity is developed during a person's lifetime; it is not inherited.

Immunity can be acquired either actively or passively. Immunity is acquired *actively* when a person is exposed to microorganisms or foreign substances, and the immune system responds by producing specialized lymphocytes and special proteins called **antibodies.** Organisms or substances that provoke such a response are called **antigens;** they are usually destroyed or inactivated by the specific antibodies and lymphocytes the body produces. Immunity is acquired *passively* when antibodies are transferred from one organism to another. This type of immunity is considered to be passive because the recipient does not actively synthesize the antibodies but has them provided by an outside source. Immunity in the recipient lasts only as long as the antibodies are present—in most cases, for several weeks or months.

Both actively acquired immunity and passively acquired immunity can be obtained by natural or artificial means.

NATURALLY ACQUIRED IMMUNITY

Naturally acquired active immunity is obtained when a person is exposed to antigens in the course of daily life and the immune system responds by producing antibodies and specialized lymphocytes. Immunity is life-long for some diseases, such as measles, chickenpox, and yellow fever. For other diseases, especially intestinal diseases, the immunity usually lasts for only a few years. Sometimes *subclinical infections* (those that produce no evident symptoms) can also confer immunity.

Naturally acquired passive immunity involves the natural transfer of antibodies from a mother to her infant. An expectant mother is able to pass some of her antibodies to her fetus across the placenta. This mechanism is called *placental transfer*. If the mother is immune to such diseases as diphtheria, rubella, or polio, the newborn infant will be temporarily immune to these diseases as well. Certain antibodies are also passed from the mother to her nursing infant in breast milk, especially in the first secretions called *colostrum*. In the infant, immunity generally lasts only as long as the transmitted antibodies are present—usually a few weeks or months. These maternal antibodies are essential for providing immunity to the infant until its own immune system matures.

ARTIFICIALLY ACQUIRED IMMUNITY

Artificially acquired active immunity results from vaccination. **Vaccination,** also called **immunization,** introduces specially prepared antigens called **vaccines** into the body. Vaccines consist of inactivated bacterial toxins (toxoids), killed microorganisms, or living but attenuated (weakened) microorganisms. These substances are no longer able to cause disease, but they are still able to stimulate an immune response, much as naturally acquired pathogens do. Vaccines are discussed in more detail in Chapter 18.

Unlike artificially acquired active immunity, which occurs in response to the injection of *antigens*, **artificially acquired passive immunity** involves the introduction of *antibodies* into the body. These antibodies come from an animal or human who is already immune to the disease. For example, a person who is bitten by a snake might be injected with antibodies from a horse that has been immunized against snake venom.

The antibodies are found in the serum of the immune animal or human. *Serum* (plural, *sera*) is the fluid remaining after blood has clotted and the blood cells and clotted matter have been removed. Because most of the antibodies remain in the serum, the word **antiserum** has become a generic term for fluids containing antibodies. When serum is subjected to an electric current in the laboratory in electrophoresis (discussed in Chapter 10), the proteins within it move at different rates, as shown in Figure 17.1. One of the components

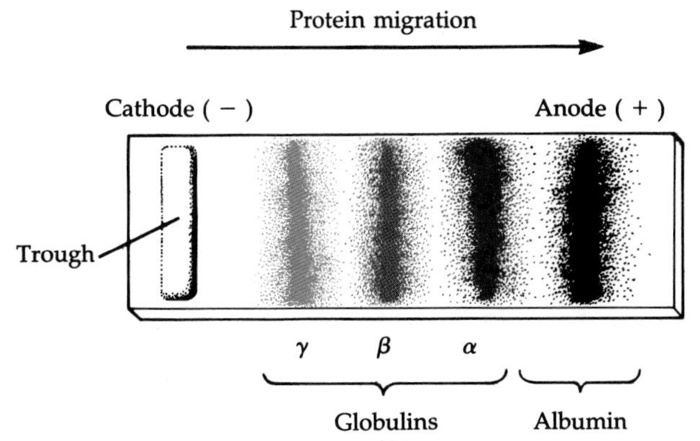

FIGURE 17.1 Separation of serum proteins by electrophoresis. In this procedure, serum is placed in a trough cut into a gel. In response to an electric current, the negatively charged proteins of the serum migrate through the gel from the negatively charged end (cathode) to the positively charged end (anode). Antibodies are concentrated mainly in the gamma (γ) fraction of globulins, hence the familiar term *gamma globulin*.

TABLE 17.1 Types of Acquired Immunity

TYPE OF IMMUNITY	HOW ACQUIRED
Naturally acquired active immunity	Antigens enter the body naturally; antibodies and specialized lymphocytes are produced
Naturally acquired passive immunity (placental and milk transfer)	Fetus or baby receives antibodies from an immunized mother
Artificially acquired active immunity	Prepared antigens in vaccines are introduced into susceptible individual, who produces antibodies and specialized lymphocytes
Artificially acquired passive immunity	Immune serum is injected into susceptible individual, who receives preformed antibodies

that is separated out by this procedure contains most of the antibodies that were present in the original sample. This antibody-rich serum component is called **gamma globulin** or **immune serum globulin.**

When immune serum globulin is injected into the body, it gives immediate protection against a disease. Immune serum can be gathered from immune persons and used to protect others. However, although artificially acquired passive immunity is immediate, it is short lived. The body produces no new antibodies. The half-life of an injected antibody is typically about three weeks.

The various types of immunity are summarized in Table 17.1.

The Duality of the Immune System

The immune system of humans, like that of all vertebrates, has two main components, the humoral immune system and the cell-mediated immune system. These components work both separately and together to protect an organism from disease.

THE HUMORAL IMMUNE SYSTEM

The **humoral immune system** involves antibodies that are dissolved in extracellular fluids, such as blood plasma, lymph, and mucus secretions—fluids once known as *humors*. This system responds when specialized lymphocytes, called **B cells (B lymphocytes),** are exposed to antigens. Presentation of antigens to B cells requires the assistance of other specialized cells in some cases, as shown in Figure 17.11. The B cells (Figure 17.2a) produce antibodies that are specifically directed against the antigens. The humoral immune response defends mostly against bacteria, bacterial toxins, and viruses in the body's fluids.

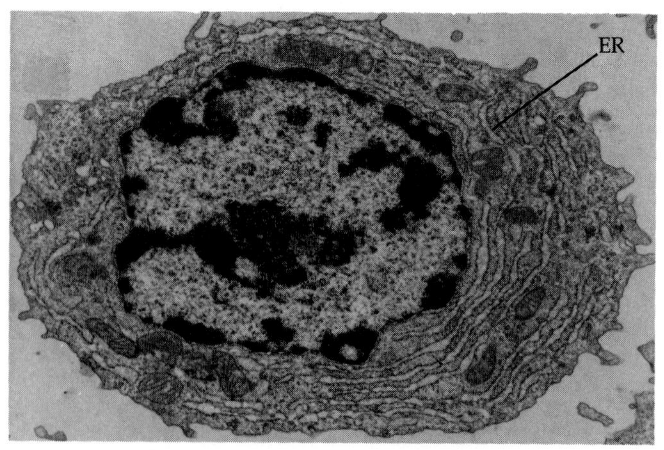

(a) B cell

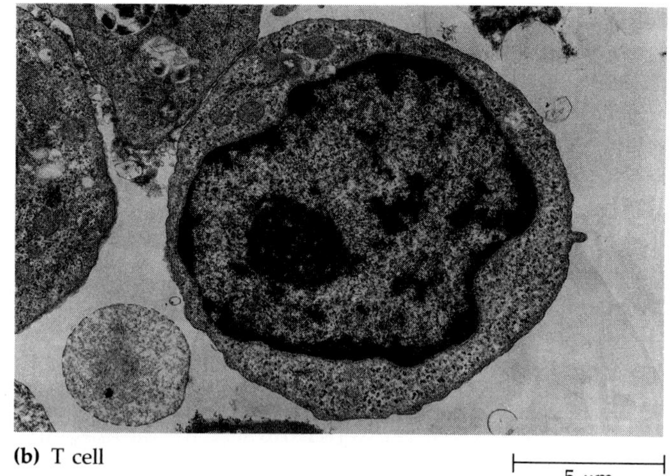

(b) T cell

FIGURE 17.2 B cells and T cells. These are the two major types of lymphocytes responsible for the immune response. **(a)** B cells produce humoral antibodies that are specifically directed against antigens. This transmission electron micrograph of a B cell shows its extensive endoplasmic reticulum (ER), characteristic of B cells, part of the protein-synthesizing machinery used to produce antibodies. **(b)** Transmission electron micrograph of a T cell. T cells play a key role in cell-mediated immunity.

THE CELL-MEDIATED IMMUNE SYSTEM

The second component of the immune system is called the **cell-mediated immune system** because it directly involves specialized lymphocytes called **T cells (T lymphocytes)** (Figure 17.2b). These cells are located in both the blood and the lymphoid tissues (see Figure 16.5). T cells do not secrete antibodies, as B cells do. Instead, they have **antigen receptors** attached to their surfaces. These receptors allow T cells to recognize and react to an antigen—different T cells being specific for different antigens. There are several different kinds of T cells, each of which carries out a specific function.

The cell-mediated immune response is most effective against bacteria and viruses located within phagocytic or infected host cells and against fungi, protozoans, and helminths. The cell-mediated immune system also responds to transplanted tissue (such as a foreign skin graft) by mounting an immune response to reject it. This system is thought to be an important defense against cancer.

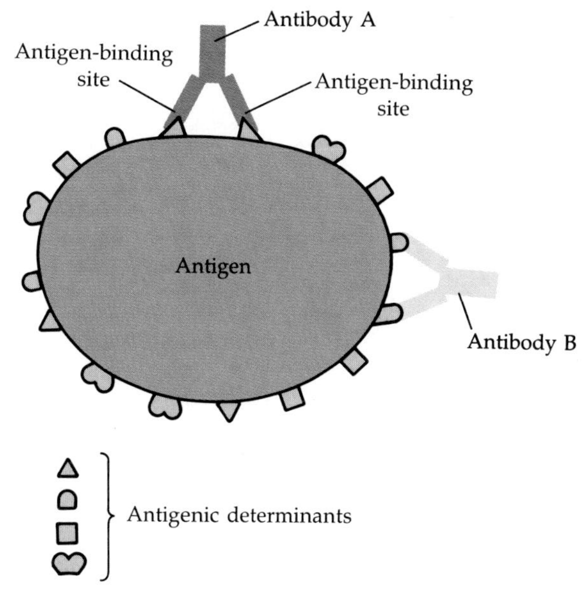

FIGURE 17.3. **Antigenic determinants.** Most antigens such as bacteria contain more than one antigenic determinant site. Each antibody molecule has at least two binding sites that can attach to a specific kind of determinant site on an antigen. It is also possible for an antibody to bind to identical sites on two different cells at the same time. This would cause neighboring antigens to aggregate if enough antibodies and antigens were present. In this mode, the Y-shaped antibodies would be in a T shape.

Antigens and Antibodies

Antigens and antibodies play key roles in the immune response system. As we mentioned previously, antigens (sometimes called **immunogens**) provoke a highly specific immune response in an organism. Usually, this response involves the formation of antibodies or highly specialized T cells.

THE NATURE OF ANTIGENS

Normally, the immune system recognizes components of the body it protects as "self" and foreign matter as "nonself." This recognition is the reason why people's defense systems do not usually produce antibodies against their own body tissues (although we will see in Chapter 19 that this sometimes occurs).

The vast majority of antigens are proteins, nucleoproteins (nucleic acid + protein), lipoproteins (lipid + protein), glycoproteins (carbohydrate + protein), or large polysaccharides. These compounds are often components of invading microbes—the capsules, cell walls, flagella, fimbriae, and toxins of bacteria, the coats of viruses, and the surfaces of many other types of cells. Nonmicrobial antigens include pollen, egg white, blood cells and serum proteins from other persons or species, and transplanted tissues and organs.

Generally, antibodies recognize and interact with specific regions on an antigen's surface, called **antigenic determinants** (Figure 17.3). The nature of this interaction depends on the size and shape of the antigenic determinant and the chemical structure of the antibody. Most antigens have many different types of determinants on their surface. Since different determinants are recognized by different antibodies, the immune system may produce several distinct antibodies against a single antigen.

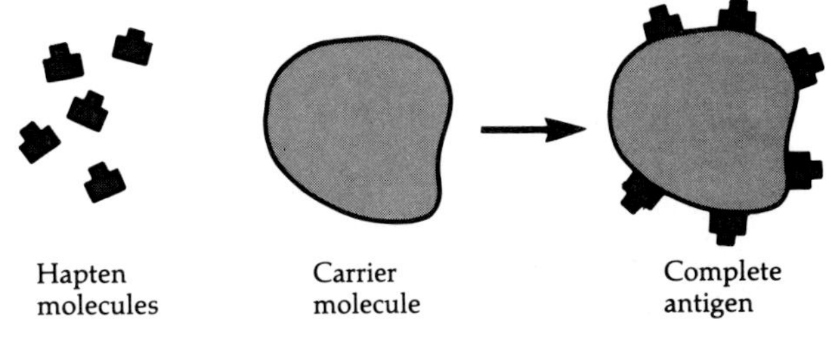

FIGURE 17.4 **Haptens.** A hapten is a molecule too small to stimulate antibody formation by itself. However, when the hapten is combined with a larger carrier molecule, usually a serum protein, the hapten and its carrier together function as an antigen and can provoke an immune response.

Most antigens have molecular weights of 10,000 or more. A foreign substance that has a low molecular weight is often not antigenic unless it is attached to a carrier molecule. These low-molecular-weight compounds are called **haptens** (from *haptein*, to grasp; see Figure 17.4). Once an antibody against the hapten has

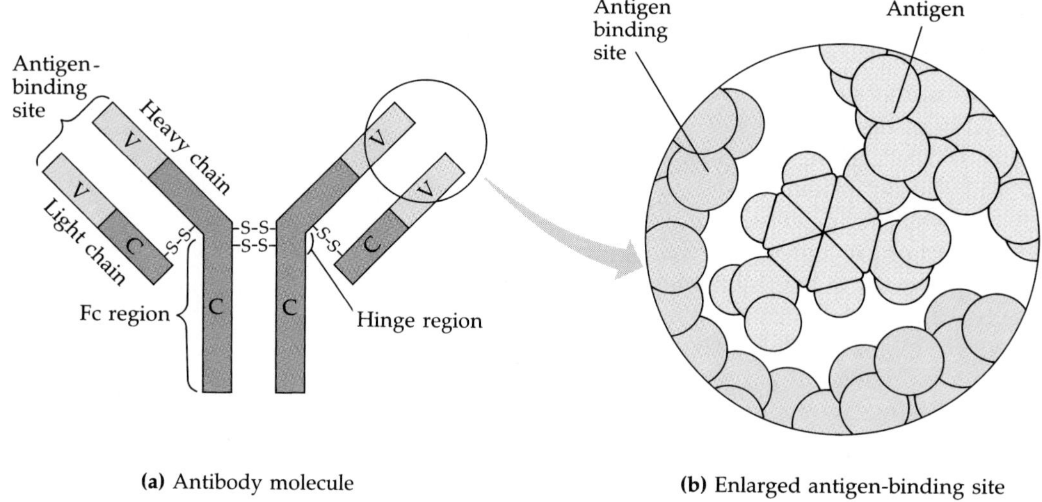

(a) Antibody molecule

(b) Enlarged antigen-binding site

FIGURE 17.5 Structure of a typical antibody molecule. **(a)** The Y-shaped molecule is composed of two light and two heavy chains linked together by disulfide bridges (S—S). Most of the molecule is made up of constant regions (C), which are the same for all antibodies of the same class. The amino acid sequences of the variable regions (V), which form the two antigen-binding sites, differ from molecule to molecule. The stem of the Y, called the Fc region, allows the antibody to bind to the surface of certain host cells. **(b)** One antigen-binding site is shown enlarged and bound to an antigen. The three-dimensional fit between the antigen-binding site and the antigen is similar to the fit between an enzyme and its substrate.

been formed, the hapten will react with the antibody independent of its carrier. Penicillin is a good example of a hapten. This drug is not antigenic by itself, and relatively few people develop allergic reactions to it. However, when penicillin combines with serum proteins present in some people, the resulting molecule initiates an immune response.

THE NATURE OF ANTIBODIES

Antibodies belong to a class of proteins called **immunoglobulins (Ig).** These proteins can recognize, bind to, and help cause the destruction of antigens. Antibodies are highly specific; they can interact with only one type of antigenic determinant on an antigen.

Each antibody has at least two identical sites that bind to antigenic determinants. These sites are known as **antigen-binding sites.** The number of antigen-binding sites on an antibody determines the **valence** of that antibody. For example, most human antibodies have two binding sites; therefore, they are bivalent.

Antibody Structure

Because a bivalent antibody is the simplest antibody type, it is called a **monomer.** A typical antibody monomer has four protein chains—two identical *light (L) chains* and two identical *heavy (H) chains* (*light* and *heavy* refer to the relative molecular weights). The chains are joined by disulfide links (see Chapter 2) and other bonds to form a Y-shaped molecule (Figure 17.5). The Y-shaped molecule is flexible and can assume a T shape (note the hinge area indicated on the antibody molecule in Figure 17.5).

Both the H and L chains have sections located at the ends of the Y's arms, called *variable (V)* regions. Together, the V regions from one L chain and one H chain form one antigen-binding site; hence, each antibody has two such sites. The amino-acid sequences of these V regions vary from one antibody to another. The resulting variations in chemical and three-dimensional structure account for the ability of different antibodies to recognize and bind with different antigens. Any single antibody molecule has only one type of L chain and one type of H chain, even if it is made up of more than one monomer. Therefore, each antibody molecule has only one type of antigen-binding site and will bind to only one type of antigenic determinant. Later in this chapter we will describe the mechanism by which the body is able to make different antibodies against the huge array of antigens we may encounter.

Figure 17.3 shows an antibody combining with antigenic determinants on an antigen. In the illustration, the antigen might be a bacterial cell with several different antigenic determinants. Note that more than one type of antibody can react with the antigen. If the two antigen-binding sites of an antibody combine with antigenic determinants on two different antigens, the antigens can aggregate into clumps, which are more easily ingested by phagocytic cells. As we will discuss in Chapter 18, this clumping can also be an important factor in the diagnosis of some diseases.

The stem of the antibody monomer and the lower parts of the Y's arms are called *constant (C)* regions (Figure 17.5). The term *constant* refers to the fact that C regions of the L and H chains of different monomers are relatively invariant in their amino acid sequence. There are five major sequences found for C regions of H chains, and there are two different sequences for L chains. Each H-chain sequence determines a different class of immunoglobulin.

The stem of the Y-shaped antibody monomer is called the *Fc region*. The antibody molecule can attach to a host cell at the Fc region. Complement (see Chapter 16) can also bind to the Fc region. The two antigen-combining sites are left free to bind to compatible antigenic determinants.

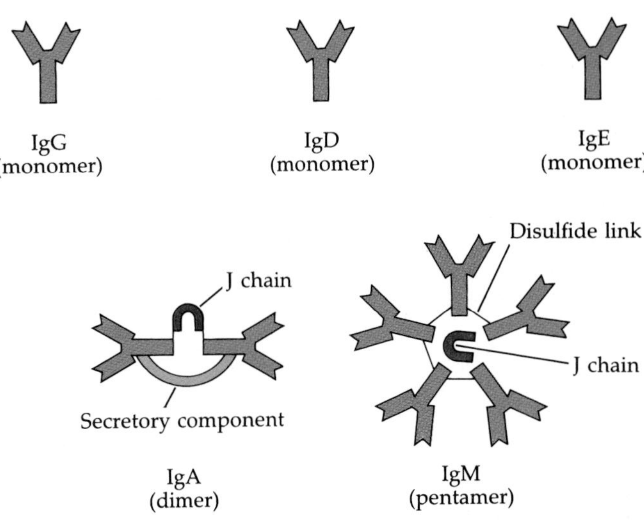

FIGURE 17.6 Human immunoglobulins. Structures of the five principal classes of human immunoglobulins are shown. Note that IgA and IgM are made of two and five monomers, respectively, in this drawing. In these cases, the monomers are held together by disulfide links, and some of these are joined by a polypeptide called the J (joining) chain. IgA is usually found attached to a protein called the secretory component; it may also occur as a monomer.

Immunoglobulin Classes

The five classes of immunoglobulins (Ig) are designated IgG, IgM, IgA, IgD, and IgE, according to the properties of their heavy chains. Each class plays a different role in the immune response. The structures of IgG, IgD, and IgE molecules resemble the structure shown in Figure 17.5a. Molecules of IgM and IgA are usually two or more monomers joined together (Figure 17.6). The characteristics of the immunoglobulin classes are summarized in Table 17.2, p. 432.

IgG. IgG antibodies account for about 80% to 85% of all antibodies in serum. These monomer antibodies readily cross the walls of blood vessels and enter tissue fluids. Maternal IgG antibodies, for example, can cross the placenta and confer passive immunity to the fetus.

IgG antibodies protect against circulating bacteria and viruses, neutralize bacterial toxins, trigger the complement system, and, when bound to antigens, enhance the effectiveness of phagocytic cells.

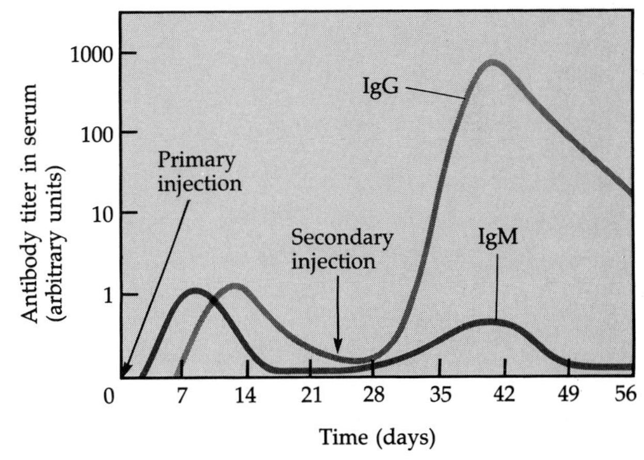

FIGURE 17.7 The primary and secondary immune response to an antigen. IgM appears first in response to the primary injection. IgG follows and provides longer-term immunity. The second (booster) injection results in a much faster and greater production of antibodies. The antibodies produced in response to the second injection are mostly IgG.

IgM. Antibodies of the IgM (from *macro*, reflecting its large size) class make up 5% to 10% of the antibodies in serum. IgM has a pentamer structure, consisting of five Y-shaped monomers that are held in position by a *J (joining) chain* (Figure 17.6).

IgM antibodies are the first ones to appear in response to the initial exposure to an antigen. However, the initially high IgM concentration in the blood rapidly declines, and the IgG concentration increases (Figure 17.7). A second exposure to antigen results mostly in the increased production of IgG.

The large size of the molecule may prevent IgM from moving about as freely as IgG does. IgM antibodies generally remain in the blood vessels and do not enter the surrounding tissues.

TABLE 17.2 Summary of Immunoglobulin Classes

CHARACTERISTICS	IgG	IgM	IgA	IgD	IgE
Percent of total serum antibody	80–85	5–10	15	0.2	0.002
Location	Blood, lymph, intestine	Blood, lymph, B-cell surface (as monomer)	Secretions (tears, saliva, mucus, intestine, milk), blood, lymph	B-cell surface, blood, lymph	Bound to mast and basophil cells throughout body, blood
Molecular weight	150,000 (monomer)	970,000 (pentamer)	405,000 (dimer with secretory component)	175,000 (monomer)	190,000 (monomer)
Half-life (days) in serum*	23	5	6	3	2
Complement fixation	Yes	Yes	No**	No	No
Placental transfer	Yes	No	No	No	No
Functions known	Enhances phagocytosis, neutralizes toxins and viruses, and protects fetus and newborn	Especially effective against microorganisms and agglutinating antigens; first antibodies produced in response to initial infection	Localized protection on mucosal surfaces	Not known, but presence on B cells may indicate function in initiation of immune response	Allergic reactions; possibly expulsion or lysis of protozoan parasites

*Time required for one-half of the antibodies to disappear.
**May be "yes" via alternate pathway.

Because of its numerous antigen-binding sites, the IgM molecule is especially effective at cross-linking particulate antigens and causing their aggregation (illustrated in Figure 18.6). IgM is the predominant antibody involved in the response to the ABO blood group antigens on the surface of red blood cells (discussed in Chapter 19). It is also especially effective in reactions involving complement. It can enhance the ingestion of target cells by phagocytic cells, as IgG does.

The fact that IgM appears first in response to a primary infection and is relatively short lived makes it uniquely valuable in disease diagnosis. If high concentrations of IgM against a disease pathogen are detected in a sick patient, then it is likely that this disease is caused by that pathogen. Detection of IgG, which is relatively long lived, may indicate only that immunity against a particular disease was acquired in the more distant past.

IgA. IgA antibodies account for about 15% of the antibodies in serum. The effective form of IgA is a dimer, consisting of two Y-shaped monomers connected by a J chain. It is produced in this form by plasma cells abundant in the mucous membranes. Each dimer then enters and passes through a mucosal cell, where it acquires a *secretory component* that protects it from enzymatic degradation. The main function of secretory IgA is probably to prevent the attachment of pathogens, particularly viruses and certain bacteria, to mucosal surfaces. Its presence in colostrum probably helps to protect the infant, especially from gastrointestinal infections.

IgD. IgD antibodies make up only about 0.2% of the total serum antibodies. Their structure resembles that of the IgG molecules. IgD antibodies are found in blood and lymph and on the surfaces of B cells. These antibodies do not fix complement and cannot cross the placenta. IgD has no known function in serum. However, its high concentration on the surface of B cells is considered an indication that it is a factor in B-cell differentiation triggered by an antigen.

IgE. Antibodies of the IgE class are slightly larger than IgG molecules, but they constitute only 0.002% of the total serum antibodies. The IgE molecules bind tightly by their Fc ends to receptors on mast cells and basophils, which are specialized cells that participate in allergic reactions (see Chapter 19). When an antigen such as pollen reacts with the IgE antibodies attached to a mast cell or basophil, that cell releases histamine and other chemical mediators. These chemicals provoke an inflammatory response—for example, an allergic reaction such as hayfever. However, the inflammation can be protective as well. It causes the attraction of IgG, complement, and phagocytic cells. This is especially useful against parasitic worms.

B Cells and Humoral Immunity

As we mentioned earlier, there are two basic kinds of immune responses. One of these, the humoral response, is carried out by a special group of cells called B cells. B cells are responsible for the production of antibodies.

B cells, and other cells, such as T cells and macrophages, develop from stem cells located in the bone marrows in adults and in the liver in embryos. By the time a B cell is mature, it may carry as many as 100,000 antibody molecules on its surface. Each B cell can produce antibodies against only one specific antigen; however, the total population of B cells in a person can produce antibodies against an enormous assortment of antigens.

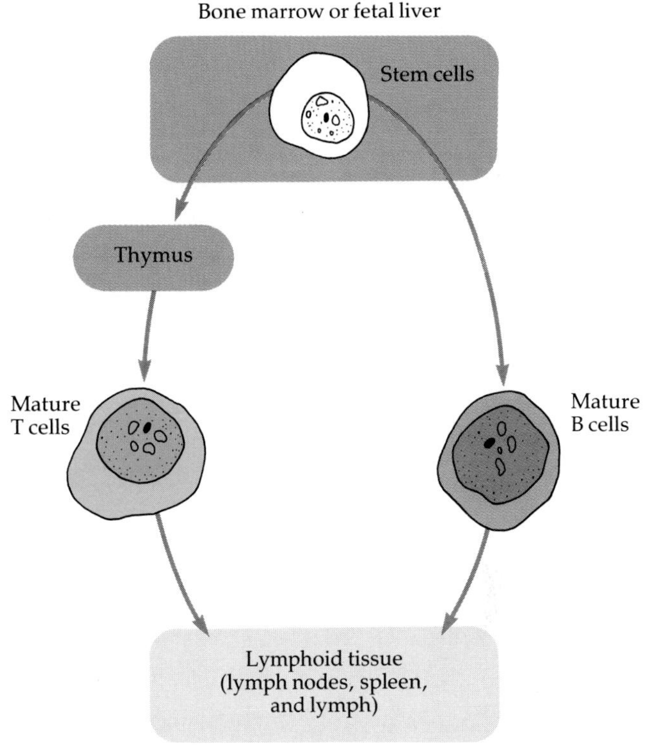

FIGURE 17.8 Differentiation of T cells and B cells. Both types of cells originate from stem cells in adult bone marrow or in the fetal liver. (Red blood cells, macrophages, neutrophils, and other white blood cells also originate from these same stem cells.) Some cells pass through the thymus and emerge as mature T cells. Other cells are influenced by the bone marrow and become B cells. Both types of cells then migrate to lymphoid tissues, such as the lymph nodes or spleen.

B-Cell and Antigen Interactions

The mature B cells migrate to lymphoid organs, where they encounter antigens (Figure 17.8). When the appropriate antigen contacts the antigen receptor antibodies on a B cell, the B cell proliferates into a large clone of cells (Figure 17.9). (This phenomenon, called clonal selection, will be discussed later.) Some of these cells differentiate into **plasma cells,** which secrete antibodies against the antigen. Each plasma cell lives for only a few days but is capable of producing about 2000 humoral antibody molecules per second.

Stimulation of a B cell by an antigen also results in the production of a population of **memory B cells.** These cells will be discussed later in the chapter.

Often, the production of antibodies by a B cell will depend on the cooperation of other cells. For example, B cells must work cooperatively with certain macrophages and T cells to produce antibodies against a type of antigen known as **T-dependent antigens.** Examples of T-dependent antigens include bacteria, foreign red blood cells, proteins, and hapten–carrier combinations that have many different antigenic determinants.

The T-dependent antigen is attacked and partly digested by macrophages or dendritic cells (cells with a highly branched appearance), which are known as *antigen-presenting cells (APC)* (Figure 17.10). Portions of the antigen, usually polypeptide fragments, are taken up by an antigen-presenting cell and are then positioned on the cell's surface. Specific T cells, the *helper T cells* (T_H), interact with antigen-presenting cells. T_H cells recognize the antigen fragments and certain "self" antigens (surface antigens that identify the cell as "self") that are also carried on the surface of antigen-presenting cells (Figure 17.11).

In fact, the T_H cell recognizes the foreign antigen only when it is combined with a self antigen on the APC. This encounter primes the T_H cell to activate the appropriate B cell. It is necessary that the B cell also carry self antigens; indeed, it is likely that the T-cell receptor recognizes a single complex of the antigen and the self antigen on the B-cell surface. The requirement that both types of antigens be matched in order to activate antibody production minimizes the chances of making antibodies against our own antigens. Helper

B cell after antigen stimulation

Clone of B cells

Plasma cells

Memory cells

Secrete antibodies into circulation

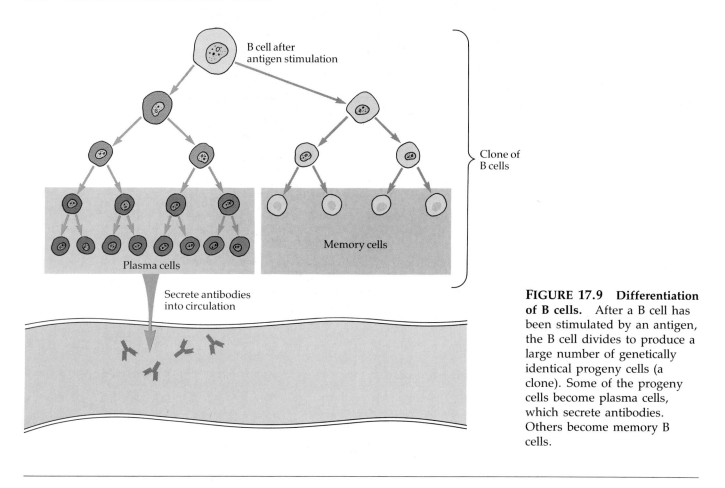

FIGURE 17.9 Differentiation of B cells. After a B cell has been stimulated by an antigen, the B cell divides to produce a large number of genetically identical progeny cells (a clone). Some of the progeny cells become plasma cells, which secrete antibodies. Others become memory B cells.

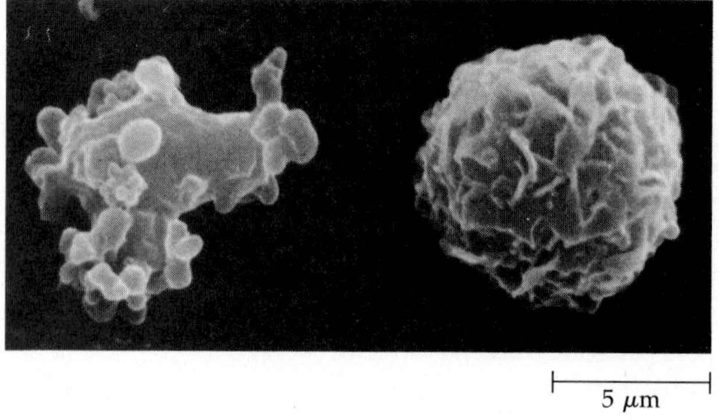

FIGURE 17.10 Antigen-presenting cells. Antigen-presenting cells are usually macrophages or dendritic cells that present antigen to helper T cells. This scanning electron micrograph shows a macrophage (right) covered with typical ridges and a dendritic cell (left) exhibiting the branching morphology that suggested their name. The antigen molecules being presented are presumably on the surface of the cells but are too small to be seen here.

T cells are involved in immune responses that produce IgG, IgA, and IgE.

The self antigens just mentioned are actually cell-surface proteins that are components of the **major histocompatibility complex (MHC).** The particular collection of MHC proteins a person carries is unique to that individual. Thus, MHC proteins are well suited to serve as self antigens—signals by which the immune system distinguishes self from nonself. The MHC will be discussed further in Chapter 19.

T-independent antigens can stimulate a response from B cells without the intervention of T cells. These T-independent antigens are usually composed of repeating polysaccharide or protein subunits. Bacterial flagella, which are composed of protein, and the lipopolysaccharide layer of gram-negative bacteria are important examples of T-independent antigens. These antigens are able to form multiple bonds with B-cell receptors, which is probably why they do not need T-cell assistance (Figure 17.12). The immune response

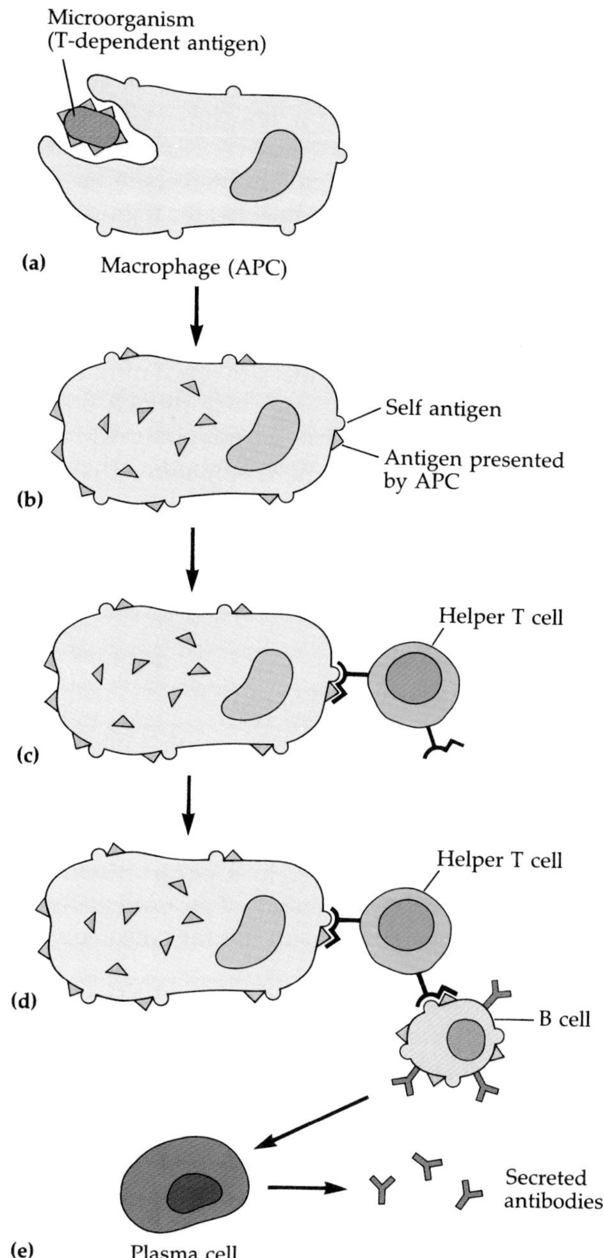

FIGURE 17.11 How helper T cells may activate B cells to make antibodies against T-dependent antigens. B cells require the help of antigen-presenting cells (APCs) and specialized helper T cells to produce antibodies against T-dependent antigen. The process is probably as follows: **(a)** The microbial antigen is attached by an APC and partially digested. **(b)** Antigen fragments are presented on the APC surface and form complexes with self (MHC) antigens. **(c)** A helper T (T_H) cell specific for the presented antigen interacts with the complex. **(d)** The helper T cell then activates an appropriate B cell, probably one that itself has both antigens on its surface, as well as antibodies to the microbial antigen. **(e)** This interaction, along with direct stimulation by the microbial antigen, triggers the B cell to differentiate into a plasma cell, which secretes antibodies.

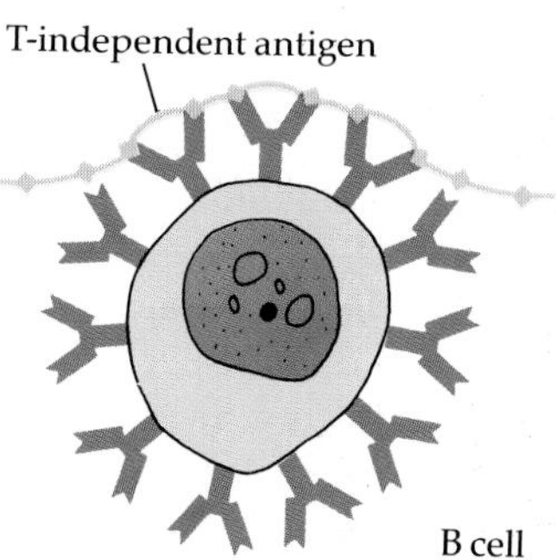

FIGURE 17.12 T-independent antigens. T-independent antigens have repeating units that form multiple bonds with a B cell. These antigens stimulate the B cell to make antibodies without the aid of helper T cells.

to T-independent antigens tends to be weaker than the response to T-dependent antigens. Once T-independent antigens are bound to B-cell receptors (surface immunoglobulins), the B cells then produce antibodies, which are almost exclusively of the IgM class. This distinction between T-dependent and T-independent antigens is clear in mice, but it is uncertain whether it can be made in humans.

ANTIBODY–ANTIGEN BINDING

Antibody recognition of antigens is based on the ability of the antibody to fit and bind to an antigen. An antigen binds to an antibody in the antigen-binding site formed by the variable regions of the light and heavy chains. The resulting combination is called an **antigen–antibody complex.** The formation of such complexes can protect a host in several ways. For example, antibodies can neutralize toxins by blocking their active sites, or they can inactivate viruses by combining with them and preventing their attachment to host cells. These complexes can also fix complement and start the lysis of invading cells.

As we will see in Chapter 19, the action of antibodies can also damage the host. For example, immune complexes of antibody, antigen, and complement can damage host tissue. Antigens combining with IgE on mast cells can initiate allergic reactions, and antibodies can react with host cells and cause autoimmune disorders.

In practice, each B-cell receptor is not always an exact fit for the antigen. When large amounts of antigen are present, some antigens attach to receptors for which they are less than a perfect fit. Because of this, some antibodies produced are poor matches for the antigen; these are said to have less *affinity*.

MOLECULAR BASIS OF THE DIVERSITY OF ANTIGEN RECEPTORS

A problem that long troubled immunologists is how the body produces such an immense variety of antibodies—even antibodies against synthesized compounds that probably do not occur in nature. Fortunately, nature has evolved a mechanism by which the immune system can respond to any conceivable antigen. According to some estimates, an individual may be able to respond to as many as 100 million different antigens; other estimates suggest 100 billion antigens.

A detailed explanation of the mechanism is beyond the scope of this book, but it is analogous to the generation of huge numbers of words from a limited alphabet. As shown in Figure 17.5a, each antibody molecule has antigen-binding sites composed of a light chain and a heavy chain. Both types of chains are divided into constant regions and variable regions. The constant regions are not directly involved in antigen binding, but they differ with the different immunoglobulin classes, i.e., IgG, IgM, etc.

The variable regions of the antibody molecule carry the antigen-binding sites, which are composed of light and heavy chains (Figure 17.5b). It is in these regions that diversity is most required. Because the light chains are produced in two major classes, kappa and lambda, three sets of genes are required to encode the information controlling the formation of the antigen-binding sites (one set for heavy chains, one for kappa light and one for lambda light). The sets of genes that control the kappa light chains contain several hundred V genes, which code for different variable regions, and a single C gene, which codes for the constant region. In addition there are four J genes, which code for the segments joining the variable and constant regions of the chain. (There is no relationship between the J genes and the J chain in Figure 17.6.) As the lymphocytes develop, the light chains are assembled by selecting a V gene from the collection of hundreds and selecting one of the four J genes to connect it to the C gene (Figure 17.13).

Therefore, we do not need extraordinarily large numbers of genes to generate huge numbers of different combinations. As an example, if there are 200 V genes and 5 J genes, then 1000 (200 times 5) combinations are possible for the chain.

The arrangement of the heavy-chain genes is similar in principle to that of the light-chain genes. However, more variation is available. Not only are there well over 100 V genes, but also there are a dozen D (diversity) segments and four J genes to make up the entire variable region of the heavy chain. The number of possible variants becomes very large. For example, 100 V genes, 12 D genes, and 4 J genes would provide for 4800 variations of heavy chains. If this were multiplied by the 1000 variations of the light chains calculated earlier, the total would be over 7 million combinations derived from only a few hundred different genes. This example is conservative; a larger library of genes is actually available. Furthermore, we have not yet considered the heavy-class constant region genes, which code for the information differentiating between antibody classes.

In addition to these gene-sorting mechanisms, antibody variability is generated by the mechanism of somatic point mutations. These are random mutations that arise as the B cell, stimulated by a suitable antigen, begins proliferating and making antibodies. Many of the new B cells will synthesize antibodies that differ slightly, by one or a few amino acids. The cells that make antibodies that are a better match to the antigen are more likely to be stimulated, with the result that the antigen-binding power of the antibodies produced tends to improve.

MONOCLONAL ANTIBODIES

Scientists have long observed that antibody-secreting B cells may become cancerous. When their proliferation is unchecked, they form tumors called myelomas. It is possible to isolate these cancer cells and propagate them in cell culture indefinitely. Cancer cells, in this sense, are "immortal," in contrast to normal B cells, which can be propagated in culture for only a few generations.

In 1975, Georges Köhler and César Milstein discovered how to fuse an "immortal" cancerous cell (a mutant that has lost the ability to form antibody) with a noncancerous antibody-secreting plasma (B) cell taken from a mouse that had been immunized with a particular antigen. The resulting hybrid cell could multiply to form a clone of cells called a **hybridoma.**

When a hybridoma is grown in culture, its genetically identical cells continue to produce the type of antibody characteristic of the ancestral B cell. The importance of the discovery is that clones of the antibody-secreting cells now can be maintained indefinitely in cell culture and can produce immense amounts of a desired pure antibody.

Because the antibodies are produced by a single hybridoma clone and are identical, they are called **monoclonal antibodies** (Figure 17.14). Monoclonal antibodies are extremely useful for three reasons—they are uniform, they are highly specific, and they can be readily produced in large quantities. Because of these qualities, monoclonal antibodies have assumed enormous importance as diagnostic tools. For example, kits have recently been developed that use monoclonal antibodies to recognize chlamydial and streptococcal

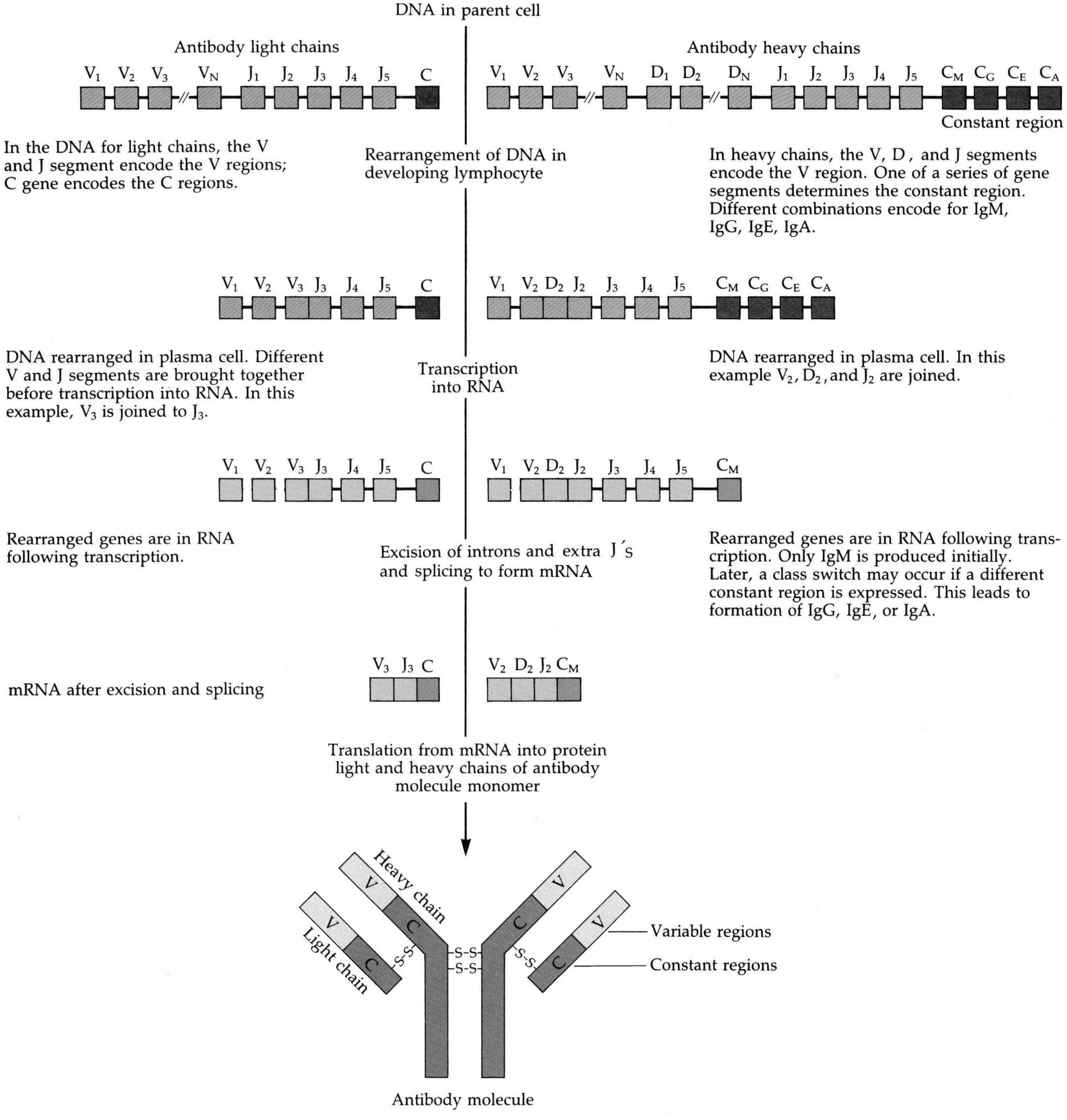

FIGURE 17.13. The genetic basis of antibody diversity. Light-chain production requires that, to code for the variable region, one V gene and one J gene be rearranged to be adjacent to each other in the B cell. These genes eventually are transcribed, along with the C gene for the constant region, into mRNA that encodes the protein to produce the light chain. The heavy gene is more complex. A V gene, a J gene, and also a D gene are sorted from the germ line DNA. The constant region of the heavy chain varies with the immunoglobulin type, e.g., IgM, IgG, IgE, or IgA. Once these genes are transcribed into mRNA, the B cell uses it to code for heavy-chain protein.

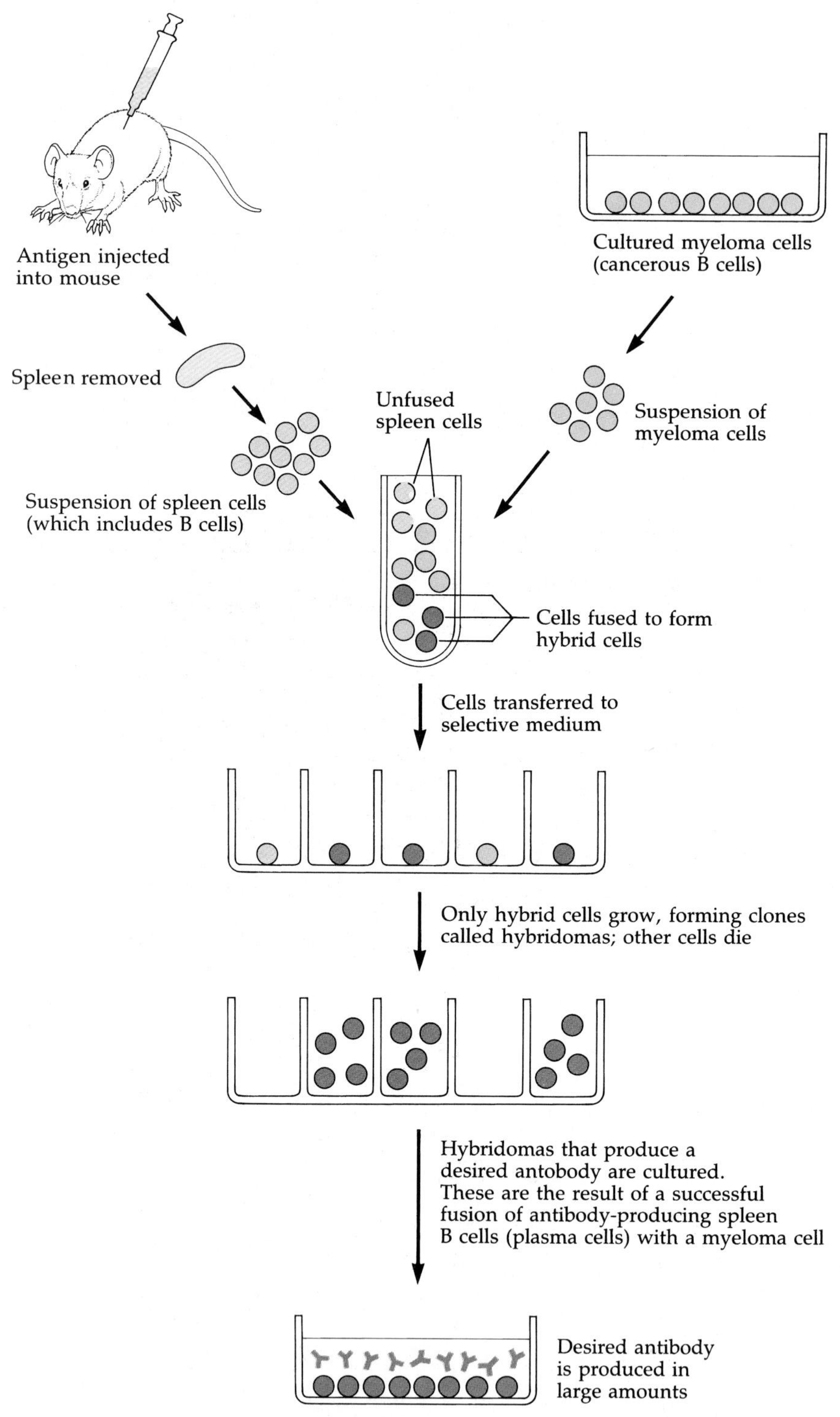

FIGURE 17.14 Production of monoclonal antibodies. A mouse injected with a specific antigen will produce antibodies against that antigen. The spleen of the mouse is removed, and a suspension is made. The spleen cells are then mixed with myeloma cells. These particular myeloma cells have usually lost their ability to produce antibody. Some of the antibody-producing spleen cells and myelomas fuse to form hybrid cells. The mixture of cells is placed in a medium that allows only successfully fused cells to proliferate into clones, called hybridomas. The hybridomas are then screened for production of the desired antibody. These hybridomas are then cultured to produce large amounts of monoclonal antibodies.

bacteria, and pregnancy tests are also available that use monoclonal antibodies to indicate the presence of a hormone excreted only in the urine of a pregnant woman.

Monoclonal antibodies also have therapeutic applications. For example, when antibiotics are used to treat blood infections by gram-negative bacteria, endotoxins are released into the blood and cause septic shock (see Chapter 15). Monoclonal antibodies have been produced to neutralize these endotoxins.

Monoclonal antibodies are also being used to overcome unwanted effects of the immune system, such as rejection of transplanted organs (see Chapter 19). In this case, antibodies are prepared that react with the T cells that are involved in rejection of the transplanted tissue; the antibodies suppress the T-cell activity. These versatile antibodies have also been proposed as a way to neutralize the receptor sites by which the AIDS virus attaches to susceptible cells.

The prospect of using monoclonal antibodies to treat cancer is also of great interest. One approach is to combine monoclonal antibodies with a toxin to make an *immunotoxin*. The general idea is that such immunotoxins could be carried throughout the body by the circulatory system, specifically attach to cancer cells wherever they might be, and kill them.

The therapeutic use of monoclonal antibodies has been limited because these antibodies are currently produced by mouse cells. The immune systems of some people have reacted against the foreign mouse proteins. Monoclonal antibodies derived from human cells would probably provoke fewer reactions. Several approaches are being taken to solve this problem. One is to construct, by methods involving recombination of DNA, an antibody with variable regions derived from mouse cells and constant regions derived from human sources. This would still make use of the large number of hybridoma cells available from mouse cells and would also make the molecule more compatible with the human immune system. These are *chimeric monoclonal antibodies*—after the mythological monster with a lion's head, a goat's body, and a serpent's tail. Genetic engineering is also proposed as a method for altering mouse antibodies so that they have characteristics that are more human.

New techniques promise to greatly expand the scale and variety of monoclonal antibody production. For example, genetic engineering can be used to produce antibodies without using hybridomas made from mammalian cells. The antibody-producing genes from animals are transplanted into bacteria such as *E. coli*. By such means, essentially the entire array of antibodies that an animal might produce are cloned into a bacterium that can produce them in large amounts. When perfected, this method could be used to make human monoclonal antibodies.

T Cells and Cell-Mediated Immunity

Early in the history of immunological experimentation, biologists learned that some forms of immunity could be transferred between animals by transferring serum from immunized to nonimmunized animals. However, other forms of immunity were not transferable with blood serum. Much later it was learned that these latter forms of immunity were transferred only when certain lymphocytes were transferred; the immunity associated with these lymphocytes is now called *cell-mediated immunity*. Immunity to tuberculosis, for example, was found to be most effectively transferred by these lymphocytes.

THE COMPONENTS OF CELL-MEDIATED IMMUNITY

T cells are the key component of cell-mediated immunity. Like B cells, they derive from precursor stem cells in bone marrow, but unlike the precursors of B cells, these precursors develop and differentiate into mature T cells within the thymus. After differentiation, the T cells migrate to lymphoid organs, such as the lymph nodes or the spleen. When T cells recognize foreign antigens, they do not secrete antibodies but instead differentiate into several types of **specialized T cells** and into long-lived memory cells. The different classes of T cells are responsible for attacking foreign antigens and regulating the immune response.

Types of T Cells

There are four main types of T cells—helper T (T_H) cells, delayed hypersensitivity (T_D) cells, suppressor T (T_S) cells, and cytotoxic T (T_C) cells. Mature T cells can be identified as to type by characteristic antigens on the cell surfaces.

Helper T (T_H) cells play a variety of roles. The cause of the pathogenicity of the AIDS virus was identified by using monoclonal antibodies to show that a certain group of T_H cells was being decimated. These T cells contain a surface antigen called CD4 and are most important in a vigorous immune response. We have already seen that certain T_H cells present T-dependent antigens to B cells. Some other T_H cells help other T cells to respond to antigens.

Delayed hypersensitivity T (T_D) cells are associated with certain allergic reactions, such as poison ivy, and with rejection of transplanted tissues. T_D cells are also important in the body's defense against cancer. T_D cells are the cells for which cell-mediated immunity was originally named. Early investigators found that the transfer of these cells between animals also transferred immunity to tuberculosis, while the transfer of

MICROBIOLOGY *IN THE NEWS*

Parasite Outwits the Immune System

How protozoans survive—and, indeed, thrive—in the inhospitable environment of the host was once explained by the belief that these organisms do not evoke an immune response, thus eliminating a primary host defense. For example, *Giardia*, an intestinal protozoan, causes chronic diarrheal illness that is long-lasting, is difficult to eliminate, and tends to recur—all signs that the immune defenses are not working. Giardiasis occurs worldwide and is common in the United States, where the carrier rate may be as high as 20% depending on the community surveyed. In its ability to resist the immune system, giardiasis resembles other protozoan diseases, such as malaria and African sleeping sickness, that are less common in the United States.

However, when researchers tested the theory that there is no immune response to protozoans, they found that the opposite is true. For example, they found that secretory IgA and IgM antibodies against *Giardia* are abundant in the small intestine of infected persons. *Giardia* normally navigates the hostile environment of the small intestine propelled by flagella, and uses attachment disks to fasten itself to the intestinal lining (see the figure). Researchers found that the body makes both anti-flagella antibodies, which interfere with the motility and cell division of *Giardia*, and anti-disk antibodies, which interfere with attachment. These discoveries, which appeared to indicate an effective immune response, only deepened the mystery of the protozoan's ability to survive.

In 1990, a team of researchers at the National Institutes of Health led by Theodore Nash sought the answer by inoculating 19 human volunteers with *Giardia* cells that were cloned from one cell and had identical surface antigens. When the researchers took samples of *Giardia* from the volunteers during the second week of infection, they found that the surface antigens had changed radically and now presented numerous variations. The researchers estimated the rate of change in surface antigens to be from 1:100 to 1:1000 per cell division. Thus, it appeared that although the body made plenty of antibodies, the organism was able to defeat the "lock-and-key" fit between antigen and antibody by continually changing locks. At present it is unclear whether the surface antigens of this organism can change indefinitely or whether there is a limited variety of antigens. It is possible to speculate that more severe and chronic infections are caused by strains with higher frequencies of change.

Mechanism of Change

Molecular biologists are intrigued by the quick-change artistry of protozoans and have been probing the mysteries of gene expression in another protozoan, *Trypanosoma*. *Trypanosoma*, the parasite that causes African sleeping sickness, changes its surface antigens more slowly than *Giardia* (at a rate of 1:10,000). The trypanosome's coat is made of a single protein, called *variable surface glycoprotein*, or VSG, so it can be entirely transformed by the alteration of a single gene. The trypanosome has genes for many variants of VSG, and these genes are switched on and off in quick succession.

A Grim Future?

Will understanding how these protozoans outwit the immune system lead to development of vaccines? The prospects for a conventional vaccine seem poor. To eliminate an organism that can interchange hundreds or thousands of surface antigens, most of which appear in seemingly random order, vaccination might have to produce hundreds or thousands of kinds of antibodies. However, recent research suggests a more practical approach. Researchers are finding evidence that, among the thousands of variable antigens, there are a few that do not change. A vaccine containing numerous copies of those few antigens might stimulate the production of enough antibodies against the selected antigens to produce effective immunity. Such research offers at least a slim ray of hope.

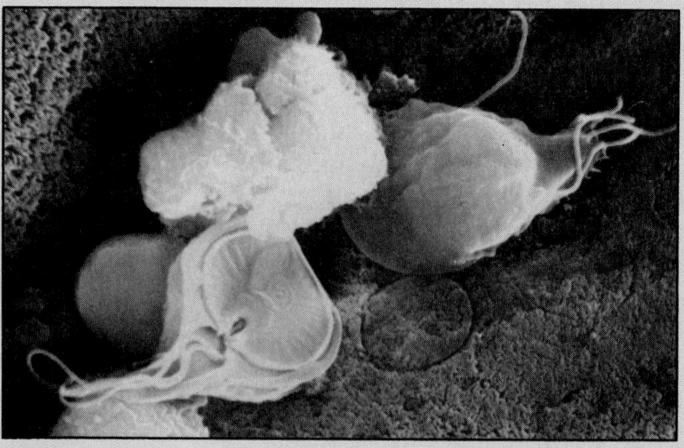

5 μm

Scanning electron micrograph of the trophozoite form of *Giardia lamblia,* the flagellated protozoan that causes giardiasis. Note the circular mark left on the intestinal wall by the ventral sucker disk the parasite uses to attach itself.

serum does not. In response to antigenic stimulation, T_D cells produce a variety of substances called *lymphokines*, which recruit defensive cells such as macrophages. We will discuss these topics more fully in Chapter 19.

Suppressor T (T_S) cells are not completely understood. It is generally accepted that they inhibit the conversion of B cells into plasma cells, which might be a way of turning off the immune response when an antigen is no longer present. They apparently also suppress the activity of some other T cells so that immune responses are not developed against host antigens. Both T_H cells and T_S cells are also known as **regulatory T cells** because they act as regulators of the immune response.

Cytotoxic T (T_C) cells destroy target cells, such as cancer cells and transplanted tissue, upon contact. Protection against viral infections and intracellular bacterial infections is the most important function of T_C cells. Because viruses (and some bacteria) reproduce within host cells, they cannot be attacked there by humoral antibodies. T_C cells recognize viral antigens on the surface of the virus-producing host cell and cause the destruction of that cell. T_C cells cause the lysis of their target cells by coming into close contact with them and releasing a protein called *perforin*. The protein forms a pore in the target cell membrane by a mechanism that is similar to the complement mechanism.

T_C cells will not react with free viruses but only with cells carrying both the target antigen and a class of self (MHC) antigens that are found on most body cells. After it contacts and kills a target cell, the T_C cell continues to live and can kill repeatedly. T_C cells also play at least some role in the defense against cancer (Figure 19.13) and a few diseases caused by protozoans and helminths.

Lymphokines and Cytokines

As mentioned previously, when T_D cells (and some other T cells) are stimulated by an antigen, they release proteins called **lymphokines.** Since it is now known that cells other than lymphoid cells produce these substances, the more general term **cytokine** is coming into use. One lymphokine produced by T_D cells is *macrophage chemotactic factor*, which attracts macrophages to the infection site. Other lymphokines are *macrophage migration inhibition factor*, which inhibits the movement of macrophages away from the infection site, and *macrophage activation factor*, which activates macrophages. Activated macrophages are much more efficient than unactivated macrophages at destroying these cellular antigens.

Another important lymphokine is *interleukin-2* (IL-2). IL-2 is responsible for the proliferation and differentiation of T cells. It is being used experimentally to treat cancer. The convention is to name lymphokines for their biological activity until their amino-acid sequence is known. They are then assigned an *interleukin number*.

Lymphokines are probably not involved in the contact destruction of target cells by T_C cells, although a lymphokine called *lymphotoxin* can destroy nonlymphocytic target cells in vitro. The functions of these lymphokines are summarized in Table 17.3

TABLE 17.3 Summary of Some Cytokines and T-Cell Lymphokines

LYMPHOKINE	*FUNCTION*
Macrophage chemotactic factor	Attracts macrophages to infection site
Macrophage migration inhibition factor	Prevents macrophages from leaving infection site
Macrophage activation factor	Activates macrophages to improve phagocytic activity
Interleukin-1 (a cytokine)	Promotes multiplication and activation of B cells and T cells; triggers fever response in body
Interleukin-2	Stimulates proliferation and differentiation of T cells and natural killer cells
Lymphotoxin	Destroys nonlymphocytic target cells in vitro; function in vivo uncertain
Interferon	Inhibits viral replication
Transfer factor	Intensifies the action of sensitized T cells

THE CELL-MEDIATED IMMUNE RESPONSE

Like B cells, individual T cells seem to have specificity for only a single antigen, which indicates the presence of T-cell receptors to recognize the antigen. These T-cell receptors are not the IgM and IgD molecules that serve this purpose on B cells; but their structures have been recently worked out, and they share many properties with immunoglobulins. However, T cells do not respond to antigens in the same way that B cells do. For one thing, T cells cannot be stimulated by soluble antigens but only by antigens displayed on cell surfaces. Before a T cell can respond to an antigen, the antigen must be processed by an antigen-presenting cell (APC) such as a macrophage (Figure 17.15a). As we mentioned earlier, the APC displays the processed antigens on its surface. T cells also respond to MHC antigens. In fact, an important difference between the response of a T cell to an antigen and the response of a

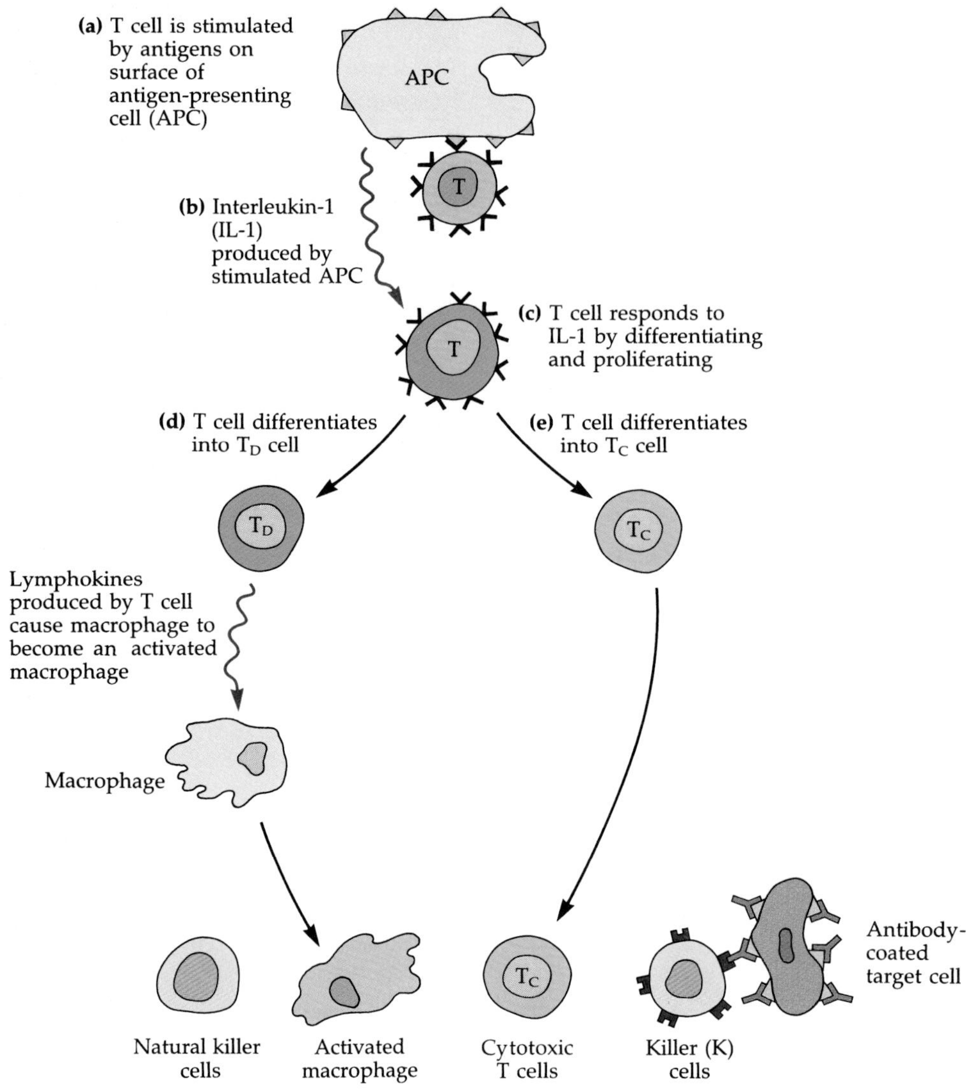
(a) T cell is stimulated by antigens on surface of antigen-presenting cell (APC)
APC
T
(b) Interleukin-1 (IL-1) produced by stimulated APC
T
(c) T cell responds to IL-1 by differentiating and proliferating
(d) T cell differentiates into T_D cell
(e) T cell differentiates into T_C cell
T_D
T_C
Lymphokines produced by T cell cause macrophage to become an activated macrophage
Macrophage
T_C
Antibody-coated target cell
Natural killer cells
Activated macrophage
Cytotoxic T cells
Killer (K) cells

FIGURE 17.15 Cell-mediated cytotoxicity. The cells responsible for cell-mediated cytotoxicity are shown in this diagram. An activated macrophage is the indirect result of antigen stimulation, but its ability to now ingest cells is not immunologically specific against that antigen. Cytotoxic T cells arise from stimulation of a T cell by an antigen and are antigenically specific against that antigen. Natural killer cells are not immunologically specific against any particular cell but attack certain foreign cells, such as tumor cells. Killer (K) cells carry determinants for the Fc region of antibody molecules. They kill antibody-coated target cells they encounter.

B cell to that antigen is that the T cell will recognize the antigen only if it is in close association with an MHC antigen. This type of recognition is known as *associative recognition*.

If the T cell were triggered by the MHC antigen alone, it would mount an immune attack on the body's own tissues. However, it is likely that any T cells with receptors that react with these self antigens are eliminated in the thymus very early in life.

Once an antigen stimulates an APC cell, the cell secretes a substance called interleukin-1 (IL-1), which is a *monokine* (Figure 17.15b). Monokines are biologically active substances secreted by macrophages. The monokine activates the T cell, which begins to synthesize interleukin-2. The T cell also synthesizes surface receptors for IL-2, which bind to the IL-2. When the receptors bind to the IL-2, the T cell begins to proliferate and differentiate into mature T cells (Figure

17.15d,e). Only T cells that have been stimulated by an antigen have receptors for IL-2. So, although the lymphokine IL-2 is nonspecific, the resulting T cells are specific to that antigen.

Killer Cells

The body's cell-mediated defense system also uses cells that are not T cells. The function of these cells might be considered an example of cell-mediated **cytotoxicity** rather than cell-mediated immunity. Lymphoid cells called **natural killer (NK) cells** are capable of destroying tumor cells. In contrast to cytotoxic T cells, they do not seem to be immunologically specific; they are simply programmed to attack a broad group of cells, most particularly tumor cells. Natural killer cells are not phagocytic, and they must contact the target cell to lyse it. Another cytotoxic cell is the **killer (K) cell.** Killer cells are also not very specific; they will attack any target cells that are coated with antibodies. Receptors on the K-cell surface are specific for the Fc region of the antibodies. Complement activity is not involved, and the precise lethal mechanism is unknown. The functions of killer cells and the other principal cells involved in cell-mediated immunity are briefly summarized in Table 17.4.

TABLE 17.4 The Principal Cells in Cell-Mediated Immunity

CELL	FUNCTION
Helper T (T_H) cell	Necessary for B-cell activation by T-dependent antigens
Delayed hypersensitivity T (T_D) cell	Provides protection against infectious agents; causes inflammation associated with tissue transplant rejection
Suppressor T (T_S) cell	Regulates immune response and helps maintain tolerance
Cytotoxic T (T_C) cell	Destroys target cells on contact
Killer (K) cell	Attacks antibody-coated target cells
Natural killer (NK) cell	Attacks and destroys target cells

Clonal Selection

We have already mentioned that an individual B or T lymphocyte is specific for only a single antigen. We have also described how the presence of an antigen leads to the proliferation and maturation (differentiation) of the B or T cell that is specific for it. Thus, an antigen in a sense *selects* the lymphocyte that will multiply to form a *clone* of cells with the identical immunological specificity. This fundamental process is called **clonal selection.**

Clonal selection was first proposed in the 1950s as a theory to explain how antigens stimulate the production of antibodies. It proved to apply equally well to both B cells and T cells. As shown in Figure 17.16, immature B cells are directly stimulated by antigen to proliferate and mature into plasma cells that secrete antibodies against the antigen, as well as into B memory cells (which will be discussed in the next section). In a similar though more complicated process, antigen that binds to a complementary receptor on the surface of an immature T cell also triggers the cell to multiply and differentiate to produce a clone of mature T cells, as well as T memory cells.

The immature B or T cell recognizes its complementary antigen by means of special cell surface molecules that fit with the antigen. What are these molecules? In the case of B cells, we know a simple and satisfying answer to our question: The receptor molecules are simply antibodies bound to the cell membrane! T-cell receptors also share many properties with immunoglobulins—in fact, they have been drawn to resemble antibodies in some diagrams in this chapter.

An important question raised by the clonal selection theory is why the immune system does not react to the body's own cells and macromolecules, a phenomenon called *tolerance*. The exact answer is still a mystery, but apparently the B and T cells that interact with self antigens are somehow destroyed during fetal development. This mechanism is called **clonal deletion.** Recently, an additional mechanism, **clonal anergy** (lack of an immune response) has been proposed. The T cells are not eliminated in the thymus but lose their ability to produce IL-2 and are not activated by antigens.

Immunological Memory

The intensity of the humoral response is reflected by the **antibody titer,** which is the amount of antibody in the serum. After the initial contact with an antigen, the exposed person's serum contains no detectable antibodies for several days. Then there is a slow rise in antibody titer; first IgM antibodies are produced, followed by IgG. Finally, a gradual decline in antibody titer occurs. This pattern is characteristic of the **primary response.** The immune responses of the host intensify after a second exposure to an antigen. This **secondary response** is also called the **memory** or **anamnestic response,** and it is explained by the clonal selection mechanism.

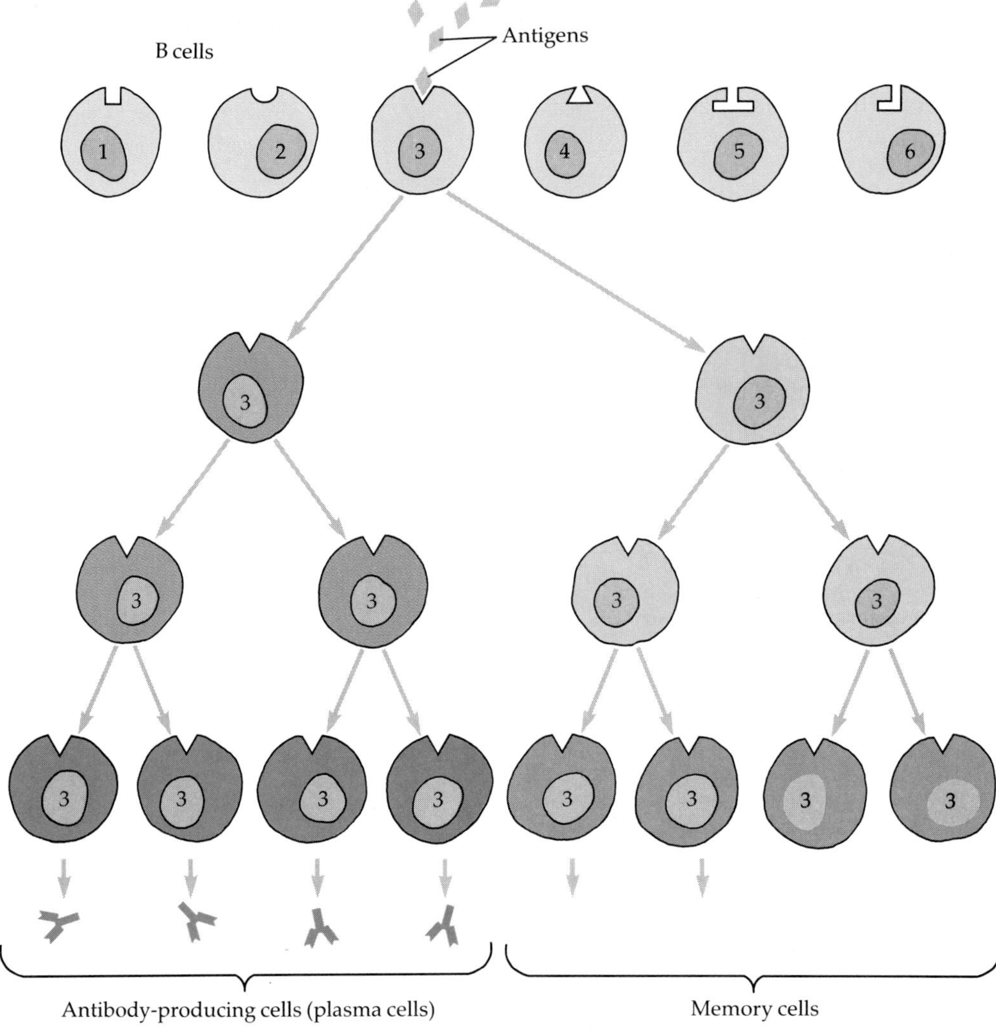

FIGURE 17.16 Clonal selection. The B cells and T cells of the body are capable of recognizing an almost infinite number of antigens, but each cell recognizes only one type of antigen. The presence of a particular antigen triggers proliferation of a cell that is specific for that antigen into a clone of cells with the same specificity. In this illustration, B cell 3 responds to the antigen by proliferating into a clone of cells, which then differentiate further into subclones of antibody-producing cells and memory cells. The antibodies that are produced will bind to the antigenic determinant that stimulated the B cell.

As we already mentioned, one component of the primary response in both humoral and cell-mediated immune responses is the production of clones of long-lived memory cells. These cells have a lifetime measured in years, or even decades. On the second or any subsequent exposure to the same antigen, these memory cells can quickly differentiate into antibody-producing plasma cells; they are thus responsible for the much faster and higher rise in antibody titer than occurred during the primary response (see Figure 17.7). The secondary response involves mainly the production of IgG antibodies.

Probably the most familiar clinical example of the anamnestic response is seen after a tetanus toxoid vaccine is injected into someone who has received a serious cut, puncture wound, or bite. Individuals who have an established immunity to tetanus (most children acquire this in their vaccination series) display a secondary response to the booster injection of tetanus toxoid. This response renews protection against tetanus quickly enough to be effective against any toxin that might be produced if *Clostridium tetani* bacteria infect the wound. For individuals without previous immunity, the response is too slow to be effective. (For these persons, immune serum globulin can provide passive, temporary immunity.)

Memory in cellular immunity is mainly important in distinguishing between self and nonself. Certain mature T cells seem to retain a memory of antigens of the host and use it to distinguish between host cells that they tolerate and foreign cells that they attempt to eliminate.

In the next chapter, we will discuss different types of immunization and explain how they stimulate the immune system to prevent disease.

STUDY OUTLINE

Introduction (p. 426)

1. An individual's genetically predetermined resistance to certain diseases is called innate resistance.
2. Individual resistance is affected by sex, age, nutritional status, and general health.
3. An individual may develop or acquire immunity after birth.
4. Immunity is the ability of the body to specifically counteract foreign organisms or substances.

Types of Acquired Immunity (pp. 427–428)

1. Acquired immunity is specific resistance to infection developed during the life of the individual.
2. Acquired immunity results from the production of antibodies and sensitized lymphocytes.

NATURALLY ACQUIRED IMMUNITY (p. 427)

1. Immunity resulting from infection is called naturally acquired active immunity; this type of immunity may be long lasting.
2. Antibodies transferred from a mother to a fetus (placental transfer) or to a newborn in colostrum results in naturally acquired passive immunity in the newborn; this type of immunity can last up to a few months.

ARTIFICIALLY ACQUIRED IMMUNITY (pp. 427–428)

1. Immunity resulting from vaccination is called artificially acquired active immunity and can be long lasting.
2. Vaccines can be prepared from attenuated, inactivated, or killed microorganisms and toxoids.
3. Artificially acquired passive immunity refers to humoral antibodies acquired by injection; this type of immunity can last for a few weeks.
4. Antibodies made by a human or other mammal may be injected into a susceptible individual.
5. Serum containing antibodies is often called antiserum.
6. When serum is separated by electrophoresis, antibodies are found in the gamma fraction of the serum (gamma globulin or immune serum globulin).

The Duality of the Immune System (pp. 428–429)

THE HUMORAL IMMUNE SYSTEM (p. 428)

1. The humoral immune system involves antibodies that are found in blood plasma and lymph.
2. Antibodies are produced by B cells in response to a specific antigen.
3. Antibodies primarily defend against bacteria, viruses, and toxins in body fluids.

THE CELL-MEDIATED IMMUNE SYSTEM (p. 429)

1. The cell-mediated immune system depends on T cells and does not involve antibody production.
2. T lymphocytes are coated with antibodylike molecules.
3. Cellular immunity is primarily a response to intracellular viruses, multicellular parasites, transplanted tissue, and cancer cells.

Antigens and Antibodies (pp. 429–433)

THE NATURE OF ANTIGENS (pp. 429–430)

1. An antigen or immunogen is a chemical substance that causes the body to produce specific antibodies or sensitized T cells, which the antigen can then combine with.
2. As a rule, antigens are foreign substances; they are not part of the body's chemistry.
3. Most antigens are proteins, nucleoproteins, lipoproteins, glycoproteins, or large polysaccharides.
4. Generally, antigens have a molecular weight greater than 10,000.
5. Antibodies are formed against specific regions on the surface of an antigen called antigenic determinant groups.
6. Most antigens have many different determinants.
7. A hapten is a low-molecular-weight substance that combines with an antibody but cannot cause the formation of antibodies unless combined with another molecule.

THE NATURE OF ANTIBODIES (pp. 430–433)

1. An antibody or immunoglobulin is a protein produced by B lymphocytes in response to the presence of an antigen and is capable of combining specifically with the antigen.
2. An antibody has at least two antigen-binding sites.
3. A single bivalent antibody unit is a monomer; multivalent antibodies are composed of monomers.
4. Most antibody monomers consist of four polypeptide chains. Two are heavy chains, and two are light chains.
5. Within each chain is a variable portion (where antigen binding occurs) and a constant portion (which serves as a basis for distinguishing the classes of antibodies).
6. An antibody monomer is Y- or T-shaped; the variable regions form the tips, and the constant regions form the base and Fc fragment.
7. The Fc region can attach to a host cell or complement.

Immunoglobulin Classes (pp. 431–433)

1. IgG antibodies are the most prevalent in serum; they provide naturally acquired passive immunity, neutralize bacterial toxins, participate in complement fixation, and enhance phagocytosis.
2. IgM antibodies are involved in agglutination and complement fixation.
3. Secretory IgA antibodies protect mucosal surfaces from invasion by pathogens.
4. IgD antibodies may help initiate the immune response in B cells.
5. IgE antibodies bind to mast cells and basophils and are involved in allergic reactions.

B Cells and Humoral Immunity (pp. 433–439)

1. Humoral immunity involves antibodies dissolved in extracellular fluids.
2. Bone marrow stem cells give rise to T cells, B cells, and macrophages.

3. Stem cells that mature in lymphoid organs become B cells.
4. A mature B cell has thousands of antibodies on its surface.

B-CELL AND ANTIGEN INTERACTIONS (pp. 433–435)

1. A B cell becomes activated when an antigen reacts with antigen receptors on its surface.
2. The activated B cell produces a clone of plasma cells and memory B cells.
3. Plasma cells secrete antibody. Memory cells recognize pathogens from previous encounters.
4. T-dependent antigens are attacked by antigen-presenting cells (APC) and then presented to helper T cells. The helper T cell then activates a B cell to produce IgG, IgA, or IgE.
5. The major histocompatibility complex (MHC) consists of cell surface proteins that are unique to each individual and provide "self" antigens.
6. T-independent antigens can directly activate a B cell and primarily elicit production of IgM.

ANTIBODY–ANTIGEN BINDING (p. 435)

1. Antibodies react with antigens to form antigen–antibody complexes, which neutralize toxins, inactivate viruses, or lyse cells.

MOLECULAR BASIS OF THE DIVERSITY OF ANTIGEN RECEPTORS (p. 436)

1. The constant region of the heavy chain of each immunoglobulin class (IgG, IgM, etc.) is identical in a person. The light chains are either a kappa or lambda class peptide.
2. The constant region of an antibody is coded for by the C gene; the variable region is coded for by hundreds of V genes and 4 J genes (and 12 D genes in heavy chains).
3. When lymphocytes develop, the V genes, a J gene (D genes in heavy chains), and the C gene recombine to form an antibody gene.
4. Somatic point mutations can occur in a formed antibody gene.
5. Recombination and mutations result in the ability to produce more than 100 million different antibody molecules.

MONOCLONAL ANTIBODIES (pp. 436–439)

1. Hybridomas are produced in the laboratory by fusing a cancerous cell with an antibody-secreting plasma cell.
2. A hybridoma cell culture produces large quantities of the plasma cell's antibody, called monoclonal antibodies.
3. Monoclonal antibodies are used in serologic identification tests, to prevent tissue rejections, and to treat septic shock.

T Cells and Cell-Mediated Immunity (pp. 439–443)

1. Cell-mediated immunity is associated with lymphocytes.

THE COMPONENTS OF CELL-MEDIATED IMMUNITY (pp. 439–441)

1. T cells are responsible for cell-mediated immunity.
2. After processing in the thymus gland, T cells migrate to lymphoid tissue.
3. T cells differentiate into several types of active cells and memory cells when they are stimulated by an antigen.

Types of T Cells (pp. 439–441)

1. Helper T (T_H) cells present T-dependent antigens to B cells.
2. Delayed hypersensitivity T (T_D) cells produce lymphokines.
3. Suppressor T (T_S) cells inhibit development of plasma cells from B cells and prevent T cells from reacting to self antigens.
4. Cytotoxic T (T_C) cells lyse target cells, such as virus-infected cells and cancer cells, by release of perforin.

Lymphokines and Cytokines (p. 441)

1. T_D cells release lymphokines, such as macrophage chemotactic factor, macrophage inhibition factor, and interleukins (IL-1 and IL-2).
2. IL-2 causes proliferation and differentiation of T cells.

THE CELL-MEDIATED IMMUNE RESPONSE (pp. 441–443)

1. T cells recognize antigens associated with MHC (self) antigens on APCs.
2. The APC secretes IL-1 to activate the T cell, which secretes IL-2.

Killer Cells (p. 443)

1. Natural killer cells lyse tumor cells, and killer cells eliminate target cells that are coated with antibodies.

Clonal Selection (p. 443)

1. Many different antigen-binding lymphocytes are produced during embryonic development.
2. When stimulated by an antigen, the B cell or T cell produces a clone of cells with the same antigen specificity.

Immunological Memory (pp. 443–444)

1. The amount of antibody in serum is called the antibody titer.
2. The response of the body to the first contact with an antigen is called the primary response. It is characterized by the appearance of IgM followed by IgG.
3. Subsequent contact with the same antigen results in very high antibody titer and is called the secondary, anamnestic, or memory response. The antibodies are primarily IgG.
4. T memory cells distinguish between self and nonself.

STUDY QUESTIONS

Review

1. Define immunity.
2. Distinguish between the following sets of terms.
   (a) Nonspecific resistance and immunity
   (b) Humoral and cell-mediated immunity
   (c) Active and passive immunity
   (d) Innate resistance and acquired immunity
   (e) Natural and artificial immunity
   (f) T-dependent and T-independent antigens
3. Classify the following examples of immunity as naturally acquired active immunity, naturally acquired passive immunity, artificially acquired active immunity, or artificially acquired passive immunity.
   (a) Immunity following injection of diphtheria toxoid
   (b) Immunity following an infection
   (c) Newborn's immunity to yellow fever
   (d) Immunity following an injection of anti-rabies serum
4. Explain what an antigen is. Distinguish between immunogen and hapten.
5. Explain what an antibody is by describing the characteristics of antibodies. Diagram the structure of a typical antibody; label the heavy chain, light chain, and constant, variable, and Fc regions.
6. Discuss the clonal selection mechanism.
7. Define each of the following on the bases of location in the body and role in the immune response: IgG, IgM, IgA, IgD, IgE.
8. By means of a diagram, explain the role of T cells and B cells in immunity.
9. Assume that the gene pool for heavy chains consists of five V, four D, and four J genes. How many different heavy chains are possible from these genes? Diagram the events leading to the production of a heavy chain consisting of $V_3D_2J_3C$.
10. Explain a function for the following types of cells: T_C, T_D, T_H, and T_S. What is a lymphokine?
11. **(a)** At time *A*, the host was injected with tetanus toxoid. At time *B*, the host was given a booster dose. Explain the meaning of the areas of the curve that are marked *a* and *b*.
    (b) Identify in the graph the antibody response of this same individual to exposure to a new antigen indicated at time *B*.

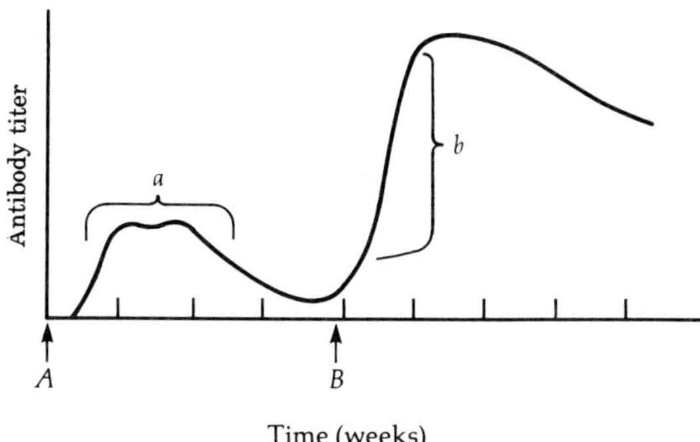

12. What effect does an antibody have on an antigen?
13. How does a T cell recognize an antigen?
14. What are natural killer cells? Killer cells?
15. How would each of the following prevent infection:
    (a) antibodies against *Neisseria gonorrhoeae* fimbriae?
    (b) antibodies against host cell mannose?
16. How are monoclonal antibodies produced?
17. Explain why a person who recovers from a disease can attend others with the disease without fear of contracting the disease.
18. Pooled human gamma globulin is sometimes administered to a patient after exposure to hepatitis. What is human gamma globulin? What type of immunity might this confer on the patient?

Challenge

1. A woman had life-threatening salmonellosis that was successfully treated with monoclonal antibodies. Why did this treatment work when antibiotics and her own immune system failed?
2. Provide an explanation for the following:
   (a) IL-2 has been used to treat pancreatic cancer.
   (b) IL-2 would exacerbate autoimmune diseases.
3. A patient with acquired immune deficiency syndrome (AIDS) has a low T_H/T_S cell ratio. What clinical manifestations would this cause?
4. A patient with chronic diarrhea was found to lack IgA in his secretions, although he had a normal level of serum IgA. What was this patient found to be unable to produce?
5. A positive tuberculin skin test shows cell-mediated immunity to *Mycobacterium tuberculosis*. How could a person acquire this immunity?
6. Newborns (under 1 year) who contract dengue have a higher chance of dying from it if their mothers had dengue prior to pregnancy. Explain why.

FURTHER READING

Ada, G.L., and G. Nossal. "The clonal selection theory." *Scientific American* 257(2):62–69, August 1987. An excellent explanation of the development and current understanding of the clonal selection theory.

Grey, H.M., A. Sette, and S. Buus. "How T cells see antigen." *Scientific American* 261(5):56–64, November 1989. Describes how antigens are processed for presentation to T cells.

Hood, L.E., I.L. Weissman, W.B. Wood, and J.H. Wilson. *Immunology*, 2nd ed. Redwood City, Calif: Benjamin/Cummings, 1984. A concise textbook emphasizing the essential concepts of immunology.

Laurence, J. "The immune system in AIDS." *Scientific American* 253(6):84–93, December 1985. A well-illustrated article on the mechanism of AIDS pathogenicity.

Lerner, R.A., and A. Tramontano. "Catalytic antibodies." *Scientific American* 258(3):58–70 March 1988. Illustrates how antigen-binding sites react with antigens and suggests new approaches to vaccines.

Pollock, R.R., J.-L. Teillaud, and M.D. Scharff. "Monoclonal antibodies: a powerful tool for selecting and analyzing mutations in antigens and antibodies." *Annual Review of Microbiology* 38:389–417, 1984. Explains production of monoclonal antibodies and discusses their use as possible vaccines.

Smith, K.A. "Interleukin-2." *Scientific American* 262(3):50–57, March 1990. Describes how interleukin-2 causes T cells to proliferate.

Tonegawa, S. "The molecules of the immune system." *Scientific American* 253(4):122–131, October 1985. Describes how billions of antibodies can be formed from a few genes.

Watson, J.D., N.H. Hopkins, J.W. Roberts, J.A. Steitz, and A.M. Weiner. *Molecular Biology of the Gene*, 4th ed. Redwood City, Calif.: Benjamin/Cummings, 1987. Chapter 23 is an excellent up-to-date summary of the molecular and genetic aspects of immunology.

Young, J.D.-E., and Z.A. Cohn. "How killer cells kill." *Scientific American* 258(1):38–44, January 1988. Shows how killer cells cause their target cancer and virus-infected cells to leak to death.

CHAPTER 18

Practical Applications of Immunology

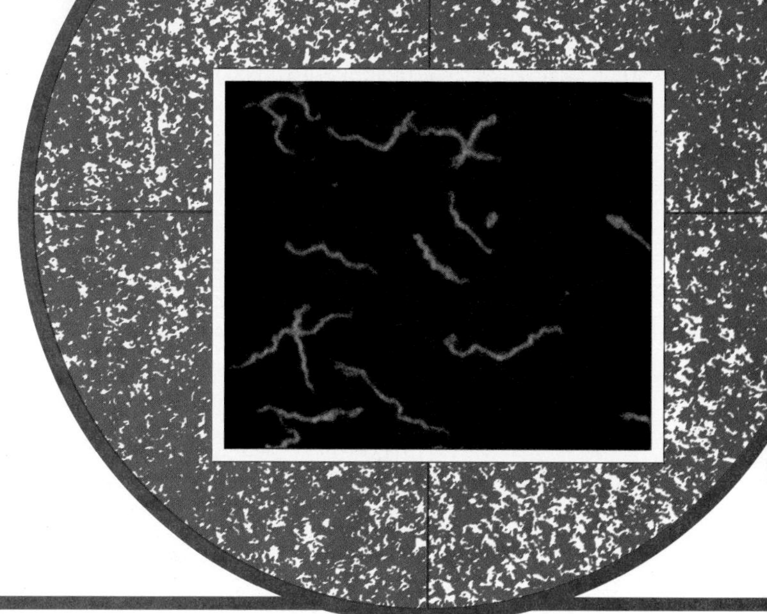

LEARNING OBJECTIVES

- Define toxoid, attenuated, subunit vaccine, acellular vaccine, conjugated vaccine, anti-idiotypic vaccine, immunodiffusion, and herd immunity.
- Explain why vaccination prevents disease.
- Differentiate between direct and indirect diagnostic tests.
- Explain how each of the following can be used in the diagnosis of a disease: precipitation, agglutination, hemagglutination-inhibition, complement fixation, neutralization, immunofluorescence, ELISA.

The science of immunology has its roots in observations about humans and disease. In this chapter, we will look at two tools that resulted from such observations—vaccines, which protect people from disease, and diagnostic tests, which tell us which microorganism is causing a disease. As we increase our understanding of the immune system at the cellular and molecular levels, scientists will be able to provide even better immunological tools.

Vaccines

Long before the invention of vaccines, it was known that people who recovered from certain diseases, such as smallpox, were immune to the disease thereafter. The Chinese may have been the first to try to exploit this phenomenon to prevent disease—they had their children inhale dried smallpox scabs. We can only speculate as to whether the results were successful or not.

In the 1700s in Turkey, Lady Montague, the wife of the British ambassador, reported that it was the practice there that ". . . the old woman comes with a nutshell full of the matter of the best sort of smallpox and asks what veins you please to have opened, . . . and puts into the vein as much venom as can lie upon the head of her needle. . . ." This practice apparently often led to only a couple of days of mild illness. The success probably depended on two factors—first, the choice of a different portal of entry for the infection, whose pathogen normally infects the body by entering through the respiratory system, and second, the use of a less virulent form of the virus. This process, which is called **variolation,** can be a useful form of protection, but it can, on occasion, backfire and kill the recipient.

One of those who received this treatment, at the age of eight, was Edward Jenner. Later in life, as a

TABLE 18.1 Principal Vaccines Used in Prevention of Bacterial Diseases in Humans

DISEASE	*VACCINE*	*RECOMMENDATION*	*BOOSTER*
Cholera	Crude fraction of *Vibrio cholerae*	For persons who work and live in endemic areas	Every 6 months as needed
Diphtheria	Purified diphtheria toxoid	See Table 18.3	Every 10 years for adults
Meningococcal meningitis	Purified polysaccharide from *Neisseria meningitidis*	For persons with substantial risk of infection	No booster effect with additional doses
Pertussis (whooping cough)	Killed *Bordetella pertussis*	Children prior to school age; see Table 18.3	For high-risk adults
Plague	Crude fraction of *Yersinia pestis*	For persons who come in regular contact with wild rodents in endemic areas	Every 6 to 12 months as needed
Pneumococcal pneumonia	Purified polysaccharide from *Streptococcus pneumoniae*	For adults with certain chronic diseases; persons over 65	No booster effect with additional doses
Tetanus	Purified tetanus toxoid	See Table 18.3	Every 10 years for adults
Tuberculosis	BCG vaccine, an attenuated strain of *Mycobacterium bovis*	For persons who are tuberculin-negative and who are exposed to tuberculosis for prolonged periods	Every 3 to 4 years as needed
Typhoid fever	Killed *Salmonella typhi*	For persons in endemic areas or areas having outbreak	Every 3 years as needed
Typhus fever	Killed *Rickettsia prowazekii*	For scientists and medical personnel in rural areas endemic for typhus	Every 6 to 12 months as needed
Hemophilus influenzae b meningitis	Capsular polysaccharide from *Hemophilus influenzae* b conjugated with protein to enhance effectiveness	See Table 18.3	At 12 or 15 months

physician, Jenner was intrigued by a dairymaid's assertion that she had no fear of smallpox because she had already had cowpox. Cowpox was a disease that caused lesions on cow udders; dairymaid's hands often became similarly infected during milking. Motivated by his childhood memory of variolation, Jenner began a series of experiments in 1798, in which he deliberately inoculated people with cowpox to prevent smallpox. This eventually led (in 1977) to the worldwide eradication of smallpox, the first disease for which this has been deliberately accomplished.

The development of conventional vaccines based on the model of the smallpox vaccine is the single most important application of immunology. Vaccines have greatly improved human health. Many pathogens transmitted by food or water can be controlled by sanitation or by antibiotics, if disease prevention fails. Viral diseases, however, are not readily treated once contracted, and transmission of viral pathogens by air or by direct contact is not easily prevented. Therefore, vaccination may be the only feasible method of controlling viral diseases. *Control* of a disease does not necessarily imply that everyone is immune to it. If most of the population is immune, outbreaks are limited to sporadic cases because there are not enough susceptible people to support the spread of epidemics. This is known as **herd immunity.**

CHARACTERISTICS OF VACCINES

A **vaccine** is a suspension of microorganisms (or some part or product of them) that will induce immunity in a host. These microorganisms may be either *inactivated* (killed) or only *attenuated.* In the latter case, they are still living but are so weakened or altered that they are no longer virulent; however, they will still provoke an immune response. **Toxoids** (inactivated bacterial toxins) will also induce immunity against their active forms.

Live, attenuated virus vaccines tend to mimic an actual infection and usually provide better immunity than that provided by inactivated viruses. Examples of live vaccines are the Sabin polio vaccines and those used against yellow fever, measles, rubella, and mumps. Many attenuated virus vaccines provide lifelong immunity without booster immunizations, and

TABLE 18.2 Principal Vaccines Used in Prevention of Viral Diseases in Humans

DISEASE	VACCINE	RECOMMENDATION	BOOSTER
Influenza	Inactivated virus	For chronically ill persons, especially with respiratory diseases, or for healthy persons over 65 years old	Annual
Measles	Attenuated virus	For infants 15 months old	Second dose before or during school years
Mumps	Attenuated virus	For infants 15 months old	*
Rubella	Attenuated virus	For infants 12 to 19 months old; for females of childbearing age who are not pregnant	*
Poliomyelitis	Attenuated or inactivated virus	For children, see Table 18.3; for adults, as risk to exposure warrants	*
Rabies	Inactivated virus	For field biologists in contact with wildlife in endemic areas; for veterinarians	Every 2 years
Yellow fever	Attenuated virus	For persons traveling to endemic areas; for military personnel	Every 10 years
Hepatitis B	Subunit vaccine	Homosexual males, intravenous drug abusers, health workers exposed to blood	*

*The duration of immunity is not known.

an effectiveness of 95% is not unusual. This long-term effectiveness probably occurs because the attenuated viruses tend to replicate in the body, and the original dose thereby increases considerably over time. One danger of such vaccines is that the live viruses can mutate to a virulent form, although this very rarely happens.

Viruses for vaccines may be inactivated by treatment with formalin or other chemicals. Heat is not used for this treatment because it is likely to alter the surface components of the virus and thus interfere with its ability to provoke an effective immune response. Commonly used inactivated virus vaccines include those used in humans against rabies (animals sometimes receive a live vaccine considered too hazardous for humans), influenza, and polio (the Salk polio vaccine).

In some vaccines, such as that for pneumococcal pneumonia, the antigens are the polysaccharide molecules of the bacterium's capsule. These vaccines must be readministered every few years, apparently because these antigens are less effective in stimulating antibody formation. Experience has also shown that vaccines against enteric bacterial pathogens, such as those causing cholera and typhoid, are not nearly as effective or long lived as those against viral diseases, such as measles and smallpox.

Vaccines that are effective against bacteria (including rickettsias and mycoplasmas) and against viruses have been produced, but to date no useful vaccines against chlamydias, fungi, protozoans, or helminthic parasites in humans are in use. However, researchers are working hard to develop a vaccine against malaria, which is caused by a protozoan parasite (see the box, p. 454).

The principal vaccines used to prevent bacterial and viral diseases in the United States are listed in Tables 18.1 and 18.2. Recommendations for the administration of some of them are given in Table 18.3. Travelers who might be exposed to cholera, yellow fever, or other diseases not endemic in this country will find that current inoculation recommendations are available from the U.S. Public Health Service and local public health agencies.

NEW VACCINE DEVELOPMENT

A basic problem with developing a new vaccine is the need for sufficient quantities of the organism. In some cases, this is very difficult—for example, when the pathogen does not grow in anything but a living human. The early successful vaccines used animal cultivation—for example, the vaccinia virus for smallpox was grown on the shaved bellies of calves, and the rabies virus was grown in the central nervous system of rabbits. The first vaccine against hepatitis B virus used viral antigens extracted from the blood of chronically infected humans because no other source was available. The successful development of cell culture methods for growing human viruses preceded the appearance in recent decades of the now familiar vaccines against polio, mumps, measles, and other

TABLE 18.3 Recommended Schedule for Active Immunization of Children

RECOMMENDED AGE	*IMMUNIZING AGENT*
2 months	DPT vaccine (diphtheria toxoid, pertussis vaccine, and tetanus toxoid)
	Hemophilus b conjugate vaccine (Hib)
	OPV (oral poliomyelitis vaccine), trivalent preparation
4 months	DPT vaccine
	Hib vaccine
	OPV, trivalent preparation
6 months	DPT vaccine
	Hib vaccine
	OPV, trivalent preparation (only in high-risk areas)
15 months	DPT vaccine
	OPV, trivalent preparation
	Mumps vaccine, Measles vaccine, Rubella vaccine } or combined MMR vaccine
	Hib vaccine booster
4–6 years (at or before school entry)	DPT vaccine
	OPV, trivalent preparation
14–16 years, and each 10 years throughout life	TD vaccine (tetanus and diphtheria toxoid)

human viral diseases. Although cell culture techniques have been successful in many cases in allowing the development of vaccines against human viruses, they have not answered the problem of a suitable test animal. For example, the AIDS virus can be grown in cell cultures, but presently not even monkeys develop AIDS. (Chimpanzees, which are an endangered species, become infected but do not develop the disease; the readily bred rhesus macaque monkey develops a similar but much more rapidly developing disease when infected by a variant of the AIDS virus called the simian immunodeficiency virus). It is unlikely that humans would be administered a killed version of such a lethal agent as the AIDS virus. Scientists remember the manufacturing defect in making the first polio vaccine in the 1950s. Not all the viruses were killed, and some vaccine batches contained infective viruses. Therefore, most believe that a successful AIDS vaccine will probably not contain any genetic information at all but will be genetically engineered to contain only antigenic subunits of the virus (see pp. 484–485).

Subunit Vaccines

A **subunit vaccine** uses only those antigenic fragments of a microorganism that are best suited to stimulating a strong immune response. The genes for these protein subunits can be introduced into the genome of a bacterium or yeast by the genetic-engineering techniques described in Chapter 9. These subunits are produced in quantity by the bacterium or yeast and are then harvested in pure form for use as a subunit vaccine. For example, the latest hepatitis B vaccine is a subunit vaccine produced in a yeast (see Table 18.2). The vaccinia virus used to control and eradicate smallpox has also been genetically engineered to express the antigens of other viruses.

These subunit vaccines are expected to produce fewer of the side-effects that are occasionally caused by extraneous elements in current whole-agent vaccines. They are also inherently safer because they do not contain any genes of the original organism and cannot reproduce in the recipient.

Acellular and Conjugated Vaccines

It is also possible to fragment a conventional vaccine and collect only those portions that contain the desired antigens. Since the complete cells are not used, these vaccines are called **acellular vaccines.** A new vaccine against pertussis is of this type. Some conventional vaccines based on polysaccharide antigens have enhanced effectiveness when combined with proteins such as toxoids for tetanus or diphtheria. These are **conjugated vaccines;** an example is that newly developed for *Hemophilus* b infections.

Anti-Idiotypic Vaccines

Anti-idiotypic vaccines are expected to be of importance in the near future. The basic idea of **anti-idiotypic vaccines** is that, rather than using an antigen as the vaccine, an antibody is used that mimics the shape of the antigen (and thus induces immunity) but is itself harmless. To produce such a vaccine, the first step is to make an antibody against the antigen. This antibody, called an idiotypic antibody (Ab-1), is then injected into a recipient, which responds by making an antibody against the antibody (anti-idiotypic antibody, or Ab-2). The Ab-2 antibody can now function as a vaccine since it contains an antigenic determinant similar to that of the original antigen (Figure 18.1). A subject receiving the vaccine containing Ab-2 responds by making yet a third antibody (anti-anti-idiotypic antibody, or Ab-3). If the immunized recipient encounters the original antigen, the Ab-3 antibodies will react with the antigen, destroying or inactivating it. Thus, Ab-3 antibodies confer immunity against the original antigen.

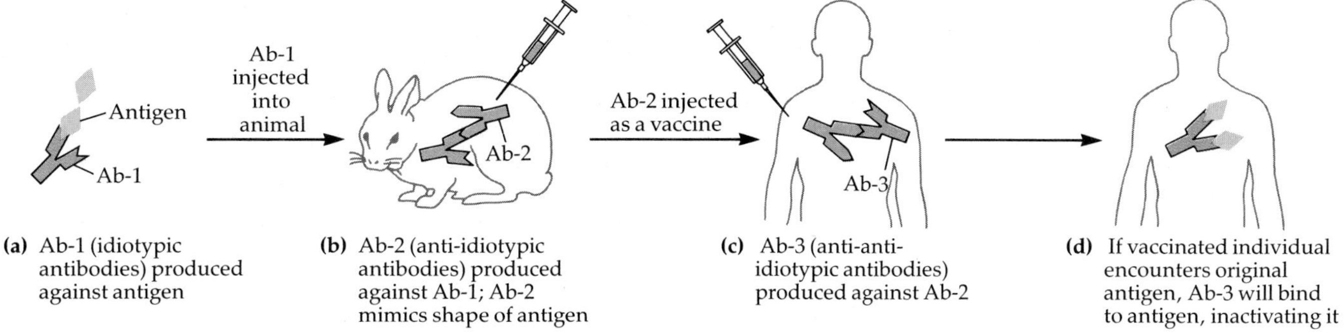
Antigen
Ab-1
Ab-1 injected into animal
Ab-2
Ab-2 injected as a vaccine
Ab-3
(a) Ab-1 (idiotypic antibodies) produced against antigen
(b) Ab-2 (anti-idiotypic antibodies) produced against Ab-1; Ab-2 mimics shape of antigen
(c) Ab-3 (anti-anti-idiotypic antibodies) produced against Ab-2
(d) If vaccinated individual encounters original antigen, Ab-3 will bind to antigen, inactivating it

FIGURE 18.1 Anti-idiotypic vaccine. **(a)** To make an anti-idiotypic vaccine against an antigen, the first step is to make an antibody against the antigen. **(b)** This first antibody, called an idiotypic antibody (Ab-1), is then injected into an animal, which produces an anti-idiotypic antibody (Ab-2). **(c)** The Ab-2 antibody is injected as a vaccine into a human, resulting in production of a third-generation anti-anti-idiotypic antibody (Ab-3). **(d)** This Ab-3 antibody will react with the original antigen if the antigen enters the immunized individual. Thus, immunization with Ab-2 provides protection against the original antigen.

This may seem like a complicated approach, but it has a number of advantages. First, no pathogenic organism, alive or dead, is injected. Second, such vaccines are highly specific because they are directed against the antigenic character of the pathogen best suited to stimulating an antigenic response. Third, certain nonprotein antigens are not able to stimulate immunity in newborn infants, but anti-idiotypic vaccines, being proteins, would be able to confer immunity upon the infant quickly. Fourth, this type of vaccine could be used to recognize the receptor sites on a cell and thus could be used to block the attachment of a virus to the cell. Such an approach has been proposed as a way of developing a potential vaccine for AIDS, but the attempt has not been successful to date.

A major problem encountered in the development of anti-idiotype technology is the source of sufficient amounts of idiotypic antibody (Ab-1). One obvious source, monoclonal antibodies, awaits the development of methods of producing these antibodies from human cells rather than from mouse cells. In Chapter 17 we discussed possible methods for doing this.

Diagnostic Immunology

More than 100 years ago, when Robert Koch was trying to develop a vaccine against tuberculosis, he observed that when tubercular guinea pigs were injected with a suspension of *Mycobacterium tuberculosis*, the site of the injection became red and slightly swollen a day or two later. Many of us will recognize this symptom as essentially the same test we use today to determine whether a person has been infected by the tuberculosis pathogen. Koch, of course, had no idea of the mechanism of cell-mediated immunity that caused this phenomenon, and he had no idea about the existence of antibodies.

Since then, immunology has given us many other invaluable diagnostic tools, most of which are based on interactions of humoral antibodies with antigens. In this section, we will discuss several of the most important techniques used to detect antigens and antibodies.

One problem our diagnostic tools must overcome is that antibodies can never be seen directly. Even at magnifications of well over 100,000×, they appear only as fuzzy, ill-defined particles. All we can do is infer their presence indirectly through a variety of reactions.

In most instances it is the presence of antibodies that is of primary interest, but the reactions are generally complementary, and known antibodies can be used to detect antigens.

PRECIPITATION REACTIONS

Precipitation reactions involve the reaction of soluble antigens with IgG or IgM antibodies (described in Chapter 17) to form large, interlocking aggregates called *lattices*. Precipitation reactions occur in two distinct stages. First, there is the rapid interaction between antigen and antibody to form small antigen–antibody complexes. This interaction occurs within

MICROBIOLOGY *IN THE NEWS*

Molecular Biology: The Latest Weapon Against Malaria

The latest microbiological and biochemical techniques of gene mapping, monoclonal-antibody production, and protein synthesis give hope that a vaccine against malaria may become available soon. Traditional techniques for creating vaccines against viruses and bacteria are ineffective against larger and more complex organisms, such as protozoans.

The need for an antimalarial vaccine has become desperate. In the early 1960s, we thought that malaria would soon disappear completely from America and other temperate-zone countries and would become relatively rare even in tropical countries. The World Health Organization planned to reduce the number of *Anopheles* mosquitoes with massive sprayings of DDT and planned to treat already infected people with a variety of drugs related to quinine.

But that battle has gone against us. The mosquitoes became resistant to DDT, and *Plasmodium* has begun to develop resistance to antimalarial drugs. About 150 million new cases of malaria occur every year throughout the world. Moreover, the number of malaria cases in the United States has increased dramatically because of increased immigration from tropical countries and increased travel to tropical countries.

Researchers headed by Ruth and Victor Nussenzweig at New York University Medical Center identified a surface antigen of the sporozoite stage of all *Plasmodium* species.

The discovery of this antigen, called circumsporozoite (CS), has led to the production of monoclonal antibodies against it, which can be used in immunologic experiments. In 1983, the CS gene was cloned, making it possible to produce large quantities of the CS protein for analysis and antibody preparation.

The first human tests of vaccines were conducted in 1987 at the University of Maryland School of Medicine and Walter Reed Army Institute of Research. Each institution tested a different vaccine preparation. The trials were concluded in 1989, and their results were disappointing. Only one-third of the volunteers produced antibodies, and the vaccines did not initiate a memory response. The Nussenzweig team is now determining characteristics of vaccines that will stimulate B cells, T_H cells, and T_C cells that produce γ interferon, which inhibits liver-stage parasites.

Research leading to a generally effective malaria vaccine is hindered by the fact that the life cycle of *Plasmodium* is so complex (see Figure 12.22). The CS vaccines were directed against the sporozoite, which is injected into the victim by the mosquito. The immune response that is triggered by the vaccine must kill all the sporozoites within 30 minutes after the mosquito bites its victim, before the sporozoites enter the protection of the liver cells. A separate vaccine would be needed to provoke an immune response against the merozoites, which destroy red blood cells and cause the typical symptoms of malaria. Yet a third vaccine would be needed to destroy the gametocytes, which continue the life cycle within the mosquito's body.

Research on vaccines to attack the merozoites and gametocytes is moving ahead at research centers worldwide. Researchers in Colombia believe they have isolated a surface antigen present on merozoites. Monoclonal antibodies against this protein did inhibit in vitro development of the parasite. The Centers for Disease Control is running trials on monkeys.

If experiments in developing a malaria vaccine are successful, it will be the first antiparasitic vaccine for humans and will provide a model for the development of vaccines against other parasitic diseases, such as trypanosomiasis and schistosomiasis.

seconds and is followed by a slower reaction, which may take minutes to hours, in which the antigen–antibody complexes form lattices that precipitate from solution. Precipitation reactions normally occur only when there is an optimal ratio of antigen to antibody. Figure 18.2 shows that no visible precipitate forms when there is an excess of either one. The optimal ratio is produced when separate solutions of antigen and antibody are placed adjacent to each other and allowed to diffuse together. In a **precipitin ring test** (Figure 18.3), a cloudy line of precipitation (ring) appears in the area in which the optimal ratio has been reached (the *zone of equivalence*).

Immunodiffusion tests are precipitation reactions carried out in an agar gel medium. In one such test, the **Ouchterlony test,** wells are cut into a purified agar gel in a Petri plate. A serum containing antibodies (antiserum) is added to one well, usually centrally located, and soluble test antigens are added to each surrounding well. A line of visible precipitate develops between the wells at the point where the optimal antigen–antibody ratio is reached. The Ouchterlony test is most useful in determining whether the antigens are identical, partially identical, or totally different (Figure 18.4).

Other precipitation tests do not depend entirely on the passive diffusion of antigen and antibody in a gel

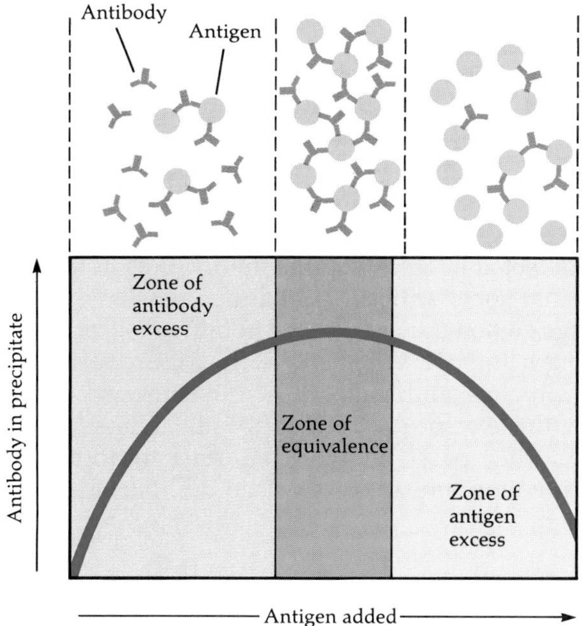

FIGURE 18.2 Precipitation curve. The curve is based on the ratio of antigen to antibody. The maximum amount of precipitate forms in the zone of equivalence, where the ratio is optimal.

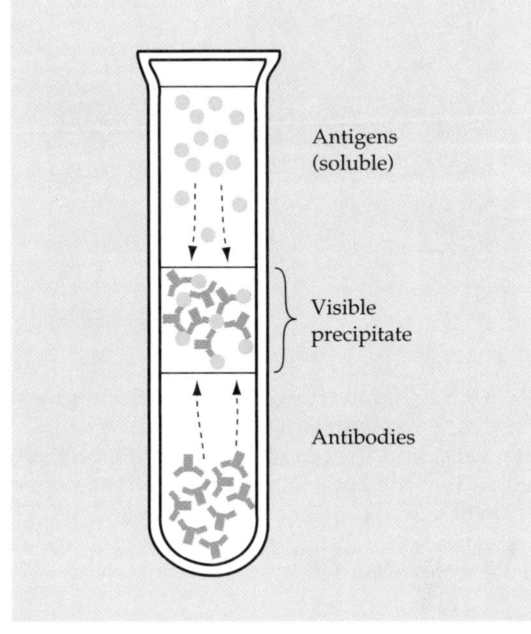

FIGURE 18.3 Precipitin ring test. A diagrammatic representation of a precipitin ring test. Antigens and antibodies diffuse toward each other in a tube of small diameter and form a visible line or ring where the zone of equivalence is reached.

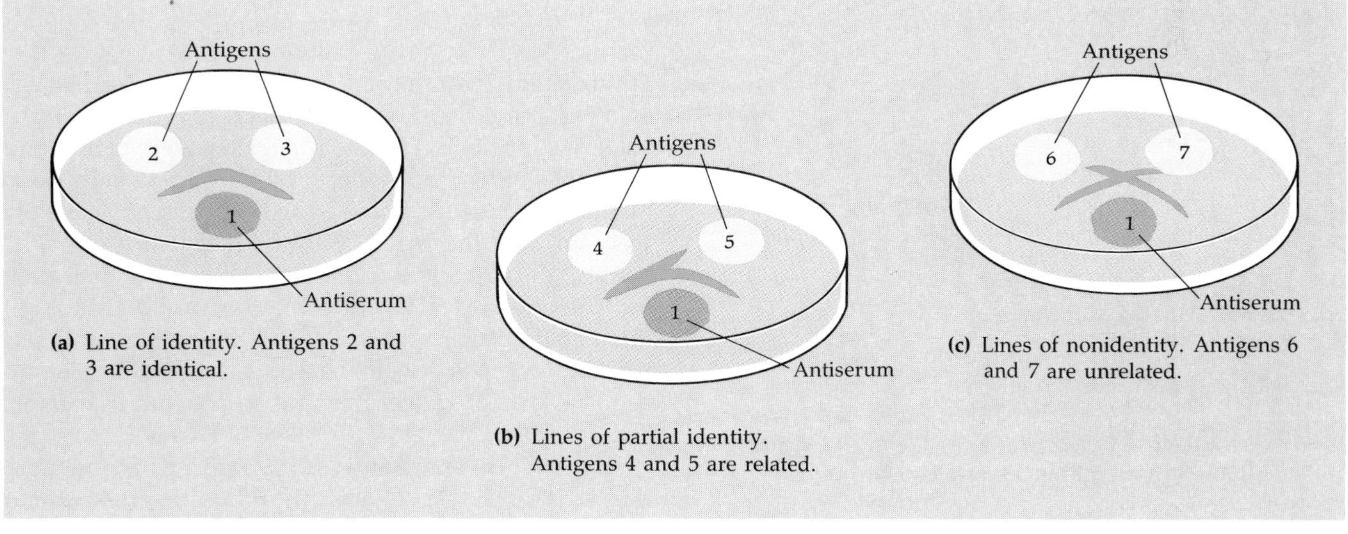

FIGURE 18.4 The Ouchterlony immunodiffusion precipitation test. **(a)** Line of identity. Neither antigen (in wells 2 and 3) nor antibody molecules (in well 1) are able to diffuse past each other because they react and precipitate to form the line (shown here in pink). Therefore, wells 2 and 3 contain the same antigen. **(b)** Lines of partial identity. The antigens are not identical, but they share many antigenic determinant sites and are therefore related. The spur shape indicates that some of the antigens did not react with antibody and diffused through the precipitation zone. **(c)** Lines of nonidentity. The antiserum contains antibodies against antigens in wells 6 and 7. The antigens are not related, because they diffused across each other's zones of precipitation.

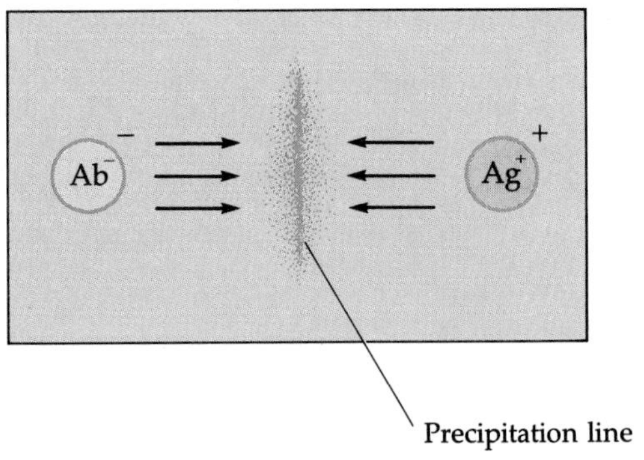

FIGURE 18.5 Countercurrent immunoelectrophoresis. Antigens (Ag^+) and antibodies (Ab^-) are placed in opposite wells, and the pH of the surrounding medium is adjusted so that the antigen and antibody have opposite charges. When an electrical current is applied, antigens and antibodies move in opposite directions. A line of precipitate forms where antigen and antibody meet.

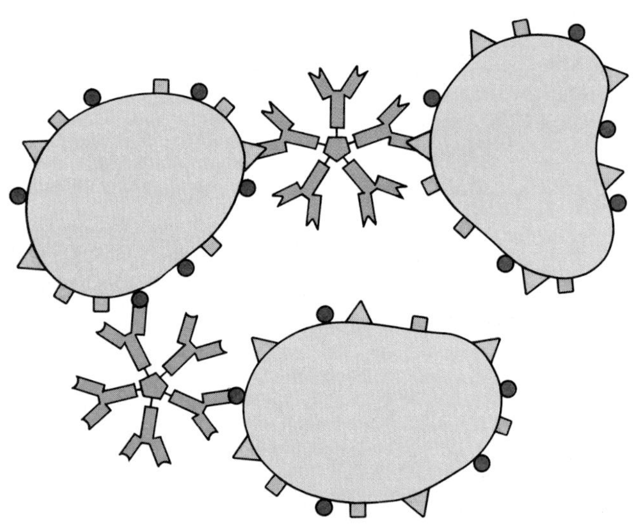

FIGURE 18.6 Agglutination reaction. When antibodies react with antigenic determinant sites on neighboring antigens, such as bacteria or red blood cells, the particulate antigens agglutinate. IgM, the most efficient immunoglobulin for agglutination, is shown here, but IgG also participates in agglutination reactions.

but instead use electrophoresis to speed up their movement. Protein mixtures can be separated rapidly, sometimes in less than an hour, with this method. A modification of the precipitation reaction combines the techniques of immunodiffusion and electrophoresis in a procedure called **immunoelectrophoresis.** This procedure is used in research to separate proteins in human serum and is the basis of certain diagnostic tests. The **countercurrent immunoelectrophoresis test,** also called **counter-immunoelectrophoresis (CIE),** can be used in the diagnosis of bacterial meningitis and other diseases. A test serum known to contain certain antibodies is used to identify an unknown antigen in some body fluid (Figure 18.5). For suspected cases of meningitis, the antigens associated with the pathogen are detected in cerebrospinal fluid. CIE is based on the fact that some antigens and antibodies have opposite charges when they are placed in buffers of correct ionic strength and pH. When an electrical current is applied, the antigens and antibodies move toward the pole with the electrical charge opposite to the one they carry. They pass through each other to do so, and a precipitation line appears within an hour if a reaction occurs.

AGGLUTINATION REACTIONS

Whereas precipitation reactions involve *soluble* antigens, agglutination reactions involve *particulate* antigens or soluble antigens adhering to particles. These antigens can be linked together by antibodies to form visible aggregates, a process called **agglutination** (Figure 18.6). Agglutination reactions are very sensitive, relatively easy to read, and available in great variety. Agglutination tests are classified as either *direct* or *indirect*.

Direct Agglutination Tests

Direct agglutination tests detect antibodies against relatively large cellular antigens, such as red blood cells, bacteria, and fungi. At one time, they were carried out in a series of test tubes, but now they are usually done in plastic *microtiter plates,* which have many shallow wells to take the place of individual test tubes. The amount of particulate antigen in each well is the same, but the amount of serum containing antibodies is diluted so that each successive well has half the antibodies of the previous well. These tests, for example, are used to test for brucellosis and to separate *Salmonella* isolates into serological types.

Clearly, the more antibody we start with, the more dilutions it will take to lower the amount to the point at which agglutination does not occur. This is the measure of **titer,** or concentration of serum antibody (Figure 18.7). For infectious diseases in general, the higher the serum antibody titer, the greater the immunity to the disease. The titer alone is of limited use in diagnosing an existing illness. There is no way of knowing whether the measured antibodies were generated in response to the immediate infection or to an earlier illness. For diagnostic purposes, a **rise in titer** is significant; that is, the titer is higher later in the disease than at its onset. If it is possible to demonstrate that the person's blood had no antibody titer before the illness but has a significant titer while the disease is progressing, this change in titer, called **seroconversion,** is also diagnostic.

Indirect (Passive) Agglutination Tests

Antibodies against soluble antigens can be detected by agglutination tests if the antigens are adsorbed onto particles such as red blood cells, bentonite clay, or minute latex spheres (Figure 18.8). Such tests, known as latex agglutination tests, are commonly used for rapid detection of antibodies against the streptococci causing sore throats. In such **indirect** or **passive agglutination tests,** the antibody reacts with the soluble antigen adhering to the particles. The particles then agglutinate with one another much as particles do in the direct agglutination tests. The same principle can be applied in reverse by using particles coated with antibodies to detect the antigens against which they are specific. A test to detect the antigen in blood that indicates infection by the hepatitis B virus makes use of antibody-coated red blood cells.

Hemagglutination

When agglutination reactions involve the clumping of red blood cells, the reaction is called **hemagglutination.** These reactions are used routinely in blood typing (see Chapter 19) and in the diagnosis of infectious mononucleosis.

Certain viruses, such as those causing mumps, measles, and influenza, have the ability to agglutinate red blood cells; this process is called **viral hemagglutination.** If a person's serum contains antibodies against such viruses, these antibodies will react with the viruses and neutralize them (Figure 18.9). For example, if hemagglutination occurs in a mixture of measles virus and red blood cells but does not occur when the patient's serum is added to the mixture, this sequence indicates that the serum contains antibodies that have bound to and neutralized the measles virus. This **hemagglutination inhibition test** is widely used in the diagnosis of influenza, measles, mumps, and a number of other viral infections.

COMPLEMENT FIXATION REACTIONS

In Chapter 16, we discussed a group of serum proteins collectively called complement. During most antigen–antibody reactions, the complement binds to the antigen–antibody complex and is used up, or fixed. This process of **complement fixation** can be used to detect very small amounts of antibody. Antibodies that do not produce a visible reaction, such as precipitation or agglutination, can be demonstrated by the fixing of complement during the antigen–antibody reaction. Complement fixation was once used in the diagnosis

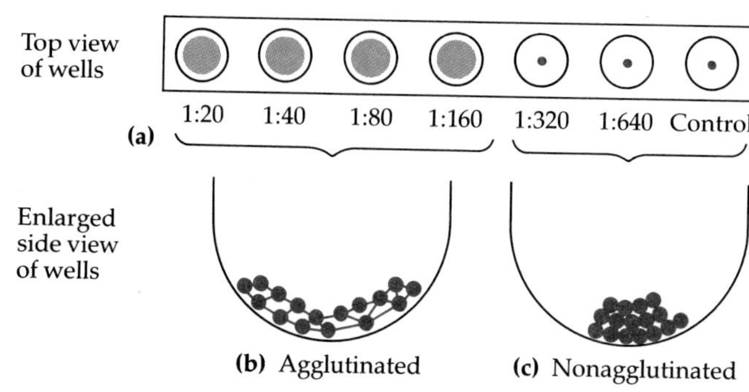

FIGURE 18.7 Direct agglutination test. (a) Each well in this microtiter plate contains, from left to right, only half the concentration of serum in the preceding well. Each well contains the same concentration of particulate antigens, such as red blood cells. **(b)** In a positive (agglutinated) reaction, sufficient antibodies are present in the serum to link the antigens together, forming an antibody–antigen mat that sinks to the bottom of the well. **(c)** In a negative (nonagglutinated) reaction, insufficient antibodies are present to cause the linking of antigens. The particulate antigens roll down the sloping sides of the well, forming a pellet at the bottom. In this example, the antibody titer is 160 since the well with a 1:160 concentration is the most dilute concentration that gives a positive reaction.

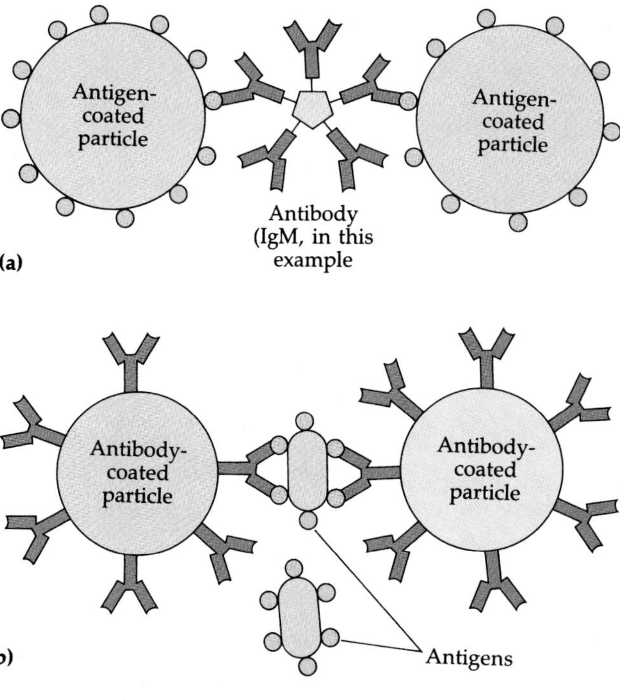

FIGURE 18.8 Indirect agglutination test. These tests are performed using antigens or antibodies coated onto particles such as red blood cells or minute latex spheres. **(a)** IgM molecule crosslinking two antigen-coated particles in a test to detect antibodies. **(b)** When particles are coated with monoclonal antibodies, agglutination indicates the presence of antigens.

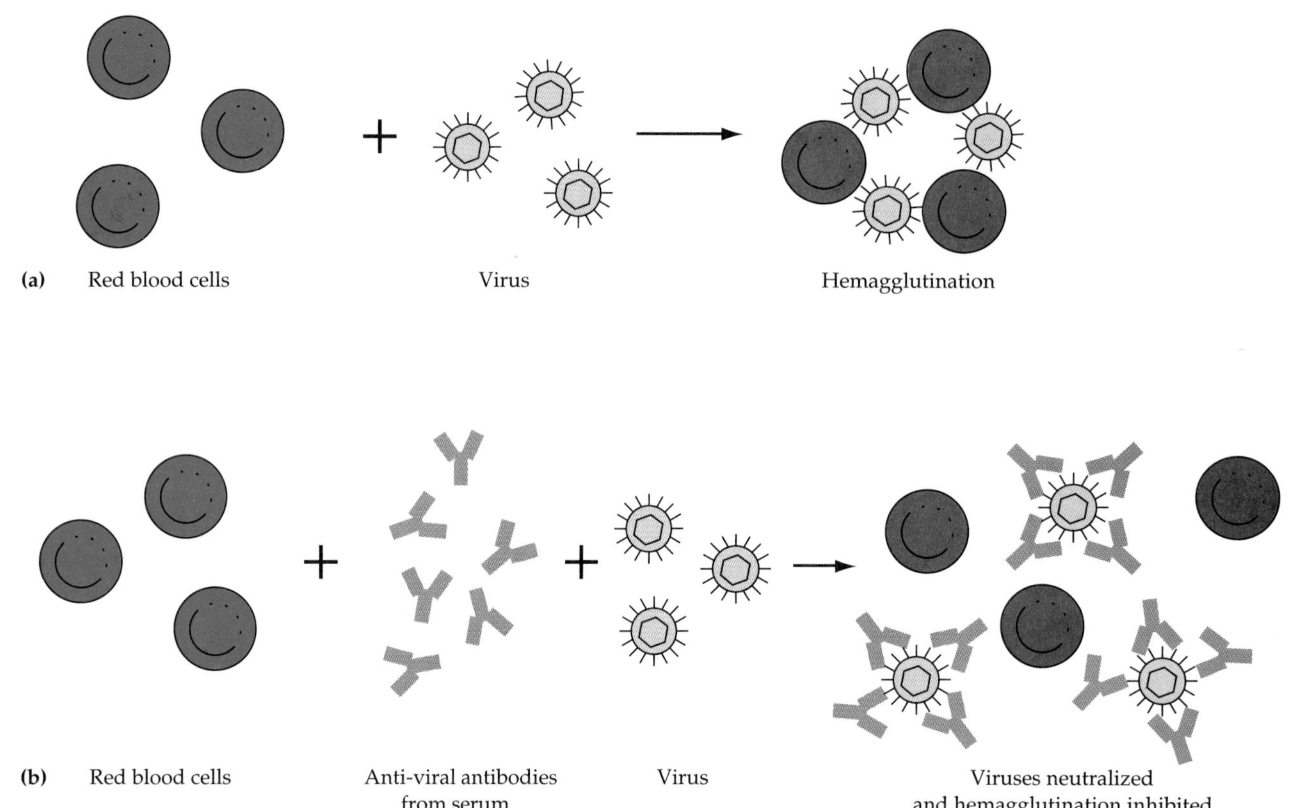

FIGURE 18.9 Viral hemagglutination. (a) Viral hemagglutination is not an antigen–antibody reaction. It simply happens that certain viruses bind to red blood cells and cause their agglutination. **(b)** If patient serum that contains antibodies to the virus is mixed with the red blood cells and the virus, the antibodies will neutralize the virus and inhibit hemagglutination—demonstrating the presence of antibodies to the virus.

of syphilis (Wasserman test) and is still used in the diagnosis of certain viral, fungal, and rickettsial diseases.

Execution of the complement fixation test requires great care and good controls. The test is performed in two stages (Figure 18.10). First, the subject's serum must be heated at 56°C for 30 minutes to inactivate its complement. The inactivated serum is then diluted and mixed with a specific amount of known antigen and fresh complement. The mixture is next incubated for about 30 minutes. An antigen–antibody reaction cannot be observed at this point. To determine whether the complement present in the mixture is fixed by an antigen–antibody reaction, which is evidence of the presence of antibodies specific for the antigen, another stage of the test must be performed.

The second stage uses an indicator system to determine whether complement is free or combined (fixed). The indicator system consists of sheep red blood cells and specific antibodies that will attach to their surfaces. The exposure of these "sensitized" cells to complement causes lysis of the red blood cells (hemolysis), which changes the color of the mixture. Therefore, if complement has been fixed by an antigen–antibody reaction during the first stage, it is not available to cause blood cell lysis in the second stage; the test is positive (Figure 18.10a). But if the complement has not been fixed during the first stage, then it is available to cause blood cell lysis in the second stage (Figure 18.10b). This negative test result means that no antibodies specific for the test antigen are present in the patient's serum.

NEUTRALIZATION REACTIONS

Neutralization reactions are antigen–antibody reactions in which the harmful effects of a bacterial exotoxin or a virus are eliminated by specific antibodies.

Neutralization reactions are used both in disease treatment and in diagnostic tests, especially for virus. These reactions were first described in 1890, when investigators observed that immune serum could neutralize the toxic substances produced by *Corynebacterium diphtheriae.* Such a neutralizing substance is called an **antitoxin.** This is a specific antibody produced by a host as it responds to a bacterial exotoxin or its corresponding toxoid (inactivated toxin). The antitoxin combines with the exotoxin to neutralize it (Figure 18.11a). Antitoxins produced in an animal can be injected into humans to provide passive immunity against a toxin. Antitoxins from horses are routinely used for prevention or treatment of diphtheria and botulism; tetanus antitoxin is usually of human origin.

Neutralization tests are used in the diagnosis of viral infections. The body responds to such infections by producing specific antibodies that bind to receptor sites on the viral surface. The binding of these antibodies prevents the virus from attaching to a host cell, and thus destroys the virus's infectivity (Figure 18.11b). Viruses that exhibit their cytopathic (cell-damaging) effects in cell culture or embryonated eggs can be used to detect the presence of neutralizing viral antibodies. If the serum to be tested contains antibodies against the particular virus, the antibodies will prevent that virus from infecting cells in the cell culture or eggs, and no cytopathic effects will be seen. The inability of a specific virus to cause cytopathic effects in the presence of immune serum can thus be used to determine the identity of a virus as well as to find the viral antibody titer.

In vitro neutralization tests are not common in modern clinical laboratories. A more frequent form of neutralization test is a skin test such as the **Schick test,** which determines the status of a person's immunity to diphtheria. A small amount of diphtheria exotoxin is inoculated into the skin. If there is sufficient serum antitoxin to neutralize the exotoxin, there is no visible reaction, indicating that the person would be immune to diphtheria. If antitoxin is insufficient, the exotoxin damages the tissues at the site of entry and produces a swollen, tender, reddish area that turns brown in four to five days.

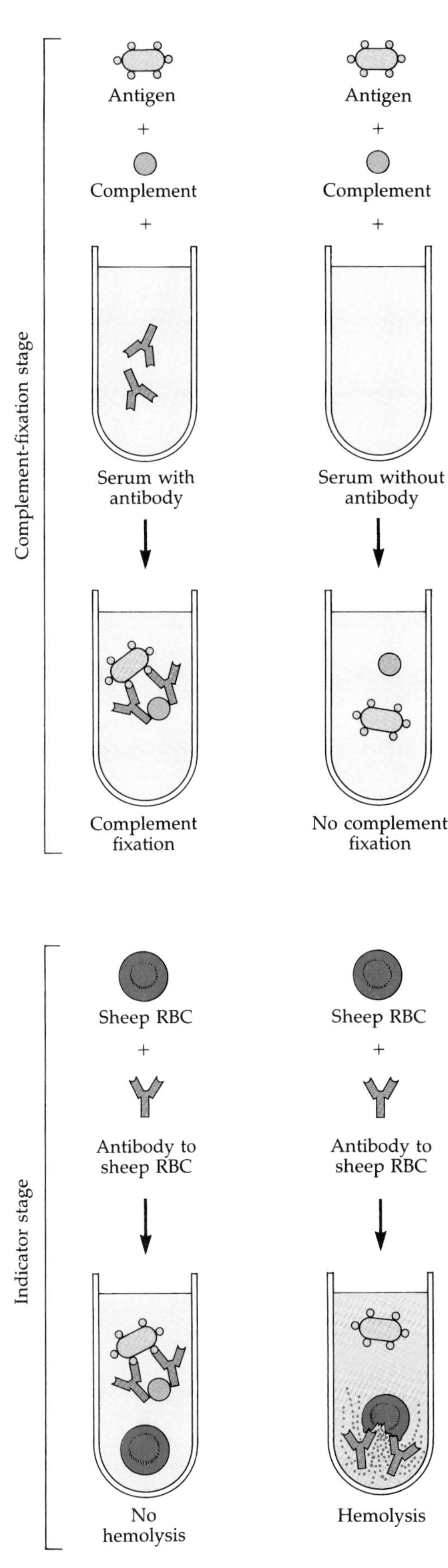

FIGURE 18.10 Complement fixation test. Complement will combine (be fixed) with the antibody that is reacting with an antigen. If all the complement is fixed in the complement-fixation stage, then none will remain to cause hemolysis of the red blood cells in the indicator stage in the diagrams. **(a)** All available complement is fixed by the antigen–antibody reaction; no hemolysis occurs, so the test is positive. **(b)** No antigen–antibody reaction occurs. The complement remains, and the red blood cells are lysed in the indicator test, so the test is negative.

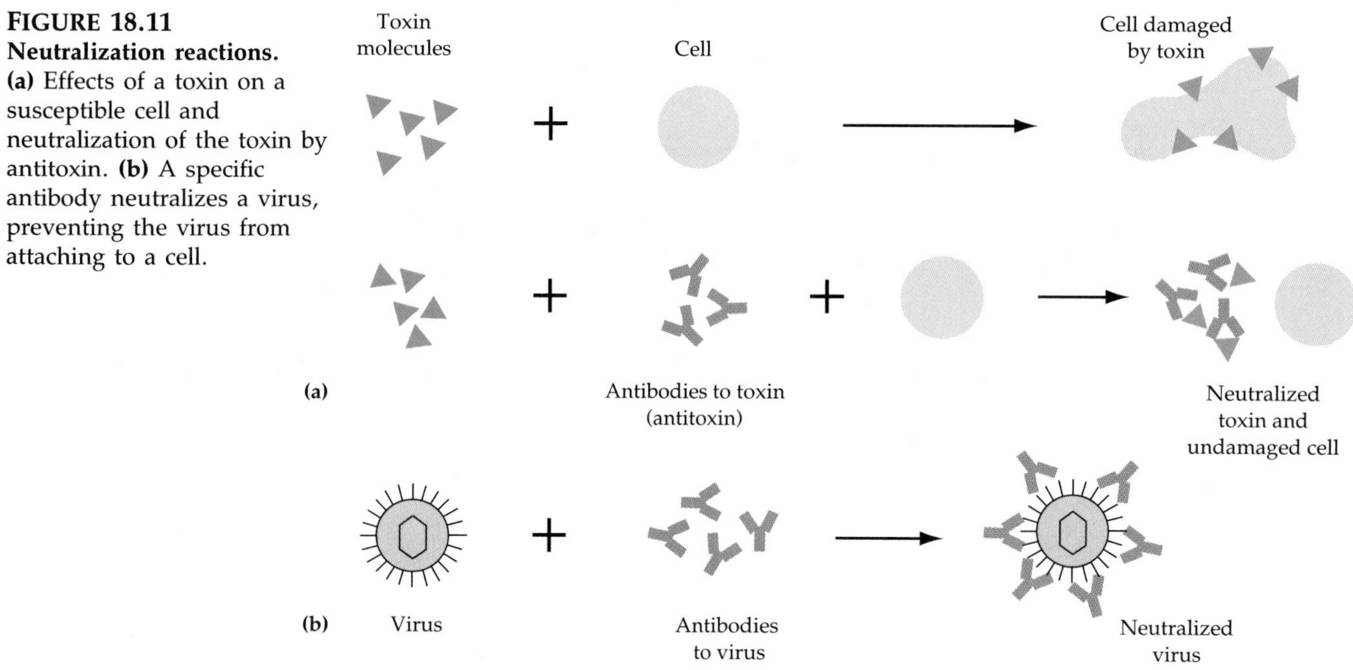

FIGURE 18.11 Neutralization reactions. **(a)** Effects of a toxin on a susceptible cell and neutralization of the toxin by antitoxin. **(b)** A specific antibody neutralizes a virus, preventing the virus from attaching to a cell.

FIGURE 18.12 Fluorescent-antibody (FA) techniques. **(a)** A direct FA test to identify group A streptococci. **(b)** In an indirect FA test, such as that used in the diagnosis of syphilis, the fluorescent dye is attached to anti-human gamma globulin, which reacts with any human immunoglobulin (such as the *Treponema pallidum*-specific antibody) that has previously reacted with antigen. **(c)** The reaction is viewed through a fluorescence microscope, and the antigen with which the dye-tagged antibody has reacted—in this case *T. pallidum*—glows in the ultraviolet illumination.

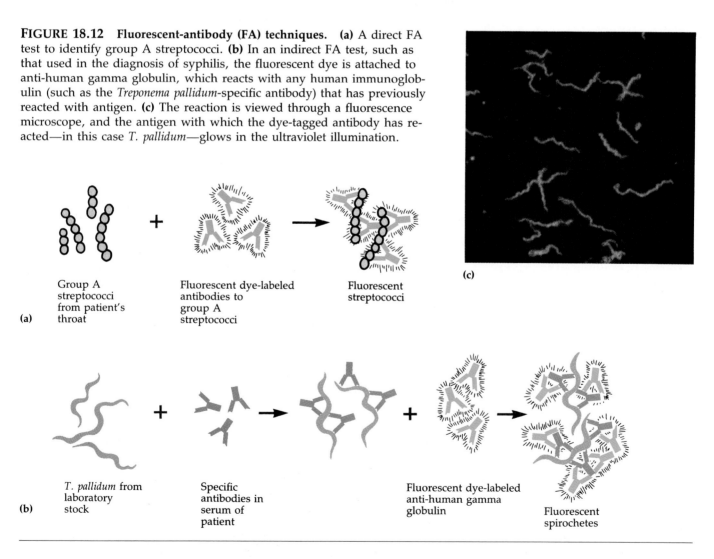

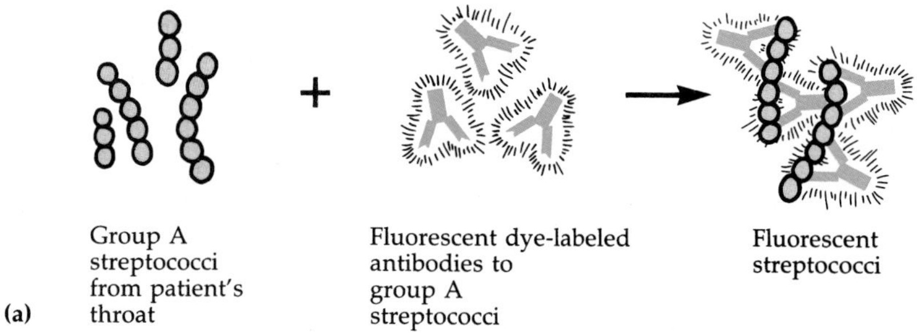

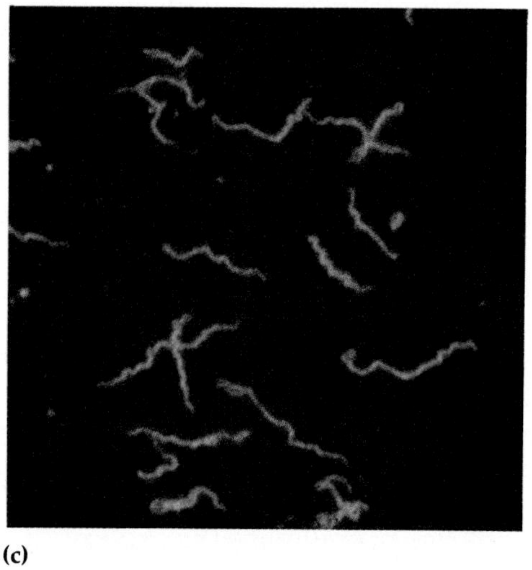

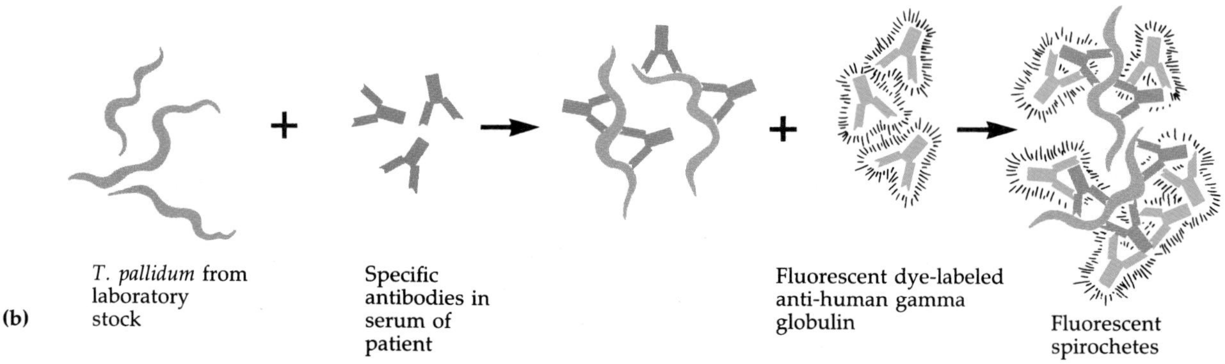

(a) Group A streptococci from patient's throat

Fluorescent dye-labeled antibodies to group A streptococci

Fluorescent streptococci

(c)

(b) *T. pallidum* from laboratory stock

Specific antibodies in serum of patient

Fluorescent dye-labeled anti-human gamma globulin

Fluorescent spirochetes

IMMUNOFLUORESCENCE AND FLUORESCENT-ANTIBODY TECHNIQUES

Fluorescent-antibody (FA) techniques (Figure 18.12) can identify microorganisms in clinical specimens and detect the presence of a specific antibody in serum. These techniques use fluorescent dyes, such as *fluorescein isothiocyanate* (FITC), which are combined with antibodies to make them fluoresce when exposed to ultraviolet light. These procedures are quick, sensitive, and very specific; the fluorescent-antibody test for rabies can be done in a few hours and has an accuracy rate close to 100%.

Fluorescent-antibody tests are of two types, direct and indirect. **Direct FA tests** are usually used to identify a microorganism in a clinical specimen. During this procedure, the specimen containing the antigen to be identified is fixed onto a slide. Fluorescein-labeled antibodies are then added, and the slide is incubated briefly. The slide is washed to remove any antibody not bound to antigen and is then examined under the ultraviolet microscope for yellow-green fluorescence (Figure 18.12a).

Indirect FA tests (Figure 18.12b) are used to detect the presence of a specific antibody in serum following exposure to a microorganism. During this procedure, a known antigen is fixed onto a slide. The test serum is then added, and, if antibody that is specific to that microbe is present, it reacts with the antigen to form a bound complex. So the antigen–antibody complex can be seen, fluorescein-labeled anti-human gamma globulin (anti-HGG), an antibody that reacts specifically with human antibody, is added to the slide. After the slide has been incubated and washed (to remove unbound antibody), it is examined under a fluorescence microscope. If the known antigen fixed to the slide appears fluorescent, the antibody specific to the test antigen is present.

An especially interesting adaptation of fluorescent antibodies is the **fluorescence-activated cell sorter (FACS).** In the previous chapter we learned that T cells carry antigenically specific receptors such as CD4 and CD8 on their surface, and these are characteristic of certain groups of T cells. The FACS is a modification of a *flow cytometer*, in which a suspension of cells leaves a nozzle as droplets containing no more than one cell each. Rapid vibration of the nozzle encourages formation of such droplets. A laser beam strikes the cell-containing droplet. The beam is then received by a detector as shown in Figure 18.13, which determines certain characteristics such as size. If the cells carry fluorescent antibody markers to identify them as CD4 or CD8 T cells, a detector can measure this fluorescence. Equipped with such a detector, the flow cytometer becomes a FACS. As the laser beam detects a cell of a preselected size or fluorescence, an electric charge, either positive or negative, can be imparted to it. As the charged droplet falls between electrically charged plates, it is moved to fall into one receiving tube or another, effectively separating cells of different types. Millions of cells can be separated in an hour, all under

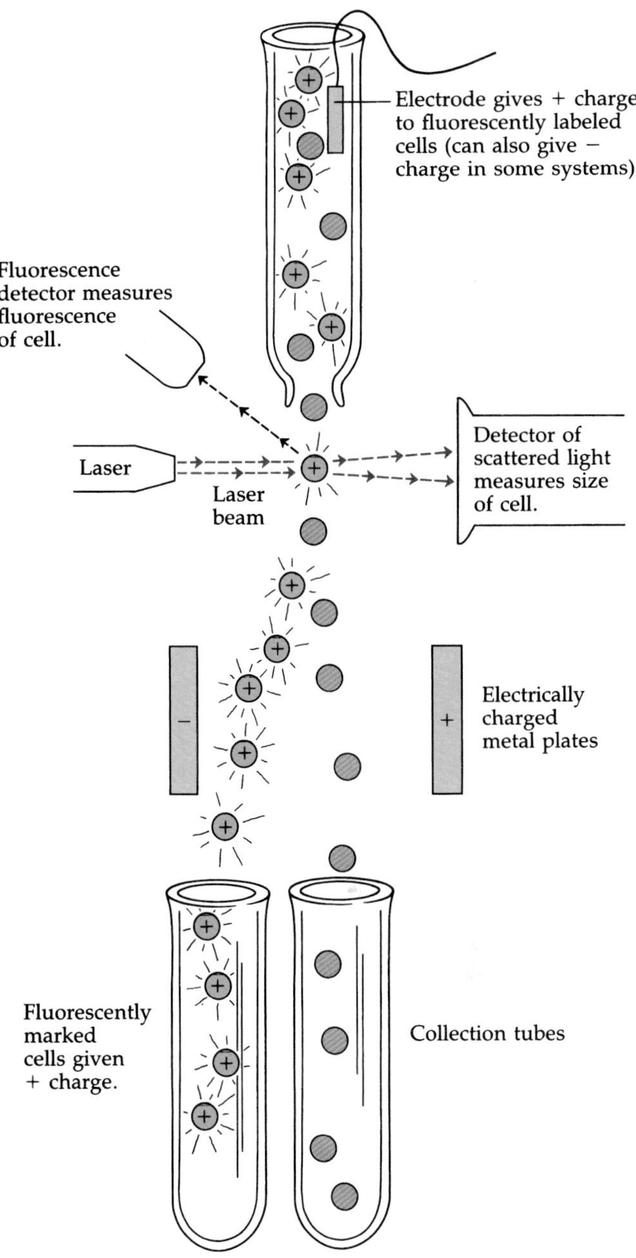

FIGURE 18.13 Fluorescence-activated cell sorter. This can be used to separate different classes of T cells. A fluorescence-labeled antibody reacts with, for example, the CD4 receptor on a T cell. When the laser beam strikes such a labeled cell, an electrical charge is applied to the droplet carrying this cell. As the charged cell falls between the charged metal plates it changes direction and is collected in an assigned tube.

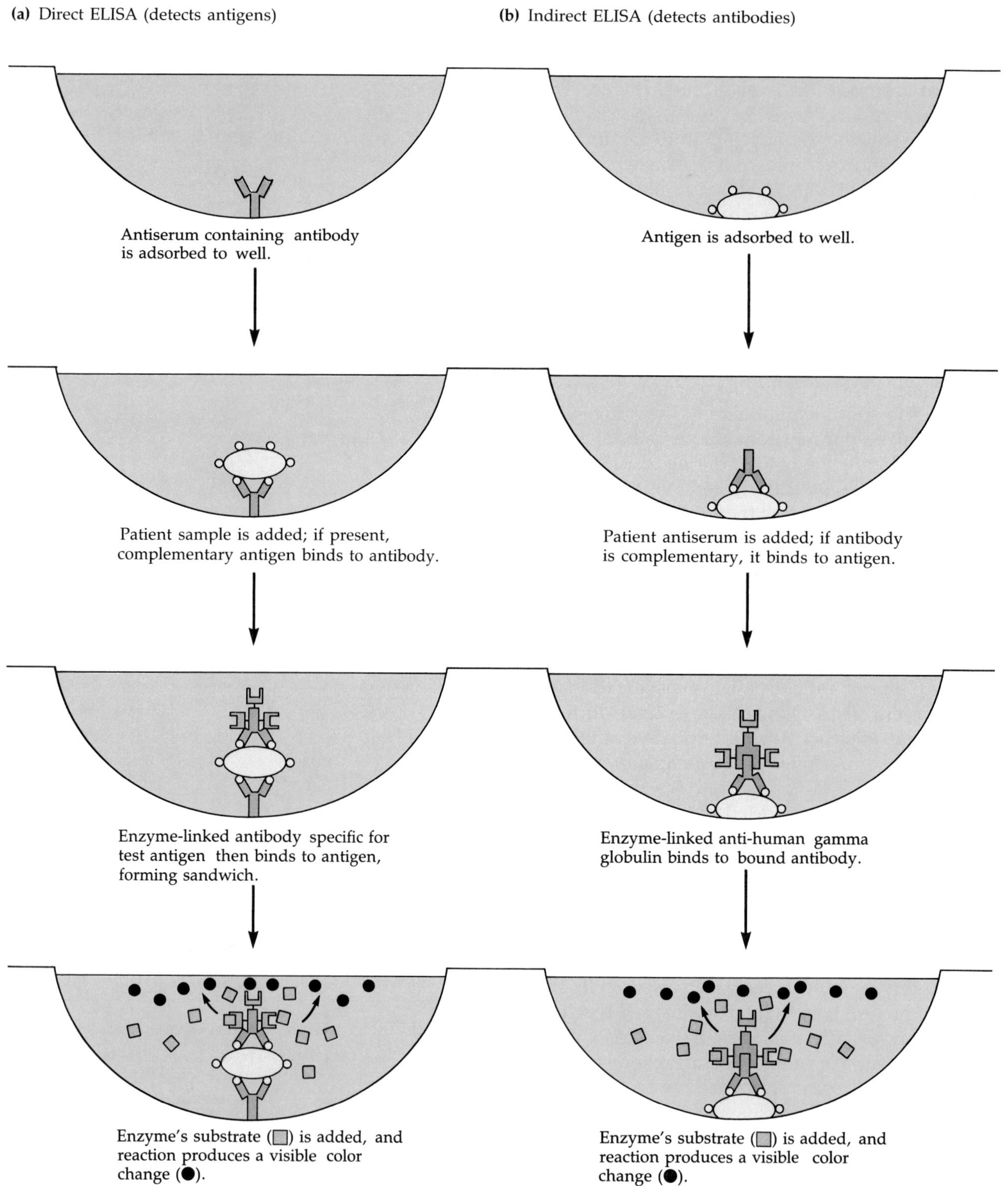

(a) Direct ELISA (detects antigens)
(b) Indirect ELISA (detects antibodies)
Antiserum containing antibody is adsorbed to well.
Antigen is adsorbed to well.
Patient sample is added; if present, complementary antigen binds to antibody.
Patient antiserum is added; if antibody is complementary, it binds to antigen.
Enzyme-linked antibody specific for test antigen then binds to antigen, forming sandwich.
Enzyme-linked anti-human gamma globulin binds to bound antibody.
Enzyme's substrate (□) is added, and reaction produces a visible color change (●).
Enzyme's substrate (□) is added, and reaction produces a visible color change (●).

FIGURE 18.14 The ELISA, or EIA, method. The components are usually contained in small wells of a microtiter plate. **(a)** The direct ELISA method is used to detect antigens. **(b)** The indirect ELISA method detects antibodies.

sterile conditions that allows them to be used in experimental work.

ENZYME-LINKED IMMUNOSORBENT ASSAY (ELISA)

The **enzyme-linked immunosorbent assay (ELISA),** also known as the **enzyme immunoassay (EIA),** has become a widely used serological technique. There are two basic methods. The **direct ELISA** detects antigens, and the **indirect ELISA** detects antibodies. A microtiter plate with numerous shallow wells is used in both procedures. The test has become highly automated in many applications, such as testing for AIDS antibodies. Variations of the test exist; for example, the reagents can be bound to tiny latex particles rather than to the surfaces of the microtiter plates.

Direct ELISA

In the first step of this method, antiserum against the antigen to be detected is adsorbed to the surface of the wells on the microtiter plate (Figure 18.14a). A patient sample containing unidentified antigen is then added to each well. If the antigen reacts specifically with the antibodies adsorbed to the well, the antigen will be retained there when the well is washed free of unbound antigen. A second antibody specific for the antigen is then added. If both the antibody adsorbed to the wall of the well and the antibody known to be specific for the antigen have reacted with the antigen, a "sandwich" will have formed, with the antigen between two antibody molecules.

This reaction is visible only because the second added antibody is linked to an enzyme, such as horseradish peroxidase or alkaline phosphatase. Unbound enzyme-linked antibody is washed from the well, and then the enzyme's substrate is added to it. Enzymatic activity is indicated by a color change that can be seen by eye or read with a special spectrophotometer. The test will be positive if the antigen has reacted with adsorbed antibodies in the first step. If the test antigen was not specific for the antibody adsorbed to the wall of the well, the test will be negative because the unbound antigen will have been washed away. For example, this form of the ELISA is used to detect a hormone secreted in the urine of pregnant women. A positive test indicates pregnancy.

Indirect ELISA

In the first step of this method, a known antigen from the laboratory, rather than the antibody, is adsorbed to the walls of the shallow wells on the plate (Figure 18.14b). To see whether a serum sample contains antibodies against this antigen, the antiserum is added to the well. If the serum contains antibody specific to the antigen, the antibody will bind to the adsorbed antigen. All unreacted antiserum is washed from the well. Anti-HGG (an antibody that reacts with any human immunoglobulin) is then allowed to react with the antigen–antibody complex. The anti-HGG, which has been linked with an enzyme, reacts with the antibodies that are bound to the antigens in the well. Finally, all unbound anti-HGG is rinsed away, and the correct substrate for the enzyme is added. A colored enzymatic reaction occurs in the wells in which the bound antigen has reacted with antibody present in the serum sample. This procedure resembles that of the indirect fluorescent-antibody test, except that the anti-HGG of the indirect assay is linked with an enzyme rather than with a fluorescing dye. This assay is used to test for antibodies against the AIDS virus.

The Future of Vaccines and Diagnostic Immunology

Vaccines and diagnostic immunology are changing rapidly in response to new knowledge and new techniques. For example, genetic engineering is certain to revolutionize current vaccine production. Most of our present vaccines will soon be considered primitive. We may eventually develop new vaccines for such diseases as AIDS, malaria (see the box on p. 454), African trypanosomiasis, and leprosy, to name just a few. There will also be safer and even more effective vaccines for such diseases as cholera and pertussis and many others.

New serological tests and tests that are more sensitive, specific, rapid, and simple than those currently used will also be developed. For example, the development of monoclonal antibodies has already made it possible, for the first time, to test for chlamydial sexually transmitted diseases. The use of DNA probes, while not serological tests (see Figure 10.15), should prove increasingly useful for detecting pathogens in tissues or in foods.

It is virtually certain that rapid new developments in the practical applications of immunology can be expected in the near future, promising dramatic improvements in human health.

STUDY OUTLINE

Vaccines (pp. 449–453)

1. The earliest vaccination procedures included inhaling dried smallpox scabs and variolation, the injection of a less virulent virus.
2. Edward Jenner developed the modern practice of vaccination when he inoculated people with cowpox virus to protect them from smallpox.
3. Herd immunity results when most of a population is immune to a disease.

CHARACTERISTICS OF VACCINES (pp. 450–451)

1. A vaccine is a suspension of infectious agents (or some part of them) that will induce a state of immunity.
2. Attenuated or inactivated microbes may be used.
3. Toxoids (inactivated bacterial exotoxins) may be used in vaccines.

NEW VACCINE DEVELOPMENT (pp. 451–453)

1. Bacteria and yeasts can be genetically engineered to produce the antigenic proteins of pathogens, and these proteins can be used for subunit vaccines.
2. Antigens separated from vaccine preparations can be used for acellular or conjugated vaccines.
3. Anti-idiotypic vaccines are composed of antibodies against an antipathogen. This antibody will stimulate antibody production against the antigenic site of the pathogen.

Diagnostic Immunology (pp. 453–463)

1. Serology is the study of antibodies in serum.
2. Many serological tests can be performed quickly to determine the presence and amounts of antibodies in serum.

PRECIPITATION REACTIONS (pp. 453–456)

1. The interaction of soluble antigens with IgG or IgM antibodies (precipitins) leads to precipitation reactions.
2. Precipitation reactions depend on the formation of lattices and occur best when antigen and antibody are present in optimal proportions. Excesses of either component decrease lattice formation and subsequent precipitation.
3. The precipitin ring test is performed in a small tube.
4. Immunodiffusion procedures such as the Ouchterlony test involve precipitation reactions carried out in an agar gel medium.
5. Such tests detect the presence of multiple antigens and provide information about their relatedness.
6. Immunoelectrophoresis combines electrophoresis with immunodiffusion for the analysis of serum proteins.
7. Countercurrent immunoelectrophoresis can be used to detect antigens in serum.

AGGLUTINATION REACTIONS (pp. 456–457)

1. The interaction of particulate antigens with antibodies leads to agglutination reactions.
2. Diseases may be diagnosed by combining the patient's serum with a known antigen.
3. Direct agglutination reactions can be used to determine antibody titer.
4. Diseases can be diagnosed by a rising titer or seroconversion (from no antibodies to the presence of antibodies).
5. Antibodies cause visible agglutination of soluble antigens affixed to latex spheres in indirect or passive agglutination tests.
6. Hemagglutination reactions involve agglutination reactions using red blood cells. Hemagglutination reactions are used in blood typing, diagnosis of certain diseases, and identification of viruses.
7. Antibodies against certain viruses can be detected by their ability to interfere with viral hemagglutination in hemagglutination inhibition tests.

COMPLEMENT FIXATION REACTIONS (pp. 457–458)

1. Complement fixation reactions are serological tests based on the depletion of a fixed amount of complement in the presence of an antigen–antibody reaction.
2. Hemolysis or its absence is used as an indicator in evaluating complement fixation.

NEUTRALIZATION REACTIONS (pp. 458–460)

1. In these reactions, the harmful effects of a bacterial exotoxin or virus are eliminated by a specific antibody.
2. The Schick test uses a toxin neutralization reaction to determine the presence of antibodies to diphtheria toxin.
3. In a virus neutralization test, the presence of antibodies against a virus can be detected by the antibodies' ability to prevent cytopathic effects of viruses in cell cultures.

IMMUNOFLUORESCENCE AND FLUORESCENT-ANTIBODY TECHNIQUES (pp. 461–463)

1. Immunofluorescence techniques use antibodies labeled with fluorescent dyes.
2. Direct fluorescent-antibody tests are used to identify specific microorganisms.
3. Indirect fluorescent-antibody tests are used to demonstrate the presence of antibody in serum.
4. A fluorescence-activated cell sorter can be used to detect and count cells labeled with fluorescent antibodies.

ENZYME-LINKED IMMUNOSORBENT ASSAY (ELISA) (p. 463)

1. ELISA techniques use antibodies linked to an enzyme such as horseradish peroxidase or alkaline phosphatase.
2. Antigen–antibody reactions are detected by enzyme activity. If the indicator enzyme is present in the test well, an antigen–antibody reaction has occurred.
3. The direct ELISA is used to detect antigens against a specific antibody bound in a test well.
4. The indirect ELISA is used to detect antibodies against an antigen bound in a test well.

The Future of Vaccines and Diagnostic Immunology (p. 463)

1. New and safer vaccines will be developed by genetic engineering.
2. Subunit, acellular, and conjugated vaccines will be developed.
3. The use of monoclonal antibodies will make serological testing faster and more accurate.
4. Monoclonal antibodies have improved diagnostic testing.

STUDY QUESTIONS

Review

1. Identify the following as a direct or indirect test:
   (a) Test for *Mycobacterium tuberculosis*
   (b) Test for antibodies against *M. tuberculosis*

2. Which test in Question 1 is proof of a disease state? Why doesn't the other test confirm a disease state? What is the disease?

3. Explain how an anti-idiotypic vaccine works. Why is this a desirable vaccine?

4. The following data were obtained from fluorescent-antibody tests for anti-*Legionella* in four people. What conclusions can you draw? What is the disease?

	ANTIBODY TITER			
	Day 1	Day 7	Day 14	Day 21
Patient A	1:128	1:256	1:512	1:1024
Patient B	0	0	0	0
Patient C	1:256	1:256	1:256	1:256
Patient D	0	0	1:128	1:512

5. Streptococcal erythrogenic toxin is injected into a person's skin in the Dick test. What results are expected if a person has antibodies against this toxin? What type of immunological reaction is this? What is the disease?

6. An antigen of the hepatitis B virus is used in the HBV vaccine. Can this vaccine cause hepatitis?

7. What do the following vaccines consist of? Which could cause the disease it is supposed to prevent?
   (a) Attenuated poliomyelitis virus
   (b) Dead *Rickettsia prowazekii*
   (c) *Vibrio cholerae* toxoid

8. Influenza virus, patient's serum, and red blood cells are mixed in a tube. If the patient has antibodies against influenza, what should be observed?

9. Measles virus, patient's serum, complement, red blood cells, and anti-red blood cells are mixed in a tube. If the patient has antibodies against measles, what should be observed?

10. How would the addition of excess complement to a complement-fixation test affect the results?

11. What is a precipitin? Explain the effects of excess antigen and antibody on the precipitation reaction. How is the precipitin ring test different from an immunodiffusion test?

12. How does the antigen in an agglutination reaction differ from that in a precipitation reaction?

13. Define the following terms, and give an example of how each reaction is used diagnostically.
    (a) Viral hemagglutination
    (b) Hemagglutination inhibition
    (c) Passive agglutination

14. Explain the complement fixation test.

15. Differentiate between a direct and indirect fluorescent-antibody test.

16. Explain the ELISA techniques.

17. Identify the direct and indirect ELISA:
    (a) Respiratory secretions to detect respiratory syncytial virus.
    (b) Blood to detect human immunodeficiency virus antibodies.

    Which of these tests provides definitive proof of disease?

18. Match the following serological tests to the descriptions.

    ____ Precipitation
    ____ Immunoelectrophoresis
    ____ Agglutination
    ____ Complement fixation
    ____ Neutralization
    ____ ELISA

    (a) Occurs with particulate antigens
    (b) Peroxidase activity indicates a positive test
    (c) Hemolysis is used as an indicator
    (d) Employs anti-human gamma globulin
    (e) Occurs with a free soluble antigen
    (f) Used to determine the presence of antitoxin

Challenge

1. What problems are associated with the use of live vaccines?

2. The World Health Organization has announced the complete eradication of smallpox and is working toward eradication of measles. Why would vaccination be more likely to eradicate a viral disease than a bacterial disease?

3. Many of the serological tests require a supply of antibodies against pathogens. For example, to test for *Salmonella*, anti-*Salmonella* antibodies are mixed with the unknown bacterium. How are these antibodies obtained?

4. A test for antibodies against *Treponema pallidum* uses an antigen called cardiolipin and the patient's serum (suspected of having antibodies). Why do the antibodies react with cardiolipin? What is the disease?

FURTHER READING

Arnon, R. "Chemically defined antiviral vaccines." *Annual Review of Microbiology* 34:593–618, 1980. Discusses immunologic responses to viral vaccines.

Kaufman, L., and P.G. Standard. "Specific and rapid identification of medically important fungi by exoantigen detection." *Annual Review of Microbiology* 41:209–225, 1987. Discusses important rapid identification and classification tests for fungal pathogens.

Kemp, D.J., R.L. Coppel, and R.F. Anders. "Repetitive proteins and genes of malaria." *Annual Review of Microbiology* 41:181–208, 1987. Discusses how the malaria parasite evades the immune system and summarizes the possibility of development of a malaria vaccine.

Kennedy, R.C., J.L. Melnick, and G.R. Dreesman. "Anti-idiotypes and immunity." *Scientific American* 255(1):48–56, July 1986. Describes how these antibodies open the way to new manipulations of the immune system.

Langer, W.L. "Immunization against smallpox before Jenner." *Scientific American* 234(1):112–117, January 1976. An interesting history of the use of smallpox itself as a vaccination against smallpox.

Lerner, R.A. "Synthetic vaccines." *Scientific American* 248(2):66–74, February 1983. Discusses use of a synthetic viral protein to produce antibodies.

Matthews, T.J. and D.P. Bolognesi. "AIDS vaccines." *Scientific American* 259(4):120–127, October 1988. Describes the AIDS vaccines currently in human trials.

Sabin, A.B. "Evaluation of some currently available and prospective vaccines." *Journal of the American Medical Association* 246:236–241, 1981. A summary of the currently available vaccines against bacterial pneumonia, poliomyelitis, measles, rubella, and influenza, with a discussion of new vaccines under investigation.

Stites, D.P., J.D. Stobo, and J.V. Wells. *Basic and Clinical Immunology*, 6th ed. East Norwalk, Conn.: Appleton & Lange, 1987. An immunology textbook with a comprehensive discussion of the serological and diagnostic aspects of immunology.

CHAPTER 19

Disorders Associated with the Immune System

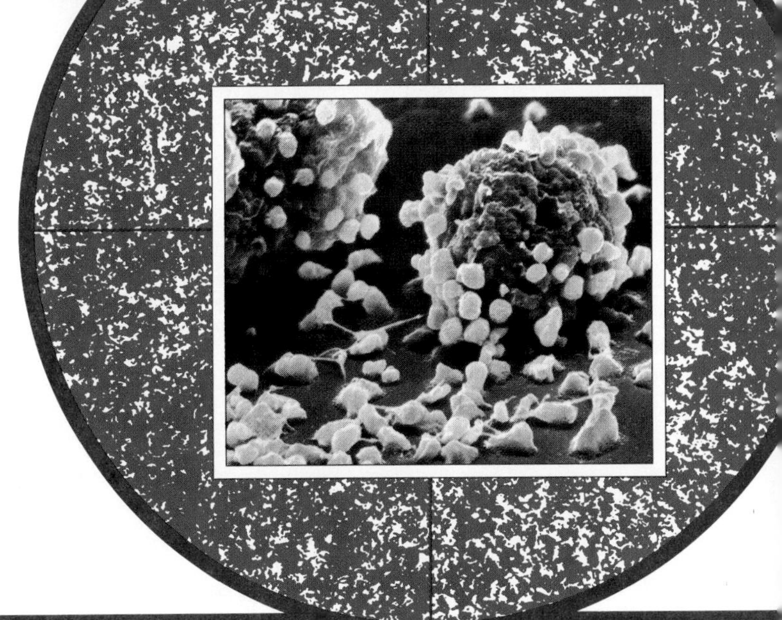

LEARNING OBJECTIVES

- Define hypersensitivity, desensitization, histocompatibility antigens, HLA, immunosuppression, and immunologic tolerance.
- Describe the mechanisms of anaphylaxis and contact dermatitis.
- Differentiate among the four types of hypersensitivity reactions.
- Describe the basis of human blood groups and their relationship to blood transfusions and hemolytic disease of the newborn.
- Explain how rejection of a transplant occurs and how rejection is prevented.
- Define and give a possible explanation for autoimmunity.
- Discuss one immune-complex disease.
- Discuss the causes and effects of immune deficiencies.
- Describe the origin of AIDS, its effect on the immune system, mode of transmission, stages, and treatments.
- Describe the immune responses to cancer and immunotherapy.

In this chapter we shall see that not all immune system responses produce a desirable result such as immunity to disease. Occasionally, the reactions of the immune system are harmful. Hayfever resulting from repeated exposures to plant pollen is a familiar example. Most of us also know that a transfusion will be rejected if the donor and recipient's blood types are incompatible and that rejection is also a problem with organ transplants. Even when a transplant is desperately needed, it is often necessary to wait until an organ of compatible tissue type can be found. Rejection of transfusions or transplants occurs because the immune system recognizes the donated cells as nonself and attacks them. Occasionally, also, the immune system will mistakenly attack one's own tissues, causing an autoimmune disease.

The host is also harmed if the immune system fails to function. For example, if the immune system is functioning properly, cancer cells are perceived as nonself as they arise, and they are eliminated. Cancer, therefore, is usually an indication of a failure of the immune system. Defective immune systems may be congenital or may be a result of damage by some agent such as the AIDS virus. Either of these conditions usually results in overwhelming microbial infections, cancers, or both.

Hypersensitivity

The term **hypersensitivity** refers to sensitivity beyond what is considered normal; the term **allergy** is probably more familiar and is essentially synonymous. Hypersensitivity responses occur in people who have been previously "sensitized" by exposure to an antigen, which in this context is sometimes called an **allergen.** Once sensitized, the immune system responds to a subsequent exposure to that antigen by reacting with it in a manner that damages the host.

TABLE 19.1 Types of Hypersensitivity

TYPE OF REACTION	TIME BEFORE CLINICAL SIGNS	CHARACTERISTICS	EXAMPLES
Type I (anaphylaxis)	30 min	IgE binds to mast cells or basophils. Causes degranulation of mast cell and release of reactive substances such as histamine	Anaphylactic shock from drug injections and insect venom; common allergic conditions, such as hay fever, asthma
Type II (cytotoxic)	5–12 hours	Antigen causes formation of antibodies that bind to target cell. Combined with action of complement destroys target cell	Transfusion reactions, Rh incompatibility
Type III (immune complex)	3–8 hours	Antibodies and antigens form complexes that cause damaging inflammation	Arthus reactions, serum sickness
Type IV (cell-mediated)	24–48 hours	No humoral antibodies involved, Antigens cause formation of cytotoxic T (T_C) cells that kill target cells	Rejection of transplanted tissues; contact dermatitis such as poison ivy

Hypersensitivity reactions are considered to be of four principal types (Table 19.1).

Type I (Anaphylaxis) Reactions

Type I reactions often occur within a few minutes after a person sensitized to an antigen is reexposed to that antigen. **Anaphylaxis** comes from the Greek word *phylaxis,* which means protection, plus the prefix *ana* (against), which makes the word mean unprotected. Anaphylaxis is an inclusive term for the reactions caused when certain antigens combine with IgE antibodies. Anaphylactic responses can be *systemic reactions,* which produce shock and breathing difficulties and are sometimes fatal, or *localized reactions,* which include common allergic conditions such as hay fever, asthma, and hives (slightly raised, often itchy and reddened areas of the skin).

The IgE antibodies produced in response to an antigen such as insect venom or pollen bind to the surfaces of cells such as mast cells and basophils. These two cell types are similar in morphology and in their contribution to allergic reactions. **Mast cells** are especially prevalent in the connective tissue of the skin and respiratory tract and in surrounding blood vessels. **Basophils** circulate in the bloodstream, where they constitute less than 1% of the leukocytes. Both are packed with granules containing a variety of chemicals called *mediators.* The Fc region of the IgE antibody (discussed in Chapter 17) attaches to a specific receptor site on the mast cell or the basophil, leaving two antigen-combining sites free.

These cells can have as many as 500,000 sites for IgE attachment, although not all attached IgE monomers are specific for the same antigen. When an antigen, such as plant pollen, binds to antigen-combining sites on two adjacent IgE antibodies and bridges the space between them, the mast cell or basophil is triggered to undergo *degranulation.* This releases the granules that pack the interior of these cells, thus also releasing the mediators contained in the granules (Figure 19.1).

The best-known mediator is **histamine.** The pharmacological effects of histamine are to increase the permeability and dilation of blood capillaries, resulting in edema (swelling) and erythema (redness). Other effects include increased mucus secretion (a runny nose, for example) and smooth-muscle contraction, which in the respiratory bronchi results in breathing difficulty.

Other mediators include **leukotrienes** of various types and **prostaglandins.** These mediators are not preformed and stored in the granules but are synthesized by the antigen-triggered cell. The most widely studied leukotriene, *SRS-A* (slow-reacting substances of anaphylaxis), is actually a combination of several leukotrienes. Because leukotrienes tend to cause prolonged contractions of certain smooth muscles, their action contributes to the spasms of the bronchial tubes that occur during asthmatic attacks. Prostaglandins affect smooth muscles of the respiratory system and cause increased mucus secretion. A number of other mediators are also released with the degranulation of mast cells and basophils.

Collectively, all these mediators serve as chemotactic agents that, in a few hours, attract neutrophils and eosinophils to the mast cell site. They then activate various factors that cause inflammatory symptoms, such as dilation of the capillaries and swelling, increased secretion of mucus, and involuntary contractions of smooth muscles.

SYSTEMIC ANAPHYLAXIS

At the turn of the century, two French biologists studied the response of dogs to the venom of the stinging jellyfish known as the Portuguese man-of-war. Large

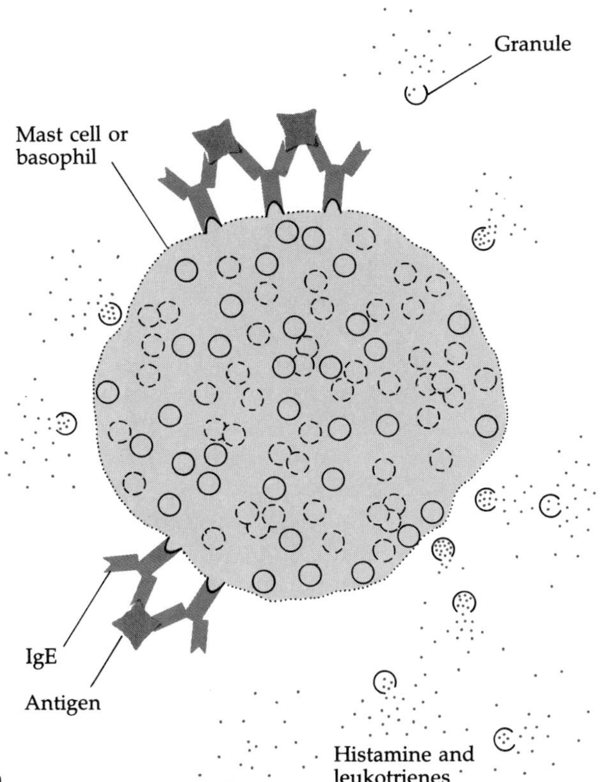

doses of the venom usually killed the dogs, but a few sometimes survived the injections. These surviving dogs were used for repeat experiments with the venom. The results were surprising. Even a very tiny dose of the venom, one that should have been almost harmless, quickly killed the dogs. They suffered difficulty in respiration, entered shock as their cardiovascular systems collapsed, and quickly died. This phenomenon was named **anaphylactic shock.**

Systemic anaphylaxis (anaphylactic shock) can result when an individual sensitized to an antigen is exposed to it again. Injected antigens are more likely to cause a dramatic response than antigens introduced via other portals of entry. The release of mediators causes peripheral blood vessels throughout the body to undergo dilation, resulting in a drop in blood pressure (shock). This reaction can be fatal within a few minutes. The drug epinephrine (adrenaline) counteracts these effects, and kits for epinephrine self-administration are available to persons allergic to insect stings. There is very little time to act once someone develops systemic anaphylaxis. In the United States, 50 to 60 persons die each year from anaphylactic shock caused by insect stings.

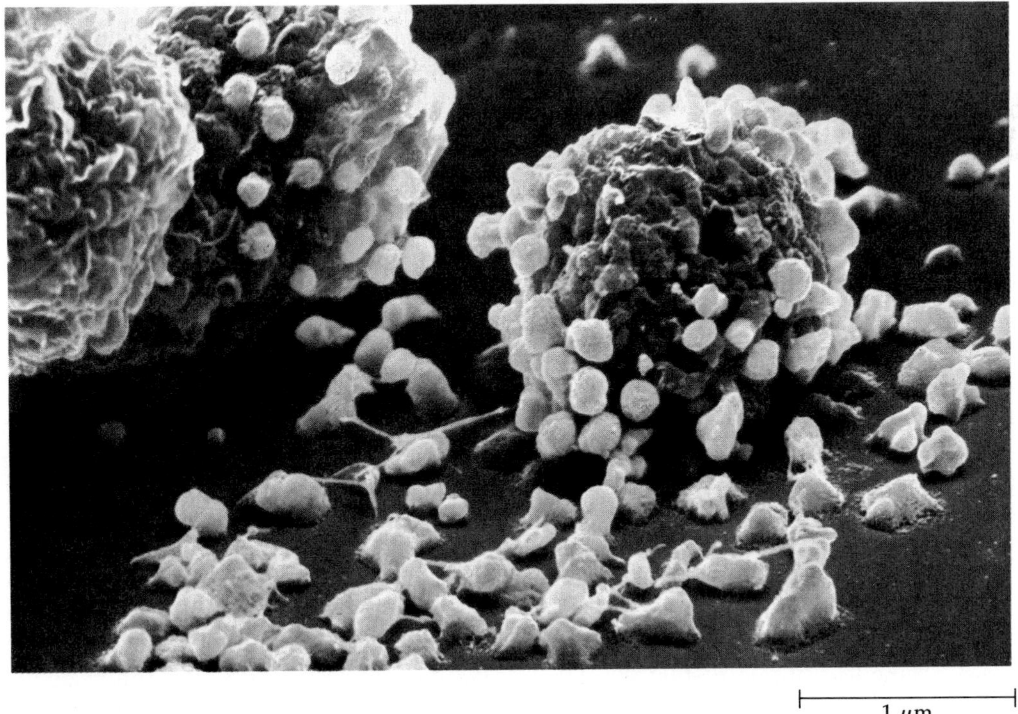

FIGURE 19.1 Mechanism of anaphylaxis. (a) In response to an antigen, IgE antibodies are produced. These antibodies coat mast cells and basophils. When an antigen bridges the gap between two adjacent antibody molecules of the same specificity, the cell undergoes degranulation and releases mediators such as histamine and leukotrienes. **(b)** Scanning electron micrograph of a degranulated mast cell that has reacted with an antigen and released granules of histamine and other reactive substances.

FIGURE 19.2 House dust mite. Many people are allergic to these tiny creatures or, more specifically, their airborne fecal pellets. House dust mites are found in bedding and house dust, where they feed on sloughed off skin cells and similar materials. They are barely visible to the naked eye, about the size of the period printed at the end of this sentence.

Penicillin is a case of special interest because many of us are acquainted with people who are sensitive to this drug. In these persons, the penicillin, which is a nonimmunogenic hapten, combines with a carrier serum protein and in that form is able to induce antibody formation. Penicillin allergy probably occurs in about 2% of the population. Anyone who has had an adverse reaction to penicillin that included generalized hives, swelling of throat tissues, or chest constriction should not receive the drug again. A childhood rash without hives is probably not a significant reaction. However, there is no completely reliable skin test for sensitivity to penicillin.

LOCALIZED ANAPHYLAXIS

Whereas sensitization to injected antigens is a common cause of systemic anaphylaxis, **localized anaphylaxis** is usually associated with antigens that are ingested (foods) or inhaled (pollen). The symptoms that develop depend primarily on the route by which the antigen enters the body.

In allergies involving the upper respiratory system, such as hay fever (allergic rhinitis), sensitization usually involves mast cells localized in the mucous membranes of the upper respiratory tract. Reexposure to the airborne antigen, which might be a common environmental material such as plant pollen, fungal spores, animal dander (scaly, dried skin), or the feces of house dust mites (Figure 19.2), results in a rapid release of large amounts of histamine, leukotrienes, and other mediators from mast cells in contact with the antigen. The typical symptoms are itchy and tearing eyes, congested nasal passages, coughing, and sneezing. Antihistamine drugs, which compete for histamine receptor sites, are often used to treat these symptoms.

Asthma is an allergic reaction that affects mainly the lower respiratory system. Such symptoms as wheezing and shortness of breath are caused by the constriction of smooth muscles in the bronchial tubes. Antihistamines are not an effective treatment for asthma since other mediators, such as leukotrienes and prostaglandins, are more important than histamine in this reaction. Treatment usually involves administration of epinephrine, which constricts the blood vessels and raises the blood pressure.

Antigens that enter the body via the gastrointestinal tract can also sensitize a host. Many of us may know someone who is allergic to a particular food. Children tend to outgrow these problems, and in fact many food allergies may not be related to hypersensitivity but are more accurately described as food intolerances. For example, many people are unable to digest the lactose in milk because they lack the necessary enzyme. The diarrhea that results from milk ingestion is a symptom of food intolerance. Gastrointestinal upset is a common symptom of food allergies, but it can also result from many other factors. Hives are more characteristic of a true food allergy, and ingestion of the antigen may result in systemic anaphylaxis. Death has even resulted when a person sensitive to fish ate french fries that had been prepared in oil previously used to fry fish. Skin tests are not reliable indicators for the diagnosis of food-related allergies, and completely controlled tests for hypersensitivity to ingested foods are very difficult to perform. Only eight foods are responsible for 97% of food-related allergies: eggs, peanuts, other nuts, milk, soy, fish, wheat, and peas.

PREVENTION OF ANAPHYLACTIC REACTIONS

Avoiding contact with the sensitizing antigen is the most obvious way to prevent allergic reactions. Unfortunately, avoidance is not always possible. Some allergic persons may never know exactly what the antigen is. In other cases, skin tests might be of use in diagnosis (Figure 19.3). These tests involve inoculating small amounts of the suspected antigen just beneath the epidermal skin layer. Sensitivity to the antigen is indicated by a rapid inflammatory reaction that features redness, swelling, and itching at the inoculation site. This small affected area is called a *wheal.*

Once the responsible antigen has been identified, the person can either try to avoid contact with it or undertake **desensitization.** This procedure consists of

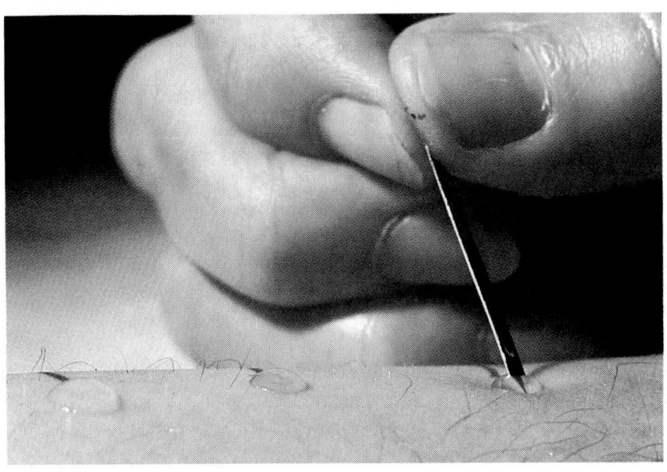

FIGURE 19.3 Skin test to identify allergens. Drops of fluid containing test substances are placed on the skin. A light scratch is made with the needle to allow the substances to penetrate the skin. If reddening and swelling occurs, this identifies the substance as a probable cause of an allergic reaction.

a series of dosages of the antigen carefully injected beneath the skin. The objective is to cause the production of IgG antibodies rather than those of the IgE class, in the hope that the circulating IgG antibodies will act as *blocking antibodies* to intercept and neutralize the antigens before they can react with cell-bound IgE. Recent evidence indicates that desensitization might also induce the production of suppressor T cells (discussed in Chapter 17). Desensitization is not a routinely successful procedure, but it is effective in 65% to 75% of persons whose allergies are induced by inhaled antigens.

Type II (Cytotoxic) Reactions

Type II reactions generally involve the activation of complement by the combination of IgG or IgM antibodies with an antigenic cell. This leads to lysis of the affected cell, which might be either a foreign cell or a host cell that carries a foreign antigenic determinant, such as a drug, on its surface. There may be additional cellular damage by the action of K cells or macrophages that attack antibody-coated cells.

When the action of the immune system is in response to self antigens and causes damage to one's own organs, this is called an **autoimmune disease.**

The most familiar cytotoxic hypersensitivity reactions are *transfusion reactions,* in which red blood cells are destroyed as a result of reaction with circulating antibodies. These involve the ABO and Rh blood group systems.

THE ABO BLOOD GROUP SYSTEM

In the early 1900s, it was discovered that human blood could be grouped into four principal types, which were designated A, B, AB, and O. This method of classification is called the **ABO blood group system.** Since then, at least fifteen other blood group systems have been discovered, but our discussion will be limited to two of the best known, the ABO and the Rh systems.

A person's ABO blood type depends on the presence or absence of two very similar carbohydrate antigens located on the cell membranes of red blood cells, or erythrocytes (Figure 19.4). Persons with type A blood possess a mosaic of antigens designated A on their red blood cells; persons of blood type B have antigens designated B on theirs. Persons with blood type AB have both A and B antigens on their red blood cells, and persons with blood type O lack both A and B surface antigens.

The serum of persons with type A antigens on their red blood cells contains antibodies against type B cells (anti-B antibodies). Persons with type B blood have antibodies against type A cells (anti-A antibodies). Type O serum contains antibodies against both A and B cells, and type AB serum contains no anti-A or anti-B antibodies at all. The presence of these antibodies is a consequence of long exposure to a vast array of antigens encountered in the environment, especially on bacteria. When a transfusion is incompatible, as when type B blood is transfused into a person with type A blood, the antigens on the type B blood cells will react with the anti-B antibodies in the recipient's serum. This antigen–antibody reaction activates complement, which in turn causes lysis of the donor's red blood cells as they enter the recipient's system. The main features of the ABO blood group system are summarized in Table 19.2.

In about 80% of the population (called **secretors**), soluble antigens of the ABO type appear in saliva and other bodily fluids. In criminal investigations, it has been possible to type saliva residues from a cigarette and to type semen in cases of rape.

THE Rh BLOOD GROUP SYSTEM

In the 1930s, the presence of a different surface antigen on human red blood cells was discovered. It was found that when rabbits were immunized with red blood cells from Rhesus monkeys, their serum soon contained antibodies that were directed against the monkey blood cells but would also agglutinate human red blood cells. This indicated that a common antigen was present on both human and monkey red blood cells. The antigen was named the **Rh factor** (*Rh* for *Rhesus*).

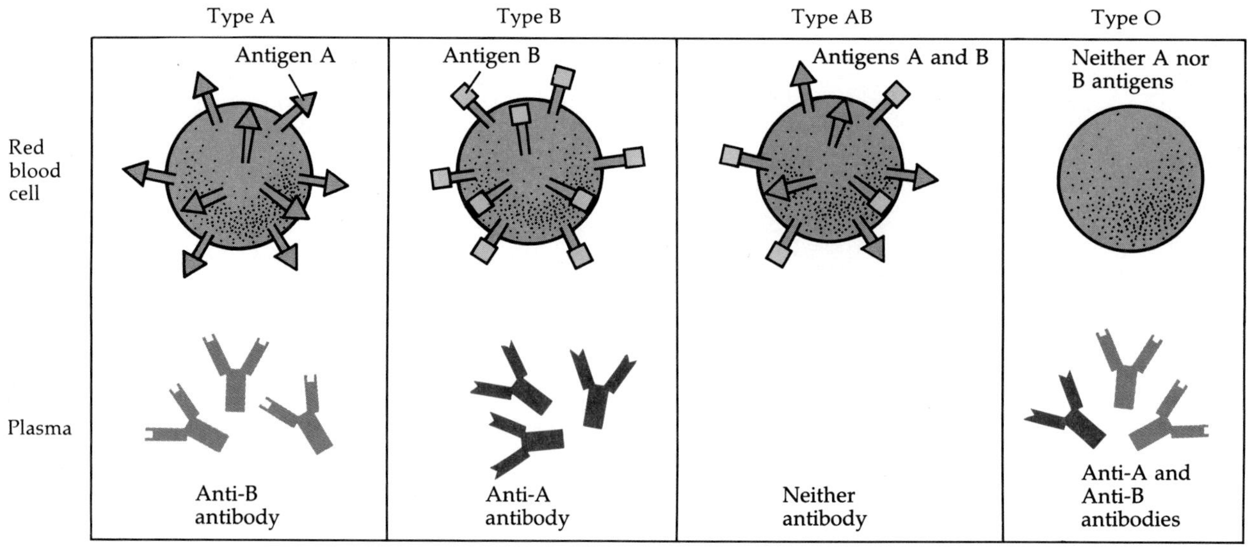

FIGURE 19.4 Relationships between the antigens on the surface of human red blood cells and the antibodies in the plasma.

TABLE 19.2 The ABO Blood Group System

	BLOOD TYPE			
CHARACTERISTIC	***A***	***B***	***AB***	***O***
Antigen present on the red blood cells	A	B	Both A and B	Neither A nor B
Antibody normally present in the plasma	anti-B	anti-A	Neither anti-A nor anti-B	Both anti-A and anti-B
Plasma causes agglutination of red blood cells of these types	B, AB	A, AB	None	A, B, AB
Percent in a Caucasian population	41	10	4	45
Percent in a black population	27	20	7	46

The roughly 85% of the population whose cells possess this antigen are called Rh^+. Those lacking this antigen (about 15%) are Rh^-. Antibodies that react with the Rh antigen do not occur naturally in the serum of Rh^- individuals, but exposure to this antigen can sensitize them to produce anti-Rh antibodies.

Blood Transfusions and Rh Incompatibility

If blood from an Rh^+ donor is given to an Rh^- recipient, the donor's red blood cells stimulate the production of anti-Rh antibodies. If the recipient receives Rh^+ red blood cells in a subsequent transfusion, a rapid hemolytic reaction will develop.

Hemolytic Disease of the Newborn

Blood transfusions are not the only way in which an Rh^- person can become sensitized to Rh^+ blood. When an Rh^- female and an Rh^+ male produce a child, the chances are 50% that the child will be Rh^+ (Figure 19.5a,b). If the child is Rh^+, the Rh^- mother can become sensitized to this antigen during birth, when the placental membranes tear and Rh^+ fetal red

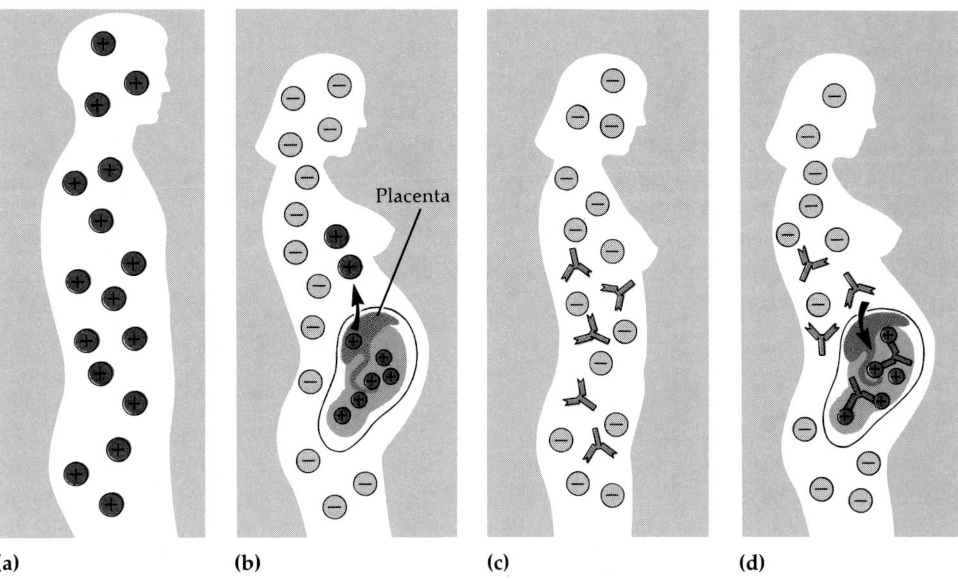

FIGURE 19.5 Hemolytic disease of the newborn. **(a)** Rh^+ father. **(b)** Rh^- mother carrying her first Rh^+ fetus. Rh antigens from the developing fetus can enter the mother's blood during delivery. **(c)** In response to the fetal Rh antigens, the mother will produce anti-Rh antibodies. **(d)** If the woman becomes pregnant again with an Rh^+ fetus, her anti-Rh antibodies will pass through the placenta into the blood of the fetus and will damage fetal red blood cells.

blood cells enter the maternal circulation, causing the mother to produce anti-Rh antibodies of the IgG type. If the fetus in a later pregnancy is Rh^+, the mother's anti-Rh antibodies will cross the placenta and destroy the fetal red blood cells (Figure 19.5c,d). The fetus responds to this immune attack by producing large numbers of immature red blood cells called erythroblasts; hence the name *erythroblastosis fetalis* was once used to describe what is now called **hemolytic disease of the newborn.** Before birth, the maternal circulation removes most of the toxic by-products of fetal red cell disintegration. After birth, however, the fetal blood is no longer purified by the mother, and the newborn develops jaundice and severe anemia.

Hemolytic disease of the newborn is usually prevented today by immediate passive immunization of the Rh^- mother with anti-Rh antibodies, which are available commercially. These anti-Rh antibodies combine with any fetal Rh^+ red blood cells that have entered the mother's circulation, so it is much less likely that she will become sensitized to the Rh antigen. If the disease is not prevented, it might be necessary that the newborn's Rh^+ blood, contaminated with maternal antibodies, be replaced by transfusion of uncontaminated blood.

DRUG-INDUCED CYTOTOXIC REACTIONS

Blood platelets (thrombocytes) are minute cell-like bodies that are essential to blood clotting. They are destroyed by antibodies and complement in the disease **thrombocytopenic purpura.** In the situation illustrated in Figure 19.6, the platelet has become coated

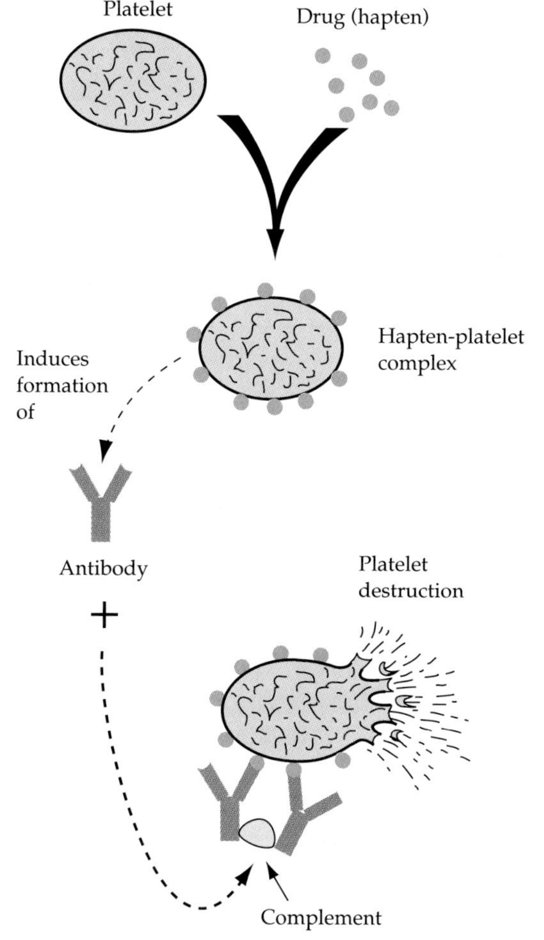

FIGURE 19.6 Thrombocytopenic purpura. Molecules of a drug such as aspirin accumulate on the surface of a platelet and stimulate an immune response that destroys the platelet.

with molecules of a drug, such as aspirin and some antibiotics, that functions as a hapten. Because platelets are necessary for blood clotting, their loss results in hemorrhages that appear on the skin as purple spots (purpura). Similarly, drugs may bind to white or red blood cells. Immune-caused destruction of granulocytic white cells is called **agranulocytosis,** and it affects our phagocytic defenses. When red blood cells are destroyed in the same manner, the condition is termed **hemolytic anemia.**

AUTOIMMUNE DISORDERS (TYPE II REACTIONS)

Graves' disease is caused by antibodies called long-acting thyroid stimulators. These antibodies attach to receptors on the thyroid gland that are the normal target of the thyroid-stimulating hormone produced by the pituitary gland. The result is that the thyroid gland is stimulated to produce increased amounts of thyroid hormones and becomes greatly enlarged. The most striking signs of the disease are goiter, which results from the enlarged thyroid, and markedly bulging, staring eyes.

Myasthenia gravis is a disease in which muscle tone becomes progressively weaker. It is caused by antibodies that coat the acetylcholine receptors at the junctions at which nerve impulses reach the muscles. Eventually, the muscles controlling the diaphragm and the rib cage fail to receive the necessary nerve signals, and respiratory arrest and death result. Both Graves' disease and myasthenia gravis are considered examples of type II immune reactions. Both diseases involve antibody reactions to cell surface antigens, although there is no cytotoxic destruction of the cells.

Type III (Immune Complex) Reactions

Type III reactions involve antibodies against antigens, often soluble, that are circulating in the serum. The antigen–antibody complexes are deposited in organs and cause inflammatory damage. In contrast, type II immune reactions are directed against antigens located on cell or tissue surfaces.

Immune complexes that cause damage form only when certain ratios of antigen and antibody occur. The antibodies involved are usually IgG. A significant excess of antibody (Figure 19.7a) leads to the formation of large complement-fixing complexes that are rapidly removed from the body by phagocytosis. When there is a significant excess of antigen (Figure 19.7b), soluble complexes form that do not fix complement and do not cause inflammation. However, when a certain antigen–antibody ratio exists, usually with a slight excess

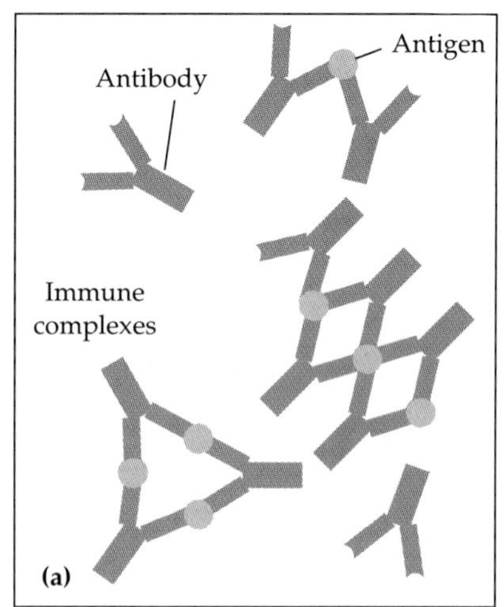

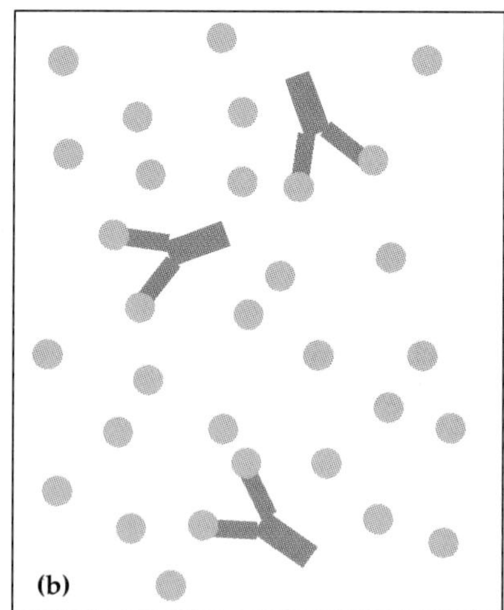

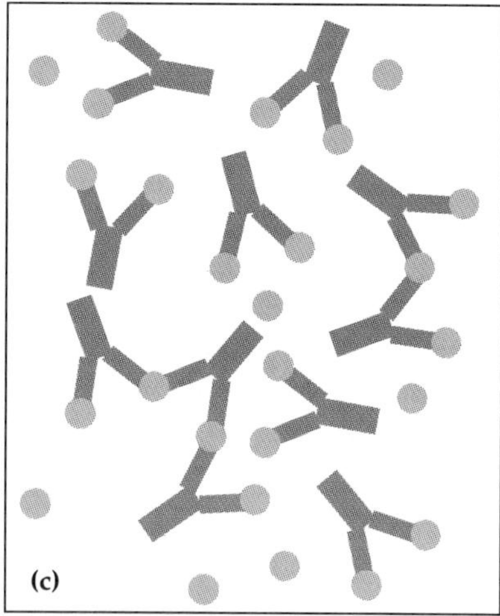

FIGURE 19.7 Formation of immune complexes. **(a)** Large crosslinked immune complexes form when antibody is in excess. Most of these are removed from the circulation by phagocytosis. **(b)** Small separate immune complexes form when antigen is in excess. These generally do not fix complement or cause inflammation. **(c)** Small, partially crosslinked complexes form when there is a certain ratio of antigen to antibody; there is usually a small excess of antigen. These complexes fix complement but are small enough to escape phagocytosis. They damage tissues when they become trapped in the basement membrane of blood vessels.

of antigen (Figure 19.7c), the soluble complexes that form are small and escape phagocytosis. These complexes circulate in the blood, pass between endothelial cells of the blood vessels, and become trapped in the basement membrane beneath the cells. In this location, they may activate complement and cause a transient inflammatory reaction. Repeated introduction of the same antigen can lead to more serious inflammatory reactions.

Such damage from formation of immune complexes was first observed when rabbits were immunized repeatedly by injections of horse serum. In the area of the injections, edema (swelling) and reddening increased with each injection, and eventually the area became necrotic (i.e., the tissue died). This response was named the **Arthus reaction,** after the researcher. In the days before antibiotics, and when only a limited number of vaccines were available, diseases caused by bacterial toxins were sometimes treated by antiserum from horses immunized with these toxins. The antibodies in the horse serum were intended to neutralize the toxins, but repeated injections often led to fever, itchy rashes, and swollen, painful joints in the recipient. This phenomenon became familiar as **serum sickness.**

Glomerulonephritis is an immune complex condition that causes inflammatory damage to the kidney glomeruli, which are sites of blood filtration. Antibodies generated in response to the M-protein of streptococci are believed to be one cause of this disease (see Chapter 15). Most people can be treated successfully with anti-inflammatory drugs, but in some, the condition progresses to fatal kidney failure.

AUTOIMMUNE DISORDERS (TYPE III REACTIONS)

Systemic lupus erythematosus is a systemic autoimmune disease that mainly affects women. The etiology of the disease is not completely understood, but afflicted persons produce antibodies directed at components of their own cells. These include antibodies against DNA, which is probably released during the normal breakdown of tissues, especially the skin. The most damaging effects of the disease result from deposits of immune complexes in the kidney glomeruli.

Crippling **rheumatoid arthritis** is a disease in which immune complexes of IgM, IgG, and complement are deposited in the joints. In fact, there is evidence that immune complexes called rheumatoid factors may be formed by IgM binding to the Fc region of normal IgG. These factors are found in 70% of persons suffering from rheumatoid arthritis. The chronic inflammation caused by this deposition eventually leads to severe damage to the cartilage and bone of the joint.

Type IV (Cell-Mediated) Reactions

Up to this point, we have discussed humoral types of immune reactions involving IgE, IgG, or IgM. Type IV reactions involve cell-mediated immune responses and are caused mainly by T cells, although macrophages may also be involved. Instead of occurring within a few minutes or hours after a sensitized individual is again exposed to an antigen, these delayed reactions are not apparent for a day or more. A major factor in the delay is the time required for the participating T cells and macrophages to migrate to and accumulate near the foreign antigens.

CAUSES OF TYPE IV REACTIONS

Type IV hypersensitivity reactions occur when certain foreign antigens, particularly of a type that bind to tissue cells, are phagocytized by macrophages and then presented to receptors on the T-cell surface. Contact between the antigenic determinant sites and the appropriate T cell causes the T cell to proliferate. The T cells involved in delayed-type hypersensitivity reactions are primarily T_D cells. In some types of hypersensitivities resulting in tissue damage, T_C cells may also participate. If a person sensitized in this way is reexposed to the same antigen, a cell-mediated hypersensitivity reaction might result. A principal factor in this reaction is the release of lymphokines by T cells reacting with the target antigen. Some lymphokines, as we saw in Chapter 17, contribute to the inflammatory reaction to the foreign antigen by attracting macrophages to the site and activating them.

CELL-MEDIATED HYPERSENSITIVITY REACTIONS OF THE SKIN

We have seen that the skin is frequently the site on which hypersensitivity symptoms are displayed. One cell-mediated hypersensitivity reaction that involves the skin is the familiar skin test for tuberculosis. Because *Mycobacterium tuberculosis* is often located within macrophages, this disease can stimulate a cell-mediated immune response. As a screening test, protein components of the bacteria are injected into the skin. If the recipient has, or has had, a prior infection by tuberculosis bacteria, an inflammatory reaction to the injection of these antigens will appear on the skin in one or two days; this interval is typical of delayed hypersensitivity reactions.

Allergic contact dermatitis, another common manifestation of type IV hypersensitivity, is usually caused by haptens that combine with proteins in the skin of some persons to produce an immune response. Reactions to poison ivy (Figure 19.8), cosmetics, and the

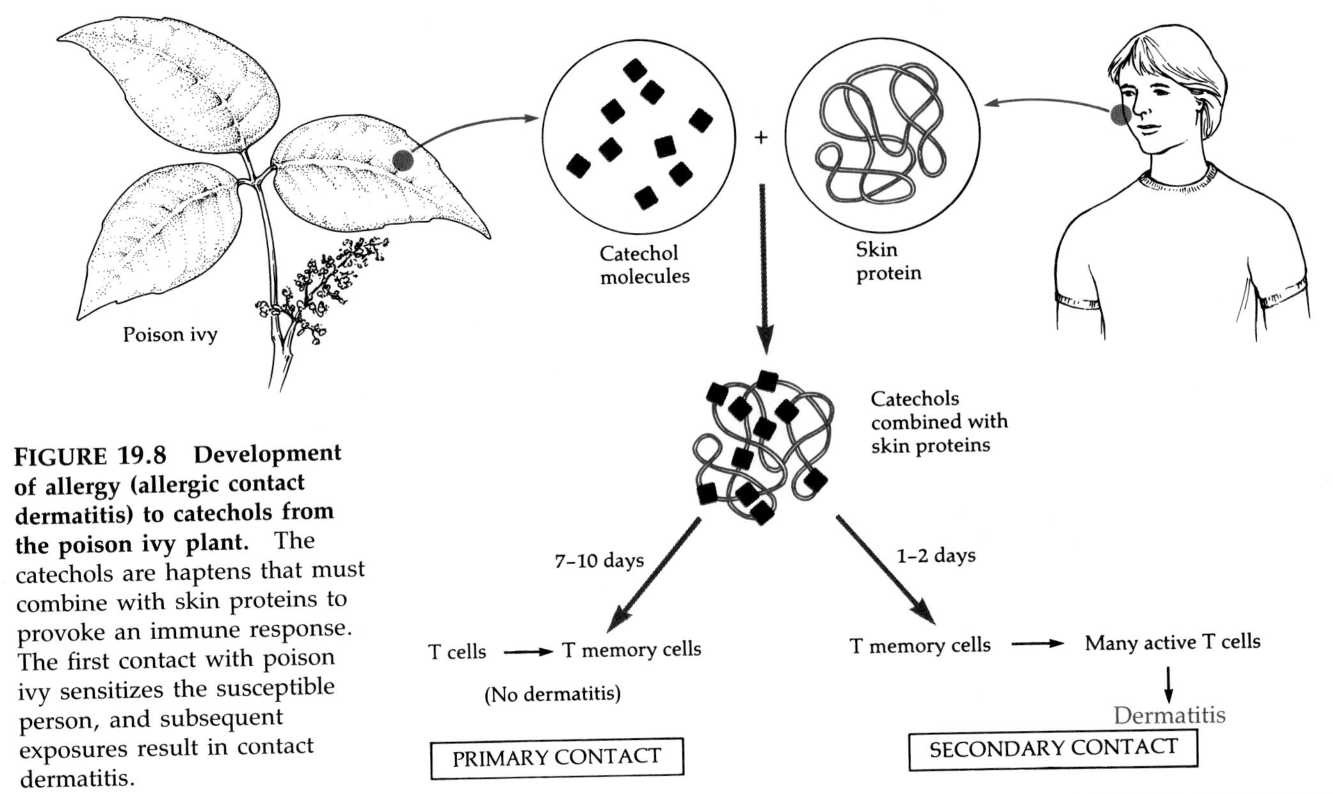

FIGURE 19.8 Development of allergy (allergic contact dermatitis) to catechols from the poison ivy plant. The catechols are haptens that must combine with skin proteins to provoke an immune response. The first contact with poison ivy sensitizes the susceptible person, and subsequent exposures result in contact dermatitis.

metals in jewelry (especially nickel) are familiar examples of these allergies. The identity of the offending environmental factor can usually be determined by a *patch test.* Samples of suspected materials are taped to the skin; after 48 hours, the area is examined for inflammation.

AUTOIMMUNE DISORDERS (TYPE IV REACTIONS)

An autoimmune disease involving the cell-mediated immune system is **lymphocytic choriomeningitis,** caused by the lymphocytic choriomeningitis virus. This virus is sometimes spread by contact with pet hamsters; it infects membranes surrounding the brain. The body responds to the viral infection by producing T cells that fail to protect, but instead cause fatal neurological damage.

There are a number of disease conditions, with less obvious etiologies, in which the cell-mediated immune system destroys the patient's tissues. Examples are **Hashimoto's thyroiditis** and **Addison's disease,** which result from T cells attacking the thyroid and adrenal glands, respectively.

LOSS OF IMMUNOLOGICAL TOLERANCE

We have seen that some hypersensitivity reactions involve autoimmunity, an immune reaction to one's own cells. Autoimmune diseases occur when there is a loss of **tolerance,** the immune system's ability to discriminate between self and nonself. Apparently tolerance arises during fetal development. If a fetus is exposed to an antigen, exposure to the same antigen after birth does not stimulate the production of antibodies or sensitized T cells. It is believed that some clones of lymphocytes (*forbidden clones*) having the potential to respond to self antigens may be produced during fetal life but are destroyed or suppressed during fetal development.

In autoimmune diseases, loss of tolerance leads to a response by antibodies or sensitized T cells against a person's own tissue antigens. Some of the major autoimmune diseases are summarized in Table 19.3.

MAJOR HISTOCOMPATIBILITY COMPLEX

The inherited genetic characteristics of individuals are expressed not only in the color of their eyes and the curl of their hair but also in differences in the antigens on their cell surfaces. Some of these are called **histocompatibility antigens.** The genes controlling the production of the most important of these antigens are known as the **major histocompatibility complex (MHC);** in humans, these genes are called the **human leukocyte antigen (HLA)** complex. We encountered these antigens in Chapter 17, where we saw that most

TABLE 19.3 Autoimmune Diseases

DISEASE	HYPERSENSITIVITY CLASS	DESCRIPTION
Addison's disease	IV	Destruction of adrenal gland cells that produce cortical hormones
Hemolytic anemia	II	Destruction of red blood cells
Diabetes mellitus, insulin dependent	IV	Destruction of pancreatic islet cells that produce insulin
Goodpasture's syndrome	II	Damage to basement membranes, but most symptoms result from attack on glomeruli of kidney
Graves' disease	II	Binding of antibodies to receptors for thyroid-stimulating hormone leads to overstimulation of thyroid activity
Hashimoto's thyroiditis	IV	Destruction of cells of the thyroid gland
Thrombocytopenia purpura	II	Destruction of blood platelets
Myasthenia gravis	II	Blockage of acetylcholinesterase receptors prevents transmission of nerve signals
Rheumatoid arthritis	III	Inflammation of joints due to deposition of complexes of IgG and anti-IgG
Systemic lupus erythematosus	III	Antigen–antibody complexes (including antibodies against RNA, IgG, red blood cells, platelets, and chromosomes) are deposited in many locations, triggering inflammation; complexes containing antibodies to DNA are often deposited in kidneys

antigens can stimulate an immune reaction only if they are associated with an MHC antigen.

An important medical application of HLA typing is in transplant surgery, where the donor and the recipient must be matched. This is done by *tissue typing.* The serological technique shown in Figure 19.9 is the one most often used. In serological tissue typing, the laboratory uses standardized antisera that are specific for particular HLAs. Lymphocytes from the person being tested are incubated with a selected specific antiserum. Complement and a dye, such as trypan blue, are then added. If antibodies in the antiserum have reacted specifically with the lymphocyte, the cell is damaged. The damaged cell will take up the dye (undamaged cells will not), thus indicating that the lymphocyte possesses a certain antigen. This method is simple and rapid. Unfortunately, sera are not always available for the HLA groups most likely to cause problems with transplants between unrelated individuals.

A second method exists for identifying such HLA groups. However, it requires about six days, which is usually too long to be useful. Research to improve this situation continues, much of it aimed at developing antisera for more HLA types. It is hoped that monoclonal antibodies for all HLA types will eventually be available.

There are two main classes of HLA antigens that are important in such matches: class I antigens (HLA-A, -B, and -C) and class II antigens (HLA-DR, -DP, and

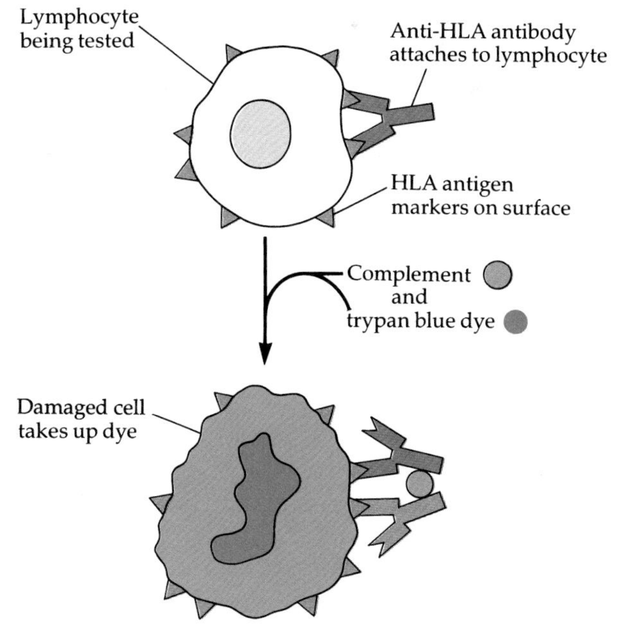

FIGURE 19.9 Tissue typing, serological method. Lymphocytes from the person being tested are incubated with laboratory test stocks of anti-HLA antibodies that are specific for a particular HLA antigen. If the antibodies react with a lymphocyte, then complement damages the lymphocyte and dye can enter the cell. A positive test result indicates that the person has the particular HLA antigen being tested for.

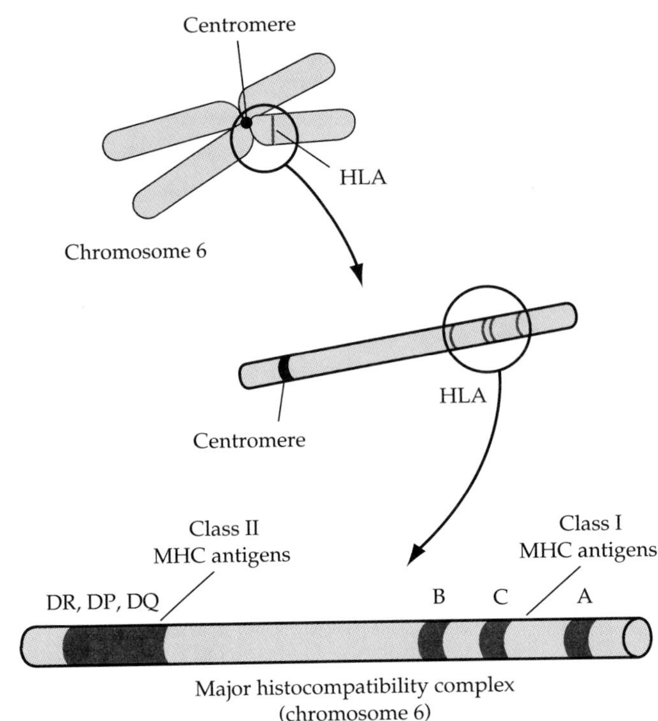

FIGURE 19.10 Genes controlling human leukocyte antigens (HLA). The genes that control HLA are located on chromosome 6. Note how the genes for class I antigens (A, B, and C) and for class II antigens (DR, DP, and DQ) are all located within a very short length of chromosome.

-DQ). The genes that control production of these antigens are located close together on the human chromosome 6, as shown in Figure 19.10. Matching for class I antigens has long been a standard procedure. These antigens, which are found on the surface of almost all nucleated cells, stimulate a strong immune response by the antibodies and T_C cells mediating transplant rejection. However, matching for class II antigens might well be more important, especially for tissue from a totally unrelated person. These MHC antigens are present, along with the processed foreign antigens, on the surface of the antigen-presenting cells (Figure 17.11) and are simultaneously recognized by T_H cells. The DR antigens are especially important; if the donor and the recipient do not share these antigens, the graft is unlikely to be accepted. The donor and the recipient must also be of the same ABO blood type.

As we mentioned previously, certain HLA antigens are related to increased susceptibility to specific diseases, which represents another medical application of HLA typing. A few of these relationships are summarized in Table 19.4.

Transplantation

In Italy in the sixteenth century, crimes were often punished by cutting off the offender's nose. A surgeon of the time, in his attempts to repair this mutilation, observed that if skin was taken from the patient, it healed properly, but if it was taken from another person, it did not. He called this a manifestation of "the force and power of individuality." Four hundred years later, as we have seen, the principles behind this phenomenon have become known.

Human organ transplants have added productive years to the lives of many individuals. Since the first kidney transplant was performed in 1954, this particular type of transplant has become a nearly routine medical procedure. Other types of transplants that are now feasible include bone marrow, thymus, heart, liver, and cornea. Tissues and organs for transplant are usually taken from recently deceased individuals, although nonessential duplicates of an organ, such as a kidney, occasionally come from a living donor.

TYPES OF TRANSPLANTS

Some transplants or grafts do not stimulate an immune response. A transplanted cornea, for example, is rarely rejected, mainly because antibodies do not circulate into the anterior chamber of the eye, which is therefore considered an immunologically **privileged site.** The brain is also an immunologically privileged site, probably because it does not have lymphatic vessels and because the walls of the blood vessels in the brain differ from blood vessel walls elsewhere in the body (the blood–brain barrier is discussed in Chapter 22). Someday, it may be possible to graft foreign nerves to replace damaged nerves in the brain and spinal cord.

TABLE 19.4 Diseases Related to Specific HLAs

	RELATED HLA ANTIGEN	*INCREASED RISK OF OCCURRENCE AS COMPARED WITH GENERAL POPULATION*	*DESCRIPTION*
Inflammatory Diseases			
Multiple sclerosis	DR2	5 times	Progressive inflammatory disease affecting nervous system
Rheumatic fever	DR (antibody 883)	4–5 times	Cross-reaction with antibodies against streptococcal antigen
Endocrine Diseases			
Addison's disease	DR3	4–10 times	Deficiency in production of hormones by adrenal gland
Graves' disease	DR3	10–12 times	Antibodies attached to certain receptors in the thyroid gland cause it to enlarge and produce excessive hormones
Malignant Disease			
Hodgkin's disease	A(A1)	1.4–1.8 times	Cancer of lymph nodes

It is also possible to transplant **privileged tissue** that does not stimulate an immune rejection. An example is replacement of a person's damaged heart valve with a valve from a pig's heart. However, privileged sites and tissues are more the exception than the rule.

It would be interesting to understand how animals tolerate pregnancy without rejecting the fetus. It has been determined that the uterus is not a privileged site, and yet in pregnancy the tissues of two genetically different individuals are in direct contact. Current theories are that there is a barrier that masks the fetal antigens and that the maternal immune system is suppressed.

When one's own tissue is grafted to another part of the body, as is done in burn treatment or in plastic surgery, the graft is not rejected. Recent technology has made it possible to use a few cells of a burn patient's uninjured skin to culture extensive sheets of new skin. The graft of this new skin is an example of an **autograft.** Identical twins have the same genetic makeup; therefore, skin or organs such as kidneys may be transplanted between them without provoking an immune response. These transplants are called **isografts.**

Most transplants, however, are made between persons who are not identical twins; these transplants do trigger an immune response. Attempts are made to match the HLA antigens of the donor and recipient as closely as possible so that the chances of rejection are reduced. Because HLA antigens of close relatives are most likely to match, blood relatives, especially siblings, are the preferred donors. Grafts between persons who are not identical twins are called **allografts.**

Since there is a shortage of available organs, it would be helpful if **xenografts** (organs from animals) could be more successfully transplanted to humans. However, the body tends to mount an especially severe immune assault on such transplants. Perhaps the most famous xenograft recipient was Baby Fae, a two-week-old baby girl who received the heart of a baboon on October 26, 1984. The girl died 20 days later from heart and kidney problems. The various types of grafts are summarized in Table 19.5.

When bone marrow is transplanted into persons with a defective immune system, the transplanted tissue may attack the host. The purpose of the transplant is to provide such persons with the B-cell and T-cell

TABLE 19.5 Types of Grafts

GRAFT	*DESCRIPTION*
Autograft (*autos*, "self")	A graft transplanted from one site to another site on the same person—for example, the skin from the thigh grafted over a burned area on the arm
Isograft (*isos*, "equal")	A graft between genetically identical persons—identical twins
Allograft (*allos*, "other")	A graft between genetically different members of the same species
Xenograft (*xenos*, "strange")	A graft between different species—for example, a transplant of a baboon heart to a human

manufacturing capability they lack. However, the result can be **graft-versus-host (GVH) disease.** The transplanted bone marrow contains immunocompetent cells that mount a mostly cell-mediated immune response against the tissue into which they have been transplanted. Because the recipients lack effective immunity, GVH disease is a serious complication and can even be fatal.

IMMUNOSUPPRESSION

To keep the problem of rejection of transplanted tissue in perspective, it is useful to remember that the immune system is simply doing its job and has no way of recognizing that its attack against the transplant is not helpful. In an attempt to prevent rejection, the recipient of an allograft usually receives treatment to suppress this normal immune response against the graft. Unfortunately, the treatments used in the past suppressed the immune response to all antigens, by both the humoral and cell-mediated immune systems. The persons so treated then become very susceptible to infectious diseases and cancer.

The drug **cyclosporine,** isolated from a mold in 1976, has revolutionized transplant medicine. Cyclosporine selectively inhibits the synthesis of certain lymphokines without interfering with the synthesis of other proteins. The drug prevents the formation of T_C cells that are responsible for transplant rejection, inhibits the release by macrophages of lymphokines that cause inflammation, and prevents the differentiation of B cells into antibody-secreting plasma cells. It does this without suppressing the ability of bone marrow to produce red blood cells and platelets.

As a result of this discovery, the success rate of organ transplants has risen dramatically. Before cyclosporine came into general use, only about 50% of kidneys from donors who were not closely related to the patient were accepted; now as many as 90% of these transplants succeed. Liver transplants became practical only after cyclosporine became available, and heart transplants have achieved unprecedented success rates.

The greatest drawback to the use of the drug appears to be an increased risk of cancer, as that disease is combatted by the same cell-mediated immune system that rejects transplanted organs. Problems with liver and kidney toxicity have also necessitated careful attention to dose levels.

A new immunosuppressant drug, FK506, has a mode of action similar to that of cyclosporine.

Natural Immune Deficiencies

The drugs we have just discussed impair the immune response artificially. Occasionally, people are born with defective immune systems. A number of inherited genes can result in impaired immunity. For example, individuals with a certain recessive trait may lack a thymus gland and therefore lack cell-mediated immunity. A different recessive trait causes lowered numbers of B cells. Some immunodeficiencies have no known cause. For example, about 1 in 700 Caucasian adults produces less IgA than other adults.

A wide variety of diseases can also impair immune function. Hodgkin's disease lowers cell-mediated immunity. Many viruses are capable of infecting and killing lymphocytes, hence lowering the immune response. Removal of the spleen decreases humoral immunity. Table 19.6 summarizes several of the better-known natural immune deficiency conditions, including AIDS.

Acquired Immunodeficiency Syndrome (AIDS)

THE ORIGIN OF AIDS

In 1981, a cluster of cases of *Pneumocystis* pneumonia appeared in the Los Angeles area. This extremely rare disease was usually seen only in persons who were immunosuppressed. Investigators soon correlated the appearance of this disease with an unusual incidence of a rare form of cancer of the skin and blood vessels called Kaposi's sarcoma. The people affected were all young homosexual men, and all showed loss of immune function. By 1983, the pathogen causing the loss of immune function had been identified as a retrovirus (see Chapter 13) that selectively infects certain T cells. This virus is now known as human immunodeficiency virus (HIV) (Figure 19.11, and see Figure 1.1a).

There have been several theories about the origin of HIV. It is now believed that it arose by mutation of a virus that had been endemic in some areas of central Africa for many years. The virus has been found in blood samples preserved from as early as 1959 in several African nations and in England.

THE HIV INFECTION

HIV infects helper T cells named T4 cells (for the CD4 antigen on their surface). Once inside the infected cell, the viral RNA is transcribed into DNA that remains incorporated into the genetic material of that cell. HIV may remain latent in the cell or begin replicating. Sometimes, replication results in continuous production of new virus particles that bud from the cell without killing it and infect other cells. In other cases, the cell is quickly killed either by the virus itself or by the action of the immune system in response to viral anti-

TABLE 19.6 Immune Deficiency Diseases

DISEASE	*CELLS AFFECTED*	*COMMENTS*
Acquired immunodeficiency syndrome (AIDS)	T cells (virus destroys T_H cells)	Allows cancer and bacterial, viral, fungal, and protozoan diseases
Common variable hypogammaglobulinemia	B, T cells (decreased immunoglobulins)	Frequent viral and bacterial infections; second most common immunodeficiency disease
Reticular dysgenesis	B, T, and stem cells (a combined immunodeficiency; deficiencies in B cells, T cells, and neutrophils)	Usually fatal in early infancy; very rare, bone marrow transplant a possible treatment
Selective dysgammaglobulinemias, may be IgA, IgG, or IgM	B cells (deficiencies in one or another immunoglobulin)	IgA deficiency is quite common, causing frequent mucosal infections; IgM deficiency rare; IgG deficiency causes frequent bacterial infections
Severe combined immunodeficiency	B, T, and stem cells (deficiency of both T cells and B cells)	Occurs in several forms; allows severe viral, fungal infections; treated with transplantation of bone marrow, fetal thymus
Thymic hypoplasia (DiGeorge syndrome)	T cells (defective thymus, causes deficiency of T cells)	Lack of cell-mediated immunity; usually fatal in infancy from *Pneumocystis* pneumonia or viral or fungal infections
Wiskott–Aldrich syndrome	B, T cells (few platelets in blood, abnormal T cells)	Frequent infections by viruses, fungi, protozoans; eczema, defective blood clotting; usually causes death in childhood
X-linked infantile (Bruton's) agammaglobulinemia	B cells (decreased immunoglobulins)	Frequent extracellular bacterial infections

gens on the surface of the cell. The virus also infects other types of cells that carry CD4 surface antigens. These cells include macrophages, antigen-presenting dendritic cells, about 40% of all monocytes, and 5% of B cells. HIV infection generally does not kill these cells but impairs their normal functioning.

The ability of the virus to remain latent intracellularly is one reason why the anti-HIV antibodies developed by infected persons fail to prevent the disease. The virus also evades immune defenses by undergoing antigenic changes, even during the course of an infection. There is evidence that a change of a single amino acid in the glycoprotein envelope of the virus may enable it to avoid antibodies against the previous antigenic configuration. HIV also evades the immune system by staying in vesicles within cells so that viral antigens are not displayed on the cell surface and T cells cannot detect the infected cells. Moreover, infected cells displaying viral antigens can fuse to uninfected cells to spread the virus.

AIDS is actually only the end stage of an infection by HIV. Shortly after the initial infection, the patient undergoes seroconversion—that is, tests positive for antibodies to HIV. This interval is almost always less than six months. The symptoms at this point are absent or resemble mononucleosis—a mild fever, swollen lymph nodes, and fatigue (see Chapter 23). Even these symptoms spontaneously disappear in a few weeks.

Chronic lymphadenopathy (swollen lymph nodes) usually announces the beginning of stage 2 and is often the first indication of illness. This stage lasts for several years; the patient does not feel especially unwell, but the number of T4 cells in the circulation declines steadily, while the number of viruses in the body increases steadily (Figure 19.12). Stage 3 is signaled by a T4 count below 400 per milliliter, and stages 4 and 5 are measured by evidence of progressive loss of effective immune response. The evidence includes loss of ability to respond to certain hypersensitivity

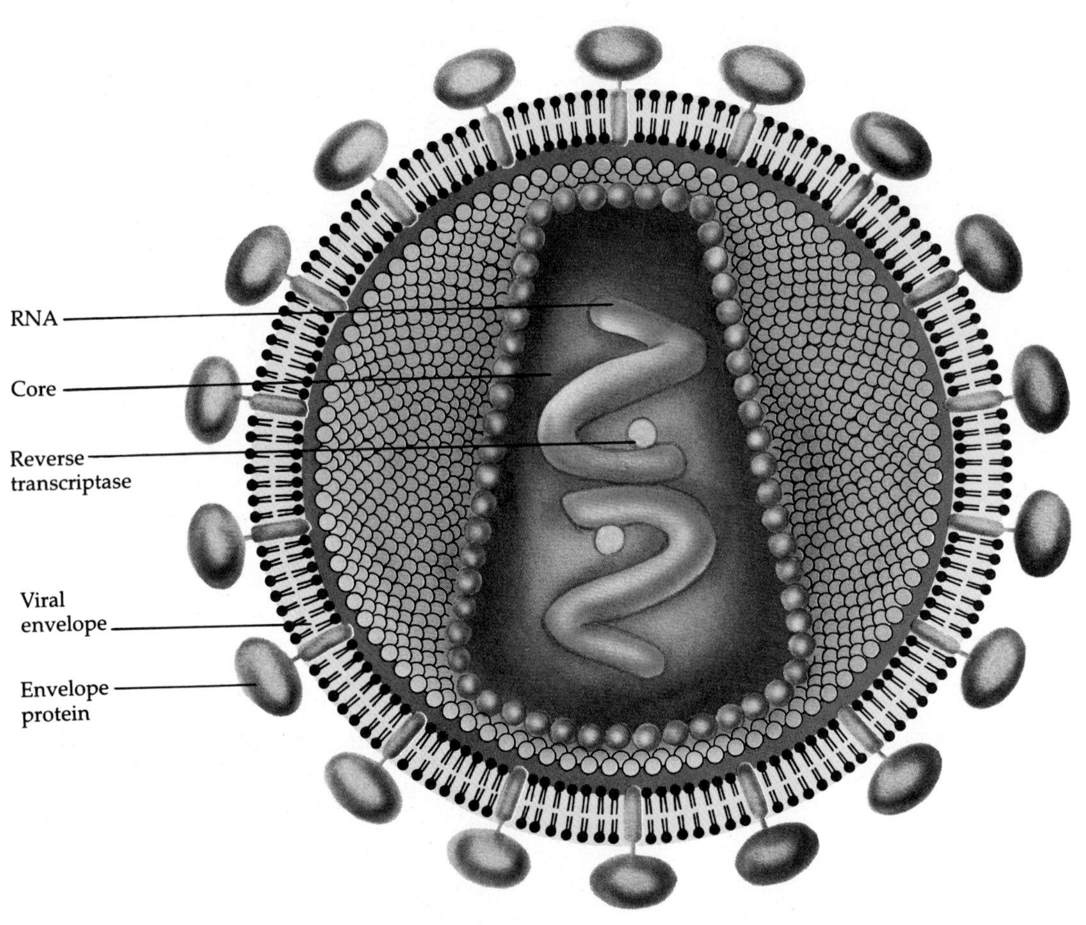

FIGURE 19.11 The HIV virion (cross-section). The protein molecules of the viral envelope form knobs that project outward. Within the spherical capsid is a cone-shaped core that contains two molecules of RNA with a molecule of reverse transcriptase attached to each. Once the RNA enters the host cell, the reverse transcriptase transcribes it into DNA.

tests and development of disease conditions such as an overgrowth by the fungus *Candida albicans*. Stage 6 of the infection is the disease condition actually called AIDS; the T4 cell count falls below 100 per milliliter and the patient falls victim to one or several opportunistic infections such as *Pneumocystis* pneumonia (Table 19.7). At the present state of therapy, most persons entering this stage will die within about two years.

The average time from infection to the development of AIDS is about 10 years. Because of this long incubation period, it is not yet known what percentage of persons infected with HIV will develop full-blown AIDS: present indications are that the great majority will.

MODES OF TRANSMISSION

Transmission of HIV usually requires transfer of bodily fluids. The most important of these are blood, semen, and vaginal secretions that contain the virus, or the transfer of cells, especially macrophages, containing the virus. It has been established that the routes of transmission include intimate sexual contact, breast milk, blood-contaminated needles, and blood-to-blood contact such as transfusions. Heterosexual contact is much more likely to transmit HIV when genital lesions are present. This is considered a very important factor in the spread of AIDS in central Africa, where sexually transmitted diseases that result in such lesions are prevalent. Heterosexual spread in Africa is so common that the male: female ratio among AIDS patients is about 1:1 as compared with 13:1 in the United States. Heterosexual transmission is increasing in the United States, however, especially via women who are drug abusers and resort to prostitution to support their addiction. The risk of sexual transmission is minimized by the use of condoms. Saliva may contain the virus, but transmission is not known to occur by kissing.

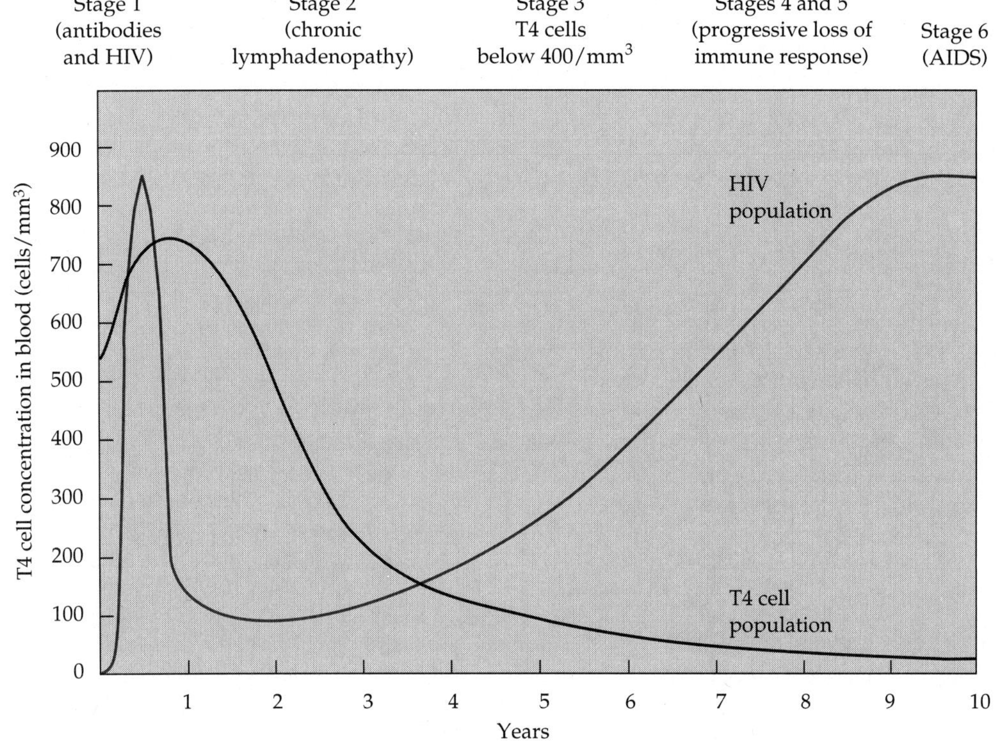

FIGURE 19.12 The stages of an HIV infection. Note how the HIV population increases rapidly upon initial infection and drops rapidly as the immune system responds. Then the HIV population slowly increases, and the T4 cell population slowly decreases. AIDS is the final stage of the process.

TABLE 19.7 Some Common Diseases Associated with AIDS

DISEASE OR AGENT	*DISEASE*
Protozoan	
Pneumocystis carinii	Life-threatening pneumonia
Cryptosporidium muris	Persistent diarrhea
Toxoplasma gondii	Encephalitis
Isospora belli	Gastroenteritis
Viral Diseases	
Cytomegalovirus	Fever, encephalitis, blindness
Herpes simplex virus	Vesicles of skin and mucous membranes
Varicella-zoster virus	Chickenpox and shingles
Epstein–Barr virus	Oral hairy leukoplakia (whitish patches on mucous membranes, probably precancerous)
Bacterial Diseases	
Mycobacterium tuberculosis	Tuberculosis
Mycobacterium avium-intracellulare	May infect many organs; gastroenteritis and other highly variable symptoms
Fungal Diseases	
Cryptococcus neoformans	Life-threatening meningitis
Candida albicans	Overgrowth on mucous membranes
Cancer	
Kaposi's sarcoma	Cancer of skin and blood vessels

MICROBIOLOGY *IN THE NEWS*

New Weapons Against AIDS

HIV's ability to destroy, evade, and even hide inside cells of the human immune system makes it extremely difficult to prevent by vaccination and to treat. Because the virus is so unusual and so deadly, it has stimulated intense research into both conventional and extremely unconventional approaches to vaccination and chemotherapy.

Vaccines

One of the most promising approaches to an HIV vaccine works by turning one of the virus's best weapons against itself. gp120 is a viral envelope glycoprotein that binds to receptors on the host's T cells. This protein enables viruses not only to infect T cells, but also to spread from cell to cell without emerging from the cells and exposing themselves to attack by humoral antibodies. The viral genome directs the synthesis of gp120 on the surface of an infected cell, and the gp120 then binds to receptors on healthy cells. The result is cell fusion and transmission of the virus.

Experimental vaccines are being made with gp120 and other envelope glycoproteins. It is thought that antibodies against gp120 would not only attack free viruses, but also block the gp120 on the surfaces of infected T cells, thus preventing cell fusion and cell-to-cell transmission. Human trials of these and other vaccines have already begun. Initially, these human tests can determine only whether the vaccines produce an immune response and whether they are safe. Determining whether they actually prevent AIDS is much more difficult. Even if an uninfected volunteer became infected with HIV early in these tests, researchers don't expect to know whether the vaccines actually prevent AIDS for 3 to 5 years, because AIDS develops so slowly.

In March 1987, French scientist Daniel Zagury tested a vaccine prepared from genetically engineered vaccinia virus into which the gene for gp120 had been inserted. The vaccinia virus causes host cells to produce gp120 proteins, which bind to the cell membranes. This should then stimulate the production of antibodies against the viral proteins. Zagury, working in Zaire, noted an antibody response after injecting himself with the vaccine. In January 1988, human trials of a similar vaccine (HIVAC-1e) were begun by Oncogen, a division of Bristol-Myers.

During September of 1987, MicroGeneSys, Inc., of West Haven, Connecticut, began human trials of a vaccine that uses a different envelope glycoprotein, gp160. The gp160 gene from a genetically engineered HIV-1 (supplied by the National Institute of Allergy and Infectious Diseases, NIAID) is inserted into baculovirus, which infects moths and butterflies. The virus is grown in cultures, and the viral protein is harvested from the cultures and used in the vaccine.

NIAID scientists are studying the effects of the vaccine in male volunteers who are healthy and free of AIDS. By 1990, one-third of the volunteers had developed anti-HIV. Cell-mediated immunity and the effectiveness of these antibodies in vivo have not been evaluated.

In 1990, the University of Washington announced plans to test HIVAC-1e and gp160 vaccines in human volunteers. This combination of vaccines successfully protected mice against HIV infection. Another vaccine consisting of a synthetic protein (gp41) carried by the Sabin vaccine attenuated poliovirus has been developed by Albert Sabin. Twenty-four infected patients have been given this vaccine.

Although these envelope glycoprotein-based vaccines hold promise, research on other types of vaccine is also under way. Interestingly, Jonas Salk and Albert Sabin, the researchers responsible for different polio vaccines, are also each developing an HIV vaccine. Jonas Salk has begun trials of a killed-HIV vaccine in infected volunteers at the University of Southern California. The virus is killed with gamma radiation and freeze dried. No adverse side-effects have been observed in the 19 human volunteers. Seven of the 19 have shown an increase in CD4 cells, indicating an improved immune condition. Another approach is being tried in England, where trials of anti-idiotypic antibodies are in progress. Finally, NIAID scientists are suggesting that passive immunization of infants born to AIDS-infected women may prevent infection of the newborns.

There have been recorded instances, though, of sexual transmission of HIV in persons whose only sexual contacts were oral-genital.

Worldwide epidemiological studies indicate three geographic patterns of transmission: (1) Transmission is primarily among homosexual or bisexual males and intravenous drug abusers in North America, Western Europe, Australia, and New Zealand. (2) Heterosexual contact is the primary mode of transmission in African and Caribbean countries. (3) Most AIDS cases in Eastern Europe, the Middle East, and Asia have occurred among people who have traveled to endemic areas and had sexual contact with infected homosexual men and female prostitutes. Lax procedures for needle sterilization have resulted in a number of hospital-associated outbreaks of AIDS in the Soviet Union and Romania.

Chemotherapy

The viral enzyme reverse transcriptase may provide a potential target for chemotherapeutic agents. In addition, several newly described genes or their products may provide potential targets. These genes include *orf*, whose product controls latency by suppressing viral replication, and *sos*, whose product is responsible for infectivity.

AZT (azidothymidine) was approved for treating AIDS in 1987 (Retrovir®, Burroughs Wellcome Co.). AZT works by competitive inhibition. Reverse transcriptase inserts it into a growing molecule of DNA where thymidine would normally go. The DNA molecule is terminated because the next nucleoside cannot be added to the AZT. The observed effects of AZT are a decrease in the incidence of opportunistic infections, Kaposi's sarcoma, and conversion to stage 3 AIDS. However, AZT helps only certain AIDS patients. It also has side-effects that include immunosuppression and anemia. Based on the success of AZT, other nucleoside analogs are currently being tested (see the table). In 1991, ddI was approved for patients who do not respond to AZT.

NIAID and 44 other medical centers are involved in testing new drugs in human volunteers. These phase I studies are intended to establish maximum tolerated doses and pharmacologic effects.

The use of immunotoxins—an antibody bound to a toxin—has been suggested. The antibody would bind to glycoproteins on the surface of infected cells, and the toxin would kill the targeted cell. Two immunotoxins are being produced, using a cytotoxic *Pseudomonas* exotoxin and ricin, a protein synthesis inhibitor.

Biochemists at the National Cancer Institute are using an antisense RNA that binds to the mRNA transcribed from HIV's *rev* gene. Products of the *rev* gene initiate viral synthesis. If antisense RNA binds to the *rev* mRNA, the proteins would not be made.

The most radical approaches to AIDS therapy have employed radiation. One patient was exposed to total-body radiation to kill every blood and bone marrow cell. He then received a bone marrow transplant and AZT. The patient died of a recurrence of cancer, but no HIV was found in his body during the remaining two weeks of his life. Five patients with early AIDS symptoms have undergone photopheresis. In this procedure, some of the patient's blood is removed once a month and exposed to ultraviolet radiation before being returned to the patient's body. The AIDS symptoms of swollen lymph nodes, fever, and night sweats were lessened in all five patients after 9 months of therapy.

Because of the rapid progress that scientists have made since the AIDS virus was discovered in 1984, many scientists are optimistic that AIDS will one day be as well controlled as measles and smallpox are today. Nevertheless, it will take years of continued research and testing to discover effective drugs and vaccines.

AIDS DRUGS CURRENTLY BEING TESTED

Nucleoside analogs Dideoxycytidine (ddc) Dideoxyadenosine (ddA) Dideoxyinosine (ddI) Ribavirin Phosphorothioate deoxynucleotides	Stop DNA synthesis by competitive inhibition
Castanospermine	Inhibits action of glucosidase necessary for HIV biosynthesis
Dextran sulfate	Blocks attachment of viruses
Rifabutin	Inhibits reverse transcriptase
Alpha interferon	Prevents virus replication
Ampligen	Induces interferon
Phosphonoformate	Inhibits reverse transcriptase
Interleukin-2	Causes proliferation of T cells
Peptide T	Blocks virus receptor sites
Soluble CD4	Virus binds to this instead of to real CD4 sites on lymphocytes

AIDS is not transmitted by insects or by casual social contact such as hugging and sharing household facilities, drinking glasses, or towels. Transmission by blood transfusion in developed countries is unlikely because blood is tested for AIDS antibodies. However, there will always be a slight risk, because blood might be donated during the interval between infection and appearance of detectable antibodies. Tests for the virus itself are also available, but they have not proved to be superior in screening blood. There is no risk at all in donating blood. AIDS has been transmitted by organ transplants and artificial insemination with donated sperm. HIV-positive women should not become pregnant because of the probability (at least 30%) of transmitting the virus to the fetus.

VACCINES AND TREATMENTS

There are two basic areas of AIDS research. Some researchers are working on vaccines to prevent the disease, and others are looking for drugs to treat AIDS.

Vaccines

There are great obstacles to production of an AIDS vaccine, among them the lack of a suitable animal host for the virus. However, researchers are now optimistic that a vaccine, once produced, could be effective. One reason for optimism is that a very few persons who were once HIV-positive have spontaneously become HIV-negative, indicating that the immune system is probably capable in rare instances of eliminating the virus. Because of the extreme virulence of the virus, many think it unlikely that any whole-virus vaccine, either killed or attenuated, would be acceptable for use on uninfected persons. However, such a vaccine would be acceptable for use in attempting to clear the virus after infection. Most efforts are directed at subunit vaccines (see Chapter 18) based on surface envelope antigens of the virus. However, such vaccines must overcome the problem of many antigenic variants, which we mentioned earlier. In addition, an effective vaccine would have to stimulate cell-mediated as well as humoral immunity to deal with HIV contained within macrophages or other cells.

Chemotherapy

A promising approach to arresting an HIV infection is to flood the body with artificially produced, soluble CD4-type molecules that would bind to circulating viruses before they could locate a CD4 receptor on a T cell. In early experiments the interceptor CD4 molecules were rapidly degraded, requiring repeated injections at an impractical rate. The soluble CD4 would have to be modified in some way so as to remain in circulation for an extended time for this approach to be practical.

Most early anti-HIV drugs such as Zidovudine® (AZT) are inhibitors of the enzyme reverse transcriptase. HIV is a retrovirus, you will recall, that copies RNA into DNA. The drugs, mostly analogs of nucleic acids, trick the enzyme into terminating the synthesis of viral DNA. Such drugs have slowed the progress of the disease but have not led to a cure. Other than the reverse transcriptase step, there are at least 13 other points at which the production of HIV could be selectively interrupted by drugs. For example, because it has no equivalent in human cells, an attractive target is the viral enzyme protease. This enzyme cuts proteins into pieces that are then reassembled into new HIV particles. Inhibiting it would prevent viral synthesis. Numerous other approaches are being intensively studied, and the most successful solutions might well be some that are not even anticipated now.

THE FUTURE OF AIDS

The Centers for Disease Control estimate that about 1 million persons in the United States are now infected with HIV and projects that 365,000 AIDS cases will have been diagnosed in the United States by the end of 1992 and 263,000 patients will have died by that time. The annual cost for the care of AIDS patients will be in the billions of dollars. Worldwide, the estimates are that there are more than 1 million total cases of AIDS and that 10 million persons are infected.

The AIDS epidemic gives clear evidence of the value of basic scientific research. It is important to reflect that without the advances of the past few decades in molecular biology, we would have been unable even to identify the agent of AIDS, to develop the tests used to screen donated blood, or to monitor the course of the infection.

Immune Response to Cancer

Cancer, like an infectious disease, represents a failure of the body's defenses, including the immune system. One of the most promising avenues for effective therapy of cancer makes use of immunological techniques.

IMMUNOLOGICAL SURVEILLANCE

The immune system's presumed responsibility for patrolling the body for cancer cells is called **immunological surveillance.** Persons who are immunosuppressed by either natural or artificial means are markedly more susceptible to cancers than the rest of the population. The conventional explanation for this susceptibility is that the immune system normally recognizes and destroys cancer cells before they become established in the body. The tendency of cancer to occur most often in the elderly, whose immune systems are thought to be less efficient, or in the very young, whose immune systems may not have developed properly, is thought to support this view.

A cell becomes cancerous when it undergoes transformation (see Chapter 13) and begins to proliferate without control. Because of transformation, the surfaces of tumor cells may acquire tumor-associated antigens that mark them as nonself to the immune system (Figure 19.13).

IMMUNOLOGICAL ESCAPE

Cancer sometimes occurs even in people with presumably normal immune systems. Once established, the cancer seems to become resistant to immune rejection,

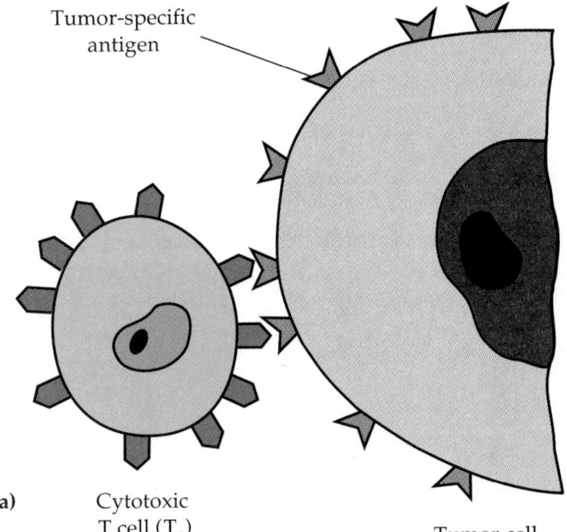

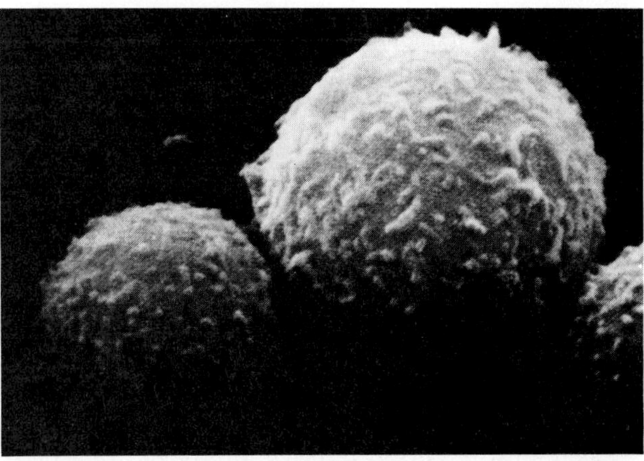

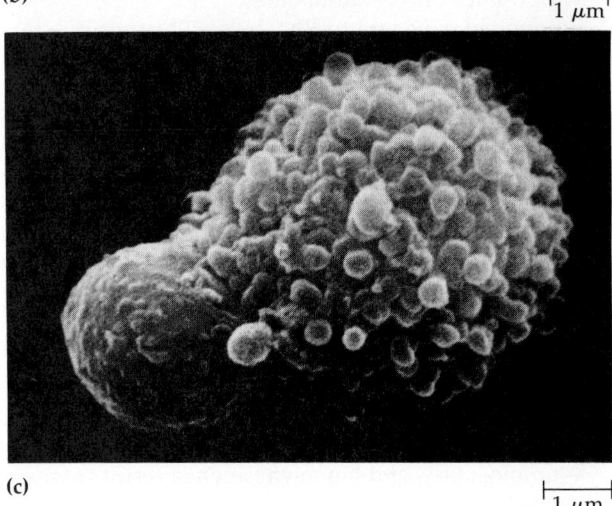

FIGURE 19.13 Interaction of a cytotoxic T (T_C) cell and a tumor cell. **(a)** Diagram of a cytotoxic T cell binding to a tumor-associated antigen on the surface of a tumor cell. **(b)** Scanning electron micrograph showing a cytotoxic T cell (smaller sphere at the lower left) attaching to the tumor cell after recognizing the tumor-associated antigens. **(c)** Lysis of the tumor cell is indicated by the deep folds and swelling in its surface membranes.

even though it is easy to demonstrate that cell-mediated and humoral immunity are directed against the tumor-associated surface antigens. This resistance to rejection is called **immunological escape,** and several mechanisms have been proposed to explain it. One possibility is *antigen modulation,* in which tumor cells are believed to shed their associated antigens, thus evading recognition by the immune system. Another proposed mechanism is that some cancers, such as Hodgkin's disease, release factors that eventually suppress the entire cell-mediated immune system.

IMMUNOTHERAPY

At the turn of the century, Dr. William B. Coley at a New York City hospital observed that if cancer patients contracted typhoid fever, their cancers often diminished noticeably. Following this lead, Coley made vaccines of killed bacteria, called Coley's toxins, to simulate a bacterial infection. Some of this work was very promising, but its results were inconsistent, and advances in surgery and radiation treatment caused it to be nearly forgotten. We now know that endotoxins from gram-negative bacteria are powerful stimulants for the production of **tumor necrosis factor** by macrophages. Tumor necrosis factor is a small protein that interferes with the blood supply of cancers in animals. There is now also much interest in cancer therapy with tumor necrosis factor and other cytokines (see Chapter 17) such as interleukin-2 and interferons.

The treatment of cancer by immunological means is termed **immunotherapy.** It seems likely to be an ef-

TABLE 19.8 Principal Types of Cancers

TYPE OF CANCER	DESCRIPTION
Carcinoma	The most common form of cancer; arises from the cells forming the skin, the glands (such as the mammary and prostate), the uterus, and the membranes of the respiratory and gastrointestinal tracts; metastasizes (spreads to other parts of the body) mainly through the lymph vessels
Leukemia	Cancer of the tissues that form blood; characterized by uncontrolled multiplication and accumulation of abnormal white blood cells
Lymphoma	Cancer of the lymph nodes
Sarcoma	Another form of malignant tumor; arises from connective tissues, such as muscle, bone, cartilage, and membranes covering muscles and fat; metastasizes through the blood vessels

fective future approach to cancer treatment. One promising technique involves immunotoxins. In Chapter 17, we defined an immunotoxin as a combination of a monoclonal antibody and a toxic agent, such as the drug ricin or a radioactive compound. The monoclonal antibody is targeted at a particular type of tumor antigen. The antibody selectively locates the cancer cell, and the attached toxic agent destroys the cell but causes little or no damage to healthy tissue. There is great hope that this approach will become analogous to the use of antibiotics in treating infectious diseases. The principal types of cancers are summarized in Table 19.8.

During the final years of this century, we can expect to see more advances in the battle against cancer than in all of previous history. Almost certainly, immunology will be one of the major weapons in that fight.

STUDY OUTLINE

Hypersensitivity (pp. 467–468)

1. Hypersensitivity reactions represent immunologic responses to an antigen (allergen) that lead to tissue damage rather than immunity.
2. Hypersensitivity reactions occur only when a person has been sensitized to an antigen.
3. Hypersensitivity reactions can be divided into four classes: Types I, II, and III are immediate reactions based on humoral immunity, and type IV is a delayed reaction based on cell-mediated immunity.

Type I (Anaphylaxis) Reactions (pp. 468–471)

1. Anaphylaxis reactions involve the production of IgE antibodies that bind to mast cells and basophils to sensitize the host.
2. Binding of two adjacent IgE antibodies to an antigen causes the target cell to release chemical mediators, such as histamine, leukotrienes, and prostaglandins, which cause the observed allergic reactions.
3. Systemic anaphylaxis may develop in minutes after injection or ingestion of the antigen; this may result in circulatory collapse and death.
4. Localized anaphylaxis is exemplified by hives, hay fever, and asthma.
5. Skin testing is useful in determining sensitivity to an antigen.
6. Desensitization to an antigen can be achieved by repeated injections of the antigen, which leads to the formation of blocking (IgG) antibodies and T_S cells.

Type II (Cytotoxic) Reactions (pp. 471–474)

1. Type II reactions are mediated by IgG or IgM antibodies and complement.
2. The antibodies are directed toward host cell antigens; complement fixation may result in cell lysis, phagocytosis, or K cell activity.

THE ABO BLOOD GROUP SYSTEM (p. 471)

1. Human blood may be grouped into four principal types, designated A, B, AB, and O.
2. The presence or absence of two carbohydrate antigens designated A and B on the surface of the red blood cell determines a person's blood type.
3. Naturally occurring antibodies are present or absent in serum against the opposite AB antigen.
4. Incompatible blood transfusions lead to the complement-mediated lysis of the donor red blood cells.

THE Rh BLOOD GROUP SYSTEM (pp. 471–473)

1. Approximately 85% of the human population possesses another blood group antigen, designated the Rh antigen; these individuals are designated Rh^+.
2. The absence of this antigen in certain individuals (Rh^-) can lead to sensitization upon exposure to it.

Blood Transfusions and Rh Incompatibility (p. 472)

1. An Rh^+ person can receive Rh^+ or Rh^- blood transfusions.
2. When an Rh^- person receives Rh^+ blood, that person will produce anti-Rh antibodies.
3. Subsequent exposure to Rh^+ cells will result in a rapid hemolytic reaction.

Hemolytic Disease of the Newborn (pp. 472–473)

1. An Rh^- mother carrying an Rh^+ fetus will produce anti-Rh antibodies.
2. Subsequent pregnancies involving Rh incompatibility may result in hemolytic disease of the newborn.
3. The disease may be prevented by passive immunization of the mother with anti-Rh antibodies.

DRUG-INDUCED CYTOTOXIC REACTIONS (pp. 473–474)

1. In the disease thrombocytopenic purpura, platelets are destroyed by antibodies and complement.
2. Agranulocytosis and hemolytic anemia result from antibodies against one's own blood cells coated with drug molecules.

AUTOIMMUNE DISORDERS (TYPE II REACTIONS) (p. 474)

1. Graves' disease results in excessive thyroid hormone production when antibodies react with the thyroid gland.
2. Myasthenia gravis results in muscle weakness when antibodies react with acetylcholine receptors.

Type III (Immune Complex) Reactions (pp. 474–475)

1. Immune complex diseases occur when IgG antibodies and soluble antigen form small complexes that lodge in the basement membranes of cells.
2. Subsequent complement fixation results in inflammation.
3. Serum sickness, glomerulonephritis, systemic lupus erythematosus, and rheumatoid arthritis are immune complex diseases.

Type IV (Cell-Mediated) Reactions (pp. 475–478)

1. Delayed hypersensitivity responses are due primarily to T_D cell proliferation.
2. Sensitized T cells secrete lymphokines in response to the appropriate antigen.
3. Lymphokines attract and activate macrophages and initiate tissue damage.
4. The tuberculin skin test and allergic contact dermatitis are examples of delayed hypersensitivities.
5. Lymphocytic choriomeningitis, Hashimoto's thyroiditis, and Addison's disease are cell-mediated autoimmune diseases.

LOSS OF IMMUNOLOGICAL TOLERANCE (p. 476)

1. Immunological tolerance represents a state of unresponsiveness to a specific antigen.
2. Tolerance to self-antigens occurs under natural conditions and develops during fetal development.
3. Autoimmunity is a humoral or cell-mediated immune response against self-antigens. Autoimmune responses frequently result in disease.

MAJOR HISTOCOMPATIBILITY COMPLEX (pp. 476–478)

1. Histocompatibility antigens located on cell surfaces express genetic differences between individuals; these antigens are coded for by MHC or HLA gene complexes.
2. To prevent rejection of transplants, HLA and ABO blood group antigens of the donor and recipient are matched as closely as possible.

Transplantation (pp. 478–480)

TYPES OF TRANSPLANTS (pp. 478–480)

1. Transplantation to a privileged site (such as the cornea) or of a privileged tissue (such as pig heart valves) does not cause an immune response.
2. Four types of transplants have been defined on the basis of genetic relationships between the donor and the recipient: autografts, isografts, allografts, and xenografts.
3. Bone marrow (with immunocompetent cells) can cause graft-versus-host disease.

IMMUNOSUPPRESSION (p. 480)

1. Successful transplant surgery often requires immunosuppressive treatment, which increases the host's susceptibility to infections and cancer.
2. Cyclosporine blocks synthesis of certain lymphokines to inhibit T-cell growth.

Natural Immune Deficiencies (p. 480)

1. Immune deficiencies may be inherited.
2. A variety of diseases can also impair the immune response.

Acquired Immunodeficiency Syndrome (AIDS) (pp. 480–486)

THE ORIGIN OF AIDS (p. 480)

1. HIV is thought to have originated in central Africa and was brought to other countries by immigration, vacationing, and imported blood products.

THE HIV INFECTION (pp. 480–482)

1. AIDS is caused by the retrovirus HIV, which kills cells with CD4 antigen, especially T_H cells.
2. HIV can remain latent or begin replication at once.
3. The virus evades host defenses by antigenic changes or latency.
4. Viral replication can kill host cells or host cells can be killed by the immune response to viral antigens on their surfaces.
5. Initial symptoms may be fever and malaise.
6. The second stage includes swollen lymph nodes, loss of appetite, fever, and rashes.
7. In the final stage, HIV infection results in opportunistic infections that eventually cause the death of the host.

MODES OF TRANSMISSION (pp. 482–485)

1. HIV is transmitted in blood, semen, vaginal secretions, and breast milk and across the placenta.
2. Blood transfusions are not a likely source of infection because blood is tested for HIV antibodies.
3. In North America and Western Europe, HIV is transmitted by homosexual activity and intravenous drug use.
4. In Africa and the Caribbean, HIV is transmitted by heterosexual activity.
5. In Eastern Europe and Asia, HIV infection occurs in people who became infected via sexual activity while traveling in endemic areas.

VACCINES AND TREATMENTS (p. 486)

1. Subunit vaccines are being tested.
2. Drugs aimed at reverse transcriptase are being used. Viral protease is another possible drug target.
3. Artificial CD4 molecules may be useful because viruses may bind to them instead of to cells.

THE FUTURE OF AIDS (p. 486)

1. The number of AIDS cases is expected to increase during the next few years, with females making up an increasing percentage.

Immune Response to Cancer (pp. 486–488)

IMMUNOLOGICAL SURVEILLANCE (p. 486)

1. Cancer cells are normal cells that have undergone transformation, fail to exhibit contact inhibition, and possess tumor-associated antigens.

2. The response of the immune system to cancer is called immunological surveillance.

IMMUNOLOGICAL ESCAPE (pp. 486–487)

1. The ability of tumor cells to escape immune responses against them is called immunological escape.
2. Antigen modulation and immunological enhancement represent two ways tumor cells escape detection and destruction by the immune system.

IMMUNOTHERAPY (pp. 487–488)

1. Tumor necrosis factor, interleukin-2, and interferons may be useful in treating some cancers.
2. Immunotoxins are chemical poisons with a monoclonal antibody; the antibody selectively locates the cancer cell for release of the poison.

STUDY QUESTIONS

Review

1. Define hypersensitivity.
2. Compare and contrast the characteristics of the four types of hypersensitivity reactions. Given an example of each type.
3. List three mediators released in anaphylactic hypersensitivities, and explain their effects.
4. Contrast systemic and localized anaphylactic reactions. Which type is more serious? Cite an example of each.
5. Explain what happens when a person develops a contact sensitivity to poison oak.
   - **(a)** What causes the observed symptoms?
   - **(b)** How did the sensitivity develop?
   - **(c)** How might this person be desensitized to poison oak?
6. Discuss the roles of antibodies and antigens in an incompatible tissue transplant.
7. What happens to the recipient of an incompatible blood type?
8. Explain how hemolytic disease of the newborn develops and how this disease might be prevented.
9. Which type of graft (autograft, isograft, allograft, or xenograft) is most compatible? Least compatible?
10. Which of the following blood transfusions is compatible? Explain your answers.

	Donor	Recipient
(a)	AB, Rh^-	AB, Rh^+
(b)	B, Rh^+	B, Rh^-
(c)	A, Rh^+	O, Rh^+

11. Define immunosuppression. Why is it used, and what problems does it cause?
12. Define autoimmunity. Present a theory that could explain autoimmune responses. Discuss one autoimmune disease in relation to this theory.
13. Summarize the causes of natural immune deficiencies. What is the effect of an immune deficiency?
14. In what ways do tumor cells differ antigenically from normal cells? Explain how tumor cells may be destroyed by the immune system.
15. If tumor cells can be destroyed by the immune system, how does cancer develop? What does immunotherapy involve?
16. Do people with AIDS make antibodies? If so, why are they said to have an immune deficiency?

Challenge

1. When and how does our immune system discriminate between self and nonself antigens?
2. The first preparations used for artificially acquired passive immunity were antibodies in horse serum. A complication that resulted from the therapeutic use of horse serum was immune complex disease. Why did this occur?
3. After working in a mushroom farm for several months, a worker develops these symptoms: hives, edema, and swelling of lymph nodes.
   - **(a)** What do these symptoms indicate?
   - **(b)** What mediators cause these symptoms?
   - **(c)** How may sensitivity to a particular antigen be determined?
   - **(d)** Other employees do not appear to have any immunological reactions. What could explain this?

   (*Note:* The allergen is conidiospores from molds growing in the mushroom farm.)
4. Physicians administering live, attenuated mumps and measles vaccines prepared in chick embryos are instructed to have epinephrine available. Since epinephrine will not treat these viral infections, what is the purpose of keeping this drug on hand?
5. What treatments are administered to a person with an immune deficiency?

FURTHER READING

"AIDS." *Science* 239:573–623, February 5, 1988. This special issue includes eight articles on AIDS: virus biology, epidemiology, and legal and ethical issues.

Atkinson, M.A., and N.K. Maclaren. "What causes diabetes?" *Scientific American* 263(1):62–71, July 1990. Describes the development of insulin-dependent diabetes and the role of T cells in destroying insulin-producing cells.

Buisseret, P.D. "Allergy." *Scientific American* 247(2):86–95, August 1982. Discusses the roles of mast cells, IgE, and antigens in hypersensitivity responses.

Cohen, I.R. "The self, the world and autoimmunity." *Scientific American* 258(4):52–60, April 1988. Describes recognition of self by T cells and suggests vaccine treatment for autoimmune diseases.

Collier, J.R., and D.A. Kaplan. "Immunotoxins." *Scientific American* 251(1):56–64, July 1984. Illustrates how cancer cells might be targeted for destruction by a monoclonal antibody carrying a toxic agent.

Feldman, M., and L. Eisenbach. "What makes a tumor cell metastatic?" *Scientific American* 259(5):30–85, November 1988. Discussion of the use of killer cells to destroy cancerous cells and experiments to turn metastatic cells into benign cells.

Lockey, R.F., and S.C. Bukantz, eds. "Primer on allergic and immunologic diseases," 2nd ed. *Journal of the American Medical Association* 258(20):2829–3034, November 27, 1987. A special issue containing 26 articles on abnormal immune responses.

Rennie, J. "The body against itself." *Scientific American* 263(6):107–115, December 1990. A discussion of current trends in immunology including treatments for autoimmune diseases.

Rodger, J.C., and B.L. Drake. "The enigma of the fetal graft." *American Scientist* 75:51–57, January-February 1987. Addresses the question of why the mother doesn't reject the fetus, which is genetically foreign tissue.

Rosenberg, S.A. "Adoptive immunotherapy for cancer." *Scientific American* 262(5):62–69, May 1990. Describes the use of cell-transfer therapy to treat cancer.

"What science knows about AIDS." *Scientific American* 259(4), October 1988. This issue is devoted entirely to AIDS. Ten articles cover a wide range of topics, including molecular biology, epidemiology, and treatments.

CHAPTER 20

Antimicrobial Drugs

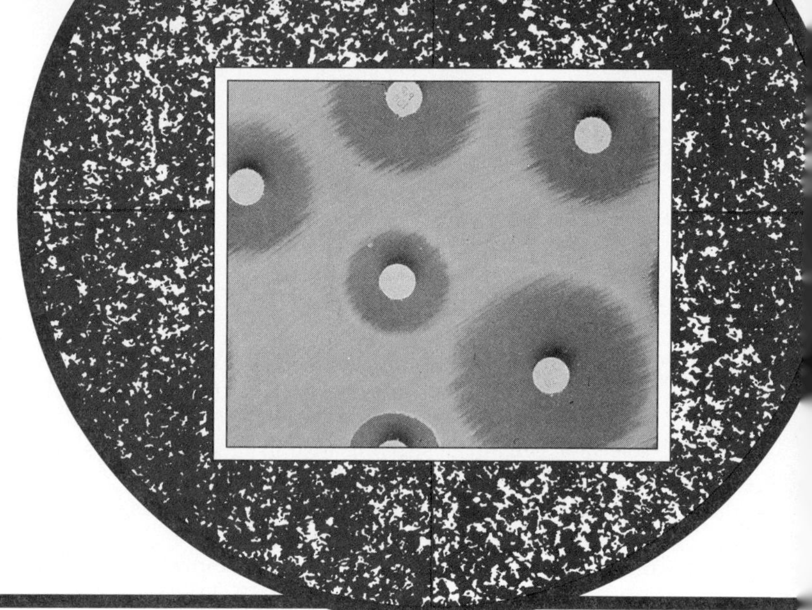

LEARNING OBJECTIVES

- Define a chemotherapeutic agent, and distinguish between a synthetic drug and an antibiotic.
- Identify the contributions of Paul Ehrlich and Alexander Fleming to chemotherapy.
- List the criteria used to evaluate antimicrobial agents.
- Identify five methods of action of antimicrobial agents.
- Describe the methods of action of each of the commonly used antibacterial drugs.
- Describe the problems of chemotherapy for viral, fungal, protozoan, and helminthic infections.
- Explain the actions of currently used antiviral, antifungal, antiprotozoan, and antihelminthic drugs.
- Describe three tests for microbial susceptibility to chemotherapeutic agents.
- Describe the mechanisms of drug resistance.

When the body's normal defenses cannot prevent or overcome a disease, it is often treated with **chemotherapy** (drugs).

In this chapter, we focus on **antimicrobial drugs,** the class of **chemotherapeutic agents** used to treat infectious diseases. Like the disinfectants discussed in Chapter 7, these chemicals act by interfering with the growth of microorganisms. Unlike disinfectants, however, they must often act *within* the host. Therefore, their effects on the cells and tissues of the host are important. The ideal antimicrobial drug kills the harmful microorganism without damaging the host, the principle of **selective toxicity.**

Drugs used in the chemotherapy of infectious disease are classified into two groups. Drugs that have been synthesized by chemical procedures in the laboratory are called **synthetic drugs.** Drugs produced by bacteria and fungi are called **antibiotics.** The distinction is often ignored in practice.

Historical Development

The birth of modern chemotherapy is credited to the efforts of Paul Ehrlich in Germany during the early part of this century. While attempting to stain bacteria without staining the surrounding tissue, he speculated about some "magic bullet" that would selectively find and destroy pathogens but not harm the host. This idea provided the basis for chemotherapy, a term he coined. Recall from Chapter 1 that Ehrlich eventually discovered a chemotherapeutic agent called salvarsan, an arsenic derivative that was effective against syphilis (by the standards of the day). Prior to Ehrlich's discovery, there had been only one effective chemotherapeutic agent in the medical arsenal—quinine, a drug used for the treatment of malaria.

In the late 1930s, the discovery of sulfa drugs, a group of synthetic drugs, touched off a new interest in chemotherapy. This discovery evolved from a systematic survey of chemicals, including many synthesized derivatives of aniline dyes. The chemical prontosil, which had been first synthesized and used as a dye many years before, was found to be an effective antimicrobial agent. Oddly enough, it worked only within living animals and was completely ineffective when exposed to the same bacteria in a test tube. The active ingredient in prontosil proved to be *sulfanilamide,* which forms as the animal metabolizes prontosil. This discovery led to the rapid development of a number of related drugs—the *sulfonamides,* or *sulfa drugs,* which are still used today.

We also mentioned in Chapter 1 that, in 1928, Alexander Fleming observed that the growth of the bacterium *Staphylococcus aureus* was inhibited in the area surrounding the colony of a mold that had contaminated a Petri plate (Figure 20.1). The mold was identified as *Penicillium notatum,* and its active compound, which was isolated a short time later, was named *penicillin.* Similar inhibitory reactions between colonies on solid media are commonly observed in microbiology, and the mechanism of inhibition is called *antibiosis.* From this term comes the term *antibiotic,* meaning a substance produced by microorganisms that in small amounts will inhibit another microorganism. There-

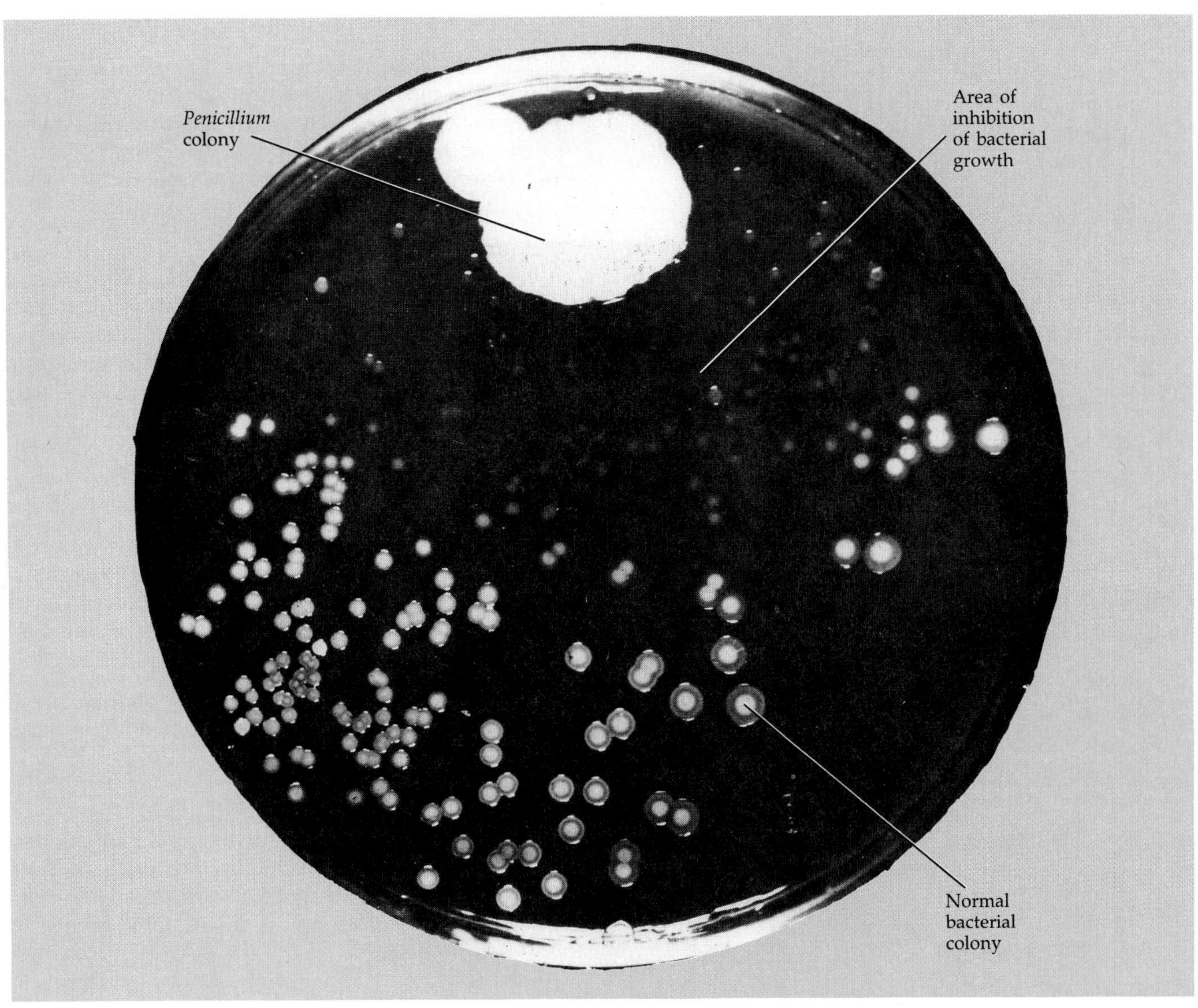

FIGURE 20.1 Discovery of penicillin. Alexander Fleming took this photograph in 1929. The colony of *Penicillium* mold accidentally contaminated the plate and is inhibiting nearby bacterial growth.

fore, the wholly synthetic sulfa drugs technically are not antibiotics.

In 1940, a group of scientists at Oxford University headed by Howard Florey and Ernst Chain succeeded in the first clinical trials of penicillin. Intensive research in the United States then led to the isolation of especially productive *Penicillium* strains for use in mass production of the antibiotic. (The most famous of these high-producing strains was originally isolated from a cantaloupe bought at a market in Peoria, Illinois.) Penicillin is still one of our most effective antibiotics, and its enormous success led to the search for others, a search that has revolutionized medicine. You might be interested to know that penicillin has a singular toxicity for guinea pigs, and it is entirely possible that this drug would never have left the experimental laboratory under the more demanding safety requirements that must be met today in the development of drugs.

Antibiotics are actually rather easy to discover, but few are of medical or commercial value. Some are used commercially other than to treat disease, for example, as a supplement in animal feed (see the box on p. 495). Many antibiotics are toxic to humans or lack any advantage over antibiotics already in use.

More than half of our antibiotics are produced by species of *Streptomyces,* filamentous bacteria that commonly inhabit soil. A few antibiotics are produced by bacteria of the genus *Bacillus,* and others are produced by molds, mostly of the genera *Penicillium* and *Cephalosporium* (sef-ä-lō-spô'rē-um). See Table 20.1 for the sources of many antibiotics in use today—a surprisingly limited group of organisms.

TABLE 20.1 Representative Sources of Antibiotics

MICROORGANISM	*ANTIBIOTIC*
Gram-Positive Rods	
Bacillus subtilis	Bacitracin
Bacillus polymyxa	Polymyxin
Actinomycetes	
Streptomyces nodosus	Amphotericin B
Streptomyces venezuelae	Chloramphenicol
Streptomyces aureofaciens	Chlortetracycline (Aureomycin®) and tetracycline
Streptomyces erythraeus	Erythromycin
Streptomyces fradiae	Neomycin
Streptomyces noursei	Nystatin
Streptomyces griseus	Streptomycin
Micromonospora purpurea	Gentamicin
Fungi	
Cephalosporium	Cephalothin
Penicillium griseofulvum	Griseofulvin
Penicillium notatum	Penicillin

Criteria for Evaluating Antimicrobial Drugs

Antimicrobial drugs can be evaluated by a number of criteria, including the following:

1. **The drug should demonstrate selective toxicity.** This means that, at the optimum concentration, the drug should be toxic for the microorganism but not for the host. In practice, some toxicity or side-effects are often considered acceptable.
2. **The drug should not produce hypersensitivity (allergy) in most hosts** (see Chapter 19). Allergic reactions, which are produced by the immune system, are not the same as toxicity. Penicillin, for example, sometimes causes allergic reactions but is one of the least toxic antibiotics for humans.
3. **A drug must be soluble in body fluids so that it can rapidly penetrate body tissues.** The rates at which the drug is broken down and excreted from the body must be low enough that the drug remains in the infected body tissue long enough to exert its effects. Certain antibiotics are especially valuable because they penetrate into the brain—later we will discuss the blood–brain barrier. Other drugs penetrate into bacteria-containing macrophages, concentrate in the urine, or have similar advantages.
4. **Microorganisms should not readily become resistant to the drug.** Resistance can be caused by a number of mechanisms and will be discussed later in this chapter.

Developers of a drug attempt to obtain the best possible combination of properties for effective human use.

Spectrum of Activity

It is comparatively simple to find or develop drugs that are effective against procaryotic cells (bacteria) and do not affect the eucaryotic cells of humans. These two cell types differ substantially in many ways, such as in the presence or absence of cell walls, the fine structure of their ribosomes, and details of their metabolism. Thus, selective toxicity has numerous targets. The problem is more difficult when the pathogen is a eucaryotic cell, such as a fungus, protozoan, or helminth. At the cellular level, these organisms resemble the human cell much more closely than a bacterial cell

MICROBIOLOGY *IN THE NEWS*

Antibiotics in Animal Feed Linked to Human Disease

Thirty years ago livestock growers began using antibiotics in the feed of closely penned animals that were being fattened for market. The drugs helped reduce bacterial infections and their spread in such ripe conditions. They also unexpectedly accelerated the animals' growth, an effect that led to wider and wider use. The growth-promoting effects are thought to be due to suppression of intestinal *Clostridium perfringens,* whose toxins may retard the animals' growth.

Meat that reaches the consumer's table is not heavily ridden with antibiotics because the Food and Drug Administration (FDA) has established limits for antibiotic residues in edible tissues. In random-sample testing, only 2.6% of veal calves exceeded the tolerance limits in 1989. Veal would be expected to have the highest residue because the animals do not live long enough to excrete much of the antibiotics. Fewer than 1% of cattle and pigs tested exceeded tolerance limits. In February 1990, the FDA found no antibiotics in milk that was tested. Nevertheless, the subtherapeutic use of antibiotics has come under attack.

The Risk to Humans

Bacteria such as *Salmonella* and *Listeria* can be transferred from animals to humans in meat or milk. The first known case of salmonellosis occurred in Germany in 1888, when 50 people became ill after eating ground beef from one cow. The use of antibiotics in animal feed preferentially allows the growth of strains of bacteria that are resistant to drugs commonly used to treat human infections. In the 1980s, researchers proved that antibiotic-resistant bacteria are being transferred from animals to humans directly in meat or milk. In 1984, Dr. Scott Holmberg and his associates at the Centers for Disease Control (CDC) traced an antibiotic-resistant *Salmonella newport* infection from South Dakota beef cattle to 18 people in four states. Antibiotic-resistant *S. typhimurium* originating from one Illinois dairy infected 16,000 people in seven states in 1985. Reports from other countries have shown similar patterns of resistance. Before 1984 in Europe, *S. dublin* was sensitive to antibiotics. Since 1984, isolates from cattle and from children with salmonellosis are resistant to chloramphenicol and streptomycin.

People taking antibiotics may have a higher risk of infection by resistant organisms. In an outbreak of antibiotic-resistant *S. newport* in California, people taking penicillin for respiratory infections had a higher rate of infection than others. Apparently, penicillin therapy gave the *S. newport* from beef a selective advantage over other bacteria, many of which were killed by the penicillin. In an outbreak of *S. typhimurium* linked to milk consumption in the Midwest, people taking antimicrobials were more rapidly infected by the bacteria. Before 1980, only one case of salmonellosis that was exacerbated by antibiotic therapy had been reported.

The number of reported cases of *Salmonella* infections of humans is increasing, and 20% to 25% of these are due to antibiotic-resistant strains of bacteria. CDC is urging public health laboratories to identify strains of *Salmonella* by studying antibiotic resistance patterns and by using serotype and molecular biology techniques to trace the origins of infection.

New Techniques for Tracing Sources

Molecular biology has provided important new tools for tracing the sources of *Salmonella* infections. In 1985, nearly 1000 cases of infection by *S. newport* resistant to multiple antibiotics were reported in Los Angeles County, California. A radioactive DNA probe was used to identify an identical plasmid carrying multiple antibiotic resistance genes in 99% of the *S. newport* isolates. A technique called plasmid profile analysis was used to track the bacterium from the sick people back to a slaughterhouse, to a meat-deboning plant, and finally to three farms.

In spite of this evidence, cattle growers, feed manufacturers, and the drug industry still argue that the link between antibiotics in feed and human disease is sketchy, and claim that the benefits derived from the use of antibiotics outweigh the potential harmful effects. However, some farmers have reported that their sick animals are not responding to penicillin and tetracycline therapy. Thus, some limitation of antibiotic use may actually benefit the farmer.

does. We shall see that our arsenal against these types of pathogens is much more limited than our arsenal of antibacterial drugs. Viral infections are particularly difficult to treat because the pathogen is within the human host's cells, and the genetic information of the virus is directing the human cell to make viruses rather than to synthesize normal cellular materials. The more we discover about the reproduction, metabolism, and structure of pathogens, the better equipped we are to discover fresh targets for antimicrobials. No doubt the search will never end.

Some antimicrobial drugs have a narrow *spectrum of activity* (range of cells) they target; penicillin, for instance, affects gram-positive bacteria but very few

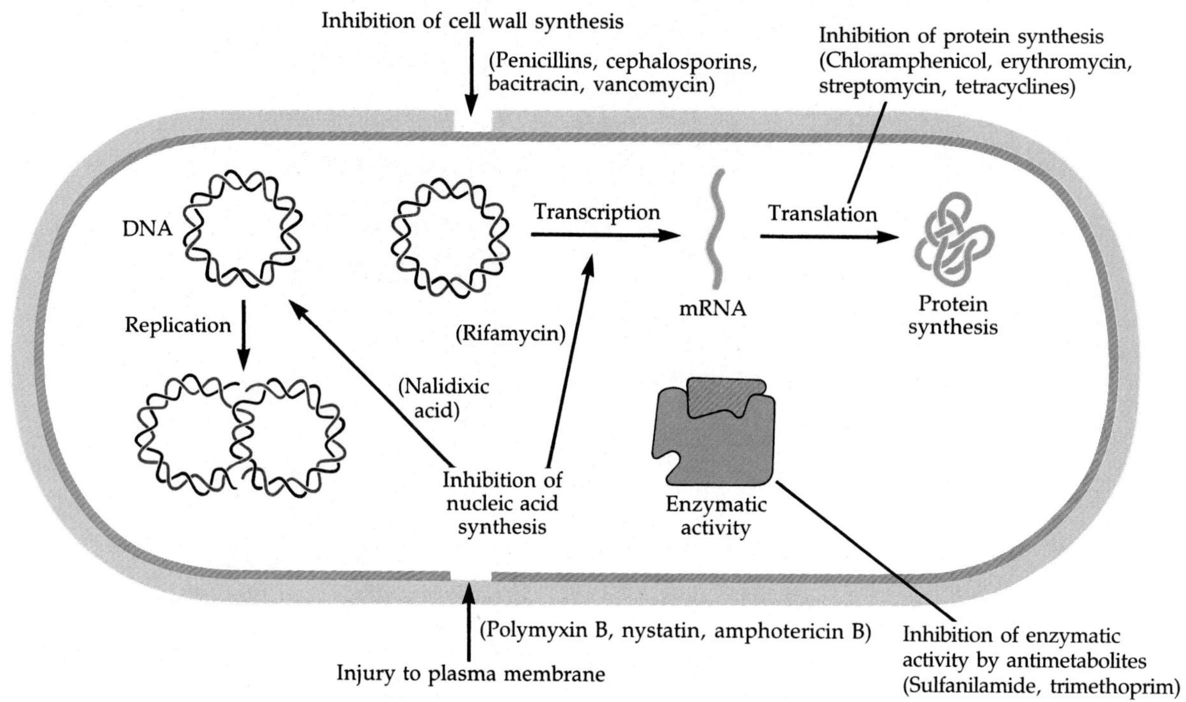

FIGURE 20.2 Summary of the actions of antimicrobial drugs in a highly diagrammatic composite of a microbial cell.

gram-negative bacteria. Other drugs affect a large number of gram-positive or gram-negative bacteria and are therefore called **broad spectrum drugs.** Since the identity of the pathogen is not always immediately known, a broad spectrum drug would seem to have an advantage in the treatment of disease by saving valuable time. The disadvantage is that many members of the normal flora of the host are destroyed by broad spectrum drugs. The normal flora ordinarily competes with and checks the growth of pathogens or other microbes. If certain organisms in the normal flora are not destroyed by the antibiotic and their competitors are destroyed, the survivors may flourish and become opportunistic pathogens. An example is the overgrowth by the yeastlike fungus *Candida albicans,* which is not sensitive to bacterial antibiotics. This overgrowth is called a **superinfection,** a term that is also applied to growth of a target pathogen that has developed resistance to the antibiotic. Such an antibiotic-resistant strain replaces the original sensitive strain, and the infection continues.

Action of Antimicrobial Drugs

We will next examine the various ways in which antibiotics exert their antimicrobial activity. Antimicrobial drugs either kill microorganisms directly **(bactericidal)** or simply prevent them from growing **(bacteriostatic).** In *bacteriostasis,* the host's own defenses, such as phagocytosis and antibody production, usually destroy the microorganisms. The major modes of action are summarized in Figure 20.2.

INHIBITION OF CELL WALL SYNTHESIS

Recall from Chapter 4 that the cell wall of a bacterium consists of a macromolecular network called peptidoglycan. Peptidoglycan is found only in bacterial cell walls. Penicillin and certain other antibiotics prevent the synthesis of intact peptidoglycan; consequently, the cell wall is greatly weakened and the cell undergoes lysis (Figure 20.3). Because penicillin affects the synthesis process, only actively growing cells are affected by these antibiotics. And, because human cells do not have peptidoglycan cell walls, penicillin has very little toxicity for the host cells. You may also recall that gram-negative bacteria have an outer membrane of lipopolysaccharides and other materials over a comparatively thin layer of peptidoglycan. Gram-negative bacteria are relatively resistant to penicillin because this outer membrane apparently prevents penicillin from reaching the site of peptidoglycan synthesis.

There are several ways in which cell wall synthesis can be affected. Two commonly used antibiotics, bacitracin and vancomycin, interfere with the synthesis of the linear strands of peptidoglycan. Penicillin and cephalosporins prevent the final crosslinking of the peptidoglycans, which interferes with the construction of the macromolecular cell wall.

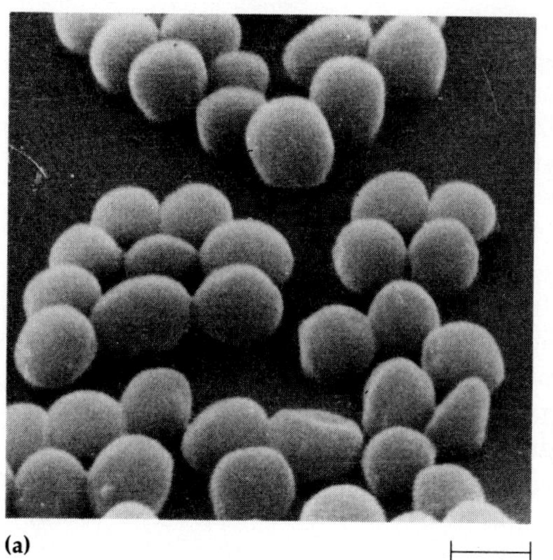

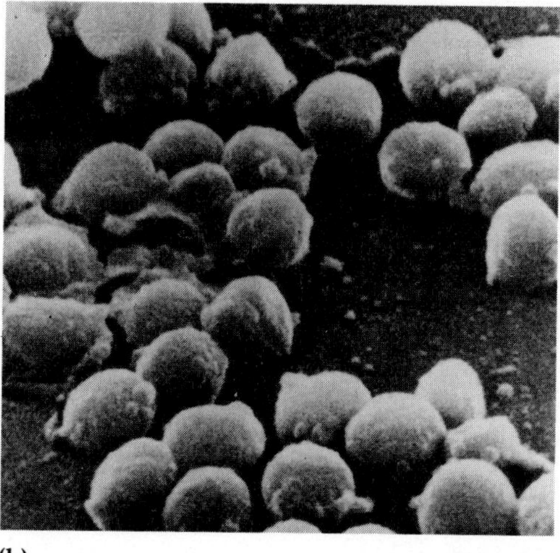

FIGURE 20.3 Effect of an antibiotic on bacterial cell walls. **(a)** *Staphylococcus aureus* before exposure to the antibiotic ampicillin. **(b)** These cells of *S. aureus* are beginning to show the effects of exposure to ampicillin, which inhibits proper formation of cell walls.

INHIBITION OF PROTEIN SYNTHESIS

Because protein synthesis is a common feature of all cells, both procaryotic and eucaryotic, it would seem an unlikely target for selective toxicity. One notable difference between procaryotes and eucaryotes, however, is the structure of their ribosomes. As we discussed in Chapter 4, eucaryotic cells have 80S ribosomes; procaryotic cells have 70S ribosomes. (The 70S ribosome is made up of a 50S and a 30S unit.) The difference in ribosomal structure accounts for the selective toxicity of antibiotics that affect protein synthesis. However, mitochondria (important eucaryotic organelles) also contain 70S ribosomes similar to those of bacteria. Antibiotics targeting the 70S ribosomes can therefore have adverse effects on the cells of the host. Among the antibiotics that interfere with protein synthesis are chloramphenicol, erythromycin, streptomycin, and tetracyclines (Figure 20.4).

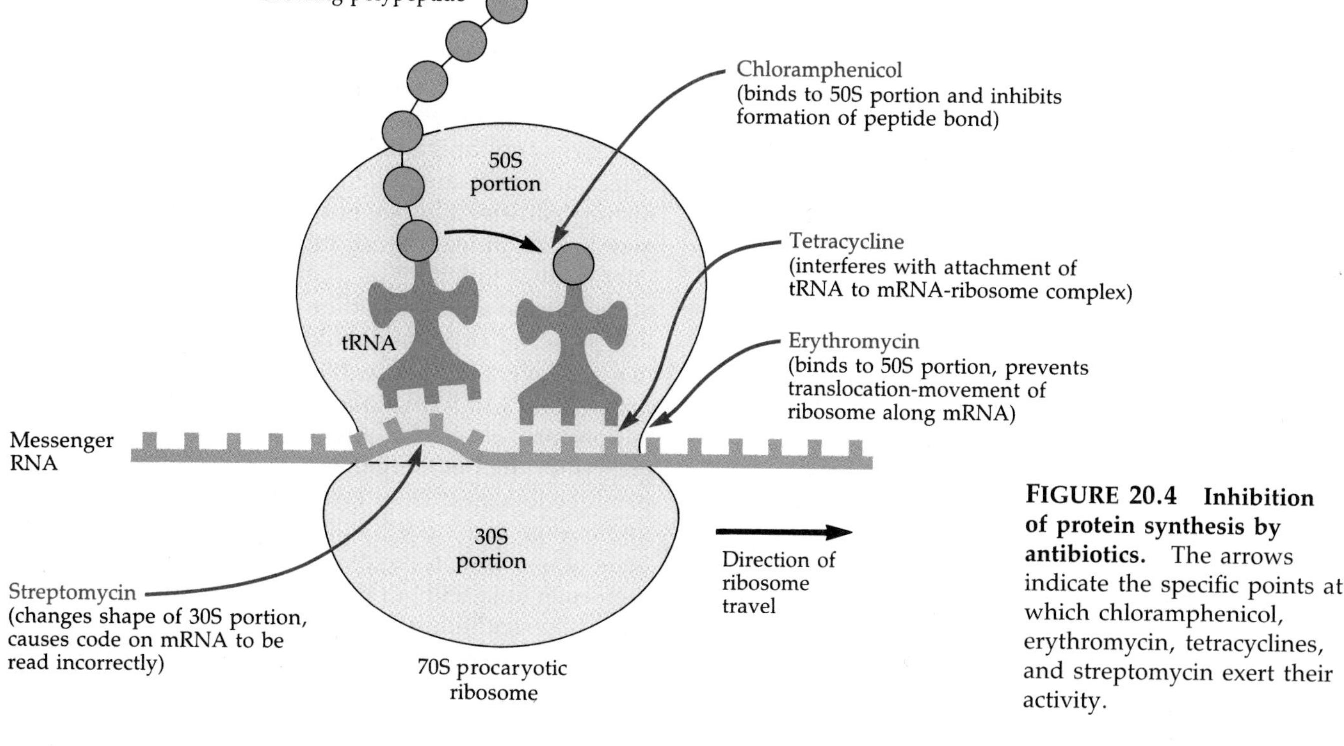

FIGURE 20.4 Inhibition of protein synthesis by antibiotics. The arrows indicate the specific points at which chloramphenicol, erythromycin, tetracyclines, and streptomycin exert their activity.

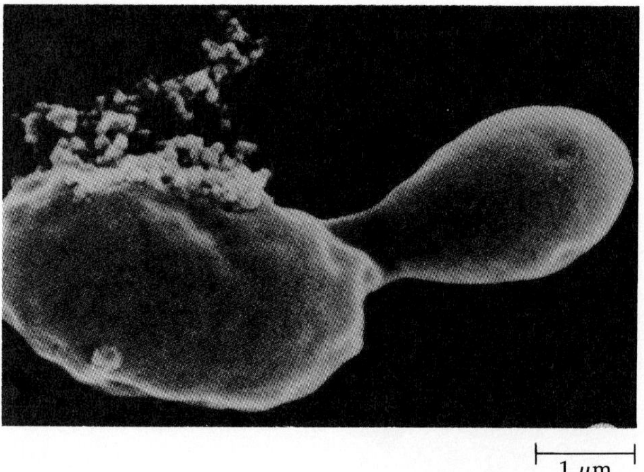

FIGURE 20.5 Disintegration of a yeast cell. The budding cell releases its cytoplasmic contents as the plasma membrane is disrupted by the antifungal drug miconazole.

Reacting with the 50S portion of the 70S procaryotic ribosome, chloramphenicol inhibits the formation of peptide bonds in the growing polypeptide chain. Erythromycin also reacts with the 50S portion of the 70S procaryotic ribosome. Most drugs that inhibit protein synthesis have a broad spectrum of activity; erythromycin is an exception. Because it does not penetrate the gram-negative cell wall, it affects mostly gram-positive bacteria.

Some other antibiotics react with the 30S portion of the 70S procaryotic ribosome. The tetracyclines interfere with the attachment of the tRNA carrying the amino acids to the ribosome, preventing the addition of amino acids to the growing polypeptide chain. Tetracyclines do not interfere with mammalian ribosomes because they do not penetrate very well into intact mammalian cells. However, at least small amounts are able to enter the host cell, as is apparent from the fact that the intracellular pathogenic rickettsias and chlamydias are sensitive to tetracyclines. The selective toxicity of the drug in this case is due to a greater sensitivity of the bacteria at the ribosomal level.

Aminoglycoside antibiotics, such as gentamicin and streptomycin, interfere with the initial steps of protein synthesis by changing the shape of the 30S portion of the 70S procaryotic ribosome. This interference causes the genetic code on the mRNA to be read incorrectly.

Injury to the Plasma Membrane

Certain antibiotics, especially polypeptide antibiotics, bring about changes in the permeability of the plasma membrane; these changes result in the loss of important metabolites from the microbial cell. One such antibiotic is polymyxin B. It causes disruption of the plasma membrane by attaching to the phospholipids of the membrane.

Some antifungal drugs, such as nystatin, amphotericin B, and ketoconazole, are effective against many systemic fungal diseases. Such drugs combine with sterols in the fungal plasma membrane to disrupt the membrane (Figure 20.5). Since bacterial plasma membranes generally lack sterols, these antibiotics do not act on bacteria. However, the plasma membranes of animal cells do contain sterols, and nystatin and amphotericin B can be toxic to the host. Fortunately, animal cell membranes have mostly *cholesterol,* and fungal cells have mostly *ergosterol,* against which the drug is most effective, so that the balance of the toxicity is tilted toward the fungus.

Inhibition of Nucleic Acid Synthesis

A number of antibiotics interfere with nucleic acid metabolism of microorganisms. Some drugs with this mode of action, such as the antiviral idoxuridine, have an extremely limited usefulness because they interfere with mammalian DNA and RNA synthesis as well. Others, such as rifamycin, nalidixic acid, and trimethoprim, are more widely used in chemotherapy because they are more selectively toxic.

Inhibition of Enzymatic Activity

In Chapter 5, we mentioned that an enzymatic activity of a microorganism can be *competitively inhibited* by a substance (antimetabolite) that closely resembles the normal substrate for the enzyme (see Figure 5.6). An example of competitive inhibition is the relationship between the antimetabolite sulfanilamide (a sulfa drug) and para-aminobenzoic acid (PABA). In many microorganisms, PABA is the substrate for an enzymatic reaction leading to the synthesis of folic acid, a vitamin that functions as a coenzyme for the synthesis of the purine and pyrimidine bases of nucleic acids. In the presence of sulfanilamide, the enzyme that normally converts PABA to folic acid combines with the drug instead of with PABA (Figure 20.6). This combination prevents folic acid synthesis and stops the growth of the microorganism. Because humans do not produce folic acid from PABA (they obtain it as a vitamin in ingested foods), sulfanilamide exhibits selective toxicity—it affects microorganisms that synthesize their own folic acid but does not harm the human host. Other chemotherapeutic agents that act as antimetabolites are the sulfones and trimethoprim, as well as many antivirals such as acyclovir.

Sulfanilamide PABA

FIGURE 20.6 Structure of sulfanilamide, a representative sulfa drug. Note the resemblance to PABA, for which it is a competitive inhibitor.

Survey of Commonly Used Antimicrobial Drugs

We will next examine the properties and activities of some commonly used antimicrobial drugs. Commonly used drugs are listed in Table 20.2.

ANTIBACTERIAL SYNTHETICS

Isoniazid (INH)

Isoniazid (INH) is a very effective synthetic antimicrobial drug against *Mycobacterium tuberculosis*. It has little effect on nonmycobacteria. The primary effect of INH is to inhibit synthesis of mycolic acids, which are components of cell walls only of the mycobacteria. When used to treat tuberculosis, INH is usually administered simultaneously with other drugs, such as rifampin or ethambutol. Because the tubercle bacillus is usually found only within macrophages or walled off in tissue, any antitubercular drug must be able to penetrate into such sites.

Ethambutol

Ethambutol is effective only against mycobacteria. The drug apparently inhibits incorporation of mycolic acid into the cell wall. It is a comparatively weak antitubercular drug; its principal use is as the secondary drug to avoid resistance problems.

Sulfonamides

As we noted earlier, **sulfonamides (sulfa drugs)** were among the first synthetic antimicrobial drugs used to treat microbial diseases. Unfortunately, many pathogens have developed resistance to them. Moreover, sulfonamides cause allergic reactions in many people. For these reasons, antibiotics have tended to diminish the importance of sulfonamides in chemotherapy. The sulfonamides continue to be used to treat certain urinary tract infections and have other specialized uses, as in the combination drug **silver sulfadiazine,** used in the control of infections in burn patients. Sulfonamides are bacteriostatic; as we mentioned, their action is due to their structural similarity to para-aminobenzoic acid (PABA) (see Figure 20.6).

Probably the most widely used sulfa today is a combination of **trimethoprim** and **sulfamethoxazole (TMP–SMZ).** This combination is an excellent example of drug *synergism* (see p. 511). When the drugs are used in combination, only about one-tenth of the concentration is needed as when each is used alone. The combination also has a broader spectrum of action and greatly reduces the emergence of resistant strains.

Figure 20.7 illustrates how the two drugs interfere with different steps of a metabolic sequence leading to the synthesis of DNA. Sulfamethoxazole inhibits the synthesis of dihydrofolic acid from PABA so that the amount made is much less than usual. Then trimethoprim, which is a structural analog of a portion of dihydrofolic acid but is not classified as a sulfa drug, blocks the conversion of any dihydrofolic acid that is present into tetrahydrofolic acid (Figure 20.8, p. 502).

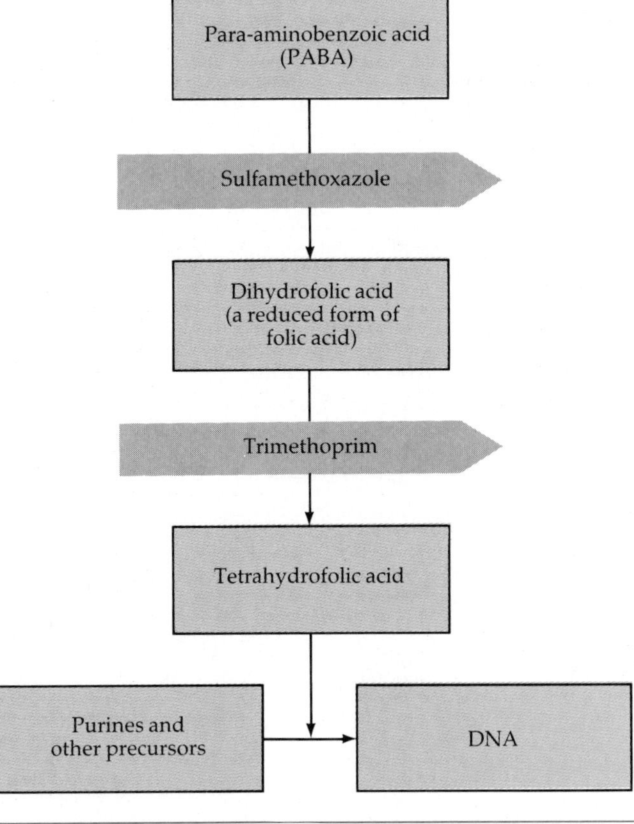

FIGURE 20.7 Trimethoprim and sulfamethoxazole. TMP–SMZ works by inhibiting different points in the synthesis of DNA. Together the drugs are synergistic.

TABLE 20.2 Commonly Used Antimicrobial Drugs

ANTIMICROBIAL DRUG	*EFFECT ON MICROORGANISMS*	*MODE OF ACTION*	*CLINICAL USE/TOXICITY*
Synthetic Drugs			
Isoniazid (INH)	Bacteriostatic	Competitive inhibition	With streptomycin and rifampin to treat tuberculosis
Ethambutol	Bacteriostatic	Competitive inhibition	With INH to treat tuberculosis
Sulfonamides	Bacteriostatic	Competitive inhibition	*Neisseria meningitidis* meningitis and some urinary tract infections
Quinolones	Bactericidal	Inhibit DNA synthesis	Urinary tract infections
Antibiotics			
Penicillin, natural penicillins G, V	Bactericidal	Inhibition of cell wall synthesis	Gram-positive bacteria
Penicillin, semisynthetic ampicillin, methicillin, oxacillin, and others	Bactericidal	Inhibition of cell wall synthesis	Broad spectrum (ampicillin); resistant to penicillinase and stomach acid (methicillin, oxacillin)
Cephalosporins	Bactericidal	Inhibition of cell wall synthesis	Similar to penicillins
Carbapenems	Bactericidal	Inhibition of cell wall synthesis	Very broad spectrum
Aminoglycosides	Bactericidal	Inhibition of protein synthesis	
Streptomycin			Gram-negative bacteria
Neomycin			Topical ointment; gram-negative bacteria
Gentamicin			Gram-negative bacteria
Tetracyclines	Bacteriostatic	Inhibition of protein synthesis	Very broad spectrum; some toxic side-effects include tooth discoloration and liver and kidney damage
Chloramphenicol	Bacteriostatic	Inhibition of protein synthesis	*Salmonella;* aplastic anemia is an occasional side-effect
Macrolides Erythromycin	Bacteriostatic	Inhibition of protein synthesis	Diphtheria; legionellosis; various infections in persons allergic to penicillin
Polypeptides Bacitracin	Bactericidal	Inhibition of cell wall synthesis	Gram-positive bacteria, primarily; toxic to kidneys; usually used as topical ointment
Polymyxin B	Bactericidal	Injury to plasma membranes	Gram-negative bacteria, including *Pseudomonas;* toxic to brain and kidneys
Vancomycin	Bactericidal	Inhibition of cell wall synthesis	Staphylococci; very toxic
Rifamycins Rifampin	Bactericidal	Inhibition of RNA synthesis	Gram-positive bacteria, some gram-negatives, chlamydias, and poxviruses; often used with INH and ethambutol for tuberculosis

(Continued)

TMP–SMZ is useful against many gram-negative pathogens of the urinary tract and intestinal tract.

Other Synthetic Antimicrobial Drugs

Nitrofurans constitute a family of drugs that are active against a large number of gram-positive and gram-negative bacteria, protozoans, and fungi. *Nitrofurantoin* is administered orally for urinary tract infections. Like nalidixic acid (see below), it usually does not reach a high enough concentration elsewhere than in

TABLE 20.2 ***(Continued)***

ANTIMICROBIAL DRUG	EFFECT ON MICROORGANISMS	MODE OF ACTION	CLINICAL USE/TOXICITY
Antifungal Drugs			
Polyenes	Fungicidal	Injury to plasma membrane	
Nystatin			*Candida* infections of skin and vagina
Amphotericin B			Histoplasmosis, coccidioidomycosis, and blastomycosis
Imidazoles	Fungicidal	Injury to plasma membrane	
Clotrimazole, miconazole			Topical use for fungal infections
Ketoconazole			Systemic fungal infections
Griseofulvin	Fungistatic	Inhibition of mitotic spindle function	Fungal infections of skin, primarily ringworm
Antiviral Drugs			
Amantadine	Antiviral	Blocks entry or uncoating	Influenza A prophylaxis and treatment
Guanine analog	Antiviral	Interferes with DNA and RNA synthesis	
Ribavirin			Viral pneumonia
Acyclovir			Herpes infections
Thymine analog Idoxuridine Zidovudine	Antiviral	Inhibition of DNA synthesis	Herpes keratitis AIDS virus
Adenine analog Vidarabine	Antiviral	Inhibition of DNA and RNA synthesis	Herpes encephalitis
Antiprotozoan Drugs			
Chloroquine	Antiprotozoan	Inhibition of DNA synthesis	Malaria
Pentamidine isethionate	Antiprotozoan	Reacts with DNA	African trypanosomiasis, *Pneumocystis* pneumonia
Diiodohydroxyquin	Antiprotozoan	Unknown	Amoebic infections; optic nerve damage
Metronidazole	Antiprotozoan	Anaerobic metabolism products damage DNA	Amoebic infections, *Trichomonas*, anaerobic bacteria
Antihelminthic Drugs			
Praziquantel	Antihelminthic	Unknown	Schistosomiasis and other fluke infestations
Niclosamide	Antihelminthic	Prevents generation of ATP	Tapeworm infestations

urine to be useful in systemic chemotherapy. Its primary mode of action is not precisely known, but it might cause breakage of DNA polymers. A nitrofuran derivative *nifurtimox* is effective against trypanosomal protozoans.

Nalidixic acid stops bacterial replication of DNA and is especially effective against gram-negative enterics. Because it accumulates in the urine, this drug is used to treat urinary tract infections in which gram-negative enteric species are implicated. The quino-

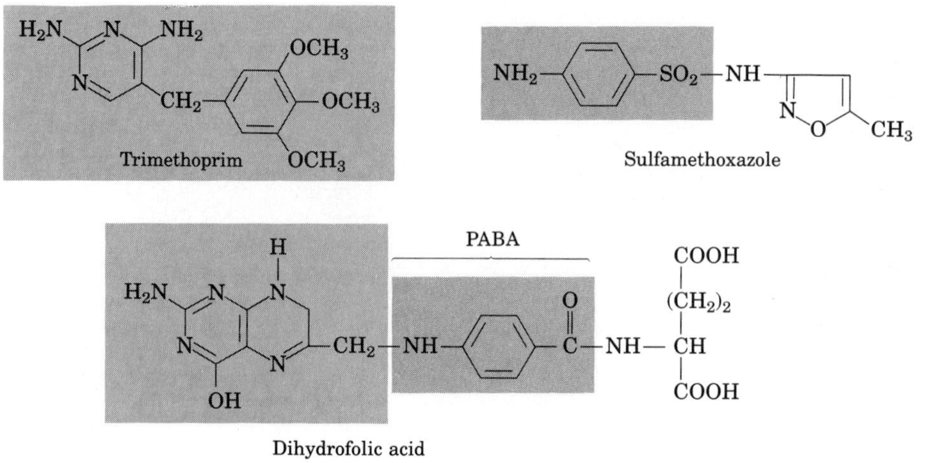

FIGURE 20.8 Trimethoprim, a structural analog of a portion of dihydrofolic acid. Note also the presence of PABA, for which sulfamethoxazole is a structural analog, in another portion of the dihydrofolic acid molecule.

lones are an emerging group of synthetic antimicrobials that are related to nalidixic acid and have the advantage of oral administration. Fluoroquinolones, introduced during the 1980s, made these drugs more effective against gram-negative bacteria and broadened their spectrum against gram-positives. Like nalidixic acid, they are especially valuable against urinary tract infections; they are also frequently prescribed for treatment of gastroenteritis. Their mode of action is to inhibit the gyrase enzyme that bacteria use to coil DNA into the double helical form. *Norfloxacin* was the first of the fluoroquinolones, but *ciprofloxacin* is considered the most potent.

ANTIFUNGAL SYNTHETICS

The antifungal drug *flucytosine* is an antimetabolite of the base cytosine that interferes with protein synthesis. It is preferentially taken up by fungal cells. The spectrum of activity is limited to only a few systemic fungal infections, and it is almost always used in combination with other antifungal drugs.

ANTIBIOTICS

Penicillin

The term **penicillin** refers to a group of chemically related antibiotics (Figure 20.9). Penicillin extracted from cultures of the mold *Penicillium* exists in several closely related forms. These are the so-called **natural penicillins** (Figure 20.9a). The prototype compound of all the penicillins is *penicillin G*. It has a narrow but useful spectrum of action and is often the drug of choice against staphylococci that do not produce penicillinase, streptococci, and several spirochetes. When injected intramuscularly, penicillin G is rapidly excreted

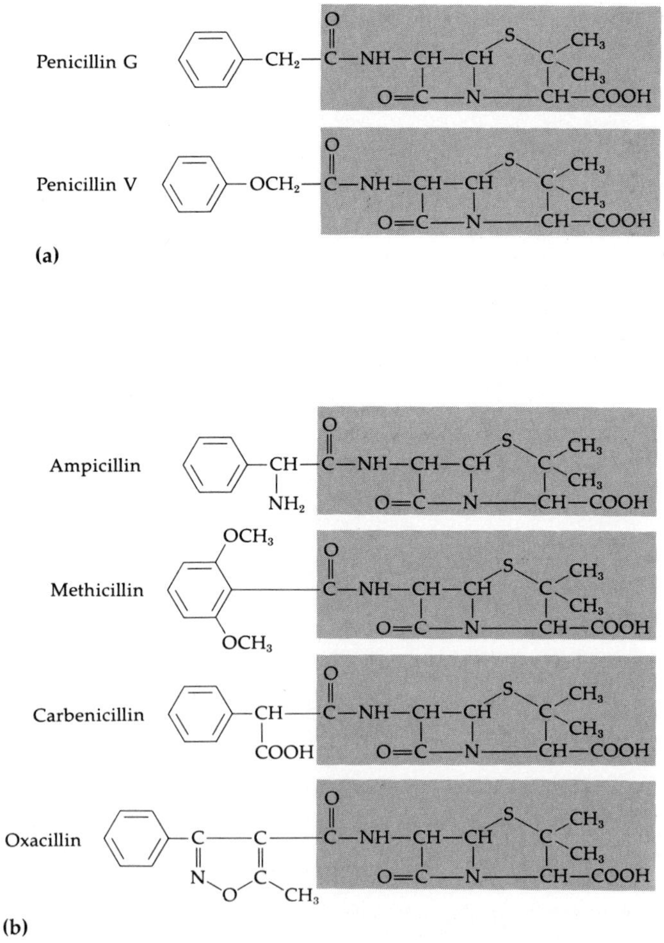

FIGURE 20.9 Structure of penicillins. The portion that all penicillins have in common, containing the β-lactam ring, is shaded in color. The unshaded portions represent the side chains that distinguish one penicillin from another. **(a)** Natural penicillins. **(b)** Semisynthetic penicillins.

from the body in three to six hours (Figure 20.10); when the drug is taken orally, the acidity of the digestive fluids in the stomach diminishes its concentration. *Procaine penicillin*, a combination of the drugs procaine and penicillin G, is retained at detectable concentrations for up to 24 hours; concentration peaks at about four hours. Still longer retention times can be achieved with *benzathine penicillin*, a combination of benzathine and penicillin G. Although retention times of as long as four months can be obtained, the concentration of the drug is so low that the organisms must be very sensitive to the drug. Penicillin V, which is acid-stable and can be best taken orally, and penicillin G are the natural penicillins most often used.

A large number of **semisynthetic penicillins** have been developed in attempts to overcome the natural penicillins' disadvantages, such as their narrow spectrum of activity and their destruction by penicillinases. In developing semisynthetic penicillins, scientists perfected a technique whereby they could either stop synthesis by *Penicillium* and obtain only the common penicillin nucleus or remove the side chains from the completed natural molecules and then chemically add other side chains. Thus the term *semisynthetic* is applied to these penicillins; part of the penicillin is produced by the mold, and part is added synthetically.

Penicillinases are enzymes produced by many bacteria, especially *Staphylococcus* spp., that cleave the β-lactam ring of the penicillin molecule (Figure 20.11). Because of this characteristic, penicillinases are sometimes called *β-lactamases*. The type of a penicillin molecule depends on the chemical side chains attached to the β-lactam ring.

Semisynthetic penicillins have been found that are more resistant to stomach acid—and therefore are more effective when taken orally—and have a spectrum of activity that is often greater than that of the natural penicillins (see Figure 20.9b). For example, *ampicillin, amoxicillin,* and *carbenicillin* have a broad spectrum and are effective against a number of important gram-negative pathogens. Except for carbenicillin, they are acid-stable enough to be taken orally.

Special attention has been given to the problem of developing variants of penicillin not destroyed by penicillinases. *Methicillin* has been in use the longest, and resistance to it has appeared. *Aztreonam* avoids the action of penicillinases by having only a single ring (therefore known as a monobactam) rather than the conventional β-lactam double ring. Its spectrum of activity is remarkable for a penicillin; it affects only gram-negative bacteria, including pseudomonads. Another approach is to combine penicillins with *potassium clavulanate* (*clavulanic acid,* a product of a streptomycete), which is a noncompetitive inhibitor of penicillinase with essentially no antimicrobial activity of its own. It has been combined with broad spectrum penicillins,

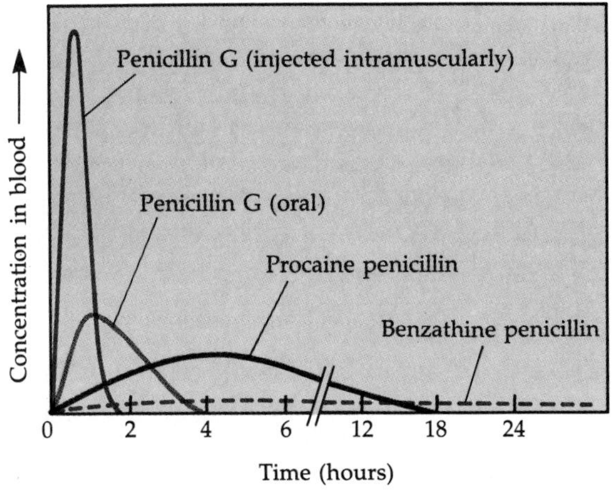

FIGURE 20.10 Penicillin G. This drug reaches different concentrations in the blood when administered by different routes and can be combined with compounds that prolong its retention in the body.

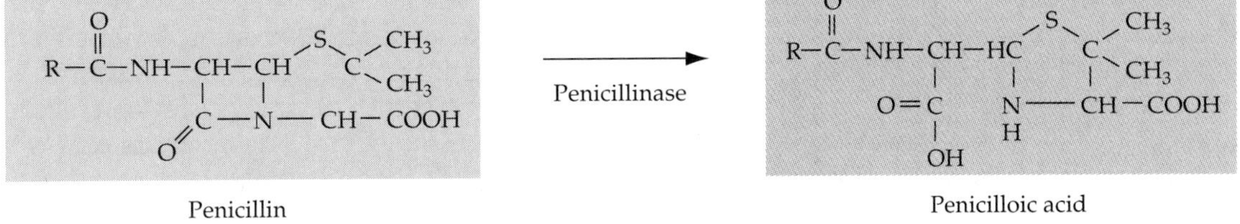

Penicillin → (Penicillinase) → Penicilloic acid

FIGURE 20.11 Penicillinase. Production of this enzyme, which is shown breaking the β-lactam ring, is by far the most common form of resistance to penicillin.

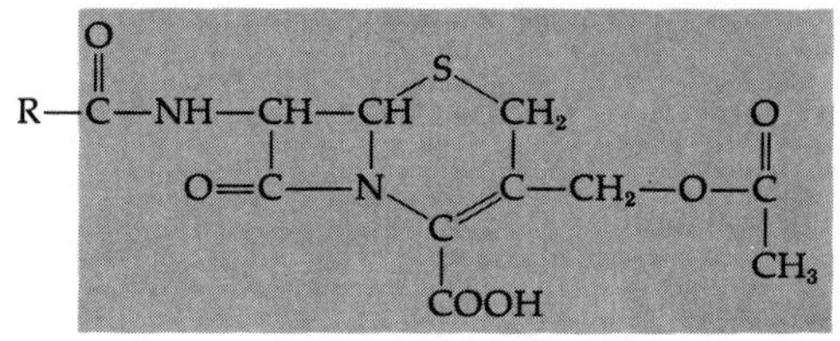

Cephalosporin nucleus

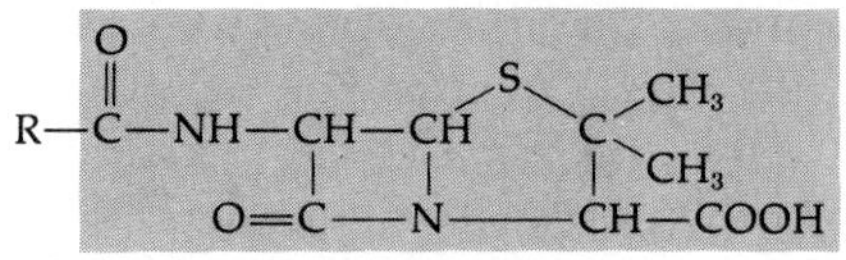

Penicillin nucleus

FIGURE 20.12 Comparison of the structure of the cephalosporin nucleus with that of penicillin. R is an abbreviation for chemical groups that make one compound different from another.

such as *amoxicillin* (*Augmentin*®) and *ticarcillin* (*Timentin*®).

Some newer, broader-spectrum variants, sometimes called *ureidopenicillins,* have also been developed. Some of these variants show antipseudomonad activity and have fewer side-effects than the available aminoglycosides currently used for these infections. Examples are *mezlocillin* and *azlocillin.*

Cephalosporins

In structure, the nuclei of **cephalosporins** resemble those of penicillin (Figure 20.12). Cephalosporins are also similar in action to penicillin; they inhibit the synthesis of cell walls. However, cephalosporins are sufficiently different from penicillin that they are resistant to penicillinases and are effective against more gram-negative organisms than the natural penicillins. However, the cephalosporins are susceptible to a separate group of β-lactamases.

The number of second- and third-generation cephalosporins has proliferated in recent years. Each generation tends to be more effective against gram-negatives and has a broader spectrum of activity than the previous generation. Some characteristic cephalosporins are *cephalothin, cefamandole,* and *cefotaxime.* Most cephalosporins are injected, but a few may be taken orally. Cephalosporins generally are more expensive than penicillins.

Carbapenems

A new class of antibiotics, the **carbapenems,** is remarkable for their extremely broad spectrum of activity. The first of this group is *Primaxin*®, a combination of a β-lactam antibiotic (*imipenem*) and *cilastatin sodium.* The cilastatin sodium prevents degradation of the combination in the kidneys. Tests have demonstrated that Primaxin is active against 98% of all organisms isolated from hospital patients. Carbapenems work by inhibiting cell wall synthesis.

Aminoglycosides

Aminoglycosides are a group of antibiotics in which amino sugars are linked by glycoside bonds. Probably the best-known aminoglycoside is *streptomycin,* which was discovered in 1944 in a culture of *Streptomyces griseus* (gri-sē'us) taken from the throat of a chicken. Historically, streptomycin is significant as the first effective antibiotic against tuberculosis and against large numbers of gram-negative bacteria. Streptomycin is still used as an alternative drug in the treatment of tuberculosis, but rapid development of resistance and the appearance of serious toxic effects have diminished its usefulness.

Aminoglycosides are bactericidal and inhibit protein synthesis. They can affect hearing by causing permanent damage to the auditory nerve, and damage to the kidneys has also been reported. Because of this, their use has been declining. *Neomycin* is present in many topical preparations. *Gentamicin* (Figure 20.13) plays a role in treatment of many enteric gram-negative infections and *Pseudomonas aeruginosa* infections.

Tetracyclines

Tetracyclines are a group of closely related broad spectrum antibiotics, produced by *Streptomyces* spp., that inhibit protein synthesis. Not only are they effective against gram-positive and gram-negative bacteria, but they are also especially valuable against the intracellular rickettsias and chlamydias. Three of the more commonly used tetracyclines are *oxytetracycline* (*Terramycin*®), *chlortetracycline* (*Aureomycin*®), and tetracycline

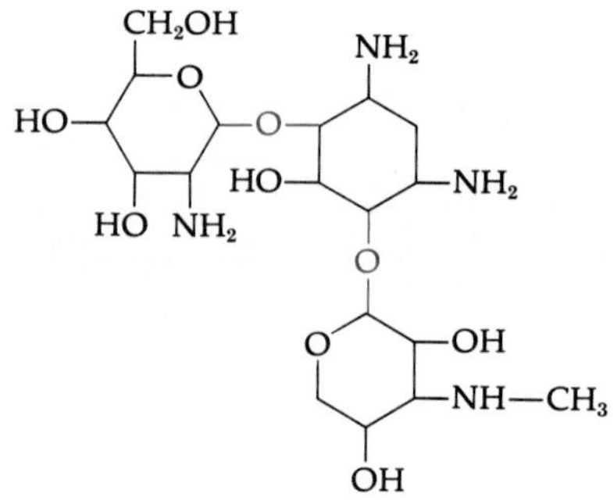

FIGURE 20.13 Structure of gentamicin, a representative aminoglycoside. Glycosidic bonds (red) are bonds between amino sugars.

itself (Figure 20.14). Some newer semisynthetic tetracyclines, such as *doxycycline* and *minocycline,* are now available. They have the advantage of being retained longer in the body.

Tetracyclines are used for many urinary tract infections and are useful in treatment of mycoplasmal pneumonia, and especially chlamydial and rickettsial infections. They are also frequently used as alternative drugs for such diseases as syphilis and gonorrhea. Tetracyclines often suppress the normal intestinal flora because of their broad spectrum, causing gastrointestinal upsets and often leading to superinfections, particularly by *Candida albicans.* They are not advised for administration to children, who might experience a brownish discoloration of the teeth, or to pregnant women, in whom they might cause liver damage. Tetracyclines are one of the most common antibiotics added to animal feeds, where their use results in significantly faster weight gains.

Chloramphenicol

Chloramphenicol is a broad spectrum bacteriostatic antibiotic that interferes with protein synthesis on the 50S ribosome (Figure 20.15). Because of its relatively simple structure, it is less expensive for the pharmaceutical industry to synthesize it chemically than to isolate it from *Streptomyces.* Its relatively small molecular size promotes its diffusion into areas of the body that are normally inaccessible to many other drugs. However, chloramphenicol has serious side-effects; most important is the suppression of bone marrow activity. This suppression affects the formation of blood cells. In about 1 in 40,000 users, the drug appears to cause aplastic anemia, a potentially fatal condition; the normal rate for this condition is only about 1 in 500,000 persons. Physicians have been advised not to use the drug for trivial conditions or ones for which suitable alternatives are available. It is still a drug of choice in the treatment of certain types of meningitis and typhoid fever.

Macrolides

Macrolides are a group of antibiotics named for the presence of a macrocyclic lactone ring. The only macrolide in common clinical use is *erythromycin* (Figure 20.16). Its mode of action is the inhibition of protein synthesis. However, erythromycin is not able to penetrate the cell walls of most gram-negative bacilli. Its spectrum of activity is therefore similar to that of penicillin G, and it is a frequent alternative drug for penicillin. Erythromycin is, however, effective against *Legionella* and some *Neisseria* spp.—important gram-negative pathogenic genera. Because it can be administered orally, an orange-flavored preparation of erythromycin is a frequent penicillin substitute for the treatment of streptococcal and staphylococcal infections in children.

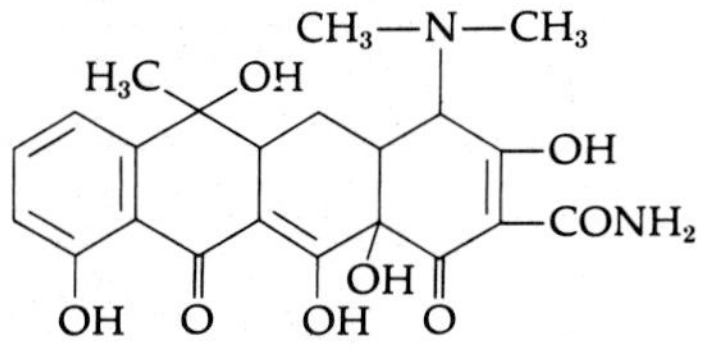

FIGURE 20.14 Structure of tetracycline. Other tetracycline-type antibiotics share the four-cyclic-ring structures of tetracycline and closely resemble it.

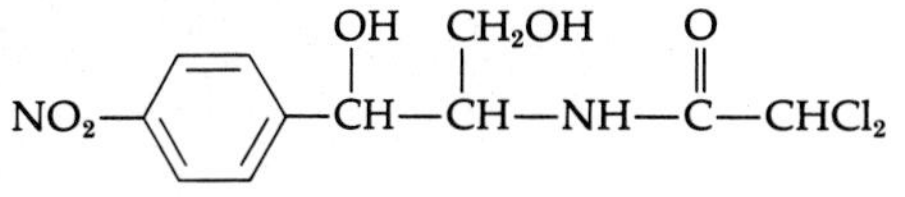

FIGURE 20.15 Structure of chloramphenicol. Note the simple structure that allows this drug to be synthesized inexpensively.

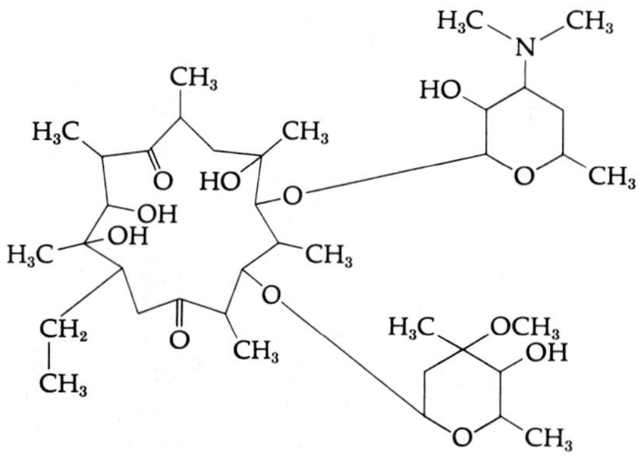

FIGURE 20.16 Structure of erythromycin, a representative macrolide.

Erythromycin is the drug of choice for the treatment of mycoplasmal pneumonia. The immunosuppressant drug FK 506 is a macrolide.

Polypeptides

A number of antibiotics are **polypeptides,** chains of amino acids linked by peptide bonds. Almost all these drugs have been isolated from the genus *Bacillus.* Two well-known examples are bacitracin and polymyxin B

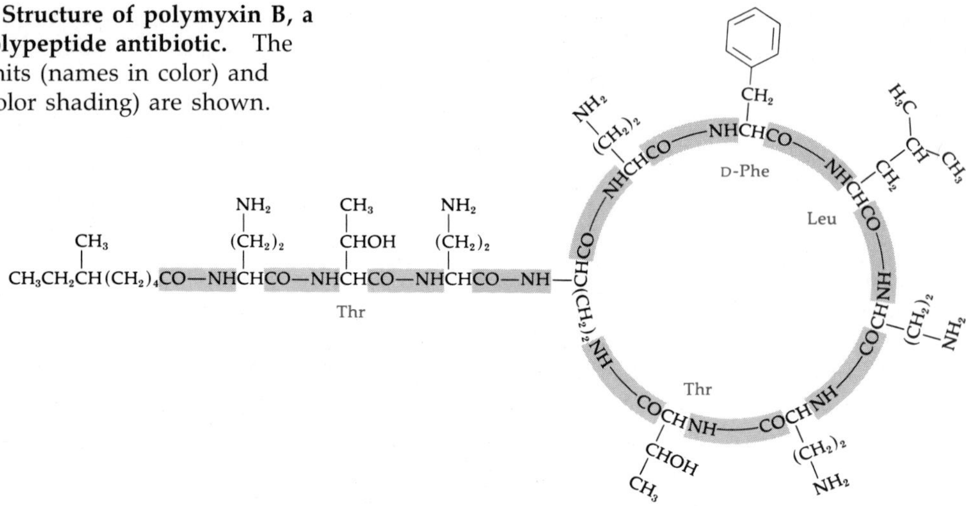

FIGURE 20.17 Structure of polymyxin B, a representative polypeptide antibiotic. The amino acid subunits (names in color) and peptide bonds (color shading) are shown.

(Figure 20.17). Both drugs are relatively toxic, and their use is mostly topical.

Bacitracin (the name is derived from its source, a *Bacillus* isolated from a wound on a girl named Tracy) is effective primarily against gram-positive bacteria, such as staphylococci and streptococci. It is also used against some important gram-negative pathogens, such as *Neisseria*. It inhibits the synthesis of cell walls. Its use is restricted to topical application for superficial infections.

Polymyxin B is a bactericidal antibiotic effective against gram-negative bacteria. For many years, it was one of very few drugs used against infections by gram-negative *Pseudomonas*. The mode of action of polymyxin B is to injure plasma membranes. Polymyxin B is seldom used today except in topical treatment of superficial infections because of toxicity when it is administered by injection. Both bacitracin and polymyxin B are available in nonprescription antiseptic ointments, in which the polypeptides are usually combined with neomycin, a broad spectrum aminoglycoside. These are rare exceptions to the requirement for a prescription for antibiotics.

Vancomycin

Vancomycin is apparently unrelated chemically to any other antibiotic. It is a very toxic drug, is difficult to administer, and has a very narrow spectrum of activity that is based on inhibition of cell wall synthesis (peptidoglycans). The drug is important because it is probably the most effective drug in clinical use against penicillinase-producing staphylococci. Vancomycin is also used in special situations, such as treatment of streptococcal endocarditis and staphylococcus infections of such devices as prosthetic heart valves.

Rifamycins

The best-known derivative of the **rifamycin** family of antibiotics is *rifampin*. These drugs inhibit the synthesis of mRNA. By far the most important use of rifampin is against mycobacteria in the treatment of tuberculosis and leprosy. A valuable characteristic of rifampin is its ability to penetrate tissues and reach therapeutic levels in cerebrospinal fluid and abscesses. This characteristic is probably an important factor in its antitubercular activity because the tuberculosis pathogen is usually located walled off in tissues or intracellularly in macrophages. An unusual side-effect of rifampin is the appearance of orange-red urine, feces, saliva, sweat, and even tears.

ANTIFUNGAL DRUGS

Polyenes

Two of the more commonly used **polyene antibiotics,** both of which are products of *Streptomyces* spp., are *nystatin* and *amphotericin B* (Figure 20.18). Both drugs

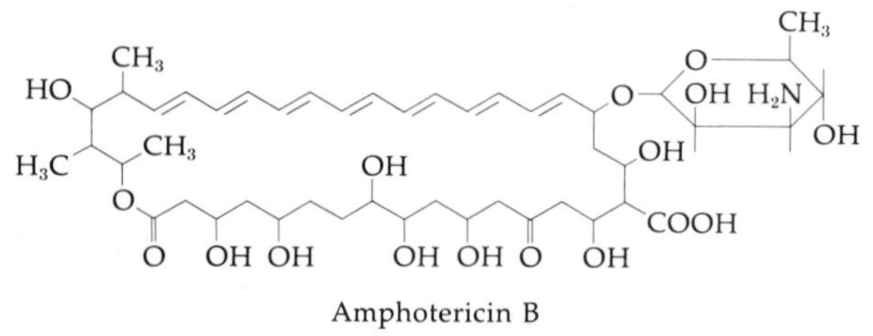

Amphotericin B

FIGURE 20.18 Structure of amphotericin B, a representative polyene antibiotic.

are fungicidal. The drugs combine with the sterols in the fungal plasma membranes, making the membranes excessively permeable and killing the cell. Bacterial plasma membranes (except those of mycoplasmas) do not contain sterols, and the drugs therefore do not affect bacteria. Nystatin is used to treat local infections of the vagina and skin and can be taken orally for *Candida* infections of the intestinal tract. It is so poorly soluble that it does not cross the intestinal lining, so it does not enter the body tissue in toxic amounts. For many years, amphotericin B has been a mainstay of clinical treatment for systemic fungal diseases, such as histoplasmosis, coccidioidomycosis, and blastomycosis. The drug's toxicity, particularly to the kidneys, is a strongly limiting factor in these uses.

Imidazoles

Imidazole antifungals, such as *clotrimazole, miconazole,* and *ketoconazole,* primarily interfere with sterol synthesis in fungi (see Figure 20.5), although they probably have other antimetabolic effects. Clotrimazole and miconazole (Figure 20.19) are generally used topically, most often in treatment of cutaneous mycoses such as athlete's foot. When taken orally, ketoconazole is effective and is a less toxic alternative to amphotericin B for many systemic fungal infections, although occasional liver damage has been reported. Ketoconazole also has an unusually wide spectrum of activity.

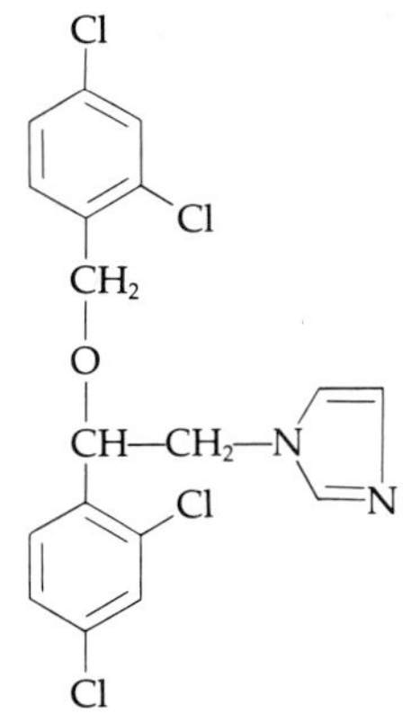

Cl

Cl

CH_2

O

$CH-CH_2-N$

Cl

N

Cl

Miconazole

FIGURE 20.19 Structure of miconazole, a representative imidazole antibiotic.

Griseofulvin

Griseofulvin is an antibiotic produced by a species of *Penicillium.* It has the interesting property of being active against superficial dermatophytic fungal infections of the hair (tinea capitis, or ringworm) and nails, even though its route of administration is oral. The drug apparently binds selectively to the keratin found in the skin, hair follicles, and nails. Its mode of action is to interfere with mitosis and thereby inhibit fungal reproduction.

Other Antifungal Drugs

Tolnaftate is a common alternative to miconazole as a topical agent for the treatment of athlete's foot. Its mechanism of action is not known. *Undecylenic acid* is a fatty acid that has antifungal activity against athlete's foot, although it is not as effective as tolnaftate or the imidazoles.

ANTIVIRAL DRUGS

In the developed parts of the world, it is estimated that at least 60% of infective illnesses are caused by viruses, and only about 15% by bacteria. Every year, at least 90% of the population of the United States suffers from some viral disease. Yet, only a handful of antiviral drugs have been approved in the United States, and these are effective against only an extremely limited group of diseases.

The first antiviral drug licensed for systemic use in the United States was *amantadine.* Although it is not known precisely how amantadine works, it prevents the virus either from entering the cell or, in at least one case, from uncoating after it enters. The drug has limited usefulness in the prevention of influenza A. It has little effect on the course of the disease once it has been contracted, but in confined situations, such as institutions or nursing homes, it can limit the spread of an outbreak. Another drug with activity against influenza is *ribavirin,* an analog of guanine, a purine in DNA and RNA. We will see that most antiviral drugs are synthetic nucleotides that interfere with DNA or RNA synthesis. Ribavirin has also been tested on infants suffering from a viral pneumonia. However, this drug must be administered by aerosol apparatus, is too expensive for general use, and is very toxic. Nevertheless, interest in ribavirin remains high because it is effective in vitro against a wide range of viruses.

The newest antiviral drugs, such as acyclovir, ganciclovir, and Zidovudine® (AZT), are not active until acted upon by enzymes that are synthesized by cells infected by a virus; this factor gives them more antiviral specificity.

Acyclovir is an analog of a guanine-containing nucleoside (Figure 20.20). It is available for administration topically as an ointment, orally, or by injection. Its best known use is in the treatment of genital herpes, but it is generally useful against infections by many herpesviruses, especially in immunosuppressed individuals. These diseases include encephalitis or eye infections caused by herpes simplex viruses and shingles or chickenpox. The mode of action of acyclovir is illustrated in Figure 20.20. The drug terminates the synthe-

FIGURE 20.20 Structure and functions of acyclovir. **(a)** Structural resemblance between acyclovir and the nucleoside 2′-deoxyguanosine. **(b)** The enzyme thymidine kinase combines phosphates with nucleosides to form nucleotides. The nucleotides are then assembled into DNA. **(c)** Acyclovir is a structural analog of the nucleoside, and the enzyme uses it to assemble a false nucleotide that cannot be assembled into DNA.

Guanine

(a) Deoxyguanosine

Acyclovir

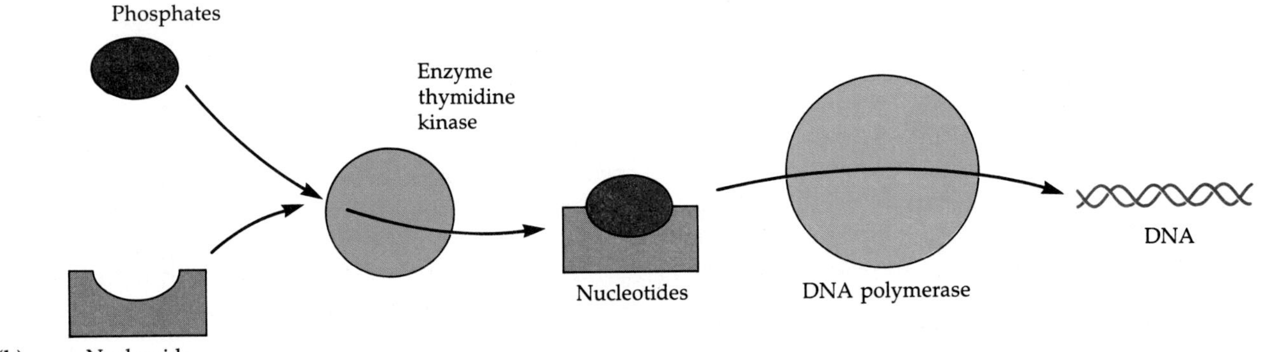

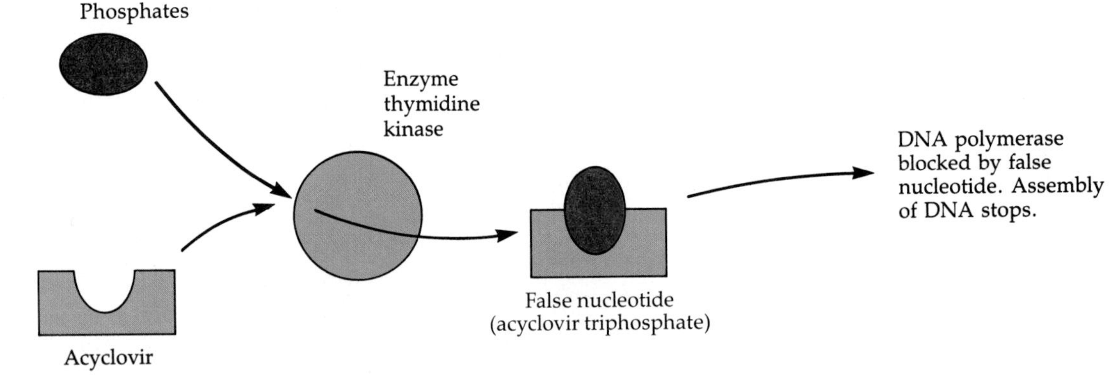

sis of viral DNA but has little effect on host cell DNA synthesis. However, it has no effect on latent herpes viruses (see Chapter 13).

Ganciclovir is a newer antiviral that, like acyclovir, is an analog of guanine. However, it has a wider spectrum of activity that includes all the herpesviruses, most important the cytomegalovirus affecting AIDS patients or transplant patients.

Several other analogs of DNA nucleosides containing purine and pyrimidine have been synthesized and have limited use against herpesviruses. *Idoxuridine* and *trifluridine* are both analogs of thymidine (the thymine-containing nucleoside), and *vidarabine* is an analog of the purine adenine. All these drugs are available as topical ointments and are useful in treating herpes simplex virus infections of the eye (herpes keratitis). Vidarabine has also been approved for intravenous use in the treatment of encephalitis caused by herpesviruses and for treatment of immunosuppressed persons infected with the varicella-zoster virus (chickenpox, shingles). *Zidovudine®* (azidothymidine, AZT) is an analog of thymidine that is used in chemotherapy of AIDS patients. Its mode of action is to interfere with the synthesis of DNA from RNA by the enzyme reverse transcriptase (see the box in Chapter 19, p. 484).

A new approach to viral chemotherapy that may bear good results involves interferon, a protein secreted by certain host cells responding to viral infections (see Chapter 16). Interferon inhibits cell infection by many kinds of viruses. Genetic-engineering techniques using bacteria have made sufficient amounts of interferon available for clinical studies of its possible effectiveness as a chemotherapeutic agent against cancer as well as against viruses.

ANTIPROTOZOAN AND ANTIHELMINTHIC DRUGS

For hundreds of years, quinine from the Peruvian cinchona tree was the only drug known to be effective for the treatment of a parasitic infection (malaria). It was first introduced into Europe in the early 1600s and was known as "Jesuit's powder." There are now many antiprotozoan and antihelminthic drugs, although many of them are still considered experimental. This does not preclude their use, however, by qualified physicians. The Centers for Disease Control (CDC) provide several of them on request when they are not available commercially.

Antiprotozoan Drugs

Quinine still has very limited use as an alternative in the control of the protozoan disease malaria, but synthetic derivatives, such as *chloroquine,* have largely replaced it. Chloroquine fits itself between nitrogenous base pairs in DNA (called *intercalation*) and prevents the DNA from serving as a template for further DNA synthesis. For prevention of malaria in areas where resistance to chloroquine is a problem, the new drug *mefloquine* is recommended. *Quinacrine* functions in a very similar manner to chloroquine and is the drug of choice for treating the protozoan disease giardiasis. *Diiodohydroxyquin* (*iodoquinol*) is an important drug prescribed for several intestinal amoebic diseases, but its dosage must be carefully controlled to avoid optic nerve damage. Its mode of action is unknown.

Metronidazole (*Flagyl®*) is one of the most widely used antiprotozoan drugs. It is unique in having activity not only against parasitic protozoans but also against obligate anaerobic bacteria. For example, as an antiprotozoan agent, it is the drug of choice for vaginitis caused by *Trichomonas vaginalis.* It is also used in the treatment of giardiasis and amoebic dysentery. The mode of action is damage to DNA by products of anaerobic metabolism.

Pentamidine isethionate is used in the early stages of African trypanosomiasis, but in the United States its most familiar clinical use is in the treatment of *Pneumocystis* pneumonia, a common complication in immunocompromised persons, particularly victims of AIDS. The drug's mode of action is unknown, but it appears to bind to DNA. *Nifurtimox* is a member of the synthetic nitrofuran drugs and is effective against the trypanosome-caused Chagas' disease. Treatment may produce side-effects of nausea and even convulsions.

Antihelminthic Drugs

With the increased popularity of sushi, a Japanese specialty often made with raw fish, the CDC began to notice an increased incidence of tapeworm infestations. To estimate the incidence, the CDC notes requests for *niclosamide,* which is the usual first choice in treatment. The drug is effective because it inhibits ATP production under anaerobic conditions. *Praziquantel* is about equally effective for treatment of tapeworms, although its mode of action is not known. Praziquantel is highly recommended for treatment of several fluke-caused diseases, especially schistosomiasis. It causes the helminths to undergo muscular spasms and apparently makes them susceptible to attack by the immune system. *Mebendazole* is an antihelminthic drug frequently used in the treatment of several of the most common intestinal helminths—ascariasis, *Ascaris lumbricoides;* pinworms, *Enterobius vermicularis;* and whipworms, *Trichuris trichiura* (trik'ėr-is trik-ē-yėr'a). The mode of action is disruption of the microtubules in the cytoplasm, which indirectly results in an inability to generate ATP from carbohydrates. *Piperazine* is used in the treatment of pinworms and ascariasis. It paralyzes the organisms, which soon pass out of the victim's body.

Tests for Microbial Susceptibility to Chemotherapeutic Agents

Different microbial species and strains have different degrees of susceptibility to different chemotherapeutic agents. Moreover, the susceptibility of a microorganism can change with time, even during therapy with a specific chemotherapeutic agent. Thus a physician must know the sensitivities of the disease-causing microorganism before treatment can be started.

Several tests can be used to indicate which chemotherapeutic agent is most likely to combat a specific pathogen. These tests, however, are not always necessary. If the organisms have been identified—for example, as *Pseudomonas aeruginosa,* beta-hemolytic streptococci, or gonococci—certain drugs can be selected without specific testing for susceptibility. These tests are necessary only when susceptibility is not predictable, or when antibiotic resistance problems develop.

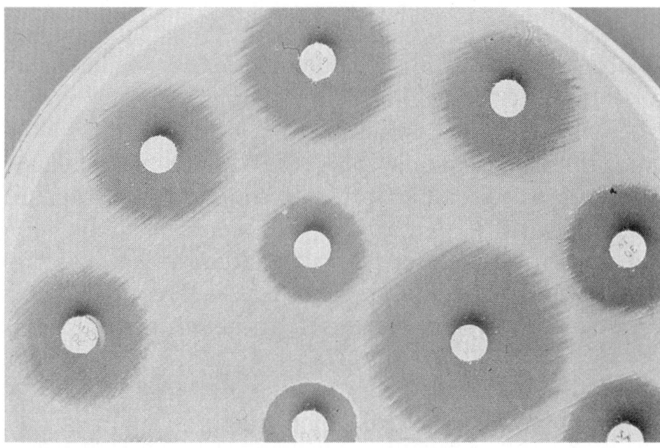

FIGURE 20.21 Disk-diffusion method for determining the activity of antimicrobials. Each disk contains a different antibiotic, which diffuses into the surrounding agar. The clear zones indicate inhibition of growth of the microorganism swabbed onto the plate surface.

DISK-DIFFUSION METHOD

Probably the most widely used—but not necessarily the best—method of testing is the **disk-diffusion method,** also known as the **Kirby–Bauer test.** A Petri plate containing an agar medium is inoculated ("seeded") uniformly over its entire surface with a standardized amount of a test organism. Next, filter-paper disks impregnated with known concentrations of chemotherapeutic agents are placed on the solidified agar surface. During incubation, the chemotherapeutic agents diffuse from the disks into the agar. The farther the agent diffuses from the disk, the lower its concentration. If the chemotherapeutic agent is effective, a *zone of inhibition* forms immediately around the disk (Figure 20.21). The diameter of the zone of inhibition can be measured. Because the size of the zone of inhibition is affected by the diffusion rate of the chemotherapeutic agent, a wider zone does not always indicate greater antimicrobial activity. The zone diameter is compared to a standard table for that drug and concentration, and the results report that the organism is *sensitive, intermediate,* or *resistant.* This result is often inadequate information for many clinical purposes. However, the test is simple and inexpensive and has considerable value in medical practice.

BROTH DILUTION TESTS

A weakness of the disk-diffusion method is that it does not determine if a drug is bactericidal and not just bacteriostatic. Some form of **broth dilution testing** is needed to determine the *minimal inhibitory concentration (MIC)* and *minimal bactericidal concentration (MBC)* of an antimicrobial. The MIC is determined by making a sequence of decreasing concentrations of the drug in a broth, which is then inoculated with the test bacteria (Figure 20.22). The lowest concentration that prevents visible growth is the MIC. The wells that do not show growth can be cultured into broth free of the antimicrobial. If growth occurs in this broth, the drug was not bactericidal, and the MBC can be determined.

Dilution tests are often highly automated. The drugs are purchased already diluted into broth in wells formed in a plastic tray. A suspension of the test organism is prepared and inoculated into all the wells simultaneously by a special inoculating device. After incubation, the turbidity may be read visually, although clinical laboratories with high work loads may read the trays with special scanners that enter the data into a computer that provides a printout of the MIC.

Effectiveness of Chemotherapeutic Agents

DRUG RESISTANCE

Bacteria become resistant to antimicrobials in a number of different ways. Some resistance is due to the ability of a particular mutant to destroy a given antibiotic. For example, as we noted earlier, some staphylococci produce the enzyme penicillinase, which destroys natural penicillins. Resistance to tetracycline is usually related to cellular changes that decrease the cell's intake of the drug. Bacteria become resistant to sulfa drugs and antibiotics when receptor sites for the antibiotic develop less affinity for it. Resistance to trimethoprim can arise when the microbe begins synthesizing very large amounts of the enzyme against which the drug is targeted. Conversely, polyene antibiotics can become less effective when resistant organisms produce smaller amounts of the sterols with which the drug produces its effects. Of particular concern is the possibility that such *resistant mutants* will increasingly replace the susceptible normal populations (see the box on p. 495). The increasing appearance of antibiotic-resistant gonorrhea is discussed in the box in Chapter 26, p. 657.

Hereditary drug resistance is often carried by extrachromosomal genetic elements called plasmids. Some plasmids, including those called **resistance (R) factors,** can be transferred between bacterial cells in a population and between different but closely related bacterial populations (see Chapter 8). R factors often contain genes for resistance to several antibiotics.

Strains of bacteria that are resistant to antibiotics are particularly common among persons who work in hospitals, where antibiotics are in constant use. *Staphylococcus aureus,* a common opportunistic pathogen that

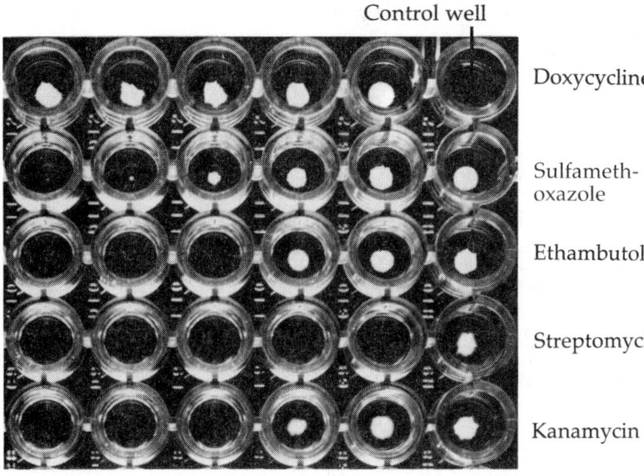

FIGURE 20.22 A microdilution MIC plate. Growth is shown by a white dot in the bottom of the well. Doxycycline had no effect at all. Kanamycin and ethambutol were about equally effective. Streptomycin was effective at all concentrations. A trailing endpoint, observed with sulfamethoxazole, is read where there is an 80% reduction in growth. The negative control well (with broth only) contained no antibiotics and no inoculum; the positive control well (not shown) contained no antibiotics but was inoculated.

is carried in the nasal passages, develops resistance to antibiotics very frequently. Every day in hospitals, surprisingly large volumes of antibiotics are dispensed as aerosols by the universal practice of clearing syringes of air before an injection is made. As these aerosols are generally inhaled in small amounts, the staphylococci harmlessly residing in a nurse's nose, for example, can quickly develop resistance. If these resistant microbes are transferred to surgical wounds and other exposed sites, they can cause nosocomial infections (see Chapter 14).

Drugs should not be used indiscriminately, and the concentrations administered should always be of optimum strength so that the chances of survival of resistant mutants are minimized. A few years ago, one medical journal reported on a survey of antibiotic use in one hospital. The results indicated that more than 60% of the antibiotics prescribed there were either given in the wrong dosage or used in applications for which they would not be effective. Therefore, many hospitals now have special monitoring committees who review the use of antibiotics for effectiveness and cost. Another way the development of resistant strains can be reduced is through the administration of two or more drugs simultaneously. If a strain is resistant to one of the drugs, the other might destroy it. The probability that the organism will acquire resistance to both drugs is roughly the product of the two individual probabilities.

Resistance to drugs is not the only problem with antimicrobial therapy. For example, antibiotics have no effect on toxins that might already have been released into the host's body fluids.

EFFECTS OF COMBINATIONS OF DRUGS

It sometimes happens that the chemotherapeutic effect of two drugs given simultaneously is greater than the effect of either given alone, a phenomenon called **synergism.** For example, in the treatment of bacterial endocarditis, penicillin and streptomycin are much more effective when taken together than when either drug is taken alone. Damage to bacterial cell walls by penicillin makes it easier for streptomycin to enter.

Other combinations of drugs can be **antagonistic.** For example, the simultaneous use of penicillin and tetracycline is often less effective than when either drug is used alone. By stopping the growth of the bacteria, the bacteriostatic drug tetracycline interferes with the action of penicillin, which requires bacterial growth.

Combinations of antimicrobial drugs should be used for the following purposes:

1. To prevent or minimize the emergence of resistant strains.
2. To take advantage of the synergistic effect.
3. To provide optimal therapy in life-threatening situations before a diagnosis can be established with certainty.
4. To lessen the toxicity of individual drugs by reducing the dosage of each in combination.

THE FUTURE OF CHEMOTHERAPEUTIC AGENTS

Antibiotics are clearly one of the greatest triumphs of medical science, although they are not without their problems and disadvantages. Our most pressing concern is with the misuse of antibiotics, which increases the likelihood of the emergence of resistant strains of pathogens. There is also renewed interest in treatments for tropical parasitic diseases.

Antiviral drugs are urgently needed, especially for AIDS. As research discovers additional aspects of the internal workings of viruses and cells, more chemicals that are effective against viruses will probably be developed. There is considerable similarity between the requirements of cancer therapy and the requirements for treatments for viral disease. It will be interesting to see whether research for one will benefit the other.

STUDY OUTLINE

Introduction (p. 492)

1. A chemotherapeutic agent is a chemical that combats disease in the body.
2. An antimicrobial drug is a chemical substance that destroys disease-causing microorganisms with minimal damage to host tissues.
3. Antimicrobial drugs may be synthetic drugs (prepared in the laboratory) or antibiotics (produced by bacteria or fungi).

Historical Development (pp. 492–494)

1. The first chemotherapeutic agent, discovered by Paul Ehrlich, was salvarsan, used to treat syphilis.
2. Sulfa drugs came into prominence in the late 1930s.
3. Alexander Fleming discovered the first antibiotic, penicillin, in 1929; its first clinical trials were done in 1940.
4. Antibiotics are produced by species of *Streptomyces, Bacillus, Penicillium,* and *Cephalosporium.*

Criteria for Evaluating Antimicrobial Drugs (p. 494)

1. Antimicrobial agents should have selective toxicity for microorganisms.
2. They should not produce hypersensitivity in most patients.
3. They should be stable during storage, be soluble in body fluids, and be retained in the body long enough to be effective.
4. Microbes should not readily become resistant to the drug.

Spectrum of Activity (pp. 494–496)

1. Antibacterial drugs affect many targets in a procaryotic cell.
2. Fungal, protozoan, and helminthic infections are more difficult to treat because these organisms have eucaryotic cells.
3. Narrow spectrum drugs affect only a select group of microbes—gram-positive cells, for example.
4. Broad spectrum drugs affect a large number of microbes.
5. Antimicrobial agents should not cause excessive harm to normal flora.
6. Superinfections occur when a pathogen develops resistance to the drug being used or when normally resistant flora multiply excessively.

Action of Antimicrobial Drugs (pp. 496–499)

1. General action is either by directly killing microorganisms (bactericidal) or by inhibition of growth (bacteriostatic).
2. Some agents, such as penicillin, inhibit cell wall synthesis in bacteria.
3. Other agents, such as chloramphenicol, erythromycin, tetracyclines, and streptomycin, inhibit protein synthesis by acting on 70S ribosomes.
4. Agents such as polymyxin B cause injury to plasma membranes.
5. Rifamycin, nalidixic acid, and trimethoprim inhibit nucleic acid synthesis.
6. Agents such as sulfanilamide act as antimetabolites by competitively inhibiting enzyme activity.

Survey of Commonly Used Antimicrobial Drugs (pp. 499–509)

ANTIBACTERIAL SYNTHETICS (pp. 499–502)

1. Isoniazid (INH) inhibits mycolic acid synthesis in mycobacteria. INH is administered with rifampin or ethambutol to treat tuberculosis.
2. The antimetabolite ethambutol is used with other drugs to treat tuberculosis.
3. Sulfonamides competitively inhibit folic acid synthesis. Sulfonamides are bacteriostatic. They are used to treat urinary tract infections and can cause allergies and drug resistance.
4. Trimethoprim and sulfamethoxazole competitively inhibit dihydrofolic acid synthesis for treatment of urinary tract and intestinal infections.
5. Nitrofurans may break DNA molecules; they are used to treat urinary tract infections.
6. Nalidixic acid and quinolones interfere with DNA synthesis; they are used to treat infections caused by gram-negative bacteria.
7. Quinolones inhibit DNA gyrase for treatment of urinary tract infections.
8. The antifungal agent flucytosine is an antimetabolite of cytosine.

ANTIBIOTICS (pp. 502–506)

1. Penicillins have low toxicity. They inhibit peptidoglycan synthesis; cell death is by osmotic lysis.
2. Penicillin G and V are natural penicillins and are effective against gram-positive cocci and spirochetes.
3. Several semisynthetic penicillins such as ampicillin are resistant to stomach acid and penicillinases and are broad spectrum.
4. Ureidopenicillins such as mezlocillin are effective against *Pseudomonas* spp.
5. Cephalosporins inhibit cell wall synthesis and are used against penicillin-resistant strains.
6. Carbapenems are broad spectrum antibiotics that inhibit cell wall synthesis.
7. Aminoglycosides include streptomycin, neomycin, spectinomycin, and gentamicin; all inhibit protein synthesis and are bactericidal.
8. Tetracyclines inhibit protein synthesis and are bacteriostatic toward many bacteria, including rickettsias and chlamydias.
9. Chloramphenicol inhibits protein synthesis and is bacteriostatic; a side-effect of prolonged use is aplastic anemia.
10. Macrolides, such as erythromycin, inhibit protein syn-

thesis and are bacteriostatic; erythromycin is used to treat legionellosis and *Neisseria* infections.

11. Polypeptides include bacitracin and polymyxin B. They are applied topically to treat superficial infections.
12. Bacitracin inhibits cell wall synthesis primarily in gram-positive bacteria.
13. Polymyxin B damages plasma membranes and is effective against gram-negative bacteria.
14. Vancomycin inhibits cell wall synthesis and may be used to kill penicillinase-producing staphylococci.
15. Rifamycins are bactericidal and are especially effective against gram-positive bacteria, including mycobacteria.

ANTIFUNGAL DRUGS (pp. 506–507)

1. Polyenes, such as nystatin and amphotericin B, combine with plasma membrane sterols and are fungicidal.
2. Imidazoles interfere with sterol synthesis and are used to treat cutaneous and systemic mycoses.
3. Griseofulvin interferes with eucaryotic cell division and is used primarily to treat skin infections caused by fungi.

ANTIVIRAL DRUGS (pp. 507–509)

1. Amantadine blocks penetration or uncoating of influenza A virus.
2. Purine and pyrimidine analogs include ribavirin, acyclovir, ganciclovir, azidothymidine (AZT), idoxuridine, trifluridine, and vidarabine.
3. Interferon inhibits cell infection by viruses.

ANTIPROTOZOAN AND ANTIHELMINTHIC DRUGS (p. 509)

1. Chloroquine, quinacrine, diiodohydroxyquin, pentamidine, and metronidazole are used to treat protozoan infections.
2. Chloroquine and quinacrine stop DNA synthesis by intercalation.
3. Antihelminthic drugs include niclosamide, mebendazole, praziquantel, and piperazine.
4. Mebendazole disrupts microtubules; piperazine paralyzes intestinal roundworms.

Tests for Microbial Susceptibility to Chemotherapeutic Agents (pp. 509–510)

1. These tests are used to determine the degree of susceptibility of different microorganisms to chemotherapeutic agents.
2. They help to determine which chemotherapeutic agent is most likely to combat a specific pathogen.
3. These tests are used when susceptibility cannot be predicted or when drug resistance arises.

DISK-DIFFUSION METHOD (p. 510)

1. In this test, also known as the Kirby–Bauer test, a bacterial culture is inoculated on an agar medium, and filter-paper disks impregnated with chemotherapeutic agents are overlayed on the culture.
2. After incubation, the absence of microbial growth around a disk is called a zone of inhibition.
3. The diameter of the zone of inhibition, when compared with a standardized reference table, is used to determine whether the organism is sensitive, intermediate, or resistant to the drug.

BROTH DILUTION TESTS (p. 510)

1. In the broth dilution test, the microorganism is grown in liquid media containing different concentrations of a chemotherapeutic agent.
2. The minimum inhibitory concentration (MIC) is the lowest concentration of chemotherapeutic agent capable of preventing microbial growth.
3. The lowest concentration of chemotherapeutic agent that kills bacteria is called the minimum bactericidal concentration (MBC).

Effectiveness of Chemotherapeutic Agents (pp. 510–511)

DRUG RESISTANCE (pp. 510–511)

1. Drug resistance refers to the ability of a microorganism to resist the antimicrobial effects of a chemotherapeutic agent.
2. Resistance may be due to enzymatic destruction of a drug or cellular or metabolic changes at target areas.
3. Hereditary drug resistance is carried by plasmids called resistance (R) factors.
4. Resistance can be minimized by the discriminate use of drugs in appropriate concentrations and dosages.

EFFECTS OF COMBINATIONS OF DRUGS (p. 511)

1. Combinations of drugs may be used to minimize the development of resistant strains, to employ a synergistic effect, to provide therapy prior to diagnosis, and to use small concentrations of each drug to lessen toxicity.
2. Some combinations of drugs are synergistic—they are more effective when taken together.
3. Some combinations of drugs are antagonistic—when taken together, both drugs become less effective than when taken alone.

THE FUTURE OF CHEMOTHERAPEUTIC AGENTS (p. 511)

1. More antiviral drugs are urgently needed. Their development might also prove to be helpful in cancer therapy.
2. Drugs to treat tropical parasites are needed.

STUDY QUESTIONS

Review

1. Fill in the following table.

Antimicrobial Agent	Synthetic or Antibiotic	Method of Action	Principal Use
Isoniazid			
Sulfonamides			
Ethambutol			
Trimethoprim			
Nitrofurans			
Penicillin, natural			
Penicillin, semisynthetic			
Cephalosporins			
Carbapenems			
Aminoglycosides			
Tetracyclines			
Chloramphenicol			
Macrolides			
Polypeptides			
Vancomycin			
Rifamycins			
Polyenes			
Griseofulvin			
Amantadine			
Acyclovir			
Chloroquine			
Niclosamide			

2. Define a chemotherapeutic agent. Distinguish between a synthetic chemotherapeutic agent and an antibiotic.
3. Paul Ehrlich discovered the first __________. Alexander Fleming discovered __________; it was the first __________.
4. List and explain five criteria used to identify an effective antimicrobial agent.
5. What similar problems are encountered with antiviral, antifungal, antiprotozoan, and antihelminthic drugs?
6. Identify three methods of action of antiviral drugs. Give an example of a currently used antiviral drug for each method of action.
7. Compare and contrast the tube dilution and agar dilution tests. Identify at least one advantage of each.
8. Describe the disk-diffusion test for microbial susceptibility. What information can you obtain from this test?
9. Define drug resistance. How is it produced? What measures can be taken to minimize drug resistance?
10. List the advantages of using two chemotherapeutic agents simultaneously to treat a disease. What problem can be encountered using two drugs?
11. Why does a cell die from the following antimicrobial-actions?
    (a) Amodiaquine intercalates into DNA.
    (b) Colistimethate binds to phospholipids.
    (c) Kanamycin binds to 70S ribosomes.
12. How is translation inhibited by each of the following?
    (a) Chloroamphenicol
    (b) Erythromycin
    (c) Tetracycline
    (d) Streptomycin
13. Dideoxyinosine (ddI) is an antimetabolite of guanine. The 3′-OH is missing from ddI. How does ddI inhibit DNA synthesis?

Challenge

1. Penicillin and streptomycin can be used together under certain circumstances, but penicillin and tetracycline cannot be used together. Offer an explanation for this.
2. Which of the following can affect human cells? Explain why or why not.
   (a) Penicillin
   (b) Nystatin
   (c) Erythromycin
   (d) Polymyxin
3. Why is idoxuridine effective if host cells also contain DNA?
4. Some bacteria become resistant to tetracycline because they don't make porins. Why can a porin-deficient mutant be detected by its inability to grow on a medium containing a single carbon source such as succinic acid?
5. Why do you suppose that *Streptomyces griseus* produces an enzyme that inactivates streptomycin? Why is this enzyme produced early in idiophase?

6. The following data were obtained from a disk-diffusion test.

Antibiotic	Zone of Inhibition
A	15 mm
B	0 mm
C	7 mm
D	15 mm

(a) Which antibiotic was most effective against the bacteria being tested?
(b) Which antibiotic would you recommend for treatment of a disease caused by this bacterium?
(c) Was antibiotic A bactericidal or bacteriostatic? How can you tell?

7. The following results were obtained from a tube dilution test for microbial susceptibility.

Tube Number	Antibiotic Concentration	Growth	Growth in Subculture
1	200 μg	–	–
2	100 μg	–	–
3	50 μg	–	+
4	25 μg	+	+

(a) The MIC of this antibiotic is ________.
(b) The MBC of this antibiotic is ________.

FURTHER READING

Abraham, E.P. "The beta-lactam antibiotics." *Scientific American* 244(6):76–86, June 1981. A discussion of semisynthetic penicillins and cephalosporins.

Baldry, P. *The Battle Against Bacteria.* Cambridge: Cambridge University Press, 1976. The history of the fight against diseases and the development of antibacterial drugs.

Cundliffe, E. "How antibiotic-producing organisms avoid suicide." *Annual Review of Microbiology* 43:207–233, 1989. Discussion of mechanisms such as enzymatic degradation and target alternation.

Hirsch, M.S., and J.C. Kaplan. "Antiviral therapy." *Scientific American* 246(4):76–85, April 1987. A review of the mechanisms by which antiviral drugs work.

Pratt, W.B., and R. Fekety. *The Antimicrobial Drugs.* New York: Oxford University Press, 1986. A very good summary of most of the commonly used antibiotics, with uses, modes of action, pharmacology, toxicity, and other information.

Sanders, C.C. "Chromosomal cephalosporinases responsible for multiple resistance to newer β-lactam antibiotics." *Annual Review of Microbiology* 41:573–593, 1987. Describes methods of action of β-lactamases and methods to detect production of these enzymes by bacteria.

Schwarcz, S.K., et al. "National surveillance of antimicrobial resistance in *Neisseria gonorrhoeae.*" *Journal of the American Medical Association* 264:1413–1417, Sept. 19, 1990. Shows drug resistance by drug and by city.

Wishnow, R.M., and J.L. Steinfeld. "The conquest of the major infectious diseases in the United States: a bicentennial retrospective." *Annual Review of Microbiology* 30:427–450, 1976. A concise summary of control of diseases, including tuberculosis, cholera, malaria, and yellow fever.

PART FOUR

Microorganisms and Human Disease

CHAPTER 21

Microbial Diseases of the Skin and Eyes

LEARNING OBJECTIVES

- Describe the structure of the skin and the ways pathogens can invade the skin.
- Provide examples of normal skin flora, and state their locations and ecological roles.
- Differentiate between staphylococci and streptococci, and list several skin infections caused by each.
- List the etiologic agent, method of transmission, and clinical symptoms of the following skin infections; acne, warts, smallpox, chickenpox, measles, rubella, and cold sores.
- Differentiate among the types of mycoses, and provide an example of each.
- Discuss the roles of bacteria, fungi, viruses, and protozoans in conjunctivitis and keratitis.
- Describe the epidemiologies of neonatal gonorrheal ophthalmia and inclusion conjunctivitis.

Part Four photo, p. 517: Recent research on the common cold indicates that aerosol transmission is the primary mode of infection (see p. 595).

In Chapter 16, we saw that the human body possesses a number of defenses that contribute to *nonspecific resistance* against many types of pathogens. These defenses include antimicrobial substances, phagocytosis, the inflammatory response, fever, and the skin and mucous membranes.

The skin, which covers and protects the body, is the body's first line of defense against pathogens. Basically, the skin is an inhospitable place for most microorganisms because the secretions of the skin are acidic and most of the skin contains little moisture. Moreover, much of the skin is exposed to radiation, which discourages microbial life. Some parts of the body, however, such as the axilla (armpit), have enough moisture to support microbial growth. Since excretions in the axillary region tend to contain more organic matter than excretions elsewhere on the body, this region can support relatively large bacterial populations, whereas other regions, such as the scalp, support rather small numbers of microorganisms. The skin is a physical as well as an ecological barrier, and it is almost impossible for pathogens to penetrate it, although some can enter through openings that are not readily apparent.

Structure and Function of the Skin

The skin of an average adult occupies a surface area of about 1.9 m^2 and varies in thickness from 0.05 to 3.0 mm. As we mentioned in Chapter 16, skin consists of two principal parts, the epidermis and the dermis (Figure 21.1). The **epidermis** is the thin, outer portion, composed of several layers of epithelial cells. The outermost layer of the epidermis, the **stratum corneum,** consists of dead cells that contain a waterproofing pro-

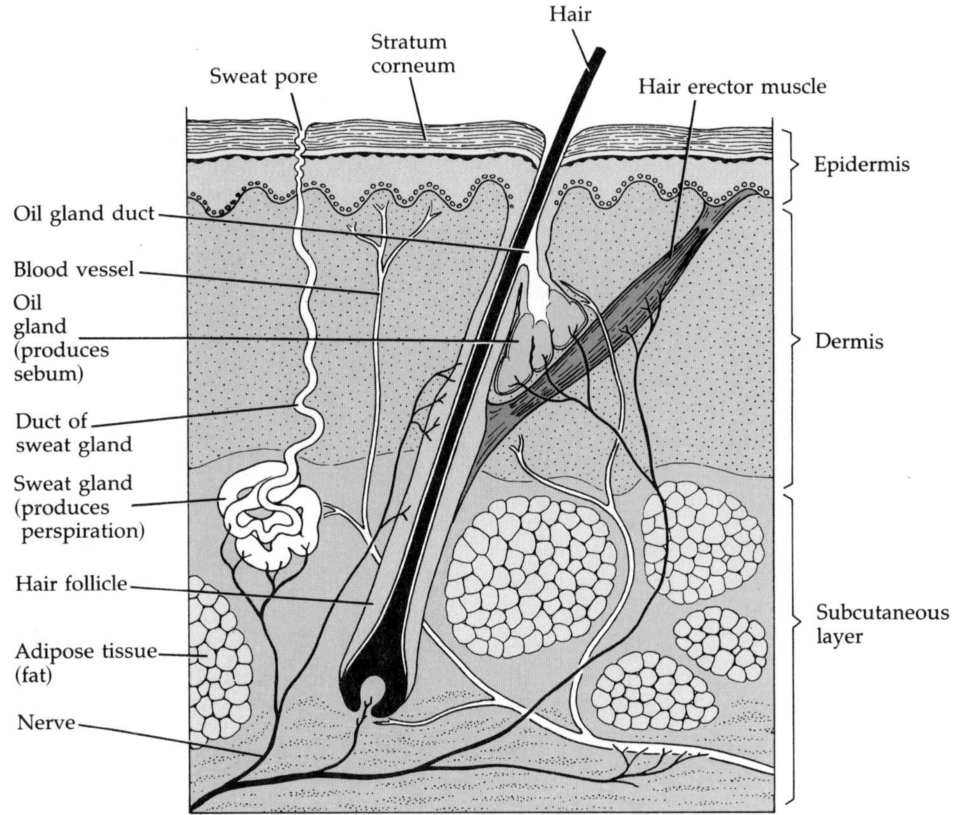

FIGURE 21.1 Structure of the skin. Note the passageways between the hair follicle and hair shaft through which microbes can penetrate the deeper tissues. Microbes can also enter the skin through sweat pores. Acne is caused by bacterial metabolism of sebum from the oil glands.

tein called **keratin.** The epidermis, when unbroken, is an effective physical barrier against microorganisms.

The **dermis** is the inner, relatively thick portion of skin that is composed mainly of connective tissue. The hair follicles, sweat-gland ducts, and oil-gland ducts contained in the dermis provide passageways through which microorganisms can enter the skin and penetrate deeper tissues.

Perspiration provides moisture and some nutrients for microbial growth. However, it contains salt, which inhibits many microorganisms, and lysozyme which is capable of breaking down the cell walls of certain bacteria.

Sebum, secreted by oil glands, is a mixture of lipids (unsaturated fatty acids), proteins, and salts that prevents skin and hair from drying out. Although the fatty acids inhibit the growth of certain pathogens, sebum, like perspiration, is also nutritive for many microorganisms, most notably the corynebacteria.

In the linings of body cavities, such as the mouth, nasal passages, lungs, and urinary, genital, and gastrointestinal tracts, the outer protective barrier is specialized in various ways. For example, some cells are ciliated. Others secrete mucus from glands beneath this outer layer of epithelial cells—hence the name *mucous membrane (mucosa).* Mucous membranes are often formed to maximize surface area to increase absorption. The total area in an average human is about 400 m^2, much more than the surface area of the skin. In the respiratory system, the mucous layer traps particles, including microorganisms, and the ciliary movement sweeps them upward out of the body (see Figure 16.3). Mucous membranes are frequently acidic, which tends to limit their microbial populations. The eyes are mechanically washed by tears, and the enzyme lysozyme in the tears destroys the cell walls of certain bacteria.

Normal Flora of the Skin

Although the skin is generally inhospitable to most microorganisms, it supports the growth of certain microbes, which are established as part of the normal flora. On superficial skin surfaces, certain aerobic bacteria produce fatty acids from sebum. These acids tend to inhibit many microbes and allow better adapted bacteria to flourish.

Microorganisms that find the skin a satisfactory environment tend to be resistant to drying and to relatively high salt concentrations. The skin's normal flora contains large numbers of gram-positive bacteria such as staphylococci and micrococci. Some of these are

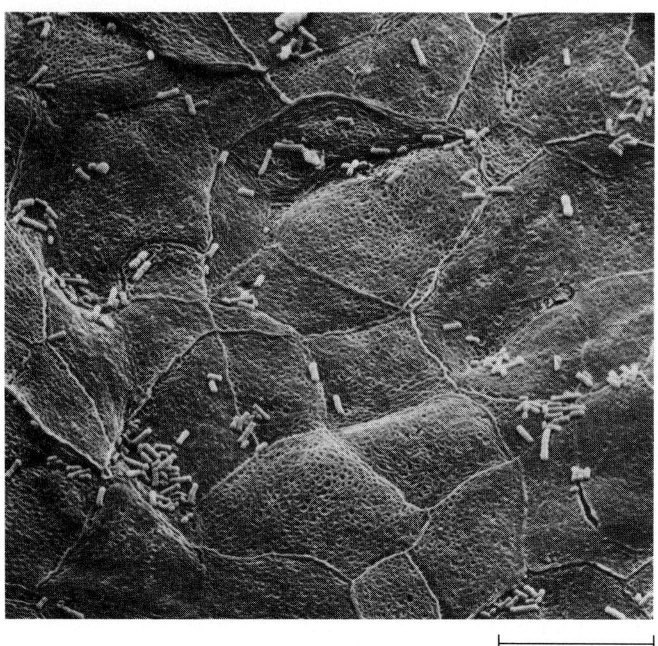

FIGURE 21.2 A section of skin showing resident bacteria.

capable of growth at sodium chloride concentrations of 7.5% or more. Scanning electron micrographs (Figure 21.2) show that bacteria on the skin tend to be grouped into small clumps. Vigorous washing can reduce their numbers but will not eliminate them. Microorganisms remaining in hair follicles and sweat glands after washing will soon reestablish the normal populations. Areas of the body with more moisture, such as the armpits and between the legs, have higher populations of microbes. These metabolize the secretions from the sweat glands and are the main contributors to what the advertising world advises us is offensive body odor.

Also part of the skin's normal flora are gram-positive pleomorphic rods called diphtheroids. Some diphtheroids, such as *Propionibacterium acnes*, are typically anaerobic and inhabit hair follicles. Their growth is supported by secretions from the oil glands (sebum), which we will see makes them a factor in acne. These bacteria produce propionic acid, which helps maintain the low pH of skin, generally between 3 and 5. This acidity has a bacteriostatic effect on many potentially harmful microbes. Other diphtheroids, such as *Corynebacterium xerosis* (ze-rō'sis) are aerobic and occupy the skin surface. A yeast belonging to the genus *Pityrosporum* (pit-i-ros'pô-rum) is capable of growing on oily skin secretions and is a frequent inhabitant of the skin. Some consider it to be responsible for the scaling skin condition known as dandruff.

DISEASES OF THE SKIN

Many important diseases manifest themselves by such symptoms as rashes, vesicles, and other lesions that mainly affect the skin. Often the focus of infection is elsewhere in the body, but it is convenient to classify these diseases by the organ most obviously affected, the skin. Table 21.1 on p. 531 summarizes the more important diseases associated with the skin.

Our discussion is primarily concerned with bacterial, viral, and fungal skin infections. The box on p. 529 describes a skin disease, scabies, caused by a parasite.

Bacterial Diseases of the Skin

Two genera of bacteria, *Staphylococcus* and *Streptococcus* (commonly called staphylococci and streptococci), are frequent causes of skin-related diseases and merit special discussion. We will also discuss these bacteria in later chapters in relation to other organs and conditions. Superficial staphylococcal and streptococcal infections of the skin are very common. The bacteria frequently come into contact with the skin and have adapted fairly well to the physiological conditions there. Both genera also produce invasive enzymes and damaging toxins that contribute to the disease process.

STAPHYLOCOCCAL SKIN INFECTIONS

Staphylococci are spherical gram-positive bacteria about 0.5 to 1.5 μm in diameter. They tend to form in irregular clusters like grapes because the cells divide at random points about their circumference and the daughter cells do not completely separate from each other (see Figures 4.1d and 11.10). For almost all clinical purposes these bacteria are divided into those that produce **coagulase,** an enzyme that coagulates (clots) fibrin in blood, and those that do not (coagulase-positive and coagulase-negative strains). The coagulase-negative strains are very common on the skin, where they may represent 90% of the normal flora. They are generally pathogenic only when the skin barrier is broken or is invaded by medical procedures such as the insertion and removal of catheters. At one time the coagulase-negative staphylococci were considered to be all one species, *Staphylococcus epidermidis*. They have since been subdivided into several species, and the name *S. epidermidis* is applied now only to the predominant species from human skin.

Staphylococcus aureus is the most pathogenic of the staphylococci. Typically, it forms golden yellow colonies. Almost all pathogenic strains of *S. aureus* are coagulase-positive. It is possible that a fibrin clot protects the microorganisms from phagocytosis and isolates them from other defenses of the host. There is a high correlation between the bacterium's ability to form coagulase and its production of damaging toxins, several of which may injure tissues. Staphylococci may also produce *leukocidin*, which destroys phagocytic leukocytes, and *exfoliative toxin*, which is responsible for scalded skin syndrome, to be discussed shortly. Some staphylococcal toxins, *enterotoxins*, affect the gastrointestinal tract and will be discussed in Chapter 25 with diseases of the digestive system.

S. aureus is a very common problem in the hospital environment. Because *S. aureus* is carried by patients, hospital personnel, and hospital visitors, the danger of infection of surgical wounds and other breaks in the skin is very high. Moreover, such infections are difficult to treat because *S. aureus* is exposed to so many antibiotics that it quickly becomes resistant to them. At one time, this organism was almost uniformly extremely susceptible to penicillin, but now only about 10% of *S. aureus* strains are sensitive. Most of the microorganism's resistance derives from the production of penicillinases.

The nasal passages provide an especially favorable environment for *S. aureus*, which is often present there in very large numbers. Its presence on unbroken skin is usually the result of transport from the nasal passages. *S. aureus* often enters the body through a natural opening in the skin barrier, the hair follicle's passage through the epidermal layer. Infections of hair follicles, or **folliculitis,** often occur as **pimples.** The infected follicle of an eyelash is called a **sty.** A more serious hair follicle infection is the **furuncle (boil),** which is a type of **abscess,** a localized region of pus surrounded by inflamed tissue. Antibiotics do not penetrate well into abscesses, which are difficult to treat. Draining pus from the abscess is frequently a preliminary step to successful treatment.

When the body fails to wall off a bacterial infection, neighboring tissue can be progressively invaded. The extensive damage is called a **carbuncle,** a hard, round deep inflammation of tissue under the skin. At this stage of infection, the patient usually exhibits the symptoms of generalized illness with fever.

Staphylococci are the primary cause of a very troublesome problem in hospital nurseries, **impetigo of the newborn.** Symptoms of this disease are thin-walled vesicles on the skin that rupture and later crust over. To prevent outbreaks, which can reach epidemic proportions, hexachlorophene-containing skin lotions are commonly prescribed (Chapter 7).

Staphylococcal infections always carry the risk that the underlying tissue will become infected or that the infection will enter the bloodstream. The circulation of toxins, such as those produced by staphylococci, is called **toxemia.** One such toxin, which is produced by staphylococci lysogenized by certain phage types, causes **scalded skin syndrome.** (See Chapters 13 and 15 for discussion of lysogeny and phage conversion.) This condition is first characterized by a lesion around the nose and mouth. The lesion develops rapidly into a bright red area, and, within 48 hours, the skin of the palms and soles peels off in sheets when it is touched (Figure 21.3). The scalded skin syndrome is frequently observed in children under the age of two, especially in newborns, as a complication of staphylococcal infections. These patients are seriously ill, and vigorous antibiotic therapy is required.

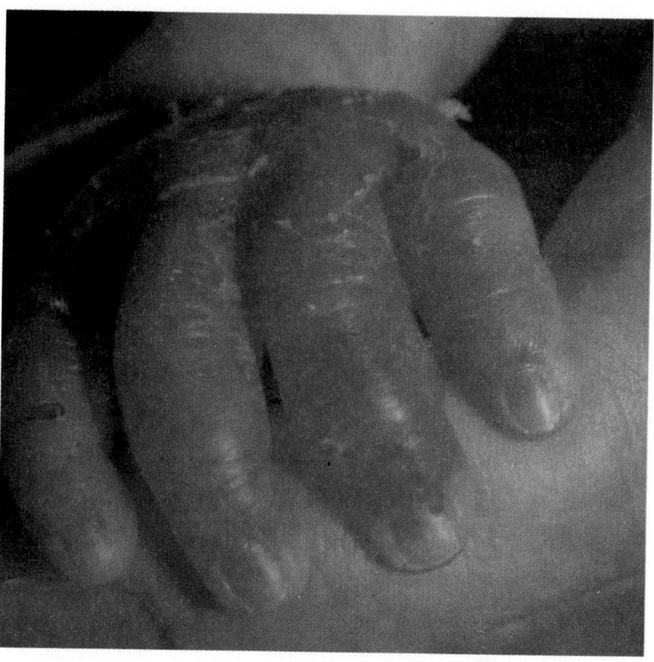

FIGURE 21.3 Scalded skin syndrome. Some staphylococci produce a toxin that causes the skin to peel off in sheets. It is especially likely to occur in children under two years of age, as shown on the hand of this infant.

The scalded skin syndrome is also characteristic of the late stages of **toxic shock syndrome (TSS).** In this potentially life-threatening condition, fever, vomiting, and a sunburnlike rash are followed by shock (a sudden drop in blood pressure). TSS originally became known as a staphylococcal growth associated with the use of a new type of highly absorbent vaginal tampons; the correlation is especially high for cases in which the tampons are retained too long. Staphylococcal toxin (mainly toxic shock syndrome toxin-1, or TSST-1) enters the bloodstream from the bacterial growth site in and around the tampon, and its circulation causes the symptoms.

Today only little more than about half of the cases of TSS are associated with menstruation. Nonmenstrual TSS occurs from staphylococcal infections that follow nasal surgery in which absorbent packing is

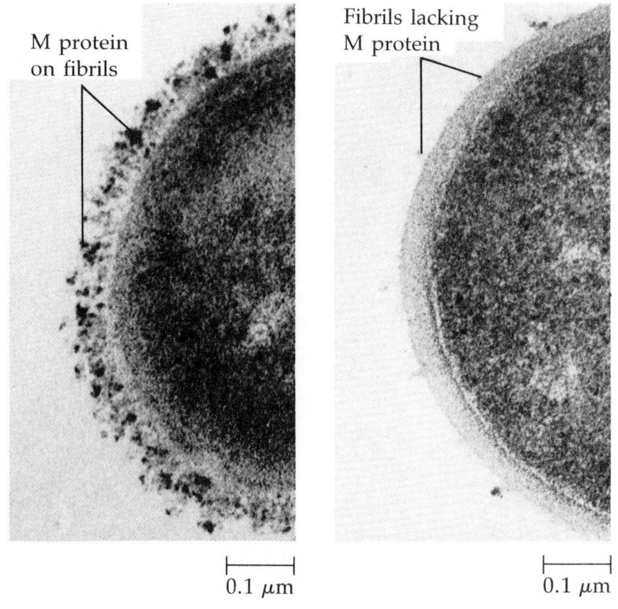

FIGURE 21.4 Electron micrographs showing portions of group A β-hemolytic streptococci. The cell on the left carries M protein on hairlike surface fibrils; the cell on the right lacks M protein.

used, after surgical incisions, and in women who have just given birth.

STREPTOCOCCAL SKIN INFECTIONS

Like staphylococci, **streptococci** are gram-positive spherical bacteria. Unlike staphylococci, the streptococcal cells grow in chains. Prior to division, the individual cocci elongate on the axis of the chain, and then the cells divide into pairs. When the dividing pairs do not separate, chaining occurs (see Figures 4.1a and 11.11). The bridges between the individual cocci in the chain consist of cell wall material that has not cleaved.

The metabolism of streptococci is strictly fermentative (they cannot use oxygen). Unlike either aerobic or facultative anaerobic bacteria, they do not produce catalase (see Chapter 6).

Many nonpathogenic streptococci are commonly found inhabiting the mouth, gastrointestinal tract, and upper respiratory system, but some streptococci are responsible for skin infections. As streptococci grow, they secrete several toxins and enzymes into the growth medium. Among these toxins are *hemolysins*, which damage red blood cells. Depending on the type of destruction caused by the hemolysins, streptococci can be divided into α-hemolytic, β-hemolytic, and γ-hemolytic streptococci (see Chapter 11). β-hemolytic streptococci have remained highly susceptible to penicillin. Unlike the staphylococci, they have no ability to produce penicillinases.

β-hemolytic streptococci are further differentiated into a number of groups, designated A through O, according to the antigenic carbohydrates in their cell walls. As causes of human disease, β-hemolytic group A streptococci are the most important. β-hemolytic group A streptococci can be further subdivided into over 55 immunologic types according to the antigenic properties of the *M protein* found in some strains (Figure 21.4). This protein is found external to the cell wall on fibrils. The M protein has antiphagocytic properties that contribute to the strain's pathogenicity. It also appears to aid the bacteria in adhering to and colonizing on mucous membranes.

In addition to hemolysins and M proteins, a number of other substances contribute to the pathogenicity of β-hemolytic group A streptococci. These substances include *erythrogenic toxin* (responsible for the scarlet fever rash), *deoxyribonucleases* (enzymes that degrade DNA), *NADase* (an enzyme that breaks down NAD), *streptokinases* (enzymes that dissolve blood clots), *hyaluronidase* (an enzyme that dissolves hyaluronic acid, the cementing substance of connective tissue), and *leukocidins* (enzymes that kill white blood cells).

β-hemolytic group A streptococci cause a wide variety of diseases, a number of which will be discussed in later chapters. The most common species of this group, *Streptococcus pyogenes*, may be implicated in some cases of impetigo. Although impetigo of the newborn is caused by staphylococci, **impetigo** is also common in children of toddler and grade-school age and is likely in these cases to be caused by streptococci. A superficial skin infection, impetigo is characterized by isolated pustules (small, round elevations containing pus) that become crusted and rupture (Figure 21.5). The disease is spread largely by contact, and the bacteria penetrate the skin through some minor abrasion or wound whose scab has been prematurely removed. Staphylococci can also be found in these lesions but are probably only secondary invaders. Fortunately, the condition is seldom serious. The drugs of choice are penicillin and erythromycin.

When streptococcal infections affect the dermis, they cause **erysipelas.** In this disease, the skin erupts into reddish patches that enlarge and thicken and swell at the margins (Figure 21.6). High fever is common. The reddening is caused by toxins produced by the streptococci as they invade new areas. Usually, the skin outbreak is preceded elsewhere in the body by a streptococcal infection, such as a streptococcal sore throat. It is not certain if the skin is infected from an external invasion or via the lymphatic system. Erysipelas is most likely to occur in the very young and the very old. The drugs of choice for the treatment of erysipelas are penicillin and erythromycin.

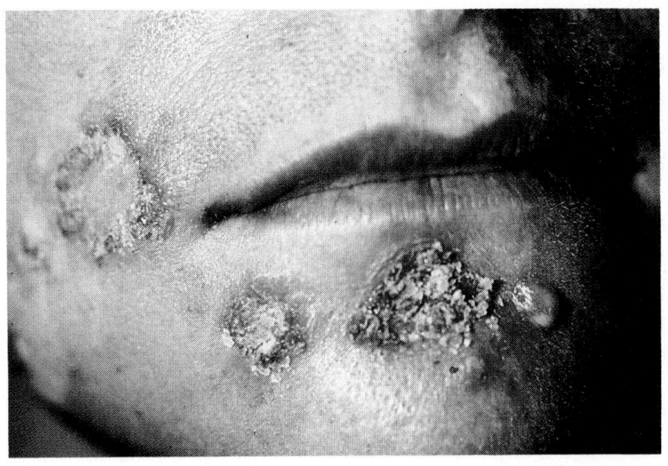

FIGURE 21.5 Impetigo. This disease is characterized by isolated pustules (round elevations on the skin containing pus) that become crusted.

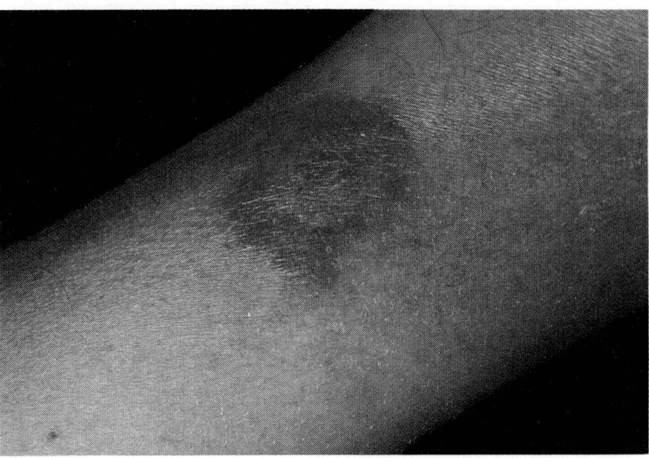

FIGURE 21.6 Erysipelas caused by group A β-hemolytic streptococci. The patches of skin have been reddened by streptococcal toxins.

INFECTIONS BY PSEUDOMONADS

Pseudomonads are aerobic gram-negative rods that are widespread in soil and water. Able to survive in any moist environment, they can grow on traces of even unusual organic matter, such as soap films or cap liner adhesives, and are resistant to many antibiotics and disinfectants. The most prominent species is *Pseudomonas aeruginosa*.

Pseudomonads frequently cause outbreaks of ***Pseudomonas* dermatitis.** This is a self-limiting rash of about two weeks' duration often associated with swimming pools and pool-type saunas and hot tubs. When large numbers of people use these facilities, the alkalinity rises and the chlorines become less effective; at the same time, the concentration of nutrients that support growth of pseudomonads increases. Hot water facilitates the entry of bacteria into hair follicles by causing the follicles to dilate. Competition swimmers are often troubled with **otitis externa,** or swimmer's ear, a pseudomonad infection of the external ear canal leading to the eardrum.

P. aeruginosa produces several exotoxins that account for much of its pathogenicity. It also has an endotoxin. Except for superficial skin infections and otitis externa, infection by *P. aeruginosa* is rare in healthy individuals. However, it often causes respiratory infections in persons already compromised by immunologic deficiencies (natural or drug induced) or by chronic pulmonary disease, especially cystic fibrosis. (Respiratory infections are discussed in Chapter 24.)

P. aeruginosa is also a very common and serious opportunistic pathogen in burn patients, particularly those with second- and third-degree burns. In such cases, infection may produce blue-green pus caused by the bacterial pigment **pyocyanin** (sometimes spelled *pyocyanine*). Of major concern in many hospitals is the ease with which *P. aeruginosa* is carried by flowers or plants sent by well-wishers.

Pseudomonads are relatively resistant to many of the commonly used disinfectants (Chapter 7) and antibiotics (see Chapter 20). Antibiotic resistance by pseudomonads is still a problem, but in recent years a number of new antibiotics have been developed, so the number of antimicrobials available for chemotherapy of these infections is not as restricted as it once was. Two antibiotics commonly used against pseudomonad infections are gentamicin and carbenicillin, often used in combination. Silver sulfadiazine is very useful in the treatment of burn infections by *P. aeruginosa*.

ACNE

Acne is probably the most common skin disease of humans, affecting an estimated 17 million people in the United States. More than 85% of all teenagers have the problem to some degree. In the United States, about 350,000 persons each year are estimated to suffer from unusually severe acne, called **cystic acne,** that produces inflamed cysts and subsequent severe scarring of the face and upper body. Although the occurrence of acne decreases after the teen years, the scarring from these severe cases often remains.

Acne begins when channels for passage of sebum to the skin surface are blocked. The sebum accumulation leads to the formation of the familiar whiteheads—(if the opening remains open, the lesions known as blackheads form). The accumulation of sebum often ruptures the lining of the hair follicle. Bacteria, especially *Propionibacterium acnes*, a diphtheroid commonly found on the skin, become involved at this stage. By metabolizing the sebum, *P. acnes* forms free fatty acids that cause an inflammatory response by the body. It is this inflammation that leads to tissue damage and subsequent acne scars. Picking or scratching the lesions—or even having tight collars or other clothing in contact with lesions—increases the incidence of scar formation. Cosmetics frequently aggravate the condition, but diet—including consumption of chocolate—has been demonstrated to have no significant effect on the disease.

Topical applications of preparations containing benzoyl peroxide are often useful, and antibiotics such as tetracycline have also been used. Severe cases should receive a dermatologist's treatment rather than household remedies. The most important recent development in treatment of severe cystic acne is isotretinoin (Accutane®). This drug, taken orally, inhibits sebum formation, and dramatic improvement follows. Isotretinoin is not recommended for use against the usual mild cases of acne because of its side-effects; for one thing, it is *teratogenic* (causes damage to the fetus) if taken by a pregnant woman for even a few days.

Viral Diseases of the Skin

Many viral diseases, while systemic in nature, are most apparent by their effects on the skin.

WARTS

Warts are caused by viruses that stimulate an uncontrolled but benign growth of skin cells. It was long known that warts can be transmitted from one person to another by contact, even sexually, but it was not until 1949 that viruses were identified in wart tissues. About 40 different types of papillomaviruses are now known to cause different types of warts, often of greatly differing appearance. After infection, there is an incubation period of several weeks before the warts appear.

The most common medical treatments for warts are applying extremely cold liquid nitrogen to them (cryotherapy), drying them with an electric current (electrodesiccation), or burning them with acids. Burning with acids is also the basis of some home remedies. Warts that do not respond to any other treatments, especially genital warts, can now be treated with injected interferon or lasers. The use of lasers results in a virus-laden aerosol. Physicians using lasers to remove warts have contracted warts themselves, especially in their nostrils. Although warts are not a form of cancer, some skin and cervical cancers seem to be associated with papillomaviruses.

SMALLPOX (VARIOLA)

It is estimated that, during the Middle Ages, 80% of the population of Europe could expect to contract **smallpox** during their lives. Those who recovered from the disease carried disfiguring scars. The disease was even more devastating to Native Americans, who had had no previous exposure and thus little resistance.

Smallpox is caused by a poxvirus known as the smallpox (variola) virus. There are two basic forms of this disease—**variola major,** which has a mortality rate of 20% or higher, and **variola minor,** which has a mortality rate of less than 1%. Recovery from one form of the disease produces effective immunity against both forms.

Transmitted first by the respiratory route, the viruses infect many internal organs before their eventual movement into the bloodstream **(viremia)** leads to infection of the skin and the production of more recognizable symptoms. The growth of the virus in the epidermal layers of the skin causes lesions (Figure 21.7).

Smallpox was the first disease to which immunity was artificially induced (see Chapters 1 and 18) and the first to be eradicated from the human population. It is believed that the last victim of a natural case of smallpox was one who recovered from variola minor in 1977 in Somalia, Africa.

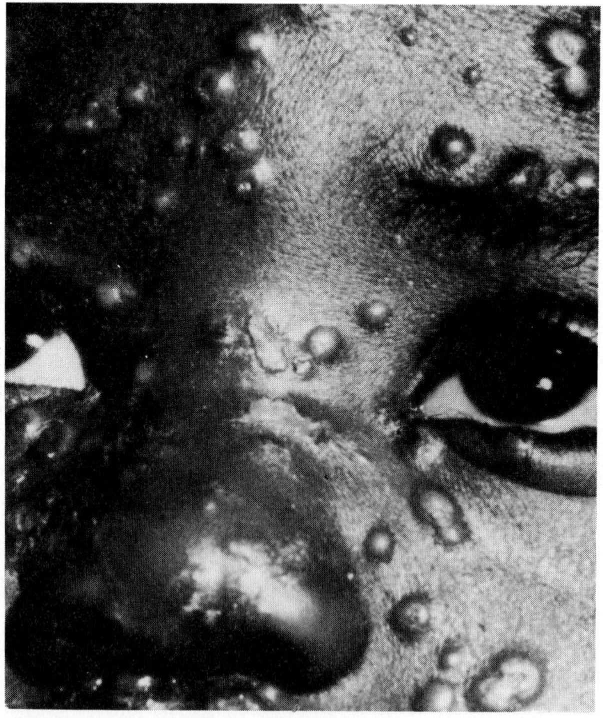

FIGURE 21.7 Lesions of smallpox. In some severe cases, the lesions nearly run together.

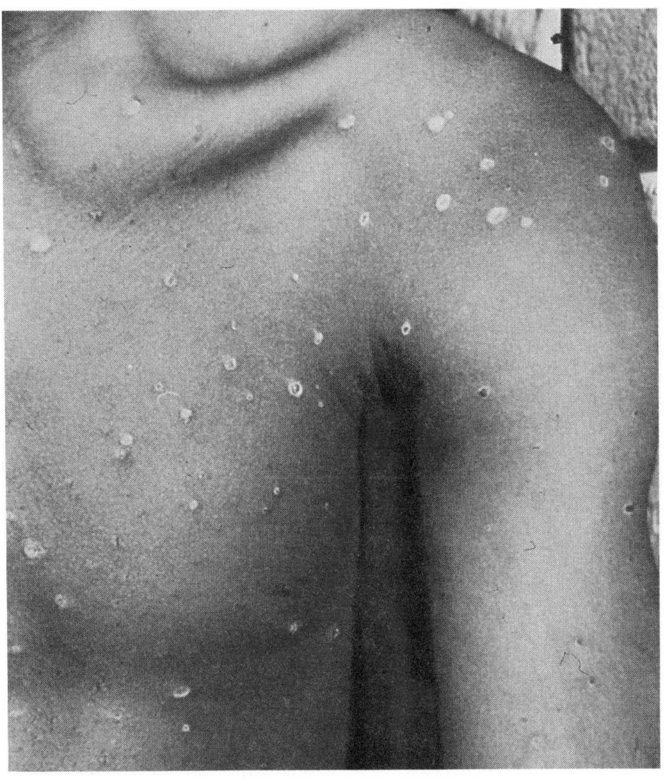

(a)

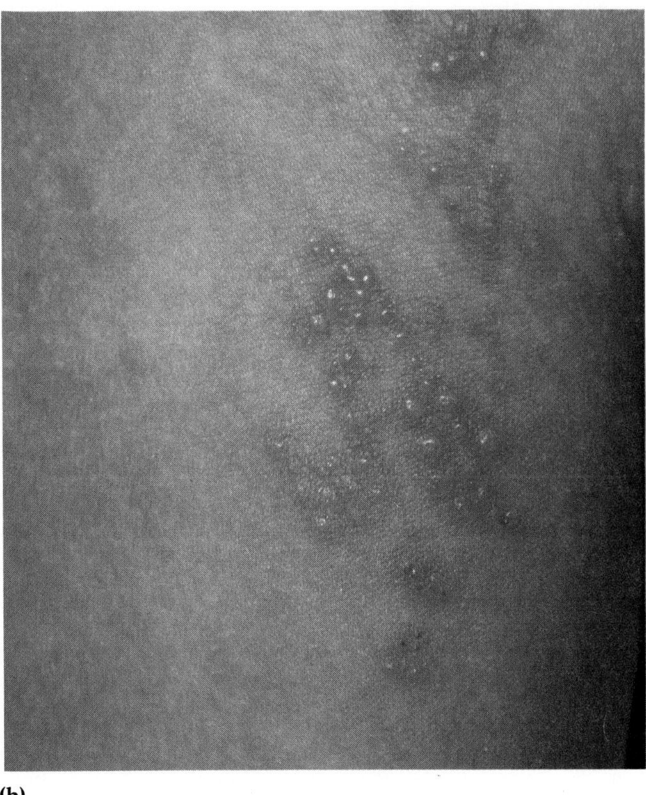

(b)

FIGURE 21.8 Typical lesions associated with (a) chickenpox and (b) shingles (herpes zoster), shown affecting the back of this patient.

The eradication of smallpox was possible because there are no animal host reservoirs for the disease. Once an effective vaccine became available, eradication was accomplished by a concerted vaccination effort coordinated by the World Health Organization.

Currently, the smallpox virus collections in laboratories have been the most likely sources of new infections. The risk of such infection is not merely a hypothetical concern as there have already been several laboratory-associated infections, one of which caused death. Today, only two sites maintain the smallpox virus, one in the United States and one in the USSR.

CHICKENPOX (VARICELLA) AND SHINGLES (HERPES ZOSTER)

Chickenpox (varicella) is a relatively mild childhood disease. After gonorrhea, it is the most common reportable infectious disease in the United States. It is probably greatly underreported, and more than 2 million cases probably occur each year in the United States. Disease summaries of the Centers for Disease Control show that about 100 deaths per year, usually from encephalitis (infection of the brain), are attributed to chickenpox.

Chickenpox is acquired by infection of the respiratory system, and the infection localizes in skin cells after about two weeks. The infected skin is vesicular for three to four days. During that time, the vesicles fill with pus, rupture, and form a scab before healing (Figure 21.8a). Lesions are mostly confined to the face, throat, and lower back. The vesicular rash can also appear in the mouth and throat. When chickenpox occurs in adults—which is not frequent because the high incidence in childhood grants immunity to most persons—it is a more severe disease with a significant mortality rate.

Reye's syndrome is an occasional severe complication of chickenpox, influenza, and some other viral diseases. A few days after the initial infection has receded, the patient persistently vomits and exhibits signs of brain dysfunction. Coma, fatty degeneration of the liver, and death can follow. Death or brain damage in survivors is from brain swelling, which prevents blood circulation. At one time, the death rate of reported cases approached 90%, but this rate has been declining with improved care and is now 30% or lower when the disease is recognized and treated in time. Reye's syndrome affects children and teenagers almost exclusively. The use of aspirin to lower fevers in chick-

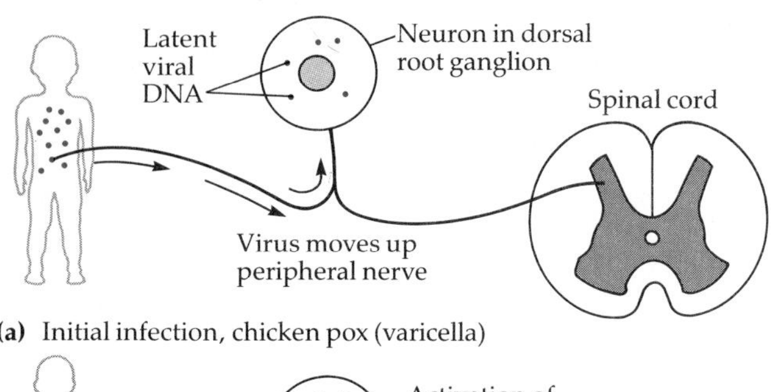

(a) Initial infection, chicken pox (varicella)

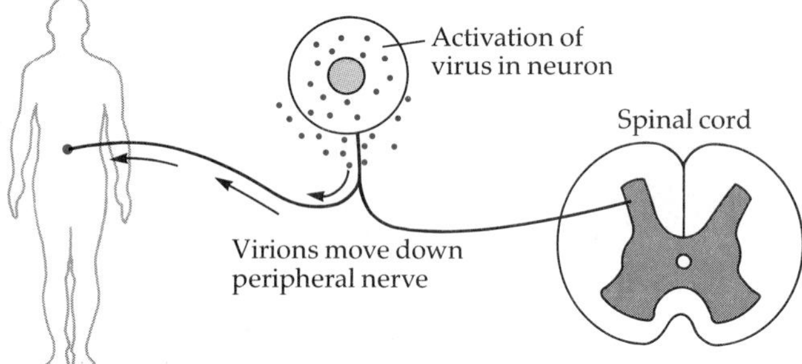

(b) Recurrence of infection, shingles (herpes zoster)

FIGURE 21.9 Latency of typical herpesviruses, varicella-zoster. After an initial infection of the varicella virus causes chickenpox **(a),** the virus moves to a central nerve ganglion, where it remains latent indefinitely. Age-related immune system depression or stress can trigger a reactivation of the virus, causing shingles **(b).**

enpox and influenza increases the chances of acquiring Reye's syndrome.

The cause of chickenpox, the varicella-zoster virus, is a herpesvirus. A characteristic of herpesviruses is *latency*. Following a primary infection, the virus enters the peripheral nerves and moves to a central nerve ganglion, where it persists as viral DNA. The DNA may or may not be transcribed, but in any event it persists, latent, indefinitely. The location of latent varicella-zoster virus is in the dorsal root ganglion near the spine. Later, perhaps as long as decades later, the virus may be reactivated. The trigger can be stress or simply the lower immune competence associated with aging. The virions produced by the reactivated DNA move along the peripheral nerves to the cutaneous sensory nerves of the skin, where they cause a new outbreak of the virus in the form of **shingles** (herpes zoster) (Figure 21.9).

In shingles, vesicles similar to those of chickenpox occur but are localized in distinctive areas. Typically, there is a distribution about the waist, although facial shingles and infections of the upper chest also occur (Figure 21.8b). The infection follows the distribution of the affected cutaneous sensory nerves and is usually limited to one side of the body at a time since these nerves are unilateral. Occasionally, such nerve infections can result in nerve damage that impairs vision or even causes paralysis. Severe pain is also frequently reported.

Shingles is simply a different expression of the virus that causes chickenpox; it expresses differently because the patient, having had chickenpox, now has partial immunity to the virus. Exposing children to shingles has led to their contracting chickenpox. Shingles seldom occurs in persons under age 20, and by far the highest incidence is in the elderly population.

Immunocompromised patients are in serious danger from infection by varicella-zoster virus—multiple organs become infected, and a mortality rate of 17% is common. In such cases, the antiviral drug *acyclovir* has proven helpful. A live, attenuated varicella vaccine has been approved and is administered at about age 15 months, with the measles–mumps–rubella (MMR) series.

HERPES SIMPLEX

Serologic surveys show that about 90% of the population of the United States have been infected with the herpes simplex virus. The initial infection usually occurs in infancy. Frequently, this infection is subclinical, but perhaps as many as 15% of the cases develop lesions known as **cold sores** or **fever blisters** (Figure 21.10). Usually occurring in the oral mucous membrane, these lesions heal as the infection subsides. Recurrences are usually associated with some trauma—excessive exposure to UV radiation from the sun is a

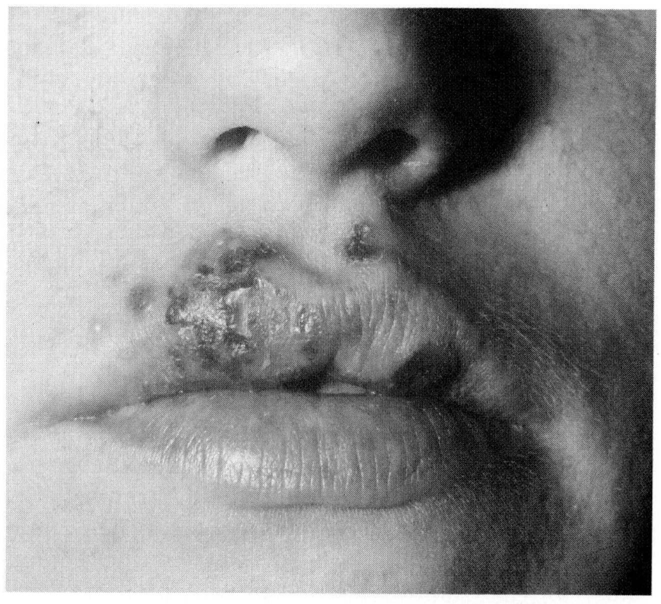

FIGURE 21.10 Cold sore or fever blisters caused by herpes simplex virus.

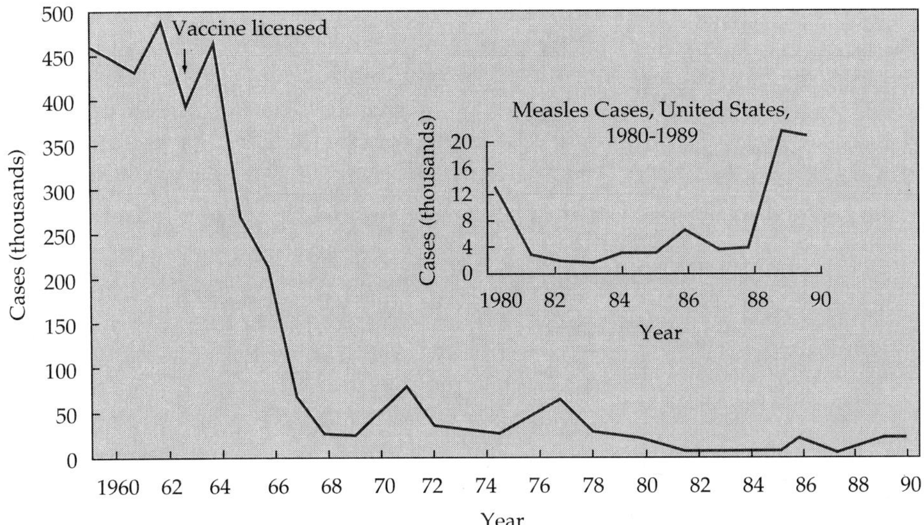

FIGURE 21.11 Number of reported measles cases in the United States, 1960 to 1990. Note the sharp decline in cases after introduction of the measles vaccine in 1963. Inset: The increase in measles cases in the last few years. [Source: CDC, *Summary of Notifiable Diseases 1989*, *MMWR* 38:54; *MMWR* 39:52 (1/4/91).]

frequent cause, as are the hormonal changes associated with menstrual periods or emotional upsets. These infections are caused by *herpes simplex type 1 virus* (HSV-1), which is transmitted primarily by oral or respiratory routes. An important complication is herpetic keratitis (discussed later in this chapter), in which the cornea of the eye becomes infected. Between recurrences, herpes simplex type 1 viruses are latent in the trigeminal nerve ganglia communicating between the face and the central nervous system. Transmission of HSV-1 infections occurs by skin contact among high school and college wrestlers **(herpes gladiatorium).** An incidence as high as 3% has been reported in high school wrestlers.

A different infection by a very similar virus, *herpes simplex type 2 virus* (HSV-2), is transmitted primarily by sexual contact. HSV-2 is differentiated from HSV-1 by its antigens and by its effect on cells in tissue cultures. We will discuss this infection in Chapter 26 in our discussion of diseases of the urinary and genital systems. Herpes simplex type 2 viruses causing genital herpes are latent in the sacral ganglia near the base of the spine.

Very rarely, either type of the herpes simplex virus may spread to the brain, causing **herpes encephalitis.** Infections by HSV-2 are more serious, with a fatality rate as high as 70% if untreated. Only about 10% of survivors can expect to lead normal lives. However, this is one of the rare viral diseases for which a chemotherapeutic treatment is reasonably effective. Tests have shown that acyclovir is almost twice as effective as vidarabine, the previously recommended drug. Even so, the mortality in the tests was still 28%, and only 38% of the survivors escaped serious neurological damage.

MEASLES (RUBEOLA)

Measles is an extremely contagious viral disease that is spread by the respiratory route. Because a person with measles is infectious before symptoms appear, quarantine is not an effective measure of prevention.

Humans are the only reservoir for measles in most parts of the world, although monkeys are also susceptible. Therefore, measles can potentially be eradicated, much as smallpox was. Since the licensing of a vaccine in 1963, the number of measles cases has declined from more than 400,000 cases per year to a record low of 1497 cases in 1983 (Figure 21.11). Unfortunately, the number of cases has been slowly creeping upward since then. Despite laws requiring immunization for school entry, the immunization rate is low in some densely populated inner-city groups, and about 40% of measles cases now occur in preschool children in these groups. In addition, although the measles vaccine is about 95% effective, cases continue to occur among those who do not develop or retain good immunity. Some of these infections are caused by contact with infected persons who come from outside the United States.

A previously developed vaccine provided poor immunity to the virus, and the practice of vaccination before the age of one year proved ineffective because of interference from inherited maternal antibodies. The new vaccine is now administered at age 15 months, frequently in combination as MMR vaccine.

The development of rubeola is similar to that of smallpox and chickenpox. Infection begins in the upper respiratory system. After an incubation period of 10 to 12 days, symptoms develop resembling those of a common cold—sore throat, headache, and cough.

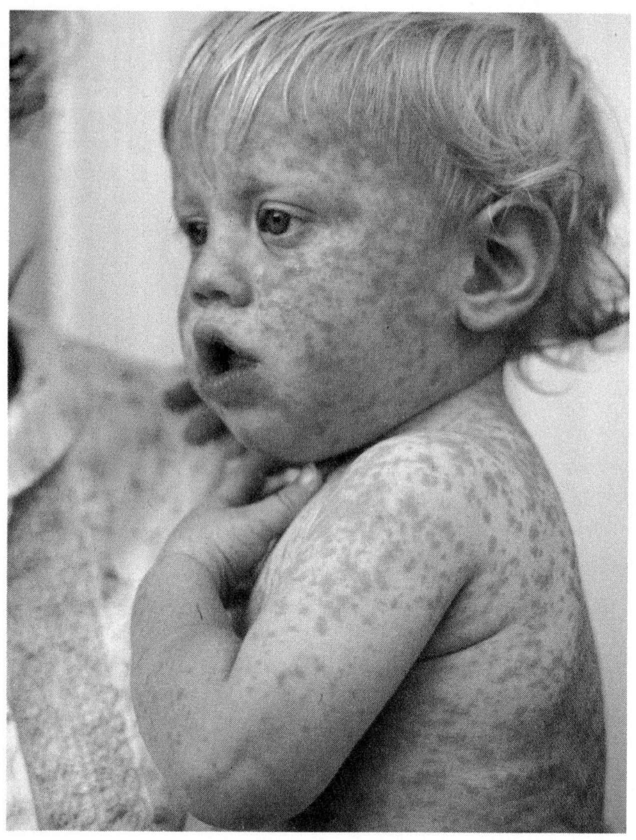

FIGURE 21.12 The rash of small raised spots typical of measles (rubeola). The rash typically begins on the face and spreads to the trunk and extremities.

Shortly thereafter, a macular rash appears on the skin, beginning on the face and spreading to the trunk and extremities (Figure 21.12). Lesions of the oral cavity include the diagnostically useful *Koplik spots* (tiny red patches with central white specks) on the oral mucosa opposite the molars.

Measles is an extremely dangerous disease, especially among very young and elderly individuals. It is frequently complicated by middle ear infection or pneumonia caused by the virus itself or by secondary bacterial infection. Encephalitis strikes approximately 1 in 1000 measles victims; its survivors are often left with permanent brain damage. As many as 1 in 3000 cases is fatal, mostly in infants. The virulence of the virus seems to vary with different epidemic outbreaks. Complications such as encephalitis occur, if at all, about a week after the rash appears.

Measles viruses can persist in the recovered patient. Apparently, a mutant form of the virus, which has lost the ability to complete its normal life cycle, sometimes remains in the patient's cells. The presence of this defective measles virus eventually causes **subacute sclerosing panencephalitis (SSPE).** This disease, which usually occurs in children or young adults with a history of measles acquired before one year of age, involves a sudden rapid and fatal degeneration of the nervous system. As a result of the decline in the number of cases of measles, SSPE has almost disappeared in the United States.

RUBELLA

Rubella, or German measles, is caused by a togavirus (rubella virus). It is a much milder disease than rubeola (measles) and often goes undetected. A rash of small red spots and a light fever are the usual symptoms (Figure 21.13). Complications are rare, especially in children, but encephalitis occurs in about 1 case in 6000, mostly in adults. Transmission is by the respiratory route, and an incubation time of two to three weeks is the norm. Recovery from clinical or subclinical cases appears to give a firm immunity.

The seriousness of rubella was not appreciated until 1941, when the association was made between serious birth defects and maternal infection during the first trimester (three months) of pregnancy—the **congenital rubella syndrome.** If a pregnant woman contracts the disease during this time, there is about a 35% incidence of serious fetal damage, including deafness, eye cataracts, heart defects, mental retardation, and death. Fifteen percent of babies with congenital rubella syndrome die during their first year. The last major epidemic of rubella in the United States was during

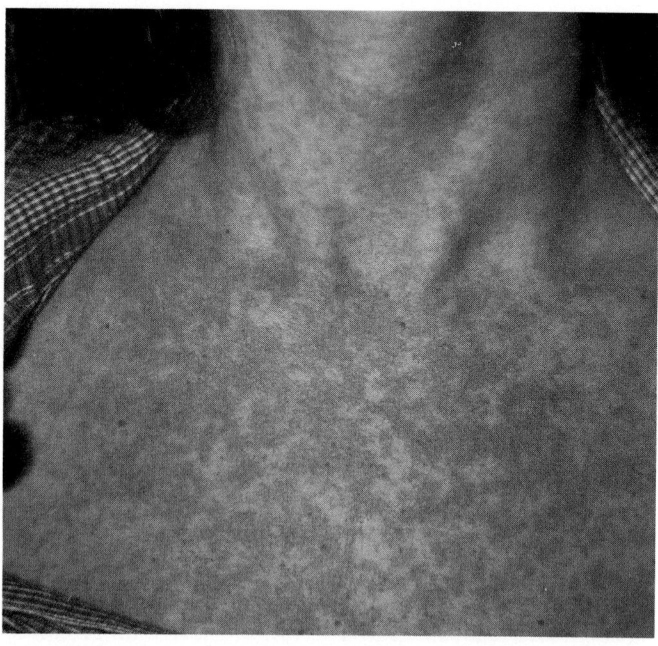

FIGURE 21.13 The rash of red spots characteristic of rubella. The spots are not raised above the surrounding skin.

MMWR

Morbidity and Mortality Weekly Report

Scabies in Health Care Facilities, Iowa

Scabies continues to occur among residents and staff of Iowa nursing homes and hospitals. During an 8-year period, the Iowa Department of Public Health confirmed scabies in 28 health care facilities and received reports of it in 11 others. The following is a report of their investigation of the problem in one nursing home.

In April 1987, an investigation revealed seven residents and three staff members with confirmed or probable scabies. All but two residents were confined to a ward of patients with Alzheimer's disease. The index patient, who had a rash of long duration, had transferred from another nursing home and probably had scabies upon arrival. Twice during 1986 the state health department had investigated the previous nursing home, which was the probable source of infestation, and had found rashes compatible with scabies but no positive scrapings. The index patient had been included in these investigations.

Comment: Scabies becomes pandemic at approximately 30-year intervals. Evidence suggests that community scabies peaked in the mid-1970s but has persisted at high levels for the past 10 years.

Scabies is a disease of the human skin caused by the mite *Sarcoptes scabiei*, a parasitic arthropod (see the figure). The disease is common among school children and is also found in adults. Sometimes it occurs as a nosocomial infection in hospital personnel treating patients with symptoms described as pruritic dermatitis. It is a major problem in nursing homes, particularly among patients who are debilitated and require extensive hands-on care.

The fingers, wrists, and elbows are the most frequent sites of infection. The mites burrow into the skin and fill the tunnel with their eggs and feces. The eggs hatch, and new mites mature, mate, and lay more eggs, perpetuating the life cycle. The symptoms of scabies are the result of hypersensitivity reactions to the mites. Symptoms first occur two to six weeks after the initial infection. The main symptom is itching, especially when the skin is warm (for example, when in bed at night). Red, raised lesions (erythematous papules) develop, which may become infected with bacteria through scratching. More advanced, chronic cases may result in generalized eczema.

Diagnosis is made by microscopic examination of scrapings from lesions for mites. Scabies is treated by topical application of gamma benzene hexachloride (Kwell®) or lindane. Intensive retreatment may be necessary. Clothing, bedding, and other personal objects that might contain mites must be thoroughly cleaned. Scabies is transmitted by direct contact with infected persons or by fomites carrying female mites.

Source: *MMWR* 37:178–179 (3/25/88).

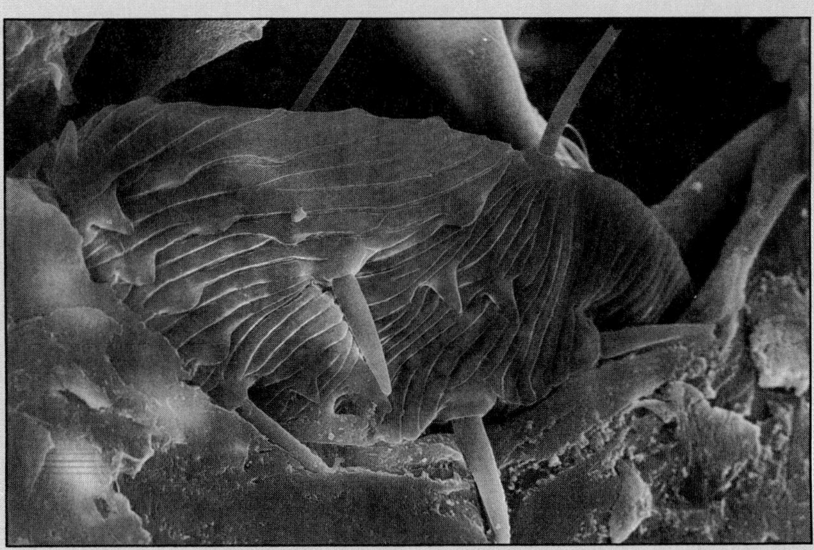

Female *Sarcoptes scabiei*.

1964 and 1965. About 20,000 severely impaired children from this epidemic are still alive. Now, fewer than 10 cases a year are reported.

It is very important to establish whether women of childbearing age are immune to rubella. Accurate diagnosis always requires laboratory tests; histories alone are very unreliable. Serum antibody against rubella can be assayed by hemagglutination inhibition. ELISA tests (see Chapter 18) are commercially available and provide comparably accurate results.

In 1969, a rubella vaccine was approved for use. It is important to know if the immunity conferred by the vaccine will be maintained from childhood through a woman's childbearing years. Follow-up studies indi-

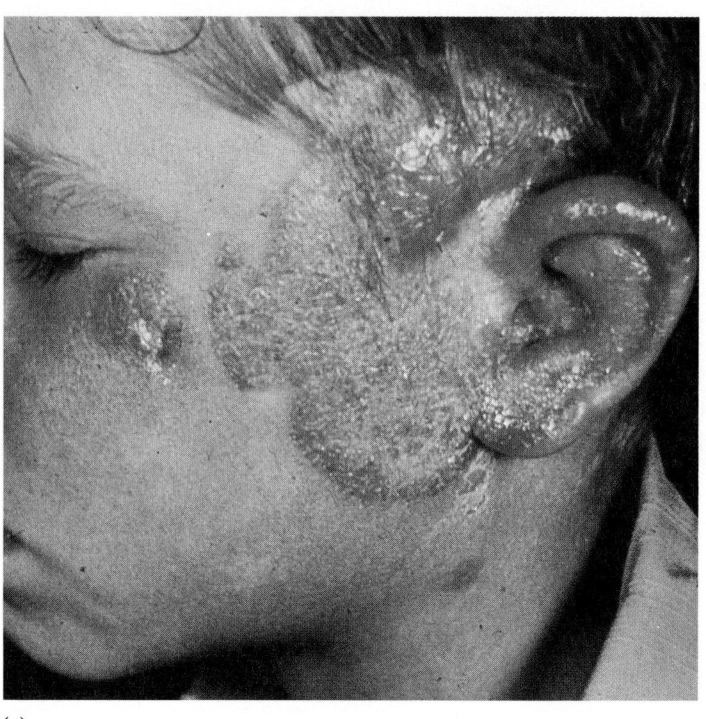

(a)

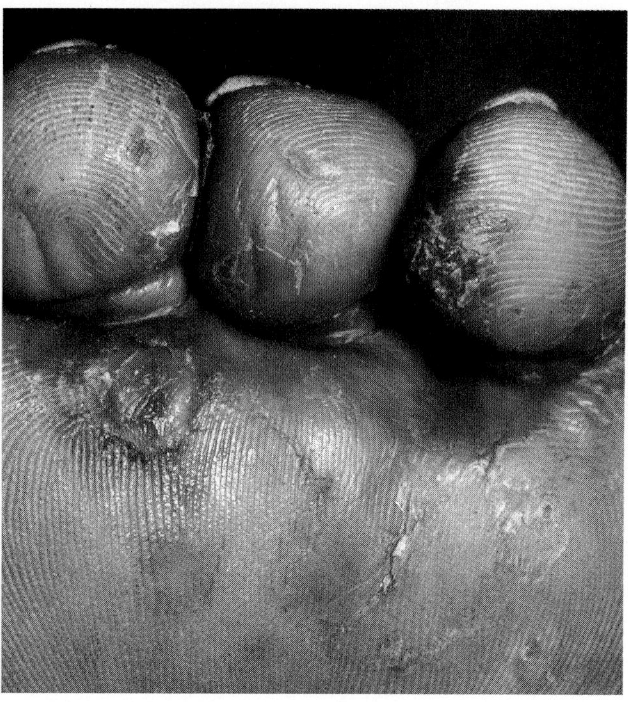

(b)

FIGURE 21.14 Dermatomycoses. **(a)** A severe case of ringworm on the side of a child's head (tinea capita). **(b)** Ringworm of the foot, or athlete's foot (tinea pedis). Moisture between the toes favors fungal infections.

cate that there has been little decline in antibody levels in the first children to receive the vaccine, who are now entering their childbearing years.

There has been concern about the possibility of fetal damage from pregnant women receiving the rubella vaccine. In hundreds of cases in which women were vaccinated three months before or three months after their presumed dates of conception, no case of congenital rubella syndrome defects has occurred. Nevertheless, the vaccine is still not recommended for pregnant women.

Fungal Diseases of the Skin

The skin is most susceptible to microorganisms that are able to resist high osmotic pressure and low moisture. It is not surprising, therefore, that fungi cause a number of skin disorders. Any fungal infection of the body is called a **mycosis.**

CUTANEOUS MYCOSES

Fungi that colonize the hair, nails, and the outer layer (stratum corneum) of the epidermis (see Figure 21.1) are called **dermatophytes,** and their infections are called **dermatomycoses.** Dermatophytes grow on the keratin present in those locations, causing infections called **tineas,** or **ringworms.** The name *ringworm* arose from the ancient Greek belief that the infections, which tend to expand circularly, were caused by a worm. The Romans incorrectly associated ringworm with lice, and the term *tinea* is derived from the Latin for worms or insect larvae. **Tinea capitis,** or ringworm of the scalp, is fairly common among elementary school children and can result in bald patches (Figure 21.14a). It is usually transmitted by contact with fomites. Dogs and cats are also frequently infected with fungi that cause ringworm in children. Ringworm of the groin, or jock itch, is known as **tinea cruris,** and ringworm of the feet, or athlete's foot, is known as **tinea pedis** (Figure 21.14b).

Three genera of fungi are involved in cutaneous mycoses. *Trichophyton* (trik-ō-fī'ton) can infect hair, skin, or nails; *Microsporum* (mī-krō-spô'rum) usually involves only the hair or skin; *Epidermophyton* (ep-i-dėr-mō-fī'ton) affects only the skin and nails. The topical drug of choice for tinea infections is usually miconazole or clotrimazole. Other topical agents found in nonprescription remedies are tolnaftate and zinc

TABLE 21.1 Diseases Associated with the Skin

	CAUSATIVE AGENT	*CHARACTERISTICS*	*TREATMENT*
Bacterial Diseases			
Impetigo	*Staphylococcus aureus;* occasionally, *Streptococcus pyogenes*	Superficial skin infection; isolated pustules	Penicillin (for *Streptococcus* infections only); erythromycin
Erysipelas	*Streptococcus pyogenes*	Reddish patches on skin; often with high fever	Penicillin, erythromycin
Pseudomonas dermatitis	*Pseudomonas aeruginosa*	Superficial rash	Usually self-limited
Otitis externa	*Pseudomonas aeruginosa*	Superficial infection of external ear canal	Gentamicin, carbenicillin
Acne	*Propionibacterium acnes*	Inflammatory lesions originating with accumulations of sebum that rupture a hair follicle	Benzoyl peroxide, tetracyclines, isotretinoin
Viral Diseases			
Warts	Papillomaviruses	A horny projection of the skin formed by proliferation of cells	May be removed by liquid nitrogen cryotherapy, electrodesiccation, acids, or lasers
Smallpox (variola)	Poxvirus (smallpox virus)	Pustules that may be nearly confluent on skin. Systemic viral infection affects many internal organs	None
Chickenpox (varicella)	Herpesvirus (varicella-zoster)	Vesicles in most cases confined to face, throat, and lower back	Acyclovir for immunocompromised patients
Shingles (zoster)	Herpesvirus (varicella-zoster)	Vesicles similar to chickenpox; typically on one side of waist, face and scalp, or upper chest	Acyclovir for immunocompromised patients
Herpes simplex	Herpesvirus (herpes simplex type 1)	Most commonly as cold sores—vesicles around mouth; can also affect other areas of skin and mucous membranes, including eyes (see herpetic keratitis, Table 21.2)	Acyclovir may modify symptoms
Measles (rubeola)	Paramyxovirus (measles virus)	Skin rash of reddish macules first appearing on face and spreading to trunk and extremities	None
German measles (rubella)	Togavirus (rubella virus)	Mild disease with a rash resembling measles, but less extensive and disappears in 3 days or less	None
Fungal Diseases			
Ringworm (tinea)	*Microsporum, Trichophyton, Epidermophyton* species	Skin lesions of highly varied appearance, on scalp may cause local loss of hair	Griseofulvin (orally); miconazole, clotrimazole, tolnaftate (topically)
Sporotrichosis	*Sporothrix schenckii*	Ulcer at site of infection spreading into nearby lymphatic vessels	Potassium iodide solution (orally)
Maduromycosis	*Allescheria boydii*	Extensive, chronic tissue damage of affected area—usually the feet or hands	Antifungals relatively ineffective
Candidiasis	*Candida albicans*	Symptoms vary with infection site. Usually affects mucous membranes or moist areas of skin	Miconazole, clotrimazole, nystatin (topically)

undecylenate. An oral antibiotic, griseofulvin, is often useful in these infections because it can localize in keratinized tissue.

Diagnosis of cutaneous mycoses is usually made by microscopic examination of scrapings of the affected areas. The dermatophytes can be readily cultured. Physicians with little specialized knowledge of fungal morphology often use a selective medium, Dermatophyte Test Medium. If, after a few days of growth, the white, fluffy fungal colonies turn red, the culture is recorded as positive for dermatophytes.

SUBCUTANEOUS MYCOSES

Subcutaneous mycoses are more serious than cutaneous mycoses. Even when the skin is broken, cutaneous fungi do not seem to be able to penetrate past the stratum corneum, perhaps because they cannot obtain sufficient iron for growth in the epidermis and the dermis. Usually these diseases are caused by fungi that inhabit the soil, especially decaying vegetation, and penetrate the skin through a small wound that allows entry into subcutaneous tissues. In this country, the disease of this type most commonly encountered is **sporotrichosis,** caused by the dimorphic fungus *Sporothrix schenckii* (spô-rō'thriks shen'kē-ē). Most cases occur among gardeners or others working with soil; the infection frequently forms a small ulcer on the hands. The fungus often enters the lymphatic system in the area and there forms similar lesions. The condition is seldom fatal and is effectively treated by ingesting a dilute solution of potassium iodide, even though the organism is not affected in vitro by even a 10% solution of potassium iodide.

Another subcutaneous mycosis is **maduromycosis,** which is caused by the fungus *Allescheria boydii* (al-lesh-er'ē-ä boi-dē-ē). Maduromycosis may be called a **mycetoma,** or fungal tumor. It destroys subcutaneous tissues and progresses slowly, eventually causing serious deformities.

Treatment of mycetomas of fungal origin is a difficult problem, since none of the standard antifungal

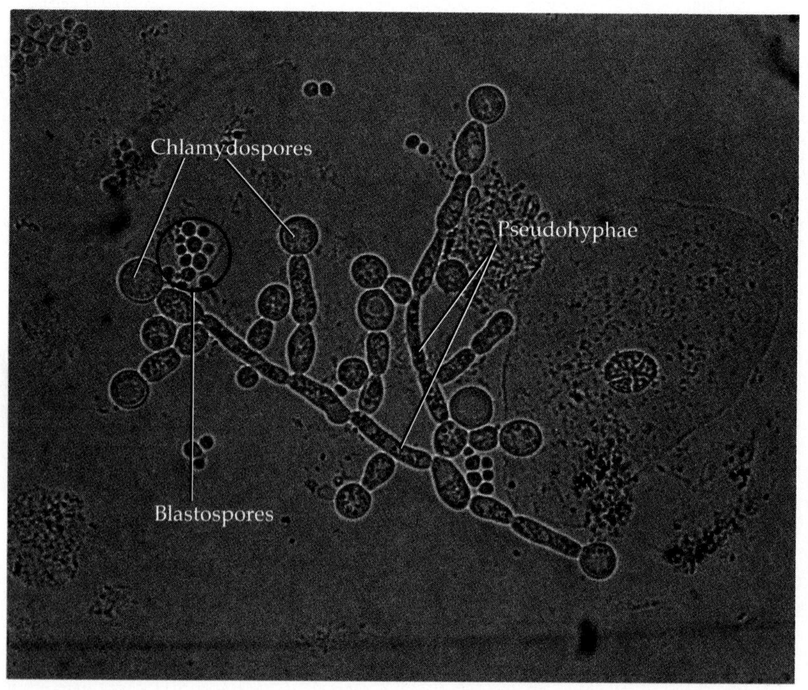

(a)

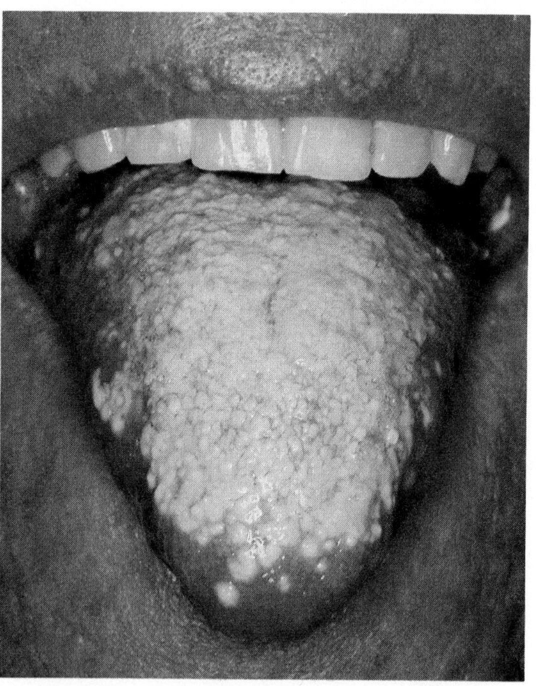

(b)

FIGURE 21.15 Candidiasis. **(a)** Photomicrograph of *Candida albicans*. Note the spherical chlamydospores (thick-walled, nonsexual spores formed from a hyphal cell), the smaller blastospores (spores produced by budding), and the pseudohyphae. *Candida* does not form filamentous hyphae but does form long cells that resemble hyphae. These pseudohyphae are relatively resistant to phagocytosis and may be a factor in pathogenicity. **(b)** Thrush, or oral candidiasis, showing a thick, creamy coating over the tongue.

drugs is outstandingly helpful. Although the condition usually does not extend beyond the foot or hand, it may persist for many years. A similar condition is caused by moldlike bacteria such as *Nocardia* and is called **actinomycete mycetoma.** A correct diagnosis is important because the bacterial type of mycetoma can be effectively treated with antibiotics.

CANDIDIASIS

The bacterial flora of the mucous membranes in the genitourinary tract and mouth usually suppresses the growth of such fungi as *Candida albicans* (Figure 21.15a). Because the fungus is not affected by antibacterial drugs, it sometimes overgrows mucosal tissue when the antibiotics suppress the normal bacterial flora. Changes in the normal mucosal pH may have a similar effect. Such overgrowths by the yeastlike *C. albicans* are called **candidiasis.** Newborn infants, whose normal flora has not become established, often suffer from a whitish overgrowth of the oral cavity, called **thrush** (Figure 21.15b). *C. albicans* is also a very common cause of vaginitis (see Chapter 26). Immunosuppressed individuals are unusually prone to candida infections of the skin and mucous membranes. On persons who are obese or diabetic, the areas of the skin with more moisture tend to become infected with this fungus. The infected areas become bright red, with lesions on the borders. Skin and mucosal infections by *C. albicans* are usually treated with topical applications of miconazole, clotrimazole, or nystatin. If candidiasis becomes systemic, as can happen in immunosuppressed individuals, fulminating disease and death can result. Oral ketoconazole is the usual treatment for systemic candidiasis.

DISEASES OF THE EYE

The epithelial cells covering the eye can be considered a continuation of the skin or mucosa. The organisms most commonly associated with the eye, *Staphylococcus epidermidis* (e-pi-dėr'mi-dis), *S. aureus*, and diphtheroids, usually originate from the skin and upper respiratory tract. A summary of diseases associated with the eye is presented in Table 21.2, p. 534.

Conjunctivitis

Many microbes can infect the eye, largely through the conjunctiva, the mucous membrane that lines the eyelids and covers the outer surface of the eyeball. **Conjunctivitis** is an inflammation of the conjunctiva. The popularity of contact lenses has been accompanied by problems with microbial infections of the eye. This is especially true of the soft-lens designs often worn for extended periods. The most common bacterial pathogens are pseudomonads, which can cause serious eye damage. Homemade saline solutions, which are a common source of infection, should not be used, and the manufacturer's recommendations for disinfection should be followed scrupulously. The most effective methods are thermal; lenses that cannot be heated can be disinfected with hydrogen peroxide, which is then neutralized.

Neonatal Gonorrheal Ophthalmia

Neonatal gonorrheal ophthalmia is a serious form of conjunctivitis caused by *Neisseria gonorrhoeae* (the cause of gonorrhea). Large amounts of pus are formed; if treatment is delayed it will frequently lead to ulceration of the cornea. It is acquired as the infant passes through the birth canal, and infection carries a high risk of blindness. Early in this century, legislation required that the eyes of all newborn infants be treated with a 1% solution of silver nitrate, which proved to be a very effective treatment in preventing this eye infection. Between 1906 and 1959, the percentage of admissions to schools for the blind that could be attributed to neonatal ophthalmia declined from 24% to only 0.3%. Silver nitrate is being replaced with antibiotics because there are frequent coinfections by the gonococcus and sexually transmitted chlamydiae, and silver nitrate is not effective against chlamydiae.

Inclusion Conjunctivitis

Chlamydial conjunctivitis, or **inclusion conjunctivitis,** is quite common today. It is caused by *Chlamydia trachomatis*, a bacterium that grows only as an obligate intracellular parasite. In infants, who acquire it in the birth canal, the condition tends to resolve spontaneously in a few weeks or months, but in rare cases it can lead to scarring of the cornea, much as in trachoma, discussed shortly. The current recommendation is to treat the eyes of newborn infants within an hour of birth (the timing is very important) with erythromycin or tetracyclines, which are usually effective against both the gonococcus and chlamydial organisms. Chlamydial conjunctivitis also appears to spread in the unchlorinated waters of swimming pools and in this context is called swimming pool conjunctivitis. Tetracycline applied as an ophthalmic ointment is an effective treatment.

Trachoma

A much more serious infection caused by *Chlamydia trachomatis* is **trachoma,** the greatest single cause of

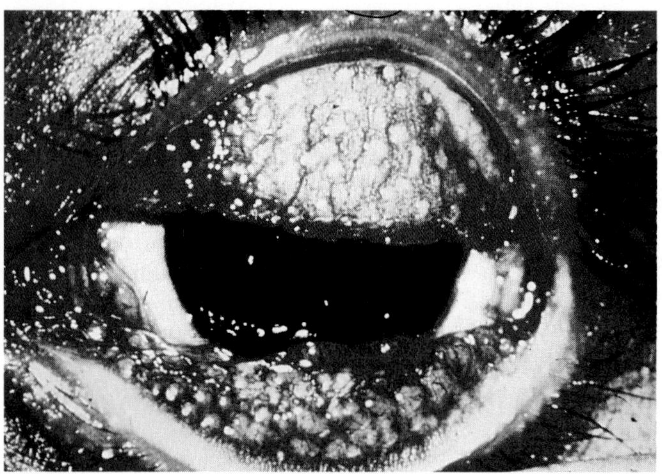

FIGURE 21.16 Trachoma. The eyelids have been pulled back to show the inflammatory nodules on the conjunctival membrane in contact with the cornea, which is damaged by the abrasion and is susceptible to secondary infections.

blindness in the world today. Millions now have the disease, and many millions have been blinded by the infection. In the arid parts of Africa and Asia, almost all children are infected early in their lives. Trachoma also occurs occasionally in the southwestern United States, especially among Native Americans. The disease is transmitted largely by hand contact or by such objects as towels. Flies may also carry the bacteria.

The disease is a conjunctivitis that eventually leads to permanent scarring. Blindness is a result of long-term mechanical abrasion of the cornea by these scars and by turned-in eyelashes (Figure 21.16). Secondary infections by other bacterial pathogens are also a factor. Antibiotic ointments, especially tetracycline, are useful in treatment. Partial immunity is generated by recovery. The disease can be controlled through sanitary practices and health education.

Herpetic Keratitis

Herpetic keratitis is usually caused by herpes simplex type 1 virus. This is a localized infection of the cornea that can recur with an epidemiology similar to that of cold sores. It is characterized by inflammation and corneal ulcers that can be quite deep. Herpetic keratitis is one of the few viral diseases for which there is an effective chemotherapeutic treatment; trifluridine is the drug of choice.

Acanthamoeba Keratitis

The first case of ***Acanthamoeba*** (a-kan-thä-mē'bä) ***keratitis*** was reported in 1973 in a Texas rancher. Since then, well over 100 cases have been diagnosed. This amoeba has been found in fresh water, hot tubs, soil, and even homemade contact lens saline solutions. Most recent cases have been associated with the wear-

TABLE 21.2 Diseases Associated with the Eye

	CAUSATIVE AGENT	CHARACTERISTICS	TREATMENT
Bacterial Diseases			
Neonatal gonorrheal ophthalmia	*Neisseria gonorrhoeae*	Acute infection with much pus formation; if treatment is delayed, ulcers form on cornea	Silver nitrate, tetracycline, or erythromycin for prevention; penicillin for treatment
Inclusion conjunctivitis	*Chlamydia trachomatis*	Swelling of eyelid; mucus and pus formation	Tetracycline
Trachoma	*Chlamydia trachomatis*	Conjunctivitis causes scarring of eyelid that mechanically damages cornea, often causing secondary infections	Tetracycline
Viral Disease			
Herpetic keratitis	Herpesvirus (herpes simplex type 1)	Initially a mild conjunctivitis, but may progress to corneal ulcers and severe damage	Idoxuridine, vidarabine, trifluridine may be effective
Protozoan Disease			
Acanthamoeba keratitis	*Acanthamoeba polyphaga*	Frequently results in severe eye damage requiring corneal transplants or removal of eye	Ketoconazole, miconazole have some moderate success; surgery required to restore vision

ing of contact lenses. Contributing factors are faulty disinfecting procedures (only heat will reliably kill the cysts), homemade saline solutions, and wearing the lenses while swimming. The infection in its early stages is only a mild inflammation, but later stages are often accompanied by severe pain. Damage is often so severe as to require a corneal transplant, or even removal of the eye. Diagnosis is confirmed by the presence of trophozoites and cysts in stained scrapings of the cornea.

STUDY OUTLINE

Introduction (p. 518)

1. The skin is a physical and chemical barrier against microorganisms.
2. Moist areas of the skin (such as the axilla) support larger populations of bacteria than dry areas (such as the scalp).

Structure and Function of the Skin (pp. 518–519)

1. The outer portion of the skin, called the epidermis, contains keratin, a waterproof coating.
2. The inner portion of the skin, the dermis, contains hair follicles, sweat ducts, and oil glands that provide passageways for microorganisms.
3. Sebum and perspiration are secretions of the skin that can inhibit growth of organisms.
4. Sebum and perspiration provide nutrients for some microorganisms.
5. Mucous membranes line the body cavities. The mucus traps microorganisms, and ciliary movement facilitates their removal from the body.
6. The eye is mechanically washed by tears that contain lysozyme.

Normal Flora of the Skin (pp. 519–520)

1. Microorganisms that live on skin are resistant to desiccation and high concentrations of salt.
2. Gram-positive cocci predominate on the skin.
3. The normal skin flora is not completely removed by washing.
4. Members of the genus *Propionibacterium* metabolize oil from the oil glands and colonize hair follicles.

DISEASES OF THE SKIN (pp. 520–533)

Bacterial Diseases of the Skin (pp. 520–524)

STAPHYLOCOCCAL SKIN INFECTIONS (pp. 520–522)

1. Staphylococci are gram-positive bacteria that often grow in clusters.
2. The majority of skin flora consists of coagulase-negative *S. epidermidis*.
3. Almost all pathogenic strains of *S. aureus* produce coagulase.
4. Pathogenic *S. aureus* can produce enterotoxins, leukocidins, and exfoliative toxin.
5. Localized infections (sties, pimples, and carbuncles) result from *S. aureus* entering openings in the skin.
6. Many strains of *S. aureus* produce penicillinase.
7. Impetigo of the newborn is a highly contagious superficial skin infection caused by *S. aureus*.
8. Toxemia occurs when toxins enter the bloodstream; staphylococcal toxemias include scalded skin syndrome and toxic shock syndrome.

STREPTOCOCCAL SKIN INFECTIONS (p. 522)

1. Streptococci are gram-positive cocci that often reproduce in chains. They are strictly fermentative and catalase negative.
2. Streptococci are classified according to their hemolytic enzymes and cell-wall (M) antigens.
3. Group A β-hemolytic streptococci (including *S. pyogenes*) are the pathogens most important to humans.
4. Group A β-hemolytic streptococci produce a number of virulence factors: erythrogenic toxin, deoxyribonuclease, NADase, streptokinases, and hyaluronidase.
5. Streptococci are susceptible to penicillin.
6. Impetigo (isolated pustules) and erysipelas (reddish patches) are skin infections caused by *S. pyogenes*.

INFECTIONS BY PSEUDOMONADS (p. 523)

1. Pseudomonads are gram-negative rods. They are strict aerobes found primarily in soil and water. They are resistant to many disinfectants and antibiotics.
2. *Pseudomonas aeruginosa* is the most prominent species.
3. *P. aeruginosa* produces an endotoxin and several exotoxins.
4. Diseases caused by *P. aeruginosa* include otitis externa, respiratory infections, burn infections, and dermatitis.
5. Infections have a characteristic blue-green pus caused by the pigment pyocyanin.
6. Carbenicillin, gentamicin, and silver sulfadiazine are useful in treating *P. aeruginosa* infections.

ACNE (pp. 523–524)

1. *Propionibacterium acnes* can metabolize sebum trapped in hair follicles.
2. Metabolic end products (fatty acids) cause an inflammatory response known as acne.
3. Treatment with benzoyl peroxide and tetracycline is somewhat effective.

Viral Diseases of the Skin (pp. 524–530)

WARTS (p. 524)

1. Papillomaviruses cause skin cells to proliferate and produce a benign growth called a wart.

2. Warts are spread by direct contact.
3. Warts may regress spontaneously or be removed chemically or physically.

SMALLPOX (VARIOLA) (pp. 524–525)

1. Variola virus causes two types of skin infections, variola major and variola minor.
2. Smallpox is transmitted by the respiratory route, and the virus is moved to the skin via the bloodstream.
3. The only host for smallpox is humans.
4. Smallpox has been eradicated as a result of vaccination effort by the WHO.

CHICKENPOX (VARICELLA) AND SHINGLES (HERPES ZOSTER) (pp. 525–526)

1. Chickenpox is caused by varicella-zoster virus.
2. Varicella-zoster virus is transmitted by the respiratory route and is localized in skin cells, causing a vesicular rash.
3. Complications of chickenpox include encephalitis and Reye's syndrome.
4. After chickenpox, the virus can remain latent in nerve cells and subsequently activate as shingles.
5. Shingles is characterized by a vesicular rash along the affected cutaneous sensory nerves.
6. The virus can be treated with acyclovir. An attenuated live vaccine is used.

HERPES SIMPLEX (pp. 526–527)

1. Herpes simplex infection of mucosal cells results in cold sores and occasionally encephalitis.
2. The virus remains latent in nerve cells, and cold sores can recur when the virus is activated.
3. Herpes simplex type 1 virus is transmitted primarily by oral and respiratory routes.
4. Acyclovir has proven successful in treating herpes encephalitis.

MEASLES (RUBEOLA) (pp. 527–528)

1. Measles is caused by measles virus (a paramyxovirus) and transmitted by the respiratory route.
2. After the virus has incubated in the upper respiratory tract, macular lesions appear on the skin, and Koplik spots appear on the oral mucosa.
3. Complications of measles include middle ear infections, pneumonia, encephalitis, and secondary bacterial infections.
4. Measles virus can persist following infection and later cause subacute sclerosing panencephalitis.
5. Vaccination provides effective long-term immunity.

RUBELLA (pp. 528–530)

1. The rubella virus (togavirus) is transmitted by the respiratory route.
2. A red rash and light fever might occur in an infected individual; the disease can be asymptomatic.
3. Congenital rubella syndrome can affect a fetus when a woman contracts rubella during the first trimester of her pregnancy.
4. Damage from congenital rubella syndrome includes stillbirth, deafness, eye cataracts, heart defects, and mental retardation.
5. Vaccination with live rubella virus provides immunity of unknown duration.

Fungal Diseases of the Skin (pp. 530–533)

CUTANEOUS MYCOSES (pp. 530–532)

1. Fungi that colonize the outer layer of the epidermis cause dermatomycoses.
2. *Microsporum*, *Trichophyton*, and *Epidermophyton* cause dermatomycoses called ringworm, or tinea.
3. These fungi grow on keratin-containing epidermis, such as hair, skin, and nails.
4. Ringworm and athlete's foot are usually treated with topical application of antifungal chemicals.
5. Diagnosis is based on microscopic examination of skin scrapings or fungal culture.

SUBCUTANEOUS MYCOSES (pp. 532–533)

1. Sporotrichosis and mycetoma result from soil fungi that penetrate the skin through a wound.
2. The fungi grow and produce subcutaneous nodules along the lymphatic vessels.

CANDIDIASIS (p. 533)

1. *Candida albicans* causes infections of mucous membranes and is a common cause of thrush (in oral mucosa) and vaginitis.
2. *C. albicans* is an opportunistic pathogen that may proliferate when normal bacterial flora are suppressed.
3. Topical antifungal chemicals may be used to treat candidiasis.

DISEASES OF THE EYE (pp. 533–535)

1. *Staphylococcus epidermidis*, *S. aureus*, and diphtheroids are normal flora of the eye.
2. These microorganisms usually originate from the skin and upper respiratory tract.

Conjunctivitis (p. 533)

1. Conjunctivitis is caused by a number of bacteria.
2. Conjunctivitis can be transmitted by improperly disinfected contact lenses.

Neonatal Gonorrheal Ophthalmia (p. 533)

1. Neonatal gonorrheal ophthalmia is caused by the transmission of *Neisseria gonorrhoeae* from an infected mother to an infant during its passage through the birth canal.
2. All newborn infants are treated with 1% silver nitrate or an antibiotic to prevent the growth of *Neisseria*.

Inclusion Conjunctivitis (p. 533)

1. Inclusion conjunctivitis is an infection of the conjunctiva caused by *Chlamydia trachomatis*. It is transmitted to infants during birth and is transmitted in unchlorinated swimming water.

Trachoma (pp. 533–534)

1. In trachoma, which is caused by *C. trachomatis*, scar tissue forms on the cornea.
2. Trachoma is transmitted by hands, fomites, and perhaps flies.

Herpetic Keratitis

(p. 534)

1. Herpetic keratitis causes corneal ulcers. The etiology is herpes simplex type 1 that invades the central nervous system and can recur.
2. Trifluridine is an effective treatment for herpes keratitis.

Acanthamoeba Keratitis

(pp. 534–535)

1. *Acanthamoeba*, transmitted via water, can cause keratitis.

STUDY QUESTIONS

Review

1. Discuss the usual mode of entry of bacteria into the skin. Compare bacterial skin infections with those caused by fungi and viruses with respect to method of entry.
2. Compare and contrast impetigo and erysipelas.
3. A teenaged male with confirmed influenza was hospitalized when he developed respiratory distress. He had a fever, rash, and low blood pressure. *S. aureus* was isolated from his respiratory secretions. Discuss the relationship between his symptoms and the etiological agent.
4. Complete the table of epidemiology below.

Disease	Etiologic Agent	Clinical Symptoms	Method of Transmission
Acne			
Pimples			
Warts			
Chickenpox			
Fever blisters			
Measles			
German measles			

5. How do mycetoma and athlete's foot differ? In what ways are they similar?
6. **(a)** Differentiate between conjunctivitis and keratitis.
   (b) Select a bacterial and viral eye infection, and discuss the epidemiology of each.
7. A laboratory test used to determine the identity of *Staphylococcus aureus* is its growth on mannitol salt agar. The medium contains 7.5% sodium chloride (NaCl). Why is it considered a selective medium for *S. aureus*?
8. Why are people immunized against rubella since the symptoms of the disease are mild or even inapparent?
9. Explain the relationship between shingles and chickenpox.
10. Why are the eyes of all newborn infants washed with an antiseptic or antibiotic?
11. What is the leading cause of blindness in the world?
12. An opportunistic dimorphic fungus that causes skin infections is ________.
13. Identify the following diseases based on the symptoms in the chart below.

Symptoms	Disease
Koplik spots	
Macular rash	
Vesicular rash	
Small, spotted rash	
Recurrent "blisters" on oral mucosa	
Corneal ulcer and swelling of lymph nodes	

14. What complications can occur from herpes simplex type 1 infections?
15. What is in the MMR vaccine? Why are children not vaccinated against smallpox?

Challenge

1. You have isolated an organism from what appears to be impetigo. The organisms are gram-positive cocci in singles, pairs, and small groups. What test would help you quickly determine whether your isolate is *Staphylococcus* or *Streptococcus*? What is the result of this test if the organism is *Staphylococcus*?
2. Is it necessary to treat a patient for warts? Explain briefly.
3. A hospitalized patient recovering from surgery develops an infection that has blue-green pus and a grapelike odor. What is the probable etiology? How might the patient have acquired this infection?
4. A 12-year-old diabetic girl using continuous subcutaneous insulin infusion to manage her diabetes developed a fever (39.4°C), low blood pressure, abdominal pain, and erythroderma. She was supposed to change the needle

insertion site every three days after cleaning the skin with an iodine solution. Frequently she did not change the insertion site more often than every ten days. Blood culture was negative, and abscesses at insertion sites were not cultured. What is the probable cause of her symptoms?

5. Analyses of nine conjunctivitis cases provided the data in the chart at right. How were these infections transmitted? How could they be prevented?

No.	Etiology	Isolated from Eye Cosmetics or Contact Lenses
5	*S. epidermidis*	+
1	*Acanthamoeba*	+
1	*Candida*	+
1	*P. aeruginosa*	+
1	*S. aureus*	+

FURTHER READING

Chesney, P.J., M.S. Bergdoll, J.P. Davis, and J.M. Vergeront. "The disease spectrum, epidemiology, and etiology of toxic shock syndrome." *Annual Review of Microbiology* 38:315–338, 1984. History and recent developments related to this disease.

Cliff, A., and P. Haggert. "Island epidemics." *Scientific American* 250(5):138–147, May 1984. An excellent review of epidemiology using measles as an example.

Fischetti, V.A. "Streptococcal M protein." *Scientific American* 264(6):58–65, June 1991. Compares the M protein fibrils on streptococcal bacteria with the defensive quills of the porcupine.

Henderson, D.A. "The eradication of smallpox." *Scientific American* 235(4):25–33, October 1976. A summary of the incidence of smallpox and investigation of the last endemic infection.

Langmuir, A.D., et al. "The Thucydides syndrome." *New England Journal of Medicine* 313:1027–1030, 1985. Descriptive epidemiology using translations of Thucydides' observations are used to understand an ancient epidemic.

Marples, M.J. "Life on the human skin." *Scientific American* 220(1):108–115, January 1969. A study of the microflora of the skin and the relationship between bacterial populations and hair, moisture, and temperature.

Sheagren, J.N. "*Staphylococcus aureus*, the persistent pathogen." *New England Journal of Medicine* 310:1368–1373, 1437–1442, 1984. Part One describes infections and immunity; Part Two discusses toxemias, treatment, and prevention.

Shepherd, M.G., R.T.M. Poulter, and P.A. Sullivan. "*Candida albicans*: biology, genetics, and pathogenicity." *Annual Review of Microbiology* 39:579–614, 1985. A thorough monograph on diseases caused by *C. albicans* and host defenses; includes an overview of antifungal drugs.

Tachibana, D.K. "Microbiology of the foot." *Annual Review of Microbiology* 30:351–375, 1976. An interesting report on the normal flora of the foot, foot infections, and foot odor.

See also Further Reading for Part Four at the end of Chapter 26.

CHAPTER 22

Microbial Diseases of the Nervous System

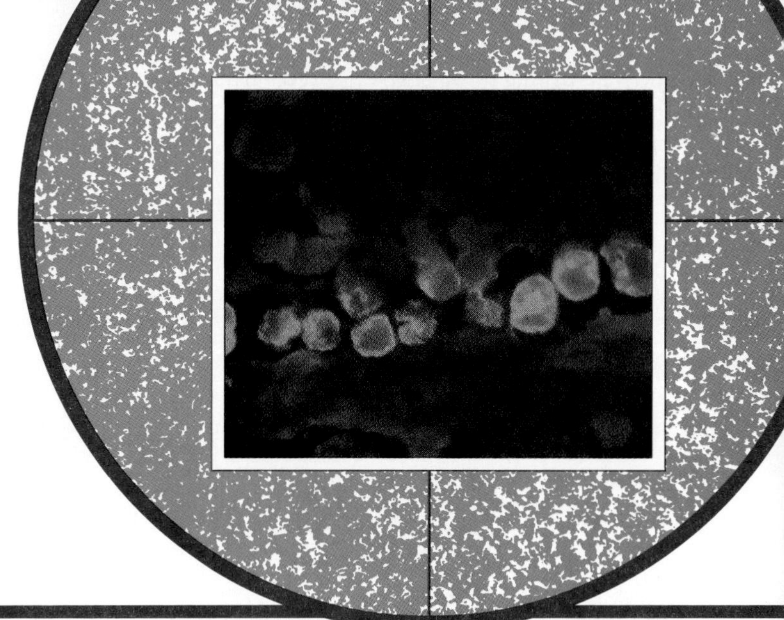

LEARNING OBJECTIVES

- Differentiate between meningitis and encephalitis.
- Discuss the epidemiology of meningococcal meningitis, *Hemophilus influenzae* meningitis, cryptococcosis, and listeriosis.
- Discuss the causative agent, symptoms, and method of transmission of tetanus.
- Provide the etiologic agent, symptoms, suspect foods, and treatment for botulism.
- Discuss the epidemiology of leprosy and poliomyelitis, including method of transmission, etiology, disease symptoms, and preventive measures.
- List the etiology, methods of transmission, and reservoirs for rabies.
- Describe how rabies vaccines are made and used.
- Identify the causative agent, vector, symptoms, and treatment for arthropod-borne encephalitis, cryptococcosis, and African trypanosomiasis.
- List the characteristics of slow virus diseases.

Structure and Function of the Nervous System

The human nervous system is organized into two divisions—the central nervous system and the peripheral nervous system (Figure 22.1a). The **central nervous system** consists of the brain and the spinal cord. As the control center for the entire body, it picks up sensory information from the environment, interprets the information, and sends impulses that coordinate the body's activities. The **peripheral nervous system** consists of all the nerves that branch off from the brain and the spinal cord. These nerves are the lines of communication between the central nervous system, the various parts of the body, and the external environment.

Both the brain and the spinal cord are covered and protected by three continuous membranes called *meninges.* These are the outermost dura mater, the middle arachnoid, and the innermost pia mater. Between the pia mater and arachnoid membranes is a space called the *subarachnoid space,* in which *cerebrospinal fluid* circulates (Figure 22.1b).

A very interesting feature of the brain is the **blood–brain barrier.** Certain capillaries permit some substances to pass from the blood into the brain but restrict others. These capillaries are less permeable than others within the body and are therefore more selective in passing materials.

Drugs cannot cross the blood–brain barrier unless they are lipid-soluble. (Glucose and many amino acids are not lipid-soluble, but they can cross the barrier because special transport systems exist for them.) The lipid-soluble antibiotic chloramphenicol enters the brain readily. Penicillin is only slightly lipid-soluble, but if it is given in very large doses, enough may cross the barrier to be effective. Inflammations of the brain tend to alter the blood–brain barrier in such a way as to

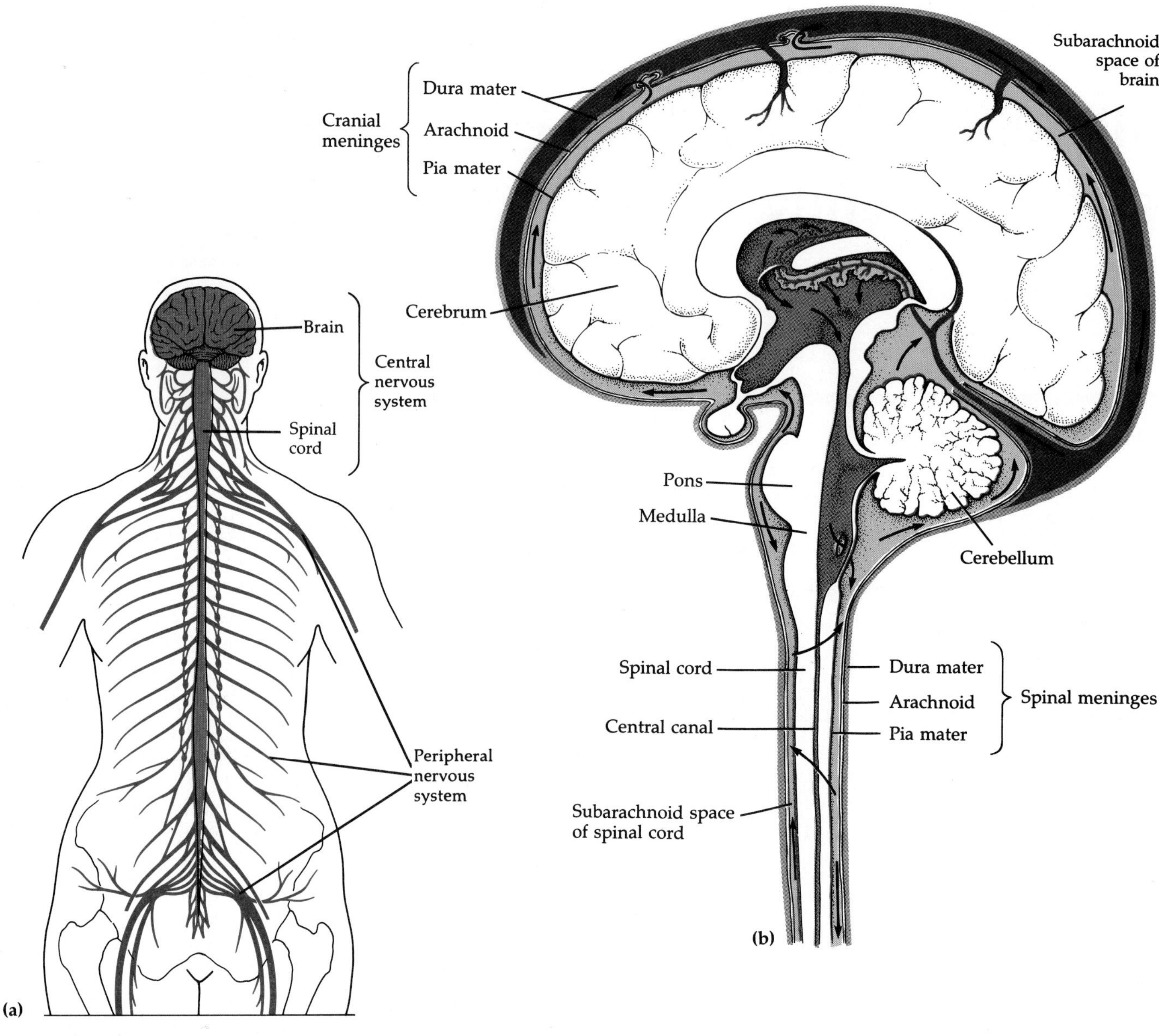

FIGURE 22.1 Human nervous system. **(a)** Organization of the nervous system. **(b)** Meninges and cerebrospinal fluid. Circulation of the fluid is indicated by black arrows in the colored zones.

allow antibiotics to cross that would not be able to cross if there were no infection.

Even though the central nervous system has considerable protection, it can still be invaded by microorganisms in several ways. For example, microorganisms can gain access through trauma, such as a skull or backbone fracture, or through a medical procedure such as a spinal tap, in which a needle is inserted into the subarachnoid space of the spinal meninges to obtain a sample of fluid for diagnosis. Some microorganisms can also move along peripheral nerves. But probably the most common routes of central nervous system invasion are the bloodstream and lymphatic system (see Chapter 23). Because cerebrospinal fluid communicates with the lymphatic system, invading microorganisms can enter the cerebrospinal fluid through it.

An infection of the meninges is called **meningitis.** An infection of the brain itself is called **encephalitis.** Table 22.4, p. 555, summarizes the main diseases associated with the nervous system.

Bacterial Diseases of the Nervous System

Microbial infections of the central nervous system are infrequent but often have serious consequences.

BACTERIAL MENINGITIS

Meningitis is an inflammation of the meninges. Most patients suffering from any type of meningitis complain of a headache and have symptoms of nausea and vomiting. Convulsions and coma accompany the infection in many cases. The mortality rate varies with the pathogen but is generally high for an infectious disease today. Many people who survive an attack suffer some sort of neurological damage.

Meningitis can be caused by different types of pathogens, including viruses, bacteria, fungi, and protozoans. In this section, we will discuss bacterial meningitis. There are three major types of bacterial meningitis—meningococcal meningitis caused by *Neisseria meningitidis*, pneumococcal meningitis caused by *Streptococcus pneumoniae*, and *Hemophilus influenzae* meningitis. These three types make up the great majority of meningitis cases. Only meningococcal meningitis is a notifiable disease (see Chapter 14). Table 22.1 presents data about these and some other types of bacterial meningitis.

TABLE 22.1 Types of Bacterial Meningitis

BACTERIUM	*PERCENTAGE OF ALL REPORTED MENINGITIS*	*COMMENTS*
Neisseria meningitidis	27%	Can occur as epidemics; fatality rate of about 5%
Hemophilus influenzae	48%	Fatality rate of about 6%
Streptococcus pneumoniae	13%	Fatality rate of 27% to 40%
Mycobacterium tuberculosis	Uncommon	Complication of tuberculosis; fatality rate of about 20%
Leptospira interrogans	Uncommon	Meningitis is a common symptom of leptospirosis
Listeria monocytogenes	Uncommon	Causes spontaneous abortions and fetal damage; meningitis in immunosuppressed persons
Nocardia asteroides	Uncommon	Meningitis is a common complication of pulmonary infections by *Nocardia*

Nearly 50 species of bacteria have been reported to be opportunistic pathogens that occasionally cause meningitis. These include such common organisms as *Escherichia coli*, *Pseudomonas aeruginosa*, and *Klebsiella pneumoniae*. Although the number of cases of meningitis caused by these gram-negative organisms is low, the fatality rate is in excess of 30%.

Neisseria Meningitis (Meningococcal Meningitis)

Meningococcal meningitis is caused by the bacterium *Neisseria meningitidis*. The neisseriae are aerobic, nonmotile, gram-negative cocci.

N. meningitidis can inhabit the portion of the throat behind the nose (nasopharynx) without causing any symptoms. This carrier state can last from several days to several months and provides the reservoir for the organism. If an individual without adequate immunity acquires the microorganism, usually through contact with a healthy carrier, the resulting throat infection can lead to bacteremia, which is followed by meningitis.

The bacteria observed in extracellular or cerebrospinal fluid are found in leukocytes that have entered the fluid in response to the infection. Most of the symptoms associated with meningitis are thought to be caused by an endotoxin produced by the bacteria.

Meningococcal meningitis usually strikes children younger than two years of age, with the highest incidence occurring during the first year of life. Children are usually born with maternal immunity and become susceptible as the immunity weakens, at about six months of age. The death rate for meningococcal meningitis has been about 5% in recent years.

Explosive epidemics of meningococcal meningitis among young adults in military training camps used to be common. However, a recently developed vaccine, consisting of purified capsular polysaccharide, has proved to be effective in protecting military recruits.

Hemophilus influenzae Meningitis

Hemophilus influenzae is a nonmotile, aerobic, gram-negative, pleomorphic bacterium (Figure 22.2). A member of the normal throat flora, it can also cause meningitis. The bacterium, which is encapsulated, is further divided into six types on the basis of its antigenic capsular carbohydrates. Only one type has real importance to medical microbiology, type b. The virulence of *H. influenzae* type b is related to its capsule.

The name *Hemophilus influenzae* was given because the microorganism was erroneously thought to be the causative agent of the influenza pandemics of 1890 and World War I. *H. influenzae* was probably only a secondary invader during those virus-caused pandemics. The

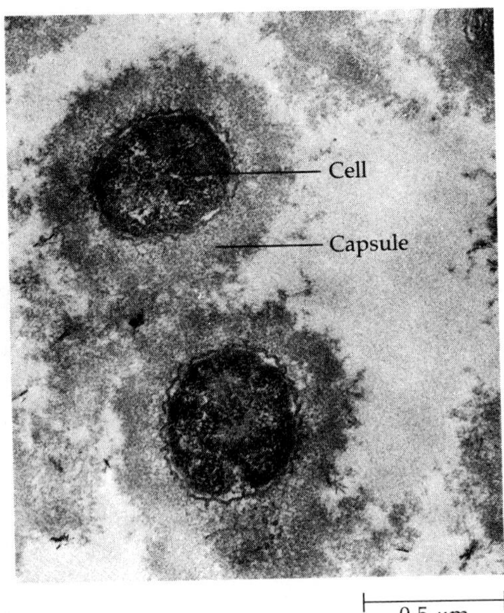

FIGURE 22.2 Electron micrograph of *Hemophilus influenzae* type b. Note the thick layer of capsular material that makes it resistant to phagocytosis.

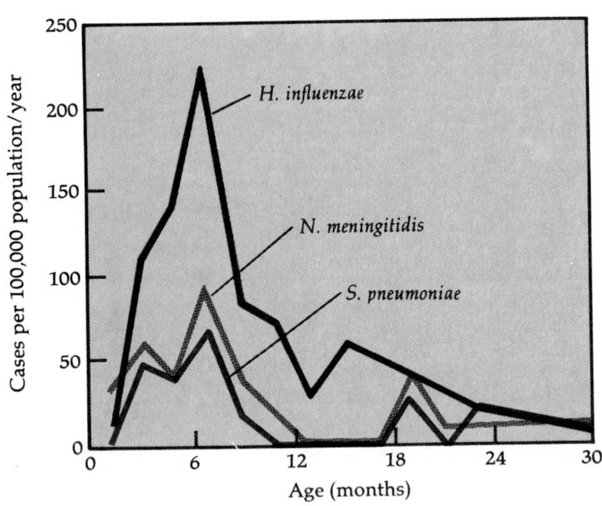

FIGURE 22.3 Reported cases of meningitis. The graph shows the incidence by age of the most common bacterial types of meningitis, those caused by *Hemophilus influenzae*, *Neisseria meningitidis*, and *Streptococcus pneumoniae*.

term *Hemophilus* refers to the fact that the microorganism requires factors in blood for growth (*hemo* means blood).

H. influenzae type b is by far the most common cause of bacterial meningitis; it produces an estimated 20,000 cases each year, mostly in children (Figure 22.3). It accounts for most of the cases of reported bacterial meningitis (48%) but has the lowest mortality rate of the common types of bacterial meningitis (6%). It occurs mostly in children under four years of age and is especially prevalent at about age six months. Residual damage to the nervous system occurs in many cases. The antibodies acquired from the mother provide protection for about two months. The carrier state in the throats of children is quite high, and transmission in daycare centers is now an increasing problem.

Invasion of the meninges by *H. influenzae* is usually preceded by a viral infection of the respiratory tract, which opens the way for the bacteria to enter the bloodstream and be transported to the meninges. Conjugated vaccines directed at the antigenic polysaccharide capsule have been developed. Current recommendations are to begin the vaccination series at 2 months. In most cases, good immunity is achieved at 6 or 7 months.

Streptococcus pneumoniae Meningitis (Pneumococcal Meningitis)

Streptococcus pneumoniae, like *H. influenzae,* is a common inhabitant of the nasopharyngeal region. The capsule on *S. pneumoniae* makes it very resistant to phagocytosis and is the most important element in its pathogenicity. About half of the cases of pneumococcal meningitis occur among children ages 1 month to 4 years. However, the hospitalized elderly are another susceptible group. For this group, some protection may be available from the vaccine for pneumococcal pneumonia, which is caused by the same organism.

Although pneumococcal meningitis accounts for only about 13% of the cases of reported bacterial meningitis, the mortality rate of 21% to 40% is very high.

Diagnosis and Treatment of Bacterial Meningitis

Prompt treatment of any bacterial meningitis is essential, so chemotherapy of suspected cases is usually initiated before identification of the pathogen is complete. Penicillin and ampicillin are frequently the first choice, but the broad spectrum third-generation cephalosporins are often alternatives. As soon as identification is confirmed, or perhaps when the antibiotic sensitivity has been determined from cultures, the antibiotic treatment may be changed. If *H. influenzae* meningitis outbreaks occur in a daycare center, other children in the center can be given antibiotics as a preventive measure. This also has the effect of lowering the carrier rate in the group.

Diagnosis of bacterial meningitis requires a sample of cerebrospinal fluid obtained by a spinal tap. A simple Gram stain is often useful; it will frequently determine the identity of the pathogen with considerable reliability. Cultures are also made from the fluids. For this purpose, prompt and careful handling are required because many of the likely pathogens are very sensitive and will not survive much storage time or even changes in temperature. The cultures are usually inoculated into an enrichment medium that helps protect them from oxygen, and they are then streaked onto media such as blood agar. The two most frequently used serological tests made on cerebrospinal fluid are counter-immunoelectrophoresis and rapid latex agglutination tests.

LISTERIOSIS

Listeria monocytogenes is a gram-positive rod associated with a great variety of animals. It is excreted in animal feces and hence is widely distributed in soil and water. The name is derived from the proliferation of monocytes (a type of leukocyte) found in some animals infected by it. The disease **listeriosis** usually is a mild, often symptomless disease in healthy adult humans, but recovering or apparently healthy persons often shed the pathogen indefinitely in their feces. *L. monocytogenes* is ingested by phagocytic cells but is not destroyed; it even proliferates within them. It also has the unusual capability of moving directly from one macrophage to an adjacent one.

As a pathogen, *L. monocytogenes* has two principal characteristics—it mainly affects adults who are immunosuppressed or pregnant or have cancer, and it has a special affinity for growth in the central nervous system and the placenta supplying the fetus with nutrients. Growth in the central nervous system is usually expressed as meningitis. When it infects a pregnant woman, the growth on the placenta leads to a high rate of spontaneous abortion or stillbirth. A surviving newborn may be acutely ill with septicemia and meningitis. The infant mortality related to this type of infection is about 60%.

In human outbreaks, the organism is mostly foodborne. It is frequently isolated from a wide variety of foods; dairy products have been involved in several outbreaks. There has been some speculation that the organism's intracellular location allows it to survive pasteurization, but the current opinion is that proper pasteurization eliminates it. *L. monocytogenes* is one of the few pathogens that is capable of growth at refrigerator temperatures. In fact, it can be isolated, in part, by *cold enrichment*, meaning incubation at a low temperature at which it outgrows most other bacteria.

Considerable effort is being made to develop methods by which *L. monocytogenes* can be better detected in foods. Considerable improvement has been made with selective growth media. However, eventually DNA probes (see Chapter 10) and serological tests employing monoclonal antibodies are expected to be the most satisfactory. Diagnosis in humans depends on isolation and culturing of the organism. Penicillin and ampicillin are the antibiotics of choice for treatment.

TETANUS

The causative agent of **tetanus,** *Clostridium tetani,* is an obligately anaerobic, endospore-forming, gram-positive rod. It is especially common in soil contaminated with animal fecal wastes.

The symptoms of tetanus are caused by an extremely potent neurotoxin *(tetanospasmin)* that is released upon death and lysis of the growing bacteria (see Chapter 15). It enters the central nervous system via peripheral nerves or the blood. An amount of tetanospasmin weighing as much as the ink in one period on this page could kill 30 people. The bacteria themselves do not spread from the infection site, and there is no inflammation.

In a muscle's normal operation, a nerve impulse initiates the contraction of the muscle. At the same time, an opposing muscle receives a signal to relax so as not to oppose the contraction. The tetanus neurotoxin blocks the relaxation pathway so that both sets of muscles contract, resulting in the characteristic muscle spasms called *spastic paralysis* (Figure 22.4). The muscles of the jaw are affected early in the disease, which prevents opening of the mouth (lockjaw). Gradually, other skeletal muscles become affected, including those involved in swallowing. Death results from spasms of the respiratory muscles.

Because the organism is an obligate anaerobe, the wound through which it enters the body must provide anaerobic growth conditions. Improperly cleaned deep puncture wounds, especially those with little or no bleeding, serve very well. Of the approximately 100 tetanus cases that have occurred each year in the United States in recent years (about 25% of which were fatal), many arise from trivial but fairly deep injuries that are thought to be too minor to bring to a physician. The death rate of intravenous drug users from tetanus is the highest of any group. The elderly often lack effective immunity and are also a susceptible group.

Most persons in this country have received the DPT (diphtheria, pertussis, tetanus) immunization, which includes the tetanus toxoid that stimulates the formation of antibodies to fight tetanus toxin. Immunization is nearly 100% effective, but recovery from the clinical disease does not confer immunity because the amount of toxin usually produced is too small to be immunogenic.

When a wound is severe enough to be brought to the attention of a physician, the decision is usually made to provide protection against tetanus, and a

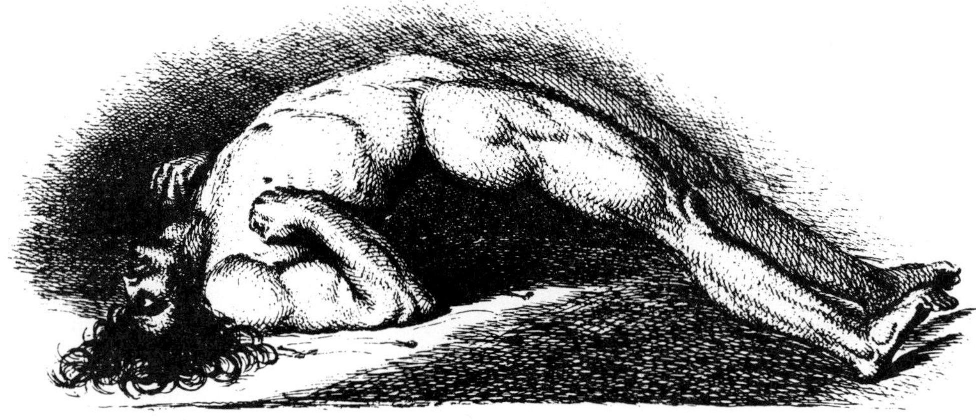

FIGURE 22.4 Tetanus. Drawing of a British soldier during the Napoleonic wars dying of the final stages of tetanus. (Drawing by Charles Bell in the Royal College of Surgeons, Edinburgh.)

booster of toxoid may be given. Table 22.2 summarizes the questions involved in this decision. A patient who has had the full sequence of three doses of toxoid and a booster every ten years should be immune. However, people often do not get boosters on schedule, and serological surveys show that at least 50% of the population does not have adequate protection. Even so, if toxoid has been given before, the anamnestic response (see Figure 17.7) to a booster given after the injury will be rapid enough to be protective. Tetanus immune globulin (TIG), which is prepared from the serum of immunized humans, is often given simultaneously to provide temporary immunity. It is injected at a different site, however, to avoid neutralizing the toxoid. To minimize production of more toxin, damaged tissue should be removed (debridement) and antibiotics should be administered. However, once the toxin has attached to the nerves, such therapy is of little use.

In less developed areas of the world, tetanus arising from the severed umbilical cords of infants is a major cause of death. In some cultures, the practice of dressing the cut umbilical cord with such materials as soil, clay, and cow dung is a major contributor to the development of tetanus. Worldwide, there are probably several hundred thousand cases of tetanus from all causes each year.

BOTULISM

Botulism, a form of food poisoning, is caused by *Clostridium botulinum,* an obligately anaerobic, endospore-forming, gram-positive rod that is found in soil and in many freshwater sediments. Ingesting the endospores usually does no harm, as will be explained shortly. However, in anaerobic environments, such as sealed cans, the microorganism produces an exotoxin that animal assays show to be the most potent of all natural toxins. This neurotoxin is highly specific for the synaptic end of the nerve, where it blocks the release of acetylcholine, a chemical necessary for transmission of nerve impulses across synapses. Persons suffering from botulism undergo a progressive *flaccid paralysis* for one to ten days and may die from respiratory and cardiac failure. Nausea, but no fever, may precede the neurological symptoms. The initial neurological symptoms vary, but nearly all sufferers have double or blurred vision. Other symptoms include difficulty in swallowing and general weakness. Incubation time

TABLE 22.2 Wound Management for Prevention of Tetanus

HISTORY OF TETANUS IMMUNIZATION	MINOR WOUNDS		EXTENSIVE WOUNDS	
	Administer Toxoid	Administer TIG	Administer Toxoid	Administer TIG
Unknown or 0 to 2 doses	Yes	No	Yes	Yes
3 or more doses	No*	No	No**	No

*Yes, if more than 10 years since last dose.
**Yes, if more than 5 years since last dose.

varies, but symptoms typically appear within a day or two. As with tetanus, recovery from the disease does not confer immunity because the toxin is usually not present in amounts large enough to be immunogenic. A vaccine is available for exposed laboratory workers and is used by veterinarians to protect valuable horse-breeding stock in some states.

Botulism was first described as a clinical disease in the early 1800s, when it was known as the sausage disease (*botulus* means sausage). Blood sausage was made by filling a pig stomach with blood and ground meats, tying shut all the openings, boiling it for a short time, and smoking it over a wood fire. The sausage was then stored at room temperature. This attempt at food preservation included most of the requirements for an outbreak of botulism—it killed competing bacteria but allowed the more heat-stable *C. botulinum* endospores to survive, and it provided anaerobic conditions and an incubation period for toxin production.

Most botulism results from attempts at food preservation that fail to eliminate the *C. botulinum* endospores. The botulinal toxin is very heat labile and will be destroyed by most ordinary cooking methods that bring the food to a boil. Sausage and bacon rarely cause botulism today, largely because of the addition of nitrites. It was discovered that nitrites prevent *C. botulinum* growth following germination of the endospores.

Botulinal toxin is not formed in acidic foods (below pH 4.7). Such foods as tomatoes therefore can be safely preserved without the use of a pressure cooker. There have been cases of botulism from acidic foods that normally would not have supported the growth of the botulism organisms. Most of these episodes are related to mold growth, which metabolized enough acid to allow the initiation of growth of the botulism organisms.

There are several serological types of the botulinal toxin produced by different strains of the pathogen. These differ considerably in their virulence and other factors.

Type A toxin is probably the most virulent. There have been deaths from type A toxin when the food was only tasted but not swallowed. It is even possible to absorb lethal doses through skin breaks while handling laboratory samples. In untreated cases, there is a 60% to 70% mortality rate. The type A endospore is the most heat resistant of all *C. botulinum* strains. In the United States, it is found mainly in California, Washington, Colorado, Oregon, and New Mexico. Some eastern states have never had a type A outbreak. The type A organism is usually proteolytic (the breakdown of proteins by clostridia releases amines with unpleasant odors), but obvious spoilage odor is not always apparent in low-protein foods, such as corn or beans.

Type B toxin is responsible for most European outbreaks of botulism and is the most common type in the eastern United States. The mortality rate in cases without treatment is perhaps 25%. Type B botulism organisms occur in both proteolytic and nonproteolytic strains. The endospores of the proteolytic strains are more heat resistant and more likely to survive in food.

Type E toxin is produced by botulism organisms that are often found in marine or lake sediments. Therefore, outbreaks often involve seafood and are especially common in the Pacific Northwest, Alaska, and the Great Lakes area. The endospore of type E botulism is less heat resistant than that of other strains and is usually destroyed by boiling. Types A and B will survive much higher temperatures. Type E is nonproteolytic, so the chance of detecting spoilage in high-protein foods such as fish is minimal. It is also capable of producing a toxin at refrigerator temperatures and requires less strictly anaerobic conditions for growth.

Botulism is not a common disease. Only a few cases are reported each year, but outbreaks from restaurants occasionally involve 20 or 30 cases. Eskimos and coastal Native Americans in Alaska probably have the highest rate of botulism in the world, mostly of type E. The problem arises from food preparation methods that reflect a cultural need to avoid use of scarce fuels for heating or cooking. One of the more common foods involved in Alaskan outbreaks of botulism is *muktuk*. This is prepared by slicing the flippers of seals or whales into strips and then drying them for a few days. To tenderize them, they are stored anaerobically in a container of seal oil for several weeks until they approach putrefaction. The 40% mortality rate for type E botulism observed in recent years among Alaskan natives reflects the difficulty in getting prompt treatment for isolated ethnic groups.

Botulism organisms do not seem to be able to compete successfully with the normal intestinal flora, so production of toxin by ingested bacteria almost never causes botulism in adults. However, the intestinal flora of infants is not well established, and they may suffer from **infant botulism.** This disease was first reported in 1976; an estimated 250 cases occur in the United States every year, many more than any other form of botulism. Infants have many opportunities to ingest soil and other materials contaminated with the endospores of the organism, but 30% of reported cases have been associated with honey. Endospores of *C. botulinum* are recovered with some frequency from honey, and a lethal dose may be as few as 2000 bacteria. The recommendation is not to feed honey to infants under one year of age; there is no problem with older children or adults who have normal intestinal flora.

The treatment of botulism relies heavily on supportive care. Antibiotics are of almost no use because the toxin is preformed. Antitoxins aimed at the neutralization of A, B, and E toxins are available and are usually administered together. This trivalent antitoxin will not affect the toxin already attached to the nerve ends and is probably more effective on type E than on types A and B. The toxin is quite firmly bound. Recovery requires that the nerve endings regenerate; it therefore proceeds slowly. Extended respiratory assistance may be needed, and some neurological impairment may persist for months.

Botulism is diagnosed by the inoculation of mice with samples from patient serum, stool, or vomitus specimens. The toxin in food can similarly be identified by mouse inoculation. Different sets of mice are immunized with type A, B, or E antitoxin. All the mice are then inoculated with the test toxin; if, for example, those protected with type A antitoxin are the only survivors, then the toxin is type A.

The botulism pathogen can also grow in wounds in a manner similar to that of clostridia causing tetanus or gas gangrene (see Chapter 23). Such episodes of **wound botulism** occur occasionally.

LEPROSY

Mycobacterium leprae is probably the only bacterium that grows in the peripheral nervous system, although it can also grow in skin cells. It is an acid-fast rod closely related to the tuberculosis pathogen, *Mycobacterium tuberculosis.* The organism was first isolated and identified about 1870 by Gerhard A. Hansen of Norway; his discovery was one of the first links ever made between a specific bacterium and a disease. **Leprosy** is sometimes called **Hansen's disease** in an effort to avoid using the dreaded name *leprosy.*

The organism has an optimum growth temperature of 30°C and shows a preference for the outer, cooler portions of the human body. A very slow generation time of about 12 days has been estimated. The microorganism has never been grown on artificial media, but in 1969 armadillos, which have a body temperature of only 30° to 35°C, were experimentally infected with *M. leprae.* It is now known that armadillos can contract a leprosylike disease in the wild, although they are not considered a source of human infection. Armadillos are used in studies of the disease itself and in evaluations of the effectiveness of chemotherapeutic agents.

Only about one person in 200 exposed to the leprosy bacillus develops detectable disease at all. Leprosy occurs in two main forms (although borderline forms are also recognized) that apparently reflect the effectiveness of the host's cell-mediated immune system. The *tuberculoid (neural) form* is characterized by regions of the skin that have lost sensation and are surrounded by a border of nodules (Figure 22.5a). This form of the disease occurs in persons with effective immune reactions. Recovery sometimes occurs spontaneously. The disease can be diagnosed by detecting acid-fast bacilli in the fluids from a slit cut in a cool site, such as an earlobe. The **lepromin test** uses an extract of lepromatous tissue injected into the skin. A visible

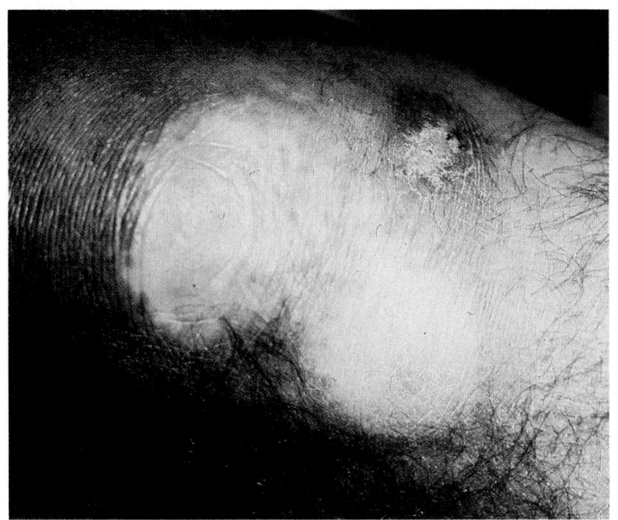

(a)

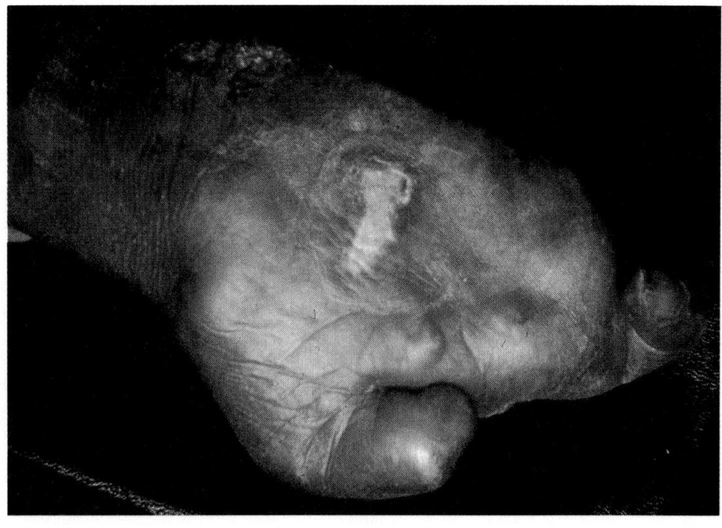

(b)

FIGURE 22.5 Symptoms of leprosy. (a) Tuberculoid leprosy. The depigmented area of skin surrounded by a border of nodules is typical of this stage. **(b)** Lepromatous leprosy. If the immune system fails to control the disease, the result is progressive tissue damage of the cooler parts of the body, as in this severely deformed hand.

skin reaction that develops at the injection site indicates that the body has developed an immune response to the leprosy bacillus. This test is negative during the later lepromatous stage of the disease.

In the *lepromatous (progressive) form* of leprosy, skin cells are infected and disfiguring nodules form all over the body. Patients with this type of leprosy have the least effective cell-mediated immune response, and the disease has progressed from the tuberculoid stage. Mucous membranes of the nose tend to become affected, and a lion-faced appearance is associated with this type of leprosy. Deformation of the hand into a clawed form and considerable necrosis of tissue can also occur (Figure 22.5b). The progression of the disease is unpredictable, and remissions may alternate with rapid deterioration.

The exact means of transfer of the leprosy bacillus is uncertain, but patients with lepromatous leprosy shed large numbers in their nasal secretions and in exudates (oozing matter) of their lesions. Most persons probably acquire the infection when secretions containing the pathogen contact their nasal mucosa. Leprosy, however, is not very contagious, and transmission usually occurs only between persons in fairly intimate and prolonged contact. The time from infection to the appearance of symptoms is usually measured in years, although children can have a much shorter incubation period. Death is not usually a result of the leprosy itself but of complications, such as tuberculosis.

Much of the public's fear of leprosy can probably be attributed to biblical and historical references to the disease. In the Middle Ages, patients with leprosy were rigidly excluded from normal European society and sometimes were even given bells to wear so that people could avoid contact with them. This isolation might have contributed to the near disappearance of the disease in Europe. But patients with leprosy are no longer kept in isolation since patients can be made noncommunicable within a few days by the administration of sulfone drugs. The National Leprosy Hospital in Carville, Louisiana, presently treats about 500 leprosy patients on an ambulatory basis. These outpatients are free to carry out their normal daily routines and return to the hospital for periodic treatment.

The number of leprosy cases in the United States has been gradually increasing. Currently, over 200 cases are reported each year, and the total number of cases is estimated at about 2600. Many of these cases are imported, for the disease is usually found in tropical climates. Millions of persons, most of them in Asia and Africa, suffer from leprosy today.

Although the disease is still difficult to cure, prolonged treatment with sulfone drugs (such as dapsone) has been effective in arresting its progress. Dapsone has been the mainstay of treatment, but resistance to the drug is a problem. Rifampin and a fat-soluble dye, clofazimine, are the other main drugs that are used, often along with dapsone.

A vaccine for leprosy is undergoing field trials. However, because of the extremely long incubation period for leprosy, it will be many years before the vaccine's effectiveness is established. An encouraging sign that a vaccine might be effective is the fact that the BCG vaccine used to provide immunity to tuberculosis (also caused by a *Mycobacterium* species) is somewhat protective against leprosy. A BCG program in Africa to control tuberculosis showed that the vaccine gave at least 50% protection against leprosy at the same time.

Viral Diseases of the Nervous System

Most viruses affecting the nervous system enter it by circulation in the blood or lymph, but some enter peripheral nerve axons and move up them.

POLIOMYELITIS

Poliomyelitis (polio) is best known as a cause of paralysis. However, only about 10% of infected people develop identifiable symptoms, and the paralytic form of poliomyelitis probably affects less than 1% of those infected with the poliomyelitis virus. The great majority of cases are asymptomatic or cause mild symptoms, such as headache, sore throat, fever, and nausea, which are often interpreted as mild meningitis or influenza. Asymptomatic or mild cases of polio are most common in the very young.

In some parts of the world, because of poor sanitary conditions, most of the population contract asymptomatic poliomyelitis as infants and develop immunity. Nonetheless, the incidence of the disease in the tropical and subtropical zones is so high that there are an estimated 250,000 cases of paralytic polio each year. As sanitation has improved in the more developed regions of the world, the chance of developing the disease in infancy has decreased, particularly for those in the upper and middle socioeconomic groups. When infection occurs in adolescence or early adulthood, the paralytic form of the disease is frequent.

Humans are the only known natural host for polioviruses. Polioviruses are more stable than most other viruses and can remain infectious for relatively long periods in water and food. The primary mode of transmission is ingestion of water contaminated with feces containing the virus.

Because the infection is initiated by ingestion of the virus, its primary areas of multiplication are the throat and small intestine. This accounts for the sore throat and nausea. Next, the virus invades the tonsils

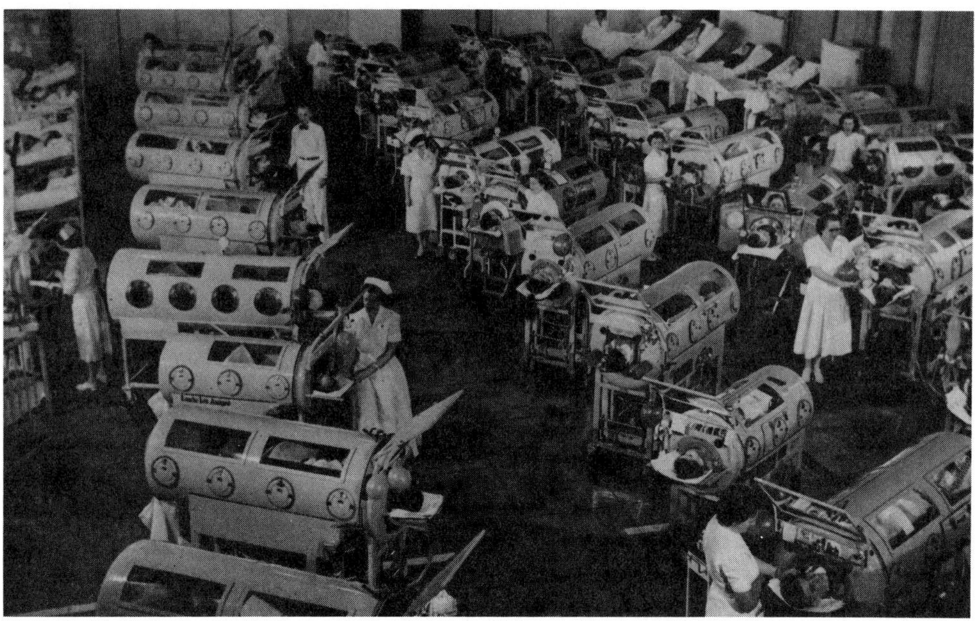

FIGURE 22.6 Iron lungs at a March of Dimes Respiratory Center in Los Angeles in the 1950s. Breathing was possible for many only with these mechanical aids. A few survivors from these epidemics still use these machines, at least part-time. Others are able to use portable respiratory aids.

and the lymph nodes of the neck and ileum (terminal portion of the small intestine). From the lymph nodes, the virus enters the blood *(viremia).* In most cases, the viremia is only transient, the infection does not progress past the lymphatic stage, and clinical disease does not result. If the viremia is persistent, however, the virus penetrates the capillary walls and enters the central nervous system. Once in the central nervous system, the virus displays a high affinity for nerve cells, particularly motor nerve cells in the upper spinal cord (anterior horn cells). The virus does not infect the peripheral nerves or the muscles. As the virus multiplies within the cytoplasm of the motor nerve cells, the cells die and paralysis results. Death results from respiratory failure (Figure 22.6).

Diagnosis of polio is usually based on isolation of the virus from feces and throat secretions. Cell cultures can be inoculated, and cytopathic effects on the cells can be observed (see Table 15.4).

The incidence of polio in the United States has decreased markedly since the availability of the polio vaccines. The last cases acquired in the United States that were attributed to a wild virus were reported in 1979. The development of the first polio vaccine was made possible by the introduction of practical techniques for cell culture, for the virus does not grow in any common laboratory animal. The polio vaccine was the prototype for the vaccines for mumps, measles, and rubella.

There are three different serotypes of the poliovirus, and immunity must be provided for all three. Two vaccines are available. The *Salk vaccine,* which was developed in 1954, uses viruses that have been inactivated by treatment with formalin. Vaccines of this type are inactivated polio vaccines (IPV). They require a series of injections. Their effectiveness is high, perhaps 90% against paralytic polio. The antibody levels decline with time, and booster shots are needed every few years to maintain full immunity. Using only IPV, several European countries have almost eliminated polio from their populations. A newer IPV has been introduced that is produced on human diploid cells. It is called *enhanced inactivated polio vaccine* (E-IPV).

The *Sabin vaccine,* introduced in 1963, contains three living, attenuated strains of the virus and is more popular in the United States than the Salk vaccine. It is less expensive to administer, and most people prefer taking a sip of orange-flavored drink containing the virus (oral polio vaccine, or OPV), to having a series of injections. The immunity achieved with the OPV resembles that acquired by natural infection. One disadvantage is that, on rare occasions—one in 560,000 first doses, one in about 2,000,000 on subsequent doses—one of the attenuated strains of the virus (type 3) seems to revert to virulence and cause the disease. These cases often occur in secondary contacts, not the person who received the vaccine. This illustrates that recipients of the vaccine can infect contacts, leading—in most cases—to immunization.

Some medical scientists have suggested that a return to the IPV might be desirable despite its disadvantages. It is currently recommended for persons with defective immune systems. As the level of polio in the population declines, the cases that can be linked to the live vaccine become a majority of the total cases. There is no reason why polio could not be eliminated in this country by an effectively administered program of immunization (Figure 22.7).

There is spirited debate in the medical community on the future course of polio vaccinations. However, approval is expected of an E-IPV combined with the DPT vaccine that is now regularly administered. If this vaccine was followed by booster doses of the OPV, the dangers of paralytic polio from the OPV would be minimized and the benefits of the live vaccine retained.

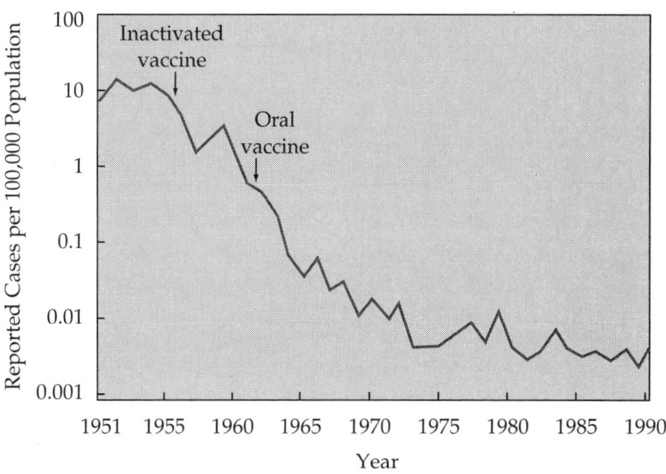

FIGURE 22.7 Poliomyelitis. Number of cases of poliomyelitis reported in the United States from 1951 to 1990. Note the decrease after introduction of the Salk and Sabin vaccines. [*Source:* CDC. *Summary of Notifiable Diseases 1989, MMWR* 38:54 (1989); *MMWR* 39:52 (1/4/91).]

RABIES

Rabies is an acute infectious disease that usually results in fatal encephalitis. The causative agent is *rabies virus* (a rhabdovirus; rhabdos—rod) that has a characteristic bullet shape (Figure 22.8). Humans usually acquire the rabies virus from the bite of an infected animal. Even the lick of such an animal is dangerous because rabies virus in its saliva can enter minute scratches. The virus can cross the intact mucous membrane lining the eyes or body openings. Scientists investigating bat behavior in caves have contracted rabies, presumably from rabies virus present in aerosols of bat secretions in the cave atmosphere. Humans have also acquired rabies from aerosols of the virus when macerating infected tissue in a laboratory blender.

Rabies is unique, as the incubation period is usually long enough to allow development of immunity from postexposure vaccination. (The amount of virus introduced into the wound is usually too small to provoke adequate natural immunity in time.) Initially, the virus multiplies in skeletal muscle and connective tissue, where it remains localized for periods ranging from days to months. Then it enters and travels along the peripheral nerves to the central nervous system, where it causes encephalitis. Bites in areas rich in nerve fibers, such as the hands and face, are especially dangerous, and the resulting incubation period tends to be short. Bites on the hands have an average incubation period of eight weeks; bites on the face, five weeks. Once the virus enters the peripheral nerves, it is no longer accessible to the immune system.

The preliminary symptoms at the onset are mild and varied—resembling several common infections. When the central nervous system becomes involved, the patient tends to alternate between periods of agitation and intervals of calm. At this time, a frequent symptom is spasms of the muscles of the mouth and pharynx that occur when liquids are being swallowed. In fact, even the mere sight or thought of water can set off the spasms—thus the common name *hydrophobia* (fear of water). The final stages of the disease result from extensive damage to the nerve cells of the brain and the spinal cord. Death in a few days is almost inevitable.

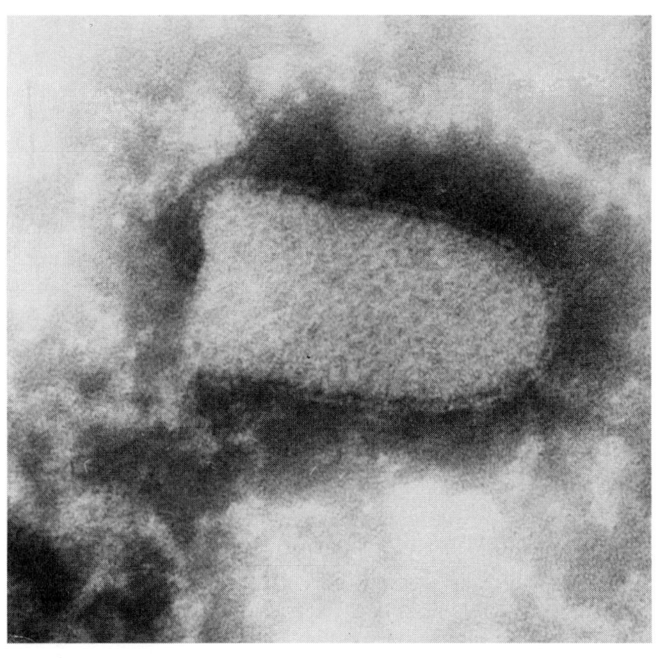

FIGURE 22.8 Rabies virus. Note the unusual bullet shape.

Animals with **furious rabies** are at first restless, then become highly excitable and snap at anything within reach. The biting behavior is essential to maintain the virus in the animal population. When paraly-

MMWR

MORBIDITY AND MORTALITY WEEKLY REPORT

Human Rabies in San Francisco

San Francisco's first case of human rabies since 1940 was reported in December 1987. The diagnosis was made postmortem in a 13-year-old Filipino immigrant boy. The patient was well until November 26, when he began experiencing painful lower back spasm. Symptoms progressed, and he presented to the emergency room on November 28, complaining of penile, buttock, and lower back pain. His neurological exam was normal, and he was afebrile, but his white blood cell (WBC) count was 20,200. He was sent home with Tylenol® for pain and Flexeril® for muscle spasm. He returned the next day, reporting priapism (penile erection) and abdominal pain causing him to scratch at his right lower abdomen. He was admitted to the hospital with a differential diagnosis of pyelonephritis or appendicitis. On admission, he had a low-grade fever (38.3°C rectal) and high blood pressure (150/90), with an elevated WBC count (17,000) and creatinine phosphokinase. An exploratory laparotomy and appendectomy were performed, but nothing abnormal was found. Over the next two days, he became increasingly agitated and combative, complaining of blurred vision and an inability to move his legs. Because no focal neurological abnormalities were identified, a psychiatric disorder was also considered in the diagnosis. Two lumbar punctures were normal. A CT scan of the head and an EEG (electroencephalogram) were unremarkable. On December 2, he experienced episodes of subnormal body temperature, abnormally low blood pressure, subnormal rate of ventilation, and slow heart action, and his oral secretions increased markedly. He became comatose, was given an oxygen tube, and was then transferred to a university hospital with a discharge diagnosis of metabolic encephalopathy.

After transfer, the absence of corneal reflexes was noted. The first abnormal lumbar puncture was on December 3 (protein = 52, WBC = 4), but MRI and CT scans of the head were nondiagnostic. The patient's condition continued to deteriorate, and he died on December 15, 19 days after onset of illness. At autopsy, multiple sections of brain tissue were positive by fluorescent rabies antibody test. No bite marks were recorded by the medical examiner. Serum specimens collected as late as December 9, 13 days after illness began, were negative for antibody, but a postmortem specimen had a titer of 1:65,536. No antibodies were detected in cerebrospinal fluid.

The patient had not traveled outside of California since his arrival from the Philippines in 1981. Except for a three-day camping trip to Sonoma County with 82 classmates and teachers in October 1987, he had not traveled outside of San Francisco in the previous year. No animal contacts were reported during the camping trip or while living in San Francisco. Interviews with relatives, classmates, friends, and teachers did not identify any likely animal contact. However, the patient had been bitten by a dog shortly before leaving the Philippines. Preliminary monoclonal antibody studies suggested that the patient's isolate was compatible with viral isolates from carnivores.

Because of the exposure to the patient and his copious secretions, 12 family members and 75 of 177 health care workers received postexposure rabies prophylaxis.

Comment: Genetic analyses of rabies viruses from the patient and from animals in the United States and the Philippines suggest that the patient acquired rabies in 1981, thus making the incubation period 6 years. Incubation times longer than one year occur in less than 3 percent of rabies cases. One 19-year incubation period has been reported.

Brain tissue samples submitted to U.S. public health laboratories were used to obtain viral RNA. Samples from bats, foxes, and skunks were obtained in the United States, and samples from rabid dogs and infected humans were obtained from the Philippines. By using reverse transcriptase, cDNA was made for each RNA sample. The polymerase chain reaction was then used to amplify the cDNA to provide enough samples for restriction enzyme analysis. Electrophoresis of the fragments shows that the patient's virus was unknown in the United States and identical to a rabies virus found in the Philippines.

Sources: California Morbidity 2, Jan. 22, 1988; *MMWR* 37:305–308 (5/20/88). Updated 1991.

sis sets in, the flow of saliva increases as swallowing becomes difficult, and nervous control is progressively lost. The disease is almost always fatal within a few days. Some animals suffer from **dumb (paralytic) rabies,** in which there is only minimal excitability. This form is especially common in cats. The animal remains relatively quiet and even unaware of its surroundings, but it might snap irritably if handled. A similar manifestation of rabies occurs in humans and is often misdiagnosed as encephalitis or other conditions. There is some speculation that the two forms of the disease may be caused by slightly different forms of the virus.

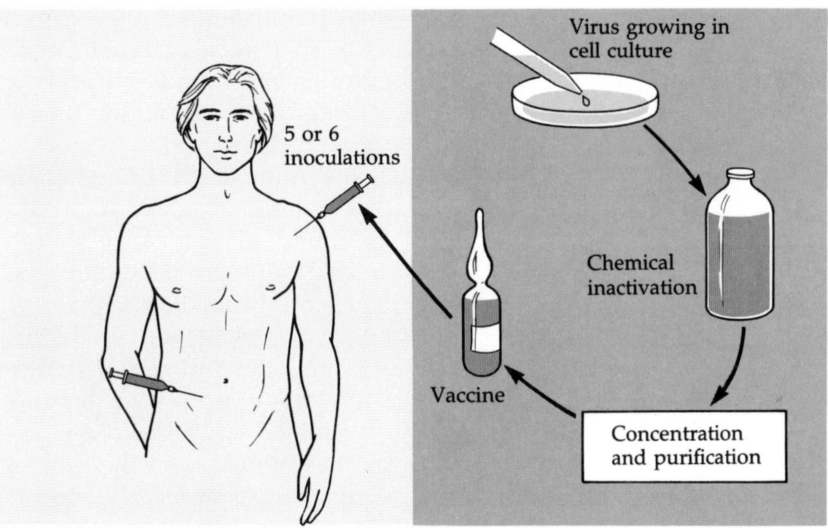

FIGURE 22.9 Preparation of the most recent rabies vaccines.

Laboratory diagnosis of rabies in humans and animals is based on several findings. When the patient or animal is alive, a diagnosis can sometimes be confirmed by immunofluorescent studies, in which viral antigens are detected in saliva, serum, or cerebrospinal fluid. After death, diagnosis is confirmed by a fluorescent-antibody test performed on the brain tissue.

Any person bitten by an animal that is positive for rabies must take antirabies treatment. Another indication for antirabies treatment is any unprovoked bite by a skunk, bat, fox, coyote, bobcat, or raccoon not available for examination. Treatment after a dog or cat bite, if the animal cannot be found, is determined by the prevalence of rabies in the area.

The original Pasteur treatment, in which the virus was attenuated by drying in the dissected spinal cords of rabies-infected rabbits, has been replaced by human diploid cell vaccine (HDCV) (Figure 22.9). The HDCV is administered in a series of injections at intervals during a 28-day period. Passive immunization is provided simultaneously by injecting human rabies immune globulin (RIG) that has been harvested from people who are immunized against rabies, such as laboratory workers.

Some problems have arisen with the HDCV. Veterinarians and laboratory workers, who must be immunized periodically, have developed allergic reactions. There have been cases of rabies in HDCV-immunized persons in Africa, causing concern that some African rabies may be caused by viruses that are antigenically different from the standard strains used to make the vaccines. Furthermore, the HDCV is much too expensive for many parts of the world; in these areas, cheaper but less effective vaccines are prepared from animal nerve tissue. However, new vaccines are being developed.

Rabies occurs all over the world. In the United States, 20,000 to 30,000 persons are given the rabies vaccine each year. Nearly twice that many *die* of rabies each year in India, where the vaccine is given to about 3 million persons. Australia, Great Britain, New Zealand, and Hawaii are free of the disease, a condition maintained by rigid quarantine. In North America, rabies is widespread among wildlife (Figure 22.10). Worldwide, dogs are the most common carriers of rabies. Rabies is seldom found in squirrels, rabbits, rats, or mice. Rabies has long been endemic in vampire bats of South America. In Europe and North America, there are ongoing experiments to immunize wild animals with live rabies vaccine produced in genetically engineered vaccinia viruses, in food dropped for the animals to find. In Pennsylvania, for example, inoculated sandwiches have been used to immunize raccoons. It is too early to assess the results of these attempts. In the United States, up to 6000 cases of rabies are diagnosed in animals each year, but in recent years, only about one case per year has been diagnosed in humans (see the box on p. 550).

ARTHROPOD-BORNE ENCEPHALITIS

Encephalitis, caused by a mosquito-borne arbovirus (encephalitis virus) is rather common in the United States. Figure 22.11 shows the incidence of encephalitis in the United States from 1975 to 1989. The increase in the summer months coincides with the proliferation

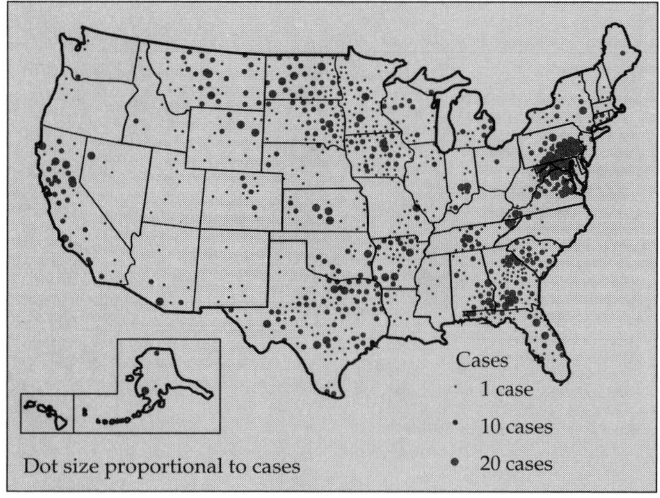

(a)

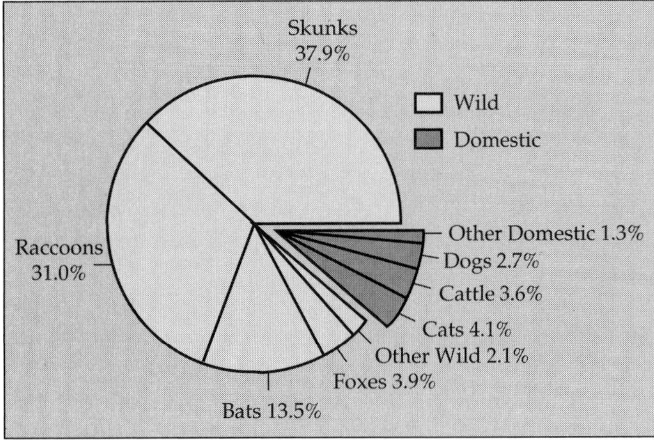

(b)

FIGURE 22.10 Rabies. (a) Cases of rabies in animals, by county in the United States, 1988. **(b)** Distribution of rabies cases in various animals, United States, 1988.

of adult mosquitoes during these months. Sentinel animals, such as caged rabbits or chickens, are tested periodically for antibodies to arboviruses. This gives health officials information on the incidence and types of virus in their area.

A number of clinical types of arthropod-borne (sometimes shortened to arbo) encephalitis have been identified; all can cause symptoms ranging from the subclinical to rapid death from encephalitis. Active cases of these diseases are characterized by chills, headache, and fever. As the disease progresses, mental confusion and coma are observed. Survivors often suffer from permanent neurological problems.

Horses as well as humans are frequently affected by these viruses, so there are strains causing Eastern equine encephalitis (EEE) and Western equine encephalitis (WEE). These two viruses are the most likely to cause severe disease in humans. EEE is the least prevalent arboencephalitis (its main mosquito vector prefers to feed on birds), but it is the most severe. The mortality rate is about 25% or more, and a high incidence of brain damage, deafness, and similar neurological problems occurs in survivors. St. Louis encephalitis (SLE) acquired its name from the location of an early major outbreak (in which the fact that mosquitoes were involved in transmission of these diseases was discovered). California encephalitis (CE) was first identified in that state, but most cases occur elsewhere. The La Crosse strain of CE is the most commonly encountered (Table 22.3).

The Far East also has endemic encephalitis. **Japanese B encephalitis** is the best known; it is a serious public health problem, especially in Japan, Korea, and

FIGURE 22.11 Arbovirus infections of the central nervous system. Cases per month in the United States, 1975 to 1990.

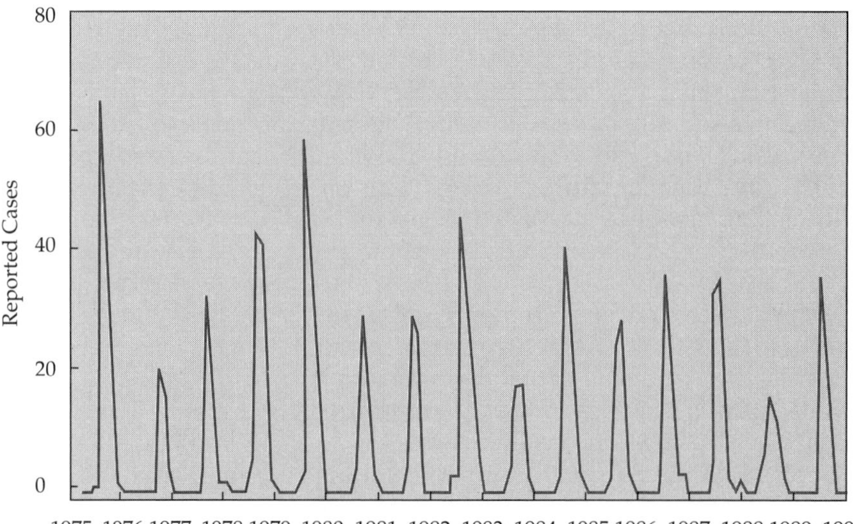

TABLE 22.3 Arthropod-Borne Encephalitis

DISEASE	MOSQUITO VECTOR	HOST ANIMALS	GEOGRAPHIC DISTRIBUTION IN U.S.	COMMENT
St. Louis encephalitis	*Culex*	Birds	Throughout country	Mostly urban outbreaks; affects all age groups
California encephalitis	*Aedes*	Small mammals	North Central states, New York state	Affects mostly children in rural or suburban areas, La Crosse strain medically most important
Western equine encephalitis	*Culex*	Birds, horses	Throughout country	Severe disease, frequent neurological damage
Eastern equine encephalitis	*Aedes, Culiseta*	Birds, horses	East coast	More severe even than WEE; affects mostly young children and younger adults; relatively uncommon in humans

China. Vaccines are used to control the disease in these countries.

Diagnosis of arthropod-borne encephalitis is usually made by serological tests, such as complement fixation. The most effective control measure is local elimination of the mosquitoes.

Fungal Disease of the Nervous System

The central nervous system is seldom invaded by fungi, but one pathogenic fungus is well adapted to growth in fluids of the central nervous system.

CRYPTOCOCCUS NEOFORMANS MENINGITIS (CRYPTOCOCCOSIS)

Fungi of the genus *Cryptococcus* are spherical cells resembling yeasts; they reproduce by budding and produce polysaccharide capsules, some much thicker than the cells themselves (Figure 22.12). Only one species, *Cryptococcus neoformans*, is pathogenic for humans, causing the disease called **cryptococcosis.** The organism is widely distributed in soil, especially soil contaminated with pigeon droppings. It is also found in pigeon roosts and nests on the window ledges of urban buildings. Most cases of cryptococcosis occur in urban areas. It is thought to be transmitted by the inhalation of dried infected pigeon droppings.

Inhalation of *C. neoformans* initially causes infection of the lungs, frequently subclinical, and often the disease does not proceed beyond this stage. However, it can spread through the bloodstream to other parts of the body, including the brain and meninges, especially in immunosuppressed individuals and persons receiving steroid treatments for major illnesses. The disease is usually expressed as chronic meningitis, which is usually progressive and fatal if untreated.

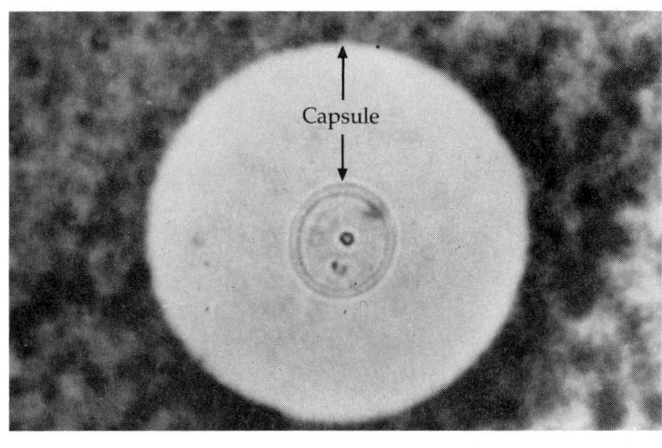

FIGURE 22.12 ***Cryptococcus neoformans.*** This yeastlike fungus has an unusually thick capsule. In this micrograph, the capsule is made visible by suspending the cells in dilute India ink.

The best serological diagnostic test is a latex agglutination test to detect cryptococcal antigens in the patients' serum or cerebrospinal fluid. The drugs of choice for treatment are amphotericin B and flucytosine.

Protozoan Diseases of the Nervous System

Protozoans capable of invading the central nervous system are rare, but their effects are devastating.

AFRICAN TRYPANOSOMIASIS

African trypanosomiasis, or sleeping sickness, is a protozoan disease that affects the nervous system (see the box in Chapter 17, p. 440). In 1907, Winston Churchill described Uganda during an epidemic of sleeping sickness as a "beautiful garden of death." Even today, the disease affects about a million people in Central and East Africa, and about 20,000 new cases are reported each year.

The disease is caused by *Trypanosoma brucei gambiense* and *Trypanosoma brucei rhodesiense,* flagellates that are injected by the bite of a tsetse fly *(Glossina* [gläs-sē'nä]). Animal reservoirs for the two trypanosomes are similar, but they occur in different habitats and are spread by different species of tsetse fly vector. Human-to-human transmission by bites of the insect vector is more likely with infections by *T. b. gambiense,* which circulates in the patient's blood during the two- to four-year course of the disease. *T. b. rhodesiense* causes a disease that is more acute, running its course in only a few months.

During the early stages of either form of the disease, a few trypanosomes can be found in the blood. During later stages, the trypanosomes move into the cerebrospinal fluid. Symptoms of the disease include decreases in physical activity and mental acuity. Untreated, the host enters a coma, and death is almost inevitable.

There are some moderately effective chemotherapeutic agents, such as suramin and pentamidine isethionate. However, the drugs produce toxic effects, and, because they do not cross the blood–brain barrier, they are effective only when the central nervous system has not become affected. A toxic arsenical, called melarsoprol, is usually chosen when the central nervous system is affected. A vaccine is being developed, but a major obstacle is that the trypanosome is able to change protein coats at least 100 times and can thus evade antibodies aimed at only one or a few of the proteins. Each time the body's immune system is successful in suppressing the trypanosome, a new clone of parasites appears with a different antigenic coat (Figure 22.13).

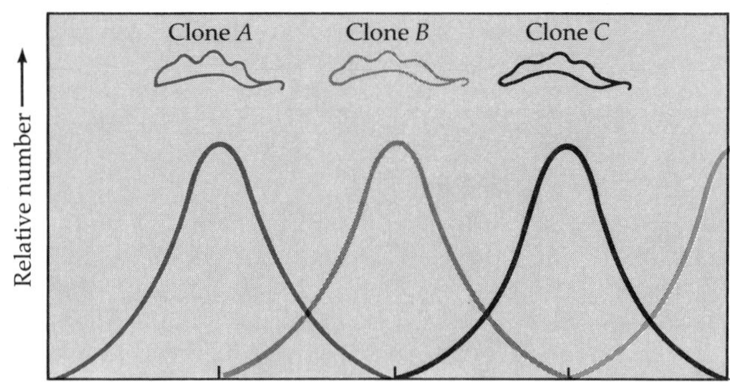

FIGURE 22.13 Trypanosomiasis. The population of each trypanosome clone drops nearly to zero as the immune system suppresses them, but a new clone with a different antigenic surface then replaces it.

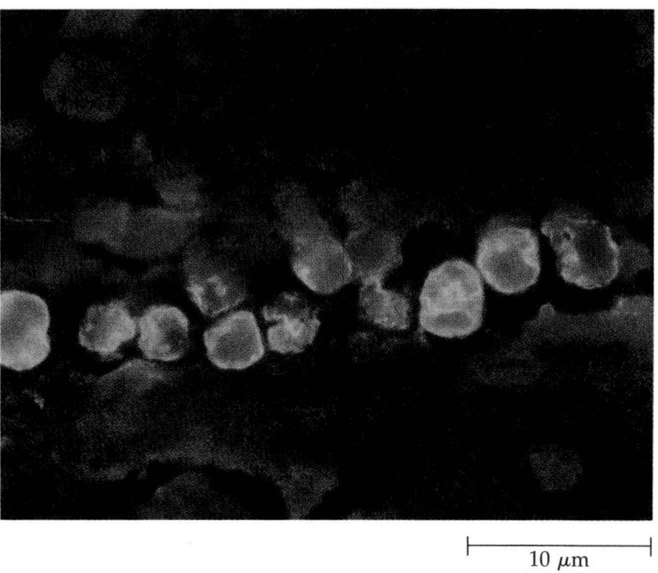

FIGURE 22.14 ***Naegleria fowleri*** **from a case of amebic meningoencephalitis.** The parasite is shown stained by the yellow-green fluorescent dye used in the fluorescent-antibody test and is surrounded by brain tissue.

NAEGLERIA MICROENCEPHALITIS

Naegleria fowleri (nī-gle'rē-ä fou'lėr-ē) is a protozoan (amoeba) that is known to cause the neurological disease ***Naegleria*** **microencephalitis.** Although cases of it are reported in most parts of the world, only a few cases per year are reported in the United States. The most common victims are children who swim in ponds or streams. The organism initially infects the nasal mucosa and later proliferates in the brain (Figure 22.14). The fatality rate is nearly 100%.

Nervous System Diseases Caused by Unconventional Agents

There are a number of fatal diseases of the human central nervous system that are thought to be caused by what have been called **slow viruses.** However, no one is absolutely sure of the causative agents, which in most cases are clearly not conventional viruses. Therefore many prefer the term **unconventional agents** (see Chapter 13). The name "slow virus" alludes to the

TABLE 22.4 Diseases Associated with the Nervous System

DISEASE	CAUSATIVE AGENT	MODE OF TRANSMISSION	TREATMENT
Bacterial Diseases			
Meningococcal meningitis	*Neisseria meningitidis*	Contact with a healthy carrier via respiratory tract	Penicillin, ampicillin, third-generation cephalosporins
Hemophilus influenzae meningitis	*Hemophilus influenzae*	Via respiratory tract	Third-generation cephalosporins
Pneumococcal meningitis	*Streptococcus pneumoniae*	Via respiratory tract	Third-generation cephalosporins
Botulism	*Clostridium botulinum*	Ingestion of toxin	Antitoxins and respiratory support
Listeriosis	*Listeria monocytogenes*	Contaminated food	Penicillins
Leprosy	*Mycobacterium leprae*	Transfer of exudates from lesions	Sulfone drugs (dapsone), rifampin, clofazimine
Tetanus	*Clostridium tetani*	Contamination of deep anaerobic wounds	Immune globulin and respiratory support
Fungal Disease			
Cryptococcosis	*Cryptococcus neoformans*	Inhalation of dried infected pigeon droppings	Amphotericin B, flucytosine
Viral Diseases			
Poliomyelitis	Poliovirus	Ingestion of virus	None
Rabies	Rabies virus	Bite of a rabid animal	Immune globulin and vaccination exposure
Arthropod-borne encephalitis	Arboviruses	Mosquitoes	None
Protozoan Diseases			
African trypanosomiasis	*T. b. rhodesiense*, *T. b. gambiense*	Tsetse flies	Suramin, pentamidine isethiouate, melarsoprol
Naeglaria microencephalitis	*Naeglaria fowlerii*	From fresh water, probably nasal passages	None
Unconventional Agents			
Creutzfeldt–Jakob disease	Unconventional agent	Ingestion or accidental wounds	None
Kuru	Unconventional agent	Ingestion or wound contamination	None

slow course of the diseases, which are characterized by long incubation times. The damage to the central nervous system is insidious and progressive, without the fever and inflammation seen in encephalitis. Autopsies show a characteristic spongiform (porous, like a sponge) degeneration of the brain. In recent years, the study of slow virus diseases has been one of the most interesting areas of microbiology in medical science.

A typical slow virus disease is **sheep scrapie.** The infected animal rubs against fences and walls until areas on its body are raw. During a period of several weeks or months, the animal gradually loses motor control and dies. The agent, whatever it is, can be passed to other animals, such as mice, by the injection of brain tissue from one animal to the next.

Humans suffer from diseases similar to scrapie. **Creutzfeldt–Jakob disease (CJD)** is an example. This disease is rare (about 200 cases per year in this country) but often occurs in families. There is no doubt that an infective agent is involved since transmission via corneal transplants and accidental scalpel nicks of a surgeon during autopsy have been reported. Several cases have been traced to injection of a growth hormone derived from human tissue.

Some tribes in New Guinea have suffered from a slow virus disease called **kuru.** Kuru is apparently related to the practice of smearing brain tissue onto the body during cannibalistic rituals. The infection results when the agent is introduced into open sores or cuts. The disease is disappearing as cannibalism dies out.

A current international public health concern is **bovine spongiform encephalopathy.** Since 1986 thousands of cattle in the British Isles have come down with this condition—known locally as mad cow disease. It causes the animals to become unmanageable, and they must be destroyed. The present theory is that the disease is due to a scrapie-like pathogen probably transmitted in feed supplements derived from animals such as sheep. There is great concern in the international medical community that the condition might be transmitted to humans via the ingestion of beef.

As we discussed in Chapter 13, the unconventional agents that cause these diseases seem to be pure protein. If nucleic acid is present, it has not been detected to date. The exact nature of these agents and their mode of replication are still mysteries. They continue to be the subject of intense research—and contention.

STUDY OUTLINE

Structure and Function of the Nervous System (pp. 539–540)

1. The central nervous system consists of the brain, which is protected by the skull bones, and the spinal cord, which is protected by the backbone.
2. The peripheral nervous system consists of the nerves that branch from the central nervous system.
3. The central nervous system is covered by three layers of membranes called meninges. Cerebrospinal fluid circulates between the inner and middle meninges and in the ventricles of the brain.
4. The blood–brain barrier normally prevents many substances, such as antibiotics, from entering the brain.
5. Microorganisms can enter the central nervous system through trauma, along peripheral nerves, and through the bloodstream and lymphatic system.
6. An infection of the meninges is called meningitis. An infection of the brain is called encephalitis.

Bacterial Diseases of the Nervous System (pp. 541–547)

BACTERIAL MENINGITIS (pp. 541–543)

1. Meningitis can be caused by viruses, bacteria, fungi, and protozoans.
2. Nearly 50 species of opportunistic bacteria can cause meningitis.

Neisseria Meningitis (Meningococcal Meningitis) (p. 541)

1. *Neisseria meningitidis* causes meningococcal meningitis. This bacterium is found in the throats of healthy carriers.
2. The bacteria probably gain access to the meninges through the bloodstream. The bacteria may be found in leukocytes in cerebrospinal fluid.
3. Symptoms are due to endotoxin. The disease occurs most often in young children.
4. Military recruits are vaccinated with purified capsular polysaccharide to prevent epidemics in training camps.
5. Encapsulated *N. meningitidis* is resistant to phagocytosis.
6. Diagnosis is based on isolation and identification of the bacteria in blood or cerebrospinal fluid.
7. *N. meningitidis* must be cultured on media containing blood and incubated in an atmosphere containing 5% to 10% carbon dioxide.

Hemophilus influenzae Meningitis (pp. 541–542)

1. *Hemophilus influenzae* requires blood factors for growth.
2. *H. influenzae* is the most common cause of meningitis in children under four years old.
3. The disease often occurs as a secondary infection following a viral infection such as influenza.
4. A moderately successful vaccine (Hib) directed against the capsular polysaccharide antigen is available.
5. Diagnosis is based on identification of the organism in spinal fluid.

Streptococcus pneumoniae Meningitis (Pneumococcal Meningitis) (p. 542)

1. Hospitalized patients and young children are most susceptible to *S. pneumoniae* meningitis. It is rare but has a high mortality rate.

Diagnosis and Treatment of Bacterial Meningitis (pp. 542–543)

1. Diagnosis is based on isolation and identification of the bacteria in cerebrospinal fluid.

2. Cultures are usually made on blood agar and incubated in an atmosphere containing reduced oxygen levels.
3. Counter-immunoelectrophoresis and rapid latex agglutination tests are made on cerebrospinal fluid.
4. Penicillin, ampicillin, or cephalosporins may be administered before identification of the pathogen.

LISTERIOSIS (p. 543)

1. *Listeria monocytogenes* causes meningitis in newborns, the immunosuppressed, pregnant women, and cancer patients.
2. It is acquired by ingestion of contaminated food.
3. Listeriosis may be asymptomatic in healthy adults.
4. *L. monocytogenes* can cross the placenta and cause spontaneous abortion and stillbirth.

TETANUS (pp. 543–544)

1. Tetanus is caused by a localized infection of a wound by *Clostridium tetani.*
2. *C. tetani* produces the neurotoxin tetanospasmin, which causes the symptoms of tetanus—spasms, contraction of muscles controlling the jaw, and death resulting from spasms of respiratory muscles.
3. *C. tetani* is an anaerobe that will grow in unclean wounds and wounds with little bleeding.
4. Acquired immunity results from DPT immunization that includes tetanus toxoid.
5. Following an injury, an immunized person may receive a booster of tetanus toxoid. An unimmunized person may receive (human) tetanus immune globulin.
6. Debridement (removal of tissue) and antibiotics may be used to control the infection.

BOTULISM (pp. 544–546)

1. Botulism is caused by an exotoxin produced by *C. botulinum* growing in foods.
2. Serological types of botulinum toxin vary in virulence, with type A being the most virulent.
3. The toxin is a neurotoxin that inhibits transmission of nerve impulses.
4. Blurred vision occurs in one to two days; progressive flaccid paralysis follows for one to ten days, resulting in respiratory and cardiac failure.
5. *C. botulinum* will not grow in acidic foods or in an aerobic environment.
6. Endospores are killed by proper canning. Addition of nitrites to foods inhibits outgrowth after endospore germination.
7. The toxin is heat labile and is destroyed by boiling (100°C) for five minutes.
8. Infant botulism results from the growth of *C. botulinum* in an infant's intestines.
9. Wound botulism occurs when *C. botulinum* grows in anaerobic wounds.
10. For diagnosis, mice protected with antitoxin are inoculated with toxin from the patient or foods.

LEPROSY (pp. 546–547)

1. *Mycobacterium leprae* causes leprosy, or Hansen's disease.
2. *M. leprae* has never been cultured on artificial media. It can be cultured in armadillos.
3. The tuberculoid form of the disease is characterized by loss of sensation in the skin surrounded by nodules. The lepromin test is positive.
4. In the lepromatous form, disseminated nodules and tissue necrosis occur. The lepromin test is negative.
5. Leprosy is not highly contagious and is spread by prolonged contact with exudates.
6. Untreated individuals often die of secondary bacterial complications, such as tuberculosis.
7. Patients with leprosy are made noncommunicable within four or five days with sulfone drugs and then treated as outpatients.
8. Leprosy occurs primarily in the tropics.
9. Laboratory diagnosis is based on observations of acid-fast bacilli (rods) in lesions or fluids.

Viral Diseases of the Nervous System (pp. 547–553)

POLIOMYELITIS (pp. 547–549)

1. The symptoms of poliomyelitis are usually headache, sore throat, fever, stiffness of the back and neck, and occasionally paralysis (less than 1% of the cases).
2. Poliovirus (picornavirus) is found only in humans and is transmitted by ingestion of water contaminated with feces.
3. Poliovirus first invades lymph nodes of the neck and small intestine. Viremia and spinal cord involvement may follow.
4. At present, outbreaks of polio in the United States are uncommon because of the use of vaccines.
5. The Salk vaccine (an inactivated polio vaccine) involves injection of formalin-inactivated viruses and boosters every few years. The Sabin (oral polio) vaccine contains three attenuated live strains of poliovirus and is administered orally.
6. Diagnosis is based on isolation of the virus from feces and throat secretions.

RABIES (pp. 549–551)

1. Rabies virus (a rhabdovirus) causes an acute, usually fatal, encephalitis called rabies.
2. Rabies may be contracted through the bite of a rabid animal, by inhalation of aerosols, or invasion through minute skin abrasions. The virus multiplies in skeletal muscle and connective tissue.
3. Encephalitis occurs when the virus moves along peripheral nerves to the central nervous system.
4. Symptoms of rabies include spasms of mouth and throat muscles followed by extensive brain and spinal cord damage and death.
5. Laboratory diagnosis may be made by direct immunofluorescent tests of saliva, serum, and cerebrospinal fluid or brain smears.
6. Reservoirs for rabies in the United States include skunks, bats, foxes, and raccoons. Domestic cattle, dogs, and cats may get rabies. Rodents and rabbits seldom get rabies.
7. The Pasteur treatment for rabies involved multiple subcutaneous injections of rabies virus grown in rabbit spinal cord tissue.

8. Current postexposure treatment includes administration of human rabies immune globulin (RIG) along with multiple intramuscular injections of human diploid cell vaccine (HDCV).
9. Preexposure immunization consists of injections of HDCV.

ARTHROPOD-BORNE ENCEPHALITIS (pp. 551–553)

1. Symptoms of encephalitis are chills, headache, fever, and eventually coma.
2. Many types of arboviruses transmitted by mosquitoes cause encephalitis.
3. The incidence of arthropod-borne encephalitis increases in the summer months when mosquitoes are most numerous.
4. Horses are frequently infected by EEE and WEE viruses.
5. Diagnosis is based on serological tests.
6. Elimination of the vector is the most effective control measure.

Fungal Disease of the Nervous System (p. 553)

***CRYPTOCOCCUS NEOFORMANS* MENINGITIS (CRYPTOCOCCOSIS)** (p. 553)

1. *Cryptococcus neoformans* is an encapsulated yeastlike fungus that causes *Cryptococcus neoformans* meningitis.
2. The disease may be contracted by inhalation of dried infected pigeon droppings.
3. The disease begins as a lung infection and may spread to the brain and meninges.
4. Immunosuppressed individuals are most susceptible to *Cryptococcus neoformans* meningitis.
5. Diagnosis is based on latex agglutination for cryptococcal antigens in patient's serum or cerebrospinal fluid.

Protozoan Diseases of the Nervous System (pp. 553–554)

AFRICAN TRYPANOSOMIASIS (p. 554)

1. African trypanosomiasis is caused by the protozoans *Trypanosoma brucei gambiense* and *T. b. rhodesiense* and transmitted by the bite of the tsetse fly *(Glossina).*
2. The disease affects the nervous system of the human host, causing lethargy and eventually coma. It is commonly called sleeping sickness.

***NAEGLERIA* MICROENCEPHALITIS** (p. 554)

1. Encephalitis caused by the protozoan *N. fowleri* is almost always fatal.

Nervous System Diseases Caused by Unconventional Agents (pp. 554–556)

1. The diseases called slow virus diseases are of uncertain etiology. The diseases progress slowly in the host.
2. Sheep scrapie and bovine spongiform encephalopathy are examples of diseases caused by unconventional agents that are transferable from one animal to another.
3. Creutzfeldt–Jakob disease and kuru are human diseases similar to scrapie. Kuru occurs in isolated groups of cannibals who eat brains.
4. The unconventional agents may be pure protein, or protein associated with undetectable nucleic acid.

STUDY QUESTIONS

Review

1. Differentiate between meningitis and encephalitis.
2. Fill in the following table.

Causative Agent of Meningitis	Susceptible Population	Method of Transmission	Treatment
N. meningitidis			
H. influenzae			
S. pneumoniae			
L. monocytogenes			
C. neoformans			

3. Briefly explain the derivation of the name *Hemophilus influenzae*.
4. If *Clostridium tetani* is relatively sensitive to penicillin, then why doesn't penicillin cure tetanus?
5. Compare and contrast the Salk and Sabin vaccines with respect to composition, advantages, and disadvantages.
6. What treatment is used against tetanus under the following conditions?
   (a) Before a person suffers a deep puncture wound
   (b) After a person suffers a deep puncture wound
7. Why is the following description used for wounds that are susceptible to *C. tetani* infection: ". . . Improperly cleaned deep puncture wounds . . . ones with little or no bleeding . . ."
8. List the following information for botulism: etiologic agent, suspect foods, symptoms, treatment, conditions necessary for microbial growth, basis for diagnosis, prevention.
9. Provide the following information on leprosy: etiology, method of transmission, symptoms, treatment, prevention, and susceptible population.
10. Provide the following information on poliomyelitis: etiology, method of transmission, symptoms, prevention. Why aren't Salk and Sabin vaccines considered treatments for poliomyelitis?

11. Provide the etiology, method of transmission, reservoirs, and symptoms for rabies.
12. Outline the procedures for treating rabies after exposure. Outline the procedures for preventing rabies prior to exposure. What is the reason for the differences in the procedures?
13. Why is the incidence of arbovirus encephalitis in the United States higher in the summer months?
14. Describe the symptoms and laboratory test results that would lead to a diagnosis of slow virus disease.
15. Provide evidence that slow virus diseases are caused by transmissible agents that resemble viruses.
16. Fill in the following table.

Disease	Etiology	Vector	Symptoms	Treatment
Arthropod-borne encephalitis				
African trypanosomiasis				

17. Why are meningitis and encephalitis generally difficult to treat?

Challenge

1. Most of us have been told that a "rusty nail" causes tetanus. What do you suppose is the origin of this adage?
2. A one-year-old infant was lethargic and had a fever. When he was admitted to the hospital, he had multiple brain abscesses with gram-negative coccobacillary rods. Identify the disease, etiology, and treatment.
3. A 40-year-old bird handler was admitted to the hospital with soreness over his upper jaw, progressive vision loss, and bladder dysfunction. He had been well two months earlier. Within weeks he lost reflexes in his lower extremeties and subsequently died. Examination of cerebrospinal fluid showed lymphocytes. What etiology do you suspect? What further information do you need?
4. A BCG vaccination will result in positive lepromin and tuberculin tests. What is the relationship between leprosy and tuberculosis?
5. One week after bathing in a hot spring, a nine-year-old girl was hospitalized after a three-day history of progressive, severe headaches, nausea, lethargy, and stupor. Examination of CSF revealed ameboid organisms. Identify the disease, etiology, and treatment.

FURTHER READING

Bingham, R. "Outrageous ardor." *Science 81* 2:54–61, September 1981. An interesting article on slow viruses and their discoverer, Carleton Gajdusek.

Chasan, D.J. "The polio paradox." *Science 86* 7:37–39, April 1986. A discussion about one little-used vaccine and another that is the leading cause of polio.

Gaylord, H., and P.J. Brennan. "Leprosy and the leprosy bacillus: recent developments." *Annual Review of Microbiology* 41:645–675, 1987. A review of new information about antigens and immunity to *M. leprae* from armadillo cultures.

Hall, S.S. "The La Crosse file." *Science 84* 5:54–62; July–August 1984. The epidemiology of California encephalitis virus.

John, D.T. "Primary amoebic meningoencephalitis and the biology of *Naegleria fowleri*." *Annual Review of Microbiology* 36:101–123, 1982. A good article on the natural history of the disease and the parasite.

Kaplan, C., G.S. Turner, and D.A. Warrell, eds. *Rabies, the Facts*, 2nd ed. New York: Oxford University Press, 1986. A summary of information on rabies.

Kristensson, K., and E. Norrby. "Persistence of RNA viruses in the central nervous system." *Annual Review of Microbiology* 40:159–184, 1986. Descriptions of viruses such as measles virus that remain in the CNS for years and of the resultant host changes.

Rosen, L. "The natural history of Japanese encephalitis virus." *Annual Review of Microbiology* 40:395–414, 1986. A good lesson in natural history that describes alternate hosts and overwintering.

Schlech, W.F., et al. "Bacterial meningitis in the United States, 1978 through 1981; The National Bacterial Meningitis Study." *Journal of the American Medical Association* 253:1649–1754, 1985. Analysis and discussion of incidence and types of bacterial meningitis.

Smith, L., and H. Sugiyama. *Botulism: The Organism, Its Toxins, the Disease*, 2nd ed. Springfield, Ill.: Thomas, 1988. A summary of information on botulism.

Spector, D.H., and D. Baltimore. "The molecular biology of poliovirus." *Scientific American* 232(5):24–31, May 1975. Information on viral multiplication is obtained from laboratory-grown poliovirus.

See also Further Reading for Part Four at the end of Chapter 26.

CHAPTER 23

Microbial Diseases of the Cardiovascular System

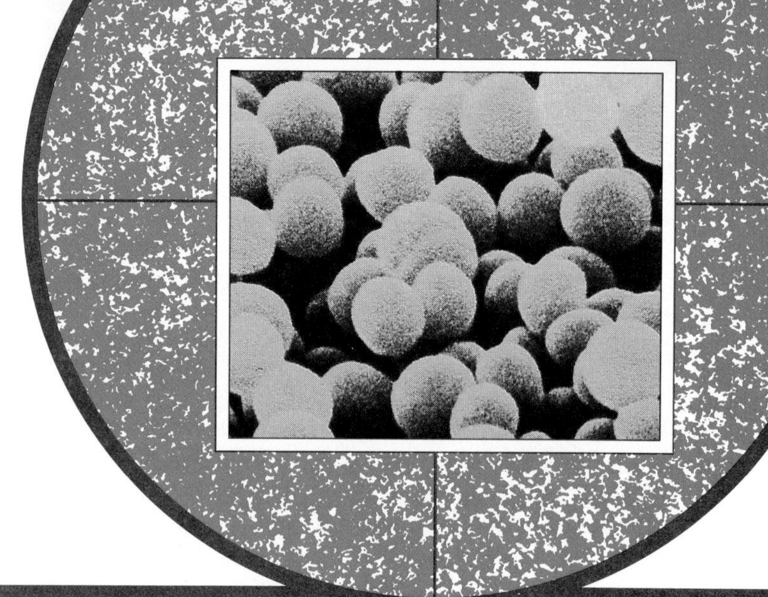

LEARNING OBJECTIVES

- List the symptoms of septicemia, and explain the importance of infections that develop into septicemia.
- Discuss the epidemiologies of puerperal sepsis, bacterial endocarditis, myocarditis, and Burkitt's lymphoma.
- Discuss the causes, treatments, and preventive measures of rheumatic fever and infectious mononucleosis.
- Describe the epidemiologies of tularemia, brucellosis, anthrax, and gas gangrene.
- Compare and contrast the causative agents, vectors, reservoirs, symptoms, and treatments of malaria, yellow fever, dengue, relapsing fever, typhus, Rocky Mountain spotted fever, and Lyme disease.
- List the causative agents and methods of transmission of toxoplasmosis, American trypanosomiasis, and schistosomiasis. Also, describe their worldwide effects on health.
- Identify the best control measures for vector-borne diseases.

The **cardiovascular system** consists of the heart, blood, and blood vessels. The **lymphatic system** consists of the lymph, lymph vessels, lymph nodes, and the lymphoid organs (tonsils, appendix, spleen, and thymus gland—see Figure 16.5). Because both systems circulate various substances throughout the body, they can serve as vehicles for the spread of infection. A summary of diseases associated with the cardiovascular system is presented in Table 23.1, pp. 580–581.

Structure and Function of the Cardiovascular System

The center of the cardiovascular system is the *heart* (Figure 23.1). The function of the heart is to circulate the blood through the body's tissues so it can deliver certain substances to cells and remove other substances from them. *Blood vessels* are the tubes that carry circulating blood throughout the body. An *artery* is a blood vessel that carries blood from the heart to different tissues of the body. When the arteries reach their destinations, they branch into smaller vessels called *arterioles,* which in turn branch into even smaller vessels called *capillaries.* Capillary walls are only one cell thick, and it is through these walls that blood and tissue cells exchange materials. The capillaries that take blood away from the tissue cells converge to form small veins called *venules.* Venules unite to form *veins,* and veins return the blood to the heart.

The *blood* itself is a mixture of formed elements and a liquid called plasma. The *plasma* transports dissolved nutrients to body cells and carries wastes away from the cells. The formed elements in blood include red blood cells, white blood cells, and platelets. Red blood cells (*erythrocytes*) carry oxygen and some carbon diox-

FIGURE 23.1 The cardiovascular and lymphatic systems. Details of circulation to the head and extremities are not shown in this simplified diagram. The blood circulates from the heart through the arterial system (lighter color) to the capillaries in the lungs and other parts of the body. From these capillaries, the blood returns through the venous system (gray) to the heart. From the capillaries, some plasma filters into the surrounding tissue and enters the lymph capillaries **(a)**. This fluid, now called lymph, returns to the heart through the lymphatic circulatory system (darker color). All lymph returning to the heart must pass through at least one lymph node, as shown in **(b)**. See also Figure 16.5.

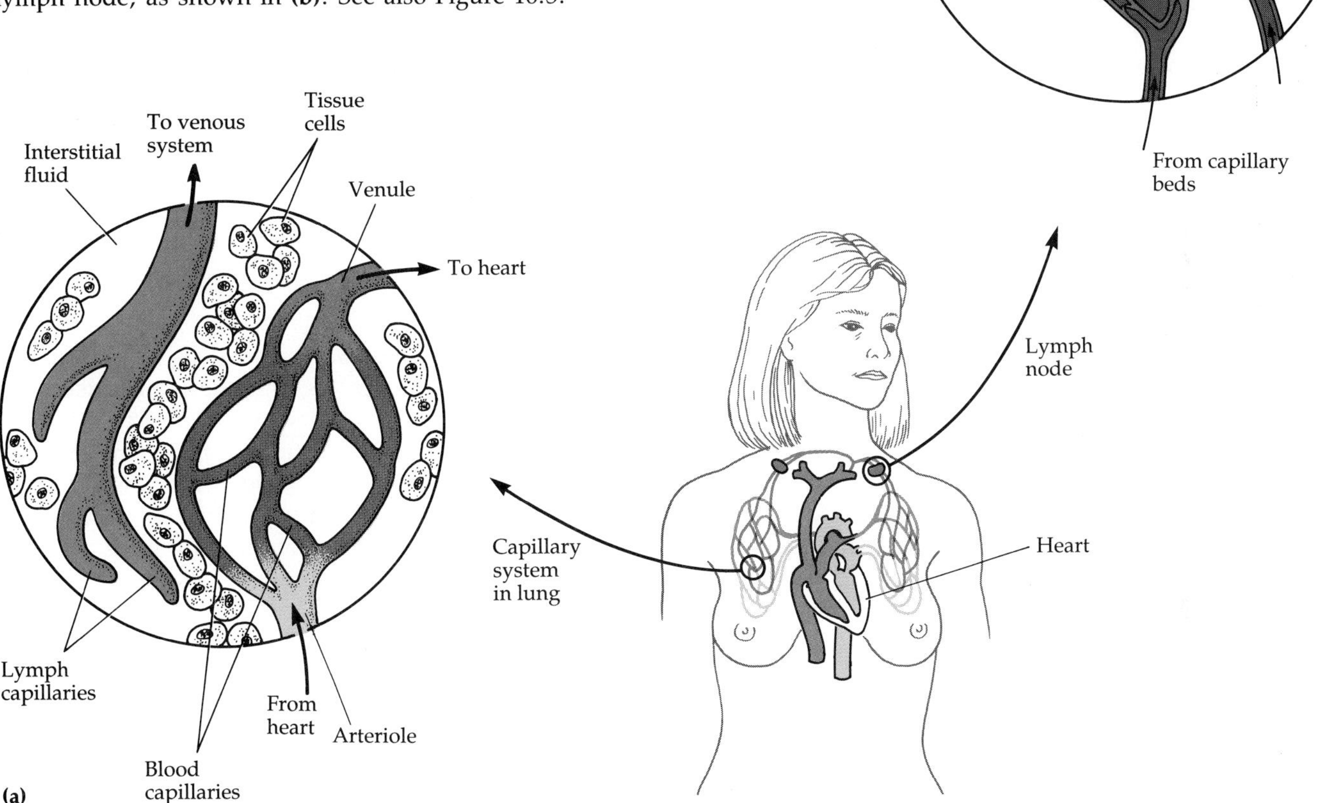

ide, although most of the carbon dioxide in blood is dissolved in the plasma. White blood cells (leukocytes) play several roles in defending the body against infection, as was discussed in Part Three. Some white cells, in particular the neutrophils, are phagocytes. B cells and T cells, two types of lymphocytes, play key roles in immunity.

Structure and Function of the Lymphatic System

As part of the overall pattern of circulation, some plasma filters out of the blood capillaries and into spaces between tissue cells. These spaces are called *interstitial spaces.* The fluid circulating around and between tissue cells is called *interstitial fluid.* Microscopic lymphatic vessels that surround tissue cells are called *lymph capillaries;* they are larger and more permeable than blood capillaries. As the interstitial fluid moves around the tissue cells, it is picked up by the lymph capillaries; the fluid is then called *lymph* (Figure 23.1).

Because lymph capillaries are very permeable, they readily pick up microorganisms or their products. From lymph capillaries, lymph is transported into larger lymph vessels called *lymphatics,* which contain valves that keep the lymph moving toward the heart. Eventually, all the lymph is returned to the blood just before the blood enters the right atrium of the heart. As a result of this circulation, proteins and fluid that have filtered from the plasma are returned to the blood.

At various points along the lymphatic system are oval structures called *lymph nodes,* through which lymph flows. Within the lymph nodes are fixed macrophages that help to clear the lymph of microorganisms. At times, the number of microorganisms circulating through lymph nodes is so great that the nodes themselves become infected and hence become enlarged and tender. The lymph nodes are also important to the body's immune reaction against invading microorganisms. Microorganisms entering the lymph nodes and other lymphoid tissues encounter two types of lymphocytes—B cells that are stimulated to produce humoral antibodies, and T cells that differentiate into mature T cells that carry out important defensive functions (see Figure 17.2).

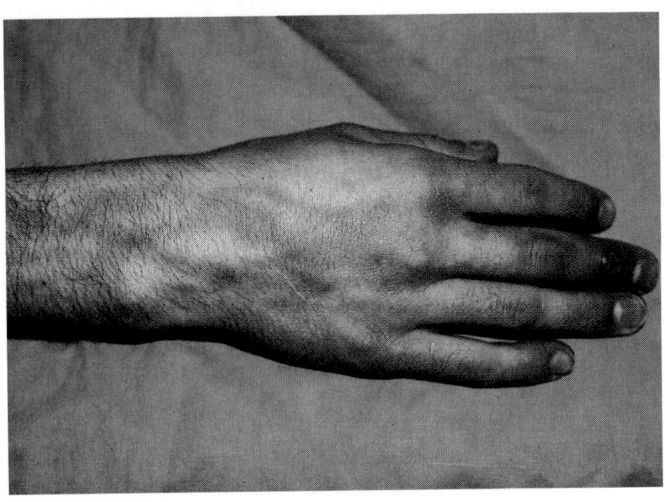

FIGURE 23.2 Lymphangitis. As the infection spreads from its original site along the lymphatic channels, the inflamed walls of the channels become visible as red streaks.

Bacterial Diseases of the Cardiovascular System

Once bacteria gain access to the bloodstream, they may become widely disseminated and sometimes are also able to reproduce rapidly.

SEPTICEMIA

Although blood is normally sterile, moderate numbers of microorganisms are usually not harmful. However, if the defenses of the blood and lymphatic systems fail, the microorganisms can undergo uncontrolled proliferation in the blood. That condition is called **septicemia,** or blood poisoning. One symptom of septicemia is the appearance of **lymphangitis** (Figure 23.2), inflamed lymph vessels shown as red streaks under the skin, running along the arm or leg from the infection site. Sometimes the streaks end at a lymph node, where the lymphocytes attempt to stop the invading microorganisms. Clinically, a person suffering from septicemia exhibits the symptoms of **septic shock.** The small arteries and veins are constricted, peripheral blood circulation decreases, and tissues served by the peripheral circulation become oxygen-starved. The patient develops rapid breathing and heartbeat, a drop in blood pressure, and mental confusion. The extremities become cool and may turn blue. Septic shock may progress to coma and death.

The organisms most frequently associated with septicemia are gram-negative rods, although a few gram-positive bacteria and fungi are also implicated. Among the gram-negative rods are *Escherichia coli, Serratia marcescens, Proteus mirabilis, Enterobacter aerogenes, Pseudomonas aeruginosa,* and *Bacteroides* species. These bacteria enter the blood from a focus of infection in the body. As you may recall, the cell walls of many gram-negative bacteria contain endotoxins that are released upon the lysis of the cell. It is the endotoxin that actually causes the symptoms. Once released, the endotoxin damages blood vessels; this damage causes the low blood pressure and subsequent shock. In some cases, antibiotics aggravate the condition by causing the lysis of large numbers of bacteria that then release damaging endotoxins. Products that will neutralize these endotoxins with monoclonal antibodies are currently being introduced. Many of these gram-negative rods are of nosocomial origin and are introduced into the bloodstream by medical procedures, such as catheters or feeding tubes, that bypass the normal barriers between the environment and the blood.

PUERPERAL SEPSIS

Puerperal sepsis, also called **puerperal fever** and **childbirth fever,** is a nosocomial infection that frequently leads to septicemia. This begins as an infection of the uterus as a result of childbirth or abortion. *Streptococcus pyogenes,* a group A β-hemolytic streptococcus, is the most frequent cause, although other organisms may cause infections of this type. Recall from Chapter 21 that group A β-hemolytic streptococci have an M protein on their surface that increases virulence by improving adherence to the mucous membranes and by increasing resistance to phagocytosis.

Puerperal sepsis progresses from an infection of the uterus to an infection of the abdominal cavity (peritonitis) and in many cases to septicemia. At one Paris hospital between 1861 and 1864, of the 9886 women who gave birth, 1226 (12%) died of such infections. At that time, the death rate from puerperal sepsis outside such hospitals was frequently twice that. These deaths were largely unnecessary. Some 20 years before, Oliver Wendell Holmes in America and Ignaz Semmelweiss in Austria had clearly demonstrated that the disease was transmitted by the hands and instruments of the attending midwives or physicians and that disinfection of hands and instruments could prevent such transmission. Yet, in 1879, Louis Pasteur still found it necessary to lecture physicians about the cause of the disease.

Antibiotics, especially penicillin, and modern hygienic practices have now made *S. pyogenes* puerperal sepsis an uncommon complication of childbirth. Infections related to improperly performed abortions—not strictly puerperal sepsis—are often of mixed bacterial types, many of them caused by anaerobic bacteria of the *Bacteroides* or *Clostridium* genera. Penicillin is also effective in these cases.

BACTERIAL ENDOCARDITIS

The wall of the heart consists of three layers. The outer layer, the *pericardium,* is a sac that encloses the heart. The middle layer, the *myocardium,* is the thickest layer and consists of cardiac muscle tissue. The inner layer is a lining of epithelium and is called the *endocardium.* This layer lines the heart muscle itself and covers the valves in the heart. An inflammation of the endocardium is called **endocarditis.**

One type of bacterial endocarditis, **subacute bacterial endocarditis,** is characterized by fever, anemia, general weakness, and heart murmur. It is usually caused by α-hemolytic streptococci, although β-hemolytic streptococci or staphylococci can be involved. About 5% to 10% of the cases are caused by enterococci. The condition probably arises from a focus of infection elsewhere in the body, such as in the teeth or tonsils. Microorganisms are released by tooth extractions or tonsillectomies, enter the blood, and find their way to the heart. Normally, such bacteria would be quickly cleared from the blood by the body's defensive mechanisms. However, in persons whose heart valves are abnormal, because of either congenital heart defects or such diseases as rheumatic fever and syphilis, the bacteria lodge in the preexisting lesions. Within the lesions, the bacteria multiply and become entrapped in blood clots that protect them from phagocytes and antibodies. As multiplication progresses and the clot gets larger, pieces of the clot break off and can occlude blood vessels or lodge in the kidneys. In time, the function of the heart valves is impaired. Left untreated, subacute bacterial endocarditis is invariably fatal.

Another type of bacterial endocarditis is **acute bacterial endocarditis** (Figure 23.3), which is caused by *Staphylococcus aureus* and *Streptococcus pyogenes.* These organisms find their way from the initial site of infection to normal or abnormal heart valves; the rapid destruction of the heart valves is frequently fatal with days or weeks. Penicillin is sometimes used prophylactically to prevent endocarditis during procedures such as tooth extractions and tonsillectomies. Streptococci can also cause **pericarditis,** inflammation of the sac around the heart (the pericardium). Infections of the heart muscle tissue, or myocardium, are called **myocarditis** and can involve a widely varied array of microbes, including bacteria, fungi, protozoa, and helminths. In the United States, the most common cause is enteroviruses of the coxsackie B type. In Central and South America, *Trypanosoma cruzi,* the protozoan causing Chagas' disease (see p. 577), is the most prevalent agent.

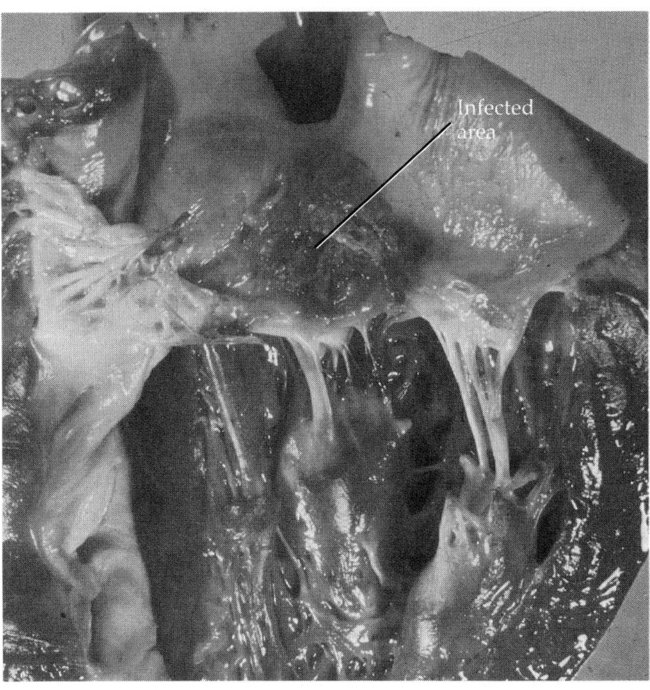

FIGURE 23.3 Bacterial endocarditis. This heart has been dissected to expose the bicuspid (mitral) valve. The cordlike structures connect the heart valve to the operating muscles. In a normal heart, the region at the upper part of the heart would all be smooth and light colored. The bacterial infection is responsible for the rough, reddened area; the inflammation resulted in a narrowing of the valve opening.

RHEUMATIC FEVER

Group A β-hemolytic streptococcal infections, such as those caused by *Streptococcus pyogenes,* sometimes lead to **rheumatic fever,** which is generally considered an autoimmune type of complication (see Chapter 19). The disease is usually expressed as an arthritis, especially in older persons. Another frequent form is an

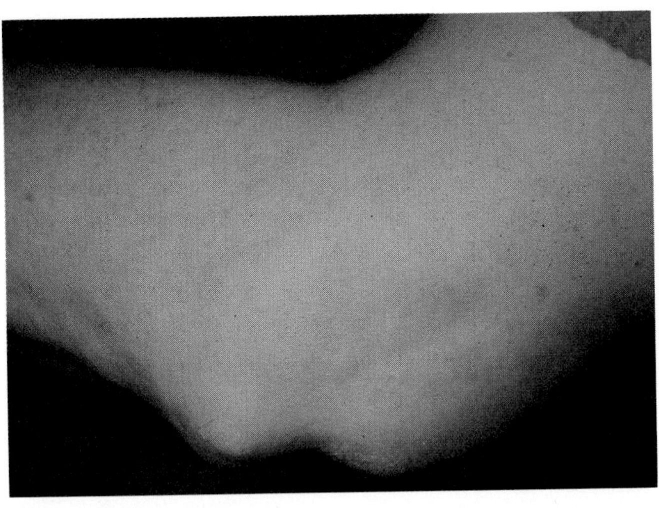

FIGURE 23.4 Rheumatic fever was named, in part, because of the characteristic subcutaneous nodules that appear at the joints, as shown here.

inflammation of the heart, which damages the valves. Subcutaneous nodules at joints often accompany this stage (Figure 23.4). Other symptoms include fever and malaise. Although rheumatic fever usually does no permanent damage to the joints, there can be permanent heart damage.

In addition, as many as 10% of people with rheumatic fever develop **Sydenham's chorea,** an unusual complication known in the Middle Ages as St. Vitus' dance. Several months following an episode of rheumatic fever, the patient (much more likely to be a girl), exhibits purposeless, involuntary movements during waking hours. Occasionally, sedation is required to prevent self-injury from flailing arms and legs. The condition disappears in a few months.

Rheumatic fever is usually precipitated by a streptococcal sore throat. One to five weeks later, probably after the original infection has disappeared, evidence of heart abnormalities appears. Reinfection with streptococci (for example, another sore throat) renews the attack and causes further damage to the heart. Perhaps as many as 3% of children with untreated β-hemolytic streptococcal infections contract rheumatic fever. However, many of these cases are essentially subclinical.

The exact mechanism by which streptococci produce rheumatic fever is still obscure, but some data point to an immunological reaction that is localized in the heart and joints. For example, infection with group A β-hemolytic streptococci may result in deposition of streptococcal antigens such as M proteins in the joints and heart. When the body produces antibodies against the antigens, the antigen–antibody reaction might cause the damage. It is also possible that a streptococcal antigen is cross-reactive with components of heart muscle and produces antibodies that subsequently damage the heart. Persons with symptoms of damage from rheumatic fever do have relatively high titers of antibody against streptococcal antigens. Genetic factors might also play a role in this disease. It has been demonstrated that persons of certain MHC types (see Chapter 19) are about 15 times more at risk than the normal population.

The number of cases of rheumatic fever has been declining in this country because of early treatment of streptococcal infections. In the 1940s, there were an estimated 250,000 cases annually; by the 1980s, the disease had become rare. Some find this difficult to explain when group A streptococci are so frequently found in the throats of schoolchildren. One possibility is that the streptococci may have mutated and lost some virulence factor. There has been concern, reinforced by recent clusters of cases, that another mutation might reestablish the virulence. The disease remains a major health problem in much of Asia, Africa, and South America.

Penicillin is administered to patients with rheumatic fever as a prophylactic. The penicillin will not alleviate the symptoms (which are treated with anti-inflammatory drugs), but it will prevent subsequent infections that could cause a recurrence of rheumatic fever.

TULAREMIA

Tularemia is a disease caused by a small, gram-negative, facultatively anaerobic, pleomorphic, rod-shaped bacterium, *Francisella tularensis.* The microorganism was named for Tulare County, California, where it was originally observed in ground squirrels in 1911.

Tularemia exhibits one of several forms, depending on whether the infection is acquired by inhalation, ingestion, bites, or, most commonly, contact through minor skin breaks. The first sign of infection is usually local inflammation and a small ulcer at the site of infection. About a week after infection, the regional lymph nodes enlarge; many will contain pockets filled with pus. If the disease is not contained by the lymphatic system, the microorganisms can produce septicemia, pneumonia, and abscesses throughout the body. Ingestion of infected, inadequately cooked meat leads to a focus of infection in the mouth or throat. The highest mortality occurs with pneumonic tularemia, which has even resulted from breathing aerosols created by handling dead but intact rabbits.

Humans most frequently acquire the infection through minor skin abrasions and by rubbing the eyes after handling small wild mammals. Probably 90% of the cases in this country are contracted from rabbits; it is estimated that 1% of all American rabbits are in-

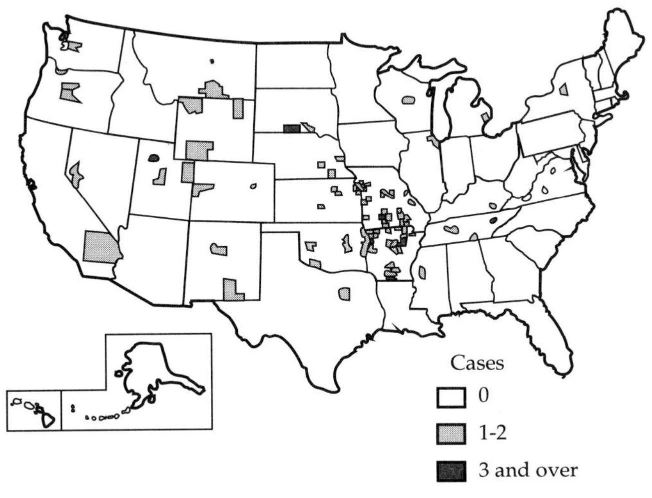

FIGURE 23.5 Tularemia distribution. The geographical distribution, by county, of 152 cases of tularemia reported in the United States during 1989. (Source: CDC. *Summary of Notifiable Diseases 1989, MMWR* 38:54.)

fected. The disease can also be spread by the bites of arthropods, such as deer flies (see Figure 12.31c), ticks, or rabbit lice, particularly in the western states; contact with small animals is a more likely cause in the central and eastern states (Figure 23.5).

F. tularensis is difficult to grow from serum samples, even on highly specialized media. Although no well-defined toxins have been identified to account for its virulence, *F. tularensis* survives for long periods within body cells, including phagocytes. In fact, resistance to phagocytosis might account for the low infective dose required. Naturally acquired immunity is usually permanent, but recurrences have been reported. An attenuated live vaccine is available for high-risk laboratory workers.

Serological diagnosis can be done with an agglutination test that becomes positive after about ten days of infection. A rising titer is diagnostic. The organism is considered so dangerous to handle (because of the ease with which aerosol infections are transmitted) that such diagnosis should not be attempted without special isolation hoods.

Streptomycin and gentamicin are the antibiotics of choice, but prolonged administration is necessary to prevent relapses. Before the days of antibiotics, mortality rates were as high as 15%. The intracellular location of the bacterium is a problem in chemotherapy.

BRUCELLOSIS (UNDULANT FEVER)

Like the bacteria that cause tularemia, the *Brucella* that cause **brucellosis** favor intracellular growth, move through the lymphatic system, and travel to organs via the bloodstream. These bacteria are very small, gram-negative, aerobic rods. They are difficult to grow on artificial media because they require that carbon dioxide be added to the atmosphere of the incubator, and they also require specially enriched media.

The most common *Brucella* species in the United States is *Brucella abortus* (ä-bôr'tus). It is found in cattle and is transmitted by unpasteurized milk products (Figure 23.6). The disease caused by this species is usually mild and often self-limited. The most common form in the rest of the world is *Brucella melitensis* (me-li-ten'sis), which has a reservoir in goats and camels and is also transmitted by unpasteurized milk. The disease it causes is severe and often results in disability or death. *Brucella suis* (sü'is) is transmitted by contact with swine. Infections by *B. suis* are distinguished by the formation of destructive abscesses.

Brucella tends to multiply in the uterus of a susceptible animal, where their growth is favored by the presence of a carbohydrate, *mesoerythritol*, that is produced in the fetus and the membranes surrounding it. When the animal is not pregnant, it excretes the bacteria in milk. About 200 cases of brucellosis are reported in the United States each year, but there are few if any deaths. Because of pasteurization, dairy products are involved in only about 10% of brucellosis cases today; most cases occur as a result of contact with

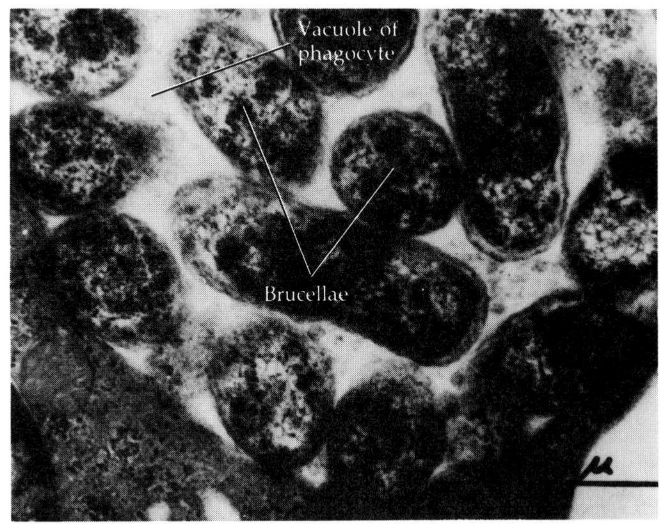

FIGURE 23.6 ***Brucella abortus.*** Phagocyte containing *B. abortus* in a vacuole. The organisms are able to survive and reproduce in phagocyte cells, where chemical and immunological defenses have difficulty reaching.

diseased animal tissue by farmers, veterinarians, and meat packers.

The organisms apparently enter the human body by passing through minute abrasions in the skin or mucous membrane of the mouth, throat, or gastrointestinal tract. *B. melitensis* and *B. suis* are the most easily transmitted, and some think that they may even be able to penetrate intact skin. Once in the body, the microorganisms are ingested by fixed macrophages, in which they multiply and travel via the lymphatic system to the lymph nodes. From there, the microorganisms can be transported to the liver, spleen, or bone marrow. The ability of the microorganisms to grow inside the phagocytes partly accounts for their virulence and resistance to antibiotic therapy and antibodies. If any immunity is achieved, it is not reliable. A vaccine is used for cattle but not for humans.

Brucellosis is highly variable in its signs and symptoms as different organisms infect different organs. Many cases are subclinical; others are acute with fever and chills. Chronic brucellosis may last more than a year and may have very low-grade symptoms. The fever, particularly with *B. melitensis,* typically spikes to about 40°C (104°F) each evening; that is why the disease is sometimes called *undulant fever.*

Because the symptoms are difficult to interpret, serological testing is important in the diagnosis of brucellosis. Several tests are in use, the most common being an agglutination test. An ELISA test is considered more specific and sensitive, however. A simple agglutination test used on serum or milk to identify brucellosis-positive cattle has been an important factor in establishing a number of brucellosis-free areas. Culture is difficult and results in many false-negative reports; it is most likely to be successful with *B. melitensis,* which tends to be present in blood samples.

Treatment must be prolonged because of the intracellular growth habit of the bacteria. Frequently, combinations of tetracycline plus streptomycin are used for several weeks.

ANTHRAX

In 1877, Robert Koch isolated *Bacillus anthracis,* the bacterium that causes **anthrax** in animals. The endospore-forming bacillus is a large, aerobic, gram-positive microorganism that is apparently able to grow slowly in soil types having specific moisture conditions. The endospores have survived in soil tests for up to 60 years. The disease strikes primarily grazing animals, such as cattle and sheep. The *B. anthracis* endospores are ingested along with grasses, causing a fulminating, fatal septicemia.

There is often less than one case of human anthrax on average in the United States each year. Persons at risk are those who handle animals, hides, wool, and other animal products from certain foreign countries. Goat hair and handicrafts containing animal hides from the Middle East have been a repeated source of infection.

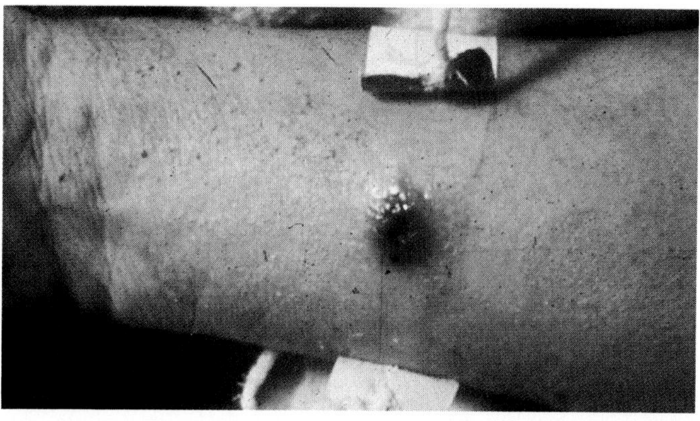

FIGURE 23.7 A cutaneous pustule of anthrax on a human arm. If this localized infection is not contained by the body's defenses, septicemia might be the next stage.

If contact is made with materials containing anthrax endospores, the organism can enter the skin through a cut or abrasion and cause a pustule there (Figure 23.7). This pustular infection is sometimes kept localized by the defenses of the body, but there is always danger of septicemia. Probably the most dangerous form of anthrax is pulmonary anthrax, contracted when the endospores are inhaled. *Woolsorters disease,* a dangerous form of pneumonia, results. It begins abruptly with high fever, difficulty in breathing, and chest pain. This disease eventually results in septicemia, and the mortality rate is high.

In humans, anthrax is best diagnosed by isolation of the bacterium. A number of morphological and biological tests can be used for precise identification. Penicillin is the drug of choice in human treatment. However, once septicemia is well advanced, antibiotic therapy may prove useless, probably because the exotoxins remain despite the death of the bacteria.

GANGRENE

If a wound causes the blood supply to be interrupted (a condition known as *ischemia*), the wound becomes anaerobic. Ischemia leads to *necrosis,* or death of the tissue. The death of soft tissue from loss of blood supply is called **gangrene** (Figure 23.8). These conditions can also occur as a complication of diabetes.

Substances released from dying and dead cells provide nutrients for many bacteria. Various species of the genus *Clostridium,* which are gram-positive, endo-

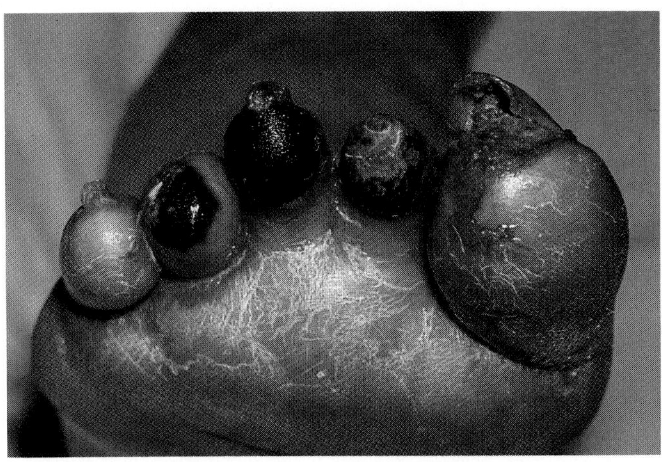

FIGURE 23.8 Toes of a patient with gangrene. This disease is caused by *Clostridium perfringens* and other clostridia. The black, necrotic tissue on the toes furnishes anaerobic growth conditions for the bacteria, which then progressively destroy adjoining tissue.

spore-forming anaerobes widely found in soil and in the intestinal tracts of humans and domesticated animals, grow readily in such conditions. *Clostridium perfringens* is the species most commonly involved in gangrene, but other clostridia and several other bacteria can also grow in such wounds.

Once ischemia and the subsequent necrosis have developed, **gas gangrene** can develop, especially in muscle tissue. As the *C. perfringens* microorganisms grow, they ferment carbohydrates in the tissue and produce gases (carbon dioxide and hydrogen) that swell the tissue. The bacteria produce toxins that move along muscle bundles, killing cells and producing necrotic tissue that is favorable for further growth. Eventually, these toxins and bacteria enter the bloodstream and cause systemic illness. Enzymes produced by the bacteria degrade collagen and proteinaceous tissue, facilitating spread of the disease. Without treatment, the condition is uniformly fatal.

One complication of improperly performed abortions is the invasion of the uterine wall by *C. perfringens,* which resides in the genital tract of about 5% of all women. This infection can lead to gas gangrene and result in a life-threatening invasion of the bloodstream.

The surgical removal of necrotic tissue is called debridement; it and amputation are the most common medical treatments for gas gangrene. When gas gangrene occurs in such regions as the abdominal cavity, the patient can be treated in a *hyperbaric chamber,* which contains a pressurized oxygen-rich atmosphere (Figure 23.9). The oxygen saturates the infected tissues and thereby prevents the growth of the obligately anaerobic clostridia. Prompt cleaning of serious wounds and precautionary antibiotic treatment are the most effective steps in the prevention of gas gangrene. Penicillin is effective against *C. perfringens.*

SYSTEMIC DISEASES CAUSED BY BITES AND SCRATCHES

Animal bites can result in serious infections. Most serious bites are by domestic animals, such as dogs and cats, because they live in close contact with humans. It has been reported that 69% of dog bites become infected unless effectively cleansed. Domestic animals often harbor *Pasteurella multocida* (pas-tyėr-el'lä multō'si-dä), a nonmotile, gram-negative rod similar to the *Yersinia* bacterium that causes plague. *P. multocida* is primarily a pathogen of animals; it causes septicemia, hence the name *multocida,* meaning many killing. Humans infected with *P. multocida* may show varied responses. For example, local infections with severe swelling and pain can develop at the site of the wound. Development of forms of pneumonia and septicemia is possible and life-threatening. Penicillin and tetracycline are usually effective in treatment of these infections.

Clostridium species and other anaerobes, such as species of *Bacteroides* and *Fusobacterium,* can also infect deep animal bites.

Rat Bite Fever

Rat bites are typically infected by organisms of the genus *Streptobacillus* (strep-tō-bä-sil'lus) or *Spirillum.* These infections occur more commonly in Asiatic

FIGURE 23.9 A multiplace hyperbaric chamber that can accommodate as many as 12 patients. Such chambers are usually available at major medical centers.

countries than in the United States. The disease, often called **rat bite fever,** is characterized by recurring fever, arthritislike symptoms (inflammation, pain, stiffness), and infections of the lymphatic vessels.

Bites by Humans

In one typical recent year, there were 210 rat bites of humans reported in New York City—and 760 cases of human bites of humans! Bites by humans are even more prone to infection and complications such as septicemia and tissue destruction than are animal bites. Bites can be self-inflicted, such as a bitten lip or one's clenched fist striking another's unoffending tooth. The infections are often mixed; most frequently isolated are *Staphylococcus aureus,* hemolytic streptococci, and anaerobes such as *Bacteroides* and *Fusobacterium.* The wounds should be thoroughly washed and debrided and, to further avoid anaerobic conditions, perhaps not closed for a few days. Penicillin is the usual initial choice for antibiotic treatment.

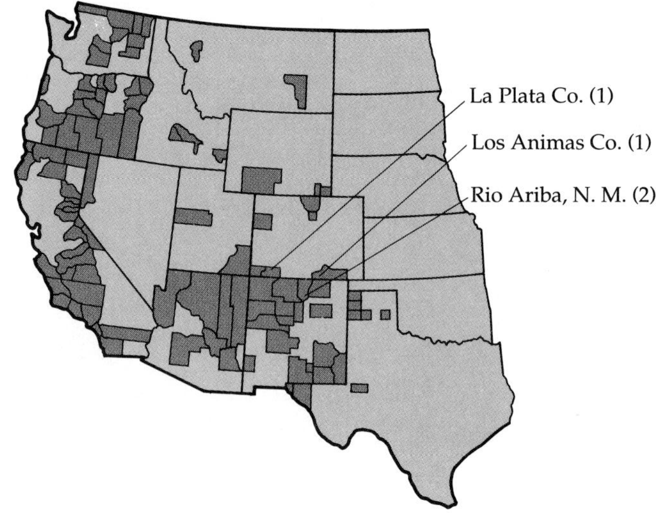

FIGURE 23.10 Plague distribution. Geographical distribution of human and animal plague in the United States in 1989. Counties reporting plague in animals are shown in brown. The human cases are indicated in parentheses following the county names at top right.

PLAGUE

Few diseases have affected human history more dramatically than **plague,** known in the Middle Ages as the Black Death. This term comes from one of its characteristics, the dark blue areas of skin caused by hemorrhages. In the fourteenth century, this disease destroyed perhaps one-fourth of the total population of Europe. Near the turn of this century, as many as 10 million people are reported to have died from plague in India in a 20-year period. The number of plague cases reported in the United States has been increasing in recent years, and there are now about 20 cases annually (Figure 23.10). A mortality rate of 25% is not unusual.

The disease is caused by a gram-negative, rod-shaped bacterium, *Yersinia pestis.* Normally a disease of rats, plague is transmitted from one rat to another by the rat flea, *Xenopsylla cheopis* (ze-nop-sil'lä kē-ō'pis) (see Figure 12.31b). European rats, introduced into the United States many years ago, are the primary reservoirs of plague. In the United States, particularly in the far West and Southwest, the disease is endemic in wild rodents, especially ground squirrels, prairie dogs, and chipmunks.

If its host dies, the rat flea seeks a replacement, which may be another rodent or a human. A plague-infected flea is hungry for a meal because the growth of the bacteria blocks the flea's digestive tract, and the blood the flea ingests is quickly regurgitated. An arthropod vector is not always necessary for plague transmission. Contact from the skinning of infected animals, scratches of domestic cats, and similar incidents have been reported to cause infection.

In parts of the world where human proximity to rats is common, infection from this source still prevails. From the flea bite, bacteria enter the human's bloodstream and proliferate in the lymph and blood. One factor in the virulence of the plague bacterium is its ability to survive and proliferate inside phagocytic cells rather than being destroyed by them. An increased number of highly virulent organisms eventually emerges, and an overwhelming infection results. The lymph nodes in the groin and armpit become enlarged, and fever develops as the body's defenses react to the infection. Such swellings are called *buboes,* which accounts for the name **bubonic plague** (Figure 23.11). The mortality rate of untreated bubonic plague ranges from 50% to 75%. Death from bubonic plague, if it occurs, is usually within less than a week after the appearance of symptoms.

A particularly dangerous condition arises when the bacteria are carried by the blood to the lungs and a form of the disease called **pneumonic plague** results. The mortality rate for this type of plague is nearly 100%. Even today, this disease can rarely be controlled if it is not recognized within 12 to 15 hours of the onset of fever. Persons with pneumonic plague usually die within three days. Like influenza, pneumonic plague is easily spread by airborne droplets from human or animal cases. Great care must be taken to prevent airborne infection of persons in contact with patients.

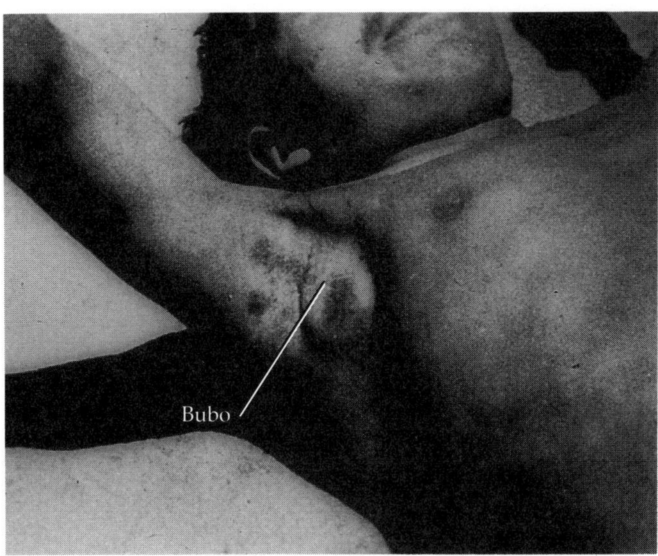

FIGURE 23.11 Bubonic plague. This photograph shows a bubo (swollen lymph node) under the arm of the patient. Swollen lymph nodes are a common indication of systemic infection.

Contacts of pneumonic plague cases should receive prophylactic antibiotics.

Diagnosis of plague is most commonly done by isolating the bacterium and then identifying it by fluorescent-antibody and phage tests. Persons exposed to infection can be given prophylactic antibiotic protection. A number of antibiotics, including streptomycin and tetracycline, are effective. Recovery from the disease gives a reliable immunity. No vaccines are available, except for persons likely to come into contact with infected fleas during field operations or for laboratory personnel exposed to the organism. The long-term effectiveness of the plague vaccine is not high.

Control of rat-based plague largely involves modern sanitation procedures. Sanitary garbage disposal eliminates a source of food for rats, and building maintenance denies rats access to hiding places in attics and walls.

Relapsing Fever

Except for the species that causes Lyme disease (discussed in the next section), all members of the spirochete genus *Borrelia* cause **relapsing fever.** In the United States, the disease is transmitted by soft ticks that feed on rodents. The incidence of relapsing fever increases during the summer months, when the activity of rodents and arthropods increases. The disease is characterized by a fever, sometimes in excess of 40.5°C (105°F), jaundice, and rose-colored spots. After 3 to 5 days, the fever subsides. Three or four relapses may occur, each shorter and less severe than the initial fever. Each recurrence is caused by a different antigenic type of the spirochete. Diagnosis is made by observation of bacteria in the patient's blood, unusual for a spirochete disease.

Lyme Disease (Lyme Borreliosis)

In 1975, a cluster of disease cases in young people that was first diagnosed as rheumatoid arthritis was reported near the city of Lyme, Connecticut. The seasonal occurrence (summer months), lack of contagiousness among family members, and descriptions of an unusual skin rash that appeared several weeks before the first symptoms suggested a tick-borne disease. The fact that penicillin alleviated the progression of symptoms suggested a bacterial pathogen. In 1983, a spirochete that was later named *Borrelia burgdorferi* was identified as the cause. **Lyme disease** may now be the most common tick-borne disease in this country. It is also found in Europe and Australia. In the United States, the disease is most prevalent on the Atlantic Coast, as shown on the map (Figure 23.12a). Deer and field mice are the most important animal reservoirs that maintain the spirochete in the tick population.

The tick feeds three times during its life cycle. The first and second feedings, as a larva and then as a nymph, are usually on a field mouse. The third feeding, as an adult, is usually on a deer. These feedings are one year apart, and the ability of the spirochete to remain viable in the disease-tolerant field mice is critical in maintaining the disease in the wild. Other animals also serve as reservoirs for the pathogen. None of these wild animals shows symptoms of being affected by the disease, although dogs and horses sometimes do.

On the Pacific Coast, the tick is the western black-legged tick *Ixodes pacificus* (iks-ō'dēs pas-i'fi-kus) (Figure 23.12b). In the rest of the country, it is mostly *Ixodes dammini* (dam'mi-nē). This latter tick is so small that it is often missed on the body (Figures 12.30a, 23.13b). On the Atlantic Coast, almost all *Ixodes* ticks carry the spirochete, but on the Pacific Coast few are infected. This is because that tick feeds on lizards that do not carry the spirochete effectively.

The first symptom of Lyme disease is usually a rash that appears at the bite site. It is a red area that clears in the center as it expands to a final diameter of about 15 cm (Figure 23.12c). This distinctive rash is seen in about 75% of the cases. Flulike symptoms appear in a couple of weeks as the rash fades. Antibiotics given during this interval are very effective in limiting the disease. During a second phase, when it occurs, there is often evidence that the heart is affected. Heartbeats may become so irregular that a pacemaker is re-

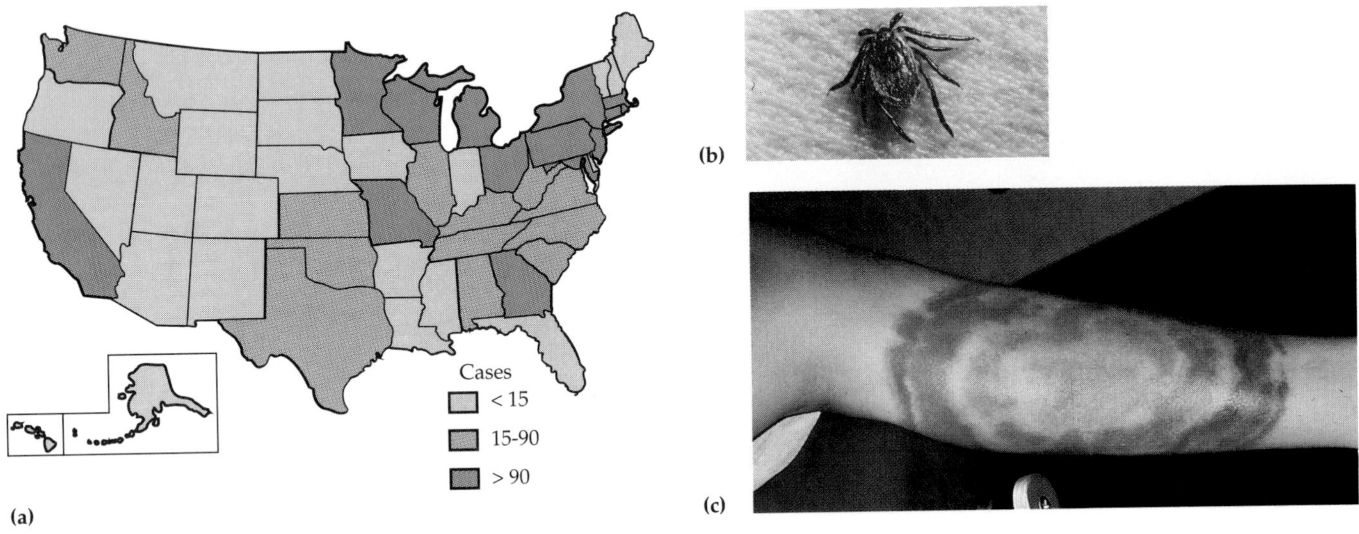

FIGURE 23.12 Lyme disease. **(a)** Map showing incidence of Lyme disease in the United States in 1989. **(b)** A tick (*Ixodes pacificus*). Ticks of this species may carry the spirochete *Borrelia burgdorferi,* which causes Lyme disease. **(c)** The classic "bull's-eye" rash of Lyme disease.

quired. Neurological symptoms such as facial paralysis, meningitis, and encephalitis may be seen. Months or years later, patients may develop arthritis that may affect them for years. The knees may be so damaged that walking is difficult. Immune responses to the presence of the bacteria are probably the cause of this joint damage. Many of the symptoms of long-term Lyme disease resemble those of the later stages of syphilis, also caused by a spirochete.

A number of antibiotics are effective in treatment of the disease, although in the later stages large amounts may need to be administered intravenously. Diagnosis of Lyme disease depends partly on the symptoms and an index of suspicion based on the prevalence of the carrier in the geographical area. For the diagnosis of Lyme disease, an almost bewildering array of serological tests have been developed, but at this time it is unclear which give the best results. False-negative and false-positive results are quite common. Physicians are cautioned that serological tests must be interpreted in conjunction with clinical symptoms and the likelihood of exposure to infection. Another disease transmitted by *Ixodes* ticks is babesiosis (see the box on p. 578).

TYPHUS

The various typhus diseases are caused by rickettsias (bacteria that are obligate intracellular parasites of eucaryotes). Rickettsias are spread by arthropod vectors. Rickettsias infect mostly the endothelial cells of the vascular system and multiply within them. The resulting inflammation causes local blockage of the small blood vessels. There are several related rickettsial diseases that differ mainly in their severity and in their arthropod vectors, including epidemic typhus, endemic murine typhus, and the spotted fevers (tick-borne typhus). Q fever, which is discussed in the next chapter (on respiratory diseases), is an atypical rickettsial disease caused by an atypical rickettsia.

Epidemic Typhus

Epidemic typhus (louse-borne typhus) is caused by *Rickettsia prowazekii* and carried by the human body louse *Pediculus humanus corporis* (ped-ik'ū-lus hü'ma-nus côr'pô-ris) (see Figure 12.31a). The pathogen grows in the gastrointestinal tract of the louse and is excreted by it. The pathogen is not transmitted directly by the bite of an infected louse but rather is transmitted when the feces of the louse are rubbed into the wound when the bitten host scratches the bite. The disease can flourish only in crowded and unsanitary conditions, when lice can transfer readily from an infected host to a new host. In recent years, a number of cases of a mild form of epidemic typhus have been traced to contact with a reservoir in flying squirrels in the eastern United States. There have been no deaths. Although the vector in these cases is unknown, it is probably something other than a louse.

Epidemic typhus disease produces a high and prolonged fever for two or more weeks. Stupor and a rash of small red spots caused by subcutaneous hemorrhag-

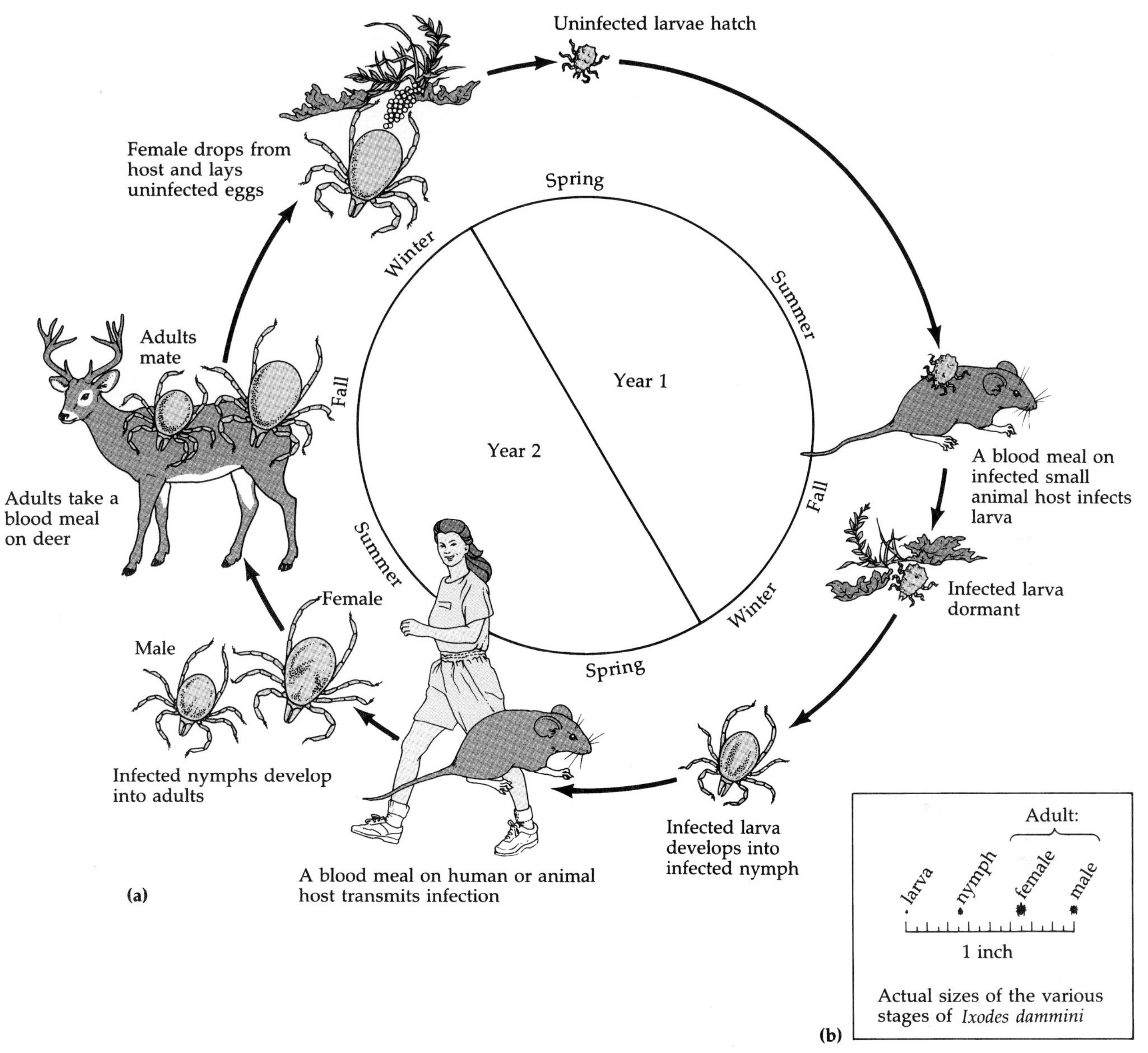

FIGURE 23.13 Life cycle of Lyme disease. (a) The tick vector of Lyme disease has a two-year life cycle in which it requires three blood meals. The larval form becomes infected from small animals, usually field mice, then enters a dormant stage until the following spring. Then it molts into a nymph, which is still infected. This is the stage at which it is most likely to infect humans. It is crucial to the maintenance of the life cycle in the wild that the spirochete be able to remain viable in small animals to reinfect the larvae the following year. The field mouse is well adapted to this. The third feeding, by the infected adults that develop from the nymph, is taken from deer. After this, the adults lay eggs, which are uninfected, that overwinter to develop into larvae the following spring. **(b)** The scale shows the small sizes of the various stages of the ticks' life cycle, which makes detection on the body so difficult.

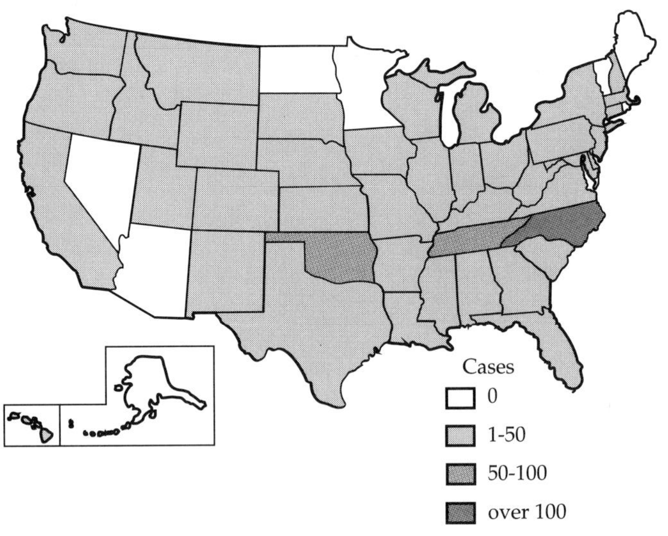
Cases
0
1-50
50-100
over 100

FIGURE 23.14 Distribution of Rocky Mountain spotted fever (tick-borne typhus). The map shows the geographical distribution of the 654 cases reported in the United States in 1990. [(Source: *MMWR* 39:52 (1/4/91).]

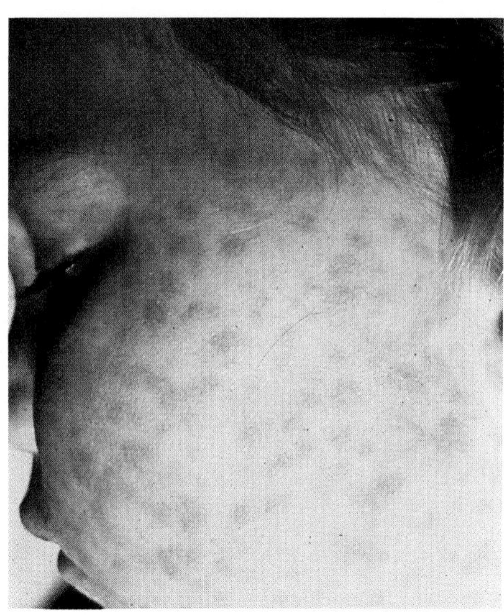

FIGURE 23.16 Rocky Mountain spotted fever. The rash is often mistaken for measles.

FIGURE 23.15 The cycle of Rocky Mountain spotted fever. Mammals are not essential to maintenance of the disease in the tick population; the pathogen is passed by transovarian passage, so new ticks are infected at birth. A blood meal is required to advance to successive stages.

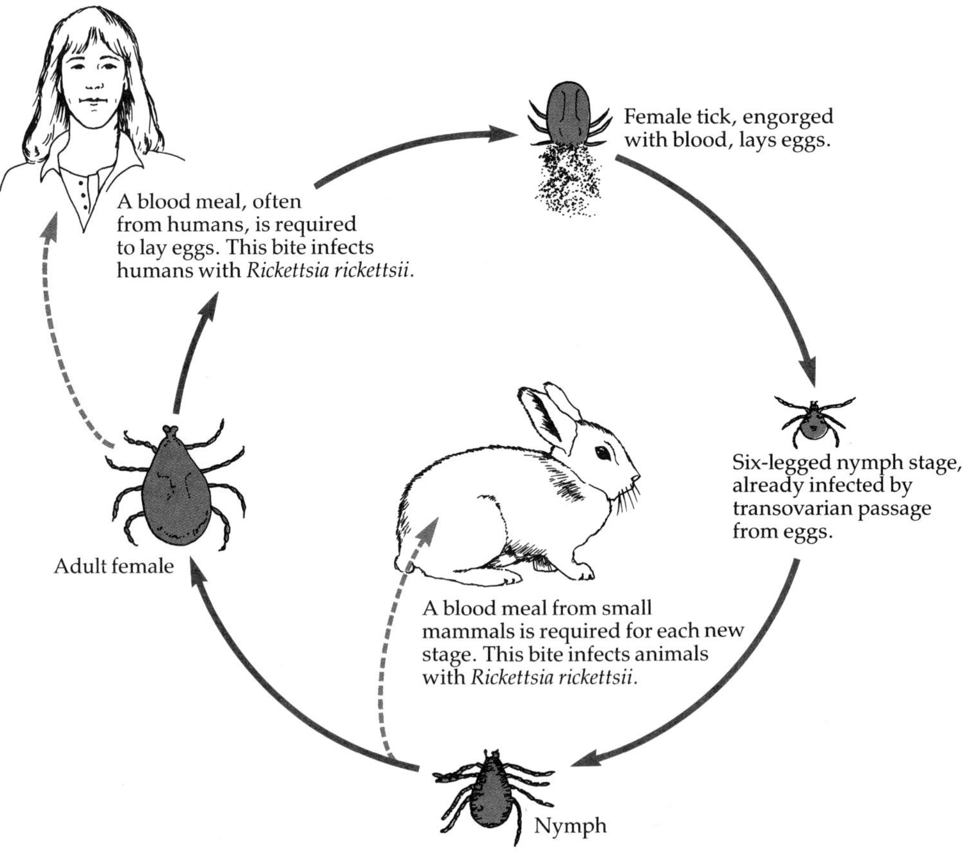
Female tick, engorged with blood, lays eggs.
A blood meal, often from humans, is required to lay eggs. This bite infects humans with *Rickettsia rickettsii*.
Six-legged nymph stage, already infected by transovarian passage from eggs.
Adult female
A blood meal from small mammals is required for each new stage. This bite infects animals with *Rickettsia rickettsii*.
Nymph

ing are characteristic as the rickettsias invade blood vessel linings. Mortality rates are very high when the disease is untreated.

Tetracycline and chloramphenicol are usually effective against epidemic typhus, but the elimination of conditions in which the disease can flourish is more important in its control. The most sensitive diagnostic test is an indirect fluorescent-antibody test, but a passive agglutination test using latex beads is also in use. Vaccines are available for military populations, which historically have been highly susceptible to the disease. Recovery from the disease gives a solid immunity and also renders a person immune to the related endemic murine typhus.

Endemic Murine Typhus

Endemic murine typhus occurs sporadically rather than in epidemics. The term *murine* (derived from Latin for mouse) refers to the fact that rodents, such as rats and squirrels, are the common hosts for this typhus. Endemic murine typhus is transmitted by the rat flea *Xenopsylla cheopis* (see Figure 12.31b), and the pathogen responsible for the disease is *Rickettsia typhi,* a common inhabitant of rats. Texas has had a number of outbreaks of murine typhus in recent years, often associated with campaigns to eliminate rodents, which caused the rat fleas to seek a new host. With a mortality rate of less than 5%, the disease is considerably less severe than the epidemic form. Except for the reduced severity of the disease, endemic murine typhus is clinically indistinguishable from epidemic typhus.

Tetracycline and chloramphenicol are effective treatments for endemic murine typhus, and the diagnostic procedure for the disease is the same as for epidemic typhus. Rat control is the best preventive measure for the disease.

Spotted Fever

Tick-borne typhus, or **Rocky Mountain spotted fever,** is probably the best-known rickettsial disease in the United States. It is caused by *Rickettsia rickettsii*. Despite its name (it was first recognized in the Rocky Mountain area), it is most common in the southeastern United States and Appalachia (Figure 23.14). This rickettsia is a parasite of ticks and is usually passed from one generation of ticks to another through their eggs, a mechanism called *transovarian passage* (Figure 23.15). Surveys show that in endemic areas perhaps 1 out of every 1000 ticks is infected. The rickettsias do not kill the tick. In different parts of the country, different ticks are involved—in the West, the wood tick *Dermacentor andersoni* (dėr-mä-sen'tôr an-der-sōn'ē), in the East the dog tick *Dermacentor variabilis* (vār-ē-a'bil-is).

About a week after the tick bites, a rash develops that is sometimes mistaken for measles (Figure 23.16); however, it often appears on palms and soles, which does not occur with viral rashes. The rash is evidence of the vascular damage, resulting in leakage of blood into surrounding tissues, that is characteristic of many rickettsial infections. The rash is accompanied by fever and headache. Death, which occurs in about 3% of the approximately 1000 cases reported each year, is usually caused by kidney and heart failure. Antibiotics such as tetracycline and chloramphenicol are very effective if administered early enough. No vaccine is available.

Serological tests are available but do not become positive until late in the illness. Therapy should be started before confirmation and then be based on clinical evidence. A misdiagnosis can be crucial—if treatment is not prompt and correct, the mortality rate is about 20%. It should be mentioned that there are a number of other tick-borne rickettsial diseases known by local names in various regions of the world.

Viral Diseases of the Cardiovascular System

BURKITT'S LYMPHOMA

In the 1950s, Denis Burkitt, a physician working in eastern Africa, noticed the frequent occurrence in young children of a fast-growing tumor of the jaw (Figure 23.17). Now well known as **Burkitt's lymphoma,** this is the commonest childhood cancer in Africa. The disease has a limited geographical distribution similar to that of malaria in central Africa. Burkitt suspected a viral cause of the tumor and a mosquito vector. At that time there was no known virus that caused human cancer, although several viruses were clearly associated with animal cancers. Intrigued by this possibility, the British virologist Tony Epstein and his student Yvonne Barr made biopsies of the tumors. A virus was cultured from this material, and the electron microscope showed a herpeslike virus in the culture cells—the Epstein–Barr (EB) virus. When used to infect certain monkeys, the EB virus causes a malignant tumor within a few weeks. Research eventually showed, however, that mosquitoes do not transmit the virus or the disease. Instead, mosquito-borne malarial infections apparently foster the development of Burkitt's

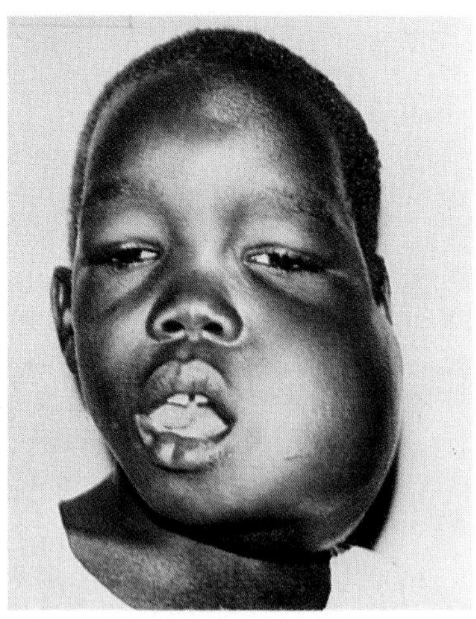

FIGURE 23.17 Burkitt's lymphoma. Cancerous tumors of the jaw caused by Epstein-Barr virus are seen mainly in children. This child was successfully treated.

lymphoma by impairing the immune response to the EB virus, which is almost universally present in human adults. In areas without endemic malaria, such as the United States, Burkitt's lymphoma is rare. The appearance of the lymphoma in AIDS patients is an indication of the importance of immune surveillance in preventing expression of the disease. A much more important medical problem that was later associated with the EB virus is **nasopharyngeal carcinoma.** This disease is a major cause of death in Southeast Asia, where it has an incidence of 10 to 20 cases per 100,000 population.

INFECTIOUS MONONUCLEOSIS

The EB virus was the subject of one of the accidental discoveries that often advance science. A technician in a laboratory investigating the EB virus was being used as the negative control in serological tests for the EB virus. While on vacation she contracted an infection, characterized by fever, sore throat, swollen lymph nodes in the neck, and general weakness—the symptoms familiar to many young people as **infectious mononucleosis,** or mono. The most interesting facet of the technician's disease was that she became serologically positive for the EB virus. It was soon confirmed that the same virus that causes Burkitt's lymphoma also causes infectious mononucleosis.

In less developed parts of the world, infection with the EB virus is acquired in early childhood and 90% of the children over four years of age have acquired antibodies. Childhood infections by the EB virus are usually asymptomatic, but if infection is delayed until young adulthood, as is often the case in the United States, the result is infectious mononucleosis. The peak incidence of the disease in this country occurs at about 15 to 25 years of age. College populations, particularly those from the upper socioeconomic strata, have a high incidence of the disease. Most college students have no immunity, and about 15% of these students can expect to contract the disease. The disease is generally self-limiting and seldom fatal. A principal cause of the rare deaths is rupture of the enlarged spleen (a common response to a systemic infection) during vigorous activity. Recovery is usually complete in a few weeks, and immunity is permanent.

The usual route of infection is by transfer of saliva by kissing or, for example, by sharing drinking vessels. Reproduction of the virus appears to occur mainly in the parotid glands, which accounts for its presence in saliva. It does not spread among casual household contacts, so aerosol transmission is unlikely. The incubation period before appearance of symptoms is from four to seven weeks.

Although reproduction is considered to occur in epithelial cells of the parotid glands, the infection is almost exclusively selective for B cells. The virus does not reproduce significantly in B cells, but the infection causes B cells to transform into plasma cells, which reproduce rapidly. These are then attacked by cytotoxic T cells of the cell-mediated immune system. The symptoms of infectious mononucleosis are associated with B-cell proliferation and the immune response. After recovery, the virus remains latent in a small number of B cells.

The disease name, *mononucleosis,* refers to lymphocytes with unusual lobed nuclei that proliferate in the blood during the acute infection. The infected B cells produce nonspecific antibodies called heterophil antibodies. Older diagnostic tests depended heavily on blood counts of the distinctive lymphocytes and on detection of the heterophil antibodies. A fluorescent-antibody test that detects IgM antibodies against the EB virus is the most specific diagnostic method.

The EB virus has been identified as one cause of **chronic fatigue syndrome (CFS).** People with this condition often have difficulty performing routine tasks and become so debilitated that they are unable to work. Symptoms include profound fatigue, fever, muscle weakness, and headaches. The disease affects females more than males, and there is no evidence of person-to-person transmission. Laboratory diagnosis of CFS caused by EB virus is based on the presence of antibodies against the virus. It is unlikely that EB virus is the only cause of this condition. Other herpes viruses that infect B cells might be responsible, and depression is thought to be a predisposing factor.

YELLOW FEVER

Arboviruses, or arthropod-borne viruses, can reproduce either in arthropods, such as mosquitoes, or in humans. **Yellow fever** is caused by an arbovirus *(yellow fever virus)* and is historically important because it was the first such virus discovered and provided the first confirmation that an insect could transmit a virus.

The yellow fever virus is injected into the skin by the mosquito. The virus then spreads to local lymph nodes, where it multiplies; from the lymph nodes, it advances to the liver, spleen, kidney, and heart, where it can persist for days. In the early stages of the disease, the person experiences fever, chills, headache, and backache, followed by nausea and vomiting. This stage is followed by jaundice, a yellowing of the skin that gave the disease its name. The jaundice is caused by the deposition of bile pigments in the skin and mucous membranes as a result of liver damage. In severe cases, the virus produces lesions in the infected organs, and hemorrhaging occurs.

Yellow fever is still endemic in many tropical areas, such as Central America, tropical South America, and Africa. Monkeys are a natural reservoir for the virus. Control of the disease depends on local control of *Aedes aegypti* (ä-e'dēz ē-jip'tē), the usual mosquito vector. Immunization of the exposed population is also an effective control.

There is no specific treatment for yellow fever. Diagnosis is usually by clinical signs, but it can be confirmed by a rise in antibody titer or isolation of the virus from the blood. The vaccine in use is an attenuated live viral strain and yields a very effective immunity with few side-effects.

DENGUE

Dengue (den'ghē) is a similar but milder viral disease also borne by the *Aedes aegypti* mosquito. This disease, endemic in the Caribbean and other tropical environments around the world, is caused by *dengue fever virus,* an arbovirus. It is characterized by fever, severe muscle and joint pain, and rash. Except for the painful symptoms, which have led to the name **breakbone fever,** classic dengue fever is a relatively mild disease and is rarely fatal.

The countries surrounding the Caribbean are reporting an increasing number of cases of dengue. In most years, more than 100 cases are imported into the United States, mostly by travelers from the Caribbean and South America. The mosquito vector for dengue is common in the Gulf states, and there is some concern that the virus will sooner or later be introduced into this region and become endemic. A special concern of health officials is the introduction into the United States of an Asian mosquito, *Aedes albopictus* (al-bō-pik'tus), that is an efficient carrier of the virus and an aggressive biter. It transmits the virus by transovarian passage and from person to person. The range of this mosquito can potentially cover much of the country (see the box in Chapter 12, p. 324).

A more serious form of dengue, **dengue hemorrhagic fever,** is characterized by bleeding from the skin, gums, and gastrointestinal tract and sometimes by circulatory failure and shock. Diagnosis is usually based on symptoms, and there are also several serological tests to detect either the virus or its antibodies. Control measures are directed at eliminating the *Aedes* mosquitoes. No specific treatment is effective against the virus.

Protozoan Diseases of the Cardiovascular System

TOXOPLASMOSIS

Toxoplasmosis, a disease of blood and lymphatic vessels, is caused by the protozoan *Toxoplasma gondii,* a small, crescent-shaped organism. *T. gondii* is a sporozoan, like the malarial parasite.

Cats are an essential part of the life cycle of *T. gondii* (Figure 23.18). Random tests on urban cats have shown that a large number of them are infected with the organism, which causes no apparent illness in the cat. The organism undergoes its only sexual phase in the intestinal tract of the cat. Oocysts are then shed in the cat's feces and contaminate food or water that can be ingested by other animals. The *oocysts* contain *sporozoites* that invade host cells and form trophozites called *tachyzoites* (about the size of large bacteria, 2×7 μm). The intracellular parasite reproduces rapidly (*tachys* means speed), and the increased numbers cause the rupture of the host cell, with release of more tachyzoites. As the immune system becomes increasingly effective, the disease enters a chronic phase in which the infected host cell develops a wall to form a *tissue cyst.* The numerous parasites within such a cyst reproduce very slowly, if at all, and persist for years, especially in the brain. Loss of immune function, AIDS being the best example, allows a reactivation of the infection from such cysts. Humans generally acquire the infection by ingestion of undercooked meats containing tachyzoites or tissue cysts, although there is a possibility of contracting the disease more directly by contact with cat feces.

The disease is a rather undefined mild illness, and the primary danger is in the congenital infection of a fetus. The effects on the fetus can be drastic, including convulsions, severe brain damage, blindness, and death. The mother is probably unaware of the disease,

FIGURE 23.18 Life cycle of *Toxoplasma gondii*. The domestic cat is the definitive host, in which the protozoan reproduces sexually. Oocysts are produced in the cat's intestinal tract and passed in its feces. Each oocyst contains four sporozoites, which are ingested by an intermediate host. There, in blood and other tissues, the sporozoites mature into tachyzoites. Eating the undercooked meat of the intermediate host or contact with cat feces causes infection in humans.

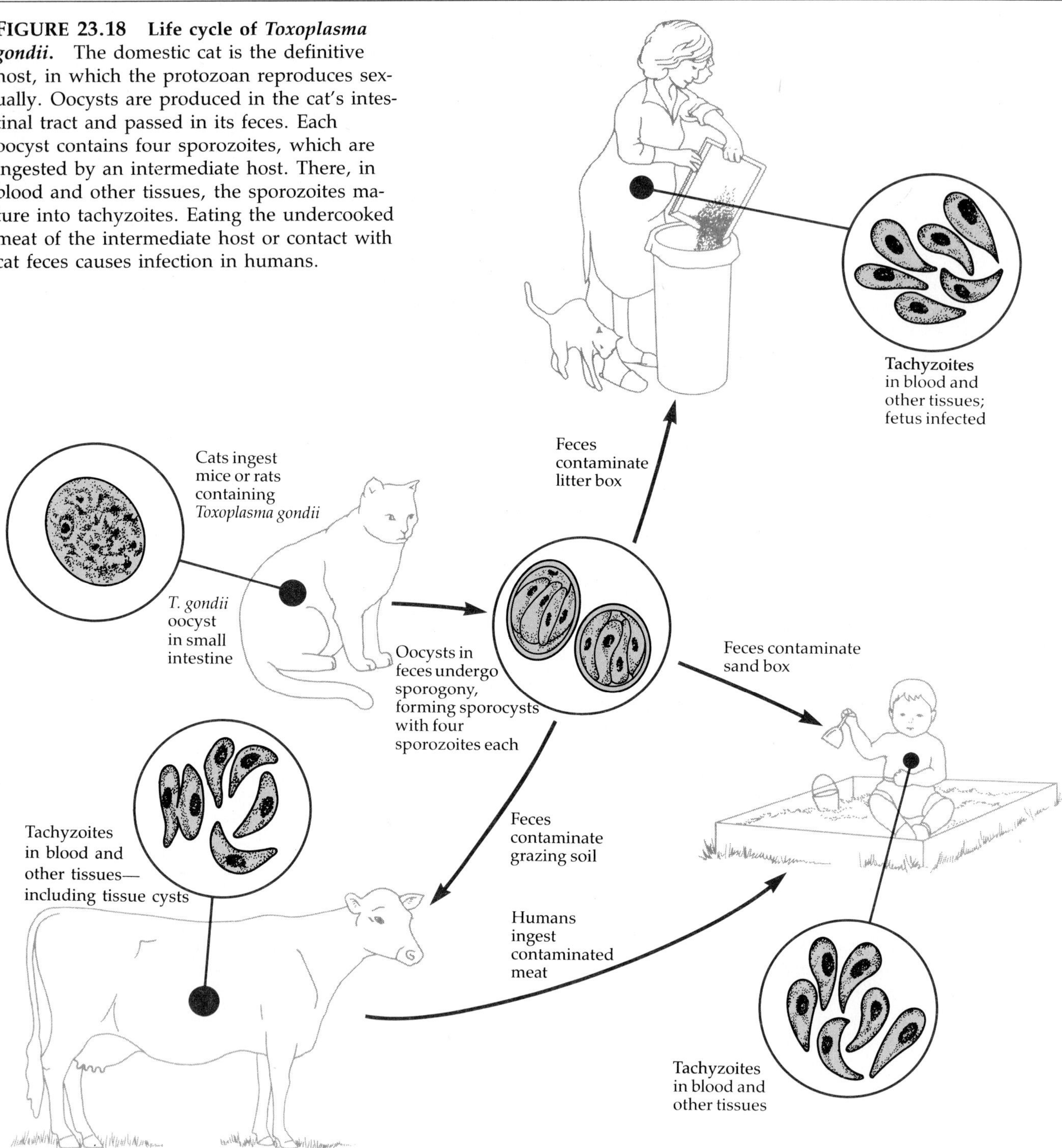

which is transmitted across the placenta. Some surveys have shown that approximately 30% of the population carries antibodies for this organism; this percentage indicates the high rate of subclinical, unrecognized infections.

T. gondii can be isolated and grown in cell cultures. The preferred serological tests use an indirect fluorescent-antibody method or an ELISA test. Interpretation of tests to justify extreme action such as a therapeutic abortion is difficult. Toxoplasmosis is treated with pyrimethamine in combination with either trisulfapyrimidines or sulfadiazine.

AMERICAN TRYPANOSOMIASIS (CHAGAS' DISEASE)

American trypanosomiasis (Chagas' disease) is a protozoan disease of the cardiovascular system. The causative agent is *Trypanosoma cruzi* (tri-pa-nō-sō'mä kruz'ē), a flagellated protozoan (Figure 23.19). The disease occurs in the extreme southern United States, Mexico, Central America, and into South America (Figure 23.20). Only a few cases have been reported in the United States, but the disease infects 40% to 50% of the population in some rural areas of South America.

The reservoir for *T. cruzi* includes a wide variety of wild animals, including rodents, opossums, and armadillos. The arthropod vector is the reduviid bug, called the "kissing bug" because it often bites persons near the lips (see Figure 12.31d). The insects live in the cracks and crevices of mud or stone huts with thatched roofs. The trypanosomes, which grow in the gut of the bug, are passed on if the bug defecates while feeding. The bitten human or animal often rubs the feces into the bite wound or other skin abrasions by scratching or into the eye by rubbing. A bizarre means of transmission occasionally occurs in remote areas of Mexico, where reduviid bugs are eaten as an aphrodisiac.

At the site of inoculation, the trypanosomes reproduce and pass through various stages of their life cycle. A swollen lesion develops at the inoculation site (often near the eye). The parasites are carried by the blood to many organs of the body, where they multiply in host cells and cause destructive inflammation of the affected tissues. The disease is most dangerous to children, whose death rate may be as high as 10%. Although many organs may be affected, the principal cause of death is damage to the heart. Growth in the brain is also a common development. If the parasites damage the nerves controlling the peristaltic action of the esophagus or colon, these organs no longer transport food and become grossly enlarged, conditions known as megaesophagus and megacolon, respectively.

Laboratory procedures for the diagnosis of American trypanosomiasis include observing the rather sparse population of trypanosomes in blood or growing the microorganisms from blood cultures. An extraordinary form of diagnosis, which is actually done with some frequency, is *xenodiagnosis*. Reduviid bugs, reared in parasite-free conditions, are allowed to feed on the arm of a suspected patient. The trypanosome is then identified in the intestinal tract of the insect 10 to 20 days later. Serological tests are not very satisfactory.

Treatment of this disease is very difficult when chronic, progressive stages have been reached. The trypanosome multiplies intracellularly and is difficult to reach chemotherapeutically. Nifurtimox is the drug of choice in treatment. The drug decreases the number of parasites but is not a cure.

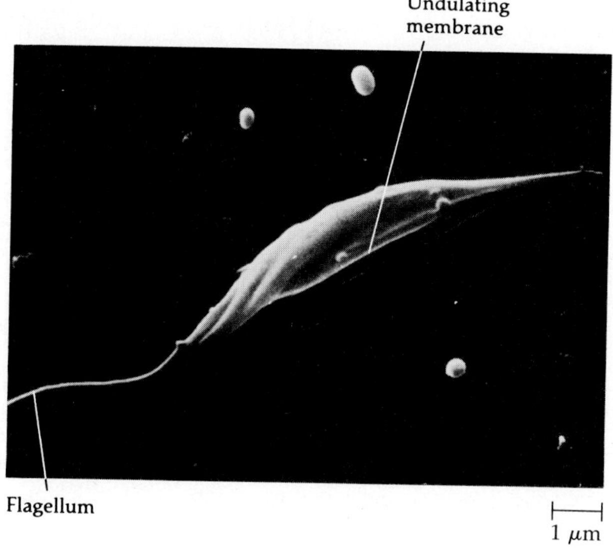

FIGURE 23.19 ***Trypanosoma cruzi*, the cause of American trypanosomiasis (Chagas' disease).** Scanning electron micrograph. The flagellum appears as an extension of the undulating membrane.

FIGURE 23.20 Distribution of American trypanosomiasis. A very small area of the United States in southern Texas is affected.

MMWR

MORBIDITY AND MORTALITY WEEKLY REPORT

Babesiosis—Connecticut

Since August 1988, six cases of babesiosis—a rare protozoan parasitic disease—have been reported to the Connecticut Department of Health Services (CDHS); only two cases thought to have been acquired in Connecticut were reported before 1988.

The first person became ill in August 1988; onset of illness in the other five persons occurred between late June and mid-August 1989. Ages ranged from 68 to 86 years; five were men. All six persons had fever, headache, and fatigue. Two of the patients were taking oral corticosteroids for chronic obstructive pulmonary disease; none was otherwise immunosuppressed. Four patients were treated with both quinine and clindamycin; one received quinine without clindamycin; the sixth received no specific therapy for babesiosis. All six are now asymptomatic, and their parasitemia has cleared.

Five of the patients lived within 3 miles of each other; the sixth lived 22 miles away. None of the patients gave a history of recent travel to areas with known endemic babesiosis, and none had received blood transfusions before becoming ill. Gardening near the home was the principal outdoor activity of four persons; the other two walked in fields near their homes. Only one person recalled being bitten by a tick before becoming ill, and all six had observed mice in the areas around their homes.

Parasites were detected on smears of peripheral blood taken from all six persons. In addition, each had IgG antibody titers to *Babesia microti* of $\geq$1:1024. *B. microti* was isolated (by hamster inoculation) from blood of two patients and from 8 of 11 mice (73%) trapped near four of the patients' homes. A statewide survey conducted in 1976 to 1977 detected *B. microti* antibodies in mice collected in 4 of 22 sites. Three of these four sites are within 20 miles of five of the patients' homes and within 45 miles of the other patient's home.

Comment: *Babesia* is a protozoan parasite of red blood cells. In the United States, babesiosis is most commonly caused by *B. microti.* Babesiosis was recognized in the Northeast in the 1960s and is endemic in Nantucket, Martha's Vineyard, Shelter Island, and parts of Long Island.

In humans, *B. microti* infection may be subclinical or may present as a febrile illness with constitutional symptoms and anemia. Manifestations are most severe in elderly, immunosuppressed, or asplenic persons.

The natural hosts for *B. microti* include the white-footed mouse and the meadow vole. Tick bite by *Ixodes dammini* is the usual source of human infection. In addition, infection can be transmitted by blood transfusion. Entomologic surveys have detected increases in *I. dammini* and its spread to new areas. Physicians should be aware that babesiosis could occur in areas where *Babesia* was not previously considered endemic.

Source: MMWR 38:649–650 (9/29/89).

MALARIA

Malaria is characterized by chills and fever and often by vomiting and severe headache. These symptoms typically appear at two- or three-day intervals that alternate with asymptomatic periods. Malaria is found wherever the mosquito vector *Anopheles* is found (see Figure 12.29b) and there are human hosts for the protozoan parasite *Plasmodium.* The disease was once common in the United States, but effective mosquito control and a reduction in the number of human carriers caused the reported cases to drop below 100 by 1960 (Figure 23.21). In recent years, however, there has been an upward trend in the number of cases in the United States, reflecting a worldwide resurgence of malaria, increased travel to malarial areas, and an increase in immigration from malarial areas. In southern California there was a recent outbreak of nearly 30 cases in which malaria was transmitted from infected persons by local mosquitoes. Occasionally, malaria has been transmitted by unsterilized syringes used by drug addicts, and blood transfusions from persons who have been in an endemic area are also a potential risk. In tropical Asia, Africa, and Central and South America, malaria is still a serious problem. In fact, as the protozoan develops resistance to antimalarial drugs and as political instability continues to interfere with control measures, malaria is more often being described as a resurgent disease. Current estimates are that it affects some 300 million persons worldwide each year and causes from 2 to 4 million deaths.

The causative organisms of malaria are the spore-forming protozoans (sporozoans) of the genus *Plasmo-*

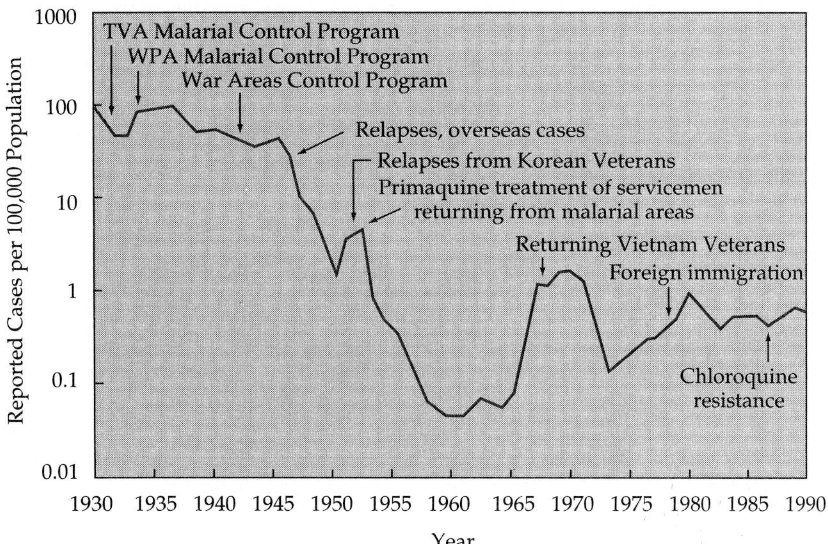

FIGURE 23.21 Reported cases of malaria in the United States, 1930 to 1990. Malaria was a common disease as recently as 1935. Its decline coincided with government control programs such as the Tennessee Valley Authority (TVA) flood control program, the Work Projects Administration (WPA) malarial control program, and war area control programs during World War II. Note the recent increases in incidence as veterans returned from Vietnam and foreign immigration increased. [Source: CDC. *Summary of Notifiable Diseases 1989, MMWR* 38:54; *MMWR* 39:52 (1/4/91).]

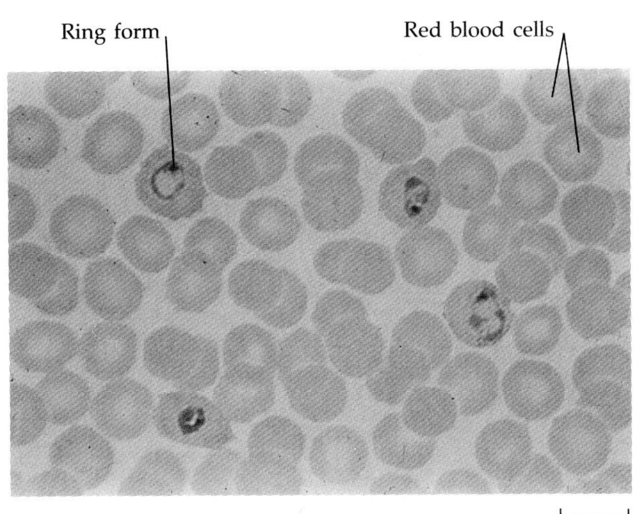

FIGURE 23.22 Blood smear from a malarial patient. The protozoan grows within infected red blood cells. The ring forms are an early stage. The dark spot of the ring is the nucleus, and the light central area within the circular ring is the food vacuole of the parasite. The food in the vacuole is the cytoplasm of the red blood cell. In later stages, just before lysis of the red blood cell, the cell becomes crowded with developing merozoites.

dium. Four pathogenic species are recognized, each of which causes a distinctive form of the disease. The most dangerous species is *Plasmodium falciparum* (fal-sip'är-um), a species that is also widespread geographically. Also widely distributed is *P. vivax* (vī'vaks). *P. malariae* (mä-lā'rē-ī) and *P. ovale* (ō-vä'lē) cause geographically restricted, relatively milder diseases.

The mosquito carries the *sporozoite* form of the *Plasmodium* protozoan in its saliva (see Figure 12.22). The sporozoite enters the bloodstream of the bitten human and within about 30 minutes enters the liver cells. The sporozoites in the liver cells undergo reproductive schizogony by a series of stages that finally results in the release of large numbers of *merozoite* forms into the bloodstream. These merozoites infect red blood cells and undergo reproductive schizogony again. Laboratory diagnosis of malaria is made by examining a blood smear for infected red blood cells (Figure 23.22). Eventually, the red blood cells rupture and release large numbers of merozoites. This rupturing of the red

TABLE 23.1 Diseases of the Cardiovascular and Lymphatic Systems

DISEASE	CAUSATIVE AGENT	MODE OF TRANSMISSION	TREATMENT
Bacterial Diseases			
Puerperal sepsis and infections related to abortions	Primarily *Streptococcus pyogenes; Clostridium* and *Bacteroides spp.* often cause postabortion infections	Unsanitary conditions during childbirth and abortion	Penicillin
Endocarditis			
Subacute bacterial	Many organisms, especially α-hemolytic streptococci	Bacteremia localizing in the heart	Varies with agent; pencillin for α-hemolytic streptococci
Acute bacterial	*Staphylococcus aureus, Streptococcus pyogenes*	Bacteremia localizing in the heart	Penicillin
Pericarditis	*Streptococcus pyogenes*	Bacteremia localizing in the heart	Penicillin
Rheumatic fever	Group A β-hemolytic streptococci	Inhalation leading to streptococcal sore throat	Penicillin (for sore throat)
Tularemia	*Francisella tularensis*	Animal reservoir (rabbits); skin abrasions, ingestion, inhalation, bites	Streptomycin or gentamicin
Brucellosis	*Brucella* species	Animal reservoir (cows); ingestion in milk, direct contact with skin abrasions	Streptomycin, tetracycline
Anthrax	*Bacillus anthracis*	Reservoir is soil or animals; inhalation or ingestion of heavy spore concentrations	Penicillin
Gangrene	*Clostridium perfringens*	Contamination of open wound by clostridial endospores	Debridement, amputation, hyperbaric chamber, penicillin
Rat bite fever	*Spirillum* or *Streptobacillus* spp.	Rat bite	Penicillin or tetracycline
Plague	*Yersinia pestis*	*Xenopsylla cheopis* (rat flea)	Tetracycline and streptomycin
Relapsing fever	*Borrelia* species	Soft ticks	Tetracycline

(Continued)

blood cells is fairly simultaneous, and the release of toxic compounds is the cause of the paroxysms (recurrent intensifications of symptoms) of chills and fever that are the characteristic symptoms of malaria. The fever reaches 40°C (104°F), and a sweating stage begins as the fever subsides. Between paroxysms, the patient feels normal. Anemia results from the loss of red blood cells, and enlargement of the liver and spleen are added complications.

When the red blood cells rupture, many of the released merozoites infect other red blood cells within a few seconds to renew the cycle in the bloodstream. If only 1% of the red blood cells contain parasites, there are an estimated 100,000,000,000 parasites in circulation at one time in a typical malarial patient! Other merozoites become male or female *gametocytes*. When these enter the digestive tract of a feeding mosquito, they pass through a sexual cycle that produces new infective sporozoites. It took the combined labors of several generations of scientists to discover this complex life cycle of the malaria parasite.

The highest mortality from malaria is in young children. Persons who survive malaria acquire a limited immunity. Although they can be reinfected, they tend to have a less severe form of the disease. This relative immunity almost disappears if the person leaves an endemic area with its periodic reinfections. Persons who have the genetic sickle-cell trait—common in many areas where malaria is endemic—are resistant to malaria.

Considerable effort is currently being expended on the search for an effective vaccine (see the box in Chap-

TABLE 23.1 ***(Continued)***

DISEASE	*CAUSATIVE AGENT*	*MODE OF TRANSMISSION*	*TREATMENT*
Bacterial Diseases			
Lyme disease	*Borrelia burgdorferi*	*Ixodes* spp. (tick)	Penicillin
Epidemic typhus	*Rickettsia prowazekii*	*Pediculus humanus corporis* body louse	Tetracycline and chloramphenicol
Endemic murine typhus	*Rickettsia typhi*	*Xenopsylla cheopis* (rat flea)	Tetracycline and chloramphenicol
Rocky Mountain spotted fever	*Rickettsia rickettsii*	*Dermacentor andersoni* and other species (tick)	Tetracycline and chloramphenicol
Viral Diseases			
Myocarditis	Several agents, especially coxsackievirus (enterovirus)	Inhalation and ingestion	None
Infectious mononucleosis	Epstein–Barr virus	Oral secretions, kissing	None
Yellow fever	Arbovirus (yellow fever virus)	*Aedes aegypti* (mosquito)	None
Dengue	Arbovirus (dengue fever virus)	*Aedes aegypti* (mosquito)	None
Protozoan Diseases			
Toxoplasmosis	*Toxoplasma gondii*	Contact with cat feces or contaminated, under-cooked meats	Pyrimethamine in combination with trisulfapyrimidines or sulfadiazine
Chagas' disease	*Trypanosoma cruzi*	Bite of reduviid bug	Nifurtimox in early stages
Malaria	*Plasmodium* species	*Anopheles* mosquito	Chloroquine, mefloquine
Helminthic Disease			
Schistosomiasis	*Schistosoma* species	Contaminated water; cercariae enter the body through the skin	Praziquantel

ter 18, p. 454). The sporozoite stage is of primary interest as a target for a vaccine because neutralizing it would prevent the initial infection from becoming well established. However, some believe that any effective vaccine would have to protect against all three stages.

Malaria was once fairly effectively treated with quinine, but newer derivatives of quinine, such as primaquine and especially chloroquine, have been the mainstays for prevention and treatment of malaria for many years. Resistance to these drugs is spreading rapidly, and current chemical alternatives are more expensive, which prohibits widespread use in the parts of the world most troubled by malaria. The latest drug of choice for many applications is mefloquine, a quinine derivative. Another approach to chemotherapy has been to use drug combinations such as Fansidar®, a combination of pyrimethamine and sulfadoxine. Both drugs are inhibitors of folic acid synthesis (see the discussion of sulfa drugs in Chapter 20); together, they have a synergistic effect. Resistance against such combinations is also emerging.

Effective control of malaria is not in sight. It will probably require a combination of chemotherapeutic and immunological approaches. The expense and the need for an effective political organization in malarial areas are probably going to be as important in controlling the disease as advances in medical research.

Helminthic Diseases of the Cardiovascular System

Many helminths find the cardiovascular system a convenient highway to travel as part of their life cycle. Schistosomes find a home there, shedding eggs that are distributed in the bloodstream.

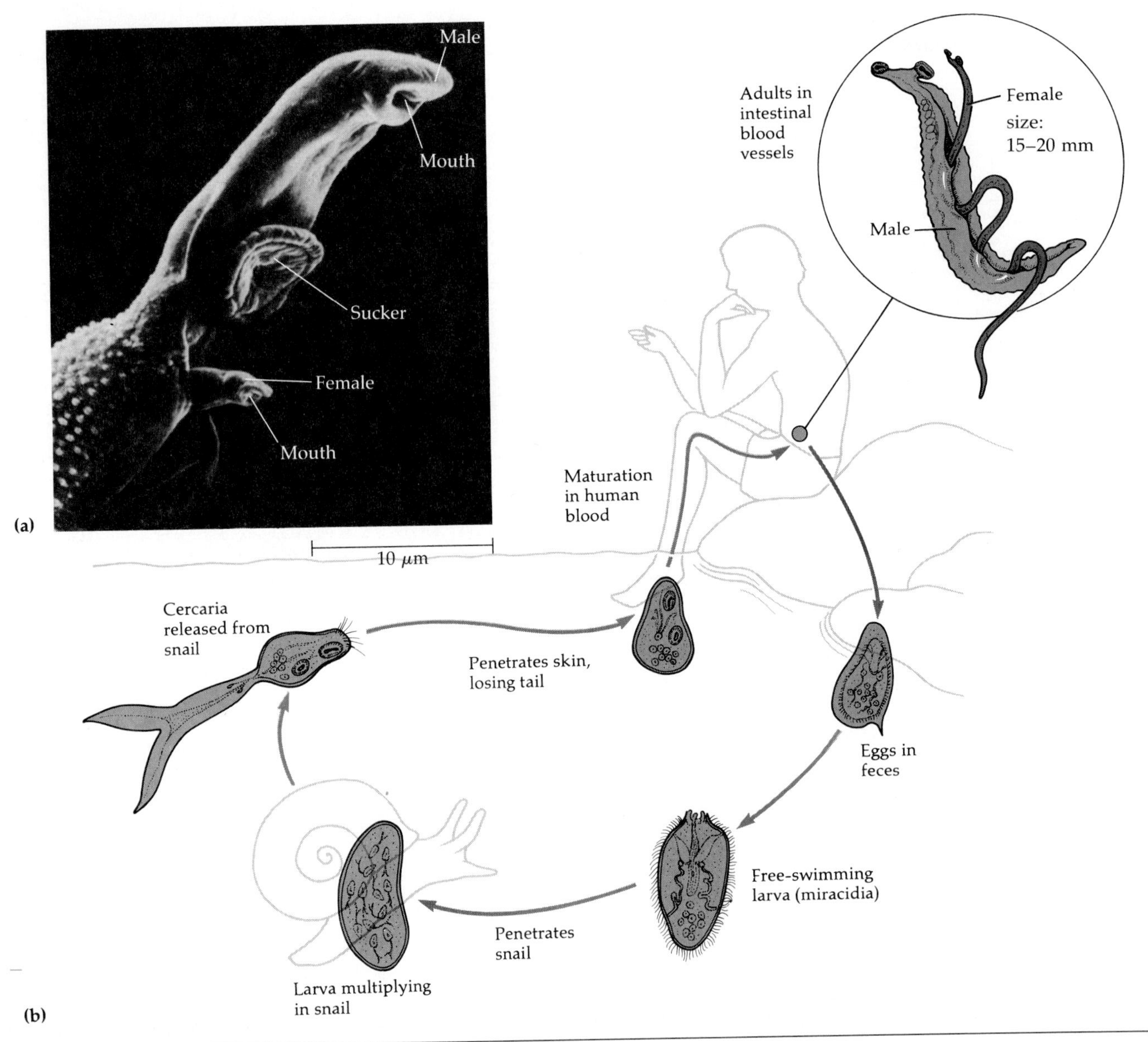

FIGURE 23.23 (a) Male and female schistosomes. The schistosomes (split-body) take their name from the life-style of the adults. The female lives in a cleft on the ventral surface of the male worm, is continuously fertilized, and continuously lays eggs. The mouths by which the worms feed are visible, as is the sucker on the male by which it attaches to the host. **(b)** Life cycle of *Schistosoma*, cause of the disease schistosomiasis.

SCHISTOSOMIASIS

Schistosomiasis is a debilitating disease caused by a flatworm parasite, a fluke. Eggs shed by the fluke lodge in body organs and initiate a damaging inflammatory reaction. The disease is not contracted in the United States, but more than 400,000 immigrants to the United States have the disease, and about 200 million persons in Asia, Africa, South America, and the Caribbean are also affected. Worms of the genus *Schistosoma* (shis-tō-sō′mä) cause the disease; they vary with geographical location, but the characteristics of the disease are similar.

When waters become contaminated with *Schistosoma* ova excreted in human wastes (Figure 23.23), a motile, ciliated larval form of *Schistosoma* called a miracidium is released from the ova and enters certain species of snails. (The lack of a suitable host snail is a primary reason why schistosomiasis is not transmitted in the United States.) Eventually, the pathogen emerges from the snail in an infective form called the *cercaria.* These cercariae have forked swimming tails. When they contact the skin of a person wading or swimming in the water, they discard the tail and enzymatically penetrate the skin. They are then carried by the bloodstream to the veins of the liver or urinary bladder, depending partly on the species of schistosome. There they mature into an adult form, in which the slender female lives in a cleft of the male. The union produces a supply of new ova, some of which cause local tissue damage *(granulomas)* from defensive body reactions (Figure 23.24). Other ova enter the water to continue the cycle.

The adult worm appears to be unaffected by the host's immune system. Apparently, it quickly coats itself with a layer that mimics the host's tissues. However, a previous infection tends to be protective to some degree against subsequent infections, which encourages work toward development of a vaccine.

Laboratory diagnosis consists of microscopic identification of the flukes or their ova in fecal and urine specimens, intradermal tests, and serological tests, such as complement-fixation and precipitin tests.

Praziquantel and oxiaminiquine are approved for use against schistosomes in the United States. Sanitation and elimination of the host snail are also useful forms of control.

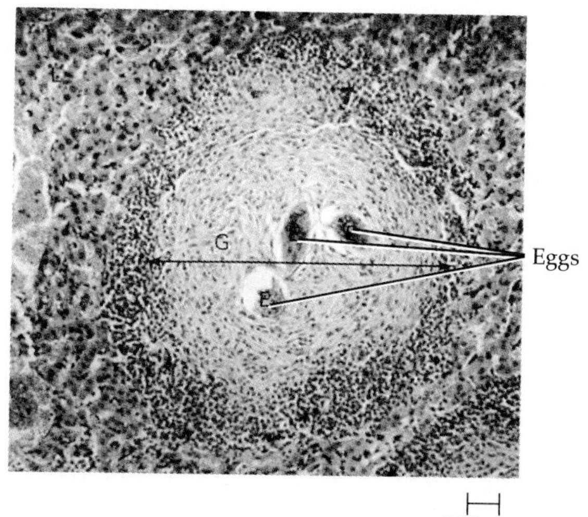

FIGURE 23.24 Granuloma from a patient with schistosomiasis. The eggs laid by the adult schistosomes lodge in tissue, and the body responds to the irritant by forming scarlike tissue surrounding it (a granuloma). This photo of a slice of liver tissue shows three eggs surrounded by a granulomatous area, whose diameter is shown by the arrow.

SWIMMER'S ITCH

Swimmers in lakes in the northern United States are sometimes troubled by **swimmer's itch.** This is a cutaneous allergic reaction to cercariae similar to that of schistosomiasis. However, these parasites mature only in wildfowl and not in humans, so infection does not progress beyond penetration of the skin and a local inflammatory response.

STUDY OUTLINE

Introduction (p. 560)

1. The heart, blood, and blood vessels make up the cardiovascular system.
2. Lymph, lymph vessels, lymph nodes, and lymphoid organs constitute the lymphatic system.

Structure and Function of the Cardiovascular System (pp. 560–561)

1. The heart circulates substances to and from tissue cells.
2. Arteries and arterioles transport blood away from the heart. Venules and veins bring blood to the heart.
3. Materials are exchanged between blood and tissue cells at capillaries. Capillaries connect arterioles to venules.
4. Blood is a mixture of plasma and cells.
5. Plasma transports dissolved substances. Red blood cells carry oxygen. White blood cells are involved in the body's defense against infection.

Structure and Function of the Lymphatic System (pp. 561–562)

1. Fluid that filters out of capillaries into spaces between tissue cells is called interstitial fluid.
2. Interstitial fluid enters lymph capillaries and is called lymph.
3. Lymphatics return lymph fluid to the blood.
4. Lymph nodes contain fixed macrophages, B cells, and T cells.

Bacterial Diseases of the Cardiovascular System (pp. 562–573)

SEPTICEMIA (p. 562)

1. The growth of microorganisms in blood is called septicemia.

2. Symptoms include lymphangitis (inflamed lymph vessels). Septicemia can lead to septic shock, characterized by decreased blood pressure.
3. Septicemia usually results from a focus of infection in the body.
4. Gram-negative rods are usually implicated. Endotoxin causes the symptoms.

PUERPERAL SEPSIS (pp. 562–563)

1. Puerperal sepsis begins as a uterine infection following childbirth or abortion; it can progress to peritonitis or septicemia.
2. *Streptococcus pyogenes* is the most frequent cause.
3. Oliver Wendell Holmes and Ignaz Semmelweiss demonstrated that puerperal sepsis was transmitted by the hands and instruments of midwives and physicians.
4. Puerperal sepsis is now uncommon because of modern hygienic techniques and antibiotics.
5. Anaerobic bacteria, including *Bacteroides* and *Clostridium*, can cause infections after improperly performed abortions.

BACTERIAL ENDOCARDITIS (p. 563)

1. The outer layer of the heart is the pericardium. The inner layer is the endocardium.
2. Subacute bacterial endocarditis is usually caused by α-hemolytic streptococci, although other gram-positive cocci can be involved.
3. The infection arises from a focus of infection, such as a tooth extraction.
4. Preexisting heart abnormalities are predisposing factors.
5. Symptoms include fever, anemia, and heart murmur.
6. Acute bacterial endocarditis is usually caused by *Staphylococcus aureus* or *Streptococcus pyogenes.*
7. The bacteria cause rapid destruction of heart valves.
8. *Streptococcus* can also cause pericarditis.
9. Inflammation of the heart muscle (myocardium) is called myocarditis.
10. Coxsackievirus (enterovirus) and *T. cruzi* are common causes.

RHEUMATIC FEVER (pp. 563–564)

1. Rheumatic fever is an autoimmune complication of group A β-hemolytic streptococcal infections.
2. Rheumatic fever is expressed as arthritis, inflammation of the heart, or Sydenham's chorea and can result in permanent heart damage.
3. Antibodies against group A β-hemolytic streptococci react with streptococcal antigens deposited in joints or heart valves or cross-react with the heart muscle.
4. Rheumatic fever can be a sequel to a streptococcal infection, such as streptococcal sore throat. Streptococci might not be present at the time of rheumatic fever.
5. The incidence of rheumatic fever in the United States has declined because of prompt treatment of streptococcal infections.
6. Rheumatic fever is treated with anti-inflammatory drugs.

TULAREMIA (pp. 564–565)

1. Tularemia is caused by *Francisella tularensis.* The reservoir is small wild mammals, especially rabbits.
2. Symptoms include ulceration at the site of entry, followed by septicemia and pneumonia.
3. Humans contract tularemia by handling diseased carcasses, eating undercooked meat of diseased animals, and being bitten by certain vectors (such as deer flies).
4. *F. tularensis* is difficult to culture in vitro. In vivo, it is resistant to phagocytosis.
5. Laboratory diagnosis is based on an agglutination test on isolated bacteria.

BRUCELLOSIS (UNDULANT FEVER) (pp. 565–566)

1. Brucellosis can be caused by *Brucella abortus*, *B. melitensis*, and *B. suis*.
2. Farm animals (cattle, pigs, goats, and camels) comprise the reservoir.
3. The bacteria enter through minute breaks in the mucosa or skin, reproduce in macrophages, and spread via lymphatics to liver, spleen, or bone marrow.
4. Symptoms include malaise and fever that spikes each evening (undulant fever).
5. A vaccine for cattle is available.
6. Diagnosis is based on serological tests.

ANTHRAX (p. 566)

1. *Bacillus anthracis* causes anthrax. In soil, endospores can survive for up to 60 years.
2. Grazing animals acquire an infection after ingesting the endospores.
3. Humans contract anthrax by handling hides from infected animals. The bacteria enter through cuts in the skin or through the respiratory tract.
4. Entry through the skin results in a pustule that can progress to septicemia. Entry through the respiratory tract can result in pneumonia.
5. Diagnosis is based on isolation and identification of the bacteria.

GANGRENE (pp. 566–567)

1. Soft tissue death from ischemia is called gangrene.
2. Microorganisms grow on nutrients released from gangrenous cells.
3. Gangrene is especially susceptible to the growth of anaerobic bacteria such as *Clostridium perfringens*, the causative agent of gas gangrene.
4. *C. perfringens* can invade the uterine wall during improperly performed abortions.
5. Debridement, hyperbaric chambers, and amputation are used to treat gas gangrene.

SYSTEMIC DISEASES CAUSED BY BITES AND SCRATCHES (pp. 567–568)

1. *Pasteurella multocida*, introduced by the bite of a dog or cat, can cause septicemia.
2. Anaerobic bacteria such as *Clostridium*, *Bacteroides*, and *Fusobacterium* infect deep animal bites.
3. Rat bite fever is caused by *Streptobacillus* or *Spirillum.*
4. *S. aureus*, *Bacteroides*, α-hemolytic streptococci, and *Fusobacterium* can infect human bites.

PLAGUE (pp. 568–569)

1. Plague is caused by *Yersinia pestis.* The vector is usually the rat flea *(Xenopsylla cheopis).*

2. Reservoirs for plague include European rats and North American rodents.
3. Symptoms of bubonic plague include bruises on the skin and enlarged lymph nodes (buboes).
4. The bacteria can enter the lungs and cause pneumonic plague.
5. Laboratory diagnosis is based on isolation and identification of the bacteria.
6. Antibiotics are effective to treat plague, but they must be administered promptly after exposure to the disease.
7. Control of the rat population is an effective deterrent to the spread and incidence of plague.

RELAPSING FEVER (p. 569)

1. Relapsing fever is caused by *Borrelia* species and transmitted by soft ticks.
2. The reservoir for the disease is rodents.
3. Symptoms include fever, jaundice, and rose-colored spots. Symptoms recur three or four times after apparent recovery.
4. Laboratory diagnosis is based on the presence of spirochetes in the patient's blood.

LYME DISEASE (LYME BORRELIOSIS) (pp. 569–570)

1. Lyme disease is caused by *Borrelia burgdorferi* and is transmitted by a tick *(Ixodes).*
2. Lyme disease is prevalent on the Atlantic Coast of the United States.
3. Deer and field mice are the animal reservoirs.
4. Diagnosis is based on serological tests and clinical symptoms.

TYPHUS (pp. 570–573)

1. Rickettsias are obligate intracellular parasites of eucaryotic cells.

Epidemic Typhus (pp. 570–573)

1. The human body louse *Pediculus humanus corporis* transmits *Rickettsia prowazekii* in feces, which are deposited while the louse is feeding.
2. Epidemic typhus is prevalent in crowded and unsanitary living conditions that allow the proliferation of lice.
3. The symptoms of typhus are rash, prolonged high fever, and stupor.
4. Tetracyclines and chloramphenicol are used to treat typhus.

Endemic Murine Typhus (p. 573)

1. Endemic murine typhus is a less severe disease caused by *Rickettsia typhi* and transmitted from rodents to humans by the rat flea.

Spotted Fever (p. 573)

1. *Rickettsia rickettsii* is a parasite of ticks (*Dermacentor* species) in the southeastern United States, Appalachia, and the Rocky Mountain states.
2. The rickettsia may be transmitted to humans, in whom it causes tick-borne typhus fever.
3. Chloramphenicol and tetracyclines are effective to treat the disease.
4. The serological tests are used for laboratory diagnosis.

Viral Diseases of the Cardiovascular System (pp. 573–575)

BURKITT'S LYMPHOMA (pp. 573–574)

1. Epstein–Barr (EB) virus causes Burkitt's lymphoma and nasopharyngeal cancer.
2. EB virus causes cancer in laboratory inoculated monkeys.
3. Burkitt's lymphoma tends to occur in patients whose immune system has been weakened by malaria.

INFECTIOUS MONONUCLEOSIS (p. 574)

1. Infectious mononucleosis is caused by EB virus.
2. The virus multiplies in the parotid glands and is present in saliva. It causes the proliferation of atypical lymphocytes.
3. The disease is transmitted by ingestion of saliva from infected individuals.
4. Diagnosis is made by indirect fluorescent-antibody technique.
5. EB virus can cause chronic fatigue syndrome (CFS).
6. CFS symptoms include profound fatigue and muscle weakness.

YELLOW FEVER (p. 575)

1. Yellow fever is caused by an arbovirus (yellow fever virus). The vector is the mosquito *Aedes aegypti.*
2. The virus multiplies in lymph nodes and then disseminates to liver, spleen, kidney, and heart.
3. Symptoms include fever, chills, headache, nausea, and jaundice.
4. Diagnosis is based on the presence of virus-neutralizing antibodies in the host.
5. No treatment is available. An attenuated, live viral vaccine is available.

DENGUE (p. 575)

1. Dengue is caused by an arbovirus (dengue fever virus) and is transmitted by the mosquito *Aedes aegypti.*
2. Symptoms are fever, muscle and joint pain, and rash.
3. Dengue hemorrhagic fever involving bleeding from the skin, gums, and gastrointestinal tract is a more serious form of the disease.
4. Laboratory identification is based on serological tests that detect either the virus or its antibodies.

Protozoan Diseases of the Cardiovascular System (pp. 575–581)

TOXOPLASMOSIS (pp. 575–577)

1. Toxoplasmosis is caused by the sporozoan *Toxoplasma gondii.*
2. *T. gondii* undergoes sexual reproduction in the intestinal tract of domestic cats, and oocysts are eliminated in cat feces.
3. Oocysts can be ingested by cattle and other animals.
4. Humans contract the infection by ingesting oocysts in undercooked meat from an infected animal or in dried cat feces.
5. Subclinical infections are probably common because the disease symptoms are rather mild.
6. Congenital infections can occur. Symptoms include convulsions, brain damage, blindness, and death.

7. *T. gondii* can be isolated and grown in all cultures and identified by serological tests.

AMERICAN TRYPANOSOMIASIS (CHAGAS' DISEASE) (p. 577)

1. *Trypanosoma cruzi* causes Chagas' disease. The reservoir includes many wild animals. The vector is the "kissing bug."
2. The infection begins at the site of inoculation and disseminates via the bloodstream to other organs. It causes destructive inflammation of infected organs.
3. Observation of the trypanosomes in blood confirms diagnosis.

MALARIA (pp. 578–581)

1. The symptoms of malaria are chills, fever, vomiting, and headache, which occur at two- or three-day cycles.
2. Malaria is transmitted by *Anopheles* mosquitoes. The causative agent is any one of four species of *Plasmodium*.
3. Sporozoites reproduce in the liver and release merozoites into the bloodstream, where they infect red blood cells and produce more merozoites.
4. Laboratory diagnosis is based on microscopic observation of merozoites in red blood cells.
5. A vaccine is being developed.
6. New drugs are being developed as the protozoans develop resistance to drugs such as chloroquine.

Helminthic Diseases of the Cardiovascular System (pp. 581–583)

SCHISTOSOMIASIS (pp. 582–583)

1. Species of the blood fluke *Schistosoma* cause schistosomiasis.
2. Eggs eliminated with feces hatch into larvae that infect the intermediate host. Free-swimming cercariae are released and penetrate the skin of a human.
3. The adult flukes live in the veins of the liver or urinary bladder in humans.
4. Adult flukes reproduce, and eggs are excreted or remain in the host.
5. Symptoms are from the host's defense to eggs that remain in the body.
6. Observation of eggs or flukes in feces, skin tests, or indirect serological tests may be used for diagnosis.
7. Chemotherapy is used to treat the disease; sanitation and snail eradication are used to prevent the disease.

SWIMMER'S ITCH (p. 583)

1. Swimmer's itch is a cutaneous allergic reaction to cercariae that penetrate the skin. The definitive host for this fluke is wildfowl.

STUDY QUESTIONS

Review

1. What are the symptoms of septicemia?
2. How can septicemia result from a single focus of infection, such as an abscess?
3. Complete the following table.

Disease	Frequent Causative Agent	Predisposing Condition(s)
Puerperal sepsis		
Subacute bacterial endocarditis		
Acute bacterial endocarditis		
Myocarditis		
Pericarditis		

4. Describe the probable cause of rheumatic fever. How is rheumatic fever treated? How is it prevented?
5. Compare and contrast epidemic typhus, endemic murine typhus, and tick-borne typhus.
6. Plot the temperature of a patient with brucellosis for a one-week period.
7. Complete the following table.

Disease	Causative Agent	Method of Transmission	Reservoir	Symptoms	Prevention
Tularemia					
Brucellosis					
Anthrax					
Lyme Disease					

8. Provide the following information on plague: causative agent; vector; U.S. reservoir; control; treatment; prognosis (probable outcome).
9. List the causative agent, method of transmission, and reservoir for schistosomiasis, toxoplasmosis, and American trypanosomiasis. Which disease are you most likely to get in the United States? Where are the other diseases endemic?

10. Complete the following table.

Disease	Causative Agent	Vector	Symptoms	Treatment
Malaria				
Yellow fever				
Dengue				
Relapsing fever				

11. Cite evidence implicating EB virus in Burkitt's lymphoma. Offer an explanation of why a large number of people who are infected with EB virus do not get Burkitt's lymphoma or nasopharyngeal cancer.

12. List four bacterial infections that might result from animal bites.

13. Why is *Clostridium perfringens* likely to grow in gangrenous wounds?

14. List the causative agents and methods of transmission of infectious mononucleosis.

15. What is the most effective control measure for mosquito-borne diseases?

16. Differentiate between the transmission and symptoms of bubonic plague and pneumonic plague.

17. What is infectious mononucleosis?

Challenge

1. Indirect fluorescent-antibody tests on the serum of three 25-year-old women, each of whom is considering pregnancy, provided the following information. Which of these women may have toxoplasmosis? What advice might be given to each woman with regard to toxoplasmosis?

Patient	Antibody Titer		
	Day 1	Day 5	Day 12
Patient A	1:1,024	1:1,024	1:1,024
Patient B	1:1,024	1:2,048	1:3,072
Patient C	0	0	0

2. A 19-year-old man went deer hunting. While on the trail, he found a partially dismembered dead rabbit. The hunter picked up the front paws for good luck charms and gave them to another hunter in the party. The rabbit had been handled with bare hands that were bruised and scratched from the hunter's work as an automobile mechanic. Festering sores on his hands, legs, and knees were noted two days later. What infectious disease do you suspect the hunter has? How would you proceed to prove it?

3. On March 30, a 35-year-old veterinarian experienced fever, chills, and vomiting. On March 31, he was hospitalized with diarrhea, left axillary bubo, and secondary bilateral pneumonia. On March 27, he had treated a cat who had labored respiration; an X ray revealed pulmonary infiltrates. The cat died on March 28 and was disposed of. Chloramphenicol was administered to the veterinarian. On April 10, his temperature returned to normal, and on April 20, he was released from the hospital. Sixty human contacts were given tetracycline. Identify the incubation and prodromal periods for this case. Explain why the 60 contacts were treated. What was the etiologic agent? How would you identify the agent?

4. Three of five patients who underwent aortic-valve replacement surgery developed bacteremia. The causative agent was *Enterobacter cloacae*. What were the patients' symptoms? How would you identify this bacterium? A manometer used in the operations was culture-positive for *E. cloacae*. What is the most likely source of this contaminant? Suggest a way of preventing such occurrences.

5. In August and September, six persons who each at different times spent a night in the same cabin developed the symptoms shown in the graph below. Three recovered after tetracycline therapy, two recovered without therapy, and one was hospitalized with septic shock. What is the disease? What is the incubation period of this disease? How do you account for the periodic temperature changes? What caused septic shock in patient no. 6?

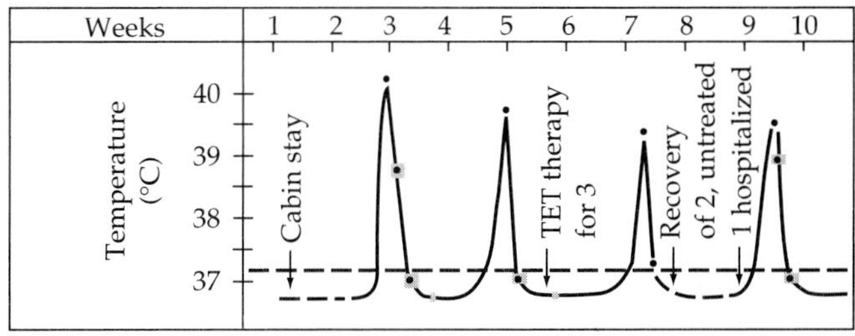

6. A 67-year-old man worked in a textile mill that processed imported goat hair into fabrics. He noticed a painless, slightly swollen pimple on his chin. Two days later he developed a 1-cm ulcer at the pimple site and a temperature of 37.6°C. He was treated with tetracycline. What is the etiology of this disease? Suggest ways to prevent it.

FURTHER READING

Barbour, A.G. "Antigenic variation of a relapsing fever *Borrelia* species." *Annual Review of Microbiology* 44:155–171, 1990. Antigenic variation may account for relapses of *Borrelia* infection.

Bisno, A.L. "Acute rheumatic fever: forgotten but not gone." *New England Journal of Medicine* 316:421–427, 476–478, 1987. A discussion of the 1985–1986 resurgence of rheumatic fever in the intermountain area.

Gregg, C.T. *Plague: An Ancient Disease in the 20th Century.* Albuquerque: University of New Mexico Press, 1985. An interesting and factual account of plague in this century.

Habicht, G.S., G. Beck, and J.L. Benach. "Lyme disease." *Scientific American* 257(1):78–83, July 1987. Describes how investigators discovered the cause of this new disease since the first report in 1975. Also describes the actions of bacterial LPS on interleukin-1 in causing symptoms.

Henle, W., G. Henle, and E.T. Lennette. "The Epstein–Barr virus." *Scientific American* 241(1):48–59, July 1979. A summary of information on one of the most common viruses infecting humans.

Kadis, S., T.C. Montie, and S.J. Ajl. "Plague toxin." *Scientific American* 220(3):92–100, March 1969. The mode of action of *Yersinia* toxin.

McDade, J.E., and V.F. Newhouse. "Natural history of *Rickettsia rickettsii*." *Annual Review of Microbiology* 40:287–309, 1986. Discusses the ecology of these bacteria and reasons for the increasing range and incidence of the disease.

Miller, L.H., et al. "Research toward a malaria vaccine." *Science* 234:1349–1355, 1986. A survey of progress in the search for a vaccine against malaria.

See also Further Reading for Part Four at the end of Chapter 26.

CHAPTER 24

Microbial Diseases of the Respiratory System

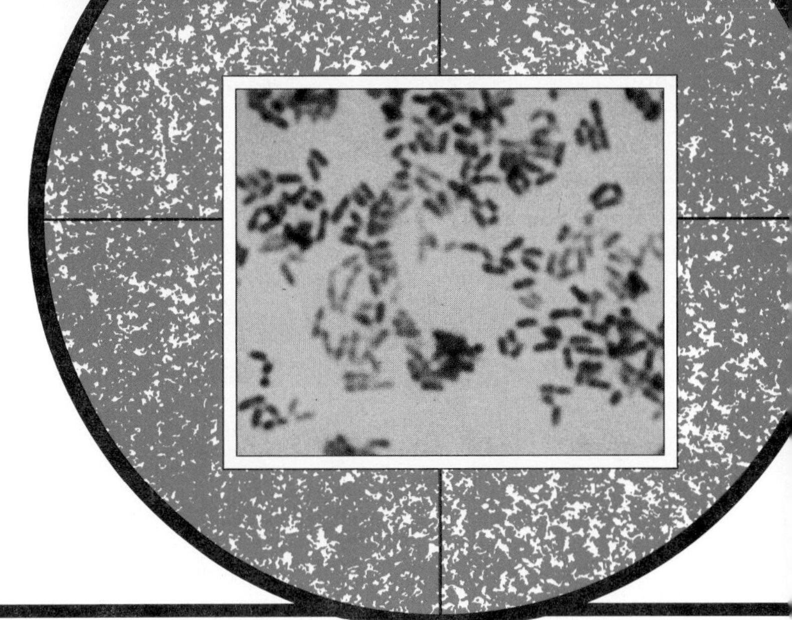

LEARNING OBJECTIVES

- Describe how microorganisms are prevented from entering the respiratory system.
- List the methods by which infections of the respiratory system are transmitted.
- Characterize the normal flora of the upper and lower respiratory systems.
- Describe the etiologic agent, symptoms, preferred treatment, and laboratory identification tests for ten bacterial diseases and three viral diseases of the respiratory system.
- Describe vaccines that are available to prevent respiratory system infections.
- Explain why vaccination against influenza and colds is impractical.
- List the etiologic agents, methods of transmission, preferred treatments, and laboratory identification tests for several fungal and protozoan diseases of the respiratory system.

With every breath, we inhale several microorganisms, so the upper respiratory system is a major portal of entry for pathogens. In fact, respiratory system infections are the most common type of infection—and among the most damaging. Some pathogens that enter the body via the respiratory system can infect other parts of the body—for example, the pathogens that cause measles, mumps, and rubella. Another very serious aspect of respiratory infections is the ease with which they are spread, both by *direct contact* with droplets emitted during sneezing, coughing, and talking and by *fomites* (contaminated objects). Transmission by droplets is especially common for highly communicable respiratory infections, such as pneumonic plague and influenza. Diseases associated with the respiratory system are summarized in Table 24.1, pp. 610–611.

Structure and Function of the Respiratory System

It is convenient to think of the respiratory system as being composed of two divisions, the upper respiratory system and the lower respiratory system. The **upper respiratory system** consists of the nose, the throat, and the structures associated with them, including the middle ear and the auditory (eustachian) tubes (Figure 24.1). Ducts from the sinuses and the nasolacrimal ducts from the lacrimal apparatus empty into the nasal cavity (see Figure 16.2). The auditory tubes from the middle ear empty into the upper portion of the throat.

The upper respiratory system has several anatomical defenses against airborne pathogens. Coarse hairs in the nose filter large dust particles from the air. The nose is lined with a mucous membrane that contains

FIGURE 24.1 Structures of the upper respiratory system.

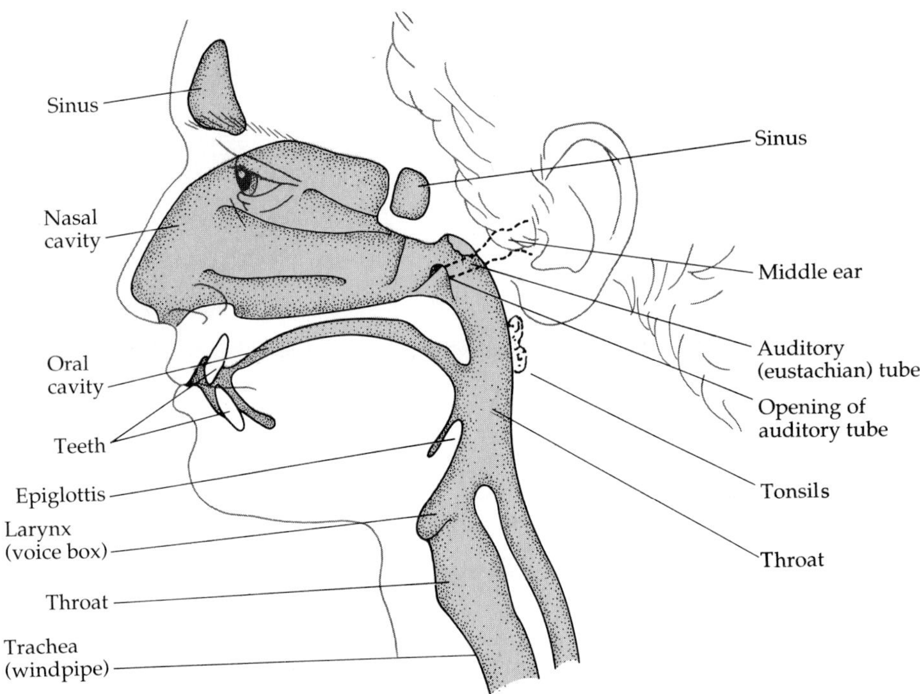

numerous mucus-secreting cells and cilia. The upper portion of the throat also contains a ciliated mucous membrane. The mucus moistens inhaled air and traps dust and microorganisms, especially particles larger than 4 to 5 μm. The cilia assist in removing these particles by moving them toward the mouth for elimination.

At the junction of the nose and throat are masses of lymphoid tissue, the tonsils and adenoids, that contribute to immunities to certain infections. Occasionally, however, these tissues become infected and help spread infection to the ears via the auditory tubes. Because the nose and throat are connected to the sinuses, nasolacrimal apparatus, and middle ear, infections commonly spread from one region to another.

The **lower respiratory system** consists of the larynx (voicebox), trachea (windpipe), bronchial tubes, and alveoli (Figure 24.2). Our lungs contain millions of alveoli, with an absorptive area of about 90 square meters. Alveoli are the air sacs that make up the lung tissue; within them, oxygen and carbon dioxide are exchanged between the lungs and blood. The double-layered membrane covering the lungs is the *pleura*. A ciliated mucous membrane lines the lower respiratory system down to the smaller bronchial tubes and helps to prevent microorganisms from reaching the lungs. As we discussed in Chapter 16, particles trapped in the larynx, trachea, and larger bronchial tubes are moved up toward the throat by a ciliary action called the ciliary escalator (see Figure 16.3b). If microorganisms actually reach the lungs, then certain phagocytic cells—the *alveolar macrophages*, or *dust cells*—usually locate, ingest, and destroy most of them. IgA antibodies in such secretions as respiratory mucus, saliva, and tears also help to protect mucosal surfaces of the respiratory system from many pathogens. Thus, the body has several mechanisms for removing the pathogens that cause airborne infections. However, if all these mechanisms fail, then the microorganism wins the host–parasite competition and a respiratory disease results.

Normal Flora of the Respiratory System

A number of potentially pathogenic microorganisms are part of the normal flora in the upper respiratory

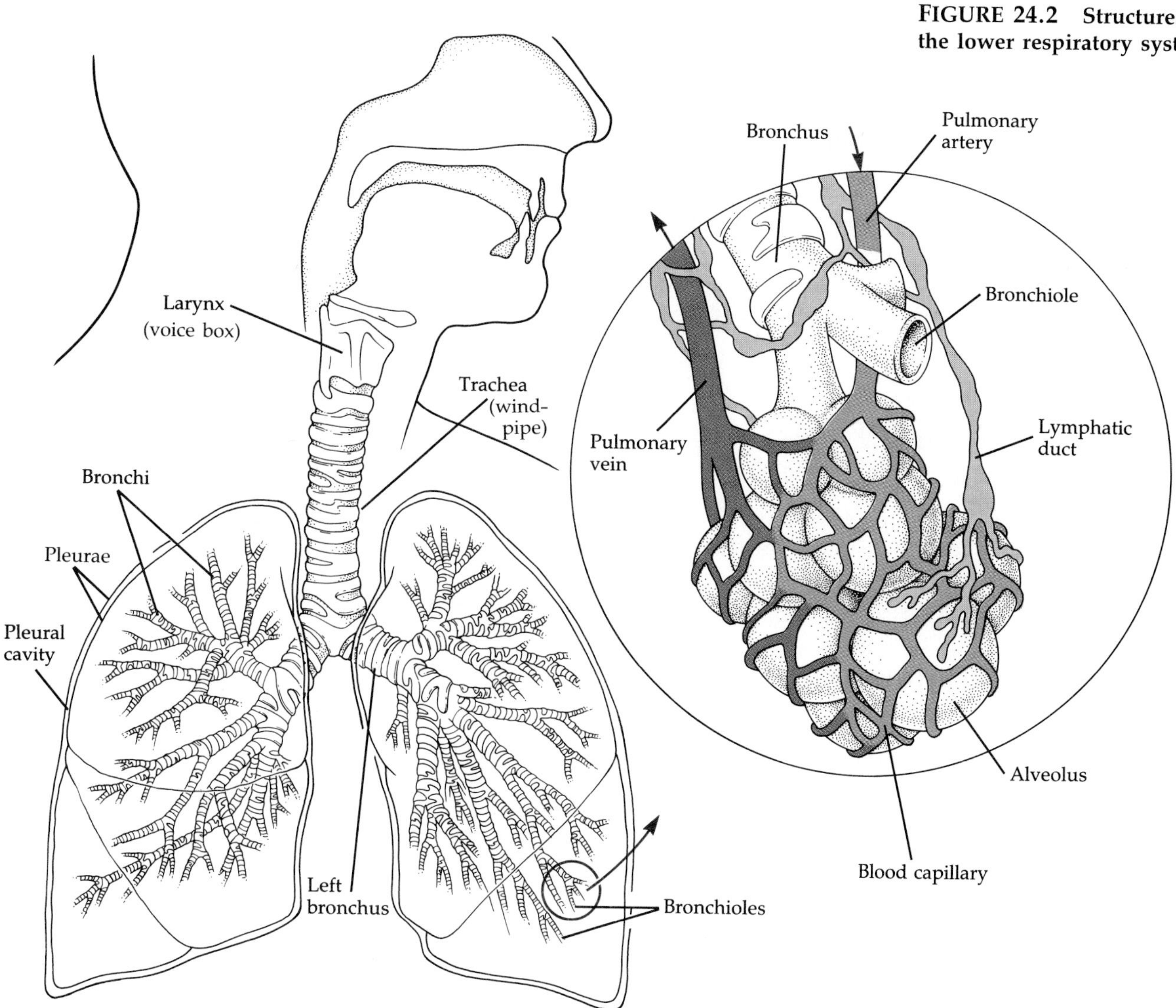

FIGURE 24.2 Structures of the lower respiratory system.

system. Despite the presence of these microorganisms, they rarely cause epidemics or morbidity because certain microorganisms of the normal flora suppress their growth by competing with them for nutrients and producing inhibitory substances.

By contrast, the lower respiratory tract is usually sterile—although the trachea may contain a few bacteria—because of the normally efficient functioning of the ciliary escalator in the bronchial tubes.

DISEASES OF THE UPPER RESPIRATORY SYSTEM

As most of us know from personal experience, the respiratory system is the site of many common infections. We will soon discuss **pharyngitis,** inflammation of the mucous membranes of the throat, or sore throat. When the larynx is the site of infection, we suffer from **laryngitis,** which affects our ability to speak. This infection is caused by bacteria, such as *Streptococcus pneumoniae* or *S. pyogenes,* or viruses, often in combination. The organisms that cause pharyngitis also can cause inflamed tonsils, or **tonsillitis.**

The nasal sinuses are cavities in certain cranial bones that open into the nasal cavity. They have a mucous membrane lining that is continuous with that of the nasal cavity. When a sinus becomes infected with such organisms as *S. pneumoniae* or *Hemophilus influenzae,* the mucous membranes become inflamed, and there is a heavy nasal discharge of mucus. This condition is called **sinusitis.** If the opening by which the mucus leaves the sinus becomes blocked, internal pressure can cause pain or a sinus headache. These diseases are almost always *self-limiting*, meaning that recovery will usually occur even without medical intervention.

Probably the most threatening infectious disease of the upper respiratory system is **epiglottitis,** inflammation of the epiglottis. The epiglottis is a flaplike structure of cartilage that prevents ingested material from entering the larynx (see Figure 24.1). Inflammation of the epiglottis is a rapidly developing disease that can result in death within a few hours. It is caused by opportunistic pathogens, usually *H. influenzae* type b.

Bacterial Diseases of the Upper Respiratory System

Airborne pathogens make their first contact with the body's mucous membranes as they enter the upper respiratory system. Many respiratory or systemic diseases initiate infections here.

STREPTOCOCCAL PHARYNGITIS (STREP THROAT)

Streptococcal pharyngitis (strep throat) is an upper respiratory infection caused by group A β-hemolytic streptococci. They are gram-positive bacteria. The most common of these is *Streptococcus pyogenes,* the same bacterium responsible for many skin and soft tissue infections, such as impetigo, erysipelas, and acute bacterial endocarditis.

The pathogenicity of group A streptococci is enhanced by their resistance to phagocytosis. They are also able to produce special enzymes that lyse fibrin clots (*streptokinase*), discussed in Chapter 15. They also produce *streptolysins* that are cytotoxic to tissue cells, red blood cells, and protective leukocytes. There is some concern that group A streptococci of unusual virulence are now in circulation.

Without some analysis, streptococcal pharyngitis cannot be distinguished from pharyngitis of other causes (mostly viral). Probably no more than half of so-called strep throats are actually streptococcal in origin. At one time, the diagnosis of strep throat was based on culturing of bacteria from a throat swab. Results took overnight or longer, and this test has been almost universally replaced by new indirect agglutination diagnostic tests for streptococcal pharyngitis that use microscopic latex particles coated with antibodies against group A streptococci. A swab of the throat is still used to collect bacteria, and the streptococcal antigens are extracted from the swab. An agglutination-type reaction is used to detect the reaction between the antigen and antibody-coated latex particles. Many of these tests take as little as ten minutes to perform and are highly specific in detecting the presence or absence of group A streptococci. In fact, *S. pyogenes* can be detected in the throats of many people who are only asymptomatic carriers. Therefore, a positive result is not a certain indication that the symptoms are caused by the detected streptococci. A rise in IgM antibody titer is the best indication of a sore throat of genuinely streptococcal origin, but this is not a practical option for day-to-day use. A negative result should be checked by culturing a throat swab.

Streptococcal pharyngitis is characterized by local inflammation and a fever. Frequently, tonsillitis occurs and the lymph nodes in the neck become enlarged and tender. Another frequent complication is otitis media (infection of the middle ear).

Nearly all streptococci are sensitive to penicillin, the drug of choice. Erythromycin is also effective. Treatment of streptococcal pharyngitis is important to prevent such complications as rheumatic fever and glomerulonephritis, both of which are caused by an immune response to streptococcal infections.

There are more than 60 serological types of group A streptococci. Since resistance to streptococcal diseases is type-specific, a person who has recovered from infection by one type is not necessarily immune to infection by another type.

Strep throat is now most commonly transmitted by respiratory secretions, but epidemics spread by unpasteurized milk were once frequent.

SCARLET FEVER

When the *S. pyogenes* strain causing streptococcal pharyngitis produces an *erythrogenic* (reddening) *toxin,* the resulting infection is called **scarlet fever.** When the strain produces this toxin, it has been lysogenized by a bacteriophage. As you may recall, this means that the genetic information of a bacteriophage (bacterial virus) has been incorporated into the chromosome of the bacterium, so the characteristics of the bacterium have been altered. The toxin causes a pinkish-red skin rash, which is probably the skin's hypersensitivity reaction to the circulating toxin, and a high fever. The tongue has a spotted, strawberrylike appearance and then, as it loses its upper membrane, becomes very red and enlarged. As the disease runs its course, the affected skin frequently peels off, as if sunburned, very similar to the scalded skin syndrome caused by *Staphylococcus aureus.*

Scarlet fever seems to vary in severity and frequency at different times and in different locations but generally has been declining in recent years. It is a communicable disease spread mainly by inhalation of infective droplets from an infected person. Classically, scarlet fever has been considered to be associated with streptococcal pharyngitis, but it might accompany a streptococcal skin infection.

Penicillin is effective in treating the original streptococcal infection.

DIPHTHERIA

Another bacterial infection of the upper respiratory system is **diphtheria,** once a major cause of death in children. The disease begins with a sore throat and fever, followed by general malaise and swelling of the neck. The pathogenesis of diphtheria is better understood than that of most other infectious diseases. The organism responsible is *Corynebacterium diphtheriae,* which is a gram-positive, non-endospore-forming, somewhat pleomorphic rod. The dividing cells are often observed to fold together to form V- and Y-shaped figures resembling Chinese characters (Figure 24.3).

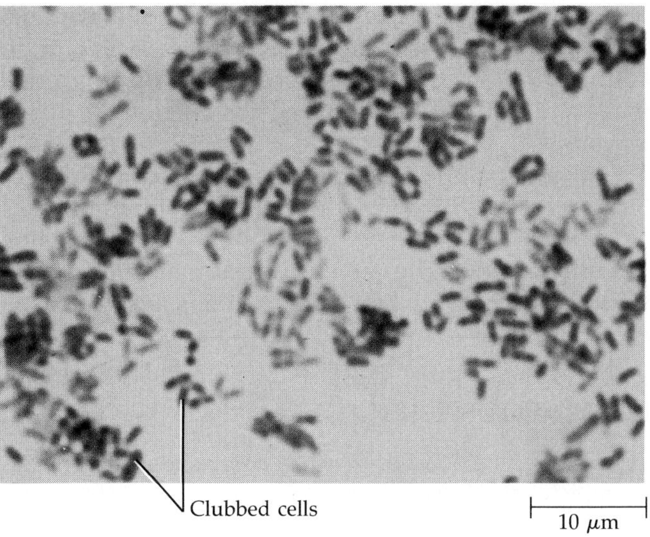

FIGURE 24.3 ***Corynebacterium diphtheriae,* the organism that causes diphtheria.** This Gram stain shows the club-shaped morphology. The dividing cells are often observed to fold together to form V- and Y-shaped figures resembling Chinese characters.

Part of the normal immunization program for children in the United States is the *DPT vaccine* (see Chapter 18). The *D* stands for diphtheria toxoid, an inactivated toxin that causes the body to produce antibodies against the diphtheria toxin.

C. diphtheriae has adapted to a generally immunized population, and relatively nonvirulent strains are found in the throats of many symptomless carriers. The bacterium is well suited to airborne transmission and is very resistant to drying. Its cell walls are very similar to those of the tubercle bacillus, which also shows high resistance to environmental stress. The bacteria are usually spread by respiratory transmission, and, once established in the upper respiratory system of a susceptible person, they increase rapidly in number.

Characteristic of diphtheria is a tough (*diphtheria,* from the Greek for leather), grayish membrane that forms in the throat in response to the infection. It contains fibrin, dead tissue, and bacterial cells and can totally block the passage of air to the lungs.

Although the bacteria do not invade tissues, those that have been lysogenized by a phage can produce a powerful exotoxin (see Chapter 15). Circulating in the bloodstream, the toxin interferes with protein synthesis. Only 0.01 mg of this highly virulent toxin is enough to kill a 91-kg (200-pound) person. Thus, if antitoxin therapy is to be effective, it must be administered before the toxin enters the tissue cells. When such organs as the heart or kidneys are affected by the toxin, the disease can rapidly be fatal. In other cases, the nerves can be involved and partial paralysis results.

Laboratory diagnosis can usually be made if the organisms are isolated from the primary lesion. Typical colonies grow on differential media (grayish colonies on Loeffler coagulated serum medium, black colonies on tellurite agar). Once isolated, the microorganisms are tested for toxigenicity by the guinea pig virulence test or the gel diffusion test. In the *guinea pig test,* the microorganisms are injected into one side of a shaved guinea pig. After four hours, the guinea pig is injected with antitoxin. Thirty minutes later, microorganisms are injected on the opposite side. If the microorganisms are toxigenic, a characteristic lesion will form between 48 and 72 hours later at the site injected before the antitoxin was administered. In the *gel diffusion test,* a strip of filter paper containing diphtheria antitoxin is placed on agar containing antitoxin-free calf serum. Then the microorganisms are streaked at a right angle to the filter paper. If the microorganisms are toxigenic, a visible line of antigen–antibody precipitate will form.

Even though antibiotics such as penicillin and erythromycin control the growth of the bacteria, they do not neutralize the diphtheria toxin. Thus, to treat diphtheria, antibiotics should be used only in conjunction with antitoxin.

The number of diphtheria cases reported in the United States each year is presently well under 100, but the death rate for respiratory cases is still from 5% to 10%, mostly among the elderly or the very young. In young children, the disease occurs mainly in groups that have not been immunized for religious or other reasons.

In the past, diphtheria was spread mainly to healthy carriers by droplet infection. Respiratory cases have been known to arise from contact with cutaneous diphtheria. With the advent of immunization, however, the carrier rate of toxigenic strains has decreased dramatically.

CUTANEOUS DIPHTHERIA

Diphtheria is also expressed as **cutaneous diphtheria.** In this form of the disease, *C. diphtheriae* infects the skin, usually at a wound or similar skin lesion, and there is minimal systemic circulation of the toxin. In cutaneous infections, the bacteria cause slow-healing ulcerations covered by a gray membrane.

Cutaneous diphtheria is fairly common in tropical countries. In the United States, it occurs mostly among Native Americans and among adults in low socioeconomic circumstances. It is responsible for most of the reported cases of this disease in persons over 30 years old. When diphtheria was more common, repeated contacts with toxigenic strains reinforced the immunity, which otherwise weakens with time. Many adults now lack immunity because routine immunization was less available during their childhood. Some surveys indicate effective immune levels in as few as 25% of the adult population. Because of this, when any wound or trauma requires tetanus toxoid, it is the usual practice to combine it with diphtheria toxoid (DT vaccine).

OTITIS MEDIA

One of the more uncomfortable complications of the common cold, or of any infection of the nose or throat, is infection of the middle ear, **otitis media,** leading to earache. The middle ear can be infected directly by contaminated water from swimming pools and by some severe trauma, such as injury to the eardrum and certain skull fractures. The infecting microorganisms cause the formation of pus, which builds up pressure against the eardrum and causes it to become inflamed and painful. The condition is most frequent in early childhood, probably because the auditory tube connecting the middle ear to the throat is small and is more easily blocked by infection (see Figure 24.1). Enlarged adenoids might be a contributing factor for children with an unusually high frequency of infection.

A number of bacteria can cause otitis media. The one most commonly isolated is *S. pneumoniae* (about 35% of cases). Other bacteria frequently involved include *H. influenzae* (25% to 30%), *Branhamella catarrhalis* (bran-ha-mel'lȧ ka-tär'al-is) (10% to 15%), *S. pyogenes* (8% to 10%), and *S. aureus* (1% to 2%). In about 3% to 5% of cases, no bacteria can be detected. Most of these cases are probably caused by obligate anaerobic bacteria. Viruses are possibly a cause, but treatment always assumes that bacteria are the causative agents. Broad spectrum penicillins, such as amoxicillin, are usually the first choice for therapy for children under eight. Penicillin is the recommended antibiotic in older patients.

Viral Disease of the Upper Respiratory System

COMMON COLD (ACUTE CORYZA)

A number of different viruses are involved in the etiology of the **common cold (acute coryza).** About 50% of all colds are caused by *rhinoviruses. Coronaviruses* probably cause another 15% or 20%. About 10% of all colds are caused by one of an assortment of other viruses. In about 40% of colds, no agent can be identified.

We tend to accumulate immunities against cold viruses during our lifetimes, which may be a reason why older persons tend to get fewer colds. It has been shown that young children have three or four colds per year, but adults at age 60 average less than one cold per year. Immunity is based on the ratio of IgA antibodies to single serotypes and has a reasonably high short-term effectiveness. Isolated populations may develop a group immunity, and their colds disappear until a new set of viruses is introduced. Altogether, there are probably more than 200 agents that cause colds. There are at least 113 serotypes of rhinoviruses alone, so a vaccine effective against so many different agents does not seem practical.

However, a new approach has been suggested. It has been found that most cold viruses, although antigenically different, often use as few as two different receptor sites on the mucosa. This may make it possible to approach the problem from a different direction. Instead of trying to produce antibodies against many different virus configurations, perhaps an antibody can be produced that will achieve the same results by blocking the relatively few receptor sites in the body.

The symptoms of the common cold are familiar to all of us. They include sneezing, excessive nasal secretion, and congestion. (In ancient times, one school of medical thought believed that the nasal discharges were waste products from the brain.) The infection can easily spread from the throat to the sinus cavities, the lower respiratory system, and the middle ear, leading to complications of laryngitis and otitis media. The

FIGURE 24.4 A card-playing experiment that included cold sufferers with healthy persons tended to support the theory of airborne transmission of cold-causing viruses.

uncomplicated cold usually is not accompanied by fever.

The rhinoviruses prefer a temperature slightly below that of normal body temperature—such as might be found in the upper respiratory system, which is open to the outside environment. No one knows exactly why the number of colds seems to increase with colder weather in temperate zones. It is not known whether closer indoor contact promotes epidemic-type transmission or whether physiological changes increase susceptibility.

A single rhinovirus deposited on the nasal mucosa is often sufficient to cause a cold. However, there is surprisingly little agreement on how the cold virus is transmitted to a site in the nose. One line of experimentation tends to show that cold sufferers deposit the viruses on doorknobs, telephones, and other surfaces, where they remain viable for hours. Healthy people can then transfer these viruses to their hands and thence to their nasal passages. This theory was supported by an experiment in which healthy persons who used virucidal iodine solutions on their hands had a much reduced incidence of colds. However, another series of experiments involving a group of card-players, half with colds and half without colds, supported a different conclusion. Half of the healthy players were restrained so that they could not use their hands to transfer to their noses viruses picked up from the playing cards, and the other half were not restrained. The players who could not touch their noses came down with as many colds as those who could—an argument for airborne transmission. Even when healthy participants in card games were placed in a room separate from anyone suffering from a cold and isolated from their airborne secretions, but handled cards literally soaked in nasal secretions, none developed colds (Figure 24.4). In a perhaps less repellent series of experiments, researchers required healthy volunteers to kiss cold sufferers for 60 to 90 seconds—only 8% of the volunteers came down with colds.

Because colds are caused by viruses, antibiotics are of no use in treatment. The common cold will usually run its course to recovery in about a week. Recovery time is not affected by over-the-counter drugs presently available, although such drugs may lessen the severity of certain signs and symptoms. Research to develop chemotherapeutic agents that interfere with viral replication is ongoing.

DISEASES OF THE LOWER RESPIRATORY SYSTEM

The lower respiratory system can be infected by many of the same bacteria and viruses that infect the upper respiratory system. As the bronchi (see Figure 24.2) become involved, **bronchitis** or **bronchiolitis** develops. *Mycoplasma pneumoniae* and a number of common respiratory viruses are suspected causes. In infants, the most common agent is probably the respiratory syncytial virus, discussed later in this chapter. Diseases such as **whooping cough** are also a form of bronchitis. A severe complication of bronchitis is **pneumonia,** in which the pulmonary alveoli become involved.

Bacterial Diseases of the Lower Respiratory System

WHOOPING COUGH (PERTUSSIS)

Infection by the bacterium *Bordetella pertussis* results in whooping cough (pertussis). *B. pertussis* is a small, nonmotile, gram-negative coccobacillus. The virulent strains possess a capsule. It is an obligate aerobe and produces an endotoxin as part of its cell wall. In the cytoplasm, the bacteria produce an exotoxin that is apparently released along with the endotoxin when the cell autolyses upon death. The precise role of the toxins in pathogenesis has yet to be determined. The organism does not invade tissues. However, it attaches to the cilia in the trachea and impedes their action, allowing mucus to accumulate. Eventually, a toxic fragment of the bacterial cell wall, *tracheal cytotoxin,* causes loss of the ciliated cells (Figure 24.5).

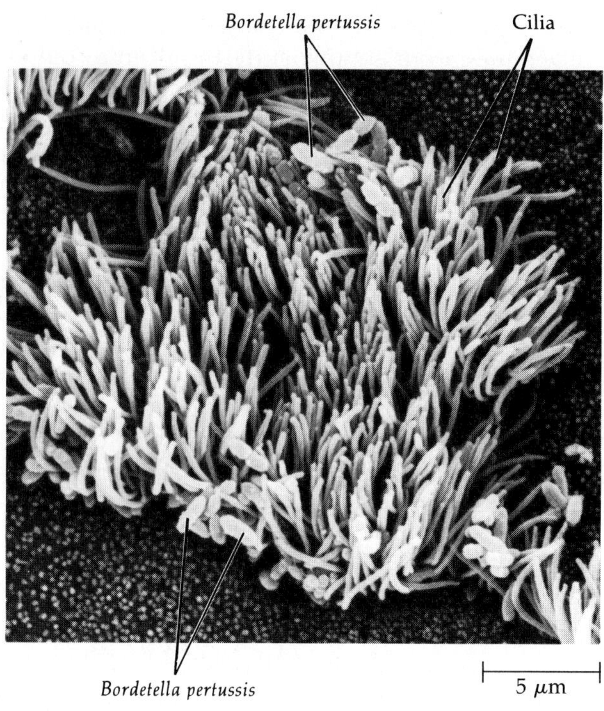

FIGURE 24.5 Ciliated cells of the respiratory system. In this scanning electron micrograph, cells of *Bordetella pertussis* can be seen growing on the cilia; they will eventually cause the loss of the ciliated cells.

Whooping cough, which is primarily a childhood disease, can be quite severe. The initial stage, which is called the *catarrhal stage,* resembles a common cold. Prolonged sieges of coughing characterize the second, *paroxysmal stage.* When ciliary action is compromised, the mucus accumulates, and the infected person desperately attempts to cough up these mucus accumulations. The violence of the coughing in small children can actually result in broken ribs. Gasping for air between coughs causes a whooping sound, hence the name of the disease. Coughing episodes occur several times a day for one to six weeks. The *convalescence stage,* the third stage, may last for months. The disease is of unusually long duration for a respiratory infection. The disease is transmitted by inhaling pathogens expelled by the coughing of infected patients. It is highly infectious, with a transmission rate of 90% among non-immune contacts.

Diagnosis of pertussis depends in large part upon clinical signs and symptoms. The organism can be cultured from a throat swab inserted through the nose on a thin wire and held in the throat while the patient coughs. The pathogen is very sensitive to environmental stresses and must be transferred to the culture medium quickly. Many laboratories still use Bordet–Gengou medium, developed in 1906, but newer media that are more selective and have a longer shelf life are also available. A direct fluorescent-antibody test is the diagnostic serological test most often used. It is not especially satisfactory, and there is considerable interest in developing improved serological tests or DNA probes for diagnosis of pertussis.

Mild cases of whooping cough require no specific treatment. The disease is most severe in infants under one year of age, and nearly half the cases occur in this age group. Severe cases, especially in infants, are usually treated with erythromycin. Although antibiotics might not result in rapid improvement, the drugs do render the patient noninfectious.

After recovery, immunity is good—at least, second attacks tend to be very mild. A vaccine prepared from heat-inactivated whole bacteria is a part of the regular immunization schedule for children (the P in DPT vaccine stands for *pertussis*). Vaccination is usually recommended at about two months of age, when the titer of maternal antibodies in the baby's circulation is dropping and the child's own antibodies have not yet replaced them. Vaccination has lowered the annual number of cases in the United States from more than 250,000 per year to fewer than 2000, with fewer than 10 deaths. There has been considerable concern about the safety of the pertussis vaccine. Because it contains higher levels of endotoxins than any other vaccine, it frequently produces feverish reactions. The main question, however, is whether it causes permanent neurological damage in recipients. The most recent consensus of the medical community is that it does not, but that concern about the possibility is leading to unacceptably low rates of vaccination in the general population. In Japan and England, adverse publicity has led to widespread avoidance of pertussis vaccination. The result was major epidemics.

Because infants are less capable of coping with the effort of coughing to maintain an airway, the mortality rate in this group is relatively high. Authorities point to these figures as indications that vaccination should continue while research on better vaccines is intensified. An acellular vaccine, with fewer nonfunctional components, has been tested in Japan, Sweden, and limited areas of the United States. Very few side-effects have been reported with this newer vaccine. It is uncertain whether the observed effectiveness is high enough for it to be licensed in the United States.

TUBERCULOSIS

Tuberculosis is an infectious disease caused by the bacterium *Mycobacterium tuberculosis,* a slender rod and an obligate aerobe. The rods grow slowly (20-hour generation time), sometimes form filaments (Figure 24.6), and tend to grow in clumps. On the surface of

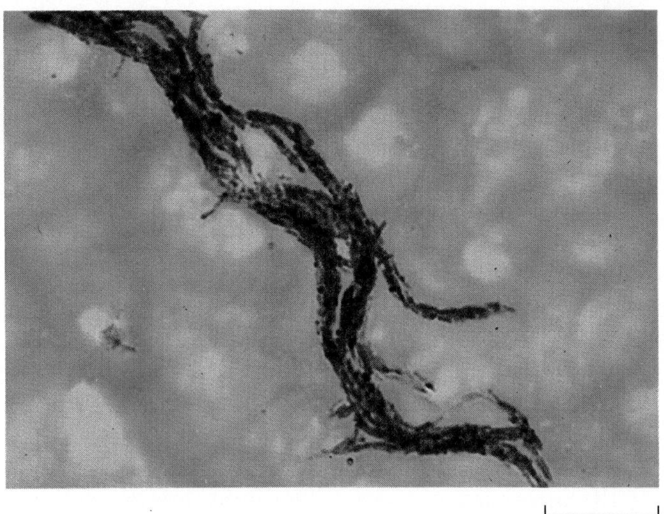

FIGURE 24.6 Mycobacterium tuberculosis, showing the cordlike growth that often occurs in liquid growth media. This filamentous, funguslike growth under some conditions is responsible for the organism's name, *myco* (fungus).

liquid media, their growth appears moldlike, which suggested the genus name *Mycobacterium,* from the Greek *mykes,* meaning fungus.

These bacteria are relatively resistant to normal staining procedures. When stained by the Ziehl–Neelson or Kinyoun technique that stains the cell with carbolfuchsin dye, they cannot be decolorized with a mixture of acid and alcohol and are therefore classified as *acid-fast.* This characteristic reflects the unusual composition of the cell wall, which contains large amounts of lipid materials. These lipids might also be responsible for the resistance of mycobacteria to environmental stresses, such as drying. In fact, these bacteria can survive for weeks in dried sputum and are very resistant to chemical antimicrobials used as antiseptics and disinfectants.

Tuberculosis is a good example of the importance of the ecological balance between host and parasite in infectious disease. Hosts are not usually aware of pathogens that invade the body and are defeated. If defenses fail, however, hosts become very much aware of the resulting disease. Several factors may affect host resistance levels—the presence of other illness and physiological and environmental factors such as malnutrition, overcrowding, and stress.

Tuberculosis is most commonly acquired by inhaling the tubercle bacillus. Only very fine particles containing one to three bacilli reach the lungs, where they are usually phagocytized by a macrophage in the alveoli (see Figure 24.2). The macrophages of a healthy individual usually destroy the bacilli. If they do not, the macrophages actually protect the microbe from the chemical and immunological defenses of the body, and many of the bacilli survive and multiply within the macrophage.

These macrophages eventually lyse, releasing an increased number of pathogens. The tubercle bacilli released from dying macrophages form a lesion. A hypersensitivity reaction against these organisms causes formation of a *tubercle,* which effectively walls off the pathogen. These small lumps are characteristic of tuberculosis and give the disease its name. Tubercles are composed of packed masses of tissue cells and the disintegration products of bacilli and leukocytes; they usually have a necrotic center. Few bacteria are present in the tubercle.

The tubercle bacillus does not produce any injurious toxins. Tissue damage is mostly from the hypersensitivity reaction. As the reaction continues, the tubercle undergoes necrosis and eventually forms a *caseous lesion* that has a cheeselike consistency. If the caseous lesions heal, they become calcified and show clearly on X rays, where they are called *Ghon complexes* (Figure 24.7). If the disease is not arrested at this point, the caseous lesions progress to liquefaction. An air-filled *tuberculous cavity* is formed from the caseous lesion. Conditions within the cavity favor the prolifera-

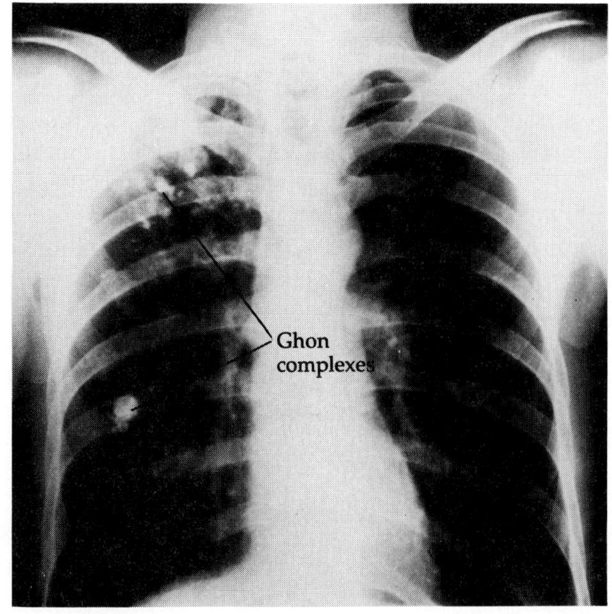

FIGURE 24.7 X ray showing Ghon complexes in a case of tuberculosis. The lower area of the right lung (left side of photo) has a large complex, and numerous smaller lesions toward the upper part of the lung are visible. The left lung is clear.

tion of the tubercle bacillus, which then grows for the first time extracellularly. Bacilli soon reach very large numbers, and eventually the lesion ruptures, releasing the microorganisms into the blood and lymphatic system. This condition of rapidly spreading infection that overwhelms the body's remaining defenses is called *miliary tuberculosis* (the name is derived from the numerous millet seed-sized tubercles formed in the infected tissues).

This condition leads to a progressive disease characterized by loss of weight, coughing (often with a show of blood), and general loss of vigor. (At one time, tuberculosis was commonly known as consumption.) Even when patients are considered cured, tubercle bacilli often remain in the lung, and the disease may be reactivated. Reactivation may be precipitated by old age, poor nutrition, or immunosuppression.

The first effective antibiotic in the treatment of tuberculosis was streptomycin. Probably the drug most commonly used today is the synthetic chemical isoniazid (INH). A combination of INH and rifampin is particularly effective. Some treatments use a capsule containing these two drugs plus pyrazinamide, which is especially effective against bacteria within macrophages. Ethambutol, pyrazinamide, and streptomycin are used in cases of intolerance or resistance to the primary drugs. To be effective, chemotherapy for tuberculosis must be continued for months—a nine-month regimen is considered a short-course treatment. Prolonged treatment is necessary, partly because the tubercle bacillus grows very slowly and many antibiotics are effective only against growing cells. Also, the bacillus is often hidden for long periods in macrophages or other locations difficult to reach with antibiotics. A major practical problem with treatment of tuberculosis is the patient's compliance with medication regimens during the long course of treatment. For preventive prophylaxis for persons exposed to tuberculosis, a 6- to 12-month course of INH is usually given alone because there are relatively few bacteria and development of resistance is not likely.

Persons infected with tuberculosis develop cell-mediated immunity against the bacterium. This form of immune response, rather than humoral immunity, is because the pathogen is located mostly within macrophages. This immunity, involving sensitized T cells, is the basis for the *tuberculin skin test.* In this test, a purified protein derivative (PPD) of the tuberculosis bacterium, derived by precipitation from broth cultures, is injected cutaneously. If the injected person has been infected with tuberculosis in the past, sensitized T cells react with these proteins and a delayed hypersensitivity reaction appears in about 48 hours. This reaction appears as an induration (hardening) and reddening of the area around the injection site. Probably the most accurate tuberculin test is the *Mantoux test,* in which dilutions of 0.1 ml of antigen are injected and the reacting area of the skin is measured. A number of similar tests are also in common use.

A positive tuberculin test in the very young is a probable indication of an active case of tuberculosis. In older persons, it might indicate only hypersensitivity resulting from a previous infection or vaccination, not a current active case. Nonetheless, it is an indication that further examination is needed, such as a chest X ray for the detection of lung lesions and attempts to isolate the bacterium.

Confirmation of tuberculosis by isolation of the bacterium is complicated by the very slow growth of the pathogen. Eight weeks might be needed for growth of an isolate, and at least one week is required for the appearance of an exceedingly minute colony. Selective media are recommended, such as the Lowenstein–Jensen medium containing glycerol and eggs.

Another species, *Mycobacterium bovis* (bō'vis), is a pathogen mainly of cattle. It is the cause of **bovine tuberculosis,** which is transmitted to humans via contaminated milk or food. Bovine tuberculosis accounts for less than 1% of tuberculosis cases in the United States. It seldom spreads from human to human, but before the days of pasteurized milk and the development of control methods, such as tuberculin testing of cattle herds, this disease was a common form of tuberculosis in humans. *M. bovis* infections cause tuberculosis that primarily affects the bones or lymphatic system. At one time, a common manifestation of this type of tuberculosis was hunchbacked deformation of the spine.

The *BCG vaccine* is a live culture of *M. bovis* that has been made avirulent by long cultivation on artificial media. (BCG stands for the *b*acillus of Calmette and Guerin, the persons who originally isolated the strain.) The vaccine is fairly effective in preventing tuberculosis and has been in use since the 1920s. In the United States, the vaccine is no longer recommended for health care workers or high-risk adults. It is used only for certain children at high risk who have negative skin tests. Apparently, the vaccine enhances cell-mediated immunity, and persons who have received the vaccine show a positive reaction to tuberculin skin tests.

After the introduction of effective antibiotics in the 1950s, the incidence of tuberculosis underwent a steady decline. This was reversed in 1985, when the number of reported cases showed a perceptible increase. Currently, more than 20,000 new cases of tuberculosis are reported in the United States each year (Figure 24.8a). The mortality, currently about 1700 per year, has continued to decrease, probably because of more effective chemotherapy. Many of the new cases are found among immigrant groups, especially Asian-born refugees. Certain ethnic groups tend to have much higher rates of tuberculosis. For example,

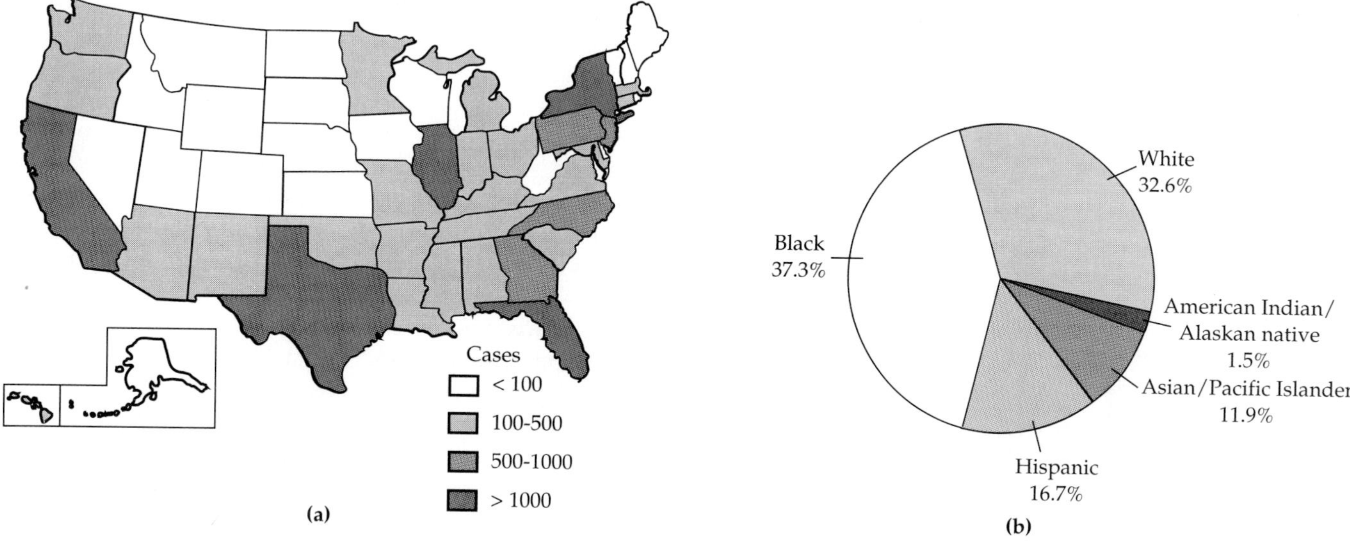

FIGURE 24.8 Tuberculosis distribution. **(a)** Geographic distribution of reported cases of tuberculosis in the United States in 1990. [*Source: MMWR* 39:52 (1/4/91).] **(b)** Percentage of cases by race or ethnic origin in the United States. [*Source:* CDC. *Summary of Notifiable Diseases 1989, MMWR* 38:54.]

blacks, Native Americans, Asians, and Hispanics account for about two-thirds of tuberculosis cases (Figure 24.8b). Most cases in the white population occur in the very elderly.

Recently, however, there has been a marked increase of tuberculosis in persons about 25 to 44 years of age. The Centers for Disease Control believe that the increase in this age group is caused primarily by increased susceptibility associated with HIV infection. Other mycobacterial diseases also affect persons in the late stages of HIV infections. A majority of the isolates are of a related group of organisms known as the *M. avium-intracellulare* (ā′vē-um in′trä-cel-ū-lār) complex. In the general population, infections by these organisms are uncommon.

In the United States, 10 to 12 million are estimated to be infected by the tubercle bacillus, many with only latent infections. Worldwide, the number of active cases in developing countries is estimated in the tens of millions; at least 3 million die of the disease each year.

BACTERIAL PNEUMONIAS

The term *pneumonia* is applied to many pulmonary infections. In the United States, the annual number of cases of the various forms of pneumonia probably exceeds 2 million and there are about 50,000 deaths. Most pneumonias are bacterial in origin. The more common pathogens are *S. pneumoniae, H. influenzae, S. aureus, Legionella pneumophila,* and *Mycoplasma pneumoniae.*

Some pneumonias caused by certain gram-negative organisms are most likely to occur when the patient's defenses are naturally lowered from such conditions as diabetes or from alcoholism or drug abuse. Among such organisms that cause pneumonias are *K. pneumoniae, E. coli, P. aeruginosa, Branhamella catarrhalis,* and several species of *Enterobacter.*

Most of the organisms that cause pneumonia are members of the normal flora of the mouth and throat that cause opportunistic infections. Viruses, fungi, and protozoans are also responsible for different pneumonias. Examples of these will be discussed later in this chapter. In practice, the causes of many pneumonia cases are never diagnosed. The physician usually makes an initial "best-guess" diagnosis based on symptoms, history, and microscopic examination of a sputum specimen. An antibiotic treatment is then directed at the suspected organisms.

Pneumococcal Pneumonia

The most common cause of bacterial pneumonia in adults is *S. pneumoniae* (the pneumococcus); the disease it causes is **pneumococcal pneumonia.** *S. pneumoniae* is a gram-positive, ovoid bacterium (Figure 24.9). This organism is also a common cause of otitis media, meningitis, and septicemia. Because it usually forms cell pairs, it was formerly called *Diplococcus pneumoniae.* It produces a dense capsule that makes the pathogen

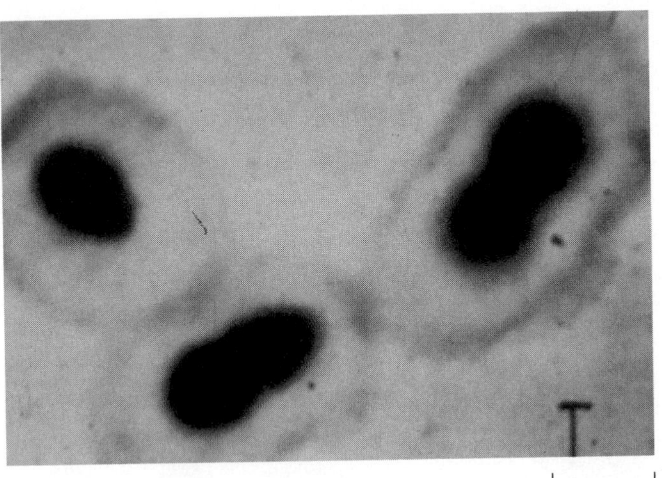

FIGURE 24.9 ***Streptococcus pneumoniae.*** Note the paired arrangement, the reason the organism was once called *Diplococcus pneumoniae.* The capsule, important in pathogenicity, has been made more apparent here by reaction with a specific pneumonococcal antiserum (quellung reaction).

resistant to phagocytosis. These capsules are also the basis of serological differentiation of pneumococci into some 83 serotypes. Before antibiotic therapy became available, antisera directed at these capsular antigens were used to treat the disease.

Pneumococcal pneumonia involves both the bronchi and the alveoli (see Figure 24.2). Symptoms include fever, difficult breathing, and chest pain. The lungs have a reddish appearance because blood vessels are dilated. In response to the infection, alveoli fill with some red blood cells, polymorphonuclear leukocytes (PMNs), and fluid from surrounding tissues. The sputum is often rust-colored from blood coughed up from the lungs. Pneumococci can invade the bloodstream, the pleural cavity surrounding the lung, and occasionally the meninges. No bacterial toxin has been clearly related to pathogenicity.

A presumptive diagnosis can be made by isolation of the pneumococci from the throat, sputum, and other fluids. Pneumococci can be distinguished from other α-hemolytic streptococci by observing the inhibition of growth next to a disk of optochin (ethylhydrocupreine hydrochloride) or by performing a bile solubility test. They can also be serologically typed. The bacteria are observed under a microscope; when they react with a positive antiserum, their capsules appear to swell. This swelling is called the *quellung reaction* (from the German word for swelling).

There are many healthy carriers of the pneumococcus. Virulence of the bacteria seems to be based mainly on the carrier's resistance, which can be lowered by stress. Many illnesses of the elderly terminate in pneumococcal pneumonia.

Recurrence of pneumococcal pneumonia is not uncommon, but the second serological type is usually different. Before chemotherapy was available, the fatality rate was as high as 25%. This has now been lowered to less than 1% for younger patients treated early in the course of their disease. Penicillin is the drug of choice, but a few strains of penicillin-resistant pneumococci have been reported. A vaccine has been developed from the purified capsular material of the 23 types of pneumococci that cause at least 80% of the pneumococcal pneumonias in the United States. This vaccine is used for the groups most susceptible to infection, the elderly and debilitated individuals.

Klebsiella Pneumonia

Klebsiella pneumoniae, the causative agent of ***Klebsiella* pneumonia,** is a member of the enteric group of bacteria. It is commonly found in the throat and mouth of healthy persons. The organism is an encapsulated gram-negative rod, and its virulence is strongly related to the presence of the capsule. This pneumonia is most commonly found in persons who are chronically debilitated, and malnutrition is often a contributing factor. Male alcoholics past 40 years of age are probably the most susceptible group. Probably 1% to 3% of bacterial pneumonias are of this type. Symptoms resemble those of pneumococcal pneumonia, but a prime difference is the formation of lung abscesses and permanent lung damage with *Klebsiella* pneumonia.

The fatality rate in untreated cases can be very high, perhaps exceeding 85%. Early treatment is vital but is usually not received by the most susceptible population. *K. pneumoniae,* like most gram-negative bacteria, is not sensitive to penicillin. In fact, the normal penicillin treatment for pneumococcal pneumonia may even be harmful because it suppresses the gram-positive members of the throat's normal flora, which afford some protection against invading pathogens. Quick identification is essential because delay allows the disease to progress without impedance, which is very dangerous. The antibiotics of choice are cephalosporins or gentamicin.

Mycoplasmal Pneumonia (Primary Atypical Pneumonia)

Typical pneumonia has a bacterial origin. If a bacterial agent is not isolated, the pneumonia is considered *atypical* and viral agents are usually suspected. However, there could be another source of infection—mycoplasmas. These are bacteria that do not form cell walls. The mycoplasmas do not grow under the conditions normally used to recover most bacterial pathogens. Because of this characteristic, pneumonias caused by mycoplasmas are often confused with viral pneumonias.

The bacterium *Mycoplasma pneumoniae* is the causative agent of **mycoplasmal pneumonia (primary atypical pneumonia).** This type of pneumonia was first discovered when such atypical infections responded to tetracyclines, indicating that the agent was nonviral.

The disease is endemic and is a fairly common cause of pneumonia in young adults and children. The pathogens are transmitted in airborne droplets. Mycoplasmas usually infect the upper respiratory tract, and relatively few cases develop into pneumonia. The mild symptoms, which persist for three weeks or longer, are low fever, cough, and headache. Death is very rare.

When isolates from throat swabs and sputum grow on a medium containing horse serum and yeast extract, they form distinctive colonies of a "fried-egg" appearance. The colonies are usually so small that they have to be observed with a hand lens or microscope. *M. pneumoniae* is β-hemolytic for guinea pig red blood cells; this trait differentiates it from other mycoplasmas. Conclusive identification is made by fluorescent-antibody methods. The mycoplasmas are highly varied in appearance because they lack cell walls (see Figure 11.9). Their flexibility allows them to pass through filters with pores as small as 0.2 μm in diameter, which is small enough to retain most other bacteria.

Diagnosis based on the recovery of organisms might not be useful in treatment because as long as two weeks may be required for the slow-growing organisms to develop. A complement fixation test, using antigens from *M. pneumoniae* cells, can be used to test for a significantly rising titer of circulating antibodies, although this procedure also takes from two to three weeks. A DNA probe test (see Figure 10.15) now available should allow reliable diagnosis within hours. Early treatment with tetracycline and erythromycin shortens the illness.

Legionellosis

A type of pneumonia identified only recently is **legionellosis (Legionnaires' disease),** which was discussed in Chapters 1 and 14. This disease first received public attention in 1976, when a series of deaths occurred among members of the American Legion who had attended a meeting in Philadelphia. A total of 182 persons contracted pulmonary disease, apparently at this meeting, 29 of whom died. Because no obvious bacterial cause could be found, the deaths were attributed to viral pneumonia. Close investigation, mostly with techniques directed at locating a suspected rickettsial agent, eventually identified an unknown bacterium, an aerobic gram-negative rod now known as *Legionella pneumophila.* Several species of *Legionella* have now been identified.

The disease is commonly characterized by a high fever of 41°C (105°F), cough, and general symptoms of pneumonia. No person-to-person transmission seems to be involved. Recent studies have shown that the bacterium can be isolated with some frequency from natural waters. Of particular interest is the fact that the organisms can grow well in the water of air-conditioning cooling towers, which might explain some epidemics in hotels, urban business districts, and hospitals as caused by airborne transmission. A recent outbreak was traced to an artificial waterfall (see the box on page 602). The organism has also been found to inhabit the water lines of many hospitals. Most hospitals keep the temperature of hot water lines relatively low (43° to 55°C) as a safety measure, and this inadvertently maintains a good growth temperature in cooler parts of the system for this organism. The bacterium is considerably more resistant to chlorine than most other bacteria and can survive for long periods in water with a low level of chlorine. One possible explanation for the resistance of *Legionella* to chlorine and heat is an interrelationship with waterborne amoebae. The bacteria are ingested by the amoebae but continue to proliferate; they may even survive within encysted amoebae.

Although the disease has only recently been identified, it seems now to have always been fairly common. Over 1000 cases are reported each year, but the actual incidence is estimated at more than 25,000. *L. pneumophila* organism is often overlooked as a possible cause of disease. One hospital that had never recorded a case of nosocomial legionellosis found upon investigation that more than 14% of nosocomial pneumonias had this cause and that most of the sites in its water distribution system contained *L. pneumophila.* Males over 50 years of age are most likely to contract the disease, especially if they are heavy smokers or alcohol abusers or have a chronic illness.

The best diagnostic method is by culture on a selective charcoal–yeast extract medium. Examination of respiratory specimens by fluorescent-antibody methods (see Chapter 18) may be used, and a DNA probe test is available (see Figure 10.15). Erythromycin, with or without rifampin, is the drug of choice for treatment.

PSITTACOSIS (ORNITHOSIS)

The term **psittacosis (ornithosis)** was derived from the disease's association with psittacine birds, such as parakeets and other parrots. It was later found that the disease can also be contracted from many other birds, such as pigeons, chickens, ducks, and turkeys. Therefore, the more general term *ornithosis* has come into use.

The causative organism is *Chlamydia psittaci,* a

MMWR

MORBIDITY AND MORTALITY WEEKLY REPORT

Legionellosis from a Decorative Fountain

In April 1988, an influenzalike illness was reported by members of two groups that held day-long conferences, two days apart, at a hotel in Santa Clara County, California. Both groups were served breakfast and lunch at tables around a decorative fountain in the lobby. The fountain was slightly more than 1 meter high and emitted a fine aerosol spray from multiple jets that gave an overall spherical appearance.

The hotel is served by the municipal water supply, and the fountain used recirculated municipal water. Before the outbreak, there was no set schedule for cleaning. When the fountain was cleaned, no disinfecting chemicals were used nor were records of maintenance kept.

A questionnaire was sent to all 56 persons who attended either of the two conferences. The attack rate was 82%, and the mean incubation period was 56 hours. Symptoms reported by more than 80% of patients in each group included headache, muscle aches, chills, fever, and malaise. Symptoms resolved spontaneously within five days. Nine sought medical care; none was hospitalized. No secondary cases occurred. No rise in antibody titers to influenza A or B virus, parainfluenza virus, respiratory syncytial virus, *Legionella pneumophila*, or *L. feeleii* was detected in the patients' sera. Water samples from the fountain grew *L. anisa*. When eight samples of serum were tested against *L. anisa*, five (63%) showed at least a fourfold rise in antibody titer.

No hotel employees reported similar illness during the outbreak period, but 10 (42%) of 24 of those tested showed *L. anisa* antibody titers of ≥ 1:256. By contrast, none of 48 healthy controls (matched to hotel workers by age, sex, and ethnic group) who had premarital blood tests at the county health department had *L. anisa* titers of at least 1:128.

The symptoms, incubation period, duration of illness, and lack of secondary cases were typical of legionellosis. *L. anisa* was first reported to be a pathogen in 1989, but this is the first outbreak due to this agent. The seven prior reported outbreaks of legionellosis were due to three *Legionella* species: *L. pneumophila, L. feeleii,* and *L. micdadei.*

Comment: It is now widely appreciated that drift aerosols from evaporative condensers and cooling towers can pose a risk of legionellosis if their water is not regularly disinfected. However, aerosolization of water from the kinds of decorative fountains found in shopping malls, hotel atria, and other enclosed spaces can also pose a risk. It is likely that sporadic cases and even undetected outbreaks of legionellosis are occurring elsewhere from similar exposures.

The water in the implicated fountain had never been disinfected, and this permitted the growth of *L. anisa*. After the outbreak, disinfection procedures were instituted. The procedures used for cooling towers can probably be recommended for decorative fountains, too, and include regular inspection, drainage, removal of accumulated dirt, scale, sludge, and slime, and systematic continuous use of biocides, along with periodic microbiologic analysis to ensure that the total bacterial count is kept under control. Maintenance workers should wear protective gear. Documentation of fountain maintenance and the application of biocides should be kept in a bound log book. These efforts should help prevent future outbreaks of legionellosis from decorative fountains.

Source: *California Morbidity* 45/46, Nov. 16, 1990.

gram-negative, obligate intracellular bacterium (see Figure 11.8b). One way chlamydias differ from the rickettsias, which are also obligate intracellular bacteria, is that chlamydias form tiny **elementary bodies** as one part of their life cycle. Unlike most rickettsias, elementary bodies are resistant to environmental stress; therefore, they can be transmitted through air and do not require a bite to transfer the infective agent directly from one host to another. The elementary bodies attach to epithelial cells of the mucous membrane of the respiratory system. They enter the cell by phagocytosis and develop into larger **reticulate bodies** that reproduce by repeated binary fission. Reticulate bodies eventually change back into infectious elementary bodies that leave the cell to infect other cells.

Psittacosis is a form of pneumonia that usually causes fever, headache, and chills. Subclinical infections are very common, and stress appears to enhance susceptibility to the disease. Disorientation and even delirium in some cases indicate that the nervous system can be involved.

The disease is seldom transmitted from one human to another but is usually spread by contact with the droppings and other exudates of fowl. One of the most common modes of transmission is inhalation of dried particles from droppings. The birds themselves usually have diarrhea, ruffled feathers, respiratory illness, and a generally droopy appearance. The parakeets and other parrots sold commercially are usually but not always free of the disease. Pet store employees and persons involved in raising turkeys are at greatest risk of contracting the disease.

Diagnosis is made by isolation of the bacterium in embryonated eggs or mice, or by cell culture. A fluorescent-antibody staining technique can then be used to identify the organism by the presence of its specific antigens. If organisms are not isolated, a complement fixation test that detects rising serum antibody titer in the patient can be used. No vaccine is available, but tetracyclines are effective antibiotics in treating humans and animals. Effective immunity does not result from recovery, even when high titers of antibody are present in the person's serum.

Most years, fewer than 100 cases and very few deaths are reported in the United States. The main danger is in late diagnosis. Before antibiotic therapy was available, mortality was about 20%.

CHLAMYDIAL PNEUMONIA

In recent years there have been outbreaks of a respiratory illness, especially common in populations of college students, that were found to be caused by a chlamydial organism. Originally it was considered a strain of *C. psittaci* (sit'tä-sē) (known by the letters *TWAR,* which were derived from laboratory markers, *TW* for Taiwan and *AR* for acute respiratory, that were being used to identify the original isolates). The pathogen has now been assigned the species name *Chlamydia pneumoniae,* and the disease is known as chlamydial pneumonia. It is apparently transmitted from person to person, probably by the respiratory route, but not as readily as infections such as influenza. Nearly half of the population is serologically positive to antibodies against the organism, an indication that this is a fairly common illness. Clinically, it resembles mycoplasmal pneumonia. Several serological tests have been developed that are useful in diagnosis; the one most used at present is a fluorescent-antibody test. The most effective antibiotic has been tetracycline.

Q FEVER

In Australia in the mid-1930s, a disease was identified that was characterized by a fever lasting one or two weeks, chills, chest pain, severe headache, and other evidence of a pneumonia-type infection. The disease was rarely fatal. In the absence of an obvious cause, the affliction was labeled **Q fever** (for *query*)—much as one might say "X fever." The causative agent was subsequently identified as the obligate, parasitic, intracellular bacterium *Coxiella burnetii,* a rickettsia (Figure 24.10). Most rickettsias are not resistant enough to sur-

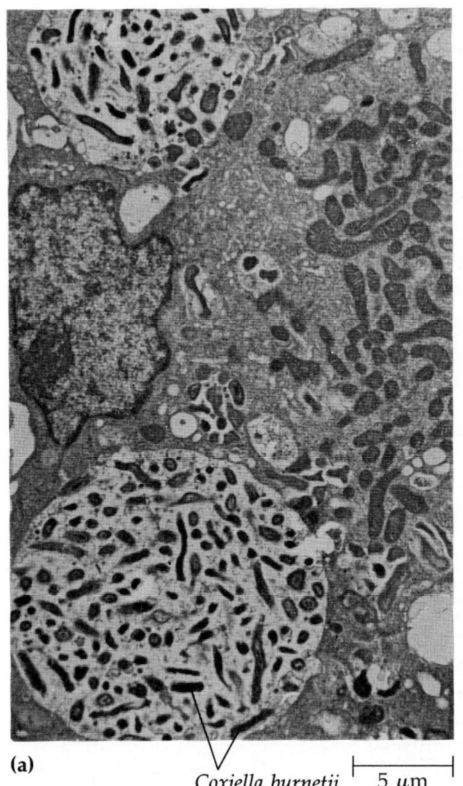

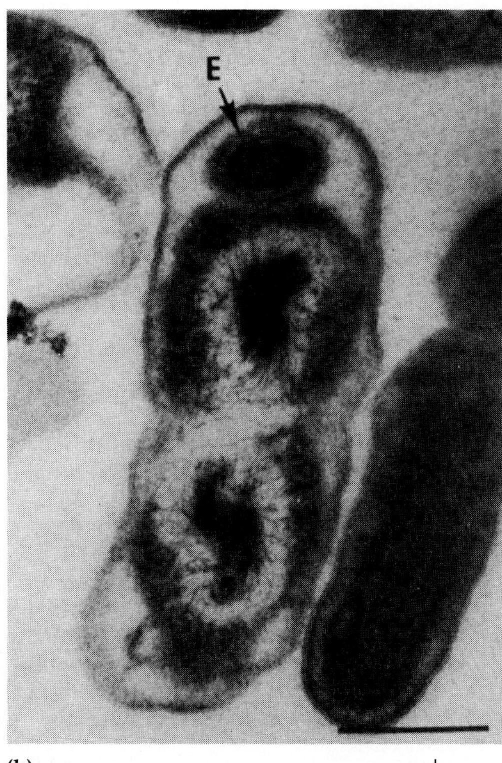

FIGURE 24.10 *Coxiella burnetii.* **(a)** Note the intracellular growth of this rickettsial organism in vacuoles of the host cell. **(b)** This cell has just undergone division. Note the endosporelike body (E), which is probably responsible for the relative resistance of the organism.

vive airborne transmission, but this organism is an exception.

The most serious complication of Q fever is endocarditis, which occurs in about 10% of cases. Five to ten years might elapse between the initial infection and appearance of endocarditis. The organisms apparently reside in the liver during this interval.

The organism is a parasite of a number of arthropods, especially cattle ticks, and is transmitted among animals by tick bites. The infection in animals is usually subclinical. The cattle ticks first spread the disease among dairy herds, and the organisms are shed in the feces, milk, and urine of infected cattle. Once the disease is established in a herd, it is maintained by aerosol transmission. The disease is spread to humans by ingestion of unpasteurized milk and inhalation of aerosols of microbes generated in dairy barns, especially from placental material at calving time. Inhalation of a single organism is enough to cause infection. Many dairy workers have acquired at least subclinical infections. Workers in meat- and hide-processing plants are also at risk. The pasteurization temperature of milk, which was originally aimed at eliminating tuberculosis organisms, was raised slightly in 1956 to ensure the killing of *C. burnetii.* In 1981, an endosporelike body was discovered, which may account for this heat resistance (Figure 24.10b).

The organism may be identified by isolation and growth in chick embryos in eggs or in cell culture. Laboratory personnel testing for *Coxiella*-specific antibodies in the patient's serum can use serological tests, mostly complement fixation and agglutination types.

Most cases of Q fever in the United States are reported from the western states. The disease is endemic to California, Arizona, Oregon, and Washington. A vaccine for laboratory workers and other high-risk personnel is available. Tetracyclines are very effective in treatment.

Viral Diseases of the Lower Respiratory System

VIRAL PNEUMONIA

Viral pneumonia can occur as a complication of influenza or even chickenpox. A number of enteric and other viruses have been shown to cause viral pneumonia, but viruses are isolated and identified in less than 1% of pneumonia-type infections because few laboratories are equipped to properly test clinical samples for viruses. In those cases of pneumonia for which no cause is determined, viral etiology can be assumed if mycoplasmal pneumonia has been ruled out.

Respiratory syncytial virus is probably the most common cause of viral respiratory disease among infants. The name is derived from its characteristic of causing cell fusion (syncytium formation) in cell culture. It is worldwide in occurrence. Epidemics in hospital nurseries are of particular concern. The symptoms are coughing and wheezing that last for more than one week. If an infant is ill enough to be hospitalized, the mortality rate is 1% or 2%. Aerosol administration of the antiviral drug ribavirin at least significantly diminishes the severity of symptoms.

INFLUENZA (FLU)

The developed countries of the world are probably more aware of **influenza (flu)** than of any other disease, except for the common cold. The flu is characterized by chills, fever, headache, and general muscular aches. Recovery normally occurs in a few days, and coldlike symptoms appear as the fever subsides. Incidentally, diarrhea is not a normal symptom of the disease, and the intestinal discomforts attributed to "stomach flu" are probably from some other cause.

The influenza virus consists of eight distinct RNA fragments of differing lengths enclosed by an inner

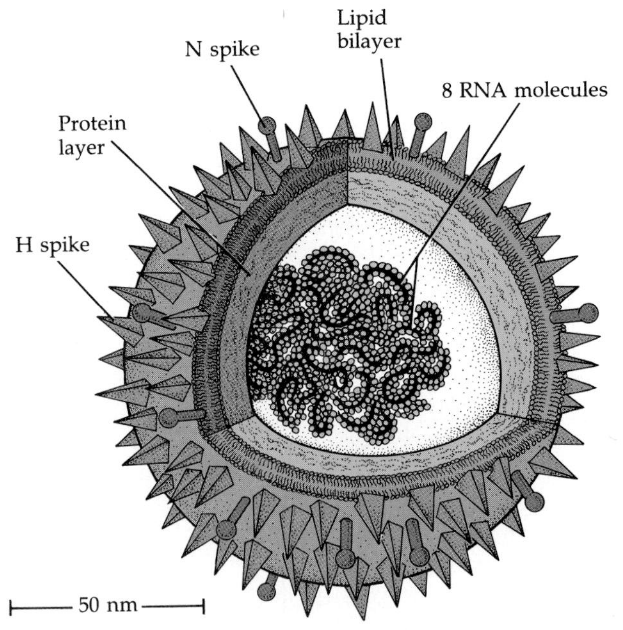

FIGURE 24.11 Detailed structure of the influenza virus. The envelope is composed of a protein layer, lipid layer, and two types of spikes. Because the eight separate RNA molecules can be readily rearranged, the virus can evade previous immunity by major changes in the protein composition of the H and N spikes. This is called antigenic shift. The outer spikes also undergo minor variations from time to time, which is called antigenic drift.

layer of protein and an outer lipid bilayer. Embedded in the lipid bilayer are numerous projections that characterize the virus (Figure 24.11). There are two type of projections, *hemagglutinin (H) spikes* and *neuraminidase (N) spikes,* both of which are highly antigenic.

The H spikes, of which there are about 500 on each virus, allow the virus to recognize and attach to body cells before infecting them. Antibodies against the influenza virus are directed mainly at these spikes. The term *hemagglutinin* refers to the agglutination of red blood cells (hemagglutination) that occurs when the viruses are mixed with them. This reaction is important in serological tests, such as the hemagglutination inhibition test used to identify influenza viruses.

The N spikes, of which there are about 100 per virus, differ from the H spikes in appearance and function. Apparently, they help the virus to separate from the infected cell as the virus exits after intracellular reproduction. N spikes also stimulate formation of antibodies, but these are less important in the body's resistance to the disease than those produced in response to the H spikes.

Viral strains are identified by variation in the H and N antigens. The different forms of the antigens are assigned numbers—for example, H_1, H_2, H_3, N_1, and N_2. Each number change represents a substantial alteration in the protein makeup of the spike. The first antigenic type was isolated in 1933 and was named H_0N_1. The Asiatic flu pandemic of 1957 was of type H_2N_2, and the Hong Kong pandemic of 1968 was designated H_3N_2 (Figure 24.12). These changes are called *antigenic shifts,* and they are great enough to evade most of the immunity developed in the human population. These antigenic shifts are probably caused by a major genetic recombination. Because influenza virus RNA occurs as eight molecules, recombination is likely in infections caused by more than one strain. Recombination between the RNA of animal viral strains (found in swine, horses, and birds, for example) and the RNA of human strains might be involved. Ducks in southern Chinese farming communities have come under suspicion as the animals most likely to be involved in genetic shifts. Wild ducks and other migratory birds then become carriers that spread the virus over large geographical areas.

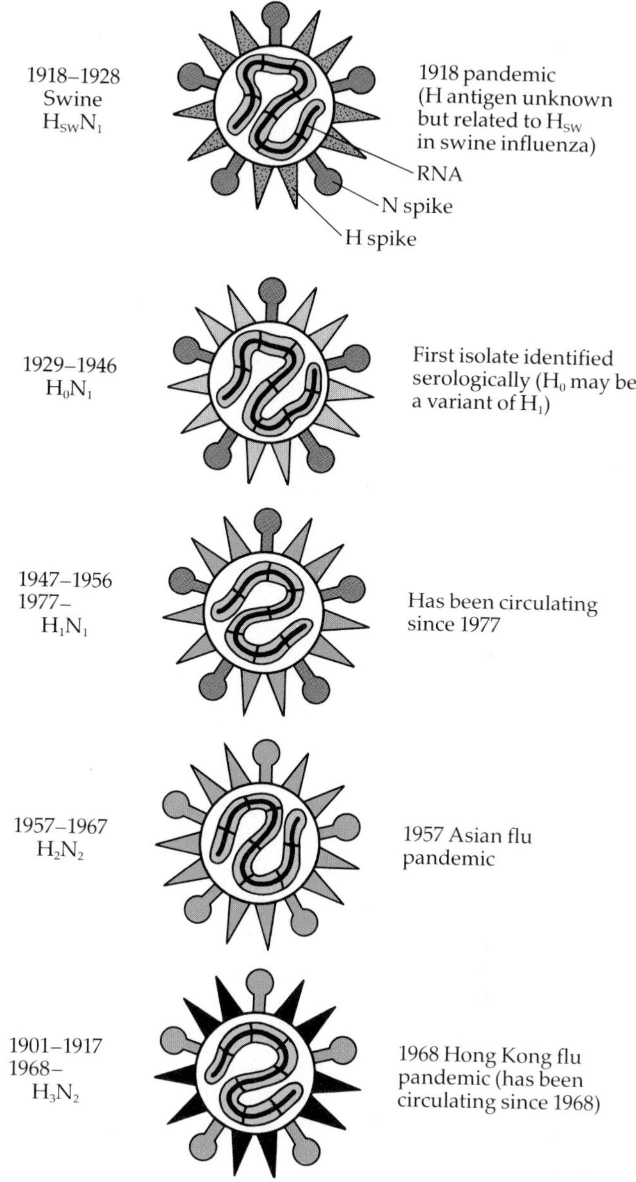

FIGURE 24.12 Differences among influenza viruses. Antigenic shifts in the influenza virus account for most of the major flu epidemics of this century.

Between episodes of such major antigenic shifts, there are minor annual variations in the antigenic makeup called *antigenic drift.* The virus might still be designated as H_3N_2, for example, but viral strains arise reflecting minor antigenic changes within the antigenic group. These strains are sometimes assigned names related to the locality in which they were first identified. They usually reflect an alteration of only a single amino acid in the protein makeup of the H or N spike. Such a minor, one-step mutation is probably a response to selective pressure by antibodies (usually IgA in the mucous membranes) that neutralizes all viruses except for the new mutations. Such mutations can be expected about once in each million multiplications of the virus.

The usual result of antigenic drift is that a vaccine effective against H_3, for example, will be less effective against H_3 isolates circulating ten years after the event. There will have been enough drift in that time that the virus can largely evade the antibodies originally stimulated by the earlier strain. In 1977, the Russian strain

H_1N_1 circulated in the United States. Because it closely resembled a strain that had circulated in the 1950s (see Figure 24.12), it primarily affected persons under the age of about 25.

The influenza viruses are also classified into major groups according to the antigens of their protein coats. These groups are A, B, and (rarely) C. The A-type viruses are the ones responsible for major pandemics. The B-type virus also circulates and mutates, but it is usually responsible for more geographically limited and milder infections.

To date, it has not been possible to make a vaccine for influenza that gives long-term immunity to the general population. Although it is not difficult to make a vaccine for a particular antigenic strain of virus, each new strain of circulating virus must be identified in time for useful development and distribution of a new vaccine. Representatives from the Centers for Disease Control recently visited mainland China to train medical workers there in the techniques needed to detect molecular changes in the influenza virus. By combining information from China, Japan, and Taiwan, we now get earlier warning of new influenza virus types. This usually gives more opportunity to develop annual vaccines directed at the likely antigenic types. Unless the strain appears to be unusually virulent, vaccine administration is usually limited to the elderly, to hospital personnel, and to similar high-risk groups. The vaccines are often *multivalent,* that is, directed at several strains in circulation at the time. At present, influenza viruses for manufacturing vaccines are grown in embryonated egg cultures. The vaccines are usually 70% to 90% effective, but the duration of protection is probably no more than three years.

One new approach to the problem of antigenic variation might be to use genetic-engineering techniques. A laboratory bacterium can theoretically be made to produce very quickly a large number of antigens of the influenza virus when genes for this purpose have been inserted into the bacterial DNA.

Almost every year, epidemics of the flu spread rapidly through large populations. The disease is so readily transmissible that epidemics are quickly propagated through populations susceptible to the newly changed strain of virus. The mortality rate from the disease is not high, usually less than 1%, and these deaths are mainly among the very young and the very old. However, so many persons are infected in a major epidemic that the total number of deaths is often large. Very often, the cause of death is not the influenza virus but secondary bacterial infections. The bacterium *H. influenzae* was named under the mistaken belief that it was the principal pathogen causing influenza rather than being a secondary invader. *S. aureus* and *S. pneumoniae* are other prominent secondary bacterial invaders.

In any discussion of influenza, the great pandemic of 1918–1919 must be mentioned. Worldwide, more than 20 million people died. No one is sure why it was so unusually lethal. Usually, the very young and very old are the principal victims, but in 1918–1919, young adults had the highest mortality rate. Normally the infection is restricted to the upper respiratory system, but some change in virulence allowed the virus to invade the lungs and frequently cause a fatal viral pneumonia. Bacterial complications also frequently accompanied the infection and, in those preantibiotic days, were often fatal. The 1918 viral strain apparently became endemic in the swine population of the United States. Occasionally, influenza is still spread to humans from this reservoir, but the disease does not propagate like the virulent disease of 1918. In 1976, however, a recruit at Fort Dix, New Jersey, died of influenza caused by the so-called swine flu strain. This precipitated a national campaign of preventive inoculations for swine influenza.

An antiviral drug, amantadine, has been found to significantly reduce the symptoms of influenza if administered promptly. In prophylactic use, it apparently reduces the rate of infection and illness by perhaps as much as 70%. The bacterial complications of influenza are amenable to treatment with antibiotics.

New techniques that use immunofluorescence testing with monoclonal antibodies have made it unnecessary to send influenza virus isolates to central laboratories to be typed. Many hospital laboratories, using this method, can type influenza viruses within 24 to 72 hours.

Fungal Diseases of the Lower Respiratory System

HISTOPLASMOSIS

Histoplasmosis superficially resembles tuberculosis. In fact, it was first recognized as a disease in the United States when X-ray surveys showed lung lesions in many persons who were tuberculin test negative. Although the lungs are the organs most likely to be initially infected, the organisms cause lesions in almost all organs of the body. Normally, symptoms of the disease are rather poorly defined and mostly subclinical, and the disease passes for a minor respiratory infection. In a few cases, histoplasmosis becomes progressive and is a severe, generalized disease. This occurs in only a small number of infections, perhaps less than 0.1%.

The causative organism, *Histoplasma capsulatum* (his-tō-plaz′mä kap-su-lä′tum), is a dimorphic fungus; that is, it has a yeastlike morphology in tissue growth,

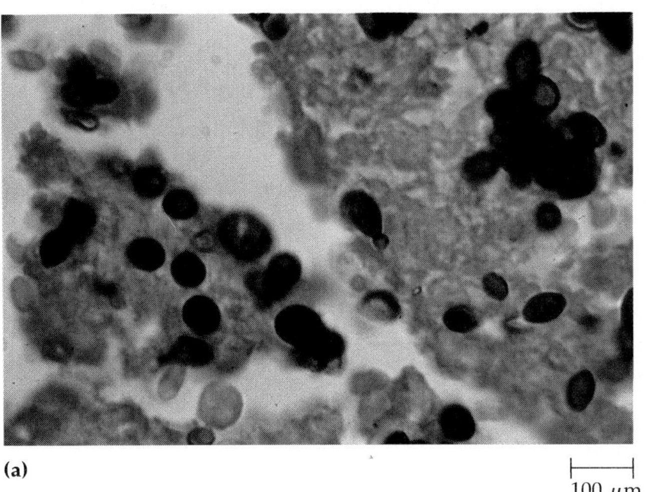

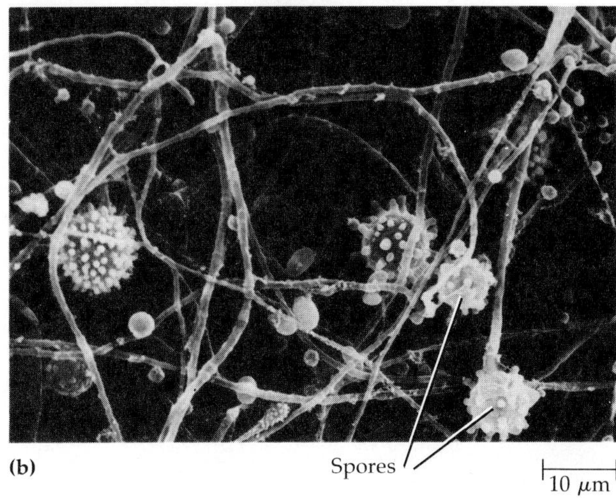

FIGURE 24.13 ***Histoplasma capsulatum,* a dimorphic fungus that causes histoplasmosis.** **(a)** Yeastlike form typical of growth in tissue or at 37°C. One cell near the center is budding. **(b)** Filamentous, spore-forming phase found in soil or at temperatures below 35°C. The spores are usually the infectious particles.

and, in soil or artificial media, it forms a filamentous mycelium carrying reproductive conidia (Figure 24.13). In the body, the yeastlike form is found intracellularly in macrophages, where it survives and multiplies.

Although histoplasmosis is rather widespread throughout the world, it has a limited geographical range in the United States (Figure 24.14). For example, of 513 cases reported in a recent year, 403 were from the states of Minnesota, Iowa, Missouri, and Alabama. In general, the disease is found in the states adjoining the Mississippi and Ohio rivers. More than 75% of the population in some of these states have antibodies against the infection. In other states—Maine, for example—a positive test is a rather rare event. Approximately 50 deaths are reported in the United States each year from histoplasmosis. The total number of infected persons may run into the millions, so the death rate is low.

Humans acquire the disease from airborne conidia produced under conditions of particular moisture and pH levels. These conditions occur particularly where droppings from birds and bats have accumulated.

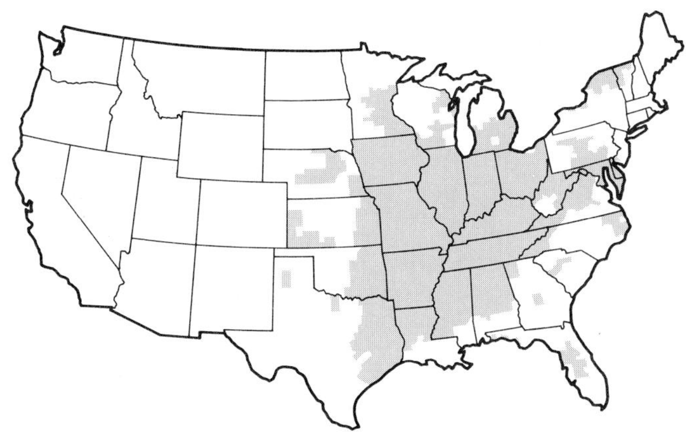

FIGURE 24.14
Histoplasmosis distribution. Geographical distribution of histoplasmosis (colored area) in the United States.

Birds themselves do not carry the disease, but their droppings provide nutrients, particularly a source of nitrogen, for the fungus. Bats, which have a lower body temperature than birds, carry the fungus, shed it in their feces, and probably infect new soil sites.

Clinical signs and history, complement fixation tests, and, most important, isolation of the organism or its identification in tissue specimens are necessary for proper diagnosis. Currently, the most effective chemotherapy is with amphotericin B or ketoconazole.

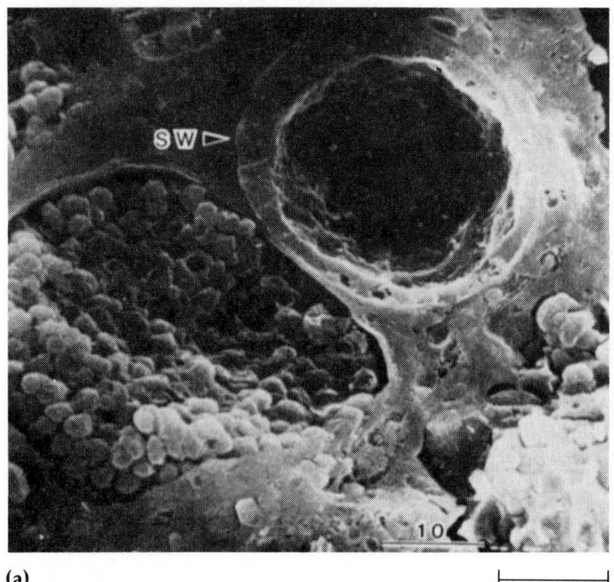

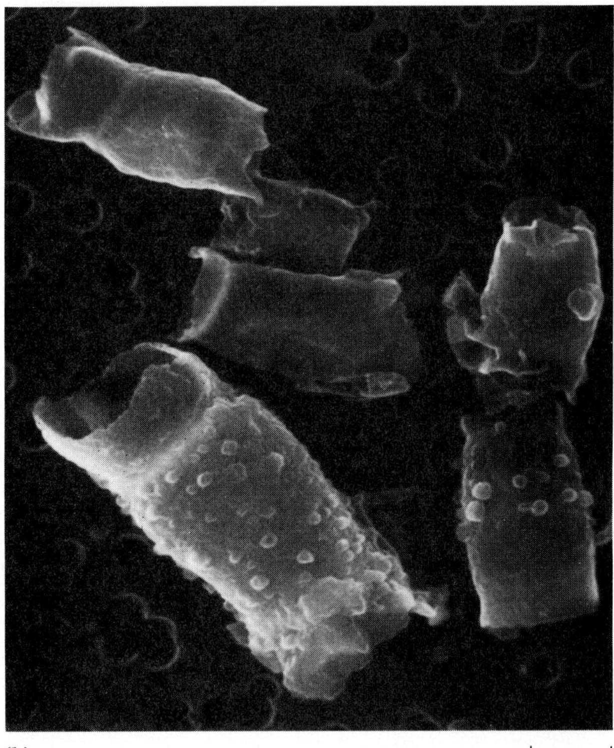

FIGURE 24.15 *Coccidioides immitis.* **(a)** Scanning electron micrograph of two spherical bodies found in mouse lung tissue. Note the thick wall (SW) of the empty spherule and the mass of spores in the other spherule. **(b)** Scanning electron micrograph of arthrospores, the airborne infectious particles that spread coccidioidomycosis. The holes in the background are about 0.45 μm.

COCCIDIOIDOMYCOSIS

Another fungal pulmonary disease, also rather restricted geographically, is **coccidioidomycosis.** The causative organism is *Coccidioides immitis,* a dimorphic fungus. The spores are found in dry, highly alkaline soils of the American Southwest and in similar soils of South America and northern Mexico. In tissues, the organism forms a thick-walled body filled with spores *(spherule)* (Figure 24.15a). In soil, it forms filaments that reproduce by formation of arthrospores (Figure 24.15b). The wind carries the arthrospores to transmit the infection. Arthrospores are often so abundant that simply driving through an endemic area can result in infection, especially during a dust storm.

The symptoms of coccidioidomycosis include chest pain and perhaps fever, coughing, and loss of weight. Most infections are inapparent, and almost all patients recover in a few weeks, even without treatment. Sound immunity remains. However, in less than 1% of the infections, a progressive disease resembling tuberculosis disseminates throughout the body. The resemblance to tuberculosis is so close that isolation of the organism is necessary to properly diagnose coccidioidomycosis. Diagnosis is most reliably made by identification of the spherules (Figure 24.15a) in tissue or fluids. The organism is also readily cultured from fluids or lesions. Laboratory workers must handle it with unusual care because of the possibility of infectious aerosols. A tuberculinlike skin test is used in screening for cases. Of those tested in California, more than half had positive skin tests, although most had not been aware of the infection. Farm workers are most likely to be infected, and predisposing factors, such as fatigue and poor nutrition, can lead to more severe disease.

Approximately 50 to 100 deaths each year are reported from this disease in the United States. Most cases are reported from southern California and Arizona desert regions, with only a scattering of cases reported from other states (Figure 24.16).

A vaccine is currently being tested. If successful, it is expected to be widely used in the areas affected by coccidioidomycosis. Amphotericin B has been used to

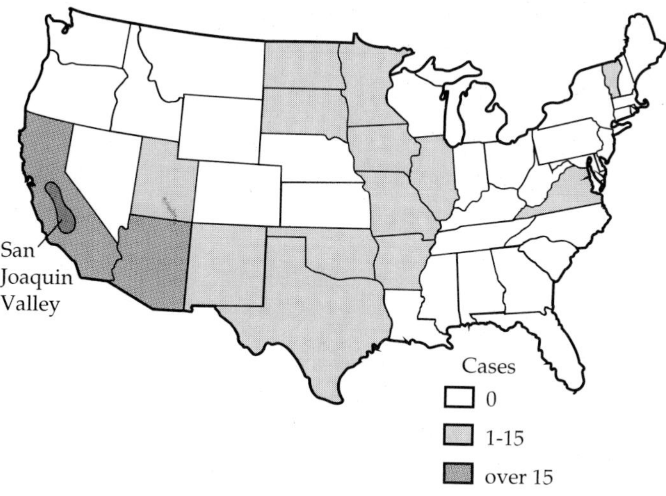

FIGURE 24.16 Cases of coccidioidomycosis in the United States in a typical year. Because of the large number of cases that occur in the San Joaquin Valley in California, this disease is sometimes called San Joaquin Valley fever. A number of the cases shown in states outside the Southwest were probably acquired during a stay in the endemic areas, mostly California and Arizona.

treat serious cases, but less toxic imidazole drugs, such as ketoconazole, might replace it.

BLASTOMYCOSIS (NORTH AMERICAN BLASTOMYCOSIS)

Blastomycosis is usually called **North American blastomycosis** to differentiate it from a similar South American blastomycosis. It is caused by the fungus *Blastomyces dermatitidis* (blas-tō-mī'sēz dėr-mä-tit'i-dis), a dimorphic fungus found most often in the Mississippi Valley, where it probably grows in soil. Approximately 30 to 60 deaths are reported each year, although most infections are asymptomatic. The infection begins in the lungs and can spread rapidly. Cutaneous ulcers commonly appear, and there is extensive abscess formation and tissue destruction. The organism can be isolated from pus and biopsy specimens. Amphotericin B is usually an effective treatment.

OTHER FUNGI INVOLVED IN RESPIRATORY DISEASE

Many other opportunistic fungi may cause respiratory disease, particularly in immunosuppressed hosts or when there is exposure to massive numbers of spores. **Aspergillosis** is an important example; it is transmitted by the spores of *Aspergillus fumigatus* (fū-mi-gä'tus) and other species of *Aspergillus*, which are fairly widespread in decaying vegetation. Compost piles are ideal sites for growth, and farmers and gardeners are most often exposed to infective amounts of such spores. Similar pulmonary infections sometimes result when individuals are exposed to spores of mold genera, such as *Rhizopus* or *Mucor*. Such diseases can be very dangerous, particularly invasive infections of pulmonary aspergillosis. Predisposing factors include an impaired immune system, cancer, and diabetes. As with most systemic fungal infections, there is only a limited arsenal of antifungal agents available; amphotericin B has proved the most useful.

Protozoan Disease of the Lower Respiratory System

PNEUMOCYSTIS PNEUMONIA

***Pneumocystis* pneumonia** is caused by *Pneumocystis carinii*. The taxonomic position of this organism has been uncertain ever since its first discovery in 1909, when it was thought to be a developmental stage of a trypanosome. Since that time there has been no universal agreement about whether it is a protozoan or a fungus. It has some characteristics of both groups of organisms. Recent analysis of the RNA shows that it is strongly related to certain yeasts. The disease occurs throughout the world and can be endemic in hospitals. The organism is found in healthy human lungs but causes disease among immunosuppressed patients. This group includes persons receiving immunosuppressive drugs to minimize rejection of transplanted tissue and those whose immunity is depressed because of cancer. Persons with acquired immunodeficiency syndrome (AIDS) are also very susceptible to this organism.

Before the AIDS epidemic, *Pneumocystis* pneumonia was an uncommon disease; perhaps 100 cases occurred each year. But a leading AIDS researcher observed in 1990 that, "during the past year, the number of deaths from *Pneumocystis carinii* pneumonitis (pneumonia) in the United States exceeded the combined number of deaths from meningococcal infections, all types of viral hepatitis and encephalitis, gonorrhea, syphilis, varicella, measles, mumps, rubella, diphtheria, tetanus, pertussis, polio, amebiasis, shigellosis, salmonellosis, typhoid fever, typhus fevers, cholera, rabies, brucellosis, anthrax, tularemia, botulism, and malaria. More patients died of *P. carinii* pneumonitis than of all types of tuberculosis."

The life cycle of *P. carinii* is not completely known. In the human lung the organisms are found mostly in

TABLE 24.1 Diseases Associated with the Respiratory System

DISEASE	CAUSATIVE AGENT	MODE OF TRANSMISSION	TREATMENT
Bacterial Diseases of the Upper Respiratory System			
Streptococcal pharyngitis (strep throat)	Streptococci, especially *Streptococcus pyogenes*	Respiratory secretions	Penicillin
Scarlet fever	Erythrogenic toxin-producing strains of *Streptococcus pyogenes*	Respiratory secretions	Penicillin and erythromycin
Diphtheria	*Corynebacterium diphtheriae*	Respiratory secretions, healthy carriers	Antitoxin and penicillin, tetracyclines, erythromycin
Cutaneous diphtheria	*Corynebacterium diphtheriae*	Respiratory secretions, healthy carriers	Antitoxin and penicillin, tetracyclines, erythromycin
Otitis media	Several agents, especially *Staphylococcus aureus, Streptococcus pneumoniae,* β-hemolytic streptococci, and *Hemophilus influenzae*	Complication of cold, or nose or throat infection	Penicillin and amoxicillin

(Continued)

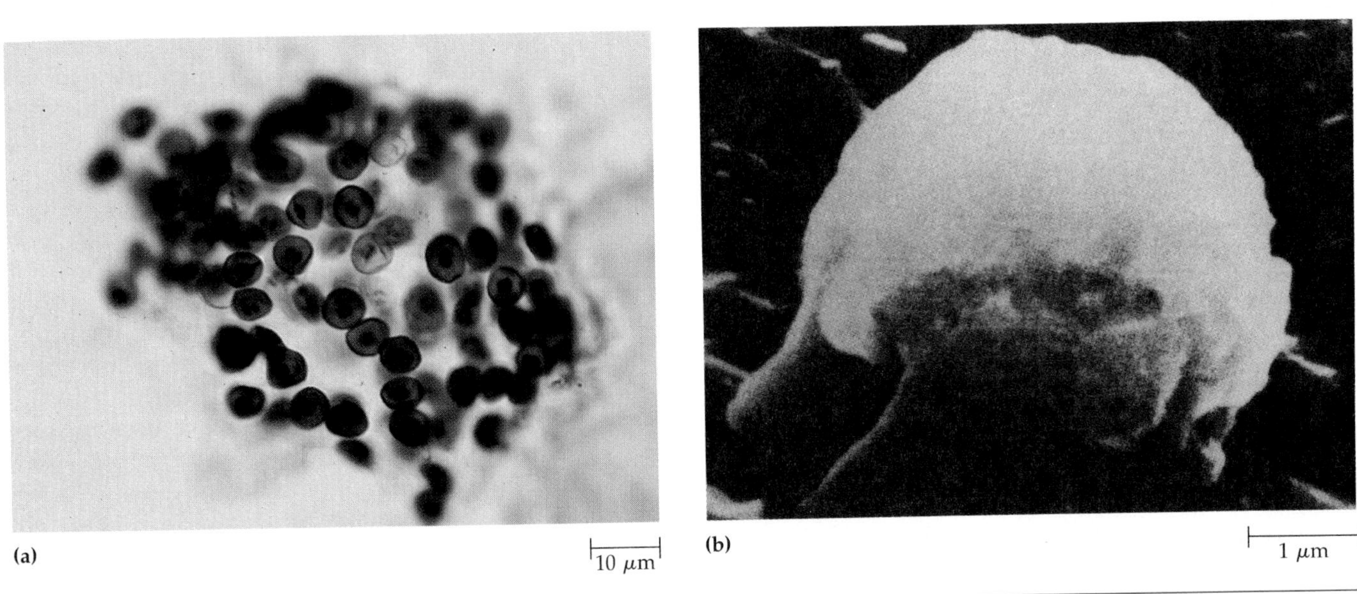

FIGURE 24.17 ***Pneumocystis carinii.*** **(a)** Stained cysts in a smear taken from the lung of a patient. **(b)** The trophozoite stage adhering to the surface of a chick embryo lung cell. In this scanning electron micrograph, note the tubular extensions through which the organism extracts nutrients from the host.

the lining of the alveoli. There they form a thick-walled cyst in which spherical intracystic bodies successively divide as part of a sexual cycle. The mature cyst contains eight such bodies (Figure 24.17a). Eventually it ruptures and releases them, each body developing into a vegetative cell (see Figure 24.17b). The vegetative cells can reproduce asexually by binary fission or budding, but may also enter the encysted sexual stage.

The drug of choice is currently trimethoprim-sulfamethoxazole, but pentamidine isethionate is a frequent alternative.

TABLE 24.1 *(Continued)*

DISEASE	*CAUSATIVE AGENT*	*MODE OF TRANSMISSION*	*TREATMENT*
Viral Disease of the Upper Respiratory System			
Common cold	Coronaviruses, rhinoviruses	Respiratory secretions	None
Bacterial Diseases of the Lower Respiratory System			
Whooping cough	*Bordetella pertussis*	Respiratory secretions	Erythromycin
Tuberculosis	*Mycobacterium tuberculosis*	Respiratory secretions	Isoniazid and rifampin
	Mycobacterium bovis	Usually milk-borne	Isoniazid and rifampin
Pneumococcal pneumonia	*Streptococcus pneumoniae*	Healthy carriers; primarily a disease following viral respiratory infection or other stress	Penicillin
Klebsiella pneumonia	*Klebsiella pneumoniae*	Primarily a disease in debilitated hosts (for example, alcoholics)	Cephalosporins or gentamicin
Mycoplasmal pneumonia	*Mycoplasma pneumoniae*	Respiratory secretions	Tetracycline, erythromycin
Legionellosis	*Legionella pneumophila*	Aerosols from contaminated water	Erythromycin, rifampin
Psittacosis (ornithosis)	*Chlamydia psittaci*	Animal reservoirs; aerosols of dried droppings and other exudates of birds; person-to-person transmission is rare	Tetracyclines
Q fever	*Coxiella burnetii*	Animal reservoirs; aerosols in dairy barns; unpasteurized milk	Tetracyclines
Chlamydial pneumonia	*Chlamydia pneumoniae*	Respiratory secretions	Tetracyclines
Viral Diseases of the Lower Respiratory System			
Viral pneumonia	Several viruses	Aerosols; complication of other viral diseases	Ribivirin diminishes symptoms
Influenza	Influenza viral strains; many serotypes	Respiratory secretions	Amantadine (prophylaxis)
Fungal Diseases of the Lower Respiratory System			
Histoplasmosis	*Histoplasma capsulatum*	Bird or bat reservoirs; airborne fungal spores from growth in soil; person-to-person transmission is rare	Amphotericin B, ketoconazole
Coccidioidomycosis	*Coccidioides immitis*	Soil organisms; aerosols containing arthrospores	Amphotericin B, ketoconazole
Blastomycosis	*Blastomyces dermatitidis*	Soil organisms; airborne spores	Amphotericin B
Other fungal pneumonias	Species of *Aspergillus, Rhizopus, Mucor,* and other genera	Airborne spores of opportunistic fungi	Amphotericin B
Protozoan Disease of the Lower Respiratory System			
Pneumocystis pneumonia	*Pneumocystis carinii*	Probably inhalation of respiratory secretions	Pentamidine isethionate

STUDY OUTLINE

Introduction (p. 589)

1. Infections of the upper respiratory system are the most common type of infection.
2. Pathogens that enter the respiratory system can infect other parts of the body.
3. Respiratory infections are transmitted by direct contact and fomites.

Structure and Function of the Respiratory System (pp. 589–591)

1. The upper respiratory system consists of the nose, throat, and associated structures, such as the middle ear and auditory tubes.
2. Coarse hairs in the nose filter large particles from air entering the respiratory tract.
3. The ciliated mucous membranes of the nose and throat trap airborne particles and remove them from the body.
4. Lymphoid tissue, tonsils, and adenoids provide immunity to certain infections.
5. The lower respiratory system consists of the larynx, trachea, bronchial tubes, and alveoli.
6. The ciliary escalator of the lower respiratory system helps prevent microorganisms from reaching the lungs.
7. Microorganisms in the lungs can be phagocytized by alveolar macrophages.
8. Respiratory mucus contains IgA antibodies.

Normal Flora of the Respiratory System (pp. 590–591)

1. The normal flora of the nasal cavity and throat can include pathogenic microorganisms.
2. The lower respiratory system is usually sterile because of the action of the ciliary escalator.

DISEASES OF THE UPPER RESPIRATORY SYSTEM (pp. 591–595)

1. Specific areas of the upper respiratory system can become infected to produce pharyngitis, laryngitis, tonsillitis, sinusitis, and epiglottitis.
2. These infections may be caused by several bacteria and viruses, often in combination.
3. *H. influenzae* type b can cause epiglottitis.
4. Most respiratory tract infections are self-limiting.

Bacterial Diseases of the Upper Respiratory System (pp. 592–594)

STREPTOCOCCAL PHARYNGITIS (STREP THROAT) (p. 592)

1. This infection is caused by group A β-hemolytic streptococci, the group to which *Streptococcus pyogenes* belongs.
2. Symptoms of this infection are inflammation of the mucous membrane and fever; tonsillitis and otitis media may also occur.
3. Rapid diagnosis is made by indirect agglutination tests.
4. Penicillin is used to treat streptococcal pharyngitis.
5. Immunity to streptococcal infections is type-specific.
6. Streptococcal sore throat is usually transmitted by droplets but has been associated with unpasteurized milk.

SCARLET FEVER (pp. 592–593)

1. Streptococcal sore throat, caused by an erythrogenic toxin-producing *S. pyogenes*, results in scarlet fever.
2. *S. pyogenes* produces erythrogenic toxin when lysogenized by a phage.
3. Symptoms include a pink rash, high fever, and a red, enlarged tongue.

DIPHTHERIA (pp. 593–594)

1. Diphtheria is caused by exotoxin-producing *Corynebacterium diphtheriae.*
2. Exotoxin is produced when the bacteria are lysogenized by a phage.
3. A membrane, containing fibrin and dead human and bacterial cells, forms in the throat and can block the passage of air.
4. The exotoxin inhibits protein synthesis, and heart, kidney, or nerve damage may result.
5. Laboratory diagnosis is based on isolation of the bacteria and appearance of growth on differential media.
6. Toxigenicity is determined with the guinea pig test or gel diffusion test.
7. Antitoxin must be administered to neutralize the toxin, and antibiotics can stop growth of the bacteria.
8. Routine immunization in the United States includes diphtheria toxoid in the DPT vaccine.

CUTANEOUS DIPHTHERIA (p. 594)

1. Slow-healing skin ulcerations are characteristic of cutaneous diphtheria.
2. There is minimal dissemination of the exotoxin in the bloodstream.

OTITIS MEDIA (p. 594)

1. Earache, or otitis media, can occur as a complication of nose and throat infections or through direct inoculation from an external source.
2. Pus accumulation causes pressure on the eardrum.
3. Bacterial causes include *Streptococcus pneumoniae, Hemophilus influenzae, Branhamella catarrhalis, Streptococcus pyogenes*, and *Staphylococcus aureus*.

Viral Disease of the Upper Respiratory System (pp. 594–595)

COMMON COLD (ACUTE CORYZA) (pp. 594–595)

1. Any one of approximately 200 different viruses can cause the common cold; rhinoviruses cause about 50% of all colds.
2. Symptoms include sneezing, nasal secretions, and congestion.
3. Sinus infections, lower respiratory tract infections, laryngitis, and otitis media can occur as complications of a cold.

4. Colds are most often transmitted by indirect contact.
5. Rhinoviruses prefer temperatures slightly lower than body temperature.
6. The incidence of colds increases during cold weather, possibly because of increased interpersonal indoor contact or physiological changes.
7. Antibodies are produced against the specific viruses.

DISEASES OF THE LOWER RESPIRATORY SYSTEM (pp. 595–611)

1. Many of the same microorganisms that infect the upper respiratory system also infect the lower respiratory system.
2. Diseases of the lower respiratory system include bronchitis and pneumonia.

Bacterial Diseases of the Lower Respiratory System (pp. 595–604)

WHOOPING COUGH (PERTUSSIS) (pp. 595–596)

1. Whooping cough is caused by *Bordetella pertussis.*
2. The initial stage of whooping cough resembles a cold and is called the catarrhal stage.
3. The accumulation of mucus in the trachea and bronchi causes deep coughs characteristic of the paroxysmal (second) stage.
4. The convalescence (third) stage can last for months.
5. Laboratory diagnosis is based on isolation of the bacteria on enrichment and selective media followed by serological tests.
6. Regular immunization for children includes dead *B. pertussis* cells as part of the DPT vaccine.

TUBERCULOSIS (pp. 596–599)

1. Tuberculosis is caused by *Mycobacterium tuberculosis.*
2. Large amounts of lipids in the cell wall account for the bacterum's acid-fast characteristic as well as its resistance to drying and disinfectants.
3. *M. tuberculosis* may be ingested by alveolar macrophages; if not killed, the bacteria reproduce in the macrophages.
4. Lesions formed by *M. tuberculosis* are called tubercles; necrosis results in a caseous lesion that might calcify and appear in an X ray as a Ghon complex.
5. New foci of infection can develop when a caseous lesion ruptures and releases bacteria into blood or lymph vessels; this is called miliary tuberculosis.
6. Miliary tuberculosis is characterized by weight loss, coughing, and loss of vigor.
7. A positive tuberculin skin test can indicate either an active case of tuberculosis or prior infection or vaccination and immunity to the disease.
8. Laboratory diagnosis is based on isolation of the bacteria and requires incubation of up to eight weeks.
9. *Mycobacterium bovis* causes bovine tuberculosis and can be transmitted to humans by unpasteurized milk.
10. *M. bovis* infections usually affect the bones or lymphatic system.
11. BCG vaccine for tuberculosis consists of a live, avirulent culture of *M. bovis.*
12. *M. avium-intracellulare* complex infects patients in the late stages of HIV infection.

BACTERIAL PNEUMONIAS (pp. 599–601)

1. Most cases of pneumonia are caused by normal flora of the mouth and throat.
2. The most common etiologic agents are *S. pneumoniae, H. influenzae, S. aureus, L. pneumophila,* and *M. pneumoniae.*

Pneumococcal Pneumonia (pp. 599–600)

1. Pneumococcal pneumonia is caused by encapsulated *Streptococcus pneumoniae.*
2. Symptoms of this disease are fever, difficult breathing, chest pain, and rust-colored sputum.
3. The bacteria can be identified by the production of α-hemolysins, inhibition by optochin, bile solubility, and the quellung reaction.
4. A vaccine consists of purified capsular material from 23 serotypes of *S. pneumoniae.*

Klebsiella Pneumonia (p. 600)

1. *Klebsiella pneumoniae* causes *Klebsiella* pneumonia.
2. *Klebsiella* pneumonia results in lung abscesses and permanent lung damage; mortality is 85%.

Mycoplasmal Pneumonia (Primary Atypical Pneumonia) (pp. 600–601)

1. *Mycoplasma pneumoniae* causes primary atypical pneumonia, or mycoplasmal pneumonia.
2. *M. pneumoniae* produces small "fried-egg" colonies after two weeks' incubation on enriched media containing horse serum and yeast extract.
3. A complement fixation test, used to diagnose the disease, is based on the rising of antibody titer. A DNA probe provides rapid identification of the bacteria.

Legionellosis (p. 601)

1. The disease is caused by the aerobic gram-negative rod *Legionella pneumophila.*
2. The bacterium can grow in water, such as air-conditioning cooling towers, and then be disseminated in the air.
3. This pneumonia does not appear to be transmitted from person to person.
4. Bacterial culture, fluorescent-antibody tests, or DNA probes are used for laboratory diagnosis.

PSITTACOSIS (ORNITHOSIS) (pp. 601–603)

1. *Chlamydia psittaci* is transmitted by contact with contaminated droppings and exudates of fowl.
2. Elementary bodies allow the bacteria to survive outside a host. Reticulate bodies develop in host cells.
3. The bacteria are isolated in embryonated eggs, mice, or cell culture; identification is based on fluorescent-antibody staining.
4. Commercial bird handlers are most susceptible to this disease.

CHLAMYDIAL PNEUMONIA (p. 603)

1. *Chlamydia pneumoniae* causes pneumonia; it is transmitted from person to person.
2. A fluorescent-antibody test is used for diagnosis.

Q FEVER (pp. 603–604)

1. Obligately parasitic, intracellular *Coxiella burnetii* causes Q fever.

2. The disease is usually transmitted to humans through unpasteurized milk or inhalation of aerosols in dairy barns.
3. Laboratory diagnosis is made with the culture of bacteria in embryonated eggs or cell culture.

Viral Diseases of the Lower Respiratory System (pp. 604–606)

VIRAL PNEUMONIA (p. 604)

1. Many pneumonia cases are thought to be caused by viruses.
2. A number of viruses can cause pneumonia as a complication of infections such as influenza.
3. The etiologies are not usually identified in a clinical laboratory because of the difficulty in isolating and identifying viruses.
4. Respiratory syncytial virus is a common cause of pneumonia in infants.

INFLUENZA (FLU) (pp. 604–606)

1. Influenza is caused by influenza virus and is characterized by chills, fever, headache, and general muscular aches.
2. Hemagglutinin (H) and neuraminidase (N) spikes project from the outer lipid bilayer of the virus.
3. Viral strains are identified by antigenic differences in the H and N spikes; they are also divided by antigenic differences in their protein coats (A, B, and C).
4. Viral isolates are identified by immunofluorescence testing with monoclonal antibodies.
5. Antigenic shifts that alter the antigenic nature of the H and N spikes make natural immunity and vaccination of questionable value. Minor antigenic changes are caused by antigenic drift.
6. Deaths during an influenza epidemic are usually from secondary bacterial infections.
7. Multivalent vaccines are available for the elderly and other high-risk groups.
8. Amantadine is an effective prophylactic and curative drug.

Fungal Diseases of the Lower Respiratory System (pp. 606–609)

1. The following mycoses can be treated with amphotericin B.

HISTOPLASMOSIS (pp. 606–608)

1. *Histoplasma capsulatum* causes a subclinical respiratory infection that only occasionally progresses to a severe, generalized disease.
2. The disease is acquired by inhalation of airborne conidia.
3. Isolation of the fungus or identification of the fungus in tissue samples is necessary for diagnosis.

COCCIDIOIDOMYCOSIS (pp. 608–609)

1. Inhalation of the airborne arthrospores of *Coccidioides immitis* can result in coccidioidomycosis.
2. Most cases are subclinical, but when there are predisposing factors such as fatigue and poor nutrition, a progressive disease resembling tuberculosis can result.

BLASTOMYCOSIS (NORTH AMERICAN BLASTOMYCOSIS) (p. 609)

1. *Blastomyces dermatitidis* is the causative agent of blastomycosis.
2. The infection begins in the lungs and can spread to cause extensive abscesses.

OTHER FUNGI INVOLVED IN RESPIRATORY DISEASE (p. 609)

1. Opportunistic fungi can cause respiratory disease in immunosuppressed hosts, especially when large numbers of spores are inhaled.
2. Among these fungi are *Aspergillus, Rhizopus,* and *Mucor.*

Protozoan Disease of the Lower Respiratory System (pp. 609–611)

1. Immunosuppressed patients and cancer patients are susceptible to an endemic parasite, *Pneumocystis carinii.*
2. *Pneumocystis* pneumonia is currently being treated with trimethoprim or pentamidine.

STUDY QUESTIONS

Review

1. Describe how microorganisms are prevented from entering the upper respiratory system. How are they prevented from causing infections in the lower respiratory system?
2. Respiratory diseases are usually transmitted by __________. List other methods by which they can be transmitted.
3. How do the normal flora of the respiratory system illustrate microbial antagonism?
4. Compare and contrast primary atypical pneumonia and viral pneumonia.
5. How is otitis media contracted? What causes otitis media? Why was otitis media included in a chapter on diseases of the respiratory system?
6. List the causative agent, symptoms, and treatment for three viral diseases of the respiratory system. Separate the diseases according to whether they infect the upper or lower respiratory system.
7. Under what conditions can the saprophytes *Aspergillus* and *Rhizopus* cause infections?
8. Why is the DPT vaccine routinely given to children in the United States? What does it consist of?

9. Complete the following table.

Disease	Causative Agent	Symptoms	Treatment
Streptococcal sore throat			
Scarlet fever			
Diphtheria			
Whooping cough			
Tuberculosis			
Pneumococcal pneumonia			
Klebsiella pneumonia			
Chlamydial pneumonia			
Legionellosis			
Psittacosis			
Q fever			
Epiglottitis			

10. A patient has been diagnosed as having pneumonia. Is this sufficient information to begin treatment with antimicrobial agents? Briefly discuss why or why not.

11. List the causative agent, method of transmission, and endemic area for the fungal diseases histoplasmosis, coccidioidomycosis, and blastomycosis.

12. Briefly describe the procedures and positive results of the tuberculin test, and indicate what is revealed by a positive test.

13. Discuss reasons for the increased incidences of colds and pneumonias during cold weather.

14. Match the bacteria involved in respiratory infections with the following laboratory test results.

Gram-positive cocci
- Catalase-positive ________
- Catalase-negative
  - β-hemolytic, bacitracin inhibition ________
  - α-hemolytic, optochin inhibition ________

Gram-positive rods
- Not acid-fast ________
- Acid-fast

Gram-negative cocci ________

Gram-negative rods
- Aerobes
  - Rods ________
  - Coccobacilli ________
- Facultative anaerobes
  - Require X and V factors ________
  - Capsule ________

Intracellular parasites
- Form elementary bodies ________
- Do not form elementary bodies ________

Wall-less ________

Challenge

1. Differentiate between the following.
   (a) *S. pyogenes* causing strep throat and *S. pyogenes* causing scarlet fever.
   (b) Diphtheroids and *C. diphtheriae*.

2. Why might vaccination against influenza be of questionable value?

3. In a two-week period, eight infants in an intensive care nursery (ICN) developed pneumonia caused by respiratory syncytial virus (RSV). Fluorescent-antibody (FA) screening and viral cultures were used to diagnose infections, and positive patients were placed in a separate room. A two-week-old girl from the newborn nursery, adjacent to the ICN, also developed an RSV infection. Toward the end of this outbreak, FA tests and viral cultures were made of 15 ICN personnel. (All 15 reported having had upper respiratory illness during the previous week.) Viral cultures were negative, but the RSV-FA test was strongly positive in one nurse and weakly positive in five others. Comment on the probable source of this outbreak. Explain the apparent discrepancy between the FA test and viral culture results. How can RSV infections in nurseries be prevented?

4. During a 6-month period in 1988, 72 clinic staff members became tuberculin-positive. A case–control study was undertaken to determine the most likely source of *M. tuberculosis* infection among the staff. A total of 16 cases and 34 tuberculin-negative controls were compared. Pentamidine is not used for TB treatment. What disease was probably being treated with this drug? What is the most likely source of infection?

	Cases	Control
Works $\geq$ 40 hr/week	100	62
In room during aerosolized pentamidine therapy in TB patients	31	3
Patient contact	94	94
Lunch eaten in staff lounge	38	35
Resident of western Palm Beach	75	65
Female	81	77
Cigarette smoker	6	15
Contact with nurse diagnosed with TB in 1987	15	12
In unventilated room during collection of TB-positive sputum samples	13	8

FURTHER READING

"*Bordetella* pertussis: pathogenesis and prevention of whooping cough." In L. Leive and D. Schlessinger, eds., *Microbiology—1984.* Washington, D.C.: American Society for Microbiology, pp. 157–183. A collection of articles on the epidemiology, toxins, and immunity of whooping cough.

Miller, J.A. "Clinical opportunities for plant and soil fungi." *BioScience* 36:656–658; 1986. Describes the rise in fungal infections from species previously known only as plant pathogens and soil saprophytes and from new genera of fungi.

Skinner, F.A., and L.B. Quesnel. *Streptococci.* New York: Academic Press, 1978. Describes the pathogenicity, immunology, and taxonomy of the genus *Streptococcus.*

Stuart-Harris, C. "The epidemiology and prevention of influenza." *American Scientist* 69:166–172, 1981. Discusses the relationship between antigenic shift in the virus and immunity to influenza.

Tyrrell, D.A.J. "Hot news on the common cold." *Annual Review of Microbiology* 42:35–47, 1988. Describes research on drugs and hot-air therapy to treat the common cold.

Weiss, R. "TB troubles." *Science News* 133:92–93, 1988. Discusses reasons for the recent increase in tuberculosis in the United States.

See also Further Reading for Part Four at the end of Chapter 26.

CHAPTER 25

Microbial Diseases of the Digestive System

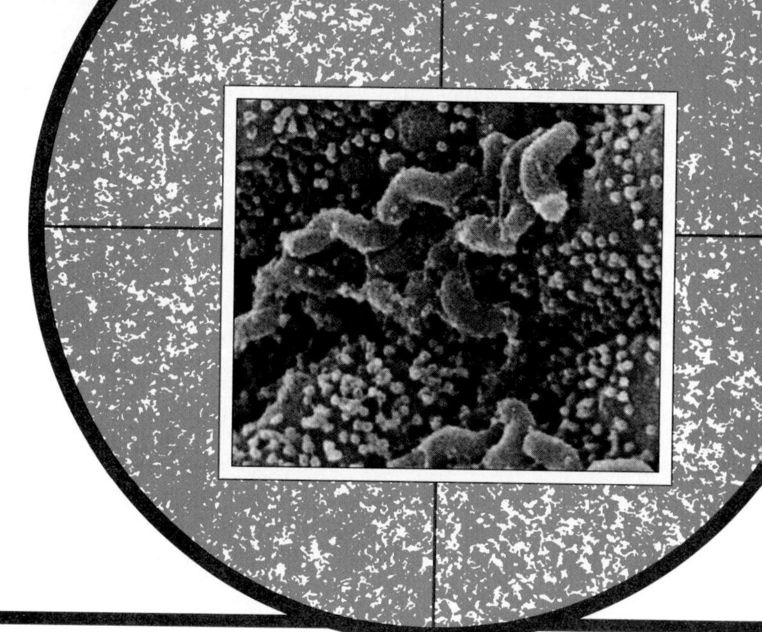

LEARNING OBJECTIVES

- Describe the antimicrobial features of the digestive system.
- Describe the events that lead to dental caries and periodontal disease.
- List the etiologic agents, suspect foods, symptoms, and treatment for staphylococcal food poisoning, salmonellosis, typhoid fever, bacillary dysentery (shigellosis), cholera, and gastroenteritis.
- List the etiologic agents, methods of transmission, sites of infection, and symptoms for gastritis, mumps, cytomegalovirus inclusion disease, infectious hepatitis, and serum hepatitis.
- Compare and contrast giardiasis, balantidiasis, and amoebic dysentery.
- Compare and contrast tapeworm and nematode infections.
- Discuss preventive measures for infections of the gastrointestinal system.

Microbial diseases of the digestive system are second only to respiratory diseases as the leading cause of illness in the United States. Most such diseases result from the ingestion of food or water contaminated with pathogenic microorganisms or their toxins. These pathogens enter the food or water supply after being shed in the feces of persons or animals infected with them. Therefore, microbial diseases of the digestive system are typically transmitted by a *fecal–oral* cycle. This cycle is interrupted by good sanitation practices in food handling as well as by modern methods of sewage treatment and disinfection of drinking water.

Table 25.2 (see pp. 640–641) is a summary of diseases associated with the mouth and digestive system.

Structure and Function of the Digestive System

The **digestive system** may be divided into two principal groups of organs (Figure 25.1). One group is the *gastrointestinal (GI) tract* or *alimentary canal,* essentially a tubelike structure that includes the mouth, pharynx (throat), esophagus (food tube), stomach, small intestine, and large intestine. The other group of organs, the *accessory structures,* consists of the teeth, tongue, salivary glands, liver, gallbladder, and pancreas. Except for the teeth and tongue, the accessory structures lie outside the tract and produce secretions that are conveyed by ducts into the tract.

The purpose of the digestive system is to digest foods—that is, to break them down into small molecules that can be taken up and used by body cells. In a process called *absorption,* these end-products of digestion then pass from the small intestine into the blood or lymph for distribution to body cells. Then the food

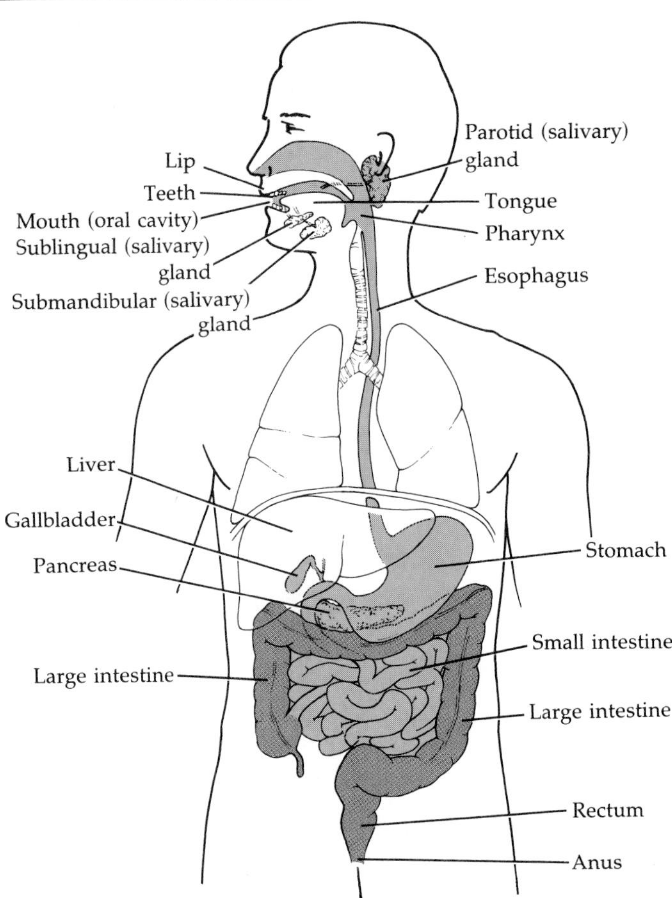

FIGURE 25.1 Anatomy of the human digestive system.

moves through the large intestine where water, vitamins, and nutrients are absorbed from it and microbial cells are added to it. The resulting undigested solids, called *feces,* are eliminated from the body through the anus. Feces carry intestinal pathogens, and it is necessary to prevent them from contaminating food and water.

Normal Flora of the Digestive System

Bacteria heavily populate most of the digestive system. In the mouth, each milliliter of saliva can contain millions of bacteria. The stomach and small intestine have relatively few microorganisms because of the hydrochloric acid produced by the stomach and the rapid movement of food through the small intestine. By contrast, the large intestine has enormous microbial populations, exceeding 100 billion bacteria per gram of feces. (Up to 40% of fecal mass is microbial cell material.) The population of the large intestine is composed mostly of anaerobes of the genera *Lactobacillus* and *Bacteroides* and facultative anaerobes, such as *E. coli, Enterobacter, Klebsiella,* and *Proteus.* Most of these bacteria assist in the enzymatic breakdown of foods, and some of them synthesize useful vitamins, such as niacin, vitamins B_1, B_2, B_6, and B_{12}, folic acid, biotin, and vitamin K.

Bacterial Diseases of the Mouth

The entrance to the digestive system provides an environment that supports a large and varied microbial population.

DENTAL CARIES (TOOTH DECAY)

The teeth are unlike any other exterior surface of the body. They are hard, and they do not shed surface cells (Figure 25.2). This allows the accumulation of masses of microorganisms and their products. These accumulations are called **plaque** (Figure 25.3b), and they are intimately involved in the formation of **dental caries,** or tooth decay (Figure 25.3c).

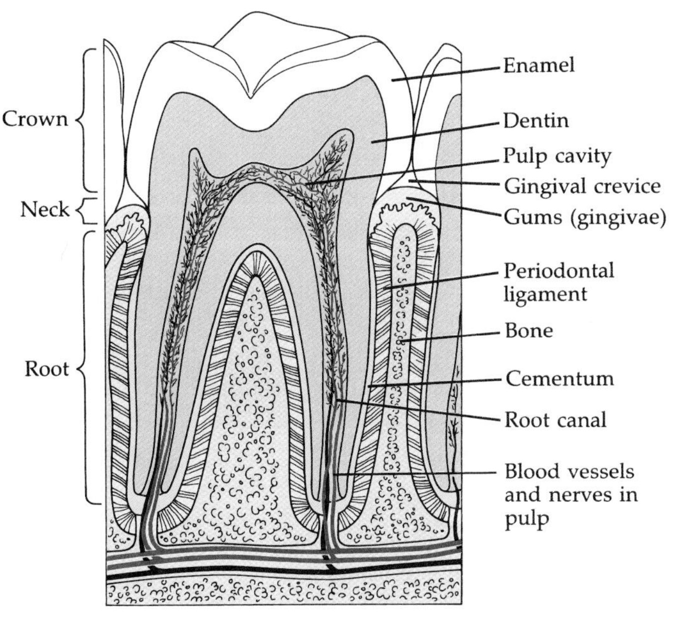

FIGURE 25.2 Normal tooth.

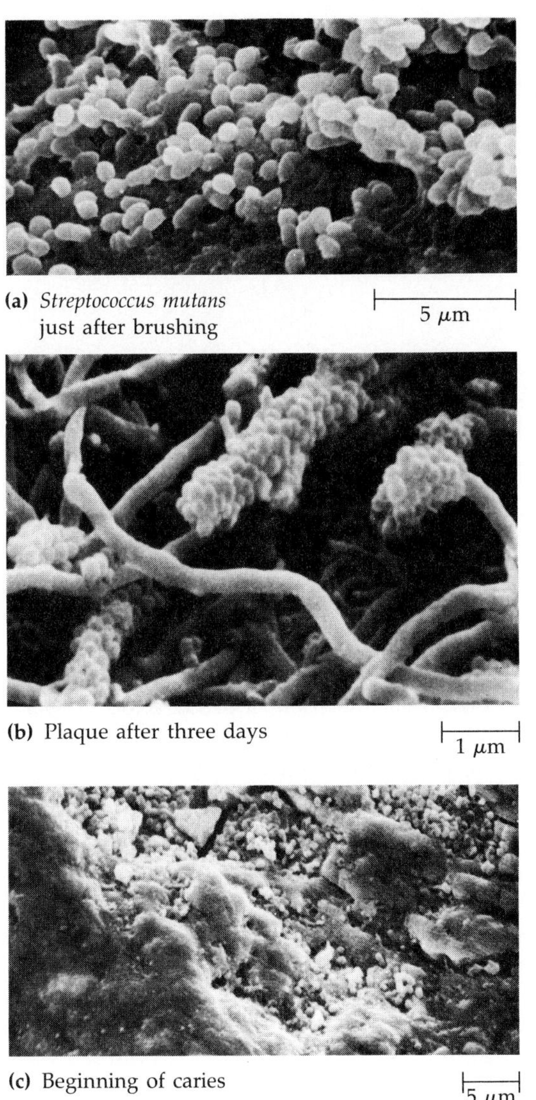

FIGURE 25.3 Initiation of dental caries. (a) Spherical cells of *Streptococcus mutans* adhere to tooth enamel shortly after brushing. **(b)** Dental plaque formed on tooth after three days without brushing. Both cocci and filamentous bacteria are present. **(c)** The beginning of a dental caries lesion. Numerous cells of *S. mutans* can be seen in the eroded area.

Oral bacteria convert sucrose and other carbohydrates into lactic acid, which in turn attacks the tooth enamel. The microbial population on and around the teeth is very complex; more than 300 species have been identified. Probably the most important *cariogenic* (causing caries) bacterium is *Streptococcus mutans,* a gram-positive coccus (Figure 25.3a). Some other species of streptococci are also cariogenic but play a lesser role in initiating caries.

The initiation of caries depends on the attachment of *S. mutans,* or other streptococci, to the tooth. These bacteria do not adhere to a clean tooth, but within minutes a freshly brushed tooth will become coated with a pellicle (thin film) of proteins from saliva. Within a couple of hours, cariogenic bacteria become established on this pellicle and begin to produce a gummy polysaccharide of glucose molecules called *dextran* (Figure 25.4). In the production of dextran, the bacteria first hydrolyze sucrose into its component monosaccharides, fructose and glucose. The enzyme glucosyltransferase then assembles the glucose molecules into dextran. The residual fructose is the primary sugar fermented into lactic acid. Accumulations of bacteria and dextran adhering to the teeth make up dental plaque. The bacterial population of plaque is predominantly streptococci and filamentous members of the genus *Actinomyces* (Figure 25.3b). (Older, calcified deposits of plaque are called *dental calculus* or *tartar.*) *S. mutans* especially favors crevices or other sites on the teeth protected from the shearing action of chewing, or the flushing action of the liter or so of saliva produced in the mouth each day. On protected areas of the teeth, plaque accumulations can be several hundred cells thick. Because plaque is not very permeable to saliva, the lactic acid produced by bacteria is not diluted or neutralized, and it breaks down the enamel of the teeth to which the plaque adheres.

While saliva contains nutrients that encourage the growth of bacteria, it also contains antimicrobial substances, such as lysozyme (see Chapter 16), that help protect exposed tooth surfaces. Some protection is also

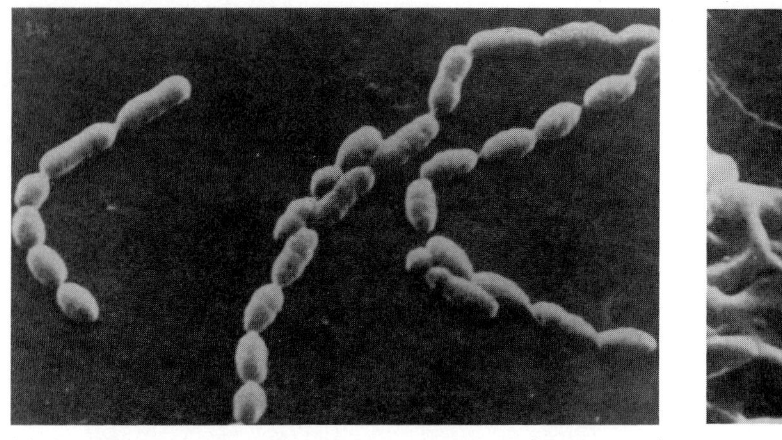

(a) *S. mutans* in glucose 5 μm

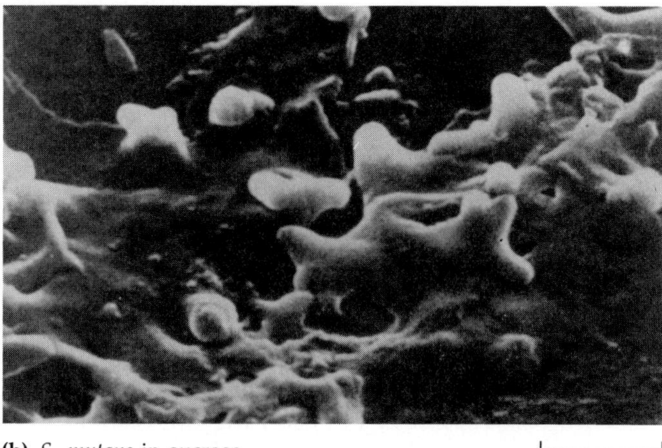

(b) *S. mutans* in sucrose 5 μm

FIGURE 25.4 Involvement of *S. mutans* and sucrose in dental caries. **(a)** *S. mutans* growing in glucose broth. **(b)** *S. mutans* growing in sucrose broth. Note the large accumulation of dextran.

provided by *crevicular fluid,* a tissue exudate that flows into the gingival crevice (Figure 25.2) and is closer in composition to serum than to saliva. It protects teeth by virtue of both its flushing action and its phagocytic cell and immunoglobulin content.

Localized acid production within deposits of dental plaque results in a gradual softening of the external *enamel* (Figures 25.3c and 25.5a, b). Enamel low in fluoride is more susceptible to the effects of the acid. This is the reason for fluoridation of water and toothpastes, which has been an important factor in the recent decline in dental caries in the United States.

If the initial penetration of the enamel by caries remains untreated, bacteria can penetrate into the interior of the tooth. The composition of the bacterial population involved in advancement of the decayed area from the enamel into the *dentin* is entirely different from that of the population initiating the decay. The dominant microorganisms are gram-positive rods and filamentous bacteria; *S. mutans* is present in small numbers only. Although once considered the cause of dental caries, *Lactobacillus* species actually play no role in initiating the process. However, these very prolific lactic acid producers are important in advancing the front of the decay once it is established.

The decayed area eventually advances to the *pulp,* which connects with the tissues of the jaw and contains the blood supply and the nerve cells (Figure

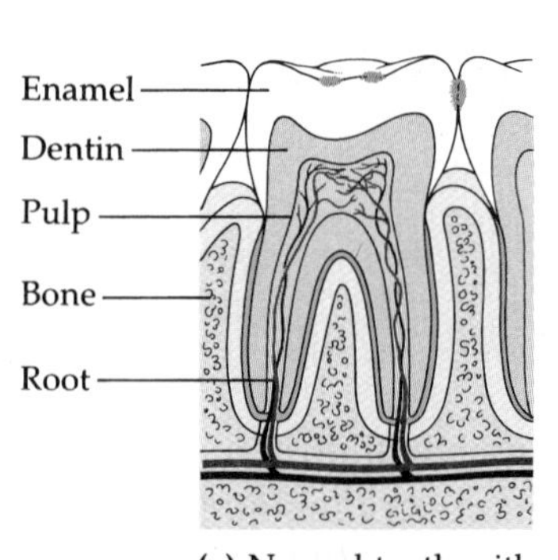

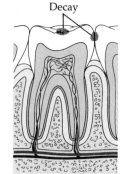

(a) Normal tooth with plaque

Decay

(b) Decay in enamel

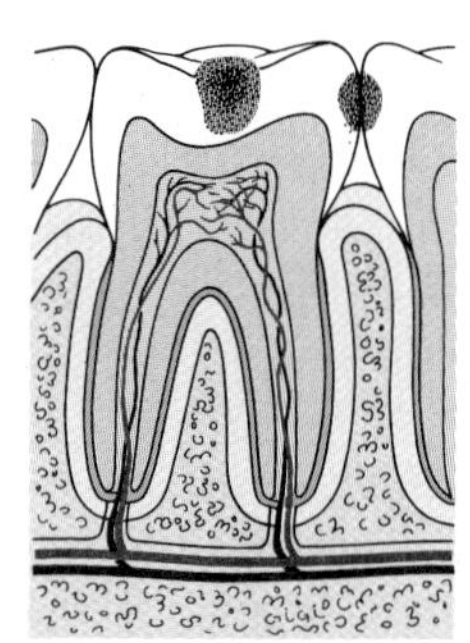

(c) Advanced decay

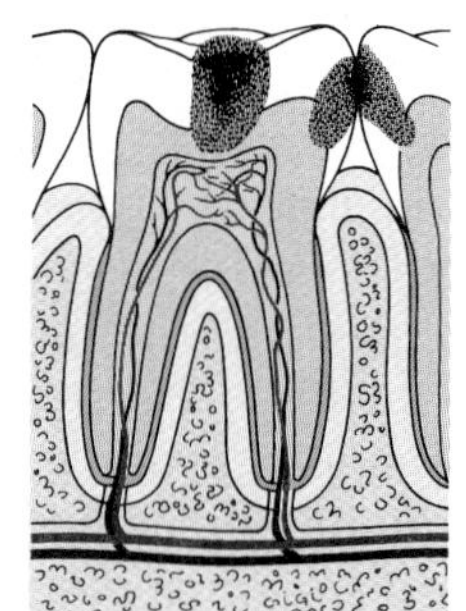

(d) Decay in dentin

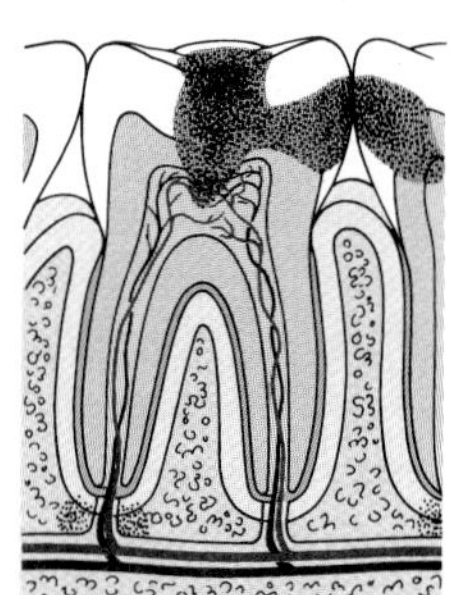

(e) Decay in pulp

FIGURE 25.5 Stages of dental caries. **(a)** Tooth with plaque accumulation in hard-to-clean areas. **(b)** Decay begins as enamel is attacked by acids formed by bacteria. **(c)** Decay advances through the enamel. **(d)** Decay advances into the dentin. **(e)** Decay enters the pulp and may form abscesses in the tissues surrounding the root.

25.5c, d, e). Almost any member of the normal flora of the mouth can be isolated from the infected pulp and roots. Once this stage is reached, root canal therapy is required to remove the infected and dead tissue and to provide access for antimicrobials that suppress renewed infection. If untreated, the infection may advance from the tooth to the soft tissues, producing dental abscesses caused by mixed bacterial populations that contain many anaerobes. These, and most other dental soft tissue infections, can be treated by penicillin and its derivatives.

Although dental caries are probably one of the more common infectious diseases in humans today, they were not common in the Western world until about the seventeenth century. In human remains from older times, only about 10% of the teeth contain caries. The introduction of table sugar, or sucrose, into the diet is highly correlated with our present level of caries in the Western world. Studies have shown that sucrose, a disaccharide composed of glucose and fructose, is much more cariogenic than either glucose or fructose individually (see Figure 25.4). People living on high-starch diets (starch is a polysaccharide of glucose) have a low incidence of caries unless sucrose is also a significant part of their diet. The contribution of bacteria to tooth decay has been shown by experiments with germ-free animals. Such animals do not develop caries even when fed a sucrose-rich diet designed to encourage their formation.

Sucrose is pervasive in the modern Western diet. It appears, however, that if sucrose is ingested only at regular mealtimes, the protective and repair mechanisms of the body are not overwhelmed. It is the sucrose that is ingested between meals that is most damaging to the teeth. In one set of experiments, it was shown that 330 grams of sucrose per day at mealtimes did not cause caries, but only 120 grams per day at frequent intervals between meals caused an increase of five to seven times in the number of caries. Although such sugars as mannitol, sorbitol, and xylitol are metabolized to acid by oral bacteria, they are not cariogenic. This is why they are used to sweeten "sugarless" candies and chewing gums.

The best strategies for the prevention of caries are minimal ingestion of sucrose; brushing, flossing, and professional cleaning to remove plaque; and the use of fluorides. As for the numerous mouthwashes that are said to prevent or reduce plaque, the Food and Drug Administration does not recognize any as being of proven effectiveness. The ancient Chinese and a number of other cultures made use of urine as a mouthwash to improve oral health. Indeed, it tends to lower acidity and may be useful, but this should not be construed as a recommendation. A vaccine based on IgA antibodies in the saliva is theoretically possible, and tests conducted on animals have given encouraging results.

PERIODONTAL DISEASE

Even persons who avoid tooth decay might, in later years, lose their teeth to **periodontal disease.** Periodontal (surrounding the tooth) disease is a collective term for a number of conditions characterized by inflammation and degeneration of structures that support the teeth (Figure 25.6).

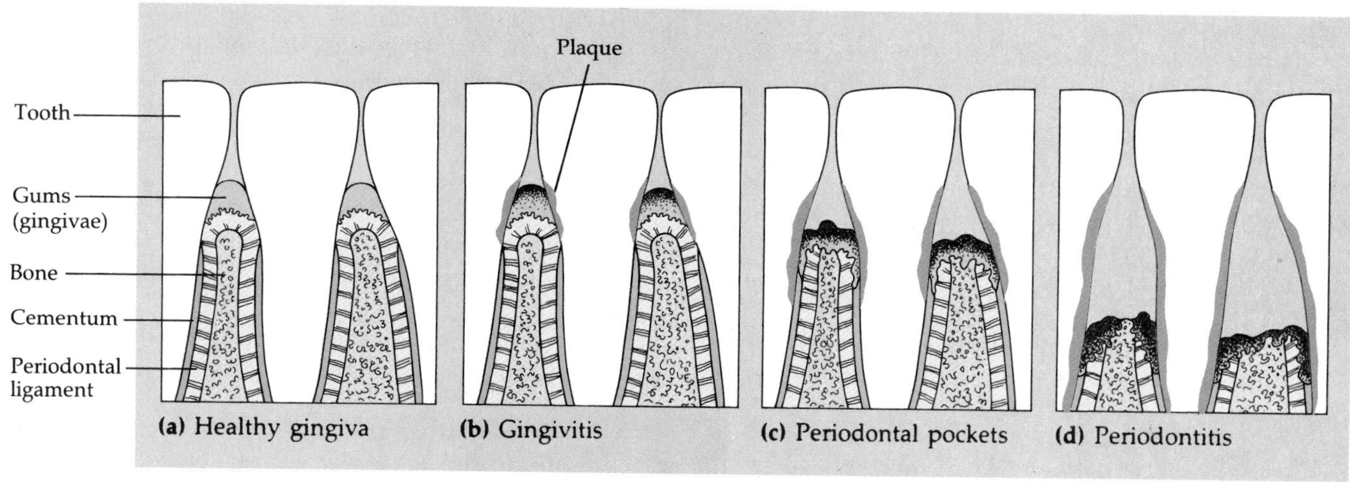

FIGURE 25.6 Periodontal disease. **(a)** Teeth firmly anchored by healthy bone and gum tissue (gingiva). **(b)** Toxins in plaque irritate gums, causing gingivitis. **(c)** Periodontal pockets form as tooth separates from gingiva. **(d)** Gingivitis progresses to periodontitis. Toxins destroy the gingiva and bone that support the tooth and the cementum that protects the root.

The roots of the tooth are protected by a covering of specialized connective tissue called cementum. As the gums recede with age, formation of caries on the cementum becomes more common.

Gingivitis

In many cases of periodontal disease, the infection is restricted to the gums, or gingivae. These infections, called **gingivitis,** are characterized by bleeding of the gums while brushing the teeth. This is a condition experienced to some degree by almost everyone (Figure 25.6b). It has been shown experimentally that gingivitis will appear in a few weeks if brushing is discontinued and plaque is allowed to accumulate. An assortment of streptococci, actinomycetes, and anaerobic gram-negative bacteria predominate in these infections.

Periodontitis

Gingivitis can progress to a chronic condition called **periodontitis,** which is responsible for nearly 10% of tooth loss in older adults. This is an insidious condition that generally causes little discomfort. The gums are inflamed and bleed easily. Sometimes pus is formed in pockets surrounding the teeth (periodontal pockets) (Figure 25.6c). As the infections continue, they progress toward the root tips. The bone and tissue that support the teeth are destroyed, leading eventually to loosening and loss of the teeth (Figure 25.6d). Numerous bacteria of many different types are involved in these infections; the damage to tissue is done by an inflammatory response to the presence of these bacteria. Periodontitis is treated either by surgically eliminating the periodontal pockets or by using specialized cleaning techniques on the tooth surfaces normally protected by the gums.

Acute necrotizing ulcerative gingivitis, also termed **Vincent's disease** or **trench mouth,** is one of the more common serious mouth infections. The disease causes enough pain that normal chewing is difficult. Foul breath (halitosis) also accompanies the infection. The bacteria usually associated with this condition are the anaerobic *Fusobacterium* and actinomycetes (see the box in Chapter 16, p. 410). Because these causative organisms are usually anaerobic, treatment with oxidizing agents, debridement, and administration of antibiotics may be temporarily effective.

Bacterial Diseases of the Lower Digestive System

Diseases of the digestive system are essentially of two types, infections and intoxications.

An **infection** occurs when a pathogen enters the gastrointestinal (GI) tract and multiplies. Microorganisms can penetrate into the intestinal mucosa and grow there or can pass through to other systemic organs. Infections are characterized by a delay in the appearance of gastrointestinal disturbance while the pathogen increases in numbers or affects invaded tissue. There is also usually a fever, one of the body's general responses to an infective organism.

Some pathogens cause disease by elaborating toxins that affect the GI tract. An **intoxication** is caused by the ingestion of such a preformed toxin. Most intoxications, such as that caused by *Staphylococcus aureus,* are characterized by a very sudden appearance (usually in only a few hours) of symptoms of a GI disturbance. Fever is less often one of the symptoms.

Both infections and intoxications often cause *diarrhea,* which most of us have experienced. Severe diarrhea, accompanied by blood or mucus, is called *dysentery.* Both types of digestive system diseases are also frequently accompanied by *abdominal cramps, nausea,* and *vomiting*. The general term *gastroenteritis* is applied to diseases causing inflammation of the stomach and intestinal mucosa. Botulism (see Chapter 22) is a special case of intoxication because the ingestion of the preformed toxin affects the nervous system rather than the GI tract.

In the developing countries, diarrhea is a major factor in infant mortality. Approximately one in every ten children dies of it before the age of five. It also affects the absorption of nutrients from their food and adversely affects the growth of the survivors. The cause of diarrhea may be any of several organisms. Most are not identified, but surveys in such countries as Bangladesh indicate that the three most common causes are enterotoxigenic *E. coli, Shigella* spp., and intestinal rotaviruses. It is estimated that mortality from childhood diarrhea could be halved by *oral rehydration therapy*. Ideally, this is a solution of sodium chloride, potassium chloride, and sodium bicarbonate. However, even a solution of a handful of table sugar and a pinch of salt in a liter of water has proved to be a very useful treatment for diarrheal diseases.

STAPHYLOCOCCAL FOOD POISONING (STAPHYLOCOCCAL ENTEROTOXICOSIS)

A very common cause of gastroenteritis is **staphylococcal food poisoning.** This is an *intoxication* caused by ingestion of an *enterotoxin* produced by *S. aureus.* Staphylococci are comparatively resistant to environmental stresses, as we discussed in Chapter 11. Staphylococci have a fairly high resistance to heat; vegetative cells can tolerate 60°C for half an hour. They are also resistant to drying and radiation, which helps them survive on skin surfaces. Resistance to high osmotic pressures helps them grow in foods, such as cured ham, in which the high osmotic pressure of salts inhibits the growth of competitors.

S. aureus is a common inhabitant of the nasal passages, from which it often contaminates the hands. It is also a frequent cause of skin lesions on the hands. From these sources, it can readily enter food. If the organisms are allowed to incubate in the food (a situation called *temperature abuse*), they reproduce and release enterotoxin into the food. These events, which lead to outbreaks of staphylococcal intoxication, are illustrated in Figure 25.7.

S. aureus produces several toxins that damage tissues or increase the organism's virulence. The enterotoxins causing food poisoning are classified as serological types A (which is responsible for most cases) through D. Production of this toxin is often correlated with production of an enzyme that coagulates blood plasma. Such bacteria are described as *coagulase-positive* (see Chapter 11). No direct pathogenic effect can be attributed to the enzyme, but it is useful in the tentative identification of types that are likely to be virulent.

Generally, a population of about 1 million bacteria per gram will produce enough enterotoxin to cause illness. The growth of the organism is facilitated if the competing organisms in the food have been eliminated—by cooking, for example. The organism is also more likely to grow if competing bacteria are inhibited by a higher-than-normal osmotic pressure or by a relatively low moisture level. *S. aureus* tends to outgrow most competing bacteria under these conditions.

Custards, cream pies, and ham are examples of high-risk foods. Competing organisms are eliminated in custards by the high osmotic pressure of sugar and by cooking. In ham they are eliminated by curing agents such as salts and preservatives. Poultry products can also harbor staphylococci if they are handled and allowed to stand at room temperatures. Hamburger, however, is seldom associated with any type of bacterial food poisoning because it contains microorganisms of the normal flora that are competitive with staphylococci, and hamburger is usually cooked immediately before consumption. Any foods that are prepared in advance and are not kept chilled are a potential source of staphylococcal food poisoning. Because contamination of foods cannot be avoided completely, the most reliable method of preventing staphylococcal food poisoning is adequate refrigeration during storage to prevent toxin formation.

The toxin itself is heat stable and can survive up to 30 minutes of boiling. So, once the toxin is formed, it is not destroyed when the food is reheated, although the bacteria will be killed.

Staphylococcal food poisoning is characterized by nausea, vomiting, and diarrhea starting one to six

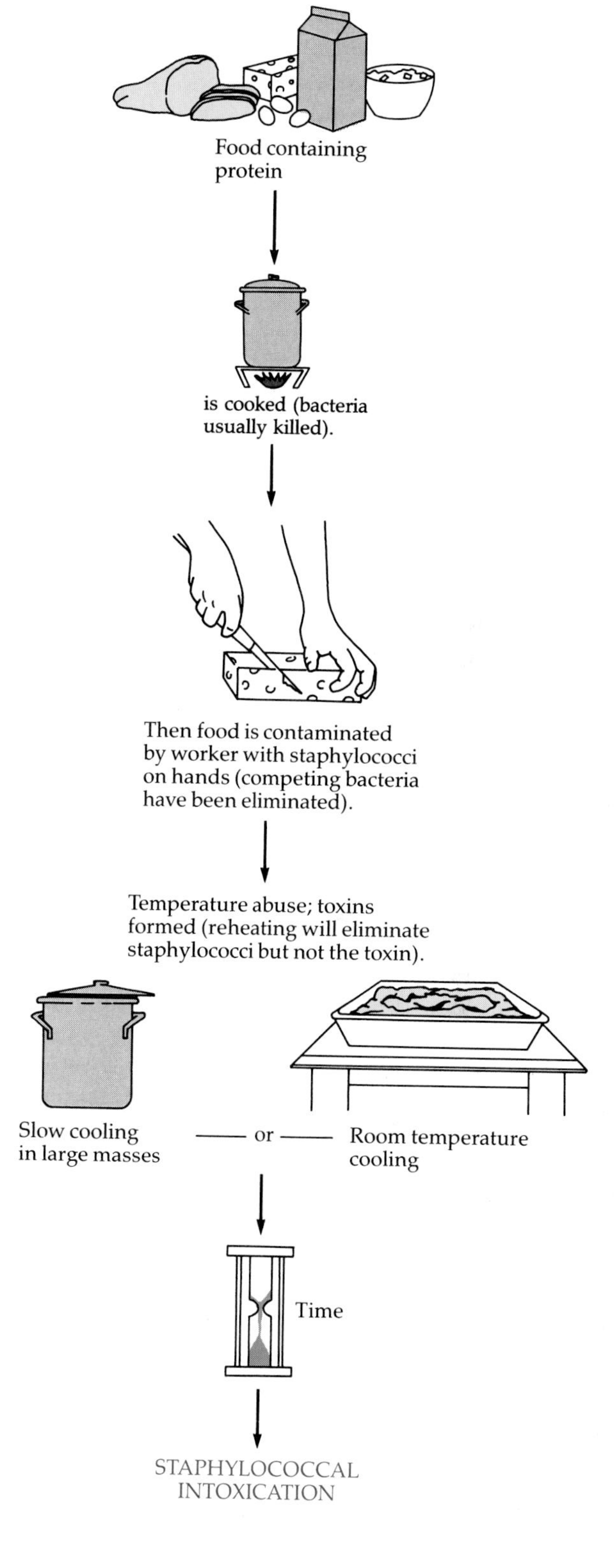

FIGURE 25.7 The sequence of events in a typical outbreak of staphylococcal food poisoning.

hours after the contaminated food is ingested. The toxin affects the vagus nerve in the upper intestinal tract and triggers the brain's vomiting reflex center. The cause of the diarrheal symptoms is not as clearly identified. The symptoms usually disappear in 24 hours. Because considerable amounts of water and electrolytes are lost during episodes of vomiting and diarrhea, treatment for the disease consists of replacing the lost water and electrolytes.

The mortality rate of staphylococcal food poisoning is almost nil among otherwise healthy individuals, but it can be significant in weakened individuals, such as residents of nursing homes. No reliable immunity results from recovery. However, there is a great deal of variation in individual susceptibility to the toxin, and it is suspected that immunity acquired from a previous exposure might account for some of this variation.

The diagnosis of staphylococcal food poisoning is usually based on the symptoms, particularly the short incubation time characteristic of intoxication. If the food has not been reheated so that the bacteria are not killed, the organism can be recovered and grown. *S. aureus* isolates can be tested by the *phage-typing* method, which is useful in tracing the source of the contamination (see Figure 10.9, p. 261). These bacteria will grow well in 7.5% sodium chloride, so this concentration is often used in media for their isolation or identification. Pathogenic staphylococci usually ferment mannitol, produce blood hemolysins and coagulase, and form golden-yellow colonies. They cause no obvious spoilage when growing in foods.

No simple procedure will identify the toxin in foods. Identification requires a complex extraction and a method of serological testing against antisera; these procedures are not yet commercially available, but kits that make use of monoclonal antibodies to detect them are in development. Food can be easily tested, however, for the presence of *thermostable nuclease*, an enzyme produced by *S. aureus* that survives heating to temperatures that will kill the cells themselves. It can be detected by a simple test, in which an extract of the food is allowed to react with DNA. The presence of detectable amounts of the enzyme usually indicates a microbial population large enough to have produced sufficient toxin to cause illness.

SALMONELLOSIS (*SALMONELLA* GASTROENTERITIS)

As we noted earlier, in bacterial infections, disease results from microbial growth in body tissues rather than from the ingestion of food and drink already contaminated by preformed toxins resulting from microbial growth. Bacterial infections, such as salmonellosis, usually have longer incubation periods (12 hours to 2 weeks) than bacterial intoxications, reflecting the time needed for the organism to grow in the tissues of the host. Bacterial infections are also often characterized by some fever, indicating the host's response to the infection.

The *Salmonella* bacteria (named for the discoverer, Daniel Salmon) are gram-negative, facultatively anaerobic, non-endospore-forming, usually motile rods that ferment glucose to produce acid and gas. Their normal habitat is the intestinal tracts of humans and many animals. All salmonellae are considered pathogenic to some degree, causing **salmonellosis,** or ***Salmonella* gastroenteritis.**

The nomenclature of the *Salmonella* organisms is confusing. According to *Bergey's Manual*, none of the present methods for naming salmonellae is satisfactory from a scientific viewpoint. Rather than recognized species, there are more than 2000 *serovars* (closely related organisms differentiated by serological testing). Only about 50 such serovars are isolated with any frequency in the United States. Some of these are named much like species—for example, *Salmonella dublin* and *Salmonella eastbourne* (ēst'bôrn) were named after the sites where they were first isolated, and *Salmonella typhimurium* (tī-fi-mûr'ē-um) causes a typhoidlike disease in mice. Others are represented by the Kauffmann–White scheme (see the discussion of *Salmonella* in Chapter 11), which is generally used in clinical laboratories. This method assigns numbers and letters to different antigens: O (somatic or body), Vi (capsular), and H (flagellar).

Salmonellosis has an incubation time of about 12 to 36 hours. The salmonellae first invade the intestinal mucosa and multiply there. Sometimes they manage to pass through the intestinal mucosa to enter the cardiovascular system, and from there they spread to eventually affect many organs. The fever associated with *Salmonella* infections might be from endotoxins released by lysed cells, but this relationship is not certain. There is usually a moderate fever accompanied by nausea, abdominal pain and cramps, and diarrhea. As many as 1 billion salmonellae per gram can be found in the infected person's feces during the acute phase of the illness.

The mortality rate is overall very low, probably lower than 1%. However, the death rate is high in infants and among the very old; death is usually from septicemia. Individual responses to the infection vary considerably. The severity and incubation time can depend on the number of *Salmonella* ingested. Normally, recovery will be complete in a few days, but many patients will continue to shed the organisms in their feces for up to six months. Antibiotic therapy is not useful in treating salmonellosis; treatment consists of oral rehydration therapy.

Salmonellosis is probably greatly underreported. The 20,000 to 30,000 cases reported each year are prob-

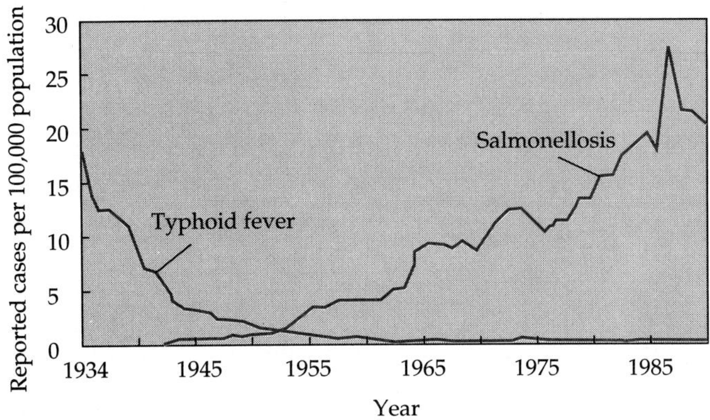

FIGURE 25.8 Incidence of salmonellosis and typhoid fever. The incidence of typhoid fever has been declining in the United States, while that of salmonellosis (gastroenteritis) has been increasing. (Source: CDC. *Summary of Notifiable Diseases* 1988, *MMWR* 37:54.)

ably only from 1% to 10% of the actual total (see Figure 25.8).

Meat products are particularly susceptible to contamination by *Salmonella* (see the box in Chapter 20, p. 495). If these products are mishandled, the bacteria can grow to infective numbers rather quickly. The sources of the bacteria are the intestinal tracts of many animals, and meats can be contaminated readily in processing plants. Poultry, eggs, and egg products are often contaminated by *Salmonella*.

Prevention of salmonellosis depends on good sanitation practices to deter contamination and on proper refrigeration to prevent increases in bacteria. The organisms are generally destroyed by normal cooking that heats the food to an internal temperature of at least 68°C (145°F). However, contaminated food can contaminate a surface, such as a cutting board. Although the food first prepared on the board might later be cooked and its bacteria killed, another food subsequently prepared on the board might not be cooked.

A recent series of outbreaks of salmonellosis have occurred in the New England states and were traced to contaminated eggs. Indications are that the bacteria are transmitted to the eggs before they are laid, although the chickens may be asymptomatic. Persons in the endemic areas, especially, have been cautioned to eat only well-cooked eggs. Eggs fried "sunny-side up" or boiled for only four minutes are not well enough cooked to reliably kill *Salmonella*. An often unsuspected factor is the presence of inadequately cooked or raw eggs in foods such as hollandaise sauce and Caesar salad dressing.

Diagnosis of salmonellosis usually depends on isolating the organism from the patient's stool or from leftover food. Isolation requires specialized selective and differential media; these methods are relatively slow. Also, the small numbers of *Salmonella* generally present in foods present a special problem in detection. Because this disease is important, much effort has been applied to improving detection and identification methods. The most promising of these are DNA probes in kit form (see Figure 10.15) and ELISA-type kits that make use of monoclonal antibodies. These have made one-day detection and identification of *Salmonella* possible and are especially useful in screening food for small numbers of the organisms.

TYPHOID FEVER

A few serovars of *Salmonella* are much more virulent than others. The most virulent species, *Salmonella typhi*, causes the bacterial infection **typhoid fever.** This pathogen crosses the intestinal wall and enters the bloodstream.

The incubation period is much longer than that of salmonellosis, normally about two weeks. The patient first suffers from a high fever of 40°C (104°F) and continual headache. Diarrhea appears only during the second or third week, and the fever tends to decline. The organism becomes disseminated in the body and can be isolated from the blood, urine, and feces. In severe cases, there can be perforation of the intestinal wall. The mortality rate is now about 1%to 2%; at one time, it was about 10%. Before the days of proper sewage disposal, water treatment, and food sanitation, typhoid was an extremely common disease. Its incidence has been declining in the United States, while that of salmonellosis has been increasing (Figure 25.8). Typhoid fever is still a frequent cause of death in parts of the world with poor sanitation.

A substantial number of recovered patients become carriers. They harbor the pathogen in the gallbladder and continue to shed bacteria for several months. A number of such carriers continue to shed the organism indefinitely. Most of us are familiar with the case of Typhoid Mary. Her name was Mary Mallon; she worked as a cook in New York state in the early part of the century and was responsible for several outbreaks of typhoid and three deaths. Her case became well known through the attempts of the state to restrain her from working at her chosen trade.

In each recent year, there have been about 400 to 500 cases of typhoid fever in the United States. Approximately 70% of these cases were acquired during foreign travel. Normally, there are fewer than three deaths each year.

The third-generation cephalosporins, such as ceftriaxone, are the usual drugs of choice for the treatment of severe typhoid fever. However, drug sensitivity testing is required because of the continuing appearance of resistant strains. Chloramphenicol, ampicillin, and trimethoprim-sulfamethoxazole are among the effective alternatives. Recovery from typhoid confers a life-long immunity.

Immunization for typhoid is normally not done in more developed countries, except for high-risk laboratory and military personnel. The vaccine that has long been in use is a killed-organism type, which must be injected. For parts of the world where the disease is endemic, an improved vaccine is a priority. Currently at least two new live oral vaccines are being tested (see the box in Chapter 8, p. 205). One of these is already licensed and available in many countries. Results today indicate that it gives only about 67% protection after three doses. However, this immunity appears to persist for at least five years, and mass immunizations seem to have led to an encouraging herd immunity in these populations.

BACILLARY DYSENTERY (SHIGELLOSIS)

Bacillary dysentery (shigellosis) is a severe form of diarrhea caused by a group of facultatively anaerobic gram-negative rods of the genus *Shigella,* named for the Japanese microbiologist Kiyoshi Shiga.

There are four species of pathogenic *Shigella: S. sonnei* (sōn′ne-ē), *S. dysenteriae* (dis-en-te′rē-ī), *S. flexneri* (fleks′nėr-ē), and *S. boydii* (boi′dē-ē). These bacteria are residents only of the human intestinal tract, and animal reservoirs are unimportant. They are closely related to the pathogenic *E. coli. S. sonnei,* the most common in the United States, causes a relatively mild dysentery. At the other extreme, infection with *S. dysenteriae* (sometimes called the Shiga bacillus) often results in a severe dysentery and prostration. The toxin responsible is unusually virulent and is known as the Shiga toxin. Fortunately, this is the species found least frequently in the United States.

Virulent shigellae produce exotoxins that inhibit protein synthesis, thereby killing cells. The bacteria proliferate to immense numbers in the small intestine, but the damage they cause is to the large intestine. There they produce tissue destruction in the intestinal mucosa, causing severe diarrhea with blood and mucus in the stools. Infected persons may have as many as 20 bowel movements in a day. The ulcerations in the intestinal mucosa eventually heal but form scar tissue. Additional symptoms of infection are abdominal cramps and fever. Except for *S. dysenteriae, Shigella* rarely invades the bloodstream. Diagnosis is usually based on recovery of the organisms from rectal swabs.

In recent years, the number of cases reported in the United States has been about 15,000 to 20,000, with 20 to 35 (about 0.2%) deaths. *S. dysenteriae* has a significant mortality rate, however, and the death rate in tropical areas can be much higher, perhaps 20%. The disease is probably more common than the reported numbers indicate. Some cases of so-called traveler's diarrhea might be mild forms of bacillary dysentery. This dysentery is especially common in institutionalized patients and in persons living on Native American reservations. Some immunity seems to result from recovery, but no satisfactory vaccine has yet been developed.

In severe cases of bacillary dysentery, antibiotic therapy and fluid and electrolyte replacement are indicated. At present, trimethoprim-sulfamethoxazole is the antibiotic of choice.

CHOLERA

During the 1800s, the bacterial infection called Asiatic cholera crossed Europe and North America in repeated epidemics. Today, **cholera** is endemic in Asia, particularly India, and has only occasional outbreaks in Western countries. These outbreaks are caused by temporary lapses in sanitation, and the number of cases is limited.

The causative organism is *Vibrio cholerae,* a slightly curved, gram-negative rod with a single polar flagellum (Figure 25.9). The serogroup 0:1 of *V. cholerae* causes the classically recognized epidemic form of the disease. It can be subdivided into two *biotypes* (differentiated other than by serology) as classical or *eltor* (named for the original culture isolated at a quarantine camp, El Tor, for pilgrims to Mecca). At present, the most widespread cholera pathogen is serogroup 0:1, biotype eltor. It grows in the small intestine and produces an enterotoxin that results in the secretion of chlorides, bicarbonates, and water. The excess water and mineral electrolytes are excreted, taking on the typical appearance of "rice water stools" from the masses of intestinal mucus, epithelial cells, and bacteria. The sudden loss of these fluids and electrolytes (from 12 to 20 liters of fluid might be lost in one day) causes shock, collapse, and often death. Because of the loss of fluid, the blood becomes so viscous that vital organs are unable to function properly. Violent vomiting sometimes occurs. The organisms are not invasive, and a fever is usually not present. The severity of the disease varies considerably, and the number of subclinical cases might be several times the number that are reported.

In the United States there have been occasional cases of cholera caused by the 0:1 serogroup. These have all been in the Gulf coast area, and it is speculated that the organism may be endemic in coastal waters. Most gastroenteritis due to *V. cholerae* in this country has been caused by the non-0:1 serogroup, usually

FIGURE 25.9 *Vibrio cholerae,* **the pathogen that causes cholera. (a)** Note the slightly curved morphology in this scanning electron micrograph. **(b)** Normal intestinal wall of an infant rabbit. **(c)** Rabbit intestine after infection with *V. cholerae.*

ingested in contaminated seafood. The waters of the Gulf and Pacific coasts support an indigenous population of these organisms. These bacteria differ from the 0:1 serotypes in several ways, and they are more likely to invade the intestinal mucosa, causing bloody stools and fever.

The cholera bacteria can be readily isolated from the feces, partly because the organisms can grow in media that are alkaline enough to suppress the growth of many other organisms. The non-0:1 serogroup is occasionally isolated from blood and wounds.

Recovery from the disease results in an effective immunity based on the antigenic activity of both the cells and the enterotoxin. However, because of antigenic differences among bacterial strains, the same person can have cholera more than once. Most victims in endemic areas are children. The available vaccines are protective to some extent but provide immunity of relatively short duration as compared with the immunity from a natural infection.

Tetracycline is used for treatment, but chemotherapy is not as effective as replacement of the lost fluids and electrolytes. Untreated cases of cholera may have a 50% mortality rate, whereas the rate for cases having proper supportive care can be less than 1%.

VIBRIO PARAHAEMOLYTICUS GASTROENTERITIS

Vibrio parahaemolyticus is found in salt water estuaries in many parts of the world. It is morphologically similar to *V. cholerae,* but it is halophilic and requires 2% or more sodium chloride for optimum growth. It is the most common cause of gastroenteritis in Japan, with thousands of cases reported annually. It is also a common cause of gastroenteritis in many nations of Southeast Asia. The organism is present in coastal waters of the continental United States and Hawaii. Crustaceans, such as shrimp and crabs, have been associated with several outbreaks caused by *V. parahaemolyticus* in the United States in recent years.

Symptoms include abdominal pain, vomiting, a burning sensation in the stomach, and watery stools, resembling those of cholera. The organism has also been known to cause cutaneous infections of cuts that have come in contact with contaminated clams and oysters. The incubation time is normally less than 24 hours. The generation time under optimum conditions is less than 10 minutes, and large numbers are reached quickly. Recovery usually follows in a few days, and the fatality rate is low.

Because *V. parahaemolyticus* has a requirement for sodium, as well as for a high osmotic pressure, isolation media containing 2% to 4% sodium chloride are used in diagnosis of the disease.

ESCHERICHIA COLI GASTROENTERITIS (TRAVELER'S DIARRHEA)

One of the most prolific microorganisms in the human intestinal tract is *E. coli*. Because it is so common and so easily cultivated, microbiologists often regard it as something of a laboratory pet. *E. coli* is normally harmless, but certain strains can be pathogenic. All pathogenic strains have specialized fimbriae that allow them to bind to certain intestinal epithelial cells. They also produce toxins that cause gastrointestinal disturbances.

There are several distinct pathogenic groups of *E. coli*. The *enterotoxigenic* form is not invasive but forms an enterotoxin that produces a watery diarrhea that resembles a mild form of cholera. This group of organisms is a primary cause of traveler's diarrhea and, in developing countries, most infant diarrhea. *Enteroinvasive E. coli* strains produce an enterotoxin that has an activity much like that of the *Shigella* toxins. These organisms penetrate the lining of the intestinal tract, resulting in inflammation, fever, and sometimes a *Shigella*-like dysentery. The *enteropathogenic E. coli* strains are not important as a cause of traveler's diarrhea but are known to cause epidemic diarrhea among small children. In recent years, the *enterohemorrhagic E. coli* strains have become well known in this country as the cause of several outbreaks of disease associated with ingestion of undercooked hamburger or raw milk. Cattle are suspected as the reservoir. These organisms produce verotoxins, which are responsible for hemorrhagic colitis (an inflammation of the colon with bleeding). The best known organism in this group is *E. coli* serotype O157:H7. Many patients, especially children, infected by this organism produce stools combined with copious amounts of blood, but without fever. This toxin may be identical to the Shiga toxin produced by the unusually virulent strains of *S. dysenteriae* previously mentioned. A dangerous complication is hemolytic uremic syndrome (blood in the urine, kidney failure), which occurs when the kidneys are affected by the toxin.

In practice, it is difficult to differentiate among the isolates of pathogenic and nonpathogenic *E. coli*. Probably 50% to 65% of traveler's diarrhea is caused by enterotoxigenic *E. coli*. Of the remaining cases, from 10% to 20% are probably caused by *Shigella*. Other enteric bacteria, such as *Salmonella* and *Campylobacter*, as well as an assortment of unidentified bacterial pathogens, viruses, and protozoan parasites, are also occasionally involved.

Traveler's diarrhea is almost impossible to avoid in parts of the world with poor sanitation practices. Prophylactic antibiotic therapy is generally not recommended, although it can be somewhat effective. An alternative preventive regimen is to take two bismuth subsalicylate (Pepto-Bismol®) tablets four times a day. One trial showed that only 14% of those following this regimen developed diarrhea during a three-week trip, whereas 40% of the control group, who took a placebo, developed traveler's diarrhea.

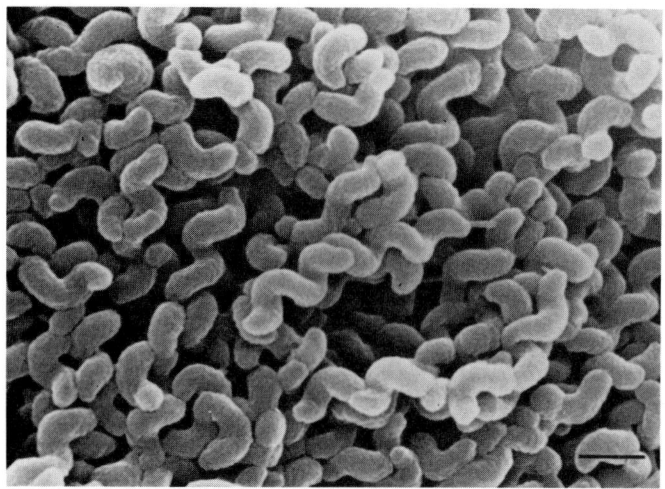

1 μm

FIGURE 25.10 ***Campylobacter jejuni.*** Scanning electron micrograph showing the spiral-shaped rod morphology.

There has been considerable research on a vaccine, which would be very popular commercially, but results have been disappointing. In adults, the disease is usually self-limiting, and chemotherapy is not attempted. Once contracted, the best treatment is the usual oral rehydration recommended for all diarrhea. In severe cases, antimicrobials may be necessary.

In Chapter 26, *E. coli* is discussed as one of the most common causes of urinary tract infections.

CAMPYLOBACTER GASTROENTERITIS

Campylobacter strains are gram-negative, microaerophilic, spirally curved rods (Figure 25.10) that were once considered only minor causes of human gastroenteritis. Like *Salmonella,* they are part of the intestinal flora of a number of animals, especially sheep and cattle, but their infective dose is much smaller than that of *Salmonella*. Cows with no evidence of illness may excrete them in their milk. After improved methods for their isolation and identification were developed in the late 1970s, it became clear that they were a common cause of human disease. In fact, they have replaced *Shigella* spp. as the second most common cause of diarrhea in the United States. In some states, they are even more frequently isolated in diarrhea cases than salmonellae.

***Campylobacter* gastroenteritis** is usually caused by *C. jejuni*. Clinically, it is characterized by fever, cramp-

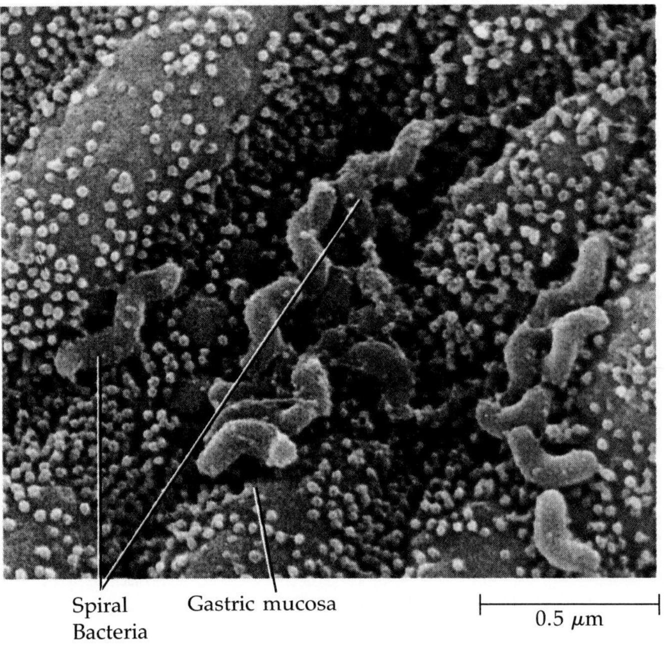

FIGURE 25.11 ***Helicobacter pylori.*** These spiral bacteria are able to grow in the acidic environment of the stomach by producing urease, which neutralizes the acidity. The bacteria can be seen in the depression formed in the surrounding gastric mucus layer in this biopsy specimen.

ing abdominal pain, and diarrhea with blood and mucus present in the stools. A billion organisms per gram of stool have been isolated at the peak of an infection. Normally, recovery follows within a week.

HELICOBACTER GASTRITIS

During the 1980s there was increasing speculation that the organism then known as *Campylobacter pylori* (pī'lô-rē) was the cause of gastric ulcers. It was found associated with ulcerated areas in the deepest parts of the mucus layer lining the stomach interior (Figure 25.11). Analysis of the RNA content later led to the conclusion that it did not belong in the genus *Campylobacter,* and in 1989 it was assigned the new name *Helicobacter pylori*. It has not yet been conclusively proven that *H. pylori* causes duodenal ulcers, although it has been proven to cause gastritis (inflammation of the mucous membrane of the stomach). Ulcers heal more rapidly and recur less often, however, when treatment eliminates the organism. For a discussion of ulcers, see the box in Chapter 15, p. 398.

The highly acidic environment of the stomach is lethal for most microorganisms. *H. pylori,* however, produces high concentrations of an especially efficient urease, an enzyme that converts urea to the alkaline compound ammonia. This results in a locally high pH in the area of growth.

Therapy based upon knowledge of the organism and its activity is not very advanced. The bacterium is quite sensitive to bismuth, and products containing bismuth have been fairly successful in several tests. It is also susceptible to metronidazole, but in practice, eradication of the organism has been difficult.

YERSINIA GASTROENTERITIS

Other enteric pathogens that are being identified with increasing frequency are *Yersinia enterocolitica* (en-tėr-ō-kōl-it'ik-ä) and *Yersinia pseudotuberculosis* (sū-dō-tü-ber-kū-lō'sis). These gram-negative organisms are intestinal inhabitants of many domestic animals and are often transmitted in meat and milk. Both organisms are distinctive in their ability to grow at refrigerator temperatures of 4°C (39°F). The organisms cause a gastroenteritis known as **yersiniosis.** Outbreaks of yersiniosis are more common in northern Europe than in the United States. The symptoms of yersiniosis are diarrhea, fever, headache, and abdominal pain. The pain is often severe enough to cause a misdiagnosis of appendicitis. Diagnosis requires culture of the organism, which can then be evaluated by serological tests.

CLOSTRIDIUM PERFRINGENS GASTROENTERITIS

Probably one of the more common, if underrecognized, forms of food poisoning in the United States is caused by *Clostridium perfringens,* a large, gram-positive, endospore-forming, obligately anaerobic rod. This organism is also responsible for human gas gangrene (see Chapter 23).

Most outbreaks are associated with meats or meat stews contaminated with intestinal contents of the animal during slaughter. The organism's nutritional requirement for amino acids is met by such foods, and when the meats are cooked, the oxygen level is lowered enough for clostridial growth. The endospores survive most routine heatings, and the generation time of the vegetative bacterium is less than 20 minutes under ideal conditions. Large populations can therefore build up rapidly when foods are being held for serving or when inadequate refrigeration leads to slow cooling.

The organism grows in the intestinal tract and produces an exotoxin that causes the typical symptoms of abdominal pain and diarrhea. Most cases are mild and self-limiting and probably are never clinically diagnosed. The symptoms usually appear from 8 to 12 hours after ingestion.

Diagnosis is usually based on isolation and identification of the organism in stool samples.

BACILLUS CEREUS GASTROENTERITIS

Bacillus cereus (se'rē-us) is a large, gram-positive, endospore-forming organism that is very common in soil and vegetation and is usually considered harmless. It has, however, been identified as the cause of outbreaks of foodborne illness. Heating the food does not always kill the spores, which germinate as the food cools. Because competing organisms have been eliminated in the cooked food, *B. cereus* grows rapidly and produces toxins. Rice dishes served in oriental restaurants seem especially susceptible. Some cases resemble *C. perfringens* infections and are almost entirely diarrheal in nature. Other episodes involve nausea and vomiting. It is suspected that different toxins are involved in the differing symptoms. Both forms of the disease are self-limiting.

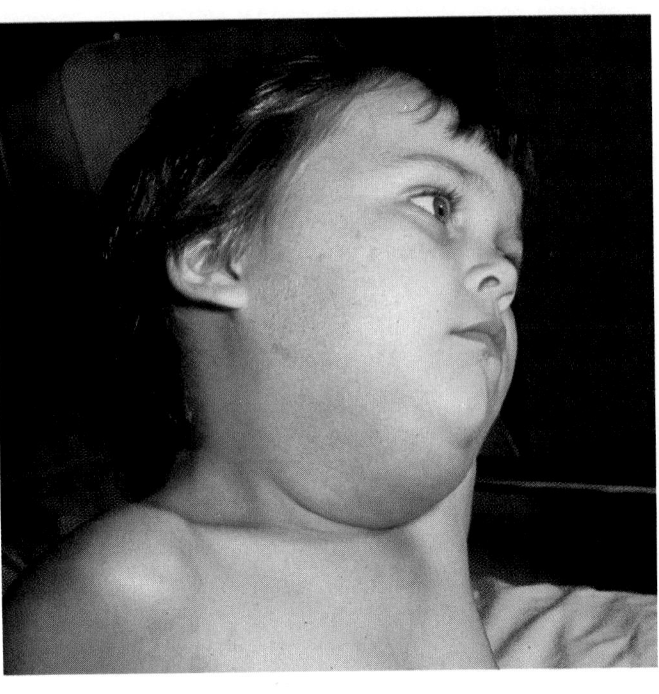

FIGURE 25.12 Mumps. A patient showing the typical swelling of mumps.

Viral Diseases of the Digestive System

Although viruses do not reproduce within the contents of the digestive system like bacteria, they invade many organs associated with the system.

MUMPS

The targets of the mumps virus, the parotid glands, are located just below and in front of the ears (see Figure 25.1). Since the parotids are one of the three pairs of salivary glands of the digestive system, it is appropriate to include a discussion of mumps in this chapter. The virus itself is a *paramyxovirus,* the group to which the measles virus belongs. The mumps virus is an enveloped virus containing single-stranded, fragmented RNA.

Mumps typically begins with painful swelling of one or both parotid glands 16 to 18 days after exposure to the virus (Figure 25.12). The virus is transmitted in saliva and respiratory secretions, and its portal of entry is the respiratory tract. An infected person is most infective to others during the first 48 hours before clinical symptoms appear. Once the viruses have begun to multiply in the respiratory tract and local lymph nodes in the neck, they reach the salivary glands via the blood. Viremia (presence of virus in the blood) begins several days before the onset of mumps symptoms and before the virus appears in saliva. The virus is present in the blood and saliva for three to five days after the onset of the disease and in the urine after ten days or so.

Mumps is characterized by inflammation and swelling of the parotid glands, fever, and extreme pain during swallowing. About four to seven days after the onset of symptoms, the testes can become inflamed (orchitis). This happens in about 20% to 35% of males past puberty. Sterility is a possible but rare consequence. Other complications of mumps include meningitis, inflammation of the ovaries, and pancreatitis. In children, mumps is less common than chickenpox or measles because the disease is less infectious.

An effective attenuated live vaccine is available and is often administered as part of the trivalent measles–mumps–rubella (MMR) vaccine. The number of cases of mumps has dropped sharply since the introduction of the vaccine in 1968. For example, in 1968, 152,209 cases of mumps were reported, whereas in 1989, there were 5611 cases, which was higher than the 2982 cases in 1985. Second attacks are rare, and cases involving only one parotid gland, or subclinical cases (about 30% of those infected), are as effective as bilateral mumps in conferring immunity.

Serological diagnosis is not usually necessary. If laboratory confirmation of a diagnosis based only on symptoms is desired, the virus can be isolated by embryonated-egg or cell culture techniques. The virus is now usually identified by ELISA tests.

CYTOMEGALOVIRUS (CMV) INCLUSION DISEASE

The cytomegalovirus (CMV) is a herpesvirus that induces a cellular swelling characterized as *cytomegaly*

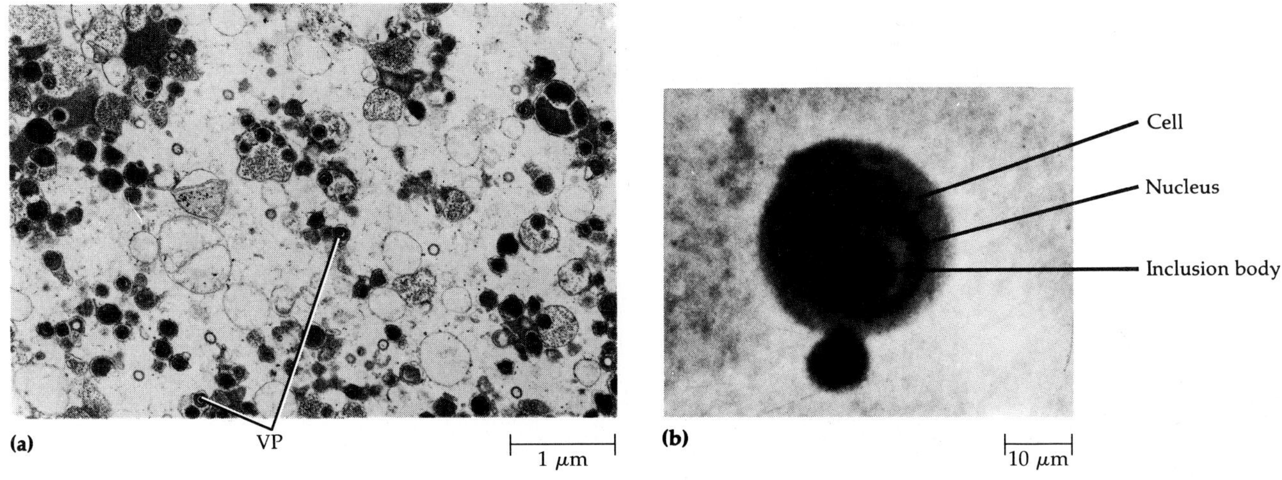

FIGURE 25.13 Cytomegalovirus (CMV). **(a)** Electron micrograph showing individual viral particles (VP) in tissue culture. **(b)** Light micrograph of a single animal cell swollen as a result of infection with cytomegalovirus. The dark, intranuclear inclusion body has the typical "owl-eyed" appearance.

(large cell). These cells are known as "owl's eyes" because of their distinctive appearance (Figure 25.13). Once a person is infected with CMV, the infection persists for life. Blood macrophages and T lymphocytes are the sites of latency. The virus is shed at intervals in such body secretions as saliva, urine, semen, cervical secretions, and breast milk.

Some estimate that 80% of the population of the United States may carry the virus. It has been said that if CMV were accompanied by a skin rash, it would be familiar as one of the common childhood diseases. In less developed parts of the world, infection may approach 100%. It can be spread by kissing and other personal contacts. There is a high incidence of the virus in the male homosexual population. In adults and older children, most infections are subclinical or, at worst, much like a mild case of infectious mononucleosis. However, immunosuppressed individuals, including those with AIDS, may develop life-threatening pneumonia. Recently, commercial products were marketed that contain antibodies that fight CMV present in donated kidneys.

A primary CMV infection acquired during pregnancy by a nonimmune mother can seriously harm the fetus. At the time of her first pregnancy, a woman of affluent background has about a 50% chance of being naturally immune to CMV infection; a woman from a disadvantaged background has about an 80% chance. Virus transmission to the fetus by an immune mother will sometimes occur, but almost no damage ensues. Tests for immune status are readily available, and young women should realize that the main source of infection is young children. Any physician should determine the immune status of a woman of childbearing age and inform any nonimmune woman of the risks of pregnancy. Each year nearly 4000 infants are born with symptomatic disease; many die, and the survivors are severely damaged. In addition, another 4500 to 6000 children, while showing no symptoms at birth, later develop handicaps such as deafness or neurological problems.

A live attenuated vaccine has been tested for several years, but not in women of childbearing years since there is concern about using a live virus that might conceivably become latent in the recipient.

The virus is susceptible in vitro to the antiviral drug ganciclovir. At this time the drug has not been approved for infected infants; however, it has had beneficial effects when used on immunocompromised adult patients. One of the leading microbial complications for AIDS patients is reactivated cytomegalovirus infections, especially affecting the eyesight and lungs.

The most reliable diagnostic method is isolation of the virus from bodily fluids during the first two weeks of life; the sample is usually shipped to a central reference laboratory. Serological tests are not very specific or sensitive, but intensive research is under way.

HEPATITIS

Hepatitis is an inflammation of the liver. Viral hepatitis is now the second most frequently reported infectious disease in the United States. There are at least

TABLE 25.1 Characteristics of Viral Hepatitis

CHARACTERISTIC	HEPATITIS A	HEPATITIS B	HEPATITIS C	HEPATITIS D	HEPATITIS E
Transmission	Fecal–oral (ingestion of contaminated food and water)	Parenteral (injection of contaminated blood or other body fluids)	Parenteral	Parenteral (host must be coinfected with hepatitis B)	Fecal–oral
Agent	Hepatitis A virus (HAV); single-stranded RNA, no envelope	Hepatitis B virus (HBV); double-stranded DNA, envelope	Hepatitis C virus (HCV); single-stranded RNA, envelope	Hepatitis D virus (HDV); single-stranded RNA; envelope from HBV	Hepatitis E virus (HEV); single-stranded RNA, no envelope
Incubation period	2 to 6 weeks	4 to 26 weeks	1 to 8 weeks	Uncertain	2 to 6 weeks
Manifestations or symptoms	Mostly subclinical; severe cases—fever, headache, malaise, jaundice	Frequently subclinical; similar to HAV, but fever, headache absent, and more likely to progress to severe liver damage	Similar to HBV	Severe liver damage, high mortality rate	Similar to HAV, but pregnant women may have high mortality rate
Chronic liver disease	No	Yes	Yes	Yes	No
Vaccines	None, but immunoglobulins give temporary protection	Two types; plasma derived, genetically engineered	None	HBV vaccine is protective because coinfection required	None

five different viruses causing hepatitis, and probably more remain to be discovered or become better known. Hepatitis is also an occasional result of infections by other viruses such as Epstein-Barr virus or cytomegalovirus. Drug and chemical toxicity can also cause acute hepatitis that is clinically identical to viral hepatitis. The characteristics of the various forms of viral hepatitis are summarized in Table 25.1.

Hepatitis A (Infectious Hepatitis)

The *hepatitis A virus (HAV)* is the etiological agent of **hepatitis A (infectious hepatitis).** The HAV contains single-stranded RNA and lacks an envelope. It can be grown in cell culture.

After a typical entrance via the oral route, HAV multiplies in the epithelial lining of the intestinal tract. Viremia eventually occurs, and the virus spreads to the liver, kidneys, and spleen. The virus is shed in the feces and can also be detected in the blood and urine. The amount of virus excreted is greatest before symptoms appear and then declines rapidly. Therefore, a food handler responsible for spreading the virus might not appear to be ill at the time. The virus is probably able to survive for several days on such surfaces as cutting boards. Contamination of food or drink by feces is aided by the resistance of HAV to chlorine disinfectants at concentrations ordinarily used in water. Mollusks, such as oysters, that live in contaminated waters are also a source of infection (see the box on p. 634).

At least 50% of infections with HAV are subclinical, especially in children. In clinical cases, the initial symptoms are anorexia (loss of appetite), malaise, nausea, diarrhea, abdominal discomfort, fever, and chills. These symptoms are more likely to appear in adults; they last from 2 to 21 days, and the mortality is low. In about two-thirds of the cases, there is also jaundice, with yellowing of the skin and the whites of the eyes and the dark urine typical of liver infections. In such cases the liver becomes tender and enlarged. Liver damage is probably caused by immunological reactions.

There is no chronic form of HAV infection, and the virus is usually shed only during the acute stage of disease, although it is difficult to detect. The incubation time averages about four weeks and ranges from two to six weeks, which makes epidemiological studies

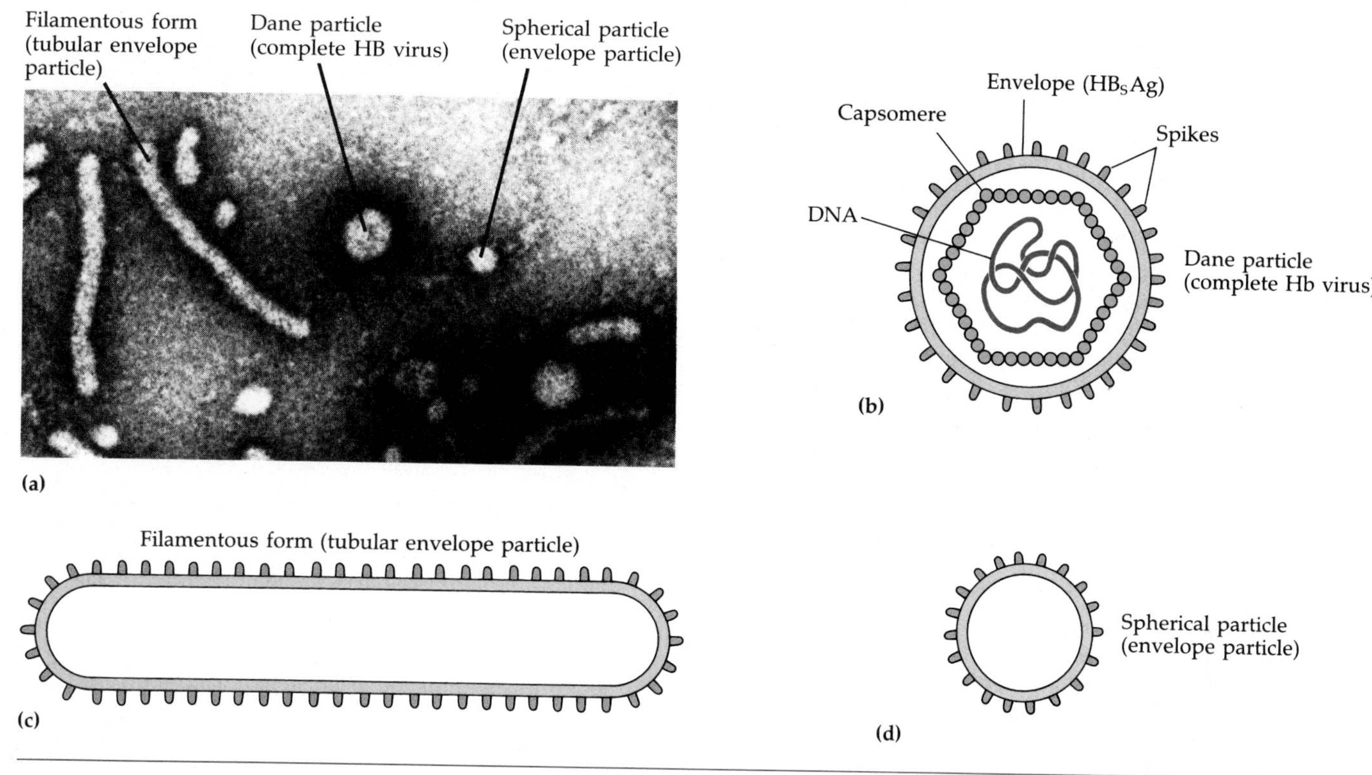

FIGURE 25.14 Hepatitis B virus (HBV). **(a)** Electron micrograph shows all three distinct types of HBV particles discussed in the text. **(b)–(d)** Diagrammatic representations.

for the source of infections difficult. There are no animal reservoirs.

In the United States, the percentage of the population with HAV is much higher among lower socioeconomic groups (72% to 88%) than among middle and upper socioeconomic groups (18% to 30%). Most cases are unreported, and the 28,919 cases reported in 1990 in the United States represent only a fraction of the actual number.

HAV antibodies can be identified by several serological tests, of which the ELISAs are the most reliable. Acute disease is diagnosed by detection of IgM anti-HAV because these antibodies appear about four weeks after infection and disappear about three to four months after infection.

No specific treatment for the disease exists, but persons at risk for exposure to hepatitis A can be given immune globulin (HAIG), which provides protection for several months. This is sufficient for travelers to high-risk areas.

Hepatitis B (Serum Hepatitis)

Hepatitis B, or **serum hepatitis,** is caused by the *hepatitis B virus (HBV)*. HBV and HAV are completely different viruses. HBV is larger, its genome is double-stranded DNA, and it is enveloped. Because HBV has often been transmitted by blood transfusions, there has been great interest in studying this virus to identify contaminated blood.

The serum from patients with hepatitis B contains three distinct particles. The largest, the *Dane particle,* is the complete virion; it is infectious and capable of replicating. There are also smaller *spherical particles,* about half the size of a Dane particle, and *filamentous particles,* which are tubular particles similar in diameter to the spherical particles but about ten times as long (Figure 25.14). The spherical and filamentous particles are unassembled components of Dane particles without nucleic acids; assembly is evidently not very efficient, and large numbers of these unassembled components accumulate. Fortunately, these numerous unassembled particles contain *hepatitis B surface antigen (HB_SAg),* which can be detected with antibodies specific against them. Such antibody tests make large-scale screening of blood for HBV possible.

Physicians, nurses, dentists, medical technologists, and others who are in daily contact with blood have a considerably higher incidence of this disease than the general population. In 1988, for example, 12,000 health care workers became infected and 300 died. Physicians have five times and dentists have two to three times the normal rate of infection. In 1990,

MMWR
MORBIDITY AND MORTALITY WEEKLY REPORT

Foodborne Hepatitis A in Seven States

From 1983 through 1989, the incidence of hepatitis A in the United States increased 58%. This report summarizes recent foodborne outbreaks of hepatitis A.

Alaska

Between June 18 and July 20, 1988, 32 serologically confirmed hepatitis A cases occurred in Peters Creek. Twenty-three additional cases occurred among household contacts of the original patients. All 32 patients had consumed an ice-slush beverage purchased from a local convenience market between May 23 and June 10.

The ice-slush was prepared daily with tap water from a bathroom sink, using utensils stored beside a toilet. Four of the five employees tested negative for IgM antibody to hepatitis A virus (IgM anti-HAV). The fifth employee, who was one of the two persons who prepared the ice-slush, refused to be tested. However, a household contact of this employee had had serologically confirmed hepatitis A in early June and reported that the employee had had jaundice concurrently with her illness.

Florida

Sixty-one cases of serologically confirmed hepatitis A occurred in five states (see map). Fifty-nine people had eaten raw oysters, one had eaten raw scallops, and one had eaten baked oysters. All the oysters and scallops were traced to the same growing area of Bay County.

The implicated oysters apparently had been illegally harvested from oyster beds outside approved coastal waters. Sources of human fecal contamination were identified near these oyster beds and included boats with inappropriate sewage disposal systems and a local sewage treatment plant with discharges containing high levels of fecal coliforms.

North Carolina

Hepatitis A cases were reported among employees of businesses located in east Greensboro. In a control study, patients were more likely than controls to have eaten at one nearby restaurant. Only consumption of iced tea was associated with illness.

All food handlers at the restaurant were tested for IgM anti-HAV; one employee, who was IgM anti-HAV positive, denied symptoms of hepatitis. However, this employee was a suspected intravenous-drug user and had prepared fountain drinks.

Comment: These outbreaks illustrate two principal modes of transmission of foodborne hepatitis A: (1) contamination of food during preparation and (2) contamination of food before it reaches the food service establishment.

Contamination of food during preparation by an infected food handler is the most common mode of transmission in foodborne outbreaks. The outbreak in North Carolina is also consistent with a nationwide phenomenon of increased reports of hepatitis A among intravenous-drug users who may become food handlers.

Contamination of food before it reaches the service establishment is less common. Mollusks filter large quantities of water during feeding and, in the process, can concentrate microorganisms.

Measures to prevent foodborne hepatitis A outbreaks include training food handlers in proper hygiene and food-handling practices and investigating food handlers who have symptoms of hepatitis or are otherwise ill. Prevention of hepatitis A outbreaks associated with mollusks relies on surveillance of water beds where mollusks are harvested to ensure that there is no evidence of fecal contamination, and thorough cooking.

Source: *MMWR* 39:228–232 (4/13/90).

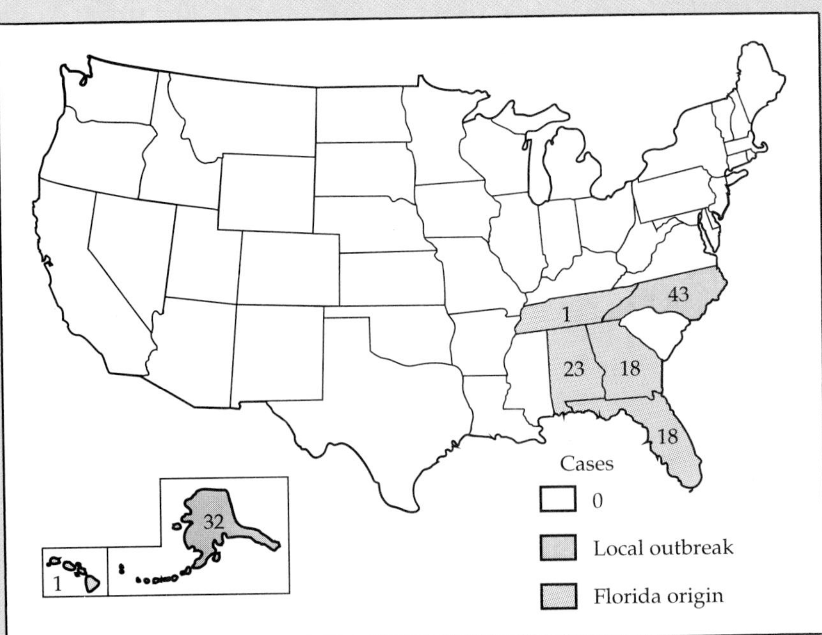

Colored areas show hepatitis A outbreaks from a common vehicle originating in Florida.

however, surveys showed that only about 40% of health care workers were vaccinated against HBV and among these only 25% of physicians were vaccinated. There have also been instances of transmission to patients by surgeons and dentists. Intravenous drug abusers often share needles and fail to practice proper sterilization; as a consequence, they also have a high incidence of hepatitis B.

In addition to transmission by virus-carrying blood, recent evidence indicates that hepatitis B can be transmitted by any secretion of bodily fluids, such as saliva, breast milk, and semen. Transmission by semen donated for artificial insemination has been documented, and semen has been implicated in transmission between heterosexuals with multiple partners and in male homosexuals. A mother who is positive for HB_SAg, especially if she is a chronic carrier, may transmit the disease to her infant, usually at birth. In most cases, this type of transmission can be prevented by giving the newborn immune globulin (HBIG) immediately after birth. These babies should also be vaccinated.

The incubation period averages about 12 weeks but ranges from 4 to 26 weeks. Because this incubation period is lengthy and of uneven duration, determination of the origin of an infection can be difficult.

The CDC estimates that 200,000 or more persons, mostly young adults, are infected with hepatitis B each year. Clinically, hepatitis B varies widely. Probably about one-half of the cases are entirely asymptomatic. Symptoms are highly variable and include, in the early stages, loss of appetite, low-grade fever, and joint pains. Later, jaundice usually appears. It is difficult to distinguish between hepatitis A and B solely on clinical grounds, although there is less likely to be fever and headache with HBV infections.

Ninety percent of acute hepatitis B infections end in complete recovery. The mortality rate is higher than that of hepatitis A but is probably less than 1% of the generally infected population. It might be as high as 2% or 3% among hospitalized patients, many of whom are elderly. Up to 10% of patients become chronic carriers. These carriers are a reservoir for transmission of the virus, and they also have a high rate of liver disease. A special concern is the strong correlation between the occurrence of liver cancer and the incidence of chronic hepatitis B infections. Chronic carriers are about 200 times more likely to get liver cancer than the general population. Liver cancer is the most common form of cancer in sub-Saharan Africa and the Far East, areas where hepatitis B is extremely common.

Researchers have been unable to cultivate HBV in cell culture. This has been the means by which vaccines for polio, mumps, measles, and rubella have been developed. However, two effective vaccines for hepatitis B have been developed by other means. One uses HB_SAg harvested from the serum of human carriers. It has been rigorously purified, and any fear that it might contain the AIDS virus is unwarranted. A newer vaccine uses HB_SAg produced by a genetically engineered yeast. Vaccination is recommended for high-risk groups; a partial listing would include medical personnel exposed to blood and blood products, persons undergoing hemodialysis, patients and staff at institutions for the mentally retarded, and homosexually active males. The incidence of hepatitis B is very high in this latter group, which is considered evidence that these cases are sexually transmitted.

In 1984, a campaign was begun in Taiwan to prevent mother-to-infant transmission of the disease. All infants delivered of HB_SAg-positive mothers are vaccinated, and infants of highly infectious carrier mothers also receive HBIG. Similar recommendations have been proposed in the United States.

Hepatitis C (Non-A, Non-B)

Screening of donated blood for HB_SAg became so effective that transmission of hepatitis B by blood transfusion declined dramatically. However, in the 1960s a previously unsuspected form of hepatitis appeared that could also be transmitted by transfusion. This form was named, by elimination, non-A, non-B **(NANB)** hepatitis. Soon, almost all cases of transfusion-transmitted hepatitis were of the NANB type. This form of hepatitis has an incubation period of about 60 days and causes symptoms that are subclinical or relatively mild, but it is likely to become chronic and cause liver damage.

In 1989 a serological test was developed to detect antibodies to NANB virus, and the virus was then named **hepatitis C virus.** This virus has a single strand of RNA and is probably enveloped.

This antibody test should be very useful in preventing hepatitis C infections from blood transfusions. However, only an estimated 10% or so of the cases arise from this source. Most cases are associated with intravenous drug use, occupational exposure to blood, kidney dialysis, or possibly heterosexual activity. A significant fraction of infected persons have no identifiable, probable source of infection. The test detects antibodies, and these may not appear in response to the infection for several months. Also, the new test is much more sensitive in detecting infections in the chronic stage than in the acute stage. Even so, the test indicates that approximately 2% of volunteer blood donors are infected with hepatitis C, many more than those infected with hepatitis B. A test that would detect the virus itself is still needed.

Finally, because some individuals are known to have had more than one episode of NANB hepatitis, there remains the strong possibility that hepatitis C virus is not the only blood-transmitted NANB hepatitis agent.

Hepatitis D (Delta Hepatitis)

In 1977, a new hepatitis virus, now known as **hepatitis D virus (HDV),** was discovered in carriers of HBV in Italy. Persons who carried this so-called *delta antigen* and were also infected with hepatitis B had a much higher incidence of severe liver damage and a much higher mortality rate than persons who had antibodies against HBV alone. With time, it became clearer that HDV can occur with either acute *(coinfection form)* or chronic *(superinfection form)* hepatitis. In persons with a case of self-limiting acute HBV, the coinfection with HDV disappeared as the HBV was cleared from the system, and the condition resembled a typical case of acute HBV. However, if the HBV infection progressed to the chronic stage, the superinfection with HDV was often accompanied by progressive liver damage and a fatality rate several times that of persons infected with HBV alone.

Epidemiologically, HDV is linked to the epidemiology of HBV. Most cases occur in the Middle East and Mediterranean area, where the infection appears to spread mostly by intimate contact. In the Far East, however, HBV is extremely common but HDV is not. Furthermore, in the Amazon Basin and northern South America the disease takes the form of sudden epidemics with a very high mortality rate. In the United States and northern Europe the disease occurs predominantly in high-risk groups such as intravenous drug abusers.

Structurally, the HDV antigen is a single strand of RNA, shorter than any other animal-infecting virus, that is protected by a capsid. This particle is not capable of causing an infection. It becomes infectious when an external envelope of HB_SAg, whose formation is controlled by the genome of HBV, covers the HDV protein core (the delta antigen) (Figure 25.14b).

Hepatitis E (Infectious NANB Hepatitis)

Another form of NANB hepatitis is spread by fecal–oral transmission, much like hepatitis A, which it clinically resembles. The agent is now known as **hepatitis E virus (HEV)** and is endemic in areas of the world with poor sanitation, especially India and southeast Asia. It resembles HAV in being an unenveloped virus with a single strand of RNA but is not related serologically to it. Like hepatitis A virus, it does not cause chronic liver disease, but for some unexplained reason it is responsible for a very high mortality rate in pregnant women.

VIRAL GASTROENTERITIS

A number of viruses, such as polioviruses, echoviruses, and coxsackieviruses, are transmitted by the fecal–oral route. However, despite the name *enteroviruses,* these viruses generally do not directly affect the digestive system. About 90% of cases of viral gastroenteritis are caused by rotavirus or by the Norwalk agent.

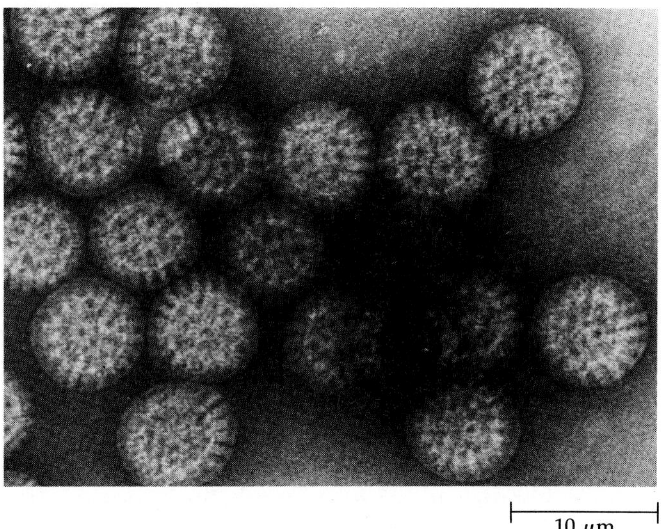

FIGURE 25.15 Rotavirus. The name of this virus comes from its wheel shape.

Rotavirus (*rota* means wheel) is the most common cause of viral gastroenteritis, mostly causing disease in small children (Figure 25.15). Following an incubation period of two or three days, the patient suffers from low-grade fever, diarrhea, and vomiting, which persist for a mean of five to eight days. There may be more than one serotype of rotavirus causing disease, and immunity is only partially protective. An ELISA and several other serological tests can be used to detect the virus in feces.

Major epidemics of viral gastroenteritis have been caused by a virus known as the *Norwalk agent* (named after an outbreak in Norwalk, Ohio, in 1968). All age groups are affected by this virus. Following a two-day incubation period, infected persons typically suffer nausea, abdominal cramps, diarrhea, and vomiting for one to three days. Symptoms are mild, as compared with those of rotavirus infections. About one-half of middle-aged Americans show evidence, by serum antibodies, that they have been infected. Susceptibility to the virus varies considerably; some persons are not affected and do not even respond with an immune reaction. Others develop gastroenteritis and an immunity that might be only short term.

The only treatment for viral gastroenteritis is oral rehydration or, in exceptional cases, intravenous rehydration.

Fungal Diseases of the Digestive System

Some fungi produce toxins called **mycotoxins,** which cause blood diseases, nervous system disorders, kidney damage, and liver damage.

ERGOT POISONING

Some mycotoxins are produced by *Claviceps purpurea,* a fungus causing smut infections on crops (see Figure 15.7). The mycotoxins produced by *C. purpurea* cause **ergot poisoning,** which results from the ingestion of rye or other cereal grains contaminated with the fungus. Ergot poisoning was very widespread during the Middle Ages. The toxin can restrict blood flow in the extremities, and gangrene results. It may also cause hallucinogenic symptoms, producing bizarre behavior similar to that caused by LSD.

AFLATOXIN POISONING

Aflatoxin is a mycotoxin produced by the fungus *Aspergillus flavus,* a common mold. Aflatoxin has been found in many foods but is particularly likely to be found on peanuts. Aflatoxin is highly toxic and can cause serious damage to livestock when their feed is contaminated with *A. flavus.* Although the risk to humans is unknown, there is strong evidence that aflatoxin contributes to cirrhosis of the liver and cancer of the liver in parts of the world, such as India and Africa, where food is subject to aflatoxin contamination.

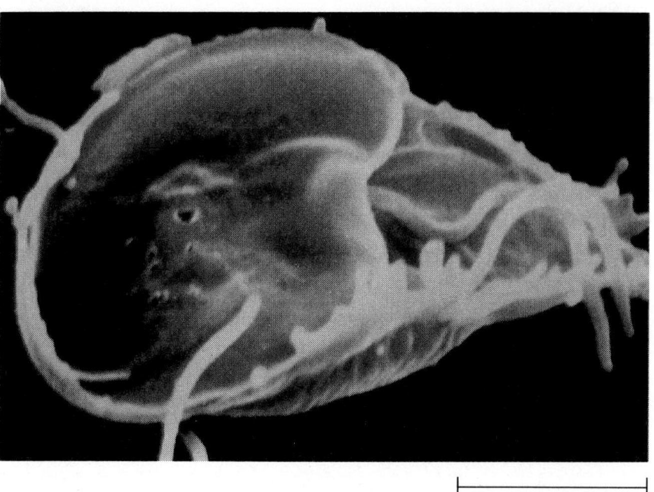

FIGURE 25.16 Scanning electron micrograph of the trophozoite form of *Giardia lamblia,* the flagellated protozoan that causes giardiasis. Note the prominent sucker disc it uses to adhere to the intestinal mucosa. The reverse side is smoothly streamlined, and the intestinal contents move easily around the attached organism.

Protozoan Diseases of the Digestive System

A number of pathogenic protozoans complete their life cycles in the human digestive tract. Usually they are ingested as resistant, infective cysts and are shed in greatly increased numbers as newly produced cysts.

GIARDIASIS

Giardia lamblia is a flagellated protozoan that is able to attach firmly to a human's intestinal wall (Figure 25.16). It is the cause of **giardiasis,** a prolonged diarrheal disease of humans. The disease sometimes persists for weeks and is characterized by malaise, nausea, flatulence (intestinal gas), weakness, weight loss, and abdominal cramps. The protozoans sometimes occupy so much of the intestinal wall that they interfere with food absorption. There are frequent outbreaks of giardiasis in the United States; it is probably the most common cause of epidemic waterborne diarrheal disease. About 7% of the population are healthy carriers and shed the cysts in their feces.

The organism is also shed by a number of wild mammals, especially beavers, and the disease is not uncommon in backpackers who drink from wilderness waters. Most outbreaks in the United States are transmitted by contaminated water supplies. In a recent national survey of surface waters serving as sources for municipalities, the organism was detected in 18% of the samples. Because the cyst stage of the protozoan is relatively insensitive to chlorine, filtration of water supplies is usually necessary to eliminate the cysts from water.

Giardia is not reliably found in stools, and the *string test* is recommended for diagnosis. In this test, a gelatin capsule packed with about 140 cm of fine string is swallowed by the patient. One end of the string is taped to the cheek. The gelatin capsule dissolves in the stomach, and an enclosed weighted rubber bag attached to the other end of the string enters the upper bowel. After a few hours, the string is drawn up through the mouth and examined for parasites. A recent development is a fluorescent-antibody test that employs monoclonal antibodies specific for antigens on the cyst. Specimens for the test are taken by rectal swab.

Treatment with metronidazole or quinacrine hydrochloride is usually effective within a week. Immunity to giardiasis is discussed in the box in Chapter 17, p. 440.

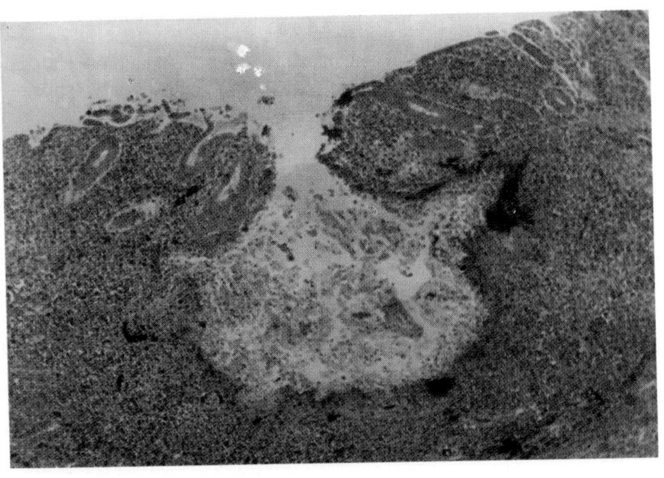

FIGURE 25.17 Typical flask-shaped ulcer in intestinal wall caused by *Entamoeba histolytica.*

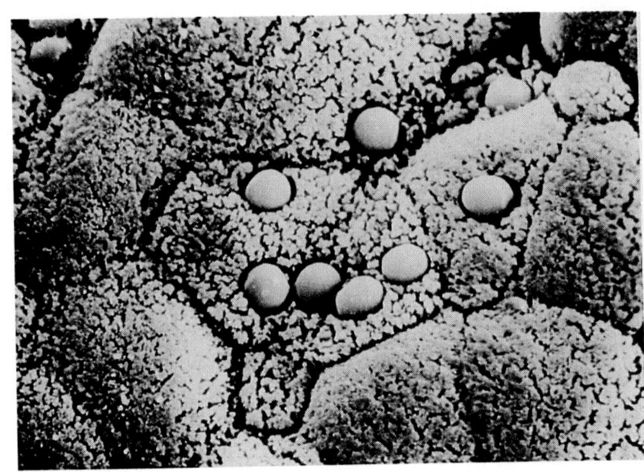

FIGURE 25.18 *Cryptosporidium.* Oocysts embedded into mucosal surface of intestine.

BALANTIDIASIS (BALANTIDIAL DYSENTERY)

Balantidium coli (bal-an-tid'ē-um kō'lī) is a ciliated protozoan that causes **balantidiasis, or balantidial dysentery.** The organism, which is the only ciliate known to cause human disease, exists in a vegetative form (trophozoite) and a cyst form. Humans acquire *B. coli* by ingesting cysts in food or water contaminated by feces. In the colon the cysts release the vegetative form, which lives in the large intestine. It typically causes a mild disease involving abdominal pain, nausea, vomiting, diarrhea, and weight loss. Occasionally it invades the epithelial lining and causes fatal dysentery.

Laboratory confirmation of balantidiasis is based on the demonstration of vegetative cells in feces. Treatment consists of tetracyclines. Metronidazole is sometimes used as an alternative.

AMOEBIC DYSENTERY (AMOEBIASIS)

Amoebic dysentery (amoebiasis) is found worldwide and is spread mostly by food or water contaminated by cysts of the protozoan amoeba *Entamoeba histolytica* (see Figure 12.19). Although stomach acid (HCl) can destroy vegetative cells, it does not affect the cysts. In the intestinal tract, the cyst wall is digested away and the vegetative forms are released. The vegetative forms then multiply in the epithelial cells of the wall of the large intestine. A severe dysentery results, and the feces characteristically contain blood and mucus. The vegetative forms feed on red blood cells and destroy tissue in the gastrointestinal tract (Figure 25.17).

Severe infections result if the intestinal wall is perforated. Abscesses might have to be treated surgically, and invasion of other organs, particularly the liver, is not uncommon. Perhaps 5% of the population in the United States are asymptomatic carriers of *E. histolytica*. It is estimated that, worldwide, one person in ten is infected, mostly asymptomatically, and that a tenth of these infections progress to the more serious stages.

Diagnosis depends largely on recovery and identification of the organisms in feces. (Red blood cells observed within the trophozoite stage of an amoeba are considered an indication of *E. histolytica*.) There are several serological tests that can also be used for diagnosis, including latex-agglutination and fluorescent-antibody tests. Such tests are especially useful if the patient is not passing amoebae.

Metronidazole plus iodoquinol are the drugs of choice in treatment.

CRYPTOSPORIDIOSIS

Cryptosporidiosis is caused by the protozoan *Cryptosporidium*, long recognized as a pathogen in calves but not known to infect humans until 1976. Since then, the number of reported cases has been increasing. This is another example of a disease with particularly serious effects on immunosuppressed individuals, including those with AIDS. In otherwise healthy individuals, the organism causes intestinal distress of short duration. Immune-deficient individuals, however, suffer from a severe life-threatening diarrhea that may last for months.

Transmission is largely fecal–oral between humans. However, a zoonosis-type transmission from calves has been observed. Diagnosis is based on microscopic examination of feces for *Cryptosporidium* oocysts (Figure 25.18). At present there is no reliable treatment.

Helminthic Diseases of the Digestive System

Helminthic parasites are very common in the human intestinal tract, especially under conditions of poor sanitation. Figure 25.19 shows the worldwide incidence of some helminthic diseases. In spite of their size and formidable appearance, they often produce few symptoms. They have become so well adapted to their human hosts, and vice versa, that when their presence is revealed it is often a surprise.

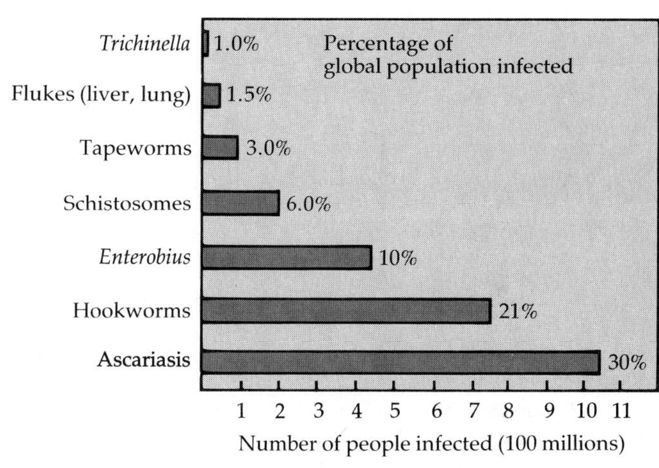

FIGURE 25.19 Incidence of selected intestinal helminths in the world population.

TAPEWORM INFESTATIONS

Taenia saginata is a typical tapeworm whose life cycle extends through three stages. The adult worm lives in the intestine of a human host, where it lays eggs that are excreted in the feces. The eggs are ingested with contaminated feed by animals, such as grazing cattle, and hatch into a larval form called a cysticercus that lodges in the animal's muscles. When a human ingests undercooked meat containing cysticerci, the cysticerci develop into adult worms. Infestations by tapeworms begin with the consumption of undercooked beef, pork, or fish containing larval forms of the tapeworm. The larvae develop into adults that attach to the intestinal wall by suckers on the scolex. The adult beef tapeworm, *T. saginata,* can live in the human intestine for 25 years and reaches a length of six meters (18 feet) or longer. Even a worm of this size seldom causes significant symptoms beyond a vague abdominal discomfort. There is, however, psychological distress when a meter or more of detached segments (proglottids) break loose and unexpectedly slip out of the anus, which happens occasionally.

Taenia solium (sō'lē-um), the pork tapeworm, has a life cycle similar to the beef tapeworm. An important difference is that *T. solium* may produce the larval stage in the human host, a disease condition called **cysticercosis.** In this disease, the infection can arise from the tapeworm egg rather than only from ingestion of undercooked meat containing cysticerci. The eggs can be ingested as a result of poor sanitary practices or possibly by autoinfection (when eggs produced by the adult tapeworm in the intestinal tract somehow enter the stomach).

Cysticerci can develop in many human organs (Figure 25.20). When this occurs in muscle tissue, the symptoms are seldom severe. Cysticerci that develop in the eyes and in the brain, however, may be more serious. Cerebral cysticercosis, which is endemic in Mexico and Central America, has become a fairly common condition in parts of the United States within Hispanic cultures. The symptoms often mimic those of a brain tumor. The number of cases reported reflects, in

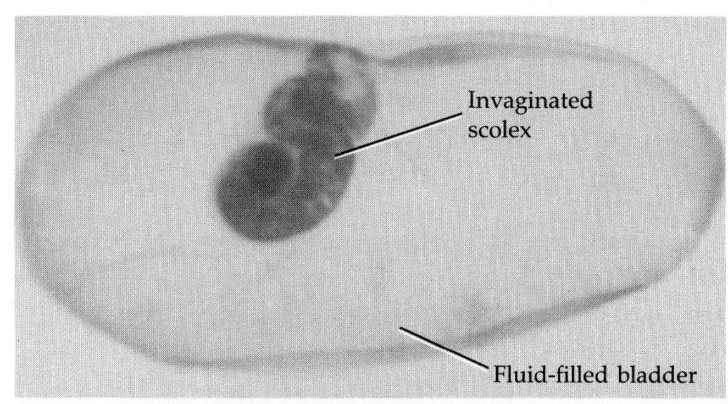

(a)

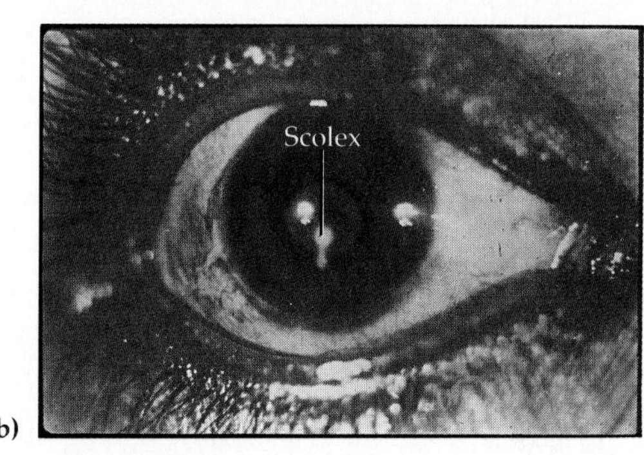

(b)

FIGURE 25.20 Cysticercosis in humans.
(a) A cysticercus of *Taenia solium.* Note the scolex of the larval tapeworm invaginated in the fluid-filled bladder. Cysticerci are typically 6 to 18 mm in length. **(b)** Cysticerci can occur in many tissues. This one has formed in an eye. The scolex has evaginated from the bladder.

TABLE 25.2 Diseases Associated with the Digestive System

DISEASE	*CAUSATIVE AGENT*	*MODE OF TRANSMISSION*	*TREATMENT*
Bacterial Diseases of the Mouth			
Dental caries	*Streptococcus mutans* is most prominent	Bacteria use sucrose to form plaque	Drilling and filling. Prevention includes restricting ingestion of sucrose; brushing, flossing, and fluoridation
Peridontal disease	Variety of organisms	Presence of bacterial plaque initiates an inflammatory response	Antibiotics such as penicillin and tetracycline; cleaning teeth may prevent occurrence
Bacterial Diseases of the Lower Digestive System			
Staphylococcal food poisoning	*Staphylococcus aureus*	Ingestion of exotoxin in food, usually improperly refrigerated	Replace lost water and electrolytes
Salmonellosis	*Salmonella* spp.	Ingestion of contaminated food and drink	Replace lost water and electrolytes
Typhoid fever	*Salmonella typhi*	Ingestion of contaminated food and drink	Third-generation cephalosporins
Bacillary dysentery (shigellosis)	*Shigella* spp.	Ingestion of contaminated food and drink	Replace lost water and electrolytes; trimethoprim-sulfamethoxazole
Cholera	*Vibrio cholerae*	Ingestion of contaminated food and drink	Replace lost water and electrolytes
Vibrio parahaemolyticus gastroenteritis	*Vibrio parahaemolyticus*	Often ingestion of contaminated shellfish	Replace lost water and electrolytes
Escherichia coli gastroenteritis	Enterotoxigenic, enteroinvasive, enteropathogenic, and enterohemorrhagic strains of *Escherichia coli*	Ingestion of contaminated food and drink	Replace lost water and electrolytes
Campylobacter gastroenteritis	*Campylobacter jejuni*	Ingestion of contaminated food and drink	Replace lost water and electrolytes
Helicobacter gastritis	*Helicobacter pylori*	Presumably ingested	Bismuth-containing drugs, metronidazole
Yersinia gastroenteritis	*Yersinia enterocolitica*	Ingestion of contaminated food and drink	Replace lost water and electrolytes
Clostridium perfringens gastroenteritis	*Clostridium perfringens*	Ingestion of contaminated food (usually meat) and drink	Replace lost water and electrolytes
Bacillus cereus gastroenteritis	*Bacillus cereus*	Ingestion of contaminated food	None, self-limiting
Viral Diseases of the Digestive System			
Mumps	Mumps virus	Saliva and respiratory secretions	None, but a vaccine is available (MMR)
Cytomegalovirus (CMV) inclusion disease	Cytomegalovirus (CMV)	Placental transfer and blood transfusion	Ganciclovir

(Continued)

part, the use of CT (computerized tomography) scanning in diagnosis. These expensive machines X ray the body in continuous "slices."

The fish tapeworm, *Diphyllobothrium latum* (dī-fil-lō-bo'thrē-um lā'tum), is found in pike, trout, perch, and salmon. The CDC has issued warnings about the

TABLE 25.2 ***(Continued)***

DISEASE	CAUSATIVE AGENT	MODE OF TRANSMISSION	TREATMENT
Hepatitis A	Hepatitis A virus	Ingestion of contaminated food or drink	None, but passive immunization available
Hepatitis B	Hepatitis B virus	Blood or blood contamination by needles, transfusions; sexual contact, especially between male homosexuals	Vaccines available
Hepatitis C	Hepatitis C virus	Blood or blood contamination by needles, transfusions	None
Hepatitis D	Hepatitis D virus	Probably same as hepatitis B	Hepatitis B vaccine effective because coinfection required
Hepatitis E	Hepatitis E virus	Ingestion of contaminated food and water	None
Viral gastroenteritis	Norwalk agent or rotavirus	Ingestion of contaminated food and water	Replace lost water and electrolytes
Fungal Diseases of the Digestive System			
Ergot poisoning	Mycotoxin produced by *Claviceps purpurea*	Ingestion of contaminated food, usually cereal grain	Antispasmodic drugs
Aflatoxin poisoning	Mycotoxin produced by *Aspergillus flavus*	Ingestion of contaminated food, usually peanuts	None
Protozoan Diseases of the Digestive System			
Giardiasis	*Giardia lamblia*	Ingestion of contaminated water	Metronidazole and quinacrine hydrochloride
Balantidiasis (balantidial dysentery)	*Balantidium coli*	Ingestion of contaminated food and water	Tetracycline or metronidazole
Amoebic dysentery (amoebiasis)	*Entamoeba histolytica*	Ingestion of contaminated food and water	Metronidazole plus iodoquinol
Cryptosporidiosis	*Cryptosporidium*	Uncertain, probably fecal–oral	None
Helminthic Diseases of the Digestive System			
Tapeworm infestations	*Taenia saginata* (beef tapeworm); *T. solium* (pork tapeworm); *Diphyllobothrium latum* (fish tapeworm)	Ingestion of contaminated meat or fish	Niclosamide
Hydatid disease	*Echinococcus granulosus*	Ingestion of eggs	Surgical removal
Pinworms	*Enterobius vermicularis*	Ingestion of eggs	Pyrantel pamoate, mebendazole
Hookworms	*Necator americanus*	Larvae in soil penetrate skin	Mebendazole
Ascariasis	*Ascaris lumbricoides*	Ingestion of eggs	Mebendazole
Trichinosis	*Trichinella spiralis*	Ingestion of larvae in meat	Thiabendazole and corticosteroids

risks of fish tapeworm infection from the increasingly popular sushi and sashimi (Japanese dishes prepared from raw fish). To relate a vivid example, about ten days after eating sushi, one person developed symptoms of abdominal distention, flatulence (intestinal gas), belching, intermittent abdominal cramping, and diarrhea. Eight days later, the patient passed a 1.2-meter (four-feet) long tapeworm identified as a species of *Diphyl lobothrium*.

Laboratory diagnosis of human infestation consists of identification of the tapeworm eggs or segments in feces. The drug of choice for eliminating tapeworms is niclosamide.

HYDATID DISEASE

Not all tapeworms are large. One of the most dangerous is *Echinococcus granulosus,* which is only a few millimeters in length (see Figure 12.26). Humans are not the definitive hosts. The adult form lives in the intestinal tract of carnivorous animals, such as dogs and wolves. Typically, humans become infected from the feces of a dog that has become infected by eating the flesh of a sheep or deer containing the cyst form of the tapeworm. Unfortunately, humans can be an intermediate host, and cysts can develop in the body.

Once ingested by a human, the egg of *E. granulosus* may migrate to various tissues of the body. The liver and lungs are the most common sites, but the brain and numerous other sites also may be infected. Once in place, the egg develops into a **hydatid cyst** that can grow to a diameter of 10 mm in a few months (Figure 25.21). In some locations, cysts may not be apparent for many years. Some become enormous, containing up to 15 liters (4 gallons) of fluid. Damage may arise from the size of the cyst in such areas as the brain or the interior of bones. Another factor in the pathogenicity of such cysts is that the fluid contains proteinaceous material, to which the host becomes sensitized. If the cyst suddenly ruptures, the result can be life-threatening anaphylactic shock. Treatment is limited to surgical removal, but care must be taken to avoid release of the fluid and potential anaphylactic shock. The disease is seen most frequently among people who raise sheep or hunt or trap wild animals.

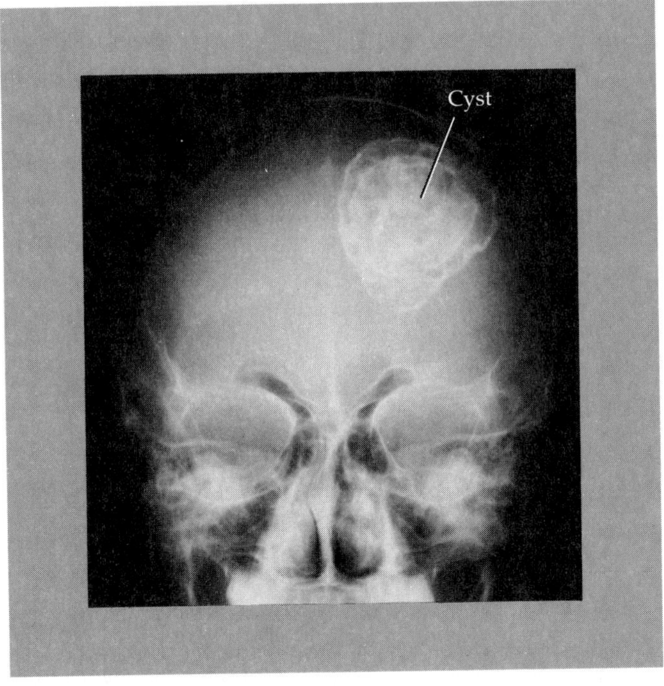

FIGURE 25.21 Hydatid cyst. A large cyst can be seen in this X ray of an infected brain.

NEMATODE INFESTATIONS

Pinworms

Most of us are familiar with the **pinworm,** *Enterobius vermicularis* (see Figure 12.27). This tiny worm migrates out of the anus of the human host to lay its eggs, causing local itching. Whole households may become infected. Such drugs as pyrantel pamoate or mebendazole are usually effective in treatment.

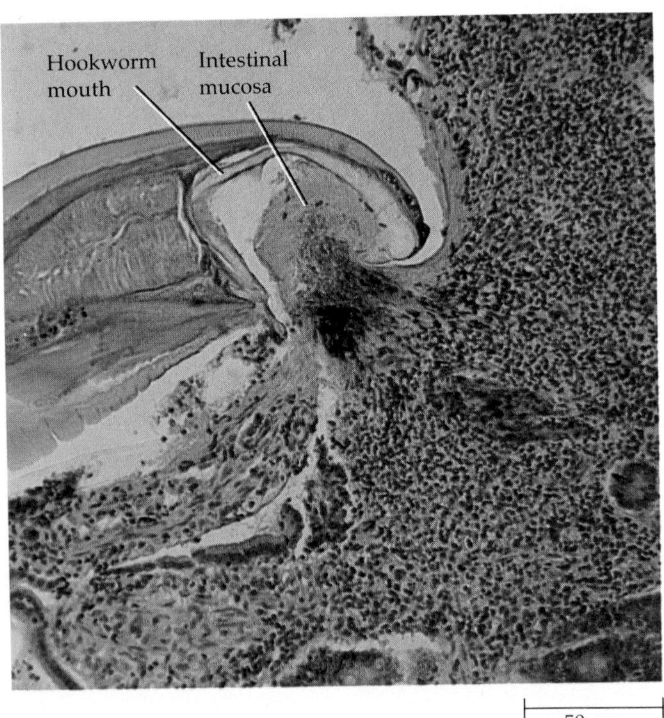

FIGURE 25.22 *Ancylostoma* hookworm attached to intestinal mucosa. Note how the mouth is allowing the worm to feed on the tissue. (See also Figure 12.28 for a different species.)

Hookworms

Hookworm infestations were once a very common parasitic disease in the southeastern states. In the United States, the species most commonly seen is *Necator americanus* (see Figure 12.28). The hookworm attaches to the intestinal wall and feeds on blood and tissue (Figure 25.22) rather than on partially digested food, so the presence of large numbers of worms can lead to anemia and lethargic behavior. Large infections can also lead to a bizarre symptom, *pica,* which is a craving for peculiar foods, such as laundry starch or soil containing a certain type of clay. Pica is a symptom of iron deficiency anemia. Because the life cycle of the hookworm requires human feces to enter the soil and bare skin to contact contaminated soil, the incidence of the disease has declined greatly with improved sanitation and the practice of wearing shoes. Hookworms can be treated effectively with mebendazole.

Ascariasis

One of the most widespread helminthic diseases is **ascariasis,** caused by *Ascaris lumbricoides.* This condition is still fairly common in the United States. It does not usually cause severe symptoms, but its presence can be manifested in distressing ways. As we described in Chapter 12, diagnosis is often made when an adult worm emerges from the anus. These worms can be quite large, up to 30 cm (about 1 foot) in length (Figure 25.23). In the intestinal tract, they live on partially digested food and cause few symptoms, but ingested eggs undergo a peculiar life cycle. In the upper intestine, they hatch into small, wormlike larva that pass into the bloodstream and then into the lungs. In the lungs, they migrate into the throat and are swallowed—all this movement, just to get back to the place they started! In the lungs, they may cause some pulmonary symptoms, and extremely large numbers cause pneumonia.

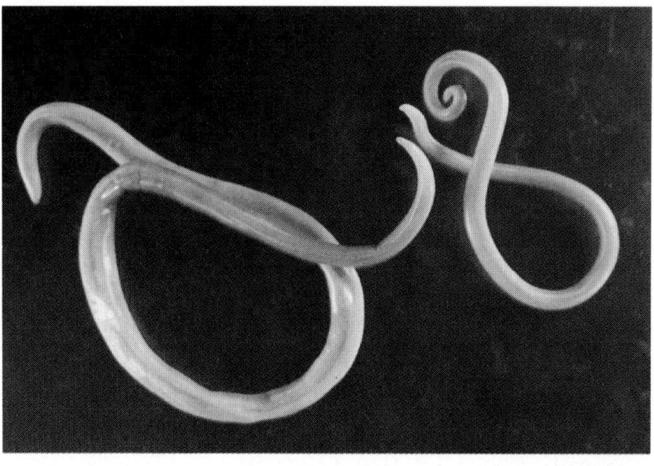

FIGURE 25.23 ***Ascaris lumbricoides.*** This photograph shows a male (the smaller worm with the coiled end) and a female. These worms are up to 30 cm (1 foot) in length.

The most dramatic consequences of infection with *A. lumbricoides* are from the migrations of adult worms. The worms are equipped with minute cutting teeth and can penetrate tissue. If they pass through the intestinal wall, they can infect the abdominal cavity. Worms have been known to leave the body of small children through the umbilicus (navel) and to escape through the nostrils of a sleeping person. Once ascariasis is diagnosed, it can be effectively treated with mebendazole.

Trichinosis

Most infestations by the small roundworm *Trichinella spiralis,* called **trichinosis,** are insignificant. The larvae, in encysted form, are located in muscles of the host. In 1970, routine autopsies of human diaphragm muscles showed that about 4% of those tested carried this parasite. Severe cases of trichinosis can be fatal—sometimes in only a few days. The severity of the disease is generally proportional to the number of larvae ingested. Ingestion of undercooked pork is the most common vehicle of infestation, but ingestion of the flesh of animals that feed on garbage (bears, for example) also causes outbreaks.

Any ground meat can be contaminated from machinery previously used to grind contaminated meats. Eating raw sausage or hamburger is a poor health habit. One person acquired trichinosis by chewing the fingernails after handling infected pork. Prolonged freezing of meats containing *Trichinella* tends to eliminate the worms, but freezing should not be considered a substitute for thorough cooking.

In the muscles of hosts such as pigs, the *T. spiralis* larvae are encysted in the form of short worms about 1 mm in length. When the flesh of an infected animal is ingested by humans, the cyst wall is removed by digestive action in the intestine. The organism then matures into the adult form. The adult worms spend only about one week in the intestinal mucosa and produce larvae that invade tissue. Eventually, the encysted larvae localize in muscle (common sites include the diaphragm and eye muscles), where they are barely visible in biopsied specimens (Figure 25.24).

Symptoms of trichinosis include fever, swelling about the eyes, and gastrointestinal upset. Small hemorrhages under the fingernails are often observed. Biopsy specimens, as well as a number of serological tests, can be used in diagnosis. Recently, a serological ELISA test that detects the parasite in meats has been developed. Treatment consists of the administration of thiabendazole to kill intestinal worms and corticosteroids to reduce inflammation.

In the past ten years, the number of cases reported annually in the Unites States has varied from 39 to 206. Deaths are rare, and in most years there are none.

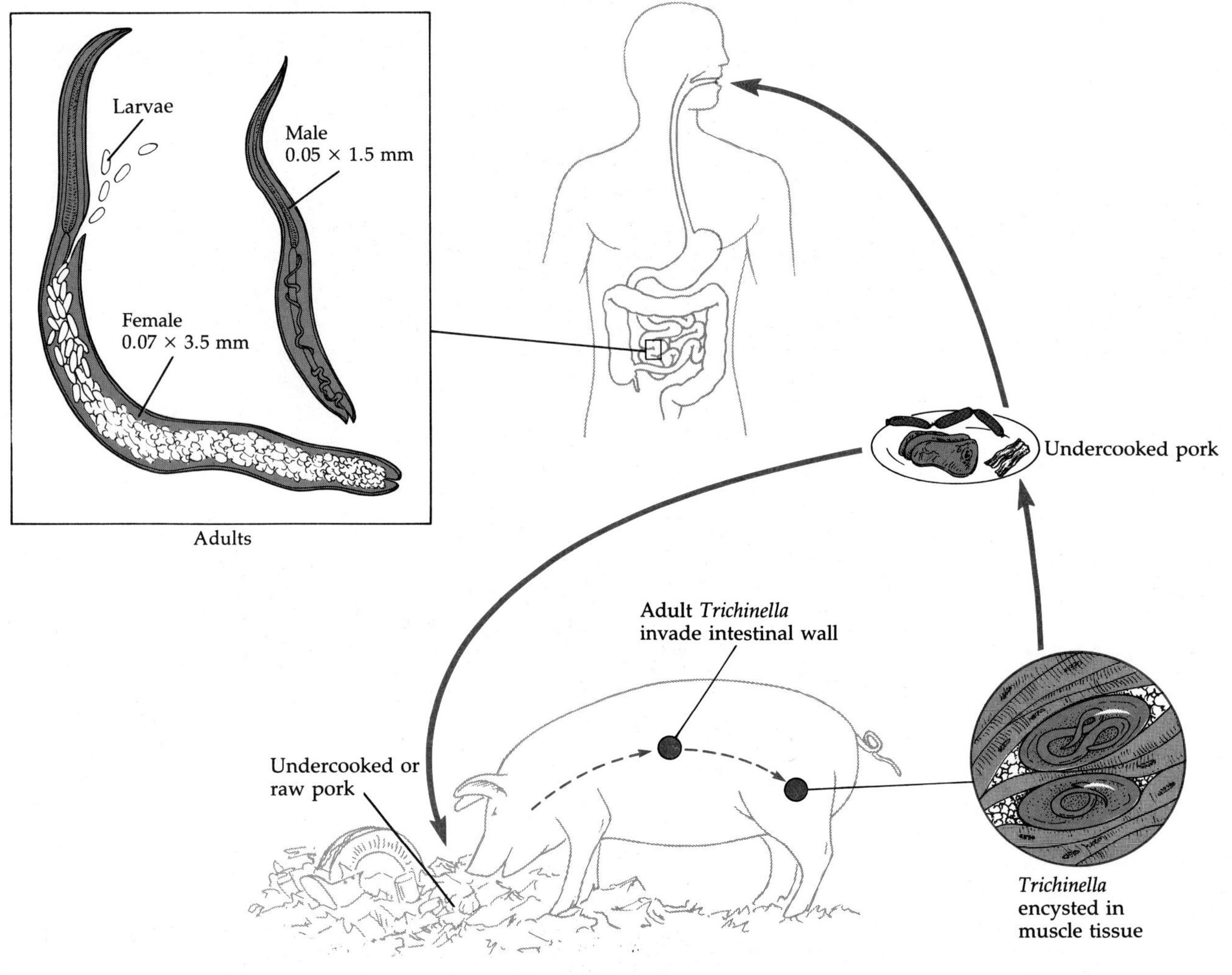

FIGURE 25.24 Life cycle of *Trichinella spiralis,* the causative agent of trichinosis.

STUDY OUTLINE

Introduction (p. 617)

1. Diseases of the digestive system are the second most common illnesses in the United States.
2. Diseases of the digestive system usually result from ingestion of microorganisms and their toxins in food and water.
3. The fecal–oral cycle of transmission can be broken by proper disposal of sewage, drinking-water disinfection, and proper preparation and storage of foods.

Structure and Function of the Digestive System (pp. 617–618)

1. The gastrointestinal (GI) tract, or alimentary canal, consists of the mouth, pharynx, esophagus, stomach, small intestine, and large intestine.
2. The teeth, tongue, salivary glands, liver, gallbladder, and pancreas are accessory structures.
3. In the GI tract, with mechanical and chemical help from the accessory structures, large food molecules are broken down into smaller molecules that can be transported by blood or lymph to cells.
4. Feces, resulting from digestion, are eliminated through the anus.

Normal Flora of the Digestive System (p. 618)

1. A wide variety of bacteria colonize the mouth.
2. The stomach and small intestine have few resident microorganisms.

3. The large intestine is the habitat for *Lactobacillus, Bacteroides, E. coli, Enterobacter, Klebsiella,* and *Proteus.*
4. Bacteria in the large intestine assist in degrading food and synthesizing vitamins.
5. Up to 40% of fecal mass is microbial cells.

Bacterial Diseases of the Mouth (pp. 618–622)

DENTAL CARIES (TOOTH DECAY) (pp. 618–621)

1. Dental caries begin when tooth enamel and dentin are eroded and the pulp is exposed to bacterial infection.
2. *Streptococcus mutans,* found in the mouth, uses sucrose to form dextran from glucose and lactic acid from fructose.
3. Bacteria adhere to teeth with a sticky dextran capsule, forming dental plaque.
4. Acid produced during carbohydrate fermentation destroys tooth enamel at the site of the plaque.
5. Gram-positive rods and filamentous bacteria can penetrate into dentin and pulp.
6. Carbohydrates such as starch, mannitol, and sorbitol are not used by cariogenic bacteria to produce dextran and do not promote tooth decay.
7. Caries are prevented by restricting ingestion of sucrose and by physical removal of plaque; a vaccine against *S. mutans* is theoretically possible.

PERIODONTAL DISEASE (pp. 621–622)

1. Caries of the cementum and gingivitis are caused by streptococci, actinomycetes, and anerobic gram-negative bacteria.
2. Chronic gum disease (periodontitis) can cause bone destruction and tooth loss; periodontitis is due to inflammatory response to a variety of bacteria growing on the gums.
3. Acute necrotizing ulcerative gingivitis is caused by *Fusobacterium* and actinomycetes.

Bacterial Diseases of the Lower Digestive System (pp. 622–630)

1. A gastrointestinal infection is caused by growth of a pathogen in the intestines.
2. Incubation times, the times required for bacterial cells to grow and their products to produce symptoms, range from 12 hours to 2 weeks. Symptoms of infection generally include a fever.
3. A bacterial intoxication results from ingestion of preformed bacterial toxins.
4. Symptoms appear from 1 to 48 hours after ingestion of the toxin. Fever is not usually a symptom of intoxication.
5. Infections and intoxications cause diarrhea, dysentery, or gastroenteritis.
6. These conditions are usually treated with fluid and electrolyte replacement.

STAPHYLOCOCCAL FOOD POISONING (STAPHYLOCOCCAL ENTEROTOXICOSIS) (pp. 622–624)

1. Staphylococcal food poisoning is caused by ingestion of an enterotoxin produced in improperly stored foods.
2. *S. aureus* is inoculated into foods during preparation. The bacteria grow and produce enterotoxin in food stored at room temperature.
3. The exotoxin is not denatured by boiling for 30 minutes.
4. Foods with high osmotic pressure and those not cooked immediately before consumption are most often the source of staphylococcal enterotoxicosis.
5. Nausea, vomiting, and diarrhea begin between one and six hours after eating, and the symptoms last approximately 24 hours.
6. Laboratory identification of *S. aureus* isolated from foods or the presence of thermostable nuclease in foods can confirm diagnosis.

SALMONELLOSIS (*SALMONELLA* GASTROENTERITIS) (pp. 624–625)

1. Salmonellosis, or *Salmonella* gastroenteritis, is caused by many *Salmonella* species.
2. Symptoms include nausea, abdominal pain, and diarrhea and begin 12 to 36 hours after ingestion of large numbers of *Salmonella*. Septicemia can occur in infants and in the elderly.
3. Fever might be caused by endotoxin.
4. Mortality is lower than 1%, and recovery can result in a carrier state.
5. Heating food to 68°C will usually kill *Salmonella*.
6. Laboratory diagnosis is based on isolation and identification of *Salmonella* from feces and foods.

TYPHOID FEVER (pp. 625–626)

1. *Salmonella typhi* causes typhoid fever.
2. Fever and malaise occur after a two-week incubation. Symptoms last from two to three weeks.
3. *S. typhi* is harbored in the gallbladder of carriers.
4. A killed-bacteria vaccine is available for high-risk persons.

BACILLARY DYSENTERY (SHIGELLOSIS) (p. 626)

1. Bacillary dysentery is caused by four species of *Shigella*.
2. Symptoms include blood and mucus in stools, abdominal cramps, and fever. Infections by *S. dysenteriae* result in ulceration of the intestinal mucosa.
3. Isolation and identification of the bacteria from rectal swabs are used for diagnosis.

CHOLERA (pp. 626–627)

1. *Vibrio cholerae* produces an exotoxin that alters membrane permeability of the intestinal mucosa; the resulting vomiting and diarrhea cause loss of body fluids.
2. The incubation period is approximately three days. The symptoms last for a few days. Untreated cholera has a 50% mortality rate.
3. Diagnosis is based on isolation of *Vibrio* from feces.
4. *Vibrio cholerae* non-0:1 causes gastroenteritis in the United States. It is usually transmitted via contaminated seafood.

VIBRIO PARAHAEMOLYTICUS GASTROENTERITIS (p. 627)

1. Gastroenteritis can be caused by the halophile *V. parahaemolyticus.*
2. Onset of symptoms begins within 24 hours after ingestion of contaminated foods. Recovery occurs within a few days.
3. The disease is contracted by ingestion of contaminated crustaceans or contaminated mollusks.

ESCHERICHIA COLI GASTROENTERITIS (TRAVELER'S DIARRHEA) (p. 628)

1. Gastroenteritis may be caused by enterotoxigenic, enteroinvasive, enteropathogenic, or enterohemorrhagic strains of *E. coli*.
2. The disease occurs as epidemic diarrhea in nurseries, as traveler's diarrhea, as endemic diarrhea in underdeveloped countries, and as hemorrhagic colitis.
3. In adults, the disease is usually self-limiting and does not require chemotherapy.

CAMPYLOBACTER GASTROENTERITIS (pp. 628–629)

1. *Campylobacter* is the second most common cause of diarrhea in the United States.
2. *Campylobacter* is transmitted in cow's milk.

HELICOBACTER GASTRITIS (p. 629)

1. *Helicobacter pylori* produces ammonia, which neutralizes stomach acid; the bacteria colonize the stomach mucosa and cause gastritis.
2. Bismuth and several antibiotics may be useful in treating gastritis.

YERSINIA GASTROENTERITIS (p. 629)

1. *Y. enterocolitica* and *Y. pseudotuberculosis* are transmitted in meat and milk.
2. *Yersinia* can grow at refrigeration temperatures.

CLOSTRIDIUM PERFRINGENS GASTROENTERITIS (p. 629)

1. A self-limiting gastroenteritis is caused by *C. perfringens*.
2. Endospores survive heating and germinate when foods (usually meats) are stored at room temperature.
3. Exotoxin produced when the bacteria grow in the intestines is responsible for the symptoms.
4. Diagnosis is based on isolation and identification of the bacteria in stool samples.

BACILLUS CEREUS GASTROENTERITIS (p. 630)

1. Ingesting food contaminated with the soil saprophyte *Bacillus cereus* can result in diarrhea, nausea, and vomiting.

Viral Diseases of the Digestive System (pp. 630–636)

MUMPS (p. 630)

1. Mumps virus (a paramyxovirus) enters and exits the body through the respiratory tract.
2. About 16 to 18 days after exposure, the virus causes inflammation of the parotid glands, fever, and pain during swallowing. About 4 to 7 days later, orchitis may occur.
3. After onset of the symptoms, the virus is found in the blood, saliva, and urine.
4. A measles–mumps–rubella (MMR) vaccine is available.
5. Diagnosis is based on symptoms or an ELISA test on viruses cultured in embryonated eggs or cell culture.

CYTOMEGALOVIRUS (CMV) INCLUSION DISEASE (pp. 630–631)

1. CMV (a herpesvirus) causes intranuclear inclusion bodies and cytomegaly of host cells.
2. CMV is transmitted by saliva, urine, semen, cervical secretions, and human milk.
3. CMV inclusion disease can be asymptomatic, a mild disease, or progressive and fatal. Immunosuppressed patients may develop pneumonia.
4. If the virus crosses the placenta, it can cause congenital infection of the fetus resulting in impaired mental development, neurological damage, and stillbirth.
5. Diagnosis is based on isolation of the virus; serological tests are not specific.

HEPATITIS (pp. 631–636)

1. Inflammation of the liver is called hepatitis. Symptoms include loss of appetite, malaise, fever, and jaundice.
2. Viral causes of hepatitis include hepatitis viruses, EB virus, and CMV.

Hepatitis A (Infectious Hepatitis) (pp. 632–633)

1. Hepatitis A virus causes infectious hepatitis. At least 50% of all cases are subclinical.
2. HAV is ingested in contaminated food or water, grows in the cells of the intestinal mucosa, and spreads to the liver, kidneys, and spleen in the blood.
3. The virus is eliminated with feces.
4. The incubation period is 2 to 6 weeks; the period of disease is 2 to 21 days; and recovery is complete in 4 to 6 weeks.
5. Diagnosis is based on serological tests for the virus or antibodies.
6. Passive immunization can provide temporary protection.

Hepatitis B (Serum Hepatitis) (pp. 633–635)

1. Hepatitis B virus causes serum hepatitis.
2. HBV is transmitted by blood transfusions, contaminated syringes, saliva, sweat, breast milk, and semen.
3. Dane particles (complete virion), spherical particles (HB_SAg), and filamentous particles (containing HB_SAg) are found in serum of patients with hepatitis B.
4. The average incubation period is three months; recovery is usually complete, but some patients develop a chronic infection.
5. A vaccine against HB_SAg is available.

Hepatitis C (Non-A, Non-B) (pp. 635–636)

1. Hepatitis C virus is transmitted via blood.
2. About 2% of blood donors may be infected with hepatitis C.

Hepatitis D (Delta Hepatitis) (p. 636)

1. Delta agent has a circular strand of RNA and uses HB_SAg as a coat.

Hepatitis E (Infectious NANB Hepatitis) (p. 636)

1. Hepatitis E virus is spread by the fecal–oral route.

VIRAL GASTROENTERITIS (p. 636)

1. Viral gastroenteritis is most often caused by rotavirus or the Norwalk agent.

Fungal Diseases of the Digestive System (p. 637)

1. Mycotoxins are toxins produced by some fungi.
2. Mycotoxins affect the blood, nervous system, kidney, or liver.

ERGOT POISONING (p. 637)

1. Ergot poisoning is caused by the mycotoxin produced by *Claviceps purpurea*.
2. Cereal grains are most often contaminated with the *Claviceps* mycotoxin.

AFLATOXIN POISONING (p. 637)

1. Aflatoxin is a mycotoxin produced by *Aspergillus flavus*.
2. Peanuts are most often contaminated with aflatoxin.

Protozoan Diseases of the Digestive System (pp. 637–639)

GIARDIASIS (p. 637)

1. *Giardia lamblia* grows in the intestines of humans and wild animals and is transmitted in contaminated water.
2. Symptoms of giardiasis are malaise, nausea, flatulence, weakness, and abdominal cramps that persist for weeks.
3. Diagnosis is based on identification of the protozoan in the small intestine.

BALANTIDIASIS (BALANTIDIAL DYSENTERY) (p. 638)

1. *Balantidium coli* causes balantidial dysentery when growing in the large intestine.
2. *B. coli* can cause ulceration of the intestinal wall and, rarely, fatal dysentery.
3. Infections are acquired by ingesting cysts in contaminated food and water.
4. Diagnosis is based on observation of trophozoites in feces.

AMOEBIC DYSENTERY (AMOEBIASIS) (p. 638)

1. Amoebic dysentery is caused by *Entamoeba histolytica* growing in the large intestine.
2. The amoeba feeds on red blood cells and GI tract tissues. Severe infections result in abscesses.
3. Diagnosis is confirmed by observation of trophozoites in feces and several serological tests.

CRYPTOSPORIDIOSIS (pp. 638–639)

1. *Cryptosporidium* causes prolonged diarrhea in immunosuppressed patients.
2. The presence of oocysts in feces confirms diagnosis.

Helminthic Diseases of the Digestive System (pp. 639–644)

TAPEWORM INFESTATIONS (pp. 639–642)

1. Tapeworms are contracted by the consumption of undercooked beef, pork, or fish containing encysted larvae (cysticerci).
2. The scolex attaches to the intestinal mucosa of humans (the definitive host) and matures into an adult tapeworm.
3. Eggs are shed in the feces and must be ingested by an intermediate host.
4. Adult tapeworms can be undiagnosed in a human.
5. Diagnosis is based on observation of proglottids and eggs in feces.
6. Cysticercosis in humans occurs when the pork tapeworm larvae encyst in humans.

HYDATID DISEASE (p. 642)

1. Humans infested with the tapeworm *Echinococcus granulosus* might have hydatid cysts in their lungs or other organs.
2. Dogs and wolves are usually the definitive hosts and sheep or deer are the intermediate hosts for *E. granulosus*.

NEMATODE INFESTATIONS (pp. 642–644)

Pinworms (p. 642)

1. Humans are the definitive host for pinworms, *Enterobius vermicularis*.

Hookworms (pp. 642–643)

1. Hookworm larvae bore through skin and migrate to the intestine to mature into adults.

Ascariasis (p. 643)

1. *Ascaris lumbricoides* adults live in human intestines.

Trichinosis (pp. 643–644)

1. *Trichinella spiralis* larvae encyst in muscles of humans, swine, and other mammals to cause trichinosis.
2. The roundworm is contracted by ingesting undercooked meat containing larvae.
3. Adults mature in the intestine and lay eggs. The new larvae migrate to invade muscles.
4. Symptoms include fever, swelling around the eyes, and gastrointestinal upset.
5. Biopsy specimens and serological tests are used for diagnosis.

STUDY QUESTIONS

Review

1. State examples of the representative normal flora, if any, and rectum.
2. What properties of *S. mutans* implicates this bacterium in the formation of dental caries? Why is sucrose, more than any other carbohydrate, responsible for the formation of dental caries?
3. Complete the following table.

Disease	Etiologic Agent	Suspect Foods	Symptoms	Treatment
Staphylococcal food poisoning				
Salmonellosis				
Bacillary dysentery				
Cholera				
Gastroenteritis				
Traveler's diarrhea				

4. Differentiate between salmonellosis and typhoid fever.
5. You probably listed *E. coli* in answer to Questions 1 and 3. Explain why this one bacterial species is both beneficial and harmful.
6. What preventive treatments are currently used for infectious hepatitis? For hepatitis B? For hepatitis C?
7. How is blood that is to be used for transfusions tested for HBV? For HCV?
8. Define mycotoxin. Provide an example of a mycotoxin.
9. Explain how the following diseases differ and how they are similar: giardiasis, balantidiasis, amoebic dysentery, and cryptosporidiosis.
10. Differentiate between amoebic dysentery and bacillary dysentery.
11. List the general symptoms of gastroenteritis. Since there are many etiologies, what is the laboratory diagnosis usually based on?
12. Differentiate among the following factors of bacterial intoxication and bacterial infection: prerequisite conditions, etiologic agents, onset, duration of symptoms, and treatment.
13. Gastritis is caused by ________. How can this bacterium tolerate stomach acid?
14. Complete the following table.

Disease	Etiologic Agent	Method of Transmission	Site of Infection	Symptoms	Prevention
Mumps					
CMV inclusion disease					
Infectious hepatitis					
Serum hepatitis					
Viral gastroenteritis					

15. Diagram the life cycle of a human tapeworm.
16. Diagram the life cycle of *Trichinella*, and include humans in the cycle.
17. How can bacterial and protozoan infections of the GI tract be prevented?
18. Look at your diagram for questions 15 and 16. Indicate sequences in the life cycles that could be easily broken to prevent these diseases.

Challenge

1. Twenty-eight kindergarten children and seven adults visited a certified raw milk (CRM) bottling plant, where they were given ice cream and CRM. Three to six days later, nine children and three adults developed gastroenteritis. The only foods eaten by all these children (ill and well) were in the school-provided lunches. No one else in the school became sick. What was the source of this gastroenteritis outbreak? Stool cultures showed one bacterium common to nine of the ill children and not present in samples from nine well children. This bacterium is a curved gram-negative rod; it neither ferments nor oxidizes glucose. What is the bacterium? How could this disease have been prevented?

2. Why is a human infection of trichinosis considered a "dead end" for the parasite?

3. Complete the following table.

Disease	Conditions Necessary for Microbial Growth	Basis for Diagnosis	Prevention
Staphylococcal food poisoning			
Salmonellosis			

4. 2130 students and employees of a public school system developed diarrheal illness on April 2. The cafeteria served chicken that day. On April 1, part of the chicken was placed in water-filled pans and cooked in an oven for two hours at a dial setting of 177°C. The oven was turned off, and the chicken was left overnight in the warm oven. The remainder of the chicken was cooked for two hours in a steam cooker and then left in the device overnight at the lowest possible setting (43°C). Two serotypes of a gram-negative, oxidase-negative, lactose-negative rod were isolated from 32 patients. What is the pathogen? How could this outbreak have been prevented?

5. Staff in one hospital ward noted an increase in the number of cases of HBV. Fifty cases occurred during a six-month period compared with four cases during the previous six months. Between January 1 and 15, all 50 patients had multiple invasive procedures as shown below:

Transfusion, fingerstick, IV catheter, heparin injection	78%
Transfusion, insulin injection, surgery, fingerstick	64%
Fingerstick, IV catheter, insulin injection, heparin injection	80%
Transfusion, heparin injection, surgery, IV catheter	2%
Heparin injection, IV catheter, insulin injection, surgery	0%

How did the patients acquire HBV? Provide an explanation for the 2% and 0%.

FURTHER READING

Betley, M.J., V.L. Miller, and J.J. Mekalanos. "Genetics of bacterial enterotoxins." *Annual Review of Microbiology* 40:577–605, 1986. Discusses the production of enterotoxins by *V. cholerae, E. coli, S. dysenteriae,* and *S. aureus.*

Dick, J.D. "*Helicobacter (Campylobacter) pylori:* a new twist to an old disease." *Annual Review of Microbiology* 44:249–269, 1990. An in-depth treatise on the general biology, diagnosis, and identification of *Helicobacter.*

Kolenbrander, P.E. "Intergeneric coaggregation among human oral bacteria and ecology of dental plaque." *Annual Review of Microbiology* 42:627–656, 1988. Describes how bacteria aggregrate on tooth surfaces to form plaque.

Prince, A.M. "Non-A, non-B hepatitis viruses." *Annual Review of Microbiology* 37:217–232, 1983. Describes ongoing research seeking the causative agents of NANB hepatitis.

Reimann, H., and F.L. Bryan. *Food-borne Infections and Intoxications,* 2nd ed. New York: Academic Press, 1979. An authoritative discussion of nearly all foodborne diseases.

Riley, L.W. "The epidemiologic, clinical, and microbiologic features of hemorrhagic colitis." *Annual Review of Microbiology* 41:383–407, 1987. Discusses intestinal diseases caused by *E. coli* that are endemic in North America.

Tiollais, P., and M. Buendia. "Hepatitis B virus." *Scientific American* 264(4):116–123, April 1991. Describes the epidemiology, reproduction, and possible eradication of this virus.

van der Waaij, D. "The ecology of the human intestine and its consequences for overgrowth by pathogens such as *Clostridium difficile.*" *Annual Review of Microbiology* 43:69–87, 1989. Describes colonization of the intestine by microbial flora and the role of the indigenous flora in causing life-threatening disease.

See also Further Reading for Part Four at the end of Chapter 26.

CHAPTER 26

Microbial Diseases of the Urinary and Reproductive Systems

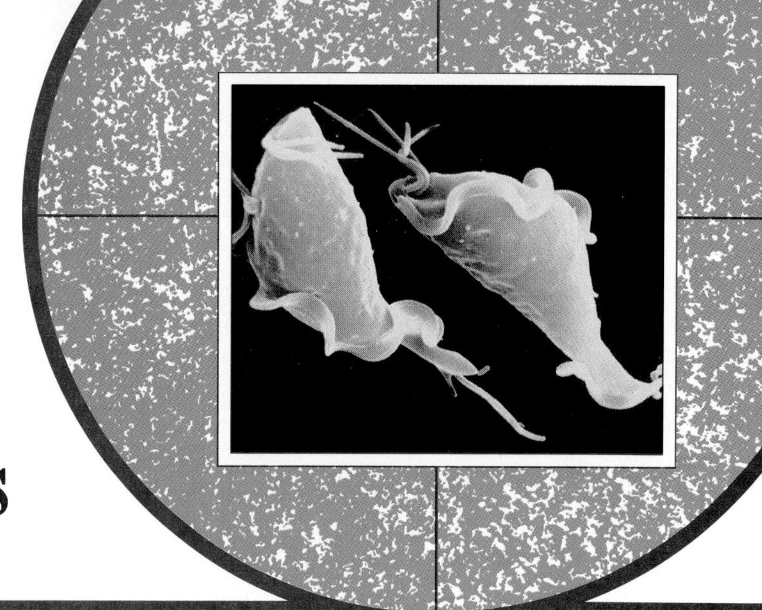

LEARNING OBJECTIVES

- List the members of the normal flora of the urinary and reproductive systems and their habitats.
- Describe methods of transmission for urinary and reproductive system infections.
- List the microorganisms that cause cystitis and pyelonephritis, and name the predisposing factors for these diseases.
- Describe the cause and treatment of glomerulonephritis.
- List the etiologic agents, symptoms, methods for diagnosis, and treatments for leptospirosis, gonorrhea, syphilis, NGU, vaginitis, lymphogranuloma venereum, chancroid, candidiasis, and trichomoniasis.
- Discuss the epidemiology of genital herpes and genital warts.
- List genital diseases that can cause congenital and neonatal infections, and explain how these infections can be prevented.

The **urinary system** is composed of organs that regulate the chemical composition and volume of the blood and excrete the waste products of metabolism. The **reproductive system** is composed of organs that produce gametes to propagate the species and, in the female, organs that support and nourish the developing embryo. Because the urinary and reproductive systems are closely related anatomically, we will discuss them in the same chapter. In fact, some diseases that affect one system also affect the other, especially in the male. Both systems open to the external environment and thus share portals of entry for microorganisms that can cause disease. The normal floras of these and other bodily systems also cause opportunistic infections of the urinary and reproductive systems. The major microbial diseases of the urinary and reproductive systems are summarized in Table 26.1 on p. 666.

Structure and Function of the Urinary System

The urinary system consists of two *kidneys,* two *ureters,* a single *urinary bladder,* and a single *urethra* (Figure 26.1). Certain wastes are removed from the blood as it circulates through the kidneys. The wastes, which include urea, uric acid, creatinine, and various salts together with water, are collectively called *urine.* The urine passes through the ureters into the urinary bladder, where it is stored prior to elimination from the body. Elimination occurs through the urethra. In the female, the urethra conveys only urine to the exterior. In the male, the urethra is a common tube for both urine and seminal fluid.

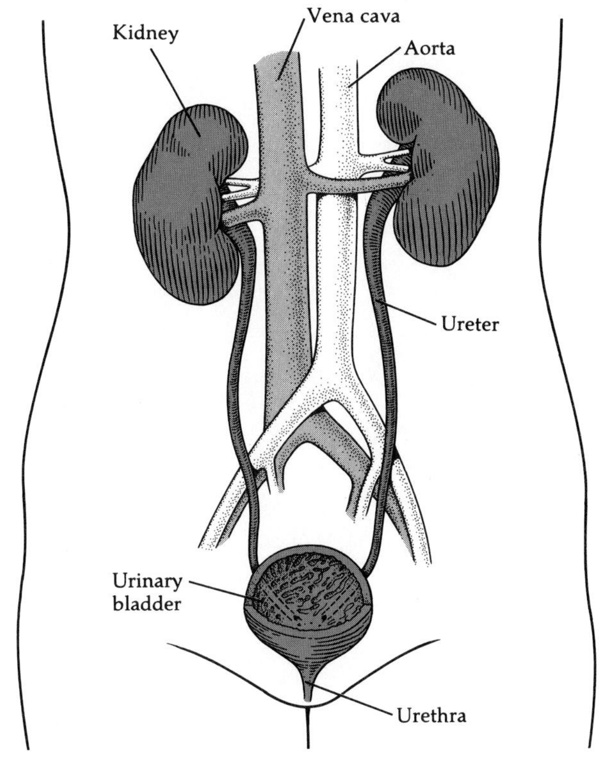

FIGURE 26.1 Organs of the urinary system.

The urinary tract has certain characteristics that help to prevent infection. Where the ureters enter the urinary bladder, there are valves that normally prevent the backflow of urine to the kidneys. This mechanism helps to shield the kidneys from lower urinary tract infections. In addition, the acidity of normal urine has some antimicrobial properties. The flushing action of urine to the exterior also aids in preventing microorganisms from setting up foci of infection.

Structure and Function of the Reproductive System

The **female reproductive system** consists of two *ovaries*, two *uterine* (*fallopian*) *tubes*, the *uterus*, the *vagina*, and *external genitals* (Figure 26.2). The ovaries produce female sex hormones and ova (eggs). When an ovum is released in the process called ovulation, it enters a uterine tube; there fertilization occurs if viable sperm are present. The fertilized ovum (zygote) descends the uterine tube and enters the uterus. It implants in the inner wall and remains there while it develops into an embryo and, later, a fetus. At birth, the infant is expelled from the uterus through the vagina. The vagina also serves as a copulatory canal. The external genitals (vulva) include the clitoris, labia, and glands that produce a lubricating secretion during copulation.

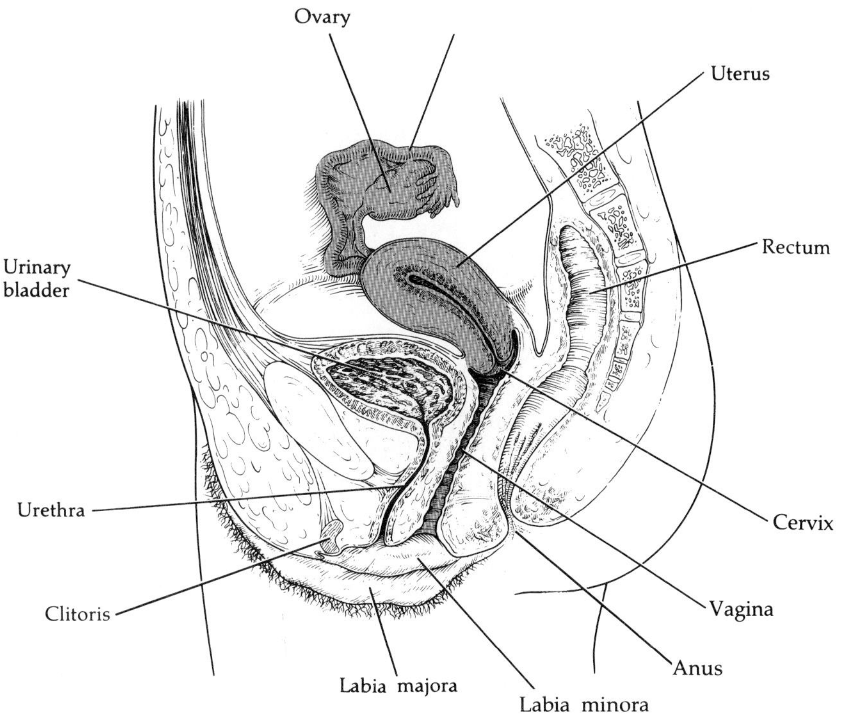

FIGURE 26.2 Female reproductive organs.

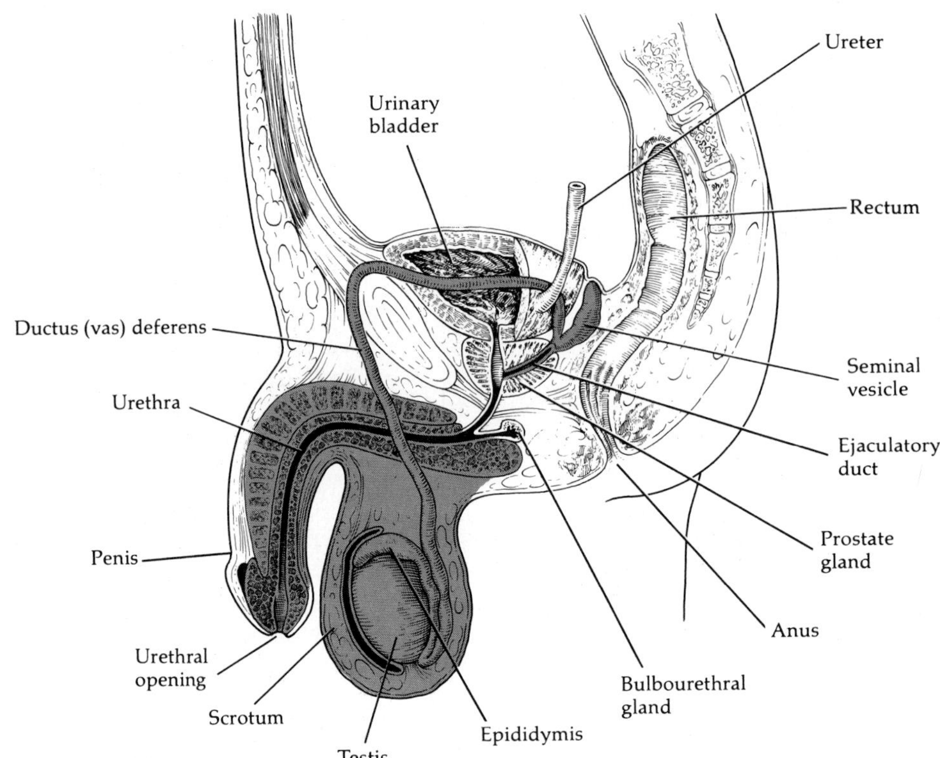

FIGURE 26.3 Male reproductive organs.

The **male reproductive system** consists of two *testes*, a system of *ducts, accessory glands*, and the *penis* (Figure 26.3). The testes produce male sex hormones and sperm. Newly produced sperm move into the epididymis, where they are stored until ejaculation. To exit from the body, the sperm cells pass through a series of ducts—epididymis, ductus (vas) deferens, ejaculatory duct, and urethra. During ejaculation, the seminal fluid leaves the body through the urethra.

Normal Flora of the Urinary and Reproductive Systems

Normal urine is sterile in the urinary bladder and the organs of the upper urinary tract. The urethra, however, does contain a normal resident flora that includes *Streptococcus, Bacteroides, Mycobacterium, Neisseria*, and a few enterobacteria. Urine becomes contaminated with members of the skin flora during its passage through the urethra.

In the female genital system, the normal flora of the vagina is greatly influenced by sex hormones. For example, within a few weeks after birth, the female infant's vagina is populated by lactobacilli. This population grows because estrogens are transferred from maternal to fetal blood and cause glycogen to accumulate in the cells lining the vagina. Lactobacilli convert the glycogen to lactic acid, and the pH of the vagina becomes acidic. This glycogen–lactic acid sequence provides the conditions under which the acid-tolerant normal flora grows in the vagina.

The physiological effects of estrogens diminish several weeks after birth, and other bacteria—including corynebacteria and a variety of cocci and bacilli—become established and dominate the flora. As a result, the pH of the vagina becomes more neutral until puberty. At puberty, estrogen levels increase, lactobacilli again dominate, and the vagina again becomes acidic. During the reproductive years, small numbers of other bacteria and fungi become part of the flora. In the adult, a disturbance of this ecosystem by an increase in glycogen (caused by oral contraceptives or pregnancy, for example) or elimination of the normal flora by antibiotics can lead to *vaginitis*, an infection of the vagina. When the female reaches menopause, estrogen levels again decrease, the flora returns to that of childhood, and the pH again becomes neutral.

Diseases of the Urinary System

The urinary system normally contains relatively few microbes, but it is often subject to opportunistic infections that can be quite troublesome.

Bacterial Diseases of the Urinary System

Many infections of the urinary system appear to be opportunistic and related to a number of predisposing factors. These factors include disorders of the nervous system, toxemia associated with pregnancy, diabetes mellitus, and obstructions to the flow of urine, such as tumors and kidney stones. Because the anus, from which feces are excreted, is located close to the urethra, it is not unusual, especially in females, for the urinary tract to become contaminated with intestinal bacteria. These are usually *Escherichia coli, Proteus* species, and other gram-negative enteric bacteria; staphylococci, enterococci, and pseudomonads are also common in such infections. These *Pseudomonas* infections are, as elsewhere, unusually troublesome to treat. The fungus *Candida albicans* is also an opportunistic agent of urinary tract infections. Moreover, a number of sexually transmitted diseases can cause inflammation in the urinary tract.

Infections usually cause inflammation of the affected tissue. Inflammation of the urethra is called *urethritis;* of the urinary bladder, *cystitis;* and of the ureters, *ureteritis.* The most significant danger from lower urinary tract infections is that they can affect the kidneys (causing *pyelonephritis*) and impair their function. The kidneys are also sometimes affected by systemic bacterial diseases, such as typhoid fever or leptospirosis; when the kidneys are thus infected, the pathogens causing these diseases can be found in excreted urine.

Many infections of the urinary tract are of nosocomial origin. In fact, about 35% of all nosocomial infections occur in the urinary tract. Operations on the urinary bladder and prostate gland and catheterization for draining the urinary bladder are procedures that can introduce bacteria into the bladder and ureters. *E. coli* causes more than half of the nosocomial infections of the urinary tract: fecal streptococci, *Proteus, Klebsiella,* and *Pseudomonas* also commonly cause such infections.

Treatment of diseases of the urinary tract depends on diagnostic isolation of the causative organism and determination of its antibiotic sensitivity. Normal urine contains fewer than 10,000 bacteria per milliliter. When more than 100,000 bacteria per milliliter are found, there is usually an infection. Dipslide kits of various designs are often used to estimate the bacterial content of urine.

CYSTITIS

Cystitis is an inflammation of the urinary bladder and is very common, especially among females. Symptoms often include *dysuria* (difficult or painful urination) and *pyuria* (the presence of leukocytes in the urine). The female urethra has many microorganisms around its opening, and it is shorter than the male urethra, so microorganisms can traverse it more readily. Sexual intercourse or careless personal hygiene facilitates such transfer. Contributing factors in females include gastrointestinal system infections and preexisting infections of the vagina, uterus, or urethra. In males, cystitis might be associated with infections of the gastrointestinal system, kidneys, or urethra. Most cases are due to infection by *E. coli;* the second most common bacterial cause is the coagulase-negative *Staphylococcus saprophyticus* (sap-rō-fit'i-kus).

Treatment of cystitis depends on the bacterium that is responsible. Nitrofurantoin, which concentrates preferentially in the urine, is often the drug of choice. Broad spectrum penicillins, such as ampicillin, are also recommended.

PYELONEPHRITIS

In 25% of untreated cases, cystitis may progress to **pyelonephritis,** an inflammation of one or both kidneys. The disease is generally a complication of infection elsewhere in the body. In females, it is often a complication of lower urinary tract infections. The causative agent in about 75% of the cases is *E. coli.* Other bacteria associated with pyelonephritis are *Enterobacter aerogenes, Proteus* species, *Pseudomonas aeruginosa, Streptococcus pyogenes,* and staphylococci. Should pyelonephritis become chronic, scar tissue forms in the kidneys and severely impairs their function. This is a potentially life-threatening condition; treatment usually begins with a broad spectrum antibiotic such as a second- or third-generation cephalosporin. An extended course of 14 days is recommended.

LEPTOSPIROSIS

Leptospirosis is primarily a disease of domestic or wild animals, but it can be passed to humans and sometimes causes severe kidney or liver disease. The causative organism is the spirochete *Leptospira interrogans* (in-tėr'rä-ganz), shown in Figure 26.4. *Leptospira* has a characteristic shape—an exceedingly fine spiral wound so tightly that it is barely discernible under the darkfield microscope. Like other spirochetes, *L. inter-*

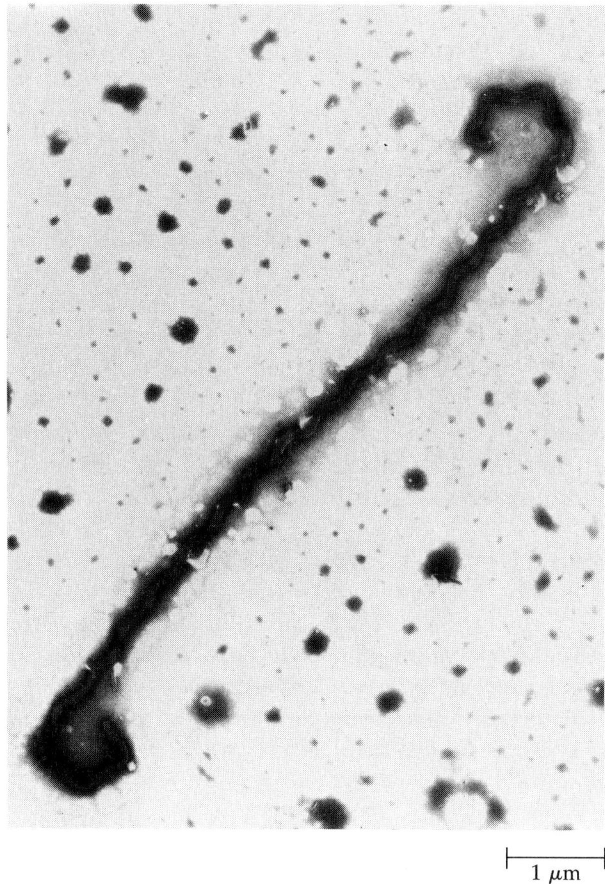

FIGURE 26.4 ***Leptospira interrogans*, the causative agent of leptospirosis.** The hooked ends are often seen in preparations of this organism. Remnants of the threadlike axial filament are barely visible on the hooks.

rogans (so named because the hooked ends suggest a question mark) stains poorly and is hard to see under a normal light microscope. It is an obligate aerobe that can be readily grown in a variety of artificial media supplemented with rabbit serum.

Animals infected with the spirochete shed the bacteria in their urine for extended periods. Humans become infected by contact with urine-contaminated water or soil, or sometimes with animal tissue. Persons whose occupations expose them to animals or animal products are most at risk. Usually the pathogen enters through minor abrasions in the skin or mucous membranes. In the United States, dogs and rats are the most common sources. Domestic dogs have a sizeable rate of infection; even when immunized, they may continue to shed leptospira.

After an incubation period of one or two weeks, symptoms of headaches, muscle aches, chills, and fever abruptly appear. Several days later the acute symptoms disappear and the temperature returns to normal. A few days later, however, a second episode of fever may occur. In a small number of cases the kidneys and liver become seriously infected; kidney failure is the most common cause of death. Recovery results in a solid immunity. There are usually about 50 cases reported each year in the United States, but because the clinical symptoms are not distinctive, many cases are probably not diagnosed correctly.

Serological tests usually must be done by central reference laboratories, which use suspensions of either killed or living leptospiras. If antibodies are present, they cause agglutination of the suspension. Most diagnoses are made by isolation of the pathogen from the blood or cerebrospinal fluids. Treatment of leptospirosis with antibiotics is often unsatisfactory because the correct diagnosis is delayed, which minimizes the effectiveness of the treatment. The drug of choice is penicillin.

GLOMERULONEPHRITIS

Glomerulonephritis (Bright's disease) is an inflammation of the glomeruli in the kidneys. The glomeruli are blood capillaries that assist in filtering blood as it passes through the kidneys. Glomerulonephritis is an immune-complex hypersensitivity disease. Most cases are a sequel to infection with β-hemolytic streptococci (*S. pyogenes*). Soluble streptococcal antigens combine with specific antibodies to form antigen–antibody complexes that interact with complement (see Chapter 16). The complexes are deposited in the glomeruli, where they cause inflammation and kidney damage.

The disease is characterized by fever, high blood pressure, and the presence of protein and red blood cells in the urine. The protein and red blood cells are in the urine because the permeability of the glomeruli increases as a result of inflammation. Although a few patients die of glomerulonephritis and some develop chronic conditions, most recover completely.

DISEASES OF THE REPRODUCTIVE SYSTEM

Microbes causing infections of the reproductive system are usually very sensitive to environmental stresses and require intimate contact for transmission.

Bacterial Diseases of the Reproductive System

Most diseases of the reproductive system are transmitted by sexual activity and are therefore called **sexually**

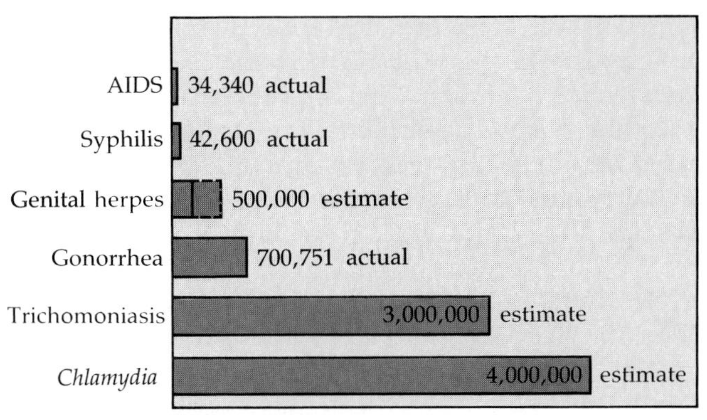

FIGURE 26.5 Incidence of sexually transmitted diseases in the United States. The data are estimates by the Centers for Disease Control for United States cases in 1989. They include reported cases plus the CDC's estimates of unreported cases. [Source: *MMWR* 36:52 (1/8/88).]

transmitted diseases (STDs). Most of these diseases can be readily cured with antibiotics, if treated early, and can largely be prevented by the use of condoms. Nevertheless, STDs are a major public health problem in the United States. Figure 26.5 illustrates the relative frequency of the more common STDs in the United States.

GONORRHEA

By far the most common reportable communicable disease in the United States is **gonorrhea,** an STD caused by the gram-negative diplococcus *Neisseria gonorrhoeae* (see Figure 26.8). (Health care providers must report "reportable" disease cases to local, state, or federal health agencies.) An ancient disease, gonorrhea was described and given its present name by the Greek physician Galen in A.D. 150. At present, the number of cases is decreasing after steadily and steeply increasing for many years (Figure 26.6). About 700,000 cases in the United States are reported to the Centers for Disease Control each year, and the true number of cases is probably much larger. More than 60% of the cases are in the 15- to 24-year-old age group.

To infect, the gonococcus must attach to the mucosal cells of the epithelial wall by means of fimbriae. One experimental vaccine is aimed at preventing these fimbriae from attaching. The organism does not infect the layered squamous cells characteristic of the external skin but invades the spaces separating mucosal cells. Mucosal cells are found in the oral–pharyngeal area, the eyes, joints, and rectum, both male and female genitals, and external genitals of prepubertal females. The invasion sets up an inflammation, and, when leukocytes move into the inflamed area, the characteristic pus formation results. In men, a single unprotected exposure results in infection with gonorrhea from 20% to 35% of the time. Women become infected from 60% to 90% of the time from a single exposure.

Males become aware of a gonorrheal infection by painful urination and discharge of pus-containing material from the urethra (Figure 26.7). About 80% of infected males show these obvious symptoms after an incubation period of only a few days; most others show symptoms in less than a week. In the days before antibiotic therapy, symptoms persisted for weeks. In

FIGURE 26.6 Incidence and distribution of gonorrhea. **(a)** Incidence of gonorrhea in the United States since 1940. **(b)** Geographical distribution of cases in 1990. [Sources: CDC. *Summary of Notifiable Diseases 1989, MMWR* 38:54 (10/5/90). *MMWR* 39:52 (1/4/91).]

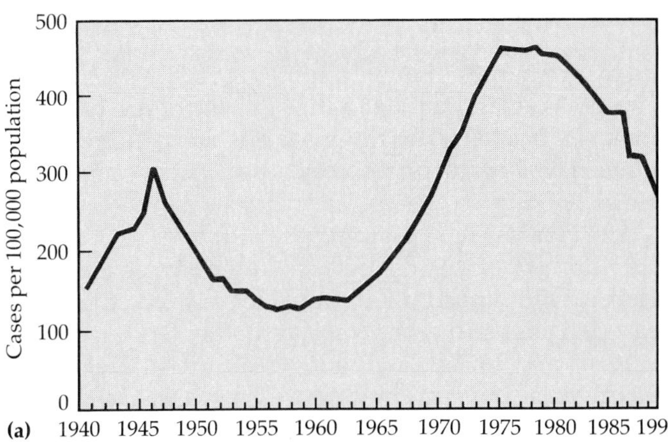

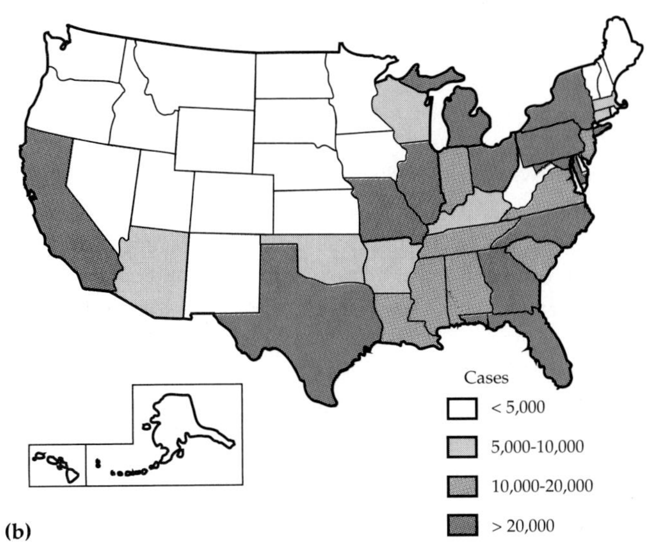

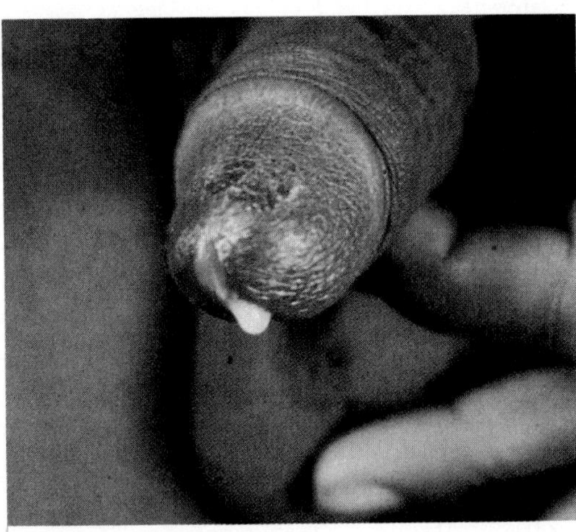

FIGURE 26.7 Pus-containing discharge from the urethra of a male with an acute case of gonorrhea.

untreated cases, recovery may eventually occur without complication, but, when complications do occur, they can be serious. In some cases, the urethra is scarred and partially blocked. Sterility can result when the testes become infected or when the ductus (vas) deferens, the tube carrying sperm from the testes, becomes blocked by scar tissue.

In females, the disease is more insidious. Very few women are aware of the early stages of the infection. Later in the course of the disease, there might be abdominal pain from **pelvic inflammatory disease (PID).** Some estimate that 20% to 30% of untreated gonorrhea cases progress to PID. Probably 1 million cases occur each year, and 150,000 of these require surgery, such as removal of the infected uterine tube.

The cause of PID is not restricted to *N. gonorrhoeae;* coinfection with chlamydial bacteria is frequent, and they are also an important agent of this type of infection. *PID* is a collective term for any extensive bacterial infection of the pelvic organs, particularly the uterus, cervix, uterine tubes, and ovaries. The most serious of these is the infection of the uterine tubes **(salpingitis).** It is theorized that the bacteria frequently attach to spermatozoa and are transported by them to the uterine tubes. Certainly, it has been demonstrated that women using barrier-type contraceptives, especially with spermicides, have a significantly lower rate of PID.

The result of salpingitis can be scarring that blocks the passage of ova from the ovary to the uterus, resulting in sterility. Tens of thousands of women have been rendered sterile by such infections in recent years. Only about 20% of such cases of sterility can be reversed surgically. Also, implantation of the fertilized ova may take place in the uterine tube rather than in the uterus. This is an *ectopic,* or *tubal, pregnancy,* a life-threatening condition. The reported cases of ectopic pregnancies have been increasing steadily, which corresponds to increasing occurrence of PID. One episode of salpingitis causes infertility in 10% to 15% of women, and 50% to 75% become infertile after three or more such infections.

Currently, there are an estimated 1 million cases of PID each year. The mortality rate is under 1%; deaths usually occur because misdiagnosis prevented treatment. The recommended treatment of PID is simultaneous administration of doxycycline (a long-acting tetracycline) and cefoxitin sodium (a cephalosporin). This combination is active against both the gonococcus and chlamydia.

In both sexes, untreated gonorrhea can become a serious, systemic infection. Complications of gonorrhea can involve the joints, heart **(gonorrheal endocarditis),** meninges **(gonorrheal meningitis),** eyes, pharynx, or other parts of the body. **Gonorrheal arthritis,** which is caused by the growth of the gonococcus in fluids in joints, occurs in about 1% of gonorrhea cases. Joints commonly affected include the wrist, knee, and ankle.

Gonorrheal eye infections occur most often in newborns. If the mother is infected, then the eyes of the infant can become infected as it passes through the birth canal. This condition, **ophthalmia neonatorum,** can result in blindness. Because of the seriousness of this condition and the difficulty of being certain that the mother is free of gonorrhea, silver nitrate in dilute solution or antibiotics such as erythromycin are placed in the eyes of all newborn infants. If the mother is known to be infected, an intramuscular injection of penicillin is also administered to the infant. Some sort of prophylaxis is required by law in most states. Gonorrheal infections can also be transferred by hand contact from infected sites to the eyes of adults.

Gonorrheal infections can be acquired at any point of sexual contact; pharyngeal and anal gonorrhea are not uncommon. The symptoms of **pharyngeal gonorrhea** often resemble those of the usual septic sore throat. **Anal gonorrhea** can be painful and accompanied by discharges of pus. In some cases, however, the symptoms are limited to itching. The incidence of anal gonorrhea has been declining sharply since more individuals have been taking precautions against transmission of AIDS.

The increase in sexual activity with a series of partners and the fact that the disease in the female may go unrecognized contributed considerably to the increased incidence of gonorrhea and other STDs during the 1960s and 1970s. The widespread use of oral contraceptives also contributed to the increase. Oral contraceptives tend to increase the moisture content and

MMWR

MORBIDITY AND MORTALITY WEEKLY REPORT

Antibiotic-Resistant Gonococci and Natural Selection

Strains of gonococci resistant to each of the agents recommended for gonorrhea therapy have emerged. Penicillinase-producing *Neisseria gonorrhoeae* (PPNG) was first reported in 1976. Although the number of reported cases of gonorrhea has declined during the last ten years, the incidence of PPNG infection has increased. In 1989, PPNG cases made up 7.4% of all gonorrhea cases reported. This represents a 131% increase over the PPNG cases reported in 1988.

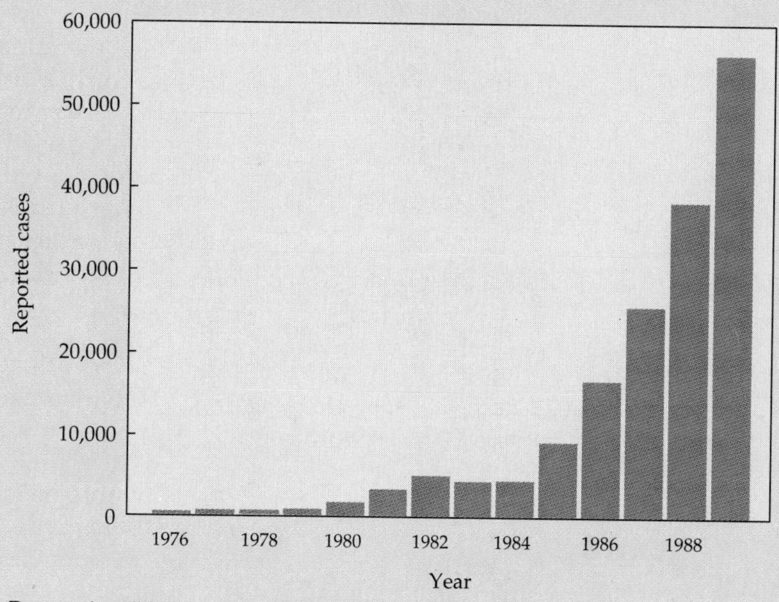

Reported antibiotic-resistant *Neisseria gonorrhoeae* cases in the United States, 1976–1989. (Source: CDC. *Summary of Notifiable Diseases, MMWR* 37:54.)

Importation from Other Countries

Over forty countries have reported cases of PPNG to the World Health Organization. The organism accounts for about 30% of all recent gonococcal isolates in the Philippines and 16% in the Republic of Singapore. (Among the factors thought to contribute to the high prevalence of PPNG in the Philippines and Singapore is the preventive use of oral penicillins, especially by prostitutes.)

Multiple-Drug Resistance

Tetracycline-resistant *N. gonorrhoeae* (TRNG) emerged in 1986, and the incidence of TRNG infection in the United States has risen each year since. It appears that *N. gonorrhoeae* acquired the *tetM* plasmid from streptococci.

In May 1981, a PPNG strain resistant to spectinomycin was reported from Travis Air Force Base, California. The patient had acquired the infection in the Philippines. The CDC now recommends that all isolates found to be PPNG also be tested for spectinomycin resistance. Patients who have infections caused by spectinomycin-resistant PPNG should be treated with ceftriaxone. Spectinomycin was used for five years before resistance was seen, and resistance to ceftriaxone was reported in 1988, which raises the question of how long the life span of these and other drugs will be.

In 1983, 56 cases of penicillin-resistant gonococcal infection occurred in North Carolina. These cases represent the first reported outbreak of gonorrhea caused by bacteria that are resistant to penicillin but do not produce penicillinase, a phenomenon called chromosomally mediated resistance (CMRNG). By 1984, 400 cases of CMRNG were reported from 22 states. These cells make a penicillin-binding protein instead of penicillinase. This is significant because the isolates are negative in a test for PPNG but are not treatable with penicillin.

Combinations of plasmid and chromosomally mediated resistance to multiple antibiotics have been isolated. Multiple-drug resistance is increasing so rapidly that, as of 1990, the CDC recommends that health departments determine antibiotic susceptibilities of all *N. gonorrhoeae* isolates by the disk-diffusion test.

Eradication of endemic antibiotic-resistant gonorrhea is extremely difficult. The once effective single dose of penicillin or tetracycline is no longer an acceptable treatment. Resistance can result in delays in treatment until an effective drug is found, and patients with inadequately treated gonorrhea are at high risk for complications.

Asked to list the reasons for the emergence of drug-resistant bacterial strains, a spokesman for the National Institutes of Health blamed the unnecessary use of prophylaxis, the use of antibiotics in animal feeds, the availability of over-the-counter antibiotics in many countries, and misuse by health professionals. It has been pointed out that the indiscriminate use of antibiotics may result in a vast majority of infections being caused by antibiotic-resistant bacteria.

Sources: *MMWR* 39:167–169, 284–287, 293 (1990).

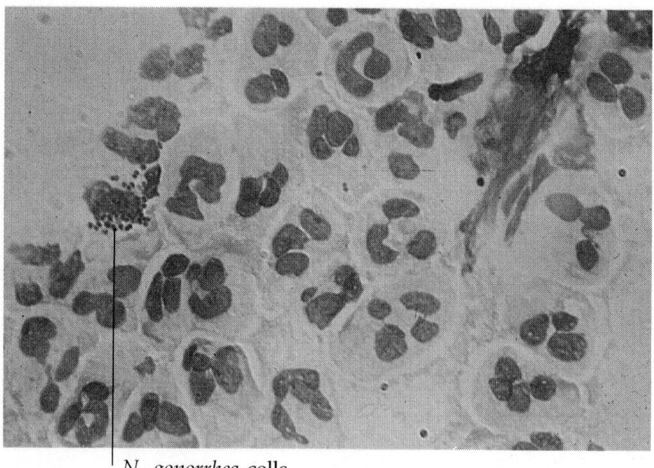

FIGURE 26.8 A smear of pus from a patient with gonorrhea. The *Neisseria gonorrhoeae* bacteria are contained within phagocytic leukocytes. They are visible as pairs of cocci and are gram-negative. The large stained bodies are the nuclei of the leukocytes.

raise the pH of the vagina while increasing the susceptibility of mucosal cells. Oral contraceptives have often replaced condoms and spermicides, which helped to prevent disease transmission. All these factors predispose the female to infection and therefore have increased the overall incidence of disease. Immunity to reinfection does not result from recovery from gonorrhea.

Penicillin has been an effective treatment for gonorrhea over the years, although the dosages have had to be much increased because of the appearance of penicillin-resistant bacteria. Penicillin resistance is usually caused by the presence of a gene for penicillinase (an enzyme that degrades penicillin). The penicillinase gene might be carried by the bacterial chromosome or by a plasmid (see Chapter 8). Gonococci having such plasmids first appeared in 1976 (see the box on p. 657).

Ceftriaxone, a third-generation cephalosporin, is currently the drug of choice, but this is subject to change. Antibiotic resistance to the gonococcus has accelerated at an alarming rate in recent years. Because of frequent coinfection with chlamydia, antibiotic treatment should include an effective antichlamydial antibiotic, usually of the tetracycline group.

Gonorrhea in men is diagnosed by finding gonococci in a stained smear of pus from the urethra. The typical gram-negative diplococci within the phagocytic leukocytes are readily identified (Figure 26.8). Gram staining of exudates is not as reliable with women. Usually, a culture is taken from within the cervix and grown on special media. Cultivation of the nutritionally fastidious bacterium requires an atmosphere containing carbon dioxide. The gonococcus is very sensitive to adverse environmental influences (desiccation and temperature) and survives poorly outside the body. It even requires special transporting media to keep it viable for short intervals before the cultivation is under way. Cultivation has the advantage of allowing determination of antibiotic sensitivity. Diagnosis of gonorrhea has been aided by the development of an ELISA test that detects *N. gonorrhoeae* in urethral pus or on cervical swabs within about three hours with high accuracy. Other rapid tests now available use monoclonal antibodies against antigens on the surface of the gonococcus. Colorimetric results are available within 40 minutes.

NONGONOCOCCAL URETHRITIS (NGU)

Nongonococcal urethritis (NGU), also known as **nonspecific urethritis (NSU),** refers to any inflammation of the urethra not caused by *Neisseria gonorrhoeae.* Symptoms include pain during urination and a watery discharge.

Nonmicrobial factors, such as trauma (passage of a catheter), can cause this condition, but at least 40% of the cases of NGU are acquired sexually. In fact, NGU might be the most common sexually transmitted disease in the United States today. Although it is not a reportable disease and exact data are lacking, the Centers for Disease Control estimate that there are 3 to 5 million new cases of NGU each year.

Because the symptoms are often mild in males, and females are usually asymptomatic, many cases go untreated. Physicians often treat NGU as a male urological disease rather than as an STD. Complications are not common but can be serious. Males may develop inflammation of the epididymis. In females, inflammation of the uterine tubes may cause sterility by scarring the tubes. It is estimated that as many as 60% of such cases of salpingitis are from chlamydial infection.

Probably the most common pathogen associated with NGU is *Chlamydia trachomatis.* A substantial number of persons suffering from gonorrhea are coinfected with *C. trachomatis. C trachomatis* is responsible for the sexually transmitted disease lymphogranuloma venereum (see p. 662) and trachoma (an eye infection). Like *N. gonorrhoeae, C. trachomatis* can infect the eyes of a newborn during birth.

Chlamydias are small gram-negative bacteria that are obligately intracellular parasites. Although chlamydial infections may be asymptomatic or cause only mild symptoms, especially in women, they have an alarming capacity to cause salpingitis and subsequent infertility.

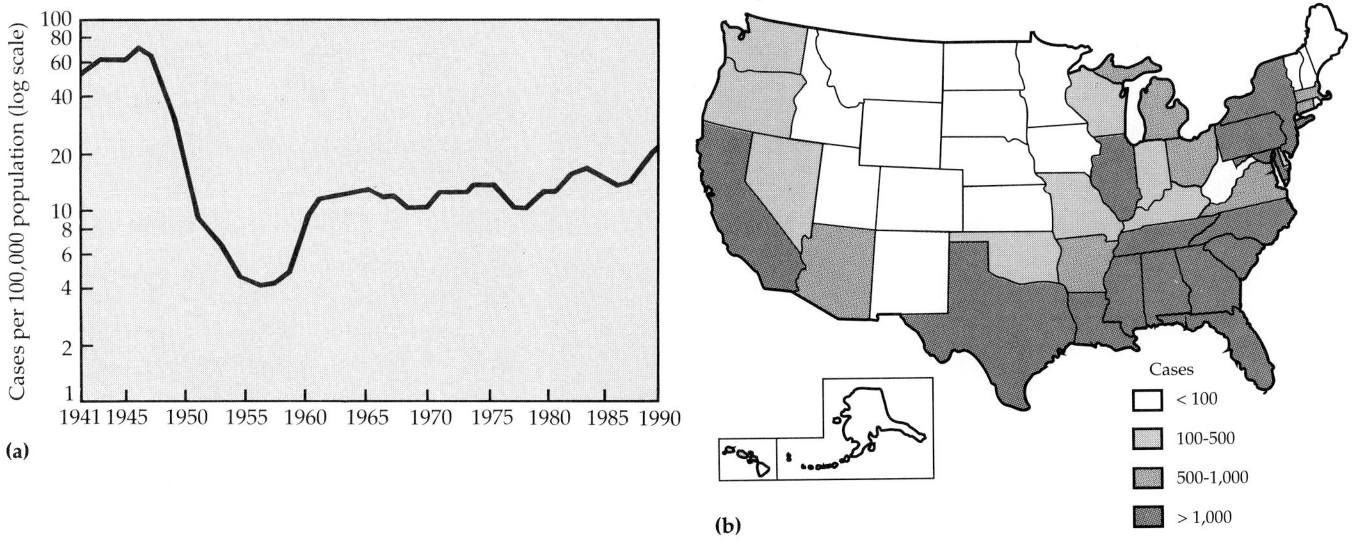

FIGURE 26.9 Incidence and distribution of syphilis. **(a)** Frequency of cases in the United States since 1941. **(b)** Geographical distribution of cases in 1990. [Sources: CDC. *Summary of Notifiable Diseases 1989, MMWR* 38:54 (10/5/90). *MMWR* 39:52 (1/4/91).]

At one time, culture of chlamydias required cell culture lines not available in all laboratories. This diagnostic picture has changed dramatically with the availability of several new rapid tests for chlamydial infections. Some test kits require no specialized equipment and give reliable results in 30 minutes.

Bacteria other than *N. gonorrhoeae* and *C. trachomatis* can also be implicated in NGU. The next most common cause of urethritis and infertility is probably *Ureaplasma urealyticum* (ū-rē-ă-lit'i-kum). This organism is a member of the mycoplasma group (bacteria without a cell wall). Another mycoplasma, *Mycoplasma hominis* (ho'mi-nis), commonly inhabits the normal vagina but can opportunistically cause salpingitis.

Both chlamydia and mycoplasma are sensitive to tetracycline-type antibiotics. Erythromycin is an acceptable alternative drug.

SYPHILIS

The earliest reports of **syphilis** date back to the end of the fifteenth century in Europe. This was coincidental with the return of Columbus from the New World and gave rise to a hypothesis that syphilis was introduced to Europe by his men. Another hypothesis is that syphilis existed in Europe and Asia before the fifteenth century but became widespread as urban living became more common. One description of the Morbus Gallicus (French disease) seems to clearly describe syphilis as early as 1547 and ascribes its transmission in these terms, ". . . it is taken when one pocky person doth synne in lechery one with another."

In contrast to the precipitous increase in the number of gonorrhea cases, the number of new syphilis cases in the United States has remained fairly stable (Figure 26.9). This relative stability as compared with gonorrhea (see Figure 26.6) is remarkable because the epidemiology of the two diseases is quite similar and concurrent infections are not uncommon.

The highest incidence of syphilis is in the 20- to 39-year-old age group, which is also the most sexually active. The incidence among male homosexuals is several times that in the general population but is declining. Until recently, syphilis had actually become an uncommon disease in the heterosexual population. Many states discontinued the requirements for pre-

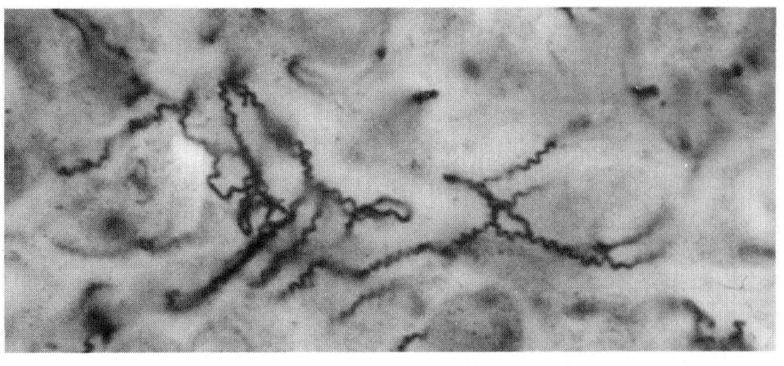

FIGURE 26.10 ***Treponema pallidum,* the causative agent of syphilis.**

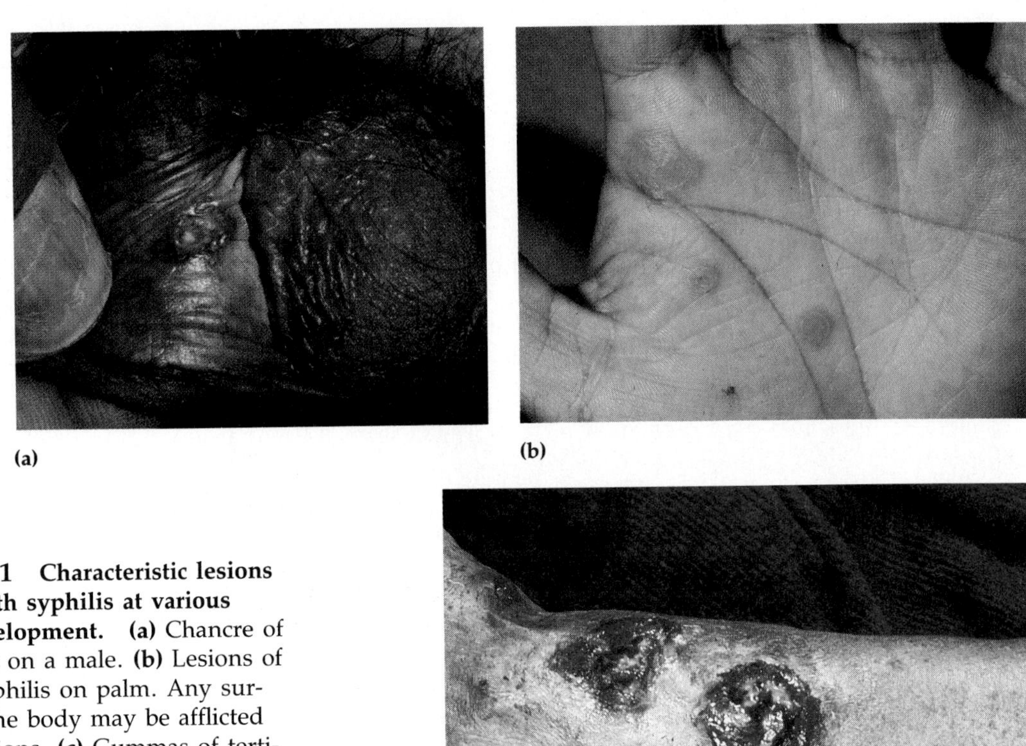

FIGURE 26.11 Characteristic lesions associated with syphilis at various stages of development. **(a)** Chancre of primary stage on a male. **(b)** Lesions of secondary syphilis on palm. Any surface area of the body may be afflicted with such lesions. **(c)** Gummas of tertiary stage on the back of an arm. Gummas such as these are rarely seen today, in the era of antibiotics.

marital syphilis tests because so few cases were detected. At present, the population most at risk is economically disadvantaged inner-city residents, especially drug-using prostitutes of both sexes.

The causative organism of syphilis is a weakly staining, gram-negative spirochete, *Treponema pallidum*. The spirochete is a thin, tightly coiled helix no more than 20 μm in length (Figure 26.10). The virulent strains of the spirochetes have been successfully cultured only in cell cultures, which is not very useful for routine clinical diagnosis. Separate strains of *T. pallidum* are responsible for certain tropically endemic skin diseases such as yaws. These diseases are not sexually transmitted.

Syphilis is transmitted by sexual contact of all kinds, via syphilitic infections of the genitals or other bodily parts. The incubation period averages about three weeks but can range from two weeks to several months. The disease progresses through several recognized stages.

In the *primary stage* of the disease, the initial symptom is a small, hard-based **chancre,** or sore, which usually appears at the site of infection (Figure 26.11a). The chancre is painless, and a serous exudate is formed in the center. This fluid is highly infectious, and examination with the darkfield microscope shows many spirochetes. In a few weeks, this lesion disappears. None of these symptoms causes any distress. In fact, many women are entirely unaware of the chancre, which is commonly on the cervix. In males, the chancre sometimes forms in the urethra and is not visible. Serological diagnostic tests are positive in about 80% of patients in the primary stage. During this stage, bacteria enter the bloodstream and lymphatic system, which distribute them widely in the body.

Several weeks after the primary stage (the exact length of time varies), the disease enters the *secondary stage,* characterized mainly by skin rashes of varying appearance (Figure 26.11b). Other symptoms often observed are the loss of patches of hair, malaise, and mild fever. The rash is widely distributed on the skin and is also found in the mucous membranes of the mouth, throat, and cervix. At this stage, the lesions of the rash contain many spirochetes and are very infectious. Dentists or other medical workers coming into contact with fluid from these lesions can easily become infected by the spirochete entering through minute breaks in the skin. Such nonsexual transmission is possible, but the organisms do not survive long on environmental surfaces and are very unlikely to be transmitted via such objects as toilet seats. Serological tests for syphilis are almost always positive at this stage.

The symptoms of secondary syphilis usually subside after a few weeks, and the disease enters a *latent period.* During this period, there are no symptoms of the disease, although serological tests are positive. After two to four years of latency, the disease is not normally infectious, except for transmission from mother to fetus. The majority of cases do not progress beyond the latent stage, even without treatment. Some resistance to reinfection is observed at this period.

Because the symptoms of primary and secondary syphilis are not disabling, persons may enter the latent period without having received medical attention. In less than half of these cases, the disease reappears in a *tertiary stage.* This stage occurs only after an interval of many years—usually at least ten—after the onset of the latent phase. *T. pallidum* has an outer layer of lipids that stimulates remarkably little effective immune response, especially from cell-destroying complement reactions. It has been described as a Teflon® pathogen. Nonetheless, most of the symptoms of tertiary syphilis are probably due to the body's immune reactions, of a cell-mediated nature, to surviving spirochetes. Inflammatory responses by phagocytic cells such as neutrophils and macrophages also contribute. This causes lesions called *gummas,* rubbery masses of tissue that appear in many organs and sometimes on the external skin (Figure 26.11c). Although many of these lesions are not very harmful, some can cause extensive tissue damage, such as deafness or blindness from lesions in the central nervous system or perforation of the palate (roof of the mouth), which interferes with speech. Few, if any, organisms are found in the lesions of the tertiary stage, and they are not considered very infectious. Today, cases of syphilis allowed to progress to this stage have become rare.

One of the most distressing and dangerous forms of syphilis, called **congenital syphilis,** is transmitted across the placenta to the unborn fetus. Damage to mental development and other neurological symptoms are among the more serious consequences. This type of infection is most common when pregnancy occurs during the latent stage of syphilis. A pregnancy during the primary or secondary stage is likely to produce a stillbirth.

Serological testing for primary syphilis is not completely reliable. Often, the best method for diagnosis is an examination of exudates removed from punctured lymph nodes in the affected area or from lesions. A darkfield microscope is necessary because the organisms do not stain well and are only about 0.2 μm in diameter, a size that approaches the limits of resolution for a brightfield microscope. Figure 26.10 shows the spirochete in such a preparation.

The **Venereal Disease Research Laboratory (VDRL)** slide flocculation test is probably the most widely used serological screening test. The **rapid plasma reagin (RPR)** card test, which is similar, is also in common use. Both of these tests are nonspecific, in that they do not detect antibodies produced against the spirochete itself but detect *reagin-type antibodies.* Reagin-type antibodies are apparently a response to lipid materials that the body forms as an indirect response to infection by the spirochete. The antigen used in such precipitation-type slide tests is thus not the syphilis spirochete but is an extract of beef heart (cardiolipin) that seems to contain lipids similar to those that stimulated the reagin-type antibody production. These antibodies are formed from three to five weeks after infection, so the VDRL and RPR tests are of only limited usefulness in detecting primary syphilis. Each of these serological tests will detect only about 70% to 80% of primary syphilis cases, but they detect 99% of secondary syphilis cases.

Most slide tests are likely to produce a significant percentage of false-positive reactions for syphilis. (These are most likely to occur in persons with immune disorders.) Therefore, positive reactions should be confirmed by a more exacting test based on spirochete-type antigens. These tests often have less than 1% false-positive reaction. One such test is the **fluorescent treponemal antibody absorption (FTA–ABS) test.** This is an indirect immunofluorescence test, in which a preparation of an avirulent strain of *T. pallidum* (one that can be cultured) is allowed to react with a sample of a patient's serum on a microscope slide (see Figure 18.12c). Antibodies in the serum will combine with the spirochetes, but this reaction is not visible. To make it visible, an antibody that will combine with any human antibody (anti-human gamma globulin, or anti-HGG) is tagged with a dye that will be visible under a fluorescence microscope. The tagged anti-HGG is added to the slide and will form a combination of spirochete–antibody and tagged anti-HGG. The slide is again washed and examined with a fluorescence microscope. The spirochetes are detected by their fluorescent glow when struck by ultraviolet light. A disadvantage of the FTA–ABS test is that it remains positive even after syphilis has been successfully treated. The nonspecific reagin-based tests tend to become negative upon successful treatment and are therefore more useful for determining the effect of therapy.

Benzathine penicillin, a long-acting formulation that remains effective in the body for about two weeks, is the usual antibiotic treatment of syphilis. The serum concentrations achieved by this formulation are low, but the spirochete has remained very sensitive to this treatment. Penicillin therapy is especially effective during the primary stage, but antibiotics are generally not effective in tertiary syphilis, probably because the organisms are usually not present.

The spirochete grows slowly, and penicillin is effective only against growing, not dormant, organisms. Therefore, antibiotic treatment might be prolonged. For penicillin-sensitive persons, a number of other antibiotics, such as erythromycin and the tetracyclines, have also proved effective. Antibiotic therapy aimed at gonorrhea and other infections is not likely to eliminate syphilis as well, because such therapy for those other diseases is usually administered for too short a period. Similarly, penicillin directed at syphilis does not reach a high enough concentration to successfully treat gonorrheal infections.

GARDNERELLA VAGINITIS

Vaginitis is most commonly caused by one of three organisms: the fungus *Candida albicans* (kan'did-ä al'bi-kans), the protozoan *Trichomonas vaginalis* (trik-ŏ-mon' as va-jin-al'is), or the bacterium *Gardnerella vaginalis.* Vaginal infections not attributed to *Trichomonas* or *Candida* were once termed *nonspecific vaginitis.* Now most of these cases are attributed to the presence of *G. vaginalis.* There is no sign of inflammation, and the condition is something of an ecological mystery. The population of *G. vaginalis* is many times higher than in uninfected women, the population of *Lactobacillus* bacteria is correspondingly decreased, and there is an overgrowth by anaerobic bacteria such as *Bacteroides* spp. It is not exactly known whether the decrease in acid-producing lactobacilli, and the higher pH characteristic of the condition, is a cause or effect of the overall microbial changes. Whether the condition is sexually transmitted is also uncertain; the organism is a common inhabitant of the vaginas of asymptomatic women. There is no corresponding disease condition in males.

Probably one-third of all vaginitis cases are of this type, which occurs when the vaginal pH is above 5 (normally it would be less than 4). The condition is characterized by a pronounced fishy odor of a frothy vaginal discharge, which is usually copious in volume. Diagnosis is based on the fishy odor, the level of vaginal pH, and the microscopic observation of *clue cells* in the discharge. These clue cells are sloughed-off vaginal epithelial cells covered with bacteria, mostly *G. vaginalis* (Figure 26.12). Treatment is primarily by metronidazole, a drug that eradicates the anaerobes essential to continuation of the disease but allows the normal lactobacilli to repopulate the vagina.

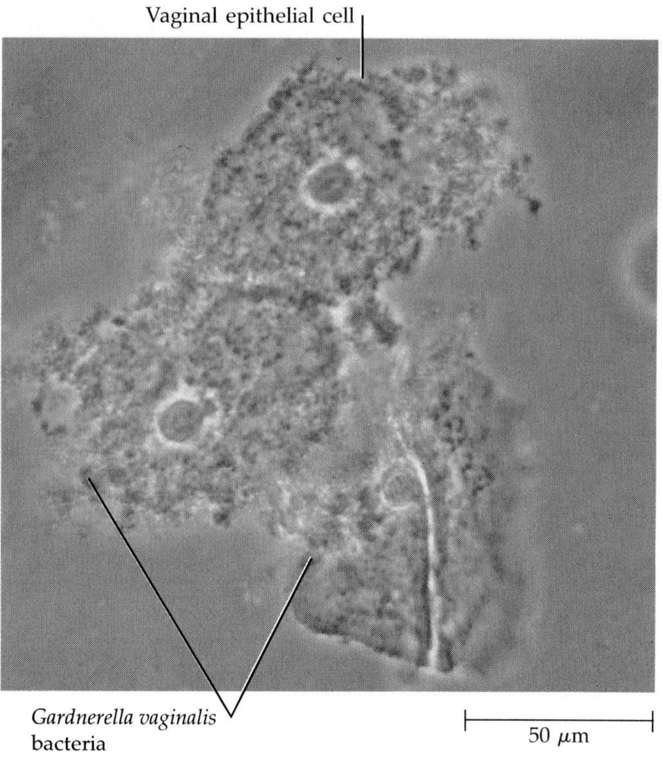

FIGURE 26.12 Clue cells. An essential part of diagnosis of *Gardnerella vaginalis* infections is the finding of clue cells, which are vaginal epithelial cells coated with gram-negative rods, mostly *G. vaginalis.*

LYMPHOGRANULOMA VENEREUM

There are a number of STDs that are uncommon in the United States but more common in the tropical areas of the world. For example, *Chlamydia trachomatis,* the cause of trachoma (see Chapter 21) and a major cause of NGU, is also responsible for **lymphogranuloma venereum,** a disease found in tropical and near-tropical regions. It is apparently caused by a strain of *C. trachomatis* that is invasive and tends to infect lymphoid tissue. In the United States, there are usually about 250 cases per year, mostly in the southeastern states.

After a latent period from 7 to 12 days, a small lesion appears at the site of infection, usually on the genitals. The lesion ruptures and heals without scarring. From one week to two months later, the microorganisms invade the lymphatic system and the regional lymph nodes become enlarged and tender. Suppuration (a discharge of pus) may also occur. The inflammation of the lymph nodes results in scarring, which occasionally obstructs the lymph vessels. This scarring leads to edema of the genital skin and massive enlargement of the external genitals in males. In females, rectal narrowing results from involvement of the lymph nodes in the rectal region. Enlarged nodes may be slow to subside, even after successful antibiotic therapy. These conditions can eventually require surgery.

For diagnosis, pus can be aspirated from infected lymph nodes. When infected cells are properly stained with an iodine preparation, the clumped, intracellular chlamydias can be seen as inclusions. The isolated organisms can also be grown in cell culture or in embryonated eggs. Several serological tests are available, but all are troubled with frequent false-positive reactions. The drug of choice for treatment is tetracycline, and alternatives are such drugs as doxycycline or erythromycin.

CHANCROID (SOFT CHANCRE)

The STD known as chancroid occurs more frequently in tropical areas, where it is often more common than syphilis. The number of reported cases in the United States has been increasing steadily, from 840 to 5000 during a recent nine-year period. Almost all occur in the states of New York, Texas, California, Florida, and Georgia.

In **chancroid (soft chancre),** a swollen, painful ulcer that forms on the genitals, involves an infection of the adjacent lymph nodes. About a week elapses between infection and appearance of the symptoms. Infected lymph nodes in the groin area sometimes even break through and discharge pus to the surface. These ulcers are highly infective, as are the other genital lesions that occur. Lesions might also occur on such diverse areas as the tongue and lips. Because chancroid is so seldom seen by some physicians, they might confuse it with primary syphilis or genital herpes infections. The causative organism is *Hemophilus ducreyi* (dü-krā'ē), a small gram-negative rod that can be isolated from exudates of lesions. The recommended antibiotics are erythromycin and trimethoprim-sulfamethoxazole.

Viral Diseases of the Reproductive System

GENITAL HERPES

A much publicized STD is genital herpes, usually caused by *herpes simplex type 2 virus.* The herpes simplex virus occurs as either type 1 or type 2. Herpes simplex type 1 virus is primarily responsible for the common cold sore or fever blister (see Chapter 21). Herpes simplex type 2 virus, and sometimes type 1, causes an STD called **genital herpes.** The lesions appear after an incubation period of about one week or less and cause a burning sensation. After this, vesicles appear (Figure 26.13). In both sexes, urination can be painful and walking is quite uncomfortable; the patient is even irritated by clothing. Usually, the vesicles heal in a couple of weeks.

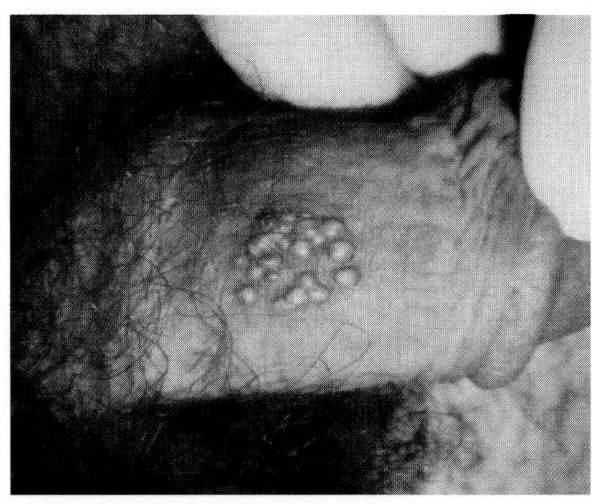

FIGURE 26.13 Vesicles of genital herpes on a penis, shown approximately natural size.

The vesicles contain fluid that is infectious. Condoms provide uncertain protection since the vesicles on women are usually on the external genitals (seldom on the cervix or within the vagina), and the vesicles on men are often at the base of the penis.

One of the most distressing characteristics of genital herpes is the possibility of recurrences. As in other herpes infections, such as cold sores or chickenpox–shingles, the virus enters a latent state in nerve cells. The site of latency is in nerve ganglia near the base of the spine. There is apparently considerable individual variation in latency. Some persons have frequent attacks; for others, recurrence is rare. Reactivation appears to be triggered by a number of factors, including menstruation, emotional stress or illness (especially if accompanied by fever, a factor that is also involved in appearance of cold sores), and perhaps just scratching the affected area. About 88% of patients with type 2 virus and about 50% of those with type 1 virus will have recurrences. The average patient will have recurrences every three or four months. If a person is going to experience recurrences, the first will usually appear within about six months after the primary infection.

In the United States, the incidence of genital herpes has increased so much that there are cumulatively probably more than 20 million infected individuals. Females are less likely to be aware that they have the disease. This lack of awareness can have serious effects because of *neonatal herpes.* The virus can cross the placental barrier and affect the fetus if the mother suffers from an acute infection early in pregnancy. How frequently this happens is not known, but the fetus is considered infected for practical purposes if the virus can be grown from the amniotic fluid. Infection can result in spontaneous abortion or serious damage to the fetus, such as mental retardation and defective sight and hearing.

If the fetus remains free of the virus, it is necessary to prevent infecting it as it passes through a herpesvirus-infected birth canal. Identifying women who are unaware of their infection is difficult; asymptomatic infections are common. Ideally, weekly viral cultures from the mother should be taken during the final six weeks before delivery; many asymptomatic women shed the virus. If results are positive, a cesarean section is usually recommended. This operation should be done before the fetal membrane ruptures. If the woman is shedding the virus, the rupture of the fetal membrane will cause the virus to spread to the uterus within a few hours. Infected infants have a high mortality rate, and survivors almost always have severe neurological damage.

At present, there is no cure for genital herpes, although research on its prevention and treatment is intensive. Acyclovir, administered topically or orally, is proving to be reasonably effective in alleviating the symptoms of a primary outbreak. There is some relief of pain and other symptoms and slightly faster healing. Continuous administration for several months has been found to lower the chances of recurrence; however, it has no effect on the frequency of recurrence after it is discontinued.

GENITAL WARTS

Warts are an infectious disease; since 1907 it has been known that they are caused by viruses known as papillomaviruses. It is probably less commonly known that warts can be transmitted sexually and that this is an increasing problem. An estimated half million new cases are reported each year. There are more than 50 serotypes of papillomaviruses, and certain serotypes tend to be linked with certain forms of genital warts. Morphologically some warts are extremely large and "warty" in appearance with multiple fingerlike projections; others tend to be relatively smooth or flat (Figure 26.14). The incubation time is usually a matter of weeks or months. The greatest danger from genital warts is their connection to cancer. Some 90% of genital warts are caused by viruses of serotypes that do not cause cancer, but there are several serotypes that are frequently associated with a progression to cancer. In women this is usually cervical cancer, and in men it is usually cancer of the penis. Genital warts in women are much more likely to be precancerous than those in men. It is hoped that serological typing of warts will eventually become a routine aid in determining which are the most dangerous, but at present this is expensive and limited to relatively few laboratories. Treatments of warts were discussed previously in Chapter 21, and none of them is completely satisfactory.

FIGURE 26.14 Genital warts on a penis. These infections are often much more extensive.

AIDS

In concluding this section on viral STDs, we should recall that AIDS is a viral disease that is frequently transmitted by sexual contact. However, its effect is on the immune system, so it was discussed in Chapter 19 and the boxes in Chapters 13 and 19, pp. 344 and 484.

Fungal Disease of the Reproductive System

CANDIDIASIS

Candida albicans is a yeastlike fungus that often grows on mucous membranes of the mouth, intestinal tract, and genitourinary tract (see Figure 21.15). Infections are usually a result of opportunistic overgrowth when the competing microflora is suppressed by antibiotics or other factors. As we discussed in Chapter 21, *C. albicans* is the cause of thrush (oral candidiasis). It is also responsible for occasional cases of NGU in males and **vulvovaginal candidiasis,** which is the most common cause of vaginitis. About 75% of all women experience at least one episode.

The lesions of vulvovaginal candidiasis resemble those of thrush but produce more irritation, severe itching, a thick, yellow, cheesy discharge, and a yeasty odor. *C. albicans* is an opportunistic pathogen. Predisposing conditions include use of oral contraceptives and pregnancy. These cause an increase of glycogen in the vagina (see the discussion of the normal vaginal flora on p. 652). Diabetes and treatment with broad spectrum antibiotics are also commonly associated with the occurrence of *C. albicans* vaginitis.

TABLE 26.1 Diseases Associated with the Urinary and Reproductive Systems

DISEASE	*CAUSATIVE AGENT*	*MODE OF TRANSMISSION*	*TREATMENT*
Bacterial Diseases of the Urinary System			
Cystitis (bladder infection)	*Escherichia coli, Staphylococcus saprophyticus*	Opportunistic infections	Broad spectrum penicillins, nitrofurantoin
Pyelonephritis (kidney infection)	Most common are *E. coli, Enterobacter* spp., *Klebsiella* spp., and *Proteus* spp.	From systemic bacterial infections or infections of the lower urinary tract	Broad spectrum cephalosporins
Leptospirosis (kidney infection)	*Leptospira interrogans*	Direct contact with infected animals, urine of infected animals, or contaminated water	Penicillin
Glomerulonephritis (kidney infection)	Immune-complex hypersensitivity reaction to certain strains of *Streptococcus pyogenes*	Sequel to infection by particular strains of *S. pyogenes* in another part of the body	None, autoimmune condition
Bacterial Diseases of the Reproductive System			
Gonorrhea	*Neisseria gonorrhoeae*	Direct contact, especially sexual contact	Penicillin for nonresistant strains; third-generation cephalosporins for resistant strains
Nongonococcal urethritis (NGU)	*Chlamydia* or other bacteria, including *Mycoplasma hominis* and *Ureaplasma urealyticum*	Sexual contact or opportunistic infections	Tetracycline or erythromycin
Syphilis	*Treponema pallidum*	Direct contact, especially sexual contact	Benzathine penicillin
Gardnerella vaginitis	*Gardnerella vaginalis*	Opportunistic pathogen, vaginal pH of 5 to 6	Metronidazole
Lymphogranuloma venereum (LVG)	*Chlamydia trachomatis*	Direct contact, especially sexual contact	Tetracycline, doxycycline, or erythromycin
Chancroid (soft chancre)	*Hemophilus ducreyi*	Direct contact, especially sexual contact	Erythromycin or trimethoprim–sulfamethoxazole
Viral Diseases of the Reproductive System			
Genital herpes	Herpes simplex type 2 virus, and occasionally type 1	Direct contact, especially sexual contact	Acyclovir may alleviate symptoms
Genital warts	Papillomavirus	Direct contact	Removal by various mechanical and chemical means; interferon injection
Fungal Disease of the Reproductive System			
Candidiasis	*Candida albicans*	Opportunistic pathogen, may be transmitted by sexual contact	Clotrimazole, miconazole
Protozoan Disease of the Reproductive System			
Trichomoniasis (usually vaginitis)	*Trichomonas vaginalis*	Usually sexually transmitted	Metronidazole

Vulvovaginal candidiasis is diagnosed by microscopic identification of the fungus in scrapings of lesions and by isolation of the fungus in culture. Treatment consists of topical application of clotrimazole or miconazole. Oral ketoconazole has also been used successfully in some difficult cases in which frequent recurrences were a problem.

Protozoan Disease of the Reproductive System

TRICHOMONIASIS

The protozoan *Trichomonas vaginalis* is frequently a normal inhabitant of the vagina in females and of the urethra in many males (Figure 26.15; see also Figure 12.20). If the normal acidity of the vagina is disturbed, the protozoan may overgrow the normal microbial population of the genital mucosa and cause **trichomoniasis.** (Males rarely have any symptoms as a result of the presence of the organism.) It is well documented as an STD and is often a coinfection with gonorrhea. In response to the protozoan infection, the body accumulates leukocytes at the infection site. The resulting purulent discharge is profuse, greenish-yellow, and characterized by a foul odor. This discharge is accompanied by irritation and itching. This type of vaginitis is usually sexually transmitted. *T. vaginalis*, *C. albicans*, and *G. vaginalis* cause the three most common types of vaginitis. Table 26.2 summarizes the characteristics of the three types.

Diagnosis is easily made by microscopic examination and identification of the organisms in the discharge. The organism can be found in semen or urine of male carriers. Treatment is by oral metronidazole, administered to both sexual partners, which readily clears the infection. The organism can also be isolated and grown on laboratory media.

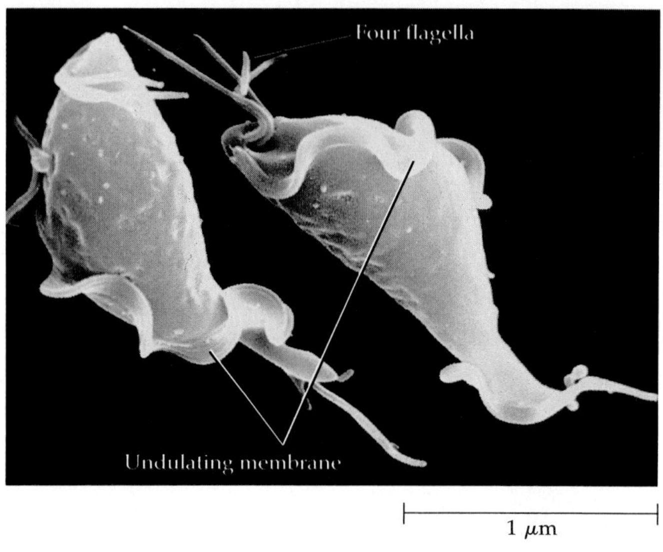

FIGURE 26.15 ***Trichomonas vaginalis*, the flagellated protozoan that causes trichomoniasis.**

TABLE 26.2 Characteristics of the Most Common Types of Vaginitis

	CANDIDA ALBICANS	*GARDNERELLA VAGINALIS*	*TRICHOMONAS VAGINALIS*
Organism	Yeastlike fungi	Coccobacillary bacteria	Flagellated protozoans
Odor	Yeasty	Fishy	Foul
Color of discharge	White	Gray-white	Greenish-yellow
Consistency of discharge	Curdy	Thin, frothy	Frothy
Amount of discharge	Varies	Copious	Profuse
Appearance of vaginal mucosa	Dry, red	Pink	Tender, red
pH	Below 4	Above 5	Above 5

Source: Adapted from P. A. Gilly, "Vaginal discharge: its causes and cures." *Postgraduate Medicine* 80(8):231, December 1986.

STUDY OUTLINE

Introduction (p. 650)

1. The urinary system regulates the chemical composition of the blood and excretes nitrogenous waste.
2. The genital system produces gametes for reproduction and, in the female, supports the growing embryo.
3. The urinary and genital systems are closely related anatomically, and both open to the external environment.
4. Microbial diseases of these systems can result from infection from an outside source or from opportunistic infection by members of the normal body flora.

Structure and Function of the Urinary System (pp. 650–651)

1. Urine is transported from the kidneys through ureters to the urinary bladder and is eliminated through the urethra.
2. Valves prevent urine from flowing back to the urinary bladder and kidneys.
3. The flushing action of urine and the acidity of normal urine have some antimicrobial value.

Structure and Function of the Reproductive System (pp. 651–652)

1. The female genital system consists of two ovaries, two uterine tubes, the uterus, the vagina, and external genitals.
2. The vagina functions as the birth canal and copulatory canal.
3. The male genital system consists of two testes, ducts, accessory glands, and the penis.
4. Seminal fluid leaves the male body through the urethra.

Normal Flora of the Urinary and Reproductive Systems (p. 652)

1. *Streptococcus, Bacteroides, Mycobacterium, Neisseria,* and enterobacteria are members of the resident flora of the urethra.
2. The urinary bladder and upper urinary tract are sterile under normal conditions.
3. Lactobacilli dominate the vaginal flora during reproductive years.

DISEASES OF THE URINARY SYSTEM (pp. 653–654)

Bacterial Diseases of the Urinary System (pp. 653–654)

1. Opportunistic gram-negative bacteria from the intestines often cause urinary tract infections, especially in females.
2. Predisposing factors of urinary tract infections include nervous system disorders, toxemia, diabetes mellitus, and obstructions to the flow of urine.
3. Urethritis, cystitis, and ureteritis are terms describing inflammations of tissues of the lower urinary tract.
4. Pyelonephritis can result from lower urinary tract infections or from systemic bacterial infections.
5. Diagnosis and treatment of urinary tract infections depend on the isolation and antibiotic sensitivity testing of the etiologic agents.
6. More than 100,000 bacteria per milliliter of urine indicates an infection.
7. About 35% of all nosocomial infections occur in the urinary system. *E. coli* causes more than half of these infections.

CYSTITIS (p. 653)

1. Inflammation of the urinary bladder, or cystitis, is common in females.
2. Microorganisms at the opening of the urethra and along the length of the urethra, careless personal hygiene, and sexual intercourse contribute to the high incidence of cystitis in females.

PYELONEPHRITIS (p. 653)

1. Inflammation of the kidneys, or pyelonephritis, is usually a complication of lower urinary tract infections.
2. About 75% of pyelonephritis cases are caused by *E. coli.*

LEPTOSPIROSIS (pp. 653–654)

1. The spirochete *Leptospira interrogans* is the cause of leptospirosis.
2. Leptospirosis is characterized by chills, fever, headache, and muscle aches.
3. Serological identification is difficult because there are many serotypes of *L. interrogans.*

GLOMERULONEPHRITIS (p. 654)

1. Glomerulonephritis, or Bright's disease, is an immune-complex disease occurring as a sequel to a β-hemolytic streptococcal infection.
2. Antigen–antibody complexes are deposited in the glomeruli and cause inflammation and damage.

DISEASES OF THE REPRODUCTIVE SYSTEM (pp. 654–666)

Bacterial Diseases of the Reproductive System (pp. 654–663)

1. Most diseases of the genital system are sexually transmitted diseases (STDs).
2. Most STDs can be prevented by the use of condoms and are treated with antibiotics.

GONORRHEA (pp. 655–658)

1. *Neisseria gonorrhoeae* causes gonorrhea.
2. Gonorrhea is the most common reportable communicable disease in the United States.
3. *N. gonorrhoeae* attaches to mucosal cells of the oral–pharyngeal area, genitals, eyes, joints, and rectum by means of fimbriae.
4. Symptoms in males are painful urination and pus discharge. Blockage of the urethra and sterility are complications of untreated cases.
5. Females might be asymptomatic until the infection spreads to the uterus and uterine tubes. Blockage of the uterine tubes, sterility, and pelvic inflammatory disease are complications of untreated cases.
6. Gonorrheal endocarditis, gonorrheal meningitis, and gonorrheal arthritis are complications that can affect both sexes if gonorrheal infections are untreated.
7. Ophthalmia neonatorum is an eye infection acquired by infants during passage through the birth canal of an infected mother.
8. Penicillin has been used effectively to treat gonorrhea. In 1976, penicillinase-producing strains of *N. gonorrhoeae* appeared.
9. Culture of gram-negative diplococci isolated from patients is used to confirm diagnosis.

NONGONOCOCCAL URETHRITIS (NGU) (pp. 658–659)

1. NGU, or nonspecific urethritis, is any inflammation of the urethra not caused by *N. gonorrhoeae.*
2. About 40% of NGU cases are transmitted sexually.
3. Most cases of NGU are caused by *Chlamydia trachomatis.*
4. Symptoms of NGU are often mild or lacking, although salpingitis and sterility may occur.
5. *C. trachomatis* can be transmitted to infants' eyes at birth.
6. *Ureaplasma urealyticum* and *Mycoplasma hominis* also cause NGU.

SYPHILIS (pp. 659–662)

1. Syphilis is caused by *Treponema pallidum,* a spirochete that has not been cultured in vitro. Laboratory cultures are grown in cell cultures.
2. *T. pallidum* is transmitted by direct contact and can invade intact mucous membranes or penetrate through breaks in the skin.
3. The primary lesion is a small, hard-based chancre at the site of infection. The bacteria then invade the blood and lymphatic system, and the chancre spontaneously heals.
4. The appearance of a widely disseminated rash on the skin and mucous membranes marks the secondary stage. Spirochetes are present in the lesions of the rash.
5. The patient enters a latent period after the secondary lesions spontaneously heal.
6. At least ten years after the secondary lesion, tertiary lesions called gummas can appear on many organs.
7. Congenital syphilis, resulting from *T. pallidum* crossing the placenta during the latent period, can cause neurological damage in the newborn.
8. *T. pallidum* is identifiable through darkfield microscopy of fluid from primary and secondary lesions.
9. Many serological tests, such as VDRL, RPR, and FTA–ABS, can be used to detect the presence of antibodies against *T. pallidum* during any stage of the disease.

GARDNERELLA VAGINITIS (p. 662)

1. Vaginitis can be caused by *Candida albicans, Trichomonas vaginalis,* or *Gardnerella vaginalis.*
2. Diagnosis of *G. vaginalis* is based on increased vaginal pH, fishy odor, and the presence of clue cells.

LYMPHOGRANULOMA VENEREUM (pp. 662–663)

1. *C. trachomatis* causes lymphogranuloma venereum, which is primarily a disease of tropical and subtropical regions.
2. The initial lesion appears on the genitals and heals without scarring.
3. The bacteria are spread in the lymph system and cause enlargement of lymph nodes, obstruction of lymph vessels, and edema of the genital skin.
4. The bacteria are isolated and identified from pus taken from infected lymph nodes.

CHANCROID (SOFT CHANCRE) (p. 663)

1. Chancroid, a swollen, painful ulcer on the mucous membranes of the genitals or mouth, is caused by *Hemophilus ducreyi.*

Viral Diseases of the Reproductive System (pp. 663–664)

GENITAL HERPES (pp. 663–664)

1. Herpes simplex type 2 virus causes an STD called genital herpes.
2. Symptoms of the infection are painful urination, genital irritation, and fluid-filled vesicles.
3. Neonatal herpes is contracted during fetal development or birth. It can result in neurological damage or infant fatalities.
4. The virus might enter a latent stage in nerve cells. Vesicles reappear following trauma and hormonal changes.
5. Genital herpes is associated with cervical cancer.
6. The drug acyclovir has proven effective in treating the symptoms of genital herpes; however, it does not cure the disease.

GENITAL WARTS (p. 664)

1. Papillomaviruses cause warts.
2. The papillomaviruses that cause genital warts have been associated with cancer of the cervix or penis.

AIDS (p. 664)

1. AIDS is a sexually transmitted disease of the immune system.

Fungal Disease of the Reproductive System (pp. 664–666)

CANDIDIASIS (pp. 664–666)

1. *Candida albicans* causes NGU in males and vulvovaginal candidiasis in females.

2. Vulvovaginal candidiasis is characterized by lesions that produce itching and irritation.
3. Predisposing factors for candidiasis are pregnancy, diabetes, tumors, and broad spectrum antibacterial chemotherapy.
4. Diagnosis is based on observation of the fungus and its isolation from lesions.

Protozoan Disease of the Reproductive System (p. 666)

TRICHOMONIASIS (p. 666)

1. *Trichomonas vaginalis* causes trichomoniasis when the pH of the vagina increases.
2. Diagnosis is based on observation of the protozoan in purulent discharges from the site of infection.

STUDY QUESTIONS

Review

1. List the normal flora of the urinary system, and show their habitats in Figure 26.1.
2. List the normal flora of the genital system, and show their habitats in Figures 26.2 and 26.3.
3. How are urinary tract infections transmitted?
4. Explain why *E. coli* is frequently implicated in cystitis in females. List some predisposing factors for cystitis.
5. Name one organism that causes pyelonephritis. What are the portals of entry for organisms that cause pyelonephritis?
6. Leptospirosis is a kidney infection of humans and other animals. How is this disease transmitted? What types of activities would increase one's exposure to this disease? What is the etiology?
7. Describe the symptoms of genital herpes. What is the etiologic agent? When is this infection least likely to be transmitted?
8. What is glomerulonephritis? How is it transmitted? How is it treated?
9. Complete the following table.

Disease	Causative Agent	Symptoms	Method of Diagnosis	Treatment
Gardnerella vaginitis				
Gonorrhea				
Syphilis				
NGU				
Lymphogranuloma venereum				
Chancroid				

10. Name one fungus and one protozoan that can cause genital system infections. What symptoms would lead you to suspect these infections?
11. List the genital infections that cause congenital and neonatal infections. How can transmission to a fetus or newborn be prevented?

Challenge

1. The tropical skin disease called yaws is transmitted by direct contact. Its causative agent, *Treponema pallidum pertenue,* is indistinguishable from *T. pallidum.* The appearance of syphilis in Europe coincides with the first importation of slaves. How might *T. pallidum pertenue* have evolved into *T. pallidum* in the temperate climate of Europe?
2. Why can frequent douching be a predisposing factor to bacterial vaginitis, vulvovaginal candidiasis, or trichomoniasis?
3. A previously well 19-year-old female was admitted to a hospital after two days of nausea, vomiting, headache, and neck stiffness. Cerebrospiral fluid and cervical cultures showed gram-negative diplococci in leukocytes; a blood culture was negative. What disease did she have? How was it probably acquired?
4. *Neisseria* is cultured on Thayer–Martin media, consisting of chocolate agar and nystatin, incubated in a 5% CO_2 environment. How is this selective for *Neisseria*?

5. The list below is a key to selected microorganisms that cause genitourinary infections. Complete this key by listing genera discussed in this chapter in the blanks that correspond to their respective characteristics.

Gram-negative bacteria
- Spirochete
  - Aerobic ________
  - Anaerobic ________
- Coccus
  - Oxidase-positive ________
- Bacillus, nonmotile
  - Requires X factor ________
  - Gram-positive wall ________
- Obligate intracellular parasite ________
- Lacking cell wall
  - Urease-positive ________
  - Urease-negative ________

Fungus
- Pseudohyphae ________

Protozoan
- Flagella ________

No organism observed/cultured from patient ________

6. Using the following information, determine what the disease is and how the infant's illness might have been prevented.

May 11: A 23-year-old woman has her first prenatal examination. She is 4-1/2 months pregnant. Her VDRL results are negative.

June 6: The woman returns to her physician complaining of a labial lesion of a few days' duration. A biopsy is negative for malignancy, and herpes test results are negative.

July 1: The woman returns to her physician because the labial lesion continues to cause some discomfort.

Sept. 15: The baby's father has multiple penile lesions and a generalized body rash.

Sept. 25: The woman delivers her baby. Her RPR is 1:32 and the infant's is 1:128.

Oct. 1: The woman takes her infant to a pediatrician because the baby is lethargic. She is told the infant is well and not to worry.

Oct. 2: The baby's father has a persistent body rash and plantar and palmar rashes.

Nov. 8: The infant becomes acutely ill with pneumonia and is hospitalized. The admitting physician finds clinical and radiologic signs of osteochondritis.

FURTHER READING

Aral, S.O., and K.K. Holmes. "Sexually transmitted diseases in the AIDS era." *Scientific American* 264(2):620–69, February 1991. Discusses the epidemiology and risk factors for STDs in the United States; includes suggestions for controlling these diseases.

Braun, M.M., W.L. Heyward, and J.W. Curran. "The global epidemiology of HIV infection and AIDS." *Annual Review of Microbiology* 44:555–577, 1990. Discusses geographic differences in the epidemiology of AIDS.

Corey, L., and P.G. Spear. "Infections with herpes simplex viruses." *New England Journal of Medicine* 314:686–691, 749–755, 1986. Part One describes the virus; Part Two describes clinical infections and treatment.

Gunby, P. "Genital herpes research: many aim to tame maverick virus." *Journal of the American Medical Association* 250:2417–2427, 1983. A summary of prospects for control of genital herpes infections.

Holmes, K.K. "The *Chlamydia* epidemic." *Journal of the American Medical Association* 245:1718–1723, 1981. A readable summary of recent information on chlamydial bacteria and their sexually transmitted diseases.

Journal of the American Medical Association 264(11):1413–1417, 1432–1437, 1451–1452, September 19, 1990. This issue includes three articles on the current rise in gonorrhea and syphilis in the United States.

Shafer, W.M., and R.F. Rest. "Interactions of gonococci with phagocytic cells." *Annual Review of Microbiology* 43:121–145, 1989. Describes the conflicts between the antiphagocytic properties of *Neisseria* and PMNs.

Further Reading for Part Four

The following medical microbiology textbooks discuss principles of microbiology with an emphasis on bacterial pathogens. Some texts also include viral, fungal, and parasitic diseases.

Baron, S., ed. *Medical Microbiology,* 3rd ed. New York: Churchill Livingstone, 1991.

Davis, B.D., R. Dulbecco, H.N. Eisen, and H.S. Ginsberg. *Microbiology,* 4th ed. New York: Harper & Row, 1990.

Freeman, B.A. *Burrow's Textbook of Microbiology,* 22nd ed. Philadelphia: Saunders, 1985.

Joklik, W.K., H.P. Willett, D.B. Amos, and C.M. Wilfert, eds. *Zinsser Microbiology,* 19th ed. Norwalk, Conn.: Appleton–Century–Crofts, 1988.

Mandell, G.L., R.G. Douglas, and J.E. Bennett, eds. *Principles and Practice of Infectious Disease,* 3rd ed. New York: Wiley, 1989.

Murray, P.R., W.L. Drew, G. Kobayashi, and J. Thompson. *Medical Microbiology.* St. Louis: C.V. Mosby, 1990.

Schaechter, M., G. Medoff, and D. Schlessinger. *Mechanisms of Microbial Disease.* Baltimore: Williams and Wilkins, 1989.

Sherris, J.C., ed. *Medical Microbiology,* 2nd ed. New York: Elsevier, 1990.

Wilson, G., A. Miles, and M.T. Parker. *Topley and Wilson's Principles of Bacteriology, Virology and Immunity,* 8th ed., 5 vols. Baltimore: Williams and Wilkins, 1990.

PART FIVE

Applied Microbiology and the Environment

CHAPTER 27

Soil and Water Microbiology

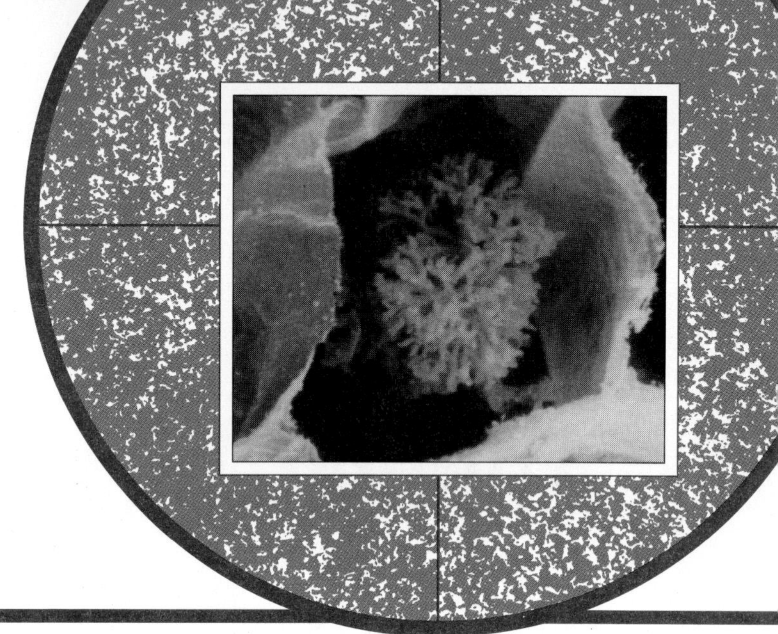

LEARNING OBJECTIVES

- Explain how the components of soil affect soil microflora.
- Outline the carbon and nitrogen cycles, and explain the roles of microorganisms in these cycles.
- Describe the freshwater and seawater habitats of microorganisms.
- Discuss the causes and effects of eutrophication.
- Explain how water is tested for bacteriologic quality.
- Compare primary, secondary, and tertiary sewage treatments.
- List some of the biochemical activities that take place in an anaerobic sludge digester.
- Define BOD, septic tank, oxidation pond, activated sludge, and trickling filter.
- List three examples illustrating the use of bacteria to remove pollutants.

Part Five photo, p. 671: Techniques for bioremediation are developed in environmental laboratories such as this one.

Many people associate bacteria with disease or food spoilage and therefore regard microorganisms as harmful things to be avoided or controlled. Actually, harmful microorganisms are only a very small fraction of the total microbial population. Bacteria and other microorganisms are in fact essential to the maintenance of life on Earth. In previous chapters, we focused mainly on disease-causing capabilities of microorganisms. In this chapter, you will learn about many of the positive functions microbes perform in the environment.

We can find microbes, especially bacteria, in the most widely varied habitats on Earth. They are found in the frozen Antarctic and in boiling hot springs. Aircraft flying miles above the Earth can recover microbes from the thin atmosphere. Exploration of the deepest ocean sediments shows large numbers of bacteria living there, in eternal darkness and subjected to incredible pressures (see the box in Chapter 6, p. 128). Microorganisms are found in the clearest mountain streams flowing from a melting glacier and in waters nearly saturated with salts, such as those of the Dead Sea (see the box in Chapter 5, p. 143).

Some microbes are *parasites*—they derive their nutrients and their ability to reproduce from other microbes. *Bdellovibrio* bacteria, for example, prey on other bacteria (see the box in Chapter 3, p. 62). Other microorganisms, such as those that make the lactic acid of cheese, pickles, or sauerkraut, minimize their competition by forming an acidic environment that is inhospitable to other organisms.

Mutualism, in which two organisms benefit, is exemplified by the association of a fungus and an alga in lichen (discussed later in this chapter). Some protozoans have even adapted to being hosts to single-celled algae that reside within their cytoplasm—the protozoan furnishes protection, and the algae provide additional nutrients for the host.

Commensalism, where one organism benefits from the relationship without affecting the other, is common among microbes. For example, the few bacteria that can degrade polymers (such as cellulose) release simpler compounds (such as glucose) that neighboring scavengers quickly utilize. Some organisms participate in *cometabolism,* which carries generosity to an extreme. The organisms degrading a particular substrate get no benefit whatsoever in terms of energy or nutrition but leave behind a product that can be used by other microbes. However, this sort of behavior is the exception—most microbes live in an intensely competitive world.

Soil Microbiology and Cycles of the Elements

Some microbial population or another fills almost every imaginable ecological niche. Many of these niches are in soil and water. The microorganisms that inhabit soil and water are the subject of this chapter.

The Components of Soil

Soil is a complex mixture of solid inorganic matter (rocks and minerals), organic matter, water, air, and living organisms.

Minerals

The weathering of rocks adds such elements as silicon, aluminum, and iron to the soil. Calcium, magnesium, potassium, sodium, phosphorus, and small amounts of other elements that are essential for life are also present in the mineral component of soil.

Organic Matter

The organic matter in soil consists of carbohydrates, proteins, lipids, and other materials. Organic matter makes up from 2% to 10% of most agriculturally important soils. Swamps and bogs, by contrast, have a higher content of organic matter, up to 95% in some peat bog soils. The waterlogged soil of swamps and bogs is an anaerobic environment, and the microbial decomposition of organic matter occurs very slowly there.

All organic matter in soil is derived from the remains of microorganisms, plants, and animals, their waste products, and the biochemical activities of various microorganisms. A great portion of the organic matter is of plant origin—mostly dead roots, wood and bark, and fallen leaves. A second source of organic matter is the vast numbers of bacteria, fungi, algae, protozoans, and viruses, which can total billions per gram of fertile soil. These organisms break down organic substances, producing and maintaining a continuous supply of inorganic substances that plants and other organisms require for growth. Much of the organic matter in soil is ultimately decomposed to inorganic substances, such as ammonia, water, carbon dioxide, and various compounds of nitrate, phosphate, and calcium.

A considerable part of the organic matter in soil occurs as **humus.** This dark material consists of partially decomposed organic matter, chiefly materials that are relatively resistant to decay.

The addition of organic matter, either completely or partially decomposed, is essential to the fertility of soil. Moreover, these spongy organic materials loosen the soil and thereby prevent the formation of heavy crusts and increase the pore spaces. This addition of pore spaces in turn increases aeration and water retention.

Water and Gases

Microbial life in soil depends on the availability of some water, which dissolves many inorganic and organic constituents in soil and makes them available to the living inhabitants. However, if the water content is too high, atmospheric oxygen cannot easily penetrate the soil, and this can be a factor in limiting growth efficiency and the types of organisms that flourish. Generally, because of respiration by soil organisms, soil gases contain a high proportion of carbon dioxide and a low proportion of oxygen. Most of these soil gases exist in pore spaces between soil particles or are dissolved in the water.

Organisms

Fertile soil contains a great many animals, ranging from microscopic forms, including numerous nematodes, to larger forms, including insects, millipedes, centipedes, spiders, slugs, snails, earthworms, mice, moles, gophers, and reptiles. Most of these animals are beneficial, in that they promote some mechanical movement of the soil and thereby help to keep the soil loose and open. All soil organisms also contribute to the organic matter of soils in the form of their waste products and eventual remains.

Soil also contains the root systems of higher plants and enormous numbers of microorganisms.

Soil Microbial Flora

The soil is one of the main reservoirs of microbial life (Table 27.1). A good agricultural soil the size of a football field usually contains a microbial population of a

TABLE 27.1 Microorganisms per Gram of Typical Garden Soil at Various Depths

DEPTH (CM)	BACTERIA	ACTINO-MYCETES	FUNGI	ALGAE
3–8	9,750,000	2,080,000	119,000	25,000
20–25	2,179,000	245,000	50,000	5,000
35–40	570,000	49,000	14,000	500
65–75	11,000	5000	6000	100
135–145	1400	—	3000	—

Source: Adapted from M. Alexander, *Introduction to Soil Microbiology*, 2nd ed. New York: Wiley, 1977.

size approaching the weight of a cow eating grass on that field. The metabolic capabilities of these vast numbers of microorganisms are probably about 100,000 times those of the grazing cow. However, measurement of the carbon dioxide from soil and other evidence indicate that these microorganisms exist in near starvation and at low reproductive rates. When usable nutrients are added to soil, the microbial populations and their activity rapidly increase until the nutrients are depleted, and then the microbial activity returns to the lower levels.

The most numerous organisms in soil are bacteria. Typical garden soil has millions of bacteria in each gram. As Table 27.1 shows, the population is highest in the top few centimeters of the soil and declines rapidly with depth. The populations are usually estimated with plate counts on nutrient media, and the actual numbers are probably greatly underestimated by this method. No single nutrient medium or growth condition can possibly meet all the myriad nutrient and other requirements of soil microorganisms.

Actinomycetes are bacteria, but they are usually considered separately in enumerations of soil populations because they differ so greatly from our conventional concept of bacteria. They typically form long filaments resembling those of molds, but of bacterial dimensions. This filamentous growth habit is a considerable advantage for a microorganism in soil. Under relatively dry conditions, the filament can bridge the gap between one particle of soil and another. This type of growth also maximizes surface area for a given weight and may give the actinomycetes a nutritional advantage. These organisms are found in soil in very large numbers and produce a gaseous substance called *geosmin*, which gives the soil its characteristic musty odor. The **biomass** (total mass of living organisms in a given volume) of the actinomycetes is probably about equal to the biomass of all other bacteria in soil. Interest in actinomycetes was greatly stimulated by the discovery that some genera, particularly *Streptomyces*, produce valuable antibiotics.

Fungi are found in soil in much smaller numbers than actinomycetes and other bacteria. Because plate counts of fungal colonies are based on the germination of asexual spores, the actual fungal population is probably lower than such counts indicate. It is estimated that the biomass of fungi probably equals the biomass of all bacteria including actinomycetes. This large biomass occurs because fungal mycelia are much larger than bacterial cells. Molds greatly outnumber yeasts in soil.

Algae and cyanobacteria are found even in dry desert soils. They sometimes form visible accumulations on the surfaces of moist soils. As photosynthetic organisms, they are found mainly on the soil surface, where sunlight is available. However, significant numbers of algae and cyanobacteria are also found more than 50 cm below the surface. The environmental contribution of these microorganisms is significant only in special cases. For example, fixation of atmospheric nitrogen (discussed shortly) by some species of cyanobacteria in grasslands and tundra regions and after rainfall in deserts adds significantly to the fertility of soils in those regions.

Protozoans are plentiful in soil. They become dormant as cysts under adverse conditions. When microbial numbers increase, many protozoans function as predators, and their numbers tend to rise and fall with the bacterial populations.

Pathogens in Soil

Human pathogens, which are mostly parasites, find the soil an alien, hostile environment. Even relatively resistant enteric pathogens, such as *Salmonella* species, have been observed to survive for only a few weeks or months when introduced into soil. Most human pathogens that can survive in soil are endospore-forming bacteria. For example, endospores of *Bacillus anthracis*, which causes anthrax in animals, can survive in certain soils for decades before finally germinating when ingested by grazing animals. Disposing of the body of an animal infected with anthrax requires considerable care so that the soil is not seeded with the endospores from the dead animal.

Clostridium tetani (the causative agent of tetanus), *Clostridium botulinum* (the causative agent of botulism), and *Clostridium perfringens* (the causative agent of gas gangrene) are also endospore-forming pathogens whose normal habitat is the soil. From the soil they are introduced into foods or wounds, where they grow and produce toxins.

Certain helminthic pathogens, such as the hookworm, spend part of their life cycle as larval worms in the soil. These larvae penetrate the skin of humans coming into contact with them. Because the larval de-

MICROBIOLOGY IN THE NEWS

Bacteria Contribute to the Greenhouse Effect

For years, scientists have been warning that the increased burning of fossil fuels is raising the carbon dioxide content of the atmosphere and that this added carbon dioxide absorbs some infrared radiation that the Earth normally radiates out into space. This absorption traps the Earth's heat, ultimately causing a greenhouse effect that can eventually warm the atmosphere enough to melt the polar ice caps and flood major coastal cities. In addition, the more pessimistic scientists have warned that the warmer air could cause continual droughts in major crop-growing areas, resulting in worldwide famine.

Other scientists, who are more optimistic, claim that nature takes care of the problem automatically. The extra carbon dioxide stimulates the growth of plants, which fix carbon dioxide into organic molecules, thus reestablishing a balance.

However, there are gases other than carbon dioxide that absorb the infrared rays leaving the Earth, thus exacerbating the greenhouse effect, and these gases, molecule for molecule, are far more active infrared absorbers than carbon dioxide. One of them is methane, which is now much more prevalent in the atmosphere than it was two or three decades ago.

One key link in the chain is *carbon dioxide fertilization,* the stimulating effect of carbon dioxide on plant growth. At first this seems to be a blessing. A Department of Energy report estimates that doubling the amount of carbon dioxide in the atmosphere could cause a 30% to 50% increase in the growth of some agricultural plants. Dr. Boyd Strain, a botanist at Duke University, suggests that much of the increase of agricultural productivity after the Industrial Revolution came not from the use of fertilizers and pesticides but from the increasing abundance of carbon dioxide in the atmosphere.

Because of the additional plant growth, however, there is a corresponding increase in the amount of plant matter that decays. According to Paul Guthrie, an atmospheric scientist at the NASA Goddard Space Flight Center, the dangerous increase in the methane content of the air may largely be a result of the decay of the extra plant matter whose growth was stimulated by the increased amount of carbon dioxide. When dead plants decay in anaerobic pockets of organic debris in forests or grasslands, in the sediment of lakes and bays, or in rice paddies, bacterial fermentation breaks down the organic matter, releasing carbon dioxide and hydrogen. Anaerobic archaeobacteria known as *methanogens* (see p. 291) then combine these into methane and water (see the figure).

The amount of methane in the atmosphere has doubled during the last 150 years. If it is true that a large percentage of this increase is a result of the action of methanogenic bacteria on products of organic decay, then the menace of the greenhouse effect is worse than the pessimists thought. The process that at first cheered the optimists—increased plant growth—may be a reason for even darker pessimism.

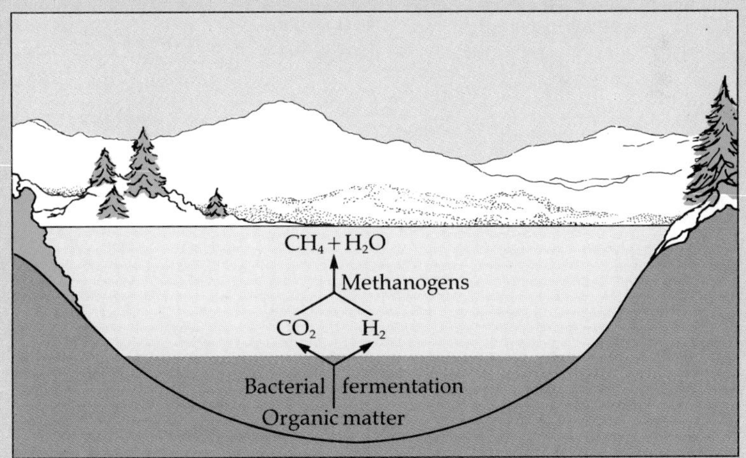

Methane is generated in the sediments of lakes and bays by the action of methanogens.

velopment takes place only in warm, moist soil, these diseases are limited to tropical or subtropical areas.

Pathogens of plants are much more likely to inhabit soil. Most plant pathogens are fungi, mainly because fungi are capable of growing at the low moisture level typical of plant surfaces. Many of the rusts, smuts, blights, and wilts affecting plants are caused by fungi that pass part of their life cycle in soil. Bacterial diseases of plants are less common, but bacterial pathogens in soil do cause plant diseases, especially rots, in which bacterial enzymes hydrolyze plant tissue.

Viruses also cause plant diseases. The mechanism by which they penetrate the thick plant cell walls is frequently via a sap-feeding insect.

Some of the microorganisms found in soil are insect pathogens, which are potentially useful for pest control. *Bacillus thuringiensis*, for example, is a soil bacterium that is pathogenic to the larvae of many insects; it is now used in their control (see Chapter 11, p. 287). A number of other insect pathogens, including viruses and fungi, are under investigation as biological pesticides.

Microorganisms and Biogeochemical Cycles

Perhaps the most important role of soil microorganisms is their participation in **biogeochemical cycles**—the recycling of certain chemical elements so that they can be used over and over again. Among these elements are carbon, nitrogen, sulfur, and phosphorus. Were it not for the activities of microorganisms in the various biogeochemical cycles, essential elements would become depleted and all life would cease.

THE CARBON CYCLE

The first biogeochemical cycle we will consider is the carbon cycle (Figure 27.1). As you know, all organic compounds contain carbon. Most of the inorganic carbon used in the synthesis of organic compounds comes from the carbon dioxide in the atmosphere. Some carbon dioxide is also dissolved in water.

In photosynthesis—the first step in the carbon cycle—carbon dioxide is incorporated, or *fixed*, into organic compounds by such photoautotrophs as cyanobacteria, green plants, algae, and green and purple sulfur bacteria. In the next step in the cycle, chemoheterotrophs consume the organic compounds—animals eat photoautotrophs, especially green plants, and may in turn be eaten by other animals. Thus, as the organic compounds of the photoautotrophs are digested and resynthesized, the carbon atoms of carbon dioxide are transferred from organism to organism up the food chain.

Some of the organic molecules are used by chemoheterotrophs, including animals, to satisfy their energy requirements. When such energy is released through the process of respiration, carbon dioxide is released into the atmosphere. This carbon dioxide immediately becomes available to start the cycle over again. However, much of the carbon remains within the organisms until they excrete wastes or die. When the organisms die, the organic compounds of their

FIGURE 27.1 The carbon cycle.

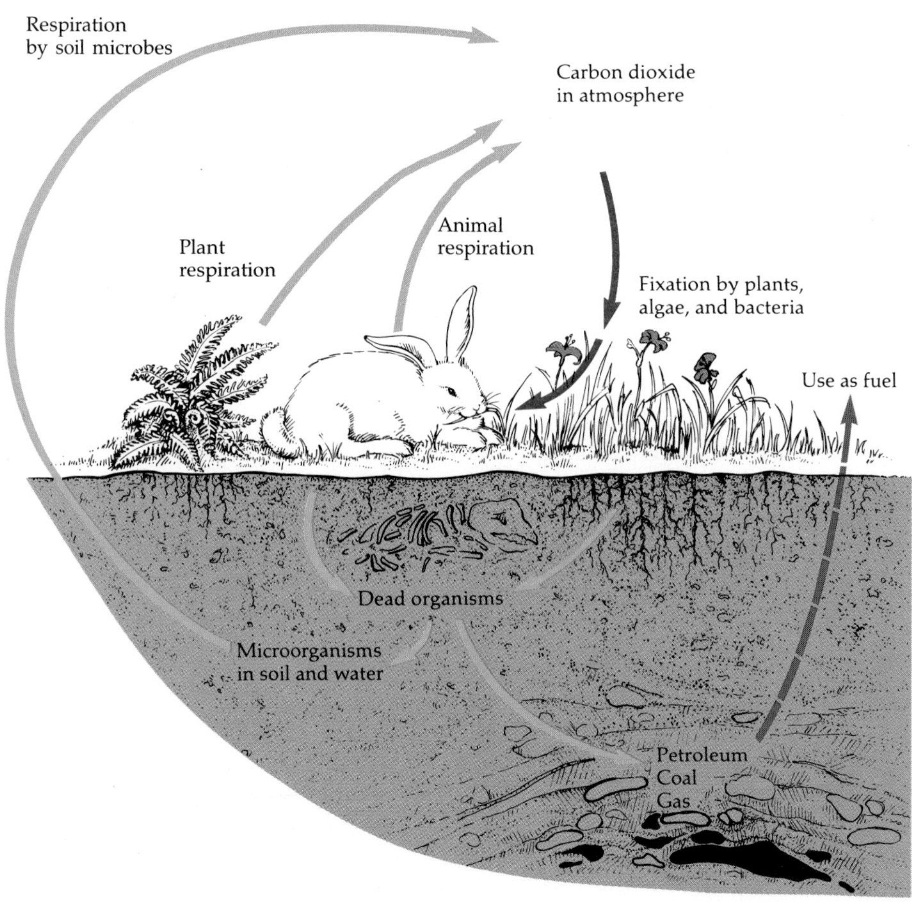

FIGURE 27.2 The nitrogen cycle.

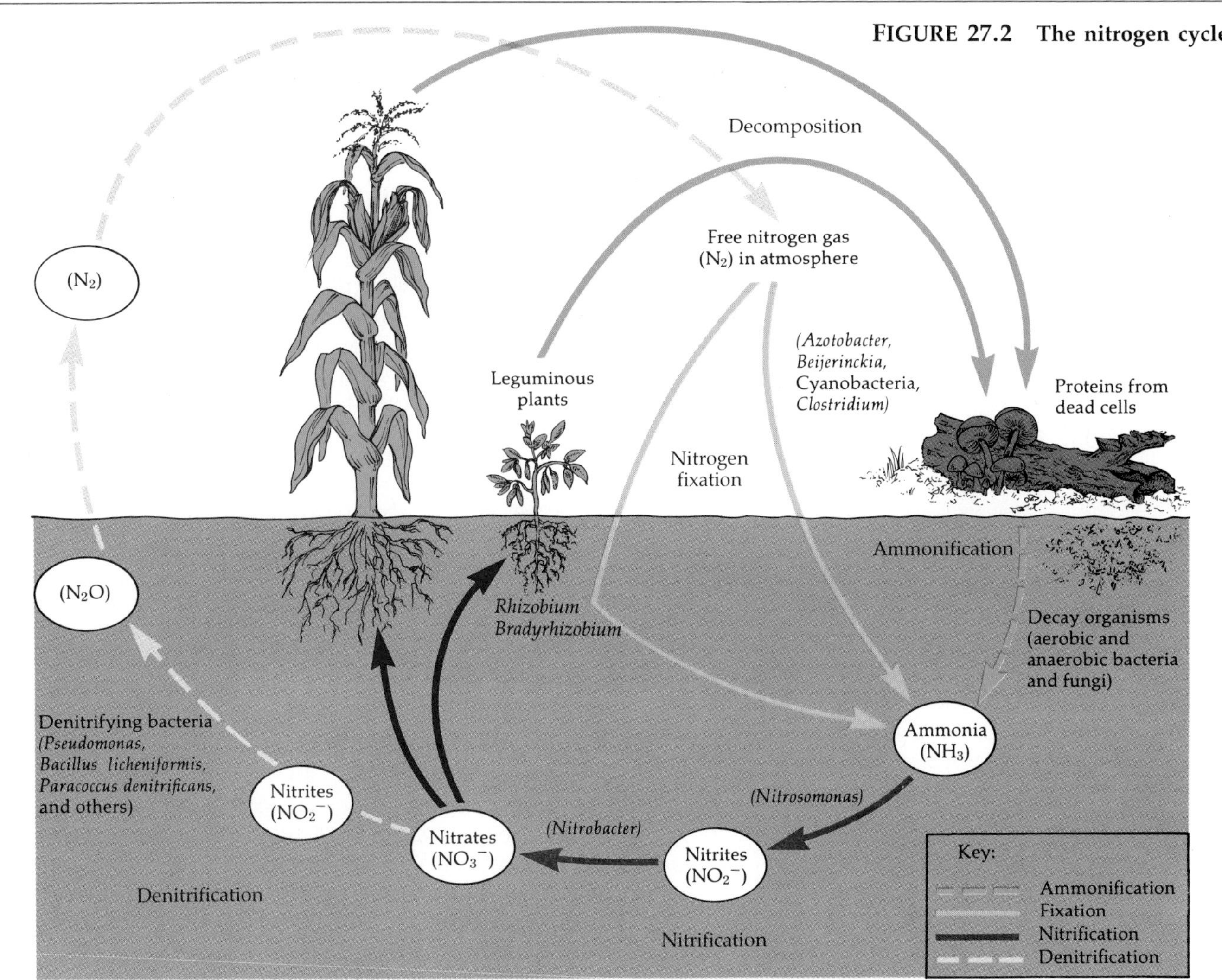

bodies are deposited in the soil and are decomposed by microorganisms, principally by bacteria and fungi. During this decomposition, carbon dioxide is returned to the atmosphere. Although the carbon dioxide of the atmosphere makes up only about 0.03% of the atmospheric gases, it is essential for the growth of plants and algae.

Carbon is stored in rocks, such as limestone ($CaCO_3$), and is dissolved as carbonate ions (CO_3^{2-}) in oceans; it is also stored in organic mineral forms, such as coal and petroleum. Burning such fossil fuels releases carbon dioxide. There is evidence that the amount of carbon dioxide in the atmosphere has been increasing as a result of widespread use of fossil fuels. Some scientists believe that this is a potential problem because the increase of atmospheric carbon dioxide could cause a warming of the Earth known as the greenhouse effect (see the box on p. 675).

THE NITROGEN CYCLE

The nitrogen cycle is shown in Figure 27.2. Nitrogen is needed by all organisms for their synthesis of protein, nucleic acids, and other nitrogen-containing compounds. Molecular nitrogen (N_2) makes up almost 80% of the Earth's atmosphere; the atmosphere over every acre of fertile soil contains more than 30,000 tons of nitrogen. Despite the abundance of nitrogen as a molecular gas, however, no eucaryote is able to make direct use of it. Instead, the nitrogen must be fixed (combined) with other elements, such as oxygen and hydrogen. The resulting compounds, such as nitrate ion (NO_3^-) and ammonium ion (NH_4^+), are then used by autotrophic organisms. The chemical and physical forces operating in the soil, water, and air, together with the activities of specific microorganisms, are important to the conversion of nitrogen to usable forms.

Ammonification

Almost all the nitrogen in the soil exists in organic molecules, primarily in proteins. When an organism dies, the process of microbial decomposition results in the hydrolytic breakdown of proteins into amino acids. In a process called deamination, the amino groups of amino acids are removed and converted into ammonia (NH_3) (see Figure 27.2). This release of ammonia is called **ammonification.** Ammonification is brought about by numerous bacteria and fungi. It can be represented as follows:

$$\text{Proteins from dead cells and waste products} \xrightarrow{\text{Microbial decomposition}} \text{Amino acids}$$

$$\text{Amino acids} \xrightarrow{\text{Microbial ammonification}} \text{Ammonia } (NH_3)$$

Microbial growth releases extracellular proteolytic enzymes that decompose proteins. The resulting amino acids are transported into the microbial cells, where ammonification occurs. The fate of the ammonia produced by ammonification depends on soil conditions. Because ammonia is a gas, it rapidly disappears from dry soil, but in moist soil it becomes solubilized in water, and ammonium ions (NH_4^+) are formed:

$$NH_3 + H_2O \longrightarrow NH_4^+OH \longrightarrow NH_4^+ + OH^-$$

Ammonium ions from this sequence of reactions are used by bacteria and plants for amino acid synthesis.

Nitrification

The next sequence of reactions in the nitrogen cycle involves the oxidation of the ammonium ion to nitrate, a process called **nitrification.** Living in the soil are autotrophic nitrifying bacteria, such as those of the genera *Nitrosomonas* and *Nitrobacter.* These organisms obtain energy by oxidizing ammonia or nitrate. In the first stage, *Nitrosomonas* oxidizes ammonium to nitrites:

$$\underset{\text{Ammonium ion}}{NH_4^+} \xrightarrow{\textit{Nitrosomonas}} \underset{\text{Nitrite ion}}{NO_2^-}$$

In the second stage, such organisms as *Nitrobacter* oxidize nitrites to nitrates:

$$\underset{\text{Nitrite ion}}{NO_2^-} \xrightarrow{\textit{Nitrobacter}} \underset{\text{Nitrate ion}}{NO_3^-}$$

Plants tend to use nitrate as their source of nitrogen for protein synthesis because nitrate is highly mobile in soil and is more likely to encounter a plant root than ammonium. Ammonium ions would actually make a more efficient source of nitrogen as they require less energy to incorporate into protein, but these positively charged ions are usually bound to negatively charged clays in the soil, whereas the negatively charged nitrate ions are not bound.

Denitrification

At various points in the cycle, atmospheric nitrogen is either added or removed. The loss of nitrogen from the cycle involves a process called **denitrification,** the conversion of nitrate to nitrogen gas. Denitrification can be represented as follows:

$$\underset{\text{Nitrate ion}}{NO_3^-} \longrightarrow \underset{\text{Nitrite ion}}{NO_2^-} \longrightarrow \underset{\text{Nitrous oxide}}{N_2O} \longrightarrow \underset{\text{Nitrogen gas}}{N_2}$$

Pseudomonas species appear to be the most important group of bacteria in denitrification in soils. A number of other genera, including *Paracoccus* (păr-ä-kok'kus), *Thiobacillus,* and *Bacillus,* also include species that are capable of carrying out denitrification reactions. Denitrifying bacteria are aerobic, but under anaerobic conditions they can use nitrate in place of oxygen as a final electron acceptor. (This process is called *anaerobic respiration;* see Chapter 5.) Thus, denitrification occurs in waterlogged soils depleted of oxygen. Because denitrifying bacteria deliver nitrogen to the atmosphere at the expense of soil nitrates, denitrification is unfavorable for soil fertility.

Nitrogen Fixation

During the initial phase of the nitrogen cycle, nitrogen gas is converted into ammonia in a process called **nitrogen fixation.** Only a few species of bacteria and cyanobacteria are capable of enacting this process. The nitrogenase enzyme responsible for nitrogen fixation is anaerobic, so it probably evolved early in the history of the planet, before the atmosphere contained molecular oxygen and before nitrogen-containing compounds were available from decaying organic matter. Nitrogen fixation is brought about by two types of organisms, nonsymbiotic and symbiotic.

Nonsymbiotic (free-living) *nitrogen-fixing bacteria* are found in particularly high concentrations in the *rhizosphere,* the region where the soil and roots make contact—especially in grasslands. Among the nonsymbiotic bacteria that can fix nitrogen are aerobic species such as *Azotobacter.* These aerobic organisms apparently shield the nitrogenase enzyme from oxygen by, among other things, having a very high rate of oxygen utilization that minimizes the diffusion of oxygen into the cell where the enzyme is located.

Another nonsymbiotic obligate aerobe that fixes nitrogen is *Beijerinckia* (bī-yė-rink'ē-ä). Some anaerobic bacteria, such as certain species of *Clostridium,* also fix nitrogen. The bacterium *Clostridium pasteurianum* (pas-tyėr-ē-ā'num), an obligately anaerobic, nitrogen-fixing microorganism, is a prominent example. Other nonsymbiotic nitrogen-fixing bacteria include certain species of the facultatively anaerobic *Klebsiella, Enterobacter,* and *Bacillus* and the anaerobic photoautotrophic *Rhodospirillum* (rō-dō-spī-ril'um) and *Chlorobium.*

There are many species of aerobic, photosynthesizing cyanobacteria that fix nitrogen. Because their energy supply is independent of carbohydrates in soil or water, they are especially useful suppliers of nitrogen to the environment. Cyanobacteria usually carry their nitrogenase enzymes in specialized structures called *heterocysts* that provide anaerobic conditions for fixation.

Most of the nonsymbiotic nitrogen-fixing organisms are capable of fixing large amounts of nitrogen under laboratory conditions. However, in the soil there is usually a shortage of usable carbohydrates to supply the energy needed for the reduction of nitrogen to ammonia, which is then incorporated into protein. Nevertheless, these nitrogen-fixing bacteria make important contributions to the nitrogen economy of such areas as grasslands, forests, and the arctic tundra.

Symbiotic nitrogen-fixing bacteria serve an even more important role in plant growth for crop production. Members of the genera *Rhizobium* and *Bradyrhizobium* infect the roots of leguminous plants, such as soybeans, beans, peas, peanuts, alfalfa, and clover. (These agriculturally important plants are only a few of the thousands of known leguminous species, many of which are bushy plants or small trees found in poor soils in many parts of the world.) Rhizobia are spe-

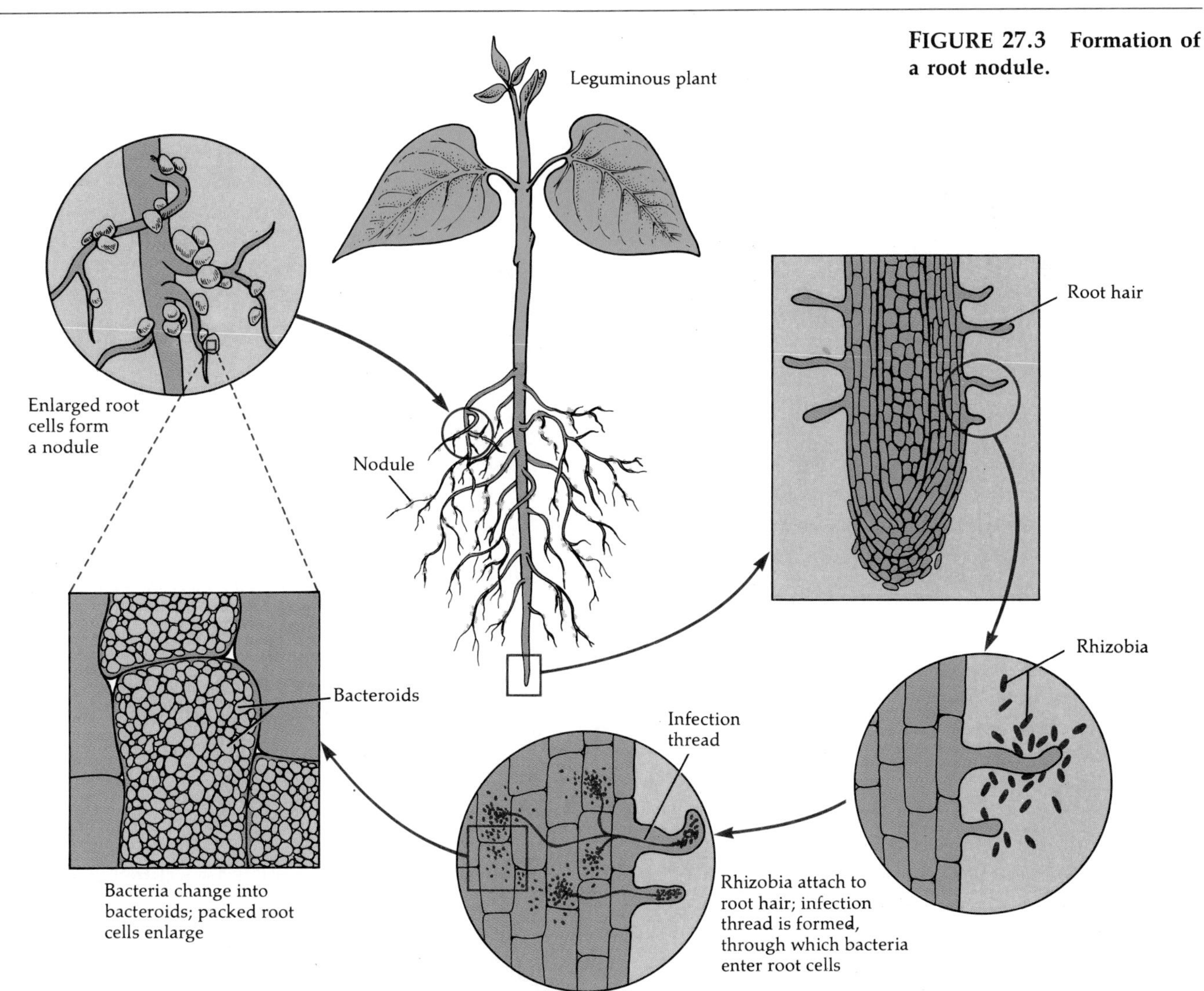
Leguminous plant
Root hair
Enlarged root cells form a nodule
Nodule
Rhizobia
Bacteroids
Infection thread
Bacteria change into bacteroids; packed root cells enlarge
Rhizobia attach to root hair; infection thread is formed, through which bacteria enter root cells

FIGURE 27.3 Formation of a root nodule.

cially adapted to particular leguminous plant species. The bacteria attach to the root of the host legume, usually at a root hair (Figure 27.3). In response to the bacterial infection, an indentation forms in the root hair and an *infection thread* synthesized by the plant passes down the root hair into the root itself. The bacteria follow this infection thread and enter the cells in the root. Inside these cells, the bacteria alter their morphology into larger forms called *bacteroids* that eventually pack the plant cell. The root cells are stimulated by this infection to form a tumorlike nodule of bacteroid-packed cells (Figure 27.3). Nitrogen is then fixed by a symbiotic process of the plant and the bacteria. The plant furnishes anaerobic conditions and growth nutrients for the bacteria, and the bacteria fix nitrogen to be incorporated into plant protein. Millions of tons of nitrogen are fixed in this way each year.

There are similar examples of symbiotic nitrogen fixation in nonleguminous plants, such as alder trees. These trees are among the first to appear in forests after fires or glaciation. The alder tree is symbiotically infected with an actinomycete (*Frankia*) and forms nitrogen-fixing root nodules. About 50 kg of nitrogen can be fixed each year by the growth of one acre of alder trees; the trees thus make a valuable addition to the forest economy.

Another important contribution to the nitrogen economy of forests is made by **lichens,** which are a symbiosis between a fungus and an alga or cyanobacterium (Figure 12.14). When one symbiont is a nitrogen-fixing cyanobacterium, the product is fixed nitrogen that eventually enriches the forest soil. Free-living cyanobacteria can fix significant amounts of nitrogen in desert soils after rains and on the surface of arctic tundra soils. Rice paddies can accumulate heavy growths of such nitrogen-fixing organisms. The cyanobacteria also form a symbiosis with a small floating fern, *Azolla,* which grows thickly in rice paddy waters. So much nitrogen is fixed by these organisms that other nitrogenous fertilizers are often unnecessary for rice cultivation.

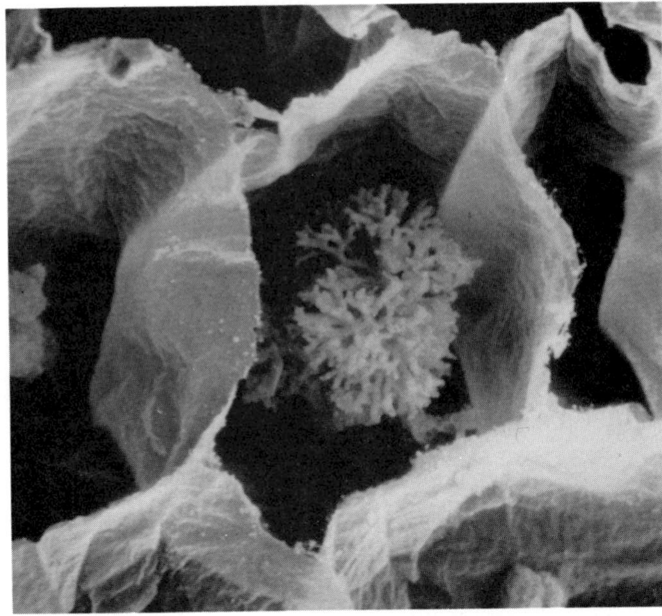

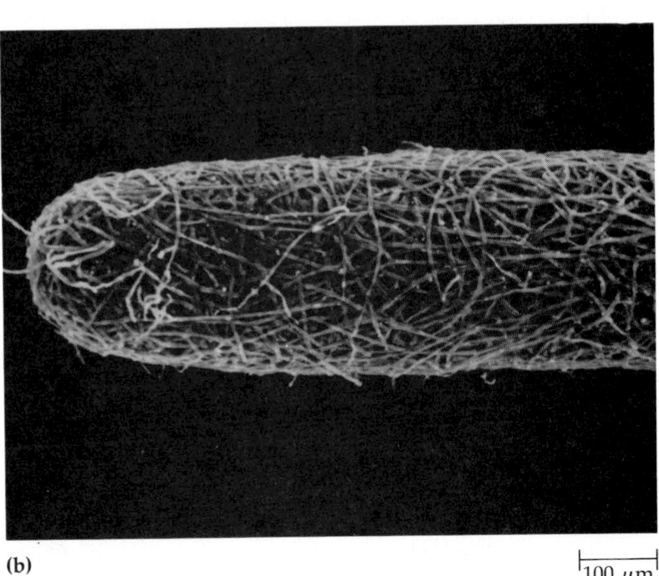

FIGURE 27.4 Scanning electron micrographs of mycorrhizae. (a) A fully developed arbuscule of an endomycorrhiza in a plant cell. (The term *arbuscule* means little bush.) As the arbuscule decomposes, it releases accumulated nutrients for the plant. **(b)** The mycelial mantle of a typical ectomycorrhizal fungus surrounding a eucalyptus tree root.

The Role of Mycorrhizae

A very important contribution to plant growth is made by **mycorrhizae** (*myco* means fungus; *rhiza* means root). There are two types of these fungi: *endomycorrhizae,* also known as *vesicular-arbuscular mycorrhizae,* and *ectomycorrhizae.* Both types function as root hairs on plants—that is, they extend the surface area through which the plant can absorb nutrients, especially phosphorus, which is not very mobile in soil. Vesicular-arbuscular mycorrhizae form large spores that can be isolated easily from soil by sieving. The hyphae from these germinating spores penetrate into the plant root and form two types of structures, vesicles and arbuscules. **Vesicles** are smooth oval bodies that are most probably storage structures. **Arbuscules** (Figure 27.4a) are formed inside plant cells. Nutrients travel from the soil through fungal hyphae to these arbuscules, which gradually break down and release the nutrients to the plants. Most grasses and other plants are surprisingly dependent on these fungi for proper growth, and their presence is nearly universal in the plant world.

Ectomycorrhizae mainly infect trees such as pine and oak. The fungus forms a mycelial *mantle* over the smaller roots of the tree (Figure 27.4b). Ectomycorrhizae do not form vesicles or arbuscules. Managers of commercial pine tree farms must take care to see that seedlings are inoculated with soil containing effective mycorrhizae. Truffles, known as a culinary delicacy, are ectomycorrhizae. In France, trained pigs are often used to find them by smell and root them up. To the female pig, the odor is that of a potential mate. In nature, the fungus is dependent upon ingestion by an animal, which distributes the undigested spores into new locations.

OTHER BIOGEOCHEMICAL CYCLES

Microorganisms are also important in cycles involving other elements, such as sulfur. In addition, there are important microbial transformations of potassium, iron, manganese, mercury, selenium, zinc, and other minerals. The various chemical reactions in these cycles are often essential to make the minerals available to plants in soluble form for their metabolism.

Degradation of Synthetic Chemicals in the Soil

We seem to take for granted that soil microorganisms will degrade materials entering the soil. Natural organic matter, such as falling leaves or animal residues, are in fact readily degraded. However, in this industrial age, there are many chemicals that do not occur in nature—such as plastics—that enter the soil in large amounts. Some plastics used for garbage bags now contain starch incorporated into their chemical structure. The starch is readily biodegradable, but after it is consumed, plastic polymers remain.

Many synthetic chemicals such as pesticides are highly resistant to degradation by microbial attack. A well-known example is the insecticide DDT. When DDT was first introduced, its property of *recalcitrance* (resistance to degradation) was considered quite beneficial because one application remained effective in the soil for an extended time. However, it was soon found that the chemical tended to accumulate and concentrate in parts of the food chain because of its solubility in fat. Eagles and other predatory birds accumulated DDT from contaminated food and suffered impaired reproductive ability—that is, their eggs had soft shells and broke during incubation.

Not all synthetic chemicals are as recalcitrant as DDT. Some are made up of chemical bonds and sub-

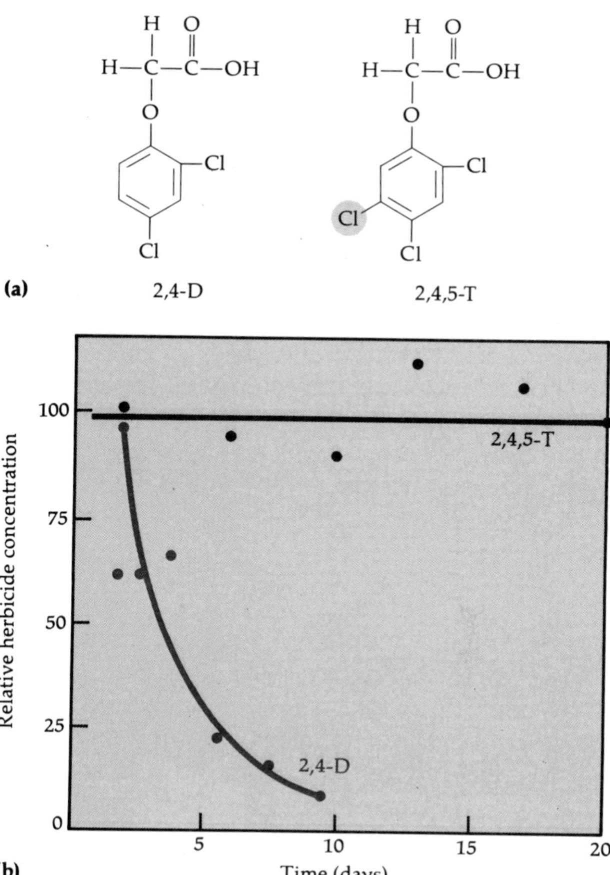

FIGURE 27.5 Slight structural differences often affect biodegradability. (a) Structures of the herbicides 2,4-D and 2,4,5-T. **(b)** Rates of microbial decomposition of 2,4-D and 2,4,5-T.

units that are subject to attack by bacterial enzymes. Small differences in chemical structure can make large differences in biodegradability. The classic example is that of two herbicides—2,4-D, the common chemical used to kill lawn weeds, and 2,4,5-T, which is used to kill shrubs. The addition of a single chlorine atom to the structure of 2,4-D extends its life in soil from a few days to an indefinite period (Figure 27.5).

A growing problem is the leaching into groundwaters of toxic materials that are not biodegradable or that degrade very slowly. The sources of these materials may include landfills, illegal industrial dumps, or pesticides applied to agricultural crops. Once groundwater becomes contaminated, the economic damage can be devastating. Researchers are developing processes and isolating bacteria that promote degradation or removal of these toxic materials. Examples of current work are in the boxes in Chapter 2, p. 38, and Chapter 9, p. 240.

Aquatic Microbiology and Sewage Treatment

Aquatic Microorganisms

Aquatic microbiology refers to the study of microorganisms and their activities in natural waters, such as lakes, ponds, streams, rivers, estuaries, and the sea. Large numbers of microorganisms in a body of water generally indicate high nutrient levels in the water. Water contaminated by inflows from sewage systems or from biodegradable industrial organic wastes is relatively high in bacterial counts. Similarly, ocean estuaries (fed by rivers) have higher nutrient levels and hence higher microbial counts than other shoreline waters.

In water, particularly in water with low nutrient concentrations, microorganisms tend to grow on stationary surfaces and on particulate matter. In this way, a microorganism has contact with more nutrients than if it were randomly suspended and floating freely with the current. Many bacteria whose main habitat is water have appendages and holdfasts that attach to various surfaces. One example is *Caulobacter* (see Figure 11.15a). Some bacteria also have gas vesicles that they can fill and empty to adjust buoyancy.

FRESHWATER MICROBIAL FLORA

Figure 27.6 shows a typical lake or pond that serves as an example to represent the various zones and the kinds of microbial flora found in a body of fresh water. The **littoral zone** along the shore has considerable

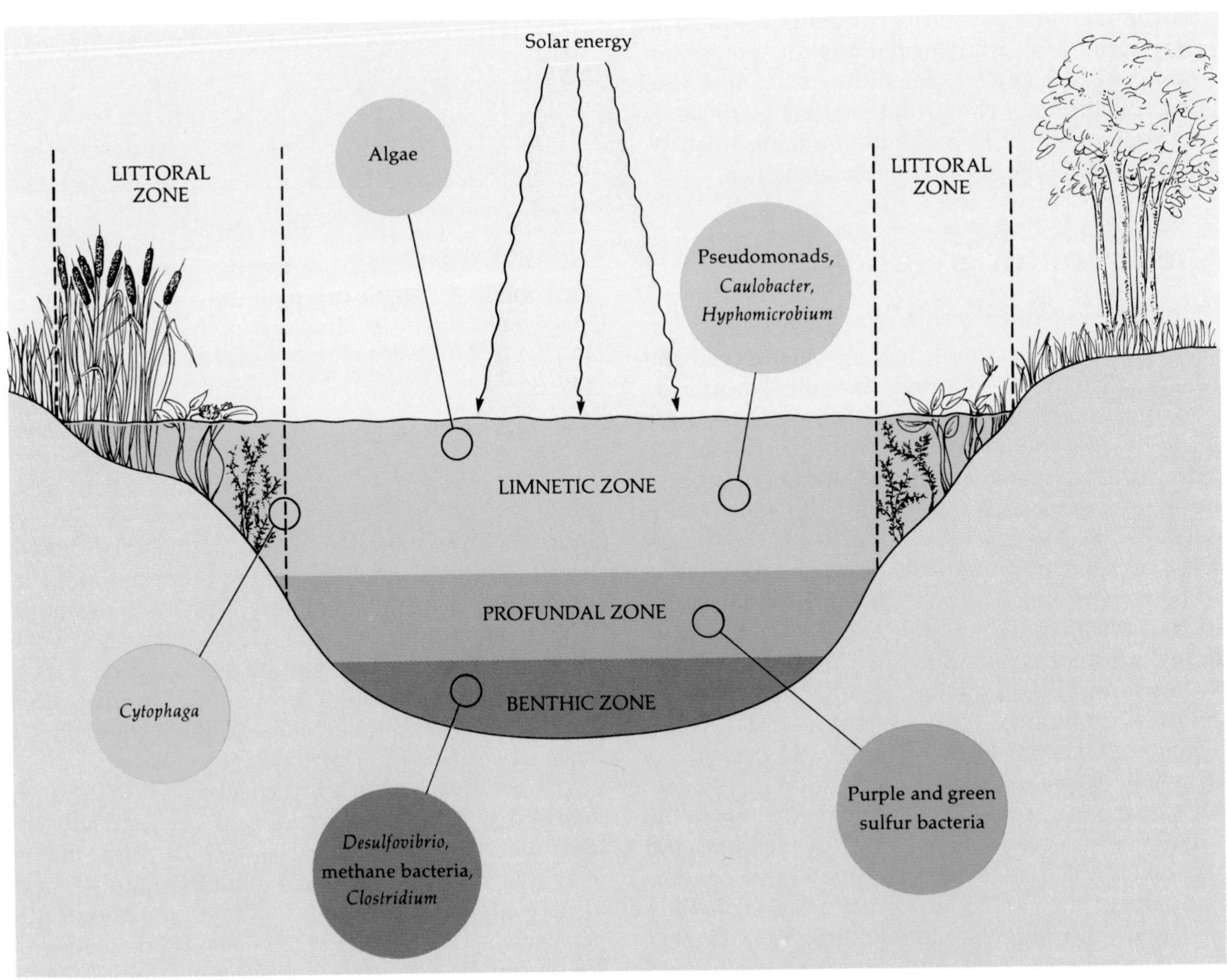

FIGURE 27.6 The zones of a typical lake or pond and some representative microorganisms of each zone. The microorganisms fill niches that vary in light, nutrients, and oxygen availability.

rooted vegetation, and light penetrates throughout it. The **limnetic zone** consists of the surface of the open water area away from the shore. The **profundal zone** is the deeper water under the limnetic zone. The **benthic zone** contains the sediment at the bottom.

Microbial populations of freshwater bodies tend to be affected mainly by the availability of oxygen and light. Light is, in many ways, the more important resource because photosynthetic algae are the main source of organic matter, and hence of energy, for the lake. These organisms are the *primary producers* of a lake that supports a population of bacteria, protozoans, fish, and other aquatic life. Photosynthetic algae are located in the limnetic zone.

Areas of the limnetic zone with sufficient oxygen contain pseudomonads and species of *Cytophaga* (sī-täf'äg-ä), *Caulobacter,* and *Hyphomicrobium.* Oxygen does not diffuse into water very well, as any aquarium owner knows. Microorganisms growing on nutrients in stagnant water quickly use up the dissolved oxygen in the water. In the oxygenless water, fish die, and odors (from hydrogen sulfide and organic acids, for example) are produced from anaerobic activity. Wave action in shallow layers, or water movement in rivers, tends to increase the amount of oxygen throughout the water and aid in the growth of aerobic populations of bacteria. Movement thus improves the quality of water and aids in the degradation of polluting nutrients.

Deeper waters of the profundal and benthic zones have low oxygen concentrations and less light. Algal growth near the surface often filters the light, and it is not unusual for photosynthetic microorganisms in deeper zones to use different wavelengths of light from those used by surface-layer photosynthesizers (see Figure 12.11). The purple and green sulfur bacteria are found in these deeper zones. These bacteria are anaerobic photosynthetic organisms that metabolize hydrogen sulfide to sulfur and sulfate in the bottom sediments of the benthic zone. The sediment also includes bacteria such as *Desulfovibrio* that use sulfate (SO_4^{2-}) as an electron acceptor and reduce it to hydrogen sulfide (H_2S), which is responsible for the rotten-egg odor of many lake muds.

Methane-producing bacteria are also part of these anaerobic benthic populations. In swamps, marshes, or bottom sediments, they produce methane gas (see the box, p. 675). *Clostridium* species are common in bottom sediments and may include botulism organisms, particularly those causing outbreaks of botulism in water-fowl.

SEAWATER MICROBIAL FLORA

The open ocean is relatively high in osmotic pressure, low in nutrients, and quite cold at great depths. The pH also tends to be higher than is optimal for most microorganisms. Bacterial populations in such waters tend to be much smaller than in estuaries and in small freshwater bodies fed by rivers and streams. Much of the microscopic life of the ocean is composed of photosynthetic diatoms and other algae. Largely independent of preformed organic nutrient sources, these microbes use energy from photosynthesis and atmospheric carbon dioxide for carbon. These organisms constitute the marine **phytoplankton** community, the basis of the oceanic food chain. Ocean bacteria benefit from the eventual death and decomposition of phytoplankton and also attach to their living bodies. Protozoans in turn feed on bacteria and the smaller phytoplankton. Krill—shrimplike crustaceans—feed on the phytoplankton and in turn are an important food supply of larger sea life. Many fish and whales are also able to feed directly on the phytoplankton.

Microbial luminescence is an interesting minor aspect of deep-sea life. Certain bacteria produce light flashes when they are agitated by wave action or by a boat's wake. Many bacteria are also luminescent, and some have established symbiotic relationships with benthic-dwelling fish. These fish sometimes use the glow of their resident bacteria as an aid in attracting and capturing prey in the complete darkness of the ocean depths. These bioluminescent organisms have an enzyme called luciferase that picks up electrons from flavoproteins in the electron transport chain and then emits some of the electron's energy as a photon of light.

The Role of Microorganisms in Water Quality

Water in nature is seldom totally pure. Even rainfall is contaminated as it falls to Earth. For example, combustion of fossil fuels puts sulfur compounds in the air. These encounter the falling rain and form sulfurous acid—the notorious acid rain. Water quality is measured by the degree of its contamination by soluble chemicals, suspended colloids such as clay particles (turbidity), and microbes.

WATER POLLUTION

The form of water pollution that will be our primary interest is microbial pollution, especially by pathogenic organisms.

Transmission of Infectious Diseases

Water that moves below the ground's surface undergoes a filtering that tends to remove microorganisms. For this reason, water from springs and deep wells is generally of good quality. The most dangerous form of water pollution occurs when human feces enter the

TABLE 27.2 Waterborne Disease Outbreaks in Public Water Systems, 1986 to 1988

AGENT	NUMBER OF OUTBREAKS	NUMBER OF CASES
Shigella spp.	4	2733
Salmonella spp.	2	70
Campylobacter spp.	1	250
Giardia lamblia (cyst-forming protozoan)	9	1169
Cryptosporidium spp. (sporeforming protozoan)	1	13,000
Acute gastrointestinal illness of unknown etiology	24	2975

Source: *Surveillance Summaries, MMWR* 39:SS-1 (March 1990).

water supply. Many diseases are perpetuated by the fecal–oral route of transmission, in which a pathogen shed in human feces contaminates water or food and is later ingested (see Chapter 25). Typical of these diseases are typhoid fever and cholera, which are caused by bacteria, and hepatitis A, which is caused by a virus. An especially troublesome waterborne disease is diarrhea caused by ingestion of cysts of the protozoan *Giardia lamblia*. This protozoan is a problem not only in municipal water systems but also in mountain streams, which are contaminated by beavers and other animals. Mollusks that feed by filtering tend to concentrate waterborne viruses and bacteria in their tissues, so mollusks from polluted water may be dangerous to eat.

Not all pathogens must be ingested to cause disease. For example, the helminthic disease schistosomiasis is spread among persons who swim or wade in waters contaminated by human wastes. The pathogens are not usually ingested but are in the form of swimming cercaria that bore through the skin.

As good sanitation practices have nearly eliminated a few diseases, such as typhoid fever and cholera, in the United States, attention has focused on other waterborne diseases. Note in Table 27.2 that most outbreaks of waterborne bacterial diseases are now caused by *Shigella* species, which cause shigellosis (bacterial dysentery). There is some anxiety about pathogens such as *Legionella*, the cause of a pneumonia often spread by inhalation of water aerosols. *Mycobacterium* spp., *Pseudomonas* spp., and other opportunistic pathogens that are dangerous to immunosuppressed individuals are being found more and more often in tap water.

Chemical Pollution

Preventing chemical contamination of water is a difficult problem. Industrial and agricultural chemicals leached from the land enter water in great amounts and in forms that are resistant to biodegradation. Many of these chemicals become biologically concentrated in some of the organisms in the food chain.

A striking example of industrial water pollution involved mercury, used in the manufacture of paper. The metallic mercury was allowed to flow into waterways as waste. It was assumed that the mercury was inert and would remain segregated in the sediments. However, bacteria in the sediments converted the mercury into a soluble chemical compound, methyl mercury, which was then taken up by fish and invertebrates in the waters. When such seafood is a substantial part of the human diet, the mercury concentrations can accumulate with devastating effects on the nervous system.

Another example of chemical pollution is the synthetic detergents developed immediately after World War II. These rapidly replaced many of the soaps then in use. Because these new detergents were not biodegradable, they rapidly accumulated in the waterways. In some rivers, large rafts of detergent suds could be seen traveling downstream; in some cities, a small head of bubbles might appear on the surface of a glass of water. These detergents were replaced in 1964 by new biodegradable formulations.

The new detergents, however, brought new problems. Substantial amounts of phosphates were added to many of these detergents to improve their effectiveness. Unfortunately, phosphates pass virtually unchanged through most sewage treatment systems and can cause **eutrophication** (*eu* means well; *troph* means nourish) of lakes and streams—an overabundance of nutrients, resulting in overgrowth of algae or cyanobacteria and the eventual death of other organisms.

To understand how phosphate pollution leads to eutrophication, we must first consider the nutritional requirements of algae and cyanobacteria. These organisms get their energy from sunlight and their carbon from carbon dioxide dissolved in the water. Most waters contain adequate amounts of the minerals required for algal growth, except for nitrogen and phosphorus. Both of these nutrients can enter water from domestic, farm, and industrial wastes when waste treatment is absent or inefficient. These additional nutrients cause dense aquatic growths called **blooms.** Since many cyanobacteria are able to fix nitrogen from the atmosphere, these photosynthesizing organisms require only traces of phosphorus to initiate blooms. Once eutrophication results in blooms of algae or cyanobacteria, the eventual effect is the same as the addition of equivalent amounts of organic matter. But in this case, blooms generate additional organic matter by photosynthesis and nitrogen fixation from the atmosphere. In the short run, these algae and cyanobacteria produce oxygen. However, they eventually die and are degraded by bacteria. During the degradation process the oxygen in the water is used up, which may kill the fish. Undegraded remnants of organic matter settle to the bottom and hasten the filling of the lake.

Municipal waste containing detergents is likely to be the main source of phosphates in lakes and streams because soil phosphorus is relatively insoluble and tends to be retained in soil. Because of this, phosphate-containing detergents are banned in some localities.

Oil spills on land and at sea represent some of the most dramatic examples of chemical pollution. Bacteria that are especially effective in degrading petroleum products can be selected and applied to such spills (Figure 27.7). Oil is rich in carbon but low in nutrients, such as nitrogen and phosphorus, that are essential for microbial growth. If oil-soluble "fertilizer" containing these nutrients is spread over oil spills along with the adapted bacteria, the oil is metabolized efficiently. Indigenous bacteria capable of utilizing the spilled oil are also stimulated into activity by the additional nutrients. Oil also requires oxygen for its metabolism; therefore, in cases when oil is spilled on land, vigorous tilling of the soil to maximize aeration is necessary.

Coal-mining wastes, particularly in the eastern United States, are very high in sulfur content, mostly iron sulfide (FeS_2). In the process of obtaining energy from the oxidation of the ferrous ion (Fe^{2+}), bacteria such as *Thiobacillus ferrooxidans* convert the sulfide into sulfate (see the box in Chapter 28, p. 715). The sulfate enters streams as sulfuric acid, which lowers the pH of the water and damages aquatic life. The low pH also promotes the formation of insoluble iron hydroxides, which form the yellow precipitates often seen clouding such polluted waters.

(a)

(b)

FIGURE 27.7 The use of bioremediation to clean up an oil spill. **(a)** A rocky beach at Green Island, Alaska, contaminated with oil spilled from the Exxon *Valdez.* **(b)** The same beach after three applications of carbon-free nutrients (fertilizer) to encourage microbial activity.

TESTS FOR WATER PURITY

Historically, most of our concern about water purity has been related to the transmission of disease. Therefore, tests have been developed to determine the safety of water; many of these tests are also applicable to foods.

It is not practical, however, to look only for pathogens in water supplies. For one thing, if we were to find the pathogen causing typhoid or cholera in the water system, the discovery would already be too late to prevent an outbreak of the disease. Moreover, such pathogens would probably be present only in small numbers and might not be included in tested samples.

The tests for water safety in use today are aimed instead at detecting particular **indicator organisms.** There are several criteria for an indicator organism. The most important criterion is that the organism be consistently present in human feces in substantial numbers so that its detection will be a good indication that human wastes are entering the water. The indicator organisms should also survive in the water at least as well as the pathogenic organisms would. The indicator organisms must also be detectable by simple tests that can be carried out by persons with relatively little training in microbiology.

In the United States, the usual indicator organisms are the **coliform bacteria.** Coliforms are defined as aerobic or facultatively anaerobic, gram-negative, nonendospore-forming, rod-shaped bacteria that ferment lactose to form gas within 48 hours of being placed in lactose broth at 35°C. Because some coliforms are not solely enteric bacteria but are more commonly found in plant and soil samples, many standards for food and water specify the determination of *fecal coliforms.* The predominant fecal coliform is *E. coli,* which constitutes a large proportion of the human intestinal population. There are specialized tests to distinguish between fecal coliforms and nonfecal coliforms. It is important to note that coliforms are not themselves pathogenic under normal conditions, although they can cause diarrhea and opportunistic urinary tract infections.

The United States Environmental Protection Agency Drinking Water Standards specify the minimum number of water samples to be examined each month and the maximum number of coliform organisms permitted in each 100 ml of water. Figure 6.16 shows the most probable number (MPN) method for determining the number of bacteria in a 100-ml sample; Appendix C provides the MPN table used for calculations. For detection and enumeration of coliforms, selective and differential media are required, and one of two specialized methods is used. These are summarized in Figure 27.8.

An increasingly popular method of detecting *E. coli* makes use of an enzyme, β-glucuronidase (GUD), produced by almost all strains. If *E. coli* is added to a medium containing 4-methylumbelliferone-β-D-glucuronide (MUG), GUD converts MUG to a product that is visible when the medium is illuminated by ultraviolet light. Results can be obtained in less than a day. A considerable number of salmonellae and shigellae are also positive for GUD, but they can be eliminated by including MUG in coliform-selective media.

Coliforms have been very useful as indicator organisms in water sanitation, but they have limitations. One problem has been the growth of coliform bacteria in *biofilms* on the inner surfaces of water pipes. Their presence in tap water has led to a number of community orders to boil water even though these coliforms are not necessarily related to fecal contamination and there is no evidence that they have a significant effect on public health. A more serious problem is that some pathogens are even more resistant than coliforms to chemical disinfection—in their order of increasing resistance to chlorination, viruses and protozoan cysts. By using sophisticated methods of detecting viruses, it has been found that chemically disinfected water samples that are free of coliforms are often contaminated with enteric viruses. The cysts of *G. lamblia* are so resistant to chlorination that eliminating them by this method is probably impractical. Mechanical methods are probably necessary, such as filtration and flocculation to remove colloidal particles.

Concern for these problems has led to recommendations for revision of drinking water regulations. These call for a dual barrier treatment of all drinking water. Surface waters should be treated first by flocculation and then by sand or diatomaceous earth filtration to remove cysts and turbidity; these processes would remove most bacteria and viruses as well. This barrier treatment would be followed by chemical disinfection to inactivate any remaining pathogens. These purification methods are described in the next section.

WATER TREATMENT

When water is obtained from uncontaminated reservoirs fed by clear mountain streams or from deep wells, it requires minimal treatment to make it safe to drink. Many cities, however, obtain their water from badly polluted sources, such as rivers that have received municipal and industrial wastes upstream. Very turbid (cloudy) water is allowed to stand in a holding reservoir for a time to allow as much particulate suspended matter as possible to settle out (Figure 27.9). The water then undergoes **flocculation treatment,** that is, removal of colloidal materials such as clay, which would otherwise remain in suspension indefinitely. A flocculant chemical, such as aluminum potassium sulfate (alum), forms aggregations of fine suspended particles called floc. As these aggregations slowly settle out, they entrap colloidal material and carry it to the bottom. Large numbers of viruses and bacteria are also removed by this treatment. Alum was used to clear muddy river water during the first half of the nineteenth century in the military forts of the American West, long before the germ theory of disease was developed. There were also observations at the time in European cities that people who used water filtered through sand to remove turbidity had a lower incidence of cholera during outbreaks.

After flocculation treatment, water is passed through beds of sand or diatomaceous earth to accomplish **sand filtration.** As we mentioned previously, some protozoan cysts, such as those of *G. lamblia,* appear to be removed from water only by such filtration treatment. The microorganisms are trapped mostly by surface adsorption in the sand beds. They do not penetrate the tortuous routing of the sand beds, even though the openings might be larger than the organisms that are filtered out. These sand filters are periodically backflushed to clear them of accumulations. Water systems of cities that have an exceptional con-

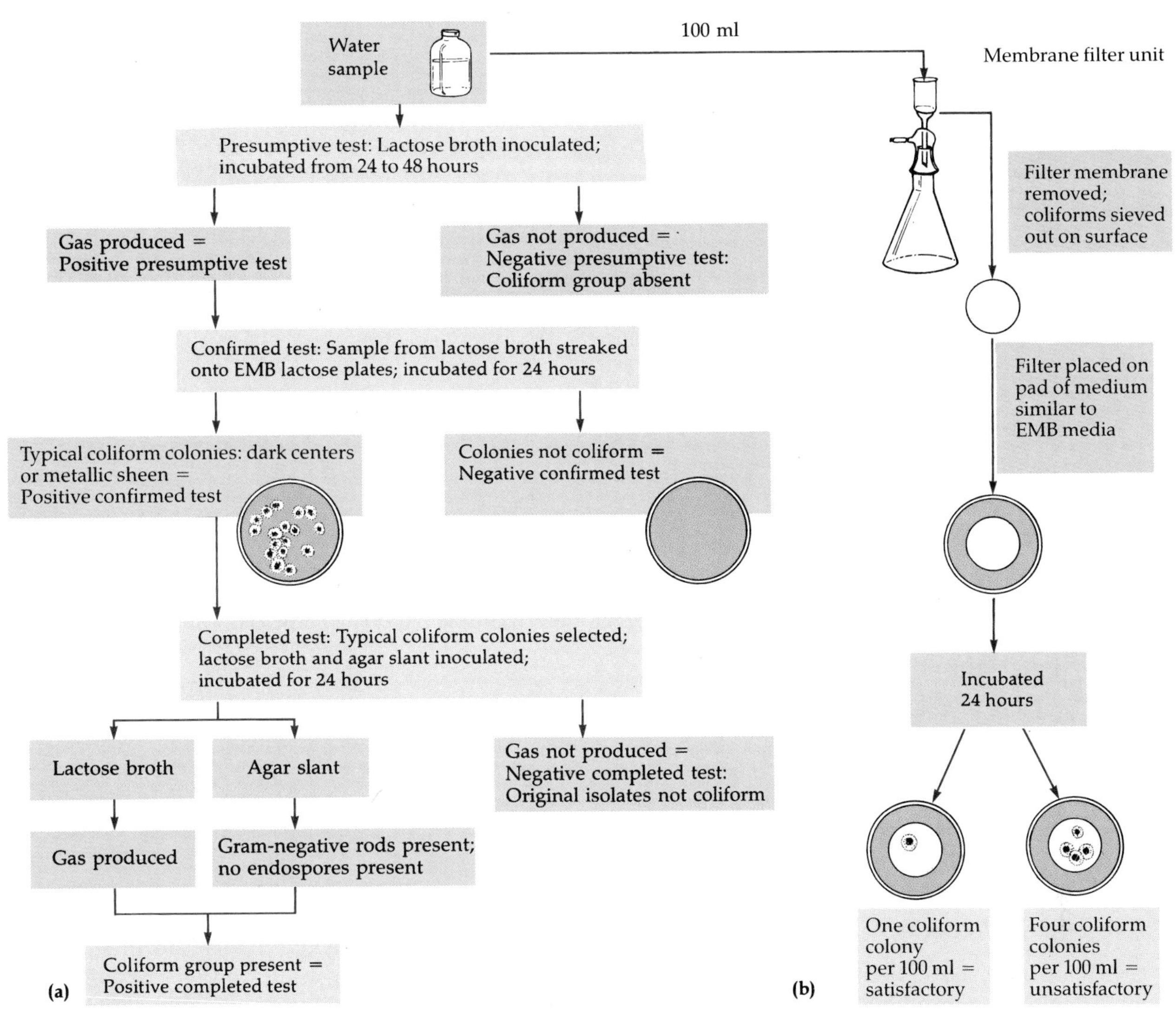

FIGURE 27.8 Analysis of drinking water for coliforms. **(a)** Note how the series of tests uses the characteristics associated with coliforms—the presence of gas from lactose fermentation and gram-negative rods. Coliform numbers are estimated by inoculating a number of tubes of lactose broth. The percentage of tubes positive for coliforms is a statistical indication of the number of coliforms in the sample. **(b)** Many laboratories use the membrane filtration method for coliforms. The coliform colonies have a distinctive appearance (dark centers or metallic sheen).

cern for toxic chemicals supplement sand filtration with filters of activated charcoal (carbon). Charcoal has the advantage of removing not only particulate matter but also some dissolved organic chemical pollutants.

Before entering the municipal distribution system, the filtered water is chlorinated. Because organic matter neutralizes chlorine, the plant operators must pay constant attention to maintaining effective levels of chlorine. There has been some concern that chlorine itself might be a health hazard, that it might react with organic contaminants of the water to form carcinogenic compounds. At present, this possibility is considered minor when compared with the proven usefulness of chlorination of water.

As we noted in Chapter 7, one substitute for chlorination is ozone treatment. Ozone (O_3) is a highly reactive form of oxygen that is formed by electrical spark discharges and ultraviolet light. (The fresh odor of air following an electrical storm or around an ultraviolet light bulb is from ozone.) Ozone for water treatment is

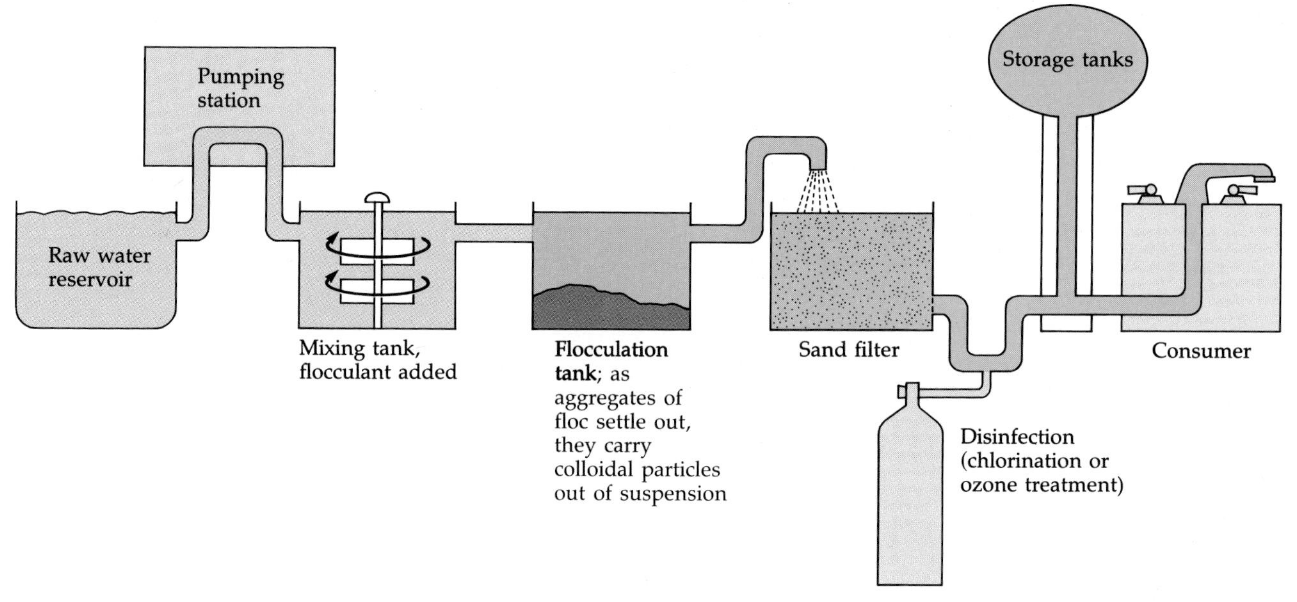

FIGURE 27.9 Steps of water treatment in a typical municipal water purification plant.

generated electrically at the site of treatment. Use of ultraviolet light is also a possible alternative to chemical disinfection. Arrays of ultraviolet tube lamps are arranged in quartz tubes so that water flows close to the lamps. This is necessary because of the low penetrating power of ultraviolet radiation.

SEWAGE TREATMENT

After water has been used, it becomes sewage. Sewage includes all the water from a household that is used for washing, as well as toilet wastes. Rainwater flowing into street drains and some industrial wastes enter the sewage systems in some cities. Sewage is mostly water and contains little particulate matter, perhaps only about 0.03%. Even so, in large cities, this solid portion of sewage can total more than 1000 tons of solid material per day.

Until environmental awareness intensified, a surprising number of large cities in this country had only rudimentary sewage treatment systems or no system at all. Raw sewage, untreated or nearly so, was simply discharged into rivers or oceans. A flowing, well-aerated stream is capable of considerable self-purification. Therefore, until increases in populations and their wastes exceeded this capability, casual treatment of municipal wastes caused little complaint. In the United States, most methods of simple discharge have been improved.

Primary Treatment

The usual first step in sewage treatment is called **primary treatment** (Figure 27.10). In this process, incoming sewage receives preliminary treatment—large floating materials are screened out, the sewage is allowed to flow through settling chambers so that sand and similarly gritty material can be removed, skimmers remove floating oil and grease, and floating debris are shredded and ground. After this step, the sewage passes through sedimentation tanks, where solid matter settles out. (The design of these primary settling tanks varies.) Sewage solids collecting on the bottom are called **sludge;** sludge at this stage is called *primary sludge.* From 40% to 60% of suspended solids are removed from sewage by this settling treatment, and flocculating chemicals that increase the removal of solids are sometimes added at this stage. Biological activity is not particularly important in primary treatment, although some digestion of sludge and dissolved organic matter can occur during long holding times. The sludge is removed on either a continuous or an intermittent basis, and the effluent (the liquid flowing out) then undergoes secondary treatment.

Biochemical Oxygen Demand

Primary treatment removes approximately 25% to 35% of the **biochemical oxygen demand (BOD)** of the sewage. An important concept in sewage treatment and in the general ecology of waste treatment, BOD is a measure of the biologically degradable organic matter in water. BOD is determined by the amount of oxygen required by bacteria to metabolize the organic matter. The classic method of measurement is to use special bottles with airtight stoppers. Each bottle is first filled with the test water or dilutions of the test water. The

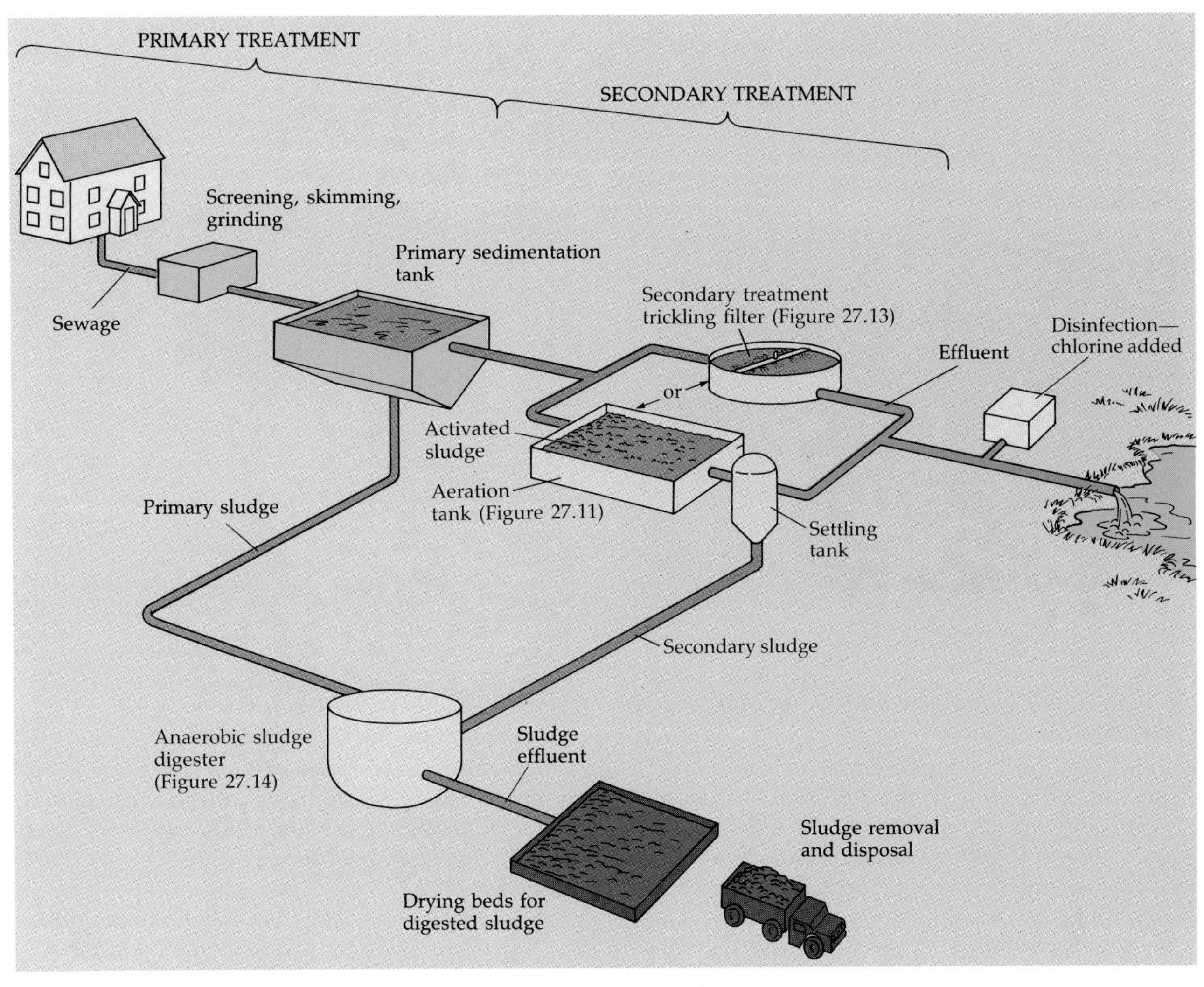

FIGURE 27.10 Steps of typical sewage waste treatment. A particular system would use either activated sludge aeration tanks or trickling filters, not both, as shown in this figure. The sludge is disposed of in landfills or agricultural land. Microbial activity occurs aerobically in trickling filters or in activated sludge aeration tanks and anaerobically in the anaerobic sludge digester.

water is initially aerated to provide a relatively high level of dissolved oxygen and is seeded with bacteria if necessary. The filled bottles are then incubated in the dark for five days at 20°C, and the decrease in dissolved oxygen is determined by a chemical or electronic testing method. The more oxygen that is used up as the bacteria degrade the organic matter in the sample, the greater the BOD—which is usually expressed in milligrams of oxygen per liter of water. The amount of oxygen that normally can be dissolved in water is only about 10 mg/liter. Typical BOD values of waste water may be twenty times this amount. If this waste water enters a lake, for example, bacteria in the lake begin to consume the organic matter responsible for the high BOD, rapidly depleting the oxygen in the lake water.

Secondary Treatment

After primary treatment, the greater part of the BOD remaining in the sewage is in the form of dissolved organic matter. **Secondary treatment,** which is primarily biological, is designed to remove most of this organic matter and reduce the BOD (Figure 27.10). In this process, the sewage undergoes strong aeration to encourage the growth of aerobic bacteria and other microorganisms that oxidize the dissolved organic

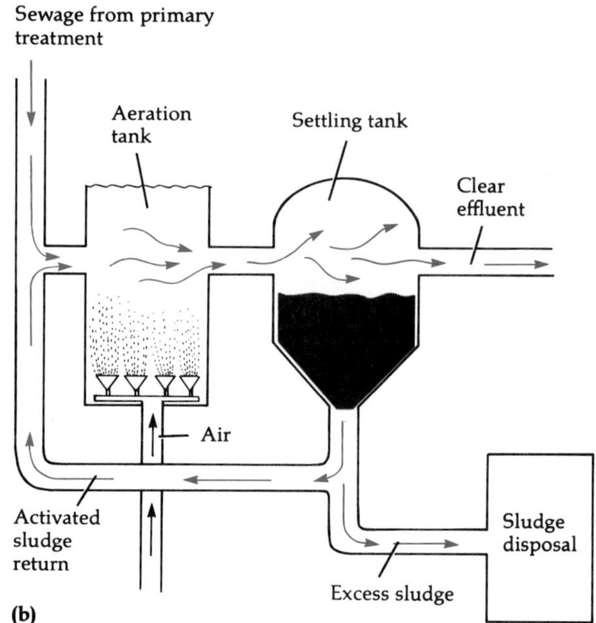

FIGURE 27.11 Activated sludge system of secondary treatment. **(a)** An aeration tank. Note that the surface is frothing from aeration. **(b)** Diagram of an activated sludge system.

matter to carbon dioxide and water. Two commonly used methods of secondary treatment are activated sludge systems and trickling filters.

In the aeration tanks of the **activated sludge system,** air or pure oxygen is added to the effluent from primary treatment (Figure 27.11). The sludge in the effluent contains large numbers of metabolizing bacteria, together with yeasts, molds, and protozoans. An especially important ingredient of the sludge are species of *Zoogloea* bacteria, which form flocculant masses (floc) in the aeration tanks (Figure 27.12). The activity of these aerobic microorganisms oxidizes much of the effluent's organic matter into carbon dioxide and water. When the aeration phase is completed, the floc (secondary sludge) is allowed to settle to the bottom, just as the primary sludge settles in primary treatment.

Soluble organic matter in the sewage is adsorbed onto the floc and is incorporated into microorganisms in the floc. As the floc settles out, this organic matter is removed with the floc and is subsequently treated in an anaerobic sludge digester. More organic matter is probably removed by this process than by the relatively short-term aerobic oxidation.

Most of the settled sludge is removed for treatment in an anaerobic sludge digester; some of the sludge is recycled to the activated sludge tanks as a starter culture for the next sewage batch. The effluent water is sent on for final treatment. Occasionally, when aeration is stopped, the sludge will float rather than settle out; this phenomenon is called *bulking*. When this happens, the organic matter in the floc flows out with the discharged effluent and often causes serious problems of local pollution. A considerable amount of research has been devoted to the causes of bulking and its possible prevention. It is apparently caused by the growth of filamentous bacteria of various types; the sheathed bacterium *Sphaerotilus natans* is often mentioned as the primary offender. Activated sludge systems are quite efficient: they remove from 75% to 95% of the BOD from sewage.

Trickling filters are the other commonly used method of secondary treatment. In this method, the sewage is sprayed over a bed of rocks or molded plastic (Figure 27.13). The rocks or other components of the bed must be large enough so that air penetrates to the bottom but small enough to maximize the surface area available for microbial activity. Actually, there is no filtering action. A slimy, gelatinous film of aerobic microorganisms grows on the rocks' surfaces. In many ways, this film is functionally similar to the organisms found in activated sludge systems. Because air circulates throughout the rock bed, these aerobic microorganisms in the slime layer are able to oxidize much of the organic matter trickling over the surfaces into carbon dioxide and water. Trickling filters remove from 80% to 85% of the BOD, so they are generally less efficient than activated sludge systems. However, trickling filters are usually less troublesome to operate and have fewer problems from overloads or toxic sewage.

Treated sewage is disinfected, usually by chlorination, before being discharged. Not all microbes are necessarily killed by chlorination. The number of coliforms allowed in the effluent depends on where it is discharged. The discharge is usually into an ocean or into flowing streams, although spray irrigation fields are sometimes used to avoid phosphorus and bacterial

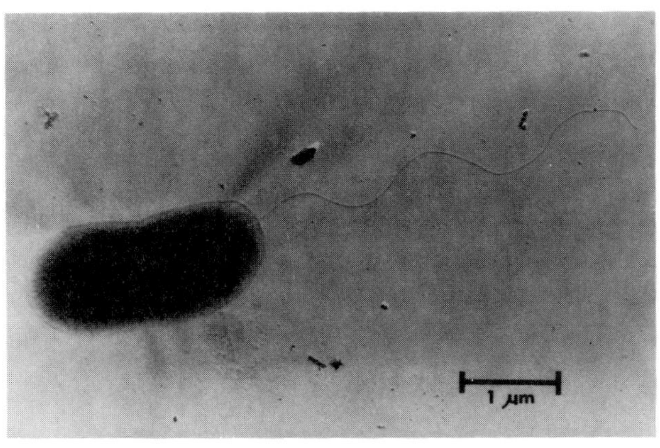

(a)

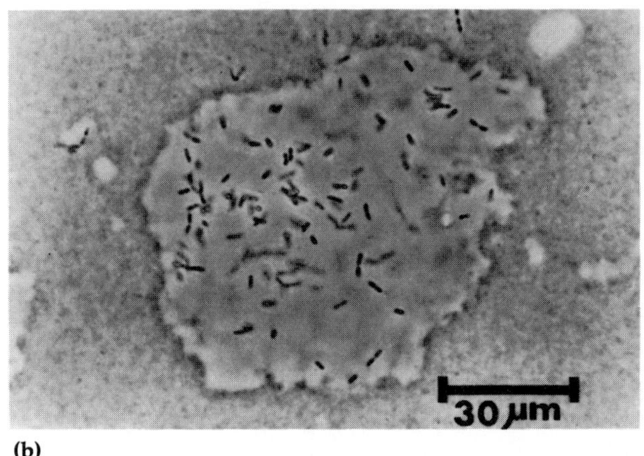

(b)

FIGURE 27.12 ***Zoogloea,* the organism that contributes to floc formation in activated sludge systems.** **(a)** *Zoogloea ramigera* cell with a single polar flagellum. **(b)** Particle of floc formed by *Z. ramigera*. Note the cells embedded in gelatinous matter synthesized by the bacteria.

contamination of waterways. For ocean discharge, 240 coliforms per 100 ml of water may be allowed, but in a smaller body of water coliforms might have to be reduced to 23 per 100 ml.

In cities with limited freshwater supplies, there is already some recycling of sewage into drinking water. This is safe and practical and will probably become commonplace in the future.

Sludge Digestion

Primary sludge accumulates in primary sedimentation tanks; sludge also accumulates in activated sludge and in trickling filter secondary treatments. For further treatment, these sludges are often pumped to **anaerobic sludge digesters** (Figure 27.14). The process of sludge digestion is carried out in large tanks, from which oxygen is almost completely excluded. In sec-

(a)

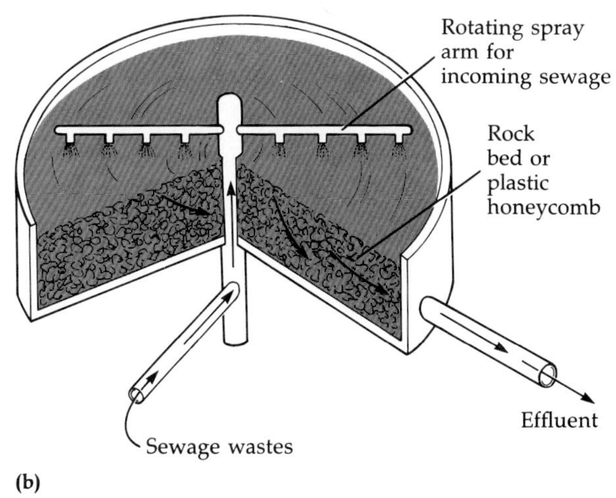

(b)

FIGURE 27.13 Trickling filter method of secondary treatment. **(a)** The sewage is sprayed from the system of rotating pipes onto a bed of rocks or plastic honeycomb designed to have a maximum surface area and allow oxygen to penetrate deeply into the bed. Microorganisms grow on the enormous surface area, forming a microbial slime layer that aerobically metabolizes the organic matter in the sewage trickling down through the bed. **(b)** Diagram of a trickling filter system.

(a)

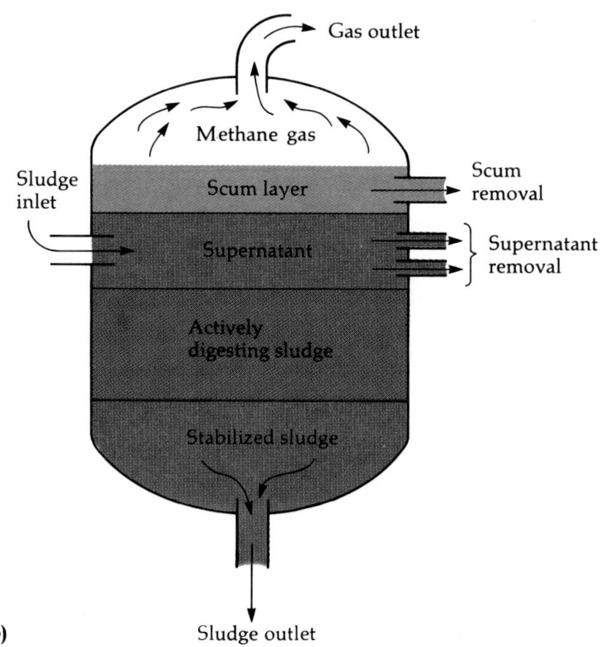

(b)

FIGURE 27.14 Sludge digestion. **(a)** An anaerobic sludge digester at a California treatment plant. Much of a typical digester is below ground level. In colder climates, the above-ground part of the tank is insulated by mounded earth. Methane from such a digester is often used to run pumps or heaters in the treatment plant. **(b)** Diagram of a sludge digester. The scum layer and supernatant layers are low in solids and are recirculated through secondary treatment.

ondary treatment, emphasis is placed on the maintenance of aerobic conditions so that organic matter is converted to carbon dioxide, water, and solids that can settle out. By comparison, anaerobic fermentations are metabolically inefficient, and the microorganisms leave large amounts of only partially digested organic materials in the form of fatty acids, alcohols, and similar products that retain much of their original BOD.

An anaerobic sludge digester is designed to encourage the growth of anaerobic bacteria, especially methane-producing bacteria that decrease organic solids by degrading them to soluble substances and gases, mostly methane (60% to 70%) and carbon dioxide (20% to 30%). The methane is routinely used as a fuel for heating the digester and is also frequently used to run power equipment in the plant.

There are essentially three stages in the activity of an anaerobic sludge digester. The first stage is the production of carbon dioxide and organic acids from anaerobic fermentation of the sludge by various anaerobic and facultatively anaerobic microorganisms. In the second stage, the organic acids are metabolized to form considerable hydrogen and carbon dioxide as well as organic acids, such as acetic acid. These products are the raw materials for a third stage, in which the methane-producing bacteria produce methane (CH_4). Most of the methane is derived from the energy-yielding reduction of carbon dioxide by hydrogen gas:

$$CO_2 + 4H_2 \longrightarrow CH_4 + 2H_2O$$

Other methane-producing microbes split acetic acid (CH_3COOH) to yield methane and carbon dioxide:

$$CH_3COOH \longrightarrow CH_4 + CO_2$$

Methane and carbon dioxide are relatively innocuous end-products, comparable to the carbon dioxide and water from aerobic treatment. However, considerable amounts of undigested sludge still remain, although it is relatively stable and inert. To reduce its volume, this sludge is pumped to shallow drying beds or filters. It is then carried away for disposal. It can be used for landfill or as a soil conditioner, or it can be incinerated. Sludge has about one-fifth of the value of normal commercial lawn fertilizers but has desirable soil-conditioning qualities, much as soil humus does.

Septic Tanks

Homes and businesses in areas of low population density that are not connected to municipal sewage systems often use a **septic tank** (Figure 27.15), a device whose operation is similar in principle to primary treatment. Sewage enters a holding tank, and suspended solids settle out. The sludge in the tank must be pumped out periodically and disposed of. The effluent flows through a system of perforated piping into a leaching (soil drainage) field. The effluent entering the soil is decomposed by soil microorganisms. These systems work well when not overloaded and when the drainage system is properly sized to the load and soil

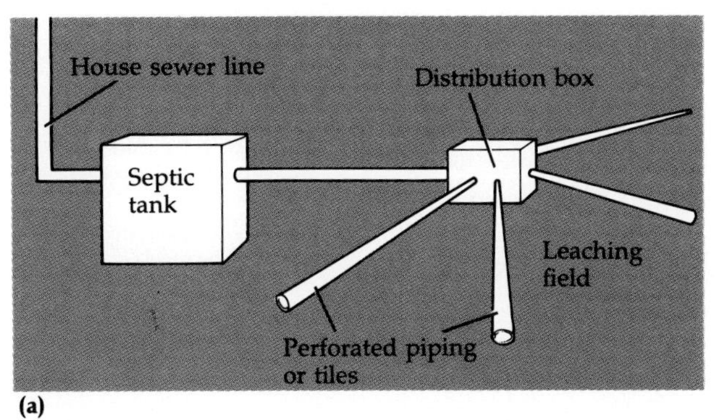

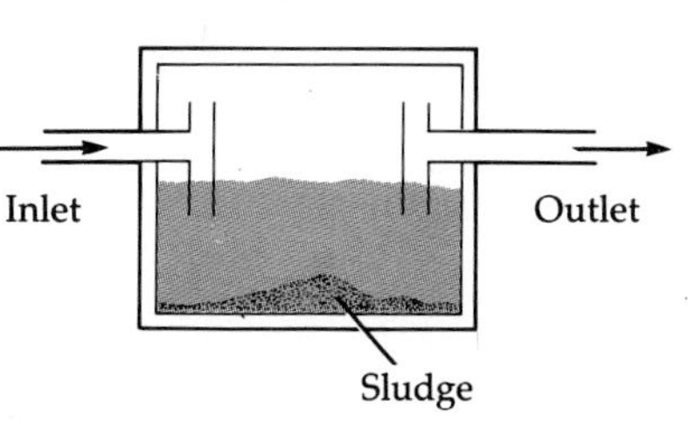

FIGURE 27.15 Septic tank system. **(a)** Overall plan. Solids settle out in the septic tank as sludge; most soluble organic matter is disposed of by percolation into soil. **(b)** The inside of a septic tank. The sludge is pumped out at periodic intervals.

type. Heavy clay soils require extensive drainage systems because of the soil's poor permeability. The high porosity of sandy soils can allow chemical or bacterial pollution of nearby water supplies.

Oxidation Ponds

Many small communities and many industries use **oxidation ponds,** also called **lagoons** or **stabilization ponds,** for water treatment. These are inexpensive to build and operate but require considerable land. Designs vary, but most incorporate two stages. The first stage is analogous to primary treatment—the sewage pond is deep enough that conditions are almost entirely anaerobic. Sludge settles out in this stage. In the second stage, which corresponds to secondary treatment, effluent is pumped into an adjoining pond or system of ponds that are shallow enough to be aerated by wave action. Because it is difficult to maintain aerobic conditions for bacterial growth in ponds with so much organic matter, the growth of algae is encouraged to produce oxygen. Bacterial action in decomposing the organic matter in the wastes generates carbon dioxide. Algae, which use carbon dioxide in their photosynthetic metabolism, grow and produce oxygen, which in turn encourages activity of aerobic microorganisms in the sewage. Considerable amounts of organic matter in the form of algae accumulate, but this is not a problem because the oxidation pond, unlike a lake, already has a large nutrient load.

Some small sewage-producing operations, such as isolated campgrounds or highway rest stop areas, use an **oxidation ditch** for sewage treatment. In this method, a small oval channel in the shape of a race track is filled with sewage water. A paddle wheel similar to that on a Mississippi steamboat propels the water in a self-contained flowing stream aerated enough to oxidize the wastes.

Tertiary Treatment

As we have seen, primary and secondary treatments of sewage do not remove all the biologically degradable organic matter. Amounts of organic matter that are not excessive can be added to a flowing stream without causing a serious problem. Eventually, however, the pressures of increased population might increase wastes beyond a body of water's carrying capacity, and additional treatments might be required. Even now, primary and secondary treatments are inadequate in certain situations, such as when the effluent is discharged into small streams or recreational lakes. Some communities have therefore developed tertiary treatment plants. Lake Tahoe in the Sierra Nevada, surrounded by extensive development, is the site of one of the best-known tertiary systems.

The effluent from secondary treatment plants contains some residual BOD. It also contains about 50% of the original nitrogen and 70% of the original phosphorus—which can greatly affect a lake's ecosystem. Tertiary treatment is designed to remove essentially all the BOD, nitrogen, and phosphorus. Tertiary treatment depends less on biological treatment than on physical and chemical treatments. Some systems encourage denitrifying bacteria to form volatile nitrogen gas. Nitrogen is converted to ammonia and evaporates into the air in stripping towers. Phosphorus is precipitated out by combining with such chemicals as lime, alum, and ferric chloride. Filters of fine sands and activated charcoal remove small particulate matter and dissolved chemicals. Finally, chlorine is added to the purified water to kill or inhibit any remaining microorganisms and to oxidize any remaining odor-producing substances.

Tertiary treatment provides water that is suitable for drinking, but the process is extremely costly. Secondary treatment is less costly, but water that has undergone only secondary treatment contains many substances that pollute water. Much work is being done to design secondary treatment plants in which the effluent can be used for irrigation. This design would eliminate a source of water pollution, provide nutrients for plant growth, and reduce the demand on already scarce water supplies. The soil would act as a trickling

(a) Nonbiodegradable materials are removed as the wastes pass on a moving belt.

(b) The degradable wastes are placed in a rotating digester that increases the efficiency of bacterial digestion and slowly moves the material through.

(c) The composted material leaves the digester, where microbial action caused temperatures to reach about 60°C (140°F).

(d) The composted material has much less mass than the original wastes, thus lessening the demand on the landfill. The compost is also useful as a soil conditioner.

FIGURE 27.16 Composting municipal wastes.

filter to remove chemicals and microorganisms before the water reaches groundwater and surface water supplies. Even now, waste water with coliform counts below 2.2/100 ml is being used to irrigate food crops, orchards, and pastures; water with coliform counts below 23/100 ml is being used to irrigate landscaping and recreational areas.

SOLID MUNICIPAL WASTE

Solid municipal waste (garbage) is most frequently placed into large compacted landfills. Conditions are largely anaerobic, and even presumably biodegradable materials such as paper are not very effectively attacked by microorganisms. It is not at all unusual to recover a 20-year-old newspaper in readable condition. Such anaerobic conditions, however, do promote the activity of the same methanogens encountered in anaerobic sludge digesters in sewage treatment. The methane they produce can be tapped with drill holes and burned to generate electricity or purified and introduced into natural gas pipeline systems. Such systems are part of the design of more than a hundred large landfills, some of which provide energy for several thousand homes.

The amount of organic matter entering landfills can be considerably lessened if it is first separated from material that is not biodegradable and *composted*. Composting is a process long used by gardeners to convert plant remains into the equivalent of natural humus. A pile of leaves or grass clippings will undergo microbial degradation. Under favorable conditions thermophilic bacteria will raise the temperature of the compost to 55 or 60°C in a couple of days. After the temperature declines, the pile can be turned to renew the oxygen supply and a second temperature rise will occur. Over time, the thermophilic microbial populations are replaced by mesophilic populations that slowly continue the conversion to a stable material similar to humus. Where space is available, municipal wastes are composted in long windrows (long, low piles) that are distributed and periodically turned over by specialized machinery. Municipal waste disposal now also makes increasing use of composting methods. Continuous-flow digesters can be used to compost solid municipal wastes very efficiently (Figure 27.16).

STUDY OUTLINE

Introduction (pp. 672–673)

1. Microorganisms live in a wide variety of habitats.
2. Most microorganisms do not cause disease.

SOIL MICROBIOLOGY AND CYCLES OF THE ELEMENTS (pp. 673–681)

The Components of Soil (pp. 673–676)

1. Soil consists of solid inorganic matter, water, air, and living organisms and products of their decay.

MINERALS (p. 673)

1. Weathering of rocks adds minerals to soil.

ORGANIC MATTER (p. 673)

1. Organic matter in soil comes from microorganisms, plants, animals and their waste products, and the biochemical activities of microorganisms.
2. Organic matter that is relatively resistant to decay is called humus.

WATER AND GASES (p. 673)

1. Microbial life in soil depends on the availability of water and oxygen.

ORGANISMS (pp. 673–676)

1. Microorganisms in the soil decompose organic matter and transform nitrogen- and sulfur-containing compounds into usable forms.
2. Bacteria are the most numerous organisms in the soil.
3. The characteristic musty odor of soil is from geosmin produced by actinomycetes.
4. Soil is not a reservoir for human pathogens, except for some endospore-forming bacteria and helminthic larvae; human pathogens find soil to be a hostile environment.
5. Insect and plant pathogens are found in soil.

Microorganisms and Biogeochemical Cycles (pp. 676–681)

1. In biogeochemical cycles, certain chemical elements are recycled.
2. Microorganisms are essential to the continuation of biogeochemical cycles.

THE CARBON CYCLE (pp. 676–677)

1. CO_2 is fixed into organic compounds by photoautotrophs.
2. These organic compounds provide nutrients for chemoheterotrophs.
3. Chemoheterotrophs release CO_2 that is then used by photoautotrophs.

THE NITROGEN CYCLE (pp. 677–681)

1. Microorganisms decompose proteins from dead cells and release amino acids.
2. Ammonia is liberated by microbial ammonification of the amino acids.
3. Ammonia is oxidized to nitrates for energy by nitrifying bacteria.
4. Denitrifying bacteria reduce nitrates to molecular nitrogen (N_2).
5. N_2 is converted into ammonia by nitrogen-fixing bacteria.
6. Nitrogen-fixing bacteria include free-living genera such as *Azotobacter*, cyanobacteria, and the symbiotic bacteria *Rhizobium* and *Frankia*.
7. Ammonium and nitrate are used by bacteria and plants to synthesize amino acids that are assembled into proteins.
8. Symbiotic fungi called mycorrhizae live in and on plant roots.
9. Mycorrhizae increase the surface area and nutrient absorption of the plant.

OTHER BIOGEOCHEMICAL CYCLES (p. 681)

1. Microorganisms are also involved in the transformation of other elements, including sulfur, potassium, iron, manganese, mercury, zinc, and selenium.
2. These reactions make minerals available in soluble form to plants for their metabolism.

Degradation of Synthetic Chemicals in the Soil (p. 681)

1. Many synthetic chemicals, such as pesticides and plastics, are recalcitrant (resistant to degradation).
2. Recalcitrance is based on the nature of the chemical bonding.

Aquatic Microbiology and Sewage Treatment (pp. 682–695)

Aquatic Microorganisms (pp. 682–683)

1. The study of microorganisms and their activities in natural waters is called aquatic microbiology.
2. Natural waters include lakes, ponds, streams, rivers, estuaries, and the sea.
3. The concentration of bacteria in water is proportional to the amount of organic material in the water.
4. Most aquatic bacteria tend to grow on surfaces rather than in a free-floating state.

FRESHWATER MICROBIAL FLORA (pp. 682–683)

1. Numbers and locations of freshwater microbial flora depend on the availability of oxygen and light.
2. Photosynthetic algae are the primary producers of a lake. They are found in the limnetic zone.
3. Pseudomonads, *Cytophaga, Caulobacter,* and *Hyphomicrobium* are found in the limnetic zone, where oxygen is abundant.
4. Microbial growth in stagnant water uses available oxygen and can cause odors and the death of fish.
5. The amount of dissolved oxygen is increased by wave action.
6. Purple and green sulfur bacteria are found in the benthic zone, which contains light and H_2S but no oxygen.
7. *Desulfovibrio* reduces SO_4^{2-} to H_2S in benthic mud.
8. Methane-producing bacteria are also found in the benthic zone.

SEAWATER MICROBIAL FLORA (p. 683)

1. The open ocean is not a favorable environment for most microorganisms because of its high osmotic pressure, low nutrients, and high pH.
2. Phytoplankton, consisting mainly of diatoms, are the primary producers of the open ocean.
3. Some algae and bacteria are bioluminescent. They possess the enzyme luciferase, which can emit light.

The Role of Microorganisms in Water Quality (pp. 683–695)

WATER POLLUTION (pp. 683–685)

1. Microorganisms are filtered from water that percolates into groundwater supplies.
2. Some pathogenic microorganisms are transmitted to humans in water supplies.
3. Recalcitrant chemical pollutants may be concentrated in animals in an aquatic food chain.
4. Mercury is metabolized by certain bacteria into a soluble compound that is concentrated in animals.
5. Nutrients such as phosphates cause eutrophication of aquatic ecosystems.
6. Eutrophication means well nourished. It is the result of the addition of pollutants or natural nutrients.
7. The growth of oil-degrading bacteria can be enhanced by the addition of nitrogen and phosphorous fertilizer.
8. *T. ferrooxidans* produces sulfuric acid at coal-mining sites.

TESTS FOR WATER PURITY (pp. 685–686)

1. Tests for the bacteriological quality of water are based on the presence of indicator organisms.
2. The most common indicator organisms are coliforms.
3. Coliforms are aerobic or facultatively anaerobic, gram-negative, non-endospore-forming rods that ferment lactose with the production of acid and gas within 48 hours of being placed in a medium at 35°C.
4. Fecal coliforms, predominantly *E. coli,* are used to indicate the presence of human feces.

WATER TREATMENT (pp. 686–688)

1. Drinking water is held in a holding reservoir long enough that suspended matter settles.
2. Flocculation treatment uses a chemical such as alum to coalesce and then settle colloidal material.
3. Sand filtration removes bacteria, viruses, and protozoan cysts.
4. Drinking water is disinfected with chlorine to kill remaining pathogenic bacteria.

SEWAGE TREATMENT (pp. 688–694)

1. Domestic wastewater is called sewage.
2. It includes household water, toilet wastes, industrial wastes, and rainwater.

Primary Treatment (p. 688)

1. Primary sewage treatment is the removal of solid matter called sludge.
2. Biological activity is not very important in primary treatment.

Biochemical Oxygen Demand (pp. 688–689)

1. Primary treatment removes approximately 25% to 35% of the biochemical oxygen demand (BOD) of the sewage.
2. BOD is a measure of the biologically degradable organic matter in water.
3. It is determined by measuring the amount of oxygen bacteria require to degrade the organic matter.

Secondary Treatment (pp. 689–691)

1. Secondary treatment is the biological degradation of organic matter in sewage after primary treatment.
2. Activated sludge and trickling filters are methods of secondary treatment.
3. Microorganisms degrade the organic matter aerobically.
4. Secondary treatment removes up to 95% of the BOD.

Sludge Digestion (pp. 691–692)

1. Sludge is placed in an anaerobic sludge digester; bacteria degrade organic matter and produce simpler organic compounds, methane, and CO_2.
2. The methane produced in the digester is used to heat the digester and to operate other equipment.
3. Excess sludge is periodically removed from the digester, dried, and disposed of as landfill or as soil conditioner or incinerated.

Septic Tanks (pp. 692–693)

1. Septic tanks can be used in rural areas to provide primary treatment of sewage.
2. They require a large leaching field for the effluent.

Oxidation Ponds (p. 693)

1. Small communities can use oxidation ponds for secondary treatment.
2. These require a large area in which to build an artificial lake.

Tertiary Treatment (pp. 693–694)

1. Tertiary treatment employs physical filtration and chemical precipitation to remove all the BOD, nitrogen, and phosphorus from water.
2. Tertiary treatment provides drinkable water, whereas secondary treatment provides water usable only for irrigation.

SOLID MUNICIPAL WASTE (pp. 694–695)

1. Municipal landfills prevent decomposition of solid wastes because they are dry and anaerobic.
2. In some landfills, methane produced by methanogens can be recovered for an energy source.
3. Composting and continuous-flow digesters can be used to promote biodegradation of organic matter.

STUDY QUESTIONS

Review

1. Write a one-sentence description of each of the following soil constituents: solid inorganic matter, water, gases, organic matter.
2. The precursor to coal is peat found in bogs. Why does peat accumulate in bogs?
3. The metabolic activities of microorganisms often produce acids. Why would microbial growth increase the solid inorganic matter in soil?
4. Compare and contrast humus and recalcitrant chemicals.
5. Diagram the carbon cycle in the presence and absence of oxygen. Name at least one microorganism that is involved at each step.
6. Fill in the following table with the information provided below.

Process	Chemical Reactions	Microorganisms
Ammonification		
Nitrification		
Denitrification		
Nitrogen fixation		

Choices:

$NO_3^- \longrightarrow N_2$
$N_2 \longrightarrow NH_3$
$-NH_2 \longrightarrow NH_3$
$NH_3 \longrightarrow NO_2^-$
$NO_2^- \longrightarrow NO_3^-$

Bacillus
Rhizobium
Nitrosomonas
Azotobacter
Nitrobacter
Proteolytic bacteria

7. The following organisms have important roles as symbionts with plants and fungi. Describe the symbiotic relationship of each organism with its host: cyanobacteria, mycorrhizae, *Rhizobium*, *Frankia*.
8. Compare and contrast the physical conditions of the ocean with those of fresh water.
9. Matching.
   (a) Methane-producing bacteria
   (b) *Desulfovibrio*
   (c) Photosynthetic bacteria

   ____ $CO_2 + H_2S \xrightarrow{\text{light}} C_6H_{12}O_6 + S^0$
   ____ $SO_4^{2-} + H^+ \longrightarrow H_2S$
   ____ $CO_2 + H_2 \longrightarrow CH_4$
10. Indicate which reactions in Question 9 require atmospheric oxygen (O_2).
11. Outline the treatment process for drinking water.
12. What is the purpose of a coliform count on water?
13. If coliforms are not normally pathogenic, why are they used as indicator organisms for bacteriological quality?
14. The following processes are used in wastewater treatment. Match the type of treatment with the processes. Each choice can be used once, more than once, or not at all.

Processes	Types of Treatment
____ Leaching field	**(a)** Primary
____ Removal of solids	**(b)** Secondary
____ Biological degradation	**(c)** Tertiary
____ Activated sludge	
____ Chemical precipitation of phosphorus	
____ Trickling filter	
____ Results in drinking water	
____ Effluent can be used for irrigation	
____ Produces methane	

15. Define BOD.
16. Why is activated sludge a more efficient means of removing BOD than a sludge digester?
17. Why are septic tanks and oxidation ponds not feasible for large municipalities?
18. Explain the effect of dumping untreated sewage into a pond on the eutrophication of the pond. The effect of sewage that has primary treatment? The effect of sewage that has secondary treatment? Contrast your previous answers with the effect of each type of sewage on a fast-moving river.
19. Bioconversion refers to the generation of energy sources by bacteria. Describe an example of bioconversion.
20. Bioremediation refers to the use of living organisms to remove pollutants. Describe three examples of bioremediation.

Challenge

1. Here are the formulas of two detergents that have been manufactured:

```
C—C—C—C—C—C—C—C—C—C— . . .

        C
        |
        C     C
        |     |
C—C—C—C—C—C— . . .
  |
  C
```

Which of these would be recalcitrant, and which would be readily degraded by microorganisms? (*Hint:* Refer to the degradation of fatty acids in Chapter 5.)

2. A sewage treatment plant in Redwood City, California, is a tertiary treatment plant. Outline the flow of water and solids through this plant. What is the final quality of the water with respect to BOD, nitrogen, and phosphorus?
3. What is the MPN of a water sample with these multiple-tube fermentation test results: 10-ml portions: 4 positive for acid and gas; 1-ml portions: 3 positive for acid and gas; 0.1-ml portions: 1 positive. (See MPN table, Appendix C.)
4. Flooding after two weeks of heavy rainfall in Tooele, Utah, preceded the high rate of diarrheal illness shown below. *G. lamblia* was isolated from 26% of the patients. A comparison study of a town 65 miles away revealed that there was diarrheal illness in 2.9% of the 103 persons interviewed. Tooele has a municipal water system and a municipal sewage treatment plant. Explain the probable cause of this epidemic and method(s) of stopping it.

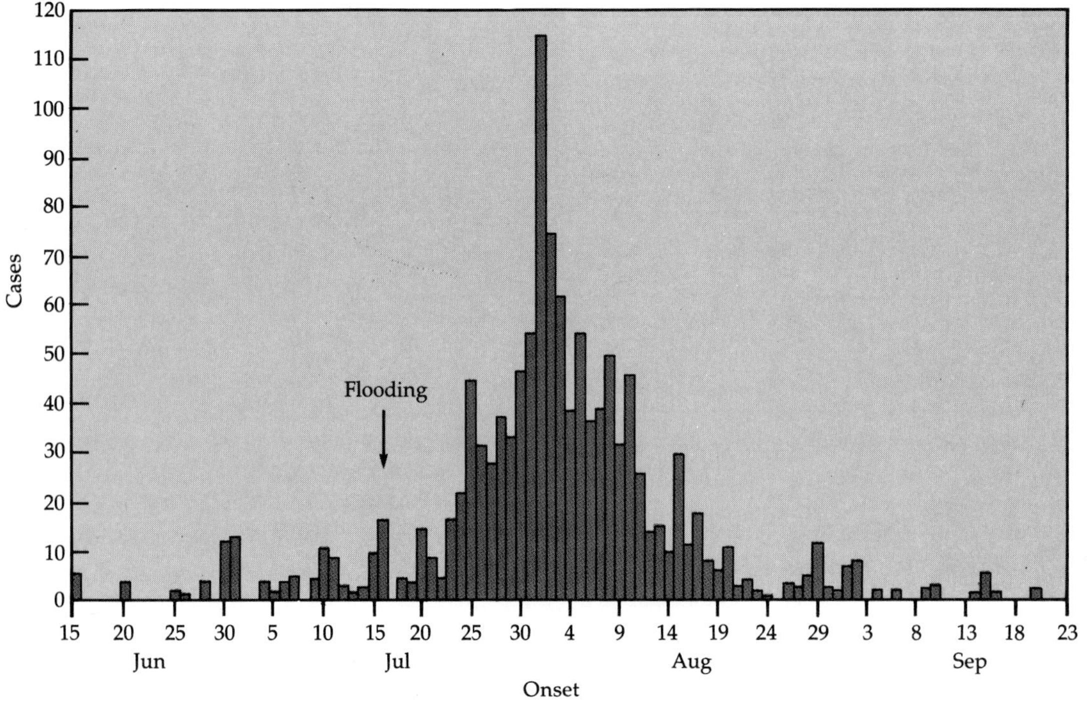

Distribution of cases of diarrheal illness.

FURTHER READING

Atlas, R.M., and R. Bartha. *Microbial Ecology: Fundamentals and Applications,* 2nd ed. Redwood City, Calif.: Benjamin/Cummings, 1987. The foremost textbook on the subject of microbial ecology.

AWWA Organisms in Water Committee. "Committee report: microbiological considerations for drinking water regulation revisions." *Journal of the American Water Works Association* 79:81–88, 1987. A discussion of the current status of the microbiology of drinking water and suggested changes for the future.

Brock, T.D., ed. *Thermophiles: General Molecular and Applied Microbiology.* New York: Wiley, 1986. Discusses the microorganisms, their metabolism, and growth requirements of the unique environment of the hot springs in Yellowstone National Park.

Childress, J.J., H. Felbeck, and G.N. Somero. "Symbiosis in the deep sea." *Scientific American* 256(5):114–120, May 1987. Describes the chemoautotrophic bacterial symbionts in worms living at hydrothermal vents.

Cloud, P. "The biosphere." *Scientific American* 249(3):176–189, September 1983. A discussion of how life has been shaped and sustained on Earth in soil, water, and the atmosphere.

Dowling, D.N., and W.J. Broughton. "Competition for nodulation of legumes." *Annual Review of Microbiology* 40:131–157, 1986. Describes the effects of the environment, other organisms, and host genes on root nodule formation.

Erlich, H.L. *Geomicrobiology,* 2nd ed. New York: Marcel Dekker, 1990. A textbook on the role of microorganisms in transformations of carbon, phosphorus, mercury, manganese, and other elements.

Margulis, L., D. Chase, and R. Guerero. "Microbial communities." *BioScience* 36:160–170, 1986. A discussion about bacteria in their natural environment. Photographs of living systems show relationships that are not apparent in vitro.

Padwa, D., ed. "Benefits to agriculturists from the studies of microorganisms." *Developments in Industrial Microbiology* 24:19–67, 1983. Papers on technological uses of nitrogen-fixing bacteria, mycorrhizae, yeasts, and agrobacteria.

Paul, E.A., and F.E. Clark. *Soil Microbiology and Biochemistry.* New York: Academic Press, 1989. This authoritative text includes chapters on microbial transformations of nitrogen and metals and methods for studying soil organisms.

Shapiro, J.A. "Bacteria as multicellular organisms." *Scientific American* 258(6):82–89, June 1988. Describes colonies of bacterial cells that function as multicellular organisms, such as predatory *Myxococcus* colonies.

Sieburth, J.M. *Sea Microbes.* New York: Oxford University Press, 1979. Discusses microorganisms and their adaptations for living in ocean waters.

CHAPTER 28

Applied and Industrial Microbiology

LEARNING OBJECTIVES

- Provide a brief history of the development of food preservation.
- Explain why sterilization of canned foods is important.
- Describe thermophilic anaerobic spoilage and flat sour spoilage by mesophilic bacteria.
- Describe low-temperature preservation and aseptic packaging.
- Define pasteurization, and explain why dairy products are pasteurized.
- Provide examples of chemical food preservatives, and explain why they are used.
- Outline at least four beneficial activities of microorganisms in food production.
- Describe the role of microorganisms in the production of alternative energy sources, industrial chemicals, and pharmaceuticals.
- Define industrial fermentation, bioreactor, and biotechnology.

We will now turn our attention from soil and aquatic microbiology to food and industrial microbiology. This chapter will discuss food spoilage and preservation, foodborne infections and food poisoning, and food production involving microorganisms. We will also look at some important aspects of industrial microbiology.

FOOD MICROBIOLOGY

Food Preservation and Spoilage

Modern civilization and its large populations could not be supported without effective methods of food preservation. In fact, civilization arose only after agriculture produced a year-round stable food supply in a single site so that people were able to give up the nomadic hunting cultures. Many of the methods of food preservation used today were probably discovered by chance in centuries past. Primitive people observed that dried meat and salted fish resisted decay. Nomads no doubt observed that soured animal milk resisted further decomposition and was still palatable. Moreover, if the curd of the soured milk was pressed to remove moisture and allowed to ripen (in effect, cheese-making), it was even more effectively preserved and tasted better. Farmers soon learned that if grains were kept dry, they did not become moldy.

All such phenomena are readily understood by anyone familiar with the factors that control the growth of microorganisms (see Chapters 6 and 7). Bacteria and fungi require a minimum amount of available moisture for growth. Drying food and adding salt or sugar minimizes the available moisture and thus prevents spoilage. Acidity created by the natural fermentation of milk and vegetable juices, as in yogurt and sauerkraut, also prevents the growth of many spoilage bacteria.

These methods of preservation are still in use today. However, a tour of any supermarket will demonstrate that heat sterilization, pasteurization, refrigeration, and freezing are now much more popular methods for the control of food spoilage. These modern methods preserve food nearer its natural state and palatability than the methods that were used in the past.

CANNING

Preservation by heat originated in the early 1800s, before microorganisms were known to cause human disease and food spoilage. The technology of canning was promoted by the military. When armies were small professional organizations, they could be supported by the land, with supplements of portable rations of wine, cheese, and dried grains. But with the rise of large armies of citizen soldiers, a development of the French Revolution, this strategy was no longer viable. Thus, the French government offered a prize to anyone who could devise a method of preserving food, particularly meat. In 1810, the prize was won by a confectioner, Nicholas Appert, who showed that food could be preserved if it was sealed in tightly stoppered containers and boiled for specified periods. He developed detailed tables of the boiling times required for different foods and container sizes. As we know today, there are many endospore-forming bacteria that will survive hours of such boiling. Failures were manageably few, however, and were usually attributed to faulty sealing.

The concept of heat preservation was followed quickly by the invention of the metal can. Temperatures above the boiling point of water were introduced into processing by oil or salt baths. Achieving these temperatures with pressurized steam in closed containers, the method used today, was introduced late in the nineteenth century, when reliable pressure controls and safety valves were developed. By then, of course, there was a better theoretical understanding of microbe control.

Commercial Sterilization

Preserving foods by heating a properly sealed container is not difficult. The problem is to use the minimum amount of heat necessary to kill spoilage organisms and dangerous microorganisms, such as the endospore-forming *Clostridium botulinum*. Because heating degrades the quality of food, much research was needed to determine the exact heat treatment that would succeed in sterilizing the food but minimize these reductions in quality.

Canned goods today undergo what is called **commercial sterilization** (Figure 28.1), which is not as rig-

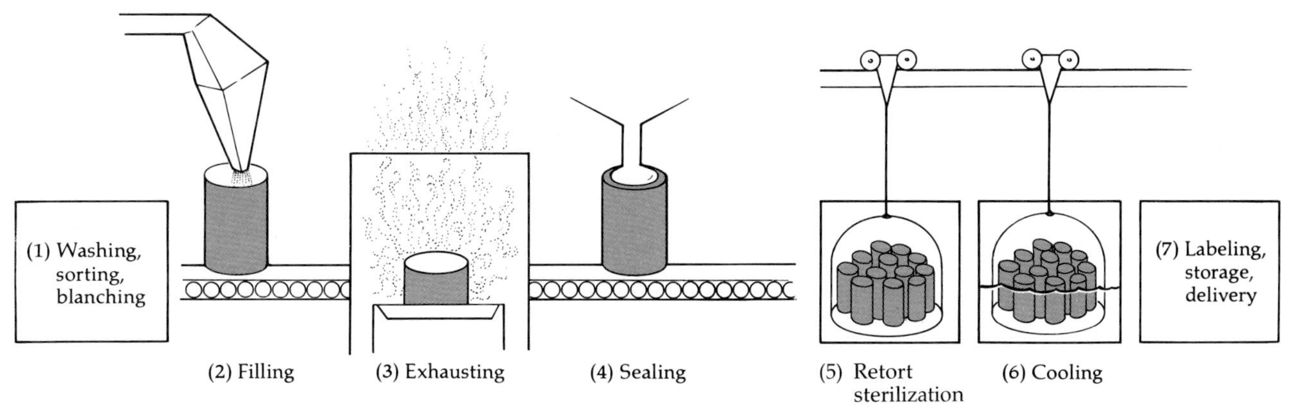

FIGURE 28.1 Industrial canning. (1) The food to be canned is washed, sorted, and blanched. Blanching is a treatment with hot water or live steam that softens the product so that the can will fill better. It also destroys enzymes that might alter the color, flavor, or texture of the product and lowers the microbial count. (2) Cans are filled to capacity, leaving as little dead space as possible. (3) Cans are placed in a steam box, where the heat exhausts (drives out) dissolved air. (4) Cans are sealed. (5) Sterilization is done in a large retort with steam under pressure. (6) Cooling is done with a water spray or by submergence. (7) Cans are then labeled, stored, and delivered.

orous as true sterilization. Commercial sterilization is intended to destroy *C. botulinum* endospores. If this is accomplished, then any other significant spoilage or pathogenic bacterium will also be destroyed. To ensure complete sterilization, enough heat is applied for the **12D treatment** (12-decimal reductions), by which a theoretical population of botulism endospores would be decreased by 12 logarithmic cycles. What this means is that if there were 10^{12} (1,000,000,000,000) botulism endospores in a can, after treatment there would be only one survivor. Because 10^{12} is an improbably large population, this treatment is considered quite safe. Certain thermophilic endospore-forming bacteria have endospores that are more resistant to heat treatment than *C. botulinum*. However, these bacteria are obligate thermophiles and generally remain dormant at temperatures lower than about 45°C (112°F). Therefore, they are not a problem at normal storage temperatures.

Spoilage of Canned Food

If canned foods are incubated at high temperatures, such as exist in a truck in the hot sun or next to a steam radiator, the thermophilic bacteria that often survive commercial sterilization can germinate and grow. **Thermophilic anaerobic spoilage** is a fairly common cause of spoilage in low-acid canned foods. The can usually swells from gas, and the contents have a lowered pH and a sour odor. A number of thermophilic species of *Clostridium* can cause this type of spoilage. When thermophilic spoilage occurs but the can is not swollen by gas production, the spoilage is termed **flat sour spoilage.** This type of spoilage is caused by thermophilic organisms such as *Bacillus stearothermophilus* (ste-rō-thėr-mä'fil-us), which is found in the starch and sugars used in food preparation. Many industries have standards for the numbers of such thermophilic organisms permitted in raw materials. Both types of spoilage occur only when the cans are stored at higher than normal temperatures, which permits the growth of bacteria whose endospores are not destroyed by normal processing.

Mesophilic bacteria can spoil canned foods if the food is underprocessed or if the can leaks. Normally, these bacteria are killed by proper processing procedures. Underprocessing is more likely to result in spoilage by endospore-formers; the presence of non-endospore-formers strongly suggests that the can leaks. Leaking cans are often contaminated during the cooling of cans after processing by heat. The hot cans are sprayed with cooling water or passed through a trough filled with water. As the can cools, a vacuum is formed inside, and external water can be sucked through a leak past the heat-softened sealant in the crimped lid (Figure 28.2). Contaminating bacteria in the cooling water are drawn into the can with the water. Spoilage from underprocessing or can leakage is likely to produce odors of putrefaction, at least in high-protein foods, and occurs at normal storage temperatures. In such types of spoilage, there is always the potential that botulism bacteria will be present.

Some acidic foods, such as tomatoes or preserved fruits, are preserved by temperatures of 100°C or below. As a rule, the only important spoilage organisms in acidic foods are molds, yeasts, and occasional species of acid-tolerant, non-endospore-forming bacte-

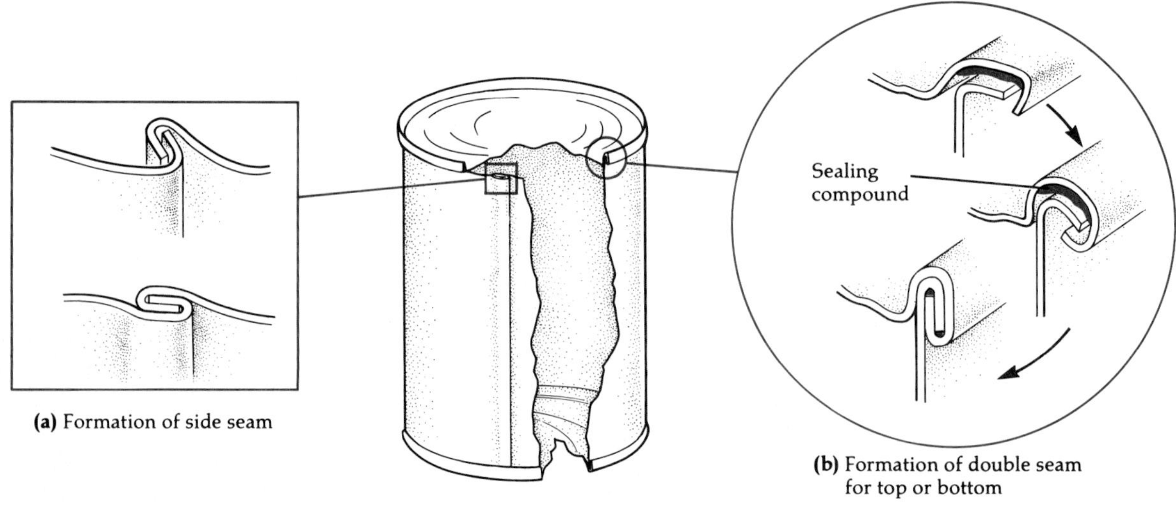

FIGURE 28.2 Construction of a metal can. Note the seam construction. During cooling after sterilization, the vacuum formed in the can may actually force the entry of contaminating organisms along with water.

TABLE 28.1 Types of Canned Food Spoilage

TYPE OF SPOILAGE	INDICATIONS OF SPOILAGE: APPEARANCE OF CAN	INDICATIONS OF SPOILAGE: CONTENTS OF CAN
Low- and Medium-Acid Foods (pH above 4.5)		
Flat sour *(Bacillus stearothermophilus)*	Possible loss of vacuum on storage	Appearance not usually altered; pH markedly lowered; sour; may have slightly abnormal odor; sometimes cloudy liquid
Thermophilic anaerobic *(Clostridium thermosaccharolyticum)*	Can swells	Fermented, sour, cheesy, or butyric acid odor
Sulfide *(Desulfotomaculum nigrificans)*	No swelling	Usually blackened; "rotten-egg" odor
Putrefactive anaerobic *(Clostridium sporogenes)*	Can swells	May be partially digested; pH slightly above normal; typical putrid odor
Aerobic endospore-formers (*Bacillus* spp.)	Usually no swelling, except in cured meats when nitrate and sugar are present	Coagulated evaporated milk; foods usually have sour taste
High-Acid Foods (pH below 4.5)		
Flat sour *(Bacillus coagulans)*	Can flat, little change in vacuum	Slight pH change; off odor and flavor
Butyric anaerobic *(Clostridium butyricum)*	Can swells	Fermented, butyric acid odor
Non-endospore-formers (mostly lactic acid bacteria)	Can swells	Acid odor
Yeasts	Can swells, may burst	Fermented; yeasty odor
Molds	Can flat	Surface growth; musty odor

Source: Data from the National Canners Association, 1950 6th Street, Berkeley, CA 94710.

ria. These organisms are capable of growing at the normal pH of these foods but are easily killed by temperatures lower than 100°C. Occasional problems in acidic foods develop from a few organisms that are both heat-resistant and acid-tolerant. Examples of heat-resistant fungi are the mold *Byssochlamys fulva* (bis-sō-klam'is fül'vä), which produces a *heat-resistant ascospore,* and a few molds, especially species of *Aspergillus,* that sometimes produce specialized resistant bodies called *sclerotia.* A spore-forming bacterium, *Bacillus coagulans* (kō-ag'ū-lans), is unusual in that it is capable of growth at a pH of almost 4.0. A summary of the types of canned food spoilage is presented in Table 28.1.

Home canning is an important means of preserving food. Because of the possibility of botulism food poisoning resulting from improper canning methods, persons involved in home canning should obtain reliable directions and follow them exactly.

PASTEURIZATION

As we discussed in Chapter 1, Louis Pasteur began his career as a microbiologist when the French brewing industry commissioned him in 1871 to investigate the causes of spoilage in beer and wine. He concluded that spoilage was caused by the growth of microorganisms, and he worked out a method of eliminating them by mild heating. Too much heat would alter the beverage's characteristics to an unacceptable degree, but killing all the microorganisms was found to be unnecessary. The principle of the selective killing of microorganisms, now called pasteurization, was later applied to other foods, milk in particular (see Chapter 7). Other products, such as ice cream, yogurt, and beer, all have individual pasteurization times and temperatures, which often differ considerably. Reasons for variation include less efficient heating in more viscous foods and the protective effects of fats.

ASEPTIC PACKAGING

A recent development in food preservation is the increasing use of **aseptic packaging.** Packages are usually made of some material that cannot tolerate conventional heat treatment, such as laminated paper or plastic. The packaging materials come in continuous rolls that are fed into a machine that sterilizes the material with a hot hydrogen peroxide solution, some-

FIGURE 28.3 Spoilage temperatures.

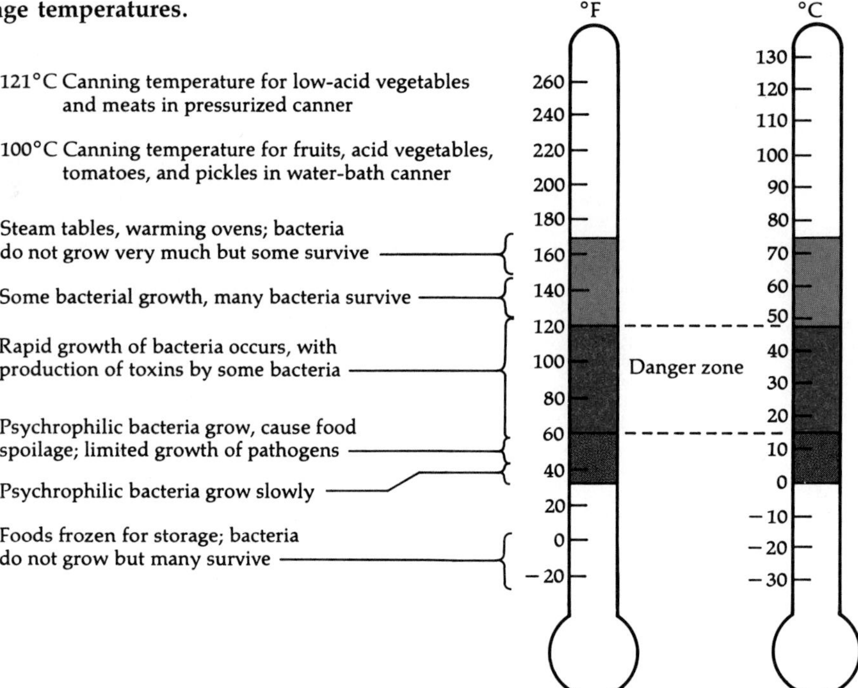
°F
°C
121°C Canning temperature for low-acid vegetables and meats in pressurized canner
100°C Canning temperature for fruits, acid vegetables, tomatoes, and pickles in water-bath canner
Steam tables, warming ovens; bacteria do not grow very much but some survive
Some bacterial growth, many bacteria survive
Rapid growth of bacteria occurs, with production of toxins by some bacteria
Psychrophilic bacteria grow, cause food spoilage; limited growth of pathogens
Psychrophilic bacteria grow slowly
Foods frozen for storage; bacteria do not grow but many survive
Danger zone
260
240
220
200
180
160
140
120
100
80
60
40
20
0
−20
130
120
110
100
90
80
70
60
50
40
30
20
10
0
−10
−20
−30

times aided by ultraviolet light. If the containers are metal, they can be sterilized with superheated steam or other high-temperature methods. High-energy electron beams can also be used to sterilize the packaging materials. While still in the sterile environment, the material is formed into packages, which are then filled with liquid foods that have been conventionally sterilized by heat. The filled package is not sterilized after it is sealed.

LOW-TEMPERATURE PRESERVATION

Low temperatures lengthen the reproduction time of microorganisms. Even so, some molds and bacteria can grow at a significant rate at temperatures below the freezing point of water, 0°C. (Food typically does not freeze solid until it is several degrees below this point.) Growth of microorganisms in foods a few degrees below 0°C is common, and growth at minus 18°C has been reported. A properly set refrigerator maintains a temperature range from 0° to 7°C. Many microorganisms grow slowly at these temperatures and will alter the taste and appearance of foods that are stored for too long.

Pathogenic bacteria, with a few exceptions, will not grow at these temperatures. Among the exceptions are the clostridia that cause type E botulism, *Yersinia enterocolitica*, and *Listeria monocytogenes*. Freezing does not immediately kill significant numbers of bacteria, but frozen bacterial populations are dormant and decline slowly with time. Some parasites, such as the roundworms that cause trichinosis, are killed by several days of freezing. Some important temperatures associated with microorganisms and food spoilage are shown in Figure 28.3.

Anyone responsible for preparing food in large amounts should be aware of the spoilage that can arise from the slow cooling of large amounts of hot food. Hot food placed in a refrigerator cools more slowly than most persons realize, and the food can spend a considerable time at the incubation temperatures of pathogenic or spoilage bacteria before its temperature drops to that of the refrigerator (Figure 28.4).

RADIATION AND FOOD PRESERVATION

There has been considerable research into the use of ionizing radiation for food preservation, especially for military applications. It is possible to sterilize food by radiation, but it is unlikely that this procedure will soon replace conventional heat sterilization for most purposes. The taste and appearance of irradiated food have been much improved in recent years. This improvement has been accomplished mainly by freezing the food in liquid nitrogen and exhausting oxygen from the package before it is irradiated.

However, a major factor affecting the adoption of this type of preservation is that the public is apprehensive about foods exposed to radiation, even though there is no evidence that harmful effects result from

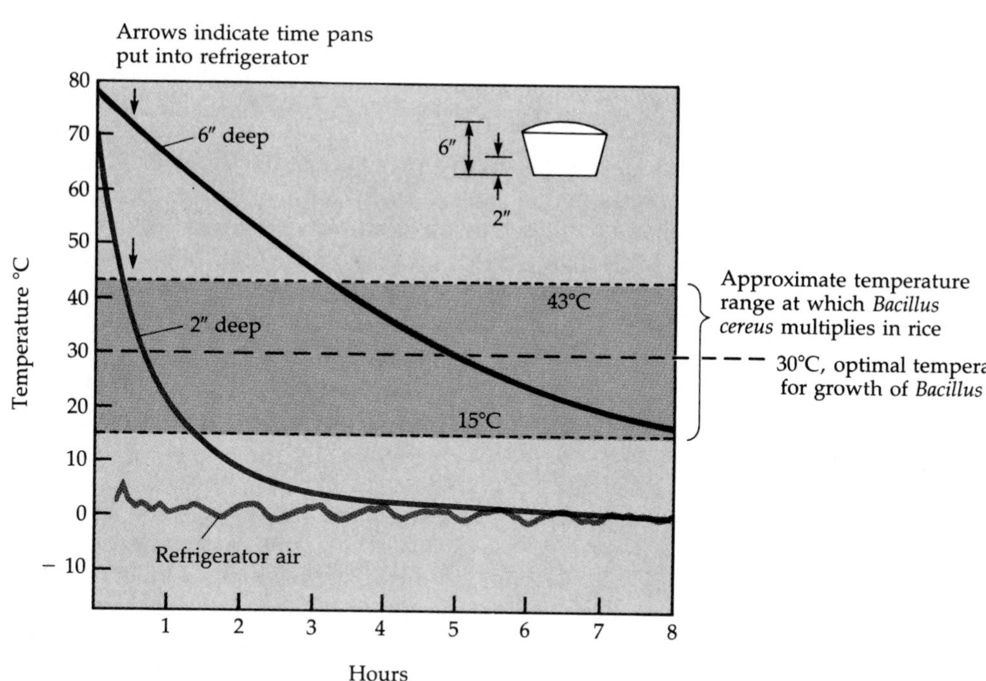

FIGURE 28.4 The effect of a food's mass on its cooling rate. Note that the two-inch-deep pan of rice cooled through the incubation temperature of the *Bacillus cereus* in about an hour, whereas the six-inch-deep pan of rice remained in this temperature range for more than five hours.

the amount and types of radiation used. Irradiation is considered to be another form of food additive. As such, it must be reported on the label and must meet the stringent requirements for safety testing required of chemical food additives. These are expensive and can be justified only when a considerable market is expected.

Pasteurization-like radiation treatments that fall short of attempts at total sterilization may prove to be commercially practical. For several years the Food and Drug Administration has allowed irradiation to decontaminate spices. More recently, it has given approval for the irradiation in moderate doses of certain foods to eliminate insects and to prevent sprouting or other undesirable maturation of certain fruits and vegetables. Pork may also be irradiated to eliminate the parasitic worms that cause trichinosis. Foods treated in this manner must be labeled with a logo (Figure 28.5) and a statement such as "Treated with gamma radiation to extend shelf life." After a few years, to allow for familiarization, only the logo will be required.

FIGURE 28.5 Irradiation logo. This logo indicates that a food has received irradiation treatment.

Many other countries have allowed the use of irradiation on foods for some time. In the United States, the greatest commercial interest is in irradiation treatment of spices and pet foods. The preferred method for irradiation is gamma rays produced by cobalt-60, when deep penetration is a requirement. However, this type of treatment requires several hours of exposure in isolation behind protective walls. High-energy electron accelerators are much faster and sterilize in a few seconds, but this treatment has low penetrating power and is suitable only for sliced meats, bacon, or similar thin products.

Microwaves, such as those used for cooking, kill bacteria only by heating foods and have little or no direct killing effect. Bacteria can survive (and are easily isolated) on the interior walls of microwave ovens.

CHEMICAL PRESERVATIVES

Chemical preservatives are frequently added to foods to retard spoilage. Among the more common additives are sodium benzoate, sorbic acid, and calcium propionate. These chemicals are simple organic acids, or salts of organic acids, which the body readily metabolizes and which are generally judged to be safe in foods. **Sorbic acid,** or its more soluble salt *potassium sorbate,* and **sodium benzoate** prevent molds from growing in certain acidic foods, such as cheese and soft drinks. Such foods, usually with a pH of 5.5 or less, are most susceptible to mold-type spoilage. **Calcium propionate** is an effective fungistat used in bread. It prevents the growth of surface molds and the *Bacillus* bacterium that causes ropy bread. These organic acids inhibit mold growth, not by affecting the pH but by interfering with the mold's metabolism or the integrity of the plasma membrane.

Sodium nitrate and **sodium nitrite** are added to many meat products, such as ham, bacon, wieners, and sausage. The active ingredient is sodium nitrite, which certain bacteria in the meats can also produce from sodium nitrate. These bacteria use nitrate as a substitute for oxygen under anaerobic conditions, much as it is used in denitrification in soil. The nitrite has two main functions—to preserve the pleasing red color of the meat by reacting with blood components in the meat and to prevent the germination and growth of any botulism endospores that might be present. There has been some concern that the reaction of nitrites with amino acids can form certain carcinogenic products known as **nitrosamines,** and the amount of nitrites added to foods has generally been reduced recently for this reason. However, the use of nitrites continues because of their established value in preventing botulism. Since nitrosamines are formed in the body from other sources, the added risk posed by limited use of nitrates and nitrites in meats might be lower than was once thought.

Foodborne Infections and Microbial Intoxications

We mentioned in Chapter 25 that illness resulting from microbial growth in food is associated with two principal mechanisms. In the mechanism known as *foodborne infection,* the contaminating microorganism infects the person who ingests contaminated food. As the pathogen grows in this host, it produces damaging toxins. Diseases caused by this mechanism include gastroenteritis, typhoid fever, and dysentery. In the other mechanism, called *microbial intoxication,* the toxin is formed in the food by microbial growth and then is ingested with the food. Diseases associated with this mechanism include botulism, staphylococcal food poisoning, and mycotoxicoses (intoxications caused by fungal toxins, such as ergot or aflatoxin). Several microbial intoxication diseases are described in Table 28.2.

Dairy products, which are often consumed without having been cooked, are particularly likely to transmit food-related diseases. Standards for sanitation in the dairy industry are therefore very stringent. Because most dairy milk is drawn mechanically and is promptly put into cooled holding tanks, most of the bacteria in milk today are gram-negative psychrophiles. Grade A pasteurized cultured products, such as buttermilk and cultured sour cream, have high counts of lactic acid bacteria as a natural result of their method of manufacture. Therefore, the standards require only that they contain fewer than 10 coliforms per milliliter. Pasteurized Grade A milk is required to have a standard plate count of fewer than 20,000 bacteria and not more than 10 coliforms per milliliter. Grade A dry milk products are required to have a standard plate count of fewer than 30,000 bacteria per gram or a coliform count of fewer than 10 per gram.

Some people prefer raw milk because of the lack of additives or because they believe it has not been altered chemically. In the United States, unpasteurized (raw) dairy products are available commercially only in California, where they carry a warning. The standard for certified raw milk is a maximum bacterial count of 10,000 per milliliter and 10 coliforms per milliliter. The microbial standards guaranteed by certification do not prevent transmission of disease in raw milk; in fact, certified raw milk has been the source of several outbreaks of salmonellosis.

The Role of Microorganisms in Food Production

In the latter part of the nineteenth century, microbes used in food production were grown in pure culture for the first time. This development quickly led to improved understanding of the relationships between specific microbes and their products and activities. This period can be considered the beginning of industrial food microbiology. For example, once it was understood that a certain yeast grown under certain conditions produced beer and that certain bacteria could spoil the beer, brewers were better able to control the quality of their products. Specific industries became active in microbiological research and selected certain microbes for their special qualities. For example, the brewing industry extensively investigated the isolation and identification of yeasts and selected those that were able to produce more alcohol.

As our knowledge of microbial genetics increases, we will continue to improve our ability to select and use desirable strains of microorganisms. Tremendous advances in molecular biology have permitted the alteration of microorganisms by techniques that are increasingly replacing the relatively hit-or-miss approaches of the past. The application of these new techniques, known as *genetic engineering,* is discussed in Chapter 9. In this section, we will discuss the role of microorganisms in the production of several common foods.

TABLE 28.2 Foodborne Microbial Intoxications

DISEASE	FOODS INVOLVED	PREVENTION	CLINICAL FEATURES	ONSET AND DURATION
Aflatoxin	Moldy grains, peanuts	Avoid eating contaminated grains, peanut products	Low doses may induce liver cancer; high doses cause general liver damage	Carcinogenic effects have indefinite onset; direct toxicity to animals usually sudden
Bacillus cereus intoxication	Custard, cereal, starchy foods	Refrigeration of foods	Cramps, diarrhea, nausea, vomiting	Onset: 8–16 hours Duration: Less than 1 day
Botulism	Canned foods	Proper canning procedures; boiling food prior to consumption	Difficulty swallowing, double vision, respiratory paralysis	Onset: 2 hours–6 days Duration: Weeks
Ciguatera	Carnivorous tropical fish in which dinoflagellate toxin is concentrated	Avoid eating large tropical fish	Tingling, rash, fever, breathing discomfort	Onset: 2–6 hours Duration: To 2 weeks
Ergotism	Moldy grains	Avoid eating contaminated grains	Burning abdominal pain, hallucinations	Onset: 1–2 hours Duration: Months
Methyl mercury poisoning	Freshwater or ocean fish	Stop dumping mercury into waters	Blurred vision, numbness, apathy, coma	Onset: 1 week Duration: May be chronic
Mushroom poisoning	*Amanita* species	Avoid eating poisonous mushrooms	Vomiting, liver necrosis, neurotoxic effects	Onset: Less than 1 day Duration: Less than 10 days
Paralytic mollusk poisoning	Bivalve mollusks during red tide (dinoflagellate blooms)	Avoid eating mollusks during red tide	Tingling, rash, fever, respiratory paralysis	Onset: Less than 1 hour Duration: Less than 12 hours
Scombroid poisoning	Histaminelike substance produced by *Proteus* growing on certain fish, such as tuna	Refrigeration of fish	Headache, cramps, hives, shock (rare)	Onset: Several minutes–1 hour
Staphylococcal intoxication	High-osmotic-pressure foods not cooked before eating	Refrigeration of foods	Nausea, vomiting, diarrhea	Onset: Less than 1 day Duration: Less than 3 days

CHEESE

The United States leads the world in the manufacture of cheese, producing more than 1.5 million tons each year. Although there are many types of cheeses, all require the formation of a *curd*, which can then be separated from the main liquid fraction, or *whey* (Figure 28.6). The curd is made up of a protein, *casein*, and is usually formed by the action of an enzyme, *rennin*, which is aided by acidic conditions provided by certain lactic acid-producing bacteria. These inoculated lactic acid bacteria also provide the characteristic flavors and aromas of fermented dairy products during the ripening process. The curd undergoes a microbial ripening process, except for a few unripened cheeses, such as ricotta and cottage cheese.

Cheeses are generally classified by their hardness, which is produced in the ripening process. The more moisture lost from the curd, and the more the curd is compressed, the harder the cheese. Romano and Parmesan cheeses, for example, are classified as *very hard* cheeses; Cheddar and Swiss are *hard* cheeses. Limburger, blue, and Roquefort cheeses are classified as *semisoft;* Camembert is an example of a *soft* cheese.

The hard Cheddar and Swiss cheeses are ripened by lactic acid bacteria growing anaerobically in the interior. Such hard, interior-ripened cheeses can be quite large. The longer the incubation time, the higher the acidity and the sharper the taste of the cheese. A *Propionibacterium* (prō-pē-on-ē-bak-ti'rē-um) species produces carbon dioxide, which forms the holes in Swiss cheese. Semisoft cheeses, such as Limburger, are rip-

(a)

(b)

(c)

FIGURE 28.6 Making cheddar cheese. (a) The milk has been coagulated (curd) by the action of rennin and is inoculated with ripening bacteria for flavor and acidity. Here the workers are cutting the curd into slabs. **(b)** The liquid whey is being drained away down the central trough from the slabs of curd. **(c)** The curd is milled to allow even more drainage of whey and is compressed into blocks for extended ripening. The longer the ripening period, the more acidic (sharper) the cheese.

ened by bacteria and other contaminating organisms growing on the surface. Blue and Roquefort cheeses are ripened by *Penicillium* molds inoculated into the cheese. The texture of the cheese is loose enough that adequate oxygen can reach the aerobic molds. The growth of the *Penicillium* molds is visible as blue-green clumps in the cheese. Camembert cheese is ripened in small packets so that the enzymes of *Penicillium* mold growing aerobically on the surface will diffuse into the cheese for ripening.

OTHER DAIRY PRODUCTS

Butter is made by churning cream until the fatty globules of butter separate from the liquid buttermilk. The typical flavor and aroma of butter and buttermilk are from diacetyls, a combination of two acetic acid molecules that is a metabolic end-product of the fermentation by some lactic acid bacteria. Today, buttermilk is usually not a by-product of butter-making but is made by inoculating skim milk with bacteria that form lactic acid and the diacetyls. The inoculum is allowed to grow for 12 hours or more before the buttermilk is cooled and packaged. Sour cream is made from cream inoculated with organisms similar to those used to make buttermilk.

A wide variety of slightly acidic dairy products—probably a heritage of a nomadic past—are found around the world. Many of them are part of the daily diet in the Balkans, Eastern Europe, and Russia. One such product is yogurt, which is also popular in the United States. Commercial yogurt is made from low-fat milk, from which much of the water has been evaporated in a vacuum pan. The resulting thickened milk is inoculated with a mixed culture of *Streptococcus thermophilus*, primarily for acid production, and *Lactobacillus bulgaricus* (bul-gă'ri-kus), to contribute flavor and aroma. The temperature of the fermentation is about 45°C for several hours, during which time *S. thermophilus* outgrows *L. bulgaricus*. Maintaining the proper balance between the flavor-producing and the acid-producing organisms is the secret of making yogurt.

Kefir and kumiss are popular beverages in Eastern Europe. The usual lactic acid-producing bacteria are supplemented with a lactose-fermenting yeast to give these drinks an alcohol content of 1% or 2%.

TABLE 28.3 Some Fermented Foods and Related Products

FOODS AND PRODUCTS	RAW INGREDIENTS	FERMENTING ORGANISMS	LOCATION PRODUCED
Dairy Products			
Cheeses (ripened)	Milk curd	*Streptococcus* spp., *Leuconostoc* spp.	Worldwide
Kefir	Milk	*Streptococcus lactis, Lactobacillus bulgaricus, Candida* spp.	Southwestern Asia
Kumiss	Raw mare's milk	*L. bulgaricus, Lactobacillus leichmannii, Candida* spp.	Russia
Taette	Milk	*S. lactis* var. *taette*	Scandinavia
Yogurt	Milk, milk solids	*Streptococcus thermophilus, L. bulgaricus*	Worldwide
Meat and Fish Products			
Country-cured hams	Pork hams	*Aspergillus, Penicillium* spp.	Southern United States
Dry sausages	Pork, beef	*Pediococcus cerevisiae*	Europe, United States
Fish sauces	Small fish	Halophilic *Bacillus* spp.	Southeast Asia
Izushi	Fresh fish, rice, vegetables	*Lactobacillus* spp.	Japan
Nonbeverage Plant Products			
Cocoa beans	Cacao fruits (pods)	*Candida krusei, Geotrichum* spp.	Africa, South America
Coffee beans	Coffee cherries	*Erwinia dissolvens, Saccharomyces* spp.	Brazil, Congo, Hawaii, India
Kimchi	Cabbage and other vegetables	Lactic acid bacteria	Korea
Miso	Soybeans	*Aspergillus oryzae, Saccharomyces rouxii*	Japan
Olives	Green olives	*Leuconostoc mesenteroides, Lactobacillus plantarum*	Worldwide
Poi	Taro roots	Lactic acid bacteria	Hawaii
Sauerkraut	Cabbage	*L. mesenteroides, L. plantarum*	Worldwide
Soy sauce (shoyu)	Soybeans	*A. oryzae* or *Aspergillus soyae; S. rouxii, Lactobacillus delbrueckii*	Japan
Tempeh	Soybeans	*Rhizopus oligosporus; Rhizopus oryzae*	Indonesia, New Guinea, Surinam
Breads			
Idli	Rice and bean flour	*Leuconostoc mesenteroides*	Southern India
Rolls, cakes, etc.	Wheat flours	*Saccharomyces cerevisiae*	Worldwide
San Francisco sourdough bread	Wheat flour	*Saccharomyces exiguus, Lactobacillus sanfrancisco*	Northern California
Sour pumpernickel	Wheat flour	*L. mesenteroides*	Switzerland, other areas

Source: J.M. Jay. *Modern Food Microbiology*, 3rd ed. New York: Van Nostrand Reinhold, 1986.

NONDAIRY FERMENTATIONS

Microorganisms are also used in baking, especially for bread. The sugars in bread dough are fermented by yeasts, much as they are in the fermentation of alcoholic beverages (discussed in the next section). Anaerobic conditions for production of ethanol by the yeasts are mandatory for production of alcoholic beverages. In baking, carbon dioxide forms the typical bubbles of leavened bread. Aerobic conditions favor carbon dioxide production and are encouraged as much as possible. This is the reason the bread dough is kneaded repeatedly. Whatever ethanol is produced evaporates during baking. In some breads, such as rye or sourdough, the growth of lactic acid bacteria gives the typical tart flavor (see the box in Chapter 1, p. 5).

Fermentation is also used in the production of such foods as sauerkraut, pickles, and olives. In Asia, extremely large amounts of soy sauce are produced by molds that form starch-degrading enzymes to produce fermentable sugars. This principle is used in making other Asian fermented foods, including sake, the Japanese rice wine. In soy sauce production, molds such as *Aspergillus oryzae* (a-spėr-jil'lus ô'ri-zī) are grown on wheat bran and then are allowed to act along with lactic acid bacteria on cooked soybean and crushed wheat mixtures. After this process has produced fermentable carbohydrates, a prolonged fermentation results in soy sauce. Table 28.3 lists many fermented foods.

ALCOHOLIC BEVERAGES AND VINEGAR

Microorganisms are involved in the production of almost all alcoholic beverages. Beer and ale are products of grain starches fermented by yeast (Table 28.4). Because yeasts are unable to use starch directly, the starch from grain must be converted to glucose and maltose, which the yeasts can ferment into ethanol and carbon dioxide. In this conversion, called **malting,** starch-containing grains, such as malting barley, are allowed to sprout and then are dried and ground. This product, called **malt,** contains starch-degrading enzymes (amylases) that convert cereal starches into carbohydrates that can be fermented by yeasts. For **distilled spirits,** such as whiskey, vodka, and rum, carbohydrates from cereal grains, potatoes, and molasses are fermented to alcohol. The alcohol is then distilled to make a concentrated alcoholic beverage.

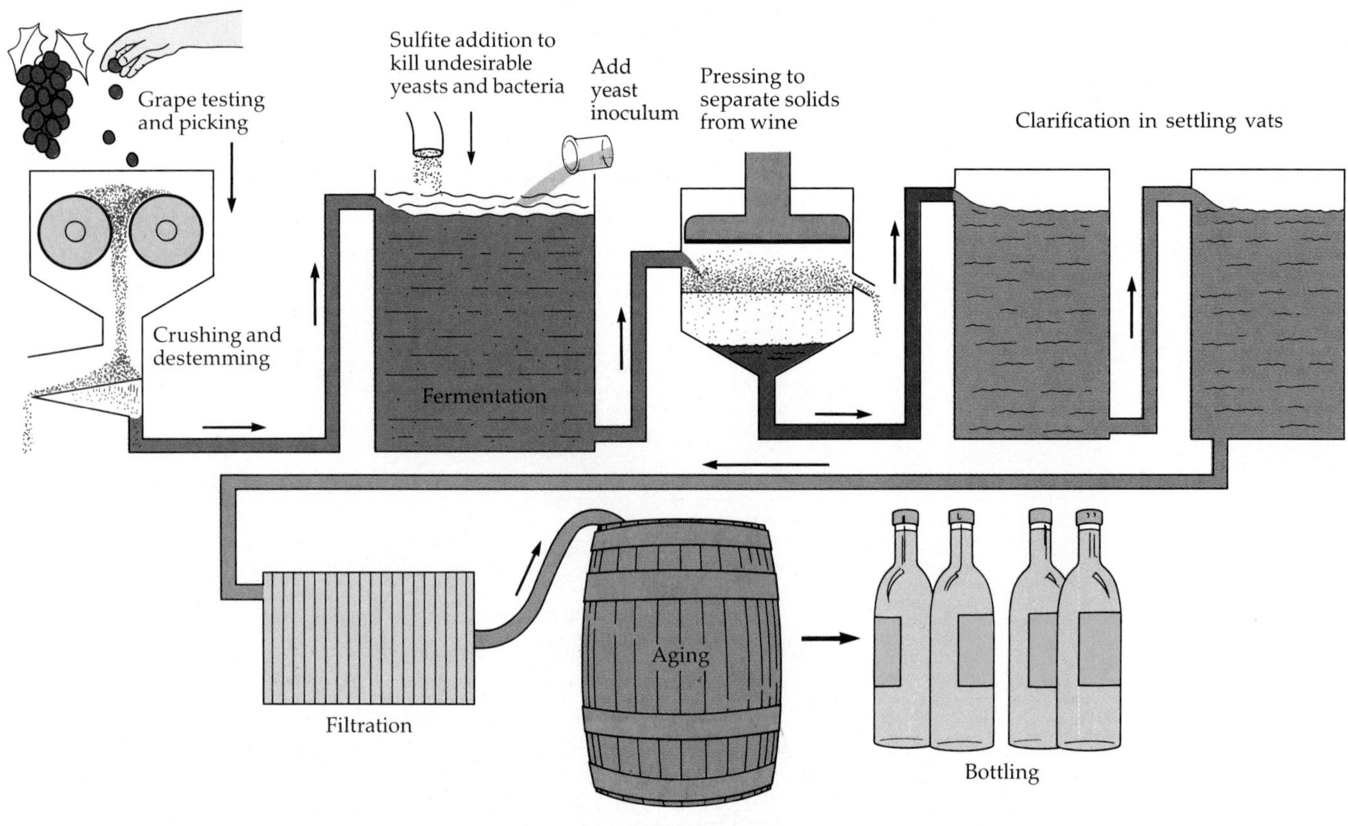

FIGURE 28.7 Basic steps in making red wine. For white wines, the pressing precedes fermentation so that the color is not extracted from the solid matter.

TABLE 28.4 Production of Alcoholic Beverages by Yeasts

BEVERAGE	YEAST	METHOD OF PREPARATION	FUNCTION OF YEAST
Root beer (pre-1905 and homemade)	*Saccharomyces cerevisiae*	Molasses, sassafras bark, wintergreen bark, and sarsaparilla root added for flavor; yeast added; incubated aerobically.	Converts sugar into carbon dioxide; by aerobic metabolism; 0.03% alcohol.
Beer	*S. carlsbergensis* (bottom yeast)	Germinated barley releases starches and amylase enzymes (malting). Enzymes in malt hydrolyze starch to fermentable sugars (mashing). Liquid (wort) sterilized. Hops (flowers) added for flavor. Yeast added, incubated at 37°–49°C.	Converts sugar into alcohol and carbon dioxide; 4.0% alcohol. Yeast grows on bottom of fermenting vessel.
Ale	*S. cerevisiae* (top yeast)	As in beer; incubated at 50°–70°C.	Converts sugar into alcohol; 6% alcohol. Yeast grows at top of fermentation vessel.
Sake	*S. cerevisiae*	*Aspergillus oryzae* converts starch in steamed rice into sugar; yeast added; incubated at 20°C.	Converts sugar into alcohol; 14%–16% alcohol.
Wine, natural	*S. cerevisiae*	Strain of grape provides various flavors and sugar concentrations. Grapes crushed into must; sulfur dioxide added to inhibit wild yeast; yeast added. Red wines: incubated at 25°C. Aged in oak for 3–5 years and in bottle for 5–15 years. White wines: incubated at 10°–15°C. Aged 2–3 years in bottle.	Converts grape sugar into alcohol; 14% or less alcohol.
Wine, sherry	*S. cerevisiae* and *S. beticus* or *S. bayanus*	As natural wine with additional surface growth (flor) at 27°C. Alcohol added to 18%–21%.	*S. beticus* grows as surface film producing aldehydes from alcohol.
Wine, sparkling (champagne)	*S. cerevisiae*	As natural wine with secondary fermentation in bottle. 2.5% sugar and yeast added to bottled wine; incubated at 15°C; bottle inverted to collect yeast in neck.	In secondary fermentation, produces carbon dioxide; yeast settles quickly.
Distilled Beverages			
Rum, Jamaica	Wild yeast	Cane molasses inoculated from previous fermentation. Oak aging adds color. Distilled to concentrate.	Converts sugar to alcohol; 50%–95% alcohol.
Brandy	*S. cerevisiae*	Fruits pressed; yeast added. Distilled to concentrate alcohol, blended with other brandies.	Converts sugar into alcohol; 40%–43% alcohol.
Whiskey	*S. cerevisiae*	Wort (see beer) is fermented by yeast. Distilled to concentrate alcohol; aged in charred oak barrels.	Converts sugar to alcohol; 50%–95% alcohol.

Wines are made from fruits, typically grapes, which contain sugars that can be used directly by yeasts for fermentation; malting is unnecessary in wine-making. Grapes usually need no additional sugars, but other fruits might be supplemented with sugars to ensure enough alcohol production. The steps of wine-making are shown in Figure 28.7. Lactic acid bacteria are important when wine is made from grapes that are especially acidic from high concentrations of malic acid. These bacteria convert the malic acid to the weaker lactic acid in a process called **malolactic fermentation.** The result is a less acidic, better-tasting wine than would otherwise be produced.

Wine producers who allowed wine to be exposed to air found that it soured from the growth of aerobic bacteria that converted the ethanol in the wine to acetic acid. The result was vinegar (*vin* means wine; *aigre* means sour). The process is now used deliberately to make vinegar. Ethanol is first produced by anaerobic fermentation of carbohydrates by yeasts. The ethanol is then aerobically oxidized to acetic acid by acetic acid-producing bacteria of the genera *Acetobacter* and *Gluconobacter*.

MICROORGANISMS AS A FOOD SOURCE

In the future, as human populations increase, if the amount of arable land does not correspondingly increase, microorganisms may become more important to the food requirements of the world. Protein is in particularly short supply. Microorganisms, which can usually double their weight in a few hours and often in less than an hour, may help solve this problem. When used as a food source, microorganisms are referred to as **single-cell protein (SCP).** Microorganisms can use as substrates many carbonaceous materials, such as cellulose, methanol, and petroleum hydrocarbons.

Experimentation is under way to develop SCP that is palatable, economical, and without harmful side-effects for humans. So far, many SCPs have failed to meet one or more of these criteria. However, one microorganism already sold as food in the United States and Mexico is the cyanobacterium *Spirulina* (spī-rü-lī'nä), which has long been grown as food in Lake Chad, Africa, and was also used by the Aztecs in Mexico. Other microorganisms are currently being used in animal feed, thus providing food for humans indirectly.

INDUSTRIAL MICROBIOLOGY

Much of industrial microbiology—the use of microbes to make industrial products—has centered on the lactic acid and ethanol fermentations that are the basis of the major food industries we just discussed. Although the lactic acid and ethanol that are produced are incidental to food fermentation, they have industrial uses. Microbes are used to make a variety of products, many of which are not related to foods. During the two World Wars, microbes were used to make chemical compounds such as glycerol and acetone. In the future, as the petroleum products from which many chemicals are synthesized become scarce, the world may come to depend on microbial fermentations of renewable substrates to produce these chemicals. The potential shortage of petroleum has also led to interest in microbes as alternative energy producers.

In the coming years, industrial microbiology will be revolutionized by the application of genetically engineered organisms. In Chapter 9 we discussed the

FIGURE 28.8 Bioreactors. (a) Drawing of a continuously stirred bioreactor. **(b)** A bioreactor tank.

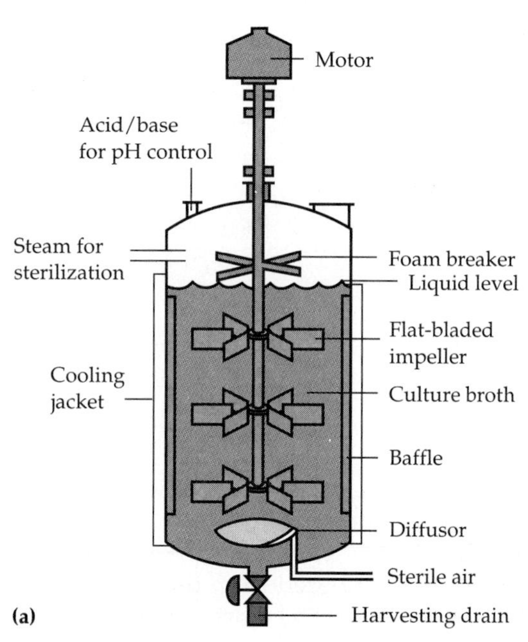

methods for making these modified organisms and some of the products derived from them—what is now known as **biotechnology.**

Fermentation Technology

Industrial production of microbial products usually involves fermentation. In industry, this term refers to large-scale cultivation of microbes or other single cells to produce a commercially valuable substance (see the box on p. 120 for other definitions of fermentation). We have just discussed the most familiar examples: the anaerobic food fermentations used in the dairy, brewing, and wine-making industries. Much of the same technology, with the frequent addition of aeration, has been adapted to make other industrial products, such as insulin and human growth hormone, from genetically engineered microorganisms, including aerobic cells. Industrial fermentation is also used in biotechnology to obtain useful products from genetically engineered plant and animal cells. For example, animal cells are used to make monoclonal antibodies (see Chapter 17).

Vessels for industrial fermentations are called **bioreactors** and are designed with close attention to aeration, pH control, and temperature control. There are many different designs, but the most widely used bioreactors are of the continuously stirred type (Figure 28.8). The air is introduced through a diffuser at the bottom (which breaks up the incoming airstream to maximize aeration) and a series of stirrer paddles and stationary wall baffles that keep the microbial suspension agitated. Oxygen is not very soluble in water, and keeping the heavy microbial suspension well aerated is difficult. Highly sophisticated designs have been developed to achieve maximum efficiency in aeration and other growth requirements, including medium formulation. The high value of the products of genetically engineered microorganisms and eucaryotic cells has stimulated development of newer types of bioreactors and computerized controls for them.

Generally speaking, the microbes in industrial fermentation produce either primary metabolites, such as ethanol, or secondary metabolites, such as penicillin. A primary metabolite is formed essentially at the same time as the cells, and the production curve follows the cell population curve almost in parallel, with only minimal lag. Secondary metabolites are not produced until the microbe has largely completed its growth phase (known as the **trophophase**) and has entered the stationary phase of the growth cycle (Figure 28.9). The following period, during which most of the secondary metabolite is produced, is known as the **idiophase.** The secondary metabolite may be a microbial conversion of a primary metabolite. Alternatively, it may be a product of metabolism of the original growth medium

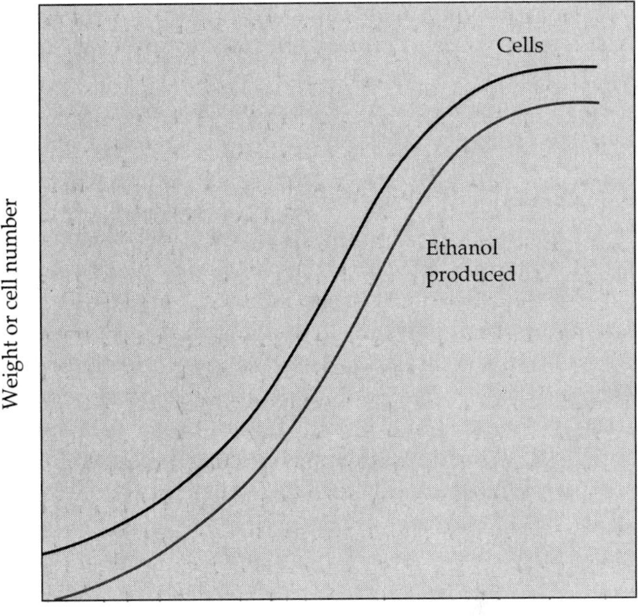

(a) Primary metabolites

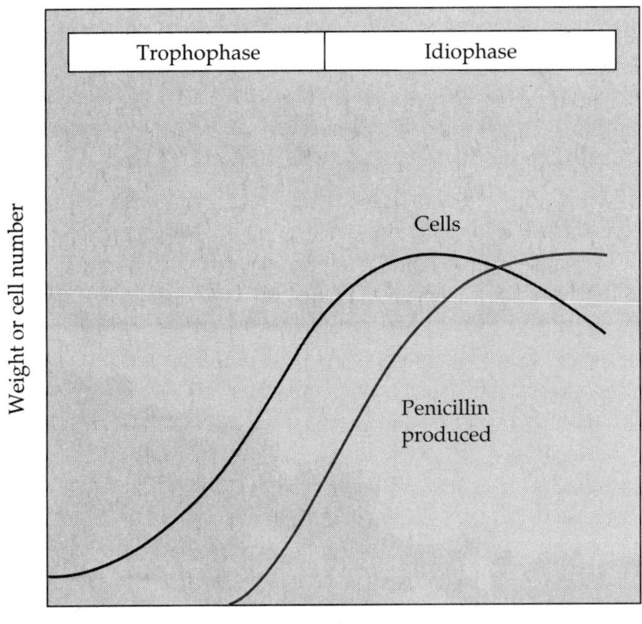

(b) Secondary metabolites

FIGURE 28.9 Primary and secondary fermentation. (a) Primary metabolite. A primary metabolite such as ethanol from yeasts has a production curve that lags only slightly behind the line showing cell growth. **(b)** Secondary metabolite. A secondary metabolite such as penicillin begins to be produced only after the main growth phase of the cell (trophophase) is completed. The main production of the secondary metabolite occurs during the stationary phase of cell growth (idiophase).

that the microbe makes only after considerable numbers of cells and a primary metabolite have accumulated.

Strain improvement is also an ongoing activity in industrial microbiology. Probably the best-known example is that of the mold used for penicillin production. The original culture of *Penicillium* did not produce penicillin in large enough quantities for commercial use. A more efficient culture was isolated from a cantaloupe from a Peoria, Illinois, supermarket. This strain was treated variously with ultraviolet light, X rays, and nitrogen mustard (a chemical mutagen). Selections of mutants, including some that arose spontaneously, quickly increased the production rates by a factor of more than a hundred. Improvements in fermentation techniques have nearly tripled even this yield.

IMMOBILIZED ENZYMES AND ORGANISMS

Microbes are, in many ways, packages of enzymes. Industries have been making limited use of free enzymes to manufacture many products such as high-fructose syrups or semisynthetic antibiotics such as ampicillin. To do this, the enzyme must be immobilized on the surface of some solid support or otherwise manipulated so that it can convert a continuous flow of substrate to product without being lost. These techniques have also been adapted to live whole cells, and sometimes even to dead cells (Figure 28.10). Whole-cell systems are difficult to aerate, and they lack the single-enzyme specificity of immobilized enzymes. However, whole cells are advantageous if the process requires a series of steps that can be carried out by one microbe's enzymes. They also have the advantage of allowing continuously running processes with large cell populations operating at high reaction rates. Immobilized cells, which are usually anchored to microscopically small spheres or fibers, are presently used to make high-fructose syrups, aspartic acid, and numerous other products of biotechnology.

FIGURE 28.10 Immobilized cells. In some industrial processes, the cells are immobilized on surfaces such as the silk fibers shown here. The substrate flows past the immobilized cells.

Industrial Products

In this section, we will discuss some of the more important commercial microbial products and the growing alternative energy industry.

AMINO ACIDS

Amino acids have become a major industrial product from microorganisms. For example, over 165,000 tons of glutamic acid, used to make the flavor enhancer monosodium glutamate, are produced every year. Certain amino acids, such as lysine, cannot be synthesized by animals and are present only at low levels in the normal diet. Therefore, the commercial synthesis of lysine and some of the other essential amino acids as food supplements is an important industry. More than 35,000 tons of lysine are produced every year. In nature, microbes rarely produce amino acids in excess of their own needs because feedback inhibition (see Chapter 8) prevents wasteful production of primary metabolites. Commercial microbial production of amino acids depends on specially selected mutants and sometimes on ingenious manipulations of metabolic pathways.

Lysine is produced by a bacterium, *Corynebacterium glutamicum* (glü-tam'i-kum), which normally produces both threonine and lysine. If both accumulate, feedback inhibition prevents further production of both amino acids, which share a common enzyme in their synthetic pathway. A mutant bacterium that lacks the enzymes necessary to make threonine is used to produce lysine in large amounts. Enough threonine is added to the medium to allow this mutant organism to grow, but not enough to trigger feedback inhibition. The organism then continues to produce lysine because lysine alone does not cause feedback inhibition.

Glutamic acid is produced by the same bacterium that produces lysine. The bacterium is induced to excrete glutamic acid when it is provided with only a minimal amount of the vitamin biotin. The plasma membrane is weakened by the vitamin deficiency, and the glutamic acid leaks into the medium. This tech-

MICROBIOLOGY HIGHLIGHTS

Microbial Miner

For thousands of years, refiners have been extracting copper from low-grade copper ore by a process called *leaching*. Until 1957, however, nobody realized that the extraction would be impossible without the bacterium *Thiobacillus ferrooxidans,* which thrives in acidic environments such as mining sites. Thanks to this microbe, millions of tons of copper are recovered from billions of tons of low-grade ore every year.

Thiobacillus ferrooxidans is one of several species of chemoautotrophic bacteria, which are becoming increasingly important in the extraction of a variety of metals. The chemoautotrophs obtain their energy from the oxidation of inorganic substances (iron and sulfur, in the case of *Thiobacillus ferrooxidans*). This allows them to stimulate reactions involving minerals that are unaffected by, and may even be poisonous to, other bacteria. Moreover, chemoautotrophic bacteria thrive in the absence of organic matter because they extract their carbon directly from the carbon dioxide in the atmosphere.

The bacteria are naturally present in the leaching solution prepared at the mining site. The refiners grind the copper-bearing rock and heap it into a pit, called a *dump* (see photo). Then they pour water mixed with sulfuric acid (the *leach solution*) through the dump. When the leach solution runs out through the bottom of the dump, it contains copper sulfate ($CuSO_4$) in solution. The refiners then add metallic iron to the leach solution. The iron reacts with the copper sulfate to release metallic copper (see Figure 28.12).

Thiobacillus frees copper from the ore in a variety of ways. Most commonly, the bacillus reacts with iron and sulfur compounds that are in the rock. The name *ferrooxidans* means "oxidizing iron compounds."

The rock is likely to contain compounds of iron and sulfur, such as iron pyrite (FeS_2), also known as "fool's gold." Note that the pyrite molecule contains a divalent *ferrous* ion (Fe^{2+}) and two monovalent sulfur ions (S^-). As the leach solution trickles down through the ore-bearing rock, *Thiobacillus ferrooxidans* oxidizes the Fe^{2+} (ferrous) ion by removing an electron from it. This turns the ferrous ion into a trivalent *ferric* (Fe^{3+}) ion.

The sulfur that was in the FeS_2 molecule can now combine with hydrogen ions and oxygen molecules to form sulfuric acid (H_2SO_4).

The equation describing the chemical reaction looks like this:

$$\underset{\text{Iron pyrite}}{FeS_2} + \underset{\text{Hydrogen ions}}{2H^+} + \underset{\text{Oxygen}}{2O_2} \longrightarrow$$

$$\underset{\text{Ferric ion}}{Fe^{3+}} + \underset{\text{Sulfuric acid}}{H_2SO_4} + \underset{\text{Sulfur}}{S^0}$$

In addition to the iron pyrite, the ore contains compounds of copper and sulfur, such as copper sulfide (CuS). The ferric ion (Fe^{3+}) oxidizes the monovalent copper ion (Cu^+) to create a divalent copper ion (Cu^{2+}), which then combines with the *sulfate* ion (SO_4^{2-}), contributed by the sulfuric acid, to form copper sulfate ($CuSO_4$), as follows:

$$\underset{\text{Copper sulfide}}{CuS} + \underset{\text{Ferric ion}}{Fe^{3+}} + \underset{\text{Sulfuric acid}}{H_2SO_4} \longrightarrow$$

$$\underset{\text{Copper sulfate}}{CuSO_4} + \underset{\text{Ferrous ion}}{Fe^{2+}} + \underset{\text{Hydrogen ions}}{2H^+} + \underset{\text{Sulfur}}{S^0}$$

Now that the leach solution has copper sulfate dissolved in it, the refiners add metallic iron (in the form of iron filings or scrap iron). The iron reacts with the copper sulfate to form ferrous sulfate and metallic copper, as in the following equation:

$$\underset{\text{Copper sulfate}}{CuSO_4} + \underset{\text{Iron}}{2Fe^0} + \underset{\text{Sulfuric acid}}{H_2SO_4} \longrightarrow$$

$$\underset{\text{Ferrous sulfate}}{2FeSO_4} + \underset{\text{Copper}}{Cu^0} + \underset{\text{Hydrogen}}{2H^+}$$

Thiobacillus can now reoxidize the ferrous ion to ferric ion.

The mining industry has begun to use chemoautotrophic bacteria to retrieve other elements that are present in low concentrations in ores. *Thiobacillus* is being used to extract cobalt at the Idaho National Laboratory, and U.S. Gold Corporation uses *Thiobacillus* to free gold from rock.

Microbial mining offers a possible solution to another problem that threatens modern industrial societies—pollution of the soil and groundwater by unextracted metallic ions in waste water. Chemoautotrophic bacteria that can take up and accumulate toxic metal ions in their cells may be useful in removing metal pollutants from waste water.

Biological leaching of copper ores.

nique of inducing the cell to form a leaky membrane is also sometimes used in antibiotic production.

Two microbially synthesized amino acids, **phenylalanine** and **aspartic acid,** have become important as ingredients in the sugar-free sweetener aspartame (NutraSweet®).

In applications in which only the L-isomer of an amino acid is wanted, microbial production, which forms only the L-isomer, has an advantage over chemical production, which forms both the D- and L-isomers.

About the only amino acids that are not produced by microbial fermentation are methionine, alanine, glycine, and cysteine. These are chemically synthesized.

CITRIC ACID

Citric acid is a constituent of citrus fruits, such as oranges and lemons, and at one time these were its only industrial source. However, over a hundred years ago, citric acid was identified as a product of mold metabolism. It can be chemically synthesized, but not cheaply enough to replace production by mold fermentation. Citric acid has an extraordinary range of uses beyond the obvious ones of giving tartness and flavor to foods. It is an antioxidant and pH adjuster in many foods, and in dairy products it often serves as an emulsifier. Most citric acid is used in foods, and only about 30% of its applications are industrial and pharmaceutical. Well over 300,000 tons of citric acid are produced every year. Much of it is produced by a mold, *Aspergillus niger* (nī'jėr), using molasses as a substrate. The mold excretes citric acid when provided with only a limited supply of iron and manganese.

ENZYMES

Enzymes are widely used in different industries. For example, amylases are used in the production of syrups from corn starch, in the production of paper sizing (a coating for smoothness, as on this page), and in the production of glucose from starch. Glucose isomerase is an important enzyme; it converts the glucose that amylases form from starches into fructose, which is

TABLE 28.5 Some Microbial Enzymes Produced Commercially

ENZYME	MICROORGANISM	USE OF ENZYME
Amylase	*Aspergillus niger*, *Aspergillus oryzae*, *Bacillus subtilis*	Baking: Flour supplement Brewing: Mashing Food: Precooked foods, syrup Pharmaceuticals: Digestive aids Starch: Cold-water laundry Textiles: Desizing agent
Cellulase	*A. niger*	Food: Liquid coffee concentrate
Dextransucrase	*Leuconostoc mesenteroides*	Pharmaceuticals: Dextran
Glucose oxidase	*A. niger*	Food: Glucose removal from egg solids Pharmaceuticals: Test papers
Invertase	*Saccharomyces cerevisiae*	Candy: Prevents granulation in soft center Food: Artificial honey
Lactase	*Saccharomyces fragilis*	Dairy: Prevents crystallization of lactose in ice cream and concentrated milk
Lipase	*A. niger*	Dairy: Flavor production in cheese
Pectinase	*A. niger*	Wine and juice: Clarification
Penicillinase	*B. subtilis*	Pharmaceuticals: Diagnostic agent
Protease	*A. oryzae*	Brewing: Beer stabilizer Baking: Bread-making Food: Meat tenderizer Pharmaceuticals: Digestive aid Textiles: Desizing
Streptodornase	*Streptococcus pyogenes*	Pharmaceuticals: Reagent, wound debridement

Source: J.R. Porter, "Microbiology and the food and energy crisis." *ASM News* 40(11):822, November 1974.

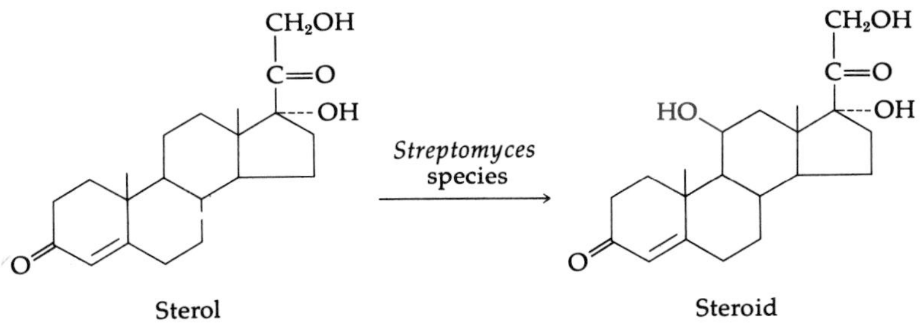

FIGURE 28.11 Production of steroids. Conversion of a precursor compound such as a sterol into a steroid by *Streptomyces*. The addition of a hydroxyl group to carbon number 11 (in color on steroid) might require more than 30 steps by chemical means, but the microorganism can make the addition in only one step.

used in place of sucrose as a sweetener in many foods. Probably half of the bread baked in this country is made with proteases, which adjust the amounts of glutens (protein) in wheat so that baked goods are improved or made uniform. Other proteolytic enzymes are used as meat tenderizers or in detergents as an additive to remove proteinaceous stains. Rennin, which is an enzyme used to form curds in milk, is usually produced commercially by fungi. Several enzymes produced commercially by microorganisms are listed in Table 28.5.

VITAMINS

Vitamins are sold in large quantities combined in tablet form. They are used as individual food supplements. Microbes can provide an inexpensive source of some vitamins. **Vitamin B_{12}** is produced by *Pseudomonas* and *Propionibacterium* spp. Cobalt is a part of the molecule, and cobalt chloride must be provided in the medium. **Riboflavin** is another vitamin produced by fermentation, mostly by fungi such as *Ashbya gossypii* (ash'bē-ä gos-sip'ē'ē).

PHARMACEUTICALS

Modern pharmaceutical microbiology developed after World War II, with the production of antibiotics. To increase the number of random mutations, antibiotic-producing organisms were subjected to ultraviolet radiation in the hope that some mutations would be useful. These and other techniques soon led to strains that produce antibiotics in amounts thousands of times greater than the original isolates did. All antibiotics were originally the products of microbial metabolism. Many antibiotics are still produced by microbial fermentations, and work continues on selection of more productive mutants by nutritional and genetic manipulations. Antibiotics are typically made industrially by inoculating a solution of growth medium with spores of the appropriate mold or streptomycete and vigorously aerating it. After the antibiotic reaches a satisfactory concentration, it is extracted by solution, precipitation, and other standard industrial procedures.

Vaccines are a product of industrial microbiology. Many antiviral vaccines are mass produced in chicken eggs or cell cultures. Production of vaccines against bacterial diseases usually requires the growth of large amounts of the bacteria. Genetic engineering techniques are increasingly important in the area of vaccine development and production (see Chapter 9).

Steroids (Figure 28.11) are a very important group of chemicals that include cortisone, used as an anti-inflammatory drug, and estrogens and progesterone, used in birth control pills. It is difficult to recover steroids from animal sources, but microorganisms can synthesize steroids from sterols or from related compounds derived from animals and plants. Certain strains of molds and bacteria (particularly streptomycetes) can produce steroids by converting chemical groups on sterols and similar molecules. Genetic engineering will probably revolutionize the microbial production of medical pharmaceuticals.

URANIUM AND COPPER

Thiobacillus ferrooxidans is used in the recovery of otherwise unprofitable grades of uranium and copper ores (see Figure 28.12 and the box on p. 715). For example, when solutions containing ferric ion (Fe^{3+}) are washed through deposits of insoluble copper compounds, the copper is oxidized to soluble compounds. In this process, Fe^{3+} is reduced to Fe^{2+}. The Fe^{2+} can be reoxidized by *T. ferrooxidans*. The soluble copper then moves out of the ores and is reclaimed.

MICROORGANISMS

Microorganisms themselves sometimes constitute an industrial product. A simple example is the sale of yeast cultures for making wines at home. Bakers' yeast is produced in large aerated fermentation tanks. At the end of the fermentation, the contents of the tank are about 4% yeast solids. The concentration can be much higher, but only at a cost in efficiency. The cells are harvested by continuous centrifuges and are pressed into the familiar yeast cakes sold in the supermarket

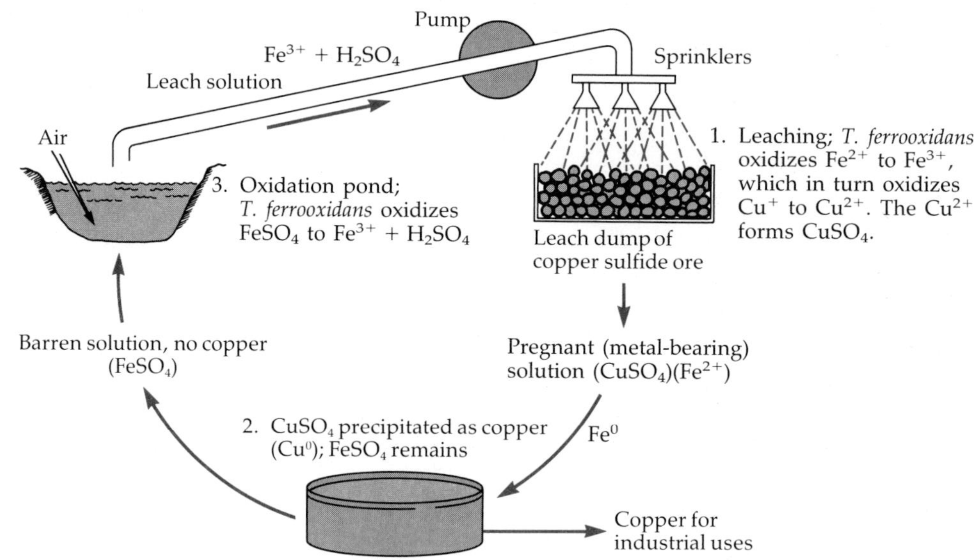

FIGURE 28.12 Biological leaching of copper ores. The chemistry of the process is considerably more complicated than shown here. The *Thiobacillus* bacteria essentially derive their energy from oxidation of sulfides to sulfates.

for home baking. Wholesale bakers purchase yeast in 50-lb boxes.

Other important microbes that are sold industrially are the symbiotic nitrogen-fixing bacteria *Rhizobium* and *Bradyrhizobium*. These organisms are usually mixed with peat moss to preserve moisture; the farmer mixes the peat moss and bacterial inoculum with the seeds of legumes to ensure infection of the plants with efficient nitrogen-fixing strains (see Chapter 27). Gardeners for many years have used the insect pathogen *Bacillus thuringiensis* to control leaf-eating insect larvae. Commercial preparations containing toxic crystals and endospores of this organism are available at almost any gardening supply store.

Alternative Energy Sources Using Microorganisms

Each year, the United States produces many hundreds of tons of organic wastes from crops, forests, and municipalities. This waste is called biomass, and so far it has been disposed of by burning or placing in a dump. Conversion of biomass into alternative fuel sources, called **bioconversion,** would help meet our energy needs as well as help dispose of troublesome accumulations of solid waste. The bioconversion of organic materials into alternative fuels by microorganisms is a developing industry.

Methane is probably one of the most convenient energy sources produced from biomass conversion. Methane was discussed in Chapter 27 as a product of the anaerobic treatment of sewage sludge. New York City operates a plant that produces 10 million cubic feet of methane per day from wastes at one of its landfill sites. Large cattle-feeding lots must dispose of immense amounts of animal manure, and considerable effort has been devoted to devising practical methods for producing methane from these wastes. A major problem with any scheme for large-scale methane production is the need to economically concentrate the widespread biomass material. If it could be concentrated, the animal and human wastes in the United States could supply much of our energy now supplied by fossil fuels and natural gas.

An interesting aspect of methane production from animal wastes is the inhibition of the process by certain antibiotics fed to cattle. The purpose of these antibiotics is to suppress methane production from the largely cellulosic foods in the animal's rumen. It is desirable that as much as possible of the nutrients in these foods be converted into meat or milk, but certain anaerobic bacteria convert some of these nutrients into methane gas. This gas tends to be lost to the atmosphere and represents an inefficient use of nutrients. The antibiotics used to inhibit these bacteria are excreted in the animal's wastes, and, unfortunately, they inhibit desired methane production from the wastes.

The agricultural industry has encouraged the production of ethanol from agricultural products. *Gasohol* (90% gasoline + 10% ethanol) is available in many parts of the United States and is used in automobiles worldwide. Corn is presently the most frequently used substrate, but eventually agricultural waste products may be used. A by-product of ethanol production is a large quantity of ethanol-producing yeasts. The Food and Drug Administration has approved use of such yeasts as a food additive in human snack food. In the future, we will see more efforts to recover the energy now lost in industrial, municipal, and rural wastes.

Industrial Microbiology and the Future

Microbes have been exceedingly useful to mankind, even when their existence was unknown. They will always remain an essential part of many basic food-processing technologies. The development of genetic engineering has further intensified interest in industrial microbiology by expanding the potential for new products and applications. As new biotechnology applications and products enter the marketplace, they will affect our lives and well-being in ways that we can only guess about today.

STUDY OUTLINE

FOOD MICROBIOLOGY (pp. 700–712)

Food Preservation and Spoilage (pp. 700–706)

1. The earliest methods of preserving foods were drying, the addition of salt or sugar, and fermentation.

CANNING (pp. 701–703)

1. In 1810, Nicholas Appert devised a canning process that involved keeping air out of sealed cans and heating food in the cans.
2. In the nineteenth century, steam under pressure (pressure cooking) was used in canning.

Commercial Sterilization (pp. 701–702)

1. Today, commercial sterilization heats canned foods to the minimum temperature necessary to destroy *C. botulinum* endospores while minimizing alteration of the food.
2. The commercial sterilization process uses sufficient heat to reduce a population of *C. botulinum* by 12 logarithmic cycles (12D treatment).
3. Endospores of thermophiles can survive commercial sterilization.

Spoilage of Canned Food (pp. 702–703)

1. Canned foods stored above 45°C can be spoiled by thermophilic anaerobes.
2. Thermophilic anaerobic spoilage is sometimes accompanied by gas production; if no gas is formed, the spoilage is called flat sour spoilage.
3. Spoilage by mesophilic bacteria is usually from improper heating procedures or leakage.
4. Acidic foods can be preserved by heat of 100°C because microorganisms that survive are not capable of growth in a low pH.
5. *Byssochlamys, Aspergillus,* and *Bacillus coagulans* are acid-tolerant and heat-resistant organisms that can spoil acidic foods.

PASTEURIZATION (p. 703)

1. Louis Pasteur determined that, because spoilage was caused by only certain microorganisms, it was not necessary to kill all microorganisms to prevent spoilage.
2. Pasteurization is the application of heat that selectively kills microorganisms without greatly altering the quality of the food.

ASEPTIC PACKAGING (pp. 703–704)

1. Presterilized materials are assembled into packages and aseptically filled with heat-sterilized liquid foods.

LOW-TEMPERATURE PRESERVATION (p. 704)

1. Many microorganisms, although few pathogens, will grow slowly at refrigeration temperatures (0°C to 7°C).
2. *Clostridium botulinum* type E, *Yersinia enterocolitica,* and *Listeria monocytogenes* are pathogens that will grow at low temperatures.
3. Freezing might not kill bacteria, but it will prevent their growth.

RADIATION AND FOOD PRESERVATION (pp. 704–705)

1. Gamma radiation can be used to sterilize food, kill insects and parasitic worms, and prevent sprouting of fruits and vegetables.
2. Microwaves kill bacteria indirectly by heating the food.

CHEMICAL PRESERVATIVES (pp. 705–706)

1. Sodium benzoate, sorbic acid or potassium sorbate, and calcium propionate are added to foods to inhibit the growth of molds.
2. These chemicals are metabolized by humans.
3. Sodium nitrate and sodium nitrite are added to meats to preserve the red color and to prevent the germination of *C. botulinum* endospores.
4. Nitrates are converted to nitrites by some bacteria during anaerobic respiration; nitrites are the active ingredient.

Foodborne Infections and Microbial Intoxications (p. 706)

1. In a foodborne infection, the pathogen grows in the infected host.
2. Gastroenteritis, typhoid fever, and dysentery are foodborne infections.

3. Microbial intoxication results from ingestion of a toxin produced by microbes growing in food.
4. Botulism, staphylococcal food poisoning, and mycotoxicoses are examples of microbial intoxications.
5. Pasteurized dairy milk might contain gram-negative psychrotrophs.
6. Pasteurized products of cultured milk (such as buttermilk and sour cream) contain large numbers of lactic acid bacteria.
7. Public health service standards for Grade A pasteurized milk are a standard plate count of fewer than 20,000 bacteria per milliliter and not more than 10 coliforms per milliliter.
8. Dry milk products must have a standard plate count of fewer than 30,000 bacteria per gram or a coliform count fewer than 10 per gram.
9. The species of bacteria in unpasteurized (raw) milk are not controlled, and raw milk can transmit disease.

The Role of Microorganisms in Food Production (pp. 706–712)

CHEESE (pp. 707–708)

1. The milk protein casein curdles because of the action by lactic acid bacteria or the enzyme rennin.
2. Cheese is the curd separated from the liquid portion of milk, called whey.
3. Hard cheeses are produced by lactic acid bacteria growing in the interior of the curd.
4. The growth of microorganisms in cheeses is called ripening.
5. Semisoft cheeses are ripened by bacteria growing on the surface.
6. Soft cheeses are ripened by *Penicillium* growing on the surface.

OTHER DAIRY PRODUCTS (p. 708)

1. Old-fashioned buttermilk was produced by lactic acid bacteria growing during the butter-making process.
2. Commercial buttermilk is made by letting lactic acid bacteria grow in skim milk for 12 hours.
3. Sour cream, yogurt, kefir, and kumiss are produced by lactobacilli, streptococci, or yeasts growing in low-fat milk.

NONDAIRY FERMENTATIONS (p. 710)

1. Sugars in bread dough are fermented by yeast to ethanol and CO_2; the CO_2 causes the bread to rise.
2. Sauerkraut, pickles, olives, and soy sauce are the products of microbial fermentations.

ALCOHOLIC BEVERAGES AND VINEGAR (pp. 710–712)

1. Carbohydrates obtained from grains, potatoes, or molasses are fermented by yeasts to produce ethanol in the production of beer, ale, and distilled spirits.
2. The sugars in fruits such as grapes are fermented by yeasts to produce wines.
3. In wine-making, lactic acid bacteria convert malic acid into lactic acid in malolactic fermentation.
4. *Acetobacter* and *Gluconobacter* oxidize ethanol in wine to acetic acid (vinegar).

MICROORGANISMS AS A FOOD SOURCE (p. 712)

1. Microorganisms can be used to produce single-cell protein (SCP) from otherwise unusable substrates such as cellulose.

INDUSTRIAL MICROBIOLOGY (pp. 712–719)

1. Microorganisms produce alcohol, lactic acid, glycerol, and antibiotics that are used in food manufacturing and other industrial processes.
2. Industrial microbiology has been revolutionized by the ability of genetically engineered cells to make many new products.
3. Biotechnology is a way of making commercial products by using living organisms.

Fermentation Technology (pp. 713–714)

1. In industry, the growth of cells on a large scale is called fermentation.
2. Industrial fermentations are carried on in bioreactors, which control aeration, pH, and temperature.
3. Primary metabolites such as ethanol are formed as the cell grows (during the trophophase).
4. Secondary metabolites such as penicillin are produced during the stationary phase (idiophase).
5. Mutant strains that produce a desired product can be selected.

IMMOBILIZED ENZYMES AND ORGANISMS (p. 714)

1. Enzymes or whole cells can be bound to solid spheres or fibers. When substrate passes over the surface, enzymatic reactions change the substrate to the desired product.

Industrial Products (pp. 714–718)

1. Most amino acids used in foods and medicine are produced by bacteria.
2. Microbial production of amino acids can be used to produce L-isomers; chemical production results in both D- and L-isomers.
3. Lysine and glutamic acid are produced by *Corynebacterium glutamicum.*
4. Citric acid, used in foods, is produced by *Aspergillus niger*.
5. Enzymes used in manufacturing foods, medicines, and other goods are produced by microbes.
6. Some vitamins used as food supplements are made by microorganisms.
7. Vaccines, antibiotics, and steroids are products of microbial growth.
8. The metabolic activities of *T. ferrooxidans* can be used to recover uranium and copper ores.
9. Yeasts are grown for wine- and bread-making; other microbes (*Rhizobium, Bradyrhizobium,* and *Bacillus thuringiensis*) are grown for agricultural use.

Alternative Energy Sources Using Microorganisms (p. 718)

1. Organic waste, called biomass, can be converted by microoorganisms into alternative fuels, a process called bioconversion.
2. Fuels produced by microbial fermentation are methane and ethanol.

Industrial Microbiology and the Future (p. 719)

1. Genetic engineering will enhance the ability of industrial microbiology to produce medicines and other useful products.

STUDY QUESTIONS

Review

1. Why are foods preserved?
2. List five methods used to preserve foods.
3. Matching.

____ Sorbic acid	(a) Sterilization
____ 5°C	(b) Pasteurization
____ 72°C for 15 seconds	(c) Low-temperature preservation
____ 121°C for 15 minutes	(d) Chemical preservation
____ 12D treatment	

4. Since many mesophilic bacteria can grow at low temperatures, what value is it to store food in a refrigerator?
5. Define pasteurization and aseptic packaging.
6. Pasteurization does not kill all microorganisms, so food can still spoil. Why, then, are dairy products pasteurized?
7. State one advantage and one disadvantage of the addition of nitrites to foods.
8. Outline the steps in the process of wine production.
9. Outline the steps in the production of cheese, and compare the production of hard and soft cheeses.
10. List two examples of whole cells as the desired product of industrial fermentation.
11. How can microorganisms provide energy sources? What metabolic processes can result in fuels?
12. Nicholas Appert's invention was the forerunner of hermetically sealed cans. Why do hermetically sealed cans prevent food spoilage?
13. Why is a can of blackberries preserved by commercial sterilization typically heated to 100°C instead of at least 116°C?
14. Under what conditions would you expect to find thermophiles spoiling foods? Psychotrophs?
15. Discuss the importance of industrial microbiology.
16. Explain the processes involved in the production of sourdough bread.

Challenge

1. Which bacteria seem to be most frequently used in the production of food? Can you guess why?
2. Why do the following processes preserve food?
   (a) Fermentation
   (b) Salting
   (c) Drying
3. Three to five days after eating Thanksgiving dinner at a restaurant, 112 people developed fever and gastroenteritis. All the food had been consumed except for five "doggie" bags. Bacterial analysis of the mixed contents of the bags (containing roast turkey, giblet gravy, and mashed potatoes) showed the same bacterium that was isolated from the patients. The gravy had been prepared from giblets of 43 turkeys that had been refrigerated for three days prior to preparation. The uncooked giblets were ground in a blender and added to a thickened, hot stock mixture. The gravy was not reboiled and was stored at room temperature throughout Thanksgiving Day. What was the source of the illness? What was the most likely etiologic agent? Was this an infection or an intoxication? (*Hint:* Refer to Chapter 25.)
4. *Methylophilus methylotrophus* can convert methane (CH_4) into proteins. Amino acids are represented by

$$H_2N-\underset{H}{\overset{R}{\underset{|}{\overset{|}{C}}}}-C\begin{matrix}\nearrow O \\ \searrow OH\end{matrix}$$

Diagram a pathway illustrating the production of at least one amino acid.

5. An outbreak of typhoid fever occurred following a family gathering of 293 people. Cultures from 17 of these persons yielded *Salmonella typhi*. Nine foods prepared by three foodhandlers were available at the event.

Foods Consumed	Percent Ill
Salad and roast beef	60
Noodles and baked beans	42
Noodles and egg salad	12
Egg salad and roast beef	0
Baked beans and fruit	0

What is the most likely source of *S. typhi?* Which tests would verify this?

6. Suppose you are culturing an organism that produces enough lactic acid to kill itself in a few days.

(a) How can the use of a bioreactor help you maintain the culture for weeks or months? The graph at right shows conditions in the bioreactor.

(b) If your desired product is a secondary metabolite, when can you begin collecting it?

(c) If your desired product is the cells themselves, and you want to maintain a continuous culture, when can you begin harvesting?

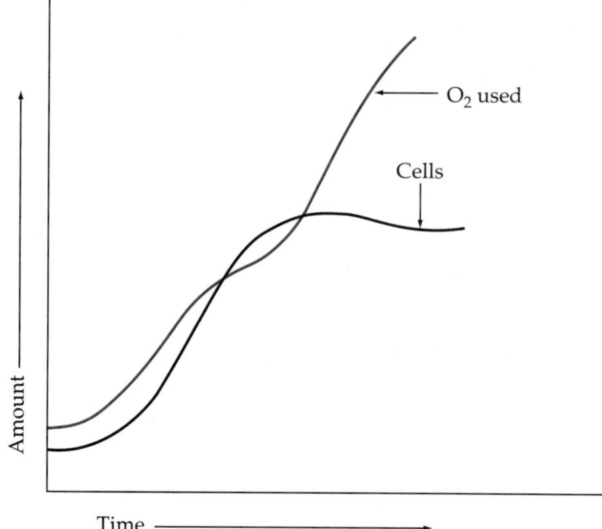

FURTHER READING

Banwart, G.J. *Basic Food Microbiology,* 2nd ed., New York: Van Nostrand Reinhold, 1989. A comprehensive textbook including foodborne illnesses, control of microorganisms, and useful microorganisms.

Brierley, C.L. "Microbiological mining." *Scientific American* 247(2):44–53, August 1982. Discusses the role of *Thiobacillus* in leaching copper from ore.

Crueger, W., and A. Crueger. *Biotechnology: A Textbook of Industrial Microbiology,* 2nd ed. Sunderland, Mass.: Sinauer, 1990. This text covers the principles and processes for production of enzymes, antibiotics, single-cell protein, and fuels.

Hesseltine, C.W. "Microbiology of oriental fermented foods." *Annual Review of Microbiology* 37:575–601, 1983. An interesting monograph on the bacteria and fungi used to prepare traditional Asian foods.

Kosikowski, F.V. "Cheese." *Scientific American* 252(5):88–99, May 1985. Describes how types of milk, species of microbes, and ripening times are used to produce 2000 cheese varieties.

Reed, G. *Prescott and Dunn's Industrial Microbiology,* 4th ed. Westport, Conn.: AVI, 1983. A standard reference book in industrial microbiology.

Rose, A.H., ed. *Economic Microbiology,* 8 vols. New York: Academic Press. Published over several years, these volumes cover food, energy, and other aspects of applied microbiology.

Sawyer, C.A.D., and J.J. Pestka. "Foodservice systems: Presence of injured bacteria in foods during food product flow." *Annual Review of Microbiology* 39:51–67, 1985. An analysis of microbiological quality of foods that are prepared and stored, both hot and cold, before serving.

Society for Industrial Microbiology. *Developments in Industrial Microbiology.* Arlington, Va.: Society for Industrial Microbiology. Proceedings of the general meeting of the Society for Industrial Microbiology, published annually. Each volume contains discussions of current advances in food, dairy, and pharmaceutical microbiology.

Zeikus, J.G. "Chemical and fuel production by anaerobic bacteria." *Annual Review of Microbiology* 34:423–464, 1980. A summary of the technology and practicality of producing hydrocarbon fuels and organic solvents from agricultural and municipal wastes.

APPENDIX A

Classification of Bacteria According to *Bergey's Manual of Systematic Bacteriology*

Kingdom Procaryotae

Division I.	**Gracilicutes**	Procaryotes with thin cell walls, implying a gram-negative type of cell wall
Division II.	**Firmicutes**	Procaryotes with thick and strong skin, indicating a gram-positive type of cell wall
Division III.	**Tenericutes**	Procaryotes of a pliable and soft nature, indicating the lack of a rigid cell wall
Division IV.	**Mendosicutes**	Procaryotes that lack conventional peptidoglycan (archaeobacteria)

VOLUME 1

Section 1 The Spirochetes

Order I. Spirochaetales
- Family I. Spirochaetaceae
  - Genera
    - I. *Spirochaeta*
    - II. *Cristispira*
    - III. *Treponema*
    - IV. *Borrelia*
- Family II. Leptospiraceae
  - Genus I. *Leptospira*

Other organisms: Hindgut spirochetes of termites and *Cryptocercus punctulatus* (wood-eating cockroach).

Section 2 Aerobic/Microaerophilic, Motile, Helical/Vibrioid Gram-Negative Bacteria

Genera
- *Aquaspirillum*
- *Spirillum*
- *Azospirillum*
- *Oceanospirillum*
- *Campylobacter*
- *Bdellovibrio*
- *Vampirovibrio*

Section 3 Nonmotile (or Rarely Motile), Gram-Negative Curved Bacteria

- Family I. Spirosomaceae
  - Genera
    - I. *Spirosoma*
    - II. *Runella*
    - III. *Flectobacillus*
- Other Genera
  - *Microcyclus*
  - *Meniscus*
  - *Brachyarcus*
  - *Pelosigma*

Section 4 Gram-Negative Aerobic Rods and Cocci

- Family I. Pseudomonadaceae
  - Genera
    - I. *Pseudomonas*
    - II. *Xanthomonas*
    - III. *Frateuria*
    - IV. *Zoogloea*
- Family II. Azotobacteraceae
  - Genera
    - I. *Azotobacter*
    - II. *Azomonas*

Family III. Rhizobiaceae
Genera
I. *Rhizobium*
II. *Bradyrhizobium*
III. *Agrobacterium*
IV. *Phyllobacterium*

Family IV. Methylococcaceae
Genera
I. *Methylococcus*
II. *Methylomonas*

Family V. Halobacteriaceae
Genera
I. *Halobacterium**
II. *Halococcus**

Family VI. Acetobacteraceae
Genera
I. *Acetobacter*
II. *Gluconobacter*

Family VII. Legionellaceae
Genus I. *Legionella*

Family VIII. Neisseriaceae
Genera
I. *Neisseria*
II. *Moraxella*
III. *Acinetobacter*
IV. *Kingella*

Other Genera
Beijerinckia
Derxia
Xanthobacter
Thermus
Thermomicrobium
Halomonas
Alteromonas
Flavobacterium
Alcaligenes
Serpens
Janthinobacterium
Brucella
Bordetella
Francisella
Paracoccus
Lampropedia

Section 5 Facultatively Anaerobic Gram-Negative Rods

Family I. Enterobacteriaceae
Genera
I. *Escherichia*
II. *Shigella*
III. *Salmonella*
IV. *Citrobacter*
V. *Klebsiella*
VI. *Enterobacter*
VII. *Erwinia*
VIII. *Serratia*
IX. *Hafnia*
X. *Edwardsiella*
XI. *Proteus*
XII. *Providencia*
XIII. *Morganella*
XIV. *Yersinia*

Other Genera of the Family Enterobacteriaceae
Obesumbacterium
Xenorhabdus
Kluyvera
Rahnella
Cedecea
Tatumella

Family II. Vibrionaceae
Genera
I. *Vibrio*
II. *Photobacterium*
III. *Aeromonas*
IV. *Plesiomonas*

Family III. Pasteurellaceae
Genera
I. *Pasteurella*
II. *Haemophilus*
III. *Actinobacillus*

Other Genera
Zymomonas
Chromobacterium
Cardiobacterium
Calymmatobacterium
*Gardnerella**
Eikenella
Streptobacillus

Section 6 Anaerobic Gram-Negative Straight, Curved, and Helical Rods

Family I. Bacteroidaceae
Genera
I. *Bacteroides*
II. *Fusobacterium*
III. *Leptotrichia*
IV. *Butyrivibrio**
V. *Succinimonas*
VI. *Succinivibrio*
VII. *Anaerobiospirillum*
VIII. *Wolinella*
IX. *Selenomonas*
X. *Anaerovibrio*
XI. *Pectinatus*
XII. *Acetivibrio*
XIII. *Lachnospira**

Section 7 Dissimilatory Sulfate- or Sulfur-Reducing Bacteria

Genera
Desulfuromonas
Desulfovibrio
Desulfomonas
Desulfococcus
Desulfobacter
Desulfobulbus
Desulfosarcina

Section 8 Anaerobic Gram-Negative Cocci

Family I. Veillonellaceae
Genera
I. *Veillonella*
II. *Acidaminococcus*
III. *Megasphaera*

Section 9 The Rickettsias and Chlamydias

Order I. Rickettsiales
Family I. Rickettsiaceae
Tribe I. Rickettsieae
Genera
I. *Rickettsia*
II. *Rochalimaea*
III. *Coxiella*

Tribe II. Ehrlichieae
Genera
IV. *Ehrlichia*
V. *Cowdria*
VI. *Neorickettsia*

Tribe III. Wolbachieae
Genera
VII. *Wolbachia*
VIII. *Rickettsiella*

Family II. Bartonellaceae
Genera
I. *Bartonella*
II. *Grahamella*

Family III. Anaplasmataceae
I. *Anaplasma*
II. *Aegyptianella*
III. *Haemobartonella*
IV. *Eperythrozoon*

Order II. Chlamydiales
  Family I. Chlamydiaceae
    Genus I. *Chlamydia*

Section 10 The Mycoplasmas

Order I. Mycoplasmatales
  Family I. Mycoplasmataceae
    Genera
     I. *Mycoplasma*
     II. *Ureaplasma*
  Family II. Acholeplasmataceae
    Genus I. *Acholeplasma*
  Family III. Spiroplasmataceae
    Genus I. *Spiroplasma*
  Other Genera
    Anaeroplasma
    *Thermoplasma**
    Mycoplasmalike organisms of plants and invertebrates

Section 11 Endosymbionts

A. Endosymbionts of Protozoa
    Genera
     I. *Holospora*
     II. *Caedibacter*
     III. *Pseudocaedibacter*
     IV. *Lyticum*
     V. *Tectibacter*

B. Endosymbionts of Insects
    Genus *Blattabacterium*

C. Endosymbionts of Fungi and Invertebrates Other Than Arthropods

VOLUME 2

Section 12 Gram-Positive Cocci

  Family I. Micrococcaceae
    Genera
     I. *Micrococcus*
     II. *Stomatococcus*
     III. *Planococcus*
     IV. *Staphylococcus*
  Family II. Deinococcaceae
    Genus I. *Deinococcus*
  Other Genera
    Streptococcus (pyogenic hemolytic streptococci, oral streptococci, enterococci, lactic acid streptococci, anaerobic streptococci, other streptococci)
    Leuconostoc
    Pediococcus
    Aerococcus
    Gemella
    Peptococcus
    Peptostreptococcus
    Ruminococcus
    Coprococcus
    Sarcina

Section 13 Endospore-Forming Gram-Positive Rods and Cocci

    Genera
    Bacillus
    Sporolactobacillus
    Clostridium
    Desulfotomaculum
    Sporosarcina
    Oscillospira

Section 14 Regular, Nonsporing, Gram-Positive Rods

    Genera
    Lactobacillus
    Listeria
    Erysipelothrix
    Brochothrix
    Renibacterium
    Kurthia
    Caryophanon

Section 15 Irregular, Nonsporing, Gram-Positive Rods

    Genus *Corynebacterium*
    Plant-pathogenic species of *Corynebacterium*
    Genera
    *Gardnerella**
    Arcanobacterium
    Arthrobacter
    Brevibacterium
    Curtobacterium
    Caseobacter
    Microbacterium
    Aureobacterium
    Cellulomonas
    Agromyces
    Arachnia
    Rothia
    Propionibacterium
    Eubacterium
    Acetobacterium
    *Lachnospira**
    *Butyrivibrio**
    Thermoanaerobacter
    Actinomyces
    Bifidobacterium

Section 16 Mycobacteria

  Family Mycobacteriaceae
    Genus *Mycobacterium*

Section 17 Nocardioforms

    Genera
    *Nocardia**
    *Rhodococcus**
    *Nocardioides**
    *Pseudonocardia**
    *Oerskovia**
    *Saccharopolyspora**
    *Micropolyspora**
    *Promicromonospora**
    *Intrasporangium**

VOLUME 3

Section 18 Anoxygenic Photosynthetic Bacteria

Purple Bacteria
  Family I. Chromatiaceae
    Genera
     I. *Chromatium*
     II. *Thiocystis*
     III. *Thiospirillum*
     IV. *Thiocapsa*
     V. *Lamprobacter*
     VI. *Lamprocystis*
     VII. *Thiodictyon*
     VIII. *Amoebobacter*
     IX. *Thiopedia*
  Family II. Ectothiorhodospiraceae
    Genus *Ectothiorhodospira*
  Purple nonsulfur bacteria
    Genera
    Rhodospirillum
    Rhodopila
    Rhodobacter
    Rhodopseudomonas
    Rhodomicrobium
    Rhodocyclus

Green Bacteria
Green sulfur bacteria
Genera
Chlorobium *Anacalochloris*
Prosthecochloris *Chloroherpeton*
Pelodictyon
Symbiotic consortia
Multicellular filamentous green bacteria
Genera
Chloroflexus *"Oscillochloris"***
Heliothrix *Chloronema*
Genera Incertae Sedis
Genera
Heliobacterium
Erythrobacter

Section 19 Oxygenic Photosynthetic Bacteria

Cyanobacteria
Order Chroococcales
Genera
I. *Chamaesiphon*
II. *Gloeobacter*
Synechococcus-group
III. *Gloethece*
Cyanothece-group
Gloeocapsa-group
Synechocystis-group
Order Pleurocapsales
Genera
I. *Dermocarpa*
II. *Xenococcus*
III. *Dermocarpella*
IV. *Myxosarcina*
V. *Chroococcidiopsis*
Pleurocapsa-group
Order Oscillatoriales
Genera
I. *Spirulina*
II. *Arthrospira*
III. *Oscillatoria*
IV. *Lyngbya*
V. *Pseudanabaena*
VI. *Starria*
VII. *Crinalium*
VIII. *Microcoleus*
Order Nostocales
Family I. Nostocaceae
Genera
I. *Anabaena*
II. *Aphanazomenon*
III. *Nodularia*
IV. *Cylindrospermum*
V. *Nostoc*
Family II. Scytonemataceae
Genus I. *Scytonema*
Family III. Rivulariaceae
Genus I. *Calothrix*
Order Stigonematales
Genera
I. *Chlorogloeopsis*
II. *Fischerella*
III. *Stigonema*
IV. *Geitleria*
Order Prochlorales
Family Prochloraceae
Genus *Prochloron*
"Prochlorothrix"

Section 20 Aerobic Chemolithotrophic Bacteria and Associated Organisms

Nitrifying Bacteria
Family I. Nitrobacteraceae
Nitrate-Oxidizing Bacteria
Genera
I. *Nitrobacter*
II. *Nitrospina*
III. *Nitrococcus*
IV. *Nitrospira*
Ammonia-Oxidizing Bacteria
Genera
I. *Nitrosomonas*
II. *Nitrosococcus*
III. *Nitrosospira*
IV. *Nitrosolobus*
V. *"Nitrosovibrio"*
Colorless Sulfur Bacteria
Genera
Thiobacterium *Thiomicrospira*
Macromonas *Thiosphaera*
Thiospira *Acidiphilium*
Thiovulum *Thermothrix*
Thiobacillus
Obligate Chemolithotrophic Hydrogen Bacteria
Genus *Hydrogenobacter*
Iron- and Manganese-Oxidizing and/or -Depositing Bacteria
Family *"Siderocapsaceae"*
Genera
I. *"Siderocapsa"*
II. *"Naumaniella"*
III. *"Siderococcus"*
IV. *"Ochrobium"*
Magnetotactic Bacteria
Genera
Aquaspirillum (A. magnetotacticum)
"Bilophococcus"

Section 21 Budding and/or Appendaged Bacteria

Prosthecate Bacteria
Budding Bacteria

Genera
Hyphomicrobium *Ancalomicrobium*
Hyphomonas *Prosthecomicrobium*
Pedomicrobium *Labrys*
Stella
Nonbudding Bacteria
Genera
Caulobacter *Prosthecobacter*
Asticcacaulis
Nonprosthecate Bacteria
Budding Bacteria
Lack Peptidoglycan
Genera
Planctomyces
"Isophaera"
Contain Peptidoglycan
Genera
Ensifer
Blastobacter
Angulomicrobium
Gemmiger
Nonbudding Stalked Bacteria
Genera
Gallionella
Nevskia
Other Bacteria
Genera
Seliberia
"Metallogenium"
"Thiodendron"
Spinate Bacteria

Section 22 Sheathed Bacteria

Genera
Sphaerotilus *"Phragmidiothrix"*
Leptothrix *Crenothrix*
Haliscominobacter *"Clonothrix"*
"Lieskeella"

Section 23 Nonphotosynthetic, Nonfruiting Gliding Bacteria

Order I. Cytophagales
Family I. Cytophagaceae
Genera
I. *Cytophaga*
II. *Capnocytophaga*
III. *Flexithrix*
IV. *Sporocytophaga*
Other Genera
Flexibacter
Microscilla
Saprospira
Chitinophaga
Order II. Lysobacterales
Family I. Lysobacteraceae
Genus *Lysobacter*
Order III. Beggiatoales
Family I. Beggiatoaceae
Genera
I. *Beggiatoa* III. *Thioploca*
II. *Thiothrix* IV. *"Thiospirillopsis"*
Others
Simonsiellaceae
Genera
I. *Simonsiella*
II. *Alysiella*
"Pelonemataceae"
Genera
I. *"Pelonema"* III. *"Peloploca"*
II. *"Achroonema"* IV. *"Desmanthus"*
Other Genera
Toxothrix *Vitreoscilla*
Leucothrix *Desulfonema*
Agitococcus
Herpetosiphon

Section 24 Gliding, Fruiting Bacteria

Order I. Myxobacterales
Family I. Myxococcaceae
Genus *Myxococcus*
Family II. Archangiaceae
Genus *Archangium*
Family III. Cystobacteraceae
Genera
I. *Cystobacter*
II. *Melittangium*
III. *Stigmatella*
Family IV. Polyangiaceae
Genera
I. *Polyangium*
II. *Nannocystis*
III. *Chondromyces*

Section 25 Archaeobacteria

Group I. Methanogenic Archaeobacteria
Order I. Methanobacteriales
Family I. Methanobacteriaceae
Genera
I. *Methanobacterium*
II. *Methanobrevibacter*
Family II. Methanothermaceae
Genus I. *Methanothermus*
Order II. Methanococcales
Family I. Methanococcaceae
Genus I. *Methanococcus*
Order III. Methanomicrobiales
Family I. Methanomicrobiaceae
Genera
I. *Methanomicrobium*
II. *Methanogenium*
III. *Methanospirillum*

Family II. Methanosarcinaceae
Genera
I. *Methanosarcina* III. *Methanothrix*
II. *Methanolobus* IV. *Methanococcoides*
Family III. Methanoplanaceae
Genus I. *Methanoplanus*
Other Genera
Methanosphaera
Group II. Archaeobacterial Sulfate Reducers
Order "Archaeoglobales"
Family "Archaeoglobaceae"
Genus *Archaeoglobus*
Group III. Extremely Halophilic Archaeobacteria
Order Halobacteriales
Family I. Halobacteriaceae
Genera
I. *Halobacterium** IV. *Halococcus*
II. *Haloarcula* V. *Natronobacterium*
III. *Haloferax* VI. *Natronococcus**
Group IV. Cell Wall-less Archaeobacteria
Genus *Thermoplasma**
Group V. Extremely Thermophilic S°-Metabolizers
Order I. Thermococcales
Family I. Thermococcaceae
Genera
I. *Thermococcus*
II. *Pyrococcus*
Order II. Thermoproteales
Family I. Thermoproteaceae
Genera
I. *Thermoproteus*
II. *Thermofilum*
Family II. Desulfurococcaceae
Genera
I. *Desulfurococcus*
Other Bacteria
Staphylothermus
Pyrodictium
Order VI. Sulfolobales
Family I. Sulfolobaceae
Genera
I. *Sulfolobus*
II. *Acidianus*

VOLUME 4

Section 26 Nocardioform Actinomycetes

Genera
*Nocardia** *Faenia*
*Rhodococcus** *Promicromonospora**
*Nocardioides** *Intrasporangium**
*Pseudonocardia** *Actinopolyspora*
*Oerskovia** *Saccharomonospora*
*Saccharopolyspora**

Section 27 Actinomycetes with Multilocular Sporangia

Genera
Geodermatophilus
Dermatophilus
Frankia

Section 28 Actinoplanetes

Genera
Actinoplanes *Dactylosporangium*
Ampullariella *Micromonospora*
Pilimelia

Section 29 *Streptomyces* and Related Genera

Genera
Streptomyces *Kineosporia*
Streptoverticillium *Sporichthya*

Section 30 Maduromycetes

Genera
Actinomadura *Planomonospora*
Microbispora *Spirillospora*
Microtetraspora *Streptosporangium*
Planobispora

Section 31 *Thermomonospora* and Related Genera

Genera
Thermomonospora *Nocardiopsis*
Actinosynnema *Streptoalloteichus*

Section 32 Thermoactinomycetes

Genus *Thermoactinomyces*

Section 33 Other Genera

Genera
Glycomyces *Kitasatosporia*
Kibdelosporangium *Saccharothrix*
Pasteuria

---

* Genera that have been placed in more than one volume: *Gardnerella, Lachnospira,* and *Butyrivibrio* have thin, gram-positive walls but stain as gram-negatives. *Halobacterium, Halococcus,* and *Thermoplasma* are archaeobacteria that are also included among the Gram-Negative Rods and Cocci or the Mycoplasmas. Nocardioforms are included with Gram-Positive Eubacteria and Actinomycetes because their relatedness to either group is uncertain.

** Names in quotation marks have not been validated by publication in the *International Journal of Systematic Bacteriology*.

APPENDIX B

Word Roots Used in Microbiology

The Latin rules of grammar pertain to singular and plural forms of scientific names.

	Gender		
	Feminine	Masculine	Neuter
Singular	-a	-us	-um
Plural	-ae	-i	-a
Examples	alga, algae	fungus, fungi	bacterium, bacteria

a-, an- absence, lack. Examples: abiotic, in the absence of life; anaerobic, in the absence of air.
-able able to, capable of. Example: viable, having the ability to live or exist.
actino- ray. Example: actinomycetes, bacteria that form star-shaped (with rays) colonies.
aer- air. Examples: aerobic, in the presence of air; aerate, to add air.
albo- white. Example: *Streptomyces albus* produces white colonies.
ameb- change. Example: ameboid, movement involving changing shapes.
amphi- around. Example: amphitrichous, tufts of flagella at both ends of a cell.
amyl- starch. Example: amylase, an enzyme that degrades starch.
ana- up. Example: anabolism, building up.
ant-, anti- opposed to, preventing. Example: antimicrobial, a substance that prevents microbial growth.
archae- ancient. Example: archaeobacteria, "ancient" bacteria, thought to be like the first form of life.
asco- bag. Example: ascus, a baglike structure holding spores.
aur- gold. Example: *Staphylococcus aureus,* gold-pigmented colonies.
aut-, auto- self. Example: autotroph, self-feeder.
bacillo- a little stick. Example: bacillus, rod-shaped.
basid- base, pedestal. Example: basidium, a cell that bears spores.
bdell- leech. Example: *Bdellovibrio,* a predatory bacterium.
bio- life. Example: biology, the study of life and living organisms.
blast- bud. Example: blastospore, spores formed by budding.
bovi- cattle. Example: *Mycobacterium bovis,* a bacterium found in cattle.
brevi- short. Example: *Lactobacillus brevis,* a bacterium with short cells.
butyr- butter. Example: butyric acid, formed in butter, responsible for rancid odor.
campylo- curved. Example: *Campylobacter,* curved rod.
carcin- cancer. Example: carcinogen, a cancer-causing agent.
-caryo, -karyo a nut. Example: eucaryote, a cell with a membrane-enclosed nucleus.
caseo- cheese. Example: caseous, cheeselike.
caul- a stalk. Example: *Caulobacter,* appendaged or stalked bacteria.
cerato- horn. Example: keratin, the horny substance making up skin and nails.
chlamydo- covering. Example: chlamydospores, spores formed inside hypha.
chloro- green. Example: chlorophyll, green-pigmented molecule.
chrom- color. Examples: chromosome, readily stained structure; metachromatic, intracellular colored granules.
chryso- golden. Example: *Streptomyces chryseus,* golden colonies.
-cide killing. Example: bactericide, an agent that kills bacteria.
cili- eyelash. Example: cilia, a hairlike organelle.
cleisto- closed. Example: cleistothecium, completely closed ascus.
co-, con- together. Example: concentric, having a common center, together in the center.
cocci- a berry. Example: coccus, a spherical cell.
coeno- shared. Example: coenocyte, a cell with many nuclei not separated by septa.
col-, colo- colon. Examples: colon, large intestine; *Escherichia coli,* a bacterium found in the large intestine.
conidio- dust. Example: conidia, spores developed at the end of aerial hypha, never enclosed.

coryne- club. Example: *Corynebacterium,* club-shaped cells.
-cul small form. Example: particle, a small part.
-cut the skin. Example: *Firmicutes,* bacteria with a firm cell wall, gram-positive.
cyano- blue. Example: cyanobacteria, blue-green pigmented organisms.
cyst- bladder. Example: cystitis, inflammation of the urinary bladder.
cyt- cell. Example: cytology, the study of cells.
de- undoing, reversal, loss, removal. Example: deactivation, becoming inactive.
di-, diplo- twice, double. Example: diplococci, pairs of cocci.
dia- through, between. Example: diaphragm, the wall through or between two areas.
dys- difficult, faulty, painful. Example: dysfunction, disturbed function.
ec-, ex-, ecto out, outside, away from. Example: excrete, to remove materials from the body.
en-, em- in, inside. Example: encysted, enclosed in a cyst.
entero- intestine. Example: *Enterobacter,* a bacterium found in the intestine.
epi- upon, over. Example: epidemic, a disease affecting all the people.
erythro- red. Example: erythema, redness of the skin.
eu- well, proper. Example: eucaryote, a proper cell.
exo- outside, outer layer. Example: exogenous, from outside the body.
extra- outside, beyond. Example: extracellular, outside the cells of an organism.
firmi- strong. Example: *Bacillus firmus* forms resistant endospores.
flagell- a whip. Example: flagellum, a projection from a cell; in eucaryotic cells, it pulls cells in a whiplike fashion.
flav- yellow. Example: *Flavobacterium* cells produce yellow pigment.
fruct- fruit. Example: fructose, fruit sugar.
-fy to make. Example: magnify, to make larger.
galacto- milk. Example: galactose, monosaccharide from milk sugar.
gamet- to marry. Example: gamete, a reproductive cell.
gastr- stomach. Example: gastritis, inflammation of the stomach.
gel- to stiffen. Example: gel, a solidified colloid.
-gen an agent that initiates. Example: pathogen, any agent that produces disease.
-genesis formation. Example: pathogenesis, production of disease.
germ, germin- bud. Example: germ, part of an organism capable of developing.
-gony reproduction. Example: schizogony, multiple fission producing many new cells.
gracili- thin. Example: *Aquaspirillum gracile,* a thin cell.
halo- salt. Example: halophile, an organism that can live in high salt concentrations.
haplo- one, single. Example: haploid, half the number of chromosomes or one set.
hema-, hemato-, hemo- blood. Example: *Hemophilus,* a bacterium that requires nutrients from red blood cells.
hepat- liver. Example: hepatitis, inflammation of the liver.
herpes creeping. Example: herpes, or shingles, lesions appear to creep along the skin.
hetero- different, other. Example: heterotroph, obtains organic nutrients from other organisms; other feeder.
hist- tissue. Example: histology, the study of tissues.
hom-, homo- same. Example: homofermenter, an organism that produces only lactic acid from fermentation of a carbohydrate.
hydr-, hydro- water. Example: dehydration, loss of body water.
hyper- excess. Example: hypertonic, having a greater osmotic pressure in comparison with another.
hypo- below, deficient. Example: hypotonic, having a lesser osmotic pressure in comparison with another.
im- not, in. Example: impermeable, not permitting passage.
inter- between. Example: intercellular, between the cells.
intra- within, inside. Example: intracellular, inside the cell.
io- violet. Example: iodine, a chemical element that produces a violet vapor.
iso- equal, same. Example: isotonic, having the same osmotic pressure when compared with another.
-itis inflammation of. Example: colitis, inflammation of the large intestine.
kin- movement. Example: streptokinase, an enzyme that lyses or moves fibrin.
lacti- milk. Example: lactose, the sugar in milk.
lepis- scaly. Example: leprosy, disease characterized by skin lesions.
lepto- thin. Example: *Leptospira,* thin spirochete.
leuko- whiteness. Example: leukocyte, a white blood cell.
lip-, lipo- fat, lipid. Example: lipase, an enzyme that breaks down fats.
-logy the study of. Example: pathology, the study of changes in structure and function brought on by disease.
lopho- tuft. Example: lophotrichous, having a group of flagella on one side of a cell.
luc-, luci- light. Example: luciferin, a substance in certain organisms that emits light when acted upon by the enzyme luciferase.
lute-, luteo- yellow. Example: *Micrococcus luteus,* yellow colonies.
-lysis loosening, to break down. Example: hydrolysis, chemical decomposition of a compound into other compounds as a result of taking up water.
macro- largeness. Example: macromolecules, large molecules.
mendosi- faulty. Example: Mendosicutes, archaeobacteria lacking peptidoglycan.
meningo- membrane. Example: meningitis, inflammation of the membranes of the brain.
meso- middle. Example: mesophile, an organism whose optimum temperature is in the middle range.
meta- beyond, between, transition. Example: metabolism, chemical changes occurring within a living organism.
micro- smallness. Example: microscope, an instrument used to make small objects appear larger.
-mnesia memory. Examples: amnesia, loss of memory; anamnesia, return of memory.

molli- soft. Example: Mollicutes, a class of wall-less eubacteria.

-monas a unit. Example: *Methylomonas,* a unit (bacterium) that utilizes methane as its carbon source.

mono- singleness. Example: monotrichous, having one flagellum.

morpho- form. Example: morphology, the study of the form and structure of organisms.

multi- many. Example: multinuclear, having several nuclei.

mur- wall. Example: murein, a component of bacterial cell walls.

mus-, muri- mouse. Example: murine typhus, a form of typhus endemic in mice.

mut- to change. Example: mutation, a sudden change in characteristics.

myco-, -mycetoma, -myces a fungus. Example: *Saccharomyces,* sugar fungus, a genus of yeast.

myxo- slime, mucus. Example: Myxobacteriales, an order of slime-producing bacteria.

necro- a corpse. Example: necrosis, cell death or death of a portion of tissue.

-nema a thread. Example: *Treponema* has long, threadlike cells.

nigr- black. Example: *Aspergillus niger,* a fungus that produces black conidia.

ob- before, against. Example: obstruction, impeding or blocking up.

oculo- eye. Example: monocular, pertaining to one eye.

-oecium, -ecium a house. Examples: perithecium, an ascus with an opening that encloses spores; ecology, the study of the relationships between organisms and between an organism and its environment (household).

-oid like, resembling. Example: coccoid, resembling a coccus.

-oma tumor. Example: lymphoma, a tumor of the lymphatic tissues.

-ont being, existing. Example: schizont, a cell existing as a result of schizogony.

ortho- straight, direct. Example: orthomyxovirus, a virus with a straight, tubular capsid.

-osis, -sis condition of. Examples: lysis, the condition of loosening; symbiosis, the condition of living together.

pan- all, universal. Example: pandemic, an epidemic affecting a large region.

para- beside, near. Example: parasite, an organism that "feeds beside" another.

peri- around. Example: peritrichous, projections from all sides.

phaeo- brown. Example: *Phaeophyta,* brown algae.

phago- eat. Example: phagocyte, a cell that engulfs and digests particles or cells.

philo-, -phil liking, preferring. Example: thermophile, an organism that prefers high temperatures.

-phore bears, carries. Example: conidiophore, a hypha that bears conidia.

-phyll leaf. Example: chlorophyll, the green pigment in leaves.

-phyte plant. Example: saprophyte, a plant that obtains nutrients from decomposing organic matter.

pil- a hair. Example: pilus, a hairlike projection from a cell.

plano- wandering, roaming. Example: plankton, organisms drifting or wandering in water.

plast- formed. Example: plastid, a formed body within a cell.

-pnoea, -pnea breathing. Example: dyspnea, difficulty in breathing.

pod- foot. Example: pseudopod, a footlike structure.

poly- many. Example: polymorphism, many forms.

post- after, behind. Example: posterior, places behind (a specific) part.

pre-, pro- before, ahead of. Examples: procaryote, a cell with the first nucleus; pregnant, before birth.

pseudo- false. Example: pseudopod, false foot.

psychro- cold. Example: psychrophile, an organism that grows best at low temperatures.

-ptera wing. Example: Diptera, the order of true flies, insects with two wings.

pyo- pus. Example: pyogenic, pus-forming.

rhabdo- stick, rod. Example: Rhabdovirus, an elongated, bullet-shaped virus.

rhin- nose. Example: rhinitis, inflammation of mucous membranes in the nose.

rhizo- root. Examples: *Rhizobium,* a bacterium that grows in plant roots; mycorrhiza, mutualism between a fungus and the roots of a plant.

rhodo- red. Example: *Rhodospirillum,* a red-pigmented, spiral-shaped bacterium.

rod- gnaw. Example: rodents, the class of mammals with gnawing teeth.

rubri- red. Example: *Clostridium rubrum,* red-pigmented colonies.

rumin- throat. Example: *Ruminococcus,* a bacterium associated with a rumen (modified esophagus).

saccharo- sugar. Example: disaccharide, a sugar consisting of two simple sugars.

sapr- rotten. Example: *Saprolegnia,* a fungus that lives on dead animals.

sarco- flesh. Example: sarcoma, a tumor of muscle or connective tissues.

schizo- split. Example: schizomycetes, organisms that reproduce by splitting, an early name for bacteria.

scolec- worm. Example: scolex, the head of a tapeworm.

-scope, -scopic watcher. Example: microscope, an instrument used to watch small things.

semi- half. Example: semicircular, having the form of half a circle.

sept- rotting. Example: aseptic, free from bacteria that could cause decomposition.

septo- partition. Example: septum, a cross-wall in a fungal hypha.

serr- notched. Example: serrate, with a notched edge.

sidero- iron. Example: *Siderococcus,* a bacterium capable of oxidizing iron.

siphon- tube. Example: Siphonaptera, the order of fleas, insects with tubular mouths.

soma- body. Example: somatic cells, cells of the body other than gametes.

speci- particular things. Examples: species, the smallest group of organisms with similar properties; specify, to indicate exactly.

spiro- coil. Example: spirochete, a bacterium with a coiled cell.

sporo- spore. Example: sporangium, a structure that holds spores.

staphylo- grapelike cluster. Example: *Staphylococcus,* a bacterium that forms clusters of cells.

-stasis arrest, fixation. Example: bacteriostasis, cessation of bacterial growth.

strepto- twisted. Example: *Streptococcus,* a bacterium that forms twisted chains of cells.

sub- beneath, under. Example: subcutaneous, just under the skin.

super- above, upon. Example: superior, the quality or state of being above others.

sym-, syn- together, with. Examples: synapse, the region of communication between two neurons; synthesis, putting together.

-taxi to touch. Example: chemotaxis, response to the presence (touch) of chemicals.

taxis- orderly arrangement. Example: taxonomy, the science dealing with arranging organisms into groups.

tener- tender. Example: Tenericutes, the phylum containing wall-less eubacteria.

thallo- plant body. Example: thallus, an entire macroscopic fungus.

therm- heat. Example: *Thermus,* a bacterium that grows in hot springs (to 75°C).

thio- sulfur. Example: *Thiobacillus,* a bacterium capable of oxidizing sulfur-containing compounds.

-thrix. See **trich-**

-tome, -tomy to cut. Example: appendectomy, surgical removal of the appendix.

-tone, -tonic strength. Example: hypotonic, having less strength (osmotic pressure).

tox- poison. Example: antitoxin, effective against poison.

trans- across, through. Example: transport, movement of substances.

tri- three. Example: trimester, three-month period.

trich- a hair. Example: peritrichous, hairlike projections from cells.

-trope turning. Example: geotropic, turning toward the Earth (pull of gravity).

-troph food, nourishment. Example: trophic, pertaining to nutrition.

-ty condition of, state. Example: Immunity, the condition of being resistant to disease or infection.

undul- wavy. Example: undulating, rising and falling, presenting a wavy appearance.

uni- one. Example: unicellular, pertaining to one cell.

vaccin- cow. Example: vaccination, injection of a vaccine (originally pertained to cows).

vacu- empty. Example: vacuoles, an intracellular space that appears empty.

vesic- bladder. Example: vesicle, a bubble.

vitr- glass. Example: in vitro, in culture media in a glass (or plastic) container.

-vorous eat. Example: carnivore, an animal that eats other animals.

xantho- yellow. Example: *Xanthomonas,* produces yellow colonies.

xeno- strange. Example: axenic, sterile, free of strange organisms.

xero- dry. Example: xerophyte, any plant that tolerates dry conditions.

xylo- wood. Example: xylose, a sugar obtained from wood.

zoo- animal. Example: zoology, the study of animals.

zygo- yoke, joining. Example: zygospore, a spore formed from the fusion of two cells.

-zyme ferment. Example: enzyme, any protein in living cells that catalyzes chemical reactions.

APPENDIX C

Most Probable Numbers (MPN) Table

MPN Index for Various Combinations of Positive and Negative Results When Five 10-ml Portions, Five 1-ml Portions, and Five 0.1-ml Portions Are Used **(see pp. 160–162)**

NO. OF TUBES GIVING POSITIVE REACTION OUT OF			*MPN Index per 100 ml*	*NO. OF TUBES GIVING POSITIVE REACTION OUT OF*			*MPN Index per 100 ml*
5 of 10 ml Each	*5 of 1 ml Each*	*5 of 0.1 ml Each*		*5 of 10 ml Each*	*5 of 1 ml Each*	*5 of 0.1 ml Each*	
0	0	0	<2	4	3	1	33
0	0	1	2	4	4	0	34
0	1	0	2	5	0	0	23
0	2	0	4	5	0	1	30
1	0	0	2	5	0	2	40
1	0	1	4	5	1	0	30
1	1	0	4	5	1	1	50
1	1	1	6	5	1	2	60
1	2	0	6	5	2	0	50
2	0	0	4	5	2	1	70
2	0	1	7	5	2	2	90
2	1	0	7	5	3	0	80
2	1	1	9	5	3	1	110
2	2	0	9	5	3	2	140
2	3	0	12	5	3	3	170
3	0	0	8	5	4	0	130
3	0	1	11	5	4	1	170
3	1	0	11	5	4	2	220
3	1	1	14	5	4	3	280
3	2	0	14	5	4	4	350
3	2	1	17	5	5	0	240
4	0	0	13	5	5	1	300
4	0	1	17	5	5	2	500
4	1	0	17	5	5	3	900
4	1	1	21	5	5	4	1600
4	1	2	26	5	5	5	≥2400
4	2	0	22				
4	2	1	26				
4	3	0	27				

Source: *Standard Methods for the Examination of Water and Wastewater,* 17th ed., New York: American Public Health Association, 1989.

APPENDIX D

Methods for Taking Clinical Samples

To diagnose a disease, it is often necessary to obtain a sample of material that may contain the disease-causing organism. Samples must be taken aseptically. The sample container should be labeled with the patient's name, room number (if hospitalized), date, time, and medications being taken. Samples must be transported to the laboratory immediately for culture. Delay in transport may result in growth of some organisms, and their toxic products may kill other organisms. Pathogens tend to be fastidious and die without their optimum environmental conditions.

In the laboratory, samples from infected tissues are cultured on differential and selective media in an attempt to isolate and identify any pathogens or organisms that are not normally found in association with that tissue.

UNIVERSAL PRECAUTIONS

The following procedures should be used by all health care workers, including students, whose activities involve contact with patients or with blood or other body fluids. These procedures were developed to minimize the risk of transmitting AIDS in a health care environment, but adherence to these guidelines will minimize transmission of *all* nosocomial infections.

1. Gloves should be worn when touching blood and body fluid, mucous membranes, and nonintact skin and when handling items or surfaces soiled with blood or body fluids. Gloves should be changed after contact with each patient.
2. Hands and other skin surfaces should be washed immediately and thoroughly if contaminated with blood or other body fluids. Hands should be washed immediately after gloves are removed.
3. Masks and protective eyewear or face shields should be worn during procedures that are likely to generate droplets of blood or other body fluids.
4. Gowns or aprons should be worn during procedures that are likely to generate splashes of blood or other body fluids.
5. To prevent needlestick injuries, needles should not be recapped, purposely bent or broken, or otherwise manipulated by hand. After disposable syringes and needles, scalpel blades, and other sharp items are used, they should be placed in puncture-resistant containers for disposal.
6. Although saliva has not been implicated in HIV transmission, mouthpieces, resuscitation bags, and other ventilation devices should be available for use in areas in which the need for resuscitation is predictable. Emergency mouth-to-mouth resuscitation should be minimized.
7. Health care workers who have exudative lesions or weeping dermatitis should refrain from all direct patient care and from handling patient-care equipment.
8. Pregnant health care workers are not known to be at a greater risk of contracting HIV infection than health care workers who are not pregnant; however, if a health care worker develops HIV infection during pregnancy, the infant is at risk of infection. Because of this risk, pregnant health care workers should be especially familiar with and strictly adhere to precautions to minimize the risk of HIV transmission.

Source: Centers for Disease Control and National Institutes for Health. *Biosafety in Microbiological and Biomedical Laboratories,* 2nd ed. Government Printing Office, 1988.

INSTRUCTIONS FOR SPECIFIC PROCEDURES

Wound or Abscess Culture

1. Cleanse the area with a sterile swab moistened in sterile saline.
2. Disinfect the area with 70% ethanol or iodine solution.
3. If the abscess has not ruptured spontaneously, a physician will open it with a sterile scalpel.
4. Wipe the first pus away.
5. Touch a sterile swab to the pus, taking care not to contaminate the surrounding tissue.
6. Replace the swab in its container, and properly label the container.

Ear Culture

1. Clean the skin and auditory canal with 1% tincture of iodine.
2. Touch the infected area with a sterile cotton swab.
3. Replace the swab in its container.

Eye Culture

This procedure is usually performed by an ophthalmologist.

1. Anesthetize the eye with topical application of a sterile anesthetic solution.
2. Wash the eye with sterile saline solution.
3. Collect material from the infected area with a sterile cotton swab. Return the swab to its container.

Blood Culture

1. Close room windows to avoid contamination.
2. Clean skin around selected vein with 2% tincture of iodine on a cotton swab.
3. Remove dried iodine with gauze moistened with 80% isopropyl alcohol.
4. Draw a few milliliters of venous blood.
5. Aseptically bandage the puncture.

Urine Culture

1. Provide the patient with a sterile container.
2. Instruct the patient to collect a midstream sample. This is obtained by voiding a small volume from the bladder before collection. This washes away extraneous bacteria of the skin flora.
3. A urine sample may be stored under refrigeration (4° to 6°C) for up to 24 hours.

Fecal Culture

For bacteriological examination, only a small sample is needed. This may be obtained by inserting a sterile swab into the rectum or feces. The swab is then placed in a tube of sterile enrichment broth for transport to the laboratory. For examination for parasites, a small sample may be taken from a morning stool. The sample is placed in a preservative (polyvinyl alcohol, buffered glycerol, saline, or formalin) for microscopic examination for eggs and adult parasites.

Sputum Culture

1. A morning sample is best as microorganisms will have accumulated while the patient is sleeping.
2. Patient should rinse his or her mouth thoroughly to remove food and normal flora.
3. Patient should cough deeply from the lungs and expectorate into a sterile glass wide-mouth jar.
4. Care should be taken to avoid contamination of personnel.
5. In cases such as tuberculosis where there is little sputum, stomach aspiration may be necessary.
6. Infants and children tend to swallow sputum. A fecal sample may be of some value in these cases.

APPENDIX **E**

Biochemical Pathways

FIGURE E.1 The Calvin–Benson Cycle for Photosynthetic Carbon Metabolism. Steps 1–3 provide for the initial fixation and reduction of carbon, generating the three-carbon compounds glyceraldehyde 3-phosphate and dihydroxyacetone phosphate, which are interconvertible (step 4). On the average, 2 of every 12 three-carbon molecules are used in the synthesis of glucose (steps A–D). Ten of every 12 three-carbon molecules are used to generate ribulose 5-phosphate by a complex series of reactions (step 5). The ribulose 5-phosphate is then phosphorylated at the expense of ATP (step 6), forming ribulose 1,5-diphosphate, the acceptor molecule with which the sequence began. (See Figure 5.23 on p. 127 for a simplified version of the Calvin–Benson cycle.)

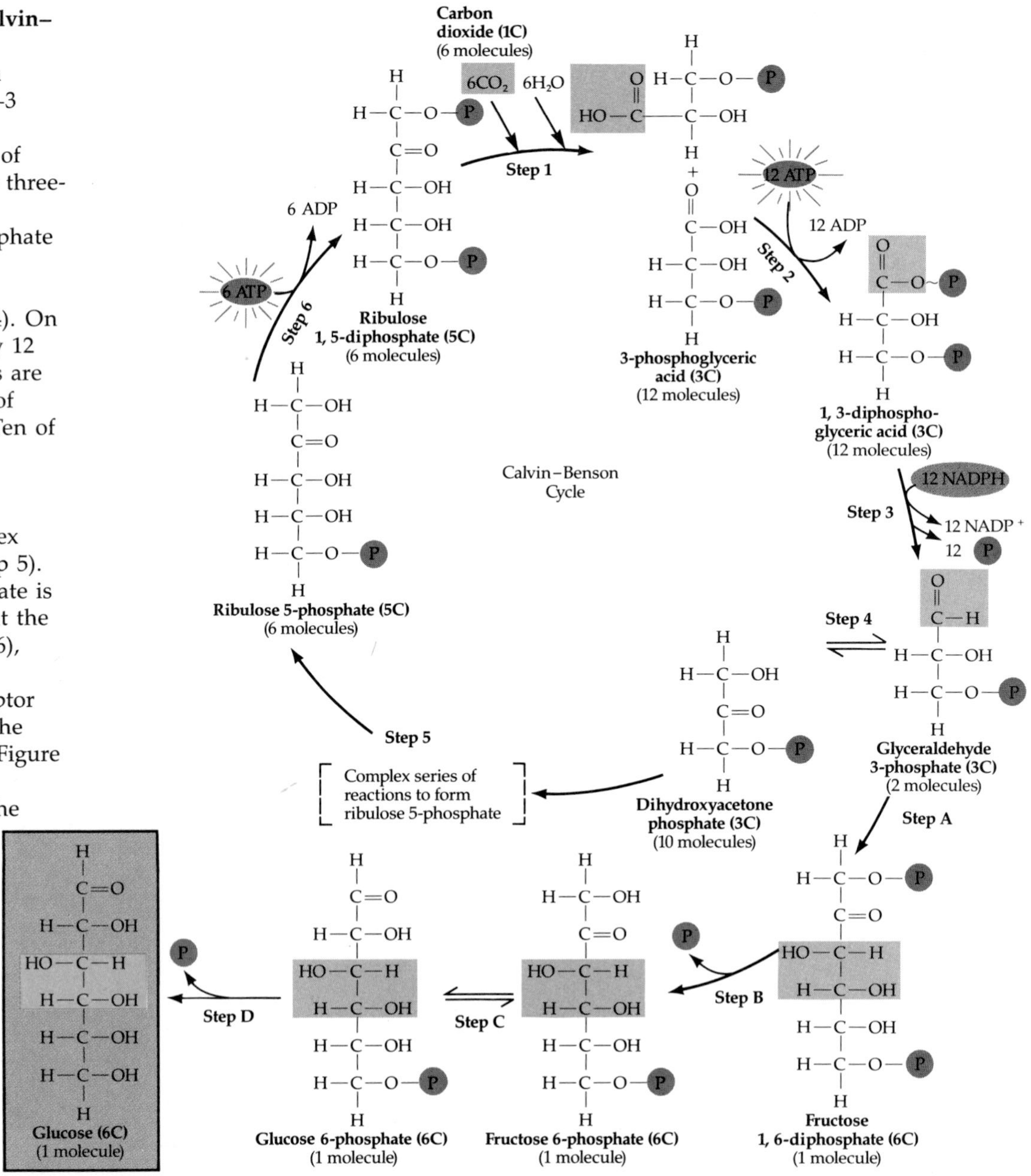

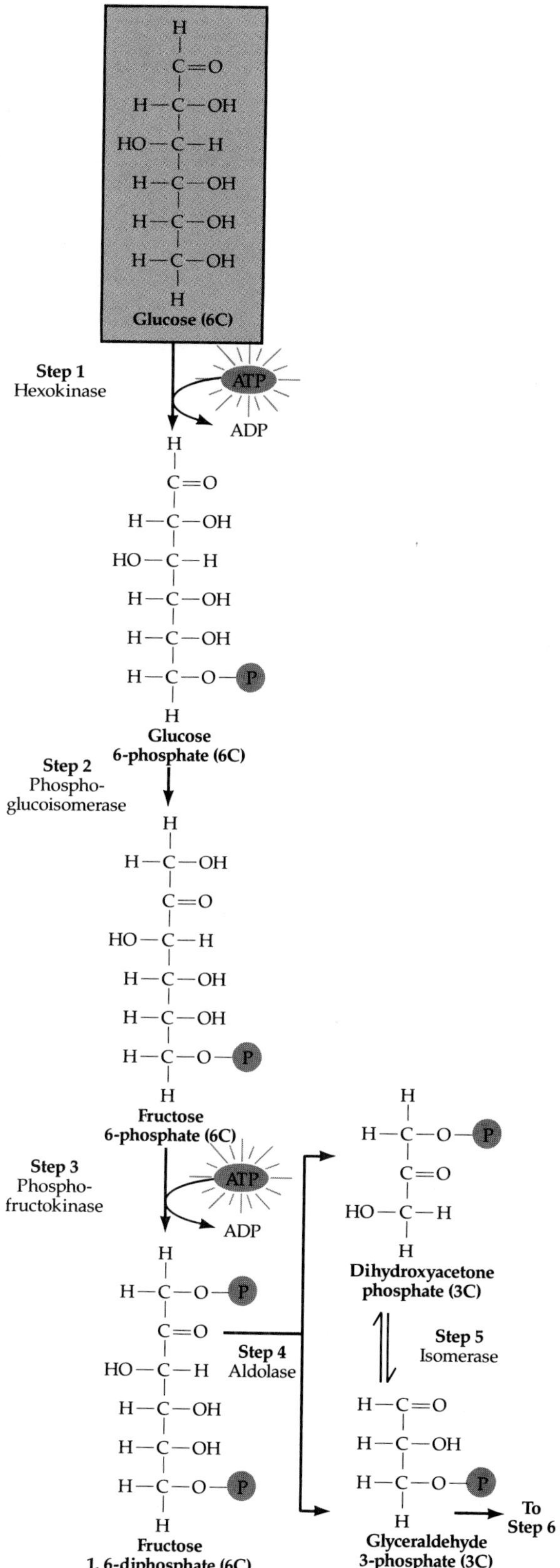

FIGURE E.2 Glycolysis (Embden–Meyerhof Pathway). Each of the ten steps of glycolysis is catalyzed by a specific enzyme, which is named under each step number. (See Figure 5.12 on p. 113 for a simplified version of glycolysis.)

Step 1: Glucose enters the cell and is phosphorylated by the enzyme hexokinase, which transfers a phosphate group from ATP to the number six carbon of the sugar. The product of the reaction is glucose 6-phosphate. The electrical charge of the phosphate group traps the sugar in the cell because of the impermeability of the plasma membrane to ions. Phosphorylation of glucose also makes the molecule more chemically reactive. Although glycolysis is supposed to *produce* ATP, in step 1, ATP is actually consumed—an energy investment that will be repaid with dividends later in glycolysis.

Step 2: Glucose 6-phosphate is rearranged to convert it to its isomer, fructose 6-phosphate. Isomers have the same number and types of atoms but in different structural arrangements.

Step 3: In this step, still another molecule of ATP is invested in glycolysis. An enzyme transfers a phosphate group from ATP to the sugar, producing fructose 1,6-diphosphate.

Step 4: This is the reaction from which glycolysis gets its name ("sugar splitting"). An enzyme cleaves fructose 1,6-diphosphate into two different three-carbon sugars: glyceraldehyde 3-phosphate and dihydroxyacetone phosphate. These two sugars are isomers.

Step 5: The enzyme isomerase interconverts the three-carbon sugars. The next enzyme in glycolysis uses only glyceraldehyde 3-phosphate as its substrate. This pulls the equilibrium between the two three-carbon sugars in the direction of glyceraldehyde 3-phosphate, which is removed as fast as it forms.

(*Continued*)

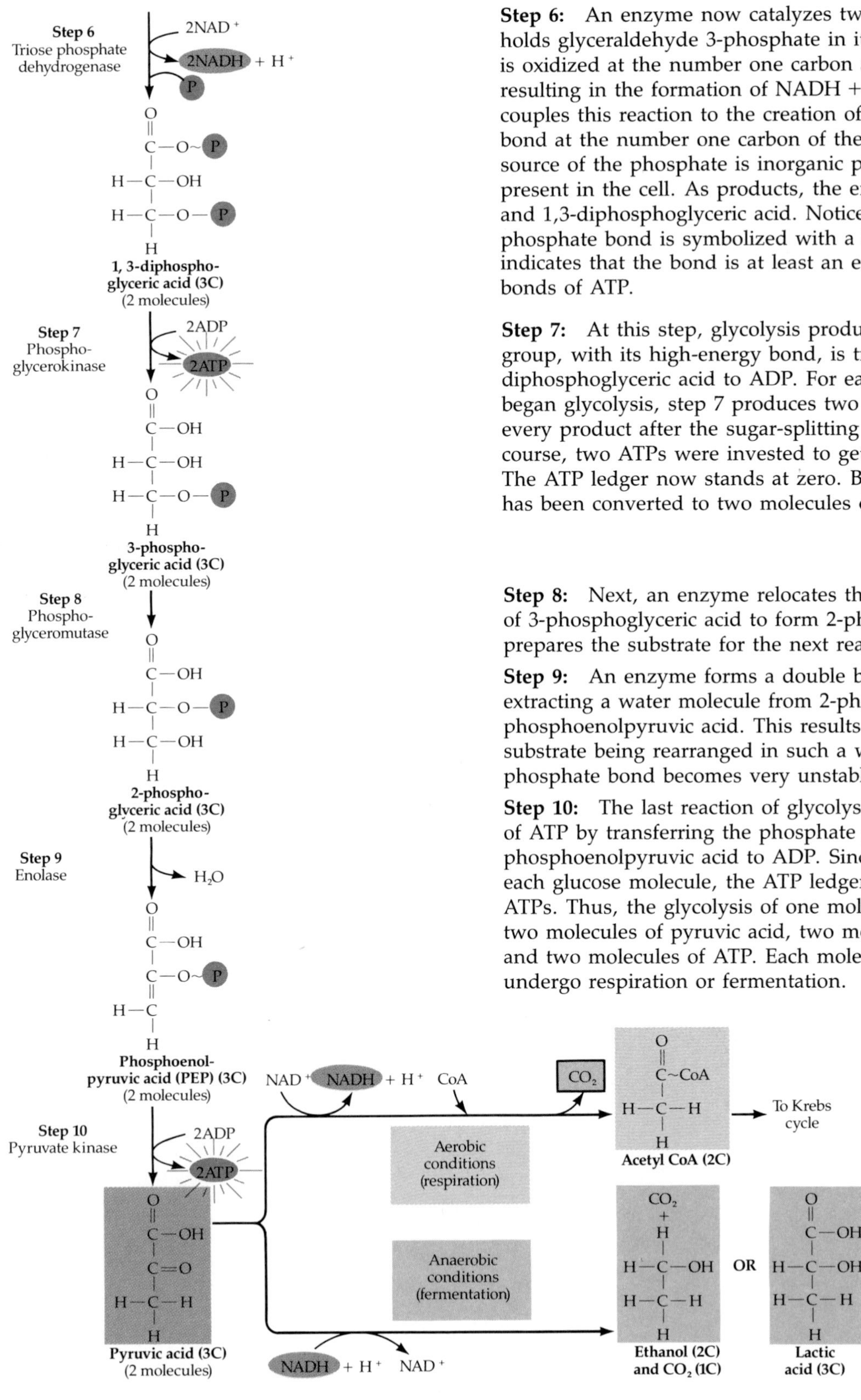

Step 6: An enzyme now catalyzes two sequential reactions while it holds glyceraldehyde 3-phosphate in its active site. First, the sugar is oxidized at the number one carbon and NAD^+ is reduced, resulting in the formation of NADH + H^+. Second, the enzyme couples this reaction to the creation of a high-energy phosphate bond at the number one carbon of the oxidized substrate. The source of the phosphate is inorganic phosphate, which is always present in the cell. As products, the enzyme releases NADH + H^+ and 1,3-diphosphoglyceric acid. Notice in the figure that the new phosphate bond is symbolized with a high-energy bond (~), which indicates that the bond is at least an energetic as the phosphate bonds of ATP.

Step 7: At this step, glycolysis produces ATP. The phosphate group, with its high-energy bond, is transferred from 1,3-diphosphoglyceric acid to ADP. For each glucose molecule that began glycolysis, step 7 produces two molecules of ATP, because every product after the sugar-splitting step (step 4) is doubled. Of course, two ATPs were invested to get sugar ready for splitting. The ATP ledger now stands at zero. By the end of step 7, glucose has been converted to two molecules of 3-phosphoglyceric acid.

Step 8: Next, an enzyme relocates the remaining phosphate group of 3-phosphoglyceric acid to form 2-phosphoglyceric acid. This prepares the substrate for the next reaction.

Step 9: An enzyme forms a double bond in the substrate by extracting a water molecule from 2-phosphoglyceric acid to form phosphoenolpyruvic acid. This results in the electrons of the substrate being rearranged in such a way that the remaining phosphate bond becomes very unstable.

Step 10: The last reaction of glycolysis produces another molecule of ATP by transferring the phosphate group from phosphoenolpyruvic acid to ADP. Since this step occurs twice for each glucose molecule, the ATP ledger now shows a net gain of two ATPs. Thus, the glycolysis of one molecule of glucose results in two molecules of pyruvic acid, two molecules of NADH + H^+, and two molecules of ATP. Each molecule of pyruvic acid can now undergo respiration or fermentation.

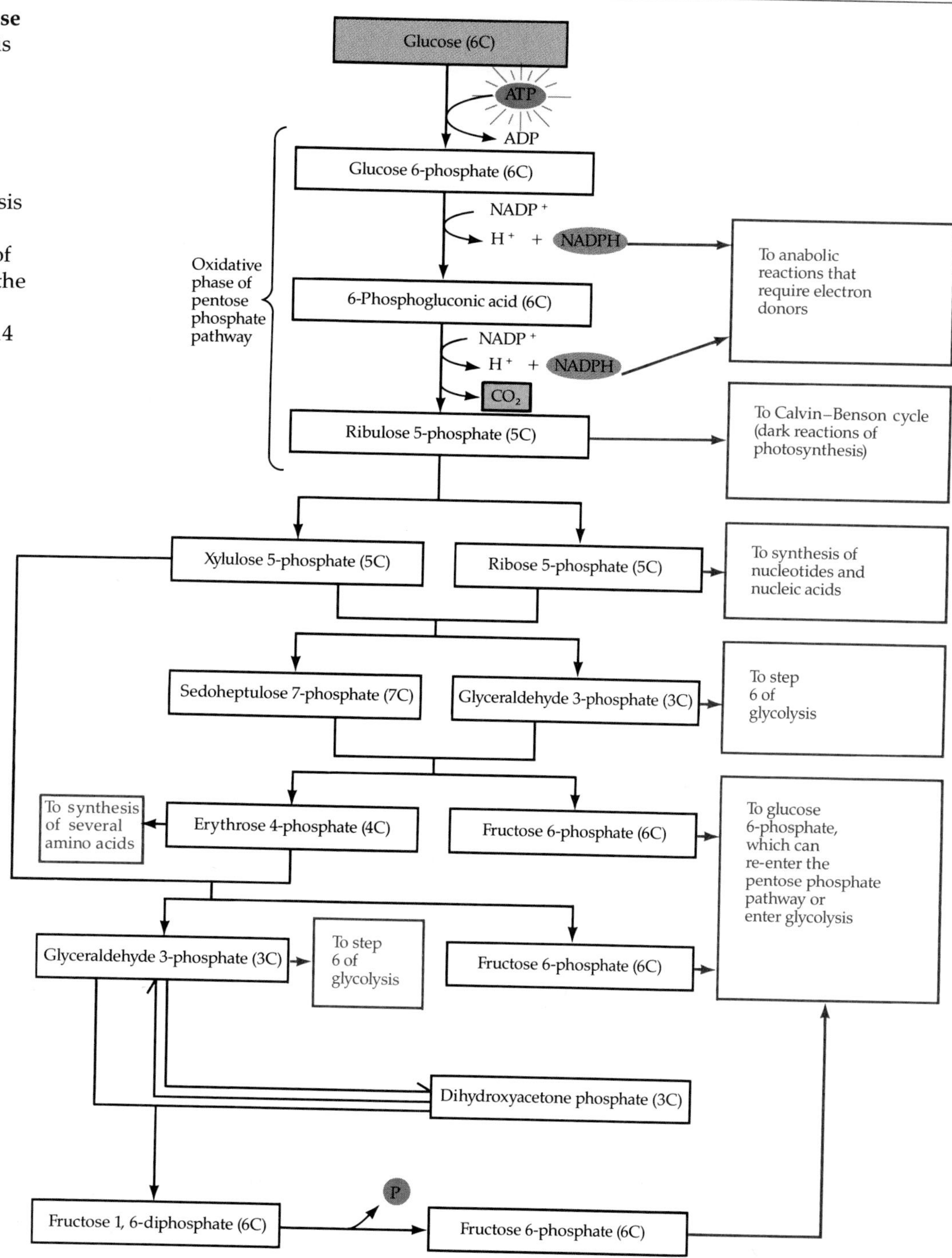

FIGURE E.3 The Pentose Phosphate Pathway. This pathway, which operates simultaneously with glycolysis, provides an alternate route for the oxidation of glucose and plays a role in the synthesis of biological molecules, depending on the needs of the cell. Possible fates of the various intermediates are shown in color. (See p. 114 for a discussion of the pentose phosphate pathway.)

FIGURE E.4 The Entner–Doudoroff Pathway. This pathway is an alternate to glycolysis for the oxidation of glucose to pyruvic acid. (See p. 114 for a discussion of this pathway.)

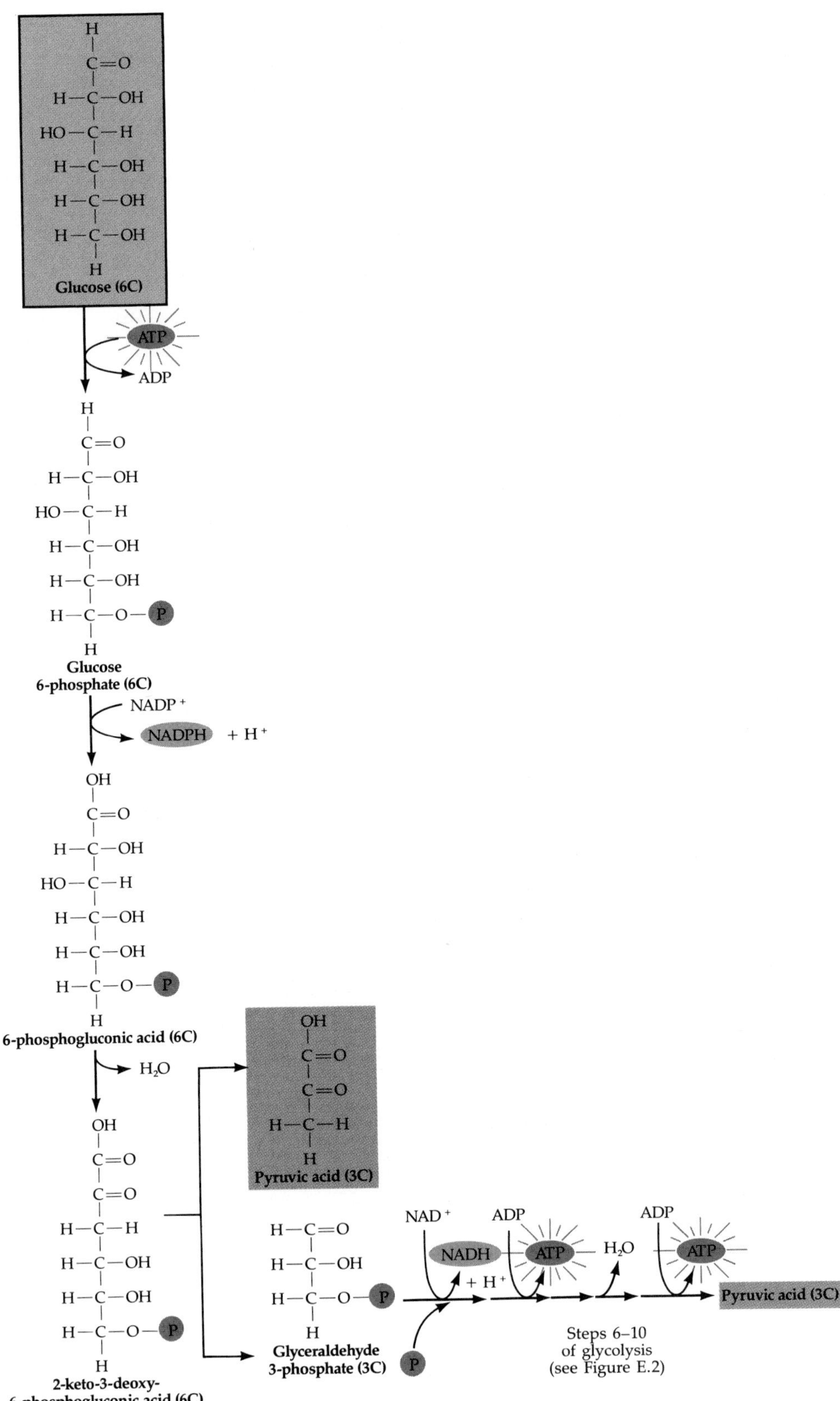

FIGURE E.5 The Krebs Cycle. (See Figure 5.13 on p. 115 for a simplified version.)

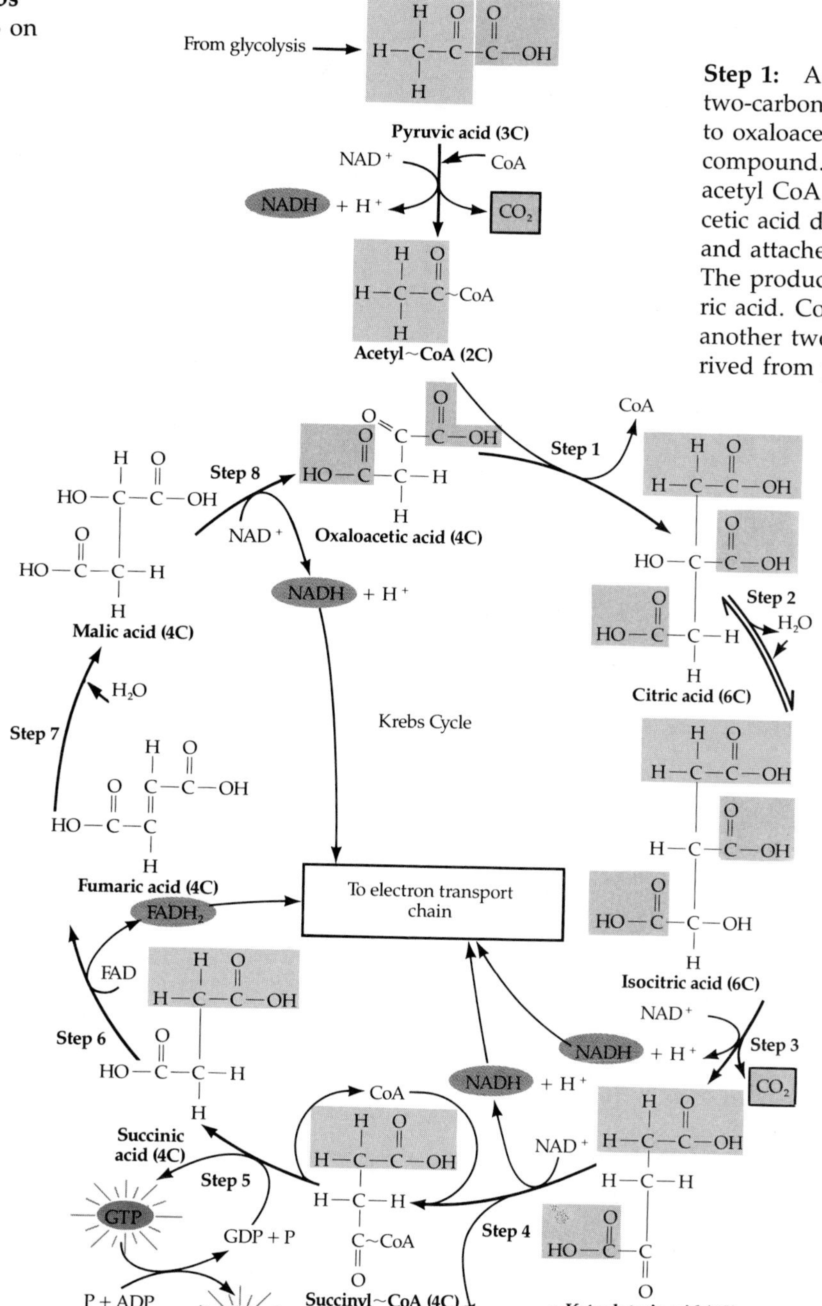

Step 1: Acetyl CoA adds its two-carbon acetyl fragment *(red)* to oxaloacetic acid, a four-carbon compound. The unstable bond of acetyl CoA is broken as oxaloacetic acid displaces the coenzyme and attaches to the acetyl group. The product is the six-carbon citric acid. CoA is then free to prime another two-carbon fragment derived from pyruvic acid.

Step 2: A molecule of water is removed, and another is added back. The net result is the conversion of citric acid to its isomer, isocitric acid.

Step 3: The substrate loses a CO_2 molecule *(gray)*, and the remaining five-carbon compound is oxidized, reducing NAD^+ to $NADH + H^+$.

Step 4: CO_2 *(gray)* is lost; the remaining four-carbon compound is oxidized by the transfer of electrons to NAD^+ to form $NADH + H^+$ and is then attached to CoA by an unstable bond.

Step 5: Substrate-level phosphorylation occurs in this step. CoA is displaced by a phosphate group, which is then transferred to GDP to form guanosine triphosphate (GTP). GTP is similar to ATP, which is formed when GTP donates a phosphate group to ADP.

Step 6: In another oxidative step, two hyrogens are transferred to FAD to form $FADH_2$. The function of this coenzyme is similar to that of $NADH + H^+$, but $FADH_2$ stores less energy.

Step 7: Bonds in the substrate are rearranged in this step by the addition of a water molecule.

Step 8: The last oxidative step produces another molecule of $NADH + H^+$ and regenerates oxaloacetic acid, which accepts a two-carbon fragment from acetyl CoA for another turn of the cycle.

APPENDIX F

Exponents, Exponential Notation, Logarithms, and Generation Time

Very large and very small numbers—such as 4,650,000,000 and 0.00000032—are cumbersome to work with. It is more convenient to express such numbers in exponential notation, that is, as a power of 10. For example, 4.65×10^9 is in **standard exponential notation,** or **scientific notation.** 4.65 is the **coefficient,** and 9 is the power or **exponent.** In standard exponential notation, the coefficient is always a number between 1 and 10, and the exponent can be positive or negative.

To change a number into exponential notation, follow two steps. First, determine the coefficient by moving the decimal point so there is only one nonzero digit to the left of it. For example,

0.0000003 2

The coefficient is 3.2. Second, determine the exponent by counting the number of places you moved the decimal point. If you moved it to the left, the exponent is positive. If you moved it to the right, the exponent is negative. In the example, you moved the decimal point 7 places to the right, so the exponent is −7. Thus

$$0.00000032 = 3.2 \times 10^{-7}$$

Now suppose we are working with a large number instead of a very small number. The same rules apply, but our exponential value will be positive rather than negative. For example,

$$4{,}650{,}000{,}000 = 4.65 \times 10^{+9} = 4.65 \times 10^{9}$$

To multiply numbers written in exponential notation, multiply the coefficients and *add* the exponents. For example,

$$(3 \times 10^4) \times (2 \times 10^3) = (3 \times 2) \times 10^{4+3} = 6 \times 10^7$$

To divide, divide the coefficient and *subtract* the exponents. For example,

$$\frac{3 \times 10^4}{2 \times 10^3} = \frac{3}{2} \times 10^{4-3} = 1.5 \times 10^1$$

Microbiologists use exponential notation in many situations. For instance, exponential notation is used to describe the number of microorganisms in a population. Such numbers are often very large (Chapter 6). Another application of exponential notation is to express concentrations of chemicals in a solution—chemicals such as media components (Chapter 6), disinfectants (Chapter 7), or antibiotics (Chapter 20). Such numbers are often very small. Converting from one unit of measurement to another in the metric system requires multiplying or dividing by a power of 10, which is easiest to carry out in exponential notation.

A **logarithm (log)** is the power to which a base number is raised to produce a given number. Usually we work with logarithms to the base 10, abbreviated **$\log_{10}$**. The first step in finding the $\log_{10}$ of a number is to write the number in standard exponential notation. If the coefficient is exactly 1, the $\log_{10}$ is simply equal to the exponent. For example,

$$\log_{10} 0.00001 = \log_{10}(1 \times 10^{-5}) = -5$$

If the coefficient is not 1, as is often the case, a logarithm table or calculator must be used to determine the logarithm.

Microbiologists use logs for calculating pH levels and for graphing the growths of microbial populations in culture (Chapter 6).

CALCULATING GENERATION TIME

As a cell divides, the population increases exponentially. Numerically this is equal to 2 (since one cell divides into two) raised to the number of times the cell divided (generations):

$$2^{\text{number of generations}}$$

To calculate the final concentration of cells:

$$\text{initial number of cells} \times 2^{\text{number of generations}} = \text{number of cells}$$

For example: if 5 cells were allowed to divide 9 times, this would result in

$$5 \times 2^9 = 2560 \text{ cells}$$

To calculate the number of generations a culture has undergone, cell numbers must be converted to logarithms. Standard logarithm values are based on 10. 0.301 is the log of 2 that is used because one cell divides into two.

$$\text{Number of generations} = \frac{\text{log number of cells (end)} - \text{log number of cells (beginning)}}{0.301}$$

To calculate the generation time for a population:

$$\frac{60 \text{ min} \times \text{hours}}{\text{Number of generations}} = \text{minutes/generation}$$

As an example we calculate the generation time if 100 bacterial cells growing for 5 hours produced 1,720,320 cells.

$$\frac{\log 1{,}720{,}320 - \log 100}{0.301} = 14 \text{ generations}$$

$$\frac{60 \text{ min} \times 5 \text{ hr}}{14 \text{ generations}} = 21 \text{ minutes/generation}$$

A practical application of the calculation is determining the effect of a newly developed food preservative on the culture. Suppose 900 of the same species were grown under the same conditions as the above example, except that the preservative was added to the culture medium. After 15 hours, there were 3,276,800 cells. Calculate the generation time, and decide whether the preservative inhibited growth.

Answer: 75 min/generation. The preservative did inhibit growth.

PRONUNCIATION OF SCIENTIFIC NAMES

RULES OF PRONUNCIATION

The easiest way to learn new material is to talk about it, and that requires saying scientific names. Scientific names may look difficult at first glance, but keep in mind that generally every **syllable** is pronounced and the primary requirement in saying a scientific name is to communicate it.

The rules for the pronunciation of scientific names depend, in part, on the derivation of the root word and its vowel sounds. We have provided some general guidelines below. Pronunciations frequently do not follow the rules because a common usage has become "accepted" or the derivation of the name cannot be determined. Furthermore, for many scientific names there are alternative correct pronunciations.

Vowels Pronounce all the vowels in scientific names. Two vowels written together and pronounced as one sound are called a **diphthong** (for example, the *ou* in *sound*). A special comment is needed about the pronunciation of the vowel endings *-i* and *-ae*: There are two alternative ways to pronounce each of these. In this book, we usually give the pronunciation of a long *e* (*ē*) to the *-i* ending and a long *i* (*ī*) to the *-ae* ending. However, the reverse pronunciations are also correct and in some cases are preferred. For example, *coli* is usually pronounced kō'lī.

Consonants When *c* or *g* is followed by *ae, e, oe, i,* or *y*, it has a soft sound. When *c* or *g* is followed by *a, o, oi,* or *u*, it has a hard sound. When a double *c* is followed by *e, i,* or *y*, it is pronounced as *ks* (e.g., cocci).

Accent The accented syllable is usually the next to last or third to last syllable.

1. The accent is on the next to last syllable:
   a. When the name contains only two syllables, for example, péstis.
   b. When the next to last syllable is a diphthong, for example, *Amoéba*.
   c. When the vowel of the next to last syllable is long, for example, *Treponéma*. The vowel in the next to last syllable is long in words ending in these suffixes:

Suffix	Example
-ales	orders such as Eubacteriáles
-ina	*Sarcína*
-anus, -anum	*pasteuriánum*
-uta	*diminúta*

   d. When the word ends in one of these suffixes:

Suffix	Example
-atus, -atum	caudátum
-ella	Salmonélla

2. The accent is on the third to last syllable in family names. Families end in *-aceae*, which is always pronounced -ā'sē-ē.

PRONUNCIATION OF ORGANISMS IN THIS TEXT

Pronunciation key:

a	hat	ē	see	o	hot	th	thin
ā	age	ė	term	ō	go	u	cup
ã	care	g	go	ô	order	u̇	put
ä	father	i	sit	oi	oil	ü	rule
ch	child	ī	ice	ou	out	ū	use
e	let	ng	long	sh	she	zh	seizure

Acanthamoeba polyphaga a-kan-thä-mē'bä pol-ē-fā'gä
Acetobacter ä-sē-tō-bak'tėr
Achinanthes minutissima ā'kin-an-thēs min-ü-tis'sē-mä
Acinetobacter a-si-ne'tō-bak-tėr

Actinobacillus actinomycemcomitans ak-tin-ō-bä-sil'lus ak-tin-ō-mī-sem-kō'mi-tanz
Actinomyces israelii ak-tin-ō-mī'sēs is-rā'lē-ē
Aedes aegypti ä-e'dēz ē-jip'tē
A. albopictus al-bō-pik'tus
Aeromonas hydrophilia ãr-ō-mō'nas hī-dro'fil-ä
Agrobacterium tumefaciens ag-rō-bak-ti'rē-um tü-me-fā'shenz
Ajellomyces ä-jel-lō-mī'sēs
Alcaligenes al'kä-li-gen-ēs
Allescheria boydii al-lesh-er'ē-ä boi'dē-ē
Amanita phalloides am-an-ī'ta fal-loi'dēz
Anabaena an-ä-bē'nä
Anopheles an-of'-el-ēz
Aquaspirillum bengal a-kwä-spī-ril'lum ben'gal
A. graniferum gra-ni'fir-um
A. magnetotacticum mag-ne-tō-tak'ti-kum
Arthroderma är-thrō-dėr'mä
Ascaris lumbricoides as'kar-is lum-bri-koi'dēz
Ashbya gossypii ash'bē-ä gos-sip'ē'ē
Aspergillus flavus a-spėr-jil'lus flā'vus
A. fumigatus fü-mi-gä'tus
A. niger nī'jėr
A. oryzae ô'ri-zī
A. soyae soi'ī
Azolla ā-zō'lä
Azomonas ā-zō-mō'nas
Azospirillum ā-zō-spī-ril'lum
Azotobacter ä-zō-tō-bak'tėr
Babesia microti ba-bē'sē-ä mī-krō'tē
Bacillus anthracis bä-sil'lus an-thrā'sis
B. cereus se'rē-us
B. coagulans kō-ag'ū-lanz
B. licheniformis lī-ken-i-fôr'mis
B. megaterium meg-ä-tėr'ē-um
B. polymyxa po-lē-miks'ä
B. sphaericus sfe'ri-kus
B. stearothermophilus ste-rō-thėr-mä'fil-us
B. subtilis su'til-us
B. thuringiensis thůr-in-jē-en'sis
Bacteroides hypermegas bak-tė-roi'dēz hī'pėr-meg-äs
Balantidium coli bal-an-tid'ē-um kō'lī (or kō'lē)
Bdellovibrio bacteriovorus del-lō-vib'rē-ō bak-tė-rē-o'vô-rus
Beauveria bō-vãr'ē-a
Beggiatoa bej-jē-ä-tō'ä
Beijerinckia bī-yė-rink'ē-ä
Bifidobacterium globosum bī-fi-dō-bak-ti'rē-um glob-ō'sum
B. pseudolongum sū-dō-lông'um
Blastomyces dermatitidis blas-tō-mī'sēz dėr-mä-tit'i-dis
Blattabacterium blat-tä-bak-ti'rē-um
Boletus edulis bō-lē'tus e'dū-lis
Bordetella pertussis bôr-de-tel'lä pėr-tus'sis
Borrelia burgdorferi bôr-rel'ē-ä burg-dôr'fėr-ē
Bradyrhizobium brad-ē-rī-zō'bē-um
Branhamella catarrhalis bran-ha-mel'lȧ ka-tär'al-is
Brucella abortus brü-sel'lä ä-bôr'tus
B. melitensis me-li-ten'sis
B. suis sü'is
Byssochlamys fulva bis-sō-klam'is fül'vä
Campylobacter fetus kam-pī-lō-bak'tėr fē'tus
C. jejuni jē-jū'nē
Candida albicans kan'did-ä al'bi-kanz
C. krusei krūs'ā-ē
C. utilis ū'til-is
Canis familiaris kānis fa-mil'yãr-is
Carpenteles kär-pen'tel-ēz
Caulobacter kô-lō-bak'tėr
Cephalosporium sef-ä-lō-spô'rē-um
Ceratocystis ulmi sē-rä-tō-sis'tis ul'mē
Chilomastix kē'lō-ma-sticks
Chlamydia pneumoniae kla-mi'dē-ä nü-mō'nē-ī
C. psittaci sit'tä-sē
C. trachomatis trä-kō'mä-tis
Chlamydomonas klam-i-dō-mō'näs
Chlorobium klô-rō'bē-um
Chloroflexus klô-rō-flex'us
Chromatium vinosum krō-mā'tē-um vi-nō'sum
Chroococcus turgidus krō-ō-kok'kus tėr'gi-dus
Chrysops krī'sops
Citrobacter sit'rō-bak-tėr
Cladosporium werneckii kla-dō-spô'rē-um wėr-ne'kē-ē
Claviceps purpurea kla'vi-seps pür-pü-rē'ä
Clonorchis sinensis klo-nôr'kis si-nen'sis
Clostridium acetobutylicum klôs-tri'dē-um a-sē-tō-bū-til'li-kum
C. botulinum bo-tū-lī'num
C. butyricum bü-ti'ri-kum
C. difficile dif'fi-sil-ē
C. pasteurianum pas-tyėr-ē-ā'num
C. perfringens pėr-frin'jens
C. sporogenes spô-rä'jen-ēz
C. tetani te'tan-ē
C. thermosaccharolyticum thėr-mō-sak-kär-ō-li'ti-kum
Coccidioides immitis kok-sid-ē-oi'dēz im'mi-tis
Corynebacterium diphtheriae kô-rī-nē-bak-ti'rē-um dif-thi'rē-ī
C. glutamicum glü-tam'i-kum
C. xerosis ze-rō'sis
Coxiella burnetii käks-ē-el'lä bėr-ne'tē-ē
Cristispira pectinis kris-tē-spī'rä pek'tin-is
Cryptococcus neoformans krip-tō-kok'kus nē-ō-fôr'manz
Cryptonectria parasitica krip-tō-nek'trē-ä par-ä-si'ti-kä
Cryptosporidium krip-tō-spô-ri'dē-um
Culex kū'leks
Culiseta kū-li'se-tä
Cytophaga sī-täf'äg-ä

Daptobacter dap′to-bak-tėr
Dermacentor andersoni dėr-mä-sen′tôr an-dėr-sōn′ē
D. variabilis vãr-ē-a′bil-is
Desulfotomaculum nigrificans dē-sul-fō-to-ma′kū-lum nī′gri-fi-kans
Desulfovibrio desulfuricans dē-sul-fō-vib′rē-ō dē-sul-fėr′i-kans
Didinium nasutum dī-di′nē-um nä-sūt′um
Diphyllobothrium latum dī-fil-lō-bo′thrē-um lā′tum
Echinococcus granulosus ē-kīn-ō-kok′kus gra-nū-lō′sus
Ectothiorhodospira mobilis ek′tō-thī-ō-rō-dō-spī-rä mō′bil-is
Emericella em-ėr-ē-sel′lä
Emmonsiella em-mon-sē-el′lä
Entamoeba histolytica en-tä-mē′bä his-tō-li′ti-kä
Enterobacter aerogenes en-te-rō-bak′tėr ā-rä′jen-ēz
E. cloacae klō-ā′kī
Enterobius vermicularis en-te-rō′bē-us ver-mi-kū-lar′is
Enterococcus faecalis en-te-rō-kok′kus fē-kā′lis
Epidermophyton ep-i-dėr-mō-fī′ton
Erwinia ėr-wi′nē-ä
E. dissolvens dis-solv′ens
Erysipelothrix rhusiopathiae ãr-i-si-pel′ō-thrix rus-ē-ō-path′ē-ī
Erysiphe grammis ãr′i-sīf gram′mis
Escherichia coli esh-ėr-i′kē-ä kō′lī (or kō′lē)
Euglena ū-glē′nä
Eurotium yėr-ō′tē-um
Filobasidiella fi-lō-ba-si-dē-el′lä
Fonsecaea pedrosoi fon-se′kē-ä pe-drō′sō-ē
Francisella tularensis fran-sis-el′lä tü-lä-ren′sis
Frankia frank′ē-ä
Fusobacterium fü-sō-bak-ti′rē-um
Gambierdiscus toxicus gam′bē-ėr-dis-kus toks′i-kus
Gardnerella vaginalis gärd-nė-rel′lä va-jin-al′is
Gelidium jel-id′ē-um
Geotrichum gē-ō-trik′um
Giardia lamblia jē-är′dē-ä lam′lē-ä
Gloeocapsa glē-ō-kap′sä
Glomus fasciculatus glo′mus fa-sik-ū-lä′tus
Glossina gläs-sē′nä
Gluconobacter glü-kon-ō-bak′tėr
Gonyaulax
Gymnoascus jim-nō-as′kus
Gymnodynium microadriaticus jim-nō-din′ē-um mī-krō-ā′drē-at-i-kus
Haloarcula hal′ō-är-kū-lä
Halobacterium halobium hal-ō-bak-ti′rē-um hal-ō′bē-um
Halococcus hal-ō-kok′kus
Helicobacter pylori hē′lik-ō-bak-tėr pī′lô-rē
Hemophilus ducreyi hē-mä′fi-lus dü-krā′ē
H. influenzae in-flü-en′zī
Histoplasma capsulatum his-tō-plaz′mä kap-su-lä′tum
Holospora hō-lo′spô-rä
Homo sapiens hō′mō sā′pē-ens
Hydrogenomonas hī-drō-je-nō-mō′näs
Hyphomicrobium hī-fō-mi-krō′bē-um
Isospora ī-so′spô-rä
Ixodes dammini iks-ō′dēs dam′mi-nē
I. pacificus pas-i′fi-kus
Klebsiella pneumoniae kleb-sē-el′lä nü-mō′nē-ī
Lactobacillus bulgaricus lak-tō-bä-sil′lus bul-gā′ri-kus
L. delbrueckii del-brük′ē-ē
L. leichmannii līk-man′nē-ē
L. plantarum plan-tä′rum
L. sanfrancisco sän-fran-sis′kō
Legionella anisa lē-jä-nel′lä a′nis-ä
L. feeleii fē′lē-ē
L. micdadei mik-dā′dē
L. pneumophila nü-mō′fi-lä
Leptospira interrogans lep-tō-spī′rä in-tėr′rä-ganz
Leuconostoc mesenteroides lü-kō-nos′tok mes-en-ter-oi′dēz
Licmorphora lik-môr′fôr-ä
Listeria monocytogenes lis-te′rē-ä mo-nō-sī-tô′je-nēz
Meniscus mē-nis′kus
Methanosarcina meth-a-nō-sär′sī-nä
Methylophilus methylotrophus meth-i-lo′fi-lus meth-i-lō-trōf′us
Microcladia mī-krō klād′ē-ä
Micrococcus luteus mī-krō-kok′kus lū′tē-us
Micromonospora purpureae mī-krō-mo-nä′spô-rä pür-pü-rē′ī
Microsporum gypseum mī-krō-spô′rum jip′sē-um
Mixotricha mix-ō-trik′ä
Moraxella lacunata mô-raks-el′lä la-kü-nä′tä
M. osloensis os′lō-en-sis
Mucor rouxii mū-kôr rō′ē-ē
Mycobacterium avium-intracellulare mī-kō-bak-ti′rē-um ā′vē-um-in′trä-cel-ū-lä-rē
M. bovis bō-vis
M. leprae lep′rī
M. smegmatis smeg-ma′tis
M. tuberculosis tü-bėr-kū-lō′sis
Mycoplasma hominis mī-kō-plaz′mä ho′mi-nis
M. pneumoniae nu-mō′nē-ī
Myxococcus fulvus micks-ō-kok′kus ful′vus
Naegleria fowleri nī-gle′rē-ä fou′lėr-ē
Nannizia nan-nī′zē-ä
Necator americanus ni-kā′tôr ä-me-ri-ka′nus
Neisseria gonorrhoeae nī-se′rē-ä go-nôr-rē′ī
N. meningitidis me-nin-ji′ti-dis
Nereocystis nē-rē-ō-sis′tis
Nitrobacter nī-trō-bak′tėr
Nitrosomonas nī-trō-sō-mō′näs
Nocardia asteroides nō-kär′dē-ä as-tėr-oi′dēz
Oocystis ō-ō-sis′tis
Ornithodorus ôr-nith-ō′dô-rus
Paracoccidioides brasiliensis pãr-ä-kok-sid′ē-oi-dēz bra-sil-ē-en′sis

Paracoccus denitrificans pãr-ä-kok′kus dē-nī-tri′fi-kanz
Paragonimus westermanni pãr-ä-gōn′e-mus we-ster-man′nē
Paramecium multimicronucleatum pãr-ä-mē′sē-um mul-tē-mī-krō-nü-klē-ä′tum
Pasteurella haemolytica pas-tyėr-el′lä hēm-ō-lit′i-kä
P. multocida mul-tō′si-dä
Pediculus humanus corporis ped-ik′ū-lus hü′ma-nus kôr′pô-ris
Pediococcus cerevisae pe-dē-ō-kok′kus se-ri-vis′e-ī
Penicillium chrysogenum pen-i-sil′lē-um krī-so′gen-um
P. griseofulvum gri-sē-ō-fůl′vum
P. notatum nō-tä′tum
Peridinium per-i-din′ē-um
Petriellidium pet-rē-el-li′dē-um
Phialophora verruscosa fē-ä-lō′fô-rä ver-rü-skō′sä
Phormidium luridum fôr-mi′dē-um ler′i-dum
Phytophthora infestans fī-tof′thô-rä in-fes′tans
Pichia pik′ē-ä
Pityrosporum pit-i-ros′pô-rum
Plasmodium falciparum plaz-mō′dē-um fal-sip′är-um
P. malariae mä-lā′rē-ī
P. ovale ō-vä′lē
P. vivax vī′vaks
Plesiomonas shigelloides ple-sē-ō-mō′nas shi-gel-loi′des
Pneumocystis carinii nü-mō-sis′tis kär-i′nē-ī (or kär-i′nē-ē)
Propionibacterium acnes prō-pē-on-ē-bak-ti′rē-um ak′nēz
P. freudenreichii froi-den-rīk′ē-ē
Proteus mirabilis prō′tē-us mi-ra′bi-lis
P. vulgaris vul-ga′ris
Prototheca prō-tō-thā′kä
Pseudomonas aeruginosa sū-dō-mō′nas ā-rü-ji-nō′sä
P. andropogonis ān′dro-po-gō-nis
P. cepacia se-pā′sē-ä
P. fluorescens flôr-es′ens
P. haemolytica hē-mō-lit′i-kä
P. multocida mul′tō-sid-ä
P. syringae sėr-in′gī
Quercus kwer′kus
Rhizobium japonicum rī-zō′bē-um jap-on′i-kum
R. meliloti mel-i-lot′ē
Rhizopus nigricans rī′zō-pus nī′gri-kans
R. oryzae ô′rī-zī
Rhodopseudomonas rō-dō-sū-dō-mō′nas
Rhodospirillum rubrum rō-dō-spī-ril′lum rūb′rum
Rickettsia prowazekii ri-ket′sē-ä prou-wä-ze′kē-ē
R. rickettsii ri-ket′sē-ē
R. typhi tī′fē
Riftia rift′ē-ä
Rosa multiflora rō′sä mul-ti-flô′rä
Saccharomyces beticus sak-ä-rō-mī′sēs bet′i-kus
S. bayanus bā′an-us
S. carlsbergensis kärls-bėr′gen′sis
S. cerevisiae se-ri-vis′ē-ī
S. ellipsoideus ē-lip-soi′dē-us
S. exiguus egz-ij′ū-us
S. fragilis fra′jil-is
S. rouxii rō-ē′ē
Salmonella choleraesuis sal-mōn-el′lä kol-ėr-ä-sü′is
S. dublin dub′lin
S. eastbourne ēst′bôrn
S. enteritidis en-tėr-ī′ti-dis
S. newport new′pôrt
S. typhi tī′fē
S. typhimurium tī-fi-mūr′ē-um
Sargassum sär-gas′sum
Sartorya sär-tô′rē-ä
Schistosoma shis-tō-sō′mä or skis-tō-sō′mä
Serratia marcescens ser-rä′tē-ä mär-ses′sens
Shigella boydii shi-gel′lä boi′dē-ē
S. dysenteriae dis-en-te′rē-ī
S. flexneri fleks′nėr-ē
S. sonnei son′nē
Sphaerotilus natans sfe-rä′ti-lus nā′tans
Spirillum ostreae spī-ril′lum ost′rā-ī
S. volutans vō′lū-tans
Spiroplasma spī-ro-plaz′mä
Spirosoma spī-rō-sō′mä
Spirulina spī-rü-lī′nä
Spongomorphora spon′jō-môr-fô-rä
Sporothrix schenkii spô-rō′-thriks shen′kē-ē
Staphylococcus aureus staf-i-lō-kok′kus ô′rē-us
S. epidermidis e-pi-dėr′mi-dis
S. saprophyticus sap-rō-fit′i-kus
Stella stel′lä
Stigmatella aurantiaca stig-mä′tel′lä ô-rän-tē′ä-kä
Streptobacillus strep-tō-bä-sil′lus
Streptococcus lactis strep-tō-kok′kus lak′tis
S. mutans mū′tans
S. pneumoniae nü-mō′nē-ī
S. pyogenes pī-äj′en-ēz
S. thermophilus thėr-mo′fil-us
Streptomyces aureofaciens strep-tō-mī′sēs ô-rē-ō-fa′si-ens
S. erythraeus ā-rith′rē-us
S. fradiae frā′dē-ī
S. griseus gri′sē-us
S. nodosus nō-dō′sus
S. noursei nôr′sē-ī
S. olivoreticuli ō-liv-ō-re-tik′ū-lē
S. venezuelae ve-ne-zü-e′lī
Sulfolobus sul-fō-lō′bus
Taenia saginata te′nē-ä sa-ji-nä′tä
T. solium sō′lē-um
Talaromyces ta-lä-rō-mī′sēs
Thermoactinomyces vulgaris thėr-mō-ak-tin-ō-mī′sēs vul-gar′is

Thermoplasma thėr-mō-plaz′mä
Thiobacillus ferrooxidans thī-ō-bä-sil′lus fer-rō-oks′i-danz
T. thiooxidans thī-ō-oks′i-danz
Thiocapsa floridana thī-ō-kap′sä flôr-i′dä-nä
Thunnus tün′nus
Toxoplasma gondii toks-ō-plaz′mä gon′dē-ē
Trachelomonas trä-kel-ō-mōn′as
Treponema pallidum trē-pō-nē′mä pal′li-dum
Triatoma trī-ä-tō′ma
Trichinella spiralis trik-in-el′lä spī-ra′lis
Trichomonas vaginalis trik-ō-mōn′as va-jin-al′is
Trichonympha sphaerica trik-ō-nimf′ä sfe′ri-kä
Trichophyton trik-ō-fī′ton
Trichosporon trik-ō-spôr′on
Trichuris trichiura trik′ėr-is trik-ē-yėr′a
Tridacna trī-dak′nä
Trypanosoma brucei gambiense tri-pa-nō-sō′mä brüs′ē gam-bē-ens′
T. brucei rhodesiense rō-dē-sē-ens′
T. cruzi kruz′ē
Ureaplasma urealyticum ū-rē-ä-plaz′mä ū-rē-ä-lit′i-kum
Veillonella vi-lo-nel′lä
Vibrio cholerae vib′rē-ō kol′ėr-ī
V. parahaemolyticus pa-rä-hē-mō-li′ti-kus
Wuchereria bancrofti vū-kėr-ār′ē-ä ban-krof′tē
Xenopsylla cheopis ze-nop-sil′lä kē-ō′pis
Yersinia enterocolitica yėr-sin′ē-ä en-tėr-ō-kōl-it′ik-ä
Y. pestis pes′tis
Y. pseudotuberculosis sū-dō-tü-bėr-kū-lō′sis
Zoogloea zō-ō-glē′ä

GLOSSARY

ABO blood group Classification of red blood cells based on the presence or absence of A and B carbohydrate antigens.

Abscess A localized accumulation of pus.

Acellular vaccine Vaccine consisting of antigenic parts of cells.

Acetyl coenzyme A (acetyl CoA) A substance composed of an acetyl group and a carrier molecule called coenzyme A (CoA); provides the means for pyruvic acid to enter the Krebs cycle. *See* Krebs cycle.

Acetyl group $CH_3{-}\overset{\displaystyle O}{\overset{\|}{C}}{-}$

Acid A substance that dissociates into one or more hydrogen ions and one or more negative ions.

Acid-fast stain A differential stain used to identify bacteria that are not decolorized by acid-alcohol.

Acidic dye A salt in which the color is in the negative ion; used for negative staining.

Acidophile A bacterium that grows below pH 4.

Acquired immunity The ability, obtained during the life of the individual, to produce specific antibodies.

Actinomycetes Gram-positive bacteria that tend to form branching filaments; may form true mycelia; may produce conidiospores; see *Bergey's Manual,* Vols. 2 and 4.

Activated sludge Sludge being digested by aerobic organisms; used in secondary sewage treatment.

Activation energy The minimum collision energy required for a chemical reaction to occur.

Actively acquired immunity Production of antibodies or specialized lymphocytes by an individual in response to an antigen.

Active site Region on an enzyme that interacts with the substrate.

Active transport Net movement of a substance across a membrane against a concentration gradient; requires energy.

Acute disease A disease in which symptoms develop rapidly but last for only a short time.

Adenine A purine nucleic acid base that pairs with thymine in DNA and uracil in RNA.

Adenocarcinoma Cancer of glandular tissue.

Adenosine diphosphate (ADP) The substance formed when ATP is split and energy is released.

Adenosine diphosphoglucose (ADPG) Glucose activated by ATP; precursor for glycogen synthesis.

Adenosine triphosphatase The enzyme that catalyzes the two reactions ADP + Ⓟ $\longrightarrow$ ATP and ATP $\longrightarrow$ ADP + Ⓟ.

Adenosine triphosphate (ATP) An important intracellular energy source.

Adherence Attachment of a microbe or phagocyte to another's plasma membrane or other surface.

Adhesin *See* Ligands.

Aerial mycelium A mycelium composed of fungal hyphae that project above the surface of the growth medium and produce asexual spores.

Aerobe An organism requiring oxygen (O_2) for growth.

Aerobic respiration Respiration in which the final electron acceptor in the electron transport chain is oxygen (O_2).

Aerotolerant anaerobe An organism that does not use oxygen (O_2) but is not affected by its presence.

Aflatoxin $C_{17}H_{10}O_6$, a carcinogenic toxin produced by *Aspergillus flavus.*

Agar A complex polysaccharide derived from a marine alga and used as a solidifying agent in culture media.

Agglutination A joining together or clumping of cells.

Airborne transmission The spread of pathogens farther than 1 meter in air from reservoir to susceptible host.

Alcohol An organic molecule with the functional group —OH.

Alcohol fermentation A catabolic process, beginning with glycolysis, that produces ethyl alcohol.

Aldehyde An organic molecule with the functional group $-\underset{\displaystyle O}{\underset{\|}{C}}-H$.

Algae A group of photosynthetic eucaryotes; some are included in the Kingdom Protista and some in the Kingdom Plantae.

Algin Sodium salt of mannuronic acid ($C_6H_8O_6$); found in brown algae.

Alkaline Having more OH^- ions than H^+ ions; pH is greater than 7.

Allergen An antigen that evokes a hypersensitivity response.

Allergy *See* Hypersensitivity.

Allograft A graft between persons who aren't identical twins.

Allosteric inhibition The process in which an enzyme's activity is changed because of binding on the allosteric site.

Allosteric site The site on an enzyme at which an inhibitor binds.

Alpha-amino acid An amino acid with —COOH and $—NH_2$ attached to the same carbon atom.

Alum Aluminum sulfate $[Al_2(SO_4)_3]$.

Amanitin Polypeptide mushroom toxin that causes liver and nerve damage.

Ames test A procedure using bacteria to identify potential carcinogens.

Amination The addition of an amino group.

Amino acid An organic acid containing an amino group and a carboxyl group.

Amino acid activating enzyme Enzyme that attaches an amino acid to tRNA.

Aminoglycoside An antibiotic consisting of amino sugars and an aminocyclitol ring; for example, streptomycin.

Amino group $—NH_2$

Ammonification Removal of amino groups from amino acids to form ammonia.

Amoeba An organism belonging to the Kingdom Protista that moves by means of pseudopods.

Amphibolic pathway A pathway that is anabolic and catabolic.

Amphitrichous Having tufts of flagella at both ends of a cell.

Amplify Increase the number of genes in a cell.

Anabolism All synthesis reactions in a living organism.

Anaerobe An organism that does not require oxygen (O_2) for growth.

Anaerobic respiration Respiration in which the final electron acceptor in the electron transport chain is an inorganic molecule other than oxygen (O_2); for example, a nitrate or sulfate.

Anaerobic sludge digester Anaerobic digestion used in secondary sewage treatment.

Anal pore A site in certain protozoans for elimination of waste.

Analytical epidemiology Comparison of a diseased group and a healthy group to determine the cause of the disease.

Anamnestic response A rapid rise in antibody titer following exposure to an antigen after the primary response to that antigen.

Anaphylaxis A hypersensitivity reaction involving IgE antibodies, mast cells, and basophils.

Angstrom (Å) A unit of measurement equal to 10^{-10} m, 10^{-4} μm, and 10^{-1} nm; no longer an official unit.

Animalia A kingdom composed of multicellular eucaryotes lacking cell walls.

Animal virus A virus that multiplies in animal tissues.

Anion An ion with a negative charge.

Anoxygenic Not producing molecular oxygen; typical of bacterial photosynthesis.

Antagonism Active opposition; for example, between two drugs or two microbes.

Antibiotic An antimicrobial agent produced naturally by a bacterium or fungus.

Antibody A protein produced by the body in response to an antigen and capable of combining specifically with that antigen.

Antibody titer The amount of antibody in serum.

Anticodon The three nucleotides by which a tRNA recognizes an RNA codon.

Antigen Any substance that, when introduced into the body, causes antibody formation and reacts only with its specific antibody.

Antigen-binding sites Sites on an antibody that bind to antigenic determinants.

Antigenic determinant A specific region on the surface of an antigen against which antibodies are formed.

Antigenic drift Minor variations in the antigenic makeup of a virus that occur with time.

Antigenic shift Major genetic changes in influenza viruses causing changes in H and N antigens.

Antigen modulation The process whereby tumor cells shed their specific antigens, thus avoiding the host's immume response.

Antigen-presenting cell Macrophage that engulfs an antigen and presents fragments to T cells.

Antigen receptors Antibodylike molecules on T cells.

Antihuman gamma globulin Antibodies that react specifically with human antibodies.

Anti-idiotypic antibody Antibody that mimics the shape of an antigen to induce immunity.

Antimetabolite Any substance that interferes with metabolism by competitive inhibition of an enzyme.

Antimicrobial agent A chemical that destroys pathogens without damaging body tissues.

Antiparallel DNA Double-stranded DNA molecule in which one strand is $5' \longrightarrow 3'$ and the complement is $3' \longrightarrow 5'$.

Antisense strand (−strand) Viral RNA that cannot act as mRNA.

Antiseptic A chemical for disinfection of the skin, mucous membranes, or other living tissues.

Antiserum A solution containing antibodies.

Antitoxin A specific antibody produced by the body in response to a bacterial exotoxin or its toxoid.

Antiviral protein Protein made in response to interferon that blocks viral multiplication.

Apoenzyme The protein portion of an enzyme, which requires activation by a coenzyme.

Arbuscule Fungal mycelia in plant root cells.

Archaeobacteria Procaryotic organisms lacking peptidoglycan.

Arthropod An animal phylum characterized by an exoskeleton and jointed legs; includes insects and ticks.

Arthrospore An asexual fungal spore formed by fragmentation of a septate hypha.

Arthus reaction Inflammation and necrosis at the site of injection of foreign serum, due to immune complex formation.

Artificially acquired active immunity The production of antibodies by the body in response to a vaccination.

Artificially acquired passive immunity The transfer of humoral antibodies formed by one individual to a susceptible individual, accomplished by injection of antiserum.

Ascospore A sexual fungal spore produced in an ascus, formed by the Ascomycetes class of fungi.

Ascus A saclike structure containing ascospores.

Asepsis The absence of contamination by unwanted organisms.

Aseptic packaging Commercial food preservation by filling sterile containers with sterile food.

Asexual reproduction Reproduction without opposite mating strains.

Asexual spores Reproductive cells produced by mitosis and cell division (eucaryotes) or binary fission (actinomycetes).

A site Site on ribosome where new tRNA binds.

Associative recognition T-cell response to a foreign antigen and MHC antigen.

Atom The smallest unit of matter that can enter into a chemical reaction.

Atomic number The number of protons in the nucleus of an atom.

Atomic weight The total number of protons and neutrons in the nucleus of an atom.

Attenuation Lessening of virulence of a microorganism. Also, a regulatory mechanism for protein synthesis.

Attenuator Site on mRNA where translation can be stopped.

Autoclave Equipment for sterilization by steam under pressure, usually operated at 15 psi and 121°C.

Autogenous hypothesis Model for evolution of eucaryotes stating that organelles are formed from internal membranes in procaryotes.

Autograft A tissue graft from one's self.

Autoimmunity An immunological response against a person's own tissue antigens.

Autotroph An organism that uses carbon dioxide (CO_2) as its principal carbon source.

Auxotroph A mutant microorganism with a nutritional requirement not possessed by the parent.

Axial filament The structure for motility found in spirochetes.

Bacillus Any rod-shaped bacterium; when written as a genus, refers to rod-shaped, endospore-forming, facultatively anaerobic, gram-positive bacteria.

Bacteremia A condition in which there are bacteria in the blood.

Bacteria All living organisms with procaryotic cells.

Bacterial growth curve A graph indicating the growth of a bacterial population over time.

Bactericidal Capable of killing bacteria.

Bacteriochlorophyll The light-absorbing pigment found in green sulfur and purple sulfur bacteria.

Bacteriocin Toxic protein produced by bacteria that kill other bacteria.

Bacteriocinogenic plasmid Plasmid containing genes for the synthesis of bacteriocins.

Bacteriophage (phage) A virus that multiplies in bacterial cells.

Bacteriorhodopsin The light-absorbing purple pigment in the *Halobacterium* cell membrane.

Bacteriostatic Capable of inhibiting bacterial growth.

Bacteroid Enlarged *Rhizobium* cells found in root nodules.

Basal body A structure that anchors flagella to the cell wall and plasma membrane.

Base A substance that accepts hydrogen ions and is capable of uniting with water to form an acid.

Base analog A chemical that is structurally similar to the normal nitrogenous bases in nucleic acids but with altered basepairing properties.

Base pairs The arrangement of nitrogenous bases in nucleic acids based on hydrogen bonding; in DNA, base pairs are A–T and G–C; in RNA, base pairs are A–U and G–C.

Base substitution The replacement of a single base in DNA by another base, causing a mutation.

Basic dye A salt in which the color is in the positive ion; used for bacterial stains.

Basidiospore A sexual fungal spore produced in a basidium, characteristic of the Basidiomycetes.

Basidium A pedestal that produces basidiospores; found in the Basidiomycetes.

Basophil A granulocyte that readily takes up basic dye.

B cells Antibodies secreting plasma cells and memory cells.

BCG vaccine A live, attenuated strain of *Mycobacterium bovis* used to provide immunity to tuberculosis.

Benign tumor A noncancerous tumor.

Benthic zone The sediment at the bottom of a body of water.

Bergey's Manual The standard taxonomic reference on bacteria.

Beta oxidation The removal of two carbon units of a long chain of fatty acid to form acetyl-CoA.

Binary fission Bacterial reproduction by division into two daughter cells.

Binomial nomenclature The system of having two names (genus and specific epithet) for each organism.

Biochemical oxygen demand (BOD) A measure of the biologically degradable organic matter in water.

Biochemical pathway A sequence of enzymatically catalyzed reactions occurring in a cell.

Biochemistry The science of chemical processes in living organisms.

Bioconversion Changes in organic matter brought about by the growth of microorganisms.

Biogenesis The concept that living cells can arise only from preexisting cells.

Biogeochemical cycles The recycling of chemical elements by microorganisms for use by other organisms.

Biological transmission Transmission of a pathogen from one host to another when the pathogen reproduces in the vector.

Bioluminescence The emission of light from the electron transport chain of certain living organisms.

Biomass Organic matter produced by living organisms and measured by weight.

Bioreactor Fermentation vessel with controls for environmental conditions, e.g., temperature and pH.

Bioremediation The use of microbes to remove an environmental pollutant.

Biotechnology The use of living cells in industry, usually to produce products such as pharmaceuticals.

Biovar A subgroup of a *Salmonella* serovar based on biochemical testing.

Bladder A bag. In brown algae, an air sac.

Blade Flat leaflike structure of multicellular algae.

Blastospore An asexual fungal spore produced by budding from the parent cell.

Blocking antibody An IgG antibody that reacts with an allergen to prevent a hypersensitivity reaction.

Blood–brain barrier Cell membranes that allow some substances to pass from the blood to the brain but restrict others.

Bloom (algal) Abundant growth of microscopic algae, producing visible colonies in nature.

Brightfield microscope A microscope that uses visible light for illumination; the specimens are viewed against a white background.

Broad spectrum antimicrobial agent A chemical that has antimicrobial activity against many infectious microorganisms.

Broth dilution test A method of determining the MIC by using serial dilutions of an antimicrobial drug.

Brownian movement The movement of particles, including microorganisms, in a suspension owing to bombardment by the moving molecules in the suspension.

Bubo An enlarged lymph node caused by inflammation.

Budding Asexual reproduction beginning as a protuberance from the parent cell that grows to become a daughter cell; also, release of an enveloped virus through the plasma membrane of an animal cell.

Buffer A substance that tends to stabilize the pH of a solution.

Bulking Condition arising when sludge floats rather than settles in secondary sewage treatment.

Burst size The number of newly synthesized bacteriophage particles released from a single cell.

Burst time The time required from bacteriophage adsorption to release.

Cachectin A polypeptide released by phagocytes in response to bacterial endotoxins; induces shock.

Calvin–Benson cycle Conversion of CO_2 into reduced organic compounds.

Cancer A malignant, invasive cellular tumor that has the capability of spreading throughout the body or body parts.

Candle jar A sealed jar containing a lighted candle, used to incubate bacterial cultures in high CO_2 atmosphere.

Capsid The protein coat of a virus that surrounds the nucleic acid.

Capsomere A protein subunit of a capsid.

Capsule An outer, viscous covering on some bacteria composed of a polysaccharide or polypeptide.

Carbapenems Antibiotics containing β-lactam and other active agents.

Carbohydrates Organic compounds composed of carbon, hydrogen, and oxygen, with the hydrogen and oxygen present in a 2:1 ratio; include starches, sugars, and cellulose.

Carbon cycle The series of processes that converts carbon dioxide (CO_2) to organic substances and back to carbon dioxide in nature.

Carbon fixation *See* Calvin–Benson cycle.

Carbon skeleton The basic chain or ring of carbon atoms in a molecule; for example,

$$-\overset{|}{\underset{|}{C}}-\overset{|}{\underset{|}{C}}-\overset{|}{\underset{|}{C}}-$$

Carboxyl group —COOH.

Carboxysome Procaryotic inclusion containing ribulose 1,5-diphosphate carboxylase.

Carcinogen Any cancer-producing substance.

Cardiolipin A beef heart extract used in the venereal disease research laboratory (VDRL) slide test to detect antibodies against syphilis.

Carrageen A galactose polymer in the cell walls of red algae.

Carrier An individual who harbors a pathogen but exhibits no signs of illness.

Carrier proteins Plasma membrane proteins that transport substances across the membrane.

Casein Milk protein.

Catabolic repression Inhibition of decomposition of alternate carbon sources by glucose.

Catabolism All decomposition reactions in a living organism.

Catalase An enzyme that catalyzes the breakdown of hydrogen peroxide to water and oxygen.

Catalyst A substance that affects the rate of a chemical reaction, usually increasing the rate, but isn't changed in the reaction.

Cation A positively charged ion.

cDNA (complementary DNA) DNA made from mRNA in vitro.

Cell The basic microscopic unit of structure and function of all living organisms.

Cell culture Animal or plant cells grown in vitro.

Cell-mediated immunity An immune response that involves the binding and elimination of antigens by T-cell lymphocytes.

Cell theory The principle that all living things are composed of cells.

Cellulose A glucose polysaccharide that is the main component of plant cell walls.

Cell wall The outer covering of most bacterial, fungal, algal, and plant cells: in eubacteria, it consists of peptidoglycan.

Centrioles Paired, cylindrical structures found in the centrosome of eucaryotic cells.

Cephalosporin An antibiotic produced by the fungus *Cephalosporium* that inhibits the synthesis of gram-positive bacterial cell walls.

Cercaria A free-swimming larva of trematodes.

Chancre A hard sore, the center of which ulcerates.

Chemical bond Attractive force between atoms forming a molecule.

Chemical element A fundamental substance composed of atoms that have the same atomic number and behave the same way chemically.

Chemical energy The energy of a chemical reaction.

Chemically defined medium A culture medium in which the exact chemical composition is known.

Chemical reaction The process of making or breaking bonds between atoms.

Chemiosmosis A proton gradient across a cytoplasmic membrane; can be used to generate ATP.

Chemistry The science of the interactions of atoms and molecules.

Chemoautotroph An organism that uses an inorganic chemical as an energy source and carbon dioxide (CO_2) as a carbon source.

Chemoheterotroph An organism that uses organic molecules as a source of carbon and energy.

Chemostat An apparatus to keep a culture in log phase indefinitely.

Chemotaxis Motion in response to the presence of a chemical.

Chemotherapy Treatment of a disease with chemical substances.

Chemotroph An organism that uses oxidation–reduction reactions as its primary energy source.

Chimeric monoclonal antibodies Genetically engineered antibodies made of human constant regions and mouse variable regions.

Chitin A glucosamine polysaccharide that is the main component of fungal cell walls and arthropod skeletons.

Chlamydospore An asexual fungal spore formed within a hypha.

Chloramphenicol A broad spectrum bacteriostatic chemical.

Chlorobium vesicle *See* Chlorosome.

Chlorophyll *a* The light-absorbing pigment in cyanobacteria, algae, and plants.

Chloroplast The organelle that performs photosynthesis in photoautotrophic eucaryotes.

Chlorosome Plasma membrane folds in green sulfur bacteria containing bacteriochlorophylls.

Choleragen A *Vibrio* enterotoxin.

Chromatin Threadlike, uncondensed DNA in an interphase eucaryotic cell.

Chromatophore An infolding in the plasma membrane where bacteriochlorophyll is located in photoautotrophic bacteria.

Chromosome The structure that carries hereditary information.

Chromosome number Diploid number of a eucaryote.

Chronic disease An illness that develops slowly and is likely to continue or recur for long periods.

Cilia Relatively short cellular projections that move in a wavelike manner.

Ciliate A member of the protozoan phylum Ciliata that uses cilia for locomotion.

Cisternae Stacked elements of the Golgi complex.

Class A taxonomic ranking between phylum and order.

Clonal deletion Elimination of B and T cells that react with self.

Clonal selection theory Development of clones of B and T cells against a specific antigen.

Clone A population of cells that are identical to the parent cell.

Clue cells Sloughed-off vaginal cells covered with *G. vaginalis*.

Coagulase A bacterial enzyme that causes blood plasma to clot.

Coccobacillus A bacterium that is an oval rod.

Coccus A spherical or ovoid bacterium.

Codon A group of three nucleotides in mRNA that specifies the insertion of an amino acid into a protein.

Coenocytic hyphae Fungal filaments that are not divided into uninucleate cell-like units because they lack septa.

Coenzyme A nonprotein substance that is associated with and that activates an enzyme.

Coenzyme A (CoA) A coenzyme that functions in decarboxylation.

Cofactor The nonprotein component of an enzyme.

Cohort method Comparison of two populations: control and experimental.

Coliforms Aerobic or facultatively anaerobic, gram-negative, nonspore-forming, rod-shaped bacteria that ferment lactose with acid and gas formation within 48 hours at 35°C.

Collagen The main structural protein of muscles.

Collision theory The principle that chemical reactions occur because energy is gained as particles collide.

Colony A clone of bacterial cells on a solid medium that is visible to the naked eye.

Commensalism A system of interaction in which two organisms live in association and one is benefited while the other is neither benefited nor harmed.

Commercial sterilization A process of treating canned goods aimed at destroying the endospores of *Clostridium botulinum*.

Communicable disease Any disease that can be spread from one host to another.

Competence The physiological state in which a recipient cell can take up and incorporate a large piece of donor DNA.

Competitive inhibition The process by which a chemical competes with the normal substrate for the active site of an enzyme.

Complement (C) A group of serum proteins involved in phagocytosis and lysis of bacteria.

Complement fixation The process in which complement combines with an antigen–antibody complex.

Complete digestive system A digestive system with a mouth and an anus.

Completed test The final test for detection of coliforms in the multiple-tube fermentation test.

Complex lipids Lipids containing P, N, and/or S.

Complex medium A culture medium in which the exact chemical composition is not known.

Complex virus A virus with a complicated structure, such as a bacteriophage.

Compost Decomposing organic matter.

Compound A substance composed of two or more different chemical elements.

Compound light microscope An instrument with two sets of lenses that uses visible light as the source of illumination.

Compromised host A host whose resistance to infection is impaired.

Condensation reaction A chemical reaction in which a molecule of water is released.

Condenser A lens system located below the microscope stage that directs light rays through the specimen.

Confirmed test The second stage of the multiple-tube fermentation test, used to identify coliforms on solid differential media.

Congenital disease A disease present at birth as a result of some condition that occurred in utero.

Conidiophore An aerial hypha bearing conidiospores.

Conidiospore An asexual spore produced in a chain from a conidiophore.

Conjugated protein Molecule consisting of amino acids and other organic or inorganic compounds.

Conjugated vaccine A vaccine consisting of the desired antigen and other proteins.

Conjugation The transfer of genetic material from one cell to another involving cell-to-cell contact.

Constitutive enzyme An enzyme that is produced regardless of how much substrate is present.

Contact inhibition The cessation of animal cell movement and division as a result of contact with other cells.

Contact transmission Spread of disease by direct or indirect contact or via droplets.

Contagious disease A disease that is easily spread from one person to another.

Continuous cell line Animal cells that can be maintained through an indefinite number of generations in vitro.

Convalescent period The period of recovery from a disease.

Corepressor The molecule (end-product) that brings about repression of a repressible enzyme.

Countercurrent immunoelectrophoresis (CIE) The movement of antigen and antibody toward each other through an electric field. *See also* Electrophoresis.

Counterstain A stain used to give contrast in a differential stain.

Covalent bond A chemical bond in which the electrons of one atom are shared with another atom.

Cresol A mixture of isomers from petroleum.

Crisis The phase of a fever characterized by vasodilation and sweating.

Cristae Foldings of the inner membrane of a mitochondrion.

Crossing over Process by which a portion of a chromosome is exchanged with a portion of another chromosome.

Culture Microorganisms that grow and multiply in a container of culture medium.

Culture medium The nutrient material prepared for growth of microorganisms in a laboratory.

Curd The solid part of milk that separates from the liquid in the making of cheese, for example.

Cutaneous mycosis A fungal infection of the epidermis, nails, and hair.

Cuticle Nonliving outer covering of helminths.

Cyanobacteria Oxygen-producing photoautotrophic procaryotes; formerly called blue-green algae.

Cyclic AMP A molecule derived from ATP, in which the phosphate group has a cyclic structure; acts as a cell messenger.

Cyclic photophosphorylation Movement of an electron from chlorophyll through a series of electron acceptors and back to chlorophyll; purple and green bacterial photophosphorylation.

Cyclosporine A drug that suppresses the T-cell response.

Cyst A sac with a distinct wall containing fluid or other material; also, a protective capsule of some protozoans.

Cysticercus Encysted tapeworm larva.

Cystitis Inflammation of the urinary bladder.

Cytochrome oxidase An enzyme that oxidizes cytochrome *c*.

Cytochromes Proteins that function as electron carriers in respiration and photosynthesis.

Cytocidal Resulting in cell death.

Cytokines Small proteins released from human cells in response to bacterial infection; directly or indirectly may induce fever, pain, or T-cell proliferation.

Cytomegaly Enlarged cells.

Cytopathic effect (CPE) Tissue deterioration caused by viruses.

Cytoplasm In a procaryote, everything inside the plasma membrane; in a eucaryote, everything inside the plasma membrane and external to the nucleus.

Cytoplasmic streaming The flowing of cytoplasm in a eucaryotic cell.

Cytosine A pyrimidine nucleic acid base that pairs with guanine.

Cytoskeleton Microfilaments and microtubules that provide support and movement for eucaryotic cytoplasm.

Cytostome The mouthlike opening in some protozoans.

Cytotoxic T (T_C) cells Cells that destroy antigens.

Cytotrophic Binding to cells; for example, IgE antibodies bind to target cells.

D- Prefix describing a stereoisomer.

Darkfield microscope A microscope that has a device to scatter light from the illuminator so that the specimen appears white against a black background.

Dark reaction *See* Calvin–Benson cycle.

Deamination Removal of an amino group.

Death phase Period of logarithmic decrease in a bacterial population.

Debridement Surgical removal of necrotic tissue.

Decarboxylation Removal of carbon dioxide (CO_2) from an amino acid.

Decimal reduction time The time (in minutes) required to kill 90% of a bacterial population at a given temperature.

Decolorization The process of removing a stain.

Decomposition reaction A chemical reaction in which bonds are broken to produce smaller parts from a large molecule.

Definitive host An organism that harbors the adult, sexually mature form of a parasite.

Degeneracy (of the genetic code) Redundancy of genetic code; that is, most amino acids are signaled by several codons.

Degranulation Release of contents of secretory granules.

Dehydration The removal of water.

Dehydration synthesis *See* Condensation reaction.

Dehydrogenation The loss of hydrogen atoms from a substrate.

Delayed hypersensitivity Cell-mediated hypersensitivity.

Delayed hypersensitivity T (T_D) cells Cells that produce lymphokines in type IV hypersensitivities.

Denaturation A change in the molecular structure of a protein.

Denitrification The reduction of nitrates to nitrites or nitrogen gas.

Dental plaque A combination of bacterial cells, dextran, and debris adhering to the teeth.

Deoxyribonucleic acid (DNA) The nucleic acid of genetic material.

Deoxyribose A five-carbon sugar contained in DNA nucleotides.

Dermatophyte A fungus that causes a cutaneous mycosis.

Dermis The inner portion of the skin.

Descriptive epidemiology Analysis of all data regarding the occurrence of a disease to determine the cause of the disease.

Desensitization The prevention of allergic inflammatory responses.

Desiccation The absence of water.

Detergent Any substance that reduces the surface tension of water.

Dextran A polymer of glucose.

Diacetyl $CH_3COCOCH_3$ produced from carbohydrate fermentation; chiefly responsible for the odor of dairy products.

Diagnosis Identification of a disease.

Diapedesis The process by which phagocytes move out of blood vessels.

Differential count The number of each kind of leukocyte in a sample of 100 leukocytes.

Differential medium A solid culture medium that makes it easier to distinguish colonies of the desired organism.

Differential stain A stain that distinguishes objects on the basis of reactions to the staining procedure.

Diffusion The net movement of molecules or ions from an area of higher concentration to an area of lower concentration.

Digestion The process of breaking down substances physically and chemically.

Digestive vacuole An organelle in which substrates are broken down enzymatically.

Dimorphism The property of having two growth forms.

Dioecious Referring to organisms in which organs of different sexes are located in different individuals.

Diphtheroid A gram-positive pleomorphic rod.

Dipicolinic acid Chemical substance found in bacterial endospores and not in vegetative cells.

Diplobacilli Rods that divide and remain attached in pairs.

Diplococci Cocci that divide and remain attached in pairs.

Diploid Having two sets of chromosomes; normal state of a eucaryotic cell.

Direct contact A method of spreading infection from one host to another through some kind of close association of the hosts.

Direct FA test A fluorescent-antibody test to detect the presence of an antigen.

Direct microscopic count Enumeration of cells by observation through a microscope.

Direct test Identification of a pathogen.

Disaccharide A sugar consisting of two monosaccharides.

Disease Any change from a state of health.

Disinfectant Any substance used on inanimate objects to kill or inhibit the growth of microorganisms.

Disk-diffusion test *See* Kirby–Bauer test.

Dissimilation plasmids Plasmids containing genes coding for the production of enzymes that catalyze the catabolism of certain unusual sugars and hydrocarbons.

Dissociation Transformation of a compound into positive and negative ions in solution.

Distilled spirits Alcoholic beverages made by distillation of wine.

Disulfide bond Two atoms of sulfur held together by a covalent bond (S—S).

Division A phylum; used in botany and microbiology.

DNA ligase Enzyme that covalently bonds the phosphate of one nucleotide with the 3′ carbon of another.

DNA probe A labeled piece of DNA used to identify the presence of bacteria in a sample.

Donor cell A cell that gives DNA to a recipient cell in recombination.

DPT vaccine A combined vaccine used to provide active immunity, containing diphtheria and tetanus toxoids and killed *Bordetella pertussis* cells.

Droplet infection The transmission of infection by small liquid droplets carrying microorganisms.

Dysentery A disease characterized by frequent, watery stools containing blood and mucus.

Eclipse period The time during viral multiplication when complete, infective virions are not present.

Edema An abnormal accumulation of fluid in body parts or tissues, causing swelling.

Electron A negatively charged particle in motion around the nucleus of an atom.

Electron acceptor An ion that picks up an electron that has been lost from another atom.

Electron donor An ion that gives up an electron to another atom.

Electronic configuration The arrangement of electrons in shells or energy levels in an atom.

Electron microscope A microscope that uses a flow of electrons instead of light to produce an image.

Electron shells Regions of an atom corresponding to different energy levels.

Electron transport chain A series of compounds that transfer electrons from one compound to another, generating ATP by oxidative phosphorylation.

Electrophoresis The separation of substances (for example, serum proteins) by their rate of movement through an electric field.

Elementary body An infectious form of *Chlamydia.*

ELISA (enzyme-linked immunosorbent assay) A group of serological tests that use enzyme reactions as indicators.

Embden–Meyerhof pathway *See* Glycolysis.

Emulsify To mix two liquids that do not dissolve in each other.

Encephalitis Inflammation of the brain.

Encystment Formation of a cyst.

Endemic disease A disease that is constantly present in a certain population.

Endergonic reaction A chemical reaction that requires energy.

Endocarditis Inflammation of the lining of the heart (endocardium).

Endocytosis The process of moving material into a eucaryotic cell.

Endogenous pyrogen Interleukin-1, causes fever.

Endomembrane system Organelles in eucaryotic cells; so called because of their interaction.

Endoplasmic reticulum A membrane network in eucaryotic cells connecting the plasma membrane with the nuclear membrane.

Endospore A resting structure formed inside some bacteria.

Endosymbiotic hypothesis Model for evolution of eucaryotes stating that organelles arose from procaryotic cells living inside a host procaryote.

Endotoxin Part of the outer portion of the cell wall of most gram-negative bacteria; lipid A.

Endotoxin shock *See* Septic shock.

End-product inhibition *See* Feedback inhibition.

Energy level The energy of an electron and its position relative to the atomic nucleus.

Enrichment culture A culture medium used for preliminary isolation that favors the growth of a particular microorganism.

Enterics The common name for bacteria in the family Enterobacteriaceae.

Enterotoxin An exotoxin that causes gastroenteritis; produced by *Staphylococcus*, *Vibrio*, and *Escherichia*.

Entner–Doudoroff pathway An alternate pathway for the oxidation of glucose to pyruvic acid.

Envelope An outer covering surrounding the capsid of some viruses.

Enzyme A protein that catalyzes chemical reactions in a living organism.

Enzyme induction The process by which a substance can cause the synthesis of an enzyme.

Enzyme poison Substance that can permanently inactivate enzymes, e.g., cyanide.

Enzyme repression The process by which a substance can stop the synthesis of an enzyme.

Enzyme-substrate complex A temporary union of an enzyme and its substrate.

Eosinophil A granulocyte whose granules take up the stain eosin.

Epidemic disease A disease acquired by many people in a given area in a short time.

Epidemiology The science dealing with when and where diseases occur and how they are transmitted.

Epidermis The outer portion of the skin.

Equilibrium The point of even distribution.

Equivalent treatments Different methods that have the same effect on controlling microbial growth.

Ergot A substance produced in sclerotia by the fungus *Claviceps purpurea* that causes contraction of arteries and uterine muscle.

Erythrogenic toxin A substance, produced by some streptococci, that causes erythema.

Ester linkage The bonding between two organic molecules (R) as $R{-}\overset{\overset{\displaystyle O}{\|}}{C}{-}O{-}R$.

Ethambutol A synthetic antimicrobial agent that interferes with the synthesis of RNA.

Etiology The study of the cause of a disease.

Eubacteria Procaryotic organisms; possess peptidoglycan cell walls.

Eucaryote A cell with DNA enclosed within a distinct membrane-bounded nucleus.

Eutrophication The addition of organic matter and subsequent removal of oxygen from a body of water.

Exchange reaction A chemical reaction that has both synthesis and decomposition components.

Exergonic reaction A chemical reaction that releases energy.

Exon A region of a eucaryotic chromosome that codes for a protein.

Exonuclease An enzyme that cuts DNA.

Exotoxins Protein toxins released from bacterial cells into the surrounding medium.

Experimental epidemiology The study of a disease using controlled experiments.

Extracellular enzyme An enzyme released from a cell to break down large molecules.

Extreme halophile An organism that requires a high salt concentration for growth.

Eyespot A pigmented area capable of detecting the presence of light.

Facilitated diffusion The transfer of a substance across a plasma membrane from an area of higher concentration to an area of lower concentration mediated by carrier proteins (permeases).

Facultative anaerobe An organism that can grow with or without oxygen (O_2).

Facultative halophile An organism capable of growth in, but not requiring, 1% to 2% salt.

Family A taxonomic group between order and genus.

Fat An organic compound consisting of glycerol and fatty acids.

Fatty acids Long hydrocarbon chains ending in a carboxyl group.

Feedback inhibition Inhibition of an enzyme in a particular pathway by the accumulation of end-product from the pathway.

Fermentation The enzymatic degradation of carbohydrates in which the final electron acceptor is an organic molecule, ATP is synthesized by substrate-level phosphorylation, and oxygen (O_2) is not required.

Fever An abnormally high body temperature.

F factor Fertility factor; a plasmid found in the donor cell in bacterial conjugation.

Fibrinolysin A kinase produced by streptococci.

Filtration Passage of a liquid or gas through a screenlike material; a 0.45-μm filter removes bacteria.

Fimbria An appendage on a bacterial cell used for attachment.

Five kingdoms Organisms are assigned to one of five kingdoms: Animalia, Plantae, Fungi, Protista, or Monera.

Fixed macrophage Macrophage that is located in a certain organ or tissue—for example, in liver, lungs, spleen, or lymph nodes.

Fixing (chemical elements) Combining elements so that a critical element can enter the food chain. *See* Nitrogen fixation; Calvin–Benson cycle.

Fixing (in slide preparation) The process of attaching the specimen to the slide.

Flagella Thin appendages that arise from one or more locations on the surface of a cell and are used for cellular locomotion.

Flagellate A member of the protozoan phylum Mastigophora that uses flagella for locomotion.

Flaming The process of sterilizing an inoculating loop by holding it in an open flame.

Flat sour Thermophilic spoilage of canned goods not accompanied by gas production.

Flatworm An animal belonging to the phylum Platyhelminthes.

Flavoprotein Protein with flavin coenzyme that functions as an electron carrier in respiration.

Flocculation Removal of colloidal material by addition of a chemical that causes the colloidal particles to coalesce.

Flora *See* Normal flora.

Fluid mosaic model A way of describing the dynamic arrangment of phospholipids and proteins comprising the plasma membrane.

Fluke A flatworm belonging to the class Trematoda.

Fluorescence The ability to give off light of one color when exposed to light of another color.

Fluorescence microscope A microscope that uses an ultraviolet light source to illuminate specimens that will fluoresce.

Fluorescent-antibody technique A diagnostic tool using antibodies labeled with fluorochromes and viewed through a fluorescent microscope.

Fluorochromes Dyes used to stain bacteria that fluoresce when illuminated with ultraviolet light.

Focal infection A systemic infection that began as an infection in one place.

Fomite A nonliving object that can spread infection.

Forespore A structure consisting of chromosome, cytoplasm, and endospore membrane inside a bacterial cell.

Frameshift mutation A mutation caused by the addition or deletion of one or more bases in DNA.

Free radical A highly reactive particle with an unpaired electron; designated $X\cdot$.

Freeze-drying *See* Lyophilization.

FTA-ABS test An indirect fluorescent-antibody test used to detect syphilis.

Functional groups Arrangement of elements in organic molecules that are responsible for most of the chemical properties of those molecules.

Fungi Organisms that belong to the Kingdom Fungi; eucaryotic chemoheterotrophs.

Gamete A male or female reproductive cell.

Gametocyte A male or female protozoan cell.

Gamma globulin *See* Immune serum globulin.

Gangrene Tissue death from loss of blood supply.

Gastroenteritis Inflammation of the stomach and intestine.

Gas vacuole A procaryotic inclusion of buoyancy compensation.

Gene A segment of DNA or a sequence of nucleotides in DNA that codes for a functional product.

Gene amplification A mechanism that causes a gene to be replicated many times.

Generalized transduction Transfer of bacterial chromosome fragments from one cell to another by a bacteriophage.

Generation time The time required for a cell or population to double in number.

Genetic engineering Manufacturing and manipulating genetic material in vitro.

Genetics The science of heredity.

Genotype The genetic makeup of an organism.

Genus The first name of the scientific name (binomial); the taxon between family and species.

Geosmin An alcohol produced by actinomycetes that has an earthy odor.

Germ Part of an organism capable of developing.

Germicidal Capable of killing microorganisms.

Germicidal lamp An ultraviolet light (wavelength = 260 nm) capable of killing bacteria.

Germination The process of starting to grow from a spore or endospore.

Germ theory The principle that microorganisms cause disease.

Globulin The protein type to which antibodies belong.

Glomerulonephritis Inflammation of the glomeruli of the kidneys, but not a result of kidney infection.

Glucan A polysaccharide component of yeast cell walls.

Glycerol An alcohol; $C_3H_5(OH)_3$.

Glycocalyx A gelatinous polymer surrounding a procaryotic cell wall.

Glycogen A polysaccharide stored by some cells.

Glycolysis The main pathway for the oxidation of glucose to pyruvic acid.

Golgi complex An organelle involved in the secretion of certain proteins.

Graft-versus-host (GVH) disease A condition that occurs when a transplanted tissue has an immune response to the tissue recipient.

Gram-negative cell wall A peptidoglycan layer surrounded by a lipopolysaccharide outer membrane.

Gram-positive cell wall Composed of peptidoglycan and teichoic acids.

Gram stain A differential stain that divides bacteria into two groups, gram-positive and gram-negative.

Granulocyte A leukocyte with visible granules in the cytoplasm; includes neutrophils, basophils, and eosinophils.

Griseofulvin A fungistatic antibiotic.

Group translocation In procaryotes, active transport in which a substance is chemically altered during transport across the plasma membrane.

Guanine A purine nucleic acid base that pairs with cytosine.

Gumma A rubbery mass of tissue characteristic of tertiary syphilis and tuberculosis.

Halogen One of the following elements: fluorine, chlorine, bromine, iodine, or astadine.

Halophile An organism that grows in high concentrations of salt.
Haploid A eucaryotic cell or organism with one of each type of chromosome.
Hapten An antigen that has reactivity and not immunogenicity.
Heavy metals Certain elements with specific gravity greater than 4 that are used as antimicrobial agents; for example, silver (Ag), copper (Cu), and mercury (Hg).
Helminth A parasitic roundworm or flatworm.
Helper T (T_H) cells Cells that interact with an antigen before B cells interact with the antigen.
Hemagglutination Clumping of red blood cells.
Hemagglutination-inhibition A process whereby an antibody inhibits viral hemagglutination.
Hemoflagellate A parasitic flagellate found in the circulatory system of its host.
Hemolysins Enzymes that lyse red blood cells.
Herd immunity The presence of immunity in most of a population.
Hermaphroditic Having both male and female reproductive capacities.
Heterocyst A large cell in certain cyanobacteria; site for nitrogen fixation.
Heterolactic Any organism producing lactic acid and other acids or alcohols as end-products of fermentation (for example, *Escherichia*).
Heterotroph An organism that requires an organic carbon source.
Hexachlorophene A chlorinated phenol used as an antiseptic.
Hexose monophosphate shunt *See* Pentose phosphate pathway.
Hfr A bacterial cell in which the F factor has become integrated into the chromosome.
High-energy bond A bond that can readily be broken to release usable energy; designated by ~.
High-temperature short-time (HTST) pasteurization Pasteurizing at 72°C for 15 seconds.
Histamine A substance released by tissue cells that causes vasodilation, capillary permeability, and smooth muscle contraction.
Histocompatibility antigens Antigens on the surface of human cells.
Histones Proteins associated with DNA in eucaryotic chromosomes.
Holdfast The branched base of an algal stipe.
Holoenzyme An enzyme consisting of an apoenzyme and a cofactor.
Homolactic An organism producing only lactic acid from fermentation (for example, *Streptococcus*).
Homologous chromosome A chromosome that has the same base sequence as another. In a diploid cell, one of a pair of chromosomes.
Host An organism infected by a pathogen.
Hot-air sterilization The use of an oven at 170°C for approximately 2 hours.
H (hemagglutination) spikes Antigenic projections from the outer lipid bilayer of influenza virus.
Human leukocyte antigens Human cell surface antigens.
Humoral immunity Immunity produced by antibodies dissolved in body fluids, mediated by B cells.
Humus Organic matter that remains in soil following partial decomposition.
Hyaluronic acid A mucopolysaccharide that holds together certain cells of the body.
Hybridoma A cell made by fusing an antibody-producing B cell with a cancer cell.
Hydrogen bond A bond between a hydrogen atom covalently bonded to oxygen or nitrogen and another covalently bonded oxygen or nitrogen.
Hydrogen ion (H^+) A proton.
Hydrolysis A decomposition reaction in which chemicals react with the H^+ and OH^- of a water molecule.
Hydroxide OH^-.
Hydroxyl $-OH$.
Hyperbaric chamber An apparatus to hold bases at pressures greater than 1 atmosphere.
Hypersensitivity Altered, enhanced immune reactions leading to pathologic changes.
Hypertonic Describing a solution that has a higher concentration of solutes than an isotonic solution.
Hyphae Long filaments of cells in fungi or actinomycetes.
Hypotonic Describing a solution that has a lower concentration of solutes than an isotonic solution.

Icosahedron A polyhedron with 20 triangular faces and 12 corners.
ID_{50} The bacterial concentration required to produce a demonstrable infection in 50% of the test host population.
Idiophase Stationary growth phase.
IgA The class of antibodies found in secretions.
IgD Antibodies found on B cells.
IgE The class of antibodies involved in hypersensitivities.
IgG The most abundant antibodies found in serum.
IgM The first antibodies to appear after exposure to an antigen.
Illuminator A light source.
Imidazoles Antifungal drugs that interfere with sterol synthesis.
Immediate hypersensitivity Allergic reactions involving humoral antibodies.
Immune complex A circulating antigen–antibody aggregate capable of fixing complement.
Immune serum globulin The serum fraction containing immunoglobulins (antibodies).
Immunity The body's defense against a particular microorganism.
Immunization A process that produces immunity.
Immunodiffusion test A test consisting of precipitation reactions carried out in an agar gel medium.
Immunoelectrophoresis Identification of proteins by electrophoretic separation and then serological testing.
Immunofluorescence Procedures using the fluorescent-antibody technique.
Immunogen *See* Antigen.
Immunoglobulin (Ig) An antibody.
Immunological enhancement The binding of antibodies to tumor antigens.

Immunological escape Resistance of cancer cells to immune rejection.
Immunological surveillance The body's immune response to cancer.
Immunological tolerance The immune system's ability to respond to foreign (nonself) antigens and not to respond to self antigens.
Immunosuppression Inhibition of the immune response.
Immunotherapy Treatment using antibodies.
Immunotoxin An immunotherapeutic agent consisting of a poison bound to a monoclonal antibody.
IMViC Biochemical tests used to identify enterics, including indole, methyl red, Voges–Proskauer, and citrate.
Inapparent infection *See* Subclinical infection.
Incidence The fraction of the population that contracts a disease during a particular length of time.
Inclusion Material inside a cell.
Incomplete digestive system A digestive system with one opening (mouth) for intake of food and elimination of waste.
Incubation period The time interval between the actual infection and first appearance of any signs or symptoms of disease.
Indirect (passive) agglutination test Indirect agglutination test using soluble antigens attached to latex or other small particles.
Indirect contact Transmission of pathogens by fomites.
Indirect FA test A fluorescent-antibody test to detect the presence of specific antibodies.
Indirect tests Diagnosis based on the presence of antibodies against a pathogen.
Inducer A substrate that brings about an increased amount of an enzyme.
Inducible enzyme *See* Enzyme induction.
Inert Inactive.
Infection Growth of microorganisms in the body.
Infection thread An invagination in a root hair that allows *Rhizobium* to infect the root.
Infectious disease A disease caused by pathogens.
Inflammation A host response to tissue damage characterized by reddening, pain, heat, and swelling.
Innate resistance Resistance of an individual to diseases that affect other species and other individuals of the same species.
Inoculate To introduce microorganisms into a culture medium or host.
Inoculating loop (needle) An instrument used to transfer bacteria from one culture medium to another.
Inorganic compounds Small molecules not usually containing carbon.
Insertion sequence *See* Transposon.
Interferon An antiviral protein produced by certain animal cells in response to a viral infection.
Interleukins Chemicals that cause T-cell proliferation. *See* Cytokines.
Intermediate host An organism that harbors the larval or asexual stage of a helminth or protozoan.
Intoxication Poisoning.
Intron A region of a eucaryotic chromosome that does not code for a protein.
Invasiveness The ability of microorganisms to establish residence in a host.
In vitro "In glass"; not in a living organism.
In vivo Within a living organism.
Iodophor A complex of iodine and a detergent.
Ion A negatively or positively charged atom or group of atoms.
Ionic bond A chemical bond formed when atoms gain or lose electrons in the outer energy levels.
Ionization Separation of a molecule into groups of atoms with electrical charges.
Ionizing radiation High-energy radiation that causes ionization; for example, X rays and gamma rays; wavelengths less than 1 nm.
Ischemia Local loss of blood supply.
Isoelectric focusing Electrophoresis of a protein until it reaches its isoelectric point.
Isoelectric point The pH at which a protein no longer migrates in electrophoresis.
Isograft Tissue graft between identical twins.
Isomers Two molecules with the same chemical formula but different structures.
Isoniazid (INH) A bacteriostatic agent used to treat tuberculosis.
Isotonic Referring to a solution in which osmotic pressure is equal across a membrane.
Isotope A form of a chemical element in which the number of neutrons in the nucleus is different from the other forms of that element.

Kelp Multicellular brown algae.
Keratin A protein found in epidermis, hair, and nails.
Killer (K) cells Effector cells that kill antibody-coated target cells.
Kinases (1) Bacterial enzymes that break down fibrin (blood clots). (2) Enzymes that remove a Ⓟ from ATP and attach it to another molecule.
Kingdom The highest category in the taxonomic hierarchy of classification.
Kinins Substances released from tissue cells that cause vasodilation.
Kirby–Bauer test An agar-diffusion test to determine microbial susceptibility to chemotherapeutic agents.
Koch's postulates Criteria used to determine the causative agent of infectious diseases.
Krebs cycle A pathway that converts two-carbon compounds to carbon dioxide (CO_2), transferring electrons to NAD^+ and other carriers.

L- Prefix describing a stereoisomer; L-amino acids are more commonly found in proteins.
Lactic acid fermentation Catabolic process, beginning with glycolysis, that produces lactic acid.
Lagging strand During DNA replication, the daughter strand synthesized discontinuously.
Lag phase The time interval in a bacterial growth curve during which there is no growth.
Larva Sexually immature stage of a helminth or arthropod.
Latent disease A disease characterized by a period of no symptoms when the pathogen is inactive.

Latent infection A condition in which a pathogen remains in the host for long periods without producing disease.

LD_{50} The lethal dose for 50% of the inoculated hosts within a given period.

Leading strand During DNA replication, the daughter strand synthesized continuously.

Lepromin test A skin test to determine the presence of antibodies to *Mycobacterium leprae*.

Leukemias Cancers characterized by abnormally large numbers of leukocytes.

Leukocidins Substances produced by some bacteria that can destroy neutrophils and macrophages.

Leukocyte White blood cell.

Leukopenia A condition in which the number of leukocytes is smaller than normal.

Leukotrienes Mediators of anaphylaxis.

L form A natural bacterium with a defective cell wall.

Lichen A symbiosis between a fungus and an alga or cyanobacterium.

Ligands Carbohydrate-specific binding proteins projecting from procaryotic cells; used for adherence.

Light reaction *See* Photophosphorylation.

Limnetic zone The surface zone of an inland body of water away from the shore.

Lipase An exoenzyme that breaks down fats into their component fatty acids and glycerol.

Lipid A molecule composed of glycerol and fatty acid.

Lipid A A component of gram-negative outer membrane; endotoxin.

Lipopolysaccharide (LPS) A molecule consisting of a lipid and a polysaccharide, forming the outer layer of gram-negative cell walls.

Lipoprotein A molecule consisting of a lipid and protein.

Liposome A fatty globule that may be used to administer chemotherapeutic agents.

Lithotroph *See* Autotroph.

Littoral zone The region along the shore of an inland body of water where there is considerable vegetation and where light penetrates to the bottom.

Local infection An infection in which pathogens are limited to a small area of the body.

Localized reaction An anaphylaxis-type reaction, such as hay fever, asthma, and hives.

Log phase Period of bacterial growth or logarithmic increase in cell numbers.

Lophotrichous Having two or more flagella at one end of a cell.

Luciferase An enzyme that accepts electrons from flavoproteins and emits a photon of light.

Lymphangitis Inflammation of the lymph vessels.

Lymphocyte A white blood cell involved in antibody production.

Lymphokines *See* Cytokines.

Lymphoma Cancer of lymphoid tissue.

Lymphotoxin A lymphokine that can destroy target cells in vitro.

Lyophilization Freeze-drying; freezing a substance and sublimating the ice in a vacuum.

Lysis Disruption of the plasma membrane. In disease, a gradual period of decline.

Lysogeny A state in which phage DNA is incorporated into the host cell without lysis.

Lysosome An organelle containing digestive enzymes.

Lysozyme An enzyme capable of lysing bacterial cell walls.

Lytic cycle A sequence for replication of phages that results in host cell lysis.

Macrolides Antibiotics that inhibit protein synthesis; for example, erythromycin.

Macromolecules Large organic molecules.

Macrophage A phagocytic cell; an enlarged monocyte.

Macrophage activation factor Increases macrophages' efficiency at destroying ingested cells.

Macrophage chemotactic factor Attacts macrophages to infection site.

Macrophage migration-inhibiting factor Prevents macrophages from leaving infection site.

Macular rash Small red spots appearing on the skin.

Major histocompatibility complex (MHC) The genes that code for the histocompatibility antigens; also known as human leukocyte antigens (HLA).

Malolactic fermentation Conversion of malic acid to lactic acid by lactic acid bacteria.

Malt Germinated barley grains containing maltose and amylase.

Malting Germination of starchy grains resulting in glucose and maltose production.

Mannan A polysaccharide component of yeast cell walls.

Mantoux test A tuberculin skin test.

Margination The process by which phagocytes stick to the lining of blood vessels.

Mast cell A type of cell found throughout the body that contains histamine and other substances that stimulate vasodilation.

Maximum growth temperature The highest temperature at which a species can grow.

Mechanical transmission The process by which arthropods transmit infections by carrying pathogens on their feet and other body parts.

Meiosis The process that leads to the formation of haploid gametes in a diploid organism.

Membrane filter A screenlike material with pores small enough to retain microorganisms.

Memory cells Long-lived B or T cells responsible for the anamnestic response.

Meningitis Inflammation of the meninges covering the central nervous system.

Merozoite A trophozoite of *Plasmodium* found in infected red blood cells or liver cells.

Mesophile An organism that grows between 25°C and 40°C.

Mesosome An irregular fold in the plasma membrane of a procaryotic cell.

Messenger RNA (mRNA) The type of RNA molecule that directs the incorporation of amino acids into proteins.

Metabolism The sum of all the chemical reactions that occur in a living cell.

Metacercaria The encysted stage of a fluke in its final intermediate host.

Metachromatic granule The intracellular volutin stored by some bacteria.

Meter (m) The standard unit of length in the metric system; one ten-millionth of the distance from the equator to the pole.

Methane Odorless, colorless, flammable gas; CH_4.

Microaerophile An organism that grows best in an environment with less oxygen (O_2) than is found in air.

Microbiota *See* Normal flora.

Micrometer (μm) A unit of measure equal to 10^{-6} m.

Microorganism A living organism too small to be seen with the naked eye; includes bacteria, fungi, protozoans, and microscopic algae; also includes viruses.

Microtubule The structure of the proteins that make up eucaryotic flagella and cilia.

Microwave An electromagnetic wave with wavelength between 10^{-1} and 10^{-3} m.

Minimal bactericidal concentration (MBC) The lowest concentration of a chemotherapeutic agent that will kill the test microorganisms.

Minimal inhibitory concentration (MIC) The lowest concentration of a chemotherapeutic agent that will prevent growth of the test microorganism.

Minimum growth temperature The lowest temperature at which a species will grow.

Miracidium The free-swimming, ciliated larva of a fluke that hatches from the egg.

Missense mutation A mutation that results in substitution of an amino acid in a protein.

Mitochondria Organelles containing the respiratory ATP-synthesizing enzymes.

Mitosis The division of the cell nucleus, often followed by division of the cytoplasm of the cell.

Mixed culture A culture containing more than one kind of microorganism.

MMR vaccine Attenuated measles, mumps, rubella viruses.

MMWR *Morbidity and Mortality Weekly Report;* weekly publication of the Centers for Disease Control containing data on notifiable diseases and topics of special interest.

Molds Fungi that form mycelia and appear as cottony tufts.

Mole An amount of a chemical equal to the atomic weights of all the atoms in a molecule of the chemical.

Molecular biology The science dealing with DNA and protein synthesis of living organisms.

Molecular weight The sum of the atomic weights of all atoms making up a molecule.

Molecule A combination of atoms forming a specific chemical compound.

Monera The kingdom to which all procaryotic organisms belong.

Monoclonal antibodies Specific antibodies produced by in vitro clones of B cells hybridized with cancerous cells.

Monocyte A phagocyte.

Monokine Substances secreted by macrophages. *See* Cytokines.

Monolayer A single layer of cells due to the cessation of cell division by contact inhibition.

Monomers The units that combine to form polymers.

Mononuclear phagocytic system A system of fixed macrophages located in the spleen, liver, lymph nodes, and bone marrow.

Monotrichous Having a single flagellum.

Morbidity The incidence of a specific disease.

Mordant A substance added to a staining solution to make it stain more intensely.

Morphology The external appearance.

Mortality The deaths from a specific notifiable disease.

Most probable number (MPN) A statistical determination of the number of coliforms per 100 ml of water or food.

Motility The ability of an organism to move by itself.

M protein A heat- and acid-resistant protein of streptococcal cell walls and fimbriae.

Multiple-tube fermentation test A method of detecting the presence of coliforms.

Murein *See* Peptidoglycan.

Mutagen An agent in the environment that brings about mutations.

Mutation Any change in the base sequence of DNA.

Mutation rate The probability of a gene mutating each time a cell divides.

Mutualism A symbiosis in which both organisms are benefited.

Mycelium A mass of long filaments of cells that branch and intertwine, typically found in molds.

Mycetoma A chronic infection caused by certain fungi and *Nocardia,* characterized by a tumorlike appearance.

Mycology The science dealing with fungi.

Mycorrhiza A fungus growing in symbiosis with plant root hairs.

Mycosis A fungal infection.

Mycotoxin A toxin produced by a fungus.

Myocarditis Inflammation of the heart muscle.

Naked virus A virus without an envelope.

Nanometer (nm) A unit of measurement equal to 10^{-9} m, 10^{-3} μm, and 10 Å.

Natural killer (NK) cells Lymphoid cells that destroy tumor cells.

Naturally acquired active immunity Antibody production in response to an infectious disease.

Naturally acquired passive immunity The natural transfer of humoral antibodies; transplacental transfer.

Necrosis Tissue death.

Negative (indirect) selection *See* Replica plating.

Negative stain A procedure that results in colorless bacteria against a stained background.

Neurotoxin A chemical that is poisonous to the nervous system.

Neutralizing antibody An antibody that inactivates a bacterial exotoxin or virus.

Neutron An uncharged particle in the nucleus of an atom.

Neutrophil Also called polymorphonuclear leukocyte; a highly phagocytic granulocyte.

Nicotinamide adenine dinucleotide (NAD^+) A coenzyme that functions in the removal and transfer of H^+ and electrons from substrate molecules.

Nicotinamide adenine dinucleotide phosphate ($NADP^+$) A coenzyme similar to NAD^+.

Nitrification The oxidation of nitrogen from ammonia to nitrites and nitrates.

Nitrofuran A synthetic antimicrobial drug.

Nitrogen cycle The series of processes that converts nitrogen (N_2) to organic substances and back to nitrogen in nature.

Nitrogen fixation The conversion of nitrogen (N_2) into ammonia.

Nitrosamine A carcinogen formed by the combination of nitrite and amino acids; nitroso-: —N═O.

Nomenclature The system of naming things.

Noncommunicable disease A disease that is not transmitted from one person to another.

Noncompetitive inhibitor *See* Allosteric inhibition.

Noncyclic photophosphorylation Movement of an electron from chlorophyll to NAD^+; plant and cyanobacterial photophosphorylation.

Nonionizing radiation Radiation that does not cause ionization; for example, ultraviolet radiation.

Nonsense codon A special terminator codon that does not code for any amino acid.

Nonsense mutation A base substitution in DNA that results in a nonsense codon.

Nonspecific resistance Host defenses that tend to afford protection from any kind of pathogen.

Normal flora Collection of microorganisms that colonize an animal without causing disease.

Nosocomial infection An infection that develops during the course of a hospital stay and was not present at the time the patient was admitted.

Notifiable disease A disease that physicians must report to the public health service.

N (neuraminidase) spikes Antigenic projections from the outer lipid bilayer of influenza virus.

Nuclear envelope The double membrane that separates the nucleus from the cytoplasm in a eucaryotic cell.

Nucleic acid A macromolecule consisting of nucleotides; for example, RNA and DNA.

Nucleic acid hybridization The process of combining single complementary strands of DNA.

Nucleoid The region in a bacterial cell containing the chromosome.

Nucleoli Areas in a eucaryotic nucleus where rRNA is synthesized.

Nucleoplasm The gel-like fluid within the nuclear envelope.

Nucleoprotein A macromolecule consisting of protein and nucleic acid.

Nucleoside A compound consisting of a purine or pyrimidine and pentose sugar.

Nucleotide A compound consisting of a purine or pyrimidine base, a five-carbon sugar, and a phosphate.

Nucleus The part of a eucaryotic cell that contains the genetic material; also, the part of an atom consisting of the protons and neutrons.

Numerical taxonomy A method of comparing organisms on the basis of many characteristics.

Nutrient broth (agar) A complex medium made of beef extracts.

Objective lens In a compound light microscope, the lens closest to the specimen.

Obligate anaerobe An organism that is unable to use oxygen.

Ocular lens In a compound light microscope, the lens closest to the viewer.

Oligodynamic action The ability of small amounts of a heavy metal compound to exert antimicrobial activity.

Oncogene A gene that can bring about malignant transformation.

Oncogenic virus A virus that is capable of producing tumors.

Oocyst An encysted sporozoan zygote in which cell division occurs to form the next infectious stage.

Operator The region of DNA adjacent to structural genes that controls their transcription.

Operon The operator site and structural genes it controls.

Opportunistic pathogen An organism that does not ordinarily cause a disease but can become pathogenic under certain circumstances.

Opsonization The enhancement of phagocytosis by coating microorganisms with certain serum proteins (opsonins).

Optimum growth temperature The temperature at which a species grows best.

Oral groove On some protozoans, the site at which nutrients are taken in.

Order A taxonomic classification between class and family.

Organelles Membrane-bounded structures within eucaryotic cells.

Organic compounds Molecules that contain carbon and hydrogen.

Organic growth factor An essential organic compound that an organism is unable to synthesize.

Organotroph *See* Heterotroph.

Osmosis The net movement of solvent molecules across a selectively permeable membrane from an area of higher concentration to an area of lower concentration.

Osmotic pressure The force with which a solvent moves from a solution of lower solute concentration to a solution of higher solute concentration.

Ouchterlony test An immunodiffusion test.

Outer membrane The outer layer of a gram-negative cell wall consisting of lipoproteins, lipopolysaccharides, and phospholipids.

Oxidase test A diagnostic test for the presence of cytochrome *c* using *p*-aminodimethylaniline.

Oxidation The removal of electrons from a molecule or the addition of oxygen to a molecule.

Oxidation pond A method of secondary sewage treatment.

Oxidation–reduction (redox) reaction A coupled reaction in which one substance is oxidized and one is reduced.

Oxidative phosphorylation The synthesis of ATP coupled with electron transport.

Oxygen cycle The processes that convert molecular oxygen (O_2) to oxides, water, and organic compounds and back to O_2.

Oxygenic Producing oxygen, as in plant and cyanobacterial photosynthesis.

PABA Para-aminobenzoic acid; a precursor for folic acid synthesis.

PAGE Polyacrylamide gel electrophoresis. *See also* Electrophoresis.

Pandemic disease An epidemic that occurs worldwide.

Papular rash A skin rash characterized by raised spots.

Parasite An organism that derives nutrients from a living host.

Parasitism A symbiosis in which one organism (the parasite) exploits another (the host) without providing any benefit in return.

Parenteral route Deposition directly into tissues beneath the skin and mucous membranes.

Passive agglutination *See* Indirect agglutination test.

Passively acquired immunity Immunity acquired when antibodies produced by another source are transferred to the individual who needs them.

Pasteurization The process of mild heating to kill particular spoilage organisms or pathogens.

Pathogen A disease-causing organism.

Pathology Study of the causes and development of disease.

Pellicle The flexible covering of some protozoans.

Penicillins A group of antibiotics produced either by *Penicillium* (natural penicillins) or by adding side chains to the β-lactam ring (semisynthetic penicillins).

Pentose phosphate pathway A metabolic pathway that can occur simultaneously with glycolysis to produce pentoses and NADH without ATP production.

Peptide A chain of two (di-), three (tri-), or more (poly-) amino acids.

Peptide bond A bond joining the amino group of one amino acid to the carboxyl group of a second amino acid with the loss of a water molecule.

Peptidoglycan The structural molecule of eubacterial cell walls consisting of the molecules N-acetylglucosamine, N-acetylmuramic acid, tetrapeptide side chain, and peptide side chain.

Peptone Short chains of amino acids produced by the action of acids or enzymes on proteins.

Perforin Protein that makes a pore in a target cell membrane, released by T_c cells.

Pericarditis Inflammation of the sac around the heart (pericardium).

Periplasmic gel The region between the outer membrane and the cytoplasmic membrane.

Peritrichous Having flagella distributed over the entire cell.

Permease A carrier protein in the plasma membrane.

Peroxidase An enzyme that breaks down hydrogen peroxide: $H_2O_2 + NADH + H^+ \longrightarrow 2H_2O + NAD^+$.

Peroxide An oxygen oxide consisting of two atoms of oxygen.

pH The symbol for hydrogen ion concentration; a measure of the relative acidity of a solution.

Phage *See* Bacteriophage.

Phage typing A method of identifying bacteria using specific strains of bacteriophages.

Phagocyte A cell capable of engulfing and digesting particles that are harmful to the body.

Phagocytosis The ingestion of solids by cells.

Phagolysosome A digestive vacuole.

Phagosome Food vacuole of a phagocyte.

Phalloidin *See* Amanitin.

Phase-contrast microscope A compound light microscope that allows examination of structures inside cells through the use of a special condenser.

Phenol C_6H_5OH; carbolic acid.

Phenol coefficient A standard of comparison for the effectiveness of disinfectants; the disinfecting action of a chemical is compared with phenol for the same length of time on the same organism under identical conditions.

Phenolic A synthetic derivative of phenol.

Phenotype The external manifestations of the genetic makeup of an organism.

Phosphate group A portion of a phosphoric acid molecule attached to some other molecule:

$$—O—\overset{\displaystyle O}{\overset{\|}{P}}—O— \\ \underset{\displaystyle OH}{|}$$

Phospholipid A complex lipid composed of glycerol, two fatty acids, and a phosphate group.

Phosphorylation The addition of a phosphate group to an organic molecule.

Photoautotroph An organism that uses light as its energy source and carbon dioxide (CO_2) as its carbon source.

Photoheterotroph An organism that uses light as its energy source and an organic carbon source.

Photophosphorylation The production of ATP by photosynthesis.

Photosynthesis The light-driven synthesis of carbohydrate from carbon dioxide (CO_2).

Phototaxis Movement in response to the presence of light.

Phototroph An organism that uses light as its primary energy source.

Phylogeny The evolutionary history of a group of organisms.

Phylum A taxonomic classification between kingdom and class.

Phytoplankton Free-floating photoautotrophs.

Pilus An appendage on a bacterial cell used for transfer of genetic material during conjugation.

Pinocytosis The engulfing of small molecules by infolding the plasma membrane.

Plankton Free-floating organisms.

Plantae A kingdom composed of multicellular eucaryotes with cellulose cell walls.

Plant virus A virus that multiplies in plant tissues.

Plaque A clearing in a confluent growth of bacteria due to lysis by phages. *See also* Dental plaque.

Plaque-forming units Visible plaques counted, perhaps due to more than one phage.

Plasma The liquid portion of blood in which the formed elements are suspended.

Plasma cells Cells produced by B lymphocytes that manufacture specific antibodies.

Plasma membrane The selectively permeable membrane enclosing the cytoplasm of a cell; the outer layer in animal cells, internal to the cell wall in other organisms.

Plasmid A small cyclic DNA molecule in bacteria replicating independently of the chromosome.

Plasmodium A multinucleated mass of protoplasm; when written as a genus, refers to the agent of malaria.

Plasmolysis Loss of water from a cell in a hypertonic environment.

Plate count A method of determining the number of bacteria in a sample by counting the number of colony-forming units on a solid culture medium.

Pleomorphic Having many shapes.

Pneumonia Inflammation of the lungs.

Point mutation *See* Base substitution.

Polar molecule A molecule with an unequal distribution of charges.

Poly-β-hydroxybutyric acid A fatty acid storage material unique to bacteria.

Polyene An antimicrobial agent that alters sterols in eucaryotic plasma membranes and contains more than four carbon atoms and at least two double bonds.

Polyhedron A many-sided solid.

Polykaryocyte A multinucleated giant cell.

Polymer A molecule consisting of a sequence of similar units or monomers.

Polymerase An enzyme that synthesizes specific polymers.

Polymorphonuclear leukocyte (PMN) A neutrophil.

Polypeptide A chain of amino acids. Also, a group of antibiotics.

Polyribosome An mRNA strand with several ribosomes attached to it.

Porins Proteins in the outer membrane of gram-negative cell walls that allow passage of small molecules.

Portal of entry The avenue by which a pathogen gains access to the body.

Portal of exit The route by which a pathogen leaves the body.

Positive (direct) selection A procedure for picking out mutant cells by growing them.

Potential energy Energy that is stored.

Pour plate method A method of inoculating a solid nutrient medium by mixing bacteria in the melted medium and pouring the medium into a Petri plate to solidify.

PPD (purified protein derivative) Proteins extracted from a boiled culture of *Mycobacterium tuberculosis;* the antigen used in the tuberculin skin test.

Precipitation reaction A reaction between soluble antigens and multivalent antibodies to form aggregates.

Precipitin ring test A precipitation test performed in a capillary tube.

Predisposing factor Anything that makes the body more susceptible to a disease or alters the course of a disease.

Presumptive test The first stage of the multiple-tube fermentation test in the detection of coliforms.

Prevalence The fraction of a population having a specific disease at a given time.

Primary cell line Human tissue cells that grow for only a few generations in vitro.

Primary infection An acute infection that causes the initial illness.

Primary producer Photoautotroph that converts light energy into chemical energy.

Primary response Antibody production to the first contact with an antigen.

Primary treatment Physical removal of solid matter from waste water.

Prion Infectious protein.

Privileged tissue Body tissue to which there is no immune response.

Procaryote A cell whose genetic material is not enclosed in a nuclear envelope. *See* Monera.

Prodromal period The time following incubation when the first symptoms of illness appear.

Product The substance formed in a chemical reaction.

Profundal zone The deeper water under the limnetic zone in an inland body of water.

Proglottid A body segment of a tapeworm containing male and female organs.

Promoter site The starting point on DNA for transcription of RNA by RNA polymerase.

Prophage Phage DNA inserted into the host cell's DNA.

Prostaglandins Hormonelike substances that are synthesized in many tissues and circulate in the blood, the exact function of which is unknown.

Prostheca A stalk or bud protruding from a procaryotic cell.

Prosthetic group A coenzyme that is bound tightly to its apoenzyme.

Protein A large molecule containing carbon, hydrogen, oxygen, and nitrogen (and sulfur); some proteins have a globular structure and others are pleated sheets.

Protein kinase Enzyme that transfers a phosphate group from ATP to an amino acid.

Proteolytic An enzyme capable of hydrolyzing a protein.

Protista The kingdom to which protozoans belong; unicellular, eucaryotic organisms.

Proton A positively charged particle in the nucleus of an atom.

Proto-oncogene An oncogene that is functioning normally.

Protoplasmic streaming Movement of protoplasm in a plasmodial slime mold.

Protoplast A gram-positive bacterium without a cell wall; a plant cell lacking a cell wall.

Protoplast fusion Joining of two cells, used in genetic engineering.

Protozoa Unicellular eucaryotic organisms belonging to the kingdom Protista.

Provirus Viral DNA that is integrated into the host cell's DNA.

Pseudohypha A short chain of cells that results from the lack of separation of daughter cells after budding.

Pseudomurein The structural molecules of archaeobacterial cell walls consisting of N-acetyltalosaminuronic acid, N-acetylglucosamine, and L amino acids.

Pseudopods Extensions of a cell that aid in locomotion and feeding.

P site Site on ribosome where tRNA is attached to peptide chain.

Psychrophile An organism that grows best at 15°C and does not grow above 20°C.

Psychrotroph An organism that is capable of growth at 4°C and above 20°C.

Pure culture A population of one strain or species of bacteria.

Purines The class of nucleic acid bases that includes adenine and guanine.

Pus An accumulation of dead phagocytes, dead bacterial cells, and fluid.

Pyocin A bacteriocin produced by *Pseudomonas aeruginosa.*

Pyocyanin Blue-green pigment produced by *Pseudomonas aeruginosa.*

Pyogenic Pus-forming.

Pyrimidines The class of nucleic acid bases that includes uracil, thymine, and cytosine.

Quaternary ammonium compound (quat) A cationic detergent with four organic groups attached to a central nitrogen atom; used as a disinfectant.
Quellung reaction Apparent swelling of a bacterial capsule in the presence of a specific antibody.
Quinine An antimalarial drug derived from the cinchona tree and effective against sporozoites in red blood cells.

Rapid plasma reagin A serologic test for syphilis.
r-determinant Genes for antibiotic resistance carried on R factors.
Reactants Substances that are combined in a chemical reaction.
Reactivity The ability of an antigen to combine with an antibody.
Reagin IgE antibodies made in response to a treponemal infection characterized by their ability to combine with lipids; reagin-lipid complex will fix complement.
Recalcitrance Being resistant to degradation.
Receptor An attachment for a pathogen on a host cell.
Recipient cell A cell that receives DNA from a donor cell in recombination.
Recombinant DNA A DNA molecule produced by recombination.
Recombinant DNA techniques *See* Genetic engineering.
Recombinant RNA technology Techniques used to make RNA molecules.
Recombination The process of joining pieces of DNA from different sources.
Redia A trematode larval stage that may reproduce asexually one or two times before developing into a cercaria.
Redox reactions Paired oxidation–reduction reactions.
Red tide A bloom of planktonic dinoflagellates.
Reducing medium A culture medium containing ingredients that will remove dissolved oxygen from the medium to allow the growth of anaerobes.
Reduction The addition of electrons to a molecule; the gain of hydrogen atoms.
Refractive index The relative velocity with which light passes through a substance.
Regulator gene The gene that codes for a repressor protein.
Regulatory T cells Helper T and suppressor T cells that regulate the immune response.
Rennin A proteolytic enzyme obtained from a calf's stomach that forms curd from milk.
Replica plating A method of inoculating a number of solid minimal culture media from an original plate of complete medium; mutant colonies that don't grow on the minimal media can be selected from the original plate.
Replication fork The point where DNA separates and new strands of DNA will be synthesized.
Replicative form A double-stranded RNA molecule produced during the multiplication of certain RNA viruses.
Reportable disease *See* Notifiable disease.
Repressible enzyme *See* Enzyme repression.
Repressor A protein that binds to the operator site to prevent transcription.
Reservoir of infection A continual source of infection.
Resistance The ability to ward off diseases through nonspecific and specific defenses.
Resistance (R) factor A bacterial plasmid carrying genes that determine resistance to antibiotics.
Resistance transfer factor A group of genes for replication and conjugation on the R factor.
Resolution The ability to distinguish fine detail with a magnifying instrument.
Respiration An ATP-generation process in which chemical compounds are oxidized and the final electron acceptor is usually an inorganic molecule; also, the process by which living organisms produce carbon dioxide (CO_2).
Restriction enzymes Enzymes that cut double-stranded DNA, leaving staggered ends.
Reticulate body An intracellular stage of *Chlamydia*.
Reverse transcriptase RNA-dependent DNA polymerase; an enzyme that synthesizes a complementary DNA from an RNA template.
Reversible reaction A chemical reaction in which the end products can readily revert to the original molecules.
R factors Plasmids carrying genes for antibiotic resistance.
Rh blood group Classification of red blood cells based on the presence or absence of Rh antigens.
Rhizosphere The region in soil where the soil and roots make contact.
Riboflavin A B vitamin that functions as a flavoprotein.
Ribonucleic acid (RNA) The class of nucleic acids that comprises messenger RNA, ribosomal RNA, and transfer RNA.
Ribose A five-carbon sugar that is part of ribonucleotide molecules and RNA.
Ribosomal RNA (rRNA) The RNA molecules that form the ribosomes.
Ribosomes The sites of protein synthesis in a cell, composed of RNA and protein.
Ricin A toxin protein from castor oil beans.
Rifamycin An antibiotic that inhibits bacterial RNA synthesis.
Ring stage A young *Plasmodium* trophozoite that looks like a ring in a red blood cell.
RNA-dependent RNA polymerase An enzyme that synthesizes a complementary RNA from an RNA template.
RNA primer A short strand of RNA used to start synthesis of the lagging strand of DNA.
Root nodule A tumorlike growth on the roots of certain plants containing a symbiotic nitrogen-fixing bacterium.
Roundworms Animals belonging to the phylum Aschelminthes.

S (Svedberg unit) Notes the relative rate of sedimentation during ultra-high-speed centrifugation.
Sabin vaccine A preparation containing three attenuated strains of polio virus administered orally.
Saccharide Sugar; general formula $(CH_2O)_n$.
Salk vaccine A preparation of a formalin-inactivated polio virus that is injected.
Salt A substance that dissolves in water to cations and anions, neither of which is H^+ or OH^-.
Sanitation Removal of microbes from eating utensils and food preparation areas.
Saprophyte An organism that obtains its nutrients from dead organic matter.

Sarcina A group of eight bacteria that remain in a packet after dividing; when written as a genus, refers to gram-positive, anaerobic cocci.

Sarcoma A cancer of fleshy, nonepithelial tissue or connective tissue.

Saturation The condition in which the active site on an enzyme is occupied by the substrate or product at all times. In a hydrocarbon, having the maximum number of hydrogen atoms.

Saxitoxin Neurotoxin produced by some dinoflagellates.

Scanning electron microscope An electron microscope that provides three-dimensional views of the specimen magnified about 10,000 times.

Schaeffer–Fulton stain An endospore stain that uses malachite green to stain the endospores and safranin as a counterstain.

Schick test A skin test to detect the presence of antibodies to diphtheria.

Schizogony The process of multiple fission, in which one organism divides to produce many daughter cells.

Sclerotia The reddish hardened ovaries of a grain that are filled with mycelia of the fungus *Claviceps purpurea.*

Scolex The head of a tapeworm, containing suckers and possibly hooks.

Secondary infection An infection caused by an opportunistic pathogen after a primary infection has weakened the host's defenses.

Secondary treatment Biological degradation of the organic matter in waste water, following primary treatment.

Secretion The production and release of fluid containing a variety of substances from a cell.

Selective medium A culture medium designed to suppress the growth of unwanted bacteria and encourage the growth of desired microorganisms.

Selective permeability The property of a plasma membrane to allow certain molecules and ions to move through the membrane while restricting others.

Selective toxicity The property of some antimicrobial agents to be toxic for a microorganism and nontoxic for the host.

Semiconservative replication The process of DNA replication in which each double-stranded molecule of DNA contains one original strand and one new strand.

Sense codon A codon that codes for an amino acid.

Sense strand (+strand) Viral RNA that can act as mRNA.

Sepsis The presence of unwanted bacteria.

Septate hypha A hypha consisting of uninucleate cell-like units.

Septicemia A condition characterized by the multiplication of bacteria in the blood.

Septic shock Sudden drop in blood pressure due to infection by gram-negative bacteria.

Septic tank A tank, built into the ground, in which waste water is treated by primary treatment.

Septum A crosswall dividing two parts.

Serial dilution The process of diluting a sample several times.

Seroconversion Change in a person's response to an antigen in a serological test.

Serology The branch of immunology concerned with the study of antigen–antibody reactions in vitro.

Serovar A subgroup of *Salmonella* based on the presence of antigens.

Serum The liquid remaining after blood plasma is clotted, which contains immunoglobulins.

Sexual dimorphism The distinctly different appearance of adult male and female organisms.

Sexual reproduction Reproduction that requires two opposite mating strains, usually designated male and female.

Signs Changes due to a disease that a physician can observe and measure.

Simple stain A method of staining microorganisms with a single basic dye.

Single-cell protein (SCP) A food substitute consisting of microbial cells.

Singlet oxygen Highly reactive O_2.

Skin test The intradermal injection of an antigen or antibody to determine susceptibility to an antigen.

Slide agglutination test A method of identifying an antigen by combining it with a specific antibody in a slide.

Slime layer *See* Capsule.

Slime mold Funguslike protists.

Slow-reacting substance of anaphylaxis (SRS-A) Leukotrienes released by target cells after being bound by IgE antibodies.

Slow virus infection A disease process that occurs gradually over a long period, caused by viruses or unconventional agents. *See also* Prion, Virino, Viroid.

Sludge Solid matter obtained from sewage.

Smear A thin film of material on a slide.

Soap A surface-active agent made from animal fats and lye (NaOH).

Solubility The ability to be dissolved, usually in water.

Solute A substance dissolved in another substance.

Solvent A dissolving medium.

Somatic Relating to the cell or body.

Specialized transduction The process of transferring a piece of cell DNA adjacent to a prophage to another cell.

Species (1) The most specific level in the taxonomic hierarchy. **Bacterial species:** A population of cells with similar characteristics. (2) The second name in scientific binomial nomenclature.

Specific channel proteins Proteins in the outer membrane of gram-negative cell walls that allow passage of specific molecules.

Specific epithet The second name in a scientific binomial.

Specific resistance *See* Immunity.

Spheroplast A gram-negative bacterium lacking a complete cell wall.

Spike A carbohydrate–protein complex that projects from the surface of certain enveloped viruses.

Spirillum A spiral or corkscrew-shaped bacterium; when written as a genus, refers to aerobic, helical bacteria with peritrichous flagella.

Spirochete A corkscrew-shaped bacterium with an axial filament.

Spontaneous generation The idea that life could arise spontaneously from nonliving matter.

Spontaneous mutation A mutation that occurs without a mutagen.

Sporadic disease A disease that occurs occasionally in a population.

Sporangiophore Aerial hypha supporting a sporangium.
Sporangiospore An asexual fungal spore formed within a sporangium.
Sporangium A sac containing one or more spores.
Spore A reproductive structure formed by fungi and actinomycetes.
Sporogenesis The process of spore and endospore formation.
Sporozoite A trophozoite of *Plasmodium* found in mosquitoes, infective for humans.
Stability The condition of not deteriorating with time.
Staphylococci A broad sheet of spherical cells.
Stationary phase The period in a bacterial growth curve when the number of cells dividing equals the number dying.
Stem cells Fetal cells that give rise to bone marrow, blood cells, and B and T cells.
Stereoisomers Two molecules consisting of the same atoms, arranged in the same manner but differing in their relative positions; mirror images.
Sterile Free of microorganisms.
Steroids A specific group of chemical substances, including cholesterol and hormones.
Sterol A lipid alcohol found in the plasma membranes of fungi and *Mycoplasma*.
Stipe A stemlike supporting structure of multicellular algae and basidiomycetes.
Strain A group of cells all derived from a single cell.
Streak plate method A method of isolating a culture by spreading microorganisms over the surface of a solid culture medium.
Streptobacilli Rods that remain attached in chains after cell division.
Streptococci Cocci that remain attached in chains after cell division.
Structural gene A gene that codes for an enzyme.
Subacute disease A disease with symptoms between acute and chronic.
Subclinical infection An infection that does not cause a noticeable illness.
Subcutaneous mycosis A fungal infection of tissue beneath the skin.
Substrate Any compound with which an enzyme reacts.
Substrate-level phosphorylation The synthesis of ATP by direct transfer of a high-energy phosphate group from an intermediate metabolic compound to ADP.
Subunit vaccine Vaccine consisting of an antigenic fragment produced by genetic engineering.
Suicide gene A gene that codes for a product that kills the cell.
Sulfa drugs Any synthetic chemotherapeutic agent containing sulfur and nitrogen. *See also* Sulfonamides.
Sulfhydral group —SH.
Sulfonamides Bacteriostatic compounds that interfere with folic acid synthesis by competitive inhibition.
Superficial mycosis A fungal infection localized in surface epidermal cells and along hair shafts.
Superinfection Growth of the target pathogen that has developed resistance to the antimicrobial drug being used; growth of an opportunistic pathogen.
Superoxide dismutase Enzyme that destroys O_2^-; $O_2^- + O_2^- + 2H^+ \longrightarrow H_2O_2 + O_2$.
Superoxide free radical O_2^-.
Suppressor T cells (T_S) Cells that inhibit an immune response.
Surface-active agent (surfactant) Any compound that decreases the tension between molecules lying on the surface of a liquid.
Susceptibility The lack of resistance to a disease.
Sylvatic Belonging in the woods; a wild animal.
Symbiosis The living together of two different organisms.
Symptom A change in body function that is felt by a patient as a result of a disease.
Syndrome A specific group of signs or symptoms accompanying a particular disease.
Synergistic effect The principle whereby the effectiveness of two drugs used simultaneously is greater than that of either drug used alone.
Synthesis reaction A chemical reaction in which two or more atoms combine to form a new, larger molecule.
Synthetic chemotherapeutic agent An antimicrobial agent that is prepared from chemicals in a laboratory.
Synthetic drug A chemotherapeutic agent that is prepared from chemicals in a laboratory.
Systemic anaphylaxis Hypersensitivity reaction causing vasodilation and resulting in shock.
Systemic (generalized) infection An infection throughout the body.
Systemic mycosis A fungal infection in deep tissues.

Tachyzoite Rapidly growing trophozoite.
T antigen An antigen in the nucleus of a tumor cell.
Tapeworm A flatworm belonging to the class Cestoda.
Target cell A body cell to which antibodies bind.
Taxis Response to an environmental stimulus.
Taxon A taxonomic category.
Taxonomy The science of classification.
T cell Stem cell processed in the thymus gland that is responsible for cellular immunity.
Teichoic acid A polysaccharide found in gram-positive cell walls.
Temperate phage A bacteriophage existing in lysogeny with a host cell.
Temperature abuse Improper food storage at a temperature that allows bacteria to grow.
Teratogenic Causing abnormal development.
Terminator site The site on DNA at which transcription ends.
Tertiary treatment Physical and chemical treatment of waste water to remove all BOD, nitrogen, and phosphorus, following secondary treatment.
Tetracyclines Broad spectrum antibiotics that interfere with protein synthesis.
Tetrad A group of four cocci.
Tetrahedron A four-sided solid structure.
T-even bacteriophage A complex virus with double-stranded DNA that infects *E. coli;* for example, T2, T4, T6.
Thallus The entire vegetative structure or body of a fungus, lichen, or alga.

Thermal death point (TDP) The temperature required to kill all the bacteria in a liquid culture in 10 minutes at pH 7.

Thermal death time (TDT) The length of time required to kill all bacteria in a liquid culture at a given temperature.

Thermoduric Heat resistant.

Thermophile An organism whose optimum growth temperature is between 50°C and 60°C.

Thermophilic anaerobic spoilage Spoilage of canned foods due to the growth of thermophilic bacteria.

Thermostable nuclease A heat-stable enzyme produced by *Staphylococcus aureus.*

Three-kingdom Organisms are assigned one of three kingdoms: eucaryota, eubacteria, or archaeobacteria.

Thylakoid A chlorophyll-containing membrane in a chloroplast.

Thymine A pyrimidine nucleic acid base in DNA that pairs with adenine.

Tincture An alcoholic or aqueous solution.

Tinea A cutaneous fungal infection; ringworm.

Ti plasmid *Agrobacterium* plasmid carrying genes for tumor induction in plants.

Tissue culture *See* Cell culture.

Titer Reciprocal of a dilution.

Tolerance A state of immunological nonresponsiveness to self antigens.

Total magnification Magnification of a specimen, determined by multiplying the ocular lens magnification by the objective lens magnification.

Toxemia Symptoms due to toxins in the blood.

Toxigenicity The capacity of a microorganism to produce a toxin.

Toxin Any poisonous substance produced by a microorganism.

Toxoid An inactivated toxin.

Trace element A chemical element required in small amounts for growth.

Transamination The transfer of an amino group from an amino acid to an organic acid.

Transcription The process of synthesizing RNA from a DNA template.

Transduction Transfer of DNA from one cell to another by a bacteriophage. *See also* Generalized transduction, Specialized transduction.

Transferrins Human iron-binding proteins that reduce iron available for a pathogen.

Transfer RNA (tRNA) The class of molecules that brings amino acids to the site where they are incorporated into proteins.

Transformation The process in which genes are transferred from one bacterium to another as "naked" DNA in solution; also, the changing of a normal cell into a cancerous cell.

Transient flora Collection of microorganisms that are present on an animal for a short time without causing a disease.

Translation The use of RNA as a template in the synthesis of protein.

Translocation Movement of a gene from one chromosomal locus to another.

Transmembrane channels Pores in target cells' membranes produced by complement.

Transmission electron microscope An electron microscope that provides high magnifications of thin sections of a specimen.

Transposon A small piece of DNA that can move from one region of the DNA molecule to another.

Transverse fission *See* Binary fission.

Transverse septum A crosswall that separates genetic material into two daughter cells in binary fission.

Trickling filter A method of secondary sewage treatment.

Trophophase *See* Log phase.

Trophozoite The vegetative form of a protozoan.

Tuberculin test A skin test used to detect the presence of antibodies to *Mycobacterium tuberculosis.*

Tumor Excessive tissue caused by uncontrolled cell growth.

Tumor necrosis factor *See* Cachectin.

Tumor-specific transplantation antigen (TSTA) A viral antigen on the surface of a transformed cell.

Turbidity The cloudiness of a suspension.

Turnover number The number of substrate molecules metabolized per enzyme molecule per second.

12D treatment A sterilization process that results in a decrease of the bacterial population by 12 logarithmic cycles.

Ubiquinones Low molecular weight, nonprotein carriers in the electron transport chain.

UDP-N-acetyl glucosamine (UDPNAC) A compound necessary for the biosynthesis of peptidoglycan.

Ultra-high-temperature (UHT) pasteurization Pasteurizing by changing the temperature from 74 to 140°C and back to 74°C in less than 5 seconds.

Ultrastructure Fine detail not seen with a compound light microscope.

Ultraviolet (UV) radiation Radiation from 10 to 390 nm.

Uncoating The separation of viral nucleic acid from its protein coat.

Unconventional agents Infectious agents that may consist of protein only.

Undulating membrane A highly modified flagellum on some protozoans.

Uracil A pyrimidine nucleic acid base in RNA that pairs with adenine.

Uridine diphosphoglucose (UDPG) Precursor for synthesis of glycogen.

Use-dilution test A method of determining effectiveness of a disinfectant using serial dilutions.

Vaccination The process of conferring immunity by using a vaccine.

Vaccine A preparation of killed, inactivated, or attenuated microorganisms or toxoids to induce artificially acquired active immunity.

Vacuole An intracellular inclusion, in eucaryotic cells, surrounded by a plasma membrane; in procaryotic cells, surrounded by a proteinaceous membrane containing gas.

Valence The combining capacity of an atom or molecule.

Vancomycin An antibiotic that inhibits cell wall synthesis.

Vasodilation Dilation or enlargement of blood vessels.

VDRL test A rapid screening test to detect the presence of antibodies against *Treponema pallidum.*

Vector An arthropod that carries disease-causing organisms from one host to another. A plasmid or virus used in genetic engineering to insert genes into a cell.

Vegetative cells Cells involved with obtaining nutrients, as opposed to reproduction or resting.

Vehicle transmission Transmission of a pathogen to a large number of people by an inanimate reservoir.

Vesicle The expanded, terminal area of the Golgi complex; also, a fluid-filled blister. In a procaryote, a protein-covered hollow cylinder in a gas vacuole. In a mycorrhizal fungus, a storage structure.

V factor NAD^+ or $NADP^+$.

Vibrio A curved or comma-shaped bacterium; when written as a genus, refers to gram-negative, motile, facultatively anaerobic curved rods.

Viral hemagglutination The ability of certain viruses to cause agglutination of red blood cells.

Viremia The presence of viruses in the blood.

Virino An infectious agent composed of a small piece of nucleic acid and a protein from the host.

Virion A fully developed complete viral particle.

Viroid An infectious piece of "naked" RNA.

Virulence The degree of pathogenicity of a microorganism.

Virus A submicroscopic, parasitic, filterable agent consisting of a nucleic acid surrounded by a protein coat.

Visible light Radiation from 400 to 700 nm, which the human eye can see.

Volutin Stored phosphate in a procaryotic cell.

Wandering macrophage A macrophage that leaves the blood and migrates to infected tissue.

Wassermann test A complement fixation test used to diagnose syphilis.

Wheal An area of edema of the skin resulting from a skin test.

Whey The fluid portion of milk that separates from curd.

Xenodiagnosis A method of diagnosis based on exposing a parasite-free normal host to the parasite and then examining the host for parasites.

Xenograft A tissue graft from another species.

X factor A precursor necessary to synthesize cytochromes.

Yeast A unicellular fungus belonging to the phylum Ascomycetes.

Zone of inhibition The area of no bacterial growth around an antimicrobial agent in the disk-diffusion test.

Zoonosis A disease that occurs primarily in wild and domestic animals but can be transmitted to humans.

Zygospore A sexual fungal spore characteristic of the Zygomycetes.

Zygote A diploid cell produced by the fusion of two haploid gametes.

PHOTOGRAPH ACKNOWLEDGMENTS

Note: CDC = Centers for Disease Control, Public Health Service, U.S. Department of Health and Human Services, Atlanta, Georgia. BPS = Biological Photo Service, Moss Beach, California.

Frontispiece and Contents i: ©T.E Adams/Visuals Unlimited. iii: A.B. Dowsett/SPL/Photo Researchers, Inc. xvi: Part One, ©Will & Deni McIntyre/Allstock. Part Two, ©Tony Ward. xvii: Part Three, ©Ted Horowitz/The Stock Market. Part Four, ©Tom Lyle/Medichrome/The Stock Shop. Part Five, ©Ed Young. xviii: top, CDC; bottom, G.L. Baron, University of Guelph/BPS. xix: top, CDC; bottom, ©Paul W. Johnson, University of Rhode Island/BPS. xx: top, A.B. Dowsett/SPL/Photo Researchers, Inc.; bottom, USDA/Science Source/Photo Researchers, Inc. xxi: Courtesy of Thomas Steitz. xxii: top, Chris Case; bottom, courtesy of Millipore/Greg Hoff. xxiii: ©Dr. Jeremy Burgess/SPL/Photo Researchers, Inc. xxiv: ©T.E. Adams/Visuals Unlimited. xxv: Chris Case. xxvi: David M. Phillips/Visuals Unlimited. xxvii: top, ©Veronika Burmeister/Visuals Unlimited; bottom, ©Lennart Nilsson, Boehringer Ingelheim International. xxviii: top, reprinted by permission of the publisher. From "Ultrastructural Observations on Giardiasis in a Murine Model," by Robert Owen, Paulina Nemanic, and David P. Stevens, *Gastroenterology,* 76:757–769. Copyright 1979 by the American Gastroenterology Association; bottom, CDC. xxx: CDC/BPS. xxxi: ©Visuals Unlimited. xxxii: top, taken by E. Calomeni and provided by M. Estes, Baylor College of Medicine, Houston, TX.; bottom, courtesy of Marietta Voge, Ph.D, from *Diagnostic Medical Parasitology,* by L.S. Garcia and D.A. Bruckner, ©Elsevier Scientific Publications. xxxiii: Steve Woit.

Part Openers 1: ©Will & Deni McIntyre/Allstock. 2: ©Tony Ward. 3: ©Ted Horowitz/The Stock Market. 4: ©Tom Lyle/Medichrome/The Stock Shop. 5: ©Ed Young.

Chapter 1 Opener: CDC. 1.1a: ©CDC. 1.1b: ©Lennart Nilsson. 1.6a: ©CNRI/SPL/Photo Researchers, Inc. 1.6b: G.L. Barron, University of Guelph/BPS. 1.6c: ©K.W. Jean/Visuals Unlimited. 1.6d: ©Cabisco/Visuals Unlimited. 1.6e: ©Lee D. Simon/Photo Researchers, Inc. 1.7: ©David Scharf/Peter Arnold, Inc.

Chapter 2 Opener: CDC. Box: ©Randy Brandon 1989/Peter Arnold, Inc.

Chapter 3 Opener: CNRI/Science Photo Library/Photo Researchers, Inc. 3.4a–c: ©David M. Phillips/Visuals Unlimited. 3.5: ©Paul W. Johnson/University of Rhode Island/BPS. 3.6b: CDC. 3.8a: ©CNRI/Science Photo Library/Photo Researchers, Inc. 3.8b: NIBSC/Science Photo Library/Photo Researchers, Inc. 3.9e: ©Leon J. Le Beau/BPS. 3.10a: ©M. Abbey 1981/Photo Researchers, Inc. 3.10b: ©Leon J. Le Beau/BPS. 3.10c: ©Eric V. Grave 1980/Photo Researchers, Inc.

Chapter 4 Opener, 4.1a: ©Custom Medical Stock Photo. 4.1b: ©Beveridge/BPS. 4.1c1: T.J. Beveridge and C. Forsberg, University of Guelph/BPS. 4.1c2: ©David Scharf 1990/SPL/Custom Medical Stock Photo. 4.2b–d, 4.3a: BPS. 4.3b: Burans Krieg, N.R. "Biology of the Chemoheterotrophic Sporilla," *Bacterial Review,* 40:55–115, 1976. ©American Society of Microbiology. 4.3c: O. Carleton, N.W. Charonn, P. Allander, and S. O'Brien, *J. Bacteriology,* 137:1413–1416, 1979. 4.4a: Dr. Heinz Schlesner, Kiel University, Germany. 4.4b: Dr. J.S. Poindexter, New York University School of Medicine. 4.6a: ©Cabisco/Visuals Unlimited. 4.6b: ©Barry Dowsett/SPL/Photo Researchers, Inc. 4.6c: ©E.C.S. Chan/Visuals Unlimited. 4.6d: USDA/Science Source/Photo Researchers, Inc. 4.7a: CNRI/SPL/Custom Medical Stock Photo. 4.8: CNRI/SPL/Photo Researchers, Inc. 4.11a: T.J. Beveridge, University of Guelph/BPS. 4.12: H.S. Pankratz and R.L. Uffen, Michigan State Univ./BPS. 4.13b: Chris Case. 4.17b: H.S. Pankratz/BPS. 4.19a&b: Reprinted by permission. Figure by Wessenberg and G.A. Antipa, 1970, "Capture and Ingestion of Paramecium by *Didinium nasutum,*" *J. Protozoology,* 17:250–270. 4:20a: R. Rodewald, Univ. of Virginia/BPS. 4.21a, 4.22a: G.E. Palade, Yale University Medical School. 4.23a: Courtesy N. Simionescu. 4.24: E.H. Newcomb and T.D. Pugh, University of Wisconsin/BPS.

Chapter 5 Opener, 5.3b–c: Thomas Steitz, Yale University, Department of Molecular Biophysics and Biochemistry.

Chapter 6 Opener, 6.7, 6.8: Chris Case. 6.9b: Janice Sheldon/Photo 20-20. 6.10b: L.D. Simon, Rutgers University. 6.15a: H.W. Jannasch, Woods Hole Oceanographic Institute, Woods Hole MA. 6.15b: Courtesy of Millipore/Greg Hoff. Box: ©D. Foster/WHOI/Visuals Unlimited.

Chapter 7 Opener: Chris Case. 7.8: ©CDC.

Chapter 8 Opener, 8.1a: Dr. Jack Griffith. 8.1b: Courtesy of H. Ris. 8.5a: J. Cairns, Imperial Research Fund Laboratory, Mill Hill, London. 8.7: O.L. Miller, Jr., B.A. Hamkalo, and C.A. Thomas, *J. Science,* 169:392, July 24, 1970. Copyright 1970 by the AAAS. 8.24: L.G. Caro and R. Curtiss. 8.28: R. Welch, University of Wisconsin Medical School. Box: Courtesy of S. Falkow, Rocky Mountain Laboratory, Hamilton, MT.

Chapter 9 Opener and 9.6: Reprinted from *American Laboratory,* 8 (5), 1990, p. 36. Copyright 1990 by International Scientific Community, Inc. 9.1: Courtesy of Keith Wood. 9.5b1: ©Dr. Jeremy Burgess/SPL/Photo Researchers, Inc. 9.9: Courtesy of Pharmacia. 9.11: ©Dr. Daniel C. Williams/ Lilly Research Laboratories. 9.14: Courtesy of USDA. 9.16: Courtesy of Mycogen Corporation. Box: Randall von Wedel, Cytoculture International.

Chapter 10 Opener, 10.4a: USM/Visuals Unlimited. 10.4b&c: Courtesy of J.W. Schopf, University of California, Los Angeles. 10.9: Courtesy of the Microbial Diseases Laboratory, Berkeley, CA. 10.11: Biavatti et al., *Int. J. Systematic Bacteriology,* 32:358–373, 1982. 10.12d&e: Courtesy of Steffen Kjelleberg. From Jouper-Jaan et al., *Applied and Environmental Microbiology,* 1986, 53:1419–21. Box: Courtesy of Marine World USA, Vallejo, CA.

Chapter 11 Opener: A.B. Dowsett/SPL/Photo Researchers, Inc. 11.1: From T.J.J. Fitzgerald et al., *J. Bacteriology,* 130:1333, 1977. Photo courtesy of Churchill-Livingstone. 11.2a: BPS. 11.2b: A.B. Dowsett/SPL/Photo Researchers, Inc. 11.3: John A. Fuerst, from *Bergey's Manual of Systematic Bacteriology* ©Williams & Wilkins Co., Baltimore. 11.4: Cheun-mo To and Charles C. Brinton, Jr. 11.5: Judith F.M. Hoeniger, *J. Gen. Microbiology,* 40:29–42, 1965. Copyright 1965 Society for General Microbiology. 11.6: London School of Hygiene and Tropical Medicine/SPL/Custom Medical Stock Photo. 11.7a: CDC. 11.7b: Sidney M. Feingold. 11.8a: W. Burgdorfer, Rocky Mountain Laboratory, Hamilton, MT. 11.8b: R.C. Cutlip, National Animal Disease Center, Ames, IA. 11.9: ©Michael Gabridge/Visuals Unlimited. 11.10: Dr. Tony Brain/SPL/Custom Medical Stock Photo. 11.11: Lin Tao, J.M. Tanzer, and T.J. MacAlister, *J. Bacteriology,* 169:2534, 1987. ©American Society for Microbiology. 11.12a: R.E. Strange and J.R. Hunter in G.W. Gould and A. Hurst, eds., *The Bacterial Spore.* New York: Academic Press, 1969, p. 461. 11.12b: C.L. Hannay and P. Fitz-James, *Canadian J. Microbiology,* 1:694, 1955. 11.12c: Courtesy of H.M. Solomon, FDA, from "Botulism," by D.A. Kautter and R.K. Lynt, Jr., in *FDA Papers,* Nov. 1971. 11.13, 11.14: ©David M. Phillips/Visuals Unlimited. 11.15a: J.S. Poindexter, *Bacteriology Review,* 28:231–295, 1965. 11.15b: H. Rechenbach and M. Dworkin, 1981, "Introduction to the Gliding Bacteria," in *The Prokaryotes,* eds. M. Stars, H. Stolp, H.G. Trupu, A. Balows, and H.G. Schlegel, Springer-Verlag, pp. 315–327. 11.15c: K. Stephens, Stanford University/BPS. 11.16a: R.L. Moore/BPS. 11.16b: Courtesy of J.L. Stokes. 11.17a: Karl O. Stetter. 11.17b: Holger Jannasch, Woods Hole Oceanographic Institute, Woods Hole, MA. 11.18: ©Paul Johnson/BPS. 11.19a&c: ©T.E. Adams/ Visuals Unlimited. 11.19b: ©Sherman Thomson/Visuals Unlimited. 11.20: Elizabeth M.H. Wellington, Neil Cresswell, and Venetia A. Saunders, "Growth and Survival of Streptomycete Inoculants," *Applied and Environmental Microbiology,* May 1990; 56:1413–1419. Copyright ©American Society of Microbiology.

Chapter 12 Opener: Chris Case. 12.3: ©David Scharf 1986/ Peter Arnold, Inc. 12.7: ©Cecil Fox/Science Source/Photo Researchers, Inc. 12.9a: ©Manfred Kage/Peter Arnold, Inc. 12.9b: G.T. Cole, University of Texas, Austin/BPS. 12.13a: Catherine M. Pringle/BPS. 12.15: ©S. Sharnoff/Visuals Unlimited. 12.18: Carolina Biological Supply. 12.19: CDC. 12.30a: CDC. 12.30b: Michael Nachtigall, M.S.

Chapter 13 Opener: CDC/Peter Arnold, Inc. 13.3b: Dr. Jack Griffith. 13.4b: R.C. Valentine and H.G. Pereira. 13.5b: F.A. Murphy, Viral Pathology Branch, CDC. 13.6b: A.K. Harrison, Viral Pathology Branch, CDC. 13.7b&c: Robley Williams. 13.8a: Courtesy of Bernard Roizman. 13.8b: C. Garon and J. Rose, National Institute of Allergy and Infectious Diseases, CDC. 13.9a: A.K. Harrison, Division of Viral Diseases, CDC. 13.9b: J.J. Cardamone, Jr., University of Pittsburgh/BPS. 13.9c: G. Smith, National Cancer Institute. 13.10: Chris Case. 13.13a&b: G. Wertz, University of North Carolina School of Medicine, Chapel Hill. 13.18a&b: C. Morgan, H.M. Rose, and B. Mednis, *J. Viology,* 2:501–516, 1968. 13.22b: D.R. Brown et al., *J. Virology,* 10:524–536, 1972. 13.23: A.K. Harrison, Division of Viral Diseases, CDC. 13.24: T.O. Diener, U.S. Department of Agriculture.

Chapter 14 Opener, 14.1a&c: ©David M. Phillips/Visuals Unlimited. 14.1b: ©Fred Hossler/Visuals Unlimited. 14.4a: ©Michael P. Gadomski 1988/Photo Researchers, Inc. 14.4b: ©Erika Stone/Peter Arnold, Inc. 14.4c: ©L. Dardelet/Photo Researchers, Inc. 14.5a: ©Sybil Shelton/Peter Arnold, Inc. 14.5b: ©Bonnie M. Freer/Peter Arnold, Inc. 14.5c: ©Scott Camazine 1990/Photo Researchers, Inc. 14.6: ©SIU 1988/ Visuals Unlimited. 14.7: M.W. Jennison, Department of Biology, Syracuse University, Syracuse, NY.

Chapter 15 Opener: J.P. Bader, National Cancer Institute. 15.1a: L.R. Inman and J.R. Cantey, *J. Clinical Investigation,* 71:1–8, 1983. 15.1b: S. Murphee, Auburn University/BPS. 15.2: S.C. Holt, University of Texas Health Science Center, San Antonio/BPS. 15.5a&b: F.A. Murphy, Viral Pathology Branch, CDC. 15.6: J.P. Bader, National Cancer Institute. Box: ©Veronika Burmeister/Visuals Unlimited.

Chapter 16 Opener, 16.1: ©Ed Reschke. 16.3a: K.E. Muse, Duke University Medical Center/BPS. 16.6: ©Lennart Nilsson, Boehringer Ingelheim International, GmbH. 16.12e: Schreiber et al., *J. Experimental Medicine,* 149:870–882, 1979.

Chapter 17 Opener: ©David M. Phillips/The Population Council/Science Source/Photo Researchers, Inc. 17.2a: Dr. Dorothea Zucker-Franklin, *Atlas of Blood Cells: Function and Pathology.* Reprinted by permission of Edi Ermes, s.r.l. Italy. 17.2b: ©David M. Phillips/The Population Council/Science Source/Photo Researchers, Inc. 17.10: R.M. Steinman et al., *J. Experimental Medicine,* 149:1–16. Box: Reprinted by permission of the publisher. From "Ultrastructural Observations on Giardiasis in a Murine Model," by Robert Owen, Paulina Nemanic, and David P. Stevens, *Gastroenterology,* 76:757–769. Copyright 1979 by the American Gastroenterological Association.

Chapter 18 Opener, 18.12: CDC.

Chapter 19 Opener, 19.1b: ©Lennart Nilsson, Boehringer Ingelheim International GmbH. 19.2: ©David Scharf. 19.3: ©Lennart Nilsson, Boehringer Ingelheim International GmbH. 19.13b&c: A. Liepins.

Chapter 20 Opener: ©Visuals Unlimited. 20.1: The Bettmann Archive. 20.3a&b: D. Greenwood and F. O'Grady, *Science,* 163:1076, 1969. Copyright ©1969 by the AAAS. 20.5: M. Bastide, S. Jouvert, and J.M. Bastide, *Canadian J. Microbiology,* 28:1119–1126, 1982. 20.21: ©Visuals Unlimited. 20.22: Courtesy of Donald Nash, from *J. Clinical Microbiology,* Dec.

1986, p. 977. Reprinted with permission of the American Society for Microbiology.

Chapter 21 Opener: CDC. 21.2: Dr. Jean-Paul Revel, California Institute of Technology. 21.3: ©Charles Stoer, M.D. 1985/Camera M.D. Studios. 21.4: P.P. Cleary, University of Minnesota School of Medicine/BPS. 21.5: ©Ken Greer/ Visuals Unlimited. 21.6: ©Carroll H. Weiss, RBP, 1982/ Camera M.D. Studios. 21.7: World Health Organization, Geneva, Switzerland. 21.8a: CDC. 21.8b: Camera M.D. Studios. 21.10: ©Carroll H. Weiss, RPB, 1980/Camera M.D. Studios. 21.12: ©Lowell Georgia/Science Source/Photo Researchers, Inc. 21.13: ©Carroll H. Weiss, RPB,1989/Camera M.D. Studios. 21.14a: CDC. 21.14b: Camera M.D. Studios. 21.15a: CDC. 21.15b: ©Science Photo Library/Photo Researchers, Inc. 21.16: AFIP/Visuals Unlimited. Box: Courtesy of Walter B. Shelley, M.D., and E. Dorinda Shelley, M.D.

Chapter 22 Opener: CDC/BPS. 22.2: F.L.A. Buckmire, The Medical College of Wisconsin. 22.4: The Bettmann Archive. 22.5a: ©Kenneth E. Greer/Visuals Unlimited. 22.5b: CDC. 22.6: Courtesy of Warren E. Collins, Inc. 22.8: A.K. Harrison/CDC. 22.12: Edward Bottone, The Mount Sinai Hospital. 22.14: CDC/BPS.

Chapter 23 Opener: Dr. Tony Brain/SPL/Custom Medical Stock Photo. 23.2: ©Ken Greer/Visuals Unlimited. 23.3: ©Camera M.D. Studios. 23.4: Courtesy of Dr. Victor Marks. 23.6: Marilyn J. Tufte. 23.7: CDC/BPS. 23.8: ©SIU/Visuals Unlimited. 23.9: ©Giselle Hosgood. 23.11: Air Force Institute of Pathology. 23.12b: ©Edward S. Ross. 23.12c, 23.16: CDC. 23.17: CDC/Visuals Unlimited. 23.19: S.G. Baum, Albert Einstein College of Medicine/BPS. 23.22: CDC. 23.23a: Wilmar Jansma, University of Southern Illinois; Harry Schafer and Bruce Wetzel, National Institute of Health. 23.24: Kreier/Mortensen, Ohio State University, from *Infection, Resistance and Immunity*, Harper & Row.

Chapter 24 Opener, 24.3: ©A.M. Siegelman/Visuals Unlimited. 24.4: Michael Kientz/*Discover* magazine. 24.5: K.E. Muse, Duke University Medical Center/BPS. 24.6: Frederick S. Nolte. 24.7: R.B. Morrison, Austin, TX. 24.9: ©Raymond B. Otero/Visuals Unlimited. 24.10a: J.A. Stuekemann and D. Paretzky, University of Kansas/BPS. 24.10b: Thomas F. McCaul and Jim C. Williams, "Developmental Cycle of *Coxiella burnetti:* Structure and Morphogenesis of Vegetative and Sporogenic Differences," *J. Bacteriology*, Sept. 1981, p. 1068. ©American Society For Microbiology. 24.13a: Visuals Unlimited. 24.13b: G.T. Cole, University of Texas, Austin/BPS. 24.15a: David Drutz and Milton Huppert, *J. Infectious Diseases*, 147:379. ©1983 University of Chicago. 24.15b: G.T. Cole, University of Texas, Austin/BPS. 24.17a: G.W. Willis, Ochsner Medical Institution/BPS. 24.17b: From "Propagation of *Pneumocystis carinii* in Vitro," by L. Pifer et al., *Pediatric Research*, 11:305–316, 1977.

Chapter 25 Opener: L.K. Ng, R. Sherburne, D.E. Taylor, and M.E. Stiles, *J. Bacteriol* 164:339, ©American Society for Microbiology. 25.3 a–c: Z. Scobel, Forsyth Dental Center/ BPS. 25.4a&b: Hutton D. Slade, *Microbiol. Review*, 14(2):331–384, June 1980. 25.9a: G.T. Cole, University of Texas at Austin/BPS. 25.9b&c: Reprinted from *Food Technology*, 36(3):93–96, 1982. ©Institute of Food Technologists. 25.10: L.K. Ng, R. Sherburne, D.E. Taylor, and M.E. Stiles, *J. Bacteriology*, 164:339. ©American Society for Microbiology. 25.11: ©Veronika Burmeister/Visuals Unlimited. 25.12b: F.A. Murphy, Viral Pathology, CDC. 25.13a: A.K. Harrison, Div. of Viral Diseases, CDC. 25.13b: CDC. 25.14a: Dr. Jack Griffith. 25.15: Taken by E. Calomeni and provided by M. Estes, Baylor College of Medicine, Houston, TX. 25.16: ©Jerome Paulin/Visuals Unlimited. 25.17: Air Force Institute of Pathology. 25.18: Courtesy of Marietta Voge, Ph.D., from *Diagnostic Medical Parasitology*, L.S. Garcia and D.A. Bruckner. ©Elsevier Scientific Publications. 25.20b: Herman Zaiman, M.D., Mercy Hospital, Valley City, ND. 25.21: Armed Forces Institute of Pathology. 25.22: ©R. Calentine/ Visuals Unlimited. 25.23: ©A.M. Siegelman/Visuals Unlimited.

Chapter 26 Opener: ©David M. Phillips/Visuals Unlimited. 26.4: D. Bromley, West Virginia University Medical Center. 26.7, 26.8: CDC. 26.10: ©M. Abbey/Photo Researchers, Inc. 26.11a&b: ©Carroll H. Weiss 1981/Camera M.D. Studios. 26.11c, 26.12:CDC. 26.13: ©Mike Remmington, University of Washington Viral Disease Clinic. 26.14: Camera M.D. Studios. 26.15: ©David M. Phillips/Visuals Unlimited.

Chapter 27 Opener, 27.4a: M.F. Brown, University of Wisconsin, Madison/BPS. 27.4b: R.L. Peterson, University of Guelph. 27.7: Courtesy of Exxon Corporation. 27.11a: ©V. Paulson 1984/BPS. 27.12a&b: R.F. Unz, *Int. J. Systematic Bacteriology*, 21:91–99, 1971. 27.13a: Douglas Munnecke/BPS. 27.14a: Carl May/BPS. 27.16a-d: Steve Woit.

Chapter 28 Opener: Chris Case. 28.6a&b: ©David R. Frazier/ Photo Researchers, Inc. 28.6c: ©Junebug Clark/Photo Researchers, Inc. 28.8b: LSL Biolafitte, Inc. 28.10: ©Manfred Kage/Peter Arnold, Inc.

INDEX

Note: A *t* following a page number indicates tabular material and an *f* following a page number indicates a figure. Drugs are listed under their generic names. When a drug trade name is listed, the reader is referred to the generic name.

A

A site, 201
Abdominal cramps, in digestive system infections and intoxications, 622
ABO blood group system, 471, 472*f*, 472*t*
  characteristics of, 472*t*
Abortion, infections after, 563, 580*t*
Abscess, 416, 521
Absorbance, in estimating bacterial numbers by turbidity, 162
Absorption, in digestion, 617
Acanthamoeba spp, 318*t*
  keratitis caused by, 534–535, 534*t*
Acanthamoeba polyphaga, keratitis caused by, 534–535, 534*t*
Accessory glands, male, 652. *See also* Reproductive system
Accessory structures, digestive, 617, 618*f*. *See also* Digestive system
Accutane. *See* Isotretinoin
Acellular slime molds, 313
Acellular vaccines, 452
Acetic acid, fermentation of, 123*t*
Acetobacter spp, 280
  in vinegar production, 712
Acetyl CoA. *See* Acetyl coenzyme A
Acetyl coenzyme A, in Krebs cycle, 114–116
N-Acetylglucosamine, in peptidoglycan, 77
N-Acetylmuramic acid, in peptidoglycan, 77
Acid-anionic detergents, antimicrobial activity of, 182, 184*t*
Acid-base balance, 33–34
Acid-fast stain, 65
  for *Mycobacterium tuberculosis*, 65, 597
Acidic dyes, 60
Acidic foods, spoilage of, 702–703
Acidic solutions, 34
Acidophiles, 144
Acids, 33
Acinetobacter spp, genetic transformation occurring in, 215
Acne, 288, 410, 523–524, 531*t*
  Propionibacterium acnes in, 520, 524
Acquired immunity, 427–428
Acquired immunodeficiency syndrome (AIDS), 2–4, 480–486, 664. *See also* Human immunodeficiency virus
  animal models and, 342
  chemotherapy for, 484, 486
  cytomegalovirus infection and, 631
  diseases associated with, 483*t*
  feline, 342
  future of, 486
  helper T cells in, 439
  incidence of, 655*f*
  monoclonal antibodies in treatment of, 439
  opportunistic infection and, 370–371
  origin of, 480
  Pneumocystis pneumonia in, 370–371, 609
  and risk to health care workers, 344
  simian, 342
  transmission of, 344, 482–485
  treatment of, 484–485, 486
  vaccine for, 452, 484, 486
Acridine dyes, as frameshift mutagens, 210
Actinobacillus actinomycetemcomitans, localized juvenile periodontitis caused by, 410
Actinomyces spp, 275*t*, 288
  in tooth decay, 395, 619
Actinomyces israelii, 288
Actinomycete mycetoma, 533
Actinomycetes, 275*t*, 293
  in acute necrotizing ulcerative gingivitis, 622
  antibiotics produced by, 494*t*
  in nitrogen fixation, 680
  phylogenetic relationships of, 258*f*
  in soil, 674
Actinomycosis, 288
Activated sludge system, 690, 691*f*
Activation energy, 30, 102
Active immunity
  artificially acquired, 427, 428*t*
  naturally acquired, 427, 428*t*
Active processes, movement across membranes by, 83
Active site, enzyme, 31, 104, 105*f*
Active transport, 86
Activity, spectrum of, for antimicrobial agent, 494–496
Acute bacterial endocarditis, 563, 580*t*
Acute coryza (common cold), 594–595, 611*t*
  adenoviruses causing, 350
Acute disease, 374
Acute necrotizing ulcerative gingivitis (Vincent's disease, trench mouth), 622
Acyclovir, 501*t*, 507–508
  for genital herpes, 664, 665*t*
  for varicella-zoster infection in immunocompromised patients, 526
Addison's disease, 476, 477*t*
  HLA antigens in, 479*t*
Adenine, 46, 190–191
Adenine analogs, 501*t*. *See also specific type*
Adenocarcinomas, virus-induced, 357
Adenosine diphosphate. *See* ADP
Adenosine triphosphate. *See* ATP
Adenoviruses, 338*t*, 350. *See also specific type*
  attachment sites of, 349
  cytopathic effects of, 403*t*
  morphology of, 336*f*, 337
  multiplication of, 350, 351*f*
  opportunistic infection and, 371
  respiratory system disease caused by, 350
  size of, 334*f*
Adherence, 393–395, 404*f*
  in phagocytosis, 414
Adhesins, 393
  plasmid gene encoding, 401
ADP, 47
Adrenaline. *See* Epinephrine
Adsorption
  in bacteriophage multiplication, 345, 346*f*
  in RNA virus multiplication, 352*f*
Aedes spp, 326*t*, 553*t*

Aedes spp (*Continued*)
  diseases transmitted by, 379*t*
Aedes aegypti, 575, 581*t*
  diseases transmitted by, 379*t*
Aedes albopictus, 324, 575
Aerial mycelium, 298
Aerobes, 146, 274*t*, 275*t*
  gram-negative rods and cocci, 274*t*, 277–280
  obligate, 146
  oxygen-related enzymes of, 147*t*
Aerobic endospore-formers, food spoilage caused by, 703*t*
Aerobic respiration, 114, 118–120
  anaerobic respiration compared with, 123*t*
  fermentation compared with, 123*t*
Aerosolization, and legionellosis, 602
Aerotolerant anaerobes, 146*f*, 147–148
  oxygen-related enzymes of, 147*t*
Affinity, in antibody-antigen binding, 435
Aflatoxin, 404, 637, 641*t*, 706, 707*t*
  as frameshift mutagen, 210
African trypanosomiasis (African sleeping sickness), 316, 440, 554, 555*t*
  arthropod vector for, 379*t*
Agammaglobulinemia, X-linked infantile (Bruton's), 481*t*
Agar, 148–149
  nutrient, 150
  from red alga, 310
Agglutination reactions, 456–457
  Agglutination tests
  direct, 456–457
  indirect (passive), 457
  slide, 261
Agranulocytosis, 474
Agricultural products, genetic engineering in production of, 237*t*
Agriculture, recombinant DNA technology applications for, 241–243
Agrobacterium spp, 241, 274*t*
  Entner-Doudoroff pathway in, 114
Agrobacterium tumefaciens, 279
  Ti plasmid in, 241
AIDS (acquired immunodeficiency syndrome), 2–4, 480–486, 664. *See also* HIV
  animal models and, 342
  chemotherapy for, 484, 486
  cytomegalovirus infection and, 631
  diseases associated with, 483*t*
  feline, 342
  future of, 486
  helper T cells in, 439
  incidence of, 655*f*
  monoclonal antibodies in treatment of, 439
  opportunistic infection and, 370–371
  origin of, 480
  Pneumocystis pneumonia in, 370–371, 609
  and risk to health care workers, 344
  simian, 342
  transmission of, 344, 482–485
  treatment of, 484–485, 486
  vaccine for, 452, 484, 486
Air filters, microorganisms removed by, 173
Airborne transmission, in spread of infection, 377
Ajellomyces spp, and reclassification of fungi, 303*t*
Alanine, 42*t*
Alarmone, *lac* operon regulation and, 206
Alcohol fermentation, 121–122
Alcoholic beverages, microorganisms in production of, 710–712
Alcohols
  antimicrobial activity of, 180, 184*t*
  ethyl (ethanol)
    antimicrobial activity of, 180, 184*t*
      effectiveness and, 183*f*
    fermentation of, 123*t*
    production of from agricultural products, 718
Aldehydes, antimicrobial activity of, 182–183, 185*t*
Ale, microorganisms in production of, 710, 711*t*
Aleutian mink disease, 360*t*
Algae, 15, 16*f*, 306–310. *See also specific type and* Microorganisms
  blooms of, 310, 685
  in carbon cycle, 17
  characteristics of, 309*f*
  comparison of, 297*t*
  classification of, 307, 309*f*
  copper sulfate for destruction of, 181, 184*t*
  filamentous, 306
  in fresh water, 683
  habitats of, 307*f*
  in lichens, 310–312
  in oxygen cycle, 17
  pathogenetic properties of, 403–404
  photosynthesis in, 129*t*
  phyla of, 307–310
  reproduction of, 310
  role of in nature, 310
  in soil, 674
  structure of, 310
Algin, 310
Alimentary canal (gastrointestinal tract), 617, 618*f*. *See also* Digestive system
  in host defense, 409
  as portal of entry, 393, 394*t*, 404*f*
Alkaline solutions, 34
Alkylation, in gas sterilization, 183
Allergen, 467
Allergic contact dermatitis, 475–476
Allergic reaction (allergy), 467. *See also* Hypersensitivity
  to house dust mites, 470
  IgE in, 432, 468
  localized, 470
  to poison ivy, 476*f*
  systemic, 468–470
Allergic rhinitis, 470
Allescheria spp, and reclassification of fungi, 303*t*
Allescheria boydii, 302*t*
  skin infections caused by, 531*t*, 532
Allograft, 479
Allosteric inhibition, of enzyme, 107
Alpha-amino acids, 41
Alpha-carbon, 41
Alpha-hemolytic streptococci, 286, 522
  subacute bacterial endocarditis caused by, 580*t*
Alpha interferon, 422
  for AIDS, 485
  genetic engineering in production of, 237*t*
Alpha virus, 338*t*
Alternative energy sources, microorganisms in production of, 718
Alternative pathway for complement activation, 418, 419*f*
Aluminum, in soil, 673
Alveolar macrophages, 413, 590
Amanita muscaria, 305*f*
Amanita mushrooms, toxins produced by, 404
Amanitin toxin, 305*f*, 404
Amantadine, 501*t*, 507
  for influenza, 606, 611*t*
American trypanosomiasis (Chagas' disease), 316, 577, 581*t*
  transmission of, 378*t*, 581*t*
Ames test, for carcinogen identification, 213–214
Amino acid activating enzyme, 199, 201*f*
Amino acid sequencing, 262
Amino acids, 41–42
  activation of, by transfer RNA, 199, 201*f*
  biosynthesis of, 132, 133*f*
  industrial microbiology in production of, 714–716
Amino group, 35, 36*t*
Aminoglycosides, 500*t*, 504
  protein synthesis affected by, 498
2-Aminopurine, as chemical mutagen, 209–210
Ammonia
  nitrogen gas converted to (nitrogen fixation), 17, 145, 678–680
  release of, 678
Ammonification, 678
Ammonium compounds, quaternary, antimicrobial activity of, 182, 184*t*
Ammonium ion, 677
  oxidation of to nitrate, 678
Amoebas, 15, 313, 314*f*, 318*t*. *See also* Protozoans
  aggregation of, 312–313
Amoebic dysentery (amoebiasis), 315, 638, 641*t*
Amoxicillin, 503
  for otitis media, 594, 610*t*
  potassium clavulanate combined with, 504
Amphibolic pathways, 135
Amphitrichous flagella, 73, 74*f*
Amphotericin B, 501*t*, 507–508
  for blastomycosis, 609, 611*t*
  for coccidioidomycosis, 608–609, 611*t*
  for histoplasmosis, 608, 611*t*
  plasma membrane affected by, 498
  sources of, 494*t*
Ampicillin, 500*t*, 503
  for cystitis, 653
  structure of, 502*f*
  for typhoid fever, 626
Amplification gene, in oncogene activation, 358
Ampligen, for AIDS, 485
Amylase, commercial production of, 716*t*
Anabaena spp, 275*t*
Anabaena spiroides, 292*f*
Anabolism (anabolic reactions), 29, 101–102, 131–132
  catabolism related to, 102*f*
Anaerobes, 274*t*
  aerotolerant, 146*f*, 147–148
    oxygen-related enzymes of, 147*t*

facultative, 146, 274*t*, 280–282
oxygen-related enzymes of, 147*t*
gram-negative cocci, 274*t*, 283
gram-negative straight, curved, and helical rods, 274*t*, 282, 283*f*
obligate, 146*f*, 147
oxygen-related enzymes of, 147*t*
Anaerobic glove boxes, 151
Anaerobic growth media and methods, 150–151
Anaerobic respiration, 114, 120, 678
Anaerobic sludge digesters, 691–692
Anal gonorrhea, 656
Anal pore, 314
Analytical epidemiology, 386
Anamnestic immune response, 443–444
Anaphylactic shock, 469
Anaphylaxis, 468–471, 468*t*
localized, 468, 470
mechanism of, 469*f*
prevention of, 470–471
systemic, 468–470
Anemia, hemolytic, 474, 477*t*
Anergy, clonal, 443
Angstrom, 53
Animal bites, systemic diseases caused by, 567–568
Animal husbandry products, genetic engineering in production of, 237*t*, 242–243
Animal reservoirs, in spread of infection, 374–375
Animal viruses, 334. *See also specific type*
growth of in laboratory, 341–343
multiplication of, 348–355
Animalia, 251, 252*f*, 254*f*
Animals, classification of, 251, 252*f*, 253*f*
Anions, 26
Anisakidae, 325*t*
Anopheles mosquito, 326*f*, 326*t*
diseases transmitted by, 379*t*
Plasmodium transmitted by, 316–317, 327, 578, 581*t*
Anoxygenic photosynthesis, 127, 292
Antagonism
in chemotherapy, 511
microbial, 370
Anthrax, 287, 566, 580*t*. *See also Bacillus anthracis*
transmission of, 378*t*, 580*t*
Antibacterial synthetics, 499–502
Antibiosis, 493
Antibiotics, 11, 492, 500*t*, 502–506. *See also specific agent and* Antimicrobial agents
cell wall synthesis affected by, 496, 497*f*
discovery of, 11–13, 493–494
microbial fermentations in production of, 717
nucleic acids affected by, 498
plasma membrane affected by, 498
protein synthesis affected by, 497–498
resistance to, 510–511. *See also* Drug resistance
sources of, 494*t*
Antibodies, 427, 430–433. *See also* Immunoglobulins; Serology
antigen binding to, 435
in artificially acquired passive immunity, 427
blocking, in desensitization, 471
diversity of, molecular basis of, 436, 437*f*
fluorescent, 57–58
in humoral immune system, 428
monoclonal. *See* Monoclonal antibodies
in naturally acquired passive immunity, 427
structure of, 430–431
Antibody-antigen binding, 435. *See also* Antigen-antibody complexes
Antibody titer, 443, 456
Anticodon, 199
Antifungal drugs, 501*t*, 502, 506–507. *See also specific type and* Antimicrobial agents
plasma membrane affected by, 498
Antigen-antibody complexes (immune complexes), 435, 474–475
in glomerulonephritis, 475, 654, 665*t*
Antigen-antibody reactions. *See also* Immune complex reactions
complement fixation during, 457–458
ionic bonds in, 26
rheumatic fever and, 564
Antigen-binding sites, 430
Antigen modulation, and immunologic escape, 487
Antigen-presenting cells, 433, 434*f*
Antigen receptors, 429
molecular basis of diversity of, 436, 437*f*
Antigenic determinants, 429
Antigenic drift, in influenza virus, 605–606
Antigenic shifts, in influenza virus, 605
Antigens, 427, 429–430
antibody binding to, 435. *See also* Antigen-antibody complexes
in artificially acquired active immunity, 427
B-cell interactions and, 433–435
in fluorescence microscopy, 57–58
histocompatibility, 476
in naturally acquired active immunity, 427
"self," 433, 434, 435*f*. *See also* Tolerance
cytotoxic T cells and, 441
surface, and immune response to protozoans, 440
T-dependent, 433
T-independent, 434–435
Antihelminthic drugs, 501*t*, 509. *See also specific type and* Antimicrobial agents
Anti-idiotypic vaccines, 452–453
Antimetabolites, and inhibition of enzymatic activity, 498
Antimicrobial agents, 492. *See also specific type and* Antibiotics; Chemotherapy
actions of, 169–170, 496–498
in animal feed, human disease and, 495
cell wall synthesis affected by, 496, 497*f*
chemical, 176–185
combinations of, 511
effectiveness of, 183*f*
enzymes affected by, 169–170, 498
evaluation of, criteria for, 494
historical development of, 492–494
nucleic acids affected by, 169–170, 498
physical, 170–176
plasma membrane affected by, 83, 169, 498
proteins affected by, 169–170, 497–498
resistance to, 510–511. *See also* Drug resistance
spectrum of activity of, 494–496
Antimicrobial substances, body producing, 418–423
Antiparallel structure of DNA, 195
Antiprotozoan drugs, 501*t*, 509. *See also specific type and* Antimicrobial agents
Antirabies treatment, 551
Antisense strand RNA viruses, 338–339*t*, 354
Antisepsis, 168*t*
Antiserum (antisera), 260–261, 427–428
Antitoxins, 397, 459
for botulism, 545–546
for diphtheria, 593, 610*t*
Antiviral drugs, 501*t*, 507–509. *See also specific type and* Antimicrobial agents
Antiviral proteins, 422
APC. *See* Antigen-presenting cells
Apoenzyme, 103
Appendaged bacteria, 275*t*, 289
Approved Lists of Bacterial Names, 258
Aquaspirillum bengal, 277*f*
Aquaspirillum magnetotacticum, 76, 276
Aquatic food chain, algae in, 310
Aquatic microbiology, 682–695
organisms in, 682–683
Arboviruses, 338*t*
arthropod-borne encephalitis caused by, 551, 555*t*
dengue fever caused by, 575, 581*t*
yellow fever caused by, 575, 581*t*
Arbuscules, 680
Archaeobacteria, 71, 275*t*, 291
cell wall of, 80
characteristics of, 253*t*
classification of, 251–254
photosynthesis without chlorophyll in, 128
phylogenetic relationships of, 258*f*
Arcyria slime mold, 312*f*
Arenaviruses, 339*t*
Arginine, 43*t*
Arsenic, 23*t*
Arterioles, 560, 561*f*. *See also* Cardiovascular system
Artery, 560. *See also* Cardiovascular system
Arthritis
gonorrheal, 656
rheumatic fever and, 563–564
rheumatoid, 475, 477*t*
septic, *Hemophilus influenzae* causing, 282
Arthroderma spp, and reclassification of fungi, 303*t*
Arthropod vectors, 325–327, 377–379. *See also specific type or disease*
diseases transmitted by, 379*t*
Arthropod-borne encephalitis, 551–553, 555*t*
vector for, 379*t*, 553*t*
Arthrospore, 300
Arthus reaction, 475
Artificially acquired immunity, 427–428
active, 427, 428*t*
passive, 427, 428*t*
As. *See* Arsenic
Ascariasis, 641*t*, 643

Ascaris lumbricoides, 322–323, 325*t*, 641*t*, 643
Aschelminthes, 321–325, 325*t*
Ascomycota, 302–303*t*, 304
ascospores produced by, 301
Ascospores, 301
heat-resistant, food spoilage and, 703
Ascus, 301
Asepsis, 168*t*
Aseptic packaging, for food preservation, 703–704
Aseptic surgery, 167
Aseptic technique, 7
Asexual reproduction
in algae, 310
in eucaryotes, 220
in protozoans, 314
Asexual spores, 299, 300
Ashyba gossypii, riboflavin produced by, 717
Asian tiger mosquito, 324
Asiatic flu, 605
Asparagine, 43*t*
Aspartame, 716
Aspartic acid, 43*t*
industrial microbiology in production of, 716
Aspergillosis, 306, 609
Aspergillus spp, 302*t*, 304. *See also specific type*
aspergillosis caused by, 306, 609, 611*t*
in food production, 709*t*
food spoilage and, 703
reclassification of, 303*t*
Aspergillus flavus
aflatoxin produced by, 404, 637, 641*t*
frameshift mutations and, 210
conidiospores produced by, 305*f*
Aspergillus fumigatus, 609
Aspergillus niger, 298*f*
in commercial citric acid production, 716
in commercial enzyme production, 716*t*
Aspergillus oryzae
in commercial enzyme production, 716*t*
in food production, 709*t*
Aspergillus soyae, in food production, 709*t*
Aspirin, and Reye's syndrome, 525–526
Assays, microbiological, chemically defined media in, 149
Associative recognition, 442
Asthma, 470
Athlete's foot, 530
Atomic number, 23
Atomic weight, 23
Atoms, 22
chemical bonds between, 24–29
molecule formation and, 24–29
structure of, 22–24
valence of, 24
ATP, 47, 118
in anabolic and catabolic reactions, 102
movement across membranes and, 83, 86
production of, 110
in aerobic respiration, 119*t*
chemiosmotic mechanism of, 110, 117–118
by halobacterial system, 128
mitochondria in, 94
summary of, 130
ATP synthase, 117*f*, 118
Attachment
in animal virus multiplication, 349, 350*f*
in bacteriophage multiplication, 345
Attenuated vaccine, 450
Attenuation, 206–207
Attenuator, 206–207
Attractant, 75
Atypical pneumonia (mycoplasmal pneumonia), 285, 600–601, 611*t*
Augmentin. *See* Amoxicillin, potassium clavulanate combined with
Auramine O, 57
Aureomycin. *See* Chlortetracycline
Autoclaving, for sterilization, 171–172, 174*t*
Autogenous hypothesis of eucaryotic evolution, 95
Autograft, 479
Autoimmune diseases, 471, 477*t*
type II reactions, 474
type III reactions, 475
type IV reactions, 476
Autolysins, teichoic acid affecting, 79
Autotrophs, 126
Auxotrophs, 212
histidine, of *Salmonella*, 213
AVPs. *See* Antiviral proteins
Axial filaments, 75–76
on spirochetes, 276
Axilla, microbial growth in, 518
Azidothymidine (AZT, zidovudine), 501*t*, 509
for AIDS, 4, 485, 486
Azlocillin, 504
Azolla, 680
Azomonas spp, 279
Azospirillum spp, 274*t*, 276
Azotobacter spp, 279
lipid inclusions in, 87
in nitrogen fixation, 678
Azotobacter vinelandii, amino acid sequencing of, 262*f*
AZT (azidothymidine, zidovudine), 501*t*, 509
for AIDS, 4, 485, 486
Aztreonam, 503

B

B cells (B lymphocytes), 428, 433–439, 561
antigen interactions and, 433–435
differentiation of, 433*f*, 434*f*
helper T cells in activation of, 433, 435*f*
in humoral immunity, 433–439
memory, 433
Babesia microti, 318*t*, 578
Babesiosis, 570, 578
arthropod vector for, 379*t*, 570, 578
Bacillaceae, phylogenetic relationships of, 258*f*
Bacillary dysentery (shigellosis), 258*f*, 280, 626, 640*t*
water pollution in transmission of, 640*t*, 684
Bacillus (bacilli), 15, 71
Bacillus spp, 275*t*, 287. *See also specific type*
antibiotics produced by, 494*t*
in denitrification, 678
endospores formed by, 88
in food production, 709*t*
food spoilage caused by, 703*t*
genetic transformation occurring in, 215
lipid inclusions in, 87
in nitrogen fixation, 679
polypeptides isolated from, 505
Bacillus anthracis, 287, 566, 580*t*. *See also* Anthrax
capsule of, 72–73, 396
fluorescein isothiocyanate staining for, 57
and Koch's postulates, 371, 372
in soil, 674
Bacillus of Calmette and Guerin, vaccine from, 598
Bacillus cereus, 287*f*, 630, 640*t*
foodborne microbial intoxications caused by, 707*t*
gastroenteritis caused by, 630, 640*t*
Bacillus coagulans, in food spoilage, 703, 703*t*
Bacillus polymyxa, antibiotics produced by, 494*t*
Bacillus stearothermophilus, flat sour spoilage caused by, 702, 703*t*
Bacillus subtilis
antibiotics produced by, 494*t*
in commercial enzyme production, 716*t*
gene products secreted by, 235
Bacillus thuringiensis, 287
industrial microbiology in production of, 718
in pest control, 18
in soil, 676
Bacitracin, 500*t*, 505, 506
cell wall synthesis affected by, 496
sources of, 494*t*
Bacteremia, 374
nosocomial, 382*t*
Bacteria, 14–15, 16*f*, 273–295. *See also specific type and* Microorganisms
axial filaments of, 75–76
binary fission in, 15, 70, 155–156
cardiovascular and lymphatic system diseases caused by, 562–573, 580–581*t*
cell wall of, 73*f*, 77–81
atypical, 79–80
composition and characteristics of, 77–79
damage to, 80–81
structures external to, 72–77
structures internal to, 81–89
chromosomes of, 86, 93, 192
classification of, 257, 273, 274–275*t*
conjugation in, 216–217
cytoplasm of, 73*f*, 86
denitrifying, 678
digestive system diseases caused by, 618–630, 640*t*
division of, 155–156
DNA replication in, 196–197
drugs for infections with, 11, 492, 500*t*, 502–506. *See also specific agent and* Antibiotics
endocarditis caused by, 563, 580*t*
endospores formed by, 88–89
fimbriae of, 73*f*, 76–77
flagella of, 15, 73–75, 73*f*
fungi compared with, 297*t*
gene expression in, regulation of, 203–207

glycocalyx of, 72–73
gram-negative. *See* Gram-negative bacteria
gram-positive. *See* Gram-positive bacteria
greenhouse effect and, 675
growth of. *See* Bacterial cultures, growth of
hydrothermal, 143
inclusions in, 73*f*, 87
lower respiratory system diseases caused by, 595–604, 611*t*
luminescent, 683
magnetotactic, 76
meningitis caused by, 541–543, 555*t*
diagnosis and treatment of, 542–543
nervous system diseases caused by, 541–547, 555*t*
nitrifying, carboxysomes in, 87
nitrogen fixing, 678–680
nonsymbiotic nitrogen-fixing, 678–679
nuclear area of, 73*f*, 86
oral diseases caused by, 618–622
phototropic, 275*t*, 291–293
pili of, 76–77
plasma (cytoplasmic) membrane of, 73*f*, 81–83
pleomorphic, 71
pneumonia caused by, 599–601, 611*t*. *See also specific type*
respiratory system diseases caused by, 592–594, 595–604, 610*t*, 611*t*
ribosomes in, 73*f*, 86–87
size/shape/arrangement of, 71–72
skin diseases caused by, 520–524, 531*t*
in soil, 674
structure of, 73*f*
"sulfur," 87
symbiotic nitrogen-fixing, 679–680
transduction in, 218
transformation in, 214–216
classification and, 267–268
upper respiratory system diseases caused by, 592–594, 610*t*
urinary tract infections caused by, 653–654, 665*t*
viruses compared with, 333*t*
Bacterial cultures, growth of, 155–163
death phase of, 157*f*, 158
indirect methods for estimation of, 162–163
lag phase of, 157
log phase of, 157
logarithmic representation and, 156–157
measurement of, 158–162
phases of, 157–158
stationary phase of, 157–158
preserving, 155
Bacterial endocarditis, 563, 580*t*
Bacterial growth curve, 157–158
logarithmic vs. arithmetic representation of, 156–157
Bacterial kinases, virulence and, 396
Bacterial meningitis, 541–543, 555*t*
diagnosis and treatment of, 542–543
Bacterial pneumonias, 599–601, 611*t*. *See also specific type*
Bacterial predator, *Bdellovibrio* as, 62
Bacterial species, 257
Bacterial viruses, 334
Bactericidal agents, 496
Bacteriochlorophylls, 125, 129
Bacteriocins, 220–221
enterics producing, 280
Bacteriophage f2, size of, 334*f*
Bacteriophage λ, lysogeny for, 347–348
Bacteriophage M13, size of, 334*f*
Bacteriophage MS2, size of, 334*f*
Bacteriophage T4, size of, 334*f*
Bacteriophages
growth of in laboratory, 341
lysogenic, 347
multiplication of, 345–348
temperate, 347
transduction by, 218
Bacteriorhodopsin, 128
Bacteriostasis, 168*t*, 496
Bacteriostatic agents, 496
Bacteroides spp, 274*t*, 282, 283*f*
bites infected with, 567
human, 568
in digestive system, 618
and postabortion infections, 563, 580*t*
septicemia caused by, 562
in urethra, 652
in vaginitis, 662
Bacteroides hypermegas, 283*f*
Bacteroids, in nitrogen fixation, 680
Bakers' yeast. *See also Saccharomyces cerevisiae*
budding of, 299*f*
gene products expressed by, 235
industrial production of, 717–718
Balantidiasis (balantidial dysentery), 316, 638, 641*t*
Balantidium coli, 316, 318*t*, 638, 641*t*
Basal body, of bacterial flagella, 73, 74*f*
Base analogs, as chemical mutagens, 209–210
Base composition of nucleic acids, and classification/identification of microorganisms, 265, 267*t*
Base sequence, and classification/identification of microorganisms, 265
Base substitution, 207, 208*f*
Bases, 33
Basic dyes, 60
Basic solutions, 34
Basidiomycota, 302–303*t*, 304, 305*f*
basidiospores produced by, 301
Basidiospores, 301, 304, 305*f*
Basidium, 301, 304
Basophils, 411*f*, 411*t*, 412
in hypersensitivity reactions, 468
BCG vaccine, 598
Bdellovibrio spp, 62, 254, 274*t*, 276–277, 673
Beef tapeworm (*Taenia saginata*), 321, 325*t*, 639, 641*t*
transmission of, 379*t*, 641*t*
Beer, microorganisms in production of, 710, 711*t*
Beggiatoa spp, 130, 275*t*, 290
Beijerinckia, in nitrogen fixation, 679
Benthic zone, microbial flora in, 682*f*, 683
Benzalkonium chloride (Zephiran), antimicrobial activity of, 180, 182, 184*t*
effectiveness and, 183*f*
Benzathine penicillin, 503
for syphilis, 661–662, 665*t*
Benzoic acid (sodium benzoate)
antimicrobial activity of, 182, 185*t*
in food preservation, 705, 706
Benzoyl peroxide
for acne, 524
antimicrobial activity of, 185
Benzpyrene, as frameshift mutagen, 210
Bergey's Manual of Determinative Bacteriology, 257–258
Bergey's Manual of Systematic Bacteriology, 257, 258, 273
Bestatin, for localized juvenile periodontitis, 410
Betadine. *See* Povidone-iodines
Beta-hemolytic streptococci, 286, 522. *See also specific type and Streptococcus* spp
in glomerulonephritis, 654
in pharyngitis, 592
in puerperal sepsis, 562, 580*t*
in rheumatic fever, 563, 580*t*
Beta interferon, 422
Beta oxidation, 122–124
BGH. *See* Bovine growth hormone
Bidirectionality, in bacterial DNA replication, 196–197
Bifidobacterium spp, PAGE patterns of, 262*f*
Bifidobacterium globosum, PAGE patterns of, 262*f*
Bifidobacterium pseudolongum, PAGE patterns of, 262*f*
Binary fission, bacterial reproduction by, 15, 70, 155–156
Binomial nomenclature, 254
Bioaugmentation, 38
Biochemical oxygen demand of sewage, 688–689
Biochemical pathway, 111
Biochemical tests
and classification/identification of microorganisms, 259–260, 267*t*
for enterics, 280
Bioconversion, 718
Biodegradation of chemicals
in soil, 17, 38, 229*f*, 240, 681
in water, 684–685
Biogenesis, 7–8
Biogeochemical cycles, 676–681
Biological leaching of copper, by *Thiobacillus ferrooxidans*, 715, 717, 718*f*
Biological transmission, in spread of infection by vectors, 379
Bioluminescent organisms, 683
Biomass, 674
conversion of into alternative fuel sources, 718
Bioreactors, for industrial fermentations, 712*f*, 713
Bioremediation, in toxic dump cleanup, 17, 38, 229*f*, 240, 681
Biosynthetic reactions, 102. *See also* Anabolism
Biotechnology, 18, 227–247. *See also* Genetic engineering
definition of, 228
Biotin, function of, 104*t*
Biovars (biotypes)
for *Salmonella*, 280
for *Vibrio*, 626
Bird's nest fungus, 305*f*
Bismuth subsalicylate, for prevention of traveler's diarrhea, 628
Bismuth sulfite agar, 152

Bites
human, 568
rat, 567–568, 580*t*
systemic diseases caused by, 567–568
Black bread mold (*Rhizopus nigricans*), 304
and scientific nomenclature, 254–255
Black Death, 568. *See also* Plague
Bladder, 650, 651*f*
bacterial infection of, 653, 665*t*
Blastomyces spp, reclassification of, 303*t*
Blastomyces dermatitidis, 302*t*, 304, 609, 611*t*
Blastomycosis, 609, 611*t*
Blastospores, 300
Blattabacterium spp, 274*t*
Bleach, as disinfectant, 179
Blindness, trachoma causing, 534
Blocking antibodies, in desensitization, 471
Blood, 560. *See also* Cardiovascular system
formed elements in, 411–413
Blood agar, 152
Blood poisoning (septicemia), 374, 562
Blood types
ABO system, 471, 472*f*
Rh system, 471–473
hemolytic disease of the newborn and, 472–473
transfusions and, 472
Blood vessels, 560. *See also* Cardiovascular system
increased permeability of in inflammation, 415–416
Blood–brain barrier, 539–540
Blooms
algal, 310, 685
of cyanobacteria, 685
BOD. *See* Biochemical oxygen demand
Body odor, and normal flora of skin, 520
Body temperature, pathogens affecting, 417–418
Boil (furuncle), 521
Boiling, for sterilization, 171, 174*t*
Boletus edulis, 305*f*
Bonds, chemical, 24–29
Bordet-Gengou medium, in whooping cough diagnosis, 596
Bordetella spp, 274*t*, 279
Bordetella pertussis, 279, 394*t*, 595, 611*t*. *See also* Pertussis
respiratory tract as portal of entry for, 394*t*
Borrelia spp, 274*t*, 276, 569, 580*t*
Borrelia burgdorferi, 4, 569, 581*t*. *See also* Borreliosis axial filaments of, 75
Borreliosis, Lyme (Lyme disease), 4, 276, 569–570, 581*t*. *See also* *Borrelia burgdorferi*
arthropod vector for, 4, 276, 379*t*, 569, 581*t*
life cycle of, 571*f*
transmission of, 4, 276, 378*t*, 569, 581*t*
Botulinum toxin, 399, 545
Botulism, 287, 399, 400*t*, 544–546, 555*t*, 706, 707*t*. *See also* *Clostridium botulinum*
food preservation and, 701–702
freezing for, 704
infant, 545
wound, 546
Bovine growth hormone, genetic engineering in production of, 237*t*, 243
Bovine spongiform encephalopathy, 556
Bovine tuberculosis, 598
Bradyrhizobium spp, 279
industrial microbiology in production of, 718
plant growth and, 679–680
and symbiosis in nitrogen fixation, 145, 679–680
Brandy, microorganisms in production of, 711*t*
Branhamella catarrhalis, otitis media caused by, 594
Bread, microorganisms in production of, 5, 709*t*, 710
Bread mold, black (*Rhizopus nigricans*), 304
and scientific nomenclature, 254–255
Breakbone fever, 575
Breed count method, for direct microscopic count of bacteria, 160
Brightfield microscopy, 55, 56*f*, 61*t*
Bright's disease (glomerulonephritis), 475, 654, 665*t*
Brilliant Green agar, 152
Brome grass mosaic virus, 361*t*
5-Bromouracil, as chemical mutagen, 209–210
Bronchiolitis, 595
Bronchitis, 595
Hemophilus influenzae causing, 282
Broth dilution tests, 510
Brown algae, 308*f*, 310
characteristics of, 309*t*
Brucella spp, 274*t*, 279, 565, 580*t*
gastrointestinal tract as portal of entry for, 394*t*
Brucella abortus, 565
Brucella melitensis, 565, 566
Brucella suis, 565, 566
Brucellosis (undulant fever), 565–566, 580*t*
transmission of, 378*t*, 580*t*
Bruton's agammaglobulinemia, 481*t*
"Bubble boy," and EB virus causing cancer, 359, 372
Buboes, 568, 569*f*
Bubonic plague, 568, 569*f*
transmission of, 378*t*
Budding
of bacteria, 156, 275*t*, 290
of enveloped virus, 355, 356*f*
of yeasts, 298
Buffers, in growth medium, 144
Bulking, and sewage treatment, 690
Burkitt's lymphoma, 358, 573–574
Burst size, 347
Burst time, 347
Butter, microorganisms in production of, 708
Buttermilk, microorganisms in production of, 708
Butyric anaerobic food spoilage, 703*t*
Byssochlamys fulva, food spoilage and, 703

C

C. *See* Carbon
C_α. *See* Alpha-carbon
C1, 418, 420*f*
deficiencies of, 421
C2, 418, 420*f*
deficiencies of, 421
C2a, 418
C2b, 418, 420*f*
C3, 418, 420*f*
deficiency of, 421
C3a, 418, 421
C3b, 418, 420*f*, 421
C4, 418
deficiencies of, 421
C4a, 418
C4b, 418, 420*f*
C5-C9, 418, 420*f*
C5a, 421
C genes, and diversity of antigen receptors, 436, 437*f*
C regions (constant regions), 431
Ca. *See* Calcium
Cachectin. *See* Tumor necrosis factor
Calcium, 23*t*
in soil, 673
Calcium hypochlorite, antimicrobial activity of, 179
Calcium ion, 26
Calcium propionate
antimicrobial activity of, 182, 185*t*
in food preservation, 705, 706
Calculus, dental, 619
California encephalitis, 552, 553*t*
Calvin–Benson cycle, 125, 127*f*
Campylobacter spp, 274*t*
gastroenteritis caused by, 628–629, 640*t*
water pollution in transmission of, 684*t*
Campylobacter fetus, 276
Campylobacter jejuni, 276, 629, 640*t*
Campylobacter pylori. *See Helicobacter pylori*
Cancer (malignant disease)
EB virus causing, 358–359, 574
genital herpes and, 359
genital warts and, 664
HLA antigens in, 479*t*
immune response to, 486–488
immunologic treatment of, 487–488
interferons for treatment of, 422–423
papovaviruses causing, 351
types of, 487*t*
viruses and, 356–359
Candida spp. *See also specific type* and Candidiasis
in food production, 709*t*
reclassification of, 303*t*
Candida albicans, 302*t*, 664, 665*t*. *See also* *Candida* spp
candidiasis caused by, 306, 531*t*, 532*f*, 664–666, 665*t*, 666*t*
chlamydospores produced by, 300
skin infections caused by, 531*t*, 533
urinary tract infections caused by, 653
Candida krusei, in food production, 709*t*
Candidiasis, 306, 531*t*, 532*f*, 533, 664–666, 665*t*, 666*t*
oral (thrush), 306, 533, 664
vulvovaginal, 664–666, 665*t*, 666*t*
Candle jars, 151, 152*f*
Canning, for food preservation, 701–703
spoilage and, 702–703
temperatures for, 704*f*
Capillaries, 560, 561*f*. *See also* Cardiovascular system
Capsid, 335
assembly of in viral maturation, 355
of naked virus, 336

Capsomeres, 335
Capsules, bacterial, 72–73, 73*f*
  and adherence in phagocytosis, 414
  negative staining of, 64*f*, 65, 72
  and penetration of host defenses, 395–396, 404*f*
Carbapenems, 500*t*, 504
Carbenicillin, 503
  structure of, 502*f*
Carbohydrates, 36–37. *See also specific type*
  catabolism of, 111–122. *See also specific aspect*
  in soil, 673
Carbolfuchsin
  for flagella staining, 66
  as simple stain, 63, 65*t*
Carbolic acid. *See* Phenol
Carbon, 23*t*
  bonding patterns of, 35
  covalent bonding and, 27, 28*f*
  electronic configuration for, 25*t*
  for microbial growth, 145
Carbon cycle, 17, 676–677
Carbon dioxide, fermentation of, 123*t*
Carbon dioxide fertilization, greenhouse effect and, 675
Carbon dioxide fixation, 676
Carbon dioxide-generating packets, 151–152
Carbon dioxide incubators, 151
Carbon skeleton, 35
Carbon synthesis, 125
Carboxyl group, 35, 36*t*
Carboxysomes, 87
Carbuncle, 521
Carcinogens, 212–214
  chemical, identification of, 212–214
Carcinoma, 487*t*
Cardiolipin, in syphilis diagnosis, 661
Cardiovascular system
  diseases of, 560–588. *See also specific type*
    bacterial, 562–573, 580–581*t*
    helminthic, 581–583, 581*t*
    protozoan, 575–581, 581*t*
    viral, 573–575, 581*t*
  structure and function of, 560–561
Caries, dental (tooth decay), 618–621, 640*t*
  adhesion in, 393–395
Cariogenic bacteria, 619
Carpenteles spp, and reclassification of fungi, 303*t*
Carrageen, from red algae, 310
Carrier protein, in facilitated diffusion, 83
Carriers, 374
Cascade, for complement activation, 418
Case reporting, in epidemiology, 386
Casein, in curd, 707
Caseous lesion, in tuberculosis, 597
Castanospermine, for AIDS, 485
Cat-scratch fever, transmission of, 378*t*
Catabolism (catabolic reactions), 30, 101–102
  anabolism related to, 102*f*
  carbohydrate, 111–122
  lipid, 122–124
  protein, 124
Catabolitic repression, 206
Catalase, 147
Catalysts, 31, 102
Catarrhal stage, of whooping cough, 596
Cationic detergents. *See* Quaternary ammonium compounds
Cations, 26
Cats, in *Toxoplasma gondii* life cycle, 575, 576*f*, 581*t*
Cauliflower mosaic virus, 361*t*
Caulobacter spp, 275*t*, 289, 683
  in fresh water, 683
CD4, soluble, for AIDS, 485, 486
CD4 T cells
  in AIDS, 439, 480–481
  fluorescence-activated cell sorter in identification of, 461
CD8 T cells, fluorescence-activated cell sorter in identification of, 461
CDC. *See* Centers for Disease Control
cDNA. *See* Complementary DNA
Cefamandole, 504
Cefotaxime, 504
Ceftriaxone
  for gonorrhea, 658
  for typhoid fever, 625
Cell clone. *See* Clone
Cell counters, electronic, for direct microscopic count of bacteria, 162
Cell cultures. *See* Cultures
Cell division
  bacterial, 156
  in procaryotic vs. eucaryotic cells, 96*t*
Cell lysis, teichoic acids affecting, 79
Cell wall
  in archaeobacteria, 253*t*
  bacterial, 73*f*, 77–81
    antimicrobial drugs affecting, 496, 497*f*
    atypical, 79–80
    composition and characteristics of, 77–79
    damage to, 80–81
    and penetration of host defenses, 396, 404*f*
    structures external to, 72–77
    structures internal to, 81–89
  in eubacteria, 253*t*
  eucaryotic, 89*f*, 90, 96*t*, 253*t*
  gram-negative, 78*f*
  gram-positive, 78*f*
  in procaryotic vs. eucaryotic cells, 96*t*, 253*t*
Cell-mediated cytotoxicity, 442*f*
Cell-mediated hypersensitivity reactions, 468*t*, 475–478
  autoimmune, 476
  causes of, 475
  major histocompatibility complex and, 476–478
  of skin, 475–476
  tolerance and, 476
Cell-mediated immunity (cell-mediated immune system), 429, 439–443
Cellular slime molds, 312
Cellulase, commercial production of, 716*t*
  genetic engineering in, 237*t*
Cellulose, 37
  in algal cell wall, 90
Cementum, caries formation on, 622
Centers for Disease Control, 386–387
Central nervous system, 539, 540*f*. *See also* Nervous system
Centrioles, 89*f*, 95
Cepacol. *See* Cetylpyridinium chloride
Cephalosporins, 500*t*, 504. *See also specific agent*
  cell wall synthesis affected by, 496
  for gonorrhea, 658, 665*t*
  for *Klebsiella* pneumonia, 600, 611*t*
  for pyelonephritis, 653, 665*t*
  structure of, 504*f*
  for typhoid fever, 625
Cephalosporium, antibiotics produced by, 494*t*
Cephalothin, 504
  sources of, 494*t*
Cercaria, 320, 583
  swimmer's itch caused by, 583
Cerebrospinal fluid, 539
Cervical cancer
  genital herpes and, 359
  genital warts and, 664
Cestodes, 320–321, 325*t*. *See also* Tapeworms
Cetylpyridinium chloride (Cepacol), antimicrobial activity of, 182, 184*t*
CFS. *See* Chronic fatigue syndrome
Chagas' disease, 316, 577, 581*t*
  transmission of, 378*t*, 581*t*
Chain of transmission. *See* Transmission, of disease
Champagne, microorganisms in production of, 711*t*
Chancre
  soft, 663, 665*t*
  syphilitic, 660
Chancroid, 663, 665*t*
Cheese, microorganisms in production of, 707–708, 709*t*
Chemical bonds, 24–29
Chemical carcinogens, identification of, 212–214
Chemical elements, 23–24
Chemical energy, 29
Chemical mutagens, 208–210
Chemical packets, carbon dioxide generating, 151–152
Chemical pollution, of water, 684–685
Chemical preservatives, in food, 705–706
Chemical reactions, 29–31
  collision theory of, 30
  decomposition, 29–30
  energy of, 29
  enzymes and, 31
  exchange, 30
  factors affecting, 30
  mechanism of, 30–31
  reversibility of, 30
Chemically defined media, 148*t*, 149, 153*t*
Chemicals, degradation of in soil, 17, 38, 229*f*, 240, 681
  in water, 684–685
Chemiosmosis, 110, 117–118
Chemistry, 22
  basic principles of, 22–51
    atomic structure, 22–24
    bonds, 24–29
    inorganic compounds and, 32–34
    molecule formation, 24–29
    organic compounds and, 34–47
    reactions, 29–31
Chemoautotrophs, 126, 129*t*, 130, 275*t*, 290–291
  at hydrothermal vents, 143
Chemoheterotrophs, 126, 129*t*, 130
  in carbon cycle, 676
Chemostat, 158
Chemosterilizers, gaseous, 183, 185*t*

Chemotactic agents, in hypersensitivity reactions, 468
Chemotaxis, 75, 414
 depressed, and neutrophil dysfunction, 410
 phagocyte migration and, 416
Chemotherapy, 11, 492. *See also* Drugs; Antimicrobial agents
 antibacterial, 499–502
 antibiotic, 502–506
 antifungal, 502, 506–507
 antihelminthic, 509
 antiprotozoan, 509
 antiviral, 507–509
 birth of, 11–13, 492–494
 effectiveness of agents in, 510–511
 future of, 511
 microbial susceptibility tests and, 509–510
Chemotrophs, 126
Chickenpox, 525–526, 531*t*
 and latent infections, 359, 526
Childbirth fever (puerperal fever, puerperal sepsis), 562–563, 580*t*
Chimeric monoclonal antibodies, 439
Chitin, in fungal cell wall, 15, 90
Chlamydia spp, 274*t*, 283–284
 pneumonia caused by, 603, 611*t*
 sexually transmitted diseases caused by
  incidence of, 655*f*
  pelvic inflammatory disease and, 656
 size of, 334*f*
Chlamydia pneumoniae, 285, 603, 611*t*
Chlamydia psittaci, 284–285, 284*f*, 601–602, 611*t*
 TWAR strain of, 603
Chlamydia trachomatis, 284
 eye infection caused by, 533–534, 534*t*, 658
 lymphogranuloma venereum caused by, 284, 662–663, 665*t*
 nongonococcal urethritis caused by, 284, 658–659, 665*t*
Chlamydomonas, 308*f*
Chlamydospores, 300
Chloramines, antimicrobial activity of, 179–180
Chloramphenicol, 500*t*, 505
 for endemic murine typhus, 573, 580*t*
 for epidemic typhus, 573, 580*t*
 protein synthesis affected by, 497*f*, 498
 for Rocky Mountain spotted fever, 573, 581*t*
 sources of, 494*t*
 for typhoid fever, 626
Chlorhexidine, antimicrobial activity of, 178–179, 184*t*
Chloride ion, 26
Chlorine, 23*t*
 as disinfectant, 179, 184*t*
Chlorobium spp, 275*t*
 in nitrogen fixation, 679
 photosynthesis in, 126
Chlorobium vesicles (chlorosomes), 129
Chloroflexus, 130
Chlorophyll
 algae colored by, 307
 in chloroplasts, 95
 in green algae, 310
 in photophosphorylation, 125
 photosynthesis without, 128
Chloroplasts, 89*f*, 95
 classification of, 253*f*
 procaryotic cells compared with, 251*t*
Chloroquine, 501*t*, 509
 for malaria, 581
Chlorosomes (chlorobium vesicles), 129
Chlortetracycline, 504
 sources of, 494*t*
Cholera, 281, 399, 400*t*, 626–627, 640*t*. *See also Vibrio cholerae*
 vaccine for, 450*t*, 627
 water pollution in transmission of, 626–627, 640*t*, 684
Choleragen, 399
Cholesterol, structure of, 40*f*
Chorea, Sydenham's, 564
Choriomeningitis, lymphocytic, 360*t*, 476
Chromatin, 93
Chromatium spp, 275*t*, 291*f*
 photosynthesis in, 129
Chromatophores (thylakoids), 82, 83*f*, 89*f*, 95
Chromium, 23*t*
Chromosome number, in eucaryotic organism, 218
Chromosomes, 93, 190
 bacterial (procaryotic), 86, 93, 192
 DNA and, 191–192
 eucaryotic, 192
 genetic transfer and recombination and, 214–222
 in procaryotic vs. eucaryotic cells, 96*t*
 viruses damaging, 356
Chronic disease, 374
Chronic fatigue syndrome, 574
Chronically mediated resistance, 657
Chroococcus spp, 275*t*
Chrysops, 327*f*
Cidex. *See* Glutaraldehyde
CIE (counter-immunoelectrophoresis). *See* Countercurrent immunoelectrophoresis test
Ciguatera, 307, 707*t*
Cilastin sodium, with imipenem, 504
Cilia
 eucaryotic, 90, 91*f*
 protozoan, 15
Ciliary escalator, in host defense, 409
Ciliata (ciliates), 313, 314*f*, 316, 318*t*
 conjugation of, 314
Cisternae, 93, 94*f*
Citric acid
 fermentation of, 123*t*
 industrial microbiology in production of, 716
Citric acid cycle, 114. *See also* Krebs cycle
Citrobacter spp, biochemical tests for, 259
CJD. *See* Creutzfeldt–Jakob disease
Cl. *See* Chlorine
Cladosporium werneckii, 302*t*
Class, 254*f*, 256
 in Procaryotae, 257*t*
Classical pathway for complement activation, 418, 419*f*
Classification of microorganisms, 250–272. *See also* Microorganisms, classification/identification of
Claviceps purpurea, 403, 637, 641*t*
Clavulanic acid, penicillin combined with, 503–504
Clofazimine, for leprosy, 547
Clonal anergy, 443
Clonal deletion, 443
Clonal selection, 443, 444*f*
Clone, 153, 228. *See also* Genetic engineering
 forbidden, and loss of tolerance, 476
 in gene library, 233–234
  identification of, 234–235
 pure, obtaining, 153–155
Cloning vectors, 228
 plasmids for, 231–232
Clostridium spp, 275*t*, 287. *See also specific type*
 bites infected with, 567
 collagenase produced by, 396–397
 endospores formed by, 88
 food spoilage caused by
  freezing and, 704
  thermophilic anaerobic, 702, 703*t*
 in fresh water, 683
 in gangrene, 566–567
 hyaluronidase produced by, 396
 in nitrogen fixation, 679
 and postabortion infections, 563, 580*t*
Clostridium botulinum, 287, 544. *See also* Botulism; *Clostridium* spp
 commercial sterilization for destruction of, 701–702
 exotoxins produced by, 399, 400*t*
 nervous system disease caused by, 555*t*
 prophage gene coding for, 348
 in soil, 674
Clostridium pasteurianum, in nitrogen fixation, 679
Clostridium perfringens, 287. *See also Clostridium* spp; Gas gangrene
 animal growth affected by, and antibiotics in animal feed, 495
 exotoxins produced by, 400*t*
 in gangrene, 567, 580*t*
 gastroenteritis caused by, 629, 640*t*
 hemolysins produced by, 396
 as phosphoprotein, 46
 portal of entry for, 394*t*
 in soil, 674
Clostridium sporogenes, putrefactive anaerobic food spoilage caused by, 703*t*
Clostridium tetani, 287, 543, 555*t*. *See also Clostridium* spp; Tetanus
 portal of entry for, 394*t*
 in soil, 674
 toxin produced by, 399, 400*t*
Clostridium thermosaccharolyticum, thermophilic anaerobic spoilage caused by, 703*t*
Clotrimazole, 501*t*, 507
 for tinea infections, 530
 for vulvovaginal candidiasis, 665*t*, 666
Club fungi (Basidiomycota), 302–303*t*, 304
Clue cells, in vaginitis, 662
CMRNG. *See* Chronically mediated resistance
CMV. *See* Cytomegalovirus
Co. *See* Cobalt
CoA. *See* Coenzyme A
Coagulase-negative staphylococci, skin infections caused by, 520
Coagulase-positive staphylococci
 food poisoning and, 623
 skin infections caused by, 520

Coagulases
  plasmid gene encoding, 401
  virulence and, 396
Coal tar derivatives. *See also* Phenolics
  antimicrobial activity of, 177–178
Coal-mining wastes, and water pollution, 685
Cobalt, 23*t*
Coccidioides immitis, 302*t*, 608, 611*t*
  arthrospores produced by, 300
  respiratory tract as portal of entry for, 394*t*
Coccidioidomycosis, 608–609, 611*t*
  vaccine for, 608
Coccobacillus (coccobacilli), 71, 279
Coccus (cocci), 15, 71, 274*t*, 275*t*
  cell arrangement of, 70*f*
  gram-negative aerobic, 274*t*, 277–280
  gram-negative anaerobic, 274*t*, 283
  gram-positive, 274*t*, 285–286
  gram-positive endospore-forming, 275*t*, 286–287
Cocoa beans, microorganisms in production of, 709*t*
Codons, 199–201
Coenocytic hyphae, 297, 298*f*
Coenzyme A, 103–104
  acetyl, in Krebs cycle, 114–116
Coenzyme Q, 116
Coenzymes, 103–104
Cofactors, 103, 104
Coffee beans, microorganisms in production of, 709*t*
Cohort groups, 386
Cohort method, in epidemiology, 386
Cold, common, 594–595, 611*t*
  adenoviruses causing, 350
Cold sores, 350, 359, 526. *See also* Herpes simplex viruses, type 1
  as latent infection, 359, 527
Coliform bacteria, 686
  fecal, 686
  testing water for, 686, 687*t*
Colitis, hemorrhagic, 628
Collagen vascular disorders, complement deficiencies causing, 421
Collagenase, virulence and, 396–397
Collision theory, 30
Colony, 153
  pure, obtaining, 153–155
Colony-stimulating factor, genetic engineering in production of, 236, 237*t*
Colostrum, antibodies passed in, 427
Cometabolism, 673
Commensalism, 370, 673
Commercial sterilization, for food preservation, 701–702
Common cold, 594–595, 611*t*
  adenoviruses causing, 350
Common variable hypogammaglobulinemia, 481*t*
Competence, in genetic transformation, 216
Competitive inhibitors, enzyme, 107
  antimicrobial agents and, 498
Complement, 418–421
  activation of
    consequences of, 418–421
    pathways of, 418, 419*f*
  components of, 418
  deficiencies of, 421
Complement fixation, 421, 457
Complement fixation reactions, 457–458
Complement fixation test, 458, 459*f*
Complementary DNA, 47, 191
  in gene library clones, 233–234
Complex lipids, 39–40
Complex media, 150, 153*t*
Complex viruses, 337
  morphology of, 339*f*
Composting, 694–695
Compound light microscope, 53–55
  images produced by, 56*f*
Compounds, 24
Compromised host, in spread of infection, 382
Concentration gradient, 83
Condensation, 36
Condenser, for compound light microscope, 53, 54*f*
Congenital rubella syndrome, 528–529
Congenital syphilis, 661
Conidiophore, 300
Conidiospore, 293, 300, 304, 305*f*
Conjugated proteins, 44–46
Conjugated vaccines, 452
Conjugation
  in bacteria, 216–217
  in protozoans, 314–315
Conjugation fungi (Zygomycota), 302–303*t*, 304
Conjunctiva, as portal of entry, 392, 404*f*
Conjunctivitis, 533
  inclusion, 533, 534*t*
Constant regions, 431
Constitutive enzymes, 203
Contact dermatitis, allergic, 475–476
Contact inhibition, loss of in transformed cells, 358, 403
Contact transmission, in spread of infection, 376–377, 589
Continuous cell lines, 343
Convalescence, 384
  and whooping cough, 596
Copper, 23*t*
  antimicrobial activity of, 180–181, 184*t*
  industrial microbiology in production of, 715, 717, 718*f*
Copper sulfate, antimicrobial activity of, 181, 184*t*
Corepressor, 206
Coronaviruses, 338*t*
  common cold caused by, 594, 611*t*
Cortex, of lichen thallus, 311
Corticosteroids, for trichinosis, 641*t*, 643
Corticosterone, microbial production of, 717
Corynebacteria. *See Corynebacterium* spp
Corynebacterium spp, 275*t*, 288
  in vagina, 652
Corynebacterium diphtheriae, 288, 593, 610*t*
  cutaneous diphtheria caused by, 594, 610*t*
  metachromatic granules in, 87
  respiratory tract as portal of entry for, 394*t*
  temperate phages and, 347–348
  toxin produced by, 397–399, 400*t*
Corynebacterium glutamicum, in industrial production of amino acids, 714–716
Corynebacterium xerosis, 288*f*
  on skin surface, 520
Coryza, acute (common cold), 594–595, 611*t*
  adenoviruses causing, 350
Countercurrent immunoelectrophoresis test (counter-immunoelectrophoresis), 456
Counterstains, 64
Covalent bonds, 27, 28*f*
  antimicrobial agents affecting, 170
Cowpox, 350, 450
  smallpox prevented by vaccination with, 450
Cowpox virus, 338*t*
Coxiella spp, 274*t*
Coxiella burnetii, 283, 603, 611*t*. *See also* Q fever
  endosporelike structures formed by, 88
Coxsackievirus, myocarditis caused by, 563, 581*t*
CPE. *See* Cytopathic effect of viruses
Cr. *See* Chromium
Cramps, abdominal, in digestive system infections and intoxications, 622
Cresols. *See also* Phenolics
  antimicrobial activity of, 177–178
Creutzfeldt–Jakob disease, 555*t*, 556
Crevicular fluid, tooth decay and, 620
Cristae, 93, 94*f*
Crop production, symbiotic nitrogen-fixing bacteria and, 679
Crossing over, 214
Crown gall disease, 241, 279
Crustose lichens, 311
Cryphonectria parasitica, 306
Cryptococcosis, 553, 555*t*
Cryptococcus spp, reclassification of, 303*t*
Cryptococcus neoformans, 302*t*, 403, 553, 555*t*
  meningitis (cryptococcosis) caused by, 553, 555*t*
  pathogenetic properties of, 403
Cryptosporidiosis, 638–639, 641*t*
Cryptosporidium spp, 318–319, 638, 641*t*
  characteristics of, 318*t*
  water pollution in transmission of, 684*t*
Crystal violet, as simple stain, 63, 65*t*
CSF. *See* Colony-stimulating factor
Cu. *See* Copper
Culex, 326*t*, 553*t*
  diseases transmitted by, 379*t*
Culiseta, in encephalitis, 553*t*
Culture media, 148–155
  for anaerobic growth, 150–151
  chemically defined, 148*t*, 149, 153*t*
  complex, 150, 153*t*
  differential, 152–153, 153*t*, 154*f*
  enrichment, 153, 153*t*
  purposes of, 153*t*
  reducing, 150, 153*t*
  selective, 152–153, 153*t*, 154*f*
  special techniques and, 151–152
Cultures, 148
  bacterial
    growth of, 155–163
    preserving, 155
  enrichment, 153
  media for. *See* Culture media
  pure, obtaining, 153–155
  for virus growth in laboratory, 342–343
Curd, in cheese production, 707
Cutaneous diphtheria, 594, 610*t*
Cutaneous infection, nosocomial, 382*t*
Cutaneous mycoses, 306, 530–532

Cyanobacteria, 275*t*, 292*f*, 293
blooms of, 685
carboxysomes in, 87
classification of, 253*f*
gas vacuoles in, 87, 293
in lichens, 310–312, 680
in nitrogen cycle, 17, 679, 680
in oxygen cycle, 17
photosynthesis in, 129*t*
phylogenetic relationships of, 258*f*
in soil, 674
Cyanocobalamin (vitamin B_{12})
commercial production of, 717
function of, 104*t*
Cyclic AMP
amoeboid aggregation and, 312
lac operon regulation and, 206
Cyclic photophosphorylation, 125, 126*f*
Cyclosporine, for immunosuppression in transplantation, 480
Cysteine, 42*t*
Cystic acne, 410, 523. *See also* Acne
Cysticerci, 321
Cysticercosis, 639
Cystitis, 653, 665*t*
Cysts
hydatid, 321, 322*f*, 641*t*, 642
transmission of, 379*t*, 641*t*
protozoans forming, 315
Cytochromes, 116
PAGE analysis of, 263
Cytokines, 441
Cytolysis, 418–421
Cytomegalovirus, 630–631, 640*t*
cytopathic effects of, 403*t*
transmission of to fetus, 631, 640*t*
vaccine for, 631
Cytomegalovirus inclusion disease, 630–631, 640*t*
Cytometry, flow
classification and, 267
and fluorescence-activated cell sorter, 461
Cytopathic effect of viruses, 343, 355, 402–403
Cytophaga spp, 275*t*, 290
in fresh water, 683
Cytoplasm
bacterial, 73*f*, 86
eucaryotic, 89*f*, 91–92
in procaryotic vs. eucaryotic cells, 96*t*
Cytoplasmic membrane (plasma membrane)
antimicrobial agents affecting, 169
bacterial, 73*f*, 81–83
antimicrobials in destruction of, 83, 498
functions of, 81–83
structure of, 81
eucaryotic, 89*f*, 90–91
in procaryotic vs. eucaryotic cells, 96*t*
Cytoplasmic streaming, 92, 313
Cytosine, 190–191
Cytoskeleton, 91
Cytostome, 309, 314
Cytotoxic reactions, 468*t*, 471–474
autoimmune, 474
drug-induced, 473–474
Cytotoxic T cells, 441
function of, 443*t*
tumor cell interaction and, 487*f*
Cytotoxicity, cell-mediated, 442*f*, 443
Cytotoxins, 397
diseases caused by, 400*t*
Streptococcus pyogenes producing, 399, 400*t*
tracheal, in whooping cough, 595

D

2,4-D, degradation of in soil, 681
D segments, and diversity of antigen receptors, 436, 437*f*
12D treatment, in food preservation, 702
D value. *See* Decimal reduction time
Dairy products
foodborne infections caused by, 706
microorganisms in production of, 708
Dandruff, *Pityrosporum* causing, 520
Dane particle, 633
Dapsone, for leprosy, 547
Daptobacter, 254
Dark reactions (light-independent reactions), in photosynthesis, 125
Darkfield microscope, 55–57, 61*t*
Darwin's theory of evolution, 250–251
ddA. *See* Dideoxyadenosine
ddc. *See* Dideoxycytidine
ddI. *See* Dideoxyinosine
DDT, degradation of in soil, 681
Deamination, 124
Death curve, microbial, 170
Death phase, 158
Death point, thermal, 170
Death rate, microbial, 170
Death time, thermal, 170
Decarboxylation, 114, 124
12-Decimal reductions treatment, in food preservation, 702
Decimal reduction time, 170
Decimeter, 52, 53*t*
Decline, period of, 384
Decolorizing agent, 63
Decomposition reactions, 29–30
Deep test tubes for microbial growth, 149
for anaerobic bacteria, 151
Deep-freezing, for bacterial culture preservation, 155, 174, 175*t*
Deer fly, 327*f*
Defenses, of host. *See* Host defenses
Definitive host
for helminths, 319
for tapeworms, humans as, 321
Degerming, 168*t*
alcohols for, 180
soap for, 181, 184*t*
Degradation of chemicals
in soil, 17, 38, 229*f*, 240, 681
in water, 684–685
Degradative reactions, 101. *See also* Catabolism
Degranulation, in hypersensitivity reactions, 468
Dehydration synthesis, 36, 37*f*
in anabolic processes, 102
peptide bond formation by, 43, 44*f*
Dehydrogenation, 114, 124
Dehydrogenation reactions, 109
Delayed-hypersensitivity T cells, 439–441
function of, 443*t*
Deletion, clonal, 443
Delta antigen, in hepatitis, 636
Delta hepatitis (hepatitis D), 636, 641*t*
Denaturation, 44, 106
Dengue, 575, 581*t*
in Americas, 324
arthropod vector for, 379*t*, 581*t*
Dengue fever virus, 575, 581*t*
Dengue hemorrhagic fever, 324, 575
Denitrification, 678
Denitrifying bacteria, 678
Dental calculus, 619
Dental caries (tooth decay), 618–621, 640*t*
adhesion in, 393–395
Dental plaque, 618, 619*f*
Dental procedures, HBV and HIV transmission during, 344
Dentin, decay in, 620
Deoxyribonucleases, 522
Deoxyribonucleic acid. *See* DNA
Deoxyribose, 46
Dermacentor spp, 326*t*, 581*t*
Dermacentor andersoni, 573, 581*t*
diseases transmitted by, 379*t*
Dermacentor variabilis, 573
Dermatitis
allergic contact, 475–476
Pseudomonas, 523, 531*t*
Dermatomycoses, 306, 530–532
Dermatophytes, 306, 530
Dermis, 408, 519
Descriptive epidemiology, 386
Desensitization, 470–471
Desiccation, microbial growth affected by, 174–175, 175*t*
Desulfotomaculum nigrificans, sulfide food spoilage caused by, 703*t*
Desulfovibrio spp, 274*t*, 282
in fresh water, 683
Desulfovibrio desulfuricans, 38
Detergents, antimicrobial activity of, 181, 184*t*
Deuteromycota, 302–304
Devescovinids, in termites, 278
Dextran, 37
in tooth decay, 619, 620*f*
Dextran sulfate, for AIDS, 485
Dextransucrase
commercial production of, 716*t*
plasmid gene encoding, 401
Diabetes mellitus, insulin dependent, 477*t*
Diagnostic immunology, 453–463
future of, 463
Diapedesis, in inflammation, 416
Diarrhea. *See also* Gastroenteritis
digestive tract infections and intoxications causing, 622
traveler's, 628
Diatoms, 309
characteristics of, 309*t*
Dideoxyadenosine, for AIDS, 485
Dideoxycytidine, for AIDS, 485
Dideoxyinosine, for AIDS, 485
Differential media, 152–153, 153*t*, 154*f*
Differential stains, 63–65, 65*t*
and classification/identification of microorganisms, 259, 267*t*
Differential white blood cell count, 411–412
Diffusion
facilitated, 83–84
simple, 83, 84*f*
DiGeorge syndrome, 481*t*
Digestion, in phagocytosis, 414–415
Digestive system. *See also specific structure or organ*
diseases of, 617–649. *See also specific type*

bacterial, 618–630, 640*t*
fungal, 637, 641*t*
helminthic, 639–643, 641*t*
protozoan, 637–639, 641*t*
viral, 630–636, 640–641*t*
normal flora of, 618
structure and function of, 617–618
Digestive vacuole (phagolysosome), 414
Diiodohydroxyquin (iodoquinol), 501*t*, 509
for amoebic dysentery, 638, 641*t*
Dilution tests, 510
Dimorphism
fungi displaying, 299
sexual, of *Ascaris lumbricoides*, 322
Dinoflagellates, 307–309
blooms of, 310
characteristics of, 309*t*
in seawater, luminescence and, 683
toxins produced by, 404
Dioecious helminths, 319
Dipeptides, 43–44
Diphtheria, 288, 397–399, 400*t*, 593–594, 610*t*
cutaneous, 594, 610*t*
vaccine for, 450*t*, 593
schedule for administration of, 452*t*
Diphtheria toxin (diphtherotoxin), 397–399, 593
Diphtheroids, on skin, 520
Diphyllobothrium latum (fish tapeworm), 640–642, 641*t*
Dipicolinic acid, in endospores, 88
Diplobacillus (diplobacilli), 71
Diplococcus (diplococci), cell arrangement of, 70*f*, 71
Diplococcus pneumoniae. See Streptococcus pneumoniae
Diploid cell, 220
Diploid cell lines, 343
Diploid number, 220
Direct agglutination tests, 456–457
Direct contact transmission
in respiratory disease, 589
in spread of infection, 376
Direct ELISA, 463
Direct FA tests, 460*f*, 461
Direct flaming, for sterilization, 173, 174*t*
Direct microscopic count, bacteria measured by, 160–162
Direct selection (positive selection), mutant
identification by, 211
Disaccharides, 36–37
Disease, 369. *See also specific type and* Infection
acute, 374
chronic, 374
development of, 383–384
endemic, 373
epidemic, 373
etiology of, 369
germ theory of, 10
infectious
classification of, 373–374
etiology of, 372–373
latent, 374
microbes and, 18–19
pandemic, 373–374
patterns of, 383–384
predisposing factors affecting, 383
principles of, 368–391
sporadic, 373
subacute, 374
transmission of, 375–379
Disinfection
definition of, 168*t*
evaluating agents for, 177
principles of, 176–177
types of agents for, 177–185
Disk-diffusion method, 510
Dissimilation plasmids, 220
Dissimilatory sulfate-reducing or sulfur-reducing bacteria, 274*t*, 282–283
Dissociation, in water, 32, 33
Distilled spirits, microorganisms in production of, 710, 711*t*
Disulfide bridges, antimicrobial agents affecting, 170
Diversity segments (D segments), and diversity of antigen receptors, 436, 437*f*
Division, 254*f*, 256
in Procaryotae, 257*t*
DNA, 46–47, 190–191
antiparallel structure of, 195
chemical structure of, 193–195
chromosomes and, 191–192
cloning, 234–235
double helix structure of, 46, 190–191, 193–195
injection of into foreign cells, 232–233
in nucleus, 92
in procaryotic vs. eucaryotic cells, 251*t*
recombinant, 14, 214, 227–247. *See also* Genetic engineering
replication of, 192–197
in bacteria, 196–197
rate of, 197
sequencing of, 239
sources of, 233–234
synthetic, 234
DNA hybridization, and classification/identification of microorganisms, 265–267, 267*t*
DNA ligase, in DNA synthesis, 196
DNA polymerases, in DNA synthesis, 195–196
DNA probes, 234–235
in identification of bacteria, 267
in medical diagnosis, 241
for mycoplasmal pneumonia, 601
in salmonellosis diagnosis, 625
DNA sequencing, 239
DNA viruses, 15, 335, 340*f*. *See also specific type*
biosynthesis for, 350–351
double-stranded, 338*t*
multiplication of, 350–351
oncogenic, 358–359
single-stranded, 338*t*
tumors caused by, 357
Dog bites, systemic diseases caused by, 567
Dolphins, bacteria in, 264
Donor cell, 214
Double covalent bond, 27, 28*f*
Double helix structure of DNA, 46, 190–191, 193–195
Double-stranded RNA viruses, 339*t*
Doxycycline, 505
for lymphogranuloma venereum, 663, 665*t*
DPT vaccine, 543, 593, 596
schedule for administration of, 452*t*
Droplet transmission
in respiratory disease transmission, 589
in spread of infection, 377
DRT. *See* Decimal reduction time
Drug resistance, 510–511
and antibiotics in animal feed, 495
chronically mediated, 657
of gonococci, 657, 658
multiple, 657
and natural selection, 657
plasmid mediated, 657
of *Staphylococcus aureus*, 286
transposons and, 222
Drugs. *See also specific type and* Chemotherapy
synthetic, 11, 492
antibacterial, 499–502
antifungal, 502
Dry heat sterilization, 173, 174*t*
Dry weight, bacterial numbers estimated by, 163
Ductal system, in male, 652
Dumb rabies, 550
Dust cells, 590
Dysentery, 280, 622
amoebic (amoebiasis), 315, 638, 641*t*
bacillary (shigellosis), 258*f*, 280, 626, 640*t*
water pollution in transmission of, 640*t*, 684
Balantidium coli causing, 316, 638, 641*t*
water pollution in transmission of, 684
Dysgammaglobulinemias, 481*t*
Dysuria, in cystitis, 653

E

E-IPV. *See* Enhanced inactivated polio vaccine
Eastern equine encephalitis, 552, 553*t*
EB virus. *See* Epstein–Barr virus
Echinococcus granulosus, 321, 322*f*, 325*t*, 641*t*, 642
Echoviruses, opportunistic infection and, 371
Eclipse period, 345
Ecology, microbial, 17, 38, 240, 681, 685
Ectomycorrhizae, 680, 681
Ectopic pregnancy, salpingitis and, 656
Ectosymbiosis, and microorganisms in termites, 278
Edema, in inflammation, 415
Edwardsiella tarda, dolphins infected by, 264
EEE. *See* Eastern equine encephalitis
EGF. *See* Epidermal growth factor
Eggs
embryonated, inoculation of for virus growth in laboratory, 342
Salmonella contamination of, 625
Ehrlich, Paul, 492
EIA (enzyme immunoassay). *See* Enzyme-linked immunosorbent assay
Electron acceptor, 26
Electron beams, high energy, microbial growth affected by, 175–176
Electron donor, 26
Electron micrographs, 58–60
Electron microscope, 58–60, 61*t*
scanning, 59*f*, 60

Electron microscope (*Continued*)
  transmission, 58–60
Electron shells, 24, 25*t*
Electron transport chain, 110, 116–117
  and chemiosmotic generation of ATP, 118
Electronic cell counters, for direct microscopic count of bacteria, 162
Electronic configurations, 24, 25*t*
Electrons, 23–26
Electrophoresis, 262
  polyacrylamide gel, 262–265
  serum proteins separated by, 427–428
Elementary body, 283, 602
Elements, chemical, 23–24
ELISA. *See* Enzyme-linked immunosorbent assay
Embden–Meyerhof pathway, 112. *See also* Glycolysis
Embryonated eggs, inoculation of, for virus growth in laboratory, 342
Emericella spp, and reclassification of fungi, 303*t*
Emmonsiella spp, and reclassification of fungi, 303*t*
Emulsification, and soap as antimicrobial agent, 181
Enamel, tooth, in dental caries, 620
Encephalitis, 540
  arthropod-borne, 551–553, 555*t*
    vector for, 379*t*, 553*t*
  California, 552, 553*t*
  equine
    eastern, 552, 553*t*
    western, 552, 553*t*
    transmission of, 378*t*
  herpes, 527
  measles and, 528
  progressive, 360*t*
  St. Louis, 552, 553*t*
Encephalitis virus, 551
Encephalopathy, 360*t*
  bovine spongiform, 556
Encystment, protozoan, 315
End-product inhibition, 108–109
Endemic disease, 373
Endemic murine typhus, 283, 573, 581*t*
  arthropod vector for, 379*t*, 581*t*
Endergonic reaction, 29
Endocarditis, 563
  bacterial, 563, 580*t*
  gonorrheal, 656
  Q fever and, 604
Endocardium, 563
Endocrine diseases, HLA antigens in, 479*t*
Endocytosis, 90–91
  and penetration in animal virus multiplication, 349–350
Endogenous pyrogen, 401. *See also* Interleukin-1
Endomembrane system, 96
Endomycorrhizae, 680–681
Endoplasmic reticulum, 89*f*, 93
Endospores, 88–89
  bacteria forming, 275*t*, 286–287
  and resistance to microbial control, 169
  staining of, 64*f*, 65–66, 65*t*, 89
Endosymbionts, 274*t*
Endosymbioses, 143, 251
Endosymbiotic hypothesis, of eucaryotic evolution, 96
Endotoxic shock (septic shock), 401, 562
Endotoxins, 79, 399–401, 404*f*
  exotoxins compared with, 401*t*
Energy
  activation, 30, 102
  for active transport, 83, 86
  alternative sources of, microorganisms in production of, 718
  chemical, 29
  for group translocation, 86
  production of in microbial metabolism, 109–110
    biochemical pathways of, 110–111
    summary of, 130
  utilization of in microbial metabolism, 131–132
Energy levels, atomic, 24
Enhanced inactivated polio vaccine, 548
Enrichment culture, 153
Entamoeba histolytica, 315, 318*t*, 638, 641*t*
Enterics. *See* Enterobacteriaceae
Enterobacter spp, 274*t*, 281
  biochemical tests for, 259
  in digestive system, 618
  in nitrogen fixation, 679
  in nosocomial infection, 281, 380*t*
  pyelonephritis caused by, 653, 665*t*
  two-dimensional polyacrylamide gel electrophoresis of, 263*f*
Enterobacter aerogenes, 281
  pyelonephritis caused by, 653, 665*t*
  septicemia caused by, 562
Enterobacter cloacae, 281
Enterobacteriaceae (enterobacteria), 280–281. *See also specific genus*
  members of, 259
  in nosocomial infection, 380*f*
  phylogenetic relationships of, 258*f*
  in urethra, 652
Enterobius vermicularis, 322, 323*f*, 325*t*, 641*t*, 642
Enterococcus, in nosocomial infection, 380*t*
Enterohemorrhagic *E. coli*, 628, 640*t*
Enteropathogenic *E. coli*, 628, 640*t*
Enterotoxicosis, staphylococcal (staphylococcal food poisoning), 286, 622–624, 640*t*, 706, 707*t*
Enterotoxigenic *E. coli*, 628, 640*t*
Enterotoxins, 397
  diseases caused by, 400*t*
  heat-labile, 399
    plasmid gene encoding, 401
  Staphylococcus aureus producing, 286, 399, 400*t*, 622–623
  Vibrio cholerae producing, 399, 400*t*
Enteroviruses, 636
Entner–Doudoroff pathway, 114
Entry, portals of, 392–395, 404*f*
  preferred, 393, 394*t*
Envelope, viral, 335
  synthesis of, 355
Enveloped viruses, 337
  attachment sites of, 349
  budding of, 355, 356*f*
  helical, 337
  polyhedral, 337
Environment, microbial growth affected by, 169
Enzyme immunoassay. *See* Enzyme-linked immunosorbent assay
Enzyme induction, 203
Enzyme poisons, 107
Enzyme-linked immunosorbent assay (ELISA), 463
  in brucellosis, 566
  direct, 463
  in gonorrhea, 658
  in hepatitis A, 633
  indirect, 463
  in mumps, 630
  in salmonellosis, 625
  in toxoplasmosis, 576
Enzyme-substrate complex, 31, 104, 105*f*
Enzymes, 103–109
  antimicrobial agents affecting, 169–170, 498
  bacterial virulence related to, 396–397, 404*f*
  and chemical reactions, 31, 103–109
  classification of, 103
  components of, 103–104
  constitutive, 203
  extracellular, in facilitated diffusion, 84
  factors affecting activity of, 105–107
  immobilized, and fermentation technology, 714
  inducible, 203
  industrial microbiology in production of, 716–717
  inhibitors of, 107
  mechanism of action of, 104–105
  microbial, commercial production of, 716–717
  oxygen-related, 147*t*
  pH affecting, 106
  in phagocytosis, 414–415
  restriction, 230*f*, 231
  substrate concentration affecting, 106*f*, 107
  temperature affecting, 105–106
Eosinophils, 411*f*, 411*t*, 412
Epidemic disease, 373
Epidemic typhus, 283, 570–573, 581*t*
  arthropod vector for, 379*t*, 581*t*
Epidemiology, 384–387
  analytical, 386
  descriptive, 386
  experimental, 386
Epidermal growth factor, genetic engineering in production of, 237*t*
Epidermis, 408, 518–519
Epidermophyton spp, 302*t*
  skin infections caused by, 530, 531*t*
Epiglottis, 590*f*, 592
Epiglottitis, 592
  Hemophilus influenzae causing, 282, 592
Epinephrine, for anaphylactic shock, 469
EPO. *See* Erythropoietin
Epstein–Barr virus, 351, 358–359
  in Burkitt's lymphoma, 573–574
  in chronic fatigue syndrome, 574
  in infectious mononucleosis, 574, 581*t*
  in nasopharyngeal carcinoma, 358–359, 574
  respiratory tract as portal of entry for, 394*t*
  tumors caused by, 358–359, 573–574
Equilibrium, simple diffusion and, 83
Equine encephalitis
  eastern, 552, 553*t*
  western, 552, 553*t*
    transmission of, 378*t*
Equivalent treatments, 173
ER. *See* Endoplasmic reticulum

Ergot, 403, 706
Ergot poisoning (ergotism), 403, 637, 641*t*, 707*t*
Erwinia spp, 281
Erwinia dissolvens, in food production, 709*t*
Erysipelas, 522, 523*f*, 531*t*
Erythroblastosis fetalis (hemolytic disease of the newborn), 472–473
Erythrocytes (red blood cells), 411*t*, 560
  in malaria, 579–580
Erythrogenic toxins, 399, 522
  and scarlet fever, 592, 610*t*
Erythromycin, 500*t*, 505
  for chancroid, 663, 665*t*
  for legionellosis, 601, 611*t*
  for lymphogranuloma venereum, 663, 665*t*
  for mycoplasmal pneumonia, 601, 611*t*
  for nongonococcal urethritis, 658
  protein synthesis affected by, 497*f*
  sources of, 494*t*
  for streptococcal pharyngitis, 592
  for syphilis, 662
  for whooping cough, 596, 611*t*
Erythropoietin, genetic engineering in production of, 237*t*
Escape, immunologic, 486–487
Escherichia coli, 274*t*, 280
  adhesins of, 395
  Bdellovibrio affecting, 62
  biochemical tests for, 259
  conjugation in, 216–217
  in cystitis, 653, 665*t*
  in digestive system, 618
  enterohemorrhagic, 628, 640*t*
  enteropathogenic, 628, 640*t*
  enterotoxigenic, 628, 640*t*
  feedback inhibition in, 108–109
  in gastroenteritis, 628, 640*t*
  insulin produced by, 229*f*, 235, 236, 237*t*
  lactose metabolism in, 203–206
    induction and, 203
    operon model and, 203–206
  lysogeny of bacteriophage λ in, 347–348
  in meningitis, 541
  in nosocomial infection, 380*t*, 381
  in opportunistic infection, 370
  in pyelonephritis, 653, 665*t*
  in septicemia, 562
  size of, 334*f*
  testing water for, 686, 687*f*
  toxins produced by, 400*t*
  in urinary tract infections, 653, 665*t*
  water contaminated by, 686
Ester linkage, 39
Estrogens, microbial production of, 717
Ethambutol, 499, 500*t*
  for tuberculosis, 598
Ethanol
  antimicrobial activity of, 180, 184*t*
    effectiveness and, 183*f*
  fermentation of, 123*t*
  production of from agricultural products, 718
Ethylene oxide, for gaseous chemosterilizers, 183, 185*t*
Etiology of disease, 369, 371–373
Eubacteria, 71. *See also* Bacteria
  characteristics of, 253*t*
  classification of, 251–254
  fungi compared with, 297*t*
Eucaryotic cells (eucaryotes), 15, 70, 89–96. *See also* Algae; Fungi; Helminths; Protozoans
  cell wall of, 89*f*, 90
  centrioles in, 89*f*, 95
  characteristics of, 253*t*
    comparison of, 297*t*
  chloroplasts in, 89*f*, 95
  chromosomes of, 192
  cilia of, 90, 91*f*
  classification of, 251–253
  cytoplasm of, 89*f*, 91–92
  endoplasmic reticulum in, 89*f*, 93
  evolution of, 95–96
  flagella of, 89*f*, 90, 91*f*
  glycocalyx of, 90
  Golgi complex in, 89*f*, 93, 94*f*
  lysosomes in, 89*f*, 95
  mitochondria in, 89*f*, 93–94
  nucleus of, 89*f*, 92–93
  organelles in, 89*f*, 90, 92
  photosynthesis in, 129*t*
  plasma (cytoplasmic) membrane of, 89*f*, 90–91
  procaryotic cells compared with, 96*t*, 251*t*
  recombination in, 219–222
    classification and, 267–268
  ribosomes in, 89*f*, 93
Euglena, 308*f*, 314
Euglenoids, 308*f*, 309
  characteristics of, 309*t*
Eurotium spp, and reclassification of fungi, 303*t*
Eutrophication, 684–685
Evolution
  Darwin's theory of, 250–251
  genes and, 222
Exchange reactions, 30
Exergonic reaction, 29
Exfoliative toxin, staphylococci producing, 521
Exit, portals of, in spread of infection, 379–380
Exoenzymes, virulence and, 396
Exons, 199
Exonuclease activity, in DNA synthesis, 195
Exotoxins, 397–399, 400*t*, 404*f*. *See also specific type*
  diseases caused by, 400*t*
  endotoxins compared with, 401*t*
  mechanism of action of, 399
Experimental epidemiology, 386
Exponential growth phase, 157
Expression, of gene, 191
Extracellular enzymes, in facilitated diffusion, 84
Extreme halophiles, 144
  classification of, 253
Extreme thermophiles, 144
Exxon *Valdez* oil spill, bacteria in cleanup of, 38, 685
Eye diseases, 533–535, 534*t*. *See also specific type*
  bacterial, 534*t*
  gonococcal, 533, 534*t*, 656
    silver nitrate for, 181, 184*t*, 533, 534*t*, 656
  protozoan, 534*t*
  viral, 534*t*
Eyepiece, for compound light microscope, 53, 54*f*
Eyespot, 309

F

F. *See* Fluorine
F^+ cells, in bacterial conjugation, 216–217
F^- cells, in bacterial conjugation, 216–217
F factor
  in bacterial conjugation, 216–217
  as plasmid, 220
FA. *See* Fluorescent-antibody techniques
Facilitated diffusion, 83–84
FACS. *See* Fluorescence-activated cell sorter
Factor VIII, genetic engineering in production of, 237*t*
Facultative anaerobes, 146, 274*t*, 280–282
  oxygen-related enzymes of, 147*t*
Facultative halophiles, 144
FAD. *See* Flavin adneine dinucleotide
FAIDS. *See* Feline AIDS
Fallopian tubes, 651
  infection of, 656
Family, 254*f*, 256
Fansidar. *See* Pyrimethamine, with sulfadoxine
Fastidious bacteria, 149
Fats, 37–39. *See also* Lipids
  saturated, 39
  unsaturated, 39
Fatty acids, 37–39
Fc region, 431
  in hypersensitivity reactions, 468
Fe. *See* Iron
Fecal coliforms, 686
Fecal-oral route of transmission
  for cryptosporidiosis, 639, 641*t*
  for digestive system diseases, 617
  for hepatitis A, 632, 634
  for hepatitis E (infectious NANB hepatitis), 636
  for viral gastroenteritis, 636
  water pollution and, 684
Feces, 618
Feedback inhibition, 108–109
Feline AIDS, 342
Feline leukemia virus, 359
FeLV. *See* Feline leukemia virus
Female reproductive system. *See also* Reproductive system
  normal flora of, 652
  structure and function of, 651
Fermentation, 8, 112*f*, 120–122
  aerobic respiration compared with, 123*t*
  alcoholic beverages and vinegar produced by, 710–712
  anaerobic respiration compared with, 123*t*
  dairy products produced by, 708, 709*t*
  food and related products produced by, 709*t*
  industrial, 123*t*, 712–714
  malolactic, 711
  nondairy, 709*t*, 710
Fermentation technology, 712–714
Fertilization, carbon dioxide, and greenhouse effect, 675
Fetus
  cytomegalovirus infection and, 631, 640*t*
  genital herpes and, 663–664
  gonococcal eye infection and, 533, 534*t*, 656

Fetus (*Continued*)
silver nitrate for, 181, 184*t*, 533, 534*t*, 656
Fever, 417–418
childbirth (puerperal fever, puerperal sepsis), 562–563, 580*t*
endotoxins causing, 400–401
Fever blisters (cold sores), 350, 359, 526. *See also* Herpes simplex viruses, type 1
as latent infection, 359, 527
Fibrinolysin. *See* Streptokinase
Filament, of bacterial flagella, 73, 74*f*
Filamentous organisms
algae, 306
measuring methods for, 163
Filamentous particles, in hepatitis B, 633
Filaments, intermediate, 91
Filobasidiella spp, and reclassification of fungi, 303*t*
Filoviruses, 339*t*
Filter paper method, disinfectant evaluated by, 177, 178*f*
Filterable viruses, 173
Filtration
bacteria counted by, 160
for sterilization, 173, 174*t*
Fimbriae, bacterial, 73*f*, 76–77
Fingerprinting, genetic, 229*f*, 236–239
Fish products, microorganisms in production of, 709*t*
Fish tapeworm (*Diphyllobothrium latum*), 640–642, 641*t*
Fission
binary, 15, 70, 155–156
multiple, 314
FITC. *See* Fluorescein isothiocyanate
Five-kingdom system, 251, 252*f*
Fixation, prior to staining, 60
Fixed macrophages, 413. *See also* Histiocytes
FK 506, 505
Flaccid paralysis, in botulism, 544
Flagellates, 313, 314*f*, 315–316, 318*t*
Flagellin, 73
Flagellum (flagella)
bacterial, 15, 73–75, 73*f*
staining of, 64*f*, 65*t*, 66
eucaryotic, 89*f*, 90, 91*f*
preemergent, 309
in procaryotic vs. eucaryotic cells, 96*t*
protozoan, 15, 313, 314*f*, 315
Flagyl. *See* Metronidazole
Flaming, direct, for sterilization, 173, 174*t*
Flat sour spoilage, 702, 703*t*
Flatworms (Platyhelminthes), 15, 319–320. *See also specific type*
Flavin adenine dinucleotide, 103
Flavin mononucleotide, 103
Flavivirus, 338*t*
Flavoproteins, 116
Fleas
diseases transmitted by, 326*t*, 379*t*
rat, as vector, 326*t*, 327*f*, 568, 580*t*, 581*t*
diseases transmitted by, 379*t*
Yersinia pestis carried by, 281
Fleming, Alexander, 11, 493
Fleshy fungi, 297–298
Flies, in disease transmission, 326*t*, 376*f*, 377
Floc
and sewage treatment, 690, 691*f*
and water treatment, 686, 688*f*
Flocculation treatment, 686, 688*f*
Flora
in fresh water, 682–683
normal, 18, 369–371
of digestive system, 618
host relationships and, 370
of reproductive system, 652
of respiratory system, 590–591
of skin, 519–520
of urinary tract, 652
of vagina, 652
in seawater, 683
in soil, 673–674
transient, 370
Flow cytometry
and classification/identification of microorganisms, 267, 267*t*
and fluorescence-activated cell sorter, 461
Flu. *See* Influenza
Flucytosine, 502
Fluid mosaic model, 81
Flukes (trematodes), 319–320, 325*t*
schistosomiasis caused by, 582–583
Fluorescein isothiocyanate, 57, 461
Fluorescence-activated cell sorter, 461–463
Fluorescence microscopy, 57–58, 61*t*
Fluorescent-antibody techniques, 57, 460*f*, 461–463. *See also* Immunofluorescence
Fluorescent treponemal antibody absorption test, for syphilis, 661
Fluorine, 23*t*
Fluorochromes, 57
Fluoroquinolones, 502
FMN. *See* Flavin mononucleotide
Focal infection, 374
Folic acid, function of, 104*t*
Foliose lichens, 311
Folliculitis, 521
Fomites
in respiratory disease transmission, 589
in spread of infection, 377
Fonsecaea, 302*t*
Food allergies, 470
Food chain, aquatic, algae in, 310
Food intolerance, 470
Food microbiology, 700–722
Food poisoning, 707*t*. *See also* Gastroenteritis
clostridia causing, 287
exotoxins causing, 400*t*
staphylococcal, 286, 622–624, 640*t*, 706, 707*t*
Food preservation, 700–706
aseptic packaging for, 703–704
canning for, 701–703
chemical preservatives for, 705–706
heat for, 701–703
low-temperature, 704
pasteurization for, 703
radiation in, 704–705
Food production, microorganisms in, 706–712
genetic engineering and, 237*t*, 706
Food source, microorganisms as, 712
Food spoilage
butyric anaerobic, 703*t*
and canned foods, 702–703
flat sour, 702, 703*t*
putrefactive anaerobic, 703*t*
sulfide, 703*t*
temperatures of, 704*f*
thermophilic anaerobic, 702, 703*t*
Foodborne hepatitis A, 634
Foodborne infections, 622, 706
Foodborne microbial intoxications, 706, 707*t*
Foodborne transmission, in spread of infection, 377
Forbidden clones, and loss of tolerance, 476
Forespore, 88
Formaldehyde gas, antimicrobial activity of, 182, 185*t*
Formalin, antimicrobial activity of, 182, 185*t*
Formed elements in blood, 411–413
Fossil fuels
alternatives to, microorganisms in production of, 718
burning
carbon dioxide released by, 677
greenhouse effect and, 675
Fountain water, legionellosis from, 602
Frameshift mutagens, 210
Frameshift mutations, 208, 209*f*
Francisella spp, 274*t*, 279
Francisella tularensis, 279, 564, 565, 580*t*
Frankia spp, 275*t*, 293
in nitrogen fixation, 680
Free radicals
hydroxyl, 147
superoxide, 147
Free-floating algae, 307
Freeze-drying, for bacterial culture preservation, 155
Freezing
for bacterial culture preservation, 155
for food preservation, 704
microbial growth affected by, 174, 175*t*
Freshwater microbial flora, 682–683
Fruticose lichens, 311
FTA-ABS test, for syphilis, 661
Fuels, fossil
alternatives to, microorganisms in production of, 718
burning
carbon dioxide released by, 677
greenhouse effect and, 675
Functional groups, 35–36
Fungal hyphae. *See* Hyphae
Fungi, 15, 16*f*, 296–306. *See also specific type and* Microorganisms
antibiotics produced by, 494*t*
bacteria compared with, 297*t*
characteristics of, 297–301, 302–303*t*
comparison of, 297*t*
classification of, 251, 252*f*, 253*f*
imperfect, 302–303
club (Basidiomycota), 302–303*t*, 304
conjugation (Zygomycota), 302–303*t*, 304
digestive system diseases caused by, 637, 641*t*
dimorphic, 299
diseases caused by, 304–306. *See also* Mycoses
drugs for treatment of, 501*t*, 502, 506–507. *See also specific type and* Antifungal drugs
economic effects of, 306
fleshy, 297–298
Imperfecti, 302–303
in lichens, 310–312

lower respiratory system diseases caused by, 606–609, 611*t*
medically important, 302–304
nervous system diseases caused by, 553, 555*t*
nutritional adaptations of, 301
pathogenetic properties of, 403–404
phyla of, 302–304
reproductive structures of, 299–301
respiratory system diseases caused by, 606–609, 611*t*
Sabouraud's dextrose agar for growth of, 152
sac (Ascomycota), 302–303*t*, 304
in septicemia, 562
skin diseases caused by, 530–533, 531*t*
in soil, 674, 675
toxins produced by, 403–404
foodborne microbial intoxications caused by, 706, 707*t*
as vegetative structures, 297–299
Furious rabies, 549–550
Furuncle (boil), 521
Fusion, penetration in animal virus multiplication by, 350
Fusobacterium spp, 274*t*, 282, 283*f*
in acute necrotizing ulcerative gingivitis, 622
bites infected with, 567
human, 568

G

β-Galactosidase
induction and, 203
operon model and, 203–206
Gambierdiscus toxicus, 309
Gametes, 218
protozoans producing, 315
Gametocytes
of *Plasmodium*, 317, 580
protozoans producing, 315
Gamma globulin, 427*f*, 428
Gamma-hemolytic streptococci, 286, 522
Gamma interferon, 422
genetic engineering in production of, 237*t*
Gamma rays
microbial growth affected by, 175–176
mutations caused by, 210
Ganciclovir, 508
for cytomegalovirus infection, 631, 640*t*
Gangrene, 566–567, 580*t*
gas, 287, 400*t*, 567. *See also Clostridium perfringens*
Garbage, 694–695
Gardnerella spp, 274*t*, 275*t*, 282
Gardnerella vaginalis, 282, 662, 665*t*, 666*t*
Gas gangrene, 287, 400*t*, 567. *See also Clostridium perfringens*
Gas vacuoles, 87, 293
Gas vesicles, 87
Gaseous chemosterilizers, 183, 185*t*
Gases, in soil, 673
Gasohol, 718
Gastric juice, in host defense, 410
Gastritis, *Helicobacter*, 398, 629, 640*t*
Gastroenteritis, 622. *See also* Food poisoning
Bacillus cereus, 630, 640*t*
Campylobacter, 628–629, 640*t*
Clostridium perfringens, 629, 640*t*
Escherichia coli, 628, 640*t*
Salmonella (salmonellosis), 280, 624–625, 640*t*
transmission of, 378*t*, 640*t*
Vibrio parahaemolyticus, 281, 627, 640*t*
viral, 636, 640*t*
Yersinia, 629, 640*t*
Gastrointestinal tract, 617, 618*f*. *See also* Digestive system
in host defense, 409
as portal of entry, 393, 394*t*, 404*f*
Gel diffusion test, for diphtheria, 593
Gel electrophoresis, in Southern blotting, 239
Gelidium, agar extracted from, 310
Gender, as predisposing factor, 383
Gene amplification, in oncogene activation, 358
Gene libraries, 233–234
clone identification in, 234–235
Gene products
making, 235–236
for medical therapy, 236, 237*t*
Gene therapy, 239
Gene-cloning vectors. *See* Cloning vectors
Generalized infection (systemic infection), 374
Generalized transduction, 218
Generation time, bacterial, 156
Genes, 191
evolution and, 222
expression of, 191
in bacteria, regulation of, 203–207
I, 204
proP, *Salmonella*, 205
stress-induced, 205
structural, in operon model, 203–204
"suicide," 243
Genetic code, 202–203
Genetic engineering (recombinant DNA technology), 6, 13–14, 18, 227–247. *See also* Genetic recombination
advent of, 227–228
agricultural applications for, 241–243
alternatives to vectors in, 232–233
basic research and, 236–241
clone identification and, 234–235
and cloning with a plasmid vector, 231–232
definition of, 228
and DNA sources, 233–234
in food production, 237*t*, 706
future of, 243
gene libraries in, 233–234
clone identification and, 234–235
gene product synthesis and, 235–236
medical applications for, 236–241
procedures used in, 228–231
products of, 236, 237*t*
restriction enzymes and, 231
safety issues in, 243
synthetic DNA for, 234
Genetic fingerprinting, 229*f*, 236–239
Genetic information, flow of, 197
Genetic recombination, 214–222. *See also* Genetic engineering
in bacteria
conjugation and, 216
transduction and, 218
transformation and, 214–216
and classification/identification of microorganisms, 267–268, 267*t*
in eucaryotes, 218–222
plasmids and, 216, 220–221
transposons and, 221–222
Genetic transformation, in bacteria, 214–216
mechanism of, 216*f*
Genetics, 190
microbial, 13, 190–226
Genital herpes, 359, 527, 663–664, 665*t*. *See also* Herpes simplex viruses, type 2
incidence of, 655*f*
Genital warts, 664, 665*t*
Genitals, external, in female, 651
Genitourinary tract. *See also* Reproductive system; Urinary tract
as portal of entry, 393, 394*t*, 404*f*
Genotype, 191
Gentamicin, 500*t*, 504
for *Klebsiella* pneumonia, 600, 611*t*
protein synthesis affected by, 498
sources of, 494*t*
for tularemia, 565
Genus (genera), 254*f*, 255
in nomenclature, 14
Geosmin, 293, 674
Geotrichum spp, in food production, 709*t*
Germ, 4. *See also* Microorganisms
Germ theory of disease, 10
German measles. *See* Rubella
Germicide, 168*t*
Germination, of endospore, 89
Ghon complexes, in tuberculosis, 597
Giant cells, 402
Giardia lamblia, 316, 318*t*, 637, 641*t*
immune response to, 440
resistance of to chlorination, 637, 686
water pollution in transmission of, 637, 641*t*, 684
Giardiasis, 637, 641*t*. *See also Giardia lamblia*
Gingivitis, 621*f*, 622
acute necrotizing ulcerative (Vincent's disease, trench mouth), 622
Gleocapsa, 292*f*
Gliding bacteria
fruiting, 275*t*, 289*f*, 290
nonfruiting, 275*t*, 290
Global warming (greenhouse effect), bacteria and, 675
Globulin, immune. *See* Immune globulin
Glomerulonephritis, 475, 654, 665*t*
Glossina (tsetse fly), 326*t*
diseases transmitted by, 379*t*
T. brucei gambiense transmitted by, 316, 554
Glove boxes, anaerobic, 151
Glucan, in yeast cell wall, 90
Gluconobacter spp, 280
in vinegar production, 712
Glucose, *lac* operon regulation and, 206
Glucose effect, 206
Glucose isomerase, commercial production of, 716
Glucose oxidase, commercial production of, 716*t*
β-Glucuronidase, *E. coli* detected with, 686
Glutamic acid, 42*t*
industrial microbiology in production of, 714–715
Glutamine, 43*t*

Glutaraldehyde, antimicrobial activity of, 182–183, 185t
Glycerol, 37, 39f
fermentation of, 123t
Glycine, 42t
Glycocalyx
bacterial, 72–73
eucaryotic, 90
in procaryotic vs. eucaryotic cells, 96t
Glycogen, 37
Glycolysis, 111–114
alternatives to, 114
reactions of, 113f
Glycoproteins, 46
Golgi complex in formation of, 93
variable surface, 440
Glycosyltransferase, and *S. mutans* in tooth decay, 393
Glyphosate, resistance to, 241
Golgi complex, 89f, 93, 94f
Gonococcal eye infection, 533, 534t, 656
silver nitrate for, 181, 184t, 533, 534t, 656
Gonococcus. *See Neisseria gonorrhoeae*
Gonorrhea, 279, 655–658, 665t. *See also Neisseria gonorrhoeae*
anal, 656
distribution of, 655f
incidence of, 655f
pharyngeal, 656
Gonorrheal arthritis, 656
Gonorrheal endocarditis, 656
Gonorrheal meningitis, 656
Gonyaulax spp
neurotoxins produced by, 404
shellfish poisoning caused by, 307
Goodpasture's syndrome, 477t
Graft-versus-host disease, 480
Grafts. *See also* Transplantation
types of, 479t
Graham stick-tape method, 322
Gram stain, 63–65, 65t
procedure for, 63f
Gram-negative bacteria, 64, 274t
aerobic rods and cocci, 274t, 277–280
aerobic/microaerophilic, motile, helical/vibrioid, 274t, 276–277
anaerobic cocci, 274t, 283
anaerobic straight, curved, and helical rods, 274t, 282, 283f
Brilliant Green agar for growth of, 152
cell wall of, 78f, 79
classification of, 253f
facultatively anaerobic rods, 146, 274t, 280–282
gram-positive bacteria compared with, 80t
phylogenetic relationships of, 258f
and resistance to microbial control, 169
in septicemia, 562
Gram-positive bacteria, 64, 274t, 275t
antibiotics produced by, 494t
cell wall of, 78f, 79
classification of, 253f
cocci, 274t, 285–286
endospore-forming rods and cocci, 275t, 286–287
gram-negative bacteria compared with, 80t
nonsporing rods
irregular, 275t, 288
regular, 275t, 287–288
phylogenetic relationships of, 258f
and resistance to microbial control, 169
in septicemia, 562
Grana. *See* Granum
Granstein cells, 408
Granules
metachromatic, 87
polysaccharide, 87
sulfur, 87
Granulocytes, 411f, 411t, 412. *See also specific type*
classification of, 413f
Granulomas, in schistosomiasis, 583
Granum (grana), 95
Graves' disease, 474, 477t
HLA antigens in, 479t
Green algae, 308f, 310
characteristics of, 309t
copper sulfate for destruction of, 181, 184t
in lichens, 310–312
Green bacteria, 291–293
nonsulfur, 129, 292–293
sulfur, 126–129, 129t, 292
in fresh water, 683
phylogenetic relationships of, 258f
Greenhouse effect, bacteria and, 675
Griffith, Frederick, 214
Griseofulvin, 501t, 507
for cutaneous mycoses, 532
sources of, 494t
Group translocation, 86
Growth, microbial, 141–166. *See also* Bacterial cultures, growth of
agents controlling. *See also* Antimicrobial agents
alcohols affecting, 180, 184t
aldehydes affecting, 182–183, 185t
anaerobic media and methods for, 150–151
of bacterial cultures, 155–163
chemical requirements for, 145–148
chemically defined media for, 149
chlorhexidine affecting, 178–179, 184t
complex media for, 150
control of, 167–189
and actions of microbial control agents, 169–170. *See also* Antimicrobial agents
chemical methods of, 176–185
conditions influencing, 168–169
physical methods of, 170–176
terminology related to, 168t
culture media for, 148–155
desiccation affecting, 174–175, 175t
differential media for, 152–153, 154f
disinfection and
evaluation of, 177
principles of, 176–177
types of, 177–185
dry heat sterilization and, 173, 174t
enrichment culture for, 153
environment affecting, 169
filtration affecting, 173, 174t
gaseous chemosterilizers and, 183, 185t
halogens affecting, 179–180, 184t
heat affecting, 170–173, 174t
heavy metals affecting, 180–181, 184t
ionizing radiation affecting, 175–176, 175t
low temperature affecting, 169, 173–174, 175t
membrane permeability alterations and, 169
microbe type affecting, 169
and microbial death rate, 170
moist heat sterilization and, 171–172, 174t
nonionizing radiation affecting, 175t, 176
nucleic acid damage and, 169–170
organic acids affecting, 182, 185t
osmotic pressure affecting, 175, 175t
oxidizing agents affecting, 183–185, 185t
pasteurization and, 172–173, 174t
phenol and phenolics affecting, 177–178, 184t
physical requirements for, 142–145
physiologic state of microbe affecting, 169
and preserving bacterial cultures, 155
in procaryotic vs. eucaryotic cells, 251t
protein damage and, 169–170
pure cultures and, 153–155
quaternary ammonium compounds affecting, 182, 184t
radiation affecting, 175–176, 175t
requirements for, 141–148
selective media for, 152–153, 154f
special culture techniques for, 151–152
surface-active agents (surfactants) affecting, 181–182, 184t
temperature affecting, 142–144, 169
Growth curve, bacterial, 157–158
logarithmic vs. arithmetic representation of, 156–157
Growth factors
epidermal, genetic engineering in production of, 237t
organic, microbial growth affected by, 148
Growth hormone, genetic engineering in production of
bovine, 237t, 243
human, 229f, 236, 237t
porcine, 237t
Guanine, 190–191
Guanine analogs, 501t. *See also specific type*
Guinea pig test, for diphtheria, 593
Gummas, in syphilis, 660f, 661
Gymnoascus spp, and reclassification of fungi, 303t

H

H. *See* Hydrogen
H chains (heavy chains), 430
genes coding for, diversity and, 436, 437f
H spikes. *See* Hemagglutinin spikes
Haemophilus spp. *See Hemophilus* spp
Halitosis, in acute necrotizing ulcerative gingivitis, 622
Haloarcula, 71
Halobacteria. *See Halobacterium* spp
Halobacterium spp, 128, 275t, 291
gas vacuoles in, 87
Halobacterium halobium
ATP synthesis by, 128
photosynthesis without chlorophyll in, 87

Halococcus spp, 291
Halogens, antimicrobial activity of, 179–180, 184*t*
Halophiles, 291
  classification of, 253*f*
  extreme, 144
    classification of, 253
  facultative, 144
  in food production, 709*t*
  phylogenetic relationships of, 258*f*
Halophilic archaeobacteria, 71, 72*f*
Hansen's disease, 546. *See also* Leprosy
Haploid cell, 218
Haptens, 429–430
Hashimoto's thyroiditis, 476, 477*t*
HAV. *See* Hepatitis A virus
Hay fever, 470
HBsAG. *See* Hepatitis B surface antigen
HBV. *See* Hepatitis B virus
HDCV. *See* Human diploid cell vaccine
HDV. *See* Hepatitis D virus
Heart, 560, 561*f*. *See also* Cardiovascular system
Heat
  for food preservation, 701–703
  microbial growth affected by, 170–173, 174*t*
Heat-labile enterotoxin, 399
  plasmid gene encoding, 401
Heat-resistant ascospore, food spoilage and, 703
Heavy chains, 430
  genes coding for, diversity and, 436, 437*f*
Heavy metals, antimicrobial activity of, 180–181, 184*t*
HeLa cell line, 343
Helical viruses, 336, 337*f*
Helical/vibrioid bacteria, 274*t*, 276–277
Helicobacter spp, 276
  adhesion of, 395, 398
  gastritis caused by, 398, 629, 640*t*
  ulcers caused by, 398, 629
Helicobacter pylori, 629, 640*t*
  adhesion of, 398
  ulcers and, 398, 629
Helminths, 15, 319–325. *See also specific type*
  biology of, 319
  cardiovascular and lymphatic system diseases caused by, 581–583, 581*t*
  characteristics of, 325*t*
    comparison of, 297*t*
  digestive system diseases caused by, 639–643, 641*t*
  dioecious, 319
  drugs for infection with, 501*t*, 509. *See also specific type and* Antihelminthic drugs
  hermaphroditic, 319
  hosts for, 319
  life cycle of, 319
  pathogenetic properties of, 403–404
  reproduction of, 319
  in soil, 674–675
Helper T cells, 433, 439
  in AIDS, 439, 480–481
  in B-cell activation, 433, 435*f*
  function of, 443*t*
Hemagglutination, 457
  by viruses, 336, 457, 458*f*
Hemagglutination inhibition tests, 457
  for influenza viruses, 605
Hemagglutinin spikes, 605
Heme fraction of blood, *Hemophilus* using, 282
Hemoflagellates, 316
Hemolysins
  streptococci producing, 522
  virulence and, 396
Hemolytic anemia, 474, 477*t*
Hemolytic disease of the newborn, 472–473
Hemolytic streptococci. *See also specific type and Streptococcus spp*
  alpha, 286, 522
    in subacute bacterial endocarditis, 580*t*
  beta, 286, 522
    in glomerulonephritis, 654
    in pharyngitis, 592
    in puerperal sepsis, 562, 580*t*
    in rheumatic fever, 563, 580*t*
  gamma, 286, 522
  in human bites, 568
Hemolytic uremic syndrome, *E. coli* and, 628
Hemophilus spp, 274*t*, 282. *See also specific type*
  genetic transformation occurring in, 215
Hemophilus ducreyi, 663
Hemophilus influenzae, 282. *See also Hemophilus* spp
  capsule of, 396
  meningitis caused by, 282, 541–542, 555*t*
  otitis media caused by, 594, 610*t*
  sinusitis caused by, 591
  type b, 542*f*
    meningitis caused by, 542
    vaccine for, 450*t*
      as conjugated vaccine, 452
      schedule for administration of, 452*t*
Hemophilus vaginalis. *See Gardnerella vaginalis*
Hemorrhagic colitis, 628
Hemorrhagic fever, dengue, 324, 575
HEPA filters. *See* High-efficiency particulate air filters
Hepadnaviruses, 339*t*
Hepatitis, 631–636, 641*t*
  characteristics of, 632*t*
  type A, 632–633, 641*t*
    characteristics of, 632*t*
    foodborne outbreaks of, 634
    water pollution in transmission of, 641*t*, 684
  type B, 633–635, 641*t*
    vaccine for, 451*t*, 635, 641*t*
  type C (non-A, non-B), 635–636, 641*t*
  type D (delta), 636, 641*t*
  type E, 636, 641*t*
Hepatitis A virus, 338*t*, 632–633, 641*t*
  gastrointestinal tract as portal of entry for, 394*t*, 632
Hepatitis B surface antigen, 633
Hepatitis B vaccine, 635, 641*t*
  genetic engineering in production of, 237*t*
Hepatitis B virus, 339*t*, 633, 641*t*
  portal of entry for, 394*t*
Hepatitis C virus, 338*t*, 635, 641*t*
Hepatitis D virus, 636, 641*t*
Hepatitis E virus, 636, 641*t*
Herd immunity, 450
Hermaphroditic helminths, 319
Herpes encephalitis, 527
Herpes gladiatorium, 527
Herpes simplex viruses, 337, 338*t*, 531*t*
  entry of into animal cell, 350*f*
  latent infection and, 359, 526
  skin infections caused by, 526–527
  type 1, 338*t*, 350, 359
    eye infection caused by, 534
    skin infections caused by, 527, 531*t*
  type 2, 338*t*, 351, 359, 663, 665*t*. *See also* Genital herpes
    fetus infected by, 663–664
    skin infections caused by, 527
Herpes zoster (shingles), 338*t*, 526, 531*t*. *See also* Varicella-zoster virus
  and latent infections, 359, 526
Herpesviruses, 338*t*, 350–351. *See also specific type*
  eye infection caused by, 534
  human, 350–351, 358–359
  and latent infection, 359, 526
  multiplication of, 350–352
Herpetic keratitis, 527, 534
Heterocysts, 293
  in nitrogen fixation, 679
Heterolactic bacteria (heterofermentative bacteria), 122
Heterotrophs, 126
Hexachlorophene, antimicrobial activity of, 178, 184*t*
Hfr cell, in bacterial conjugation, 216–217
HHV 1. *See* Herpes simplex viruses, type 1
HHV 2. *See* Herpes simplex viruses, type 2
HHV 3. *See* Herpes zoster; Varicella-zoster virus
HHV 5. *See* Epstein–Barr virus
HHV (human herpesviruses). *See* Herpesviruses, human
Hierarchy, for classification
  phylogenetic, 256–257
  taxonomic, 255–256
High-efficiency particulate air filters, 173
High-energy electron beams, microbial growth affected by, 175–176
High-energy phosphates, in group translocation, 86
High frequency of recombination cell, in bacterial conjugation, 216–217
High-temperature short-time pasteurization, 172
Histamine
  in hypersensitivity reactions, 468
  in inflammation, 415
Histidine, 43*t*
Histidine auxotrophs, of *Salmonella*, Ames test and, 213
Histiocytes, 413
Histocompatibility antigens, 476. *See also* HLA antigens
Histones, 92
  in procaryotic vs. eucaryotic cells, 251*t*
Histoplasma spp, reclassification of, 303*t*
Histoplasma capsulatum, 302*t*, 606–607, 611*t*
  respiratory tract as portal of entry for, 394*t*
Histoplasmosis, 606–608, 611*t*
HIV (human immunodeficiency virus), 3, 355. *See also* AIDS
  and animal models of infection, 342
  cytopathic effects of, 403*t*

HIV (*Continued*)
infection with, 480–482
stages of, 483*f*
and risk to health care workers, 344
transmission of, 482–485
virion of, 482*f*
HLA antigens, 477–478
diseases related to, 479*t*
genes controlling, 478*f*
HLA complex, 476–478
HLA typing, 477–478
H_0N_1 strain, of influenza virus, 605
H_2N_2 strain, of influenza virus, 605
H_3N_2 strain, of influenza virus, 605
Hodgkin's disease
EB virus in, 359
HLA antigens in, 479*t*
Holdfasts, 310, 311
Holoenzyme, 103, 104*f*
Holospora spp, 274*t*
Homolactic bacteria (homofermentative bacteria), 121
Hong Kong flu, 605
Hook, of bacterial flagella, 73, 74*f*
Hooke, Robert, 6, 53
Hookworms, 323–325, 641*t*, 642–643. *See also Necator americanus*
in soil, 674–675
Hormones, genetic engineering in production of, 236
Hospital, microorganisms in, 380–382
Hospital-acquired infections (nosocomial infections), 380–383
control of, 383
Enterobacter causing, 281, 380*t*
principal kinds of, 382*t*
Host cells, damage to, 397–401, 404*f*
Host defenses, 407–424. *See also specific type and* Immunity
nonspecific, 407–424
pathogen penetration of, 395–397, 404*f*
specific, 426–448
Host range, of virus, 334–335
Hosts
compromised, in spread of infection, 382
defenses of. *See* Host defenses
extent of involvement of in disease, classification and, 374
for helminths, 319
infectious dose for 50% of, 393
lethal dose for 50% of, 393
normal flora in, 370
portals of entry for, 392–395, 404*f*
portals of exit for, 379–380
Hot-air sterilization, 173, 174*t*
House dust mite, allergy to, 470
Houseflies, in disease transmission, 326*t*, 376*f*, 377
HSV-1. *See* Herpes simplex viruses, type 1
HSV-2. *See* Herpes simplex viruses, type 2
HTLV. *See* Human T cell leukemia viruses
HTST pasteurization. *See* High-temperature short-time pasteurization
Human bites, 568
Human diploid cell vaccine, 551
Human genome project, 243
Human growth hormone, genetic engineering in production of, 229*f*, 236, 237*t*
Human immunodeficiency virus (HIV), 3, 355. *See also* Acquired immunodeficiency syndrome
and animal models of infection, 342
cytopathic effects of, 403*t*
infection with, 480–482
stages of, 483*f*
and risk to health care workers, 344
transmission of, 482–485
virion of, 482*i*
Human insulin, genetic engineering in production of, 229*f*, 235, 236, 237*t*
Human leukocyte antigen complex, 476–478. *See also* HLA antigens
Human rabies, 550
Human reservoirs, in spread of infection, 374
Human T cell leukemia viruses, 359
Humoral immunity (humoral immune system), 428, 433–439
Humus, 673
Hyaluronidase
therapeutic use of, 396
virulence and, 396, 522
Hybridization, nucleic acid, and classification/identification of microorganisms, 265–267, 267*t*
Hybridoma, 436
Hydatid cyst, 321, 322*f*, 641*t*, 642
transmission of, 379*t*, 641*t*
Hydrochloric acid, formation of, 27
Hydrogen, 23*t*
covalent bonding and, 27, 28*f*
electronic configuration for, 25*t*
hydrogen bonding and, 27, 29*f*
Hydrogen bonds, 27, 29*f*
antimicrobial agents affecting, 169–170
Hydrogen ions, 27
in acid-base balance, 33–34
Hydrogen molecule, formation of, 28*f*
Hydrogen peroxide, antimicrobial activity of, 183–185, 185*t*
Hydrogenomonas, 130
Hydrolase class of enzymes, 103*t*
Hydrolysis, 37
Hydrolytic enzymes, in phagocytosis, 415
Hydrophobia, in rabies, 549
Hydrothermal bacteria, 143
Hydroxyl free radicals, 147
Hydroxyl group, 35, 36*t*
Hyperbaric chamber, for gas gangrene treatment, 567
Hypersensitivity, 467–478. *See also* Allergic reaction
tissue damage in tuberculosis from, 597–598
type I (anaphylaxis), 468–471, 468*t*
type II (cytotoxic), 468*t*, 471–474
type III (immune complex), 468*t*, 474–475
type IV (cell-mediated), 468*t*, 475–478
Hypertonic solution, 85
Hypha (hyphae), 297, 298*f*
in lichen thallus, 311
Hyphomicrobium spp, 275*t*, 290
in fresh water, 683
Hypogammaglobulinemia, common variable, 481*t*
Hypothermic factors, virulence and, 397
Hypotonic solution, 85

I

I. *See* Iodine
I gene, 204
Icosahedron, viruses in shape of, 336–337
ID_{50}, 393
Idiophase, and fermentation technology, 713
Idli, microorganisms in production of, 709*t*
Idoxuridine, 501*t*, 508
IFNs. *See* Interferons
Ig. *See* Immunoglobulins
IgA, 432
characteristics of, 432*t*
deficiency of, 481*t*
secretory component of, 431*f*, 432
structure of, 431*f*
IgD, 432
characteristics of, 432*t*
structure of, 431*f*
IgE, 432–433
characteristics of, 432*t*
in hypersensitivity reactions, 432, 468
structure of, 431*f*
IgG, 431
characteristics of, 432*t*
deficiency of, 481*t*
structure of, 431*f*
IgM, 431–432
characteristics of, 432*t*
deficiency of, 481*t*
structure of, 431*f*
IL-1. *See* Interleukin-1
IL-2. *See* Interleukin-2
Illness, period of, 384
Illuminator, for compound light microscope, 53, 54*f*
Imidazoles, 501*t*, 507
for coccidioidomycosis, 609
Imipenem, with cilastin sodium, 504
Immobilized enzymes and organisms, and fermentationm technology, 714
"Immortal" cell lines, 343
Immune complex reactions, 468*t*, 474–475. *See also* Antigen-antibody reactions
autoimmune, 475
Immune complexes (antigen-antibody complexes), 435, 474–475
in glomerulonephritis, 475, 654, 665*t*
Immune deficiency. *See* Immunodeficiency
Immune globulin
for hepatitis A, 633
for hepatitis B, 635
rabies, 551
tetanus, 544
Immune response. *See also* Immune system; Immunity; Immunology
cancer and, 486–488
cell-mediated, 429, 439–443
humoral, 428, 433–439
parasites evoking, 440
primary, 443
secondary (anamnestic), 443–444
Immune serum globulin (gamma globulin), 427*f*, 428
Immune system, 426–448. *See also* Immune response; Immunity; Immunology
AIDS affecting, 2–4

and opportunistic infection, 370–371
cell-mediated, 429
disorders associated with, 467–490. *See also specific type*
in dolphins, weakened, 264
duality of, 428–429
humoral, 428
Immunity, 426–448. *See also* Immune response; Immune system; Immunology
acquired, 427–428
active, 427, 428*t*
artificially acquired, 427–428
herd, 450
humoral, 428, 433–439
naturally acquired, 427, 428*t*
passive, 427, 428*t*
Immunization (vaccination), 10–11, 427. *See also* Vaccines
schedule of for children, 452*t*
Immunodeficiency
acquired. *See* Acquired immunodeficiency syndrome
natural, 480, 481*t*
severe combined, 481*t*
Immunodiffusion tests, 454–456
Immunofluorescence, 57, 58*f*, 461–463
Immunogens, 429. *See also* Antibodies; Antigens
Immunoglobulins, 430. *See also specific type under* Ig *and* Antibodies
classes of, 431–433
structures of, 431*f*
Immunologic escape, 486–487
Immunologic memory, 443–444
Immunologic surveillance, 486
Immunologic tolerance, 443. *See also* "Self" antigens
loss of, 476
Immunology, 13. *See also* Immune response; Immune system; Immunity
diagnostic, 453–463
future of, 463
practical applications of, 449–466
Immunosorbent assay, enzyme-linked. *See* Enzyme-linked immunosorbent assay (ELISA)
Immunosuppression, 480
Immunotherapy, 487–488
Immunotoxins, 439
for AIDS, 485
Impetigo, 522, 523*f*, 531*t*
of newborn, 521
Inactivated polio vaccines, 548, 549*f*
enhanced, 548
Inactivated vaccine, 450
Inapparent infection (subclinical infection), 374
immunity conferred by, 427
Incidence, disease, 373
Incineration, for sterilization, 173, 174*t*
Inclusion conjunctivitis, 533, 534*t*
Inclusions (inclusion bodies)
bacterial, 73*f*, 87
in magnetotactic bacteria, 76
lipid, 87
viral, 356, 357*f*, 402
Incubation period, 383–384
Incubators, carbon dioxide, 151
India ink, for negative staining, 64*f*, 65
Indicator organisms, in water purity tests, 685
Indirect agglutination tests (passive agglutination tests), 457
Indirect contact transmission, in spread of infection, 377
Indirect ELISA, 463
Indirect FA tests, 460*f*, 461
Indirect selection (negative selection), mutant identification by, 211–212
Inducers, 203
Inducible enzymes, 203
Induction, 203
enzyme, 203
Industrial canning, for food preservation, 701–702
Industrial fermentations, 713–714
Industrial microbiology, 712–719
energy sources and, 718
future of, 719
genetic engineering and, 18
products of, 714–718
Industrial products, 714–718
Infant botulism, 545
Infection, 369. *See also specific type and* Disease
digestive system, 622
focal, 374
foodborne, 622, 706
inapparent. *See* Infection, subclinical
local, 374
nosocomial (hospital-acquired), 380–383
control of, 383
Enterobacter causing, 281, 380*t*
principal kinds of, 382*t*
postabortion, 563, 580*t*
postsurgical, aseptic techniques and, 381
primary, 374
secondary, 374
spread of, 375–380
subclinical, 374
immunity conferred by, 427
subcutaneous, 408
systemic (generalized), 374
Infection control committee, 383
Infection thread, 680
Infectious diseases
classification of, 373–374
etiology of, 371–373
foodborne, 706
water pollution in transmission of, 683–684
Infectious dose for 50% of hosts, 393
Infectious hepatitis. *See* Hepatitis, type A
Infectious mononucleosis, 359, 574, 581*t*
Infectious NANB hepatitis (hepatitis E), 636, 641*t*
Inflammation, 415–417
complement causing, 421
Inflammatory diseases, HLA antigens in, 479*t*
Influenza, 604–606, 611*t*
transmission of, 378*t*, 611
vaccine for, 451*t*, 606
Influenza viruses, 337, 604–606, 611*t*
type A, 338*t*, 606
type B, 338*t*, 606
type C, 338*t*, 606
Ingestion, in phagocytosis, 414
INH. *See* Isoniazid
Inhibition
contact, loss of in transformed cells, 358, 403
feedback (end-product), 108–109
zone of, 510
Inhibitors, enzyme, 107
antimicrobial agents and, 498
Innate resistance, 426
Inner membrane. *See* Plasma membrane
Inoculation, for virus growth in laboratory
animal, 342
of embryonated eggs, 342
Inorganic compounds, 31, 32–34. *See also specific type*
and acid–base balance, 33–34
acids, 33
bases, 33
salts, 33
water, 32–33
Insects, microorganisms for control of, 17–18
Bacillus thuringiensis for, 287
genetic engineering and, 229*f*, 242
Insertion sequences, 221
Insulin, genetic engineering in production of, 229*f*, 235, 236, 237*t*
Integral proteins, 81
Intercalation, chloroquine and, 509
Interferons, 421–423
function of, 441*t*
genetic engineering in production of, 236, 237*t*
and viral chemotherapy, 509
virus-infected cells producing, 356
Interleukin-1, 442
in fever, 417–418
function of, 441*t*
in pyrogenic response, 400–401
Interleukin-2, 441, 442–443
for AIDS, 485
function of, 441*t*
genetic engineering in production of, 237*t*
Interleukin number, 441
Intermediate filaments, 91
Intermediate hosts
for helminths, 319
for tapeworms, humans as, 321
Interstitial fluid, 561
Interstitial spaces, 561
Intoxications, digestive system, 622
foodborne, 706, 707*t*
Intracytoplasmic membranes, bacteriochlorophylls in, 129
Introns, 199
Invertase, commercial production of, 716*t*
Iodide ion, 26
Iodine, 23*t*
antimicrobial activity of, 179, 184*t*
effectiveness and, 183*f*
Iodophor solutions
antimicrobial activity of, 179
contamination of, 179, 180
Iodoquinol (diiodohydroxyquin), 501*t*, 509
for amoebic dysentery, 638, 641*t*
Ionic bonds, 26–27
Ionization, 33, 175–176
Ionizing radiation. *See* Radiation
IPV. *See* Inactivated polio vaccines
Iris moss, carrageen from, 310
Iron, 23*t*
in soil, 673
Iron oxide, bacterial synthesis of, 76
Irradiation. *See* Radiation
IS. *See* Insertion sequences
Ischemia, gangrene and, 566
Isodine. *See* Povidone-iodines

Isoelectric focusing, in two-dimensional PAGE, 263
Isoelectric point, in two-dimensional PAGE, 263
Isograft, 479
Isoleucine, 42*t*
Isomerase class of enzymes, 103*t*
Isomers, 37
  amino acid, 41–42
Isoniazid, 499, 500*t*
  for tuberculosis, 598, 611*t*
Isopropanol, antimicrobial activity of, 180, 184*t*
Isospora, 318*t*
Isotonic solution, 85
Isotopes, 23–24
Isotretinoin, for acne, 410, 524
Ixodes spp, 326*t*, 569, 580*t*. *See also specific type*
  diseases transmitted by, 379*t*
Ixodes dammini, 327*f*, 571*f*
  in babesiosis, 578
  diseases transmitted by, 379*t*
  in Lyme disease, 569
Ixodes pacificus, 327*f*
  in Lyme disease, 569
Izushi, microorganisms in production of, 709*t*

J

J chain
  genes coding for, diversity and, 436, 437*f*
  IgA, 431*f*, 432
  IgM, 431
J genes, diversity of antigen receptors and, 436, 437*f*
Janssen, Zaccharias, 53
Japanese B encephalitis, 552–553
Jenner, Edward, 10, 449–450
"Jesuit's powder," 509
Jock itch, 530
Joining chain (J chain)
  genes coding for, diversity and, 436, 437*f*
  IgA, 431*f*, 432
  IgM, 431

K

K. *See* Potassium
K cells. *See* Killer cells
Kappa light chains, genes coding for, diversity and, 436
Kauffmann-White scheme, 280, 624
Kefir, microorganisms in production of, 708, 709*t*
Kelp (brown algae), 308*f*, 310
  characteristics of, 309*t*
Keratin, 408, 519
Keratinase, 306
Keratitis
  Acanthamoeba, 534–535, 534*t*
  herpetic, 527, 534
Ketoconazole, 501*t*, 507
  for coccidioidomycosis, 609, 611*t*
  for histoplasmosis, 608, 611*t*
  plasma membrane affected by, 498
  for vulvovaginal candidiasis, 666
Kidneys, 650, 651*f*
  bacterial infection affecting, 653, 654, 665*t*. *See also* Glomerulonephritis
Killer cells, 443
  function of, 443*t*
  natural, 443
    function of, 443*t*
Kilometer, 53*t*
Kimchi, microorganisms in production of, 709*t*
Kinases, bacterial, virulence and, 396
Kingdoms, 256
  five, for classification, 251, 252*f*
  three, for classification, 251–254
Kinins, in inflammation, 415
Kinyoun technique, for staining *Mycobacterium tuberculosis*, 597
Kirby–Bauer test, 510
"Kissing bug" (*Triatoma*, reduviid bug), 326*t*, 327*t*
  diseases transmitted by, 379*t*
  T. cruzi transmitted by, 316, 577, 581*t*
Klebsiella spp, 274*t*, 281. *See also Klebsiella pneumoniae*
  in digestive system, 618
  in nitrogen fixation, 679
  in nosocomial infection, 380*t*
    of urinary tract, 653
Klebsiella pneumoniae, 281. *See also Klebsiella* spp
  biochemical tests for, 260
  capsule of, 73, 396, 414
  meningitis caused by, 541
  pneumonia caused by, 600, 611*t*
  and scientific nomenclature, 254–255
Koch, Robert, 10, 371, 566
Koch's postulates, 10, 371–372
  exceptions to, 372
Koplik spots, 528
Krebs cycle, 114–116
Kumiss, microorganisms in production of, 708, 709*t*
Kupffer's cells, 413
Kuru, 555*t*, 556

L

L chains (light chains), 430
  genes coding for, diversity and, 436, 437*f*
L forms, 80
  mycoplasmas compared with, 285
La Crosse strain, of California encephalitis, 552
Lac operon, 204–206, 207*t*
Lacrimal apparatus, 408*f*
  as host defense, 408–409
Lactase, commercial production of, 716*t*
Lactic acid bacteria, 121
  in cheese production, 707
  in dairy products, 706, 708
  in food production, 708, 709*t*
  in wine production, 711
Lactic acid fermentation, 120–121, 122*f*, 123*t*
Lactobacilli. *See Lactobacillus* spp
Lactobacillus spp, 275*t*, 287–288
  in bread baking, 5
  chemically defined medium for growth of, 149
  in digestive system, 618
  in food production, 5, 709*t*
  lactic acid fermentation and, 121
  phylogenetic relationships of, 258*f*
  in tooth decay, 620
  in vagina, 652, 662
Lactobacillus bulgaricus, in food production, 708, 709*t*
Lactobacillus delbrueckii, in food production, 709*t*
Lactobacillus leichmannii, in food production, 709*t*
Lactobacillus plantarum, in food production, 709*t*
Lactobacillus sanfrancisco, in food production, 5, 709*t*
Lactose, *lac* operon regulation and, 206
Lactose fermenters
  differentiation of from nonfermenters, 152–153
  MacConkey agar for growth of, 152–153
Lactose metabolism, in *E. coli*
  genes for, 203
  operon model and, 203–206
Lag phase, 157
Lagging strand, in DNA synthesis, 195
Lagoons, for water treatment, 693
Lambda light chains, genes coding for, diversity and, 436
Landfills, for solid municipal waste, 694–695
Langerhans cells, 408
Laryngitis, 591
Latent disease, 374
  syphilis as, 661
Latent viral infections, 359. *See also specific type*
Lattices, in precipitation reactions, 453
Lauric acid, 39*f*
"Lawn," 341
LD_{50}, 393
Leaching, of copper from ore, microorganisms in, 715, 717, 718*f*
Leading strand, in DNA synthesis, 195
Lecithinase, virulence and, 397
van Leeuwenhoek, Anton, 6, 53
Legionella spp, 274*t*, 279, 602
  water pollution in transmission of, 602, 684
Legionella anisa, 602
Legionella feeleii, 602
Legionella pneumophila, 4, 601, 602, 611*t*
Legionellosis (Legionnaires' disease), 4, 601, 602, 611*t*
  as exception to Koch's postulates, 373
Lenses, for compound light microscope, 53, 54*f*
Lentiviruses, 339*t*
Lepromatous leprosy (progressive leprosy), 546*f*, 547
Lepromin test, 546–547
Leprosy, 288, 546–547, 555*t*. *See also Mycobacterium leprae*
  vaccine for, 547
Leptospira spp, 274*t*, 276, 653
  axial filaments of, 75*f*
Leptospira interrogans, 653–654, 665*t*
  portal of entry for, 394*t*, 654
Leptospirosis, 276, 653–654, 665*t*
  transmission of, 378*t*, 654, 665*t*
Lethal dose for 50% of hosts, 393
Leucine, 42*t*
Leuconostoc spp, in food production, 709*t*
Leuconostoc mesenteroides
  in commercial enzyme production, 716*t*
  in food production, 709*t*
Leukemia, 487*t*
  virus-induced, 356–357
Leukemia virus

feline, 359
human T cell, 359
Leukocidins
staphylococci producing, 521
streptococci producing, 522
virulence and, 396
Leukocytes (white blood cells), 411, 411*t*, 561
polymorphonuclear, 412. *See also* Neutrophils
principal types of, 411*f*
Leukocytosis, 411
Leukoencephalopathy, progressive multifocal, 360*t*
Leukopenia, 411
Leukotrienes
in hypersensitivity reactions, 468
in inflammation, 415
Lice, 326*t*, 327*f*, 570, 581*t*. *See also Pediculus* spp
diseases transmitted by, 379*t*
Lichens, 310–312
and nitrogen fixation, 680
Ligands, for adherence, 393
Ligase class of enzymes, 103*t*
Light chains, 430
genes coding for, diversity and, 436, 437*f*
Light-independent reactions (dark reactions), in photosynthesis, 125
Light microscope, compound, 53–55
images produced by, 56*f*
Light reactions, in photosynthesis, 125
Limestone, carbon in, 677
Limnetic zone, microbial flora in, 682*f*, 683
Lipases, 122
commercial production of, 716*t*
Lipid A, 79
as endotoxin, 400
Lipid catabolism, 122–124
Lipid inclusions, 87
Lipids, 37–41. *See also specific type*
biosynthesis of, 131, 132*f*
in soil, 673
Lipopolysaccharides, in outer membrane of gram-negative cell, 79
Lipoproteins, in outer membrane of gram-negative cell, 79
Lipoteichoic acid, 78*f*, 79
Lister, Joseph, 10, 167
Lister, Joseph Jackson, 53
Listeria spp, 275*t*
flow cytometry in identification of, 267
Listeria monocytogenes, 288, 543, 555*t*
food spoilage caused by, freezing and, 704
Listeriosis, 543, 555*t*
Lithotrophs, 126
Littoral zone, microbial flora in, 682*f*, 682–683
Live vaccines, 450
Local infection, 374
Localized anaphylaxis, 468, 470
Localized juvenile periodontitis, 410
Lockjaw, 287, 399, 543. *See also* Tetanus
Log phase, 157
Logarithmic decline phase, 158
Lophotrichous flagella, 73, 74*f*
Louse. *See* Lice
Louse-borne typhus. *See* Epidemic typhus
Low-temperature food preservation, 704, 705*f*
Low-temperature microbial control, 169, 172–174, 175*t*
Lower respiratory system
diseases of, 595–610, 611*t*. *See also specific type and* Pneumonia
bacterial, 595–604, 611*t*
fungal, 606–609, 611*t*
nosocomial, 382*t*
protozoan, 609–610, 611*t*
viral, 604–606, 611*t*
structures of, 590, 591*f*
LPJ. *See* Localized juvenile periodontitis
LPS. *See* Lipopolysaccharides
LSD, *Claviceps purpurea* as source of, 403
Luminescence, microbial, in seawater, 683
Lupus erythematosus, systemic, 475, 477*t*
LVG. *See* Lymphogranuloma venereum
Lyase class of enzymes, 103*t*
Lyme disease (Lyme borreliosis), 4, 276, 569–570, 581*t*. *See also Borrelia burgdorferi*
arthropod vector for, 4, 276, 379*t*, 569, 581*t*
life cycle of, 571*f*
transmission of, 4, 276, 378*t*, 569, 581*t*
Lymph, 561
Lymph capillaries, 561
Lymph nodes, 561*f*, 562
Lymphangitis, 562
Lymphatic system. *See also* Cardiovascular system
components of, 412*f*, 413
structure and function of, 561–562
Lymphatics, 561
Lymphocytes, 411*f*, 411*t*, 412–413. *See also* B cells; T cells
Lymphocytic choriomeningitis, 360*t*, 476
Lymphogranuloma venereum, *C. trachomatis* causing, 284, 662–663, 665*t*
Lymphokines, 441
delayed-hypersensitivity T cells producing, 441
genetic engineering in production of, 236
Lymphoma, 487*t*
Burkitt's, 358, 573–574
Lymphotoxin, 441
function of, 441*t*
Lyophilization, for bacterial culture preservation, 155, 175*t*
Lysine, 43*t*
industrial microbiology in production of, 714
Lysis
in bacteriophage multiplication, 346–347
cell, teichoic acids affecting, 79
osmotic, 81
Lysogenic cells, 347
Lysogenic phages, 347
Lysogeny, 347–348
pathogenicity and, 401–402
Lysosomes
eucaryotic, 89*f*, 95
in phagocytosis, 415
Lysozymes
cell wall damaged by, 80–81, 345, 346
in host defense, 410
in phage multiplication, 345, 346
in phagocytosis, 414–415
in saliva, 619
Lytic cycle, 347

M

M protein, 396, 522
in glomerulonephritis, 475
phagocytic adherence affected by, 415
rheumatic fever and, 564
MacConkey agar, 152–153
Macrolides, 500*t*, 505
Macromolecules, 36
Macrophage activation factor, 441
function of, 441*t*
Macrophage chemotactic factor, 441
function of, 441*t*
Macrophage migration inhibition factor, 441
function of, 441*t*
Macrophages, 411*f*, 412, 413
alveolar, 413, 590
classification of, 413*f*
fixed, 413. *See also* Histiocytes
wandering, 413
Maduromycosis, 531*t*, 532–533
"Magic bullet," and modern chemotherapy, 11, 492
Magnesium, 23*t*
electronic configuration for, 25*t*
in soil, 673
Magnetite, bacterial synthesis of, 76
Magnetosomes, 76
Magnetotactic bacteria, 76
Magnetotaxis, 76
Magnification, calculation of, for compound light microscope, 53
Major histocompatibility complex, 434, 476–478. *See also* "Self" antigens
Malachite green, in endospore staining, 66
Malaria, 316, 317–318, 578–581, 581*t*. *See also Plasmodium* spp
arthropod vector for, 379*t*
transmission of, 378*t*, 581*t*
vaccine for, 454, 580–581
Male reproductive system. *See also* Reproductive system
structure and function of, 652
Malignant disease. *See* Cancer
Malolactic fermentation, 711
Malt, 710
Mammalian cells, in genetic engineering, 235–236
Manganese, 23*t*
Mannan, in yeast cell wall, 90
Mannitol salt agar, 152
Mantle, ectomycorrhizae forming, 681
Margination, in inflammation, 416
Mast cells, in hypersensitivity reactions, 468
Mastigophora, 313, 315–316, 318*t*
Matrix, mitochondrial, 93, 94*f*
Maturation
in animal virus multiplication, 355
in bacteriophage multiplication, 345–346
in RNA virus multiplication, 352*f*
Maximum growth temperature, 142
MBC. *See* Minimal bactericidal concentration
Measles, 527–528, 531*t*
German. *See* Rubella

Measles (*Continued*)
vaccine for, 451*t*, 526, 527
schedule for administration of, 452*t*
Measles virus, 338*t*, 531*t*
cytopathic effects of, 403*t*
Meat products, microorganisms in production of, 709*t*
Mebendazole, 509
for nematode infections, 641*t*, 642, 643
Mechanical transmission, in spread of infection by vectors, 376*f*, 377
Mediators, in hypersensitivity reactions, 468. *See also specific type*
Medicine, recombinant DNA technology applications for, 236–241
Medulla, of lichen thallus, 311
Mefloquine, for malaria, 581
Meiosis, 219*f*, 220
Melarsoprol, for African trypanosomiasis, 554
Membrane, undulating, 315
Membrane attack complex, 418, 420*f*
Membrane filtration
for sterilization, 173
in water testing, 687*f*
Memory, immunologic, 443–444
Memory B cells, 433
Meninges, 539, 540*f*
Meningitis, 540
bacterial, 541–543, 555*t*
diagnosis and treatment of, 542–543
organisms causing, 541*t*, 542*f*, 555*t*
gonorrheal, 656
Hemophilus, 282, 541–542, 555*t*
vaccine for, 450*t*
meningococcal, 279, 401, 541, 555*t*
vaccine for, 450*t*
pneumococcal, 541*t*, 542, 555*t*
Meningococcal meningitis, 279, 401, 541, 555*t*. *See also Neisseria meningitidis*
vaccine for, 450*t*
Meniscus spp, 274*t*
Mercuric chloride, antimicrobial activity of, 181
Mercurochrome, antimicrobial activity of, 181, 184*t*
Mercury
antimicrobial activity of, 180–181, 184*t*
bacteria in cleanup of contamination by, 38
Merozoites, of *Plasmodium*, 316–317, 579
Merthiolate, antimicrobial activity of, 181, 184*t*
effectiveness and, 183*f*
Mesoerythritol, *Brucella* growth and, 565
Mesophiles, 142, 144
Mesosomes, 82–83
Messenger RNA, 47
in bacteriophage multiplication, 345
for making complementary DNA, 233, 234*f*
in protein synthesis, 197–198, 199–201, 202*t*
for retrovirus virions, 355
Metabolic activity, bacterial numbers estimated by, 163
Metabolism (microbial), 101–140
anabolism, 101–102, 131–132
ATP generation and, 110
carbohydrate catabolism and, 111–122
catabolism, 101–102, 111–124
energy production in, 109–110
biochemical pathways of, 110–111
summary of mechanisms for, 130
energy utilization in, biochemical pathways of, 131–132
enzymes in, 103–109
feedback inhibition in, 108
integration of, 132–135
lipid catabolism and, 122–124
oxidation–reduction in, 109–110
photosynthesis, 125–130
protein catabolism and, 124
Metacercaria, 320
Metachromatic granules, 87
Metals, heavy, antimicrobial activity of, 180–181, 184*t*
Meter, 52, 53*t*
Methane. *See also* Methanogens
biomass conversion producing, 718
fermentation of, 123*t*
Methane molecule, formation of, 27, 28*f*
Methane-producing bacteria. *See* Methanogens
Methanobacterium spp, 275*t*
Methanogens, 291, 675
classification of, 253
in landfills, 694
phylogenetic relationships of, 258*f*
Methicillin, 500*t*, 503
structure of, 502*f*
Methionine, 42*t*
Methyl mercury poisoning, 707*t*
Methylene blue, as simple stain, 63, 65*t*
Methylparaben. *See also* Parabens
antimicrobial activity of, 182
Metric system, for measurement of microbes, 52–53
Metronidazole, 501*t*, 509
for amoebic dysentery, 638, 641*t*
for balantidiasis, 638, 641*t*
for giardiasis, 637, 641*t*
for trichomoniasis, 665*t*, 666
for vaginitis, 662, 665*t*
Mezlocillin, 504
Mg. *See* Magnesium
MHC. *See* Major histocompatibility complex
MIC. *See* Minimal inhibitory concentration
Mice, AIDS infection in, 342
Miconazole, 501*t*, 507
for tinea infections, 530
for vulvovaginal candidiasis, 665*t*, 666
Microaerophilic bacteria, 146*f*, 148
Microbes. *See* Microorganisms
Microbial antagonism, 370
Microbial control agents. *See* Antimicrobial agents
Microbial death curve, 170
Microbial death rate, 170
Microbial ecology, 17, 38, 240, 681, 685
Microbial enzymes, commercial production of, 716–717
Microbial genetics, 13, 190–226
Microbial growth, 141–166. *See also* Growth, microbial
control of, 167–189
Microbial intoxications, foodborne, 622, 706, 707*t*
Microbial metabolism, 101–140. *See also* Metabolism (microbial)
Microbial mining, 715
Microbial susceptibility tests, 509–510
Microbiological assays, chemically defined media in, 149
Microbiology
aquatic (water), 682–695
food, 700–722
Golden Age of, 8–11
history of, 6–14
industrial, 712–719
energy sources and, 718
future of, 719
and genetic engineering, 18
products of, 714–718
milestones in, 9*f*
modern developments in, 13–14
Nobel prizes in, 12–13*t*
pharmaceutical, 717
soil, 673–681
veterinary, 264
Microbiota, normal (normal flora), 18, 369–371
of digestive system, 618
host relationships and, 370
of reproductive system, 652
of respiratory system, 590–591
of skin, 519–520
of urinary tract, 652
of vagina, 652
Microcladia, 308*f*
Micrococcaceae, phylogenetic relationships of, 258*f*
Microdilution MIC plate, 511
Microencephalitis, *Naegleria*, 554, 555*t*
Microfilaments, 91
Microglial cells, 413
Micrographs, electron, 58–60. *See also* Microscopy
Micrometer, 53
Micromonospora spp, 275*t*
Micromonospora purpurea, antibiotics produced by, 494*t*
Micron. *See* Micrometer
Microorganisms (microbes), 4–6, 14–15, 16*f*. *See also specific type*
aquatic, 682–683
biogeochemical cycles and, 676–681
classification/identification of, 14, 250–272
amino acid sequencing in, 262, 267*t*
base composition of nucleic acids and, 265, 267*t*
biochemical tests in, 259–260, 267*t*
criteria for, 257–269
differential staining for, 259, 267*t*
five-kingdom system and, 251, 252*f*
flow cytometry in, 267, 267*t*
genetic recombination and, 267–268, 267*t*
morphologic characteristics in, 258–259, 267*t*
nucleic acid hybridization and, 265–267, 267*t*
numerical taxonomy and, 267*t*, 268–269
phage typing in, 261, 267*t*
phylogenetic hierarchy and, 257
phylogenetic relationships and, 251–254
protein analysis in, 262–265, 267*t*
scientific nomenclature and, 254–255

serology in, 260–261, 267*t*
taxonomic hierarchy and, 255–256
three-kingdom system and, 251–254
commercial production of, 717–718
disease and, 18–19
in food production, 706–712
as food source, 712
growth of, 141–166. *See also* Growth, microbial
in hospital, 380–382
human welfare and, 17–18, 38, 240, 681, 685
industrial production of, 717–718
measurement of, 52–53
metabolism of, 101–140. *See also* Metabolism (microbial)
microscopy for observation of, 53–66. *See also* Microscopy
naming, 14. *See also* Microorganisms, classification/identification of
portals of entry for, 392–395, 404*f*
portals of exit for, 379–380
in soil, 673–676
types of, 14–15, 16*f*
in water, 682–683
water quality affected by, 683
Microscopy, 53–66
brightfield, 55, 56*f*, 61*t*
compound light, 53–55
images produced by, 56*f*
darkfield, 55–57, 61*t*
electron, 58–60, 61*t*
fluorescence, 57–58, 61*t*
instruments for, 53–60, 61*t*
phase-contrast, 56*f*, 57, 61*t*
preparation of specimens for, 60–66
scanning electron, 59*f*, 60
transmission electron, 58–60
Microsporida, 318*t*
Microsporum spp, 302*t*
reclassification of, 303*t*
skin infections caused by, 530, 531*t*
Microtiter plates, for direct agglutination tests, 456
Microtrabecular lattice, 91
Microtubules, 91
Microwaves, microbial growth affected by, 176
Middle ear infection (otitis media), 594, 610*t*
Migration, phagocyte, in inflammation, 416–417
Miliary tuberculosis, 598
Milk, foodborne infections and, 706
Millimeter, 52, 53*t*
Millimicron. *See* Nanometer
Minerals, in soil, 673
Minimal bactericidal concentration, 510
Minimal inhibitory concentration, 510
microdilution plate for determination of, 511*f*
Minimum growth temperature, 142
Mining
microorganisms in, 715
wastes from, and water pollution, 685
Minocycline, 505
Miracidium, 320
Miso, microorganisms in production of, 709*t*
Missense mutation, 207, 209*f*
Mites, 326*t*
house dust, allergy to, 470
Mitochondrion (mitochondria), 89*f*, 93–94
classification of, 253*f*
procaryotic cells compared with, 251*t*
Mitosis, 219*f*, 220
Mixotricha, 278
MMR vaccine, 630, 640*t*
schedule for administration of, 452*t*, 526
MMWR. *See Morbidity and Mortality Weekly Report*
Mn. *See* Manganese
Mo. *See* Molybdenum
Moist heat sterilization, 171–172, 174*t*
Molds, 15, 16*f*, 297–298. *See also specific type and* Fungi; Microorganisms
antibiotics produced by, 494
black bread (*Rhizopus nigricans*), 304
and scientific nomenclature, 254–255
in food spoilage, 703*t*
slime, 312–313
Molecular biology, 13–14. *See also* Genetic engineering
and malaria vaccine development, 454
Salmonella infections traced with, 495
Molecular weight, 27–29
Molecules, 22
formation of, 24–29
Moles, 27–29
Mollusk poisoning, paralytic, 707*t*. *See also* Paralytic shellfish poisoning
Molybdenum, 23*t*
Monera, 251, 252*f*, 255*f*. *See also* Cyanobacteria
divisions and classes in, 257*t*
Monoclonal antibodies, 436–439
chimeric, 439
production of, 438*f*
genetic engineering in, 237*t*
viral attachment site and, 349
Monocytes, 411*f*, 411*t*, 412
Monokines, 442
Monomers, 36
antibody, 430
Monomorphic bacteria, 71
Mononuclear phagocytic system (reticuloendothelial system), 413
Mononucleosis, infectious, 359, 574, 581*t*
Monosaccharides, 36
Monotrichous flagella, 73, 74*f*
Moraxella spp, 279
Moraxella lacunata, 279
Morbidity, CDC data on, 386–387
Morbidity and Mortality Weekly Report, 386
Mordant, in staining, 60–63
Mortality, CDC data on, 386–387
Mosquitoes, as vectors, 326*f*, 326*t*, 575, 578, 581*t*
diseases transmitted by, 379*t*
Most probable number method
bacteria counted by, 160, 161*f*
in water testing, 686, 687*f*
Motive force, proton, 118
Mouth, bacterial diseases of, 618–622. *See also specific type*
MPN method. *See* Most probable number method
mRNA. *See* Messenger RNA
Mucor spp, 302*t*
mucormycosis caused by, 306
pulmonary infections caused by, 609, 611*t*
Mucormycosis, 306
Mucous membranes (mucosa)
as host defense, 407–410, 519
chemical factors and, 409–410
mechanical factors and, 408–409
as portals of entry, 392–393, 404*f*
of trachea, 409*f*
Mucus, in host defense, 408
Muktuk, botulism caused by, 545
Multiple fission, 314
Multiple sclerosis, HLA antigens in, 479*t*
Multiplication, 345–355. *See also* Reproduction
of animal viruses, 348–355
of bacteriophages, 345–348
lysogeny and, 347–348
T-even, 345–347
Multivalent vaccines, for influenza, 606
Mumps, 630, 640*t*
vaccine for, 451*t*, 526, 630, 640*t*
schedule for administration of, 452*t*
Mumps virus, 338*t*, 630, 640*t*
Municipal waste, solid, 694–695
Murein (peptidoglycan), 15, 77–79
Murine typhus, endemic, 283, 573, 581*t*
arthropod vector for, 379*t*, 581*t*
Mushroom poisoning, 404, 707*t*
Mutagens
chemical, 208–210
frameshift, 210
Mutants
identification of, 211–212
resistant, 510
Mutation, 191, 207–214. *See also specific type*
chemicals causing, 208–210
identification of, 212–214
and diversity of antigen receptors, 436
frequency of, 210–211
identification of, 211–212
in oncogene activation, 358
radiation causing, 210
types of, 207–210
Mutation rate, 210–211
Mutualism (mutualistic relationship), 370
in lichens, 310
Myasthenia gravis, 474, 477*t*
myc oncogene, translocation and, 358
Mycelium (mycelia), 15, 16*f*, 293
reproductive (aerial), 298
vegetative, 298
Mycetoma, 532–533. *See also* Maduromycosis
actinomycete, 533
N. asteroides causing, 288
Mycobacteria. *See Mycobacterium* spp
Mycobacterium spp, 275*t*, 288. *See also specific type*
acid-fast staining for, 65
ethambutol for infection with, 499
lipid inclusions in, 87
and resistance to microbial control, 169
in urethra, 652
water pollution in transmission of, 684
Mycobacterium avium-intracellulare, 599
Mycobacterium bovis, 598
Mycobacterium leprae, 288, 546, 555*t*. *See also* Leprosy; *Mycobacterium* spp

Mycobacterium leprae (*Continued*)
acid-fast staining for, 65
as exception to Koch's postulates, 372
special culture techniques for, 151
Mycobacterium tuberculosis, 288, 596–597, 611*t*. *See also Mycobacterium* spp; Tuberculosis
acid-fast staining for, 65, 597
auramine O staining for, 57
cell wall of, and resistance to phagocytosis, 396
isoniazid for infection caused by, 499, 598, 611*t*
and Koch's postulates, 372, 373
lipids in, 39–40
mucous membrane penetrated by, 408
phagocytosis and, 415
resistance of to microbial control, 169
respiratory tract as portal of entry for, 394*t*
skin test for, 598
and cell-mediated hypersensitivity reactions, 475
Mycology, 296–306. *See also* Fungi
Mycoplasma spp, 274*t*, 285
cell wall of, 79–80
classification of, 253*f*
phylogenetic relationships of, 258*f*
pneumonia caused by, 285, 600–601, 611*t*
Mycoplasma hominis, nongonococcal urethritis caused by, 659, 665*t*
Mycoplasma pneumoniae, 285, 595
pneumonia caused by, 285, 600–601, 611*t*
Mycoplasmal pneumonia, 285, 600–601, 611*t*
Mycoplasmas. *See Mycoplasma* spp
Mycorrhizae, 680–681
Mycoses, 304–306
cutaneous, 306, 530–532
subcutaneous, 306, 532–533
superficial, 306
systemic, 304–306
Mycotoxicoses, 706, 707*t*
Mycotoxins, 404, 637, 641*t*
Myeloperoxidase, in phagocytosis, 415
Myocarditis, 563, 581*t*
Myocardium, 563
Myxobacteria, 290
Myxococcus spp, 275*t*
Myxococcus fulvus, 289*f*
Myxospores, 290
Myxoviruses, respiratory tract as portal of entry for, 394*t*

N

N. *See* Nitrogen
N spikes. *See* Neuraminidase spikes
Na. *See* Sodium
NAD^+. *See* Nicotinamide adenine dinucleotide
NADase, 522
$NADP^+$. *See* Nicotinamide adenine dinucleotide phosphate
Naegleria fowleri, 318*t*
microencephalitis caused by, 554, 555*t*
NAG. *See* N-Acetylglucosamine
Naked viruses, 335*f*, 336
release of, 355
Nalidixic acid, 501–502
NAM. *See* N-Acetylmuramic acid
NANB hepatitis. *See* Non-A, non-B hepatitis
Nannizia spp, and reclassification of fungi, 303*t*
Nanometer, 53
Nasal passages, *S. aureus* in, 521
Nasopharyngeal carcinoma, Epstein–Barr virus in, 358–359, 574
Natural killer cells, 443
function of, 443*t*
Natural penicillins, 500*t*, 502
Natural selection, drug resistance and, 657
Naturally acquired immunity, 427
active, 427, 428*t*
passive, 427, 428*t*
Necator americanus, 323, 325*t*, 641*t*, 642. *See also* Hookworms
skin as portal of entry for, 393, 641*t*
Necrosis, 566. *See also* Gangrene
Necrotizing factors, virulence and, 397
Needham, John, 7
Negative selection (indirect selection), mutant identification by, 211–212
Negative staining, 60, 64*f*, 65
Negri bodies, 356
Neisseria spp, 274*t*, 279
genetic transformation occurring in, 215
in urethra, 652
Neisseria gonorrhoeae, 279, 655, 658*f*, 665*t*. *See also* Gonorrhea; *Neisseria* spp
adhesion of, 395
antibiotic-resistant, 657, 658
eye infection caused by, 533, 534*t*, 656
silver nitrate for, 181, 184*t*, 533, 534*t*, 656
genitourinary tract as portal of entry for, 394*t*, 665*t*
pelvic inflammatory disease and, 656
penicillinase-producing, 657, 658
tetracycline-resistant, 657, 658
transformation in identification of, 267–268
Neisseria meningitidis, 279. *See also Neisseria* spp
endotoxins produced by, 401
meningitis caused by (meningococcal meningitis), 279, 401, 541, 555t
vaccine for, 450*t*
opportunistic infection and, 371
respiratory tract as portal of entry for, 394*t*
Nematodes, 322–325, 325*t*, 642–643
and eggs infective for humans, 322–323
infections caused by, 642–643
and larvae infective for humans, 323–325
Neomycin, 500*t*, 504
sources of, 494*t*
Neonatal gonorrheal ophthalmia, 533, 534*t*, 656
silver nitrate for, 181, 184*t*, 533, 534*t*, 656
Neonatal herpes, 663–664
Nereocystis, 308*f*
Nervous system
diseases of, 539–559. *See also specific type*
bacterial, 541–547, 555*t*
fungal, 553, 555*t*
protozoan, 553–554, 555*t*
unconventional agents causing, 554–556, 555*t*
viral, 547–553, 555*t*
structure and function of, 539–540
Neuraminidase spikes, 605
Neurotoxins, 397
algae producing, 404
Clostridium botulinum producing, 399, 400*t*
Clostridium tetani producing, 399, 400*t*
diseases caused by, 400*t*
Gonyaulax spp producing, 307
Neutral solutions, 34
Neutralization reactions, 458–459, 460*f*
Neutralization tests, 459
Neutrons, 23
Neutrophils, 411*f*, 411*t*, 412, 561
defects of, host defense affected by, 410
Newborn
gonococcal eye infection in, 533, 534*t*, 656
silver nitrate for, 181, 184*t*, 533, 534*t*, 656
hemolytic disease of, 472–473
NGU. *See* Nongonococcal urethritis
Ni. *See* Nickel
Niacin (nicotinic acid), function of, 104*t*
Nickel, 23*t*
Niclosamide, 501*t*, 509
for tapeworms, 641*t*, 642
Nicotinamide adenine dinucleotide, 103
Hemophilus using, 282
Nicotinamide adenine dinucleotide phosphate, 103
Nicotinic acid (niacin), function of, 104*t*
Nifurtimox, 501, 509
for Chagas' disease, 577, 581*t*
Nigrosin, for negative staining, 65
Nitrate ion, 677
Nitrates, in food preservation, 706
Nitrification, 678
Nitrifying bacteria, 678
carboxysomes in, 87
Nitrites, in food preservation, 706
Nitrobacter spp, 130, 275*t*, 290
in nitrification, 678
Nitrofurans, 500–501
Nitrofurantoin, 500–501
for cystitis, 653, 665*t*
Nitrogen, 23*t*
covalent bonding and, 28*f*
electronic configuration for, 25*t*
for microbial growth, 145
Nitrogen cycle, 17, 677–681
Nitrogen fixation, 17, 145, 678–680
and genetic engineering in plants, 241–242
Nitrogen-fixing bacteria
nonsymbiotic, 678–679
symbiotic, 679–680
Nitrogen molecule, 677
formation of, 28*f*
Nitrosamines, and nitrites in food preservation, 706
Nitrosomonas spp, 130, 275*t*, 290
in nitrification, 678
NK cells. *See* Natural killer cells
Nobel prizes in microbiology, 12–13*t*
Nocardia spp, 275*t*, 288
acid-fast staining for, 65
skin infection caused by, 533

Nocardia asteroides, 288
Nocardioforms, 275*t*, 288
Nocardiosis, 288
Nomenclature, 14, 254–255
 binomial, 254
Non-A, non-B hepatitis
 hepatitis C, 635–636, 641*t*
 infectious (hepatitis E), 636, 641*t*
Noncompetitive inhibitors, enzyme, 107
Noncyclic photophosphorylation, 125, 126*f*
Nondairy fermentations, 709*t*, 710
Non-endospore formers, food spoilage caused by, 703*t*
Nonenveloped viruses (naked viruses), 335*f*, 336
 release of, 355
Nongonococcal urethritis, 658–659, 665*t*
 C. trachomatis causing, 284, 658–659, 665*t*
Nonhemolytic streptococci, 286
Nonionization radiation, microbial growth affected by, 175*t*, 176
Nonliving reservoirs, in spread of infection, 375
Nonpolar molecules, lipids as, 37
Nonsense codon, 201
 nonsense mutation and, 207–208
Nonsense mutation, 208, 209*f*
Nonspecific resistance, 407–425, 518
Nonspecific urethritis (nongonococcal urethritis), 658–659, 665*t*
 C. trachomatis causing, 284, 658–659, 665*t*
Nonspecific vaginitis, 662
Nonsymbiotic nitrogen-fixing bacteria, 678–679
Norfloxacin, 502
Normal flora (normal microbiota), 18, 369–371
 of digestive system, 618
 host relationships and, 370
 of reproductive system, 652
 of respiratory system, 590–591
 of skin, 519–520
 of urinary tract, 652
 of vagina, 652
North American blastomycosis, 609, 611*t*
Norwalk agent, gastroenteritis caused by, 636, 641*t*
Nose, *S. aureus* in, 521
Nosocomial infections (hospital-acquired infections), 380–383
 control of, 383
 Enterobacter causing, 281, 380*t*
 principal types of, 382*t*
Nuclear area, bacterial, 73*f*, 86
Nuclear envelope, 92
Nucleases, thermostable, testing food for, 624
Nucleic acid hybridization, and classification/identification of microorganisms, 265–267, 267*t*
Nucleic acids, 46–47
 antimicrobial agents affecting, 169–170, 498
 base composition of, and classification/identification of microorganisms, 265, 267*t*
 viral, 335
  biosynthesis of in bacteriophage multiplication, 345, 346*f*
  classification of animal viruses by, 338–339*t*
Nucleoid (nuclear area), bacterial, 73*f*, 86
Nucleolus (nucleoli), 89*f*, 92
Nucleoplasm, 92
Nucleoproteins, 46
Nucleoside analogs. *See also specific type*
 for AIDS, 485
Nucleosides, 46
Nucleosomes, 92–93
Nucleotides
 biosynthesis of, 132, 133*f*
 DNA, 46, 190–191
Nucleus
 eucaryotic, 89*f*, 92–93
 in procaryotic vs. eucaryotic cells, 96*t*
Numerical taxonomy, in classification, 268–269
Nutrasweet. *See* Aspartame
Nutrient agar, 150
Nutrient broth, 150
Nutritional patterns, 125–130
Nystatin, 501*t*, 506–507
 plasma membrane affected by, 498
 sources of, 494*t*

O

O. *See* Oxygen
O polysaccharides, 79
Objective lens, for compound light microscope, 53, 54*f*
Obligate aerobes, 146
Obligate anaerobes, 146*f*, 147
 oxygen-related enzymes of, 147*t*
Obligatory intracellular parasites, viruses as, 333
Ocean, microbial flora in, 683
Ocular lens, for compound light microscope, 53, 54*f*
OD. *See* Optical density
Oil spills
 bacteria in cleanup of, 38, 685
 water pollution and, 685
Oligodynamic action of heavy metals, 181
Olives, microorganisms in production of, 709*t*
Oncogenes, 357, 402
 activation of, 358
Oncogenic viruses, 357
 DNA, 358–359
 RNA, 359
One-step growth experiment, 347
Oocysts, of *Toxoplasma gondii*, 575, 576*f*
Operator site, in *lac* operon, 204
Operon, 204
 lac, 204–206
Operon model, 203–206
Ophthalmia neonatorum (neonatal gonorrheal ophthalmia), 533, 534*t*, 656
 silver nitrate for, 181, 184*t*, 533, 534*t*, 656
Opportunists (opportunistic pathogens), 306, 370
Opsonins, 414
Opsonization, 414
 complement in, 421
 in phagocytosis, 414
Optical density, in estimating bacterial numbers by turbidity, 162
Optimum growth temperature, 142
OPV vaccine, 548, 549*f*
 schedule for administration of, 452*t*
Oral candidiasis (thrush), 306, 533, 664. *See also Candida albicans*
Oral disease, bacterial, 618–622. *See also specific type*
Oral polio vaccine (OPV), 548, 549*f*
 schedule for administration of, 452*t*
Order, 254*f*, 256
Organelles, 89*f*, 90, 92–95. *See also specific type*
 in procaryotic vs. eucaryotic cells, 96*t*, 251*t*
Organic acids, antimicrobial activity of, 182, 185*t*
Organic compounds, 31–32, 34–47. *See also specific type*
 ATP, 46–47
 carbohydrates, 36–37
 functional groups and, 35–36
 lipids, 37–41
 macromolecules, 36
 nucleic acids, 46–47
 proteins, 41–46
 in soil, 673
Organic growth factors, microbial growth affected by, 148
Organisms, immobilized, and fermentation technology, 714
Organotrophs, 126
Ornithodorus spp, 326*t*
 diseases transmitted by, 379*t*
Ornithosis (psittacosis), 601–603, 611*t*
 transmission of, 378*t*, 611*t*
Orphan viruses, reoviruses as, 355
Orthomyxoviruses, 338*t*
 attachment sites of, 349
Osmosis, 84–86
Osmotic lysis, 81
Osmotic pressure, 85
 microbial growth affected by, 144–145, 175, 175*t*
Otitis externa, 523, 531*t*
Otitis media, 594, 610*t*
Ouchterlony test, 454, 455*f*
Outer membrane, of gram-negative cell, 79
Ovaries, 651
"Owl's eyes" cells, 631
Oxacillin, 500*t*
 structure of, 502*f*
Oxiaminiquine, for schistosomiasis, 583
Oxidation, 109, 110*f*
 beta, 122–124
Oxidation ditch, for sewage treatment, 693
Oxidation ponds, for sewage treatment, 693
Oxidation-reduction, 109–110
Oxidative phosphorylation, 110
Oxidizing agents, antimicrobial activity of, 183–185, 185*t*
Oxidoreductase class of enzymes, 103*t*
Oxygen, 23*t*
 covalent bonding and, 27, 28*f*
 electronic configuration for, 25*t*
 for microbial growth, 146–148
 singlet, 147
 in soil, 673
 toxic forms of, 147
Oxygen cycle, 17
Oxygen demand, biochemical, of sewage, 688–689
Oxygen molecule, formation of, 28*f*
Oxygenic photosynthesis, 127
Oxytetracycline, 504

Ozone
  antimicrobial activity of, 183, 185t
  for water treatment, 687–688

P

P. *See* Phosphorus
P site, 201
PAGE. *See* Polyacrylamide gel electrophoresis
Pandemic disease, 373–374
Panencephalitis, subacute sclerosing, 360, 528
Pantothenic acid, function of, 104t
Papillomaviruses, 338t
  genital warts caused by, 664, 665t
  tumors caused by, 358, 524
  warts caused by, 524, 531t
Papovaviruses, 338t, 351. *See also specific type*
  cytopathic effects of, 403t
  multiplication of, 351
  tumors caused by, 358
Para-aminobenzoic acid
  structure of, 499f
  sulfanilamide as competitive inhibitor for, 498
Parabens, antimicrobial activity of, 182, 185t
Paracoccidioides brasiliensis, 302t
Paracoccus spp, in denitrification, 678
Paragonimus westermani, 325t
  life cycle of, 320
Paralysis
  flaccid, in botulism, 544
  in poliomyelitis, 547
  spastic, in tetanus, 543, 544f
Paralytic mollusk poisoning, 707t
Paralytic rabies, 550
Paralytic shellfish poisoning, 307
Paramecium, conjugation of, 315f
Paramyxoviruses, 338t. *See also specific type*
  gastrointestinal tract as portal of entry for, 394t
  mumps caused by, 630
  polykaryocytes produced by, 356
  respiratory tract as portal of entry for, 394t
  skin infections caused by, 531t
Parasites, 15, 130, 672. *See also specific type*
  and freezing for food preservation, 704
  immune response to, 440
  obligatory intracellular, viruses as, 333
Parasitism, 370
Parenchyma, 417
Parenteral route, as portal of entry, 393, 394t, 404f
Parotid glands, in mumps, 630
Paroxysmal stage, of whooping cough, 596
Parvoviruses, 338t
Passive agglutination tests (indirect agglutination tests), 457
Passive immunity, 427
  artificially acquired, 427, 428t
  naturally acquired, 427, 428t
Passive processes, movement across membranes by, 83
Pasteur, Louis, 7–8, 8–10, 172, 703
Pasteurella spp, 274t, 282
Pasteurella multocida, 282
  bites infected with, 567
Pasteurellaceae, 281–282. *See also specific genus*
Pasteurization, 10, 172–173, 174t, 703
Patch test, for allergic contact dermatitis, 476
Pathogenesis of disease, 369
Pathogenicity. *See also* Virulence
  adherence and, 393–395, 404f
  lysogeny affecting, 401–402
  mechanisms of, 392–406
  and numbers of invading microbes, 393, 404f
  plasmids affecting, 401–402
  and portals of entry, 392–395, 404f
Pathogens, 368
  opportunistic, 306, 370
  in soil, 674–676
Pathology, 369
PCBs. *See* Polychlorinated biphenyls
PCR. *See* Polymerase chain reaction
Pectinase, commercial production of, 716t
Pediculus spp, 326t, 327f
Pediculus humanus, diseases transmitted by, 379t
Pediculus humanus corporis, 570, 581t
Pediococcus cerevisiae, in food production, 709t
Pellicle, 90, 309, 314
Pelvic inflammatory disease, 656
Penetration
  in animal virus multiplication, 349–350
  in bacteriophage multiplication, 345, 346f
  in RNA virus multiplication, 352f
Penicillin, 500t, 502–504
  allergy to, 470
  for anthrax, 566
  benzathine, 503
    for syphilis, 661–662, 665t
  cell wall synthesis affected by, 496
  for cystitis, 653, 665t
  discovery of, 13, 493–494
  for gas gangrene, 567
  gonococci resistant to, 657, 658
  for gonorrhea, 658, 665t
  for leptospirosis, 654, 665t
  natural, 500t, 502
  for otitis media, 594, 610t
  for pneumococcal pneumonia, 600, 611t
  procaine, 503
  prophylactic
    bacterial endocarditis and, 563
    rheumatic fever and, 564, 580t
  for puerperal sepsis, 562, 580t
  for scarlet fever, 593, 610t
  semisynthetic, 500t, 503
  sources of, 494t
  for streptococcal pharyngitis, 592, 610t
  structure of, 502f, 504f
  for syphilis, 661–662, 665t
Penicillin G, 500t, 502–503
  structure of, 502f
Penicillin V, 500t, 503
  structure of, 502f
Penicillinase-producing *Neisseria gonorrhoeae*, 657, 658
Penicillinases, 503
  commercial production of, 716t
  and gonococcal resistance, 657, 658
Penicillium spp, 304
  cheeses ripened by, 708
  conidiospores produced by, 300, 305f
  in food production, 708, 709t
  reclassification of, 303t
Penicillium chrysogenum, 11
  conidiospores produced by, 305f
Penicillium griseofulvum, antibiotics produced by, 494t
Penicillium notatum, 13, 493
  antibiotics produced by, 494t
Penis, 652
Pentamidine isethionate, 501t, 509
  for African trypanosomiasis, 554
PEP. *See* Phosphoenolpyruvic acid
Peptidases, 124
Peptide bonds, 42–44
Peptide cross bridge, 78f, 79
Peptide T, for AIDS, 485
Peptidoglycan, 15, 77–79
  antibiotics affecting, 496
  structure of, 78f
Pepto-Bismol. *See* Bismuth subsalicylate
Peptones, 150
Percentage of transmission, in estimating bacterial numbers by turbidity, 162
Perforin, 441
Pericarditis, 563, 580t
Pericardium, 563
Peridinium, 308f
Period of convalescence, 384
Period of decline, 384
Period of illness, 384
Period of incubation, 383–384
Periodontal disease, 410, 621–622, 640t
Periodontal pockets, 621f, 622
Periodontitis, 621f, 622
  localized juvenile, 410
Peripheral nervous system, 539, 540f
Peripheral proteins, 81
Peritrichous flagella, 73, 74f
Permeability
  blood vessel, in inflammation, 415–416
  plasma membrane, 81–82
Permeases, in facilitated diffusion, 83
Peroxidase, 147
"Persistent viral infection," 360
Person-to-person transmission, in spread of infection, 376
Perspiration, 519
  as host defense, 410, 519
Pertussis (whooping cough), 279, 595–596, 611t. *See also Bordetella pertussis*
  vaccine for, 450t, 596
    schedule for administration of, 452t
Pest control
  Bacillus thuringiensis for, 287
  microorganisms in, 17–18
    genetic engineering and, 229f, 242
Pesticides, degradation of in soil, 681
Pestivirus, 338t
Petri plates, 149
  anaerobic container for, 150
Petriellidium spp, reclassification of, 303t
Petroff–Hausser counter, for direct microscopic count of bacteria, 160–162
Pfu. *See* Plaque-forming units
pH, 33, 34f
  enzyme activity affected by, 106
  of human body fluids and common substances, 34t

microbial growth affected by, 144
of skin, in host defense, 409, 520
pH buffers, 34
Phage lysozyme. *See also* Lysozymes
in bacteriophage multiplication, 345, 346
Phage typing
in classification/identification of microorganisms, 261, 267*t*
and staphylococcal food poisoning, 624
Phages. *See* Bacteriophages
Phagocytes, 410, 561
actions of, 413
classification of, 413*f*
migration of, in inflammation, 416–417
Phagocytic vacuole (phagosome), 414
Phagocytosis, 72, 90, 410–415
mechanism of, 414–415
Phagolysosome (digestive vacuole), 414
Phagosome (phagocytic vacuole), 414
Phalloidin, 404
Pharmaceutical microbiology, 717
Pharyngeal gonorrhea, 656
Pharyngitis (sore throat), 591
Streptococcus causing, 286, 592, 610*t*
rheumatic fever and, 564, 580*t*
scarlet fever and, 592–593
"Phase halo," 56
Phase-contrast microscope, 56*f*, 57, 61*t*
Phenol
antimicrobial activity of, 177–178, 184*t*
structure of, 178*f*
Phenol coefficient, 177
Phenol coefficient test, 177
Phenolics, antimicrobial activity of, 177–178, 184*t*
Phenotype, 191
Phenylalanine, 43*t*
industrial microbiology in production of, 716
o-Phenylphenol, antimicrobial activity of, 178, 184*t*
Phialophora, 302*t*
pHisoHex. *See* Hexachlorophene
Phosphates, water polluted by, eutrophication and, 684–685
Phosphoenolpyruvic acid, in group translocation, 86
Phospholipid bilayer, 81, 82*f*
Phospholipids, 39, 40*f*
in outer membrane of gram-negative cell, 79
Phosphonoformate, for AIDS, 485
Phosphoproteins, 46
Phosphorothioate deoxynucleotides, for AIDS, 485
Phosphorus, 23*t*
electronic configuration for, 25*t*
for microbial growth, 145–146
in soil, 673
Phosphorylation, 110
oxidative, 110
substrate-level, 110
Photoautotrophs, 126–129, 129*t*
in carbon cycle, 676
Photoheterotrophs, 126, 129–130, 129*t*
Photophosphorylation, 110, 125, 126*f*
Photosynthesis, 125–130
anoxygenic, 127, 292
in carbon cycle, 676
without chlorophyll, 128
chloroplasts in, 95
in eucaryotes vs. procaryotes, 129*t*
oxygenic, 127
Phototaxis, 75
Phototrophs, 126
Phototropic bacteria, 275*t*, 291–293
Phylogenetic hierarchy, 256–257
Phylogenetic relationships, 250–254
of procaryotes, 258*f*
Phylum, 254*f*, 256
Phytophthora infestans, 306
Phytoplankton, 683
Pica, in hookworm infection, 643
Pichia spp, and reclassification of fungi, 303*t*
Picornaviruses, 338*t*, 351–354. *See also specific type*
multiplication of, 351–354
PID. *See* Pelvic inflammatory disease
Pili
bacterial, 76–77
sex, 77, 216
Pilin, 77
Pimples, 521
Pinocytosis, 91
Pinworms, 322, 323*f*, 641*t*, 642. *See also Enterobius vermicularis*
Piperazine, 509
Pityrosporum spp, on skin, 520
Placental transfer, immunity and, 427
Plague, 568–569, 580*t*. *See also Yersinia pestis*
arthropod vector for, 379*t*
bubonic, 568, 569*f*
transmission of, 378*t*
pneumonic, 568–569
transmission of, 378*t*
vaccine for, 450*t*
Planktonic algae, 307
Plant cells, in genetic engineering, 236
Plant products, nonbeverage, microorganisms in production of, 709*t*
Plant viruses, 334, 360–361
Plantae, 251, 252*f*, 255*f*
Plants, classification of, 251, 252*f*, 253*f*
Plaque (dental), 618, 619*f*
Plaque-forming units, 341
Plaque method, for detecting and counting viruses, 341
Plaques
formation of, 341
in phage typing, 261
Plasma, 411, 560
Plasma cells, B-cell differentiation to, 433, 434*f*, 435*f*
Plasma membrane (cytoplasmic membrane)
antimicrobial agents affecting, 169
bacterial, 73*f*, 81–83
antimicrobials in destruction of, 83, 498
functions of, 81–83
structure of, 81
eucaryotic, 89*f*, 90–91
in procaryotic vs. eucaryotic cells, 96*t*
Plasmid-mediated resistance, 657
Plasmids, 220–221
bacterial, 73*f*, 86, 220–221
in conjugation, 216
as cloning vectors, 231–232
dissimilation, 220
pathogenicity and, 401–402
Ti, 241
Plasmodial slime molds, 313
Plasmodium spp, 313, 316–318, 318*t*, 578–579, 581*t*. *See also* Malaria
portal of entry for, 394*t*
vaccine against, 454
vector for, 327
Plasmodium falciparum, 579
Plasmodium malariae, 579
Plasmodium vivax, 579
Plasmolysis, 145*f*
Plate counts, for bacterial growth measurement, 158–160
Platelets (thrombocytes), 411*t*
and drug-induced cytotoxic reactions, 473–474
Platyhelminthes, 319–320, 325*t*. *See also specific type*
Pleomorphic bacteria, 71
PMNs. *See* Polymorphonuclear leukocytes
Pneumococcal meningitis, 541*t*, 542, 555*t*
Pneumococcal pneumonia, 286, 599–600, 611*t*
vaccine for, 450*t*, 600
Pneumococcus. *See Streptococcus pneumoniae*
Pneumocystis carinii, 302*t*, 304*f*, 318*t*, 609–610, 611*t*
classification of, 303–304, 609
life cycle of, 609–610
and opportunistic infection in AIDS patients, 370–371, 609
Pneumocystis pneumonia, 303–304, 609–610, 611*t*
in AIDS patients, 2, 370–371, 609
Pneumonia, 595
atypical (mycoplasmal), 285, 600–601, 611*t*
bacterial, 599–601, 611*t*
chlamydial, 285, 603, 611*t*
Hemophilus influenzae, 282
Klebsiella, 600, 611*t*
mycoplasmal, 285, 600–601, 611*t*
pneumococcal, 286, 599–600, 611*t*
vaccine for, 450*t*, 600
Pneumocystis, 303–304, 609–610, 611*t*
in AIDS patients, 2, 370–371, 609
progressive, 360*t*
viral, 604, 611*t*
Pneumonic plague, 568–569
transmission of, 378*t*
Poi, microorganisms in production of, 709*t*
Point mutations, 207, 208*f*
and diversity of antigen receptors, 436
Poisoning, food. *See* Food poisoning
Poisons, enzyme, 107
Polar molecule, water as, 32
Poliomyelitis (polio), 547–549, 555*t*
vaccine for, 451*t*, 548–549
schedule for administration of, 452*t*
Poliovirus, 338*t*, 547–548, 555*t*
cytopathic effects of, 403*t*
gastrointestinal tract as portal of entry for, 394*t*, 547
multiplication of, 351–354
size of, 334*f*
Pollution
in toxic dumps, microbes in cleanup of, 17, 38, 229*f*, 240, 681
water, 683–685

Polyacrylamide gel electrophoresis, 262–265
two dimensional, 263–265
Polychlorinated biphenyls
bioremediation of, 240
in dolphins, 264
Polyenes, 501*t*, 506–507
Polyhedral viruses, 336–337
Poly-β-hydroxybutyric acid, in lipid inclusions, 87
Polykaryocytes, 356
Polymerase chain reaction, 239–241
Polymerases
DNA, in DNA synthesis, 195–196
RNA
in DNA synthesis, 195
in transcription, 198
Polymers, 36
Polymorphonuclear leukocytes, 412. *See also* Neutrophils
Polymorphs. *See* Polymorphonuclear leukocytes
Polymyxin B, 500*t*, 505, 506
structure of, 506*f*
Polymyxins
plasma membrane destroyed by, 83
sources of, 494*t*
Polyomaviruses, 338*t*
cytopathic effects of, 403*t*
tumors caused by, 358
Polypeptides, 44, 500*t*, 505–506
Polyribosome, 201
Polysaccharide granules, 87
Polysaccharides, 37
biosynthesis of, 131
Porcine growth hormone, genetic engineering in production of, 237*t*
Pores, in nuclear membrane, 92
Porins, 79
Pork, *Trichinella spiralis* in, 643
Pork tapeworm (*Taenia solium*), 639–640, 641*t*
Portals of entry, 392–395, 404*f*
preferred, 393, 394*t*
Portals of exit, 379–380
Positive selection (direct selection), mutant
identification by, 211
Postabortion infection, 563, 580*t*
Postsurgical infection, 382*t*. *See also specific organ or structure affected*
aseptic techniques and, 381
Potassium, 23*t*
in soil, 673
Potassium clavulanate, penicillin combined with, 503–504
Potassium ion, 26
Potassium sorbate (sorbic acid)
antimicrobial activity of, 182, 185*t*
in food preservation, 705, 706
Potato spindle tuber viroid, 361*f*
Potato yellow dwarf virus, 361*t*
Pour plate method, for plate count, 158, 159*f*
Povidone-iodine solutions
antimicrobial activity of, 179
contamination of, 179, 180
Poxviruses, 337, 338*t*, 350. *See also specific type*
multiplication of, 350
respiratory tract as portal of entry for, 394*t*
PPNG. *See* Penicillinase-producing *Neisseria gonorrhoeae*
Praziquantel, 501*t*, 509
for schistosomiasis, 583
Precipitation curve, 454, 455*f*
Precipitation reactions, 453–456
Precipitin ring test, 454, 455*f*
Precursors, in amino acid synthesis, 132
Predator, *Bdellovibrio* as, 62
Predisposing factors, 383
Preemergent flagellum, 309
Preferred portal of entry, 393, 394*t*
Pregnancy
cytomegalovirus infection and, 631, 640*t*
ectopic (tubal), salpingitis and, 656
genital herpes and, 663–664
gonococcal eye infection and, 533, 534*t*, 656
silver nitrate for, 181, 184*t*, 533, 534*t*, 656
rubella and, 528–529
Preservation, food, 700–706
Prevalence, disease, 373
Primaquine, for malaria, 581
Primary atypical pneumonia (mycoplasmal pneumonia), 285, 600–601, 611*t*
Primary cell lines, 343
Primary immune response, 443
Primary infection, 374
Primary metabolites, and fermentation technology, 713
Primary producers, 683
Primary sludge, in sewer treatment, 688
Primary stain, 63
Primary structure of protein, 44, 45*f*
Primary treatment of sewage, 688
Primaxin. *See* Imipenem, with cilastin sodium
Primer, RNA, in DNA synthesis, 195
Prion, 360
Privileged site, 478
Privileged tissue, transplantation of, 479
Probes, 234–235
in identification of bacteria, 267
in medical diagnosis, 241
for mycoplasmal pneumonia, 601
for *Salmonella*, 625
Procaine penicillin, 503
Procaryotae, 251, 252*f*, 255*f*
divisions and classes in, 257*t*
Procaryotic cells (procaryotes), 15, 69–70, 71–89. *See also* Bacteria
axial filaments of, 75–76
cell wall of, 73*f*, 77–81
atypical, 79–80
composition and characteristics of, 77–79
damage to, 80–81
structures external to, 72–77
structures internal to, 81–89
classification of, 251, 252*f*
cytoplasm of, 73*f*, 86
endospores formed by, 88–89
eucaryotic cells compared with, 96*t*, 251*t*
fimbriae of, 73*f*, 76–77
flagella of, 73–75, 73*f*
glycocalyx of, 72–73
inclusions in, 73*f*, 87
movement across membranes in, 83–86
nuclear area of, 73*f*, 86
photosynthesis in, 129*t*
phylogenetic relationships of, 258*f*
pili of, 76–77
plasma (cytoplasmic) membrane of, 73*f*, 81–83
ribosomes in, 73*f*, 86–87
size/shape/arrangement of, 71–72
structure of, 73*f*
Prodromal period, 384
Profundal zone, microbial flora in, 682*f*, 683
Progesterone, microbial production of, 717
Proglottids, 321
Progressive encephalitis, 360*t*
Progressive multifocal leukoencephalopathy, 360*t*
Progressive pneumonia, 360*t*
Proline, 43*t*
Promoter site, 198
in *lac* operon, 204
Prontosil, 493
proP gene, *Salmonella*, 205
Prophage, 347
Prophylactic penicillin administration
bacterial endocarditis and, 563
rheumatic fever and, 564, 580*t*
Propionibacterium spp, 275*t*
in cheese production, 708
vitamin B_{12} produced by, 717
Propionibacterium acnes, 288, 531*t*
on skin, 520, 524
Propionic acid, fermentation of, 123*t*
Propylparaben. *See also* Parabens
antimicrobial activity of, 182
Prospective study, in epidemiology, 386
Prostaglandins
in hypersensitivity reactions, 468
in inflammation, 415
Prosthecae, 288
Proteases, 124
commercial production of, 716*t*, 717
virulence and, 397
Protein analysis, 262–265
Protein catabolism, 124
Protein kinases, oncogenes producing, 357
Proteins, 41–46. *See also specific type*
antimicrobial agents affecting, 169–170
antiviral, 422
conjugated, 44–46
genes coding for, 191
repressor, in *lac* operon, 204–206
simple, 44
single cell (SCP), 18, 712
in soil, 673
structure of, 44–46
synthesis of, 132
antibiotics affecting, 497–498
RNA in, 197–202
Proteus spp, 281
in digestive system, 618
endotoxins produced by, 401
in nosocomial infection, 380*t*
pyelonephritis caused by, 653, 665*t*
urinary tract infections caused by, 653, 665*t*
Proteus mirabilis, septicemia caused by, 562
Protista, 251, 252*f*
Protists, classification of, 252*f*, 253*f*
Proto-oncogenes, 357
Proton motive force, 118
Proton pumps, 117
Protons, 23
Protoplast, 80
Protoplast fusion, 232

Protozoans, 15, 16*f*, 313–319. *See also specific type and* Microorganisms
biology of, 313–315
cardiovascular and lymphatic system diseases caused by, 575–581, 581*t*
characteristics of, comparison of, 297*t*
digestive system diseases caused by, 637–639, 641*t*
drugs for infection with, 501*t*, 509. *See also specific type and* Antiprotozoan drugs
encystment of, 315
immune response to, 440
lower respiratory system diseases caused by, 609–610, 611*t*
medically important, 315–319
nervous system diseases caused by, 553–554, 555*t*
nutrition of, 314
pathogenetic properties of, 403–404
phyla of, 315–319
reproduction of, 314–315
respiratory system diseases caused by, 609–610, 611*t*
in soil, 674
Prourokinase, genetic engineering in production of, 237*t*
Provirus, 355
Pseudohypha, 298
Pseudomonadaceae, phylogenetic relationships and, 258*f*
Pseudomonads. *See Pseudomonas* spp
Pseudomonas spp, 274*t*, 277–279. *See also specific type*
in denitrification, 678
dermatitis caused by, 523, 531*t*
Entner–Doudoroff pathway in, 114
flow cytometry in identification of, 267
in fresh water, 683
povidone-iodine solutions contaminated by, 179, 180
resistance of, 169, 182, 523
skin infections caused by, 523, 531*t*
in toxic cleanup, 38, 240
urinary tract infections caused by, 653
Vitamin B_{12} produced by, 717
water pollution in transmission of, 684
Pseudomonas aeruginosa, 277. *See also Pseudomonas* spp
amino acid sequencing of, 262*f*
meningitis caused by, 541
in nosocomial infection, 380*t*, 381–382
povidone-iodine solutions contaminated by, 180
pyelonephritis caused by, 653
septicemia caused by, 562
skin infections caused by, 523, 531*t*
Pseudomonas andropogonis, 277*f*
Pseudomonas cepacia, povidone-iodine solutions contaminated by, 180
Pseudomonas fluorescens, genetic engineering of, 237*t*, 242
Pseudomonas syringae, ice-minus bacterium, genetic engineering in production of, 237*t*
Pseudomurein, 80
Pseudopods, 15, 313
Psittacosis (ornithosis), 601–603, 611*t*
transmission of, 378*t*, 611*t*
PSTV. *See* Potato spindle tuber viroid
Psychrophiles, 142
Psychrotrophs, 142
Puerperal sepsis (childbirth fever, puerperal fever), 562–563, 580*t*
Pulmonary anthrax, 566
Pulp, of tooth, decay of, 620–621
Pumps, proton, 117
Pure cultures, obtaining, 153–155
Purines, 46
biosynthesis of, 132, 133*f*
Purity, of water, tests for, 685–686
Purple bacteria, 291–293
nonsulfur, 130, 292–293
sulfur, 129, 129*t*, 291*f*, 292
in fresh water, 683
Pus, 416–417
Putrefactive anaerobic food spoilage, 703*t*
Pyelonephritis, 653, 665*t*
Pyocyanin (pyocyanine), 523
Pyrantel pamoate, for pinworms, 641*t*, 642
Pyrazinamide, for tuberculosis, 598
Pyridoxine (vitamin B_6), function of, 104*t*
Pyrimethamine
with sulfadoxine, for malaria, 581
for toxoplasmosis, 576–577
Pyrimidines, 46
biosynthesis of, 132, 133*f*
Pyrogen, endogenous, 401. *See also* Interleukin-1
Pyrogenic response. *See also* Fever
endotoxins causing, 400–401
Pyuria, in cystitis, 653

Q

Q. *See* Coenzyme Q
Q fever, 283, 603–604, 611*t*. *See also Coxiella burnetii*
transmission of, 378*t*, 611*t*
Quaternary ammonium compounds, antimicrobial activity of, 182, 184*t*
Quaternary structure of protein, 44, 45*f*
Quats. *See* Quaternary ammonium compounds
Quellung reaction, 600
Quinacrine, 509
for giardiasis, 637, 641*t*
Quinine, 509
discovery of, 11
for malaria, 581
Quinolones, 500*t*

R

R-determinant, 220*f*, 221
R factors (resistance factors), 220*f*, 221, 401, 510
nosocomial infections and, 382
R group, 41
Rabbits, in tularemia transmission, 564–565, 580*t*
Rabies, 549–551, 553*f*, 555*t*
dumb (paralytic), 550
furious, 549–550
human, 550
transmission of, 378*t*
vaccine for, 451*t*, 551
Rabies immune globulin, 551
Rabies virus, 338*t*, 355, 549, 555*t*
inclusion bodies produced by, 356
Radiation
for AIDS, 485
for food preservation, 704–705
microbial growth affected by, 175–176, 175*t*
mutations caused by, 210
Rapid plasma reagin test, for syphilis, 661
Rapid-identification tests, 259–260
ras oncogene, mutation and, 358
Rat bite fever, 567–568, 580*t*
Rat flea, as vector, 326*t*, 327*f*, 568, 580*t*, 581*t*
diseases transmitted by, 379*t*
Rats, *Yersinia pestis* carried by, 281
Reaction rate, 30
Reagin-type antibodies, in syphilis diagnosis, 661
Recalcitrance, 681
Recipient cell, 214
Recombinant DNA, 14, 214, 228. *See also* Recombinant DNA technology
Recombinant DNA technology (genetic engineering), 6, 13–14, 18, 227–247
advent of, 227–228
agricultural applications for, 241–243
alternatives to vectors in, 232–233
basic research and, 236–241
clone identification and, 234–235
and cloning with a plasmid vector, 231–232
definition of, 228
and DNA sources, 233–234
in food production, 237*t*, 706
future of, 243
gene libraries in, 233–234
clone identification and, 234–235
gene product synthesis and, 235–236
medical applications for, 236–241
procedures used in, 228–231
products of, 236, 237*t*
restriction enzymes and, 231
safety issues in, 243
synthetic DNA for, 234
Recombination, genetic, 214–222. *See also* Genetic engineering
in bacteria
conjugation and, 216
transduction and, 218
transformation and, 214–216
and classification/identification of microorganisms, 267–268, 267*t*
in eucaryotes, 218–222
plasmids and, 216, 220–221
transposons and, 221–222
Recycling, microbes in, 17, 38, 240, 681, 685. *See also* Bioremediation
Red algae, 308*f*, 310
characteristics of, 309*t*
Red blood cells (erythrocytes), 411*t*, 560
in malaria, 579–580
Red tide, 307, 310, 404
Redi, Francesco, 6
Rediae, 320
Reducing media, 150, 153*t*
Reduction, 109
Reduviid bug ("kissing bug"), 326*t*, 327*t*
diseases transmitted by, 379*t*. 581*t*
T. cruzi transmitted by, 316, 577
Refractive index, 55
Refrigeration
for bacterial culture preservation, 155
microbial growth affected by, 169, 173–174, 175*t*
Regulatory sites, in *lac* operon, 204

Regulatory T cells, 441
Rejection, transplant, monoclonal antibodies in prevention of, 439
Relapsing fever, 276, 569, 580*t*
 arthropod vector for, 379*t*, 580*t*
Release
 in animal virus multiplication, 355
 in RNA virus multiplication, 352*f*
 of virions from host cell, in bacteriophage multiplication, 346–347
Rennin
 in casein formation, 707
 commercial production of, 717
 genetic engineering in production of, 237*t*
Reoviruses, 339*t*, 355. *See also specific type*
 multiplication of, 355
Repair of tissue, 417
Repair synthesis, in radiation damage, 210, 211*f*
Repellent, 75
Replica-plating, for mutant identification, 211–212
Replication, DNA, 192–197
 in bacteria, 196–197
 rate of, 197
Replication forks, 193, 195*f*
 multiple, 197
Repression, 203
Repressor protein, in *lac* operon, 204–206
Repressors, 203
Reproduction
 asexual
  in algae, 310
  in eucaryotes, 220
  protozoan, 314
 of helminths, 319
 sexual
  in algae, 310
  in eucaryotes, 218–220
   vs procaryotes, 96*t*
  protozoan, 314–315
 viral, 345–355
Reproductive mycelium, 298
Reproductive system
 diseases of, 654–666. *See also specific type*
  bacterial, 654–663, 665*t*
  fungal, 664–666, 665*t*
  protozoan, 665*t*, 666
  viral, 663–664, 665*t*
 female, 651
 male, 652
 normal flora of, 652
 as portal of entry, 393, 394*t*, 404*f*
 structure and function of, 651–652
Reservoirs, in spread of infection, 374–375
Residual body, 415
Resistance, 19, 407
 drug, 510–511
  and antibiotics in animal feed, 495
  chronically mediated, 657
  of gonococci, 657, 658
  multiple, 657
  and natural selection, 657
  plasmid mediated, 657
  of *Staphylococcus aureus*, 286
  transposons and, 222
 innate, 426
 nonspecific, 407–425, 518
 specific, 407. *See also* Immunity
Resistance factors, 220*f*, 221, 401, 510
 nosocomial infections and, 382
Resistance transfer factor, 221
Resistant mutants, 510
Resolution, 53–55
Resolving power, 53–55
Respiration, 112*f*, 114
 aerobic, 114, 118–120
  anaerobic respiration compared with, 123*t*
  fermentation compared with, 123*t*
 anaerobic, 114, 120, 678
  aerobic respiration compared with, 123*t*
  fermentation compared with, 123*t*
Respiratory syncytial virus, 604
Respiratory system. *See also* Lower respiratory system; Upper respiratory system
 diseases of, 589–616. *See also specific type*
  adenoviruses causing, 350
  bacterial, 592–594, 595–604, 610*t*, 611*t*
  fungal, 606–609, 611*t*
  nosocomial, 382*t*
  protozoan, 609–610, 611*t*
  viral, 594–595, 604–606, 611*t*
 function of, 589–590
 in host defense, 409
 normal flora of, 590–591
 as portal of entry, 392–393, 394*t*, 404*f*
 structure of, 589–590
Restriction enzymes, 230*f*, 231
Reticular dysgenesis, 481*t*
Reticulate body, 284, 602
Reticuloendothelial system (mononuclear phagocytic system), 413
Retorts, 171
Retrospective study, in epidemiology, 386
Retroviruses, 339*t*, 355. *See also specific type*
 cancer caused by, 359
 multiplication of, 354, 355*f*
rev gene products, in AIDS treatment, 485
Reverse transcriptase
 for making complementary DNA, 233, 234*f*
 in retrovirus multiplication, 355
Reverse transcription viruses, 339*t*
Reversible reactions, 30
Reye's syndrome, 525–526
Rh blood group system, 471–473
 hemolytic disease of the newborn and, 472–473
 transfusions and, 472
Rh factor, 471–472
Rh incompatibility, 472
Rhabdoviruses, 338*t*, 355. *See also specific type*
 cytopathic effects of, 403*t*
 multiplication of, 355
 portal of entry for, 394*t*
Rheumatic fever, 563–564, 580*t*
 HLA antigens in, 479*t*
Rheumatoid arthritis, 475, 477*t*
Rhinitis, allergic, 470
Rhinoviruses, 338*t*. *See also specific type*
 common cold caused by, 594–595, 611*t*
Rhizobium spp, 274*t*, 279
 Entner–Doudoroff pathway in, 114
 and genetic engineering for enhanced nitrogen fixation, 242
 industrial microbiology in production of, 718
 plant growth and, 679–680
 and symbiosis in nitrogen fixation, 145, 679–680
Rhizobium meliloti, genetic engineering in production of, 237*t*
Rhizoids, 311
Rhizopus spp, 302*t*
 mucormycosis caused by, 306
 pulmonary infections caused by, 609, 611*t*
 sporangiospores produced by, 300
Rhizopus nigricans, 304
 and scientific nomenclature, 254–255
Rhizopus oligosporus, in food production, 709*t*
Rhizopus oryzae, in food production, 709*t*
Rhizosphere, nonsymbiotic nitrogen-fixing bacteria in, 678
Rhodopseudomonas, 130
Rhodospirillum spp, 275*t*
 in nitrogen fixation, 679
Ribavirin, 501*t*, 507
 for AIDS, 485
 for viral pneumonia, 604, 611*t*
Riboflavin (vitamin B_2)
 commercial production of, 717
 function of, 104*t*
Ribonucleic acid. *See* RNA
Ribosomal RNA, 47
 in protein synthesis, 191, 197, 199, 202*t*
Ribosomes
 eucaryotic, 89*f*, 93, 254
 procaryotic, 73*f*, 86–87
 in procaryotic vs. eucaryotic cells, 96*t*, 251*t*
Rice, and symbiotic relationship with microorganisms, 290
"Rice water stools," in cholera, 626
Rickettsia spp, 274*t*, 283, 284*f*. *See also specific type*
Rickettsia prowazekii, 283, 284*f*, 570, 581*t*
Rickettsia rickettsii, 283, 573, 581*t*
 portal of entry for, 394*t*
Rickettsia typhi, 283, 573, 581*t*
Rifabutin, for AIDS, 485
Rifampin, 500*t*, 506
 for legionellosis, 601, 611*t*
 for leprosy, 547
 for tuberculosis, 598, 611*t*
Rifamycins, 500*t*, 506
Riftia, hydrothermal bacteria and, 143
RIG. *See* Rabies immune globulin
Ring stage, 316
Ringworm, 530, 531*t*
 transmission of, 378*t*
Rise in titer, 456
RNA, 46, 47
 messenger. *See* Messenger RNA
 processing of in eucaryotic cells, 198–199
 and protein synthesis, 197–202
 ribosomal. *See* Ribosomal RNA
 transfer. *See* Transfer RNA
RNA polymerases
 in bacteriophage multiplication, 345
 in DNA synthesis, 195
 RNA-dependent, 354

in transcription, 198
RNA primer, in DNA synthesis, 195
RNA viruses, 15, 335, 340*f*. *See also specific type*
antisense strand, 338–339*t*, 354
biosynthesis for, 351–355
double-stranded, 339*t*
multiplication of, 351–355
oncogenic, 359
sense strand, 338*t*, 351–354
tumors caused by, 357
RNA-dependent RNA polymerase, 354
Rocky Mountain spotted fever, 283, 572*f*, 573, 581*t*
arthropod vector for, 379*t*, 581*t*
cycle of, 572*f*
distribution of, 572*f*
transmission of, 378*t*, 581*t*
Rods
gram-negative aerobic, 274*t*, 277–280
gram-negative anaerobic, straight, curved, and helical, 274*t*, 282, 283*f*
gram-negative facultatively anaerobic, 146, 274*t*, 280–282
gram-positive, antibiotics produced by, 494*t*
gram-positive endospore-forming, 275*t*, 286–287
gram-positive nonsporing
irregular, 275*t*, 288
regular, 275*t*, 287–288
Roll tubes, 151
Root beer, microorganisms in production of, 711*t*
Rotavirus, gastroenteritis caused by, 636, 641*t*
Rough endoplasmic reticulum, 89*f*
Roundworms (Aschelminthes), 15, 321–325, 325*t*
Route of transmission. *See* Transmission, of disease
RPR test, for syphilis, 661
rRNA. *See* Ribosomal RNA
RTF. *See* Resistance transfer factor
Rubella, 338*t*, 528–530, 531*t*
pregnancy and, 528–529
vaccine for, 451*t*, 526, 529–530
schedule for administration of, 452*t*
Rubella virus, 528, 531*t*
Rubeola (measles), 527–528, 531*t*
vaccine for, 451*t*, 526, 527
schedule for administration of, 452*t*
Rum, microorganisms in production of, 711*t*
"Run," 74*f*, 75

S

S. *See* Sulfur
Sabin vaccine, 548, 549*f*
Sabouraud's dextrose agar, 152
Sac fungi (ascomycota), 302–303*t*, 304
Saccharomyces spp. *See also specific type*
fermentation by, 299
in food production, 709*t*
Saccharomyces bayanus, in alcoholic beverage production, 711*t*
Saccharomyces beticus, in alcoholic beverage production, 711*t*
Saccharomyces carlsbergensis, in alcoholic beverage production, 711*t*
Saccharomyces cerevisiae, 5. *See also Saccharomyces* spp
in alcoholic beverage production, 711*t*
asexual reproduction in, 220
budding of, 299*f*
in commercial enzyme production, 716*t*
in food production, 5, 709*t*
gene products expressed by, 235
Saccharomyces exiguus, in food production, 5, 709*t*
Saccharomyces rouxii, in food production, 709*t*
Safety, of genetic engineering of microbes, 243
Safranin
in endospore staining, 66
in negative staining, 65
as simple stain, 63, 65*t*
SAIDS. *See* Simian AIDS
St. Louis encephalitis, 552, 553*t*
Sake, microorganisms in production of, 711*t*
Saliva, as host defense, 409
Salk vaccine, 548
Salmonella spp, 274*t*, 280, 624. *See also specific type and* Salmonellosis
antibiotic-resistant, and antibiotics in animal feed, 495
Bdellovibrio affecting, 62
biochemical tests for, 259, 280
Brilliant Green agar for growth of, 152
DNA probes in identification of, 266*f*, 267
foods contaminated by, 625
gastroenteritis caused by, 624–625, 640*t*
and glyphosate resistance, 241
mutant, Ames test for, 213–214
stress-induced genes and, 205
water pollution in transmission of, 684*t*
Salmonella choleraesuis, gastrointestinal tract as portal of entry for, 394*t*
Salmonella dublin, 280, 624
antibiotic-resistant, and antibiotics in animal feed, 495
Salmonella eastbourne, 624
Salmonella enteritidis, gastrointestinal tract as portal of entry for, 394*t*
Salmonella newport
antibiotic-resistant, and antibiotics in animal feed, 495
molecular biology in tracing infections with, 495
Salmonella typhi, 280, 625, 640*t*. *See also Salmonella* spp; Typhoid fever
bismuth sulfite agar for growth of, 152
endotoxins produced by, 401
gastrointestinal tract as portal of entry for, 394*t*
phage typing of, 261*f*
vaccine for, 205
Salmonella typhimurium, 280, 624
antibiotic-resistant, and antibiotics in animal feed, 495
gastrointestinal tract as portal of entry for, 394*t*
Salmonellosis, 280, 624–625, 640*t*. *See also Salmonella* spp
transmission of, 378*t*, 640*t*
Salpingitis, 656
Salts, 33. *See also* Sodium
high concentrations of in water, *Halobacterium* and, 128
Salvarsan, discovery of, 11
Sand filtration, for water purification, 686
Sanitization, 168*t*
Saprophytes, 130
Sarcinae, cell arrangement of, 70*f*, 71
Sarcodina, 313, 315, 318*t*
Sarcoma, 487*t*
virus-induced, 357
Sarcoma viruses, 359
Sarcoptes scabiei, 529
Sargassum, 306
Sartorya spp, and reclassification of fungi, 303*t*
Saturated enzyme, 106*f*, 107
Saturated fat, 39
Sauerkraut, microorganisms in production of, 709*t*
Saxitoxin, 404
Scabies, 529
Scalded skin syndrome, 400*t*, 521
Scalp, ringworm of, 530
Scanning electron microscope, 59*f*, 60
Scarlet fever, 286, 400*t*, 522, 592–593, 610*t*
Schaeffer–Fulton endospore stain, 64*f*, 66, 89
Schick test, 459
Schistosoma spp, 325*t*, 581*t*, 582–583
Schistosomiasis, 581*t*, 582–583
water pollution in transmission of, 684
Schizogony, 314
Scientific nomenclature, 14, 254–255
Sclerotia
of *Aspergillus*, food spoilage and, 703
of *Claviceps purpurea*, 403
Scolex, 320, 321*f*
Scombroid poisoning, 707*t*
SCP. *See* Single-cell protein
Scrapie, 360, 555–556
Scratches, systemic diseases caused by, 567–568
Se. *See* Selenium
Seawater, microbial flora in, 683
Sebum, 519
in acne, 523–524
in host defense, 409–410, 519
Propionibacterium acnes supported by, 520, 524
Secondary immune response, 443–444
Secondary infection, 374
Secondary metabolites, and fermentation technology, 713–714
Secondary structure of protein, 44, 45*f*
Secondary treatment of sewage, 689–691
Secretors, blood type and, 471
Secretory component, IgA, 431*f*, 432
Selection, clonal, 443, 444*f*
Selective media, 152–153, 153*t*, 154*f*
Selective toxicity, 492
Selectively permeable membrane, 81–82
Selenium, 23*t*
"Self" antigens, 433, 434, 435*f*. *See also* Tolerance
cytotoxic T cells and, 441
Semen
AIDS transmission and, 344, 482
hepatitis B transmission and, 635, 641*t*
Semiconservative replication, 193
Semipermeable membrane, 81–82

Semisynthetic penicillins, 500*t*, 503
Semmelweis, Ignatz, 167
Sense strand RNA viruses, 338*t*, 351–354
Sepsis, puerperal (childbirth fever, puerperal fever), 562–563, 580*t*
Septate hyphae, 297, 298*f*
Septic shock, 401, 562
Septic tanks, 692–693
Septicemia, 374, 562
Septum (septa), of hyphae, 297
Serial dilutions, for bacterial growth measurement, 158, 159*f*
Serine, 42*t*
 in oncogenic transformation, 357
Seroconversion, 457
Serology
 in classification/identification of microorganisms, 260–261, 267*t*
 for viral identification, 343
Serovars (serotypes), for *Salmonella*, 280, 624
Serratia spp, 281
Serratia marcescens, 281
 in nosocomial infection, 380*t*
 septicemia caused by, 562
Serum hepatitis. *See* Hepatitis, type B
Serum (sera), 427. *See also* Antiserum (antisera)
Serum sickness, 475
Severe combined immunodeficiency, 481*t*
Sewage ponds, 693
Sewage treatment, 688–694
 microbes in, 17
 primary, 688
 secondary, 689–691
 tertiary, 693–694
 Zoogloea in, 280, 690, 691*f*
Sex pili, 77, 216
Sexual dimorphism, of *Ascaris lumbricoides*, 322
Sexual reproduction
 in algae, 310
 in eucaryotes, 218–220
 in procaryotes, 96*t*
 protozoan, 314–315
Sexual spores, 299, 300–301
Sexually transmitted diseases, 654–663, 665*t*. *See also specific type*
 incidence of, 655*f*
Sheathed bacteria, 275*t*, 290
Sheep scrapie, 360, 555–556
Shellfish poisoning, paralytic, 307. *See also* Paralytic mollusk poisoning
Sherry, microorganisms in production of, 711*t*
Shiga bacillus, 626. *See also Shigella* spp
Shigella spp, 274*t*, 280, 400*t*, 626
 adhesion of, 395
 biochemical tests for, 259
 dysentery caused by, 626, 640*t*
 gastrointestinal tract as portal of entry for, 394*t*
 water pollution in transmission of, 684
Shigella boydii, 626
Shigella dysenteriae, 626
Shigella flexneri, 626
Shigella sonnei, 626
Shigellosis (bacillary dysentery), 258*f*, 280, 626, 640*t*
 water pollution in transmission of, 640*t*, 684
Shingles (herpes zoster), 338*t*, 526, 531*t*. *See also* Varicella-zoster virus
 and latent infections, 359, 526
Shock
 anaphylactic, 469
 septic (endotoxic), 401, 562
Shoyu, microorganisms in production of, 709*t*
Si. *See* Silicon
Side chains, tetrapeptide, 78–79
Siderophores, virulence and, 397
Silicon, 23*t*
 in soil, 673
Silver, antimicrobial activity of, 180–181, 184*t*
Silver nitrate
 antimicrobial activity of, 181, 184*t*
 in neonatal gonorrheal ophthalmia, 181, 184*t*, 533, 534*t*, 656
Silver sulfadiazine, 499
Simian AIDS, 342
Simian virus 40, tumors caused by, 358
Simian viruses, 338*t*
Simple diffusion, 83, 84*f*
Simple lipids, 37–39
Simple proteins, 44
Simple stains, 60–63, 65*t*
Single-cell protein (SCP), 18, 712
Single covalent bond, 27, 28*f*
Single-stranded DNA viruses, 338*t*
Singlet oxygen, 147
Sinus headache, 592
Sinusitis, 591
Skin
 cell-mediated hypersensitivity reactions of, 475–476
 diseases of, 520–533. *See also specific type*
  bacterial, 520–524, 531*t*
  fungal, 530–533, 531*t*
  pseudomonads causing, 523
  staphylococcal, 520–522
  streptococcal, 522, 523*f*
  viral, 524–530, 531*t*
 function of, 518–519
 as host defense, 407–410, 518
  chemical factors and, 409–410
  mechanical factors and, 408–409
 normal flora of, 519–520
 as portal of entry, 393, 394*t*, 404*f*
 structure of, 408*f*, 518–519
Skin tests
 allergens identified by, 470, 471*f*
 for *Mycobacterium tuberculosis*, 598
  and cell-mediated hypersensitivity reactions of skin, 475
Slants, 149
SLE. *See* St. Louis encephalitis
Sleeping sickness, African (African trypanosomiasis), 316, 440, 554, 555*t*
 arthropod vector for, 379*t*
Slide agglutination test, 261
Slime layer, 72
Slime molds, 312–313
Slow-reacting substances of anaphylaxis, in hypersensitivity reactions, 468
Slow viral infections, 359–360
 of nervous system, 554–556
Sludge, 688
 activated, 690, 691*f*
 digestion of, 691–692
 primary, 688
Slug, 313
Smallpox, 350, 524–525, 531*t*. *See also* Smallpox virus
Smallpox vaccination, 524–525
 discovery of, 10–11, 450, 524
Smallpox virus (variola virus), 338*t*, 524, 531*t*
Smears, preparing, 60
Smooth endoplasmic reticulum, 89*f*
Sn. *See* Tin
Snomax, genetic engineering in production of, 229*f*, 237*t*
Snow, John, 384
Soap, antimicrobial activity of, 181, 184*t*
 effectiveness and, 183*f*
Sodium, 23*t*. *See also* Salts
 in soil, 673
Sodium benzoate (benzoic acid)
 antimicrobial activity of, 182, 185*t*
 in food preservation, 705, 706
Sodium chloride, formation of, 26
Sodium hypochlorite, antimicrobial activity of, 179
Sodium ion, 26
Sodium nitrate, in food preservation, 706
Sodium nitrite, in food preservation, 706
Soft chancre, 663, 665*t*
Soil
 components of, 673–676
 gases in, 673
 microorganisms in, 673–676
 minerals in, 673
 as nonliving reservoir in spread of infection, 375
 organic matter in, 673
 organisms in, 673–676
 pathogens in, 674–676
 water in, 673
Soil microbiology, 673–681
Solid municipal waste, 694–695
Solute, water and, 32
Solvent, water as, 32
Somatotropin (human growth hormone), genetic engineering in production of, 229*f*, 236, 237*t*
Sorbic acid (potassium sorbate)
 antimicrobial activity of, 182, 185*t*
 in food preservation, 705, 706
Sorbose, fermentation of, 123*t*
Sore throat (pharyngitis), 591
 Streptococcus causing, 286, 592, 610*t*
  rheumatic fever and, 564, 580*t*
  scarlet fever and, 592–593
Sour cream, microorganisms in production of, 708
Southern blotting, 238*f*, 239
Soy sauce, microorganisms in production of, 709*t*
Spallanzani, Lazaro, 7
Spastic paralysis, in tetanus, 543, 544*f*
Special culture techniques, 151–152
Special stains, 65–66, 65*t*
Specialized transduction, 218, 348, 349*f*
Species (specific epithet)
 bacterial, 257
 in nomenclature, 14, 254
Specific channel proteins, 79
Specific resistance, 407. *See also* Immunity
Specificity, enzyme, 104–105
Spectinomycin, PPNG resistant to, 657
Spectrum of activity, of antimicrobial agent, 494–496
Sphaerotilus spp, 275*t*

Sphaerotilus natans, 290
  and bulking in sewage treatment, 690
Spherical particles, in hepatitis B, 633
Spheroplast, 81
Spherule, of *Coccidioides immitis*, 608
Spicules, 321
Spikes
  hemagglutinin, 605
  neuraminidase, 605
  viral, 335–336
Spiral bacteria, 15, 71, 72*f*
Spirillum spp, 71, 72*f*, 274*t*
  lipid inclusions in, 87
  rat bites infected by, 567, 580*t*
Spirillum volutans, 276
Spirochetes, 71, 72*f*, 274*t*, 275–276, 278. *See also specific type*
  axial filaments of, 75–76
  phylogenetic relationships of, 258*f*
Spiroplasma spp, 285
Spirosoma spp, 274*t*
Spirulina, as food source, 712
Spliceosome, 199
Spoilage, food
  butyric anaerobic, 703*t*
  and canned foods, 702–703
  flat sour, 702, 703*t*
  putrefactive anaerobic, 703*t*
  sulfide, 703*t*
  temperatures of, 704*f*
  thermophilic anaerobic, 702, 703*t*
Spontaneous generation, 6–8
Spontaneous mutations, 208
Sporadic disease, 373
Sporangia, 300, 313*f*
Sporangiophores, 300
Sporangiospores, 300
Spore coat, 88
Spore septum, 88
Spore staining, 64*f*, 65–66, 65*t*
Spores
  asexual, 299, 300
  reproduction in fungi by, 299–301
  sexual, 299, 300–301
Sporothrix schenckii, 302*t*
  skin infections caused by, 531*t*, 532
Sporotrichosis, 531*t*, 532
Sporozoa, 316–319, 318*t*
Sporozoites
  of *Plasmodium*, 316, 317*f*, 579
  of *Toxoplasma gondii*, 575, 576*f*
Sporulation (sporogenesis), 88
Spotted fever, Rocky Mountain, 283, 572*f*, 573, 581*t*
  arthropod vector for, 379*t*, 581*t*
  cycle of, 572*f*
  distribution of, 572*f*
  transmission of, 378*t*, 581*t*
Spread plate method, for plate count, 159*f*, 160
Square cells, 71, 72*f*
SRS-A. *See* Slow-reacting substances of anaphylaxis
SSPE. *See* Subacute sclerosing panencephalitis
Stabilization ponds, for water treatment, 693
Staggered cuts, 230*f*, 231
Stains (staining), 60–66. *See also specific type*
  acid-fast, 65
  differential, 63–65, 65*t*
    in classification/identification of microorganisms, 259, 267*t*
  for endospores, 64*f*, 65–66, 65*t*, 89
  for flagella, 64*f*, 65*t*, 66
  Gram, 63–65, 65*t*
  negative, 60, 64*f*, 65
  primary, 63
  simple, 60–63, 65*t*
  special, 65–66
Staphylococci. *See Staphylococcus* spp
Staphylococcus spp, 274*t*, 285–286. *See also specific type*
  cell arrangement of, 70*f*, 71
  coagulases produced by, 396
  food poisoning (foodborne microbial intoxications) caused by, 622–624, 640, 707*t*
  genetic transformation occurring on, 215
  hemolysins produced by, 396
  leukocidins produced by, 396
  penicillinases produced by, 503
  pyelonephritis caused by, 653
  skin infections caused by, 520–522
  toxins produced by, 399, 521
    foodborne microbial intoxications and, 622–624, 640, 707*t*
    phagocytosis and, 415
    plasmid gene encoding, 401
  urinary tract infections caused by, 653, 665*t*
Staphylococcus aureus, 285–286. *See also Staphylococcus* spp
  acute bacterial endocarditis caused by, 563, 580*t*
  adhesion of, 395
  culture media for growth of, 152
  food poisoning caused by, 622–624, 640*t*
  in human bites, 568
  in nosocomial infection, 380*t*, 381
  otitis media caused by, 594, 610*t*
  resistance of, 286, 521
  selective and differential media for growth of, 152
  skin infections caused by, 520–522, 531*t*
  staphylokinase produced by, 396
  toxic shock syndrome caused by, 4, 521–522
  toxins produced by, 399, 400*t*
Staphylococcus epidermidis, skin infections caused by, 520
Staphylococcus saprophyticus, cystitis caused by, 653, 665*t*
Staphylokinase, virulence and, 396
Star-shaped cells, 71, 72*f*
Starch, 37
Start codon, 199
Stationary phase, 157–158
STDs. *See* Sexually transmitted diseases
Steam, for sterilization, 171–172, 174*t*
Stella, 71, 72*f*
Stereoisomers, amino acid, 41–42
Sterile medium, 148
Sterility, salpingitis causing, 656
Sterilization, 168*t*, 170–173, 174*t*
  commercial, for food preservation, 701–702
  gas, 183, 185*t*
Steroids, 40–41
  microbial production of, 717
Sterols, 40–41, 90
  in mycoplasma, 80
Sticky ends, 230*f*, 231
Stigmatella aurantiaca, 289*f*
Stipes, 310
Stop codon, 201
  nonsense mutation and, 207–208
Strain improvement, and fermentation technology, 714
Stratum corneum, 518–519
Streak plate method, for pure culture, 153–155
Streaming, cytoplasmic, 92, 313
Strep throat (streptococcal pharyngitis), 286, 592, 610*t*
  rheumatic fever and, 564, 580*t*
  scarlet fever and, 592–593
Streptobacilli, 71
Streptobacillus spp, rat bites infected by, 567, 580*t*
Streptococci. *See Streptococcus* spp
Streptococcus spp, 274*t*, 286. *See also specific type*
  α-hemolytic, 286, 522
    in subacute bacterial endocarditis, 580*t*
  β-hemolytic, 286, 522
    in glomerulonephritis, 654
    in pharyngitis, 592
    in puerperal sepsis, 562, 580*t*
    in rheumatic fever, 563, 580*t*
  cell arrangement of, 70*f*, 71
  in endocarditis, 563, 580*t*
  fluorescent-antibody test in identification of, 460*f*
  in food production, 709*t*
  γ-hemolytic, 286, 522
  genetic transformation occurring on, 215
  in glomerulonephritis, 475, 654, 665*t*
  hemolytic, 286, 522
    in human bites, 568
  hyaluronidase produced by, 396
  lactic acid fermentation and, 121
  leukocidins secreted by, 396
  in pericarditis, 563
  in pharyngitis, 286, 592, 610*t*
    rheumatic fever and, 564, 580*t*
    scarlet fever and, 592–593
  in puerperal sepsis, 562–563, 580*t*
  in rheumatic fever, 563, 580*t*
  in skin infections, 522
  streptolysins produced by, 592
  temperate phages and, 348
  in urethra, 652
  in urinary tract infections, 653, 665*t*
Streptococcus lactis, in food production, 709*t*
  var. *taette*, 709*t*
Streptococcus mutans, 286*f*
  adhesins of, 393
  capsule of, 73
  in tooth decay, 393–395, 619–620, 640*t*
Streptococcus pneumoniae, 286. *See also Streptococcus* spp
  capsule of, 72, 396, 414
  genetic transformation of, 214–215
  in laryngitis, 591
  in meningitis, 541*t*, 542, 555*t*
  mucous membrane penetrated by, 408
  opportunistic infection and, 371
  in otitis media, 594, 610*t*
  in pneumonia, 286, 599–600, 611*t*
  respiratory tract as portal of entry for, 394*t*
  in sinusitis, 591
Streptococcus pyogenes. *See also Streptococcus* spp

Streptococcus pyogenes (*Continued*)
in acute bacterial endocarditis, 563, 580*t*
blood agar for growth of, 152
in commercial enzyme production, 716*t*
exotoxins produced by, 399, 400*t*
in glomerulonephritis, 654, 665*t*
impetigo caused by, 522, 523*f*
and Koch's postulates, 373
in laryngitis, 591
M protein produced by, 396, 414
in otitis media, 594
in pericarditis, 580*t*
in pharyngitis, 592, 610*t*
in puerperal sepsis, 562–563, 580*t*
in pyelonephritis, 653
in rheumatic fever, 563
in scarlet fever, 592, 610*t*
in skin infections, 531*t*
streptokinase produced by, 396
Streptococcus thermophilus, in food production, 708, 709*t*
Streptodornases, commercial production of, 716*t*
Streptokinase (fibrinolysin)
streptococci producing, 522, 592
therapeutic use of, 396
virulence and, 396
Streptolysins, streptococci producing, 592
Streptomyces spp, 275*t*, 293
antibiotics produced by, 494
tetracyclines produced by, 504
Streptomyces aureofaciens, antibiotics produced by, 494*t*
Streptomyces erythraeus, antibiotics produced by, 494*t*
Streptomyces fradiae, antibiotics produced by, 494*t*
Streptomyces griseus
antibiotics produced by, 494*t*
streptomycin produced by, 504
Streptomyces nodosus, antibiotics produced by, 494*t*
Streptomyces noursei, antibiotics produced by, 494*t*
Streptomyces olivoreticuli, Bestatin from, 410
Streptomyces venezuelae, antibiotics produced by, 494*t*
Streptomycetes, in steroid production, 717
Streptomycin, 500*t*, 504
for brucellosis, 566
protein synthesis affected by, 497*f*, 498
sources of, 494*t*
for tuberculosis, 598
for tularemia, 565
Stress, bacteria affected by, 205
String test, for giardiasis diagnosis, 637
Stroma, 417
Sty, 521
Subacute bacterial endocarditis, 563, 580*t*
Subacute disease, 374
Subacute sclerosing panencephalitis, 360, 528
Subarachnoid space, 539, 540*f*
Subclinical infection, 374
immunity conferred by, 427
Subcutaneous infection, 408
Subcutaneous mycoses, 306, 532–533
Substrate, 31, 102
enzyme activity affected by concentration of, 106*f*, 107
Substrate-level phosphorylation, 110
Subunit vaccines, 236, 452
Subunits, 36
Sucking lice, 326*t*
Sucrose, in dental caries, 620*f*, 621, 640*t*
Sugar-phosphate backbone, 191
"Suicide genes," 243
Sulfa drugs (sulfonamides), 499–500, 500*t*
discovery of, 11, 492–493
Sulfadiazine, for toxoplasmosis, 577
Sulfadoxine, with pyrimethamine, for malaria, 581
Sulfamethoxazole and trimethoprim, 499–500
for bacillary dysentery, 626, 640*t*
for chancroid, 663, 665*t*
for typhoid fever, 626
Sulfanilamide, 493
competitive inhibition in action of, 498
structure of, 499*f*
Sulfate-reducing bacteria, dissimilatory, 274*t*, 282–283
Sulfide, in food spoilage, 703*t*
Sulfide ion, 26
Sulfolobus spp, 275*t*, 291
Sulfonamides (sulfa drugs), 499–500, 500*t*
discovery of, 11, 492–493
Sulfone drugs, for leprosy, 547
Sulfur, 23*t*
biogeochemical cycles involving, 681
electronic configuration for, 25*t*
for microbial growth, 145–146
"Sulfur bacteria," 87, 291*f*. *See also* Green bacteria; Purple bacteria
Sulfur granules, 87
Sulfur-reducing bacteria, dissimilatory, 274*t*, 282–283
Sunlight, microbial growth affected by, 176
Superficial mycoses, 306
Superoxide dismutase, 147
genetic engineering in production of, 237*t*
Superoxide free radicals, 147
Suppressor T cells, 441
function of, 443*t*
Suramin, for African trypanosomiasis, 554
Surface antigens, changing, and immune response to protozoans, 440
Surface-active agents (surfactants), antimicrobial activity of, 181–182, 184*t*
Surgery, aseptic, 167
Surgical wound infections
aseptic techniques and, 381
nosocomial, 382*t*
Staphylococcus aureus causing, 286
Surveillance, immunologic, 486
Susceptibility, 407
SV40. *See* Simian virus 40
"Swarm," 75
Sweat. *See* Perspiration
"Swim" ("run"), 74*f*, 75
Swimmer's itch, 583
Sydenham's chorea, 564
Symbiosis, 672
between host and normal flora, 370
in microbial growth, 145
and microorganisms in termites, 278
in nitrogen fixation, 679–680
Symbiotic nitrogen-fixing bacteria, 679–680
Syncytia, 402
Synergism, in chemotherapy, 499, 511
Synthesis reactions, 29
Synthetic chemicals, degradation of in soil, 681
Synthetic DNA, 234
Synthetic drugs, 11, 492
antibacterial, 499–502
antifungal, 502
Syphilis, 4, 276, 659–662, 665*t*. *See also Treponema pallidum*
congenital, 662
distribution of, 659*f*
incidence of, 655*f*, 659*f*
latent period and, 661
primary stage of, 660
secondary stage of, 660–661
tertiary stage of, 660*f*, 661
tests for, 661
Systemic anaphylaxis, 468–470
Systemic infection (generalized infection), 374
Systemic lupus erythematosus, 475, 477*t*
Systemic mycoses, 304–306

T

2,4,5-T, degradation of in soil, 681
T antigen, 358
T cell antigen receptors, 429
T cells (T lymphocytes), 428*f*, 429, 439–443, 561
in B-cell activation, 433, 435*f*
CD4
in AIDS, 439, 480–481
fluorescence-activated cell sorter in identification of, 461
CD8, fluorescence-activated cell sorter in identification of, 461
in cell-mediated immunity, 439–443
cytotoxic, 441
function of, 443*t*
and tumor cell interaction, 487*f*
delayed hypersensitivity, 439–441
function of, 443*t*
differentiation of, 433*f*
helper, 433, 439
in AIDS, 439, 480–481
in B-cell activation, 433, 435*f*
function of, 443*t*
regulatory, 441
specialized, 439
suppressor, 441
function of, 443*t*
T_C cells. *See* Cytotoxic T cells
T_D cells. *See* Delayed-hypersensitivity T cells
T_H cells. *See* Helper T cells
T_S cells. *See* Suppressor T cells
T-dependent antigens, 433
T-DNA, 241
T-even bacteriophages, multiplication of, 345–347
T-independent antigens, 434–435
t-PA. *See* Tissue plasminogen activator
Tachyzoites, of *Toxoplasma gondii*, 318, 575, 576*f*
Taenia saginata (beef tapeworm), 321, 325*t*, 639, 641*t*
transmission of, 379*t*, 641*t*

Taenia solium (pork tapeworm), 639–640, 641*t*
Taette, microorganisms in production of, 709*t*
Talaromyces spp, and reclassification of fungi, 303*t*
Tapeworms, 320–321, 325*t*, 639–642, 641*t*
  beef, 321, 325*t*, 639, 641*t*
    transmission of, 379*t*, 641*t*
  fish, 640–642, 641*t*
  pork, 639–640, 641*t*
Tartar, 619
Taxa, arrangement of organisms into, 251
Taxis, 75
Taxonomic hierarchy, 255–256
Taxonomy, 250. *See also* Microorganisms, classification/identification of
  numerical, 268–269
TCA cycle. *See* Tricarboxylic acid cycle
TD vaccine, schedule for administration of, 452*t*
TDP. *See* Thermal death point
TDT. *See* Thermal death time
Tears, as host defense, 408–409, 519
Teeth. *See under Tooth*
Teichoic acids, 78*f*, 79
Tempeh, microorganisms in production of, 709*t*
Temperate phages, 347
Temperature
  body, pathogens affecting, 417–418
  enzyme activity affected by, 105–106
  food spoilage and, 704, 704*f*, 705*f*
  microbial growth affected by, 142–144, 169
    heat and, 170–173, 174*t*
    low temperature, 169, 172–174, 175*t*
Temperature abuse, in staphylococcal food poisoning, 622
Terminator region, 198
Termites, microorganisms used by, 278
Terramycin. *See* Oxytetracycline
Tertiary structure of protein, 44, 45*f*
Tertiary treatment of sewage, 693–694
Testes, 652
Tetanospasmin, 399, 543
  plasmid gene encoding, 401
Tetanus, 287, 399, 400*t*, 543–544, 555*t*. *See also Clostridium tetani*
  vaccine for, 450*t*, 543–544
    schedule for administration of, 452*t*
Tetanus immune globulin, 544
Tetanus toxin, 399
tetM plasmid, drug resistance and, 657
Tetracycline-resistant *Neisseria gonorrhoeae*, 657
Tetracyclines, 500*t*, 504–505
  for acne, 524
  for balantidiasis, 638, 641*t*
  for brucellosis, 566
  for chlamydial pneumonia, 603, 611*t*
  for cholera, 627
  for endemic murine typhus, 573, 581*t*
  for epidemic typhus, 573, 581*t*
  for gonorrhea, 658
  for lymphogranuloma venereum, 663, 665*t*
  for mycoplasmal pneumonia, 601, 611*t*
  for nongonococcal urethritis, 658, 665*t*
  protein synthesis affected by, 497*f*, 498
  for Q fever, 604, 611*t*
  for Rocky Mountain spotted fever, 573, 581*t*
  sources of, 494*t*
  structure of, 505
  for syphilis, 662
Tetrad, cell arrangement, 70*f*, 71
Tetrahedron, 35
Tetrapeptide side chains, 78–79
Thallus (thalli)
  of algae, 306, 310
  of lichens, 311
  of mold or fleshy fungus, 297
  of red algae, 310
Theory of evolution, Darwin's, 250–251
Thermal death point, 170
Thermal death time, 170
Thermoacidophiles
  classification of, 253
  phylogenetic relationships of, 258*f*
Thermoactinomyces vulgaris, endospores of, 88
Thermoduric organisms, pasteurization and, 172
Thermophiles, 142, 144
  extreme, 144
Thermophilic anaerobic spoilage, 702, 703*t*
Thermoplasma spp, 254, 285
Thermostable nuclease, testing food for, 624
Thiabendazole, for trichinosis, 641*t*, 643
Thiamine (vitamin B_1), function of, 104*t*
Thiobacillus spp, 275*t*, 290–291. *See also specific type*
  carboxysomes in, 87
  in denitrification, 678
  in mining, 715, 717
  sulfur granules in, 87
Thiobacillus ferrooxidans, 130, 685
  in biological leaching of copper, 715, 717, 718*f*
Thiobacillus thiooxidans, 130
Three-kingdom system, 251–254
Threonine, 42*t*
  in oncogenic transformation, 357
Thrombocytes (platelets), 411*t*
  and drug-induced cytotoxic reactions, 473–474
Thrombocytopenic purpura, 473–474, 477*t*
Thrush (oral candidiasis), 306, 533, 664. *See also Candida albicans*
Thylakoids (chromatophores), 82, 83*f*, 89*f*, 95
Thymic hypoplasia, 481*t*
Thymine, 46, 190–191
Thymine analogs, 501*t*. *See also specific type*
Thymine dimers
  microbial growth and, 176
  in radiation damage, 210, 211*f*
Thyroiditis, Hashimoto's, 476, 477*t*
Ti plasmid, 241, 242*f*
Ticarcillin, potassium clavulanate combined with, 504
Tick-borne typhus (Rocky Mountain spotted fever), 283, 572*f*, 573, 581*t*
  arthropod vector for, 379*t*, 581*t*
  cycle of, 572*f*
  distribution of, 572*f*
  transmission of, 378*t*, 581*t*
Ticks, as vectors, 326*t*, 327*f*, 580*t*
  diseases transmitted by, 379*t*
  in Lyme disease, 4, 276, 379*t*, 569, 581*t*
TIG. *See* Tetanus immune globulin
Timentin. *See* Ticarcillin, potassium clavulanate combined with
Tin, 23*t*
Tincture of iodine, antimicrobial activity of, 179
  effectiveness and, 183*f*
Tinea, 530, 531*t*
Tinea capitis, 530
Tinea cruris, 530
Tinea pedis, 530
Tissue cultures. *See* Cultures
Tissue cyst, in toxoplasmosis, 575
Tissue plasminogen activator, genetic engineering in production of, 229*f*, 237*t*
Tissue typing, 477
Titer (antibody), 443, 456
TMP-SMZ. *See* Trimethoprim-sulfamethoxazole
TMV. *See* Tobacco mosaic virus
TNF. *See* Tumor necrosis factor
Tobacco mosaic virus, 332, 336*f*, 361*t*
  size of, 334*f*
Togaviruses, 338*t*, 354–355. *See also specific type*
  multiplication of, 354–355
  portal of entry for, 394*t*
  skin infections caused by, 531*t*
Tolerance, 443. *See also* "Self" antigens
  loss of, 476
Tolnaftate, 507
Tonsillitis, 591
Tooth
  extraction of, HBV and HIV transmission during, 344
  normal, 618, 619*f*
Tooth decay (dental caries), 618–621, 640*t*
  adhesion in, 393–395
Toxemia, 374, 397
  staphylococci causing, 521
Toxic dumps, microbes in cleanup of, 17, 38, 229*f*, 240, 681
Toxic shock syndrome, 4, 286, 399, 400*t*, 521–522
Toxic shock syndrome toxin-1, 521
Toxicity, selective, 492
Toxigenicity, 397
Toxins, 397–401, 404*f*. *See also specific type*
  exfoliative, staphylococci producing, 521
  foodborne microbial intoxications and, 622, 706, 707*t*. *See also* Food poisoning; Gastroenteritis
  fungi producing, 403–404
  staphylococcal, 521, 622–623
  toxic shock syndrome caused by, 4, 399, 400*t*, 521
  in vaccines, 427
Toxoids, 397
  in vaccines, 427, 450
Toxoplasma gondii, 318, 575, 581*t*
  characteristics of, 318*t*
  life cycle of, 575, 576*f*
Toxoplasmosis, 575–577, 581*t*
  transmission of, 379*t*, 581*t*

Trace elements, for microbial growth, 146
Trachea, mucous membranes of, 409*f*
Tracheal cytotoxin, in whooping cough, 595
Trachoma, 284, 533–534, 534*t*
Transamination, 132, 133*f*
Transcription, 191, 197–199, 202*t*
Transduction
  in bacteria, 218
  of oncogenes, 358
  specialized, 218, 329*f*, 348
Transfer factor, function of, 441*t*
Transfer RNA, 47
  in protein synthesis, 191, 197, 199–201, 202*t*
Transferase class of enzymes, 103*t*
Transferrins, in host defense, 410
Transformation
  in bacteria, 214–216
    classification and, 267–268
    mechanism of, 216*f*
  cancerous, and cytotoxic T cell-tumor cell interaction, 486, 487*f*
  and continuous cell lines, 343
  of normal cells to tumor cells, 357–358, 403
Transfusion reactions, 471
  and ABO blood group system, 471
  and Rh blood group system, 471–472
Transient flora, 370
Translation, 191, 198*f*, 199–202, 202*t*
Translocation
  group, 86
  of oncogenes, 358
Transmembrane channels, 420*f*, 421
Transmission
  of disease, 375–379
    for nosocomial infections, 382–383
  percentage of, in estimating bacterial numbers by turbidity, 162
Transmission electron micrograph, 58–59
Transmission electron microscope, 58–60
Transovarian passage, in Rocky Mountain spotted fever, 572*f*, 573
Transplantation, 478–480
  bone marrow, 479–480
  rejection of, monoclonal antibodies in prevention of, 439
Transport, active, 86
Transport chain, electron, and chemiosmotic generation of ATP, 118
Transposons (transposable genetic elements), 221–222
  antibiotic resistance and, 222
  bacterial, 222
Traveler's diarrhea, 628
  exotoxins causing, 400*t*
Trematodes, 319–320, 325*t*
Trench mouth (acute necrotizing ulcerative gingivitis, Vincent's disease), 622
Treponema spp, 274*t*, 276
Treponema pallidum, 276, 659*f*, 660, 665*t*. *See also* Syphilis
  axial filaments of, 75
  darkfield microscopy in identification of, 57, 659*f*, 661
  as exception to Koch's postulates, 373
  fluorescent-antibody test in identification of, 460*f*, 661
  genitourinary tract as portal of entry for, 393, 394*t*, 660, 665*t*
  mucous membrane penetrated by, 408
Triatoma ("kissing bug," reduviid bug), 326*t*, 327*f*
  diseases transmitted by, 379*t*
  Trypanosoma cruzi transmitted by, 316, 577, 581*t*
Tricarboxylic acid cycle, 114. *See also* Krebs cycle
Trichinella spiralis, 323–325, 325*t*, 641*t*, 643
  gastrointestinal tract as portal of entry for, 394*t*, 641*t*, 643
Trichinosis, 323, 325, 641*t*, 643
  transmission of, 379*t*, 641*t*
Trichomonas vaginalis, 315*f*, 316, 318*t*, 665*t*, 666
  characteristics of vaginitis caused by, 666*t*
Trichomoniasis, 665*t*, 666
  incidence of, 655*f*
Trichonympha sphaerica, 278
Trichophyton spp, 302*t*
  reclassification of, 303*t*
  skin infections caused by, 530, 531*t*
Trichosporon, 302*t*
Trickling filters, for sewage treatment, 690–691
Triclocarban, antimicrobial activity of, 181, 184*t*
Trifluridine, 508
Trimethoprim, 502*f*
Trimethoprim-sulfamethoxazole, 499–500
  for bacillary dysentery, 626, 640*t*
  for chancroid, 663, 665*t*
  for typhoid fever, 626
Tripeptides, 44
Triple covalent bond, 27, 28*f*
Trisulfaprimidines, for toxoplasmosis, 577
tRNA. *See* Transfer RNA
TRNG. *See* Tetracycline-resistant *Neisseria gonorrhoeae*
Trophophase, and fermentation technology, 713
Trophozoites
  of *G. lamblia*, 316
  of *Plasmodium*, 316
Truffles, as ectomycorrhizae, 681
Trypanosoma spp, 316
  immune response to, 440, 554
Trypanosoma brucei gambiense, 316, 318*t*, 554, 555*t*
Trypanosoma brucei rhodesiense, 318*t*, 554, 555*t*
Trypanosoma cruzi, 316, 318*t*, 563, 577, 581*t*
Trypanosomiasis
  African (African sleeping sickness), 316, 440, 554, 555*t*
    arthropod vector for, 379*t*
  American (Chagas' disease), 316, 577, 581*t*
    transmission of, 378*t*, 581*t*
Tryptophan, 43*t*
  attenuator affected by, 207
Tsetse fly, 326*t*
  diseases transmitted by, 379*t*
  T. brucei gambiense transmitted by, 316, 554
TSS. *See* Toxic shock syndrome
TSST-1. *See* Toxic shock syndrome toxin-1
TSTA. *See* Tumor-specific transplantation antigen
Tubal pregnancy, salpingitis and, 656
Tubercle, 597
Tuberculin skin test, 598
  and cell-mediated hypersensitivity reactions of skin, 475
Tuberculoid leprosy (neural leprosy), 546
Tuberculosis, 288, 596–599, 611*t*. *See also Mycobacterium tuberculosis*
  bovine, 598
  distribution of, 598–599
  miliary, 598
  skin test for, 598
    and cell-mediated hypersensitivity reactions of skin, 475
  vaccine for, 450*t*
Tuberculous cavity, 597–598
Tubeworms, hydrothermal bacteria and, 143
Tularemia, 279, 564–565, 580*t*. *See also Francisella tularensis*
  transmission of, 378*t*, 580*t*
"Tumble," 74*f*, 75
Tumor cells
  cytotoxic T-cell interaction and, 487*f*
  transformation of normal cells to, 357–358, 403
Tumor necrosis factor, 487
  genetic engineering in production of, 237*t*
  in septic shock, 401
Tumor-specific transplantation antigen, 358
Turbidity, bacterial numbers estimated by, 162
Turnover number, 103
TWAR strain, 603
Two-dimensional PAGE, 363–365
Typhoid fever, 280, 401, 625–626, 640*t*. *See also Salmonella typhi*
  vaccine for, 205, 450*t*, 626
  water pollution in transmission of, 640*t*, 684
Typhus, 570–573, 581*t*
  endemic murine, 283, 573, 581*t*
    arthropod vector for, 379*t*, 581*t*
  epidemic, 283, 570–573, 581*t*
    arthropod vector for, 379*t*, 581*t*
  transmission of, 378*t*
  vaccine for, 450*t*
Typing, phage, 261
Tyrosine, 43*t*
  in oncogenic transformation, 357

U

Ubiquinones, 116
UHT sterilization. *See* Ultra-high-temperature sterilization
Ulcers, as infectious disease, 398, 629
Ultra-high-temperature sterilization, 172
Ultrastructure, 53
Ultraviolet light
  microbial growth affected by, 176
  mutations caused by, 210, 211*f*
Uncoating
  in animal virus multiplication, 350
  in RNA virus multiplication, 352*f*
Unconventional agents, 360, 554–555

nervous system disease caused by, 554–556, 555*t*
Undecylenic acid, 507
Undulant fever (brucellosis), 565–566, 580*t*
transmission of, 378*t*, 580*t*
Undulating membrane, 315
Unsaturated fat, 39
Upper respiratory system
diseases of, 591–595, 610–611*t*. *See also specific type*
bacterial, 592–594, 610*t*
viral, 594–595, 611*t*
normal flora of, 591
structures of, 589–590
Uranium, industrial microbiology in production of, 717
Ureaplasma spp, 285
Ureaplasma urealyticum, nongonococcal urethritis caused by, 659, 665*t*
Ureidopenicillins, 504
Ureteritis, 653
Ureters, 650, 651*f*
Urethra, 650, 651*f*
Urethritis, 653
nongonococcal, 658–659, 665*t*
C. trachomatis causing, 284, 658–659, 665*t*
Urinary bladder, 650, 651*f*
bacterial infection of, 653, 665*t*
Urinary tract
bacterial diseases of, 653–654, 665*t*
normal flora of, 652
as portal of entry, 393, 394*t*, 404*f*
structure and function of, 650–651
Urinary tract infections, 653–654, 665*t*
Enterobacter causing, 281, 665*t*
nosocomial, 382*t*, 653
Urine, 650
in host defense, 409
normal flora in, 652
Use-dilution test, 177
Uterine tubes, 651
infection of, 656
Uterus, 651
UV light. *See* Ultraviolet light

V

V. *See* Vanadium
V factor, *Hemophilus* using, 282
V genes, and diversity of antigen receptors, 436, 437*f*
V regions (variable regions), 430
Vaccination (immunization), 10–11, 427. *See also* Vaccines
schedule of for children, 452*t*
Vaccines, 427, 449–453. *See also specific type or disease*
acellular, 452
anti-idiotypic, 452–453
characteristics of, 450
conjugated, 452
development of, 451–453
embryonated eggs for growth of viruses for, 342
future of, 463
industrial microbiology in production of, 717
live, 450
recombinant DNA techniques in making, 235, 236
subunit, 236, 452
Vaccinia virus, 338*t*
and recombinant DNA techniques in making vaccines, 235, 236
size of, 334*f*
Vacuoles
digestive (phagolysosome), 414
gas, 87
phagocytic (phagosome), 414
in protozoan nutrition, 314
Vacuum heap treatment, 240
Vagina, 651
normal flora of, 652
Vaginitis, 652, 662, 665*t*, 666*t*
candidal, 664–666, 665*t*, 666*t*
characteristics of, 666*t*
Gardnerella vaginalis, 282, 662, 665*t*, 666*t*
nonspecific, 662
Trichomonas, 665*t*, 666
characteristics of, 666*t*
Valence, 24
antibody, 430
Valine, 42*t*
van Leeuwenhoek, Anton, 6, 53
Vanadium, 23*t*
Vancomycin, 500*t*, 506
cell wall synthesis affected by, 496
Variable regions, 430
Variable surface glycoprotein, 440
Varicella, 525–526, 531*t*
and latent infections, 359, 526
Varicella vaccine, 526
Varicella-zoster virus. *See also* Herpes zoster
latent infection caused by, 359, 526
respiratory tract as portal of entry for, 394*t*
Variola, 338*t*, 524, 531*t*. *See also* Smallpox
Variola major, 524
Variola minor, 524
Variola virus (smallpox virus), 338*t*, 524, 531*t*
Variolation, 449
Vasodilation, in inflammation, 415–416
VDRL test, for syphilis, 661
Vectors, 228, 325, 376*f*, 377–379. *See also specific type and disease*
Aedes albopictus as, 324
arthropods as, 325–327, 376*f*, 377–379, 553*t*
diseases transmitted by, 379*t*
plasmids for, 231–232
Vegetative mycelium, 298
Vegetative structures, 297–299
Vehicle transmission, in spread of infection, 376*f*, 377
Veillonella spp, 274*t*, 283
Veins, 560. *See also* Cardiovascular system
Venereal Disease Research Laboratory test, for syphilis, 661
Venules, 560, 561*f*. *See also* Cardiovascular system
Verotoxins, in hemorrhagic colitis, 628
Vesicles, 680
gas, 87
Vesicular-arbuscular mycorrhizae, 680–681
Veterinary microbiology, 264
Vibrio spp, 71, 72*f*, 274*t*, 281
dolphins infected by, 264
two-dimensional polyacrylamide gel electrophoresis of, 263*f*
Vibrio cholerae, 281, 626–627, 640*t*. *See also* Cholera
gastrointestinal tract as portal of entry for, 394*t*
toxin produced by, 399, 400*t*
Vibrio enterotoxin, 399, 400*t*
Vibrio parahaemolyticus, 281
gastroenteritis caused by, 627, 640*t*
Vibrioids, 276–277
Vibrionaceae, 281. *See also Vibrio* spp
Vidarabine, 501*t*, 508–509
Vincent's disease (acute necrotizing ulcerative gingivitis, trench mouth), 622
Vinegar, microorganisms in production of, 712
Viral components, biosynthesis of, in bacteriophage multiplication, 345, 346*f*
Viral gastroenteritis, 636, 641*t*
Viral hemagglutination, 336, 457, 458*f*
Viral hepatitis, 631–636, 641*t*. *See also* Hepatitis
Viral inclusion bodies. *See* Inclusion, viral
Viral isolates, identification of, 343–345
Viral plaques. *See* Plaques
Viral pneumonia, 604, 611*t*
Viremia
poliovirus causing, 548
smallpox virus causing, 524
Virino, 360
Virion, 335
Viroids, plant diseases caused by, 361
Virology, 13
Virulence. *See also* Pathogenicity
capsules and, 65, 72–73, 395–396
enzymes affecting, 396–397
Viruses, 15, 16*f*, 332–365. *See also specific type and* Microorganisms
animal, 334
growth of in laboratory, 341–343
bacteria compared with, 333*t*
bacterial, 334
cancer and, 356–359
cardiovascular and lymphatic system diseases caused by, 573–575, 581*t*
characteristics of, 333–335
classification of, 257, 338–339*t*, 339–340, 361*t*
base sequence and, 265
complex, 337
morphology of, 339*f*
cultivation of, 340–345
digestive system diseases caused by, 630–636, 640–641*t*
DNA. *See* DNA viruses
enveloped, 337
budding of, 355, 356*f*
filterable, 173
gastroenteritis caused by, 636, 641*t*
gene products expressed by, 235
helical, 336, 337*f*
hepatitis caused by, 631–636, 641*t*. *See also* Hepatitis
host range of, 334–335
identification of, 340–345
infection caused by, 355–356
drugs for, 501*t*, 507–509. *See also specific type and* Antiviral drugs
latent, 359. *See also specific type*
slow, 359–360, 554–556
isolation of, 340–345
laryngitis caused by, 591
latent infection caused by, 359. *See also specific type*

Viruses (*Continued*)
lower respiratory system diseases caused by, 604–606, 611*t*
morphology of, 336–337, 339*f*
multiplication of, 345–355
mutant, 336
naked (nonenveloped), 335*f*, 336
release of, 355
nervous system diseases caused by, 547–553, 555*t*
orphan, reoviruses as, 355
pathogenetic properties of, 402–403
plant, 334, 360–361
pneumonia caused by, 604, 611*t*
polyhedral, 336–337
respiratory system disease caused by, 594–595, 604–606, 611*t*
reverse transcription, 339*t*
RNA. *See* RNA viruses
size of, 334*f*, 335
skin infections caused by, 524–530, 531*t*
slow infection caused by, 359–360, 554–556
in soil, 675
structure of, 335–337, 339*f*
taxonomy of, 339–340
unconventional, 360
upper respiratory system diseases caused by, 594–595, 611*t*
Visna, 360*t*
Vitamin B_1 (thiamine), function of, 104*t*
Vitamin B_2 (riboflavin)
commercial production of, 717
function of, 104*t*
Vitamin B_6 (pyridoxine), function of, 104*t*
Vitamin B_{12} (cyanocobalamin)
commercial production of, 717
function of, 104*t*
Vitamin E, function of, 104*t*
Vitamin K, function of, 104*t*
Vitamins
coenzymes derived from, 103, 104*t*
commercial production of, 717
function of, 104*t*
Volutin, 87
VSG. *See* Variable surface glycoprotein
Vulva, 651
Vulvovaginal candidiasis, 664–666, 665*t*
characteristics of, 666*t*

W

Wall teichoic acid, 78*f*, 79
Wandering macrophages, 413
Warming, global (greenhouse effect), bacteria and, 675
Warts, 524, 531*t*
genital, 664, 665*t*
papovaviruses causing, 351, 524
Water, 32–33
fresh, microbial flora in, 682–683
hydrogen bond formation and, 27, 29*f*
as nonliving reservoir in spread of infection, 375
purification of, 686–688
quality of, microorganisms affecting, 683–695
recycling, microbes in, 17
seawater, microbial flora in, 682–683
in soil, 673
treatment of, 686–688
Water microbiology, 682–695
Water pollution, 683–685
chemical, 684–685
and infectious disease transmission, 683–684
Water purity, tests for, 685–686, 687*f*
Waterborne transmission, in spread of infection, 377
WEE. *See* Western equine encephalitis
Western equine encephalitis, 552, 553*t*
transmission of, 378*t*
Wheal, in allergy skin test, 470
Whey, in cheese production, 707
Whiskey, microorganisms in production of, 711*t*
White blood cell count, differential, 411–412
White blood cells (leukocytes), 411, 411*t*, 561
polymorphonuclear, 412. *See also* Neutrophils
principal types of, 411*f*
Whooping cough (pertussis), 279, 595–596, 611*t*. *See also Bordetella pertussis*
vaccine for, 450*t*, 596
schedule for administration of, 452*t*
Windpipe. *See* Trachea
Wine, microorganisms in production of, 710*f*, 711–712, 711*t*
Wiskott–Aldrich syndrome, 481*t*
Woolsorters disease, 566
Wound botulism, 546
Wound tumor virus, 361*t*
Wuchereria bancrofti, pathogenetic properties of, 404

X

X factor, *Hemophilus* using, 282
X-linked infantile agammaglobulinemia, 481*t*
X rays, mutations caused by, 210
Xenodiagnosis, in Chagas' disease, 577
Xenograft, 479
Xenopsylla, as vector, 326*t*, 327*f*, 568, 573, 580*t*, 581*t*
diseases transmitted by, 379*t*

Y

Yeasts, 15, 298–299. *See also specific type and* Fungi; Microorganisms
alcoholic beverages produced by, 710–712
asexual reproduction in, 220
bakers'. *See also Saccharomyces cerevisiae*
budding of, 299*f*
gene products expressed by, 235
industrial production of, 717–718
bread making and, 5, 709*t*, 710
fermentation by, 299
in food spoilage, 703*t*
sexual reproduction in, 219*f*
Yellow fever, 575, 581*t*
arthropod vector for, 379*t*, 581*t*
transmission of, 378*t*, 581*t*
vaccine for, 451*t*
Yellow fever virus, 575, 581*t*
Yersinia spp, 274*t*, 281
gastroenteritis caused by, 629, 640*t*
Yersinia enterocolitica
food spoilage caused by, freezing and, 704
gastroenteritis caused by, 629, 640*t*
Yersinia pestis, 281, 568, 580*t*. *See also* Plague
capsule of, 396
portal of entry for, 394*t*
Yersinia pseudotuberculosis, 629
Yersiniosis, 629
Yogurt, microorganisms in production of, 708, 709*t*

Z

Zephiran. *See* Benzalkonium chloride
Zidovudine (AZT, azidothymidine), 501*t*, 509
for AIDS, 4, 485, 486
Ziehl–Neelsen technique, for staining *Mycobacterium tuberculosis*, 597
Zinc, 23*t*
Zinc chloride, antimicrobial activity of, 181
Zinc oxide, antimicrobial activity of, 181
Zinc peroxide, antimicrobial activity of, 183–185
Zn. *See* Zinc
Zone of equivalence, in precipitin ring test, 454
Zone of inhibition, 510
Zoogloea spp, in sewage treatment, 280, 690, 691*f*
Zoonoses, 374–375, 378–379*t*. *See also specific disease*
Zoster. *See* Herpes zoster (shingles)
Zygomycota, 302–303*t*, 304
zygospores produced by, 301
Zygospores, 301
Zygote, 218–220
in *Plasmodium* reproduction, 317